The chemistry of
alkanes and cycloalkanes

THE CHEMISTRY OF FUNCTIONAL GROUPS

A series of advanced treatises under the general editorship of
Professor Saul Patai

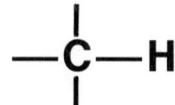

The chemistry of
alkanes and cycloalkanes

Edited by

SAUL PATAI

and

ZVI RAPPOPORT

The Hebrew University, Jerusalem

1992

JOHN WILEY & SONS
CHICHESTER–NEW YORK–BRISBANE–TORONTO–SINGAPORE

An Interscience® Publication

Other Wiley Editorial Offices

John Wiley & Sons, Inc., 605 Third Avenue,
New York, NY 10158-0012, USA

Jacaranda Wiley Ltd, G.P.O. Box 859, Brisbane,
Queensland 4001, Australia

John Wiley & Sons (Canada) Ltd, 22 Worcester Road,
Rexdale, Ontario M9W 1L1, Canada

John Wiley & Sons (SEA) Pte Ltd, 37 Jalan Pemimpin #05-04,
Block B, Union Industrial Building, Singapore 2057

Library of Congress Cataloging-in-Publication Data

The Chemistry of alkanes and cycloalkanes / edited by Saul Patai.
    p.   cm.—(The Chemistry of functional groups)
'An Interscience publication.'
Includes bibliographical references and indexes.
ISBN 0 471 92498 9 (cloth)
    1. Alkanes. 2. Cycloalkanes. I. Patai, Saul. II. Rappoport, Zvi.
III. Series.
QD305.H6C46   1992   90-26878
547.411—dc20   CIP

British Library Cataloguing-in-Publication Data

The chemistry of alkanes and cycloalkanes.
—(The chemistry of functional groups)
I. Patai, Saul. II. Rappoport, Zvi. III. Series.
547

ISBN 0 471 92498 9

Typeset by Thomson Press (India) Ltd., New Delhi and
Printed in Great Britain by Biddles Ltd, Guildford, Surrey

Contributing authors

W. Adam

Department of Organic Chemistry, University of Würzburg, D-8700 Würzburg, Germany

Zeev Aizenshtat

Department of Organic Chemistry and Casali Institute of Applied Chemistry, The Hebrew University of Jerusalem, Jerusalem 91904, Israel

William C. Agosta

The Rockefeller University, 1230 York Avenue, New York, NY 10021-6399, USA

J. E. Anderson

Chemistry Department, University College London, Gower Street, London WCIE 6BT, UK

Richard F. W. Bader

Department of Chemistry, McMaster University, Hamilton, Ontario L8S 4MI, Canada

Derek V. Banthorpe

Department of Chemistry, University College London, 20 Gordon Street, London WC1H 0AJ, UK

S. W. Benson

Loker Hydrocarbon Research Institute, University of Southern California, Los Angeles, California 90089, USA

Stefan Berger

Department of Chemistry, Philipps-Universität Marburg, Hans-Meerwein-Strasse, D-3550 Marburg, Germany

James H. Brewster

Department of Chemistry, Purdue University, West Lafayette, Indiana 47907-1393, USA

N. Cohen

The Aerospace Corporation, P.O. Box 92957, Los Angeles, California 90009, USA

Robert H. Crabtree

Department of Chemistry, Yale University, P.O. Box 6666, New Haven, Connecticut 06511-8118, USA

Tino Gäumann

Institute of Physical Chemistry, Federal School of Technology, CH-1015 Lausanne, Switzerland

Lowell H. Hall

Department of Chemistry, Eastern Nazarene College, Quincy, MA 02170, USA

Contributing authors

E. Heilbronner — Grütstrasse 10, CH-8704 Herrliberg, Switzerland

A. C. Hopkinson — Department of Chemistry, York University, Downsview, Ontario M3J IP3, Canada

A. Hummel — Interfaculty Reactor Institute, Delft University of Technology, Mekelweg 15, 2629 JB Delft, The Netherlands

Marianna Kańska — Department of Chemistry, The University of Warsaw, Warsaw, Poland

Lemont B. Kier — Department of Medicinal Chemistry, Virginia Commonwealth University, Richmond, VA 23298, USA

Goverdhan Mehta — School of Chemistry, University of Hyderabad, Central University PO, Hyderabad-500 134, India

George A. Olah — Katherine B. and Donald P. Loker Hydrocarbon Research Institute and Department of Chemistry, University of Southern California, Los Angeles, CA 90089–1661, USA

T. Oppenländer — Fachhochschule Furtwangen, Abteilung Villingen-Schwenningen, Fachbereich Verfahrenstechnik, D-7730 VS-Schwenningen, Germany

Eiji Ōsawa — Department of Knowledge-Based Information Engineering, Toyohashi University of Technology, Tempaku-cho, Toyohashi 441, Japan

G. K. Surya Prakash — Katherine B. and Donald P. Loker Hydrocarbon Research Institute and Department of Chemistry, University of Southern California, Los Angeles, CA 90089-1661, USA

H. Surya Prakash Rao — Department of Chemistry, Pondicherry University, Pondicherry-605 014, India

Glenn A. Russell — Department of Chemistry, Iowa State University, Ames, Iowa 50011, USA

Hans J. Schäfer — Organisch-Chemisches Institut der Westfälischen Wilhelms-Universität, Correns-Strasse 40, 4400 Münster, Germany

G. Zang — Department of Organic Chemistry, University of Würzburg, D-8700 Würzburg, Germany

Mieczysław Zieliński — Isotope Laboratory, Faculty of Chemistry, Jagiellonian University, Cracow, Poland

Foreword

In nearly thirty years the series *The Chemistry of Functional Groups* has treated all the major groups of organic chemistry. During all this time we were aware of the omission of the treatment of the C—H group, which may be regarded as the parent of all these groups even though most authors do not include it in the list of functional groups.

For a long time we did not dare to deal with the C—H group, but in 1987 the publication of a volume on the cyclopropyl group served as a natural link to approach the 'non-functional' C—H group and the present volume is the result of this approach.

Two chapters of the originally planned list of contents did not materialize: these were on '*Combustion and pyrolysis*' and on '*Bioinorganic oxidation*'.

Again, we succeeded in securing worldwide cooperation of authors. In the present volume these include contributors from ten countries, among them from Canada, Germany, India, Israel, Japan, The Netherlands, Poland, Switzerland, the UK and USA.

The literature coverage in most chapters in until 1991.

We would be grateful to readers who would draw our attention to mistakes or omissions in this volume as well as in all other volumes of the series.

SAUL PATAI
ZVI RAPPOPORT

Jerusalem
November 1991

The Chemistry of Functional Groups
Preface to the series

The series 'The Chemistry of Functional Groups' was originally planned to cover in each volume all aspects of the chemistry of one of the important functional groups in organic chemistry. The emphasis is laid on the preparation, properties and reactions of the functional group treated and on the effects which it exerts both in the immediate vicinity of the group in question and in the whole molecule.

A voluntary restriction on the treatment of the various functional groups in these volumes is that material included in easily and generally available secondary or tertiary sources, such as Chemical Reviews, Quarterly Reviews, Organic Reactions, various 'Advances' and 'Progress' series and in textbooks (i.e. in books which are usually found in the chemical libraries of most universities and research institutes), should not, as a rule, be repeated in detail, unless it is necessary for the balanced treatment of the topic. Therefore each of the authors is asked not to give an encyclopaedic coverage of his subject, but to concentrate on the most important recent developments and mainly on material that has not been adequately covered by reviews or other secondary sources by the time of writing of the chapter, and to address himself to a reader who is assumed to be at a fairly advanced postgraduate level.

It is realized that no plan can be devised for a volume that would give a complete coverage of the field with no overlap between chapters, while at the same time preserving the readability of the text. The Editor set himself the goal of attaining reasonable coverage with moderate overlap, with a minimum of cross-references between the chapters. In this manner, sufficient freedom is given to the authors to produce readable quasi-monographic chapters.

The general plan of each volume includes the following main sections:

(a) An introductory chapter deals with the general and theoretical aspects of the group.

(b) Chapters discuss the characterization and characteristics of the functional groups, i.e. qualitative and quantitative methods of determination including chemical and physical methods, MS, UV, IR, NMR, ESR and PES—as well as activating and directive effects exerted by the group, and its basicity, acidity and complex-forming ability.

(c) One or more chapters deal with the formation of the functional group in question, either from other groups already present in the molecule or by introducing the new group directly or indirectly. This is usually followed by a description of the synthetic uses of the group, including its reactions, transformations and rearrangements.

(d) Additional chapters deal with special topics such as electrochemistry, photochemistry, radiation chemistry, thermochemistry, syntheses and uses of isotopically labelled compounds, as well as with biochemistry, pharmacology and toxicology. Whenever applicable, unique chapters relevant only to single functional groups are also included (e.g. 'Polyethers'. 'Tetraaminoethylenes' or 'Siloxanes').

This plan entails that the breadth, depth and thought-provoking nature of each chapter will differ with the views and inclinations of the authors and the presentation will necessarily be somewhat uneven. Moreover, a serious problem is caused by authors who deliver their manuscript late or not at all. In order to overcome this problem at least to some extent, some volumes may be published without giving consideration to the originally planned logical order of the chapters.

Since the beginning of the Series in 1964, two main developments occurred. The first of these is the publication of supplementary volumes which contain material relating to several kindred functional groups (Supplements A, B, C, D, E and F). The second ramification is the publication of a series of 'Updates', which contain in each volume selected and related chapters, reprinted in the original form in which they were published, together with an extensive updating of the subjects, if possible, by the authors of the original chapters. A complete list of all above mentioned volumes published to date will be found on the page opposite the inner title page of this book.

Advice or criticism regarding the plan and execution of this series will be welcomed by the Editor.

The publication of this series would never have been started, let alone continued, without the support of many persons in Israel and overseas, including colleagues, friends and family. The efficient and patient co-operation of staff members of the publisher also rendered me invaluable aid. My sincere thanks are due to all of them, especially to Professor Zvi Rappoport who, for many years, shares the work and responsibility of the editing of this Series.

The Hebrew University   SAUL PATAI
Jerusalem, Israel

Contents

Contents

List of abbreviations used

Ac	acetyl (MeCO)
acac	acetylacetone
Ad	adamantyl
Alk	alkyl
All	allyl
An	anisyl
Ar	aryl
Bz	benzoyl (C_6H_5CO)
Bu	butyl (also t-Bu or Bu^t)
CD	circular dichroism
CI	chemical ionization
CIDNP	chemically induced dynamic nuclear polarization
CNDO	complete neglect of differential overlap
Cp	η^5-cyclopentadienyl
DBU	1,8-diazabicyclo[5.4.0]undec-7-ene
DME	1,2-dimethoxyethane
DMF	N,N-dimethylformamide
DMSO	dimethyl sulphoxide
ee	enantiomeric excess
EI	electron impact
ESCA	electron spectroscopy for chemical analysis
ESR	electron spin resonance
Et	ethyl
eV	electron volt
Fc	ferrocene
FD	field desorption
FI	field ionization
FT	Fourier transform
Fu	furyl(OC_4H_5)
Hex	hexyl(C_6H_{13})
c-Hex	cyclohexyl(C_6H_{11})
HMPA	hexamethylphosphortriamide
HOMO	highest occupied molecular orbital

i-	iso
Ip	ionization potential
IR	infrared
ICR	ion cyclotron resonance

LCAO	linear combination of atomic orbitals
LDA	lithium diisopropylamide
LUMO	lowest unoccupied molecular orbital

M	metal
M	parent molecule
MCPBA	*m*-chloroperbenzoic acid
Me	methyl
MNDO	modified neglect of diatomic overlap
MS	mass spectrum

n	normal
Naph	naphthyl
NBS	*N*-bromosuccinimide
NMR	nuclear magnetic resonance

Pen	pentyl(C_5H_{11})
Pip	piperidyl($C_5H_{10}N$)
Ph	phenyl
ppm	parts per million
Pr	propyl (also *i*-Pr or Pr^i)
PTC	phase transfer catalysis
Pyr	pyridyl (C_5H_4N)

R	any radical
RT	room temperature

s-	secondary
SET	single electron transfer
SOMO	singly occupied molecular orbital

t-	tertiary
TCNE	tetracyanoethylene
THF	tetrahydrofuran
Thi	thienyl(SC_4H_3)
TMEDA	tetramethylethylene diamine
Tol	tolyl(MeC_6H_4)
Tos or Ts	tosyl(*p*-toluenesulphonyl)
Trityl	triphenylmethyl(Ph_3C)

Xyl	xylyl($Me_2C_6H_3$)

In addition, entries in the 'List of Radical Names' in *IUPAC Nomenclature of Organic Chemistry*, 1979 Edition. Pergamon Press, Oxford, 1979, p. 305–322, will also be used in their unabbreviated forms, both in the text and in formulae instead of explicitly drawn structures.

CHAPTER **1**

Atomic and group properties in the alkanes

RICHARD F. W. BADER

Department of Chemistry, McMaster University, Hamilton, Ontario, L8S 4M1, Canada

I. ATOMS IN CHEMISTRY

A. A Theory of Molecular Structure

The title of this series of books, 'Chemistry of the Functional Groups', is a mirroring of the cornerstone of experimental chemistry, that atoms and functional groupings of atoms exhibit characteristic sets of properties (static, reactive and spectroscopic) which, in general, vary between relatively narrow limits. In chemistry we recognize the presence of a group in a given system and predict its effect upon the static and dynamic properties of the system in terms of a set of properties assigned to that group. Thus, the knowledge of chemistry is classified, understood and predicted in terms of the chemistry of the functional groups. The functional group is the central concept in the working hypothesis of chemistry, that a molecule is a collection of atoms linked by a network of bonds, the molecular structure hypothesis.

The purpose of theory is to provide the conceptual framework in our pursuit of the prediction and understanding of observation. Thus it should be possible to recover from theory the concepts that make up the molecular structure hypothesis, since this hypothesis is itself a synthesis of ideas developed to account for the observations of chemistry. It has been demonstrated that quantum mechanics can be generalized to a particular class of open systems, i.e. to pieces of some total system, and that these open systems faithfully recover the characteristics that are ascribed to the chemical atom[1]. The resulting theory of atoms in molecules, since it defines the concept central to the molecular structure hypothesis, recovers as well the structural aspects of the hypothesis, namely the network of bonds which link the atoms and defines a molecular structure, together with a theory of structural stability which describes the changes in structure resulting from the making and breaking of chemical bonds[2]. This chapter describes the application of this theory of molecular structure[3] to a study of the atoms and their groupings which determine the chemistry of the alkanes.

B. Additivity and Transferability of Group Properties

The finding that the variations in group properties through certain series of molecules can be so slight as to enable one to establish a group additivity scheme played an important role in establishing the concept of the functional group in chemistry. A group additivity scheme requires not only that a molecular property be additive over the groups in a given molecule, but that the contribution from each group, and hence the group itself be transferable amongst a series of molecules. The earliest and best studied of the systems exhibiting such group additivity schemes were the alkanes, both acyclic and cyclic. The study of the molar volumes of the normal alkanes at their boiling point by Kopp in 1855[4a] provided the earliest example of the transferability and additivity of group properties. This was followed by the demonstration of characteristic and additive group contributions to the molar refraction[4b] and the related property of molar polarization[5], as made possible by the introduction of the Abbe refractometer in 1874. The molar refraction provided the first example of a measured physical property being used to establish the presence of a group in a given molecule through its additive contribution to the total property.[5]

Even a property as sensitive as the energy can exhibit additive group contributions as was once again first demonstrated in the experimentally determined heats of formation of the normal hydrocarbons[6-9]. Benson[10] has demonstrated how this and other thermodynamic properties of molecules and the changes in these properties upon chemical reaction can be usefully tabulated and predicted in terms of group contributions.

The properties of the alkanes, both acyclic and cyclic, are so well ordered and classified in terms of atomic and group contributions that the use of the theory of atoms in molecules as the basis for the theoretical discussion of alkane chemistry is most appropriate. The theory, in addition to predicting these properties from fundamental principles, leads directly to an understanding of their additive and constitutive nature.

The discussion of atomic properties will focus on the transferability of the volumes, energies and polarizabilities of the methyl and methylene groups, as these are important properties from both the historical and operational points of view. The discussion of a group's contribution to the molecular polarizability, a property important to the understanding of intermolecular interactions and chemical reactivity, requires a study of the atomic charges and first moments, as a basis for obtaining an understanding of how the atomic charge distributions respond to an externally applied field. One finds that the response of the charge distribution to an electric field is a characteristic group property, whether the field be externally applied or internally generated by a vibrational displacement of the nuclei.

Having defined the transferable methylene group one may inquire as to how its properties, in particular its energy, change when it is subjected to the geometrical constraints found in the cyclopropane molecule. It is shown that a small amount of charge is transferred from the hydrogens to carbon when the methylene group is introduced into cyclopropane, causing a correspondingly small increase in its energy, an increase which equals the measured and so-called strain energy. Through the use of the theory of atoms in molecules one may use quantum mechanics to predict not only the properties of a total system, but to predict as well the properties of the atoms which comprise it. In this manner one obtains a direct link between chemical concepts such as strain energy and the underlying physics.

On the other hand, it is at the limit of near-perfect transferability of group properties as exhibited by the alkanes that one can detrmine whether or not a proposed theory meets the requirements of experiment, for at this limit one may experimentally determine to high precision the properties of atoms in molecules via their additive contributions. By demonstrating that the atoms defined by quantum mechanics account for and recover the experimentally measured properties of atoms in molecules, one establishes that the atoms of theory are the atoms of chemistry. There is no test of theory other than the demonstration that it predicts what can be measured.

There have been many previous attempts to define certain properties of atoms in molecules, particularly atomic charges. Arbitrary definitions of an atom or of some of its properties do not account for the characteristics ascribed to the atom within the molecular structure hypothesis and hence play no operational role in the prediction and understanding of chemical observations. Aside from the obvious statement that all definitions should follow directly from quantum mechanics using only information contained in the state function, it is instructive to review what are the essential requirements imposed upon the definition of an atom in a molecule by the observations of chemistry. The reader may judge for himself the correctness of these criteria and whether any definition other than the quantum definition of an atom is able to fulfill them.

The first requirement stems from the necessity of two identical pieces of matter exhibiting identical properties, with the important understanding that the form of a substance in real space is determined by its distribution of charge. Thus two systems, macroscopic or microscopic, are identical only if they have identical charge distributions. This requires that atoms be defined in real space and that their properties be directly determined by their distribution of charge. Thus if an atom has the same form in the real space of two systems, i.e. the same distribution of charge, then it contributes the same amount to every property in both systems. This condition demands in turn that the atomic contributions $M(\Omega)$ to some property M are additive over the atoms in a molecule to yield the molecular average $\langle M \rangle$ (equation 1). The boundary of the atom,

$$\langle M \rangle = \sum_{\Omega} M(\Omega) \tag{1}$$

as defined by a partitioning which necessarily exhausts all of real space, must be such as to maximize the transferability of an atom between systems, i.e. maximize the transfer of chemical information. These three requirements must be met if one is to predict that atoms and functional groupings of atoms contribute characteristic and measurable sets of properties to every system in which they occur. It is because of the direct relationship between the spatial form of an atom and its properties that we are able to identify them in different systems. The existence of an additivity scheme is a limiting situation wherein not only equation 1, that each atom make an additive contribution to every property, must hold, but requiring in addition that the atom be transferable amongst a set of molecules without change. The definition of an atom which meets these requirements will recover those properties of atoms in molecules which are directly measurable in the limiting situation wherein the properties for a series of molecules follow an additivity relationship.

There is no known relationship between the energy and the distribution of charge in a molecular system of the form outlined above as being a requirement for a theory of atoms in molecules, although such a relationship is thought to exist[11]. Yet the atoms of theory, as is to be demonstrated in this chapter, exhibit just this requirement, a constancy in their contribution to the total energy of a system paralleling the constancy in their charge distribution. This property, together with the definition of an atom's energy which sums to the total energy of a system according to equation 1, are unique to the quantum atom. The reader is invited to consider how one might go about devising a non-arbitrary partitioning into atomic contributions of the electron–electron, electron–nuclear and nuclear–nuclear interactions, all of which contribute to the total energy of a molecule, and do so in such a way that when the atom is transferred to a second system with differing numbers of electrons and nuclei, thereby causing all of the individual contributions to the energy to change, its form and energy remain unaltered. This is the substance of chemistry. The total energy of a methyl group is in excess of twenty-five thousand kcal mol^{-1}. Yet it is an experimental fact that it can be transferred between members of the homologous series of saturated alkanes with a change in energy of less than one kcal mol^{-1}, and this in spite of changes of thousands of kcal mol^{-1} in the individual potential energy contributions to its total energy. The energy is singled out only because it is a non-trivial and fundamental property. A theory which properly accounts for the transferability of an atom's energy will so account for all properties.

It is hoped that a critical reading of this account, which describes how the theory of atoms in molecules is able to account for the properties of the alkanes in a manner which parallels the historical role which the experimental study of this series of molecules played in the development of the concept of a functional group, will convince the reader that the predictions of quantum mechanics have been extended to the realm of the atom. It is time to forego the arbitrary and subjective definitions of chemical concepts and bring the full predictive power of theory to bear on chemistry. The concepts of chemistry, of atoms and bonds, of structure and changes in structure, evolved from experimental chemistry, and they are part of the common vocabulary of chemistry. Unfortunately, while chemistry has a single vocabulary, every individual has his or her own dictionary, the term atom or bond leading to different images in different minds. Theory offers the opportunity of greatly extending the operational usefulness of this common vocabulary by providing a single dictionary, one based on physics.

C. Atoms and the Charge Distribution

For the reasons discussed above, an atom is to be defined in real space. Thus, while the theory of atoms in molecules is obtained from quantum mechanics, its vehicle of expression is the electronic charge density, the distribution function which determines an atom's form in real space. Because the charge density must exhibit a cusp at the position of a nucleus and satisfy a particular condition there, the information contained in the electronic charge density serves also to determine the positions of and the charges on the nuclei in a molecule.

Because of the dominance of the nuclear–electron force, the electronic charge distribution exhibits local maxima at the positions of the nuclei with the result that recognizable atomic forms are created within a molecular system. These forms are so dominant in determining the structure of the charge distribution that their individual properties make characteristic contributions to the properties of the total system. Thus the atomic basis for the classification of chemistry could and did evolve as the working model of chemistry before its underlying physical basis was known.

Schrödinger's state function Ψ contains all the information that can be known about a quantum system. Among the properties of a system it predicts is the distribution of charge. The quantity $\Psi^*\Psi dx_1 dx_2 \cdots dx_n$, where dx_i denotes an infinitesimal volume element $(d\tau_i)$ and the spin component for electron i, is the probability that each of the N electrons in a system will be in some particular volume element with a particular spin component. If this quantity is summed over all spins and integrated over the spatial coordinates of all electrons but those of electron number one, what remains is the probability that electron number one is in some particular volume element $d\tau_1$. That is,

$$\text{probability that electron one is in } d\tau_1 = \left\{ \int d\tau' \Psi^* \Psi \right\} d\tau_1 \qquad (2)$$

where the summation over all spins and integration over all spatial coordinates but one is denoted by the symbol $\int d\tau'$. Since Ψ is an antisymmetrized function, all electrons are equivalent and multiplication of the result given in equation 2 by N will give the total probability of finding electronic charge in $d\tau_1$. The probability per unit volume, the electronic charge density $\rho(r)$, is therefore given by

$$\rho(r) = N \int \Psi^* \Psi d\tau' \qquad (3)$$

The electronic charge density is responsible for scattering the X-rays in a crystal structure determination, a determination made possible by the charge density exhibiting its maximum values at the positions of the nuclei. Such experiments can be used to determine the charge density, averaged over the thermal motions of the nuclei. Rather than a probability, $\rho(r)$ can be and, for the present purposes, is more usefully interpreted as providing a description of the static (or nuclear motion averaged) distribution of negative charge throughout real space. The model of matter employed here is one wherein the distribution of negative charge is measurably finite over a relatively large volume of space and is most highly concentrated at the positions of the point-like nuclei which are embedded in it.

The atomic forms defined by the topology of the charge density are open systems and their boundaries, as defined in real space, satisfy the quantum condition for an open system. Thus all of the properties of an atom in a molecule or a crystal employed in the atomic classification of the properties of matter are predicted by quantum mechanics. The same topological properties of the charge density which define the atom also lead to the definition of bonds, structure and structural stability and the whole of the molecular structure hypothesis is given a basis in physics[3].

This chapter, which describes the application of the theory of atoms in molecules to the chemistry of the alkanes, gives only a curtailed account of the quantum mechanical basis of the theory, as full accounts have been presented elsewhere[1,3]. The development of the theory of molecular structure is presented in somewhat more detail and the application begins with a review of the topological features of molecular charge distributions and the associated definitions of atoms, bonds and molecular structure.

II. TOPOLOGY OF THE CHARGE DENSITY

A. The Dominant Form in the Charge Density

The dominant topological feature of the electronic charge density—that it exhibits local maxima at the positions of the nuclei—is illustrated in Figure 1a which gives a display of $\rho(r)$ for the tetrahedrane molecule, C_4H_4. The plane shown contains two carbon nuclei and their associated protons. The density exhibits a maximum at the position of a nucleus for any plane containing the nucleus. This behaviour of the charge density is to be contrasted with that exhibited at the mid-point of a C—C internuclear axis. The charge density has the appearance of a saddle at this point for the two carbon nuclei in the plane of Figure 1a, but appears as a maximum in ρ at the corresponding point between the two carbon nuclei lying above and below the plane shown in the

FIGURE 1. (a) Relief map of the electronic charge density ρ, for tetrahedrane in a plane containing two carbon and two hydrogen nuclei with an arbitrary cut-off in ρ at the positions of the carbon nuclei. The rounded maximum in the foreground lies at the mid-point of the axis between the two out-of-plane carbon nuclei. (b) Map showing the trajectories traced out by the gradient vectors of the charge density for the same plane as in (a). Most of the trajectories terminate at the positions of the nuclei whose positions are encompassed by large open circles. A set of trajectories also terminates as the critical point denoting the two-dimensional maximum in the foreground of (a). This is the bond critical point between the out-of-plane carbons, and the associated set of trajectories defines their interatomic surface. The critical point in the interior of the molecule, a cage critical point, serves as an origin for trajectories. Bond or $(3, -1)$ critical points are denoted by black dots. The pairs of trajectories which originate at each of these points and terminate at the neighbouring nuclei are drawn in heavy bold and they denote the bond paths. The pairs of bold trajectories which, in this plane, terminate at each such critical point, denote the intersections of the corresponding interatomic surfaces with the plane of the diagram. (c) Contour map of the charge density for the same plane overlaid with bond paths and the intersections of the interatomic surfaces. The contour values in atomic units (au) increase in the order 2×10^n, 4×10^n, 8×10^n with n beginning at -3 and increasing in steps of unity (open crosses mark the projected positions of out of plane nuclei)

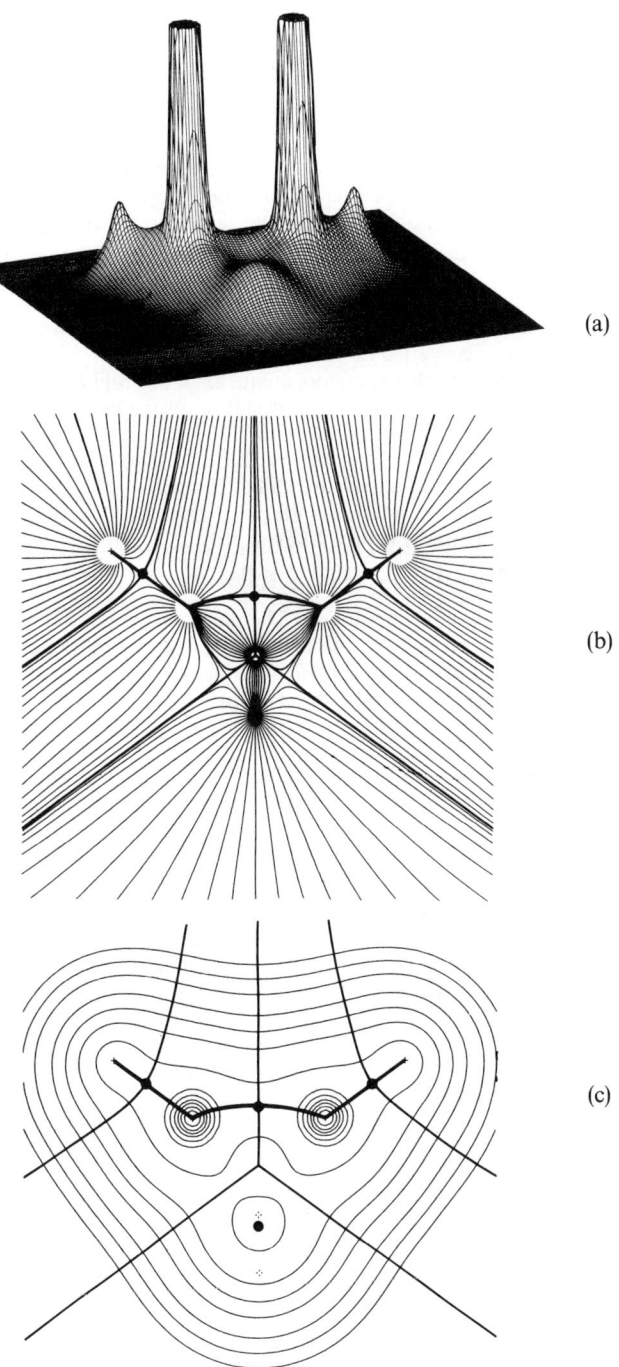

(a)

(b)

(c)

FIGURE 1. (*caption opposite*)

figure. Knowledge of ρ in one or two dimensions is insufficient to characterize its three-dimensional form. What is needed is a method for summarizing in a succinct manner the principal topological features of a charge distribution.

B. Critical Points and Their Classification

Each topological feature of $\rho(r)$, whether it be a maximum, a minimum or a saddle, has associated with it a point in space called a *critical point*, where the first derivatives of $\rho(r)$ vanish. Thus at such a point, denoted by the position vector $\mathbf{r}_c$, $\nabla\rho(\mathbf{r}_c) = 0$, where $\nabla\rho$ denotes the operation

$$\nabla\rho = \mathbf{i}\partial\rho/\partial x + \mathbf{j}\partial\rho/\partial y + \mathbf{k}\partial\rho/\partial z \tag{4}$$

Whether a function is a maximum or a minimum at an extremum is, of course, determined by the sign of its second derivative or curvature at this point.

In general, for an arbitrary choice of coordinate axes, one will encounter nine second derivatives of the form $\partial^2\rho/\partial x\partial y$ in the determination of the curvatures of ρ at a point in space. Their ordered 3×3 array is called the *Hessian matrix* of the charge density, or simply, the Hessian of ρ. This is a real, symmetric matrix and as such it can be diagonalized. The new coordinate axes are called the principal axes of curvature. The trace of the Hessian matrix, the sum of its diagonal elements, is invariant to a rotation of the coordinate system. Thus the value of the quantity $\nabla^2\rho$, called the Laplacian of ρ,

$$\nabla^2\rho = \nabla\cdot\nabla\rho = \partial^2\rho/\partial x^2 + \partial^2\rho/\partial y^2 + \partial^2\rho/\partial z^2 \tag{5}$$

is invariant to the choice of coordinate axes. The principal axes and their corresponding curvatures at a critical point in ρ are obtained as the eigenvectors and corresponding eigenvalues in the diagonalization of the Hessian matrix of ρ. Thus the pairs of names 'curvature and eigenvalue' and 'axes of curvature and eigenvectors' can be used interchangeably in describing the properties of a critical point in ρ.

While all of the eigenvalues of the Hessian matrix of ρ at a critical point are real, they may equal zero. The *rank* of a critical point, denoted by ω, is equal to the number of non-zero eigenvalues or non-zero curvatures of ρ at the critical point. The *signature*, denoted by σ, is simply the algebraic sum of the signs of the eigenvalues, i.e. of the signs of the curvatures of ρ at the critical point. The critical point is labelled by giving the duo of values (ω, σ).

With relatively few exceptions, the critical points of charge distributions for molecules at or in the neighbourhood of energetically stable geometrical configurations of the nuclei are all of rank three. The near ubiquitous occurrence of critical points with $\omega = 3$ in such cases is another general observation regarding the topological behaviour of molecular charge distributions. It is in terms of the properties of critical points with $\omega = 3$ that the elements of molecular structure are defined. A critical point with $\omega < 3$, i.e. with at least one zero curvature, is said to be degenerate. Such a critical point is unstable in the sense that a small change in the charge density, as caused by a displacement of the nuclei, causes it to either vanish or to bifurcate into a number of non-degenerate or stable ($\omega = 3$) critical points. Since structure is generic in the sense that a given structure or arrangement of bonds persists over a range of nuclear configurations, the observed limited occurrence of degenerate critical points is not surprising. One correctly anticipates that the appearance of a degenerate critical point in a molecular charge distribution denotes the onset of structural change.

There are just four possible signature values for critical points of rank three. They are:

$(3, -3)$ all curvatures are negative and ρ is a local maximum at $\mathbf{r}_c$;

$(3, -1)$ two curvatures are negative and ρ is a maximum at $\mathbf{r}_c$ in the plane defined by their corresponding axes. ρ is a minimum at $\mathbf{r}_c$ along the third axis which is perpendicular to this plane;

(3, +1)   two curvatures are positive and ρ is a minimum at $\mathbf{r}_c$ in the plane defined
            by their corresponding axes. ρ is a maximum at $\mathbf{r}_c$
            along the third axis which is perpendicular to this plane;
(3, +3)   all curvatures are positive and ρ is a local minimum at $\mathbf{r}_c$.

C. Critical Points of Molecular Charge Distributions

The coulombic potential becomes infinitely negative when an electron and a nucleus coalesce and, because of this, the state function and charge density for an atom or molecule must exhibit a cusp at a nuclear position. However, this is not a problem of practical import and the nuclear positions behave topologically as do $(3, -3)$ critical points in the charge distribution, and hereafter they will be referred to as such. The local maxima in the charge density at the positions of the nuclei are also illustrated for the cyclopropane molecule in Figure 2. The charge density exhibits a maximum at a carbon nucleus in the plane of the ring and in another perpendicular plane, showing that all three curvatures at a nuclear critical point are negative. For the two carbon nuclei in the plane of Figure 1a, or each such pair in Figure 2a, the charge density exhibits one positive and one negative curvature at the C–C critical point and it has the appearance of a saddle. For the out-of-plane carbon nuclei in C_4H_4 the corresponding critical point exhibits two negative curvatures and appears as a maximum in this one particular plane. This same property of a C–C critical point is evident in Figure 2d for C_3H_6. Thus the critical points between the carbon nuclei and between a carbon and its adjacent protons are $(3, -1)$ critical points. It is important to note that since the charge density is a maximum at a $(3, -1)$ critical point in a plane perpendicular to the internuclear axis, electronic charge is accumulated in the region between the nuclei (Figure 1a and Figure 2d). Such a critical point is found between every pair of nuclei which are considered to be linked by a chemical bond. The two saddles in the charge density in Figure 1a which link the in-plane carbon nuclei to the maximum at the $(3, -1)$ critical point between the out-of-plane carbon nuclei are $(3, +1)$ critical points, the third curvature at each such point being positive. A $(3, +1)$ critical point is found in the centre of the ring in cyclopropane. Figure 2a shows the plane containing the two positive curvatures of this critical point and the density is a minimum in the plane of the ring nuclei. Such a critical point has a unique plane or surface for which ρ is a minimum at the position of the critical point, a behaviour opposite to that of a $(3, -1)$ critical point. The plane shown in Figure 2d contains the one negative and one of the two positive curvatures of ρ at the $(3, +1)$ critical point, and it appears as a saddle in this plane. The critical point in ρ at the centre of the tetrahedrane molecule which appears as a minimum in Figure 1a is indeed a local minimum in the charge density, a $(3, +3)$ critical point.

The $(3, +1)$ and $(3, +3)$ critical points are associated with the presence of rings and cages respectively, just as the $(3, -1)$ critical points are associated with the presence of chemical bonds which link the $(3, -3)$ local maxima at the positions of the nuclei.

These qualitative associations of topological features of ρ with elements of molecular structure can be replaced with a complete theory, one which recovers all of the elements of structure in a manner that is totally independent of any information other than that contained within the charge density[2,12]. The underlying structure of the charge density is brought to the fore in its associated gradient vector field. The boundary condition of a quantum subsystem is also stated in terms of this field[13].

D. Gradient Vector Field of the Charge Density

The gradient vector field of the charge density is represented through a display of the trajectories traced out by the vector $\nabla\rho$. A trajectory of $\nabla\rho$, also called a gradient path, starting at some arbitrary point $\mathbf{r}_0$ is obtained by calculating $\nabla\rho(\mathbf{r}_0)$, moving a

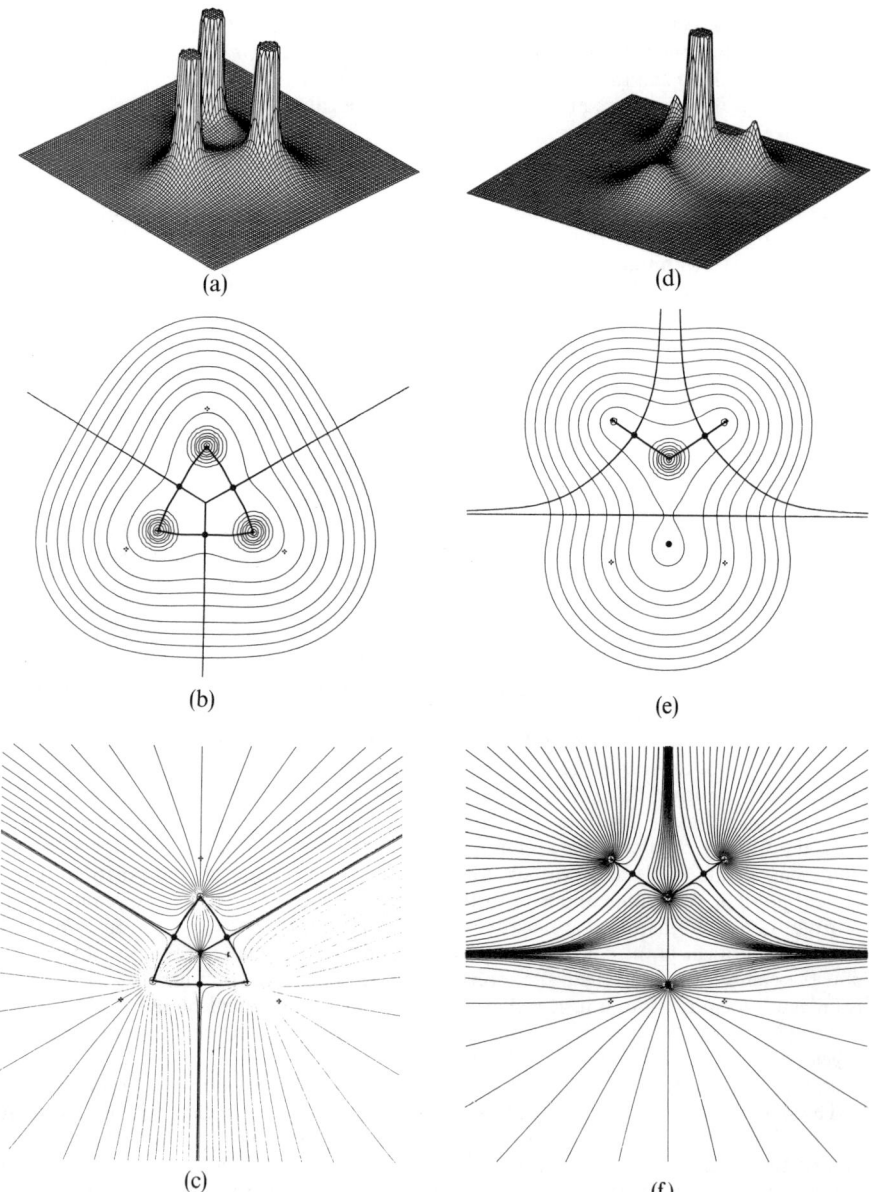

FIGURE 2. (a) and (d). Relief maps of ρ in cyclopropane in a plane containing carbon nuclei (a) and in a symmetry plane containing a set of CH_2 nuclei (d). (b) and (e). Contour maps of ρ for the same two planes. The outer contour is 0.001 au, the remainder having the values given for Figure 1. Bond paths and interatomic surfaces are indicated by bold lines and bond critical point by black dots. The lower bond critical point in (e) corresponding to the two-dimensional maximum in the foreground of (d) is the bond critical point for the two out-of-plane carbons. (c) and (f). Trajectories of the gradient vectors of the charge density for the same two planes. The trajectories which terminate at the lower bond critical point in (f) define the interatomic surface between the out-of-plane carbons

distance $\Delta\mathbf{r}$ away from this point in the direction indicated by the vector $\nabla\rho(\mathbf{r}_0)$ and then repeating this procedure until the path so generated terminates.

The gradient vector field of the charge density in tetrahedrane is exhibited in Figure 1b for the same plane as shown in Figure 1a. The gradient paths which terminate at a given nucleus are shown as ending at the boundary of a circle of arbitrary radius. Because ρ is a local maximum at each nuclear position, each nucleus or $(3, -3)$ critical point serves as the terminus of all the gradient paths starting from and contained in some neighbourhood of the nucleus and it behaves as an *attractor* in the gradient vector field of the charge density. Associated with each attractor is a *basin*, the space traversed by all the gradient paths which terminate at the attractor. Gradient vector fields of the charge density for two planes of cyclopropane are shown in Figures 2c and f. They show carbon nuclei acting as attractors in three dimensions.

Since $(3, -3)$ critical points in a many-electron charge distribution are generally found only at the positions of the nuclei, the nuclei act as the attractors of the gradient vector field of $\rho(\mathbf{r})$. The result of this identification is that the space of a molecular charge distribution, real space, is partitioned into disjoint regions, the basins, each of which contains one point attractor or nucleus. This fundamental topological property of a molecular charge distribution is illustrated in Figure 1b, and Figure 2c and f. It is to be emphasized that because ρ is a local maximum at a nucleus, a $(3, -3)$ critical point, the basin of an attractor is a region of three-dimensional space and the partitioning so clearly indicated in these Figures extends throughout all of space. *An atom, free or bound, is defined as the union of an attractor and its associated basin.*

Alternatively, an atom can be defined in terms of its boundary. The atomic basin is separated from neighbouring atoms by interatomic surfaces. The existence of an interatomic surface S_{AB} denotes the presence of a $(3, -1)$ critical point between neighbouring nuclei A and B. The presence of such a critical point between certain pairs of nuclei was noted above as being a general topological property of molecular charge distributions. Their presence now appears as providing the boundaries between the basins of neighbouring atoms. Figures 1a and 2d illustrate that the charge density is a maximum in the surface defined by the eigenvectors associated with the two negative curvatures of ρ at a $(3, -1)$ critical point. Such a critical point acts as a two-dimensional attractor since it serves as the terminus for the gradient paths which lie in this surface. This behaviour is illustrated by the critical point between the out-of-plane carbon nuclei in Figure 1b or the corresponding point in Figure 2f (the critical point at the bottom of each diagram). In each case, the plane of the diagram coincides with the surface between the out-of-plane carbon nuclei and the gradient paths which terminate at the critical point define a surface in three dimensions, the interatomic surface. The interatomic surfaces associated with each of the $(3, -1)$ critical points, which appear as saddles in Figures 1 or 2, are perpendicular to the plane of the diagram. Hence only two trajectories terminate at these critical points, as denoted by bold lines, and they mark the intersection of the interatomic surface with the plane of the diagram.

An atomic surface S_A comprises the union of a number of interatomic surfaces separating two neighbouring basins and some portions which may be infinitely distant from the attractor. The atomic surface of a carbon atom in cyclopropane, for example, consists of four interatomic surfaces, two with the other carbon atoms and two with hydrogen atoms.

If the topological property which defines an atom is also one of physical significance, then it should be possible to obtain from quantum mechanics an equivalent mechanical definition. This can be accomplished through a generalization of the quantum action principle to obtain a statement of this principle which applies equally to the total system or to an atom within the system. The generalization of the action principle to a subsystem of some total system is unique, as it applies only to a region that satisfies a particular constraint on the variation of its action integral. The constraint requires that the subsystem be bounded by a surface of zero flux in the gradient vector of the charge density as stated in equation 6. In order for the scalar product of $\mathbf{n}$, the vector normal

$$\nabla\rho(\mathbf{r})\cdot\mathbf{n}(r) = 0 \text{ for all points on the surface of an atom} \qquad (6)$$

to the surface, with $\nabla\rho$ to vanish, it is necessary that the atomic surface not be crossed by any trajectories of $\nabla\rho$ and as such it is referred to as a *zero flux surface*. A total isolated system is also bounded by a surface satisfying equation 6. Since the generalized statement of the action principle applies to any region bounded by such a surface, the zero flux surface condition places the description of the total system and the atoms which comprise it on an equal footing.

Because of the dominant topological property exhibited by a molecular charge distribution—that it exhibits local maxima only at the positions of the nuclei—the imposition of the quantum boundary condition of zero flux leads directly to the topological definition of an atom. Indeed the interatomic surfaces, along with the surfaces found at infinity, are the only closed surfaces in a three dimensional space which satisfy the zero flux surface condition of equation 6.

E. Chemical Bonds and Molecular Graphs

Reference to the gradient vector fields shown for tetrahedrane and cyclopropane illustrates that no trajectories cross from one atomic basin to another across an interatomic surface. The reader is asked to recall that $\nabla\rho$ vanishes at a critical point and thus the bold lines shown linking neighbouring nuclei are not trajectories which cross the surface, but rather pairs of trajectories each of which originates at a $(3, -1)$ critical point where $\nabla\rho = 0$. Their significance is discussed below. Another way of seeing that an interatomic surface is one of zero flux in $\nabla\rho$ is to recall that such a surface is defined by the gradient paths which terminate at a $(3, -1)$ critical point, as illustrated by the lower such point in Figure 1b or 2f. Since gradient paths do not cross, an interatomic surface must obey equation 6. This condition is equivalent to the statement that the vector $\nabla\rho(\mathbf{r})$ is tangent to its trajectory at each point $\mathbf{r}$.

Figure 1b and Figures 2c and f also show the pairs of gradient paths which originate at each $(3, -1)$ critical point and terminate at the neighbouring attractors. Each such pair of trajectories is defined by the eigenvector associated with the single positive eigenvalue of a $(3, -1)$ critical point. These two unique gradient paths define a line through the charge distribution linking the neighbouring nuclei along which $\rho(r)$ is a maximum with respect to any neighbouring line. Such a line is found between every pair of nuclei whose atomic basins share a common interatomic surface, and in the general case it is referred to as an *atomic interaction line*.

The existence of a $(3, -1)$ critical point and its associated atomic interaction line indicates that electronic charge density is accumulated between the nuclei that are so linked. This is made clear by reference to the displays of the charge density for such a critical point, as given in Figures 1a and 2d for example, which show that the charge density is a maximum in an interatomic surface at the position of the critical point. This is the point where the atomic interation line intersects the interatomic surface and charge is so accumulated between the nuclei along the length of this line. Both theory and observation concur that the accumulation of electronic charge between a pair of nuclei is a necessary condition if two atoms are to be bonded to one another. This accumulation of charge is also a sufficient condition when the forces on the nuclei are balanced and the system possesses a minimum energy equilibrium internuclear separation. Thus the presence of an atomic interaction line in such an equilibrium geometry satisfies both the necessary and sufficient conditions that the atoms be bonded to one another. In this case the line of maximum charge density linking the nuclei is called a *bond path* and the $(3, -1)$ critical point referred to as a *bond critical point*.

For a given configuration **X** of the nuclei, a *molecular graph* is defined as the union of the closures of the bond paths or atomic interaction lines. Pictorially the molecular graph is the network of bond paths linking pairs of neighbouring nuclear attractors. The molecular graph isolates the pair-wise interactions present in an assembly of atoms which dominate and characterize the properties of the system, be it at equilibrium or in a state of change.

A molecular graph is the direct result of the principal topological properties of a system's charge distribution: that the local maxima, $(3, -1)$ critical points, occur at the positions of the nuclei thereby defining the atoms, and that $(3, -1)$ critical points are found to link certain, but not all pairs of nuclei in a molecule. The network of bond paths thus obtained is found to coincide with the network generated by linking together those pairs of atoms which are assumed to be bonded to one another on the basis of chemical considerations. Molecular graphs are shown in Figure 3 for a sampling of alkanes in equilibrium geometries with widely different structures include acyclic, cyclic, bicyclic, cages and the very strained propellanes[14].

The recovery of a chemical structure in terms of a property of the system's charge distribution is a most remarkable and important result. A great deal of chemical knowledge goes into the formulation of a chemical structure and, correspondingly, the same information is successfully and succinctly summarized by such structures. The demonstration that a molecular structure can be faithfully mapped onto a molecular graph imparts new information to them—that nuclei joined by a line in the structure are linked by a line through space along which electronic charge density, the glue of chemistry, is maximally accumulated.

The small ring propellanes such as [1.1.1]propellane possess unusual structures with 'inverted' geometries at the bridgehead carbon atoms. A bond path links the bridgehead nuclei in the propellane molecules even though this results in structures that have four bond paths, all to one side of a plane, terminating at each of the bridgehead nuclei. While the model of hybridized orbitals cannot describe such a situation, even assuming bent bonds, the molecular graphs for the propellanes demonstrate that the charge distribution of a carbon atom can be so arranged as to yield bond paths—lines of maximum charge density—which correspond to an inverted structure. The propellanes and their bicyclic congeners exemplify the contrasting behaviour exhibited by the charge density between a pair of nuclei in situations where the atoms are and are not bonded to one another. Figure 4 gives relief maps of the charge density in the plane which bisects and is perpendicular to the bridgehead internuclear axis for [2.2.2]propellane and its bicyclic analogue. Such a plane contains the critical point in the interatomic surface between the bridgehead atoms. It also contains the critical point in the interatomic surface between the methylenic carbon atoms linked by a peripheral bond. The reader will recall that the charge density is a maximum at a bond critical point in the interatomic surface. Thus the diagrams for the [2.2.2] systems both exhibit three maxima in the charge density corresponding to the three peripheral bonds between the methylene groups. The purpose of the diagram is to contrast the behaviour of the charge density at the bridgehead critical point for the propellanes with that found for the bicyclic molecules. *In the propellanes which possess a bridgehead bond, the charge density is a maximum at this point, while in the bicyclic molecules which do not possess a bridgehead bond, the charge density is a minimum at this same point.* In the former molecules there is a line of maximum charge density linking the nuclei while in the latter such a line is absent and the charge density is instead a local minimum at the central critical point in the bicyclic molecules. There is, therefore, an essential, qualitative difference in the manner in which electronic charge is distributed along a line linking a pair of bonded

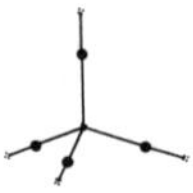

1

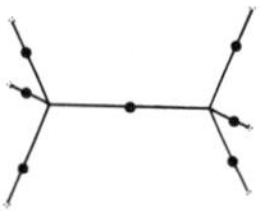

2

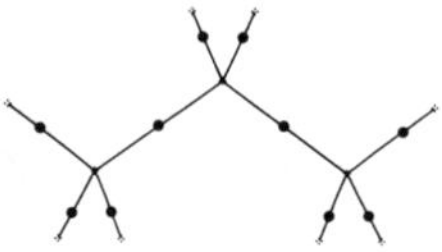

3

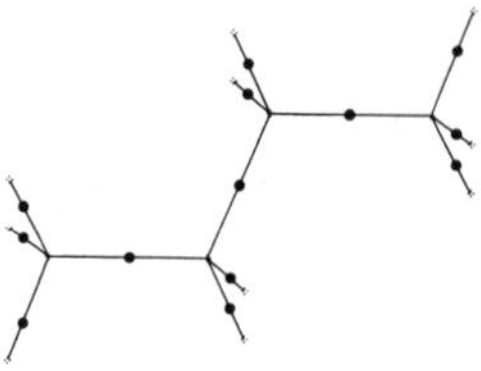

4

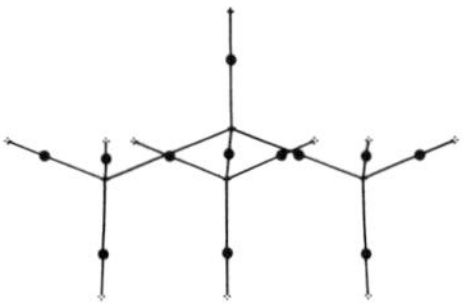

5

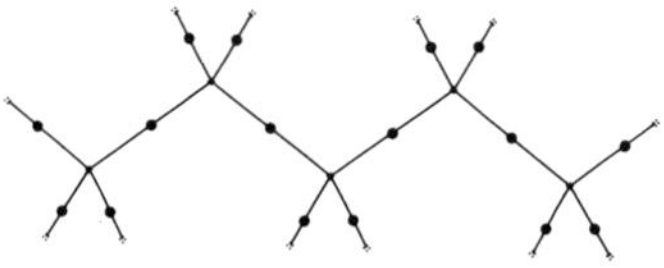

6

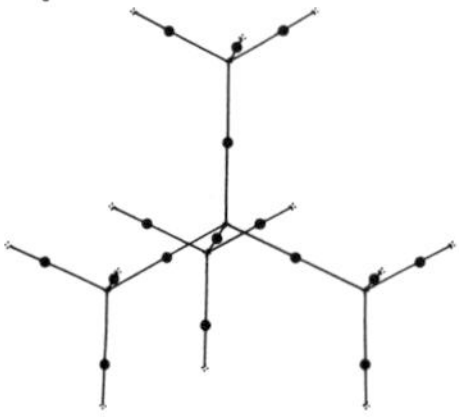

7

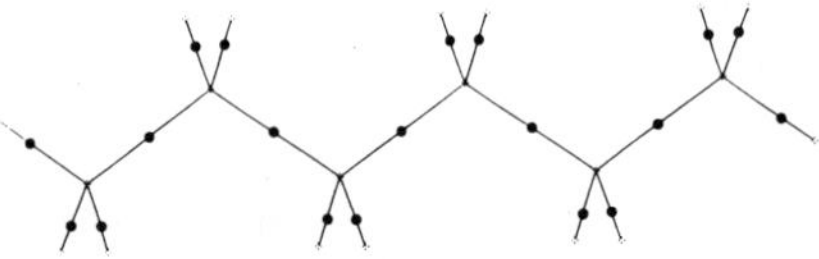

8

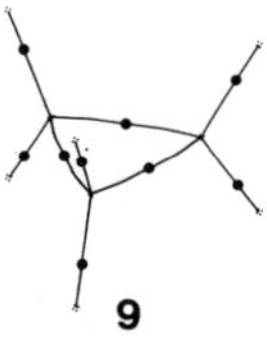

9

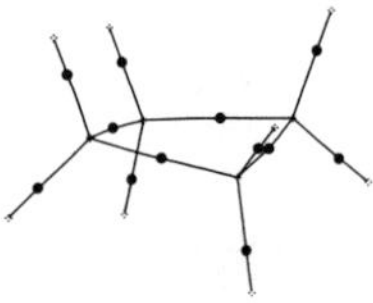

10

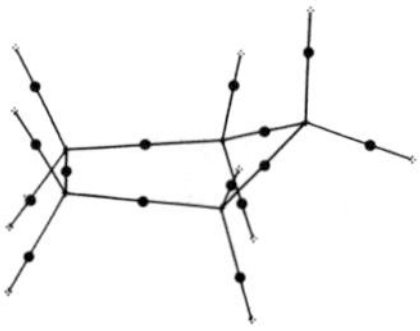

11

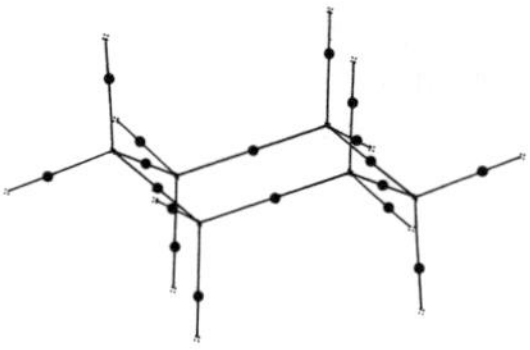

12

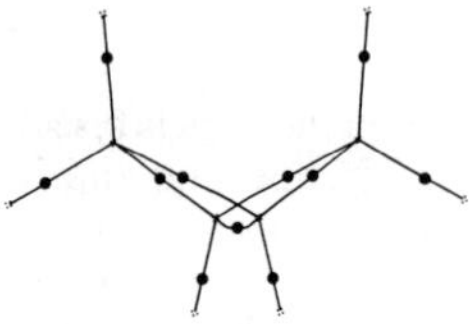

13

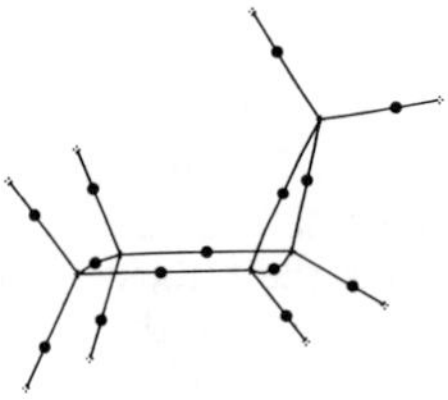

14

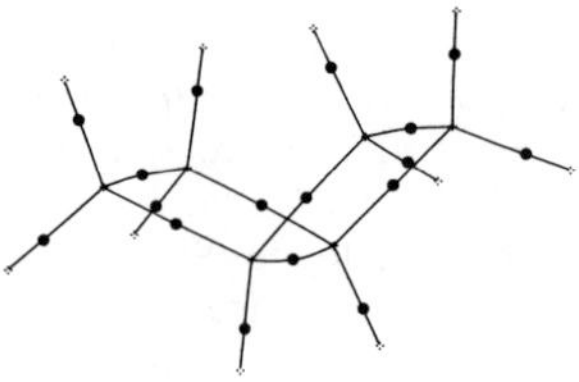

15

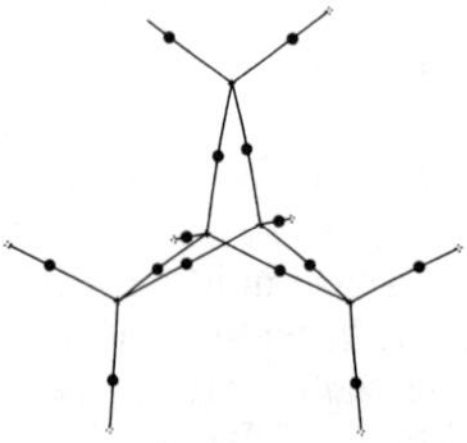

16

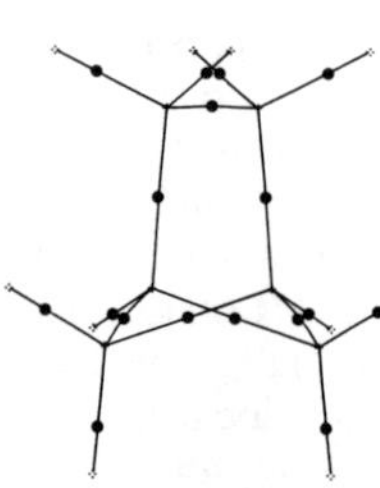

17

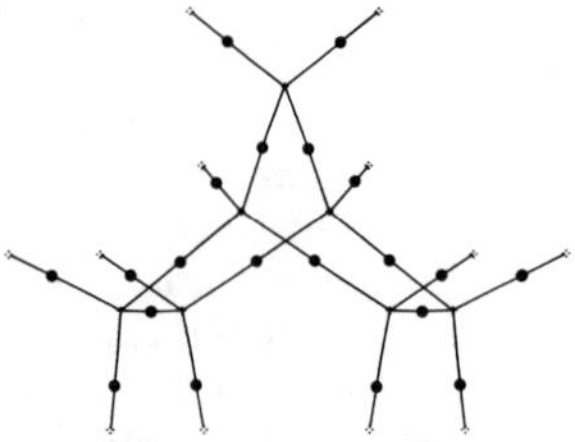

18

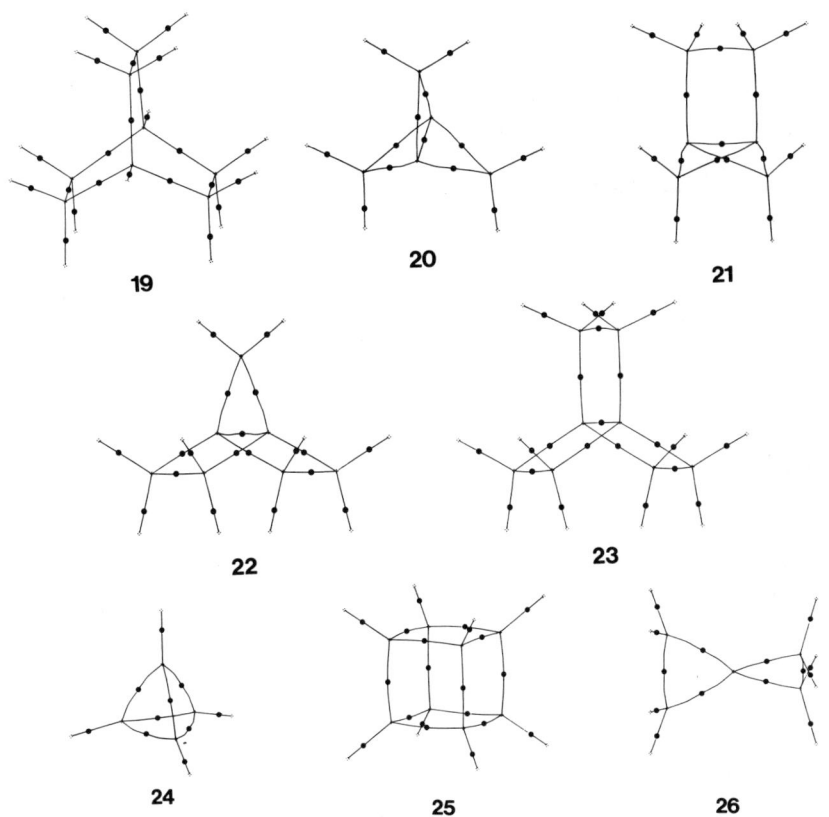

FIGURE 3. Molecular graphs for hydrocarbon molecules. The same numbering system is used in Tables 1 and 2. The bond critical points are denoted by black dots. The structures depicted in this figure are predicted by the quantum mechanical charge distributions using the theory of atoms in molecules

nuclei (as between the bridgehead nuclei in the propellanes) and along a line linking two nuclei that are not bonded (as between the nuclei in the corresponding bicyclic structures).

A calculation that includes electron correlation and yields a bridgehead internuclear separation in [1.1.1]propellane which is in agreement with the experimental value of 1.596 Å, still shows the presence of a bridgehead bond critical point and hence the presence of a bridgehead bond.

The fundamental difference in the manner in which electronic charge is distributed in the bridgehead internuclear region, between the propellanes and the corresponding bicyclic molecules, is not made apparent in density difference maps. Such deformation maps show a region of charge depletion between these nuclei for the propellanes as well as for the bicyclic molecules[15,16]. First, there is no physical basis for demanding that a density difference map in a polyatomic molecule show a charge buildup between a pair of nuclei if the nuclei are to be considered bonded to one another. Second, the reference density, in addition to being physically non-realizable, is arbitrary in its construction. Different results are obtained and different conclusions are reached depending on whether

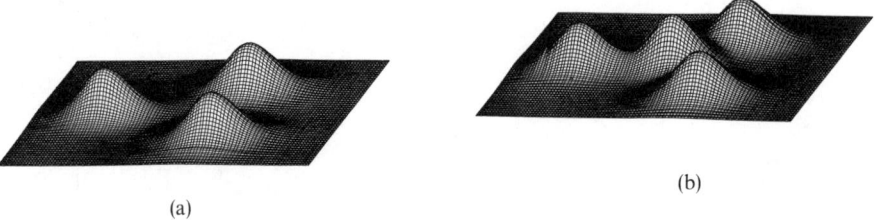

(b)

(a)

FIGURE 4. Relief maps of the charge density for bicyclo[2.2.2]octane (a) and [2.2.2]propellane (b) in the symmetry plane bisecting the bridgehead internuclear axis and the peripheral H_2C—CH_2 bonds. The three maxima for (a) correspond to the presence of an interatomic surface and hence a bond path between each of the peripheral C—C bonds. The charge density at the midpoint of the bridgehead axis is a local minimum and there is no accumulation of electronic charge between the bridgehead carbon nuclei. The map for the propellane molecule exhibits the same maxima denoting the presence of the peripheral bonds and, in addition, an even larger maximum at the midpoint of the bridgehead axis. There is an accumulation of electronic charge between the bridgehead nuclei in this molecule, they are linked by a bond path and they are bonded to one another

one employs spherical atom densities or densities of atoms in prepared valence states in the construction of the promolecule density. Third, in performing a comparison between density difference maps for pairs of molecules, as for example Jackson and Allen[16] do for [1.1.1]propellane and various cyclic and bicyclic molecules, one is actually comparing four different charge distributions. It is clear that the spherical atom promolecule density for bicyclo[1.1.1]pentane will, because of the larger bridgehead separation of 1.87 Å, subtract considerably less charge from the bridgehead symmetry plane than will be corresponding promolecule density for the equilibrium geometry of [1.1.1]propellane where the corresponding separation is calculated to be 1.54 Å. Thus their observation of an 'essential similarity' in the deformation density distributions for these two molecules, in particular the lack of a charge buildup between the bridgehead nuclei, is an artifact of the promolecule distributions and is not a reflection of the relative properties of the two charge distributions of interest. Qualitatively and of fundamental importance, the propellane molecule does accumulate charge at the bond midpoint and along the resultant bond path, while the bicyclic compound exhibits a local minimum in ρ in this same region. In addition, as demonstrated by a direct quantitative comparison of their charge distributions, the bicyclic molecule has significantly less charge in this region than does the propellane molecule. The value of $\rho(\mathbf{r})$ at the bridgehead bond critical point in [1.1.1]propellane is 0.203 au while its value at the corresponding cage critical point in bicyclo[1.1.1]pentane is 0.098 au. The corresponding values of $\rho(\mathbf{r})$ for [2.2.2] propellane and bicyclooctane are 0.288 au and 0.021 au, respectively. These essential observations are lost in a comparison of the deformation densities because the reference density for propellane removes more charge from the critical bridgehead region than does that for the bicyclic molecule. Why complicate a comparison of two distributions through the introduction of two more distributions which are arbitrary in their definition and are of no direct physical interest?

F. Rings and Cages

The remaining critical points of rank three occur as consequences of particular geometrical arrangements of bond paths and they define the remaining elements of molecular structure—rings and cages. If the bond paths are linked so as to form a ring of bonded atoms, as in cyclopropane for example, whose molecular graph is shown in Figure 3, then a (3, +1) critical point is found in the interior of the ring. As discussed above and illustrated in Figure 2c, the eigenvectors associated with the two positive eigenvalues of the Hessian matrix of ρ at this critical point generate an infinite set of gradient paths which originate at the critical point and define a surface, called the ring surface. The bond paths which form the perimeter of the ring are noticeably outwardly curved away from the geometrical perimeter of the ring, a behaviour characteristic of strained hydrocarbons. *A ring, as an element of structure, is defined as a part of a molecular graph which bounds a ring surface.*

If the bond paths are so arranged as to enclose the interior of a molecule with ring surfaces, then a (3, +3) or cage critical point is found in the interior of the resulting cage. The charge density is a local minimum at a cage critical point. The phase portrait in the vicinity of a cage critical point is shown in Figure 1 for C_4H_4, whose molecular graph is shown in Figure 3. Trajectories only originate at such a critical point and terminate at nuclei, and at bond and ring critical points, thereby defining a bounded region of space. *A cage, as the final element of molecular structure, is a part of a molecular graph which contains at least two rings, such that the union of the ring surfaces bounds a region of R^3* (i.e., three dimensional space) *which contains a (3, +3) critical point.*

G. Calculation of the Charge Density

The molecular graphs shown in Figure 3 are calculated from single determinant SCF wave functions using the 6-31G* basis at the corresponding optimized geometries, calculations denoted as 6-31G*/6-31*[17]. In a few cases, namely structures **11, 15, 17** and **22**, the 6-31G* basis was used in conjunction with the 4-31G basis in 6-31G*/4-31G calculations. These basis sets give bond lengths and bond angles for the unstrained hydrocarbons in good agreement with experiment and recover the important trends in these parameters[17]. The SCF geometries of the highly strained systems do change significantly with the addition of correlation as illustrated above for the bridgehead bond in [1.1.1]propellane which increases by 0.06 Å. The trends in geometrical parameters throughout the various series of molecules represented by structures **1** to **26** are properly described by the calculations reported here, as are the topological properties of the charge density. Even much smaller basis sets recover the topological features of the saturated hydrocarbon molecules which define the molecular graph and structure of a system.

The group additivity of the energy, polarizability and other properties is demonstrated anew here using a larger basis set than those initially used to demonstrate additivity in the hydrocarbons[18-20]. The basis used here is a Dunning [5s4p/3s] contraction[21] of the Huzinaga (11s6p/5s) set[22] for carbon, complemented with two d functions for carbon and two p functions for the hydrogens. While the absolute energies change with basis set, the differences in the group energies in changed environments change by less than $0.5\,kcal\,mol^{-1}$ between the 6-31G*, the 6-31G** bases employed previously and the larger Dunning–Huzinaga basis described above. The variations in geometrical, bond and group properties through series of molecules presented here are basis set independent.

III. STRUCTURE AND STRUCTURAL STABILITY

A. The Molecular Structure Concept

The finding that the charge distribution contains the information required to define a molecular graph for every geometrical arrangement of the nuclei allows for much more than a pedestrian recovery of the notion of structure as a set of atoms linked by a network of bonds. Instead, the dynamical behaviour of the molecular graphs as induced by a motion of the system through nuclear configuration space is shown to be describable within the framework of a relatively new field of mathematics, one which demonstrates that the notions of structure and structural stability are inseparable. The application of these ideas to the dynamical properties of the molecular graph leads to a complete theory of structure and structural stability. For a full account the reader should turn to References 2 or 3. We give here only a summary of the principal findings.

The essential idea of structure in a molecular system and its original intent, as opposed to its geometry, is that it be generic. Thus the network of bonds linking the atoms are assumed to persist over a range of nuclear displacements until some geometrical parameter attained a critical value, at which point bonds were assumed to be made and/or broken to yield a new structure. A molecular geometry, on the other hand, as well as being a classical idea associated with the concept of a potential energy surface, changes with every displacement of the nuclei. This notion of structure, including the making and breaking of chemical bonds, is simply and faithfully recovered in the changing topology of a molecular charge distribution as induced by nuclear displacements.

B. The Conflict and Bifurcation Mechanisms of Structural Change

A simple isomerization will be used to illustrate the definition of structure and its change. Consider the transfer of a methyl group from one methylene group to the other in an open structure on the potential energy surface of protonated cyclopropane:

$$CH_3—CH_2—CH_2{}^+ \longrightarrow {}^+CH_2—CH_2—CH_3 \qquad (7)$$

Consider first the process denoted by the sequence of molecular graphs in Figures 5a to 5c. Any displacement of the nuclei from the geometry depicted in the first of the molecular graphs leaves the graph unchanged; the same nuclei remain linked by the same set of bond paths. Molecular graphs with this property are said to be equivalent and such an equivalent set of molecular graphs defines a molecular structure. The structure is stable as it persists for arbitrary nuclear displacements in the neighbourhood of any geometry lying within the region of nuclear configuration space yielding an equivalent molecular graph. A structure is stable if its network of bond paths persists over arbitrary nuclear motions.

The central molecular graph depicted in Figure 5b is not equivalent to those which preceded it, nor to those which follow. It exists for just one geometry along this reaction path. It defines a new structure, one which is intermediate between the open structures characteristic of the reactants and products. In it, the carbon of the methyl group is not linked to either carbon of the methylene groups, but rather to the bond critical point of the C—C bond path. A bond or (3, −1) critical point, it will be recalled, acts as an attractor in the two-dimensional interatomic surface and in this case the surface between the two methylenic carbon atoms coincides with the symmetry plane of the molecule. For just this one geometry can the unique, downwardly directed gradient path originating at the methyl carbon critical point intersect with, i.e. become part of, the set of gradient paths forming the interatomic surface between the methylenic carbon atoms. Mathematically, such an intersection of the one- and two-dimensional manifolds of two bond critical points is unstable and, consequently, so are the molecular graph and the structure it defines. They are changed by any motion of the nuclei which destroys the plane of symmetry. The unstable structure is called a conflict structure, as it represents a state of metastable balance in the competition of the two methylenic carbon nuclei for the line of maximum electronic charge density extending from the methyl carbon.

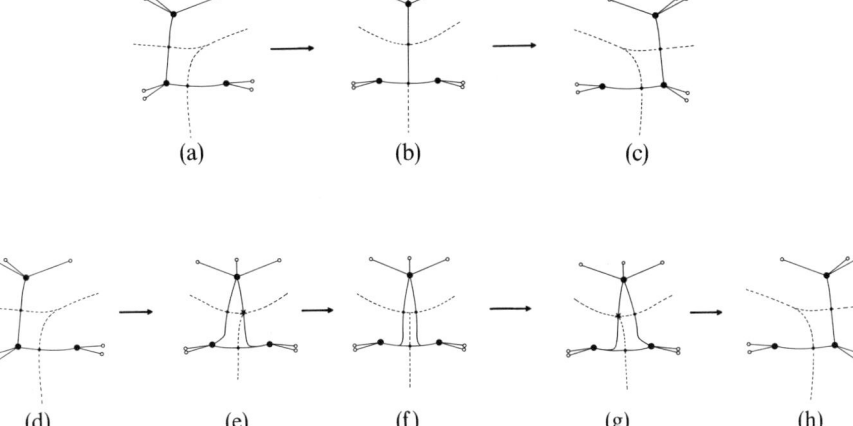

FIGURE 5. Molecular graphs (a) to (c) depict the conflict mechanism of structural change. Carbon nuclei are denoted by black circles, protons by open circles, bond critical points by dots and interatomic surfaces by dashed lines. The network of bond paths, i.e. the structure denoted by (a), is stable to nuclear motions until the motion of the methyl carbon yields the symmetrical geometry in (b). The attainment of this geometry causes an abrupt change in structure, the bond path from the methyl carbon switching from the methylenic carbon nucleus to the critical point of the H_2C—CH_2 bond. This structure is unstable and an infinitesimal continuation of the motion of Me causes the bond path to switch to the second methylenic carbon nucleus to yield the product structure. The graphs (d) to (h) illustrate the bifurcation mechanism. The unstable critical point is denoted by a star in the unstable structures (e) and (g)

The conflict mechanism, which corresponds to an abrupt switching of nuclear attractors, is one of only two possible mechanisms for bringing about a change in structure.

The other mechanism is illustrated by the second sequence of molecular graphs in Figure 5. In this case the change in structure is brought about by following a lower-energy reaction pathway. As discussed above, the first of the molecular graphs represents a stable structure as it persists for arbitrary displacements of the nuclei in the neighbourhood of its geometry. However, the interatomic surfaces of the two carbon–carbon bonds are close to merging and continued motion of the methyl group eventually results in the formation of a new critical point as represented by the star in Figure 5e. The abrupt appearance of this critical point in the charge distribution signals the formation of a bond between the methyl carbon and the second methylene group and a new structure is formed at this point along the reaction pathway. The starred critical point is unstable. Unlike the critical points of rank three which define the elements of molecular structure, this critical point is of rank two as it exhibits a zero curvature in the plane of the newly formed ring. Such a critical point is mathematically unstable to perturbations of the charge density and thus so is the structure it defines. Further infinitesimal motion along the reacton pathway causes the starred critical point to bifurcate into a ring and a bond critical point to yield the intermediate ring structure (Figure 5f), a structure which is stable to arbitrary nuclear displacements. As in the conflict mechanism, two stable structural regimes are separated by an unstable structure. Continued motion of the methyl group causes the newly formed ring critical point to migrate towards the bond critical point of the original C—C bond. In a process which is the reverse of the formation of the new C—C bond to methyl, these two critical points

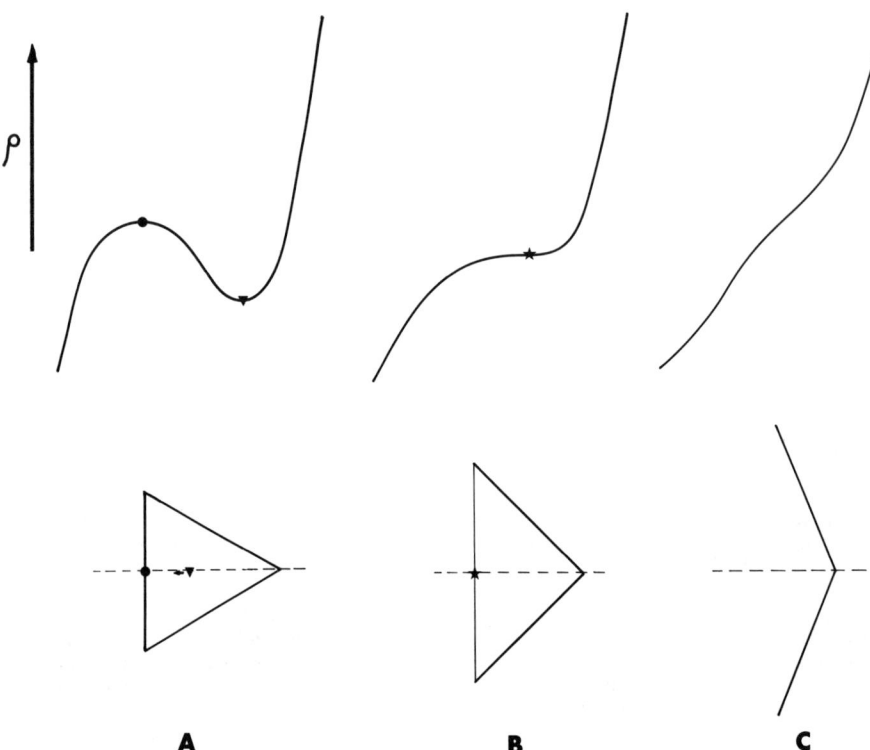

FIGURE 6. Profiles of the charge density along a C_2 symmetry axis for the opening of a ring structure, as caused by the coalescence of a bond (dot) and a ring (triangle) critical point. The profile for the ring structure in A exhibits a maximum at the bond critical point and a minimum at the ring critical point. The two critical points coalesce to yield an inflexion point in ρ and the unstable structure B. It is for this geometry of the nuclei that the ring bond is broken (or formed in the reverse reaction). The profile for the resulting open structure C exhibits a simple monotonic decrease in ρ from the position of the apex nucleus

eventually coalesce to yield an unstable critical point whose subsequent disappearance signals the breaking of the original C—C bond and the formation of the open, product structure. The annihilation of a negative curvature of the bond critical point by a positive curvature of the ring critical point to yield an unstable critical point of rank two is illustrated by the series of profiles of the charge density shown in Figure 6. This method of inducing a change in structure is called the bifurcation mechanism.

C. A Theory of Molecular Structure

A brief, quantitative summary of these ideas can now be readily given. The topological properties of a system's charge distribution enable one to assign a molecular graph to each point X in the nuclear configuration space of a system. This assignment corresponds to defining a unique network of atomic interaction lines to each molecular geometry. A gradient vector field of the charge density, the field $\nabla\rho(\mathbf{r}, \mathbf{X})$ where $\mathbf{X}$ denotes a set of nuclear coordinates, exists for every geometry X. The definition of molecular structure makes use of the mathematical device of an equivalence relation of vector fields over real, three-dimensional space R^3. The equivalence relation is defined as follows: two vector fields $\mathbf{V}$ and $\mathbf{V}'$ over R^3 are said to be equivalent if and only if the trajectories of $\mathbf{V}$ can be mapped one-to-one onto the trajectories of $\mathbf{V}'$. By applying this definition to the gradient vector fields $\nabla\rho(\mathbf{r}, X)$, one obtains an equivalence relation operating in nuclear configuration space R^Q—a space of Q dimensions. An equivalence relation is then obtained for the molecular graph defined by each $\nabla\rho(\mathbf{r}, X)$ and *a molecular structure is defined as an equivalence class of molecular graphs.*

The result of applying the equivalence relationship to the field $\nabla\rho(\mathbf{r}, X)$ is a partitioning of nuclear configuration space R^Q into a finite number of non-overlapping regions, each of which is characterized by a unique molecular structure. These structurally stable, open regions are separated by boundaries, hypersurfaces in the space R^Q. A point on a boundary possesses a structure which is different from, but transitional to, the structures characteristic of either of the regions it separates. Since a boundary is of dimension less than R^Q, arbitrary motions of the nuclei will carry a point on the boundary into neighbouring stable structural regions and its structure will undergo corresponding changes. The boundaries are the loci of the structurally unstable configurations of a system. In general, the trajectory representing the motion of a system point in R^Q will carry it from one stable structural region through a boundary to a neighbouring stable structural region. The result is an abrupt and discontinuous change in structure. A change in structure is catastrophic and for this reason the set of unstable structures is called the catastrophe set. A point in a structurally stable region of nuclear configuration space is termed a *regular point*, and a point on one of the structurally unstable boundaries is termed a *catastrophe point*. This definition of molecular structure associates a given structure with an open neighbourhood of the most probable nuclear geometry, and removes the need of invoking the Born–Oppenheimer approximation for the justification or rationalization of structure in a molecular system.

By appealing to the theorem of structural stability of Palis and Smale[23] one can show that only two kinds of structural instabilities or catastrophe points can arise and that there are therefore only two basic mechanisms for structural change in a chemical system.

Palis and Smale's theorem of structural stability when used to describe structural changes in a molecular system predicts a configuration $X \in R^Q$ to be structurally stable if $\rho(\mathbf{r}, X)$ has a finite number of critical points such that:

(a) each critical point is non-degenerate, and
(b) the stable and unstable manifolds of any pair of critical points intersect transversely.

The immediate consequence of the theorem is that a structural instability can be established through only one of two possible mechanisms which correspond to the bifurcation and conflict catastrophes. A change in molecular structure—the making and breaking of chemical bonds—can only be caused by the formation of a degenerate critical point in the electronic charge distribution or by the attainment of an unstable intersection of the submanifolds of bond and ring critical points.

IV. CLASSIFICATION OF CHEMICAL BONDS

A. Bond Order

The interaction of two atoms results in the formation of a $(3, -1)$ or bond critical point in the charge density. The trajectories of the gradient vector field of ρ which originate and terminate at this critical point define the bond path and interatomic surface, respectively. Thus the properties of the charge density at such a critical point serve to

TABLE 1. Bond properties in hydrocarbons calculated by 6-31G*/6-31G*[a]

Formula	Structure	Bond	R_e	R_b	r_A	r_B (Å)	n	ρ_b (e/a$_0^3$)	$\nabla^2\rho_b$	λ_1	λ_2	λ_3 (e/a$_0^5$)	ε	ρ_r and ρ_c (e/a$_0^3$)
a. Acyclic compounds														
CH$_4$ (1)		C—H	1.0836		0.6814	0.4022		0.2772	−0.9801	−0.7185	−0.7185	0.4569	0.0000	
C$_2$H$_6$ (2)		C—C	1.5274		0.7637	0.7637	1.00	0.2520	−0.6604	−0.4766	−0.4766	0.2928	0.0000	
		C—H	1.0856		0.6808	0.4048		0.2787	−0.9838	−0.7250	−0.7182	0.4595	0.0095	
C$_3$H$_8$ (3)		C—C	1.5283		0.7677	0.7606	1.01	0.2540	−0.6868	−0.4845	−0.4798	0.2957	0.0098	
		C1—H4	1.0858		0.6806	0.4051		0.2784	−0.9824	−0.7231	−0.7170	0.4576	0.0085	
		C1—H5	1.0866		0.6808	0.4058		0.2777	−0.9754	−0.7195	−0.7139	0.4580	0.0078	
		C2—H	1.0873		0.6806	0.4068		0.2800	−0.9886	−0.7282	−0.7222	0.4617	0.0083	
n-C$_4$H$_{10}$ (4)		C1—C2	1.5277		0.7673	0.7605	1.01	0.2541	−0.6694	−0.4842	−0.4803	0.2951	0.0082	
		C2—C3	1.5302		0.7651	0.7651	1.02	0.2555	−0.6735	−0.4906	−0.4813	0.2985	0.0192	
		C1—H5	1.0847		0.6798	0.4049		0.2790	−0.9867	−0.7250	−0.7188	0.4571	0.0086	
		C1—H6	1.0858		0.6802	0.4057		0.2782	−0.9787	−0.7211	−0.7155	0.4579	0.0079	
		C2—H	1.0880		0.6803	0.4077		0.2792	−0.9813	−0.7232	−0.7181	0.4600	0.0070	
i-C$_4$H$_{10}$ (5)		C—C	1.5302		0.7707	0.7595	1.02	0.2551	−0.6727	−0.4873	−0.4831	0.2976	0.0086	
		C1—H5	1.0874		0.6809	0.4065		0.2769	−0.9682	−0.7149	−0.7104	0.4571	0.0062	
		C1—H6	1.0861		0.6803	0.4058		0.2781	−0.9788	−0.7203	−0.7150	0.4565	0.0075	
		C2—H	1.0884		0.6804	0.4080		0.2813	−0.9945	−0.7289	−0.7289	0.4634	0.0000	
n-C$_5$H$_{12}$ (6)		C1—C2	1.5269		0.7666	0.7603	1.02	0.2546	−0.6714	−0.4853	−0.4815	0.2953	0.0079	
		C2—C3	1.5281		0.7638	0.7642	1.03	0.2565	−0.6786	−0.4926	−0.4843	0.2983	0.0017	
		C1—H6	1.0848		0.6799	0.4049		0.2789	−0.9870	−0.7250	−0.7191	0.4572	0.0082	
		C1—H7	1.0863		0.6806	0.4058		0.2779	−0.9769	−0.7201	−0.7148	0.4581	0.0074	
		C2—H	1.0885		0.6806	0.4079		0.2790	−0.9793	−0.7220	−0.7174	0.4600	0.0064	
		C3—H	1.0895		0.6806	0.4090		0.2780	−0.9703	−0.7160	−0.7125	0.4582	0.0049	
neo-C$_5$H$_{12}$ (7)		C—C	1.5332		0.7732	0.7601	1.02	0.2552	−0.6719	−0.4853	−0.4853	0.2987	0.0000	
		C—H	1.0866		0.6803	0.4063		0.2778	−0.9752	−0.7178	−0.7131	0.4559	0.0066	
n-C$_6$H$_{14}$ (8)		C1—C2	1.5284		0.7677	0.7606	1.01	0.2538	−0.6678	−0.4834	−0.4794	0.2950	0.0083	
		C2—C3	1.5298		0.7648	0.7650	1.02	0.2554	−0.6736	−0.4900	−0.4815	0.2979	0.0176	
		C3—C4	1.5296		0.7648	0.7648	1.02	0.2554	−0.6738	−0.4894	−0.4817	0.2973	0.0160	
		C1—H7	1.0857		0.6805	0.4052		0.2784	−0.9823	−0.7229	−0.7168	0.4574	0.0086	
		C1—H8	1.0864		0.6807	0.4058		0.2778	−0.9761	−0.7199	−0.7142	0.4580	0.0079	
		C2—H	1.0880		0.6803	0.4077		0.2793	−0.9818	−0.7234	−0.7185	0.4601	0.0069	

b. Cyclic compounds

Compound	Bond												ρ_r
C_3H_6 (9)	C—C	1.4974	1.5069	0.98	0.7511	0.7511	0.2490	-0.5331	-0.4892	-0.3284	0.2846	0.4896	$\rho_r = 0.2044$
	C—H	1.0759			0.6802	0.3958	0.2849	-1.0383	-0.7604	-0.7412	0.4632	0.0260	
C_4H_8 (10)	C—C	1.5485	1.5507	0.97	0.7751	0.7751	0.2467	-0.6269	-0.4616	-0.4610	0.2958	0.0014	$\rho_r = 0.0847$
	C1—H5	1.0848			0.6818	0.4030	0.2806	-1.0034	-0.7355	-0.7312	0.4633	0.0059	
	C1—H6	1.0848			0.6815	0.4034	0.2796	-0.9922	-0.7303	-0.7243	0.4624	0.0082	
C_5H_{10} (11)[a]	C1—C2	1.5377		1.00	0.7688	0.7689	0.2517	-0.6504	-0.4755	-0.4730	0.2981	0.0053	$\rho_r = 0.0359$
	C2—C3	1.5449		0.97	0.7722	0.7728	0.2482	-0.6349	-0.4672	-0.4643	0.2966	0.0063	
	C3—C4	1.5515		0.96	0.7758	0.7758	0.2451	-0.6212	-0.4596	-0.4567	0.2952	0.0064	
	C1—H6	1.0836			0.6783	0.4053	0.2813	-1.0020	-0.7314	-0.7281	0.4575	0.0046	
	C1—H7	1.0841			0.6772	0.4070	0.2812	-0.9964	-0.7299	-0.7240	0.4575	0.0081	
	C2—H8	1.0839			0.6772	0.4067	0.2816	-0.9989	-0.7320	-0.7250	0.4581	0.0096	
	C2—H9	1.0826			0.6776	0.4050	0.2820	-1.0068	-0.7343	-0.7299	0.4575	0.0061	
	C3—H10	1.0832			0.6779	0.4053	0.2820	-1.0043	-0.7348	-0.7282	0.4587	0.0090	
	C3—H11	1.0838			0.6778	0.4059	0.2819	-1.0010	-0.7341	-0.7261	0.4592	0.0109	
C_6H_{12} (12)	C—C	1.5325		1.02	0.7663	0.7663	0.2543	-0.6673	-4837	-0.4807	0.2971	0.0062	$\rho_r = 0.0176$
	C1—H7	1.0870			0.6799	0.4071	0.2798	-0.9882	-7248	-0.7210	0.4576	0.0053	
	C1—H8	1.0892			0.6803	0.4090	0.2780	-0.9700	-7161	-0.7121	0.4582	0.0057	

c. Bicyclic compounds

Compound	Bond												ρ_r
C_4H_6 (13)	C1—C3	1.4658	1.4819	0.98	0.7376	0.7376	0.2488	-0.3791	-0.3890	-0.2692	0.2792	0.4450	$\rho_r = 0.2120$
	C1—C2	1.4886	1.4966	1.01	0.7538	0.7384	0.2541	-0.5341	-0.4944	-0.3232	0.2836	0.5295	
	C1—H5	1.0697			0.6841	0.3856	0.2854	-1.0630	-0.7756	-0.7485	0.4611	0.0362	
	C2—H6	1.0781			0.6826	0.3956	0.2854	-1.0485	-0.7682	-0.7480	0.4678	0.0270	
	C2—H7	1.0832			0.6855	0.3977	0.2828	-1.0210	-0.7534	-0.7396	0.4719	0.0187	
C_5H_8 (14)	C1—C4	1.5129	1.5261	0.95	0.7602	0.7602	0.2440	-0.4877	-0.4748	-0.3059	0.2930	0.5523	$\rho_r(3) = 0.2063$
	C1—C2	1.5282	1.6504	1.02	0.7757	0.7542	0.2559	-0.6668	-0.4935	-0.4746	0.3012	0.0399	
	C1—C5	1.4935	1.5033	1.00	0.7496	0.7485	0.2528	-0.5472	-0.4973	-0.3348	0.2849	0.4856	$\rho_r(4) = 0.0872$
	C2—C3	1.5578	1.5601	0.94	0.7798	0.7798	0.2425	-0.6100	-0.4600	-0.4478	0.2979	0.0273	
	C1—H6	1.0753			0.6812	0.3941	0.2849	-1.0451	-0.7656	-0.7421	0.4626	0.0317	
	C3—H9	1.0834			0.6809	0.4025	0.2819	-1.0094	-0.7368	-0.7349	0.4623	0.0026	
	C3—H10	1.0856			0.6821	0.4036	0.2805	-0.9984	-0.7352	-0.7282	0.4649	0.0096	
	C5—H12	1.0768			0.6806	0.3962	0.2842	-1.0341	-0.7594	-0.7360	0.4614	0.0318	
	C5—H13	1.0795			0.6814	0.3980	0.2833	-1.0162	-0.7480	-0.7312	0.4629	0.0229	

TABLE 1. (continued)

Formula	Structure	Bond	R_e (Å)	R_b (Å)	r_A (Å)	r_B (Å)	n	ρ_b (e/a$_0^3$)	$\nabla^2\rho_b$	λ_1 (e/a$_0^5$)	λ_2	λ_3	ε	ρ_r and ρ_c (e/a$_0^3$)
C_6H_{10}	**(15)**[a]	C1—C4	1.5734	1.5985	0.7888	0.7888	0.92	0.2398	-0.5893	-0.4489	-0.4443	0.3040	0.0103	$\rho_r = 0.0816$
		C1—C2	1.4593	1.5629	0.7761	0.7749	0.97	0.2479	-0.6326	-0.4687	-0.4645	0.3006	0.0091	
		C1—C6	1.5524	1.5545	0.7775	0.7766	0.96	0.2466	-0.6262	-0.4654	-0.4606	0.2998	0.0104	
		C2—C3	1.5600		0.7807	0.7811	0.94	0.2418	-0.6070	-0.4546	-0.4486	0.2962	0.0133	
		C1—H7	1.0782		0.6763	0.4014		0.2851	-1.0376	-0.7492	-0.7481	0.4597	0.0015	
		C2—H8	1.0817		0.6782	0.4035		0.2822	-1.0097	-0.7368	-0.7320	0.4592	0.0066	
		C2—H9	1.0817		0.6779	0.4038		0.2821	-1.0088	-0.7380	-0.7310	0.4602	0.0095	
		C3—H10	1.0814		0.6781	0.4033		0.2824	-1.0116	-0.7382	-0.7326	0.4592	0.0077	
		C3—H11	1.0818		0.6781	0.4037		0.2820	-1.0073	-0.7373	-0.7302	0.4602	0.0097	
C_5H_8	**(16)**	C—C	1.5457	1.5507	0.7740	0.7757	0.96	0.2457	-0.5984	-0.4502	-0.4499	0.3017	0.0007	$\rho_r = 0.1020$
		C1—H6	1.0823		0.6839	0.3984		0.2826	-1.0337	-0.7511	-0.7511	0.4685	0.0000	$\rho_c = 0.0983$
		C2—H7	1.0846		0.6812	0.4034		0.2819	-1.0111	-0.7404	-0.7323	0.4616	0.0111	
C_6H_{10}	**(17)**[a]	C1—C5	1.5583	1.5615	0.7772	0.7837	0.94	0.2421	-0.5919	-0.4460	-0.4442	0.2982	0.0041	$\rho_r = 0.0468$
		C1—C2	1.5418	1.5421	0.7741	0.7679	1.00	0.2517	-0.6444	-0.4760	-0.4677	0.2994	0.0178	
		C2—C3	1.5639	1.5641	0.7821	0.7821	0.92	0.2397	-0.5933	-0.4470	-0.4392	0.2929	0.0178	$\rho_r(1,1) = 0.0916$
		C1—H7	1.0770		0.6775	0.3995		0.2865	-1.0566	-0.7597	-0.7574	0.4605	0.0031	
		C2—H8	1.0822		0.6777	0.4045		0.2827	-1.0120	-0.7366	-0.7337	0.4583	0.0039	
		C5—H13	1.0816		0.6772	0.4044		0.2826	-1.0161	-0.7369	-0.7364	0.4571	0.0007	
		C5—H14	1.0807		0.6777	0.4045		0.2827	-1.0120	-0.7366	-0.7337	0.4583	0.0039	
C_7H_{12}	**(18)**	C1—C7	1.5392	1.5399	0.7656	0.7741	1.00	0.2525	-0.6454	-0.4737	-0.4719	0.3001	0.0038	$\rho_r(5) = 0.0418$
		C1—C2	1.5425	1.5427	0.7691	0.7735	1.00	0.2517	-0.6481	-0.4753	-0.4718	0.2990	0.0074	
		C2—C3	1.5577	1.5578	0.7789	0.7789	0.94	0.2424	-0.6071	-0.4524	-0.4478	0.2930	0.0103	
		C1—H	1.0853		0.6809	0.4044		0.2815	-1.0080	-0.7336	-0.7326	0.4582	0.0015	
		C7—H9	1.0869		0.6801	0.4068		0.2789	-0.9839	-0.7202	-0.7180	0.4543	0.0030	
		C2—H9	1.0856		0.6793	0.4063		0.2806	-0.9930	-0.7278	-0.7233	0.4580	0.0062	
		C2—H10	1.0853		0.6794	0.4059		0.2806	-0.9931	-0.7278	-0.7236	0.4582	0.0058	
C_8H_{14}	**(19)**	C1—C2	1.5352	1.5353	0.7639	0.7714	1.02	0.2550	-0.6672	-0.4839	-0.4820	0.2986	0.0039	$\rho_r = 0.0216$
		C2—C3	1.5510	1.5511	0.7755	0.7755	0.96	0.2450	-0.6236	-0.4580	-0.4576	0.2921	0.0008	$\rho_c = 0.0211$
		C1—H	1.0855		0.6779	0.4076		0.2824	-1.0052	-0.7310	-0.7310	0.4569	0.0000	
		C2—H	1.0857		0.6783	0.4074		0.2806	-0.9900	-0.7264	-0.7205	0.4569	0.0082	

d. Propellanes and others

Formula	No.	Bond												ρ
C₅H₆	(20)	C1—C3	1.5430		0.7715	0.7715	0.73	0.2030	+0.0253	−0.1089	−0.1089	0.2432	0.0000	ρ = 0.1990
		C1—C2	1.5020	1.5112	0.7398	0.7632	1.00	0.2515	−0.5248	−0.4914	−0.3176	0.2843	0.5472	
		C2—H	1.0750		0.6862	0.3888		0.2898	−1.0924	−0.7864	−0.7799	0.4739	0.0084	
C₆H₈	(21)	C1—C4	1.5944	1.5955	0.7976	0.7976	0.70	0.1971	−0.0674	−0.2820	−0.0347	0.2493	7.1237	ρ(3) = 0.1961
		C1—C2	1.5468	1.5483	0.7956	0.7522	0.96	0.2461	−0.6216	−0.4666	−0.4488	0.2938	0.0396	
		C1—C5	1.4926	1.5022	0.7382	0.7558	1.01	0.2530	−0.5355	−0.4975	−0.3176	0.2796	0.5664	ρ(4) = 0.0883
		C2—C3	1.5387	1.5413	0.7704	0.7704	1.00	0.2526	−0.6555	−0.4813	−0.4747	0.3006	0.0140	
		C2—H7	1.0820		0.6819	0.4001		0.2833	−1.0217	−0.7477	−0.7396	0.4656	0.0109	
		C5—H12	1.0781		0.6877	0.3904		0.2874	−1.0685	−0.7747	−0.7690	0.4751	0.0075	
		C5—H11	1.0775		0.6857	0.3918		0.2868	−1.0661	−0.7716	−0.7665	0.4721	0.0067	
C₇H₁₀	(22)ᵃ	C1—C4	1.5343	1.5388	0.7673	0.7673	0.98	0.2482	−0.4922	−0.4914	−0.3013	0.3005	0.6312	ρ(3) = 0.2130
		C1—C2	1.5452	1.5482	0.7874	0.7600	0.95	0.2438	−0.6033	−0.4625	−0.4344	0.2936	0.0647	
		C1—C7	1.4973	1.5049	0.7490	0.7508	0.97	0.2477	−0.4959	−0.4887	−0.2816	0.2744	0.7355	ρ(4) = 0.0869
		C2—C3	1.5808	1.5832	0.7914	0.7914	0.89	0.2341	−0.5720	−0.4376	−0.4283	0.2939	0.0217	
		C2—H8	1.0819		0.6811	0.4008		0.2834	−1.0234	−0.7505	−0.7387	0.4658	0.0161	
		C2—H9	1.0804		0.6802	0.4003		0.2836	−1.0253	−0.7473	−0.7410	0.4630	0.0085	
		C7—H	1.0714		0.6798	0.3916		0.2887	−1.0749	−0.7762	−0.7636	0.4648	0.0165	
C₈H₁₂	(23)	C1—C4	1.5122		0.7561	0.7561	1.26	0.2878	−0.8303	−0.5821	−0.5821	0.3339	0.0000	ρ = 0.0861
		C1—C2	1.5508	1.5545	0.7760	0.7758	0.95	0.2441	−0.6083	−0.4615	−0.4458	0.2978	0.0381	
		C2—C3	1.5745	1.5768	0.7883	0.7883	0.90	0.2360	−0.5810	−0.4393	−0.4353	0.2936	0.0090	
		C—H	1.0859		0.6825	0.4034		0.2795	−0.9899	−0.7308	−0.7215	0.4624	0.0129	
C₄H₄	(24)	C—C	1.4634	1.4842	0.7378	0.7378	1.02	0.2551	−0.4672	−0.3813	−0.3733	0.2874	0.0213	ρ = 0.2013
		C—H	1.0635		0.6890	0.3745		0.2852	−1.0965	−0.7747	−0.7747	0.4528	0.0000	ρ = 0.1833
C₈H₈	(25)	C—C	1.5632	1.5677	0.7834	0.7834	0.95	0.2440	−0.6079	−0.4571	−0.4546	0.3038	0.0057	ρ = 0.0814
		C—H	1.0852		0.6858	0.3994		0.2786	−0.9989	−0.7307	−0.7307	0.4635	0.0000	ρ = 0.0118
C₅H₈	(26)	C1—C2	1.5127	1.5213	0.7586	0.7586	0.94	0.2425	−0.5005	−0.4722	−0.3138	0.2855	0.5049	ρ = 0.2036
		C1—C3	1.4789	1.4923	0.7369	0.7491	1.03	0.2571	−0.5719	−0.5027	−0.3623	0.2931	0.3874	
		C—H	1.0772		0.6817	0.3955		0.2839	−1.0322	−0.7578	−0.7394	0.4650	0.0249	

ᵃThe data for these structures are from a 6-31G*/4-31G calculation.

characterize the interaction. These properties include a bond order, a bond ellipticity, a bond path length and angle as well as a classification of the interaction in terms of the stresses induced in the charge density both parallel and perpendicular to the bond path. These parameters, as well as providing a physical basis for the classification of the bonding present in a molecule, enable one to translate the predicted electronic effects of the orbital-based concepts of hybridization, aromaticity, homoaromaticity, conjugation and hyperconjugation into observable consequences in the charge density. Bond properties for the hydrocarbons 1 to 26 are given in Table 1.[14]

For bonds between a given pair of atoms one may define a bond order whose value is determined by ρ_b, the value of the charge density at the bond critical point[24]. The extent of charge accumulation in the interatomic surface and along the bond path increases with the assumed number of electron pair bonds and this increase is faithfully monitored by the value of ρ_b. When applied to the ρ_b values of the C—C bonds in hydrocarbons one obtains bond orders of 1.0, 1.6, 2.0 and 3.0 for ethane, benzene, ethylene and acetylene, respectively. The value of ρ_b provides a useful measure of bonds strength for all types of bonds. In a hydrogen bond AH—BX, obtained when an acid AH binds to atom B of a base BX, the value of ρ_b is relatively small and only slightly greater than the sum of the unperturbed densities of the H and B atoms of the acid and base molecules at the degree of penetration found in the adduct. Nevertheless, the strength of the hydrogen bond is found to parallel this degree of penetration of the van der Waals envelopes of the acid and base molecules and to again increase with an increase in ρ_b. A related property is the bonded radius of an atom, r_A or r_B in Table 1, the distance from the nucleus to the associated bond critical point, a quantity closely paralleling the relative electronegativity of two bonded atoms.

For densities obtained using the 6-31G* basis, the bond order n is given by[14]

$$n = \exp[6.458(\rho_b - 0.2520)] \tag{8}$$

This definition yields values of 1.00, 1.62, 2.05 and 2.92 for the C—C bond orders in ethane, benzene, ethylene and acetylene, respectively. This empirical bond order is meant to provide a convenient measure of the extent to which electronic charge is accumulated between pairs of bonded nuclei relative to a set of standard values. In the acyclic hydrocarbons, interior bonds and bonds in the branched structures have the highest bond orders, with $n = 1.02$–1.03. Bond orders slightly less than unity are found for C—C bonds in the cyclic and bicyclic molecules. The smallest values are found for the bridgehead bonds in [1.1.1] and [2.1.1]propellanes. The bridgehead bond in [2.2.2]propellane has a relatively high value for ρ_b, yielding an order of 1.3. Bond orders are not simply correlated with bond length, and a smaller R_e does not necessarily imply a greater accumulation of charge along the bond path. For example, a C—C bond is longer in a branched alkane than in ethane, but its order is greater. The ring bonds of cyclopropane and certain of those in the three-membered rings of 13 and 14 are considerably shorter than the C—C bond in ethane but are of unit order. The [2.1.1] and [2.2.1]propellanes also have three-membered ring bonds with R_e values less than 1.50 Å, and in these very strained molecules the bond orders are less than unity.

While the values of ρ_b for the C—H bonds exhibit a much smaller variation—they are all formally of order one—there are important trends. In the acyclic alkanes the value of ρ_b increases in the order $CH_4 < CH_3 < CH_2 < CH$. The value of ρ_b is greater still for the C—H bond in cyclopropane and decreases in value as the ring size increases, exhibiting the acyclic value in cyclohexane. Still larger values are found for the C—H bonds in the more strained bicyclic structures, particularly for hydrogens bonded to bridgehead carbons. The largest values of all are found for the C—H bonds of the three-membered rings of the propellanes. The trend in the values of ρ_b parallel the increase in the percent s character of the C—H bond as discussed in a subsequent section. A C—H bond in cyclopropane has a greater percent s character, a shorter bond length, a larger stretch frequency and a larger dissociation energy than any C—H bond in an acyclic alkane, in agreement with the larger value of ρ_b in the cyclic molecule. The decrease in ρ_b with increase in ring size to the values found in cyclopentane and cyclohexane correlates with corresponding decreases in percent s character, C—H stretch frequency and bond dissociation energy. For cyclohexane itself, ρ_b is greater for equatorial than for axial hydrogen and again, the C—H stretch frequency and dissociation energy exhibit identical trends.

B. Bond Path Angle

The bond paths are noticeably curved in those structures where the geometry of the molecule precludes the possibility of tetrahedrally directed bonds between carbon nuclei. The *bond path angle* α_b, the limiting value of the angle subtended at a nucleus by two bond paths, when compared to the corresponding geometrical or so-called bond angle α_e, is important in quantifying the concept of strain in these structures. Another related quantity is the *bond path length*, R_b, which for a curved bond path exceeds the corresponding internuclear separation or so-called bond length, R_e (Table 1). The C—H bond paths do not exhibit any significant curvature.

Values of the bond path angle and the difference $\Delta\alpha = \alpha_b - \alpha_e$ are listed in Table 2 for the cyclic, bicyclic and propellane molecules together with their strain energies as calculated using Franklin's group equivalents. For most molecules $\Delta\alpha > 0$ and in these cases the bonds are less strained than the geometrical angles α_e would suggest. In general, $\Delta\alpha$ provides a measure of the degree of relaxation of the charge density away from the geometrical constraints imposed by the nuclear framework. In cyclopropane, for example, $\Delta\alpha = 18.8°$ and the deviation of the angle formed by the bond paths from the normal tetrahedral angle is correspondingly less. This difference decreases to 6.1° in cyclobutane and becomes small and negative for five- and six-membered rings. The negative value of $\Delta\alpha$ for cyclohexane brings the bond path angle to within 0.5° of the tetrahedral angle. The strain enrgy of cyclopropane is nearly the same as that for cyclobutane in spite of its bond angle being smaller by 30°. This similarity in strain energies can be accounted for, at least in part, by the much greater relaxation of angle strain between the bonds in cyclopropane than between those in cyclobutane. A quantitative discussion of the strain energies in these molecules is given in Section VII in terms of the relative energies of the carbon atoms.

The value of $\Delta\alpha$ exceeds 10° only for three-membered rings. The largest values are found for the three-membered rings in tetrahedrane, $\Delta\alpha = 21°$, and for the angle of the three-membered ring formed by the bonds terminating at the central carbon in spiropentane where $\Delta\alpha = 23°$. In spite of these large relaxations, the strain energies for these two molecules are large relative to any comparison with cyclopropane as a prototype. The reason for this is that, in tetrahedrane, each carbon is common to three three-membered rings, and in spiropentane the central carbon is common to two such rings. These additional constraints result in an increase in the strain energy and they are present to an even greater extent in bicyclo[1.1.0]butane and in the propellanes. The fusing of three rings to a common bridgehead bond inhibits the relaxation of the bond paths in these latter molecules relative to that found to occur in cyclopropane. This is particularly true for the apex angles of the three-membered rings in [1.1.1] and [2.1.1]propellane. In the former case, $\Delta\alpha$ is slightly negative and the bond path angle is less than 60°. Both angles in the three-membered ring of [2.2.1]propellane show larger relaxations than in either of the former molecules. The relatively large values of $\Delta\alpha$ for the angles made by the bond paths of the three-membered rings that terminate at a bridgehead carbon in [1.1.1] and [2.1.1]propellane result in bond path angles that differ

TABLE 2. Geometric and bond path angles[a]

Molecule	Angle	Geometric angle α_e	Bond path angle α_b	$\Delta\alpha$ $\alpha_b - \alpha_e$	Strain energy (kcal mol^{-1})
 Cyclopropane (9)	1	60.0	78.84	18.84	27.5
 Cyclobutane (10)	1	89.01	95.73	6.72	26.5
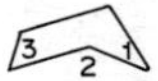 Cyclopentane (11)	1	104.55	104.02	−0.53	6.2
	2	105.36	104.43	−0.93	
	3	106.50	105.03	−1.54	
 Cyclohexane (12)	1	111.41	110.08	−1.34	0.0
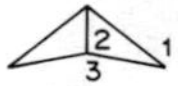 Bicyclo[1.1.0]butane (13)	1	58.99	72.78	13.79	63.9
	2	60.50	76.62	16.12	
	3	97.91	105.07	7.16	
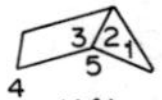 Bicyclo[2.1.0]pentane (14)	1	60.86	79.28	18.42	54.7
	2	69.57	78.30	18.72	
	3	90.84	96.76	5.92	
	4	89.16	95.77	6.61	
	5	110.03	109.72	−0.31	
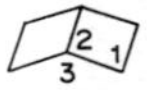 Bicyclo[2.2.0]hexane (15)	1	90.17	97.01	6.85	51.8
	2	89.91	96.59	6.67	
	3	114.43	112.64	−1.79	

Compound	Structure	Atom				
Bicyclo[1.1.1]pentane	(16)	1 2	74.44 87.20	84.72 95.85	10.27 8.65	68.0
Bicyclo[2.1.1]hexane	(17)	1 2 3 4	99.20 101.77 86.11 82.60	100.69 103.46 94.21 90.95	1.48 1.69 8.10 8.35	37.0
Bicyclo[2.2.1]heptane	(18)	1 2 3 4	94.37 101.51 108.43 103.13	97.42 102.90 108.69 103.57	3.05 1.39 0.26 0.44	14.4
Bicyclo[2.2.2]octane	(19)	1 2	109.30 109.64	108.68 108.56	−0.62 −1.08	7.4
[1.1.1]Propellane	(20)	1 2 3	61.81 95.98 59.09	59.37 107.99 69.09	−2.44 12.01 9.99	98.0
[2.1.1]Propellane	(21)	1 2 3 4 5 6	88.97 91.03 112.30 57.72 97.27 64.57	93.65 97.91 116.3 67.96 111.45 65.27	4.69 6.88 4.00 10.25 14.18 0.71	104.0
[2.2.1]Propellane	(22)	1 2 3 4 5 6	59.18 61.64 112.38 90.86 89.14 128.49	75.40 74.32 116.43 92.97 96.43 126.57	16.22 12.6 4.06 2.11 7.30 −1.91	105.0

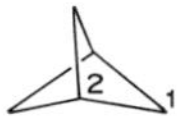

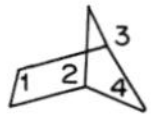

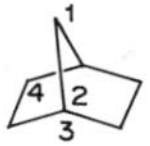

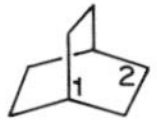

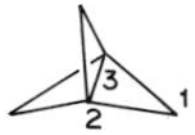

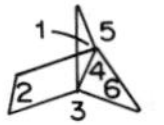

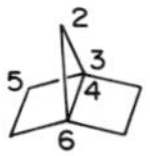

TABLE 2. (*continued*)

Molecule	Angle	Geometric angle α_e	Bond path angle α_b	$\Delta\alpha$ $\alpha_b - \alpha_e$	Strain energy (kcal mol^{-1})
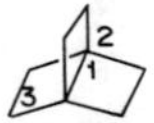 [2.2.2]Propellane (23)	1 2 3	91.15 119.96 88.85	97.05 118.51 96.81	5.90 −1.44 7.97	89.0
 Tetrahedrane (24)	1	60.00	81.36	21.36	140.0
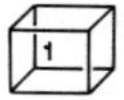 Cubane (25)	1	90.00	97.43	7.43	154.7
 Spiropentane (26)	1 2 3	59.24 61.51 137.60	79.04 84.84 123.02	19.80 23.33 −14.58	63.2

[a] Angles in degrees.

by only $\pm 1°$ from the tetrahedral angle of 109°. The corresponding bond path angles made by the bond paths from three- and four-membered rings or from two four-membered rings are larger by *ca* 8° and exhibit smaller relaxations from their geometrical angles. These bridgehead bond path angles can be compared with those in the corresponding bicyclic molecules. The value of 96° for α_b in bicyclo[1.1.1]pentane indicates a greater degree of strain for these bonds than is found in its propellane analogue.

C. Bond Ellipticity

In a bond with cyclindrical symmetry the two negative curvatures of ρ at the bond critical point are of equal magnitude. However, if electronic charge is preferentially accumulated in a given plane along the bond path (as it is for a bond with π-character, for example), then the rate of fall-off in ρ from its maximum value ρ_b in the interatomic surface is less along the axis lying in this plane than along the one perpendicular to it and the magnitude of the corresponding curvature of ρ is smaller. If λ_2 is the value of the smallest curvature, then the quantity $\varepsilon = (\lambda_1/\lambda_2 - 1)$, the *ellipticity of the bond*, provides a measure of the extent to which charge is preferentially accumulated in a given plane[24]. The axis of the curvature λ_2, the major axis, determines the relative orientation of this plane within a molecule (see Figure 7). For the 6-31G* basis set, the ellipticities of the C—C bonds in ethane, benzene and ethylene are 0.0, 0.23 and 0.45, respectively, and the major axes of the ellipticities in the latter two cases are perpendicular to the plane of the nuclei as anticipated for molecules with π bonds.

The chemistry of a three-membered ring is very much a consequence of the high concentration of charge in the interior of the ring relative to that along its bond paths[24]. The values of ρ_r, the value of ρ at a ring critical point, are generally only slightly less than, and in some cases almost equal to, the values of ρ_b for the peripheral bonds in the case of a three-membered ring. In four-membered and larger rings the values of ρ_r are considerably smaller as the geometrical distance between the bond and ring critical points is greater than in a three-membered ring. [Values of the charge density at ring (ρ_r) and cage (ρ_c) critical points are listed in Table 1.] Because electronic charge is concentrated to an appreciable extent over the entire surface of a three-membered ring, the rate of fall-off in the charge density from its maximum value along the bond towards the interior of the ring is much less than its rate of decline in directions perpendicular to the ring surface. Thus the C—C bonds have substantial ellipticities and their major axes lie in the plane of the ring. In hydrocarbons, this property is unique to three-membered ring. It accounts for their well-documented ability to act as an unsaturated system with a π-like charge distribution in the plane of the ring that is able to conjugate with a neighbouring unsaturated system[24]. This is rationalized within molecular orbital theory through the choice of a particular set of orbitals, the so-called Walsh orbitals. The ellipticity of a C—C bond in cyclopropane is actually greater than that for the double bond in ethylene, indicating that the extent to which charge is preferentially accumulated in the plane of the ring in the former compound is greater than that accumulated in the π plane of ethylene. Large ellipticities are found for bonds in three-membered rings of the bicyclic and propellane molecules. The very large value, in excess of seven, found for the bridgehead bond in [2.1.1]propellane indicates that it is potentially structurally unstable. The relationship between the value of ε and the susceptibility of a bond to rupture is nicely illustrated by the properties of the propellanes.

It was discussed in Section III that the structure of a system, as characterized by its molecular graph, is determined by the number of bond, ring and cage critical points. These are stable critical points in the sense that they retain their properties and hence the charge density retains its corresponding form as the nuclei are displaced. A structure which contains only stable critical points will thus persist over some range of all possible

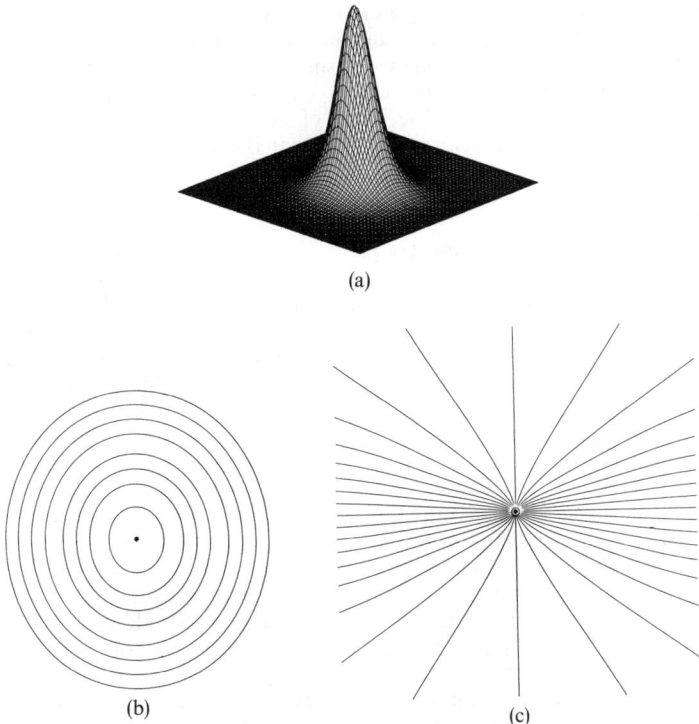

(a)

(b)   (c)

FIGURE 7. Relief map, (a) contour map (b) and associated gradient vector field map (c) of the charge density in the C—C interatomic surface of the ethylene molecule, the symmetry plane bisecting the C—C internuclear axis. The contour map illustrates the elliptical nature of the ethylenic charge distribution with the major axis perpendicular to the plane of the nuclei. This figure again emphasizes that the charge density is a maximum in the interatomic surface at the position of the bond critical point.

nuclear motions and is, therefore, a stable structure. When two or more stable critical points coalesce, however, the properties of the resulting critical point are unstable with respect to nuclear motions, as is the structure it defines. The coalescence of a bond and a ring critical point, such as occurs in the opening of a ring structure, was discussed in Section III.B. The same bifurcation mechanism is used here in a discussion of the relative stabilities of the propellanes. When a bond of a ring structure is extended, the resulting change in ρ causes the critical point of the ring to migrate towards that of the bond. The charge density has a positive curvature at the ring critical point (in the ring surface) and a negative curvature at the bond critical point along their direction of approach, as illustrated in Figure 6. The densities for the [2.1.1] and [2.2.1]propellanes, shown in Figure 8, indicate that in these two molecules the three-membered ring critical points are relatively close to the critical point of the bridgehead bond in their equilibrium geometries. The diagram for the [2.2.1] molecule, in particular, shows in profile the maximum of the bridgehead bond linked to the minimum of the ring critical point of the three-membered ring. The values of ρ at these two critical points are nearly equal

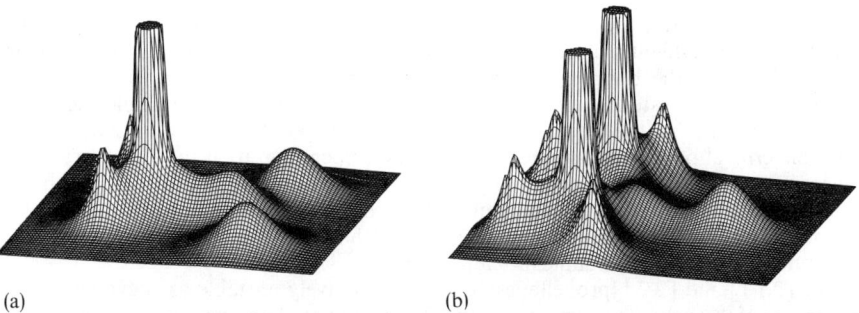

(a)   (b)

FIGURE 8. Relief maps ρ for [2.2.1] (a) and [2.1.1]propellane (b) in the symmetry plane bisecting the bridgehead internuclear axis. Each map shows the maxima at the bond critical points for the bridgehead bond and the peripheral C—C bonds, two for [2.2.1] and one for [2.1.1]propellane. Also appearing in this plane are the maxima in ρ associated with the C and H nuclei of a methylene group. The saddle between the bridgehead bond maximum and the maximum at the carbon nucleus is the critical point of the three-membered ring referred to in the text. The values of ρ at these bond and ring critical points are nearly equal and they are in close proximity. An extension of the bridgehead bond causes the positive curvature of the ring point to coalesce with and annihilate a negative curvature of the bond point (see profiles in Figure 6) and thus break the bridgehead bond

(Table 1) and they are separated by only 0.46 Å. Consequently, as is evident from the figure, λ_2, the negative curvature of the bond in the direction of the ring critical point, is smaller in magnitude than that of its perpendicular counterpart λ_1, whose magnitude is essentially unaltered from that of a normal C—C bond. Thus because of the proximity of the three-membered ring critical point, one of the negative curvatures of the bridgehead bond is decreased in magnitude and the bond has a large ellipticity.

Upon coalescence of the bond and ring critical points, which will occur for some particular extension of the bridgehead bond, the curvatures of the two critical points, one positive and the other negative, must be equal and hence the curvature of the new critical point formed by their merger must be zero. This new critical point with one zero curvature is unstable—it exists only for this particular value of the bond extension and at this extension the bond is broken. Since the magnitude of the negative curvature λ_2 decreases during the approach of the critical points and ultimately equals zero on their coalescence, the ellipticity of the bond undergoes a dramatic increase, becoming infinite when the bond is broken. Gatti and coworkers[25] have discussed the relative stability of the substituted [10]annulene versus the dinorcaradiene structures as a function of the ellipticity of the C_1—C_6 bond which links the bridgehead nuclei in the latter structures.

If the unstable critical point is formed in an isolated ring system, a further extension of the bond causes it to vanish and the ring structure is changed into an open structure, corresponding to the loss of a bond and a ring critical point as illustrated in Figure 6. The rupture of the bridgehead bond in [1.1.1]propellane transforms the structure into a cage structure similar to that of the related bicyclic compound.

The bridgehead bond in [2.1.1]propellane, with an ellipticity of 7.1, is potentially the most susceptible to rupture. As is clear from Figure 8 and the data in Table 1, the values of ρ at the bond critical point and at the two three-membered ring critical points are of almost equal value and they are separated by only 0.06 Å. The magnitude of the bond curvature λ_2 is close to zero, the value which signals the formation of an unstable structure through the breaking of the bridgehead bond. Both the bridgehead bond and a bond of the three-membered ring in [2.2.1]propellane have relatively high ellipticities.

Because of symmetry, the bridgehead bond in [1.1.1]propellane does not exhibit any ellipticity. As is evident from the data in Table 1, however, the ring critical points are only 0.25 Å from the bond critical point and their values differ by only 0.004 au. As previously demonstrated[2], an extension of the bridgehead bond in this molecule results in the coalescence of the bond and three ring critical points and the formation of an unstable critical point with two zero curvatures. Depending upon the symmetry of the nuclear displacement away from the geometry of the unstable structure, a number of stable structures, including a cage, can be formed. This is illustrated in Figure 9. The bonds of the [2.2.2]propellane molecule, since it has only four-membered rings in its structure, do not exhibit significant ellipticities.

The [2.1.1] and [2.2.1]propellanes are indeed relatively unstable as suggested by the properties of their charge distributions and they readily undergo polymerization of 50 K. The [1.1.1] molecule is the most stable of these propellanes and considerably more stable than the [2.2.2] propellane with respect to thermolysis. Stability is not a function of just the static properties of the equilibrium charge distribution, but also depends upon how the distribution changes with the possible nuclear motions. A comparison of the relative stabilities of the propellane structures from this dynamic point of view has been given in terms of a study of the charge relaxations accompanying their nuclear motions[14].

The sign and magnitude of the Laplacian of the charge density at the bond critical point, the quantity $\nabla^2\rho_b$, provide a classification of the mechanics of the atomic interactions involved in chemical bond formation[26]. This aspect of the classification of the bonds in hydrocarbon molecules is now presented following an introduction to the properties of the Laplacian of a molecular charge distribution.

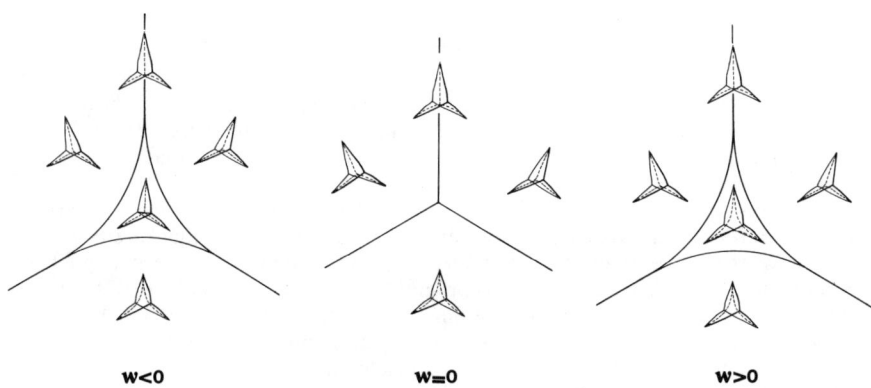

w<0   w=0   w>0

FIGURE 9. A sampling of nuclear configuration space illustrating the structures predicted and obtained by displacements from the geometry of the unstable structure of [1.1.1]propellane, resulting from the coalescence of the bridgehead bond critical point with all three ring critical points. The parameter w is a measure of the change in the bridgehead internuclear separation from its value in the unstable geometry, the geometry located at the intersection of the three axes in the case w = 0. The lines denote the partitioning of 'geometry space' into different structural regions. For w > 0, the bridgehead bond is broken and a cage structural region results, bounded by symmetrically equivalent open double-ring structures. For w < 0, the system enters the propellane structural region. Any geometry lying on a partitioning surface is unstable, as exemplified by the conflict structures in all three cases. Such a structure results from an unstable intersection of the two manifolds of two ring critical points

V. LAPLACIAN OF THE CHARGE DENSITY

A. The Physical Basis of the Electron Pair Model

Cenral to the models of molecular geometry and chemical reactivity is the concept of localized pairs of electrons, bonded and non-bonded[27]. In general, aside from core regions, electrons are not spatially localized[28]. Neither the properties of the pair density nor the topology of the total charge distribution offer evidence of localized pairs of electrons, bonded or non-bonded. One may arbitrarily select a set of orbitals which individually show some degree of spatial localization. However, only the total density has physical meaning, and every set of orbital densities for a system, no matter how the orbitals are chosen, sum to the same total density. What is indeed remarkable, is that when one sums the individual orbital densities, their nodes and all suggestions of spatially localized patterns of charge dissolve to yield the relatively simple topology exhibited by the total charge density. This topology, as discussed above, provides a faithful mapping of the chemical concepts of atoms, bonds and structure, but it must be recognized that it does not provide any indication of maxima that would correspond to the electron pairs of the Lewis model. The physical basis of this most important of all chemical models is one level of abstraction above the visible topology of ρ and appears instead in the topology of the Laplacian of ρ[29].

The Laplacian of ρ is the sum of its three principal curvatures at each point in space, the quantity $\nabla^2\rho(\mathbf{r})$ (cf equation 5). When $\nabla^2\rho(\mathbf{r}) < 0$, the value of the charge density at the point $\mathbf{r}$ is greater than the value of $\rho(\mathbf{r})$ averaged over all neighbouring points in space, and when $\nabla^2\rho(\mathbf{r}) > 0$, $\rho(\mathbf{r})$ is less than this averaged value. Thus a maximum (a minimum) in $-\nabla^2\rho(\mathbf{r})$ means that electronic charge is locally concentrated (locally depleted) in that region of space even though the charge density itself exhibits no corresponding maximum (minimum). One may think of $\nabla^2\rho$ as providing a measure of the extent to which the charge density is locally compressed or expanded. This property of ρ must be distinguished from local maxima and minima in ρ itself.

The Laplacian of ρ recovers the shell structure of an atom by displaying a corresponding number of shells of charge concentration and charge depletion[26]. For a spherical atom, the outer or valence shell of charge concentration, the VSCC, contains a sphere over whose surface elecctronic charge is maximally and uniformly concentrated. Upon entering into chemical combination, this valence shell charge concentration is distorted and maxima and saddles appear on the surface of the sphere[29]. The maxima correspond in number, location and size to the localized pairs of electrons assumed in the Lewis model[27] and in the extension of this model, as contained in the VSEPR model of molecular geometry[30,31]. All of the properties postulated in the VSEPR model for bonded and non-bonded pairs of electrons are recovered by the maxima in the VSCC for the central atom and the Laplacian of the charge density thus provides a physical basis for this most successful of models of molecular geometry[32,33]. The shell structure and the maxima and minima in the VSCC of the carbons in cyclopropane and [1.1.1]propellane are illustrated in Figure 10.

The Lewis model encompasses chemical reactivity as well, through the concept of a generalized acid–base reaction. Complementary to the local maxima or charge concentrations in the VSCC of an atom for the discussion of reactivity are its local minima or charge depletions. A local charge concentration is a Lewis base or nucleophile, while a local charge depletion is a Lewis acid or electrophile. A chemical reaction corresponds to the combination of a 'lump' in the VSCC of the base with a 'hole' in the VSCC of the acid[29]. The positions of the local charge concentrations and depletions together with their magnitudes are determined by the positions of the corresponding critical points in the VSCCs of the respective base and acid atoms. This information

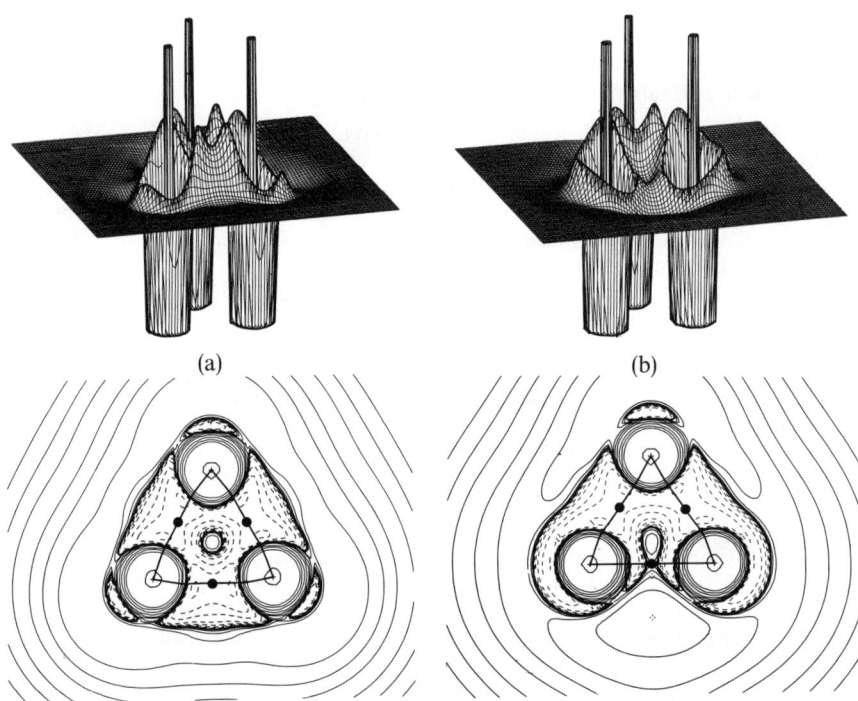

FIGURE 10. Relief maps of the negative of the Laplacian of the charge density for a plane containing a three-membered ring in cyclopropane (a) and in [1.1.1]propellane (b). Also shown are contour maps overlaid with bond paths of the same function for the same planes. The spike-like charge concentration at each carbon nucleus has been terminated at -2.0 au. Dashed contours denote negative values of $\nabla^2 \rho$, regions of charge concentration, while solid contours denote positive values, regions of charge depletion. The values of $\nabla^2 \rho$ at the bond critical points are given in Table 1. The value at the non-bonded charge concentration on a bridgehead carbon in propellane is -0.40 au. What appears to be a corresponding non-bonded maximum on each carbon in cyclopropane is in fact not a local maximum (three negative curvatures) but rather a $(3, -1)$ critical point (two negative curvatures and one positive curvature) in the negative of the Laplacian

enables one to predict the positions of attack within a molecule and the relative geometry of approach of the reactants. The properties of the Laplacian have been used to predict the approach of nucleophiles to keto, amidic and ethylenic carbon atoms, to predict the preferred sites of electrophilic attack in substituted benzenes and to locate the points of attachment and resulting geometries of hydrogen bonded complexes[29,34]. The propensity of a given site through a series of molecules to electrophilic or nucleophilic attack has been shown to be proportional to the size of the charge concentration or to the extent of charge depletion, respectively[34].

The Laplacian of ρ plays a dominant role in the theory of atoms in molecules[3]. The quantum condition of zero flux in the gradient vector of ρ that defines the atom, when integrated over the surface of an atom Ω, yields a constraint on the atomic average of the Laplacian of ρ:

$$\oint dS \nabla \rho \cdot \mathbf{n} = \int_\Omega \nabla^2 \rho \, d\tau = 0 \qquad (9)$$

According to equation 9, if charge is concentrated in some regions of an atom it must be depleted to a corresponding extent in others, since the integral of the Laplacian over an atom must vanish. This property is common to all atoms, free or bound. The Laplacian appears in the local expression for the virial theorem:

$$(\hbar^2/4m)\nabla^2\rho(\mathbf{r}) = \mathscr{V}(\mathbf{r}) + 2G(\mathbf{r}) \tag{10}$$

The quantity $\mathscr{V}(\mathbf{r})$ when integrated over an atom or over all space yields the corresponding value of the potential energy and is, therefore, the potential energy density. Correspondingly, $G(\mathbf{r})$ is the kinetic energy density since its integration over an atom or over all space yields the kinetic energy T. Since the integral of the Laplacian of ρ over an atom Ω or over the total system vanishes, corresponding integrations of equation 10 yield the virial theorem for an atom, $\mathscr{V}(\Omega) + 2T(\Omega) = 0$, or for the total system, $\mathscr{V} + 2T = 0$. Since $G(\mathbf{r}) > 0$ and $\mathscr{V}(\mathbf{r}) < 0$, equation 10 demonstrates that the lowering of the potential energy dominates the energy in those regions of space where electronic charge is concentrated or compressed, i.e. where $\nabla^2\rho < 0$.

B. Characterization of Atomic Interactions

The charge density is a maximum in an interatomic surface at the bond critical point (Figure 7). The two curvatures of ρ perpendicular to the bond path at the bond critical point, λ_1 and λ_2 in Table 1, are therefore negative and charge is concentrated locally in the surface at this point. The charge density is a minimum at the same point along the bond path, the third curvature of ρ at the bond critical point λ_3 being positive. Charge is locally depleted at the critical point with respect to neighbouring points on the bond path. Thus the formation of an interatomic surface and a chemical bond is the result of a competition between the perpendicular contractions of ρ which lead to a contraction or compression of charge along the bond path, and the parallel expansion of ρ which leads to its depletion in the surface and to its separate concentration in the basins of the neighbouring nuclei. The sign of $\nabla^2\rho$ at the bond critical point determines which of the two competing effects is dominant.

When $\nabla^2\rho_b < 0$ and large in magnitude, the perpendicular contractions of ρ dominate the interaction and electronic charge is concentrated between the nuclei along the bond path. The result is a sharing of electronic charge between the atoms as is found in covalent or polar bonds. This is the situation for all of the C—H and all of the C—C bonds of the hydrocarbons with the exception of the bridgehead bonds in [2.1.1] and [1.1.1]propellane. Figure 11a for ethane or Figure 10a for cyclopropane illustrates the manner in which the valence electronic charge is concentrated between the nuclei in shared interactions, a sharing which bridges the basins of the neighbouring atoms. According to equation 10, these bonds achieve their stability through the lowering of the potential energy because of the electronic charge that is concentrated between the nuclei and shared by both atoms.

The value of $\nabla^2\rho_b$ for the C—H bonds parallels ρ_b in its behaviour, becoming more negative as the value of ρ_b increases. The relatively large value of ρ_b found for H bonded to a geometrically strained carbon as in cyclopropane, tetrahedrane or [1.1.1]propellane is a result of an increase in the degree of contraction of the charge density in the interatomic surface towards the bond path.

The values of $\nabla^2\rho_b$ and its individual components show only minor variations for the majority of the C—C bonds. The magnitude of $\nabla^2\rho_b$ for bonds with significant ellipticities, as found in cyclopropane and in three-membered rings of bicyclic structures, is reduced as a result of the softening of one of the negative curvatures. As discussed above, electronic charge is delocalized over the ring surface as a consequence of this softening. A comparison of the contour maps of $\nabla^2\rho$ for a four-membered ring and that for

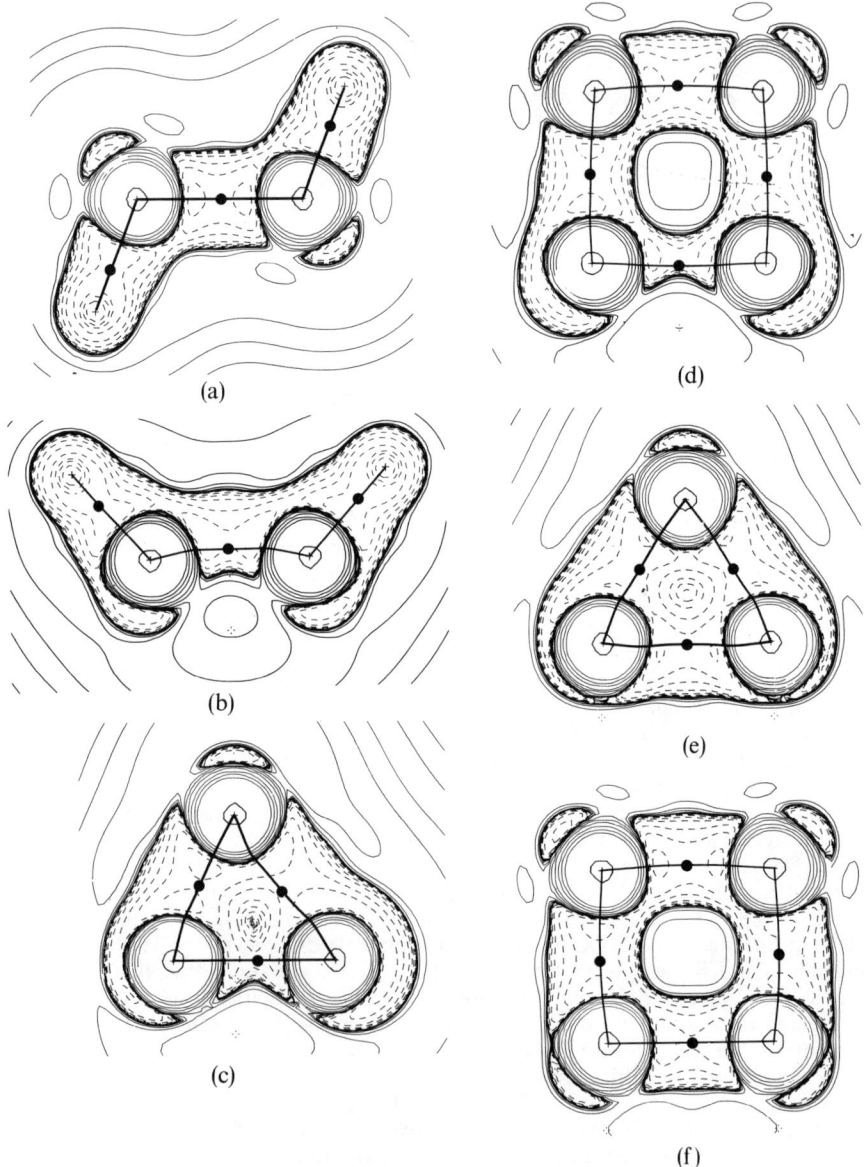

FIGURE 11. Contour maps of the Laplacian of the charge density overlaid with bond paths for the (a) H—C—C—H plane of ethane, (b) H—C—C—H symmetry plane containing the bridgehead bond in bicyclo[1.1.0]butane, (c) distorted geometry of [1.1.1]propellane in which the bridgehead nuclei and the bridging methylene groups are displaced in opposite directions, (d) plane of the four-membered ring in [2.1.1]propellane, (e) plane of the three-membered ring in [2.2.1]propellane, and (f) plane of the four-membered ring in [2.2.2]propellane. The values of $\nabla^2 \rho_b$ are given in Table 1. The value of the bridgehead non-bonded charge concentration in (d) is -0.51 and in (b) is -0.37

cyclopropane (Figure 10 and 11) shows that, in the latter, all three nuclei are bound by what is essentially a common charge concentration in the ring interior as opposed to the former case, where the dominant concentrations are localized separately along the bonds. Thus, the significant ellipticities of the perimeter bonds in cyclopropane lead to a delocalization of charge over the surface of the ring. The extent of this delocalization and the accompanying concentration of charge in the ring interior are reflected in the outward curvature of the C—C bond paths and in their unusually short bond lengths. The concentration of charge exerts attractive forces on all three ring nuclei and simultaneously lowers their potential energy. Therefore, while the small bond angles of a three-membered ring result in strain, the amount of strain energy is less than anticipated on this basis alone because of the presence of the π-like delocalized charge over the ring surface and its sharing by all of the ring nuclei unique to a three-membered ring.

It has been shown that the enhanced electrophilic reactivity of cyclopropanes over cyclopentanes and similar hydrocarbons is not related to the release of strain energy which accompanies the rupture of a C—C bond on acetolysis[35]. Cyclopropane and cyclobutane have similar strain energies and enthalpies of acetolysis, yet the latter is essentially inert while the former has a moderate reactivity. The susceptibility of cyclopropanes to protonation has the same origin as that for an olefin—bonds with large ellipticities and resulting concentrations of less tightly bound electronic charge, π-density in an olefin and in-plane delocalized charge for the three-membered rings.

When $\nabla^2 \rho_b > 0$, one has the other limiting type of atomic interaction—an interaction dominated by contractions of the charge density towards each of the nuclei as reflected in the dominance of the positive curvature of ρ, λ_3 along the bond path. They are called closed-shell interactions as they typify interactions between closed-shell atoms as found in noble gas repulsive states, ionic bonds, hydrogen bonds, van der Waals molecules and in the relatively long bonds found between what are formally closed-shell atoms in compounds such as S_4N_4 and S_8^{2+}. The dominance of the positive curvature of ρ along the atomic interaction line is a consequence of the Pauli exclusion principle which leads to a separate concentration of charge in each atomic basin. The electronic energy of these systems is lowered by the charge concentrated in the basin of each atom. One also observes a progressive change from an interaction dominated by the perpendicular contractions to one dominated by the parallel contraction of ρ for the bond between identical but progressively more electronegative atoms. Thus $\nabla^2 \rho_b$ is negative for the C—C bond in ethane and for the N—N bond in H_2NNH_2, less negative for the O—O bond in HOOH and positive for F—F. The positive, parallel curvature λ_3 increases dramatically in this isoelectronic series from 0.29 in ethane to 1.88 au in F_2. The charge density of a bound fluorine atom is both tightly bound and strongly localized within its boundaries. This is a favourable trait for an ionic interaction, but the antithesis of that required for a shared homopolar interacton, and the F_2 bond is relatively weak. The charge density of each fluorine atom in F_2 is polarized towards the other nucleus and the molecule is bound by the net accumulation of charge in the binding region of each atom.

The trend in $\nabla^2 \rho_b$ towards positive values for bonds between atoms of increasing electronegativity parallels the observed decrease in the amount of charge accumulated in the binding regions of such bonds as determined by a map of the deformation density, $\Delta \rho$[36]. Thus $\Delta \rho > 0$ in the binding region of a normal C—C bond where the interaction is dominated by a shared accumulation of charge between the nuclei. Its value becomes smaller and eventually negative as the bonded atoms become more electronegative and the interaction is increasingly dominated by the separate localization of charge in each of the atomic basins, as, for example, in bonds between oxygen atoms. Dunitz and Seiler[37] also ascribe this trend in the behaviour of $\Delta \rho$ in the binding region to the increasing importance of the exclusion principle and its effect of separately localizing charge on each of the atoms. It is found that an experimental density difference map does not show a build-up of charge between the bridgehead nuclei in a [3.1.1]propellane[15] and, in agreement with the above observation, the value of $\nabla^2\rho_b$ is also positive or close to zero for the bridgehead bond in a propellane with at least two three-membered rings (Table 1).

The bridgehead bonds in the propellanes exhibit the greatest variation in properties of the bonds encountered in this study (Figures 10 and 11). The value of $\nabla^2\rho_b$ at the bridgehead bond critical point is negative and large in magnitude for [2.2.2]propellane, as anticipated for a bond of order greater than unity. This bond differs from a normal bond with $n > 1$ in that it possesses the largest value for the positive curvature of ρ, λ_3. Usually, this measure of the curvature of ρ along the bond path decreases—the curvature is softened—as more charge is accumulated between the nuclei. Thus there is a relatively large tension in the charge density along the bridgehead bond path tending to separately concentrate the density in the two atomic basins. The value of $\nabla^2\rho_b$ for the bridgehead bond becomes increasingly less negative through the series as the number of three-membered rings increases. Its value is close to zero in [2.1.1]propellane and slightly positive in [1.1.1]propellane. The bridgehead bonds in these two molecules are intermediate between a shared interaction and one dominated by the separate localization of charge in each of the atomic basins. The resulting pattern of charge concentration and removal is different from that for a normal C—C bond and the chemistry of the bridgehead carbon atoms are altered correspondingly.

C. The Laplacian and Reactivity

The valence shell of charge concentration exhibits local maxima equal in number to the number of bonded and non-bonded Lewis pairs and their relative orientation and sizes are as anticipated on the basis of the VSEPR model of molecular geometry. Carbon in methane, for example, has four tetrahedrally directed maxima, nitrogen in ammonia has three bonded maxima and one larger, broader non-bonded maximum while oxygen in water has two bonded and two non-bonded maxima. This parallelism between the model of bonded and non-bonded pairs and the local maxima in the valence shell of charge concentration as defined by the Laplacian of the charge density is always found[33]. Thus the chlorine atom in ClF_3 has three bonded maxima, two axial and one equatorial, and two larger, broader equatorially placed non-bonded maxima as anticipated on the basis of its T-shaped geometry.

In addition to the maxima, as outlined above, local minima exist on the surface of the sphere of charge concentration in the valence shell of a bound atom. The maxima are linked by lines originating at intervening saddle points in $\nabla^2\rho$, similar to the bond paths as determined by the topology of ρ itself. In the case of a carbon atom in methane or ethane, the result is a tetrahedron with curved faces. In the centre of each face and opposite to a bonded maximum there is a local minimum in $\nabla^2\rho$. This arrangement of maxima linked by lines which encompass local minima is called an atomic graph. The relief and contour maps for the plane of the carbon nuclei in cyclopropane (Figure 10a) show two of the four bonded maxima on each carbon where $\nabla^2\rho$ is very negative, one for each C—C bond, and opposite each is a minimum where $\nabla^2\rho$ is positive. Although slightly distorted from a regular tetrahedron, this pattern is similar to that for a carbon in ethane (Figure 11a).

In [2.2.2]propellane the four bonded maxima of a bridgehead carbon atom lie approximately on the corresponding internuclear axes. While the atomic graph of this atom is distorted from the regular tetrahedral arrangement, the basic pattern survives. The most noticeable difference is that $\nabla^2\rho$ is very slightly negative, rather than positive, at the centre of the face opposite the bridgehead bond. Because the arrangement of the four bonded maxima on the bridgehead carbon in [1.1.1]propellane reflects its inverted geometry, the structure and nature of the atomic graph for this atom is greatly altered from the regular pattern as can been seen by comparing its Laplacian plot (Figure 10b) with that for cyclopropane (Figure 10a). The shift of all four bonded maxima to one side of a plane causes the three saddle points whose lines define the face opposite the bridgehead bond to coalesce with the local minimum at the face centre to produce a broad local maximum in $-\nabla^2\rho$—a non-bonded concentration of charge. The magnitude of the Laplacian is greater at this poins in [1.1.1]propellane than it is for the bonded maximum directed at the opposing bridgehead carbon. Thus the valence shell of a bridgehead carbon atom in [1.1.1]propellane exhibits five local charge concentrations, four bonded and one non-bonded. The exposed bridgehead non-bonded charge concentration is a site of electrophilic attack. Its ease of protonation explains the pronounced susceptibility of [1.1.1]propellane to acetolysis[38]. A display of the Laplacian of ρ for [1.1.1]propellane that has been displaced in the manner of an antisymmetric stretch of the carbon skeletal framework is shown in Figure 11c. This is a very intense low-frequency motion and is the most facile of the framework motions for this molecule. Its effect is to shorten the bridgehead bond by concentrating more charge between these nuclei and to greatly enhance the non-bonded concentration of charge on a bridgehead carbon. Thus this dynamic property of the charge density further enhances the reactivity of [1.1.1]propellane towards electrophiles.

The bridgehead carbons in [2.1.1]propellane and bicyclo[1.1.0]butane also exhibit inverted geometries and they possess a relatively large charge concentration as opposed to the normal depletion, in the centre of the face opposite the bridgehead bond [Figures 11b and d)]. The principal concentration of charge in the non-bonded regions of these carbons does not lie on the bridgehead axis as it does in [1.1.1]propellane. While not as accessible, these charge concentrations serve as centres of electrophilic attack and [2.1.1]propellane should be, and bicyclo[1.1.0]butane is, susceptible to acetolysis[39]. This concentration is gone in [2.2.1]propellane, which approaches the more regular pattern found in [2.2.2]propellane.

The presence of the non-bonded charge concentrations on the bridgehead carbons of [1.1.1]propellane (Figure 10b) does not imply that the net forces exerted on these nuclei by the bridgehead density are antibinding rather than binding. The dipolar polarization of a bridgehead carbon atom in [1.1.1]propellane is large and is directed away from the second nucleus, being dominated by the diffuse concentration of non-bonded charge. The force exerted on the bridgehead nucleus by its own charge density is, however, binding, being dominated by density closer to the nucleus[14]. This force is directed at the other bridgehead nucleus to which it is linked by a bond path. This pattern of atomic moments—dipolar directed away from and electric field directed towards the bonded nucleus—is not unique to the bridgehead carbons but is the normal pattern for the atoms in C_2, N_2, O_2 and F_2[40]. The largest atomic moments in this set are for the nitrogen atom which possesses an axial 'lone pair' in N_2 and the strongest bond. Thus the bridgehead carbons in [1.1.1]propellane differ not in kind but only in the degree of the polarizations, the dipole being larger and the binding force smaller in magnitude than those found for atoms in more strongly bound systems. In [2.2.2]propellane the dipolar moment is greatly reduced in magnitude and the binding force is slightly larger, whilst the bridgehead carbons do not exhibit the special properties they do in[1.1.1]propellane.

VI. ATOMIC PROPERTIES

A. Quantum Mechanics of an Atom in a Molecule

It is said that not since the time of Berzelius has it been possible for one mind to grasp all of the knowledge of chemistry. Today, not only is one forced to focus one's endeavours on just a portion of the broad spectrum of chemical knowledge and observation, one has the added responsibility of attempting to understand what physics has to say about accounting for the observations of chemistry. Faced with the impossible task of mastering all relevant knowledge, one is forced to concentrate on concepts. Readers of this chapter may be unfamiliar with the reformulation of quantum mechanics by Feynman and Schwinger which occurred around 1950. It is their work which enables one to extend the predictions of quantum mechanics to an atom in a molecule and this application of their ideas provides the opportunity to introduce such readers to the concepts involved.

In 1926 Schrödinger derived his wave equation by requiring that the value of a certain integral be a minimum[41]. This integral expressed the total energy of a system as an average over a 'wave function' and the minimization was carried out under the condition that the wave function remain normalized to unity. Denoting the integral and its constraint by $G[\Psi]$, Schrödinger's equation

$$H\Psi = E\Psi \tag{11}$$

is obtained by demanding that the condition

$$\delta G[\Psi] = 0 \tag{12}$$

be fulfilled, i.e. that the function Ψ be such that the first-order change in the value of $G[\Psi]$ caused by small variation in Ψ be zero. Using equation 11 one easily obtains the so-called hypervirial theorem

$$\langle \Psi | [H, F] | \Psi \rangle = 0 \tag{13}$$

which states that the average of the commutator of the Hamiltonian operator H with any other operator F equals zero. Through the proper choice of F one may derive the important theorems of quantum mechanics, such as the Ehrenfest relations, the virial theorem and, with a slight modification, the Hellmann–Feynman theorem. These equations suffice to determine the physics of a molecule in a stationary state.

In 1948 and 1951 respectively, Feynman[42] and Schwinger[43] each gave a new formulation of quantum mechanics. Both formulations, which were later shown to be equivalent, are based upon novel generalizations of the principle of least action. This principle, which has a long history in physics and can be used as the basis for classical mechanics, states that the passage of a system from one time to another is predicated by the requirement that a quantity known as the 'action' be minimized. Both of the new formulations are built upon parallelisms between quantum and classical mechanics first pointed out by Dirac. In Schwinger's case, this led to his extension of the action principle through the introduction of 'generators of infinitesimal unitary transformations'. Through the action of such generators, which are operators that act on the state function or on another operator, one may describe all possible changes in a system. Through their use Schwinger was able to demonstrate that all of quantum mechanics can be obtained from a single principle, the quantum action principle. This principle yields Schrödinger's equation which describes the time evolution of the system, it defines the observables of a quantum mechanical system together with their equations of motions, the Heisenberg equation, and it predicts the commutation relationships.

We are interested in molecules in stationary states and in this case Schwinger's principle takes on a particularly simple form. In this instance, it yields Schrödinger's equation for a stationary state, equation 11, and equation 14 for the variation in Schrödinger's energy functional $G[\Psi]$,

$$\delta G[F\Psi] = \langle \Psi | [H, F] | \Psi \rangle = 0 \tag{14}$$

where now the operator F denotes the generator which causes the infinitesimal change, the variation, in the wave function Ψ. That is, one identifies the mathematical variations in Ψ with the action of operators on Ψ and in this manner one defnes the observables, their average values and their equations of motion which, for a stationary state, reduce to the hypervirial theorem (equation 13), Equation 14 is in effect a variational derivation of the hypervirial theorem and from it one obtains a complete descripton of the properties of a stationary state.

It is through a generalization of Schwinger's principle that one obtains a prediction of the properties of an atom in a molecule[3]. The generalization is possible only if the atom is defined to be a region of space bounded by a surface which satisfies the zero flux boundary condition, a condition repeated here as equation 15,

$$\nabla \rho(\mathbf{r}) \cdot \mathbf{n}(r) = 0 \text{ for all points on the surface of the atom } \Omega \qquad (15)$$

By averaging over an atom in a molecule in a stationary state in the manner determined by the generalization of Schwinger's principle, Schrödinger's energy integral $G[\Psi]$ becomes the atomic integral $G[\Psi, \Omega]$, and Schwinger's principle, equation 14, assumes the form of equation 16:

$$\delta G[F\Psi, \Omega] = \tfrac{1}{2}\{\langle \Psi | [H, F] | \Psi \rangle_\Omega + \text{complex conjugate}\} \qquad (16)$$

where the subscript Ω implies that the commutator is averaged over the atom. Equation 16 is identical in form and content to Schwinger's principle for a total system in a stationary state (equation 14), but as presented in equation 16 it applies to any region of space bounded by a zero flux surface (equation 15). This includes the total system whose boundary is at infinity, as a special case. When Ω refers to the total system, the variation $\delta G[F\Psi, \Omega]$ equals the variation $\delta G[F\Psi]$ which vanishes, and equation 16 reduces to equation 14. The boundary of an atom in a molecule, however, is comprised of surfaces with neighbouring atoms and in this case the commutator average in equation 16 does not, in general, vanish. It is instead given by the flux in J_F, the current of the generator F, through the surface $S(\Omega)$ of the atom, a necessary consequence of an atom being an open system, one which can exchange charge and momentum across its boundaries with neighbouring atoms. Equation 16 determines the mechanics of an atom in a molecule and predicts its properties, just as equation 14 does for the total system.

B. Definition and Calculation of Atomic Properties

For a particular choice of the generator F, equations 14 and 16 both yield statements of the virial theorem. The virial theorem for an atom in a molecule obtained from equation 16 states that twice the average kinetic energy of the electrons in atom Ω, $T(\Omega)$, equals the negative of the virial of the forces exerted on the electrons, $\mathscr{V}(\Omega)$:

$$2T(\Omega) = -\mathscr{V}(\Omega) \qquad (17)$$

This equation is identical in form to the expression for the total system:

$$2T = -\mathscr{V} \qquad (18)$$

The virial $\mathscr{V}(\Omega)$ equals the potential energy of the electrons, and the electronic energy of an atom in a molecule is accordingly defined by

$$E_e(\Omega) = T(\Omega) + \mathscr{V}(\Omega) \qquad (19)$$

an energy which satisfies the various statements of the virial theorem:

$$E_e(\Omega) = -T(\Omega) \quad \text{and} \quad E_e(\Omega) = \tfrac{1}{2}\mathscr{V}(\Omega) \qquad (20)$$

Like all atomic properties, summation of $E_e(\Omega)$, $T(\Omega)$ or $\mathscr{V}(\Omega)$ over all the atoms in a molecule yields the corresponding molecular average. This is a direct consequence of the mode of definition of the action integral for an atom in the extension of Schwinger's principle to an open sysem. When the molecule is in an equilibrium geometry and no forces are exerted on the nuclei, the electronic virial $\mathscr{V}$ is identical to the usual total potential energy of the molecule and in this case the sum of the atomic energies equals E, the total energy of the molecule:

$$E = \sum_\Omega E_e(\Omega) \tag{21}$$

Another important result governing the mechanics of an atom in a molecule is obtained from equation 16 when the operator F is set equal to the momentum for an electron. The result in this case is an expression for the force acting on the electrons in an atom, the Ehrenfest force, a force not to be confused with the Hellmann–Feynmann force acting on a nucleus. The expression for the Ehrenfest force is equivalent to having Newton's equation of motion for an atom in a molecule, as it determines all of the mechanical properties of the atom. The force $F(\Omega)$ is determined entirely by the pressure acting on the surface of the atom $S(\Omega)$,

$$F(\Omega) = -\oint dS(\Omega)\sigma(\mathbf{r})\cdot\mathbf{n}(\mathbf{r}) \tag{22}$$

where the pressure, or force per unit area, is determined by the quantum mechanical stress tensor σ. What is important about this result is that the atomic surface $S(\Omega)$ is composed of a number of interatomic surfaces, there being one such surface for each atom linked to the atom in question by a bond path, i.e. bonded to the atom Ω. Thus

$$F(\Omega) = -\sum_{\Omega'}\oint dS(\Omega,\Omega')\sigma(\mathbf{r})\cdot\mathbf{n}(\mathbf{r}) \tag{23}$$

where $S(\Omega,\Omega')$ denotes the interatomic surface between atoms Ω and Ω'. Equation 23 states that all of the mechanical properties of an atom are determined by the pressure exerted on each of the interatomic surfaces it shares with its bonded neighbours. Thus the properties of a carbon atom in cyclopropane, for example, are determined by the flux in the forces through the two interatomic surfaces it shares with hydrogen atoms and through the two surfaces it shares with the other carbon atoms.

All atomic properties are determined by an integration of a corresponding density over the basin of the atom. This is possible even for a property like the energy which involves two-electron interactions because the potential as well as the kinetic energy are expressible in terms of $\sigma(\mathbf{r})$, the quantum mechanical stress tensor, a quantity which in turn in completely determined by the one-electron density matrix.

An atomic population $N(\Omega)$ is determined by an integration of the charge density $\rho(\mathbf{r})$ over the atomic basin,

$$N(\Omega) = \int_\Omega \rho(\mathbf{r})d\tau \tag{24}$$

and the net charge on an atom, $q(\Omega)$, is given by

$$q(\Omega) = Z_\Omega - N(\Omega) \tag{25}$$

where Z_Ω is the nuclear charge. The dipolar polarization of an atomic charge distribution is determined by an atom's first moment $\mathbf{M}(\Omega)$,

$$\mathbf{M}(\Omega) = -\int_\Omega \mathbf{r}_\Omega\rho(\mathbf{r})d\tau \tag{26}$$

with the nucleus serving as the origin of the position vector $\mathbf{r}_\Omega$. Higher atomic moments are also useful. The atomic quadrupole moment provides a model-independent measure of an atom's π character by determining the extent to which electronic charge is removed from a given plane and accumulated along an axis perpendicular to the plane[3].

The idea of defining the shape of a molecule with respect to its non-bonded interactions with other molecules, its 'van der Waals envelope', in terms of an outer contour of the charge density was made some time ago[40,44]. It has been shown that the 0.002 au contour of the charge density provides a good measure of the size and shape of a molecule in a crystal, while the slightly larger envelope defined by the 0.001 au contour reproduces the equilibrium diameters of gas-phase molecules determined by second virial coefficient or viscosity data. Closely associated with the notion of a molecule's shape is a measure of the volume it occupies. An atomic volume $v(\Omega)$ is a measure of the space enclosed by the atom's interatomic surfaces and by the intersection of these surfaces with a particular envelope of the charge density for those open regions not bounded by an interatomic surface. A group or molecular volume then equals the corresponding sum of atomic volumes. The hydrocarbons are used to illustrate the transferability of group volumes and explain the relationship of these volumes to changes in hybridization and both steric and geometric strain. The mean molecular polarizability and its group contributions are also shown to be linearly proportional to corresponding sums of atomic volumes, making possible the prediction of molecular polarizabilities from tabulated atomic volumes, as determined by the theory of atoms in molecules.

The integration of a property density over the basin of an atom is accomplished numerically using the program PROAIM developed in this laboratory. Table 3 lists the values of the atomic populations and energies for structures 1 to 26 for the 6-31G*/6-31G* calculations. In general, the error in a total electron population is less than ± 0.01 e and less than $\pm 2.0\,\mathrm{kcal\,mol^{-1}}$ in the total energy, as determined by a comparison of the sums of the integrated atomic values with the corresponding molecular values. A test of the goodness of the integration over the basin of an atom is provided by the condition that the atomic average of the Laplacian of ρ should vanish, as indicated in equation 9. The quantity $L(\Omega)$ listed in Table 3 is $-1/4$ the atomic average of $\nabla^2\rho$ and, in general, its value is less than 1×10^{-4} or $0.06\,\mathrm{kcal\,mol^{-1}}$ for a hydrogen atom. The term $L(\Omega)$ is to vanish in the integration of equation 10 to obtain the atomic statement of the virial

TABLE 3. Atomic properties in hydrocarbons[a]

Molecule	Ω	$N(\Omega)$	$L(\Omega)$ (au)	$T(\Omega) = -E(\Omega)$ (au)
Methane (1)	C1	5.9668	−0.000 17	37.739 36
	H2	1.0083	0.000 07	0.613 95
Ethane (2)	C1	5.9249	0.000 40	37.733 23
	H3	1.0248	0.000 07	0.627 11
Propane (3)	C1	5.9373	−0.000 70	37.750 04
	*C2	5.8924	−0.004 77	37.724 85
	H4	1.0248	−0.000 02	0.626 40
	H5	1.0275	0.000 13	0.627 35
	H7	1.0367	0.000 06	0.638 26
n-Butane (4)	C1	5.9350	−0.000 12	37.749 48
	*C2	5.9043	0.000 53	37.738 27
	H5	1.0251	0.000 17	0.627 34
	H6	1.0281	0.000 10	0.628 04
	H8	1.0397	0.000 20	0.639 02
Isobutane (5)	*C1	5.9433	−0.000 58	37.763 37
	C2	5.8706	−0.000 41	37.711 28
	H5	1.0275	0.000 09	0.627 76

(continued)

TABLE 3. (*continued*)

Molecule	Ω	$N(\Omega)$	$L(\Omega)$ (au)	$T(\Omega) = -E(\Omega)$ (au)
	H6	1.0296	0.000 10	0.627 59
	H7	1.0452	0.000 39	0.648 25
Pentane (**6**)	C1	5.9343	0.000 24	37.747 25
	*C2	5.9047	0.000 72	37.740 62
	C3	5.9142	0.000 69	37.757 32
	H6	1.0251	0.000 08	0.627 02
	H7	1.0280	0.000 09	0.627 90
	H9	1.0399	0.000 11	0.638 84
	H11	1.0430	0.000 11	0.639 49
Neopentane (**7**)	C1	5.9460	0.000 05	37.774 88
	*C2	5.8626	−0.001 63	37.692 91
	H6	1.0295	−0.000 08	0.628 45
Hexane (**8**)	C1	5.9357	0.000 51	37.749 40
	*C2	5.9044	0.002 23	37.738 37
	C3	5.9136	0.000 86	37.755 98
	H7	1.0249	0.000 08	0.626 60
	H8	1.0280	0.000 09	0.627 75
	H10	1.0400	0.000 10	0.639 26
	H12	1.0427	0.000 11	0.639 77
Cyclopropane (**9**)	*C1	6.0102	0.000 35	37.782 34
	H4	0.9949	0.000 09	0.618 66
Cyclobutane (**10**)	*C1	5.9434	0.000 28	37.759 30
	H5	1.0289	0.000 08	0.632 20
	H9	1.0277	0.000 09	0.632 52
Cyclopentane (**11**)	*C1	5.9160	0.001 85	37.758 49
	C2	5.9241	0.000 77	37.756 20
	C3	5.9308	0.000 40	37.748 61
	H6	1.0326	0.000 09	0.635 51
	H7	1.0416	0.000 44	0.642 24
	H8	1.0407	0.000 10	0.642 00
	H9	1.0386	0.000 13	0.636 91
	H10	1.0338	0.000 04	0.638 61
	H11	1.0369	0.000 10	0.640 88
Cyclohexane (**12**)	*C1	5.9202	0.000 24	37.758 03
	H7	1.0361	0.000 09	0.637 08
	H8	1.0437	0.000 10	0.639 65
Bicyclo[1.1.0]butane (**13**)	C1	6.1210	0.000 17	37.860 24
	*C2	5.9283	0.002 08	37.741 52
	H5	0.9463	0.000 06	0.592 96
	H6	0.9976	0.000 08	0.618 78
	H7	1.0068	0.000 11	0.622 33
Bicyclo[2.1.0]pentane (**14**)	C1	6.0433	0.001 36	37.827 17
	C2	5.9160	0.000 01	37.744 29
	C5	6.0063	−0.000 34	37.793 96
	H6	0.9893	−0.000 03	0.614 40
	H9	1.0288	0.000 09	0.631 50
	H10	1.0211	−0.000 07	0.630 70
	H12	0.9941	0.000 07	0.617 07
	H13	1.0027	0.000 09	0.619 85
Bicyclo[2.2.0]hexane (**15**)	C1	5.9696	−0.001 23	37.786 65
	C2	5.9408	0.001 00	37.760 71
	C3	5.9444	−0.000 20	37.760 75
	H7	1.0284	0.000 20	0.635 93

TABLE 3. (*continued*)

Molecule	Ω	$N(\Omega)$	$L(\Omega)$ (au)	$T(\Omega) = -E(\Omega)$ (au)
	H8	1.0275	0.000 08	0.634 20
	H9	1.0317	0.000 10	0.635 27
	H10	1.0269	0.000 12	0.633 91
	H11	1.0307	0.000 09	0.635 06
Bicyclo[1.1.1]pentane (16)	*C1	5.9644	−0.002 86	37.767 00
	C2	5.9570	−0.000 02	37.778 70
	H6	1.0108	0.000 06	0.621 35
	H7	1.0298	0.000 09	0.632 15
Bicyclo[2.1.1]hexane (17)	C1			
	C2			
	C5	5.9722	−0.004 31	37.789 35
	H7	1.0233	0.000 78	0.632 24
	H8	1.0341	0.000 10	0.637 81
	H13	1.0365	−0.000 04	0.636 54
	H14	1.0257	0.000 09	0.632 32
Bicyclo[2.2.1]heptane (18)	C1	5.9264		37.769 91
	C2	5.9418	0.001 20	37.767 87
	C7	5.9641	−0.001 74	37.808 57
	H8	1.0326	−0.000 08	0.637 34
	H9	1.0356	0.000 10	0.638 08
	H10	1.0349	0.000 10	0.636 90
	H13	1.0344	0.000 11	0.633 34
Bicyclo[2.2.2]octane (19)	*C1	5.8991	0.004 25	37.765 59
	C2	5.9389	−0.002 41	37.764 53
	H9	1.0441	0.000 11	0.646 31
	H10	1.0400	0.000 15	0.640 99
[1.1.1]Propellane (20)	*C1	6.1076	0.001 36	37.806 19
	C2	5.9744	−0.000 84	37.799 98
	H6	0.9770	0.000 10	0.613 13
[2.1.1]Propellane (21)	*C1	6.1494	0.001 59	37.858 21
	C2	5.8969	−0.000 17	37.729 70
	C5	5.9640	−0.000 63	37.789 28
	H7	1.0138	0.000 07	0.628 13
	H11	0.9835	0.000 10	0.613 73
	H12	0.9786	0.000 10	0.613 76
[2.2.1]Propellane (22)	C1	6.1523		37.915 03
	C2	5.9009	0.000 44	37.717 59
	C7	5.9584	0.000 90	37.776 66
	H8	1.0232	0.000 09	0.631 98
	H9	1.0168	0.000 10	0.629 53
	H16	0.9867	0.000 11	0.618 22
[2.2.2]Propellane (23)	*C1	6.0156	0.001 92	37.900 48
	C2	5.9494	0.001 10	37.744 31
	H9	1.0227	0.000 09	0.628 55
Tetrahedrane (24)	C1	6.1112	−0.000 02	37.836 69
	H5	0.8889	−0.000 09	0.562 98
Cubane (25)	*C1	5.9966	0.001 02	37.809 32
	H9	1.0034	0.000 06	0.614 81
Spiropentane (26)	*C1	5.9969	0.001 20	37.782 13
	C3	6.0718	0.001 15	37.856 33
	H6	0.9926	0.0001 14	0.616 58

*a*Numbering of structures and of atoms is as given in Table 1.

theorem. More details concerning the determination of the atomic properties are given in Reference 18.

C. Atomic Charges and the Hybridization Model

The atomic populations exhibit a basis set dependence. Because the atom is defined in terms of a property of the charge density, its population $N(\Omega)$ and other properties change in an ordered and understandable way with a change in the basis. The atomic charges tabulated in Table 3 are obtained from 6-31G*/6-31G* calculations, for which the basis includes polarization functions on carbon but not on hydrogen. The charges obtained using this basis differ by a fixed amount from those obtained using a larger basis, one which in particular includes polarization functions on hydrogen. Thus, for example, the values of $N(H)$ and $N(C)$ are brought into agreement with those obtained using the Dunning–Huzinaga set augmented with polarization functions previously described in Section II.G and hereafter referred to as the D—H basis, through the addition of 0.037e to each $N(H)$ value and adjusting the value of $N(C)$ to which the hydrogens are bonded, accordingly. This has the important consequence that there is no change in the net charge of any of the functional groups, CH_3, CH_2 or CH, with a change in basis, all of the changes in population occurring across the C—H interatomic surfaces within a given functional group. It further implies that when the group is just the carbon atom, such as the bridgehead carbon in a propellane, the value of $N(C)$ is independent of basis. For example, the values of $Q(C)$ for the bridgehead carbon in [1.1.1]propellane are $-0.108e$ and $-0.102e$, respectively, for the 6-31G* and 6-31G** basis sets, the existence of a small difference of 0.006e in the two populations in this case being a result of the very strained nature of the system.

The same behaviour is found for the energies of the atoms. Thus not only are the trends in atomic populations and energies obtained using larger basis sets reproduced by the data given in Table 3, the *changes in the values of the group properties through the series of molecules are quantitatively recovered.*

Reference to Table 3 shows hydrogen to be more electronegative than carbon in all of the acyclic hydrocarbons. Hydrogen has its most negative charge in the CH group of isobutane, followed by its values in CH_2, and exhibits its least negative values in CH_3 of normal alkanes. However, a hydrogen atom becomes less electronegative relative to carbon as the carbon is subjected to increasing geometric strain. Thus the hydrogen in cyclopropane has a small positive charge, a slightly larger one when attached to the bridgehead carbon of bicyclobutane and its largest net charge is in tetrahedrane. This trend of a decreasing population of H with increasing geometric strain is identical to that obtained using larger basis sets, but the net charge on hydrogen is positive only in the most highly strained systems because of the greater populations for hydrogen predicted by the larger basis sets.

Since hydrogen withdraws charge from carbon in the acyclic hydrocarbons, the relative group electron-withdrawing ability is in the order $CH_3 > CH_2 > CH > C$. A methyl group linked to a methylene group withdraws an almost constant amount of charge from methylene, as $q(Me) = -0.0165 \pm 0.0005e$ in propane through to hexane. This number changes slightly to $-0.0175e$ for the 6-31G** and D—H basis sets. The reader is referred to Table 4 for the net charges on the Me and methylene groups obtained using the D—H basis. The methylene group in propane is linked to two methyls and its net charge of $+0.034e$ is, within integration error, twice that of a methylene linked to a methyl and a second methylene, as found in butane through to hexane. A methylene linked only to other methylenes as found in pentane and hexane has a zero net charge (Table 4). Thus the inductive transfer of charge from methylene to methyl is a nearly

TABLE 4. Net charges in methyl and methylene groups and their energies relative to standard values[a]

Molecule	$q(CH_3)$	$q(CH_2)$ (e)	$q(CH_2)^b$	$\Delta E(CH_3)$	$\Delta E(CH_2)$ (kcal mol^{-1})	$\Delta E(CH_2)^b$
Ethane	+0.000					
Propane	−0.017	+0.034		−10.7	+21.1	
n-Butane	−0.018	+0.018		−10.4	+10.5	
n-Pentane	−0.018	+0.018	+0.002	−10.3	+10.4	−0.6
n-Hexane	−0.017	+0.017	+0.001	−10.4	+10.6	−0.3

[a] $E(CH_3)$ in ethane $= A^0 = -39.62953$ au, $E(CH_2) = B^0 = -39.04743$ au using the D—H basis set.
[b] This CH_2 is bonded only to other methylenes.

constant quantity, equal to 0.017e per methyl, and this transfer is damped by a single methylene group, as a CH_2 group once removed has a zero net charge.

The net charges on the individual atoms in these groups also exhibit a similar constancy. The H atom of methyl in the plane of the carbon framework has a charge of −0.025e, 0.003e smaller in magnitude than that of the two other H atoms of methyl. The H of CH_2 linked to a single methyl bears a charge of −0.040e, while that for a H on a carbon linked only to other methylenes has a charge of −0.043e. One correctly anticipates, as illustrated in Figure 12, that these constancies in average populations are a result of essentially constant distributions of charge over corresponding atoms in these molecules and that this will in turn result in these atoms making essentially constant contributions to the total energies. The two central methylene groups in hexane and the single such group in pentane have essentially zero net charge and their energies are found to be equal to the additive contribution per methylene group in the energy additivity scheme for the normal alkanes. These groups will be taken to define the properties of the 'standard' CH_2 group for comparison with the methylene groups of the cyclic and bicyclic hydrocarbons.

When three methyls are attached to a single carbon as in isobutane, each withdraws significantly more electronic charge than when bonded to methylene and this effect is further enhanced in neopentane. The net charges on the carbon atoms in CH_3, CH_2 (linked only to methylenes), CH_2 (linked to one methyl), CH and C are, respectively, +0.065, +0.085, +0.097, +0.129 and +0.137e. There is a dramatic decrease in the average population of a hydrogen atom when it is linked to a carbon atom in a system with ring strain. Figure 13 gives net atomic and group charges for all cyclic structures. By symmetry, a CH_2 group of the three-, four- and six-membered cyclic molecules has zero net charge as does the standard methylene group of the acyclics defined above. Compared to the hydrogens of this standard group, the hydrogens of cyclopropane have 0.048 fewer electrons, the same decrease in population as predicted using the D—H basis. In cyclobutane this difference has decreased to 0.014e and in cyclohexane the average of the axial and equatorial H populations is only 0.003e less than that in the standard methylene group. Correspondingly, the average electron population of a carbon atom in these molecules increases with increasing strain, the net charge of a C atom in the six-, five-, four- and three-membered cyclics being +0.086, +0.077 (average), +0.057 and −0.010e, respectively. The value for cyclohexane, which may be considered to be strain free, is within 0.006e of the standard value. The bridgehead carbon atoms in bicyclobutane 13 and in the propellanes 21 and 22 bear the largest negative charges of all the carbon atoms and the order of electron-withdrawing ability found in the unstrained acyclic molecules is reversed. In the strained molecules the CH and C groups

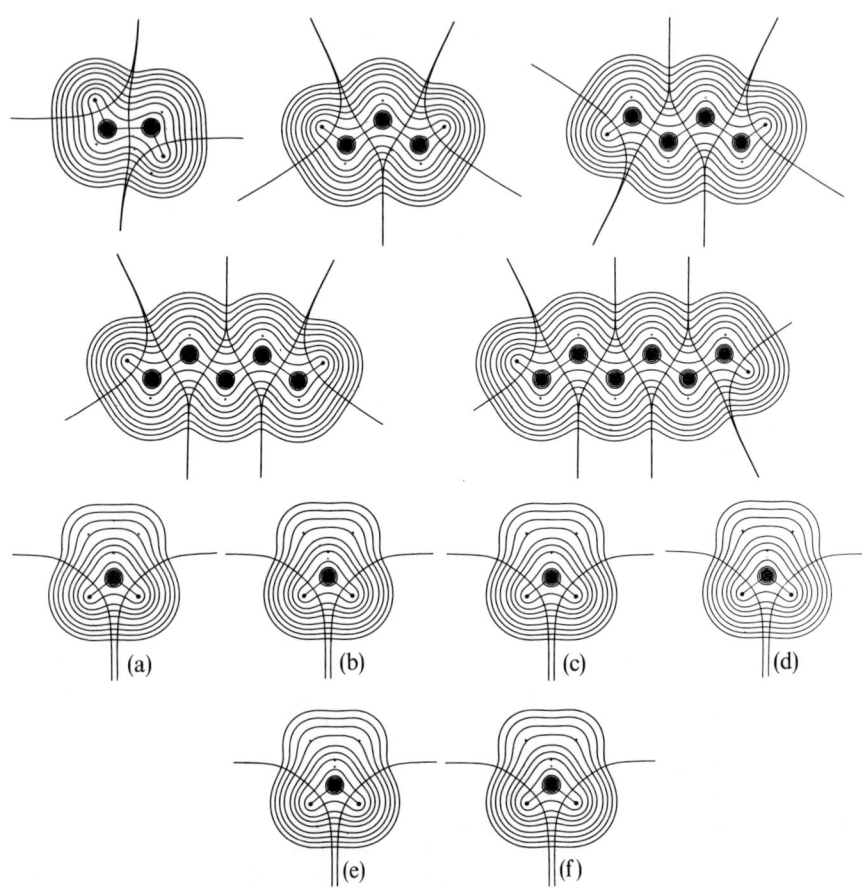

FIGURE 12. Contour plots of the charge distributions in the normal alkanes, ethane to hexane, for the plane containing the carbon nuclei and the in-plane protons of the Me groups. Also shown are the charge distributions for the methylene groups for a plane containing the C and H nuclei: (a) in propane, (b) in butane, (c) in pentane, (d) in hexane, (e) the central group in pentane, (f) an equivalent group in hexane. The atomic boundries and bond paths are shown. The plots for the Me groups in propane to hexane are superimposable, as are plots (b), (c) and (d) as well as (e) and (f) for the methylene groups. The outer contour has the value 0.001 au with remaining values being the same as those given in Figure 1

withdraw charge from the CH_2 groups. These observations indicate that a carbon becomes more electronegative as it is subjected to more angular strain and this fact is of primary importance in understanding the gross effects of geometrical constraints on the charge distribution and its energy.

Orbital theories relate an increase in the electronegativity of a carbon atom relative to that of a bonded hydrogen to an increase in the s character of its bonding hybrid orbital. Thus the decrease in bond length and pK_a and the increase in force constant and bond dissociation energy for C—H through the series cyclohexane, ethylene and

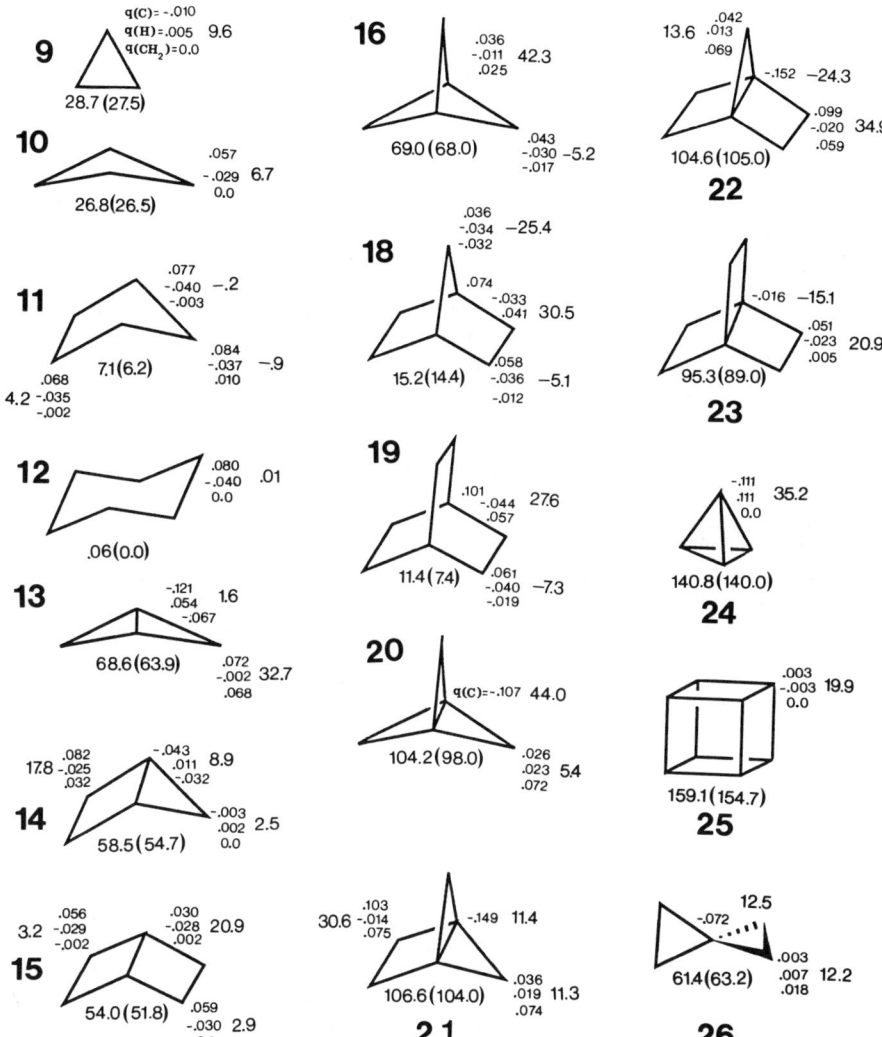

FIGURE 13. Structure of strained hydrocarbons showing opposite each CH_2, CH or C group the net charge on carbon, the average net charge on hydrogen, the net charge for the group and the strain energy of the group in kcal mol^{-1}. The total calculated strain energy followed by the value in parentheses is given beneath each structure.

acetylene is accounted for by a change in hybridization on carbon from sp^3 to sp^2 to sp, respectively. The increasing electron populations of these carbons to yield net charges of $+0.080e$, $-0.035e$ and -0.177e, respectively, quantify the accompanying increase in electronegativity. An empirical expression has been proposed to relate the C—H coupling constant $J(^{13}C—H)$ to the percent s character of the bond[45] and the values of this constant for cyclohexane, ethylene and acetylene (123, 156 and 249 Hz, respectively) yield values of 25, 31 and 50% for the s character.

The hybridization model also predicts that the smaller bond angles found in a system with angular strain should result in an increase in the p character of the strained C—C bonds and hence to an increase in the s character of the associated C—H bonds[46]. In this way one can account for the increase in the electronegativity of carbon relative to hydrogen which, as noted above, accompanies the introduction of strain into a cyclic system. The validity of this model can be demonstrated by showing that the properties of the CH_2 group in cyclopropane resemble those for ethylene, a point emphasized by Coulson and Moffitt[47]. The C—H bond length and HCH bond angle, both observed and calculated at the 6-31G* level, are similar for the CH_2 groups in cyclopropane and ethylene, the observed values being 1.082 and 1.090 Å, and 116.6° and 116.5°, respectively. The similarity in properties extends to the C—H coupling constants, the value for cyclopropane being 161 Hz, to the C—H force constant and bond dissociation energies as well as to their pK_a values[48] of 46 and 44, ethylene being more acidic. The carbon is negatively charged in both molecules, slightly more so in ethylene. Similarly, the properties for the C—H bridgehead group in bicyclo[1.1.0]butane and the still larger net negative charge on this carbon are understood by noting that its properties approach those observed for the same group in acetylene. A value of 205 Hz for $J(^{13}C—H)$ correlates with 41%s character. The C—H bond length of 1.070 Å, while larger than the value of 1.057 Å in acetylene, is less than that found in ethylene. Since bicyclobutane reacts with phenyl lithium whose pK_a is 43, the pK_a of this bicyclic molecule is probably less than 40 which makes it more acidic than ethylene or cyclopropane but less acidic than acetylene itself, which has a pK_a of 25. The atomic charges for the cyclic molecules are considered in detail below in the discussion of the origin of strain energy and its variation with structure throughout this series of molecules. The ultimate argument for an increase in electronegativity with increasing s character is, of course, based on energy, an s electron being more tightly bound than a p electron. As described below, one indeed finds that the energy of a carbon atom decreases as the s character of its hybrids to dissimilar neighbouring atoms increases. The bridgehead carbon atom in bicyclobutane, for example, is found to be only 9.5 kcal mol^{-1} less stable than the carbon in acetylene.

D. Atomic and Group Volumes

The study of the molar volumes of hydrocarbons by Kopp in 1855[4a] provided the earliest example of the addivitity of group properties. Table 5 and 6 list the molecular volume and the contribution of each atom and functional group to this volume as determined by both the 0.002 and 0.001 au envelope of the charge density for a number of hydrocarbons obtained from 6-31G**/6-31G* calculations. Table 5 also lists the fraction of the total electronic charge of the atom contained within the stated envelope[44].

The volumes of the methyl and methylene groups in the homologous series of normal hydrocarbons exhibit the same pattern of transferable values noted above for the group populations. The volume of the methyl group in ethane is slightly larger than that of the methyl groups in the remaining members whose volumes vary by only ±0.02 au from their average value. The volume of the methylene group bonded to two methyls in propane is slightly greater than the volumes of the methylene bonded to one methyl, their volumes varying by ±0.02 au about the average value. Finally, the methylene groups bonded only to other methylenes also have a characteristic volume. The small differences between the latter two kinds of methylene groups are evident in Figure 12. The methylene bonded to one methyl is less compressed than one bonded only to other methylenes and this is reflected in this larger volume. This effect is enhanced for propane where the single methylene is larger still. The volume varies in the reverse order to the population and energy of the methylene group: the more contracted the group, the larger its population and the lower its energy. The same is true for the methyl group, the

TABLE 5. Atomic energies, populations and volumes[a]

Molecule	Atom[b]	$-E(\Omega)$	$N(\Omega)$	$v(\Omega)$ 0.001	$v(\Omega)$ 0.002	$\%N(\Omega)$ 0.001	$\%N(\Omega)$ 0.002
CH$_4$	C	37.6097	5.754	76.90	64.94	99.7	99.4
	H	0.6480	1.062	52.31	38.61	97.8	96.0
C$_2$H$_6$	C	37.6328	5.762	64.66	57.90	99.8	99.6
	H	0.6621	1.079	52.30	38.79	98.0	96.1
C$_3$H$_8$	C	37.6502	5.774	63.45	57.31	99.9	99.7
	H	0.6615	1.079	52.44	38.87	98.0	96.1
	H	0.6625	1.082	52.59	39.45	98.1	96.4
	C	37.6539	5.778	54.75	51.37	98.0	97.9
	H	0.6744	1.092	52.10	38.83	98.0	96.3
n-C$_4$H$_{10}$	C	37.6470	5.773	63.36	57.29	99.9	99.7
	H	0.6622	1.079	52.38	38.80	97.9	96.2
	H	0.6632	1.083	52.52	39.44	98.1	96.4
	C	37.6695	5.791	53.79	50.72	99.9	99.9
	H	0.6750	1.095	52.20	39.43	98.2	96.6
n-C$_5$H$_{12}$	C	37.6469	5.773	63.64	57.30	99.9	99.7
	H	0.6621	1.080	52.37	38.85	97.9	96.1
	H	0.6629	1.082	52.41	39.49	98.1	96.4
	C	37.6716	5.791	53.64	50.69	99.9	99.9
	H	0.6747	1.095	52.09	39.38	98.2	96.6
	C	37.6910	5.803	52.36	49.98	100.0	99.9
	H	0.6753	1.098	52.13	40.03	98.3	96.8
n-C$_6$H$_{14}$	C	37.6480	5.775	63.57	57.29	99.9	99.7
	H	0.6615	1.079	52.40	38.89	97.9	96.1
	H	0.6627	1.082	55.59	39.44	98.1	96.3
	C	37.6707	5.790	53.60	50.55	99.9	99.9
	H	0.6751	1.095	52.16	39.42	98.3	96.6
	C	37.6884	5.804	52.76	50.30	100.0	99.9
	H	0.6756	1.098	52.21	39.99	98.4	96.8
iso-C$_4$H$_{10}$	C	37.6649	5.781	61.29	56.23	99.9	99.8
	H	0.6625	1.084	52.21	39.99	99.9	98.2
	H	0.6627	1.082	52.41	39.38	98.1	96.4
	C	37.6759	5.808	47.28	45.74	100.0	99.9
	H	0.6845	1.102	51.49	38.65	98.3	96.6
neo-C$_5$H$_{12}$	C	37.6746	5.786	59.75	55.25	99.9	99.8
	H	0.6634	1.084	52.10	39.76	98.3	96.6
	C	37.6938	5.851	40.97	40.97	100.0	100.0
c-C$_3$H$_6$	C	37.7115	5.894	70.84	62.11	99.8	99.6
	H	0.6557	1.053	50.21	37.47	97.9	96.2
c-C$_4$H$_8$	C	37.6905	5.830	60.83	55.34	99.9	99.7
	H	0.6683	1.086	52.64	38.93	98.0	96.2
	H axial	0.6686	1.084	52.12	38.66	98.1	96.3
c-C$_6$H$_{12}$	C	37.6896	5.809	53.70	50.67	99.9	99.9
	H	0.6729	1.092	52.25	38.85	98.0	96.3
	H axial	0.6754	1.099	52.24	39.99	98.3	96.7
Bicyclo[1.1.0]butane	C	37.6702	5.814	68.14	59.62	99.8	99.6
	H axial	0.6577	1.062	51.49	38.51	98.1	96.3
	H	0.6557	1.055	50.72	37.54	97.9	96.1
	C	37.8253	6.061	78.69	68.22	99.8	99.5
	H	0.6322	1.009	48.05	36.10	97.8	96.1
[1.1.1]Propellane	C	37.7330	5.864	65.86	58.08	99.8	99.6
	H	0.6494	1.033	48.39	36.83	98.2	96.5

(*continued*)

TABLE 5. (continued)

Molecule	Atom[b]	$-E(\Omega)$	$N(\Omega)$	$v(\Omega)$		$\%N(\Omega)$	
				0.001	0.002	0.001	0.002
	C	37.8041	6.102	93.77	76.83	99.7	99.2
C_2H_4	C	37.7250	5.919	92.37	77.27	99.6	99.3
	H	0.6473	1.041	50.30	37.31	97.8	96.0
C_2H_2	C	37.8418	6.121	121.06	98.46	99.4	98.8
	H	0.5691	0.879	41.35	31.18	97.7	96.1
CH_3^+	C	37.6826	5.757	95.46	77.01	99.5	99.0
	H	0.5179	0.748	35.70	26.61	97.5	95.8
$C_4H_9^+$	C	37.7136	5.796	68.22	60.39	99.8	99.6
	H[c]	0.6170	0.969	45.18	34.91	98.1	96.6
	H	0.5896	0.934	44.53	34.21	97.9	96.4
	C	37.9288	6.099	64.58	60.21	99.9	99.8

[a]All results in atomic units and calculated from 6-31G**/6-31G*.
[b]The carbon atom of a methyl group, if present, is listed first. This is followed by the unique H of a methyl and then by one of the two equivalent hydrogens of a methyl. Given next are the carbon and hydrogen atoms of methylene. The final entries are for carbon and hydrogen of methine and finally a single carbon as in neopentane.
[c]In plane.

transferable methyl having a smaller volume, but larger population and lower energy than the methyl of ethane. It has also been shown that the methyl and methylene group dipole moments and their correlation energies as determined by a density functional exhibit the identical pattern of transferable values[20]. The group contributions to the molecular dipole are discussed in Section VIII.

The methyl groups in isobutane and neopentane have still larger net charges and still lower energies since they withdraw charge from CH and C, respectively. Correspondingly they have smaller volumes, the smallest being found for the most stable, that in neopentane. One notes that the volume and contained charge of the central carbon in this latter molecule is independent of the choice of envelope as it is interior to both the 0.001 or 0.002 au contour. This observation is borne out by the fact that both choices of envelope recover 100% of the charge of this atom. The stability and electron population of a carbon atom increases with the extent of methyl substitution, as methyl is less electron-withdrawing than is hydrogen. The volume of a substituted carbon, however, exhibits the opposite order, its value decreasing in the order ethane > propane > isobutane > neopentane.

What might at first seem surprising from the next entries in Table 5 is that the volume of a carbon atom increases as it is subjected to increased geometric strain. The extent of geometric strain present in a molecule is measured by comparing the bond path angle with the normal bond angle (see Table 2). The bond path angle of 78.8° in cyclopropane, for example, is considerably less than the normal tetrahedral C—C—C bond angle. The strain in this molecule is, however, less than that suggested by the geometric angle which is less than the bond path angle by 18.8°. The strain is still less in cyclobutane and, correspondingly, the bond path angle exceeds the geometric angle of 89.0° by only 6.7°. In the strain-free cyclohexane molecule the bond path angle is actually closer to the tetrahedral value than is the geometric one since the latter, equal to 111.4°, *exceeds* the former by 1.3°. Correspondingly, the volume of the carbon in cyclopropane is greater than that for the carbon in cyclobutane which in turn is greater than that for cyclohexane. Unlike the variation in volume of a carbon with extent of methyl substitution discussed above, one finds both the electron population and stability of a strained carbon to parallel

TABLE 6. Molar volumes of molecules and functional groups

Molecule and group	$V_M(cm^3\,mol^{-1})$		$N - \Sigma_\Omega N(\Omega)$
	0.001	0.002	
CH_4	25.53	19.58	+0.0003
CH_3	20.87	16.13	
C_2H_6	39.54	31.10	+0.0006
CH_3	19.77	15.55	
C_3H_8	53.64	42.76	+0.002
CH_3	19.73	15.62	
CH_2	14.18	11.51	
$n\text{-}C_4H_{10}$	67.64	44.13	+0.002
CH_3	19.70	15.61	
CH_2	14.12	11.56	
$n\text{-}C_5H_{12}$	81.56	65.96	+0.004
CH_3	19.71	15.63	
CH_2	14.08	11.55	
CH_2^a	13.98	11.61	
$n\text{-}C_6H_{14}$	95.71	72.49	+0.003
CH_3	19.74	15.62	
CH_2	14.09	11.56	
CH_2^a	14.03	11.63	
$i\text{-}C_4H_{10}$	67.34	54.36	-0.002
CH_3	19.51	15.61	
CH	8.81	7.53	
$neo\text{-}C_5H_{12}$	80.78	65.95	+0.002
CH_3	19.28	15.57	
C	3.66	3.66	
$c\text{-}C_3H_6$	45.85	36.69	+0.002
CH_2	15.28	12.23	
$c\text{-}C_4H_8$	59.11	47.45	+0.001
CH_2	14.78	11.86	
$c\text{-}C_6H_{12}$	84.25	69.35	+0.002
CH_2	14.04	11.53	
Bicyclo[1.1.0]butane	53.02	42.83	+0.001
CH_2	15.20	12.02	
CH	11.31	9.31	
[1.1.1]Propellane	60.28	48.98	+0.002
CH_2	14.51	11.76	
C	8.37	6.86	
C_2H_4	34.44	27.11	-0.00007
CH_2	17.22	13.56	
C_2H_2	28.99	23.24	+0.00006
CH	14.49	11.62	
CH_3^+	18.14	14.00	-0.001
$t\text{-}C_4H_9^+$	59.97	49.19	+0.002
CH_3	18.07	14.61	
C	5.76	5.37	

^{a}These methylenes are bonded only to other methylenes.

the increase in its volume. These observations, together with an understanding of strain energy itself, can be rationalized in terms of the hybridization model. As illustrated above in the discussion of the atomic and group populations, the smaller C—C—C bond angles found in hydrocarbons with angular strain should result in an increase of the p character of the strained C—C bonds and hence in an increase in the s character of the associated C—H bonds. Since s electrons are bound more tightly than are p electrons, the effect of the angular strain is to increase the stability of a strained carbon atom and increase its electronegativity relative to its bonded hydrogen atoms.

The atomic volume of carbon is another property like its population $N(C)$, whose value increases with increasing s character, the values of $v(C)$ in ethane, ethylene and acetylene being 65, 92 and 121 au, respectively, paralleling the trend in $N(C)$ values which at this basis set in the same order as 5.76, 5.92 and 6.12e. On the basis of these results one anticipates that the atomic volume of carbon will also exhibit an increase with increasing geometric strain, as this also causes an increase in the s character of the carbon atom.

The increase in stability, charge and volume of a carbon atom with increasing geometric strain is illustrated by the results for the three cyclic molecules cyclohexane, cyclobutane and cyclopropane[44]. It is important to note that the theory of atoms in molecules, as illustrated in a subsequent section, again recovers the experimental findings regarding the values of the strain energies for these molecules. The energy of the methylene group in cyclohexane differs from the energy of the standard methylene group (a value fixed independently by the partitioning of the energies of the normal hydrocarbons) by only $0.06\,\mathrm{kcal\,mol^{-1}}$. The charges of the carbon and hydrogen atoms in this group and in the transferable methylene group in pentane and hexane are also very similar. Thus in agreement with experiment, cyclohexane has a zero strain energy. The strain present in the four- and three-membered rings results in a transfer of charge from hydrogen to carbon relative to their values in the standard methylene group. The extent of this charge transfer increases with an increase in strain. The result is an increase in the stability of the carbon atom, but an even greater decrease in the stability of the two hydrogen atoms relative to their values in the standard methylene group, and the net result is a strain energy. The strain energy per methylene group is greater for cyclopropane than for cyclobutane, the values being 9.3 and $6.5\,\mathrm{kcal\,mol^{-1}}$, respectively. In agreement with experiment, the predicted total strain energies of 27.9 and $26.0\,\mathrm{kcal\,mol^{-1}}$ for the two molecules are very similar.

The volume changes for the hydrogens attached to the strained carbon atoms parallel the shifts in charge, their volumes being slightly less than the volumes of the hydrogen atoms in the standard methylene group. The increase in volume of the carbon atom dominates the volume change accompanying the introduction of geometric strain into a cyclic molecule. The reason for this can be seen from Figure 14, which shows the boundaries of the carbon atom of the standard methylene group superimposed on the carbon atom of cyclopropane in the plane of the carbon nuclei. The effect of decreasing the angle between the carbon–carbon bond paths subtended at the nucleus of the standard methylene carbon to its value in cyclopropane is a fan-like opening up of the carbon atom and a corresponding increase in its breadth. There is a loss in its relative volume where the boundaries of the three ring atoms meet, but this is small compared to the gain in the outer regions of the atom. It is also important to note that this result of an increase in volume of the methylene group with increase in strain is not an artifact of the partitioning method. Thus the incremental increase in the total molecular volumes, propane through to hexane, is 14.0 au. This increment per added methylene group is less than one-third and one-quarter, respectively, of the total molecular volumes of cyclopropane and cyclobutane. Gavezzotti has noted that the surface area of a methylene group increases with geometric strain and decreases with steric crowding[49].

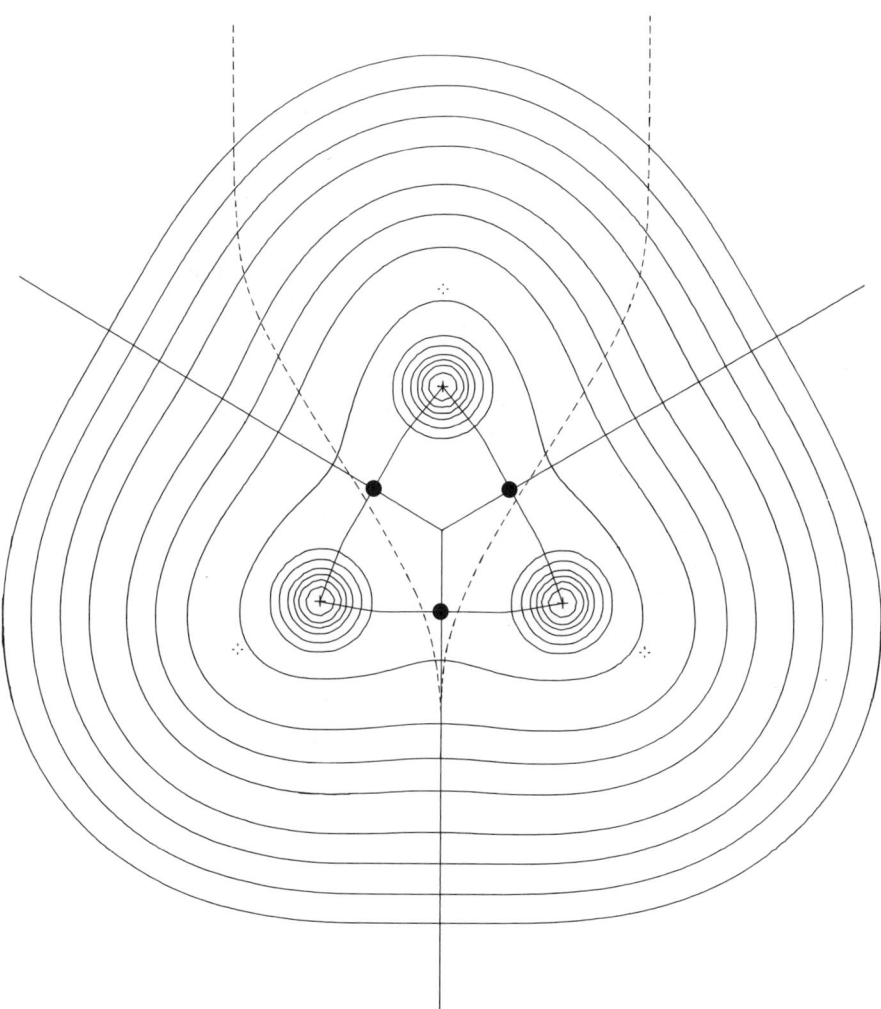

FIGURE 14. Contour plot of the charge distribution of the cyclopropane molecule in the ring plane showing the bond paths and the C—C interatomic surfaces. Also shown for one carbon (and its out-of-plane hydrogens) is the corresponding boundary of the standard methylene group in the acyclic hydrocarbons. The outer contour value is 0.001 au, the remainder as in Figure 1

Also included in Tables 5 and 6 are the results for bicyclo[1.1.0]butane and [1.1.1]-propellane. In both these molecules all four bonds to a bridgehead carbon atom lie on one side of a plane through its nucleus. In keeping with the considerable geometrical strain present at the bridgehead positions, these atoms are the most stable, and possess the largest net charges and volumes of all the saturated carbon atoms considered here. The bridgehead carbon in bicyclo[1.1.0]butane is over one and one-half times larger than the methine carbon in isobutane. It is only 10.4 kcal mol^{-1} less stable than the sp-hydridized carbon atom of acetylene with a net charge one-half as great, suggesting a considerable degree of s character in its bonds to neighbouring groups from which it withdraws charge. The volume of a bridgehead carbon in [1.1.1]propellane is nearly 50% larger than that of a methyl group carbon atom and more than twice as large as the central carbon in neopentane. The bridgehead carbons are large and exposed in these two molecules. In keeping with this observation, both molecules undergo acetolysis by protonation of the bridgehead carbons, as previously discussed in Section V.C. in terms of the Laplacian of the charge density.

In the normal and branched alkanes the general behaviour, as noted above, is for the volume of a carbon atom to decrease as its stability and electron population increase. This behaviour is what is observed for atomic volumes of the free atoms across a short row of the periodic table, from an alkali metal to an inert gas. In the strained hydrocarbons, on the other hand, just the opposite behaviour is observed, with the volume increasing along with an increase in population and stability of the carbon.

The parallel increases in electron population, stability and volume of a carbon atom which accompany an increase in its s character is further illustrated by the series ethane, ethylene and acetylene. The idea extends to carbocations as well (Table 5), the sp^2 carbons of CH_3^+ and $C(CH_3)_3^+$ having volumes in excess of those for an sp^3 carbon in the hydrocarbons from which they are derived. Correspondingly, they also have greater electronegativities as reflected in their net charges and stabilities. Thus the positive charge in CH_3^+ is shared equally over all four nuclei. As anticipated in terms of hyperconjugation model, the electron population of the central carbon increases with increasing methyl substitution, the central carbon in tertiary butyl carbon bearing a net negative charge. While $v(C)$ is greater for the sp^2 carbon of a carbocation than it is for the corresponding carbon in the saturated hydrocarbon from which it is derived, the value of $v(C)$ for the sp^2 carbon is found to decrease, rather than increase, with $N(C)$. The width of an sp^2 carbon in the plane perpendicular to the plane of the nuclei does indeed increase with the increase in hyperconjugative electron release to this carbon which accompanies an increase in methyl substitution, from 5.7 au in CH_3^+ to 6.7 au in $C(CH_3)_3^+$. In the plane of the nuclei, however, the C of CH_3^+ extends between the H atoms while in $C(CH_3)_3^+$ the central carbon is totally enclosed by the interatomic surfaces with other carbons. In addition to the net charges on the carbon in CH_3^+ and in methane being the same, the former is more stable than the latter by 45.7 kcal mol^{-1}. The stability of the carbon which changes from sp^3 to sp^2 on formation of a carbocation increases with the extent of methyl substitution, equalling 147 kcal mol^{-1} in the case of *tert*-butyl cation, thereby accounting for the observation that the acidity of a branched hydrocarbon increases with the extent of methyl substitution.

The volume of a methylene group decreases along with a decrease in the number of adjacent methyl groups. This is evident in Figure 12, where the methylene carbon adjacent to the methyl carbon appears less confined than the one next to it which is bonded only to methylenic carbons. The volume of the carbon atom in a methyl group and of the group itself decrease as the number of methyl groups substituted on a single carbon increases. The volume of a carbon atom also decreases with the extent of methyl substitution. In all of these cases, the volume decrease of the carbon atom is accompanied by an increase in its electron population and stability. If these are examples of what could be termed increases in *steric strain*, then such strain and its consequences must be sharply distinguished from the *geometric strain* found in small ring systems as described above. The two types of strain exhibit opposite behaviour with respect to the volume change accompanying increases in the stability and electron population of a carbon atom.

This study of the properties of atoms in molecules shows that a carbon atom subjected to geometrical strain, an unsaturated carbon atom and an sp^2 carbon in a carbocation exhibit similar properties with respect to changes in their atomic populations, energies and volumes. For example, reactions leading to a reduction in geometric strain, in unsaturation or in positive charge will all proceed at a faster rate with an increase in pressure, for all these reactions lead to a decrease in the volumes of the associated carbon atoms. (The atomic volume decrease associated with the loss of charge in a reaction of a carbocation could be offset to some extent by an accompanying decrease in the electroconstriction of a solvent.) Most important, a carbon atom subjected to an increasing degree of geometric strain behaves increasingly like an unsaturated carbon atom with respect to the paralleling behaviour of its charges, stability and volume. The hydrocarbons also demonstrate that the volumes of atoms and of functional groups can be a characteristic property, one that is transferable between systems.

E. Acyclic Hydrocarbons and Additivity of the Energy

It is clear from Figure 12 that to within the accuracy of the plots shown in that figure, the charge densities of the chemically equivalent methyl and methylene groups are superimposable, as are the interatomic surfaces which define the atoms and the boundaries of the groups. This suggests that the groups are transferable among the molecules. This is further borne out by the corresponding constancies in the group populations and volumes discussed above. The energy is the most sensitive of properties and it is shown here that the energies of these same equivalent groups change by less than one kcal mol^{-1} upon transfer among members of the series and thus the charge distributions and all of the properties of each group, including its polarizability, are predicted to be transferable to within the same experimental accuracy, as is indeed observed. The transferability of the polarizability of these same groups is demonstrated in a subsequent section.

This constancy in properties and charge distributions requires that the equilibrium geometries of the groups remain essentially unchanged throughout the series of molecules, as predicted by SCF calculations at the different levels of basis set. Table 1 permits a comparison of bond lengths at the 6-31G* level of basis. Examples of the constancy in geometrical parameters using the D—H basis set are as follows: the bond lengths of the in-plane (H') and out-of-plane (H) hydrogens in the transferable methyl groups in propane to hexane all equal 1.0841 and 1.0850 Å, respectively, while the corresponding bond angles vary by 0.02° about the value of 107.76° for HCH and by 0.01° about the value 107.67° for HCH', except for propane where it has the value 107.80°. A methylene bonded to a methyl group has a C—H bond length of 1.0867 Å and an HCH angle of $106.25 \pm 0.01°$.

It is possible to fit the experimental heats of formation of the members of the homologous series $CH_3(CH_2)_mCH_3$ starting with $m = 0$, with the expression $\Delta H_f^\circ = 2A + mB$ where A is the contribution from the methyl group and B that from the methylene group. The generally accepted value of B at 25°C is -4.93 kcal mol^{-1} while $A = -10.12$ kcal mol^{-1}. Wiberg[50] and Schulman and Disch[51] have shown that the correlation energy correction, the zero point energies and the change in ΔH_f° on going from 298 to 0 K are well represented by group equivalents. This is borne out by the fact that the additivity of the energy is recovered by single determinant SCF calculations which refer to the vibrationless molecules at 0 K. Thus the total calculated molecular energies E can be fitted to a similar relationship, with $E = 2A^\circ + mB^\circ$ for all three basis sets with a maximum error of 0.05 kcal mol^{-1}. This means that the calculated state functions, energies and charge distributions contain the necessary information to account for the additivity observed in this homologous series of molecules. For the D—H basis set $A^\circ = -39.62953$ au and $B^\circ = -39.04743$ au, where A° is the energy of a methyl group in ethane, equal to $\frac{1}{2}E(C_2H_6)$.

It is clear from the group populations given in Table 4 that the methyl group in ethane is not identical to the corresponding group in the other members of the homologous series, and similarly, the methylene group differs slightly depending on whether it is bonded to one methyl or to two other methylenes. The small differences in populations found for these groups are to be anticipated as their environments change by correspondingly small amounts in these cases. In ethane, methyl is bonded to methyl, while in the other molecules it is bonded to methylene, from which it withdraws charge. Table 4 also lists the energies of the methyl groups relative to the constant $A°$, the energy of a methyl group in ethane. To within the accuracy of the numerical integrations, the energy and populations of methyl are constant when it is bonded to methylene, and the Me group is the same in all members of the homologous series past ethane. The Me group in these molecules is more stable relative to methyl in ethane by an amount $\Delta E = -10.4 \pm 0.4\,\mathrm{kcal\,mol^{-1}}$ and its electron population is greater by an amount $\Delta N = 0.0176 \pm 0.0005e$. The charges and energies obtained using the results of the 6-31G*/6-31G* and 6-31G**/6-31G* calculations behave in the identical manner, their values of ΔE and ΔN being $-10.7 \pm 0.9\,\mathrm{kcal\,mol^{-1}}$ and $0.0165 \pm 0.005e$ for the former and $-10.5 \pm 0.5\,\mathrm{kcal\,mol^{-1}}$ and $0.018 \pm 0.001e$ for the latter.

The charge and energy gained by methyl is taken from the methylene groups. What is most remarkable, and what accounts for the additivity observed in these molecules is that the energy gained by methyl is equal to the energy lost by methylene. Table 4 lists the energies and charges of the methylene groups relative to the energy increment $B°$. One finds that in propane, where it is bonded to two methyls, the energy of the methylene group is $B° - 2\Delta E$ and it transfers $2\Delta N\,e$ to the methyl groups, where ΔE and ΔN are the differences obtained above for the Me group bonded to a methylene. In butane, a methylene is bonded to a single methyl, its energy is given by $B° - \Delta E$ and its population decreases by ΔN to give a net charge $q(\mathrm{CH_2}) = +0.018$. The corresponding $\mathrm{CH_2}$ groups in pentane and hexane, those bonded to a single methyl, have the same properties as the $\mathrm{CH_2}$ groups in butane. Thus the central methylene in pentane and the two such groups in hexane, those bonded only to other methylenes, should have an energy equal to the increment $B°$ and a zero net charge. This is what is found to within the uncertainties of the integrated values. Therefore, methylene groups bonded only to other methylenes, as found in pentane and in all succeeding members of the series, possess zero charge and contribute the standard increment $B°$ to the total energy of the molecule.

The group additivity scheme for the energy in hydrocarbons is not the result of methyl and methylene groups having the same energies in every molecule in spite of small changes in their environments. Instead, their properties do change with changes in environment to give two kinds of methyl groups and three kinds of methylene groups. There are only three different $\mathrm{CH_2}$ groups because the change in environment is damped by a single such group. The underlying reason for the observation of additivity in the face of these small differences is the fact that the change in energy for a change in population, the quantity $\Delta E/\Delta N$, is the same for both the methyl and methylene groups. Thus the small amount of charge shifted from methylene to methyl makes the same contribution to the total energy. Reference to Table 1 shows that nearly all of this small shift in charge and its accompanying change in energy are restricted to the carbon atoms of the methyl and methylene groups, the properties of corresponding hydrogen atoms remaining essentially unchanged. This is to be expected, as it is the carbons that are bonded and share a common interatomic surface and the charge transfer is accomplished by a small shift in this surface. This shift is reflected in the movement of the C—C bond critical point 0.003 or 0.004 Å in the counter direction of the charge transfer and the bonded radius of the methyl carbon is slightly greater than that for the methylenic carbon. Since the transfer of charge occurs between chemically similar atoms, the resulting change in energy is zero. It is also necessary that the change in correlation energy for a change in population be the same for both groups if one is to account for the experimental observation of group additivity of the energy.

It is to be emphasized that there is no *a priori* reason why the energy of the central transferable methylene group in pentane or hexane, as defined by the surfaces of zero flux in the gradient vector field of the charge density, should have an energy equal to the constant $B°$ appearing in the expression $E = 2A° + mB°$ for the total calculated energies, an equation which mirrors the expression for the experimental heats of formation. It is straightforward to use quantum mechanics to relate a spectroscopically measured energy to the theoretically defined difference in energy between two states of a system. In a less direct, but no less rigorous manner, quantum mechanics also relates the difference in the experimentally determined heats of formation of pentane and hexane to the corresponding, theoretically defined energy of the methylene group.

F. Physical Basis For Transferability of Group Properties

According to equation 23, which determines the mechanics of an atom in a molecule, an atom responds only to the total force $\sigma \cdot d\mathbf{S}$ exerted on every element of its atomic surface. If it were not for this property of an atom and its properties being determined by the total force exerted on it, rather than by the individual contributions to this force, there would be no chemically recognizable atoms or functional groups. The virial of this force determines both the kinetic and potential energies of an atom. Reference 18 gives a table of the individual electronic and nuclear contributions to the virials of these forces for the atoms in the transferable methyl groups. These contributions differ by large amounts between successive members of the series of the normal hydrocarbons, but the total virial and total potential energy, as determined by the Ehrenfest force, remain essentially constant.

One finds that the total atomic populations and kinetic energies, the latter equalling minus the atomic energy, are all remarkably constant for each kind of atom. If the distribution of charge is the same for each type of atom, then the quantity $V°_{en}(\Omega)$, the potential energy of interaction of the nucleus of atom Ω with its own charge distribution, should also be the same. This is indeed the case, the largest difference being $0.4\,\mathrm{kcal\,mol^{-1}}$ for a hydrogen and the carbons exhibiting a variation of $\pm 2.6\,\mathrm{kcal\,mol^{-1}}$ about the mean. The interaction of *all* the nuclei in the molecule with the charge in atom Ω, the quantity $V_{ne}(\Omega)$, increases by very large amounts with the removal of each CH_2 group, changing by $1.5 \times 10^3\,\mathrm{kcal\,mol^{-1}}$ for H and by $9.2 \times 10^3\,\mathrm{kcal\,mol^{-1}}$ for carbon from hexane to propane. The repulsion of the electrons in atom Ω by the other electrons in the molecule, the quantity $V_{ee}(\Omega)$, and the contribution of the nuclear–nuclear repulsion energy to the energy of Ω, the quantity $V_{nn}(\Omega)$, on the other hand, decrease by large amounts with the removal of each CH_2 group. The sum of all three contributions to give $\mathscr{V}(\Omega)$, the total potential energy of the atom (equation 19), is, however, the same for each kind of atom through the series. This must be, since $\mathscr{V}(\Omega)$ is equal to twice the total energy of the atom by the virial theorem. The quantity $\mathscr{V}(\Omega)$ is the total virial, the virial of all the forces—electron–nuclear, electron–electron and nuclear–nuclear—acting on atom Ω and this quantity is conserved along with the kinetic energy when the distribution of charge of the atom remains unchanged.

VII. THE ORIGIN OF STRAIN ENERGY

A. Definition of Strain Energy

To discuss and compare theoretically defined strain energies with those defined in terms of the experimental heats of formation for the cyclic, bicyclic and propellane molecules, we shall define group energies in the same manner as is done using the experimental heats of formation except that they will not be referenced to the standard states of the elements. The strain energy of a given group in a molecule is obtained by subtracting from its energy, the energy of the corresponding standard group. The energies of the standard methyl and methylene groups are the constants A° and B° reported above in the expression for the calculated energy E. The energies of the standard methyne and C groups are obtained by subtracting three and four times the value of A°, respectively, from the total energies of isobutane and neopentane. The strain energies calculated in this manner using the results of the 6-31G*/6-31G* calculations are given in kcal mol^{-1} opposite the CH$_2$, CH and C groups for the structures shown in Figure 13. The sum of the group strain energies for a given structure is also given, followed in brackets by the experimental value determined using the Franklin group equivalents. While there are small differences between the theoretical and experimental values for the strain energy, the overall agreement is rather good, the largest errors being found for [1.1.1] and [2.2.2]propellane.

The results presented in this section make use of the 6-31G*/6-31G* calculations, since only for this basis has such a large set of molecules been treated in a consistent manner. However, some cyclic and bicyclic molecules have been calculated using the larger basis sets and, as demonstrated in the text, the results given here quantitatively mirror those obtained using the larger basis sets.

B. Cyclic Molecules

As noted above, the relative order of the electronegativities of H and C is reversed from that found in the standard methylene group when carbon is subjected to the geometrical strain encountered in the formation of the small ring cyclic compounds because of the accompanying increase in s character of the C—H bonds. Thus, relative to its population in the standard CH$_2$ group, each hydrogen in cyclopropane transfers 0.048e to carbon. This leads to an increase of 15.5 kcal mol^{-1} in the stability of the carbon but to a 12.6 kcal mol^{-1} decrease in the stability of each of the two hydrogens relative to their values in the standard group. Overall, the transfer of charge within the CH$_2$ group leads to a 9.6 kcal mol^{-1} decrease in its stability relative to the standard value and to a total strain energy three times this, or 28.8 kcal mol^{-1}, in good agreement with the generally accepted value. The use of the double star and D—H basis sets leads not only to the same group charges, but also to the same charge transfers within a group and to the same changes in energy relative to the standard groups. This was noted above for the methyl and methylene groups in the acyclic molecules and remains true for the strained systems. The double star results for cyclopropane yield a charge transfer of 0.044e from H to C relative to the standard group and a contribution of 9.2 kcal mol^{-1} to the strain energy, while the D—H basis yields corresponding values of 0.048e and 9.2 kcal mol^{-1}. The CH$_2$ group in cyclopropane is only 2.3 kcal mol^{-1} more stable than the same group in ethylene and, in terms of the charge transfer within the group and its energy, it resembles the ethylene fragment more than it does the standard methylene group.

If one takes the s character in the C—H bonds to carbon in the standard methylene group to be 25%, in ethylene to be 33.3% and in acetylene to be 50%, then the energy of this carbon varies in a linear manner with the s character of this bond, the energy decreasing by 3.0 kcal mol^{-1} with each percent increase in s character. In all of these molecules the charge transfer is between C and H and the functional group has no net charge. The population of carbon in cyclopropane is slightly smaller than that of the carbon in ethylene and its energy is correspondingly greater, by 7.9 kcal mol^{-1}, yielding ca 31% s character. While the carbon of ethylene is more stable than the carbon in the standard CH$_2$ group, the CH$_2$ fragment in ethylene is less stable than the standard, as is the same fragment in cyclopropane. In both cases the hydrogens are destabilized more than the carbon is stablized by the transfer of charge within the CH$_2$ group.

In the less strained cyclobutane, the transfer of charge from H to C relative to the populations in the standard CH_2 group is reduced to 0.014e. Correspondingly, carbon has an energy only slightly greater than the standard value and, in terms of the above linear relationship, a 26% s character. In agreement with this, the value of the C—H coupling constant, 134 Hz, is only slightly greater than the value for methylene in cyclohexane and the relationship between J(C—H) and s character referred to previously[45] yield a value of 27%. Similarly, the bond length, stretching force constant and the HCH bond angle for the CH_2 group in this molecule suggest only a small increase in s character over that found in the standard CH_2 group. Thus the stability of C is increased by only 1.0 kcal mol^{-1} over the standard value while the stability of each hydrogen is decreased by 3.9 kcal mol^{-1} and each CH_2 group in cyclobutane is 6.7 kcal mol^{-1} less stable than the standard. The total strain energy is calculated to be 26.8 (experimental, 26.5) kcal mol^{-1}.

The populations and energies of the atoms in the methylene group of cyclohexane differ little from their values in the standard group and the small differences which are present lead to a total calculated strain energy of 0.06 kcal mol^{-1} compared to an anticipated value of zero for a system with no geometrical strain, the same value obtained using the double star basis. The results for cyclopentane refer to but one configuration of a molecule which undergoes a rapid ring-puckering motion. The results are, however, intermediate between those for its neighbours. Two of the methylene groups have almost zero net strain while the third is less stable by 4.2 kcal mol^{-1}. The total strain energy for this one geometry is calculated to be 7.1 kcal mol^{-1} compared to the conformationally averaged experimental value of 6.2 kcal mol^{-1}.

The relative degree of geometric strain in these molecules can be measured in terms of the departure of the C—C—C bond path angles from the tetrahedral value (Table 2). A carbon atom in a CH_2 group subjected to an increasing amount of geometric strain, as determined in this way, has an increasing amount of charge transferred to it from H relative to the populations in the standard methylene group. This redistribution of charge within the methylene group leads to a stabilization of the carbon atom, but to an even greater destabilization of the hydrogen atoms. The overall result is a strain energy equal to the increase in the energy of the group relative to that of standard methylene.

As discussed above, these results obtained using the theory of atoms in molecules can be rationalized in terms of the hybridization model, which predicts the s character of the carbon hybrid orbitals in the C—H bonds to increase with increasing geometric strain. This model of associating an increase in the stability and electronegativity of an atom with an increase in the s character of its hybrid orbitals gives results in agreement with theory in other systems as well. The inversion barrier in ammonia or phosphine and the barrier to the linearization of water have the same mechanical origin as does the strain energy in a cyclic hydrocarbon. The s character of the bonds from hydrogen to the N, P and O atoms increases in attaining the geometry associated with these barriers. In all three cases, this increase in the s character of the A—H bonds results in the transfer of electronic charge from the hydrogen atoms to the central atom A, and the stability of A is increased. As in the cyclic hydrocarbons, this transfer of charge to the A atom results in a shortening of the bonds to the hydrogen atoms and, to complete the analogy, the resulting decrease in the stability of the hydrogen atoms exceeds the increase in the stability of the central atom and the energy of the molecule is increased. The same mechanical factors operate to yield a planar geometry for the amide group, HN—C=O. A rotation about the C—N bond in planar formamide, where the N is presumed to be sp^2 hybridized, results in its pyramidalization and a change to sp^3 hybridization. As anticipated on the basis of the above examples, the theory of atoms in molecules demonstrates that the electronegativity of nitrogen relative to carbon is decreased by this rotation, which is accompanied by a transfer of charge to the carbon and hydrogen atoms and by a lengthening of the C—N and C—H bonds with the largest changes being observed for the carbon atom. The decrease in stability of the nitrogen caused by the loss of charge is greater than the increase in the stability of the carbon and hydrogen atoms and the planar geometry is favoured. Theory shows, as opposed to the results anticipated on the basis of the resonance model, that in addition to the amide nitrogen atom losing rather than gaining charge upon rotation, the oxygen atom undergoes only minor changes in its energy and population.

C. Bicyclic Molecules

Bicyclo[1.1.0]butane **13** is the most strained of the bicyclic molecules. The magnitude of the net charge of $-0.121e$ on the bridgehead carbon of this molecule, which possesses an inverted geometry, is exceeded only by that for the bridgehead carbons in [2.1.1] and [2.2.1]propellane. This carbon withdraws charge from its bonded hydrogen and from the two methylene groups. Its energy is only $9.5\,kcal\,mol^{-1}$ above that of carbon in acetylene, suggesting a considerable degree of s character in its hybrid bonds to the groups from which it withdraws charge. Orbital analyses indicate that the bridgehead bond is essentially pure p and hence all the s character is in the hybrids directed at H and the CH_2 groups. The ^{13}C coupling constant for the bridgehead C—H bond yields 41% s character leaving $ca\,30\%$ for each of the hybrids to the methylene groups. The coupling constants for the C—H bonds in these CH_2 groups average to the same value as found for cyclopropane, giving them a 32% s character. This implies that each of the hybrids directed at the bridgehead carbons have 18% s character, considerably less than that of the hybrids directed from the bridgehead carbons. While admittedly approximate, the analysis indicates that in the bonds between a bridgehead and a methylenic carbon there is considerably more s character in the hybrids from the former than from the latter. Therefore, the bridgehead carbons should withdraw charge from the methylene carbons as well as from the bridgehead hydrogens. Thus what at first might appear as an anomaly—the most strained carbon atom possessing the most negative total energy—follows directly from the orbital model. It predicts the hybrids from the bridgehead carbons to be relatively rich in s character and for these atoms to be the most electronegative, and hence the most stable in the molecule.

As noted above, in assigning the group energy to CH it is assumed that there is no transfer of charge from this group to the three attached methyl groups, whereas it bears a charge of $+0.084e$ in isobutane. As a result, the energy for the standard CH group is $60\,kcal\,mol^{-1}$ more stable than its actual energy in the isobutane molecule. Even so, because of the very electronegative nature of the bridgehead carbon and the similarity of its energy with the energy of carbon in acetylene, the energy of the bridgehead CH group differs from the standard value by only $1.6\,kcal\,mol^{-1}$. Thus essentially all of the strain energy in this molecule, when computed using standard group energies, has its origin in the charge distributions of the methylenic groups. The charges and energies of the methylenic hydrogens are very similar to those for cyclopropane, as is the average of the ^{13}C coupling constants for the C—H bonds. Thus the increase in the strain energy for the methylenic groups in this molecule over that found in cyclopropane comes from the further transfer of charge from the methylenic to the bridgehead carbons. Each CH_2 group has a net charge of $+0.068e$ and is calculated to be $32.7\,kcal\,mol^{-1}$ less stable than the electrically neutral standard CH_2 group. The total calculated strain energy is $68.6\,(63.9)\,kcal\,mol^{-1}$.

In the progression through the bicyclic series from the most strained bicyclobutane **13** to the least strained bicyclo[2.2.2]octane **19**, the populations of the bridgehead carbons undergo a continuous decrease and the CH group becomes positively charged. The populations of the bridging methylene groups, on the other hand, increase through the

This term, together with the sum of the atomic polarization terms

$$\mu_p(CH_3) = \sum_j M(H_j) + M(C) \tag{34}$$

determines the contribution of the Me group to the molecular dipole moment if there is no transfer of charge from the Me group to the remainder of the system, i.e. if $q(Me) = 0$. If the group charge is different from zero, then a term $q(Me)(X_{Me} - X_R)$, which represents the contribution resulting from the transfer of charge from Me to the rest of the system, is included in the group contribution.

Table 7 lists the magnitudes of the charge transfer and atomic polarization contributions defined as in equations 33 and 34, as well as their sum, for the methyl and methylene groups in the normal hydrocarbons and for the methylene group in cyclopropane. The weak field resulting from the small transfer of charge from C to H does not determine the direction of the polarization of the atomic densities in the hydrocarbons. Each hydrogen of a methyl or a methylene group in an acyclic hydrocarbon is polarized away from carbon, in the direction of the charge transfer, while the carbon is polarized to a lesser extent away from the hydrogens. The magnitude of a hydrogen moment is three to four times greater than that for carbon and thus both μ_c and μ_p are directed away from carbon. The fact that the sum of these two contributions is nearly equal to the magnitude of the group moment even in those cases where symmetry does not require such an equality shows that μ_c, μ_p and their sum μ are essentially parallel vectors. The group moment for methyl is directed along the C—C axis linking the group to the remainder of the molecule to within 0.7° or less, while the moment for a methylene group is directed along the C_2 axis which bisects the group to within ca 1°.

The group moments exhibits the same pattern of values as found for other group properties in this series of molecules, with the methyl group in ethane and the methylene group in propane being distinct. The charge transfer contribution is only slightly larger than the atomic polarization term for the methyl group, but it is ca 1.7 times larger for the methylene group.

There is a reduction in the transfer of charge from carbon to hydrogen in cyclopropane and the charge transfer contribution to the methylene group in this molecule is correspondingly reduced in value from the standard group found in pentane or hexane. Even so, the group moment is larger in this case because of a considerable increase in the atomic polarization contribution. There is an accumulation of charge in the interior of the ring in the cyclic molecule causing a reversal in the direction of the polarization of the charge density on carbon. In cyclopropane, both the C and H atoms are polarized

TABLE 7. Group contributions to molecular dipole moments[a]

Molecule	$\mu_c(CH_3)$	$\mu_p(CH_3)$	$\mu(CH_3)$	$\mu_c(CH_2)$	$\mu_p(CH_2)$	$\mu(CH_2)$
C_2H_6	0.135	0.126	0.261			
C_3H_8	0.140	0.114	0.254	0.183	0.125	0.308
$n\text{-}C_4H_{10}$	0.140	0.114	0.254	0.190	0.112	0.302
$n\text{-}C_5H_{12}$	0.141	0.115	0.256	0.191	0.114	0.305
				0.198^b	0.104^b	0.302
$n\text{-}C_6H_{14}$	0.134	0.119	0.256	0.191	0.116	0.306
				0.198^b	0.106^b	0.304^b
$c\text{-}C_3H_6$				0.071	0.290	0.361

[a]In atomic units, 1 au = 2.542 debyes.
[b]Methylene bonded only to other methylenes.

in the direction of the transfer of charge from C to H, with the moment for carbon being approximately twice that for hydrogen.

The relatively large, outwardly directed polarization of the carbon atom density in a three-membered ring is responsible for the negative end of the observed dipole of cyclopropene, ca 0.5 debye, being directed away from the double bond[20]. In this cyclic molecule the charge transfer contributions to the dipole from the double bond and from the methylene group nearly cancel, and the final moment is determined by the polarization of the methylene group, and this primarily by the polarization of the carbon which, in this case, is 1.5 times larger than that found in cyclopropane itself.

Corresponding to equation 28 for the dipole moment, the expression for the atomic contributions to a change in the dipole moment, as caused by a displacement of the nuclei or by an applied field, is

$$\Delta\boldsymbol{\mu} = \sum_{\Omega}\{[\Delta q(\Omega)\mathbf{X}_{\Omega} + \mathbf{Z}_{\Omega}\Delta\mathbf{X}_{\Omega}] + \Delta\mathbf{M}(\Omega)\} \tag{35}$$

$$= \Delta\boldsymbol{\mu}_{c} + \Delta\boldsymbol{\mu}_{p}$$

The charge transfer term can be changed by a change in the atomic charges or by a displacement of the nuclei. The change $\Delta\boldsymbol{\mu}_{c}$ can be expressed in terms of origin-independent group contributions in the same manner as described above in equations 31 and 32 for $\boldsymbol{\mu}_{c}$ itself. Equation 35 is used to determine the group contributions to the molecular polarizability and to vibrational intensities.

B. Definition and Additivity of Group Polarizabilities

The polarizability of a molecule is the proportionality constant $\boldsymbol{\alpha}$ which relates the first-order change in the dipole moment to an applied electric field:

$$\Delta\boldsymbol{\mu} = \boldsymbol{\alpha}\cdot\mathscr{E} \tag{36}$$

The change in the dipole moment is related to the first-order correction to the charge density $\rho'(r)$ by

$$\Delta\boldsymbol{\mu} = -\int \mathbf{r}\rho'(\mathbf{r})d\tau\cdot\mathscr{E} \tag{37}$$

and in terms of this expression one may define an electric polarizability density $\boldsymbol{\alpha}(\mathbf{r})$:

$$\boldsymbol{\alpha}(\mathbf{r}) = -\mathbf{r}\rho'(\mathbf{r}) \tag{38}$$

The integration of this density over the basin of an atom gives the corresponding atomic contribution to the polarizability, and the polarizability, like all properties, is determined by a sum of atomic contributions. The polarizability density includes the interaction of the density at a point $\mathbf{r}$ with the effects of the external field at all points within the molecule. The mean polarizability $\bar{\alpha}$ is defined as the average of the three principal components of the polarizability tensor:

$$\bar{\alpha} = (\alpha_{xx} + \alpha_{yy} + \alpha_{zz})/3 \tag{39}$$

The computational details of determining the atomic contributions to α are given in Reference 52, where the theory is applied to a number of molecules covering a wide range of atomic interactions. The extent to which an applied external field changes the polarization of an atomic density or the degree of interatomic charge transfer is readily related to the static properties of the atomic charge distributions and the corresponding contributions to the polarizability may be understood and predicted. This is illustrated here by a discussion of the polarizabilities of the methyl and methylene groups.

Placing a molecule in an external electric field has two effects: it causes electronic charge to be transferred across the interatomic surfaces in the direction of the (positive) field, and it changes the polarization of the atomic densities, inducing atomic polarizations, which can be in the direction of the applied field or counter to it. The first effect is described by the term $\Delta\mu_c$ and the second by $\Delta\mu_p$ in equation 35. Division of these changes by the field $\mathcal{E}$ yields the proportionality constant between $\mathcal{E}$ and $\Delta\mu$ i.e. the polarizability α (equation 36). Since the application of the electric field perturbs the electronic charge density, the interatomic surfaces are also shifted by the field. For a tumbling molecule, however, this effect averages to zero and the atomic contributions are correspondingly averages.

The interatomic charge transfers and atomic polarizations induced in the hydrocarbons by an applied electric field are characteristic and quite novel. In each case, for a field applied parallel to the extended chain, there is an apparent transfer of electronic charge over the length of the molecule between the terminal methyl groups, as illustrated in Figure 15. In this figure, the area enclosed by the 0.001 au density envelope for each molecule is partitioned into group contributions by the interatomic surfaces between bonded carbon atoms, and the charge lost or gained per unit field, applied parallel to the major axis or perpendicular to it, is indicated for each group. There is no transfer of charge between the groups for a field applied perpendicular to the plane of the diagram. In propane the parallel field leads to a transfer of charge across the intervening methylene group from one methyl group to the other, while the perpendicular field causes charge to be transferred from both methyl groups to the methylene group. In butane, the application of a field parallel to the two terminal C—C bonds in the *trans* conformation leads to an alternation in the sign of the intergroup charge transfers, the direction of the charge transfer between the methylene groups being counter to a larger transfer of charge between the terminal methyl groups in the direction of the applied field. This alternation in sign is not observed in pentane for the parallel field because of its orientation with respect to the C—C bond axes. Even here, however, the charge transfer between the methylene groups, while in the same direction as that of the applied field, is close to zero. In this molecule one observes an alternation in the signs of the intergroup charge transfers for the perpendicular field, with the terminal methyls and central methylene groups gaining charge at the expense of the two equivalent intervening methylene groups. In butane, the orientation of the perpendicular field is such as to cause a transfer of charge from the downfield methyl and methylene groups to their upfield counterparts.

Most of the charge transferred as a result of the application of a field in any direction occurs between the hydrogen atoms, their changes in population being, in general, twenty to one hundred times larger than those for the carbon atoms. The application of a field perpendicular to the major axis of the molecule induces a transfer of electronic charge from the downfield to the upfield hydrogen atoms of the methyl and methylene groups when the atoms are so oriented with respect to the field. A perpendicular field in the plane of the chain of carbon nuclei in propane also induces a transfer of charge from the methyl groups to the methylene group, while in butane the same field, as well as transferring charge between the hydrogen atoms in each methyl group, results in a transfer of charge between the two methylene groups. The transfer of charge caused by the application of a field perpendicular to the plane containing the chain of carbon nuclei is by symmetry restricted to occur between pairs of hydrogen atoms which are equivalent with respect to the symmetry plane and within a given group.

The charge transfer and polarization contributions to the methyl group in ethane are given in Table 8 to illustrate the general response of the charge distribution of a hydrocarbon group to an external electric field. The terms labelled α_c are the contributions to the polarizability arising from the change in the populations of the hydrogens, referenced to the carbon nucleus. This term is calculated in the manner prescribed in

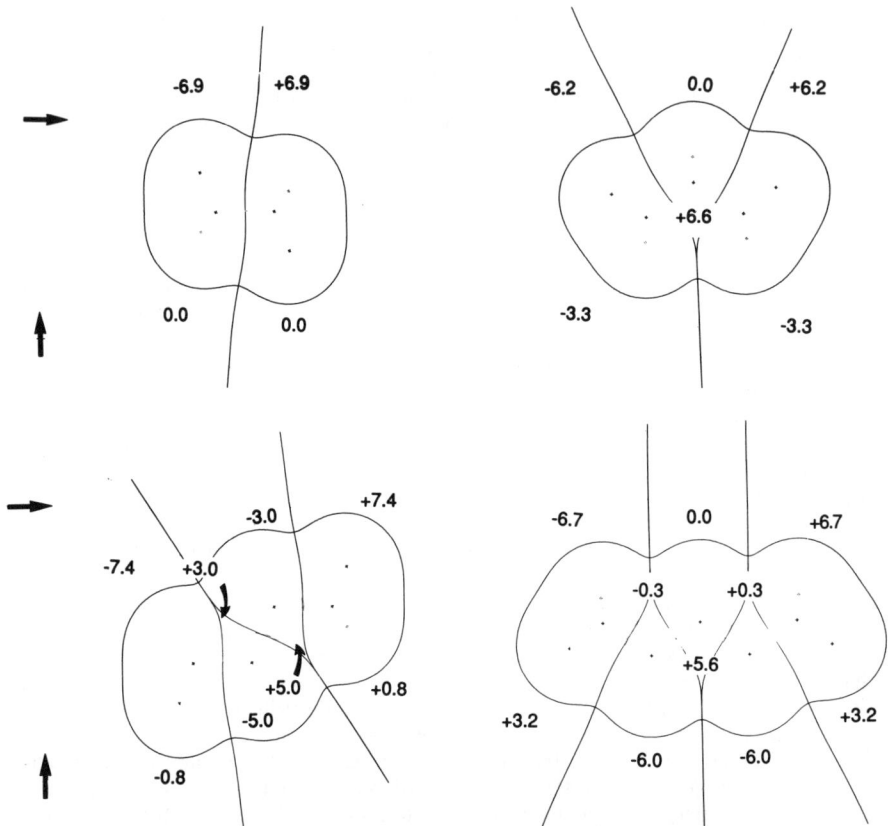

Figure 15. The 0.001 au outer density envelope for the hydrocarbons ethane to pentane showing the group boundaries. The numbers shown are the transfers of charge induced between equivalent groups per unit field, by fields applied in the directions indicated by the arrows on the left-hand side of the figure

equation 33 with q(H) replaced by Δq(H) and the resultant $\Delta\mu_c$ divided by the field strength. Since the applied field also causes a change in the net charge of the methyl group, there is an additional charge transfer contribution to the polarizability, labelled the intergroup charge transfer, which is given by $\Delta q(CH_3)/\mathscr{E}$ multiplied by the distance of the carbon nucleus to the critical point of the bond path linking the carbon nucleus to the carbon of the neighbouring group. The contribution arising from the changes in the atomic polarizations is calculated as in equation 34 with $\mathbf{M}(\Omega)$ replaced by $\Delta\mathbf{M}(\Omega)$ and the resultant $\Delta\mu_p$ divided by $\mathscr{E}$ to give α_p. The hydrogen polarization contribution is listed separately from that for carbon.

The charge distributions of the hydrogen atoms always polarize in the direction of the applied field. Since a hydrogen is polarized away from carbon, the net polarization of a downfield hydrogen is decreased, while that of an upfield atom is increased. The charge distributions of the carbon atoms consistently polarize in a direction counter to that of the applied field, in an apparent response to the oppositely directed internal field

TABLE 8. Atomic and group contributions to molecular polarizability[a]

Atomic contributions in ethane: all contributions per unit applied field

Component	$\Delta M(H)$	$\Delta M(C)$	$\alpha_p(CH_3)$	$\alpha_c(CH_3)$	Intergroup charge transfer	$\alpha_{ii}(CH_3)$
$\parallel$	1.55	−5.15	−0.48	5.10	9.87	14.49
$\perp$	1.07	−3.28	−0.06	12.93	0.00	12.87

Group contributions to mean polarizability

Molecule	$\bar{\alpha}(CH_3)$	$\bar{\alpha}(CH_2)$	Molecular polarizability				
			calc.	exp.	$\alpha_\parallel$	$\alpha_{\perp i}$	$\alpha_{\perp 0}$
C_2H_6	13.43		26.87	30.51	29.06	25.77	25.77
C_3H_8	13.43	11.25	38.11	43.30	41.67	37.21	35.47
$n\text{-}C_4H_{10}$	13.48	11.26	49.47	~55.8	55.17	48.34	44.90
$n\text{-}C_5H_{12}$	13.59	11.26	60.90		71.03	57.49	54.24
		11.16^b					
$c\text{-}C_3H_6$		11.38	34.51	38.51	35.60	35.60	31.23

[a]All quantities in atomic units.
[b]For the central methylene.

created by the transfer of charge between the hydrogen atoms. The same behaviour is found in other molecules for atoms shielded from the external field, the carbon in carbon dioxide or ethylene, for example, when the field is applied along the major axis and for the carbon in methane for any field direction.

The counter polarization induced in a carbon atom by an external field is much greater than is the polarization induced in a hydrogen. The field-induced polarizations in the methyl group of ethane are typical (Table 8). For a parallel field, the magnitude of the induced moment per unit field strength, i.e. the contribution to the polarizability, is greater for the carbon atom than it is for all three hydrogens. The net induced polarization of the methyl group is opposed to the applied field and decreases the group's polarizability. For perpendicularly applied fields, the opposing polarizations of the hydrogens and the carbon approximately cancel for a methyl group. In a methylene group, the counter polarization of the carbon atom always dominates the group's net polarization and the polarizability of the group is correspondingly decreased.

The group contributions to the mean molecular polarizabilities calculated in this manner are given in Table 8. The methylene group in propane and those bonded to a single methyl group in butane and pentane are found to exhibit a single value for their mean polarizabilities, to within the uncertainties of the calculation. The contribution of the central methylene group to the mean polarizability of pentane is slightly less. There is no experimental value with which to compare the calculated polarizability for pentane, and the slight increase in the contribution from the methyl group in this case over the values for propane and butane cannot be verified. However, even here the change is less than 1 percent, a variation within the bounds of the experimentally determined additivity of group polarizabilities.

Using the experimental results one may determine group contribution in an empirical manner, by subtracting the contribution of ethane (which is itself assumed to equal twice the mean polarizability of a methyl group) from that of propane to obtain the contribution of the methylene group. Following this procedure one obtains $\bar{\alpha}(CHH_3) = 15.26$ au and

$\bar{\alpha}(CH_2) = 12.79$ au, values which reproduce the mean polarizabilities of the other acyclic hydrocarbons. The theoretical values underestimate the experimental values by an almost constant amount through the series of molecules. Thus the present basis yields values which, when multiplied by a factor of 1.135, recover the experimental values for ethane, propane and butane with an error of less than 0.1 au.

The ratio $\bar{\alpha}(CH_3)/\bar{\alpha}(CH_2)$ equals 1.198 using the theoretically determined group values, while the same ratio determined from the experimental values equals 1.195. Thus the total molecular polarizability is partitioned between the methyl and methylene groups in a manner determined by the quantum definition of an atom in a molecule. The partitioning of space resulting from the imposition of the zero flux boundary condition yields methyl and methylene groups whose predicted properties recover the experimentally observed group additivity of both the energy and polarizability.

It was noted in the preceding sections that there is a small but significant change in the distribution of charge of a methylene group when it is transferred from an acyclic hydrocarbon to the cyclopropane ring. Thus, as anticipated on this basis, the experimental mean polarizability of cyclopropane differs slightly from the value obtained by taking three times the mean polarizability of the same group determined from the experimental values for the acyclic hydrocarbons, the predicted value being 38.37 au compared to the observed value of 38.51 au. The theoretical values for the group contributions recover this experimental trend with $\bar{\alpha}(CH_2)$ for the standard methylene group equalling 11.25 au compared to 11.38 au for the same group in cyclopropane.

Table 8 also includes the calculated principal components of the molecular polarizability tensors; $\parallel$ denotes the component parallel to the carbon chain, while $\perp$ denotes the perpendicular components, $\perp_i$ in the plane containing the carbon nuclei, $\perp_0$ perpendicular to this plane as well. The parallel component is the largest in each case and it exhibits the largest homologous increase, a result of the contribution to this component from the transfer of charge over the length of the molecule between the terminal methyl groups.

A linear correspondence exists between the mean polarizability of a molecule and its volume, as determined by an outer envelope of the charge density within the theory of atoms in molecules. This relationship holds not only for the hydrocarbons but for molecules covering the spectrum of atomic interactions from shared to closed-shell.[52] The theory demonstrates that one can rigorously define a group contribution to a molecular polarizability. This, together with the ability of the theory to assign volumes to individual atoms and functional groups, not only places the correlations between these two properties on a firm theoretical basis, but makes possible the assignment of a mean polarizability to a molecule through a knowledge of the volumes of its constituent groups, quantities which are readily determined and tabulated, as they are here for the hydrocarbons.

C. Vibrationally Induced Molecular Dipole Moments

Knowledge of the characteristic features of the charge distribution of a given functional group enables one to understand and correlate its properties. This is further illustrated for the hydrocarbons by demonstrating that their response to an electric field internally generated by a displacement of the nuclei is similar to that obtained in the presence of an externally applied field. Both responses are similarly correlated with the polarizations and atomic charges characteristic of a hydrocarbon charge distribution.

This correlation of properties is illustrated for the vibrationally induced dipole moments for the infrared active modes of methane and ethylene. The details of the calculation are given in Reference 20. Table 9 lists the charges on the hydrogen and carbon atoms in methane together with their first moments determined in an SCF

TABLE 9. Vibrationally and field-induced dipole moments in methane[a]

Atom Ω	Equilibrium geometry $q(\Omega)$	$\mathbf{M}(\Omega)$	Stretched geometry $\Delta q(\Omega)$	$\Delta\mathbf{M}(\Omega)$	Bent geometry $\Delta q(\Omega)$	$\Delta\mathbf{M}(\Omega)$	External field $\Delta q(\Omega)$	$\Delta\mathbf{M}(\Omega)$
H_0	−0.062	−0.065	+0.238	+0.010	+0.084	+0.037	+2.92	+1.20
H_i	−0.062	+0.065	−0.303	+0.008	−0.061	+0.036	−2.96	+1.21
C	+0.248	0.000	+0.130	−1.540	−0.046	−0.605	+0.04	−2.98

Contribution from			
$\Delta\mu_c$	+1.133	+0.544	+13.88
$\Delta\mu_p$	−1.538	−0.749	+1.84
$\Delta\mu$	−0.405	−0.205	+15.72

[a]All changes are pet unit displacement or per unit field in atomic units.

6-31G**/6-31G** calculation. Also listed are the changes in these quantities, per unit displacement, for the antisymmetric stretching and bending motions depicted in the table.

The electronic population of hydrogen increases when the C—H bond is shortened or the HCH angle is decreased. This has the effect of creating a charge transfer moment in the direction of the shortened bonds or in the direction of the hydrogens with the smaller bond angle, resulting in the net charges on the atoms labelled H_i becoming more negative. In the antisymmetric stretching motion, the change in population on the carbon atom is quite small compared with the changes for the hydrogen atoms. Just the reverse is true for the atomic first moments, the values for the hydrogens being close to zero while the charge distribution of the carbon atom is strongly polarized to yield a large first moment whose direction is counter to the charge transfer moment. This relaxation of the atomic charge distribution within the boundaries of the carbon atom is the largest single change caused by the antisymmetric motion of the nuclei and it dominates the induced dipole moment for this mode (Table 9). Thus there are two opposing charge transfers induced by the antisymmetric stretch: charge in the outer region of the distribution is transferred from the outwardly to the inwardly displaced protons, while in the interior of the molecule an even larger amount of charge is transferred in the opposite direction within the basin of the carbon atom.

The extent of charge transfer between the hydrogens caused by the antisymmetric bending motion in methane is smaller by a factor of 10, while the polarizations of the hydrogens are larger by about the same factor compared to the corresponding changes found for the stretching mode. The changes in the charge transfer moment are determined by the changes in the hydrogen charges as they are in the stretching mode and also, as for the stretching mode, the overall change in the dipole is dominated by a counter-polarization of the charge density of the carbon atom, and the direction of the final moment corresponds to a flow of negative charge in the direction of the opened bond angle.

The displacements of the nuclei in the antisymmetric stretching and bending modes create an electric field within the molecule in the direction of the inwardly displaced protons, H_i. The reader will recognize the pattern of interatomic charge transfers and atomic polarizations resulting from these vibrationally generated fields as being similar in all respects to those resulting from a similarly directed externally applied field. Included in Table 9 are the changes in net charges and atomic first moments induced by an

76   Richard F. W. Bader

external field applied in the direction of the negative z axis. There is a transfer of charge
between the hydrogen atoms in the direction of the applied field, and their atomic
densities are polarized in the same direction. The carbon atom, on the other hand, while
undergoing a smaller change in population, exhibits a much larger polarization whose
direction is counter to that of the applied field. The only difference in the vibrationally
and field-induced dipole moments is that the former are dominated by the counter
polarizations of the carbon charge distribution, while the latter is dominated by a much
larger charge transfer component (Table 9).

IX. CONCLUSION

The introductory section emphasized the important historical role that the study of the
alkanes has played in establishing the concept of a functional group in chemistry. This
chapter has illustrated how quantum mechanics can be used to better understand and
predict the chemistry of the alkanes because of its ability to define the relevant groups
and characterize their properties. Atoms and functional groupings of atoms do make
additive contributions to the properties of the system in which they occur, the constancy
in these contributions for a given group, as it occurs in difference molecules, mirroring
the constancy in its distribution of charge. Thus a knowledge of the charge distribution
of a given group and of how it reacts to a changed environment enables one to predict
and understand its characteristic properties, both static and dynamic, and their
susceptibility to change.

   The study of the alkanes has also played a definitive role in the development of the
structural aspects of chemistry, the notion that a molecule is a collection of atoms linked
by a three-dimensional network of bonds. Figure 3 provides dramatic evidence that
theory recovers all of the molecular structure hypothesis, by predicting the network of
bonds which define a given structure, as well as the atoms they link. These predictions
are possible not only for alkanes but for all systems, including the metallic state. Theory
demonstrates that the notion of structure is inseparable from that of structural stability,
thereby extending the operational usefulness of the molecular structure hypothesis
through the definition of what is meant by the making and breaking of chemical bonds
and a delineation of the mechanisms whereby this can be accomplished. Theory provides
new ways of asking questions concerning the stability of chemical bonds, by restating
them in terms of the dynamic topology of the charge density. The propensity and location
of a system to chemical attack is in turn shown to be determined by the concentrations
and depletions of charge determined by the Laplacian of the charge density.

   Chemists think of problems in synthesis, structure and reactivity in terms of the
properties of the functional groups involved. The theory of atoms in molecules makes
possible not only the quantification of this idea by providing the chemist with a complete
prediction of group properties but also, as a consequence of being able to apply physics
to an atom in a molecule, it imbues him with a deeper physical insight into chemistry.

X. REFERENCES

1. R. F. W. Bader and T. T. Nguyen-Dang, *Adv. Quantum Chem.*, **14**, 63 (1981).
2. R. F. W. Bader, T. T. Nguyen-Dang and Y. Tal, *Rep. Prog. Phys.*, **44**, 893 (1981).
3. R. F. W. Bader, *Atoms In Chemistry—A Quantum Theory*, Oxford University Press, Oxford, 1990.
4. cf. (a) S. Glasstone, *Textbook of Physical Chemistry*, 2nd edition, D. Van Nostrand, Co. New York, 1946, p. 524.
   (b) J. W. Bruhl, *Justus Leibigs Ann. Chem.*, **200**, 139 (1880); **203**, 1, 255 (1880).
5. R. J. W. Le Fevre, in *Advances in Physical Organic Chemistry*, Vol. 3 (Ed. V. Gold), Academic Press, London, 1965, pp. 1–90.
6. F. D. Rossini, *J. Res. Natl. Bur. Stand.*, **6**, 37 (1931); **7**, 329 (1931).

7. F. D. Rossini, *J. Res. Natl. Bur. Stand.*, **13**, 21 (1934).
8. E. J. Prosen, W. H. Johnson and F. D. Rossini, *J. Res. Natl. Bur. Stand.*, **37**, 51 (1946).
9. J. L. Franklin, *Ind. Eng. Chem.*, **41**, 1070 (1949).
10. S. W. Benson, F. R. Cruickshank, D. M. Golden, G. R. Haugen, H. E. O'neal, A. S. Rodgers, R. Shaw and R. Walsh, *Chem. Rev.*, **69**, 279 (1969).
11. P. Hohenberg and W. Kohn, *Phys. Rev.*, **136**, B864 (1964).
12. R. F. W. Bader, T. T. Nguyen-Dang and Y. Tal, *J. Chem. Phys.*, **70**, 4316 (1979).
13. R. F. W. Bader and P. M. Beddall, *J. Chem. Phys.*, **56**, 3320 (1972).
14. K. B. Wiberg, R. F. W. Bader and C. D. H. Lau, *J. Am. Chem. Soc.*, **109**, 985 (1987).
15. P. Chakrabarti, P. Seiler and J. D. Dunitz, *J. Am. Chem. Soc.*, **103**, 7378 (1981).
16. J. E. Jackson and L. C. Allen, *J. Am. Chem. Soc.*, **106**, 591 (1984).
17. W. J. Hehre, L. Radom, P. v. R. Schleyer and J. A. Pople, *Ab Initio Molecular Orbital Theory*, Wiley, New York, 1986.
18. K. B. Wiberg, R. F. W. Bader and C. D. H. Lau, *J. Am. Chem. Soc.*, **109**, 1001 (1987).
19. R. F. W. Bader, *Can. J. Chem.*, **64**, 1036 (1986).
20. R. F. W. Bader, A. Larouche, C. Gatti, M. Carroll, P. J. MacDougall and K. B. Wiberg, *J. Chem. Phys.*, **87**, 1142 (1987).
21. T. H. Dunning, *J. Chem. Phys.*, **53**, 2823 (1970); **55**, 716 (1971).
22. S. Huzinaga, *Gaussian Basis Sets for Molecular Calculations*, Elsevier, Amsterdam, 1984.
23. J. Palis and S. Smale, *Pure Math.*, **14**, 223 (1970).
24. R. F. W. Bader, T. S. Slee, D. Cremer and E. Kraka, *J. Am. Chem. Soc.*, **105**, 5061 (1983); D. Cremer, E. Kraka, T. S. Slee, R. F. W. Bader, C. D. H. Lau, T. T. Nguyen-Dang and P. J. MacDougall, *J. Am. Chem. Soc*, **105**, 5069 (1983).
25. C. Gatti, M. Barzaghi and M. Simonetta, *J. Am. Chem. Soc.*, **107**, 878 (1985).
26. R. F. W. Bader and H. Essen, *J. Chem. Phys.*, **80**, 1943 (1984).
27. G. N. Lewis, *J. Am. Chem. Soc.*, **38**, 762 (1916).
28. R. F. W. Bader and M. E. Stephens, *J. Am. Chem. Soc.*, **97**, 7391 (1975).
29. R. F. W. Bader, P. J. MacDougall and C. D. H. Lau, *J. Am. Chem. Soc.*, **106**, 1594 (1984).
30. R. J. Gllespie and R. S. Nyholm, *Quart. Rev. Chem. Soc.*, **11**, 239 (1957).
31. R. J. Gillespie, *Molecular Geometry*, Van Nostrand Reinhold, London, 1972.
32. R. F. W. Bader, R. J. Gillespie and P. J. MacDougall, *J. Am. Chem. Soc.*, **110**, 7392 (1988).
33. R. F. W. Bader, R. J. Gillespie and P. J. MacDougall, in *From Atoms To Polymers* (Eds. J. F. Liebman and A. Greenberg), VCH Publ. Inc., New York, 1989.
34. M. T. Carroll, C. Chang and R. F. W. Bader, *Mol. Phys.*, **63**, 387 (1988); M. T. Carroll, J. R. Cheeseman, R. Osman and H. Weinstein, *J. Phys. Chem.*, **93**, 5120 (1989); R. F. W. Bader and C. Chang, *J. Phys. Chem.*, **93**, 2946 (1989); T. G. Tang, W. J. Hu, D..Y. Yan and Y. P. Cui, *J. Mol. Struct. (Theochem)*, **207**, 327 (1990).
35. K. B. Wiberg and S. R. Kass, *J. Am. Chem. Soc.*, **107**, 988 (1985).
36. J. D. Dunitz, *X-ray Analysis and the Structure of Organic Molecules*, Cornel University Press, Ithaca, NY, 1979, pp. 391ff.
37. J. D. Dunitz and P. Seiler, *J. Am. Chem. Soc.*, **105**, 7056 (1983).
38. K. B. Wiberg, W. P. Dailey, F. H. Walker, S. T. Wadell, L. S. Crocker and M. D. Newton, *J. Am. Chem. Soc.*, **107**, 7247 (1985).
39. K. B. Wiberg and G. Szeimies, *J. Am. Chem. Soc.*, **92**, 571 (1970).
40. R. F. W. Bader, W. H. Henneker and P. E. Cade, *J. Chem. Phys.*, **46**, 3341 (1967).
41. E. Schrödinger, *Ann. Phys.*, **79**, 361 (1926).
42. R. P. Feynman, *Rev. Mod. Phys.*, **20**, 367 (1948).
43. J. Schwinger, *Phys. Rev.*, **82**, 914 (1951).
44. R. F. W. Bader, M. T. Carroll, J. R. Cheeseman and C. Chang, *J. Am. Chem. Soc.*, **109**, 7968 (1987).
45. J. N. Shoolery, *J. Chem. Phys.*, **31**, 1427 (1959); N. Muller and D. E. Pritchard, *J. Chem. Phys.*, **31**, 1471 (1959).
46. A. D. Walsh, *Nature (London)*, **159**, 167, 712 (1947).
47. C. A. Coulson and W. E. Moffitt, *Philos. Mag.*, **40**, 1 (1949).
48. J. March, *Advanced Organic Chemistry*, 3rd ed., Wiley–Interscience, New York, 1985, p. 222.
49. A. Gavezzotti, *J. Am. Chem. Soc.*, **105**, 5220 (1983); **107**, 962 (1985).
50. K. B. Wiberg, *J. Comput. Chem.*, **5**, 197 (1984).
51. J. M. Schulman and R. L. Disch, *Chem. Phys. Lett.*, **113**, 291 (1985).
52. K. E. Laidig and R. F. W. Bader, *J. Chem. Phys.*, **93**, 7213 (1990).

CHAPTER **2**

Structural chemistry of alkanes—a personal view

EIJI ŌSAWA

Department of Knowledge-Based Information Engineering, Toyohashi University of Technology, Toyohashi 441, Japan

I. INTRODUCTION

This chapter describes *static* structural aspects of alkanes and cycloalkanes as seen through average atomic positions[1]. Neither conformational and electronic structures nor energetic features are explicitly dealt with here, since other chapters take care of these. Because of the limited space, the type of molecules reviewed here must be restricted as well: we focus only on the unstrained, not-so-large and saturated hydrocarbons except for a very few cases. As the result of these restrictions, such favorite topics for this kind of review, like cyclopropane[2], cyclobutane[3], cubane[4], inverted carbon[5], catenane[6], pagodane[7], dodecahedrane[8] and cage molecules in general[9,10], are not included. Fortunately, these have recently been reviewed elsewhere. We place some emphasis on the aspects of computational chemistry related to structural problems, which may not necessarily be approved by every chemist. This is the reason for adding a subtitle.

The Chemistry of Alkanes and Cycloalkanes
Edited by S. Patai and Z. Rappoport © 1992 John Wiley & Sons Ltd

80   E. Ōsawa

II. EXPERIMENTAL STRUCTURES

A. Vapor Phase

Every discussion on the structure of organic molecules traditionally starts from the bond distances and valence and torsional angles of small and saturated hydrocarbons in the vapor phase. Table 1 presents an exemplary set of such structures obtained by the gas-phase electron diffraction (ED)[11]. Only the r_g values, namely structural parameters containing effects of thermal vibration[12], are collected mainly because the electron diffraction results are usually given in terms of r_g. However, as Rasmussen noted[13], the r_g structure is suitable only when bond lengths are compared among closely related structures. On the other hand, r_α° structures, which Rasmussen recommends[13] and which are based on average nuclear positions in the vibrational ground state, are somewhat artificial due to the assumptions made for the thermal correction. Although structural chemists are very keen on these definitions, the differences in the structural parameters among various definitions are actually very small, less than 0.01 Å in bond length and 1° in valence angles. This is true even including r_e values which correspond to completely motionless structures frozen at 0 K and are obtained by ab initio theoretical calculations. Since these differences are of the same order of magnitude as the experimental uncertainties, no serious practical problem arises even if the definitions are ignored, as do Hehre and coworkers in their book[14]. For this reason, the recent trend in the experimental structure determination of free molecules seems to be to combine the results of electron diffraction, microwave spectroscopy and computational methods (such as molecular mechanics and ab initio calculations) to reach the best compromise[15]. Norbornane in Table 1 is a typical example, which even includes infrared/Raman and X-ray information.

Table 1 was used, together with several additional molecules, as the reference dataset in the construction of an empirical molecular mechanics force field[16]. A similar but more extensive collection of dataset served as the structural standard for Allinger's new force field MM3[17]. However, fitting these temperature-dependent, dynamically averaged values to a static structural model creates a lot of problems, principally because thermal molecular vibration varies depending on individual structure, and simple averaging is basically wrong. Instead, more and more attempts are being made to use the results of high level ab initio calculations for the standard[18]. It may be added here that, in the near future, the utility of r_g structures as given in Table 1 may revive as the ultimate goal for the theoretical predictions of hydrocarbon structures starting from r_e values.

B. Solid Phase

Cambridge Structural Database (CSD) has been playing an increasingly important role in many aspects of structural organic chemistry, already containing the results of more than 80,000 X-ray crystallographic analyses of organic compounds[19]. The old Sutton Table of standard bond lengths is extensively revised based on the statistical analysis of CSD data taking into account influences from the vicinity of the bond in question[20]. Table 2 reproduces a part of the average distances of C_{sp^3}—C_{sp^3} bonds substituted with H and/or C_{sp^3} atoms. Note that the X-ray bond length is determined as the distance between the points of highest electron density. Variations in the distances due to the substitution pattern are surprisingly large. It remains to be seen how to account quantitatively for these effects which include steric, hybridization and other miscellaneous factors. The X-ray diffraction method is much more versatile and flexible in molecular structure determinations than other experimental methods. When higher precision regarding the proton positions is required, one can turn to the neutron diffraction method[21].

Chemists have long been warned that the closely packed molecules in crystals are deformed by lattice forces, hence the derived structural parameters may differ from the vapor phase values. Arguments trying to explain molecular properties measured in the gas or solution in terms of X-ray structure generally stumble upon this remark. For saturated hydrocarbons, one must consider two effects: First, packing forces are much weaker than for the polar molecules, hence the deformations arising from such weak interactions can be largely neglected. Second, the effects of crystal packing must have been averaged out if statistically processed values like the bond lengths given in Table 2 are used. Conceptually a more attractive correction for the packing force will be to evaluate the effect by some computational approach. The potential energy calculation based on a rigid body keeps the molecular structure fixed at what was obtained by X-ray analysis or at some standard structure, and optimizes intermolecular van der Waals (and electrostatic interaction) energies over a range of up to 10 Å. Program WMIN[22] based on this approach has been extensively used[23], for example for comparing the three solid modifications of ethane at low temperatures[24]. In this particular case, however, the ethane molecule is fixed at the standard geometry (C—C, 1.532, C—H, 1.095 Å; C—C—H, 109.5 or 111.5°), hence no new information on the molecular structure can be obtained.

Removal of the rigid-body constraint provides insights into the ways packing force deforms molecules. Our preliminary observations[25] illustrate the point. When the crystal of n-hexane (space group P1) was optimized with regard to intra- and intermolecular MM2'[26] steric energies, it gave unit cell parameters close to those observed, as expected (Table 3). An interesting point is that the observed C—C bond lengths seemed to *decrease* as one goes to the inside of a molecule (see bond lengths in Table 3). Although the low precision of X-ray measurements in this instance precludes conclusive statements, this trend contradicts the well-known trend for n-alkanes in the vapor phase, according to which the bond lengths *increase* as one goes to the inside of a chain[27]. *Ab initio* calculations also support this tendency[28]. Is the trend in the crystal a new effect of the crystal packing force? According to our preliminary calculations (Table 3), the trend of C—C bond length variation in the optimized crystal (entry K) follows that in the vapor phase and *ab initio* calculation, a rational result for a nonpolar molecule.

When a compound fails to give crystals suitable for diffraction experiments, solid-phase NMR spectroscopy sometimes give important information on the structure. For example, the lengths of the central bond in two derivatives of hexaphenylethane (1 and 2)[29,30] have been determined by the nutation NMR method, which measures splitting due to magnetic dipole–dipole coupling in the solid state ^{13}C-NMR spectrum of a non-reorienting sample[31]. The splitting pattern of 2, ^{13}C-enriched at the central ethane carbons, fits best to a bond length between 1.64 and 1.65 Å. Similar results are obtained for the ^{13}C-enriched 1.

There is a small story behind these measurements. Initially, 1 gave poor crystals, the X-ray analysis of which showed an abnormally *short* central bond distance of 1.47(2)Å[29] and aroused some controversy, since theoretical computations consistently predict it to be longer than 1.6 Å [30–32]. The solid-state NMR experiments mentioned above led to a conclusion that the previously reported X-ray distance of 1 was grossly in error[31].

(1) R=C_6H_5

(2) R=H

82   E. Ōsawa

TABLE 1. Bond lengths r (r_g, Å), valence angles θ(r_α, deg), torsional angles ϕ(deg) and intermolecular nonbonded distances d(Å) of small hydrocarbons as determined by the vapor-phase electron diffraction method (experimental uncertainties are given in parentheses, the last digit coinciding with that of the measured value)

Hydrocarbon	Quantity	Value	Reference
Methane	r(CH)	1.1068(10)	a
	d(H···H)	1.8118(70)	
	θ(HCH)	107.30(30)	
Ethane	r(CC)	1.5323(20)	b
	r(CH)	1.1017(20)	
	θ(HCH)	107.30(30)	
	d(C···H)	2.1964	
Propane	r(CC)	1.5323(30)	c
	r(CH)	1.1073(51)	
	θ(CCC)	112.4(12)	
	θ(HCH)$_{CH_3}$	107(3)	
	θ(HCH)$_{CH_2}$	106.1(assumed)	
Isobutane	r(CC)	1.535(1)	d
	r(C^{t}H)	1.122(6)	
	r(CH)$_{CH_3}$	1.113(2)	
	θ(CCC)	110.8(2)	
	θ(^{t}HCC)	108.1(2)	
	θ(CCH$_1$)	111.4(4)	
	θ(CCH$_2$)	110.1(3)	
	θ(CCH)$_{av}$	111.4(2)	
	θ(H$_1$CH$_3$)	108.7(11)	
	θ(H$_1$CH$_2$)	106.5(17)	
n-Butane (average for *trans* and *gauche*)	r(CC)	1.531(2)	e
	r(CH)	1.119(2)	
	θ(CCC)	113.3(4)	
	θ(CCH)$_{av}$	110.7(4)	
	ϕ(CCCC)	72.4(48)	
	%*trans*	64(7)	
Cyclopentane	r(CC)	1.546(1.2)	f
	r(CH)	1.1135(15)	
Neopentane	r(CC)	1.537(3)	g
	r(CH)	1.114(8)	
	θ(CCH)	112.2(28)	
Cyclohexane	r(CC)	1.536(2)	h
	r(CH)	1.121(4)	
	θ(CCC)	111.4(2)	
Methylcyclohexane	r(CC)$_{av}$	1.536(2)	i
	r(CH)$_{av}$	1.124(4)	
	θ(CCC)$_{ring}$	111.4(5)	
	θ(CCC)$_{exo}$	112.1(16)	
	θ(CCH)$_{av}$	109.3(5)	
1,1-Dimethylcyclohexane	r(CC)$_{endo,av}$	1.532(2)	j
	r(CC)$_{exo,av}$	1.536(2)	
	θ(CCH)$_{av}$	110.2(5)	
	ϕ(C$_6$C$_1$C$_2$C$_3$)	50.4(10)	
	ϕ(C$_1$C$_2$C$_3$C$_4$)	51.3(10)	
	ϕ(C$_2$C$_3$C$_4$C$_5$)	53.2(10)	
Norbornane	r(C$_1$C$_2$)	1.536(15)	k
	r(C$_1$C$_7$)	1.546(24)	

TABLE 1. (continued)

Hydrocarbon	Quantity	Value	Reference
	$r(C_2C_3)$	1.573(15)	
	$\theta(C_1C_2C_3)$	102.7	
	$\theta(C_2C_1C_6)$	109.0	
	$\theta(C_2C_1C_7)$	102.0(1)	
	$\theta(C_1C_7C_4)$	93.4	
	$\phi(C_6C_1C_2C_3)$	71.6	
	$\phi(C_2C_1C_7C_4)$	56.3	
	$\phi(C_7C_1C_2C_3)$	35.8	
	$\phi(C_1C_2C_3C_4)$	0.0	
Bicyclo[3.3.1]nonane (65 °C)	$r(CC)_{av}$	1.536(2)	l
	$r(CH)_{av}$	1.107(7)	
	$\theta(HCH)_{av}$	106.5(45)	
	$\theta(C_1C_9C_5)$	107.3(56)	
	$\theta(C_2C_3C_4)$	113.0(41)	
	%BC^p	5(4)	
Cyclooctane (BC)^p	$r(CC)_{av}$	1.540(1)	m
	$r(CH)_{av}$	1.116(2)	
	$\theta(C_2C_1C_8)$	120.2(10)	
	$\theta(C_1C_2C_3)$	116.1(18)	
	$\theta(C_2C_3C_4)$	117.0(20)	
	$\theta(C_3C_4C_5)$	116.1(15)	
	$\theta(HCH)_{av}$	104.5(11)	
	$\phi(C_8C_1C_2C_3)$	63.1(41)	
	$\phi(C_1C_2C_3C_4)$	98.4(35)	
Cyclodecane ([2323], C_{2h})	$r(CC)_{av}$	1.545(3)	n
	$r(CH)_{av}$	1.115(3)	
	$\theta(CCC)_{av}$	116.1(11)	
	$\theta(HCC)_{av}$	108.7(10)	
Cyclododecane([3333], D_4)	$r(CC)_{av}$	1.540(1)	o
	$r(CH)_{av}$	1.114(3)	
	$\theta(HCH)$	104.2(18)	
	$\theta(C_1C_2C_3)$	115.0(7)	
	$\phi(C_1C_2C_3C_4)$	159.5(12)	

^{a}L. S. Bartell, K. Kuchitsu and R. de Neui, J. Chem. Phys., 35, 1211 (1961).
^{b}T. Iijima, Bull. Chem. Soc. Jpn., 46, 2311 (1973). See also K. Kuchitsu, J. Chem. Phys., 49, 4456 (1968); L. S. Bartell and H. K. Higginbotham, J. Chem. Phys., 42, 851 (1965).
^{c}T. Iijima, Bull. Chem. Soc. Jpn., 45, 1291 (1972).
^{d}R. L. Hilderbrandt and J. D. Wieser, J. Mol. Struct., 15, 27 (1973).
^{e}R. K. Heenan and L. S. Bartell, J. Chem. Phys., 78, 1270 (1983). See also F. W. Bradford, S. Fitzwater and L. S. Bartell, J. Mol. Struct., 38, 185 (1977).
^{f}W. J. Adams, H. J. Geise and L. S. Bartell, J. Am. Chem. Soc., 92, 5013 (1970).
^{g}L. S. Bartell and W. F. Bradford, J. Mol. Struct., 37, 113 (1977).
^{h}O. Bastiansen, L. Fernholt, H. M. Seip, H. Kambara and K. Kuchitsu, J. Mol. Struct., 18, 163 (1973). See also J. D. Ewbank, G. Kirsch and L. Schäfer, J. Mol. Struct., 31, 39 (1976).
^{i}A. Tsuboyama, A. Murayama, S. Konaka and M. Kimura, J. Mol. Struct., 118, 351 (1984).
^{j}H. J Geise, F. C. Mijlhoff and C. Altona, J. Mol. Struct., 13, 211 (1972).
^{k}L. Doms, L. Van den Enden, H. J. Geise and C. Van Alsenoy, J. Am. Chem. Soc., 105, 158 (1983).
^{l}V. S. Mastryukov, M. V. Popik, O. V. Dorofeeva, A. V. Golubinskii, L. V. Vilkov, N. A. Belikova and N. L. Allinger, J. Am. Chem. Soc., 103 1333 (1981).
^{m}O. V. Dorofeeva, V. S. Mastryukov, N. L. Allinger and A. Almenningen, J. Phys. Chem., 89, 252 (1985). For ab initio calculation, see K. Siam, O. V. Dorofeeva, V. S. Mastryukov, J. D. Ewbank, N. L. Allinger and L. Schäfer, J. Mol. Struct., 164, 93 (1988).
^{n}R. L. Hilderbrandt, J. D. Wieser and L. K. Montgomery, J. Am. Chem. Soc., 95, 8598 (1973).
^{o}E. G. Astavin, V. S. Mastryukov, N. L. Allinger, A. Almenningen and R. Seip, J. Mol. Sruct., 212, 87 (1989).
^{p}BC = boat-chair.

TABLE 2. Statistical analysis of X-ray lengths of C_{sp^3}—C_{sp^3} bonds substituted with H or C_{sp^3} atoms[a]

	Mean	50%	σ	n
RCH_2—CH_3	1.513	1.514	0.014	192
R_2CH—CH_3	1.524	1.526	0.015	226
R_3C—CH_3	1.534	1.534	0.011	825
RCH_2—CH_2R	1.524	1.524	0.014	2459
R_2CH—CH_2R	1.531	1.531	0.012	1217
R_3C—CH_2R	1.538	1.539	0.010	330
R_2CH—CHR_2	1.542	1.542	0.011	321
R_3C—CHR_2	1.556	1.556	0.011	215
R_3C—CR_3	1.588	1.580	0.025	21
C—C(overall)	1.530	1.530	0.015	5777

[a]Excerpted from Reference 20. The unit is Ångstrom. Mean = unweighted sample mean. 50% = average over the median quartile of sample data. σ = standard deviation, n = number of data.

TABLE 3. n-Hexane crystal optimized with regard to inter- and intramolecular steric energies by the use of the KESSHOU program[25]

Unit-cell dimensions		X-ray	KESSHOU
	a	4.17(2)Å	4.03
	b	4.70(2)	4.33
	c	8.57(2)	8.46
	α	96.6(3)°	96.4
	β	87.2(3)	87.9
	γ	105.0(3)	102.5

Bond lengths (Å)

	C1—C2	C2—C3	C3—C4	C4—C5	C5—C6
X-ray[a,b]	1.530(10)	1.526(10)	1.524(10)	1.526(10)	1.530(10)
K[c]	1.531	1.534	1.534	1.534	1.531
MM2'	1.534	1.537	1.537	1.537	1.534
6–31G*[d]	1.528	1.529	1.529	1.529	1.528

[a]N. Norman and H. Mathisen, Acta Chem. Scand., 15, 1755 (1961). At −115 °C.
[b]Calculated based on published atomic coordinates.
[c]KESSHOU.
[d]Reference 28.

The ^{13}C–^{13}C dipolar coupling method was used for the determination of C—C bond lengths in the fascinating carbon cluster C_{60} (3), although this is not a *hydro*carbon. This giant molecule had failed to give clear X-ray results due to extremely rapid rotation even at low temperatures. Two types of bonds, the one fusing five- and six-membered rings and the other between two six-membered rings, are determined to be 1.45(0.015) and 1.40(0.015)Å, respectively[33]. These values are in close agreement with the AM1-calculated bond distances of 1.464 and 1.385 Å [34,35].

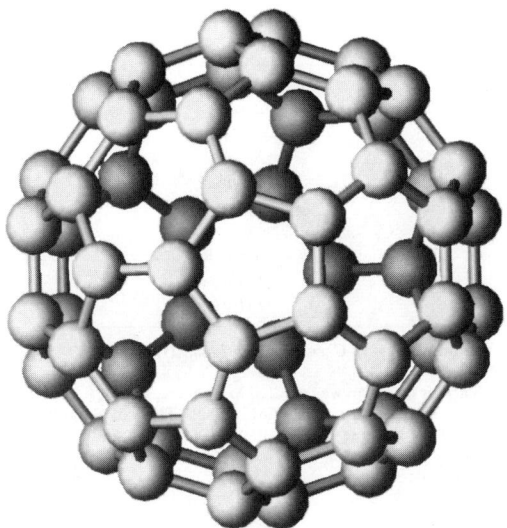

C_{60} molecule, **3**

C. Other Phases

Information obtainable from experimental determination of molecular structure in other phases than those mentioned above is severely limited for two reasons: the scarcity of available means except for the NMR technique, and the complications arising from heterogeneous environments. Recent intensive interest in the structure of molecules forming monolayers or other types of aggregates absorbed on the surface spur exploration of new methods for observing surface structures such as scanning tunneling microscope (STM). Dynamic structural aspects obtainable from these studies are, however, outside of this review[36].

Among various techniques of NMR, the nuclear Overhauser effect (NOE) and the coupling constant are the most frequently used for deriving microscopic information on the solution structures of molecules, especially bioactive substances like macrolide antibiotics[37]. Vicinal coupling constants give the angle of rotation around a bond via the Karplus equation[38-40], whereas NOE provides distances of close, nonbonded protons[37]. Using these data one can generally deduce 'average' structures in solution. For hydrocarbons, however, the application of these techniques requires partially deuterated samples and has been reported only occasionally.

Here we deviate a little from the topic of this chapter to mention an interesting possibility of stereochemical analysis. Since diastereoisomers show different vicinal coupling constants, the configuration of a partially deuterated hydrocarbon molecule having multiple asymmetric centers can be assigned by measuring these constants. This is the simplest way of determining the 3D stereochemistry, but has been used only for conformationally restricted cases like chair six-membered ring compounds[38]. Flexible molecules, like partially deuterated n-alkanes, cannot be subjected to this analysis unless the conformer distribution is known. No practical experimental means is available to determine the distribution of conformational isomers of flexible molecules. Hence, it seems worthwhile to test available computational techniques for estimating the conformer distribution and to expand the spectroscopic assignment of 3D stereochemistry to flexible molecules.

E. Ōsawa

TABLE 4. Vicinal H/H coupling constants (Hz) across the C2—C3 bond of partially deuterated n-alkanes 4–7

Compound		Vicinal protons	Observed[a]	Calculated[b]
meso-n-Butane	**4a**	*erythro*	8.76	9.42
dl-n-Butane	**4b**	*threo*	6.11	4.68
n-Pentane	**5**	*erythro*	9.13	9.87
n-Hexane[c]	**6a**	*erythro*	9.11	9.79
	6c	*erythro*	9.15	9.79
	6b	*threo*	5.97	4.52
	6b, 6c[c]	*erythro*	9.08	10.26
n-Heptane	**7**	*erythro*	9.13	9.76

[a] In tetramethylsilane: G. Schrumpf, *Angew. Chem., Int. Ed. Engl.*, **21**, 146 (1982); U. Hartge and G. Schrumpf, *J. Chem. Res. (S)*, 189 (1981).
[b] A version of the modified Karplus equation was used: S. Masamune, P. Ma, R. E. Moore, T. Fujiyoshi, C. Jaime and E. Ōsawa, *J. Chem. Soc., Chem. Commun.*, 261 (1986). The coupling constant given above is the population-weighted average of all possible C—C rotamers. Structure and relative energy were obtained by geometry-optimizing with MM2′ ($\varepsilon = 1.5$), Ref. 26. For the detailed methodology of calculating $^3J_{HH}$ values of flexible molecules, see Reference 40.
[c] Across the C3—C4 bond.

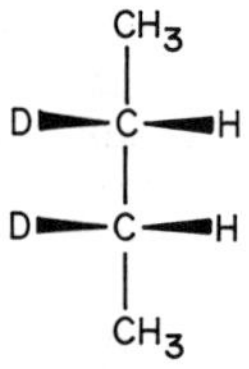

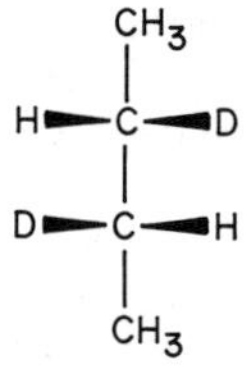

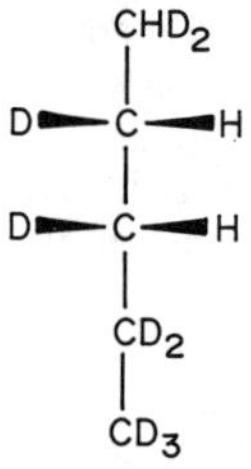

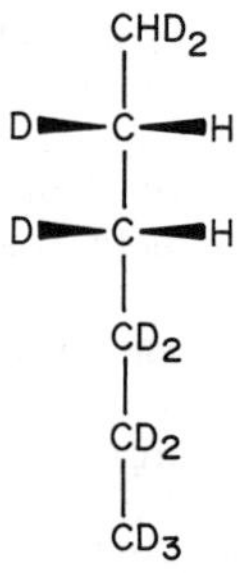

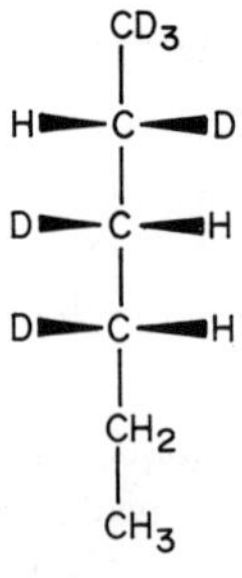

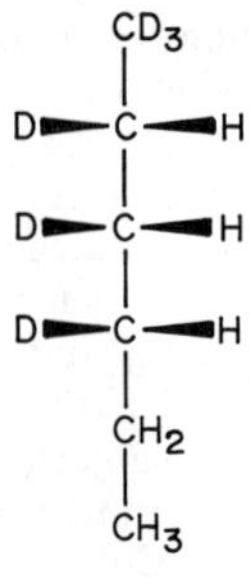

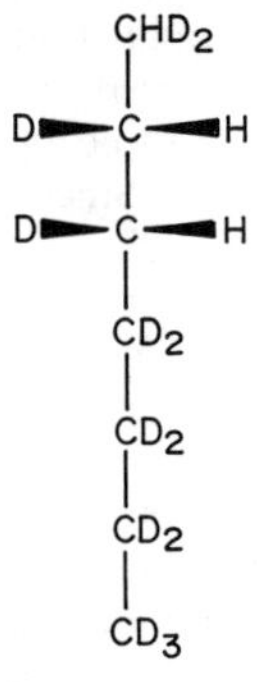

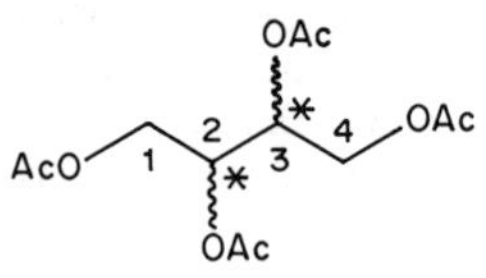

Table 4 shows the results of such an attempt, where the molecular mechanics method was used to estimate the relative population of equilibrating rotamers and to compute population-weighted vicinal H/H coupling constants by means of a Karplus-type equation. The calculated coupling constants are weight-averaged over three C—C rotamers for **4**; 27 for **5**; 81 for **6a**; 27 for **6b** and **6c**, and 243 for **7**. It can be seen that the preliminary results are promising, the calculated values agreeing moderately well with the observed values.

The sources of error can be ascribed to both the Karplus-type equation and the molecular mechanics method used. The Karplus equation was further improved to a precision of about 0.4 Hz[39], while MM2 was used instead of MM2', when the same strategy was applied to peracetylated alditols like **8**. Since such acetoxy derivatives lie outside the scope of this review, only a brief explanation will be given here. Instead of deuterium in **4–7**, the acetoxy group now imparts chirality to the carbon atoms. Test calculations were performed on tetra- (**8**) to hexaalditols and the error in the calculated vicinal H/H coupling constants was about 0.7 Hz in terms of standard deviation. By root-mean-square (rms) comparison of the observed vicinal coupling constants with the calculated constants of all magnetically unique diastereoisomers, it was possible to assign relative stereochemistry of asymmetric carbon atoms with a probability of 92%[40].

III. COMPUTATIONAL STRUCTURES

From the arguments presented in the previous sections, it is clear that informations on the exact molecular structure that are obtainable by experimental methods are limited and sometimes even ambiguous: limited because of technical difficulties and conditions inherent to the methods (e.g. the electron diffraction technique is applicable only to symmetric molecules, X-ray analysis only to crystals, etc.)[12], ambiguous because of imposed constraints and also because of thermal vibration effects.

The static structures of simpler hydrocarbons have been (as mentioned above) the most frequently used checkpoints for the accuracy of theoretical calculations. If the results of comparison indicate that the accuracies of theoretical computation of molecular structure are good enough, then we should be able to supplement or even replace experimental structural determination. Despite the well-known difficulties in reproducing energies by *ab initio* calculations, such as heats of formation of small molecules[41], the answer to the question on structure is clearly, yes. Early in the 1970s, Boggs[42] and Schäfer and coworkers[43] recognized the power of Pulay's gradient method, which can be used to automatically optimize molecular geometry during *ab initio* SCF MO calculations. The optimized geometry under the double-zeta level of theory turned out to be close to *real* geometry. They have since been systematically accumulating the examples and providing valuable structural informations, even those not amenable to experiments.

Continuing exponential improvements in the performance/price rates of high-speed computers[44] have made it possible for practising chemists to perform gradient geometry optimizations at levels higher than 4-21G for molecules of moderate size, a fact surprising in a sense, since such calculations have been monopolized by a handful of theoreticians having access to very large computers, until only a few years ago. The general merits of utilizing computational structures may be summarized as follows:

(1) Unstable, short-lived structures including energy maxima and saddle points can be obtained as readily as the ground state structures.

(2) Minute details outside of experimental uncertainties, like variations of C—H bond lengths under various circumstances, can be easily assessed (*vide infra*).

(3) Molecules under study need not be present as chemicals; it is often faster and easier to compute than to synthesize.

(4) The quality of the computed results is consistent and unaffected by the ability of the individual researcher as long as the same and well-maintained program and methodology are used.

A. The 4-21G/Gradient Method

The Arkansas and Texas computationist groups use program TEXAS equipped with Pulay's gradient technique of geometry optimization to obtain the molecular structures of small to medium-sized organic molecules on a 4-21G basis set[45,46]. Thus far, 600 structures of 150 molecules have been optimized, among them the largest hydrocarbon being a cage dimer of 7-*tert*-butylnorbornadiene (9)[47].

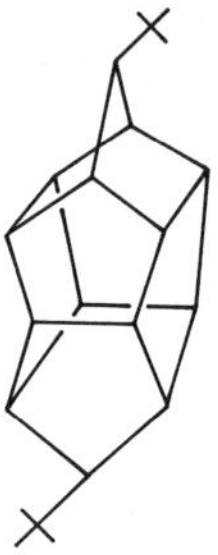

(9)

How well do the calculated geometries reproduce the observed ones? Here one must be cautious since, as mentioned above, definitions are different. Experimental geometries, e.g. r_g values, involve thermal vibration effects, whereas the *ab initio*-derived values refer to the potential minimum in the absence of vibrational motion (r_e, θ_e). With the basis set of double-zeta quality like 3-21G and 4-31G, C—C bond lengths are reproduced quite well, errors being typically less than 0.01 Å, while the mean absolute error in the bond angle calculation is about $1.0°$[14]. These error ranges cover variations of structural parameters with definitions (or experimental methods). Among a number of studies that have been made in the course of accumulating 4-21G/gradient structures, some pertaining to this chapter will be mentioned below.

Structural parameters, such as those given in Tables 1 and 2, change from molecule to molecule since they are the result of a number of intramolecular interactions, acting simultaneously within the molecular force field. Changes are significantly dependent on local environments, but they are so subtle that their magnitudes are usually comparable to the experimental errors. Hence it is difficult to accurately assess the cause of small geometrical variations. Being free from experimental uncertainties, computational results are an ideal vehicle with which to study the trends. For example, in the n-alkane series, the local symmetry of methyl groups is twofold rather than threefold, due to differences between the planar and nonplanar C—H bonds and H—C—H angles[48]. From n-propane to n-pentane, the C1—H2 bond (see 10) is calculated to be slightly (< 0.001 Å) but consistently shorter than the C1—H3 (= C1—H4) bond, whereas the C5—C1—H2 angle is about 0.5° larger than the C5—C1—H3 (= C5—C1—H4) angle. As mentioned above in connection with Table 3, the inner C—C (and C—H) bond lengths are about 0.001 Å longer than those of terminal bonds. The changes are marginal but the trend

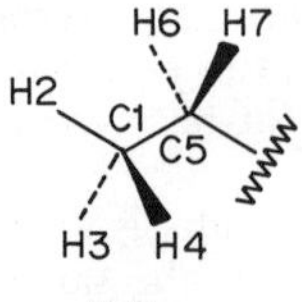

(10)

persists. The 4-21G C—C bonds for these small n-alkanes are about 0.01 Å longer than the r_g lengths.

Similarly, for cyclohexanes, the 4-21G/gradient calculation confirmed that equatorial C—H bonds are consistently shorter than axial bonds by 0.002 Å. Since equatorial C—H bonds are *trans* to the ring C—C bond, a general rule including n- and cycloalkanes is that, within the same CH_x group, a C—H bond antiperiplaner to C—C is consistently *shorter* than those antiperiplanar to C—H. The relative lengths between terminal and internal C—C bond mentioned above are also general: in dimethylcyclohexanes, for example, the equatorial Me—C_α(ring) bond is always *shorter* than the adjacent endocyclic C—C bonds by 0.005–0.01 Å[49].

In the course of these studies, it was realized that the rotation of, for example, a backbone C—C bond in n-alkane produces periodical variation in the C—C bond length as well as in the C—C—X (X = H, C) valence angles[46]. These observations are the most striking evidence for the *molecular mechanics* principle, wherein the dynamics of molecules is assumed to respond to a mechanical model[50].

B. Higher Levels of *ab initio* Calculations

The success of the 4-21G/gradient method prompted several authors to compare various computational methods for calculating geometries of small, unstrained hydrocarbons[28,51,52]. Among them, Häfelinger and coworkers[52] reported a systematic study about the dependence of bond lengths on the basis set. They found, among other things, that the differences between the observed and calculated bond distance depend linearly on bond distances. For this reason he proposed, instead of adding a single correction, using a corrective equation,

$$r^{exp} = m \cdot r^{calc} + a \tag{1}$$

where m and a are constants which vary depending on the definition of experimental lengths and basis set. Within the eight basis sets examined, r_g values of hydrocarbons are best reproduced by using the 6-31G basis set ($m = 0.948$, $a = 0.081$ Å).

The dependence of the calculated error in the bond length (within the Hartree–Fock limit) on bond lengths probably arises from the neglect of electron correlation. The primary reason for this guess is that the electron correlation is required to reach the true potential minimum[51], which in turn affects the geometry. Because of the very long computer time needed for the electron correlation calculations, no truly systematic study has been conducted on the effect of including electron correlation on the equilibrium geometry after the pioneering work by Raghavachari[53].

The merits of using the 6-31G level of calculation are well illustrated by the successful correlation between the observed and calculated *deuterium-isolated* CH stretching frequencies (v_{CH})[54]. Partial deuteration of a hydrocarbon, for example, conversion of a CH_3 group into CHD_2, isolates the CH stretch vibration from coupling and greatly increases the precision of frequency measurements to within 1 cm^{-1}[55]. It was by this means that the in-plane (ip) C1—H2 (see **10**) stretching (2950 cm^{-1}) of the methyl group in propane was demonstrated to absorb at 13 cm^{-1} higher frequency than the out-of-plane (op) C1—H3 stretching (2937 cm^{-1})[54].

It is at the moment not possible for *ab initio* calculations to reproduce the absolute values of these frequencies. Indeed, in absolute terms v_{CH} values are too high by about 10%, when the frequencies are calculated on the 6-31G-optimized geometry by the usual normal coordinate analysis. However, these two quantities *correlate* quite well; when relative shifts from v_{CH} of methane are considered, they agree within 3 cm^{-1}.

The ip and op C—H bond lengths of propane (cf. **10**) have been calculated to be 1.0845 and 1.0854 Å, respectively, at the HF/6-31G level[54]. This bond length variation

E. Ōsawa

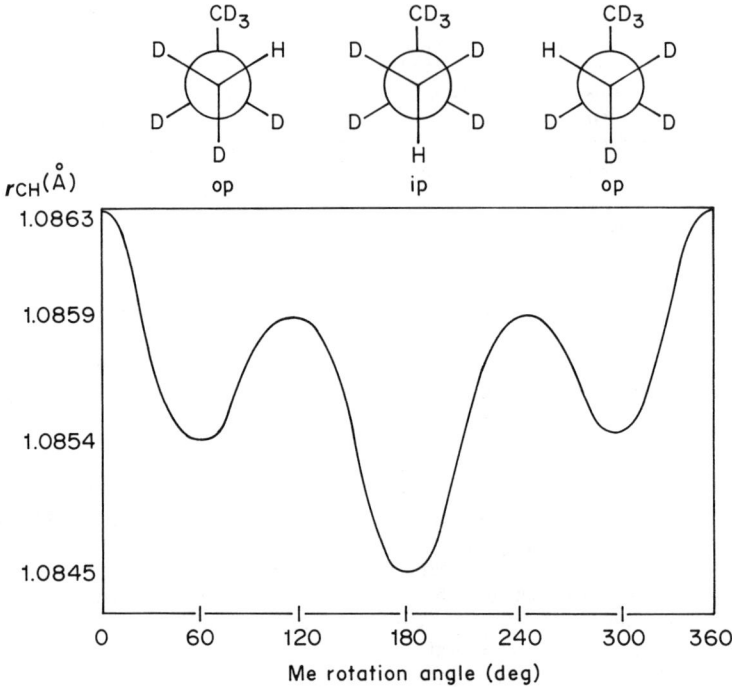

FIGURE 1. Schematic redrawing of the calculated variation in the C—H bond lengths of the methyl group of propane as a function of the methyl rotation angle

of 0.0009 Å corresponds to the frequency difference of 13 cm^{-1} as mentioned above, or, more simply, 10^{-4} Å corresponds to 1 cm^{-1}. Comparison of the observed frequency of the *isolated* v_{CH} mode with the calculated CH bond length for a large number of partially deuterated alkanes gave very good linear relations[55]. These relations are among the most successful and precise correspondences between experiment and calculation. From these results it follows that, in the course of the rotation of the methyl group of propane, the C—H bond length should vary periodically as shown in Figure 1. This figure is *real*, in that the data consist of quantities reproduced by experiments.

Some interesting points are a few notable deviations (not included in the above regression analysis) in the linear correlation between the calculated and observed v_{CH} frequencies (equation 1). One of the op CH bonds in *gauche* butane (**11**) is under

(11) *gauche* butane

steric crowding and was assigned to that observed at 2949 cm^{-1}, which is 43 cm^{-1} down from 2992 cm^{-1} of methane. The 6-31G/gradient calculation gave the ν_{CH} absorption of this bond also at a very low frequency of 3241 cm^{-1}, or 30 cm^{-1} down from 3271 cm^{-1} of methane. Both of the observed and calculated red shifts are exceptionally large and deviate significantly from the correlation line of equation 1 of uncrowded C—H bonds. Although these two shifts do not agree satisfactorily (inclusion of the electron correlation effect with the MP2/6-31G level of theory reduced the disagreement only by 2 cm^{-1}), such a distinctive deviation from equation 1 is a very clear indication of some special environment, namely steric crowding.

C. Empirical Approaches

Even though the *ab initio* gradient technique of geometry optimization produces accurate molecular geometries, the cost of such calculations is still too high for cursory purposes and for very large molecules like polymers or molecular aggregates. Hence less expensive methods are sometimes sought.

The latest version of Dewar's semiempirical SCF-MO method is concisely packaged into the popular program MOPAC[56]. According to Stewart's review[57], average errors of MNDO, AM1 and PM3 methods in reproducing molecular geometries are 0.04–0.05 Å for bond lengths, 3–4° for valence angles and 13–21° for torsion angles. These figures may seem to indicate only modest performance of MOPAC in the calculation of structures. However, one must consider the fact that the averaging covers all types of molecules examined. Actually, we have confirmed that the MNDO method reproduced the X-ray distance of C—C bonds in hydrocarbons with a standard deviation of 0.013 Å[58]. Stewart later indicated that the average error in calculating bonds involving a carbon atom by the PM3 method is 0.005 Å[59]. It is likely that the accuracy in the calculation of hydrocarbon geometries is very high, especially when closely related structures are compared[58].

Molecular mechanics, on the other hand, are generally constructed with considerable emphasis on reproducing molecular geometries as well as energies and other molecular properties. Especially for hydrocarbons, performance with regard to structure can be judged as generally good enough for most practical purposes like molecular modeling (Table 5).

TABLE 5. Standard deviations of errors in several molecular mechanics schemes in reproducing molecular geometries

	MM2[a]	DREIDING[b]	COSMIC(90)[c]	TRIPOS$_{v5.2}$[d]
Bond length (Å)	0.0093	0.035	0.009	0.025
Valence angle (deg)	1.3	3.2	—	2.5
Torsion angle (deg)	6.6	8.9	—	9.5
Number of molecules	22	76[e]	18[f]	76[e]
Experimental method[g]	ED	X	diverse	X

[a] Taken from Reference 26.
[b] S. L. Mayo, B. D. Olafson and W. A. Goddard, III, *J. Phys. Chem.*, **94**, 8897 (1990).
[c] S. D. Morley, R. J. Abraham, I. S. Haworth, D. E. Jackson, M. R. Saunders and J. G. Vinter, *J. Comput. -Aided Mol. Design* (in press).
[d] M. Clark, R. D. Cramer, III and N. Van Opdenbosch, *J. Comput. Chem.*, **10**, 982 (1989).
[e] Includes general types of molecules.
[f] Aromatic hydrocarbons.
[g] ED = electron diffraction , X = X-ray analysis.

Allinger's MM2, one of the most popular molecular mechanics force fields[50], has recently been updated to MM3[17]. This extensively improved force field is claimed to perform better in many respects than its previous version. With regard to structure, a remarkable improvement was achieved by introducing a stretch–torsion cross term, which works to elongate the eclipsed bond. A good example is norbornane (Table 1), for which the C2—C3 bond [1.573(15) Å] is observed to be longer than the other two kinds of C—C bonds [C1—C7, 1.546(24); C1—C2, 1.536(15) Å]. MM2 did not reproduce this order (1.541, 1.538, 1.542 Å), but MM3 does (1.557, 1.548, 1.540 Å). This new device works well also in reproducing C—C bond lengths of cyclopentane and dodecahedrane[60].

IV. CONCLUDING REMARKS AND PERSPECTIVES

The static geometries of simple and unstrained hydrocarbons, one of the most fundamental areas of chemistry, are briefly reviewed above, with one serious question in mind: can the experimental determination of precise nuclear dispositions in a molecule be replaced with computational methods? My personal answer to this question is very positively affirmative. As a matter of fact, a considerable proportion of recent papers report the use of some computational method to obtain static and dynamic structural informations when the experimental determination is difficult. The best advisable method seems to be the MP2/6-31G strategy for small molecules and the MM3 computation for larger molecules[61]. Computational methods are still under intensive development but they will merge in the future into a few, truly good methods after critical evaluation by the general users.

A few conceptual insights into the future of computational techniques may be worth commenting on before concluding this chapter. Up to this point, we have equated the term structure to geometry, namely disposition of atomic nuclei in a molecule. If the meaning of structure is expanded to include some of the usual chemical concepts that relate to structural properties of a molecule, the second most important concept next to positions of the nuclei will be the bond, which in turn should be closely related to electronic charge density. Bader[62] has put forward a theory of interpreting structure in terms of a gradient vector of the molecular charge density to give new definitions of chemical bonds. Here again, it is more convenient to obtain the charge density distribution from all the occupied molecular orbitals, rather than from experiments. This new interpretation of bonds is real in the sense that electron densities are the measurable quantities (e.g. by X-ray diffraction), but practical application of the theory has been carried out using computed molecular orbitals. In this respect, the Bader theory resembles the present widespread use of computed geometry.

Kutzelnigg[63] addressed fundamental problems concerning the *ab initio* calculation of molecular properties, including those of higher-order ones resulting from some kind of perturbation to the wave function, such as NMR chemical shift and hyperpolarizability. Properties like equilibrium geometry represent the simplest quantity in this category. His work suggests enormous potential of computational approaches to a wide range of molecular structural properties.

V. ACKNOWLEDGEMENTS

I thank Ms Claudia Marquard for preparing Table 1, Mr M. Yoshida for drawing structure **3** and Mrs M. Murakami for typing and drawing.

VI. REFERENCES

1. A recent review on the dynamic molecular structures: I. Hargittai, *Pure Appl. Chem.*, **61**, 651 (1989).

2. T. S. Slee, in *Modern Models of Bonding and Delocalization* (Eds. J. F. Liebman and A. Greenberg), VCH Publ., New York, 1986, p. 63.
3. F. H. Allen, *Acta Crystallogr.*, **B40**, 64 (1984).
4. G. W. Griffin and A. P. Marchand, *Chem. Rev.*, **89**, 987 (1989).
5. K. B. Wiberg, *Acc. Chem. Res.*, **17**, 379 (1984).
6. J.-P. Sauvage, *Nouv. J. Chim.*, **9**, 299 (1985).
7. H. Prinzbach, W.-D. Fessner, B. A. R. C. Murty, J. Worth, D. Hunkler, H. Fritz, W. D. Roth, P. V. R. Schleyer, A. B. McWen, W. F. Maier, P. R. Spurr, J. Mortensen, J. Heinze, G. Gescheidt and F. Gerson, *Angew. Chem., Int. Ed. Engl.*, **26**, 451, 452, 455, 457 (1987).
8. L. A. Paquette, *Chem. Rev.*, **89**, 1051 (1989).
9. G. A. Olah (Ed.), *Cage Hydrocarbons*, Wiley, New York, 1990.
10. E. Ōsawa and O. Yonemitsu (Eds.), *Chemistry of Three-Dimensional Polycyclic Hydrocarbons*, VCH Publ., New York (in preparation).
11. I. Hargittai and M. Hargittai (Eds.), *Stereochemical Applications of Gas-Phase Electron Diffraction*, VCH Publ., New York, 1988. For other methods for vapor phase, see A. P. Cox, *J. Mol. Struct.*, **87**, 61 (1983).
12. I. Hargittai and M. Hargittai, in *Molecular Structure and Energetics, Vol. 2, Physical Measurements* (Eds. J. F. Liebman and A. Greenberg), VCH Publ., New York, 1987, p. 1.
13. K. Rasmussen, *Potential Energy Functions in Conformational Analysis*, Springer-Verlag, Berlin, 1985, p. 82.
14. W. J. Hehre, L. Radom, P. v. R. Schleyer and J. A. Pople, *Ab Initio Molecular Orbital Theory*, Chap. 6, Wiley, New York, 1986.
15. J. E. Boggs, *J. Mol. Struct. (Theochem)*, **97**, 1 (1983).
16. E. Ōsawa and C. Marquard, unpublished results.
17. N. L. Allinger, Y. H. Yuh and J.-H. Lii, *J. Am. Chem. Soc.*, **111**, 8551 (1989).
18. J. R. Maple, U. Dinur and A. T. Hagler, *Proc. Natl. Acad. Sci. USA*, **85**, 5350 (1988); U. Dinur and A. T. Hagler, *J. Am. Chem. Soc.*, **111**, 5149 (1989); K. Palmo, L.-O. Peitila and S. Krimm, *J. Comput. Chem.*, **12**, 385 (1991); T. Hirano and E. Ōsawa. *Croat. Chem. Acta*, **57**, 1633 (1984); F. A. Momany, V. J. Klimkowski and L. Schäfer, *J. Comput. Chem.*, **11**, 654 (1990).
19. R. Taylor and O. Kennard, *J. Chem. Inf. Comput. Sci.*, **26**, 28 (1986); F. H. Allen and O. Johnson, *Acta Crystallogr.*, **B47**, 62 (1991); F. H. Allen, M. J. Doyle and R. Taylor, *Acta Crystallogr.*, **B47**, 29, 41, 50 (1991).
20. F. H. Allen, O. Kennard, D. G. Watson, L. Brammer, A. G. Orpen and R. Taylor, *J. Chem. Soc., Perkin Trans. 2*, S1 (1987).
21. H. Fuess, *Chem. Brit.*, **14**, 37 (1978). G. A. Jeffrey, *J. Mol. Struct. (Theochem)*, **108**, 1 (1984).
22. W. R. Busing, *WMIN, A Computer Program to Model Molecules and Crystals in Terms of Potential Energy Functions*, Oak Ridge National Laboratory, Contract No. W-7405-eng-26 (1981); W. R. Busing, *Acta Crystallogr.*, **A28**, 5252 (1972).
23. I. Tickle, J. Hess, A. Vos and J. B. F. N. Engberts, *J. Chem. Soc., Perkin Trans. 2*, 460 (1978); R. J. J. Visser, A. Vos and J. B. F. N. Engberts, *J. Chem. Soc., Perkin Trans. 2*, 634 (1978); S. C. De Sanctis, *Acta Crystallogr.*, **B39**, 366 (1983); H. Kimoto, K. Saigo and M. Hasegawa, *Chem. Lett.*, 711 (1990); G. Brink and L. Glasser, *J. Mol. Struct.*, **160**, 357 (1987); I. Simon, L. Glasser, H. A. Scheraga and R. St. John Manley, *Macromolecules*, **21**, 990 (1988).
24. A. W. M. Braam and A. Vos, *Acta Crystallogr.*, **B36**, 2688 (1980).
25. E. Ōsawa, P. M. Ivanov and J. M. Rudzinski, unpublished results.
26. C. Jaime and E. Ōsawa, *Tetrahedron*, **39**, 2769 (1983).
27. S. Fitzwater and L. S. Bartell, *J. Am. Chem. Soc.*, **98**, 8338 (1976).
28. K. B. Wiberg and M. A. Murcko, *J. Am. Chem. Soc.*, **110**, 8029 (1988).
29. M. Stein, W. Winter and A. Rieker, *Angew. Chem., Int. Ed. Engl.*, **17**, 692 (1978); W. Winter, *Fresenius' Z. Anal. Chem.*, **304**, 279 (1980).
30. B. Kahr, D. Van Engen and K. Mislow, *J. Am. Chem. Soc.*, **108**, 8305 (1986).
31. N. Yannoni, B. Kahr and K. Mislow, *J. Am. Chem. Soc.*, **110**, 6670 (1988).
32. E. Ōsawa, Y. Onuki and K. Mislow, *J. Am. Chem. Soc.*, **103**, 7475 (1981).
33. C. S. Yannoni, P. P. Bernier, D. S. Bethune, G. Meijer and J. R. Salem, *J. Am. Chem. Soc.*, **113**, 3190 (1991).
34. J. M. Rudzinski, Z. Slanina, M. Togashi and E. Ōsawa, *Thermochim. Acta*, **125**, 155 (1988).
35. There are more computed bond distances for C_{60}: J. Feng, J. Li, A. Wong and M. C. Zerner, *Int. J. Quantum Chem.*, **37**, 599 (1990).

36. H. Ringsdorf, B. Schlarb and J. Venzmer, *Angew. Chem., Int. Ed. Engl.,* **27**, 113 (1988); J. Sauer, *Chem. Rev.,* **89**, 199 (1989).
37. J. C. Hempel and F. K. Brown, *J. Am. Chem. Soc.,* **111**, 7323 (1989); F. K. Brown, J. C. Hempel, J. S. Dixon, S. Amoto, L. Muller and P. W. Jeffs, *J. Am. Chem. Soc.,* **111**, 7328 (1989).
38. R. E. Moore, G. Bartolini, J. Barchi, A. A. Bothner-By, J. Dadok and J. Fond, *J. Am. Chem. Soc.,* **104**, 3776 (1982).
39. K. Imai and E. Ōsawa, *Magn. Reson. Chem.,* **28**, 668 (1990).
40. E. Ōsawa, K. Imai, T. Fujiyoshi-Yoneda, C. Jaime, P. Ma and S. Masamune, *Tetrahedron,* **47**, 4579 (1991).
41. Recently developed (albeit expensive) Gaussian-1 theory gives energies within 0.1 eV: J. A. Pople, M. Head-Gordon, D. J. Fox, K. Raghavachari and L. A. Curtiss, *J. Chem. Phys.,* **90**, 5622 (1989); L. A. Curtiss, C. Jones, G. W. Trucks, K. Raghavachari and J. A. Pople, *J. Chem. Phys.,* **93**, 2537 (1990); M. P. Mcgrath and L. Radom, *J. Chem. Phys.,* **94**, 511 (1991).
42. See also J. E. Boggs. *Pure Appl. Chem.,* **60**, 175 (1988).
43. L. Schäfer, S. Q. Newton, F. A. Momany and V. J. Klimkowski, *J. Mol. Struct.* (in press).
44. G. C. Levy, *J. Chem. Inf. Comput. Sci.,* **28**, 167 (1988). According to one of the figures given in this paper, the speed increases roughly 10 times per year, predicted to reach 1000 GFLOPS (1 TFLOPS) by the year 2000. Perhaps a more interesting prediction that derives from the same figure is that the price of the computer, which keeps decreasing at a rate of 300% per year in terms of $/MIPS, will become zero in the year 2010.
45. L. Schäfer, *J. Mol. Struct.,* **100**, 51 (1983).
46. L. Schäfer, J. D. Ewbank, V. J. Klimkowski, K. Siam and C. Van Alsenoy, *J. Mol. Struct.* (*Theochem*), **135**, 141 (1986).
47. K. Siam, C. Van Alsenoy, K. Wolinski, L. Schäfer, A. P. Marchand, P. W. Jin and M. N. Deshpende, *J. Mol. Struct.,* **204**, 209 (1990).
48. J. N. Scarsdale, C. Van Alsenoy and L. Schäfer, *J. Mol. Struct.* (*Theochem*), **86**, 277 (1982).
49. V. J. Klimkowski, J. P. Manning and L. Schäfer, *J. Comput. Chem.,* **6**, 570 (1985).
50. U. Burkert and N. L. Allinger, *Molecular Mechanics,* ACS, Washington, D.C., 1982.
51. S. Tsuzuki, L. Schäfer, H. Goto, E. D. Jemmis, H. Hosoya, K. Siam, K. Tanabe and E. Ōsawa, *J. Am. Chem. Soc.,* **113**, 4665 (1991).
52. G. Häfelinger, C. U. Regelmann, T. M. Krygowski and K. Wozniak, *J. Comput. Chem.,* **10**, 329 (1989).
53. K. Raghavachari, *J. Chem. Phys.,* **81**, 1383 (1984). See also Reference 28.
54. A. L. Aljibury, R. G. Snyder, H. L. Strauss and K. Raghavachari, *J. Chem. Phys.,* **84**, 6872 (1986).
55. D. C. McKean, *Chem. Soc. Rev.,* **7**, 399 (1978).
56. MOPAC 6.0 (QCPE #455, JCPE #44, 49). Inquiry addresses: QCPE, Creative Arts Bldg. 181, Indiana University, Bloomington, IN 47405, USA; JCPE, c/o Japan Association for International Chemical Information, Gakkai Center Bldg., 2-4-16 Yayoi, Bunkyo-ku, Tokyo 113, Japan.
57. J. J. P. Stewart, *J. Comput. Chem.,* **10**, 221 (1989).
58. D. A. Dougherty, C. S. Choi, G. Kaupp, A. B. Buda, J. M. Rudzinski and E. Ōsawa, *J. Chem. Soc., Perkin Trans. 2,* 1063 (1986).
59. J. J. P. Stewart, *J. Comput.-Aided Mol. Design,* **4**, 1 (1990).
60. N. L. Allinger, H. J. Geise, W. Pyckhout, L. A. Paquette and J. C. Gallucci, *J. Am. Chem. Soc.,* **111**, 1106 (1989).
61. One other force field which aims at improving MM2 has been announced: J. L. M. Dillen, *J. Comput. Chem.,* **11**, 1125 (1990).
62. K. B. Wiberg, R. F. W. Bader and C. D. H. Lau, *J. Am. Chem. Soc.,* **109**, 985 (1987). R. F. W. Bader, *Acc. Chem. Res.,* **9**, 18 (1985).
63. W. Kutzelnigg, *J. Mol. Struct.* (*Theochem*), **202**, 11 (1989).

## CHAPTER 3

Conformational analysis of acyclic and alicyclic saturated hydrocarbons

J. E. ANDERSON

Chemistry Department, University College London, Gower Street, London, WC1E 6BT, England

The Chemistry of Alkanes and Cycloalkanes
Edited by S. Patai and Z. Rappoport © 1992 John Wiley & Sons Ltd

I. INTRODUCTION

A. Introduction

The conformational analysis of saturated hydrocarbons has important aspects which are peculiar to itself. For other groups, analysis largely concerns their interactions with hydrocarbon frameworks, whereas here the alkyl groups are the framework. Hydrogen is very slightly electronegative compared to carbon[1] so carbon–hydrogen bonds may have small dipoles, but in the usual terms of conformational analysis, dipole–dipole repulsions and attractions in saturated hydrocarbons are subsumed into steric repulsions. Some attempts to demonstrate induced dipole–induced dipole (i.e. attractive van der Waals) interactions will be discussed later but make up a rather small canon.

A further peculiarity of saturated hydrocarbons is that they can be branched. At the simplest level methyl, ethyl, isopropyl and *tert*-butyl groups need to be considered as examples of primary to quaternary carbon centres, whose increasing branching has important consequences. They have an internal conformational analysis—about each carbon–methyl bond—as well as the analysis of the bond to the rest of the molecule.

As a result there is much that can be said about hydrocarbons that does not apply to other groups, and the reader may occasionally have to recall that the statements of this review refer implicitly to saturated hydrocarbons. Likewise, the reader should keep in mind that this is not an account of the development of the subject, which can be found elsewhere[2,3].

In a conformational analysis, chemists want to know which conformations are significantly populated, but beyond that there are various facts they can hope to learn experimentally and discuss about each conformation:

1. The dihedral relationships along bonds, both generally in terms of *gauche*, *anti* and eclipsed conformations, and specifically in terms of the angles actually found which are more or less close to 60° and 180°. For this review I define a positive dihedral angle as follows. If in a fragment ABCD, looking along B towards C, AB has to be rotated clockwise ($< 180°$) to eclipse CD, the ABCD dihedral angle is positive. The DCBA dihedral angle is also positive, of course. If clockwise is replaced by *anti*clockwise, positive is replaced by negative.

2. The longer-range relationships between individual alkyl groups, beyond the two ends of a given bond.

3. The spectroscopic observables like vibration frequencies or spin–spin coupling constants and other indirect conformational indicators, which are eventually used to determine the conformation adopted but which must be related to known conformations.

4. The kinetics of interconversion of conformations.

5. The bond angles C—C—C, C—C—H and H—C—H at each carbon atom in the molecule and the length of carbon–carbon bonds. One might add the length of carbon–hydrogen bonds, but these are less readily, less regularly and less surely determined, and then less interesting anyway. These bond angles and bond lengths will often be considered by only the most enthusiastic conformational analyst, who might also be interested in knowledge of conformations not significantly populated.

A point which should be addressed here, that is not peculiar to hydrocarbon conformational analysis, is the meaning of a conformational minimum. The organic chemist tends to write structures in terms of the conformers situated at the bottom of potential minima, so that a few very well defined conformational states describe the system. It is desirable that he think or be aware of the less certain nature of things in reality, that aside from zero-point energy, thermal energy implies that a series of vibrational and torsional states, close to the minimum, are significantly populated. Even an adjacent minimum separated by a small barrier may be so readily accessible that the distinct identity of the two minima is subverted.

This reservation having been brought forward, it will not arise again, since the illustration of the important features of conformational analysis is best done by treating effects as being clear-cut even when they are very small. This is part of the attraction of molecular mechanics calculations which most readily give minimum enthalpy conformations. It is an important goal of this review to show that marked conformational effects strikingly demonstrated by highly branched hydrocarbons can be seen in embryonic form in very simple hydrocarbons. Thermal energy effects are trivial in the striking examples, but may be relatively more important in simple molecules.

B. Experimental Methods

A detailed review of the experimental methods used to determine hydrocarbon conformations and dynamics will not be given, since two fine accounts are available. In their 1965 textbook *Conformational Analysis*[4] Eliel, Allinger, Angyal and Morrison give a sixty page survey of physical methods which had been used until then. Berg and Sandstrom[5] have given an up-to-date account of experimental techniques in a recent review.

For only a few compounds does the experimental observation consists of separate signals (in the broadest sense) for each conformation, with signal intensity directly related to conformational population. The techniques which give such direct results are often not applicable to a wide range of hydrocarbons. As a result, indirect methods have to be used and one problem that must be faced is the reliability of the results offered by the various experimental methods of this kind. The classical question of the relative stability of the *qauche* and *anti* conformations of 2,3-dimethylbutane discussed further below illustrates the point. While there is no doubt that *gauche* and *anti* conformations are about equally populated, two indirect methods using refractive indices[6] and ultrasonic relaxation[7] suggest that the *anti* conformation is markedly the more stable. The reviews cited above[4,5] give some help in evaluating such indirect techniques some of which, like the NMR coupling constant method, are widely accepted as reliable.

Direct determinations of the structure by diffraction or vibrational/rotational spectroscopy are restricted to small or highly symmetrical hydrocarbons[8]. The reports of such determinations are occasionally short of one or two dihedral angles, which are of interest

98   J. E. Anderson

for the conformational analysis. Direct results may be obtained from dynamic NMR investigations of more highly branched hydrocarbons, such as 2,3-dimethylbutane[9,10], for which a separate set of signals is observed for *gauche* and *anti* conformations. In many cases, however, each observed set of signals arises not from a single conformation, but from a rapidly interconverting family of conformations, as in the case of cyclooctane discussed in Section VI.E[11,12]. In such direct but ambiguous determinations, where some, but not all, conformational processes are slow on the NMR timescale, molecular mechanics calculations have been used as a valuable aid to elucidating the experimental observations.

Very similar comments might be made about the techniques for investigating conformational dynamics—the barriers to rotation or to more complicated interconversion processes, and again reviews[4,5,13] give much background information. NMR spectroscopy is the technique of choice for barriers greater than about 4.5 kcal mol^{-1} and, at best, yields values of the rate of interconversion at several temperatures, which can lead to the activation parameters for the interconversion[14]. Often, the rate of conformational interconversion is measured at only one temperature, and thence a free energy of activation for rotation at that temperature results.

Most other methods are appropriate for measuring lower barriers than those obtained by the NMR method, and many involve extrapolations and assumptions which make them somewhat less reliable for barriers of intermediate size, say greater than 3 kcal mol^{-1}.

Such uncertainties are not a major problem now since, for many molecules, several techniques shed some light on the dynamics, and molecular mechanics calculations are also reliably illuminating as discussed below. Consequently, an overall clear picture has emerged in most cases.

C. Medium Effects

Conformational analysis should not be discussed without some attention being paid to any differences in results in the solid, in the gas phase and in various types of solvents. It may be reasonable to assume that with saturated hydrocarbons, conformational results peculiar to the solid state reflect some property of that state rather than of the conformation.

Organic chemists would like to hope that in molecules as nonpolar as saturated hydrocarbons there is no significant difference between solution and the gas phase. This is implicit in the wide acceptance of the gas-phase value of 0.8–0.9 kcal mol^{-1}[15,16] for the *gauche–anti* energy difference in butane in discussions of results for other molecules which are almost invariably in solution. It is therefore striking that the energy difference for butane in solution or as a neat liquid is about 0.55 kcal mol^{-1}[15,16]. The effect of external pressure is to increase even further the proportion of the *gauche* conformation[17]. Calculations[18–20] support the experimental observation of the greater population of *gauche* butane in the liquid.

From time to time the effect of the internal pressure of a solution on a conformational equilibrium is discussed[21] but convincing separation of this from other effects is often difficult. Hydrocarbons, and particularly butane, give a favourable opportunity for discussing this effect, for neither the *gauche* nor the *anti* conformation of butane has a dipole moment[22]. The former conformation appears to be more compact than the *anti*, and there is less of the latter in the liquid state.

Internal pressure is an unfortunate term when concentrating on a single molecule of butane in the neat liquid, for it seems to imply compression and repulsion when in fact the molecule experiences net attractive stabilizing interactions with its neighbours. Interaction presumably comprises a few repulsive interactions with the nearest atoms of the nearest neighbours and attractive interactions with all more distant atoms. The more compact *gauche* butane should pack better with an assembly of solvent molecules than the *anti* conformation, and this appears to favour attractive interactions more than repulsive ones since the intrinsic preference for *anti* is somewhat offset on coming into solution.

It is not clear whether for hydrocarbons larger than butane the more compact conformation is always[17] relatively more favoured in the liquid state. The phase-dependence of hydrocarbon conformational equilibria merits further experimental investigaton.

It is calculated[18-20] that in aqueous solution, as simple ideas of hydrophobicity would predict, there is proportionately more of the *gauche* conformation of butane. However, it is simplistic even when discussing a butane molecule in neat butane, and far less so in water, to talk in terms of enthalpy and to ignore entropy.

Entropy *differences* between conformations and entropies of activation are often not discussed in the course of a conformational analysis, which is paradoxical since calorimetric measurements of *absolute* entropies[23] did much to initiate interest in the conformational behaviour of simple hydrocarbons[2,3]. Some organic chemists prefer to hope that entropy differences between conformations are not significant. This has caused problems, for example, in trying to reconcile differences between calculated enthalpies of activation for internal rotation and experimental free energies of activation as determined by dynamic NMR spectroscopy[24-26]. Lii and Allinger[26] have discussed this matter and view a large part of the entropy relevant to conformational equilibria as arising from methyl group rotational freedom. Since a large part of the conformational analysis of alkyl groups comes down to methyl groups interacting in a *gauche* arrangement or molecules distorting to alleviate this, the hope may be misplaced. Substantial negative entropies of activation for rotation are predicted[26] and help to reconcile the calculated and experimental barriers.

Another obvious contributor to entropy is the steepness of the conformational energy minimum. Vibrational energy levels are sparser in a steep-sided energy well than in a broad flat-bottomed one. Calculations certainly suggest that potential minima do have different shapes, so the entropy predictions of the calculations should be given attention comparable to the enthalpy predictions.

D. Molecular Mechanics and Other Calculations

It is characteristic of the organic chemist to seek simple explanations of observed facts, often with a view to extending the explanation to cases where there are no facts to serve. An explanation may require the combining of several factors, giving much scope for favouring one factor over all others. In a sense, molecular mechanics calculations are a way of applying quite rigorously a set of explanations which are rather less rigorous. Such calculations are particularly helpful in the conformational analysis of saturated hydrocarbons, but I do not intend to give a detailed account of their application although I will use results appropriately. Some limited account of the method is needed, and what follows is based on the Allinger MM2 and MM3 programs as described in the original literature[27-29] and experienced by me. Several monographs and reviews[30-33] give accounts and evaluations of other programs. Anet[34] has commented on the preponderant use of MM2 when different, equally satisfactory programs are available.

It is assumed that, beyond the bonding energy, any particular structure and conformation has a steric energy, made up of several components depending on how greatly various molecular parameters differ from their ideal values. These components and their magnitudes are calculated and listed individually and correspond quite closely to the factors that the organic chemist expects to be important. There is energy due to bond lengthening or shortening (compression), bond-angle opening or closing (bending), the dihedral arrangement of the six bonds at either end of a given carbon–carbon bond (torsion), the van der Waals interactions of atoms at both ends of a carbon–carbon bond (1,4-interactions), to which are added longer-range attractive and repulsive van der Waals interactions. Finally, there are some usually small contributions from combinations of distortions, stretch–bend, bend–bend, torsion–stretch and torsion–bend, which have some physical reality. These have been introduced over time[28a,29] as a means of correcting systematic shortcomings[28b] of calculations based on only the more obvious major contributors to the steric energy.

The usefulness of calculating the steric energy of a specific structure is crucially enhanced by energy minimization procedures, which move downhill in energy terms from the initial structure to find the local conformational energy minimum, and a driving procedure by which the molecule can be moved to other points in conformational space, whence new minima may be located. The driving procedure usually involves changing a single dihedral angle in a controlled way. If done judiciously in a series of suitably small steps, it produces a steric energy graph which may be a good representation of the rotational potential for the bond in question.

One important point that must be considered is what aspects of the molecular structure are most important when judging how a calculated structure matches the known experimental structure. Clearly, calculated bond lengths, bond angles and dihedral angles should match well when these are known, but Allinger and coworkers[27–29] have placed emphasis on matching calculated and experimental heats of formation. They have deprecated empirical modifications of their program for specific purposes of matching rotational barriers, partly because the resultant match to experimental heats of formation is much poorer, and partly for another important reason. Calculations lead to enthalpies, while some experimental results, particularly those from NMR studies, lead to free energies and the two are directly comparable only when the entropy is known or can safely be assumed to be zero. Lack of care for entropy depreciates the value of some work in the literature. However, the calculated values of bond angles, bond lengths, dihedral angles and relative steric energies of conformations usually agree well with experiments, and trends along homologous series are usually well matched.

An important problem in parametrizing comes with van der Waals attractions and repulsions. With the gaugeable terms, namely bond lengths, bond angles and dihedral angles, what has been calculated can be measured experimentally, and parameters can be adjusted appropriately. However, atom–atom interactions are never measured individually. The sum of all pairwise interactions contributes, along with other terms, to the total calculated steric energy of the molecule, which may also be gaugeable from the heat of formation, or estimated relative to another structure from some chemical or conformational equilibrium. Any lack of agreement in the experimental and calculated gaugeable terms makes the calculated van der Waals interactions much more uncertain. A consequence of this is that different programs put different values on such interactions and care should be taken not to overemphasize individual pairwise interactions, or the sum of $(m) \times (n)$ pairwise interactions of an n-atom group with an m-atom group. A different force field may produce a strikingly different result in a way that it would not if calculations of a gaugeable term were being compared. A result may tell more about the program than about the molecule being studied.

One problem of calculations should more reasonably be attributed to the calculator, and is that of finding best local minima. Procedures do exist[35–38] for searching widely for these, but may produce large numbers of minima which require much analysis by the chemist[34] of the kind he sought to avoid. A simpler minimization procedure and chemical intuition may be a more satisfactory approach but, as we show in Section II.C, chemical intuition is not necessarily strongly developed.

It also must not be forgotten that the calculations show what obtains once the minimum energy conformation has been attained. Unless asked specifically, the calculations do not show what obtained, in, say, the perfectly staggered conformation. The strain which is the driving force for moving from the symmetrical conformation to the minimum energy conformation, and which is often the chemist's particular interest, may not be apparent if only the latter conformation is calculated.

One very useful aspect of calculations is the information they give on experimentally unobservable conformations, and this is particularly true of conformational minima which are only sparsely populated. Such conformations, while no doubt embellishing a primary account of populated conformations of a given molecule, hardly deserve to be reviewed here.

With all the above reservations, the chemist interested in conformations of saturated hydrocarbons is fortunate in having molecular mechanics calculations as an auxiliary to his experimental investigations.

Ab initio calculations relevant to saturated hydrocarbon conformational analysis have been carried out, but do not extend much beyond simple compounds[39-43]. Satisfactory results require the inclusion of electron correlation effects and big basis sets[41] so that even simple problems require extensive calculation.

II. MOLECULAR DISTORTIONS

The conformational response of a molecular fragment when substitution introduces additional strain takes one of four general forms. The first and most drastic one is to change to a different kind of conformation. The others are distortions of bond lengths, of bond angles and of dihedral angles, which are dealt with successively here, minding that real molecules undergo a combination of these distortions. Some distortion may be described more fully in the chapter on structure elsewhere in this book, but, at the risk of some repetition, its importance in conformational analysis will be illustrated here.

A. Bond Lengths

The various definitions of bond lengths depend on how they are measured, on vibrational states etc., and, among others, Rasmussen[33] has given an account of these. Lengthening and shortening of bonds is not a very efficient way of relieving steric strain compared with changing bond and dihedral angles but, where it is appropriate, marked bond lengthening is found. Ruchardt, Beckhaus and their coworkers[44-49] have discussed calculations and experimental observations showing that, while tri- or less-substituted ethane bonds usually have more efficient means of relieving strain, unusually long bonds are increasingly likely as the branching at either end of the bond and the total strain of a hydrocarbon molecule increase. Bonds can be grouped together depending on whether they link two tertiary carbon atoms, one tertiary to a quaternary, or two quaternary ones, and there is a fairly direct relationship between the overall strain calculated for the molecule and the calculated length of the central bond. The longest, experimentally determined bond in a saturated hydrocarbon is 1.647 Å in 2,3-bis-(1-adamantyl)-2,3-dimethylbutane[47], but since rather more strained molecules can be imagined, longer bonds may yet be found.

Calculations[50] of rotation about sp^3–sp^3 carbon–carbon bonds suggest that the bond length is somewhat greater in the eclipsed transition state than in the ground state. It remains to be seen whether in rotationally free molecules with eclipsed ground-state conformations (see below) the bond concerned is unusually long. It is difficult to imagine conformational reasons for bond shortening, so that such examples as do occur for saturated hydrocarbons[51] are better discussed in terms of hybridization.

B. Bond Angles

Opening of the C^1—C^2—C^3 bond angle, when C^1 and C^3 are parts of sterically demanding groups, is a common feature of saturated hydrocarbons, particularly when one or both of the remaining substituents on C^2 are hydrogen atoms, for compensating closing of the C—C^2—H bonds meets little steric resistance. The C—C—C bond-angle opening in propane can be ascribed to C—C bond repulsion being greater than that of C—H (see below) rather than to repulsion of the methyl group hydrogens. However, with substituents larger than methyl, repulsive forces between atoms become important.

The biggest conformational consequence of bond-angle opening is that, in the Newman projection, the C^1 and C^3 substituents are forced nearer to their neighbours at the opposite end of the bond (see 1). If there is only one substituent R^5 at the opposite end of the bond, the tendency of R^5 to rotate towards the methine hydrogen depending on the size of C^1 or C^3 will be accentuated (see 2, and the discussion of eclipsed conformations below).

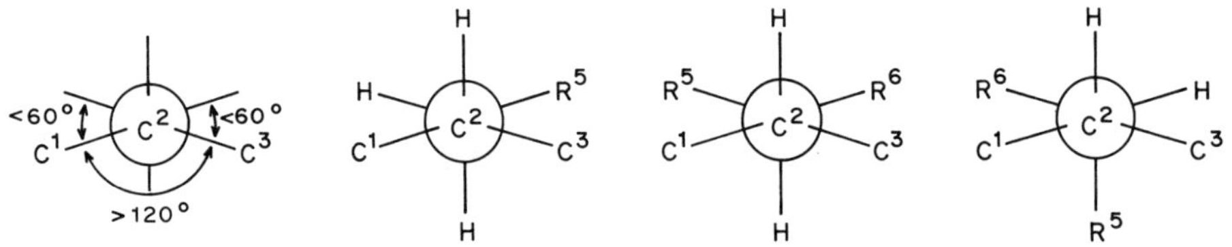

If there are two substituents at the opposite end of the bond, the conformation 3 with only two *gauche* interactions will be destabilized by interactions along the central bond resulting from opening of both the C^1—C^2—C^3 and R^5—C^4—R^6 bond angles. The conformation 4 with formally three *gauche* interactions is stabilized compared to 3, and the consequences of this for the range of compounds from tetramethylethane (2,3-dimethylbutane) to the tetra-*tert*-alkylethanes are discussed below.

In a pentasubstituted ethane with three substituents at the opposite end of the bond, opening of the C^1—C^2—C^3 angle has unpredictable consequences, difficult to distinguish from other effects, while in an ethane bond with six alkyl substituents bond-angle opening cannot be readily matched by low-strain bond-angle closing. Hence in both these cases no useful generalizations can be made. Overall, bond-angle opening certainly leads to, but is not the only reason for, distorted dihedral angles, as the next section shows.

C. Dihedral Angles, Skewing, Libration and Steric Repulsion

There is great scope for reducing steric strain in saturated hydrocarbons by distortion of dihedral angles away from perfectly staggered 60° and 180° values, i.e. by skewing. Co-operative skewing distortions of group after group often extend throughout the most stable conformation of a crowded molecule[52], and can be detected in even very simple saturated hydrocarbons[53]. They are little appreciated by practising organic chemists and are not mentioned in general organic chemistry texts.

When a methyl group is introduced at *gauche*-position +g or −g in propane, as in 5, hydrogen–methyl parallel 1,3-interactions result. These can be reduced somewhat by the *anti*clockwise rotation shown in 6 or by the clockwise rotation shown in 7 (thick arrows). This skewing is no more than methyl groups moving apart, but the strain can be further reduced by the concomitant rotation of the methyl groups *in the same sense* as shown by the small arrows in **6** and **7**. These diagrams represent the *gauche* conformations of butane, which are known[54,55] to have methyl—C—C—methyl dihedral angles of about 70° and H—CH$_2$—C—X dihedral angles averaging 64.5° by calculation[56] (see below). Any *gauche* 1,2-dialkyl ethane fragment shows comparable rotation in the same sense of both the ethane bond and the alkyl groups, and in the enantiomeric −g and +g conformations the adjustment is clockwise and anticlockwise, respectively.

(5)   (6)   (7)

Even at a bond with an alkyl group apparently symmetrically *gauche* to *two* methyl groups, interactions can be reduced and better minima found by skewing away from the perfectly staggered conformation in either sense, provided the alkyl and methyl groups skew in the same sense about the bond joining them to the framework of the molecule. The alkyl group rapidly librates between these skewed minima by way of the perfectly staggered conformation as a transition state, but there is also libration within the alkyl and methyl groups, from one side of perfect staggering to the other. Calculation and experiment on more highly branched molecules suggest that libration has a lower barrier if all groups move at the same time[57].‡

The degree of skewing away from perfect staggering is eventually limited by increasing eclipsing and other short-range interactions. If the alkyl group is *gauche* to two different alkyl groups, the two skewed minima are or different energy. If the gain in stabilization by skewing is small compared with the loss due to eclipsing, a minimum may not exist, but a trace of the relief skewing offers may appear as a point of inflection in the calculated potential profile.[58]

All saturated hydrocarbons with *gauche* alkyl substituents have skewing and libration as part of their conformational analysis. Hexamethylethane is a fine example, seeming ideal at first for the perfectly staggered conformation **8**. This, however, has substantial parallel 1,3-interactions (see **9**), and if rotation about all bonds (the six methyl–carbon bonds *and* the central ethane bond) takes place in an *anti*clockwise direction (see **10**), these long-range interactions are relieved. The same could be achieved by rotation about all bonds in clockwise direction (see **11**) and the minima occur at about 15° on either side of the perfectly staggered conformation[59,60]. Hexamethylethane therefore exists as a mixture of skewed conformations **10** and **11** with a barrier[57,61–63] of about 0.4 kcal mol^{-1} to libration between these minima through the *staggered* transition state, and of course

‡In the 360° rotational cycle for a bond, two populated minima may be separated by a low barrier while in the opposite direction each is separated from the rest of the cycle by a high barrier. Passage over the high barrier will usually involve a large change in dihedral angle and will be infrequent compared with interconversion over the low barrier, which will usually involve a smaller dihedral angle change. If the high barrier process is called a rotation then in between each rotation molecules will go back and forth many times over the low barrier and this movement is called a libration. The barrier to libration in a highly branched molecule may be greater than the barrier to rotation in a simple molecule, but no matter. The name libration is a useful concept in a molecule where there is already a classical rotation. If the libration barrier is low, thermal energy may blur the integrity of the adjacent minima as indicated in Section I.A.

a much larger barrier, calculated[61] and measured[64] to be 9.4 kcal mol^{-1}, to rotation through an *eclipsed* transition state to another double minimum. There are thus six equivalent minimum energy conformations during 360° of rotation.

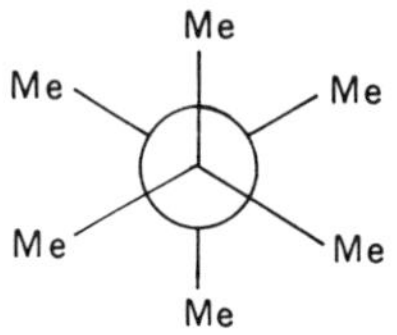

(8) 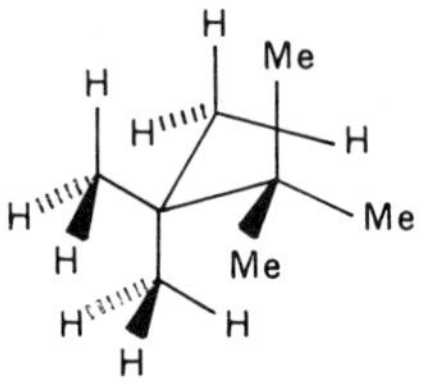   (9) 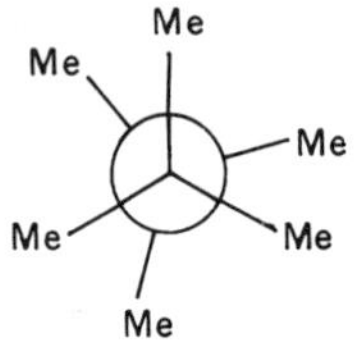   (10) 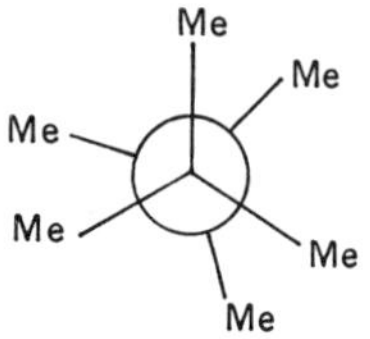   (11)

Much use is made of molecular mechanics to explore rotational conformations, and it is easy to see that if each staggered conformation is a double minimum, minimization from an approximate structure, or an approach from a different staggered conformation by driving the dihedral angle, may arrive at one or other minimum. Unless the second local minimum is sought and evaluated explicitly, there can be no certainty that the conformational analysis is correctly based. An analysis of an unsymmetrical acyclic alkyl compound which does not discuss all six conformations about each bond, even if only to propose a simplification, is liable to be erroneous in a way which may be trivial in simple cases, but critically important when the two local minima are quite different.

The most dramatic demonstration of the sixfold nature of the rotational potential in branched hydrocarbons is provided by tri-*tert*-butylmethane[8,65,66] whose results are complemented by subsequent work[67,68] on the similar tri-*tert*-alkylmethanols. Calculations show that minimum energy conformations are located between 13° and 20° on either side of perfect staggering[66,68] (experiment[8] suggests 11°), and dynamic NMR shows the slowing down of two processes in such molecules. The barrier to rotation through the eclipsed state in tri-*tert*-butylmethanol, for example[67], is 10.0 kcal mol^{-1} while the barrier to libration through the perfectly staggered conformation is 9.2 kcal mol^{-1}.

It has already been emphasized that rotation about the central bond and of alkyl groups attached to the central bond should be in the same sense for optimal release of steric strain. If one alkyl group is skewed in the opposite sense to all others, a slightly less stable minimum may result. This has been called incoherent as opposed to coherent skewing[69], and introduces a further family of less stable but no doubt accessible conformations. A particularly striking example of incoherent skewing, 14 kcal mol^{-1} less stable than the coherently skewed conformation, has been calculated for tetra-*tert*-butylmethane[60]. Evidence for such conformations other than from calculations will be difficult to come by[70], but when inspection of a calculated conformation arrived at by energy minimization shows that it is not coherently skewed, the nearby conformation which is coherently skewed should be sought and can be expected to be more stable.

Thus the approximately 120° of rotational conformational space between two eclipsed arrangements, which holds a single staggered conformation according to the simple analysis, may in fact contain two coherently skewed conformations and an indeterminate number of easily accessible, incoherently skewed ones. Such a proliferation of conformational states may have entropic consequences.

The minimum-energy path for libration between skewed conformations appears to involve a concerted movement about all skewed bonds, whereas rotation through an eclipsed conformation happens independently for all groups involved. This has been demonstrated experimentally in the case of adamantyl-di-*tert*-butylcarbinol[67] where the barriers to rotation of the adamantyl and *tert*-butyl groups are different (10.8 and 11.7 kcal mol^{-1}, respectively), while the libration barrier measured by observing NMR signals of both groups is the same (10.3 kcal mol^{-1}).

Rotation and libration in such systems cannot be unrelated, however[56], for when one group in a coherently skewed molecule rotates to a new staggered position, it becomes incoherently skewed compared with all the other groups, so additionally they or the rotating group must librate to give finally a new coherently skewed stable conformation. This means that libration will almost invariably have a lower barrier than rotation, for the latter comprehends a very large measure of the former.

Skewing, coherence, libration and rotation are clearly of great importance in discussing the real examples of tri-*tert*-butylmethane and tri-*tert*-alkylmethanols. In the case of hexamethylethane, some of their consequences are significant, and in as simple as example as *gauche*-butane they can be traced vestigially. Any simple molecule with adjacent alkyl groups is affected by skewing and libration, and any simplifier who deems these effects to be negligible should be able to justify such neglect.

D. Eclipsed Ground States

Any systematic study of the conformation of a 1,1,2-trisubstituted ethane $R^1CH_AH_B$—$CH_CR^2R^3$ (12) leads on to the question of eclipsed ground-state conformations as the groups R become more demanding of space. A conformation like 13, whether skewed or not, is unlikely to be significantly populated and conformations 14 or 15 are likely to be skewed away from perfect staggering, to reduce interactions between groups R. When presented this way it is not surprising that, as these groups R become larger, the likelihood of a conformation close to eclipsed increases. It has been suggested[65] on the basis of molecular mechanics calculations and coupling constant measurements that in 1,1,2-tri-*tert*-butylethane the preferred conformation is more or less eclipsed with the R^1—C—C—H_C dihedral angle at 5° (see 16).

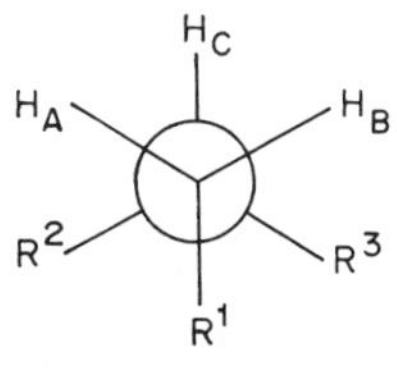

(13) 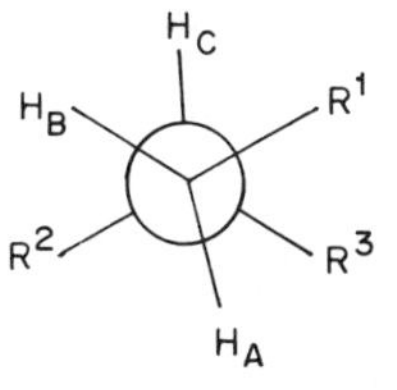   (14) 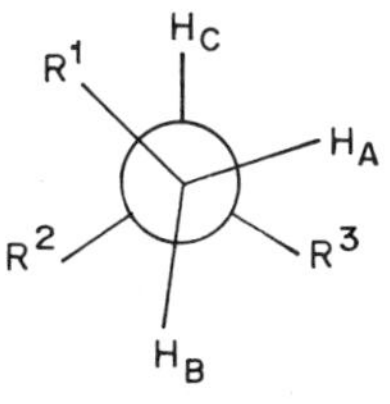   (15)

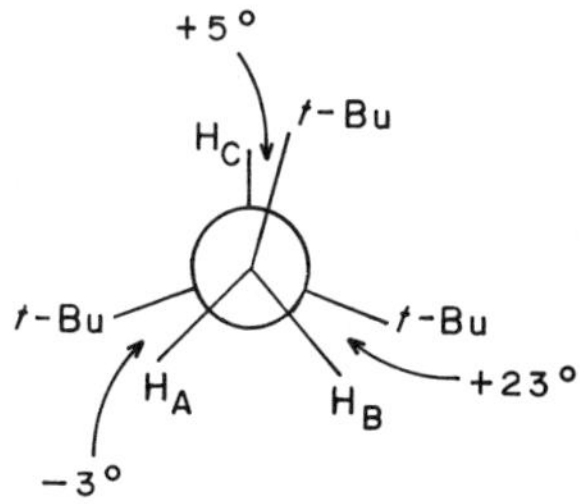

(16)

J. E. Anderson

TABLE 1. Vicinal proton–proton coupling constants (Hz) in acyclic alkanes
$R^1CH^AH^B$—$CH^CR^2R^3$

R^1	R^2	R^3	J_{A-C}	J_{B-C}	$\sum J$	Reference
t-Bu	t-Bu	t-Bu	3.6	3.6	7.2	69
Me	t-Bu	t-Bu	3.7	3.7	7.4	69
Me	Et	t-Bu	3.3	6.5	9.8	71
t-Pentyl	Me	Me	5.1	5.1	10.2	72
Me	i-Pr	i-Pr	5.1	5.1	10.2	55
t-Bu	Me	i-Pr	8.8	1.7	10.5	72
t-Bu	Me	Me	5.3	5.3	10.6	72
t-Bu	Me	CH_2Bu-t	7.1	3.7	10.8	55
Me	i-Pr	t-Bu	4.2	7.0	11.2	55
Me	Et	i-Pr	4.8	6.6	11.4	75
i-Pr	Me	Me	7.1	7.1	14.2	72

NMR can be used to follow what happens with simpler groups R. From the NMR spectrum of compounds like 12 it should be possible to measure two coupling constants, J_{A-C} and J_{B-C}, which should be equal if $R^2 = R^3$. The sum of these, $\sum J$, indicates the degree of eclipsing. When the conformation is near to staggered, like 14 or 15, $\sum J = 14.2$ Hz, if 2,4-dimethylpentane can be taken as an example. In the tri-tert-butylethane, which is a good model for the eclipsed conformation, $\sum J = 7.2$ Hz. Table 1 shows these values and several more for acyclic alkanes which seem to span the range of conformations intermediate between staggered and eclipsed, with the sum of the coupling constants quantifying the intermediacy.

When in a molecule 12 the groups R^2 and R^3 are part of a ring system, there is less scope for distortion than in an acyclic system, and eclipsed conformations may be more likely. For example, 2,6-dimethylneopentylcyclohexane in the structure where all groups are equatorial is calculated to be most stable when the tert-butyl group eclipses the 1-axial hydrogen[55].

While focussing on the eclipsing of R^1 and a hydrogen, it should not be forgotten that R^2 and R^3 are also quite near to being eclipsed. However, the closing down of the H—C—H bond angle and the opening up of the R^2—C—R^3 bond angle to produce a non-alternating conformation[69,73], which is calculated to occur in the most extreme of the cases quoted above, does mean that the eclipsing may be less than perfect. A crystal structure confirmation of such eclipsed conformations is obviously desirable.

E. Attractive Steric Interactions

Attractive steric interactions are a feature of all molecules which can be analysed conformationally. The total enthalpy of any conformation will thus include an attraction term, which in a polyatomic molecule may be considerable. Often, of course, repulsive steric interactions will be considerably greater and will be analysed as determining the molecular conformation. It is not uncommon, however, that molecular mechanics calculations for a molecule suggest that the sum of all interactions greater than 1,4 with respect to each other is attractive, which is inherently reasonable when a little consideration is given to the interatomic distances involved.

Such weak forces are difficult to quantify and will not be further discussed here, except to consider whether the position of a conformational equilibrium in saturated hydrocarbons is ever determined by attractive steric interactions. The best evidence that can be adduced comprises the known results[74] for the series of compounds $RCH_2C(Me)_2$—$C(Me)_2CH_2R$. The groups R will no doubt be *anti* to the group —$C(Me)_2CH_2R$ as shown in **17**, but it is the experimentally observed equilibrium along the central bond of the molecule that is of interest. The conformation **18** with RCH_2 groups *gauche* to each other (or its enantiomer) is more stable than the *anti*-conformation **19** by 0.08–0.2 kcal mol^{-1}. Since methyl groups and CH_2R groups as in **17** are similar in shape (seen from the opposite end of the bond), it is reasonable to attribute the conformational preference for **18** and its enantiomer to attractive steric interactions between groups R, which are greater when these groups are nearer, i.e. in either *gauche* conformation. Since the groups R are beyond repelling each other in all conformations, molecular mechanics inevitably confirms that R—R attractions are greater in *gauche* conformations[55].

R —CH$_2$

    C —— C(Me)$_2$CH$_2$R
      ''''Me
Me

(17)

R
|
CH$_2$
Me   R
   \   /
    \   CH$_2$
Me   Me

Me   Me
    Me

(18)

R
|
CH$_2$
Me   Me

Me   Me
    CH$_2$
      |
      R

(19)

    There are one or two consequences of long-range attractive interactions that are worth commenting on. There is a natural tendency to think in terms of the extended structure of a molecule, since saturated carbon–carbon bonds are known to prefer an *anti*-conformation, and a clear representation of the overall structure emerges if groups are drawn in a linear sequence. If any longish molecule is folded up, with some care taken not to involve drastically unfavourable conformations, large sections of the molecule involving many atoms may be brought near to the optimum structure for attractive steric interactions. Molecular mechanics suggests that attractive steric interactions between carbon and hydrogen atoms may be 0.05 kcal mol^{-1} at a maximum. Two cyclohexyl groups C_6H_{11} near each other have 289 pairwise interactions, and if these are on average only 10% of the maximum, the net attraction is over 1.4 kcal mol^{-1}.

    Well-judged folding thus has enthalpic attractions. It may correctly be argued that opportunities for intermolecular attraction are reduced in a folded conformation in the liquid phase, but sets of fluctuating random intermolecular interactions are unlikely to match a favourable folded conformation sitting in a conformational potential-energy well.

    In molecules with little steric strain but a large number of atoms, attractive steric interactions contribute significantly to the net steric strain and offer an extreme test of whether a molecular mechanics program gives a correct balance of repulsive and attractive van der Waals forces. Molecules which may fall into this class are the medium- to large-ring cycloalkanes.

    Helical conformations have been observed in polymethylene, the epitome of a long saturated hydrocarbon[75]. Insofar as a helical structure brings distant atoms closer together than in an extended linear conformation, attractive steric interactions offer an enthalpic contribution towards an explanation of this helicity.

III. SUBSTITUTED METHANES AND ETHANES

A. Introduction

Any single C_n saturated hydrocarbon is a substituted methane n times over and a substituted ethane $(n-1)$ times over (or n times over if cyclic), so simple methyl-susbstituted methanes and ethanes are well worth considering. Discussing the significant conformational points of each and considering their use and the limits on their use as models for more highly branched hydrocarbons is therefore important.

B. Propane and Dialkylmethanes

The experimental[76] C—C—C bond angle of 112.4° and the central H—C—H bond angle of 106.1° in propane show that repulsion of geminal C—C bonds is greater than that between C—H bonds. The nearest hydrogens of different methyl groups do not appear to repel each other, since such repulsion would be relieved by concerted methyl group rotation(skewing), as well as by bond-angle opening, as discussed above, but there is no evidence for this[77-79]. The methyl group rotational barrier in propane[80], 3.17 kcal mol^{-1}, which is 0.29 kcal mol^{-1} higher than in ethane, represents a simple steric augmentation of the barrier.

In propane analogues CH_2RR^1 with more branched substituents which cannot avoid parallel 1,3-interactions, R—R^1 repulsion becomes important. A small increase in the C—C—C bond angle moves substituents apart significantly because of divergence, and the necessary closing down of H—C—C and H—C—H bond angles at the same centre is not too demanding a distortion. Consequently, further opening of C—C—C bond angles is a common and favourable way of relieving this strain. Relief can also be achieved by skewing of the dihedral angles in the two alkyl groups to reduce 1,3-interactions. Both features should become more marked as these latter interactions increase, and di-tert-butylmethane has received detailed consideration[79,81]. An electron diffraction study[81] shows that the central C—C—C, H—C—H and H—C—C bond angles are 125–128°, 105° and 112°, respectively, while the dihedral angles of the CH$_2$-tert-butyl and C–methyl bonds average about 15° and 12° from perfectly staggered, respectively. Calculation[79] agrees well with experiment.

2,2-Dimethylbutane is halfway between propane and di-tert-butylmethane, for its methylene group carries one methyl and one tert-butyl group. Molecular mechanics calculations of the ground-state conformation[82] show substantial C—C—C bond-angle opening to 116° but surprisingly little skewing (1.2°). This molecule with intermediately large 1,3-interactions is clearly able to maximize strain relief without skewing, which emphasizes how bond- angle opening is the preferred reliever in a disubstituted methane.

The threefold barrier to tert-butyl group rotation in this molecule has been determined by dynamic NMR spectroscopy[82,83] to be 4.9 kcal mol^{-1} at 92 K. This observation implies that the barrier to rotation about the CH$_3$—CH$_2$Bu-t bond is less than 4.9 kcal mol^{-1}, so that the energy associated with a tert-butyl group eclipsing a hydrogen atom represents less, maybe much less, than 2.0 kcal mol^{-1}. This is not surprising, since the two hydrogen–hydrogen distances in the perfectly staggered ground state **20** are shorter than the two equivalent distances in the perfectly eclipsed state **21**. These symmetrical conformations are not expected to be the actual ground and transition states for this rotation, but the comparison does make an important point. For a branched alkyl substituent, the 1,2-interactions in an eclipsed transition state may be significantly offset by the parallel 1,3-interactions of a staggered ground state. Herein lies part of the explanation of low rotational barriers in molecules with extensive 1,3-interactions, discussed elsewhere in this review.

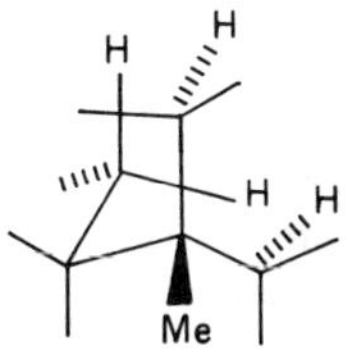

(20) 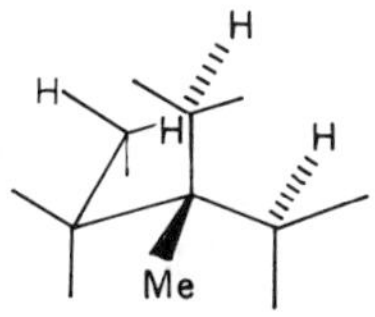   (21)

C. 2-Methylpropane and Trialkylmethanes

Again C—C—C bond angles are opened[84], to 111.2°, in 2-methylpropane. Dihedral angles are not clearly different from 60° and 180°. The rotational barrier is 3.9 kcal mol^{-1} [85-98], even greater than that in propane. Trialkylmethanes can still compensate opening of three C—C—C bonds with closing of three H—C—C bonds and skewing to reduce parallel 1,3-interactions. These distortions reach the extreme in tri-tert-butylmethane[8,65,66] discussed in detail in Section II.C. Even in isopropyl-di-tert-butylmethane through to triisopropylmethane[52,89,90] there is nothing like the extreme behaviour of tri-tert-butylmethane, although interaction between the three alkyl groups does lead to considerable distortion of dihedral angles. Molecular mechanics calculations and NMR coupling constants permit quite firm conclusions about conformations, and how three groups around a single point interact with each other, which should allow a useful discussion of simpler molecules.

D. 2,2-Dimethylpropane and Tetraalkylmethanes

Bond angles cannot open up in 2,2-dimethylpropane except as a vibration, since there are no others to close down in compensation. The structure has been determined[81] and calculated[91], and the rotational barrier[88,92] is 4.3 kcal mol^{-1}. Tetraalkylmethanes are inevitably much more strained than their trialkylmethane structural isomers[52] due to the restriction on bond-angle opening. Significant skewing is to be expected, but this is also somewhat reduced by the presence of four substituents at the central methane carbon.

Iroff and Mislow[60] have carried out force field calculations on tetra-tert-butylmethane. The striking features to emerge are the compression within the tert-butyl groups seen in methyl—C—methyl bond angles of about 100° and restricted skewing of the methyl groups by only 4.8° due to all-round congestion. The skewing of the tert-butyl groups is 14–16°. A brief report has recently appeared[93] of calculations of the conformations of tetraethylmethane and methane with any four primary alkyl groups. Not unexpectedly, out of the range of combinations of conformations, only two are enthalpically preferred, those that avoid all methyl–methyl parallel 1,3-interactions.

This last investigation is connected to one aspect of conformational analysis of tetraalkylmethanes that has been well known for a long time. The gem-dimethyl effect[94] can be generalized as follows in conformational terms. In a molecular fragment R—CH$_2$—CX$_2$—CH$_2$—R when X is hydrogen, there is an enthalpic preference for each R to be anti to CH$_2$R along a CH$_2$—CX$_2$ bond, so the groups R are far apart in space. When X = methyl (or, more generally, when X is not hydrogen), the likelihood that R is gauche to CH$_2$—R along a CH$_2$—CX$_2$ bond is greater, so the groups R may be close together. The consequence of the effect may simply be conformational as described, or may be seen as greater reactivity of the groups R to give ring closure or other reactions. The effect should be seen, if less markedly, even when only one X is not hydrogen.

In more highly branched tetraalkylmethanes, the drive of all groups to minimize 1,3-interactions (although there is limited scope for doing so) dominates the conformational analysis. Lengthening of bonds to the quaternary centre is to be expected, but it buys small relief at a high price in enthalpy. Hence unusual distortion away from the quaternary centre, as calculated for tetra-*tert*-butylmethane above, may be common. A high strain energy compared with its structural isomers is to be expected.

E. Butane and 1,2-Dialkylethanes

Butane exists in one of three minimum energy conformations, two of which are enantiomeric and *gauche* and one that is *anti* and most stable by about 0.89 kcal mol^{-1} in the gas phase[15,16]. This is a particularly notable number as a model for the strain in a *gauche* arrangement in any —(CH$_2$)$_4$— fragment, for half the strain in a methyl axial arrangement in cyclohexane and for the strain in methyl groups vicinal in other saturated fragments. The comments in Section I.B on the liquid-phase value of this number and in Section II.C on skewing should not be forgotten.

The *gauche* conformation of butane already shows conformational distortions in that the C—C—C—C dihedral angle is opened[53,54] to about 70°, while the methyl group hydrogens are calculated[55] to be skewed coherently by about 4.6°. The rotational barrier to *anti–gauche* interconversion derived spectroscopically[95] is 3.63 kcal mol^{-1}, and as to the methyl eclipsing–methyl barrier, there has been considerable speculation described recently by Allinger and coworkers[42], who conclude it to be 4.89 kcal mol^{-1}.

There seems to have been little work on the corresponding *anti–gauche* equilibrium in 1,2-dialkylethanes. 1,2-Di-*tert*-butylethane provides a reasonable model for an all-*anti* conformation[96]. Calculations of the strain in many compounds in the series have been reported without a conformational analysis[97,98] and in fact many have been prepared in pure form and had their density, refractive index and heat of combustion discussed in terms of the average number of *gauche* interactions[6,99,100].

F. 2-Methylbutane and 1,1,2-Trialkylethanes

2-Methylbutane is a simple model for an alkyl chain with a single methyl substituent H(CH$_2$)$_n$-CH(CH$_3$)—CH$_2$(CH$_2$)$_m$H where $m = n = 1$. The conformation with the C-4 methyl group *gauche* to both the geminal methyl groups is less stable than the enantiomeric conformations, where it is *anti* to one by 0.81 and 0.64 kcal mol^{-1} in the gas[101] and in the liquid[17] phase, respectively (0.92 kcal mol^{-1} by calculation[41]). 2-Methylbutane would serve reliably as a student's first textbook exercise in conformational analysis, since the enthalpy difference equates very well with the value expected on counting *gauche* interactions, but it is to be hoped that teaching will not stop there. *Ab initio* calculations suggest libration by 3° on either side of perfect staggering in the less stable conformation and skewing of *gauche* methyl groups by 4° in the preferred conformations[41].

1,1,2-Trisubstituted ethanes have been discussed in Section II.E, where the tendency towards more skewed and eventually eclipsed conformations was noted.

G. 2,3-Dimethylbutane and 1,1,2,2-Tetraalkylethanes

The resistance of organic chemists to teaching an acceptable conformational picture of 2,3-dimethylbutane is remarkable. Studies by many techniques[17,58,101] and calculations[29,41,102] all suggest that the *anti* conformation 23 has very similar enthalpy to the *gauche* conformations 22 and 24, with the *gauche* forms slightly more favoured in solution than in the gas phase[17]. Tyro students, however, are usually expected to deduce that the *anti*-conformation 23 with two *gauche* methyl–methyl interactions is

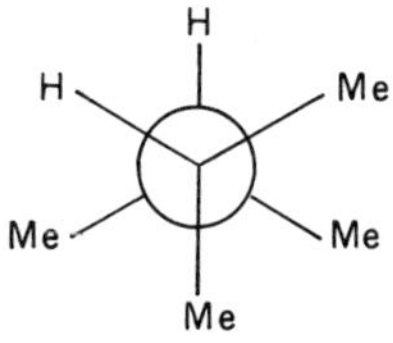

(22) 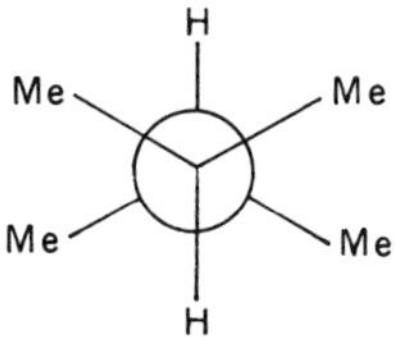   (23) 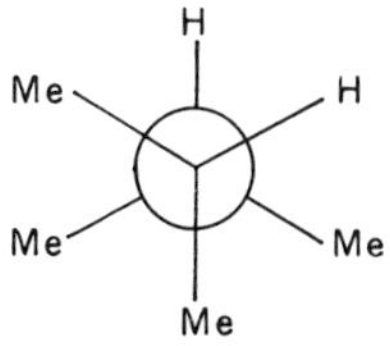   (24)

more stable than the *gauche* conformations **22** and **24** with three such interactions. In spite of this wrong conclusion there is a quite patent determination to limit students' knowledge of the subject of this review to *gauche* interactions as the only determinant, and some may progress uncommonly far knowing no better. A slightly more complicated description introducing bond-angle opening and skewing, with 2,3-dimethylbutane and hexamethylethane as exemplars, allows a correct appreciation of the basic conformational analysis of saturated organic compounds.

In the *anti*-conformation **23** with two *gauche* methyl–methyl interactions, opening of the methyl—C—methyl bond angles at both ends of the central bond causes vicinal methyl groups to be closer to each other as shown in the exaggerated diagram **25**. Any skewing to relieve this for one pair of methyls accentuates the interaction of the other pair. In the *gauche*-conformations **22** and **24**, methyl—C—methyl bond-angle opening and skewing in one direction (but not in the other) moves all methyl groups quite far apart, as shown in the exaggerated diagram **26**. Molecular mechanics calculations[102] suggest that the two methyl–methyl *gauche*-dihedral angles of **25** are 57.5°, while the three angles in **26** are 64.7°, 64.7° and 61.4°. *Ab initio* calculations[41] suggest that the *anti*-conformation librates 3° on either side of perfect staggering while dihedral angle distortions in the *gauche* isomer are from 2° to 15°. The net result of these distortions is suggested by calculation and by experiment to make the three possible 'staggered' conformations for 2,3-dimethylbutane of simlar enthalpy.

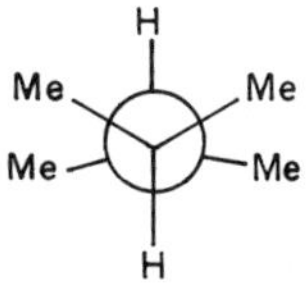

(25) 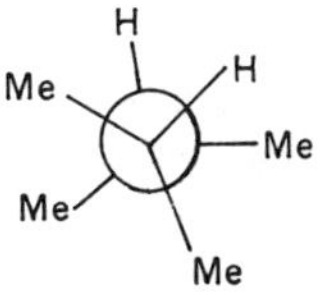   (26)

The barrier to interconversion of *gauche* and *anti* forms of 2,3-dimethylbutane has been determined[9,16] by dynamic NMR spectroscopy to be 4.3 kcal mol^{-1} at $-177\,°C$. Direct interconversion of *gauche* forms involves two more-or-less simultaneous methyl–methyl eclipsing interactions and a higher barrier[7].

1,1,2,2-Tetraalkylethanes show their relationship to 2,3-dimethylbutane in that their analysis is dominated by the C—C—C bond angles opening while H—C—C bond angles close down at both ends of the ethane bond. This disfavours the *anti* conformation and favours the two *gauche* conformations like **27** and **28**, in which rotation about the ethane bond to move alkyl groups towards the remaining hydrogens has taken place.

Compounds **29–31** show the consequences of extreme branching in such molecular situations, Tetra-*tert*-butylethane **29**[103,104] and its less symmetrical analogues **30** and

TABLE 2. Vicinal coupling constants (in H_Z) between protons H^1 and H^2 in $(CH_3)_2CH^1—CH^2AB$

A	Me	Me	Me	Me	Et	Et	Et	i-Pr	i-Pr	t-Bu
B	Me	Et	i-Pr	t-Bu	Et	i-Pr	t-Bu	i-Pr	t-Bu	t-Bu
$^3J_{H^2-H^2}$	5.4	3.9	5.5	2.0	4.5	5.1	1.8	4.5	1.4	0.8
Ref.	55	72	55	89	72	55	89	90	52	89

31^{49} exist only in distorted versions of the enantiomeric *gauche* conformations **27** and **28**, with a barrier to interconversion of the two forms by rotation about the central bond so high that dissociation[105] of the molecule takes place more readily. The two enantiomeric conformations of **30** can be separated chemically by picking crystals, and this provides the first example of stable rotational isomerism in aliphatic hydrocarbons[104]. A crystal structure determination of the (S, S)-configuration of **30** showed that both methine hydrogens more or less eclipse a *tert*-alkyl group, and that the H—C—C—H dihedral angles are 107° and 109° in the two rotational isomers[104]. Several groups[63,73,106] have carried out molecular mechanics calculations on **29** with results which are somewhat force-field dependent.

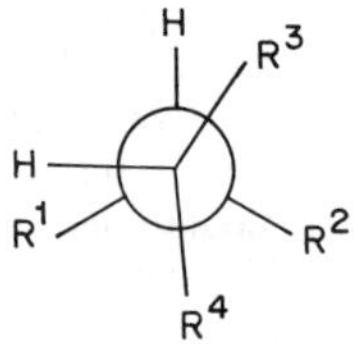

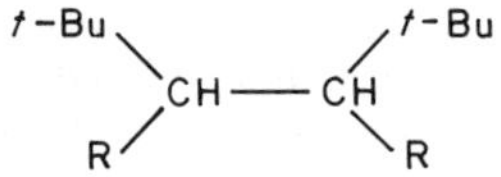

(29) R = t-Bu

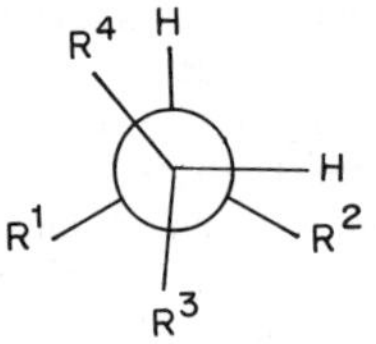

(30) R = 1-adamantyl (d,l configuration)

(31) R = 1-adamantyl (meso-configuration)

(27)   (28)

It is interesting to observe what happens in intermediate situations in a series of compounds $(CH_3)_2CH^1—CH^2AB$ as exemplified by the $H^1—H^2$ vicinal coupling constant. If only two perfectly staggered *gauche* conformations are populated, the coupling constant should be about 3 Hz. In general terms a larger coupling constant indicates the presence of significant amounts of the *anti* conformation, and a smaller coupling constant indicates significant skewing towards an H—C—C—H dihedral angle of 90° or greater. Table 2 shows such coupling constant and demonstrates that, as substituents vary, a full range of conformations can be encountered. A full discussion of conformations in some of the compounds, including molecular mechanics calculations, has been given[52,89,90]. The corresponding coupling constant in tetra-*tert*-butylethane **29** is 2.0 Hz[106], rather more than some values in Table 2, but this is an indication that the H—C—C—H dihedral angle is rather greater than 90° in this case.

H. 2,2,3-Trimethylbutane and Pentaalkylethanes

For 2,2,3-trimethylbutane (pentamethylethane **32**), there is presumably a sixfold rotational potential with six equivalent conformations skewed on either side of perfect staggering. There is a barrier of 6.9 kcal mol^{-1} to rotation about the central ethane bond[63,107]. Pentaalkylethanes can also be analysed as trialkylmethanes, remembering

32 R = Me

RCH(Me)—C(Me)₃   33 R = Et

34 R = i-Pr

the skewing about the central bond. Surprisingly, the rotational barriers in ethyltetramethyl- and isopropyl tetramethylethane (**33** and **34**), 6.3 and 6 kcal mol^{-1} respectively[83], are smaller than in **32**. This suggests increased strain in the ground state of the more substituted compounds due to both bond-angle compression at the methine carbon and 1,3-interactions.

I. Hexamethylethane and Hexaalkylethanes

The conformational analysis of hexaalkylethanes, including hexamethylethane which has been discussed in Section II.C, is dominated by the sixfold rotational potential with skewing on either side of perfect staggering to attain the most stable minimum conformation. There are barriers (which may often be more than one kcal mol^{-1}[57]) to the concerted passage through perfectly staggered conformation to reach the opposite skewed conformation, but barriers to rotation through the eclipsed conformation are as low as the 8.4 kcal mol^{-1} for hexamethylethane[64] discussed above and seldom[89] more than about 11 kcal mol^{-1}, reflecting substantial steric interactions of the 1,3-type in the ground state. A structural feature of such molecules is lengthening of the central bond which helps minimize these interactions. Certain aspects of the conformational analysis of such molecules is well covered in papers by Ruchardt and his collaborators and by Anderson and coworkers[44,45,52,57,89].

IV. DYNAMIC STEREOCHEMISTRY

A. Introduction

The rotational barrier for all carbon–carbon sp^3–sp^3 bonds is greater than that in ethane[13,108-110], and to the extent that barriers broadly increase with steric congestion, hexasubstituted ethanes have higher barriers than pentasubstituted ethanes, which are in turn higher than tetrasubstituted ethanes and so on. It is attractive to think of rotational barriers *in saturated hydrocarbons* in terms of an intrinsic barrier[111] of 2.9 kcal mol^{-1} overlaid by a steric contribution that can be analysed in terms of the alkyl substituents on the bond, but some caveats are necessary.

In highly-branched alkanes, experimental and calculated barriers are often for the interconversion of stable populated conformations, which may for convenience be described as a rotation about one significant bond, although some rotation about several bonds and other distortions are involved. Focussing thus on one bond is acceptable as long as any rationalization considers other bonds adequately.

The ground-state conformation about a carbon–carbon bond is perfectly staggered only in the simplest cases as discussed above, and the analyst may find that the most interesting aspect of the interpretation of a barrier or the comparison of two barriers is what is learnt about ground states.

The status of the 'eclipsed conformation' as the rotational transition state needs considering. When C^1 and C^4 in a C^1—C^2—C^3—C^4 fragment rotate past each other, the high-energy point may not occur when the dihedral angle is 0°, but rather at that point nearby where the inward pointing substituents on C^1 and C^4 have their maximum interaction. Furthermore, as Mislow has pointed out[73], there is no reason for all three pairs of substituents along the bond to have maximum interactions at the same time. A little distortion can allow the high-energy points of each pairwise interaction to occur consecutively so as to reduce the maximum energy somewhat. Such points are particularly true in highly branched molecules, so the terms 'eclipsed conformation' and 'the rotational transition state', if apparently used interchangably have some implied qualifications.

TABLE 3. Dihedral angle changes in some rotational and librational processes

Compound showing bond involved	Nature of the process	Change in dihedral angle (deg)	References
t-BuCH$_2$—CH(Bu-t)$_2$	Rotation through nearly eclipsed ground state	10	69
	Rotation through syn state	350	69
Me$_3$C—CMe$_3$	Libration through perfectly staggered state	31	57
Me$_3$C—CMe$_3$	Rotation through eclipsed state	89	57
(t-BuMe$_2$C—)$_2$	Libration through the $anti$ conformation	32	57
CH$_3$CH$_2$—CH$_2$CH$_3$	Rotation between $gauche$ conformations	100	53, 54
CH$_3$CH$_2$—CH$_2$CH$_3$	Rotation between $gauche$ and $anti$ conformations	130	53, 54
(t-Bu)$_2$CH—Pr-i	Rotation through the H—C—C—H = 0° conformation	173	89
	Rotation through the H—C—C—H = 180° conformation	187	89
(t-Bu$_2$CH—)$_2$	Rotation through the H—C—C—H = 0° conformation	191	63
	Rotation through the H—C—C—H = 180° conformation	169	63

In a range of different molecules passing from one significant conformational minimum to another, the change in dihedral angle may vary from a very small amount for some librations and even some rotations through an eclipsed conformation, through a range of intermediate values, with libration reaching as high as 38°, to the normal value for a simple substituted ethane about 120°. In 1,1,2,2-tetrasubstituted ethanes with a twofold rotational potential discussed above, stable conformations are likely to interconvert by somewhat more than 180° of rotation rather than the more hindered rotation of somewhat less than 180°. Occasionally there is only one significantly populated conformation, so rotation means a change of close to 360°. Table 3 lists examples of the range that can be found as indicated by experiment and by calculation.

B. Normal Behaviour

Since it is important to be aware of both libration and rotation, normal behaviour is best represented by hexamethylethane[56,60-64] with its six minima during rotation through 360° and two separate processes, of libration through about 31°, and rotation of about 89° through an eclipsed conformation. All hydrocarbons described in Section III can be treated as modifications of this.

In simpler substituted ethanes in symmetrical conformations, the librational minima will be nearer to each other and they coincide at the perfectly staggered position in ethane, which has no librational barrier. If the simpler ethane has an unsymmetrical conformation, e.g. $gauche$ butane, one librational minimum may disappear or be merely a point of inflection, so that the sixfold potential becomes an unsymmetrical threefold one with more substituted carbon–carbon bonds with other than simple methyl substituents, the six possible minima should be considered as three librational pairs. Whether any librational pair is so unsymmetrical that it can be considered as a single minimum, and thus simplified, can be decided by calculation, or perhaps by intuition. The libration may not be suitable for experimental study, but awareness of the

TABLE 4. Calculated barriers[52] to libration^a (kcal mol^{-1}) in gauche and anti conformations of compounds $RCMe_2$—CMe_2R

R	gauche Libration	anti Libration
Methyl	0.4	0.4
Ethyl	1.2	1.4
Isopropyl	b	1.5
tert-Butyl	2.2	2.1
Neopentyl	1.7	2.1

^{a}From more stable, skewed conformation when there are two.
^{b}Not straightforward; see original paper[52].

phenomenon allows the chemist to be surer about the location and interpretation of minima and of barriers. The conformational possibilities of each alkyl substituent on the bond can be analysed analogously.

C. Concerted Motions and Libration

As discussed in Section II.C, a coherently skewed ground-state conformation is expected in a molecule with several branched groups around a central carbon. If libration or rotation of one group takes place, an unstable, incoherently skewed state is reached, and only after libration of other groups at the centre (in the same sense if the first group librated, in the opposite sense if it rotated) is a new stable, coherently skewed populated state reached. A coherently skewed conformation also results if the rotating group rather than the other groups subsequently librates, so while the other groups no doubt move during the process, rotation of one group can happen independently, leaving all other groups in their original position.

In seeking to demonstrate that concerted motion or libration is involved in a conformational interconversion, the experimental evidence is usually the barrier to interconversion of conformational minima, and the molecular structure in the minima. We have to speculate as to the means of achieving the exchange process, or use molecular mechanics to illuminate this. In the case of the tri-tert-alkylcarbinols[67,68] the frequency of libration of different alkyl groups is the same and, while this does not show that they rotate at the same time, calculations for a less restricted molecule tri-tert-butylsilane[73] suggest that groups do rotate together.

The bulk of knowledge of libration in hydrocarbons is derived from calculations, which show how unusually high the librational barrier for tri-tert-butylmethane is[65]. In a set of 1,2-dialkyltetramethylethanes, calculations[52] show that all libration barriers are less than 2.2 kcal mol^{-1} (Table 4), with different values for the gauche and anti conformations.

D. Rotation

It is a general rule for a carbon–carbon bond that the greater the dihedral angle change between conformational minima, the higher the barrier is likely to be, so a onefold or twofold rotational barrier should be higher than a sixfold. The exceptions tend to be the most memorable, but generally the very factors which produce minima at only one or two positions in the rotational cycle are likely to increase the potential between the two minima.

The onefold rotational barrier in the trisubstituted 1,1,2-tri-tert-butylethane is calculated[69] to be about 17 kcal mol^{-1}. In the series of 1,1,2,2-tetrasubstituted ethanes with a twofold rotational potential, the exceptional tetra-*tert*-alkyl compounds with barriers certainly over 23 kcal mol^{-1}[103] and probably over 40 kcal mol^{-1}[104] have already been discussed in Section II.G. Table 5 shows results for tetraalkylethanes with twofold and threefold rotational barriers.

It has long been recognized[112,113] that in highly-branched hydrocarbons parallel 1,3-interactions lead to considerable destabilization of ground states and to low rotational barriers. In this context it is well worth considering rotation about hexaalkylethane bonds for which there may be up to six rotational minima during the 360° rotation, as discussed for hexamethylethane. As long as libration is easy, these may be considered as three double minima, although the origin of the librational barrier, and of steric reduction of rotational barriers, is the same.

Table 6 shows two sets of results, for *tert*-butyl—C(Me)R^{1}R^2 and for R^{1}CH$_2$C(Me)$_2$—C(Me)$_2$CH$_2$R^2. It is noteworthy that in the first series, although barriers rise somewhat with increased branching, which thus affects the rotational transition state more than the ground state, there appears to be an upper limit of about 12 kcal mol^{-1} on *tert*-butyl rotational barriers. Outside the hydrocarbon field, where substituents can be bulky without branching, barriers to *tert*-butyl rotation are markedly higher as, for example, that of 13.6 kcal mol^{-1} for dibromo-di-*tert*-butylmethane[114]. Higher *tert*-butyl group rotational barriers are also found[115] when the group is attached to a quaternary carbon of a rigid bicyclic system.

The striking feature of the second series is the effect of a β-substituent on the rotational

TABLE 5. Some twofold and threefold rotational barriers (kcal mol^{-1}) in tetrasubstituted ethanes

				Barrier		
R^1	R^2	R^3	R^4	expt.	calc.	Reference
Twofold barriers in R^{1}R^{2}CH—CHR^{3}R^4						
i-Pr	*i*-Pr	Me	Me	6.6	5.5	52
t-Bu	*t*-Bu	Me	Me	11.0	9.0	90
t-Bu	*t*-Bu	*t*-Bu	*t*-Bu	>23		103
Threefold barriers in R^{1}R^{2}R^{3}C—CH$_2$R^4						
Me	Me	Me	Me	4.9		83
Me	Me	Me	*t*-Bu		4.9	79

TABLE 6. Rotational barriers (kcal mol^{-1}) in hexaalkylethanes

Compounds	R^1	R^2	expt.	References
Me$_3$C—CMeR^{1}R^2	Me	Me	8.4	64
	Me	Et	9.4	61, 83
	Me	CMe$_2$Bu-*t*	11.7	61
	i-Pr	*i*-Pr	8.7	90
	i-Pr	*t*-Bu	11.0	89
R^{1}CH$_2$C(Me)$_2$—C(Me)$_2$CH$_2$R^2	H	H	8.4	64
	Me	Me	10.4–10.7	57, 61, 83
	i-Pr	*i*-Pr	10.3	57
	t-Bu	*t*-Bu	13.8	57, 63

TABLE 7. Rotational barriers (in $kcal\,mol^{-1}$) in compounds *tert*-butyl—$C(X)R^1R^2$

R^1	R^2	t-Bu—$C(Me)R^1R^2$	t-Bu—CHR^1R^2	References
		Rotational barrier		
Me	Me	8.4	6.9	64, 83, 107
Me	Et	9.4	6.3	61, 83
Me	i-Pr	8.7	6.0	83
i-Pr	i-Pr	11.0	7.3	52
t-Bu	i-Pr		7.6, 8.4	89
t-Bu	t-Bu		8.0	117
—$(CH_2)_3$—			6.0	118
—$(CH_2)_4$—			6.3	118
—$(CH_2)_5$—			7.4	118
—$(CH_2)_6$—			7.8	118
—$(CH_2)_7$—			7.3	118
—$(CH_2)_8$—			7.3	118

barrier[57,113,116]. It is reasonable to expect R^1 and R^2 in such molecule to prefer a position *anti* to the $C(Me)_2R$ group, and thus to have no interactions with groups at the other end of the central bond either in the ground or the transition state for rotation. The undeniable affect of R on the barrier must therefore be indirect, and a buttressing of the CH_2 group to which it is attached has been suggested[63,113].

With one fewer substituent, barriers to *tert*-butyl group rotation in pentasubstituted ethanes are rather smaller. Table 7 shows the reduction in rotation barriers in *tert*-butyl—$C(Me)R^1R^2$ when the methyl group is replaced by a hydrogen atom, and barriers for some other pentaalkylethanes.

The measured *tert*-butyl rotational barriers in compounds t-Bu—CH_2R are 4.9 and $4.3\,kcal\,mol^{-1}$ for $R = CH_3$ and $R = H$, respectively[82,83,88,92]. The values are also for a straightforward rotation, although in the former case there is undoubtedly some slight skewing away from perfect staggering in the ground state.

Further examples exist in highly branched hydrocarbons which involve a considerable degree of co-operative rotation of several groups. Such conformational interconversions are normal when alkyl groups, particularly the unsymmetrical ethyl and isopropyl groups, interact with each other around a centre or other simple structural feature. The subject has been reviewed recently[5] in a slightly wider context than that of hydrocarbons.

An example, *tris*-isopropylmethane, is now given in some detail.[90] The preferred conformation is calculated to be one in which two of the groups have the isopropyl hydrogen *gauche* in the same sense to the unique hydrogen while the third group is *anti*, although skewing makes dihedral angles quite different from the perfectly staggered values implied by conformation 35 (and methyl group hydrogens are skewed as well). This conformation is only slightly more stable than a conformation like 36 with all three groups *gauche* in the same sense, and there is a barrier to interconversion of 35 and 36 which is calculated to be $3.4\,kcal\,mol^{-1}$. NMR coupling constants agree with an equilibrium involving these structures, but the picture has to be extended somewhat, because there are clearly three equivalent versions of 35 depending on which of the three isopropyl groups is *anti*, and these with 36 make up one set of conformations. There is a second set of conformations equivalent to the first except that all *gauche* groups are *gauche* in the opposite sense, e.g. 37 and 38. The barrier to interconversion of the sets is calculated to be $5.3\,kcal\,mol^{-1}$, and this is borne out by NMR observations of a dynamic process with a barrier of $6.6\,kcal\,mol^{-1}$.

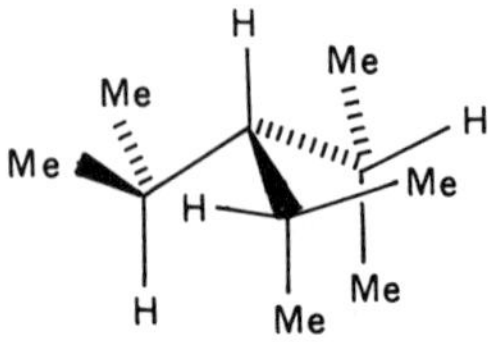

J. E. Anderson

(35) 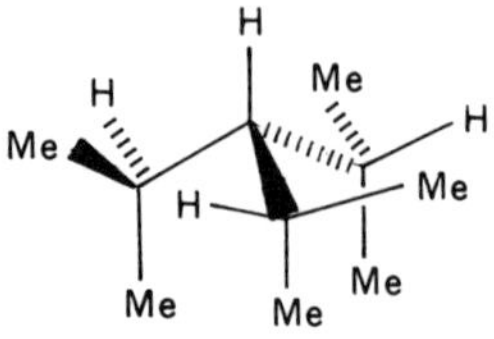 (36) 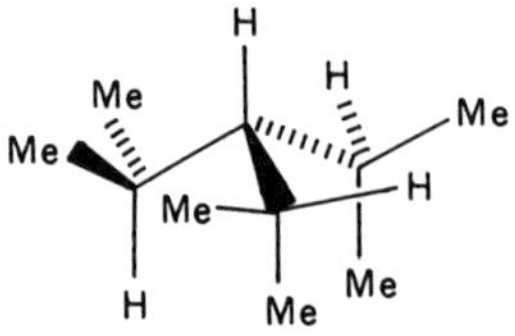 (37)

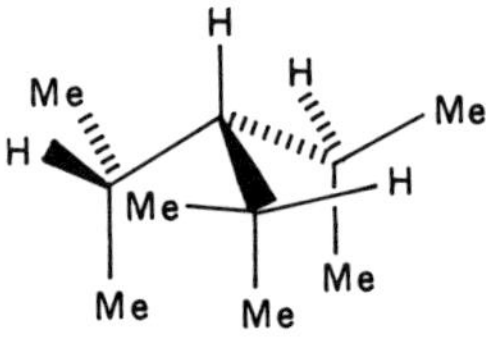

(38)

The analyst may well feel that in a molecule like this, interactions along any one bond and rotation about that bond are only a small part of the picture. The methyl hydrogens in adjacent isopropyl groups thus help define the molecular conformation by interacting through space (their relationship through bonds is 1,7). Immediate interactions along any one bond and rotation about that bond are only a small part of the picture so that comparison of the size of 'barriers' in different molecules, as made earlier in this section with simpler molecules are not particularly helpful. When all groups are branched, every molecule is a special case.

V. CYCLOHEXANE AND ALKYLCYCLOHEXANES

A. Cyclohexane

The conformation of cylcohexane is a chair, as described in the chapter on structure. The twist-boat conformation has been measured[119] to be $5.2\,kcal\,mol^{-1}$ less stable than the chair and the barrier to interconversion of the chair and twist boat is about $10.1\,kcal\,mol^{-1}$ in the liquid phase[120,121] and slightly higher in the gas phase[112] ($10.4\,kcal\,mol^{-1}$), while pressure increases the rate of the process[123]. A summary of a considerable body of experimental work on this topic is given in these last references[122,123].

Many calculations have been made of cyclohexane, for the success of such calculations is a measure of the success of a program. Some accounts of these have been given[29,124].

B. Monoalkylcyclohexanes

The conformational questions here are not only of axial or equatorial preference (A value[125,126]), but also of the rotational conformation about the alkyl-to-cyclohexyl bond. The ethyl, isopropyl and tert-butyl groups serve as models for any primary, secondary or tertiary alkyl group, and the methyl and tert-butyl groups have a rotational isotropy not present in the ethyl and isopropyl groups. The A value is a treacherous universal measure of steric size because of the restraints to reducing steric strain by rotation or other distortions that the cyclohexyl ring must obey. Much of the interest of the acyclic alkane conformational analysis comes from the complexity of additional choices in such molecules compared with cyclohexane.

There is nothing remarkable about the conformation of the exocyclic bond in methylcyclohexane[127,128] and, in both axial and equatorial ethylcyclohexane, the conformation with the methyl group *gauche* to both ring-CH_2 groups is less stable than those where it is *gauche* to one ring-CH_2 group and to the methine hydrogen[102]. In the equatorial conformation of isopropylcyclohexane, behaviour quite analogous to that of 2,3-dimethylbutane is observed. The opening-up of the methyl-C-methyl and CH_2—C—CH_2 bond angles at either end of the exocyclic bond means that enantiomeric conformations **39** and **40**, and conformation **41** are of apparently equal energy, as shown by observation[125] of the ^{13}C NMR spectrum at $-165\,^\circ$C, and by calculation[102].

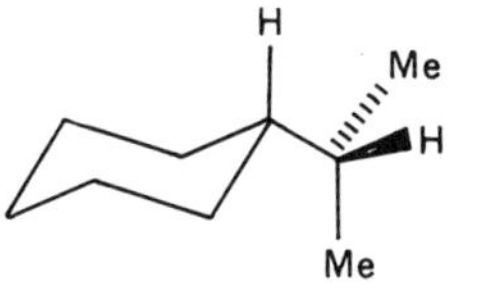

(**39**)

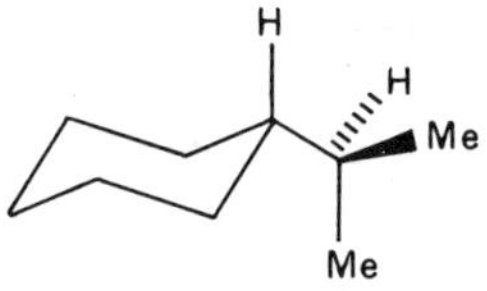

(**40**)

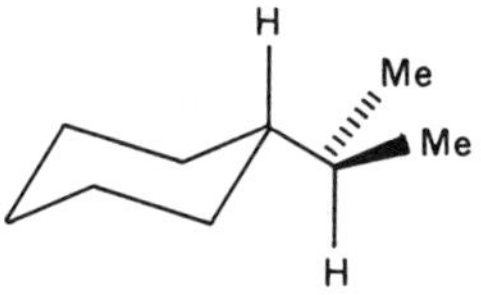

(**41**)

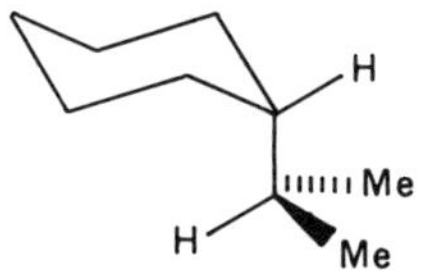

(**42**)

On the question of how cyclohexane accommodates a *tert*-butyl group substituent, calculation suggests[130] that torsion about the exocyclic bond relieves strain and the perfectly staggered arrangement is a transition state between conformations skewed by 6–10° in the equatorial and 15–27° in the axial conformation, where parallel 1,3-interactions are no doubt greater. How the ring distorts in these monoalkylcyclo-hexanes has also been studied[102,130,131].

The *A* value, the free-energy difference between the two chair conformations with axial and equatorial alkyl groups, is widely used as a measure of steric size[125,126]. For all alkyl groups there is less than 7% of the axial conformation present at ambient temperature and direct quantitative observation of the equilibrium is not possible. Historically, indirect examination of conformational equilibria in disubstituted cyclohexanes has been used to predict equilibria in monoalkyl compounds (assuming additivity of substituent effects), and reviews of early work in this field and of the pitfalls encountered have been given[126,132].

More recently, direct observation of the conformational mixture by NMR spectroscopy at low temperatures[131,133,134], sometimes using enrichment of the axial isomer population by rapid cooling of a high-temperature equilibrium mixture, has been possible. Along with this, the reliability of molecular mechanics calculations of saturated hydrocarbons has led to a clear picture of the equilibrium.

One important point to emerge is that, while the enthalpy of methyl, ethyl and isopropyl groups is similar, there are important differences in entropy between some axial and equatorial conformations as well as between substituents[102,131,134]. Some of this is due to easily understood conformational entropy. For isopropylcyclohexane there is only one likely axial conformation **42**, but three equatorial conformations which have been measured experimentally to be of equal enthalpy[131]. There is therefore expected to be three times as much of the equatorial conformation than is predicted on considerations of enthalpy alone. What is less clear is whether other terms such as vibrational entropy differ between conformations. In the case of methylcyclohexane, which has been much studied and in which there is no possibility of conformational entropy, there does appear to be an entropy term favouring the equatorial conformation slightly, so that steric strain not only disfavours the axial conformation enthalpically, but also by restricting higher frequency vibrations and torsions it has an effect on the entropy term. Calculated values for the enthalpy difference[131] are 1.78, 1.81 and 1.71 kcal mol^{-1} for methyl, ethyl and isopropyl axial vs equatorial and 4.8 kcal mol^{-1} for tert-butyl[130]. Other references[102,134] will lead to earlier calculations which are not greatly different. Various experimental measurements agree reasonably well with these[32,133,134].

C. Polyalkylcyclohexanes

It is not difficult to imagine that there is considerable scope for conformational analysis of dialkylcyclohexanes and polyalkylcyclohexanes, many of which have been prepared in a state of high stereochemical purity[135]. In the case of 1,3- and 1,4-disubstituted and 1,3,5-trisubstituted cyclohexanes, conformational effects are expected to be nearly additive but there may be some interest in the extent to which they are not exactly so. Dialkylcyclohexanes with tert-butyl as alkyl groups have received detailed consideration as discussed in Section V.E below. I am unaware of systematic work on 1,2-dialkycyclohexanes, although the interaction between substituents in such cases should be interesting.

1,2,3,4,5,6-Hexaalkyl cyclohexanes have received some attention, the ring inversion of the all-cis hexamethyl compound being discussed below. A crystal structure determination of all-equatorial hexaethylcyclohexane[136] gains interest from the fact that hexaisopropylcyclohexane with all isopropyl groups equatorial is less stable than its conformational isomer with all isopropyl groups axial, reflecting interactions of isopropyl groups[137]. An alkyl group flanked by four methyl groups in cyclohexane might have little reason to prefer an equatorial position to an axial one, since its interactions with methyl groups when equatorial resemble its transannular diaxial interactions when it is axial[89a], though in fact the opposite is observed with a chloro, bromo or hydroxy substituent[89b].

Isotope effects on conformational equilibria have been studied using cyclohexane as a test bed[138-140]. In monodeuterocyclohexane[140a] and monotritiocyclohexane[140b] deuterium and tritium are more stable in the equatorial position than in the axial position by 8.2 and 11.2 cal mol^{-1}, respectively.

D. Chair–Chair Interconversions

One much-studied subject is the ring inversion of cyclohexane and polymethylcyclohexanes, as shown in Table 8. The subject has been reviewed[148] but, very generally, compounds with flattened chair conformations due to methyl–methyl 1,3-diaxial interactions[141,143] have lowered barriers. Simply substituted compounds where rotation about substituted bonds can take place in the boat–twist manifold of conformations have barriers similar[135,142] to that of cyclohexane, while more highly substituted compounds, where rotation about substituted bonds must take place on the way to the transition state[143-147], have markedly higher barriers than in cyclohexane.

Various studies have been reported of cyclohexane rings with one or more hydrocarbon rings attached spiro-fashion[144b,149-151]. In compound **43**[150], there is an equilibrium

TABLE 8.   Barriers (kcal mol^{-1}) to ring inversion of polymethylcyclohexanes

Substituents	Barrier	References
None	10.3	120, 121, 122
1,1-Dimethyl	10.3	135, 142
cis-1,2-Dimethyl	10.3	142
trans-1,3-Dimethyl	9.8	142
cis-1,4-Dimethyl	9.8	142
1,1,3,3-Tetramethyl	8.7	143
1,1,4,4-Tetramethyl	11.4	143, 144
1,1,3,3,5,5-Hexamethyl	<8	141
all-cis 1,2,4,5-Tetramethyl	12.5	145
1r, 2c, 3t, 4t-Tetramethyl	11.5	146
all-cis 1,2,3,4,5,6-Hexamethyl	17.4	145
Dodecamethyl	16.4	147

between 88% of a chair conformation with a barrier to interconversion to twist conformations of 22 kcal mol^{-1}, and 12% of a twist–boat conformation with a pseudorotation barrier of about 10 kcal mol^{-1}. Cyclohexanes with various combinations of five spiro-linked rings have barriers between 13 and 20 kcal mol^{-1}, while with six such spiro-linked rings the barriers are between 22 and 33 kcal mol^{-1} [151].

E. Non-chair Conformations of Alkylcyclohexanes

The developing interest in conformational analysis in the 1950s diverted attention from the cyclohexane ground state **44** to non-chair conformations, and early work is summarized in sections[152,153] of books published in 1965, and in greater detail in a review of 1974[154]. The boat conformation **45**, except in a few constrained bicyclic compounds, is a transition state between conformations **46**, described as twist–boat or twist, so the description 'non-chair' should usually be taken to mean this latter conformation.

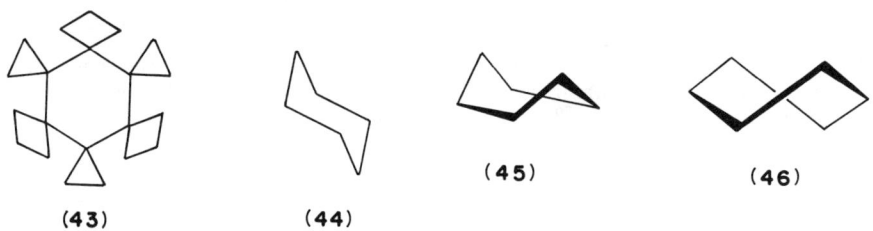

(43)   (44)   (45)   (46)

Non-chair conformations of cyclohexane itself have been investigated in an elegant matrix-isolation experiment by Anet, Chapman and their coworkers[155]. Rapid cryogenic trapping from 1073 K down to 20 K produced a sample whose infrared spectrum showed several absorptions additional to those for cyclohexane chair conformation. Their intensity and rate of disappearance at several low temperatures allowed the determination of the stability of the twist conformation relative to the chair, viz $\Delta H_0 = 5.5$ kcal mol^{-1} and $\Delta S_0 = 4$ eu.

Certain alkylcyclohexanes have a non-chair conformation that is more stable than any chair, and the first experimental determination[156] of the relative enthalpy and entropy of non-chair conformations in 1960 was based on a determination of the *cis–trans* equilibrium for 1,3-di-*tert*-butylcyclohexane, the *trans* isomer being largely, if not completely, in a non-chair conformation. The Delft school[157–160] has made a systematic study of all di-*tert*-butylcyclohexanes, and this and more recent work[161,162] suggest that while *trans*-1,4- and *cis*-1,3-di-*tert*-butylcyclohexane exist preferentially as diequatorial chair conformations, all other isomers exist as conformational mixtures including large amounts of non-chair conformations. All conformations are, of course, distorted from ideal symmetrical structures to accommodate the large substituents.

Thus, *cis*-1,2-di-*tert*-butylcyclohexane exists as two interconverting chair conformations for preference, with one axial and one equatorial substituent[158,159,162]. The *trans*-isomer could exist as a chair with two equatorial substituents, but the close proximity of the substituents disfavours this by several kcal mol^{-1} compared with the diaxial chair conformation and the most favoured twist conformation, which are of similar energy according to calculations. There is some historical irony here, for the improbability of a chair with an axial *tert*-butyl group was the starting assumption for significant early work[152,153,156]. *trans*-1,3-Di-*tert*-butylcyclohexane, the object of Allinger and Freiberg's experiments[156] discussed above, shows NMR spectral features[158,163,164] which confirm the dominance of non-chair conformations. *cis*-1,4-Di-*tert*-butylcyclohexane is concluded from experiment and calculation to exist as a mixture of conformations of which twist ones are more stable, but there is a significant amount of a chair conformation with an axial substituent[158].

When a cyclohexane ring has methyl groups diaxial in a 1,3-relationship, most readily achieved in 1,1,3,3-tetramethylcyclohexane, the chair conformation is destabilized with respect to some twist conformations. The *syn*-diaxial repulsion leads to opening of the bond angle at the 2-position and a complementary closing down of bond angles at the other end of the molecule (the reflex effect)[165]. The preferred conformation is nonetheless this distorted chair[143,144], and even 1,1,3,3,5,5-hexamethylcyclohexane is calculated to have a chair as its lowest energy conformation[166,167]. This last example and the di-*tert*-butylcyclohexanes discussed above make it clear how extreme substitution of a simple cyclohexane ring must be, before non-chair conformations become important.

The other circumstance in which non-chair conformations occur is in saturated fused-ring cycloalkanes[168]. It is sufficient to examine models to recognize that in one stereoisomer each of perhydrophenanthrene **47** and perhydroanthracene **48** it is not possible for all three rings to exist in a chair conformation. This was recognized quite early, and was used as a second basis for estimating the chair/non-chair energy difference[169], since the heat of combustion of the *trans–syn–trans* and *trans–anti–trans* configurations of perhydroanthracene differ by 5.39 kcal mol^{-1}. Perhydrophenalene has since been considered as well[170,171] and has two stereoisomers **49** and **50** which cannot have all rings in chair conformations.

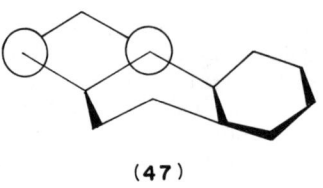

(47)

(48)

(49)

(50)

F. Decalin and Other Fused-ring Hydrocarbons

Bicyclo[$m.n.0$]alkanes are epitomised by decalin and their conformational analysis is particularly clear when a six-membered ring is involved (m or $n = 4$).

Unsubstituted compounds divide into two sets depending on whether the ring junction is of *cis* or *trans* configuration, and ring 2 can be treated as a pair of substituents on ring 1, or *vice versa*. In the *trans* case one conformation with 'substituents' pseudoaxial is likely to be much less stable (if not geometrically impossible) than its ring-inverted form with pseudo-equatorial 'substituents'. In the *cis* configuration, one 'substituent' is pseudoequatorial and the other pseudoaxial or *vice versa* in two equivalent conformations.

The ring inversion of the six-membered ring in *cis*-decalin[172-174] and *cis*-perhydroindane[174-176] has barriers of about 12.4 and 7.7 kcal mol^{-1}, respectively. The barrier is particularly high in the decalin case, since in the middle of the conformational interconversion both rings must be in unstable twist conformations at the same time.

Perhydrophenanthrenes and anthracenes with three fused rings have received considerable attention because of their interesting dynamic and static conformational properties, which have long been known and understood[168]. Non-chair conformations have been discussed in Section V.E and recent work on perhydroanthracenes[29,177,178] and perhydrophenanthrenes[177] extends this understanding. Compounds with four saturated fused rings, such as perhydrochrysenes, have been considered[179].

VI. SMALL, MEDIUM AND LARGE-SIZED RING HYDROCARBONS

A. Four- and Five-membered Rings[180]

Cyclobutane has a non-planar carbon framework with a puckering angle of 28° as determined by electron diffraction[181]. The barrier to ring inversion is about 1.45 kcal mol^{-1} [182]. In methylcyclobutane, puckering is reduced to 21° and 16° in the equatorial and axial conformations with the latter less stable[183] by 0.84 kcal mol^{-1}, and a barrier to ring inversion from the latter conformation[184] of 1.12 kcal mol^{-1}.

Cyclopentane exists in two conformations with free (i.e. zero-barrier) pseudorotation among their various degenerate forms[181]. The bent, or C_s, or envelope conformation with four carbon framework atoms in the same plane has a puckering angle of 25° while the twisted, or C_2, or butterfly conformation with carbon atoms 1 and 3 above and below the plane of atoms 2, 4 and 5 has 26° for the angle of twist[185]. The planar conformation is considerably less stable, 4.8 kcal mol^{-1} by calculation[186]. Cyclopentane is calculated to prefer to locate methyl substituents in pseudoequatorial or isoclinal positions[187,188], and calculations of a wide range of polymethyl and alkylcyclopentanes have been briefly reported[189].

B. Medium-ring Conformations

In saturated cyclic hydrocarbons with seven or more methylene groups, the conformational possibilities are much more diverse and a reviewer's best service to his reader may lie in presenting a judiciously simplified picture. A convincingly full set of possible conformations for an n-membered ring can be determined using molecular mechanics calculations linked to a search procedure for local minima of any kind. Rings up to seventeen-membered have recently been taken[35-39] as a suitable test bed for such search procedures. Evaluation of minima may be simplified by rejection of all those which are 'high' in enthalpy above the global minimum, but many of those remaining may be dimples separated by a very small barrier from a better defined minimum and demonstrably[34] part of that minimum for most purposes.

Often the chemist may prefer to use experimental evidence gained principally from dynamic NMR studies to limit this vast calculated total picture. Usually, even with most efficient slowing of conformational processes on the NMR timescale, only one or two sets of signals are observed, so each cycloalkane can be considered in terms of one or two conformational types, with different positions that a CH_2 group can occupy within each type. Calculations may show slightly less stable local minima nearby, but the simplification subsumes these by treating the conformational type as a dynamic equilibrium fast on the NMR timescale.

Only a few of these populated conformational types are symmetrical and in the absence of all symmetry there will be two sets of n conformations for each type. In each set a labelled CH_2 group will occupy successively each of the n different positions in the conformation, and interconversion of the n-members of the set is a pseudorotation. The second pseudorotating set comprises the enantiomers of each member of the first set and the interconversion of the two sets is a different pseudorotation invariably called a ring inversion. The simplification is complicated enough and pseudorotation and ring inversion are discussed further in the next section.

Something needs to be said about the representation of conformations in medium-sized rings. Any conformation of an n-membered ring alkane is defined by the n internal $CH_2—CH_2—CH_2—CH_2$ dihedral angles φ_1 to φ_n listed in turn. In a recognizably different kind of conformational minimum, several of the set of dihedral angles will have changed sign or their value will have changed significantly. A conformation can be represented by a polygon, even a regular polygon **51**, with a sign and a number indicating the ring $C—C—C—C$ dihedral angle along each bond, and is totally defined thereby. Such a picture almost inevitably requires recourse to models, and will not be used here.

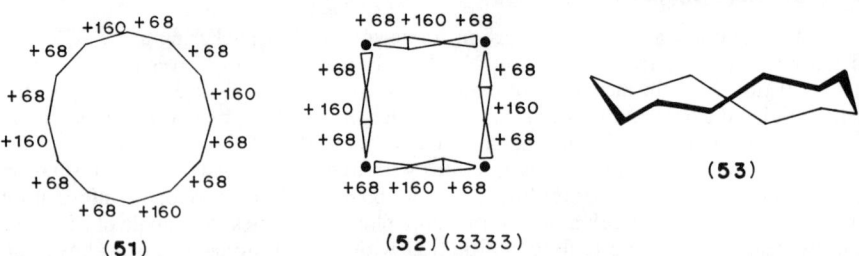

(51)   (52)(3333)   (53)

A view of the conformation from above the mean plane, gives a better impression. Conformations appear to have sides and corners, the latter occurring where two bonds in succession have acute dihedral angles of the same sign and a side occurring when successive bonds have different signs[190].

Numbers may once again be written alongside bonds to show the dihedral angles involved, and wedges may be used to represent the position with respect to the mean plane of atoms in the framework. Such representations (52) are particularly associated with Dale[191] and can be described cryptically in the form [abcde], which means that the plan view of the $(a + b + c + d + e = n)$-membered ring conformation appears as a set of sides successively a, b, c, d, e bonds long. Models help, of course, and, as expected, calculations usually find that there are several similar stable minima with slightly different sets of dihedral angles all of which qualify for the description [abcde]. The analyst will often be grateful to have this set of minima grouped under one such description and will take particular note of the lowest-energy form in the set.

A schematic perspective drawing 53 is particularly useful, with a view from the side and somewhat above the plane, to avoid overlap of atoms. This is usually a version of the true perspective drawing that a computer might produce, modified to clarify significant aspects of the conformation. Models help. The diagrams 51–53 illustrate the preferred conformation of cyclododecane in these conventions[192,193].

C. Pseudorotation and Ring Inversion

Any process which moves a molecule from one potential energy minimum conformation to another by changing one or more dihedral angles is an interconversion. If the two minima are of the same type e.g. boat-chairs, the process is called a pseudorotation. A complete pseudorotation cycle comprises all the degenerate forms in the set. In the second, enantiomeric, ring-inverted set mentioned above, each member differs from a member of the first set only in the sign of the dihedral angles – their magnitudes are more or less[194] the same.

Diagrams 54 and 55 show such a change of sign from a positive to a negative dihedral angle in the familiar ring inversion of cyclohexane, noting the definition of + and − given in Section I.A.

(54)   (55)

In the course of pseudorotation and ring inversion of boat–chair conformations, intermediate conformational minima of a different type, e.g. a chair, may be visited. This chair conformation will have its own pseudorotation cycle and ring inversion process, which is thus linked to the boat-chair cycle. There are likely to be further different, less stable conformational types, some of them little populated no doubt, but linked into the total conformational network by an interconversion process.

The three terms interconversion, pseudorotation and ring inversion thus seem distinct, the first including both the latter, but every interconversion between *different* conformational types is part of some pseudorotation cycle perhaps not one of the obvious ones. For this reason I prefer to call all interconversions except ring inversions by the same name pseudorotation, to direct attention to the multitude of possibilities beyond those that are NMR visible. The reader may prefer to reserve pseudorotation for the interconversion of conformations of the same type. If aware of the alternatives, it is easy to recognise which usage any particular writer has adopted.

In practice the expression 'ring inversion of a cycloalkane' is usually applied to ring inversion of the most stable conformational type of that alkane and the justification for this comes from dynamic NMR observations of such medium-ring compounds.

Two distinct sets of NMR spectral changes are often observed because the n-membered ring being studied exists as a pseudorotating set of n equivalent forms of a single conformation *or* as two sets of different pseudorotating conformations (the sets interconverting by a different pseudorotation) *and* a ring-inverted version of all these. Ring inversion will usually be the process with the high barrier, since *all* dihedral angles have to change sign. When ring inversion is slow on the NMR timescale but other pseudorotations are fast, skeletal carbon atoms of a cycloalkane will give a single signal, since all populated positions in the skeleton can be visited by pseudorotation. There will, however, be two signals for the two hydrogens H_a and H_b attached to a skeletal atoms since, by pseudorotation alone, H_a cannot reach the set of positions occupied by H_b.

Anet and Anet[195] reserve the name pseudorotation for interconversion of the same conformational type and have given a detailed description of ring inversion and pseudorotation in their review. It is likely that the discussion of ring inversion and pseudorotation in any particular case will describe and illustrate the processes being referred to, so the author's meaning will be clear.

What calculations may tell beyond the NMR result is that either or both of these kinds of 'conformation' is in fact a low barrier, NMR-invisible interconversion of two different but similar conformations, e.g. twist–chair–chair with true chair–chair, and also that the 'pseudorotations' which have become slow on the NMR timescale actually take place in a serious of steps during which other kinds of conformations like boat–boat, not significantly populated, nonetheless must be visited to achieve the conformational interconversion. These steps and the NMR-invisible interconversion are further pseudorotation processes.

The multiplicity of these choices brings home the simplicity of cyclohexane systems. Only one kind of conformation, the chair, is found, except in rare and quite well understood circumstances, and it has only one kind of skeletal position, and only two kinds of position for a hydrogen or a substituent.

D. Alkyl-substituted Cycloalkanes

The conformational questions of unsubstituted cycloalkanes are overlaid in alkylcycloalkanes with the conformational requirements and influence of the substituent. In a single boat–chair conformation of cyclooctane, for example, there are ten different kinds of position that a methyl group can occupy. Anet[196] has discussed the meaning of the terms axial and equatorial for distinguishing the two locations for a substituent at each position in a conformation. I suggest that a substituent at a carbon labelled 3 in a ring is quintessentially equatorial if it is perfectly antiperiplanar to the C—1 to C—2 bond and the C—5 to C—4 bond. Its geminal companion is then quintessentially axial. In all other cases the substituent is more or less equatorial or axial depending on whether it approaches the quintessential, and the terms pseudoequatorial and pseudoaxial may turn out to be useful. The substituent in a monoalkylcycloalkane in a given conformation will prefer the position where it is nearest to being quintessentially equatorial. Hendrickson[197] has calculated the least hindered position for a methyl group in various conformations of several cycloalkanes, and the same position may of course be preferred by other substituents.

1,1-Dimethylcycloalkanes and compounds with more than one geminal dimethyl group are an interesting group for study. It is a manifestation of the *gem*-dimethyl effect[94] that such groups occupy a position at a corner in the conformation. This limits the conformational possibilities somewhat, and clearly strategic placing of a second *gem*-dimethyl group in the ring might limit the ring to a single conformation and its ring-inverted form[198]. If, for example, a ring prefers a $[33xy]$ conformation, the 1,1,7,7-tetramethyl derivative may be restricted in this way.

E. Some Medium-sized Rings

This subject has been reviewed[12,31,195,199] well and extensively. Conformational isomerization of cycloheptane is particularly easy, since neither ring inversion nor pseudorotation has been seen to be slow on the NMR timescale[200] and so both have barriers less than about 5kcal mol^{-1}. The chair and twist–chair conformations interconvert by a pseudorotaion, as do the boat and twist–boat, although the latter are calculated to be more than 2kcal mol^{-1} less stable[201]. Cyclooctane has been much studied by experiment and calculation, and illustrates well the complexity of the conformational possibilities which are present in cycloalkanes. Several detailed accounts of the possible conformations and how they interconvert have been given[202–205]. An electron diffraction study[206] of cyclooctane gave results that could be explained in terms of a single kind of conformation in the gas phase, the boat–chair. NMR studies[202,207] taken in conjunction with molecular mechanics calculations[29,208] show that cyclooctane exists in solution as an equilibrium of two sets of conformations, principally a pseudorotating and ring-inverting set of boat–chairs like **56**, which set interconverts with a small percentage of a pseudorotating and ring inverting set of conformations like **57** and **58** that make up the twist chair–chair/crown family. The equilibrium is temperature-dependent, reflecting the differences in symmetry of the conformations, among other things[29]. The highest conformational barrier in cyclooctane ($\Delta H^{\ddagger} = 10.5 \text{kcal mol}^{-1}$)[207] is that passing from the boat–chair set to the crown set, and thus is a pseudorotation. There is a barrier to ring inversion of the boat–chair conformation[208] of 7.7kcal mol^{-1}, and the ring-inversion barrier in the crown set of conformations is similarly large[207].

| (56) | (57) | (58) |

Pseudorotation within both sets of conformations is rather easy with barriers estimated to be less than 5kcal mol^{-1}, and so these are not measured experimentally. In simple substituted cyclooctanes the barrier to pseudorotation is sufficiently large to be measurable. Examination of models and calculations permits some convincing speculation on the likely pathway for all these interconversions[202].

The preferred conformation of cyclononane and the work leading to its elucidation have been fully described[209]. A [333] twist boat chair is found both by calculation and experiment. The [144] twist chair boat and [225] twist chair–chair are calculated to be 1.15 and $0.84 \text{kcal mol}^{-1}$ less stable, respectively[29], but both are favoured by entropy. Experimentally, the 95% population of the [333] conformation at $-173\,^{\circ}\text{C}$ becomes about 40% at room temperature[209], with about 50% of [225] and 10% of [144], while calculations[29,191] and electron diffraction experiments[210] agree with this. The molecule has recently served as the object of various methods for locating conformational minima[34,36,37]. The ring-inversion barrier has been determined[199,209] and it increases dramatically with the introduction of gem-dimethyl substitution. The values are 6 and 9kcal mol^{-1} for cyclononane and 1,1-dimethylcyclononane[191], respectively, and about 20kcal mol^{-1} for 1,1,4,4-tetramethylcyclononane[199].

Beyond nine-membered rings it is decreasingly helpful to use words like boat and chair to describe conformations. The investigation of the conformation of cyclododecane by various techniques has been summarized recently[193]. The highly symmetrical [3333] conformational description applies to two quite different conformational minima of widely different energy, indicating a shortcoming of the numerical nomenclature.

Cyclohexadecane is an interesting and instructive example. From both X-ray diffraction[213] and vibrational spectroscopy[211-213] the low-temperature crystal form has the square [4444] conformation[211], which is also found for derivatives[199]. This is reckoned most stable for the gas phase by molecular mechanics calculations[199,214,215], and best explains[215] the dynamic NMR results for a solution at $-152\,°C$. What seems to be a consistent picture is perturbed by the fact that there is a high-temperature crystal form which, like the liquid at room temperature, has a vibrational spectrum that indicates the presence of little of the [4444] conformation[211]. The experimental heat of formation and that calculated for the [4444] conformation are also discrepant[214]. As yet there is no explanation for these various facts[211].

Aside from these last cycloalkanes that I have singled out for explicit discussion, there have been analyses of many of the other rings of similar size, from which no particularly useful generalizations seem to me to emerge. However, some of the original papers that seem most likely to help the reader to form his own conclusions are now cited: cyclodecane[39,216-218], cycloundecane[39,199,219], cyclotridecane[39,219,220], cyclotetradecane[39,211,218], cyclopentadecane[219], cycloheptadecane[35], cyclooctadecane[221].

A surprisingly large number of cycloalkanes with greater than twenty-membered rings have been prepared, mainly by polymer chemists as models for polyethylene[222]. These have been studied by X-ray diffraction[223,224], solid state ^{13}C NMR spectroscopy[222-224], differential scanning calorimetry and other techniques[222,225,226]. In general the conformational preference seems to be for two long parallel chains linked at either end, viz [4n4m].

VII. REFERENCES

1. K. B. Wiberg, R. F. W. Bader and C. D. H. Lau, J. Am. Chem. Soc., 109, 1001 (1987).
2. E. L. Eliel, N. L. Allinger, S. J. Angyal and G. A. Morrison, Conformational Analysis, Interscience, New York, 1965.
3. E. L. Eliel, Stereochemistry of Carbon Compounds, McGraw-Hill, New York, 1962.
4. Reference 2, Chap. 3.
5. U. Berg and J. Sandstrom, Adv. Phys. Org. Chem., 25, 1 (1989).
6. G. Mann, M. Muhlstadt, J. Brabant and E. Doering, Tetrahedron, 23, 3393 (1967).
7. J. Lamb, Z. Elektrochem., 64, 135 (1960).
8. H. B. Burgi and L. S. Bartell, J. Am. Chem. Soc., 94, 5236 (1972).
9. L. Lunazzi, D. Macciantelli, B. Bernardi and K. U. Ingold, J. Am. Chem. Soc., 99, 4573 (1977).
10. W. Ritter, W. Hull and H. J. Cantow, Tetrahedron Lett., 3093 (1978).
11. F. A. L. Anet and V. J. Basus, J. Am. Chem. Soc., 95, 4424 (1973).
12. F. A. L. Anet, Top. Curr. Chem., 45, 169 (1974).
13. J. P. Lowe, Prog. Phys. Org. Chem., 6, 1 (1968).
14. J. Sandstrom, Dynamic NMR Spectroscopy, Academic Press, London, 1982.
15. L. Rosenthal, J. F. Rabolt and J. Hummel, J. Chem. Phys., 76, 817 (1982) and earlier work cited therein.
16. K. B. Wiberg and M. A. Murcko, J. Am. Chem. Soc., 110, 8029 (1988) and earlier work cited therein.
17. J. Devaure and J. Lascombe, Nouv. J. Chim., 3, 579 (1979).
18. W. L. Jorgensen, Acc. Chem. Res., 22, 184 (1989) and earlier work cited therein.
19. R. O. Rosenberg, R. Mikkilineni and B. J. Berne, J. Am. Chem. Soc., 104, 7647 (1982).
20. D. J. Tobias and C. L. Brooks III, J. Chem. Phys., 92, 2582 (1990) and work cited therein.
21. R. A. Ford and N. L. Allinger, J. Org. Chem., 35, 3178 (1970).
22. J. R. Durig and D. A. C. Compton, J. Phys. Chem., 83, 265 (1979).
23. J. G. Aston, in Determination of Organic Structures by Physical Methods (Eds. E. A. Braude and F. C. Nachod). Vol. 1, Academic Press, New York, 1955.
24. C. Jaime and E. Osawa, Tetrahedron, 39, 2769 (1983).
25. J. E. Anderson, H. Pearson and D. I. Rawson, J. Am. Chem. Soc., 107, 1447 (1985).

26. J.-H. Lii and N. L. Allinger, *J. Am. Chem. Soc.*, **111**, 8566 (1989).
27. N. L. Allinger, *J. Am. Chem. Soc.*, **99**, 8127 (1977).
28. (a) J. L. M. Dillen, *J. Comput. Chem.*, **11**, 1125 (1990).
    (b) K. B. Lipkowitz and N. L. Allinger, *QCPE Bull.*, **7**, 19 (1987).
29. N. L. Allinger, Y. H. Yuh and J.-H. Lii, *J. Am. Chem. Soc.*, **111**, 8551 (1989).
30. E. M. Engler, J. D. Andose and P. v. R. Schleyer, *J. Am. Chem. Soc.*, **95**, 8005 (1973).
31. U. Burkert and N. L. Allinger, *Molecular Mechanics* A.C.S. Monograph 177, Washington 1982. Chapter 4 gives a particularly fine account of calculations of hydrocarbons.
32. E. Osawa and H. Musso, *Top. Stereochem.*, **13**, 117 (1982).
33. K. Rasmussen, *Potential Energy Functions in Conformational Analysis*, Springer-Verlag, Berlin, 1985.
34. F. A. L. Anet, *J. Am. Chem. Soc.*, **112**, 7172 (1990).
35. M. Saunders, K. N. Houk, Y.-D. Wu, W. C. Still, M. Lipton, G. Chang and W. C. Guida, *J. Am. Chem. Soc.*, **112**, 1419 (1990) and earlier work cited therein.
36. D. M. Ferguson, W. A. Glauser and D. J. Raber, *J. Comput. Chem.*, **10**, 903 (1989).
37. M. Lipton and W. C. Still, *J. Comput. Chem.*, **9**, 343 (1988).
38. M. Saunders, *J. Comput. Chem.*, **10**, 203 (1989).
39. H. Goto and E. Osawa, *J. Am. Chem. Soc.*, **111**, 8950 (1989).
40. Reference 31, pp. 10–15.
41. Reference 5, p. 7.
42. N. L. Allinger, R. S. Grev, B. F. Yates and H. F. Schaefer III, *J. Am. Chem. Soc.*, **112**, 114 (1990).
43. D. A. Dougherty and K. Mislow, *J. Am. Chem. Soc.*, **101**, 1401 (1979).
44. C. Ruchardt and H.-D. Beckhaus, *Angew. Chem., Int. Ed. Engl.*, **19**, 429 (1980) and references cited therein.
45. C. Ruchardt and H.-D. Beckhaus, *Angew. Chem., Int. Ed. Engl.*, **24**, 529 (1985).
46. R. Winiker, H.-D. Beckhaus and C. Ruchardt, *Chem. Ber.*, **113**, 3456 (1980).
47. G. Hellmann, S. Hellmann, H.-D. Beckhaus and C. Ruchardt, *Chem. Ber.*, **115**, 3364 (1982).
48. S. Hellmann, H.-D. Beckhaus and C. Ruchardt, *Chem. Ber.*, **116**, 2219 (1983).
49. M. A. Flamm-ter Meer, H.-D. Beckhaus, K. Peters, H. G. von Schnering, H. Fritz and C. Ruchardt, *Chem. Ber.*, **119**, 1492 (1986).
50. R. F. W. Bader, J. R. Cheeseman, K. E. Laidig, K. B. Wiberg and C. Breneman, *J. Am. Chem. Soc.*, **112**, 6530 (1990) and other work cited therein.
51. O. Ermer, P. Bell, J. Schafer and G. Szeímies, *Angew. Chem., Int. Ed. Engl.*, **28**, 473 (1989).
52. J. E. Anderson and B. R. Bettels, *Tetrahedron*, **46**, 5353 (1990).
53. R. K. Heenan and L. S. Bartell, *J. Chem. Phys.*, **78**, 1270 (1983).
54. K. Kuchitsu, *Bull. Chem. Soc. Jpn.*, **32**, 748 (1959).
55. J. E. Anderson, unpublished results.
56. W. D. Hounshell, L. D. Iroff, R. J. Wroczynski and K. Mislow, *J. Am. Chem. Soc.*, **100**, 5212 (1978).
57. H.-D. Beckhaus, C. Ruchardt and J. E. Anderson, *Tetrahedron*, **38**, 2299 (1982).
58. L. S. Bartell and T. L. Boates, *J. Mol. Struct.*, **32**, 379 (1976).
59. E. J. Jacob, H. B. Thompson and L. S. Bartell, *J. Chem. Phys.*, **47**, 3736 (1967).
60. L. D. Iroff and K. Mislow, *J. Am. Chem. Soc.*, **100**, 2121 (1978).
61. C. H.. Bushweller, W. G. Anderson, M. J. Goldberg, M. W. Gabriel, L. R. Gilliom and K. Mislow, *J. Org. Chem.*, **45**, 3880 (1980).
62. H. Braun and W. Luttke, *J. Mol. Struct.*, **31**, 97 (1976).
63. E. Osawa, H. Shirahama and T. Matsumoto, *J. Am. Chem. Soc.*, **101**, 4824 (1979).
64. J. E. Anderson, unpublished results for 2,2-dimethyl-3,3-*bis*-trideuteromethylbutane.
65. R. J. Wroczynski and K. Mislow, *J. Am. Chem. Soc.*, **101**, 3980 (1979).
66. A. T. Hagler, P. S. Stern, S. Lifson and S. Ariel, *J. Am. Chem. Soc.*, **101**, 813 (1979).
67. J. E. Anderson, P. A. Kirsch and J. S. Lomas, *J. Chem. Soc., Chem. Commun.*, 1065 (1988).
68. J. E. Anderson and P. A. Kirsch, unpublished results.
69. J. E. Anderson, *J. Chem. Soc., Perkin Trans.* 2, 299 (1991).
70. A search for an incoherently skewed conformation by calculation may produce a point of inflection rather than a minimum in the potential energy profile.
71. S. Hellmann, Thesis, University of Freiburg, 1982.
72. American Petroleum Institute, Project 44.
73. W. D. Hounshell, D. A. Dougherty and K. Mislow, *J. Am. Chem. Soc.*, **100**, 3149 (1978).

74. The results are from Reference 57, although there is no discussion of attractive steric interactions in this paper.
75. C. X. Cui and M. Kertesz, J. Am. Chem. Soc., 111, 4216 (1989) and references therein.
76. D. R. Lide, J. Chem. Phys., 33, 1514 (1960).
77. Lide states[76] that the distances from a methylene hydrogen to its two gauche hydrogen neighbours is the same within 0.3 pm.
78. H. Dreizler, private communication.
79. H.-B. Burgi, W. D. Hounshell, R. D. Nachbar and K. Mislow, J. Am. Chem. Soc., 105, 1427 (1983).
80. G. Bestmann, W. Lalowski and H. Dreizler, Z. Naturforsch., A, 40A, 221 (1985).
81. L. S. Bartell and W. F. Bradford, J. Mol. Struct., 37, 113 (1977).
82. M. R. Whalon, C. H. Bushweller and W. G. Anderson, J. Org. Chem., 49, 1184 (1984).
83. C. H. Bushweller and W. G. Anderson, Tetrahedron Lett., 1811 (1972).
84. D. R. Lide. J. Chem. Phys., 33, 1519 (1960).
85. D. R. Lide Jr. and D. E. Mann, J. Chem. Phys., 29, 914 (1958).
86. K. S. Pitzer, J. Chem. Phys., 5, 473 (1937).
87. J. G. Aston, R. M. Kennedy and S. C. Schumann, J. Am. Chem. Soc., 62, 2059 (1940).
88. J. R. Durgi, M. S. Craven and J. Bragin, J. Chem. Phys., 53, 38 (1970) and references cited therein.
89. (a) J. E. Anderson, B. R. Bettels, H. M. R. Hoffmann, D. Pauluth, S. Hellmann, H. D. Beckhaus and C. Ruchardt, Tetrahedron, 44, 3701 (1988).
    (b) H.-J. Schneider and W. Freitag, Chem. Ber., 112, 16 (1979)
90. J. E. Anderson, K. H. Koon and J. E. Parkin, Tetrahedron, 41, 561 (1985).
91. S. Fitzwater and L. S. Bartell, J. Am. Chem. Soc., 98, 5107 (1976).
92. K. S. Pitzer and J. E. Kilpatrick, Chem. Rev., 39, 435 (1946).
93. R. W. Alder, C. M. Maunder and A. G. Orpen, Tetrahedron Lett., 31, 6717 (1990).
94. N. L. Allinger and V. Zalkow, J. Org. Chem., 25, 701 (1960).
95. D. A. C. Compton, S. Montero and W. F. Murphy, J. Phys. Chem., 84, 3587 (1980).
96. G. M. Whitesides, J. P. Sevenair and R. W. Goetz, J. Am. Chem. Soc., 89, 1135 (1967).
97. D. F. DeTar and C. J. Tempas, J. Am. Chem. Soc., 98, 4567 (1976).
98. D. F. DeTar and C. J. Tempas, J. Org. Chem., 41, 2009 (1976).
99. G. Mann, Tetrahedron, 23, 2475 (1967).
100. G. Mann, Tetrahedron, 23, 3375 (1967).
101. L. A. Verma, W. F. Murphy and H. J. Bernstein, J. Chem. Phys., 60, 540 (1974).
102. E. Osawa, J. B. Collins and P. v. R. Schleyer, Tetrahedron, 33, 2667 (1977).
103. S. Brownstein, J. Dunogues, D. Lindsay and K. U. Ingold, J. Am. Chem. Soc., 99, 2073 (1977).
104. G. D. Mendelhall, D. Griller, D. Lindsay, T. T. Tidwell and K. U. Ingold, J. Am. Chem. Soc., 96, 2441 (1974).
105. C. Ruchardt, H.-D. Beckhaus, G. Hellmann, S. Weiner and R. Winiker, Angew. Chem., Int. Ed. Engl., 16, 875 (1977).
106. H.-D. Beckhaus, G. Hellmann and C. Ruchardt, Chem. Ber., 111, 72 (1978).
107. J. E. Anderson and H. Pearson, Tetrahedron Lett., 2779 (1972).
108. B. Starck, Landolt–Bornstein New Series Numerical Data and Functional Relationships in Science and Technology, Group 2, Volume 4, Springer-Verlag, Berlin, 1967, p. 202.
109. S. Sternhell, in Dynamic Nuclear Magnetic Resonance Spectroscopy (Eds. L. M. Jackman and F. A. Cotton) Chap. 6, Academic Press, New York, 1975.
110. Tables 4 and 5 in Reference 89.
111. E. B. Wilson, Chem. Soc. Rev., 1, 293 (1972).
112. B. L. Hawkins, W. Bremser, S. Borcic and J. D. Roberts, J. Am. Chem. Soc., 93, 4472 (1971).
113. J. E. Anderson and H. Pearson, J. Chem. Soc. Chem. Commun., 908 (1972).
114. H. O. Kalinowski, E. Rocker and G. Maier, Org. Magn. Reson., 21, 64 (1983).
115. H. Duddeck, M. A. McKervey and D. Rosenblum, Tetrahedron Lett., 31, 4061 (1990).
116. J. E. Anderson and H. Pearson, J. Am. Chem. Soc., 97, 764 (1975).
117. Barrier value calculated[67] from results reported previously.[65]
118. F. A. L. Anet, M. St. Jacques and G. N. Chmurny, J. Am. Chem. Soc., 90, 5243 (1968).
119. M. Squillacote, R. S. Sheridan, O. L. Chapman and F. A. L. Anet, J. Am. Chem. Soc., 97, 3244 (1975).

3. Conformational analysis of acyclic and alicyclic saturated hydrocarbons   131

120. F. A. L. Anet and A. J. R. Bourn, *J. Am. Chem. Soc.*, **89**, 760 (1967).
121. D. Hofner, S. A. Lesko and G. Binsch, *Org. Magn. Reson.*, **11**, 179 (1978).
122. B. D. Ross and N. S. True, *J. Am. Chem. Soc.*, **105**, 4871 (1983).
123. T. L. Hasha, T. Eguchi and J. Jonas, *J. Am. Chem. Soc.*, **104**, 2290 (1982).
124. Reference 31, pp. 91–93.
125. S. Winstein and N. J. Holness, *J. Am. Chem. Soc.*, **77**, 5562 (1955).
126. J. A. Hirsch, *Top. Stereochem.*, **1**, 199 (1967).
127. H. J. Geise, H. R. Buys and F. C. Mijlhoff, *J. Mol. Struct.*, **9**, 447 (1971).
128. A. Tsuboyama, A. Murayama, S. Konaka and M. Kimura, *J. Mol. Struct.*, **118**, 351 (1984).
129. M. E. Squillacote, *J. Chem. Soc., Chem. Commun.*, 1406 (1986).
130. B. van de Graaf, J. M. A. Baas and B. M. Wepster, *Recl. Trav. Chim. Pays-Bas*, **97**, 268 (1978).
131. M. E. Squillacote and J. M. Neth, *Magn. Reson. Chem.*, **25**, 53 (1987) and references cited therein.
132. F. R. Jensen and C. H. Bushweller, *Adv. Alicyclic Chem.*, **3**, 139 (1971).
133. F. A. L. Anet and M. E. Squillacote, *J. Am. Chem. Soc.*, **97**, 3243 (1975).
134. H. Booth and J. R. Everett, *J. Chem. Soc., Perkin Trans. 2*, 255 (1980) and references cited therein.
135. G. Mann, M. Muhlstadt and J. Braband, *Tetrahedron*, **24**, 3607 (1968).
136. A. Imirzi and E. Torti, *Atti. Accad. Naz. Lincei Cl. Sci. Fis., Mat. Nat., Rend.*, 4498 (1968).
137. O. Golan, Z. Goren and S. E. Biali, *J. Am. Chem. Soc.*, **112**, 9300 (1990).
138. F. A. L. Anet, V. J. Basus, A. P. W. Havett and M. Saunders, *J. Am. Chem. Soc.*, **102**, 3945 (1980).
139. S. R. Ellison, M. S. Fellows, M. J. T. Robinson and M. J. Widgery, *J. Chem. Soc., Chem. Commun.*, 1069 (1984) and references cited therein.
140. (a) F. A. L. Anet and D. J. O'Leary. *Tetrahedron Lett.*, **30**, 1059 (1989) and references cited therein.
    (b) F. A. L. Anet, D. J. O'Leary and P. G. Williams, *J. Chem. Soc., Chem. Commun.*, 1427 (1990).
141. H. Friebolin, H. Schmidt, S. Kabuss and W. Faisst, *Org. Magn. Reson.*, **1**, 147 (1969).
142. D. K. Dalling, D. M. Grant and L. F. Johnson, *J. Am. Chem. Soc.*, **93**, 3678 (1971).
143. M. St. Jacques, M. Bernard and C. Vaziri, *Can. J. Chem.*, **48**, 2386 (1970).
144. (a) R. W. Murray and M. L. Kaplan, *Tetrahedron*, **23**, 1575 (1967).
    (b) J. B. Lambert, J. L. Gosnell Jr. and D. S. Bailey, *J. Org. Chem.*, **37**, 2814 (1972).
145. G. Mann *Tetrahedron Lett.*, 1917 (1975).
146. H. Jancke, G. Engelhardt, R. Radeglia, H. Werner and G. Mann, *Z. Chem.*, **15**, 310 (1975).
147. L. Fitjer, H. J. Scheuermann and D. Wehle, *Tetrahedron Lett.*, **25**, 2329 (1984).
148. J. E. Anderson, *Top. Curr. Chem.*, **45**, 139 (1974).
149. H. A. P. deJongh and H. Wynberg, *Tetrahedron*, **20**, 2553 (1964).
150. L. Fitjer, U. Klages, W. Kuhn, D. S. Stephenson, G. Binsch, M. Noltemeyer, E. Egert and G. M. Sheldrick, *Tetrahedron*, **40**, 4337 (1984).
151. (a) L. Fitjer, M. Giersig, D. Wehle, M. Dittmer, G. W. Koltermann, W. Schormann and E. Egert, *Tetrahedron*, **44**, 393 (1988).
    (b) L. Fitjer, U. Klages, D. Wehle, M. Giersig, N. Schormann, W. Clegg, D. S. Stephenson and G. Binsch, *Tetrahedron*, **44**, 405 (1988).
152. M. Hanack, *Conformational Theory*, Chap. 6, Academic Press, New York, 1965.
153. Reference 2, pp. 469–486.
154. G. M. Kellie and F. G. Riddell, *Top. Stereochem.*, **8**, 225 (1974).
155. M. Squillacote, R. S. Sheridan, O. L. Chapman and F. A. L. Anet, *J. Am. Chem. Soc.*, **97**, 3244 (1975).
156. N. L. Allinger and L. A. Freiberg, *J. Am. Chem. Soc.*, **83**, 5028 (1961).
157. J. D. Reminsje, H. van Bekkum and B. M. Wepster, *Recl. Trav. Chim. Pays-Bas*, **93**, 93 (1974).
158. B. van de Graaf, H. van Bekkum, H. van Koningsfeld, A. Sinnema, A. van Veen, B. M. Wepster and A. M. van Wijk, *Recl. Trav. Chim. Pays-Bas*, **93**, 135 (1974).
159. B. van de Graaf, J. M. A. Baas and B. P. Wepster, *Recl. Trav. Chim. Pays-Bas*, **97**, 268 (1978).
160. J. M. A. Baas, B. van de Graaf, A. van Veen and B. M. Wepster, *Recl. Trav. Chim. Pays-Bas*, **99**, 228 (1980) and earlier work cited in References 157–160.
161. M. Askari, D. L. Merrifield and L. Schaefer, *Tetrahedron Lett.*, 3497 (1976) and earlier work cited therein.
162. R. J. Unwalla, S. Profeta Jr. and F. A. vanCatledge. *J. Org. Chem.*, **53**, 5658 (1988).
163. D. J. Loomes and M. J. T. Robinson, *Tetrahedron*, **33**, 1149 (1977).

164. H. Jancke and H. Werner, *J. Prakt. Chem.*, **322**, 247 (1980).
165. J.-F. Biellmann, R. Hanna, G. Ourisson, C. Sandris and B. Waegell, *Bull. Soc. Chim. France*, 1429 (1960) and earlier work cited therein.
166. H. G. Schmidt, A. Jaeschke, H. Friebolin, S. Kabuss and R. Mecke, *Org. Magn. Reson.*, **1**, 163 (1969).
167. H. Friebolin, H. G. Schmid, S. Kabuss and W. Faisst, *Org. Magn. Reson.*, **1**, 147 (1969).
168. Reference 2, pp. 232–236.
169. J. L. Margrave, M. V. Frisch, R. G. Bautista, R. L. Clarke and W. S. Johnson, *J. Am. Chem. Soc.*, **85**, 546 (1963).
170. J. L. M. Dillen, *J. Org. Chem.*, **49**, 3800 (1984).
171. P. Gund and T. M. Gund, *J. Am. Chem. Soc.*, **103**, 4458 (1981).
172. D. Tavernier and M. J. O. Anteunis, *Org. Magn. Reson.*, **18**, 109 (1982).
173. J. M. A. Baas, B. van de Graaf, D. Tavernier and P. Vanhee, *J. Am. Chem. Soc.*, **103**, 5014 (1981).
174. Reference 148 gives a description of early work on decalin.
175. H. J. Schneider and N. Nguyen-Ba, *Org. Magn. Reson.*, **18**, 38 (1982).
176. R. E. Lack and J. D. Roberts, *J. Am. Chem. Soc.*, **90**, 6997 (1968).
177. P. Vanhee, B. van de Graaf, D. Tavernier and J. M. A. Buys, *J. Org. Chem.*, **48**, 648 (1983) and work cited therein.
178. D. K. Dalling and D. M. Grant, *J. Am. Chem. Soc.*, **96**, 1827 (1974).
179. M. Farina, G. diSilvestro, E. Mantica, D. Botta, F. Morandi and S. Bruckner, *Isr. J. Chem.*, **20**, 182 (1980).
180. For a review of structural studies of four- and five-membered rings with an exhaustive listing of investigations up to 1980 see A. C. Legon, *Chem. Rev.*, **80**, 231 (1980).
181. T. Egawa, T. Fukuyama, S. Yamamoto, F. Takabayashi, H. Kambara, T. Ueda and K. Kuchitsu, *J. Chem. Phys.*, **86**, 6018 (1987).
182. T. B. Mallory and W. J. McLafferty, *J. Mol. Spectrosc.*, **54**, 20 (1975).
183. J. R. Durig, T. J. Geyer, T. S. Little and V. F. Kalasinski, *J. Chem. Phys.*, **86**, 545 (1987).
184. J. R. Durig, L. A. Carreira and J. N. Willis Jr., *J. Chem. Phys.*, **57**, 2755 (1972).
185. R. L. Rosas, C. Cooper and J. Laane, *J. Phys. Chem.*, **94**, 1830 (1990).
186. K. S. Pitzer and W. E. Donath, *J. Am. Chem. Soc.*, **81**, 3213 (1959).
187. (a) N. L. Allinger, J. A. Hirsch, M. A. Miller, I. J. Tyminski and F. A. van Catledge, *J. Am. Chem. Soc.*, **90**, 1199 (1968).
     (b) N. L. Allinger, H. J. Geise, W. Pyckhout, L. A. Paquette and J. C. Gallucci, *J. Am. Chem. Soc.*, **111**, 1106 (1989).
188. H. J. Schneider, G. Schmidt and F. Thomas, *J. Am. Chem. Soc.*, **105**, 3556 (1983).
189. D. S. Egolf and P. C. Jurs, *Anal. Chem.*, **59**, 1586 (1987).
190. These are not rigid definitions. Two successive *anti* bonds, for example, can both have positive or negative dihedral angles and still be part of a side.
191. J. Dale, *Acta Chem. Scand.*, **27**, 1115 (1973).
192. J. Dale, *Acta Chem. Scand.*, **27**, 1130 (1973).
193. E. G. Atavin, V. S. Mastryukov, N. L. Allinger, A. Almenningen and R. Seip. *J. Mol. Struct.*, **212**, 87 (1989).
194. 'More or less' is used, since clearly dihedral angles will not be exactly the reverse of each other, e.g. in axial and equatorial methylcyclohexane.
195. F. A. L. Anet and R. Anet, in *Dynamic Nuclear Magnetic Resonance Spectroscopy* (Eds. L. M. Jackman and F. A. Cotton), Chap. 14, Academic Press, New York, 1975. See also Reference 11.
196. F. A. L. Anet, *Tetrahedron Lett.*, **31**, 2125 (1990).
197. J. B. Hendrickson, *J. Am. Chem. Soc.*, **89**, 7043 (1967).
198. G. Borgen and J. Dale (a) *J. Chem. Soc., Chem. Commun.*, 447 (1969). (b) *J. Chem. Soc., Chem. Commun.*, 1105 (1970).
199. J. Dale, *Top. Stereochem.*, **9**, 199 (1976) and results quoted therein.
200. J. D. Roberts, *Chem. Brit.*, **2**, 529 (1966).
201. J. B. Hendrickson, *J. Am. Chem. Soc.*, **89**, 7047 (1967).
202. J. Dale, *Angew. Chem., Int. Ed. Engl.*, **5**, 1000 (1966).
203. J. B. Hendrickson, *J. Am. Chem. Soc.*, **86**, 4854 (1964).
204. F. A. L. Anet, in *Conformational Analysis of Medium-sized Heterocycles* (Ed. R. S. Glass), Chap. 3, VCH, Weinheim, 1988.

205. J. E. Anderson, E. D. Glazer, D. L. Griffith, R. Knorr and J. D. Roberts, *J. Am. Chem. Soc.*, **91**, 1386 (1969).
206. O. V. Dorofeeva, V. S. Mastryukov, N. L. Allinger and A. Almenningen, *J. Phys. Chem.*, **89**, 252 (1975).
207. See Reference 11 and other work cited therein.
208. F. A. L. Anet and J. S. Hartman, *J. Am. Chem. Soc.*, **85**, 1204 (1963).
209. F. A. L. Anet and J. Krane, *Isr. J. Chem.*, **20**, 72 (1980).
210. O. V. Dorofeeva, V. S. Mastryukov, N. L. Allinger and A. Allmenningen, unpublished results quoted as reference 66 in Reference 29.
211. V. L. Shannon, H. L. Strauss, R. G. Snyder, C. A. Elliger and W. L. Mattice, *J. Am. Chem. Soc.*, **111**, 1947 (1989) and results quoted therein.
212. S. L. Bjornstad, G. Borgen and J. Dale, *Acta Chem. Scand.*, **B29**, 320 (1975).
213. E. Billeter and H. Gunthard, *Helv. Chim. Acta*, **41**, 338 (1958).
214. N. L. Allinger, B. Gorden and S. Profeta Jr., *Tetrahedron*, **36**, 859 (1980).
215. F. A. L. Anet and A. K. Cheng, *J. Am. Chem. Soc.*, **97**, 2420 (1975).
216. O. Ermer, *Tetrahedron*, **31**, 1849 (1975).
217. R. L. Hilderbrandt, J. D. Wieser and L. L. Montgomery, *J. Am. Chem. Soc.*, **95**, 8598 (1973).
218. G. Chang, W. C. Guida and W. C. Still, *J. Am. Chem. Soc.*, **111**, 4379 (1989).
219. F. A. L. Anet and T. N. Rawdah, *J. Am. Chem. Soc.*, **100**, 7810 (1978).
220. B. H. Rubin, M. Williamson, M. Takeshita, F. M. Meyer, F. A. L. Anet, B. Bacon and N. L. Allinger, *J. Am. Chem. Soc.*, **106**, 2088 (1984).
221. S. L. Bjornstad, G. Borgen and G. Gaupset, *Acta Chem. Scand.*, **B28**, 821 (1974).
222. M. Moeller, *Adv. Polym. Sci.*, **66**, 59 (1985) and work cited therein.
223. P. Groth, *Acta Chem. Scand.*, **A33**, 199 (1979).
224. T. Yamanobe, T. Sorita, T. Komoto, E. Ando and H. Sato, *J. Mol. Struct.*, **131**, 267 (1985) and earlier work cited therein.
225. H. P. Grossmann, *Polym. Bull. (Berlin)*, **5**, 131 (1981).
226. H. P. Grossmann and H. Boelstler, *Polym. Bull. (Berlin)*, **5**, 175 (1981).

CHAPTER **4**

Chiroptical properties of alkanes and cycloalkanes[†]

JAMES H. BREWSTER

Department of Chemistry, Purdue University, West Lafayette, Indiana 47907-3699

[†]In this chapter the stereochemical configuration is given as hatched (not dashed) lines and as thick lines (not wedges).

The Chemistry of Alkanes and Cycloalkanes
Edited by S. Patai and Z. Rappoport © 1992 John Wiley & Sons Ltd

I. INTRODUCTION

Many chiral saturated hydrocarbons are known in nonracemic form. Some are natural products or derivatives thereof; others are products of synthesis. Absolute configurations have been established by chemical correlations with substances of known configuration by changes not affecting the stereogenic atoms[1-5]. In many cases the enantiomeric composition can be estimated by comparisons with samples that have been fully resolved or by extended correlations with substances that have been analyzed by modern methods for the assessment of enantiomeric purity. This permits calculation of *maximum rotations* and these are, where possible, reported here. *Observed rotations*, α, have been corrected for tube length ($l = 1$ dcm) when necessary and are reported as such when corrections for density or concentration (ρ in g ml^{-1}) have not been made by authors. *Specific rotations*:

$$[\alpha] = \frac{\alpha}{l \times \rho} \tag{1}$$

have been converted to *molecular rotations*:

$$[M] \text{ or } [\Phi] = \frac{[\alpha] MW}{100} \tag{2}$$

to facilitate comparisons with other substances (MW is molecular weight). The units of [M] are: deg m^{-1}(mol l^{-1})$^{-1}$. Most of the lower molecular weight compounds considered here are liquids and were prepared in amounts permitting polarimetric measurements without solvent (homogeneous, neat), but in some cases solvents have been used. Temperatures of the measurements were in the range 20–27 °C.

II. CHIROPTICAL PROPERTIES

A. Effects of Temperature, Solvent and Concentration

In general, the observed rotations of organic compounds decrease with increasing temperature as a consequence of reduced density. Even when this is accounted for in the calculation of specific rotation, many substances show a decrease in rotation as the relative abundances of chiral conformers, some of which may have opposed rotations, become more nearly equal[6]. However, the specific rotations of (+)-3-methylhexane[7] (1) and (+)-3-methylheptane[8] (2) are nearly constant with temperature, especially in solution[7].

There have been a number of attempts to relate specific rotation to the refractive index of the solvent or of the solution. Theoretical models suggest that 'rotivity', Ω, may be a more fundamental property than specific rotation:

$$\Omega = \frac{[M]}{(n^2 + 2)} \tag{3}$$

but it would appear that other effects of solvents may be more important[6]. Thus, 1 and 2 seem to follow the relationship:

$$[M] = \frac{k}{(n - 1)} \tag{4}$$

for solvents such as n-hexane (n_D^{20} 1.3835), methylene chloride (n_D^{20} 1.4245), benzene (n_D^{20} 1.4992) and carbon disulfide (n_D^{20} 1.6246)—not to mention themselves as solvent: 1 (n_D^{20} 1.3890)[7], and 2 (n_D^{20} 1.4002)[8], while solutions of (+)-*cis*-pinane (3) follow a wholly different relationship[9]:

$$[M] = k(n^2 + 2)^2 \tag{5}$$

The rotations of solutions of $(-)$-isocamphane (**4**) increase even more rapidly with n, while those of $(-)$-dibornyl (**5**) are irregular and do not rise with increasing n and those of $(-)$-dimenthyl (**6**) decrease[10]. It seems clear that further work will be required before this behavior is understood.

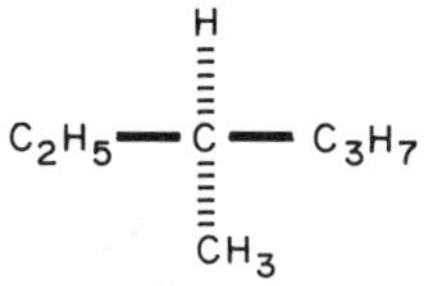

[M]$_D$ + 9.9° (neat)

(**1**)

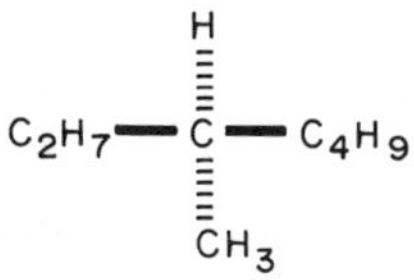

[M]$_D$ + 11.4° (neat)

(**2**)

[M]$_D$ + 31.5° (neat)

(**3**)

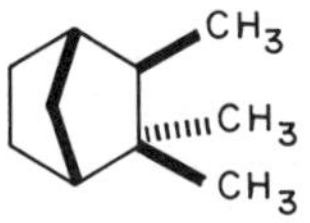

[M]$_D$ − 7.5° (C$_6$H$_6$)

(**4**)

[M]$_D$ − 53.2° (C$_6$H$_6$)

(**5**)

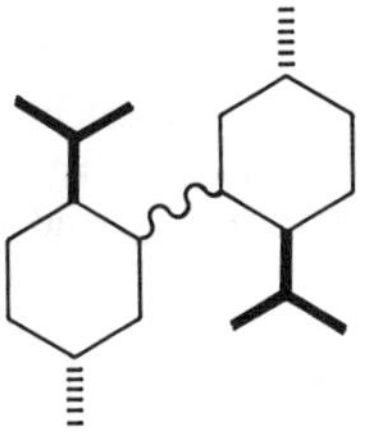

[M]$_D$ − 137° (C$_6$H$_6$)

(**6**)

Several old, but extensive and scholarly, reviews should be consulted for details on the increasingly neglected precautions that should be taken in making polarimetric measurements[11-13].

B. Effects of Wavelength of Light

1. Electronic optical activity (EOA) (visible and ultraviolet)

The most widely used light source for simple polarimetry has been the sodium flame or the sodium lamp, which produces a close doublet at 5896 and 5890 Å (usually reported as 5893 Å or 589 nm), but some rotations have been measured with a mercury lamp, which produces spectral lines at 5461 and 4358 Å.

Optical Rotatory Dispersion (ORD), or measurement of rotation as a function of wavelength[13-16], gives *plain dispersion curves* (no sharp maxima or minima) in the visible and quartz ultraviolet with saturated hydrocarbons because they do not absorb in these regions of the spectrum. With many saturated compounds the rotation increases with decreasing wavelength in accord with a single-term Drude expression[11]:

$$[M]_\lambda = \frac{k}{\lambda^2 - \lambda_o^2} \tag{6a}$$

or

$$[M]_\lambda = \frac{k}{\lambda^2} \tag{6b}$$

Here, k is an empirical constant characteristic of the substance (at a particular temperature, in a particular solvent), λ is the wavelength of light used in the observation and λ_o is the wavelength of an 'optically active absorption band'.

Simple dispersion, in which only one Drude term is required, is often observed. This implies that one particular transition in the ultraviolet controls rotation at longer wavelengths, but this is usually an artefact of arithmetic since the sum of several Drude terms can often be approximated by one, in which λ_o is averaged and k is summed. In those cases where the 'optically active absorption bands' make oppositely signed contributions, the data may fit best a two-term Drude expression:

$$[M] = \frac{k'}{\lambda^2 - \lambda_o'^2} - \frac{k''}{\lambda^2 - \lambda_o''^2} \tag{6c}$$

For $(+)$-3-methylhexane, where measurements over the range $\lambda = 643.8 - 252.9$ nm were made[7], a single-term Drude expression can be calculated from the data reported[7]:

$$[M] = \frac{3.021 \times 10^6}{\lambda^2 - \lambda_o^2} \tag{7}$$

λ in nm; $\lambda_o = 157.8$ nm

Similar ORD measurements for several steroids with alkyl side-chains at C_{17} gave similar results[17]: 5α-pregnane (7), 5β-pregnane (8), 5α-cholestane (9) and 5β-cholestane (10) (λ in nm). On the other hand, the λ_o value for 5α-androstane (11) points to complex dispersion (two terms of opposite sign). The ORD curve of 5β-androstane (12) is anomalous, being positive at long wavelengths and negative at short, but close to zero throughout the range studied (600–280 nm). This behavior is typical of systems with two Drude terms of opposite sign. Additional ORD data for saturated hydrocarbons have been reported by Levene and Rothen[1].

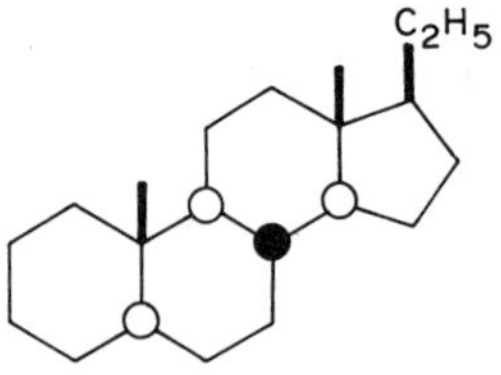

$$[M] = \frac{1.6 \times 10^7}{\lambda^2 - (167)^2}$$

$[M]_D = 50.1°$

(7)

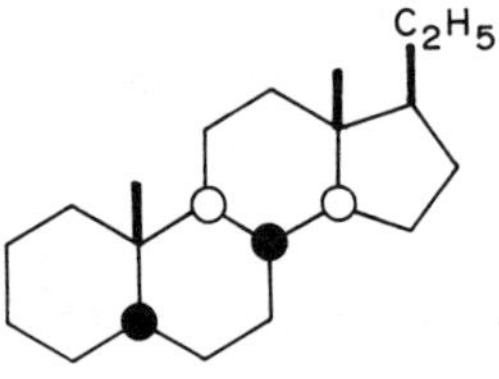

$$[M] = \frac{1.5 \times 10^7}{\lambda^2 - (135)^2}$$

$[M]_D = 45.6°$

(8)

Absorption spectroscopy in the vacuum ultraviolet[18] shows a strong band at 141–143 nm with normal paraffins and another at about 157 nm in isobutane and neopentane. A third band at 128–134 nm is prominent in all of these hydrocarbons as well. Cyclopropane shows bands at 162, 145 and, very strongly, at 124 nm, while cyclobutane shows end absorption up to 185 nm and peaks at 152, 142 and 126 nm. There is a reasonable probability that these are the bands indicated by the Drude expressions and, therefore, that they control the optical activity of saturated hydrocarbons in the quartz ultraviolet and visible. For trans-1,2-dimethylcyclopropane CD measurements down to 150 nm show one Cotton effect at 190 nm and two of opposite

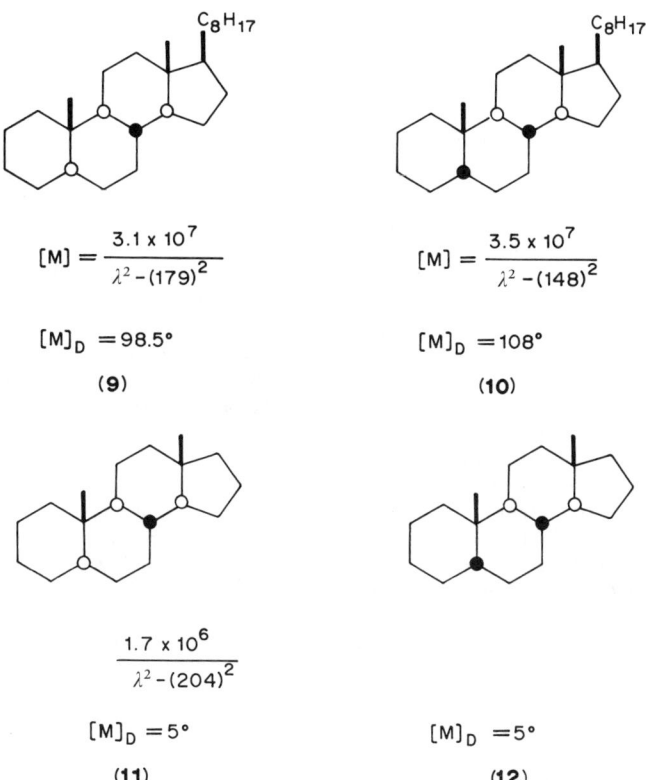

$$[M] = \frac{3.1 \times 10^7}{\lambda^2 - (179)^2}$$

$$[M]_D = 98.5°$$

(9)

$$[M] = \frac{3.5 \times 10^7}{\lambda^2 - (148)^2}$$

$$[M]_D = 108°$$

(10)

$$\frac{1.7 \times 10^6}{\lambda^2 - (204)^2}$$

$$[M]_D = 5°$$

(11)

$$[M]_D = 5°$$

(12)

sign at 180 and 160 nm (see below)[19]. In this case, it would appear that the latter, or perhaps a 'resonance giant' near 120 nm[18], actually controls long-wavelength rotation. ('Resonance giants' are especially intense ultraviolet absorption bands found with some transition metal ions and in some hydrocarbons. They have been attributed to transitions to outer Rydberg orbitals that have the same symmetry as unoccupied inner orbitals and overlap them enough that resonance between them is possible. In alkanes and cycloalkanes there will be as many empty non-bonding or anti-bonding σ orbitals as there are C—C single bonds. In nuclei of high symmetry the nodal surfaces of such orbitals may make them resemble d, f or higher orbitals to which Rydberg transitions are possible. 'Giant resonances' have been suggested as being responsible for strong bands in neopentane, cyclopropane and cyclohexane. This argument suggests that they would also be observed in caged compounds.)

It is important to realize that the observation of a Cotton effect does not establish that it is the prime source of long-wavelength optical activity. Transitions lying deeper in the ultraviolet might be stronger and more influential. It may be some time before we can say with assurance which ultraviolet Cotton effects rule the visible. In the meantime more empirical studies may provide clues as to which elements of structure do seem to do so. Saturated hydrocarbon residues may play a significant role in generating the 'background rotation' that is a prominent feature in the ORD curves of steroidal ketones and other complex compounds[14-16], but short-wavelength bands of functional groups may also contribute. Thus ketones have a strong chiroptically active absorption band at 190 nm[20].

2. Vibrational optical activity (VOA) (infrared and Raman)

ORD techniques do not appear to be useful in the infrared. In the first place, optical rotation is a *refraction* phenomenon which is not easily studied within absorption bands because of slit-width effects; these have given rise to spurious Cotton effects in infrared ORD[21]. Secondly, the multiplicity of bands in the middle IR would, at best, give ORD curves of incredible complexity. More telling is the failure of attempts to demonstrate that infrared bands could produce optical rotation in nearby regions of transparency; it was found that optical rotation between 4000 and 2000 cm^{-1} was still dominated by contributions from remote electronic transitions[21]. Background rotation of this nature would only make the observation of contributions from infrared bands difficult to discern.

Circular dichroism (CD), a measure of the differential absorption of right- and left-circular polarized light[16,22], is of some value for saturated hydrocarbons in far ultraviolet, where such compounds show absorption bands (see above). It is also becoming increasingly important in the infrared, where many of the vibrational transitions are optically active (VCD). Of particular interest is the fact that the C—H bonds and the carbon skeleton can act as chromophores which, if optically active, would provide stereochemical information not otherwise available. Different conformers give different infrared spectra and might, in principle, be studied in the presence of one another by temperature variation. The first authentic Cotton effects were observed in 1973[23]. (S) − (+)-2,2,2-Trifluoro-1-phenylethanol (13) and its enantiomer gave, respectively, positive and negative Cotton effects at the C*—H stretch (2920 cm^{-1}) and (R)-(−)-neopentyl-1-d chloride (14) gave a negative Cotton effect at a C*—D stretch (2204 cm^{-1}). Any C*—H Cotton effect in 14 was obscured by other C—H vibrations. With 13 it has been suggested that the magnetic moment change requisite for optical activity resulted from mixing with the C—O—H bend.

$(S)-(+)$ C_6H_5 ——— $\overset{\overset{H}{\vdots}}{\underset{\overset{\vdots}{OH}}{C}}$ ——— CF_3 $(R)-(-)$ $(CH_3)_3C$ ——— $\overset{\overset{D}{\|}}{\underset{\overset{\|}{H}}{C}}$ ——— Cl

(13)   **(14)**

There have been many advances in instrumentation since 1973 and it is now possible to measure VCD down to 600 cm^{-1} [24,25]. Fourier transform infrared (FTIR) technology has provided increased resolution and convenience[26,27]. This technique is becoming routinely accessible to organic chemists and we may anticipate an increasing number of applications[28,29].

Raman optical activity (ROA)[30] provides a means for studying vibrational transitions via differential scattering of right- and left-circular polarized light:

$$\Delta = \frac{I^R - I^L}{I^R + I^L} \tag{8}$$

where I is the scattered intensity of right- and left-circular polarized light. Although it is a more difficult technique than VCD, ROA has some distinct advantages. The intensity of signal increases with frequency so that the use of visible light for scattering has advantages over use of the infrared. ROA measurements are easier than VCD in the region of skeletal vibrations so that these techniques provide access to a different set of chromophores. The correlation of ROA with chiral features of structure may, however, prove difficult.

III. ALKANES

A. Preparation, Configuration, Enantiomeric Purity

During the 1920s and 1930s, Levene and his coworkers engaged in a broad program to establish relative configurations at noncyclic stereogenic centers having one hydrogen atom. Chiral alkanes played a key role in a number of these correlations:

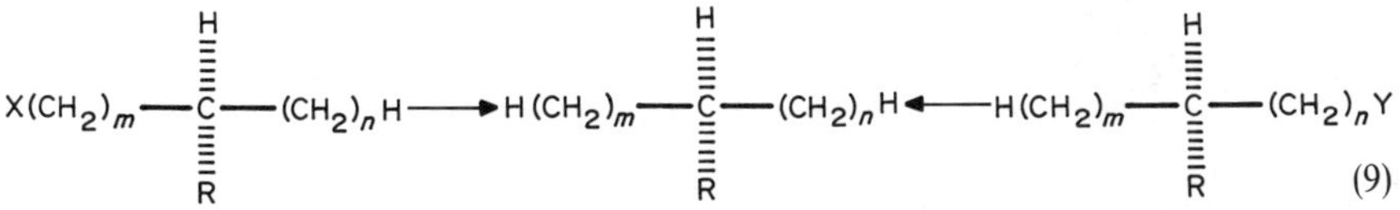

$$X(CH_2)_m \overset{H}{\underset{R}{-C-}}(CH_2)_nH \longrightarrow H(CH_2)_m \overset{H}{\underset{R}{-C-}}(CH_2)_nH \longleftarrow H(CH_2)_m \overset{H}{\underset{R}{-C-}}(CH_2)_nY \qquad (9)$$

An effort was made to prepare all of these compounds enantiomerically pure or, at least, from substances of known enantiomer composition so that maximum molecular rotations could be calculated. This work was summarized in a review[1]. Absolute configurations could not, of course, be assigned until Bijvoet established the absolute configuration (15) of (+) tartaric acid[31] showing that Fischer's convention for the carbohydrates was, fortuitously, correct. At that time it became clear that the pictorial representations in that review, which did not adhere to Fischer's convention, are the reverse of those actually obtaining. This does not diminish the value of this review, provided that due care is used in reading the figures. Tables 1 and 2 show the absolute configurations and molecular rotations of these alkanes and many others prepared since then. In addition, the configurations and rotations of several alkanes with both hydrogen and deuterium at the stereogenic center are shown in Table 3 (reviewed elsewhere[32]). Several trends are

TABLE 1. Molecular rotations of alkanes with unbranched groups at the stereogenic center

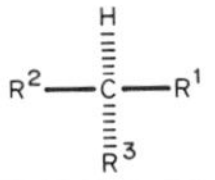

R^1	R^2	R^3	Configuration	$[M]_D$	Reference
C_3H_7	C_2H_5	CH_3	S	$+9.9°$	1,33
C_4H_9	C_2H_5	CH_3	S	$+11.1$	1
C_5H_{11}	C_2H_5	CH_3	S	$+13.3$	1,33
C_6H_{13}	C_2H_5	CH_3	S	$+13.8^a$	34
C_7H_{15}	C_2H_5	CH_3	S	$+14.4$	33
C_8H_{17}	C_2H_5	CH_3	S	$+13.2^b$	35
C_9H_{19}	C_2H_5	CH_3	S	$+14.7$	33
$C_{11}H_{23}$	C_2H_5	CH_3	S	$+14.6$	33
$C_{13}H_{27}$	C_2H_5	CH_3	S	$+14.7$	33
C_4H_9	C_3H_7	CH_3	S	$+1.7$	1
C_5H_{11}	C_3H_7	CH_3	S	$+2.4$	1
C_6H_{13}	C_3H_7	CH_3	S	$+3.2^a$	34
$C_{29}H_{59}$	C_3H_7	CH_3	S	$+3$	36
C_5H_{11}	C_4H_9	CH_3	S	0.9	1

^{a}Calculated for sample estimated to have enantiomeric purity of 68%.
^{b}Calculated for sample estimated to have enantiomeric purity of 96%.

TABLE 2. Molecular rotations of alkanes with branched alkyl groups at the stereogenic center

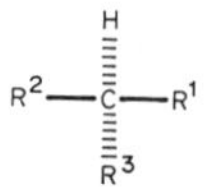

$$R^2 \!-\!\! \overset{\overset{\displaystyle H}{\vert}}{\underset{\underset{\displaystyle R^3}{\vert}}{C}} \!\!-\! R^1$$

R^1	R^2	R^3	Configuration	$[M]_D$	Reference
a. Branching at C_δ					
$(CH_2)_3CH(CH_3)_2$	C_2H_5	CH_3	S	$+13.5°$	37
b. Branching at C_γ					
$(CH_2)_2CH(CH_3)_2$	C_2H_5	CH_3	S	$+11.9$	1
$(CH_2)_2CH(CH_3)_2$	C_3H_7	CH_3	S	$+3.5$	1
$(CH_2)_2CH(CH_3)_2$	C_4H_9	CH_3	S	$+1.5$	1
$(CH_2)_2CH(CH_3)_2$	C_5H_{11}	CH_3	S	$+0.2$	1
$(CH_2)_2c\text{-}C_6H_{11}$	C_2H_5	CH_3	S	$+16.8$	1
	C_8H_{17}	CH_3	S	$+5.3$	38
c. Branching at C_β					
$CH_2CH(CH_3)_2$	C_2H_5	CH_3	S	$+21.3$	1
$CH_2CH(CH_3)_2$	C_3H_7	CH_3	S	$+14.9$	1
$CH_2CH(CH_3)_2$	C_4H_9	CH_3	S	$+11.9$	1
$CH_2CH(CH_3)_2$	C_5H_{11}	CH_3	S	$+9.3$	1
$CH_2c\text{-}C_6H_{11}$	C_2H_5	CH_3	S	$+19.3$	1
d. Branching at C_α					
$CH(CH_3)_2$	C_2H_5	CH_3	S	-26.4	1
				-11.1	42
$CH(CH_3)_2$	C_3H_7	CH_3	S	-5.5^a	1
$CH(CH_3)_2$	C_7H_{15}	CH_3	S	-1.1^a	1
	$CH_2CH(C_2H_5)_2$	CH_3	S	-60.2	39
$CH(C_2H_5)_2$	C_2H_5	CH_3	S	$+15$	40
$c\text{-}C_6H_{11}$	C_2H_5	CH_3	S	-3.0^a	41
$c\text{-}C_6H_{11}$	C_3H_7	CH_3	S	-3.8^a	44
$c\text{-}C_6H_{11}$	C_4H_9	CH_3	S	-4.5^a	44
$c\text{-}C_6H_{11}$	C_3H_7	C_2H_5	S	-1.5^a	1
$c\text{-}C_6H_{11}$	C_4H_9	C_2H_5	S	-2.5^a	44
$c\text{-}C_6H_{11}$	C_4H_9	C_3H_7	S	-0.5^a	44
$C(CH_3)_3$	C_2H_5	CH_3	S	-49.8	42
$C(CH_3)_3$	$CH_2CH(C_2H_5)_2$	CH_3	S	-132	39
$C(CH_3)_3$	$(S)\text{-}CH_2\underset{\underset{\displaystyle CH_3}{\vert}}{C}HC_2H_5$	CH_3	S	-97.5	39
$C(CH_3)_3$	$(R)\text{-}CH_2\underset{\underset{\displaystyle CH_3}{\vert}}{C}HC_2H_5$	CH_3	S	-137.8	39
$C(CH_3)_3$	$(R)\text{-}CH_2\underset{\underset{\displaystyle CH_3}{\vert}}{C}HCH(CH_3)_2$	CH_3	S	-79.8	43
$C(CH_3)_3$	$(S)\text{-}CH_2\underset{\underset{\displaystyle CH_3}{\vert}}{C}HCH(CH_3)_2$	CH_3	S	-119.1	43

^{a}Not maximum rotation.

TABLE 3. Molecular rotations of alkanes with hydrogen and deuterium at the stereogenic center

$$R^2 \underset{\overset{\displaystyle |}{R^1}}{\overset{\displaystyle H}{-}} C - D$$

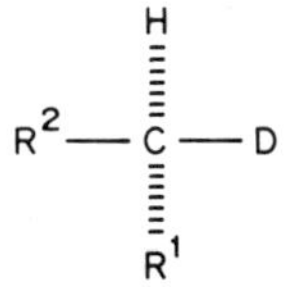

R^1	R^2	Configuration	$[M]_D$	References
CH_3	CH_2CH_3	S	$-0.36°$	32,45–48
CH_3	$CH_2CH_2CH_3$	S	-0.38	32,49,50
CH_3	$CH(CH_3)_2$	S	-0.62	51,52
CH_3	$C(CH_3)_3$	S	-0.075	53
CH_3	$(S) - \underset{\overset{\displaystyle |}{H}}{\overset{\displaystyle D}{C}} - CH_3$	S	-0.61	46

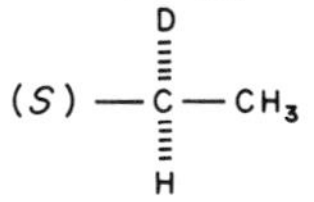

noteworthy. (S)-3-methylalkanes tend to a maximum molecular rotation of about $+15°$ as the chain length increases. Branching at an atom once removed from the stereogenic atom produces larger rotations. Branching next to the asymmetric atom, as with i-propyl, cyclohexyl and t-butyl groups, *generally* reverses the sign of rotation and increases the magnitude of rotation—except with 3-pentyl[39].

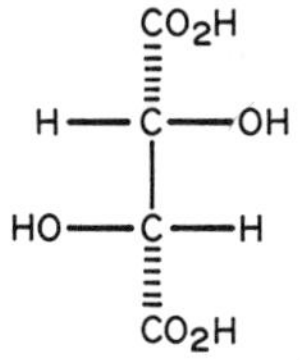

$$HO - \underset{\overset{\displaystyle |}{CO_2H}}{\overset{\displaystyle CO_2H}{\underset{\displaystyle |}{C}}} \begin{array}{l} H - C - OH \\ \quad | \\ - C - H \end{array}$$

(+)

(15)

B. Interpretations of Optical Rotation

1. Atomic asymmetry

Russell Marker, who had been one of Levene's principal lieutenants, suggested that one could rank substituents in such a way that their spatial ordering about the stereogenic center would reflect the sign of rotation while differences in an empirical rank would control the magnitude of rotation[54]. This may be termed the *atomic asymmetry* approach since it does not distinguish rigid and flexible substituents nor the steric interactions among them. Using the relative configurations and the maximum rotations accumulated by Levene and coworkers[1], he assigned ordinal numbers to the groups as shown in Table 4. The smaller the number, the higher the rank. For *C*ABCH, a clockwise arrangement: A > B > C on the face opposite the H atom corresponds to the levorotatory isomer (cf. **16**).

Marker's work contains ideas that prefigure aspects of the Cahn–Ingold–Prelog (CIP) notation[56]; in many series there will be a relationship between RS configuration and sign of rotation, as seen in Tables 1 and 2a–c (but note 2d). The 'Marker Rules' remain

TABLE 4. Maximum molecular rotations of configurationally related alkanes and simple derivatives[1,54] (abridged)

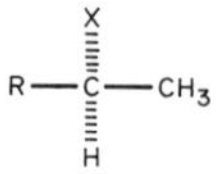

$$R-\overset{\overset{X}{\vdots}}{\underset{\underset{H}{\vdots}}{C}}-CH_3$$

X	$R = C_2H_5$	$n\text{-}C_3H_7$	$n\text{-}C_4H_9$	$n\text{-}C_5H_{11}$	Ordinal rank of X
$(CH_2)_2Br$	38.8°	21.3°	16.8°	14.7°	1
$i\text{-}C_4H_9$	21.3	14.9	11.9	9.3	2
$(CH_2)_3Br$	21.9	14.5	8.3	6.2	3
$(CH_2)_4Br$	14.9	7.8	5.3	4.0	4
$(CH_2)_2CO_2H$	13.6	6.9	4.1	1.9	5
$(CH_2)_3CO_2Et$	13.0	5.9	2.8	1.7	6
$(CH_2)_2CO_2Et$	12.7	5.5	2.3	0.5	7
$(CH_2)_3CO_2H$	11.1	3.7	1.7	0.8	8
$n\text{-}C_5H_{11}$	12.5	2.4	0.8	0	10
$n\text{-}C_4H_9$	11.4	1.7	0	−0.8	11
$(CH_2)_3OH$	12.0	1.7	0	−1.9	12
$n\text{-}C_3H_7$	10.0	0	−1.5	−2.4	14
$(CH_2)_2NH_2$	10.6	−0.4	−1.9	−3.6	15
CH_2CO_2Et	11.4	−0.7	−2.9	−4.2	16
$CH_2C_6H_5$	9.2				17
$(CH_2)_2OH$	9.0	−2.1	−4.0	−6.1	18
CH_2CO_2H	10.3	−3.6	−6.1	−8.1	19
C_2H_5	0	−10.0	−11.4	−12.5	20
CH_2OH	−5.2		levo		21
$i\text{-}C_3H_7$	−28.3		levo		22
CH_3					23
OH	10.3	12.1	12.2	12.5	24
CO_2H	18.0	21.4	24.3	25.0	25
CO_2Et	22.9	27.5	30.7	dextro	26
C_6H_5	36.6	38.0	39.0	dextro	27
Br	48.8[55]	63			28
I	70[55]	92	80.9	115	29

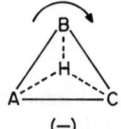

(−)

(16)

(R)-(−)

a useful summary of Levene's results but must be used with circumspection in cases where two or more hetero atoms and/or unsaturated groups are present at the stereogenic center or where those groups may interact with one another, as by hydrogen bonding. They do not apply at all when the stereogenic center is part of a ring. These rules epitomize the approach of organic chemists of the time to the problem of optical activity, namely, that it should be a function of inherent and invariant attributes of the substituents, *per se*, and of their order of attachment to the stereogenic atom[57]. This model found an appropriate expression as a means of characterizing stereogenicity in the CIP notation[56] but is severely limited as a means of accounting for optical rotatory power. Polarizability theory indicates that substituents having cylindrical symmetry will not produce optical rotation simply by virtue of their distribution in space around a point[58], although they might do so by perturbing the orbitals of the central atom[59]. The resolution and determination of enantiomer purity for bromochlorofluoromethane (**17**)[60] showed it to have a maximum rotation $[M]_D^{max}$ 1.7°. It seems likely, on this basis, that compounds having two or more saturated carbon atoms at the stereogenic center will show little or no optical rotation attributable to atomic asymmetry.

$$H \overline{\quad\quad} \overset{\displaystyle F}{\underset{\displaystyle Br}{C}} \overline{\quad\quad} Cl \quad \text{(configuration unknown)}$$

(**17**)

2. The conformational dissymmetry model

Werner Kuhn, over a period of thirty years[61–64], advocated the view that a noncoplanar system of two oscillators could, if coupled, give rise to optical rotation in a predictable way. If the oscillators are chromophores, exciton activation can give rise to the two kinds of coupling shown in Figure 1. The rotatory effects would be equal and opposite and occur at the same wavelength, unless one of the couplings were energetically favored over the other. If the transitions of the chromophore are electrically allowed there will be some displacement of charge along the arrows. For groups linked as shown in Figure 1, coulombic factors would favor mode *b*, which would then produce the long-wavelength member of a bisignate Cotton effect couplet. This concept has been developed as a powerful method for determining the stereochemistry of polyhydroxy compounds via circular dichroism measurements on their benzoate esters[65]. When the coupled oscillator concept is extended to deal with optical rotation in spectral regions of transparency the operative electronic phenomenon is not excitation but deformation of orbitals (a function of the polarizability of atoms) which, again, would be expected to be favored if cooperative. On this basis, coupling after pattern *b* in Figure 1 would be dominant. This model was developed explicitly by Kirkwood[66] and has been applied to several specific small molecules[67,68].

Whiffen[69] reduced this model to the practical level required for use by organic chemists by suggesting that the skew ethylene glycol unit **18** will make a regular contribution of 45° to the molecular rotation of polyhydroxycyclohexanes and pyranoses. This was then generalized as the Conformational Dissymmetry Model[70] for cases where the hydroxy groups are replaced by other atoms, including hydrogen. In the original treatment, the size of the chiral units was limited to four atoms, which made additive contributions, and a binary all-or-nothing conformational analysis was used to simplify application to flexible compounds (Section III.B.2.a). Subsequent refinements have included treating the chiral unit as a helix, which allows extension to larger units (Section III.B.2.b) and use

(−)   (+)

(a)   (b)

FIGURE 1. Coupled oscillators[61]. Symmetrical coupling, as shown in a, will produce levorotation; unsymmetric coupling, as shown in **b**, will produce dextrorotation

of more realistic methods of conformational analysis (Section III.B.2.c). This model is also readily applied to cyclic systems (Section IV).

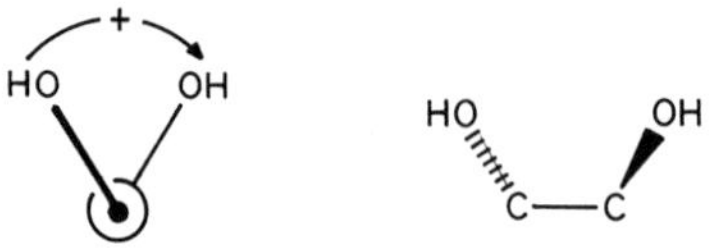

$$[M]_D = +45°$$

(18)

a. Four-atom units; binary conformational analysis. In the original form of the Conformational Dissymmetry Model[70] it was suggested that the optical rotations of saturated compounds are due primarily to contributions from skew four-atom systems: X—C—C—Y (**19**). It was suggested that the magnitude of the rotatory effects was related to some function of the polarizabilities of the terminal atoms X and Y (which functions were simply symbolized 'X' and 'Y') and a constant of proportionality, k. On this basis, the optical rotation of one enantiomer of the chiral compound: ZCH_2—C*HXY would derive from the relative amounts of the three conformers (**20**).

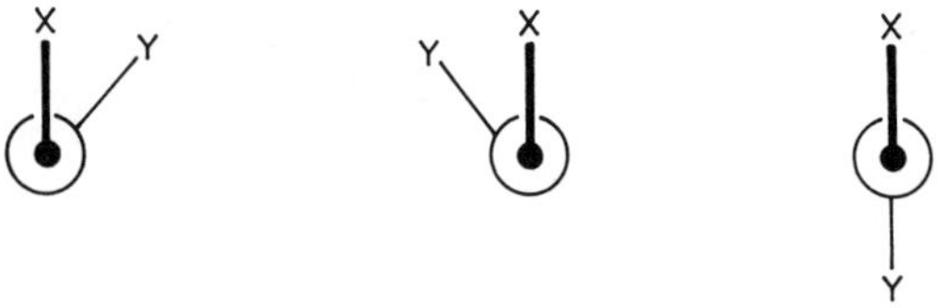

$+kXY$	$-kXY$	0
(a)	**(b)**	**(c)**

(19)

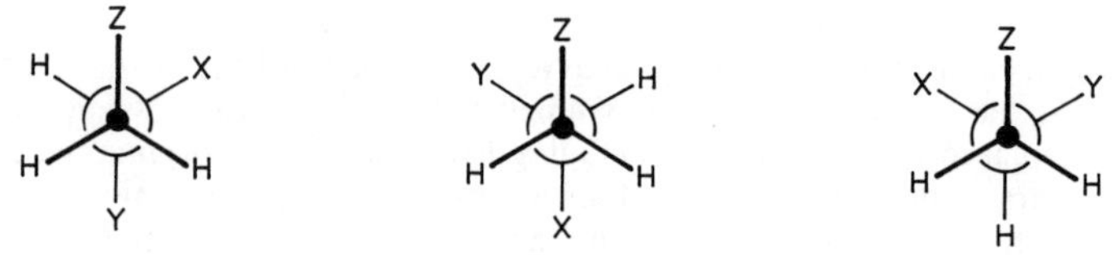

$k(XZ - XH + YH$	$k(ZH - HH + XH$	$k(ZY - YH + HH$
$-HH + XH - ZX)$	$-XH + YH - YZ)$	$-HH + XH - ZX)$
$= +k(Z—H)(X—H)$	$= -k(Z—H)(Y—H)$	$= +k(Z-H)(X-H)$
(a)	**(b)**	**(c)**

(20)

It is evident that the rotation would be zero if the three conformers were present in exactly equal amounts. This is why the rotations of many flexible compounds decrease with increasing temperature. In the original model it was considered that the *digauche bisected* conformer **20c** would be 'forbidden' if lower energy forms such as **20a** and **20b** were available. Conformations were thus given a binary classification: 'forbidden' or 'allowed', in the latter case without distinction of the steric requirements of X, Y or Z. On this basis, the rotation of **20** would be

$$[M]_D = \tfrac{1}{2}k(Z—H)(X—H) - \tfrac{1}{2}k(Z—H)(Y—H) \qquad (10)$$
$$\text{or} \quad \tfrac{1}{2}k(Z—H)(X—Y)$$

This enantiomer would be dextro or levo as the rotational contributions of X and Y differ and would, of course, be zero if the two groups were identical. The term $\tfrac{1}{2}$ derives from the fact that the mol fraction of each conformer would be: $n_a = n_b = 0.5$. Conformational rotatory powers, relative to $Y = CH_3$, were assigned (Table 6) to give reasonable fits to the data for various 2-substituted alkanes (Table 6). The values in Table 5 decrease in accord with the square root of the atomic refractions of the atoms (a measure of polarizability).

It will be noted that the rotations observed for alcohols and amines are positive, while those calculated are negative. It was suggested at that time that this reflected the conformational consequences of the smaller sizes of the hydroxy and amino groups. This is supported by subsequent computer modeling[71] (see Section III.B.2.c).

In order to cope with the more complex situation obtaining in the alkanes and substituted alkanes, a further all-or-nothing rule was adopted, namely, that the five-atom *meso digauche* conformation **21** would also be 'forbidden'. The steric strains occurring here would resemble those present in diaxial *cis*-1,3-dimethylcyclohexane (but see Section III.B.2.c). With conformations **20c** and **21** 'forbidden', the 27 possible chain conformers of 3-methylhexane reduce to six 'allowed' forms which are considered to be

TABLE 5. Conformational rotational contributions[70]

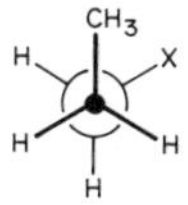

X	$R_D{}^a$	$\Delta[M]_D = k(C—H)(X—H)^b$
I	13.954	250
Br	8.741	180
SH	7.729	no data
Cl	5.844	170
CN	5.459^c	160
C_6H_5	6.757^c	140
CO_2H	4.680^c	90
CH_3	2.591	60
NH_2	2.382	55
OH	1.518	50
H	1.028	0
D	1.004^d	-1.13^e

^{a}Attachment atom only, except where noted. Values of A. I. Vogel, *J. Chem. Soc.*, 1833 (1948).
^{b}Empirical values, except where noted; rounded to nearest $10°$, except for NH_2 and D.
^{c}Whole group, not just attachment atom.
^{d}From data of C. K. Ingold, C. G. Raisin and C. L. Wilson, *J. Chem. Soc.*, 915 (1936); C_6D_6 has a molar refraction 0.144 lower than C_6H_6.
^{e}From the expression: $[M]_D = 160(R_C^{1/2} - R_H^{1/2})(R_X^{1/2} - R_H^{1/2})$.

TABLE 6. Molecular rotations of (S)-2-substituted alkanes[70]

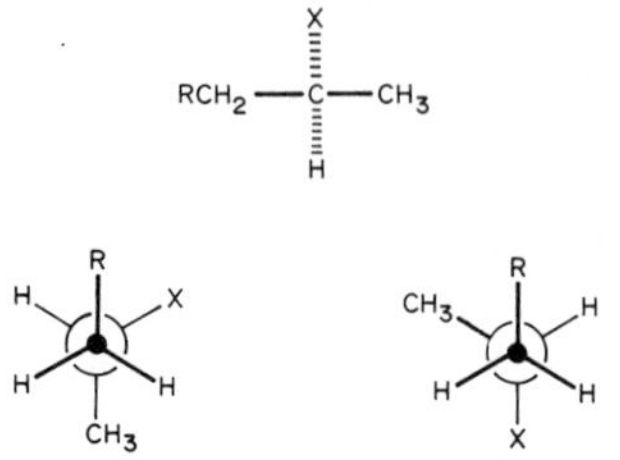

$$RCH_2 \overset{X}{\underset{H}{\overset{|}{-}\!C\!-}} CH_3$$

$$[M]_D = \tfrac{1}{2}k[(C\!-\!H)(X\!-\!H) - (C\!-\!H)(C\!-\!H)]$$

X =	I	Br	Cl	CN	C$_6$H$_5$	CO$_2$H	OH	NH$_2$	NH$_3^+$
Calcd	**95°**	**60°**	**55°**	**50°**	**40°**	**15°ᵃ**	**−5°ᵇ**	**−3°ᵇ**	**−3°ᵇ**
R = CH$_3$	70[55]	49[55]	45[55]	25	37	18	10	5	−1
C$_2$H$_5$	92	63	46	49	38	21	12		
n-C$_3$H$_7$	81			51	39	24	12	5	−3
n-C$_4$H$_9$				43		26	12		
n-C$_5$H$_{11}$	115	75	54				13	7	−10
n-C$_6$H$_{13}$				40		27	13		

ᵃ A value of: $k(C\!-\!H)(CO_2H\!-\!H) = 110°$ would give better results.
ᵇ Values calculated as though these groups have the same steric requirements as CH$_3$ (see Section III.B.2.c)

TABLE 7. Analysis of (S)-3-methylhexane (Simple Conformational Dissymmetry)[70]

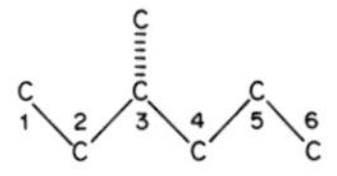

C$_2$—C$_3$	Conformations at: C$_3$—C$_4$	C$_4$—C$_5$	Δ[M]$_D$
		+	$+3k(C\!-\!H)^2$
+	+		
		o	$+2k(C\!-\!H)^2$
		+	$+k(C\!-\!H)^2$
−	+		
		o	0
		−	$-3k(C\!-\!H)^2$
−	−		
		o	$-2k(C\!-\!H)^2$

$$\text{Net } [M]_D = \frac{1k(C\!-\!H)^2}{6} = +10° \quad (\text{Obsd. } +9.9°)$$

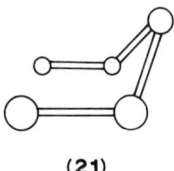

(21)

equally abundant (Table 7). There is a net excess of one dextro unit at an abundance of 1/6, so that

$$[M]_D = + k(C—H)^2/6 = + 10° \quad (Obsd. = 9.9°) \quad (11)$$

The simple arithmetic of the binary conformational analysis indicates that the rotations of tertiary alkanes and their derivatives, of the general structure **22**, should be, simply, the difference of radical rotations for the groups —CH$_2$X and —CH$_2$Y. Radical rotations for hydrocarbon groups are shown in Table 8[70]; both series appear to converge on a value of about 34.14°. Rotations for alkanes calculated as differences of these values are shown in Table 9. It is significant that this method is not limited to alkanes. Radical rotations for terminally substituted alkyl groups —(CH$_2$)$_n$X are shown in Tables 10 and 11 and rotations for compounds containing those groups, calculated and observed, are shown in Table 12.

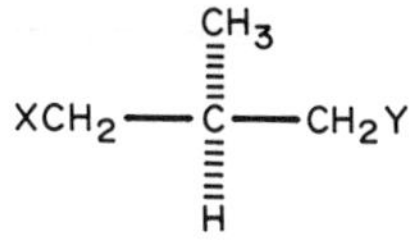

$$XCH_2—\overset{\overset{\displaystyle CH_3}{|}}{\underset{\underset{\displaystyle H}{|}}{C}}—CH_2Y$$

(22)

It is a consequence of this analysis that alkanes having four alkyl groups at the stereogenic center should show very low (predicted, zero) rotation. Thus, for $C(CH_2A)(CH_2B)(CH_2C)(CH_2D)$, the three conformations of group —CH$_2$A are equally 'forbidden' and hence, since no alternative is available, equally allowed. Since there is

TABLE 8. Calculation of radical rotations for alkyl groups[70]

n	$R = (CH_2)_nCH_3$		$R = (CH_2)_nCH(CH_3)_2$[a]	
	$\dfrac{\Delta[M]_D}{k(C—H)^2}$	$\Delta[M]_D$[b]	$\dfrac{\Delta[M]_D}{k(C—H)^2}$	$\Delta[M]_D$[b]
1	1/3	20°	2/3	40°
2	3/6	30	5/9	33.33
3	8/15	32	12/21	34.29
4	20/36	33.3	29/51	34.12
5	49/87	33.8	70/123	34.15
6	119/210	34.0	169/297	34.14
7	288/507	34.1	408/717	34.14

[a]These values should apply for $(CH_2)_n$c-C_6H_{11} groups as well.
[b]Letting $k(C—H)^2 = 60°$.

TABLE 9. Calculated molecular rotations $[M]_D{}^a$ for (S)-alkanes[70]

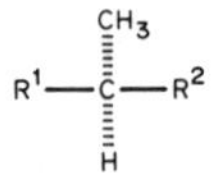

$$R^1\!-\!\overset{\overset{CH_3}{|}}{\underset{\underset{H}{|}}{C}}\!-\!R^2$$

R¹	R²							
	CH₂CH₃		(CH₂)₂CH₃		(CH₂)₃CH₃		(CH₂)₄CH₃	
$(CH_2)_2CH_3$	10°	**9.9°**						
$(CH_2)_3CH_3$	12	**11.4**	2°	**1.7°**				
$(CH_2)_4CH_3$	13	**12.5**	3.3	**2.4**	1.3°	**0.8°**		
$CH_2CH(CH_3)_2$	20	**21.3**	10	**14.9**	8	**11.9**	6.7°	**9.3°**
$(CH_2)_2CH(CH_3)_2$	13	**11.9**	3.3	**3.5**	1.3	**1.5**	0	**0.2**

^{a}Observed rotations are in boldface (See Table 1).

TABLE 10. Radical rotations $(\Delta[M]_D)$ for terminally substituted alkyl groups $-(CH_2)_nX$[70]

n	$\Delta[M]_D$
1	$\dfrac{k(C\!-\!H)(X\!-\!H)}{3}$
2	$\dfrac{2k(C\!-\!H)^2 + k(C\!-\!H)(X\!-\!H)}{6}$
3	$\dfrac{7k(C\!-\!H)^2 + k(C\!-\!H)(X\!-\!H)}{15}$
4	$\dfrac{19k(C\!-\!H)^2 + k(C\!-\!H)(X\!-\!H)}{36}$
5	$\dfrac{48k(C\!-\!H)^2 + k(C\!-\!H)(X\!-\!H)}{87}$

TABLE 11. Radical rotations $(\Delta[M]_D)$ for terminally substituted alkyl groups $(CH_2)_nX$[70]

X	CH_2X	$(CH_2)_2X$	$(CH_2)_3X$	$(CH_2)_4X$
Br	60°	50°	40°	36.7°
C_6H_5	46.7	43.3	37.3	35.6
CO_2H	30	35	34.0	34.2
NH_2	18	29.2	31.7	33.2
OH	16.7	28.3	31.3	33.0

TABLE 12. Calculated molecular rotations $[M]_D^a$ for compounds with two flexible saturated chains $R^1 = (CH_2)_mX$ and $R^2 = (CH_2)_nCH_3$ [70]

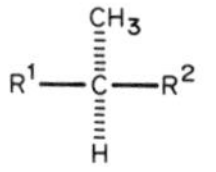

	R^2							
R^1	CH_2CH_3		$(CH_2)_2CH_3$		$(CH_2)_3CH_3$		$(CH_2)_4CH_3$	
$(CH_2)_2Br$	30°	**38.8°**	20°	**21.3°**	18°	**16.8°**	16.7°	**14.7°**
$(CH_2)_3Br$	20	**21.9**	10	**14.5**	8	**8.3**	6.7	**6.2**
$(CH_2)_4Br$	16.7	**14.9**	6.7	**7.8**	4.7	**5.3**	3.4	**4.0**
$CH_2C_6H_5$	26.7	**9.5**						
$(CH_2)_2C_6H_5$	23.3	**35.1**						
CH_2CO_2H	10	**10.3**	0	**−3.6**	−2	**−6.1**	−4.1	**−8.1**
$(CH_2)_2CO_2H$	15	**13.6**	5	**6.9**	3	**4.1**	1.7	**1.9**
$(CH_2)_3CO_2H$	14	**11.1**	4	**3.7**	2	**1.7**	0.8	**0.8**
$(CH_2)_2NH_2$	9.2	**10.6**	−0.8	**−0.4**	−2.8	**−1.9**	−4.1	**−3.6**
CH_2OH	−3.3	**−5.2**	−13.3	**−6.8**	−15.3	**−7.9**		
$(CH_2)_2OH$	8.3	**9.0**	−1.7	**−2.1**	−3.7	**−4.0**	−5.0	**−6.1**
$(CH_2)_3OH$	11.3	**11.9**	−1.3	**0**	−0.7	**−0.7**	−2.0	**−2.6**
$(CH_2)_4OH$	13.0	**12.0**	3.0	**1.7**	1.0	**0**	−0.3	**−1.9**

^{a}Observed rotations in boldface.

no steric bias in the environment each chain is constrained either to the *trans* conformation, which is achiral, or to equal amounts of the two *gauche* conformations. Several such alkanes have been prepared and found to have vanishing optical rotations: **23**[72], **24**[73], **25**[74].

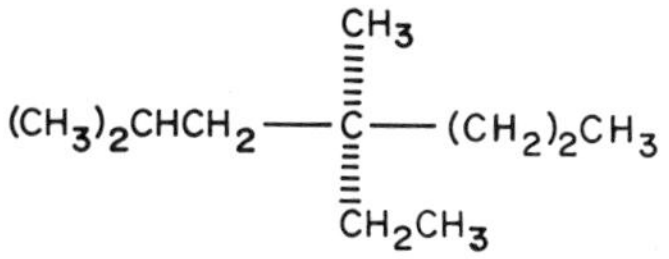

(23)[72]

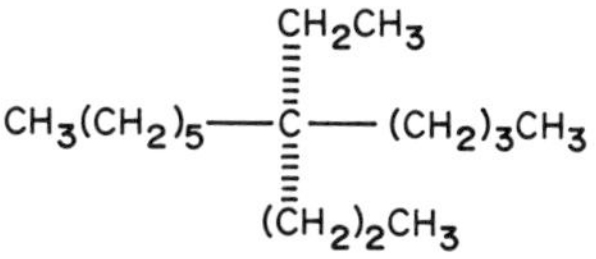

(24)[73]

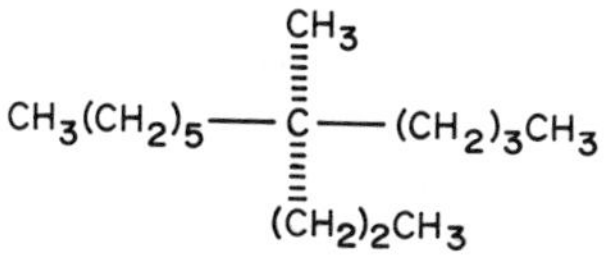

(25)[74]

The basic four-atom unit analysis is widely applicable to cyclic systems, including pyranoses, terpenes and carbohydrates[75] and alkaloids[76] and can be extended to cyclic unsaturated compounds as well[77], in what amounts to an incorporation of the Mills Rules[78] into this model. ORD and CD studies of cyclic alkenes showed that the long-wavelength rotations of these compounds are controlled by the η–π^* transitions of the ethylenic chromophore near 200 nm[79] and could be accommodated by an octant rule[80]. Drake and Mason[81] found that there are two optically active absorption bands in the 200–165 nm region of the spectrum of such alkenes, giving oppositely signed Cotton effects. A Rydberg band, with sharp vibronic structure in the vapor phase, and subject to shifting with changes in solvent and temperature was also found. (This observation is of general significance in suggesting a cause for the failure of simple 'rotivity' corrections with alkanes; see Section II.A). The presence of these additional transitions may cause difficulty in extensions of the simple Ethylene Octant Rule, especially with twisted alkenes. Drake and Mason[81] also found appreciable low-wavelength CD activity with saturated systems such as cholestane and twistane and stressed the point that large contributions from saturated portions of a molecule might mask weaker olefinic contributions. Although the Conformational Dissymmetry Model and the Ethylene Octant Rule give generally concordant results, they differ in mechanism in ways that bear upon more general applications of these two approaches to the understanding of chiroptical activity. The Conformational Dissymmetry Model is, inherently, a 'through-bond' model in which the chromophore itself may be twisted and in which additional perturbations are transmitted along bonds or through bond-centered molecular orbitals. Sector rules, in general, posit 'through-space' interactions.

Some significant exceptions to the Conformational Dissymmetry Model have been noted in the [2.2.1]-bicycloheptane series[82], with compounds having carboxy or phenyl groups at the stereogenic center, or with highly congested hydroxy groups. These findings indicate that difficulties may be expected when chromophores are held in environments that may prevent them from acting, by rotation, as essentially cylindrically symmetrical groups.

b. The helical conductor model. i. The classical version[83]. In what might now be termed the 'classical' concept of organic structure, covalent single bonds are pictured as pairs of electrons localized between atoms and having relatively little interaction with the rest of the molecule. In a chiral molecule each bond could be considered to be a *symmetrical chromophore in a chiral environment*[84] and each would make its own contribution to chiroptical properties. Perturbation of a pair of electrons (excitation by light of a suitable wavelength or simply a small adjustment in response to 'tickling' by radiation too low in energy to cause absorption) would produce a displacement of charge along the bond. This would induce smaller cooperative shifts of the electrons of neighboring bonds. A twisted chain of bonds would behave like a loop of wire with a current flowing through it and tend to produce a magnetic flux parallel or antiparallel to the electric flux. It has long been understood[85–87] that this would lead to differences in response to right- and left-circular polarized light and that this is what produces chiroptical effects. A right-handed helix produces parallel electric and magnetic moment changes and this gives a positive Cotton effect at the absorption band and dextrorotation at longer wavelength. With a left-handed helix the electric and magnetic moment changes are antiparallel and this would give a negative Cotton effect and levorotation[88,89].

The linear four atom system, X—C—C—Y, is the smallest structural unit that can produce optical activity by this mechanism. Any perturbation of the electrons of the central bond would provoke displacements of charge in the C—X and C—Y bonds that would be some function of the polarizabilities of X and Y. The helicity of the three-bond chain[90] would influence the sign and magnitude of the chiroptical response. An end-to-end projection of the dextrorotatory conformation (**26**) (dihedral angle, θ) shows that it can be regarded as a right helix. With constant bond angles, the helicity of this unit varies as the sine of the torsion angle. The localized electron model of structure suggests that the optical rotation of a complex compound would be expressed by the sum of all such contributions. This is, of course, the Conformational Dissymmetry

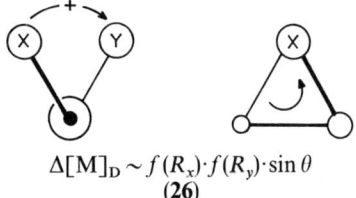

$$\Delta[M]_D \sim f(R_x) \cdot f(R_y) \cdot \sin\theta$$
(26)

Model, as presented above. A more detailed consideration[83] leads to the expression

$$\Delta[M]_\lambda = 652 \frac{LA}{D^2}\left[\sum R_D \frac{(\lambda_D^2 - \lambda^2)}{(\lambda^2 - \lambda_0^2)}\right] f(n)$$
(12)

where L = end-to-end distance (Å), A = area projected on a plane perpendicular to L (Å)^2, D = path length (a sum of bond lengths) (Å), R_D = atomic refractions, λ_D = 5893 Å, λ = wavelength of measurement, λ_0 = wavelength of the absorption band, and $f(n) = (n^2 + 2)/3$.

The last term should replace the term $f(n) = (n^2 + 3)^2/3n$ used earlier[83].

As mentioned above, the effect of refractive index is more complex than the Lorenz correction can deal with; it varies in different ways with different compounds and may reflect the properties of Rydberg transitions. Since the refractive index of many compounds is of the order of $(2)^{1/2}$, we approximate this term as 4/3 and pay it no further attention. A right-handed helix illustrating these terms is shown in Figure 2. For simple four-atom systems, the term LA/D^2 can be calculated from bond lengths (d_1, d_2, d_3), bond angles (α and β) and the dihedral angle (θ):

$$\frac{LA}{D^2} = \frac{1}{2}\frac{d_1 \cdot d_2 \cdot d_3}{(d_1 + d_2 + d_3)^2}\sin\alpha \times \sin\beta \times \sin\theta$$
(13)

This form of the expression can be used when the bond angles differ from tetrahedral and the dihedral angle is other than 60°:

$$\Delta[M]_D = \Delta[M]_D^\circ \frac{\sin\alpha \times \sin\beta \times \sin\theta}{\sin\alpha^\circ \times \sin\beta^\circ \times \sin 60^\circ}$$
(14)

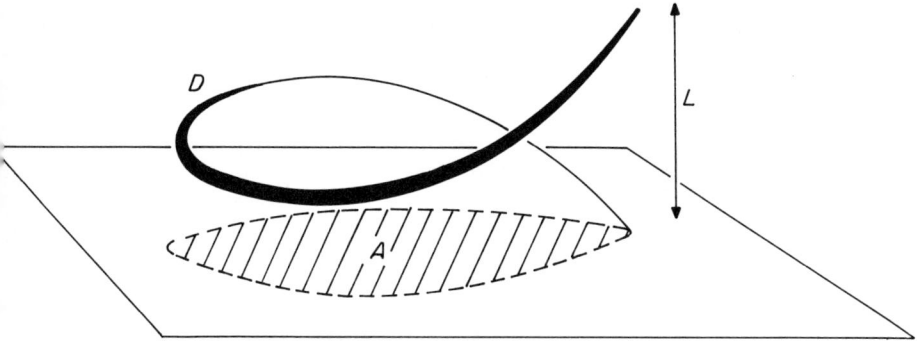

FIGURE 2. A uniform helical conductor: L is the end-to-end length, A the projected area on a plane perpendicular to L and D the length of chain (a sum of bond lengths)

TABLE 13. 'Rotivity' of skew conformational units

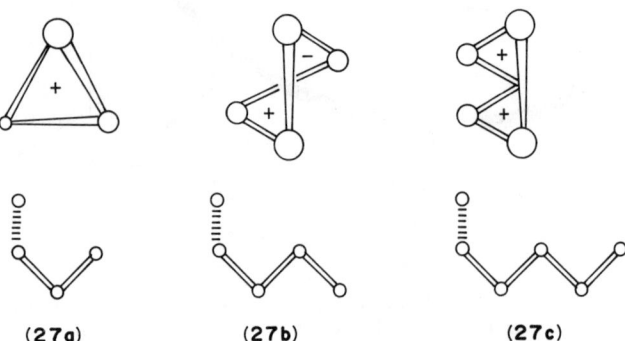

X	Calculated $\Delta[M]_D/f(n)$	Empirical values $\Delta[M]_D/1.33^a$
F	+22°	no data
OH	+35	+38°
NH$_2$	+47	+41
CH$_3$	+58	+45
Cl	+75	+127
SH	+94	no data
Br	+101	+135
I	+146	+188

$^a f(n) = (n^2 + 2)/3;\ n = 2^{1/2}$.

where $\Delta[M]_D^\circ$ is the standard rotatory effect $(\theta = 60°)$ and $\sin \alpha° = \sin \beta° = 2\sqrt{2}/3$, $\sin 60° = \sqrt{3}/2$, whence

$$[M]_D = 0.75\sqrt{3} \times \Delta[M]_D^\circ \times \sin \alpha \times \sin \beta \times \sin \theta \qquad (15)$$

'Rotivity' contributions for some common conformational units were calculated by the use of equation 12 and are compared with those used with the empirical Conformational Dissymmetry Model (Table 13).

This analysis suggested that longer chains could also contribute to optical rotation in addition to, or instead of, the four-atom units. Helical structures were identified among the staggered conformations possible to five- and six-atom chains[90]. Some of them showed large projected areas (*A* in equation 12) but contain several *gauche* or crowded conformations that would tend to make them less abundant than others in flexible compounds. They can, however, be prominent in rigid caged compounds where, indeed, they do seem to be responsible for large rotations. Mono-*gauche* conformations are interesting because they might be present in less strained conformers. As seen in **27a, b** and **c**, a *gauche* conformation at the end of an otherwise planar zigzag system of bonds will make a significant rotatory contribution under this model only when the entire

(27a)   (27b)   (27c)

chain carries an even number of atoms (and, thus, an odd number of bonds). A similar expectation, based on quite different principles, is anticipated from a molecular orbital treatment of saturated compounds (next section).

ii. A qualitative molecular orbital version. In the molecular orbital model of structure, pairs of electrons are considered to be spread over the entire molecule in large and often diffuse orbitals. The Hückel model for π systems is familiar to most organic chemists; a number of theoreticians have proposed that this model can be extended to saturated, or σ, systems[91-93]. This concept is not easily implemented in complex compounds and finds relatively little application, but pictorial representations of these orbitals provide an important alternative way of looking at organic structures[94]. What is especially important from our point of view is the notion that the electrons that are excited by ultraviolet light or perturbed by visible light are not localised in bonds but may be spread through extensive chiral structures. The chiroptical properties of saturated compounds then become those that might be expected of *inherently chiral chromophores*. We would stress here the important conclusion that especially strong through-bond interactions occur in planar zigzag chains[93]. Such units are conspicuous features of some of the most favored conformations of chain compounds and of polycyclics composed of fused chair-form cyclohexane rings; although not chiral as such, they might behave as more highly polarizable segments of skew conformational units such as **28a** and **28b**. Planar zigzag chains appear to play an important role in determining the intensity of ketonic n-π^* absorption bands[95] and the sign and magnitude of the Cotton effects associated with those transitions[96].

(28a)   **(28b)**

c. Toward a more realistic conformational analysis. For (S)-3-methylhexane, each of the bonds $C_{(2)}$—$C_{(3)}$, $C_{(3)}$—$C_{(4)}$ and $C_{(4)}$—$C_{(5)}$ would have three staggered conformations that should be taken into account (Figure 3). Those conformations can be identified by the twist they put in the main chain (T = *trans*, P = Plus *gauche*, M = Minus *gauche*) or by the sign of the net rotational contribution ($+$, $-$, 0) each would be expected to make (see Table 7, Section III.B.2.a). (Note that the two notations are not interchangeable; thus T can be $+$, $-$ or 0, depending on what other groups are present. The TPM notation is noncommittal as to models of optical activity and is preferable as a purely descriptive notation. Where applicable, the $+ -0$ notation is useful in summing up rotatory contributions.) There are twenty-seven conformers that result from different combinations of these bond conformations; they may be identified and distinguished by symbols representing their bond conformations, reading along the chain in the customary way. Thus, the conformer shown at the top of Figure 3 would be identified as TTT or ($-$, $+$, 0) the latter revealing that the expected rotation of this species would be expected to be zero.

In the original Conformational Dissymmetry Model (above), the digauche conformations **20c** and **21** were considered to be 'forbidden'; all others were equally 'allowed'. It would be more realistic to consider that conformers are neither forbidden nor allowed,

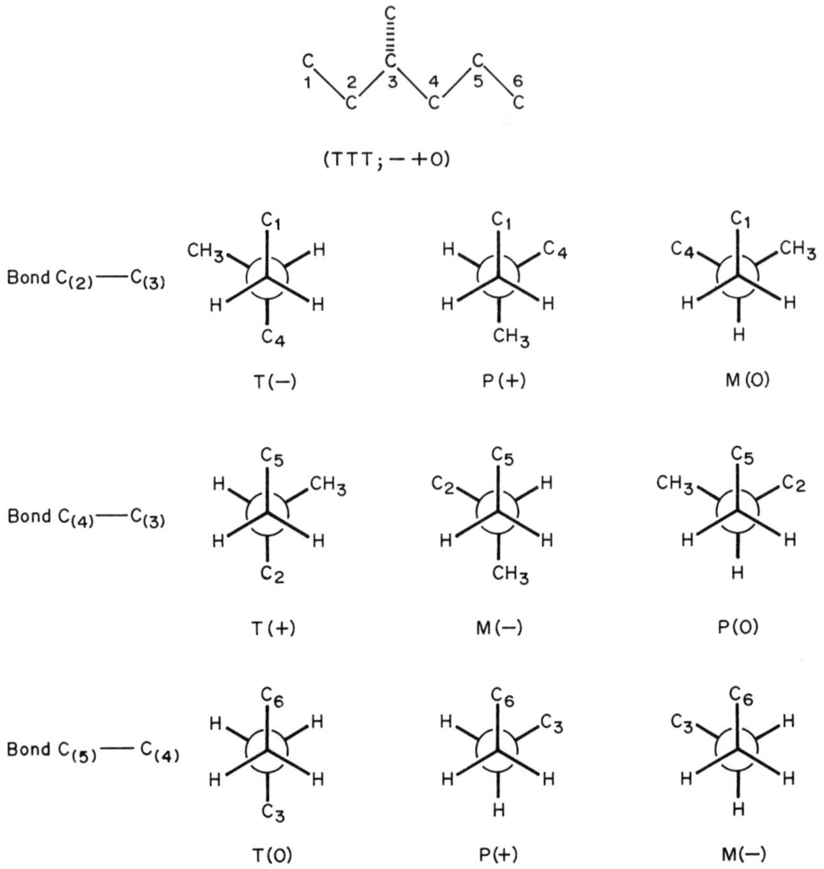

(TTT; − +0)

FIGURE 3. Bond conformations for (S)-3-methylhexane

but simply to have different energies and, therefore, to be present in different amounts. The required conformational census can be carried out by considering that every conformer (mol fraction n_a, n_b, etc.) is in direct equilibrium with the most stable one (mol fraction n_0), whence

$$K_a = n_a/n_0; \qquad K_b = n_b/n_0; \qquad K_0 = 1$$

Since $K_a \times n_0 + K_b \times n_0 + \cdots + K_0 \times n_0 = 1$, we have

$$n_0 = 1/(K_a + K_b + \cdots + 1)$$

$$n_a = K_a \times n_0; \qquad n_b = K_b \times n_0$$

If every conformer in a molecule without symmetry is counted, there will be no entropy terms to contend with. With compounds having rotational symmetry it is necessary to determine which sets of corresponding conformers can arise in different ways. Thus, both of the conformers PTTTT and TTTTP should be considered to be in equilibrium with TTTTT (see Scheme 1). Corresponding rotations about bonds $C_{(2)}$—$C_{(3)}$ and

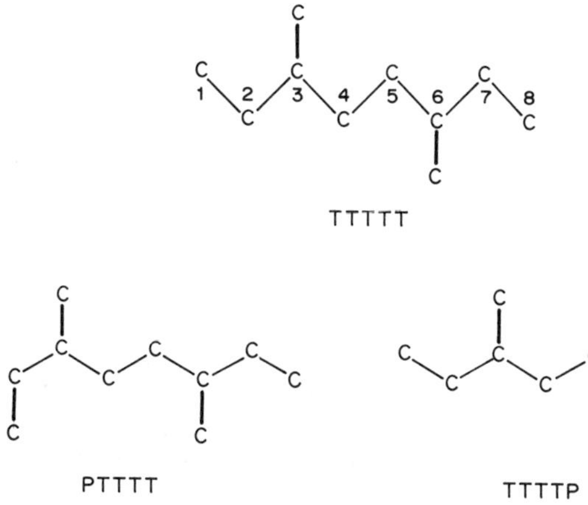

TTTTT

PTTTT   TTTTP

SCHEME 1

$C_{(6)}$—$C_{(7)}$ produce the same conformer, which therefore has a doubled probability. In such cases a branching diagram showing all possible ways of attaining each conformer will take all symmetry effects into account. Rotations of methyl groups will be ignored. Then, K_a can be calculated from enthalpy differences:

$$\Delta H = \Delta G = -RT \ln K = -2.303 \, RT \log K \qquad (16)$$

with $R = 1.9872 \, \text{cal deg}^{-1} \text{mol}^{-1}$, $T = 298 \, \text{K}$ and $\Delta H = -1.3638 \log K_a$, whence

$$K_a = 10^{-\Delta H/1.3638}$$

Some years ago a computer-assisted continuum analysis was made using units such as those shown in Figure 3 but elaborated to include substituents on the terminal atoms treated as rigid rotors. Enthalpies, and thus probabilities, were estimated for clusters ('septs') of bond conformers lying within conformational energy wells, with torsion angles varied in steps of 1°. These, in turn, were used in calculations of probabilities for sets of conformers formed by combinations of 'septs'. The rotatory contribution of each member of each sept can be calculated with the aid of equation 14 and adjusted for probability. These calculations[97], and others like it[98], gave good results with simple alkanes, but are clearly too cumbersome for general use. A simpler calculation, taking into account only those conformers with normal torsion angles, gave comparable results[97].

This method is still not fully realistic, since it does not allow for relaxation of the more congested conformers and thus tends to minimize their contribution. We have used PC Model (4.0)[99] to calculate MMX energies and hence populations for all 27 conformers of (S)-(+)-3-methylhexane. These results are shown in detail in Table 14; the conformational results are compared with those obtained by the previous methods (Table 15). [It is an interesting sidenote to these calculations that those systems containing the *meso*-digauche conformation **21** can relieve some of the van der Waals strain by twisting in either of two directions to assume torsion angles of about 90°. In one of the conformers of (S)-3-methylhexane (+, −, 0) the relief of strain is substantial but the abundance of that conformer is still only 0.0140.]

An approximate analysis is still needed. It is seen from the results presented in Table 15 that it will remain useful to 'forbid' the *meso*-digauche bond conformation **21**, since

J. H. Brewster

TABLE 14. MMX energies (kcal mol^{-1}) and mol fraction $(n_x)^a$ of individual conformers of (S)-(+)-3-methylhexane

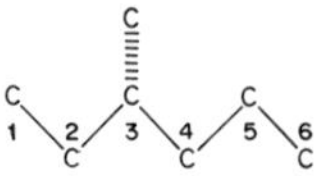

Sign of bonds		Sign of bond $(C_{(2)}-C_{(3)})$		
$C_{(3)}-C_{(4)}$	$C_{(4)}-C_{(5)}$	+	−	0
+	0	6.22 (0.1500)	6.22 (0.1500)	6.64 (0.0738)
−	0	7.61 (0.0144)	6.24 (0.1450)	6.50 (0.0935)
0	0	6.50 (0.0935)	6.67 (0.0702)	9.00 (0.0014)^b
+	+	6.85 (0.0518)	6.92 (0.0460)	7.51 (0.0170)
−	+	Not a minimum	8.66 (0.0025)	8.95 (0.0015)
0	+	8.91 (0.0016)	9.04 (0.0013)	11.16 (0.00003)^b
+	−	8.59 (0.0027)	8.63 (0.0026)	9.05 (0.0013)
−	−	8.54 (0.0030)	6.90 (0.0476)	7.23 (0.0273)
0	−	9.11 (0.0011)	9.14 (0.0011)	12.09 (0.00001)^b

^{a}Values in parentheses are mol fractions.
^{b}Not shown in Table 15.

the twelve conformers containing that unit together make up about 3.3% of the mixture. All other conformations would be allowed, but each additional *gauche* interaction beyond the two that must be present would reduce the probability by a fraction corresponding to an equilibrium constant $K_{g/t}$ (*gauche/trans*). In the original Conformational Dissymmetry Model, $K_{g/t} = 1$, since all 'allowed' conformations are considered to be present in equal amounts. PC Model calculations[99] give enthalpy differences for n-butane: $\Delta H = 0.87$ kcal mol^{-1}, $K_{g/t} = 0.230$; for 2-methylbutane: $\Delta H = 0.74$ kcal mol^{-1}, $K_{g/t} = 0.287$; and for 2,3-dimethylbutane: $\Delta H = 0.15$ kcal mol^{-1}, $K_{g/t} = 0.776$, presumably reflecting values input, since vapor phase Raman results give similar values[100]. On this basis, $K_{g/t}$ may be approximated as 1/4, except when tertiary atoms are adjacent, in which case $K_{g/t} = 1$. Calculations made on this basis ('$K_{g/t} = 0.25$') are shown in Table 15. They accord well with the rigid-rotor 'sept' calculations, but less well with those made directly by use of the PC Model. Those latter are better approximated by letting $K_{g/t} = 0.50$, which is more permissive of highly twisted chains (Table 15). This approximation seems to be more useful; it is illustrated in detail in the branching diagram for (S)-(+)-3-methylhexane shown in Table 16. Here the most stable conformers have only two *gauche* interactions. They are assigned a relative abundance of 1. With $K_{g/t} = 1/2$, those having three *gauche* interactions would have a relative abundance of 1/2 while those with four would have an abundance of 1/4. For the twelve conformers allowed, the sum of the relative abundances is 7, whence the individual abundances are 1/7, 1/14

TABLE 15. Comparison of methods for the conformational analysis of (S)-$(+)$-3-methylhexane

Conformer bonds 2-3	3-4	4-5	Gauche inter-actions[a]	Approximate methods[b] $K_{g/t} = 1$[70]	$K_{g/t} = 0.5$	$K_{g/t} = 0.25$	Computer methods Sept[90]	PC Model[99]
−	+	0	2	0.1667	0.1429	0.2051	0.1904	0.1500
		+	3	0.1667	0.0714	0.0513	0.0610	0.0460
		−	*				0.0004	0.0026
+	+	0	2	0.1667	0.1429	0.2051	0.1904	0.1500
		+	3	0.1667	0.0714	0.0513	0.0610	0.0518
		−	*				0.0004	0.0027
0	+	0	3		0.0714	0.0513	0.0502	0.0738
		+	4		0.0357	0.0128	0.0161	0.0170
		−	*				0.0001	0.0013
−	−	0	2	0.1667	0.1429	0.2051	0.1904	0.1450
		+	*				0.0004	0.0025
		−	3	0.1667	0.0714	0.0513	0.0610	0.0476
+	−	0	*				0.0015	0.0144
		+	*				0	0
		−	*				0.0004	0.0030
0	−	0	3		0.0714	0.0513	0.0573	0.0935
		+	*				0.0001	0.0015
		−	4		0.0357	0.0128	0.0200	0.0273
−	0	0	3		0.0714	0.0513	0.0462	0.0702
		+	*				0.0001	0.0013
		−	*				0.0001	0.0011
+	0	0	3		0.0714	0.0513	0.0519	0.0935
		+	*				0.0002	0.0016
		−	*				0	0.0011

[a] * contains one or more interactions of the type 21.
[b] The approximate methods are discussed in detail in the immediately following pages.

and 1/28 (Tables 15 and 16). The rotation of each conformer is shown in terms of units of $\pm k$ for each four-atom unit. There is a general cancellation of rotations but a net excess of $k/14$. Taking a value $k = 60°$, this gives a rotation of the correct sign, but about half the magnitude ($[M]_D = +4.3°$ calcd.) of that observed ($+9.9°$). Two of the allowed conformers ($+ +0$, 1/7; $- + +$, 1/14) contain the six-atom monogauche chain conformation, PTT (27c), and one ($0 + 0$, 1/14) contains the enantiomeric group, MTT. If these units make up the difference between observed and calculated rotations, then PTT must contribute 30–40° of rotation to any conformer in which it is found. A similar effect is observed in the steroids (see Section IV.B). A similar treatment of (S)-3-methyl-heptane shows that 31 of a possible 81 conformers are 'allowed', with individual abundances of 1/13, 1/26, 1/52 and 1/104. Again there is an extensive cancellation of rotatory power but a net excess of $+k/13$ or $+4.6°$ ($[M]_D + 11.4°$, obsd). There are 23 six-atom monogauche conformations for a net rotatory effect of $(2.75)/13 \times 35°$, or $+7.4°$, for a total of $+12°$ calcd. Additional calculations have been made in this way (four-atom units only!) for a number of chiral alkanes (Table 17). The correct sign of rotation is obtained in every case, including the somewhat surprising one of 3-methyl-4-ethylheptane, where the (S) isomer is dextrorotatory. It is clear that this model is only approximate when it comes to predicting magnitude of rotation, although it does identify those with large $[M]_D$. These discrepancies may be related to the question of whether this approximate model can deal with the ways in which highly congested molecules can adjust to steric strain.

J. H. Brewster

TABLE 16. (S)-$(+)$-3-Methylhexane

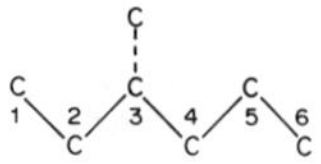

| Conformer at bond: | | | Abundance[a] | | |
C$_{(2)}$—C$_{(3)}$	C$_{(3)}$—C$_{(4)}$	C$_{(4)}$—C$_{(5)}$	relative	net	Rotation[b]
			1/2	1/14	$+3k$
			1	1/7	$+2k^{+c}$
			1/2	1/14	$+1k$
			1/2	1/14	$+1k^{2c}$
			1	1/7	0
			1/2	1/14	$-3k$
			1	1/7	$-2k$
			1/2	1/14	$-1k$
			1/4	1/28	$+2k$
			1/2	1/14	$+k^{-c}$
			1/4	1/28	$-2k$
			1/2	1/14	$-1k$
			$\sum = 7$	1/14	$\sum = +1k$

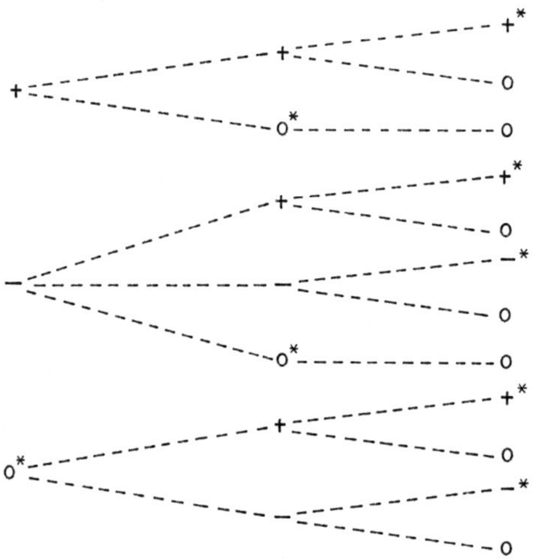

[a] $K_{g/t} = 1/2$.
[b] $k = \Delta[M]_D = 60°$.
[+c] PTT (27c) present; [-c] MTT (27c) present.

d. *Deuteroalkanes.* A relationship was suggested between rotatory power and atomic refraction[70] (A). The molar refraction for deuterium (1.004) is somewhat less than that

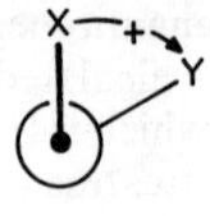

$$+kXY = 160 R_x^{1/2} \cdot R_y^{1/2}$$
(A)

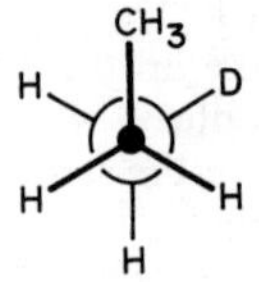

(B)

for hydrogen (1.028)[70]. Using these values with that for a saturated carbon atom (2.591), the rotatory contribution of the basic deuterium-containing unit (B) can be estimated:

$$\begin{aligned}
[M]_D &= k(C—H)(D—H) \\
&= 160(2.591^{1/2} - 1.028^{1/2})(1.004^{1/2} - 1.028^{1/2}) \\
&= -1.13°
\end{aligned}$$
(17a)

TABLE 17. Calculated molecular rotations for alkanes ($K_{g/t} = 1/2$; $k(C-H)^2 = 60°$) (**21** forbidden)

Structure: (S) $R^2 \cdots\!\!\!-\!C\!-\!\!\!\cdots R^1$ with H above and R^3 below.

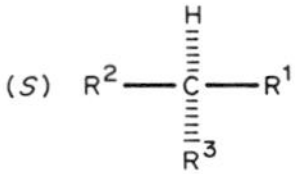

R^1	R^2	R^3	'Allowed' conformers	$\sum$ Relative abundances	$\dfrac{[M]_D}{k(C-H)^2}$	Calcd. $[M]_D$	Obsd.
n-C_3H_7	C_2H_5	CH_3	12	7	1/14	+4.3°	+9.9°
n-C_4H_9	C_2H_5	CH_3	31	13	1/13	+4.6	+11.4
n-C_5H_{11}	C_2H_5	CH_3	74	23	15/184	+4.9	+13.3
$CH_2CH(CH_3)_2$	C_2H_5	CH_3	5	4	1/4	+15	+21.3
$CH_2CH(CH_3)_2$	n-C_3H_7	CH_3	8	11/2	2/11	+10.9	+14.9
$CH_2CH(CH_3)_2$	n-C_4H_9	CH_3	21	41/4	7/41	+10.2	+11.9
$CH(CH_3)_2$	C_2H_5	CH_3	5	9/2	-2/9	-13.3	-26.4
$CH(CH_3)_2$	n-C_3H_7	CH_3	12	17/2	-3/17	-10.6	
$CH(C_2H_5)_2$	$CH_2CH(C_2H_5)_2$	CH_3	16	10	-5/4	-75	-60.2
$CH(C_2H_5)_2$	C_2H_5	CH_3	7	9/2	+2/9	+13.3	+15
$C(CH_3)_3$	C_2H_5	CH_3	1	1	-1	-60	-49.8
$C(CH_3)_3$	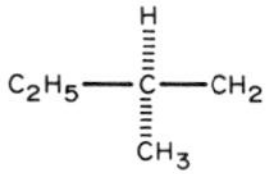 $C_2H_5\text{---}\overset{CH_3}{\underset{\;}{C}}\text{---}CH_2$ (H wedge)	CH_3	2	3/2	-8/3	-160	-97.5
$C(CH_3)_3$	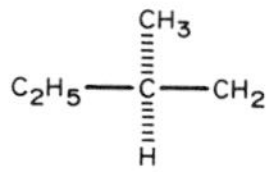 $C_2H_5\text{---}\overset{CH_3}{\underset{\;}{C}}\text{---}CH_2$ (H)	CH_3	3	5/2	-2	-120	-137.8
$C(CH_3)_3$	$(C_2H_5)_2CH\text{---}CH_2$	CH_3	4	5/2	-3	-180	-132
$C(CH_3)_3$	$(CH_3)_2CH\text{---}\overset{CH_3}{\underset{\;}{C}H}\text{---}CH_2$	CH_3	1	1	-3	-180	-79.8
$C(CH_3)_3$	$(CH_3)_2CH\text{---}\overset{CH_3}{\underset{\;}{C}}\text{---}CH_2$ (H)	CH_3	3	3	-2	-120	-119.1

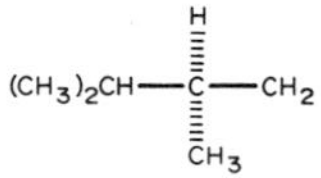

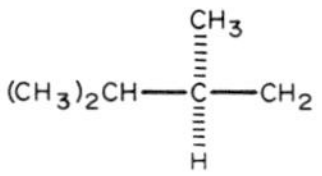

These expressions can be manipulated algebraically:

$$\begin{aligned}
k(\text{C—H})(\text{C—D}) &= k(\text{C—H})[(\text{C—H}) - (\text{D—H})] \\
&= k(\text{C—H})^2 - k(\text{C—H})(\text{D—H}) \\
&= 60° + 1.13° = 61.1°
\end{aligned} \tag{17b}$$

and

$$k(\text{C—D})^2 = \frac{[k(\text{C—H})(\text{C—D})]^2}{k(\text{C—H})(\text{C—H})} = 62.28° \tag{17c}$$

while

$$k(\text{D—H})^2 = \frac{[k(\text{C—H})(\text{D—H})]^2}{k(\text{C—H})(\text{C—H})} = +0.021° \tag{17d}$$

These values, and $K_{g/t} = 1/2$, have been used to estimate the rotations of the deutero compounds shown in Table 3 (Table 18). The results are in reasonable agreement with those observed. Rotations for other flexible deuterium compounds were estimated in a similar way[101].

These calculations implicitly impute these rotational effects to *electronic* transitions in the far ultraviolet, presumably for the C—H and C—D chromophores. A controlling role has been proposed for C—H bonds in the CD of cyclopentanones[102] and it may well be that the contributions of those bonds have been slighted here and elsewhere. Against this view, Fickett[103] has calculated a rotation of $[\text{M}]_\text{D} + 0.64°$ for *R*-2-d-butane *due to vibrations of the C—H and C—D bonds*, on what amounts to an atomic asymmetry approach using Kirkwood theory. However, ORD studies of several deuterium compounds give simple plain curves, *increasing as wavelength decreases*[52,53], suggesting strongly that this should be considered a case of electronic, rather than vibrational, optical activity.

TABLE 18. Molecular rotations of deuteroalkanes

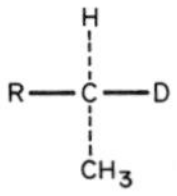

(**21** forbidden; $K_{g/t} = 1/2$; $k(\text{C—H})(\text{D—H}) = -1.13°$)

R	Allowed conformations	Σ Relative abundances	$[\text{M}]_\text{D}$	Calcd.	$[\text{M}]_\text{D}$ Obsd.[a]
C_2H_5	3	2	$\frac{1}{4}k(\text{C—H})(\text{D—H})$	$-0.28°$	$-0.36°$
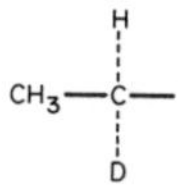	3	2	$\frac{1}{4}[k(\text{C—H})^2 - k(\text{C—D})^2]$	-0.57	-0.61
$CH(CH_3)_2$	3	5/2	$\frac{1}{5}k(\text{C—H})(\text{D—H})$	-0.23	-0.62
$n\text{-}C_3H_7$	7	7/2	$\frac{5}{14}k(\text{C—H})(\text{D—H})$	-0.40	-0.38
$C(CH_3)_3$	1	1	0	0	-0.075

[a]See Table 3.

IV. CYCLOALKANES

Cyclic compounds with three to six members are relatively rigid, compared to the chain compounds discussed above. Conformational analysis is much more simple, except perhaps for cyclopentanes. Bridged and cage compounds may display unusual torsion angles and contain long twisted chains that would be expected to produce large rotations. The detailed analysis of many such systems is complex and not definitive enough to present at this time. Some of the concepts currently being worked on seem worth illustrating and are offered for consideration here.

A. Monocyclic Compounds

(S, S)-$(+)$-1,2-d_2-Cyclopropane (29) has been prepared by a route that allows assignment of absolute configuration and estimation of the maximum rotation[104]. A special pressure polarimeter tube ($l = 1$ dcm), used in a cold room at 3 °C, allowed measurements of rotation to be made with the neat liquid in the near ultraviolet: $\alpha_{365}^{max} = +0.298°$. As shown above, a value of $\Delta[M]_D + 0.021°$ can be calculated for $k(D—H)(D—H)$. The dihedral angle here (145°; PC Model[99]) would reduce this value considerably: $0.021° \times \sin 145°/\sin 60° = 0.014°$, but the change in wavelength of light would increase it strongly. The optical rotation at 365 nm was large enough to allow a study of thermal stereomutation[104]. It has since been shown that VCD is better because it allows an accurate estimate of mixtures of chiral and racemic *trans* isomers[105]. Measurements of vibrational optical activity have also been made with chiral 2,3-d_2-oxirane[106], (R)-$(+)$-methyloxirane[107] and *trans*-2,3-dimethyloxirane[108]. These studies have provided useful insights into the structures that contribute to VOA.

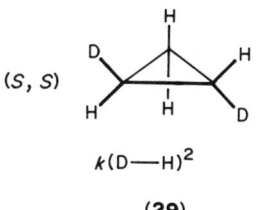

$$k(D—H)^2$$

(29)

(R, R)-$(-)$-1,2-Dimethylcyclopropane (31) has been prepared (equation 18) from (S)-$(-)$-2-methylbutanol (30) by a process that establishes the absolute configuration[109]. A mixture of products was obtained, from which the *trans* isomer could be isolated pure by glc; $[M]_D - 32.2°$. Repetition of this preparation gave a product with $[M]_D - 29.9°$ [110]. However, this hydrocarbon has also been prepared from $(-)$-*trans*-1,2-cyclopropane dicarboxylic acid (32) of 9.3% enantiomeric purity; the hydrocarbon product had a rotation which, assuming that no racemization had occurred, indicated a maximum rotation of only $-18°$ [111]. The ORD and CD of $(-)$ dimethylcyclopropane show that the first Cotton effect, at 185 nm, is positive[110]. There apears to be no absorption band in this region, only end absorption from stronger bands at lower wavelengths. Deeper penetration into the UV (to *ca* 150 nm)[19] reveals two more Cotton effects of opposite sign. Although it is not clear that they are strong enough to control rotation in the visible, *some* band in the far UV must be strong enough to outweigh the first Cotton effect in influence. None of these CD measurements approach the domain of the 'resonance giant' observed in the UV spectrum at about 115 nm[18].

$(2R, 3R)$-1,1,2,3-Tetramethylcyclopropane (35) (note change in CIP rank) has been prepared (equation 19) from methyl $(+)$-*trans*-chrysanthemate (34)[112], as has the dicarboxylic acid 33[113]. The visible rotation for 35 was not reported, but a strong positive

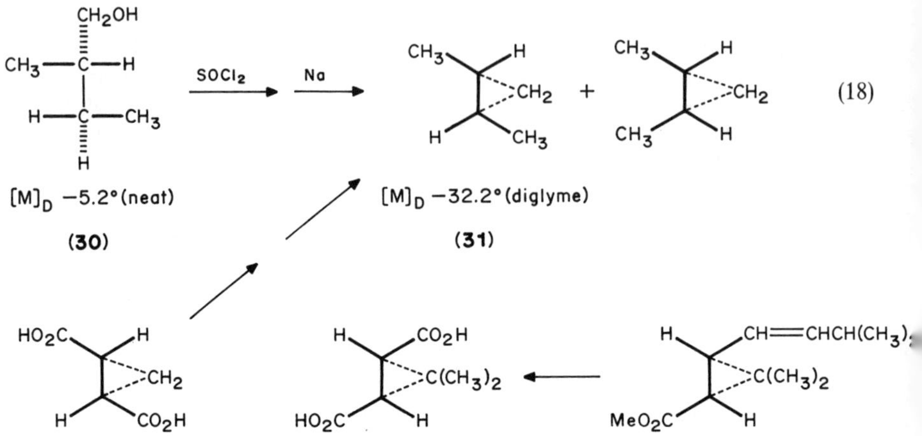

$[M]_D$ −5.2° (neat)

(30)

$[M]_D$ −32.2° (diglyme)

(31)

(−) (−) (+) **(34)**

$$[M]_D = -k(CO_2H{-}H)^2 = -135^\circ \text{ (calcd)} = -406^\circ \text{ (obsd)}^{111}$$

(32)

$$[M]_D = +k(CO_2H{-}H)^2 - 2k(CO_2H{-}H)(CH_3{-}H) = -55^\circ \text{ (calcd)} = -53^\circ \text{ (obsd)}^{113}$$

(33)

Cotton effect was observed at about 185 nm. S. F. Mason[112] pointed out that this might reflect interactions of the isolated methyl groups with the geminal pair, which should be approximately twice as strong as those of the isolated methyl groups with one another, but opposite in sign. A similar interpretation of the corresponding dichotomy in the visible rotations of the *trans* acids, **32** and **33**, was proposed earlier[75].

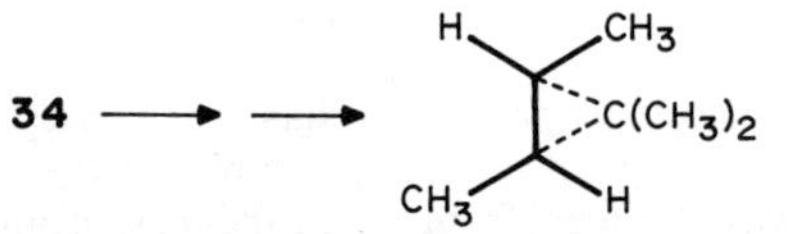

(35)

To apply the Conformational Dissymmetry Model to **31**, the bond angle and torsion angle corrections of the Helix Model are required:

$$[M]_D = \frac{-60^\circ \times (\sin 117)^2 \times (\sin 143)}{(\sin 109.28)^2 \times (\sin 60)}$$
$$= -37^\circ \text{ (Obsd.: } -18^{\circ\,111},\ -29^{\circ\,110},\ -32^{\circ\,109})$$

At a higher computational level, the Kirkwood model[67,12] gave $[M]_D + 11.8^\circ$ for the enantiomer. At a still higher level, Random Phase Approximation calculations were used to obtain an extensive catalog of transitions[114] for (R,R)-(−)-**31**, which reproduced the observable absorption and CD bands fairly well. There is an extensive clustering of transitions in bands but the lowest-energy Cotton effect should be positive and the next, negative. The deeper bands could not be sorted out.

The VCD of nonracemic **31** has been interpreted as arising from coupled C—H vibrations[115].

Nonracemic *trans*-1,2-dimethylcyclopentane (**36**) and -cyclohexane (**37**) have been prepared from (−)-*trans*-4-cyclohexene-1,2-dicarboxylic acid via the dimethylcyclohexene (equation 20). The latter was degraded to (−)-3,4-dimethyladipic acid, which has been correlated with (S)-(−)-2-methylbutanol (**30**), establishing the R,R-configuration of the cyclic compounds. It is believed that these substances are of known enantiomeric purity, on which basis the rotations shown are thought to be maximal. The dihedral angle between the methyl groups of **36** will, on average, be larger than that in **37**; this is reflected in the relative magnitude of rotation. It would be expected that **37** would have a rotation of $-k(\text{C—H})^2$. The lower value observed might reflect some loss of isomer identity by hydrogen exchange during catalytic reduction of the cycloalkene.

(R, R)-(+)-1,3-*trans*-Dimethylcyclohexane (**39**) was prepared (equation 21) from (R)-(+)-3-methylcyclohexanone (**38**) via Grignard reaction, dehydration and reduction[118]; the preparation has been confirmed[119]. The fact that the less stable of the diastereomeric 1,3-dimethylcyclohexanes could be prepared in optically active form (and was therefore *trans*) was an early clue to the preference of cyclohexane for the chair conformation. Both the simple four-atom model and the helix model predict a rotation of zero for **39**.

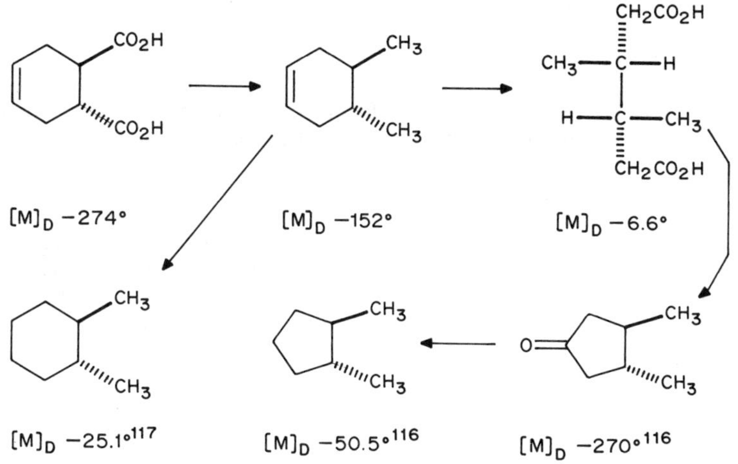

$[M]_D$ −274° $[M]_D$ −152° $[M]_D$ −6.6° (20)

$[M]_D$ −25.1°[117] $[M]_D$ −50.5°[116] $[M]_D$ −270°[116]

(**37**) (**36**)

(21)

$[M]_D$ +14.7° $[M]_D$ +1.41°

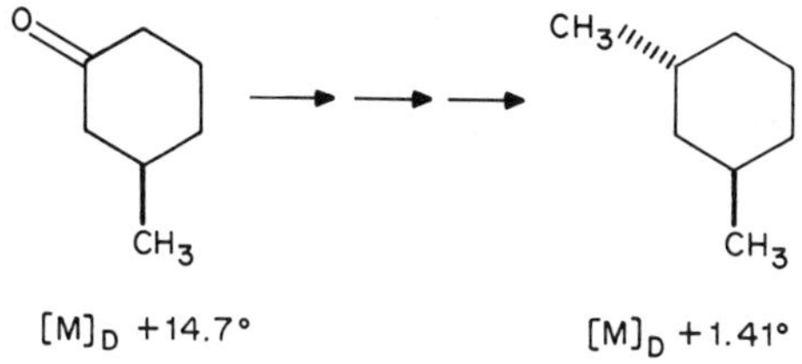

(**38**) (**39**)

The *cis*- and *trans*-*m*-menthanes (**40, 41**) have been prepared by way of thermal cleavage of *cis*- and *trans*-Δ^4-carene (equations 22 and 23)[120]. The models of optical activity so

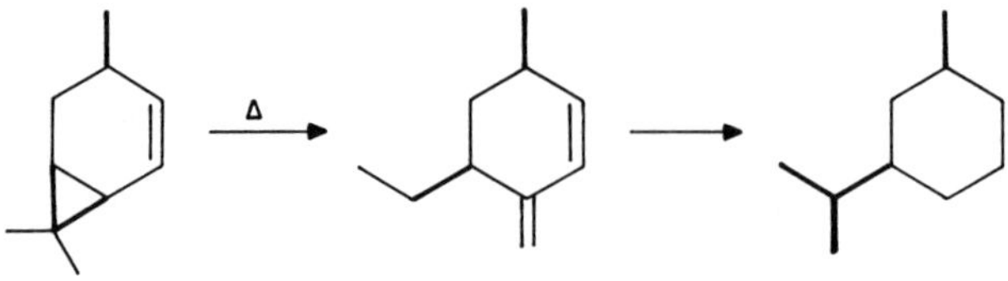

$$[M]_D -220° \qquad [M]_D -147° \qquad [M]_D +17.8°$$

(40)

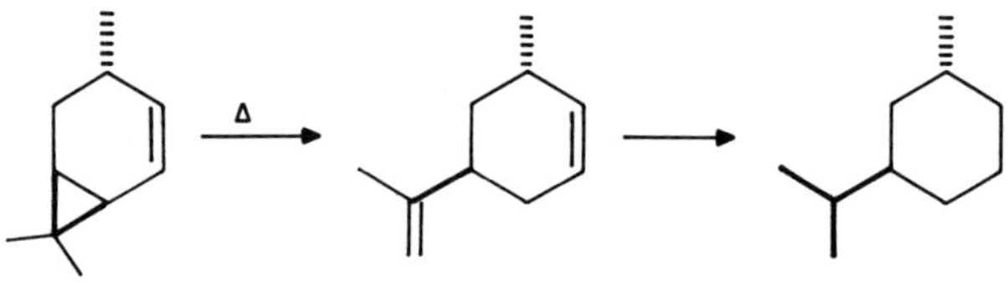

(22)

$$[M]_D -228° \qquad [M]_D -234° \qquad [M]_D +5.75°$$

(41)

(23)

far discussed would predict zero rotation for these compounds. Molecular modeling[99] shows that the isopropyl group can occupy the three staggered conformations **42** with equal ease. The six-atom monogauche conformations shown there in heavy lines are present, overall, in equal amounts so that their contributions should cancel. But if the whole molecule is taken into account, it is seen that the right-twisted chain in **42a** and the left-twisted chain in **42b** are in different environments. If this affects their rotatory power, then their effects would not cancel out. This implies that the whole notion of simple pairwise additivity[121] is an oversimplification. If the effect is small, it would show up most clearly in compounds expected to have zero rotation under simpler models.

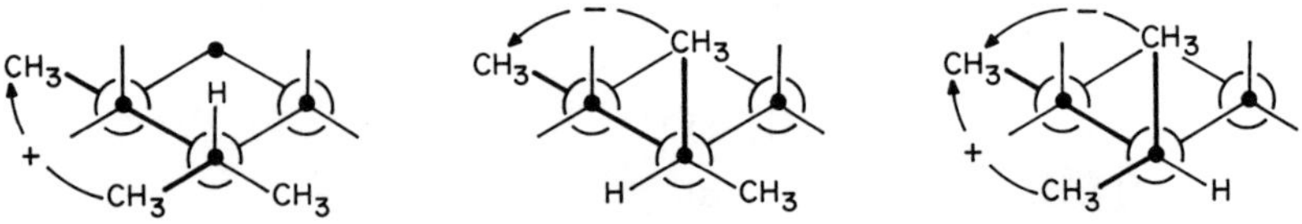

(a)   **(b)**   **(c)**

(42)

B. Valence-bridged Polycyclic Compounds

Unsubstituted cis-(m.n.0]-bicycloalkanes will be achiral as such or by equilibrium of enantiomeric conformers, as with cis-decalin (**43**), but unsubstituted trans-[m.n.0]-bicycloalkanes will be chiral if $m \neq n$. Many bicyclic compounds owe their chirality to substitution. Only a few of these can be treated in any detail.

The thujanes (cis-, **44**; trans-, **45**) are the parent hydrocarbons of a significant series of terpenes, the chemistry of which has been reviewed[122]. The bicyclic system, as such, is achiral. Molecular modeling[99] indicates that in this case the isopropyl group does not occupy all conformations equally, but resides mainly in that one in which it is

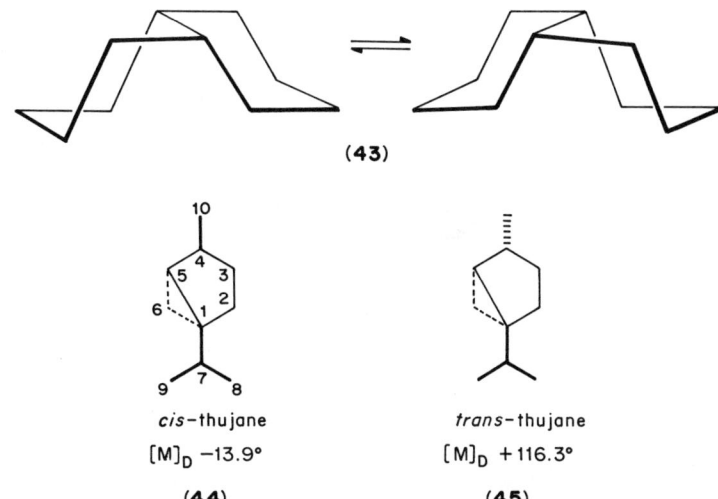

(43)

cis-thujane
[M]$_D$ −13.9°

(44)

trans-thujane
[M]$_D$ +116.3°

(45)

symmetrical relative to the cyclopropyl group (**46**). It should not make a major contribution to optical rotation. It would appear that the largest effect derives from the relationship of the methyl group to the cyclopropyl group, which in these situations may be expected to resemble an ethylenic group both in its conformational demands and as a chromophore. In *trans*-thujane, which has a large dextrorotation, the methyl group has a strong right *gauche* orientation to the cyclopropane ring (**47a**); in the *cis*-isomer, which

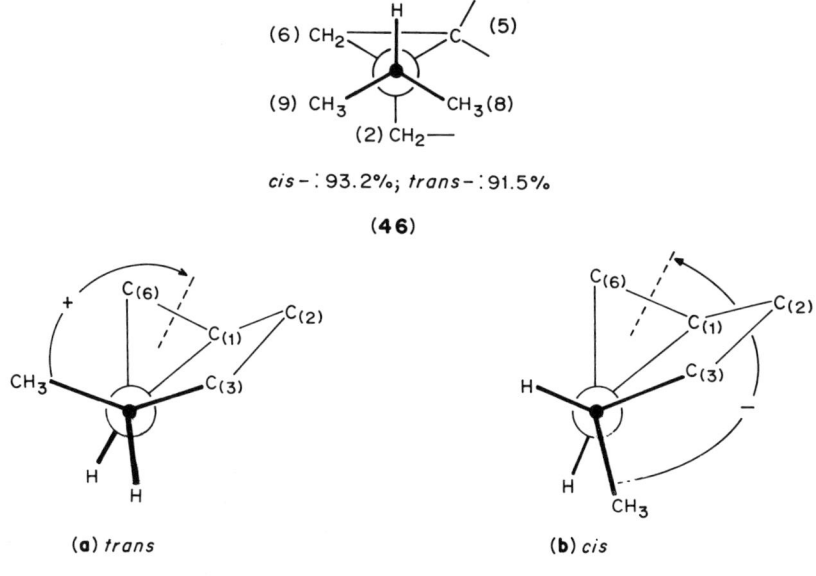

(6) CH$_2$ H   C (5)
(9) CH$_3$ CH$_3$(8)
(2) CH$_2$

cis−: 93.2%; *trans*−: 91.5%

(46)

(**a**) *trans*   (**b**) *cis*

Projections along bonds: C$_{(4)}$—C$_{(5)}$

(47)

has a small levorotation, the methyl group is *anti* and only slightly off the bisector in an obtuse left relationship (**47b**). This may be the main effect controlling sign and magnitude of rotation in these hydrocarbons.

The carane series has been extensively studied and the parent hydrocarbons have been prepared. (+)-Car-3-ene (**48**) was hydroboronated to give the 4-ol (**49**), the tosylate of which was reduced to give pure *cis*-carane (**50**) ($[M]_D - 52.0°$)[123] (equation 24). (−)-2-Carone (**51**) (from (−)-carvone, $[M]_D - 92°$) gave *trans*-carane (**52**) ($[M]_D + 87.6°$) on Wolff–Kishner reduction[123] (equation 25). Conformationally the [4.1.0]-bicyclo-heptane nucleus resembles cyclohexane in being puckered at $C_{(3)}$ and $C_{(4)}$. The methyl

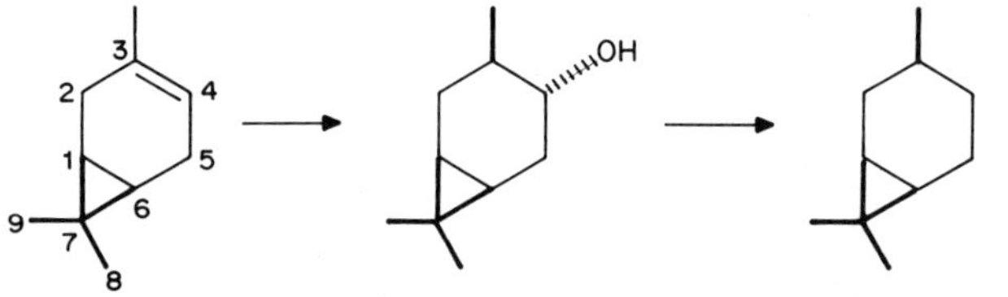

$$\begin{array}{ccc} [M]_D + 27° & [M]_D - 112.7° & [M]_D - 52.0° \\ (48) & (49) & (50) \end{array}$$

(24)

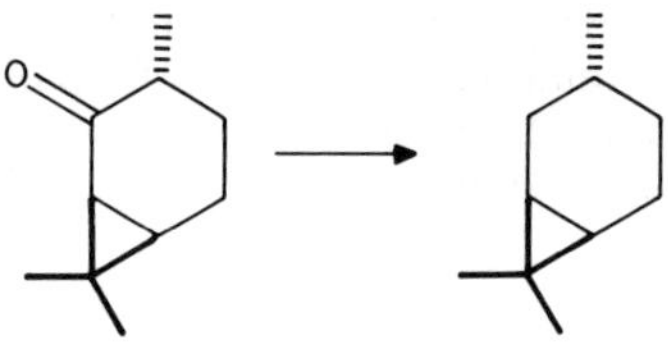

$$\begin{array}{cc} [M]_D - 225° & [M]_D + 87.6° \\ (51) & (52) \end{array}$$

(25)

TABLE 19. Conformation angles in the caranes

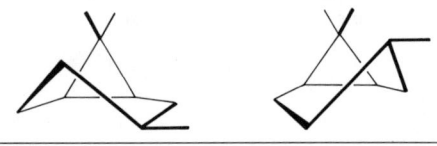

Bond conformation	Conformation angle	
	cis-Carane ($[M]_D - 52°$)	*trans*-Carane ($[M]_D + 88°$)
6–1–2–3	+32.80°	−5.21°
7–1–2–3	+102.48	+65.82
1–2–3–4	−60.76*	+34.18*
2–3–4–5	+60.61	−61.02
3–4–5–6	−34.16*	+60.89*
4–5–6–1	+5.67	−30.98
4–5–6–7	−65.00	−100.45

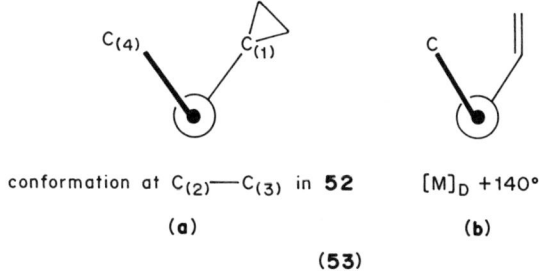

conformation at $C_{(2)}$—$C_{(3)}$ in **52** $[M]_D$ +140°

(a)   (b)

(53)

group determines the handedness of this puckering by its preference for an equatorial orientation. Ring atoms 3 and 4 are then in two kinds of relationship with the cyclopropane nucleus. PC Modeling[99] gives the torsion angles shown in Table 19. The starred values correspond to conformations of the type shown in **53**. In the corresponding situations in the cyclohexene series, it was concluded[77] that the conformations at one remove from the double bond (**53b**) were dominant. On this basis the conformers at $C_{(2)}$—$C_{(3)}$ (**53a**) (and at $C_{(4)}$—$C_{(5)}$) (see the starred items in Table 19) should be dominant in the caranes. This relationship is most easily seen by comparing *trans*-carane (**52**) with limonene (**54**).

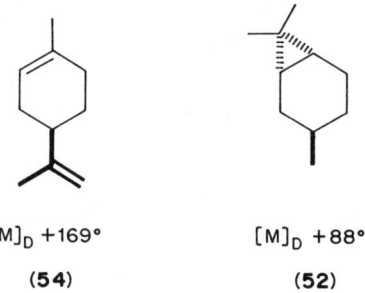

$[M]_D$ +169° $[M]_D$ +88°

(54)   (52)

The sign of the Cotton effect in cyclopropano-*cis*-decalins such as (**55**) appears to depend strongly on the twist of the cyclohexane ring, just as the sign of an ethylenic Cotton efffect would if there were a double bond in the same position in the ring[124].

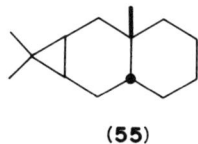

(55)

(−)-*trans*-Hydrindanone (**57**) ($[M]_D$-429° in benzene) has been prepared by pyrolysis of (+)-*trans*-1,2-cyclohexanediacetic acid (**56**) ($[M]_D$ + 96.7°)[125,126] (equation 26). **57** gives an unusually large negative Cotton effect, as do 16-ketosteroids[127]. This has been attributed to twisting of the cyclopentane skeleton or, alternatively, to a dissignate effect of quasi-axial hydrogen atoms[102]. Reduction of **57** gives (−)-2-hydrindanol (**58**) ($[M]_D$ − 17.4°; neat) and (R, R)-(−)-*trans*-hydrindane (**59**) ($[M]_D$ − 7.4°; neat). It is to be

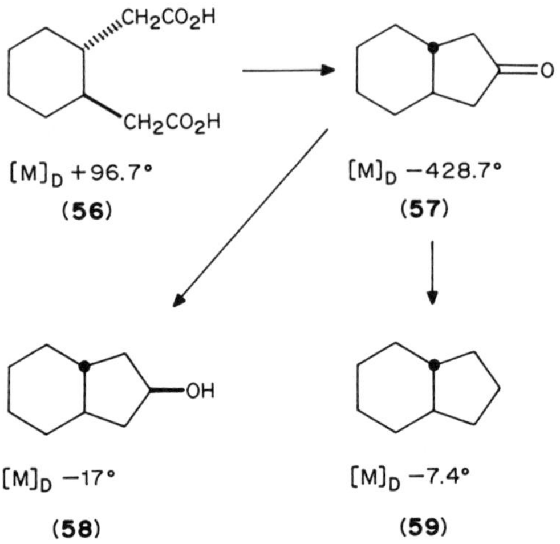

(26)

expected that the hydroxy group of **58** would have little additional rotatory effect. Molecular modeling[99] gave the torsion angles shown in Table 20. The sum of the sines of these angles is -0.2944, whence the calculated rotation is

$$[M]_D = -(0.2944/\sin 60°)k(C—H)^2 = -20.4°$$

It is noteworthy that the sum of the sines of the internal (acute) angles is almost exactly zero (-0.007). The fact that the calculated rotation is larger than that observed might suggest that obtuse angles have lower rotatory power than acute angles.

Since *trans*-decalin is achiral, the rotations of the decalols serve as a measure of the change in molecular rotation to be expected on insertion of a hydroxy group in other compounds containing *trans*-fused cyclohexane rings. The two possible equatorial

TABLE 20. Torsion angles for (R, R)-$(-)$-*trans*-hydrindane

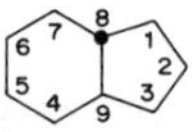

Skew unit	Torsion angle	Sine	Symmetry multiplier
6–7–8–9 = 5–4–9–8	+58.62°	0.85373	×2
6–7–8–1 = 5–4–9–3	+175.13	0.08490	×2
5–6–7–8 = 6–5–4–9	−55.00	−0.81915	×2
4–5–6–7	+54.36	+0.81269	
7–8–9–4	−61.98	−0.88278	
7–8–9–3 = 1–8–9–4	+172.28	0.13433	×2
1–8–9–3	+46.54	0.72585	
2–1–8–9 = 2–3–9–8	−37.30	−0.60599	×2
2–1–8–7 = 2–3–9–4	−158.30	−0.36975	×2
1–2–3–4 = 8–1–2–3	+14.29	0.24683	×2

secondary alcohols of this series (**60, 61**) have been prepared via resolutions expected to give optically pure materials and their configurations have been defined by their relationships to monocyclic compounds[128]. The corresponding ketones give Cotton effects as expected under the Octant Rule. The rotation of the *trans*-1*eq*-decalol (**60**) ($[M]_D + 62°$) is attributable to the right-*gauche* conformational relationship of the hydroxy group with $C_{(8)}$. The hydroxy group of the *trans*-2*eq*-decalol (**61**) is *anti* to both $C_{(4)}$ and $C_{(9)}$ and would be expected to make essentially no contribution to optical rotation ($[M]_D + 1.9°$; ethanol). These rotational shifts correspond well with those observed in the steroid series[129] (see **62**, which is configurationally related to **60**, and **63**, rings A and B of which are enantiomeric to **61**).

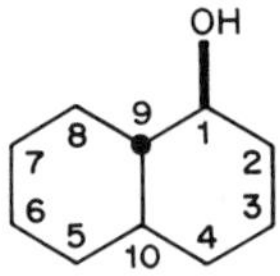

[M]$_D$ +62°

(60)

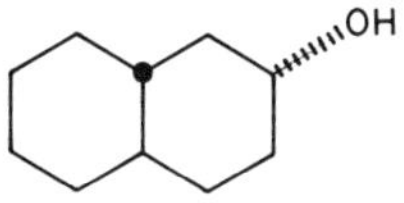

[M]$_D$ +1.9°

(61)

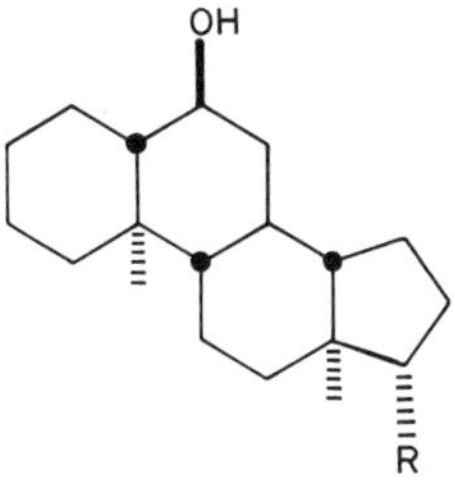

Δ[M]$_D$ +55°

(62)

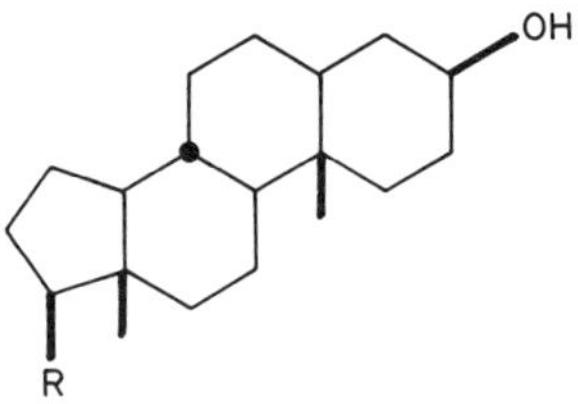

Δ[M]$_D$ −2°

(63)

When the rotational shifts of hydroxy groups in the decalols are applied to the *trans–transoid–trans*–perhydrophenanthrols (**64–68**), it becomes possible to deduce the rotatory contribution of the parent hydrocarbon (PHP) (**69**). (Note that the configuration of the ring in **67** is opposite to that in the other compounds of this series) Granting that there is considerable spread among these values, the average rotation for **69** is found to be +55.7°, which is quite close to the value $[+k(C—H)^2]$ expected from the simple four-atom analysis seen in **69**. It might be argued that **68** contains a crowded hydroxy group and should not be included in the average; if so, the average value becomes +62.4°.

Perhydrotriphenylene **70** is of interest because it forms inclusion compounds[133]. It is chiral and has been obtained in nonracemic form by peroxide decarboxylation of the acid (**71**), which could be resolved to apparent enantiomeric purity. A sample of the (+) acid (40% optical purity) was degraded to the ketone (**72**) ($[M]_D^{max} + 332°$)[134]. This showed a positive Cotton effect in the ORD, at 308–272 nm, allowing assignment of configuration by use of the Octant Rule (**72b**). A simple four-atom analysis indicates that **70** should have a rotation of $-3k(C—H)^2$ or $-180°$. In contrast with the value for *trans*-1,2-

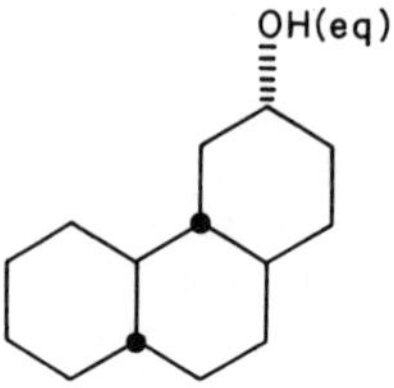

$[M]_D -79°^{130}$

PHP = -81°

(64)

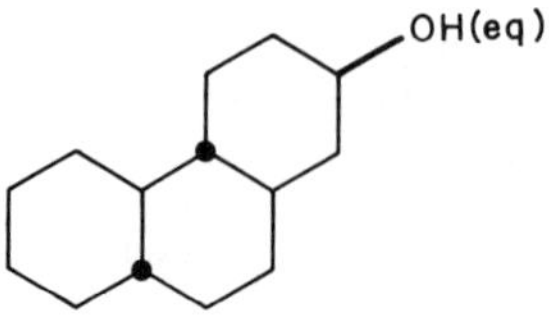

$[M]_D -56°^{131}$

PHP = -58°

(65)

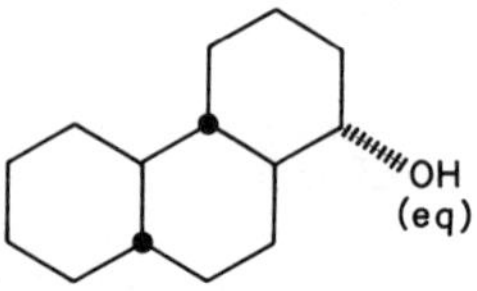

$[M]_D +10.4°^{132}$

PHP = -51.8°

(66)

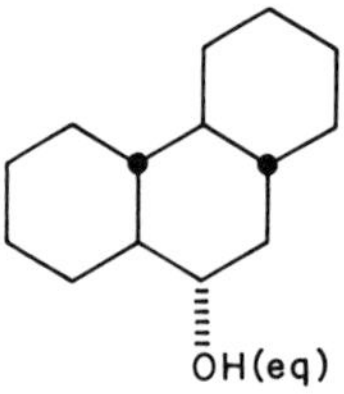

$[M]_D +120.6°^{130}$

PHP = +58.6°

(67)

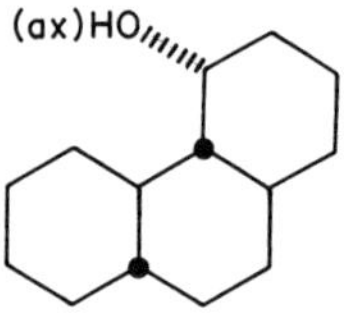

$[M]_D -91.5°^{132}$

PHP = -29°

(68)

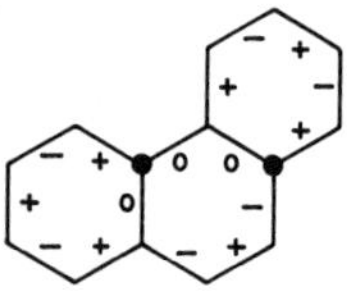

$[M]_D$ (est) +55.7°

(69)

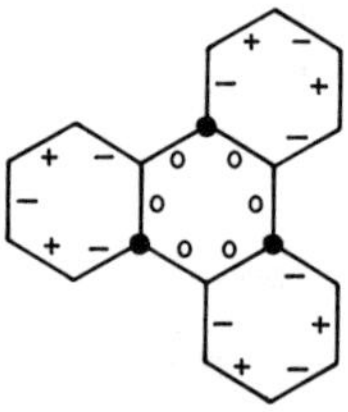

$[M]_D -229°$

(70)

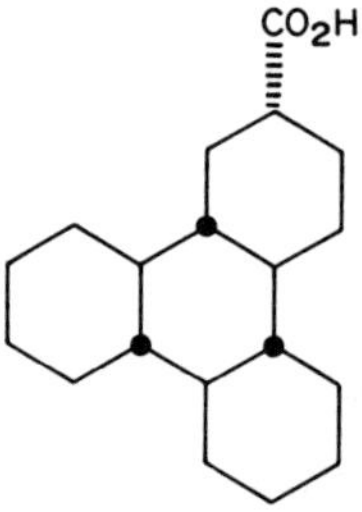

$[M]_D -216°$

(71)

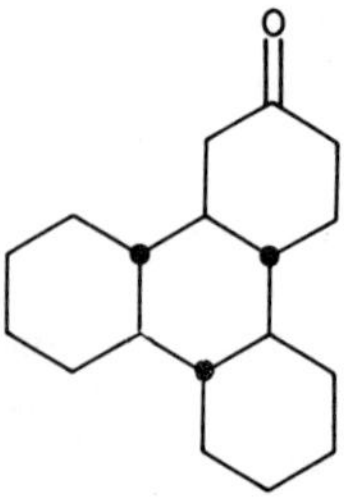

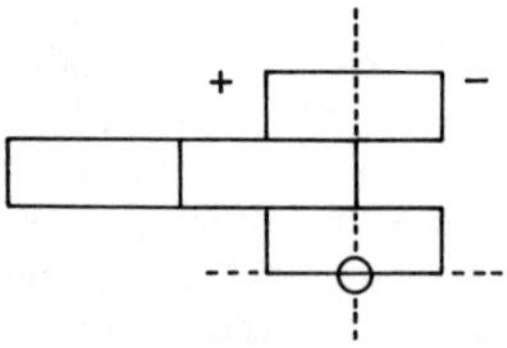

(a)

$[M]_D^{max} +332°$

(b)

Positive Cotton effect

(72)

dimethylcyclohexane (37) ($[M]_D - 25.1°$), this result suggests that the value $k(C—H)^2 = 60°$ is somewhat too small.

The diterpenes provide persuasive evidence that much remains to be learned about optical rotation as an additive property. One set of hydrocarbons in this series appears to provide useful insight into the chiroptical properties of axial methyl groups. Fichtelite (18-norabietane) (73), 19-norabietane (74) and abietane (75) differ in the number and orientation of methyl group at C_4[135]. If additivity of rotations be presumed, then the equality shown in equation 2 should hold and the molecular rotation of 18,19-norabietane (76) should be in the range $+64°$ to $+81°$. But a simple four-atom calculation (shown in 76) indicates that this nucleus should have a rotation of about zero. This discrepancy

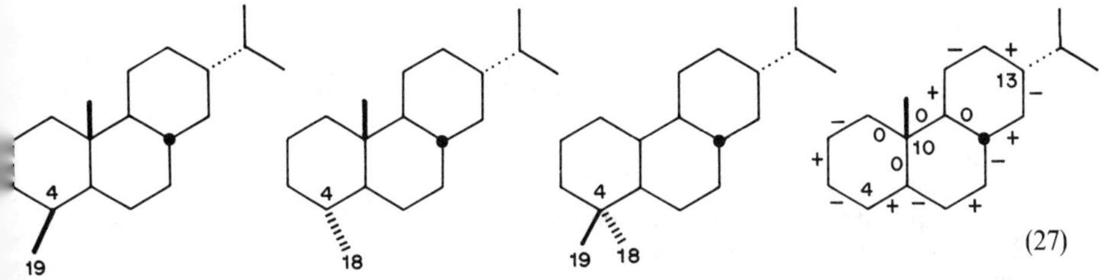

$$[M]_D +49.8° \quad + \quad [M]_D\ 0°;\ +8° \quad = \quad [M]_D -23.5°(CCl_4) \quad + \quad [M]_D\ estd$$
$$-13.8°\ (C_6H_{14}) \qquad +64° - +81°$$

(73)   (74)   (75)   (76)

(27)

does not appear to involve the equatorial methyl group ($C_{(18)}$) since it appears to make a rotatory contribution of between $-60°$ and $-70°$, which is what would be expected in the four-atom analysis. The axial methyl group ($C_{(19)}$), however, makes a contribution of about $-15°$ to $-30°$, whereas a positive contribution of about $+60°$ would have been anticipated. Preliminary PC modeling[99] does not suggest that this is the result of deformations due to 1,3-diaxial interactions. $C_{(19)}$ does make a left-*gauche* interaction with a planar zigzag chain as shown in 77, but it is required that this seven-atom system make a rotational contribution some $15°–30°$ stronger than the simple four-atom effect. In the hypothetical 18,19-bisnor-compound (76) the large rotation indicated by equation 27 would be consistent with the absence of the angular methyl group at $C_{(10)}$. This suggests that that methyl group makes little or no net contribution to optical rotation, although the four-atom approach would require it to contribute $-60°$. The molecular orbital approach

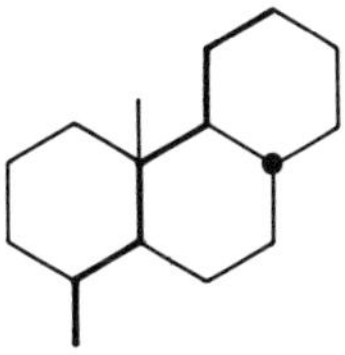

(77) 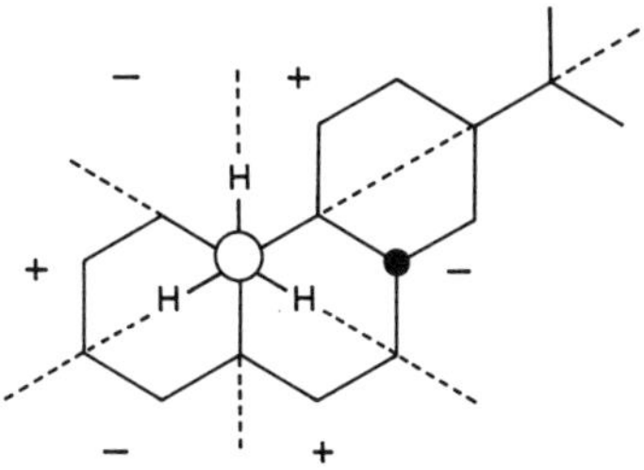   (78)

suggests that an angular methyl group might well constitute a distinctive chromophore—one that might follow a sextant rule as indicated in **78**. If so, its contribution in this way might be opposite to its four-atom contribution, in which case the sector rule would be signed as shown in **78**. It will be seen that a similar anomaly occurs in the steroid series (below). Our studies of this aspect of the rotations of hydrocarbons continue.

The steroid nucleus is uniquely rich as a source of information about the effects of structure on optical rotation. Although it is old, the compilation of Jacques, Kagan and Ourisson[136] covers the time when research on simple steroids was at its peak and will serve as a principal reference. It was in this area that the principle of additivity of molecular rotations was established[129], though a careful scrutiny of the data indicates that there are sometimes significant difficulties with quantitative aspects. The major hydrocarbons of this series are shown in Table 21. We concern ourselves here with compounds having a *trans*- A/B ring fusion, in which the ring atoms are more or less coplanar so that the bridgehead (axial) methyl groups and ring substituents (notably side chains at $C_{(17)}$) are the main variables. *Cis*-ring fusions produce major 'kinks' and 'bends' in the ring structure with which our methods do not always work well.

A preliminary analysis of the 5α-steroid nucleus indicates that an interaction of axial methyl groups with planar zigzag structures may have to be taken into account as well as the contributions they may make in simpler four-atom interactions. Thus the D-homogonane nucleus (**79**) has a center of symmetry and is, therefore, achiral. Contraction of ring D to form gonane (**80**) should be rotationally equivalent to converting *trans*-decalin (achiral, $[M]_D \, 0°$) to *trans*-hydrindane (**59**) ($[M]_D - 7.4°$). For this to be

TABLE 21. Steroidal hydrocarbons

Name	β-Substituent at $C_{(17)}$	[M]$_D$ 5α Series	[M]$_D$ 5β Series
Androstane	—	+5°	+11°
Methylandrostane	CH_3	(+28)	
Pregnane	C_2H_5	+52	+58
Dinorcholane	$CH(CH_3)_2$	+27	
Norcholane	$CH(CH_3)C_2H_5$		
Cholane	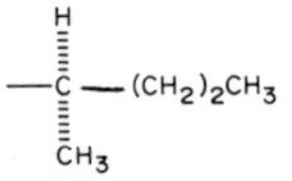 $-\overset{H}{\underset{CH_3}{C}}-(CH_2)_2CH_3$		+99
Cholestane	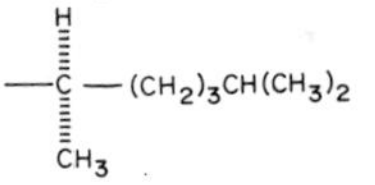 $-\overset{H}{\underset{CH_3}{C}}-(CH_2)_3CH(CH_3)_2$	+91	+97
24-$β_F$-Methylcholestane	$-\overset{H}{\underset{CH_3}{C}}-(CH_2)_2-\overset{H}{\underset{CH_3}{C}}-CH(CH_3)_2$	+66	
24-$α_F$-Ethylcholestane	$-\overset{H}{\underset{C}{C}}-(CH_2)_2-\overset{Et}{\underset{H}{C}}-CH(CH_3)_2$	+104	
24-$β_F$-Ethylcholestane	$-\overset{H}{\underset{CH_3}{C}}-(CH_2)_2-\overset{H}{\underset{Et}{C}}-CH(CH_3)_2$	+80	

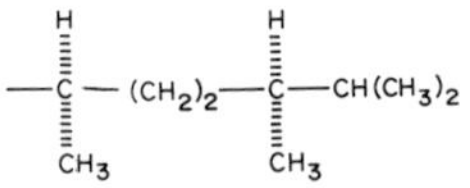

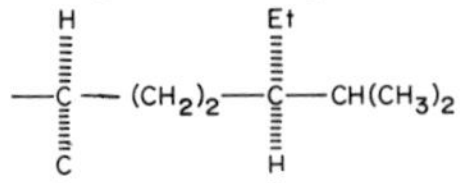

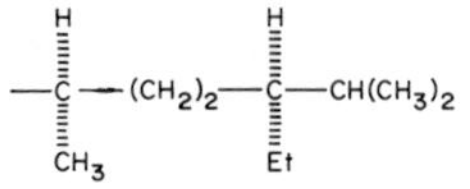

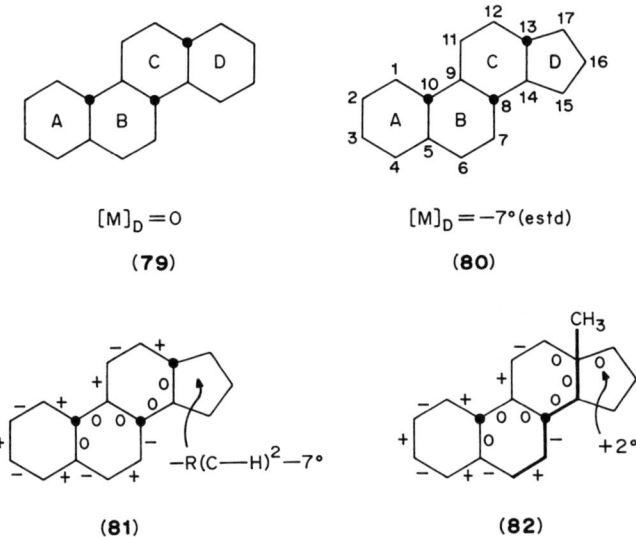

$$[M]_D = 0$$

(79)

$$[M]_D = -7° \text{(estd)}$$

(80)

$$-R(C-H)^2 - 7°$$

(81)

$$+2°$$

(82)

true, the net contribution of ring D must be $-k(C-H)^2 - 7°$, to balance the excess dextrorotation of ring A (**81**) $[+k(C-H)^2]$ (rings B and C cancel each other).

The insertion of an angular methyl group ($C_{(18)}$) at $C_{(13)}$ to form estrane (**82**) produces more or less compensatory four-atom effects along bonds $C_{(13)}-C_{(12)}$ and $C_{(13)}-C_{(17)}$, with the latter making the larger contribution because of its 90° dihedral angle:

$$\Delta M_{(13-12)} = -k(C-H)^2$$
$$\Delta M_{(13-17)} = (\sin 90/\sin 60) \times k(C-H)^2$$
$$= 1.1547k(C-H)^2$$
$$\text{Net} = +0.1547k(C-H)^2 = +9.3°$$

The net effect of this substitution, taking $k(C-H)^2 = 60°$, would be a dextro shift of 9.3°, giving **82** an estimated rotation of $+2°$. In fact, a much stronger shift is observed, as may be deduced by employing the method of molecular rotations[129] (Table 22). The difference between the observed and expected rotation ($+42°$) must be attributed to effects going beyond those due to four-atom groups. The planar zigzag chain running from $C_{(6)}$ to $C_{(13)}$ is right-*gauche* to $C_{(18)}$ and so might provide a source for the extra dextro-rotation. Any contribution from $C_{(18)}$ acting as an axial chromophore (**83**) would be expected to be small and negative since only $C_{(6)}$ and $C_{(7)}$ are uncompensated and they are at some distance. If there is any such contribution, it would mean only that the zigzag contribution must be larger.

The insertion of a second methyl group ($C_{(19)}$) at $C_{(10)}$ to form androstane (**84**) has a negative rotatory effect of $-39°$ ($M_{84}-M_{82}$). This is within the range of that to be expected from four-atom effects alone: $-k(C-H)^2$, resulting from the elimination of the positive contribution of bond $C_{(10)}-C_{(1)}$. The planar zigzag effects should be small since the counterpoised chains at $C_{(10)}-C_{(12)}$ and $C_{(10)}-C_{(16)}$ both involve odd numbers of atoms. When this axial methyl group is regarded as a chromophore (**85**) it is seen that the carbon atoms of the first shell ($C_{(2)}$, $C_{(4)}$, $C_{(6)}$, $C_{(8)}$ and $C_{(11)}$) make a net dextro contribution, those of the second shell ($C_{(12)}$ and $C_{(14)}$) cancel out and those of the outer shell ($C_{(15)}$ and $C_{(16)}$) make negative contributions. It is therefore clear that the contributions to be attributed to these extra effects cannot be evaluated for $C_{(19)}$.

TABLE 22. Molecular rotations of estrane derivatives

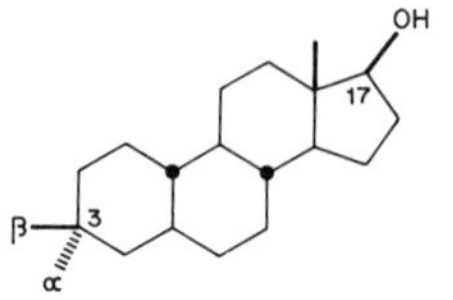

Substituents	$[M]_D$	$\Delta[M]_D$ of Substituents[129] at		Calculated $[M]_D$ of **82**
		$C_{(3)}$	$C_{(17)}$	
17-βOH	+65.6°		+20	+45.6°
3-αOH, 17-βOH	+66.8	+5	+20	+41.8
3-βOH, 17-βOH	+63.1	−2	+20	+45.1
				Average + 44.2

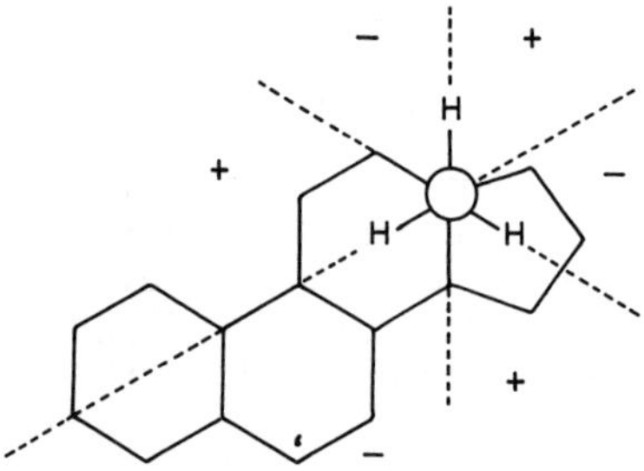

(83)

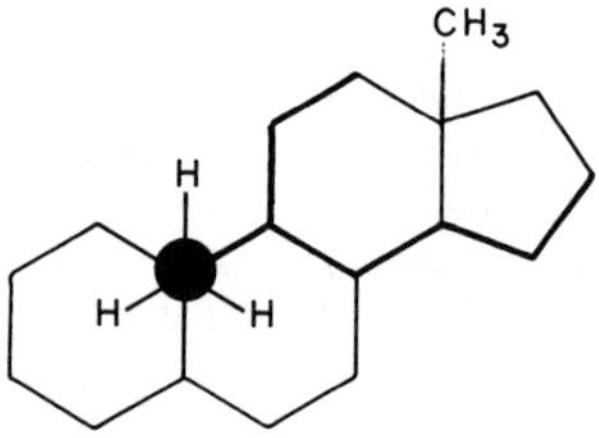

(84)

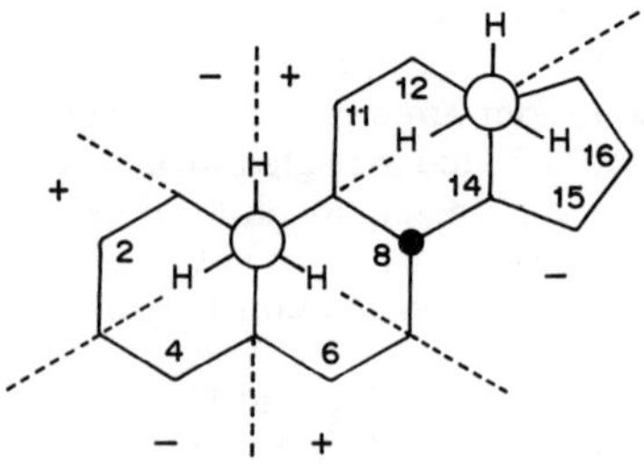

(85)

The rotatory contributions of the $C_{(17)}$ side chains can be evaluated by use of the methods described above for the alkanes. A 17β-methyl group is flanked at $C_{(13)}$ by three alkyl groups and at $C_{(16)}$ by one, with torsion angles as shown in Table 23. The 17β-ethyl group of pregnane can adopt two conformations. The most abundant of these (82%) is dextrorotatory, the less abundant (18%) levorotatory (PC Model[99]). A

TABLE 23. Torsion angles and $\Delta[M]_D$ for 17β-methylandrostane

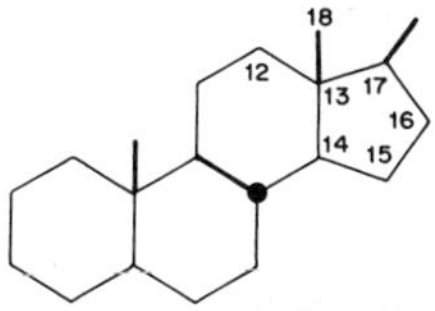

Bond	Group	Torsion angle[99]	$\Delta[M]_D$ (calcd)
$C_{(17)}$—$C_{(13)}$	$C_{(12)}$	+78.66°	+67.9°
	$C_{(14)}$	−167.23	−15.3
	$C_{(18)}$	−46.76	−50.5
$C_{(17)}$—$C_{(16)}$	$C_{(15)}$	+149.06	+35.6
			$\Delta[M]_D$ +37.7

TABLE 24. Conformational dissymmetry analysis of steroid side chains at $C_{(17)}$

Side chain	Conformers	Σ Relative abundances	$\dfrac{\Sigma \Delta[M]_D}{k(C-H)^2}$	$\Delta[M]_D$ Calcd[a]	Obsd
CH_3	1	1		+37.7°	+30°
CH_2CH_3	2			+76.1	+52
	(+82%; −18%)				
$CH(CH_3)_2$	1	1	0	+37.7	+35
—$\overset{H}{\underset{CH_3}{C}}$—$CH_2CH_3$	2	3/2	+1	+77.7	
—$\overset{H}{\underset{CH_3}{C}}$—$(CH_2)_2CH_3$	3	2	+2	+97.7	+87
—$\overset{H}{\underset{CH_3}{C}}$—$(CH_2)_3CH_3$	8	15/4	+15/4	+97.7	
—$\overset{H}{\underset{CH_3}{C}}$—$(CH_2)_3CH(CH_3)_2$	14	27/4	+7	+100	+86
—$\overset{H}{\underset{CH_3}{C}}$—$(CH_2)_2$—$\overset{H}{\underset{CH_3}{C}}$—$CH(CH_3)_2$	9	15/2	+4	+69	+66
—$\overset{H}{\underset{CH_3}{C}}$—$(CH_2)_2$—$\overset{CH_3}{\underset{H}{C}}$—$CH(CH_3)_2$	10	13/2	+11	+138.7	+111

[a]Shifts from corresponding androstanes. $K_{g/t} = 1/2$; $k(C-H)^2 = 60°$.

17β-isopropyl group is not free to rotate and possesses only the *meso* conformation (Table 24). The conformational mobility of the side chain increases with length, but is strongly biased by the configuration at $C_{(20)}$. The influence of alkyl groups at $C_{(24)}$ is quite strong. On the whole these rotations agree well with those found by the method of rotation shifts (from androstane).

C. Atom-bridged Polycyclic Compounds

The terpenes of the pinane series have been reviewed[137]. Hydroboronation of (+)-α-pinene (**86**) and protonolysis gave (+)-*cis*-pinane (**87**) (equation 28), while hydroboronation of (−)-β-pinene followed by thermal equilibration and protonolysis gave the (−)-*trans*-isomer (the enantiomers of these two substances are shown as **88** and **89** in equation 29). Catalytic hydrogenation of either pinene probably gives a mixture of the two diastereomers having the same ring configuration (as **87** and **89**); they are difficult to separate and have the same sign of rotation.

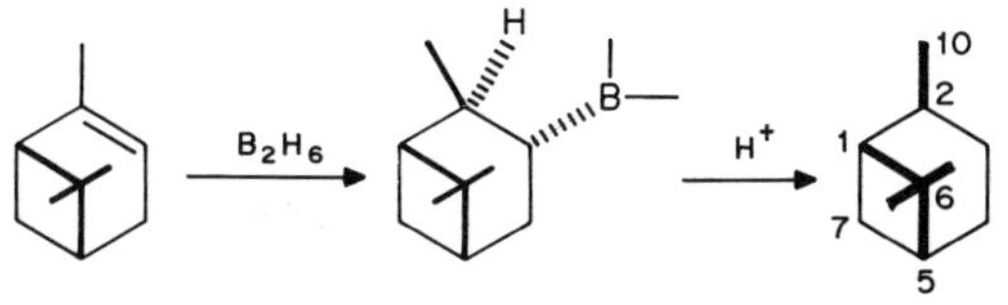

$$[M]_D +71.3°$$
(**86**)

$$[M]_D +38.1°$$
(**87**)

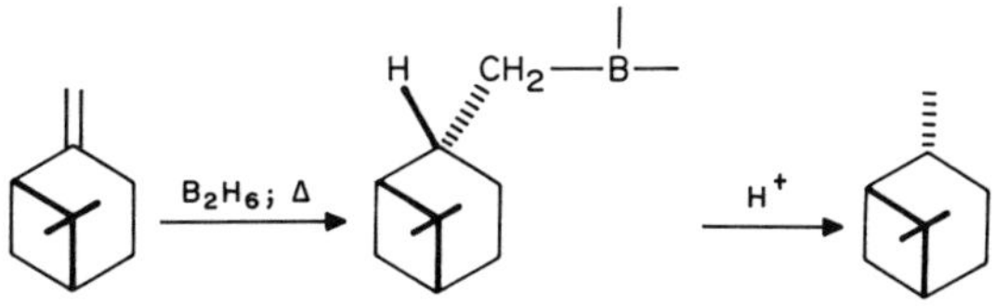

$$\text{(28)}$$

$$[M]_D +30.9°$$
(**88**)

$$[M]_D +25.3°$$
(**89**)

$$\text{(29)}$$

The pinanes are of particular importance because they make it clear that the simple four-atom analysis is, at best, incomplete. Very simply, under that model, the $C_{(10)}$ methyl group at $C_{(2)}$ should be the prime contributor to sign and magnitude of rotation, since it destroys the symmetry of the rest of the molecule and occupies essentially enantiomeric equatorial conformations in **87** and **89**. The conformation $C_{(10)}-C_{(2)}-C_{(1)}-C_{(7)}$ is clearly dextrorotatory in the *trans*-isomer (**89**) (which is dextrorotatory), but the conformation $C_{(10)}-C_{(2)}-C_{(1)}-C_{(6)}$ is equally clearly levorotatory in the *cis*-isomer which, in fact, has a stronger dextrorotation. It is difficult to estimate the relative magnitudes of the contributions of longer twisted chains which might involve pairs of methyl groups.

An alternative, suggested by the molecular orbital model but needing further study, is that the *gem*-dimethyl group should be considered, as a whole, to be a chromophore. If so, it should follow a sector rule centered on the central atom, here $C_{(6)}$ (**90**). The two methyl groups would lie in one plane (here, XY) and the other two carbon atoms (here, $C_{(1)}$ and $C_{(5)}$) would lie in another (here, XZ). The YZ plane would not be needed in

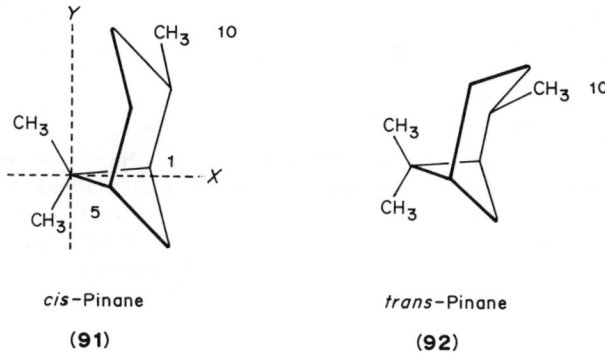

$(Z$ up)

(90)

most applications but seems required to distinguish the paired carbon atoms. In the two pinanes under consideration here the $C_{(10)}$ methyl group would lie in the same sector whether it be *cis-* or *trans-* **(91, 92)**. This proposal is tentative but it appears to merit more detailed consideration.

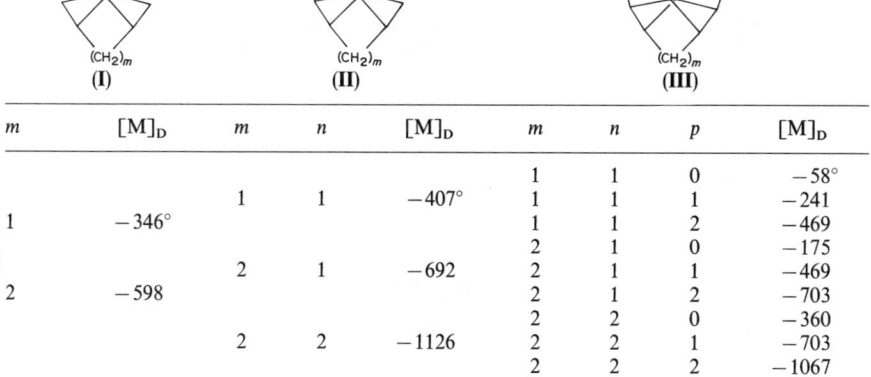

cis-Pinane	*trans*-Pinane
(91)	**(92)**

D. Caged Polycyclic Compounds

Caged compounds contain several valence bridges. They can show very large rotations, as exemplified by the *'triblattanes'* prepared by Nakazaki and coworkers (reviewed[138]).

TABLE 25. Molecular rotations of the triblattanes[138]

(I)			(II)			(III)	
m	$[M]_D$	m	n	$[M]_D$	m	n	p	$[M]_D$
					1	1	0	$-58°$
		1	1	$-407°$	1	1	1	-241
1	$-346°$				1	1	2	-469
					2	1	0	-175
		2	1	-692	2	1	1	-469
2	-598				2	1	2	-703
					2	2	0	-360
		2	2	-1126	2	2	1	-703
					2	2	2	-1067

180   J. H. Brewster

These materials can be considered to be derived from a twisted form of [2.2.2]-bicyclo-octane by successive introduction of bridges containing 0,1 or 2 carbon atoms (Table 25). It is seen that there is a regular increase in rotation with length of the bridges. This observation would indicate that those bridges are parts of left-handed helices. This, in turn, suggests a major role for the largest helices to be seen in these compounds. The major helix increases in size as a second bridge is inserted, but does not do so with a third bridge. Indeed, the third bridge may be considered to be part of a dextrorotatory helix. This may account for the fact that the third bridge may actually decrease the optical rotation.

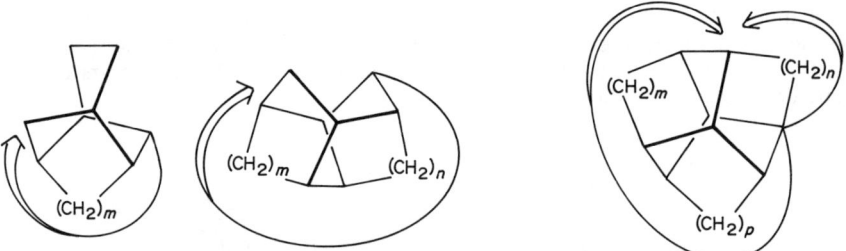

An alternative analysis of twistane (93) (Table 25, I, $n = 2$) involved projection along the three major axes of rotational symmetry (94). Helix calculations for each of these projections gave reasonably good calculated values for the rotation of twistane (here, $[M]_D - 484°$; the (+) enantiomer was used for the calculations)[139]. The projection 94a may, in fact, be the most significant if the molecular orbital model proves to be the most appropriate method for handling complex systems of this nature.

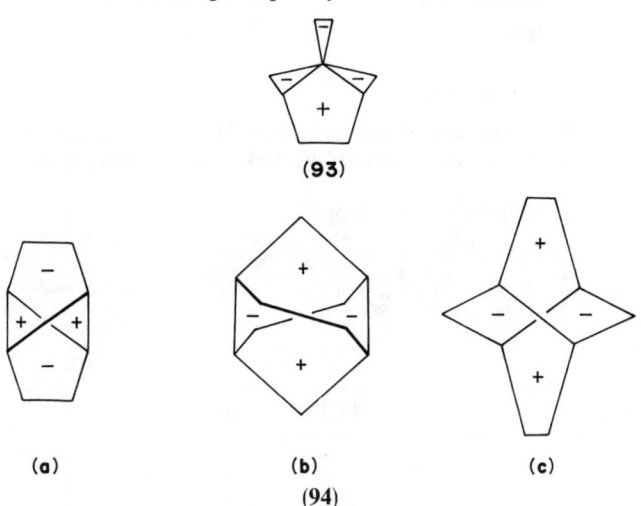

(93)

(a)   (b)   (c)

(94)

V. CONCLUSION

It would appear that a rigorous treatment of the chiroptical properties of hydrocarbons in terms of their far-ultraviolet transitions will require some method, such as the Random Phase Approximation[114], which allows for consideration of all molecular orbitals.

Optical rotation at long wavelengths represents a summation or, perhaps, a synthesis of all the contributions of those transitions. It may be that a helical conductor model, informed by insights from molecular orbital theory, can approximate such a summation but it is likely that particular chromophoric groups may make unusually strong contributions that would have to be taken into account as special cases. The cyclopropane ring clearly appears to be one such special chromophore. The quaternary carbon atom, particularly in such manifestations as the axial bridgehead methyl group of terpenoid compounds or as the *gem*-dimethyl group, which is also abundantly represented in natural products, may be another. For these cases, sector rules may prove useful. In the meantime, the simple four-atom Conformational Dissymmetry Model provides approximate results which are generally of the correct sign, but may be off in magnitude. Failures of this model are valuable in pointing to some of the situations in which a molecular-orbital-based approach may be called for.

VI. ACKNOWLEDGMENT

The author wishes to thank Dr Alfred Bader, of the Aldrich Chemical Company, for personal support which allowed purchase of the computer and the software that have been used in modeling many of the compounds considered here.

VII. REFERENCES

1. P. A. Levene and A. Rothen, *J. Org. Chem.*, **1**, 76 (1936) (the absolute configurations of the compounds shown in this paper should all be reversed); P. A. Levene and A. Rothen, *in Organic Chemistry* (Ed. H. Gilman), 1st ed., Chap. 12, Wiley, New York, 1938.
2. J. A. Mills and W. Klyne, in *Progress in Stereochemistry*, Vol. I (Ed. W. Klyne), Butterworths, London, 1954, pp. 177–222.
3. J. H. Brewster, in *Techniques of Chemistry* (Ed. A. Weissberger), Vol. IV, *Elucidation of Structures by Physical and Chemical Methods* (Eds. K. W. Bentley and G. W. Kirby), Part III, Wiley–Interscience, New York, 1972, pp. 1–249.
4. J. Jacques, C. Gross and S. Boucher, in *Stereochemistry, Fundamentals and Methods* (Ed. H. Kagan), Vol. 4 (*Absolute Configurations of Compounds with One Asymmetric Carbon Atom*), G. Thieme, Stuttgart, 1977.
5. W. Klyne and J. Buckingham, *Atlas of Stereochemistry*, Vols. I and II, 2nd ed., Oxford University Press, New York, 1978.
6. W. E. Kauzmann, J. E. Walter and H. Eyring, *Chem. Rev.*, **26**, 339 (1940).
7. B. C. Easton and M. K. Hargreaves, *J. Chem. Soc.*, 1413 (1959).
8. J. Kenyon and B. C. Platt, *J. Chem. Soc.*, 633 (1939).
9. H. G. Rule and A. R. Chambers, *J. Chem. Soc.*, 145 (1937).
10. A. W. H. Pryde and H. G. Rule, *J. Chem. Soc.*, 345 (1940).
11. T. M. Lowry, *Optical Rotatory Power*, Longmans, Green, London, 1935; reprinted by Dover Publications, New York, 1964.
12. W. Heller and D. D. Fitts, in *Technique of Organic Chemistry*, Vol. I (*Physical Methods*), Part III (Ed. A. Weissberger), 3rd ed., Wiley–Interscience, New York, 1960, pp. 2147–2333.
13. W. Heller and H. G. Curmè, in *Techniques of Chemistry*, Vol. I (*Physical Methods of Chemistry*), Part IIIc (*Polarimetry*) (Ed. A. Weissberger and B. W. Rossiter), Wiley–Interscience, New York, 1972, pp. 51–181.
14. C. Djerassi, *Optical Rotatory Dispersion*, McGraw-Hill Book Company, New York, 1960.
15. W. Klyne and A. C. Parker, in *Technique of Organic Chemistry*, Vol I (*Physical Methods*), Part III (Ed. A. Weissbeger), 3rd ed., Wiley–Interscience, New York, 1960, pp. 2335–2385.
16. P. Crabbé and A. C. Parker, in *Techniques of Chemistry*, Vol. I (*Physical Methods of Chemistry*), Part IIIC (*Polarimetry*) (Eds. A. Weissberger and B. W. Rossiter), Wiley–Interscience, New York, 1972, pp. 183–270.
17. P. M. Jones and W. Klyne, *J. Chem. Soc.*, 871 (1960).
18. M. B. Robin, *Higher Excited States of Polyatomic Molecules*, Vol. I, Academic Press, New York, 1974, pp. 105–155; Vol. III, 1985, pp. 79–106.

182 J. H. Brewster

19. A. Gedanken and O. Schnepp, *Chem. Phys.*, **12**, 341 (1976); A. Gedanken and M. Levy, *Rev. Sci. Instrum.*, **48**, 1661 (1977).
20. D. N. Kirk, *J. Chem. Soc., Perkin Trans. 1*, 787 (1980).
21. H. R. Wyss and Hs. H. Günthard, *J. Opt. Soc. Am.*, **56**, 888 (1966); H. R. Wyss and Hs. H. Günthard, *Helv. Chim. Acta*, **49**, 660 (1966).
22. L. Velluz, M. Legrand and M. Grosjean, *Optical Circular Dichroism*, Verlag Chemie, Weinheim; Academic Press, New York, 1965.
23. E. C. Hsu and G. Holzwarth, *J. Chem. Phys.*, **59**, 4678 (1973); G. Holzwarth, E. C. Hsu, H. S. Mosher, T. R. Faulkner and A. Moscowitz, *J. Am. Chem. Soc.*, **96**, 251 (1974); T. R. Faulkner, A. Moscowitz, G. Holzwarth, E. C. Hsu and H. S. Mosher, *J. Am. Chem. Soc.*, **96**, 252 (1974).
24. F. Devlin and P. J. Stevens, *Appl. Spectrosc.*, **41**, 1142 (1987).
25. P. Malon and T. A. Keiderling, *Appl. Spectrosc.*, **42**, 32 (1988).
26. E. D. Lipp and L. A. Nafie, *Appl. Spectrosc.*, **38**, 20 (1984).
27. P. L. Polavarapu, *Appl. Spectrosc.*, **38**, 26 (1984).
28. P. J. Stephens and R. Clark, in *Optical Activity and Chiral Discrimination* (Ed. S. F. Mason), Reidel, Dordrecht, 1978, pp. 263–288; P. J. Stephens, *Croat. Chim. Acta*, **62**, 429 (1989).
29. T. B. Freedman and L. A. Nafie, in *Topics in Stereochemistry*, Vol. 17 (Eds. E. L. Eliel and S. H. Wilen), Wiley–Interscience, New York, 1987, pp. 113–206.
30. L. D. Barron, *Optical Activity and Chiral Discrimination* (Ed. S. F. Mason), Reidel, Dordrecht, 1978, pp. 219–262; L. D. Barron, *Molecular Light Scattering and Optical Activity*, Cambridge University Press, Cambridge, 1982.
31. J. M. Bijvoet, A. F. Peerdeman and A. J. van Bommel, *Nature*, **168**, 271 (1951); J. M. Bijvoet, *Endeavour*, **14**, 71 (1955).
32. D. Arigoni and E. L. Eliel, in *Topics in Stereochemistry*, Vol. 4 (Eds. N. L. Allinger and E. L. Eliel), Wiley–Interscience, New York, 1969, pp. 127–243.
33. L. Lardicci, P. Salvadori and P. Pino, *Ann. Chim. (Rome)*, **52**, 652 (1962).
34. R. L. Letsinger and J. G. Traynham, *J. Am. Chem. Soc.*, **72**, 849 (1950).
35. G. S. Gordon, III and R. L. Burwell, Jr., *J. Am. Chem. Soc.*, **71**, 2355 (1949).
36. S. Ställberg-Stenhagen and E. Stenhagen, *J. Biol. Chem.*, **183**, 223 (1950).
37. G. W. O'Donnell and M. D. Sutherland, *Aust. J. Chem.*, **19**, 525 (1966).
38. P. A. Levene and S. H. Harris, *J. Biol. Chem.*, **111**, 735 (1935).
39. S. Pucci, M. Aglietto and P. L. Luisi, *Gazz. Chim. Ital.*, **100**, 159 (1970); S. Pucci, P. L. Luisi, M. Aglietto and P. Pino, *J. Am. Chem. Soc.*, **89**, 2787 (1967).
40. V. Prelog and E. Zalan, *Helv. Chim. Acta*, **27**, 535, 545 (1944).
41. P. J. Cram and J. Tadanier, *J. Am. Chem. Soc.*, **81**, 2737 (1959).
42. L. Lardicci, R. Menicagli, A. M. Caporusso and G. Giacomelli, *Chem. Ind. (London)*, 184 (1973).
43. U. Azzena, P. L. Luisi, U. W. Suter and S. Gladiali, *Helv. Chim. Acta*, **64**, 5821 (1981).
44. P. A. Levene and R. E. Marker, *J. Biol. Chem.*, **97**, 563 (1932).
45. G. K. Helmkamp, C. D. Joel and H. Sharman, *J. Org. Chem.*, **21**, 844 (1956).
46. G. K. Helmkamp and B. F. Rickborn, *J. Org. Chem.*, **22**, 479 (1957).
47. G. K. Helmkamp and N. Schnautz, *Tetrahedron*, **2**, 304 (1958).
48. A. Streitwieser, Jr., J. R. Wolfe, Jr. and W. D. Schaeffer, *Tetrahedron*, **6**, 338 (1959).
49. A. Streitwieser, Jr. and M. R. Granger, *J. Org. Chem.*, **32**, 1528 (1967).
50. A. Streitwieser, Jr. I. Schwager, L. Verbit and H. Rabitz, *J. Org. Chem.*, **32**, 1532 (1967).
51. W. A. Sanderson and H. S. Mosher, *J. Am. Chem. Soc.*, **88**, 3671 (1966).
52. G. Solladie, M. Muskatirovic and H. S. Mosher, *J. Chem. Soc., Chem. Commun.*, 809 (1968).
53. P. H. Anderson, B. Stephenson and H. S. Mosher, *J. Am. Chem. Soc.*, **96**, 3171 (1974).
54. R. E. Marker, *J. Am. Chem. Soc.*, **58**, 976 (1936).
55. D. G. Goodwin and H. R. Hudson, *J. Chem. Soc. (B)*, 1333 (1968).
56. R. S. Cahn, C. Ingold and V. Prelog, *Angew. Chem., Int. Ed. Engl.*, **5**, 385 (1966); V. Prelog and G. Helmchen, *Helv. Chim. Acta*, **55**, 2581 (1972); *Angew. Chem., Int. Ed. Engl.*, **21**, 567 (1982).
57. cf., e.g., L. F. Fieser and M. Fieser, *Natural Products Related to Phenanthrene*, 3rd ed., Reinhold, New York, 1949, pp. 211–219.
58. J. G. Kirkwood, *J. Chem. Phys.*, **5**, 479 (1937).
59. A. Julg, *Tetrahedron*, **12**, 146 (1961).
60. S. H. Wilen, K. A. Bunding, C. M. Kaschere and M. J. Wieder, *J. Am. Chem. Soc.*, **107**, 6997 (1985); J. Canceil, L. Lacombe and A. Collet, *J. Am. Chem. Soc.*, **107**, 6993 (1985).
61. W. Kuhn, *Z. Phys. Chem.*, **4B**, 14 (1929); **22B**, 406 (1933).

4. Chiroptical properties of alkanes and cycloalkanes 183

62. W. Kuhn, Z. Elektrochem., **56**, 506 (1952).
63. W. Kuhn, Angew. Chem., **68**, 93 (1956).
64. W. Kuhn, Ann. Rev. Phys. Chem., **9**, 417 (1958).
65. N. Harada and K. Nakanishi, Circular Dichroism Spectroscopy: Exciton Coupling in Organic Stereochemistry, University Science Books, Mill Valley, CA, 1983.
66. J. G. Kirkwood, J. Chem. Phys., **5**, 479 (1937).
67. W. W. Wood, W. Fickett and J. G. Kirkwood, J. Chem. Phys., **20**, 561 (1952).
68. H. J. Bernstein and E. E. Pedersen, J. Chem. Phys., **17**, 885 (1949).
69. D. H. Whiffen, Chem. Ind. (London), 964 (1956).
70. J. H. Brewster, J. Am. Chem. Soc., **81**, 5475 (1959).
71. J. H. Brewster, work in progress.
72. A. Streitwieser, Jr. and T. R. Thomson, J. Am. Chem. Soc., **77**, 3921 (1955).
73. H. Wynberg and L. A. Hulshof, Tetrahedron, **30**, 1775 (1974).
74. L. A. Hulshof and H. Wynberg, Stud. Org. Chem. (Amsterdam), **3** (New Trends in Heterocycl. Chem.), 373 (1979); Chem. Abstr., **92**, 75,738e (1980).
75. J. H. Brewster, J. Am. Chem. Soc., **81**, 5483 (1959).
76. J. H. Brewster, Tetrahedron, **13**, 106 (1961).
77. J. H. Brewster, J. Am. Chem. Soc., **81**, 5493 (1959).
78. J. A. Mills, J. Chem. Soc., 4976 (1952).
79. A. I. Scott and A. D. Wrixon, Tetrahedron, **26**, 3695 (1970).
80. A. I. Scott and A. D. Wrixon, Tetrahedron, **27**, 4787 (1971).
81. A. F. Drake and S. F. Mason, Tetrahedron, **33**, 937 (1977).
82. J. A. Berson, J. S. Walia, A. Remanick, S. Suzuki, P. Reynolds-Warnhoff and D. Wilner, J. Am. Chem. Soc., **83**, 3986 (1961).
83. J. H. Brewster, in Topics in Stereochemistry, Vol. 2 (Eds. N. L. Allinger and E. L. Eliel), Wiley–Interscience, New York, 1967, pp. 1–72.
84. A. Moscowitz, Tetrahedron, **13**, 48 (1961).
85. J. W. Gibbs, Am. J. Sci., (3), **23**, 460 (1882); Collected Works, Vol. II, part 2, Longmans, Green, New York, 1928, p. 195.
86. P. Drude, Theory of Optics (transl. by C. R. Mann and R. A. Millikan) (publ. 1900), Dover Reprint, 1959, pp. 400–407.
87. E. U. Condon, Rev. Mod. Phys., **9**, 432 (1937).
88. E. Charney, The Molecular Basis of Optical Activity, Wiley–Interscience, New York, 1979.
89. S. F. Mason, Quart. Rev., **17**, 20 (1963); S. F. Mason, Molecular Optical Activity and the Chiral Discriminations, Cambridge University Press, Cambridge, 1982.
90. J. H. Brewster, Topics in Current Chemistry, **47**, 29 (1974).
91. C. Sandorfy, Can. J. Chem., **33**, 1337 (1955); C. Sandorfy, in Sigma Molecular Orbital Theory (Eds. O. Sinanoglu and K. B. Wiberg), Yale University Press, New Haven, 1970, pp. 130–136.
92. K. Fukui, H. Kato and T. Yonezawa, Bull. Chem. Soc. Jpn., **34**, 442, 1111 (1961); K. Fukui, H. Kato, T. Yonezawa, R. Morokuma, A. Imamura and C. Nagata, Bull. Chem. Soc. Jpn., **35**, 38 (1962).
93. R. Hoffmann, J. Chem. Phys., **39**, 1397 (1963); R. Hoffmann, A. Imamura and W. J. Hehre, J. Am. Chem. Soc., **90**, 1499 (1968); R. Hoffmann, Acc. Chem. Res., **4**, 1 (1971).
94. W. L. Jorgensen and L. Salem, The Organic Chemist's Book of Orbitals, Academic Press, New York, 1973.
95. J. Hudec, J. Chem. Soc., Chem. Commun., 829 (1970); M. T. Hughes and J. Hudec, J. Chem. Soc., Chem. Commun., 805 (1971); G. P. Powell and J. Hudec, J. Chem. Soc., Chem. Commun., 806 (1971).
96. D. N. Kirk and W. Klyne, J. Chem. Soc., Perkin Trans. 1, 1076 (1974).
97. J. H. Brewster, Tetrahedron, **30**, 1807 (1974).
98. R. Colle, U. W. Suter and P. Luisi, Tetrahedron, **37**, 3727 (1981).
99. Serena Software, Box 3076, Bloomington, IN, 47402-3076. This uses the MM2 (QCPE-395, 1977) force field of N. L. Allinger with the graphic interface MODEL of P. C. Still, as adapted for the IBM-PC by M. M. Midland.
100. A. L. Verma, W. F. Murphy and H. J. Bernstein, J. Chem. Phys., **60**, 1540 (1974).
101. J. H. Brewster, Tetrahedron Lett., Number 20, 23 (1959).
102. D. N. Kirk, J. Chem. Soc., Perkin Trans. 1, 2171 (1976).
103. W. Fickett, J. Am. Chem. Soc., **74**, 4204 (1952).

184   J. H. Brewster

104. J. A. Berson and L. D. Pedersen, *J. Am. Chem. Soc.*, 97, 238 (1975); J. A. Berson, L. D. Pedersen and B. K. Carpenter, *J. Am. Chem. Soc.*, 97, 240 (1975); 98, 122 (1976).
105. S. J. Cianciosi, K. M. Spencer, T. B. Freedman, L. A. Nafie and J. E. Baldwin, *J. Am. Chem. Soc.*, 111, 1913 (1989); K. M. Spencer, S. J. Cianciosi, J. E. Baldwin, T. B. Freedman and L. A. Nafie, *Appl. Spectrosc.*, 44, 235 (1990); S. J. Cianciosi, N. Rangunathan, T. B. Freedman, L. A. Nafie and J. E. Baldwin, *J. Am. Chem. Soc.*, 112, 8204 (1990).
106. J. M. Schwab and C.-K. Ho, *J. Chem. Soc., Chem. Commun.*, 872 (1986); T. B. Freedman, M. G. Paterlini, N. S. Lee, L. A. Nafie, J. M. Schwab and T. Ray, *J. Am. Chem. Soc.*, 109, 4727 (1987); Calculations: R. Dutler and A. Rauk, *J. Am. Chem. Soc.*, 111, 6957 (1989).
107. P. K. Bose, P. L. Polavarapu, L. D. Barron and L. Hecht, *J. Phys. Chem.*, 94, 1734 (1990).
108. T. M. Black, P. K. Bose, P. L. Polavarapu, L. D. Barron and L. Hecht, *J. Am. Chem. Soc.*, 112, 1479 (1990).
109. W. von E. Doering and W. Kirmse, *Tetrahedron*, 11, 272 (1960).
110. W. R. Moore, H. W. Anderson, S. D. Clark and T. M. Ozretich, *J. Am. Chem. Soc.*, 93, 4932 (1971).
111. Y. Inouye, T. Sugita and H. Walborsky, *Tetrahedron*, 20, 1695 (1964).
112. L. Crombie, D. A. Findlay, R. W. King, I. M. Shirley, D. A. Whiting, P. M. Scopes and B. M. Tracey, *J. Chem. Soc., Chem. Commun.*, 474 (1976).
     (A suggestion by S. F. Mason is discussed in this paper.)
113. H. Staudinger and L. Ruzicka, *Helv. Chim. Acta*, 7, 201 (1924).
114. S. Bohan and T. D. Bouman, *J. Am. Chem. Soc.*, 108, 3261 (1986).
115. R. D. Amos, N. C. Handy, A. F. Drake and P. Palmieri, *J. Chem. Phys.*, 89, 7287 (1988).
116. W. C. M. C. Kokke and F. A. Varkevisser, *J. Org. Chem.*, 39, 1535 (1974).
117. H. M. Walborsky, L. Barash and T. C. Davis, *Tetrahedron*, 19, 2333 (1963).
118. M. Mousseron, R. Richaud and R. Granger, *Bull. Soc. Chim. Fr.*, 224 (1946).
119. S. Siegel and G. V. Smith, *J. Am. Chem. Soc.*, 82, 6082 (1960).
120. K. Gollnick and G. Schade, *Justus Liebigs Ann. Chem.*, 721, 133 (1969).
121. W. Kauzmann, F. B. Clough and I. Tobias, *Tetrahedron*, 13, 57 (1961).
122. D. Whittaker and D. V. Banthorpe, *Chem. Rev.*, 72, 305 (1972).
123. W. Cocker, P. V. R. Shannon and P. A. Staniland, *J. Chem. Soc.* (C), 946 (1966).
124. F. Fringuelli, A. Tatticchi, F. Fernandez, D. N. Kirk and P. M. Scopes, *J. Chem. Soc., Perkin Trans. 1*, 1103 (1974).
125. W. Huckel and H. Sowa, *Chem. Ber.*, 74, 57 (1941).
126. J. W. Barrett and R. P. Linstead, *J. Chem. Soc.*, 1069 (1935).
127. P. M. Bourne and W. Klyne, *J. Chem. Soc.*, 2044 (1960).
128. F. Fernàndez, D. N. Kirk and M. Scopes, *J. Chem. Soc., Perkin Trans. 1*, 18 (1974).
129. W. Klyne, in *Determination of Organic Structures by Physical Methods*, Vol. I (Eds. E. A. Braude and F. C. Nachod), Academic Press, New York, 1955, pp. 73–130; see also, earlier tables: W. Klyne, *Chem. Ind. (London)*, 755 (1948).
130. B. Alcaide, M. P. Tarazona and F. Fernàndez, *J. Chem. Soc., Perkin Trans. 1*, 2117 (1982).
131. B. Alcaide and F. Fernàndez, *J. Chem. Soc., Perkin Trans. 1*, 1665 (1983).
132. B. Alcaide and F. Fernàndez, *J. Chem. Soc. Perkin Trans. 1*, 2250 (1981).
133. G. Allegra, M. Farina, A. Immirzi, A. Colombo, U. Rossi, R. Bioggi and G. Natta, *J. Chem. Soc. (B)*, 1020 (1967).
134. M. Farina and G. Audisio, *Tetrahedron*, 26, 1827 (1970).
135. A. W. Burgstahler and J. N. Marx, *J. Org. Chem.*, 34, 1562, 1566 (1969).
136. J. Jacques, H. Kagan and G. Ourisson, *Tables of Constants and Numerical Data*, 14, Selected Constants. Optical Rotatory Power. 1a. Steroids (Ed. S. Allard), Pergamon Press, Oxford, 1965.
137. D. V. Banthorpe and D. Whittaker, *Chem. Rev.*, 66, 643 (1966).
138. M. Nakazaki, in *Topics in Stereochemistry*, Vol. 15 (Eds. E. L. Eliel, S. H. Wilen and N. L. Allinger), Wiley–Interscience, New York, 1983, pp. 200–244.
139. J. H. Brewster, *Tetrahedron Lett.*, 4355 (1972).

## CHAPTER 5

Enumeration, topological indices and molecular properties of alkanes

LOWELL. H. HALL

Department of Chemistry, Eastern Nazarene College, Quincy, MA 02170, USA

and

LEMONT B. KIER

Department of Medicinal Chemistry, Virginia Commonwealth University, Richmond, VA 23298, USA

The Chemistry of Alkanes and Cycloalkanes
Edited by S. Patai and Z. Rappoport © 1992 John Wiley & Sons Ltd

I. INTRODUCTION

Alkanes have a central place in the structure theory of organic molecules. Most nonaromatic organic molecules may be considered as substituted alkanes. For this reason one important approach to an understanding of the structure of any organic molecule class begins with an investigation of the structure and properties of alkanes. Further, the alkanes provide a rich arena for developing models of structure since there is such an enormous variety of skeletal arrangements without additional factors such as heteroatoms or unsaturation.

Alkanes fall broadly into two classes: cyclic and noncyclic. In each class there is a range of structure complexity relating to the degree of branching in the molecular skeleton. It is observed that alkane properties are generally related to two major aspects of alkane structure: molecule size and degree of branching. In the development of relationships between molecular properties and molecular structure, it is relatively easy to represent the molecular size. Quantities such as the number of carbon atoms and molecular weight are used, while an experimental quantity such as molar volume and computed quantities such as surface area or volume have an information potential. These quantities and other might be used to represent molecular size in quantitative structure–activity relations (QSAR).

On the other hand, representation of structure complexity, including branching and cyclization, has been more elusive. Early efforts in this area include the pioneering work of Wiener[1,2] and Platt[3,4]. For more information on these developments, see Rouvray[5,6], Trinajstic[7] and Kier and Hall[8,9]. The elusiveness of this problem arises in part because of vague definitions or lack of definition of these attributes. We will start with one specific definition and then introduce ways to approach the problem using what has become known as the chemical graph.

For a working definition of *molecular structure*, we have adopted the one given by Eliel: 'The structure of a molecule is completely defined by the number and kind of atoms and the linkages between them'[10]. We have applied the term *molecular topography* to the three-dimensional arrangement of the atoms in a particular conformation; the term *configuration* is also used. The term *molecular topology* refers to the set of atoms and connections within the molecular skeleton, that is, the molecular structure as defined by Eliel.

II. THE CHEMICAL GRAPH

The structure of an alkane consists of the carbon atoms connected to each other by the network of bonds and connected to an appropriate number of hydrogen atoms. This network of connections, called the *molecular skeleton*, consists of chemical bonds. From the valence bond viewpoint, the bond may be considered to be made up of the overlap of σ-type orbitals, although such a bonding interpretation is not necessary to our approach. It is this network of connections which determines the structure features of the molecule. It is this network which distinguishes among the structures of six-carbon alkanes such as hexane, 3-methylpentane, cyclohexane and methylcyclopentane. Having specified the molecular skeleton, one can deduce much of the structure information for a given molecule.

The usual graphical approach to alkane structure focuses on the network of skeletal bonds, the connections between the carbon atoms. Once this network is established and understood, the bonds (connections) to the hydrogen atoms are determined precisely. That is, the hydrogen atom content may be considered to follow from the carbon atom

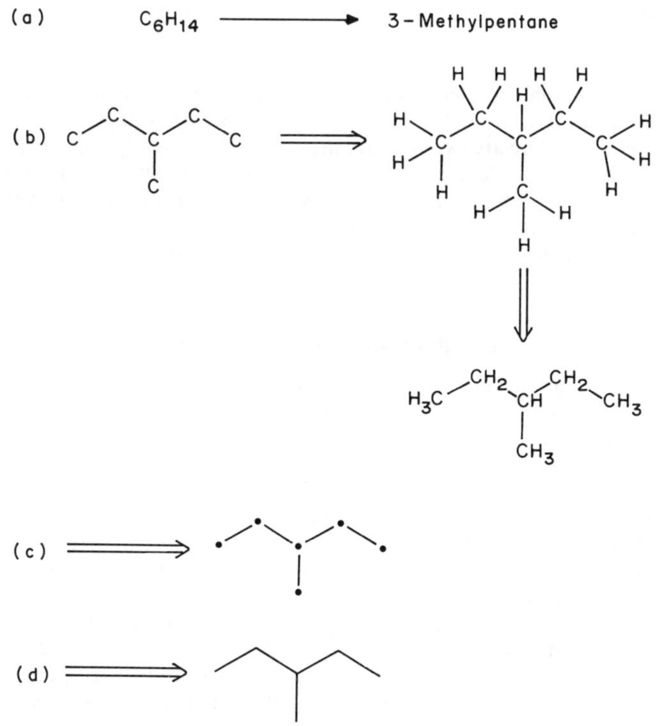

FIGURE 1. Progression of representations from structural formula to chemical graph. (a) A molecular formula is made specific with selection of 3-methylpentane. (b) The carbon skeleton is specified by the isomer selection, the hydrogen atoms are added to the skeleton and the structure written in skeleton form. (c) The graph representation using a dot for each vertex. (d) The chemical graph as usually written without a dot to signify the vertex

bonding scheme, since all alkane carbon atoms are quadrivalent. The molecular formula is insufficient to describe the structure; there are many isomeric possibilities. Once the particular isomer is named, as suggested in Figure 1(a), the carbon skeleton is established. Then, the hydrogen atoms may be added to complete the structure. In this sense, each carbon atom may be considered as part of a specific structure unit, i.e. CH_n, which is connected to other such units by the molecular skeleton. Indeed, chemists usually write alkane structures in that way, as shown in Figure 1(b). It is a relatively short step to rewrite the skeleton structure in the form of a chemical graph, as shown in Figure 1(c).

This entity, the chemical graph for 3-methylpentane (3M5 in a useful shorthand for alkanes), represents the necessary information for the development of structure information which, in turn, becomes the basis for the QSAR of alkane properties for 3M5. The juncture of each line (*edge*) in the graph is called a vertex, shown as a bold dot in Figure 1(c), or, more simply, in Figure 1(d). The vertex represents a carbon-hydride group, CH_n. Each edge represents a single skeletal bond, that is, the connection to an adjacent hydride group. In this way the value of n can be determined at each vertex. The number of adjacent neighbors for vertex i is called the vertex degree or delta value, δ_i. The value of n for a given vertex is directly related to δ_i and the usual valence of carbon, 4, which is also the number of valence electrons, Z^v (equation 1).

$$n = 4 - \delta_i = Z^v - \delta_i \tag{1}$$

In this manner, a graph representation of the alkane structure is developed. The way in which this chemical graph becomes the basis for QSAR development is discussed in subsequent sections. There are two primary aspects to this development: (a) enumeration of isomers, the number of distinct structures with the same number of atoms; and (b) enumeration of subgraph fragments, the number of distinct fragments of various types found in a given graph. Subgraph enumeration leads to a basis for numerical representation of chemical graphs which, in turn, leads to mathematical models for QSAR.

III. ENUMERATION OF ALKANES

One of the earliest applications of chemical graphs was the enumeration of alkane isomeric structures. Both Rouvray[5,6] and Biggs and coworkers[11] recount these early investigations. It is clear that counting alkane isomers is an important procedure since one approach to counting organic molecules structures is the counting of substituted alkanes. From the early work of Cayley[12], followed by the work of Polya[13,14] and later Henze and Blair[15], the chemical graph has been the basis of enumeration schemes. In this paper no attempt will be made to be comprehensive nor will an overall review be attempted. The reader is referred to the excellent work of Trinajstić for a description of methods and summary of references[7].

Cayley was the first to develop a systematic mathematical approach to isomer enumeration[12]. He listed alkanes and alkyl radicals up to those with 13 carbon atoms. Although his counts for $n = 12$ and $n = 13$ are now known to be incorrect, his pioneering work led the way. He also indicated that he could not reduce his scheme to a single formula for enumeration of alkane isomers.

It was Henze and Blair[15] who led the way to a method which could ultimately be computerized[16]. Using the concept of rooted trees, they developed a recursion scheme for alkyl alcohols. A table of their results is given by Trinajstić[7]. Using the alcohol results, Henze and Blair were able to develop recursion formulas for alkanes. They published the enumeration up to 20 carbon atoms. Later, Perry[17] extended their work to $n = 30$.

Trinajstić gives a table of rooted graphs which becomes the basis for enumeration of substituted alkanes. Alcohols and alkanes are enumerated up to 31 carbon atoms. For $n = 10$ there are 75 alkanes; 211 primary alcohols, 194 secondary alcohols and 102 tertiary alcohols for a total of 507 isomers. The number of isomers increases rapidly with the number of carbon atoms; for alkanes with $n = 5$ the isomer count is 3, for $n = 10$ it is 75, for $n = 15$ it is 4347, for $n = 20$ it is 366,319 and for $n = 30$ it is 4,111,846,763.

This enumeration work has been extended and formalized in a more general sense by utilization of the work of Polya[13,14]. The Polya counting polynomial, based on earlier work of Redfield[18] and Lunn and Senior[19], was the basis for useful approaches to enumeration of many chemical classes, including cyclic structures, as demonstrated by Trinajstić[7]. Some detailed examples are given by Balaban and coworkers[20] as well as an extensive set of references.

IV. ENUMERATION OF SUBGRAPHS FROM CHEMICAL GRAPHS

Representation of molecular structure via chemical graphs requires the characterization of the graph in a numerical sense. In a mathematical model such as a QSAR equation, one or more numerical variables of structure must be entered into the equation for each molecule in the data set. These numerical quantities must distinguish among the various structures in the data. For example, if the hexane isomers are involved in the data, the structure variables must represent the differences among hexane, 3-methylpentane, 2-methylpentane, 2,2-dimethylbutane and so on.

The hexane isomers all possess the same number of atoms, hence this crude size measure is an insufficient basis for representation. The differences among the isomers lie in the branching patterns within the different molecular skeletons. In chemical graph theory these differences are represented by decomposition of the whole graph into various fragments of a given size and type. These distinct fragments are called subgraphs[7-9].

Each of the hexane isomers possesses five edges, but they possess a different number of two-edge fragments and three-edge fragments. The counts of these higher-order fragments are a basis of differentiation among the isomers. Discussion of the fragment counts in isomers and their basis for numerical representation are given in the next section. See Figures 4 and 5 and Tables 4 and 5 for examples of heptanes.

The counting, identification and characterization of all the distinct subgraphs of molecular skeletons is a nontrivial task, especially as the number of atoms and the skeletal complexity increases. For this reason computer programs have been developed to facilitate the computation. The ALLPATH program of Randić and coworkers[21] and the Molconn-X program of Hall and Kier[22] are available for this task.

V. MOLECULAR CONNECTIVITY METHOD

A. Description of Molecular Structure

At the lowest level of information, the quantitation of structure may be a count of the number of atoms in a molecule. The count of the number of carbon atoms in a normal alkane series provides a useful description of structure which correlates closely with most physical properties. The number of carbon atoms in a series of normal alkyl-substituted molecules correlates very well with a biological activity for some cases[8,9]. Of course, such simple cases are not of great significance nor of great challenge for modeling.

At the other end of a spectrum of rigor and information content, a quantum-mechanical description of a molecule provides quantitation of electron distribution, electron energy levels and dependence on conformation, and a profile of dynamic events which the molecule can experience. Semiempirical implementation of quantum-mechanical principles, as in various forms of molecular orbital theory, have come into use in describing molecules in QSAR studies[23].

There has evolved an intuition about structural differences among molecules which is based on the structure formulas commonly used by chemists. Representation of molecular structure might be viewed along a spectrum; these intuitions may be placed between these two extremes of rigor, counts of atoms and quantum mechanics. Molecular formulas have facilitated written communication in a profound way among chemists. In a book or a paper, we see the familiar depiction of a molecule, and instantly, to the trained mind, a plethora of information is conveyed. The structural formulas of pentane, isopentane and neopentane suggest to the chemist the awareness that here are three different molecules, possessing properties which are different, and which are capable of undergoing different phenomena.

There is significant information encoded in structural formulas, however, the information is essentially qualitative. It is true that molecules described by different structures have different properties and that a ranking of expected properties may be predicted from the formulas of isomers; however, the structural formula lacks a quantitative character. How numerically different are pentane, isopentane and neopentane? The structural formula representation has usually been considered outside the realm of quantitative structure description, bounded by quantum mechanics and a count of atoms.

The success in using the qualitative information in the structural formula is a motivation to translate the structural formula into a numerical representation. A few efforts have been made in earlier years to develop a basis for numerical values assigned to structures or fragments of structures, whereby a quantitation for use in QSAR might be achieved. These efforts often lack generality; they are based on assignment of numerical values to fragments or structures, based on agreement with some specific physical property. Starting with the most elementary attributes of molecular structure, some investigators have attempted to quantitate structure based upon familiar characteristics of branching[7]. We have briefly reviewed several of these[8,9].

Although some early work provided interesting results[1-4], they were not readily extended to QSAR work. A later suggestion, an alkane branching index of Randić[24], appeared to be the best possibility to develop a widely useful quantitative description of molecular structure for the kinds of molecules of interest to the structure–activity analyst such as the medicinal chemist[8,9]. This method, developed by Kier and Hall in a series of papers[25-29], has been named molecular connectivity. The development, information content and applications are described in the rest of this section.

B. Developing an Index of Branching

As a useful starting point, let us represent the carbon atoms in Figure 2 as a skeleton structural description. The hydrogen atoms are not written explicitly because they are implied on what we know to be primary, secondary, tertiary or quaternary carbon atoms. We will use the skeleton structure, known as the *hydrogen-suppressed graph*, throughout our discussion. To develop the principles of molecular connectivity, consider two alkanes in Figure 2: (a) 2,4-dimethylpentane and (b) 3,3-dimethylpentane. We ask the question: What is the structure difference between these, expressed in quantitative terms? One approach to describing the differences in these structures is to record at each atom the number of neighboring atoms bonded to it in the skeleton. Figure 2 shows these numbers. We call these numbers δ (delta) values. It is immediately obvious that the set of δ numbers for the two molecules is different. We are apparently on a profitable course, the generation of numbers expressing structural differences between these two molecules. However, if we sum these δ values for each molecule, we arrive at 12 for both. This does not lead to differentiation. This is a general result for graphs called the handshake theorem.

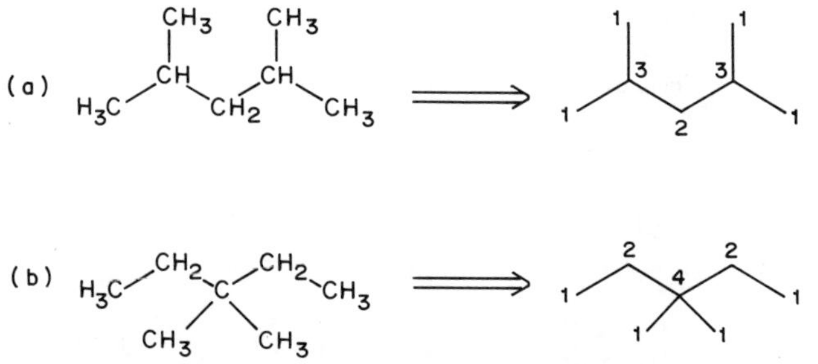

FIGURE 2. Structures and skeleton formulas of (a) 2,4-dimethylpentane and (b) 3,3-dimethylpentane. The numbers associated with each vertex in the chemical graph give the number of (adjacent) bonded skeletal neighbors

Our approach now is to develop a quantitative basis for representing molecular structure, beginning with a numerical representation for each bond. We dissect the skeleton into constituent bonds (graph edges), using the δ values to characterize each edge. Skeleton a in Figure 2 decomposes into the following pairs of delta values: (1,3), (1,3), (3,2), (2,3), (3,1), (3,1), whereas skeleton b decomposes to (1,2), (2,4), (1,4), (1,4), (4,2), (2,1).

Examination of these sets of edges (skeletal bonds), defined by the δ values, shows that they are clearly different. We will explore a basic question: how can these bond descriptors be converted into a number which will be different for these two molecules and may quantitate the structural difference?

Perhaps the first idea which might occur is to compute a product of each edge-pair of δ values, as shown within each set of parentheses below, and then sum the products. The two molecules a and b, represented by the graphs in Figure 2, can now be described by the following combinations:

$$a = (1 \times 3) + (1 \times 3) + (3 \times 2) + (2 \times 3) + (1 \times 3) + (1 \times 3)$$

$$b = (1 \times 2) + (2 \times 4) + (1 \times 4) + (1 \times 4) + (4 \times 2) + (2 \times 1)$$

Carrying out the indicated algebra leads to the following: a = 24, b = 28. The difference in these two numbers suggests that this approach may be used as a basis for differentiation among graphs.

This δ value representation for the sets of edges may be viewed as the basis for developing a set of inequalities. Thus, the operation on the set of pairs of δ values yields a number for each molecule. However, based on our experience with molecular properties, there is a certain ordering expected among property values for isomers. For example, the boiling point of the normal heptane isomer is higher than for 3-ethylpentane, which is in turn higher than more branched isomers, and so on. With this expectation in mind, we examine other combinations of the pairs of δ values to find an operation which yields a calculated series of numbers paralleling certain property values. For example, the heat of formation and the boiling point of the hexane isomers form the following set of inequalities: 6 > 3M5 > 2M5 > 23MM4 > 22MM4.

The sum-of-numbers scheme leads to different results for the two heptane isomers. A test must now be made as to whether the sum of products of edge δ values yields unique numbers for a larger group of alkanes, say, the nine ismers of heptane. Examination of the first column of Table 1 reveals two pairs of identical values. Thus, the summation

TABLE 1. Index values for heptane isomers

Chemical graph	$\sum(\delta_i\delta_j)$	$\sum(\delta_i\delta_j)^{1/2}$	$\sum(\delta_i\delta_j)^{-1/2}$
1	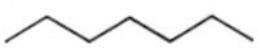 20	10.828	3.414
2	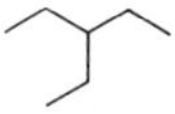 24	11.591	3.346
3	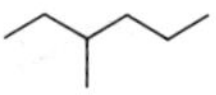 23	11.459	3.308
4	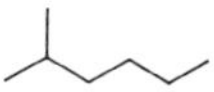 22	11.328	3.270
5	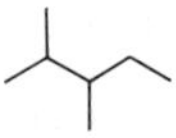 26	12.060	3.181
6	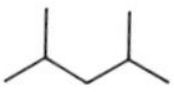 24	11.827	3.126
7	 28	12.485	3.121
8	 26	12.243	3.061
9	 30	12.928	2.943

of the product of the two adjacent δ values, yielding a single number, is not a scheme which successfully satisfies the inequalities. Another step is necessary. Let us consider an additional operation. One mathematical operation is to take the square root of the product of bonded δ values and then to sum these values. The results of this additional operation on molecules a and b of Figure 2 yield:

$$a = 1.732 + 1.732 + 2.449 + 2.449 + 1.732 + 1.732 = 11.827$$

$$b = 1.414 + 2.828 + 2.000 + 2.000 + 2.828 + 1.414 = 12.485$$

This combination of operations for the two examples from Figure 2 leads to different numbers. We can test the uniqueness of values from these operations by calculating values for the heptane isomers shown in Table 1. All values are unique. It appears that we can derive numbers for each molecule based on the number of neighbors on each atom (δ values). These are unique for heptane isomers and may carry information about relative structure. It is this quality of uniqueness which suggests that these graph-derived numbers may be a basis for developing QSARs.

There is, however, one more consideration that must be made concerning the numerical bond values. In Table 2 are listed the 10 types of bonds found in alkane skeletons. The product of δ values are computed, following by the square root of the products. It is notworthy that the (1,1) bond, which may be considered to comprise the largest part (all) of the ethane skeleton, has the lowest numerical value for $\Sigma(\delta_i\delta_j)$. In contrast, the (4,4) bond, which intuition tells us should comprise the smallest (spatial) part of, say 2,2,3,3-tetramethylbutane, has the largest numerical $\Sigma(\delta_i\delta_j)$ value. The consequence of this inverted numerical order relative to the spatial contribution of the bond type of the molecule is an inverted order of the $\Sigma(\delta_i\delta_j)$ values for the heptanes, relative to their branching and size (see column 2, Table 2).

In view of this disharmony with spatial contributions of bonds, an additional operation on the original bond description seems warranted. One approach is to reciprocate the

TABLE 2. Alkane bond types and numerical descriptors

Bond type	(δ_i, δ_j)	$(\delta_i, \delta_j)^{1/2}$	$(\delta_i, \delta_j)^{-1/2}$
(1,1)	1	1.000	1.000
(1,2)	2	1.414	0.707
(1,3)	3	1.732	0.577
(1,4)	4	2.000	0.500
(2,2)	4	2.000	0.500
(2,3)	6	2.449	0.408
(2,4)	8	2.828	0.354
(3,3)	9	3.000	0.333
(3,4)	12	3.464	0.289
(4,4)	16	4.000	0.250

$(\delta_i \delta_i)$ values for each. The results for the 10 alkane bonds can be seen in the last column of Table 2. The bond types contributing a larger portion to the size of a molecule—(1,2), (1,3), etc.—now have larger numerical values. The small spatial contributions of the (3,4) and (4,4) type bonds are now appropriately reflected by smaller numerical values.

The consequences of this additional operation can now be seen in column 3 of Table 1. The more elongated, less branched molecules like heptane, 2-methylhexane and 3-methylhexane have larger numbers. These molecules are larger, as is reflected by computed cavity surface areas[30]. At the other extreme, more branched, smaller molecules, such as the last three heptane isomers, have lower numerical values.

Based on a preliminary analysis, this algorithm appears to satisfy two criteria: (1) the generation of unique numbers for selected groups of molecules, and (2) numbers with some relationship to notions of branching and size. Thus we have, in this algorithm, a potential for the generation of quantitative information describing molecular structure.

C. Analysis of Structure Information

The reciprocal-square-root algorithm developed above generates highly discriminating numbers for alkane graphs. We now proceed to investigate the information content of the numbers generated by this algorithm. The sum of bond terms for molecules shown in Figure 2 for a and b is an index, in essence, a weighted count: the count of bonds, each weighted by its environment in the skeleton formula. The weighting is based on the adjacency relationships of both atoms forming the bond, specifically by the δ values for the bonded atoms.

The two heptanes a and b each have six bonds; these bond-count numbers generally reflect molecular size. Each term in the summation is based on two adjacency numbers (δ values). The larger these δ values, the smaller the term, as is evident in Table 2. A large δ value means the atom is highly branched. Thus, the 4 in a (1,4) bond indicates and (partially) characterizes a fragment with a structure as shown in Figure 3a. A (2,4) bond is a fragment typified by Figure 3b. The bond indices for these fragments are 0.500 and 0.354, respectively.

The encircled fragment in Figure 3b contributes a smaller portion to any structure of which it is a part, than does the encircled fragment in Figure 3a. This is obvious because the (2,4) bond contribution is made up of the space between the atoms (the overlap region) and 1/2 and 1/4, respectively, of the total contribution of the two atoms forming the bond. The (1, 4) bond, on the other hand, contributes the intervening overlap region and 1/1 and 1/4, respectively, of the total contribution of the two atoms composing this bond.

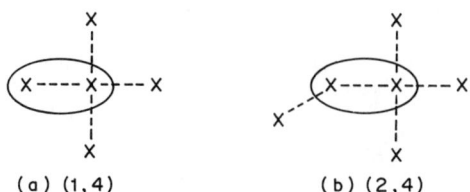

(a) (1,4)   (b) (2,4)

FIGURE 3. Structural fragments encoded by (a) a
(1,4) bond descriptor and (b) a (2,4) bond descriptor

To this point the development has relied on a parallel between bond composition and bond contribution to the overall structure of a molecule. If we examine the δ values in our alkane examples, we can readily see that they bear a relationship to the structure of the group of atoms implied in the skeleton structure, and hence to the bonds composing that group.

Two graphs of molecular structure are shown in Figure 3. Consider the (1,4) bond in Figure 3a. The value of 1 for the terminal atom denotes a structure formed from a carbon atom and from three hydrogen atoms (suppressed in the skeleton structure). The structure of the atom with an adjacency of 4 makes a smaller contribution to the overall structure, since there are no hydrogen atoms attached. The structural contribution comes from the extension of the molecule in all directions around this atom.

Compare the (1,4) bond with the (2,4) bond from Figure 3. The atom with an adjacency of 2 contributes two hydrogens to the structure of the bond. The contribution of the other atom in the (2,4) bond is the same as for the atom with an adjacency of 4, described above. Thus, the index for the (1,4) bond reflects a larger structural contribution to the molecule than does the (2,4) bond index, primarily due to the larger relative contribution of the atom with an adjacency of 1 compared to the carbon with an adjacency of 2.

Based on this development, we expect that the indices for the alkanes may carry information relevant to the structure of these compounds. If we wish to relate structure to geometry, it must be borne in mind that the volume and the surface area of a molecule are not sharply defined geometrical properties. There is no outer boundary to an atom; hence some particular basis must be used to assign the radii used to define some approximate surface.

An important test of the information content, as described thus far, arises from examination of the relationship of these indices for a series of molecules to a physical property, which is dependent on the structure. Such a comparison is made in the next section.

D. Relationship to Cavity Surface Area

For several years use has been made of a computed geometric property of molecules. This property, which can be examined to see if these indices carry relevant structure information, is the cavity surface area[30]. This is a computer-calculated property which is based on a spherical shape for each atom in a molecule and takes into account a suitable overlapping of these spheres in the formation of bonds. To each radius of an atom is added a radius for a solvent water molecule which would be in closest contact. The radii for the atoms, including the water molecule (taken as a sphere), are the van der Waals radii. The surface area is then computed over this composite of interpenetrating spheres.

Data for the 10 alkanes in Table 3 indicate a clear relationship between this computed

TABLE 3. Cavity surface area and branching index

Alkane	Computed cavity surface area[a]	$\sum(\delta_i, \delta_j)^{-1/2}$ $^1\chi$
Isobutane	249.1	1.732
Butane	255.2	1.914
Neopentane	270.1	2.000
2-Methylbutane	274.6	2.270
Pentane	287.0	2.414
2,2-Dimethylbutane	290.8	2.561
3-Methylpentane	300.0	2.807
Hexane	319.0	2.914
2,4-Dimethylpentane	324.7	3.125
Heptane	351.0	3.414

[a]Data taken from Reference 30.

surface area and the pattern of branching, expressed here in the branching index $^1\chi$. The correlation coefficient of the linear regression obtained for this data set is $r = 0.984$. We can say that the index provides enough information to permit a relative description of structure features governing this calculated molecular property. This is a major achievement in view of the quality of the relationship and the simplicity of the algorithm leading to the indices. The index is a dimensionless quantity derived by a calculation from numbers (*that do not depend upon adjustable parameters*) which reflect the adjacency relationship of atoms in the carbon skeletons.

This branching index is derived from a mathematical operation on a skeleton formula which conveys no quantitation of bond lengths or angles; thus, some information is lost. Structural characteristics in three dimensions, such as conformation, are beyond the ability of this kind of treatment to quantitate. Nevertheless, at this point, enough quantitative information is developed about alkanes to permit a development of a formal algorithm which will lead to modeling of a QSAR with a property.

E. Development of a Formal Algorithm

The development described above in general terms is an algorithm which leads to a structure index. There are certain advantages in employing formal mathematical expressions to convey an understanding of the entire procedure and to permit extension of the method.

The general name for the structural description method utilizing the adjacency relationships of molecular skeletons was selected to be *molecular connectivity*[9]. The number assigned to a skeleton atom describing its adjacency relationship is called the simple connectivity value (or *simple delta value*) of the atom. In the development the δ values were used for the first-order subgraph (or bond) between atoms i and j. The index for the entire molecule, in this case, the molecular connectivity index of the first order, is designated by the Greek letter chi, $^1\chi$, and is computed as in equation 2,

$$^1\chi = \Sigma(\delta_i\delta_j)^{-1/2} \qquad (2)$$

where the summation is over all bonds ($n - 1$ edges for noncyclic alkanes; n is the number of carbon atoms). In the absence of any further designations associated with $^1\chi$, it is understood that the index is based on δ_i values for simple connectivity of atoms in the skeleton structure. It is customary to refer to this index for the molecule as the χ index ('chi index'), or, more specifically, the simple $^1\chi$ index ('chi-one index').

F. Molecular Connectivity Indices over Larger Subgraphs

The results recorded in Table 3 suggest that the $^1\chi$ index is parallel to the cavity surface area. It is unreasonable to think that a single numerical index for each molecule could capture the complete essence of the structure and provide an entire basis for describing a complex property like the cavity surface area or the many properties of a given molecule. Several indices describing the many facets of molecular structure must be necessary to relate to any single property and, most certainly, to the many properties of a molecule.

An analogy can be found in plane geometry. One property of a triangle is its area. In order to compute the area we must know two structural features, the base length and height. Similarly, the cavity surface area value must be a result of more than one structural feature contributing to this property. We will consider ways to increase the amount of structure information which can be obtained from the molecular skeleton (chemical graph). Our considerations up to this point have focused on individual atoms and pairs of atoms connected by the graph edges. Let us now consider groups of edges within the skeleton. We shall find that each skeleton is characterized by a different profile of counts of various groupings of edges. This distinctive profile is the basis for what has become known as the molecular connectivity method[8,9].

G. Molecular Connectivity

In Figure 4, the graph of two heptane isomers along with a decompostion of the graphs into subgraph fragments are shown. These particular subgraphs (contiguous three-atom, two-edge paths) are shown below each skeleton formula. In this case each molecule has six two-bond subgraphs. For molecule (a) the subgraphs are (1,3,1), (1,3,2), (1,3,2), (3,2,2), (2,2,2) and (2,2,1) using the bond designation convention previously

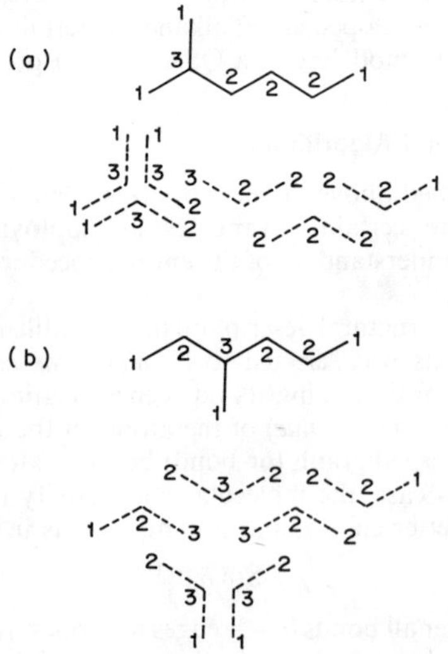

FIGURE 4. Decomposition of (a) 2-methylhexane and (b) 3-methylhexane into second-order (two-bond) fragments

employed. Molecule b is composed of the following two-bond subgraphs: (1,2,3), (1,3,2), (2,3,2), (2,3,1), (3,2,2) and (2,2,1).

Using the reciprocal-square-root algorithm developed above, we convert the two-bond descriptions to reciprocal-square-roots of the δ-value products and sum for each molecule. For molecule a the process is as in equation 3, where the [$\cdots$] means 'function of'.

$$a = \mathbf{f}[(1,3,1), (1,3,2), (1,3,2), (3,2,2), (2,2,2), (2,2,1)]$$
$$= \mathbf{f}[3, 6, 6, 12, 8, 4]$$
$$= \mathbf{f}[1.732, 2.449, 2.449, 3.464, 2.828, 2.000]$$
$$= 0.577 + 0.408 + 0.408 + 0.289 + 0.354 + 0.500$$
$$= 2.536 \tag{3}$$

This calculation leads to an index value a = 2.536. Since this index is calculated from two-bond length fragments (two-path length subgraphs), we call this index a second-order index, or $^2\chi$. The $^2\chi$ value for skeleton b is 2.302.

It is noteworthy that the first-order χ indices, $^1\chi$, for both molecules a and b are identical: 3.808. Thus the simple $^1\chi$ for these isomers conveys the same information, or, putting it another way, this index does not provide a quantitative differentiation between these molecules. These $^2\chi$, however, does differentiate between the two structures. Structure information based on the skeletal edges represent these two molecules in an identical manner whereas the information based on edge-pairs does discriminate between the two molecules.

In what ways does the information encoded in $^1\chi$ differ from $^2\chi$? To reveal this, we can examine these two indices for the heptane isomers in Table 4. The most striking observation is that the first-order index, $^1\chi$, is computed from the same number of

TABLE 4. First- and second-order χ indices for heptanes

	Chemical graph	No. of 1^0 subgraphs	$^1\chi$	No. of 2^0 subgraphs	$^2\chi$
1	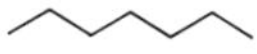	6	3.414	5	2.061
2	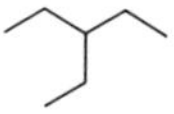	6	3.346	6	2.091
3	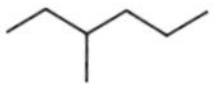	6	3.308	6	2.302
4	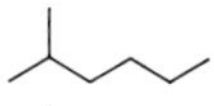	6	3.270	6	2.536
5	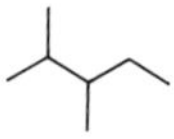	6	3.181	7	2.630
6	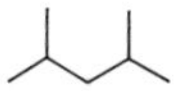	6	3.126	7	3.023
7		6	3.121	8	2.871
8	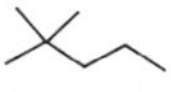	6	3.061	8	3.311
9		6	2.943	9	3.521

first-order subgraphs, 6, whereas $^2\chi$ is computed from a variable number of 2^0 (second-order) subgraphs. Also, the increase in $^2\chi$ numerical values is inverse to the $^1\chi$ order. Finally, we note that, in general, an increasing $^2\chi$ parallels an increase in the degree of branching of the heptanes.

The $^2\chi$ index appears at first glance to encode information of a similar nature to the $^1\chi$ index for these alkanes. A statistical analysis has shown that $^1\chi$ and $^2\chi$ are covariant to an extent $r = 0.66$ for a set of 75 alkane isomers. It must be emphasized that this relationship is true for alkane structures but not necessarily for all types of molecules. Even within the alkane structures, there is enough difference between the information in $^1\chi$ and $^2\chi$ that one could find that both indices, in concert, reflect structure features contributing in a somewhat different way to the numerical value of a property.

The basis of the information content in $^2\chi$ is twofold: (1) the number of second-order subgraphs, which varies with degree of branching, and (2) the variation in the weighting as the structure varies. There is an increase in the number of 2^0 subgraphs at points of branching in the branched isomer compared to the normal isomer. We compare in Figure 5 isomers with (a) a terminal ethyl, (b) a terminal isopropyl and (c) a terminal tertiary butyl group. It is apparent that these branch points contribute to an increasing number of subgraphs. Compound a contributes two subgraphs involving the first

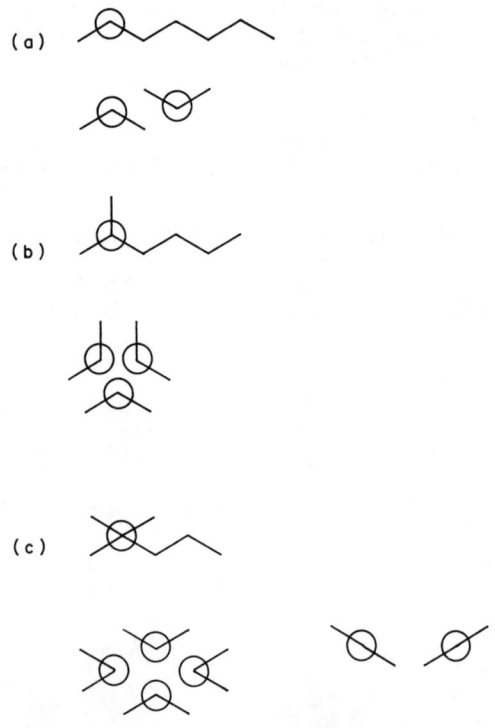

FIGURE 5. The increasing number of subgraphs of second order generated by increasing the degree of branching. Illustrated are a terminal (a) ethyl, (b) isopropyl and (c) tertiary butyl group

branched carbon (encircled), compound b contributes three subgraphs involving the first branched carbon while compound c contributes six subgraphs involving the branched carbon. Thus, the branching attribute of structure manifests itself as an increased number of subgraphs, contributing to the magnitude of $^2\chi$. In the case of the first-order χ index, the branching in the alkanes leads to more modest numerical differences due only to changes in the δ values from 1 to a maximum of 4. As a consequence the range of $^1\chi$ indices among heptanes is only 0.471, while the range for $^2\chi$ indices is much larger at 1.460.

A consequence of this difference in the structural attribute emphasis, due to δ values and due to the number of subgraphs, is that the $^1\chi$ index is more size-dependent, while the $^2\chi$ index encodes more information about branching. It is therefore expected that the quality of QSARs can be increased by combining the $^1\chi$ with the $^2\chi$ index.

Since there is a gain in structural information by considering an extended index such as $^2\chi$, let us consider more extended indices by computing the $^3\chi$ indices from 3^0 (three-path) subgraphs. These would be derived from subgraphs of three contiguous bonds (four atoms) within the skeleton structure. The $^3\chi$ indices along with the number of 3^0 subgraphs for the heptane isomers are shown in Table 5. As we shall see shortly, there is more than one type of third-order subgraph. For purposes of clear identification it is necessary to give an additional designation to the chi index: a subscript representing the type of subgraph on which it is based. For this third-order path-based index, the subscript P is used, as in $^3\chi_P$[8,9].

The structure variation among the heptanes shows that the number of 3^0 subgraphs ranges from four to six while the numerical range of the $^3\chi_P$ values is less than 1. Neither the number of 3^0 subgraphs nor the $^3\chi_P$ values appear to be related to what is commonly called branching. Thus, the last four molecules are multiply branched, yet the number

TABLE 5. Third- and fourth-order subgraph counts and χ indices for heptanes

Chemical graph	No. of 3^0 subgraphs	$^3\chi_P$	No. of 4^0 subgraphs	$^4\chi_P$
1	 4	1.207	3	0.677
2	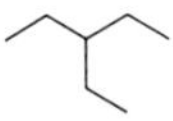 6	1.732	3	0.866
3	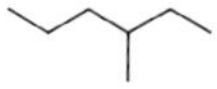 5	1.478	3	0.697
4	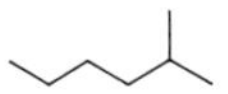 4	1.135	3	0.612
5	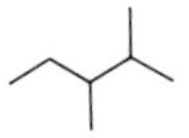 6	1.782	2	0.471
6	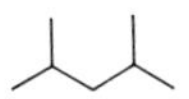 4	0.943	4	0.943
7	 6	1.914	1	0.250
8	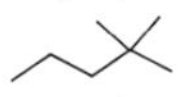 4	1.000	3	0.750
9	 6	1.732	0	0.000

of subgraphs alternate between four and six and the $^3\chi_P$ numerical values span the full range of the values for all heptanes.

We can conclude that less obvious structural characteristics are encoded in the $^3\chi_P$ values for heptanes. For example, there are more subgraphs and hence higher $^3\chi_P$ values for molecules which (a) are branched on adjacent atoms (molecules 5 and 9), (b) have a branch longer than one atom (molecule 2) or (c) have midchain quaternary carbon atom (molecule 7). In contrast, low values of $^3\chi_P$ are found for more elongated molecules or those with only one branched carbon (molecules 1,4,6 and 8). To the extent that these structural attributes may contribute significantly to the numerical value of a property, the $^3\chi_P$ index carries information of a relative numerical magnitude among heptanes. These computed quantities may be related to structure attributes other than branching, such as compactness, details of skeletal arrangement or shape. Some of these considerations will be described in a later section.

Another higher-order chi index can also be defined. The $^4\chi_P$ index, derived from subgraphs of four contiguous bonds in a manner described above, is also shown in Table 5. Molecule number 9 has a value $^4\chi_P = 0$ for this index, since there are no contiguous four-bond subgraphs. Thus the information content of $^4\chi_P$ is not large for this molecule. Five of the molecules in Table 5 have close $^4\chi_P$ values due to the common occurrence of three 4^0 subgraphs. We can conclude that this index, in the case of the heptanes, embraces too large a fragment to carry definitive information about structure, as we have previously analyzed for lower-order indices. Higher-order indices, $^5\chi_P$, $^6\chi_P$ and beyond, are obviously possible and may convey significant information about molecules larger and more complex than the heptane isomers. QSARs have been reported using these higher-order indices[8,9].

So far we have only considered subgraphs of contiguous edges, called paths; it is possible to consider other types of simple fragments of structure. For example, we may dissect the heptane isomers into 3^0 subgraphs in which three bonds join at a common atom[9]. We call this a 3^0 cluster subgraph. Figure 6 shows this decomposition and the calculation of the corresponding chi index, $^3\chi_C$. The subscript C indicates a cluster subgraph. Thus, a $^3\chi_C$ index is derived from three bonds with one atom common to all three[8,9].

The information from the $^3\chi_C$ indices for these heptanes deals largely with branching and the multiplicity at branched atoms. The number of subgraphs for molecules with

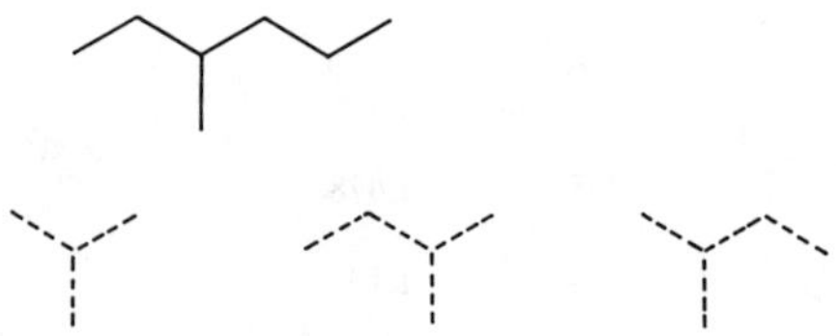

(2,3,1,2)   (1,2,3,1,2)   (2,3,1,2,2)

(2·3·1·2)   (1·2·3·1·2)   (2·3·1·2·2)

12   12   24

0.289   0.289   +   0.204

$^3\chi_C = 0.289$ $^4\chi_{PC} = 0.493$

FIGURE 6. The $^3\chi_C$ and $^4\chi_{PC}$ subgraphs of 3-methylhexane with calculated indices

quaternary carbons is four in contrast to a single subgraph for every tertiary carbon. Among the molecules with tertiary carbons, those with terminal branching yield larger $^3\chi_C$ values than those with midchain branching. Thus, this index is quite specific for deriving relative numerical structural information about the branching feature. To distinguish the $^3\chi_C$ cluster index from the $^3\chi$ index previously described, we append the subscript P in order to indicate a path-based index: $^3\chi_P$.

At least two more subgraphs are known to carry information. One is the path/cluster subgraph shown in Figure 6. This is a fourth-order (4^0) subgraph leading to an index designated as $^4\chi_{PC}$. An analysis of this index and its information content reveals a role in the structural description of substituted benzene rings[31]. The $^4\chi_{PC}$ index arises from each branch point in a skeleton and from a ring substituent. Each ring substituent generates at least two path/cluster subgraphs of order 4. With substituents longer than one atom length, the number of subgraphs per substituent is at least three. The $^4\chi_{PC}$ index also encodes information about the orientation of ring substituents. In general, the greater the adjacency or crowding of single vertex substituents, the larger the numerical value of the $^4\chi_{PC}$ index[31].

Further considerations of subgraph types leads to fragments containing a ring or a circuit. For the organic chemist the term ring is well known. Naphthalene possesses two six-membered rings. The graph theorist also recognizes the presence of another cyclic feature, the ten-membered ring which is usually called a circuit. Although the specific terminology is a matter of convention, several indices can be generated and shown to contain structural information, which have been called the chain subgraphs[9]. This structure feature is a cyclic arrangement (or chain) of atoms. The indices, designated CH as in $^6\chi_{CH}$ for six-membered rings, are calculated in the manner previously described. The fragment designations are converted to products and the reciprocal of the square root computed to yield the index value. This index is a specific descriptor for the cyclic structural feature, including presence or absence as reflected by a value of zero for the chain index. The smaller the chain index, the more substituted the positions are on the ring[8,9].

H. General Formalism of Molecular Connectivity: $^m\chi_t$ and $^m\chi_t^v$

For subgraph orders greater than two ($m > 2$), there are several types of subgraphs: path, cluster, path/cluster and chain as described above. For the third order, however, there is no path/cluster type. Each type of index is defined in a manner analogous to the lower-order indices as a product of reciprocal square roots. The general relation is given in equation 4.

$$^mc_S = \prod(\delta_i)_S^{-1/2} \tag{4}$$

The product is over the $m + 1$ delta values in the subgraph of order m, except for the chain (or ring) type in which m is the number of atoms in the ring. The chi index of order m and based on subgraph type t is obtained as defined in equation 5.

$$^m\chi_t = \sum {}^m c_s \tag{5}$$

Complete enumeration of each type of subgraph becomes computationally intense as structure complexity increases. All the necessary computations are carried out by computer software along with calculations for many other quantities of interest in QSAR[22].

The chi indices represent the whole graph. Each is composed of terms from the various parts of the graph, but the summation covers the whole graph. In this manner each index encodes in a particular manner the structure information resident in the molecular skeleton.

I. The Zero-order Index

In the development up to this point, we have dealt with structural fragments or subgraphs of various bond (edge) compositions. It is also possible to decompose a molecular structure into subgraphs of single atoms. Thus the subgraph value would be described by a single δ for the atom. Computing a connectivity index in the usual way, we would compute the reciprocal square root of each δ_i and sum these to form an index we can call a zero-order index, $^0\chi$[8,9].

This index is directly related to atoms, not bonds. It describes the number of atoms in a molecule and the branching in the molecule. For alkanes, this index is highly correlated with numerical values for $^1\chi$ indices, and more highly correlated with the number of atoms than are the other chi indices.

VI. MOLECULAR SHAPE: KAPPA INDICES

A. Background

The shape of an alkane molecule is a structure feature often discussed but difficult to quantitate. It is generally believed that molecular shape influences chemical events, however the encoding of this feature has not been well developed up to this time. One approach to quantifying shape has been to measure the effect of its presence upon some standard chemical reaction. From this effort has come the steric index of Taft[32] and later contributions by Charton[33], Hancock[34], Dubois and coworkers[35] and DeTar[36]. Other approaches have regarded molecules as hard-sphere geometric objects from which standardized dimensions can be calculated (Verloop and coworkers[37]) or size estimated from radii (Pearlman[38], Hermann[30] and Bondi[39]). Austel and coworkers[40] have proposed a scheme using branching and atom differences.

The definition of molecular shape in numerical terms for quantitative structure–activity analysis is plagued by a number of problems. A molecule has no real boundary surface but only probability domains of electrons. Geometric models are thus approximations of limited value. Conformational variation in some molecules can be great enough to make shape quantitation based on hard-sphere models impractical. Shape effect on chemical events may not be easily factored out from electronic effects, hence indexes based on reaction measurements may not contain pure shape information. Finally, a molecule is often a highly irregular object when viewed as a hard-sphere model. Many measurements in various directions may have to be made to encode anything approaching valuable shape information.

B. A Graph Model of Shape

Recognizing these problems and limitations, Kier has approached shape quantitation by breaking away from conventional geometric concepts of molecules as hard spheres[41-43]. Specifically, a molecule is regarded as an array of atoms with a bonding or no-bonding relationship to other atoms. This description is synonymous with a graph. That part of physical phenomena dependent upon shape is assumed to arise from structural features embedded in the molecular graph, representing the molecule.

We define the shape of a molecule as a summation of many structural attributes, each attribute describable by the quantitation of some feature of the graph. We define the graph as a representation of all nonhydrogen atoms in the molecule. The graph features selected to define the shape attributes are the counts of paths of various bond lengths. Thus the count of 2-bond fragment (second-order attribute) contains information about one attribute of structure which is part of the whole. This count is designated ^2P, or, in general, ^mP of order m.

Since shape is an attribute of all molecules, it is necessary to transform ^mP into an index which carries information for any number of atoms in the molecule. To accomplish this, we define a particular shape as having an intermediate relationship between two extreme shapes, each of which is easily defined both pictorially and numerically. These extreme shapes must be common to subsets of molecules of any number of atoms. The extremes selected for any order of attribute, m, are the maximum, $^mP_{max}$, and minimum, $^mP_{min}$, counts of paths in the molecular graph. The shape attribute of a particular molecule, i, is therefore given by equation 6:

$$^mP_{max} \geqslant {}^mP_i \geqslant {}^mP_{min} \tag{6}$$

where the number of atoms, A, is the same for all three structures. This intermediate numerical relationship must now be transformed into a single number for each attribute. To accomplish this, we must examine each and derive a suitable algorithm for the shape information of that attribute.

C. First-order Shape Attribute

For this attribute of shape, described by the counts of one-bond fragments, ^1P, Kier selected for $^1P_{max}$ the complete graph, where all atoms are bonded to each other. For any number of atoms, A, the value of $^1P_{max} = (A-1)A/2$. For the $^1P_{min}$ structure he has selected the linear graph where the value is $^1P_{max}$. In Table 6, entry no. 1 is the graph of $^1P_{max}$ where $A = 5$. Entry no. 2 is the graph of $^1P_{min}$ where $A = 5$.

The shape attribute of the first order for a molecule, i, lies somewhere between the complete graph and the linear graph. This is the extent of the definition of this shape attribute. This definition does not numerically define spheres, ellipsoids or other geometric figures; instead, the approach speaks of, and numerically defines, relationships to graphs which define extreme structures.

It is necessary to derive an algorithm which yields a numerical index for any molecule with A atoms and with ^1P_i edges. Based on the general expression (equation 6), the ratio of ^1P_i to the values for both extremes, $^1P_{max}$ and $^1P_{min}$, is suggested[39]. An index of shape of order one, $^1\kappa$, is given in equation 7.

$$^1\kappa = 2\,{}^1P_{max}\,{}^1P_{min}/({}^1P_i)^2 \tag{7}$$

The use of 2 in the numerator is a scaling factor which ensures the value $^1\kappa = A$ when

TABLE 6. Graphs of $^mP_{max}$ and $^mP_{min}$

No.	Graph	A	Counts of fragments		
			^1P	^2P	^3P
1		5	10		
2		5	4		
3		5		6	
4		5		3	
5		8			9
6	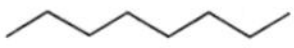	8			5

there are no cycles in the graph of the molecule. The index $^1\kappa$ can be expressed in terms of the count of atoms. A (equation 8).

$$^1\kappa = A(A - 1)^2/(^1P_i)^2 \tag{8}$$

This expression may be readily evaluated for any chemical graph from the atom count together with the count of edges, subgraphs of first order.

D. Second-order Shape Attribute

The second-order shape attribute is defined by the count of 2-bond paths, ^2P_i, and is related to the shape extremes $^2P_{max}$ and $^2P_{min}$. For $^2P_{max}$ Kier adopted for the extreme the nonlinear graph called the star graph, in which all atoms are adjacent to a common atom[38]. The star graph for $A = 5$ is shown in Table 6, no. 3. The numerical value of $^2P_{max}$ for any count of A is $^2P_{max} = (A - 1)(A - 2)/2$. For the other second-order shape attribute extreme, $^2P_{min}$, he used the linear graph, shown in Table 6, no. 4 where $A = 5$. In general, for any value of A, $^2P_{min} = A - 2$.

An algorithm expressing this second-order attribute can now be written down and a second-order shape index, $^2\kappa$, calculated (equation 9):

$$^2\kappa = 2\,^2P_{max}\,^2P_{min}/(^2P_i)^2 \tag{9}$$

The scaling value of 2 in the numerator ensures the value $^2\kappa = A - 1$ for all linear graphs. This is the same as $^2\kappa = $ (number of bonds). Equation 9 can be expressed in terms of the count of atoms, A (equation 10):

$$^2\kappa = (A - 1)(A - 2)2/(^2P_i)^2 \tag{10}$$

This expression may be readily evaluated for any chemical graph from the atom count together with the count of edge-pairs, subgraphs of second order. Since the counting of edge pairs is nontrivial for complex graphs, use is usually made of computer programs[21,22].

E. Third-order Shape Attribute

The count of paths of three contiguous bonds, ^3P, forms the basis for the quantitation of another shape attribute. This structure is compared to two extreme structures, $^3P_{max}$ and $^3P_{min}$. For the third-order attribute $^3P_{max}$ is chosen to be a twin star structure shown in Table 6, no. 5. For $^3P_{min}$, the linear graph, Table 6, no.6 is the representation when $A = 8$. In general, for any odd value of A, $^3P_{max} = (A - 1)(A - 3)/4$, and for any even value of A, $^3P_{max} = (A - 2)^2/4$. In general, $^3P_{min} = A - 3$. A suitable algorithm in which third-order shape information can be encoded from ^3P_i is given in equation 11.

$$^3\kappa = 4\,^3P_{max}\,^3P_{min}/(^3P_i)^2 \tag{11}$$

The scaling factor 4 is used in the numerator to bring the $^3\kappa$ values into rough equivalence with the other kappa values.

The $^3\kappa$ values can be expressed in terms of A using equations 12 and 13:

$$^3\kappa = (A - 1)(A - 3)^2/(^3P_i)^2 \qquad \text{when A is odd} \tag{12}$$

$$^3\kappa = (A - 3)(A - 2)^2/(^3P_i)^2 \qquad \text{when A is even} \tag{13}$$

These expressions may be readily evaluated for any chemical graph from the atom count together with the count of triple-edges, subgraphs of third order. Since the counting of paths of length three is nontrivial for complex graphs, use is usually made of computer programs[21,22].

F. Zero-order Shape Attribute

After introducing $^1\kappa$, $^2\kappa$ and $^3\kappa$ as shape indices, a possible extension arises in the form of a $^0\kappa$ index. It follows from the previous developments that $^0\kappa$ should be derived from ^0P fragments. These are logically the atoms of a graph treated in isolation. One attribute of an atom which should influence the shape of a molecule is the topological uniqueness of that atom within the molecule. The collective effect of topological uniqueness is the symmetry of the molecule.

Symmetry is certainly a shape attribute which must play some role in the influence of structure on function. One approach to the quantitation of symmetry is through the use of the Shannon equation for information content[44]. This subject has been studied quite thoroughly by Brillouin[45] and Bonchev[46]. Kier has made use of it to encode molecular symmetry and to relate it to physical properties[47].

Two numerical values may be derived from the Shannon equation. The first of these is i, the information content per atom (equation 14),

$$i = -\sum p_i \log(p_i) \qquad (14)$$

where p is the probability of randomly selecting an atom from the whole. The information content in the entire molecule with A atoms is iA. Brillouin[45] has modified the expression to give the redundancy of the molecule, R (equation 15):

$$R = 1 - (i/\log A) \qquad (15)$$

A shape index, $^0\kappa$, could be made equal to either of these symmetry/redundancy indices in an effort to encode information about this shape attribute. Computation of these zero-order indices depends upon proper classification of the atom (vertex) groups. For simple molecules, the chemist readily accomplishes this task by eye. For complex molecules and for automated computation use is made of computer programs[22].

G. Shape Information in the Kappa Values

By selecting the models for each order of shape attribute as we have, we can quantitate these attributes from their relationship to two extreme structures associated with each model. The first-order shape attribute of a molecule with a certain value of $^1\kappa$ lies between a linear graph and a complete graph, all with the same number of atoms.

TABLE 7. Range of $^m\kappa$ values and shape information encoded

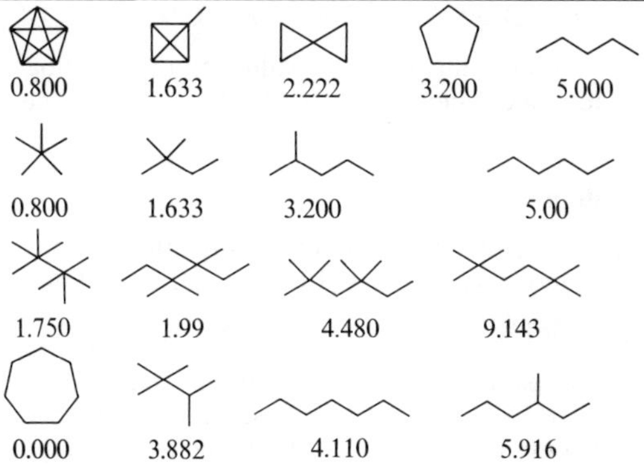

$A = 5$ $^1\kappa$					
	0.800	1.633	2.222	3.200	5.000
$A = 6$ $^2\kappa$					
	0.800	1.633	3.200		5.00
$A = 10$ $^3\kappa$					
	1.750	1.99	4.480	9.143	
$A = 7$ $^0\kappa$					
	0.000	3.882	4.110	5.916	

In Table 7, a range of $^1\kappa$ values is shown for graphs ranging from a linear structure to one with a maximum number of cycles. The structural information encoded in $^1\kappa$ is related to the complexity, or, more precisely, the cyclicity of a molecule. All noncyclic structures with five atoms would have $^1\kappa = 5.000$, or, in general, $^1\kappa = A$ for noncyclic structures.

In the case of $^2\kappa$, the structures and index values in Table 7 for $A = 6$ reveal that the encoded information relates to the degree of star graph-likeness and linear graph-likeness. Stated in more general terms, $^2\kappa$ encodes information about the spatial density of atoms in a molecule.

The information in $^3\kappa$ indices can be illustrated by the set of real and hypothetical structures in Table 7. The $^3\kappa$ values are larger when branching is nonexistent or when it is located at the extremities of a graph. We can say that $^3\kappa$ encodes information about the centrality of branching.

Finally, in Table 7 are shown the influence of several structures on the value of $^0\kappa$, which we compute as $^0\kappa = iA$ from the carbon skeleton. The greater the symmetry, the lower the $^0\kappa$ value. Stated another way, the greater the uniqueness of atoms, information content or negentropy, the larger the value of $^0\kappa$.

VII. SELECTED ALKANE STRUCTURE–PROPERTY MODELS

In this section a selection of alkane properties will be presented along with QSAR models based on topological methods. For more examples see Rouvray[5,6], Kier and Hall[8,9] and Trinajstić[7].

A. Heat of Atomization and Formation

A basic property relating to molecule stability is the heat of atomization or the heat of formation. Kier and Hall have developed models for both[8,9]. For 44 compounds with the best experimental data, heat of atomization is modeled on the number of carbon atoms, A, and the $^1\chi$ index as in equation 16[48].

$$\Delta H_a = 283.33A - 6.321\,^1\chi + 116.19, \qquad r = 0.9999, \quad s = 0.96, \quad n = 44 \qquad (16)$$

This data set includes propane through some nonanes.

In this and the other examples given here, r is the correlation coefficient, s the standard deviation and n the number of observations in the data set. The statistical quantities are determined by standard multiple linear regression methods.

For a larger data set on the gas-phase heat of formation, equation 17 is obtained:

$$\Delta H_a = 7.649A - 3.268\,^1\chi + 11.70, \qquad r = 0.9971, \quad s = 1.13, \quad n = 67 \qquad (17)$$

These high-quality relations illustrate the value of the topological chi indices. The addition of other topological indices to the model equations decreases the standard error[48].

B. Heat of Vaporization

The quantity which measures the enthalpy change upon vaporization is the heat of vaporization. A useful model can be based on the $^1\chi$ index alone (equation 18)[49].

$$\Delta H_v = 2.381\,^1\chi + 0.486, \qquad r = 0.9981, \quad s = 0.163, \quad n = 47 \qquad (18)$$

This data set includes propane through nonanes and up to hexadecane in normal alkanes. This model, based on $^1\chi$, is superior to one based on the number of carbon atoms for which the standard error is 0.476. The model may be further improved by the addition of higher-order chi indices, leading to an error near the experimental level, estimated by

Somayajulu and Zwolinski to be $0.03\,\text{kcal}\,\text{mol}^{-1}$ [50]. This QSAR, obtained by Kier and Hall[8,9], is useful for estimation of values for other alkanes.

C. Molar Refraction

Another property of wide interest for alkanes is a measure of polarizability, the molar refraction R_m, which is based upon the experimental measurement of liquid density and refractive index. For a data set of high experimental quality, the following QSAR is obtained (equation 19)[51].

$$R_m = 7.756\,^1\chi + 2.207\,^2\chi + 3.707, \qquad r = 0.9997, \quad s = 0.122, \quad n = 46 \qquad (19)$$

Addition of the $^3\chi_P$ and $^4\chi_C$ indices lowers the standard error to 0.045. Molar refraction determined in this manner is dependent upon liquid density measurement, which has also been modeled by Kier and Hall[52]. This present data set includes pentanes through some decanes.

D. Boiling Point

A widely used indicator of intermolecular forces is the boiling point (T_b). Kier and Hall obtained the following QSAR (equation 20)[53].

$$T_b = 55.69\,^1\chi + 4.71\,^4\chi_{PC} - 96.13, \qquad r = 0.9969, \quad s = 2.53, \quad n = 51 \qquad (20)$$

The data set includes pentane through nonanes. Over a wider range of compounds and boiling points, the dependence of boiling point on atom count is nonlinear[54].

E. Liquid Density

The liquid density, a measure of molecular packing and molar volume, has been modeled by Kier and Hall[52]. For normal alkanes the density is inversely proportional to the count of carbon atoms. Hence in this model, an inverse $^1\chi$ index is used. When d_4^{20} is plotted against $^1\chi/A$, a set of convergent lines is obtained for the various classes of branched alkane isomers, such as 3-methyl, 2-methyl, 2,3-dimethyl, etc. The focus of these convergent lines is a point on the density axis, 0.83 [52].

When the $^3\chi_P$ index is added to reciprocal $^1\chi$, the following QSAR equation is obtained (equation 21)[52]:

$$d_4^{20} = 0.7348 - 0.2929/^1\chi + 0.0030\,^3\chi_P, \qquad r = 0.9889, \quad s = 0.0046, \quad n = 82 \qquad (21)$$

The data set includes pentanes through nonanes.

F. Water Solubility

A limited data set is available for the solubility (S) of alkanes in water. The following model was obtained by Kier and Hall[55]:

$$\ln(S) = -1.505 - 2.533\,^1\chi, \qquad r = 0.958, \quad s = 0.511, \quad n = 18 \qquad (22)$$

The equation is established between $^1\chi$ and the natural logarithm of the water solubility. The data set includes butane through octane. QSAR models using chi indices have also been developed for other classes of organic molecules[9].

G. Octanol–Water Partition Coefficient

A quality related to the water solubility is the partition coefficient, P, for which a large quantity of data is available. QSAR models are established using the logarithm of the partition coefficient. Kier and Hall have developed QSAR models for several classes of organic molecules, including alkanes, as follows[56]:

$$\log(P) = 0.884^1\chi + 0.406, \qquad r = 0.975, \quad s = 0.160, \quad n = 45 \qquad (23)$$

This data set includes alkanes, alkenes, alkynes and aromatic hydrocarbons.

H. CNDO/2 Atomic Partial Charge

Hall and Kier investigated the possible relationship between topological indices and the partial charge computed by semiempirical MO methods[57]. The following model was obtained in a comparison of atomic indices from MO theory (partial charge) and a modified version of the chi indices. In this case the chi indices were partitioned into atomic contributions. Of course, the $^0\chi$ index is already defined as an atom index; the others are given the strike-through symbol, χ (equation 24):

$$q_C = 0.0162 + 0.049^0\chi + 0.0493^1\chi + 0.0217^2\chi, \qquad r = 0.998, \quad s = 0.0014, \quad n = 19$$

$$(24)$$

This model indicates that for alkanes there is a significant relation between the computed partial charge and the molecular topology as represented in the chi indices. The computed partial charge depends equally, in this model, upon the intrinsic atom state and the effect of the bonded atoms. Atoms, one bond removed, have a smaller effect.

I. Chromatographic Retention Index

Millership and Woolfson have investigated the relation between chi indices and the chromatographic retention index (RI). They reported a QSAR based on two chi indices (equation 25)[58].

$$RI = 0.719^1\chi + 0.125^2\chi - 0.242, \qquad r = 0.998, \quad s = 0.045, \quad n = 18 \qquad (25)$$

This data set included ethane through octanes.

J. Pitzer Acentric Factor

A quantity often used in calculations on real gases is the Pitzer acentric factor, ω. Pitzer defined the factor as a means of characterizing deviation from spherical symmetry for use in corresponding state model[59]. The acentric factor is obtained from experimental data, as follows: $\omega = \log(P'_r) - 1.0$ in which P'_r is the reduced pressure P/P_c at the reduced temperature of $0.7°C$, P_c being the critical pressure. This definition is consistent with acentric factor values of zero for rare gases.

Passut and Danner[60] have presented a list of acentric factors obtained for hydrocarbons. Kier[31] obtained an excellent correlation for 51 alkanes with the first- and second-order kappa indices (equation 26):

$$\omega = 0.019^1\kappa + 0.023^2\kappa + 0.083, \qquad r = 0.995, \quad s = 0.014, \quad F = 2277, \quad n = 51 \qquad (26)$$

Only two residuals exceed twice the standard deviation. This equation may be of sufficient quality to predict other alkane values since the average residual is less than 3%.

K. Taft Steric Parameter

Taft has determined a steric parameter from experimental hydrolysis rate data[32]. This parameter, denoted by E_s, was designed to embody the influences of substituents or groups with respect to shape, size or bulk, that is, their steric influence on reaction rate and equilibria. The Taft steric factor was obtained from the ratios of basic and acidic rates of hydrolysis of ethyl acetate as compared with those of other esters in which the substituent groups in the acyl group are varied.

For a shape analysis using the kappa indices, Kier selected the alkyl groups. It is assumed that the effect is entirely due to steric effects for alkyl groups, whereas there may be a residual electronic effect when heteroatoms are involved. Many of the groups are small, having fewer than four or five carbon atoms. For such a set, the higher-order kappa indices may be small or zero, or cover a narrow range. For this reason Kier resorted to a technique in which a dimer is constructed for each alkyl group. One-half of the computed value is used as the corresponding kappa index[31].

For the 25 alkyl groups in the Taft data, Kier found the following QSAR:

$$-E_s = -0.395 {}^0\kappa + 0.780 {}^1\kappa - 0.372 {}^3\kappa - 0.698,$$
$$r = 0.976, \quad s = 0.27, \quad F = 142, \quad n = 25 \qquad (27)$$

In this model the zero-order index is based on equation 14 as ${}^0\kappa = iA$, where A is the number of atoms in the alkyl group and symmetry classification is based only on the alkyl group, not on the dimer version. The standard deviation is small enough to permit estimation of Taft steric values for groups for which no experimental data are available.

The role of the kappa indices in equation 27 indicates that higher symmetry, as shown in the ${}^0\kappa$ index, leads to a higher E_s value; increased size, reflected in ${}^1\kappa$, favors smaller Taft values and branching in the R group, revealed in ${}^3\kappa$, at the attached carbon, leads to larger values whereas more distant branching has the opposite effect.

VIII. ELECTROTOPOLOGICAL STATE: GRAPH-INVARIANT ATOM INDEX

The preceding discussion deals exclusively with indices of the whole molecule, that is, indices based on various structures and summed over the whole molecule. These indices are said to be graph-invariant, because they are independent of the manner in which the graph is labeled or numbered. For this reason, the indices are said to represent the information resident in the structure rather than some unique feature of the labeling scheme.

It is worthwhile to pursue the development of a graph-invariant index of each atom (vertex) in the molecular skeleton (graph). Chemists often need to compare the nature and/or the effect of one atom with another within a given molecule or among molecules.

In a recent series of papers, Kier and Hall have presented the development of such a set of indices. In these papers each skeletal atom is considered to be characterized by a topological (and electrotopological) state, determined by the character of the atom itself as well as its relationship to the whole of the molecule. In the first paper, an atom index was developed as the basis for the determination of topological equivalence within a given molecule[61]. The topological equivalence index for each atom, T_i, possesses the same numerical value for all topologically equivalent atoms. Further, the ratio of atom values within a molecule, such as the methyl to methylene value, is also an index characteristic of the specific set of atoms. Examples were given to show how to characterize an ethyl group, an isopropyl group and the bay region in polyaromatic hydrocarbons[61]. Further, the sum of the T_i values for a molecule is a highly discriminating topological index called the total topological equivalence index, τ.

Hall and Kier showed, however, that this index is largely topological and extended their analysis to increase the amount of electronic information. This work led to a series of papers in which the electrotopological index was developed[62-65].

In the development of the electrotopological state index, Kier and Hall viewed each atom as possessing an intrinsic state which is modified through interaction with the field of every other atom within the molecular graph. The topological aspect of the graph vertex is derived from its relative structural status in the graph, that is, being on the surface of the molecule or buried within the molecule. For example, the methyl groups in neopentane are on the surface, possessing a mantle status. On the other hand, the central quaternary atom is a buried atom. This topologial attribute of the atom can be represented by the simple δ value, the number of skeletal neighbors. For carbon atoms in an alkane, δ ranges from 1 to 4. Terminal atoms, methyl groups, have a mantle status whereas atoms with a δ of 3 or 4 have a more buried status. The topological aspect of the intrinsic state value is selected to be this simple delta value (δ).

The electronic aspect to be encoded into the intrinsic state requires reference to the electronic structure. For alkanes this is accomplished rather easily because there is a little variation in the electronic character among the carbon atoms, at least in the intrinsic sense. In their development, Kier and Hall had in mind all general classes of organic molecules and developed the following relation for the intrinsic state, I, of a given atom:

$$I = (\delta^v + 1)/\delta \tag{28}$$

The definition may be applied to saturated carbon-hydride groups. From this definition the intrinsic state values for methyl, methylene, methine and quaternary carbons are 2.000, 1.500, 1.333 and 1.200, respectively.

The final step in the derivation of the electrotopological state is the combination of the atom intrinsic state expression with the experession for the perturbation by all other skeletal atoms. The perturbing influence on atom i by atom j was assumed to be proportional to the difference between the atom intrinsic values, $I_i - I_j$. Further, the perturbation falls off as the square of the distance between the two atoms. The distance, in this case, is taken to be the number of atoms in the (shortest) path between atoms i and j, namely r_{ij}. The overall perturbation on atom i, ΔI_i, is taken as a sum over all other atoms in the skeleton (molecular graph):

$$\Delta I_i = \sum (I_i - I_j)/r_{ij}^2 \tag{29}$$

The electrotopological state for atom i is defined by equation 30:

$$S_i = I_i + \Delta I_i \tag{30}$$

Note that, since ΔI_i may be positive or negative, the S value may be greater or lower than the corresponding I value. In general, terminal methyl groups tend to have S values greater than 2.000 whereas methine groups and quaternary carbon atoms tend to have smaller values.

Figure 7 shows several normal alkanes along with the S values for the skeletal groups. Lengthening the chain results in the S values of methyl groups approaching 2.32 whereas the midchain methylene groups approach 1.50, the intrinsic state value. The only methylene group to depart significantly from the intrinsic state value is the group immediately adjacent to the terminal methyl group.

In Figure 8 are shown several branched alkanes in which significant perturbations are indicated. Among the butane isomers, the methyl groups have S values near 2.2 whereas the methylene group has the typical penultimate carbon value, near 1.32. The tertiary-group S value is below 1.00. A similar pattern appears among the pentane isomers in which the methine group in 2-methylbutane has the value 0.88. The quaternary atom in 2,2-dimethylpropane has the lowest value encountered among this set of molecules, 0.50.

Inspection of these S values suggests a quantization of values for skeletal groups. Methyl groups lie above 2.16 whereas methylene groups lie in the approximate range of 1.25 to 1.50. The methine groups tend to be in the range from 0.80 up to 0.95 and

1.34
2.21 1.39
1.36 1.44
2.25 1.42
1.37 1.46
2.26 1.43 1.46
1.37 1.47 1.48
2.29 1.44 1.48 1.48
1.37 1.47 1.49 1.49
2.30 1.44 1.48 1.49

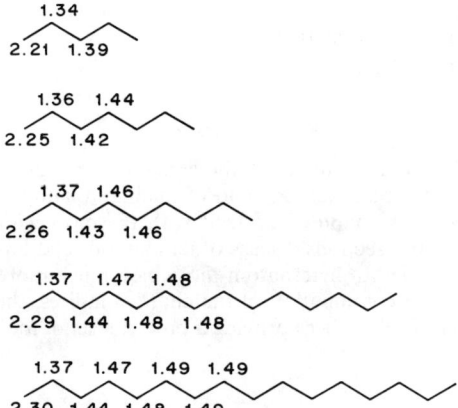

FIGURE 7. Electrotopological state values for the methyl and methylene groups in several normal alkanes. Values are omitted for the symmetrically related groups on the right side of the molecule

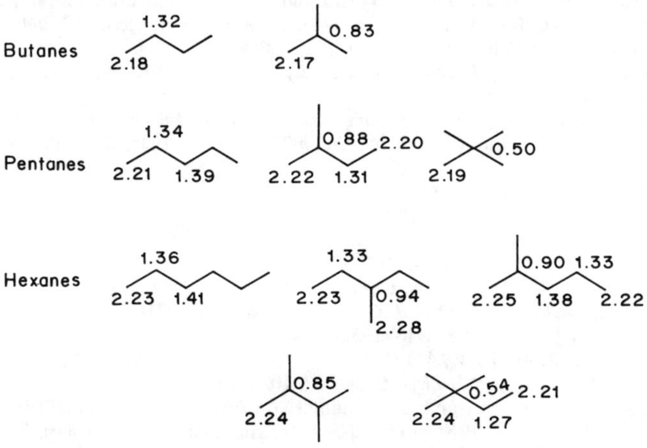

Butanes
1.32
2.18
0.83
2.17
Pentanes
1.34
2.21 1.39
0.88 2.20
2.22 1.31
0.50
2.19
Hexanes
1.36
2.23 1.41
1.33
2.23
0.94
2.28
0.90 1.33
2.25 1.38 2.22
0.85
2.24
0.54 2.21
2.24 1.27

FIGURE 8. Electrotopological state values for the skeletal groups in the butane, pentane and hexane isomers. Values are given only for the symmetry independent groups within each molecule

the quaternary atoms from 0.40 up to 0.60. On this basis the electrotopological *S* values can be used as a basis for the classification of skeletal groups in alkanes.

A small electronic effect is also revealed in this form of analysis. Examination of graphs similar to those in Figure 8 shows the influence on a methyl group by the ethyl, isopropyl and tertiary groups as −0.13, −0.17 and −0.19. The order of electron donation to a methyl group is *tert*-butyl > isopropyl > ethyl > methyl. This corresponds to the ordering of electronegativity as follows: methyl > ethyl > isopropyl > *tert*-butyl. This effect has been described by Fliszar[66].

IX. CONCLUSIONS

The selection of QSAR models presented above gives a general indication of the types of relationships which can be developed and of the high quality obtained with the use of topological indices. This chapter has dealt with the most widely used indices, but references given in the early sections discuss other methods and indices. Although much more effort has been applied to heteroatom molecules, much more remains to be done on alkane properties. The combination of the chi $(^m\chi_t)$ indices, the kappa $(^m\kappa_a)$ indices and the electrotopological (S_i) indices provide a powerful set of tools in the development of QSARs.

X. REFERENCES

1. H. Wiener, *J. Am. Chem. Soc.*, **69**, 17, 2636 (1947).
2. H. Wiener, *J. Phys. Chem.*, **52**, 425,1082 (1948).
3. J. R. Platt, *J. Phys. Chem.*, **56**, 328 (1952).
4. J. R. Platt, in *Encyclopedia of Physics* (Ed. S. Flügge), Vol. 37, Springer-Verlag, Berlin, 1961, p. 173.
5. D. H. Rouvary, *Am. Sci.*, **61**, 729 (1973).
6. D. H. Rouvary, *Sci. Am.*, **255**, 40 (1986).
7. N. Trinajstić, *Chemical Graph Theory*, Vols. I and II, CRC Press, Boca Raton, FL, 1983.
8. L. B. Kier and L. H. Hall, *Molecular Connectivity in Structure–Activity Analysis*, Research Studies Press, Wiley, Letchworth, England, 1986.
9. L. B. Kier and L. H. Hall, *Molecular Connectivity in Chemistry and Drug Research*, Academic Press, New York, 1976.
10. E. L. Eliel, *Stereochemistry of Carbon Compounds*, McGraw-Hill, New York, 1965, p. 1.
11. N. L. Biggs, E. K. Lloyd and R. J. Wilson, *Graph Theory 1736–1936*, Clarendon Press, Oxford, 1976, Chapter 4.
12. E. Cayley, *Phil. Mag.*, **67**, 444 (1874).
13. G. Polya, *Acta Math.*, **68**, 145 (1937).
14. G. Polya and R. C. Read, *Combinatorial Enumeration of Groups, Graphs, and Chemical Compounds*, Springer-Verlag, Berlin, 1987.
15. H. R. Henze and C. M. Blair, *J. Am. Chem. Soc.*, **53**, 3042, 3077 (1931).
16. C. C. Davis, K. Cross and M. Ebel, *J. Chem. Educ.*, **48**, 675 (1971).
17. D. Perry, *J. Am. Chem. Soc.*, **54**, 2918 (1932).
18. J. H. Redfield, *Am. J. Math.*, **49**, 433 (1927).
19. A. C. Lunn and J. K. Senior, *J. Phys. Chem.*, **33**, 1027 (1929).
20. A. T. Balaban, J. W. Kennedy and L. V. Quintas, *J. Chem. Educ.*, **65**, 304 (1988).
21. M. Randić, G. M. Brissey, R. B. Spenser and C. L. Wilkins, *Computers and Chem.*, **3**, 5 (1979).
22. L. H. Hall, unpublished results. Information about the computer program Molconn-X is available from the first author, L. H. Hall.
23. L. B. Kier, *Molecular Orbital Theory in Drug Research*, Academic Press, New York, 1971.
24. M. Randić, *J. Am. Chem. Soc.*, **97**, 6609 (1975).
25. L. B. Kier, L. H. Hall, W. J. Murray and M. Randić, *J. Pharm. Sci.*, **64**, 1971 (1975).
26. L. B. Kier, W. J. Murray, M. Randić and L. H. Hall, *J. Pharm. Sci.*, **65**, 1226 (1976).
27. L. B. Kier and L. H. Hall, *J. Pharm Sci.*, **65**, 1806 (1976).
28. L. B. Kier and L. H. Hall, *Eur. J. Med. Chem*, **12**, 307 (1977).
29. L. H. Hall and L. B. Kier, *J. Pharm. Sci.*, **67**, 1743 (1978).
30. R. B. Hermann, *J. Phys. Chem.*, **76**, 2754 (1972).
31. L. B. Kier, *J. Pharm. Sci.*, **69**, 1034 (1980).
32. R. W. Taft, Jr, in *Steric Effects in Organic Chemistry*, (Ed. M. S. Newman), Wiley, New York, 1965, p. 556.

33. M. Charton, *J. Am. Chem. Soc.*, **91**, 615 (1969).
34. C. K. Hancock, E. A. Meyers and B. J. Yager, *J. Am. Chem. Soc.*, **83**, 4211 (1961).
35. J. A. MacPhee, A. Panaye and J. E. Dubois, *Tetrahedron*, **34**, 3553 (1978).
36. D. F. DeTar, *J. Org. Chem.*, **45**, 5166 (1980).
37. A. Verloop, W. Hoogenstraaten and J. Tipker, in *Drug Design*, Vol. III (Ed. E. J.Ariens), Academic Press, New York, 1976.
38. R. S. Pearlman, in *Physical Chemical Properties of Drugs* (Eds. S. H. Yalkowsky, A. A. Sinkula and S. C. Valvani), Dekker, New York, 1980.
39. A. Bondi, *J. Chem. Phys.*, **68**, 441 (1964).
40. V. Austel, E. Kutter and W. Kalbflersch, *Arnzeimforsch*, **29**, 585 (1979).
41. L. B. Kier, *Quant. Struct.-Act. Relat.*, **4**, 109 (1985).
42. L. B. Kier, *Quant. Struct.-Act. Relat.*, **5**, 1,7 (1986).
43. L. B. Kier, *Acta Pharm. Jugosl.*, **36**, 171 (1986).
44. C. E. Shannon and W. Weaver, *The Mathematical Theory of Communication*, Univ. of Illinois Press, Urbana, IL, 1949.
45. L. Brillouin, *Science and Information Theory*, Academic Press, New York, 1962.
46. D. Bonchev, *Information Theoretic Indices for Characterization of Chemical Structure*, Wiley, Chichester, 1983.
47. L. B. Kier, *J. Pharm. Sci.*, **69**, 807 (1980).
48. See Reference 9, pp. 82–85.
49. See Reference 9, pp. 126–129.
50. G. R. Somayajulu and B. J. Zwolinski, *Trans. Faraday Soc.*, **68**, 1971 (1972).
51. See Reference 9, pp. 108–110.
52. See Reference 9, pp. 141–145.
53. See Reference 9, pp. 131–134.
54. C. Tsonopoulos, *AIChE J.*, **125**, 97 (1989).
55. See Reference 9, pp. 147–149.
56. See Reference 9, pp. 159–165.
57. L. H. Hall and L. B. Kier, *Tetrahedron*, **33**, 1953 (1977).
58. J. S. Millership and A. D. Woolfson, *J. Pharm. Pharmacol.*, **29**, 75 (1977).
59. K. Pitzer, D. Z. Lippmann, R. F. Curl, C. M. Huggins and D. E. Petersen, *J. Am. Chem. Soc.*, **77**, 3433 (1955).
60. C. A. Passut and R. P. Danner, *Ind. Eng. Chem., Process Res. Dev.*, **12**, 365 (1973).
61. L. H. Hall and L. B. Kier, *Quant. Struct.-Act. Relat.*, **9**, 115 (1990).
62. L. B. Kier and L. H. Hall, *Pharm. Res.*, **7**, 801 (1990).
63. L. H. Hall and L. B. Kier, *J. Chem. Inf. Comput. Sci.*, **31**, 76 (1991).
64. L. B. Kier and L. H. Hall, *J. Math. Chem.*, **47**, 229 (1991).
65. L. H. Hall and L. B. Kier, *Quant. Struct.-Act. Relat.*, **10**, 43 (1991).
66. S. Fliszar, *Charge Distributions and Chemical Effects*, p. 83, Springer-Verlag, New York, 1983.

CHAPTER **6**

The thermochemistry of alkanes and cycloalkanes

N. COHEN

The Aerospace Corporation, P.O. Box 92957, Los Angeles, California 90009, USA

and

S. W. BENSON

Loker Hydrocarbon Research Institute, University of Southern California, Los Angeles, California 90089, USA

The Chemistry of Alkanes and Cycloalkanes
Edited by S. Patai and Z. Rappoport © 1992 John Wiley & Sons Ltd

I. INTRODUCTION

'The fascination of a growing science lies in the work of the pioneers at the very borderland of the unknown, but to reach this frontier one must pass over well travelled roads; of these one of the safest and surest is the broad highway of thermodynamics'[1].

In the first modern textbook on thermodynamics, from the dedication of which the above sentence is extracted, Gilbert N. Lewis and Merle Randall cited words of over a century ago by the great French chemist, Henri Louis Le Chatelier, on the importance of the science of thermodynamics:

'It is known that in the blast furnace the reduction of iron oxide is produced by carbon monoxide, according to the reaction

$$Fe_2O_3 + 3CO \rightleftharpoons 2Fe + 3CO_2$$

but the gas leaving the chimney contains a considerable proportion of carbon monoxide, which thus carries away an important quantity of unutilized heat. Because this incomplete reaction was thought to be due to an insufficiently prolonged contact between carbon monoxide and the iron ore, the dimensions of the furnaces have been increased. In England they have been made as high as 30 meters. But the proportion of CO escaping has not diminished, thus demonstrating, by an experiment costing several hundred thousand francs, that the reduction of iron oxide by CO is a limited reaction. Acquaintance with the laws of chemical equilibrium would have permitted the same conclusion to be reached more rapidly and far more economically'[2].

Today no one professionally involved in chemical science or engineering has any doubts about the importance of thermodynamics and thermochemistry. Now, however, the danger is that the field is being taken for granted; it is assumed that all the problems have been solved, all the important data measured, and what cannot be measured has been adequately calculated. In this chapter we review the thermochemistry of the alkane and cycloalkane hydrocarbons, with some attention to the current status of our knowledge in this important field and to the problems that remain to be solved.

We discuss the basic thermochemistry and thermodynamic fundamentals, describe important data sources and experimental techniques, criticize the reliability of the database, and discuss in detail methods for calculating or estimating thermochemical properties (heat capacities, enthalpies, entropies, free energies) at temperatures from 298 to 1500 K for gaseous and liquid phase alkanes and cycloalkanes that have not been studied in the laboratory.

The method of group additivities, developed by Benson and coworkers over the past thirty years, is explained and shown to be the most reliable semiempirical technique that has been explored. Group additivity parameters for alkanes and cycloalkanes previously developed by Benson and coworkers have been revised to take into account recent literature and extended to higher temperatures, to the liquid phase, to phase change processes and to solutions. Free radicals as well as stable species are considered. The latter aspect of the study allows a reassessment of alkane bond strengths, for which current best estimates are provided. A particularly useful feature of the work is the derivation of polynomial expansions for the various thermochemical properties as a function of temperature. Until now, extrapolating thermochemical or kinetic calculations to high temperatures required the use of cumbersome tables of temperature-dependent group values. Polynomial expansions simplify these calculations considerably.

In several instances, group additivity calculations have highlighted experimental measurements that are almost certainly in error. The technique thus not only allows prediction of unmeasured thermochemical properties, but provides a simple standard by which to judge the probable accuracy of published data.

II. THERMOCHEMICAL AND STATISTICAL THERMODYNAMIC RELATIONSHIPS[3]

A. Thermochemical Fundamentals

For any gas-phase chemical reaction

$$\sum_i n_i R_i \rightleftharpoons \sum_j n_j P_j$$

at equilibrium, the equilibrium constant (expressed in pressure units) is related to the species partial pressures by

$$K_p = \frac{\prod_j [P_j]^{n_j}}{\prod_i [R_i]^{n_i}} \tag{1}$$

The Gibbs free energy change for the reaction, $\Delta G°$, is related to the equilibrium constant by

$$\Delta G_T° = -RT \ln K_p \tag{2}$$

$$= \sum_j n_j \Delta_f G_T°(P_j) - \sum_i n_i \Delta_f G_T°(R_i) \tag{3}$$

The quantities $\Delta_f G_T°(P_j)$ and $\Delta_f G_T°(R_i)$ represent the free energies of formation of the species P_j and R_i from the elements in their standard states (at a pressure of 1 atm) at temperature T. (Formerly, the Gibbs free energy was designated just 'free energy', and denoted by F rather than G.) The constant R is the universal gas constant (1.987 cal mol^{-1} K^{-1} or 8.314 joule mol^{-1} K^{-1} when G is expressed in units of cal mol^{-1} or joule mol^{-1}, respectively).

Frequently, thermochemical tables list the quantity $\log_{10} K_{fT}$ for each species as a function of temperature. This is defined by

$$2.303 \log_{10} K_{fT} = \ln_e K_{fT} = -\Delta_f G_T°/RT \tag{4}$$

(Henceforth we use *log*, without the subscript, to mean the decadic logarithm, and *ln* to mean the natural logarithm.) Using these quantities, the reaction equilibrium can be expressed conveniently as

$$K_{p,T} = \frac{\prod_{j} K_{fT}(P_j)}{\prod_{i} K_{fT}(R_i)} \tag{5}$$

The units of $K_{p,T}$ are $(atm)^{\Delta n}$ where $\Delta n = \sum_j n_j - \sum_i n_i$. If species concentrations are used rather than pressures, then the equilibrium constant is written $K_{c,T}$ and it has the units of $(concentration)^{\Delta n}$, where concentration is expressed either in $mol\,liter^{-1}$ or $mol\,cm^{-3}$. The two equilibrium constants are related by the equation

$$K_p = K_c \times (R'T)^{\Delta n} \tag{6}$$

where R' is the gas constant, but now in units of $1\,atm\,mol^{-1}\,K^{-1}$ (0.08205) or $cm^3\,atm\,mol^{-1}\,K^{-1}$ (82.05).

The Gibbs free energy represents the driving force of a reaction: a large negative value for ΔG means a large value for K_p or K_c, which means equilibrium favors products of the reaction. A large positive value for ΔG means a small equilibrium constant, which means that reagents are favored. Note that whether a reaction will actually proceed to equilibrium or not—and at what rate—may be controlled completely by kinetics, not thermochemistry. The driving force of a reaction represents the net result of two opposing tendencies: equilibrium is favored by a lower potential energy (or enthalpy) state, but a higher entropy state. That is, potential energy (or enthalpy) tends to decrease while entropy tends to increase. Historically, before the principles of thermodynamics were thoroughly understood it was thought that the energy alone was the driving force of a reaction. These ideas are expressed quantitatively by the relationships

$$G = E + PV - TS = H - TS \tag{7}$$

where E, H and S are energy, enthalpy (also called heat content) and entropy, respectively. (Formal derivation of the fundamental relationships of thermodynamics is outside the scope of this chapter; our purpose in this section is merely to gather all the equations necessary for an understanding of the remainder of the discussion.) For most practical purposes, enthalpy is more useful a concept than energy, and we will not discuss E further. Both enthalpy and entropy are related to the heat capacity of a substance by relationships such as

$$\Delta H_{T_2} - \Delta H_{T_1} = \int_{T_1}^{T_2} C_p\,dT \tag{8}$$

$$S_{T_2} - S_{T_1} = \int_{T_1}^{T_2} C_p\,d\ln T \tag{9}$$

Here, C_p is the heat capacity at constant pressure. It is related to the heat capacity at constant volume, C_v, by

$$\begin{aligned} C_p &= C_v + PV/T \\ &= C_v + R \text{ (for ideal gases)} \end{aligned} \tag{10}$$

Since equilibrium constants are related to Gibbs free energy, and the latter in turn is related to H and S, it is apparent that in some sense, heat capacities are the fundamental quantities from which all the others can be derived.

Equations 8 and 9 provide recipes for calculating changes with temperature in the enthalpy and entropy of a substance. Complete evaluations still require knowing these properties at $T = 0$ (or some other base temperature) and what ΔH and ΔS are for each phase change, such as melting or sublimation. The third law of thermodynamics, namely that the entropy of a pure crystalline substance approaches zero as T approaches absolute zero, anchors the entropy scale. Nevertheless, difficulties in measuring or calculating heat capacities at very low temperatures make it particularly useful to have in hand a value of the entropy at some temperature. There is no corresponding law regarding enthalpies: i.e. even at absolute zero a substance has a residual heat (or energy) content. While there are procedures for calculating the heat capacity of a substance (and hence, changes in enthalpy or entropy with temperature), which we will present later in this chapter, there is no practical way to calculate $\Delta_f H_0$ given our current knowledge. Hence, apart from very approximate estimating schemes, one is forced to rely on experimental measurements to anchor the enthalpy scale. Similarly, the enthalpy changes associated with phase changes must be measured experimentally (or estimated only approximately).

From the above relationships, it should be apparent that, if there are no phase changes involved, we can characterize the thermochemistry of a substance if we know $\Delta_f H$ and S at some temperature (tables and compilations list either $\Delta_f H_0$, $\Delta_f H_{273}$ or $\Delta_f H_{298}$) and C_p as a function of temperature. From the thermodynamic quantities for the reagents and products of a reaction, it is possible to derive the corresponding quantities for the reaction itself (i.e. the net change in properties in the course of a reaction, defined by equations of the form of equation 3).

For the practitioner concerned with chemical processes, whether in a catalytic cracking plant or a pharmaceutical laboratory, the equilibrium constants, K_p or K_c, are the quantities of primary concern. If the kinetics are favorable, can the reaction proceed as written? What is the maximum product yield? Will the yield increase if the temperature is raised? The equilibrium constant indicates the maximum possible yield, and its temperature dependence reveals how the yield can change with temperature. The thermodynamic quantities (G, H and S) used to calculate K relate the equilibrium constant to fundamental physical quantities. Thus, while tables of equilibrium constants may appear to be arrays of arbitrary numbers, tables of thermodynamic quantities reveal structural relationships that can be used to predict equilibrium constants for processes never previously studied.

Hence, the scope of this chapter is to provide the reader with an appreciation for the data and techniques available (and their accuracy) for evaluating fundamental thermochemical quantities that are necessary for understanding the chemistry of the hydrocarbons.

B. Statistical Thermodynamic Relations

Statistical thermodynamics provides exact formulas for the calculation of the fundamental quantities of heat capacities, enthalpies and entropies, provided that certain assumptions are valid. Among these assumptions are:

(1) Translational, vibrational, rotational and electronic degrees of freedom are completely separable for the molecule of interest.
(2) The rigid-rotor harmonic oscillator approximation provides an adequate description of molecular properties.
(3) The gaseous species obeys the ideal gas equation of state.

A fundamental equation of statistical mechanics is that for the partition function, Q, for a collection of identical molecules:

$$Q = \sum_i p_i \exp(-\varepsilon_i/RT) \tag{11}$$

where p_i is the number of discrete states (the degeneracy) that have energy E_i in a system at equilibrium at temperature T. The various thermodynamic properties are defined in terms of the partition function. In this discussion, we are interested in only four—heat capacity at constant volume, entropy, internal energy and enthalpy:

$$C_v/R = (\delta \ln Q/\delta \ln T)_v + (\delta^2 \ln Q/\delta(\ln T)^2)_v \tag{12}$$
$$= (\delta^2 \ln Q/\delta(1/T)^2)_v$$

$$S/R = \ln Q - (\delta \ln Q/\delta(1/T))/T \tag{13}$$

$$E/R = -(\delta \ln Q/\delta(1/T))_v \tag{14}$$

$$H/R = -(\delta \ln Q/\delta(1/T))_v + T \tag{15}$$

If the various degrees of freedom are separable (assumption 1 above), then the partition function can be factored into the product of partial partition functions:

$$Q = Q_{\text{trans}} \times \prod Q_{\text{vib},i} \times Q_{\text{rot}} \times Q_{\text{elec}} \tag{16}$$

The factors on the right-hand side of equation 16 can be expressed in terms of various molecular properties:

$$Q_{\text{trans}} = V \times (2\pi MkT/h)^{3/2} \tag{17}$$

$$Q_{\text{vib},i} = (1 - \exp(-h\nu_i/kT))^{-1} \tag{18}$$

for the ith vibrational mode. The total partition function includes one such factor for each vibrational mode. The rotational partition function (for external rotation, or rotation of the molecule as a whole) depends on whether the molecule is linear (2 degrees of freedom) or nonlinear (3 degrees of freedom):

$$Q_{\text{rot-2D}} = (8\pi^2 IkT/h^2)/\sigma_e \tag{19}$$

$$Q_{\text{rot-3D}} = (\pi^{1/2}/\sigma_e)(8\pi^2 I_m kT/h^2)^{3/2} \tag{20}$$

$$Q_{\text{elec}} = \sum g_i \exp(-\varepsilon_i/kT) \tag{21}$$

All the linear (i.e. noncyclic) alkanes have internal rotations about the C—C bonds. For each internal rotation, if there is no energy barrier to rotation, the partition function for free internal (one-dimensional) rotation is

$$Q_{\text{f.i.r}} = (8\pi^3 I_{\text{i.r.}} kT/h^2)^{1/2}/\sigma_i \tag{22}$$

In equations 17–22, h is Planck's constant, k Boltzmann's constant, V the molar volume, M the molecular weight, σ_e and σ_i are the symmetry numbers for external or internal rotation, respectively; I is the moment of inertia and I_m is the product of principal moments of inertia (for external, or overall, rotation), $I_{\text{i.r.}}$ is the moment of inertia for internal rotation, g_i and ε_i are the degeneracy and energy level of the ith electronic level.

In general, hydrocarbons have no low-lying excited electronic states, and the degeneracy of the ground state is unity; hence $Q_{\text{elec}} = 1$. (For alkyl free radicals it is generally assumed that the ground state has spin $\frac{1}{2}$ and hence $Q_{\text{elec}} = 2$.) Since there are no linear (i.e. two-dimensional) alkanes or cycloalkanes, we do not need equation 19 for the two-dimensional rotor.

For computational purposes, it is more convenient to work in molar rather than molecular units. In the following expressions, we have evaluated all the numerical constants in molar units. The Boltzmann constant, k, is thus replaced by the universal gas constant, $R = N_0 k = 1.987 \text{ cal mol}^{-1} \text{K}^{-1}$, where N_0 is Avogadro's number.

We can evaluate equations 12–15 for the thermodynamic functions using the specific formulas of equations 17–22. The results are summarized below.

C. Translation

$$C^\circ_{v(trans)}/R = \tfrac{3}{2} \tag{23}$$

$$C^\circ_{p(trans)}/R = C^\circ_{v(trans)}/R + 1 = \tfrac{5}{2} \tag{24}$$

$$E^\circ_{trans}/RT = \tfrac{3}{2} \tag{25}$$

$$(H^\circ_T - H^\circ_0)_{trans}/RT = \tfrac{5}{2} \tag{26}$$

$$S^\circ_{trans}/R = -1.16 + (\tfrac{3}{2})\ln M + (\tfrac{5}{2})\ln T + \ln n$$
$$= 13.08 + (\tfrac{3}{2})\ln M + (\tfrac{5}{2})\ln(T/298) + \ln n \tag{27}$$

where, in the last expressions, M is molecular weight in daltons and n is the number of optical isomers (discussed further below).

D. External Rotation (three dimensions)

$$C^\circ_{p(e.r.)}/R = \tfrac{3}{2} \text{ (rigid-rotor harmonic oscillator)} \tag{28}$$

$$E^\circ_{e.r.}/RT = \tfrac{3}{2} \tag{29}$$

$$(H^\circ_T - H^\circ_0)_{e.r.}/RT = \tfrac{5}{2} \tag{30}$$

$$S^\circ_{e.r.}/R = -0.034 + \tfrac{3}{2}\ln T + \tfrac{1}{2}\ln I_m - \ln \sigma_e$$
$$= 11.5 + \tfrac{3}{2}\ln(T/298) + \tfrac{1}{2}\ln(I_m/\sigma_e^2) \tag{31}$$

where, in the last expression, I_m, the product of principal moments of inertia, is in units of $(\text{dalton cm}^2)^3$. The subscript 'e.r.' signifies external rotation contribution.

Because the translational functions (equations 23–27) depend only on parameters that are known precisely, they can be calculated exactly. It is useful to remember this, since the translational contribution is generally by far the largest of the several degrees of freedom. The moments of inertia of the molecule depend on its large-scale spatial configuration—whether it is approximately linear, tightly coiled or somewhere in between (i.e. the rigid-rotor assumption may not be a good one). For example, the difference in rotational entropy between *gauche*- and *trans*-butane is approximately 0.3 eu (1 eu = 1 entropy unit = 1 gibbs mol^{-1} = 1 cal mol^{-1} K^{-1}). For decane, the entropy difference between stretched and coiled configurations is approximately 0.7 eu. Thus, for a large alkane, there is some uncertainty in the rotational contributions to thermodynamic functions. In principle, one would perform some sort of time-averaged calculation over the various spatial configurations. In practice, one should expect an uncertainty on the order of 0.5 eu in the calculation.

E. Internal Rotations

Internal rotations are generally treated by considering the mode as a free rotation (no potential energy barrier), and then by making approximate corrections for the effect of the barrier. Calculating the properties of free rotation is handicapped by the difficulty in evaluating the moments of inertia for the internal rotation. Most computational schemes rely on an approximate procedure developed by Herschbach and coworkers[4]. This involves calculating the moments of inertia of each of the two moieties attached to the bond that is the axis of rotation, I_a and I_b. Then,

$$1/I_{ir} = 1/I_a + 1/I_b \tag{32}$$

The separate moments I_a and I_b are estimated by assuming that each moiety is a top rotating about an axis parallel to the bond joining the two moieties and passing through

N. Cohen and S. W. Benson

TABLE 1. Approximate moments of inertia
for some free rotors

Rotor	I_{ir}^a
CH_3	3.0
C_2H_5	20
$n\text{-}C_3H_7$	28
$i\text{-}C_3H_7$	56
$n\text{-}C_4H_9$	80
$t\text{-}C_4H_9$	100

[a] Units are (dalton $\text{Å}^2)^3$, calculated assuming the rotor
is connected to an infinite mass and rotates about
an axis through its own center of mass and parallel
to the C—C bond connecting the rotor to that
infinite mass.

its own center of gravity. In Table 1 we list approximate moments of inertia for the
rotors commonly encountered in evaluating properties of alkanes.

For a free rotor with symmetry σ_i,

$$C_p^\circ/R = \tfrac{1}{2} \tag{33}$$

$$E^\circ/RT = \tfrac{1}{2} \tag{34}$$

$$(H_T^\circ - H_0^\circ)_{\text{f.i.r.}}/RT = \tfrac{1}{2} \tag{35}$$

$$S^\circ = -1.1 + R\ln(I_{\text{i.r.}}^{1/2}/\sigma_i) + \tfrac{1}{2}R\ln T$$
$$= 4.6 + R\ln(I_{\text{i.r.}}^{1/2}/\sigma_i) + \tfrac{1}{2}R\ln(T/298) \tag{36}$$

In equation 36, the units of $I_{\text{i.r.}}$ are (dalton $\text{Å}^2)^3$.

Frequently in the discussion that follows we refer to the intrinsic entropy, S_{int}°. This
is the entropy when the effects of symmetry and optical isomerism are neglected:
$S_{\text{int}}^\circ = S_{\text{observed}}^\circ + R\ln(\sigma/n)$, where $\sigma = \sigma_e \times \sigma_i$. The determination of symmetry number and
optical isomerism is discussed further below.

If there is a potential energy barrier to rotation, the rotor is hindered. If the barrier
is very low, the hindrance becomes inconsequential as temperature increases. If the
barrier is very high, the rotor can be treated approximately as a torsional vibration.
The spectroscopic vibrational frequency of a hindered rotor is given by

$$\nu(\text{in cm}^{-1}) = 360(T/298)(1/Q_f)(V/RT)^{1/2} \tag{37}$$

For $V/RT > 10$, the thermodynamic functions S°, G°/T and H°/T of the rotor are
accurately represented (within $0.05\,\text{cal mol}^{-1}\,\text{K}^{-1}$) by the functions of the vibration with
frequency given by equation 37. Only the intermediate case $(V/RT \simeq 1)$ presents
problems. The principal difficulty in this case is that of ascertaining with any precision
the barrier height. This is a consequence of the fact that several different factors determine
the barrier height, and all are highly sensitive to the interatomic distances involved.
Furthermore, while most treatments to derive thermodynamic properties assume the
barriers for an n-fold rotation (e.g. 3-fold rotation for a methyl rotor) are symmetric,
this is not in general true. For example, for rotation about the central C—C bond in
butane, the *trans → gauche* and *gauche → cis* barriers are approximately 4 and
6 kcal mol^{-1} respectively[5].

The effect of a barrier to internal rotation is to reduce the heat capacity, and hence
the enthalpy, energy and entropy, of the free rotor. The numerical magnitude of the

TABLE 2. Approximate torsional barriers to internal rotation

Bond	V (kcal mol^{-1})	Reference
CH_3—CH_3	2.9	Pitzer[6]
CH_3—C_2H_5	3.3	Chao and coworkers[7]
CH_3—n-C_3H_7	3.3	Chen and coworkers[8]
CH_3—i-C_3H_7	3.8	Chen and coworkers[8]
CH_3—t-C_4H_9	4.7	Sackwild and Richards[9]
C_2H_5—C_2H_5	3.5	Chen and coworkers[8]
C_2H_5—n-C_3H_7	4.2	Sackwild and Richards[9]
C_2H_5—neo-C_2H_{11}	[6.5]	Zirnet and Sushinskii[10a]
C_3H_7—t-C_4H_9	[10.9]	Zirnet and Sushinskii[10]
i-C_3H_7—i-C_3H_7	4.3	Lunazzi and coworkers[11], Jaime and Osawa[12]
i-C_3H_7—i-C_4H_9	[8.9]	Zirnet and Sushinskii[10]
i-C_3H_7—t-C_4H_9	7.0	Anderson and Pearson[13], Jaime and Osawa[12]
t-C_4H_9—t-C_4H_9	9.8	Anderson and Pearson[13], Jaime and Osawa[12]
i-C_4H_9—t-C_4H_9	[10.8]	Zirnet and Sushinskii[10]
2,4,4-trimethyl-2-C_5H_{11}—2,4,4-trimethyl-2-C_5H_{11}	13.8	Anderson and Pearson[13], Jaime and Osawa[12]

[a] The values deduced by Zirnet and Sushinskii are reasonable, but the procedures by which they were derived are questionable; hence they should be regarded as only approximate.

effect depends on two parameters: V/RT, where V is the barrier height; and Q_f, the partition function for free rotation. Some barrier heights pertinent to alkanes are listed in Table 2.

The contributions to thermodynamic functions from hindered internal rotation were worked out in a series of papers by Pitzer and coworkers[14] in the 1930s, '40s and '50s, and are tabulated by Pitzer and Brewer[15], to which the reader is referred for further details. Here, we simply note that an uncertainty in V/RT of 4.0 ± 1.5 (corresponding to an uncertainty of 900 cal mol^{-1} at 298 K—which is similar to the uncertainty for most barriers for rotors larger than CH_3) produces an uncertainty in rotational entropy of approximately ± 0.5 eu. Since a C_n alkane without rings has $n-1$ internal rotations, this puts a lower limit to the uncertainty in any calculation of the entropy of an alkane of approximately $0.5(n-3)$ eu for $n \geqslant 5$.

F. Vibration

For a harmonic oscillator,

$$C_p^\circ/R = u^2 e^{-u}/(1 - e^{-u})^2 \tag{38}$$

$$(H_T^\circ - H_0^\circ)_{vib}/RT = ue^{-u}/(1 - e^{-u}) \tag{39}$$

$$S^\circ/R = ue^{-u}/(1 - e^{-u}) - \ln(1 - e^{-u}) \tag{40}$$

where $u = 1.438\,\omega_i/T$ and ω_i is the fundamental wave number of the harmonic oscillator in cm^{-1}.

Vibrational frequencies are not easily deconvoluted from raw infrared spectral data, especially for a molecule with more than 4 or 5 atoms. Alkane vibrational frequencies fall into a few classes (Table 3).

The high C—H stretching frequencies are inactive except at very high temperatures; they contribute little to the C_p (and hence to other functions), and errors in their values are inconsequential. Only errors in low frequencies are significant. An error of 25 cm^{-1}

TABLE 3. Characteristic frequencies for vibrational modes

Vibrational mode	Approximate frequency (cm^{-1})
C—H stretch	2850–3000
C—C stretch	900–1100
H—C—H bend (CH_3 deformation, CH_2 scissors)	1400–1500
CH_2 twist, wag	1200–1400
C—C—H bend (CH_3 rock)	900–1300
C—C—H bend (CH_2 rock)	750–800
C—C—C bend	250–500
Ring deformations (in-plane)	400–600
Ring deformations (out-of-plane)	200–400
CH_3—CH_2R torsions (internal rotations)	200–225
RCH_2—CH_2R torsions (internal rotations)	75–100

in a $200 \, cm^{-1}$ frequency contributes approximately 0.2 eu error to the entropy at any temperature from 200 to 1500 K. The last two entries in Table 3 cover all the internal rotations of alkanes, which have characteristic frequencies just as do vibrations. The first of the two includes a methyl group rotating against the rest of the molecule (R is any alkyl group); the second includes all rotations where both moieties are ethyl groups or larger.

All routine statistical mechanical calculations for hydrocarbons assume vibrations are adequately described by the harmonic oscillator approximation. We can get a rough indication of the magnitude of the possible errors this approximation causes by examining the isolated CH and C_2 molecules, for both of which the required spectroscopic constants are available. For CH, the entropy correction is 0.012 eu at 298 K and 0.05 eu at 1000 K. For C_2, which is more nearly harmonic, the corresponding values are 0.004 and 0.02 eu. An alkane, C_nH_{2n+2}, will have $2n + 2$ C—H stretches and $n - 1$ C—C stretches, each of which can contribute to the anharmonicity. Bending modes are more complex, but we can assess their importance from a study of the thermodynamics of C_3 by Strauss and Thiele[16]. For this triatomic molecule, they found that detailed quantum corrections (including anharmonicity) to the rigid-rotor harmonic oscillator (RRHO) approximation decreased the entropy by 0.02 and 0.002 eu at 298 and 1000 K, respectively. These results suggest that near 298 K the RRHO approximation will underestimate S by about 0.01 eu per C atom, and by about 0.1 eu per C atom near 1000 K. For most purposes these are small errors, but certainly not negligible for decanes or larger alkanes.

G. Symmetry Numbers, Optical Isomers, Entropy of Mixing and Some Kinetic Considerations

If there are two or more physically distinguishable isomeric forms, the entropy of a mixture must include a term for the entropy of mixing. For i different species, the mole fraction of each of which is n_i, the entropy of mixing (per mole of final mixture) is given by

$$\Delta S_{\text{mixing}} = - R \sum n_i \ln (n_i) \qquad (41)$$

If the different species have different enthalpies of formation, the n_i will be temperature-dependent and hence so will ΔS_{mixing}. If the species all have the same enthalpy and the same statistical probability, then the n_i are all equal (and independent of temperature), and $\Delta S_{\text{mixing}} = R \ln n$, where n is the number of such species. This is the origin of the final term on the right-hand side of equation 27.

Optical isomerism will result if there is a *chiral center*, when the molecule will have the property of rotating a beam of polarized light. The two forms, or enantiomers, are designated R and S (from the Latin *rectus* and *sinister*, signifying right and left, respectively; formely, the designations *dextro-* and *levo-*, depending which one rotates the light beam clockwise and which one counterclockwise, were used). If there are n chiral centers, there can be as many as 2^n optical isomers. An exception occurs when there are two identical chiral centers, as in 3,4-dimethylhexane, in which case there are only three optical isomers, since one of the apparent four is its own mirror image. This latter form is the *meso*-compound. (See Example 1 in Section XIII for more details.) Optical isomerism can also result without a chiral center: the most stable configuration of cyclopentane is nonplanar, so that the molecule and its mirror image are not superposable and $n = 2$.

The symmetry number is easiest calculated by first treating the molecule as rigid and calculating the external symmetry number, σ_e. For CH_4, $\sigma_e = 12$; for any other linear unbranched alkane, $\sigma_e = 6$. For most branched alkanes, σ_e can be 1, 2, 3, 4 or 6. Each internal CH_3 rotor contributes a factor of 3 to σ_i; each *t*-Bu rotor contributes a factor of 3^4: 3 for each CH_3 group and 3 for overall symmetry of the three CH_3 groups.

Accounting properly for the effects of the symmetry number on both thermochemical and kinetic properties has often led to computational errors in the past, and practitioners need to take care in their bookkeeping. Cycloalkanes present ample opportunities for such pitfalls. The simplest ring, cyclopropane, has an unambiguous symmetry number of 6. Cyclobutane would have $\sigma_e = 8$ if it were planar, but there is some slight deviation from planarity, and actually $\sigma_e = 4$. Since the energy barrier to ring puckering is only about $1 \, \text{kcal mol}^{-1}$, the deformation from one side of the plane to the other is very rapid, and we could consider the molecule as dynamically planar and compute σ_e accordingly; it would simply be a matter of adjusting our bookkeeping (namely, the ring correction of Table 13) accordingly. However, we chose to make our calculations with the symmetry number of 4. (A precise calculation of entropy would take into account the fact that cyclobutane is a mixture of puckered and planar molecules, the former more stable by about $1 \, \text{kcal mol}^{-1}$. In the limit of very low temperature, the mixture consists entirely of puckered molecules; at high temperatures, $\frac{2}{3}$ of the molecules are planar. The entropy of an equilibrium mixture will thus include a contribution from ΔS_{mixing} that approaches zero at low temperatures and approximately 1.4 eu at high temperatures. But between 298 and 1500 K, ΔS_{mixing} varies by less than 0.1 eu, and the mixing aspect can be ignored without serious consequence.)

Cyclohexane exists in several nonplanar structural configurations, of which the 'chair' is the most stable. In this configuration, for which $\sigma_e = 6$, there are two different kinds of H atoms: equatorial (approximately in the plane of the skeletal ring) and axial (approximately perpendicular to the ring). A second configuration, considerably more rigid, is the 'boat' form, in which two opposite H atoms are very close to one another (the 'bowsprit–flagpole' interaction). The more stable boat form is the 'twist boat' or 'skew boat' form, in which the boat deforms slightly to increase the separation of the bowsprit–flagpole interaction. The twist boat is about $5.5 \, \text{kcal mol}^{-1}$ less stable than the chair, and the boat form is about $1.6 \, \text{kcal mol}^{-1}$ less stable yet. One chair form converts into the other (interchanging axial and equatorial H atoms) by passing through the 'boat' and 'twist boat' forms. The energy barrier to the transition is approximately $10 \, \text{kcal mol}^{-1}$. Thus, at room temperature or below, $> 99\%$ of the molecules are in the chair configuration—although the lifetime for *a particular molecule* to transform from one chair form into another, interconverting axial and equatorial H atoms, is on the order of microseconds.

Cyclopentane, as noted above, is also nonplanar, with two different configurations more stable than the planar: an 'envelope', with $\sigma = 1$ and $n = 1$, and a 'half-chair', with

$n = 2$ and $\sigma = 2$. The envelope form is the more stable. Cycloheptane has no symmetries ($\sigma = 1$), thus consisting of fourteen nonequivalent stereoisomers.

(Although kinetics problems are not directly in our present purview, we cannot entirely overlook them as one important area of application of thermochemistry. In predicting rate coefficients, one needs to know the number of reaction paths—or, what is equivalent, the change in symmetry number in a reaction. The fact that cycloheptane is puckered poses the interesting question of whether the different H atoms (since *at any given instant* they are not all equivalent) have different reactivities. Strictly speaking, a separate rate coefficient calculation should be made for each of the 14 nonequivalent H atoms; the total rate coefficient is then the sum of the 14 individual ones—each for a reagent of $\sigma = 1$. In practice, the differences in reactivities of the 14 H atoms are smaller than the uncertainties in the appropriate kinetic calculations. The single calculation can thus be carried out for a single reagent with $\sigma = 14$. In the case of cyclohexane, it is known that the preferred conformation for substituents is invariably the equatorial position. This does not necessarily mean that the equatorial H atom is more reactive, but only that once substitution takes place, the equatorial position is the more stable.)

H. Deviations from Ideality: Real Gases

All the statistical mechanical formulas presented so far were evaluated assuming that the gas obeyed the ideal gas law

$$P\underline{V} = RT \tag{42}$$

where $\underline{V}$ represents the volume per mole of the gas at pressure P and temperature T. Real gases deviate from this behavior to a slight but measurable degree, and that deviation increases as pressure increases and temperature decreases. (As $P/T \to 0$, molecules become less and less sensitive to the presence of one another, and the behavior of a real gas approaches very nearly that of the ideal gas.) The behavior of a real gas is often characterized by a power series in $\underline{V}$, namely

$$P\underline{V}/RT = 1 + B/\underline{V} + C/\underline{V}^2 + D/\underline{V}^3 + \cdots \tag{43}$$

Such equations (sometimes the expansion is written in powers of P) are called virial equations of state, and the coefficients $B, C, D, \ldots$ are the second, third, fourth,... virial coefficients. The thermodynamic functions C_p, H and S of the real gas deviate from the expressions already presented by correction factors that are functions of the virial coefficients. For example, the deviation of C_p from the ideal gas heat capacity, C_p°, is given by[17]

$$(C_p - C_p^\circ)_T = -T[(\delta^2 B/dT^2)P + (\delta^2 C/dT^2)P^2/2 + \cdots] \tag{44}$$

To a first approximation (i.e. if the molecules behave like hard spheres with no other intermolecular forces), the virial coefficients are simply related to the rigid-sphere collision diameter, σ, of the molecule:

$$B = \tfrac{2}{3}\pi N_0 \sigma^3 \tag{45}$$

$$C = \tfrac{5}{18}(\pi N_0 \sigma^3)^2 \tag{46}$$

etc., and since σ is temperature- and pressure-independent, there are no corrections for nonideality to be made to thermodynamic functions. (B and C are also respectively identical to K_2 and K_3, the equilibrium constants for dimers and trimers in a gas consisting of rigid hard spheres.) More sophisticated molecular models (e.g. Lennard-Jones, Sutherland, Buckingham) all lead to temperature- and pressure-variant virial coefficients. (Even for xenon, which is as near to an ideal gas as any molecule, B varies from $-130 \, \text{cm}^3 \, \text{mol}^{-1}$ at 298 K to $-23 \, \text{cm}^3 \, \text{mol}^{-1}$ at 573 K[18].)

The magnitude of the corrections for nonideality is illustrated by data of Waddington, Todd and Huffman[19] for the measured heat capacity of n-heptane. They found $\delta C_p / \delta P = 1.45$ and $0.27 \, \text{cal} \, \text{mol}^{-1} \, \text{K}^{-1} \, \text{atm}^{-1}$ at 375 and 466 K, respectively. Since C_p° values for this alkane are approximately 47 and $56 \, \text{cal} \, \text{mol}^{-1} \, \text{K}^{-1}$ at these temperatures[20], the corrections are 3 and 0.5%, respectively, at those temperatures and 1 atm pressure.

III. HYDROCARBON THERMOCHEMISTRY: HOW MEASUREMENTS ARE MADE; THEIR RELIABILITY; THE DATABASE

From equations such as 7, 8 and 9, it is evident that the thermochemical properties and functions of interest—including C_p°, S°, $-(G^\circ - H_{298}^\circ)/T$, $H^\circ - H_{298}^\circ$, $\Delta_f H^\circ$, $\Delta_f G^\circ$ and Log K_f—can all be reduced to $C_{p,T}^\circ$ over the relevant temperature range and S° and $\Delta_f H^\circ$ at one specific temperature (usually, but not necessarily, 298 K). (If the temperature range involves phase transitions, one must know in addition the appropriate enthalpy change: ΔH_v for vaporization, ΔH_s for sublimation or ΔH_m for melting.) All the other thermodynamic properties can be calculated if these are known. The basic quantities can be obtained by (1) experimental measurement, (2) calculation from the principles of statistical thermodynamics (C_p and S only) or (3) estimation by approximate semiempirical techniques. We will discuss and compare all of these briefly in the following paragraphs. In some cases, we refer the reader to more extensive discussions by Cox and Pilcher[21] or Stull, Westrum and Sinke ('SWS')[22].

A. Experimental Methods

1. Enthalpy of formation

There are four principal methods used for measuring the enthalpy of formation of a hydrocarbon: (1) determination of the heat of combustion of the liquid or solid in a bomb calorimeter, (2) determination of the heat of combustion of the gas or volatile liquid in a flame calorimeter, (3) determination of the heat of reaction from other species whose heats of formation are well-known and (4) determination of the equilibrium constant for some reaction involving other species all of whose heats of formation are well-known.

a. Bomb calorimetry. If a sample of solid (usually in a compressed pellet; in an ampoule if hygroscopic, volatile or reactive) or liquid (in a glass or plastic ampoule) is placed in an atmosphere of oxygen in a bath of fluid inside a thermally insulated container and ignited (usually by electric spark), the measured temperature rise of the bath can be used to determine the heat of reaction for the process

$$C_n H_{2n+2} + (3n+1)/2 \, O_2 \rightarrow n CO_2 + (n+1) H_2 O$$

from

$$\Delta_r H = \langle C_p \rangle \times \Delta T$$

The value of $\Delta_f H(C_n H_{2n+2})$ can be determined from $\Delta_r H$ (the heat of reaction), the heats of formation of the other species and the enthalpies of the phase transitions (enantiomorphic transitions, melting, sublimation, vaporization) that take place. The experimental problems include insuring purity of the reagent; complete combustion;

perfect insulation, so no heat is lost; calibration for the heat capacities of the temperature-measuring device, the ampoule, the bath fluid, the walls of the calorimeter and anything else involved; and accounting for the energy input from the spark. The heat capacities of the products must be known through the entire temperature range of combustion. (This may be very small if the unknown is in a large diluent bath.)

According to SWS (Reference 22, p. 70), heat evolution in combustion calorimetry can be measured with a precision of 0.01%. Consider the combustion of liquid n-octane, for which $\Delta_r H$ at 298 K is approximately 1220 kcal mol^{-1}. The aforementioned precision means an uncertainty in heat evolution of 0.12 kcal mol^{-1}. The heats of formation of CO_2 and $H_2O_{(g)}$ are[23] -94.054 ± 0.011 and -57.10 ± 0.010 kcal mol^{-1}, respectively, at 298 K. Since 8 moles of CO_2 and 9 moles of H_2O are produced per mole of octane burned, the total uncertainty in enthalpy of formation can be as large as 0.3 kcal mol^{-1} not taking into account uncertainties in the enthalpies of phase transitions. (In general, the uncertainty will be 0.03 kcal mol^{-1} per C atom in the alkane.) Cox and Pilcher[21] (Chapter 4) have compared the accuracy of various schemes for estimating ΔH_v; their comparison suggests that careful estimations are good to a few tenths of a kcal mol^{-1} for two representative C_7 alkanes. Thus, based on these considerations alone, we can expect experimental measurements by bomb calorimetry to have uncertainties of at least ± 0.05 kcal mol^{-1} per C atom. (Most tabulated data rely on older measurements with somewhat larger uncertainties.)

It is essential (though not always done) that the products of combustion be analyzed; the assumption of complete combustion is not always an accurate one. Alkane combustion proceeds in a series of steps, in which CO is always predecessor to the final product CO_2. If only 0.1% of the carbon content of an alkane is oxidized only as far as CO, the resulting error in $\Delta_f H$ will be 0.07 kcal per C atom. If an equivalent amount of carbon is reduced to graphite (solid soot/graphite is often observed after bomb combustion), the error is 0.1 kcal per C atom.

Another limitation is imposed by the presence of isomers. For example, butane vapor consists of 72% *trans* and 28% *gauche* conformers at 298 K[24]. (Conformers are discussed in more detail in Section V.) Since the *gauche* form is higher in energy by approximately 1.0 kcal mol^{-1}, there is a conformational contribution to enthalpy of 0.28 kcal mol^{-1}. Further complications arise in the case of pentanes or larger alkanes: though they exist in higher-energy curled configurations in the gas phase, in the liquid phase their free energy is lowered by assuming more nearly linear configurations. Thus, the distribution between *gauche* and *trans* conformers will be different in the two phases. This effect needs to be taken into account in calculating the enthalpy change of combustion for the gaseous species when starting from a liquid. (If ΔH_{vap} is measured this effect is automatically taken into account.)

b. Flame calorimetry. This procedure, though useful only for the smaller alkanes, avoids at least some of the complications of phase transitions that occur in bomb calorimetry. However, the technique is inherently less precise, so that the overall accuracy of enthalpy measurements is comparable for the two techniques. In a definitive flame calorimetric study of the heats of combustion of CH_4, C_2H_6, C_3H_8, C_4H_{10} and $i\text{-}C_4H_{10}$, Pittam and Pilcher[25] estimated their uncertainties in the five cases to be between ± 0.06 and ± 0.015 kcal mol^{-1}—a good indication of the current state of the art in this field. All the alkanes were in the gaseous state, so this represents the most favorable experimental situation.

In order to appreciate the accuracy of thermochemical measurements, it is instructive to compare the results of Pittam and Pilcher on the aforementioned five alkanes with earlier results of Rossini and coworkers[26], formerly considered the most reliable available.

Alkane	Rossini and coworkers	Pittam and Pilcher
Methane	-17.89 ± 0.07	-17.80 ± 0.10
Ethane	-20.24 ± 0.12	-20.04 ± 0.07
Propane	-24.83 ± 0.14	-25.02 ± 0.12
n-Butane	-30.36 ± 0.16	-30.03 ± 0.16
i-Butane	-32.41 ± 0.13	-32.07 ± 0.15

While the agreement is good, what is noteworthy is that in three of the five cases, the differences between the two measurements exceed the sum of the estimated uncertainties. These figures suggest that the accuracy (as distinct from the precision) of a combustion measurement of $\Delta_f H$ is probably no better than ± 0.1 kcal per C atom.

Gas phase flame calorimetry has been used for alkanes up through C_5; liquid combustion has been used for determining enthalpies of formation of the alkanes from C_4 through C_{16}, and solid combustion, for C_{18} and higher.

c. Heats of reaction. In principle, any reaction, not just combustion, can be used to determine $\Delta_f H$; the principal requirement is that the enthalpies of all other species involved be well-known. A frequently used reaction is the hydrogenation of the corresponding alkene; also used is the isomerization—e.g. the conversion of n-butane to isobutane. Other reactions include formation from H_2 and the corresponding monohalide (F, Br, Cl or I) or, in the cases of methane and ethane, the dihalides. The latter techniques are generally less precise than the direct combustion. Since all three of the above techniques depend only on the First Law of Thermodynamics (any change in the internal energy of a system is equal to the sum of the heat input less the work done by the system), they are referred to as First Law methods.

d. Equilibrium constant measurements. Differentiation with respect to T of equation 2 and substitution with equation 7 gives van't Hoff's relation,

$$(\delta \ln K_p / \delta T)_p = \Delta_r H^\circ / RT^2 \qquad (47)$$

From this, we see that the slope of a plot of $R(\ln K_p)$ vs $1/T$ gives $\Delta_r H^\circ$. (Since $\Delta_r H^\circ$ is not constant, the plot will exhibit some curvature.) As already discusssed, from the heat of reaction the enthalpy of formation of one species can be determined if those for the other species are known. This procedure, involving changes with temperature of enthalpy and entropy, is referred to as a Second Law method. This method has not been used for the alkanes because other, more accurate techniques are available.

2. Heat capacity and entropy

The measurement of entropy $S^\circ(T)$ requires a measurement of heat capacity from absolute zero up to T, the temperature of interest. Heat capacities are measured in a calorimeter in which the temperature change pursuant to an input of a known quantity of heat (e.g. electrical energy) is monitored. Calculating C_p requires knowing the mass, the temperature increment and energy input, and the heat capacity of the empty apparatus. (In the case of liquids, sometimes what is measured is C_{sat}, the saturation heat capacity, defined by $C_{sat} = T(\delta S/\delta T)_{sat}$, which differs from C_p by the relationship

$$C_{sat} = C_p - T(\delta V/\delta T)_p (dp/dT)_{sat}$$

In measurements of C_{sat}, the pressure changes with temperature along the vapor–liquid saturation curve. The difference between C_{sat} and C_p is small except at temperatures far above the normal boiling point.) Since measurements near 0 K become increasingly difficult with diminishing temperature, some sort of extrapolation is necessary over the last few kelvins. Entropy is determined by integrating the C_p function with respect to temperature. Since the method depends on the fact that the residual entropy at absolute zero is 0, this is called a Third Law procedure.

Entropy can also be calculated by the Second Law method of the preceding paragraph: if the enthalpy is known as a function of temperature, then a measurement of an equilibrium constant over that temperature range will allow a determination of S.

Heat capacity can also be determined from a measurement of the velocity of sound in the gas. For an ideal gas,

$$C_v/R = RT/(Mu^2 - RT) \qquad (48)$$

where u is the gas velocity and M the molecular weight. For a real gas, the equation must take into account the nonzero value of the second virial coefficient, B, and its derivative, dB/dT.

B. Statistical Thermodynamic Calculations

In Section II above we presented all the required formulas for carrying out statistical mechanical calculations of the thermodynamic properties and discussed their limitations. These were the procedures pioneered in the 1930s and 40s, mainly by Pitzer and his associates. This approach culminated in the compilation of SWS[22], but is no longer widely used because of the unavailability of the fundamental spectroscopic constants (vibrational frequencies and torsional barriers) for most hydrocarbons. The numerous published attempts to reconcile calculated properties of ethane with experimental values attest to the difficulties inherent in this approach. Nonexperimental studies have instead pursued approximate semiempirical methods, with very good results.

C. Approximate Semiempirical Methods

The tenor of much of our preceding discussion adumbrated the attractiveness of semiempirical methods for evaluating thermodynamic functions. For example, if all C—H stretching frequencies are in the vicinity of 2900 to 3000 cm^{-1}, and if many are not known to better than 100 cm^{-1}, why not simply assume that all C—H stretches in an alkane of interest have a mean frequency of 2950 cm^{-1} and calculate accordingly? This approach of Pitzer's was carried out in great detail by Scott and later Alberty and colleagues in the papers listed below. The procedure entails selecting the fundamental global values (vibrational frequencies, torsional barriers, next-nearest-neighbor interactions, etc.) to give good agreement with the available experimental data. Over the years, since the initial compilations of Pitzer, Rossini and colleagues, successive compilations have become more and more indistinct on the differences between measured and estimated properties, and often correlations have in effect been fits to previous estimations rather than to reliable experimental numbers. For example, in the compilation of SWS[22], it is difficult to ascertain which of the tables of data are based on measurements and which on the estimations of Pitzer and collaborators.

A different semiempirical approach has been to try to generalize to a level beyond individual vibrations and rotations and assume, for example, that all methyl groups in a molecule contribute approximately the same amount to the various thermodynamic functions. We discuss this viewpoint in great detail below in the section on Additivity.

D. Polynomial Expressions for Thermodynamic Functions

The growing use of high-speed computers for various chemical engineering applications has created a demand for thermochemical data expressed more succinctly than traditional tabular presentations allow. Currently, the favored means of presentation is to express the various functions—C_p°, S°, H°, G° and K_f—as polynomials in T. One widespread format is the NASA Chemical Equilibrium program documented by Gordon and McBride[27]. Fourteen constants are needed for each species, 7 for low temperatures and 7 for high temperatures (generally divided at 1000 K):

$$C_p^\circ/R = \alpha_1 + \alpha_2 T + \alpha_3 T^2 + \alpha_4 T^3 + \alpha_5 T^4 \tag{49}$$

$$H^\circ/RT = \alpha_1 + \alpha_2 T/2 + \alpha_3 T^2/3 + \alpha_4 T^3/4 + \alpha_5 T^4/5 + \alpha_6/T \tag{50}$$

$$S^\circ/R = \alpha_1 \ln T + \alpha_2 T + \alpha_3 T^2/2 + \alpha_4 T^3/3 + \alpha_5 T^4/4 + \alpha_7 \tag{51}$$

The coefficients are chosen so that at the matching temperature both C_p and $\delta C_p/\delta T$ are equal. Equation 49 is integrated in accordance with equations 8 and 9 in order to obtain equations 50 and 51; α_6 and α_7 are the constants of integration, the former taking into account $\Delta_f H^\circ$ and the latter, the third law of thermodynamics ($S \to 0$ as $T \to 0$). The reason two sets of coefficients are necessary is that the curve of C_p vs T has a sigmoid shape, and cannot be expressed accurately by a single polynomial in T over very wide temperature ranges. Applications of these polynomials are discussed further by Burcat[28]. In our utilization of equation 49, we obtain sufficient accuracy with only four terms ($\alpha_5 = 0$), and use only a single temperature range of 298–1500 K. Furthermore we incorporate the universal gas constant R into the coefficients so as to give C_p°, S° and H° directly.

These polynomials can be manipulated appropriately in accordance with equations 2–7 to obtain polynomial expressions for the Gibbs free energy and $\ln K_f$ (for formation from the elements) or, more generally, $\ln K_p$ (for any reaction). It can be shown that $\ln K_p$ can be written in the form

$$\ln K_p = A + B/T + C \ln T + DT + ET^2 + FT^3 \tag{52}$$

where the constants $A, \ldots, F$ are related to the constants of equations 49–51 for the products and reagents:

$$A = \Delta_r S_{298}^\circ/R - \Delta\alpha_1 - \Delta\alpha_1 \ln(298) - 298\Delta\alpha_2 - (298)^2\Delta\alpha_3/2 - (298)^3\Delta\alpha_4/3 \tag{53}$$

$$B = -\Delta_r H_{298}^\circ/R + 298\Delta\alpha_1 + (298)^2\Delta\alpha_2/2 + (298)^3\Delta\alpha_3/3 + (298)^4\Delta\alpha_4/4 \tag{54}$$

$$C = \Delta\alpha_1 \tag{55}$$

$$D = \Delta\alpha_2/2 \tag{56}$$

$$E = \Delta\alpha_3/6 \tag{57}$$

$$F = \Delta\alpha_4/12 \tag{58}$$

where

$$\Delta\alpha_i = \sum \alpha_i \text{ (products)} - \sum \alpha_i \text{ (reagents)} \tag{59}$$

and the α_i for each species are as defined in equation 49.

The constants of equation 52 for $\ln K_p$ or $\ln K_f$ can also be derived simply by fitting tabulated data for the equilibrium constant. In Table 4 below we list the appropriate constants for $\log K_f$, the equilibrium constant for formation from the elements, for alkanes from CH_4 through the octanes. For the temperature range of the data (298–1500 K), only the first five terms are necessary to give adequate fits. The numerical data used to derive the constants are taken from Alberty and Gehrig[29]. Alkane names are abbreviated for convenience according to the scheme of Somayajulu and Zwolinski, as cited in Reference 29. The last number is the length of the longest carbon chain; side chains are indicated by m (methyl), e (ethyl), b (butyl) or d (decyl). However, we modify

TABLE 4. Polynomial expansions for K_f for some alkanes[a]

$$\log K_f = A + B/T + C \log T + DT + ET^2$$

Alkane	A	$10^{-3}B$	C	$10^3 D$	$10^7 E$
1	9.03	3.361	−4.696	0.6492	0.3957
2	14.09	3.443	−8.310	1.596	0.5681
3	19.79	4.158	−12.27	3.103	−1.758
4	22.11	4.979	−14.98	4.051	−3.275
2m3	22.47	5.404	−15.44	4.334	−3.404
5	24.49	5.775	−17.71	4.912	−4.242
2m4	26.12	6.061	−18.46	5.342	−4.771
22mm3	20.88	7.030	−17.30	4.822	−2.810
6	29.09	6.507	−21.32	6.382	−6.658
2m5	29.38	6.842	−20.60	6.578	−6.535
3m5	29.84	6.683	−21.69	6.438	−6.130
22mm4	26.87	7.393	−21.03	5.962	−4.049
23mm4	30.98	6.880	−22.49	6.994	−7.162
7	30.68	7.293	−24.13	7.269	−7.727
2m6	32.49	7.613	−24.59	7.556	−7.851
3m6	34.10	7.445	−25.11	7.913	−8.721
3e5	30.80	7.389	−24.11	7.239	−7.339
22mm5	30.38	8.195	−24.38	7.647	−7.227
23mm5	32.24	7.567	−24.55	7.253	−6.517
24mm5	29.45	8.064	−23.99	7.709	−8.191
33mm5	32.58	7.839	−25.11	8.006	−7.764
223m³4	32.59	8.056	−24.98	7.656	−6.643
8	36.83	8.007	−27.95	8.910	−10.632
2m7	34.38	8.419	−27.12	8.262	−8.512
3m7	37.87	8.199	−28.38	9.127	−10.658
4m7	36.34	8.251	−27.97	8.996	−10.374
3e6	34.07	8.192	−27.08	8.400	−9.207
22mm6	34.57	8.932	−27.79	8.855	−9.006
23mm6	39.82	8.234	−29.42	9.823	−11.788
24mm6	32.32	8.733	−26.66	8.409	−8.815
25mm6	36.56	8.680	−28.34	9.082	−10.109
33mm6	35.99	8.650	−28.30	9.549	−10.649
r-34mm6[b]	39.34	8.071	−29.31	9.540	−10.803
m-34mm6[b]	39.25	8.079	−29.30	9.610	−11.071
3e2m5	29.93	8.389	−25.73	7.474	−6.671
3e3m5	35.47	8.315	−28.32	9.226	−9.738
223m³5	28.41	8.811	−25.40	6.481	−14.066
224m³5	32.96	8.927	−27.34	8.448	−7.447
233m³5	35.34	8.488	−28.18	8.976	−8.522
234m³5	34.04	8.474	−27.73	9.095	−10.012
2233m⁴4	35.96	8.824	−29.20	9.626	−9.279

[a]Polynomials fitted to data tabulated by Alberty and Gehrig[29] from 298 to 1500 K. These data are not always based on experimental measurements; some are calculated based on a semiempirical scheme, discussed further in the text.
[b]We use the shorthand prefixes r and m to designate the racemic mixture of enantiomers and the *meso* compound, respectively. The currently accepted nomenclatures are 3RS4RS-dimethylhexane for the racemate and 3R4S-dimethylhexane for the *meso* stereoisomer.

their nomenclature by using 'm³' for 'mmm', 'm⁴' for 'mmmm' and 'm⁵' for 'mmmmm'. (For example, '22mm3' is 2,2-dimethylpropane; '2233m⁴4' is 2,2,3,3-tetramethylbutane and so on.)

Some sources tabulate $\log K_f$ or $\ln K_f$ as a different polynomial in T:

$$\log K_f = A + B/T + C/T^2 + D/T^3 \tag{60}$$

The constants of equation 60 have no resemblance to those for equation 52, such as are given in Table 4. For example, the polynomial for $\log K_f(CH_4)$ is

$$\log K_f(CH_4) = -5.93 + 5210/T - 31761/T^2 + 2.376 \times 10^7/T^3 \tag{61}$$

Note in particular that even the sign of the leading term is reversed. We prefer polynomials of the form of equation 52 because they are more directly related to conventional polynomial expansions of C_p.

Note also that although polynomials with only five terms (as in the above table) have sufficient precision, should one have tabulations with six terms it is not advisable simply to discard the last term. The six-term polynomial for heptane is

$$\log K_f(C_7H_{16}) = 27.07 + 7398/T - 22.17\log T + 0.005014 \times T$$
$$+ 3.965 \times 10^{-7}T^2 - 2.717 \times 10^{-10}T^3$$

at 1000 K, neglecting the last term would produce an error in $\log K_f$ of 0.272, or nearly a factor of 2 in K_f.

Bibliography of useful compilations and references

The literature on the thermochemistry of the hydrocarbons is extensive—not only in numbers of experimental and theoretical studies, but also in numbers of data evaluations and compilations. In an Appendix to this chapter we list several useful compilations and references.

IV. ADDITIVITY LAWS: GROUP ADDITIVITY

By the early 20th century the accumulated body of experimental thermochemical data was so extensive that a scheme for organizing and relating it was highly desirable. Could thermochemical data—heat capacities, enthalpies, entropies and free energies—be related directly to chemical structure? If so, could one formulate rules for predicting these properties from chemical structure, obviating the impossible chore of measuring the hundreds of thousands of organic compounds whose numbers were growing steadily through new syntheses or identifications?

Since the efforts of Fajans and Kharasch[30] in the 1920s to relate the heats of combustion of hydrocarbons to a set of bond and group energies, there have been many efforts to reduce thermochemical properties to a set of fundamental atomic or molecular constants. The more important approaches have been reviewed in some detail by Cox and Pilcher[21], and we do not repeat the details of their useful discussion; a few general remarks are sufficient for our purposes. The basic question we are concerned with is: can one, conceptually, subdivide an arbitrary chemical compound into a set of smaller units in such a way that the thermochemical properties of that compound can be calculated from constants associated with the smaller units? That is, if we are concerned with one of these thermochemical properties, Φ, can we calculate this property by a relation such as

$$\Phi(\text{Compound}) = \sum_i \varphi(u_i) \tag{62}$$

where the u_i are the fundamental units, each of which contributes an invariant amount,

φ_i, to Φ? Such relationships are called *additivity* laws, and are fundamental in many areas of physical chemistry. An additive property is one whose value is determined by adding up the contributions from smaller units. An almost trivial example (though it wasn't a century and a half ago) is the relationship between the molecular weight of a compound and the atomic weights of its constituent atoms. Molecular weights and molecular volumes are additive; melting points or refractive indexes are not.

Thus, for thermochemical properties also, the first obvious scheme might be to consider the atoms of a compound as the fundamental units u_i. In 1958, Benson and Buss[31] discussed a natural hierarchy of additivity schemes, of which an atomic additivity was the first; the second and third were bond additivity and group additivity, respectively. The conceptual basis for this approach arises from a consideration of the laws of dilute multicomponent solutions. For many properties, in the limit of infinite dilution the contributions of the several components become additive because intermolecular interactions become negligible at sufficiently large separation (and hence, at sufficiently high dilution). Raoult's Law for vapor pressures of ideal solutions is one example of such a relationship. Benson and Buss proposed an analogous law for molecules: If RNR and SNS are molecules containing the groups (or atoms) R and S, separated by a common molecular unit N, and φ is some molecular property, then as the separation between R and S becomes large (i.e. as N becomes large), $\Delta\varphi$ for the following disproportionation reaction approaches zero except for symmetry effects:

$$RNR + SNS \rightleftharpoons 2RNS$$

In the zeroth-order approximation, N vanishes, and the statement that $\Delta\varphi = 0$ for the above reaction is equivalent to an atom additivity scheme—equation 62 with the u_i equal to atoms. As N gets larger (atoms, groups, etc.) we obtain a hierarchy of additivity laws (bond additivity, group additivity, etc.). Since the interactions of the groups R and S will surely diminish as N increases, we expect increasing accuracy as we proceed up the ladder of this hierarchy of laws. (Nonbonded interactions, discussed in Section V, may become important when N is large but the shape of the molecule is such that R and S can still be in proximity with one another.)

In a bond additivity scheme, a property is related to contributions from each of the bonds in the molecule. For example, the heat of formation of isooctane (2,2,4-trimethylpentane), whose structure is shown in Figure 1, would be calculated from its bond units as

$$\Delta_f H°(\text{isooctane}) = 18\Delta_f H°(\text{C—H}) + 7\Delta_f H°(\text{C—C})$$

where $\Delta_f H°(\text{C—H})$ and $\Delta_f H°(\text{C—C})$ are, respectively, the contributions to heat of formation from each C—H or C—C bond in the compound. If this approach were strictly valid, then the heats of formation of all the octane isomers, each of which contains 18 C—H bonds and 7 C—C bonds and no other bonds, should be identical. In Table 5 below we list $\Delta_f H°_{298}$ and $S°_{int,298}$ for several octanes.

Table 5 shows that the enthalpies and entropies of these octanes, while not identical, differ by only 4 kcal mol^{-1} and 5.3 eu, respectively, from smallest to largest. In general, bond additivity schemes for thermochemical properties can provide order-of-magnitude

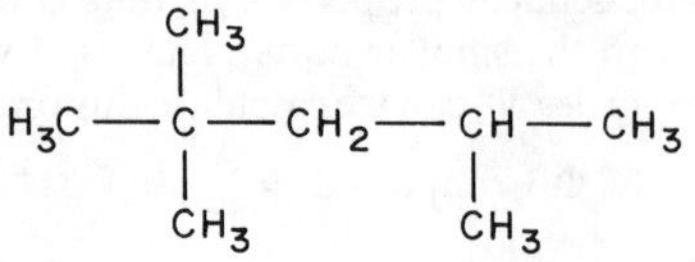

$$\begin{array}{c}
\qquad\qquad CH_3 \\
\qquad\qquad | \\
H_3C \text{——} C \text{——} CH_2 \text{——} CH \text{——} CH_3 \\
\qquad\qquad | \qquad\qquad\quad | \\
\qquad\qquad CH_3 \qquad\qquad CH_3
\end{array}$$

FIGURE 1

TABLE 5. Comparison of $\Delta_f H^\circ_{298}$ and $S^\circ_{int,298}$ of some octane isomers

Compound	$\Delta_f H^\circ_{298}$ (from Reference 53)	$S^\circ_{int,298}$ (from Reference 22)
n-C_8H_{18}	−49.86	117.29
3-Methylheptane	−50.79	115.49
2,2-Dimethylhexane	−53.68	113.97
2,3-Dimethylhexane	−51.10	113.46
2,2,4-Trimethylpentane	−53.54	114.25
2,2,3-Trimethylpentane	−52.58	113.34
3-Ethyl-3-methylpentane	−51.34	114.39
2,2,3,3-Tetramethylbutane	−53.92	111.90

predictions or better of values for arbitrary compounds. (Benson[32] has given a table of partial bond contributions for heat capacity, entropy and heat of formation for most bonds of interest in organic molecules.) In the belief that one can do better than order-of-magnitude estimates, many workers have moved on to examine *group* additivity laws. Three important schemes—since shown to be mathematically equivalent[33]—have been developed by Laidler[34], Allen[35] and Benson and Buss[31]. Specialized schemes for hydrocarbon enthalpies, also described and compared in Reference 21, have been developed by Platt[36], Greenshields and Rossini[37] and Somayajulu and Zwolinski[38]. Joshi[39] proposed another additivity scheme (based on Laidler's) for alkanes and compared his calculated results with those of seven earlier schemes[39–44] for reproducing the enthalpies of 66 alkanes. Several other approaches are outlined and compared by Reid and coworkers[45].

A group is defined as 'a polyvalent atom (ligancy $\geqslant 2$) in a molecule together with all of its ligands'[46]. A group is written as X—$(A)_i(B)_j(C)_k(D)_l$, where X is the central atom attached to i A atoms, j B atoms, etc. In hydrocarbons, X is always carbon and A, B, C and D are always carbons or hydrogens. For example, the groups of isooctane (see Figure 1) are:

$$5\,C—(C)(H)_3$$
$$1\,C—(C)_2(H)_2$$
$$1\,C—(C)_3(H)$$
$$1\,C—(C)_4$$

In contrast, the groups of another octane isomer, 2,2,3,3-tetramethylbutane (Figure 2), are

$$6\,C—(C)(H)_3$$
$$2\,C—(C)_4$$

A group additivity scheme would thus provide the basis for differentiating between these two isomers. However, the groups of some octane isomers—e.g. 2-methylheptane,

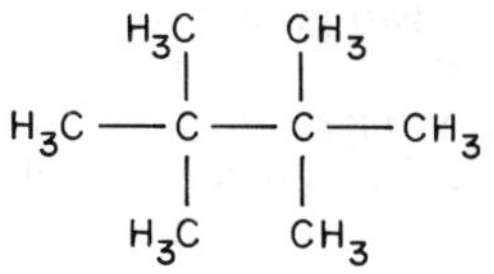

FIGURE 2

3-methylheptane and 4-methylheptane—are the same:

$$3\,C—(C)(H)_3$$
$$4\,C—(C)_2(H)_2$$
$$1\,C—(C)_3(H)$$

We could not, then, expect to differentiate between these isomers on the basis of groups alone, as defined above. However, the differences among the enthalpies of these three isomers are less than $1\,kcal\,mol^{-1}$. (An error of $1\,kcal\,mol^{-1}$ leads to an error of a factor of 5.4 in K_{eq} at 298 K and a factor of 1.4 at 1500 K.) Can we develop a simple scheme that is precise enough to predict thermochemical properties to $1\,kcal\,mol^{-1}$ or better? More importantly,, are the experimental measurements accurate to $1\,kcal\,mol^{-1}$ or better? To this latter question we shall return shortly. Now, we continue our discussion of group additivities with the observation that the species of interest in this chapter— linear and cyclic alkanes—are comprised of only the four groups already introduced: a carbon atom attached to either one, two, three or four other carbon atoms, with the remaining ligands being H atoms. (Methane, CH_4, falls in a class by itself, but since this $C—(H)_4$ group occurs in no other alkanes, the values we assign to its thermochemical properties have no direct bearing on the properties of any other compounds, and we do not consider it further.)

In the remainder of this section we discuss the derivation of group values for heat capacities, enthalpies and entropies and provide several examples to illustrate their application. In Section V we will discuss refinements to this simple group additivity scheme. There, the aim is to see if, by a simple expansion beyond the four parameters required for group additivity, we can improve significantly upon its accuracy in predicting thermochemical properties.

Heuristically, the simplest way to derive the values for the four fundamental groups is to begin with experimental values for the smallest alkanes and see what group values are necessary to match experiment. Thus, ethane, containing only two $C—(C)(H)_3$ groups, determines group values for that group:

$$\Delta_f H^\circ_{298}(C_2H_6) = -20.03\,kcal\,mol^{-1}$$
$$\Rightarrow H_f[C—(C)(H)_3] = -10.015\,kcal\,mol^{-1};$$
$$S^\circ_{298}(C_2H_6) = 54.8\,eu$$
$$\Rightarrow S_{int,298}(C_2H_6) = 54.8 + R\ln(18) = 60.5$$
$$\Rightarrow S[C—(C)(H)_3] = 30.25\,eu;$$
$$C^\circ_{p298}(C_2H_6) = 12.7\,cal\,mol^{-1}\,K^{-1}$$
$$\Rightarrow C_p[C—(C)(H)_3] = 6.35\,cal\,mol^{-1}\,K^{-1}$$

Since propane contains $2\,C—(C)(H)_3$ and $1\,C—(C)_2(H)_2$ groups, we can calculate term values for the $C—(C)_2(H)_2$ group by subtracting from the experimental quantities the values just calculated for the $C—(C)(H)_3$ group. All the straight-chain alkanes consist of just these two groups:

$$C_nH_{2n+2} \equiv 2\,C—(C)(H)_3 + (n-1)C—(C)_2(H)_2 \text{ groups}^{47}$$

The contributions of the other two groups are derived from a consideration of the branched alkanes. The simplest one, 2-methylpropane or isobutane, consists of $3\,C—(C)(H)_3$ and $1\,C—(C)_3(H)$ groups: hence the values for the latter group can be obtained from experimental data for isobutane. Similarly, the values for the $C—(C)_4$ group can be obtained from data for 2,2-dimethylpropane (neopentane).

When the group values thus derived are applied to various alkanes, we find less than perfect agreement between experimental and calculated thermochemical values.

Consequently, we choose to determine the group values by minimizing the average deviation between calculated and experimental quantities for a larger set of alkanes. In 1969, Benson and coworkers[48] reviewed the thermochemical literature and, applying this procedure to the best data, derived group values for enthalpy of formation, entropy and heat capacity. Their results are listed in Table 6. We have updated the enthalpy and entropy evaluations and present new group values in Section V[49].

The tabulated contributions to C_p° given in Table 6 are helpful for understanding the effect of temperature on the thermochemical quantities, but inconvenient to use in practical applications. Following the form of equation 49 (except that we use only four constants and set $a_5 = 0$), we can express the heat capacity contribution of each group by a polynomial in T. The coefficients are given below in Table 7.

Henceforth we shall refer to the four basic groups by the less cumbersome designations of P ($\equiv$ C—(C)(H)$_3$), S($\equiv$ C—(C)$_2$(H)$_2$), T($\equiv$ C—(C)$_3$(H)) and Q($\equiv$ C—(C$_4$)), where the initials are suggested by the standard nomenclature: primary, secondary, tertiary and quaternary for a carbon atom attached to 1, 2, 3 or 4 other carbon ligands, respectively.

When this procedure is applied to enthalpies of sets of isomers, some discrepancies emerge. For example, consider the experimental values for the following octanes, all of which have the same groups (4P, 2S, 2T):

Octane isomer	$\Delta_f H_{298}^\circ$	$S_{int,298}^\circ$
2,3-Dimethylhexane	−51.10	113.46
2,5-Dimethylhexane	−53.18	113.66
3,4-Dimethylhexane	−50.86	113.13
3-Ethyl-2-methylpentane	−50.43	112.78

TABLE 6. 1969 group values for $\Delta_f H_{298}^\circ$, $S_{int,298}^\circ$ and C_{pT}° [a]

Group	$\Delta_f H_{298}^\circ$	$S_{int,298}^\circ$	C_{pT}°						
			298	400	500	600	800	1000	1500
C—(C)(H)$_3$	−10.08	30.41	6.19	7.84	9.40	10.79	13.02	14.77	17.58
C—(C)$_2$(H)$_2$	−4.95	9.42	5.50	6.95	8.25	9.35	11.07	12.34	14.25
C—(C)$_3$(H)	−1.90	−12.07	4.54	6.00	7.17	8.05	9.31	10.05	11.17
C—(C)$_4$	0.50	−35.10	4.37	6.13	7.36	8.12	8.77	8.76	8.12

[a]From Reference 48, Table 33; the same data, with a few revisions, appear in Reference 32, Table A.1.

TABLE 7. Polynomial expressions for gas phase C_{pT}° for alkane groups[a]

$$C_{pT}^\circ = a_1 + a_2 T + a_3 T^2 + a_4 T^3$$

Group	a_1	$100 \times a_2$	$10^5 \times a_3$	$10^9 \times a_4$
C—(C)(H)$_3$	−0.10	2.442	−1.18	2.26
C—(C)$_2$(H)$_2$	−2.55	2.288	−1.32	2.97
C—(C)$_3$(H)	−1.72	2.665	−2.05	5.67
C—(C)$_4$	−3.90	3.719	−3.47	10.16

[a]Note that these a_i coefficients are related to the α_i coefficients of equation 49 by $a_i = \alpha_i \times R\,cal\,mol^{-1}\,K^{-1}$.

Because these four compounds have the same eight groups, we would calculate the same entropies and enthalpies for them. The above data show that this is a very good approximation for the entropies (< 0.9 eu spread), but not for the enthalpies (> 2.7 kcal mol^{-1} spread; an error of ΔS eu produces the same error in K_{eq} as an error in ΔH of $-T\Delta S$ cal mol^{-1}). One reason for the discrepancy lies in interactions between nonbonding groups, which are not accounted for in our group additivity scheme. Before we can derive satisfactory group values for enthalpies we need to take these nonbonding interactions into account. This is done in the next section.

V. SECOND-ORDER EFFECTS: NONBONDED INTERACTIONS

A. *Gauche* Interactions and Strain

By virtue of its tetrahedral symmetry, the HCH bonds in CH_4 are all $109.5°$. In this configuration, the $H \cdots H$ distances are all 1.78 Å. The van der Waals radius of an atom is defined such that whenever the distance between two atoms equals the sum of their van der Waals radii, their interatomic repulsion is RT. Since the H-atom van der Waals radius is 1.2 Å, any distance less than 2.4 Å between H-atom centers will produce significant repulsion between the electron clouds of the two H atoms. There have been many theoretical attempts to quantify this repulsion; its magnitude is given approximately by an empirical formula proposed by Huggins[50]:

$$V_{H \cdots H} = 4 \times 10^5 \exp(-5.4r) - 47r^{-6} - 98r^{-8} - 205r^{-10} \qquad (63)$$

where r is in Å and V is in kcal mol^{-1}. In ethane, the distance between any H atom and the H atoms of the other CH_3 group will be approximately 2.2 Å when the pairs are eclipsed (one H atom behind another when viewed along the C—C axis), but 2.5 Å when they are staggered; according to Huggins's equation, this makes the latter conformation some 3 kcal mol^{-1} more stable than the former—which is the reason that the staggered conformation is thermodynamically the preferred one. In any alkane, whenever possible, adjacent CH_xC_{3-x} groups rotate so that the ligands are staggered. When they are not constrained to tetrahedral symmetry, the H atoms will move apart from one another to reduce the interatomic repulsion. This can happen by increasing the CCC angle in alkanes with three or more carbons or by decreasing the HCH angle of a CH_3 group. In such an alkane, the CCC bond is found to be not $109.5°$ but $112.4°$ [51] and the HCH bond is $107.7°$. In ethane in its most stable conformation (with the two sets of H atoms staggered with respect to each other), the second of these effects increases the nonbonded $H \cdots H$ distance from the ideal tetrahedral structural value of 2.26 to 2.27 Å; using Huggins's formula (equation 63), we can calculate that this increase in distance decreases the $H \cdots H$ repulsion by 0.09 kcal mol^{-1} per H—H pair, or 0.5 kcal mol^{-1} for the entire molecule. In larger alkanes, the $H \cdots H$ distance on alternate CH_2 groups is increased (by both angle changes) from the tetrahedral 2.49 to 2.59 Å, which, according to equation 63, lowers the repulsion from 0.29 to 0.12 kcal mol^{-1} per $H \cdots H$ pair. This energy saving is balanced by the energy required to spread the CCC bond angle.

Consider next a normal butane molecule. Its end methyl groups can assume three spatial conformations: (1) they can both be at 12 o'clock positions relative to the central (interior) C—C bond (*cis* rotamer); (2) one can be at 12, the other at 6 o'clock (*trans* rotamer); or (3) one can be at 12 and the other at 2 or 10 o'clock (mirror-image *gauche* rotamers). This can be visualized by means of Newman diagrams, which are projections of the ligands onto a plane perpendicular to the central C–C axis, the nearer of the central C atoms being indicated by a point, the farther, by a circle (Figure 3).

If we measure the distances between H atoms on the two end methyl groups in the

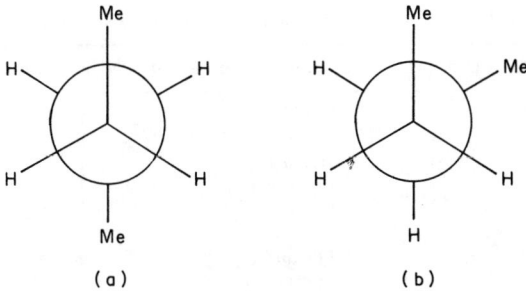

(a)   (b)

FIGURE 3. Rotamers of n-butane: (a) *trans*; (b) *gauche* (one of two mirror images). The diagram is a projection of the molecule on a plane perpendicular to the central C—C bond. The three ligands at the 12, 4 and 8 o'clock positions are attached to the number 2 carbon; the ligands and 2, 6 and 10 o'clock are attached to the number 3 carbon.

cis conformation, we find that, in the fully tetrahedral geometry, one pair of H atoms must approach within about 1.8 Å of each other—about 2.0 Å allowing for the non-tetrahedral effects just discussed. (As the CH_3 groups rotate, H atoms will approach each other considerably closer, which is the source of the potential energy barrier to internal rotation.) In the *gauche* conformation, they can be some 0.4 Å farther apart, and in the *trans* they are more than 4 Å apart. According to equation 63, any approach closer than 2.75 Å between H-atom centers will produce some repulsion between the electron clouds of the two H atoms. This repulsion manifests itself thermochemically by raising the enthalpy of formation of the compound. (Heat capacity and entropy are affected only weakly by this and other repulsive interactions discussed below.) Thus, the *trans* rotameric conformation is the stablest conformation, and the end groups will rotate into that conformation in order to minimize electron repulsions[52]. Experimental evidence indicates that the *gauche* butane conformation is 0.8 kcal mol^{-1} less stable than the *trans* form. The *cis* form is substantially less stable.

Now, consider 2-methylbutane, whose Newman diagrams are shown in Figure 4. It is impossible for the end methyl groups to configure themselves so that all are *trans* to one another. In the two *gauche* conformations, two methyls are *gauche* to one another,

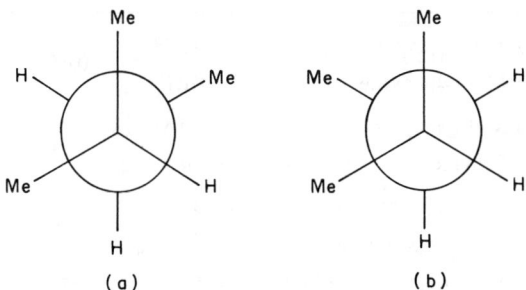

(a)   (b)

FIGURE 4. Rotamers of 2-methylbutane: (a) *gauche* (one of two mirror images); (b) *syn*.

TABLE 8. Number of *gauche* interactions for a C_a—C_b bond

C_a atom group	C_b atom group	*gauche* interactions
P	P, S, T or Q	0
S	S (*trans*)	0
S	S (*gauche*)	1
S	T (*gauche*)	1
S	T (*syn*)	2
S	Q	2
T	T ('*trans*')	2
T	T ('*cis*')	3^a
T	Q	4
Q	Q	6

[a] In the special case that there are three T groups in a row (as in 234m³5), one pair will be '*trans*' and the other '*cis*', giving a total of five *gauche* interactions.

but in the *syn* conformation, one methyl is at the *gauche* distance from two other methyl groups. Since each such *gauche* conformation contributes some $0.8 \, \text{kcal mol}^{-1}$ of destabilization, the most stable orientation is that which minimizes the number of such interactions, namely the *gauche*, rather than the *syn*, conformation. Because the interacting methyl groups are always positioned with two intervening C atoms, *gauche* interactions are also called 1,4 interactions. Any alkane with side branches will have some 1,4 interactions. The total number can, in most cases, be calculated easily by a simple procedure described by Alberty and Gehrig[29]. The number of 1,4 interactions for each C—C bond depends on: (1) how many other carbon atoms are attached to each of the carbon atoms forming the bond; and (2) what the exact stereometry is. For example, 2-methylbutane in the *gauche* arrangement will have two 1,4 interactions, but three in the *syn* arrangement. The number of interactions is tabulated in Table 8. Generally, one is interested only in the minimum number, but there are instances where the other case is important. For example, consider 3,4-dimethylhexane: there will be two *gauche* interactions if the two T groups are on opposite sides of main carbon atom chain—as, for example, in the case of *r*-34mm6; there will be an additional interaction if they are on the same side, as in *m*-34mm6.

The effect of 1,4 interactions on enthalpies is illustrated by the following pairs of alkane isomers. In each pair, the second alkane differs from the first only by an additional 1,4 interaction; the other groups are all the same:

Alkane	$\Delta_f H^\circ$	$\Delta\Delta_f H^\circ$
2-Methylpentane	−41.78	0.65
3-Methylpentane	−41.13	
4-Methylheptane	−50.67	0.31
3-Ethylhexane	−50.36	
2,3-Dimethylhexane	−51.10	0.24
3,4-Dimethylhexane	−50.86	
2,2,3-Trimethylpentane	−52.58	0.88
2,3,3-Trimethylpentane	−51.70	

Similar comparisons can be made for alkylcycloalkanes (discussed in more detail in Section VI, below). The following pairs of dimethylcyclohexanes all differ only in that the second has two more *gauche* interactions than the first:

	$\Delta_f H^\circ$	$\Delta\Delta_f H^\circ$
trans-1,2	−43.00	1.87
cis-1,2	−41.13	
cis-1,3	−44.12	1.94
trans-1,3	−42.18	
trans-1,4	−44.10	1.89
cis-1,4	−42.21	

On the basis of comparisons such as these the destabilization energy associated with each 1,4 interaction was proposed by Allen[35] to be 0.5 kcal mol^{-1}, Benson and coworkers[31,48] favored 0.7 kcal mol^{-1}, later increased[32] to 0.8 kcal mol^{-1}; and Cox and Pilcher[21] preferred 0.6 kcal mol^{-1}; we find best results with a value between 0.75 and 0.8 kcal mol^{-1}. A slightly higher value is suggested by the experimental measurement, using Raman spectroscopy, of the difference in enthalpy at low temperatures between *trans* and *gauche* n-butane—approximately 0.96 kcal mol^{-1} [24].

B. 1,5 Interactions

One other type of nonbonded interaction must be taken into account. Consider the molecule 2,2,4,4-tetramethylpentane. The H atoms on the side methyl groups (on the 2nd and 4th carbons) can be as close as 0.3 Å from one another, and it is difficult to configure the molecule so that no such H atoms are less than 1.5 Å apart. Because the interacting methyl groups can be numbered 1 and 5 with respect to one another, this is called a 1,5 interaction, and one would expect this repulsion to be even greater than that resulting from a 1,4 interaction. Note that if there is only one pair of side chains in the 1,5 configuration (e.g. 2,4-dimethylpentane) then the alkane can rotate about the central C—C bond to keep the side methyls far enough apart from one another to avoid 1,5 repulsions. But if there are two methyls (or larger groups) on one C atom (e.g. 2,2,4-trimethylpentane) then close 1,5 interaction between one pair of methyls is unavoidable. Thus, 1,5 enthalpy corrections are required for alkanes of the form $n,n,n + 2$-trialkylalkane (1 correction) or $n,n,n + 2, n + 2$-tetraalkylalkane (2 corrections).

Interactions involving centers even farther apart are probably smaller than the two just considered, but may not be completely negligible. Nevertheless, because of the flexibility of the alkane chain, it is difficult to assign any quantitative magnitude to more distant interactions, and we must therefore neglect them for the present. We will designate the 1,4 and 1,5 interactions by the letters G and F, respectively.

With these six parameters P, S, T, Q, G and F we are then ready to characterize any linear alkane. In Table 9, we list the experimental enthalpies for the 64 alkanes for which data are available and compare them with calculations. The 3rd through 8th columns list the number of P, S, T, Q, G and F contributions, respectively; the 9th column gives the calculated enthalpy using term values for P, S,... etc. that minimize the average error. (These parameters are thus slightly improved over the values given in Table 6.) The experimental values are all taken from Pedley, Naylor and Kirby (PNK)[53].

N. Cohen and S. W. Benson

TABLE 9. Experimental and calculated enthalpies of formation (298 K) for some alkanes

Alkane	$\Delta_f H°(exp)^a$	#P	#S	#T	#Q	#G	#F	$\Delta_f H°(calc)^b$	Error[c]
2	−20.03	2	0	0	0	0	0	−20.00	−0.03
3	−25.02	2	1	0	0	0	0	−25.00	−0.02
4	−30.02	2	2	0	0	0	0	−30.00	−0.02
2m3	−32.07	3	0	1	0	0	0	−32.40	0.33
5	−35.11	2	3	0	0	0	0	−35.00	−0.11
2m4	−36.74	3	1	1	0	1	0	−36.60	−0.14
22mm3	−40.18	4	0	0	1	0	0	−39.90	−0.28
6	−39.94	2	4	0	0	0	0	−40.00	0.06
2m5	−41.78	3	2	1	0	1	0	−41.60	−0.18
3m5	−41.13	3	2	1	0	2	0	−40.80	−0.33
22mm4	−44.48	4	1	0	1	2	0	−43.30	−1.18
23mm4	−42.61	4	0	2	0	2	0	−43.20	0.59
7	−44.86	2	5	0	0	0	0	−45.00	0.14
2m6	−46.51	3	3	1	0	1	0	−46.60	0.09
3m6	−45.72	3	3	1	0	2	0	−45.80	0.08
22mm5	−49.21	4	2	0	1	2	0	−48.30	−0.91
23mm5	−47.54	4	1	2	0	3	0	−47.40	−0.14
24mm5	−48.21	4	1	2	0	2	0	−48.20	−0.01
33mm5	−48.09	4	2	0	1	4	0	−46.70	−1.39
3e5	−45.32	3	3	1	0	3	0	−45.00	−0.32
223m³4	−48.88	5	0	1	1	4	0	−49.10	0.22
8	−49.86	2	6	0	0	0	0	−50.00	0.14
2m7	−51.48	3	4	1	0	1	0	−51.60	0.12
3m7	−50.79	3	4	1	0	2	0	−50.80	0.01
4m7	−50.67	3	4	1	0	2	0	−50.80	0.13
22mm6	−53.68	4	3	0	1	2	0	−53.30	−0.38
23mm6	−51.10	4	2	2	0	3	0	−52.40	1.30
24mm6	−52.39	4	2	2	0	3	0	−52.40	0.01
25mm6	−53.18	4	2	2	0	2	0	−53.20	0.02
33mm6	−52.58	4	3	0	1	4	0	−51.70	−0.88
34mm6[d]	−50.86	4	2	2	0	5	0	−50.80	−0.06
3e6	−50.36	3	4	1	0	3	0	−50.00	−0.36
3e3m5	−51.34	4	3	0	1	6	0	−50.10	−1.24
223m³5	−52.58	5	1	1	1	5	0	−53.30	0.72
224m³5	−53.54	5	1	1	1	3	1	−53.30	−0.24
233m³5	−51.70	5	1	1	1	6	0	−52.50	0.80
234m³5	−51.94	5	0	3	0	5	0	−53.20	1.26
2233m⁴4	−53.92	6	0	0	2	6	0	−55.00	1.08
3e2m5	−50.43	4	2	2	0	4	0	−51.60	1.17
9	−54.54	2	7	0	0	0	0	−55.00	0.46
33ee5	−55.52	4	4	0	1	8	0	−53.50	−2.02
2233m⁴5	−56.67	6	1	0	2	8	0	−58.40	1.73
2234m⁴5	−56.62	6	0	2	1	6	1	−58.30	1.68
2334m⁴5	−56.43	6	0	2	1	8	0	−58.30	1.87
2244m⁴5	−57.74	6	1	0	2	4	2	−58.40	0.66
10	−59.63	2	8	0	0	0	0	−60.00	0.37
2m9	−62.12	3	6	1	0	1	0	−61.60	−0.52
5m9	−61.81	3	6	1	0	2	0	−60.80	−1.01
11	−64.75	2	9	0	0	0	0	−65.00	0.25
2255m⁴7	−72.28	6	3	0	2	6	0	−70.00	−2.28
3355m⁴7	−66.13	6	3	0	2	8	2	−65.20	−0.93
22445m⁵6	−67.16	7	1	1	2	8	2	−67.60	0.44
12	−69.24	2	10	0	0	0	0	−70.00	0.76

TABLE 9 (continued)

Alkane	$\Delta_f H°(\text{exp})^a$	#P	#S	#T	#Q	#G	#F	$\Delta_f H°(\text{calc})^b$	Error^c
3366m⁴8	−76.05	6	4	0	2	8	0	−73.40	−2.65
4466m⁴9	−74.86	6	5	0	2	6	2	−76.80	1.94
35ee35mm7	−72.01	6	5	0	2	12	2	−72.00	−0.01
5577m⁴11	−87.62	6	7	0	2	4	2	−88.40	0.78
46ee46mm9	−82.96	6	7	0	2	12	2	−82.00	−0.96
16	−89.58	2	14	0	0	0	0	−90.00	0.42
18	−99.09	2	16	0	0	0	0	−100.00	0.91
5b22	−140.44	3	22	1	0	2	0	−140.80	0.36
11b22	−141.83	3	22	1	0	2	0	−140.80	−1.03
11d21	−168.69	3	27	1	0	3	0	−165.00	−3.69
32	−166.63	2	30	0	0	0	0	−170.00	3.37

^{a}Experimental data tabulated by PNK; all values in kcal mol^{-1} [53].
^{b}Calculated by group additivities using the group values: P = −10.00, S = −5.00, T = 2.40, Q = 0.10, G = 0.80, F = 1.60.
^{c}Error = experimental minus calculated value.
^{d}34mm6 exists in the racemic, or r-, form, which has 4 gauche interactions, and the meso form, with 5 gauche interactions (calculated here). The experimental value presumably applies to an equilibrium mixture, with $\Delta_f H°$ somewhere between the two. For further discussion see the first Example in Section XIII.

The average error in the predicted enthalpy for these 64 compounds is 0.77 kcal mol^{-1}; this compares favorably with the average experimental uncertainty of 0.60 kcal mol^{-1}. Nevertheless, there are some disconcertingly large discrepancies. Some of these may reflect experimental errors: e.g. those for dotriacontane (32) and 11-decylheneicosane (11d21). For both of these large, not highly branched alkanes, the experimental value is probably in error. (If we omit these two, our average calculated error is reduced to 0.65 kcal mol^{-1}.) Other discrepancies occur in patterns that suggest systematic errors. Consider the following alkanes, all of which have geminal side chains (i.e. on the same carbon atom):

Alkane	#G	#F	Error in $\Delta_f H°$
22mm4	2	0	−1.18
33mm5	4	0	−1.39
3e3m5	6	0	−1.24
33ee5	8	0	−2.02
2255m⁴7	6	0	−2.28
3366m⁴8	8	0	−2.65

Our group additivity methods underestimate the stability of these molecules. On the other hand, several large alkanes with vicinal, or adjacent side chains have errors in the opposite direction (stability overestimated):

Alkane	#G	#F	Error in $\Delta_f H°$
3e2m5	4	0	1.17
234m³5	4	0	2.06
2234m⁴5	6	1	1.68
2233m⁴5	8	0	1.73
2334m⁴5	8	0	1.87

One possible corrective step that seems physically reasonable is to assume that successive *gauche* interactions do not induce the same enthalpy increment. Consider again the Newman projection for 2-methylbutane (Figure 2a). By rotating slightly around the C—C bond, the distance between the *gauche* methyl side groups can be increased, thus decreasing the repulsive force. However, in 2,2-dimethylbutane any rotation to increase the separation between one pair of *gauche* methyls decreases the separation between the others; hence it is likely that the enthalpy increment for the second *gauche* interaction is numerically larger than that for the first. A different sort of evidence suggesting that successive *gauche* interactions do not diminish the stability by a constant amount comes from the measured Raman spectrum of 2,3-dimethylbutane, which indicates that the *gauche* and *trans* forms (with, respectively, two and three 1,4 interactions) differ in energy by only $54 \pm 30 \, \text{cal mol}^{-1}$ [24]. There are various possible schemes that can be applied to achieve improved agreement. We obtain the best results by assuming an additional $0.8 \, \text{kcal mol}^{-1}$ of repulsion for T–T and T–Q pairs and $1.6 \, \text{kcal mol}^{-1}$ for Q–Q pairs. This adjustment can be made within the framework of the additivity philosophy simply by altering the method of counting *gauche* interactions. Instead of using the conventional counting procedure given above, one counts the total number of possible 1,4 carbon–carbon interactions not on the main chain. This has the effect of increasing the destabilizing effect of multiple *gauche* interactions. It also enables us to reduce the value of the parameter for Q (from 0.1 to -0.1), which gives better agreement for 2,2-dimethylpropane, the only alkane with a quaternary carbon atom and no *gauche* interactions. The revised 1,4 interaction counting scheme is shown in Table 10.

The calculation of entropies by group additivity is not, to first approximation, affected by *gauche* interactions. (At a more detailed level, *gauche* interactions can affect entropy calculations because they can affect barrier heights to internal rotations, which in turn affects the entropy of hindered internal rotation.) In Table 11 below we have optimized the group values for entropy contributions given in Table 6 above to give best overall agreement with the entropies tabulated by SWS down through the decanes[22]. (Note that these values, like most similar compilations of gas phase entropies, are not experimental values, but are derived, either from spectroscopic data or from semiempirical methods. 'Experimental' entropies of unbranched alkanes up to $C_{16}H_{34}$ were tabulated by Person and Pimentel[54]; these values were derived from calorimetric liquid-phase entropy measurements, combined with entropies of vaporization and vapor pressure measurements. Their values, up through decane, differ from those of SWS in Table 11 by no more than 0.15 eu.) For these 150 alkanes the average discrepancy is 0.61 eu. Note that in several instances there is a considerable discrepancy between the values of SWS

TABLE 10. Revised number of *gauche* interactions for a C_a—C_b bond

C_a atom group	C_b atom group	*gauche* interactions
P	P, S, T or Q	0
S	S	0
S	T	1
S	Q	2
T	T	3[a]
T	Q	5[a]
Q	Q	8[a]

[a]These are the values that differ from the simple counting scheme of Table 8.

TABLE 11. Comparison of tabulated and calculated alkane entropies

Alkane	$S°(SWS)^a$	$S°(AG)^b$	σ	n	$S°(calc)^c$	Error[d]
2	54.85	54.79	18	1	54.86	−0.01
3	64.51	64.54	18	1	64.26	0.25
4	74.12	74.20	18	1	73.66	0.46
2m3	70.42	70.44	81	1	69.87	0.55
5	83.40	83.55	18	1	83.06	0.34
2m4	82.12	82.16	27	1	81.45	0.67
22mm4	73.23	73.14	972	1	72.53	0.70
6	92.83	92.94	18	1	92.46	0.37
2m5	90.95	91.06	27	1	90.85	0.10
3m5	90.77	91.55	27	1	90.85	−0.08
22mm4	85.62	85.65	243	1	84.69	0.93
23mm4	87.42	87.46	162	1	86.49	0.93
7	102.27	102.32	18	1	101.86	0.41
2m6	100.38	100.51	27	1	100.25	0.13
r-3m6[e]	101.37	101.83	27	2	101.63	−0.26
3e5	98.35	98.39	81	1	98.07	0.28
22mm5	93.90	93.86	243	1	94.09	−0.19
r-23mm5	98.96	99.11	81	2	98.65	0.31
24mm5	94.80	94.89	162	1	95.89	−1.09
33mm5	95.53	95.19	81	1	94.89	0.64
223m³4	91.61	91.65	729	1	91.10	0.51
8	111.55	111.70	18	1	111.26	0.29
2m7	108.81	109.87	27	1	109.65	−0.84
r-3m7	110.32	111.29	27	2	111.03	−0.71
4m7	108.35	109.35	27	1	109.65	−1.30
3e6	109.51	109.46	27	1	109.65	−0.14
22mm6	103.06	103.45	243	1	103.49	−0.43
r-23mm6	106.11	108.05	81	2	108.05	−1.94
r-24mm6	106.51	107.01	81	2	108.05	−1.54
25mm6	104.93	105.75	162	1	105.29	−0.36
33mm6	104.70	104.77	81	1	105.67	−0.97
r-34mm6	107.15[f]	106.62	81	2	108.05	
m-34mm6[e]	107.15	106.54	162	1	105.29	
3e2m5	105.43	106.09	81	1	106.67	−1.24
3e3m5	103.48	101.51	243	1	103.49	−0.01
224m³5	101.15	101.12	729	1	100.50	0.65
234m³5	102.31	102.42	243	1	103.69	−1.38
r-223m³5	101.62	101.34	729	2	101.88	−0.26
233m³5	103.14	102.09	243	1	102.69	0.45
2233m⁴4	93.06	93.09	13122	1	92.96	0.10
9	120.86	121.07	18	1	120.66	0.20
2m8	118.52	119.25	27	1	119.05	−0.53
r-3m8	119.90	120.65	27	2	120.43	−0.53
r-4m8	119.90	120.18	27	2	120.43	−0.53
3e7	118.52	118.90	27	1	119.05	−0.53
4e7	118.52	118.34	27	1	119.05	−0.53
22mm7	113.07	112.79	243	1	112.89	0.18
r-23mm7	116.79	117.68	81	2	117.45	−0.66
r-24mm7	116.79	115.81	81	2	117.45	−0.66
r-25mm7	116.79	118.04	81	2	117.45	−0.66
26mm7	114.03	115.11	162	1	114.69	−0.66
33mm7	115.25	114.40	81	1	115.07	0.18
r-34mm7	117.48	118.26	81	4	118.82	−1.34

(continued)

TABLE 11. (*continued*)

Alkane	$S°(SWS)^a$	$S°(AG)^b$	σ	n	$S°(calc)^c$	Error^d
r-35mm7	116.10	115.31	81	2	117.45	
m-35mm7	116.10^f	115.05	162	1	114.69	
44mm7	113.87	111.60	81	1	113.69	0.18
r-3e2m6	116.79	117.73	81	2	117.45	−0.66
4e2m6	115.41	114.62	81	1	116.07	−0.66
3e3m6	115.25	111.89	81	1	115.07	0.18
r-3e4m6	116.79	116.28	81	2	117.45	−0.66
33rr5	110.31	104.36	972	1	110.13	0.18
r-223m^{3}6	111.34	110.34	729	2	111.28	0.06
r-224m^{3}6	111.34	111.78	729	2	111.28	0.06
225m^{3}6	109.96	110.37	729	1	109.90	0.06
233m^{3}6	112.14	110.27	243	1	112.90	0.05
r-234m^{3}6	114.37	114.06	243	4	115.84	−1.47
r-235m^{3}6	112.30	113.44	243	2	114.46	−2.16
244m^{3}6	112.14	111.97	243	1	112.09	0.05
r-334m^{3}6	113.52	111.79	243	2	113.46	0.06
3e22mm5	109.96	110.21	729	1	119.90	0.06
3e23mm5	112.14	108.84	243	1	112.09	0.05
3e24mm5	112.30	112.62	243	1	113.09	−0.79
2233m^{4}5	106.69	104.38	2187	1	105.92	0.77
2334m^{4}5	107.65	104.45	1458	1	107.73	−0.08
2244m^{4}5	103.13	102.94	13122	1	102.36	0.77
r-2234m^{4}5	108.23	107.87	2187	2	108.30	−0.07
10	130.17	130.45	18	1	130.06	0.11
2m9	127.74	128.62	27	1	128.45	−0.71
3m9	129.12	130.02	27	2	129.83	−0.71
4m9	129.12	129.54	27	2	129.83	−0.71
5m9	127.74	128.26	27	1	128.45	−0.71
3e8	127.74	128.27	27	1	128.45	−0.71
4e8	129.12	129.19	27	2	129.83	−0.71
22mm8	122.29	122.17	243	1	122.29	0.00
23mm8	126.01	127.04	81	2	126.85	−0.84
24mm8	126.01	125.26	81	2	126.85	−0.84
25mm8	126.01	126.94	81	2	126.85	−0.84
26mm8	126.01	127.27	81	2	126.85	−0.84
27mm8	123.25	124.49	162	1	124.09	−0.84
33mm8	124.47	123.74	81	1	124.47	0.00
34mm8	126.70	127.89	81	4	128.22	−1.52
35mm8	126.70	126.86	81	4	128.22	−1.52
r-36mm8	125.32	126.06	81	2	126.85	−1.53
m-36mm8		126.06	162	1		
44mm8	124.47	122.61	81	1	124.47	0.00
r-45mm8	125.32	124.53	81	1	126.85	−1.53
m-45mm8		124.46	162	1	124.09	
4p7	125.56	125.07	81	1	126.27	−0.71
4ip7	124.63	123.87	81	1	125.47	−0.84
3e2m7	126.01	125.80	81	2	126.85	−0.84
4e2m7	126.01	124.80	81	2	126.85	−0.84
5e2m7	124.63	125.66	81	1	125.47	−0.84
3e3m7	124.47	121.52	81	1	124.47	0.00
4e3m7	126.70	126.55	81	4	128.22	−1.52
3e5m7	126.01	125.54	81	2	126.85	−0.84
3e4m7	126.01	125.21	81	2	126.85	−0.84
4e4m7	124.47	120.09	81	1	124.47	0.00
223m^{3}7	120.56	119.76	729	2	120.68	−0.12
224m^{3}7	120.56	120.66	729	2	120.68	−0.12

TABLE 11. (*continued*)

Alkane	$S°(SWS)^a$	$S°(AG)^b$	σ	n	$S°(calc)^c$	Error^d
225m^{3}7	120.56	121.13	729	2	120.68	−0.12
226m^{3}7	119.18	119.39	729	1	119.30	−0.12
233m^{3}7	121.36	119.85	243	1	121.49	−0.13
234m^{3}7	123.59	122.94	243	4	125.24	−1.65
235m^{3}7	123.59	124.32	243	4	125.24	−1.65
236m^{3}7	122.90	124.26	243	2	123.86	−0.96
244m^{3}7	121.36	120.16	243	1	121.49	−0.13
245m^{3}7	123.59	123.59	243	4	125.24	−1.65
246m^{3}7	121.52	119.74	243	1	122.49	−0.97
255m^{3}7	121.36	121.60	243	1	121.49	−0.13
334m^{3}7	122.74	120.79	243	2	122.86	−0.12
335m^{3}7	122.74	122.74	243	2	122.86	−0.12
344m^{3}7	122.74	119.97	243	2	122.86	−0.12
345m^{3}7^g	123.59	121.64	243	2	123.86	−0.27
345m^{3}7-pseud		121.50	243	2	123.86	
3ip2m6	121.52	121.45	243	1	122.49	−0.97
33ee6	122.29	115.29	243	1	122.29	0.00
34ee6	123.25	121.85	162	1	124.09	−0.84
3e22mm6	120.56	120.45	729	2	120.68	−0.12
4e22mm6	119.18	119.11	729	1	119.30	−0.12
3e23mm6	122.74	118.28	243	2	122.86	−0.12
4e23mm6	122.90	121.11	243	2	123.86	−0.96
3e24mm6	123.59	123.50	243	4	125.24	−1.65
4e24mm6	121.36	118.95	243	1	121.49	−0.13
3e25mm6	122.90	121.67	243	2	123.86	−0.96
4e33mm6	121.36	120.61	243	1	121.49	−0.13
3e34mm6	122.74	118.44	243	2	122.86	−0.12
2233m^{4}6	115.91	112.55	2187	1	115.32	0.59
2234m^{4}6	118.14	117.91	2187	4	119.07	−0.93
2235m^{4}6	117.45	115.79	2187	2	117.70	−0.25
2244m^{4}6	115.91	115.10	2187	1	115.32	0.59
2245m^{4}6	117.45	118.14	2187	2	117.70	−0.25
2255m^{4}6	112.35	112.37	13122	1	111.76	0.59
2334m^{4}6	119.63	115.46	729	2	119.88	−0.25
2335m^{4}6	118.25	117.49	729	1	118.50	−0.25
2344m^{4}6	119.63	118.22	729	2	119.88	−0.25
r-2345m^{4}6	119.10	116.25	729	2	120.88	−1.78
m-2345m^{4}6	119.10	115.92	1458	1	118.13	0.97
3344m^{4}6	116.71	112.90	729	1	117.50	−0.79
3ip24mm5	116.23	115.37	2187	1	117.32	−1.09
33ee2m5	119.18	116.12	729	1	119.30	−0.12
3e223m^{3}5	117.29	112.85	2187	1	115.32	1.97
3e224m^{3}5	115.91	117.51	2187	2	117.70	−1.79
3e234m^{3}5	118.25	114.29	729	1	118.50	−0.25
22334m^{5}5	112.80	109.89	6561	1	112.34	0.46
22344m^{5}5	110.62	108.76	19683	1	110.15	0.47

^{a}Entropy tabulated by SWS (Reference 22).
^{b}Entropy tabulated by Alberty and Gehrig (Reference 29).
^{c}Calculated using the following group values: P = 30.30, S = 9.40, T = − 12.30, Q = − 35.00.
^{d}Error is difference between group additivity calculation and values tabulated by SWS (Reference 22).
^{e}The prefix 'r-' designates the racemic mixture of optical isomers; the prefix 'm-' designates the *meso* form.
^{f}SWS do not distinguish between the stereoisomers of 3,4-dimethylhexane or of 3,5-dimethylheptane; presumably the values they quote are for equilibrium mixtures.
^{g}The stereoisomerism of 3,4,5-trimethylheptane is complicated by the presence of a pseudoasymmetric atom; see discussion in Reference 29.

and of Alberty and Gehrig (AG)[29]. In some cases, this discrepancy is surprisingly large ($\geqslant 2$ eu): 3e3m6, 33ee5, 3e23mm5, 2334m^{4}5, 3e23mm6, 4e24mm6, 3e34mm6, 2233m^{4}6, 2334m6, 3344m^{4}6, 33ee2m5, 3e223m^{3}5, 3e234m^{3}5 and 22334m^{5}5. Possibly one of the evaluations is based on an incorrect value of σ or n. Another oddity is the discrepancy in the values of AG for 23mm7, 24mm7 and 25mm7, all of which have the same groups and should have very similar entropies.

Occasionally one may be interested in calculating thermodynamic properties for a large compound that differs only slightly from one for which data are already known. (This is more likely to occur when dealing with substituted alkanes than with unsubstituted ones.) In such a case, it may save some calculations to begin with the known properties of the reference compound and then make appropriate corrections for the groups that have to be added or subtracted to get to the compound of interest. For example, if one needs S°_{298} for 3-ethyl-4,5-dimethyloctane but knows the value for 3,4,5-trimethyloctane, one simply adds the entropy contribution for an S $[= C—(C)_2(H)_2]$ group.

The contribution of *gauche* interactions to the evaluation of $\Delta_f H^\circ$ and S° poses an interesting problem that is often not taken into account. Consider first n-butane. At room temperature, the gas consists of an equilibrium mixture of *trans* and *gauche* isomers. The relative concentrations of these two forms depends on the energy difference between the two (the *gauche* is 960 cal mol^{-1} higher in energy, according to experimental results of Verma and coworkers[24]) and on the entropy difference (S_{int} is the same for both, but for the *trans* $n = 1$, while for the *gauche* $n = 2$). Thus, at equilibrium $[t]/[g] = 0.5 \exp(-960/RT) = 2.53$ at 298 K. Thus, the gas consists of 72% *trans* and 28% *gauche*. The entropy of one mole of mixture is

$$S^\circ_{298} = 0.72 \times S(t) + 0.28 \times S(g) + S_{mix}$$
$$= 0.72 \times [S_{int} + R \ln(1/18)] + 0.28 \times [S_{int} + R \ln(2/18)]$$
$$\quad - R[0.72 \ln(0.72) + 0.28 \ln(0.28)]$$
$$= S^\circ_{int} - R \times [0.72 \ln(0.72 \times 18) + 0.28 \ln(0.28 \times 9)]$$
$$= S_{int} - 2.10R$$

while the entropy of pure *trans* would be

$$S^\circ_{298} = S_{int} + R \ln(1/18)$$
$$= S_{int} - 2.89R$$

The entropy of the mixture is thus 1.57 eu larger than that of pure *trans*. Similarly, the enthalpy of one mole of mixture is

$$\Delta_f H^\circ_{298} = 0.72 \Delta_f H^\circ_{298}(t) + 0.284 \Delta_f H^\circ_{298}(g)$$
$$= \Delta_f H^\circ_{298}(t) + 0.28 \times 960$$

So the enthalpy of the mixture is 0.27 kcal mol^{-1} larger than that of pure *trans*.

Symmetry and optical isomer numbers are calculated as described earlier in section II.G. We remind the reader that highly branched alkanes can have quite large symmetry numbers ($> 10^4$), since each CH_3 group contributes a factor of 3 to σ.

We can carry out similar calculations for n-pentane, but there are now four conformers to consider: *trans*, *gauche* (with one *gauche* interaction) and two with two *gauche* orientations—call them *gauche–gauche–cis* and *gauche–gauche–trans*. The value of n/σ for the four are, respectively, 1/8, 2/9, 2/9 and 2/18. If we assume that *gauche* is approximately 900 cal mol^{-1} higher in energy than *trans*, that *gauche–gauche–trans* is 900 cal mol^{-1} higher still and that *gauche–gauche–cis* is substantially higher (say 5 kcal mol^{-1}), than at 298 K an equilibrium mixture consists of 59%, 38%, 3% and 0%

of the four conformers. We find that S_{298}° is approximately 3 eu higher than for pure *trans*, and $\Delta_f H_{298}^\circ$ is approximately 0.49 kcal mol^{-1} higher. For larger alkanes the computations rapidly get very complicated, but for n-hexane the corresponding quantities are approximately 4.5 eu and 0.77 kcal mol^{-1}.

We should then expect that our group additivity estimations, carried out on the assumption of pure *trans* alkane in each case, should be slightly in error. We tabulate the errors as follows. Recall that propane, having no *gauche* interactions, would have no calculated error.

Alkane	Error in $\Delta_f H_{298}^\circ$	Error in S_{298}°
Propane	0.0	0.0
Butane	0.27	1.57
Pentane	0.49	3.05
Hexane	0.77	4.5

We see that, past propane, the errors are approximately 0.25 kcal mol^{-1} and 1.5 eu per added CH_2 group. Since we derived the group additivity constants from experimental values, and since the experimental values are for the equilibrium mixtures, we have already implicitly taken into account the contribution of *gauche* isomers in an approximate way. But why do we not, then miscalculate the properties of propane, which has no *gauche* contributions? The answer must be that the CH_2 group in propane is sufficiently different from the CH_2 groups in larger alkanes that the errors cancel out. The possibility of different CH_2 groups is discussed in more detail in Section VII on component additivity schemes. However, it is interesting to note that this discussion provides a physical basis for the revised *gauche* counting scheme earlier. Since the empirical enthalpies apply to equilibrium mixtures and are therefore affected by the presence of all possible conformers, the scheme that counts all possible 1,4 interactions is a more accurate representation of the empirical enthalpy. Since, for branched alkanes, the counting becomes complex and there is no assurance that the empirical *gauche* contribution will continue to obey the apparent simple additive law suggested by the butane–pentane–hexane calculation, we should not be surprised that group additivity leads to some errors in those cases. What is perhaps most surprising is that it works as well as it does.

VI. CYCLOALKANES

Since cycloalkanes without side chains consist only of CH_2 groups, we should be able to calculate their thermochemical properties using the S group values only. Table 12 shows that is not the case.

Not only are the discrepancies often very large, but they are not obviously systematic. Three interesting arguments account for the sources of error in the simple group additivity calculations: (1) neglect of ring strain, which lowers the enthalpy of formation; (2) the fact that the CH_2 group in an open-chain alkane has built into its values contributions from the normal internal rotations of the molecule. Since the ring compounds have no internal rotations, ring CH_2 groups are qualitatively different from their linear counterparts. In place of internal rotations, rings have vibrational frequencies representing oscillations (or pseudorotations) of the entire ring. This difference affects both entropy and enthalpy. (3) But the most important difference becomes apparent after considering in more detail the nature of the group additivity approach to estimating thermochemical properties. From Table 9 we see that $S(298)$ for C_2H_6 is approximately

TABLE 12. Properties of selected cycloalkanes (at 298 K)

$C_nH_{2n}(\sigma)$	$\Delta_f H(\text{exp})^a$	$\Delta_f H(\text{calc})^b$	$S_{int}(\text{exp})^c$	$S_{int}(\text{calc})^d$
Cyclopropane (6)	12.74	−15.0	60.35	28.20
Cyclobutane (4)	6.79	−20.0	66.18	37.60
Cyclopentane (1)	−18.26	−25.0	70.00	47.00
Cyclohexane(6)	−29.49	−30.0	74.84	56.40
Cycloheptane(1)	−28.23	−35.0	81.82	65.80
Cyclooctane(1)	−29.73	−40.0	87.66	75.20

[a] Experimental values taken from PNK[53], Table 1.1.
[b] Calculated by group additivities using group values given in Table 9.
[c] 'Experimental' values, tabulated by SWS in Reference 22, Chapter 14, are actually calculated from experimental data for liquid cycloalkanes.
[d] Calculated by group additivities using group values given in Table 11.

54.8 eu; hence, $S_{int}(298)$ is $54.8 + R\ln(18) = 60.4$ eu. From equations 27 and 31 we see that 54 eu comes from constants that are the same for any alkane. The remaining 5.6 eu are determined by molecular weight, vibrational frequencies and moments of inertia. By our procedure of evaluating the group contributions, we determine the CH_3, or P, contribution by observing that ethane consists only of two P groups. Thus, the 54 eu constant is divided between two groups. Each successive group in a molecule will make only a small contribution according as it affects M, I and vibrational frequencies. We might then expect the P contribution to entropy to be large, but the S, T and Q contributions to be all comparable and all small. This is not quite the case: the contributions for P, S, T and Q are 30.3, 9,4, −12.3 and −35 eu, respectively. How can groups decrease the entropy? The answer is that whenever there is a T or Q group there are more P groups in the molecule than just the two present in any unbranched alkane. One T group means a 3rd P group. But the P group carries a numerical contribution of 27 eu, which have already been taken into account by the first two P groups in the alkane. Hence, the T group contribution must be small enough to cancel out this added 27 eu. We would thus expect that the T contribution would be 27 eu less than the S contribution and the Q contribution to be 54 eu less. In fact, the two groups are 21.7 and 44.4 eu smaller, respectively. Now, suppose we try to apply our group values for entropy to cycloalkanes, C_nH_{2n}. There are no P groups, which have built in the 54 eu constant contribution. This contribution must be shared among the n identical S groups. Therefore, the cyclo S groups represent something physically quite different from the linear S groups. (The effect of vibrational and internal rotations discussed in the preceding point makes a smaller contribution to the difference between the two types of S groups.) What is more, if we try to distribute the 54 eu constant among the first three S groups (the minimum number in any ring), then we have no different kinds of groups to use to cancel out the unnecessary 18 eu contributed from each successive S group. To a first approximation, we could solve this problem by calculating the entropy from the relation such as

$$S_{int}(\text{c-}C_nH_{2n}) = n \times (CH_2 \text{ contribution}) + \text{constant}$$

where the constant is approximately 18 and the CH_2 contribution is approximately 7 eu. This would still not be very accurate, because of the appearance of very low frequency torsional ring modes at certain values of n. In other words, not only is the 'constant' not constant, but does not even vary in a regular way.

Similarly, the fact that the enthalpy errors do not diminish monotonically suggests that there are certain cyclic conformations that are more stable than those with one more or one less carbon atom. Except for the highly strained cyclopropane and cyclobutane, the molecules fall into two groups: those with ring strain of 10–$13\,\text{kcal mol}^{-1}$ ($n = 8, 9, 10, 11, 14$) and those with significantly less strain of approximately 1–$6\,\text{kcal mol}^{-1}$ ($n = 7, 12, 13, 15$). There seems to be no simple modification to group additivity rules that will accommodate the cycloalkanes; consequently, Benson and coworkers (Reference 50, Table 33; Reference 34, p. 273) adopted the expedient of simply applying an additive correction to the group additivity–calculated thermochemical properties of ring compounds. (A more conceptually rigorous presentation of the argument accounting for the uniqueness of the cyclic groups is given by Benson and Buss[31].) For example, the correction for $\Delta_f H$(cycloheptane) is $6.4\,\text{kcal mol}^{-1}$; this would apply not only to cycloheptane, but to any cycloheptane derivative with one or more side chains. In Table 13 we list their ring corrections.

In Table 14, we list all the cycloalkanes and alkylcycloalkanes for which gas-phase experimental data for $\Delta_f H^\circ_{298}$ are available and compare them with calculated values using the methods already outlined. In Table 15, we compare values for S°_{298} calculated by our group additivity methods with values tabulated by SWS. (It is worth remembering that all tabulations of gas-phase entropies are based largely on semiempirical or spectroscopically derived values, and not directly on experimental results.) The disubstituted cyclopentanes will need *ortho* corrections to the enthalpy if the substituents are configured 1,2-*cis*. Because the H–H separations are slightly less than in the case of *gauche* conformations (by a few hundredths of an Å for tetrahedral geometries), the *ortho* correction should be slightly larger than $0.8\,\text{kcal mol}^{-1}$; we assume a value of $1.0\,\text{kcal mol}^{-1}$ destabilization for each such correction (in both gas and liquid phase). The disubstituted cyclohexanes requires *gauche* corrections for the enthalpy and symmetry corrections for entropy. To take these into account properly, it is necessary to understand some features about the substitution positions in cyclohexane. As noted earlier, cyclohexane exists in different conformations, of which the chair is the most stable. In this geometrical arrangement, there are six *equatorial* H atoms (located approximately in the plane of the ring) and six *axial* H atoms, alternately above and below the plane of the ring. Substitution at an equatorial position is the more stable; thus, the first ring substituent is almost always located equatorially. The second

TABLE 13. Corrections to be applied to ring compound estimates[a]

Cycloalkane (σ)	$\Delta_f H^\circ$	ΔS_{int}	$C^\circ_p(T)$						
			300	400	500	600	800	1000	1500
Cyclopropane (6)	27.7	32.1	−3.05	−2.53	−2.10	−1.90	−1.77	−1.62	−1.5
Cyclobutane (4)	26.8	28.6	−4.61	−3.89	−3.14	−2.64	−1.88	−1.38	−0.67
Cyclopentane (1)	7.1	27.8	−6.50	−5.5	−4.5	−3.80	−2.80	−1.9	−0.37
Cyclohexane (6)	0.7	18.6	−5.8	−4.1	−2.9	−1.3	1.1	2.2	3.3
Cycloheptane (1)	6.8	15.9							
Cyclöoctane (1)	10.3	12.5							
Cyclononane	13.3								
Cyclodecane	13.1								
Cycloundecane	12.1								
Cyclododecane	5.0								
Cyclotridecane	6.1								
Cyclotetradecane	12.8								
Cyclopentadecane	3.0								

[a]Data from Reference 32, Table A.1, except enthalpy and entropy data, have been updated to optimize agreement with experimental data for alkylcycloalkanes.

TABLE 14. Comparison of calculated and experimental $\Delta_f H^\circ_{298}$ for cycloalkanes and alkycyclo-alkanes[a,b]

Cycloalkane	#P	#S	#T	#Q	#G	$\Delta_f H^\circ$(exp)[c]	$\Delta_f H^\circ$(calc)	Error
C_3	0	3	0	0	0	12.74	12.7	0.04
11mmC_3	2	2	0	1	0	−1.96	−2.2	0.24
C_4	0	4	0	0	0	6.79	6.8	−0.01
eC_4	1	4	1	0	0	−6.29	−5.6	−0.69
C_5	0	5	0	0	0	−18.26	−17.9	−0.36
mC_5	1	4	1	0	0	−25.38	−25.3	−0.08
11mmC_5	2	4	0	1	0	−33.03	−32.8	−0.23
cis-12mmC_5	2	3	2	0	0[d]	−30.95	−30.7	−0.25
trans-12mmC_5	2	3	2	0	0	−32.65	−32.7	0.05
cis-13mmC_5	2	3	2	0	0	−32.48	−32.7	0.22
trans-13mmC_5	2	3	2	0	0	−31.93	−32.7	0.77
eC_5	1	5	1	0	0	−30.33	−30.3	−0.03
C_6	0	6	0	0	0	−29.49	−29.2	−0.19
mC_6	1	5	1	0	0	−36.97	−36.7	−0.27
11mmC_6	2	4	2	0	0	−43.24	−44.1	0.86
cis-12mmC_6	2	4	2	0	3	−41.13	−41.7	0.57
trans-12mmC_6	2	4	2	0	1	−43.00	−43.3	0.30
cis-13mmC_6	2	4	2	0	0	−44.12	−44.1	−0.02
trans-13mmC_6	2	4	2	0	2	−42.18	−42.5	0.32
cis-14mmC_6	2	4	2	0	2	−42.21	−42.5	0.29
trans-14mmC_6	2	4	3	0	0	−44.10	−44.1	0.00
c,c-135m$^3C_6$	3	3	3	0	0	−51.48	−51.5	0.02
c,t-135m$^3C_6$	3	3	3	0	2	−49.37	−49.9	0.53
eC_6	1	6	1	0	1	−41.04	−40.9	−0.14
pC_6	1	7	1	0	1	−46.01	−45.9	−0.11
bC_6	1	8	1	0	1	−50.98	−50.9	−0.08
PentylC_6	1	9	1	0	1	−55.88	−55.9	0.02
HexylC_6	1	10	1	0	1	−60.80	−60.9	0.10
HeptylC_6	1	11	1	0	1	−69.12	−65.9	−3.22
DecylC_6	1	14	1	0	1	−81.14	−80.9	−0.24
DodecylC_6	1	16	1	0	1	−90.51	−90.9	0.39

[a]Enthalpies calculated using group values of Table 9.
[b]Ring correction of Table 13 applied.
[c]Experimental values from PNK or SWS.
[d]Ortho correction of 2.0 kcal mol^{-1} applied.

susbstituent can be either axial or equatorial. If the substituents are adjacent (1,2 substitutions) or opposite (1,4 substitutions) one another and both are either equatorial or axial, they are *trans* to one another (i.e. on opposite sides of the ring plane). If one is equatorial and the other axial, they are *cis* to one another. If the substituents are in the 1,3 positions, the opposite is true: they are *cis* if both axial or both equatorial, and *trans* if one of each. Thus, while there is only one form each of *cis*-1,2, *trans*-1,3 and *cis*-1,4 (one CH$_3$ axial, the other equatorial), there are two possible forms for *trans*-1,2, *trans*-1,4 and *cis*-1,3 (both CH$_3$ groups axial, or both equatorial), with the e,e form being considerably more stable because of the fewer *gauche* interactions. In Table 14 the calculated enthalpy is for the conformation with the fewer *gauche* interactions. In the case of the enthalpies, the average error is 0.4 kcal mol^{-1}; for the calculated entropies, the average error is 0.52 eu. Some of the experimental values—e.g. the enthalpy of heptylcyclohexane—are clearly in error.

TABLE 15. Comparison of calculated and tabulated S°_{298} for cycloalkanes and alkylcycloalkanes[a,b]

Cycloalkane	σ	n	S°(SWS)	S°(calc)	Error
C_3	6	1	56.75	56.74	0.01
C_4	4	1	63.43	63.45	-0.02
C_5	10	1	70.00	70.22	0.22
mC_5	3	1	81.24	81.22	0.02
$11mmC_5$	18	1	85.87	85.26	0.61
cis-$12mmC_5$	9	1	87.51	87.63	-0.12
$trans$-$12mmC_5$	18	2	87.67	87.63	0.04
cis-$13mmC_5$	9	1	87.67	87.63	0.04
$trans$-$13mmC_5$	18	2	87.67	87.63	0.04
eC_5	3	1	90.42	90.62	-0.20
C_6	6	1	71.28	71.44	-0.16
mC_6	3	1	82.06	81.42	0.64
$11mmC_6$	9	1	87.24	87.83	-0.59
cis-$12mmC_6$	9	2	89.51	89.21	0.30
$trans$-$12mmC_6$	18	2	88.65	87.83	0.82
cis-$13mmC_6$	9	1	88.54	87.83	0.71
$trans$-$13mmC_6$	9	2	89.92	89.21	0.71
cis-$14mmC_6$	9	1	88.54	87.83	0.71
$trans$-$14mmC_6$	18	1	87.19	86.46	0.73
c,c-$135m^3C_6$	81	1	93.30	92.07	1.23
c,t-$135m^3C_6$	27	1	95.60	94.25	1.35
eC_6	3	1	91.44	90.82	0.62
pC_6	3	1	100.27	100.22	0.05
bC_6	3	1	109.58	109.62	-0.04
$PentylC_6$	3	1	118.89	119.02	-0.13
$HexylC_6$	3	1	128.20	128.42	-0.22
$HeptylC_6$	3	1	137.51	137.82	-0.31
$DecyclC_6$	3	1	165.43	166.02	-0.59
$DodecylC_6$	3	1	184.05	184.82	-0.77
C_7	1	1	81.82	81.70	0.12

[a] Entropies calculated using group values of Table 11.
[b] Ring correction of Table 13 applied.

VII. COMPONENT ADDITIVITY SCHEMES

Suppose we want to move up a level of complexity to extend additivity beyond the second-order form of group additivities (we designate atomic and bond additivities as zeroth- and first-order approximations, respectively). We can consider not only the immediate ligands attached to the central atom, but also the bonds of those ligands as well. These units are called *components*; they are the next hierarchical level beyond atoms, bonds and groups. A molecule has the same number of components as groups, but the designation of the components take into account the next-nearest neighbors of the central carbon atom. For example, in isooctane we have the following eight components:

3 P(4): three primary C atoms each of which is attached to a quaternary C atom;
2 P(3): two primary C atoms each of which is attached to a tertiary C atom;
1 S(34): a secondary C atom attached to a tertiary and quaternary:
1 T(112): a tertiary C atom attached to two primaries and a secondary; and
1 Q(1112): a quaternary C atom attached to three primaries and a secondary.

There are four types of primary components, ten types of secondary components, 20 types of tertiary components and 31 types of quaternary components in alkanes. The simple nomenclature of P, S, T and Q is now expanded to a series of P(i), S(ij), T(ijk) and Q(ijkl), where the indices vary from 1 to 4 and indicate whether the indexed carbon ligand is itself primary, secondary, tertiary or quaternary. The number of basic building units has now grown from 4 to 65 at this next hierarchy. Obviously, with this many more units we can distinguish between many more isomers than group additivity permits.

There has been only one significant effort published so far to extend additivity to the level of component additivity, and it is worth examining in some detail. In their 1986 detailed survey of thermochemical data of organic compounds, Pedley, Naylor and Kirby[53] developed an elaborate scheme of component additivity relationships. In order to calculate enthalpies for the 64 linear alkanes for which experimental data were available, they found that they needed not all 65 components, but only 35. These components, with their contributions to enthalpy, are listed in Table 16. The numbers in parentheses designate the type of carbon ligand: primary (1), secondary (2), tertiary (3) or quaternary (4).

The P(1), S(11), T(111) and Q(1111) components are those in unbranched alkanes. The values for the other secondary and tertiary components represent the amount of destabilization resulting from the strain associated with branching. (An increase in the enthalpy means a decrease in stability.) Qualitatively, the component value increases with greater branching, as one would expect from the cumulative effects of greater strain. Quantitatively, however, there are some surprises—for example, the large jumps in going from S(34) to S(44) or T(124) to T(134).

PNK add a further level of complexity to their analysis by relating their component values to two- and three-center interaction contributions. Thus, each S(ij) component is related to S(11) by a sum of contributions representing each two-center interaction (central atom with each carbon ligand) and a single term for the three-center nonbonding interactions between all pairs of groups attached to the central atom. Each T(ijk) component is related to T(111) by contributions from each two-center interaction and

TABLE 16. Component additivity parameters for linear alkanes (298 K)[a]

Component	$\Delta_f H_{component}$	$\Delta_f H_{group}$[b]	Component	$\Delta_f H_{component}$	$\Delta_f H_{group}$[b]
P(1)–P(4)	−10.01	−10.00	T(123)	−0.36	
S(11)	−5.00	−5.00	T(223)	0.57	
S(12)	−4.97		T(133)	0.72	
S(13)	−4.83		T(114)	−0.38	
S(14)	−4.68		T(124)	0.79	
S(22)	−4.97		T(134)	3.23	
S(23)	−4.80		Q(1111)	−0.12	−0.10
S(24)	−4.02		Q(1112)	0.19	
S(33)	−4.47		Q(1113)	1.53	
S(34)	−1.79		Q(1114)	3.06	
S(44)	0.96		Q(1122)	1.2	
T(111)	−2.03	−2.40	Q(1123)	3.44	
T(112)	−1.86		Q(1124)	4.92	
T(122)	−1.31		Q(1133)	4.42	
T(222)	−0.91		Q(1222)	2.77	
T(113)	−1.29		Q(2222)	3.27	

[a] Extracted from Pedley, Naylor and Kirby (Reference 53), Table 2.4.
[b] Group additivity values derived in the previous section.

the term representing the sum of three-center interactions; and similarly for the relation between each $Q(ijkl)$ component and $Q(1111)$. For example, the enthalpy contribution of $S(24)$ is calculated as follows:

$$hS(24) = hS(11) + b(22) + b(24) + c2(24)$$

$hS(11)$ is the value given in Table 16; $b(22)$ and $b(24)$ represent contributions from a secondary–secondary pair and a secondary–quaternary pair, respectively; $c2(24)$ represents the sum of nonbonding interactions between the two ligands of the central secondary carbon. The enthalpy contribution of $Q(1124)$ is

$$hQ(1124) = hQ(1111) + b(44) + b(24) + 2b(14) + c4(1124)$$

Here $c4(1124)$ represents the sum of three-center interactions for all the nonbonding pairs associated with the central carbon: 1–2–1, 1–2–2, 1–2–4, 1–2–4 and 2–2–4. However, $c4(1124)$ is not expressible as the sum of simpler terms. Pursuing this analysis, PNK find that they need 7 nonzero b terms and 22 c terms to describe all the compounds of interest. Three of these do not occur in the analysis of only alkanes. Thus, we see that they have replaced the 32 values of Table 16 with 29 other parameters. While this level of detail may demonstrate more advantages when extended to other classes of compounds (their analysis extends to alkenes, alkynes and various organic functional groups), it clearly gains us little advantage in the treatment of alkanes alone, and we do not discuss it further.

With their large number of assigned parameters, PNK are able to calculate the enthalpies for the 64 alkanes for which experimental data are available with an average error of $0.38 \, \text{kcal mol}^{-1}$; this compares with the average error of $0.76 \, \text{kcal mol}^{-1}$ using the six-parameter scheme outlined in Section V above and the standard way of counting *gauche* interactions. Since the average experimental error is $0.60 \, \text{kcal mol}^{-1}$, any scheme that uses additional parameters to get below $0.60 \, \text{kcal mol}^{-1}$ is probably unnecessarily elaborate.

We might note that *gauche* interactions are not specifically counted in the PNK scheme. However, since this scheme takes into account next-nearest neighbor interactions, it implicitly counts *gauche* interactions. For example, any $S(23)$ component will have at least three *gauche* interactions: one from the secondary–secondary pair and two from the secondary–tertiary pair. But by the time all components in a molecule have been tallied, each *gauche* interaction will have been counted twice. Therefore, we should recompute the component values with $0.4 \, \text{kcal mol}^{-1}$ subtracted for each *gauche* interaction to see how the adjusted values then compare with the simpler group values. This is done in Table 17.

The discrepancy between the adjusted component values and the group values is considerably reduced; the fact that components with large numbers of *gauche* interactions still have enthalpy contributions larger than that for the corresponding group indicates that, on average, the successive *gauche* interactions contribute in a nonadditive way to increasing the enthalpy.

The success of four different approaches in estimating/predicting $\Delta_f H_{298}$ is illustrated in Table 18. The five columns of enthalpy values include: (1) experimental numbers, taken from PNK (also in Table IV); (2) values calculated by the group additivity methods discussed in Section IV and listed in Table IV; (3) group additivity with the revised method of counting *gauche* interactions, as discussed above; (4) values tabulated by SWS[22] which are based on a combination of experimental data and calculations (see above); (5) values calculated by component additivity from PNK; and (6) values calculated by the rather elaborate statistical mechanical scheme of Scott[55] and modified by Alberty and Gehrig (AG)[29]. The list of 64 alkanes for which experimental values are available has been reduced slightly (to 48) to include only those (C_2 through C_{10}) for which

TABLE 17. Component enthalpy values adjusted for *gauche* interactions

Component	#Gauche interactions	$\Delta_f H_{Component}$ original[a]	$\Delta_f H_{Component}$ adjusted[b]	$\Delta_f H_{Group}$
P(1)–P(4)	0	−10.01	−10.01	−10.00
S(11)	0	−5.00	−5.00	−5.00
S(12)	0	−4.97	−4.97	
S(13)	1	−4.83	−5.23	
S(14)	2	−4.68	−5.48	
S(22)	0	−4.97	−4.97	
S(23)	1	−4.80	−5.20	
S(24)	2	−4.02	−4.82	
S(33)	2	−4.47	−5.27	
S(34)	3	−1.79	−2.99	
S(44)	4	0.96	−0.64	
T(111)	0	−2.03	−2.03	−2.40
T(112)	1	−1.86	−2.26	
T(122)	2	−1.31	−2.11	
T(222)	3	−0.91	−2.11	
T(113)	2(3)	−1.29	−2.09(−2.49)	
T(123)	3(4)	−0.36	−1.56(−1.96)	
T(223)	4(5)	0.57	−1.03(−1.43)	
T(133)	4(6)	0.72	−0.88(−1.68)	
T(114)	4(5)	−0.38	−1.98(−2.38)	
T(124)	5(6)	0.79	−1.21(−1.61)	
T(134)	6(8)	3.23	0.83(0.03)	
Q(1111)	0	−0.12	−0.12	−0.10(0.10)
Q(1112)	2	0.19	−0.61	
Q(1113)	4(5)	1.53	−0.07(−0.47)	
Q(1114)	6(8)	3.06	0.66(−0.14)	
Q(1122)	4	1.2	−0.40	
Q(1123)	6(7)	3.44	1.04(0.64)	
Q(1124)	8(10)	4.92	1.72(0.92)	
Q(1133)	8(10)	4.42	1.22(0.42)	
Q(1222)	6	2.77	0.37	
Q(2222)	8	3.27	0.07	

[a]Component values derived by PNK (Reference 53), Table 2.4.
[b]Adjusted values obtained by subtracting $0.4 \, kcal \, mol^{-1}$ for each *gauche* interaction (listed in column 2) from the component value. Parenthetical values reflect the revised *gauche* counting scheme of Table 10.

all the above references carried out evaluations. The average errors of the four sources that rely on calculations are: group additivity: $0.59 \, kcal \, mol^{-1}$; modified group additivity: $0.36 \, kcal \, mol^{-1}$; AG: $0.22 \, kcal \, mol^{-1}$ PNK: $0.16 \, kcal \, mol^{-1}$ Most of these compare favorably with the average of PNK's estimated experimental uncertainties of $0.31 \, kcal \, mol^{-1}$.

Although the component additivity scheme of PNK gives, on average, a better set of calculated enthalpies than the modified group additivity scheme, it is our view that for alkanes the scheme is not justified for the following reasons. First, the average error of the simple group additivity approach is comparable to the average experimental error; it thus seems illusory to pursue a scheme that predicts the experimental values with an average precision that exceeds the average experimental accuracy. Second, if we consider the following sequences of alkanes, in which each differs from the preceding by one S_{22} component, we see that the variability in the contribution from that *component* is no

TABLE 18. Comparison of experimental enthalpies of formation (298 K) for some alkanes with several calculations

Alkane	$\Delta_f H°(\exp)^a$	Gp. Add.[b]	Gp. Add. Mod[c]	AG[d]	PNK[e]	SWS[f]
2	−20.03	−20.00	−20.00	−20.00	−20.03	−20.24
3	−25.02	−25.00	−25.00	−25.07	−25.02	−24.82
4	−30.02	−30.00	−30.00	−30.04	−29.97	−30.15
2m3	−32.07	−32.40	−32.40	−32.17	−32.07	−32.15
5	−35.11	−35.00	−35.00	−34.99	−34.94	−35.00
2m4	−36.74	−36.60	−36.60	−36.59	−36.73	−36.92
22mm4	−40.18	−39.90	−40.10	−40.25	−40.18	−39.67
6	−39.94	−40.00	−40.00	−39.91	−39.91	−39.96
2m5	−41.78	−41.60	−41.60	−41.54	−41.68	−41.66
3m5	−41.13	−40.80	−40.80	−40.94	−41.01	−41.02
22mm4	−44.48	−43.30	−43.50	−43.95	−44.55	−44.35
23mm4	−42.61	−43.20	−42.40	−42.11	−42.64	−42.49
7	−44.86	−45.00	−45.00	−44.84	−44.88	−44.88
2m6	−46.51	−46.60	−46.60	−46.49	−46.65	−46.59
3m6	−45.72	−45.80	−45.80	−45.91	−45.96	−45.96
22mm5	−49.21	−48.30	−48.50	−49.00	−48.85	−49.27
23mm5	−47.54	−47.40	−46.60	−46.39	−46.53	−47.62
24mm5	−48.21	−48.20	−48.20	−48.16	−48.25	−48.28
33mm5	−48.09	−46.70	−46.90	−47.66	−48.23	−48.17
3e5	−45.32	−45.00	−45.00	−45.29	−45.44	−45.33
223m³4	−48.88	−49.10	−48.50	−48.69	−48.92	−48.95
8	−49.86	−50.00	−50.00	−49.76	−49.86	−49.82
2m7	−51.48	−51.60	−51.70	−51.41	−51.63	−51.50
3m7	−50.79	−50.80	−50.80	−50.84	−50.93	−50.82
4m7	−50.67	−50.80	−50.80	−50.88	−50.91	−50.69
22mm7	−53.68	−53.30	−53.50	−53.92	−53.82	−53.71
23mm6	−51.10	−52.40	−51.60	−51.36	−51.48	−51.13
24mm6	−52.39	−52.40	−52.40	−52.53	−52.53	−52.44
25mm6	−53.18	−53.20	−53.20	−53.04	−53.39	−53.21
33mm6	−52.58	−51.70	−51.90	−52.70	−52.53	−52.61
34mm6	−50.86	−51.60	−50.80	−50.67	−50.43	−50.91
3e6	−50.36	−50.00	−50.00	−50.29	−50.38	−50.40
3e3m5	−51.34	−50.10	−50.30	−51.29	−51.34	−51.83
223m³5	−52.58	−53.30	−52.70	−52.80	−52.58	−52.61
224m³5	−53.54	−53.30	−53.50	−53.78	−53.54	−53.57
233m³5	−51.70	−52.50	−51.90	−52.10	−51.70	−51.73
234m³5	−51.94	−54.00	−52.40	−51.72	−51.94	−51.97
3e2m5	−50.43	−51.60	−50.80	−50.74	−50.43	−50.48
2233m⁴4	−53.92	−55.00	−53.80	−53.94	−53.97	−53.99
9	−54.54	−55.00	−55.00	−54.71	−54.83	−54.74
2233m⁴5	−56.67	−58.40	−57.20	−56.67	−56.79	−56.70
2234m⁴5	−56.62	−58.30	−56.90	−56.19	−56.62	−56.64
2244m⁴5	−57.74	−58.40	−58.80	−57.74	−58.75	−57.83
2334m⁴5	−56.43	−58.30	−56.90	−56.24	−56.43	−56.46
33ee5	−55.52	−53.50	−53.70	−54.97	−55.52	−55.44
10	−59.63	−60.00	−60.00	−59.63	−59.80	−59.67
2m9	−62.12	−61.60	−61.60	−61.28	−61.57	−61.38
5m9	−61.81	−60.80	−61.80	−60.73	−60.85	−60.70

[a]Experimental values tabulated by PNK[53].
[b]Calculated using standard *gauche* counting; group additivity values: P = −10.00, S = −5.00, T = −2.40, Q = −0.10, G = 0.80, F = 1.60.
[c]Calculated using revised *gauche* counting scheme; the same group additivity values except for Q (= 0.10).
[d]Values tabulated by Alberty and Gehrig[29].
[e]Values calculated by the component additivity scheme of PNK[53].
[f]Values tabulated by SWS[22].

smaller than the variability of the more general S *group* itself:

$$4 \longrightarrow 5 \longrightarrow 6 \longrightarrow 7 \longrightarrow 8$$

$\Delta\Delta_f H^\circ_{298}$ 5.09   4.83   4.92   5.00

$$2m5 \longrightarrow 2m6 \longrightarrow 2m7 \qquad 3m6 \longrightarrow 3m7$$

$\Delta\Delta_f H^\circ_{298}$ 4.73   4.97   5.07

Thus, the gain from the much greater effort of using a component scheme is negligible considering the ease and reliability of the group additivity scheme. When experimental thermochemical quantities have been measured with an uncertainty that is consistently smaller than the error in group additivity predictions, it will be appropriate to reconsider replacing group additivity with a more elaborate estimation scheme.

Apart from this issue, it is important to remember that tabulated enthalpies for alkanes other than those listed in Table 18 are all based on approximative calculations, and are subject to some error. Consider for example, the following 7 branched decanes, and the values of $\Delta_f H^\circ_{298}$ listed by SWS and AG:

Alkane	$\Delta_f H^\circ_{298}$(SWS)	$\Delta_f H^\circ_{298}$(AG)	$\Delta_f H^\circ_{298}$(Gp. Add. Rev.)
4ip7	− 60.02	− 60.71	− 60.8
3ip2m6	− 61.11	− 59.25	− 61.6
4e24mm6	− 60.43	− 61.11	− 60.1
3e234m³5	− 60.09	− 57.58	− 60.1
3344m⁴6	− 60.37	− 59.37	− 60.2
22344m⁵5	− 59.04	− 58.77	− 61.0
22334m⁵5	− 59.08	− 58.27	− 60.5

The differences between the three sets of values range up to $2.5\,\mathrm{kcal\,mol^{-1}}$. Bearing in mind the earlier statement that experimental uncertainties are of the order of $0.1\,\mathrm{kcal\,mol^{-1}}$ per atom, we cannot expect to measure the enthalpies for these decanes to better than $\pm 1\,\mathrm{kcal\,mol^{-1}}$ uncertainty. The differences in the three sets of values are thus barely large enough to be significant, and the group additivity values are as reasonable as any.

VIII. ALKYL RADICAL GROUPS

A. Enthalpies

Establishing values for bond strengths in the alkanes is equivalent to determining the heats of formation of the various alkyl radicals; for this reason we consider also radicals in a chapter that should be dealing exclusively with alkanes. In order to apply group additivity rules to alkyl radicals, seven new groups have to be defined and assigned enthalpy values:

$$a = \Delta_f H^\circ [\mathrm{C\cdot\!-\!(C)(H)_2}]$$
$$b = \Delta_f H^\circ [\mathrm{C\cdot\!-\!(C)_2(H)}]$$
$$c = \Delta_f H^\circ [\mathrm{C\cdot\!-\!(C)_3}]$$
$$d = \Delta_f H^\circ [\mathrm{C\!-\!(C\cdot)(H)_3}]$$
$$e = \Delta_f H^\circ [\mathrm{C\!-\!(C)(C\cdot)(H)_2}]$$
$$f = \Delta_f H^\circ [\mathrm{C\!-\!(C)_2(C\cdot)(H)}]$$
$$g = \Delta_f H^\circ [\mathrm{C\!-\!(C)_3(C\cdot)}]$$

TABLE 19. Experimental radical enthalpies of formation at 298 K

Radical	$\Delta_f H^\circ_{298}(\text{kcal mol}^{-1})$	Notes
Methyl	35.1 ± 0.15	a
Ethyl	28.4 ± 0.5	b
n-Propyl	23.4 ± 1.0	c
i-Propyl	20.0 ± 0.5	d
s-Butyl	15.0 ± 1.0	e
t-Butyl	9.4 ± 0.5	f
i-Butyl	$(16.0 \pm 1.0?)$	g
neo-Pentyl	8.9 ± 1.0	h

[a] Dobis and Benson[57] determined a value of 35.06 ± 0.1 kcal mol^{-1} based on the kinetics of the reaction, $Cl + CH_4 \rightleftharpoons HCl + CH_3$. Pacey and Wimalasena[58] obtained 35.1 ± 0.5 from an ethane pyrolysis experiment. A slightly smaller value of 34.8 ± 0.3 was reported by Russell and coworkers[59] based on the same technique. The latter result is in good agreement with the older CH_4 photoionization spectral data upon which was based the value used by Chase and collaborators[60]. We accept the results of Reference 57. The uncertainty in $\Delta_f H^\circ$ for CH_3 is now limited by the uncertainty in the corresponding quantity for CH_4.
[b] Two recent studies[58,61] agree on the value shown, with stated uncertainties of 0.4–0.5 kcal mol^{-1} in each case resulting from the experimental uncertainties. A third experimental study[62] reported 28.3 ± 0.4 kcal mol^{-1}. Additionally, the calculated enthalpy depends on the value assumed for the entropy of the radical. An uncertainty of ± 0.2 gibbs mol^{-1} in $S^\circ(C_2H_5)$ produces an additional uncertainty in $\Delta_f H^\circ$ of ± 0.06 kcal mol^{-1}, with the likely direction of the uncertainty in S° such as to result in underestimating $\Delta_f H^\circ$.
[c] Marshall and Rahman recommended 22.6 kcal mol^{-1} [63]. Castelhano and Griller[64] reported 22.8, but relative to an assumed $\Delta_f H^\circ(298)$ for CH_3 of 34.4 kcal mol^{-1}; with the recommended value for the latter of 35.1, their $\Delta_f H^\circ(298)$ for C_3H_7 becomes 23.5. McMillen and Golden[65] had recommended 21.0 in Table 2 of their review, based on experiments published prior to 1969. We prefer the value of 23.4 ± 1.0 kcal mol^{-1}; this makes the enthalpy contribution of the e group the same as that of the S group.
[d] Castelhano and Griller[64] measured 19.2 kcal mol^{-1} relative to the value for CH_3, which, when corrected for the latter (see preceding note), becomes 19.9, slightly larger than the value of 19.0 ± 0.5 obtained by Baldwin and coworkers[66]. Doering[67] had recommended 20.0, while McMillen and Golden[65] recommended 18.2 kcal mol^{-1}. The values for $\Delta_f H^\circ$ of n-Pr, i-Pr, Et and s-Bu radicals reported by Castelhano and Griller are consistent with the requirement of equation 67, which the recommendations of Reference 65 are not. In a recent reevaluation, Tschuikow-Roux and Chen[68] recommended 21.0 ± 0.5 kcal mol^{-1}, consistent with a small barrier of approximately 1 kcal mol^{-1} for internal rotation. Luo and Benson[69] derived a value of 20.0 ± 0.5 by linear BDE relationships involving several molecules; we accept this value.
[e] Castelhano and Griller[64] reported 13.9 kcal mol^{-1}; correcting for the better value for the $\Delta_f H(CH_3)$ gives 14.6. McMillen and Golden[65] had recommended 13.0 based on experiments published prior to 1969. To be consistent with equation 67, we recommend 15.0 kcal mol^{-1}.
[f] The enthalpy of t-butyl has been the subject of considerable controversy, as reviewed elsewhere[65,70]. Values published in the past 25 years range from 7.5 to 12.5 kcal mol^{-1}, and even recent studies do not seem to be converging on a consistent number. Thus, Müller-Markgraf, Rossi and Golden[71] recently determined a value of 9.2 ± 0.5 from rate coefficients for the reactions of t-Bu with DBr and DI, whereas in contrast, Russell and coworkers[72] had recently reported a value of 11.6 ± 0.4 from measurements of the kinetics of the forward and reverse reactions $t\text{-Bu} + HBr \rightleftharpoons i\text{-}C_4H_{10} + Br$. Our choice of the value of 9.4 is based on evidence presented by Benson[73].
[g] This quantity has not been measured. The value assumed has been chosen to make the group contribution of f the same as that of the T group, which was also the basis for the assignment of O'Neal and Benson[56]. Schultz and collaborators[74] argue for a value of 15.0, based on the assumption that the primary C—H bond strength is the same as in C_3H_6. Fliszar and Minichino[75] find that 14.9 gives the best consistency in their study of charge distributions and bond dissociation energies in alkanes.
[h] The bond dissociation energy of neopentane was determined[76] to be 3.9 ± 1 kcal mol^{-1} smaller than that of methane, which is now accepted to be 105.1, making the neopentane bond 101.2. With $\Delta_f H^\circ(\text{neo-}C_5H_{12}) = -40.2$ we obtain $\Delta_f H^\circ(\text{neo-}C_5H_{11}) = 8.9$ kcal mol^{-1}. McMillen and Golden[65] recommended 8.7.

Finally, there is the contribution from *gauche* interactions involving the $\cdot CH_2$ group if it is off the main chain, which we do not discuss further here.

As O'Neal and Benson have pointed out[56], only six of the group values are linearly independent; one can be assigned arbitrarily. As they did, we choose to set $d = P$. The other six terms can be calculated from experimentally determined enthalpies of formation of the radicals ethyl, n-propyl, *i*-propyl, *t*-butyl, *i*-butyl and neo-pentyl. The enthalpy of *s*-butyl can be substituted for that of either n-propyl or *i*-propyl. The relationships are as follows:

$$a = \Delta_f H°(\text{Et}) - P \qquad (64)$$

$$b = \Delta_f H°(i\text{-Pr}) - 2P \qquad (65)$$

$$c = \Delta_f H°(t\text{-Bu}) - 3P \qquad (66)$$

$$e = \Delta_f H°(\text{n-Pr}) - \Delta_f H°(\text{Et}) \qquad (67)$$

$$[\text{or } e = \Delta_f H°(s\text{-Bu}) - 2P - b)] \qquad (68)$$

$$f = \Delta_f H°(i\text{-Bu}) - \Delta_f H°(\text{Et}) - P \qquad (69)$$

$$g = \Delta_f H°(\text{neo-Pn}) - \Delta_f H°(\text{Et}) - 2P \qquad (70)$$

In order to be consistent, the two relationships involving e (equations 67 and 68) require that

$$\Delta_f H°(s\text{-Bu}) = \Delta_f H°(\text{n-Pr}) + \Delta_f H°(i\text{-Pr}) - \Delta_f H°(\text{Et}) \qquad (71)$$

The values for the radical enthalpies of formation are listed in Table 19, together with references.

With the enthalpy values of Table 19, and using equations 45–48, 50 and 51, we can derive the group values for the free radical groups as follows:

$$a = 38.4 \pm 0.5$$
$$b = 40.2 \pm 1.0$$
$$c = 39.4 \pm 0.5$$
$$d = -10.0$$
$$e = -5.0 \pm 1.5$$
$$f = -2.4 \pm 1.5$$
$$g = 0.5 \pm 1.5$$

The values given previously by Benson in Reference 34 for these seven parameters are 35.82, 37.45, 38.0, -10.08, -4.95, -1.9 and 1.5kcal mol^{-1}, respectively. (Earlier, O'Neal and Benson[56] had selected 37.0 for c.)

B. Entropies and Heat Capacities

O'Neal and Benson[56] have carried out a detailed analysis of the entropies and heat capacities of alkyl free radicals, and we refer the reader to their work. It is worth noting that entropies and heat capacities for free radicals are not measured directly, but either calculated by one of the techniques outlined earlier or inferred from experimental kinetic measurements in conjunction with some calculations. Principal sources of uncertainty in such calculations have been, and continue to be, questions of structural symmetry and barriers to internal rotations in the radicals. The calculation of $S°_{298}$ for the *t*-butyl radical illustrates both of these uncertainties. O'Neal and Benson[56], and later Benson[32], assumed that the radical site in planar, and thus has a symmetry number $\sigma = 162$ (6 for the overall symmetry and an additional factor of 3 for each of the methyl rotors). Benson[32] further assumed a rotational barrier of $3.0\,\text{kcal}\,\text{mol}^{-1}$, down from $4.7\,\text{kcal}\,\text{mol}^{-1}$ in the parent alkane, neopentane. With these assumptions, he calculated a value of 72.1 eu for S°_{298}. Pacansky and Chang[77], on the basis of infrared matrix isolation studies, concluded that the radical has C_{3v} symmetry (thus $\sigma = 81$), and the rotational barrier is only $0.5\,\text{kcal}\,\text{mol}^{-1}$; Pacansky and Yoshimine[78] later revised the latter value to $1.5\,\text{kcal}\,\text{mol}^{-1}$. With these changes, S°_{298} is 74.6 eu. This value is compatible with a value for $\Delta_f H^\circ_{298}$ of $10.1\,\text{kcal}\,\text{mol}^{-1}$—disconcertingly larger than the value recommended in Table 19 above. Entropies of other radicals are subject to similar uncertainties.

It should be stressed that rotational barrier heights are the largest source of uncertainty in calculated entropies and heat capacities of free radicals—and, in some cases, of stable molecules. For example, if there is a high barrier to an internal rotation in a radical, it has no effect on thermodynamic properties near room temperature—where measurements are generally made. That is, a low-temperature measurement will not be sensitive to a high barrier. (Kinetic methods of determining a barrier height, such as NMR and EPR, measure the rate at which a molecule or radical undergoes transition from one conformation to another. If there are two pathways, over barriers of different heights, the path over the lower barrier will be the faster and will be measured. Transition over the higher barrier will not be seen, and hence that barrier height cannot be determined.) But at sufficiently high temperatures, the effects of the barrier become significant; thus the difficulty in extrapolating thermochemical properties from low to high temperatures.

IX. ALKANE BOND STRENGTHS

The *bond dissociation enthalpy* for the bond A—B is defined as $\Delta_r H^\circ$ for the reaction

$$A\text{—}B \rightarrow A + B$$

and is related to the *bond dissociation energy* (BDE), $\Delta_r E$, by the equation

$$\Delta_r E = E^\circ(A) + E^\circ(B) - E^\circ(AB)$$
$$= [\Delta_f H^\circ(A) - RT] + [\Delta_f H^\circ(B) - RT] - [\Delta_f H^\circ(AB) - RT]$$
$$= \Delta_r H^\circ - RT$$

For many years, BDEs for all primary C—H bonds were regarded as equal, and similarly for secondary and tertiary C—H bonds. To the limits of experimental accuracy, this was generally true. According to group additivity (or component additivity) rules, this need not be the case. The group additivity values P, S, T and Q and a through g determine the bond dissociation energies (BDEs) for various C—H and C—C bonds in alkanes. For example, the P(1) bond (C_2H_5—H) would be predicted to be 100.6 ± 0.5; the P(2) bond (RCH_2CH_2—H), 100.5 ± 1; the P(3) bond (R_2CHCH_2—H), 100.1 ± 1; and the P(4) bond (R_3CCH_2—H), $101.2 \pm 1\,\text{kcal}\,\text{mol}^{-1}$. Unfortunately, the usefulness of these predictions is considerably reduced by the fact that the uncertainties exceed the mean differences; it is quite possible (on this information alone) that all four bond types have the same BDEs. In Table 20 we list experimental (or deduced) BDEs for several alkane C—H bonds. The *bond type* for a C—H bond is identified by the component classification of the C atom; for a C—C bond, by the classification of the pair of C atoms.

X. GROUPS FOR LIQUIDS; GROUPS FOR ΔH_{vap}

All of our discussion so far has been devoted to hydrocarbons in the gaseous phase. At room temperature, alkanes from C_4 through C_{16} are normally in the liquid phase; hence,

TABLE 20. Bond dissociation energies in alkanes

Bond	Bond type	$BDE_{experimental}$
H_3C—H	P(0)	105.1 ± 0.2^a
CH_3CH_2—H	P(1)	100.6 ± 0.5^a
$CH_3CH_2CH_2$—H	P(2)	100.5 ± 1^a
$(CH_3)_2CHCH_2$—H	P(3)	100.1 ± 1^a
Neopentyl—H	P(4)	101.2 ± 1^a
$(CH_3)_2CH$—H	S(11)	97.1 ± 1^a
$CH_3CH_2CH(CH_3)$—H	S(12)	97.1 ± 1^a
c-C_3H_5—H	S(22)	106.3 ± 1^c
c-C_4H_7—H	S(22)	96.5 ± 1^b
c-C_5H_9—H	S(22)	94.5 ± 1^b
c-C_6H_{11}—H	S(22)	95.5 ± 1^b
c-C_7H_{13}—H	S(22)	92.5 ± 1^b
Spiropentyl—H	S(24)	$[98.8 \pm 2]^{b,e}$
t-Butyl—H	T(111)	93.6 ± 0.5^a
H_3C—CH_3	P(1)—P(1)	90.2 ± 0.4^a
H_3C—$CH_2CH_2CH_3$	P(2)—S(12)	87.9 ± 1.1^a
H_3CCH_2—CH_2CH_3	S(12)—S(12)	87.0 ± 1^a
C—C in c-C_3H_6	S(22)—S(22)	65 ± 2^d
C—C in c-C_4H_8	S(22)—S(22)	63 ± 2^d

^{a}Derived from data in Tables 9 and 19.
^{b}From J. A. Kerr and A. F. Trotman-Dickenson 'Strengths of Chemical Bonds', in *Handbook of Chemistry and Physics*, 67th edn. (Ed. R. C. Weast), CRC Press, Boca Raton, Florida, 1986, pp. F-233ff, and references cited therein.
^{c}From Reference 65.
^{d}From data in S. W. Benson and H. E. O'Neal, *Kinetic Data on Gas Phase Unimolecular Reactions*, U.S. Dept of Commerce National Bureau of Standards (now National Institute of Standards and Technology) NSRDS-21, Washington, D.C., 1970.
^{e}This value, taken by Kerr and Trotman-Dickenson from Reference 65, comes ultimately from a kinetic study[79] and was derived using an Evans–Polanyi activation energy–BDE relationship. In the same study, a value for the cyclopropane C—H bond of 100.7 was reported. The latter value is now known to be wrong, and we suspect the spiropentane value, which should not be very different from that for cyclopropane, is also; a value near 106 ± 2 seems more reasonable.

it is often important to be able to estimate thermochemical properties for alkanes in this state. Group additivities for liquid-phase enthalpies at 298 K are easy to derive from the experimental data collected by PNK. Since $\Delta H_{vap} = \Delta_f H_{liq} - \Delta_f H_{gas}^\circ$ group values for the enthalpy of vaporization follow directly from the groups for the liquids and gases. In Table 21 we list the experimental enthalpies for liquid alkanes, together with the values calculated with the best set of group additivity parameters (i.e. the parameters chosen to minimize the average error of calculation of the 62 measured alkanes).

Alicyclic compounds, as noted earlier, present the difficulty that there are necessary nonsystematic corrections to be applied to the calculated thermochemical properties in order to bring the group additivity values into agreement with the measurements. In Table 22, we present a comparison of experimental liquid-phase enthalpies for cycloalkanes with values calculated by group additivity, using the group values of the preceding table. As was done in the case of gas-phase enthalpies, the ring correction

TABLE 21. Enthalpies of formation of liquid alkanes (298 K)

Alkane	$\Delta_f H°(\exp)^a$	$\Delta_f H°(\text{calc})^b$	Error
4	−35.04	−35.34	0.30
2m3	−36.69	−36.88	0.19
5	−41.47	−41.45	−0.02
2m4	−42.66	−42.44	−0.22
22mm3	−45.46	−45.49	0.03
6	−47.49	−47.56	0.07
2m5	−48.90	−48.55	−0.35
3m5	−48.37	−48.00	−0.37
22mm4	−51.10	−50.50	−0.60
23mm4	−49.57	−48.99	−0.58
7	−53.59	−53.67	0.08
2m6	−54.85	−54.66	−0.19
3m6	−54.11	−54.11	0.00
22mm5	−56.96	−56.61	−0.35
23mm5	−55.71	−54.55	−1.16
24mm5	−56.07	−55.65	−0.42
33mm5	−55.98	−55.51	−0.47
3e5	−53.73	−53.56	−0.17
223m³4	−56.52	−56.50	−0.02
8	−59.78	−59.78	0.00
2m7	−60.95	−60.77	−0.18
3m7	−60.30	−60.22	−0.08
4m7	−60.13	−60.22	0.09
2233m⁴4	−64.27	−63.46	−0.81
22mm6	−62.60	−62.72	0.12
23mm6	−60.37	−60.66	0.29
24mm6	−61.42	−61.21	−0.21
25mm6	−62.24	−61.76	−0.48
34mm6	−60.18	−60.11	−0.07
3e6	−59.85	−59.67	−0.18
33mm6	−61.54	−61.62	0.08
3e2m5	−59.66	−60.11	0.45
3e3m5	−60.42	−60.52	0.10
223m³5	−61.40	−62.06	0.66
224m³5	−61.95	−61.64	−0.31
233m³5	−60.59	−61.51	0.92
234m³5	−60.95	−61.10	0.15
9	−65.65	−65.89	0.24
223m³6	−67.57	−68.17	0.60
224m³6	−67.59	−67.20	−0.39
225m³6	−70.10	−69.82	−0.28
233m³6	−67.18	−67.62	0.44
235m³6	−67.88	−67.76	−0.12
244m³6	−66.97	−66.65	−0.32
334m³6	−66.32	−67.07	0.75
33ee5	−65.82	−65.53	−0.29
3e22mm5	−65.18	−67.62	2.44
3e24mm5	−64.46	−66.66	2.20
2233m⁴5	−66.52	−68.47	1.95
2244m⁴5	−66.92	−67.63	0.71
2234m⁴5	−66.37	−66.54	0.17

(*continued*)

TABLE 21. (*continued*)

Alkane	$\Delta_f H°(\text{exp})^a$	$\Delta_f H°(\text{calc})^b$	Error
2334m⁴5	−66.42	−67.51	1.09
10	−71.92	−72.00	0.08
2m9	−74.04	−72.99	−1.05
5m9	−73.59	−72.44	−1.15
11	−78.20	−78.11	−0.09
2255m⁴7	−83.94	−82.89	−1.05
3355m⁴7	−77.84	−77.65	−0.19
22445m⁵6	−78.63	−78.64	0.01
12	−83.87	−84.22	0.35
3366m⁴8	−89.10	−87.90	−1.20
4466m⁴9	−88.67	−89.87	1.20

[a]Experimental values from PNK[53].
[b]Calculated using group values P = −11.56, S = −6.11, T = −2.20, Q = 0.75, G = 0.55 and F = 2.07; and revised *gauche* counting.

terms have been chosen so as to make the calculated enthalpies of the unbranched ring compounds equal to the experimental values; the same corrections are then assumed to apply to alkylcycloalkanes. The *ortho* correction for *cis*-1,2 substitutions on cyclopropane and cyclopentane is assumed to be the same as in the gas phase (1 kcal mol⁻¹).

The average calculated error for these cycloalkanes is 0.68 kcal mol⁻¹. The errors for three compounds—methylcyclobutane, 1,1,2-trimethylcyclopropane and 1,1,2,2-tetramethylcyclopropane—are so large that the experimental values are almost surely wrong. If these problematic cycloalkanes are omitted, the average error is reduced to 0.26 kcal mol⁻¹. The ring corrections for c-C₇H₁₄, c-C₈H₁₆ and c-C₉H₁₈ are given for comparison purposes only; they are not used to correct any other ring compound values.

Similarly, we can derive group values for calculating the entropy of the liquid alkanes at 298 K. We use as our database all the alkanes tabulated by SWS[22] for which experimental entropy measurements are available. The group parameters were chosen to minimize the average error between the calculated and tabulated values. The resulting average error is 0.34 eu. The data are listed in Table 23. The one outstanding discrepancy is in the case of octadecane, and we suspect an experimental error; there is no good explanation for the near-perfect agreement for the other large linear alkanes to be followed by such a large discrepancy.

Calculations of entropies of liquid cycloalkanes require ring corrections, just as in the case of gas-phase cycloalkanes, but the corrections are numerically different. As before, we derive the corrections by comparing calculated entropies with experimental values for the unsubstituted cycloalkanes—cyclobutane, cyclopentane, cyclohexane, cycloheptane and cyclooctane. Experimental values are listed in Table 24 and compared with the calculated ones. The average error is 0.83 eu. The fits are not good for several of the substituted ring compounds. In the case of the dimethylcyclohexanes, all of which were examined in the same study[80], the discrepancy can be traced to very large differences in the enthalpies of fusion (or melting) among the compounds; this quantity makes a contribution to the experimental liquid entropy equal to $\Delta H_m/T_m$. For example, $\Delta H_m(11\text{mmC}_6)$ was measured to be 0.48 kcal mol⁻¹ (at 239 K), much smaller than the value of 2.95 kcal mol⁻¹ for ΔH_m (*trans*-14mmC₆) (at 236 K). The former contributes 2.02 eu, the latter 12.48 eu. (In addition, it appears that the evaluation of Reference 79 did not account properly for the entropy contribution of optical isomerism.) Recall that

TABLE 22. Enthalpies of formation of liquid cycloalkanes at 298 K

Alkane[a]	$\Delta_f H°(\text{exp})^b$	$\Delta_f H°(\text{calc})^c$	Error[d]
mC_3	0.41	0.41	0.00
$11mmC_3$	−7.96	−8.20	0.24
cis-$12mmC_3$	−6.29	−6.24	−0.05
$trans$-$12mmC_3$	−7.34	−7.24	−0.09
$112m^3C_3$	−22.99	−14.85	−8.14
$11mm2eC_3$	−21.56	−21.41	−0.14
$1122m^4C_3$	−28.61	−22.91	−5.70
eC_3	−5.93	−5.70	−0.22
cis-$12eeC_3$	−19.10	−18.46	−0.63
$trans$-$12eeC_3$	−19.91	−19.46	−0.45
$11mm2pC_3$	−27.72	−27.07	−0.65
C_4	0.88	0.88	0.00
mC_4	−10.64	−6.77	−3.87
eC_4	−14.10	−12.88	−1.23
C_5	−25.12	−25.12	0.00
eC_5	−39.05	−38.88	−0.17
pC_5	−45.12	−44.99	−0.13
$11mmC_5$	−41.11	−41.38	0.27
cis-$12mmC_5$	−39.51	−39.42	−0.09
$trans$-$12mmC_5$	−40.92	−40.42	−0.50
$trans$-$13mmC_5$	−40.18	−40.42	0.24
C_6	−37.38	−37.38	0.00
mC_6	−45.43	−45.03	−0.40
eC_6	−50.65	−50.59	−0.05
$11mmC_6$	−52.27	−52.54	0.27
cis-$12mmC_6$	−50.62	−51.03	0.41
$trans$-$12mmC_6$	−52.15	−52.13	−0.02
cis-$13mmC_6$	−53.27	−52.68	−0.59
$trans$-$13mmC_6$	−51.55	−51.58	0.03
cis-$14mmC_6$	−51.53	−51.58	0.05
$trans$-$14mmC_6$	−53.15	−52.68	−0.47
$1e1mC_6$	−57.41	−57.55	0.14
cis-$1e2mC_6$	−56.45	−56.59	0.14
$trans$-$1e2mC_6$	−57.41	−57.59	0.28
cis-$1e3mC_6$	−59.06	−58.24	−0.82
cis-$1e4mC_6$	−57.10	−57.14	0.04
$trans$-$1e4mC_6$	−58.89	−58.24	−0.65
C_7	−37.43	−37.43	0.00
C_8	−40.08	−40.08	0.00
C_9	−43.31	−43.31	0.00

[a]The notation C_n designates an n-cycloalkane; the prefixes m, e and p indicate methyl, ethyl or propyl side chains, respectively.
[b]Experimental values from tabulation of PNK[53], Table 1.1.
[c]Calculated by group additivity; P = − 11.5, S = − 6.11, T = −2.2, Q = 0.75, G = 0.55; $ortho$ correction = 1.0; ring corrections; C_4 = 25.32, C_5 = 5.43, C_6 = − 0.72, C_7 = 5.34, C_8 = 8.8, C_9 = 11.68.
[d]The error is the experimental value minus the calculated one. Apparent discrepancies in this column are due to round-off error.

N. Cohen and S. W. Benson

TABLE 23. Experimental and calculated entropies of
liquid alkanes (298 K)

Alkane	$S(exp)^a$	$S(calc)^b$	Error
4	55.20	55.66	0.46
2m3	52.09	51.99	−0.10
5	62.78	63.36	0.58
2m4	62.24	61.87	−0.37
6	70.76	71.06	0.30
2m5	69.45	69.57	0.12
7	78.53	78.76	0.23
2m6	77.28	77.27	−0.01
223m³4	69.85	69.92	0.07
8	86.23	86.46	0.23
224m³5	78.40	77.62	−0.78
234m³5	78.71	79.25	0.54
9	94.09	94.16	0.07
10	101.79	101.86	0.07
2m9	100.40	100.37	−0.03
3m9	102.10	101.75	−0.35
4m9	101.70	101.75	0.05
5m9	101.30	100.37	−0.93
11	109.49	109.56	0.07
12	117.26	117.26	−0.00
13	124.97	124.96	−0.01
14	132.74	132.66	−0.08
15	140.41	140.36	−0.05
16	148.09	148.06	−0.03
18	166.50	163.46	−3.04

[a]Experimental values taken from the tabulation of SWS,
Chapter 14.
[b]Calculated by group additivity using the group values
P = 23.00, S = 7.7, T = −8.28, Q = −23.70.

the group additivity-derived gaseous entropies for these same compounds agreed well
with experiment (see Table 15); this points out one of the pitfalls in a third-law entropy
calculation, which requires knowing accurately the enthalpies of all phase transitions.

As in the case of gas-phase compounds, calculating thermodynamic functions for
liquid alkanes at temperatures above 298 K is greatly facilitated with polynomial
expansions for heat capacities. Shaw[81] has evaluated the group contributions to C_p at
298 K for hydrocarbons as well as other liquids; however, rather than evaluate the group
contributions at a series of discrete temperatures, we proceed directly to a polynomial
expansions for C_p. In Table 25 are listed the polynomial coefficients for each of the four
alkane group contributions to C_p. The polynomials agree well with experimental data
in the range of 100–400 K, but are probably unreliable near the boiling point or above.

A. Vaporization Processes

Calculations for processes that include changes of phase (fusion, vaporization,
sublimation or their reverses) must take into account the contributions to the thermodynamic properties associated with the phase change.

Hirschfelder, Curtiss and Bird[83] have shown that Eyring's equation of state for dense

TABLE 24. Entropies of liquid cycloalkanes (298 K)

Alkane	σ	n	$S(exp)^a$	$S(calc)^b$	Error
C_4	8	1	43.44	43.47	0.03
C_5	10	1	48.82	49.92	1.10
mC_5	3	1	59.26	59.34	0.08
pC_5	3	1	74.29	74.74	0.45
bC_5	3	1	82.18	82.44	0.26
$11mmC_5$	18	1	63.34	63.36	0.02
cis-$12mmC_5$	9	1	64.33	64.17	-0.16
$trans$-$12mmC_5$	18	2	64.86	64.17	-0.69
$trans$-$13mmC_5$	18	2	66.90	64.17	-2.73
dC_5	3	1	128.71	128.64	-0.07
C_6	6	1	48.84	51.04	2.20
mC_6	3	1	59.26	59.43	0.18
$11mmC_6$	9	1	63.87	64.83	0.96
cis-$12mmC_6$	9	2	65.52	65.65	0.13
$trans$-$12mmC_6$	18	2	65.30	64.27	-1.03
cis-$13mmC_6$	9	1	65.16	64.27	-0.89
$trans$-$13mmC_6$	9	2	66.03	65.65	-0.38
cis-$14mmC_6$	9	1	64.80	64.27	-0.53
$trans$-$14mmC_6$	18	1	64.06	62.90	-1.16
eC_6	3	1	67.14	67.14	0.0
pC_6	3	1	74.54	74.84	0.30
bC_6	3	1	82.45	82.54	0.09
$DecylC_6$	3	1	129.10	128.74	-0.36
$DodecylC_6$	3	1	147.1	144.14	-2.96
C_7	1	1	57.97	58.00	0.03
$HexylC_7$	3	1	106.80	101.34	-5.46
C_8	8	1	62.62	62.67	0.05

[a]Values are based on experimental measurements as tabulated by SWS, Chapter 14.
[b]Calculated using group values of Table 23 and ring corrections: $C_4 = 16.8$, $C_5 = 16.0$, $C_6 = 8.4$, $C_7 = 4.1$, $C_8 = 5.2$.

TABLE 25. Polynomial expressions for liquid phase C°_{pT} for alkane groups[a]

$$C^{\circ}_{pT} = a_1 + a_2 T + a_3 T^2 + a_4 T^3$$

Group	a_1	$100 \times a_2$	$10^5 \times a_3$	$10^7 \times a_4$
C—(C)(H)$_3$	8.459	0.211	-5.61	1.723
C—(C)$_2$(H)$_2$	-1.383	7.049	-20.63	2.269
C—(C)$_3$(H)	2.489	-4.617	31.81	-4.565
C—(C)$_4$	9.116	-23.54	128.7	-19.06

[a]M. Luria and S. W. Benson[82]. Note that these a_i coefficients are related to the coefficients of equation 49 by $a_i = \alpha_i \times R \, cal \, mol^{-1} K^{-1}$.

gases leads to the following relationship between vapor pressure and ΔH_{vap}:

$$p_{\text{vap}} = (RT/8\underline{V})[(\Delta H_{\text{vap}}/RT) - 1]^3 \exp[-\Delta H_{\text{vap}}/RT] \tag{72}$$

If the factor $RT/8\underline{V}$ is approximated by a universal constant of 35 (obtained by setting the liquid molar volume to $82\,\text{cc}\,\text{mol}^{-1}$ and T to 298 K), the preceding equation can be solved by setting $T = T_{\text{b}}$, the normal boiling point, and p_{vap} by the vapor pressure at that temperature, namely 1 atm, to give Trouton's rule,

$$[\Delta H_{\text{vap,b.p.}}/RT_{\text{b.p.}}] = 10.4 \tag{73}$$

And since, for a phase change, $\Delta G = 0$, the above gives

$$\Delta S_{\text{vap,b.p.}}/R = 10.4 \tag{74}$$

Chickos and coworkers[84] have developed semiempirical procedures for estimating both $\Delta H_{\text{vap,298}}$ and $\Delta H_{\text{sub,298}}$. The former quantity, however, is easily calculated from our group additivity values for the gas- and liquid-phase groups, since

$$\Delta H_{\text{vap}} = \Delta_{\text{f}} H_{\text{g}}^{\circ} - \Delta_{\text{f}} H_{\text{l}} \quad \text{and} \quad \Delta S_{\text{vap}} = S_{\text{g}}^{\circ} - S_{\text{l}}$$

Thus, the group contributions to ΔH_{vap}, ΔS_{vap} and $\Delta C_{\text{p,vap}}$ are determined simply by the differences in the corresponding quantities for the liquid and gas phase.

Column 2 of Table 26 provides the constants that are used to compute ΔH_{vap} of an alkane at 298 K. To carry out the same calculation at another temperature, for example, the boiling point, one can calculate $\Delta_{\text{f}} H_{\text{g}}$ and $\Delta_{\text{f}} H_{\text{l}}$ by using the appropriate polynomial expansions of equation 50, or use the set of $\Delta a_{i,\text{vap}}$, also tabulated in Table 26, which are the differences in the corresponding $a_{i,\text{g}}$ and $a_{i,\text{l}}$. (The former are given in Table 7 and the latter in Table 25.)

There are, however, easier ways to estimate ΔH_{vap} and ΔS_{vap} at the boiling point. If $T_{\text{b.p.}}$ is known, then Trouton's rule, equations 73 and 74, provides the quickest procedure. (If T_{c} and P_{c}, the critical temperature and pressure, are known in addition to $T_{\text{b.p.}}$, there are several methods for estimating $\Delta H_{\text{vap,b.p.}}$, many of which have been reviewed by Reid and collaborators[45].) For 56 alkanes from C_4 through C_{20}, equation 73 gives $\Delta H_{\text{vap,b.p.}}$ with an average error of $0.26\,\text{kcal}\,\text{mol}^{-1}$. If $T_{\text{b.p.}}$ is not known, then one can use a group additivity approach. For the same 56 alkanes, the following equation gives

TABLE 26. Group contributions to ΔH and ΔS (298 K), and ΔC_{p} of vaporization[a]

Group	ΔH_{vap}^b	ΔS_{vap}^b	$\Delta a_{1,\text{vap}}^c$	$100 \times \Delta a_{2,\text{vap}}$	$10^5 \times \Delta a_{3,\text{vap}}$	$10^7 \times \Delta a_{4,\text{vap}}$
P	1.56	7.30	−8.63	2.23	4.44	−1.701
S	1.11	1.70	1.13	−4.76	19.30	−2.239
T	−0.20	−4.02	−4.21	7.27	−33.85	4.621
Q	−0.65	−11.30	−13.19	27.35	−132.3	19.17
G	0.35	−0.47				
C_4 ring correction	1.5	13.1	−14.02	4.49	23.63	8.374
C_5 ring correction	1.3	13.3	−45.23	39.88	−117.43	−11.142
C_6 ring correction	1.2	12.4	−24.94	16.97	−28.91	−0.316
C_7 ring correction	1.1	11.8				
C_8 ring correction	1.1	11.3				
C_9 ring correction	1.1					

[a]These coefficients are valid for calculations below the boiling point.
[b]Derived from values in Table 27.
[c]Derived from values in Table 28.

$\Delta H_{\text{vap,b.p.}}$ with an average error of $0.34\,\text{kcal mol}^{-1}$:

$$\Delta H_{\text{vap,b.p.}} = 3.0n_{\text{P}} + 0.43n_{\text{S}} - 2.5n_{\text{T}} - 5.4n_{\text{Q}} \tag{75}$$

where n_{P}, n_{S}, n_{T} and n_{Q} are the numbers of primary, secondary, tertiary and quaternary groups. The method is best for larger ($> C_5$) alkanes; for C_7s and higher, the maximum error is $0.58\,\text{kcal mol}^{-1}$.

We can develop a rough relationship between $\Delta H_{\text{vap,T}}$ and $\Delta H_{\text{vap,b.p.}}$ following an argument put forth by Benson and Garland[85]. They start with the general relationship

$$\Delta H_{\text{vap,T}} = \Delta H_{\text{vap,b.p.}} + \int_{T_{\text{b.p.}}}^{T} \Delta C_{\text{p,vap}} dT \tag{76}$$

The integration is approximated by making use of the following assumptions and relationships: (1) for many hydrocarbons (especially with branching), $\Delta C_{\text{p,vap,b.p.}} \simeq 11$; (2) according to the Guldberg–Guye rule[†], $T_{\text{b.p.}}/T_{\text{c}} \simeq 0.625$ for most regular liquids; (3) at the critical point (and, approximately, at T_{c}), $\Delta C_{\text{p,vap},T_{\text{c}}} = 0$; and (4) above the boiling point, $\Delta C_{\text{p,vap}}$ can be assumed to be a linear function of temperature. From these conditions, it follows that

$$\Delta C_{\text{p,vap}} \simeq -28.0 + 17.6(T/T_{\text{b.p.}}) \tag{77}$$

whence equation 76 can be integrated to give

$$\Delta H_{\text{vap,T}} \simeq \Delta H_{\text{vap,b.p.}} + 28.0(T - T_{\text{b.p.}}) + 8.8[(T^2 - T_{\text{b.p.}}^2)/T_{\text{b.p.}}] \tag{78}$$

which gives the heat of vaporization at any temperature near the boiling point from its value at the boiling point. We can then use Trouton's rule, equation 73, to give

$$\Delta H_{\text{vap,T}} \simeq 2\Delta H_{\text{vap,b.p.}} - 28T + 170T^2/\Delta H_{\text{vap,b.p.}} \tag{79}$$

At temperatures well below the boiling point, it is more accurate to use Table 26 to calculate $\Delta C_{\text{p,vap}}$ as a function of temperature and then integrate equation 76 to obtain $\Delta H_{\text{vap,T}}$.

XI. SOLUTIONS

When two liquids are mixed, there is in general a nonzero free energy change associated with the process. This free energy change is due to contributions from both enthalpy, ΔH_{mix}, and entropy, ΔS_{mix}, of mixing. However, for a particular type of solution—a *perfect*, or *ideal* solution—the enthalpy of mixing is zero. An ideal solution is defined as one for which the fugacity, f, of each component is proportional to its molar fraction:

$$f_i = f_i^{\circ} x_i \tag{80}$$

where f_i° is a constant at a given temperature and pressure for each substance x_i. Physically, two components form an ideal, or nearly ideal, solution when their structures are very similar or when they have very weak intermolecular forces. The hydrocarbons satisfy one or both of these requirements, and consequently form nearly ideal solutions; the enthalpies of mixing are typically only a few tens of calories per mole[86]. In general,

$$G_i = G_i^{\circ} + RT \ln(x_i) \tag{81}$$

If $\Delta H_{\text{mix}} = 0$, then

$$H_i - H_i^{\circ} = 0 \tag{82}$$

[†]See J. R. Partington, *An Advanced Treatise on Physical Chemistry*, Vol. 1, Wiley, New York, 1949, p. 646, for more background on this empirical rule and pertinent references.

and

$$T(S_i - S_i^\circ) = G_i - G_i^\circ = RT \ln(x_i) \qquad (83)$$

whence

$$\Delta S_{\text{mix}}/R = \sum x_i(S_i - S_i^\circ) = -\Sigma x_i \ln(x_i) \qquad (84)$$

which is analogous to equation 41.

For practical purposes, one is generally concerned with entropies of mixing of given *volumes* of liquids rather than of *moles*. (In the case of ideal gases, the two are identical, and in the case of equation 41 no distinction was necessary.) Hence, we can modify the last equation to express the entropy of mixing on a volume basis:

$$\Delta S_{\text{mix}}/R = -\Sigma(v_i\rho_i/M_i)\ln(v_i\rho_i/M_i) \qquad (85)$$

where v_i, M_i and ρ_i are the volume, molecular weight and density of the ith species, all in consistent units. (Liquid alkane densities are all in the range of $0.7 \pm 0.1\,\mathrm{g\,cc}^{-1}$ near room temperature; consequently, the dependence of ΔS_{mix} on density may be insignificant.)

XII. SUMMARY OF IMPORTANT GROUP VALUES; THEIR RELIABILITY

A. Summary of Group Values

In Table 27 we summarize all the group additivity values necessary for estimating enthalpies and entropies of any alkane and cycloalkane, either in gas or liquid phase, at 298 K.

As indicated previously, for computational purposes, it is convenient to express the group contributions to C_p as polynomials in T rather than as tabulated values. The values in Tables 28 and 29 are taken directly from Tables 7 and 25 above, except that polynomial expressions for cycloalkane corrections have been added.

Because of the particular utility of K_f, the equilibrium constant for formation from the elements in their standard states, it is useful to develop group additivity contributions that can be used to calculate K_f directly, using equations 52–59. The necessary constants are listed in Tables 30 and 31. These values are calculated from the group additivity constants previously given, but incorporating contributions from H_2(gas) and C(graphite), the elements in their standard states. The constants in the tables permit calculation of what we will define as $K_{f,\text{int}}$, the intrinsic equilibrium constant, which is related to the standard equilibrium constant by the relation

$$K_f = K_{f,\text{int}} \times n/\sigma$$

where n and σ are the number of optical isomers and symmetry number, respectively, of the compound whose equilibrium constant of formation is being calculated. Alternatively, $S = S_{\text{int}} \times R\ln(n/\sigma)$ can be used to give K_f directly. The use of these tables is illustrated by the Examples in Section XIII.

B. Reliability of Group Additivity Schemes

Throughout this chapter we have argued for the use of group additivity methods for estimating thermodynamic and thermochemical properties for hydrocarbons when appropriate measurements are not available. We maintain our conviction in the advisability of this approach, even though there are more elaborate schemes that give better agreement with tabulated data, for the following reasons:

(1) The simplicity of the group additivity method and its straightforward relationship

TABLE 27. Summary of group values (298 K)

Group	Gas			Liquid		
	$\Delta_f H^{\circ a}_g$	$S^{\circ b}_{int,g}$	$C^{\circ c}_{pg}$	$\Delta_f H^{\circ d}_l$	$S^{\circ e}_{int,l}$	$C^{\circ\ f}_{pl}$
P = C—(C)(H)$_3$	−10.00	30.30	6.19	−11.56	23.00	8.80
S = C—(C)$_2$(H)$_2$	−5.00	9.40	5.50	−6.11	7.70	7.26
T = C—(C)$_3$(H)	−2.40	−12.30	4.54	−2.20	−8.28	5.00
Q = C—(C)$_4$	0.10	−35.00	4.37	0.75	−23.70	1.76
G = *gauche* correction	0.80			0.55		
F = 1 − − −5 correction	1.60			2.07		
a = [C·—(C)(H)$_2$]	38.4	30.7	5.76			
b = [C·—(C)$_2$(H)]	39.4	10.7	6.06			
c = [C·—(C)$_3$]	39.4	−10.8	4.06			
d = [C—(C·)(H)$_3$]	−10.00	30.4	6.19			
e = [C—(C)(C·)(H)$_2$]	−5.6	9.4	5.50			
f = [C—(C)$_2$(C·)(H)]	−2.4	−12.1	4.54			
g = [C—(C)$_3$(C·)]	0.5	−35.1	4.37			
C$_3$ ring correction	27.7	32.1	−3.05			−2.34
C$_4$ ring correction	26.8	28.6	−4.61	25.3	16.8	−3.63
C$_5$ ring correction	7.1	27.8	−6.50	5.4	16.0	−3.80
C$_6$ ring correction	0.7	18.6	−5.80	−0.7	8.4	−6.30
C$_7$ ring correction	6.4	15.9		5.3	4.1	−7.62
C$_8$ ring correction	9.9	16.5		8.8	5.2	−6.81
C$_9$ ring correction	12.8			11.7		

[a] Values from Tables 9 and 13.
[b] Values from Tables 11 and 13, and Table 2.14 of Reference 34.
[c] Values from Tables 2.14 and A.1 of Reference 34, except those for a and b groups, have been recomputed using experimental vibrational frequencies reported by Pacansky and Scharder[67] and by Pacansky and Coufal[68] for ethyl and isopropyl, respectively. Note that C_ps, for d, e, f and g are assumed equal to those for P, S, T and Q, respectively.
[d] Values from Tables 21 and 22.
[e] Values from Tables 23 and 24.
[f] Group values from Shaw[81]; ring corrections based on experimental data compiled by Domalski and coworkers[69].

to fundamental physicochemical quantities and properties makes it easy to apply manually. This offers the possibility of making simple checks of suspect experimental results to determine the likelihood of their being in error. [For example, the accepted experimental value for $\Delta_f H^\circ(C_{32}H_{66})$, listed in Table 9, is in very poor agreement with the value predicted by group additivity, and is, we believe, almost certainly wrong.] Admittedly, the group additivity approach is a simplified description of reality: in fact, nonnearest-neighbor interactions are not completely negligible. The discussion in the text regarding the best scheme for counting the number of *gauche* interactions implicitly concedes this fact. But the accuracy of the present database is insufficient to refine the group additivity method with any degree of confidence.

(2) The growing use of high-speed computers and specialized chemical software programs may make the application of more complex schemes, such as component additivity, equally convenient. However, there is the danger that complacent reliance on a complicated data-processing program will obscure the occasional generation of nonsensical results—which can happen when the inputs are nonsensical or when something happens that the software writers did not anticipate.

TABLE 28. Polynomial expressions for gas-phase C°_{pT} for alkane and alkyl^a groups^b

Group	$C^\circ_{pg,T} = a_1 + a_2 T + a_3 T^2 + a_4 T^3$			
	a_1	$10^2 \times a_2$	$10^5 \times a_3$	$10^9 \times a_4$
C—(C)(H)$_3$	−0.10	2.442	−1.18	2.26
C—(C)$_2$(H)$_2$	−2.55	2.288	−1.32	2.97
C—(C)$_3$(H)	−1.72	2.665	−2.05	5.67
C—(C)$_4$	−3.90	3.719	−3.47	10.16
[C·—(C)(H)$_2$]	1.134	1.855	−1.111	2.695
[C·—(C)$_2$(H)]	2.912	1.331	−0.994	2.754
[C·—(C)$_3$]	0.465	1.608	−1.453	4.341
Cycloalkane corrections				
Cyclopropane ($T < 1000$)	−6.54	1.70	−2.05	8.42
Cyclobutane	−7.96	1.38	−0.98	2.53
Cyclopentane	−10.97	1.85	−1.33	3.77
Cyclohexane	−11.92	2.29	−0.89	0.246

^{a}Values for [C—(C·)(H)$_3$], [C—(C)(C·)(H)$_2$], [C—(C)$_2$(C·)(H)] and [C—(C)$_3$(C·)] groups are assumed the same as for C—(C)(H)$_3$, C—(C)$_2$(H)$_2$, C—(C)$_3$(H) and C—(C)$_4$, respectively.
^{b}Note that the a_i coefficients in Tables 28 and 29 are related to the coefficients of equation 49 by $a_i = \alpha_i \times R$ cal mol^{-1} K^{-1}.

TABLE 29. Polynomial expressions for liquid-phase C°_{pT} for alkane groups

Group	$C^\circ_{pl,T} = a_1 + a_2 T + a_3 T^2 + a_4 T^3$			
	a_1	$100 \times a_2$	$10^5 \times a_3$	$10^7 \times a_4$
C—(C)(H)$_3$	8.459	0.211	−5.605	1.723
C—(C)$_2$(H)$_2$	−1.383	7.049	−20.63	2.269
C—(C)$_3$(H)	2.489	−4.617	31.81	−4.565
C—(C)$_4$	9.116	−23.54	128.7	−19.06
Cycloalkane corrections				
Cyclopropane	28.469	−26.96	65.34	1.636
Cyclobutane	6.060	−3.114	−24.61	−8.349
Cyclopentane	34.261	−38.03	116.1	11.18
Cyclohexane	13.021	−14.68	28.02	0.318

TABLE 30. Gas-phase alkane group additivity contributions required to calculate $\ln K_{f,int}(T)$

Group	$\Delta_r S_{298}/R$	$\Delta_r H_{298}/R$	Δa_1	$10^3 \Delta a_2$	$10^6 \Delta a_3$	$10^{10} \Delta a_4$
P	−8.99	−5.03	−4.61	5.13	−0.846	−3.30
S	−11.66	−2.52	−2.93	4.42	−1.63	1.12
T	−14.73	−1.21	−1.94	6.29	−5.24	15.0
Q	−18.30	0.05	−1.40	12.2	−13.1	40.9
G		0.40				
F		0.80				

TABLE 31. Liquid-phase alkane group additivity contributions required to calculate $\ln K_{f,int}(T)$

Group	$\Delta_r S_{298}/R$	$\Delta_r H_{298}/R$	Δa_1	$10^3 \Delta a_2$	$10^5 \Delta a_3$	$10^8 \Delta a_4$
P	−12.66	−5.82	−0.27	−6.09	−2.32	−8.8
S	−12.52	−3.07	−3.50	28.38	−9.88	−11.6
T	−12.71	−1.11	0.177	−30.28	16.50	22.8
Q	−12.61	0.38	5.24	−125.4	65.30	95.8
G		0.28				
F		1.04				

(3) The oft-touted greater accuracy of more elaborate semiempirical estimation schemes is generally illusory because of present limitations on the accuracy of the experimental database. We have suggested in the text that some of these limitations are not temporary, but are in fact inherent in what can physically be measured. What is thus claimed as greater accuracy of such computational schemes is more appropriately described (at best) as greater precision, not accuracy.

XIII. EXAMPLES OF APPLICATIONS

In the following examples, all S and C_p values are in $cal\,mol^{-1}\,K^{-1}$, while H and G values are in $kcal\,mol^{-1}$; however, H/R is in Kelvins and S/R is dimensionless. Thus, if H is given as 60.00, H/R will be $60{,}000/1.987 = 30{,}196$.

Example 1. Use group additivities to calculate (a) $\Delta_f H°$ and (b) $S°$ for the stereoisomers of 3,4-dimethylhexane (3,4-dmh) at 298 K. (c) Calculate the equilibrium distribution and (d) the entropy of the equilibrium mixture at 298 K.

(a) From an examination of the structure of 3,4-dmh we see that there are 4 primary, 2 secondary and 2 tertiary groups.

r-3,4-dmh:

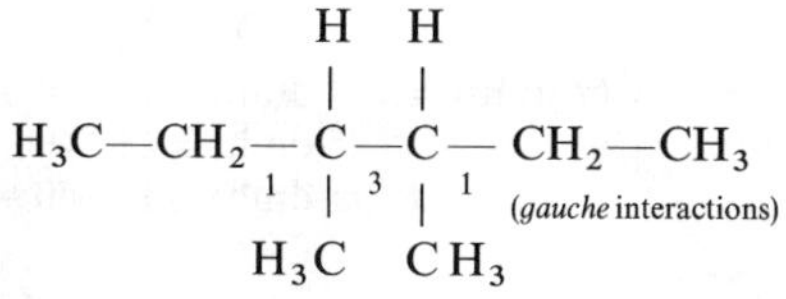

$$H_3C\text{---}CH_2\text{---}\overset{\displaystyle \overset{H}{|}}{\underset{\displaystyle \underset{H_3C}{|}}{C}}\text{---}\overset{\displaystyle \overset{H}{|}}{\underset{\displaystyle \underset{CH_3}{|}}{C}}\text{---}CH_2\text{---}CH_3$$

(gauche interactions)

m-3,4-dmh:

$$H_3C\text{---}CH_2\text{---}\overset{\displaystyle \overset{H}{|}}{\underset{\displaystyle \underset{H_3C}{|}}{C}}\text{---}\overset{\displaystyle \overset{CH_3}{|}}{\underset{\displaystyle \underset{H}{|}}{C}}\text{---}CH_2\text{---}CH_3$$

(gauche interactions)

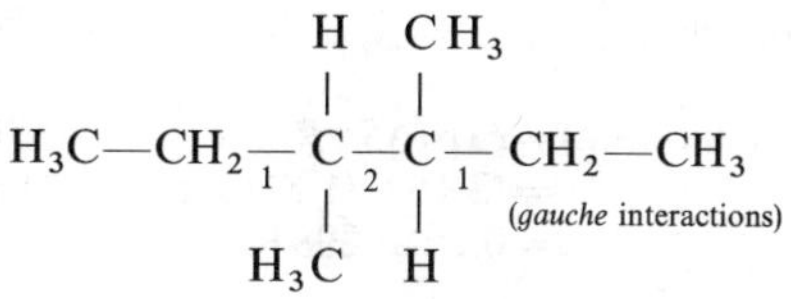

Whenever there is a tertiary group adjacent to a secondary or tertiary group, there will be some *gauche* interactions; from Table 9 we see that there can be either two or three such interactions between the two T groups, depending on the spatial arrangement, and one between each adjacent pair of S and T groups. The *meso* form has the two CH_3

side chains on opposite sides of the main chain, so it has the fewer *gauche* interactions. The values for $\Delta_f H°$(3,4-dmh) are therefore:

r-3,4-dmh:
P $4 \times -10.00 = -40.00$
S $2 \times -5.00 = -10.00$
T $2 \times -2.4 \; = -4.80$
G $5 \times \quad 0.8 \; = \quad 4.00$
$$-50.80$$

m-3,4-dmh:
P $4 \times -10.00 = -40.00$
S $2 \times -5.00 = -10.00$
T $2 \times -2.4 \; = -4.80$
G $4 \times \quad 0.8 \; = \quad 3.20$
$$-51.60$$

(b) To calculate $S°$, in addition to the properties already tallied, we need to know the symmetry number, σ, and the number of optical isomers, n. The general rule for optical isomers is that each chiral center gives rise to $n = 2$ isomers; thus, two chiral centers would give $2 \times 2 = 4$ optical isomers. However, an exception to this simple rule occurs when there are two identical chiral centers, as in 3,4-dmh, in which case there are only three optical isomers, since one of the apparent four is its own mirror image. This latter form is the *meso*-compound. There are thus three forms of 3,4-dmh: the two optically active enantiomers—for convenience, label them *d*-dmh and *l*-dmh (using an obsolescent terminology), which form a (generally inseparable) 50:50 racemic mixture—call it *r*-dmh; and the *meso*-form—call it *m*-dmh.

For either *d*-dmh or *l*-dmh, $\sigma = 3^4 \times 2$: each CH_3 group contributes a factor of 3 to the total symmetry, and there is an overall symmetry of 2. In the racemic mixture the presence of the two inseparable nonsuperposable stereoisomers makes $n = 2$. For *m*-dmh, there is no overall symmetry, so $\sigma = 81$. Because the compound is its own mirror inverse, there is no optical activity and $n = 1$.

For both forms, $S_{int}°$ is the same:

P $4 \times \quad 30.30 = \quad 121.20$
S $2 \times \quad 9.40 = \quad 18.80$
T $2 \times -12.30 = -24.60$
$$S_{int} = 115.40$$

$S°$ is given by: $S° = S_{int}° + R \ln(n/\sigma)$

so that $S°(d\text{-dmh}) = S°(l\text{-dmh}) = 115.40 + R \ln(1/162) = 105.29$
and for the racemic mixture $S°(r\text{-dmh}) = 115.40 + R \ln(2/162) = 106.67$
For the *meso* form, $S°(m\text{-dmh}) = 115.40 + R \ln(1/81) = 106.67$

(c) The equilibrium constants K_f for *r*-dmh and *m*-dmh reflect both the entropy and enthalpy contributions. Since $\ln K_f = S°/R - \Delta_f H°/RT$,

$$K_f(r\text{-dmh})/K_f(m\text{-dmh}) = \exp(\Delta S°/R) \times \exp(-\Delta\Delta H°/RT)$$
$$= (2/162)/(1/81) \times \exp(-800/RT)$$
$$= 0.26 \text{ at } 298 \text{ K}$$

which is also the equilibrium constant for the process

$$m\text{-dmh} \rightleftharpoons r\text{-dmh}$$

Thus, $K_{eq} = 0.26$ and the mixture will consist of 20.6 mol% *r*-dmh (10.3% of each optically active form) and 79.4 mol% *m*-dmh.

(d) The entropy of the mixture is the sum of the entropies of each component (multiplied by its mole fraction) plus the entropy of mixing, given by equation 41:

$$S_{\text{total}} = 0.103S°(d\text{-dmh}) + 0.103S°(l\text{-dmh}) + 0.794S°(m\text{-dmh}) + \Delta S_{\text{mixing}}$$
$$= 2 \times 0.103[S_{\text{int}}(\text{dmh}) + R\ln(1/162)] + 0.794[S_{\text{int}}(\text{dmh}) + R\ln(1/81)]$$
$$\qquad - R\Sigma n_i \ln(n_i)$$
$$= S_{\text{int}}(\text{dmh}) + R[0.206\ln(1/162) + 0.794\ln(1/81) - 2 \times 0.103\ln(0.103)$$
$$\qquad - 0.794\ln(0.794)]$$
$$= S_{\text{int}}(\text{dmh}) - 5.189R$$
$$= 105.09\,\text{eu}$$

We could also have calculated S_{total} by assuming we had a mixture of two components—20.6% racemic and 79.4% meso.

Example 2. According to experiments of Verma, Murphy and Bernstein[24], the enthalpy difference between the *trans* and *gauche* isomers of n-butane is 960 cal mol^{-1} favoring the former. Calculate the distribution of n-butane between the two isomers and the entropy of the mixture (a) at 298 K; (b) at 1000 K.

(a) 298 K. The equilibrium constant for the reaction

$$gauche\text{-}C_4H_{10} \rightleftharpoons trans\text{-}C_4H_{10}$$

is $K_{\text{eq}} = \exp(\Delta S/R) \times \exp(-\Delta\Delta_f H/RT)$. The *trans* form has $\sigma = 18$ and $n = 1$; the *gauche* form has $\sigma = 18$ and $n = 2$. (The *gauche* form is a racemic mixture of two optically active enantiomers.) Since $S = S_{\text{int}} + R\ln(n/\sigma)$ and S_{int} for the two forms is the same, $\exp(\Delta S/R) = (1/18)/(2/18) = 0.5$; $\exp(-\Delta\Delta_f H/RT) = \exp(960/RT)$. Hence, $K_{\text{eq}} = [trans]/[gauche]_{\text{eq}} = 0.5\exp(960/RT)$. At 298 K, $K_{\text{eq}} = 2.53$ and the mixture consists of 72% *trans* and 28% *gauche* (which in turn consists of an equal mixture of the two optical isomers).

The entropy of the mixture is given by

$$S_{\text{total}} = S(trans) + S(gauche) + S_{\text{mixing}}$$
$$= 0.72S°(trans) + 0.28S°(gauche) - R\Sigma(n_i \ln n_i)$$
$$= 0.72[S_{\text{int}} + R\ln(1/18)] + 0.28[S_{\text{int}} + R\ln(2/18)]$$
$$\qquad - R[(0.72\ln(0.72) + 0.26\ln(0.26)]$$
$$= S_{\text{int}} - 4.17$$

(b) 1000 K. The entropy contribution to K_{eq} is independent of temperature, but the enthalpy contribution is different. Following the above procedure but with $T = 1000$ we obtain $K_{\text{eq}} = 0.81$, $S_{\text{mixing}} = 1.34$ and $S_{\text{total}} - S_{\text{int}} = 3.62$.

[All the linear alkanes consist of mixtures of *trans* and *gauche* isomers, with each conformation possible around each internal C—C bond. Thus, in *n*-pentane we find t–t, g–t, g–g(*cis*) and g–g(*trans*). If the alkane consisted only of the *trans* isomer, S would be $S_{\text{int}} + R\ln(1/18) = S_{\text{int}} - 5.74$. The above calculation appears to suggest an error of $5.74 - 4.17 = 1.57$ eu in the entropy calculation resulting from neglecting the various isomers. Similar calculations for *n*-pentane and n-hexane give corresponding values of approximately 3.3 and 4.2 eu, while in the case of propane there would be no discrepancy. To a good approximation, our scheme for evaluating the contributions of S groups to the entropy can take this complication into account implicitly. The errors can amount to several tenths of an eu, however, and this can be one important limitation on the accuracy of any additivity scheme.]

Example 3. Use group additivities to calculate $C_p°$, $S°$, $\Delta_f H°$, $\Delta_f G°$ and $\log K_f$ for gaseous isooctane (224m^{3}5) at 1 atm pressure (a) at 298 K and (b) at 1000 K.

(a) 298 K. *Procedure.* First write out the structure and determine the number of different kinds of groups (P, S, T and Q), nonnearest-neighbor interactions (G and F), symmetry and optical isomers. (These are listed for isooctane in Tables 9 and 11). The contributions from each of these at 298 K to $\Delta_f H°$, $S°$ and $C_p°$ are given in Table 27. To calculate $\Delta_f G°$ and K_f we need the entropy of reaction for the process of formation from the elements, $8C(\text{solid}) + 9H_2(g) \rightarrow i\text{-}C_8H_{18}(g)$. This in turn requires the entropies of C(s) and $H_2(g)$. The enthalpy of reaction is just $\Delta_f H$ for isooctane, since the enthalpy of formation of the elements is by definition zero. The entropy of the elements can be taken from some standard tabulation [we use SWS (Reference 22) throughout these examples for auxiliary data; for these quantities we use their Table Nos. 1 and 2].

Group	P	S	T	Q	G	F	σ	Total
# of gps.	5	1	1	1	3	1	729	
$\Delta_f H°$	-10	-5	-2.4	0.1	0.8	1.6		-53.30
$S°$	30.3	9.4	-12.3	-35	0	0	13.0	100.50
$C_p°$	6.19	5.5	4.54	4.37	0	0		45.36
$S°[\text{C(solid)}]$								1.36
$S°(H_2)$								31.21
$\Delta S° = S° - 8S°(C) - 9S°(H_2) = 100.50 - 8 \times 1.36 - 9 \times 31.21$								-191.28
$\Delta_f G° = \Delta_f H° - T\Delta S° = -53.30 - 298 \times (-191.28/1000)$								3.703
$\log K_f = -\Delta_f G°/2.303RT = -3703/(4.575 \times 298)$								-2.716

For comparison: SWS (Table No. 137) gives $C_p° = 45.14$, $\Delta_f H°(298) = -53.57$, $S° = 101.15$, $\Delta G_f(298) = 3.27$ and $\log K_f = -2.396$. The error in K_f of about a factor of 2 is due primarily to the group additivity error in estimating $\Delta_f H$ of isooctane, as Table 9 shows.

(b) 1000 K. Thermodynamic properties are state properties; that means that the properties of a chemical/physical process are independent of the path taken. The reaction yielding isooctane from the elements at 1000 K can be written with two separate pathways:

$$(8C + 9H_2)_{1000} \xrightarrow{\text{a}} (i\text{-}C_8H_{18})_{1000}$$
$$\downarrow b \qquad \uparrow c$$
$$(8C + 9H_2)_{298} \xrightarrow{\text{d}} (i\text{-}C_8H_{18})_{298}$$

The entropy or enthalpy associated with the required pathway a is related to the others by

$$\Delta S_a = \Delta S_d + \Delta S_c + \Delta S_b; \quad \Delta H_a = \Delta H_d + \Delta H_c + \Delta H_b$$

In part (a) we calculated the functions associated with path d; we need next to calculate the entropy/enthalpy changes associated with cooling the elements from 1000 K to 298 K (path b), and with heating the isooctane from 298 to 1000 K (path c). We calculate C_p, S and $\Delta_f H$ at 1000 K by the use of equations 49–51. This requires the polynomial expansion coefficients for C_p, which are given in Table 26. Equation 49 gives $C_{p,1000}$ directly. Equation 51 is applied at both 1000 and 298 K; the difference gives the entropy change associated with path c which is then added to the value of $S(298)$ calculated in part (a). Equation 50 is similarly applied at both temperatures; the difference gives $(H°_{1000} - H°_{298})$, the enthalpy change associated with path c. This difference is added to the value of $\Delta_f H_{298}$ calculated in part (a). We then subtract the corresponding difference for the elements C and H_2 to get the contribution from path b. Values of the latter are also taken from Table Nos. 1 and 2 in SWS[22].

Groups	P	S	T	Q	Total
# of gps	5	1	1	1	
a_1	-0.1679	0.2554	-1.723	-4.078	-6.39
a_2	0.02443	0.02288	0.02654	0.03812	0.21
a_3	-1.2E-05	-1.3E-05	-2.0E-05	-3.6E-05	-1.3E-04
a_4	2.18E-09	2.97E-09	5.64E-09	1.07E-08	3.0E-08

$C_p^\circ = a_1 + a_2 \times T + a_3 \times T^2 + a_4 \times T^3$		105.44
$S^\circ - a_7 = a_1 \times \ln T + a_2 \times T + a_3 \times T^2/2 + a_4 \times T^3/3$	@ 298	20.69
	@ 1000	111.62
$S_{1000}^\circ = S_{298}^\circ + [S^\circ - a_7]_{1000} - [S^\circ - a_7]_{298}$		191.43
$H^\circ - a_6 = a_1 \times T + a_2 \times T^2/2 + a_3 \times T^3/3 + a_4 \times T^4/4$	@ 298	6.337
	@ 1000	63.323
$H_1 = H_{1000}^\circ - H_{298}^\circ = [H^\circ - a_6]_{1000} - [H^\circ - a_6]_{298}$		56.985
$H_2 = H_{1000}^\circ - H_{298}^\circ$ for H_2 (from SWS)		4.942
$H_3 = H_{1000}^\circ - H_{298}^\circ$ for C (from SWS)		2.823
$\Delta_f H_{1000}^\circ = \Delta_f H_{298}^\circ + H_1 - 8H_3 - 9H_2$		-63.376
S° [C(solid)] (from SWS)		5.846
$S^\circ (H_2)$ (from SWS)		39.704
$\Delta_r S^\circ = S^\circ (i - C_8H_{18}) - 8S^\circ(C) - 9S^\circ(H_2)$		-212.67
$\Delta_f G^\circ = \Delta_f H^\circ - T\Delta S^\circ$		149.296
$\log K_f = -\Delta_f G^\circ / RT$		-32.63

For comparison, SWS (Table No. 137) gives $C_p^\circ = 103.60$, $S^\circ = 189.83$, $\Delta_f H^\circ = -65.04$, $H^\circ - H_{298}^\circ = 55.59$, $\Delta_f G^\circ = 149.23$ and $\log K_f = -32.613$.

Alternative method using Table 30: Table 30 is designed to simplify the calculation of K_f for alkanes from the elements at any temperature by automatically including the contributions from C and H_2. The calculations use equations 52–59.

Groups	P	S	T	Q	G	F	σ	Total
$\Delta S_{298}^\circ / R$	-8.99	-11.66	-14.73	-18.30	0	0	729	-96.23
$\Delta H_{298}^\circ / R$	-5.03	-2.52	-1.21	0.05	0.40	0.80		$-26{,}817$
Δa_1	-4.61	-2.93	-1.94	-1.4				-29.32
Δa_2	0.00513	0.00442	0.00629	0.0122				4.86E-02
Δa_3	-8.5E-07	-1.6E-06	-5.2E-06	-1.3E-05				-2.42E-05
Δa_4	-3.3E-10	1.12E-10	1.50E-09	4.09E-09				4.05E-09

$A = \Delta S_{298}^\circ / R - \Delta a_1 \times \ln(298) - 298 \times \Delta a_2 - 298^2 \times \Delta a_3/2 - 298^3 \times \Delta a_4/3$	86.70
$B = -\Delta H_{298}^\circ / R + 298\Delta a_1 + 298^2 \times \Delta a_2/2 + 298^3 \times \Delta a_3/3 + 298^4 \times \Delta a_4/4$	20,030.23
$C = \Delta a_1$	-29.32
$D = \Delta a_2/2$	2.43E-02
$E = \Delta a_3/6$	-4.03E-06
$F = \Delta a_4/12$	3.38E-10
$\ln K_f = A + B/T + C \ln T + DT + ET^2 + FT^3$	-75.23
$\log K_f = \ln K/2.303$	-32.66

Example 4. Calculate C_p, S, $\Delta_f H$, $\Delta_f G$ and K_f of liquid isooctane (224m^{3}5) at 298 K.

The procedure is the same as before, except that group values for the liquid phase are used; these are given in Table 27 together with the gas-phase values.

Groups	P	S	T	Q	G	F	σ	Total
# of gps.	5	1	1	1	3	1	729	
$\Delta_f H^\circ$	-11.56	-6.11	-2.20	0.75	0.55	2.07		-61.640
S°	23.00	7.70	-8.28	-23.70	0	0	13.0	77.62
C_p°	8.80	7.26	5.00	1.76	0	0		58.02
$S^\circ[C(\text{solid})]$					(from SWS)			1.361
$S^\circ(H_2)$					(from SWS)			31.211
$\Delta_r S^\circ = S^\circ(i-C_8H_{18}) - 8S^\circ(C) - 9S^\circ(H_2)$								-214.16
$\Delta_f G^\circ = \Delta_f H^\circ - T\Delta S^\circ$								2.181
$\log K_f = -\Delta_f G^\circ/2.303RT$								-1.600

Example 5. Calculate $\Delta_f H$ for both liquid and gaseous *i*-octane at the boiling point (372.4 K); calculate ΔH_{vap} and ΔS_{vap} at that temperature.

(a) Gaseous state

Groups	P	S	T	Q	Total
# of gps.	5	1	1	1	
a_1	-0.1679	0.2554	-1.723	-4.078	-6.39
a_2	0.02443	0.02288	0.02654	0.03812	0.21
a_3	$-1.2E-05$	$-1.3E-05$	$-2.0E-05$	$-3.6E-05$	$-1.3E-04$
a_4	2.18E-09	2.97E-09	5.64E-09	1.07E-08	3.0E-08
$C_p = a_1 + a_2 \times T + a_3 \times T^2 + a_4 \times T^3$				@ 372	55.504
$S^\circ - a_7 = a_1 \times \ln T + a_2 \times T + a_3 \times T^2/2 + a_4 \times T^3/3$				@ 298	20.69
				@ 372	31.93
$S_{372}^\circ = S_{298}^\circ + [S^\circ - a_7]_{372} - [S^\circ - a_7]_{298}$					111.74
$H^\circ - a_6 = a_1 \times T + a_2 \times T^2/2 + a_3 \times T^3/3 + a_4 \times T^4/4$				@ 298	6.337
				@ 372	10.103
$H_1 = H_{372}^\circ - H_{298}^\circ = [H^\circ - a_6]_{372} - [H^\circ - a_6]_{298}$					3.765
$H_2 = (H_{372}^\circ - H_{298}^\circ)(H_2)$	(from SWS: interpolated)				0.520
$H_3 = (H_{372}^\circ - H_{298}^\circ)(C)$	(from SWS: interpolated)				0.188
$\Delta_f H_{372}^\circ = \Delta_f H_{298}^\circ + H_1 - 8H_3 - 9H_2$					-55.759

(b) Liquid state

a_1	8.459	-1.383	2.489	9.116	52.52
a_2	0.00211	0.0749	-0.04617	-0.2354	-0.20
a_3	$-5.6E-05$	$-2.1E-04$	3.2E-04	1.3E-03	1.1E-03
a_4	1.7E-07	2.3E-07	$-4.6E-07$	$-1.9E-06$	$-1.3E-06$
$C_p = a_1 + a_2 \times T + a_3 \times T^2 + a_4 \times T^3$				@ 372	68.769
$S^\circ - a_7 = a_1 \times \ln T + a_2 \times T + a_3 \times T^2/2 + a_4 \times T^3/3$				@ 298	279.17
				@ 372	293.47
$S_{372}^\circ = S_{298}^\circ + [S^\circ - a_7]_{372} - [S^\circ - a_7]_{298}$					91.93
$H^\circ - a_6 = a_1 \times T + a_2 \times T^2/2 + a_3 \times T^3/3 + a_4 \times T^4/4$				@ 298	14.294
				@ 372	19.083
$H_{372}^\circ - H_{298}^\circ = [H^\circ - a_6]_{372} - [H^\circ - a_6]_{298}$					4.788
$\Delta_f H_{372}^\circ = \Delta_f H_{298}^\circ + (H_{372}^\circ - H_{298}^\circ) - 8H_3 - 9H_2$					-63.076

(c) $\Delta H_{vap} = \Delta_f H^\circ(g) - \Delta_f H(l)$　　　　　　　　　　　　　　　7.316
　　　$\Delta S_{vap} = S^\circ(g) - S(l) = 111.74 - 91.93$　　　　　　　　19.81

For comparison, SWS gives $\Delta H_{vap}(372) = 7.410$. Note that $\Delta H_{vap}/T_{b.p.} = 19.67$; for a phase change, $\Delta G = 0$, so this quantity should equal ΔS_{vap}. According to Trouton's rule, ΔS_{vap} at the boiling point is approximately $20.7 \, \text{cal mol}^{-1} \, \text{K}^{-1}$.

Example 6. Estimate C_p for liquid hexacosane ($C_{26}H_{54}$) at the melting point, 332 K, and at 358 K. From these, calculate ΔS and ΔH on heating the liquid from 332 to 358 K.

We use the coefficients listed in Table 29 to calculate C_p at the two temperatures. We then use equations 50 and 51 to calculate the enthalpy and entropy. Note that the coefficients in Table 29 give C_p directly, not C_p/R.

Groups	P	S	Total	$a_i T^{i-1}$		Δ
				@ 332K	@358K	
# of gps	2	24	26			
a_1	8.459	1.383	-16.274	-16.27	-16.27	
a_2	0.00211	0.07049	1.69598	563.39	608.11	
a_3	-5.605×10^{-5}	-2.063×10^{-4}	-0.00506	-558.75	-650.98	
a_4	1.723×10^{-7}	2.269×10^{-7}	5.79×10^{-6}	212.25	266.92	
$C_p(T)$				200.62	207.78	15.38
$S(T)$				260.17	275.54	
H/T				132.23	137.52	5330.0
$H(T)$				43901.66	49231.63	

Andon and Martin[90] measured the heat capacity of hexacosane over the temperature range of 13 to 358 K. At 332 K they obtained $203.37 \, \text{cal mol}^{-1}$; at 358, $210.06 \, \text{cal mol}^{-1}$. The error in the group additivity calculation is 1.3% at the lower temperature and 1.1 at the higher. They also tabulated calculated thermodynamic functions over a range of temperatures. Interpolating from their tables, we find the entropy change on heating from 332 and 358 to be $15.1 \, \text{cal mol}^{-1} \, \text{K}^{-1}$ and the enthalpy change to be $5370 \, \text{cal mol}^{-1}$, both in good agreement with the group additivity estimates.

Example 7. Calculate K_f for gaseous C_2H_6 at 1000 K, using the alternative method of Table 30.

Groups	P	S	T	Q	G	H	σ	Total
# of gps	2	0	0	0	0	0	18	
$\Delta S^{\circ}_{298}/R$	-8.99	-11.66	-14.73	-18.30	0	0		-20.87
$\Delta H^{\circ}_{298}/R$	-5.03	-2.52	-1.21	0.05	0.40	0.80		$-10,060$
Δa_1	-4.61	-2.93	-1.94	-1.40				-9.22
Δa_2	0.00513	0.00442	0.00629	0.0122				0.010
Δa_3	$-8.5E-07$	$-1.6E-06$	$-5.2E-06$	$-1.3E-05$				$-1.69E-06$
Δa_4	$-3.3E-10$	$1.12E-10$	$1.50E-09$	$4.09E-09$				$-6.60E-10$
$A = \Delta S^{\circ}_{298}/R - \Delta a_1 \times \ln(298) - 298 \times \Delta a_2 - 298^2 \times \Delta a_3/2 - 298^3 \times \Delta a_4/3$								37.90
$B = -\Delta H^{\circ}_{298}/R + 298\Delta a_1 + 298^2 \times \Delta a_2/2 + 298^3 \times \Delta a_3/3 + 298^4 \times \Delta a_4/4$								7751.78
$C = \Delta a_1$								-9.22
$D = \Delta a_2/2$								5.13E-03
$E = \Delta a_3/6$								$-2.82E-07$
$F = \Delta a_4/12$								$-5.50E-11$
$\ln K = A + B/T + C \ln T + DT + ET^2 + FT^3$								-13.24
$\log K$								-5.75

Example 8. (a) Calculate the gas-phase K_{eq} between ethane and isooctane at 1000 K.

(b) What is the partial pressure of isooctane produced starting from ethane at a pressure of 1 atm? (c) 10 atm?

The equilibrium constant is related to the various K_f values for products and reagents by equation 5. Note that K_f for the elements is unity at all temperatures.

(a) $4C_2H_6 \rightleftharpoons C_8H_{18} + 3H_2$

$K_{eq} = K_f(i\text{-}C_8H_{18}) \times [K_f(H_2)]^3/[K_f(C_2H_6)]^4$

$\log K_{eq} = \log K_f(i\text{-}C_8H_{18}) - 4\log K_f(C_2H_6) = -32.63 - 4 \times (-5.75) \quad -9.63$

(b) The equilibrium constant (with the standard state of 1 atm) for this reaction is related to the species partial pressures at equilibrium by

$$K_p = \frac{(H_2)^3(i\text{-}C_8H_{18})}{(C_2H_6)^4}$$

If P_0 = original C_2H_6 pressure, then at equilibrium,

$$(i\text{-}C_8H_{18}) = x$$
$$(C_2H_6) = P_0 - 4x$$
$$(H_2) = 3x$$

and

$$\frac{(3x)^3(x)}{(P_0 - 4x)^4} = 10^{-9.63}$$

whence

$$x/(P_0 - x) = 10^{-2.77}$$

or

$$x \simeq 10^{-2.77} P_0$$

Thus, at an initial pressure of $P_0 = 1$ atm, $x = 10^{-2.77}$ atm.
(c) At an initial pressure of 10 atm, $x = 10^{-1.77}$ atm.

Example 9. Calculate ΔH_{comb} and K_{eq} for oxidation of gaseous i-octane to give gaseous products (a) at 1000 K; (b) at 500 K; (c) at 500 K, assuming that the water is produced in the liquid state.

The heat of combustion of a compound is the heat of rection for the complete oxidation process. For any alkane, the products of combustion are CO_2 and H_2O, the latter generally assumed to be in the liquid state (However, at high temperatures thermodynamic functions for water are generally tabulated in the gaseous state.) We start by writing the balanced equation for the reaction.

(a) 1000 K.

$C_8H_{18} + 12.5O_2 \rightarrow 8CO_2(g) + 9H_2O(g)$

1000 K	C_8H_{18}	O_2	$CO_2(g)$	$H_2O(g)$	Reaction
$\Delta_f H°$	-63.376	0	-94.320	-59.240	$-1,224$
$\log K_f$	-32.63	0	20.67	10.06	288.53

(b) 500 K.

$C_8H_{18} + 12.5O_2 \rightarrow 8CO_2(g) + 9H_2O(g)$

500 K	C_8H_{18}	O_2	$CO_2(g)$	$H_2O(g)$	Reaction
$\Delta_f H°$	− 59.250	0	− 94.090	− 58.276	− 1,218
$\log K_f$	− 18.99	0	41.25	22.88	554.91

(c) 500 K, giving liquid water as product.

$C_8H_{18} + 12.5O_2 \rightarrow 8CO_2(g) + 9H_2O(l)$

500 K	C_8H_{18}	O_2	$CO_2(g)$	$H_2O(l)$	Reaction
$\Delta_f H°$	− 59.250	0	− 94.090	− 59.240	− 1,227
$\log K_f$	− 18.99	0	41.25	10.06	439.53

Part (a) and (b) show that the difference in energy liberated by completely burning isooctane at 1000 K is only about 6 kcal mol^{-1} larger, or 0.5%, than the corresponding value at 500 K. The difference in heats of combustion found in parts (b) and (c) should equal to ΔH_{vap} of water at 500 K. Most thermochemical tables list properties of gaseous water at high temperatures; the more practical calculation, involving liquid water product, can therefore be carried out using ΔH_{vap} at the appropriate temperature. Note that the tabulated values are generally at 1 atm pressure, which may not correspond to the physical case of interest.

Example 10. Chickos and coworkers[84] have proposed a simple empirical relationship for predicting $\Delta H_{vap,298}$ for alkanes:

$$\Delta H_{vap,298} = 0.31 n_Q + 1.12 \tilde{n}_C + 0.71 \tag{a}$$

where n_Q is the number of quaternary carbons and $\tilde{n}_C$ is the number of nonquaternary carbons. Compare this equation's predictions with group additivity for i-C_4H_{10}, n-C_8H_{18} and i-C_8H_{18}.

From equation a above,

$$\Delta H_{vap,298} = 0.31 n_Q + 1.12 \tilde{n}_C + 0.71$$
$$= 0.31Q + 1.12(P + S + T) + 0.71$$

where we have substituted our usual nomenclature for the number of primary, secondary, tertiary and quaternary carbons. For any branched alkane without rings,

$$P = 2 + T + 2Q \tag{b}$$

whence

$$\Delta H_{vap,298} = 2.55Q + 2.24T + 1.12S + 2.95 \tag{c}$$

From Table 26,

$$\Delta H_{vap,298} = 1.56P + 1.11S - 0.2T - 0.65Q + 0.35G \tag{d}$$

Substituting from equation b we obtain

$$\Delta H_{vap,298} = 2.47Q + 1.36T + 1.11S + 3.12 + 0.35G \tag{e}$$

Comprison of equations c and e shows that the expression of Chickos and coworkers is not very different from the group additivity prediction, especially if there are no tertiary carbon atoms and no *gauche* interactions. The predictions of equation a/c and e are compared with the experimental values (from PNK) below:

Alkane	equation a/c	equation e	ΔH_{vap}(exp)
Isobutane	5.19	4.48	4.61
Octane	9.67	9.78	9.92
Isooctane ($224m^35$)	8.86	9.11	8.40

Example 11. Calculate, using group additivity, ΔH_{vap} for 1,1-dimethylcyclopentane (a) at 298 K; (b) at the boiling point, 361 K.

(a) All the numerical constants needed are in Table 26. The constitutent groups of 11mmC$_5$ are: 2Ps, 4 Ss, 1 Q, and 1 C$_5$ ring correction. From the second column of the Table we find the respective contributions to ΔH_{vap} at 298 K:

$$P: \quad 2 \times 1.56 \quad = \quad 3.12$$
$$S: \quad 4 \times 1.11 \quad = \quad 4.44$$
$$Q: \quad 1 \times -0.65 = \quad -0.65$$
$$C_5 \text{ ring corr: } 1 \times 1.3 \quad = \quad 1.3$$

$$8.21 \text{ kcal mol}^{-1}$$

(b) We calculate the change in ΔH_{vap} in going from 298 to 361 K by using equation 50 twice (first with $T = 298$, then with $T = 361$), except that instead of a_i coefficients we use the Δa_i coefficients of Table 26.

Group contributions to $\Delta C_{p,vap}$ needed to calculate the variation of ΔH_{vap} with temperature are as follows:

Groups	Δa_1	$100 \times \Delta a_2$	$10^5 \times \Delta a_3$	$10^7 \times \Delta a_4$
each P	−8.63	2.23	4.44	−1.701
each S	1.13	−4.76	19.30	−2.239
each Q	−13.19	27.35	−132.3	19.167
C$_5$ ring corr	−45.23	39.88	−117.4	11.218
Total	−71.16	52.65	−163.62	18.027

$$\Delta\Delta H_{vap}(361 - 298): \Delta a_1 \times (361 - 298) + \Delta a_2 \times (361^2 - 298^2)/2$$
$$+ \Delta a_3 \times (361^3 - 298^3)/3 + \Delta a_4 \times (361^4 - 298^4)/4$$
$$= -71.16 \times 63 + 0.5265 \times 20758 - 0.0016362 \times 6860763$$
$$+ 18.027 \times 10^{-7} \times 2.274 \times 10^9$$
$$= -680.2$$

Therefore, ΔH_{vap} at 361 K $= 8210 - 680 = 7530$ cal mol^{-1}.
From SWS, Table No. 349, we see that the experimental value is 7239 cal mol^{-1}. Trouton's rule, equation 73, predicts a value of 7460 cal mol^{-1}.

XIV. APPENDIX. BIBLIOGRAPHY OF USEFUL COMPILATIONS AND REFERENCES

The following selective list begins with three particularly useful book-length studies and then lists (in chronological order) significant monographs and journal references that deal extensively with the alkanes and cycloalkanes that are the subject of this Volume.

D. R. Stull, E. F. Westrum, Jr. and G. C. Sinke, *The Chemical Thermodynamics of Organic Compounds* (SWS), Krieger Publ. Co., Malabar, Florida, 1987. The first 200 pages include a review of basic thermodynamic concepts and thermochemistry, the evaluation of entropy, the estimation of thermodynamic quantities and applications to industrial problems. The bulk of the volume consists of ideal gas tables [including C_p°, S°, $-(G^\circ - H_{298}^\circ)/T$, H_{298}°, $\Delta_f H^\circ$, $\Delta_f G^\circ$, and $\log K_f$ for $298 < T$ (K) < 1000] for hydrocarbons, C—H—O compounds, C—H—N—O compounds, organic halogen compounds and organic sulfur compounds. A separate table lists experimental data at 298 K.

J. B. Pedley, R. D. Naylor and S. P. Kirby, *Thermochemical Data of Organic Compounds* (PNK), Chapman and Hall, London & New York, 1986. The tabulated heats of reaction and heats of formation update the tables of Cox and Pilcher. Introductory chapters discuss additivity schemes and develop a complex component additivity formalism for predicting enthalpies of organic compounds.

J. D. Cox and G. Pilcher, *Thermochemistry of Organic and Organometallic Compounds*, Academic Press, London & New York, 1970. The major feature is a tabulation of experimental measurements of $\Delta_r H^\circ$, heats of reaction, for organic and organometallic compounds. Values of $\Delta_f H^\circ$ are derived for each experimental result; where phase changes are involved measurements of ΔH_{vap}° (heat of vaporization) are tabulated. Experimental uncertainties are assessed. Introductory chapters discuss the basics of thermochemistry, the types of experimental measurements involved, their accuracies and limitations. Additional chapters examine theoretical aspects of thermochemistry and various schemes for relating thermochemical properties to structural parameters.

A. Labbauf, J. B. Greenshields and F. D. Rossini, 'Heats of formation, combustion, and vaporization of the 35 nonanes and 75 decanes', *J. Chem. Eng. Data*, **6**, 261 (1961). Tabulates calculated values of ΔH_{vap} and of ΔH_f and ΔH_{comb} for both vapor and liquid phase, all at 298 K.

R. C. Wilhoit, B. J. Zwolinski G. K. Estok, G. B. Mahoney, and S. M. Dregne, *Handbook of Vapor Pressures and Heats of Vaporization of Hydrocarbons and Related Compounds*, Thermodynamics Research Center, Dept. of Chemistry, College Station, Texas, Texas A&M University, 1971. Includes table of $\Delta H_{vap}(298)$ and ΔH_{vap}(b.p.) for many hydrocarbons, including alkanes and cycloalkanes.

J. Chao, R. C. Wilhoit and B. J. Zwolinski, 'Ideal gas thermodynamic properties of ethane and propane', *J. Phys. Chem. Ref. Data*, **2**, 427 (1973). Review and evaluation of structural parameters (including vibrational frequencies and internal rotation properties); tabulation of thermodynamic properties $[C_p^\circ, S^\circ, (H^\circ - H_0^\circ), (H^\circ - H_0^\circ)/T, -(G^\circ - H_0^\circ)/T, \Delta_f G^\circ, \Delta_f H^\circ, \log K_f]$ for $0 < T$ (K) < 1500 calculated by statistical thermodynamic methods [rigid-rotor harmonic oscillator (RRHO) approximation].

D. W. Scott, 'Correlation of the chemical thermodynamic properties of alkane hydrocarbons', *J. Chem. Phys.*, **60**, 3144 (1974). Includes 'working formulas, tables, and equations ... whereby thermodynamic properties for selected temperatures between 200 and 1500 K can be calculated for any alkane hydrocarbon with a tractably small number of molecular conformations (say $\leqslant 3^4 = 81$). ... Values so obtained for the entropy, heat capacity, and enthalpy of formation at 298.15 K are tabulated for all alkanes C_3–C_{10}. Calculated values are compared with measured ones'. The procedure is updated from the pioneering work of Pitzer[91] as modified by Person and Pimentel[54], incorporating newer experimental data and elaborate computational schemes for taking account explicitly of conformational isomers (e.g. *trans–gauche*). This paper represents perhaps the most detailed scheme for predicting alkane thermochemical properties, but the accompanying drawback is that it is very difficult for the reader to apply the procedures to other cases of interest.

S. S. Chen, R. C. Wilhoit and B. J. Zwolinski, 'Ideal gas thermodynamic properties and isomerization of n-butane and isobutane', *J. Phys. Chem. Ref. Data*, **4**, 859 (1975). Review and evaluation of structural parameters (including vibrational frequencies and internal rotation properties); tabulation of thermodynamic properties $[C_p^\circ, S^\circ, (H^\circ - H_0^\circ)/T, (G^\circ - H_0^\circ)/T]$ for $0 < T$ (K) < 1500 calculated by statistical thermodynamic methods (RRHO approximation).

K. M. Pamidimukkala, D. Rogers and G. B. Skinner, 'Ideal gas thermodynamic properties of CH_3, CD_3, CD_4, C_2D_2, C_2D_4, C_2D_6, C_2H_6, $CH_3N_2CH_3$ and $CD_3N_2CD_3$', *J. Phys. Chem. Ref. Data*, **11**, 83 (1982). Includes ideal gas thermodynamic properties $[C_p^\circ, S^\circ, (H^\circ - H_{298}^\circ)/T, -(G^\circ - H_{298}^\circ)/T, \Delta_f G^\circ, \Delta_f H^\circ,$ $\log K_f]$ for $0 < T$ (K) < 3000 calculated by statistical thermodynamic methods (RRHO approximation) for CH_3, C_2H_6 and perdeuteriated analogs.

D. F. McMillen and D. M. Golden, 'Hydrocarbon bond dissociation energies', *Ann. Rev. Phys. Chem.*, **33**, 493 (1982). A review and recommendation of best values of bond dissociation energies for various classes of hydrocarbon compounds.

W. M. Haynes and R. D. Goodwin, *Thermophysical Properties of Normal Butane from 135 to 700 K at Pressures to 70 MPa*, U.S. Dept. of Commerce, National Bureau of Standards Monograph 169, 1982, 192 pp. Tabulated data include densities, compressibility factors, internal energies, enthalpies, entropies, heat capacities, fugacities and more. Equations are given for calculating vapor pressures, liquid and vapor densities, ideal gas properties, second virial coefficients, heats of vaporization, liquid specific heats, enthalpies and entropies.

R. D. Goodwin and W. M. Haynes, *Thermophysical Properties of Propane from 135 to 700 K at Pressures to 70 MPa*, U.S. Dept. of Commerce, National Bureau of Standards Monograph 170, 1982, 244 pp. Tabulated data include densities, compressibility factors, internal energies, enthalpies, entropies, heat capacities, fugacities and more. Equations are given for calculating vapor pressures, liquid and vapor densities, ideal gas properties, second virial coefficients, heats of vaporization, liquid specific heats, enthalpies and entropies.

E. S. Domalski, W. H. Evans and E. D. Hearing, 'Heat Capacities and Entropies of Organic Compounds in the Condensed Phase', *J. Phys. Chem. Ref. Data*, **13** (1984), Supplement No. 1, 286 pp. Data for over 1400 organic compounds are tabulated, with single-temperature (as near to 298 K as possible) values for C_p and S, and ΔH and ΔS for any phase changes. Each original reference constitutes a separate entry, and while the compilers provide letter grade evaluations (A through D) of the estimated reliability of the work, there is no consolidation or intercomparison of data when there are more than one source.

R. A. Alberty and C. A. Gehrig, 'Standard chemical thermodynamic properties of alkane isomer groups', *J. Phys. Chem. Ref. Data*, **13**, 1173 (1984). Tabulations of $C_p^\circ, S^\circ, \Delta_f H^\circ$ and $\Delta_f G^\circ$ for $0 < T$ (K) < 1500 for alkanes with 10 or fewer carbons. Also tabulated are group contributions and equilibrium fractions within alkane isomer groups. The values are updated from the study by D. W. Scott (1974).

R. A. Alberty and Y. S. Ha, 'Standard chemical thermodynamic properties of alkylcyclopentane isomer groups, alkylcyclohexane isomer groups, and combined isomer groups', *J. Phys. Chem. Ref. Data*, **14**, 1107 (1985). Ideal gas thermodynamic properties $(C_p^\circ, S^\circ, \Delta_f H^\circ, \Delta_f G^\circ)$ for $298 < T$ (K) < 1000 are calculated and tabulated for alkylcyclopentanes through C_9H_{18} and alkylcyclohexanes through $C_{10}H_{20}$. Group incremental contributions for thermochemical properties are tabulated and group additivities used to calculate properties of individual species for which literature data are not available. Equilibrium mole fractions are calculated within alkylcycloalkane isomer groups for $298 < T$ (K) < 1000.

O. V. Dorofeeva, L. V. Gurvich and V. S. Jorish, 'Thermodynamic properties of twenty-one monocyclic hydrocarbons', *J. Phys. Chem. Ref. Data*, **15**, 437 (1986). Structural parameters for cycloalkanes (C_3 to C_8) and other compounds are reviewed and evaluated; structural constants were estimated where necessary; and ideal gas thermodynamic properties [C_p°, S°, $(H^\circ - H_0^\circ)$, $-G^\circ - H_0^\circ/T$, log K_f] for $100 < T(K) < 1500$ calculated by statistical thermodynamic methods (RRHO approximation).

E. S. Domalski and E. D. Hearing, 'Estimation of the thermodynamic properties of hydrocarbons at 298.15', *J. Phys. Chem. Ref. Data*, **17**, 1637 (1988). Application of group additivity techniques to calculated thermodynamic properties for liquid and solid phases at 298 K for straight chain (C_1–C_{20}) and branched alkanes, cycloalkanes and alkylcycloalkanes, and other compounds. Group values for $\Delta_f H^\circ$, C_p° and S° at 298 K are tabulated for gas, liquid and solid phase. Corrections for *gauche* interactions in gas-phase branched alkanes are reevaluated using a questionable, and certainly more complex, scheme. Sources of 'experimental' data are not indicated, and seem to include previously estimated values as well as experimentally measured ones.

XV. REFERENCES

1. G. N. Lewis and M. Randall, *Thermodynamics and the Free Energy of Chemical Substances*, McGraw-Hill, New York and London, 1923.
2. G. N. Lewis and M. Randall, Reference 1, p. 13 [from *Ann. Mines*, **13**, 157 [1888)].
3. C_p, S and H are considered thermodynamic quantities; all the properties derived from them—including G and K—are thermochemical quantities.
4. D. R. Herschbach, H. S. Johnston, K. S. Pitzer and R. E. Powell, *J. Chem. Phys.*, **25**, 736 (1956).
5. See A. Veillard, '*Ab initio* calculations of barrier heights', Chap. 11 in *Internal Rotation in Molecules* (Ed. J. Orville-Thomas), Wiley, New York, 1974.
6. R. M. Pitzer, *Acc. Chem. Res.*, **16**, 207 (1983).
7. J. Chao, R. C. Wilhoit and B. J. Zwolinski, *J. Phys. Chem. Ref. Data*, **2**, 427 (1973).
8. S. S. Chen, R. C. Wilhoit and B. J. Zwolinski, *J. Phys. Chem. Ref. Data*, **4**, 859 (1975).
9. V. Sackwild and W. G. Richards, *J. Mol. Struct. (Theochem)*, **89**, 269 (1982).
10. U. A. Zirnet and M. M. Sushinskii, *Optics and Spectroscopy*, **16**, 489 (1964).
11. L. Lunazzi, D. Macciantelli, F. Bernardi and K. U. Ingold, *J. Am. Chem. Soc.*, **99**, 4573 (1977).
12. C. Jaime and E. Osawa, *Tetrahedron*, **39**, 2769 (1983).
13. J. E. Anderson and H. Pearson, *J. Am. Chem. Soc.*, **97**, 764 (1975).
14. K. S. Pitzer, *J. Chem. Phys.*, **5**, 469 (1937); K. S. Pitzer and W. D. Gwinn, *J. Chem. Phys.*, **10**, 428 (1942); K. S. Pitzer, *J. Chem. Phys.*, **14**, 239 (1946); J. E. Kilpatrick and K. S. Pitzer, *J. Chem. Phys.*, **17**, 1064 (1949); J. C. M. Li and K. S. Pitzer, *J. Phys. Chem.*, **60**, 466 (1956).
15. G. N. Lewis and M. Randall, *Thermodynamics* (revised by K. S. Pitzer and L. Brewer), Chap. 27, 2nd edn., McGraw-Hill, New York, 1961.
16. H. L. Strauss and E. Thiele, *J. Chem. Phys.*, **46**, 2473 (1967).
17. See, for example, Reference 15, Chap. 16; or J. O. Hirschfelder, C. F. Curtiss and R. B. Bird, *Molecular Theory of Gases and Liquids*, Chap. 1, Wiley, New York, 1954.
18. See J. O. Hirschfelder, C. F. Curtiss and R. B. Bird, *Molecular Theory of Gases and Liquids*, Wiley, New York, 1954, p. 167.
19. G. Waddington, S. S. Todd and H. M. Huffman, *J. Am. Chem. Soc.*, **69**, 22 (1947).
20. Reference 22, p. 249.
21. J. D. Cox and G. Pilcher, *Thermochemistry of Organic and Organometallic Compounds* Chap. 7, Academic Press, London and New York, 1970.
22. D. R. Stull, E. F. Westrum, Jr., and G. C. Sinke, *The Chemical Thermodynamics of Organic Compounds*, Krieger Publ. Co., Malabar, Florida, 1987.
23. M. W. Chase, Jr., C. A. Davies, J. R. Downey, Jr., D. J. Frurip, R. A. McDonald and A. N. Syverud, JANAF Thermochemical Tables, 3rd Edition, *J. Phys. Chem. Ref. Data*, **14**, Supplement 1 (1985).
24. A. L. Verma, W. F. Murphy and H. J. Bernstein, *J. Chem. Phys.*, **60**, 1540 (1974).
25. D. A. Pittam and G. Pilcher, *J. Chem. Soc., Faraday Trans 1*, **68**, 2224 (1972).

26. F. D. Rossini, *J. Res. Natl. Bur. Stand.*, **6**, 37 (1931); **7**, 329 (1931); **12**, 735 (1934); **15**, 357 (1935); E. J. Prosen, F. W. Maron and F. D. Rossini, *J. Res. Natl. Bur. Stand.*, **46**, 106 (1951).
27. S. Gordon and B. J. McBride, 'Computer Program for Calculation of Complex Chemical Equilibrium Compositions, Rocket Performance, Incident and Reflected Shocks and Chapman–Jouguet Detonations', NASA SP-273 (1971).
28. A. Burcat, 'Thermochemical data for combustion calculations', Chap. 8 in *Combustion Chemistry* (Ed. W. C. Gardiner, Jr.), Springer-Verlag, New York, 1984.
29. R. A. Alberty and C. A. Gehrig, *J. Phys. Chem. Ref. Data*, **13**, 1173 (1984).
30. K. Fajans, *Chem. Ber.*, **53**, 643 (1920); **55**, 2826 (1922); M. S. Kharasch, *J. Res. Natl. Bur. Stand.*, **2**, 359 (1929).
31. S. W. Benson and J. H. Buss, *J. Chem. Phys.*, **29**, 546 (1958).
32. S. W. Benson, *Thermochemical Kinetics*, 2nd edn., Wiley, New York, 1976, p. 25.
33. Reference 21 Chap. 7 shows the equivalence of the group additivity approach and the schemes proposed by Laidler and by Allen.
34. K. J. Laidler, *Can. J. Chem.*, **34**, 626 (1956).
35. T. L. Allen, *J. Chem. Phys.*, **31**, 1039 (1959).
36. J. R. Platt, *J. Chem Phys.*, **15**, 419 (1947); *J. Phys. Chem.*, **56**, 328 (1952).
37. J. B. Greenshields and F. D. Rossini, *J. Phys. Chem.*, **62**, 271 (1958).
38. G. R. Somayajulu and B. J. Zwolinski, *Trans. Faraday Soc.*, **62**, 2327 (1966).
39. R. M. Joshi, *J. Macromol. Sci., Chem.*, **A4**, 1819 (1970).
40. Reference 37, as modified by H. A. Skinner, *J. Chem. Soc.*, 4396 (1962).
41. J. D. Overmars and S. M. Blinder, *J. Phys. Chem.*, **68**, 1801 (1964).
42. M. Souders, Jr., C. S. Matthews and C. O. Hurd, *Ind. Eng. Chem.*, **41**, 1048 (1949).
43. V. M. Tatevskii, V. A. Benderskii and S. S. Yarovoi, *Rules and Methods for Calculating the Physico-Chemical Properties of Paraffinic Hydrocarbons* (transl. B. P. Mullins), Pergamon, Oxford, 1961,
44. K. K. Verma and K. K. Doraiswamy, *Ind. Eng. Chem. Fundam.*, **4**, 389 (1965).
45. R. C. Reid, J. M. Prausnitz and B. E. Poling, *The Properties of Gases and Liquids*, 4th edn., Chap. 6, McGraw-Hill, New York, 1987.
46. Reference 32, p. 26.
47. The contribution of the C—$(C)_2(H)_2$ group is of particular historical interest. Since the CH_2 group contribution was easily derivable from the heats of formation of successive homologs in various series, it was early examined for constancy, as P. Sellers, G. Stridh and S. Sunner [*J. Chem. Eng. Data*, **23**, 250 (1978)] have pointed out. In 1886, J. Thomsen (*Thermochemische Untersuchungen*, Vol. 4, Barth, Leipzig, 1886) deduced that a CH_2 group added 158 kcal mol^{-1} to the enthalpy of combustion of a compound, whence we derive a value of 6.15 for the contribution to ΔH_f. F. D. Rossini [*J. Res. Natl. Bur. Stand.*, **13**, 21(1934)], using the then-most-recent experimental data, obtained a value of 5.15 kcal mol^{-1}, a value later refined to 4.926 [E. J. Prosen, W. H. Johnson and F. D. Rossini, *J. Res. Natl. Bur. Stand.*, **37**, 51 (1946)]. Sellers and coworkers determined separate constants for CH_2 increments in different homologous series. For n-alkanes, they found 4.95 ± 0.01; for 2-methylalkanes, 5.01 ± 0.05; for 1-cyclohexylalkanes, 4.96 ± 0.03. We discuss the question of the constancy of this 'constant' later in the text.
48. S. W. Benson, F. R. Cruickshank, D. M. Golden, G. R. Haugen, H. E. O'Neal, A. S. Rodgers, R. Shaw and R. Walsh, *Chem. Rev.*, **69**, 279 (1969).
49. It is at least of historical interest to note that M. Souders, Jr., C. S. Matthews and C. O. Hurd, Ref. 42, derived group values for hydrocarbon groups which, at least in the case of enthalpies, were not very different from current best values. Their enthalpy values, corresponding to column 2 of Table 6, were − 10.05, − 4.95, − 0.88 (except for a group on the 2nd carbon, for which the value was − 1.57) and 2.45 (0.85 for the 2nd carbon).
50. M. L. Huggins, in *Structural Chemistry and Molecular Biology* (Eds. A. Rich and N. Davidson), W. H. Freeman, San Francisco, 1968, p. 761.
51. See the discussion in S. W. Benson and M. Luria, *J. Am. Chem. Soc.*, **97**, 704 (1975).
52. S. W. Benson, Reference 32, p. 30, presented arguments supporting the supposition that the *gauche* interaction is attributable to the repulsions of the 1,4 H atoms. These arguments were elaborated in S. W. Benson and M. Luria, *J. Am. Chem. Soc.*, **97**, 704 (1975).
53. J. B. Pedley, R. D. Naylor and S. P. Kirby, *Thermochemical Data of Organic Compounds*, 2nd edn, Chapman and Hall, London & New York, 1986.

54. W. B. Person and G. C. Pimentel, *J. Am. Chem. Soc.*, **75**, 532 (1953).
55. D. W. Scott, *J. Chem. Phys.*, **60**, 3144 (1974).
56. H. E. O'Neal and S. W. Benson, *Int. J. Chem. Kinet.*, **1**, 221 (1969).
57. O. Dobis and S. W. Benson, *Int. J. Chem. Kinet.*, **19**, 691 (1987).
58. P. D. Pacey and J. H. Wimalasena, *J. Phys. Chem.*, **88**, 5657 (1984).
59. J. J. Russell, J. A. Seetula, S. M. Senkan and D. Gutman, *Int. J. Chem. Kinet.*, **20**, 759 (1988).
60. M. W. Chase, Jr., C. A. Davies, J. R. Downey, Jr., D. J. Frurip, R. A. McDonald and A. N. Syverud, 'JANAF Thermochemical Tables, 3rd Edition', *J. Phys. Chem. Ref. Data*, **14**, Supplement 1 (1985).
61. M. Brouard, P. D. Lightfoot and M. J. Pilling, *J. Phys. Chem.*, **90**, 445 (1986).
62. S. S. Parmar and S. W. Benson, *J. Am. Chem. Soc.*, **111**, 57 (1989).
63. R. M. Marshall and L. Rahman, *Int. J. Chem. Kinet.*, **9**, 705 (1977).
64. A. L. Castelhano and D. Griller, *J. Am. Chem. Soc.*, **104**, 3655 (1982).
65. D. F. McMillen and D. M. Golden, *Ann. Rev. Phys. Chem.*, **33**, 493 (1982).
66. R. R. Baldwin, G. R. Drewery and R. W. Walker, *J. Chem. Soc., Faraday Trans. 1*, **80**, 2827 (1984).
67. W. v. E. Doering, *Proc. Natl. Acad. Sci. U.S.A.*, **78**, 5279 (1981).
68. E. Tschuikow-Roux and Y. Chen. *J. Am. Chem. Soc.*, **111**, 9030 (1989).
69. Y.-R. Luo and S. W. Benson, *J. Phys. Chem.*, **93**, 3304 (1989).
70. J. E. Baggott and M. J. Pilling, *Ann. Rep. Chem. Soc., Sect. C*, 199 (1982).
71. W. Müller-Markgraf, M. J. Rossi and D. M. Golden, *J. Am. Chem. Soc.*, **111**, 956 (1989).
72. J. J. Russell, J. A. Seetula, R. S. Timonen, D. Gutman and D. F. Nava, *J. Am. Chem. Soc.*, **110**, 3084 (1988).
73. S. W. Benson, *J. Chem. Soc., Faraday Trans. 2*, **83**, 791 (1987); T. S. A. Islam and S. W. Benson, *Int. J. Chem. Kinet.*, **16**, 995 (1984).
74. J. C. Schultz, F. A. Houle and J. L. Beauchamp, *J. Am. Chem. Soc.*, **106**, 3917 (1984).
75. S. Fliszar and C. Minichino, *Can. J. Chem.*, **65**, 2495 (1987).
76. C. W. Larson, E. A. Hardwidge and B. S. Rabinovitch, *J. Chem. Phys.*, **50**, 2769 (1969).
77. J. Pacansky and J. S. Chang, *J. Chem. Phys.*, **74**, 5539 (1981).
78. J. Pacansky and M. Yoshimine, *J. Phys. Chem.*, **90**, 1980 (1986).
79. S. H. Jones and E. Whittle, *Int. J. Chem. Kinet.*, **2**, 479 (1970).
80. H. M. Huffman, S. S. Todd and G. D. Oliver, *J. Am. Chem. Soc.*, **71**, 584 (1949).
81. R. Shaw, *J. Chem. Eng. Data*, **14**, 461 (1969).
82. M. Luria and S. W. Benson, *J. Chem. Eng. Data*, **22**, 90 (1977).
83. Reference 18, p. 283.
84. J. S. Chickos, A. S. Hyman, L. H. Ladon and J. F. Liebman, *J. Org. Chem.*, **46**, 4294 (1981); J. S. Chickos, R. Annunziata, L. H. Ladon, A. S. Hyman and J. F. Liebman, *J. Org. Chem.*, **51**, 4311 (1986).
85. S. W. Benson and L. J. Garland, *J. Phys. Chem.*, **95**, 4915 (1991).
86. Data for hexane–cetane ($C_{16}H_{34}$) mixtures typify the magnitude of ΔH_{vap} and illustrate its dependence on composition and temperature:

T (K)	(Cetane mol%)	ΔH_{vap} (cal mol^{-1})
293	21	21
	50	31
	69	27
323	37	7.4
	54	8.8

(Taken from *Landolt–Börnstein Numerical Data and Functional Relations in Science and Technology, New Series*, Group IV, Vol. 2, *Heats of Mixing and Solution*, p. 492).

87. J. Pacansky and B. Schrader, *J. Chem. Phys.*, **78**, 1033 (1983).
88. J. Pacansky and H. Coufal, *J. Chem. Phys.*, **72**, 3298 (1980).
89. E. S. Domalski, W. H. Evans and E. D. Hearing, 'Heat Capacities and Entropies of Organic Compounds in the Condensed Phase', *J. Phys. Chem. Ref. Data*, **13**, Supplement 1 (1984).
90. R. J. L. Andon and J. F. Martin, *J. Chem. Thermodyn.*, **17**, 1159 (1976).
91. K. S. Pitzer, *J. Chem. Phys.*, **8**, 711 (1940).

CHAPTER **7**

The analysis of alkanes and cycloalkanes

ZEEV AIZENSHTAT

Department of Organic Chemistry and Casali Institute of Applied Chemistry, The Hebrew University of Jerusalem, Jerusalem 91904, Israel

ABBREVIATIONS

ACE	alternating chemical and electron impact ionization
CI	chemical ionization

The Chemistry of Alkanes and Cycloalkanes
Edited by S. Patai and Z. Rappoport © 1992 John Wiley & Sons Ltd

CZE	capillary zone electrophoresis
DC	dielectric constant
DCD	dielectric constant detector
1D-NMR	one-dimensional NMR (regular)
2D-NMR	two-dimensional NMR
ECD	electron capture detector
EI	electron impact
eV	electron volt
FI	field ionization
FID	flame ionization detector
FIMS	field ionization-MS
FPD	flame photometric detector
FSCOT	fused silica-SCOT
FT	Fourier transform
FTIR	Fourier transform infra-red
GC	gas chromatography
GC-C-isoMS	GC-combustion-isotope measurement MS
GC-MI	GC-matrix isolation
GC/MS	gas chromatography/mass spectrometry
GC/MS/C	GC/MS/computerized
GPC	gel permeation chromatography—size exclusion
GTA	group type analysis
HECTOR	heteronuclear shift-correlated-spectrum
HRC	high resolution chromatography
HR-MS	high resolution-MS
HPLC	high pressure liquid chromatography
INEPT-INADEQUATE	insensitive nuclear enhancement by polarization transfer
IP	ionization potential
LC	liquid chromatography
MI	matrix isolation
MI/IR	matrix isolation/IR
MI/FTIR	matrix isolation/FTIR
MPLC	medium pressure liquid chromatography
MSD	mass spectrometer detector
NBS	National Bureau of Standards
NOE	nuclear Overhauser effect
NOESY	NOE and exchange spectrometry
PFK	perfluorinated kerosene
PID	photoionization detector
PTC	programming temperature controller
RF	retention factor or response factor (as marked)
RID	refractive index detector
RRF	relative retention factor
RP	reversed phase or resolution power (MS)
SCOT	surface coated open tubular
SFC	super-critical fluid chromatography
SIM	selective ion monitoring
SIR	selected ion recording
TCD	thermoconductivity detector
TG	thermogravimetry
TIC	total ion current
TLC	thin-layer chromatography

I. INTRODUCTION

In the early stages of preparation of the present manuscript I was confronted with a surprise: 'The analysis of alkanes and cycloalkanes' has never been reviewed as such. Although the C—C and C—H bonds are the most abundant bonds in organic chemistry, the analysis of compounds constructed solely from these bonds cannot be found in analytical reviews of 'functional' groups. Therefore, I could not relate to previous secondary or tertiary sources of literature. Moreover, the reader of this chapter will find that, although I attempted to stay as objective as possible in presenting the various analytical techniques, most of the advanced applications come from petroleum and other fossil fuels studies. Since the choice of analytical techniques relates to structures of the whole alkane or cycloalkane, the various isomers, enantiomers or conformations are discussed.

In general, the scheme of the present chapter follows the analysis of single compounds by the various techniques and then tackles the more difficult problem of the separation and chromatographic analysis of mixtures. Some basic spectral properties of alkanes and cycloalkanes are mentioned briefly and the reader is directed to other chapters of this book.

Most basic methods are referred to the literature and, in cases where figures help to demonstrate the analytical concept, they are presented. However, in more advanced instrumental analyses, much of the information is in the 'grey literature': manufacturers' information data sheets. The analysis of families of alkanes and cycloalkanes has to take into account the fact that the homologous members may range from C_1 to C_{70}. Hence, the petroleum industry is still using the refining of crudes to separate it into fractions of different boiling point ranges. This separation by physical properties was followed by all the developed chromatography techniques.

Another analytical challenge presented to the researcher involved with thermal stability and catalytic reactivity of hydrocarbons relates to cracking of heavy fractions of petroleum. Thermal analysis is not discussed in this chapter. However, one should be aware that, although paraffins are considered to be chemically inactive, the application of certain analytical techniques may cause cracking, rearrangement and isomerization.

II. ALKANES

Almost every textbook on organic chemistry starts with a chapter on alkanes, also called saturated hydrocarbons or paraffins. The word paraffin, which comes from the Latin *parum affinis* (slight affinity), indicates that, already at the early stages of organic chemistry, it was recognized that non-substituted paraffins are rather inert. In petroleum analysis the mixture of the higher alkanes is named wax.

Due to the low reactivity of alkanes, their analysis by derivatization is very difficult. Moreover, most of the literature on analysis of functional groups does not include analytical methods for separation and identification of alkanes.

The diversity of alkane isomers is discussed in detail in the chapter by Kier and Hall in this volume. However, we should mention that the C_nH_{2n+2} open-chain alkanes can have from a single structural isomer (C_1—C_3) up to 366,319 isomers for $C_{20}H_{42}$. Hence, the analysis of alkanes is more and more complicated with increasing number of C atoms.

A. Normal Alkanes

Normal alkanes (straight-chain alkanes or normal paraffins) are the simplest homologous series of the group.

The C_1—C_5 n-alkanes are gases at room temperature whereas the C_5—C_{16-17} are

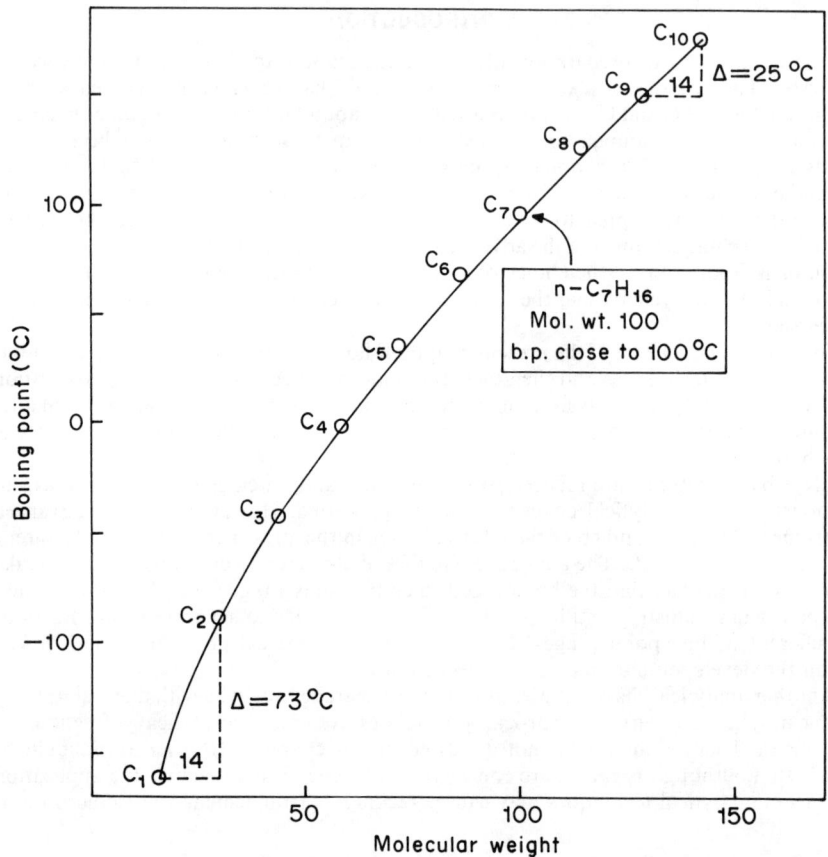

FIGURE 1. Boiling points of n-alkanes[1]

liquids and the longer ones are solids. The boiling points of n-alkanes from C_1 up to C_{10} (Figure 1)[1] allow their separation by a good distillation column. The question of refining will be addressed when we deal with mixtures.

The specific gravity of the n-alkanes (Table 1)[1] shows considerable increases from 0.424 (liquid CH_4) to 0.626 (n-pentane) and 0.794 for C_{50}.

B. Branched Alkanes

For each carbon number there is only one normal alkane, while the branched alkanes include all other isomers. The first branched isomer is 2-methylpropane (isobutane) (Table 2). The most important C_5 unit, isoprene (alkene) and the saturated isoprane are for the most part photosynthetically produced and are the building blocks of the open-chain and cyclic terpenoids. The importance of the analysis of the mono-, di-, etc. cyclic terpanes will be discussed in the cycloalkanes section.

Most terpene (C_{10}) hydrocarbons have double bonds, hence their analysis can be approached as the analysis of cycloalkenes. However, some saturated terpanes were found in petroleum and bitumens.

TABLE 1. Normal alkanes[1]

Name	Formula, C_nH_{2n+2}	m.p. (°C)	b.p. (°C)	Sp. gr. (as liquids)[a]	
Methane	CH_4	−182.6	−161.7	0.4240	gases
Ethane	C_2H_6	−172.0	−88.6	0.5462	
Propane	C_3H_8	−187.1	−42.2	0.5824	
n-Butane	C_4H_{10}	−135.0	−0.5	0.5788	
n-Pentane	C_5H_{12}	−129.7	36.1	0.6264	liquid
n-Hexane	C_6H_{14}	−94.0	68.7	0.6594	
n-Heptane	C_7H_{16}	−90.5	98.4	0.6837	
n-Octane	C_8H_{18}	−56.8	125.6	0.7028	
n-Nonane	C_9H_{20}	−53.7	150.7	0.7179	
n-Decane	$C_{10}H_{22}$	−29.7	174.0	0.7298	
n-Undecane	$C_{11}H_{24}$	−25.6	195.8	0.7404	
n-Dodecane	$C_{12}H_{26}$	−9.6	216.3	7.7493	
n-Tridecane	$C_{13}H_{28}$	−6	(230)	0.7568	
n-Tetradecane	$C_{14}H_{30}$	5.5	251	0.7636	
n-Pentadecane	$C_{15}H_{32}$	10	268	0.7688	
n-Hexadecane	$C_{16}H_{34}$	18.1	280	0.7749	
n-Heptadecane	$C_{17}H_{36}$	22.0	303	0.7767	
n-Octadecane	$C_{18}H_{38}$	28.0	308	0.7767	solids
n-Nonadecane	$C_{19}H_{40}$	32	330	0.7776	
n-Eicosane	$C_{20}H_{42}$	36.4		0.7777	
n-Heneicosane	$C_{21}H_{44}$	40.4		0.7782	
n-Docosane	$C_{22}H_{46}$	44.4		0.7778	
n-Tricosane	$C_{23}H_{48}$	47.4		0.7797	
n-Tetracosane	$C_{24}H_{50}$	51.1		0.7786	
n-Pentacosane	$C_{25}H_{52}$	53.3			
n-Triacontane	$C_{30}H_{62}$	66			
n-Pentatriacontane	$C_{35}H_{72}$	74.6		0.7814	
n-Tetracontane	$C_{40}H_{82}$	81			
n-Pentacontane	$C_{50}H_{102}$	92		0.7940	
n-Hexacontane	$C_{60}H_{122}$	99			
n-Dehexacontane	$C_{62}H_{126}$	101			
n-Tetrahexacontane	$C_{64}H_{130}$	102			
n-Heptacontane	$C_{70}H_{142}$	105	300 at 0.00001 mm		

[a]Specific gravities reported in this table refer to the liquid state. With substances liquid at 20 °C, values for this temperature are given where available. The data given for more volatile substances are for temperatures close to the boiling point, those for less volatile compounds are for temperatures just above the melting points.

The most studied open-chain diterpenoid is phytol ($C_{20}H_{40}O$) which was isolated from chlorophyll hydrolysate by Willstatter in 1909[2], and shown to be an allylic alcohol by Fischer[3]. Since then, both phytane ($C_{20}H_{42}$) and pristane ($C_{19}H_{40}$) have been identified in all fossil fuels (petroleum, coals, asphalts, bitumens, etc.).

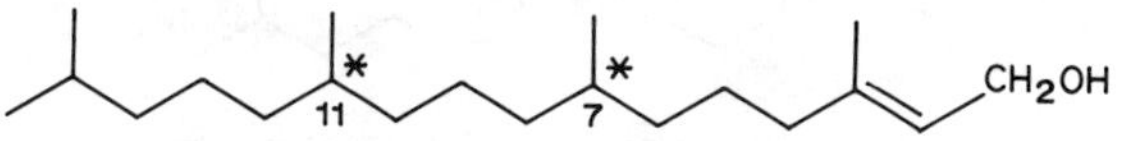

Phytol contains two asymmetric carbons (positions 7 and 11) whereas phytane ($C_{20}H_{42}$) has three such chiral centers (3, 7 and 11).

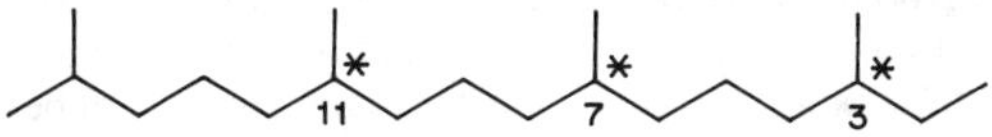

Due to the three chiral centres, phytane has diastereoisomers with different physical properties. Fischer's suggestion that phytol is biogenically produced from geraniol and farnesol was followed by Burrell[6,7], and the stereochemistry of phytol was established as *trans*-3,7(R),11(R).

Formerly, it was suggested that the isopranes found in petroleum are synthesized by nature as isoprenoids. During the early part of the 1960s a very intensive effort was made to isolate the isoprane, which appears close to the n-C_{17} hydrocarbon. For reference, 2,6,10,14-tetramethylpentadecane ($C_{19}H_{40}$ pristane) was analysed. Spectroscopic methods (IR, NMR and MS) showed very close similarity of the isolated fraction to synthetic pristane. The physical properties compared were boiling point, freezing point, specific gravity and refractive index. Since the pioneering studies by Erdman[4] and Bendoraitis and coworkers[5], single alkanes are still identified by the same methods; the isolation method is no longer used. Carbons 6 and 10 in pristane can have stereoisomeric R or S configuration. Determination of the stereochemistry can support or disprove the suggestion that some of the acyclic isopranes C_{18}, C_{15}, C_{14} derive from the phytol structure.

The stereochemical determination of alkanes and cycloalkanes is always more difficult than that of 'functionalized' organic molecules. For most methods, such as optical rotation measurements or X-ray crystallography, an isolated adequate amount of a *single* compound is required. It was also shown that for acyclic isopranes (terpanes) the specific optical rotation values[7] are very low, virtually invalidating the use of polarimetry.

For the stereochemical determination, it seemed theoretically that the most efficient method of separation for the three pristane isomers **A**, **B** and **C** would be direct gas chromatography on an optically active stationary phase. Such a method was successfully employed for amino acids and amine enantiomers[8]. However, this method was found ineffective for pristane, and hence various gas-chromatographic 'inactive' stationary phases were employed[9] for the separation of the diastereoisomers. The *relative* stereochemistry could be determined, provided standards with known respective stereochemistry are available.

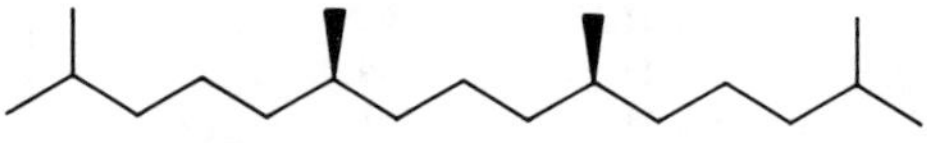

(A)

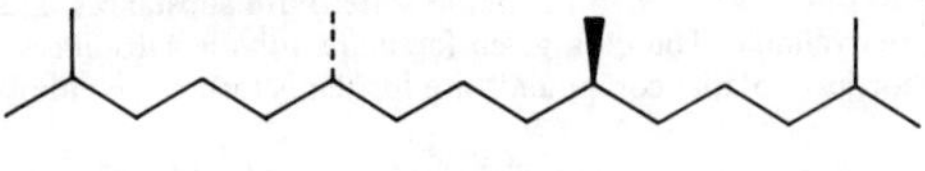

(B)

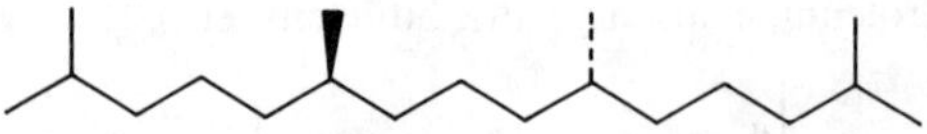

(C)

Since pristane is an 'inactive' alkane, the direct analysis will yield two peaks and not three; the 6(R),10(S) ≡ 6(S),10(R) [**A**] is symmetric and 6(R),10(R) will not separate from 6(S),10(S) [**B** + **C**]. Hence two isomers can be separated. If the examined sample is the *meso* form 6(R),10(S) then the absolute stereochemistry of the alkane can be determined[10].

Another alternative is to functionalize the alkane and form diastereoisomeric derivatives of the functional group (e.g. after oxidation to acids), which are easier to separate. Since 1971 Maxwell's group at Bristol used these methods to separate and identify the stereochemistry of isoprenoidic acyclic alkanes[10]. Whereas most acyclic isoprenoids in nature appear functionalized (alcohols, acids, alkenes, etc.), they are found in the geoenvironment as fossil alkanes. Hence it is important not only to analyse the structural isomers, but also to determine the stereochemistry.

Isoprenoids (terpanes), with a longer chain than pristane, have been detected in fossil fuels (see below the discussion on the analysis of petroleum). C_{25}—C_{30} isoprane hydro-carbons were analysed in a number of methanogenic bacteria[11]. The C_{40} isoalkane, which was isolated from the saturated fraction of the bitumen of the Green River Shale (USA), is claimed to be the product of total saturation of carotenoids. The analysis of these polyhydrogenated carotenes (PHC) requires special GC high-temperature inert columns (see GC of high carbon number wax hydrocarbons).

C. Cycloalkanes

The monocyclic hydrocarbons (C_nH_{2n}) found in nature contain from 3 to 30 carbons, although in principle there is no limit to the ring size. Even so, cyclopentanes and cyclohexanes are the most abundant in nature. The analysis of single compound monocyclic alkanes is in many respects similar to the analysis of the open-chain alkanes. However, for small rings (cyclopropane, cyclobutane) some added methods can be employed because of their higher chemical activity, due to the distortion of the C—C—C angles. The olefinic properties of cyclopropanes were reviewed by Charton[12].

The stereochemistry of cycloalkanes is discussed extensively in the chapter by Anderson, and it is important for an understanding of the analyses offered for these compounds to distinguish between isomers and conformers. Because of the low energetic barrier for equilibration between axial and equatorial conformers, only *low-temperature* ^{1}H NMR or ^{13}C NMR as well as IR will freeze the equilibration and allow the spectral analysis of the conformers. However, disubstituted cycloalkanes will exhibit, in addition to structural isomerism, also *cis*, *trans* isomerism. Both *cis* and *trans* isomers of 1,4-dimethylcyclohexane undergo, at room temperature, a conformer equilibration. In the case of the *trans* isomer the conformer with two equatorial CH_3 groups is more stable, while for the *cis* isomer both conformers are equally stable and cannot be separated.

In general the boiling points of the cycloalkanes are slightly higher than those of the corresponding alkanes and this is also true for the freezing point (see Tables 2a and 2b)[13].

Monocycloalkanes and their alkylated derivatives are named in the petroleum industry naphthenes. Most of them distil between 50–120 °C, and so these compounds are an important part of the naphtha fraction. For example, in the monocyclic terpane, *p*-menthane, the *cis* and *trans* isomers are assigned arbitrarily. As will be discussed later, only at low temperatures could the bandwidth of the Matrix Isolation FT/IR ≡ MI/FT/IR[14] analyse the differences. The MS of the two isomers (Figure 2) are essentially the same (the NBS library for *p*-menthane indicates the same spectrum for both). Figure 3 shows the fingerprint region, which is dominated by the C—H bending absorption (see IR of alkanes), and here the two isomers show different characteristics. The gem-dimethyl group absorptions are located around 1387–1369 cm^{-1} (a and b, Figure 3). However, one can see the differences which allow isomer assignment. All the above is true only if one has authentic isolated isomers, hence quantification of a mixture might be problematic. These results are claimed to be comparable to ^{13}C NMR with the advantage of the use of nanograms vs miligrams needed for NMR[14]. Since one would expect such isomers to be also separable by high-quality GC (capillary columns), the use of GC/MI/FT/IR

Z. Aizenshtat

is possible. This of course will require fast cryogenic trapping and flush heat release (see Section VI). It is also possible to identify isomers by X-ray analysis of the solid.

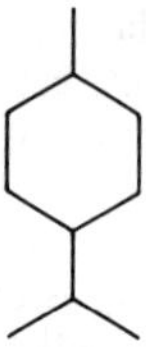

p-menthane
(b.p. 172 °C)

D. Polycyclic Alkanes

The analytical challenge for structure determination is increased with the number of rings in the cycloalkanes studied.

Classification of the polycyclic alkanes can be supported by the general formulae C_nH_{2n} for monocyclics, C_nH_{2n-2} for the dicyclics, C_nH_{2n-4} for tricyclics, etc. Later in

TABLE 2a. Paraffin hydrocarbons

Name	Formula	Melting point		Boiling point[a]		Sp. gr. (at °C)
		(°F)	(°C)	(°F)	(°C)	
Methane	CH_4	− 300	− 184	− 258.5	− 161.4	0.415(liq.-164)
Ethane	C_2H_6	− 278	− 172	− 126.4	− 88.3	0.446 (liq. 0)
Propane	C_3H_8	− 309.8	− 189.9	− 48.1	− 44.5	0.536 (liq. 0)
Butanes						
n-Butane	C_4H_{10}	− 211.8	− 135	31	− 0.6	0.60(0)
Isobutane	C_4H_{10}	− 229	− 145	13.6	− 10.2	0.559 (20)
Pentanes						
n-Pentane	C_5H_{12}	− 201.7	− 129.9	96.8	36	0.626(20)
Isopentane	C_5H_{12}	− 254.8	− 159.7	87.8	28	0.619(20)
(2-methylbutane)						
2,2-Dimethylpropane	C_5H_{12}	− 4	− 20	49.1	9.5	0.613(0)
Hexanes						
n-Hexane	C_6H_{14}	− 139.5	− 95.3	155.7	68.7	0.660(20)
Isohexane	C_6H_{14}	− 244.6	− 153.7	140.4	60.2	0.654(20)
(2-methylpentane)						
3-Methylpentane	C_6H_{14}	− 180.4	− 118	145.8	63.2	0.668(20)
2,2-Dimethylbutane	C_6H_{14}	− 144.8	− 98.2	121.5	49.7	0.649(20)
2,3-Dimethylbutane	C_6H_{14}	− 211.2	− 135.1	136.6	58.1	0.662(20)
Heptanes						
n-Heptane	C_7H_{16}	− 130.9	− 90.5	209.1	98.4	0.684(20)
Isoheptane	C_7H_{16}	− 182.4	− 119.1	194	90	0.679(20)
(2-methylhexane)						
3-Methylhexane	C_7H_{16}	− 182.9	− 119.4	197.2	91.8	0.687(20)
3-Ethylpentane	C_7H_{16}	− 181.8	− 118.8	199.9	93.3	0.698(20)
2,2-Dimethylpentane	C_7H_{16}	− 194.1	− 125.6	174	78.9	0.674(20)
2,3-Dimethylpentane	C_7H_{16}			193.5	89.7	0.695(20)
2,4-Dimethylpentane	C_7H_{16}	− 182.9	− 119.4	177.4	80.8	0.675(20)
3,3-Dimethylpentane	C_7H_{16}	− 211	− 135	186.8	86	0.693(20)
2,2,3-Trimethylbutane	C_7H_{16}	− 13	− 25	177.6	80.9	0.690(20)

TABLE 2a (continued)

Name	Formula	Melting point (°F)	Melting point (°C)	Boiling point[a] (°F)	Boiling point[a] (°C)	Sp. gr. (at °C)
Octanes						
n-Octane	C_8H_{18}	− 70.4	− 56.9	258.1	125.6	0.703 (20)
Isooctane	C_8H_{18}	− 168.3	− 111.3	243	117.2	0.698 (20)
(2-methylheptane)						
2,2,4-Trimethylpentane[b]	C_8H_{18}	− 160.3	− 107.4	210.7	99.3	0.692 (20)
n-Nonane	C_9H_{20}	− 64.7	− 53.7	303.3	150.7	0.718 (20)
n-Decane	$C_{10}H_{22}$	− 21.5	− 29.7	345.2	174	0.730 (20)
n-Undecane	$C_{11}H_{21}$	− 14.3	− 25.7	384.4	195.8	0.740 (20)
n-Dodecane	$C_{12}H_{26}$	14.4	− 9.7	420.6	216.2	0.749 (20)
n-Tridecane	$C_{13}H_{28}$	21.2	− 6	453.2	234	0.772 (0)
n-Tetradecane	$C_{14}H_{30}$	41.9	5.5	486.5	252.5	0.774 (at m.p.)
n-Pentadecane	$C_{15}H_{32}$	50	10	518.9	270.5	0.776 (at m.p.)
n-Hexadecane (cetane)	$C_{16}H_{34}$	64.4	18	549.5	287.5	0.775 (at m.p.)
n-Heptadecane	$C_{17}H_{36}$	72.5	22.5	577.4	303	0.777 (at m.p.)
n-Octadecane	$C_{18}H_{38}$	82.4	28	602.6	317	0.777 (at m.p.)
n-Nonadecane	$C_{19}H_{40}$	89.6	32	626	330	0.777 (at m.p.)
n-Eicosane	$C_{20}H_{42}$	97.7	36.5	401	205 (15 mm)	0.778 (at m.p.)
n-Heneicosane	$C_{21}H_{44}$	104.9	40.5	419	215 (15 mm)	0.778 (at m.p.)
n-Docosane	$C_{22}H_{46}$	111.9	44.4	436.1	224.5 (15 mm)	0.778 (at m.p.)
n-Tricosane	$C_{23}H_{48}$	117.8	47.7	453.2	234 (15 mm)	0.778 (at m.p.)
n-Tetracosane	$C_{24}H_{50}$	123.8	51	615.2	324	0.779 (at m.p.)
n-Pentacosane	$C_{25}H_{52}$	129.2	54	761	405	0.779 (20)
n-Hexacosane	$C_{26}H_{54}$	134.6	57	784.4	418	0.779 (20)
n-Heptacosane	$C_{27}H_{56}$	139.1	59.5	518	270 (15 mm)	0.780 (at m.p.)
n-Octacosane	$C_{28}H_{58}$	143.6	62	834.8	446	0.779 (20)
n-Nonacosane	$C_{29}H_{60}$	146.3	63.5	896	480	0.780 (20)
n-Triacontane	$C_{30}H_{62}$	150.8	66	861.8	461	0.780 (20)
n-Hentriacontane	$C_{31}H_{64}$	154.6	68.1	575.6	302 (15 mm)	0.781 (at m.p.)
n-Dotriacontane (dicetyl)	$C_{32}H_{66}$	158	70	590	310 (15 mm)	0.773 (80)
n-Tritriacontane	$C_{33}H_{68}$	161.6	72	622.4	328 (15 mm)	0.780 (at m.p.)
n-Tetratriacontane	$C_{34}H_{70}$	163.4	73	908.6	487	0.780 (at m.p.)
n-Pentatriacontane	$C_{35}H_{72}$	166.5	74.7	627.8	331 (15 mm)	0.782 (at m.p.)
n-Hexatriacontane	$C_{36}H_{74}$	168.8	76	509	265 (1.0 mm)	0.782 (at m.p.)
n-Tetracontane	$C_{40}H_{82}$	177.8	81	465.8	241 (0.3 mm)	
n-Pentacontane	$C_{50}H_{102}$	199.4	93	789.8	421 (15 mm)	0.794 (at m.p.)
nHexacontane	$C_{60}H_{122}$	210.2	99			
n-Dohexacontane	$C_{62}H_{126}$	213.8	101			
n-Tetrahexacontane	$C_{64}H_{130}$	215.6	102			
n-Heptacontane	$C_{70}H_{142}$	221	105			

[a] At 760 mm mercury, unless otherwise specified.
[b] This chemical is used as standard in octane rating.

TABLE 2b. Naphthene hydrocarbons

Name	Formula		Melting point (F°)	(°C)	Boiling point at 760 mm (°F)	(°C)	Sp. gr. (at °C)
Cyclopropane	C_3H_6	C_3H_6	-195.9	-126.6	-29.9	-34.4	0.720(-79)
Methylcyclopropane	C_4H_8	$CH_3 \cdot C_3H_5$			41	5	0.691(-20)
1,1-Dimethylcyclopropane	C_5H_{10}	$(CH_3)_2 \cdot C_3H_4$			69.8	21	0.660(20)
1,1,2-Trimethylcyclopropane	C_6H_{12}	$(CH_3)_3 \cdot C_3H_3$			127	52.8	0.695(20)
1,2,3-Trimethylcyclopropane	C_6H_{12}	$(CH_3)_3 \cdot C_3H_3$			149	65	0.692(22)
Cyclobutane	C_4H_8	C_4H_8	-58	-50	55.4	13	0.703(0)
Methylcyclobutane	C_5H_{10}	$CH_3 \cdot C_4H_7$			107.6	42	0.694(20)
Ethylcyclobutane	C_6H_{12}	$C_2H_5 \cdot C_4H_7$			161.6	72	0.745(20)
3-Cyclobutylpentane	C_9H_{18}	$C_2H_5 \cdot CH(C_4H_7) \cdot C_2H_5$			303.8–309.2	151	0.795(19)
Cyclopentane	C_5H_{10}	C_5H_{10}	-135.9	-93.3	122	50	0.751(20)
Methylcyclopentane	C_6H_{12}	$CH_3 \cdot C_5H_9$	-220.9	-140.5	161.6	72	0.750(20)
1,1-Dimethylcyclopentane	C_7H_{14}	$(CH_3)_2 \cdot C_5H_8$			189.5	87.5	0.755(20)
1,2-Dimethylcyclopentane	C_7H_{14}	$(CH_3)_2 \cdot C_5H_8$			197.6	92	0.753(20)
1,3-Dimethylcyclopentane	C_7H_{14}	$(CH_3)_2 \cdot C_5H_8$			195.8	91	0.754(20)
1-Methyl-2-ethylcyclopentane	C_8H_{16}	$(CH_3)(C_2H_5) \cdot C_5H_8$			255.2	124	0.764(20)
1-Methyl-3-ethylcyclopentane	C_8H_{16}	$(CH_3)(C_2H_5) \cdot C_5H_8$			249.8	121	0.779(20)
Cyclohexane	C_6H_{12}	C_6H_{12}	43.7	6.5	178.5	81.4	0.779(20)
Methylcyclohexane	C_7H_{14}	$CH_3 \cdot C_6H_{11}$	-195.3	-126.3	213.4	100.8	0.770(20)
1,1-Dimethylcyclohexane	C_8H_{16}	$(CH_3)_2 \cdot C_6H_{10}$	-71.5	-57.5	248	120	0.779(20)
1,2-Dimethylcyclohexane	C_8H_{16}	$(CH_3)_2 \cdot C_6H_{10}$			254.2	124	0.771(20)
1,3-Dimethylcyclohexane	C_8H_{16}	$(CH_3)_2 \cdot C_6H_{10}$	-122.8	-86	249.8	120.5	0.769(20)
1,4-Dimthylcyclohexane	C_8H_{16}	$(CH_3)_2 \cdot C_6H_{10}$			248	120	0.766(20)
Ethylcyclohexane	C_8H_{16}	$C_2H_5 \cdot C_6H_{11}$			266	130	0.777(20)
1,1,3-Trimethylcyclohexane	C_9H_{18}	$(CH_3)_3 \cdot C_6H_9$			280.4	138	0.790(20)
1,2,4-Trimethylcyclohexane	C_9H_{18}	$(CH_3)_3 \cdot C_6H_9$			284–287.6	140–2	0.778(20)
1,3,5-Trimethylcyclohexane	C_9H_{18}	$(CH_3)_3 \cdot C_6H_9$			278.6–282.2	137–9	0.772(20)
1-Methyl-2-ethylcyclohexane	C_9H_{18}	$(CH_3)(C_2H_5) \cdot C_6H_{10}$			303.8	151	0.784(20)
1-Methyl-3-ethylcyclohexane	C_9H_{18}	$(CH_3)(C_2H_5) \cdot C_6H_{10}$			300.2	149	0.799(20)
1-Methyl-4-ethylcyclohexane	C_9H_{18}	$(CH_3)(C_2H_5) \cdot C_6H_{10}$			302	150	0.804(0)
Propylcyclohexane	C_9H_{18}	$C_3H_7 \cdot C_6H_{11}$			301.1	149.5	0.767(20)
Isopropylcyclohexane	C_9H_{18}	$(CH_3)_3 \cdot C_6H_{11}$			302	150	0.787(20)
1-Methyl-4-isopropylcyclohexane	$C_{10}H_{20}$	$(CH_3)(C_3H_7) \cdot C_6H_{10}$			336.2–338	169–70	0.793(20)
1,3-Diethylcyclohexane	$C_{10}H_{20}$	$(C_2H_5)_2 \cdot C_6H_{10}$			336.2–339.8	169–71	0.796(22)
Cycloheptane	C_7H_{14}	C_7H_{14}	10.4	-12	244.6	118.1	0.811(20)
Ethylcycloheptane	C_9H_{18}	$C_2H_5 \cdot C_7H_{13}$	<-22	<-30	390.2	199	0.952(20)
Cyclooctane	C_8H_{16}	C_8H_{16}	57.7	14.3	298.4–300.2	148–9	0.835(20)

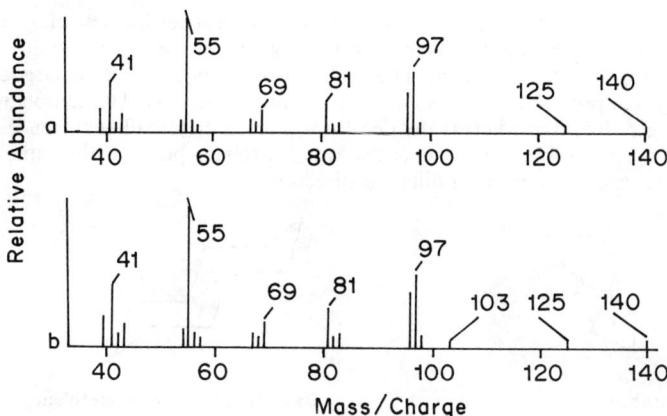

FIGURE 2. Mass spectra of (a) one isomer of *p*-menthane and (b) second isomer of *p*-methane. Reproduced by permission of the Society for Applied Spectroscopy from Reference 14

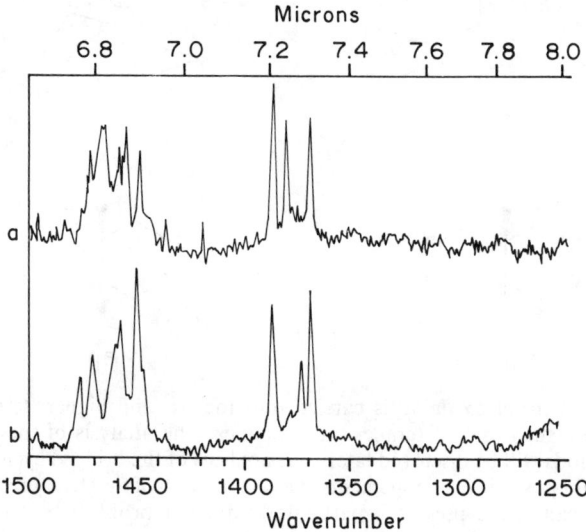

FIGURE 3. MI/FT-IR spectra, fingerprint region, of (a) one isomer of *p*-menthane and (b) second isomer of *p*-menthane, 2 ng each. Reproduced by permission of the Society for Applied Spectroscopy from Reference 14

our discussion of the use of MS (CI mode or GTA [group type analysis]) we will see that, in mixtures, this simplified approach might cause errors.

The majority of the polycyclic alkanes in nature derive from the terpenoid family (C_5 isoprene units) with the principal structural variations of six and five fused rings, formed by cyclization of the polyene open terpenoids (C_{10}, C_{15}, C_{20}, C_{25} ··· up to C_{40}).

300   Z. Aizenshtat

As good examples for the analytical approach for dicyclic terpanes we chose camphane and decalin. Camphane derives from its ketone precursor (i.e. camphor) and decalin is the fully hydrogenated naphthalene. The two bicyclic compounds exhibit large differences in their physical properties and also in their chemical reactivity. The camphane belongs to the 'bridged' bicyclics whereas the decalin is a fused dicycloalkane. Hence in decalin we have *cis, trans* isomers. In the *cis* the two hydrogens point in the same direction, whilst in the *trans* they point in different directions.

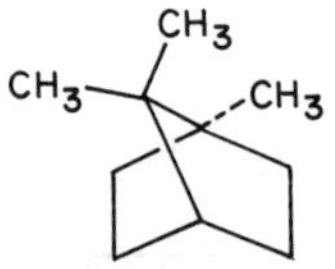

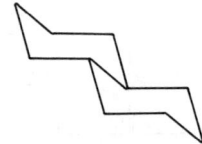

Camphane
C$_{10}$H$_{18}$
m.p. 156 °C, b.p. 161 °C

Decalin (decahydronaphtalene)
C$_{10}$H$_{18}$
cis m.p. −43 °C; *trans* m.p. −31.5 °C

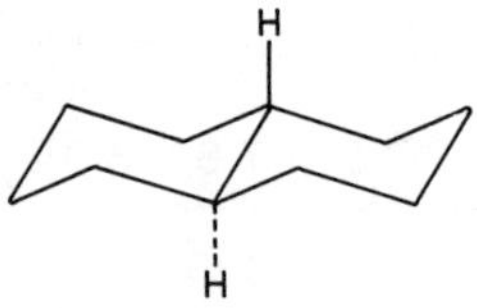

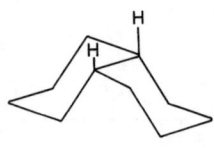

trans

cis

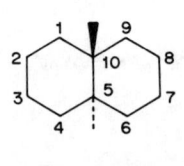

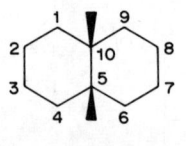

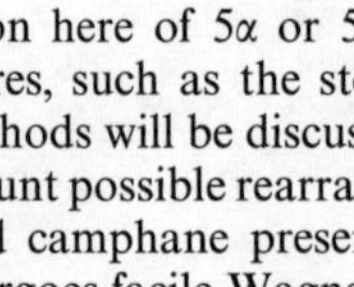

5α

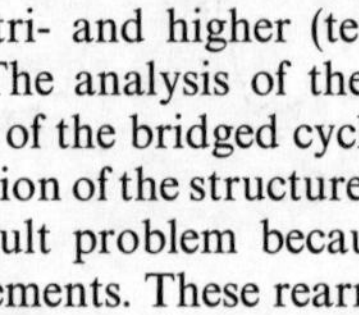

5β

The notation here of 5α or 5β is carried into the tri- and higher (tetra and penta) cyclic structures, such as the steranes and hopanes. The analysis of these by GC/MS and other methods will be discussed later. The analysis of the bridged cycloalkanes must take into account possible rearrangements. The elucidation of the structures of derivatives of pinane and camphane present a particularly difficult problem because the carbon skeleton undergoes facile Wagner–Meerwein rearrangements. These rearrangements are catalysed by acids or could occur also in the mass spectrometer, causing difficulties in fragmentation pattern recognition.

Analysis of tricyclic alkanes has been less reported in the literature, in comparison with mono-, di-, tetra- and penta-cyclics. However, since we intend to review also separation of fossil fuels, we have to mention compounds such as fichtelite (C$_{19}$H$_{34}$) and other diterpenoid tricyclics derived from abietic acid. Abietic acid and levopimaric acid belong to the resin acids, which exude from incisions of bark or trunk of high plants. The derived hydrocarbons, such as fichtelite, were found in the saturated fraction of peat bed extractions[3]. The identification and structure elucidation of fichtelite was based on m.p. 46.5 °C and optical activity $[\alpha]_D = 18°$, as well as on its resistance to chemical attack by various reagents. Early X-ray analysis[15] and mass spectra (MS) showed $m/z = 264$ and very precise elemental analysis yielded $\%H = 13.062$. Pyrolysis of the compound gave $C_{18}H_{18}$ (alkylaromatic). This method of aromatization for the analysis of polycyclic (fused) alkanes was used in many other studies.

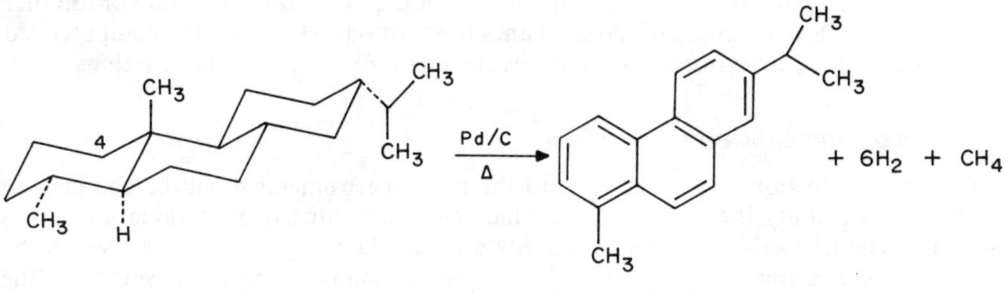

A completely different tricyclic family of compounds found in petroleum is that of derivatives of adamantanes ($C_{10}H_{16}$). The parent compound was found in petroleum, and some analytical chemists include the adamantanes in the terpene (terpanes) fraction, because alkyladamantanes could be synthesized from cholesterol, cholestane, abietic acid and cedrene by $AlBr_3$ or $AlCl_3$ 'sludge' catalysis[16]. This acidic catalysis also explains the fact that the amount of adamantanes is increased in crude oils by clay-oil heating experiments[17]. Despite its very high melting point (268 °C), adamantane and its alkylated derivatives distill at 70–80 °C/3 mmHg, and on cooling it forms cubic crystals.

III. SINGLE COMPOUND ANALYSIS

The chemical characterization of alkanes and cycloalkanes by general methods is very difficult. No suitable chemical methods for the conversion of alkanes to analytically useful derivatives have been found; they are typically inert or undergo indiscriminate reactions which produce inseparable mixtures.

In each homologous alkane or cycloalkane family, the physical constants such as boiling point and refractive index are very useful. Specific gravity measurements even in the capillary mode still need relatively large quantities of compound. The most employed method for tentative identification and relative quantification of hydrocarbons is gas chromatography (GC). This method will be discussed in detail for separation of mixtures.

Among the spectroscopic methods, MS is perhaps the most useful for the identification of alkanes and cycloalkanes. A simple computer search of reference spectra of thousands of compounds greatly enhances the utility of the method. The NMR method for single compounds is very helpful in structure determination for small molecules. However, the larger the saturated hydrocarbon, the less absolute information can be obtained by this spectroscopical method. Both two-dimensional ^1H NMR and ^{13}C NMR can furnish decisive information as to the structure of cycloalkanes.

As we have seen, cyclopropanes are exceptional in their chemical behaviour and also the hydrogen in the three-membered ring is uniquely high-field.

Despite the overall similarity of the IR spectra of most alkanes, a very careful examination of the 2900–3000 cm^{-1} and 1300–1500 cm^{-1} ranges typical for C—H bonds can yield certain structural information.

For the solid, mostly polycyclic alkanes, X-ray structural analysis can be employed for pure compounds and combined with either very accurate elemental analysis or MS study to produce very important information.

A. Mass Spectrometry of Alkanes and Cycloalkanes

The application of mass spectrometry, including electron impact (EI) studies, to the analysis of hydrocarbon fractions obtained from petroleum is as old as the earliest exploitation of the technique. Almost all of the ionization methods known have been applied to saturated hydrocarbons. The number of papers dealing with one or another aspect of the MS of alkanes and cycloalkanes is enormous. However, the computerized compilation of the mass spectra in libraries such as NBS [18] makes the search easier[19].

1. Ionization and appearance potential

By the introduction of a molecule into the mass spectrometer it will be charged. In most cases we study the positive ions, which can be produced by various ionization methods. The EI method is most commonly used for the analysis of alkanes. When the energy of the electron beam is greater than the ionization potential of the substrate, the following processes may occur:

Ionization: $\quad M + e^- \longrightarrow M^{\cdot+} + 2e^-$

Fragmentation: $\quad M^{\cdot+} \longrightarrow A^+ + B\cdot radical$

$$A^+ \longrightarrow C^+ + D \text{ neutral}$$

Rearrangement: $\quad M^{\cdot+} \rightarrow \text{Rearranged } M^{\cdot+} + \text{(neutral fragment)}$

Fragmentation could happen also directly from M. The lowest energy for which the ionization occurs is the ionization potential of M, whereas the electron beam energy necessary to cause a particular fragment to appear is the appearance potential.

Table 3 gives IP values and heats of formation[21] for selected alkanes and cycloalkanes. It is clear that the IP differences are not large enough to render these as diagnostic tools for the analysis of alkanes and cycloalkanes. Since the IP measures the energy needed

TABLE 3. Gas-phase ion and neutral theromochemistry[21]

Ion/neutral	Ionization potential (IP) (eV)	$\Delta H_f(\text{ion})$ (kcal mol^{-1})	$\Delta H_f(\text{neutral})$ (kcal mol^{-1})
CH_4^+/CH_4	12.51	271.0	-17.8
$C_2H_6^+/C_2H_6$	11.52	245.6	-20.1
$C_3H_6^+$/Cyclopropane	9.86	240	$+12.7$
$CH_3CH{=}CH_2$	9.73	229	$+4.8$
$C_3H_8^+/C_3H_8$	10.97	227	-25.0
$C_4H_8^+$/Cyclobutane	(9.92)	(235)	$(+6.8)$
$CH_2{=}CHCH_2CH_3$ E or Z	9.1	207–208[a]	-1.9
			-2.9
$C_4H_{10}^+$/ n-C_4H_{10}	10.53	213	-30.0
iso-C_4H_{10}	10.57	212	-32.1
$C_5H_{10}^+$/Cyclopentane	10.51	224	-18.7
$C_5H_{12}^+$/n-C_5H_{12}	10.35	204–211[a]	$-35.0 - 27.3$[a]
iso-C_5H_{12}	<10.22	199–207[a]	$-36.7 - 28.4$[a]
Neopentane	<10.21	195–203[a]	$-40.0 - 32.4$[a]
$C_6H_{12}^+$/Cyclohexane	9.86	198	-29.5
C_6H_{14}/n-$C_6H_{14}^+$	10.13–10.1	188–194[a]	$-43.0 - 31.0$[a]
iso-$C_6H_{14}^+$ all isomers			
$C_{10}H_{22}^+$/n-$C_{10}H_{22}$	9.65	163	$-40 - 50$[a]

[a]Range of several experimentally determined values.

for the expulsion of an electron from the alkane molecular orbitals, it does not change with the expected isomeric structure-stability of the carbocation (see C_5H_{12} isomers Table 3). However, the information is valid for an understanding of ionization processes such as photoionization (see Section VI, PID—photoionization detector). Different appearance potentials can be recorded for different fragment ions of the same molecule. The energy required for fragmentation may originate in the electron beam. Since MS are usually obtained at 70 eV (EI), considerable excess energy is imparted to the initially formed ion, causing fragmentations and rearrangements, depending on the alkane or cycloalkane structure.

The connection between fragmentation patterns and the formation of carbocation structures can be related to their heat of formation (see Table 3). The heat of formation of C_1—C_4 alkylcarbonium ions was calculated in good agreement with the experimental values[22,23]. The most stable forms of $C_3H_7^+$, $C_4H_9^+$ and $C_5H_{11}^+$ ions are predicted to have a bridged protonated cyclopropane structure:

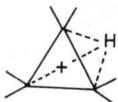

This stability of the 'non-classical carbonium ion' or its open $C_3H_7^+$ fragment makes the m/z 43 fragment the base peak even for higher normal or branched alkanes. The assumption that alkanes, which are the simplest of all organic compounds, will hence produce easy-to-interpret MS was proved erroneous[20].

2. Fragmentation patterns as related to structure

Saturated hydrocarbons produce fragmentation patterns which could be correlated with their structures only with some difficulty. A frequently cited example is the generation of the ethyl ion from isobutane [$CH_3CH(CH_3)_2$]; the $CH_3CH_2^+$ ion cannot be formed by simple bond cleavage but must involve one or more rearrangements[20]. At high energies (>70 eV), various theories were proposed to explain such drastic cleavage of bonds or even milder rearrangements.

Normal mass spectra of alkanes obtained in the EI mode (70 eV, and heated source) exhibit a characteristic and seemingly simple profile. The fragmentation pattern shows groups of peaks (m/z) spaced by 14 mass units corresponding to the CH_2 group. The fragments formed will be C_nH_{2n+1} ions having increasing abundance with decreasing carbon number (see Figure 4). The molecular M^+ ion under these conditions is quite low, and in most n-alkanes C_3 is the base peak. Unfortunately, most papers do not show the C_3 peak, which may be four to five times higher than, e.g., the '100%' C_4 peaks shown in Figures 4a and b (see also the chapter on MS by T. Gaumann in this volume). For long-chain normal alkanes the $C_6H_{11}^+$ and lower fragments are the highest (20–25% relative intensity to the $C_3H_7^+ = 100\%$). The use of these fragments in mass fragmentometry in computerized gas chromatograph MS (GC/MS/C) will be discussed later in this review.

In the case of MS study of n-alkanes, very little use has been made of other modes of ionization such as chemical ionization (CI) or photoionization (PI), field ionization (FI) and others. This is because the sensitivity of these methods and the quantity of the produced ions are very low and the molecular information is minimal. The MS of branched alkanes is more structure-dependent, and the fragmentation pattern follows the position of branching. The abundance of C_nH_{2n+1} fragments can be related to the

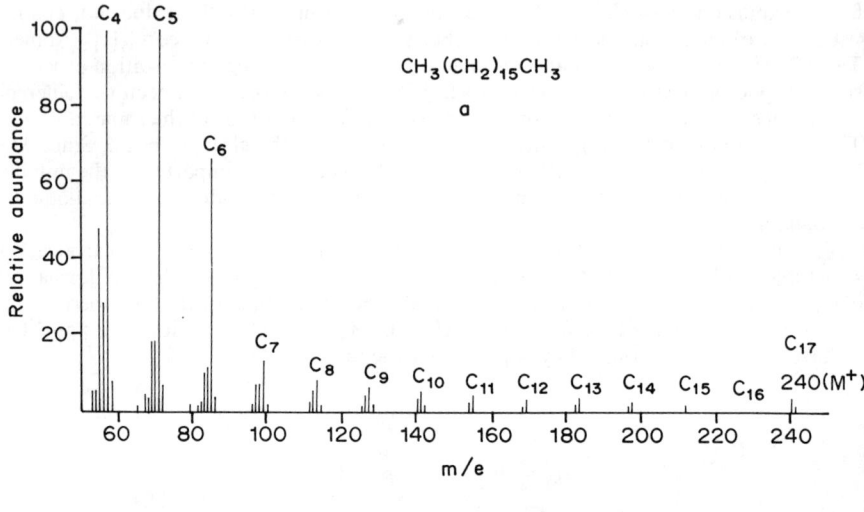

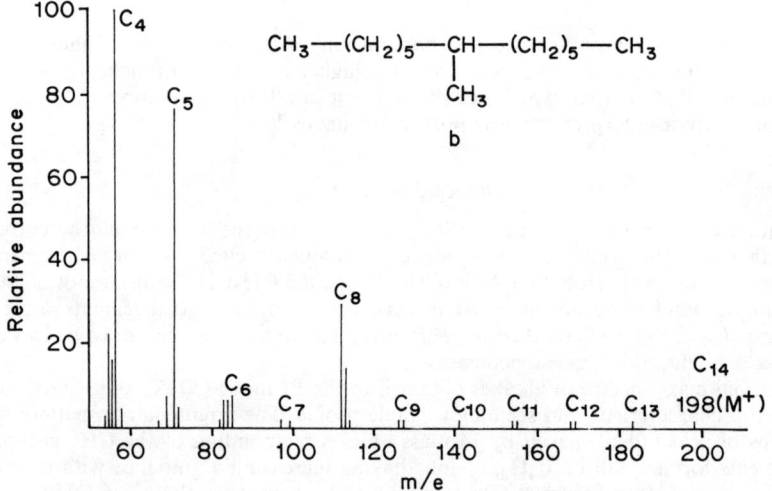

FIGURE 4. (a) Mass spectrum of n-$C_{17}H_{36}$ (MW = 240). (b) MS of 7-methyl-tridecane (branched $C_{14}H_{30}$, MW = 198). Note the m/z 14 units fragmentation for the normal alkane in comparison to the domination of C_8, C_5 in (b)[3]

relative stability of the carbocation.

$$[R_3C\!-\!CR_3]^{\cdot+} \longrightarrow R_3C^+ + \cdot CR_3$$

$$R = H, \text{ or alkyl}$$

stability: tertiary > secondary > primary

Hence, branching changes the pattern of fragmentation favouring the tertiary ions (or quaternary ions from neoalkanes) in accordance with the stability of the carbocation

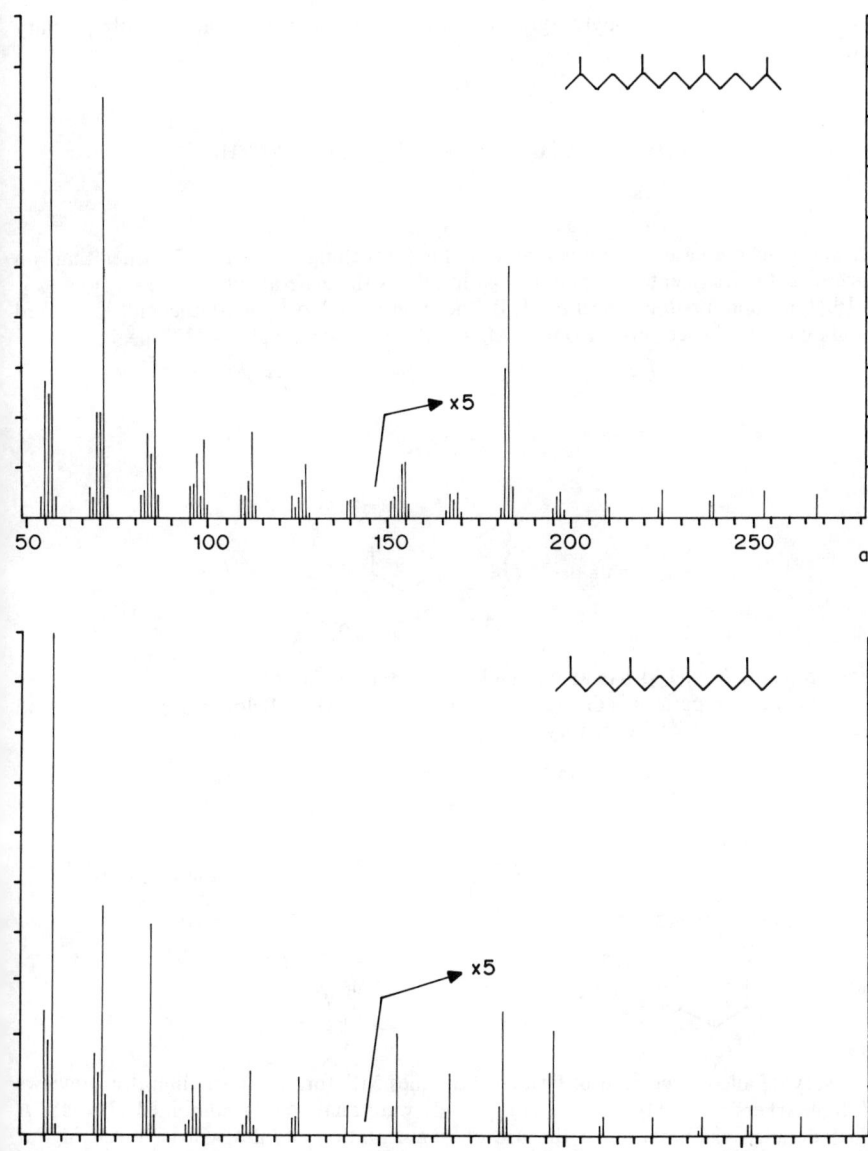

FIGURE 5. (a) Mass spectra of 2,6,10,14-tetramethylpentadecane (pristane), MW = 268, and (b) 2,6,10,14-Tetramethylhexadecane (phytane), MW = 282. Both spectra were taken under 70 eV EI[26]

formed. It is thus possible to deduce the position of branching from different fragmentation distributions from different structural isomers of the same molecular weight. An apparent exception are the 2-methylalkanes (iso-branched), which show preferred loss of an isopropyl fragment and retain the charge on the formally primary ion[24]:

$$C_nH_{2n+1} - CH_2CH \overset{CH_3}{\underset{CH_3}{\diagup\diagdown}} \xrightarrow{-e} C_nH_{2n+1}CH_2{}^+ + C_3H_7\cdot$$

However, in the case of ante-iso compounds (branching on carbon 3 rather than on carbon 2) the fragmentation pattern again follows the general rules.

Phytane and pristane (Section II.B and Figure 5)[26] belong to the multibranched isoalkanes and hence show a small $[M-15]^+$ and a strong $[M-43]^+$ peak:

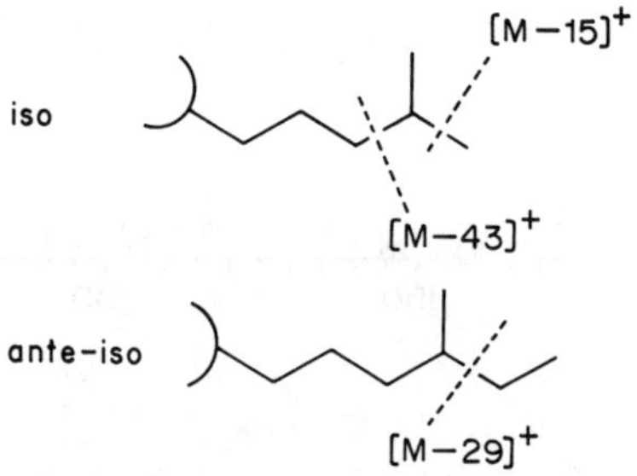

Since phytane ($C_{20}H_{42}$) has at one end an iso and at the other an ante-iso structure, its fragmentation pattern is different from pristane ($C_{19}H_{40}$). Below mass (m/z) 113 both fragmentation patterns seem very similar.

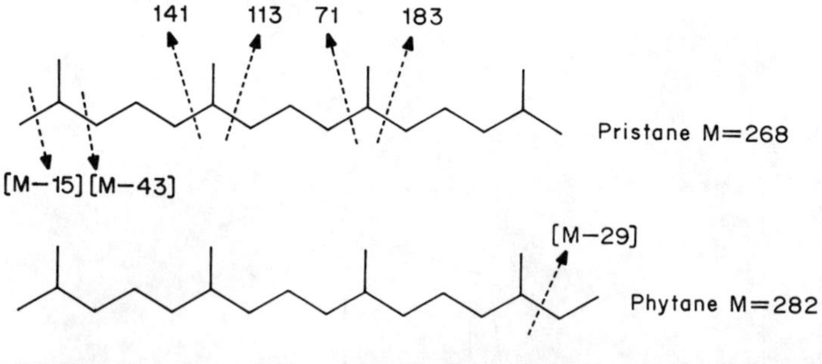

A very detailed discussion of branched alkanes MS appears in the literature reviews of 'biomarkers'[25,26]. Most of the compounds were analysed by combined GC/MS. A more detailed examination of the use of fragmentation profiles will be given in our discussion of separation methods and petroleum analysis (Section VII. A).

3. Cycloalkanes

The analysis of cycloalkanes by MS gives information not only about the number of rings but also relates to the mode of connection (fused or isolated) and stereochemistry. The structure of cyclic alkanes cannot be derived based solely on the fragmentograms (m/z profile) without consideration of eliminations and rearrangements. Even when we consider monocyclics such as cyclopentane[27] or cyclohexane[28], the complexity of the

MS fragmentation becomes evident. In order to be able to understand the fragmentation, ^{13}C and D labelling of these cycloalkanes was investigated. Also, low eV spectra were taken and alkyl derivatives measured. It is therefore suggested that the loss of ethylene is not directly from the cycloalkane but rather an elimination from the open-ring radical ion formed ($M^{\cdot +}$), so that only 6% of the ethylene is formed by the external CH_3 group[28]. This suggested model would also explain the various migrations of hydrogen (or D, if labelled).

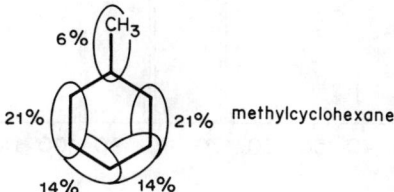

The same phenomenon was studied also in alkylated (R = Me, Et) cyclopentanes and 80% of the ethylene formed was found to originate in the ring carbons.

However, in the case of long-chain alkylated cyclopentanes, the chain fission dominates the spectra. In fossil fuels the monoalkylated cyclohexanes are characterized by the m/z 83 fragment[26]. One should be aware of the problem that fragmentation of alkylated cycloalkanes could be followed by rearrangement of the ions formed; hence m/z 83 could also be formed from the methylcyclopentyl ion[25]. Of course we must bear in mind the fact that the molecular weight of monocyclic alkanes is the same as for mono-unsaturated alkenes and, when investigating mixtures, this point is of importance (see also the discussion on fossil fuels, Section VII.A).

Cis, trans isomers of disubstituted monocyclics generally exhibit similar MS profiles. However, subtle differences in fragmentation (relative abundance) profiles were found, e.g. in cis- and trans-1,2-dimethylcyclohexane[29].

Bicyclic and higher polycyclic alkanes exhibit much more complex MS fragmentation. Budzikiewic and coworkers[20] caution that great prudence, even in the use of published spectra, must be exercised for comparison and identification purposes of polycyclic alkanes. As examples they show the mass spectrum of cis-hydrindane, in which m/e 95, 96, 81 are major fragments which can be assigned only in retrospect to the structure once known. However, given only the MS it will be impossible to reconstruct the structure of the molecule.

In Section II.D we discussed the decalines. Though the cis and trans isomers show quite similar MS, there are very noticeable quantitative differences in abundances[29] (Figure 6). The MS becomes still less characteristic in more complex structures, such as terpanes (Figure 7a,b[30], Figure 8[31]). All three sesquiterpanes shown have molecular weights of 208 but the fragmentation profile is quite different, e.g. the base peaks are m/z 109, 123 and 193, respectively. The mass spectrum of adamantane is unlike most saturated hydrocarbons; the base peak at m/z 136 is the molecular peak. This reflects the inherent rigidity of the interlocking rings, where effective fragmentation of the $M^{\cdot +}$

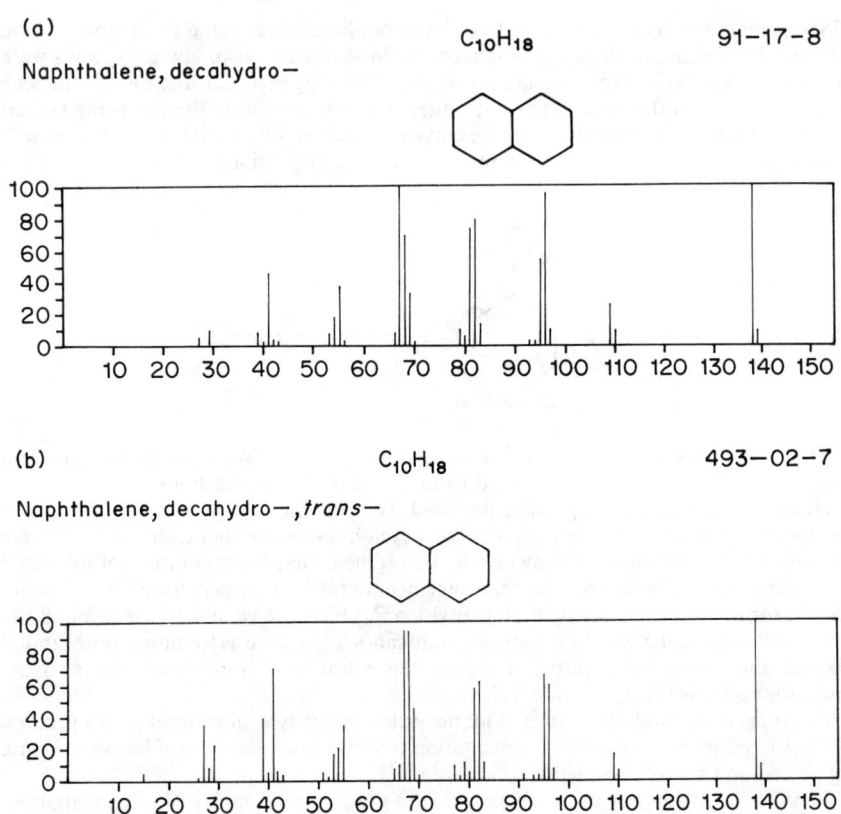

FIGURE 6. Mass spectra of probably *cis* (a) [Chemical Abstract Search number 91-17-8] and *trans* (b) [CAS 493-02-7] decahydronaphthalene (decaline), as available in the MS Library (NBS)[18]

requires that three C—C bonds be cleaved. Hence, the M − 15 and M − 28(29) which would have been abundant in acyclic or monocyclic alkanes are *not* preferred. Figure 9[32] presents the MS of adamantane showing C_3 and C_4 losses as the favoured pathways[33,34]. It is not easy to explain the fragmentation patterns of adamantane. However, in 1-hydroxyadamantane the key to the fragmentation mechanism is the formation of a protonated phenol[32]. A more direct suggestion for understanding the influence of structure on MS fragmentation was given for diamantane, $C_{19}H_{20}$. This compound is pentacyclic with m.p. 244–245 °C (see Figure 9[32]). The proposed fragmentation pathway[32] shows that m/z 173, 146, 106, 105 and 91 are all fragments containing a benzene ring. This aromatization during fragmentation is suggested for all diamonoids[32].

The mass spectra of 4-, 3- and 1-methyl-diamantanes, which were found and identified in petroleum, exhibit the M˙+ m/z 202 molecular ion to be strongest for the 3-methyl. However, for all three isomers the $[M − 15]^+$ m/z 187 is the base peak[35].

The $C_{20}H_{36}$ tricyclic diterpanes are also called diterpane resins, derived from the abietane skeleton. Fichtelite ($C_{19}H_{34}$), which belongs to the same family, is presumed to be the result of decarboxylation of the abietic acid. The MS of these relate to individual structures[36] (see Figure 10a–e[37−39]).

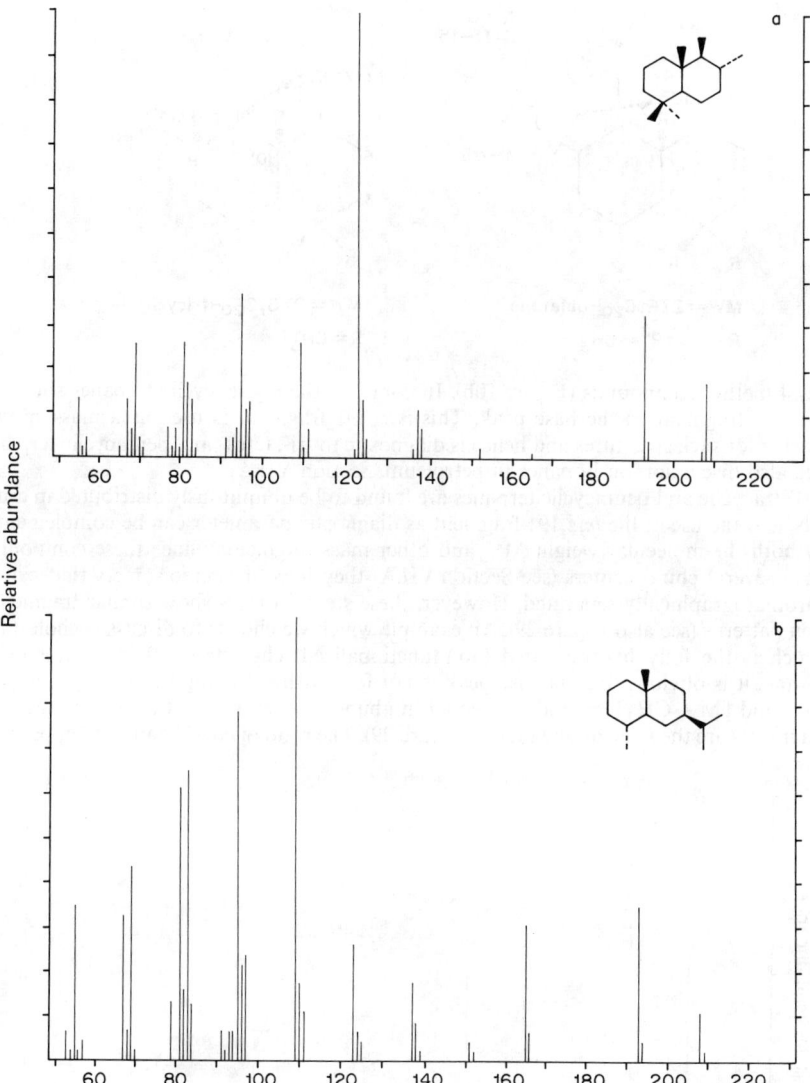

FIGURE 7. Mass spectra of two bicyclic sesquiterpanes ($C_{15}H_{28}$), (a) $8\beta(H)$-Drimane (MW = 208) with base peak m/z 123 and (b) $4\beta(H)$-eudesmane (MW = 208) with base peak m/z 109[30,26]

The mass spectra of tricyclic alkanes relate to the basic structure of the biogenic precursor. The abietane fragmentation shows both m/z 191 and 163 fragments as well as the cleaving of the isopropyl group [M − 15] and [M − 43] (m/z 261 and 233) whereas the norpimarane shows the base peak at m/z 233 [M − 29] due to ethyl group cleavage (Figure 10c). The formation of the m/z 123 and m/z 109 fragments is enhanced for

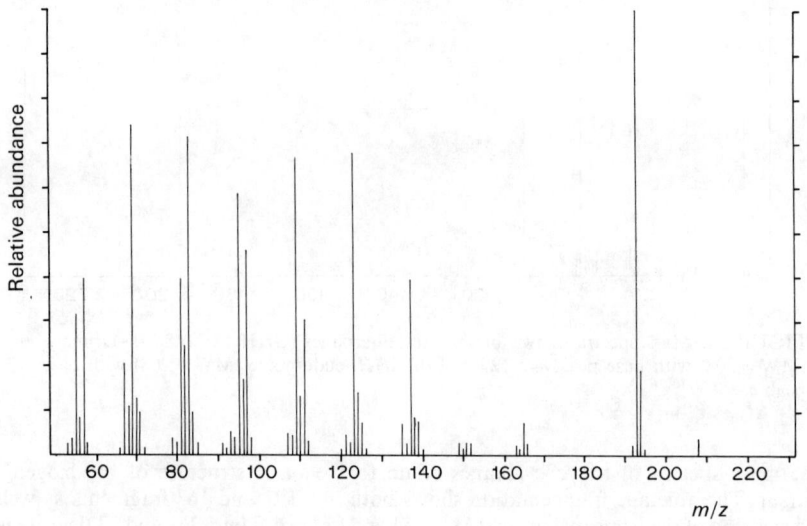

MW = 276, C$_{20}$-abietane   MW = 276, C$_{20}$-tricyclic terpane

R = R^1 = R^2 = CH$_3$ R = CH$_3$

des-4-methyl compounds (Figure 10b). In contrast, the C$_{20}$-tricyclic terpanes show the m/z 191 fragment as the base peak. This m/z 191 fragment is used as a mass spectra marker for such structures and hence is diagnostic to tri-, tetra- and pentacyclic terpanes (see also discussion on hopanes in petroleum, Section VII.A).

Tetracyclic and pentacyclic terpanes are found to be ubiquitously distributed in crude oils, and the use of the m/z 191 fragment as diagnostic parameter can be complemented by both the molecular weight M$^+$ and other mass fragments. Since these compounds have several chiral centers (see Section VII.A) they have diastereoisomers that can be chromatographically separated. However, these stereoisomers show similar fragmentation patterns (see also Figure 29). An example which we choose to discuss is cholestane, which is the fully hydrogenated (non-functionalized) cholesterol. Being a tetracyclic alkane, it is obvious that its base peak is not formed by cleaving the alkyl group. The M$^{.+}$ and [M − CH$_3$]$^+$ are almost equal in abundance. However, the m/z 218 and 217 fragments are the most abundant (see Figure 29). The ratio of m/z 218 and 217 apparently

FIGURE 8. MS of crude oil (Loma Nova) unknown C$_{15}$ bicyclic sesquiterpane with base peak m/z 193 and MW = 208[31,26]

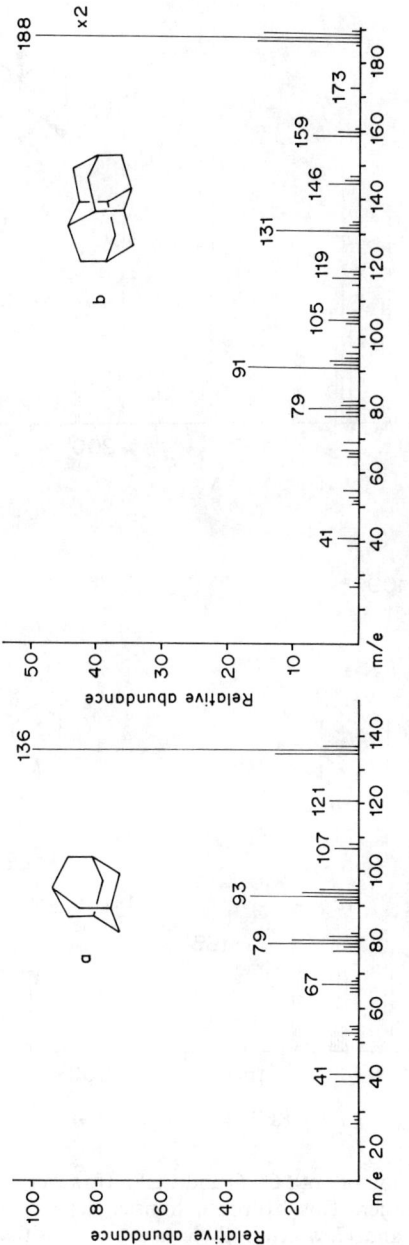

FIGURE 9. The 70 eV MS of adamantane (a) and diamantane (b)[32]

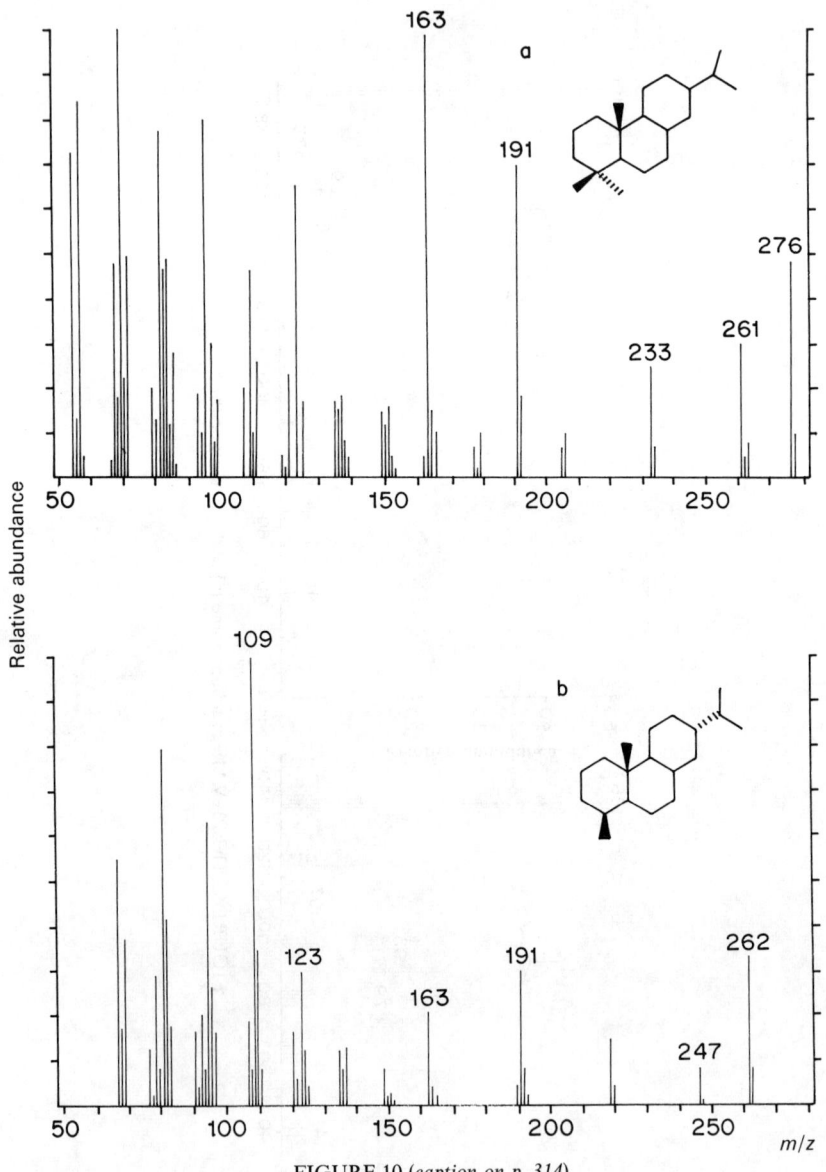

FIGURE 10 (*caption on p. 314*)

relates to the conformation of both C(16) and C(18). However, C(18) can be oriented to facilitate hydrogen transfer. This hydrogen transfer depends on the conformational positions of carbons 14 and 17, which in cholestane are α-positioned. Hence the m/z 217 dominates, whereas in the cases of 14β and 17β the m/z 218 fragment will dominate. This influence of the structure of diastereoisomers on the mass fragmentometric profile can be utilized in the study of mixtures (see Figure 29a–e).

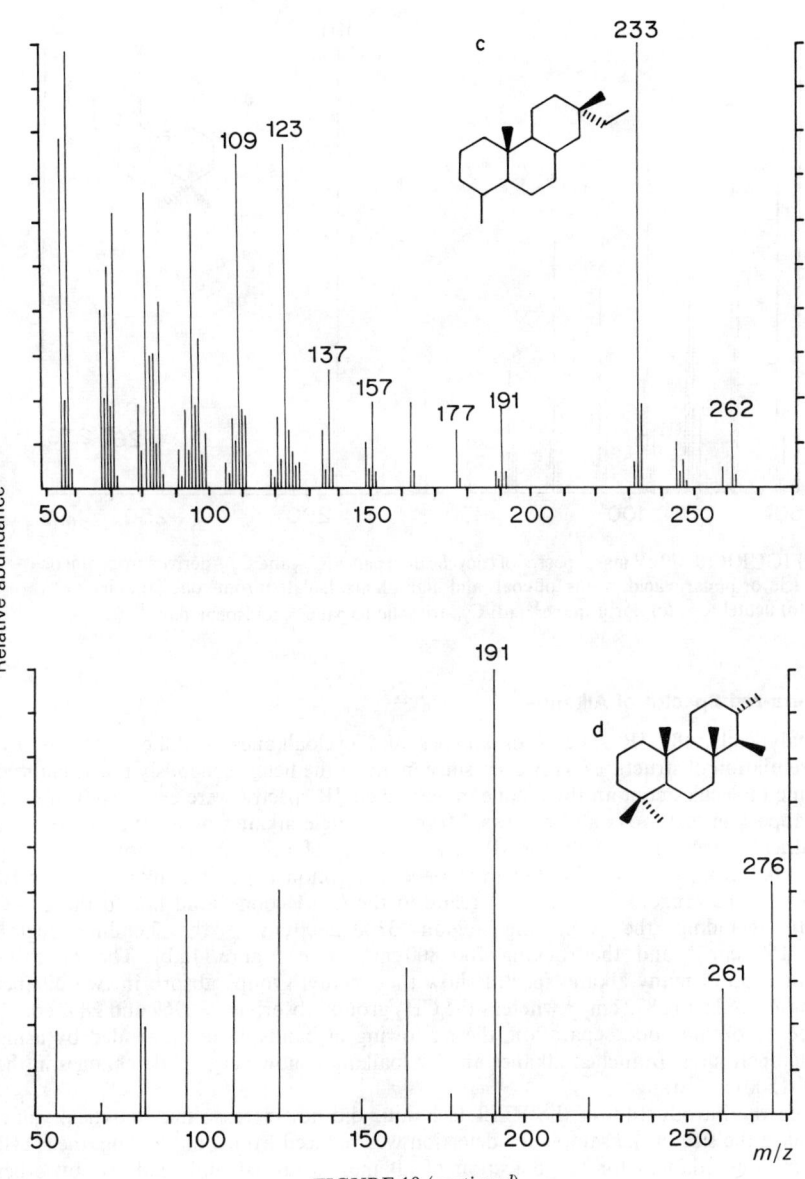

FIGURE 10 (*continued*)

Steranes are not found as such in living organisms but carry the biogenic information by the variability of structures and enantiomers. The MS, and especially the mass fragmentogram produced by a combination with other analytical methods (GC, HPLC, etc.), was proven to be a very powerful tool for both source and maturation determination of fossil fuels. The various analytical methods will be discussed when considering hydrocarbon mixtures, especially petroleum and sedimentary organic extracts.

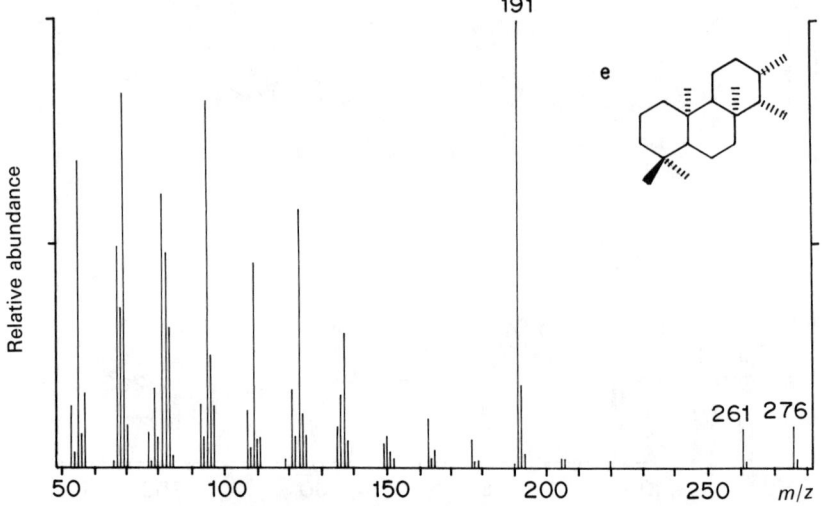

FIGURE 10. 70 eV mass spectra of tricyclic diterpanes (C_{20} and C_{19}) derived from abietic acid or pimaric acid, resins, of coals and high plants. Isolated from coal: (a) abietane[36]; (b) fichtelite[37]; (c) norprimane[36]; (d) C_{20}-tricyclic terpane[38], (e) isocopalane[39,26]

B. Infra-red Spectra of Alkanes

Analytically, the IR spectra of alkanes and cycloalkanes is difficult to use for differentiation of structures. However, since many of the heterogeneously functionalized organic molecules contain the alkane moiety, their IR spectra were extensively studied and appear in textbooks and reviews. Moreover, single alkanes or mixtures were used as either matrix or solvents for IR measurements of other compounds, containing functional groups. Most of the characteristic absorption bands for alkanes in the IR spectra in the range 4000–400 cm^{-1} relate to the C—H bond (and not to the C—C bond), including the stretching region 3300–2500 cm^{-1}, the bending region 1300–1500 cm^{-1} and the rocking 700–800 cm^{-1} (see Figure 11a,b). The extensive examination of many alkane spectra show that methyl groups absorb in two distinct bands at 2962 and 2872 cm^{-1} whereas the CH_2 groups absorb at ≈ 2962 and 2853 cm^{-1}. Hence, to obtain good separation, the narrowing of bands is recommended by using low temperatures. Branched alkanes and cycloalkanes show very subtle changes in the 1400–1500 cm^{-1} range.

Even the introduction of the FTIR technique did not increase the analytical value, but since the size of the sample for detection was reduced from 1 mg to 1 ng, the FTIR method was adapted for the detection of alkanes produced and analysed by other analytical methods, i.e. pyrolysis-FTIR[40] (see also Sections IV and V). Specific gas alkane molecules such as: CH_4, C_2H_6 and C_3H_8 are individually identified, therefore the use of FTIR gas cell permits continuous monitoring of these, in flight. This system, although designed for flow through cell detection, can be employed for remote sensors.

Since lowering of temperature will narrow the absorption bands, it will produce better resolution, which will allow one to identify single compounds and isomers. This leads to the technique of matrix isolation infrared spectrometry (MI/IR) which was used extensively to study the IR properties of molecules[41,42]. In this technique the molecules

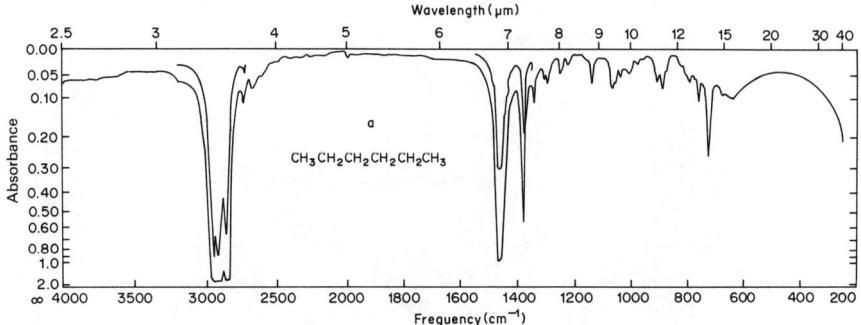

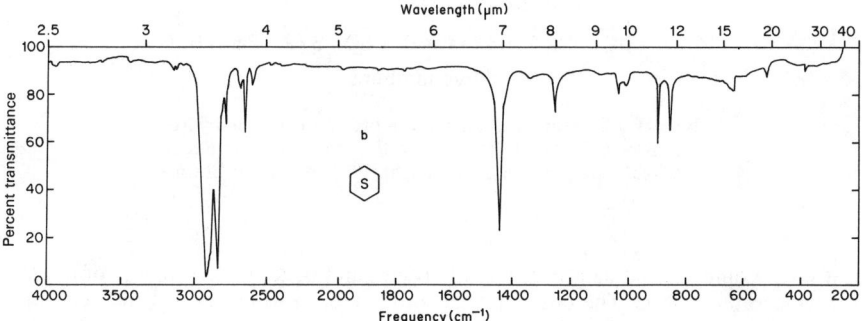

FIGURE 11. Infra-red low-resolution 'textbook' spectra of (a) a typical normal alkane (n-hexane) and (b) a cycloalkane (cyclohexane)

to be studied are co-condensed with an excess of inert gas, i.e. argon, and trapped in the argon matrix (10–15 K). Different molecules do not interact with each other and very small interaction with the argon is expected.

The MI/FTIR method can be combined with GC[45]. This was successfully done[43] even with capillary columns (see Sections II.C and III.B). Reedy and coworkers[43] concluded that straight-chain alkanes, for which vapour or liquid phase IR alone are not sufficient for positive identification, show well-resolved and unique bands in MI/FTIR. Alkanes, being non-polar, show low levels of infra-red absorption, and in the homologous series the spectral differences are very small. Figure 12[43] shows the 1400–696 cm^{-1} region for n-heptane and n-octane produced by GC/MI/IR. The claim that positive identification can be made is based on the shifts in the 900 cm^{-1} range, made possible due to MI narrowing of the bands. However, the separation of n-heptane from n-octane is easily carried out by GC even when equipped only with a packed column.

The real test of single compound MI/FTIR was to differentiate between p-menthane-based conformational isomers[14,44]. Most of the acyclic and cyclic terpenoids examined by this method were functionalized (terpanols, terpenes) but p-menthane was also studied. The isomers of each compound were investigated also by NMR and MS. The conclusion of this study is that the MI/FTIR method is very good for the identification of positional, conformational and configurational isomers. The lowest level of detection is in the picogram range. Moreover, future developments are forecast, such

Z. Aizenshtat

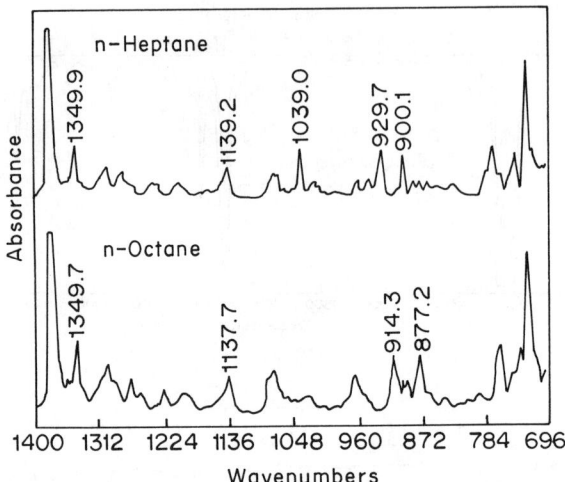

FIGURE 12. Discrimination between two straight-chain hydrocarbons by GC/MI-IR. Reprinted with permission from Reedy
et al. Anal. Chem., **57**, 1602. Copyright (1985). American Chemical Society

as a wider computerized data base for MI spectra and better understanding of the MI
low-temperature effects. The one point the authors failed to mention is the very high
cost of this analytical equipment.

C. ¹H and ¹³C NMR

Only the analytical aspects of this method will be discussed in this section, since the
other aspects of NMR are treated in another chapter of this volume. The basic
considerations for the structure determination of cyclic alkanes by NMR are as old as
the method itself. However, with the development of dynamic NMR measurements,
conformational analysis became available[46]. As an example, this method was employed
to find the cyclohexane inversion parameters ($9.5\,\text{kcal}\,\text{mol}^{-1}$ for C_6H_{12} and
$11.3\,\text{kcal}\,\text{mol}^{-1}$ for C_6HD_{11}). These energy barriers were found to depend also on the
solvent. For example, NMR spectroscopy established that only *trans*-decalin exists in a
single rigid form, whereas the *cis* form is 'highly mobile' even at quite low temperatures[47].

As an analytical tool, NMR spectroscopy of ¹H or ¹³C is best used for single compound
structure determination. Despite the introduction of the FT multispectra unit, the sample
size is only in the mg range. With the continued development of high-resolution NMR,
more and more complex structures could be elucidated[48,49]. The elucidation of the stereostructure of the 5α,14α androstane by 2D-NMR data is an excellent example of the study
of a complex polycyclic alkane of the steroid family[50]. Croasmun and Carlson[50] review
the possible use of the various 2D-NMR methods showing ¹H–¹H spin coupling, NOE
and 2D-NMR NOESY spectra (see Figures 13 and 14, and Table 4). One can see that
in the 1D (regular) NMR spectrum of androstane, even when obtained at 200 MHz
(Figure 13b), the resolution is too low. Therefore, for sterane 2D-NMR data we can
apply only 300 or even 500 MHz spectra, which has a good enough resolution.

The structure of any solid crystal-forming compound can be examined also by X-ray
analysis. The conformation is true for the solid and not for the compound in solution. The

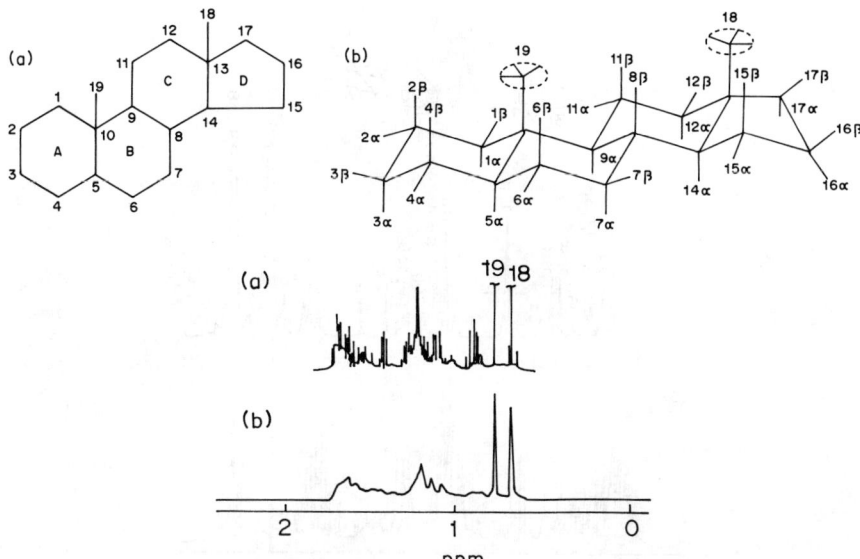

FIGURE 13. Top. Chemical structure–2D NMR connectivity diagrams: (a) skeletal chemical structure of androstane corresponds to connectivities revealed in its 2D INADEQUATE spectrum; (b) 5α,14α-androstane with H–C connectivity revealed by its ^{1}H-^{13}C heteronuclear shift-correlated, HETCOR, spectrum. Bottom. ^{1}H NMR spectra of 5α,14α-androstane: (a) measured at 500 MHz with resolution enhancement; (b) measured at 200 MHZ (adapted from Bax and Morris). Reprinted by permission of VCH Publishers, Inc., from Reference 50

TABLE 4. Reported ^{1}H chemical shifts for resonances within the steroid spectral 'hump' (B-, C- and D-ring assignments 2D NMR). Reprinted by permission of VCH Publishers, Inc., from Reference 50.

	Proton	Steroid							
		1	2	3	4	5	6	7	9
		^{1}H chemical shifts (ppm)							
	6α	2.36	2.23	na	2.80	2.58	1.75	1.59	2.28
	6β	2.47	2.48	1.82	2.80	2.44	1.30	1.83	2.41
B-ring[a]	7α	1.01	1.06	0.40	2.10	1.14	1.05	1.26	1.07
	7β	1.96	2.00	1.34	1.68	1.95	1.79	1.59	1.87
	8(β)	1.67	1.98	1.14	1.73	1.52	1.30	1.27	1.57
	9(α)	1.04	1.00	0.53	na	0.93	0.79	1.52	0.99
	11α	1.77	—	1.22	2.58	2.00	1.96	1.91	1.65
C-ring[a]	11β	1.68	na	0.98	1.71	1.40	1.31	1.29	1.46
	12α	1.09	1.65	1.75	2.25	1.86	1.84	1.93	1.46
	12β	1.18	2.21	1.29	1.88	1.73	1.69	1.74	2.08
	14(α)	9.95	1.11	1.47	2.44	1.66	1.66	1.78	1.81
	15α	1.61	1.75	1.40	1.73	1.77	1.74	1.76	1.72
D-ring[a]	15β	1.33	1.33	1.01	1.38	1.45	1.40	1.41	1.27
	16α	2.07	1.69	1.88	1.99	2.30	2.28	2.29	1.68
	16β	1.47	2.17	—	2.32	2.04	na	na	2.19

[a]See structure in Figure 13 (top).

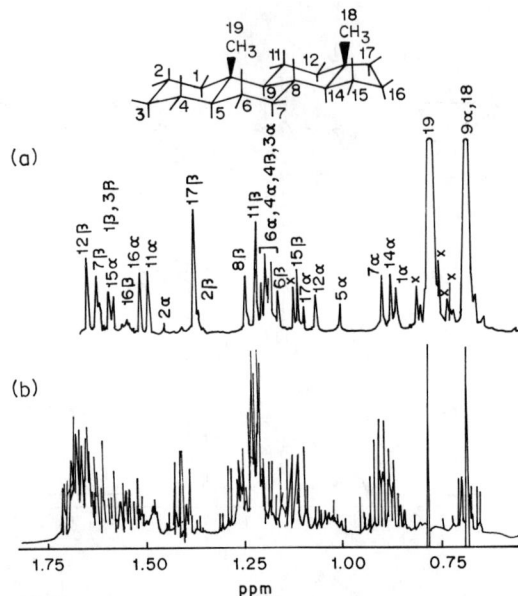

FIGURE 14. (a) 2D J-projection of 5α,14α-androstane with final assignments noted. (b) The corresponding 1D resolution-enhanced spectrum of androstane. Both spectra were run at 500 MHz with CDCl₃ as solvent. Reprinted by permission of VCH Publishers, Inc., from Reference 50

3D-NMR data provide the connectivity and long-range space interaction between methyl hydrogens and ring hydrogens. Both methods are limited to isolated and purified single compounds.

X-ray analysis is discussed in the chapter on structural chemistry in this volume.

IV. SEPARATION OF SATURATED HYDROCARBONS FROM MIXTURES

The major source of saturated hydrocarbons (alkanes and cycloalkanes) in the geosphere is petroleum. Other fossil fuels, i.e. coal and bitumens, may have small fractions of alkanes and cycloalkanes.

The C_1 to C_5 gaseous fraction of oil is mostly comprised of alkanes. Natural gas reservoirs are rich in methane whereas gas in contact with oil can contain up to 40% C_2–C_5 gases ('wet gas'). The refinery industry is interested in distilation fractions rather than the separation of the hydrocarbon families into saturates, aromatics, N, S and O containing compounds, etc. The boiling range of the components of petroleum is shown in Figure 15[51]. It is obvious that in each temperature-range 'cut' there is a mixture of hydrocarbons. Moreover, distillation at atmospheric pressure of the high boiling fractions is accompanied by cracking, which produces alkenes (mostly 1-ene) and, at higher temperatures, produces aromatics by pyrolysis. This is the reason that distillates of petroleum are limited to 390 °C and that hydrocarbon group-type analysis (determination of saturates, aromatics, etc.) is now carried out by various chromatographic methods. These techniques and the various detectors, qualitative and quantitative modes of

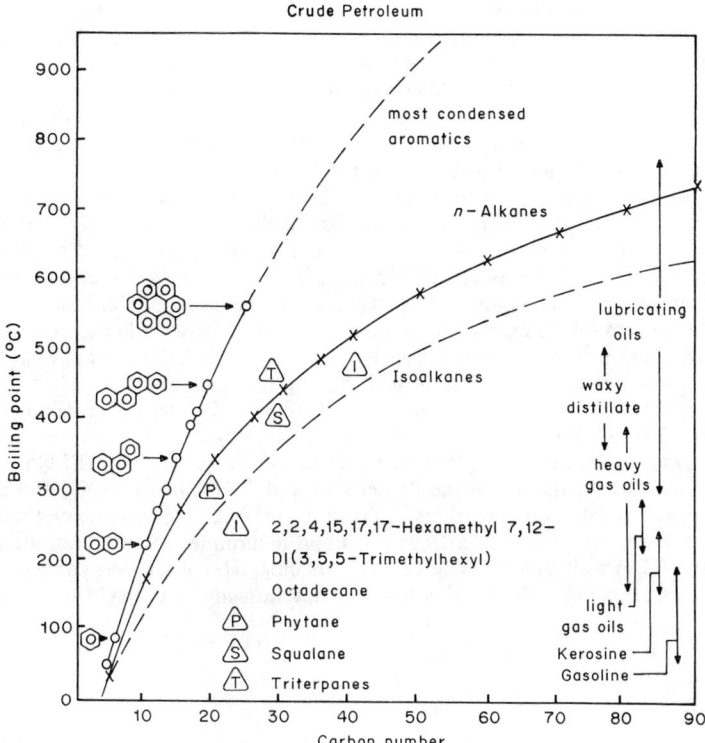

FIGURE 15. Hydrocarbon boiling point vs carbon number (see also Figures 1 and 20)[51]. Reproduced by permission of Springer-Verlag

identification and calculation, were studied and reported in hundreds of papers, and are also reviewed, every two years, by *Analytical Chemistry Review*, Petroleum Section[52]. For this reason we do not intend to review these in detail but only highlight the various approaches.

Basically, the chromatographic methods are gas phase chromatography (GC) and liquid chromatography (LC). Thin-layer chromatography (TLC) can be regarded as one of the LC methods. In the various techniques of LC, the separation of saturated hydrocarbons is based on the fact that these are non-polar molecules. In general, the use of any eluent restricts the LC method, if evaporation of the solvent is essential for the procedure.

The GC method (Section VI) is employed for crude oils, but is not the method of choice for group-type separation. This is due to the fact that its analytical power is limited by the technically feasible number of theoretical plates for distillation. Selective detectors or methods of detection will render the GC method to be group selective. On the other hand, the various LC methods are marked by the group-type separation capability, but are much less efficient in homologous separations. The oldest LC method is the gravitation-driven standing-column chromatography, using silica or alumina (or both) as the solid phase. The use of a low boiling alkane as eluent enabled the separation of the alkanes as a family from complex mixtures. With the development of smaller particles (2–5 μm size) and more active surface areas, increased pressure had to be applied to complete the analysis in a shorter time. Hence the application of the MPLC (medium pressure)[53] and HPLC (high performance)[54,55] techniques. In both methods the back-flush[53] method is used to further shorten the analysis time.

The quantification of the isolated alkane–cycloalkane fraction depends on the type of detector used and its calibration. Since the RI (refractive index) detector has low sensitivity and the alkanes absorb only in the IR region, which is difficult to quantify, it was imperative to find a better detector which would be accurate and sensitive to hydrocarbon GTA (group-type analysis). The dielectric constant (DC) detector is suggested as being accurate for GTA[56]. However, to be able to detect small changes, one must use Freon 1,2,3 as eluent. This method employs the isocratic mode with 'olefin-selective' 5-μm silica bonded to a strong cation exchange phase. The ion exchange resin is loaded with Ag[+]. This method was used earlier to separate alkanes from alkenes.

The introduction of the reversed phase (RP) columns to HPLC brought about attempts to separate alkanes by its use, e.g. in high boiling fractions of crude oil[57a]. The more polar fractions elute very fast and within 3–4 minutes C_{20} to C_{44} n-alkanes appear (Figure 16). Though the separation within the alkane–cycloalkane families is superior by GC techniques, some attempts were made to use size exclusion HPLC (GPC); see Figure 17. Because of the use of the RI detector and THF as solvent, C_6 and C_7 show reversed peaks. Burda and coworkers[57b] claim that HPLC is a suitable method for the separation of various types of hydrocarbons and determines the retention data for 54 alkanes (C_5–C_{11}) with different degrees of branching. This work was carried out on a reversed-phase column with 7000 theoretical plates, showing for each C_x that the decrease

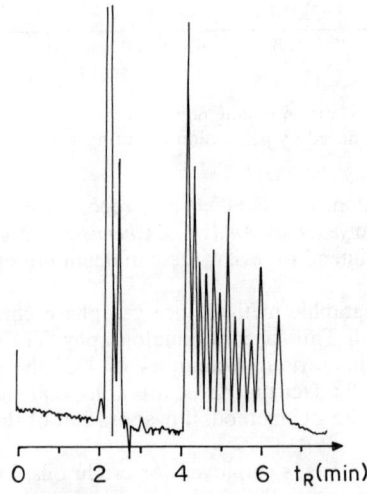

FIGURE 16. Reversed-phase separation of n-alkane standards on an LC-18-DB (5 μm) column. The mobile phase was THF–water (85:15) (0.5 ml/min[−1]) at 50 °C. The sample size was 5 μl and contained equal weights of each C_{20}, C_{22}, C_{24}, C_{26}, C_{28}, C_{30}, C_{32}, C_{34}, C_{36}, C_{38}, C_{40} and C_{44}[57a]

Separation of hydrocarbons

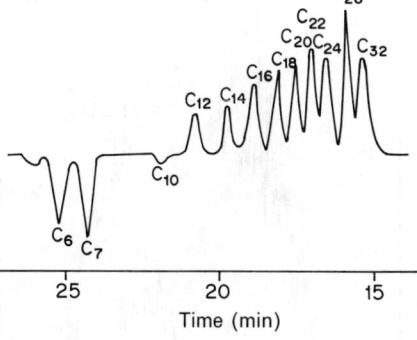

Conditions

Column: Micropak TSK GEL, 1000H

Mobile phase: THF

Flow rate: 1 ml/min^{-1}

Detector: RI

FIGURE 17. GPC (gel permeation chromatography) of n-alkanes: the C_6–C_{32} range

in branching increases retention. However, the mono-methyl branched isomers could not be separated.

The various LC methods should be examined closely before adopting any of them. If the aim is to separate for GTA, then all methods are applicable, but caution should be exercised in cases where the method is to be applied for separations within a homologous family. However, the one *distinct advantage* that both MPLC and HPLC have over GC is the use of non-destructive detectors and the low temperature of analysis (see also Section VI).

V. SEPARATION OF THE SATURATED HYDROCARBON FRACTION INTO HOMOLOGOUS FAMILIES

The methods discussed for separation of alkanes and cycloalkanes from alkenes, aromatics, resins and other polar constituents of petroleum can be employed also for synthetic mixtures, asphalts, bitumens, etc. However, for the quantification and identification of such homologous families as n-alkanes, branched alkanes, etc., further analysis is needed. Whereas the GTA relates to functionality and differences in polarity, the separation within the alkane family cannot be based on these characteristics.

The separation of n-alkanes by clathration techniques is well documented. Methods such as urea and thiourea adduction as well as molecular sieves have been employed[58–60]. The 5 Å molecular sieves effectively separate n-alkanes from other alkanes, since larger-diameter molecules do not enter the zeolitic structure[61].

Various groups have developed methods of activation of the 5 Å molecular sieves for more efficient inclusion of n-alkanes, e.g. high temperature (250 °C), vacuum and repeated N_2 'washing'. Others groups investigated larger-diameter synthetic zeolites without much success. Today the use of 5 Å resins in isooctane, followed by either thiourea adduction or first urea and then thiourea treatment, is the technique of choice to concentrate steranes and hopanes for GC/MS/C analysis[62]. In crystalline urea adducts[63] of compounds with straight carbon chains (longer than C_6), the alkanes enclosed in the hexagonal lattice are kept firmly in position. However, urea adducts can be utilized also for fatty acids (longer than C_8) whereas 5 Å molecular sieves are specific for alkanes. An impressive example for the power of the 5 Å sieving method is given in Figure 18[65].

The use of adducts with thiourea on the branched cyclic fraction is demonstrated in Figure 19[64].

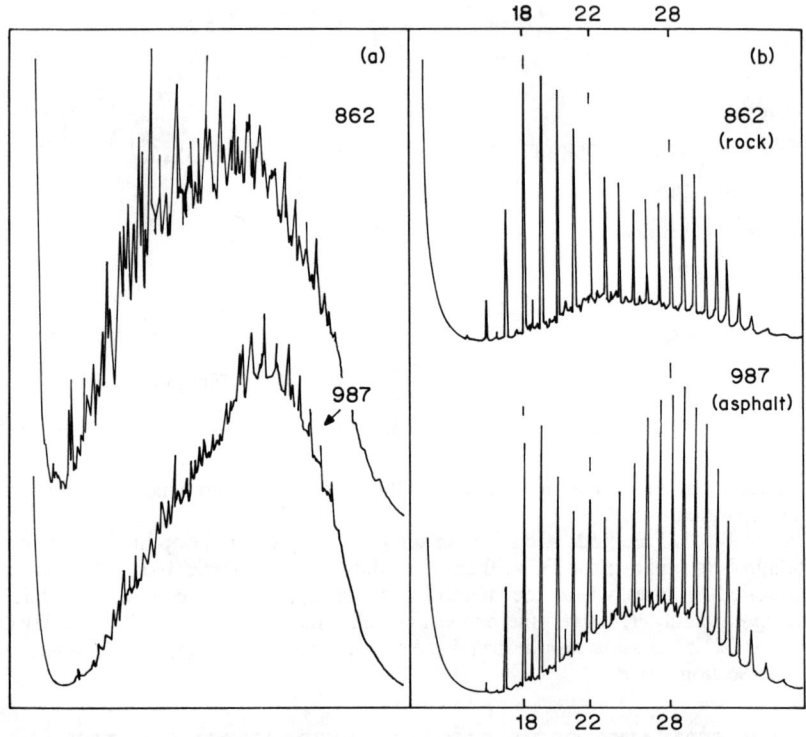

FIGURE 18. Gas chromatograms of C_{15+} saturated hydrocarbons, extracted from Senonian bituminous rocks (sample 863) and from dolomite impregnated with asphalt (sample 987), in Ein Said-1 drillhole. Asphalt-source rock correlation by n-alkane distributions (after 5 Å molecular sieving) is presented in (b). Reprinted with permission from *Org. Geochem.*, **8**, 181. Copyright (1985) Pergamon Press PLC

Figures 18 and 19 demonstrate that the separation of n-alkanes from a complex mixture is more clean-cut than their differentiation by thiourea adduction.

It is, of course, possible to combine GTA and the various inclusion techniques with mild vacuum distillation to form fractions of less complexity. One can apply selective detectors in each chromatographic method, tracing a 'pseudo-GTA' profile. These methods, sometimes called electronical sieving, will be discussed in Section VI. It is also possible to use high-resolution MS (HR-MS) for type analysis of complex hydrocarbon mixtures. Teeter[66] reviewed the methods employed until 1985, showing that the relative amount of classes of compounds in a mixture can be obtained by HR-MS.

VI. GAS CHROMATOGRAPHY, AS USED FOR ANALYSIS OF ALKANES AND CYCLOALKANES

The concept of separation by partitioning a mixture between gas and liquid phases was proposed in 1941[67] and many changes in GC theory, and consequently improvements in instrumentation, have been introduced since then.

In general, the analysis of alkanes and cycloalkanes by GC is based on boiling point

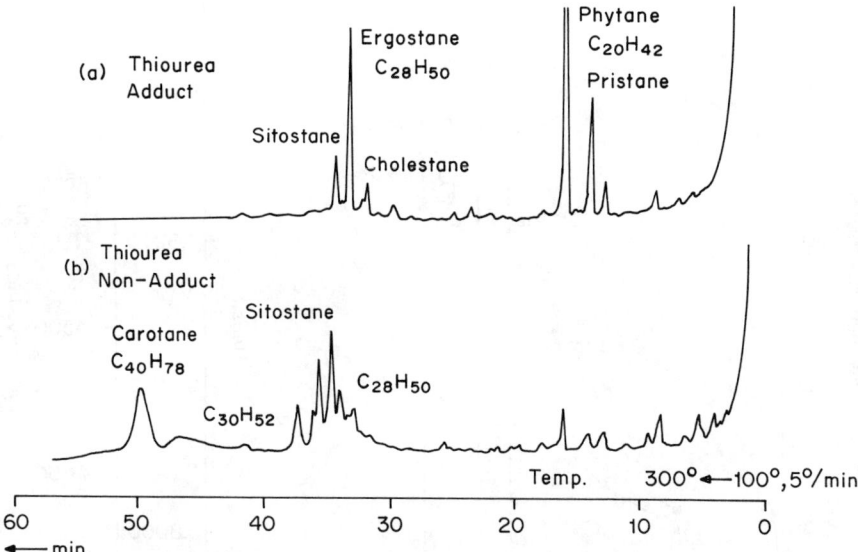

FIGURE 19. Gas–liquid chromatograms of (a) thiourea adduct and (b) a non-adduct portion of branched-cyclic hydrocarbon fraction from Green River shale. Conditions: 1.5 m × 2 mm column 4% JXR on Chromosorb G. Programmed 100–300 °C, ca 5° min^{-1} [62,64]. Reproduced by permission of Springer-Verlag

differences. Hence for the separation of n-alkanes, packed columns of about 2 m length with ca 5000 total effective plates are sufficient. The packing of these columns is based on a solid phase (diatomaceous earth, chromosorb W or any equivalent material) coated by a high-boiling liquid phase (silicon grease, a polar wax or a non-polar wax). The liquid phase (3 to 10% w/w) should be evenly distributed to form a uniform coat. The use of a 1/2″ O.D. column for the GC quantitative separation of pristane from the 294–309 °C distillation cut of petroleum is shown in Figures 20 and 21[5], however, a 22′ length column is needed.

The C_1–C_4 hydrocarbon gases are easily separated by a variety of packed columns. Most modern packing materials for this are based on synthetic porous polymers, or on porous silica beads of controlled texture (Chromosorbs 102, 104; Carbosphere, HeySep polymers, porapacks polymers, etc.). Both the analytical literature and the commercial catalogues offer many alternative materials which can be applied for gas–liquid and gas–solid chromatography.

Cyclopropane and cyclobutane should be discussed in this subsection, but since they do not appear in natural gases their GC separation is hardly mentioned in the literature.

The analysis of n-alkanes is important in the oil industry, hence the linearity of retention on various columns was studied extensively. The range of C_1 to C_7 n-alkanes was examined for linearity of the retention time vs carbon number[68]. However, the resolution power (effective plates) of packed columns is not sufficient if all isomer mixtures of the C_2–C_9 alkanes and cycloalkanes are to be studied. The introduction of the capillary GC column 'SCOT' (Surface Coated Open Tubular), as early as 1960[69], was followed immediately by changes in analytical methods[70,71]. The SCOT columns were made from either stainless steel or glass, presenting the analytical chemist with ca 150,000 effective

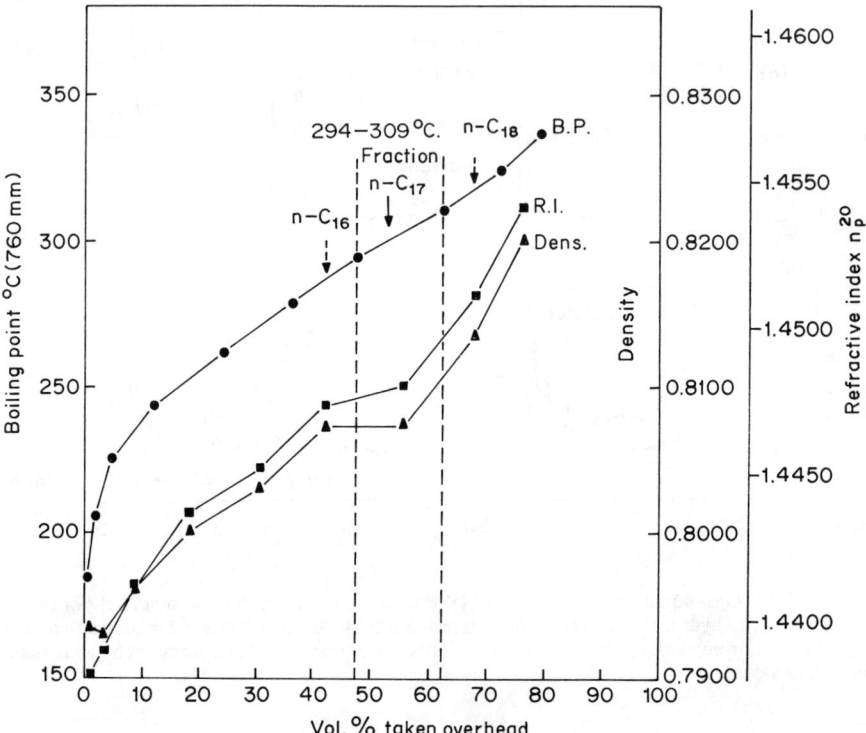

FIGURE 20. Properties of East Texas nonaromatic distillate fractions. Reprinted with permission from *Anal. Chem.*, **34**, 49. Copyright (1962) American Chemical Society

plates for 50 m. The development of their GC use, up to 1969, especially for hydrocarbon analysis, was reviewed by Douglas[72] giving 159 references.

The application of capillary columns became more abundant since the introduction of the fused silica polymer-enforced columns. Another development produced the chemi-bonded coat, which reduced 'bleeding' and solvent stripping, giving the capillary columns longer life-time. For most analyses of alkanes the carrier gas used is He; however, to optimize separation H_2 is used. For very high temperature analyses 'super pure' gases are required and the presence of traces of oxygen is detrimental to the life span of the column.

The molecular range of alkanes above C_{10} was studied extensively by GC techniques. In most cases head-space samples were taken at $\approx 100\,^{\circ}$C and injected into the gas phase. However, for very small samples ($<10^{-9}$ g) degassed from rock samples, a method described in Figure 22[73] was developed. This demonstrates some basic concepts in the new generation of GC equipment as well as high-resolution capillary columns. The use of two columns and the backflush mode operation with cold traps produce the capability of identifying more than 40 compounds in the range C_2–C_9[73]. The same range of alkanes was also analysed by HPLC, indicating faster elution of the highly branched compounds than of the n-alkanes[57b]. This general trend holds true also for the GC analysis of various branched alkanes in the C_{10} to C_{30} range, especially for lower carbon

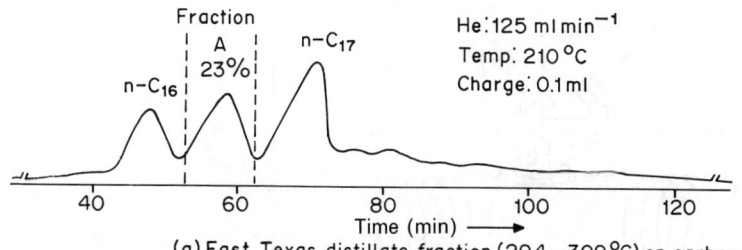

(a) East Texas distillate fraction (294—309 °C) on carbowax 20m

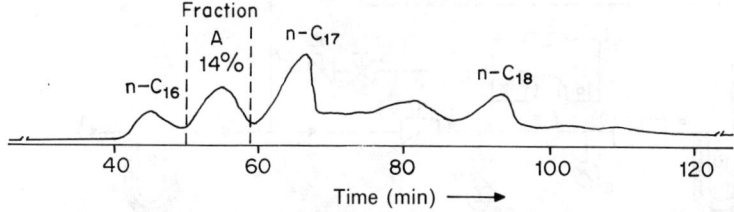

(b) Mid—continent distillate fraction (295—310 °C) on carbowax 20m

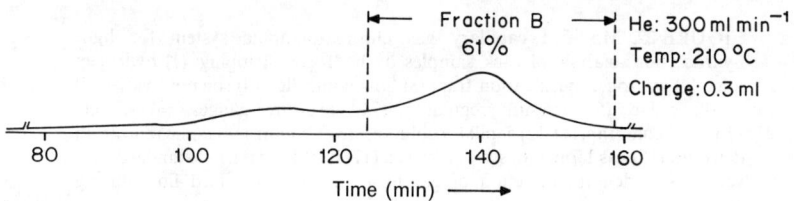

(c) East Texas carbowax fraction A on silicone grease

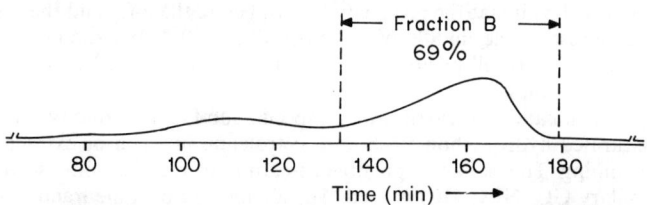

(d) Mid—continent carbowax fraction A on silicone grease

FIGURE 21. GC: (a and b) quantitative separation on Carbowax 20 M; (c and d) reinjection of fraction A on silicon grease obtaining fraction B for other analyses. Reprinted with permission from *Anal. Chem.*, **34**, 49. Copyright (1962) American Chemical Society

numbers[74]. Though most alkyl-substituted paraffins in the $C_{10}-C_{30}$ range in crude oils are 2-methyl and 3-methyl[75], it is important analytically to establish the peak assignments for the other isomers. The use of a 50 m (0.2 mm i.d.) FSCOT (Fused Silica, SCOT), capillary (0.50 μm coat of cross-linked methyl silicon) column showed: (a) all branched alkanes elute before the normal alkane of the same carbon number; (b) the longer the branching group, the faster this alkane moves, i.e. it has a lower relative retention factor (RRF); (c) identical branching groups (e.g. ethyl, propyl, butyl, etc.) show RRF according to this position on the straight-chain structure; i.e. 4,5,6,7-alkylated compounds will

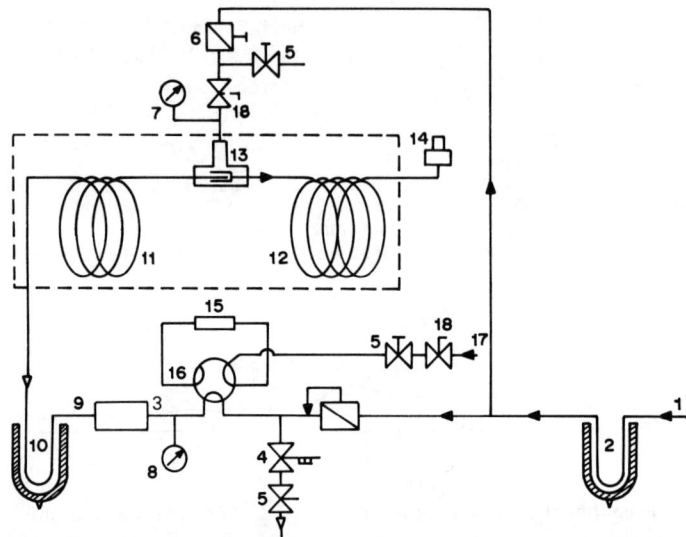

FIGURE 22. Modified capillary gas chromatographic system for light
hydrocarbon analysis of rock samples by hydrogen stripping: (1) hydrogen
inlet, (2) hydrogen purification trap, (3) flow controller, (4) solenoid valve, (5)
needle valve, (6) pressure regulator, (7,8) pressure gauges, (9) sample
tube, (10) cold trap, (11,12) capillary column, (13) T-union, (14) flame ionization
detector, (15) gas loop, (16) six-port valve, (17) inlet for external standard, (18)
valve; →— denotes direction of gas flow and ⇸— reversed flow during
backflush[73]

elute as a 'group' some 1 to 1.5 min (at the given conditions) before the n-C_x and the first
is the one substituted closest to the middle of the chain $(7,6,5,\ldots)^{74}$. We should note
that the last conclusion is true for all positions from 4 on. However, 3-methyl isomers
appear after the 2-methyl alkanes.

Detailed investigation of alkanes was carried out on specific sub-families, among which
the analysis of the saturated hydrocarbon fraction of petroleum, bitumen or extracted
coals is the most demanding. This is due to possible co-elution of various isomers even
if high-resolution capillary GC (HRC GC) is used. The alkane–cycloalkane fraction of
petroleum contains gas, liquid and wax; some groups used for the separation one column
with programmed temperature from −20 °C up to 325 °C, and others a combination of
two columns in sequence. The two columns could be placed either in one oven, the first
as a concentrator to a cold trap and the second for the detailed chromatography, or in
two ovens, allowing better temperature control on each column. Most information is
contained in commercial data bulletins and instrumentation guides, since several
companies offer various GC, FSCOT capillary columns (15 m up to 100 m in length)
with different internal diameters, with different thicknesses of coat and a variety of liquid
phases. Prospective users should take into consideration that for the analysis of alkanes
and cycloalkanes, the choice of the proper column is the heart of the GC procedure.

Until two or three years ago, the GC of hydrocarbons was restricted by the highest
temperature at which the liquid phase did not 'bleed'. Hence with the exception of an
inorganic salt 'Eutectic' packed column[76] (which could be used up to 400 °C), all standard
columns were restricted to 325–350 °C maximum temperature. This temperature limit

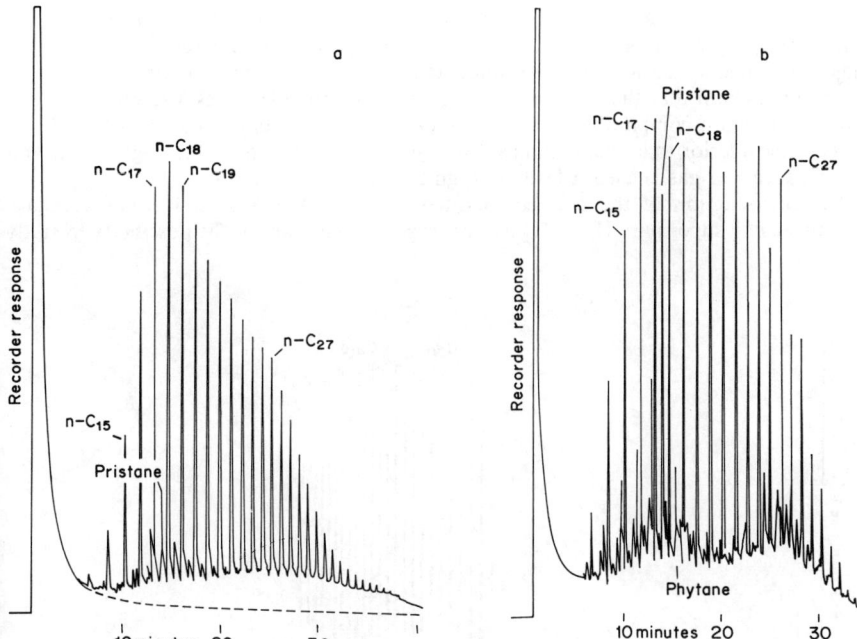

FIGURE 23. Gas chromatograms of (a) a whole crude oil from the Altamont Bluebell Field, Unita Basin, and (b) saturated hydrocarbons from a Canadian Arctic Island sedimentary rock, GC on 'eutectic' column[76]

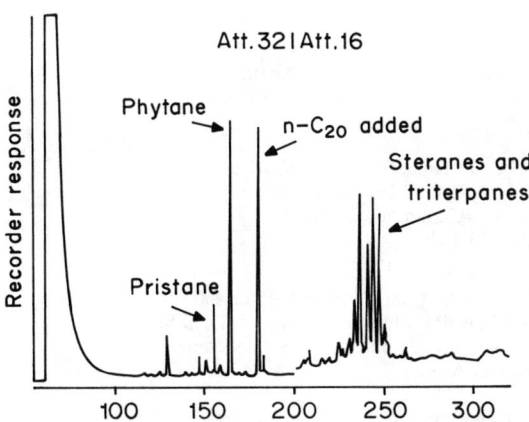

FIGURE 24. Gas chromatogram of branched and cyclic saturated hydrocarbons from the Green River oil shale. Same column as in Figure 23[76]

did not permit the analysis of waxy fractions (C_{33}–C_{36}). The use of the 'eutectic' column with rapid programming ($10\,^\circ C\,min^{-1}$ up to $400\,^\circ C$) allowed analyses of crude oils for major saturated hydrocarbons to be conducted in about 30 min (see Figures 23 and 24[76]). Despite the claim that the separation of steranes and hopanes gives a resolution in the C_{27}–C_{30} fraction comparable to a 15 m SCOT capillary column (coated with SE30), it is of course inferior to longer columns. Moreover, the column is highly hygroscopic and if impure carrier gas is used it leads to high-temperature oxidations.

The introduction of the bonded high-temperature (420–$425\,^\circ C$) FSCOT capillary columns using super pure He or H_2 as carrier gas brought about the possibility to study

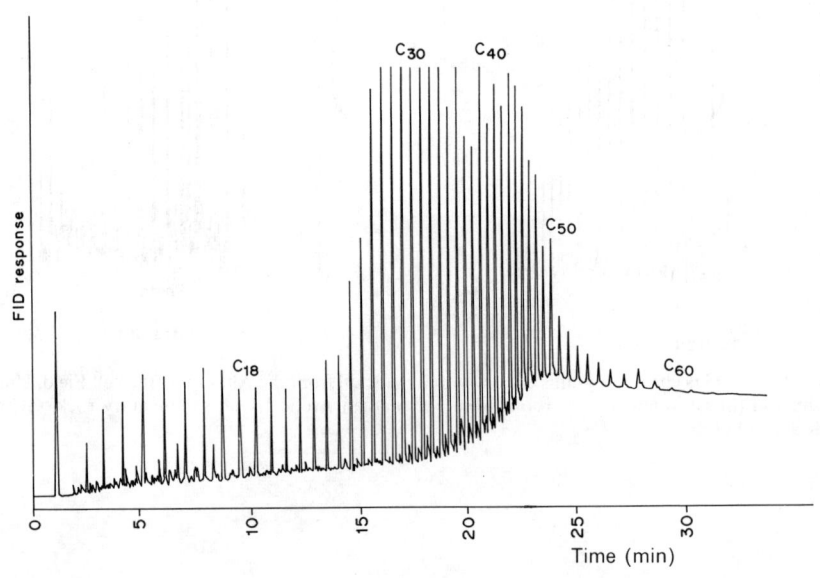

Time (min)

CANADIAN WAX

Column	SUPERCAP 'High Temperature', Al clad Fused Silica Bonded Methyl 5% Phenyl Silicone 15 Meter × 0.25 mm id, 0.1 μm film 400-2HT-15-0.1F

Press:	10 p.s.i. Helium
Temp:	60–400°C at 10 °C min^{-1} then hold at 400 °C
Injection:	Perkin-Elmer PTV Injector-60-400 °C; 1.0 μl Split 50:1
Sample	20 mg dissolved in 1 ml of 1:1 CS_2:isooctane
Detector:	420 °C FID Attn: X32
Chart:	60 cm h^{-1}
Flow Rate:	He © 1.5 ml/min

FIGURE 25. GC obtained for Canadian wax using a Supercap high temperature column, injected on 60–400 °C programmed PTV, program 60–400 °C at 10 °C min^{-1} and holds at 400 °C (oven temperature)

waxes. Figure 25 shows hydrocarbons up to C_{60}! Such analyses of waxes focus our attention on the mode of introduction of the sample into the column. Various injection methods were employed, such as simple injection ports (separate temperature control), programmed injection units (PTC) and on-column, all glass injectors. For low temperatures the use of a septum presents no problem, but at temperatures over 400 °C the bleed from the septa used is problematic. Again, the main sources of information about injectors are the commercial catalogues.

Whereas the power of resolution depends on the quality of the column, quantification and identification of the separated compounds is a function of the detector. The oldest GC detector, still used for hydrocarbon gases, is the TCD (Thermal Conductivity Detector). The TCD response is proportional to the molar fraction and the specific heat of the compound detected[77]. Its sensitivity is in the μg (10^{-6} g) range, but even for the same homologous family, this changes with molecular weight[77,78].

The FID (flame ionization detector) is ideal for hydrocarbon analysis. It is very sensitive (10^{-9}–10^{-12} g range) and responds in direct proportion to the amount of carbon combusted[77,78]. It is possible to bring the end of the separatory column very close to the flame, thus avoiding 'dead volume' and loss of resolution. The FID is simple to construct and operate, but has one major fault: the destruction of the sample. This can be overcome by splitting the eluent at the end of the column, using only part in the FID for detection and quantitation, and collecting the rest for further analyses. The technique is reported extensively in the literature, especially for the characterization of n-alkanes in petroleum fractions[80]. Lately, some companies have introduced a capillary column and quantified standard (≈ 150 compounds), using the same FID conditions and integration mode which enables characterization of petroleum. One such application for cycloalkanes in petroleum was recently demonstrated[81].

MS will be discussed in Section VII.A. Other detectors for GC which have been used for the analysis of hydrocarbons are usually more specific for molecules containing heteroatoms (e.g. ECD, FPD, etc.)[79]. Similarly, the use of FTIR as a GC detector, though used also for hydrocarbons, is applied more for functionalized compounds. The high-resolution gas chromatography using matrix isolation FTIR needs a very elaborate interface[43] whereas a 'normal' FTIR cell can be employed as detector for thermogravimetric (TG) analysis. The application of TG-FTIR for the study of hydrocarbon structure is also utilized for hydrocarbon analysis to characterize solid fossil fuels[82]. Quantification or integration of the data obtained relates to specific regions of the spectra (see also Section III.B).

VII. MASS SPECTROMETER DETECTOR (MSD)

The use of MS as a detector for both HPLC and GC added a new dimension to the analysis of alkanes, and even more so to the identification of complex cycloalkanes.

The basic concepts have already been presented in this review. However, the study of single compounds does not require an interface between the chromatographic unit and the MS. The analysis of mixtures by MS relates not only to GC/MS or HPLC/MS[84]. The popularity of GC/MS systems can be examined by the number of relevant publications. These were less than 100 in 1968, rose to a peak of 2000 papers in 1979, dropped to ≈ 1500–1750 yearly until 1988, when they rose again to 2000. The use of MSD for HPLC (also called LC/MS) also rose from a few publications in 1975 to a few hundred in 1988. A method, less used for hydrocarbon analysis, SFC (Super-critical Fluid Chromatography)/MS, was initiated only in 1985 and is gaining interest slowly. These data are given by Evershed[83]. Obviously it will be impossible to review all these developments in the use of MSD[84], even for the analysis of alkanes and cycloalkanes.

In the analysis of mixtures, it is possible to use one mass spectrometer as the separator and use a second as a MSD (MS/MS system). This method is very quick and dispenses with the need for long chromatographic columns. However, the instrumentation is expensive and for many cases, such as separation of isomers (see Figure 30), it cannot be used.

The use of MS in addition to GC/MS includes methods such as capillary zone electrophoresis (CZE/MS), thin layer chromatography (TLC/MS) and others. The major problem in employing these separation-MSD methods is the interface. The very high vacuum required for MS operation requires complete removal of the carrier gas or liquid[85]. Since the present review treats specifically the use of the MSD for the analysis of alkanes and cycloalkanes, we concentrate on the GC/MSD/C (gas chromatography/ mass spectrometry/computerized) method. The MS used can be quadrupole or magnetic and its configuration will, of course, control the power of resolution (PR), the mass range, etc. The ionization mode is also important, e.g. whether EI or CI (see Sections, III.A.1 and III.A.2).

A. Gas Chromatography/Mass Spectrometry/Computerized (GC/MS/C) System

The independent development of both GC and MS to a high degree of sophistication led to their combination in one instrument, and the need to process vast amounts of data required computerization of the system. The applications of GC/MS in organic geochemistry were reviewed by Burlingame and Schnoes[25]. Obviously, in fossil fuels the main effort is on the analysis of hydrocarbons. As an example, we can employ pioneering work using a capillary column to analyse alkanes up to C_{11} by MS[86], applying conventional MS as a detector for GC. In the early days of GC/MS application to the analysis of alkanes and cycloalkanes, quantitation of the MS-identified compounds was carried out by FID (see the previous discussion on GC). It took twenty-five years to substantiate the statement: 'GC/MS has been widely accepted for quantitative analysis of individual target compounds in organic mixtures because of its sensitivity and specificity'[87]. The capability of monitoring one or several selected ions characteristic of branched or cyclic n-alkanes allows analyses which are free of interference from chromatographic co-elutes. In general, the use of internal standards in quantitative MS is ideal, but this standard should have the same MS behaviour as the mixture to be analysed. Hence, reliable GC/MS quantitation is best performed by isotope-labelled compounds of the same structure[88]. The main reason for the need for an internal standard is the reliability and reproducibility of the MS response, which depends heavily on specific instrumental factors. These performance criteria, which need to be controlled, are different for different MS devices, and include, just to mention a few, the filament condition, electron beam energy, repeller voltage and so on. The quantitative response factors (RF) of 14 long-chain n-alkanes, using the EI mode and CI, were determined showing that RF/mol ratio is similar to TIC (total ion current)[87].

The author's group in Jerusalem is involved in both GC and GC/MS/C of fossil fuels. The availability of MS (VG ZABIIF) linked to a GC allowed various operation modes. Some of the data presented in this review have not been published previously, except in industrial reports. The main advantage in the use of GC/MS/C for saturated hydrocarbons lies in the ability to identify the branched and cyclic 'bio'-alkanes, also called 'biomarkers'. The basic analytical approach to the study of these compounds was discussed in Section III.A.3. However, the GC/MS/C method provides the possibility to work on the whole saturated hydrocarbon fraction, without employing extensive

FIGURE 26. GC profiles of (a) bitumen of the Green River formation sample as extracted, (PHC) denotes perhydrogenated carotene and (b) hydrocarbons formed by pyrolysis at high temperature, in the presence of H_2O. Both were analysed on the same column (30m SE 30 FSCOT)

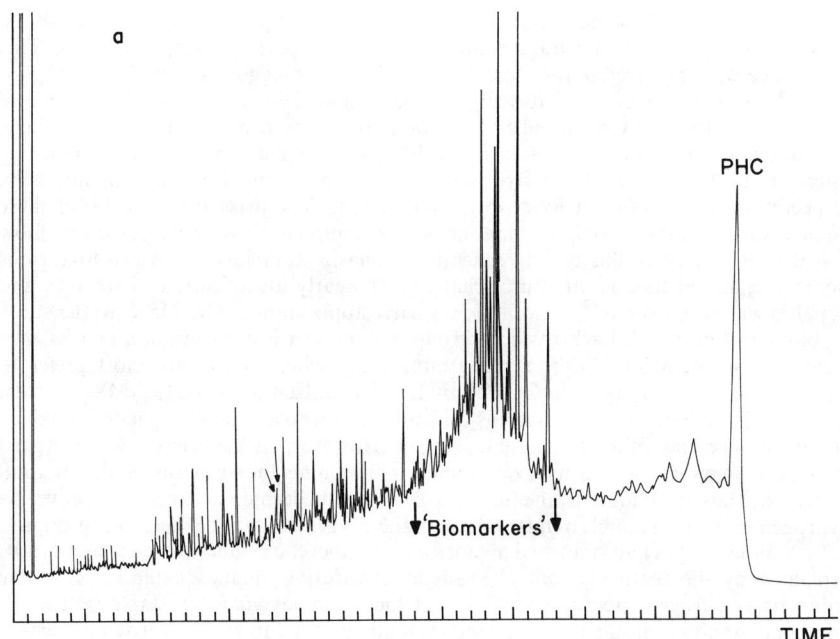

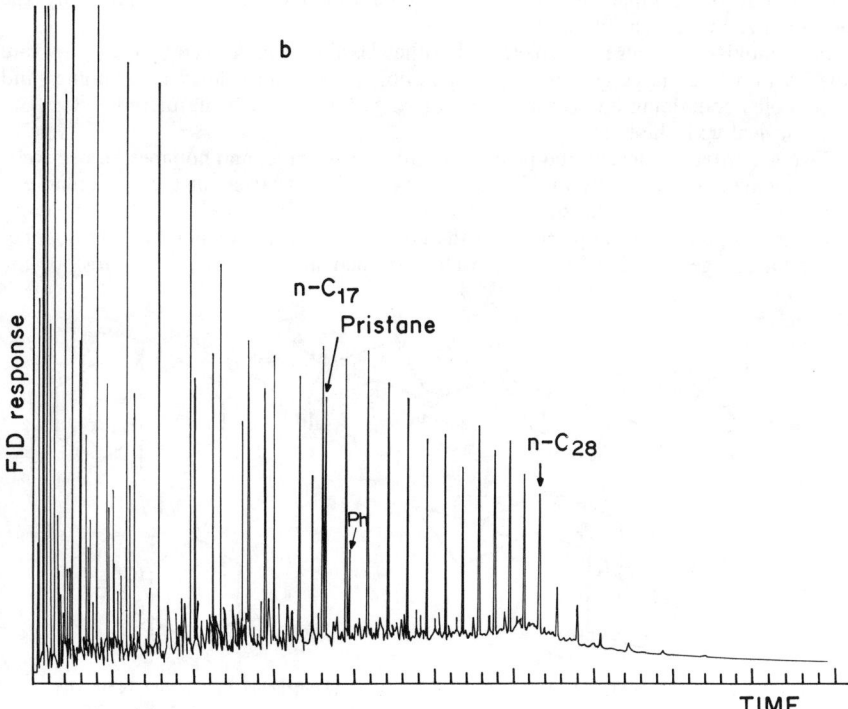

FIGURE 26. (*caption opposite*)

separation methods. The mass spectrometry of phytane and pristane, as discussed before, revealed very characteristic fragmentation patterns, typical for isoprane units. This knowledge was employed in the identification of the homologous C_5- to C_{40}-perhydrogenated carotenes (PHC) (see also Figure 26a). Figure 26b shows the whole range of hydrocarbons thermally produced and, by comparison, one can see that the 'biomarkers' that dominate the bitumen as well as the PHC (Figure 26a) appear above the n-C_{28} range. Both bitumen and the hydrous pyrolysis products (Figures 26a and 26b), respectively) are of a Green River formation sample. The mass spectra (MS) of these cyclic alkanes are more complex than the acyclic compounds. We have chosen to show first the GC profile of this boiling range to emphasize complexity due also to various isomers, enantiomers and structural changes. The early identification of steranes and hopanes was very tedious[89], hence the very early application of GC/MS/C to the study of 'biological markers'. Each research group has its own instrumentation and its own technique of GC/MS/C. While some groups study whole crude oils, most prefer to separate the petroleum prior to GC/MS by LC. The utilization of the GC/MS/C system for natural products includes steroids and other terpenoids (mostly functionalized by double bonds and other functional groups). However, in geochemistry we regard 'biological markers' as organic compounds maintaining the skeleton of the original molecule. Thus to become a 'chemical fossil' it has to become chemically inactive, by hydrogenation of the double bonds and loss of the oxygen-containing functional groups.

Low molecular weight branched alkanes and monocyclic alkanes in petroleum can be explained by the thermally-controlled 'depolymerization' (named catagenesis) of the sedimentary kerogen. Hence, we will expect the thermodynamically stable isomers to dominate. Analyses of single alkanes and cycloalkanes by MS were discussed in Section III.A.2, but the complex mixtures have to be examined by GC/MS utilizing the computerized treatment of the data[90].

Isoprenoids-isopranes in petroleum and other fossil fuels as studied by GC/MS/C are divided into three large groups: (C_{10}–C_{40}) aliphatic isopranes, alicyclic isopranes and isoprenoids containing an aromatic ring or rings (moieties). In the present review we will not deal with the latter.

Two groups of tetracyclic and pentacyclic alkanes, steranes and hopanes, respectively, are presented schematically in Figures 27 and 28. The analysis of these 'biomarkers' requires the optimal combined techniques of GC and MS, as well as a large memory and fast computers. The application of the GC/MS/C analysis of steranes and hopanes in petroleum geochemistry was extensively studied and reviewed[91,92]. Most of the

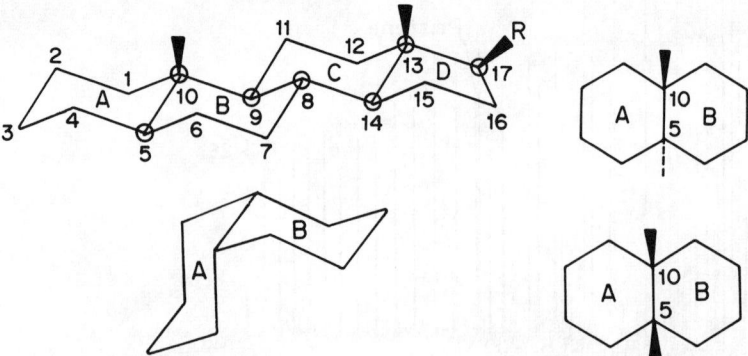

FIGURE 27. Schematic stereostructure of the sterane tetracyclic skeleton. Note that positions 5,8,9,10,13,14,17 are asymmetric carbons. Bond cleavage and change of 5α to 5β will cause the A and B rings to be in the *cis* form (see also Figure 9 and Table 6 below)

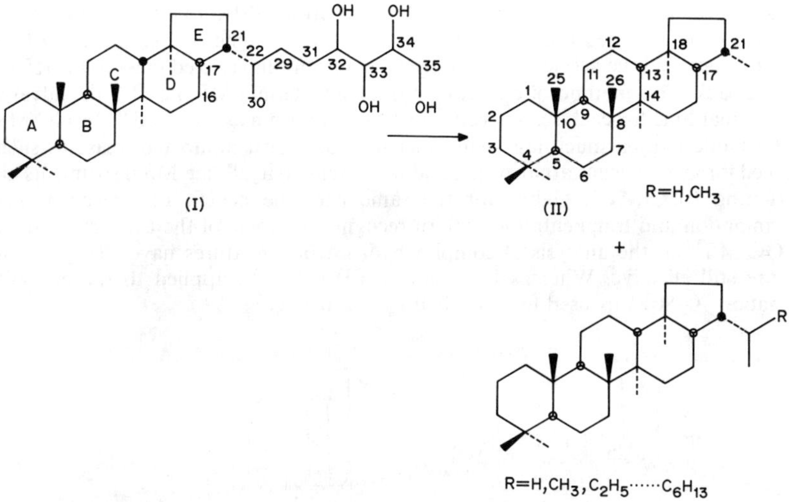

FIGURE 28. Structure and numerical assignment of hopanes (pentacyclic C_{35} terpanes). Compound I (tetrahydroxybacteriohopane), THBH is suggested as the precursor of all hopanes in the geoenvironment. Compounds II represent structures of the derived hopanes (see also Table 5 below)

modifications introduced in the GC section for the analysis of higher-boiling-point compounds and for the enhancement of the resolution at elevated temperatures were discussed previously. However, the mode of injection is important in order to obtain reliable results. The three accepted methods (splitless, split and cooled on column) were investigated by Killops[93] and by the author's group[117], leading to the conclusion that the on-column injection is superior to the other two, providing very small quantities are injected. The optimization of the MS sectors and the selection of the most suitable operational mode for the GC/MS/C of 'biomarkers' depend on the type of instrument and computerizing system. Specific detection of certain fragments and pattern recognition relate to the resolution power (RP) of the mass spectrometer and the computer's capability to match the reference file with the analysed mixture. Hence the identification of a given compound or isomer could be performed by comparison with a synthetic standard or by MS statistical analogy. The statistical approach allows GTA either by selection of typical groups of fragments or by statistical whole MS soft independent modeling of class analogy (SIMCA)[94,95].

The use of capillary columns in the GC/MS/C method produced a new challenge to the operation of the MS arising from the need for the short interval of time (*ca* 7 s at half height of the peak). Hence the MS data have to be produced and collected in 0.5–1.2 s to produce a true TIC. This, in turn, requires either fast scanning of the MS or preselection of one or several m/z ions (fragmentometry). Most of the information relating to technical changes in the hardware and software was published in the data sheets of the various GC/MS/C manufacturers.

The most popular MS operation mode is the EI (see Section III.A) and the most popular GC/MS/C recording mode is selective ion monitoring (SIM). The latter mode can be operated for low, medium or high resolution MS, in various combinations of factors. All SIM modes are similar to those described in 1967 by Gallegos and collaborators[96]. This method in practice requires a very large data storage memory and pushes the magnetic sector to its 'speed limit'[97]. Therefore, it was suggested that, for comparative runs, restricted information should be monitored, still using high resolution

334   Z. Aizenshtat

and accurate mass (AM). For this selected ion recording (SIR) one has to predetermine
the ions to be recorded and, instead of scanning, the magnet is 'parked' on a known
accurate mass and the accelerating voltage jumps from this selected mass to the reference
file and back. This method of operating a double focussing MS has also the advantage
that the real time for each scan is very short for each ion mass selected[98]. In GC/MS/C,
the CI mode is used much less than other methods for a number of reasons, such as
the need for an active gas atmosphere, and lower sensitivity. Some MS instruments allow
alternating EI, CI, ACE giving, for the same run, the benefit of molecular weight
determination and fragmentation pattern recognition. Some of the early methods used
for GC/MS[35] in the analysis of complex hydrocarbon mixtures have changed names,
but are still effective. Whereas low voltage (EI) could be applied to the analysis of
aromatics, 70–80 eV is used for saturated hydrocarbons.

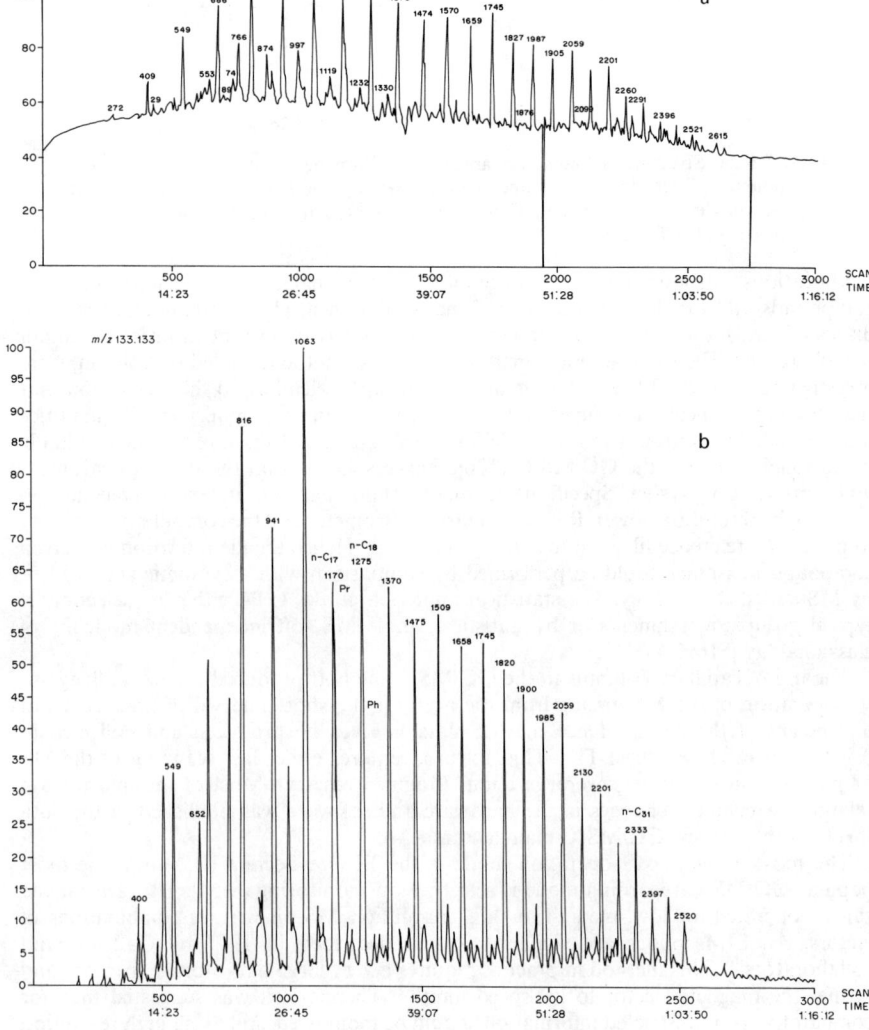

FIGURE 29. (*caption on p. 336*)

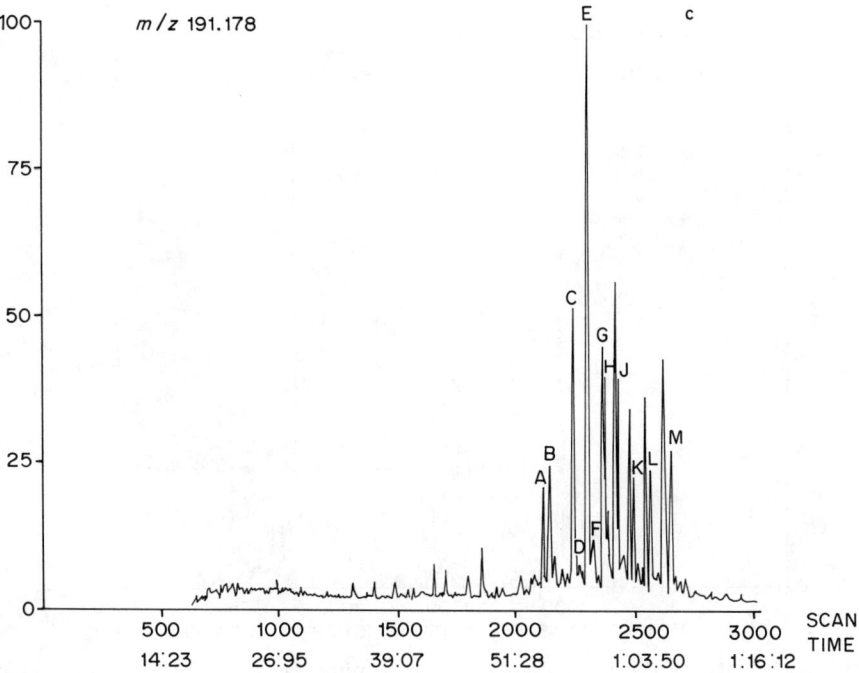

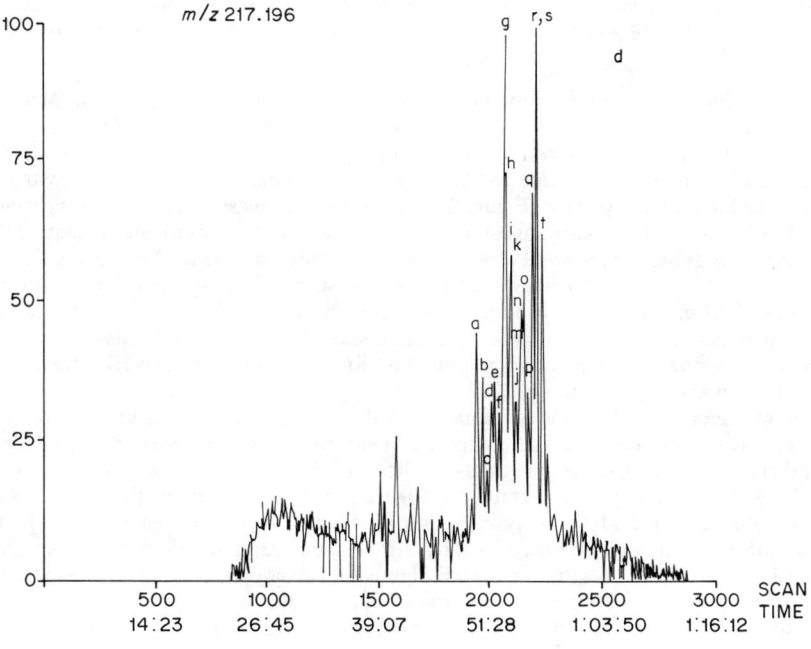

FIGURE 29. (*continued*)

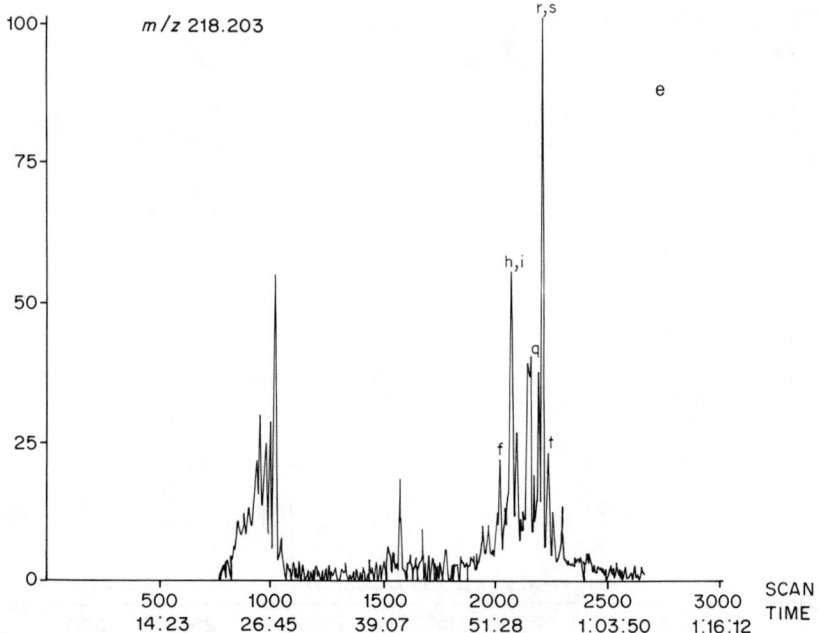

FIGURE 29. The GC/MS/C analysis of the saturated hydrocarbon fraction of an Israeli crude oil: (a) total ion current (TIC); (b) trace of fragment m/z 133.133, typical of open-chain alkanes; (c) trace of fragment (mass fragmentogram) m/z 133.133, typical for most tri-, tetra- and pentacyclic terpanes, dominated by the hopanes: (d) the m/z 217.196 mass fragmentogram, steranes, and (e) the m/z 218.203 fragmentogram, also steranes[91,92]. Traces b–e are SIM of accurate mass (AM) with RP of 5000

Figures 29a–e and 30a,b show the analyses by various GC/MS/C and MS–MS techniques, as employed for an Israeli oil (coastal plain, Heletz), using RP *ca* 5000 as described before[99–101]. The results brought in Figure 29a–e show that the TIC profile is obviously interfered with due to the reference file 'background'; PFK is constantly introduced for mass reference (Figure 29a). However, the mass fragmentogram for m/z 133.133 (saturated hydrocarbons) shows the domination of the n-alkanes (Figure 29b). The hopanes dominate the m/z 191.178 mass fragmentogram (Figure 29c). In Figure 30a the same distribution of hopanes is shown for the same oil using the SIR mode rather than the SIM method which produced the Figure 29c profile. The m/z 217.196 and m/z 218.204 peaks monitored in Figures 29d and e show the steranes with high selectivity. The m/z 218 fragment is much more abundant for the 14β(H) and 17β(H) isomers (see also Table 6 and Figure 27).

As was described before, the SIR mode provides very fast scan time and very accurate profiles, which can be stored and compared. Hence this method is recommended if large numbers of samples have to be studied[98]. The MS–MS method shown in comparison to the SIR monitoring (same sample) in Figure 30b does not require the use of a GC separation, hence the whole analysis is completed within *ca* 10 minutes. In this case the same MS apparatus can be used for HR SIR (Figure 30a) and MS–MS by the CAD (collision activated dissociation) method (Figure 30b) demonstrating the advantages and disadvantages for both methods. The direct introduction by CAD MS–MS is rapid, handles mixtures and has no molecular weight restriction, but it gives no separation of

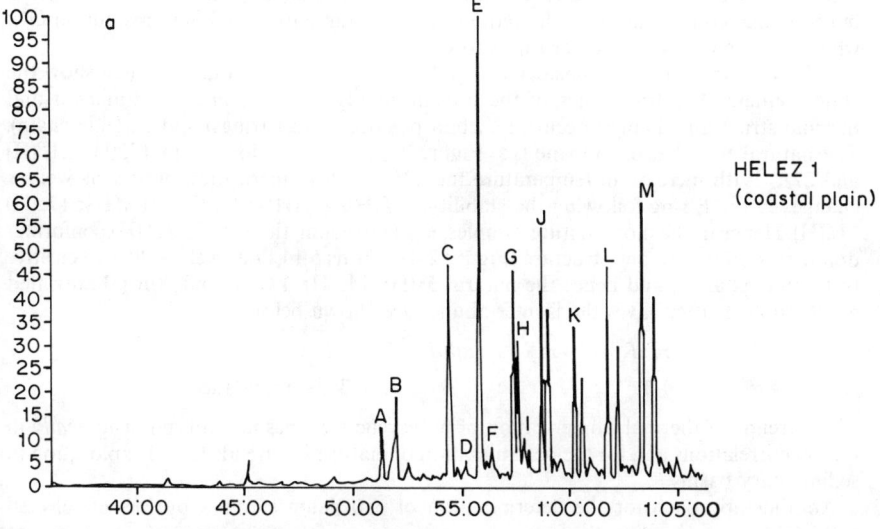

m/z 191 Triterpane chromatogram

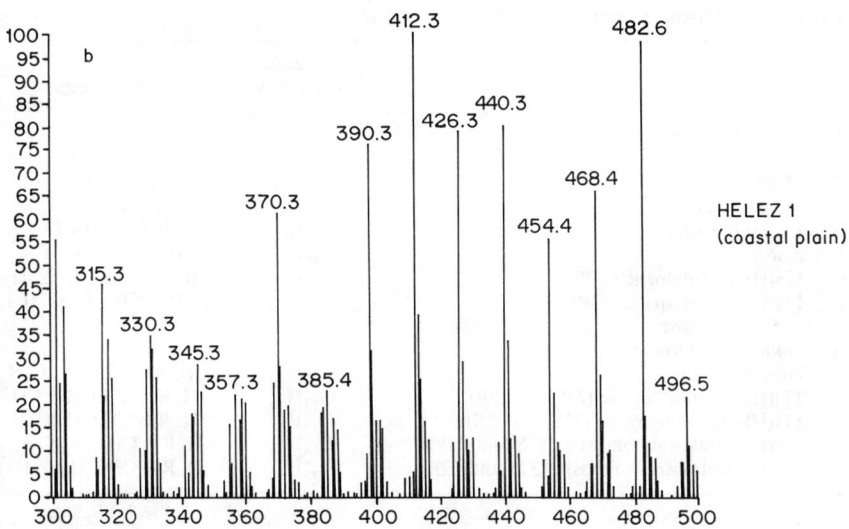

Averaged mass spectra of parents

of m/z 191 by CAD MS—MS

FIGURE 30. Comparative analysis of an Israeli oil sample (the same as in Figure 29) by GC/MS/C:
(a) SIR mode and (b) CAD (collision activated dissociation) MS–MS. The SIR for m/z 191 (hopanes)
is compared with SIM; see Figure 29 (c)[100,101] (a) Masses selected for monitoring: m/z 177, 191,
217, 218, 231 and 259. (see text for SIR mode). (b) was run under the following conditions: collision
energy of 20 eV, daughter ion selected by voltage switching, parent ion by quadropole scanning
between m/z 150–800, 1 s per scan, probe programme 50–320 °C at 60 °C/min

structural isomers and requires expensive MS–MS hardware. The GC/MS/C method is better in separation of individual components and in relative quantification of isomers, but it is time-consuming (1.5–2 h), requires data manipulation and high-resolution GC, which also limits the molecular range to C_{38} max.

Tables 5 and 6 must be considered together with Figures 27 and 28, which show the chiral centers. The triterpanes of the hopane family have the m/z 191 fragment and thermal structural changes occur at carbon positions 17,21 (rings) and 22 (side chain). The natural tetrahydroxyhopane (see Figure 28) has the conformation $17\beta(H)$, $21\beta(H)$ and 22R. With increase in temperature the $22R \rightarrow 22S$ isomerization occurs, as well as changes in the E ring following the stability: $17\beta(H)$, $21\beta(H) < 17\beta(H)$, $21\alpha(H) < 17\alpha(H)$ $21\beta(H)$. Hence in the more mature samples, e.g. petroleum, the $17\alpha(H)$, $21\beta(H)$ conformer dominates. In the sterane structure (Figure 27) carbons 5,14,17 as well as 20 are sensitive to thermal changes and hence the natural $5\alpha(H)$, $14\alpha(H)$, $17\alpha(H)$ and 20R (abbreviated $\alpha\alpha\alpha R$) conformation gives the isomer abundance shown below:

$\alpha\alpha\alpha R$	:	$\alpha\alpha\alpha S$	:	$\alpha\beta\beta R$	:	$\alpha\beta\beta S$	structure
1	:	1	:	3	:	3	abundance

These trends of thermal isomerization of polycyclic terpanes are utilized *analytically* in oil–oil correlations and for the determination of maturation trends for oil exploration in sedimentary basins.

Another approach for the determination of petroleum steranes by MS of selected metastable ions[102] utilizes the spontaneous unimolecular fragmentation of sterane parent ions, occurring in the first field-free region of a double focusing MS. The sterane

TABLE 5. Triterpane identification and structures[99]

Compound		Elemental composition	Structure
A	$18\alpha(H)$-trisnorneohopane	$C_{27}H_{46}$	I
B	$17\alpha(H)$-trisnorhopane	$C_{27}H_{46}$	II, R = H
C	$17\alpha(H)$-norhopane	$C_{29}H_{50}$	II, R = C_2H_5
D	normoretane	$C_{29}H_{50}$	III, R = C_2H_5
E	$17\alpha(H)$-hopane	$C_{30}H_{52}$	II, R = $CH(CH_3)_2$
F	moretane	$C_{30}H_{52}$	III, R = $CH(CH_3)_2$
G	$17\alpha(H)$-homohopane (22S)	$C_{31}H_{54}$	II, R = $CH(CH_3)C_2H_5$
H	$17\alpha(H)$-homohopane (22R)	$C_{31}H_{54}$	II, R = $CH(CH_3)C_2H_5$
	+ gammacerane?	$C_{30}H_{52}$	IV
X	unknown triterpane		
I	homomoretane	$C_{31}H_{54}$	III, R = $CH(CH_3)C_2H_5$
J	$17\alpha(H)$-bishomohopane (22S and 22R)	$C_{32}H_{56}$	II, R = $CH(CH_3)C_3H_7$
K	$17\alpha(H)$-trishomohopane (22S and 22R)	$C_{33}H_{58}$	II, R = $CH(CH_3)C_4H_9$
L	$17\alpha(H)$-tetrakishomohopane (22S and 22R)	$C_{34}H_{60}$	II, R = $CH(CH_3)C_5H_{11}$
M	$17\alpha(H)$-pentakishomohopane (22S and 22R)	$C_{35}H_{62}$	II, R = $CH(CH_3)C_6H_{13}$

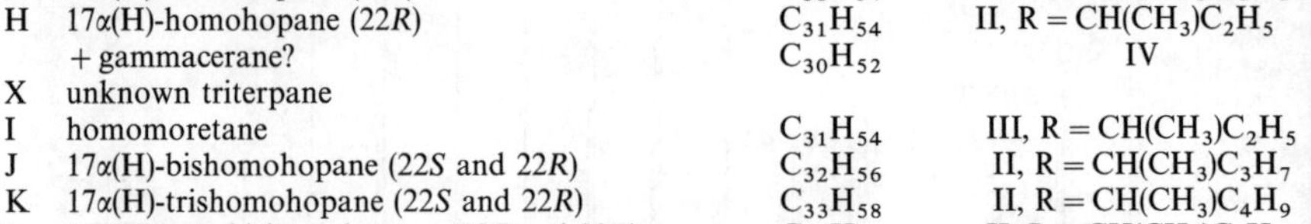

(I)   (II)   (III)   (IV)

TABLE 6. Sterane identification and structures[99]

Compound		Elemental composition	Structure
a	13β(H), 17α(H)-diacholestane (20S)	$C_{27}H_{48}$	I, R = H
b	13β(H), 17α(H)-diacholestane (20R)	$C_{27}H_{48}$	I, R = H
c	13α(H), 17β(H)-diacholestane (20S)	$C_{27}H_{48}$	II, R = H
d	13α(H), 17β(H)-diacholestane (20R)	$C_{27}H_{48}$	II, R = H
e	24-methyl-13β(H), 17α(H)-diacholestane (20S)	$C_{28}H_{50}$	I, R = CH$_3$
f	24-methyl-13β(H), 17α(H)-diacholestane (20R)	$C_{28}H_{50}$	I, R = CH$_3$
g	24-methyl-13α(H), 17β(H)-diacholestane (20S)	$C_{28}H_{50}$	II, R = CH$_3$
	+ 14α(H), 17α(H)-cholestane (20S)	$C_{27}H_{48}$	III, R = H
h	24-ethyl-13β(H), 17α(H)-diacholestane (20S)	$C_{29}H_{52}$	I, R = C$_2$H$_5$
	+ 14β(H), 17β(H)-cholestane (20R)	$C_{27}H_{48}$	IV, R = H
i	14β(H), 17β(H)-cholestane (20S)	$C_{27}H_{48}$	IV, R = H
	+ 24-methyl-13α(H), 17β(H)-diacholestane (20R)	$C_{28}H_{50}$	II, R = CH$_3$
j	14α(H), 17α(H)-cholestane (20R)	$C_{27}H_{48}$	III, R = H
k	24-ethyl-13β(H), 17α(H)-diacholestane (20R)	$C_{29}H_{52}$	I, R = C$_2$H$_5$
l	24-ethyl-13α(H), 17β(H)-diacholestane (20S)	$C_{29}H_{52}$	II, R = C$_2$H$_5$
m	24-methyl-14α(H), 17α(H)-cholestane (20S)	$C_{28}H_{50}$	III, R = CH$_3$
n	24-ethyl-13α(H), 17β(H)-diacholestane (20R)	$C_{29}H_{52}$	II, R = C$_2$H$_5$
	+ 24-methyl-14β(H), 17β(H)-cholestane (20R)	$C_{28}H_{50}$	IV, R = CH$_3$
o	24-methyl-14β(H), 17β(H)-cholestane (20S)	$C_{28}H_{50}$	IV, R = CH$_3$
p	24-methyl-14α(H), 17α(H)-cholestane (20R)	$C_{28}H_{50}$	III, R = CH$_3$
q	24-ethyl-14α(H), 17α(H)-cholestane (20S)	$C_{29}H_{52}$	III, R = C$_2$H$_5$
r	24-ethyl-14β(H), 17β(H)-cholestane (20R)	$C_{29}H_{52}$	IV, R = C$_2$H$_5$
s	24-ethyl-14β(H), 17β(H)-cholestane (20S)	$C_{29}H_{52}$	IV, R = C$_2$H$_5$
t	24-ethyl-14α(H), 17α(H)-cholestane (20R)	$C_{29}H_{52}$	III, R = C$_2$H$_5$

(I)   (II)   (IV)

(III)

metastable parent ion transitions, corresponding to $372^+ \rightarrow 217^+$, $386^+ \rightarrow 217^+$ and $400^+ \rightarrow 217^+$, can be observed separately during a single GC/MS run. The authors claim that this method is superior to the SIM (3000 RP) for the nominal 217 m/z fragmentogram since it separates much more clearly the C_{27} isomer from the C_{28} isomers[102]. A year later the same method was employed to extract by means of factor analysis information on various hopanes and steranes[103]. In Figure 31 the use of MIM (metastable ion monitoring) is demonstrated for the C_{30} series of steranes and hopanes. To some extent the SIM for m/z 191, 217, 231 nominal masses or SIR might give similar profiles. However, since C_{30} steranes relate to marine sources, the exact profile restricted to these is of importance.

A restricting factor in GC/MS/C analysis of alkanes and cycloalkanes is the scanning time, which is much shorter for quadrupole than for magnetic MS, and the mass range scanned. Quadrupole MS was restricted in the past to m/z 600, but in the last few years the range reached m/z 1200 at low resolution. In magnetic sector MS, fast scanning for large range (m/z 50–1000) causes magnetic hysteresis. For heavy bitumens and waxy

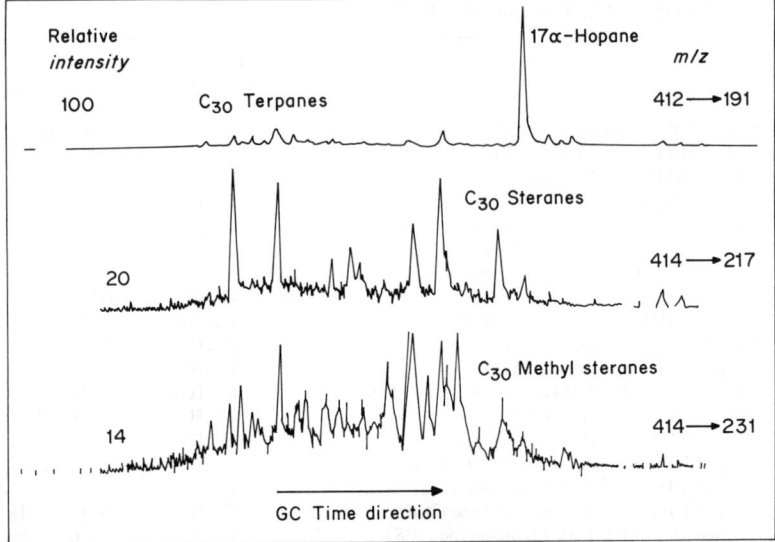

FIGURE 31. Daughter-ion scans for C_{30} steranes in Sadlerochit, Alaska (marine) crude[118] (B/E linked scan: MS is scanned in such a mode that the ratio of magnetic (B) to electric (E) selector fields is maintained constant as both are scanned, with accelerating voltage held constant

matter another problem is the volatilization of the introduced compounds, hence the adoption of the GC-field ionization MS (GC-FIMS)[104]. This method was employed by using packed GC columns connected through a separator to FI modified MS and mass range scan m/z 80–800 5 s per decade, giving a total scan time of only 11 s. This method allowed the identification of n-alkanes up to C_{40}, although with much lower GC resolution than capillary GC/MS. The authors[104] predicted that GC-FIMS has the 'potential of being one of the most powerful and valuable tools available for research in this field', but actually the major effort in GC/MS/C did not shift in this direction.

B. Isotope Analysis of Alkanes and Cycloalkanes

Both carbon and hydrogen have stable isotopes and the isotopic composition is measured as δ, i.e. the ratio of the heavier isotope to the lighter one for the measured sample compared with an elected standard compound:

$$\delta^{13}C = \left[\frac{(^{13}C/^{12}C)\,\text{sample}}{(^{13}C/^{12}C)\,\text{standard}} - 1 \right] \times 1000$$

The same basic formula is used for δD and results are given in ‰ values.

Since saturated hydrocarbons appear in nature in mixtures with other compounds, in order to otain their $\delta^{13}C$ and δD values one has to separate them from the mixture.

It is beyond the scope of the present review to discuss the use of stable carbon and hydrogen isotopes, even as used on whole fractions[105–107]. The point of interest is that the saturated hydrocarbon fraction, for all oils examined, shows an enrichment of ^{12}C as compared with aromatics, resins and asphaltenes[107]. Already in 1939 it was well established[108] that carbon compounds of biological pedigree are markedly enriched in the light isotope, while most carbonates are relatively enriched in ^{13}C. Within the hydrocarbons, in liquid petroleum, $\delta^{13}C$ values range around $-28‰ \pm 3‰$, while for methane $\delta^{13}C$ values were from $-21‰$ to $-75‰$. There are deviations from these ranges in specific cases. The genetic characterization of natural gases, mostly C_1-C_4 hydrocarbons, is usually based on concentration of the homologues and on carbon and hydrogen isotope variations[109]. The natural gases studied (*ca* 500) show methane of biogenic, thermogenic and mixed sources. The possibility to determine $\delta^{13}C$ and δD separately for each hydrocarbon became possible by the use of GC separation.

Analytically, the separated gases have to be combusted to form CO_2 for $\delta^{13}C$ measurements and the H_2O formed is turned into H_2, which is measured for δD. The standard for carbon isotopes is *usually* PDB (Peedee Belemnite, a carbonate) though the oil industry also uses the NBS_{22} (an oil fraction), whereas the δD standard is SMOW (Standard Mean Ocean Water).

A vast amount of $\delta^{13}C$ and δD information was collected on various petroleum fractions and on general concepts relating to the isotopic signature. Also, much effort was invested to minimize the amount of CO_2 needed and the possibility to interface the GC-isoMS. This method, allowing direct connection of the separating analytical method with the isotope MS, expanded the possibility to determine $\delta^{13}C$ for single compounds. Matthews and Hayes investigated the possibility to improve the technique of continuous isotopic analysis of the effluent of a gas chromatograph[110]. Until 1987[111], the isotopic carbon composition was still given for fractions isolated and sealed in quartz tubes to produce CO_2 (see Table 7)[111]. Within the last two years, two new analytical approaches to the compound-specific isotope analyses were suggested, both based on the early work by Sano and coworkers[112], recognizing the importance of interposing a combustion furnace between the GC and the MS in order to determine isotopic abundance data from small quantities of GC effluents. Thus studies of $\delta^{13}C$ of n-alkanes were performed by Gilmour and collaborators[113]. This still requires trapping of the CO_2 formed in the furnace and its transfer to the MS, but the reproducibility of the method was very good and $\delta^{13}C$ of n-alkanes could be determined with $\pm 0.2‰$ precision.

For better separation and more universal use a GC/IRMS instrument interface (Isochrom II), VG-Isogas Co., was introduced. From the applications of this system we

TABLE 7. Isotopic compositions of carbon fractions in the Messel shale[111] [a]

Sample identification	$\delta^{13}C_{PDB}(‰)$
Total carbonate	$+7.34 \pm 0.12$
Total organic carbon, kerogen	-28.21 ± 0.03
Total extractable organic material[b]	-29.72 ± 0.10
Total extract fractionated on SiO_2 column[c]	
Hexane eluent	-33.66 ± 0.14
Toluene eluent	-29.66 ± 0.03
Methanol eluent	-28.30 ± 0.07
Alkyl porphyrin fraction (Strasbourg)	-22.60 ± 0.08
Total porphyrins (Bloomington)	-23.43 ± 0.06
Acid prophyrin fraction (Strasbourg)	-23.91 ± 0.04

[a]Carbon dioxide for isotopic analysis prepared by combustion of organic material in sealed quartz tubes. Carbon-isotopic compositions reported in parts per thousand relative to the PDB standard.
[b]Soxhlet extraction, 50/50 (v/v) dichloromethane–methanol, 72 h.
[c]'Baker-10' disposable extraction column.

342

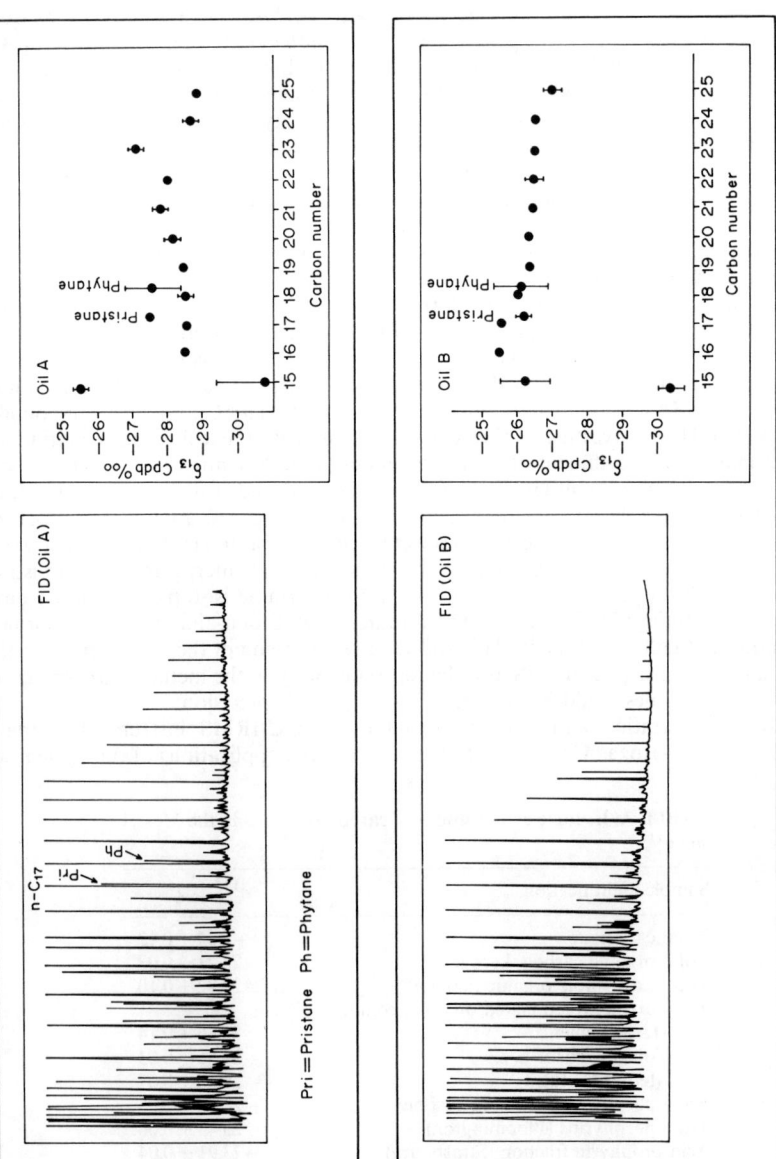

FIGURE 32. Gas chromatograms of oils A and B as detected by FID (flame ionization detector) and $\delta^{13}C$ values for selected hydrocarbons obtained by GC-C-MS instrument VG-isogas[114]

343

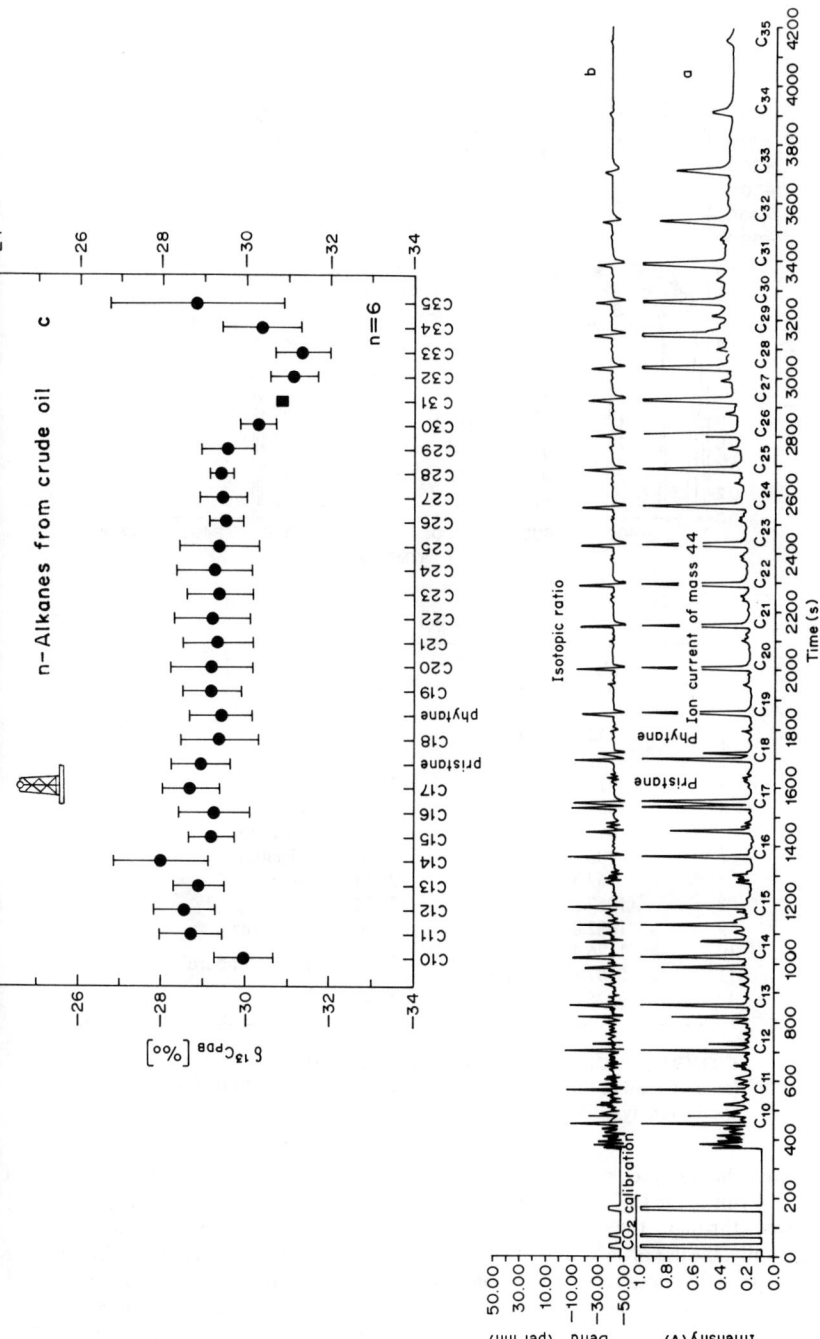

FIGURE 33. The GC-C-iso MS results of n-alkanes from crude oil (Finnigan-Mat): (a) m/z 44 trace, (b) δ^{13}C relative ratio (‰), (c) actual δ^{13}C PDB and range of experimental error for each alkane

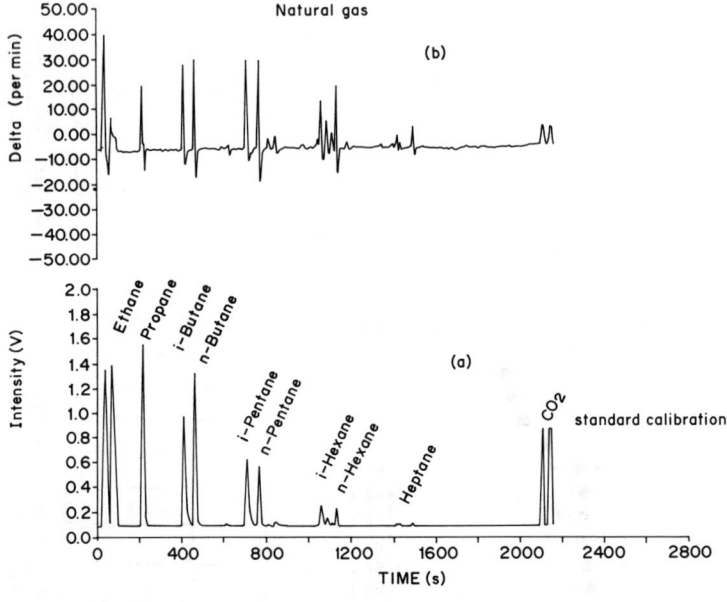

| | t_R | | | |
PEAK	(s)	nmol	$\delta^{13}C$	
1	34.7	16.3	−31.27	
2	67.0	12.6	−30.85	ethane
3	210.9	11.3	−28.54	propane
4	407.3	6.5	−30.99	i-butane
5	457.5	8.1	−28.96	i-pentane
6	705.2	5.0	−27.39	n-pentane
7	762.3	3.2	−28.14	i-hexane
8	1057.5	0.8	−26.24	
9	1082.8	0.3	−24.76	n-hexane
10	1126.4	0.7	−24.51	CO$_2$ standard
11	2197.1	7.2	−27.52	
12	2225.6	6.3	−27.90	

FIGURE 34. Natural gas analysis by GC-C-MS: (a) m/z 44 trace (quantification), (b) $\delta^{13}C$ relative to standard, (c) $\delta^{13}C$ PDB for each alkane (analyzed by Finnigan-Mat)

will discuss the analyses of two oils (A and B) as shown in Figure 32[114], which shows the normal run of the GC as recorded by the FID. During the 'heart-cut' the GC effluent is allowed through the furnace into the trap and MS (Figure 35 and Table 8). Conceptually, the use of $\delta^{13}C$ and δD data of single compounds is based on the knowledge that isotopic differences relate to the biogenic processes at the source and chemical and geochemical controlled processes at the sediment. As an example, it is claimed that the isotopic ($\delta^{13}C$) composition of n-alkanes is covariant with the total organic carbon while the δ values for pristane and phytane are covariant with those of porphyrins[116]. For those 'biomarkers' with $-60‰$ to $-73‰$ $\delta^{13}C$ values it is suggested that they relate to methane utilizing bacteria (chemoautotrophic and chemolithotrophic microorganisms). However, we cannot exclude heterotrophic bacteria as a source for the ^{13}C-depleted hopanes because of the possibility of bacterial reworking of primary matter.

In some methods, relating to isotope studies, the ratios measured provide information on biogenic and chemical processes. The introduction of the GC-C-iso-MS method allows the replacement of ^{14}C- or ^{33}S- or ^3H-labelled radioactive compounds by non-hazardous molecules.

A somewhat different interface and operational mode is offered by the GC-C-MS instrument of the Finnigan-Mat company in which the water removal is controlled by absorption, and both the FID and the MS may serve as retention and quantitation monitors and the m/z 44 total ion current is used. Calibration by CO_2 is carried out either prior to the run (Figure 33) or by using one of the known compounds within the mixtures. With this instrument, the direct analysis of the C_2–C_6 fraction could be performed (Figure 34).

We have stressed the application of this new method to the geosciences; however, it is also suggested for use in medicine, food control and nutrition. Whereas in the n-alkanes the deviation from normal was very small, in branched and alicyclic 'biomarkers' the diversity is from $ca-21‰$ to $-73.4‰$ $(\delta^{13}C)$[115]. Because of the low concentration of the

TABLE 8. Carbon-isotopic compositions of individual compounds[115]

Peak No. in Figure 35	$t_R{}^a$ (s)	Amount^b (nmol C)	$\delta^{13}C^c$ ‰ vs PDB	Identification
1	1679	1.1	-22.7 ± 1.0	norpristane
2	1722	1.0	-30.2 ± 0.3	C_{19} acyclic isoprenoid
3	1812	0.7	-25.4 ± 1.0	pristane
4	2040	2.0	-31.8 ± 0.8	phytane
5	2602	1.0	-29.1 ± 0.6	C_{23} acyclic isoprenoid
6	3161	1.3	-23.9 ± 0.6	$10\beta(H)$-des-A-lupane
7	3571	1.3	-24.9 ± 1.0	mixture of hydrocarbons
8	3688	2.6	-73.4 ± 1.3	C_{32} acyclic isoprenoid
9	3883	0.9	-24.2 ± 1.2	isoprenoid alkane
10	3957	6.8	-49.9 ± 1.1	$17\beta(H)$-22,29,30-trisnorhopane
11	3977	2.0	-60.4 ± 1.8	isoprenoid alkane
12	4100	1.6	-43.5 ± 1.0	$17\alpha(H), 21\beta(H)$-30-norhopane
13	4156	2.0	$ca-45$	$17\beta(H), 21\alpha(H)$-30-norhopane^d
14	4210	2.9	$ca-34$	$17\alpha(H), 21\beta(H)$-hopane^e
15	4256	6.2	-65.3 ± 1.4	$17\beta(H), 21\beta(H)$-30-norhopane
16	4364	1.8	-39.4 ± 0.8	$17\alpha(H),21\beta(H)$-homohopane
17	4392	1.3	-35.2 ± 1.4	$17\beta(H), 21\beta(H)$-homohopane
18	4552	4.2	-36.6 ± 0.5	$17\beta(H), 21\beta(H)$-homohopane
19	4692	15.4	-20.9 ± 1.1	lycopane^f
20	5010	0.5	-27.0 ± 0.4	unknown hydrocarbon
21	5408	0.8	-28.8 ± 1.0	unknown hydrocarbon

^{a}Chromatographic retention time (Figure 35).
^{b}Amounts of carbon represented by peaks in Figure 35.
$^c\delta \equiv 10^3\,[(R_x - R_s)/R_s]$, where $R \equiv {}^{13}C/{}^{12}C$, x designates sample, s designates PDB standard, and $R_s = 0.0112372$. Entries are mean and 95% confidence interval for two to eight replicate measurements.
^{d}Peak contains additional component, apparently an isoprenoid alkane $(\delta \sim -22‰)$.
^{e}Peak contains additional component, probably a C_{30} 4-methyl sterane. In contrast with the situation in peak 13, variations in the ratio plot suggest that δ values of the unresolved components do not differ by more than 10‰.
^{f}Peak also contains minor component, apparently an acyclic isoprenoid.

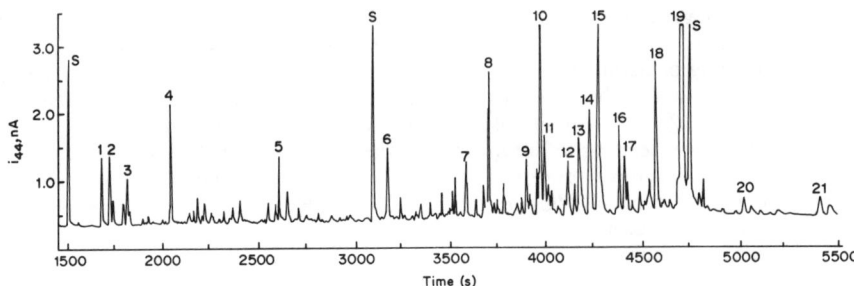

FIGURE 35. The 1500–5500 scan range of the m/z 44 trace of 'biomarkers' isolated.
Identification and $\delta^{13}C$ for each compound 1–19 in Table 8[115].

hydrocarbons studied, this fraction was concentrated to enhance also the
chromatographic separation. In a recent paper, Hayes and coworkers[116] discussed the
use of $\delta^{13}C$ for compound-specific analysis as a novel tool.

VIII. REFERENCES

1. L. F. Fieser and M. Fieser, *Advanced Organic Chemistry*, Reinhold, New York, 1961.
2. R. Willstatter, as cited in Reference 3.
3. F. G. Fischer, Ph.D. University of Freiburg, Wurzburg, 1928, as quoted by L. F. Fieser and M. Fieser, *Top. Org. Chem.*, **1**, 191 (1963).
4. J. E. Erdman, E. M. Marlett and W. E. Hanson, Division of Petroleum Chemistry, ACS Meeting, Chicago, III, No. 4, Sep, C-25, 1958.
5. J. G. Bendoraitis, B. L. Brown and L. S. Hapner, *Anal. Chem.*, **34**, 49 (1962).
6. J. W. K. Burrell, L. M. Jackman and B. C. L. Weedon, Stereochemistry and synthesis of phytol, geraniol and nerol, *Proc. Chem. Soc.* 263 (1952).
7. J. W. K. Burrell, R. F. Garwood, L. M. Jackman, E. Oskay and B. C. L. Weedon, *J. Chem. Soc. (C)*, 2144 (1966).
8. G. E. Gil-Av, B. Feibush and R. Charles-Sigler, in *Gas Chromatography* (Ed. A. B. Littlewood), Institute of Petroleum, Pittsburgh, PA, 1966, pp. 227–239.
9. M. C. Simmons, D. B. Richardson and I. Dvoretzky, in *Gas Chromatography* (Ed. R. P. W. Scott), Butterworths, Washington, 1960, pp. 211–223.
10. J. R. Maxwell, R. E. Cox, R. G. Ackman and S. N. Hooper, *Advances in Organic Geochemistry* (Ed. E. Ingerson), **33**, Earth Sciences 277, Pergamon Press, Oxford, 1972.
11. G. Holtzer, J. Ore and T. G. Tornabene, *J. Chromatogr.*, **186**, 795 (1979).
12. M. Charton, Olefinic properties of cyclopropanes, in *The Chemistry of Alkenes*, Vol. 2 (Ed. J. Zabicky), Wiley Interscience, London, 1970.
13. V. A. Kalichevsky and B. A. Stagner, *Chemical Refining of Petroleum*, Reinhold, New York, 1942.
14. W. M. Coleman III, B. M. Gordon and B. M. Lawrence, *Appl. Spectrosc.*, **43**, 298 (1989).
15. L. F. Fieser and M. Fieser, *Top. Org. Chem.*, Reinhold, New York, 1963, p. 199.
16. M. P. Nomura, P. von R. Schleyer and A. A. Arz, *J. Am. Chem. Soc.*, **89**, 3657 (1967).
17. A. I. Bogomolov, *Tr. Vses. Neftegazov. Naucho-Issled., Geologorazved. Inst.*, 10 (1964).
18. *NBS (EPA/NIH) Mass Spectra Database*, Natl. Stand. Res. Data Ser., 1987 (updated).
19. *Catalog of Mass Spectral Data*, American Petroleum Institute Research Project 44, Carnegie Institute of Technology, Pittsburg, Pennsylvania (1960).
20. H. Budzikiewic, C. Djerassi and D. H. Williams, *Mass Spectrometry of Organic Compounds,* Holden-Day, San Francisco, 1967.
21. S. G. Lias, J. E. Bartmess, J. F. Liebman, J. L. Holmes, R. D. Levin and W. G. Mallard, Gas-phase ion and neutral thermochemistry, *J. Phys. Chem. Ref. Data*, Natl. Stand. Res. Data Ser., Natl. Bur. Stand. (Library of Congress Cat. No. 88-70606), New York (1988).

22. L. Radom, J. A. Pople and P. v. R. Schleyer, *J. Am. Chem. Soc.*, **95**, 8193 (1973).
23. N. Bodor, M. J. S. Dewar and D. H. Lo, *J. Am. Chem. Soc.*, **94**, 5303 (1972).
24. A. Herlan, *Brennstoffchemie*, **45**, 244 (1964).
25. A. L. Burlingame and H. K. Schnoes, in *Organic Geochemistry* (Eds. G. Eglinton and M. T. J. Murphy), Chap. 4, Longman, Springer-Verlag, London, 1984.
26. R. P. Philp, *Fossil Fuel Biomarkers, Application and Spectra*, Elsevier, Amsterdam, 1985.
27. B. J. Millard and D. F. Shaw, *J. Chem. Soc. (B)*, 664 (1966).
28. S. Meyerson, T. D. Nevitt and P. N. Rylander, in *Advances in Mass Spectrometry*, Vol. 2 (Ed. R. M. Elliot), Pergamon Press, Oxford, 1963, pp. 313–336.
29. L. D'Or, J. Momigny and P. Natalis, in *Advances in Mass Spectrometry*, Vol. 2 (Ed. R. M. Elliot), Pergamon Press, Oxford, 1963; P. Natalis, in *Mass Spectrometry* (Ed. R. I. Reed), Academic Press, London, 1965, pp. 379–399.
30. R. Alexander, R. Kagi and R. Noble, *J. Chem. Soc., Chem. Commun.*, 226 (1983).
31. J. G. Bendoraitis, in *Advances in Organic Geochemistry* (Eds. B. Tissot and F. Bimmer), Technip, Paris, 1973, pp. 209–224.
32. R. J. Waltman and A. Campbell Ling, *Can. J. Chem.*, **58**, 2189 (1980).
33. R. C. Fort, Jr. and P. v. R. Schleyer, *Chem. Rev.*, **64**, 277 (1964).
34. R. C. Fort, Jr., *Adamantane, The Chemistry of Diamond Molecules*, Marcel Dekker, New York, 1976.
35. M. Kuras and S. Hala, *J. Chromatogr.*, **51**, 45 (1970).
36. A. L. Choffee, *The Organic Geochemistry of Australian Coals*, Ph.D. Thesis, University of Melbourne, 1983.
37. R. C. Barrick and J. I. Hedges, *Geochim. Cosmochim. Acta*, **45**, 381 (1981).
38. F. R. Aquino Neto, F. R. Restle, A. Connan, J. Albrecht and G. Ourisson, *Tetrahedron Lett.*, **23**, 2027 (1982).
39. D. S. de Miranda, G. Brendolan, P. M. Imamura, M. G. Sierra, A. J. Maraioli and E. A. Ruveda, *J. Org. Chem.*, **46**, 4851 (1981).
40. R. M. Carangelo, P. R. Solomon and D. J. Gerson, *Fuel*, **66**, 960 (1987).
41. A. J. Barnes, W. J. Orville-Thomas, A. Muller and R. Gaufres (Eds.), *Matrix Isolation Spectroscopy*, Reidel, Boston, 1981.
42. H. E. Hallam (Ed.), *Vibrational Spectroscopy of Trapped Species*, Wiley, New York, 1973.
43. G. T. Reedy, D. G. Ettinger and J. F. Schneider, *Anal. Chem.*, **57**, 1602 (1985).
44. W. M. Coleman III and B. M. Gordon, *Appl. Spectrosc.*, **41**, 886 (1987); V. Formmak and K.-H. Kubeczka, *Essential Oil Analysis by Capillary Gas Chromatography and ^{13}C NMR Spectroscopy*. John Wiley & Sons, New York, 1982.
45. P. R. Griffiths, S. L. Pentoney, Jr., A. Gorgetti and K. H. Shafer, *Anal. Chem.*, **58**, 1349 (1986).
46. W. A. Thomas, in *Annual Review of NMR Spectroscopy* (Ed. E. F. Mooney), Academic Press, London, 1968, pp. 43–89.
47. N. Muller and W. C. Tosch, *J. Chem. Phys.*, **37**, 1167 (1962).
48. E. D. Becker, *High Resolution NMR, Theory and Chemical Applications*, Academic Press, New York, 1980.
49. G. A. Gray, in *Two-Dimensional NMR Spectroscopy: Application for Chemists and Biochemists* (Eds. W. R. Croasmun and R. M. K. Carlson), Chap. 1, VCH, New York, 1987, pp. 1–64.
50. W. R. Croasmun and R. M. K. Carlson, in *Two-Dimensional NMR Spectroscopy: Application for Chemists and Biochemists* (Eds. W. R. Croasmum and R. M. K. Carlson), Chap. 7, VCH, New York, 1987, pp. 387–424.
51. G. C. Speers and E. V. Whitehead, in *Organic Geochemistry* (Eds. G. Eglinton and M. T. J. Murphy), Longman, Springer-Verlag, New York, 1969, pp. 638–675.
52. T. R. McManus, *Anal. Chem.*, **61**, 165R (1989).
53. M. Radke, H. Willsch and D. H. Welte, *Anal. Chem.*, **52**, 406 (1980).
54. J. C. Suatoni and R. E. Swab, *J. Chromatogr. Sci.*, **14**, 535 (1976).
55. J. C. Suatoni and H. R. Garber *J. Chromatogr. Sci.*, **14**, 546 (1975).
56. P. C. Hayes, Jr. and S. D. Anderson, *Anal. Chem.*, **58**, 2384 (1986).
57. (a) E. Lundanes and T. Greibrokk, *J. Chromatogr.*, **322**, 347 (1985).
    (b) J. Burda, M. Kuras, J. Kriz and L. Vodicka, *Fresenius Z. Anal. Chem.*, **321**, 549 (1985).
58. W. Meinschein and G. Kenny, *Anal. Chem.*, **29**, 1153 (1957).
59. M. Baron, Analytical application of inclusion compounds, in *Physical Methods in Analytical Chemistry*, Vol. 4 (Ed. W. Berl), Academic Press, New York, 1961, p. 226.

60. N. Nicolaides and F. Laves, *J. Am. Oil Chem. Soc.*, **40**, 400 (1963).
61. T. Thomas and R. Mays, Separation with molecular sieves, in *Physical Methods in Analytical Chemistry*, Vol. 4 (Ed. W. Berl), Academic Press, New York, 1961, p. 45.
62. M. T. J. Murphy, in *Organic Geochemistry* (Eds. G. Eglinton and M. T. J. Murphy), Chap. 3, Longman, Springer Verlag, New York, 1969, p. 74.
63. W. Schlenk, *Ann. Chem.*, **565**, 204 (1949).
64. Sr. M. Murphy, A. McCormick and G. Eglinton, *Science*, **157**, 1040 (1967).
65. E. Tannenbaum and Z. Aizenshtat, *Org. Geochem.*, **8**, 181 (1985).
66. R. M. Teeter, *Mass Spectrometry Reviews*, **4**, 123 (1985).
67. A. J. P. Martin and R. L. M. Synge, *Biochem. J.*, **35**, 1358 (1941).
68. J. K. Haken, M. S. Wainright and D. Srisukh, *J. Chromatogr.*, **257**, 107 (1983) and papers cited therein.
69. D. H. Desty, J. N. Harensnape and B. H. F. Whyman, *Anal. Chem.*, **32**, 302 (1960).
70. L. S. Ettre, *Open Tubular Columns in Gas Chromatography*, Plenum Press, New York, 1965.
71. D. H. Desty, Capillary columns, trials, tribulations and triumphs, in *Advances in Chromatography*, Vol. 1 (Eds. J. C. Giddings and R. A. Keller), Marcel Dekker, New York, 1966, pp. 199–228.
72. A. G. Douglas, *Gas Chromatography in Organic Geochemistry* (Eds. G. Eglinton and M. T. J. Murphy), Chap. 5, Longman-Springer, New York, 1969, pp. 162–180.
73. R. G. Schaefer, B. Weiner and D. Lethaeuser, *Anal. Chem.*, **50**, 1848 (1978).
74. Y. V. Kissin and G. P. Feulmer, *J. Chromatogr. Sci.*, **24**, 53 (1986).
75. B. P. Tissot and D. H. Welte, *Petroleum Formation and Occurrence*, Springer-Verlag, New York, 1978.
76. L. R. Snowdon, *Anal. Chem.*, **50**, 379 (1978).
77. R. L. Grob (Ed.), *Modern Practice of Gas Chromatography*, 2nd ed., Wiley, New York, 1977.
78. H. M. McNair, M. W. Ogden and J. L. Hensley, Recent advances in gas chromatography, *Am. Lab.*, **17**, 15 (1985).
79. S. A. Borman, *Anal. Chem.*, **55**, 726A (1983).
80. R. K. Kuchhal, B. Kumar, H. S. Malthur and G. C. Joshi, *J. Chromatogr.*, **361**, 269 (1986).
81. Y. V. Kissin, *Org. Geochem.*, **15**, 575 (1990).
82. R. M. Carangelo, D. R. Solomon and D. J. Gerson, *Fuel*, **66**, 961 (1987).
83. R. P. Evershed, in *Mass Spectrometry*, Vol. 10 (Ed. M. E. Rose), Royal Society of Chemistry, London, 1989, p. 181.
84. M. A. Grayson, *J. Chromatogr. Sci.* **24**, 529 (1986).
85. D. A. Catlow and M. E. Roe, in *Mass Spectrometry*, Vol. 10 (Ed. M. E. Rose), Royal Society of Chemistry, London, 1989.
86. L. P. Lindemann and J. J. Annis, *Anal. Chem.*, **32**, 1742 (1960).
87. H. Y. Tong, T. S. Thompson and F. W. Karasek, *J. Chromatogr.*, **350**, 27 (1985).
88. B. J. Millard, *Quantitative Mass Spectrometry*, Heyden & Son, Philadelphia, 1978.
89. D. E. Anders and W. E. Robinson, *Geochim. Cosmochim. Acta*, **35**, 661 (1971).
90. F. W. McLafferty, *J. Am. Soc. Mass Spectrom.*, **1**, 1 (1990) and references cited therein.
91. A. S. Mackenzie, Application of biological markers in petroleum geochemistry, in *Advances in Petroleum Geochemistry*, Vol. 1, Academic Press, London, 1984, p. 115.
92. A. A. Petrov and N. N. Abryutina, *Russ. Chem. Rev. (Engl. Transl.)*, **58**(6), 575 (1989); translated from *Usp. Khim.*, **58**, 983 (1989).
93. S. D. Killops, *Anal. Chem. Acta*, **183**, 105 (1986).
94. S. Wold and C. H. J. Christie, *Anal. Chem. Acta*, **165**, 51 (1984).
95. D. R. Scott, *Anal. Chem.*, **58**, 881 (1986).
96. E. J. Gallegos, J. W. Green, L. P. Lindman, R. Le Tourneall and R. M. Teeter, *Anal. Chem.*, **39**, 1833 (1967).
97. S. Zadro, J. K. Haken and W. V. Pinczewski, *J. Chromatogr.*, **323**, 305 (1985).
98. A. Stoler, E. Grushka and Z. Aizenshtat, submitted (1991); A. Stoler, Ph.D. Thesis, The Hebrew University of Jerusalem, 1991.
99. A. S. Mackenzie, U. Disko and J. Rullkotter, *Org. Geochem.*, **5**, 57 (1983) and references cited therein.
100. S. Feinstein, P. W. Brooks, M. G. Fowler, L. R. Snowdon, M. Goldberg and Z. Aizenshtat, submitted (1991).

101. M. G. Fowler, L. R. Snowdon, P. W. Brook and J. S. Hamilton, *Org. Geochem.*, **13**, 715 (1988).
102. G. A. Warburton and J. E. Zumberg, *Anal. Chem.*, **55**, 123 (1983).
103. O. H. J. Christie, T. Meyer and P. W. Brooks, *Anal. Chim. Acta*, **161**, 75 (1984).
104. J. D. Payzant, I. Rubinstein, A. M. Hogg and O. P. Straustz, *Chem. Geol.*, **29**, 73 (1980).
105. E. M. Galimov, *The Biological Fractionation of Isotopes*, Academic Press, New York, 1985.
106. W. J. Stahl, *Chem. Geol.*, **20**, 121 (1977).
107. W. J. Stahl, *Geochim. Cosmochim. Acta*, **42**, 1573 (1978).
108. A. O. Nier and E. A. Gulbransh, *J. Am. Chem. Soc.*, **61**, 697 (1939).
109. M. Schoell, *Am. Assoc. Pet. Geol. Bull.*, **67**, 2325 (1983).
110. D. E. Matthews and J. M. Hayes, *Anal. Chem.*, **50**, 1465 (1978).
111. J. M. Hayes, R. Takigku, R. Campo, H. J. Callot and P. Albrecht, *Nature*, **329**, 48 (1987).
112. M. Sano, Y. Yotsui, H. Abe and S. Sasaki, *Biomed. Mass Spectrom.*, **3**, 1 (1976).
113. I. Gilmour, P. K. Swart and C. T. Pillinger, *Org. Geochem.*, **6**, 665 (1984); and in *Advances in Organic Geochemistry*, Pergamon Press, Oxford, 1983.
114. *Preliminary GC-IRMS Studies of Hydrocarbon Mixtures*, Application Note No. 11, VG-Isogas, 1991.
115. K. H. Freeman, J. M. Hayes, J. M. Trendel and P. Albrecht, *Nature*, **343**, 254 (1990).
116. J. M. Hayes, K. H. Freeman, B. N. Popp and C. H. Hoham, *Org. Geochem.*, **16**, 1115 (1989); and in *Advances in Organic Geochemistry*, 14th EAOG Meeting, Pergamon Press, Oxford, 1991.
117. Z. Aizenshtat, unpublished results (see also Reference 98).
118. J. M. Moldowan, W. K. Seifert and E. J. Gallegos, *Am. Assoc. Pet. Geol. Bull.*, **69**, 1255 (1985).

CHAPTER **8**

NMR spectroscopy of alkanes

STEFAN BERGER

Department of Chemistry, Philipps University, D-3550 Marburg, Germany

I. INTRODUCTION

Alkanes form a class of simple molecules which provide a large structural variety, although they consist only of hydrogen and sp^3-hybridized carbon atoms. Due to both

The Chemistry of Alkanes and Cycloalkanes
Edited by S. Patai and Z. Rappoport © 1992 John Wiley & Sons Ltd

352   S. Berger

their simplicity and their structural variety, they have served as basic test molecules for
theoretical and experimental work in NMR spectroscopy. The first incremental systems,
and the first considerations of the structural dependence of chemical shifts, spin–spin
coupling constants and relaxation times have been worked out on alkanes. It is, however,
not the aim of this chapter to discuss in detail the historical development of, for example,
chemical shift theory, but rather to give an account of the current knowledge of NMR
parameters for alkanes and their application in structural chemistry. The most important
results are presented in tabular form. As one might expect, most work has been done
by ^{13}C NMR spectroscopy, since in ^{1}H NMR alkanes very often give only one broad
line with little structural information. Only recently, with the advent of NMR spectrometers of very high field, has it become possible to resolve some of the ^{1}H signals of the
higher alkanes. Therefore, this chapter is primarily devoted to ^{13}C NMR, although useful
^{1}H NMR data are also included.

II. THE CHEMICAL SHIFT

A. Chemical Shift Theory

As mentioned above, alkanes, especially methane, ethane and propane, are the prime
test molecules to check the validity of theoretical models and programs for chemical
shift calculation. Therefore, nearly each new approach in chemical shift theory, starting
from early attempts using valence bond theory[1] or a simple CNDO/2 formalism[2], was
applied to the chemical shifts of these molecules. A comparison of the results for methane
and ethane obtained by different methods can be found elsewhere[3]. Although for alkanes
the charge variation at the carbon atoms is considerably small, it was shown that with
STO-3G calculations a correlation exists between the chemical shifts and the calculated
charges for acyclic alkanes[4], cyclohexanes[5] and adamantanes[6]. The current state of the
art in ^{13}C NMR chemical shift calculation is represented by the IGLO method[7], which
furnishes chemical shift values for a large variety of alkanes and cycloalkanes showing
very good agreement (standard deviation of about 1 ppm) with experimental data. In a
very recent and detailed study it was shown[8] that the IGLO method not only gives
excellent agreement with the experimental values, but is also apt to help in understanding
the conformational effects which were earlier examined in detail with the finite
perturbation theory at the INDO level[9]. The authors were able to calculate the α, β and
γ effects used in the popular increment systems (see below). However, it was noted that
the steric γ-effect widely used in stereochemical assignments is more complex than usually
assumed. For butane, changes in the paramagnetic contributions for the C1—C2 bond
are important. The absolute shielding constant σ was calculated to be 190.6 ppm for
methane[10], the diamagnetic contribution σ_D 295.1 ppm and the paramagnetic contribution σ_P −104.5 ppm[11].

B. The Experimental Chemical Shift

1. Acyclic alkanes

The ^{13}C chemical shifts in linear and branched unsubstituted alkanes extend over a
range of about 60 ppm. The shifts, referred to TMS, increase in the order
$CH_3 < CH_2 < CH < C$. The chemical shifts are given in Table 1[3,12-14]. For the higher
linear alkanes from n-nonane upwards, the δ values become almost constant. Under
conditions of very high resolution, all signals of the carbon atoms of n-hexadecane,
2-methyltridecane and 7-methylhexadecane had been resolved[15].

TABLE 1. ^{13}C NMR chemical shifts of acyclic alkanes[a]

Compound	δ_c						
	C-1	C-2	C-3	C-4	C-5	C-6	C-7
CH_4	−2.3						
H_3C—CH_3	6.5						
H_3C—CH_2—CH_3	16.1	16.3					
H_3C—CH_2—CH_2—CH_3	13.1	24.9					

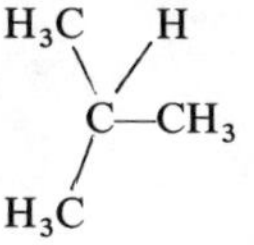 (isobutane) C—CH₃	24.6	23.3					
H_3C—CH_2—CH_2—CH_2—CH_3	13.5 (13.8)[b]	22.2 (22.8)	34.1 (34.7)				

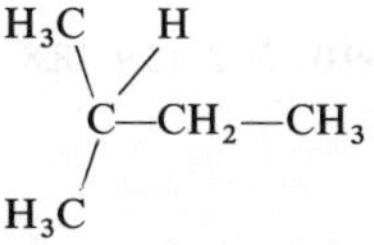 C—CH₂—CH₃	21.9	29.9	31.6	11.5			

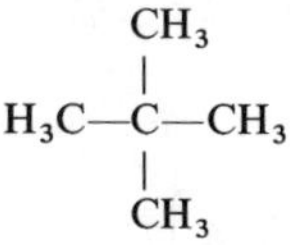 H_3C—C(CH₃)₂—CH₃	27.4	31.4					
H_3C—CH_2—CH_2—CH_2—CH_2—CH_3	13.7 (14.1)[b]	22.7 (23.2)	31.7 (32.2)				

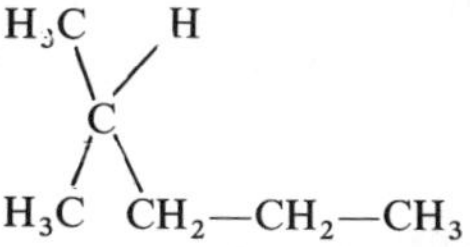 C, CH₂—CH₂—CH₃	22.7	27.9	41.9	20.8	14.3		
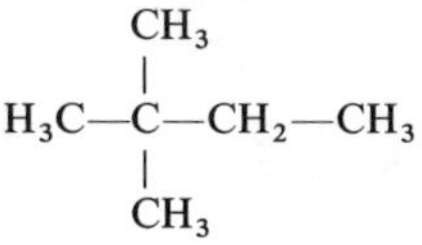 H_3C—C(CH₃)₂—CH₂—CH₃	28.7	30.3	36.1	8.5			
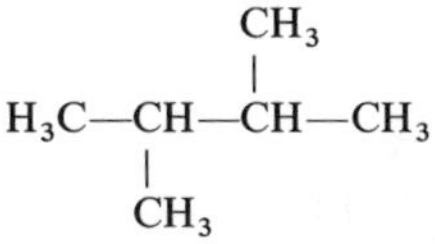 H_3C—CH(CH₃)—CH(CH₃)—CH₃	19.2	34.0					
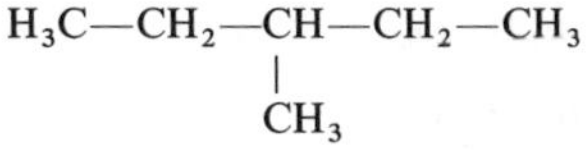 H_3C—CH_2—CH(CH₃)—CH_2—CH_3	11.4	29.4	36.8			18.7	
H_3C—CH_2—CH_2—CH_2—CH_2—CH_2—CH_3	13.7 (14.2)[b]	22.6 (23.3)	32.0 (32.6)	29.0 (29.8)			

(continued)

TABLE 1. (*continued*)

Compound	δ_c						
	C-1	C-2	C-3	C-4	C-5	C-6	C-7
$H_3C-CH_2-CH_2-CH_2-CH_2-(CH_2)_3-CH_3$	13.8	22.7	32.0	29.4	29.6		
$H_3C-CH_2-CH_2-CH_2-CH_2-(CH_2)_4-CH_3$	14.0	22.8	33.3	29.8	30.1		
$CH_3-CH_2-CH_2-CH_2-CH_2-CH_2-(CH_2)_5CH_3$	14.1	22.8	32.0	29.5	29.7	29.8	

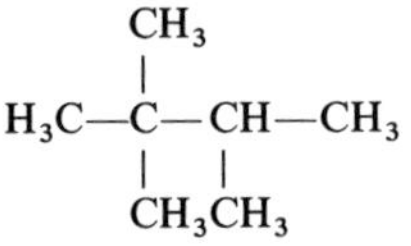

$$H_3C-\underset{\underset{CH_3}{|}}{\overset{\overset{CH_3}{|}}{C}}-\underset{\underset{CH_3}{|}}{CH}-CH_3$$

27.0 32.7 37.9 17.7

$$H_3C-\underset{\underset{CH_3}{|}}{CH}-CH_2-CH_2-CH_2-CH_3$$

22.4 28.1 38.9 29.7 23.0 13.6

$$H_3C-CH_2-\underset{\underset{{}^7CH_3}{|}}{CH}-CH_2-CH_2-CH_3$$

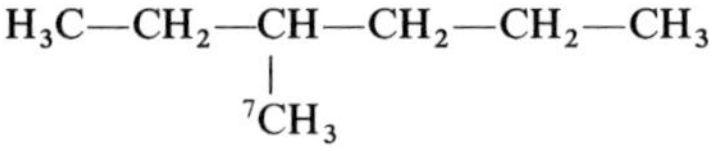

10.6 29.5 34.3 39.0 20.2 13.9 18.8

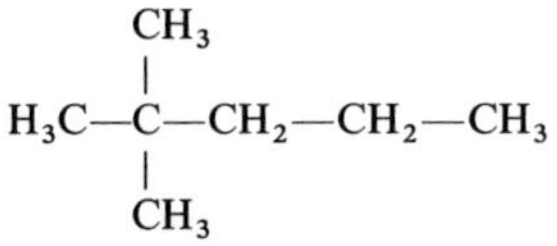

$$H_3C-\underset{\underset{CH_3}{|}}{\overset{\overset{CH_3}{|}}{C}}-CH_2-CH_2-CH_3$$

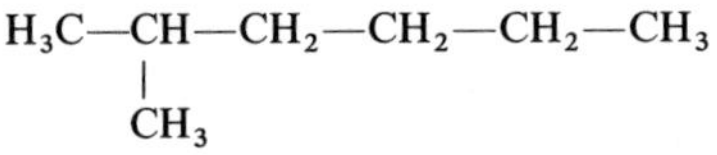

29.5 30.6 47.3 18.1 15.1

$$\overset{1}{H_3C}-\underset{\underset{H_3C}{|}}{\overset{2}{CH}}-\underset{\underset{CH_3}{|}}{\overset{3}{CH}}-\overset{4}{CH_2}-\overset{5}{CH_3}$$

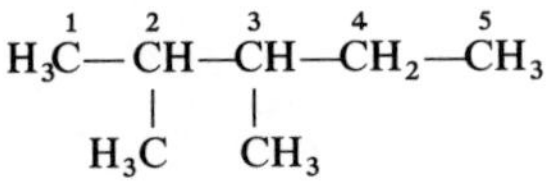

20.0 31.9 40.6 26.8 11.6 17.7 14.5

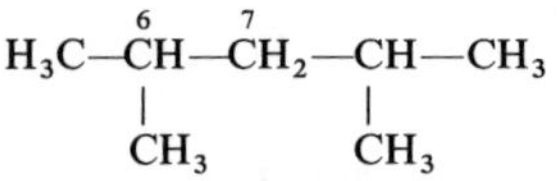

$$H_3C-\underset{\underset{CH_3}{|}}{\overset{6}{CH}}-\overset{7}{CH_2}-\underset{\underset{CH_3}{|}}{CH}-CH_3$$

22.7 25.7 49.0

$$H_3C-CH_2-\underset{\underset{{}_5CH_3}{|}}{\overset{\overset{{}^5CH_3}{|}}{C}}-CH_2-CH_3$$

7.7 32.2 33.4 25.6

$$H_3C-CH_2-\underset{\underset{C_2H_5}{|}}{\overset{\overset{H}{|}}{C}}-CH_2-CH_3$$

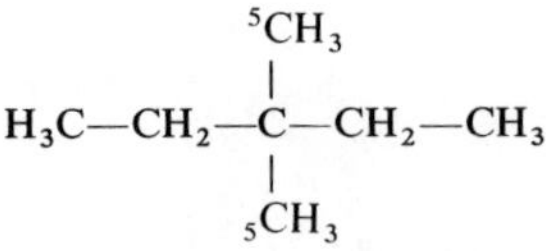

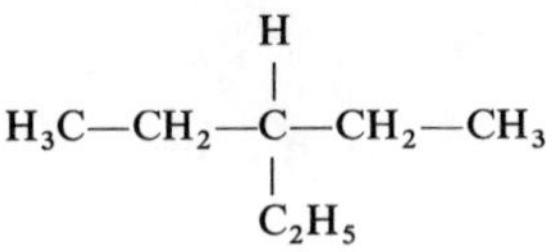

10.5 25.2 42.4

$$H_3C-CH_2-CH_2-CH_2(CH_2)_3-CH_3$$

13.6 22.7 32.1 29.4

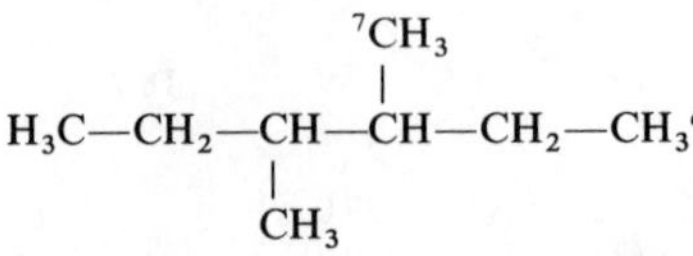

$$H_3C-CH_2-\underset{\underset{CH_3}{|}}{CH}-\overset{\overset{{}^7CH_3}{|}}{CH}-CH_2-CH_3{}^c$$

11.8 27.6 38.5 13.8

$$H_3C-CH_2-\underset{\underset{{}^7H_3C}{|}}{CH}-\underset{\underset{CH_3}{|}}{CH}-CH_2-CH_3{}^d$$

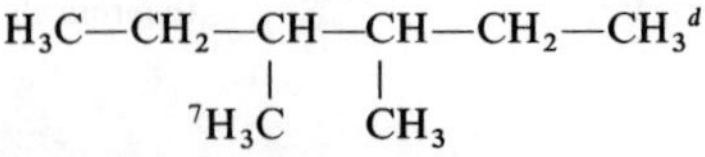

11.8 25.8 39.5 15.8

TABLE 1. (*continued*)

Compound	δ_c						
	C-1	C-2	C-3	C-4	C-5	C-6	C-7
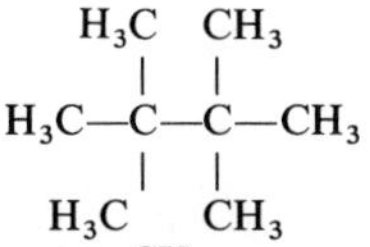 H₃C—C—C—CH₃ (with H₃C, CH₃ substituents)	25.6	35.0					
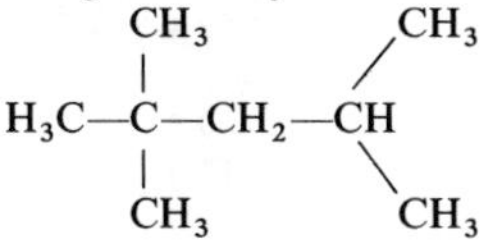 H₃C—C—CH₂—CH (with CH₃ substituents)	29.9	30.9	59.3	25.3	24.7		
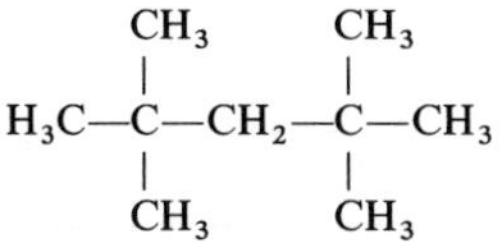 H₃C—C—CH₂—C—CH₃ (with CH₃ substituents)	31.8	32.4	56.5				
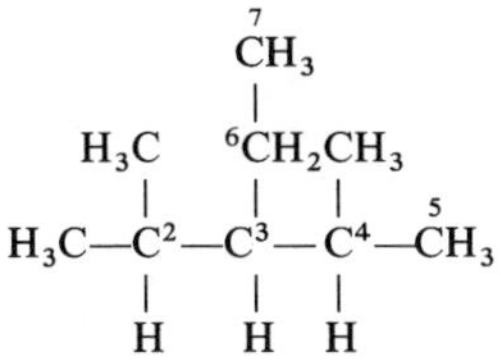 branched heptane (CH₃ at 7, ⁶CH₂CH₃, H₃C—C²—C³—C⁴—CH₃⁵)	20.0	29.0	56.8			21.1	14.5
[(CH₃)₃C]CH	65.0	38.6	34.9				

[a] Measured on a 1:1 mixture of alkane with 1.4-dioxan.
[b] Values given in parentheses for pentane, hexane and heptane are for the neat liquid, giving an indication of the effect of the solvent on the chemical shifts for the alkanes.
[c] Racemic form.
[d] *Meso* form.

2. Increment systems

Soon after the first ^{13}C NMR measurements empirical systems were devised to reproduce the values given in Table 1. Two somewhat different approaches have been published, one by Grant and Paul[16] and the other by Lindeman and Adams[17]. The former has been more often applied and its principal form was subsequently used for many other classes of compounds. Its components are given in Table 2 along with an example for 2,2′,4,4′-tetramethylpentane.

The relative contributions of the paramagnetic shielding and the diamagnetic shielding constants within this increment system have been discussed[11]. Empirical systems of this kind allow an estimation of chemical shifts with a mean deviation of less than one ppm. Modern data bank systems, e.g. SPECINFO[18] or CSEARCH[19], are able to reproduce the chemical shifts of any given alkane even more accurately within a few seconds. In a slightly different approach, a substituent constant σ_a for 34 different alkyl groups was defined, and it was shown that the chemical shift of a nucleus X in a compound XR_4 obeys equation 1, where m and b are constants.

$$\delta(X) = m\sigma_a + b \tag{1}$$

TABLE 2. Increment system for ^{13}C NMR chemical shifts of alkanes

Additive parameters for shifts caused by carbon atoms at position l relative to observed carbon (k)		Steric correction term (S_{kl}) for branched alkanes[a]	
l	A_l ppm	kl	S_{kl}
α	9.1	1° (3)	-1.1
β	9.4	1° (4)	-3.4
γ	-2.5	2° (3)	-2.5
δ	0.3	2° (4)	-7.5
ε	0.1	3° (2)	-3.7
		3° (3)	-9.5
Constant term (chemical shift for methane)		3° (4)	-15.0
$B = -2.3$ ppm		4° (1)	-1.5
		4° (2)	-8.4
		4° (3)	-15.0
		4° (4)	-25.0

$$\delta_{C(k)} = B + \sum A_k n_{kl} + S_{kl}$$

[a]1° = Primary, 2° = secondary, 3° = tertiary, 4° = quaternary carbon atom.

Example: Calculation of ^{13}C chemical shifts for

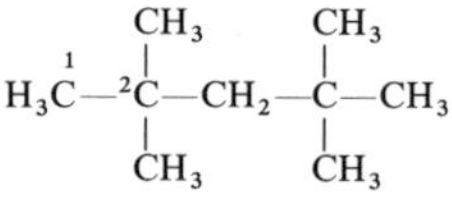

$$H_3C \overset{1}{-} \overset{2}{C} - CH_2 - C - CH_3$$

(with CH_3, CH_3 groups above and CH_3, CH_3 below)

$$\delta_{C-1} = -2.3 + A_\alpha + 3A_\beta + A_\gamma + 3A_\delta + S_{1°(4°)}$$
$$= -2.3 + 9.1 + 28.2 - 2.5 + 0.9 - 3.4 = 30.0$$
$$Exp. = 31.8$$

$$\delta_{C-2} = -2.3 + 4A_\alpha + A_\beta + 3A_\gamma + 3S_{4°(1°)}$$
$$= -2.3 + 36.4 + 9.4 - 7.5 - 4.5 = 31.5$$
$$Exp. = 32.4$$

$$\delta_{C-3} = -2.3 + 2A_\alpha + 6A_\beta + 2S_{2°(4°)}$$
$$= -2.3 + 18.2 + 56.4 - 16.8 = 55.5$$
$$Exp. = 56.5$$

This equation not only holds for alkanes, where X is carbon, but for a large variety of compounds, where X is a heteroatom[20]. A similar system using defined increments for alkyl groups had been published earlier[21]. In another interesting approach[22], the geometry of many rigid cyclic and polycyclic alkanes was calculated by the MM2 method. The authors then tried to find a minimum set of their MM2 topological and geometrical parameters which describe the ^{13}C NMR chemical shifts and were able to predict from these increments the chemical shifts of a steroid-like molecule, e.g. 4,4-dimethylpodocarpane, in good agreement with the experimental value. Thus, they conclude that from a known molecular geometry the ^{13}C chemical shifts of alkanes and polycyclic alkanes can be calculated[23]. In yet another unusual approach, it was shown that the sum of chemical shifts for alkanes obeys predictions of graph theory[24]. Based on similar earlier work[25] the authors discovered that the sum of chemical shifts is related to the molecular path counts obtained by graph theory. As an example, the alkanes 1–6 have all the same graph theory index $L = 15$ and a chemical shift sum of about 225 ppm. However, the predictive power of this system for molecules with steric crowding of the methyl groups seems to have difficulties[26].

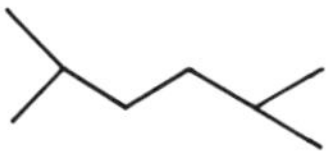

$\Sigma \delta_C$ 220.2   223.6   225.4

(1)   (2)   (3)

$\Sigma \delta_C$ 228.8   225.1   229.1

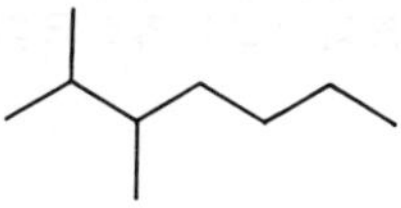

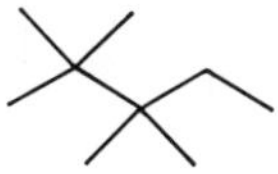

(4)   (5)   (6)

3. Cycloalkanes

The ^{13}C NMR chemical shifts of several unsubstituted cycloalkanes[27] are given in Table 3, and with few exceptions they appear in a narrow range. Due to the nature of their special bonding, cyclopropane and partly also cyclobutane deviate considerably. The signals of the carbon atoms in the C_{10} to C_{14} cyclic hydrocarbons appear at higher field than the average of the other cycloalkanes due to steric hindrance in the medium-sized rings. In the large-ring systems the chemical shifts become almost constant.

TABLE 3. ^{13}C NMR chemical shifts of cycloalkanes

Compound	δ_C
Cyclopropane	-2.8
Cyclobutane	22.4
Cyclopentane	25.8
Cyclohexane	27.0
Cycloheptane	28.7
Cyclooctane	26.8
Cyclononane	26.0
Cyclodecane	25.1
Cycloundecane	26.3
Cyclododecane	23.8
Cyclotridecane	26.2
Cyclotetradecane	25.2
Cyclopentadecane	27.0
Cyclohexadecane	26.9
Cyclooctadecane	27.5
$C_{20}H_{40}$	28.0
$C_{24}H_{48}$	28.7
$C_{30}H_{60}$	29.3
$C_{40}H_{80}$	29.4
$C_{72}H_{144}$	29.7

TABLE 4. ^{13}C NMR chemical shifts of alkyl-substituted cycloalkanes

Compound	δ								
	C-1	C-2	C-3	C-4	C-5	C-6	C-7	CH₃	—C—
Methylcyclopropane	4.9	5.6						19.4	
1,1-Dimethylcyclopropane	11.3	13.7						25.5	
cis-1,2-Dimethylcyclopropane	9.3	13.1						12.8	
trans-1,2-Dimethylcyclopropane	13.9	14.2						18.9	
1,1,2-Trimethylcyclopropane	15.1	18.4	20.9					27.4(trans) 19.5(cis) 14.0(2)	
Methylcyclobutane	31.2	30.2	18.3					22.1	
1,1-Dimethylcyclobutane	35.9	34.9	14.8					29.4	
cis-1,2-Dimethylcyclobutane	32.2		26.6					15.4	
trans-1,2-Dimethylcyclobutane	39.2		26.8					20.5	
cis-1,3-Dimethylcyclobutane	26.9	38.5						22.5	
trans-1,3-Dimethylcyclobutane	26.1	36.4						22.0	
Octamethylcyclobutane	41.1							22.3	
Methylcyclopentane	35.8	35.8	26.4	24.3				21.4	
1,1-Dimethylcyclopentane	40.1	42.3	25.8	24.2				30.0	
trans-1,2-Dimethylcyclopentane	43.7		36.0	36.2				19.7	
cis-1,2-Dimethylcyclopentane	38.6		34.2	35.3				16.1	
trans-1,3-Dimethylcyclopentane	34.2	48.2		26.4				22.4	
cis-1,3-Dimethylcyclopentane	36.3	46.0		26.7				22.1	
Methylcyclohexane	33.1	35.8	26.6	26.4				22.7	
1,1-Dimethylcyclohexane	30.0	39.8	22.6	26.7				28.8	
trans-1,2-Dimethylcyclohexane	39.6		36.0	26.9				20.2	
cis-1,2-Dimethylcyclohexane	34.4		31.5	23.7				15.7	
trans-1,3-Dimethylcyclohexane	27.1	41.5		33.9	20.8			20.5	
cis-1,3-Dimethylcyclohexane	32.8	44.7		35.5	26.5			22.8	
trans-1,4-Dimethylcyclohexane	30.1	30.9						20.1	
cis-1,4-Dimethylcyclohexane	32.6	35.6						22.6	

359

Compound								
1,1,3-Trimethylcyclohexane	30.8	49.4	28.3	35.7	22.6	39.5		24.9(trans) 33.7(cis) 23.1(3) · 32.5 32.1 32.8
t-Butylcyclohexane	48.7	28.0	26.7	27.0				27.5
trans-1,4-Di-t-Butylcyclohexane	48.4	28.1						27.7
cis-1,4-Di-t-Butylcyclohexane	42.8	23.8						27.8
Methylcycloheptane	34.9	37.5	26.9	28.9				24.4
1,1-Dimethylcycloheptane	33.3	42.6	23.8	30.8				30.8
trans-1,2-Dimethylcycloheptane	41.3		35.8	26.7	29.7			22.6
cis-1,2-Dimethylcycloheptane	37.5		34.0	26.6	29.2			17.9
trans-1,3-Dimethylcycloheptane	31.1	44.8		37.5	29.1			24.3
cis-1,3-Dimethylcycloheptane	34.2	46.9[a]		37.4	26.5			24.9
trans-1,4-Dimethylcycloheptane	35.2	36.6[a]			36.8[a]	24.1		24.3
cis-1,4-Dimethylcycloheptane	34.2	33.5			38.4	27.0		24.3
1,1,2-Trimethylcycloheptane	35.6	43.2	32.4	29.7[a]	30.4[a]	22.8	43.9	29.7(trans) 23.7(cis) 18.8(2)
1,1,3-Trimethylcycloheptane	33.6	52.6	30.7	40.6	31.3	24.7	30.7	30.7(trans) 33.6(cis) 26.9(3)
1,1,4-Trimethylcycloheptane	34.5	41.8	33.2	38.1	41.2	23.7	43.8	32.4(trans) 31.8(cis) 25.3(4)

[a]Assignments uncertain.

4. Substituted cycloalkanes

Substitution of cycloalkanes with methyl or other alkyl groups causes electronic and conformational changes which are reflected in considerable displacements of the chemical shifts. These were measured for a large body of compounds and the results are given in Table 4.

A special increment system for methyl-substituted cycloalkanes requires for each substituent a positional parameter which is different for axial and equatorial substitution. Furthermore, correction factors for vicinal methyl pairs have to be included[28]. This increment system is presented in Table 5 along with an example of its use. Conformational

TABLE 5. Increment system for ^{13}C NMR chemical shifts of methyl-substituted cycloalkanes

$$\delta_{C(k)} = 26.5 + \sum A_l n_{kl} + \sum S_l$$

Shift increment A_l due to a methyl group at position l relative to observed carbon (k)		Correction S_l for geminal (e.g. $\alpha_a \alpha_e$) and vicinal (e.g. $\alpha \beta$) methyl pairs	
l	A_l	Type	S_l
α_e	5.96	$\alpha_a \alpha_e$	-3.80
α_a	1.40	$\beta_a \beta_e$	-1.27
β_e	9.03	$\gamma_a \gamma_e$	2.02
β_a	5.41	$\alpha_e \beta_e$	-2.45
γ_e	0.05	$\alpha_e \beta_a$	-2.91
γ_a	-6.37	$\alpha_a \beta_e$	-3.43
δ_e	-0.22	$\beta_e \gamma_a$	-0.80
δ_a	-0.06	$\beta_a \gamma_e$	1.57

Example:

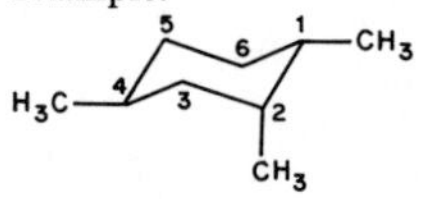

$\delta_{C-1} = 26.5 + \alpha_e + \beta_a + \beta_e + \alpha_e \beta_a$
$\qquad = 26.5 + 5.96 + 5.41 - 0.22 - 2.91 = 34.74$
$\delta_{C-2} = 26.5 + \alpha_a + \beta_e + \gamma_e + \alpha_a \beta_e = 33.55$
$\delta_{C-3} = 26.5 + \beta_e + \beta_a + \gamma_e + \beta_a \gamma_e = 42.56$
$\delta_{C-4} = 26.5 + \alpha_e + \gamma_a + \delta_e = 25.87$
$\delta_{C-5} = 26.3 + \beta_e + \gamma_e + \delta_a + \beta_e \gamma_a = 34.45$
$\delta_{C-6} = 26.5 + \beta_e + \gamma_a + \gamma_e + \beta_e \gamma_a = 28.41$

$$\delta_{CH_3(k)} = B + \sum A_l n_{kl}$$

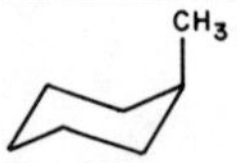

$B = 18.8$ ppm	$B = 23.1$ ppm
α 6.4	10.4
$\beta_e - 6.8$	-2.8
$\beta_a - 2.8$	-2.8
γ_a 2.0	0.0
γ_e 0.0	0.0

δ-effects in 1,3-diaxial substituted methylcyclohexanes have been reported[29]. Other papers address the chemical shifts of methylcycloheptanes[30], alkylcyclohexanes[31], tetramethylcyclohexanes[32,33], methylcyclobutanes[34] and methylcyclopentanes[35].

5. Bicyclic alkanes

In Table 6[3], which was compiled from work on bicyclo[n.1.0]alkanes[36], bicyclo[1.1.0]-butanes[37], bicyclo[2.1.0]hexanes[38], bicyclo[2.2.1]heptanes[39,40], bicyclo[3.1.1]heptanes[41], bicyclo[2.2.2]octanes[42], bicyclo[3.3.0]octanes[43], bicyclo[4.n.0]alkanes[44], bicyclo[3.3.1]-nonanes[45-48], spiro[4.5]alkanes[49] and trans-decalins[50] the chemical shifts of several bicyclic alkanes are tabulated. The values were mainly explained by the steric effects present in the different structural environments, with the exception of fused cyclopropane rings (see below).

To illustrate these effects a comparison of the disubstituted methylnorbornanes **7–9**

TABLE 6. ^{13}C NMR chemical shifts of bicyclic alkanes

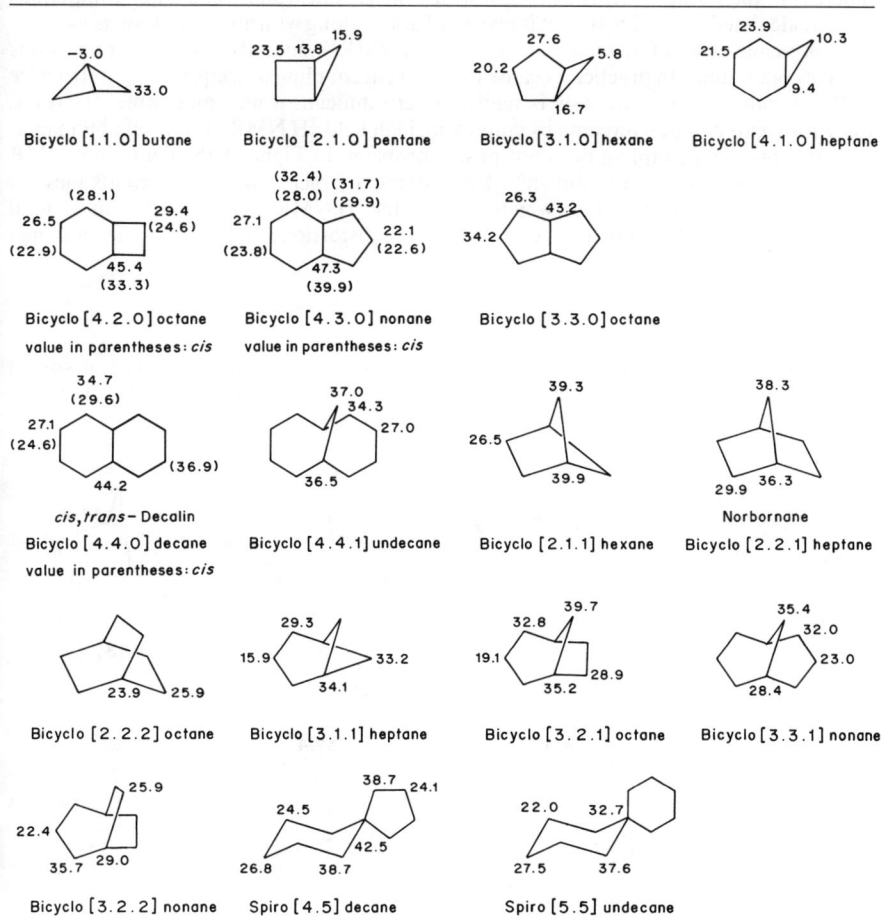

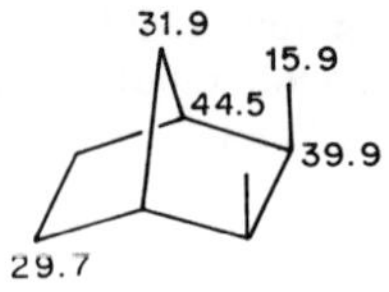

(7)

31.9
15.9
44.5
39.9
29.7

(8)

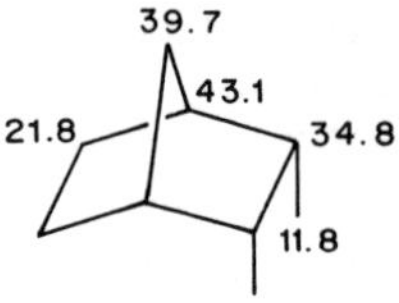

36.9   21.3
44.4
30.3
45.2
42.0   44.0
21.2
15.9

(9)

39.7
43.1
21.8
34.8
11.8

is shown[39]. Here the influence of steric hindrance of the methano bridge on the ^{13}C chemical shift, as well as the repulsion between the methyl groups themselves and the interaction between the methyl groups and the C-5 and C-6 atoms, can clearly be seen. These three compounds serve as a classical example for demonstrating steric effects in ^{13}C NMR spectroscopy.

Methyldecalins, perhydrophenanthrenes and perhydroanthracenes are models for steroids. Therefore an incremental system was developed which includes several features inherent in these ring systems and from which the chemical shifts of similar compounds can be calculated[51,52]. This system is given in Table 7 along with an example of its use.

In principle, a set of incremental rules for ^1H NMR should also be applicable to this class of compounds. In practice, because of spin–spin coupling overlap and band structure of the signals a similar analysis is rendered very difficult, if not impossible. However, these problems can be overcome by employing high field ^2H NMR at natural abundance, wherein spin–spin coupling does not pose a problem. In Figure 1 the deuterium NMR spectrum of *trans*-decalin is shown[53]. By studying a larger number of cycloalkanes the authors derived an increment set, which describes the ^1H NMR chemical shift within a total shift range of only 1.4 ppm, and accounts for ring distortions and other conformational features.

6. Polycyclic alkanes

For this class of compounds the effects of fused cyclopropane rings are of special interest, since in addition to steric effects, electronic interaction can also be discussed. For comparison one should inspect compounds **10–17** where the already known steric

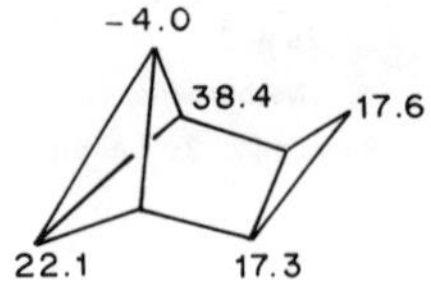

(10)

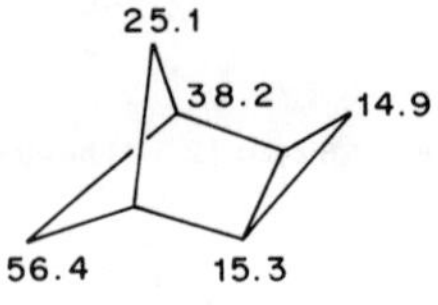

−4.0
38.4   17.6
22.1   17.3

(11)

25.1
38.2   14.9
56.4   15.3

(12)

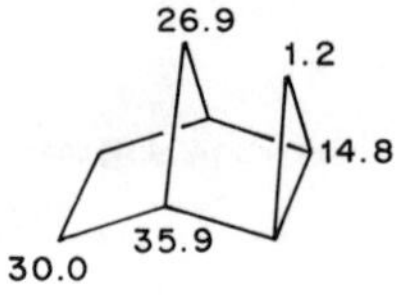

26.9   1.2
14.8
30.0   35.9

(13)

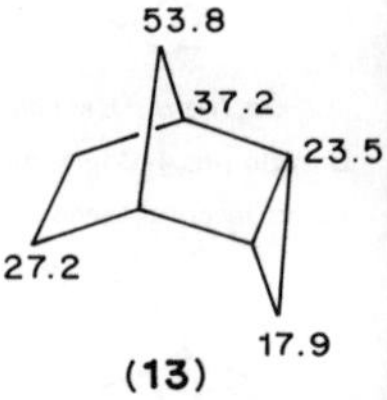

53.8
37.2
23.5
27.2
17.9

(14)

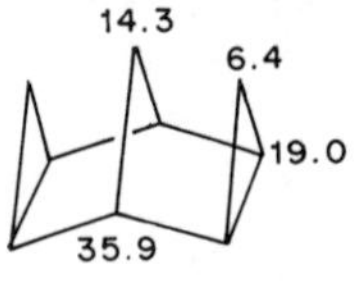

14.3
6.4
19.0
35.9

(15)

1.9   41.4
35.2
21.4
12.4
15.1

(16)

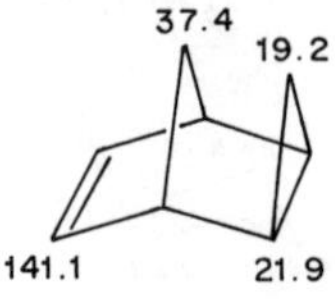

37.4   19.2
141.1   21.9

(17)

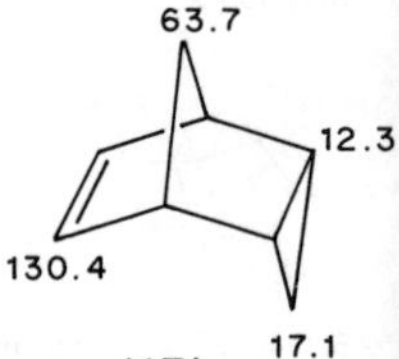

63.7
12.3
130.4
17.1

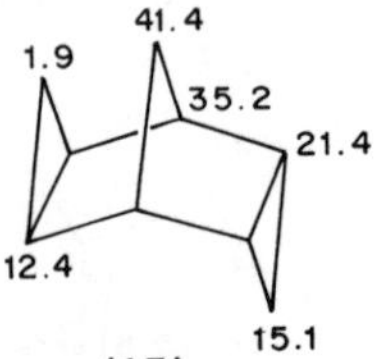

TABLE 7. Increment system for ^{13}C NMR chemical shifts in decalins and related compounds

$\delta_{C(k)} = B + \sum A_l n_{kl}$ (n_{kl} = no. of groups)

		Methyldecalin	Perhydroanthracene,-naphthacene,-phenanthrene,-pyrene
Base value		−3.07	−2.49
Molecular fragment		Shift increment A_l for position l relative to observed carbon k	
		A_l	A_l
α		9.94	9.43
β		8.49	8.81
T		−2.91	−1.13
Q		−9.04	
V_g		−3.50	−3.38
V_{tr}			−0.79
V_e			−8.19
$\beta_g\gamma_{tr}$		+1.91	
$\gamma_{H\cdots H}$	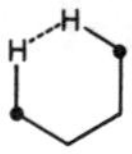	−4.56	−5.53
$\gamma_{2H\cdots H}$	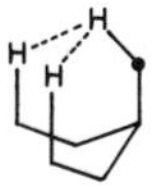	−3.95	−3.01
γ_p	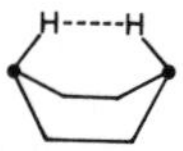		−11.01
$\delta_{syn\text{-}axial}$	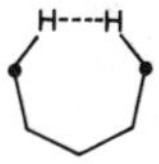		−3.65

Example: To calculate the ^{13}C chemical shifts for 2-methyl-*trans*-decalin:

$\delta_{C-2} = B + 3\alpha + 2\beta + T + 2V_g$
$= -3.07 + 29.82 + 16.98 - 2.91 - 7.00 = 33.82$
Exp. = 33.1

$\delta_{C-9} = B + 3\alpha + 4\beta + T + 4V_g$
$= -3.07 + 29.82 + 33.96 - 2.91 - 14.00 = 43.80$
Exp. = 43.5

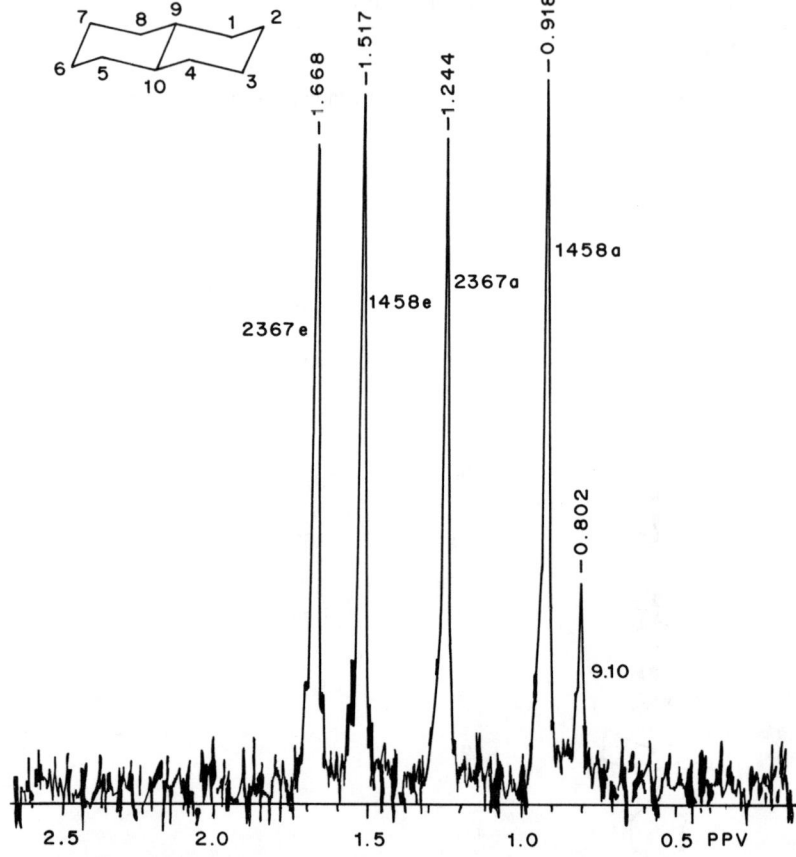

FIGURE 1. ^{2}H NMR spectrum of *trans*-decalin. Reprinted with permission from Curtis *et al.*, *J. Am. Chem. Soc.*, **111**, 7711. Copyright (1989) American Chemical Society

effects (cf compounds **7–9**) can be studied. Thus, in **10**, the C-3 carbon atom *syn* to the fused cyclopropane ring absorbs at -4 ppm, whereas the C-4 atom *anti* to the cyclopropane ring resonates at 22.1 ppm. Similar high-field steric effects can clearly be seen in the other compounds. However, the methano bridge carbon atoms in **13**, **15** and **17** show marked downfield shifts which are explained by the electron-accepting effects of the cyclopropane ring. In these cases, donation from the HOMO orbital of the methano bridge into the LUMO Walsh orbital of the cyclopropane moiety has been proposed[36,54].

This *γ-anti* effect is significantly reduced in the cyclobutane or cyclopentane fused ring systems **18–23**, since the *p* character of these orbitals is decreased markedly.

The chemical shifts of several other polycyclic alkanes are collected in Table 8. The most recent example is tricyclo[4.2.0.0.^{1,3}]octane, in which three-, four- and five-membered rings are fused[55].

The ^{13}C NMR data for selected substituted bicyclic and polycyclic alkanes are listed in Table 9. Noteworthy is the upfield shift for the quaternary carbon atom of the *tert*-butyl group in **24**[56], which can be attributed to the special bonding character in the tetrahedrane

TABLE 8. ^{13}C NMR chemical shifts of tricyclic and polycyclic alkanes

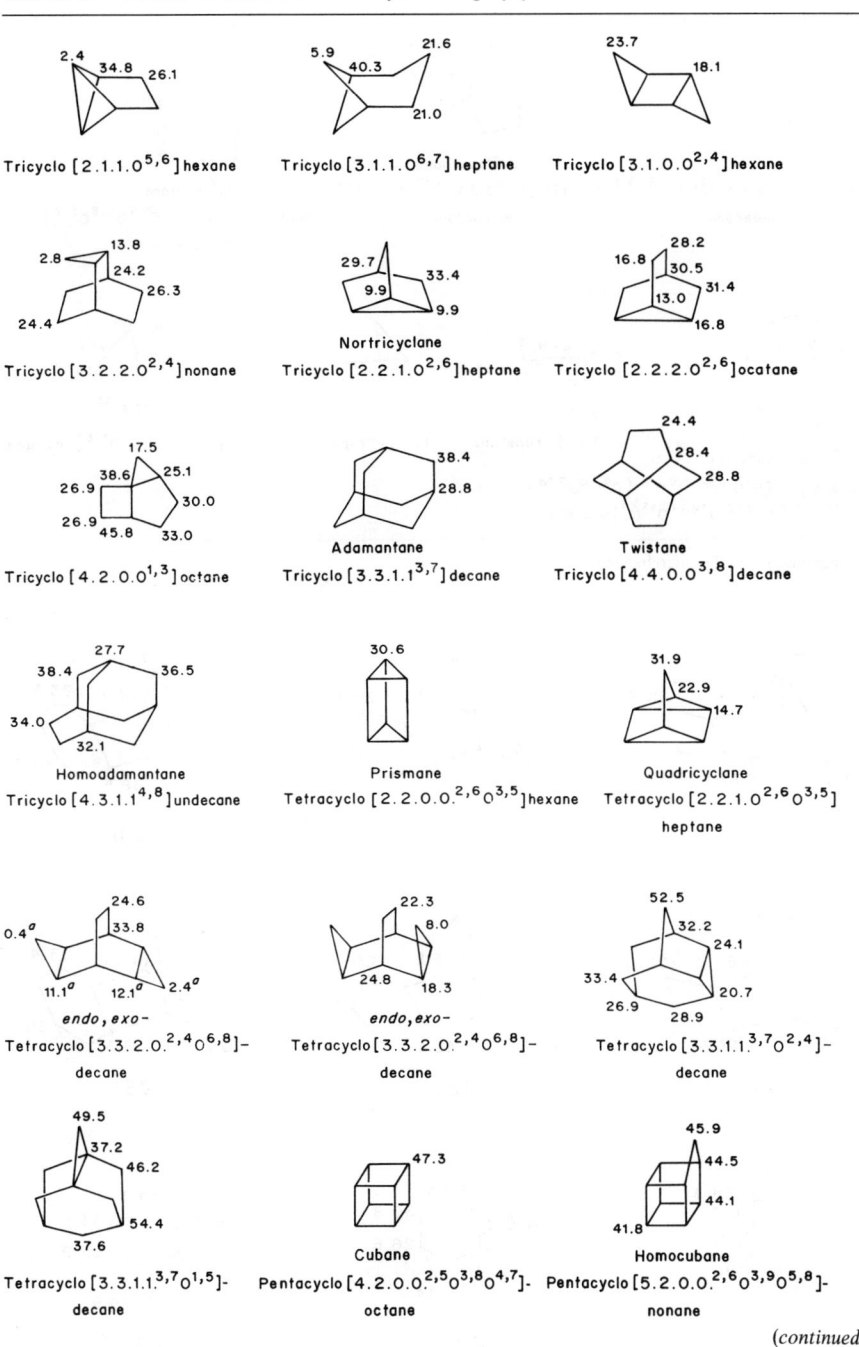

Tricyclo [2.1.1.0^{5,6}] hexane

Tricyclo [3.1.1.0^{6,7}] heptane

Tricyclo [3.1.0.0^{2,4}] hexane

Tricyclo [3.2.2.0^{2,4}] nonane

Nortricyclane
Tricyclo [2.2.1.0^{2,6}] heptane

Tricyclo [2.2.2.0^{2,6}] ocatane

Tricyclo [4.2.0.0^{1,3}] octane

Adamantane
Tricyclo [3.3.1.1^{3,7}] decane

Twistane
Tricyclo [4.4.0.0^{3,8}] decane

Homoadamantane
Tricyclo [4.3.1.1^{4,8}] undecane

Prismane
Tetracyclo [2.2.0.0^{2,6}0^{3,5}] hexane

Quadricyclane
Tetracyclo [2.2.1.0^{2,6}0^{3,5}] heptane

endo, exo-
Tetracyclo [3.3.2.0^{2,4}0^{6,8}]- decane

endo, exo-
Tetracyclo [3.3.2.0^{2,4}0^{6,8}]- decane

Tetracyclo [3.3.1.1^{3,7}0^{2,4}]- decane

Tetracyclo [3.3.1.1^{3,7}0^{1,5}]- decane

Cubane
Pentacyclo [4.2.0.0^{2,5}0^{3,8}0^{4,7}]- octane

Homocubane
Pentacyclo [5.2.0.0^{2,6}0^{3,9}0^{5,8}]- nonane

(continued)

TABLE 8. (continued)

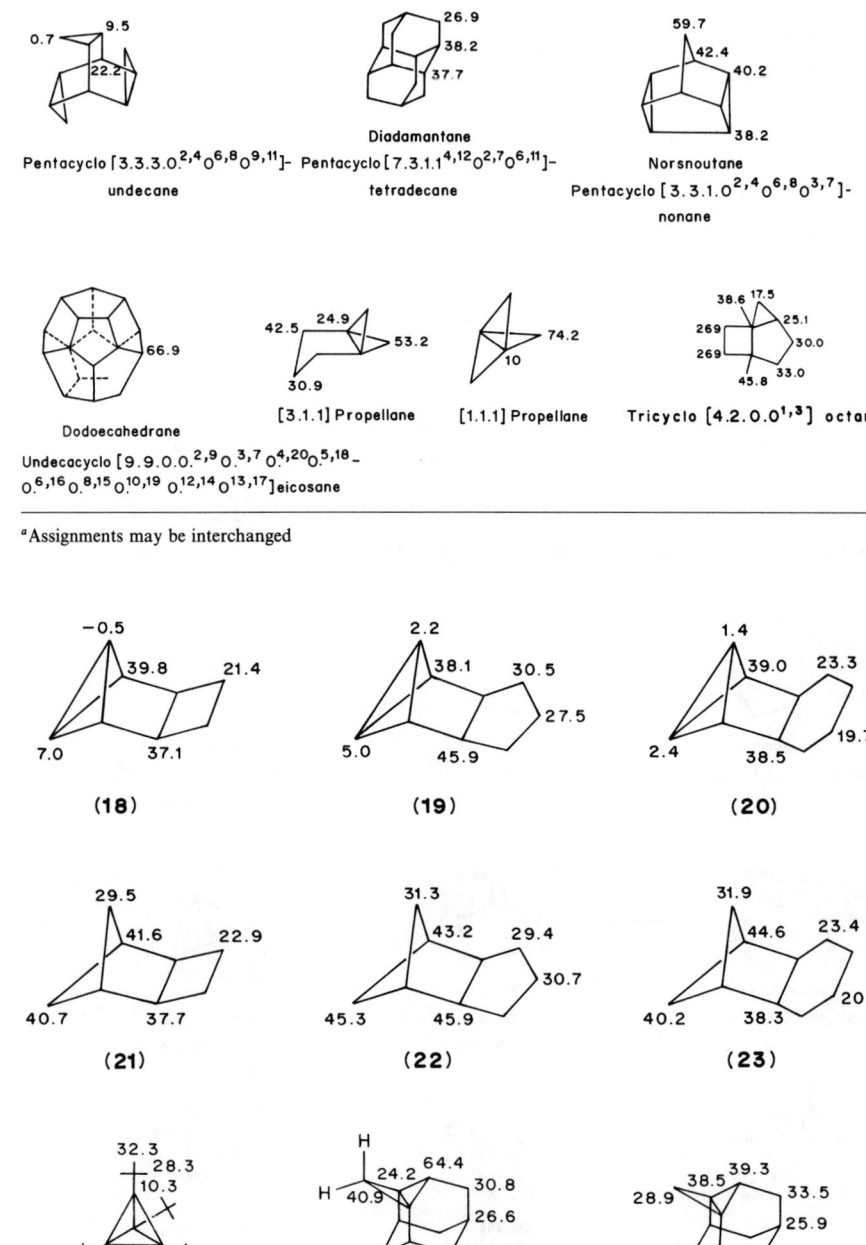

Pentacyclo [3.3.3.0.^{2,4}0^{6,8}0^{9,11}]- Pentacyclo [7.3.1.1^{4,12}0^{2,7}0^{6,11}]-
undecane   tetradecane
Diadamantane

Norsnoutane
Pentacyclo [3.3.1.0^{2,4}0^{6,8}0^{3,7}]-
nonane

Dodecahedrane
Undecacyclo [9.9.0.0.^{2,9}0^{3,7}0^{4,20}0^{5,18}_-
0.^{6,16}0^{8,15}0^{10,19}0^{12,14}0^{13,17}]eicosane

[3.1.1] Propellane

[1.1.1] Propellane

Tricyclo [4.2.0.0^{1,3}] octane

^{a}Assignments may be interchanged

(18)

(19)

(20)

(21)

(22)

(23)

(24)

(25)

(26)

TABLE 9. ^{13}C NMR chemical shifts of substituted polycyclic alkanes

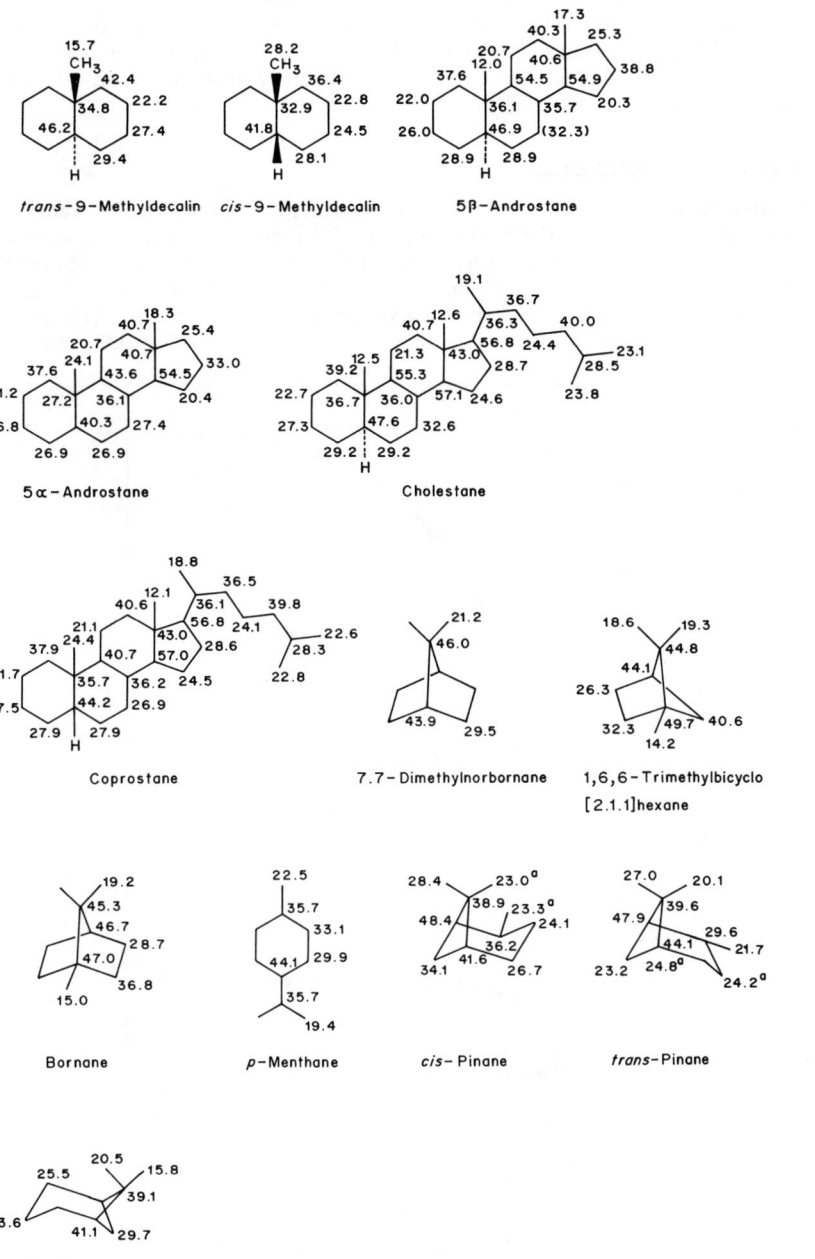

trans-9-Methyldecalin   cis-9-Methyldecalin   5β-Androstane

5α-Androstane   Cholestane

Coprostane   7.7-Dimethylnorbornane   1,6,6-Trimethylbicyclo[2.1.1]hexane

Bornane   p-Menthane   cis-Pinane   trans-Pinane

Norpinane

^aAssignment uncertain.

system. Inverted carbon atoms such as in dehydroadamantane **25** exhibit signals at
considerably higher fields than their structural analogs (cf **26**[57]).

Table 9[3] was compiled from work on methyldecalin steroids[58], tetracyclic compounds[59],
propellanes[60,61], cyclopropane fused bicyclo[2.2.2]octanes[62,63], cyclopropanodecalins[64],
tricyclooctanes[65,66], homoadamantanes[67], 1,3-dehydroadamantancs[68], 2,4-dehydro-
adamantanes[57] and diamantanes[69].

C. Medium and Solvent Effects

The aromatic solvent-induced shift (ASIS) dependence of saturated hydrocarbons was
systematically investigated for the ^{1}H resonance of the methyl groups, by determining
their chemical shifts in CCl$_4$ and benzene solutions[70]. The magnitude of the ASIS effect,
typically between -0.05 and $+0.04$ ppm, could be mainly correlated with the length of
the molecule measured along the axis of largest extension. More recent ASIS data for
a series of n-alkanes are illustrated in Figure 2[71]. The chemical-shift differences and

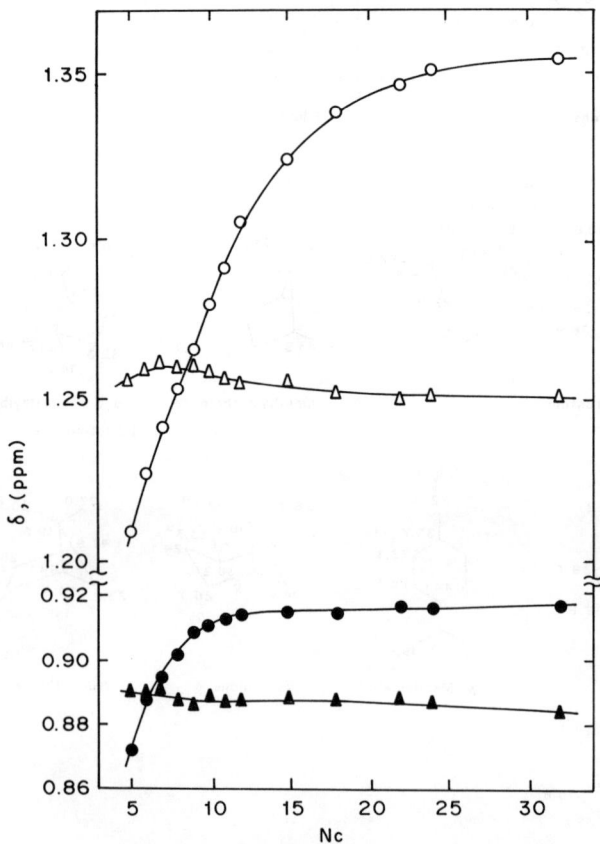

FIGURE 2. ^{1}H NMR chemical shifts for n-alkanes in benzene-d$_6$
(open circles, CH$_2$; filled circles, CH$_3$) and in CCl$_4$ (open triangles,
CH$_2$; filled triangles, CH$_3$); N_C is the number of carbon atoms.
Reproduced by permission of John Wiley & Sons from Ref. 72

their dependence on chain length were interpreted in terms of increased folding present for the higher alkanes.

A very peculiar ASIS effect was reported in 1967[72] and has been a controversial topic since then. n-Alkanes with chains longer than 15 carbon atoms in 1-chloronaphthalene solutions show a doublet splitting of their proton methylene group resonances. At higher temperature the signals broaden somewhat but do not coalesce. Some authors attributed this effect to the non-interconversion of protons in *gauche* and *trans* bonds in planar aromatic solvents[73,74]. Investigations of the pressure dependence and Monte Carlo simulations were carried out to support these interpretations[75,76]. Closely related studies in liquid crystals also reveal this doublet splitting[77]. For nonane, in various solvents, this effect was not found[78]. Other authors explain this behaviour by different solvation at the ends of the hydrocarbon chains[79]. Recent studies employing 2D-^{13}C,^{1}H correlation techniques seem to confirm this line of reasoning[80].

A ^{13}C NMR study in the gas phase for methane, ethane and propane detected significant chemical-shift changes when compared to values obtained in solution, both in the absolute sense as well as for the internal shift differences in propane[81]. Another report gives the chemical shifts of C_4 alkanes in the gas phase[82].

Deuterium NMR spectroscopy has been employed to probe the interaction between perdeuteriated n-alkanes and host molecules such as liquid crystals, urea crystals, lipid bilayers or zeolites. In general, the temperature-dependent quadrupolar splitting of the deuterium signals is interpreted in terms of conformation and ordering of the alkane chains. For lipid bilayers these studies are of interest in connection with the anaesthetic properties of alkanes. Phospatidylcholine bilayer membranes were chosen as a model. The solubilities of n-alkanes as determined by ^{2}H NMR were found to be dependent on both the membrane- and alkane-chain length[83,84]. The complex signal patterns which show the dynamic processes of perdeuteriated hexane in a multilayer is reproduced in Figure 3[85].

In nematic solvents, alkanes are found to acquiesce and conform to the nematic constraints, while retaining considerable freedom. The deuterium NMR spectra of perdeuteriated alkanes were simulated quantitatively and complete agreement was found between a motional model and the experiment[86,87].

Since their discovery in 1949[88], n-alkane urea inclusion compounds (UIC) have fascinated chemists and spectroscopists alike[89]. However, there is disagreement about the interpretation of the deuterium NMR spectra of UICs. Originally, it was believed that in these channel clathrates the alkanes exist in an all-*trans* conformation. Recently, it was shown that significant amounts (ca 30%) near the end of the chains are present in a *gauche* conformation. An illustration of the possible alkane conformations based upon MM2 calculations is shown in Figure 4[90].

Other authors[91,92] interpret these findings as decomposition artifacts of the UICs under NMR conditions. For the nonadecane-d$_{40}$ UIC, evidence was found by temperature-dependence studies[93] for two phases. In the high-temperature phase the alkane is freely rotating, while below 160 K a more rigid phase exists. Considerably larger amplitude motion toward the chain ends was also detected.

The deuterium NMR spectra of perdeuteriated n-hexane in zeolites[94-96] indicate that the hexane molecule 'reptates' within the zeolite and imply a significant interaction between the hydrocarbon and the zeolite cage wall. For cyclohexane the data suggest[94] that the chair to chair interconversion rate is damped by the zeolite.

A special medium is the magnetic field itself. An application of very high field NMR spectroscopy was recently used to calculate the magnetic susceptibility tensors of alkanes[97]. In a very high magnetic field molecules tend to align along the field axis, which leads to signal splitting in the high-resolution spectra of a quadrupolar nucleus such as deuterium. In Figure 5 the 95.2-MHz ^{2}H NMR spectrum of 2-methylpropane-

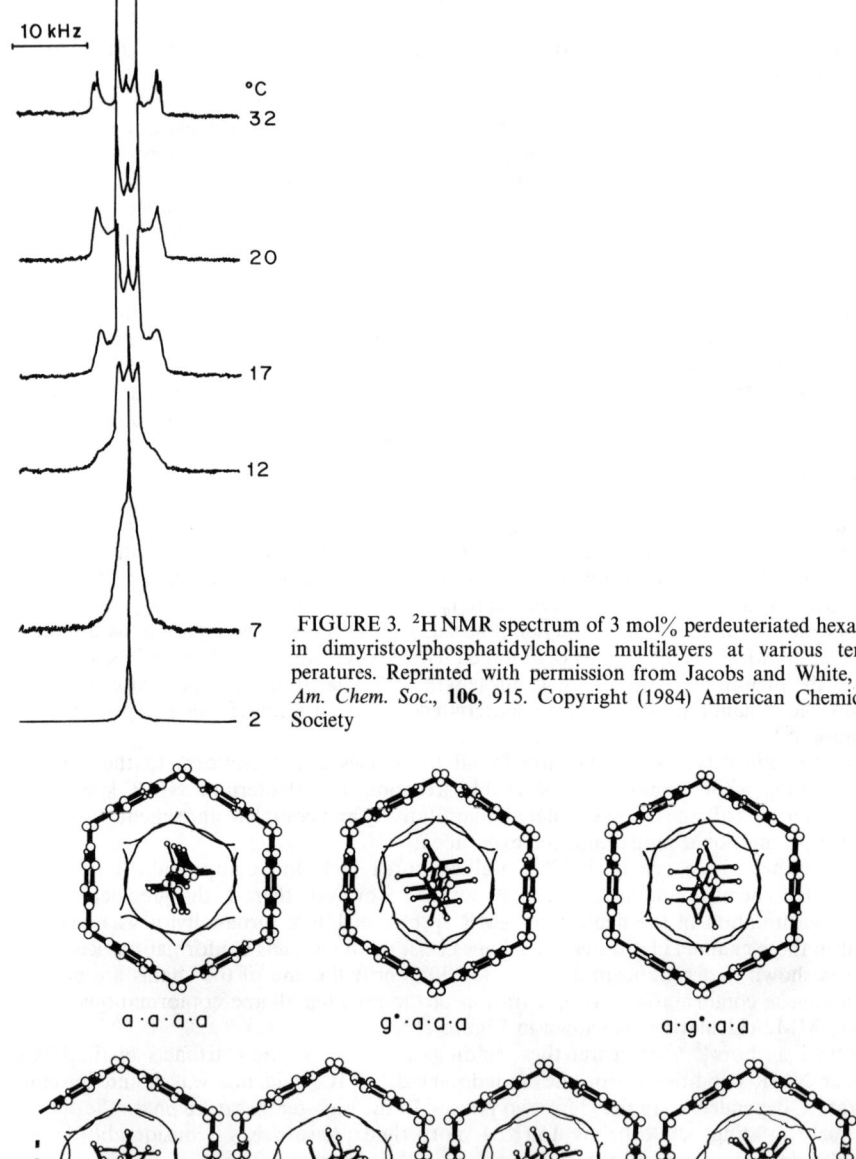

10 kHz

°C

32

20

17

12

7

2

FIGURE 3. ^{2}H NMR spectrum of 3 mol% perdeuteriated hexane in dimyristoylphosphatidylcholine multilayers at various temperatures. Reprinted with permission from Jacobs and White, *J. Am. Chem. Soc.*, **106**, 915. Copyright (1984) American Chemical Society

a·a·a·a

g*·a·a·a

a·g*·a·a

g*·a·g*·a

g*·a·g⁻·a

g*·a·a·g*

g*·a·a·g⁻

FIGURE 4. Channel axis projection of the calculated shapes of n-heptane molecules with seven conformations in the urea channel. Reproduced by permission of the American Institute of Physics from Ref. 90

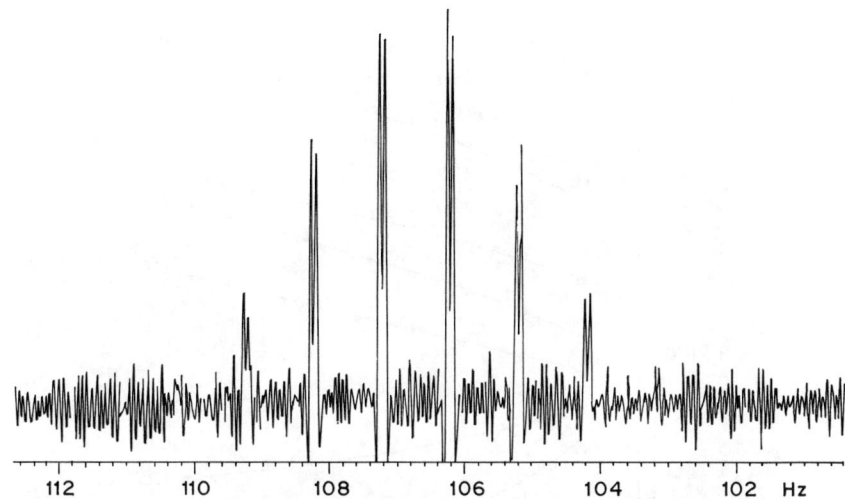

FIGURE 5. 95.2-MHz ^{2}H NMR spectrum of 2-methylpropane-2-d_1 in cyclohexane. Reproduced by permission of Academic Press, Inc., from Ref. 97

2-d_1 is shown. A decet due to ^3J (H, D) spin–spin coupling is revealed although its outer lines disappear in the noise. From the additional small doublet splitting of each decet line the magnetic susceptibility can be calculated[98].

D. Conformational and Dynamic Studies

1. Open-chain compounds

The ^{13}C chemical shifts of saturated hydrocarbon chains are the average of the chemical shifts in all molecular conformations, weighted by the probability of occurrence of each conformation. In two detailed studies[99,100] a statistical weight was assigned to several specified rotamer populations and for these rotamers chemical-shift parameters have been calculated. Using this method, it is possible to reproduce the experimental chemical shifts with a standard deviation of 1.59 ppm. These derived formulas also apply to molecules with a fixed conformation. Very recently, a microcomputer program based on these conformational increments was reported[101]. Conversely, the rotamer populations can be estimated from the ^{13}C chemical shifts. Naturally, these populations are temperature-dependent. Experimentally, both upfield and downfield shifts with increasing temperature were observed, as shown in Figure 6 for n-pentylcyclohexane[102].

Besides an intrinsic temperature dependence, *gauche/trans* transitions were discussed for the contrasting directions of shift. Other authors[103] felt that the situation is even more complex and tried to fit the values for n-alkanes to a larger set of empirical increments.

Very low temperature studies on 2,3-dimethylbutane **27** revealed that the concept of the *gauche* conformations **27b** and **27c** as the energy minima is correct[104,105]. In Figure 7, the well-resolved low-temperature spectrum is reproduced and the assignments given. The barriers of rotation were estimated to be of the order of 4.5 kcal mol^{-1}[104].

Similarly, the *gauche* conformations as lowest-energy forms were observed in 2,3-dimethylpentane and 3,4-dimethylhexane[105]. For the methyl carbon atoms of the isopropyl group in 2,4-dimethylhexane **28** a chemical shift difference of 1.5 ppm is

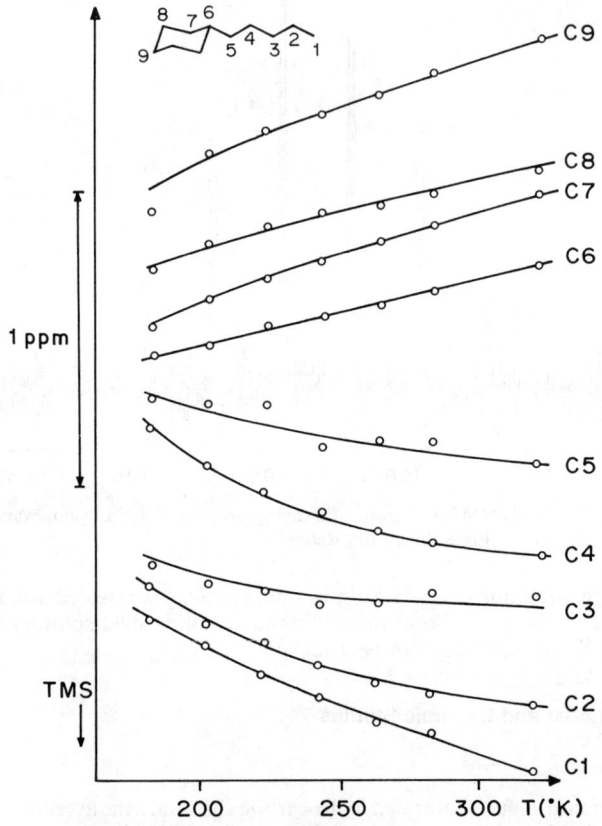

FIGURE 6. Temperature dependence of ^{13}C chemical shifts in n-pentylcyclohexane. Reprinted with permission from Schneider and Freitag, *J. Am. Chem. Soc.*, **98**, 478. Copyright (1976) American Chemical Society

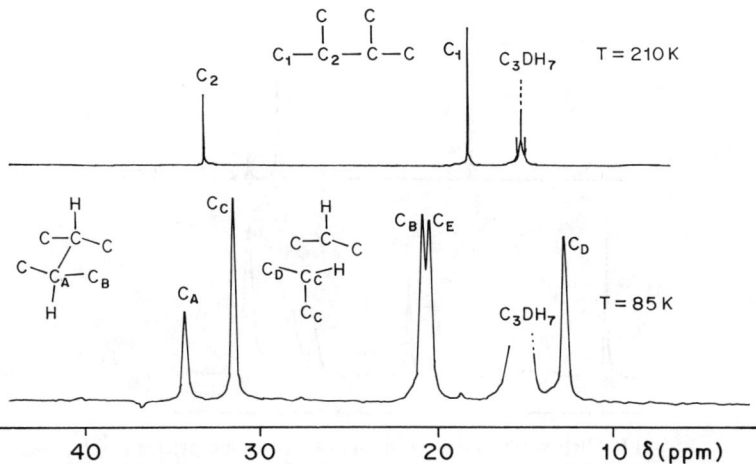

FIGURE 7. ^{13}C NMR spectra of 2,3-dimethylbutane in 1-monodeuteriopropane at 210 and 85 K. Reprinted with permission from Ritter *et al.*, *Tetrahedron Letters*, 3093 (1978). Copyright (1978) Pergamon Press PLC

observed. Using a γ-*gauche* effect of -5 ppm and a rotational isomeric state model[106], this difference and similar values for a large series of alkanes were calculated and found to be in reasonable agreement with the experiment[107].

A multitude of studies exists on sterically hindered ethanes **29**, where X, Y and Z as well as X', Y' and Z' are alkyl groups. The effect of steric hindrance on rotational barriers and rotamer population has been investigated systematically. As these studies first employed ^{1}H NMR, the *tert*-butyl group was of great value[108,109], because it shows only one singlet in the ^{1}H NMR spectrum and decoalescence was easily observed due to its steric bulk.

$$X \qquad X'$$
$$Y \rule{1cm}{0.4pt} C \rule{1cm}{0.4pt} C \rule{1cm}{0.4pt} Y'$$
$$Z \qquad Z'$$

(29)

If the steric bulk is further increased, stable rotamers can be isolated, when two adamantyl groups and two *tert*-butyl groups are attached to the ethane skeleton (X, X' = *tert*-butyl, Y, Y' = adamantyl)[110]. The sym.-tetra-*tert*-butylethane (X, X', Y, Y' = *tert*-butyl)[111] shows non-equivalent methyl groups which is attributed to an equilibrium of distorted conformers. The conformational analysis of tri-*tert*-butylethane has just recently been published[112]. Other examples are triisopropylmethane[113], and tetramethylpentane and -hexane[114]. With modern low-temperature ^{13}C NMR techniques even more complicated conformational equilibria can be assessed. As an example, the spectrum of 3-isopropyl-2,3,4-trimethylpentane (1,1,1-triisopropylethane) **30** is shown in Figure 8 for which a detailed analysis of the conformational equilibria was given[115].

Also worthy of mention is a study of *tert*-butyl-substituted cycloalkanes **31**, wherein rotation about the tertiary carbon-ring carbon bond was examined[116]. The barrier of *tert*-butyl rotation was observed to be largest (7.8 kcal mol^{-1}) in *tert*-butylcyclooctane, where the widest CCC bond angles prevail.

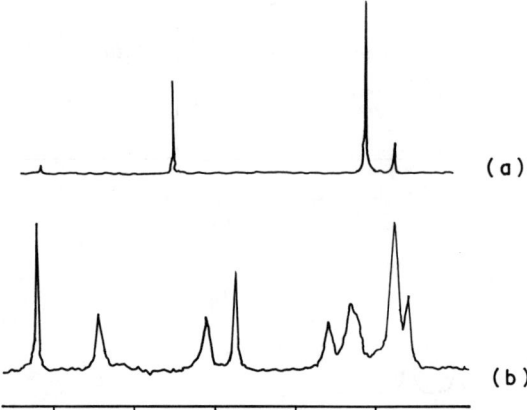

FIGURE 8. ^{13}C NMR spectrum of 1,1,1-triisopropylethane
(a) at room temperature and (b) at $-145\,°$C. Reprinted with
permission from Anderson, *Tetrahedron*, **46**, 5353. Copyright
(1990) Pergamon Press PLC

2. Dynamic processes in cyclic systems

The temperature-dependent NMR spectra of cyclohexane itself were investigated more
than 25 years ago by ^{1}H NMR using several methods and solvents[117–120]. The barrier
for chair-to-chair interconversion was determined to be 10.25 kcal mol^{-1}. Subsequently,
simple alkyl-substituted derivatives were examined by ^{13}C NMR, for example
methylcyclohexane[121] and 1,2-dimethylcyclohexane[122] (Figure 9). The chemical shifts
observed in the low-temperature region are in line with the steric interactions and the
conformational increments discussed in Section II.B.4.

Other examples are 1,2,4,5-tetramethylcyclohexane[123], hexamethylcyclohexane[124],
cis-decalin[125], *cis-syn-cis*-perhydroanthracene **32**[126] and tetracyclohexylethane[127]. After

(32)

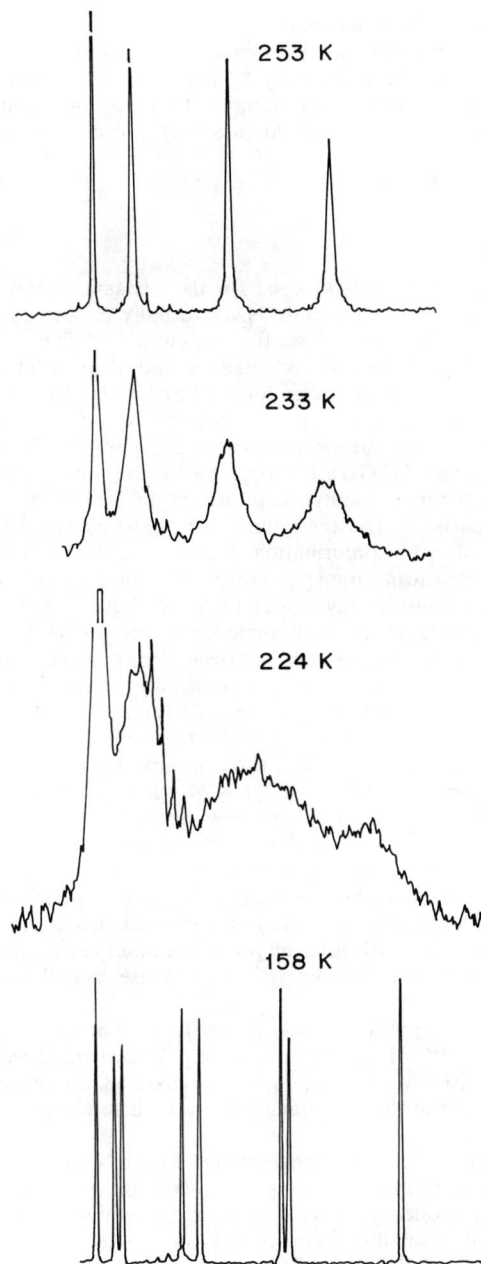

FIGURE 9. ${}^{13}C$ NMR spectra of *cis*-1,2-dimethyl-cyclohexane at various temperatures. The δ values were not given. Reproduced by permission of VCH Verlagsgesellschaft mbH from Ref. 122

an early ^{1}H NMR study of cyclooctane-d$_{14}$[128] cyclooctane was reinvestigated by ^{13}C NMR[129,130]. It was found that the boat–chair conformation is in equilibrium with a small amount of a crown form[130]. Only by high-field ^{13}C NMR was it possible to investigate the low-temperature conformations of the medium-sized cycloalkanes such as cyclononane[131], cyclodecane[132], cyclododecane[132], cyclotetradecane[132] and cyclohexadecane[133].

E. Solid State NMR

The conformational assignments based on the dynamic NMR studies above are confirmed by high-resolution solid state NMR, usually performed with the CP/MAS technique. These results were recently reviewed[134]. Thus, the results from low-temperature studies of higher cycloalkanes and their solid state spectra were compared directly[135–137]. The spectrum for solid cyclotetraeicosane, for example, splits at 248 K into five resonances, which were assigned to various *trans* (T) *gauche* (G) relations for a chain of four carbon atoms (e.g. 35.4 ppm for the sequence TTTT or 23.6 ppm for the sequence GTGG). For open-chain compounds, such as polyethylene, the chemical shift of the inner methylene groups in n-C$_{17}$H$_{36}$ was used as a model for the non-crystalline parts[138]. The crystalline parts show chemical shifts at lower field since they exist in an all-*trans* conformation. A further distinction between the two forms can be achieved by measuring the spin–lattice relaxation times (see below), since the crystalline parts relax considerably slower than the "liquid" parts of a polymer. The higher cycloalkanes crystallize in the triclinic form, whereas the higher n-alkanes such as C$_{22}$H$_{46}$, C$_{32}$H$_{66}$ and C$_{44}$H$_{90}$ as well as polyethylene crystallize in the orthorhombic form. It was proposed that the 1 ppm shift difference between the main peak of the higher cycloalkanes and the higher n-alkanes stems from this difference[139]. In a recent paper, the temperature-dependent solid state NMR spectra of the cycloalkanes allowed their calorimetric classification[140] into three groups according to phase transition behaviour. For ring sizes n, where $12 \leq n \leq 24$, all compounds have a solid–solid transition into a mesomorphic phase well below the melting transition. For $24 \leq n \leq 48$, no such transition could be detected. For $n > 48$ the phase behaviour depends strongly on the crystallization procedure. A similar study was performed for nonadecane, hexatriacontane, octahexacontahectane [CH$_3$(CH$_2$)$_{166}$CH$_3$] and polyethylene[141]. The situation seems to be more complex than in the cycloalkanes, although high-resolution solid state NMR can distinguish between inner and outer chain segments thus yielding detailed insight into the molecular organization of these crystalline and semicrystalline systems.

Solid state NMR was applied to study the radiation damage on alkanes, which is of importance with respect to radiation stability of polyethylene. It was found that, upon irradiation (low dose 0.09 Mgy), compounds like eicosane form glassy regions and that the remaining molecules in the crystalline parts near these glassy regions have a higher degree of mobility[142].

By a special NMR technique (dipole–dipole driven NMR with field cycling), the tunnelling frequency of the methyl groups of a series on n-alkanes was determined at 4.2 K. The tunnelling frequency differs for odd- and even-numbered alkanes and can be related to the different crystalline forms of these molecules[143].

Solid state static ^{13}C-NMR was applied to alkanes in order to address the fundamental issue of the anisotropy and alignment of the chemical-shift tensor of a CH$_2$ group. The spectra of cyclopropane, spiropentane, cyclooctane, cyclopentane and various other compounds[144] were obtained. It was shown that the σ_{CC} component lying perpendicular with respect to the CCC plane of the molecules is related to the CCC bond angle. The high isotropic shielding of cyclopropane was explained by this behaviour. Calculations by the IGLO method confirmed the interpretation of the experimental results.

F. Analytical Applications

^{13}C NMR and to some extent ^{1}H NMR spectroscopy are very important analytical tools to measure the content and composition of the alkanes and other hydrocarbons in crude oils[145-147], naphtha, bitumenous rocks, liquified coal products[148], gasoline[149], diesel fuels[150] and the like. Rapid and reliable detection of the main components is of high practical importance, and gave rise to numerous reports on analytical procedures and results. Quantitative methods involving NMR have been developed, and the results checked by statistical methods such as principal component analysis[151].

Another field of analytical applications is connected with polymers. Polypropylene can have different microstructures described as atactic, syndiotactic or isotactic and in one sample of the polymer combinations of these forms may be present. A detailed study of the ^{13}C NMR spectra can distinguish between these forms, although high-resolution spectra are usually needed to reveal all the necessary details[152].

A quite different analytical approach employing deuterium NMR of alkanes was demonstrated in a study of rhodium and platinum catalysts which catalyse deuterium-hydrogen exchange[153].

G. Deuterium Isotope Effects

Once again, alkanes have been employed as basic test molecules to study deuterium isotope effects on carbon chemical shifts[154,155] which can help in understanding substituent effects, shed light on hyperconjugational effects and may be useful in investigating conformational problems. The 500-MHz ^{1}H NMR spectra of the four isotopomers of methane, CH_4, CH_3D, CH_2D_2 and CHD_3, have been obtained recently to clarify the question of additivity of isotope effects. Indeed, only a small amount of non-additivity was found[156]. Careful investigations of the simple alkanes from ethane-d$_1$ through decane-d$_1$[157] showed a close parallelism between $^1\Delta$, $^2\Delta$ and $^3\Delta$ isotope effects and the chemical-shift changes in these compounds, where $^n\Delta$ refers to an isotope effect over n bonds. Comparison of the isotope effects in methane-d$_1$ (33), ethane-d$_1$ (34), propane-2-d$_1$ (35) and isobutane-2-d$_1$ (36) showed that the $^1\Delta$ values are incremental with respect to the number of methyl groups attached to the carbon atom in question, as are the chemical shifts.

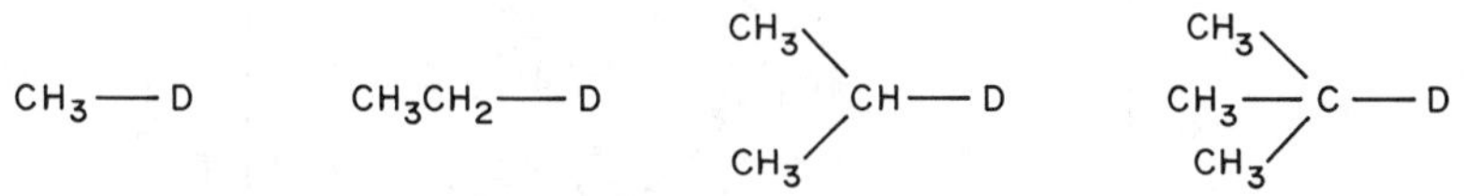

For the monodeuteriated cycloalkanes 37–41 the isotope effects shown in the scheme have been measured[158]. Remarkably, in 39 and 41 some sign changes for long-range deuterium effects were revealed. The $^1\Delta$ values correlate with the chemical shifts.

For conformationally fixed cyclohexanes 42a and 42b a strong dependence of the deuterium isotope effect on the axial or equatorial positioning was found[159].

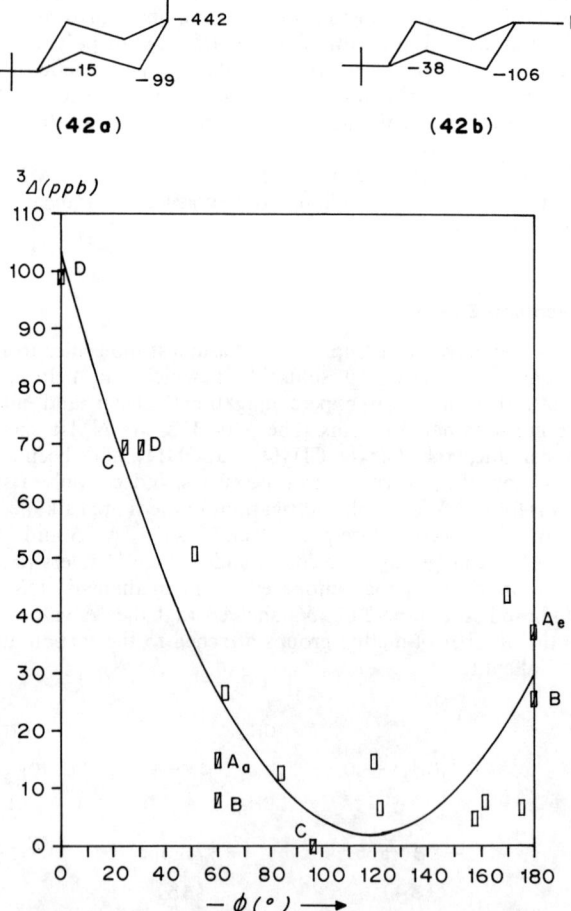

$^n\Delta$ values ($n = 1,2,3$) are given in ppb

FIGURE 10. Correlation between the absolute value of the $^3\Delta$ deuterium isotope effects on ^{13}C chemical shifts and the dihedral angle. (A) Axial and equatorial 4-tert-butylcyclohexane-d_1 (B) Adamantane-2-d_1. (C) Dicyanocobyrinic acid heptamethyl ester. (D) endo-Fenchol-2-exo-d_1. Reproduced by permission of John Wiley & Sons from Ref. 161

By the study of 12 isotopomers of protoadamantane, the stereochemical dependence of $^3\Delta$ isotope effects was also observed. A Karplus-type relationship similar to that for ^3J spin–spin coupling constants was proposed[160]. Recently, the first quantitative stereochemical dependence between $^3\Delta$ isotope effects and dihedral angle was reported for a series of deuteriated norbornanes, as shown in Figure 10[161].

Observations of the influence of substitution with deuterium on conformational equilibria led to a new method in physical organic chemistry called isotopic perturbation of equilibrium. Details can be found in a recent review[162]. The effect was first observed for the chair-to-chair interconversion of deuteriated 1,3-dimethylcyclohexane **43**[163] and later in 4-ethyl-1-methylcyclohexane[164] as well as in 1,1,4,4-tetramethylcyclohexane[165]. In contrast, the ^{13}C isotope effect on conformational equilibrium in related systems turned out to be too small to be observable[166].

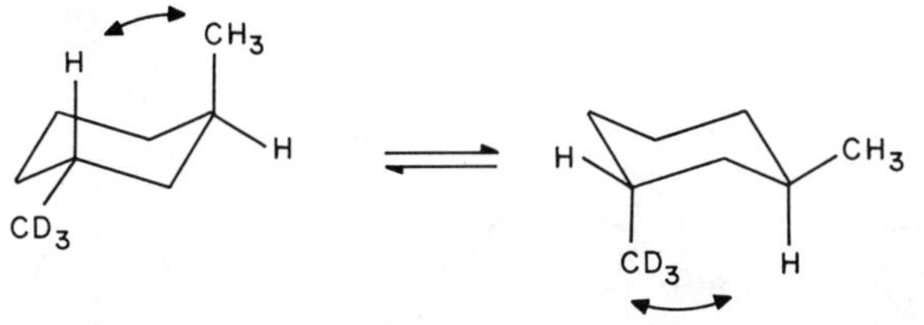

(43)

A question which has led to disagreement in the literature addresses the preference of deuterium for axial or equatorial positions in mono-deuteriocyclohexane **44**. Because of the shorter C—D bond, deuterium should be 'smaller' than hydrogen and thus prefer the axial position. However, theoretical calculations predict the opposite[167]. By integration of the low-temperature ^{13}C NMR spectra the A value for deuterium was determined to be 22 cal mol^{-1}[168], thus experimentally deuterium prefers the equatorial position. More recent measurements showed that this value is considerably smaller[169] (8.2 cal mol^{-1}) than originally reported.

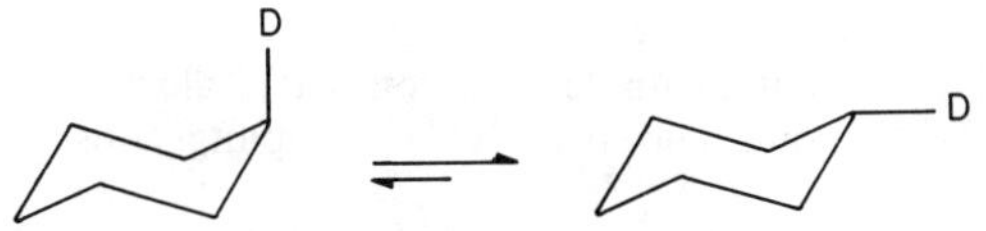

(44)

III. SPIN–SPIN COUPLING CONSTANTS

As in the case for chemical shifts the theory for spin–spin coupling constants was tested first on the alkane molecules, especially for the $^1H,^1H$ spin coupling in methane or cyclopropane. The theoretical developments have been reviewed[170,171]. The principal factors influencing the size of the couplings, e.g. hybridization, bond angle or dihedral angle, are by now common knowledge. In addition to the quantum-mechanical description of spin–spin coupling[170,171], there exists an attempt to relate the spin coupling constants in cyclic and bicyclic alkanes to the strain energy of molecules as calculated by molecular mechanics[172]. The commonly accepted nomenclature for spin–spin coupling constants is $^nJ(X,Y)$, where n is the number of bonds between the coupling nuclei X and Y.

A. ¹H,¹H Spin–Spin Coupling Constants

Geminal and vicinal proton–proton spin coupling constants in alkanes have been determined long ago and are described in any standard NMR textbook[173]. The geminal coupling constant $^1J(^1H,^1H)$ in alkanes is negative and averages to $-13\,Hz$, whereas for cyclopropane the value ($-4.3\,Hz$) is considerably different due to the larger HCH bond angle[174]. Vicinal coupling constants, which are extremely important for stereochemical assignments, especially in cyclohexane derivatives, can in pure cycloalkanes only be observed in specifically deuterated compounds[175]. Although many values of $^3J(^1H,^1H)$ in bicyclic and tricyclic alkanes have been reported in synthetic publications, there seems to be no systematic evaluation.

Noteworthy are $^4J(^1H,^1H)$ spin–spin coupling constants which, for saturated systems, can be resolved in the case of a W-arrangement[176]. Instructive examples are given for compounds **45–47**.

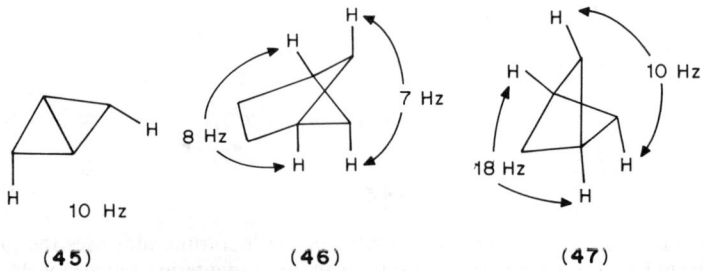

(45)   (46)   (47)

B. ¹³C,¹H Spin–Spin Coupling Constants Over One Bond

Values for $^1J(C,H)$ span a very small range in linear and branched alkanes, with deviations of only up to $2\,Hz$ from the value of $125\,Hz$ measured in methane. Some representative examples are listed in Table 10[3,177]. In contrast, the coupling constants in cycloalkanes[178,179] vary considerably. As can be seen in Table 11, there is a marked increase in the values going from cyclohexane towards the smaller ring systems.

Theoretically, a linear relationship derived from FPT calculations[180] between the %s character of the C—H bond and the $^{13}C,^1H$ coupling constant was found[180,181] (equation 2).

$$^1J(C,H) = 570\%s - 18.4 \qquad (2)$$

TABLE 10. $^1J(^{13}C,^1H)$ values for acyclic alkanes

CH₄	125
H₃C—CH₃	124.9
H₃C—CH₂—CH₃	124.4
H₃C—CH₂—CH₃	125.4
$\begin{array}{c} H \\ \vert \\ H_3C-C-CH_3 \\ \vert \\ CH_3 \end{array}$	124.0
H₃C—C(CH₃)₃	123.3

TABLE 11. $^1J(^{13}C,^1H)$ values of cycloalkanes

Cyclopropane	160.3
Cyclobutane	133.6
Cyclopentane	128.5
Cyclohexane	125.1
Cycloheptane	123.6
Cyclooctane	124.5
Cyclononane	124.0
Cyclodecane	124.3

Alternatively, this equation can be used to calculate the $\%s$ character of a specific C—H bond. One can use this formalism to describe the hybridization of a carbon atom by defining a hybridization index[182] λ (equation 3).

$$\lambda^2 = (1 - s)/s \qquad (3)$$

For example, for the carbon atoms of cyclobutane a hybridization of $sp^{2.7}$ is calculated.

Extremely high $^{13}C,^1H$ coupling constants were measured in polycyclic alkanes containing fused three-membered rings. Several examples are given in Table 12[3,36–38,183].

TABLE 12. $^1J(^{13}C,^1H)$ values of bicyclic and polycyclic alkanes

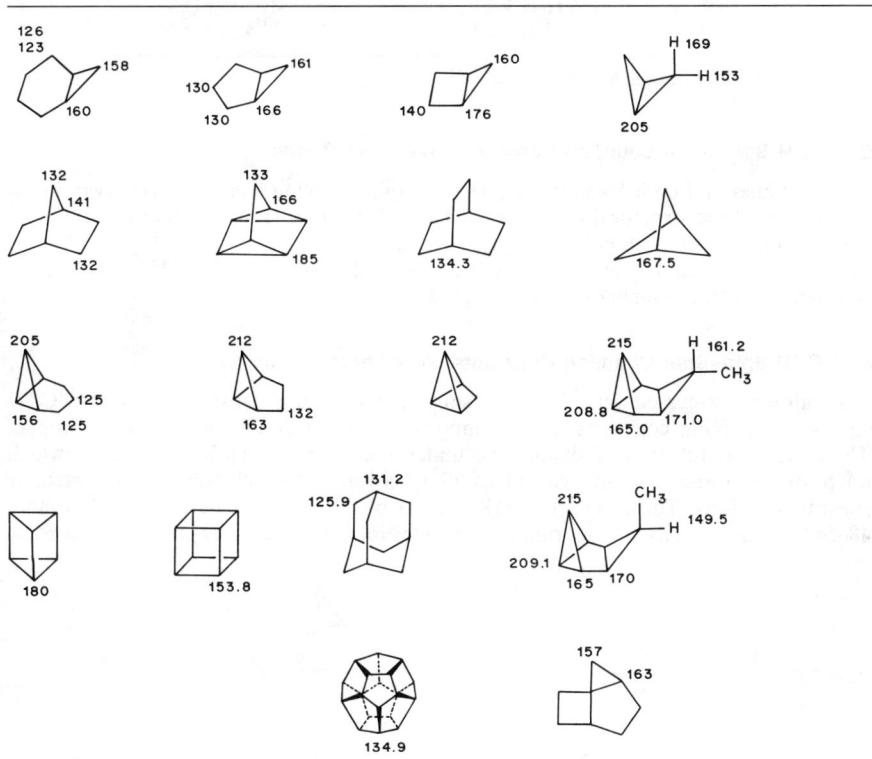

TABLE 13. $^2J(^{13}C,^1H)$ values of alkanes

H₃C—CH₂—CH₃   CH₃   H₃C—CH₃
                              |
                       H₃C—C—CH₃
                              |
                             CH₃

^2J(C-1,H-2) = −4.3
^2J(C-2,H-1) = −4.4   (−)3.9

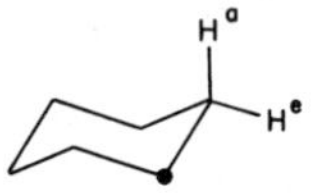

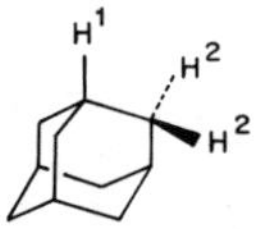

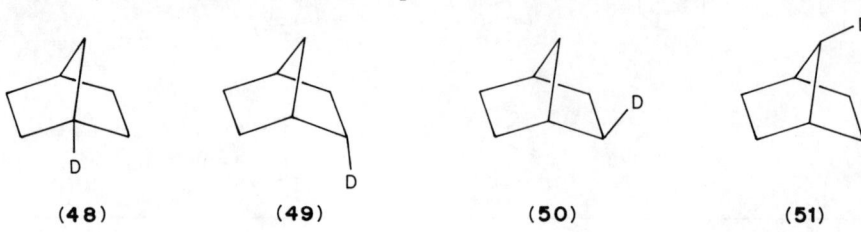

$^2J(C,H^e) = (−)3.7$ ^2J(C-1,H-2) = (−)3.3
$^2J(C,H^a) = (−)3.9$ ^2J(C-2,H-1) = (−)3.3

(−)2.6   (−)3.5   (−)3.0   (−)3.7   (−)3.7

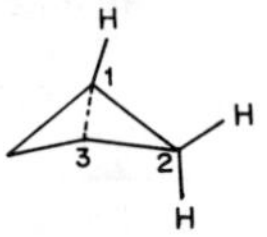

^2J(C-1,H-2) = 3.3^a
^2J(C-1,H-2) = 0^a
^2J(C-2,H-1) = 5.3
^2J(C-3,H-1) = 3.3

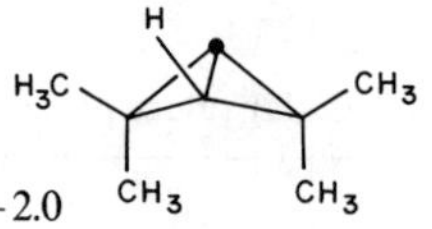

−2.0

^{a}Assignments can be interchanged.

C. ^{13}C,^{1}H Spin–Spin Coupling Constants Over Two Bonds

For alkanes and cycloalkanes, ^{13}C,^{1}H spin–spin coupling constants over two bonds provide very little structural information. The values are invariably small and negative being about −4 Hz. Some examples are given in Table 13[3,37,177–179,183–191]. Interestingly, cyclopropane displays no unusual two-bond ^{13}C,^{1}H coupling in comparison with the higher cyclic analogues.

D. ^{13}C,^{1}H Spin–Spin Coupling Constants Over Three Bonds

As already discussed for $^3J(^1H,^1H)$ spin–spin coupling constants, vicinal ^{13}C,^{1}H spin–spin coupling constants are an important tool in determining dihedral angles. Therefore, a careful study of alkanes was undertaken to provide a Karplus curve (which relate dihedral angles φ and vicinal coupling constants) which was not distorted by substituent effects. Thus, the ^{13}C NMR spectra of selectively deuteriated norbornanes **48–51**[192], where the fixed conformation provides reliable dihedral angles, were recorded.

(48)   (49)   (50)   (51)

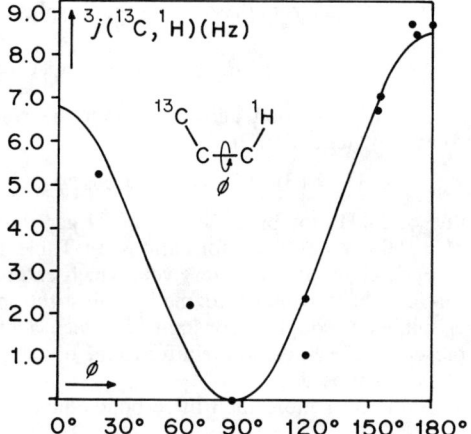

FIGURE 11. Dihedral angle dependence of $^3J(C, H)$
in norbornanes. Reproduced by permission of VCH
Verlagsgesellschaft mbH from Ref. 192

TABLE 14. $^3J(^{13}C,^1H)$ values of alkanes

$H_3CCH_2CH_3$
5.8

$H_3CCH_2CH_2CH_3$
5.4

$H_3CCH_2CH_2CH_3$
4.0

H_3C
$\diagdown$
$CHCH_2CH_3$
H_3C
4.7

$\begin{matrix} & CH_3 \\ & | \\ H_3CCH_2 - C - H \\ & | \\ & CH_3 \end{matrix}$
3.7

$\begin{matrix} H_3CCHCH_3 \\ | \\ CH_3 \end{matrix}$
5.3

$\begin{matrix} & CH_3 \\ & | \\ H_3C - C - CH_3 \\ & | \\ & CH_3 \end{matrix}$
4.7

H 2.1
3 · 2 · 1 · H 8.1

3 2 1 H
H 4
$^3J(C\text{-}3,H\text{-}1) = 5.6$
$^3J(C\text{-}2,H\text{-}4) = 7.2$

H 5.3
H 16.0

The curve is shown in Figure 11, and from these values the Karplus equation (equation 4) was derived.

$$^3J(C, H) = 3.81 - 0.9 \cos \varphi + 3.83 \cos 2\varphi \tag{4}$$

The experimental curve is in very good agreement with the results of INDO-PFT calculations for propane[193,194] (equation 5).

$$^3J(C, H) = 4.3 - 1.0 \cos \varphi + 3.6 \cos 2\varphi \tag{5}$$

The experimental value of 5.4 Hz for propane is a bit larger than the average value calculated from the 60°, 180° and 300° conformations. In Table 14[3,37,177,188,189,193], some other values for open-chain alkanes are given, which exhibit smaller couplings than for propane. This can be attributed to methyl substitution, which apparently decreases the $^3J(C,H)$ spin–spin coupling constant[195]. When a methyl, methylene or methine group were present in the γ-position relative to the observed carbon nucleus[196] a steric γ-gauche effect was proposed.

$^{13}C,^1H$ coupling constants over more than three bonds have not been reported for simple alkanes.

E. $^{13}C,^{13}C$ Spin–Spin Coupling Constants Over One Bond

With one exception (see below) $^1J(C, C)$ spin coupling constants in alkanes are all positive. According to a theoretical treatment by the INDO-FPT method[197], the Fermi contact mechanism for $^{13}C,^{13}C$ spin coupling dominates, whereas orbital and dipolar contributions are very small. For ethane, an experimental value of 34.6 Hz was found, and little variation was observed in other open-chain alkanes.

In Table 15[3,198–204,231] $^1J(^{13}C,^{13}C)$ values of several cyclic and bicyclic alkanes are collected. For symmetric compounds like cyclohexane, deuteriated species are usually investigated because the deuterium label removes the equivalence of the carbon atoms[205] and allows observation of the coupling constant.

In analogy to $^{13}C,^1H$ spin–spin coupling constants, in cyclopropane and methylcyclopropane the values deviate considerably from those of larger rings. One interpretation of the values in cyclopropane[198] discusses the %s character of the C—C bond. This is calculated by the product $s_A s_B$, where s is the fractional s-character of the carbon atoms involved. In cyclopropane $s_A s_B$ amounts to 0.18 and gives a good reproduction of the experimental values by using equation 6, which is analogous to equation 2.

$$^1J(C, C) = 658 s_A s_B - 7.9 \tag{6}$$

Decrease of coupling constants in endocyclic bonds was related to an increase in exocyclic bond coupling constants, as may be seen in methylcyclopropane.

A different line of argument maintains that in small rings or bicyclic compounds spin–spin coupling constants are transmitted by multiple pathways[206–208]. Thus, for cyclopropane, the observed $^1J(C,C)$ spin coupling constant is really the sum of $[^1J + ^2J]$. In small-membered rings, $^2J(^{13}C,^{13}C)$ spin–spin coupling constants (see below) can be

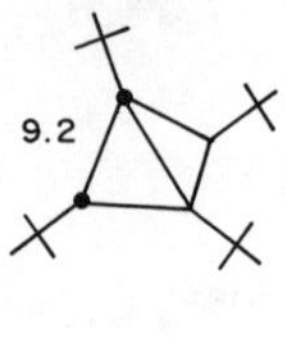

(52) 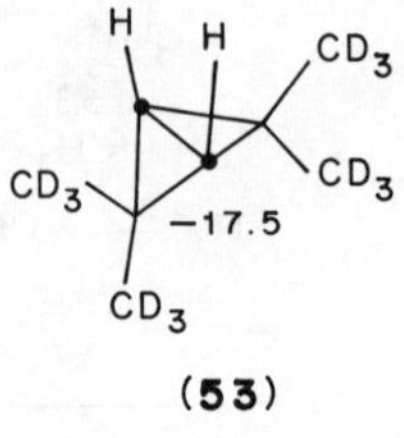   (53)

TABLE 15. $^1J(^{13}C,^{13}C)$ values of alkanes

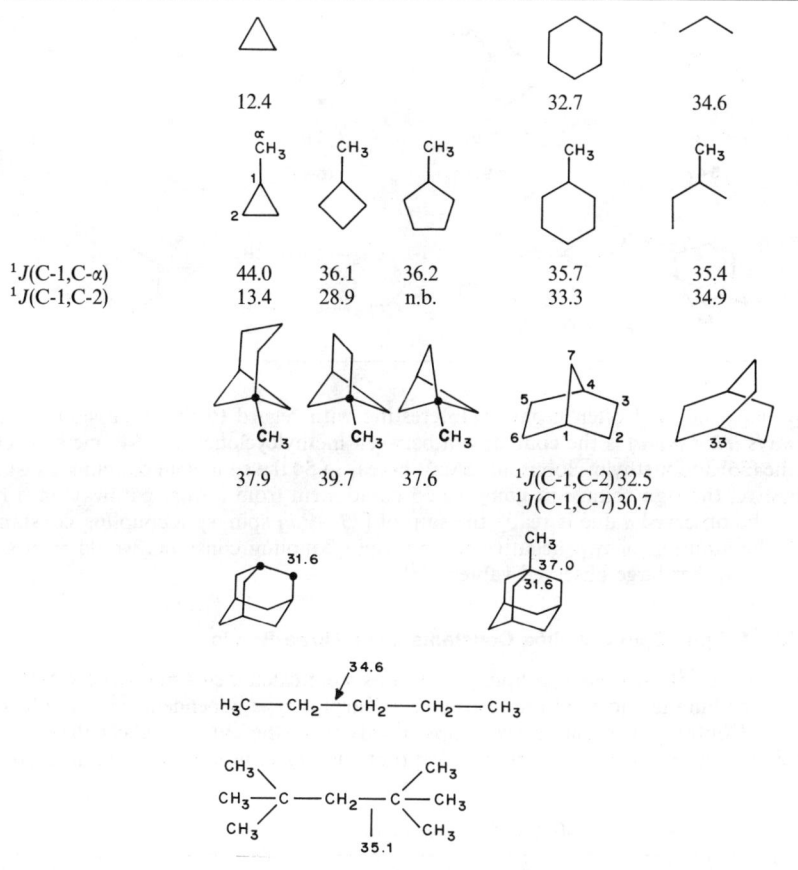

12.4	32.7	34.6

	44.0	36.1	36.2	35.7	35.4
$^1J(\text{C-1,C-}\alpha)$	44.0	36.1	36.2	35.7	35.4
$^1J(\text{C-1,C-2})$	13.4	28.9	n.b.	33.3	34.9

37.9   39.7   37.6 $^1J(\text{C-1,C-2})\,32.5$
                                          $^1J(\text{C-1,C-7})\,30.7$

31.6

CH$_3$
37.0
31.6

34.6
H_3C—CH_2—CH_2—CH_2—CH_3

CH$_3$ \
CH$_3$—C—CH$_2$—C—CH$_3$
CH$_3$ /   CH$_3$
        35.1

large and negative, which can also explain the anomaly for cyclopropane. In the stable tetra-*tert*-butyltetrahedrane **52**, a $^1J(C,C)$ value of only 9.2 Hz was found[199]. A very special case was observed in the bicyclobutane skeleton[190,209] **53**, where indeed negative $^1J(C,C)$ values have been measured. Application of equation 6 in this case would lead to a physically meaningless result for the hybridization of the carbon atoms.

Still another line of interpretation of the $^1J(C,C)$ data compares the values for a number of cyclic and acyclic alkanes including cyclopropane and relates the coupling constants to the CCC bond angle. A quadratic equation describes the relationship between $^1J(^{13}C,^{13}C)$ values and the bond angle, provided the central atom is a CH$_2$ group[204].

F. $^{13}C,^{13}C$ Spin–Spin Coupling Constants Over Two Bonds

In Table 16[3,204,206,207,210-215] some selected values of geminal C—C spin–spin coupling constants are given. As can be seen, for alkanes and cycloalkanes these values

TABLE 16. $^2J(^{13}C,^{13}C)$ values of alkanes

(−)1,6	(+)2.8	(−)8.1
(54)	(55)	(56)

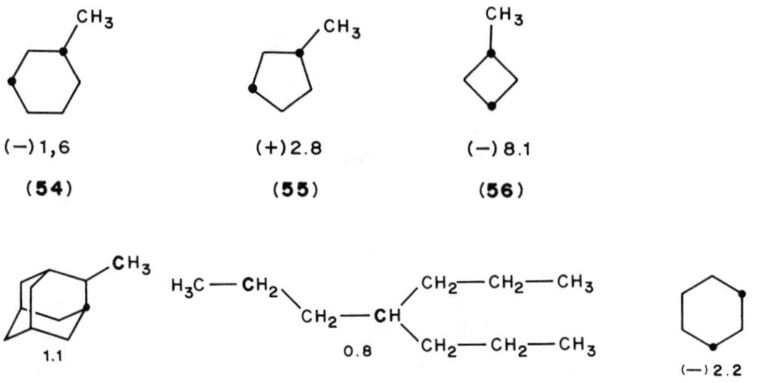

are rather small and often negative. Interesting with regard to the discussion of dual pathways (*vide supra*) is the comparison between methylcyclohexane (**54**), methylcyclopentane (**55**) and methylcyclobutane (**56**). Whereas in **54** the spin–spin coupling constant is negative, the sign change in going to **55** could stem from a dual pathway; in other words, the observed value is really the sum of $[^2J + ^3J]$ spin–spin coupling constants. In **56**, the addition of two negative ^2J spin–spin coupling constants would then give rise to the rather large observed value[206-208].

G. $^{13}C,^{13}C$ Spin–Spin Coupling Constants Over Three Bonds

Vicinal $^{13}C,^{13}C$ spin–spin coupling constants were calculated using the INDO-FPT method for butane and were found to give a Karplus-type dependence[216] on dihedral angle φ. Contrary to other relationships of this type, the curve predicts that the *cis* vicinal spin–spin coupling constant is larger than the *trans* coupling. This was interpreted

TABLE 17. $^3J(^{13}C,^{13}C)$ values of alkanes

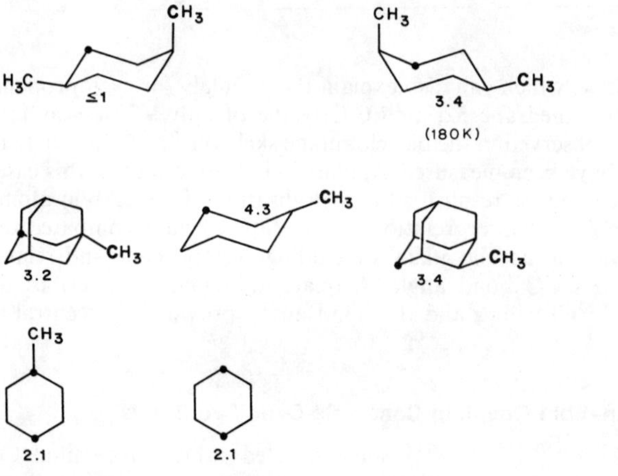

TABLE 18. Comparison of $^3J(^{13}C,^{13}C)$ values in even- and odd-membered alkanes

Alkane	$^3J(C\text{-}1,C\text{-}4)$	$^3J(C\text{-}2,C\text{-}5)$
n-Pentane	3.95	
n-Hexane	3.80	
n-Heptane	3.24	3.67
n-Octane	3.82	3.85
n-Nonane	3.73	3.66
n-Decane	3.85	3.75
n-Undecane	3.79	3.79
n-Dodecane	3.85	3.80
n-Tetradecane	3.57	3.63
n-Hexadecane	3.73	3.65

as stemming from a steric γ-effect[196]. However, in bicyclic compounds and methylcyclo-hexanes, effects of β substitution also apply[217]. Experimentally, a Karplus equation for $^3J(C,C)$ for pure alkanes has not yet been determined, although the curve obtained for cycloalkanone derivatives[218] seems to have general application[219]. In Table 17, selected values of $^3J(^{13}C,^{13}C)$ couplings are given.

The $^3J(^{13}C,^{13}C)$ spin–spin coupling in undecane (57) between carbon atoms 2 and 5 was used to assess conformational equilibria of hydrocarbons in different solvents[220]. For this purpose the compound was labelled specifically with ^{13}C at the 2 and 5 positions. However, no difference in coupling between chloroform, ether, ethanol and 13% water in ethanol was observed. In all solvents, the coupling constant was 3.6 Hz, which corresponds to 24% gauche and 76% trans conformation. Also reported was a coupling constant of 2.8 Hz between carbon atoms 7 and 10 in 8-methylhexadecane. Studies of this kind are conducted very seldom, since isotopic labelling is usually required and 2D-INADEQUATE measurements often do not give the necessary resolution. With a selective modification of the INADEQUATE sequence, dubbed SELINQUATE[221], it is now possible to measure long-range C—C spin–spin coupling constants with high resolution. Thus, a series of n-alkanes was investigated in acetone-d$_6$[222] to evaluate whether there are systematic conformational changes between even- and odd-membered n-alkanes. However, as the values in Table 18[223] demonstrate, this is not the case. For smaller alkanes the values scatter considerably, with mean coupling constants between carbon atoms 1 and 4 of 3.68 Hz and between carbon atoms 2 and 5 of 3.73 Hz. No marked concentration dependence was found.

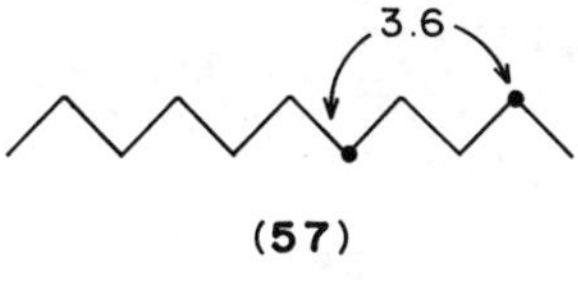

(57)

IV. SPIN–LATTICE RELAXATION

In contrast to chemical shifts and spin–spin coupling constants, spin–lattice relaxation times provide no direct structural information. Spin–lattice relaxation is a function of molecular motion, and thus, to the extent to which it is dependent on molecular size and shape, spin–lattice relaxation can provide important information not obtainable

TABLE 19. ^{13}C NMR T_1 relaxation times for
cycloalkanes $(CH_2)_n$

n	T_{1-exp}	$NOE_{(n+1)}$	T_{1-DD}
3	36.7	2.0	72.2
4	35.7	2.4	50.7
5	29.2	2.5	38.2
6	19.6	2.9	20.5
7	16.2	3.0	16.2
8	10.3	3.0	10.3
10	6.8		
15	3.2		
20	1.9		
30	1.3		
40	1.2		

by other NMR methods. Alkanes and cycloalkanes are again the class of compounds where basic observations have been made. For a series of cycloalkanes the ^{13}C spin–lattice relaxation times and the NOE effects were measured[224,225]. The dipolar relaxation times were then calculated and are given in Table 19 along with the values of some other compounds.

It can be seen that full NOE is present upwards from cyclohexane, whereas the smaller cycloalkanes, especially cyclopropane, relax via dipolar and spin–rotation mechanisms. For the higher cycloalkanes the T_1 values become constant.

Thus, although for the smaller molecules the T_1 values reflect the molecular size, for higher cycloalkanes the correlation time governing relaxation is no longer the reorientation of the molecule as a whole, but only the individual CH_2 segmental motions.

Whereas for cycloalkanes reorientation can safely be assumed to be isotropic, this condition is not valid for longer open-chain alkanes. In addition, segmental motion can be investigated. Thus, the data shown for 10-methylnonadecane **58**[226] imply a higher degree of motion towards the end of the chain. Detailed analysis of the data, including temperature effects, shows that the reorientation involves about six carbon atoms as a unit. In more recent work the T_1 data for the n-alkanes as pure liquids from C_6H_{14} through $C_{14}H_{30}$ are compared at two different temperatures. In these rod-like compounds it seems that free rotation as well as the transverse motions are severely restricted, and only a longitudinal motion is allowed[227].

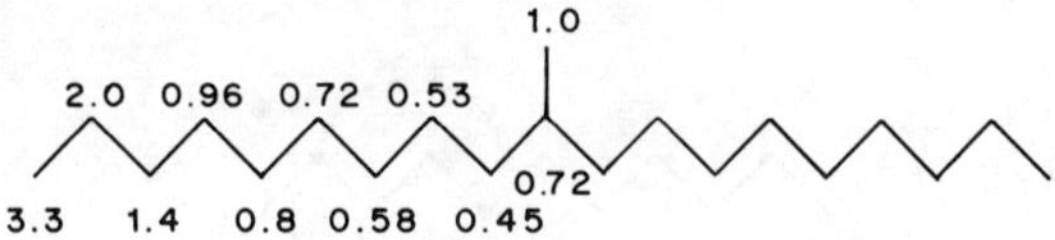

(58)

Carbon relaxation with proton spin–spin coupling was investigated for several n-alkanes, e.g. in selectively deuteriated nonane $C_4D_9CH_2C_4D_9$ with one isolated CH_2 group. For this isolated AX_2 spin system, dipolar relaxation was analysed in terms of spectral densities J_{CH}, J_{HCH}, J_{CHH} and J_{HH}. The results indicate that motion at each segment of the chain has local prolate symmetric-top character, and correlation times point to greater motional anisotropy at the middle of the chain rather than at the ends[228].

Solid state proton relaxation data in the rotating frame $(T_{1\rho})$ have been obtained at various temperatures for a series of n-alkanes from $C_{22}H_{46}$ to $C_{41}H_{84}$. A model in which the end of an alkyl chain jumps between the normal all-*trans* configuration and a defect *gauche* configuration was proposed[229]. The energy difference between configurations is chain-length dependent.

Spin–spin relaxation measurements by the spin echo method with an applied field gradient are used for determining diffusion constants. When applied to alkanes it was shown that these measurements are able to determine the grade of dispersity in polymer melts. The method was tested on paraffin mixtures containing eicosane as the main component[230].

V. CONCLUSION

The study of alkanes by NMR provided the basic knowledge on the structural dependencies of the three important NMR parameters: chemical shift, spin–spin coupling constants and spin–lattice relaxation. This information served, and still serves, to aid in understanding NMR spectroscopic results of all other organic molecules. At present, workers in this field have consolidated basic chemical-shift and spin-coupling data. Current emphasis certainly lies in analytical applications, and in the study of physicochemical interaction of alkanes in various phases (solid, liquid, gas) with a variety of host molecules.

VI. REFERENCES

1. B. V. Cheney and D. M. Grant, *J. Am. Chem. Soc.*, **89**, 5319 (1967).
2. K. Takaishi, I. Ando, M. Kondo, R. Chujo and A. Nishioka, *Bull. Soc. Chem. Jpn.*, **47**, 1559 (1974).
3. H. O. Kalinowski, S. Berger and S. Braun, *Carbon-13 NMR Spectroscopy*, Wiley, Chichester, 1988.
4. S. Fliszar, A. Goursot and H. Dugas, *J. Am. Chem. Soc.*, **96**, 4358 (1974).
5. S. Fliszar, G. Kean and R. Macaulay, *J. Am. Chem. Soc.*, **96**, 4353 (1974).
6. G. Kean, D. Gravel and S. Fliszar, *J. Am. Chem. Soc.*, **98**, 4749 (1976).
7. M. Schindler and W. Kutzelnigg, *J. Am. Chem. Soc.*, **105**, 1360 (1983).
8. M. Barfield and S. H. Yamamura, *J. Am. Chem. Soc.*, **112**, 4747 (1990).
9. K. Seidman and G. E. Maciel, *J. Am. Chem. Soc.*, **99**, 659 (1977).
10. A. B. Strong, D. Ikenberry and D. M. Grant, *J. Magn. Reson.*, **9**, 145 (1973).
11. J. Mason, *J. Chem. Soc., Perkin Trans. 2*, 1671 (1976).
12. P. Lachance, S. Brownstein and A. M. Eastham, *Can. J. Chem.*, **57**, 367 (1978).
13. A. R. N. Wilson, L. J. M. van de Ven and J. W. de Haan, *Org. Magn. Reson.*, **6**, 601 (1974).
14. R. M. Schwarz and N. Rabjohn, *Org. Magn. Reson.*, **13**, 9 (1980).
15. L. B. Alemany, *Magn. Reson. Chem.*, **27**, 1065 (1989).
16. D. M. Grant and E. G. Paul, *J. Am. Chem. Soc.*, **86**, 2984 (1964).
17. L. P. Lindeman and J. Q. Adams, *Anal. Chem.*, **43**, 1245 (1971).
18. Available from Chemical Concepts, VCH-Verlag, Weinheim.
19. Available from W. Robein, Institute Organic Chemistry, University of Vienna.
20. B.-L. Poh, *Magn. Reson. Chem.*, **24**, 816 (1986).
21. J.-E. Dubois and M. Carabedian, *Org. Magn. Reson.*, **14**, 264 (1980).
22. D. H. Smith and P. C. Jurs, *J. Am. Chem. Soc.*, **100**, 3316 (1978).
23. J. M. Bernassau, M. Fetizon and E. R. Maia, *J. Phys. Chem.*, **90**, 6129 (1986).
24. Y. Miyashita, T. Okuyama, H. Ohsako and S. Sasaki, *J. Am. Chem. Soc.*, **111**, 3469 (1989).
25. M. Randic, *J. Magn. Reson.*, **39**, 431 (1980); M. Randic and N. Trinajstic, *Theor. Chim. Acta*, **73**, 233 (1988); N. Randic, *Int. J. Quantum Chem.*, **23**, 1707 (1983).
26. Y. Miyashita, H. Ohsako, S. Sasaki and M. Randic, *Magn. Reson. Chem.*, **29**, 362 (1991).
27. J. J. Burke and P. C. Lauterbur, *J. Am. Chem. Soc.*, **86**, 1870 (1964).
28. D. K. Dalling and D. M. Grant, *J. Am. Chem. Soc.*, **89**, 6612 (1967).

29. G. Mann, E. Kleinpeter and H. Werner, *Org. Magn. Reson.*, **11**, 561 (1978).
30. M. Christl and J. D. Roberts, *J. Org. Chem.*, **37**, 3443 (1972).
31. D. J. Loomes and M. J. T. Robinson, *Tetrahedron*, **33**, 1149 (1977).
32. H. Jancke, G. Engelhardt, R. Radeglia, H. Werner and G. Mann, *Z. Chem.*, **15**, 310 (1975).
33. H.-J. Schneider and W. Freitag, *Chem. Ber.*, **112**, 16 (1979).
34. E. L. Eliel, and K. M. Pietrusiewicz, *Org. Magn. Reson.*, **13**, 193 (1980).
35. M. Christl. H. J. Reich and J. D. Roberts, *J. Am. Chem. Soc.*, **93**, 3463 (1971).
36. M. Christl, *Chem. Ber.*, **108**, 2781 (1975).
37. K. Wüthrich, S. Meiboom and L. C. Snyder, *J. Chem. Phys.*, **52**, 230 (1970).
38. M. Christl and R. Herbert, *Org. Magn. Reson.*, **12**, 150 (1979); M. Christl, H. Leininger and E. Brunn, *J. Org. Chem.*, **47**, 661 (1982).
39. J. B. Stothers, C. T. Tan and K. C. Teo, *Can. J. Chem.*, **51**, 2893 (1973).
40. R. Bicker, H. Kessler and G. Zimmermann, *Chem. Ber.*, **111**, 3200 (1978).
41. E. F. Weigand and H.-J. Schneider, *Org. Magn. Reson.*, **12**, 637 (1979).
42. J. B. Stothers and C. T. Tan, *Can. J. Chem.*, **54**, 917 (1976).
43. J. K. Whitesell and R. S. Matthews, *J. Org. Chem.*, **42**, 3878 (1977).
44. P. Metzger, E. Casadevall and M. J. Pouet, *Org. Magn. Reson.*, **19**, 229 (1982).
45. J. R. Wiseman and H. O. Krabbenhoft, *J. Org. Chem.*, **42**, 2240 (1977).
46. H.-J. Schneider and W. Ansorge, *Tetrahedron*, **33**, 265 (1977).
47. S. F. Nelsen, G. R. Weisman, E. L. Clennan and V. E. Peacock, *J. Am. Chem. Soc.*, **98**, 6893 (1976).
48. J. A. Peter, J. M. van der Toorn and H. van Bekkum, *Tetrahedron*, **33**, 349 (1977).
49. R. Kutschan, L. Ernst and H. Wolf, *Tetrahedron*, **33**, 1833 (1977).
50. W. A. Ayer, L. M. Browne, S. Fung and J. B. Stothers, *Org. Magn. Reson.*, **11**, 73 (1978).
51. D. K. Dalling, D. M. Grant and E. G. Paul, *J. Am. Chem. Soc.*, **95**, 3718 (1973).
52. D. K. Dalling and D. M. Grant, *J. Am. Chem. Soc.*, **96**, 1827 (1974).
53. J. Curtis, D. M. Grant and R. J. Pugmire, *J. Am. Chem. Soc.*, **111**, 7711 (1989).
54. M. Christl and R. Herbert, *Chem. Ber.*, **112**, 2022 (1979).
55. W. B. Smith and U. H. Brinker, *Magn. Reson. Chem.*, **29**, 465 (1991).
56. G. Maier, S. Priem, U. Schäfer and R. Matusch, *Angew. Chem.*, **90**, 552 (1978).
57. Z. Majerski, K. Mlinaric-Majerski and Z. Meic, *Tetrahedron Lett.*, **21**, 4117 (1980).
58. H. Beierbeck and J. K. Saunders, *Can. J. Chem.*, **53**, 1307 (1975); **55**, 2813 (1977).
59. M. Christl and W. Buchner, *Org. Magn. Reson.*, **11**, 461 (1978).
60. K. B. Wiberg, and F. H. Walker, *J. Am. Chem. Soc.*, **104**, 5239 (1982).
61. P. G. Gassman and G. S. Proehl, *J. Am. Chem. Soc.*, **102**, 6862 (1980).
62. A. de Meijere, O. Schallner, C. Weitemeyer and W. Spielmann, *Chem. Ber.*, **112**, 908 (1979).
63. H. Günther, W. Herig, H. Seel, S. Tobias, A. de Meijere and B. Schrader, *J. Org. Chem.*, **45**, 4329 (1980).
64. F. Fringuelli, E. W. Hagaman, L. N. Moreno, A. Tatiochi and E. Wenkert, *J. Org. Chem.*, **42**, 3168 (1977).
65. A. K. Cheng and J. B. Stothers, *Org. Magn. Reson.*, **9**, 355 (1977).
66. R. M. Cory and J. B. Stothers, *Org. Magn. Reson.*, **11**, 252 (1978).
67. S. A. Grodelski, W. D. Graham. T. W. Bentley, P. v. R. Schleyer and E. Liang, *Chem. Ber.*, **107**, 1257 (1974).
68. R. E. Pincock and F. N. Fung, *Tetrahedron Lett.*, **21**, 19 (1980).
69. M.-L. Dheu, D. Gagnaire, H. Duddeck, F. Hollowood and M. A. McKervey, *J. Chem. Soc., Perkin Trans. 2*, 357 (1979).
70. T. Winkler and W. V. Philipsborn, *Helv. Chim. Acta*, **51**, 183 (1968).
71. K. Nikki, *Magn. Reson. Chem.*, **28**, 385 (1990).
72. K.-J. Liu, *J. Polym. Sci., Part A-2*, **5**, 1209 (1967).
73. I. Ando, and A. Nishioka, *Makromol. Chem.*, **152**, 7 (1972); **160**, 145 (1972); **171**, 195 (1973).
74. I. Ando, A. Nishioka and M. Kondo, *Bull. Chem. Soc. Jpn.*, **47**, 1097 (1974).
75. I. Ando and A. Nishioka, *Makromol. Chem.*, **176**, 3089 (1975).
76. I. Ando and Y. Inoue, *Makromol. Chem., Rapid Commun.*, **4**, 753 (1983).
77. I. Ando, T. Hirai, Y. Fujii, A. Nishioka and A. Shoji, *Makromol. Chem.*, **184**, 2581 (1983).
78. K. Nikki and N. Nakagawa, *Org. Magn. Reson.*, **21**, 552 (1983).
79. M. A. Winnik, A. Mar, W. F. Reynolds, P. Dais, B. Clin and B. Caussade, *Macromolecules*, **12**, 257 (1979).

80. W. F. Reynolds, M. A. Winnik and R. G. Enriquez, *Macromolecules*, **19**, 1105 (1986).
81. L. J. M. van de Ven and J. W. de Haan, *J. Chem. Soc., Chem. Commun.*, 94 (1978).
82. I. D. Gay and J. F. Kriz, *J. Phys. Chem.*, **82**, 319 (1978).
83. J. M. Pope and D. W. Dubro, *Biochim. Biophys. Acta*, **858**, 243 (1986).
84. J. M. Pope, L. A. Littlemore and P. W. Westerman, *Biochim. Biophys. Acta*, **980**, 69 (1989).
85. R. E. Jacobs and S. H. White, *J. Am. Chem. Soc.*, **106**, 915 (1984).
86. B. Janik, E. T. Samulski and H. Toriumi, *J. Phys. Chem.*, **91**, 1842 (1987).
87. D. J. Photinos, E. T. Samulski and H. Toriumi, *J. Phys. Chem.*, **94**, 4694 (1990).
88. W. Schlenck, *Ann. Chem.*, **565**, 204 (1949).
89. G. M. Cannarozzi, G. H. Meresi, R. L. Vold and R. R. Vold, *J. Phys. Chem.*, **95**, 1525 (1991).
90. F. Imashiro, D. Kuwahara, T. Nakai and T. Terao, *J. Chem. Phys.*, **90**, 3356 (1989).
91. H. L. Casal, *J. Phys. Chem.*, **94**, 2232 (1990).
92. K. A. Wood, R. G. Snyder, H. L. Strauss, *J. Chem. Phys.*, **91**, 5255 (1989).
93. H. L. Casal, D. G. Cameron. and E. C. Kelusky, *J. Chem. Phys.*, **80**, 1407 (1984).
94. B. G. Silbernagel, A. R. Garcia, J. M. Newsam and R. Hulme, *J. Phys. Chem.*, **93**, 6506 (1989).
95. L. G. Davydov and A. G. Lundin, *Zh. Strukt. Khim.*, **30**, 76 (1989).
96. H. Lechert and W. D. Basler, *J. Phys. Chem. Solids*, **50**, 497 (1989).
97. A. A. Bothner-By, C. Stluka and P. K. Mishra, *J. Magn. Reson.*, **86**, 441 (1990).
98. E. W. Bastian, C. MacLean, P. C. M. van Zijl and A. A. Bothner-By, *Annu. Rep. NMR Spectrosc.*, **19**, 35 (1987).
99. H. Beierbeck and J. K. Saunders, *Can. J. Chem.*, **55**, 771 (1977).
100. H. Beierbeck and J. K. Saunders, *Can. J. Chem.*, **58**, 1258 (1980).
101. M. Graselli and A. C. Olivieri, *Anal. Chim. Acta*, **233**, 315 (1990).
102. H.-J. Schneider and W. Freitag, *J. Am. Chem. Soc.*, **98**, 478 (1976).
103. H. N. Cheng and F. A. Bovey, *Org. Magn. Reson.*, **11**, 457 (1978).
104. L. Lunazzi, D. Macciantelli, F. Bernardi and K. U. Ingold, *J. Am. Chem. Soc.*, **99**, 4573 (1977).
105. W. Ritter, W. Hull and H.-J. Cantow, *Tetrahedron Lett.*, 3093 (1978).
106. J. E. Mark, *J. Chem. Phys.*, **57**, 2541 (1972).
107. A. E. Tonelli, F. C. Schilling and F. A. Bovey, *J. Am. Chem. Soc.*, **106**, 1157 (1984).
108. C. H. Bushweller and W. G. Anderson, *Tetrahedron Lett.*, 1811 (1972).
109. H. Kessler, V. Gusowski and M. Hanack, *Tetrahedron Lett.*, 4665 (1968).
110. C. Rüchardt and H.-D. Beckhaus, *Angew. Chem.*, **97**, 531 (1985).
111. S. Brownstein, J. Dunogues, D. Lindsay and K. U. Ingold, *J. Am. Chem. Soc.*, **99**, 2073 (1977).
112. J. E. Anderson, *J. Chem. Soc., Perkin Trans. 2.*, 229 (1991).
113. J. E. Anderson, K. H. Koon and J. E. Parkin, *Tetrahedron*, **41**, 561 (1985).
114. C. H. Bushweller, W. G. Anderson, M. J. Goldberg, M. W. Gabriel, L. R. Gilliom and K. Mislow, *J. Org. Chem.*, **45**, 3880 (1980).
115. J. E. Anderson and R. E. Bettels, *Tetrahedron*, **46**, 5353 (1990).
116. F. A. L. Anet, M. St. Jacques and G. N. Chmurny, *J. Am. Chem. Soc.*, **90**, 5243 (1968).
117. F. R. Jensen, D. S. Noyce, C. H. Sederholm and A. J. Berlin, *J. Am. Chem. Soc.*, **82**, 1256 (1960); **84**, 386 (1962).
118. F. A. L. Anet, M. Ahmad and L. D. Hall, *Proc. Chem. Soc. (London)*, 145 (1964).
119. F. A. L. Anet, in *Dynamic NMR Spectroscopy* (Eds. L. M. Jackman and F. A. Cotton), Chap. 14, Academic Press, New York, 1975, p. 579.
120. D. Höfner, S. A. Lesko and G. Binsch, *Org. Magn. Reson.*, **11**, 179 (1978).
121. F. A. L. Anet, C. H. Bradley and G. W. Buchanan, *J. Am. Chem. Soc.*, **93**, 258 (1971).
122. H.-J. Schneider, R. Price and T. Keller, *Angew. Chem.*, **83**, 759 (1971).
123. H. Booth and J. R. Everett, *J. Chem. Soc., Perkin. Trans. 2*, 255 (1980).
124. H. Werner, G. Mann, H. Jancke and G. Engelhardt, *Tetrahedron Lett.*, 1917 (1975).
125. D. K. Dalling, D. M. Grant and L. F. Johnson, *J. Am. Chem. Soc.*, **93**, 3678 (1971); B. E. Mann, *J. Magn. Reson.*, **21**, 17 (1976).
126. A. H. Fawcett and F. Heatley, *J. Chem. Res. Synop.*, 294 (1981).
127. S. G. Baxter, H. Fritz, G. Hellmann, B. Kitschke, H. J. Lindner, K. Mislow, C. Rüchardt and S. Weiner, *J. Am. Chem. Soc.*, **101**, 4493 (1979).
128. F. A. L. Anet and J. S. Hartman, *J. Am. Chem. Soc.*, **85**, 1204 (1963).
129. H.-J. Schneider, T. Keller and R. Price, *Org. Magn. Reson.*, **4**, 907 (1972).
130. F. A. L. Anet and V. J. Basus, *J. Am. Chem. Soc.*, **95**, 4424 (1973).
131. F. A. L. Anet and J. J. Wagner, *J. Am. Chem. Soc.*, **93**, 5266 (1971).

132. F. A. L. Anet, A. K. Cheng and J. J. Wagner, *J. Am. Chem. Soc.*, **94**, 9250 (1972).
133. F. A. L. Anet and A. K. Cheng, *J. Am. Chem. Soc.*, **97**, 4220 (1975).
134. H. Saito, *Magn. Reson. Chem.*, **24**, 835 (1986).
135. I. Ando, T. Yamanobe, T. Sorita, T. Komoto, H. Sato, K. Deguchi and M. Imanari, *Macromolecules*, **17**, 1955 (1984).
136. I. Ando, T. Sorita, T. Yamanobe, T. Komoto, H. Sato, K. Deguchi and M. Imanari, *Polymer*, **26**, 1864 (1985).
137. M. Möller, W. Gronski, H.-J. Cantow and H. Höcker, *J. Am. Chem. Soc.*, **106**, 5093 (1984).
138. W. L. Earl and D. L. Vanderhart, *Macromolecules*, **12**, 762 (1979).
139. T. Sorita, T. Yamanobe, T. Komoto and I. Ando, *Makromol. Chem., Rapid Commun.*, **5**, 657 (1984).
140. H. Drotloff, D. Emeis, R. F. Waldron and M. Möller, *Polymer*, **28**, 1200 (1987).
141. M. Möller, H.-J. Cantow, H. Drotloff, D. Emeis, K.-S. Lee and G. Wegner, *Makromol. Chem.*, **187**, 1237 (1986).
142. M. Okazaki and K. Toriyama, *Chem. Phys. Lett.*, **160**, 21 (1989).
143. K. J. Abed and S. Clough, *Chem. Phys. Lett.*, **142**, 209 (1987).
144. J. C. Facelli, A. M. Orendt, A. J. Beeler, M. S. Solum, G. Depke, K. D. Malsch, J. W. Downing, P. S. Murthy, D. M. Grant and J. Michl. *J. Am. Chem. Soc.*, **107**, 6749 (1985).
145. O. M. Kvalheim, D. W. Akasnes, T. Brekke, M. O. Eide, E. Sletten and N. Telnaes, *Anal. Chem.*, **57**, 2858 (1985).
146. T. A. Holak, D. W. Aksnes and M. Stöcker, *Anal. Chem.*, **56**, 725 (1984).
147. A. Misbah-ul-Hasan, M. Farhat and M. Arab, *Fuel*, **68**, 801 (1989).
148. F. P. Miknis, T. F. Turner, L. W. Ennen and D. A. Netzel, *Fuel*, **67**, 1568 (1988).
149. J. Muhl. V. Srica and M. Jednacak, *Fuel*, **68**, 201 (1989).
150. D. J. Cookson and B. E. Smith, *Anal. Chem.*, **57**, 864 (1985).
151. T. Brekke, O. M. Kvalheim, and E. Sletten, *Anal. Chim. Acta*, **223**, 123 (1989).
152. A. E. Tonelli and F. C. Schilling, *Acc. Chem. Res.*, **14**, 233 (1981).
153. R. Brown, C. Kemball, J. A. Oliver and I. H. Sadler, *J. Chem. Res. (S)*, 274 (1985)
154. P. E. Hansen, *Prog. NMR Spectrosc.*, **20**, 207 (1988).
155. S. Berger, *NMR, Basic Principles and Progress*, **22**, 1 (1990).
156. F. A. L. Anet and D. J. O'Leary, *Tetrahedron Lett.*, **30**, 2755 (1989).
157. J. R. Wesener, D. Moskau and H. Günther, *J. Am. Chem. Soc.*, **107**, 7307 (1985).
158. R. Aydin and H. Günther, *J. Am. Chem. Soc.*, **103**, 1301 (1981).
159. R. Aydin, J. R. Wesener, H. Günther, R. L. Santillan, M. E. Garibay and P. Joseph-Nathan, *J. Org. Chem.*, **49**, 3845 (1984).
160. Z. Majerski, M. Zuanic, B. Metelko, *J. Am. Chem. Soc.*, **107**, 1721 (1985).
161. R. Aydin, W. Frankmölle, D. Schmalz and H. Günther, *Magn. Reson. Chem.*, **26**, 408 (1988).
162. H.-U. Siehl, *Adv. Phys. Org. Chem.*, **23**, 63 (1987).
163. K. W. Baldry and M. J. T. Robinson, *Tetrahedron*, **33**, 1663 (1977).
164. H. Booth and J. R. Everett, *Can. J. Chem.*, **58**, 2714 (1980).
165. S. L. R. Ellison, M. J. T. Robinson and J. G. Wright, *Tetrahedron Lett.*, **26**, 2585 (1985).
166. H. Booth and J. E. Everett, *Can. J. Chem.*, **58**, 2709 (1980).
167. I. H. Williams, *J. Chem. Soc., Chem. Commun.*, 627 (1986).
168. R. Aydin and H. Günther, *Angew. Chem.*, **93**, 1000 (1981).
169. F. A. L. Anet and D. J. O'Leary, *Tetrahedron Lett.*, **30**, 1059 (1989).
170. J. Kowalewski, *Prog. NMR Spectrosc.*, **11**, 1 (1977).
171. R. H. Contreras, M. A. Natiello and G. Scuseria, *Magn. Reson. Rev.*, **9**, 239 (1985).
172. M. P. Kozina, V. S. Mastryokov and E. M. Milvitskaya, *Russ. Chem. Rev.*, **51**, 765 (1982).
173. J. W. Emsley, J. Feeney and L. H. Sutcliffe, *High Resolution NMR Spectroscopy*, Chap. 10, Pergamon Press Oxford, 1966.
174. V. F. Bystrov, *Russ. Chem. Rev.*, **41**, 281 (1972).
175. J. L. Marshall, S. R. Walter, M. Barfield, A. P. Marchand, N. W. Marchand and A. L. Serge, *Tetrahedron*, **32**, 537 (1976).
176. M. Barfield and B. Chakrabarti, *Chem. Rev.*, **69**, 757 (1969).
177. T. Spoormaker and M. J. A. de Bie, *Recl. Trav. Chim. Pays-Bas*, **97**, 135 (1978); **98**, 380 (1979).
178. R. Aydin and H. Günther, *J. Am. Chem. Soc.*, **103**, 1301 (1981).
179. M. P. Kozina, V. S. Mastryukov and E. M. Mil'vitskaya, *Russ. Chem. Rev.*, **51**, 765 (1982).
180. M. D. Newton, J. M. Schulman and M. M. Manus, *J. Am. Chem. Soc.*, **96**, 17 (1974).

181. N. Muller and D. E. Pritchard, *J. Chem. Phys.*, **31**, 768 (1959).
182. W. A. Bingel and W. Lüttke, *Angew. Chem.*, **93**, 944 (1981).
183. R. Aydin and H. Günther, *Z. Naturforsch.*, **34b**, 528 (1979); **36b**, 398 (1981).
184. K. Tori, T. Tsushima, H. Tanida, K. Kushida and S. Satoh, *Org. Magn. Reson.*, **6**, 324 (1974).
185. H. P. Figeys, P. Geerlings, P. Raeymaekers, G. van Lommen and N. Defay, *Tetrahedron*, **31**, 1731 (1975).
186. T. J. Katz and N. Acton, *J. Am. Chem. Soc.*, **95**, 2738 (1973); R. J. Ternansky, D. W. Balogh and L. A. Paquette, *J. Am. Chem. Soc.*, **104**, 4503 (1982).
187. E. W. Della, P. T. Hine and H. K. Patney, *J. Org. Chem.*, **42**, 2940 (1977); M. Barfield, J. C. Facelli, E. W. Della and P. E. Pigou, *J. Magn. Reson.*, **59**, 282 (1984).
188. V. A. Chertkov and N. M. Sergeyev, *J. Am. Chem. Soc.*, **99**, 6750 (1977).
189. R. E. Wasylishen and T. Schaefer, *Can. J. Chem.*, **52**, 3247 (1974).
190. H. Finkelmeier and W. Lüttke, *J. Am. Chem. Soc.*, **100**, 6261 (1978).
191. T. Spoormaker and M. J. A. de Bie, *Recl. Trav. Chim. Pays-Bas*, **99**, 194 (1980).
192. R. Aydin, J.-P. Loux and H. Günther, *Angew. Chem.*, **94**, 451 (1982).
193. R. Wasylishen and T. Schaefer, *Can. J. Chem.*, **51**, 961 (1973).
194. M. Barfield, *J. Am. Chem. Soc.*, **102**, 1 (1980).
195. T. Spoormaker and M. J. A. de Bie, *Recl. Trav. Chim. Pays-Bas*, **98**, 59 (1979).
196. M. Barfield, J. L. Marshall and E. D. Canada, *J. Am. Chem. Soc.*, **102**, 7 (1980).
197. A. C. Blizzard and D. P. Santry, *J. Chem. Soc., Chem. Commun.*, 87 (1970).
198. J. Wardeiner, W. Lüttke, R. Bergholz and R. Machinek, *Angew. Chem.*, **94**, 873 (1982).
199. T. Loerzer, R. Machinek, W. Lüttke, L. H. Franz, K.-D. Malsch and G. Maier, *Angew. Chem.*, **95**, 914 (1983).
200. M. Stöcker and M. Klessinger, *Org. Magn. Reson.*, **12**, 107 (1979).
201. M. Stöcker, *Org. Magn. Reson.*, **20**, 175 (1982).
202. H. Günther and W. Herrig, *Chem. Ber.*, **106**, 3938 (1973).
203. V. V. Krishnamurthy, P. S. Iyer and G. A. Olah, *J. Org. Chem.*, **48**, 3373 (1983).
204. (a) M. Pomerantz and S. Bittner, *Tetrahedron Lett.*, **24**, 7 (1983).
     (b) E. W. Della and P. E. Pigou, *J. Am. Chem. Soc.*, **104**, 862 (1984).
205. V. A. Roznyatovsky, N. M. Sergeyev and V. A. Chertkov, *Magn. Reson. Chem.*, **29**, 304 (1991).
206. M. Klessinger and M. Stöcker, *Org. Magn. Reson.*, **17**, 97 (1981).
207. M. Klessinger, H. von Megen and K. Wilhelm, *Chem. Ber.*, **115**, 50 (1982).
208. J.-H. Cho, M. Klessinger, U. Tecklenborg and K. Wilhelm, *Magn. Reson. Chem.*, **23**, 95 (1985).
209. J. M. Schulman and M. D. Newton *J. Am. Chem. Soc.*, **96**, 6295 (1974).
210. J. L. Marshall, L. G. Faehl and R. Kattner, *Org. Magn. Reson.*, **12**, 163 (1979).
211. P. E. Hansen and J. J. Led, *Org. Magn. Reson.*, **15**, 288 (1981).
212. P. E. Hansen, O. K. Poulsen and A. Berg, *Org. Magn. Reson.*, **7**, 405 (1975).
213. F. J. Weigert and J. D. Roberts, *J. Am. Chem. Soc.*, **94**, 6021 (1972).
214. J. L. Marshall and D. E. Miiller, *Org. Magn. Reson.*, **6**, 395 (1974).
215. G. D. Andrews and J. E. Baldwin, *J. Am. Chem. Soc.*, **99**, 4851 (1977).
216. M. Barfield, I. Burfitt and D. Doddrell, *J. Am. Chem. Soc.*, **97**, 2631 (1975).
217. V. Wray, *J. Am. Chem. Soc.*, **100**, 768 (1978).
218. S. Berger, *Org. Magn. Reson.*, **14**, 65 (1980).
219. E. W. Della and P. E. Pigou, *J. Am. Chem. Soc.*, **106**, 1085 (1984).
220. F. M. Menger and L. L. D'Angelo, *J. Am. Chem. Soc.*, **110**, 8241 (1988).
221. S. Berger, *Angew. Chem.*, **100**, 1198 (1988).
222. S. Berger and M. Ochs, to appear.
223. M. Ochs, Dissertation, University of Marburg, 1990.
224. S. Berger, F. R. Kreissl and J. D. Roberts, *J. Am. Chem. Soc.*, **96**, 4348 (1974).
225. H. Fritz, P. Hug, H. Sauter and E. Logemann, *J. Magn. Reson.*, **21**, 373 (1976).
226. J. R. Lyerla, T. T. Horikawa and D. E. Johnson, *J. Am. Chem. Soc.*, **99**, 2463 (1977).
227. M. Iwahashi, Y. Yamaguchi, Y. Ogura and M. Suzuki, *Bull. Chem. Soc. Jpn.*, **63**, 2154 (1990).
228. M. S. Brown, D. M. Grant, W. J. Horton, C. L. Mayne and G. T. Evans, *J. Am. Chem. Soc.*, **107**, 6698 (1985).
229. I. Basson and E. C. Reynhardt, *J. Chem. Phys.*, **93**, 3605 (1990).
230. E. V. Meerwall and K. R. Bruno, *J. Magn. Reson.*, **62**, 417 (1985).
231. E. W. Della and P. E. Pigou, *J. Am. Chem. Soc.*, **104**, 862 (1984).

CHAPTER **9**

The mass spectra of alkanes

TINO GÄUMANN

Institute of Physical Chemistry, Federal School of Technology, Lausanne, Switzerland

I. INTRODUCTION

Hydrocarbons stand out as the most studied series of compounds in the early days of organic mass spectrometry dating back to the years after World War II. Two important reasons account for this fact: (i) there was a high demand from the petroleum industry for hydrocarbon mixtures analysis; (ii) gas chromatography was not yet known at this time and thus mass spectrometry was the only rapid and reliable analytical method available. As a consequence mass spectra of alkanes became the reference material for improvement of the theory of mass spectrometric fragmentation. The spectra were (and sometimes still are) believed to correspond to simple fragmentation reactions such as C—C and C—H bond splits. By the end of the fifties the mass spectrometer came into a more general use for structure elucidation and fragmentation mechanisms of organic compounds. Most of them include one or several heteroatoms that allow localization of the positive charge and thus to simplify the understanding of the fragmentation mechanisms. The study of alkanes became of minor interest because almost the only possibility of making progress consisted in the use of the labels ^{13}C and D. But the mass

The Chemistry of Alkanes and Cycloalkanes
Edited by S. Patai and Z. Rappoport © 1992 John Wiley & Sons Ltd

resolution needed to separate the doublet ^{13}C–CH at $m/z = 100$ exceeds 20,000 and for H_2–D, 70,000. Since it is not easy to get reproducible quantitative results at such high resolutions with a sector instrument, there exists only few—often conflicting—results. The Fourier-transform ion cyclotron resonance mass spectrometry (FT/ICR) changed the situation, but these instruments are not yet very widespread.

In most cases the neutral alkane will be ionized by electrons (equation 1).

$$C_nH_{2n+2} + e^- \rightarrow C_nH_{2n+2}^{+\cdot} + 2e^- \tag{1}$$

The ionization energy $IE(C_nH_{2n+2})$ is the difference between the heats of formation of the ion and the neutral (equation 2).

$$IE(C_nH_{2n+2}) = \Delta H_f(C_nH_{2n+2}^{+\cdot}) - \Delta H_f(C_nH_{2n+2}) \tag{2}$$

The heat of formation of the cation can be estimated by a simple equation proposed by Holmes and coworkers[1] (equation 3),

$$\Delta H_f(C_nH_{2n+2}^{+\cdot}) = 936 - 9.2N + 1246/N \quad \text{(in kJ mol}^{-1}) \tag{3}$$

where N is the total number of atoms in the ion. (For branched alkanes a small correction must be applied.) Another empirical equation has been proposed by Bachiri and coworkers[2] for prediction of ionization energies. The estimated data from these two expressions have been compared with the corresponding experimental values by Lias and colleagues (Table 2.5.1.3 in Reference 3). The ionization energy for a few simple alkanes is given in Table 1. It is highest for methane (12.6 eV), the first member of the series, and decreases to a plateau value of around 9 eV for higher alkanes.

Electrons with 70 to 100 eV energy are usually used in order to increase the sensitivity and to induce fragmentation. Following a proposal by Hurst, Platzman[7] has shown that with electrons of such a high energy part of the molecules are not directly ionized, but are 'superexcited'. These superexcited molecules either dissociate into two neutral fragments or pre-ionize to a molecular ion. Therefore there are at least two distinct mechanisms to form molecular ions. The simplified picture of a well-defined molecular ion, as it is often considered in mass spectrometry, is thus by no means justified.

Among others de Heer and colleagues have demonstrated in several publications the existence of excited neutrals and ions (see e.g. Reference 8). It seems to be a general rule for alkanes that excited molecular and atomic fragments are mainly formed from superexcited states that are optically forbidden with respect to the molecular ground state[9]. Last but not least, the possibility of an ion pair production (equation 4)

$$C_nH_{2n+2} + e^- \rightarrow C_xH_y^+ + C_{2n-x}H_{2n+2-y}^- + e^- \tag{4}$$

should be taken into consideration. However, for alkanes this reaction can be assumed to be of less importance.

In Figure 1 the spectra for n-triacontane obtained by Spiteller and coworkers[10] are shown for two different source temperatures. Maccoll[11] presents in a synoptic view some general features of alkanes. Several things are to be noted: (i) most of the spectrum consists of low mass fragments. (ii) Two fragmentation mechanisms seem to be favoured, one giving an even electron alkyl radical ion (equation 5) and the other an odd-electron olefin ion (equation 6).

$$C_nH_{2n+2}^{+\cdot} \rightarrow C_{n-m}H_{2(n-m)+1}^+ + C_mH_{2m+1} \tag{5}$$

$$C_nH_{2n+2}^{+\cdot} \rightarrow C_{n-m}H_{2(n-m)}^{+\cdot} + C_mH_{2m+2} \tag{6}$$

This general reaction scheme has been proposed by Schug[12]. He realized the disagreement of results obtained with ^{13}C labelling with this scheme, but it will be shown later that this is not necessarily the case. (iii) There seems to be a strong temperature dependence

TABLE 1. Ionization energies for some alkanes and appearance energies of some of their fragment ions by electron impact (in eV)[3-6]

Neutral	IE	CH_3^+ 9.8	$C_2H_4^{+\cdot}$ 10.5	$C_2H_5^+$ 8.1/7.4[a]	$C_3H_7^+$ 8.1	$C_4H_9^+$ 8.0/7.2/7.9/6.7[b]	$C_5H_{11}^+$ 7.9/7.4/6.6/6.9[b]	$C_6H_{13}^+$ 7.9/7.0[c]	$C_7H_{15}^+$ 7.0[d]
						AE			
CH_4	12.6	14.3(13.5)[e]							
C_2H_6	11.5	14.0		12.6(12.0)[e]					
C_3H_8	10.95		12.1	11.9					
$n\text{-}C_4H_{10}$	10.6		11.5		11.6(11.0)[e]	10.9			
$n\text{-}C_5H_{12}$	10.4	13.1[f]	11.6	13.8[f]	11.1	11.0			
$n\text{-}C_7H_{16}$	9.9				11.1	11.2	10.4	10.9	
$n\text{-}C_8H_{18}$	9.8				11.0	11.1	11.2	10.9	10.9

[a] n-/iso-.
[b] n-/sec-/iso-/t-.
[c] n-/sec-.
[d] sec-.
[e] AE for heterolytic fragmentation $RH \rightarrow R^+ + H^-$.
[f] Values for $neo\text{-}C_5H_{12}$.

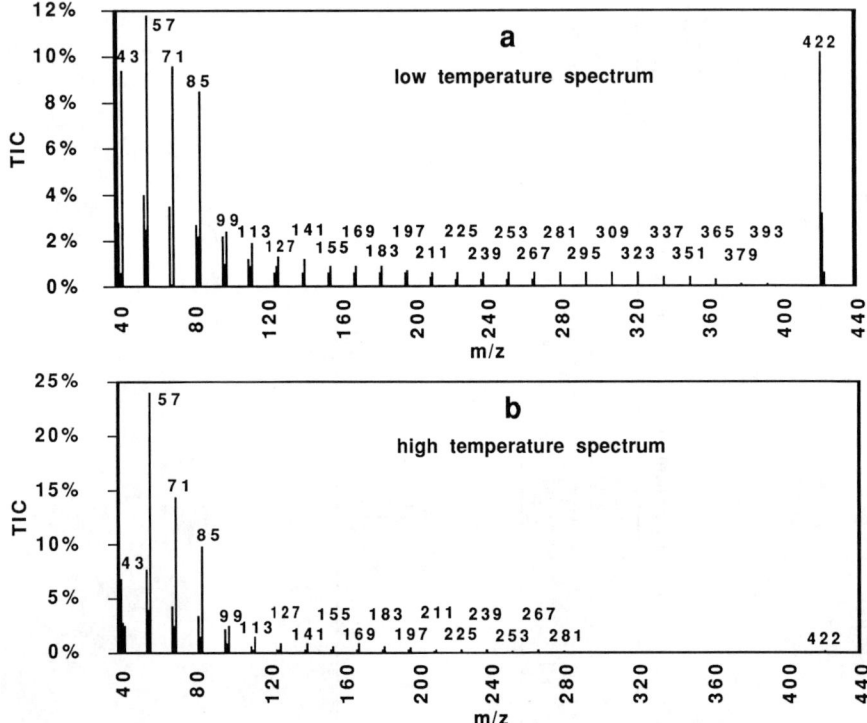

FIGURE 1. The mass spectrum of n-triacontane as a function of source temperature: (a) 70 °C, (b) 340 °C (after Spiteller and coworkers[10]. Reproduced with permission)

of the mass spectral fragmentation; the importance of higher mass fragments and, in particular, that of the molecular ion increasing with decreasing temperature, i.e. decreasing internal energy content of the neutral. This has also be shown in detail for n-hexane by Komarov and Novosel'tsev[13]. By not only decreasing the temperature within the ion source, but also the energy of the ionizing electrons, a further decrease in the intensity of the mass fragments is observed, as was demonstrated by Bowen and Maccoll[14]. The latter author[11] showed also that the spectra of several n-alkanes at low temperature and eV were nearly identical with the photoionization mass spectra[15].

In Table 1 the appearance energy (AE) for a few fragments is given. For methane there is a gap of nearly 2 eV between ionization and the appearance of the first fragment (CH_3^+), for ethane H and H_2 loss sets in practically simultaneously with ionization, but it takes nearly 2 eV for a C—C split. With increasing chain length the $C_nH_{2n+1}^+$ fragment disappears completely in favour of C—C bond splits that occur within less than one eV above ionization. To a first approximation it can be assumed that all C—C bonds have about the same energy and entropy requirements. The fact that low mass fragments prevail in the spectra of longer alkanes, in spite of their higher heats of formation, prove that a series of consecutive reactions must lead to the final observed products. Since the residence time in a sector instrument is a few μs, these fragmentations must be very fast and the ions thus contain excess internal energy. This demonstrates also the difficulty to isolate a single fragmentation.

The cross section σ for ionization by electrons of 75 eV has been determined by Vogt and coworkers[16] to be a linear function of the number n of C atoms in the molecule (equation 7).

$$\sigma = 2.03 + 2.44n \qquad (\text{in } 10^{-16}\,\text{cm}^2) \qquad (7)$$

The structure of the neutral alkane is generally known (unless one wants to determine it). However, this is by no means true for the molecular ion. In an n-alkane molecular ion it is not possible to localize the position of the charge; branched alkanes show some preference to fragment at the position of the tertiary carbon atom. One might thus assume that the charge is located at this position. Inspection of the data of Table 1 indicate that chain fragments with a higher molecular weight have a somewhat smaller appearance energy, demanding less energy for their formation. This is the basis of the so-called Stevenson rule[17], which states that the charge will be preferentially on the fragment with the lower ionization energy, i.e. on the longer alkyl or alkene ion in the case of paraffins[12,18]. In a simple C—C bond split, primary alkyl ions that have a higher ionization energy than their isomers are produced. In the formation of olefinic ions a hydrogen has to be transferred to the neutral fragment. Thus it cannot *a priori* be assumed that the molecular ion has the same structure as the neutral; even the simultaneous presence of several isomeric structures is feasible.

In Figure 1 it can be observed that low-molecular-weight ions prevail. As indicated above, this means that alkyl and olefin ions which result from primary fragmentation of the molecular ion will easily decompose further. It can be assumed that these fragments have already less excess internal energy; it means that at least some of them will undergo a slower decomposition. The multisector mass spectrometer has one or several regions where the decomposition of a selected metastable ion can be studied in the time window of several tens of μs. By special techniques the time window can be extended even to the ms range, as has been demonstrated, e.g., by Lifshitz and coworkers[19-21]. The experiments show that the part of the metastable decomposition in paraffins amounts to around 1% of the total fragmentation. The study of the metastable ions with different techniques has become very popular. However, it should not be forgotten that they correspond only to a small fraction of the decomposing ions and their behaviour is not necessarily representative either of the structure of the molecular ion or of the main reaction of the ion under study. However, it is the best one can do with present-day instrumentation to look at isolated processes.

Field ionization (FI), developed by Beckey[22,23], presents a time window on the short time scale. In this technique the molecules are ionized by a very strong inhomogeneous field in the vicinity of a sharp edge. Field strengths of several MV cm^{-1} are used. By assuming a field distribution around the edge, the time of decomposition and thus an average rate constant $\langle k \rangle$ as a function of time can be measured for the ps and ns range[24-26]. The rate constant $k(E)$ used in MS should not be confused with the rate constant that is generally used in kinetics and assumes an equilibrium distribution of the internal energies corresponding to the reaction temperature. It corresponds to some integration of the differential rate constant used in MS. In Figure 2a the logarithmic time dependence of the rate constant, determined by Beckey and colleagues[27], is shown together with a rate constant measured for EI by Hertel and Ottinger[28]. In Figure 2b a temperature dependence is given. The decrease in the value of $\langle k \rangle$ is about one power of ten by the same relative increase in time.

The molecular ions fragment according to reactions 8 and 9 to form alkyl and olefin ions, respectively. Since the primary fragmentation of the molecular ion M yields mainly ions with a molecular weight *larger* than $M/2$, but in the spectra of hydrocarbons most fragments are at much lower m/z, the consecutive fragmentations of these secondary ions furnish an important contribution to the observed spectra. Because these secondary ions have less internal energy than the fragmenting molecular ion, their fragmentation

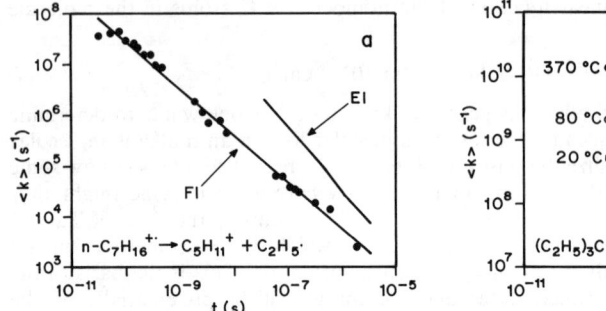

FIGURE 2. (a) The average rate constants for decomposition of n-heptane as a function of the ion residence time for FI (after Beckey and coworkers[27]) and EI (after Hertel and Ottinger[28]). (b) The average rate constants for fragmentation of 3-ethylpentane as a function of the FI emitter temperature (after Beckey and coworkers[27])

could be considered to be more specific in choosing the reaction path(s) lowest in energy. It can also be assumed that their contribution to the metastable range is more important.

Williams and coworkers[29] proposed a more detailed general scheme for fragmentations of ions generated from hydrocarbons, based on metastable measurements on n-alkanes from propane to heptane. They studied the fine structure of the metastable peaks and distinguished the following reactions (equations 8–20):

molecular ions $C_nH_{2n+2}^{+\cdot}$ $\to$ alkyl ion + alkyl radical   [20]   (8)
                                    $\to$ alkene ion + alkane   [13]   (9)
alkyl ions $C_nH_{2n+1}^{+}$ $\to$ lower alkyl ion + alkene   [29]   (10)
                                    $\to$ alkenyl ion + alkane   [11]   (11)
                                    $\to$ alkene ion + alkyl radical [1]   (12)
alkene ions $C_nH_{2n}^{+\cdot}$ $\to$ alkene ion + alkene   [8]   (13)
                                    $\to$ alkenyl ion + alkyl radical [13] (14)
                                    $\to$ alkyne ion + alkane   [5]   (15)
alkenyl ions $C_nH_{2n-1}^{+}$ $\to$ alkenyl ion + alkene   [5]   (16)
                                    $\to$ $C_3H_3^{+}$ + CH_4 [5]   (17)
                                    $\to$ alkyl ion + C_2H_2 [9]   (18)
'alkyne' ions $C_nH_{2n-2}^{+\cdot}$ $\to$ $C_{n-1}H_{2n-5}^{+}$ ions + CH_3 [8] (19)
$C_nH_{<2n-2}^{+}$ ions $\to$ loss of C_2H_2 [14]   (20)

The reactions corresponding to a loss of a hydrogen atom or molecule are not included. The numbers in brackets are the number of times the metastable reaction was observed by the authors[29], by Schug[12] and by Meyerson[30]. These numbers have no meaning as to the relative importance of the different competing fragmentation reactions, since many of these reactions are only observed on small fragments present in all mass spectra. Reactions 8 and 9 are the main reactions to be observed in the mass spectrum of an alkane. In both cases the neutral fragment contains generally not more than half of the number of carbon atoms originally present. This rule seems to be true for most of the fragmentation reactions, thus reactions 11 and 12 correspond mostly only to an elimination of CH_4 or CH_3. In addition to these reactions a loss of one or several molecules of hydrogen is often observed. In order to illustrate this reaction scheme a fragmentation diagram for n-hexane is given in Scheme 1 as proposed by Liardon and Gäumann[31]. It is based on metastable and high resolution measurements with a series of hexanes labeled with D in different positions. It is divided into two parts, the left side

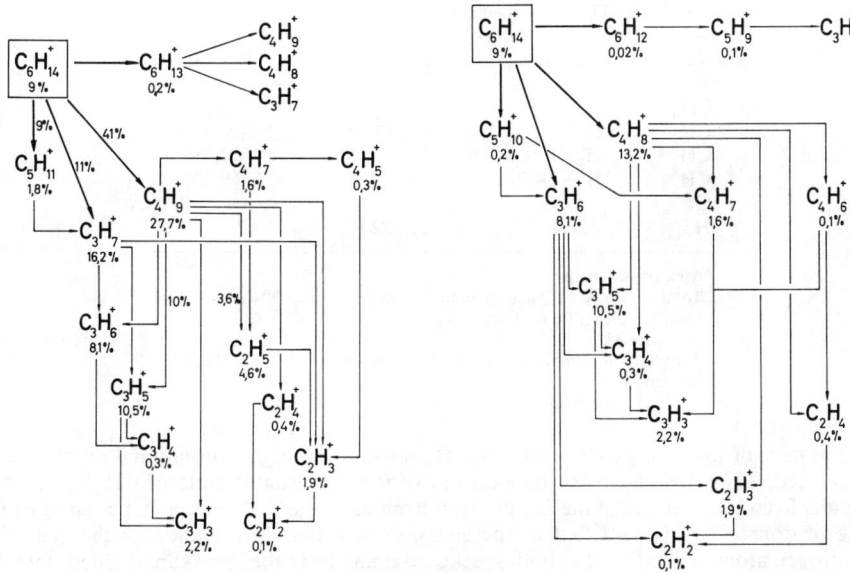

SCHEME 1

corresponding to reaction 8 and the right side to reaction 9 as primary fragmentations.
The abundance of each ion is given as % of the total ion current (TIC) for the source
spectrum. Numbers above arrows of some reactions represent an estimate of their relative
contributions to the TIC. These estimates are based on quite a few assumptions and
should be considered as approximate indications of their importance. However, this
diagram illustrates the relative importance to be expected for reactions 8–20 in the
fragmentation of an alkane.

Only in the metastable decays can the proportion of a reaction path be assessed with
some precision. However, it is not always easy to determine quantitatively the importance
of a type of reaction in the metastable decay for instrumental reasons: the limited mass
resolution when labelling is used, and contributions by CAD due to collisions with the
residual gases.

The formation of the ions H^+, $H_2^{+\cdot}$ and H_3^+ is energetically not favoured. It has
been observed in ethane and propane[32], but its importance seems to diminish with
increasing chain length. Since these ions are in a mass range that is seldom scanned, it
could also be that they are often overlooked.

In the next sections a few typical examples were selected in order to demonstrate
some of the problems encountered in mass spectrometry of alkanes. They should be
considered as typical for their class and stand for many similar samples that could not
be mentioned within this review. They cover the topics that are best known in this field.

II. METHANE—THE ISOTOPE EFFECT

Methane has the electronic structure $(1a_1)^2(2a_1)^2(1t_2)^6$, the last orbital being triply
degenerate. Its mass spectrum is given in Table 2. The ionization and appearance energies
for electron impact have been determined and reviewed by Plessis, Marmet and Dutil[38].
To a first approximation, it looks very simple. However, it should not be forgotten that
superexcited states and their fragmentation (see above) seem to play an important role[8].

TABLE 2. The mass spectrum of methane

Ion	Neutrals	AE^a(eV)	TIC^b(%)
$CH_4^{+\cdot}$	—	12.51	51/49(53)
CH_3^+	$H^{\cdot}$	14.3; 13.50^d	38/40(39)
$CH_2^{+\cdot}$	H_2	15.2	6.7/7.7(5.0)
CH^+	$H^{\cdot} + H_2(?)$	22.4	2.8/2.9(2.4)
$C^{+\cdot}$	—	$\leqslant 25.2$	0.64/1.0(0.7)
H^+	$CH_3^{\cdot}$	$22.3^c, 22.7^e$	$2.7/?^f$

^{a}Appearance energy[3].
^{b}Total ion current: Quadrupole mass spectrometer at $50\,eV^{33}$/FT/ICR at $70\,eV$ for CH_4 (% for CD_4)[34].
^{c}Reference 35.
^{d}Threshold for ion-pair formation. Electron affinity of H is $0.75\,eV^{36}$.
^{e}Reference 37.
$^f(H_2^+)$ in a quadrupole mass spectrometer is 0.48% TIC.

By impact of low energy electrons on CH_4, protons of high translational energy are produced. They arise from the dissociation of the first excited state of $CH_4^{+\cdot}$, which results from the removal of the $2a_1$ electron from methane[35]. The appearance energy of the proton coincides with that of the emission of β-Balmer radiation of the neutral hydrogen atom, indicating that both species originate from the same superexcited state[8]. Some emission from excited carbon-containing fragments has also been observed, but only little work is being done in this difficult field. Since methane contains only C—H bonds, it has been the molecule of choice for studying the primary H/D isotope effect in the fragmentation of ions.

This isotope effect has been defined by Evans and colleagues[39]. Two fragmentations can be easily defined, viz. the loss of a H or D atom and the loss of the hydrogen molecules H_2, HD and D_2 under the condition that we assume a simple one-step elimination. Although this must be true for energetic reasons near the appearance energy, its validity has not been proven for higher energies of the ionizing electrons. If we define $P(X)$ as the probability for the loss of X in a given isotopically substituted methane, corrected for its statistical probability within this molecule, a factor $\Pi_i(D/H)$ can be defined as in equation 21.

$$\Pi_i(D/H) \equiv P(D)/P(H) \qquad \text{for } CD_iH_{4-i} \tag{21}$$

This factor defines the (smaller) probability for a D atom to be lost as neutral fragment compared to an H atom within the same molecule. Another factor (equation 22)

$$\Gamma_i(H) \equiv [P(H) \text{ in } CD_iH_{4-i}]/[P(H) \text{ in } CH_4] \tag{22}$$

shows the (increased) probability to lose an H atom in an isotopically substituted methane compared to the same reaction in CH_4. An analogous probability can be defined for a D-atom elimination (equation 23):

$$\Gamma_i(D) \equiv [P(D) \text{ in } CD_iH_{4-i}]/[P(D) \text{ in } CD_4] \tag{23}$$

Several similar definitions are possible for the loss of a hydrogen molecule. Some of them are given in equations 24–26.

$$\Pi_i(HD/H_2) \equiv P(HD)/P(H_2) \qquad \text{in } CD_iH_{4-i} \tag{24}$$

$$\Gamma_i(H_2) \equiv [P(H_2) \text{ in } CD_iH_{4-i}]/[P(H_2) \text{ in } CH_4] \qquad i \leqslant 2 \tag{25}$$

$$\Gamma_i(D_2) \equiv [P(D_2) \text{ in } CD_iH_{4-i}]/[P(D_2) \text{ in } CD_4] \qquad i \geqslant 2 \tag{26}$$

TABLE 3. The isotope effect in the fragmentation of methane

		CH_4	CH_3D	CH_2D_2	CHD_3	CD_4
$\Gamma(H)$	MS^a	1	1.13	1.51	2.17	
	FT^b	1	1.24	1.58	2.17	
	QET^c	1	1.14	1.32	1.65	
$\Gamma(D)$	MS		0.36	0.54	0.75	1
	FT		0.52	0.70	0.73	1
	QET		0.62	0.70	0.80	1
$\Gamma(H_2)$	MS	1	2.4	4.1		
	FT	1	1.11	1.4		
	QET	1	1.15	1.55		
$\Gamma(HD)$	FT		1.60	1	0.71	
$\Gamma(D_2)$	MS			0.62	0.82	1
	FT			0.61	0.75	1
	QET			0.78	0.91	1
$\Pi(D/H)$	MS		0.31	0.37	0.36	
	FT		0.41	0.38	0.34	
	QET		0.56	0.54	0.49	
$\Pi(HD/H_2)$	MS		0.20	0.20		
	FT		0.50	0.25		
	QET		0.69	0.66		

[a] Measured with a double-focusing instrument[40].
[b] Measured with a FT/ICR instrument[34].
[c] Calculation based on the quasi-equilibrium theory[40].

In Table 3 values are presented for the isotope effect. One set of data has been obtained by Futrell and collaborators[40] with a high-resolution double-focusing two-sector instrument. A second set was determined with a FT/ICR mass spectrometer. The double-focusing instrument integrates over a time scale ranging from very short times to microseconds. Thus the ions must have some excess internal energy in order to be able to fragment in a sufficiently short time. In the FT/ICR, the integration extends into the millisecond range, including also the metastable range. It will be shown below that metastable fragmentations are not very important in methane, so the two instruments should not give very different results. But in addition, precise high-resolution measurements always suffer from the problem of discrimination within a mass doublet, an effect that is difficult to check. A third set of values is calculated by means of the quasi-equilibrium theory (QET). In this case some assumptions had to be made[24]. The data of Table 3 demonstrate a rather good agreement between the three sets. It can be seen that the relative probability to lose an H atom increases with a decrease in the number of H atoms within the molecule. This is even more pronounced for the loss of H_2. The contrary is true for the loss of D or D_2. This is true whether one compares with the loss within the molecule (Π-effect) or with CH_4 and CD_4 respectively (Γ-effect). The overall fragmentation of methane shows also some isotope effect: the molecular ion is a little more stable when fully deuterated, the formation of the methyl ion goes through a maximum (43%) for $CHD_3^{+\cdot}$ and the loss of molecular hydrogen is steadily decreasing[34]. A decrease of 20% in the dissociative fragmentation of superexcited states is observed for CD_4 compared with CH_4[9]. This is explained by the smaller zero-point energy of the C—D vibrations[40]. Very often H/D substitution is used to elucidate a particular fragmentation mechanism. However, an isotope effect of this order is often large enough to favour another mechanism, thus favouring a reaction path with a smaller probability in the reaction scheme of the unlabelled ion. This is particularly true in the metastable range, where the excess energy of the ion and the difference in zero-point energies of the vibration are often of the same order of magnitude. The similarity in the fragmentation patterns of low-molecular-weight alkanes led Lenz and Conner[41] to construct a general programme for calculating fragmentation patterns of deuterated methane and ethane, using a single adjustable parameter. It uses a statistical approximation and can be used for analyses of deuterated mixtures. However, it is based on an empirical approach and gives no insight into the fragmentation mechanisms.

The temperature dependence of the fragmentation of CH_4 and CD_4 has been determined by Komarov and Tikhomirov[42]. In the temperature range of 200 to 800 °C the relative intensity of the molecular ion decreases by only 12% and 18% for $CH_4^{+\cdot}$ and $CD_4^{+\cdot}$, respectively; the methyl ions increase by about 20%, CH^+ and CD^+ by 20% and 40%, respectively; $C^{+\cdot}$ shows also increase of 40%. The change corresponds again to the prediction of the QET. All possible fragment ions are formed in the decomposition of $CH_4^{+\cdot}$. This does not mean that all possible reactions leading to the products are important. Two kinds of experiment lend themselves to unravel the reaction paths: photoionization and the metastable decay. The former has the advantage of a very high energy resolution and elaborate setups such as photoelectron–photoion coincidence (PEPICO) and photoelectron–photoion–photoion coincidences (PEPIPICO, for doubly charged ions[43]) are feasible. The study of the metastable decay has the advantage that a given ion can be isolated and its fragmentation examined under the condition of small internal excess energy; the limited mass resolution is a definite disadvantage. Photoionization experiments by Chupka[36], Brehm[44] and by Dibeler and coworkers[45] disclose that the onset of ionization of the molecular ion has a complicated structure that originates from vibrational fine structure (hot bands) and a very low transition probability to the ground state of the molecular ion. Calculations by Radom and coworkers[46] indicate that the most stable structure of $CH_4^{+\cdot}$ has a C_{2v} symmetry with a pair of long $C \cdots H$ bonds due to a strong Jahn–Teller distortion. This could explain the hot bands. The $C_{2v}(^2B_1)$ ground state structure of the methane cation has been experimentally proven for $CH_4^{+\cdot}$ and $CH_2D_2^{+\cdot}$ by ESR in a solid Ne matrix at 4 K[47]. The process with the lowest energy requirement for CH_3^+ is the ion-pair formation (equation 27).

$$CH_4 + h\nu \rightarrow CH_3^+ + H^- \tag{27}$$

The analogous production of H^- by photoionization in ethane, propane and n-butane has also been observed[48]. The above process is lower in energy by the electron affinity of H (0.75 eV) than that in equation 28.

$$CH_4 + h\nu \rightarrow CH_3^+ + H^\cdot + e^- \tag{28}$$

The corresponding photoionization yield curve shows a pronounced curvature that Chupka explains by the thermal rotational energy of CH_4[36]. Chesnavich[49] summarized the different calculations for the transition state leading to this reaction. The alternative reaction (equation 29)

$$CH_4 + e^- \rightarrow CH_3^\cdot + H^+ + 2e^- \tag{29}$$

is less important and has been discussed above[35]. The methylene cation can have two precursors (equations 30 and 31).

$$CH_4^{+\cdot} \rightarrow CH_2^{+\cdot} + H_2 \tag{30}$$

$$CH_3^+ \rightarrow CH_2^{+\cdot} + H^\cdot \tag{31}$$

Again some curvature, explainable by the rotational energy of methane, is observed. The hydrogen molecule produced in reaction (30) must be in the ground state, at least at the threshold. Photoionization measurements give accurate threshold energies; the structure of the cross section as a function of photon energy sometimes allows one to discern the onset of reactions with higher energy requirement, but it does not allow one to draw a reaction scheme for ionization by electrons of higher kinetic energies. Cross sections for the different fragments in methane as a function of the electron energy have been determined by Mackenzie Peers and Milhaud[50] and extended to higher energies by Chatham and coworkers[33], but these measurements do not reveal reaction paths either.

Metastable fragmentations have the advantage that the relation between parent and fragment ion are defined and that—to a certain limit—the kinetic energy released can be determined. Rosenstock and coworkers, after an unsuccessful attempt[51], could only find a weak metastable reaction in CD_4 [52] (equation 32).

$$CD_4^{+\cdot} \rightarrow CD_3^+ + D^{\cdot} \quad (0.080\%) \tag{32}$$

The intensity of the metastable transitions in $CH_4^{+\cdot}$ is apparently several orders of magnitude weaker than in the higher alkane ions. By a careful extrapolation to low pressures, Ottinger[53] was able to find the corresponding reactions of the other isotopomers (equations 33–35).

$$CH_3D^{+\cdot} \rightarrow CH_2D^+ + H^{\cdot} \quad (0.099\%) \tag{33}$$

$$CH_2D_2^{+\cdot} \rightarrow CHD_2^+ + H^{\cdot} \quad (0.133\%) \tag{34}$$

$$CHD_3^{+\cdot} \rightarrow CD_3^+ + H^{\cdot} \quad (0.184\%) \tag{35}$$

With the exception of reaction (35) no deuterium loss was ever found. He was also unable to detect the reaction shown in equation 36.

$$CH_4^{+\cdot} \rightarrow CH_3^+ + H^{\cdot} \quad (0.061\%) \tag{36}$$

This reaction was detected by Futrell and coworkers[40]. The numbers given in parentheses are the values determined by these authors in % of the total ion current for the time interval from 1.8 μs to 3 ms, i.e. the total part of $CH_4^{+\cdot}$ ions that fragment after 1.8 μs[54]. The fraction is very small indeed, but Futrell and coworkers[24] were able to show that with a few reasonable assumptions such low figures could be explained by the quasi-equilibrium theory. They also proposed a breakdown graph for $CH_2D_2^{+\cdot}$. By extrapolation to zero pressure they observed also a hydrogen molecule loss from methylene ion, but the values are very small and especially with isotopically substituted ions the low mass resolution of the metastable spectra poses a problem. Cooks and coworkers[55] measured a very small translational energy release for reaction 36 and its temperature effect. It correlates well with the variation in centrifugal barrier height with rotational energy derived from a simple form of the Langevin collision theory. However, it is also in very good agreement with the more rigorous treatment of Klots[56,57] which allows tunnelling through the centrifugal barrier. In conclusion, it can be said that the metastable fragmentation and the photoionization studies reveal very detailed information about reaction 36 and indicate the presence of molecular hydrogen loss from the molecular and of atomic hydrogen from the methyl ion (reactions 30 and 31). Vestal[58] constructed a fragmentation scheme, based on the information available in 1968. It is astonishing that the fragmentation reactions for such a simple molecule are not better understood.

The fragmentation of methane can also be studied by allowing collisions between the ions and a colliding target gas. This technique, the collision-activated dissociation (CAD), is very popular in mass spectrometry in order to be able to differentiate among different isomeric metastable or stable ions. A detailed study of the CAD of methane has been undertaken by Ouwerkerk and colleagues[59], using ions with a kinetic energy of several keV and He, Ar and Xe as target gases. The authors measured the decomposition as a function of pressure, i.e. the number of collisions per unit time. Contrary to the fragmentation of a molecule ionized and excited by an electron or a photon, CAD is a rather brutal fragmentation method where higher electronically excited states can be occupied and, e.g., any number of hydrogens can be lost. The authors made this assumption and calculated by a least-squares procedure a fragmentation probability per collision for all ten possible reactions for hydrogen loss in methane and its fragments (4 in $CH_4^{+\cdot}$, 3 in CH_3^+, 2 in $CH_2^{+\cdot}$ and 1 in CH^+) as a function of the number of collisions. The method has the disadvantage that it is very difficult to assess any systematic errors in the results, but at least it gives some idea of the weight of the different reactions possible. The data are—as expected—rather similar for He and Ar as target gas. The most important reactions are given in equations 37–43 together with the relative probability of this reaction to take place in one collision. The first number refers to a collision in He, the second in Xe. It is evident that the simplest reaction, the loss of hydrogen from the molecular ion, is also the most important one. Its significance increases with the difference in relative molecular weight between the ion and the target gas. The loss of two hydrogens from the molecular ion is more important than the simplest reaction from the methyl ion. The probability for losing several hydrogen atoms in a collision decreases with the number of hydrogens lost in one collision and the atomic weight of the target gas. Although such measurements give no clue to the primary fragmentation of an ion after electron impact, it very often allows one to get important conclusions about the structure(s) of the ion in the moment of the collision, since this process is too fast ($\sim 10^{-15}$ s) to allow rearrangements. However, it should be remembered that it will be the structure(s) that resists fragmentation in the μs time scale. They do not always correspond to the initial structure of the ion, since it had sufficient time to rearrange to the most stable ionic structure, if it can be attained with an activation energy smaller than the difference between the ionization energy and the appearance energy of the first fragment ion.

$$CH_4^{+\cdot} + M \rightarrow CH_3^+ + H^\cdot + M \qquad 25\%/47\% \qquad (37)$$

$$CH_4^{+\cdot} + M \rightarrow CH_2^{+\cdot} + (H,H) + M \qquad 18\%/14\% \qquad (38)$$

$$CH_4^{+\cdot} + M \rightarrow CH^+ + (H,H,H) + M \qquad 8\%/4\% \qquad (39)$$

$$CH_3^+ + M \rightarrow CH_2^{+\cdot} + H^\cdot + M \qquad 16\%/12\% \qquad (40)$$

$$CH_3^+ + M \rightarrow CH^+ + (H,H) + M \qquad 9\%/4\% \qquad (41)$$

$$CH_2^{+\cdot} + M \rightarrow CH^+ + H^\cdot + M \qquad 8\%/6\% \qquad (42)$$

$$CH^+ + M \rightarrow C^{+\cdot} + H^\cdot + M \qquad 5\%/7\% \qquad (43)$$

The doubly charged ion CH_4^{2+}, the determination of its ionization energy and its decomposition have been described by Stahl and coworkers[60] and by Hanner and Moran[61]. Extended molecular orbital calculations were in agreement with their experimental finding and confirmed the ground state structure for $CH_4^{+\cdot}$ (C_{2v}) and CH_4^{2+} (D_{4h}). Wong and Radom[62] calculated that the dication is planar, but not square. Hatherly and coworkers[43] reviewed the doubly charged ion and added kinetic data obtained by a triple coincidence technique, using the high-energy photon flux of a synchrotron. Only five exit channels were found: $CH_3^+ + H^+$; $CH_2^{+\cdot} + H_2^+$; $CH_2^{+\cdot} + H^+$; $CH^+ + H^+$; $C^{+\cdot} + H^+$.

$CH_4^{+\cdot}$ and CH_3^+ have been the ions of choice to study ion/molecule reactions. A description of these results would not only exceed the framework of this review, but it is not even possible to cite all relevant publications[25,63–81]. It is evident that also the isotope effect of these reactions has received much attention[26,82–85]. The charge transfer to atoms and small molecules has also been investigated[86-88], as well as the neutralization/re-ionization behaviour[89]. It should be mentioned that ions also play an imporant role in radiation chemistry and quite a few good ideas have been transferred from irradiation experiments to mass spectrometry[58,90-92]. The methane ion and its fragments have been test substances for many experimental and theoretical ideas in mass spectrometry.

III. BUTANE—THE STRUCTURE OF THE MOLECULAR ION

Butane is the shortest n-alkane that can isomerize (to 2-methylpropane). In a molecule, the activation barrier for an isomerization is usually rather high. In the molecular ion, where one bonding electron is missing, this barrier might be lower. Further, the appearance energy for the first fragment is often one or several eV, i.e. a couple of $100\,kJ\,mol^{-1}$, above the first ionization energy. Therefore an isomerization can probably take place easily before fragmentation and some examples are known (see e.g. Lorquet[93]). The question is how to prove the structure of the cation. This is a question that is often a topic of heated discussion, not only because it is a difficult subject, but also because the problem is badly defined. It is general chemical knowledge that two stable structures are known for the formula C_4H_{10}: n-butane and isobutane. They correspond to two minima in the hyperplane defined by the coordinates of the nuclei, the transition from one minimum to the other requiring an activation energy that is large compared to the thermal energy. The ionization process results in a change of the structure of these molecules, since photoionization measurements do not allow one to determine the adiabatic ionization energy (see Section IV); from the photoelectron spectrum Heilbronner and collaborators[94] estimated it to be below 10.6 eV and Meot-Ner, Sieck and Ausloos[95] obtained from equilibrium measurements a value of 10.35 eV. Thus in such a process some additional energy is given to the ion; it might be little in mass spectrometric terms, but it is easily forgotten that an excess of 0.1 eV corresponds to an equilibrium temperature of $\sim 1000\,°C$. In terms of the quasi-equilibrium theory the excess energy is distributed randomly among the available degrees of freedom; at a given moment sufficient energy might concentrate on a given bond to dissociate it. Only a few organic molecules survive a temperature of 1000 °C!

The temperature of the molecules within the ion source will be in most cases above 100 °C. The fragmentation of butane has been used by Amorebieta and Colussi[96,97] to determine the temperature of butane molecules after collision with a hot surface. The calibration curves given by these authors can be used to estimate the effective temperature of the neutrals within the ion source. Small molecules have relatively little internal energy, but also a small number of degrees of freedom, whereas the internal thermal energy in large molecules may attain rather high values, distributed among many degrees of freedom. The first question is whether n-butane, when ionized, will form one (and only one) ionic structure, and if so, the second question will be whether such a structure is different from the one formed from isobutane. These questions can only be answered correctly with thermodynamic equilibrium measurements, since ions are unstable and spectroscopic measurements, particularly at low temperatures, are difficult, albeit not impossible. However, this would not solve the problem in mass spectrometry. If there are activation barriers between different ionic structures, are they lower than the lowest barrier for dissociation or does an isomerization take place in the fragmenting transition state? Beside the question of the activation barrier between different ionic structures it should be remembered that these ions can easily be produced—especially by electrons of 70 eV—in an excited electronic state. According to the forbiddenness of a state, its lifetime can vary between 10 ns and 1 s, but the fragmentations in mass spectrometry are measured between 10 ps and 1 s. Thus it is by no means astonishing that different experimental techniques may give different answers. Many of them are correct, but they are answers to different questions! Commonly the 'standard' mass spectrometric conditions are assumed, i.e. 70 V electrons, a temperature of the ion source somewhere between 100 and 200 °C, and a residence time in the ion source of μs. Thus the question boils down to two parts: firstly, the structure of decomposing ions within the ion source, and secondly, the structure of the ions that survived a few μs and that are either in the electronic ground state or in a metastable excited state. The research concentrates mainly on these ions in spite of the fact that they represent rarely more than one or two percent of the ions initially formed (for a review see Reference 98). The reason is that the states involved and their energies are much better defined. However, it should not be overlooked that the conclusions are only valid for a few percent of the reacting ions!

Ab initio calculations have been a considerable help in determining the different minima of the energy surface. The impact of new, calculated structures of cations in organic chemistry cannot be overestimated[99]. However, for the moment, the proof for a given structure or a mixture of different isomeric ions must be undertaken experimentally. The heat of formation of n-butane is 8 kJ mol^{-1} higher than that of 2-methylpropane; the latter has a somewhat higher ionization energy (10.57 vs 10.53 eV)[3], thus the difference in stability in favour of the cation of 2-methylpropane that Vestal[58] estimated to be 14 kJ mol^{-1} is probably too high by 10 kJ mol^{-1} and the small difference of 4 kJ mol^{-1} is more realistic[3].

Bouma, Poppinger and Radom[46] made a detailed analysis of the possible structures of the ions of n- and isobutane (among other hydrocarbons) using ab initio calculations of different levels. They reviewed also older calculations. The HOMO orbital of n-butane is the $7a_g[C_{2h}]$ orbital. It is sufficiently separated from the next lower MO, such that one can assume that the electron is eliminated from the HOMO. The corresponding ionized state does not correspond to a minimum and the authors calculate a difference between the vertical and the adiabatic ionization energy of 1.0 eV, compared with an experimentally determined value of 0.7 eV[94,100,101]. The HOMO of n-butane includes the central C—C and two C—H bonds. Therefore the authors tried to find the minimum of a structure with an elongated 2–3 bond. Accidentally this turned out to be the structure with the lowest minimum with a bond length of 2.02 Å. The proposal of a complex between an ion and radical made by Bowen and Williams[102–105] finds some confirmation in this structure, but the calculated structure is more strongly bound. However, Radom and collaborators found within a few tenths of an eV a few other minima with other bonds lengthened that are well within the excess energy furnished by the vertical ionization process. The ion has therefore to chose between several possible geometries. In isobutane the two highest orbitals are rather close, allowing an ionization from two different orbitals. Again several minima were found, allowing a facile interchange among the three methyl groups. No structure was proposed that would permit a facile isomerization between the two butane ions. Takeuchi and coworkers[106,107] repeated some of the calculations of Radom and extended them to larger values of the dissociating C—C bond. They demonstrate that for the loss of Me a barrier has to be crossed formed by a 1,2-hydrogen shift leading to the isopropyl ion, but that this barrier is lower than the energy needed to dissociate the bond leading to the formation of the Et$^+$.

The main ions in the mass spectrum of n-butane are given in Table 4. This is probably the most often measured substance in mass spectrometry, but the relative values vary according to the type of instrument and the instrumental conditions. The loss of methyl is the most important reaction; $C_3H_5^+$ results from a consecutive fragmentation, $C_2H_5^+$ can be either a direct loss of $C_2H_5^{\cdot}$ or the result of consecutive reactions. It should be noted that in the metastable time range the loss of CH_4 and $CH_3^{\cdot}$ are not just the only reactions observable, but they have similar probabilities. The photofragmentation of butane as a function of the photon energy has been determined by Steiner, Giese and

TABLE 4. Major ions in the mass spectrum of butane in % TIC

Ion	m/z	Probable neutral fragments	FT/ICR a	Source b	Metastable c	d
$C_4H_{10}^{+\cdot}$	58		5.9	3.5		
$C_4H_9^+$	57	$H^{\cdot}$	1.0	0.8		3
$C_4H_8^{+\cdot}$	56	H_2	0.3	0.2		
$C_3H_7^+$	43	$CH_3^{\cdot}$	40.9	29.9	55	37
$C_3H_6^{+\cdot}$	42	CH_4	6.0	3.8	45	60
$C_3H_5^+$	41	$H^{\cdot}, CH_4; (CH_4, H^{\cdot})$	12.9	9.5		
$C_2H_5^+$	29	$C_2H_5^{\cdot}; (H^{\cdot}, C_2H_4)$	9.6	13.2		
$C_2H_4^{+\cdot}$	28	C_2H_6	7.0	10.1		
$C_2H_3^+$	27	—	9.0	12.8		

[a]In the 0–10 ms time range[34].
[b]In the 0–2 μs range.
[c]In the ~10 μs to ~20 μs time range[108].
[d]In the 5 μs to 4 ms time range[109].

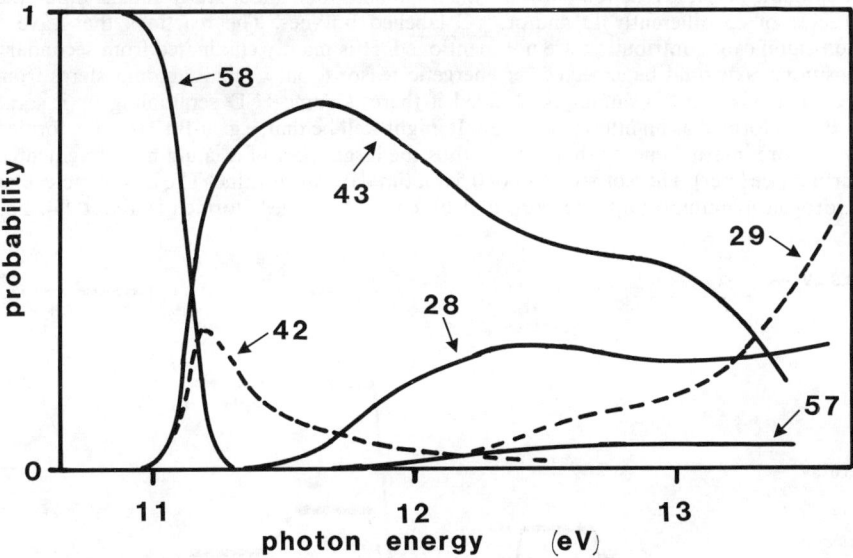

FIGURE 3. The photofragmentation of n-butane as a function of the photon energy (after Chupka and Berkowitz[48]. Reproduced with permission)

Inghram[15] and Chupka and Berkowitz[48] and is shown in Figure 3. The ionization-efficiency curve for mass 57 indicates the occurrence of the ion-pair formation (equation 4) beside the (higher energy) process of splitting off a hydrogen atom. The threshold for the formation of $C_3H_7^+$ corresponds to the energy of the secondary propyl (see later). The threshold for formation of $C_3H_6^{+\cdot}$ is, within the limits of error, identical to $C_3H_7^+$. The curve for $C_2H_5^+$ approaches the energy axis asymptotically, typical for a process that must compete with several others. A reliable threshold cannot be determined. The apparent threshold for $C_2H_4^{+\cdot}$ is 11.65 eV, higher than the calculated value of 10.45 eV. The difference may be due either to excess kinetic shift caused by competing processes of lower energy or to excess activation energy which is often required for fragmentation with rearrangement[48]. The ions produced in the primary fragmentation (the ions of the 'PIMS' as they are called by Maccoll[110]) are probably the following: $C_4H_9^+$ (+H), $C_4H_8^{+\cdot}$ (+H_2), $C_3H_7^+$ (+$CH_3^\cdot$), $C_3H_6^{+\cdot}$(+CH_4), $C_2H_5^+$ (+$C_2H_5^\cdot$), $C_2H_4^{+\cdot}$ (+C_2H_6) and correspond to the general fragmentation scheme cited in Section I. Flesch and Svec[111] arrive at a slightly different classification. Isobutane has a similar spectrum, but with reduced intensities of $C_2H_5^+$ and $C_2H_4^{+\cdot}$ in favour of $C_3H_6^+$. This seems to reflect the different structures of the molecular ions. The question is whether, and if so at which moment in the time scale of fragmentation, can there or must there be an isomerization of the molecular ion. This depends critically on the distribution of the internal energy at the moment of ionization. Since n-butane has been the workhorse for the evolution and tests of the quasi-equilibrium theory, this point is discussed in numerous publications that have become classical papers in the field (see e.g. References 58, 112–114) and has sometimes advanced the experimental confirmations.

In Scheme 2 an energy diagram is shown for different fragments in the mass spectrum of butane. It is based on the data given in References 115–117. The results of an attempt to determine the contribution of the different mechanisms involved are presented in Tables 5 and 6. The data were obtained by using a least-squares procedure containing reasonable possible reaction paths for the fragmentations under study and, as input, the spectra of 23 differently D and/or ^{13}C labelled butanes. The reactions that gave a non-significant contribution are not mentioned. H is mainly eliminated from secondary positions, as would be expected for energetic reasons, but a small amount stems from terminal positions; it cannot be decided if there is some H/D scrambling or if some n-Bu$^+$ is formed as an intermediate ion. It might well be that, e.g., n-Bu$^+$ is being formed with more internal energy than s-Bu$^+$, thus the larger part of it could have fragmented further (see later). The isotope effect of 0.5 is a time-averaged value! The loss of molecular hydrogen is unimportant; the ratio of 1-butene$^+$/2-butene$^{+\cdot}$ formed is about 1:4. For

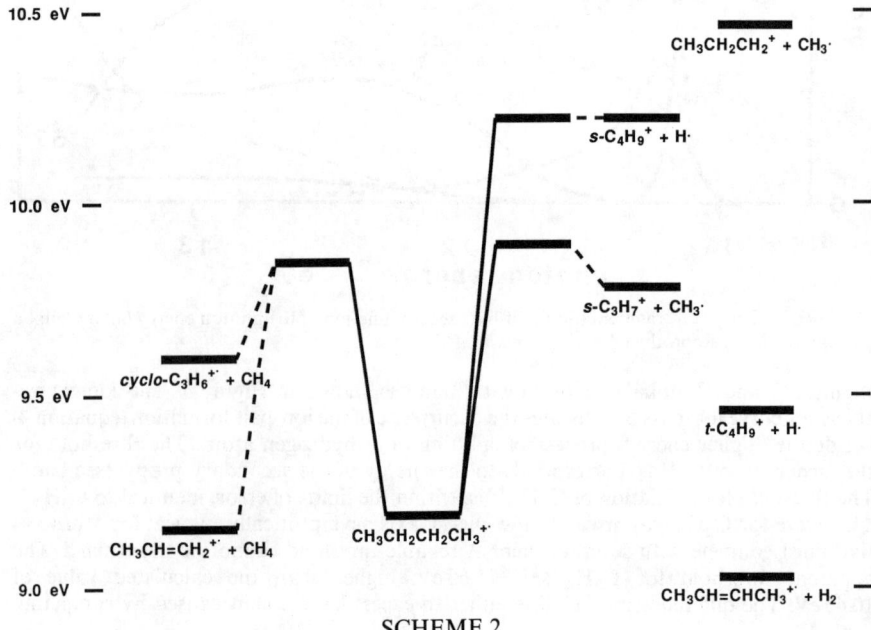

SCHEME 2

TABLE 5. The selective loss of hydrogen from molecular ion of n-butane (in %, source spectra, FT/ICR, 70 eV)[34]

Reaction	Position[a]	Probability	Isotope effect[b]
$M^{+\cdot} - (H, D)$	1	6	0.5
	2	94	0.48
$M^{+\cdot} - (H, D)_2{}^c$	1,2	20	
	2,3	80	
$M^{+\cdot} - (H, D)_3 = [M - (H, D)]^+ - (H, D)_2$	random[d]		0.96
$M^{+\cdot} - (H, D)_5 = [M - (H, D) - (H, D)_2]^+ - (H, D)_2$	random		0.92

[a]H, D from this position lost to the neutral.
[b]k_D/k_H.
[c]Weak signal.
[d]All positions have the same probability.

TABLE 6. Results from the fragmentation of labelled n-butane (in % contribution for the reaction at 70 eV; 12 eV similar)

Reaction	Position of the label lost to the neutral fragment		FT/ICR[34] (see also Reference 116)	Metastable[108]
	[13]C	D		
$M^{+\cdot} - CH_3{}^\cdot$	1	1,1,1	93	92
	2	1,2,2	2	
	2	2,2,3	2	8
	2	2,2,4	3	
$M^{+\cdot} - CH_4$	1	1,1,1,2 or 3	47	
	1			87
	1	1,1,1,4	41	
	2	1,2,2,4	3	
	2			13
	2	2,2,3,4	9	
$M^{+\cdot} - [CH_5]$		$H^\cdot, CH_4 (rnd)^a$	20	
		$CH_3, H_2 (rnd)^b$	80	
$M^{+\cdot} - C_2H_5{}^\cdot$	1,2		80	
	1,3		10	
	1,4		3	
	2,3		7	
$M^{+\cdot} - C_2H_6$	1,2		85[c]	

[a]Loss of H according to Table 5, random loss of CH_4.
[b]Loss of CH_3 according to this Table, random loss of H_2.
[c]Other reactions unknown.

the further loss of molecular hydrogen no specific reaction can be recognized, and a random process must be assumed as is often observed in the fragmentation of secondary fragments in hydrocarbons.

Several authors have realized that Me is not only lost from terminal positions[104,109,116,118–120], although different interpretations are offered. 7% of Me lost originates from a position within the chain. The ratio of internal to terminal loss does not seem to depend on time, but in the metastable range the relative importance of the two sources of the additional hydrogen atom for the internal loss seems to change, since in the metastable decay of 1,4-D_6-butane no loss of $CH_2D^\cdot$ could be observed[108]. The reaction proceeds within 0.05 eV of the calculated threshold of the most stable product, the 2-propyl ion[116]. An analogous situation is seen with higher alkane ions, showing that either prior to or simultaneously with the fragmentation, a 1,2-hydrogen shift has to take place, at least for the low-energy metastable fragmentation. Coggeshall[121] claimed that the metastable decay can be split-off in three processes with different litetimes, but Schug[122] demonstrated that the same experimental data can be represented as well with only two lifetimes. One might question the validity of results of this kind that have been obtained with relatively simple instrumentation, although there is no *a priori* reason that the fragmentation scheme can be due to populations with different lifetimes as has been demonstrated by elaborated photoionization experiments, e.g. by Brand and Baer[123]. The Me loss has been discussed in detail and an explanation by an isomerization process has been proposed by Wendelboe, Bowen and Williams[104], but it does not predict the observed isotope distribution.

The loss of methane can again be divided into an internal and a terminal loss that does not seem to depend upon time. In the latter reaction the additional hydrogen can be abstracted from position 2, 3 or 4. The origin of the additional D has been determined by different authors over different time scales. Some data are collected in Table 7. The data are remarkably similar, since the metastable data are from measurements where an error due to a superposition with the Me loss cannot be separated rigorously and no correction was applied for an internal loss. The results are equal within a reasonable error limit and do not seem to depend on the time scale of the experiment. If one assumes an equal isotope effect of 1.3 for the internal and terminal abstraction, a probability of 53% is obtained for an abstraction of a hydrogen in a primary position. It is not quite evident that the probability be divided for an internal loss between positions 2 and 3, but it seems that both positions have an equal participation of 24%. If a random distribution is assumed, the participation should be 42.8% and 28.6% for a terminal and an internal position. These values seem to be outside the error limit; a random scrambling of the seven hydrogen atoms is not observed. What is the structure of the resulting $C_3H_6^{+\cdot}$ ion, propene or cyclopropane? Appearance energy measurements indicate that sufficient energy is available for either isomer (0.69 and 0.28 eV, respectively)[116]. The activation energy for isomerization has been determined to be 1.4 eV above cyclopropane[124] and the two structures are indistinguishable by their metastable fragmentations. Also their CAD spectra are very similar, but Bowen, McLafferty and collaborators[125] were to distinguish the molecular ions of propene and cyclopropane over the fragmentation of the doubly charged ions. By comparing the $C_3H_6^{+\cdot}$ ion originating from butane, these authors concluded that this ion must have the propene structure. When measuring the released translational energy for the loss of CH_4 from butane and assuming the validity of the Haney–Franklin equation[126] that relates the internal energy of a fragmenting ion with its translational energy, Wendelboe, Bowen and Williams[104] arrive at the conclusion that the formation of cyclopropane is more probable. The two-step reaction $C_4H_{10}^{+\cdot} \rightarrow C_4H_9^+ \rightarrow C_3H_6^{+\cdot}$ can probably be excluded as a source of 'internal' loss, since the latter reaction has not been observed as a metastable fragmentation of $C_4H_9^+$ (see later).

$C_3H_5^+$ is one of the major ions in the mass spectrum of butane, but it is a tertiary ion. Loss of CH_4 from $C_4H_9^+$, H_2 from $C_3H_{7+}^{\cdot}$ and H from $C_3H_6^{+\cdot}$ are all common fragmentations. If Bu^+ would be the major precursor, the probability for the loss of the labels, in particular ^{13}C, should be random, i.e. 25%, independent of its initial position (see later). In the other cases the distribution should be the same as in the C_3 ion, i.e. ∼ 45% for a ^{13}C in position 1 of butane and ∼ 5% for position 2, nearly identical for $C_3H_7^+$ and $C_3H_6^{+\cdot}$. The latter is the case; it can thus be concluded that $C_4H_9^+$ is not an important ion for the formation of $C_3H_5^{+\cdot}$; it follows that Bu^+ cannot be an important primary fragment in the mass spectrometric fragmentation of butane.

$C_2H_5^+$ and $C_2H_4^{+\cdot}$ could be primary as well as secondary fragment ions, since ∼ 80% seem to come from a 1,2-loss to the neutral, without or with a hydrogen transfer, respectively. The reaction path appears evident, but it turns out that in the scission of

TABLE 7. The participation of position 4 to furnish H/D to a methane loss from position 1 in n-butane (in %)

Position of label in butane		Reference 116[a]	Reference 108[a]	Reference 109[b]	Reference 34[c]	Theoretical[d]
1,4-D_6		40	44	—	35; 39[e]	53
2,3-D_4		62	62	60	56; 57[f]	47
1,2-D_5	$CH_3D(CH_4)$[i]	65				77
	$CD_3H(CD_4)$[i]	80				77
	$CD_3H(CH_3D)$			44[g]	45[e], 44[h]	

[a] Metastable; first field-free region.
[b] Metastable; tandem instrument; $3\mu s$—4 ms.
[c] FT/ICR; 0–10 ms.
[d] See text.
[e] $1,^{13}CD_3$-butane ($^{13}CD_3H$–CH_3D).
[f] $1,^{13}C$-2,3-D_4-butane.
[g] 1,2,3-D_7-butane.
[h] $1,2$-$^{13}C_2$-4-D_3-butane (CD_3H–$^{13}CH_3D$).
[i] The neutral corresponding to one internal loss is given in parentheses.

the 2,3-bond about 25% of the deuterium will change place. The process is apparently not as simple as it looks. The rest can be somewhat arbitrarily distributed among the other possibilities for ^{13}C; a random process gives similar results. Nothing is known in detail about possible fragmentation routes.

Isobutane has been carefully studied by Derrick and coworkers[127,128] (among others; see e.g. References 116 and 129). These authors measured the fragmentation from the very short time scale offered by FI and FI kinetics up to the metastable range as a function of the energy of the electrons and using several compounds labelled with D. The photoionization was also studied[130]. The main fragmentation process, particularly in the metastable range, is the loss of methane. At very short time the rate constant for loss of Me is about a factor of two faster than for $-CH_4$; they cross at 20 ps and, in the long time range, loss of methane becomes more important up to the μs range, where only the loss of the latter can be observed. Up to 0.3 ns the loss of CD_4 from $(CD_3)_3CH$ is faster than the loss of CH_4 from $(CH_3)_3CH$; in the metastable range the difference is about a factor of two in favour of $-CH_4$. Different labelled products show that there is no scrambling between the methyl groups and that the additional hydrogen atom is never taken from the tertiary position, independent of the energy of the electrons. As usual, an isotope effect is observed between a CH_3 and a CD_3 group. CH_3^+ amounts to a few percent of the base peak, $C_3H_7^+$ (the C_3 peaks making up for nearly 90% of all fragments). Contrary to the other fragments, about 20–25% of this ion contained the hydrogen of the tertiary position. The formation of this ion (that appears only at high electron energies) is not known, so it is difficult to draw any conclusions.

The AE for the loss of Me and CH_4 is, within the limits of error, the same in n-butane$^{+\cdot}$, but in isobutane$^{+\cdot}$ it is 30–40 kJ mol^{-1} lower for CH_4 than for Me[115,116]. Propene$^{+\cdot}$ as well as cyclopropane$^{+\cdot}$ are energetically possible products when methane is lost from $C_4H_{10}^{+\cdot}$, although for isobutane the reaction is barely thermoneutral for the latter. Derrick and coworkers[128] offered an interesting explanation for the fact that at very short times the loss of Me is the faster reaction: the relative rate constant depends on the excess energy needed for a given time scale and the density of states within this energy window. For the loss of Me all states, including the internal rotation of the Me groups, are available, whereas for the loss of CH_4 a transition state via a 1,3-hydrogen transfer has to be formed that blocks some of the rotational degrees of freedom. At very short times this slows down the elimination of CH_4, since the excess energy is relatively large. As lower energies are considered, the difference in the energy window increases and favours the exothermic reaction.

All the results with labelled compounds seem to indicate that the proposal by Vestal[58] for an isomerization of the molecular ion of n-butane into isobutane is not correct, at least at low internal energies. This might be astonishing, since already the ionization of p orbitals can take place within an energy range of about 4 eV, as has been shown by the photoelectron spectra of n-butane[131] and isobutane[132]. The s orbitals extend up to about 25 eV, albeit with a smaller cross section[133]. But also the experimental breakdown graphs obtained by Sunner and Szabó[129] indicate that, in spite of a large excess energy, no (or at least very little) isomerization is observed, even though the structure of the molecular ions is badly defined (see above). This example also shows very clearly the advantage of the use of labelled compounds for fragmentation studies in alkanes. Even when the reaction paths are not known, the result can often furnish enough information to exclude certain hypotheses.

IV. C_7H_{16}—A MODEL FOR HIGHER HOMOLOGUES

Since it seems that the butane molecular ion does not necessarily isomerize before fragmentation, one could ask to what extent an isomerization is of importance for the higher-molecular-weight alkanes. The next member in the series, C_5H_{12}, has been the subject of a detailed study almost simultaneously by two research groups, both concentrating on the metastable decay of labelled compounds. It has the advantage that should an isomerization occur, it would most probably show up at low excess energies, i.e. in the metastable range. But this region has the disadvantage of allowing only unit mass resolution; thus some arithmetic is necessary to unravel the data, a procedure that is sometimes doubtful because of the isotope effect, particularly for H/D in the metastable range and the superposition of different isobaric fragments. Williams and coworkers[104] labelled n-pentane and 2-methylbutane with D in several positions. In 3-D_2-pentane they observed an important loss of CH_2D_2, whereas the ratio of C_2H_6/C_2H_5D was about 2/1. For 2,4-D_4-pentane CH_4 and $C_2H_3D_3$ were the main isotopomers eliminated. This and similar results with other labelled n-pentanes and 2-methylbutanes led the authors to propose a model where they assumed an intermediate $C_5H_{12}{}^{+\cdot}$ species involving an incipient carbonium ion complexed to a radical. This species is formed by stretching the appropriate bond in the ionized pentane. Subsequent isomerizations of the carbonium ion can give rise to rearranged structures where a 1,2-alkyl shift has occurred. Holmes and coworkers[134] studied, in addition to D-labelled compounds, a few pentanes labelled with ^{13}C. They realized that 3-^{13}C-pentane loses also an appreciable fraction of $^{13}CH_4$, but not 2-^{13}C-pentane. They were able to show that the proposal by Williams and coworkers could not explain all observed results. However, both groups assumed an isomerization of $C_5H_{12}{}^{+\cdot}$ prior to fragmentation. It will be shown later that the expulsion of a fragment from internal positions of the chain is a standard reaction of all linear alkanes, so it may well be that the importance of a concerted reaction for the fragmentation has been underestimated. The $C_5H_{12}{}^{+\cdot}$ ion occupies an intermediate position in the homologous series of the alkanes where an isomerization of the molecular ion seems probable, but not proven beyond doubt.

The situation in higher-molecular-weight compounds is complicated by the fact that already Thikhomirov and coworkers[135] demonstrated that in various deuterated octanes and 5-^{13}C-nonane neutral fragments containing the label were expelled. Gol'denfel'd and Korostyshevskii[136] measured the same samples under the conditions of field ionization, where it can be assumed that only fragmentations faster than a few tens of ps will be observed. Within this time window no loss of fragments from within the chain could be observed. Although the loss of Me from higher members of linear n-alkanes is a minor reaction, it attracted the attention of several groups using labelled compounds[31,101,137,138]. The general observation was that Me is lost from non-terminal positions. The most complete study was undertaken by Liardon and Gäumann[139], who were able to show that the ratio of internal (I) to terminal (T) loss for 70 eV ionizing electrons and reaction within the source followed approximately a linear relation (equation 44) for the n-alkanes C_nH_{2n+2} with $5 \leqslant n \leqslant 12$. It is shown in Table 8 that the

$$I/T = -0.41 + 0.17n \qquad (44)$$

percentage of terminal loss does not follow this relation in the metastable decay. Thus, it might be possible that it reflects only the difference in internal energy for different chain length. Decreasing the electron energy increases this ratio, although the accuracy of the measurements decreases also! They also state that Me is not eliminated from position 2. Already in n-hexane only one-third of the methane is eliminated from a terminal position. It has at least an equal probability of abstracting the additional hydrogen atom needed from another position rather than the vicinal position 2. The elimination of a Me from an internal position without an isomerization could be explained by equation 45. No measurements were conducted on simultaneous labelling with ^{13}C and D under the high resolution necessary to prove such a mechanism; thus the proposal is as hypothetical as most of these reaction schemes. The origin of the $CH_3{}^\cdot$ and CH_4

416

TABLE 8. Probability (in %) of a given position to contribute to the loss of CH_3·, CH_4, C_2H_5· and C_2H_6 in the metastable decay of n-alkanes[140]

Number of labelled compounds used						
neutral lost / position	Hexane 8	Heptane 11	Nonane 17	Decane 12	Dodecane 14	Tetradecane 26
CH_3·/CH_4[a]						
1	46/38	48/32	74/84	58/62	56/40	47/50
2		3/0	4/0	8/0		0/10
3	54/62	38/48	18/16	4/10	4/20	5/10
4		11/20c	4/0c	12/12		5/10
5				18/16	9/0	15/20
6					31/40	28/10
7						
C_2H_5·/C_2H_6[b]						
1,2	90/92	88/90	98/94	97/88	94/75	90/76
2,3				2/0		
3,4	10/8c	12/10		0/2		
4,5			2/6	0/4		
5,6				7/6c		
6,7					6/20	7/20
7,8					0/5c	3/7c

[a]These fragments contribute between 1.2 and 0.02% to the total ion current.
[b]These fragments contribute between 12 and 2% to the total current.
cThese reactions are only counted once (central position).

(45)

groups in the metastable decay of n-alkanes is given in Table 8. It can be seen that the expulsion comes mainly from the central positions within the chain. The reason for this is unknown. On the average about one-half of the C_1 fragments has a terminal origin. n-Nonane seems to be an exception. Since the loss of $CH_3^{\cdot}$ and CH_4 in this ion amount to only 0.4% of the total ion current, the precision should not be overestimated, but, on the other hand, the results were obtained with a least-squares calculation with many labelled compounds, and metastable measurements using the MIKE technique are usually rather well reproducible.

Do higher alkanes isomerize before fragmentation or do they retain their original structure? In a classical paper on the photoionization of alkanes, Steiner, Giese and Inghram[15] showed that alkanes, contrary to alkenes, do not allow the determination of the adiabatic ionization energy since the structural changes are too large in the process of ionization. Beside the impossibility to get accurate adiabatic ionization energies, this is an important statement because it means that the stable structure of the ion is different from that of the neutral. Monahan and Stanton[141] postulated, on the basis of the change of the mass spectra with the energy of the ionizing electrons (up to the keV range), that in the process of ionization electronically excited states are also formed. Toriyama and coworkers[142] studied the electron spin resonance spectra of a series of alkane cations at low temperatures in a solid matrix. They claim that in ions of a linear alkane the unpaired electron is delocalized over the in-plane C—H and C—C σ bonds forming delocalized σ radicals, whereas in highly branched alkane ions it is more confined to one of the C—C bonds forming localized σ radicals. In the solid matrix no isomerization could be observed; this could be due to the rigid matrix. Meot-Ner, Sieck and Ausloos[95] determined the thermodynamic values of the n-alkane ions by measuring ionic equilibria. This should yield adiabatic values, but they have to base their measurements on a suitable standard, which was cyclohexane in this case. Their measurements range from n-butane to n-undecane and include some deuterated, cyclic and branched isomers. They made the astonishing observation that the entropy change in ionization from n-heptane onwards to higher alkanes becomes noticeably negative by about $10\,J\,K^{-1}\,mol^{-1}$ per CH_2 unit, the change from hexane to heptane being even more than double this value. A corresponding change in the ionization enthalpies was also observed. In n-heptane such a change corresponds, e.g., to a replacement of the internal hindered rotors with a rotational barrier of about $12\,kJ\,mol^{-1}$ by six frequencies of $71\,cm^{-1}$! Based on the observation of a dimer formation of $C_2D_6^+$ with C_2D_6 by Sieck and coworkers[143], the authors make the interesting proposal that n-heptane ion 'dimerizes' its Et end groups intramolecularly, forming a pseudocyclic compound. It should be mentioned that the dimerization of ethane is very temperature dependent, its rate constant decreasing to nearly zero at the usual ion source temperatures.

It is thus not astonishing that there is some controversy about the subject. It has been reviewed by Levsen[144]. In the same publication he presents results on the CAD spectra of ten isomeric octanes and their fragments. The CAD spectra of the different molecular ions are distinguishable. Since the collision energy in this process is large compared to the internal energy and identical for all ions of the same mass, it can be concluded that with a very high probability the ions that did not decompose after $10\,\mu s$ have retained a structure that is characteristic of its neutral predecessor. On the other hand, all alkyl ions of the same mass had identical CAD spectra. Since the time of the collision process at these high translational energies is short compared to the time for rearrangements, this means that all alkyl fragments of the same mass isomerize to a common structure (or a mixture thereof) before colliding. Since also at short times no isomerization can be detected for the molecular ion, it can be assumed that higher alkane ions under normal mass spectrometric conditions do not isomerize. Levsen and coworkers[145] strengthened this point by measuring the spectra of labelled n-dodecane. This finding is confirmed by studies of the metastable decay; e.g. Krenmayer and coworkers[146] showed for some paraffin isomers that the metastable spectrum of the molecular ion is easier to interpret than that of the source and recommend it for analytical applications, although the observed reaction in the metastable decay is often of minor importance in the total fragmentation scheme (see later). The fact that fragments are apparently expelled from within the chain cannot be used as an argument for isomerization, since they seem to originate mostly from the very central position of the chain; any isomerization would destroy such a symmetry. The proposal by Meot-Ner and colleagues (see above) would be perfectly suited to explain such a reaction. Wolkoff[147] tried to explain the expulsion of fragments from non-terminal positions by a 1,2-alkyl shift. Based on measurements on metastable decays of labelled n-hexanes and literature data, he proposed the following priorities for a 1,2-alkyl shift: 1,2-Et > 1,2-Pr ≫ 1,2-Me. His proposals do not correspond to the known data.

Considering all arguments, it seems to be sure that molecular ions of n-alkanes with more than six carbon atoms do not isomerize. The same may be true for n-pentane and n-hexane. The terminal hydrogen atoms do not scramble, since they keep their identity even in the fragmentation of the secondary alkyl fragments (see Section VI). It is not evident that one can exclude mixing of the hydrogen atoms within the chain. The study of the metastable loss of, e.g., the C_2 fragments in n-nonanes, labelled with D in different positions, is indicative of a non-mixing[148], but metastable studies with labelled molecular ions are not easy because of the superposition of the labelled fragments and the isotope effects (see Table 8). Heptane is taken as a case study for understanding the fragmentation of the higher alkanes. In Table 9 the relative contributions of different fragments in the source spectrum of heptane are given. It is a typical mass spectrum of an n-alkane, where

TABLE 9. The most important peaks[a] in the source spectrum of n-heptane and the probabilities (in %) of a given position to be lost to the neutral[149]

			Position of ^{13}C					
Ion	Neutral lost[b]	% TIC	1	2	3	4	rnd[c]	$\sum^{d}$
$C_5H_{11}^{+}$	$C_2H_5^{\cdot}$	12	46	44	7	7	28.6	102%
$C_5H_{10}^{+\cdot}$	C_2H_6	6	43	42	10	10	28.6	111%
$C_4H_9^{+}$	$C_3H_7^{\cdot}$	10	49	47	49	10	42.9	105%
$C_4H_8^{+\cdot}$	C_3H_8	7	49	46	48	14	42.9	115%
$C_3H_7^{+}$	$C_4H_9^{\cdot}$	21	65	62	46	54	57.1	99%
$C_3H_6^{+\cdot}$	C_4H_{10}	5	66	56	43	69	57.1	120%
$C_3H_5^{+}$	$C_4H_{11}^{\cdot}$	9	77	71	77	59	71.4	99%
$C_2H_5^{+}$	$C_5H_{11}^{\cdot}$	6	78	73	58	80	71.4	108%

[a]Only fragments that contribute $\geqslant 5\%$ to the total ionization are given.
[b]The 'neutral lost' may (or has to) correspond to several species, lost either in one step or in consecutive reactions.
[c]Corresponding to a random distributions of all positions.
[d]Sum of the probabilities of all positions divided by the number of carbon atoms lost to the neutral.

alkyls and alkenes form the majority of the ions in a ratio of about 2:1. In the same table the probability of a given position to be lost to the neutral is also given as they were determined by Lavanchy, Houriet and Gäumann[149]. Two things can be seen: every position has a non-zero probability to be lost to the neutral fragment and this probability corresponds in no case to a random distribution. The last column gives the sum of all probabilities divided by the number of carbon atoms for the neutral. Neglecting an isotope effect, it should be 100%; the deviations give some indication of the reliability of the results. The general rule, that the fragment with the largest number of hydrogen atoms within a group (the alkyl ion) yields the most precise results, can be well seen. The authors used these (and other) data to construct a fragmentation scheme for the alkyl ions. They made the assumption that an alkyl ion can be formed either by direct fragmentation of the molecular ion (equation 46),

$$C_nH_{2n+2}^{+\cdot} \rightarrow C_mH_{2m+1}^+ + C_{n-m}H_{2(n-m)+1}^{\cdot} \quad (46)$$

or by an olefin loss from an alkyl ion with $n \geqslant l > m+1$ (equation 47)

$$C_lH_{2l+1}^+ \rightarrow C_mH_{2m+1}^+ + C_{l-m}H_{2(l-m)} \quad (47)$$

They made two hypotheses for the fragmentations: (i) reactions of the type shown in equation 46 involve only the loss of terminal positions (they call these reactions type A reactions); (ii) the formation an alkyl ion by reaction (47) (type B) involve *all* positions in a random manner. Using the data from many ^{13}C-labelled n-heptanes, they were able to calculate the probabilities for the two types of reactions that are given in Scheme 3. It can be seen that according to this scheme the fragmentations yielding alkyl ions containing more than $n/2$ atoms are mainly ($\sim 85\%$) formed by direct cleavage, where the contrary is true for smaller alkyl ions. Since the higher-molecular-weight alkyl ions will fragment further, such a fragmentation scheme allows one also to estimate the relative importance of the primary fragmentations. The authors were also able to show that the figures of Scheme 3 do not depend strongly on the energy of the ionizing electrons, as would be expected for such an overall scheme.

What is the value of such a scheme? The data were obtained from unit resolution spectra by the usual arithmetic procedures. Such calculations are difficult to check for their internal consistency. This, together with an insufficient knowledge of the fragmentations of olefins, precluded a similar calculation for the olefins. The second hypothesis (random fragmentation of alkyl ions) is shown to be true only for the lower alkyl ions (which works in this case). The postulated loss of C_3H_6 from hexyl ions is not observed in the metastable fragmentation of hexane (but it is not forbidden neither; see Reference 150). The importance of the $[M-1]^+$ ion is astonishing. It was assumed to account for the observation of an internal loss of $C_2H_5^{\cdot}$. However, this reaction, which also had been observed for the metastable decay (see Table 8), must be explained by other mechanisms. The weakness of such calculations is that they can furnish figures only for reactions that were put into the calculations; thus they depend on the knowledge of possible reaction schemes. The inclusion of many reactions asks for more data than are possibly available and has the danger that it will yield for every reaction included some small probability. In summary, it can be said that such schemes may give a reasonable semi-quantitative picture of the fragmentation that can be used as a basis for discussion, the figures for the main reactions being relatively reliable; the contributions for minor reactions must be considered with caution and are subject to later modifications when improved experimental techniques are available.

The data of Table 10 indicate the importance of the loss of neutral fragments from within the chain for all n-alkanes. Although in higher alkanes the metastable loss of larger fragments can be observed, this type of internal loss is only seen for neutral fragments containing less than five carbon atoms. However, it must be admitted that

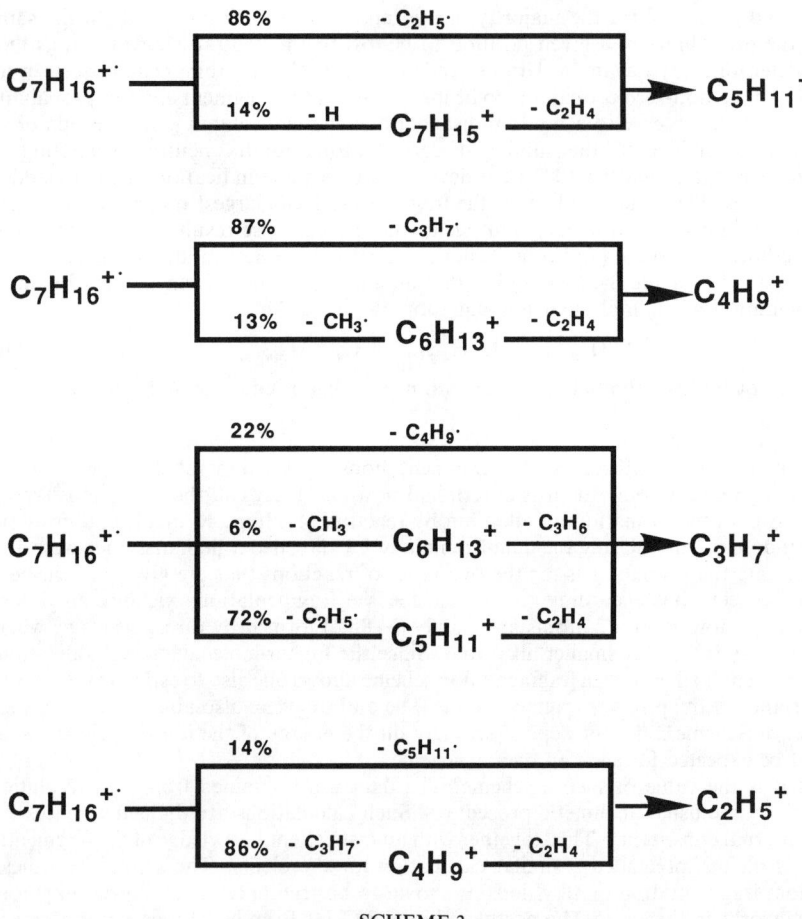

SCHEME 3

reliable metastable measurements on higher-molecular-weight alkanes are difficult to perform. The fact that the internal loss is a generally observed phenomenon led Lavanchy and coworkers[151] to extend their measurements and calculations to higher alkanes, using the same assumptions. In Section VI it will be shown that these assumptions are only approximately valid. Their principal results are shown in Table 11. The same picture is seen: the fragments with less than one-half of the carbon atoms of the molecular ion include in 81% ($s = \pm 8\%$) of the cases the terminal positions, the others in 15% ($s = \pm 4\%$). This seems to be a general picture for the fragmentation of n-alkanes within the ion source.

2-Methylhexane, an isomer of n-heptane, may serve as an illustration of the difficulty in interpreting labelled data. The most important products of primary fragmentations are $C_6H_{13}^+$ ($-CH_3\cdot$), $C_4H_9^+$ ($-C_3H_7\cdot$) and perhaps $C_3H_7^+$ ($-C_4H_9\cdot$), the latter being a possible counterpart of $C_4H_9^+$. The first two ions are also the important ions in the metastable decay beside the loss of CH_4, C_2H_6 and C_3H_6, which gain in importance in the metastable range. In the ion source fragmentation the main ion $C_4H_9^+$ can be

TABLE 10. Probability (in %) of losing a neutral fragment from the terminal positions in the metastable decay of alkanes[140]

Fragment lost	Hexane	Heptane	Nonane	Decane	Dodecane	Tetradecane
CH_3	46^a	48	72	58	56	44
CH_4	38	72	84	62	36	50
C_2H_5	90	88	98	96	94	90
C_2H_6	92	90	94	88	76	74
C_3H_7	100	78	100	100	100	94
C_3H_8	100	92	90	88	80	70
C_4H_9			80	100	100	96
$C_4H_{10}{}^b$			36	90	86	78

^{a}The complement to 100 corresponds to internal loss. It rarely includes position 2.
^{b}The higher-molecular-weight fragments in dodecane and tetradecane (C_5–C_8) have 100% terminal loss.

TABLE 11. Probability (in %) of forming the fragment alkyl ions by losing the terminal carbon atoms within the ion source (type A process)[151]

Fragment lost	C_2H_5	C_3H_7	C_4H_9	C_5H_{11}	C_6H_{13}
$C_6H_{14}^{+\cdot}$	90	72	11		
$C_7H_{16}^{+\cdot}$	86	87	22	14	
$C_9H_{20}^{+\cdot}$	88	92	82	14	14
$C_{14}H_{30}^{+\cdot}$	71	79	86	77	69 [a]

[a]Line separating the neutral fragments ($C_mH_{2m+1}^{\cdot}$) with $m \leqslant n$ from $m > n$ for $C_nH_{2n+2}^{+\cdot}$ ions.

TABLE 12. Time dependence of the positional loss of C_3 fragments from 2-methylhexane[154]

Condition Time scale Fragment lost	Position lost	FI < 10 ps	FIK 0.5 ns	source 0–2 μs	meta 20 μs	FT/ICR 0–1 s
$C_3H_7^{\cdot}$	1,2,1	100	33	36		20
	4,5,6		67	35	100	29
	1-CH$_3$, C$_2$H$_4$ rnd			29		47
C_3H_8	1,2,1	—	100	74		56
	4,5,6			18	100	32
	1-CH$_4$, C$_2$H$_4$ rnd			9		9

formed in three ways (equations 48–50).

$$C_7H_{16}^{+\cdot} \rightarrow s\text{-}C_4H_9^+ + i\text{-Pr(pos. 1,2,1')} \qquad (\Delta H_r = 66 \, \text{kJ mol}^{-1}) \quad (48)$$
$$C_7H_{16}^{+\cdot} \rightarrow t\text{-}C_4H_9^+ + n\text{-Pr(pos. 4,5,6)} \qquad (\Delta H_r = 20 \, \text{kJ mol}^{-1}) \quad (49)$$
$$C_7H_{16}^{+\cdot} \rightarrow s\text{-}C_6H_{13}^+ + CH_3^{\cdot}\text{(pos. 1)} \rightarrow t\text{-}C_4H_9^+ + C_2H_4 \quad (\Delta H_r = 119 \, \text{kJ mol}^{-1}) \quad (50)$$

Without hydrogen shift the corresponding values for reactions 48 and 49 are identical (140 kJ mol^{-1}). Methylalkanes have been discussed by Holmes and coworkers[134,152] concerning the loss of Me. The authors conclude that in the metastable Me loss from 2-methylpentane, a mixture of s- and t-pentyl ions are formed, in contrast to 2-methylhexane, where only s-hexyl ions appear to be formed, because the kinetic energy release is small and independent of the observation time.

The question which reaction of the set (48)–(50) is the most important one, is therefore legitimate. Howe[153] argued from the heat of formation for these reactions that the t-Bu must be formed. By measuring the metastable decay of $(CD_3)_2C_5H_{10}$ and $(CH_3)_2CDC_4H_9$ he found only loss of $C_3H_7^{\cdot}$ (pos. 4–6) and concluded from the isotope effect that a 1,2-shift must take place. For this purpose Gäumann and coworkers[154] prepared a series of labelled methylhexanes and studied the positional loss of different fragments over the available time scale. Some preliminary results have been published[155]. In Table 12 the data for the loss of $C_3H_7^{\cdot}$ and C_3H_8 are given as a function of fragmentation time. In the very short time range of the FI spectrum, $C_3H_7^{\cdot}$ is uniquely lost from positions 1,2,1' as the only product in this group. In the sub-nanosecond range, Pr loss from positions 4–6 is increasing in importance, together with a loss of C_3H_8 from positions 1,2,1'; it becomes the main product in the FIK range. In the time range of the metastable decay

C_3H_7 as well as C_3H_8 are only lost from positions 4–6. These findings are integrated in the source spectrum, where all fragmentations occurring up to about 2 μs are integrated: both types of fragmentation are observed, together with a contribution from reaction 50. The FT/ICR spectrum corresponds to the integrated spectrum over all fragmentations possible. Because of the high resolution, no special arithmetics are neeeded to resolve the isobaric multiplets. The integration is extended over the metastable range, as can also be seen from the contributions yielded by the three reactions. This example demonstrates with extreme clarity the basic problem in mass spectrometry: what is a mass spectrometric fragmentation? The question is somewhat academic! It is an integration over the fragmentation of molecular ions with an unknown distribution in internal energy. The results vary according to the experiment, as the case may be.

The mass spectrum of an alkane must not necessarily be complicated. The source spectrum of 2,2-dimethylpentane is conveniently interpreted by a loss of Me from position 1 (and equivalent) and C_3H_7 and C_3H_8 from positions 3–5, and the tertiary reactions from the fragment ions formed[148]. In the metastable only two fragments can be observed, loss of CH_4 from position 1 and C_3H_8 from positions 3–5. No hydrogen exchange between positions 1 and 3,4 or 5 could be observed. It is evident that the presence of a tertiary carbon atom simplifies the spectrum. A similar observation can be made when labelling 2,2,7,7-tetramethyloctane in several positions: The molecular ion loses only CH_3, CH_4, C_4H_{10} and C_8H_{18} in clear-cut C—C bond splits without any scrambling of positions[148]. A particularly nice example of how the fragmentation of branched alkane 'concentrates' on the branching points has been shown by Spiteller and coworkers[156] with the spectrum of pristane (2,6,10,14-tetramethylpentadecane) where, under suitable conditions, practically only ions due to the scission at the tertiary carbon atoms are seen as fragments.

V. METHYLCYCLOHEXANE—THE DANGER OF INSUFFICIENT LABELLING

The fragmentation of an alkane can be understood by assuming that a C—C bond is broken, sometimes accompanied by a hydrogen shift in order to minimize the energy requirements; this is, to a first approximation, a rather satisfactory model. In the case of a branched alkane the bond to a tertiary or quaternary carbon atom is more easily broken, a fact that allows the determination of the branching point(s) with a good level of confidence. Except for small alkanes, the cleavage of C—H bonds seems to be a rare event.

Non-substituted cyclic hydrocarbons cannot lose a neutral fragment just by breaking a bond. For cycles with less than eight or nine carbon atoms, three types of reaction can be envisaged:

(i) The scission of a C—C bond is preceded by a change of the ring size, thus forming a branched cyclic hydrocarbon in an intermediate step before a C—C bond split at a newly formed tertiary carbon atom.

(ii) The ring is first opened, yielding probably an olefinic compound by a hydrogen transfer. In a second step, the olefin undergoes further fragmentation.

(iii) A neutral fragment, preferably an olefin, is eliminated in a concerted reaction.

The latter reaction also seems to take place in paraffins; however, its contribution to the total fragmentation remains modest. It is typical for this reaction that consecutive carbon atoms are lost. Except perhaps in the case of a favourable steric conformation, there is no direct evidence that this type of reaction is preferred in cyclic compounds.

The ring opening, giving rise to an olefinic compound, is an attractive reaction path that is difficult to prove. It is shown in the section on olefins that the fragmentation of an olefin is a very complex reaction that is not yet understood. The saying that mass spectrometrists have a tendency to explain the fragmentation of a linear compound by ring formation (see e.g. Reference 157) and the fragmentation of a ring by its opening as a first reaction explains rather well the uncertainty of the situation. The elimination of an olefinic neutral compound is an important reaction in the mass spectra of olefins, in particular in the slower, metastable decay. According to substitution and the localization of the double bond, non-neighbouring positions can be lost together to the neutral fragment in a characteristic, however nonunderstood, reaction. If the formation of an olefinic intermediate compound is really important in the fragmentation of cyclic ions, this opens the interesting possibility that the structure of the intermediate could be traced back from its consecutive fragmentations. But this assumes a much better knowledge of the fragmentation of olefins than is available today. Levsen[158] tried to get some insight into the stability of isomeric molecular ions of alkenes and cycloalkanes. He concluded that they do not isomerize completely to a common structure with the exception of n-propylcyclopentane, the structure of which is identical to that of the normal octene ions. However, the author concentrated on CAD without labelling, a procedure with pitfalls in hydrocarbon mass spectrometry!

The change of the ring size is a reaction that is expected to take place easily in cyclic ions, but preferably in even-electron ions. Again it is very difficult to prove as a reaction path.

It is reasonable to assume that the bond to a tertiary carbon atom is preferably broken, since the resulting charge localization is rather favourable. Nevertheless, it will be shown later that even this simple reaction is not always as evident as it may look.

The safest way to trace a reaction path is to use D and ^{13}C labelling, if possible even multiple labelling. Beside the problem of interpreting the experimental data, the difficulty of the chemical syntheses of multiply labelled compounds explains the scarcity of reliable results. Until the advent of high-resolution FT/ICR methods, all data with labelled compounds were obtained from spectra taken with unit resolution, since it is extremely difficult to get reproducible results with high-resolution double-focussing instruments (this has been shown, e.g., with toluene[159,160]). Assuming no important isotope effect for the (fast) fragmentation within the ion source (since a rather large amount of excess energy is needed for a fast fragmentation) and reproducible spectra, this method gives good results for a single label within an ion, but is at its limit for multiple labelling. The study of metastable decays allows one to select a specific ion that yields usually fewer fragment ions; the reproducibility that can be obtained with a double focussing instrument is excellent, but—at least for D atoms—the isotope effect can be rather large because of the lower excess energy that is needed for fragmentation in this time range. The possibility to study metastable decays does not exist in FT/ICR, but it can be replaced, within limits, by the study of the photofragmentation induced by IR lasers[161]. It is fair to assume a reproducibility of $\sim 2\%$ for intensities measured from source spectra (this may introduce large uncertainties for less important fragments), whereas a reproducibility of better than 1% is obtained in metastable studies.

Substituted cyclopropanes are often postulated as intermediates in mass spectrometric fragmentation. Ausloos and collaborators[162] demonstrated by studying ion/molecule reactions that ring opening may occur in the cyclopropane cation. The fraction of the ions that undergo this reaction depends on its internal energy content. They estimated an energy barrier of about 0.7 eV for the conversion of cyclopropane to propene. However, these findings were not confirmed by several authors, in particular in studies of the CAD spectra. This point has been discussed by McLafferty and coworkers[163]. This demonstrates a phenomenon that is not only valid for the fragmentation of cyclic cations: the reaction paths and their respective contributions may well depend on how the ion is being formed, its internal energy content and the time scale of the experiment. Chang and Franklin[164] postulated, for example, that the cyclopropane does not open the ring, since the proton affinity measured for this ion is different from that of propene. Schwarz[165] comes to the conclusion on the basis of extended MNDO calculations that the halogen loss from bromo- or chlorocyclopropane cation is preceded by a hidden internal 1,2-hydrogen transfer within the ring before the ring opening occurs in the cyclopropyl ion. Such proposals are very difficult to prove experimentally, but they demonstrate the difficulty of proving a given reaction scheme.

Radom and coworkers[46] showed by *ab initio* MO calculations that the potential energy surfaces connecting the various isomeric structures are generally very flat. Facile wide-amplitude distortions and scrambling processes are therefore to be expected in these ions. The large structural changes which accompany ionization are manifested in large differences between vertical and adiabatic ionization energies. The same authors predicted also that the high symmetry of the smaller rings is easily lost due to Jahn–Teller distortions in the ions[166]. Isotopic substitution, particularly of H by D, may therefore have a profound influence[167]. Cyclopentane and substituted cyclopentanes were studied by several authors. The fragmentation of methylcyclopentane with a labelled methyl group yielded the astonishing result that only half of the eliminated methyl radicals originated from the labelled methyl group, as was shown by Stevenson[168] and Meyerson and colleagues[169]. This was probably one of the first indications that mass-spectrometric fragmentation of even a simple compound may be a mixture of different complex processes, a fact that is too often forgotten! Falick and Burlingame[170] measured the field-ionization kinetics of the Me loss from methylcyclopentane. They were able to differentiate two reactions for the Me loss: a fast process (< 0.1 ns), where the Me group is lost, and a slower reaction (> 10 ns), where the Me groups originate from the ring; the authors assumed for this second path that ring opening precedes the Me loss. Hirota and Niwa[171] calculated scission probabilities of skeletal C—C bonds for the monocyclic unsubstituted alkanes containing five to eight carbon atoms. The cations of two isomeric hydrocarbons may isomerize and form a mixture of two (or more) structures, separated by barriers of different heights, but lower than the energy needed for the appearance of the first fragment. Komarov and collaborators[172] demonstrated that in these cases the spectra of two isomeric neutrals may be the same if they are measured at different temperatures in order to compensate for the difference in internal energy of the ions when formed by electron impact, e.g. 500 °C for toluene and 220 °C for cycloheptatriene. For cyclopentane and 2-pentene, a mutual approach of the intensities in the two spectra is not observed with an increase in temperature up to even 1000 °C. The authors conclude that the structures of the two molecular ions must therefore be different. Brand and Baer[123] used the photoelectron photoion coincidence technique (PEPICO) to study the fragmentation of the different isomers of C_5H_{10}. They were able to show that all isomers, with the exception of 2-methyl-2-butene, exhibited a two-component decay as a function of time, indicating that dissociation occurs from at least two distinct forms of the molecular ion. The assumption of more than two forms improved somewhat the fitting of the results at the expense of additional parameters. Only 2-methyl-2-butene, the ion with the lowest heat of formation, shows a single exponential decay. The authors propose that every molecular ion isomerizes partially to the lowest-energy ion 2-methyl-2-butene before fragmentation. Ausloos and collaborators[173] show for cyclopentane and different alkylcyclopentane cations that the degree of ring opening observed depends on the internal energy of the molecular ion. It can thus be assumed that cyclic cations will undergo ring opening before fragmentation, but only to an extent that is determined by the internal energy of the ion. This is not an encouraging fact for the study of the fragmentation of these ions.

Even-electron cyclic alkyl cations have received much attention, in particular carbonium ions and protonated cyclopropanes[14,157,174,175]. The general experience is that the cyclic ion is often the most stable form and thus is preferentially formed. However, an important skeletal isomerization is always observed with a corresponding randomization of the labels in given initial positions, indicating the ease with which the structures interconvert. As an example the study by Schwarz and colleagues[176] with $C_5H_9^+$ ions may be mentioned. The authors synthesized several C_5H_9Br compounds, doubly labelled with $^{13}C_2$ in such a way that the two labels were separated by at least one unlabelled carbon atom. They studied the metastable loss of C_2H_4 from $C_5H_9^+$ for bromocyclopentane and 5-bromo-1-pentane as precursors and observed for all isomers that the probability of losing a label to the neutral fragment corresponded exactly to a random loss of two ^{13}C among five carbon atoms.

Methylcyclohexane (MCH) is taken as an example to demonstrate the possibilities and fallacies of the use of isotopic labels. The Me group in MCH has the advantage of fixing the positions within the ring and offering a favourable location on position 1 for localizing the charge. The relative importance of the different fragments in the mass spectrum of MCH for ionization with 70 eV electrons is given in Table 13 for the source spectrum in a FT/ICR instrument and the metastable MIKE spectrum in a ZAB-2F double-focussing mass spectrometer, operated with an ion energy of 8 keV. The loss of a CH_3 group yields the most important ion, in particular in the metastable spectrum. This is somewhat astonishing, since the loss of a Me group corresponds to a simple C—C bond breaking and more complicated rearrangement processes usually prevail in the metastable time range of several tens of μs. The loss of a propyl group as neutral fragment is of equal importance. The difference with the spectra of C_7H_{16} alkanes is apparent: the contributions to the total ionization by the molecular ion and by fragments with four to six carbon atoms is much more important; low-molecular-weight C_1–C_3 ions are nearly missing. This may reflect the fact that for fragmentations an additional step is needed, thus decreasing the time available for consecutive fragmentations. As in the case of alkanes of the same size, the loss of a hydrogen atom from the molecular ion is a minor reaction, unless the further fragmentation is so fast that the intermediate ion cannot be seen. This is very improbable since the fragmentation of the $(M - 1)^+$ cycloalkyl ions can be observed in the metastable spectrum.

Meyerson and collaborators[169] labelled different positions with one D atom and measured the probability of losing the label to the neutral fragment. Amir-Ebrahimi and Gault[177] prepared ^{13}C monolabelled MCH where every position was labelled. The results of the two studies are collected in Table 14. All these data present source spectra results obtained with unit resolution. In parentheses are the results obtained with a

TABLE 13. The main fragments in the mass spectrum of methylcyclohexane

m/z	Ion	Neutral lost	Source[a]	Metastable[a]	Reaction in Scheme 4
98	$C_7H_{14}^{+\cdot}$	—	15		
97	$C_7H_{13}^{+}$	$H^{\cdot}$	0.6		a
83	$C_6H_{11}^{+}$	$CH_3^{\cdot}$	24	61	
82	$C_6H_{10}^{+\cdot}$	CH_4	5	13	
70	$C_5H_{10}^{+\cdot}$	C_2H_4	6	5.6	b,c
69	$C_5H_9^{+}$	$C_2H_5^{\cdot}$	5	4.0	d
68	$C_5H_8^{+\cdot}$	C_2H_6	2	3.6	
56	$C_4H_8^{+\cdot}$	C_3H_6	5	4.0	e,f,g
55	$C_4H_7^{+}$	$C_3H_7^{\cdot}$	22	8.3	
54	$C_4H_6^{+\cdot}$	C_3H_8	1		
42	$C_3H_6^{+\cdot}$	$C_4H_8^{\cdot}$	5		h
41	$C_3H_5^{+}$	$C_4H_9^{\cdot}$	8		
40	$C_3H_4^{+\cdot}$	C_4H_{10}	—		

[a] In % of the total ion current.

TABLE 14. Percentage loss of ^{13}C[177] and D[169] to the neutral fragment in methylcyclohexane

	Position of ^{13}C						Position of D^b					
	Methyl	1	2	3	4	Σ^a	Methyl	1	2	3	4	Σ
$C_6H_{11}^+$	100(97)^c					100	100(98)	(0)				100
$C_6H_{10}^{+\cdot}$	100(99)					100	(100)	(6)				
$C_5H_{10}^{+\cdot}$	6(12)	7(13)	34(33)	43	28	98	6(13)	(25)	21(33)	40	27	85(101)
$C_5H_9^+$	62(65)	64(63)	12(22)	16	17	100	72(65)	(49)	37(28)	23	18	108(98)
$C_5H_8^{+\cdot}$	69(91)	84(85)										
$C_4H_8^{+\cdot}$	38(39)	40(39)	46(46)	36	60	101	41(36)	(40)	43(44)	51	43	95(98)
$C_4H_7^+$	68(71)	30(30)	38(42)	41	41	99	68(68)	(40)	1(43)	43		
$C_3H_6^{+\cdot}$	26(29)	29(29)	45(58)	77	76	94	26(24)	(47)	51(61)	71	67	93(98)
$C_3H_5^+$	65(66)	61(54)	42(57)	54	73	98	73(71)	(61)	57(60)	57	57	82(95)

$^a\Sigma$ = sum of the probabilities for all positions divided by the number of C/D atoms lost to the neutral.
^{b}Data for one D atom.
^{c}Data in parenthesis from Reference 178 (FT/ICR).

FT/ICR instrument operated with sufficient resolution (> 250,000) to resolve all possible multiplets containing ^{13}C and D[178]. The possibilities and the performance of the method have been demonstrated[160]. It can be seen that the results agree fairly well. Some differences may be expected, since in a sector instrument the ion intensity is integrated over $\sim 2 \mu s$ and in a FT/ICR over a few ms, including the metastable range. The formation of the $C_6H_{11}^+$ and $C_6H_{10}^{+\cdot}$ ions corresponds to the loss from the methyl group, although a very small participation of other positions seems to be real. This is rather different from methylcyclopentane (see above). For all other fragment ions all positions contribute to the neutral in different amounts. This is particularly true for D atoms: if a scrambling should take place, it is by no means complete! This somewhat contradicts theoretical predictions[46]. A check of the quality of the results is obtained by adding the probability of losing a position to the neutral over all positions. The sum, divided by the number of carbon (or deuterium) atoms lost to the neutral fragment, should give 100%, aside from a possible isotope effect. These figures are also given in Table 14. Even for D atoms, where an isotope effect could be more important, satisfactory agreement is observed, demonstrating the quality of the measurements.

Amir-Ebrahimi and Gault used the data to construct a fragmentation scheme for MCH. They defined a number of possible reactions and used the label distribution to calculate the contribution of a given reaction to yield a given ion. When the number of data exceeded the number of equations, they did not use a least-squares calculation, but used the additional data to check the validity of their assumptions. If the number of equations exceeded the number of data, they also used the data for D-labelling in order to obtain enough equations. They had to make a choice for the fragmentation reactions assumed. Firstly they had to limit the number of different fragmentations to a reasonable figure. The choice of such reactions is by no means evident, since very little is known about fragmentation of rings and the predictive power of theoretical calculations is in such systems rather nil. They further assumed that only neighbouring C atoms within the ring are lost to the neutral fragment. With these hypotheses they were able to obtain a fragmentation diagram whose main features are shown in Scheme 4. The following

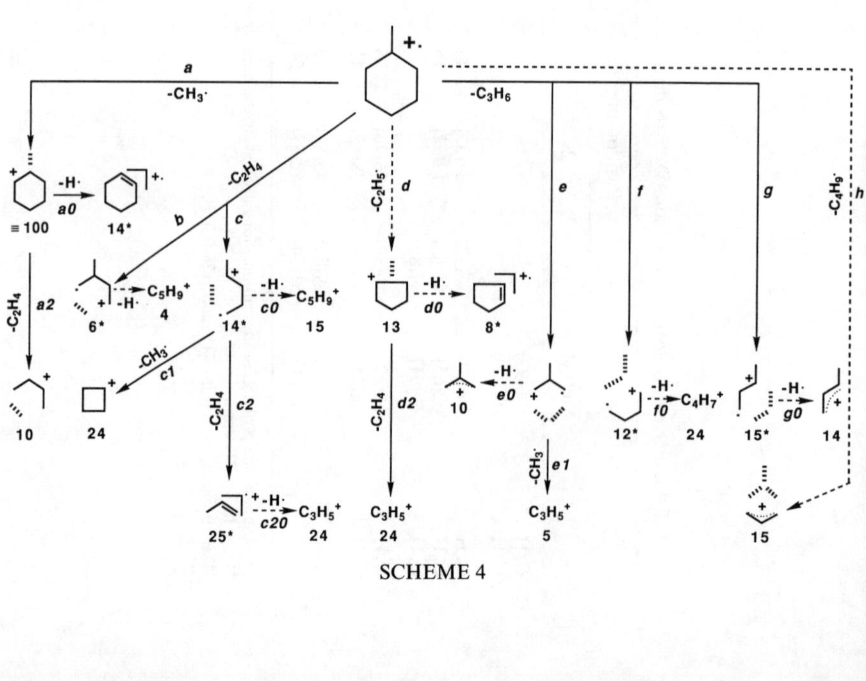

SCHEME 4

TABLE 15. The importance of the different primary fragmentation in the mass spectrum of methylcyclohexane (see Scheme 4)[177]

Reaction	Neutral lost	Positions lost	% Total ion current
a	$CH_3\cdot$	Me	30
b	C_2H_4	3,4	2
c	C_2H_4	2,3	24
d	$C_2H_5\cdot$	Me,1	11
e	C_3H_6	3,4,5	4
f	C_3H_6	Me,1,2	9
g	C_3H_6	2,3,4	7
h	$C_4H_9\cdot$	Me,1,2,6	4
Sum			91

conventions are used in this scheme:

(i) Italic letters mean a primary fragmentation path. They are summarized in Tables 13 and 15.

(ii) xi (xij) means a secondary (tertiary) fragmentation of the ion formed in the primary fragmentation x; i and j are the number of carbon atoms lost in the secondary and tertiary fragmentation, respectively.

(iii) The number below an ion is its intensity compared to that of $C_6H_{11}^+$ which is taken equal to 100, (As an indication, the sum of all ions is roughly 400.) An asterisk corresponds to an odd-electron radical ion.

(iv) A solid line corresponds to a reaction that was also observed in the metastable spectrum of the unlabelled ion. For obvious reasons one cannot distinguish between isobaric ions of different origin. Hashed lines correspond to reactions that are *not* observed in the metastable range, either because they were too fast or because the assumption of a one-step fragmentation is not valid in this case.

The percentage that can be attributed to the different primary fragmentations is given in Table 15. More than 90% of the observed ions is explained by this scheme. The loss of $CH_3\cdot$ or C_2H_4 makes up for more than 50% of the ions. The main contribution to the latter is from positions 2, 3, possibly in a concerted reaction. $C_2H_5\cdot$ originates from positions Me, 1. The elimination of $C_3H_7\cdot$ in a one-step reaction from well-defined positions is not observed, in spite of the fact that the corresponding metastable decay is seen and $C_4H_7^+$ is a main ion in the spectrum. C_3H_6 comes in roughly equal amounts from the positions Me, 1,2, 2–4 and 3–5, the only reaction paths admitted by the authors for this elimination. The elimination of $C_4H_9\cdot$ as a neutral fragment in a one-step reaction, although not very important, is a rather surprising proposal. The elimination of atomic hydrogen in reactions of the type $x0$ and $xi0$ makes up for about one-third of the ions observed. This is very astonishing, as it is rarely observed in a metastable decay and is usually energetically not favoured. This may be one of the weaknesses of the proposed scheme.

What is the value of such a scheme? To the best of our knowledge it is the only complete fragmentation scheme proposed for a cyclic hydrocarbon ion. A citation by Benson[179] is particularly appropriate for such a case: 'It is an unwritten law of chemical kinetics that all proposed mechanisms are inherently suspect, so that detailed data derived from proposed mechanisms always have somewhat of a tentative quality'. Mass spectrometrists have a tendency to concentrate on specific mechanisms, i.e. they prefer the trees to the forest. It is evidently not reasonable to try to make a complete fragmentation scheme without at least some knowledge of the detailed reactions possible in the system under study. On the other hand, such fragmentation schemes have the advantage of stimulating the directions of further research and they are finally one of the ultimate goals of mass spectrometry. There is no doubt that the quantitative statements of the scheme are at the best very approximate, because with the small number of labelled compounds available and the method of calculation used, the errors have a tendency to accumulate the further one proceeds down the scheme.

What about the qualitative aspects of such an attempt? The output can only reflect the reactions the authors put in their calculations and this corresponds to the knowledge of such systems fifteen years ago. This is particularly true for the consecutive reactions. The cyclohexyl cation, a main ion in this system, may serve as an example. Its stability has been studied by Schwarz and coworkers[180]. From measurements of the kinetic energy release in the metastable decay and CAD spectra they concluded that the cyclohexyl cation isomerizes immediately to 1-Me-cyclopentyl. Cacace and colleagues[181] tried to measure quantitatively the conversion of the cyclohexyl cation to the methylcyclo-pentyl cation using radiation chemical techniques in the gas phase. They found a pre-exponential factor of $10^{12 \pm 1.3}$ s^{-1} and an activation energy of 31 ± 4 kJ mol^{-1}. The latter value is very small for mass-spectrometric systems and confirms the conclusion by Schwarz. The conversion is well known in the liquid state from NMR studies. Olah and Lukas[182] observed the spontaneous transformation of cyclohexyl cation into methylcyclopentyl cation even at $-110\,^{\circ}$C, whereas Saunders and Rosenfeld[183] showed that the label of the Me-labelled methylcyclopentyl cation equilibrates at $-25\,^{\circ}$C over cyclohexyl cation into the five-membered ring. The conformation of the ring also has influence[184]. This seems to be the general behaviour, as has been demonstrated in NMR studies of Kirchen and Sorensen[185] for cycloalkyl cations with 8, 9 and 11 carbon atoms that convert immediately to the corresponding 1-methyl-7-, 8- and 10-carbon tertiary rings. The ten-membered ring forms an exception, as is often observed, since in this case the decalyl cation is formed.

In an attempt to be in a better position to judge the validity of Scheme 4, several labelled MCH have been synthesized and measured with high resolution[178]. The results for the methyl and 1 positions are resumed in Table 16 (several data are averages obtained for different labelled compounds). These data allow one to check to which extent the loss of a D atom in a given position is coupled with the loss of a ^{13}C in the same or another position. The following observations can be made. $C_6H_{11}^+$: Most of the Me eliminated consists of the Me group of the neutral, but a small amount stems from positions within the ring. In this case one hydrogen of the methyl group will participate in 14% of the cases (excluding an isotope effect). $C_6H_{10}^+$: The CH$_4$ comes exclusively from the Me group. Position 1 participates with 10% to the additional hydrogen atom needed, the rest being furnished mainly by positions 2 and 6. $C_5H_{10}^+$: The ethylene eliminated originates in 10% of the cases from positions 1 and Me together. Up to two hydrogen atoms in the Me position are exchanged before the elimination, demonstrating the complexity of this reaction. $C_5H_9^+$: In two-thirds of the eliminations $C_2H_5^{\cdot}$ comes from positions 1 and Me together. Little exchange of the hydrogen from the methyl group, but a large exchange from position 1, is observed. The rest stems mainly from positions 2–6 with a limited participation of a hydrogen from positions 1 and methyl. $C_5H_8^+$: The majority of C_2H_6 originates from positions 1 and Me together, coupled with some hydrogen exchange from the methyl group and a large exchange from position 1. $C_4H_8^+$: The loss of C_3H_6 is again a reaction where positions 1 and Me are either eliminated together (40%) or not at all. Again, little scrambling from the methyl group and much exchange from the position 1 is observed. $C_4H_7^+$: The formation of this ion must be a mixture of different processes, contrary to the prediction of Scheme 4. The elimination of positions 1 and Me and their hydrogen atoms together, as is proposed, corresponds to 25% or less. In 50% the Me group with all of its hydrogen atoms is lost, but with little contribution from position 1. It is fair to assume that this reaction

TABLE 16. The isotopic composition[a] of fragment ions from methylcyclohexanes labelled in positions 1 and Me (in %)[b,c,178]

Position of label	Labels lost to neutral 13C	D[b]	$C_6H_{11}^+$ $-CH_3$	$C_6H_{10}^{+·}$ $-CH_4$	$C_5H_{10}^{+·}$ $-C_2H_4$	$C_5H_9^+$ $-C_2H_5$	$C_5H_8^{+·}$ $-C_2H_6$	$C_4H_8^{+·}$ $-C_3H_6$	$C_4H_7^+$ $-C_3H_7$	$C_4H_6^{+·}$ $-C_3H_8$	$C_3H_6^{+·}$ $-C_4H_8$	$C_3H_5^+$ $-C_4H_9$
1-^{13}C	1		100	100	13	63	85	39	30	37	29	54
	0				87	37	15	61	70	63	71	46
Me-^{13}C	1		97	99	12	65	91	39	71	58	29	66
	0		3	1	88	35	9	61	29	42	71	34
1,Me-^{13}C$_2$	2		100	100	10	63	86	40	27	37	26	48
	1				5	6	4	4	51	27	13	33
	0				85	31	10	56	22	36	61	19
Me-^{13}CD$_3$	1	3	99	100	59	96	100	84	92	36	37	75
		2			28			16			42	13
		1			14						14	10
	0	0									8	6
Me-^{13}C/1-D	1	1	14	6	13	9	13	7	7	68	10	24
	0		86	94	84	89	87	93	21	32	86	45
		1	100		25	55	65	45	69	54	58	26
		0	100		75	45	35	55	38	46	42	63
					26	39	52	37	45		43	37
					74	61	48	63	55		57	56
												44

[a]data corrected for isotopic impurities.
[b]Values <5% are not given.
[c]In % of the group.

corresponds to an elimination of C_2H_4 from $C_6H_{11}{}^+$. We would expect that in this case the other positions participate in a random fashion with the neutral, the carbon as well as the hydrogen atoms. The lower-molecular-weight fragments show an increased mixing of ^{13}C and D, as can be expected for products that result from consecutive reactions. Since the further fragmentation of the intermediate ions is often accompanied by either a (hidden) hydrogen transfer or a random scrambling, no conclusions can be drawn from the present results. The data show that Scheme 4 definitely needs corrections, the importance of $C_6H_{11}{}^+$ being under-estimated and the loss of hydrogen atoms from fragments being over-estimated. From the data of Table 16 it is evident that if there is any scrambling within the molecular ion, it does not include the methyl group. But it should not be forgotten that this is no argument against ring-opening preceding fragmentation, because terminal hydrogens in olefins have generally little tendency to be exchanged. This is also expected for energetic reasons. The hydrogen in position 1 very often undergoes an exchange! As a *caveat*: it is this hydrogen that is mainly lost in the formation of $C_7H_{13}{}^+$, an ion of little intensity in the spectrum!

The fragmentation of the methylcyclohexane cation is an excellent demonstration of the smallness of our knowledge of the fragmentation of cyclic alkanes, although it is the most thoroughly studied compound in this series. The field is definitely fascinating, but progress can only be obtained with the help of a larger set of labelled compounds, a synthetic challenge! However, it would be worthwhile to have a much better knowledge of the fragmentation of at least a few cyclic alkanes because too much work in this field is based on assumptions that have no solid base.

VI. ALKYL IONS—A MYSTERY

Although the term 'alkyl ion' will soon celebrate its centenary anniversary[186], it was in solution chemistry that it was first postulated and—thanks to NMR—solution chemistry is still the main source of information about alkyl ions[99,187]. This is astonishing, since even-electron alkyl ions are by far the main fragments in the mass spectrum of most alkanes (see Section I). When measuring 1722 metastable transitions in the mass spectra of 70 n-paraffins and 2-, 3-, 4- and 5-methylparaffins, Herlan[188] classified 1019 transitions in the 'alkyl$^+ \to$ alkyl$^+$ + olefin' group and 139 were fragmentations of the molecular ion. More than forty years ago Johnson and Langer[180] labelled neopentane with ^{13}C in the central position and observed that $C_3H_5{}^+$ contained only 90% and Et$^+$ 47% of the label. Rearrangement had taken place. Grubb and Meyerson[190] proposed an intermediate with a cyclopropane structure in the fragmentation of t-Bu$^+$ to $C_3H_5{}^+$ and CH_4. The technique of labelling with ^{13}C and D has since become very important for studies in the gaseous (MS) and liquid (NMR) phase.

Bowen and Williams[150] tried to rationalize the metastable decays of alkyl ions with the reasoning that after 20^8 vibrations the ions will have a low internal energy, and have complete freedom to isomerize to any isomeric structure compatible with their internal energy. The basic assumption is further that only the lowest activation energy processes will be observed to give abundant metastable transitions. Weak metastable ions are often superimposed by CAD produced ions. Competition between two possible pathways to give abundant metastable transitions will require similar activation energies. The authors compare experimental data with thermochemical considerations. Et$^+$ and Pr$^+$ lose only H_2, yielding a vinyl ion and an allyl ion, respectively. The main loss for Bu$^+$ is methane, together with some C_2H_4 which is probably often a product of CAD. $C_5H_{11}{}^+$ and $C_6H_{12}{}^+$ are the first members in this series to lose only olefins, in this case C_2H_4 (a loss of C_3H_6 for the latter ion could not be confirmed[148]). The higher members $C_nH_{2n+1}{}^+$ of the series lose the olefins C_mH_{2m} with $3 \leqslant m \leqslant n/2 + 1$. Relevant data are presented in Table 17. They are in good agreement with the thermochemical considerations of the authors.

TABLE 17. The relative probability (in %) for a metastable loss from alkyl ions[a,b][148]

Fragment	Heptyl+	Octyl+	Nonyl+	Decyl+	Undecyl+	Dodecyl+
C_3H_6	100	75	46	28	20	15
C_4H_8		25	47	41	36	30
C_5H_{10}			7	24	30	30
C_6H_{12}				3	12	17
C_7H_{14}						6

[a]Values < 5% might be due to collision-activated dissociations. The data were obtained in the second field-free region of a ZAB-2F instrument.
[b]The alkyl ions were fragments in the dissociation of n-alkanes.

Radicals are the counterpart in the fragmentation of the molecular ion to radical ions. Thus a detailed knowledge of their structure would be of interest in the study of alkyl ions for thermodynamic reasons and for an understanding of the fragmentation mechanisms, but they are difficult to measure at low pressures since they carry no charge and cannot be trapped physically. Though much work has been performed to obtain mass spectra of the neutral fragments, mainly trivial results were obtained. Some progress was made, e.g, by using chemical trapping, in particular with substances such as TCNQ (7,7,8,8-tetracyanoquinodimethane) by the group of McEwen[191-194], but even this promising technique has so far not yielded much new information. A review of the use of radicals in analytical mass spectrometry has been published[195].

A. Butyl Ions

The Bu ion loses methane and ethylene in its metastable fragmentation. Because the first measurements by Johnson and Langer[189] already showed a positional scrambling of ^{13}C, the thermochemical stability of the isomers of $C_4H_9^+$, and the kinetics of their formation, interconversion and fragmentation has raised much interest. However, three different categories of problems should be clearly distinguished, since they are sometimes confused:

(i) The thermochemical properties of the different isomers and *ab initio* calculations of their structure are applicable to ions with a minimum of internal energy. It is not evident to produce experimentally the ions with a defined amount of excess internal energy in a mass spectrometer.

(ii) The study of the interconversion of these isomers requires a stringent control of the energy given to the stable ions. Under mass spectrometric conditions this is not always an easy task.

(iii) Bu^+ is an important fragment in the spectra of alkanes and related compounds. Thus its formation and decomposition is of interest. In these reactions, fragments are almost always formed with sufficient internal energy to isomerize, thus not allowing any definite conclusions about the path of their formation.

In Scheme 5 the energy diagram for Bu ions is given (the values are taken from Reference 3). t-Bu^+ and s-Bu^+ are in energy minima and form stable ions. A charge exchange between the positions 2 and 3 and a scrambling of H/D has been observed for the latter in the liquid state at low temperatures (see, e.g., p. 330 of Reference 99). The activation energy for its conversion to t-Bu^+ is $76\,kJ\,mol^{-1}$. The similarity of this value with the activation energy for interconversion between n-Pr^+ and s-Pr^+ suggests that the conversion passes through i-Bu^+. Whether it should be regarded as a transition state or an intermediate appears to be conjectural. The fact that (at least in the liquid state) H/D scrambling can occur without any conversion to the more stable

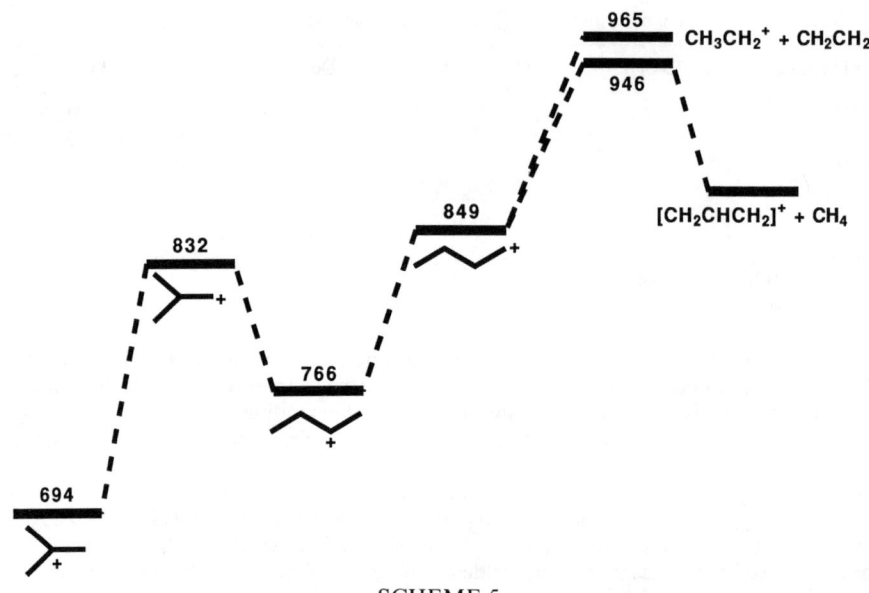

SCHEME 5

t-Bu$^+$ shows that the scheme is by no means complete. Schleyer and colleagues[196] determined the structures of 2-Bu$^+$ by an *ab initio* calculation. They found the difference between a Me-bridged and a H-bridged form to be small, somewhat in favour of the latter form. An interconversion, and thus a scrambling, seems to be an easy process. n-Bu$^+$ is unstable (or metastable). It could be considered as an intermediate in the elimination of CH$_4$ and C$_2$H$_4$, reactions for which another 97 and 116 kJ mol^{-1} are needed. Liardon and Gäumann[197] proposed protonated cyclobutane as an intermediate for this elimination. They studied the scrambling of ^{13}C and D for the fragmentation of Bu$^+$ formed in the ionization of bromobutane isomers in the ion source and in the metastable decay; contrary to earlier work the authors worked with substances that contained both labels in the same molecules. Since the carbon atoms were totally scrambled even in the source spectra of 1-bromobutane and bromoisobutane, whereas the scrambling of the deuterium process was apparently slower, they assumed a protonated cyclobutane ion where the carbon atoms are immediately equivalent, whereas the hydrogen atoms change place with the proton at a rate constant of 2×10^9 s^{-1}. Davis and colleagues[198] estimated this rate constant in the range of 10^5–10^6 s^{-1}, but because this estimation is based only on metastable transitions, it can only be a crude approximation. McLafferty, Barbalas and Tureček[163] also observed an equivalence of the carbon atoms without excessive scrambling of the hydrogens in the CAD of the C$_3$H$_6$$^{+\cdot}$, favouring in this case a cyclic transition state for an odd-electron ion. Following the randomization process, the probability of losing a given number of carbon or hydrogen atoms in a particular decomposition reaction can be calculated. For the decomposition reaction considered, viz. loss of CH$_4$, it can be assumed that the number of rearrangement steps is proportional to the number of combinations possible, i.e. 4 for ^{13}C and 126 for D. Thus it might be expected that the randomization of a D atom is anyhow a slower process than the randomization of a ^{13}C, since a larger number of steps is required. For t-butyl bromide-1-^{13}C Liardon and Gäumann had to assume two species: one with enough internal energy to fragment immediately without scrambling, whilst the other goes through an intermediate that allows an exchange of the labels. The ratio of $^{13}CC_2H_5{}^+/C_3H_5{}^+$ decreased to lower values when the energy of the electrons was lowered. A similar result was found by Shold and Ausloos[199]. When studying the detailed peak shape of the metastable transitions, Holmes, Osborne and Weese[200] found a superposition of a flat-top and a Gaussian peak, attributed to t- and s-Bu$^+$ respectively. Lias, Rebbert and Ausloos[201] observed the isomerization of s-Bu$^+$ to t-Bu$^+$ as an energy-dependent hydrogen scrambling process in the gas-phase radiolysis of the appropriate alkanes. Protonated cyclobutane isomerizes mainly to the s-Bu$^+$ structure. Fiaux, Smith and Futrell[202] observed also an incomplete scrambling of D, when the decomposing Bu$^+$ was formed by ion–molecule reaction. Parker and Bernstein[203] studied the multiphoton ionization (MPI) fragmentation pattern for n-Bu and t-Bu and other alkyl iodides in the 400–360 nm region. The MPI-MS spectra of these to isomers are significantly different, as for s-Pr and n-Pr iodide. The difference in the absorption spectra could be used for producing specific structures with labelled compounds. The difference in the MPI spectra for the different isomers was confirmed by Kühlewind, Neusser and Schlag[204]. In addition they were able to show that the metastable decays for the different Bu ions were again identical. Levine and collaborators[205] used these results to construct fragmentation pathways with the help of the maximum entropy method.

All these results demonstrate clearly that in most cases the Bu ions are formed with a specific structure. With sufficient excess energy they will fragment from this structure without any rearrangement coupled with an atom scrambling. When the excess energy is lower, the fragmentation is slower and rearrangement and isomerization reactions will compete. The fast fragmentation can be superimposed on the reaction (cf. equation 51).

$$(CH_3)_3CX^+ \rightarrow (CH_3)_2CX^+ \rightarrow C_3H_5{}^+ + HX \ (X = Cl \ or \ Br) \tag{51}$$

as was realized by Harrison and coworkers[206]. The CAD spectra of Bu ions of different precursors may vary, as has been shown by Maquestiau, Flammang and Meyrant[207]. The authors explain the difference by a ratio of s-Bu$^+/t$-Bu$^+$ that changes according to the precursor ion. This finding was put in quantitative form by Shold and Ausloos[199], who used CH_3COOH as a proton acceptor for s-Bu$^+$. By assuming that only s-Bu$^+$ and t-Bu$^+$ are present in the millisecond and second time range, they were able to measure this ratio as a function of different alkanes as the precursor ion and its internal energy at pressures of about 10^{-6} torr. In n-alkanes, 2-methylbutane, 3-methylpentane, n-butyl halides and s-butyl halides, both s-Bu$^+$ and t-Bu$^+$ ions are observed, the s-Bu$^+$ surviving without rearrangement for at least 0.1 s, the measuring time in a ICR/MS. Isobutane, neopentane, 2,2-dimethylbutane, isobutyl halides and t-butyl halides form uniquely t-Bu$^+$. They confirmed their findings by radiolytic measurements at different pressures. It is evident from these experiments that the percentage of t-Bu$^+$ from n-hexane is becoming substantially smaller with increasing pressure, whereas the i-Bu$^+$ from isobutyl bromide rearranges to t-Bu$^+$ even at atmospheric pressures. The authors propose that about 90% of the i-Bu$^+$ ions can rearrange to t-Bu$^+$ ions by simple 1,2-hydride shift. Using deuterated compounds, they were also able to determine isotope effects. Howe[153] showed that the heat of formation data are consistent with the formation of the t-Bu$^+$ when n-Pr is eliminated from the molecular ion of 2-methylhexane. The 1,2-hydrogen shift that must accompany this reaction is confirmed via a deuterium isotope effect upon the metastable decomposition. Meot-Ner[208] measured and discussed the secondary isotope effect for t-Bu due to hyperconjugation by the CD_3 groups. The fact that Bu$^+$ ion fragments from electron impact do not necessarily isomerize was also shown by Stahl and collaborators[140,209], who showed that t-butyl chloride labelled in central position by ^{13}C lost 10% of the label in the source, 16% in the CAD and 25% in the metastable spectrum. They also made the interesting observation that the induction

436 T. Gäumann

time for fragmentation induced by photons of $10\,\mu$m wavelength in a FT/ICR[161] was significantly longer for t-Bu$^+$ than for s-Bu$^+$, indicating that either more photons are needed to reach the transition state or the emission probability for IR photons is larger for t-Bu$^+$.

In conclusion it can be said that in fragmentations of alkanes—at least in the time scale observable by these instruments—t-Bu$^+$ as well as s-Bu$^+$ can be formed. s-Bu$^+$ can exchange the charge between positions 2 and 3 by a hydride shift that has a small activation energy leading to a H/D scrambling without changing its structure. Ottinger showed[210,211] that they fragment with different kinetic energies if their precursor is n-butane or n-heptane. This fact may well explain the sometimes different results obtained by different authors, but it may also be that the choice of precursors is not of general validity, since Stahl, Gäumann and coworkers[140] measured a large number of metastable Bu$^+$ fragmentations originating from n-alkanes with 7–12 carbon atoms, labelled in different positions with ^{13}C. For monolabelled Bu$^+$ they determined a probability of $23.8 \pm 0.2\%$, and for doubly labelled Bu$^+$ $48.5 \pm 0.1\%$, to be lost to the neutral. Thus, beside a small isotope effect, all positions in Bu$^+$ formed as a fragment in n-alkanes are equivalent. It is not an easy task to determine which Bu ion(s) is formed in a fragmentation, since neither the time nor the energy scale can be changed continuously. One has the impression that even a given structure, such as t-Bu$^+$, has several pathways for fragmentation. Labels within the fragment can be exchanged, its rate constant depending—among other variables—on the internal energy of the ion. This means that it is generally not feasible to draw conclusions about consecutive fragmentations from labelling experiments.

B. Heptyl Ions

Beside Bu$^+$, heptyl ions are the alkyl ions that were studied the most, in particular by labelling with D and ^{13}C. When trying to explain the fragmentation mechanism of this ion, two questions come to mind:
  (i) Where is the charge localized?
  (ii) How can heptyl ions in defined states be produced?
It is reasonable to assume that the localization of the charge in a primary position of a heptyl ion corresponds to an unstable form as in Bu$^+$. By a 1,2-hydride shift the charge will be moved immediately to a secondary, energetically more favourable, position. This reaction is probably very fast, but no measurements of its rate constant in the gas phase are known. The charge can be easily shifted over all secondary positions in the ion, thus causing a scrambling of the hydrogens in secondary positions, whereas the atoms in a primary position will remain untouched. A rate constant of $4 \times 10^8\,\text{s}^{-1}$ has been estimated for the migration frequency of hydrogen[76]. As in Bu$^+$, there is the possibility of skeletal rearrangements of the carbon atoms. Among the many different isomers possible, t-Bu-dimethylcarbenium ions have been shown by NMR measurements to be the most stable form in the liquid phase[153,212]. The fact that in the metastable spectra only one neutral fragment is lost (C_3H_6) is an additional argument in favour of the formation of this structure. This is probably also true in the gas phase. It is reasonable to assume that the energy needed to interconvert the different structures is smaller than the activation energy needed for further fragmentation. However, these processes take time and perhaps may not compete with the fast dissociation of an ion with a large amount of excess energy. Thus the time scale and the excess energy of the ion may be important. In alkanes, the fragmentation of an alkyl ion is a secondary or tertiary reaction, where it may be expected that the ion has already lost a large part of its excess energy.

The fragmentation of alkanes produces alkyl ions in different, unknown states. In order to study the reactions of a heptyl ion, it should be generated in a better-defined state. Little study has been conducted on ionization of heptyl radicals. Dearden and Beauchamp[213] studied the isomerization and decomposition of 1-pentyl, 1-hexyl and 1-heptyl radicals by photoelectron spectroscopy. The results are consistent with a mechanism involving isomerization via an intramolecular hydrogen shift to produce sec-alkyl radicals whose isomers could not be distinguished in the spectra. Further, the adiabatic and vertical ionization potentials were determined.

The most often used method is to start with a functional group in a defined position that is easily lost upon ionization. Halogen atoms are mostly used; for the production of heptyl ions, the iodides are particularly useful since the intensity of the molecular ion is weak, the heptyl parent ion is relatively abundant and the ions with iodine or fragments thereof contribute to less than 1% of the total ion current[214]. They can be considered a clean source of heptyl (or any alkyl) ions. Another possibility are ion/molecule reactions occurring in chemical ionization, either by adding a proton to an olefin, by hydride (or another anion) abstraction with, e.g., Et^+, or by proton transfer, e.g., by CH_5^+ to an alkane to form an unstable protonated alkane ion (equations 52–56).

$$AH^+ + C_7H_{14} \rightarrow A + C_7H_{15}^+ \tag{52}$$

$$C_2H_5^+ + C_7H_{16} \rightarrow C_2H_6 + C_7H_{15}^+ \tag{53}$$

$$CH_5^+ + C_7H_{16} \rightarrow CH_4 + [C_7H_{17}]^+ \tag{54}$$

$$[C_7H_{17}]^+ \rightarrow C_7H_{15}^+ + H_2 \tag{55}$$

$$[C_7H_{17}]^+ \rightarrow C_{7-m}H_{15-2m}^+ + C_mH_{2m+2} \tag{56}$$

In reaction 52 the difference in proton affinity can be changed. The fragmentation spectrum is very simple (mainly m/z 57 and 99) when water is used as reagent for chemical ionization; more fragments are observed with methane (m/z 57, 55, 97)[73]. The hydride transfer reaction in alkanes has been very extensively studied by Lias, Eyler and Ausloos[215] and, for labelled alkanes, by Houriet and Gäumann[75,78]. By using selectively deuterated compounds, these authors determined a ratio of k_s/k_p, i.e. the probability to transfer a hydrogen from a secondary or a primary position, of 3.3 for n-hexane, 1.7 for n-pentane and 1.9 for propane, respectively, and for the first substance an isotope effect of 1.14 that seems to affect equally the primary and the secondary position. All secondary positions seem to be equivalent for the hydride transfer. Chemical ionization analogous to reaction 53 has also been used as a convenient way to produce heptyl ions from heptyl halides[168].

Reaction 54 is of a more complex nature, since the intermediate heptonium is not known. For lower alkanes Hiraoka and Kebarle[216] were able to demonstrate the existence of two species, one corresponding probably to a complex between an alkyl ion and a H_2 molecule, the other to a C—C protonated alkane. Different mechanisms have been discussed by Houriet, Parisod and Gäumann[75]. In trying to rationalize the observations, the authors were led to propose two paths for the formation of heptyl ion[77]. Reaction 56 is particularly intriguing, since it is the only case where they observed a clean C—C bond split without a prior positional scrambling (vide supra). Unfortunately, it cannot be decided if this elimination of an H_2 and an olefin is so fast that there is no time for a skeletal randomization, or if the loss of an alkane neutral moiety is a rather clean, unknown elimination.

The spectrum of the heptyl ion is given in Table 18. The results for the source spectra are obtained from the spectrum of 1-iodo-n-heptane after correction for the fragments containing iodine. Bu^+ is by far the most abundant ion. The formation of propyl ion in the source might be astonishing, since it does not correspond to Stevenson's rule[17]; it could be formed by successive eliminations of two ethylenes, but we will show later

438   T. Gäumann

TABLE 18. The fragmentation spectrum of $C_7H_{15}^+$ [214] [a]

	$C_6H_{11}^+$	$C_5H_{11}^+$	$C_5H_{10}^{+\cdot}$	$C_5H_9^{+\cdot}$	$C_4H_9^+$	$C_4H_8^{+\cdot}$	$C_4H_7^+$	$C_3H_7^+$
Source[b]	0.1	0.8	0.5	1.0	64	1.1	7.5	25
Metastable[c]	<0.1	0.1	0.1	1.1	98	<0.1	0.1	0.4

[a] In % of total ion current.
[b] Data for 1-iodoheptane; the contribution by the molecular ion and the fragments containing iodide (1% of total ionization) are subtracted.
[c] Data for $C_7H_{15}^+$ from 1-iodoheptane.

that the elimination of butene can be rationalized. The elimination of an olefin from an alkyl ion is a rather complex reaction, as has been shown by many measurements using labelling with D or ^{13}C. The fragmentation of heptyl and other alkyl ions formed from the corresponding halides was studied by Fiaux and collaborators[6,214]. Houriet and collaborators studied the results of chemical ionization of halides[217] and alkenes[73]. Parisod and Gäumann[77] tried to distinguish between protonation and hydride transfer; Stahl and collaborators[218,219] concentrated on the long-lived ions by measuring the metastable decay and the CAD spectra. A few illustrative figures are collected in Table 19; the n-heptyl precursors were selectively labelled with ^{13}C or D. It is already evident from these data that any position can contribute to the neutral fragment and that neighbouring positions do not have a greater probability of being lost together than separated ones. The probability for a given position to be lost to the neutral, albeit never zero, does not correspond to a random scrambling, in particular not for the metastable decay, where the probability for a terminal position decreases further compared to the source spectrum. The metastable decay and CAD yield similar distributions: the structure of the ion after 20 μs seems to be independent of its internal energy. Deuterium is strongly scrambled, but again the figure does not correspond to a random distribution. This might come from the fact that an exchange with a terminal D atom is a rarer event.

A summary of the results obtained for heptyl is given in Figure 4. Here the probability of a given position to be lost to the C_3H_6 neutral in the metastable decay is presented for every position within the chain. The figures are an average obtained from the study of the metastable decomposition of $C_7H_{15}^+$ obtained in the fragmentation of C_9-C_{16} n-alkanes, labelled in different positions. It was assumed that the elimination of an alkyl fragment from the molecular ion corresponds to the loss of the terminal positions originating from a C—C-split[221]. The curve was obtained by a least-squares treatment of the results from 56 different alkanes with up to four ^{13}C labels. For this, it had to be assumed that the probability of losing several positions to the same neutral is obtained by multiplying their respective probabilities. The set, that could have been enlarged by including data from heptyl ions with other precursors without any change in the distribution, shows the validity of the assumptions. A corresponding distribution for fragmentation within the source cannot be obtained in this way, because it is not possible to isolate a given ion in order to study its fragmentation. The distribution obtained from the fragmentation of the n-heptyl iodides is more flat, somewhat less symmetric and seems to depend on the internal energy of the ion (vide supra).

The symmetric distribution of Figure 4 is rather strange: the terminal positions having the smallest probabilities, followed by position 4. Positions 2 and 6 have the highest values. Stahl and Gäumann[220] tried a computer simulation in order to explain the observed distribution. They supposed a number of rearrangements that proceed via protonated substituted cyclopropanes to arrive eventually at the most stable structure, t-Bu-dimethylcarbenium ion. This reaction corresponds to an approximate gain in

TABLE 19. The probability (in %) to lose a given label to the neutral for 1-heptyl ions[214,218,219]

fragment lost	conditions	Position of label								
		$1\text{-}^{13}C$	$2\text{-}^{13}C$	$3\text{-}^{13}C$	$1,2\text{-}^{13}C_2{}^a$	$2,3\text{-}^{13}C_2{}^a$	$1,3\text{-}^{13}C_2{}^a$	$1,2,3\text{-}^{13}C_3{}^a$	$1,3\text{-}D_4{}^a$	$1,5\text{-}D_4{}^a$
C_3H_6	source	39	46	46	43/20	52/19	49/18	41/30/10	32/26/18/5	35/36/15/3
	calc.[b]	43	43	43	44/20	47/20	47/20	36/33/9		
	metastable	33	50	46	56/12	60/18	56/12	51/35/3	37/24/12/5	37/35/14/4
	CAD	34	44	38	58/9	65/11	58/7			
	random[c]	42.9			57.1/14.3			51.4/34.3/2.9	36.9/39.6/13.2/1.1	
C_4H_8	source	63	69	56	24/48	34/45	37/39	20/29/33		
	calc.	61	60	60	30/45	37/42	36/32	24/28/34		
	random	57.1			57.1/28.6			34.3/51.4/11.4		

[a]For multiple labelling probability for losing one/two/three etc. labels simultaneously.
[b]Calculated for source spectra; see text and Reference 220.
[c]Calculated for random loss, all positions having the same probability (= complete scrambling).

T. Gäumann

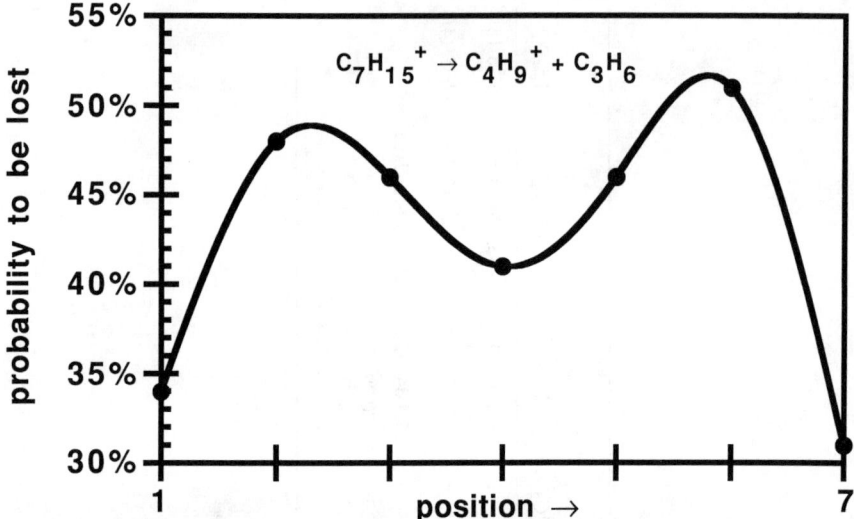

FIGURE 4. The probability of the different positions in the n-heptyl ion to be lost to the neutral C_3H_6 (after Stahl and Gäumann[221])

internal energy of $140\,\mathrm{kJ\,mol^{-1}}$. It is the central C—C bond of this ion that is supposed to fragment, either preceded by a hydride transfer to give $\mathrm{Bu^+}$ or a methide transfer for $\mathrm{Pr^+}$. Such shifts have been observed in liquid-state NMR[222]. This would explain the presence of this ion in the source spectrum (see Table 18). The ring-forming ring-opening process is simulated by a Monte Carlo method, assuming an equal probability of charge localization either on a primary, secondary or tertiary position. This is certainly wrong and could explain the overestimation of the probabilities of the terminal positions by this model. A weight function allowing one to differentiate between these cases would certainly lead to an improvement of the calculations at the expense of introducing additional arbitrary parameters. Some results are included in Table 19. Guenat, Stahl and Gäumann[223] were able to improve the model by introducing a few more assumptions; this is especially true for terminal positions and branched precursors, but the basic weakness of the model, that because of the central limit theorem the distribution will become random after a large number of rearrangements, is not solved. Certainly the metastable decay does not correspond to an infinite number of rearrangements, but the distributions of the metastable decay are so reproducible for different instrumental conditions that a basic improvement of the model is needed.

In Table 20 some data are shown for heptyl ions originating from a highly branched precursor. Analogous observations can be made as for the data of Table 19, but the evolution as a function of the time of fragmentation is even more pronounced for the branched compound. Also, in this case CI with $\mathrm{CH_5^+}$ simulates the source data, whereas $\mathrm{C_2H_5^+}$ gives results identical to the metastable case and the CAD spectrum, a strange behaviour! The evolution of the probability for position 1 is dramatic, since it changes from the highest value to the lowest as a function of time. Since a faster fragmentation corresponds to a higher excess energy, the probabilities of several labelled compounds were measured as a function of the electron energy for the ionization of the iodides[224]. The distribution starts to change when lowering the electron energy to about 15 eV to approach the metastable values near the appearance energy of $\mathrm{Bu^+}$.

TABLE 20. The probability (in %) to lose a given label to the neutral C_3H_6 for $C_7H_{15}^+$ originating from 1-iodo-4,4-dimethylpentane[219,224]

Position of label	1-^{13}C	2-^{13}C	3-^{13}C	4-^{13}C	5-^{13}C	1,3-^{13}C$_2$[c]	1,4-^{13}C$_2$[c]	5,5'-^{13}C$_2$[c]	1-D$_3$[c,d]
Conditions									
Source[a]	68	77	77	25	19	52/37	76/12		5/2/63
CH$_5^+$[b]	63			29	12				
MIKES	33(31)[e]	55(51)	55(52)	53(52)	33(52)	58/15	57/19	54/5	
CAD[a]	32	55	55	53	29	56/15	56/16	54/3	16/6/28
C$_2$H$_5^+$[b]	34			49	32				
Calc.[223]	33.3	55.6	55.6	55.5	33.3	57.7/15.6			

[a] Produced by electron impact from $C_7H_{15}I$.
[b] Produced by chemical ionization from the alkane.
[c] Probability for losing one/two/three labels simultaneously.
[d] 1-D$_3$-4-dimethylpentane.
[e] $C_7H_{15}^+$ ions from 2,2,7,7-tetramethyloctane[148].

The positional probabilities in the metastable data show again the same behaviour as for the 'linear' heptyl ions: a small value for the loss of a terminal group, which is compensated for by a higher probability for a position that was initially within the chain; all three positions (2–4) have—contrary to the linear case—the same probability. The values presented in parentheses are presented for a heptyl ion that is the primary fragment of 2,2,7,7-tetramethyloctane[148]. It is formed by an elimination of a terminal butyl group as the neutral fragment. One would expect that exactly the same ion is formed as from the 1-iodo-4,4-dimethylpentane. This is nearly true, except that the 'terminal' position the iodopentane has become an 'in-chain' position with a probability of 52%: positions 2–5 have become equivalent. This must be due to its formation: in the split of the carbon–iodine bond a hydrogen transfer apparently takes place simultaneously, producing a Me group in position 1 of the iodide, whereas the loss of a t-Bu group can take place without this transfer. How far this rule can be used to test the existence of a terminal Me group in the first step of the isomerization process remains to be tested.

Photodissociation experiments with IR photons corresponding to a wavelength of 10 μm long-lived heptyl ions from different precursors have also been conducted[140]. For all isomers the probability for all positions measured correspond to a random distribution within very narrow limits. However, this does not necessarily mean that all the ions have experienced a complete scrambling of their positions in the ms time range. Since the energy of a photon is small compared to the activation energy of the fragmentation, the internal energy content of the ion increases slowly in this experiment and allows the ion any number of internal rearrangements. Further experiments in this field are undoubtedly necessary.

C. Higher Alkyl Ions

It has been shown above that Bu ions, at least on a long time scale, scramble the carbon and hydrogen atoms randomly, whereas heptyl ions seem to follow some unknown distribution law directed by the way the lowest-energy structure(s) is attained. Practically only olefins are eliminated. The metastable fragmentation of the dodecyl ion, formed in the fragmentation of n-tetradecane (6% C_7H_{14}, 17% C_6H_{12}, 30% C_5H_{10}, 30% C_4H_8, 15% C_3H_6 [221]) may serve as an example. Bowen and Williams[225a] tried to estimate the intensities in the metastable decay by thermochemical reasoning, but their predictions do not explain the findings with labelled substances. The isomerization of alkyl ions before losing an olefin ion is a reaction that has often been postulated. Cole and Williams[225b] observed it for hexyl and octyl ions, generated from isomeric bromides, and Liardon and Gäumann[31] and coworkers[149] in the fragmentation of alkanes, Long and Munson[226] and Büker, Kuck and Grützmacher[11] for pentyl ions, Levsen and coworkers[144,145] for octanes and n-dodecane, the list not being complete. What conclusions can be drawn from these results? Although an isomerization to (probably) the lowest-energy isomer(s) has been proven by several procedures, such as thermochemical considerations, CAD spectra, labelling etc., only little work has been done with a complete set of labelling. Such measurements were undertaken on a large scale by Stahl and Gäumann[221] in the same manner as was cited above for heptyl ions. The result on a very large set of measurements of the metastable decay of ^{13}C-labelled alkanes is that alkyl radicals with less than seven carbon atoms seem to fragment only after complete scrambling. Except for a small isotope effect no deviation could be found. The distribution of the fragmentation probability to form alkyl ions with more than seven carbon atoms is very similar to that observed for heptyl ions, i.e. a low value for terminal positions, a maximum for positions 2 and $\omega - 1$ and a minimum in the center of the initial fragment. This is demonstrated for three fragmentations in Figure 5. In all

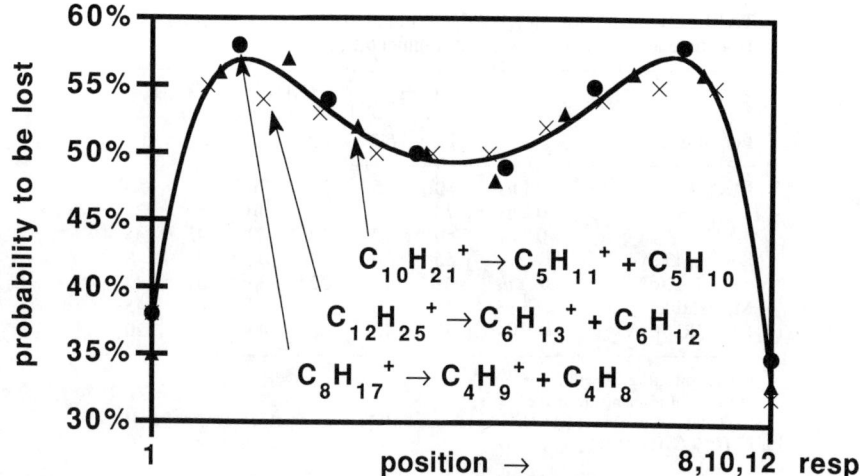

FIGURE 5. The probability of the different positions in n-alkane ions to be lost to the neutral fragment (after Stahl and Gäumann[221])

three fragmentations half of the parent ion is lost; the probability for random fragmentation is 50% in all three cases. The same deviation from a random distribution is observed. The probability for the simultaneous loss of several labels is again the product of the probabilities for the positions involved. The result is striking indeed! It shows that to a first approximation the randomization of the positions is nearly complete, but to a second approximation an identical distribution law is strictly followed, an observation that remains a mystery for the time being.

VII. OLEFINIC FRAGMENTS—WHAT IS THEIR STRUCTURE?

The loss of an alkane from an alkane cation to yield an olefin ion is one of the most important classes of reactions in the fragmentation of paraffins. It has been shown (see above) that in the majority of cases a terminal entity is eliminated. Formally, the simplest reaction would consist of a 1,3-hydrogen shift to the neutral alkane moiety, resulting in the formation of a 1-olefin ion. This reaction is difficult to prove in the source spectrum, since the subsequent fragmentation of an olefin, an odd-electron ion, is rather complex. Thus it is not possible to draw any conclusions from the source spectrum of the alkane because of too many overlaps. Such a study is further complicated by the fact that part (although minor) of the eliminated alkane stems from the inner part of the alkane ion (see above). The only way to check this proposal is to label the inner part of an alkane with D and the terminal(s) with ^{13}C. Measurements with, e.g., 2-^{13}C-1,1,4-D_4[34] nonane showed that the elimination of $^{13}CCH_2D_4$ is much less probable than $^{13}CCH_3D_3$. One has to suppose either a strong isotope effect (which is not very probable for source spectra) or hydrogen atoms being transferred from different positions within the chain. Unfortunately, it is difficult to draw any precise information from the metastable spectrum because of the overlap with the more abundant $C_2H_5^-$ fragment.

The procedure is to see how fragments are eliminated from model olefins and to compare the results with the metastable dissociation from an olefin fragment of an alkane ion. In this section we will limit ourselves mainly to the C_7H_{14} ions formed either from different isomeric C_7H_{14} or from alkanes with nine or more carbon atoms. A particularly

TABLE 21. Evolution (in %) of the terminal loss of different neutrals from 1-heptene as a function of time of fragmentation[230-234]

		C_2H_4		C_2H_5		C_3H_5	
Position lost		1,2	6,7	1,2	6,7	1-3	5-7
FIK[a]	0.1 ns	80	5	20	70	95	5
	0.2 ns	70	5	40	40	60	35
	0.7 ns	60	10	40	40	45	45
Source, 17 eV[b]	0–2 μs	55	0	29	44	90	8
10 eV[b]	0–2 μs	22	15	32	33	57	33
Metastable[c]	20 μs	17	16	33	36	45	45
FT/ICR, 70 eV[b,c]	0–10ms	35	13	30	45	74	20

[a] Field ionization kinetics.
[b] Energy of ionizing electrons.
[c] % Total ion current in FT/ICR (the metastable) spectrum $C_5H_{10}^{+\cdot}$ 17 (57)%, $C_5H_9^+$ 25 (7)%, $C_4H_9^+$ 5 (6)%.

interesting study has been made by Brand and Baer[123] with energy-selected $C_5H_{10}^{+\cdot}$ ions using the photoelectron photoion coincidence technique that was discussed in Section V. Although much work has been done with C_3–C_5 alkanes (see, e.g., References 104, 116, 134, 163, 227, 228), we think that the information obtained cannot be extrapolated to higher alkanes.

In Table 21 the probability of the elimination of the three fragments C_2H_4, C_2H_5 and C_3H_5, each a typical case for a class of neutrals, is presented for the terminal eliminations from 1-heptene as a function of the time of fragmentation. The following observations can be made:

(i) There is a strong time dependence. Within the shortest time interval observable (0.1 ns), the fragments are eliminated from the terminal and one would expect C_2H_4 and C_3H_5 from the original location of the double bond and C_2H_5 from the saturated end, respectively.

(ii) The two terminal ends become eventually identical, but the speed of equilibration is drastically different for the different fragments.

(iii) The majority of the C_2H_5 and C_3H_5 neutrals are terminal losses. Thus, one might assume that 1-heptane will suffer a shift of the double bond, but not a scrambling of the carbon skeleton.

(iv) C_2H_4 comes from anywhere within the molecule, but only neighbouring carbon atoms are eliminated together, possibly with a substituted cyclobutane ion as an intermediate.

(v) Table 21 gives a good demonstration how FT/ICR spectra yield results where the metastable decays are integrated. On the other hand the FI kinetics furnishes an interesting time dependence; however, it should be noted that the ions are produced differently, and the calculation of the time dependence requires a few assumptions and should be extrapolated with care[229-231].

(vi) When lowering the energy of the ionizing electrons, results of the source spectrum approach the label distribution observed for the metastable decay.

In Table 22 the probabilities for losing a given position to the neutral lost from heptene- or nonene-fragment ions of alkanes are compared with the corresponding probabilities for 1-heptene and 1-nonene. The data were obtained from 38 (25) selected C_{10} (C_{11})–C_{16} alkanes, labelled in such a manner that the heptene (nonene) fragment ion had the label(s) in a predictable position [beside a possible small contribution by a

TABLE 22. Probability (in %) to lose a given position to the neutral from olefin fragments in the metastable decay[234]

neutral lost	olefin	%[b]	1/1,2[a]	2/2,3	3/3,4	4/4,5	5/5,6	6/6,7	7/7,8	8/8,9	9
CH$_3$·	C$_7$H$_{14}$$^{+·}$	14	25(26)[c]	6(4)	13(14)	12(13)	13(12)	5(4)	26(25)	—	—
	C$_9$H$_{18}$$^{+·}$	6	18(28)	5(1)	17(17)	10(3)	2(1)	3(2)	17(18)	2(0)	26(28)
C$_2$H$_4$	C$_7$H$_{14}$$^{+·}$	69	17(17)[11][15][d]	5(4)[11]	28(30)[23]	29(30)[23]	5(4)[11]	16(16)[14]	—	—	—
	C$_9$H$_{18}$$^{+·}$	26	9(6)	10(0)	6(14)	27(30)	25(30)	14(14)	6(0)	3(6)	—
C$_2$H$_5$·	C$_7$H$_{14}$$^{+·}$	15	25(33)	6(16)	11(0)	18(0)	1(17)	39(36)	3(0)	—	—
	C$_9$H$_{18}$$^{+·}$	52	39(48)	3(0)	2(1)	5(1)	1(1)	0(1)	3(0)	47(48)	—
C$_2$H$_6$	C$_9$H$_{18}$$^{+·}$	16	38(48)	4(0)	4(2)	4(0)	0(0)	0(2)	0(0)	51(48)	—

[a] The first number given the position eliminated for C$_1$ fragments and the second for C$_2$ fragments.
[b] Contribution of the neutral lost to all neutrals in this list.
[c] In parentheses, data for 1-heptene and 1-nonene, respectively.
[d] In brackets, data for IR photofragmentation of olefins listed under b.

loss from the interior of the alkanes (see above)]. The elimination of four fragments are compared: CH_3, C_2H_4, C_2H_5, C_2H_6 Me is eliminated from all positions, but the probability to lose it from positions 2, 6 and 8, respectively, is very small. The same behaviour is observed for alkanes (see above). The distribution is symmetric about the centre of the ion. The C_2 fragments are always eliminated from two neighbouring positions, contrary to what is observed for even-electron ions. Most of the C_2H_4 originates from central positions, the contribution of the terminal positions decreasing in importance with increasing chain length. Similar to the methyl elimination only small amounts are lost from positions 2,3 and 5,6 for heptene and 7,8 for nonene, respectively. The slow fragmentation of 1-heptene, induced by IR photons, yields results similar to the metastable decay. One gets the impression that the olefin has reached its final structure(s) already after a few microseconds. The saturated neutral fragments are eliminated symmetrically, preferably from both ends.

The comparison of the data for the 'alkane produced' olefins with the figures for 1-heptene and 1-nonene shows that the 'fingerprints' offered by labelling are as identical as can be expected. It must be concluded that, at least in the metastable range, the olefin ions produced from higher alkanes and the corresponding olefin molecular ions fragment in the same selective and unique way: all positions within the ion participate in the neutral loss, but with different values, and the order within the chain is maintained, since neighbouring atoms are always lost. The fragmentation scheme of olefins is therefore fundamentally different from the fragmentation of alkyl ions. Levsen and coworkers[145] concluded also, on the basis of CAD spectra of the olefinic fragments of labelled n-dodecane, that the olefinic fragments have a linear structure, preferably of a 1-alkene.

The simplest explanation of the observed fact is that a shift of the double bond is possible by a successive 1,3-hydride transfer. Two additional arguments are in favour of this proposal: The metastable spectra of 2- and 3-heptene are practically indistinguishable from the label data for 1-heptene, and terminal D atoms have a much smaller tendency (if any) to scramble, contrary to atoms within the inner positions. Since the latter observation is also made for alkyl fragmentations, this is therefore not a strong argument for an absence of a skeletal isomerization. However, finer details, such as the different time dependence for different fragments, are not explained by such a model.

The mass spectra of several branched $C_7H_{14}^{+ \cdot}$ isomers have been studied. They differ from linear olefins. A particularly interesting olefin is 4,4-dimethyl-1-pentene[235]. Me is mainly lost from positions 5 in the source spectra, but the situation changes in the metastable decay since positions 1 and 3 gain in importance for this loss. The metastable loss of C_2H_4 is especially astonishing: positions 1 or 3 are lost from either position 2 or 5 with exactly the same probability of 12%, and positions 1 and 3 together with half of this value. One gets the impression that a CH_2 group of position 1 or 3 circles around a hypothetical $C(CH_2)_5^+$ ion and selects at random one CH_2 group from positions 1,3,2,5',5'' to form ethylene! When the double bond is shifted in the neutral to 4,4-dimethyl-2-pentene[234], position 3 loses its special character and Me is only lost from 1 and 5 with 25% probability for each position. In Table 23 data are presented for the loss of C_2H_4—the main metastable loss of these ions—for several branched $C_7H_{14}^{+ \cdot}$ ions. The result is astonishing: first the loss of non-neighbouring positions becomes a rule, pointing to some isomerized structure, and second, the location of the double bond within the neutral olefin determines the mechanism of the loss of C_2H_4, which is very specific and unique for each isomer. On the one hand, the olefins keep their structural identity, but on the other hand, they lose fragments that are difficult to explain without complicated—however specific—rearrangement. From this it can be assumed that there is no skeletal isomerization in olefins derived from n-alkane ions. Unfortunately, no spectra of labelled branched alkanes are known that would allow one to determine precisely the structure of the olefins derived from such alkanes.

TABLE 23. Probability (in %) to lose a given position to the neutral ethylene for branched $C_7H_{14}^{+ \cdot}$ [234-237]

Olefin	Condition	1,2	1,3	1,4	1,5	2,3	3,4	3,5	3,6	4,6	5,6
5-Me-2-hexene	source	6	6	17			71		0		
	metastable	11	8	22			54		4[a]		
	photofragm.	0	25	26			50		0		
2-Me-2-hexene	source			16[a]			19		14		45
	metastable			6			28		10		56
	photofragm.			6			16		8		72
2-Me-1-hexene[c]	source			0			64		18	10	5
	metastable			4[a]			38		25	17	14
	photofragm.			0			32		32	32	0
4,4-Di-Me-1-pentene	source	5	10			42[b]	8	33[b]			
	metastable	11	6			36	11	36			
	photofragm.	0	0			50	0	50			

[a]This position has been counted twice.
[b]This position has been counted three times.
[c]The 1 and Me positions are equivalent[236].

How typical are the mass spectra of the heptene isomers? Only little is known about other olefins that would give an insight into the details of fragmentation. Nishishita and McLafferty[238] studied the metastable and collisional activated dissociations of a large series of C_5H_{10} and C_6H_{12} isomers in order to get an answer to the question: do the molecular ions have a well-defined structure that would allow a structural analysis? The metastable spectra do not yield much information, but the CAD seems to be typical for the structure of the neutral, although the differences are not very prominent. The problem of all this kind of work is that there are not many different possibilities of losing a neutral. Bowen and Williams[239] applied enthalpy considerations to predict the most prominent fragments. The basic idea is definitely correct, but for some unknown reasons the olefins do not respect energetic calculations for reasons that we do not know. But these problems do show up only when the mechanisms are followed with isotopic labelling. This is evident from the work of several laboratories, where the isomerization of all linear octene isomers was studied by using labelling, high-resolution mass spectrometry, metastable and CAD decays and FI kinetics[240]. The authors show that at the very short times of the FI spectrum the isomers are still specific, but all decomposing ions isomerize within 1 ns to a common structure, while the isomerization of the non-decomposing ions is again not complete, in particular for 1-octene. There seems to be a complex balance between excess energy and time scale, that results in an overall scheme that is difficult to describe quantitatively.

VIII. CONCLUSIONS

Half a century ago a mass spectrometer was a single focusing magnetic sector instrument; its console contained a clock to facilitate reproducible measurements, since it was mainly used for analyses of hydrocarbon mixtures. Very often an analog computer was associated with the instrument in order to permit the solution of the linear equations needed to calculate the concentrations of the different hydrocarbons in the mixture by determining the intensity of selected peaks. The interpretation of a spectrum was left to the intuition of the operator.

What progress has been made since this period? The instrumentation evolved to multisector medium-resolution mass spectrometers. Quadrupole instruments have taken the place of unit-resolution instruments. The latest development has been the introduction of radiofrequency spectrometers such as FT/ICR and Quistors, the former allowing very high resolutions and the advantage of pulsed Fourier transform techniques, the latter the use of (relatively) high pressures, and both the trapping of ions over period of seconds. The number of ionization methods has been enlarged from simple electron-impact ionization of neutrals in the gas phase to more elaborate schemes allowing one to advance in the mass range of several 10^5 dalton.

Analog computers have been replaced by digital machines. Not only is the computer used to control the data acquisition, making the clock obsolete, but it also took over the process of interpretation of the data, allowing the search for similar spectra among a data base of 10^5 compounds. The computer enables one also to calculate structures of fragments containing not too many electrons by *ab initio* calculations. Since these calculations—like the spectral search systems—are inherently 'brute-force' methods, their potentially increases proportionally with the available computer capacity. Last but not least, the introduction of gas chromatography made the use of mass spectrometers for the analysis of hydrocarbon mixtures obsolete.

How did our understanding of the mass spectral fragmentations of alkanes improve with this revolution of the instrumental capacities? One has to take note that most of these improvements went into an enlargement of the number of classes of substances that could be measured with these instruments, the hydrocarbons having become somewhat the 'poor parent' of mass spectrometry, the fragments containing two to four carbon atoms forming a notable exception because of their widespread abundance in all mass spectra and the possibility to make *ab initio* or semi-empirical calculations of new, hitherto unknown structures.

What is the reason for this loss of interest? Certainly it was believed for a long time that the mass spectra of alkanes were simple and that their fragmentation could be explained by relatively simple theories, such as the quasi-equilibrium theory[241,242]. Although these (and other) theories were improved over the years (for a review see References 93 and 243), they were not able to correctly predict mass spectral fragmentations of alkanes. There may be two reasons for this: first, it is dangerous to predict a fragmentation on the basis of at best approximately known vibration frequencies, since for a fragmentation the amplitude of the vibrations under study has to access high quantum numbers. Because of the anharmonicity of the vibrations, the notion of orthogonal normal vibrations somewhat loses its sense (see, e.g., Ha and Günthard[244]). This is particularly true for paraffins. Secondly, detailed experiments have shown that the fragmentations observed are more complex than was assumed and belong to the most complicated reaction schemes observed in mass spectrometry, among other reasons probably because of difficulties in localizing the charge.

Our present-day knowledge of the fragmentation of higher alkanes can be summarized as follows. The molecular ion does not seem either to isomerize or to scramble randomly the positions of the carbon and hydrogen atoms. Approximately 85% of all fragments that are formed in a first step result in a C—C bond split, coupled with a hydrogen transfer. If this transfer is within the remaining ionic fragment, an alkyl radical is lost and a secondary alkyl ion is formed. This transfer is not needed for a bond split at a tertiary or quaternary carbon atom, making these carbon atoms particularly favourable as a point of fragmentation. The hydrogen transfer can also go to the neutral: the result is an olefin ion and an alkane neutral. All C—C bonds can participate, but the probability for a split of the 1-2 position is smaller, as is true for a C—H bond (in higher alkanes). Around 15% of the primary fragments are expelled from within the chain. The mechanism is unknown. It seems that it becomes more important when the molecular ion is given more time to fragment, whereas a high excess energy favours simple fragmentations. This means that the mass spectrum depends strongly on the mechanism of ionization. The secondary alkyl ions fragment further by loss of an olefin ion. The probability for the loss of a given position to the neutral corresponds to a random distribution for ions with less than seven carbon atoms; for higher alkyl ions this probability is close to random, but sufficiently different, that a general—but unknown—scheme seems to govern the fragmentation.

The olefins formed in the primary fragmentation have the structure one would expect them to have with the fragmentation rules given above. The further fragmentation of olefins is unfortunately still a black box; no general fragmentation scheme is discernible for the time being, contrary to the alkyl ions. Linear olefins $C_nH_{2n}^{+\cdot}$, in conformity with alkanes, have a small probability of eliminating the penultimate positions, again contrary to alkyl ions, where these positions have the highest probability of being lost. This means that in order to calculate a complete fragmentation scheme for a higher alkane, the fragmentations of all possible alkenyl ions had to be studied, a task that exceeds the capacities of present-day research groups in the field and is perhaps not justified.

In a very simplifying scheme one would say that odd-electron ions exhibit a smaller tendency to isomerize than even-electron ions. Levsen and Heimbrecht[245] explain this by the fact that the decomposition threshold of radical ions is lower than those of even-electron ions, since odd-electron ions are less stable than even-electron ions; the barrier for fragmentation of radical ions is apparently higher than the threshold of isomerization. However, the term 'isomerization' needs to be better defined. Very often isomerization can only be detected by selective labelling. Further, isomerization is often stipulated for fragmentations that are in reality concerted reactions. An isomerization does not mean that all hydrogen or carbon atoms are randomly scrambled, because very often the terminal or the penultimate positions within an ion exhibit particular properties. The observation that 'small' ions have a larger tendency to isomerize is not necessarily true: these ions can only lose similar fragments, but this is not a proof for an isomerization. Last but not least, the question may be raised how often the 'isomerization' takes place only in the fragmenting transition state, i.e. that the same transition state can be reached from different energy minima. The energy hyperplane of a hydrocarbon ion is very complex, as has been stated on several occasions by Lorquet[93,246]. The remark by Meyerson and coworkers[247] in 1959: 'The mass spectral behaviour of the simplest class of organic compounds, the n-alkanes, is perhaps the most complex', is still true!

IX. ACKNOWLEDGEMENTS

This work is dedicated to H. H. Günthard on his 75th birthday as thanks for a lifelong friendship and encouragement. I wish to thank my collaborators mentioned in the text, in particular D. Stahl, for many fruitful discussions. This work was made possible by generous grants from the Swiss National Science Foundation and the Federal Commission for the Encouragement of Research.

X. REFERENCES

1. J. L. Holmes, M. Fingas and F. P. Lossing, *Can. J. Chem.*, **59**, 80 (1981).
2. M. Bachiri, G. Mouvier, P. Carlier and J. E. Dubois, *J. Chim. Phys.*, **77**, 899 (1980).
3. S. G. Lias, J. E. Bartmess, J. F. Liebman, J. L. Holmes, R. D. Levin and W. G. Mallard, *J. Phys. Chem. Ref. Data*, **17**, Suppl. 1 (1988).

450	T. Gäumann

4. F. P. Lossing and J. L. Holmes, *J. Am. Chem. Soc.*, **106**, 6917 (1984).
5. R. D. Levin and S. G. Lias, *Ionization Potential and Appearance Potential Measurements*, *1971–1981*, NSRDS-NBS 71, National Bureau of Standards, Washington, USA, 1982.
6. A. Fiaux, B. Wirz and T. Gäumann, *Helv. Chim. Acta*, **57**, 708 (1974).
7. R. L. Platzman, *The Vortex*, **23**, 372 (1962); *J. Chem. Phys.*, **38**, 2775 (1963); *J. Phys. Radium*, **21**, 853 (1960).
8. C. I. M. Beenakker and F. J. de Heer, *Chem. Phys.*, **7**, 130 (1975).
9. F. J. de Heer, *Int. J. Radiat. Phys. Chem.*, **7**, 137 (1975).
10. M. Spiteller-Friedmann, S. Eggers and G. Spiteller, *Monatsh. Chem.*, **95**, 1740 (1964).
11. A. Maccoll, *Org. Mass Spectrom.*, **21**, 601 (1986).
12. J. C. Schug, *J. Chem. Phys.*, **38**, 2610 (1963).
13. V. N. Komarov and A. M. Novosel'tsev, *Russ. J. Phys. Chem.*, **48**, 1380 (1974).
14. R. D. Bowen and A. Maccoll, *Org. Mass Spectrom.*, **18**, 576 (1983).
15. B. Steiner, C. F. Giese and M. G. Inghram, *J. Chem. Phys.*, **34**, 189 (1961).
16. R. Alberti, M. M. Genoni, C. Pascual and J. Vogt, *Int. J. Mass Spectrom. Ion Phys.*, **14**, 89 (1974).
17. D. P. Stevenson, *J. Chem. Soc., Discuss. Faraday Soc.*, **10**, 35 (1951).
18. A. G. Harrison, C. D. Finney and J. A. Sherk, *Org. Mass Spectrom.*, **5**, 1313 (1971).
19. N. Ohmichi, Y. Malinovich, J. P. Ziesel and C. Lifshitz, *J. Phys. Chem.*, **93**, 2491 (1989).
20. C. Lifshitz and N. Ohmichi, *J. Phys. Chem.*, **93**, 6329 (1989).
21. C. Lifshitz, *Mass Spectrom.*, **10**, 1 (1989).
22. H. D. Beckey, *Principles of Field Ionization and Field Desorption Mass Spectrometry*. Pergamon Press, Oxford, 1977.
23. H. D. Beckey, *Field Ionization Mass Spectrometry*, Pergamon Press, Oxford, 1971.
24. A.-M. Falick, P. J. Derrick and A. L. Burlingame, *Int. J. Mass Spectrom. Ion, Phys.*, **12**, 114 (1973).
25. P. J. Derrick, R. P. Morgan, J. T. Hill and M. A. Baldwin, *Int. J. Mass Spectrom. Ion, Phys.*, **18**, 393 (1975).
26. F. Okuyama, F. W. Rollgen, and H. D. Beckey, *Z. Naturforsch.*, **A28**, 60 (1973).
27. H. D. Beckey, H. Hey, K. Levsen and G. Tenschert, *Int. J. Mass Spectrom. Ion Phys.*, **2**, 101 (1969).
28. I. Hertel and C. Ottinger, *Z. Naturforsch.*, **20a**, 1708 (1967).
29. P. Goldberg, J. A. Hopkinson, A. Mathias and A. E. Williams, *Org. Mass Spectrom.*, **3**, 1009 (1970).
30. S. Meyerson, *J. Chem. Phys.*, **42**, 2181 (1965).
31. R. Liardon and T. Gäumann, *Helv. Chim. Acta*, **52**, 1042 (1969).
32. R. Fuchs, *Int. J. Mass Spectrom. Ion Phys.*, **8**, 193 (1972).
33. H. Chatham, D. Hils, R. Robertson and A. Gallagher, *J. Chem. Phys.*, **81**, 1770 (1984).
34. T. Gäumann and J. Rapin, unpublished results.
35. J. Appell and C. Kubach, *Chem. Phys. Lett.*, **11**, 486 (1971).
36. W. A. Chupka, *J. Chem. Phys.*, **48**, 2337 (1968).
37. J. G. Smith, *Phys. Rev.*, **51**, 263 (1937).
38. P. Plessis, P. Marmet and R. Dutil, *J. Phys. B*, **16**, 1283 (1983).
39. M. W. Evans, N. Bauer and Y. J. Beach, *J. Chem. Phys.*, **14**, 701 (1946).
40. R. P. Clow and J. H. Futrell, *Int. J. Mass Spectrom. Ion Phys.*, **4**, 165 (1970). *J. Chem. Phys.*, **59**, 4061 (1973).
41. D. H. Lenz and W. M. C. Conner, Jr., *Anal. Chem. Acta*, **173**, 227 (1985).
42. V. N. Komarov and M. V. Tikhomirov, *Russian J. Phys. Chem.*, **46**, 766 (1972).
43. P. A. Hatherly, M. Stankiewicz, L. J. Frasinski, K. Codling and M. A. Macdonald, *Chem. Phys. Lett.*, **159**, 355 (1989).
44. B. Brehm, *Z. Naturforsch.*, **21a**, 196 (1966).
45. H. Dibeler, M. Kraus, R. M. Reese and F. N. Harlee, *J. Chem. Phys.*, **42**, 3791 (1965).
46. W. J. Bouma, D. Poppinger and L. Radom, *Isr. J. Chem.*, **23**, 21 (1983).
47. L. B. Knight, Jr., J. Steadman, D. Feller and E. R. Davidson, *J. Am. Chem. Soc.*, **106**, 3700 (1984).
48. W. A. Chupka and J. Berkowitz, *J. Chem. Phys.*, **47**, 2921 (1967).
49. W. J. Chesnavich, *J. Chem. Phys.*, **84**, 2615 (1986).
50. A. Mackenzie Peers and J. Milhaud, *Int. J. Mass Spectrom. Ion Phys.*, **15**, 145 (1974).
51. C. E. Melton and H. M. Rosenstock, *J. Chem. Phys.*, **26**, 568 (1957).
52. V. H. Dibeler and H. M. Rosenstock, *J. Chem. Phys.*, **39**, 1326 (1963).

53. C. Ottinger, Z. Naturforsch., **20a**, 1232 (1965).
54. R. D. Smith and J. H. Futrell, Int. J. Mass Spectrom. Ion Phys., **20**, 4257 (1976).
55. B. H. Solka, J. H. Beynon and R. G. Cooks, J. Phys. Chem., **79**, 859 (1975).
56. C. E. Klots, J. Phys. Chem., **75**, 1526 (1971).
57. C. E. Klots, Chem. Phys. Lett., **10**, 422 (1971).
58. M. L. Vestal, in Fundamental Processes in Radiation Chemistry (Ed. P. Ausloos), Wiley–Interscience, New York, 1968, pp. 67–69.
59. C. E. D. Ouwerkerk, S. A. McLuckey, P. G. Kistemaker and A. J. H. Boerboom, Int. J. Mass Spectrom. Ion Proc., **56**, 11 (1984).
60. D. Stahl, F. Maquin, T. Gäumann, H. Schwarz, P.-A. Carrupt and P. Vogel, J. Am. Chem. Soc., **107**, 5049 (1985).
61. A. W. Hanner and T. F. Moran, Org. Mass Spectrom., **16**, 512 (1981).
62. M. W. Wong and L. Radom, J. Am. Chem. Soc., **111**, 1155 (1989).
63. F. H. Field, J. L. Franklin and M. S. B. Munson, J. Am. Chem. Soc., **85**, 3575 (1963).
64. G. A. W. Derwish, A. Galli, A. Giardini-Guidoni and G. G. Volpi, J. Chem. Phys., **40**, 12 (1964).
65. F. H. Field and M. S. B. Munson, J. Am. Chem. Soc., **87**, 3289 (1965).
66. M. S. B. Munson and F. H. Field, J. Am. Chem. Soc., **87**, 3294 (1965).
67. M. Inoue and S. Wexler, J. Am. Chem. Soc., **91**, 5730 (1969).
68. M. S. B. Munson and F. H. Field, J. Am. Chem. Soc., **91**, 3413 (1969).
69. S.-L. Chong and J. L. Franklin, J. Chem. Phys., **55**, 641 (1971).
70. W. T. Huntress Jr., J. B. Laudenslager and R. F. Pinizzotto Jr., Int. J. Mass Spectrom. Ion Phys., **13**, 331 (1974).
71. A. Fiaux, D. L. Smith and J. H. Futrell, Int. J. Mass Spectrom. Ion Phys., **15**, 9 (1974).
72. M. French and P. Kebarle, Can. J. Chem., **53**, 2268 (1975).
73. R. Houriet and T. Gäumann, Helv. Chim. Acta, **59**, 107 (1976).
74. D. Smith and N. G. Adams, Int. J. Mass Spectrom. Ion Phys., **23**, 123 (1977).
75. R. Houriet, G. Parisod and T. Gäumann, J. Am. Chem. Soc., **99**, 3599 (1977).
76. R. Houriet and T. Gäumann, Int. J. Mass Spectrom. Ion Phys., **28**, 93 (1978).
77. G. Parisod and T. Gäumann, Adv. Mass Spectrom., **7**, 1402 (1978).
78. R. Houriet and T. Gäumann, Lecture Notes in Chemistry, **7**, 174 (1978).
79. R. N. Abernathy and F. W. Lampe, Int. J. Mass Spectrom. Ion Phys., **41**, 7 (1981).
80. J. K. Kim, V. G. Anicich and W. T. Huntress Jr., J. Phys. Chem., **81**, 1798 (1977).
81. M. Henchmann, D. Smith, N. G. Adams, J. F. Paulson and Z. Herman, Int. J. Mass Spectrom. Ion Proc., **92**, 15 (1989).
82. Z. Herman, P. Hieri, A. Lee and R. Wolfgang, J. Chem. Phys., **51**, 454 (1969).
83. M. D. Sefcik, J. M. S. Henis and P. P. Gaspar, J. Chem. Phys., **61**, 4321 (1974).
84. J. M. S. Henis and M. K. Tripodi, J. Chem. Phys., **61**, 4863 (1974).
85. W. T. Huntress, Jr., J. Chem. Phys., **56**. 5111 (1972).
86. E. V. Aparina, N. V. Kir'yakov, M. I. Markin and V. L. Tal'roze, Khimiya Vysokikh Energii, Engl. transl., **9**, 1 (1975).
87. E. V. Aparina, N. V. Kir'yakov, M. I. Martin and V. L. Tal'roze, Khimiya Vysokikh Energii, Engl. transl., **9**, 98 (1975).
88. G. C. Shields, L. Wennberg, J. B. Wilcox and T. F. Moran, Org. Mass Spectrom., **21**, 137 (1986).
89. E. E. C. A. Hop, J. L. Holmes, M. W. Wong and L. Radom, Chem. Phys. Lett., **159**, 580 (1989).
90. H. Arai, S. Nagai, K. Matsuda and M. Hatada, Radiat. Phys. Chem., **17**, 151 (1981).
91. H. Arai, S. Nagai, K. Matsuda and M. Hatada, Radiat. Phys. Chem., **17**, 217 (1981).
92. D. Bhattacharya, H.-Y. Wang and J. E. Willard, J. Phys. Chem., **85**, 1310 (1981).
93. J. C. Lorquet, Org. Mass Spectrom., **16**, 11 (1981).
94. G. Bieri, F. Burger, E. Heilbronner and J. P. Maier, Helv. Chim. Acta, **60**, 2213 (1977).
95. M. Meot-Ner (Mautner), L. W. Sieck and P. Ausloos, J. Am. Chem. Soc., **103**, 5342 (1981).
96. V. T. Amorebieta and A. J. Colussi, J. Phys. Chem., **88**, 4284 (1984).
97. V. T. Amorebieta and A. J. Colussi, J. Phys. Chem., **89**, 4664 (1985).
98. J. L. Holmes, Org. Mass Spectrom., **20**, 169 (1985).
99. P. Vogel, Carbocation Chemistry, Elsevier, Amsterdam, 1985.
100. K. Watanabe, J. Chem. Phys., **26**, 242 (1957).
101. N. Dinh-Nguyen, R. Ryhage, S. Ställberg-Stenhagen and E. Stenhagen, Arkiv Kemi, **18**, 393 (1961).
102. R. D. Bowen and D. H. Williams, Int. J. Mass Spectrom. Ion Phys., **29**, 47 (1979).

T. Gäumann

103. R. D. Bowen and D. Williams, *J. Am. Chem. Soc.*, **102**, 2752 (1980).
104. J. F. Wendelboe, R. D. Bowen and D. H. Williams, *J. Am. Chem. Soc.*, **103**, 2333 (1981).
105. J. F. Wendelboe and R. D. Bowen, *Org. Mass Spectrom.*, **17**, 439 (1982).
106. T. Takeuchi, M. Yamamoto, K. Nishimoto, H. Tanaka and K. Hirota, *Int. J. Mass Spectrom. Ion Phys.*, **52**, 139 (1983).
107. T. Takeuchi, M. Yamamoto and K. Nishimoto, *Adv. Mass Spectrom.*, **10B**, 783 (1986).
108. J. L. Holmes, P. Wolkoff and R. T. B. Rye, *J. Chem. Soc., Chem. Commun.*, 544 (1979).
109. R. D. Smith and J. H. Futrell, *Org. Mass Spectrom.*, **1**, 309 (1976).
110. A. Maccoll, *Org. Mass Spectrom.*, **17**, 1 (1982).
111. G. D. Flesch and H. J. Svec, *J. Chem. Soc., Faraday Trans. 1*, **69**, 1187 (1973).
112. W. A. Chupka and M. Kaminsky, *J. Chem. Phys.*, **35**, 1991 (1961).
113. H. M. Rosenstock, *Adv. Mass Spectrom.*, **4**, 523 (1967).
114. G. Khodadadi, R. Botter and H. M. Rosenstock, *Int. J. Mass Spectrom. Ion Phys.*, **3**, 397 (1969).
115. H. M. Rosenstock, K. Draxl, B. W. Steiner and J. T. Herron, *J. Phys. Chem., Ref. Data*, Suppl. 1, **6** (1977).
116. P. Wolkoff and J. L. Holmes, *J. Am. Chem. Soc.*, **100**, 7346 (1978).
117. J. L. Holmes, F. P. Lossing, and A. Maccoll, *J. Am. Chem. Soc.*, **110**, 7339 (1988).
118. D. P. Stevenson and C. D. Wagner, *J. Chem. Phys.*, **19**, 11 (1951).
119. J. R. Nesby, C. M. Drew and A. S. Gordon, *J. Phys. Chem.*, **59**, 988 (1955).
120. W. H. McFadden and A. L. Wahrhaftig, *J. Am. Chem. Soc.*, **78**, 1572 (1956).
121. N. D. Coggeshall, *J. Chem. Phys.*, **37**, 2167 (1962).
122. J. C. Schug, *J. Chem. Phys.*, **40**, 1283 (1964).
123. W. A. Brand and T. Baer, *J. Am. Chem. Soc.*, **106**, 3154 (1984).
124. J. L. Holmes and J. K. Terlouw, *Org. Mass Spectrom.*, **10**, 787 (1975).
125. R. D. Bowen, M. P. Barbalas, F. P. Pagano, P. J. Todd and F. W. McLafferty, *Org. Mass Spectrom.*, **15**, 51 (1980).
126. M. A. Haney and J. L. Franklin, in *Recent Developments in Mass Spectrometry* (Eds. K. Ogata and T. Hayakawa), p. 909, University Park Press, Baltimore, 1971.
127. P. J. Derrick, A.-M. Falick and A. L. Burlingame, *J. Chem. Soc., Perkin Trans. 2*, 98 (1975).
128. P. J. Derrick, A.-M. Falick and A. L. Burlingame, *J. Chem. Soc., Faraday Trans. 1*, 1503 (1975).
129. J. Sunner and I. Szabó, *Int. J. Mass Spectrom. Ion Phys.*, **25**, 241 (1977).
130. P. T. Mead, K. F. Donchi, J. C. Traeger, J. R. Christie and P. J. Derrick, *J. Am. Chem. Soc.*, **102**, 3364 (1980).
131. A. D. Baker, D. Betteridge, N. P. Kemp and R. E. Kirby, *J. Mol. Struct.*, **8**, 75 (1971).
132. J. N. Murrell and W. Schmidt, *J. Chem. Soc., Faraday Trans. 2*, **68**, 1709 (1972).
133. A. W. Potts and D. G. Streets, *J. Chem. Soc., Faraday Trans. 2*, **70**, 875 (1974).
134. J. L. Holmes, P. C. Burgers, M. Y. A. Mollah and P. Wolkoff, *J. Am. Chem. Soc.*, **104**, 2879 (1982).
135. M. V. Gur'ev, M. V. Tikhomirov and N. N. Tunitskii, *Dokl. Akad. Nauk SSSR*, **123**, 120 (1958); *Zh. Fiz. Khim.*, **32**, 2731 (1958).
136. I. V. Gol'denfel'd and I. Z. Korostyshevskii, *Russ. J. Phys. Chem.*, **43**, 1453 (1969).
137. C. Corolleur, S. Corolleur and F. G. Gault, *Bull. Soc. Chim. France*, 158 (1970).
138. J. H. Beynon and R. G. Cooks, *Adv. Mass Spectrom.*, **6**, 835 (1974).
139. R. Liardon and T. Gäumann, *Helv. Chim. Acta*, **52**, 528 (1969).
140. M. Bensimon, T. Gäumann, C. Guenat and D. Stahl, unpublished results.
141. J. E. Monahan and H. E. Stanton, *J. Chem. Phys.*, **37**, 2654 (1962).
142. K. Toriyama, K. Nunome and M. Iwasaki, *J. Chem. Phys.*, **77**, 5891 (1982).
143. S. K. Searles, L. W. Sieck and P. Ausloos, *J. Chem. Phys.*, **53**, 849 (1970).
144. K. Levsen, *Org. Mass Spectrom.*, **10**, 43 (1975).
145. K. Levsen, H. Heimbach, G. J. Shaw and G. W. A. Milne, *Org. Mass Spectrom.*, **12**, 663 (1977).
146. P. Krenmayer, F. Ksica and K. Varmuza, *Monatsh. Chem.*, **109**, 823 (1978).
147. P. Wolkoff, *Adv. Mass Spectrom.*, **10B**, 783 (1986).
148. T. Gäumann, unpublished results.
149. A. Lavanchy, R. Houriet and T. Gäumann, *Org. Mass Spectrom.*, **13**, 410 (1978).
150. R. D. Bowen and D. H. Williams, *J. Chem. Soc., Perkin Trans. 2*, 1479 (1976).
151. A. Lavanchy, R. Houriet and T. Gäumann, *Org. Mass Spectrom.*, **14**, 79 (1979).
152. P. Wolkoff, J. L. Holmes and F. P. Lossing, *Org. Mass Spectrom.*, **20**, 14 (1985).

153. I. Howe, *Org. Mass Spectrom.*, **10**, 767 (1975).
154. J. C. Antunes Marques, T. Gäumann, A. Heusler, R. Houriet and A. Lavanchy, to appear.
155. J. C. Antunes Marques, T. Gäumann and R. Houriet, *Adv. Mass Spectrom.*, **10B**, 761 (1986).
156. G. Remberg, E. Remberg, M. Spiteller-Friedmann and G. Spiteller, *Org. Mass Spectrom.*, **1**, 87 (1968).
157. D. G. Hall and T. H. Morton, *J. Am. Chem. Soc.*, **102**, 5686 (1980).
158. K. Levsen, *Org. Mass Spectrom.*, **10**, 55 (1975).
159. M. A. Baldwin, F. W. McLafferty and D. M. Jering, *J. Am. Chem. Soc.*, **97**, 6169 (1975).
160. M. Poretti, J. Rapin and T. Gäumann, *Int. J. Mass Spectrom. Ion Proc.*, **72**, 187 (1986).
161. M. Bensimon, J. Rapin and T. Gäumann, *Int. J. Mass Spectrom. Ion Proc.*, **72**, 125 (1986).
162. L. W. Sieck, R. Gordon Jr. and P. Ausloos, *J. Am. Chem. Soc.*, **94**, 7157 (1972).
163. F. W. McLafferty, M. P. Barbalas and F. Tureček, *J. Am. Chem. Soc.*, **105**, 1 (1983).
164. S.-L. Chong and J. L. Franklin, *J. Am. Chem. Soc.*, **94**, 6347 (1972).
165. H. Schwarz, *Org. Mass Spectrom.*, **15**, 10 (1980).
166. W. J. Bouma, D. Poppinger and L. Radom, *J. Mol. Struct.*, **103**, 209 (1983).
167. K. Matsura, K. Nunome, M. Okazaki, K. Toriyama and M. Iwasaki, *J. Phys. Chem.*, **93**, 6642 (1989).
168. D. P. Stevenson, *J. Am. Chem. Soc.*, **80**, 1571 (1958).
169. S. Meyerson, T. D. Nevitt and P. N. Rylander, *Adv. Mass Spectrom.*, **2**, 313 (1963).
170. A. M. Falick and A. L. Burlingame, *J. Am. Chem. Soc.*, **97**, 1525 (1975).
171. K. Hirota and Y. Niwa, *J. Phys. Chem.*, **72**, 5 (1968).
172. V. N. Komarov, M. V. Tikhomirov and N. M. Tunitskii, *Khim. Vysokikh Energii*, Engl. transl., **3**, 374 (1969); *Russ. J. Phys. Chem.*, **47**, 439 (1973).
173. L. W. Sieck, M. Meot-Ner (Mautner) and P. Ausloos, *J. Am. Chem. Soc.*, **102**, 6866 (1980).
174. H. Schwarz, W. Franke, J. Chandrasekhar and P. v. R. Schleyer, *Tetrahedron*, **35**, 1969 (1969)
175. M. Colosimo and R. Bucci, *J. Chem. Soc., Chem. Commun.*, 659 (1981).
176. W. Franke, H. Schwarz, H. Thies, J. Chandrasekhar, P. v. R. Schleyer, W. J. Hehre, M. Saunders and G. Walker, *Angew. Chem., Int. Ed. Engl.*, **19**, 6 (1980).
177. V. Amir-Ebrahimi and F. G. Gault, *Org. Mass Spectrom.*, **10**, 711 (1975).
178. W. Feng, A. Heusler and T. Gäumann, unpublished results.
179. S. W. Benson, *The Foundations of Chemical Kinetics*, McGraw-Hill, New York, 1960, p. 263.
180. C. Wesdemiotis, R. Wolfschütz and H. Schwarz, *Tetrahedron*, **36**, 275 (1979).
181. M. Attina, F. Cacace and A. di Marzio, *J. Am. Chem. Soc.*, **111**, 6004 (1989).
182. G. A. Olah and J. Lukas, *J. Am. Chem. Soc.*, **90**, 933 (1968).
183. M. Saunders and J. Rosenfeld, *J. Am. Chem. Soc.*, **91**, 7756 (1969).
184. R. N. Rej, E. Bacon and G. Eadon, *J. Am. Chem. Soc.*, **101**, 1668 (1979).
185. R. P. Kirchen and T. S. Sorensen, *J. Am. Chem. Soc.*, **101**, 3240 (1979).
186. J. Stieglitz, *J. Am. Chem. Soc.*, **21**, 101 (1899).
187. J. G. Traynham, *J. Chem. Educ.*, **66**, 6 (1989).
188. A. Herlan, *Org. Mass Spectrom.*, **4**, 425 (1970).
189. C. P. Johnson and A. Langer, *J. Phys. Chem.*, **61**, 1010 (1957).
190. H. M. Grubb and S. Meyerson, in *Mass Spectrometry of Organic Ions* (Ed. F. W. McLafferty), Chap. 10, Academic Press, New York, 1963.
191. C. N. McEwen and M. A. Rudat, *J. Am. Chem. Soc.*, **101**, 6470 (1979).
192. C. N. McEwen and M. A. Rudat, *J. Am. Chem. Soc.*, **103**, 4343 (1981).
193. M. A. Rudat and C. N. McEwen, *J. Am. Chem. Soc.*, **103**, 4349 (1981).
194. C. N. McEwen and M. A. Rudat, *J. Am. Chem. Soc.*, **103**, 4355 (1981).
195. C. N. McEwen, *Mass Spectrometric Reviews*, **5**, 521 (1986).
196. J. W. de M. Carneiro, P. v. R. Schleyer, W. Koch and K. Raghavachari, *J. Am. Chem. Soc.*, **112**, 4064 (1990).
197. R. Liardon and T. Gäumann, *Helv. Chim. Acta*, **54**, 1968 (1971).
198. B. Davis, D. H. Williams and A. N. H. Yeo, *J. Chem. Soc. (B)*, 81 (1970).
199. D. M. Shold and P. Ausloos, *J. Am. Chem. Soc.*, **100**, 7915 (1978).
200. J. L. Holmes, A. D. Osborne and G. M. Weese, *Org. Mass Spectrom.*, **10**, 867 (1975).
201. S. G. Lias, R. E. Rebbert and P. Ausloos, *J. Am. Chem. Soc.*, **92**, 6430 (1970).
202. A. Fiaux, D. L. Smith and J. H. Futrell, *Int. J. Mass Spectrom. Ion Phys.*, **25**, 281 (1977).
203. D. H. Parker and R. B. Bernstein, *J. Phys. Chem.*, **86**, 60 (1982).

454   T. Gäumann

204. H. Kühlewind, H. J. Neusser and E. W. Schlag, *J. Phys. Chem.*, **89**, 5600 (1985).
205. J. Silberstein, N. Ohmichi and R. D. Levine, *J. Phys. Chem.*, **89**, 5606 (1985).
206. H.-W. Leung, Ch. W. Tsang and A. G. Harrison, *Org. Mass Spectrom.*, **11**, 668 (1974).
207. A. Maquestiau, R. Flammang and P. Meyrant, *Int. J. Mass Spectrom. Ion Phys.*, **44**, 267 (1982).
208. M. Meot-Ner (Mautner), *J. Am. Chem. Soc.*, **109**, 7947 (1987).
209. M. Bensimon, T. Gäumann, C. Guenat and D. Stahl, *Adv. Mass Spectrom.*, **10B**, 807 (1986).
210. O. Osberghaus and C. Ottinger, *Phys. Lett.*, **16**, 121 (1965).
211. C. Ottinger, *Z. Naturforsch.*, **22a**, 20 (1967).
212. G. A. Olah and A. M. White, *J. Am. Chem. Soc.*, **78**, 5799 (1956).
213. D. V. Dearden and J. L. Beauchamp, *J. Phys. Chem.*, **89**, 5359 (1985).
214. A. Fiaux, B. Wirz and T. Gäumann, *Helv. Chim. Acta*, **57**, 525 (1974).
215. S. G. Lias, J. R. Eyler and P. Ausloos, *Int. J. Mass Spectrom. Ion Phys.*, **19**, 219 (1976).
216. H. Hiraoka and P. Kebarle, *J. Am. Chem. Soc.*, **98**, 6119 (1976).
217. R. Houriet and T. Gäumann, *Helv. Chim. Acta*, **59**, 119 (1976).
218. D. Stahl and T. Gäumann, *Org. Mass Spectrom.*, **12**, 761 (1977).
219. D. Stahl, C. Guenat and T. Gäumann, *Adv. Mass Spectrom.*, **8A**, 748 (1980).
220. D. Stahl and T. Gäumann, *Adv. Mass Spectrom.*, **7**, 1190 (1978).
221. D. Stahl and T. Gäumann, unpublished results.
222. M. Saunders and M. R. Kates, *J. Am. Chem. Soc.*, **100**, 7082 (1978).
223. C. Guenat, D. Stahl and T. Gäumann, *Proc. 34th Ann. Conf. Am. Soc. Mass Spectrom.*, 554 (1986).
224. D. Stahl, R. Houriet and T. Gäumann, *Z. Phys. Chem.*, **113**, 231 (1978).
225. a.) R. D. Bowen and D. H. Williams, *Org. Mass Spectrom.*, **12**, 453 (1970); b.) W. G. Cole and D. H. Williams, *J. Chem. Soc., Chem. Comm.*, 784 (1969).
226. a.) J. Long and M. S. B. Munson, *J. Am. Chem. Soc.*, **94**, 3339 (1972); b.) H.-H. Büker, D. Kuck and H.-Fr. Grützmacher, *Adv. Mass Spectrom.*, **11A**, 888 (1989).
227. W. H. McFadden, *J. Phys. Chem.*, **67**, 1074 (1963).
228. C. E. Hudson and D. J. McAdoo, *Int. J. Mass Spectrom. Ion Proc.*, **59**, 325 (1984).
229. P. Tecon, D. Stahl and T. Gäumann, *Int. J. Mass Spectrom. Ion Phys.*, **27**, 83 (1978).
230. P. Tecon, D. Stahl and T. Gäumann, *Int. J. Mass Spectrom. Ion Phys.*, **29**, 363 (1979).
231. D. Stahl and T. Gäumann, *Int. J. Mass Spectrom. Ion Phys.*, **28**, 267 (1978).
232. P. Tecon, D. Stahl and T. Gäumann, *Adv. Mass Spectrom.*, **8A**, 843 (1980).
233. J. C. Antunes Marques, D. Stahl and T. Gäumann, *Int. J. Mass Spectrom. Ion Phys.*, **47**, 101 (1983).
234. M. Bensimon, J. C. Antunes Marques, D. Stahl and T. Gäumann, unpublished results.
235. J. C. Antunes Marques, A.-M. Falick, A. Heusler, D. Stahl, P. Tecon and T. Gäumann, *Helv. Chim. Acta*, **67**, 425 (1984).
236. A.-M. Falick, T. Gäumann, A. Heusler, H. Hirota, D. Stahl and P. Tecon, *Org. Mass Spectrom.*, **15**, 440 (1980).
237. M. Bensimon, T. Gäumann, and J. Rapin, *Adv. Mass Spectrom.*, **10B**, 977 (1986).
238. T. Nishishita and F. W. McLafferty, *Org. Mass Spectrom.*, **12**, 75 (1977).
239. R. D. Bowen and D. H. Williams, *Org. Mass Spectrom.*, **12**, 453 (1970).
240. F. Borchers, K. Levsen, H. Schwarz, Ch. Wesdemiotis and H. U. Winkler, *J. Am. Chem. Soc.*, **99**, 6359 (1977).
241. H. M. Rosenstock, Ph.D. thesis, University of Utah, 1952.
242. C. Lifshitz, *Adv. Mass Spectrom.*, **11A**, 713 (1989).
243. R. G. Gilbert and S. C. Smith, *Theory of Unimolecular and Recombination Reactions*, Blackwell Scientific Publications, Oxford, 1990.
244. T.-K. Ha and H. H. Günthard, *Chem. Phys.*, **134**, 203 (1989).
245. K. Levsen and J. Heimbrecht, *Org. Mass Spectrom.*, **12**, 131 (1977).
246. J. C. Lorquet, *Discuss. Faraday Soc.*, **35**, 83 (1963).
247. S. Meyerson, T. D. Nevitz and P. N. Rylander, *Adv. Mass Spectrom.*, **1**, 313 (1959).

CHAPTER **10**

The photoelectron spectra of saturated hydrocarbons

E. HEILBRONNER

Grütstrasse 10, CH-8704 Herrliberg, Switzerland

The Chemistry of Alkanes and Cycloalkanes
Edited by S. Patai and Z. Rappoport © 1992 John Wiley & Sons Ltd

E. Heilbronner

I. INTRODUCTION

It's déjà vu, all over again.

Yogi Berra

The development of a heuristically and semi-quantitatively useful quantum-chemical model for saturated hydrocarbons took considerably longer than for unsaturated and aromatic π systems. For the latter such models were available by the mid-1930s in the form of Pauling's version of the VB ($=$ Valence Bond) treatment[1] and of Hückel's HMO ($=$ Hückel Molecular Orbital) model[2]. The reason is that the physical and chemical properties of such π systems are strongly topography-dependent, i.e. they are governed by cooperative effects which are critical functions of the connectivity between the 2p centres. Thus, the thermal, spectroscopic and chemical properties of the two C_6H_6 isomers benzene and fulvene differ dramatically, as do those of the $C_{10}H_8$ isomers naphthalene and azulene[3], although the latter both satisfy Hückel's '$2+4n$ rule'. Furthermore, these properties are easy to observe in the case of unsaturated and aromatic molecules, and a wealth of data was available, waiting to be interpreted.

This was not the case for saturated hydrocarbons C_nH_m, whose properties—as far as they were known—follow mainly additive schemes which are by and large topography-independent. A good example are their standard molar enthalpies of formation $\Delta_f H_m^{\ominus}$. For saturated hydrocarbons in the gas phase and at 298 K, the $\Delta_f H_m^{\ominus}(C_nH_m(g))$ values are, in a first approximation, the sum of molar bond increments[†]. Their electronic excitation gives rise to absorption bands in the far-UV region of the spectrum which, at the time, were not only difficult to observe but also to assign, because they are not well defined intra-valence transitions, such as the $\pi^* \leftarrow \pi$ transitions of π systems, but exhibit considerable Rydberg character. Their chemical behaviour showed none of the distinctive topography-dependent behaviour of aromatic molecules, e.g. ease and position specificity of substitution reactions, and was rather well summed up by '*parum affinis*'. Finally, the first ionization energies corresponding to the generation of the radical cations $C_nH_m^{+\cdot}$ in their electronic ground state were known to depend primarily on the size of the molecule, i.e. on the number n of carbon atoms, as shown in Figure 1, where ionization energies known up to 1969[6] from photoionization experiments have been used.

This striking difference in behaviour lead to the conclusion that electrons in π systems are strongly 'delocalized', whereas σ systems are best described in terms of 'localized' two-centre σ-bond orbitals, each occupied by an electron pair. For an understanding of the properties of unsaturated and aromatic molecules such theoretical models as the

---

[†]$\Delta_{bond} H_m^{\ominus}(CC) \approx -344 \text{ kJ mol}^{-1}$ and $\Delta_{bond} H_m^{\ominus}(CH) \approx -415 \text{ kJ mol}^{-1}$, i.e. $\Delta_f H_m^{\ominus}(C_nH_m(g)) = \frac{1}{2}(4n-m)\Delta_{bond} H_m^{\ominus}(CC) + m\Delta_{bond} H_m^{\ominus}(CH) - n\Delta_f H_m^{\ominus}(C(g)) - m\Delta_f H_m^{\ominus}(H(g))[4]$. (This additivity scheme can be improved by taking into account the slight dependence on local connectivity, i.e. by using molar group increments. $\Delta_{group} H_m^{\ominus}(CH_3) \approx -42.2 \text{ kJ mol}^{-1}$, $\Delta_{group} H_m^{\ominus}(CH_2) \approx -20.7 \text{ kJ mol}^{-1}$, $\Delta_{group} H_m^{\ominus}(CH) \approx -8.0 \text{ kJ mol}^{-1}$, $\Delta_{group} H_m^{\ominus}(C) \approx +2.1 \text{ kJ mol}^{-1}$, the sum of which yields $\Delta_f H_m^{\ominus}(C_nH_m(g))[5]$.

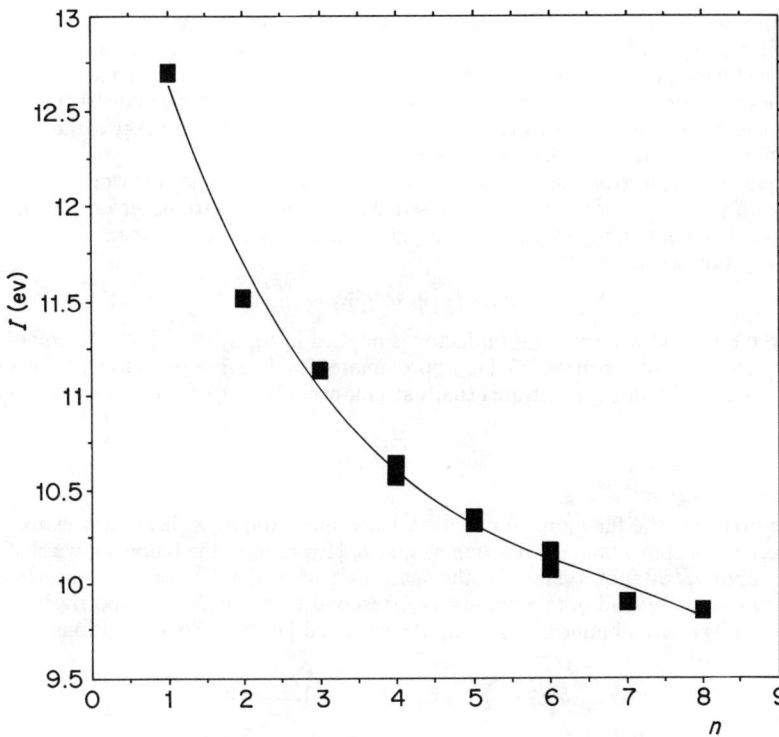

FIGURE 1. Dependence of the first ionization energies I_1 (in eV)[6] of saturated, linear hydrocarbons C_nH_{2n+2} on the number n of carbon atoms: $n = 1$, methane; $n = 2$, ethane; $n = 3$, propane; $n = 4$, butane, isobutane; $n = 5$, pentane, isopentane, neopentane; $n = 6$, hexane, isohexane, 3-ethylbutane; $n = 7$, heptane; $n = 8$, isooctane

VB or the HMO treatment were necessary, in contrast to the saturated molecules, which could be discussed satisfactorily within purely additive schemes. Although this opinion had been challenged by theoreticians rather early in the game, the realization that electrons occupying σ orbitals in saturated molecules are as 'delocalized' as those in π orbitals of unsaturated molecules gained ground mainly after photoelectron (PE) spectroscopic studies had provided a more detailed insight into the orbital structure of saturated molecules, in particular of hydrocarbons.

II. MOLECULAR ORBITAL MODELS

A. General Remarks

Nowadays excellent many-electron treatments are available in computer-ready form for the calculation of the electronic structure of molecules, in particular of saturated hydrocarbons. As far as our topic is concerned, they are of sufficient precision for a heuristically useful prediction of their ionization energies, and in many cases for a relevant interpretation and assignment of their PE spectra. The only input needed for such treatments, which are either of the *ab initio*[7] or semi-empirical[8] type, is the geometric structure of the hydrocarbon. For most purposes this information can be obtained with sufficient precision from standard molecular models, or—e.g. in the case of polycyclic, strained hydrocarbons—from molecular force field calculations. Only in particular cases, e.g. for [1.1.1]propellane, is it necessary to use either exact experimental structural data, or to determine the structure by complete minimization of the total energy of the molecule within a reliable many-electron treatment.

These many-electron, self-consistent-field (SCF) treatments provide us with a reasonably good description of the closed-shell (singlet) electronic ground state of a hydrocarbon containing $2N$ electrons, in the form of a Slater determinant, i.e. a ground configuration written as

$$^1\Psi_0 = \| \eta_1 \bar{\eta}_1 \cdots \eta_j \bar{\eta}_j \cdots \eta_N \bar{\eta}_N \| \tag{1}$$

where the necessary normalization factor is implied in the symbol $\| \cdots \|$. In equation 1 the η_j are molecular orbitals (MO), approximated by linear combinations of a set of $M \geq N$ basis orbitals ϕ_μ, which are usually atomic orbitals (AO) or Gaussian functions

$$\eta_j = \sum_{\mu=1}^{M} \phi_\mu C_{\mu j} \tag{2}$$

More precisely, the functions in equation 1 are spin orbitals, η_j being associated with an electron of spin α, and $\bar{\eta}_j$ with one of spin β. However, in the following we shall use the symbol η_j (without a dash) for the space part only of a spin orbital. According to this convention, η_j and $\bar{\eta}_j$ in equation 1 correspond to $\eta_j\alpha$ and $\eta_j\beta$, respectively.

The MOs η_j are obtained by solving the so-called Hartree–Fock equations

$$\mathscr{F}\eta_j = \sum_{k=1}^{N} \eta_k F_{\eta,kj}, \qquad j = 1, 2, \ldots, N \tag{3}$$

where $\mathscr{F}$ is the Fock operator underlying the calculation, and where the $F_{\eta,kj}$ are defined below in equations 4 and 5. The N MOs (formula 2) form an orthonormal set, i.e. $\langle \eta_j | \eta_k \rangle = \delta_{jk}$ (δ_{jk} = Kronecker's delta; $\delta_{jj} = 1$, $\delta_{jk} = 0$ if $j \neq k$). With respect to the Fock operator $\mathscr{F}$, an energy (self-energy)

$$F_{\eta,jj} = \langle \eta_j | \mathscr{F} | \eta_j \rangle \tag{4}$$

is associated with each MO η_j, and a crossterm

$$F_{\eta,jk} = \langle \eta_j | \mathscr{F} | \eta_k \rangle \tag{5}$$

with each MO pair $\eta_j, \eta_k, j \neq k$. The matrix elements 4 and 5 can be collected in a square $N \times N$ Hartree–Fock matrix

$$\mathbf{F}_\eta = (F_{\eta,jk}) \tag{6}$$

Apart from the Slater determinant (formula 1), the treatment yields values for observable quantities, such as the total electron energy E_{el} of the hydrocarbon, its charge distribution, or its dipole and multipole moments. In addition, if the total energy E_{tot}, which includes E_{el} and the core repulsion energy, has been minimized with respect to changes in geometry, it will also provide estimates of the structural parameters of the hydrocarbon. Note that all these results refer to the molecule in its electronic ground state, the latter approximated by the configuration 1, and to a fixed, rigid structure.

It is important to realize that for a given molecule the set of N MOs η_j (equation 2), obtainable by solving the Hartree–Fock equations 3, is by no means unique. If we collect a particular set in a $1 \times N$ row vector $\boldsymbol{\eta} = (\eta_1 \cdots \eta_j \cdots \eta_N)$, we are allowed to transform it into a set of N new MOs $\boldsymbol{\chi} = (\chi_1 \cdots \chi_j \cdots \chi_N)$ by multiplication with an orthogonal

$N \times N$ matrix $\mathbf{U}$ according to the equation

$$\chi = \eta \mathbf{U} \tag{7}$$

The matrix $\mathbf{U}$ is an orthogonal matrix, if it satisfies the condition

$$\mathbf{U}^T\mathbf{U} = \mathbf{U}\mathbf{U}^T = \mathbf{1} \tag{8}$$

where $\mathbf{1}$ is the unit matrix of order $N \times N$. Here and in the following we assume that the MOs η_j and χ_j are real functions. More generally, if η_j and χ_j are complex, $\mathbf{U}$ would be a unitary matrix.

If condition 8 is fulfilled, it can be shown that the transformation 7 has the following important properties:

(1) The MOs χ_j of the new set $\chi = (\chi_1 \cdots \chi_j \cdots \chi_N)$ are also solutions of the Hartree–Fock equations 3.

(2) If the original MOs η_j formed an orthonormal set, then the new MOs χ_j form also an orthonormal set, i.e. $\langle \chi_j | \chi_k \rangle = \delta_{jk}$.

(3) If the Slater determinant 1, describing the electronic ground state of the hydrocarbon, is now written in terms of the new set of MOs χ_j, i.e.

$$^1\Psi_0 = \| \chi_1 \bar{\chi}_1 \cdots \chi_j \bar{\chi}_j \cdots \chi_N \bar{\chi}_N \| \tag{9}$$

then the observable quantities that can be calculated from determinant 9 are identical to those obtained from determinant 1. This means that they are invariant with respect to an orthogonal transformation (equation 7) of the MOs. On the other hand, the Hartree–Fock matrix $\mathbf{F}_\chi$, calculated in analogy to equations 4–6 on the basis of the new MOs χ_j, i.e.

$$\mathbf{F}_\chi = (F_{\chi,jk}) \quad \text{with } F_{\chi,jk} = \langle \chi_j | \mathscr{F} | \chi_k \rangle \tag{10}$$

differs from $\mathbf{F}_\eta$, $\mathbf{F}_\chi \neq \mathbf{F}_\eta$, but is related to it according to

$$\mathbf{F}_\chi = \mathbf{U}^T \mathbf{F}_\eta \mathbf{U} \tag{11}$$

In other words, all or part of the new self-energies $F_{\chi,jj}$, and of the new crossterms $F_{\chi,jk}$ with $j \neq k$, will differ from the previous ones: $F_{\chi,jj} \neq F_{\eta,jj}$, $F_{\chi,jk} \neq F_{\eta,jk}$.

Apart from the condition 8, which ensures that the new MOs χ_j form an orthonormal set, and that the observables remain invariant, we are completely free in the choice of the orthogonal transformation matrix $\mathbf{U}$. This means that for a given molecule, there is in principle a great ambiguity of equally acceptable MO sets.

However, all available computer-ready MO programs are designed to yield MOs φ_j, called canonical molecular orbitals (CMO), which have the unique property that all of their crossterms $F_{\varphi,jk}$ vanish:

$$F_{\varphi,jk} = \langle \varphi_j | \mathscr{F} | \varphi_k \rangle = 0 \quad \text{if } j \neq k \tag{12}$$

The diagonal terms (self-energies)

$$F_{\varphi,jj} = \langle \varphi_j | \mathscr{F} | \varphi_j \rangle \equiv \varepsilon_j \tag{13}$$

are then called (molecular) orbital energies, and are given the symbol ε_j. In other words, it looks as if the program contained implicitly a transformation matrix $\mathbf{U}$, such that transformation 11 yields a diagonal matrix $\mathbf{F}_\varphi = \mathbf{diag}(F_{\varphi,jj}) \equiv \mathbf{diag}(\varepsilon_j)$. The matrix $\mathbf{F}_\varphi$ is therefore given the symbol $\varepsilon = \mathbf{diag}(\varepsilon_j)$. The orbital energies ε_j are defined by

$$\varepsilon_j = h_j + \sum_{k=1}^{N} (2J_{jk} - K_{jk}) \tag{14}$$

where h_j includes the kinetic energy of the electron moving in the CMO φ_j, and its coulombic potential energy with respect to the positive core. The terms under the summation take care of the electron–electron interaction. The individual terms J_{jk} and K_{jk} are the coulombic repulsion and the exchange energy between the particular electron moving in φ_j and an electron moving in the same orbital φ_j, yielding $J_{jj} = K_{jj}$, or in one of the other CMOs φ_k of the valence shell, in which case $J_{jk} \neq K_{jk}$. (For further details the reader is referred to standard texts on quantum chemistry, e.g. those listed in References 7 and 8.)

Apart from the fact that ε is diagonal, the advantage of working with CMOs φ_j is that they are symmetry adapted, belonging to one or other of the irreducible representations of the molecular point group of the molecule, and that they are a convenient starting point for discussing 'one-electron' properties, such as the PE spectra. Conventionally, the electronic ground state (more precisely, the ground configuration) of the molecule is therefore written in terms of the CMOs φ_j as

$$^1\Psi_0 = \| \varphi_1 \bar{\varphi}_1 \cdots \varphi_j \bar{\varphi}_j \cdots \varphi_N \bar{\varphi}_N \| \tag{15}$$

Again the observable quantities that can be derived from determinant 15 are identical to those obtained from determinants 1 or 9.

From our pragmatic point of view, wishing to obtain as simple a model as possible which is still heuristically useful, it is a major disadvantage that these conventional CMOs φ_j are usually expressed as linear combinations of a rather large set of $M \gg N$ non-orthogonal basis orbitals ϕ_μ; cf linear combination 2 which, with reference to the CMOs φ_j, is now given by

$$\varphi_j = \sum_{\mu=1}^{M} \phi_\mu c_{\mu j} \tag{16}$$

This makes them rather unwieldy for the type of qualitative or semi-quantitative perturbation arguments one has learned to appreciate from dealing with HMO treatments of π systems.

Before we show how this shortcoming can be remedied by a judicious application of the transformations 7 and 11, we discuss first simple Hückel-type, independent electron treatments, that have been proposed for saturated hydrocarbons.

B. Equivalent Bond Orbital Models

The extension of independent electron treatments—e.g. of the type proposed by Hückel for π systems[2]—to sigma systems, and in particular to hydrocarbons, has a long and well-known history. The early treatments used an orthonormal basis of atomic or bond orbitals ϕ_μ with parametrized coulomb energies α_μ and interaction matrix elements $\beta_{\mu\nu}$ restricted to nearest neighbours only. The most attractive approximation of this kind, proposed by Hall and Lennard-Jones in 1951[9], is the equivalent bond orbital (EBO) model, which has been used extensively since, with variations due mainly to Lorquet, Brailsford and Ford, Herndon, Murrell and Schmidt, and Gimarc[10]. The conceptual consequences of such a treatment, in particular the phenomenon of 'σ-conjugation' in saturated hydrocarbons, have been discussed in detail by Dewar[11].

The basis functions ϕ_μ of such an EBO model are orthonormal two-centre equivalent bond orbitals (EBO) which we designate by $\phi_\mu \equiv \lambda_\mu$. Their self-energies $H_{\mu\mu}$ and interaction crossterms $H_{\mu\nu}$ are defined as follows:

$$\begin{aligned} H_{\mu\mu} &\equiv A_\mu = \langle \lambda_\mu | \mathscr{H} | \lambda_\mu \rangle \\ H_{\mu\nu} &= \langle \lambda_\mu | \mathscr{H} | \lambda_\nu \rangle \end{aligned} \tag{17}$$

with respect to an unspecified 'Hückel' hamiltonian $\mathscr{H}$. The $H_{\mu\mu} \equiv A_\mu$ and the $H_{\mu\nu}$ play the same role in an EBO treatment as the coulomb integrals α_μ and the resonance integrals $\beta_{\mu\nu}$ in the familiar π-system HMO models. In complete analogy to the traditional handling of the latter HMO parameters, the numerical values of the quantities in equations 17 are determined by calibration with respect to an appropriate set of experimental data. The calibrated matrix elements in equations 17 are collected in the $N \times N$ matrix $\mathbf{H}_\lambda = (H_{\mu\nu})$, N being the number of EBOs in the basis set, which is equal to the number of bonds in our saturated hydrocarbon C_nH_m. (This number is $N = n + m - 1 + r = 2n + m/2$, where r is the number of rings in the molecule.)

We now assume that EBOs of the same type, say the EBOs $\lambda_{CH,\mu}$ for the CH bonds in positions μ of a hydrocarbon, are equivalent in the sense that they are assigned the same self-energies $A_{CH,\mu} = A_{CH}$ independent of the location μ of the CH bond within the molecule. The same is postulated for the CC bond orbitals $\lambda_{CC,\nu}$, which are assigned a position-independent self-energy $A_{CC,\nu} = A_{CC}$. Furthermore, nearest-neighbour interaction terms $H_{\mu\nu}$ between two geminal EBOs λ_μ and λ_ν issuing from a common carbon centre, are again assumed to be fixed quantities, depending only on the type of interacting EBOs, e.g. $H_{\mu\nu} = \langle \lambda_{CH,\mu} | \mathcal{H} | \lambda_{CH,\nu} \rangle = B_{CH,CH}$, and in analogy $B_{CH,CC}$, or $B_{CC,CC}$, as long as the angle between the bonds in question does not depart too much from a tetrahedral one. Similar assumptions are made for crossterms between more distant EBOs. In other words, all the EBOs of the same type are equivalent, hence the name EBO model.

If we make the—for the moment—arbitrary, additional assumption that $A_{CH} = A_{CC} = A$, and if we restrict the crossterms $H_{\mu\nu}$ to the geminal ones, postulating $B_{CH,CH} = B_{CH,CC} = B_{CC,CC} = B$, setting all others equal to zero, then our EBO model would be formally identical to the standard two-parameter (α, β) treatment of π systems.

As a simple illustration we choose methane CH_4, which belongs to the point group T_d, so that the equivalence of the four CH EBOs λ_1, λ_2, λ_3 and λ_4 is strictly enforced by symmetry (drawing 18). If we assign a dot to each of the EBOs λ_μ, and if we link

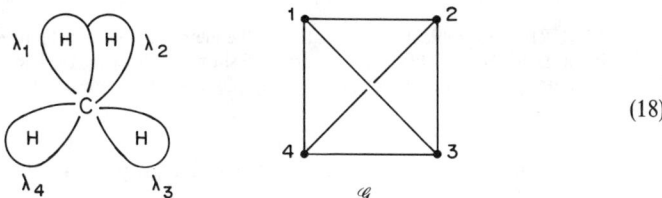

(18)

two dots (which correspond in this case necessarily to two geminal bonds) by a line, then we obtain for methane the graph $\mathcal{G}$ shown on the right side of drawing 18. As in a standard HMO treatment, we associate each dot with a self-energy $H_{\mu\mu} \equiv A_{CH,\mu} = A_{CH}$, and each line with a crossterm $H_{\mu\nu} \equiv B_{\mu\nu} = B_{CH,CH}$. If we write for short $A_{CH} \equiv A$, and $B_{CH,CH} \equiv B$, then the Hückel matrix $\mathbf{H}_\lambda = (H_{\mu\nu})$ corresponding to the methane graph $\mathcal{G}$ of drawing 18 is given by

$$\mathbf{H}_\lambda = \begin{pmatrix} A & B & B & B \\ B & A & B & B \\ B & B & A & B \\ B & B & B & A \end{pmatrix} \tag{19}$$

Diagonalization of $\mathbf{H}_\lambda$ yields the orbital energies $\varepsilon_1 = A + 3B$, $\varepsilon_2 = \varepsilon_3 = \varepsilon_4 = A - B$. Because of the high T_d symmetry of methane, the corresponding CMOs φ_j are completely predetermined, i.e. they are independent of the explicit values of the parameters A and B. One of the many possible, real representations of the methane CMOs in terms of the EBOs λ_μ is given by equations 20. These CMOs, which belong to the irreducible representations A_1 and (triply degenerate) T_2 of the point group T_d, are graphically

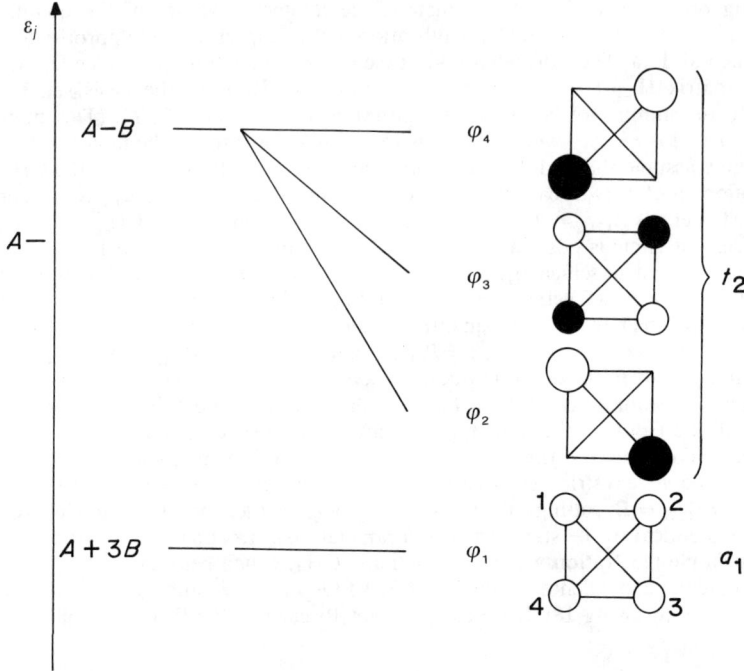

FIGURE 2. Graphical representation of the methane valence orbitals φ_j (equations 20) in terms of EBOs λ_μ, based on the graph $\mathscr{G}$ shown in drawing 18. The triply degenerate orbitals t_2 are given in one of the many possible real representations

displayed in Figure 2, based on the representative graph $\mathscr{G}$ shown in drawing 18.

$$
\begin{aligned}
\varphi_1 &= (\lambda_1 + \lambda_2 + \lambda_3 + \lambda_4)/2 \;\ldots\ldots\; a_1 \\
\varphi_2 &= (\lambda_1 \qquad - \lambda_3 \qquad)/\sqrt{2} \\
\varphi_3 &= (\lambda_1 - \lambda_2 + \lambda_3 - \lambda_4)/2 \qquad\Bigg\} \;\ldots\ldots\; t_2 \\
\varphi_4 &= (\qquad \lambda_2 \qquad - \lambda_4)/\sqrt{2}
\end{aligned}
\tag{20}
$$

In this simple form, the EBO model of a hydrocarbon C_nH_m is of interest only for the rationalization of its one-electron properties, such as the ionization energies, for the interpretation of PE spectra, and also for a transparent representation of the CMOs φ_j in terms of linear combinations of EBOs λ_μ:

$$
\varphi_j = \sum_{\mu=1}^{N} \lambda_\mu c_{\mu j}
\tag{21}
$$

as shown in equations 20 and in Figure 2 for the particular example of methane.

A closed-shell saturated hydrocarbon, with $2N$ electrons in its valence shell of N CMOs φ_j, will necessarily have N EBOs λ_μ as basis functions, as long as we do not want to include antibonding EBOs λ_μ^*. Each of the N EBOs λ_μ is occupied by two electrons to begin with. Collecting the N CMOs and N EBOs in row vectors $\boldsymbol{\varphi} = (\varphi_1 \cdots \varphi_j \cdots \varphi_N)$ and $\boldsymbol{\lambda} = (\lambda_1 \cdots \lambda_\mu \cdots \lambda_N)$, the two sets are related according to

$$
\boldsymbol{\varphi} = \boldsymbol{\lambda} \mathbf{C}
\tag{22}
$$

where $\mathbf{C}$ is the $N \times N$ orthogonal matrix of the expansion coefficients $c_{\mu j}$ of the linear combinations given in equation 21. That $\mathbf{C}$ is orthogonal follows from the fact that the CMOs (21) form an orthonormal set, and that $\langle \lambda_\mu | \lambda_\nu \rangle = \delta_{\mu\nu}$. Because transformation 22 is formally similar to transformation 7, it follows that the ground state descriptions 15 and 23 are completely equivalent within an EBO model. Consequently, we are faced

$$^1\Psi_0 = \| \lambda_1 \bar{\lambda}_1 \cdots \lambda_\mu \bar{\lambda}_\mu \cdots \lambda_N \bar{\lambda}_N \| \tag{23}$$

with the perhaps unexpected result that all observable quantities that could be derived from them are the same, and thus completely independent of the type and amount of mixing between the EBOs. We give two examples.

Because an independent electron model, such as our EBO treatment, neglects the explicit repulsion between the electrons, meaning that all J_{jk} and K_{jk} in definition 14 for the orbital energies ε_j are set equal to zero, it follows that the total electron energy E_{el} of the hydrocarbon is given by

$$E_{\mathrm{el}} = 2 \sum_{j=1}^{N} \varepsilon_j = 2 \sum_{\mu=1}^{N} A_\mu \tag{24}$$

This relationship between the set of orbital energies ε_j and the set of self-energies A_μ is known as the 'sum rule'. Within the validity of an independent electron model, it corresponds—*cum grano salis*—to the additivity scheme for enthalpies of formation $\Delta_f H_m^{\ominus}$, mentioned in the introduction.

If—in analogy to the standard HMO treatment of π systems—we wanted to define a generalized 'bond order'

$$P_{\mu\nu} = 2 \sum_{j=1}^{N} c_{\mu j} c_{\nu j} \tag{25}$$

it follows, because $\mathbf{C}$ is an orthogonal matrix, that $P_{\mu\mu} = 2$ in all cases. (The summation $\sum_{j=1}^{N} c_{\mu j} c_{\nu j}$ is the matrix element μ, ν of the product $\mathbf{C}\mathbf{C}^{\mathrm{T}} = 1$. Hence $\sum_{j=1}^{N} c_{\mu j} c_{\nu j} = \delta_{\mu\nu}$). In other words, the bond population $P_{\mu\mu}$ (which corresponds to the HMO charge order q_μ for π systems) is completely invariant, and equal to the population of the basis EBO λ_μ before any mixing has occurred. On the other hand, the quantities $P_{\mu\nu}$ between pairs of EBOs λ_μ and λ_ν (which would correspond to bond orders $p_{\mu\nu}$ in a standard Hückel π treatment) all vanish, $P_{\mu\nu} = 0$. (Both results can be checked for methane by using equations 20.)

Consequently, for the computation of many-electron properties, such as the total electron energy E_{el}, the charge distribution or the bond orders, the EBO model is rather useless, at least in the simple form presented above.

Although we shall not use them within the context of this chapter, one must mention the independent electron models of hydrocarbons, which use an AO basis. References to earlier work can be found in the papers by Klopman and Herndon[12], and attention is drawn to Hoffmann's extended Hückel theory, which is by far the most important independent electron treatment available[12]. The latter has proved extremely useful for the interpretation of PE spectra, but is a bit cumbersome for perturbation treatments, because it makes use of a non-orthogonal basis of AOs.

C. Localized Molecular Orbitals

1. Computation of localized molecular orbitals from SCF CMOs

In a recent review[13] the use of the EBO model for the interpretation of PE spectra has been discussed in detail. (Its application to the PE spectra of saturated hydrocarbons has also been described[14].) In particular, it has been shown[13] how such EBO models can be obtained by a recalibration procedure, via localized molecular orbitals derived from SCF *ab initio* calculations. The principle of this procedure can be summarized as follows, using saturated hydrocarbons as examples.

The starting point is an *ab initio* SCF calculation of a saturated hydrocarbon C_nH_m. We recall that—because of equations 12 and 13—the underlying Hartree–Fock equations take the canonical form

$$\mathscr{F}\varphi_j = \varepsilon_j\varphi_j, \qquad j = 1, 2, \ldots, N \tag{26}$$

the solution of which yields the N CMOs φ_j of the valence shell, and their corresponding orbital energies ε_j. With few execptions, we shall neglect the n inner CMOs φ_j of a hydrocarbon C_nH_m, which are, for all practical purposes, linear combinations of the pure 1s AOs of the n carbon atoms. The CMOs φ_j define the configuration 15 as an adequate approximation for the closed-shell, singlet ground state $^1\Psi_0$ of the hydrocarbon. As before, we collect the N CMOs in a row vector $\varphi = (\varphi_1 \cdots \varphi_j \cdots \varphi_N)$, and the orbital energies in a diagonal $N \times N$ matrix $\varepsilon = \mathbf{diag}(\varepsilon_1 \cdots \varepsilon_j \cdots \varepsilon_N)$.

The next step consists in transforming the CMOs φ_j into localized molecular orbitals (LMO) Λ_j, making use of an orthogonal transformation of the type defined in 7, i.e.

$$\Lambda = \varphi\mathbf{L} \tag{27}$$

where $\mathbf{L} = (L_{ij})$ is an orthogonal transformation matrix of order $N \times N$, and Λ the row vector $\Lambda = (\Lambda_1 \cdots \Lambda_j \cdots \Lambda_N)$ of the orthonormal set of the N LMO Λ_j,

$$\Lambda_j = \sum_{k=1}^{N} \varphi_k L_{kj} \tag{28}$$

The transformation matrix $\mathbf{L}$ is chosen in such a way as to yield normalized LMOs Λ_j which are as strongly localized as is possible under the restricting condition of mutual orthogonality, $\langle \Lambda_j | \Lambda_k \rangle = 0$, if $j \neq k$. This means that each LMO Λ_j is concentrated in as small a region of space as possible. Consequently, two electrons occupying such an LMO, say Λ_j, move in very close proximity to each other, with the result that their average coulombic repulsion J_{jj} becomes very large. As Ruedenberg and his coworkers have shown[15], this suggests that maximizing the sum of these repulsion terms, i.e.

$$\sum_{j=1}^{N} J_{jj} = \text{Maximum} \tag{29}$$

should be an excellent criterion for constructing a matrix $\mathbf{L}$ which yields, according to transformation 27, a set Λ of LMOs. Such a localization criterion is known as 'intrinsic', because it does not invoke any preconceived idea about the shape and/or the place of the LMOs Λ_j within the molecule. A similar localization criterion, proposed by Foster and Boys[16], is often preferred for practical, computational reasons. From our (limited) point of view both criteria yield the same LMOs Λ_j, with negligible differences.

In the case of saturated hydrocarbons (and, in fact, for saturated molecules in general), it is found that the LMOs Λ_j correspond closely to the two-centre bond orbitals which chemists have learned to associate with the line joining two atoms in their formulae. As expected on the basis of this analogy, one of the advantages of the LMOs is their high degree of transferability[15] which compares favourably with that of so-called natural orbitals[17].

Concerning the shape of such LMOs, we note briefly that the region between the two bonded centres is free of nodes, but that they have little tails of opposite sign in the region of other bonds, which ensure orthogonality with the other LMOs of the molecule. Symbolically, such an LMO could be represented as:

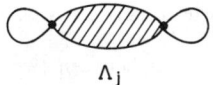

$$\Lambda_j$$

In analogy to transformation 7 and 11, it follows from equation 27 that the Hartree–Fock matrix $\mathbf{F}_\Lambda = (F_{\Lambda,ij})$ in LMO basis is given by

$$\mathbf{F}_\Lambda = \mathbf{L}^T \boldsymbol{\varepsilon} \mathbf{L} \qquad (30)$$

or explicitly by

$$F_{\Lambda,ij} = \sum_{k=1}^{N} L^T_{ik} \varepsilon_k L_{kj} = \sum_{k=1}^{N} L_{ki} \varepsilon_k L_{kj} \qquad (31)$$

where L^T_{ik} stands for the matrix element i,k of the transposed matrix $\mathbf{L}^T$, i.e. $L^T_{ik} = L_{ki}$. The transformation 30 is depicted symbolically in Figure 3 for the particular example of propane, C_3H_8, which has 20 valence electrons, occupying either $N = 10$ CMOs φ_j or the same number of LMOs Λ_j. In contrast to $\boldsymbol{\varepsilon}$ which is a diagonal matrix, $\mathbf{F}_\Lambda$ is now a full 10×10 (in general $N \times N$) matrix. In analogy to the conventions used in Section II.B, we abbreviate the matrix elements $F_{\Lambda,ij}$ of $\mathbf{F}_\Lambda$ as follows:

(1) $A_j \equiv F_{\Lambda,jj}$. These diagonal elements are the self-energies of the LMOs Λ_j (designated by ■ in Figure 3).

(2) $B_{ij} \equiv F_{\Lambda,ij}$. These are the crossterms between geminal LMOs Λ_i, Λ_j (designated by B in Figure 3).

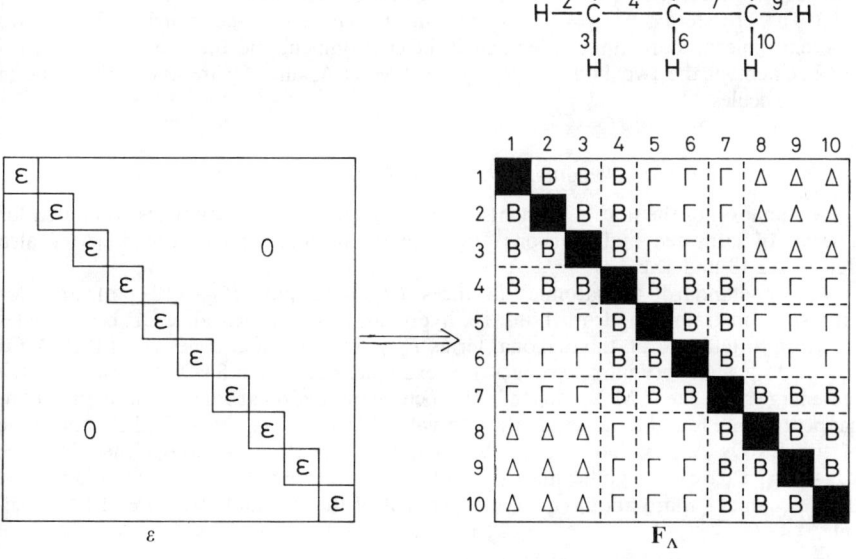

FIGURE 3. Graphical representation of the orthogonal transformation 30. The matrix ε symbolizes the eigenvalue (orbital energy) matrix pertaining to the CMOs φ_j, i.e. $\mathbf{F}_\varphi = \varepsilon$. The Hartree–Fock matrix $\mathbf{F}_\Lambda$ in localized basis $\Lambda = (\Lambda_j)$ contains the self-energies A_j (black squares), the geminal crossterms B_{ij}, the vicinal crossterms Γ_{ij} and crossterms Δ_{ij} for LMOs Λ separated by two bonds. The LMOs are numbered as shown in the structure formula

(3) $\Gamma_{ij} \equiv F_{\Lambda,ij}$. Interaction crossterm between vicinal LMOs Λ_i, Λ_j (designated by Γ in Figure 3).

(4) $\Delta_{ij} \equiv F_{\Lambda,ij}$. Interaction crossterm between two LMOs Λ_i, Λ_j separated by two bonds (designated by Δ in Figure 3).

Because of $\mathbf{LL}^{\mathrm{T}} = \mathbf{L}^{\mathrm{T}}\mathbf{L} = \mathbf{1}$, the expression 30 can be written as

$$\mathbf{F}_{\Lambda}\mathbf{L}^{\mathrm{T}} = \mathbf{L}^{\mathrm{T}}\boldsymbol{\varepsilon} \tag{32}$$

from which it follows that the eigenvalues ε_j of the Hartree–Fock matrix $\mathbf{F}_{\Lambda}$ in a localized basis are those of the CMOs φ_j. However, these CMOs can now be expressed (cf transformation 27) as linear combinations of the LMOs Λ_k, i.e.

$$\boldsymbol{\varphi} = \mathbf{\Lambda}\mathbf{L}^{\mathrm{T}}; \qquad \varphi_j = \sum_{k=1}^{N} \Lambda_k L^{\mathrm{T}}_{kj} = \sum_{k=1}^{N} \Lambda_k L_{jk} \tag{33}$$

It is evident that all this is formally identical to the EBO treatment discussed in Section II.B, the matrix $\mathbf{F}_{\Lambda}$ playing the role of $\mathbf{H}_{\lambda}$ (e.g. matrix 19), and $\mathbf{L}^{\mathrm{T}}$ the role of $\mathbf{C}$ in transformation 22.

If one computes the matrices $\mathbf{L}$, $\mathbf{\Lambda}$ and $\mathbf{F}_{\Lambda}$ for a series of saturated hydrocarbon molecules, starting with the CMOs from an *ab initio* SCF calculation, one can make the following very general observations:

(a) The LMOs Λ_j are in an obvious one-to-one correspondence to the chemical valence lines in the structural formulae. Labeling the CC and CH bonds in the formula of a hydrocarbon molecule labels the corresponding LMOs Λ_j.

(b) Two LMOs of same type and surrounded by similar bonds, but located in two different molecules, are practically identical, which means among other things that their self-energies $F_{\Lambda,jj} = A_j$ can be transferred quantitatively, almost without change, from the Hartree–Fock matrix $\mathbf{F}_{\Lambda}$ of one molecule to that of the other.

(c) The crossterms $F_{\Lambda,jk} = \langle \Lambda_j | \mathscr{F} | \Lambda_k \rangle$ in the Hartree–Fock matrices $\mathbf{F}_{\Lambda}$ of two different molecules are almost identical, if the environment and the relative geometrical arrangement of the two bonds carrying the LMOs Λ_j and Λ_k are about the same in both molecules.

2. Rules governing the elements of the matrix $\boldsymbol{F_{\Lambda}}$ for saturated hydrocarbons

Examination of the matrix elements A_j, B_{ij}, Γ_{ij}, etc. of the $\mathbf{F}_{\Lambda}$ matrices computed for a series of saturated hydrocarbons[14] has led to the following observations (cf. also References 13 and 18):

(1) For standard interatomic distances ($R_{CC} = 154\,\mathrm{pm}$, $R_{CH} = 109\,\mathrm{pm}$) and for saturated, unstrained hydrocarbons, i.e. hydrocarbons in which all CCC bond angles are close to tetrahedral, the diagonal terms $F_{\Lambda,jj}$, i.e. the self-energies A_j of the LMOs Λ_{CC} or Λ_{CH}, span very narrow ranges. For example, based on a STO-3G treatment, the self-energies of the LMOs Λ_{CC} of all open-chain hydrocarbons from methane to neopentane are found to lie within the interval $-17.84\,\mathrm{eV} < A_{CC} < -17.73\,\mathrm{eV}$, and those of the LMOs Λ_{CH} within $-16.97\,\mathrm{eV} < A_{CH} < -16.92\,\mathrm{eV}$[18]. Although the difference $A_{CC} - A_{CH} \approx 0.8\,\mathrm{eV}$ is significant, it is small enough to suggest as a first approximation that A_{CC} and A_{CH} are about equal, independent of the size and structure of the molecule:

$$A_{CC} \approx A_{CH} \tag{34}$$

(2) For tetrahedral bond angles ($\theta_{ij} \approx 109.5°$) the crossterms B_{ij} between two geminal LMOs Λ_i, Λ_j of a saturated, unstrained hydrocarbon are quite independent of the nature

of the two LMOs involved, and independent of the size and structure of the hydrocarbon molecule. Thus, one finds for the above set of molecules (using again an STO-3G treatment for a start), $-2.98\,\text{eV} < B_{\text{CC,CC}} < -2.94\,\text{eV}$; $-2.92\,\text{eV} < B_{\text{CC,CH}} < -2.91\,\text{eV}$; $-2.88\,\text{eV} < B_{\text{CH,CH}} < -2.86\,\text{eV}$. It follows that:

$$B_{\text{CC,CC}} \approx B_{\text{CC,CH}} \approx B_{\text{CH,CH}} \tag{35}$$

(3) The crossterms Γ_{ij} between two vicinal LMOs Λ_i, Λ_j in a saturated, unstrained hydrocarbon (all bond angles close to tetrahedral) are again independent of the nature of the two LMOs involved, and independent of the size and structure of the molecule, if the torsion angle τ_{ij} defining the relative, local conformation of the two LMOs Λ_i, Λ_j is the same. Using the same set of molecules and the same treatment as above, one finds for $\tau_{ij} = 180°$ (anti-planar conformation) $+0.94\,\text{eV} < \Gamma_{ij}(180°) < +1.01\,\text{eV}$, and for $\tau_{ij} = 60°$ (*gauche* conformation) $-0.57\,\text{eV} < \Gamma_{ij}(60°) < -0.55\,\text{eV}$, independent of the type of the LMOs Λ_i, Λ_j involved. In other words:

$$\Gamma_{\text{CC,CC}}(\tau_{ij}) \approx \Gamma_{\text{CC,CH}}(\tau_{ij}) \approx \Gamma_{\text{CH,CH}}(\tau_{ij}) \tag{36}$$

(4) It is found that the absolute values of matrix elements A_j and B_{ij}, and of the mean absolute values of the conformation-dependent crossterms Γ_{ij} and Δ_{ij}, follow the rule $|A_j| \gg |B_{ij}| > |\overline{\Gamma_{ij}}| > |\overline{\Delta_{ij}}|$.

The important conclusion to be drawn from these observations is that the assumption of equivalent bond orbitals λ_j as used in an EBO treatment can be validated by the properties of LMOs Λ_j derived from *ab initio* calculations. The mathematical formalism for handling LMOs according to equations, 32 and 33 is exactly the same as that of an EBO model, which implies that this Hückel-type treatment is capable of yielding excellent approximations to the SCF orbital energies ε_j and the corresponding CMOs φ_j of saturated hydrocarbons.

3. Derivation of EBOs from LMOs

Acceptance of the approximations expressed by approximations 34, 35 and 36 leads to the conclusion that all these matrix elements can be calibrated through multilinear regression treatments, with reference to observed ionization energies of a calibration set of hydrocarbons[14], if we wish to obtain a treatment appropriate for the prediction of such quantities. We shall explore this in more detail in Section V. Assuming the validity of Koopmans' theorem (cf. Section III.A) one obtains adjusted parameters A_{CC}, A_{CH}, $B_{\text{CC,CC}}$, $B_{\text{CC,CH}}$, $B_{\text{CH,CH}}$, $\Gamma_{\text{CC,CC}}(\tau)$, $\Gamma_{\text{CC,CH}}(\tau)$ etc. Thus one is led back to an EBO formalism, having replaced implicitly the LMOs by EBOs, $\Lambda_j \approx \lambda_j$, for which the calibrated matrix elements (collected in $\mathbf{H}_\lambda$) are valid. As we are going to use the EBOs λ_j as basis functions, we shall index them with greek letters, e.g. λ_μ, λ_ν etc., the indices μ, ν referring to the bonds in question.

a. The AB-model. The simplest EBO model that can be derived in this fashion consists in taking only geminal interactions into account, neglecting all others. In addition, on the basis of approximations 34 and 35 we make the simplifying assumptions $A_{\text{CC}} = A_{\text{CH}} = A$ for the self-energies, and $B_{\text{CC,CC}} = B_{\text{CC,CH}} = B_{\text{CH,CH}} = B$ for the crossterms. Under these conditions the model is described by only two parameters, namely A and B, so that the matrix $\mathbf{H}_\lambda$ is defined—in complete analogy to the standard HMO treatment of π systems—by

$$\mathbf{H}_\lambda = A\mathbf{1} + B\mathbf{A} \tag{37}$$

468   E. Heilbronner

where $1 = (\delta_{\mu\nu})$ is the unit matrix, and $A = (a_{\mu\nu})$ the adjacency matrix of the graph $\mathscr{G}^\dagger$ representing the geminal interaction pattern of the hydrocarbon. This is shown in Figure 4 for the example of propane C_3H_8, whose structural formula (a) includes the labelling of the EBOs λ_μ. In the graph $\mathscr{G}$ (b) each vertex μ stands for the EBO λ_μ, and each edge $e_{\mu\nu}$ for a geminal interaction $B_{\mu\nu}$. It is amusing to note that a three-dimensional representation of this graph $\mathscr{G}$ resembles the van't Hoff model of the corresponding hydrocarbon, i.e. in case of propane, three tetrahedra joined by their vertices, as shown in Figure 4c.

As in standard Hückel treatments, it is of advantage to introduce an abbreviation

$$-x = (A - \varepsilon)/B \tag{38}$$

which reduces the eigenvalue problem $\det(H_\lambda - \varepsilon 1) = 0$ connected with H_λ (definition 37) to the diagonalization of the adjacency matrix A:

$$\det(A - x1) = 0 \tag{39}$$

The result for propane is shown in Figure 5. The ten eigenvalues x_j are those of the adjacency matrix A underlying the matrix H_λ shown in Figure 4, and the corresponding CMO diagrams are based on graph (b) of Figure 4. Each circle refers to the coefficient $c_{\mu j}$ of the EBO λ_μ in the linear combination φ_j (cf equation 21). The radius of the circles is proportional to the absolute value $|c_{\mu j}|$, and full or open circles correspond to coefficients $c_{\mu j}$ of opposite sign. The labels of the CMOs refer to the point group C_{2v} of propane in its lowest-energy conformation, and the numbering takes only the valence orbitals into account.

This AB-model suffers from quite a few defects, some of which are already apparent from the results for propane, shown in Figure 5. The neglect of vicinal (and higher) interaction terms leads necessarily to a model which is conformation-independent. Furthermore, the AB-model yields large sets of accidentally degenerate orbitals, e.g. orbitals φ_4 to φ_8 in the case of propane. This will prove to be a very poor approximation for interpreting the low-energy part of the PE spectra of hydrocarbons, as we shall see in Section V. On the other hand, the lowest n orbital energies ε_j of the CMOs $\varphi_j, j = 1$ to n, obtained through an AB-model for a saturated hydrocarbon C_nH_m, will turn out to yield a very satisfactory interpretation of the high-energy part of its PE spectrum.

b. The $A\Gamma$-model. The above EBO AB-model can be vastly improved by including vicinal interaction terms $\Gamma_{\mu\nu}$, which leads to the $A\Gamma$-model. Because the matrix elements $\Gamma_{\mu\nu}(\tau_{\mu\nu})$ depend on the local twist angles $\tau_{\mu\nu}$, their inclusion will make the model conformation-dependent. This will also lift most of the accidental degeneracies, as can be deduced by inspecting the orbital diagrams of Figure 5. Note that the interactions would correspond to additional edges in the graph $\mathscr{G}$ (Figure 4b) between vertices separated by two consecutive edges.

---

†A graph $\mathscr{G}$ is a set of N vertices (points) $\{\cdots \mu, \nu, \cdots\}$ joined by edges (lines) $e_{\mu\nu}$, e.g. in Figure 4 the graph (a) representing the structure of propane, or the graph (b) depicting the geminal interactions between the 10 EBOs in the same molecule. Such a graph $\mathscr{G}$ can be characterized by a $N \times N$ adjacency matrix $A = (a_{\mu\nu})$ whose matrix elements are $a_{\mu\nu} = 1$, if the vertices μ and ν are joined by an edge, and zero otherwise. For example, the adjacency matrix A of the graph $\mathscr{G}$ of the methane EBO model depicted in drawing 18 is

$$A = \begin{pmatrix} 0 & 1 & 1 & 1 \\ 1 & 0 & 1 & 1 \\ 1 & 1 & 0 & 1 \\ 1 & 1 & 1 & 0 \end{pmatrix}$$

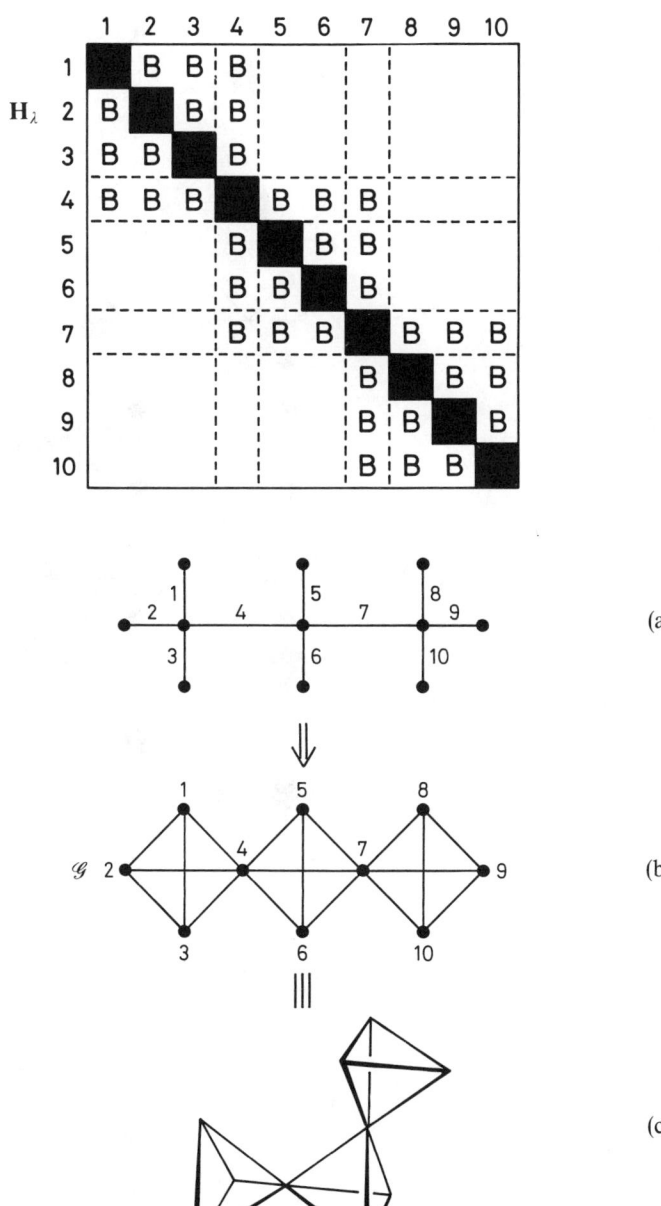

FIGURE 4. The structure of the $\mathbf{H}_\lambda$ matrix for the AB-model of propane. The black squares are the self-energies A for the CC and CH EBOs, assumed to be equal. The geminal crossterms B are the same for CH/CH, CH/CC and CC/CC interactions. All other crossterms are set equal to zero. Graph (a) shows the numbering of the bonds in propane. In graph $\mathcal{G}$ (b), each vertex (point) refers to an EBO, and each edge (line) to a geminal interaction. Graph (c) is a three-dimensional representation of (b)

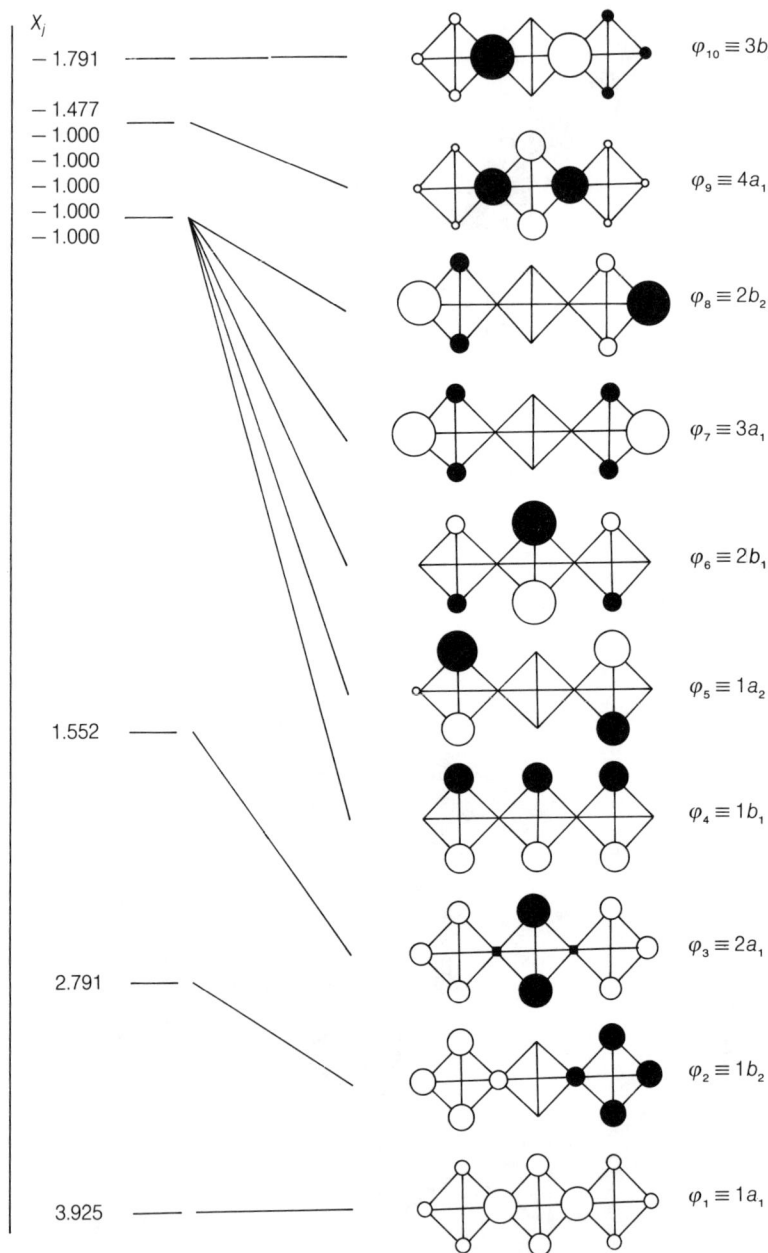

FIGURE 5. Scheme of the valence orbitals of propane obtained by diagonalizing the matrix $\mathbf{H}_\lambda$ shown in Figure 4. The orbital energies ε_j are given in terms of the parameter x_j, i.e. $\varepsilon_j = A + x_j B$ (cf equation 38) and the orbitals φ_j as linear combinations $\varphi_j = \sum_\mu \lambda_\mu c_{\mu j}$ of the EBOs λ_μ with reference to graph (b) of Figure 4. The orbital labels refer to the point group C_{2v}. Note that the orbitals φ_4 to φ_8, which are accidentally degenerate, are given in one of the many possible symmetry adapted, real representations

In its simplest form the $A\Gamma$-model makes use of the observations expressed in approximations 35 and 36, by postulating

$$B_{CC,CC} = B_{CC,CH} = B_{CH,CH} = B$$
$$\Gamma_{CC,CC}(\tau) = \Gamma_{CC,CH}(\tau) = \Gamma_{CH,CH}(\tau) = \Gamma_0 \cos(\tau) \qquad (40)$$

where Γ_0 is a negative energy parameter. Accordingly, the vicinal crossterms $\Gamma_0 \cos(\tau)$ are negative for a local *syn*-periplanar and positive for an *anti*-periplanar conformation of the two vicinal bonds. A simple mnemonic for the sign of $\Gamma_0 \cos(\tau)$ is given by the following two diagrams, which show a positive, bonding overlap, and hence a negative crossterm for the *syn*-periplanar arrangement, and a negative, antibonding overlap associated with a positive crossterm for the *anti*-periplanar one.

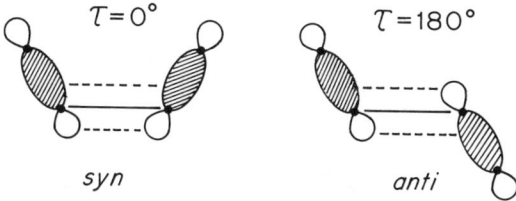

$$\tau = 0° \qquad \qquad \tau = 180°$$

$$syn \qquad \qquad \qquad anti$$

As before, the approximation in equation 40 is acceptable under the implicit assumption of (approximately) tetrahedral bond angles. The self-energies (equation 34) can be treated as either independent, $A_{CC} \approx A_{CH}$, or as identical, $A_{CC} = A_{CH} = A$, depending on the required precision.

We shall discuss this model in more detail in Section V. (Concerning the inclusion of higher interaction matrix elements, the reader is referred to References 13 and 14.)

III. PHOTOELECTRON (PE) SPECTROSCOPY

A. General Remarks

1. PE spectrum and ionization energies

The theoretical and experimental principles of PE spectroscopy have been reviewed extensively[19-22]. In particular, the book by Eland[21] provides an elegant and compact introduction. In this section we summarize only the essentials necessary to follow the arguments presented in Section V.

As far as the present review is concerned, we shall only be concerned with the primary process consisting in the ejection, by a photon of energy $h\nu$, of an electron e^- from a closed-shell hydrocarbon C_nH_m in its electronic ground state $^1\Psi_0$ (cf configuration 15). The product is a hydrocarbon radical ion $C_nH_m^+$ in one of its doublet electronic states $^2\tilde{\Psi}_j$. (In the context of PE spectroscopy it is customary to write $C_nH_m^+$ for the open-shell, radical cation, rather than $C_nH_m^{\cdot+}$). We shall not be concerned with subsequent rearrangements and/or fragmentation reactions of the radical cation $C_nH_m^+$, in contrast to its geometric relaxations which conserve the original connectivity of the parent hydrocarbon C_nH_m.

Typical photon sources used in standard PE spectroscopy are helium atoms, He(I), or helium ions, He(II), undergoing the following transitions shown in display 41.

$$\text{He(Iα):(1s)}^1(2s)^1 \rightarrow (1s)^2; \quad \lambda = 58.4\,\text{nm}; \quad hv = 21.2\,\text{eV}$$
$$\text{He(IIα):(2p)}^1 \quad\;\; \rightarrow (1s)^1; \quad \lambda = 30.4\,\text{nm}; \quad hv = 40.8\,\text{eV} \tag{41}$$

If these photon energies are larger than the ionization energies I_j needed to reach the radical ion doublet states $^2\tilde{\Psi}_j$ in which we are interested, the excess energy is carried off by the photoelectron e^- as kinetic energy $T_j = hv - I_j$. Accordingly the primary process can be written as

$$C_nH_m(^1\Psi_0) + hv \rightarrow C_nH_m^+(^2\tilde{\Psi}_j) + e^-(T_j) \tag{42}$$

The quantity measured by the PE spectrometer is the number $Z(e^-)$ of electrons ejected per unit time from a gaseous sample of fixed volume (measured in cps = counts per second) as a function of the kinetic energy T of the ejected electrons. The result is presented as a plot of $Z(e^-)$ vs ionization energies $I = hv - T$, as shown in Figure 6 for a hypothetical molecule, not necessarily a hydrocarbon.

The low-energy part of this PE spectrum consists of four bands, numbered ① to ④ with increasing ionization energy, which—for simplicity—we assume not to overlap. The bands ① and ② exhibit resolved vibrational fine structure, which is rather the exception in the case of PE spectra of saturated hydrocarbons. In contrast, the fine structure of ③ and ④ is unresolved, which could be due to the limited resolution of the recording (which is of the order of $\Delta I/I \approx 1/200$ in standard PE spectrometers), to random noise and/or to intrinsic line broadening, which would prevent a resolution even in the absence of the other factors. The latter effect is by far the most important cause for the occurrence of wide, unresolved bands in the PE spectra of saturated hydrocarbons.

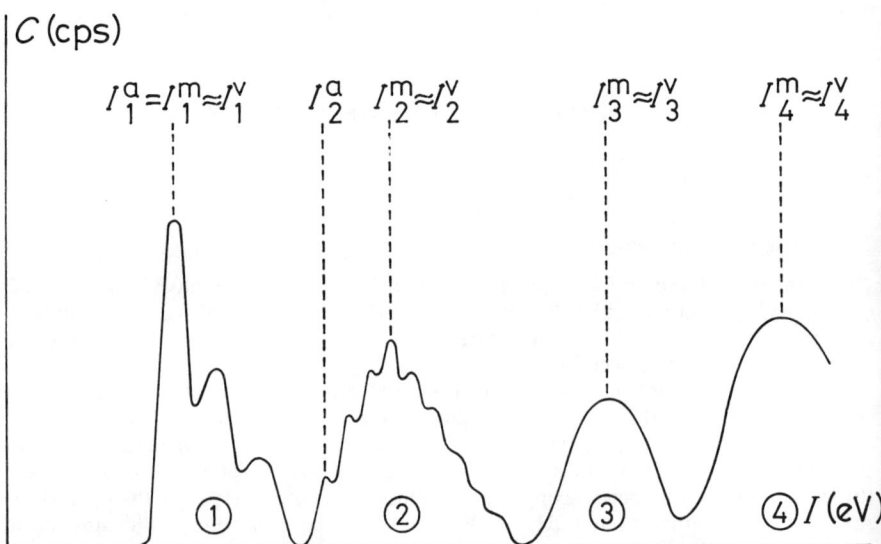

FIGURE 6. Schematic representation of the PE spectrum of a hypothetical molecule. Abscissa: Ionization energy I in eV. Ordinate: Intensity C in counts per second (cps). I_j^a = adiabatic ionization energy, I_j^v = vertical ionization energy, I_j^m = ionization energy corresponding to the maximum of the band

The position of a band $\textcircled{b}$ on the abscissa I (eV) of a PE spectrum can be characterized by:

(1) I_b^m, corresponding to the maximum of the contour of band $\textcircled{b}$. For our purposes I_b^m is usually a sufficiently close approximation for the so-called vertical ionization energy I_b^v (see below) in which we are interested primarily. As I_b^m is both easier to measure and much better defined from an experimental point of view, we shall use the approximation $I_b^m \approx I_b^v$ throughout this chapter.

(2) I_b^a, the adiabatic ionization energy, corresponds to the position of the first vibrational component of a PE band (disregarding possible hot bands). This quantity is usually not accessible with the desired precision, because of the lack of fine structure (e.g. bands $\textcircled{3}$ and $\textcircled{4}$ of Figure 6), and because of strongly overlapping bands. In a crude approximation one may assume that I_b^a coincides with the onset of the band $\textcircled{b}$, a quantity usually uncertain by about 0.2 eV.

2. Interpretation of the PE spectrum

With regard to the MO models discussed in Section II, the ionization process (reaction 42) can be visualized as the ejection of an electron from one of the CMOs of the neutral hydrocarbon C_nH_m, say from the CMO φ_j. This point of view is especially appropriate for independent electron models, such as the standard HMO treatment of π systems, or the EBO AB-model of hydrocarbons. Both are two-parameter models, in which the orbital energies $\varepsilon_j = \alpha + \beta x_j$ or $\varepsilon_j = A + Bx_j$ are given with respect to the edge of the continuum, $\varepsilon_{Continuum} = 0$, corresponding to a free electron of zero kinetic energy, $T = 0$. Within such an independent electron approximation, the ionization energy needed to remove an electron from orbital φ_j is therefore $I_j^v = -\varepsilon_j$.

The situation is more involved if we use a many-electron treatment. Making the crude (and in fact unreasonable) assumption that the removal of an electron from the CMO φ_j will not affect its shape, nor the shapes of the other CMOs φ_k, $k \neq j$, used for the description (determinant 15) of the electronic ground state $^1\Psi_0$ of the hydrocarbon C_nH_m, the resulting electronic state $^2\tilde{\Psi}_j$ of the radical cation $C_nH_m{}^+$ can be written as

$$^2\tilde{\Psi}_j = \begin{cases} \| \varphi_1 \bar{\varphi}_1 \cdots \varphi_{j-1} \bar{\varphi}_{j-1} \varphi_j \varphi_{j+1} \bar{\varphi}_{j+1} \cdots \varphi_N \bar{\varphi}_N \| \\ \| \varphi_1 \bar{\varphi}_1 \cdots \varphi_{j-1} \bar{\varphi}_{j-1} \bar{\varphi}_j \varphi_{j+1} \bar{\varphi}_{j+1} \cdots \varphi_N \bar{\varphi}_N \| \end{cases} \tag{43}$$

where the upper and lower Slater determinants correspond to the doublet components (configurations) of the radical cation state $^2\tilde{\Psi}_j$, in which the odd electron remaining in the CMO φ_j has either spin α or spin β, respectively.

It has been shown by Koopmans[23] that under the above assumption of conservation of orbital shape (i.e. using the identical orbitals φ_j in configurations 15 and 43), the vertical ionization energy $I_j^v \approx I_j^m$ is still given by

$$I_j^v = -\varepsilon_j \tag{44}$$

albeit with the important difference that the orbital energy ε_j is now the orbital energy defined according to equation 14 for the neutral, closed-shell molecule, e.g. of the parent hydrocarbon C_nH_m. This is known as Koopmans' theorem[23]. Although this theorem has severe limitations, stemming from the neglect of electron reorganization and electron correlation energies (as well as of the effects of vibronic mixing), it turns out to be a remarkably useful tool for the interpretation of PE spectra. Bearing in mind the approximations involved, we shall almost exclusively rely on it in the following sections.

If the molecule C_nH_m were completely rigid, then its PE spectrum would simply consist of a set of sharp lines (i.e. narrow peaks) at positions I_j^v. This hypothetical spectrum would then be (up to sign) a faithful copy of the orbital energy diagram, such as the one shown on the left of Figure 5. However, both the parent molecule C_nH_m in its electronic ground state $^1\Psi_0$ and the radical cation $C_nH_m{}^+$ in its different electronic states $^2\tilde\Psi_j$ undergo vibrations described by $z_Q = 3(n+m) - 6$ normal modes Q_s. Depending on the vibrational quantum numbers v_s'' and v_s' of each of these z_Q normal modes Q_s in, respectively, C_nH_m and $C_nH_m{}^+$, the ionization energy is changed by $\sum_s h v_s(v_s'' - v_s')$, $s = 1, 2, \ldots, z_Q$, assuming in a first approximation that the frequency v_s of the normal mode Q_s is the same in both C_nH_m and $C_nH_m{}^+$. As those changes differ from molecule to molecule, one obtains a whole series of lines which merge to form the broad band contours, called Franck–Condon envelopes. The relative intensities of the different lines forming a given band depend mainly on the overlap of the relevant normal-mode wave functions, which in turn are functions of the changes in internal structure parameters (e.g. bond lengths and bond angles) which accompany the transition from C_nH_m in its electronic ground state $^1\Psi_0$, to $C_nH_m{}^+$ in its structurally relaxed electronic state $^2\tilde\Psi_j$.

As a rule of thumb the Franck–Condon envelopes of the bands ⓑ can be interpreted as follows:

(1) If $I_b^a \approx I_b^m \approx I_b^v$ as shown for band ① in Figure 6, then the geometric equilibrium structure of the radical cation $C_nH_m{}^+$ is close to that of the neutral molecule C_nH_m.

(2) If $I_b^a \ll I_b^m \approx I_b^v$, as shown for band ② in Figure 6, then the geometric equilibrium structure of the radical cation $C_nH_m{}^+$ differs significantly from that of the neutral molecule C_nH_m. With rare exceptions the latter will usually be the case for saturated hydrocarbons C_nH_m.

Whereas the bands ⓑ in the PE spectrum of a hydrocarbon, and the corresponding ionization energies I_b^m, are numbered $b = 1, 2, \ldots$ according to increasing ionization energies, the corresponding CMOs φ_j are numbered $j = 1, 2, \ldots, N$ according to increasing orbital energies ε_j. Accordingly, band ① corresponds to CMO φ_N, band ② to φ_{N-1}, etc. In other words, the lower indices are related by $b = N - j + 1$, and Koopmans' theorem (equation 44) should therefore be written as $I_b^v = -\varepsilon_{N-b+1}$. However, we shall always use the simpler form (equation 44), it being understood that the lower index j refers to either the CMO quantum number or the PE band, depending on the context.

B. He(IIα) PE Spectra of Saturated Hydrocarbons

The interpretation of the PE spectra of saturated hydrocarbons owes much to the pioneering work of Price and his coworkers[24,25]. In Figure 7 are shown the He(IIα) PE spectra (cf list 41) of the simplest linear and branched hydrocarbons taken from their work. These authors have also introduced the use of He(IIα) radiation for the recording of PE spectra, thereby extending the useful range accessible to $I \approx 30\,\mathrm{eV}$.

He(IIα) PE spectra of saturated hydrocarbons are of interest because they span the complete range of valence ionization energies, i.e. those which are due to ejection from CMOs φ_j adequately represented by linear combinations 16 of carbon 2s, $2p_x$, $2p_y$, $2p_z$ and hydrogen 1s AOs, or alternatively by linear combinations 33 of the LMOs Λ_{CC} and Λ_{CH}, or the EBOs λ_{CC} and λ_{CH}. (The contribution of carbon 1s AOs may safely be neglected, because of their low basis energy of $-290\,\mathrm{eV}$ to $-295\,\mathrm{eV}$[26].) Similar PE spectra have been obtained for the linear alkanes using monochromatized X-rays[27].

Price and his coworkers noted that these PE spectra present two clearly separated regions (as indicated in Figure 7 by a dashed line), the first region terminating around $I \approx 15\,\mathrm{eV}$ to $18\,\mathrm{eV}$, and the second one extending from this value to $I \approx 30\,\mathrm{eV}$. The former series of bands is due to the ejection of an electron from outer-valence CMOs dominated by the carbon 2p and the hydrogen 1s AOs, whereas the second series of bands is similarly associated with inner-valence orbitals dominated by the carbon 2s AOs. Therefore it seemed appropriate to name the two spectral regions the C2p and the C2s region, respectively[24,25]. By analogy we call the bands in these regions the

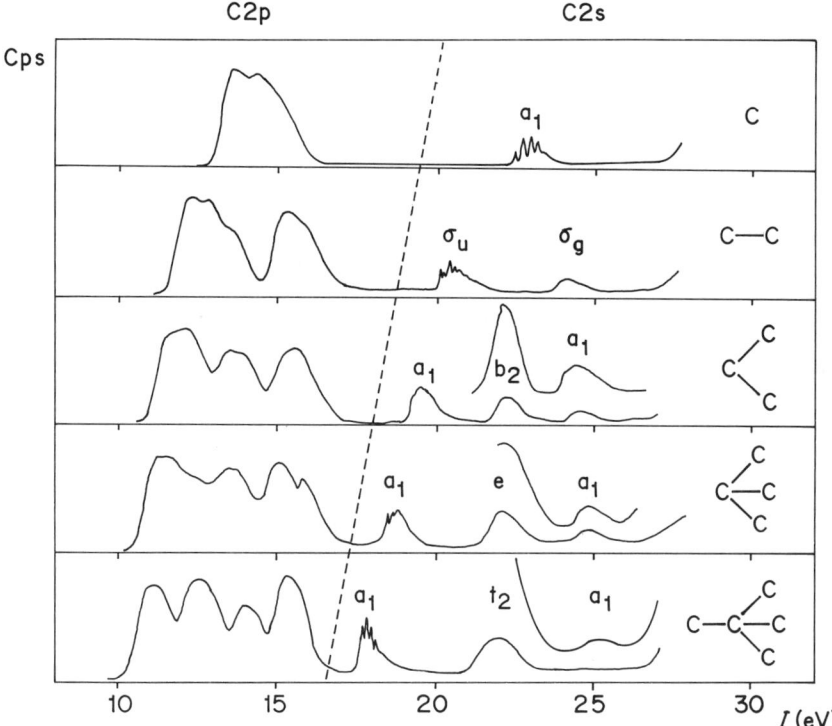

FIGURE 7. He(IIα) PE spectra of methane (C), ethane (CC), propane (CCC), isobutane $(C(C)_3)$ and neopentane $(C(C)_4)$. The dashed line separates the C2p- from the C2s-band systems. Redrawn from References 24 and 25

p-type and s-type bands, or briefly the C2p and C2s bands. The corresponding sets of CMOs φ_j are then designated as the C2p and the C2s manifold, respectively.

It is obvious from Figure 7 that the number of C2s bands equals the number n of carbon atoms in the hydrocarbon C_nH_m, as one would have deduced from the results presented in Figure 5, where the C2s manifold of CMOs φ_j is characterized by positive x_j values, the C2p manifold by negative ones. (The features labeled e and t_2 in the spectra of isobutane and neopentane are due to two and three overlapping bands. These features correspond to the removal of an electron from doubly or triply degenerate CMOs, respectively). The total number of valence-shell CMOs being $2n + m/2$, one is left with $n + m/2$ outer-valence CMOs, and correspondingly with $n + m/2$ C2p bands, which form the strongly overlapping C2p-band system squeezed into the interval from $I = 10$ to 18 eV. Thus the C2p-band system of neopentane C_5H_{12} (see Figure 7) contains eleven C2p bands, which overlap in such a way as to yield only four discernible maxima. This factor, i.e. the strong overlap between individual PE bands and the lack of detail in the Franck–Condon envelope of the resulting C2p-band systems, prevents in many cases the obtention of a unique C2p-band assignment on the basis of the PE spectrum alone[14].

He(IIα) PE spectra of saturated hydrocarbons can be found in the publications of Price and his coworkers[24,25], in work by Bieri and Åsbrink[28], and in Reference 29.

IV. A CATALOGUE OF PE SPECTRA

This section consists of a compilation of references pertaining to He(Iα) and He(IIα) gas-phase PE spectra of saturated, closed-shell hydrocarbons C_nH_m (Table 1). (With few exceptions, such PE spectra consist of broad, strongly overlapping, and usually unresolved bands. For this reason, only a few PE spectra will be shown explicitly in Section V.)

The earliest tables of PE spectra and of ionization energies derived from PE spectra are those of Turner, Baker, Baker and Brundle[19] and of Dewar and Worley[30]. However, the latter data have been obtained using a grid-type spectrometer which records the integrated intensities as a function of the ionization energies I, instead of the intensities $Z(e^-)$ themselves. These data are now of historical interest, and have not been included in our Table 1.

If in search of PE spectroscopic data, the reader is strongly urged to begin by consulting the compilation of first ionization energies (I_1^v, I_1^a, I_1^m) by Levin and Lias[31], which contains all leading references up to 1981. (See also the earlier compilation in Reference 6.) However, no PE spectra and higher ionization energies are given. For the simpler hydrocarbons the He(Iα) PE spectra and leading references are presented in the handbook authored by Kimura, Katsumata, Achiba, Yamazaki and Iwata[32], and a fair number of PE spectra and/or ionization energies of hydrocarbons are contained in References 24, 25 and 29. Finally, many PE spectra and assigned ionization energies of (saturated) hydrocarbons containing one or more three- and/or four-membered rings are to be found in the excellent review of Gleiter[33].

TABLE 1. A catalogue of references to He(Iα) and He(IIα) PE spectra of saturated hydrocarbons

m, n		Name	References
Linear and Branched Hydrocarbons			
1	4	Methane	19, 20, 21, 28, 29, 32, 34, 35, 36, 37
2	6	Ethane	19, 20, 25, 28, 29, 32, 38, 39
		Perdeuterioethane, CD_3CD_3	38
3	8	Propane	20, 25, 28, 32, 38, 39, 40, 41, 42
		2,2-Dideuteriopropane, $C_3H_6D_2$	43
		1,1,1,3,3,3-Hexadeuteriopropane, $C_3H_2D_6$	43
		Perdeuteriopropane, C_3D_8	43
4	10	Butane	14, 20, 28, 29, 32, 39, 41
		Isobutane	25, 29, 32, 42
5	12	Pentane	20, 25, 29, 32, 41, 44
		Isopentane	29, 32
		Neopentane	25, 29, 32, 42, 45, 46, 47
6	14	Hexane	20, 29, 41, 44, 45
		2,2-Dimethylbutane	29, 45
		2,3-Dimethylbutane	29
		3-Methylpentane	45
8	18	2,2,3,3-Tetramethylbutane	32, 48
9	20	Nonane	20, 27
13	28	Tridecane	20
28	58	Octacosane	49
36	74	Hexatriacontane	20, 27, 49

TABLE 1. (*continued.*)

m, n	No.	Name	References
Monocyclic Hydrocarbons			
3 6		Cyclopropane	19, 21, 29, 32, 33, 40, 50, 51, 52, 53, 54, 55, 56
4 8		Cyclobutane	29, 33, 57
		Methylcyclopropane	58, 59, 60
5 10		Cyclopentane	29, 32, 40, 44, 61
		1,1-Dimethylcyclopropane	62
		Methylcyclobutane	58
6 12		Cyclohexane	29, 32, 40, 44, 45, 50, 61, 63
		Methylcyclopentane	29
		1,1-Dimethylcyclobutane	64
7 14		Cycloheptane	61
8 16		Cyclooctane	61, 65
		trans-1,2-Diethylcyclobutane	58
		trans-1,3-Diethylcyclobutane	33
		1,1,3,3-Tetramethylcyclobutane	64
		cis-1,2-Dimethylcyclohexane	29
		trans-1,2-Dimethylcyclohexane	29
		cis-1,3-Dimethylcyclohexane	29
		trans-1,3-Dimethylcyclohexane	29
		cis-1,4-Dimethylcyclohexane	29
		trans-1,4-Dimethylcyclohexane	29
10 20		Cyclodecane	29, 61
11 22		Cycloundecane	61
Bicyclic Hydrocarbons			
4 6	1	Bicyclo[1.1.0]butane	29, 66
5 8	10	Bicyclo[1.1.1]pentane	67
	12	Bicyclo[2.1.0]pentane	29
	103	Spiro[2.2]pentane	68
6 10	13	Bicyclo[2.1.1]hexane	69
	16	Bicyclo[2.2.0]hexane	29, 70
	105	Spiro[2.3]hexane	71
	118	Bicyclopropyl	72
7 12	19	Bicyclo[2.2.1]heptane	29, 73
	14	Bicyclo[3.1.1]heptane	33
	39	Bicyclo[4.1.0]heptane	74
8 14	30	Bicyclo[2.2.2]octane	29, 73
	15	Bicyclo[4.1.1]octane	75
	2	2,2,4,4-Tetramethylbicyclo[1.1.0]butane	33
	106	Spiro[4.3]octane	55, 76
	107	Spiro[5.2]octane	77
9 16	37	Bicyclo[3.2.2]nonane	78
	38	Bicyclo[3.3.1]nonane	79
	59	Bicyclo[6.1.0]nonane	33
10 18	41	*cis*-Bicyclo[4.4.0]decane, *cis*-Decalin	29, 79, 80
	42	*trans*-Bicyclo[4.4.0]decane, *trans*-Decalin	29, 79, 80
	40	3,7,7-Trimethylbicyclo[4.1.0]heptane, Carane	81
12 22	17	1,2,3,4,5,6-Hexamethylbicyclo[2.2.0]hexane	70

(*continued*)

TABLE 1. (*continued*)

m, n	No.	Name	References
Tricyclic Hydrocarbons			
5 6	**49**	[1.1.1]Propellane	82
6 8	**53**	Tricyclo[3.1.0.0^{2,4}]hexane	33
	3	Tricyclo[3.1.0.0^{2,6}]hexane	83
7 10	**85**	Tricyclo[2.2.1.0^{2,6}]heptane, Nortricyclene	84, 85
	18	Tricyclo[3.1.1.0^{3,6}]heptane	86
	4	Tricyclo[3.1.1.0^{6,7}]heptane	83
	104	Tricyclo[4.1.0.0^{1,3}]heptane	68
8 12	**89**	Tricyclo[2.1.1.1^{2,5}]octane, D_{2d}-Dinoradamantane	87
	20	*endo*-Tricyclo[3.2.1.0^{2,4}]octane	60
	21	*exo*-Tricyclo[3.2.1.0^{2,4}]octane	60
	115	Tricyclo[3.3.0.0^{2,6}]octane	58
	54	*syn*-Tricyclo[4.2.0.0^{2,5}]octane	88
	55	*anti*-Tricyclo[4.2.0.0^{2,5}]octane	88
	60	*cis*-Tricyclo[5.1.0.0^{2,4}]octane	74
	61	*trans*-Tricyclo[5.1.0.0^{2,4}]octane	74
	22	Tricyclo[3.2.1.0^{3,7}]octane	89
	8	Tricyclo[5.1.0.0^{2,8}]octane	90
	108	Dispiro[2.0.2.2]octane	71
	109	Dispiro[2.1.2.1]octane	71
	5	1-Methyltricyclo[3.1.1.0^{6,7}]heptane	91
	6	6-Methyltricyclo[3.1.1.0^{6,7}]heptane	91
9 14	**91**	Tricyclo[3.2.1.1^{2,6}]nonane, Twist-brendane	87
	33	Tricyclo[3.2.2.0^{2,4}]nonane	92
	26	*endo*-Tricyclo[4.2.1.0^{2,5}]nonane	93
	23	Spiro[cyclopropane-1,2'-norbornane]	94
	24	Spiro[cyclopropane-1,7'-norbornane]	95
10 16	**92**	Tricyclo[3.3.1.1^{2,6}]decane, Twistane	87
	86	Tricyclo[3.3.2.0^{2,8}]decane	84
	35	Tricyclo[4.2.2.0^{2,5}]decane	92
	116	Tricyclo[5.3.0.0^{2,8}]decane	87, 96
	95	Adamantane	29, 97, 98, 99, 100
	31	Spiro[bicyclo[2.2.2]octane-2,1'-cyclopropane]	94
	111	Dispiro[2.0.2.4]decane	77
	112	Dispiro[2.2.2.2]decane	101
	90	2,5-Dimethyltricyclo[2.2.1.1^{2,5}]octane	87
11 18	**96**	1-Methyladamantane	102
	7	1-*tert*-Butyltricyclo[4.1.0.0^{2,7}]heptane	103
12 20	**117**	Tricyclo[5.5.0.0^{2,8}]dodecane	104
	97	1,3-Dimethyladamantane	79
14 24	**45**	Dodecahydro-2-methylphenalene	80
	43	*cis-syn-cis*-Perhydroanthracene	80
	44	*trans-syn-trans*-Perhydroanthracene	80
16 28	**119**	[2.2]-Perhydroparacyclophane	105
20 36	**79**	Tetra-*tert*-butyltetrahedrane	106

TABLE 1. (*continued*)

m, n	No.	Name	References
Tetracyclic Hydrocarbons			
7 8	**50**	Tetracyclo[3.2.0.0^{1,6}.0^{2,6}]heptane	107
	74	Tetracyclo[3.2.0.0^{2,7}.0^{4,6}]heptane, Quadricyclene	29, 108
8 10	**69**	Tetracyclo[3.2.1.0^{2,7}.0^{4,6}]octane	109, 110, 111
	76	Tetracyclo[3.3.0.0^{2,4}.0^{3,6}]octane	112
	51	Tetracyclo[4.2.0.0^{1,7}.0^{2,7}]octane	107
	9	Tetracyclo[4.1.0.1^{1,6}.0^{2,5}]octane	33
	81	Seco-cubane	33
9 12	**62**	*cis,trans*-Trihomobenzene	113
	88	Triasterane	85
	110	Dispiro[cyclopropane-1,2'-bicyclo[2.1.0]pentane-3',1''-cyclopropane]	114
	113	[3]Rotane	115
10 14	**52**	Tetracyclo[4.4.0.0^{1,7}.0^{2,7}]decane	107
	77	Tetracyclo[5.3.0.0^{2,6}.0^{3,10}]decane	116
	27	Tetracyclo[5.2.1.0^{2,6}.0^{3,5}]decane	93
	66	*syn-syn*-Tetracyclo[7.1.0.0^{2,4}.0^{5,7}]decane	117
	67	*syn-anti*-Tetracyclo[7.1.0.0^{2,4}.0^{5,7}]decane	117
	68	*anti-anti*-Tetracyclo[7.1.0.0^{2,4}.0^{5,7}]decane	117
	99	2,4-Dehydroadamantane	33
	11	1,1'-Bi(bicyclo[1.1.1]pentane)	67
11 16	**25**	Dispiro[cyclopropane-1,2'-bicyclo[2.2.1]heptane-3',1''-cyclopropane]	94
	100	2,4-Dehydrohomoadamantane	33
12 18	**58**	Tetracyclo[4.2.2.2^{2,5}.0^{1,6}]dodecane	118
	28	Tetracyclo[5.2.1.0^{2,6}.2^{3,5}]dodecane	119
	29	Tetracyclo[5.2.1.0^{2,6}.2^{3,5}]dodecane	120
	32	Dispiro[cyclopropane-1,2'-bicyclo[2.2.2]octane-3',1''-cyclopropane]	94
13 20	**36**	Tetracyclo[5.2.2.0^{2,6}.2^{3,5}]tridecane	119
15 24	**63**	Hexamethyl-*cis,trans*-trihomobenzene	113
16 26	**46**	Hexadecahydropyrene	80, 74
19 32	**47**	5α-Androstane	121
	48	5α, 14α-Androstane, Etioallocholane	122

Polycyclic Hydrocarbons with more than Four Rings

8 8	**80**	Cubane	109, 123, 124
	70	Cuneane	109
9 10	**71**	Pentacyclo[4.3.0.0^{3,4}.0^{3,8}.0^{5,7}]nonane	110, 111
	82	Homo-cubane	33
10 10	**65**	Diademane	74
	84	Pentaprismane	123
10 12	**72**	Pentacyclo[4.4.0.0^{2,4}.0^{3,8}.0^{5,7}]decane	110, 111
	56	Pentacyclo[5.3.0.0^{2,6}.0^{3,5}.0^{8,10}]decane	125
	57	Pentacyclo[5.3.0.0^{2,6}.0^{3,5}.0^{8,10}]decane	125
	83	Basketane	33
	64	Decahydrotricyclopropa[*cd,f,hi*]indene	74
11 12	**73**	Octahydrospiro[cyclopropane-1,3'-[1,2]-methanodicyclopropa[*cd,gh*]pentalene	126
11 14	**34**	Pentacyclo[3.3.3.0^{2,4}.0^{6,8}.0^{9,11}]undecane	127
	98	2,4-Methano-2,4-dehydroadamantane	128

(*continued*)

TABLE 1. (*continued*)

m, n	No.	Name	References
12 14	**78**	Hexacyclo[6.4.0.0^{2,7}.0^{3,12}.0^{4,6}.0^{9,11}]dodecane	116
	114	[4]Rotane	115
13 16	**93**	Trispiro[cyclopropane-(1,3′)-tricyclo[2.2.1.0^{2,6}]-heptane-(5′,1″)-cyclopropane-(7′,1″)-cyclopropane]	129
	120	Trishomotriquinacene	33
	87	Trishomobullvalene	33
14 20	**101**	Congressane	79
15 14	**75**	Dispiro[tetracyclo[3.2.0.0^{2,7}.0^{4,6}]heptane-3,1′-cyclopropane-2′,3″-tetracyclo-[3.2.0.0^{2,7}.0^{4,6}]heptane]	130
15 18	**94**	Trispiro[cyclopropane-(1,3′)-tetracyclo-[3.3.1.0^{2,8}.0^{4,6}]nonane-(7′,1″)-cyclopropane-(7′,1‴)-cyclopropane]	129
20 30	**102**	1,1′-Diadamantane	79

Concerning Table 1, the following remarks are in order:

(1) The list of references is not complete. In particular, some of the early work on ionization energies of hydrocarbons is not mentioned. However, it can easily be traced through the leading references given in the Table.

(2) Table 1 is subdivided into linear, mono-, bi-, tri-, tetra- and higher cyclic hydrocarbons $C_n H_m$, arranged according to n, m within each group. Deuteriated and Protiated derivatives appear together. The reader is reminded that in a saturated hydrocarbon $C_n H_m$ the number c of cycles is given by $c = n + 1 - m/2$.

(3) Although the names given are unambiguous, they do not always satisfy the IUPAC rules. We have used trivial names whenever convenient.

(4) The arrangement according to n, m necessarily separates the parent compounds from their alkyl-substituted derivatives. To compensate for this shortcoming, related molecules have been grouped together in the scheme of structural formulae. The bold numbers (in the column headed **No.**) refer to this formula scheme, where the hydrocarbons have been grouped according to either a common parent hydrocarbon (e.g. a given bicyclo[*h.k.l*]alkane), or their belonging to the same category (e.g. highly symmetrical hydrocarbons). However, it should be borne in mind that such groupings are largely subjective.

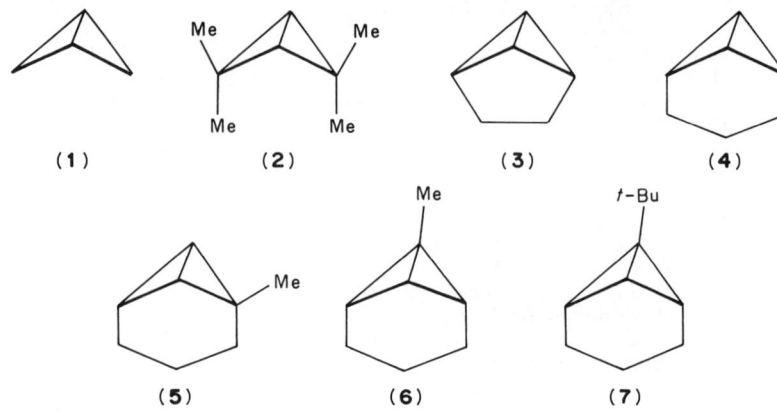

(1) (2) (3) (4)

(5) (6) (7)

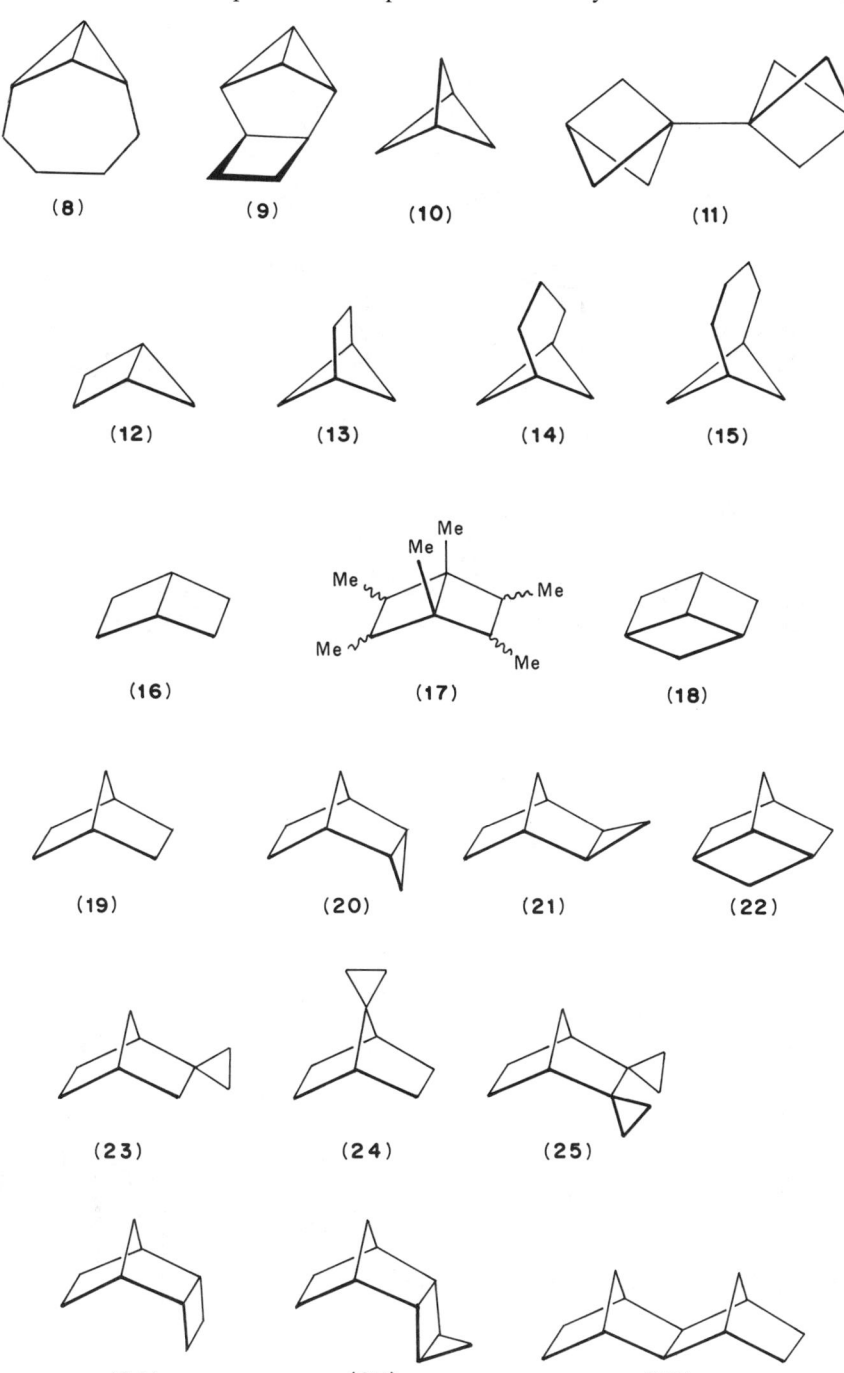

(8)   (9)   (10)   (11)

(12)   (13)   (14)   (15)

(16)   (17)   (18)

(19)   (20)   (21)   (22)

(23)   (24)   (25)

(26)   (27)   (28)

(29)   (30)   (31)   (32)

(33)   (34)   (35)   (36)

(37)   (38)   (39)

(40)

(41)   (42)   (43)   (44)

(45)   (46)   (47)   (48)

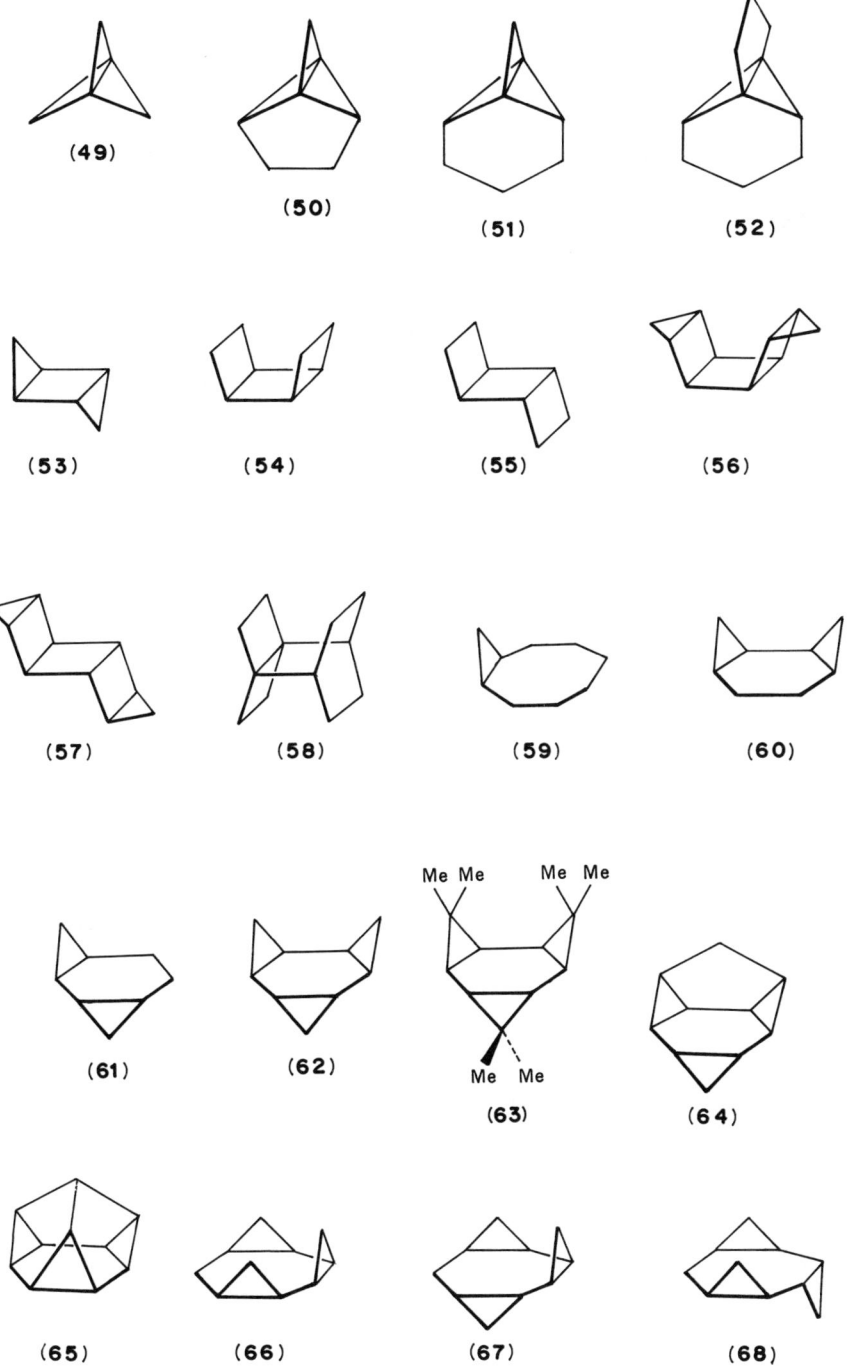

(49)

(50)

(51)

(52)

(53)

(54)

(55)

(56)

(57)

(58)

(59)

(60)

(61)

(62)

Me Me   Me Me

Me   Me

(63)

(64)

(65)

(66)

(67)

(68)

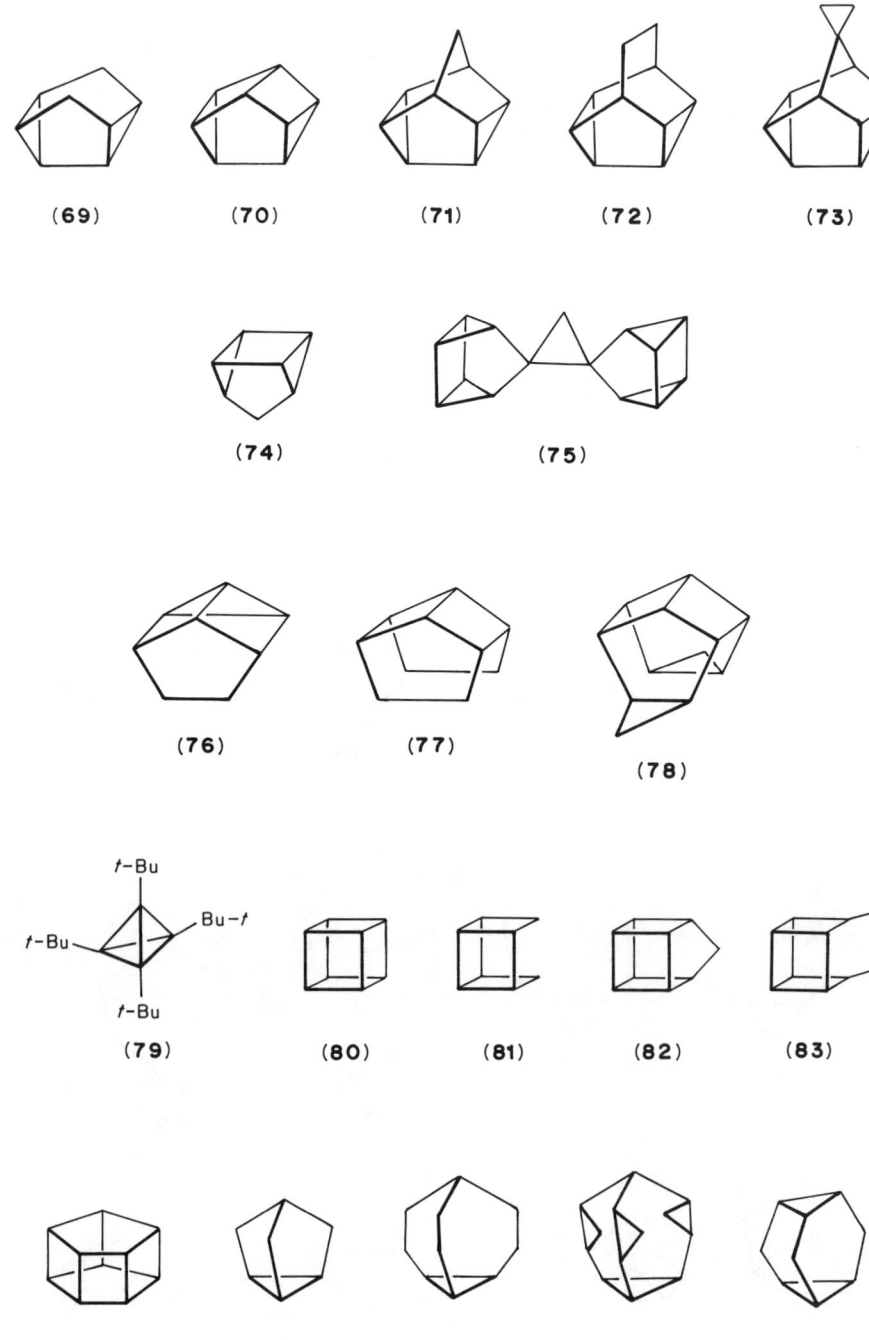

(69)   (70)   (71)   (72)   (73)

(74)   (75)

(76)   (77)   (78)

(79)   (80)   (81)   (82)   (83)

(84)   (85)   (86)   (87)   (88)

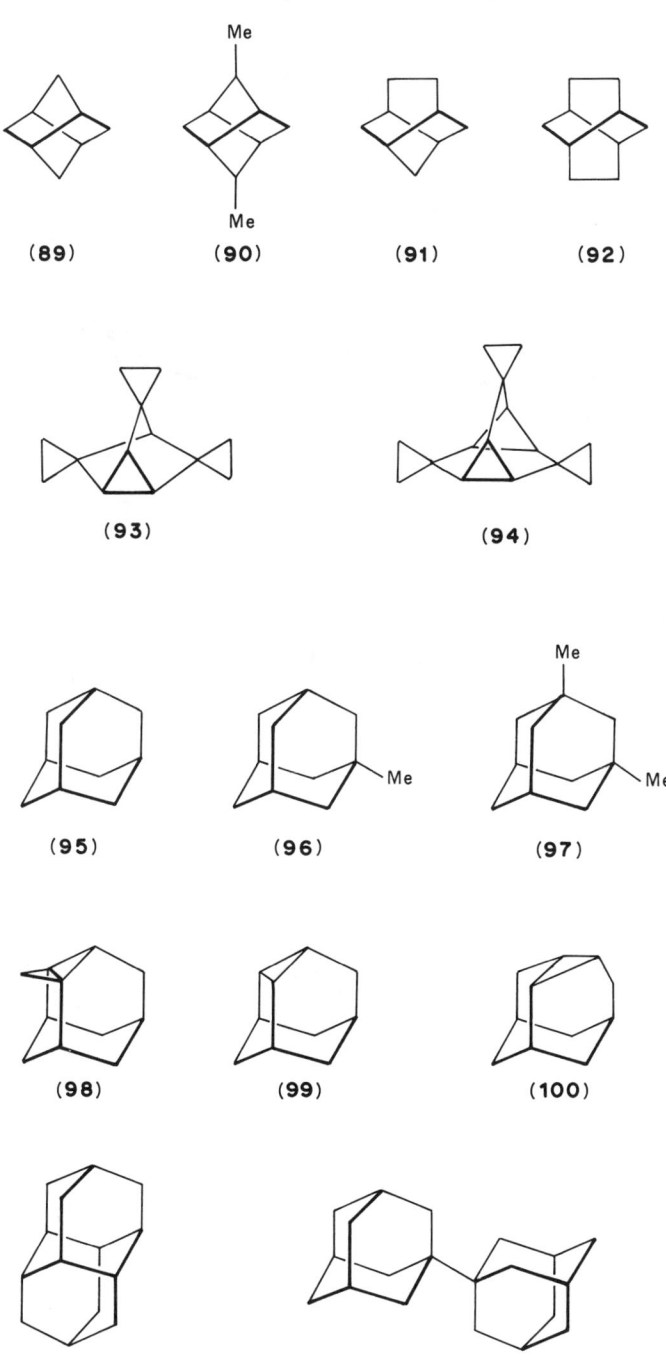

(89)   (90)   (91)   (92)

(93)   (94)

(95)   (96)   (97)

(98)   (99)   (100)

(101)   (102)

 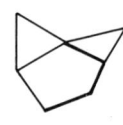 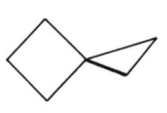 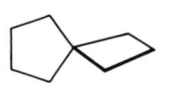 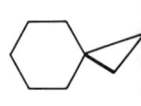

(103)   (104)   (105)   (106)   (107)

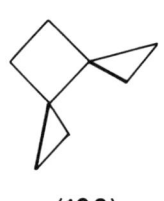 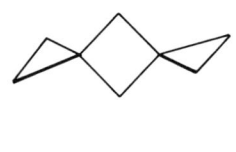 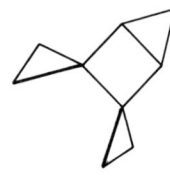

(108)   (109)   (110)

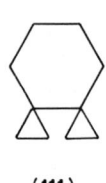 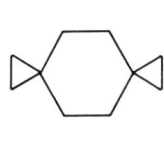 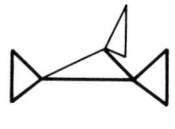 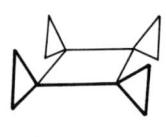

(111)   (112)   (113)   (114)

 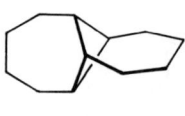

(115)   (116)   (117)

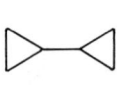 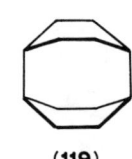 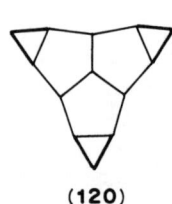

(118)   (119)

(120)

V. DISCUSSION OF PE SPECTRA

A. Methane and Ethane

The PE spectra of methane and ethane are rather special cases, because of their high symmetry. The point groups of methane and of ethane, the latter in its staggered conformation, are T_d and D_{3d} respectively, with the consequence that some of their valence shell orbitals are degenerate. They are indicated in bold type in the following list:

$$CH_4 : \mathbf{1t_2}, 2a_1 \qquad\qquad (1a_1)$$
$$C_2H_6 : 3a_{1g}, \mathbf{1e_g}, \mathbf{1e_u}, 2a_{2u}, 2a_{1g} \quad (1a_{2u})(1a_{1g}) \qquad\qquad (45)$$

$$\xrightarrow{\hspace{4cm}}$$
Descending order of orbital energy

The numbering of the orbitals (of 45) includes those of carbon 1s origin, given in parentheses. The corresponding peaks can only be observed in the ESCA spectra[27], and are found at $I(1a_1) = 290.8\,eV$ for CH_4, and at $\bar{I}(1a) = 290.7\,eV$ for C_2H_6.

The PE spectra of methane and ethane are redrawn in Figure 8 from the He(IIα) spectra recorded by Bieri and Åsbrink[28] (cf also Figure 7). They present some special features, which are due to the Jahn–Teller distortion[131] of the radical cation when the photoelectron has been ejected from one of the degenerate orbitals. Although the PE spectra of methane and ethane have been much studied, both experimentally and theoretically, following the pioneering work by Al-Joboury and Turner[132], we shall restrict our discussion to only a few points, relevant for the ensuing discussion of the PE spectra of higher alkanes.

1. The PE spectrum of methane

The ionization energies[28] of methane and the corresponding assignments are presented in Table 2.

Ejection of an electron from the triply degenerate HOMO $1t_2$ of methane gives rise in its PE spectrum to the complicated feature between 13 and 16 eV (see Figure 8), rather than the simple, single band one might have expected naively, by applying Koopmans' theorem (equation 44) to either the naive EBO results shown in Figure 2, or to those obtained by any other SCF MO treatment. This is a consequence of the Jahn–Teller instability of the orbitally degenerate methane radical cation CH_4^+ of T_d symmetry, reached vertically by removal of an electron from one of the $1t_2$ HOMOs of CH_4. The radical cation CH_4^+ can lower its total energy by undergoing a T-type distortion, whereby one of the HCH bond angles opens up and the opposite HCH bond angle closes down, so that the geometrically relaxed radical cation assumes a C_{2v} structure, as shown qualitatively in the diagrams 46.

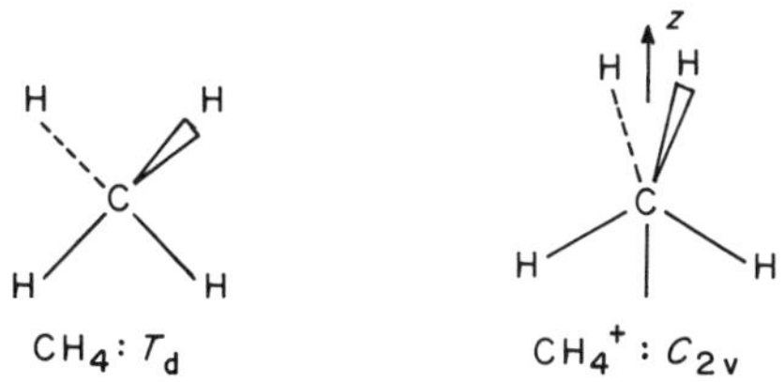

$$CH_4 : T_d \qquad\qquad CH_4^+ : C_{2v} \qquad\qquad (46)$$

Both semi-classical and quantum-mechanical calculations show[133,134] that the resulting Franck–Condon envelope of the band system ①, ②, ③ should consist of three (overlapping) maxima. The position I_2^m of the second maximum is predicted to be close

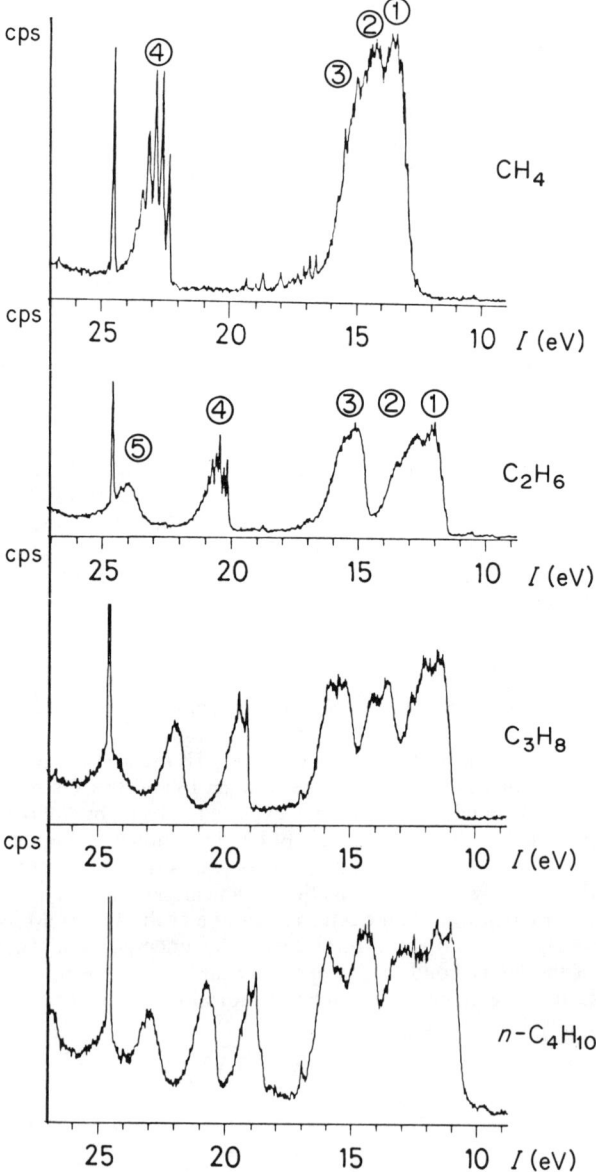

FIGURE 8. He(IIα) PE spectra of methane, ethane, propane and butane, redrawn from Reference 28

TABLE 2. Valence ionization energies of methane and ethane

Methane			Ethane		
Orb.[a]	I_j^m (eV)[b]	I_j^a (eV)[c]	Orb.	I_j^m (eV)[b]	I_j^a (eV)[c]
$1t_2$	$\begin{cases}13.6\\14.4\\15.0\end{cases}$	12.7	$3a_{1g}$	12.0	11.6
			$1e_g$	$\begin{cases}12.7\\13.5\end{cases}$	
$2a_1$	22.9	22.4	$1e_u$	$\begin{cases}15.0\\15.8\end{cases}$	
S.U.[d]	28		$2a_{2u}$	20.4	20.2
S.U.[d]	31		$2a_{1g}$	23.9	

[a] Orb. = Vacated orbital.
[b] I_j^m = position of band maximum, in eV.
[c] I_j^a = adiabatic ionization energy, in eV.
[d] The two bands of methods labelled 'S.U.' are so-called 'shake-up' bands; cf text.

to the true vertical ionization energy, $I_2^v \approx I_2^m$, which corresponds to the creation of the radical cation CH_4^+ having exactly the same T_d structure as the parent hydrocarbon CH_4. The two side bands ① and ③ should be placed symmetrically with respect to the central band ②. This prediction is in reasonable agreement with observation (cf Figure 8), especially if one considers that changes in the Franck–Condon envelope due to the neglected vibronic interactions are presumably important in this case[21]. A more detailed discussion can be found in References 20, 21, 133 and 134.

The band at 22.9 eV, labeled ④ in Figure 8, corresponds to electron ejection from orbital $2a_1$, its vibrational fine structure, with a spacing of $2190\,cm^{-1}$, being due to the breathing vibration of the radical cation. Band ④ is followed at higher ionization energies by two 'shake-up' bands at 28 and 31 eV, not shown in Figure 8. They are due to a 'non-Koopmans' ionization process, i.e. one in which an electron is ejected from the $2a_1$ orbital, and one of the remaining electrons of CH_4^+ is promoted simultaneously to a higher, antibonding (i.e. previously unoccupied) orbital[135]. (The dependence of the intensities of the methane $1t_2$ and $2a_1$ PE bands on photon energy[136,137], and the angular distribution of the corresponding photoelectrons[137,138], have been studied.)

With reference to the results of the naive, two-parameter EBO model of methane shown in Figure 2, we are now in a position to reach a first (crude) estimate of the self-energy A_{CH} of an EBO λ_{CH} and the geminal crossterm $B_{CH,CH}$ between two such EBOs, using Koopmans' theorem (equation 44), i.e. $\varepsilon(1t_2) = A_{CH} - B_{CH,CH} = -I^v(1t_2) \approx -I^m(1t_2) = -14.4\,eV$, and $\varepsilon(2a_1) = A_{CH} + 3B_{CH,CH} = -I^v(2a_1) \approx -I^m(2a_1) = -22.9\,eV$. This yields

$$\text{Methane EBO:} \quad A_{CH} = -16.5\,eV, \quad B_{CH,CH} = -2.1\,eV \tag{47}$$

However, it should be remembered that methane is a rather special case, because of its high symmetry.

2. The PE spectrum of ethane

The assignment of the PE spectrum of ethane, given in list 45 and in Table 2, is due to Richartz and his coworkers[139], in particular as far as the first three maxima are concerned. It differs from the assignments given in, e.g., References 32 and 140, where the first two maxima have been assigned to the $1e_g$ orbitals.

The problem associated with the interpretation of the low-energy part of the ethane

PE spectrum stems from the extreme closeness, i.e. the (almost) accidental degeneracy of the $1e_g$ and $3a_{1g}$ CMO energies, and their dependence on small changes of molecular geometry, which affect their relative sequence. In addition, it should be remembered that in such cases the sequence predicted, by applying Koopmans' theorem to the CMO sequence of the parent molecule, does not necessarily reflect the state sequence of the radical cation.

To illustrate the first part of the problem, we use again simple EBO models. The two-parameter AB-model yields the $3a_{1g}$ as HOMO at $\varepsilon(3a_{1g}) = A - 1.65B$. This CMO is dominated by λ_{CC}. The two CMO pairs $1e_g$ and $1e_u$ are accidentally degenerate with $\varepsilon(1e_g) = \varepsilon(1e_u) = A - B$, and entirely centred on the λ_{CH}. This suggests that lowering the self-energy A_{CC} of the EBO λ_{CC} will lower $\varepsilon(3a_{1g})$, and that introducing vicinal interactions $\Gamma_{CH,CH}$ between the EBOs λ_{CH} will move $\varepsilon(1e_g)$ towards higher energies, thereby closing up the gap between these two orbital energies. As an example we now use parameters that have been chosen to mimic an STO-3G calculation[14], i.e. $A_{CH} = -17.0\,\text{eV}$, $A_{CC} = -17.5\,\text{eV}$, $B_{CH,CH} = B_{CH,CC} = -2.9\,\text{eV}$ and $\Gamma_{CH,CH}(\tau) = \Gamma_0 \cos \tau = (-1.0\,\text{eV}) \cos \tau$, which yields $\Gamma_{CH,CH}(60°) = -0.5\,\text{eV}$ for two CH bonds in *gauche* conformation and $\Gamma_{CH,CH}(180°) = +1.0\,\text{eV}$ for two antiplanar ones. Diagonalization of the resulting 7×7 matrix $\mathbf{H}_A$ leads to $\varepsilon(1e_g) = -12.60\,\text{eV}$ and $\varepsilon(3a_{1g}) = -12.57\,\text{eV}$, two values which are identical within the significance of the EBO approximation, but well separated from the next ones, e.g. $\varepsilon(1e_u) = -15.60\,\text{eV}$. Similarly close values are obtained by other, much more sophisticated SCF models. It follows that the assignment of the first three maxima can only be obtained by taking into account geometric changes in the radical cation, electron relaxation and electron correlation effects[139] and also vibronic mixing.

Ethane is an instructive example of the inherent limitations of our EBO approach, which this simple model shares with more realistic SCF procedures, e.g. the calculation of ethane CMOs by extensively parametrized semi-empirical models[52], or by *ab initio* procedures, such as the floating Gaussian orbital model[141]. Obviously, as shown by Richartz and his coworkers[139], configuration interaction methods yielding reliable cation state energies, including their dependence on the cation geometry, are required.

The vibrational fine structure of the PE bands in the spectrum of ethane has been analysed by Rabalais and Katrib[140].

B. The C2s Manifold

1. Preliminary comments

There are good reasons for starting a discussion of the PE spectra of saturated hydrocarbons C_nH_m with the C2s bands found above 15 eV. This part of the spectrum consists—at least for the smaller hydrocarbons—of only n well-spaced bands, whereas the C2p system contains $n + m/2$ overlapping bands within a narrow range of *ca* 5 eV, as can be seen in Figures 7 and 8. This suggests that the C2s bands should be much simpler to analyse and assign than the C2p manifold, at least if we want to think in terms of simple MO models and Koopmans' theorem. As we shall see, this is indeed the case.

The reader should be warned that this seeming simplicity of the C2s band system of hydrocarbon PE spectra is spurious, and largely a consequence of the low (inherent and instrumental) resolution of the experiment. Theoretical treatments including the effects of electronic relaxation and electron correlation[135] show that the observed shapes of the individual C2s bands are the result of the superposition of a whole series of Franck–Condon envelopes, each corresponding to one of many cation states involving multiply excited configurations. However, the positions of these individual bands tend to cluster around the one band (dominantly) associated with the Koopmans' state due to simple electron ejection from a C2s CMO. Thus, from a naive point of view, one can discuss the C2s part of the PE spectrum of a saturated hydrocarbon C_nH_m, as if it corresponded to simple Koopmans' states of the radical cation $C_nH_m{}^+$.

Within the framework of an EBO model, the essential difference between the two parts of a hydrocarbon PE spectrum can be easily rationalized, e.g. with reference to the results stemming from an AB-model. Taking propane, C_3H_8, as an example, we draw from Figure 5 the following conclusions, which prove to be generally valid for all saturated hydrocarbons:

(a) The three C2s-type CMOs φ_j ($j = 1, 2, 3$) are *bonding* with respect to the self-energies $A_{CC} = A_{CH} = A$. As we shall show in the next section, both the shape of these C2s orbitals φ_j and their orbital energies ε_j change very little if higher crossterms ($\Gamma_{\mu\nu}, \Delta_{\mu\nu}$) are included in their calculation, e.g. if we implement the AB-model to obtain an $A\Gamma$-model.

(b) The seven C2p-type CMOs φ_j ($j = 4$ to 10) are *antibonding* with respect to the self-energies $A_{CC} = A_{CH} = A$. Not only do their orbital energies ε_j fall into a small range, but five of the CMOs φ_j ($j = 4$ to 8) are accidentally degenerate with orbital energies $\varepsilon_j = A - B$ ($j = 4$ to 8). In contrast to the C2s manifold, the introduction of higher crossterms ($\Gamma_{\mu\nu}, \Delta_{\mu\nu}$) has a profound effect on all the C2p orbital energies ε_j, and it will lead—symmetry allowing—to considerable mixing of the corresponding CMOs φ_j. In particular, the crossterms yield first-order splits between the formerly degenerate orbitals, with the result that the CMO sequence will depend critically on the size and sign of these crossterms. However, the resulting C2p-type CMOs will still be close in energy, with the usual consequence that inclusion of electronic reorganization and electron correlation into the calculation can (and often does) change the predicted order of the close-lying C2p states. Finally, vibronic mixing could further invalidate an assignment derived from many-electron SCF treatments, and *a fortiori* from a naive EBO model.

We conclude that an adequate and heuristically useful discussion of the C2s part of the PE spectra of saturated hydrocarbons should be possible with reference to simple EBO models, in contrast to the C2p part where the applicability of such models will presumably be limited to only the simplest cases.

2. Linear and monocyclic hydrocarbons

The top row of the correlation diagrams of Figure 9[13,18] show the observed PE band positions I_j^m ($j = 1, 2, \ldots, n$) for the lower members of the homologous series of linear and branched hydrocarbons C_nH_{2n+2}, and of the monocyclic hydrocarbons C_nH_{2n}. The bottom row of the correlation diagrams refers similarly to the eigenvalues x_j—defined in equation 38—of the adjacency matrix $\mathbf{A}$ of the graph $\mathscr{G}$ of the particular hydrocarbon, which defines the corresponding eigenvalue problem (equation 39). As an example, the graph $\mathscr{G}$ was given in Figure 4b for propane.

The similarity between the two sets of correlation diagrams, showing either the dependence of the ionization energies I_j^m or of the eigenvalues x_j on the size n and the topography of the hydrocarbons, is rather striking. Obviously the simple AB-model accounts surprisingly well for even fine details in the relative spacings of the observed I_j^m values. This suggests that a parametrization of the matrix elements of the matrices $\mathbf{H}_\lambda$ (shown in Figure 4 for the particular example of propane), i.e. of the self-energy $A_{CC} = A_{CH} = A$ and the geminal interaction parameter $B_{CC,CC} = B_{CC,CH} = B_{CH,CH} = B$ by a simple linear regression calculation of I_j^m on x_j, should yield excellent quantitative agreement. Using 58 paired data, I_j^m, x_j stemming from a calibration set consisting of the saturated linear and branched hydrocarbons C_nH_m with $n \leqslant 5$ and the monocyclic ones with $n \leqslant 8$, one obtains the parameters 48[18]. The errors quoted for A and B are

492

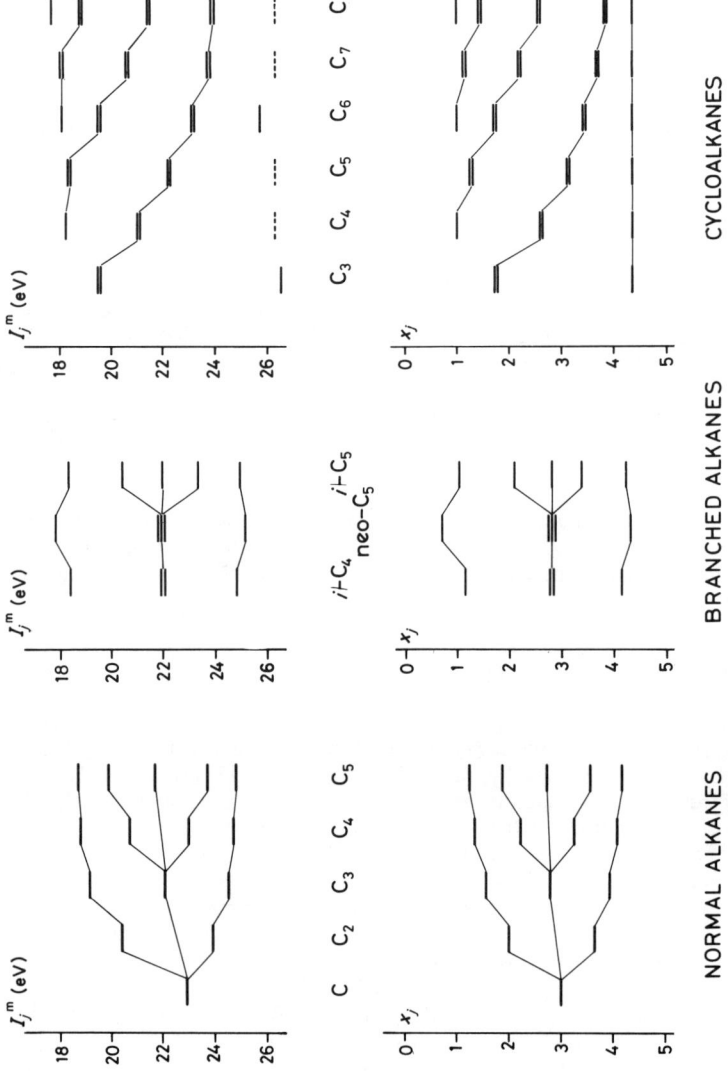

FIGURE 9. Comparison of the observed ionization energies I_j^m of the C2s bands in the He(IIα) PE spectra of linear, branched and monocyclic hydrocarbons with the lowest eigenvalues x_j (equation 38) obtained by diagonalizing the adjacency matrix A of the corresponding AB-models (cf equation 39). Normal alkanes: C_n, $n = 1$ to 5, methane to pentane. Branched alkanes: i-C_4 = isobutane, i-C_5 = isopentane, neo-C_5 = neopentane. Cycloalkanes: C_n, $n = 3$ to 8, cyclopropane to cyclooctane

AB-Model:

$$A = -15.83 \pm 0.07\,\text{eV}, \quad B = -2.17 \pm 0.03\,\text{eV} \tag{48}$$

$$r = 0.9961$$

standard deviations, based on 56 degrees of freedom. (This result should be compared to the one given in equation 47 for methane, i.e. $A = -16.5\,\text{eV}$, $B = -2.1\,\text{eV}$.) Accordingly, our calibrated EBO model yields predictions for the C2s ionization energies according to regression 49. As an example we present the results for propane in Table 3 (cf Figure 5).

$$I_j^m(\text{calc.}) = (15.9 + 2.2x_j)\,\text{eV} \tag{49}$$

$$x_j < 0$$

TABLE 3. Calculated and observed ionization energies of propane

j	Eigenvalue x_j (Figure 5)	$I_j^m(\text{calc.})$ (eV) (equation 49)	$I_j^m(\text{obs.})$ (eV) from Reference		
			28	25	27
3	1.552	19.2	19.5	19.2	19.6
2	2.791	21.9	22.1	22.1	22.0
1	3.925	24.3	24.7	24.5	24.6

The overall quality of the regression 49 is demonstrated in Figure 10, where the observed ionization energies $I_j^m(\text{obs.})$ of linear, branched and cyclic saturated hydrocarbons have been plotted vs the x_j values stemming from the diagonalization of the corresponding adjacency matrices **A**. The quality of these regressions is quite remarkable, and it is worthwhile to examine in a bit more detail why this is so.

Using again propane as a particular example, we first examine the influence of the inclusion of the vicinal crossterm $\Gamma_{\mu\nu} = \Gamma_0 \cos(\tau_{\mu\nu})$ (list 40) into the original matrix $\mathbf{H}_\lambda$ on the eigenvalues x_j, still neglecting the $\Delta_{\mu\nu}$ elements. The relative conformations of the labeled CH bonds (cf Figure 4) are shown in drawing 50. As we shall see later,

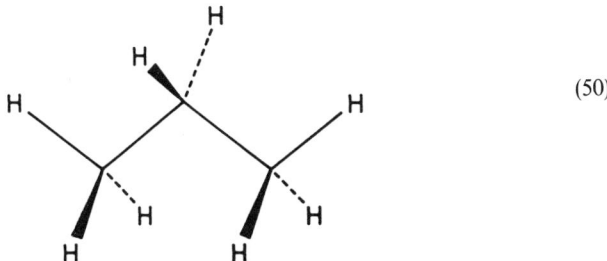

(50)

the value $\Gamma_0 = -1.0\,\text{eV}$, or in units of B values, $\Gamma_0 = 0.455B$, is an appropriate parameter. This allows us to write $\Gamma_{\mu\nu} = \Gamma_0 \cos(\tau_{\mu\nu}) = B \times 0.455 \cos(\tau_{\mu\nu}) = BG_{\mu\nu}$, with $G_{\mu\nu} = 0.455 \cos(\tau_{\mu\nu})$. This yields for a pair μ, ν of *gauche* CH bonds $\Gamma(60°) = -0.227B$, and for an anti-planar pair μ, ν, $\Gamma(180°) = \Gamma_0 \cos(180°) = 0.455B$, or $G(60°) = -0.227$ and $G(180°) = 0.455$. Making use of the abbreviation 38, our problem is reduced to the diagonalization of an augmented, symmetric matrix $\mathbf{A} + \mathbf{G}$, where **A** is the adjacency matrix underlying $\mathbf{H}_\lambda = A\mathbf{1} + B\mathbf{A}$ given in Figure 4, and **G** the matrix of the crossterms $G_{\mu\nu}$, i.e. $\mathbf{G} = (G_{\mu\nu})$. For propane, the matrix $\mathbf{A} + \mathbf{G}$ is given in 51.

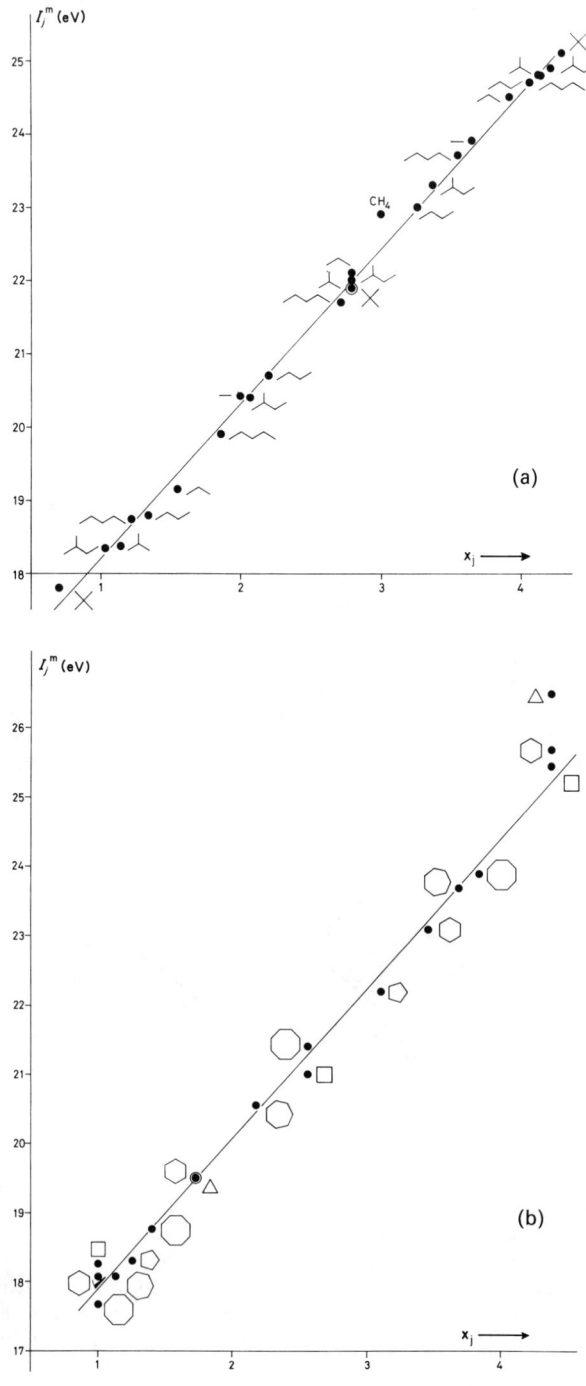

FIGURE 10. (*caption opposite*)

$$
\begin{pmatrix}
0 & 1 & 1 & 1 & 0.23 & 0.23 & -0.46 & 0 & 0 & 0 \\
 & 0 & 1 & 1 & 0.23 & -0.46 & 0.23 & 0 & 0 & 0 \\
 & & 0 & 1 & -0.46 & 0.23 & 0.23 & 0 & 0 & 0 \\
 & & & 0 & 1 & 1 & 1 & -0.46 & 0.23 & 0.23 \\
 & & & & 0 & 1 & 1 & 0.23 & 0.23 & -0.46 \\
 & & & & & 0 & 1 & 0.23 & -0.46 & 0.23 \\
 & & & & & & 0 & 1 & 1 & 1 \\
 & & \text{id} & & & & & 0 & 1 & 1 \\
 & & & & & & & & 0 & 1 \\
 & & & & & & & & & 0
\end{pmatrix} \tag{51}
$$

The lowest three eigenvalues of matrix 51 differ only marginally from those of the adjacency matrix $\mathbf{A}$ (cf Figure 5, where $\mathbf{H}_\lambda = A\mathbf{1} + B\mathbf{A}$ is shown):

	x_1	x_2	x_3
from A :	3.925	2.791	1.552
from matrix 51:	3.927	2.806	1.584

$$\tag{52}$$

and the corresponding linear combinations $\varphi_1, \varphi_2, \varphi_3$ derived from matrix 51 (cf 52) are practically identical to those depicted in Figure 5. If we are allowed to generalize this particular example, we conclude that vicinal crossterms are—by and large—irrelevant for the bonding orbitals φ_j of a hydrocarbon EBO model, i.e. for those whose eigenvalues x_j are positive, which means that their corresponding orbital energies ε_j are below the self-energy A of the EBOs.

There is a good reason for this observation. It is obvious from the diagrams of the bonding orbitals $\varphi_1, \varphi_2, \varphi_3$ of propane, shown in Figure 5, that they can be described to a good approximation in terms of the partial linear combinations 53, each of which is totally symmetric with respect to the local (almost) T_d symmetry at the respective carbon atoms, and could therefore be called 'pseudo-2s' orbitals. The numbering of the

$$
\begin{aligned}
\phi_a(2s) &= (\lambda_1 + \lambda_2 + \lambda_3 + \lambda_4)/2 \\
\phi_b(2s) &= (\lambda_4 + \lambda_5 + \lambda_6 + \lambda_7)/2 \\
\phi_c(2s) &= (\lambda_7 + \lambda_8 + \lambda_9 + \lambda_{10})/2
\end{aligned} \tag{53}
$$

EBOs and of the carbon centres refers to structure 50. The bonding CMOs $\varphi_1, \varphi_2, \varphi_3$ shown in Figure 5 can now be written to a good approximation as where—for the

$$
\begin{aligned}
\varphi_1 &= a\phi_a(2s) + b\phi_b(2s) + a\phi_c(2s) \\
\varphi_2 &= (\phi_a(2s) - \phi_c(2s))/\sqrt{2} \\
\varphi_3 &= c\phi_a(2s) - d\phi_b(2s) + c\phi_c(2s)
\end{aligned} \tag{54}
$$

FIGURE 10. Linear regressions of the ionization energies I_j^m of the C2s bands in the He(IIα) spectra of linear, branched (top) and monocyclic (bottom) hydrocarbons on the lowest eigenvalues x_j (equation 38) obtained by diagonalizing the adjacency matrix $\mathbf{A}$ of the corresponding AB-models (cf equation 39)[18]. Linear and branched hydrocarbons: $I_j^m = [(16.10 \pm 0.08) + (2.11 \pm 0.03)x_j]\,\text{eV}$, $r = 0.9980$. Monocyclic hydrocarbons: $I_j^m = [(15.67 \pm 0.09) + (2.19 \pm 0.03)x_j]\,\text{eV}$, $r = 0.9970$

moment—the only restriction governing the coefficients a, b, c and d is that of the orthogonality of the CMOs φ_i.

To calculate the orbital energies ε_j corresponding to the CMOs 54 and the values of the coefficients a, b, c and d, one must first compute the necessary matrix elements. It follows from definitions 53 that the *pseudo*-2s orbitals $\phi_\mu(2s)$ do *not* form an orthogonal set (cf equations 55). If A are the self-energies and B the geminal crossterms

$$S_{ab} = \langle \phi_a(2s)|\phi_b(2s) \rangle = 1/4$$
$$S_{bc} = \langle \phi_b(2s)|\phi_c(2s) \rangle = 1/4 \tag{55}$$
$$S_{ac} = \langle \phi_a(2s)|\phi_c(2s) \rangle = 0$$

for the EBOs λ_μ of the $A\Gamma$-model underlying the matrix 51, then the self-energies $\alpha_\mu(2s)$ and the crossterms $\beta_{\mu\nu}(2s)$ for the '*pseudo*-2s' orbitals (definition 54) are given by equations 56.

$$\alpha_\mu(2s) = \langle \phi_\mu(2s)|\mathscr{H}|\phi_\mu(2s) \rangle = A + 3B, \qquad \mu = a, b, c$$
$$\beta_{ab}(2s) = \langle \phi_a(2s)|\mathscr{H}|\phi_b(2s) \rangle = 3B/2 + A/4$$
$$\beta_{bc}(2s) = \langle \phi_b(2s)|\mathscr{H}|\phi_c(2s) \rangle = 3B/2 + A/4 \tag{56}$$
$$\beta_{ac}(2s) = \langle \phi_a(2s)|\mathscr{H}|\phi_c(2s) \rangle = B/4$$

The important point is that $\beta_{ab}(2s)$, $\beta_{bc}(2s)$ and $\beta_{ac}(2s)$ do *not* depend on the vicinal crossterms $\Gamma(\tau)$, because $2\Gamma(60°) + \Gamma(180°) = 0$, assuming tetrahedral bond angles throughout. (This can be verified by inspection of matrix 51, where the elements of the submatrices referring to vicinal interactions, e.g. the submatrix consisting of the intersection of rows 1,2,3 with columns 5,6,7, add up to zero.) Under these conditions one has obtained for the C2s manifold of CMOs an independent electron model with fixed overlap $S = 1/4$ between bonded basis functions $\phi_\mu(2s)$. Such a model had originally been proposed by Wheland[142], to include a constant overlap S between bonded 2p AOs in the Hückel treatment of π systems. If 1,3-interactions are neglected, because $B/4 \ll 3B/2 + A/4$, then the graph $\mathscr{G}_{2s}$ is simply the one representing the connectivity between the carbon atoms in the structural formula of the hydrocarbon. If $\mathbf{A}_{2s}$ is its adjacency matrix, then the eigenvalue problem to be solved is defined in equations 57.

$$\det(\mathbf{H}_{2s} - \varepsilon\mathbf{S}_{2s}) = 0$$
$$\mathbf{H}_{2s} = \alpha(2s)\mathbf{1} + \beta(2s)\mathbf{A}_{2s} \tag{57}$$
$$\mathbf{S}_{2s} = \mathbf{1} + S\mathbf{A}_{2s} \ (= \mathbf{1} + \tfrac{1}{4}\mathbf{A}_{2s})$$

The above analysis vindicates the application of this treatment by Potts, Streets and coworkers[25] for the rationalization of the C2s part of the PE spectra of saturated hydrocarbons. (Note that the calibration proposed by these authors[25] assumes that $\alpha_\mu(2s), \beta_{\mu\nu}(2s)$ and $S_{\mu\nu}$ can be treated as independent parameters, which is not the case.)

As an example we apply the Wheland treatment to propane, for which the determinant 57 takes the form given in equation 58.

$$\det \begin{pmatrix} \alpha - \varepsilon & \beta - S\varepsilon & 0 \\ \beta - S\varepsilon & \alpha - \varepsilon & \beta - S\varepsilon \\ 0 & \beta - S\varepsilon & \alpha - \varepsilon \end{pmatrix} = 0 \tag{58}$$

Using the A and B values given in the list 48, one obtains from equations 56 $\alpha_\mu(2s) \equiv \alpha = -22.4\,\text{eV}$, $\beta_{ab}(2s) = \beta_{bc}(2s) \equiv \beta = -7.25\,\text{eV}$ $[\beta_{ac}(2s) \approx 0\,\text{eV}]$, and according to equations 55, $S_{ab} = S_{bc} \equiv S = 0.25$ and $S_{ac} = 0$. Inserting these values into the determinant 58 and solving this general eigenvalue problem yields for propane the C2s orbital energies

$\varepsilon_1 = -24.1\,\mathrm{eV}$, $\varepsilon_2 = -22.4\,\mathrm{eV}$ and $\varepsilon_3 = -18.8\,\mathrm{eV}$, which compare reasonably well with those given in the list 52, considering the simplicity of the model (cf Table 3).

Of course, agreement with the observed ionization energies can be improved by a recalibration of the Wheland procedure on the basis of the C2s ionization energies of the linear (branched) hydrocarbons C_nH_{2n+2} and of the monocyclic hydrocarbons C_nH_{2n}. Assuming $S = 0.25$, as obtained above, a linear regression of the experimental I_j^m values on the computed orbital energies ε_j yields the parameters $\alpha = -22.50\,\mathrm{eV}$ and $\beta = -7.00\,\mathrm{eV}$. The residual mean square about the regression is $s^2 = 0.683\,\mathrm{eV}^2$ (for 44 degrees of freedom), and the correlation coefficient is $r = 0.949$. However, it is obvious by inspection of the plot of the 46 data pairs that the regression departs from linearity. This is confirmed by performing a quadratic regression calculation, which leads to a very significant reduction of the residual mean square to $s^2 = 0.202\,\mathrm{eV}^2$. The observed curvature corrects for the fact that a linear extrapolation of the data points referring to the lowest C2s orbitals would predict much too high orbital energies of the upper orbitals of the C2s manifold, i.e. too low ionization energies for the first bands in the C2s-band system. The reason is that our Wheland-type model neglects, necessarily, the interaction of the C2s-mainfold orbitals with the remaining, higher-lying orbitals of the C2p manifold. These interactions mix C2s-type orbitals with C2p-type orbitals of the same symmetry behaviour, which leads to depressions of the orbital energies ε_j of the C2s dominated hybrid orbitals, affecting mainly the upper ones, which are closer to the C2p manifold.

It is a curious and amusing observation that this effect can be compensated on the level of our simplistic model by setting all overlaps to zero, $S = 0$, which means that our Wheland model reverts to a standard Hückel model based on a graph $\mathscr{G}_{2s}$ in which each vertex corresponds to a carbon centre and each edge to a CC bond. Using the calibration set mentioned above, this leads to the linear regression

$$I_j^m(\mathrm{calc.}) = (21.72 \pm 0.12)\,\mathrm{eV} + (1.88 \pm 0.09)\,\mathrm{eV}\,x_j \tag{59}$$

shown in Figure 11. The uncertainties given for the constant and the slope of regression 59 refer to the 95% confidence limits. The correlation coefficient r is 0.989, and the residual mean square $s^2 = 0.152\,\mathrm{eV}^2$. Note that $I_j^m(\mathrm{calc.})$ values derived from regression 59 are affected with standard errors $s = 0.39\,\mathrm{eV}$, in addition to those due to the uncertainties of the parameters. It is obvious from Figure 11 that the quality of the regression 59 is much lower than that of the standard AB-model, depicted in Figure 10. However, the regression 59 is quite sufficient as a reasonably reliable guide for qualitative purposes, as demonstrated by the following example[14].

Assume that we wish to correlate the seven bands in the C2s band systems of norbornane **19** and quadricyclene **74**, which are observed at the values given in the list 60.

	1	2	3	4	5	6	7	
19:	17.7,	17.8,	19.8,	22.3,	23.0,	23.3,	25.9 eV	(60)
74:	17.2,	17.9,	18.1,	20.8,	22.7,	23.7,	26.9 eV	

The graphs $\mathscr{G}_{2s}$ for norbornane and quadricyclene are given in drawing 61. Diagonalization of the adjacency matrices $A_{2s}(\mathbf{19})$ and $A_{2s}(\mathbf{74})$ of the graphs $\mathscr{G}_{2s}(\mathbf{19})$ and $\mathscr{G}_{2s}(\mathbf{74})$ yields the eigenvalues x_j, and from the corresponding eigenvectors c_j the C2s CMOs φ_j. Their symmetry labels with respect to the point group C_{2v}, together with the corresponding ionization energies I_j^m obtained via the regression 59, are presented in display 62, and the resulting correlation diagram[143] in Figure 12. The relative spacings of the eigenvalues x_j are in respectable agreement with the observed values. This suggests that the assignment 62 and the correlation offered in Figure 12 are at least potentially useful working hypotheses.

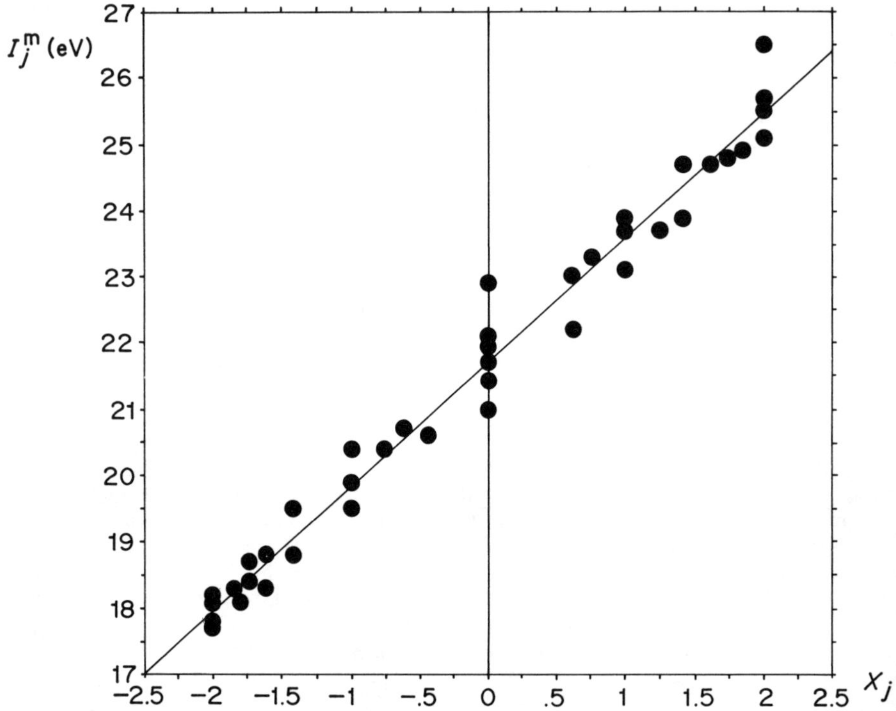

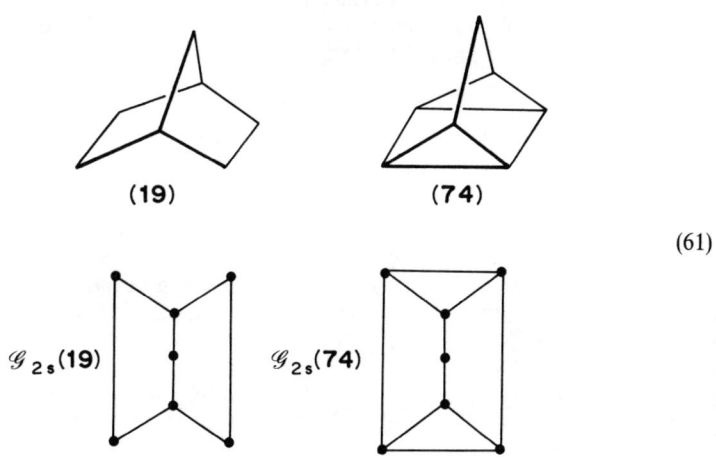

FIGURE 11. Linear regression 59 of the ionization energies I_j^m of the C2s bands in the He(IIα) PE spectra of linear, branched and monocyclic hydrocarbons on the eigenvalues x_j of the corresponding 'pseudo-2s' HMO model. The calibration set of hydrocarbons is the same as for Figure 9

(19)   (74)

$\mathscr{G}_{2s}$(19) $\mathscr{G}_{2s}$(74)

(61)

	1	2	3	4	5	6	7	
19:	$2b_1$	$3a_1$	$1a_2$	$2a_1$	$1b_2$	$1b_1$	$1a_1$	
I_j^m	18.0,	18.3,	19.8,	22.6,	23.6,	23.6,	26.1 eV	(62)
74:	$1a_2$	$3a_1$	$2b_1$	$1b_2$	$2a_1$	$1b_1$	$1a_1$	
I_j^m	18.0,	18.5,	19.1,	21.7,	23.2,	24.6,	27.2 eV	

The simplicity of this treatment, and its seeming success in correlating C2s bands in divers hydrocarbon PE spectra, should not be construed to mean that this model yields a faithful insight into the electronic mechanisms controlling the interactions of localized carbon 2s AOs. As usual, there is more juice in a grapefruit than meets the eye.

3. C2s band structure of long polymethylene chains

On top of Figure 13 are shown the PE spectra[27] of n-nonane, C_9H_{20}, and of n-hexatriacontane, $C_{36}H_{74}$, the latter being practically identical to that of an infinitely long polymethylene chain —$(CH_2)_\infty$—, i.e. of polyethylene. In the nonane PE spectrum, the C2s-band system extending from ca 17 eV to ca 25 eV is partially resolved into five broad bands, whereas in the case of hexatriacontane the C2s-band system in the same range remains completely unresolved. The electronic structure of polyethylene has been studied repeatedly, e.g. by Dewar and his coworkers (Reference 144 and references cited therein). In the following we discuss briefly the application of the AB-model to long chains of methylene groups, i.e. to the higher alkane homologues, and in particular to polyethylene[145].

Using again the abbreviation $-x = (A - \varepsilon)/B$, the problem is reduced once more to the search for the eigenvalues x_j of the adjacency matrix $A(n)$ of the graph $\mathscr{G}(n)$ representing the molecule H—$(CH_2)_n$—H, e.g. for pentane, $n = 5$ (drawing 63). It can be shown[146] that the eigenvalues x_j of $A(n)$, for arbitrary n, can be obtained (up to small corrections[147] which furthermore decrease as $1/n$, and can safely be neglected for $n > 10$) by solving the determinant 64 for $k = 1, 2, \ldots, n$, where κ stands for $\kappa = \exp[2\pi i/(2n + 2)]$. This yields $3 \times n$ eigenvalues $x_{k,i}$ ($i = 1, 2, 3$), of which we need

$$G(5) \quad \quad (63)$$

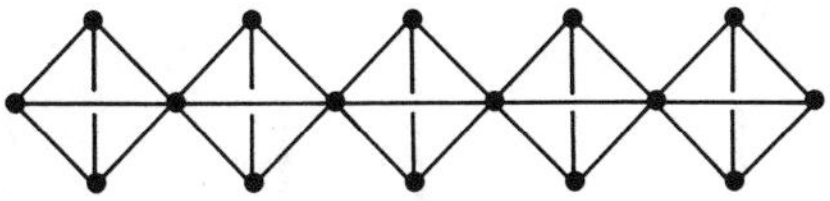

$$\mathbf{det} \begin{pmatrix} -x_k & 1 + \kappa^{-k} & 1 \\ 1 + \kappa^k & -x_k + 2\cos[2\pi k/(2n+2)] & 1 + \kappa^k \\ 1 & 1 + \kappa^{-k} & -x_k \end{pmatrix} = 0 \qquad (64)$$

only the n positive ones, which pertain to the C2s CMOs φ_j with $\varepsilon_j = A + Bx_j$, i.e. those which are bonding with respect to the self-energy A.

To derive the expected contour of the C2s-band system, one has to fold each of the n individual components in the spectrum $\{x_j\}$ with a shape function representing the band shape. This folding function could be either a normalized Gaussian $G(x)$ or a normalized Lorentzian $L(x)$, given in equations 65, where σ and τ characterize the width of the components.

$$\begin{aligned} G(x) &= (1/\sigma\sqrt{2\pi})\exp[-(x - x_j)^2/2\sigma^2] \\ L(x) &= (\tau/\pi)/[(x - x_j)^2 + \tau^2] \end{aligned} \qquad (65)$$

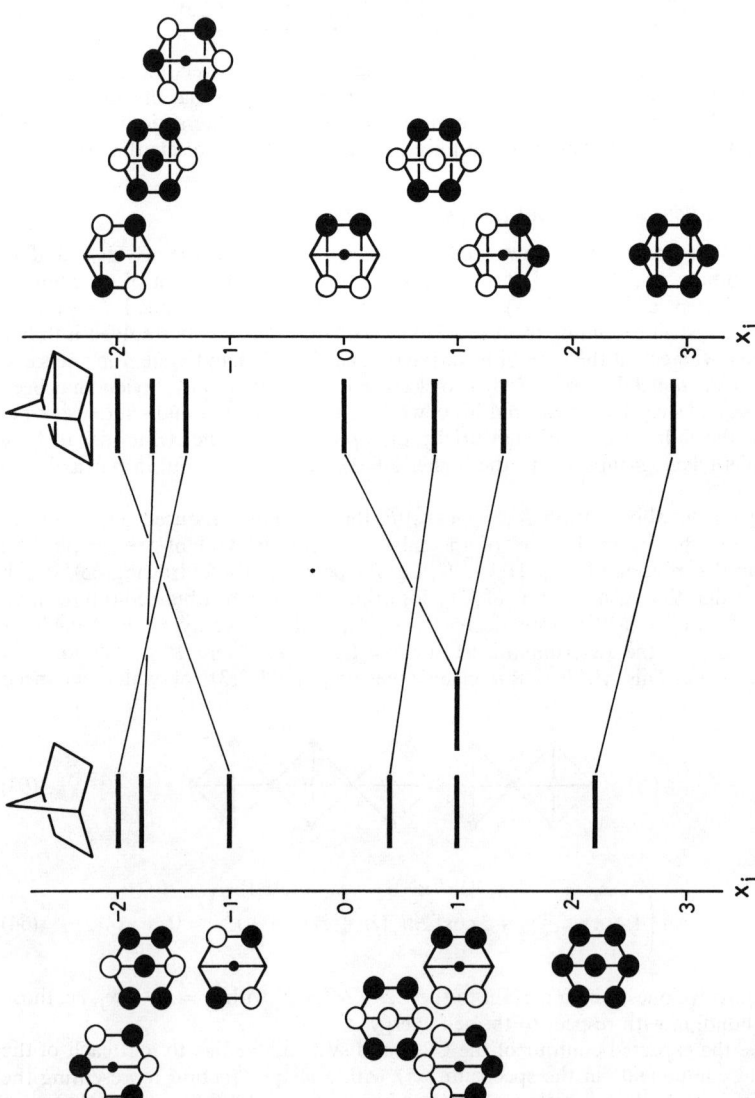

FIGURE 12. Schematic representation of the C2s HMOs of norbornane **19** and quadricyclene **74** obtained by using linear combinations of 'pseudo-2s' basis functions. Open and closed circles represent only the relative phases of the '2s' orbitals in the CMOs. Their size has not been adjusted to represent their relative contributions to the linear combinations

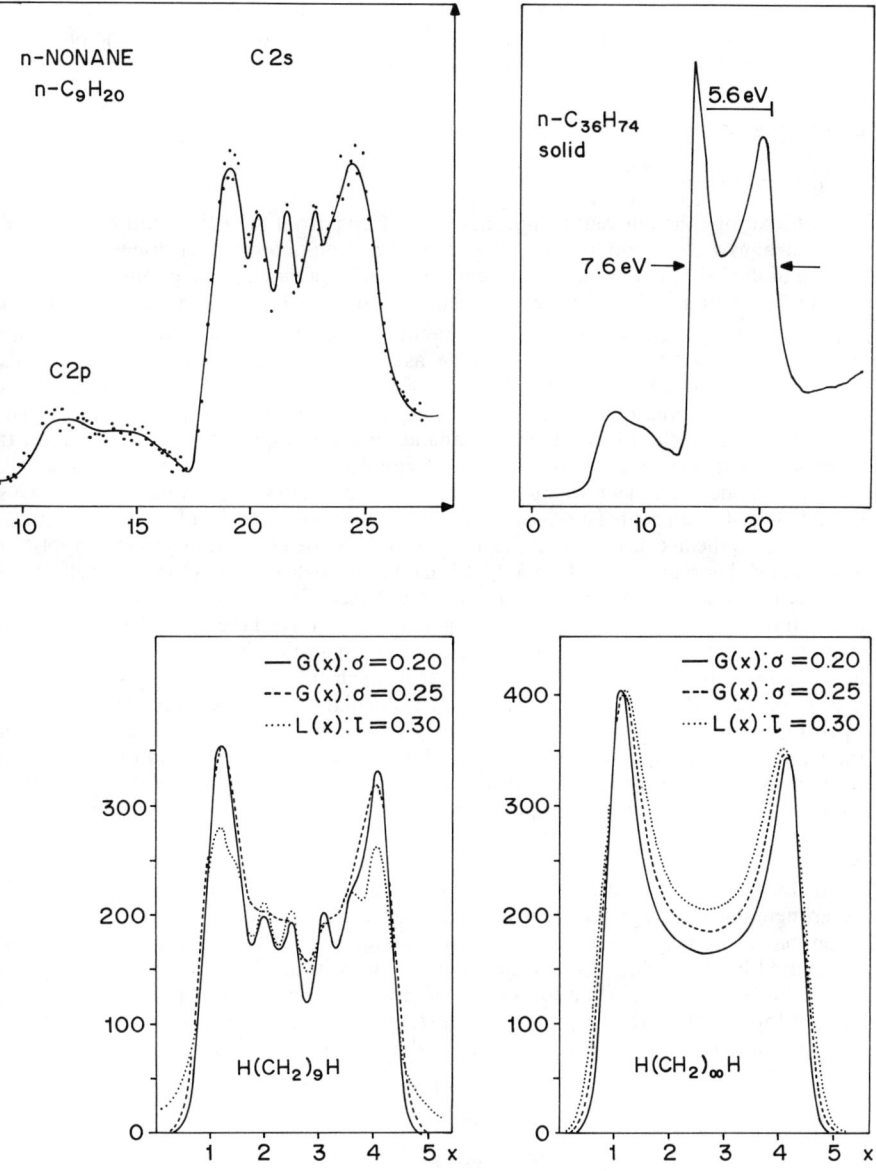

FIGURE 13. Top: PE spectra of the normal alkanes nonane C_9H_{20} and hexatriacontane $C_{36}H_{74}$ taken from Reference 27. Bottom: Calculated contours[144] of the C2s-band systems of nonane and polyethylene using the EBO model and the empirical line shape functions $G(x)$ and $L(x)$ (equations 65). The intensities are in arbitrary units. The different contours have been calculated with the half-widths τ or σ-indicated in the Figure

502   E. Heilbronner

The results^{145} are shown for nonane, $n = 9$, and for polyethylene, $n = \infty$, at the bottom of Figure 13. It is found that our simple model accounts extremely well for the observed shape of the C2s-band system, in particular if the folding function is Gaussian.

C. The C2p Manifold

1. Preliminary comments

Paradoxically, the inherent complexity of the C2p part of the PE spectra of saturated hydrocarbons C_nH_m renders their qualitative (or semi-quantitative) discussion rather easy, mainly for two reasons. To begin with the experimental determination of the individual locations I_j^m of strongly overlapping and poorly resolved bands is almost impossible by standard PE spectroscopic methods, once the hydrocarbon contains more than five or six carbon atoms. For example, as shown in Figure 8, the seven C2p bands of propane give rise to only three features, of which the first, at 11 to 13 eV, contains three bands. The situation gets worse, the bigger the hydrocarbon, and the unresolved, broad C2p part of the PE spectrum of nonane, shown in Figure 13, covers 19 bands. It is obvious that their individual positions I_j^m can only be determined roughly, if at all, so that a model treatment which produces crude estimates of the ionization energies should be quite sufficient. The second difficulty is that the sequence of the doublet states $^2\tilde{\Psi}_j$ of the radical cation $C_nH_m^+$ can no longer be determined by simply applying Koopmans' theorem to SCF CMOs φ_j of the parent hydrocarbon C_nH_m. The effects of electronic reorganization, electron correlation and of configurational and vibronic mixing can—and usually do—become very important. None of these are taken care of in standard calculations not involving (extended) configuration interaction, and a fortiori in our EBO model, which necessarily yields a much too simple, and thus very crude, picture of the real situation, except in some particular cases to be discussed below. On the other hand, if the close-lying overlapping bands, which yield the broad maxima of the C2p system, cannot be deconvoluted, then the sequence of these individual bands may not be needed explicitly. An obvious exception is the assignment of the first partial band at I_1^m, i.e. the one corresponding to the electronic ground state $^2\tilde{\Psi}_0$ of the hydrocarbon radical cation $C_nH_m^+$, the symmetry of which is relevant for its reactive behaviour, and for some reactions of the neutral molecule C_nH_m.

To illustrate this situation we take again propane C_3H_8 as an example. The only experimental fact we can deduce from its PE spectrum presented in Figure 8 is that the seven bands corresponding to Koopmans' states of $C_3H_8^+$, in which a CMO φ_j dominated by the carbon 2p and the hydrogen 1s AOs has been vacated, are grouped by 3, 2 and 2, with increasing ionization energy. Furthermore, we can guess^{23,28} from the envelopes of these three features that the ionization energies I_j^m are presumably those given in the list 66. Not much else can be deduced from this PE spectrum, in particular

PE spectrum of propane C_3H_8:

1st feature: 11.5, 12.1, 12.6 eV

2nd feature: 13.5, 14.1 eV   (66)

3rd feature: 15.2, 16.0 eV

about the sequence of states, for which one has to rely on theoretical computations.

If the coordinate system of propane, with all partial conformations staggered (point group C_{2v}), has been oriented with the x axis perpendicular to the plane containing the three carbon atoms and to the $C_2(z)$ axis, and if the numbering of the orbitals refers to the valence shell only, then the band assignments proposed by different authors compare as in diagram 67. Note that Kimura and his coworkers^{32} proposed two assignments:

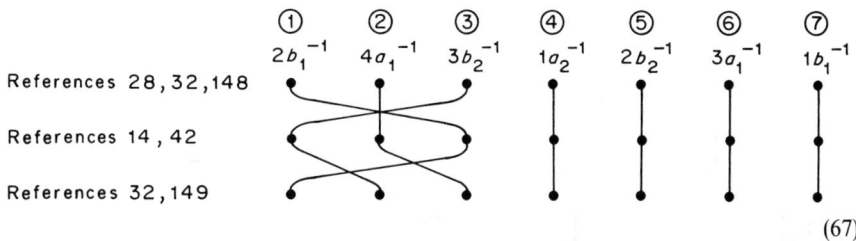

(67)

the first one (top line) stems from a SCF [6-31G], the second (bottom line) from a CI [6-32G] calculation. If anything, one concludes from diagram 67 that the proposed orbital (or state) sequences are strongly model-dependent. This situation has been analysed in detail by Müller, Nager and Rosmus[148], who came to the conclusion that the top-row assignment given in diagram 67 is presumably the correct one.

Our EBO model provides some qualitative insight as to where part of the difficulties of reaching an assignment may come, as has already been mentioned briefly in Section V.B.1. Application of the simplest EBO treatment, i.e. of the AB-model to propane, yields the five accidentally degenerate C2p-type CMOs φ_4 to φ_8 (cf Figure 4). In contrast to the C2s-type orbitals, the inclusion of vicinal $\Gamma_{\mu\nu}$ crossterms is therefore absolutely essential to obtain a rough, halfway acceptable approximation. Diagonalizing matrix 51, derived from the $A\Gamma$-model, yields eigenvalues x_j (list 68) for the seven C2p CMOs of propane, that have also been transformed into vertical ionization energies $I_j^v = -\varepsilon_j$ using the parameters given in the list 48, as included in the regression the list 49. It is immediately obvious that the orbital sequence (which incidentally corresponds

C2p orbitals of propane, according to the $A\Gamma$-model 51:

$$x_j: \quad -2.041; \; -2.015; \; -1.964; \; -1.000; \; -0.765; \; -0.496; \; -0.036$$
$$\varphi_j: \quad 3b_2 \quad\; 4a_1 \quad\; 2b_1 \quad\; 1a_2 \quad\; 2b_2 \quad\; 3a_1 \quad\; 1b_1 \qquad (68)$$
$$I_j^v: \quad 11.40; \quad 11.45; \quad 11.57; \quad 13.66; \quad 14.17; \quad 14.75; \quad 15.75$$

to the middle row of diagram 67) depends strongly on the assumed values of the vicinal crossterms $\Gamma_{\mu\nu}$, and that other effects neglected in our model can easily lead to an interchange of orbitals close in energy, in particular of the three first ones, $\varphi_1 \equiv 3b_2$, $\varphi_2 \equiv 4a_1$ and $\varphi_3 \equiv 2b_1$, depicted schematically in Figure 14. According to the results presented in list 68, the orbital energies I_1^v to I_3^v span a range of only 0.17 eV. From the phase relationship of the EBOs shown in Figure 14, it can be deduced that the relative sequence of the orbital energies of these three CMOs must depend critically on the

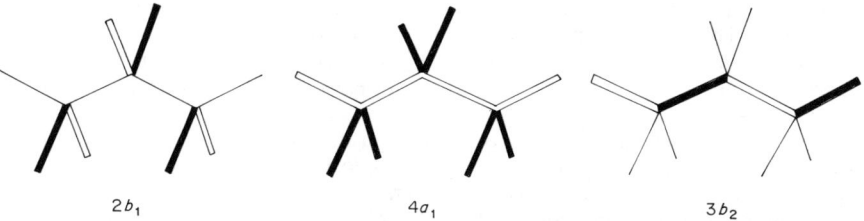

$2b_1$ $4a_1$ $3b_2$

FIGURE 14. Molecular orbital diagrams of the three highest occupied CMOs $2b_1$, $4a_1$ and $3b_2$ (not necessarily in this order; cf text) of propane, using an EBO representation. Full and empty bonds are used to represent EBOs of opposite phases. Thin lines correspond to zero contribution

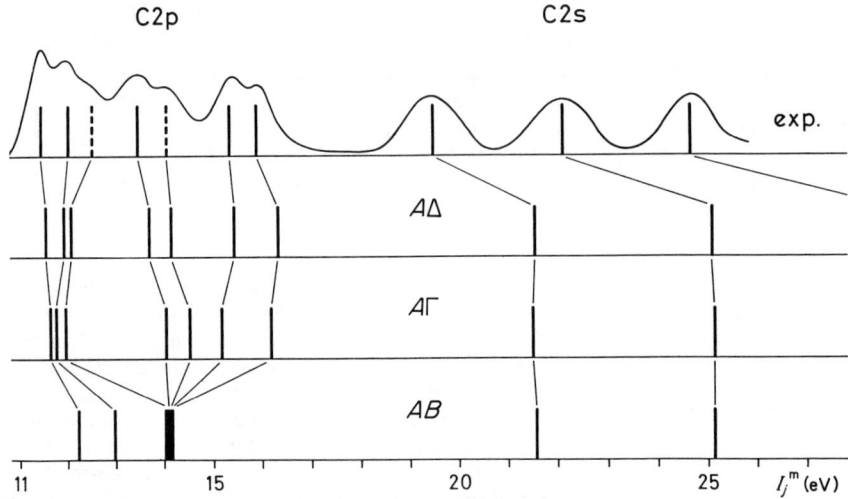

FIGURE 15. Comparison of the He(IIα) PE spectrum of propane with the orbital energies stemming from LMO models of AB, $A\Gamma$ and $A\Delta$ type, assuming the validity of Koopmans' theorem. The values of the matrix elements used in these calculations have been obtained from an STO-3G SCF calculation, using the Foster–Boys localization criterion[16]. They have not been recalibrated by comparison with experimental data, which explains the systematic shift towards higher ionization energies of the C2s-band positions

relative sizes of $B_{\mu\nu}$ and $\Gamma_{\mu\nu}$. It is noteworthy that the inclusion of $\Delta_{\mu\nu}$ crossterms will not materially change the results derived from an EBO model, as exemplified by the results shown in Figure 15, which are derived from a treatment using parameters pertaining to LMOs Λ_μ derived by the Foster–Boys localization procedure[16] from *ab initio* STO-3G LMOs of list 69[13,14]. Note that these parameters, which have not been

$$A^\circ_{CC} = -17.50\,\text{eV}, \qquad A^\circ_{CH} = -16.95\,\text{eV}$$
$$B^\circ_{CC,CC} = B^\circ_{CC,CH} = B^\circ_{CH,CH} = B = -2.89\,\text{eV}$$
$$\Gamma^\circ_{CC,CC} = \Gamma^\circ_{CC,CH} = \Gamma^\circ_{CH,CH} = \Gamma^\circ_\mu = \Gamma^\circ_0 \cos\tau_{\mu\nu} \qquad (69)$$
$$\Gamma^\circ_0 = -1.00\,\text{eV}$$

calibrated with respect to observed ionization energies, serve only to mimic an STO-3G calculation. This explains why the C2s I^m_j values are somewhat too high. One observes that the introduction of vicinal crossterms $\Gamma_{\mu\nu}$ corrects satisfactorily for the crude AB approximation, but that inclusion of $\Delta_{\mu\nu}$ does not yield a significant, further improvement.

   The take-home lession is that the EBO model can only yield crude assignments of the C2p-band system in the PE spectra of saturated hydrocarbons, especially if the predicted band positions I^v_j are very close together. Unfortunately this shortcoming is shared by other, more sophisticated SCF treatments.

   Within these limitations, an EBO model of at least $A\Gamma$ quality can be easily parametrized with respect to the observed ionization energies of a set of saturated hydrocarbons C_nH_m[150]. With respect to the parameters in the list 69, a least-squares adjustment would result in the corrections in equation 70[13,14], of which δA must be added to the self-energies A°_{CC} and A°_{CH} given in 69, whereas all crossterms of this latter list have to be multiplied by the factor f. The correction δA depends on the numbers

$$\delta A = -3.10\,\text{eV} - 0.25(N_{CC}A^\circ_{CC} + N_{CH}A^\circ_{CH})/N$$
$$f = 0.75 \tag{70}$$

N_{CC} and N_{CH} of CC and CH bonds, respectively, with $N = N_{CC} + N_{CH}$. The reader is referred to Reference 14 for further details of the calibration procedure.

2. Ribbon orbitals in polycyclic alkanes

The two highest occupied σ CMOs of benzene are the pair of degenerate e_{2g} orbitals (in D_{6h} symmetry), the symmetry behaviour of which is shown in drawing 71, on the left in a notation due to Jonsson and Lindholm[151] and on the right in terms of carbon–carbon σ LMOs Λ_{CC} or EBOs λ_{CC}. These σ CMOs, which have been called 'ribbon orbitals'[152], are not peculiar to benzene, but recur as the two highest occupied CMOs φ_{HOMO} and φ_{HOMO-1} in any saturated hydrocarbon containing one or more six-membered C_6 rings, e.g. in cyclohexane (drawing 72), where they form the degenerate e_g pair (under D_{3d} symmetry), or in polycyclic hydrocarbons such as *trans-syn-trans*-perhydroanthracene **44** (drawing 73), where $\varphi_{HOMO} = b_g$ and $\varphi_{HOMO-1} = a_g$ (point group C_{2h}) are no longer degenerate, but still close together.

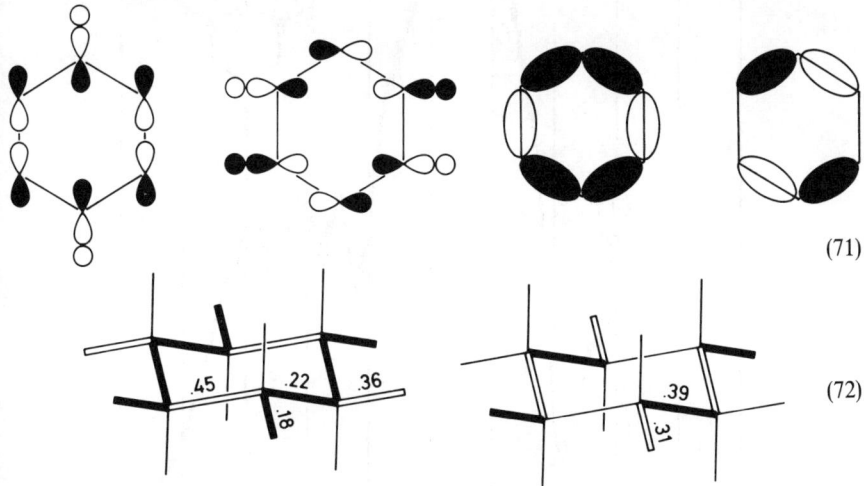

(71)

(72)

Calculations by the $A\Gamma$ model suggest that these two frontier 'ribbon orbitals' φ_{HOMO} and φ_{HOMO-1} of polycyclic hydrocarbons consisting of all-*trans* connected six-membered rings, e.g. **42**, **44**, **45** or **46**, are well separated in energy from the remaining manifold of σ orbitals by a gap which increases with increasing size n of the hydrocarbon C_nH_m. Assuming the validity of Koopmans' theorem, a PE spectroscopic study of such hydrocarbons should therefore yield a low ionization energy band, detached from the remaining C2p manifold at higher energies. It has been shown that this is indeed the case[80]. As an example Figure 16 shows the PE spectrum of perhydropyrene **46**, in which the double band ①, ② associated with the φ^{-1}_{HOMO} and φ^{-1}_{HOMO-1} ionization processes is well separated by about 1 eV from the manifold of the other σ bands. According to the $A\Gamma$ model these 'ribbon orbitals' φ_{HOMO} and φ_{HOMO-1} are evenly delocalized over the whole of the molecular frame, as shown for **46** on top of Figure 16. It can be seen that

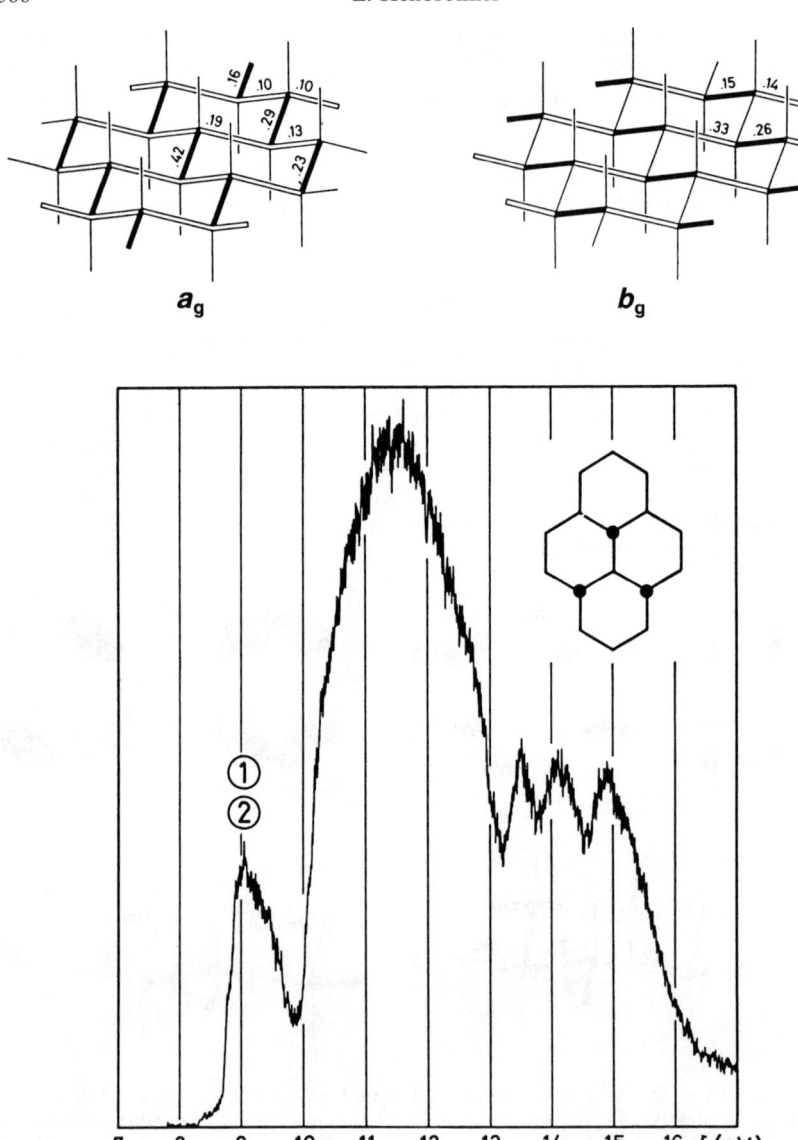

FIGURE 16. Top: The CMOs $\varphi_{\text{HOMO}} = a_{\text{g}}$ and $\varphi_{\text{HOMO}-1} = b_{\text{g}}$ of perhydropyrene **46** according to the $A\Gamma$ model, expressed as a linear combination of EBOs λ_{CC} and λ_{CH}. Open and full bonds refer to EBOs λ_μ of opposite phase. The numbers given are the absolute values $|c_{\mu j}|$ of the corresponding expansion coefficients $c_{\mu j}$. EBOs λ_μ affected with coefficients $|c_{\mu j}| > 0.1$ are indicated by thin lines. Bottom: The PE spectrum of perhydropyrene **46**. $I_j^{\text{m}}(a_{\text{g}}) \approx I_j^{\text{m}}(b_{\text{g}}) \approx 9.0 \, \text{eV}$

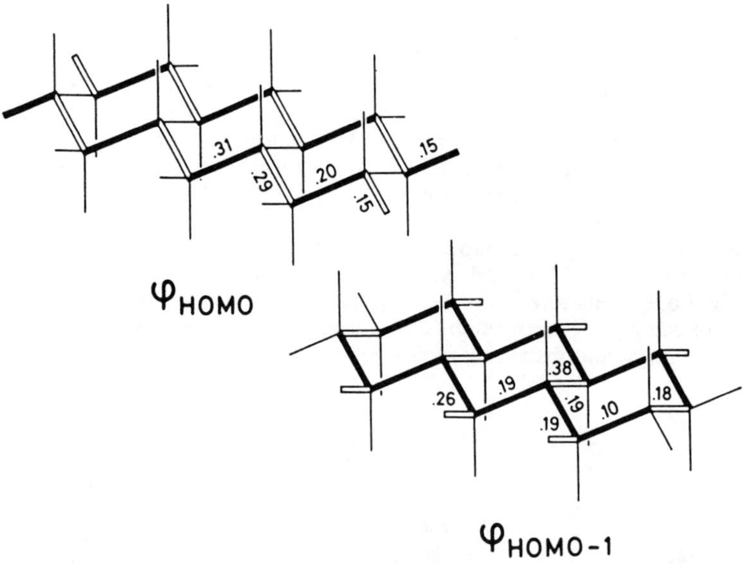

(73)

φ_{HOMO}

φ_{HOMO-1}

they conserve the characteristic phase relationship between the EBOs, exhibited by φ_{HOMO} and φ_{HOMO-1} of the e_g pair (orbitals 72) of cyclohexane, or of *trans-syn-trans*-perhydroanthracene **44** (orbitals 73).

There is little doubt that the high-lying 'ribbon orbitals' φ_{HOMO} and φ_{HOMO-1} play an important role for the 'through bond' interaction[153] of two (semi-)localized orbitals imbedded in a large σ frame[154], because they are (a) close in energy to the usual target orbitals (double-bond π or lone-pair orbitals), and (b) because they are ideally delocalized over the whole of the connecting σ frame. This makes them ideal 'relay orbitals'[155], provided that they yield significant crossterms with the target orbitals. Presumably such high-lying frontier orbitals also play an important role in 'σ conjugation', as defined by Dewar[11].

D. Small-ring Hydrocarbons

1. Cyclopropane

The chemical and physical properties of cyclopropane and of cyclopropyl groups differ significantly from those of other (cyclo)alkanes or (cyclo)alkyl groups, e.g. by the reactive behaviour of the parent hydrocarbon[156] or the almost double-bond-like character of cyclopropyl moieties in conjugation with each other, or with a π system[157]. In particular, the first ionization energy $I_1^m \approx 10.6\,\text{eV}$ of cyclopropane is smaller than the first ionization energies of cyclobutane ($I_1^m \approx 10.7\,\text{eV}$)[29,57] and cyclopentane ($I_1^m \approx 11.1\,\text{eV}$)[29,32], in contrast to the normal alkanes, where $I_1^m \approx 11.5\,\text{eV}$ of propane is greater than the first ionization energies of the higher homologues, e.g. $I_1^m(\text{butane}) \approx 11.1\,\text{eV}$, $I_1^m(\text{pentane}) \approx 10.9\,\text{eV}$[32].

To explain the properties of cyclopropane, of the cyclopropyl group and of other three-membered rings in terms of their electronic structure, two models were proposed in 1947. One model, due to Walsh[158], postulates for the three-membered ring of carbon atoms a set of three bonding, symmetry-adapted orbitals (scheme 74) based on tangential 2p AOs τ_μ and radial sp^2 AOs ρ_μ ($\mu = 1,2,3$) centred at each of the three carbon atoms. The other model, proposed by Coulson and Moffit[159], describes the electronic structure of cyclopropane in terms of three bent 'banana' EBOs $\lambda_1, \lambda_2, \lambda_3$ (scheme 75), in analogy to an earlier proposal by Förster[160].

The orbital pairs ω_S, ω_A in scheme 74 and ζ_S, ζ_A in the set 75 span the irreducible E' representation of the point group D_{3h} of cyclopropane, whereas ω_0 and ζ_0 belong to A_1.

Of the two, the Walsh model (scheme 74) has become by far the preferred one, especially if one wants to explain the conjugative properties of the cyclopropyl group in terms of overlap, e.g. between ω_A and a neighbouring p-type AO or group orbital, attached to the cyclopropane ring in position 1. For such a purpose, the model shown

$$\omega_A = (2\tau_1 - \tau_2 - \tau_3)/\sqrt{6}$$
$$\omega_S = (\quad -\tau_2 + \tau_3)/\sqrt{2}$$
$$\omega_0 = (\rho_1 + \rho_2 + \rho_3)/\sqrt{3}$$

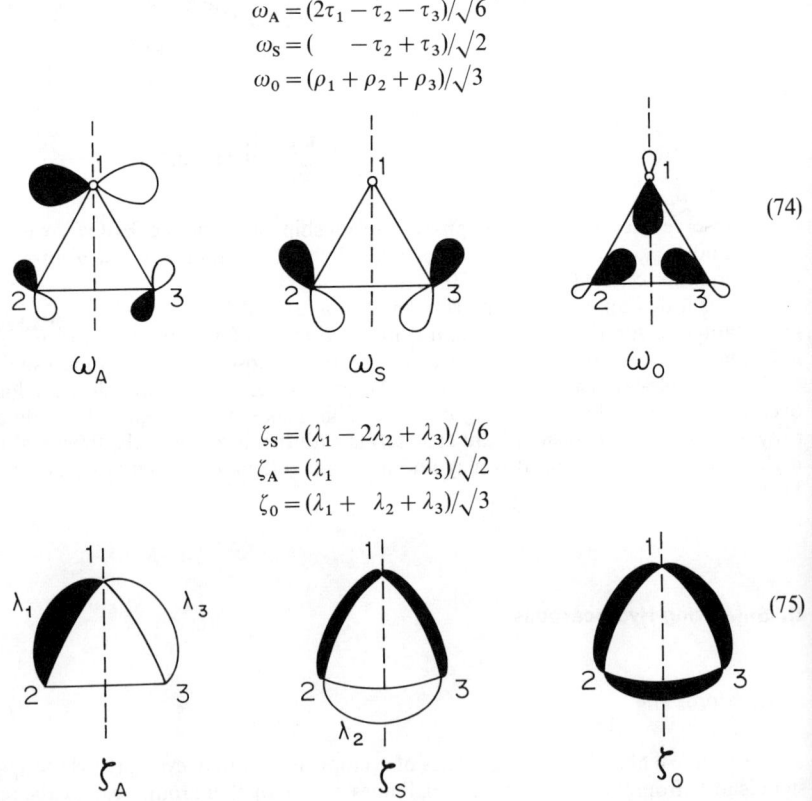

$$\zeta_S = (\lambda_1 - 2\lambda_2 + \lambda_3)/\sqrt{6}$$
$$\zeta_A = (\lambda_1 \quad\quad - \lambda_3)/\sqrt{2}$$
$$\zeta_0 = (\lambda_1 + \lambda_2 + \lambda_3)/\sqrt{3}$$

in scheme 74 is certainly easier to handle qualitatively than the model given by equation 75. By and large, Walsh orbitals have also proved to be rather useful constructs for the rationalization of the PE spectrum of cyclopropane[51,161] and of the PE spectra of cyclopropane-ring containing molecules[33,162,163]. However, if only the set of orbitals 74 is taken into consideration, difficulties have often been encountered when trying to assign the PE spectra of such molecules. These difficulties could only be remedied by including the set of antibonding Walsh orbitals ω_μ^*, $\mu = 1,2,3$ (linear combinations 76)[33] in the discussion.

$$\omega_A^* = (2\rho_1 - \rho_2 - \rho_3)/\sqrt{6}$$
$$\omega_S^* = (\qquad -\rho_2 + \rho_3)/\sqrt{2} \qquad\qquad (76)$$
$$\omega_0^* = (\tau_1 + \tau_2 + \tau_3)/\sqrt{3}$$

The PE spectra of molecules containing three-membered rings have been reviewed by Gleiter[33], and certain aspects by Klessinger and Rademacher[164].

The predilection for the set of 74 of Walsh orbitals, with or without the inclusion of the antibonding set 76, over the Förster–Coulson–Moffitt (FCM) set 75, is understandable because the former are based on the more familiar AOs, rather than on EBOs. This seemed the more reasonable, as a paper published by Bernett[165] has been widely (mis)quoted as proving that the two sets of orbitals 74 and 75 are fully equivalent. But this is not the case. In fact there exists no unitary transformation of the doubly occupied CMOs φ_j of cyclopropane, which would project out the traditional Walsh orbitals ω_j (diagram 74)[166]. In contrast, the FCM orbitals ζ_j are accessible directly via the LMOs Λ_μ of cyclopropane, which have been obtained by some localization technique[15,16] from SCF CMOs φ_j[166,167]. (For practical reasons, the FCM orbitals ζ_j in scheme 75 are then written in terms of EBOs λ_μ, as shown in scheme 75.) The FCM orbitals are therefore a much better and safer starting set for qualitative or semi-quantitative interpretations of the electronic properties of cyclopropane and cyclopropane-ring containing molecules, especially of their PE spectra[168], without the necessity of having to include antibonding orbitals.

As an illustration, we discuss first the PE spectrum of the parent hydrocarbon cyclopropane[51], shown in Figure 17. The band positions I_j^m are listed in Table 4.

If the CMOs of cyclopropane, stemming from an *ab initio* SCF calculation, are subjected to a localization routine, one obtains six LMOs Λ_{CH} and three 'banana' LMOs Λ_{CC}. From these one can form the symmetry-adapted linear combinations shown in Figure 18, which belong, under D_{3h} symmetry, to the irreducible representations A_1', E', A_2'' and E''. Of these the orbitals $3a_1'$ and the pair $3e'$ are the FCM orbitals depicted in diagram 75. They can interact with the Λ_{CH} linear combinations $2a_1'$ and the pair $2e'$, respectively, to form the valence-shell CMOs of corresponding symmetry. In contrast the linear combinations $1a_2''$ and $1e''$ are uniquely determined by symmetry and identical to the CMOs. The linear combinations $1a_1'$ and $1e'$ of the carbon 1s AOs are very much lower in energy than the other ones, and practically identical to the CMOs.

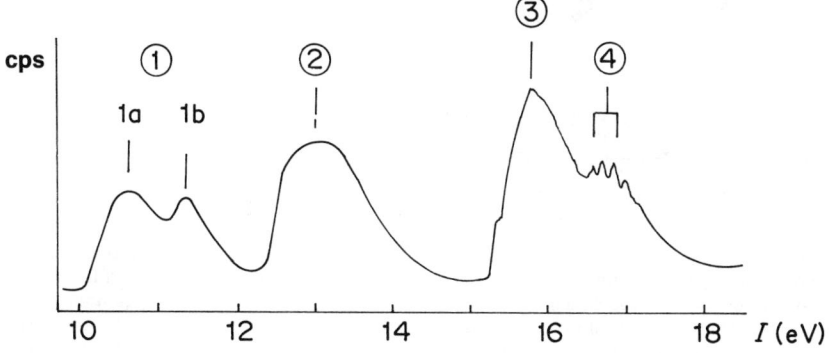

FIGURE 17. PE spectrum of cyclopropane C_3H_6 redrawn from Reference 32

TABLE 4. Observed and calculated band positions of cyclopropane

Band	CMO	I_j^m (eV)	x_j	AB-parameter[a] I_j^v (eV)	Regression[b] I_j^v (eV)	
①	$3e'$	10.95^c	−2.453	10.5	11.4	
②	$1e''$	13.0^d	−1.480	12.6	13.4	C2p
③	$3a_1'$	15.7	−0.691	14.3	15.0	
④	$1a_2''$	16.6^e	−0.040	15.7	16.3	
⑤	$2e'$	19.5	1.753	19.6	19.9 ⎫	C2s
⑥	$2a_1'$	26.5	5.091	26.9	26.5 ⎭	

[a]Parameters A and B according to equations 48 and 49.
[b]Recalibrated parameters according to equation 78.
[c]Mean of the two maxima at $I_{1,a}^m = 10.6\,\text{eV}$ and $I_{1,b}^m = 11.3\,\text{eV}$. The difference $I_{1,b}^m - I_{1,a}^m = 0.7\,\text{eV}$ corresponds to the Jahn–Teller split.
[d]The Jahn–Teller split of this band does not lead to two distinct maxima, but only to a band broadening.
[e]The vibrational fine-structure corresponds to the HCH bending vibration of the methylene groups.

To compute the energies belonging to the linear combinations of the LMOs Λ_{CH} and Λ_{CC} shown in Figure 18, and the energies of their crossterms, one has to take into account the special conditions which prevail in small rings, i.e. the dependence of the relevant matrix elements on the bond angle θ_{CCC}. Sample calculations, applying the *ab initio* STO-3G procedure to cyclopropane (D_{3h}, $\theta_{CCC} = 60°$), cyclobutane (D_{4h} assumed, $\theta_{CCC} = 90°$) and cyclopentane (D_{5h} assumed, $\theta_{CCC} = 108°$) yield the values in display 77

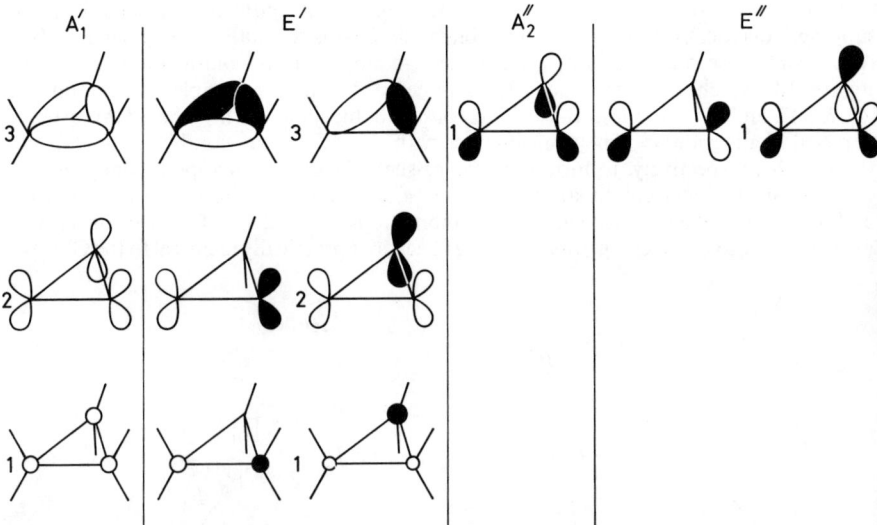

FIGURE 18. Symmetry-adapted linear combinations of the LMOs Λ_{CC} and Λ_{CH} of cyclopropane (point group D_{3h}) belonging to the irreducible representations A_1', E', A_2'' and E''. Of these, the Λ_{CC} combinations $3a_1'$ and $3e'$, which form the FCM orbital set, mix with the Λ_{CH} combinations $2a_1'$ and $2e'$ to form the CMOs of the same symmetry behaviour. The orbitals $1a_1'$ and $1e'$ are of carbon 1s parentage, and have orbital energies around − 290 eV. Note that only the relative phases and not the absolute contributions of the individual LMOs Λ_{CC} and Λ_{CH} are implied

for the self-energy A_{CC}, the geminal $(B_{CC,CC})$ and vicinal $(\Gamma_{CC,CC})$ crossterms between the 'banana' LMOs Λ_{CC}.

θ_{CCC}	A_{CC}(eV)	$B_{CC,CC}$(eV)	$\Gamma_{CC,CC}$(eV)	
60°	−16.9	−4.8		
90°	−17.2	−3.5	−2.0	(77)
108°	−17.5	−3.0	−1.0	

Apart from a slight upward shift of the self-energies A_{CC} with decreasing bond angle θ_{CCC}, the most important change is the strong increase in absolute value of the geminal crossterm $B_{CC,CC}$ as the bond angle θ_{CCC} gets smaller. This turns out to be the crucial factor making the cyclopropyl group such an excellent partner for conjugation with π systems.

As before, the matrix elements pertaining to the LMOs Λ_μ can be recalibrated by comparison with observed ionization energies to yield the parameters for the corresponding EBOs λ_μ. In a crude approximation, one can first assign relative values, suggested by the results of SCF calculations. Using the geminal crossterm $B_{CH,CH} \equiv B$ as reference, one finds $B_{CC,CH} \approx B$ and $B_{CC,CC} \approx 1.7\,B$. In addition $\Gamma_{CC,CH} \approx -0.14\,B$, $\Gamma_{CH,CH}(0°) \approx 0.24\,B$ and $\Gamma_{CH,CH}(120°) \approx -0.24\,B$ are reasonable approximations. Finally we assume that $A_{CH} \approx A_{CC} = A$, which is acceptable in view of the results shown in display 77. Under these simplifying assumptions one sets up the H_λ matrix of the $A\Gamma$-model, to obtain the eigenvalues x_j given in the fourth column of Table 4. These determine the corresponding orbital energies according to $\varepsilon_j = A + x_j B$, and thus the vertical ionization energies $I_j^v = -\varepsilon_j = -(A + x_j B)$. Using in a first approximation the parameters $A = -15.83\,\text{eV}$ and $B = -2.17\,\text{eV}$ (equation 48), one obtains the predictions listed in the fifth column of Table 4, which are not at all bad, considering all the simplifications involved. A linear regression of the experimental ionization energies I_j^m (Table 4, third column) on the x_j values (Table 4, fourth column) yields

$$I_j^v (\text{eV}) = (16.34 \pm 0.44) + (2.00 \pm 0.18)\,x_j \tag{78}$$
$$r = 0.996$$

where the error limits correspond to a 90% confidence level. Although the correlation coefficient looks rather good, it must be viewed with caution, because the major part of the correlation is due to the ordering of the data[169]. The parameters $A = -16.34\,\text{eV}$ and $B = -2.00\,\text{eV}$ are only slightly changed relative to those in equation 48. Remember that the strong increase in geminal interaction $B_{CC,CC}$ as a consequence of the small bond angle θ_{CCC} has been taken care of by postulating $B_{CC,CC} = 1.7\,B$.

As can be seen from Figure 17, the shapes of bands ① and ② are strongly influenced by the Jahn–Teller effect, i.e. by the distortion of the cyclopropane radical cation in its electronic ground state $^2E'$ and its first excited state $^2E''$. For a discussion of this effect the reader is referred to work by Haselbach[170].

It can be shown that an analogous EBO treatment of molecules containing one or more cyclopropane moieties leads to predictions of PE band positions, which allow at least a heuristically useful assignment of the PE spectra[168]. The reason for the success of this treatment can be summarized as follows:

Subjecting the occupied, bonding CMOs φ_j of cyclopropane to a localization procedure yields six CH LMOs Λ_{CH} and three CC LMOs Λ_{CC}, from which one can form the twelve symmetry-adapted linear combinations shown in Figure 18. These contain, apart from the combinations $1a_1'$ and $1e'$ of the carbon 1s AOs and six combinations of the LMOs Λ_{CH}, the FCM orbitals (diagram 75). They are part of the occupied manifold of orbitals, differing from the CMOs only by an orthogonal transformation. Consequently they, or the corresponding linear combinations of EBOs, are an ideal starting set for computing

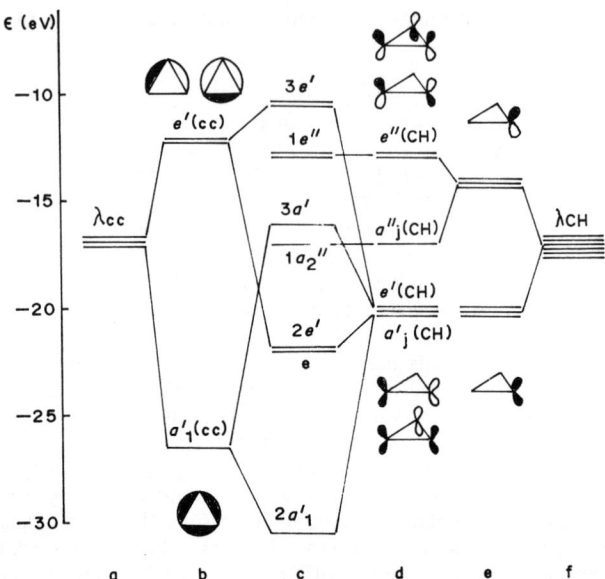

FIGURE 19. Correlation diagram showing the genesis of the cyclopropane CMOs φ_j (centre column c) from the linear combinations presented in Figure 18. Column a: self-energies A_{CC} of the EBOs λ_{CC}. Column f: self-energies A_{CH} of the EBOs λ_{CH}. Column b: FCM orbitals, split according to $B_{CC,CC}$. Column e: symmetric and antisymmetric combination of geminal λ_{CH} under the influence of $B_{CH,CH}$. Column d: symmetry-adapted linear combinations $2a_1'$ and $2e'$ of the λ_{CH} as given in Figure 18. Column c: result of mixing the linear combinations of columns b and d under the influence of $B_{CC,CH}$ and vicinal ($\Gamma_{CC,CH}$ and $\Gamma_{CH,CH}$) terms

orbital CMO energies ε_j and hence ionization energies using Koopmans' theorem (diagram 44). Walsh orbitals (equations 74) cannot be obtained in this fashion, because one would have to use also the antibonding SCF orbitals (equations 76) of cyclopropane. There is no orthogonal transformation of the above kind which would project out the orbitals τ_μ and ρ_μ needed to form the set 74 of Walsh orbitals.

Figure 19 shows the genesis of the cyclopropane valence-shell CMOs φ_j starting from EBOs λ_{CH} and λ_{CC}, the self-energy levels of which are shown in columns a and f. From the λ_{CH} one forms first the local in-phase and out-of-phase combinations (column e), and then the symmetry-adapted ones (column d), already shown in Figure 18. These λ_{CH} combinations interact, symmetry permitting, with the FCM orbitals (column b) to form the CMOs φ_j given in the central column c.

2. Conjugative property of the cyclopropyl group

The (conjugative) interaction of cyclopropane FCM orbitals with the other σ orbitals of the same hydrocarbon molecule, in particular with other FCM orbitals if it is polycyclic and contains further three-membered rings, depends strongly on the topography of the system, e.g. on the twist angle between two directly linked cyclopropane moieties, such as those in bicyclopropyl 118[72]. Therefore the individual references pertaining to the numerous different molecules of this type contained in Table 1 should be consulted[33]. Here, we shall only comment briefly on the peculiar double-bond-like character of the cyclopropyl group in conjugation with unsaturated centres. Thus a cyclopropyl group attached to a double or triple bond will reduce the ionization energy corresponding to electron ejection from the π orbital by practically the same amount as a vinyl group[167,171].

With reference to the Walsh orbitals (diagram 74), this effect has been rationalized as being mainly due to the prominent 2p character of the Walsh orbital ω_A in position 1, i.e. to the large coefficient $2/\sqrt{6} = 0.817$ of the tangential 2p AO τ_1. As a result the corresponding value of the resonance integral $\beta(\omega_A/\pi)$ between the Walsh orbital ω_A and the 2p AO of a π orbital at the point of attachment of the cyclopropyl group, is 0.817β, if β is the standard resonance integral between two neighbouring 2p AOs of a π system, and if the cyclopropyl group and the π system are in the so-called bisected conformation. Necessary consequences of this model are: (a) the crossterm $\beta(\omega_A/\pi)$ (resonance integral) is twist-angle dependent, being maximal for the bisected conformation, and zero for the perpendicular one; (b) there will be no conjugative interaction between the symmetric component ω_S of the Walsh orbitals (diagram 74) and a π system, because the AO τ_1 of ω_S has a zero coefficient in position 1. In other words $\beta(\omega_S/\pi)$ will be zero, whatever the local conformation. However, as has been shown by Bruckmann and Klessinger[171], and been confirmed by later investigations of the PE spectra of cyclopropyl-substituted aromatic molecules[168], this is not the case. Indeed one finds the ratio $\beta(\omega_A/\pi)/\beta(\omega_S/\pi) \approx 1.4$ instead of infinity, where $\beta(\omega_A/\pi)$ refers to the bisected, and $\beta(\omega_S/\pi)$ to the perpendicular conformation of the cyclopropyl group relative to the substituted π system.

If we accept the notion that the FCM orbitals (diagram 75) are indeed a better starting point for representing the E'-type CMOs of cyclopropane, in particular its HOMO $3e'$ where the contribution from the LMOs Λ_{CH} is small, we must also accept the fact that the LMOs Λ_{CC}, and thus the EBOs λ_{CC}, are not much different in self-energy A_{CC} from those of normal alkanes. Necessarily this raises the question why a cyclopropyl group does not behave like a normal alkyl group, e.g. an isopropyl group, with respect to (hyper)conjugation with a higher-lying lone pair p or π orbital, say with the HOMO of a π system.

Let us assume that the orbital energy of the HOMO $\varphi_{HOMO}(\pi)$ of a π system is $\varepsilon_{HOMO}(\pi)$, and that of the conjugating group orbital $\varphi(R)$ of an alkyl substituent R is $\varepsilon(R)$, the latter lying below the former. If the crossterm between $\varphi_{HOMO}(\pi)$ and $\varphi(R)$ is $\beta_{\pi R}$, then the upwards displacement of the orbital energy $\varepsilon_{HOMO}(\pi)$ is given by second-order perturbation in the form

$$\Delta\varepsilon_{HOMO}(\pi) = \beta_{\pi R}^2/[\varepsilon_{HOMO}(\pi) - \varepsilon(R)] \tag{79}$$

The analysis of the PE spectra of a large number of unsaturated and aromatic molecules containing alkyl and/or cyclopropyl groups R leads to the result that the resonance integral $\beta_{\pi R}$ between the relevant orbitals $\varphi(R)$ of the groups R and the orbital $\varphi_{HOMO}(\pi)$ of a π system are all essentially of the same size, i.e. $-2.4\,\text{eV} < \beta_{\pi R} < -1.9\,\text{eV}$. It follows that the reason why substitution by a cyclopropyl group leads to a larger destabilization of a π orbital than substitution by a normal alkyl group is simply that the orbital energy $\varepsilon(R)$ of the cyclopropyl FCM orbitals ζ_A and ζ_S (diagram 75) is so much closer to the orbital energy $\varepsilon_{HOMO}(\pi)$ of a π orbital, than the orbital energies $\varepsilon(R)$ of normal alkyl groups R. The reason why $\varepsilon(\zeta_A)$ and $\varepsilon(\zeta_S)$ are found at such high energies is obvious from the values given in display 77. The geminal crossterms $B_{CC,CC}$ between two banana LMOs in a cyclopropane ring is very large, giving rise to a substantial upwards shift of $\varepsilon(\zeta_A)$ and $\varepsilon(\zeta_S)$ relative to the self-energy A_{CC} of one of the LMOs Λ_{CC}. The marked tendency of cyclopropyl groups for conjugation therefore owes much more to the strong intra-ring crossterms $B_{CC,CC}$, which push the orbital energies $\varepsilon(\zeta_A)$ and $\varepsilon(\zeta_S)$ close to the HOMO energies $\varepsilon_{HOMO}(\pi)$ of π systems, rather than to some extra large coupling term $\beta_{\pi R}$ between the cyclopropyl and the π orbitals.

A rather special case concerns the influence of conjugation between the two cyclopropyl groups on the PE spectrum of bicyclopropyl **118**, which has been discussed in great detail by Gleiter[33]. The interesting feature is the presence of different conformers in the gas phase[172], which renders the analysis of the bicyclopropyl PE spectrum[72,89] more difficult, especially because of the almost free internal rotation over a wide range of the twist angle[172]. Under such conditions some bands in the PE spectrum can have complicated, and sometimes misleading, Franck–Condon envelopes[173]. This type of problem is absent in the rotanes, e.g. [3]rotane **113** and [4]rotane **114**[115], in which the cyclopropyl moieties are aligned ideally for maximum interaction[174].

3. Cyclobutane

The PE spectrum of cyclobutane, the next homologue of cyclopropane, is shown in Figure 20. In analogy to cyclopropane one can also postulate Walsh-type orbitals for cyclobutane[175], from which one would deduce that the cyclobutyl group should also be available for strong hyperconjugation, albeit to a lesser extent than the cyclopropyl group. If we first use an AB-model to derive the FCM-type orbitals of cyclobutane using only the EBOs λ_{CC}, and assuming either the real D_{2d} or an idealized D_{4h} symmetry, one obtains diagram 80. However, as is obvious from the results presented in display 77,

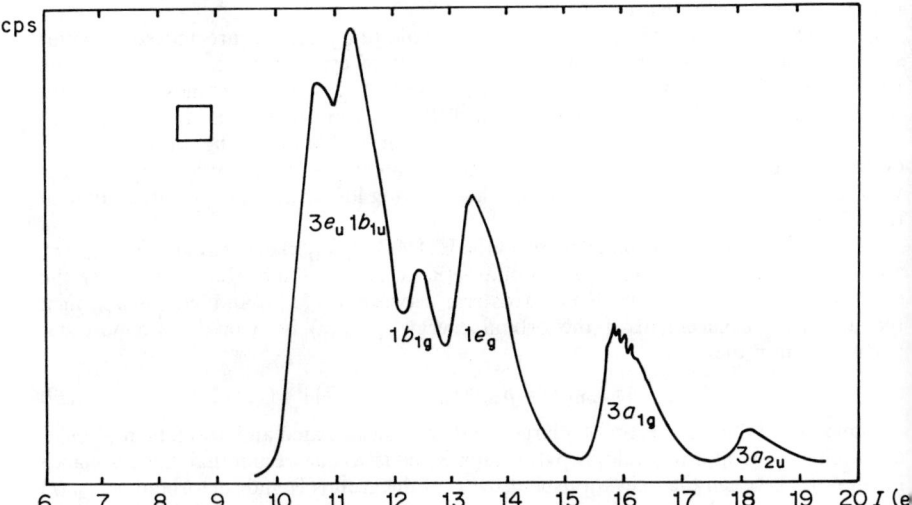

FIGURE 20. PE spectrum of cyclobutane, redrawn from Reference 33. The orbital labels refer to CMOs under the assumption of a hypothetical D_{4h} structure

j	D_{4h}	D_{2d}	I_j^m (eV)
1	$3e_u$	$4e$	10.7, 11.3[a]
2	$1b_{1u}$	$4a_1$	11.7
3	$1b_{1g}$	$1b_1$	12.5
4	$1e_g$	$3e$	13.4, 13.6[a]
5	$3a_{1g}$	$3a_1$	15.9

[a]Jahn–Teller split.

the vicinal crossterms $\Gamma_{CC,CC}$ between opposite CC EBOs become rather large, when the bond angles $\theta_{CC,CC}$ are $90°$. As deduced by inspection of the orbital diagrams in display 80, these vicinal interactions will lower the energy of the b_{1g} orbital and raise the energies of the e_u orbitals (D_{4h} assumed). In addition, only the e_u orbitals can interact with a combination of the same symmetry behaviour of the lower-lying CH EBOs λ_{CH}, which leads to an additional upwards shift. As a result the $3e_u$ CMO of cyclobutane is found above the $1b_{1g}$ CMO, as indicated in the PE spectrum shown in Figure 20. As also indicated in the spectrum, there is an additional CMO $1b_{1u}$ (between $3e_u$ and $1b_{1g}$), which is entirely localized on the CH bonds, i.e. it is a pure combination of the EBOs λ_{CH}.

The first band in the PE spectrum of cyclobutane is split into two maxima, distant $0.6\,\text{eV}$ apart, which is indicative of a strong Jahn–Teller effect. This proves that the

Orbitals		Orbital energy	
D_{2d}	D_{4h}		
b_1	b_{1g}	$A - 2B$	
e	e_u	A	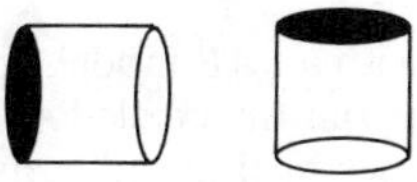 (80)
a_1	a_{1g}	$A + 2B$	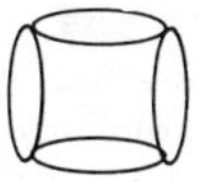

HOMOs of this molecule are indeed the degenerate pair $3e_u$. The distortion of the cyclobutane radical cation leads presumably to a D_{2h} ground-state geometry[57].

The investigation of the PE spectra of cyclopropyl and cyclobutyl halides[176] shows that hyperconjugative interactions are smaller in the latter compounds than in the former, but still larger than in non-cyclic alkyl halides.

E. Highly Symmetrical, Polycyclic Hydrocarbons

1. Tetrahedrane, cubane and pentaprismane

The PE spectra of saturated, highly symmetrical, polycyclic hydrocarbons containing three- and/or four-membered rings are of interest, because of the small CCC angles θ_{CCC}, which lead to a noticeable increase of individual geminal and/or vicinal crossterms $B_{CC,CC}$, $\Gamma_{CC,CC}$, as already shown for cyclopropane and cyclobutane in display 77. In addition, unexpected effects may result as a consequence of the high symmetry of such molecules, and their compact structure. The first factor will prevent interactions between symmetry-adjusted, linear combinations of LMOs $\Lambda_{\mu\nu}$ or EBOs $\lambda_{\mu\nu}$ belonging to different irreducible representations of the relevant point group. On the other hand, the second factor will lead to increased interaction between vicinal, or topologically more distant, but spatially close lying LMOs $\Lambda_{\mu\nu}$ or EBOs $\lambda_{\mu\nu}$. As these two factors are usually linked in a complicated fashion, depending on the topography, symmetry and local conformations of the molecules, each molecule has to be discussed in its own right, and no general rules can be given which transcend those we have already encountered.

In addition, the CMOs φ_j of such molecules are still delocalized all over the whole

516   E. Heilbronner

molecular frame, with the result that no particular, semi-localized 'target orbital' is available that could be correlated with an easily observable, detached band in the PE spectrum, as is the case for molecules containing, e.g., π orbitals and/or lone pair orbitals. Even the FCM orbitals ζ_j (diagram 75) or the Walsh orbitals ω_j (diagram 74) of three- or four-membered rings are usually mixed to such an extent with the remaining C2p orbitals, that electron ejection from a C2p CMO φ_j with a high percentage of ζ_j or ω_j will not lead to bands in the PE spectra which can be easily located.

As examples we discuss briefly the PE spectra of tetra-t-butyl-tetrahedrane 79[106], of cubane 80[123,124], of pentaprismane 84[123] and of bicyclo[1.1.1]pentane 10[67], which provide nice examples for the value of the EBO model as a guide to understanding—or more modestly, rationalizing—their electronic structure and PE spectra. That three of these examples stem from the Basel group is not supposed to imply that these are more noteworthy than many of the fine examples studied elsewhere.

Whereas the PE spectrum of the parent hydrocarbon tetrahedrane C_4H_4[177] is still unknown, its tetra-t-butyl derivative 79 has been studied in detail[106]. Its PE spectrum, which is shown in Figure 21, consists of a first band ① around 8 eV, whose double-humped Franck–Condon envelope $[I_I^m(a) = 7.5\,\text{eV}, I_I^m(b) = 8.2\,\text{eV}]$ is due to a Jahn–Teller effect. The very large band system extending from 10 eV to 16 eV is the result of electron ejection from CMOs φ_j located mainly on the four t-butyl groups, and thus of little interest for our purposes.

As usual, we start with an AB-model, applying it to the unsubstituted parent molecule tetrahedrane C_4H_4. From the labeled structural formula one obtains first the graph $\mathscr{G}$, and then the eigenvalues x_j of the adjacency matrix $\mathbb{A}$ corresponding to $\mathscr{G}$ (diagram 81). As can be seen, this model predicts that the HOMOs 1e of tetrahedrane C_4H_4 are degenerate, belonging to the irreducible representation E of the point group T_d. The crucial question is whether the introduction of vicinal crossterms $\Gamma_{\mu\nu}$ between the EBOs λ_{CC}, located on opposite edges of the tetrahedron, would alter the orbital sequence or not. To answer this open question one subjects the CMOs φ_j of tetrahedrane, obtained from an *ab initio* SCF calculation (here the STO-3G model) to the Foster–Boys[16] localization procedure, which yields LMOs $\Lambda_{\mu\nu}$ of the Λ_{CC} and Λ_{CH} type, and the crossterms given in diagram 82. As expected in view of the small angel $\theta_{CCC} = 60°$ (cf

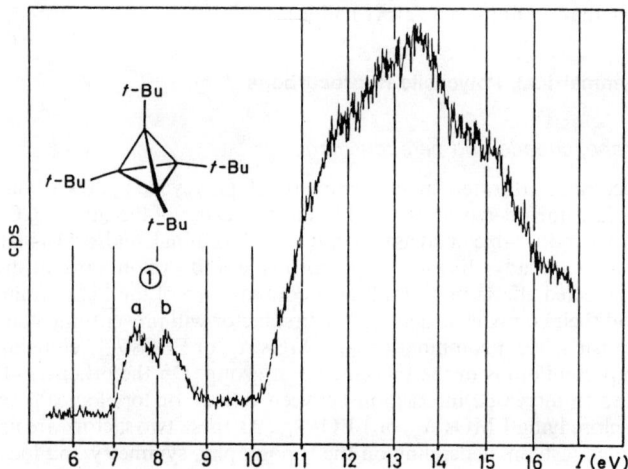

FIGURE 21. PE spectrum of tetra-t-butyltetrahedrane 79

display 77), the geminal interaction $B_{\text{CC,CC}}$ between two banana LMOs Λ_{CC} are again much larger that those for hydrocarbons with tetrahedral CCC bond angles. On the other hand, it may come as a surprise that the vicinal crossterms $\Gamma_{\text{CC,CC}}$ are not only small but zero, a result which is due to the particular twist angle $\tau_{\text{CC,CC}} = 70.5°$ prevailing in the tetrahedrane skeleton, and the departure of θ_{CCC} from 109.5°. We conclude that inclusion of vicinal crossterms would not change the orbital sequence as far as the highest occupied orbitals are concerned, a result supported by the SCF calculation. The influence of the four t-butyl groups on the CMO orbital energies ε_j consists in an upward shift, with corresponding reduction of the ionization energy. However, it can be shown that this will not lead to a change in orbital sequence, as far as the HOMO and HOMO-1 are concerned[106].

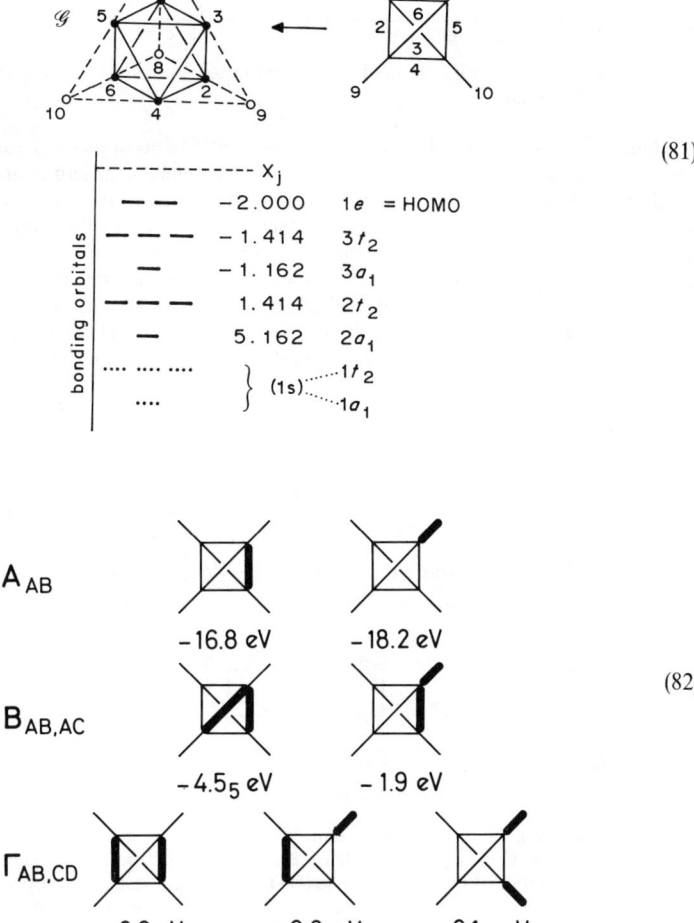

(81)

(82)

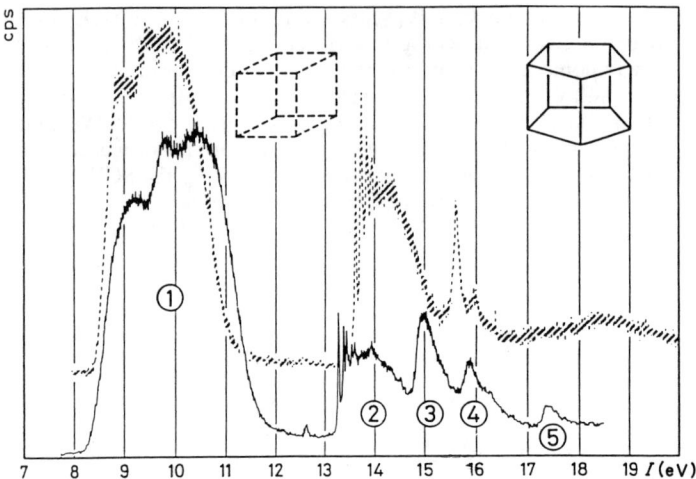

FIGURE 22. PE spectra of cubane **80** (dotted lines) and of pentaprismane **84** (solid lines)

Figure 22 presents the PE spectra of cubane $\mathbf{80}^{124}$ and of pentaprismane $\mathbf{84}^{123}$, which differ from those of other hydrocarbons by an extremely large gap of about 3 eV between the first bands, collectively numbered ① in Figure 22, which correspond to the degenerate CMOs (t_{2g}, t_{2u} of cubane, $e_2'', e_2'e_1'', e_1'$ of pentaprismane), and the band system extending from roughly 13.5 eV towards higher ionization energies. In contrast to the situation prevailing in the tetrahedrane carbon skeleton, this peculiar spacing of the CMO energies ε_j can no longer be explained by using an AB-model. For its rationalization we have to refer to the very large value of the vicinal terms $\Gamma_{CC,CC}$ between pairs of vicinal EBOs λ_{CC} opposed to each other within a cyclobutane moiety, coupled with the small value

$$A_{CH} = -17.32 \qquad A_{CC} = -16.88$$

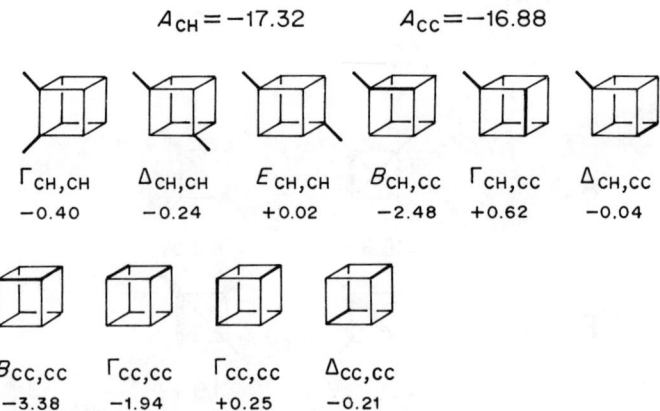

$\Gamma_{CH,CH}$	$\Delta_{CH,CH}$	$E_{CH,CH}$	$B_{CH,CC}$	$\Gamma_{CH,CC}$	$\Delta_{CH,CC}$
−0.40	−0.24	+0.02	−2.48	+0.62	−0.04

$B_{CC,CC}$	$\Gamma_{CC,CC}$	$\Gamma_{CC,CC}$	$\Delta_{CC,CC}$
−3.38	−1.94	+0.25	−0.21

FIGURE 23. Pictorial representation of the matrix elements (in eV) of the Hartree–Fock matrix of cubane **80** in a localized basis. The values have been obtained by subjecting the STO-3G *ab initio* results to a Foster–Boys localization procedure[16]. The crossterms refer to the LMOs Λ_μ indicated by bold lines and to the symmetry-equivalent pairs

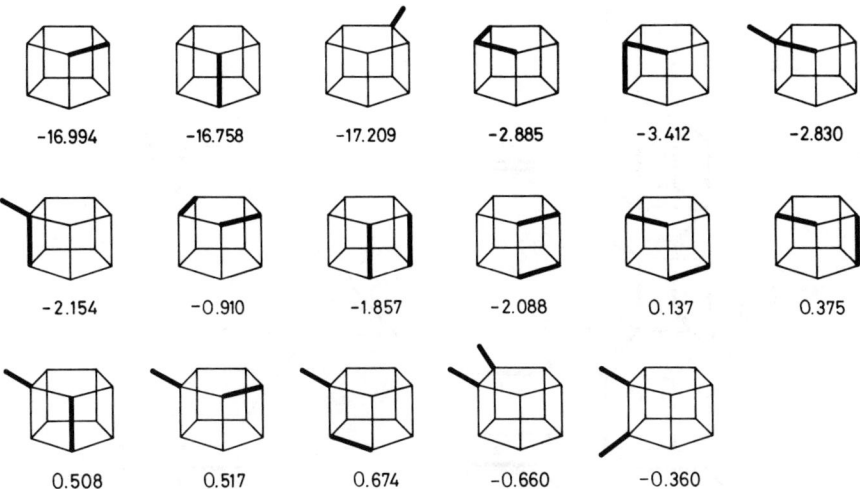

-16.994	-16.758	-17.209	-2.885	-3.412	-2.830
-2.154	-0.910	-1.857	-2.088	0.137	0.375
0.508	0.517	0.674	-0.660	-0.360	

FIGURE 24. Pictorial representation of the matrix elements (in eV) of the Hartree–Fock matrix of pentaprismane **84** in a localized basis. See legend to Figure 23

of $\Gamma_{CC,CC}$ if the vicinal EBOs λ_{CC} belong to two different rings. This is illustrated by the matrix elements for LMOs stemming from an STO-3G calculation of cubane **80**[124] (Figure 23) and of pentaprismane **84**[123] (Figure 24). The consequences of introducing vicinal interactions $\Gamma_{\mu\nu}$ into the AB-model, to yield the $A\Gamma$-model, are shown in Figure 25 for the particular example of the orbital energies of cubane **80**. The net separation in energy of the two sets of triply degenerate orbitals t_{2g} and t_{2u} from the remaining CMOs e_g, t_{1u}, a_{2u} and a_{1g} of the C2p manifold has interesting consequences for the hyperconjugative properties of the cubyl group[178].

2. [1.1.1]Propellane

Because of the extreme deviation from a tetrahedral situation of the bridge-head atoms in [1.1.1]propellane **49**, *ab initio* treatments using an extended polarized basis, e.g. an extended, polarized 6-31G basis[179], or treatments of DZ + P quality, such as a floating Gaussian basis[180], must be used to obtain reliable results. The PE spectrum of **49**[82], shown in Figure 26, agrees qualitatively and quantitatively with the orbital sequence HOMO = $3a'_1, 1e'', 3e', 1a'_2, 2e'$, predicted by such treatments. However, the resulting CMOs can again be expressed in terms of EBOs λ_μ as shown at the bottom of Figure 26.

Ejection of an electron from the HOMO $3a'_1$ of C_5H_8, **49**, yields the radical cation $C_5H_8^+$ in its electronic ground state $^2A'_1$, to which corresponds band ① in the PE spectrum of **49** shown in Figure 26. The Franck–Condon shape of this first band, and in particular its very narrow width at half-height, indicate that both the parent molecule **49** and the radical cation **49**$^+$ in its electronic ground state $^2A'_1$ differ very little in geometry. This can be interpreted in qualitative orbital language as a strong indication that the HOMO $3a'_1$ of **49** is essentially non-bonding[82].

The rather peculiar electronic structure of **49**, as compared to cyclopropane and bicyclo[1.1.0]butane **1**, is strongly reflected in the correlation diagram of Figure 27, in particular by the fact that the first ionization energy of **49** cannot be obtained by a linear extrapolation of the first ionization energies of cyclopropane and of **1**.

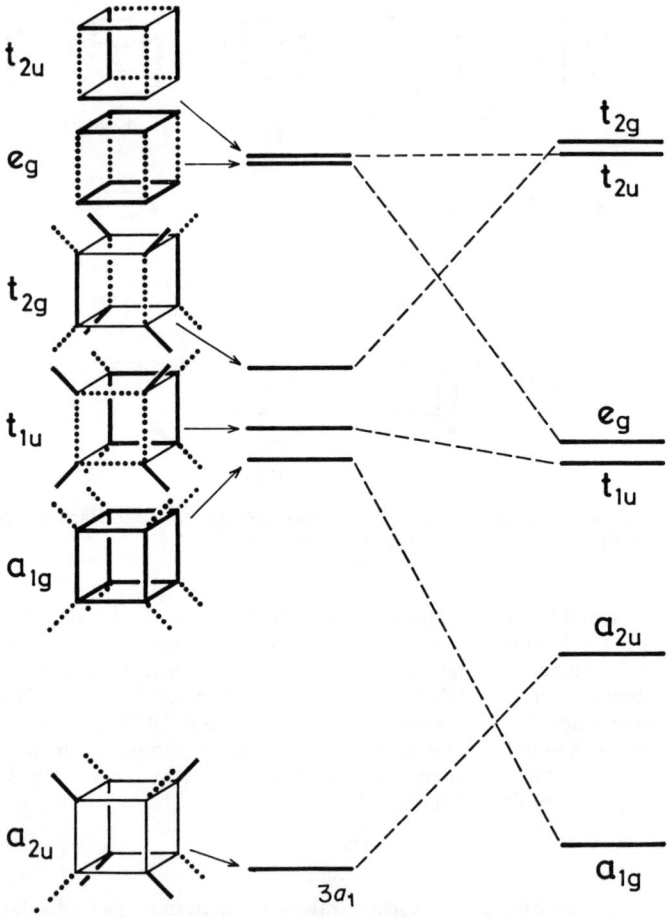

FIGURE 25. Schematic representation of the CMOs φ_j of cubane **80**, obtained by an EBO AB-model (left) and an $A\Gamma$-model (right), to demonstrate the important influence of the vicinal crossterms $\Gamma_{\mu\nu}$ in this case

3. Bicyclo[1.1.1]pentane and di-bicyclo[1.1.1]pentyl

Figure 28 presents the PE spectra[67] of bicyclo[1.1.1]pentane **10** and of di-bicyclo-[1.1.1]pentyl **11**, the latter in its D_{3d} conformation. These molecules are again examples of the fact that vicinal interaction terms must be included in an EBO model, for even a qualitative discussion of their electronic structure, and thus of their PE spectra. This

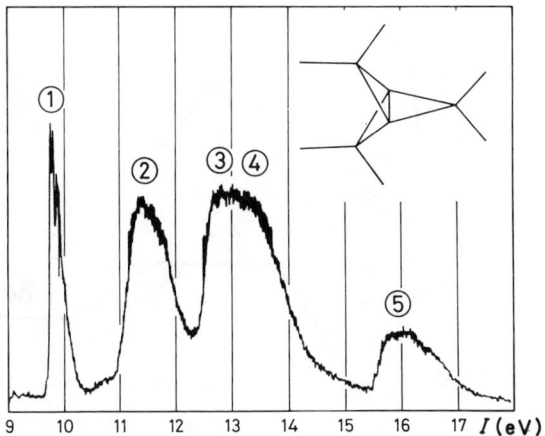

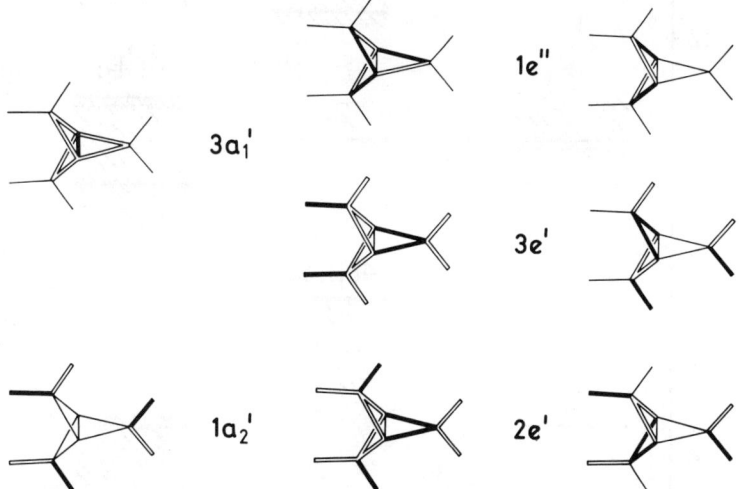

FIGURE 26. Top: The PE spectrum of [1.1.1]propellane **49**. Bottom: Valence-shell CMOs φ_j of **49** in a qualitative EBO representation. The CMOs φ_j given correspond to the first five bands in the spectrum of **49**:

Band	I_j^m (eV)	CMO
①	9.74	$3a_1'$
②	11.3_5	$1e''$
③	12.6	$3e'$
④	(13.4_5)	$1a_2''$
⑤	15.7 (16.1)	$2e'$

The diagrams show qualitatively the phase relationship between the EBOs λ_μ

E. Heilbronner

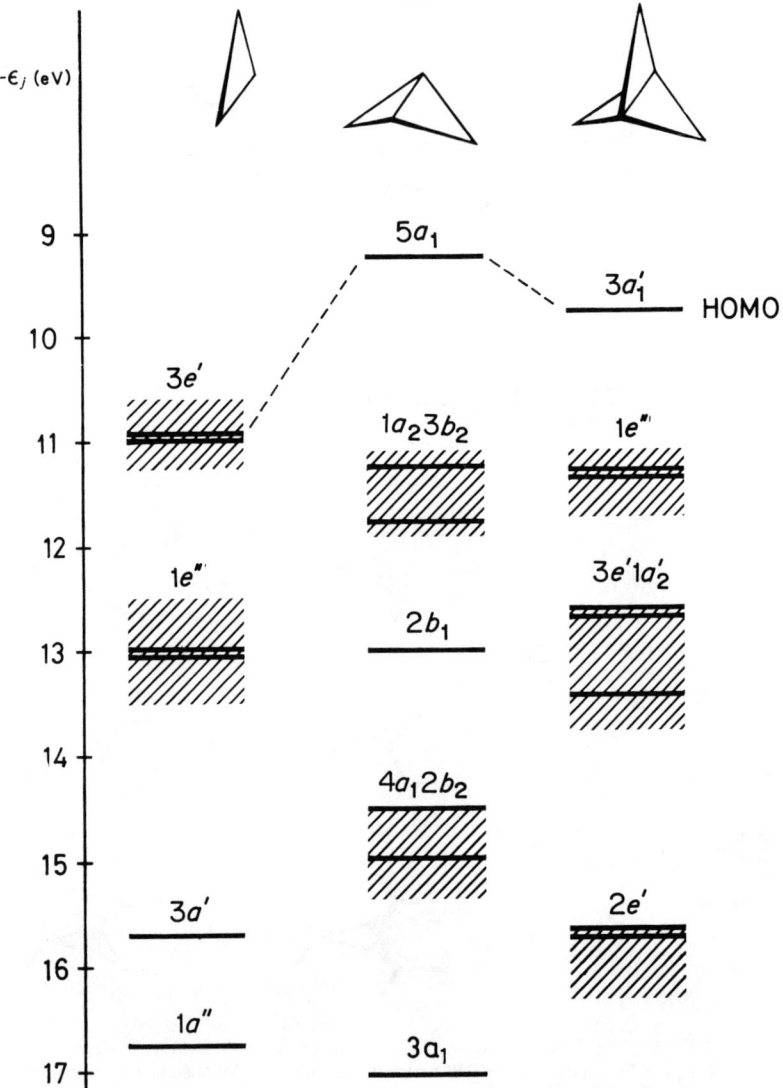

FIGURE 27. Orbital correlation diagram for cyclopropane, bicyclo[1.1.0]butane **1** and [1.1.1]propellane **49**. The orbital energy values given correspond to the observed positions of the bands in the PE spectra, i.e. $\varepsilon_j = -I_j^m$. The shaded areas indicate overlapping bands, or bands broadened by the Jahn–Teller effect

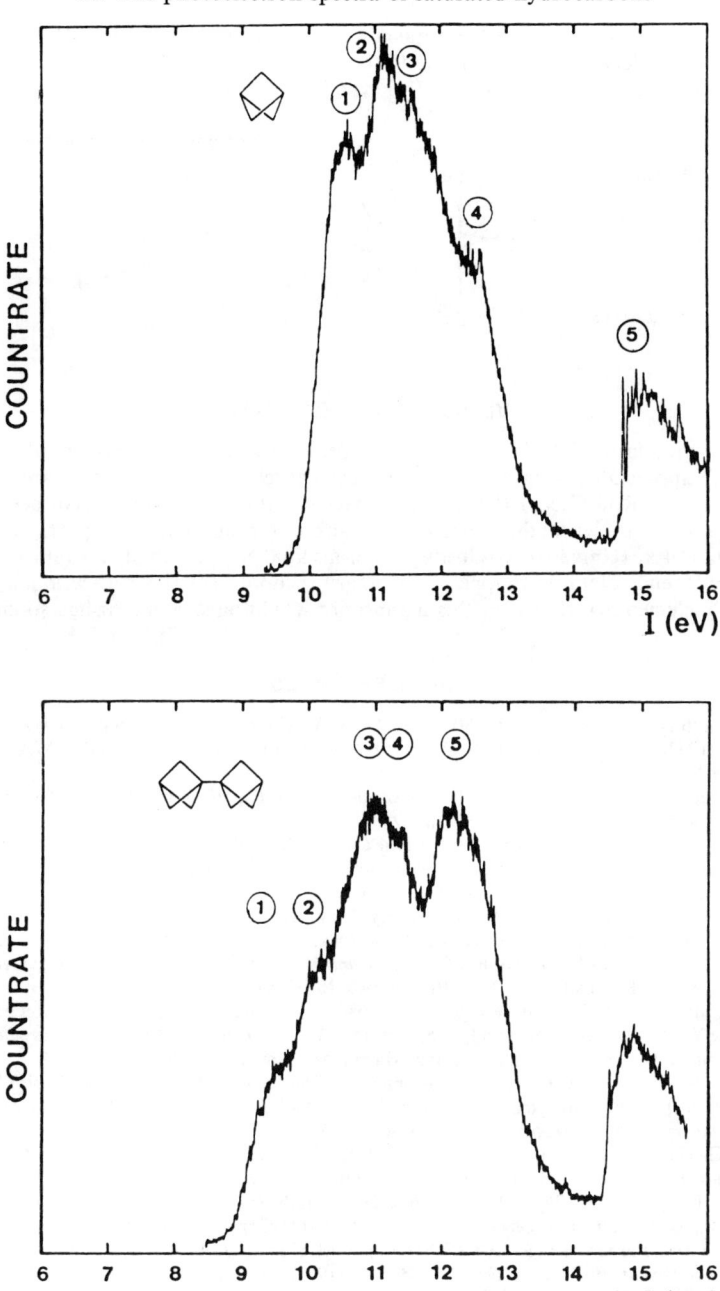

FIGURE 28. The PE spectra of bicyclo[1.1.1]pentane **10** and of di-bicyclo-[1.1.1]pentyl **11**

524   E. Heilbronner

is shown by a comparison of the sequence of the highest occupied CMOs stemming
from an AB-model and from an $A\Gamma$-model (diagram 83). The latter sequence corresponds
to that obtained from STO-3G or 4-31G type SCF calculations[67].

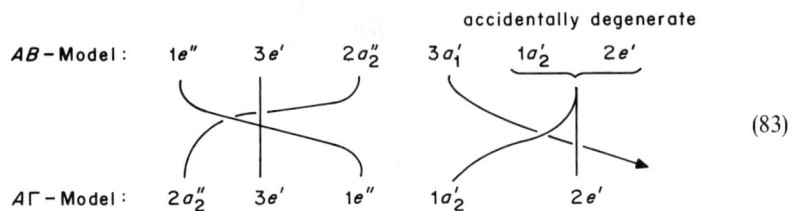

(83)

VI. ACKNOWLEDGEMENTS

I am very grateful to Dr. Engelbert Zass (Federal Institute of Technology, Zürich) for
his much appreciated help by providing a comprehensive computer search of the
literature. Prof. Rolf Gleiter (University of Heidelberg)—who would have been much
better qualified to write this review—is thanked for his active support, and Prof.
Wolfgang Lüttke (University of Göttingen) for his kind advice. Finally I want to mention
my little friend Natalie Puippe[181] (Primary School, Herrliberg), without whose
interest in elementary chemistry this manuscript would have been finished sooner.

VII. REFERENCES

1. L. Pauling, *J. Chem. Phys.*, **1**, 280 (1933); G. W. Wheland, and L. Pauling, *J. Chem. Phys.*, **1**,
   362 (1933); L. Pauling, and E. B. Wilson, *Introduction to Quantum Mechanics*, McGraw-Hill,
   New York, 1935.
2. E. Hückel, *Z. Phys.*, **70**, 204, 310 (1931); *Grundzüge der Theorie ungesättigter und aromatischer
   Verbindungen*, Verlag Chemie, Berlin, 1938.
3. D. Ginsburg (Ed.), *Non-Benzenoid Aromatic Compounds*, Wiley–Interscience, New York, 1959.
4. L. Pauling, *The Nature of the Chemical Bond and the Structure of Molecules and Crystals*,
   3rd ed., Cornell University Press, Ithaca, 1960.
5. S. Benson, *Thermochemical Kinetics*, Wiley, New York, 1968.
6. J. L. Franklin, J. G. Dillard, H. M. Rosenstock, J. T. Herron, K. Draxl and F. H. Field,
   Ionization Potentials, Appearance Potentials, and Heats of Formation of Gaseous Positive Ions,
   Natl. Stand. Ref. Data Ser., Natl. Bur. Stand., **26** (1969).
7. J. A. Pople and D. L. Beveridge, *Approximate Molecular Orbital Theory*, McGraw-Hill,
   New York, 1970; A. Szabo and N. S. Ostlund, *Modern Quantum Chemistry: Introduction to
   Advanced Electronic Structure Theory*, Macmillan, New York, 1982.
8. J. N. Murrell and A. J. Harget, *Semiempirical, Self-Consistent Molecular Orbital Theory of
   Molecules*, Wiley–Interscience, London, 1972; J. Sadlej, *Semiempirical Methods of Quantum
   Chemistry*, Ellis Horwood Ltd., Chichester, 1985.
9. G. G. Hall, *Proc. R. Soc. London, Ser. A*, **205**, 541 (1951); J. Lennard-Jones and G. G. Hall,
   Trans. Faraday Soc., **48**, 581 (1952); G. G. Hall, *Trans. Faraday Soc.*, **50**, 319 (1954).
10. J. C. Lorquet, *Mol. Phys.*, **9**, 101 (1965); D. F. Brailsford and B. Ford, *Mol. Phys.*, **18**, 621
    (1970); W. C. Herndon, *Chem. Phys. Lett.*, **10**, 460 (1971); J. N. Murrell and W. Schmidt, *J.
    Chem. Soc., Faraday Trans. 2*, **68**, 1709 (1972); B. M. Gimarc, *J. Am. Chem. Soc.*, **95**, 1417 (1973).
11. M. J. S. Dewar, *Bull. Soc. Chim. Belg.*, **88**, 957 (1979); M. J. S. Dewar and M. L. McKee, *Pure
    Appl. Chem.*, **52**, 1431 (1980); M. J. S. Dewar, *J. Am. Chem. Soc.*, **106**, 669 (1984).
12. G. Klopman, *Tetrahedron*, **19**, Suppl. 2, 111 (1963); W. C. Herndon, *J. Chem. Educ.*, **56**, 448
    (1979); R. Hoffmann, *J. Chem. Phys.*, **39**, 1397 (1963); cf. C. J. Ballhausen and H. B. Gray,
    Molecular Orbital Theory, Benjamin, New York, 1964.

13. E. Honegger and E. Heilbronner, 'The equivalent bond orbital model and the interpretation of PE spectra', in *Theoretical Models of Chemical Bonding*, Vol. 3 (Ed. Z. B. Maksic), Springer-Verlag, Berlin, 1991.
14. E. Honegger, Z.-Z. Yang and E. Heilbronner, *Croat. Chem. Acta*, **57**, 967 (1984).
15. C. Edmiston and K. Ruedenberg, *Rev. Mod. Phys.*, **35**, 457 (1963); *J. Chem. Phys.*, **43**, 597 (1965); K. Ruedenberg, in *Modern Quantum Chemistry*, Vol. 1, Academic Press, New York, 1965; W. England L. S. Salmon and K. Ruedenberg, *Top. Curr. Chem.*, **23**, 31 (1971).
16. S. F. Boys, *Rev. Mod. Phys.*, **32**, 296 (1960); J. M. Foster and S. F. Boys, *Rev. Mod. Phys.*, **32**, 300 (1960).
17. J. E. Carpenter and F. Weinhold, *J. Am. Chem. Soc.*, **110**, 368 (1988).
18. G. Bieri, J. W. Dill, E. Heilbronner and A. Schmelzer, *Helv. Chim. Acta*, **60**, 2234 (1977).
19. D. W. Turner, C. Baker, A. D. Baker and C. R. Brundle, *Molecular Photoelectron Spectroscopy*, Wiley–Interscience, London, 1970.
20. H. Siegbahn and L. Karlsson, *Photoelectron Spectroscopy*, in *Handbuch der Physik*, Vol. 31, *Corpuscles and Radiation in Matter*, Part 1 (Ed. W. Mehlhorn), Springer-Verlag, Berlin, 1982.
21. J. H. D. Eland, *Photoelectron Spectroscopy*, Butterworths, London, 1984.
22. E. Heilbronner and J. P. Maier, 'Some aspects of organic photoelectron spectroscopy', in *Electron Spectroscopy: Theory, Techniques, and Applications* (Eds. C. R. Brundle and A. D. Baker), Vol. 1, Academic Press, London, 1977.
23. T. Koopmans, *Physica*, **1**, 104 (1939).
24. W. C. Price, A. W. Potts and D. G. Streets, in *Electron Spectroscopy* (Ed. D. A. Shirley), North-Holland, Amsterdam, 1972, p. 187.
25. A. W. Potts, T. A. Williams and W. C. Price, *Faraday Discuss. Chem. Soc.*, **54**, 104 (1972); cf also Reference 40.
26. K. Siegbahn, C. Nordling, G. Johansson, J. Hedman, P. F. Hedén, K. Hamrin, U. Gelius, T. Bergmark, L. O. Werme, R. Manne and Y. Baer, *ESCA Applied to Free Molecules*, North-Holland, Amsterdam, 1969.
27. J. J. Pireaux, S. Svensson, E. Basilier, P.-Å. Malmqvist, U. Gelius, R. Caudano and K. Siegbahn, *Phys. Rev. A*, **14**, 2133 (1976).
28. G. Bieri and L. Åsbrink, *J. Electron Spectrosc.*, **20**, 149 (1980).
29. G. Bieri, F. Burger, E. Heilbronner and J. P. Maier, *Helv. Chim. Acta*, **60**, 2213 (1977).
30. M. J. S. Dewar and S. D. Worley, *J. Chem. Phys.*, **50**, 654 (1969).
31. R. D. Levin and S. G. Lias, *Ionization Potential and Appearance Potential Measurements 1971–1981*, U.S. Department of Commerce, Natl. Bur. Stand., Ref. Data 71, Washington, D.C., 1982.
32. K. Kimura, S. Katsumata, Y. Achiba, T. Yamazaki and S. Iwata, *Handbook of the He 1 Photoelectron Spectra of Fundamental Organic Molecules*, Japan Scientific Soc. Press, Tokyo, 1981.
33. R. Gleiter, *Top. Curr. Chem.*, **86**, 197 (1979).
34. A. W. Potts and W. C. Price, *Proc. R. Soc. London, Ser. A*, **326**, 165 (1971).
35. L. L. Coatsworth, G. M. Bancroft, D. K. Creber, R. J. D. Lazier and P. W. M. Jacobs, *J. Electron Spectrosc.*, **13**, 395 (1978).
36. B. P. Pullen, T. A. Carlson, W. E. Moddeman, G. K. Schweitzer, K. George, W. E. Bull and F. Grimm, *J. Chem. Phys.*, **53**, 768 (1970).
37. T. Bergmark, J. W. Rabalais, L. O. Werme, L. Karlsson and K. Siegbahn, in *Electron Spectroscopy* (Ed. D. A. Shirley), North-Holland, Amsterdam, 1972.
38. B. Narayan, *Mol. Phys.*, **23**, 281 (1972).
39. A. D. Baker, D. Betteridge, N. R. Kemp and R. E. Kirby, *J. Mol. Struct.*, **8**, 75 (1971).
40. A. W. Potts and D. G. Streets, *J. Chem. Soc., Faraday Trans. 2*, **70**, 875 (1974).
41. D. F. Brailsford and B. Ford, *Mol. Phys.*, **18**, 621 (1970).
42. J. N. Murrell and W. Schmidt, *J. Chem. Soc., Faraday Trans. 2*, **68**, 1709 (1972).
43. R. Stockbauer and M. G. Inghram, *J. Chem. Phys.*, **65**, 4081 (1976).
44. S. Ikuta, K. Yoshihara, T. Shiokawa, M. Jinno, Y. Yokoyama and S. Ikeda, *Chem. Lett.*, 1237 (1973).
45. K. Seki and H. Inokuchi, *Bull. Chem. Soc. Jpn.*, **56**, 2212 (1983).
46. W. Schmidt and B. T. Wilkins, *Angew. Chem., Int. Ed. Engl.*, **11**, 221 (1972); R. Boschi, M. F. Lappert, J. B. Pedley, W. Schmidt and B. T. Wilkins, *J. Organomet. Chem.*, **50**, 69 (1973).

526   E. Heilbronner

47. A. E. Jonas, G. K. Schweitzer, F. A. Grimm and T. A. Carlson, *J. Electron Spectrosc.*, 1, 29 (1972).
48. J. Dyke, N. Jonathan, E. Lee, A. Morris and M. Winter, *Phys. Scr.*, 16, 197 (1977); L. Szepes, T. Koranyi, G. Naray-Szabo, A. Modelli and G. Distefano, *J. Organomet. Chem.*, 217, 35 (1981).
49. N. Ueno, T. Fukushima, K. Sugita, S. Kiyono, K. Seki and H. Inokuchi, *J. Phys. Soc. Jpn.*, 48, 1254 (1980).
50. A. W. Potts, W. C. Price, D. G. Streets and T. A. Williams, *Faraday Discuss. Chem. Soc.*, 54, 168 (1972).
51. H. Basch, M. B. Robin, N. A. Kuebler, C. Baker and D. W. Turner, *J. Chem. Phys.*, 51, 52 (1969).
52. E. Lindholm, C. Fridh and L. Åsbrink, *Faraday Discuss. Chem. Soc.*, 54, 127 (1972).
53. A. Schweig and W. Thiel, *Chem. Phys. Lett.*, 21, 541 (1973).
54. S. Evans, P. J. Joachim, A. F. Orchard and D. W. Turner, *Int. J. Mass Spectrom. Ion Phys.*, 9, 41 (1972).
55. F. J. Leng and G. L. Nyberg, *J. Electron Spectrosc.*, 11, 293 (1977).
56. P. R. Keller, J. W. Taylor, T. A. Carlson, T. A. Whiteley and F. A. Grimm, *Chem. Phys.*, 99, 317 (1985).
57. P. Bischof, E. Haselbach and E. Heilbronner, *Angew. Chem.*, 82, 952 (1970); *Angew. Chem., Int. Ed. Engl.*, 9, 953 (1970).
58. P. Bischof, R. Gleiter, M. Kukla and L. A. Paquette, *J. Electron Spectrosc.*, 4, 177 (1974).
59. J. P. Gilman, T. Hsieh and G. G. Meisels, *Int. J. Mass Spectrom. Ion Phys.*, 51, 313 (1983).
60. P. Bischof, E. Heilbronner, H. Prinzbach and H. D. Martin, *Helv. Chim. Acta*, 54, 1072 (1971).
61. J. P. Puttemans, *Ing. Chim. (Brussels)*, 56, 64 (1974).
62. V. V. Plemenkov, J. Villem, N. Villem, I. G. Bolesov, I. G. Surmina, N. I. Yakushkina and A. A. Formanovskii, *Zh. Obshch. Khim.*, 51, 2076 (1981); V. V. Plemenkov, M. M. Latypova, I. G. Bolesov, V. V. Redchenko, N. Villem and J. Villem, *Zh. Org. Khim.*, 18, 1888 (1982).
63. Y. L. Sergeev, M. E. Akopyan, F. I. Vilesov and Y. V. Chizhov, *Khim. Vys. Energ.*, 7, 418 (1973).
64. P. Bruckmann and M. Klessinger, *Chem. Ber.*, 111, 944 (1978).
65. C. Batich, P. Bischof and E. Heilbronner, *J. Electron Spectrosc.*, 1, 333 (1973).
66. R. Bombach, J. Dannacher and J. P. Stadelmann, *Helv. Chim. Acta*, 66, 701 (1983).
67. R. Gleiter, K.-H. Pfeifer, G. Szeimies and U. Bunz, *Angew. Chem.*, 102, 418 (1990); *Angew. Chem., Int. Ed. Engl.*, 29, 413 (1990).
68. R. Gleiter, G. Krennrich and U. H. Brinker, *J. Org. Chem.*, 51, 2899 (1986).
69. E. W. Della, P. E. Pigou, M. K. Livett and J. B. Peel, *J. Electron Spectrosc.*, 33, 163 (1984).
70. G. Bieri, E. Heilbronner, T. Kobayashi, A. Schmelzer, M. J. Goldstein, R. S. Leight and M. S. Lipton, *Helv. Chim. Acta*, 59, 2657 (1976).
71. P. Hemmersbach and M. Klessinger, *Tetrahedron*, 36, 1337 (1980).
72. P. Asmus and M. Klessinger, *Angew. Chem.*, 88, 343 (1976).
73. P. Bischof, J. A. Hashmall, E. Heilbronner and V. Hornung, *Helv. Chim. Acta*, 52, 1745 (1969).
74. E. Heilbronner, R. Gleiter, T. Hoshi and A. de Meijere, *Helv. Chim. Acta*, 56, 1594 (1973).
75. R. Gleiter, P. Bischof, W. Volz and L. A. Paquette, *J. Am. Chem. Soc.*, 99, 8 (1977).
76. P. Bischof, R. Gleiter, A. de Meijere and L. Lueder, *Helv. Chim. Acta*, 57, 1519 (1974).
77. A. de Meijere, *Chem. Ber.*, 107, 1684 (1974).
78. M. J. Goldstein, S. Natowsky, E. Heilbronner and V. Hornung, *Helv. Chim. Acta*, 56, 294 (1973).
79. G. J. Edwards, S. R. Jones and J. M. Mellor, *J. Chem. Soc., Perkin Trans. 2*, 505 (1977).
80. E. Heilbronner, E. Honegger, W. Zambach, P. Schmitt and H. Guenther, *Helv. Chim. Acta*, 67, 1681 (1984).
81. V. A. Chuiko, E. N. Manukov, Y. V. Chizhov and M. M. Timoshenko, *Khim. Prir. Soedin.*, 639 (1985).
82. E. Honegger, H. Huber, E. Heilbronner, W. P. Dailey and K. B. Wiberg, *J. Am. Chem. Soc.*, 107, 7172 (1985).
83. P. Bischof, R. Gleiter and E. Mueller, *Tetrahedron*, 32, 2769 (1976).
84. P. Bischof, R. Gleiter, E. Heilbronner, V. Hornung and G. Schroeder, *Helv. Chim. Acta*, 53, 1645 (1970).
85. E. Haselbach, E. Heilbronner, H. Musso and A. Schmelzer, *Helv. Chim. Acta*, 55, 302 (1972).

86. R. S. Abeywickrema, E. W. Della, P. Pigou, M. Livett and J. B. Peel, *J. Am. Chem. Soc.*, **106**, 7321 (1984).
87. R. Gleiter, J. Spanguet-Larsen, 'Photoelectron Spectra of some Strained Hydrocarbons' in *Advances in Strain Energy in Organic Chemistry*, (Ed. B. Halton), J.A.I. Press Inc., Greenwich (Connecticut), to be published in 1992.
88. R. Gleiter, E. Heilbronner, M. Heckmann and H. D. Martin, *Chem. Ber.*, **106**, 28 (1973).
89. N. Bodor, M. J. S. Dewar and S. D. Worley, *J. Am. Chem. Soc.*, **92**, 19 (1970).
90. R. Gleiter, P. Bischof and M. Christl, *J. Org. Chem.*, **51**, 2895 (1986).
91. P. Bischof, R. Gleiter, R. T. Taylor, A. R. Browne and L. A. Paquette, *J. Org. Chem.*, **43**, 2391 (1978).
92. P. Bruckmann and M. Klessinger, *Angew. Chem., Int. Ed. Engl.*, **11**, 524 (1972).
93. H. D. Martin, S. Kagabu and R. Schwesinger, *Chem. Ber.*, **107**, 3130 (1974).
94. P. Asmus and M. Klessinger, *Justus Liebigs Ann. Chem.*, 2169 (1975).
95. F. Brogli, E. Heilbronner and J. Ipaktschi, *Helv. Chim. Acta*, **55**, 2447 (1972).
96. R. Gleiter, W. Sander and I. Butler-Ransohoff, *Helv. Chim. Acta*, **69**, 1872 (1986).
97. G. D. Mateescu and S. D. Worley, *Tetrahedron Lett.*, 5285 (1972).
98. W. Schmidt, *Tetrahedron*, **29**, 2129 (1973); R. Boschi, W. Schmidt, J. R. Suffolk, B. Wilkins, H. Lempka and J. N. A. Ridyard, *J. Electron Spectrosc.*, **2**, 377 (1973).
99. S. D. Worley, *J. Electron Spectrosc.*, **6**, 157 (1975); W. Schmidt, *J. Electron Spectrosc.*, **6**, 163 (1975).
100. B. Kovać and L. Klasinc, *Croat. Chem. Acta*, **51**, 55 (1978).
101. P. Asmus, M. Klessinger, L. U. Meyer and A. de Meijere, *Tetrahedron Lett.*, 381 (1975).
102. S. D. Worley, G. D. Mateescu, C. W. McFarland, R. C. Fort Jr. and C. F. Sheley, *J. Am. Chem. Soc.*, **95**, 7580 (1973).
103. R. T. Taylor and L. A. Paquette, *J. Org. Chem.*, **43**, 242 (1978).
104. R. Gleiter, A. Toyota, P. Bischof, G. Krennrich, J. Dressel, P. D. Pansegrau and L. A. Paquette, *J. Am. Chem. Soc.*, **110**, 5490 (1988).
105. R. Gleiter, J. Spanget-Larsen, H. Hopf and C. Mlynek, *Chem. Ber.*, **117**, 1987 (1984).
106. E. Heilbronner, T. B. Jones, A. Krebs, G. Mayer, K. D. Malsch, J. Pocklington and A. Schmelzer, *J. Am. Chem. Soc.*, **102**, 564 (1980).
107. R. Gleiter, K. H. Pfeifer, G. Szeimies, J. Belzner and K. Lehne, *J. Org. Chem.*, **55**, 636 (1990).
108. H. D. Martin, C. Heller, E. Haselbach and Z. Lanyjova, *Helv. Chim. Acta*, **57**, 465 (1974).
109. K. Hassenrueck, H. D. Martin and B. Mayer, *Chem. Ber.*, **121**, 373 (1988).
110. H. D. Martin, C. Heller, R. Haider, R. W. Hoffmann, J. Becherer and H. R. Kurz, *Chem. Ber.*, **110**, 3010 (1977).
111. H. D. Martin, C. Heller and J. Werp, *Chem. Ber.*, **107**, 1393 (1974).
112. G. Jonkers, W. J. Van der Meer, C. A. De Lange, E. J. Baerends, J. Stapersma and G. W. Klumpp, *J. Am. Chem. Soc.*, **106**, 587 (1984).
113. J. Spanget-Larsen, R. Gleiter, A. de Meijere and P. Binger, *Tetrahedron*, **35**, 1385 (1979).
114. P. Blickle, H. Hopf, M. Bloch and T. B. Jones, *Chem. Ber.*, **112**, 3691 (1979).
115. R. Gleiter, R. Haider, J.-M. Conia, J.-P. Barnier, A de Meijere and W. Weber, *J. Chem. Soc., Chem. Commun.*, 130 (1979).
116. J. Spanget-Larsen, R. Gleiter, G. Klein, C. Doecke and L. A. Paquette, *Chem. Ber.*, **113**, 2120 (1980).
117. J. Spanget-Larsen, R. Gleiter, M. R. Detty and L. A. Paquette, *J. Am. Chem. Soc.*, **100**, 3005 (1978).
118. R. Gleiter, G. Krennrich, P. Bischof, T. Tsuji and S. Nishida, *Helv. Chim. Acta*, **69**, 962 (1986).
119. H. D. Martin and R. Schwesinger, *Chem. Ber.*, **107**, 3143 (1974).
120. N. M. Paddon-Row, H. K. Patney and R. S. Brown, *Aust. J. Chem.*, **35**, 293 (1982).
121. L. Klasinc, B. Rušcić, N. S. Bhacca and S. P. McGlynn, *Int. J. Quantum Chem., Quantum Biol. Symp.*, **12**, 161 (1986).
122. N. Bhacca and L. Klasinc, *Z. Naturforsch., A*, **40**, 706 (1985).
123. E. Honegger, P. E. Eaton, B. K. Ravi Shankar and E. Heilbronner, *Helv. Chim. Acta*, **65**, 1982 (1982).
124. P. Bischof, P. E. Eaton, R. Gleiter, E. Heilbronner, T. B. Jones, H. Musso, A. Schmelzer and R. Stober, *Helv. Chim. Acta*, **61**, 547 (1978).
125. J. Spanget-Larsen, R. Gleiter, L. A. Paquette, M. J. Carmody and C. R. Degenhardt, *Theor. Chim. Acta*, **50**, 145 (1978).

528 E. Heilbronner

5 528 E. Heilbronner

126. P. Bischof, M. Boehm, R. Gleiter, R. A. Snow, C. W. Doecke and L. A. Paquette, *J. Org. Chem.*, **43**, 2387 (1978).
127. R. Gleiter, M. C. Boehm, A. de Meijere and T. Preuss, *J. Org. Chem.*, **48**, 796 (1983).
128. M. Eckert-Maksić, K. Mlinarić-Majerski and Z. Majerski, *J. Org. Chem.*, **52**, 2098 (1987).
129. A. de Meijere, K. Michelsen, R. Gleiter and J. Spanget-Larsen, *Isr. J. Chem.*, **29**, 153 (1989).
130. B. Kovać, E. Heilbronner, H. Prinzbach and K. Weidmann, *Helv. Chim. Acta*, **62**, 2841 (1979).
131. H. A. Jahn and E. Teller, *Proc. R. Soc. London, Ser. A*, **161**, 220 (1937).
132. M. I. Al-Joboury and D. W. Turner, *J. Chem. Soc., B*, 373 (1967).
133. R. N. Dixon, *Mol. Phys.*, **20**, 113 (1971).
134. W. Rabalais, T. Bergmark, L. O. Werme, L. Karlsson and K. Siegbahn, *Phys. Scr.*, **3**, 13 (1971); W. Meyer, *J. Chem. Phys.*, **58**, 1017 (1977); K. Takeshita, *J. Chem. Phys.*, **86**, 329 (1987).
135. L. S. Cederbaum, W. Domcke, J. Schirmer, W. von Niessen, G. H. F. Diercksen and W. P. Kraemer, *J. Chem. Phys.*, **69**, 1591 (1978); L. S. Cederbaum, W. Domcke and W. von Niessen, *Int. J. Quantum Chem.*, **14**, 593 (1978).
136. A. Schweig and W. Thiel, *J. Electron Spectrosc.*, **3**, 27 (1974).
137. S. Iwata and S. Nagakura, *Mol. Phys.*, **27**, 425 (1974).
138. J. T. Huang and F. O. Ellison, *Chem. Phys. Lett.*, **29**, 565 (1974).
139. A. Richartz, R. J. Buenker, P. J. Bruna and S. D. Peyerimhoff, *Mol. Phys.*, **33**, 1345 (1977); A. Richartz, R. J. Buenker and S. D. Peyerimhoff, *Chem. Phys.*, **28**, 305 (1978); A. Richartz, *Prog. Theor. Org. Chem.*, **2**, 64 (1977).
140. J. W. Rabalais and A. Katrib, *Mol. Phys.*, **27**, 923 (1974).
141. M. Jungen, *Theor. Chim. Acta*, **22**, 255 (1971).
142. G. W. Wheland, *J. Am. Chem. Soc.*, **63**, 2025 (1941); B. Pullman and A. Pullman, *Les Théories Électroniques de la Chimie Organique*, Masson et Cie., Paris, 1952.
143. E. Heilbronner, 'Organic chemical photoelectron spectroscopy', in: *Molecular Spectroscopy* (Sixth Conference on Molecular Spectroscopy), Institute of Petroleum, London, 1976.
144. M. J. S. Dewar, Y. Yamaguchi and S. H. Suck, *Chem. Phys. Lett.*, **50**, 259 (1977).
145. E. Heilbronner, *Helv. Chim. Acta*, **60**, 2248 (1977).
146. E. Heilbronner, *Helv. Chim. Acta*, **37**, 921 (1954).
147. R. Pauncz, *Acta Phys. Acad. Sci. Hung.*, **6**, 15 (1956).
148. W. Müller, C. Nager and P. Rosmus, *Chem. Phys.*, **51**, 43 (1980).
149. A. Richartz, R. J. Buenker and S. D. Peyerimhoff, *Chem. Phys.*, **31**, 187 (1978).
150. W. C. Herndon, *Prog. Phys. Org. Chem.*, **9**, 99 (1972); W. C. Herndon, M. L. Ellzey Jr. and K. S. Raghuveer, *J. Am. Chem. Soc.*, **100**, 2645 (1978); W. C. Herndon and M. L. Ellzey, Jr., *Chem. Phys. Lett.*, **60**, 510 (1979).
151. B.-Ö. Jonsson and E. Lindholm, *Ark. Fys.*, **39**, 65 (1969); *Chem. Phys. Lett.*, **1**, 501 (1967).
152. R. Hoffmann, P. D. Mollère and E. Heilbronner, *J. Am. Chem. Soc.*, **95**, 4860 (1973).
153. R. Hoffmann, *Acc. Chem. Res.*, **4**, 1 (1971); R. Gleiter, *Angew. Chem.*, **86**, 770 (1974); H.-D. Martin and B. Mayer, *Angew. Chem.*, **95**, 281 (1983).
154. F. S. Jørgensen, M. N. Paddon-Row and H. K. Partney, *J. Chem. Soc., Chem. Commun.*, 573 (1983); M. N. Paddon-Row, H. K. Partney, J. B. Peel and G. D. Willett, *J. Chem. Soc., Chem. Commun.*, 564 (1984).
155. E. Heilbronner and A. Schmelzer, *Helv. Chim. Acta*, **58**, 936 (1975).
156. C. H. De Puy, *Top. Curr. Chem.*, **40**, 73 (1973); D. Wendisch, 'Carbocyclische Dreiring-Verbindungen', in *Methoden der organischen Chemie*, Houben-Weyl-Müller, Vol. IV/3, G. Thieme Verlag, Stuttgart, 1971; M. Charton, 'Olefinic properties of cyclopropanes', in *The Chemistry of Alkenes* (Ed. J. Zabicky), Vol. 2, Interscience, New York, 1970.
157. A. de Meijere, *Angew. Chem.* **91**, 867 (1979); *Angew. Chem., Int. Ed. Engl.*, **18**, 809 (1979).
158. A. D. Walsh, *Nature*, **159**, 167, 712 (1947); T. M. Sugden, *Nature*, **160**, 367 (1947).
159. C. A. Coulson and W. E. Moffit, *J. Chem. Phys.*, **15**, 151 (1947): *Philos. Mag.*, **40**, 1 (1949).
160. Th. Förster, *Z. Phys. Chem.*, **B43**, 58 (1939).
161. P. R. Keller, J. W. Taylor, A. Thomas, T. A. Whitley and T. A. Grimm, *Chem. Phys.*, **99**, 317 (1985).
162. J. A. Hashmall and E. Heilbronner, *Angew. Chem.*, **82**, 320 (1970).
163. P. Bischof, R. Gleiter, E. Heilbronner, V. Hornung and G. Schröder, *Helv. Chim. Acta*, **53**, 1645 (1970).
164. M. Klessinger and P. Rademacher, *Angew. Chem.*, **91**, 885 (1979); *Angew. Chem., Int. Ed. Engl.*, **18**, 826 (1979).

165. W. A. Bernett, *J. Chem. Educ.*, **44**, 17 (1967).
166. E. Honegger, E. Heilbronner, A. Schmelzer and J.-Q. Wang, *Isr. J. Chem.*, **22**, 3 (1982).
167. E. Honegger, E. Heilbronner and A. Schmelzer, *Nouv. J. Chim.*, **6**, 519 (1982).
168. E. Honegger, E. Heilbronner and J.-Q. Wang, *Sci. Sin.*, **25**, 236 (1982).
169. E. Heilbronner and A. Schmelzer, *Nouv. J. Chim.*, **4**, 23 (1980).
170. E. Haselbach, *Chem. Phys. Lett.*, **7**, 428 (1970).
171. P. Bruckmann and M. Klessinger, *J. Electron Spectrosc.*, **2**, 341 (1973).
172. O. Bastiansen and W. Lüttke, *Acta Chem. Scand.*, **20**, 516 (1966); H. Braun and W. Lüttke, *J. Mol. Struct.*, **28**, 391 (1975).
173. E. Honegger and E. Heilbronner, *Chem. Phys. Lett.*, **81**, 615 (1981).
174. T. Prangé, C. Pascard, A. de Meijere, U. Behrens, J.-P. Barnier and J. M. Conia, *Nouv. J. Chim.*, **4**, 321 (1980).
175. R. Hoffmann and R. B. Davidson, *J. Am. Chem. Soc.*, **93**, 5699 (1971).
176. F. Brogli and E. Heilbronner, *Helv. Chim. Acta*, **54**, 1423 (1971).
177. G. Maier, *Angew. Chem.*, **100**, 317 (1988).
178. E. Honegger, E. Heilbronner, Th. Urbanek and H.-D. Martin, *Helv. Chim. Acta*, **68**, 23 (1985).
179. P. C. Harihan and J. A. Pople, *Theor. Chim. Acta*, **28**, 213 (1973).
180. H. Huber, *Theor. Chim. Acta*, **55**, 117 (1980).
181. N. Puippe (assisted by D. Lehmann), *Experiment Chemie*, Lecture with demonstrations at Herrliberg Primary School, 1991.

CHAPTER **11**

Acidity and basicity

A. C. HOPKINSON

Department of Chemistry, York University, Downsview, Ontario M3J1P3, Canada

I. INTRODUCTION

The common name for alkanes, the paraffins, is derived from the Latin *parum affinis*, meaning slightly reactive. The lack of reactivity stems from the type of bonding in saturated hydrocarbons. Strong bonding between C and H $(90–110\,\text{kcal mol}^{-1})$ and between C and C $(ca\ 85–90\,\text{kcal mol}^{-1})$ and the absence of lone pairs means the most available electrons are in tightly bound σ-orbitals. As a result reactions occur only with very aggressive electrophiles and the saturated hydrocarbons have extremely low basicity towards both Lewis and Bronsted acids. The LUMOs of saturated hydrocarbons are high-lying σ^*-orbitals and nucleophilic attack on an unactivated hydrocarbon is an even more difficult process, i.e. hydrocarbons do not easily function as Lewis bases. The alternative reaction with a powerful nucleophile, removal of a proton, leaves a saturated alkyl anion in which there is no mechanism for delocalizing the excess negative charge and the resulting instability often results in the loss of an electron, i.e. many alkyl radicals have negative electron affinities. This simple analysis of the molecular orbital description of alkanes then leads to the correct conclusion that saturated hydrocarbons have extremely low acidities and basicities.

The Chemistry of Alkanes and Cycloalkanes
Edited by S. Patai and Z. Rappoport © 1992 John Wiley & Sons Ltd

II. BASICITY

A. General

Alkanes are extremely weak bases. Nevertheless interactions with acids of both the Bronsted (proton donor) and Lewis (electron acceptor) types have been widely studied. Acid-catalysed reactions of alkanes, particularly cracking and isomerization to optimize the number of branched isomers in the C_6–C_8 range, are important industrially in the upgrading of saturated hydrocarbons for fuels.

Protonated alkanes CH_5^+, $C_2H_7^+$, $C_3H_9^+$, etc. are well characterized species in the gas phase[1-16], but they appear to be only transient species in solution[17] undergoing loss of hydrogen, probably by reduction of the acid[17-21] to form the more common trivalent carbocations. These latter ions then undergo rearrangements to generate the more stable secondary and tertiary carbocations and quenching of these ions by hydride abstraction forms branched alkanes[17,22-26]. Alternatively the carbocation may undergo β-scission (equation 1) usually to form a more stable tertiary carbocation and an olefin, a reaction which is widely exploited in the cracking of larger hydrocarbons.

$$R''-\underset{\underset{R'''}{|}}{\overset{\overset{R'}{|}}{C}}-CH_2-\overset{+}{C}\overset{R}{\underset{H}{\diagup}} \longrightarrow \underset{\underset{R'''}{|}}{\overset{R''\diagdown \diagup R'}{\overset{+}{C}}} + H_2C=C\overset{R}{\underset{H}{\diagup}} \qquad (1)$$

Protonated saturated hydrocarbons contain a carbon atom which is formally 5-coordinate. There are insufficient electrons to describe such a carbocation by a classical structure and it is necessary to invoke a 3-centre 2-electron bond. This family of non-classical alkonium ions are generally known as carbonium ions. Removal of H_2 from an alkonium ion leaves a carbocation containing a 3-coordinate carbon. These trivalent alkyl cations are known as carbenium ions. The interrelation between the two classes of carbocations is shown for the parent ions in equation 2.

$$CH_5^+ \longrightarrow CH_3^+ + H_2 \qquad (2)$$
$$\text{methonium ion} \qquad \text{methenium ion}$$
$$\text{(or methyl cation)}$$

B. Structures of Alkonium Ions

Methonium ion, CH_5^+. The structure of the methonium ion has been the subject of many theoretical studies[27-37] and there is general agreement that the global minimum structure, 1, has C_s symmetry, with two of the hydrogens much more loosely bound to carbon than the other three, i.e. the structure has considerable $CH_3^+ \cdots H_2$ character. (The experimental dissociation energy for formation of $CH_3^+ + H_2$ is 40.0 kcal mol^{-1}[14].) As the level of ab initio molecular orbital calculations used in the structure optimization is improved to include correlation energy, the three-centre bond between CH_3^+ and H_2 is strengthened[32]. The structural parameters given on structure 1[33] were optimized using configuration interaction with all double substitutions (CID) and with a 6-31G(d, p) basis set[39]. Several other possible structures have been optimized but for these other stationary points (which are close in energy to 1) the analytic second derivatives of the energy with respect to the nuclear coordinates show them to be saddle points (one or more imaginary frequencies)[32] and not minima (zero imaginary frequencies). Two of these saddle points, structures 2 and 3, are of particular interest. 2 has C_s symmetry and is achieved via a rotation by 30° about the bisector of the $C^+ \cdots H_2$ in 1. At several high levels of theory, **2** is consistently only $0.1\,kcal\,mol^{-1}$ above **1**[32]. Such a small energy difference indicates that there is essentially free rotation of the CH_3 group about the axis perpendicular to the H_2 bond.

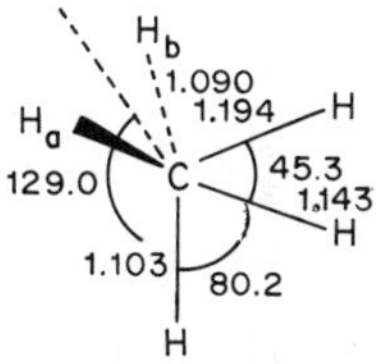

(1) $\angle H_a CH_b = 118.2$

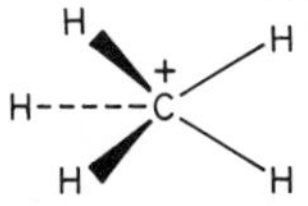

(2) 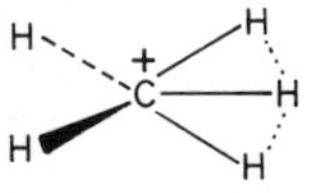   **(3)**

Structure **3** has C_{2v} symmetry and corresponds to protonation of methane along an HCH bisector. At the MP4(SDQ)/6-311G(d, p)//MP2/6-31G(d) level of theory, **3** is $1.1\,kcal\,mol^{-1}$ above **1**[32]. It has two pairs of equivalent hydrogens and the closeness in energies of **1** and **3** indicates that the barrier for scrambling of hydrogens in CH_5^+ is very low.

Ethonium ion, $C_2H_7^+$. The ethonium ion is formed in the gas phase by addition of H_2 to $C_2H_5^{+\,14}$ (equation 3). In the temperature range $-100\,°C$ to 40, $C_2H_7^+$ is formed with a small activation energy ($1.2\,kcal\,mol^{-1}$); in the temperature range $-100\,°C$ to $-130\,°C$, there is no reaction; and below $-130\,°C$, $C_2H_7^+$ is again formed with a rate of formation which *increases* with a *decrease* in temperature. This unusual behaviour is explained in terms of two isomeric forms of $C_2H_7^+$, one essentially a complex $C_2H_5^+\cdots H_2$ (structure **4**) with bonding similar to that in CH_5^+ and the other, **5**, a C—C protonated isomer. Isomer **4** is formed at very low temperature with no barrier, and at higher temperatures the negative ΔS for reaction 3 overcomes the small negative enthalpy ($\Delta H = -4 \pm 0.5\,kcal\,mol^{-1}$) and **4** dissociates. Formation of **5** disrupts the C—C bond which results in a small barrier and the reaction does not occur at very low temperatures. At higher temperatures the more stable isomer, **5**, is formed (ΔH for formation of **5** by reaction 3 is $-11.8\,kcal\,mol^{-1}$). The energy profile for this reaction is shown in Figure 1.

$$H_2 + C_2H_5^+ \rightleftharpoons C_2H_7^+ \tag{3}$$

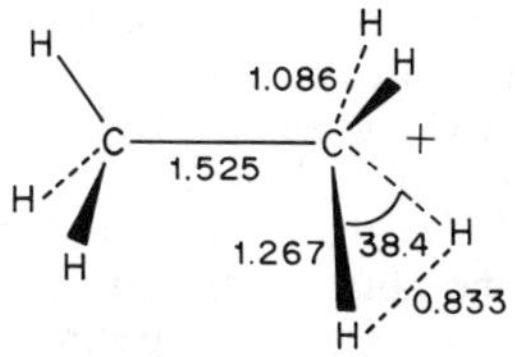

(4) 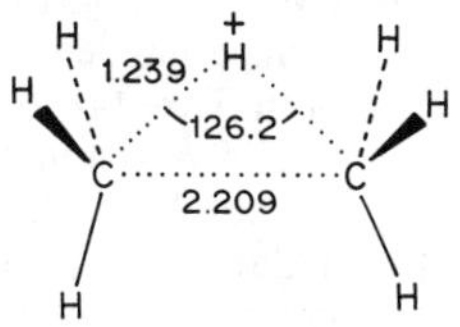   **(5)**

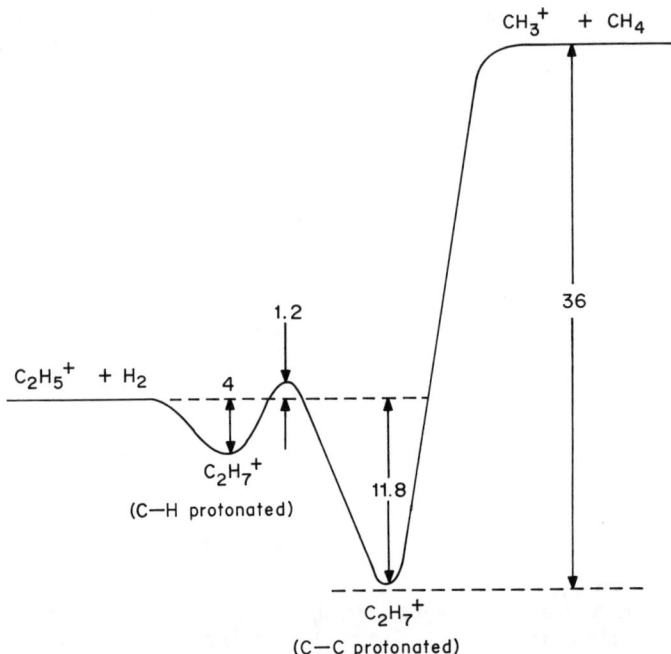

FIGURE 1. Enthalpy surface (in kcal mol^{-1}) for $C_2H_7^+$ derived from
experimental data (see Reference 14)

Ab initio molecular orbital calculations give both **4** and **5** as minima on the
$C_2H_7^+$ potential energy surface. At all levels of theory **5** is found to be lower in energy
than **4**; at the MP4(SDQ)/6-31G(d, p)//MP2/6-31G(d) level Schleyer and coworkers[32]
found the bridging structure to be lower by 6.8 kcal mol^{-1} while Hirao and Yamabe[38]
using symmetry-adapted clusters found the difference to be 4 kcal mol^{-1}. These compare
with the experimental value of 8 kcal mol^{-1}.

Some key structural parameters taken from the calculations of Hirao and Yamabe[38]
are of interest. The C—H bond lengths of the $CH_3CH_2^+$ fragment of **4** (C_s symmetry)
are not shown but are 1.081 and 1.082 Å, i.e. slightly shorter than those in CH_3CH_3
(1.096 Å) as expected for a positive ion. The distances around C^+ and H_2 reveal complex
4 to have a long C—H distance (1.267 Å) and a H—H distance of 0.833 Å, which is
considerably longer than that of H_2 (0.742 Å). Clearly the 3-centre 2-electron bonding
is quite weak.

The bridged ion **5** has the hydrogen atom symmetrically between the two carbon
atoms. There is a very low barrier to rotation[35] for the methyl groups and this flat
potential energy surface has led to some disagreement over the symmetry of this ion.
The structural data in **5** are taken from the C_{2v} structure of Hirao and Yamabe[38]. The
C—C distance of 2.209 Å is much too long for a C—C single bond (H_3C—CH_3 is
1.531 Å). The C—H bond lengths in the CH_3 group (not shown) range from 1.073 to
1.078 Å and are similar to those in CH_3^+ [1.078 Å[34] at the HF/6-31(d, p) level].

Recently the infrared spectrum of $C_2H_7^+$ in the range 2500–4000 cm^{-1} has been
obtained using laser excitation of trapped ions in a tandem mass spectrometer[15]. Two
sets of spectral features were observed and have been assigned to isomers **4** and **5**. For

5 there is excellent agreement between the experimental vibrational frequencies and scaled frequencies from *ab initio* calculations, and for **4** the agreement is also satisfactory.

Higher alkonium ions. Hiraoka and Kebarle[14] also examined the next two homologues of the alkonium ions, $C_3H_9^+$ and $C_4H_{11}^+$. As in $C_2H_7^+$ there is the possibility of either C—H or C—C protonation with the added complication that not all C—H bonds (or C—C bonds with butane) are alike. Attempts to make $C_3H_9^+$ from $(CH_3)_2CH^+$ and H_2 at temperatures as low as $-170\,^{\circ}C$ were unsuccessful, showing that the very stable isopropyl cation will, at best, form a very loose complex with H_2, i.e. the C—H protonated isomer, **6**, is very weakly bound[16]. $C_3H_9^+$ made from the addition of CH_4 to $C_2H_5^{+10}$ is believed to have the C—C protonated structure **7**, but at higher temperature it decomposes into the more stable dissociation products, $(CH_3)_2CH^+$ and H_2, with a barrier of *ca* 2.5 kcal mol^{-1} above the energy of the reactants $CH_4 + C_2H_5^+$. Furthermore, in the gas-phase protonation of labelled propane, the proton from the acid (H_3^+ or CH_5^+) always appears in the neutral products and not in the ions, i.e. there is no scrambling of hydrogens[40-42] (equations 4a and 4b). All this behaviour is consistent with a very weakly bound C—H protonated isomer **6** having lower energy than the C—C protonated isomer **7**.

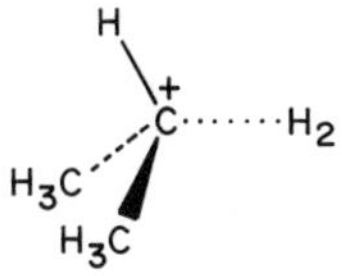

(6)

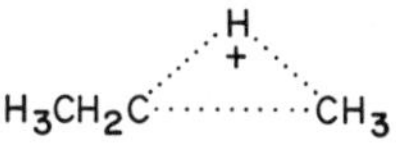

(7)

$$C_3H_8 + H_3^+ \longrightarrow C_3H_7^+ + 2H_2 \tag{4a}$$

$$C_3H_8 + H_3^+ \longrightarrow C_2H_5^+ + CH_4 + H_2 \tag{4b}$$

Isopropyl cation reacts with CH_4 to give $C_4H_{11}^+$ with an enthalpy of -3.4 kcal mol^{-1}[14]. This ion is assumed to have the C—C protonated form **8** and since its heat of formation is higher than that of t-$C_4H_9^+ + H_2$, then the C—H protonated isomer, **9**, is again lower in energy than its C—C protonated isomer.

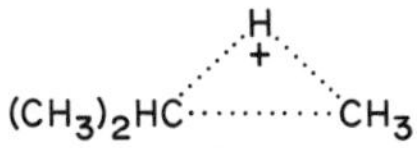

(8)

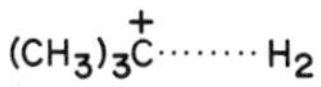

(9)

C. Structure of Protonated Cyclopropane

Protonated cyclopropane has been reported in the gas phase[43-45] to be *ca* 8 kcal mol^{-1} in energy above the isopropyl cation. The 'bent bonds' of the cyclopropane ring are susceptible to electrophilic attack leading to the expectation that cyclopropane will be more basic than saturated alkanes and that protonation will occur on the C—C bond, i.e. the edge-protonated isomer will have the lowest energy. There is, however, considerable evidence from solution chemistry that corner-protonated cyclopropanes exist as intermediates in 1,2-alkyl shifts in carbocations[46]. There have been several reviews of protonated cyclopropanes[46-48] and, in the current work, only the very recent theoretical work[49] will be reviewed.

Unambiguous determination of the site of protonation of cyclopropane from molecular orbital calculations[49-52] has proven rather difficult largely because the potential energy surface is flat. The most recent theoretical study[49], at the MP4(FC)/6-311G**//MP2/6-311G* level and including zero-point energy appears to be quite definitive. Structure **10** (Figure 2), a corner-protonated cyclopropane with one hydrogen eclipsing a C—C bond (C_s symmetry), has the lowest energy, and is at a minimum, but is only 0.1 kcal mol^{-1} lower than rotamer **11** (also C_s symmetry). The calculated CH_2—CH_3^+ distances are long (1.695 Å and 1.790 Å in **10**, 1.761 Å in **11**) while the CH_2—CH_2 distances are short (1.398 Å in **10** and 1.392 Å in **11**). These compare with experimental bond lengths of 1.51 Å in cyclopropane and structures **10** and **11** are then perhaps best thought of as a methyl cation tightly solvated by an ethylene molecule. Structure **11** has one imaginary frequency and is the transition structure for rotation of the methyl group.

Edge-protonated cyclopropane **12** (C_{2v} symmetry) is also at a minimum on the potential energy surface but lies 1.4 kcal mol^{-1} above **10**. The transition structure for formation of **12** from **10** is very close in energy to **12** and, when the zero-point energy is included, the transition state is *lower* than **12**. The existence of **12** as a minimum is then

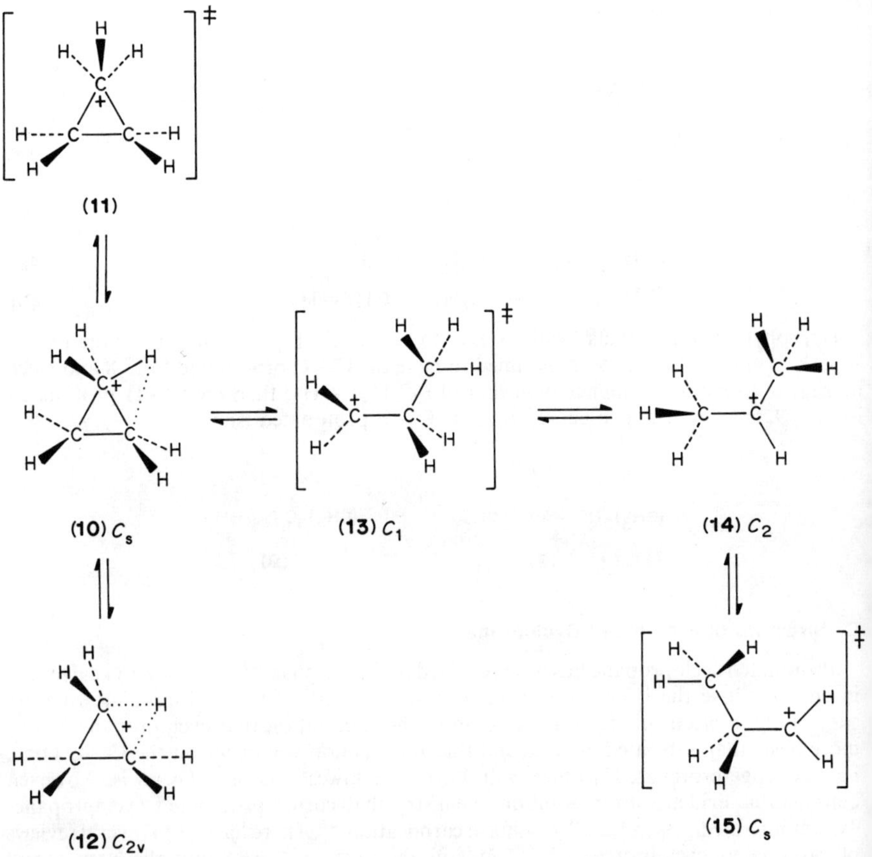

FIGURE 2. Minima and transition structures on the $C_3H_7^+$ potential energy surface (see Reference 49)

questionable. What is clear, however, is that scrambling of hydrogens in protonated cyclopropane via structure **12** requires very little activation.

The isopropyl cation **14** is the global minimum on the $C_3H_7^+$ potential energy surface (7.2 kcal mol^{-1} below **10** at 0 K[49]) but isomerization of **10** into **14** requires considerable rearrangement. Ring-opening of **10** accompanied by rotation of a CH_2 group leads to a transition structure **13** which is essentially a 1-propyl cation, and which is 13.3 kcal mol^{-1} above **10**. The hydride shift required to convert **13** into **14** occurs without barrier.

There is no minimum structure for the 1-propyl cation but one other critical point, the bisected 1-propyl cation **15**, is 1.2 kcal mol^{-1} lower in energy than **13** and is the transition structure for scrambling of hydrogens in the isopropyl cation.

D. Proton Affinities

The gas-phase proton affinity (PA) of a molecule is defined as the negative of the enthalpy of the protonation reaction at room temperature, i.e. PA $= -\Delta H$ for reaction 5.

$$B + H^+ \longrightarrow BH^+ \tag{5}$$

Relative proton affinities are determined by measuring the enthalpy of proton transfer from one base B^1 to a second base B^2 (equation 6), using high-pressure mass spectrometry, the SIFT/flow afterglow or ion cyclotron resonance techniques. This overlap method achieves an ordering of proton affinities, but to obtain absolute values these must be anchored on some independent measurement.

$$B^1H^+ + B^2 \longrightarrow B^1 + B^2H^+ \tag{6}$$

During the last decade two different reactions have been used to anchor the acidity scale for very weakly basic compounds (including methane). Bohme and coworkers[55] have used the appearance potential of OH^+ from H_2O, H_2O_2, and OH to obtain an absolute proton affinity for oxygen atom (PA $= 116.1$ kcal mol^{-1}), while McMahon and Kebarle[57] used the proton affinity of ethylene (162.6 kcal mol^{-1}) derived from photo-ionization studies on $C_2H_5^+$. Unfortunately these two scales differ by 4.2 kcal mol^{-1}, with the more recent evaluation, McMahon's, being higher. As a result of this difference, proton affinities for bases in the range 120–145 kcal mol^{-1} were revised upwards and these are the values quoted in the latest National Bureau of Standards (NBS) tables[56].

Very recent work[53,54] based on the proton affinity of Br and the spectroscopically known dissociation energy of HBr^+ has resulted in a further re-evaluation of the proton affinity of methane and the currently accepted value of 130.0 kcal mol^{-1} is very close to that of Bohme[55]. State-of-the-art *ab initio* molecular orbital calculations including correlation energy and corrected for zero-point energy and thermal energies, are capable of evaluating proton affinities to within at least 2 kcal mol^{-1} and possibly to within 0.5 kcal mol^{-1}[59]. Significantly the most accurate calculations give values of 129.0[36] and 129.5 kcal mol^{-1}[33], in excellent agreement with the most recent experiments.

From the limited amount of experimental data in Table 1 each additional carbon atom increases the proton affinity by approximately 10 kcal mol^{-1}. Cyclopropane has a higher proton affinity than propane by *ca* 20 kcal mol^{-1}, indicating considerable relief of strain on protonation. The calculated proton affinity of cyclopropane is at a very high level of theory and is 1.9 kcal mol^{-1} lower than the experimental value, leading to the suggestion that perhaps the experimental value is slightly too high[49].

The high strain energy of cubane results in a very high proton affinity, and low-level theory predicts that protonation occurs at a corner[58]. The high basicity of the dodecahedranes cannot be entirely due to relief of strain and is attributed to the high polarizability of this molecule[58].

TABLE 1. Proton affinities[a] of alkanes

	Experimental	Theoretical
Methane	130.0 ± 1^b, 130.3 ± 1^c	129.0^h, 127.7^i
	130.5 ± 2^d, 131.6^e	129.5^j
	134.7^f	
Ethane	143.6^e	139.7^i
Propane	150.0^e	
Isobutane	163.3^e	
Cyclopropane	179.8^e	178.0^k
Cyclohexane	$ca\ 169^e$	
Cubane	200.7 ± 3^g	
Dodecahedrane	196.6 ± 3^g	
Methyldodecahedrane	197 ± 3^g	
1,16-Dimethyldodecahedrane	202 ± 3^g	

[a]Energies in kcal mol^{-1}. [b]Reference 53. [c]Reference 54. [d]Reference 55. [e]Reference 56.
[f]Reference 57. [g]Reference 58. [h]Reference 36. [i]Reference 38. [j]Reference 33. [k]Reference 49.

E. Basicities in Solution

The isomerization and cracking of saturated hydrocarbons in strongly acidic media are extremely important reactions in industry. The initial discovery of the $AlCl_3$-catalysed isomerization of cyclohexane to methylcyclopentane[60] was followed by the discovery that alkanes can also be cracked by $AlCl_3$[61]. The reactions proved to be quite general for alkanes in the presence of a wide variety of acids and were quickly recognized for their ability to 'upgrade' alkanes[17,25,62]. The general concensus is that the common step which initiates all these reactions is removal of a hydride ion from the hydrocarbon. The ease with which this occurs depends upon the type of C—H bond with the order of reactivity being tertiary C—H > secondary C—H > primary C—H[17,63] consistent with the stability of the resultant carbenium ions. Standard potentials for the redox reaction $RH \rightarrow R^+$ in HF follow the same order, tertiary C—H < secondary C—H < primary C—H $\approx$ C—C[19,63].

In the reaction initiated by Bronsted acids the first step is believed to be addition of a proton to the alkane,

$$RH + H^+A^- \rightleftharpoons RH_2^+ + A^- \qquad (7a)$$
$$ \longrightarrow R^+ + H_2 \qquad (7b)$$

followed by loss of H_2 (equation 7), but the intermediate alkonium ions are not sufficiently long-lived to be characterized by spectroscopy. The concentration of CH_5^+ in superacids[64] is too low to be detected by NMR but ESCA spectra on matrices of FSO_3H/SbF_5 and of HF/SbF_5 saturated with methane at $-180\,°C$ have been interpreted as indicating the existence of CH_5^+[17]. Also, deuterated methane undergoes exchange with the solvent in superacids[65], presumably through a protonation–deprotonation mechanism. Isobutane in FSO_3D/SbF_5 undergoes rapid hydrogen deuterium exchange initially only at the tertiary carbon[17,18].

One of the problems with the simple mechanism in equation 7 is that, in the oxidation of most alkanes, much less than the stoichiometric amount of hydrogen is observed[17,62,63]. This has led to the suggestion that when the alkane is oxidized, then the solvent undergoes reduction[17,18-20,66-69] and sulphuric acid has been found to be reduced to SO_2 by alkanes[66,67]. In the commonly used superacid systems containing

SbF_5, reduction to SbF_3 has long been postulated[17-20,68,69]. An analysis of the thermodynamics for the reduction of SbF_5 to SbF_3 (equation 8) estimated this reaction to be exothermic by $33\,kcal\,mol^{-1}$[68] but, under the usual experimental conditions of SbF_5/FSO_3H and SbF_5/HF solutions, hydrogen does not react with SbF_5[19,70-72],

$$SbF_5 + H_2 \longrightarrow SbF_3 + 2HF \qquad (8)$$

Recently the Lewis acid-catalysed oxidation of isobutane has been re-examined with the rigorous exclusion of any Bronsted acid and *stoichiometric* volumes of hydrogen were collected[63]. When the temperature of a solution of isobutane in SbF_5/SO_2ClF was raised to $-30\,°C$, hydrogen evolved and the *t*-butyl cation was produced quantitatively. Analysis of a white precipitate showed SbF_3 to have been formed (equation 9). When the reaction was carried out in the presence of acetone, then only traces of hydrogen were detected and equimolar quantities of the *t*-butyl cation and protonated acetone formed (equation 10).

$$2(CH_3)_3CH + 3SbF_5 \longrightarrow 2(H_3C)_3C^+ + SbF_3 + 2SbF_6^- + H_2 \qquad (9)$$

$$(CH_3)_3CH + 3SbF_5 + (CH_3)_2CO \longrightarrow (CH_3)_3C^+ + SbF_3 + 2SbF_6^- + (CH_3)_2COH^+ \qquad (10)$$

Equation 9 can be simplified to two half-reaction, a 2-electron reduction of SbF_5 (reaction 11) with the F^- ions being absorbed subsequently by other SbF_5 molecules, and a 2-electron oxidation of the C—H bond (reaction 12). In the absence of any other base, H^+ attacks another alkane molecule to form RH_2^+ and hydrogen is evolved. If a more basic molecule like acetone is present, then the proton adds to this base and no further reaction occurs.

$$2e + SbF_5 \longrightarrow SbF_3 + 2F^- \qquad (11)$$

$$R{-}H - 2e \longrightarrow R^+ + H^+ \qquad (12)$$

The evolution of almost quantitative amounts of hydrogen in the reaction of Bronsted acids with an alkane have also been reported recently[73,74]. *In,out*-bicyclo[4.4.4]tetra-decane **16** in trifluoromethanesulphonic acid at $0\,°C$ was converted to the *in*-bicyclo-[4.4.4]-1-tetradecyl cation **17** in one hour with more than 90% of the theoretical

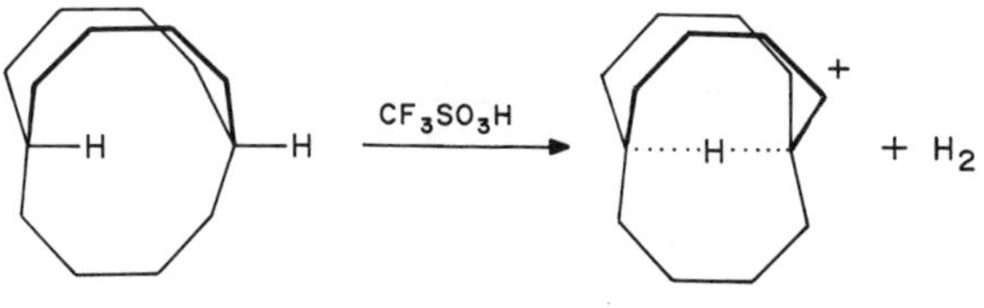

$$(13)$$

(16)   **(17)**

amount of hydrogen collected (equation 13). Deuterium labelling of the trifluoromethane-sulphonic acid and of the two bridgehead hydrogens (separately) gave kinetic isotope effects k_H/k_D of 7 with labelled solvent, 1.9 when the *out* hydrogen was labelled and 1.0 with the *in* hydrogen labelled. These isotope effects are consistent with a rate-determining step in which OH and C—H *out* bonds are being broken, but with no participation from the *in* hydrogen, i.e. formation of the symmetric μ-hydrido-bridged ion **17** is not concerted. The kinetic data are consistent with a rate-determining protonation of the C—H bond followed by rapid loss of H_2.

The very stable cation **17** has a strong band in the infrared at $2113\,cm^{-1}$ attributed to asymmetric stretch of C—H—C. *Ab initio* molecular orbital calculations at the 6-31G level show the 3-centre 2-electron bond in **17** to be linear (the ion has D_3 symmetry)[75]. This contrasts with the C—H—C bonding in the more flexible CH_3—H—$CH_3{}^+$ (angle 126°)[38] but the C—H distance of 1.251 Å for **17** is close to that from the higher-level calculations on $C_2H_7{}^+$ (1.239 Å)[38].

F. Basicities Towards Transition-metal Ions

Like protons, transition-metal ions are strongly acidic and they can, in principle, add to both the C—H and C—C bonds of alkanes. As already noted in the section on proton affinities (Table 1) strained cycloalkanes are intrinsically more basic than open-chain alkanes, and the reaction of cyclopropane with Pt((II) to form a platinacyclo-butane (equation 14) was the first reaction of a formally saturated hydrocarbon with a transition-metal ion[76-80]. The driving force in this reaction is relief of the strain associated with the small ring. The resulting metallacyclobutane is essentially free of ring strain. Many low-valent transition-metal complexes have been found to react with cycloalkanes. Metal ions convet the strained hydrocarbons quadricyclane[79,80], cubane[81,82], bicyclo-[2.1.0]pentanes[83,84], bicyclo[3.1.0]hexanes[85], bicyclo[4.1.0]heptanes[85] and bicyclo-butanes[86,87] into less strained isomers (usually cyclohexanes).

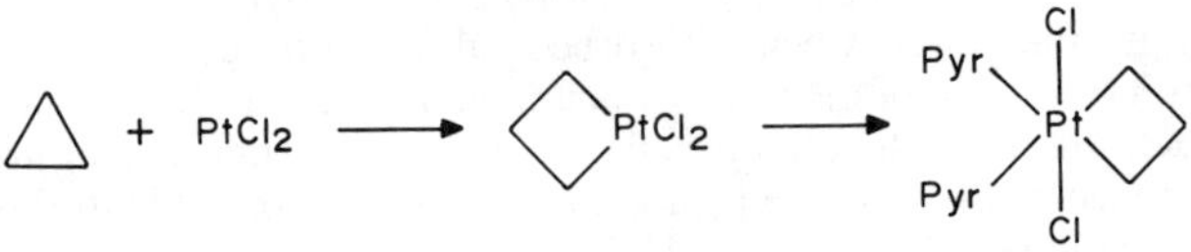

$$\triangle + PtCl_2 \longrightarrow \langle\rangle PtCl_2 \longrightarrow \underset{Pyr}{\overset{Pyr}{\underset{Cl}{\overset{Cl}{\Big|}}}}Pt\langle\rangle \tag{14}$$

The initial indication that open-chain alkanes can also be activated by transition-metal ions came in 1969 when $PtCl_4{}^{2-}$ was found to catalyse the deuteration of alkanes in CH_3COOD/D_2O[88]. This observation stimulated a search for complexes containing simultaneously M—R and M—H bonds, but initial failures to observe such compounds and the observation that C—M bonds in transition-metal complexes have low dissociation energies (*ca* 20–30 kcal mol^{-1}[89]) led to the belief that equilibria 15 and 16 were so strongly to the left that isolation of the oxidative addition products would be impossible[90].

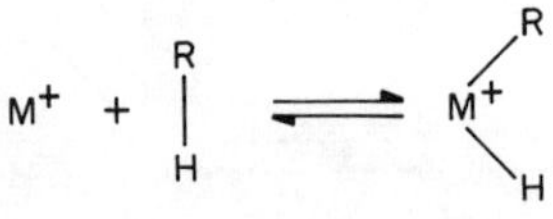

$$M^+ + \underset{H}{\overset{R}{\Big|}} \rightleftharpoons M^+\underset{H}{\overset{R}{<}} \tag{15}$$

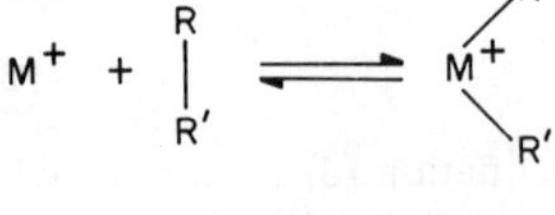

$$M^+ + \underset{R'}{\overset{R}{\Big|}} \rightleftharpoons M^+\underset{R'}{\overset{R}{<}} \tag{16}$$

Stable alkyl hydride complexes of the type shown in equation 15 were eventually synthesized by photolysis of hydride[91] and carbonyl[92] complexes of iridium in hydrocarbon solvents. These alkyl hydride complexes are surprisingly stable (alkane is eliminated at 110 °C).

$$(\text{c-}C_5Me_5)Ir(PMe_3)H_2 \xrightarrow[\text{RH}]{h\nu} (\text{c-}C_5Me_5)Ir(PMe_3)(R)(H) + H_2 \tag{17}$$

The reaction in equation 17 shows little selectivity. The 16-electron intermediate produced by elimination of H_2 in the photolysis is very reactive and attacks all C—H bonds but with a slight preference for primary carbons, probably due to steric effects. The CH_4 complex has been formed in a matrix at 12 K showing that there is essentially no barrier for addition to C—H[93].

The gas-phase addition of monopositive transition-metal ions to both C—H and C—C bonds of alkanes is currently a very active area of research[94–104]. In the gas-phase ions MR^+, the alkyl groups assist in delocalizing the positive charge[105,106] and the M—R bond energy is considerably larger than in the valence saturated $L_nM(R)(H)^+$ ions where steric interactions with the ligands L weaken the M—R bond[89,107]. Accurate 'intrinsic' bond energies are now available for many M—R bonds[103,108] in gas-phase ions, and the reactions of transition-metal ions M^+ with alkanes as a function of the kinetic energy of the ions have been studied extensively.

With methane, transition-metal ions undergo oxidative addition to give the methyl hydride. The subsequent chemistry depends upon the initial energy of the ion. At low energies H_2 is eliminated through a 4-centre reaction (equation 18a) while at higher energies homolytic bond fission occurs and H or, more commonly, CH_3 is lost (equation 18b).

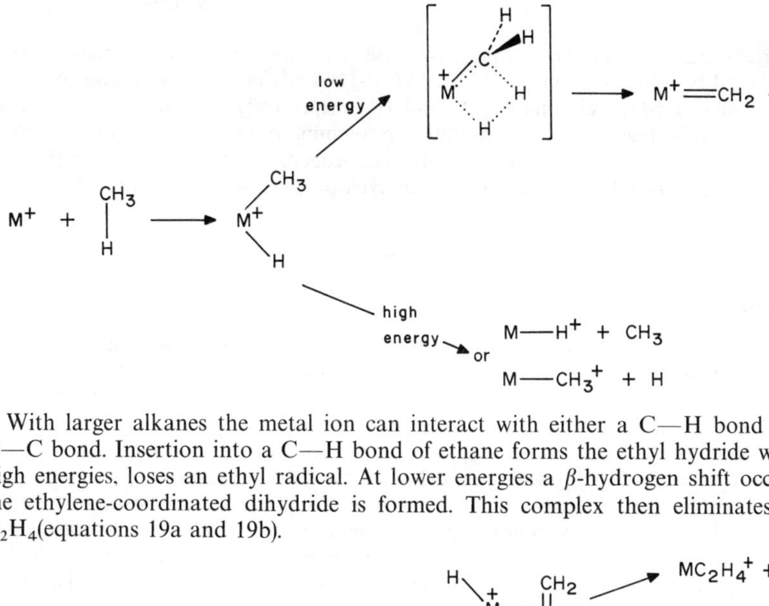

$$M^+ + \;\;\overset{CH_3}{\underset{H}{|}}\;\; \longrightarrow \;\; M^+\overset{CH_3}{\underset{H}{\diagup}}$$

low energy:
$$\left[\; M^+ \cdots C(H)(H) \cdots H \; \right] \longrightarrow M^+{=}CH_2 + H_2 \qquad (18a)$$

high energy:
$$M{-\!\!-}H^+ + CH_3 \quad \text{or} \quad M{-\!\!-}CH_3^+ + H \qquad (18b)$$

With larger alkanes the metal ion can interact with either a C—H bond or with C—C bond. Insertion into a C—H bond of ethane forms the ethyl hydride which, at high energies, loses an ethyl radical. At lower energies a β-hydrogen shift occurs and the ethylene-coordinated dihydride is formed. This complex then eliminates H_2 or C_2H_4 (equations 19a and 19b).

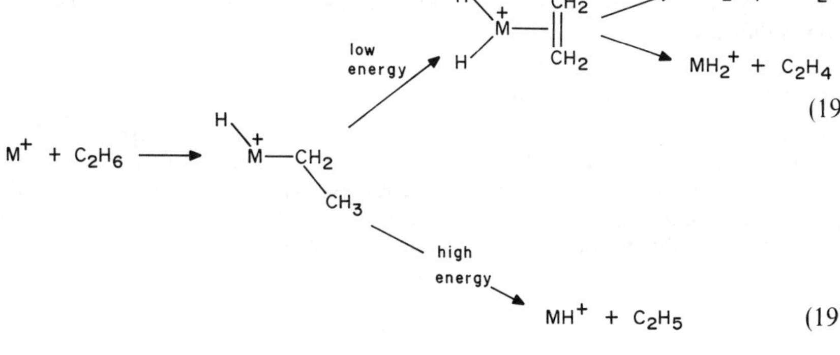

$$M^+ + C_2H_6 \longrightarrow \overset{H}{\underset{CH_3}{\diagup}}M^+{-}CH_2$$

low energy:
$$\overset{H}{\underset{H}{\diagup}}M^+\overset{CH_2}{\underset{CH_2}{|\!|}} \longrightarrow \begin{cases} MC_2H_4^+ + H_2 \\ MH_2^+ + C_2H_4 \end{cases} \qquad (19a)$$

high energy:
$$MH^+ + C_2H_5 \qquad (19b)$$

Addition to the C—C bond results in elimination of methane through a 4-centre reaction. At higher energies the M^+—CH_3 bond is broken (equation 20a and 20b).

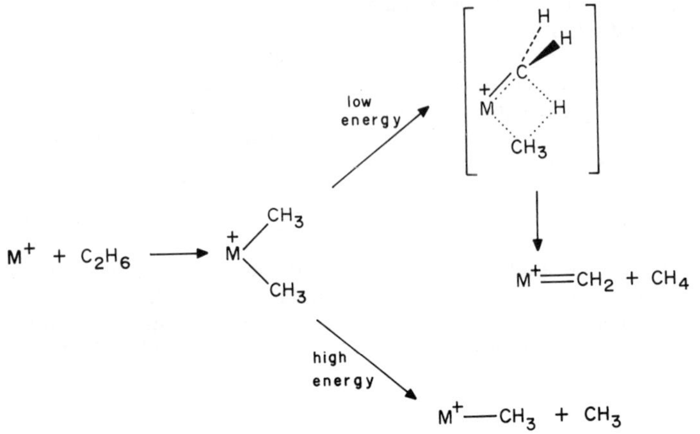

$$M^+ + C_2H_6 \longrightarrow$$

$$M^+{=}CH_2 + CH_4 \quad (20a)$$

$$M^+{-}CH_3 + CH_3 \quad (20b)$$

Higher alkanes undergo *exothermic* gas-phase reactions with some transition-metal ions. Co^+ and Ni^+ insert into the C—C and C—H bonds of propane to give oxidative addition products which eliminate CH_4 and H_2, respectively. The relative amount of methane formed increases with the amount of branching in the alkane[103]. $LaFe^+$ reacts with alkanes (except methane) giving mainly dehydrogenation products, but with some C—C cleavage, particularly in reactions with cyclopropane and cyclobutane[104].

III. ACIDITY

A. General

Alkanes are extremely weak acids. They are saturated molecules and, since in the case of first-row elements, to possess more than an octet of electrons occurs only in very unusual circumstances, they are not prone to nucleophilic attack by even the strongest of bases, i.e. they do not function as Lewis acids. They do function as Bronsted acids, but only under extreme conditions. The non-polarity of the C—H bond makes alkanes very ineffective at hydrogen bonding and loss of the proton leaves a carbanion which is unable to delocalize the excess negative charge. In the gas phase this instability results in many of the smaller alkyl anions spontaneously losing an electron to form alkyl radicals[109-112]. The extremely short lifetimes of these anions in the gas phase makes measurement of proton-transfer equilibria impossible and accurate assessment of acidities becomes difficult.

B. Acidities of Alkanes in Solution

Metal salts of alkanes, e.g. RLi, have long been used as sources of carbanions R^- in synthetic organic chemistry[114-120] and metalation of a carbon acid R'H by an organolithium compound RLi gives a measure of the relative acidities of alkanes RH and RH' (equation 21). Unfortunately this metalation reaction is not quite so simple because both RLi and R'Li usually exist as aggregates, even in coordinating solvents[117,121-125] and in the gas phase[126].

$$RLi + R'H \longrightarrow RH + R'Li \quad (21)$$

Relative stabilities of carbanions can be assessed from exchange reactions of organo-metallic compounds containing these anions[127,128]. From NMR measurements of equilibria involving R_2Mg and R'_2Hg and from the polarographic reduction of mercury alkyls[127], it has been deduced that the stability of the carbanion increases with the amount of s-character in the orbital formally holding the lone pair, i.e. $sp > sp^2 > sp^3$. For ions in which the carbanionic carbon is formally sp^2 and sp^3 hybridized, the order of stability is phenyl $\approx$ vinyl $>$ cyclopropyl $>$ methyl $>$ ethyl $>$ isopropyl $>$ t-butyl[117,127,128]. This order of stability is illustrated by the basicity of the butyllithiums[118]. sec-Butyllithium removes a proton from allyl ethers whereas n-butyllithium does not[129]; and t-butyllithium is even more basic and will remove a proton from methyl vinyl ether[130].

There are two solvent systems commonly used in measuring the relative acidities of weak acids, leading to two acidity scales. In cyclohexylamine cesium salts R^-Cs^+ exist as contact ion pairs of the type usually used in organic reactions and the transmetalation reaction in equation 21 (using cesium as the metal rather than lithium) is used to measure an ion-pair acidity scale[131]. In DMSO there is little or no ion pairing and equilibria measured in this solvent lead to an ionic acidity scale[131–133]. Unfortunately the pK_a values of cyclohexylamine (41.6) and DMSO (35) are much lower than the pK_a values of alkanes (estimated to be ca 50) and direct measurements of equilibria involving alkanes are not possible. Differences in gas-phase acidities are similar to those in DMSO and, by comparing the gas-phase acidities of $CH_n(CN)_m$, where $m + n = 4$ and m takes values from 0 to 2, the pK_a of methane has been estimated to be 56[134]. Aqueous pK_a values of methane and ethane have been estimated to have values ranging from 40 to 60[114,135–137]. By using ab initio molecular orbital calculations to calculate the deprotonation energy and calculating solvation energies from Monte Carlo statistical mechanics or molecular dynamic simulations, Jorgensen and coworkers[138,139] have calculated the aqueous pK_a of ethane to be 50.6 ± 0.2 and, in a different study, 52 ± 2.

C. Gas-phase Acidities

It is customary to characterize the acidities of molecules in the gas phase by using enthalpies, ΔH_{acid}, rather than free energies or pK_a values. The acidity of an acid RH is given by equation 22 where $\Delta H_{acid}(RH) = \Delta H_f(R^-) + \Delta H_f(H^+) - \Delta H_f(RH)$. Data for many organic acids have recently been tabulated[56,140,141] but there are few data for the unsubstituted alkanes. Standard methods of generating anions, dissociative electron attachment and chemical ionization reactions involving deprotonation, substitution and elimination are useful in generating delocalized carbanions[140,142], but in general produce little (with CH_4) or no alkyl anions[143]. This stems in part from the difficulty of finding bases sufficiently strong to remove a proton from an unactivated C—H bond[144] and in part from the instability of the alkyl anions towards loss of an electron (equation 23). The ethyl, isopropyl, isobutyl, t-butyl, cyclobutyl, cyclopentyl and cyclohexyl anions are all believed to be produced by collision-induced decarboxylation of $RCOO^-$ [111]. However, these anions were not detected and, on the basis of calculated flight times for an ion to the ion multiplier, an upper limit of $25\,\mu s$ has been set for the lifetimes of these carbanions[110,111].

$$RH \longrightarrow R^- + H^+ \qquad (22)$$

$$R^- \longrightarrow R + e \qquad (23)$$

The methyl radical has a small electron affinity ($1.8 \pm 0.7\,kcal\,mol^{-1}$ [145]) and this has been combined with the bond dissociation energy (BDE) of methane and the ionization potential of H atoms to give the enthalpy for the deprotonation of methane using the thermochemical cycle (equation 24). For most alkyl radicals the electron affinities are probably negative and are not known, and the thermochemical cycle approach cannot be used.

$$CH_4 \longrightarrow CH_3 + H \qquad BDE = 105\,kcal\,mol^{-1}$$
$$CH_3 + e \longrightarrow CH_3{}^- \qquad -EA = -2\,kcal\,mol^{-1}$$
$$H \longrightarrow H^+ + e \qquad IP = 314\,kcal\,mol^{-1}$$

$$CH_4 \longrightarrow CH_3{}^- + H^+ \qquad \Delta H_{acid} = 417\,kcal\,mol^{-1} \tag{24}$$

Alkyl anions have been implicated as intermediates stabilized by a neutral molecule. Alkoxide ions when photolysed in a pulsed ICR spectrometer dissociate into alkanes and enolate anions[146]. The intermediate 19 in equation 25 can be represented by two possible extremes. In 19a the alkyl anion R^- is 'solvated' by a ketone and in 19b the radical anion of the ketone is solvated by the radical R. The structure of this intermediate will then depend on the relative electron affinities of the alkyl group R and the ketone. Brauman and collaborators[147,148] photolysed a series of 2-substituted-2-propoxides (18 with $R' = CH_3$, $R'' = H$ and R varied). For substituents $R = CF_3$, H, Ph and $H_2C=CH$, the C—R bond dissociation energies for homolytic fission are larger than the C—CH_3 bond energy, i.e. if the intermediate complex has the structure 19b then methane would be expected to be produced. Conversely, since these R groups form more stable anions than CH_3, decomposition via 19a should result in RH. The experimental observation that only RH is formed led to the conclusion that 19 is best described by the solvated alkyl anion structure 19a.

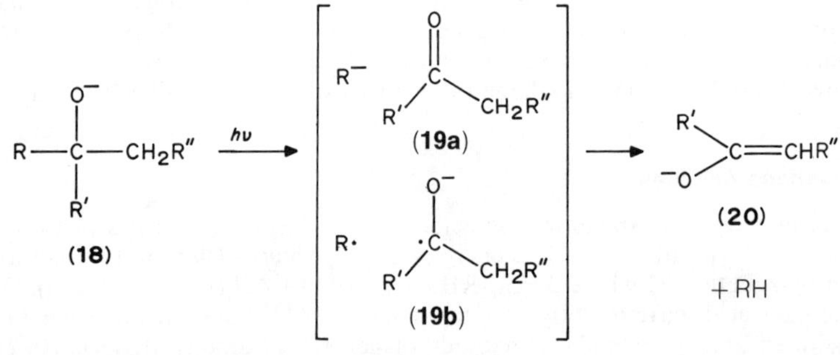

The relative leaving abilities as determined by the decomposition of alkoxides are $CF_3 > Ph > H > t\text{-}Bu > Me > i\text{-}Pr > Et$. Assuming that the intermediate has structure 19a, this order reflects the relative stabilities of the (R^-·ketone) complexes and provides an *estimate* of the relative acidities of the alkanes.

DePuy and coworkers[113,149] have established the same order of acidity for alkanes by examining the preferred modes of decomposition of the 5-coordinate silicon anions[150,151] formed by the reaction of OH^- with $(CH_3)_3SiR$. By analogy with the reaction of F^-[152] the initial step is believed to be formation of the complex 21 which can either expel $CH_3{}^-$ or R^-. The exothermicity of the initial addition reaction is insufficient to allow expulsion of an alkyl anion and heterolysis of an Si—C bond results in the solvated ion 22a or 22b (Scheme 1). Rapid proton transfer within this ion–molecule complex leads to the irreversible formation of RH or CH_4. Molecular orbital calculations[150,151] show the five-coordinate silicon anion 21 ($R = CH_3$) to be at a minimum and the transition structure for decomposition of 21 has $CH_3{}^-$ almost completely removed while the proton of the OH group is only slightly perturbed.

The branching ratios found for the decomposition of the five-coordinate anions provide a kinetic method for estimating alkane acidities. In principle this method can be applied

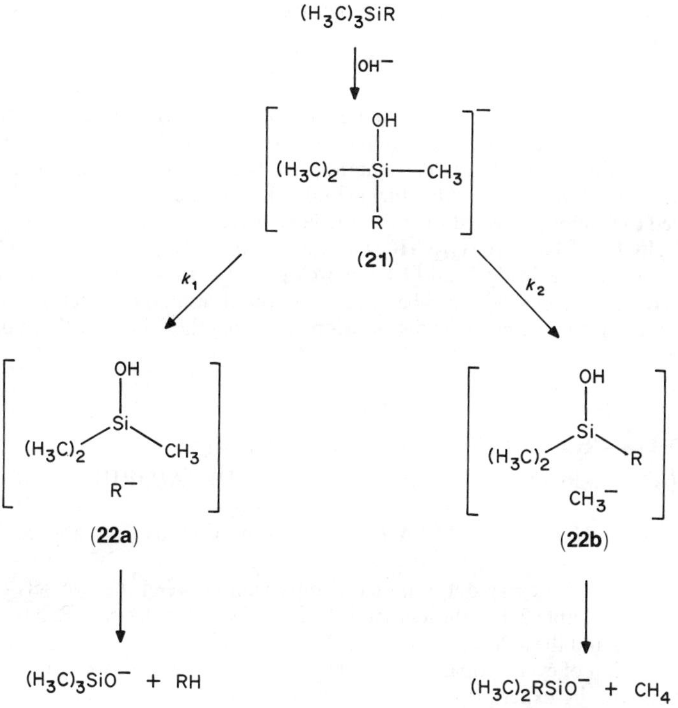

$(H_3C)_3SiR$

$\downarrow OH^-$

$$\left[(H_3C)_2\!-\!\underset{\underset{R}{|}}{\overset{\overset{OH}{|}}{Si}}\!-\!CH_3 \right]^-$$

(21)

k_1 k_2

$$\left[(H_3C)_2\!\underset{R^-}{\overset{\overset{OH}{|}}{Si}}\!{\diagdown}_{CH_3} \right] \qquad \left[(H_3C)_2\!\underset{CH_3^-}{\overset{\overset{OH}{|}}{Si}}\!{\diagdown}_{R} \right]$$

(22a)   (22b)

$\downarrow$ $\downarrow$

$(H_3C)_3SiO^- + RH$ $(H_3C)_2RSiO^- + CH_4$

SCHEME 1

to any hydrogen in a hydrocarbon, providing the appropriately substituted $RSiMe_3$ can be synthesized, although several assumptions have to be made in the analysis. The decomposition of **21** into **22** can be considered as a two-step reaction, homolytic bond dissociation followed by electron transfer to R or CH_3. Assuming that the neutral fragments Me_3SiOH and Me_2RSiOH have the same electron affinities, then the *difference* in enthalpies $\Delta\Delta H_{A\rightarrow B}$ by the two pathways is given by equation 26. The branching ratio for formation of RH and CH_4 is then assumed to be related to $\Delta\Delta H_{A\rightarrow B}$ by a linear free energy relationship, $\ln(k_1/k_2) \propto \Delta\Delta H_{A\rightarrow B}$.

$$\Delta\Delta H_{A\rightarrow B} = [BDE(R\!-\!Si) - BDE(Me\!-\!Si)] - [EA(R) - EA(CH_3)] \qquad (26)$$

The difference in acidities $\Delta\Delta H_{acid}$, taken from the thermochemical cycle in equation 24, is given by difference in $CH_3\!-\!H$ and $R\!-\!H$ bond dissociation energies and the difference in the electron affinities. Equations 26 and 27 are similar and, assuming that C—Si dissociation energies very in a similar way to C—H dissociation energies, then $\Delta\Delta H_{A\rightarrow B} \propto \Delta\Delta H_{acid}$, and $\ln(k_1/k_2) = \beta\Delta\Delta H_{acid}$, i.e. $\log(RH/CH_4)$ can be used to calculate the relative acidities. The proportionality constant, β, was calculated from the branching ratio from Me_3SiPh and from independent acidity measurements on methane and benzene. The kinetic acidities calculated using the value of β are listed in Table 2. These will be discussed shortly after describing a third gas-phase technique in which *free* alkyl anions are generated[110-112].

$$\Delta\Delta H_{acid} = [BDE(R\!-\!H) - BDE(Me\!-\!H)] - [EA(R) - EA(CH_3)] \qquad (27)$$

Alkyl anions in the gas phase have been generated by collision-induced decarboxylation of carboxylate anions.

$$RCO_2^- \longrightarrow R^- + CO_2 \tag{28}$$

Many of the alkyl anions are not detected as they immediately lose an electron. In fact these ions may never exist; the electron may be lost as the CO_2 is breaking away from the hydrocarbon fragment. Many carbanions have been detected from reaction 28 but most of them are stabilized by substituents and will not be considered here[111]. Among the saturated carbanions detected were the fragile methyl anion, and primary alkyl anions: $CH_3CH_2CH(CH_3)CH_2^-$, $(CH_3)_2CHCH_2CH_2^-$ and $(CH_3)_3CCH_2^-$. Cyclopropyl, methylcyclopropyl and bridgehead [1.1.1]bicyclopentyl anions were also observed.

Acidities of the hydrocarbons can be calculated from threshold energies (E_T), assuming there is essentially no barrier to the dissociation and using the following thermochemical cycle:

$$
\begin{array}{ll}
RCO_2^- \longrightarrow R^- + CO_2 & \Delta H \approx E_T \\
RCOOH \longrightarrow RCO_2^- + H^+ & \Delta H_{acid}(RCOOH) \\
RH + CO_2 \longrightarrow RCOOH & \Delta H = \Delta H_f(RCOOH) - \Delta H_f(RH) - \Delta H_f(CO_2) \\
\hline
RH \longrightarrow R^- + H^+ & \Delta H_{acid}(RH) = \text{sum of terms from above} \tag{29}
\end{array}
$$

It is only possible to carry out a limited comparison between the two sets of experimental acidities in Table 2. For the less acidic hydrocarbons the radicals, R, have negative electron affinities and the anions, R^-, are not observed in the decarboxylation experiments. For deprotonation of cyclopropane, neopentane and methylcyclopropane (secondary H) agreement is within experimental error, but for removal of the methyl proton of methylcyclopropane there is a large disagreement. Proton affinities calculated from *ab initio* molecular orbital calculations[109] using the isodesmic reaction in equation 30 and assuming the proton affinity of CH_3^- to be $416.6 \, kcal \, mol^{-1}$ are generally in good agreement with the experimental values and are closer to those from the silicon anion work.

$$RH + CH_3^- \longrightarrow R^- + CH_4 \tag{30}$$

The overall order of alkane acidities deduced from the decomposition of silicon anions is the same as that obtained from the decomposition of alkoxides[147,148]. The ethyl anion has the highest proton affinity, i.e., ethane is the weakest acid. The acidities of propane (secondary hydrogen) and of cyclobutane are both lower than that of methane, but all other alkanes are more acidic. The acidity of methane is particularly enhanced by a cyclopropyl substituent and by a *t*-butyl group.

It is interesting to examine the effect of α- and β-methyl substitution on the acidity of methane. One α-methyl group decreases the acidity of methane, two α-methyl groups also decrease the acidity but by a smaller amount and the three α-methyl groups of the isobutane actually increase the acidity. (This last result is at odds with the theoretical results.)

The effects of β-methyl groups are more systematic. Ethane has the largest ΔH_{acid} value ($420.1 \, kcal \, mol^{-1}$) and introduction of successive β-methyl groups decrease this value by decrements of 4.5, 2.7 and $4.0 \, kcal \, mol^{-1}$. Alcohols are intrinsically much more acidic than the alkanes, but their acidities follow a similar pattern on β-substitution by methyl groups. ΔH_{acid} values (in $kcal \, mol^{-1}$) are methanol 380.6, ethanol 377.4, isopropanol 375.4 and *t*-butanol 374.6[56], i.e. each β-methyl group decreases ΔH_{acid} by approximately $2 \, kcal \, mol^{-1}$. The effect of β-methyl groups is larger in the carbanions

TABLE 2. ΔH_{acid} values for alkanes[a]

| Anion generated | Experimental ΔH_{acid}(RH) from | | Calculated proton affinities[e] |
	silicon anion[b]	carboxylate anion[c]	
Ethyl	420.1	no anion	422.2
Isopropyl	419.4	no anion	422.5
Cyclobutyl	417.4	other products	421.2
Methyl	(416.6)[d]	418 ± 3.5	417.8[f], 418.5[g]
Cyclopentyl	416.1	no anion	—
sec-Butyl	415.7	—	421.9
n-Propyl	415.6	no anion	419.3
t-Butyl	413.1	no anion	419.6
iso-Butyl	412.9	no anion	—
Cyclopropyl	411.5	408 ± 5	412.6
[1.1.1]Bicyclopentyl	—	411 ± 3.5	—
Cyclopropylmethyl	410.5	391 ± 6	413.9
1-Methylcyclopropyl	409.2	413 ± 3.5	—
2-Methylcyclopropyl	—	411 ± 5	—
Neopentyl	408.9	411 ± 10	—
Phenyl	(400.7)[d]	401 ± 10	—

[a] All values in kcal mol^{-1}.
[b] Reference 113; errors are ± 1 kcal mol^{-1}.
[c] Reference 112.
[d] Used as reference acids in constructing acidity scale.
[e] Reference 109.
[f] Reference 33.
[g] Reference 153.

due to the greater inability of CH_2^- to carry the negative charge, and there is more delocalization by negative hyperconjugation[154] onto the β-methyl groups.

Finally, it is necessary to assess the validity of the theoretical acidities. In order to describe the loosely bound outer electrons of carbanions correctly, it is necessary to augment the usual gaussian basis sets with additional s and p functions on carbon[155,156]. The 4-31 + G basis set used to calculate the proton affinities in Table 2 is then about the minimum acceptable by today's standards. Theory agrees with experiment in predicting that ethane and cyclobutane are the two least acidic alkanes. However, theory also predicts *all* the open-chain alkyl anions to have higher proton affinities than CH_3^-, in disagreement with experiment[113,147,148]. One possible reason for this difference is that, in the experimental work, anions R^- are never generated as free ions but exist only in ion–molecule complexes and that their complexed ions have different relative stabilities than the free ions. If this is the situation, then it is not clear why 'solvation' of the more substituted ions is more exothermic than solvation of the unhindered parent ion CH_3^-. Furthermore, the qualitative agreement between the relative acidities from decomposition of very different classes of anions among alkoxide ions[148] and silicon anions[113] lends credence to the conclusion on relative acidities. The HF/4-31 + G level of theory is small and extension of the basis set and inclusion of correlation energy, which supposedly cancels in isodesmic reactions like equation 30, may significantly change the overall conclusion. Also, some of the larger carbanions examined theoretically have negative electron affinities[111] and adequate theoretical description of these unbound

548 A. C. Hopkinson

anions may be difficult, if not impossible. The theoretical results should then be treated with some caution.

IV. REFERENCES

1. V. L. Talroze and A. L. Lyubimova, *Dokl. Akad. Nauk SSSR*, **86**, 509 (1952).
2. F. H. Field and M. S. B. Munson, *J. Am. Chem. Soc.*, **87**, 3294 (1965).
3. S. Wexler and N. Jesse, *J. Am. Chem. Soc.*, **84**, 3425 (1962).
4. A. Henglein and G. Muccini, *Z. Naturforsch.*,**17**, 452 (1962).
5. F. H. Field, J. L. Franklin and M. S. B. Munson, *J. Am. Chem. Soc.*, **85**, 3575 (1963).
6. P. Kebarle and E. W. Godbole, *J. Chem. Phys.*, **39**, 1131 (1963).
7. K. Hiraoka and P. Kebarle, *J. Am. Chem. Soc.*, **97**, 4179 (1975).
8. K. Hiraoka and P. Kebarle, *J. Chem. Phys.*, **62**, 2267 (1975).
9. K. Hiraoka and P. Kebarle, *Can. J. Chem.*, **53**, 970 (1975).
10. K. Hiraoka and P. Kebarle, *J. Chem. Phys.*, **63**, 394 (1975); **63**, 1689 (1975).
11. P. Kebarle, *Ann. Rev. Phys. Chem.*, **28**, 445 (1977).
12. M. French and P. Kebarle, *Can. J. Chem.*, **53**, 2268, (1975).
13. G. I. Mackay, H. I. Schiff and D. K. Bohme, *Can. J. Chem.*, **59**, 1771 (1981).
14. K. Hiraoka and P. Kebarle, *J. Am. Chem. Soc.*, **98**, 6119 (1976).
15. L. I. Yeh, J. M. Price and Y. T. Lee, *J. Am. Chem. Soc.*, **111**, 5597 (1989).
16. D. K. Bohme, in *Interactions Between Ions and Molecules* (Ed. P. Ausloos), Plenum, New York, 1975, p. 489.
17. G. A. Olah, G. K. S. Prakash and J. Sommer, *Superacids*, Wiley, Chichester, 1985.
18. G. A. Olah, Y. Halpern, J. Shen and Y. K. Mo, *J. Am. Chem. Soc.*, **93**, 1251 (1971).
19. G. A. Olah, Y. Halpern, J. Shen and Y. K. Mo, *J. Am. Chem. Soc.*, **95**, 4960 (1973).
20. J. Lucas, P. A. Kramer and A. P. Kouwenhoven, *Recl. Trav. Chim. Pays-Bas*, **92**, 44 (1973).
21. R. Bonnifay, B. Torck and J. M. Hellin, *Bull. Soc. Chim. Fr.*, 808 (1977).
22. C. D. Nenitzescu and A. Dragan *Ber. Dtsch. Chem. Ges.*, **66**, 1892 (1933).
23. H. Pines, N. E. Joffman, in *Friedel–Crafts and Related Reactions*, Vol. II (Ed. G. A. Olah), Wiley–Interscience, New York, 1964, p. 1211.
24. C. D. Nenitzescu, in *Carbonium Ions*, Vol. II (Eds. G. A. Olah and P. v. R. Schleyer), Wiley-Interscience, New York, 1970, p. 490.
25. F. Asinger, *Paraffins*, Pergamon Press, New York, 1968, p. 695.
26. R. Bonnifay, B. Torck and J. M. Hellin, *Bull. Soc. Chim. Fr.*, 1097 (1977); 36 (1978).
27. V. Dyczmons, V. Staemler and W. Kutzelnigg, *Chem. Phys. Lett.*, **5**, 361 (1970).
28. W. A. Lathan, W. J. Hehre and J. A. Pople, *Tetrahedron Lett.*, 2699 (1970).
29. W. A. Lathan, W. J. Hehre and J. A. Pople, *J. Am. Chem. Soc.*, **93**, 808 (1971).
30. P. C. Hariharan, W. A. Lathan and J. A. Pople, *Chem. Phys. Lett.*, **14**, 385 (1972).
31. V. Dyczmons and W. Kutzelnigg, *Theor. Chim. Acta*, **33**, 239 (1974).
32. K. Raghavachari, R. A. Whiteside, J. A. Pople and P. v. R. Schleyer, *J. Am. Chem. Soc.*, **103**, 5649 (1981).
33. D. J. DeFrees and A. D. McLean, *J. Comput. Chem.*, **7**, 321 (1986).
34. J. A. Pople and L. A. Curtiss, *J. Phys. Chem.*, **91**, 155 (1987).
35. H. J. Kohler and H. Lischka, *Chem. Phys. Lett.*, **58**, 175 (1978).
36. A. Komornicki and D. A. Dixon, *J. Chem. Phys.*, **89**, 5625 (1987).
37. Results of M. Dupuis, reported in Reference 15.
38. K. A. Hirao and S. Yamabe, *Chem. Phys.*, **86**, 237 (1984).
39. P. C. Hariharan and J. A. Pople, *Theor. Chim. Acta*, **28**, 213 (1973).
40. P. Ausloos, S. G. Lias and R. Gorden Jr., *J. Chem. Phys*, **29**, 3341 (1963).
41. V. Aquilanti, A. Galli, A. Gardini-Guidoni and G. G. Volpi, *J. Chem. Phys.*, **48**, 4310 (1968).
42. B. H. Solka, A. Y. K. Lam and A. G. Harrison, *Can. J. Chem.*, **52**, 1796 (1974).
43. S. L. Chong and J. L. Franklin, *J. Am. Chem. Soc.*, **94**, 6347 (1972).
44. D. Viviani and J. B. Levy, *Int. J. Chem. Kinet.*, **11**, 1021 (1979).
45. M. Attina, F. Cacace and P. Giacomello, *J. Am. Chem. Soc.*, **102**, 4768 (1980).
46. M. Saunders, P. Vogel, E. L. Hagen and J. Rosenfeld, *Acc. Chem. Res.*, **6**, 53 (1973).
47. J. M. Coxon and M. A. Battiste, in *The Chemistry of the Cyclopropyl Group* (Ed. Z. Rappoport), Chap. 6, Wiley, Chichester, 1987.

11. Acidity and basicity 549

48. P. Vogel, *Carbocation Chemistry*, Elsevier, Amsterdam, 1985.
49. W. Koch, B. Liu and P. v. R. Schleyer, *J. Am. Chem. Soc.*, **111**, 3479 (1989).
50. H. Lischka and H. J. Kohler, *J. Am. Chem. Soc.*, **100**, 5297 (1978).
51. W. Hehre, L. Radom, P. v. R. Schleyer and J. A. Pople, *Ab Initio Molecular Orbital Theory*, Wiley–Interscience, New York, 1986.
52. M. J. S. Dewar, E. A. Healy and J. M. Ruiz, *J. Chem. Soc., Chem. Commun.*, 943 (1987).
53. N. G. Adams, D. Smith, M. Tichy, G. Javahery, N. D. Twiddy and E. E. Ferguson, *J. Chem. Phys.*, **91**, 4037 (1989).
54. M. Tichy, G. Javahery, N. D. Twiddy and E. E. Ferguson, *Int. J. Mass. Spectrom. Ion Phys.*, **93**, 165 (1989).
55. D. K. Bohme, G. I. Mackay and H. I. Schiff, *J. Chem. Phys.*, **73**, 4976 (1980).
56. S. G. Lias, J. E. Bartmess, J. F. Liebman, J. L. Holmes, R. D. Levin and G. W. Mallard, *J. Phys. Chem. Ref. Data*, **17**, Suppl. 1 (1988).
57. T. B. McMahon and P. Kebarle, *J. Am. Chem. Soc.*, **107**, 2612 (1985).
58. I. Santos, D. M. Balogh, C. W. Doecke, A. G. Marshall and L. Paquette, *J. Am. Chem. Soc.*, **108**, 8183 (1986).
59. D. A. Dixon and S. G. Lias, in *Molecular Structure and Energetics* (Eds. J. F. Liebman and A. Greenberg), Vol. 2, Chap. 7, VCH Publishers, Inc., New York, 1987.
60. O. Aschan, *Justus Liebigs Ann. Chem.*, **1**, 324 (1902).
61. A, M. McAffee, *Ind. Eng. Chem.*, **7**, 737 (1915).
62. P. L. Fabre, J. Devynck and B. Tremillon, *Chem. Rev.*, **82**, 591 (1982).
63. J. C. Culmann and J. Sommer, *J. Am. Chem. Soc.*, **112**, 4057 (1990).
64. The widely accepted definition of a superacid is an acid system that is stronger than 100% H_2SO_4. R. J. Gillespie and T. E. Peel, *Adv. Phys. Org. Chem.*, **9**, 1 (1972).
65. H. Hogeveen and C. J. Gaasbeek, *Recl. Trav. Chim. Pays-Bas*, **87**, 319 (1969).
66. J. W. Otvos, D. P. Stevenson, C. D. Wagner and O. Beeck, *J. Am. Chem. Soc.*, **73**, 5741 (1951).
67. D. P. Stevenson, C. D. Wagner, O. Beeck and J. W. Otvos, *J. Am. Chem. Soc.*, **74**, 3269 (1952).
68. J. W. Larsen, *J. Am. Chem. Soc.*, **99**, 4379 (1977).
69. T. H. Ledford, *J. Org. Chem.*, **44**, 23 (1979).
70. R. J. Gillespie and G. P. Pez, *Inorg. Chem.*, **8**, 1233 (1969).
71. H. Hogeveen, C. J. Gaasbeek and A. F. Bickel, *Recl. Trav. Chim. Pays-Bas*, **88**, 703 (1969).
72. J. W. Larsen, P. A. Buis, C. R. Watson Jr. and R. M. Pagni, *J. Am. Chem. Soc.*, **96**, 2284 (1974).
73. J. E. McMurry and T. Leckta, *J. Am. Chem. Soc.*, **112**, 869 (1990).
74. J. E. McMurry, T. Leckta and C. N. Hodge, *J. Am. Chem. Soc.*, **111**, 8867 (1989).
75. Calculations by E. R. Vorpagel, reported in Reference 74.
76. C. F. H. Tipper, *J. Chem. Soc.*, 2043 (1955).
77. D. M. Adams, J. Chatt, R. Guy and N. Sheppard, *J. Chem. Soc.*, 738 (1961).
78. M. Keeton, R. Mason and D. R. Russell, *J. Organomet. Chem.*, **33**, 259 (1961).
79. H. Hogeveen and J. F. Nusse, *Tetrahedron Lett.*, 159 (1974).
80. L. Cassar and J. Halpern, *J. Chem. Soc., Chem. Commun.*, 1082 (1971).
81. L. Cassar, P. E. Eaton and J. Halpern, *J. Am. Chem. Soc.*, **92**, 3315 (1970); **92**, 6366 (1970).
82. L. A. Paquette, R. A. Boggs, W. B. Farnham and R. S. Beckley, *J. Am. Chem. Soc.*, **97**, 1112 (1975).
83. P. G. Gassman, T. J. Atkins and J. T. Lumb, *J. Am. Chem. Soc.*, **94**, 7757 (1972).
84. K. B. Wiberg and R. C. Bishop, *Tetrahedron Lett.*, 2727 (1973).
85. K. C. Bishop, *Chem. Rev.*, **76**, 461 (1976).
86. M. Sakai and S. Masamune, *J. Am. Chem. Soc.*, **93**, 4610 (1971).
87. L. A. Paquette and G. Zon, *J. Am. Chem. Soc.*, **96**, 203 (1974).
88. N. F. Goldschleger, M. B. Tyabin, A. E. Shilov and A. A. Shteinman, *Zh. Fiz. Khim.*, **43**, 2147 (1969).
89. J. Halpern, *Acc. Chem. Res.*, **15**, 238 (1982) and references cited therein; J. Halpern, *Inorg. Chim. Acta*, **100**, 41 (1985).
90. For an excellent review see R. H. Crabtree, *Chem. Rev.*, **85**, 245 (1985).
91. A. H. Janowicz and R. G. Bergman, *J. Am. Chem. Soc.*, **104**, 352 (1982); **105**, 3929 (1983).
92. J. K. Hoyano and W. A. G. Graham, *J. Am. Chem. Soc.*, **104**, 3273 (1982).
93. A. J. Rest, I. Whitwell, W. A. G. Graham, J. K. Hoyano and A. D. McMaster, *J. Chem. Soc., Chem. Commun.*, 624 (1984).

94. M. A. Tolbert and J. L. Beauchamp, *J. Am. Chem. Soc.*, **106**, 8117 (1984).
95. L. S. Sunderlin and P. B. Armentrout, *J. Phys. Chem.*, **92**, 1209 (1988).
96. L. S. Sunderlin, N. Aristor and P. B. Armentrout, *J. Am. Chem. Soc.*, **109**, 78 (1987).
97. Y. Huang, M. B. Wise, D. B. Jacobson and B. S. Frieser, *Organometallics*, **6**, 346 (1987).
98. N. Aristor and P. B. Armentrout, *J. Phys. Chem.*, **91**, 6178 (1987).
99. R. Georgiadis and P. B. Armentrout, *J. Phys. Chem.*, **92**, 7060 (1988).
100. R. H. Schultz, J. L. Elkind and P. B. Armentrout, *J. Am. Chem. Soc.*, **110**, 411 (1988).
101. J. B. Schilling and J. L. Beauchamp, *J. Am. Chem. Soc.*, **110**, 15 (1988).
102. L. S. Sunderlin and P. B. Armentrout, *J. Am. Chem. Soc.*, **111**, 3845 (1989).
103. R. Georgiadis, E. R. Fisher and P. B. Armentrout, *J. Am. Chem. Soc.*, **111**, 4251 (1989).
104. Y. Huang and B. S. Freiser, *J. Am. Chem. Soc.*, **110**, 387 (1988).
105. P. Armentrout and J. L. Beauchamp, *J. Am. Chem. Soc.*, **103**, 784 (1981).
106. M. L. Mandich, L. F. Halle and J. L. Beauchamp, *J. Am. Chem. Soc.*, **106**, 4403 (1984).
107. F. T. T. Ng, G. L. Rempel and J. Halpern, *Inorg. Chim. Acta*, **77**, L165 (1983).
108. P. B. Armentrout and J. L. Beauchamp, *Acc. Chem. Res.*, **22**, 315 (1989).
109. P. v. R. Schleyer, G. W. Spitznagel and J. Chandrasekhar, *Tetrahedron Lett.*, **27**, 4411 (1986).
110. S. T. Graul and R. R. Squires, *J. Am. Chem. Soc.*, **110**, 607 (1988).
111. S. T. Graul and R. R. Squires, *J. Am. Chem. Soc.*, **112**, 2506 (1990).
112. S. T. Graul and R. R. Squires, *J. Am. Chem. Soc.*, **112**, 2517 (1990).
113. C. H. DePuy, S. Gronert, S. E. Barlow, V. M. Bierbaum and R. Damrauer, *J. Am. Chem. Soc.*, **111**, 1968 (1989).
114. D. J. Cram, *Fundamentals of Carbanion Chemistry*, Academic Press, New York, 1965.
115. E. Buncel, *Carbanions: Mechanistic and Isotopic Aspects*, Elsevier, Amsterdam, 1975.
116. E. Buncel and T. Durst (Eds.), *Comprehensive Carbanion Chemistry, Part A, Structure and Reactivity*, Elsevier, Amsterdam, 1980.
117. R. B. Bates and C. A. Ogle, *Carbanion Chemistry*, Springer-Verlag, Berlin, 1983.
118. J. C. Stowell, *Carbanions in Organic Synthesis*, Wiley–Interscience, New York, 1979.
119. F. Hartley and S. Patai (Eds.), *The Chemistry of the Carbon-Metal Bond*, Wiley–Interscience, Chichester, 1985, Vols. 1 and 2.
120. B. J. Wakefield, *The Chemistry of Organolithium Compounds*, Pergamon, Oxford, 1974.
121. P. West and R. Waack, *J. Am. Chem. Soc.*, **89**, 4395 (1967).
122. L. M. Seitz and T. L. Brown, *J. Am. Chem. Soc.*, **88**, 2174 (1966).
123. K. C. Williams and T. L. Brown, *J. Am. Chem. Soc.*, **88**, 4134 (1966).
124. L. D. McKeever, R. Waack, M. A. Doran and E. B. Baker, *J. Am. Chem. Soc.*, **91**, 1057 (1969).
125. E. Kaufmann, K. Raghavachari, A. E. Reed and P. v. R. Schleyer, *Organometallics*, **7**, 1597 (1988).
126. J. Berkowitz, D. A. Bafus and T. L. Brown, *J. Phys. Chem.*, **65**, 1380 (1961).
127. R. E. Dessy, W. Kitching, T. Psarras, R. Salinger, A. Chen and T. Chivers, *J. Am. Chem. Soc.*, **88**, 460 (1966).
128. D. E. Applequist and D. F. O'Brien, *J. Am. Chem. Soc.*, **85**, 743 (1963).
129. D. A. Evans, G. C. Andrews and B. Buckwalter, *J. Am. Chem. Soc.*, **96**, 5560 (1974).
130. J. E. Baldwin, G. A. Höfle and O. W. Lever, Jr., *J. Am. Chem. Soc.*, **96**, 7125 (1974).
131. A. Streitwieser, Jr., E. Juaristi and L. L. Nebenzahl, in Reference 116.
132. F. G. Bordwell, *Acc. Chem. Res.*, **21**, 456 (1988).
133. R. W. Taft and F. G. Bordwell, *Acc. Chem. Res.*, **21**, 463 (1988).
134. F. G. Bordwell and D. J. Algrim, *J. Am. Chem. Soc.*, **110**, 2964 (1988).
135. R. Stewart, *The Proton: Applications to Organic Chemistry*, Academic Press, New York, 1985.
136. B. Juan, J. Schwarz and R. Breslow, *J. Am. Chem. Soc.*, **102**, 5741 (1980).
137. K. P. Butin, I. P. Beletskaya, A. N. Kashin and O. A. Reutov, *J. Organomet. Chem.*, **10**, 197 (1967).
138. W. L. Jorgensen, J. M. Briggs and J. Gao, *J. Am. Chem. Soc.*, **109**, 6857 (1987).
139. W. L. Jorgensen and J. M. Briggs, *J. Am. Chem. Soc.*, **111**, 4190 (1989).
140. J. E. Bartmess and R. T. McIver, in *Gas Phase Ion Chemistry* (Ed. M. T. Bowers), Vol. 2, Chap. 11, Academic Press, New York, 1979.
141. G. Boand, R. Houriet and T. Gaümann, *J. Am. Chem. Soc.*, **105**, 2203 (1983).
142. C. H. DePuy and V. M. Bierbaum, *Acc. Chem. Res.*, **14**, 146 (1981).
143. J. G. Dillard, *Chem. Rev.*, **73**, 589 (1973).

144. D. K. Bohme, E. Lee-Ruff and L. B. Young, *J. Am. Chem. Soc.*, **94**, 5153 (1972).
145. G. B. Ellison, P. C. Engelking and W. C. Lineberger, *J. Am. Chem. Soc.*, **100**, 2556 (1978).
146. E. M. Arnett, L. E. Small, R. T. McIver and J. S. Miller, *J. Org. Chem.*, **43**, 815 (1978).
147. W. Tumas, R. F. Foster, M. J. Pellerite and J. I. Brauman, *J. Am. Chem. Soc.*, **105**, 7464 (1983).
148. W. Tumas, R. F. Foster and J. I. Brauman, *J. Am. Chem. Soc.*, **106**, 4053 (1984).
149. C. H. DePuy, V. M. Bierbaum and R. Damrauer, *J. Am. Chem. Soc.*, **106**, 4051 (1984).
150. J. C. Sheldon, R. N. Hayes, J. H. Bowie and C. H. DePuy, *J. Chem. Soc., Perkin Trans. 2*, 275 (1987).
151. L. P. Davis, L. W. Burggraf and M. S. Gordon, *J. Am. Chem. Soc.*, **110**, 3056 (1988).
152. C. H. DePuy, V. M. Bierbaum, R. Damrauer and J. A. Soderquist, *J. Am. Chem. Soc.*, **107**, 3385 (1985).
153. M. R. F. Siggel, T. D. Thomas and L. J. Saethre, *J. Am. Chem. Soc.*, **110**, 91 (1988).
154. P. v. R. Schleyer and A. J. Kos, *Tetrahedron*, **39**, 1141 (1983).
155. J. Chandrasekhar, J. G. Andrade and P. v. R. Schleyer, *J. Am. Chem. Soc.*, **103**, 5609 (1981).
156. G. W. Spitznagel, T. Clark, J. Chandrasekhar and P. v. R. Schleyer, *J. Comput. Chem.*, **3**, 363 (1982).

CHAPTER **12**

Alkanes: modern synthetic methods

GOVERDHAN MEHTA

School of Chemistry, University of Hyderabad, Hyderabad–500 134, India
and Jawaharlal Nehru Centre for Advanced Scientific Research, Indian Institute
of Science Campus, Bangalore–560 012, India

and

H. SURYA PRAKASH RAO

Department of Chemistry, Pondicherry University,
Pondicherry–605 014, India

The Chemistry of Alkanes and Cycloalkanes
Edited by S. Patai and Z. Rappoport © 1992 John Wiley & Sons Ltd

I. INTRODUCTION

Alkanes were the favoured targets of synthesis during the early phase of the evolution of synthetic organic chemistry, as they were considered to be the simplest organic molecules. Indeed, many of the classical reactions in organic chemistry, discovered during the last century and early part of this century, dealt with alkane syntheses. However, as synthetic organic chemistry developed, alkanes as target structures for synthesis lost their appeal and their preparation began to be motivated more and more for testing new reactions and reagents for functional group transformations rather than due to any intrinsic interest in them. This situation began to change in the 1950s as complex alkanes, particularly polycycloalkanes like tetrahedrane, cubane, dodecahedrane etc., were predicted as realizable targets by theoreticians and thus caught the imagination of many contemporary synthetic chemists. Syntheses of cubane and adamantane in the 1960s, during the exploding phase of modern organic synthesis, helped to rekindle interest in polycycloalkanes. This interest in polycycloalkanes has persisted ever since and is in the forefront of synthetic activity at present.

There is a vast literature available on alkane syntheses, particularly with regard to functional group transformations leading to alkanes[1-10]. Indeed, almost any functional group can be transformed to an alkane with the presently available repertoire of synthetic methods. However, emphasis in this chapter will be on methods that directly lead to an alkane from a given functional group. For the most part, only one-step transformations leading to alkanes will be described. Thus, in the case of cycloalkanes, the last key step leading to the target hydrocarbon and not the framework construction will be the focus of attention. In the various equations and tables that follow, only such examples and reactions will be cited, wherein the end product is an alkane. To duly recognize the importance of polycycloalkanes and to reflect the growing interest in their syntheses, the overall account is biased in their favour. Most of the examples have been drawn from the literature that appeared in the 1975–1990 period.

II. FUNCTIONAL GROUP TRANSFORMATIONS

A. Alkenes, Arenes, Alkynes and Strained Molecules

1. General aspects

Direct and ready access to alkanes is through chemical and catalytic reduction of alkenes, arenes, alkynes, cyclopropanes and strained polycycloalkanes. Among them alkenes and particularly cycloalkenes are available through a wide variety of general synthetic procedures, e.g. Diels–Alder reactions, thermal and photochemical electro-cyclizations, etc., and are therefore considered as versatile precursors of the corresponding alkanes. Additionally, alkenes are readily accessible from synthons containing functional groups like hydroxyl, carbonyl, halide, or alkyne which can be easily transformed into alkanes. Arenes are available quite abundantly from various organo-chemical feedstocks and they serve as extremely useful precursors of cycloalkanes.

2. Catalytic hydrogenation

The most convenient and widely applied method for the preparation of alkanes from alkenes is catalytic hydrogenation, which can be carried out in either heterogeneous or homogeneous media, with a wide choice of catalysts ranging from finely divided noble metals to transition metal complexes. The variation in the choice of solvent, temperature, pressure and time provides additional latitude for dealing with structurally diverse alkenes and for achieving selectivity. The general area of catalytic hydrogenation has been extensively reviewed and updated[11].

a. Hydrogenation under heterogeneous conditions. For heterogeneous hydrogenations of alkenes, catalysts based on palladium, platinum, nickel, rhodium and ruthenium have been traditionally used as finely divided metals directly or on supports like charcoal, alumina, silica gel, norit, graphite, barium sulphate, calcium and strontium carbonate

G. Mehta and H. S. P. Rao

among others[11]. In addition, metal oxides serve as excellent catalysts through *in situ* reduction to active metals during hydrogenation and platinum oxide (Adams Catalyst), palladium on charcoal and Raney nickel are some of the popular catalysts that find extensive use even in modern syntheses of alkanes. Catalytic activity of rhodium, iridium and ruthenium catalysts is considered higher than that of platinum and palladium, and the former are used in the hydrogenation of alkenes when the latter do not work well[12]. Table 1 presents examples of several interesting polycyclic alkanes that have been synthesized by the catalytic reduction of the corresponding alkenes[13]. In the preparation

TABLE 1. Alkanes from the catalytic hydrogenation of alkenes

Starting alkene	Product	Reference
$C_{104}H_{194}$	$C_{104}H_{210}$ (tetrahectane)	13a
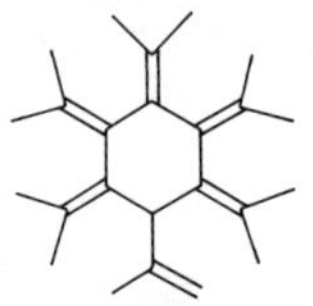	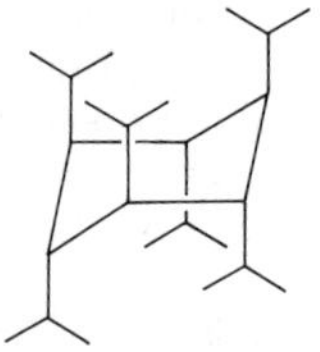	13b
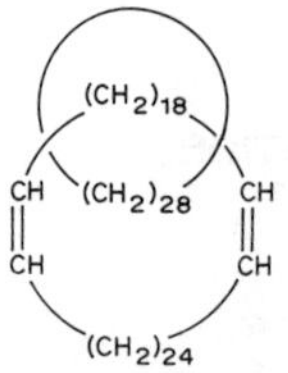	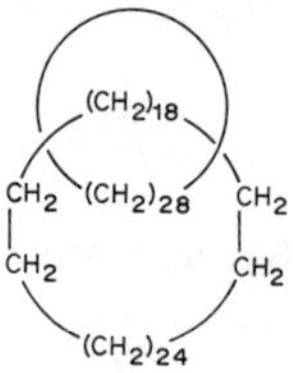	13c
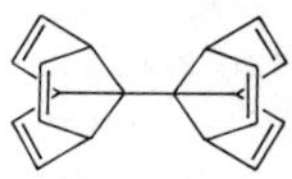	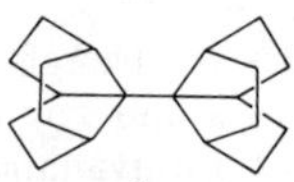	13d
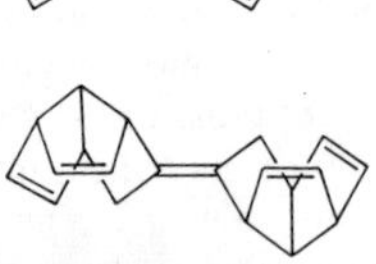	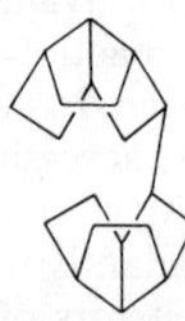	13e
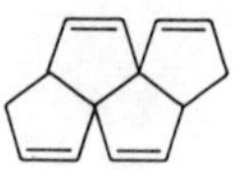	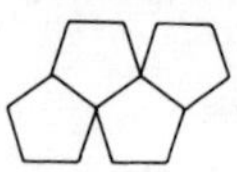	13f
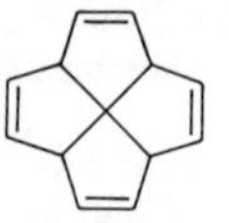	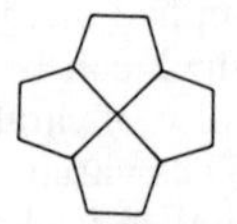	13f
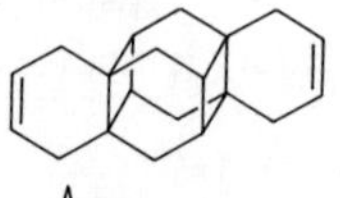	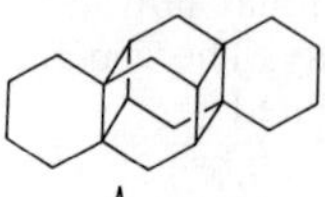	13g
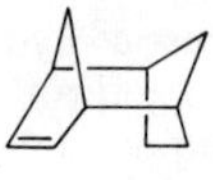	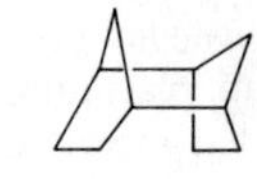	13h

(*continued*)

TABLE 1. (continued)

Starting alkene	Product	Reference
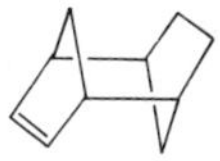	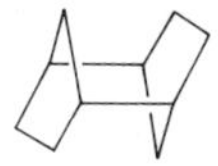	13h
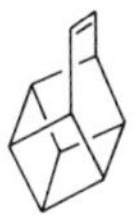	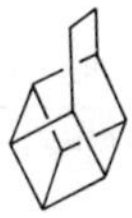	13i
	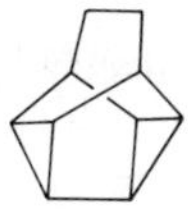	13i
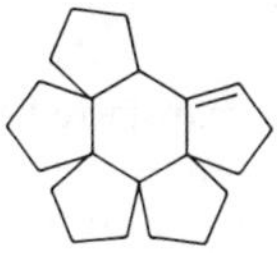	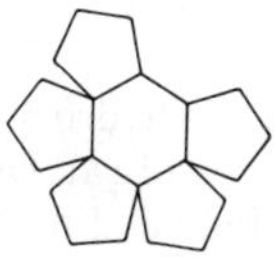	13j

of complex polycycloalkanes, catalytic reduction is usually the final step in the synthesis. Recently, a novel method for the reduction of alkenes with palladium dispersed over a siloxane matrix has been reported[14]. Interestingly, the reaction is done in aqueous THF where water may be the source of hydrogen (equation 1).

$$CH_3(CH_2)_3CH=CH(CH_2)_3 \xrightarrow[\substack{CH_3COCOOCH_3 \\ THF-H_2O\,100\%}]{Pd(OAc)_2(EtO)_3SiH} n\text{-}C_{10}H_{22} \qquad (1)$$

b. *Hydrogenation under homogeneous conditions.* The discovery of chlorotris(triphenyl-phosphine)rhodium by Wilkinson heralded the era of homogeneous hydrogenations. Reduction of alkenes in homogeneous media can be generally accomplished under ambient conditions and in a variety of organic solvents. Besides temperate conditions, the main virtue of homogeneous reductions resides in the realization of high degrees of regio- and chemoselectivity. For example, terminal alkenes and *cis*-alkenes exhibit higher rates of reduction compared to internal alkenes and *trans*-alkenes, respectively. While Wilkinson's catalyst has been used very frequently[11c-g] for alkene reductions, many catalysts like $(PPh_3)_3RuCl_2$[15], $(PPh_3)_3IrH_2Cl$[16] and $(PPh_3)_3OsHCl$[17] have also found useful applications. As an extension of the utility of these catalysts, it has been observed that homogeneous catalysts, e.g. $(PPh_3)_3Rh \cdot X$ (where $X = Cl$, HCl_2, H_2Cl, Cl_3), $(PPh_3)_3NiCl_2$, $(PPh_3)_2PdCl_2$ and $(PPh_3)_2NiBr_2$, immobilized or anchored on silica, are more effective for hydrogenation of alkenes and arenes[18]. Thus, $RhCl(NOR)_2$ on phosphinilated polystyrene or silica reduces benzene, toluene and *t*-butylbenzene to the corresponding alkane in quantitative yield (equation 2)[19]. The nickelocene–LiAlH_4 combination has proved quite useful as a homogeneous hydrogenation catalyst for the reduction of alkenes to alkanes (equation 3)[20]. Iridium and rhodium based catalysts, e.g. $[Ir(COD)(Py)(PCy_3)]PF_6$[21] and $[Rh(NOR)Ph_2P(CH_2)_4PPh_2]BF_4$[22], have proved very efficacious in the directed stereoselective delivery of hydrogen to alkenes and found many applications in natural product syntheses[23]. A significant development of far-reaching consequence is the development of several catalysts based on Rh and Ru

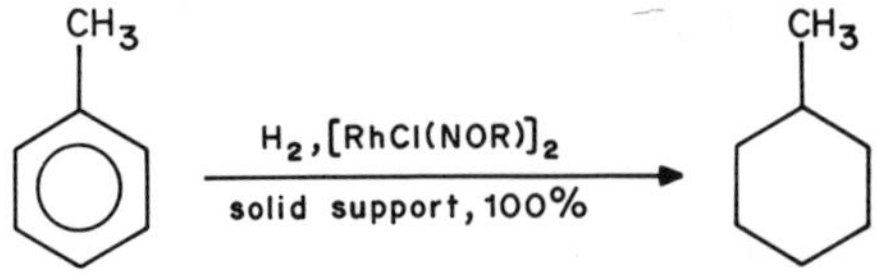

$$(2)$$

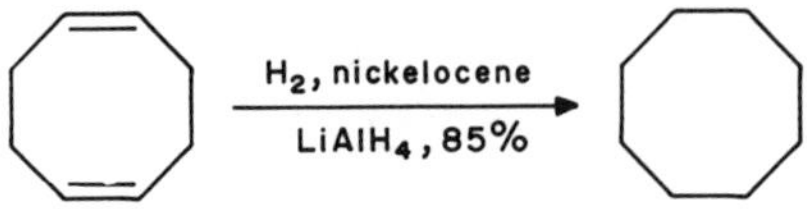

$$(3)$$

for enantioselective hydrogenation of alkenes[24]. The most notable of them is Noyori's Ru(OAc)$_2$(BINAP) catalyst[25].

3. Transfer hydrogenation

In this method of hydrogenation, the hydrogen molecule is transferred to an alkene from a hydrogen donor in the presence of catalysts like platinum, palladium or Raney Nickel (equation 4)[26]. The methodology and mechanism involved in this reaction have been reviewed recently with extensive coverage[11g]. Reduction of hindered double bonds

$$CH_3(CH_2)_5CH{=}CH_2 \xrightarrow[\substack{100\%}]{\text{d-Limonene, 10\% Pd–C}} \text{n-}CH_3(CH_2)_6CH_3 \qquad (4)$$

with diimide (N$_2$H$_2$) can be viewed as a type of transfer hydrogenation[11f,g,27]. The reaction of diimide with alkenes results in *syn* addition of hydrogen across the double bond. Diimide is generated *in situ* via decomposition of potassium azodicarboxylate, anthracene-9,10-diimine and hydrazine and its derivatives[11f]. Examples of polycyclic alkanes synthesized from the corresponding alkene precursors by diimide reductions are gathered in Table 2[28].

TABLE 2. Alkanes from the reduction of alkenes with diimide

Starting alkene	Product	Reference
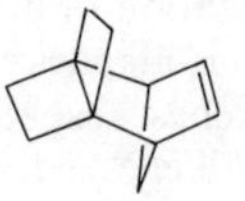	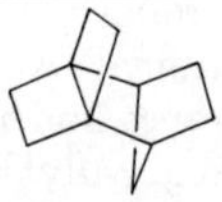	28a
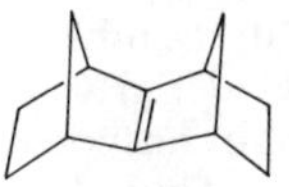	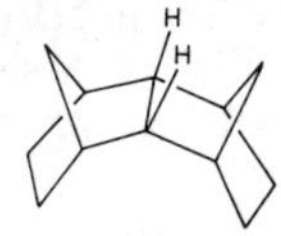	28b
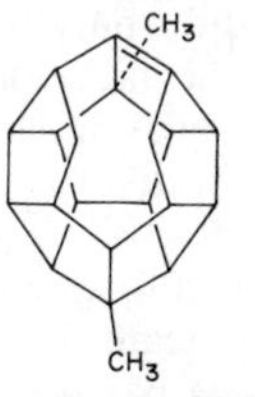	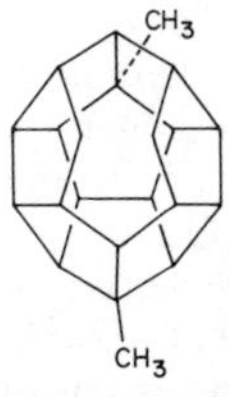	28c

4. Ionic hydrogenation

1,1-Di-, tri- and tetrasubstituted alkenes can be conveniently reduced by ionic hydrogenation procedures[29-31], which involve the sequential delivery of a proton and a hydride ion. The proton source is usually trifluoroacetic acid or trifluromethanesulphonic acid and the hydride source is a trialkylsilyl hydride. Thus, 3-methyl-3-hexene is quantitatively and rapidly reduced in high yield at $-78\,^{\circ}\text{C}$ (equation 5)[31]. A limiting aspect of ionic reductions is that they are non-stereoselective. Reduction of 1,2-dimethylcyclohexene with Et_3SiH and CF_3COOH leads to a mixture of *cis*- and *trans*-1,2-dimethylcyclohexanes[29]. However, an advantage is that hindered and tetraalkyl substituted double bonds can be preferentially reduced through ionic hydrogenations[32].

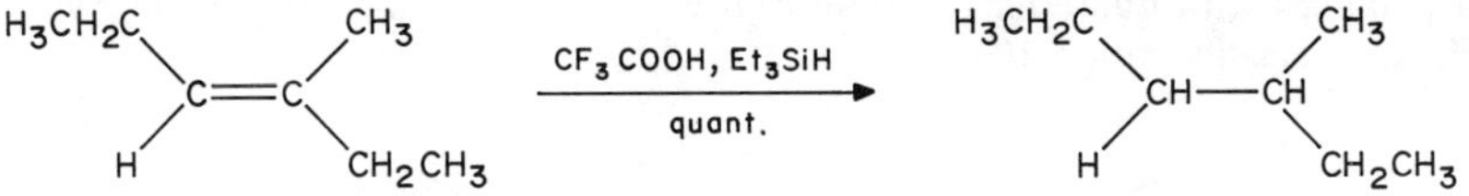

$$
\begin{array}{c}
H_3CH_2C \\
 \\
H
\end{array}
C{=}C
\begin{array}{c}
CH_3 \\
 \\
CH_2CH_3
\end{array}
\xrightarrow[\text{quant.}]{CF_3COOH,\ Et_3SiH}
\begin{array}{c}
H_3CH_2C \\
 \\
H
\end{array}
CH{-}CH
\begin{array}{c}
CH_3 \\
 \\
CH_2CH_3
\end{array}
\tag{5}
$$

5. Hydroboration – protonolysis

Conversion of alkenes to trialkylboranes and subsequent protonolysis by an organic acid affords alkanes in good yields[33]. Hexene could be converted to n-hexane as shown in equation 6. However, this reaction is not very effective for hindered alkenes like 2,2,4,4-tetramethylpentene which furnishes only 5% of alkane on protonolysis.

$$
CH_3(CH_2)_3CH{=}CH_2 \xrightarrow[91\%]{B_2H_6, CH_3CH_2COOH} \text{n-}C_6H_{14}
\tag{6}
$$

6. Borides and aluminides

Many combinations of transition metal salts with sodium borohydride or lithium aluminium hydride promote the reduction of alkenes and alkynes to alkanes. While the exact nature of the catalytic species involved and the mechanism of the reduction is shrouded in uncertainty, the procedure has found many useful applications in alkane synthesis[34]. For example, mono-, di- and trisubstituted alkenes are quantitatively reduced by an equimolar mixture of lithium aluminium hydride and transition metal halides[35-37]. Some of the transition metal halides which have been found to effect the conversion of alkenes to alkanes are $TiCl_3$, $TiCl_4$, VCl_3, $ZrCl_4$, $CrCl_3$, $FeCl_2$, $FeCl_3$, $CoCl_2$ and $NiCl_2$ (equation 7). It has been further observed that $CoCl_2$, $NiCl_2$ and $TiCl_3$ can also be used in catalytic quantities with LAH. Alkynes can also be reduced directly to alkanes by this procedure. The best reagent combination is $LAH-NiCl_2$ from the point of view of product selectivity[35].

$$
CH_3CH_2CH{=}CHCH_2CH_3 \xrightarrow[\substack{TiCl_4\ \text{or}\ ZrCl_4 \\ 96\%}]{LiAlH_4} \text{n-}C_6H_{14}
\tag{7}
$$

Nickel boride (P − 2) generated from nickel acetate and $NaBH_4$ in ethanol exhibits notable selectivity in reduction depending on the substitution pattern of the alkene[34]. Sodium borohydride with Co(II) has also proved to be of general utility for the reduction of various alkenes and alkynes (equation 8)[38,39].

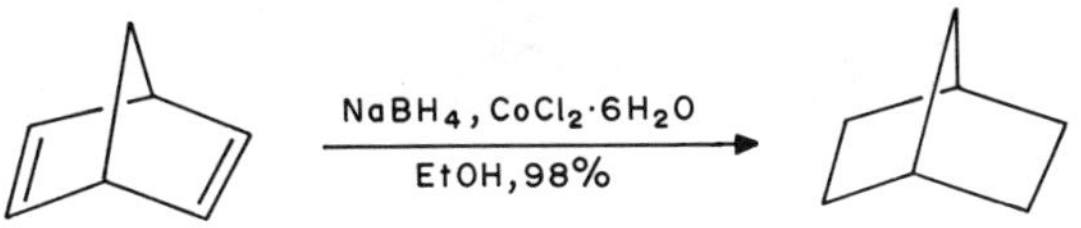

$$
\xrightarrow[\text{EtOH, 98\%}]{NaBH_4,\ CoCl_2{\cdot}6H_2O}
\tag{8}
$$

7. Chemical reduction

Alkenes can be reduced to alkanes by dissolving metal reductions. Sodium/t-BuOH reduces terminal alkenes to alkanes in fairly good yield in the presence of N,N-diethylacetamide (DEA) (equation 9)[40], or hexamethylphosphoric triamide[41]. Di- and trisubstituted olefins are resistant to reduction under these conditions.

$$CH_3(CH_2)_3CH{=}CH_2 \xrightarrow[\text{DEA or HMPT, 98\%}]{\text{Na/}t\text{-BuOH}} \text{n-}C_6H_{14} \tag{9}$$

Tetrasubstituted alkenes can be reductively alkylated with tetramethylsilane or tetramethylstannane and trifluoroacetic acid in the presence of Lewis acids, e.g. $AlBr_3$[42]. Thus, alkanes with quaternary carbon centres can be generated by this procedure in moderate yields (equation 10).

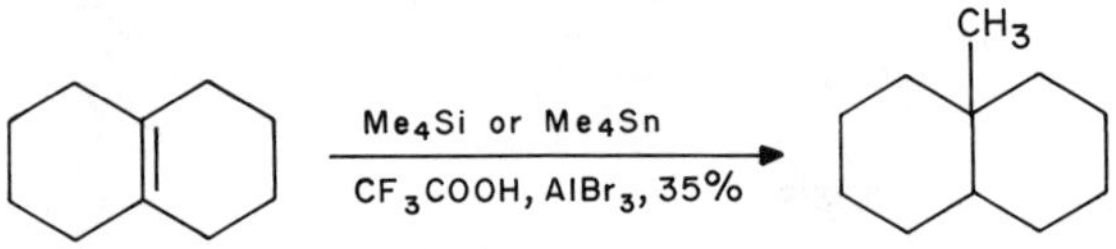

$$\xrightarrow[\text{CF}_3\text{COOH, AlBr}_3,\ 35\%]{\text{Me}_4\text{Si or Me}_4\text{Sn}} \tag{10}$$

Alkynes can be directly reduced to alkanes by the treatment of terminal alkynes with diisobutylaluminium hydride (DIBALH) in the presence of metallocenes like Cl_2TiCp_2 or Cl_2ZrCp_2. On quenching the 1,1-dimetalloalkane intermediates with D_2O, terminally deuteriated alkanes are obtained (equation 11)[43].

$$CH_3(CH_2)_4C{\equiv}CH \xrightarrow[\text{D}_2\text{O, 47\%}]{\text{DIBALH, Cl}_2\text{TiCp}_2} \text{n-}C_5H_{11}CH_2CHD_2 \tag{11}$$

TABLE 3. Reductive opening of cyclopropanes to sterically crowded alkanes and polycycloalkanes

Starting alkane	Product	Reference
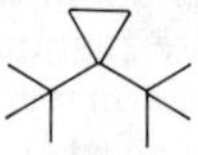	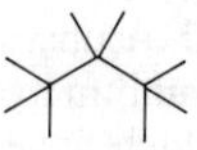	44a
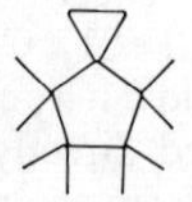	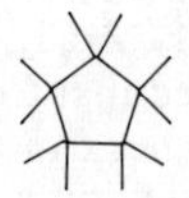	44b
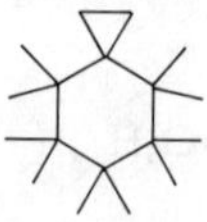	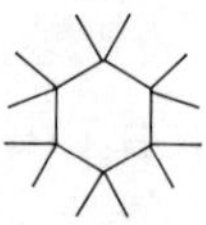	44b
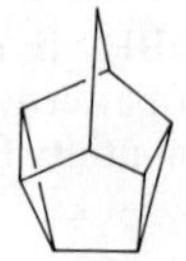	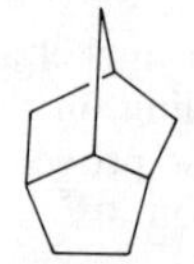	44c
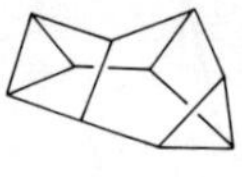	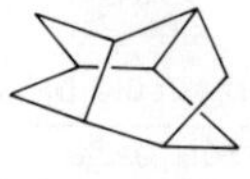	44d

8. Hydrogenolysis of cyclopropanes and other strained molecules

The strain release associated with cleavage of cyclopropanes and cyclobutanes offers the possibility of construction of many theoretically interesting alkanes. In this context, reductive opening of cyclopropanes is a frequently used protocol for assembling diverse polycyclic alkanes. Syntheses of a few sterically crowded alkanes from their cyclopropane precursors are shown in Table 3[44]. Hydrogenolysis of spiro-fused cyclopropane rings constitutes a simple procedure for the generation of cycloalkanes with a *gem*-dimethyl group (equation 12). Hydrogenolysis of strained polycycloalkanes with multiple cyclobutane rings, viz. cubane, homocubane, basketane, lead to many interesting polycyclic alkanes. A few examples are shown in Table 4[45].

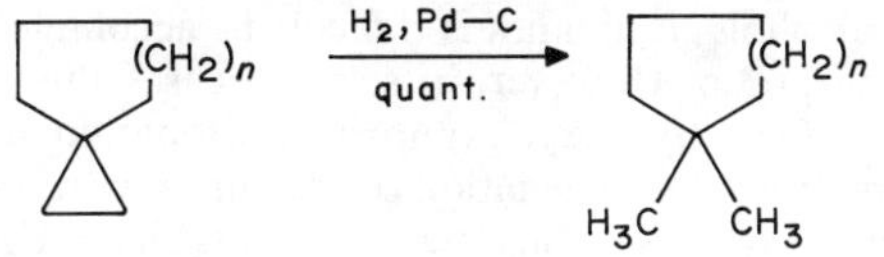

$$(12)$$

TABLE 4. Alkanes through hydrogenolysis of strained C—C bonds

Starting alkane	Products	Reference
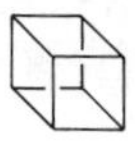		45a
	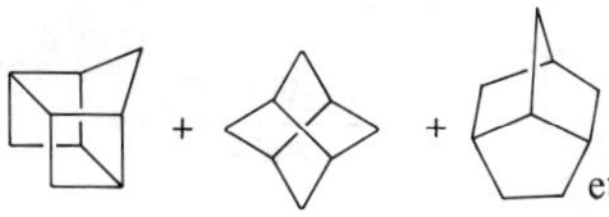	45b
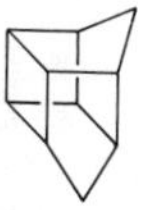	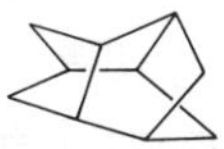	45c
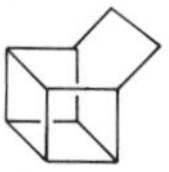	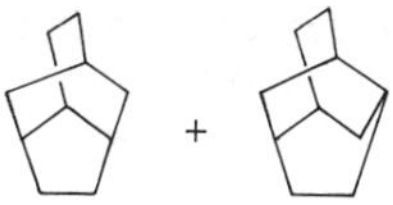	45a

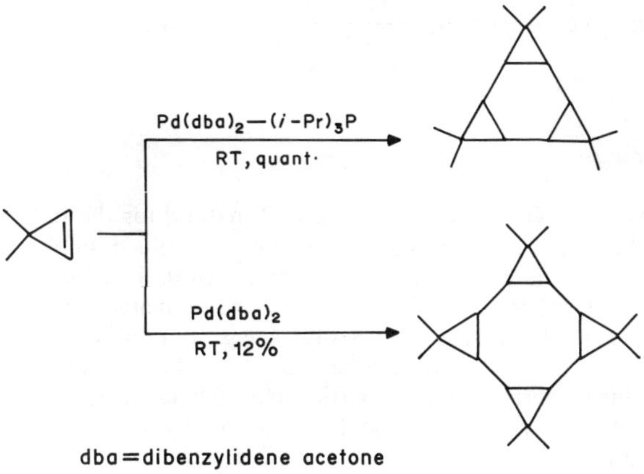

$$(13)$$

dba = dibenzylidene acetone

9. Oligomerization

Structurally novel cyclopropane-fused cycloalkanes can be constructed by oligomerization of cyclopropenes. Cycloaddition across the double bond reduces ring strain in cyclopropenes enormously. When catalysed by transition metals, dimethylcyclopropene tri- or tetramerizes stereoselectively to novel polycycloalkanes (equation 13)[46,48].

B. Alcohols and Related Derivatives

1. Direct reduction

Direct reduction of alcohols to alkanes is difficult to accomplish since the hydroxyl group is a poor leaving group. However, in selected cases this can be accomplished through the activation of the hydroxyl group by protonation or complexation with Lewis acids. For example, ionic hydrogenation conditions have been found to be effective for the direct conversion of alcohols to alkanes. Thus, $Et_3SiH-CF_3COOH$ combination has been used for the reduction of tertiary alcohols[47]. Use of boron trifluoride in the reaction medium even permits reduction of secondary aliphatic alcohols as, e.g., of 2-octanol to n-octane and of 2-adamantanol to adamantane (equation 14)[48]. Tertiary aliphatic alcohols can also be hydrogenated to alkanes in the presence of trifluoroacetic acid, but the reaction perhaps proceeds through prior dehydration to form alkenes[49].

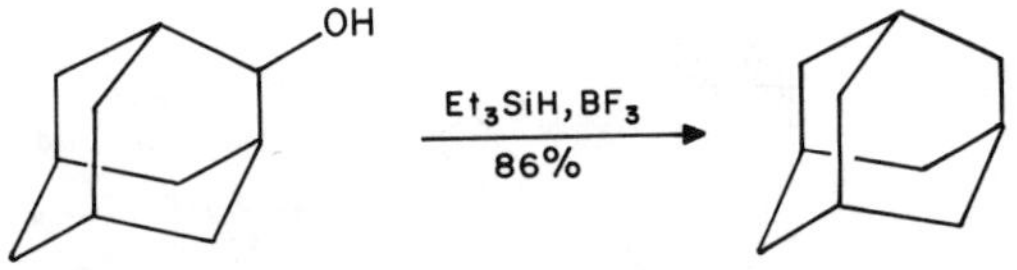

$$\text{(structure with OH)} \xrightarrow[\text{86\%}]{Et_3SiH, BF_3} \text{(adamantane structure)} \qquad (14)$$

Raney Ni has been successfully employed for the direct deoxygenation of tertiary alcohols to alkanes (equation 15). Under the reaction conditions other sensitive functionalities like t-butyldimethylsilyl ethers, acetates and alkyl chlorides are stable[50].

$$n\text{-}C_4H_9\text{---}\underset{\underset{CH_3}{|}}{\overset{\overset{OH}{|}}{C}}\text{---}n\text{-}C_6H_{13} \xrightarrow[\text{80\%}]{\text{Raney Ni}} n\text{-}C_4H_9\text{---}\underset{\underset{CH_3}{|}}{\overset{\overset{H}{|}}{C}}\text{---}n\text{-}C_6H_{13} \qquad (15)$$

2. Indirect reduction

a. Sulphonate esters. The commonly employed method for the reduction of primary and secondary alcohols consists of their conversion to p-toluenesulphonate or methanesulphonate derivatives and subsequent treatment with a reducing agent. Lithium aluminium hydride has been extensively used for the reductive elimination of tosylates as shown in Table 5[51]. However, other hydrides, e.g. sodium borohydride and sodium cyanoborohydride in HMPA, have also been used for this purpose[52]. Brown's lithium triethylborohydride (superhydride) is a particularly effective reagent for this purpose[53a]. For example, cyclooctyl mesylate could be reduced to cyclooctane with $LiEt_3BH$ in high yield as compared to the poor yield obtained with $LiAlH_4$ (equation 16)[53]. This

TABLE 5. Alkanes through the reduction of *p*-toluenesulphonate esters

Starting tosylate	Product	Reference
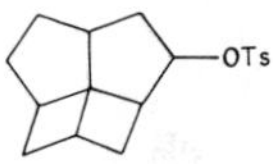 (structure with —OTs)	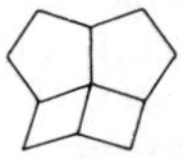 (structure)	51a
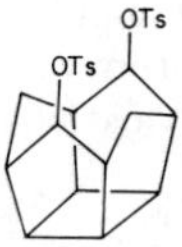 (structure with OTs, OTs)	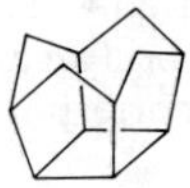 (structure)	51b
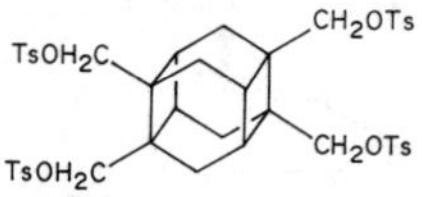 (structure with TsOH$_2$C, CH$_2$OTs, CH$_2$OTs, TsOH$_2$C)	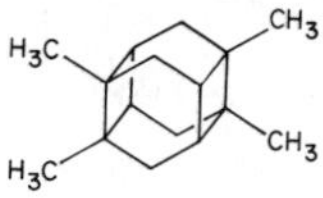 (structure with H$_3$C, CH$_3$, CH$_3$, H$_3$C)	13g

$$\text{(cyclooctane-OMs)} \xrightarrow[93\%]{\text{LiEt}_3\text{BH}} \text{(cyclooctane)} \qquad (16)$$

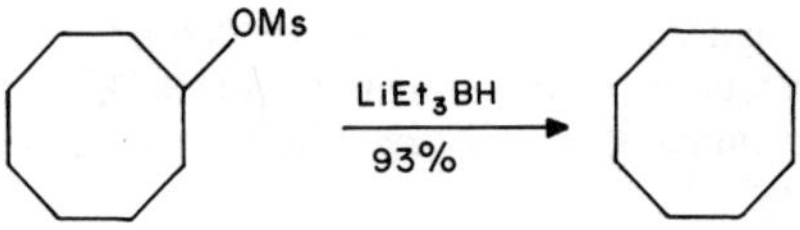

time-tested method continues to be routinely used for the two-step deoxygenation of alcohols.

Another convenient procedure for the reductive removal of the mesylate or tosylate group is through reaction with a NaI/Zn system[54]. The iodide intermediate which is formed in the reaction is reduced to the alkane with zinc powder. Following this procedure 2-octyl tosylate could be reduced to n-octane in high yield (equation 17)[54a].

$$\underset{\displaystyle \text{CH}_3\text{CH(CH}_2)_5\text{CH}_3}{\overset{\displaystyle \overset{\text{OTs}}{|}}{}} \xrightarrow[\text{glyme, 89\%}]{\text{NaI–Zn}} \text{CH}_3(\text{CH}_2)_6\text{CH}_3 \qquad (17)$$

b. Intermediate iodides. A one-pot method for the reductive deoxygenation of hydroxy and ether functionalities is via *in situ* generated iodides. The reagent system involved is chlorotrimethylsilane/sodium iodide (trimethylsilyl iodide, TMSI) and subsequent treatment with zinc and acetic acid[55]. This method is applicable to a wide variety of primary and secondary alcohols, methyl ethers and trimethylsilyl ethers. Thus, n-decanol could be readily converted to n-decane following this method (equation 18)[55].

$$\text{CH}_3(\text{CH}_2)_8\text{CH}_2\text{OH} \xrightarrow[\text{Zn–AcOH, 80\%}]{\text{Me}_3\text{SiCl–NaI, CH}_3\text{CN}} \text{CH}_3(\text{CH}_2)_8\text{CH}_3 \qquad (18)$$

Tertiary alcohols are seldom amenable to direct deoxygenation. However, sodium cyanoborohydride in the presence of zinc iodide reduces tertiary alcohols to the corresponding alkanes in high yields[56]. Primary and secondary alcohols are not reduced with this reduction reagent and thus a very desirable chemoselectivity in deoxygenation is attainable.

c. Carboxylic acid esters. Acetates generated from primary, secondary and tertiary alcohols serve as good intermediates for deoxygenation reactions. Several preparative methods for this one-step transformation, generally involving radical intermediates, are available. A method that has been used with considerable success is the reductive deacetoxylation under photochemical conditions. Irradiation of acetates is usually carried out in aq. HMPA and yields, particularly in the case of polycycloalkanes, are quite satisfactory. Several examples are shown in Table 6[57].

Acetates derived from tertiary and sterically hindered alcohols can be reduced to alkanes by treatment with lithium in ethylamine or potassium in *t*-butylamine in the presence of [18]-crown-6[58]. Thus, a bridgehead acetate derived from natural product caryophyllene was transformed to the corresponding alkane in high yield (equation 19)[58a].

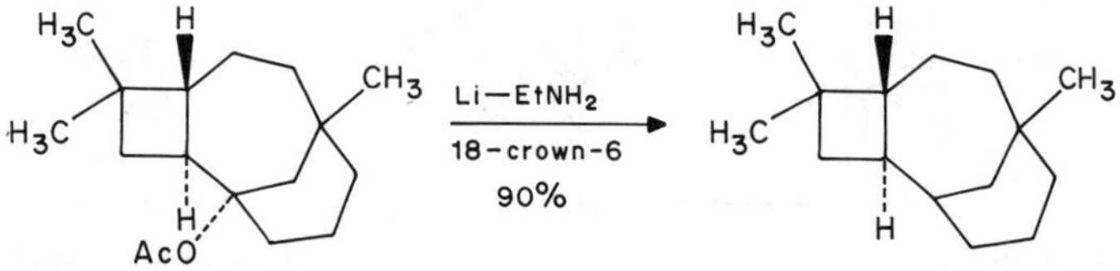

$$\text{(19)}$$

Reductive elimination of acetates from secondary alcohols has been shown to proceed in good yield, employing sodium–potassium alloy and tris(3,6-dioxaheptyl)amine (TDA), with the latter also functioning as an effective phase transfer catalyst (equation 20)[59].

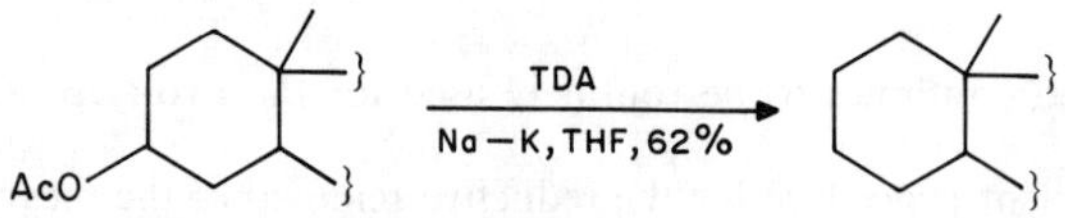

$$\text{(20)}$$

A similar reduction of acetates with sodium in hexamethylphosphoric triamide (HMPA) has been reported[60]. Following this procedure, acetates of primary and

TABLE 6. Alkanes through photochemical reductive deacetoxylation

Starting acetate	Product	Reference
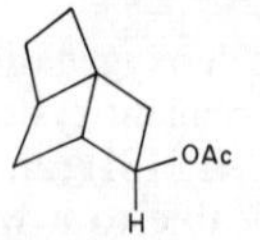	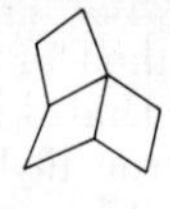	57a
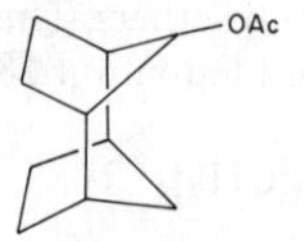	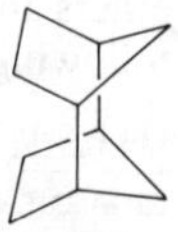	57b
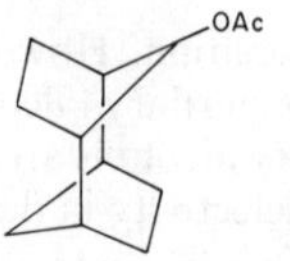	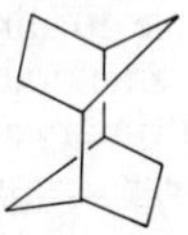	57b

secondary alcohols can be converted to the corresponding alkanes in moderate yields. Preparation of the parent hydrocarbon (laurenane) from the diterpene natural product laurenene is one such example (equation 21)[61]. In the case of highly hindered acetates alkane yields can be near-quantitative.

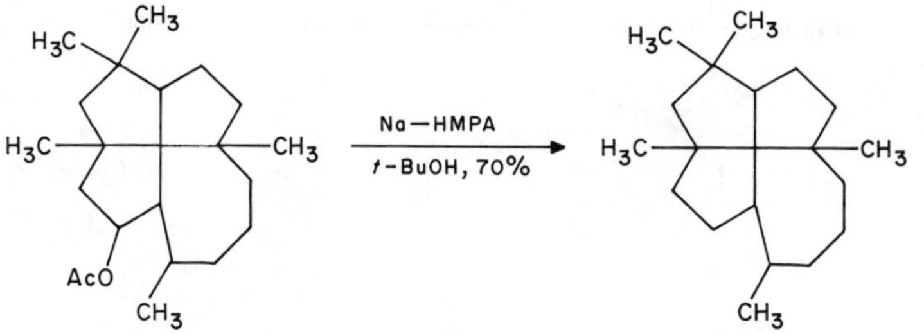

$$\xrightarrow[\text{\textit{t}-BuOH, 70\%}]{\text{Na—HMPA}} \quad (21)$$

Reduction of primary, secondary and tertiary acetates can also be conveniently performed by reaction with triphenylsilyl hydride or p-(bis-diphenylhydrosilyl)benzene[62] under radical conditions with the latter reagent being more effective (equations 22 and 23). It is to be noted that AIBN and benzoyl peroxide are not effective as radical initiators in this reduction[62].

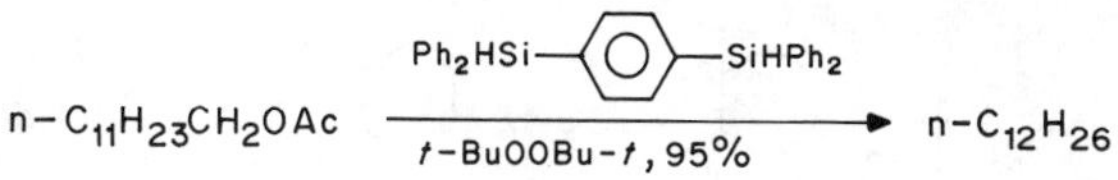

$$\text{n}-C_{11}H_{23}CH_2OAc \xrightarrow[\text{\textit{t}-BuOOBu-\textit{t}, 95\%}]{\text{Ph}_2HSi-\bigcirc-SiHPh_2} \text{n}-C_{12}H_{26} \quad (22)$$

$$\xrightarrow[\text{\textit{t}-BuOOBu-\textit{t}, 82\%}]{\text{Ph}_3SiH} \quad (23)$$

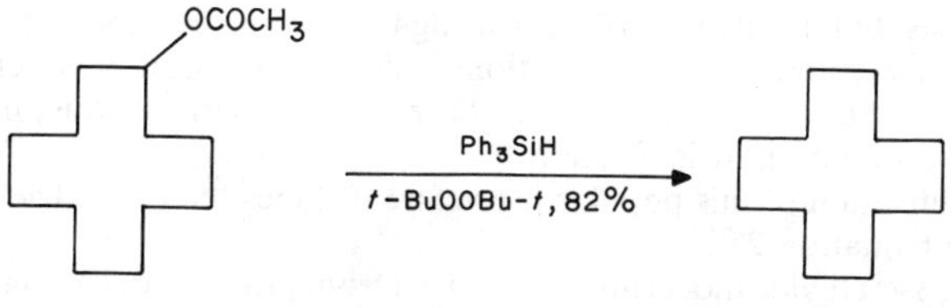

Deoxygenation of primary alcohols can also be performed by treating chloroformate esters with tripropylsilane in the presence of di-t-butyl peroxide (equation 24)[63].

$$\text{n-}C_7H_{15}CH_2O-\overset{\overset{\displaystyle O}{\|}}{C}Cl \xrightarrow[\text{\textit{t}-BuOOBu-\textit{t}, 85\%}]{\text{Pr}_3SiH} \text{n-}C_8H_{18} \quad (24)$$

d. Thiocarbonates (Barton–McCombie reaction)[64] and phosphate esters. Barton and his group have invented a very versatile methodology for the deoxygenation of alcohols through radical intermediates[65]. This approach essentially involves the treatment of derived S-methyldithiocarbonates[64,66a,b], thionobenzoates[66a], thiocarbonylimidazolides[66c], N,N-dialkylaminothiocarbonates[66d], thionoformic acid esters[66e], phenylthiocarbonate[66f] etc., with tri-n-butyltin hydride (TBTH) in the presence of the radical

G. Mehta and H. S. P. Rao

TABLE 7. Alkanes through Barton–McCombie-type radical deoxygenations

Radical precursor	Product	Reference
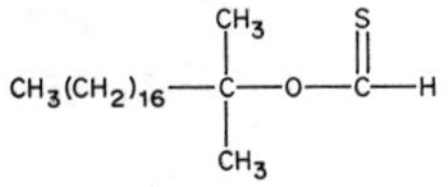 $CH_3(CH_2)_{16}$—C(CH₃)₂—O—C(=S)—H	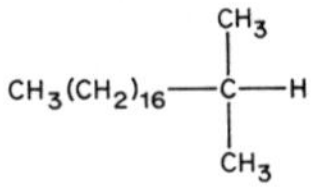 $CH_3(CH_2)_{16}$—CH(CH₃)—CH₃	66e
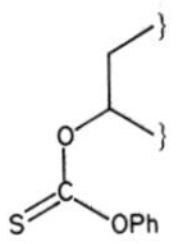	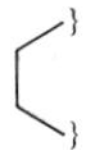	66f
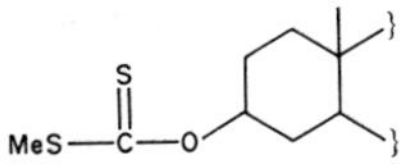 MeS—C(=S)—O—	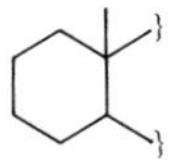	64, 66a, b
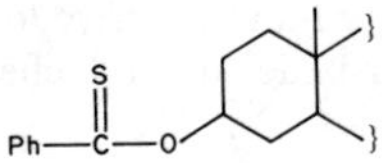 Ph—C(=S)—O—	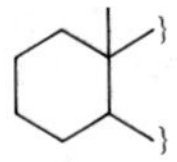	66a
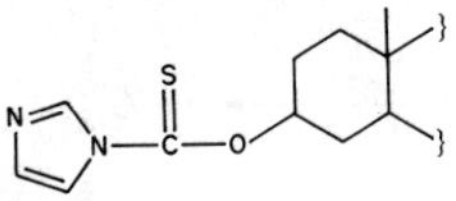 (imidazole)N—C(=S)—O—	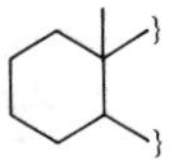	66c

generator azobisisobutyronitrile (AIBN). Through judicious selection of the thioester or thiocarbonate moiety and reaction conditions, efficient reduction of either the primary or secondary alcohol can be achieved. Several examples of this deoxygenation methodology are summarized in Table 7.

A recent modification of this popular procedure includes the use of triethyl boride as radical initiator (equation 25)[67].

Alternatively, S-methyldithiocarbonates derived from primary and secondary alcohols can be converted to alkanes using Et_3SiH with di-t-butyl peroxide and dodecanethiol in good yields (equation 26)[68].

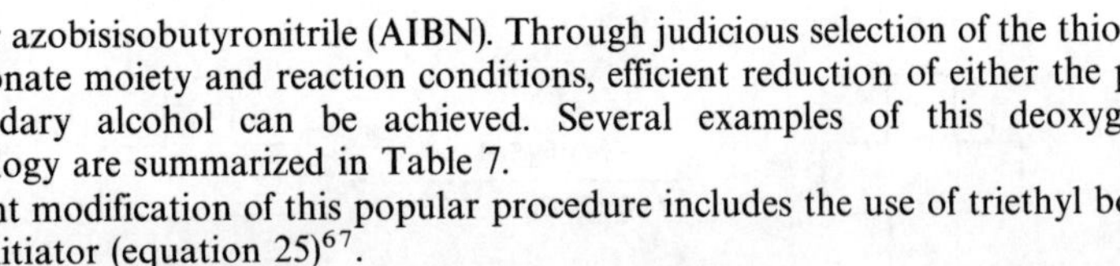

$$\text{cyclododecyl—O—C(=S)—SMe} \xrightarrow[C_6H_6,\ 93\%]{n\text{-}Bu_3SnH,\ Et_3B} \text{cyclododecane} \quad (25)$$

$$\text{MeSC(=S)—O—(menthyl)} \xrightarrow[C_{12}H_{25}SH,\ 89\%]{Et_3SiH,\ t\text{-}BuOOBu\text{-}t} \text{(menthane)} \quad (26)$$

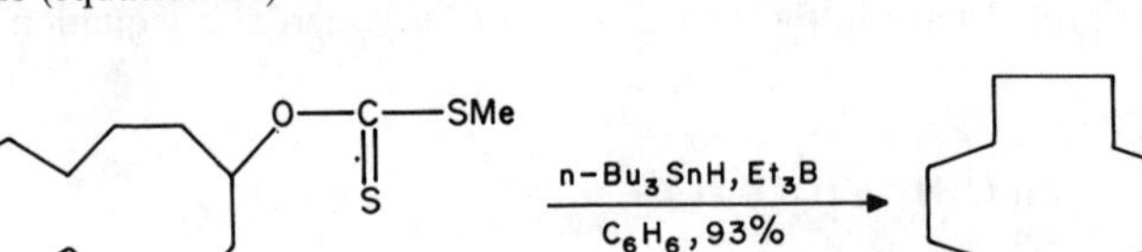

Tertiary alcohols can be conveniently deoxygenated in high yields through thiohydroxamic-O-esters. In practice, an oxalic acid half ester is prepared and decomposed in the presence of t-butylthiol or 1,1-diethylpropanethiol under radical conditions (equation 27)[69]. Diethyl phosphates and N,N,N',N'-tetramethylphosphorodiamidates of alcohols are readily reduced by lithium–ethylamine solutions. The method works well with both secondary and tertiary alcohols, the latter shown by the conversion of 1-adamantanol to adamantane (equation 28)[70].

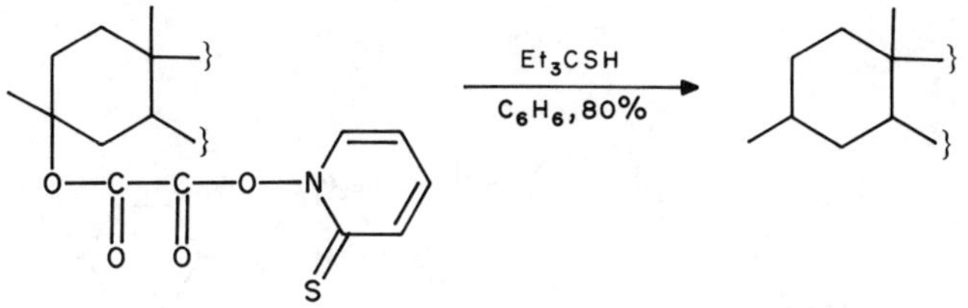

(27)

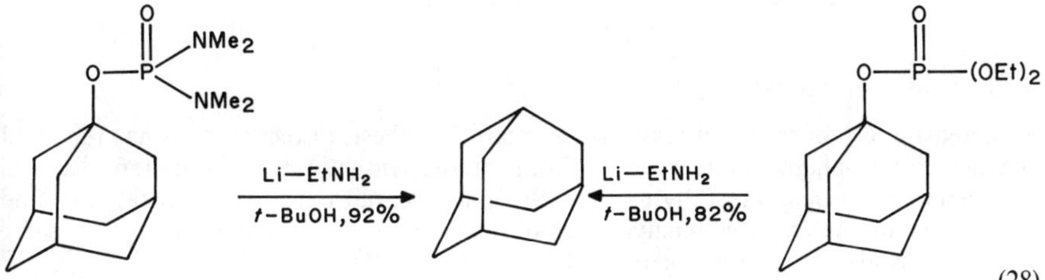

(28)

3. Sulphur compounds

Thiols and thioethers are the most commonly encountered sulphur derivatives which can be directly reduced to alkanes. For this purpose, Raney Ni has been traditionally used. An example of its use is the synthesis of 7α- and 7β-eremophilanes from a tricyclic thiophene precursor (equation 29)[71]. However, more recently nickel complex reducing agents ("NiCRA") have been profitably and effectively used for this purpose, as indicated by the reduction of dodecanethiol to dodecane (equation 30)[72]. Similarly, nickelocene in the presence of LAH has been used to effect desulphurization of thiols[73]. A combination of NaEt$_3$BH and FeCl$_2$ also achieves the thiol-to-alkane transformation quite satisfactorily[74]. The increasing utility of nickel boride has also been exploited for the transformation of sulphides to alkanes as indicated in equation 31[75].

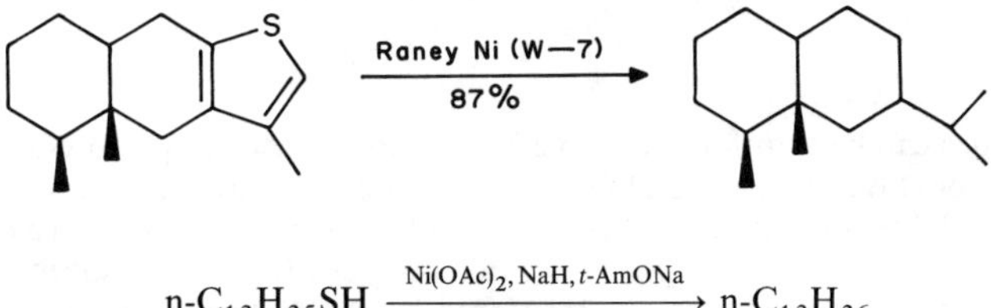

(29)

$$n\text{-C}_{12}\text{H}_{25}\text{SH} \xrightarrow[\text{THF},94\%]{\text{Ni(OAc)}_2,\,\text{NaH},\,t\text{-AmONa}} n\text{-C}_{12}\text{H}_{26}$$

(30)

$$n\text{-C}_{11}\text{H}_{23}\text{SPh} \xrightarrow[\text{MeOH},96\%]{\text{NiCl}_2\cdot6\text{H}_2\text{O},\,\text{NaBH}_4} n\text{-C}_{11}\text{H}_{24}$$

(31)

An extremely pleasing example of reductive desulphurization is the synthesis of [2]-staffane, wherein lithium in ethylamine is employed as the reducing agent (equation 32)[76]. Finally, an example of the use of radical chemistry for the conversion of an alcohol to an alkane via a sulphide intermediate has been reported (equation 33)[77].

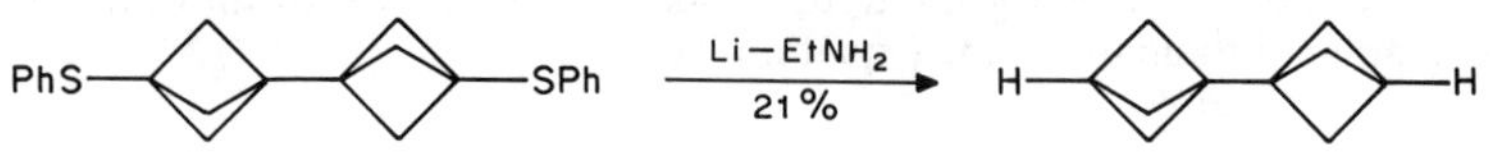

$$PhS \overline{} \cdots \overline{} SPh \quad \xrightarrow[21\%]{Li-EtNH_2} \quad H \overline{} \cdots \overline{} H \qquad (32)$$

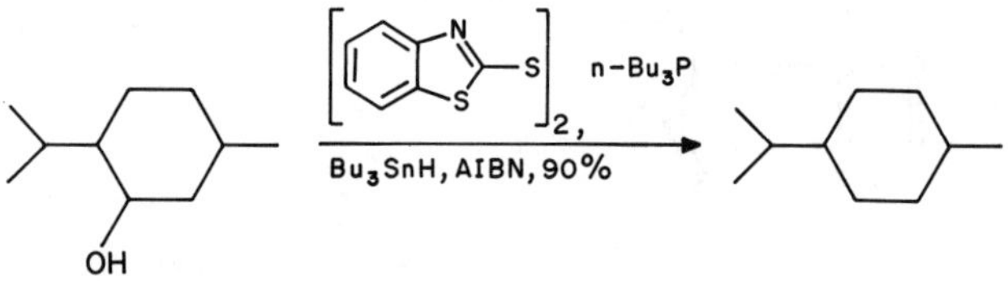

$$(33)$$

4. Selenium compounds

Increasing use of selenium reagents in organic synthesis in recent years has rendered the selenide-to-alkane conversion an important synthetic transformation. Radical chemistry has been successfully used for this purpose and both n-tributylstannane and triphenylstannane reduce phenyl selenides to alkanes (equation 34)[78]. The same procedure is also applicable for tellurides (equation 35)[78].

$$n\text{-}C_{12}H_{25}SePh \xrightarrow[73\%]{Ph_3SnH, \text{toluene}} n\text{-}C_{12}H_{26} \qquad (34)$$

$$C_{10}H_{21}-\underset{\underset{TePh}{|}}{CHCH_3} \xrightarrow[77\%]{Ph_3SnH, \text{toluene}} n\text{-}C_{12}H_{26} \qquad (35)$$

Nickel boride has also been found to be effective for reductive deselenation. The method is convenient for the preparation of mono-deuteriated alkanes (equation 36)[79].

$$n\text{-}C_{11}H_{23}CH_2SePh \xrightarrow[\text{MeOD}, 98\%]{NiCl_2, NaBD_4} n\text{-}C_{11}H_{23}CH_2D \qquad (36)$$

C. Alkyl Halides

1. General aspects

Alkyl halides can be transformed into alkanes through two-step procedures involving dehydrohalogenation to an alkene followed by reduction. However, we will only discuss those methodologies that directly deliver alkanes from organic halides through reductive dehalogenations or hydrogenolyses[80]. For the purpose of dehalogenations, either chlorides, bromides or iodides can be employed but the last two are generally preferred because of greater ease of reduction. Alkyl fluorides are usually resistant to reductive dehalogenation but a few reagent systems are available for the conversion of primary alkyl fluorides to alkanes (equations 37 and 38)[81].

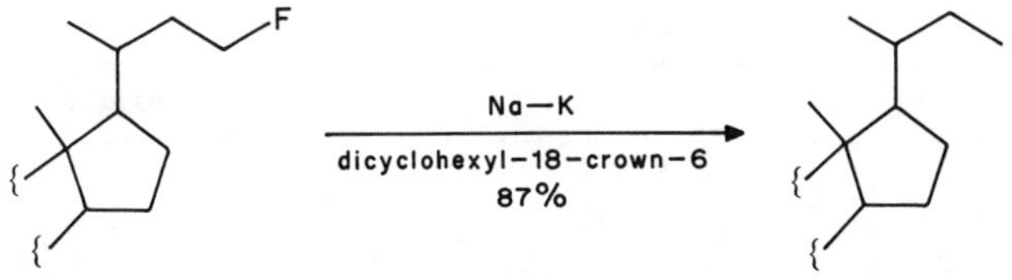

$$\text{CH}_3(\text{CH}_2)_{10}\text{CH}_2\text{F} \xrightarrow[99\%]{\text{LiAlH}_4-\text{CeCl}_3} \text{CH}_3(\text{CH}_2)_{10}\text{CH}_3 \qquad (38)$$

2. Hydrogenolysis

Catalytic hydrogenolysis of alkyl halides is one of the classical methods of alkane syntheses. Among other catalysts, Raney Ni has proved to be quite effective for the hydrogenolysis of iodides (equation 39)[82]. A relatively recent discovery is the reductive dehalogenation of alkyl halides under the conditions of ionic hydrogenation.

$$\text{CH}_3(\text{CH}_2)_4\text{CH}_2\text{I} \xrightarrow[99\%]{\text{H}_2, \text{Raney Ni}} \text{CH}_3(\text{CH}_2)_4\text{CH}_3 \qquad (39)$$

Alkyl halides can be reduced conveniently by trialkylsilanes in the presence of catalytic amounts of BF_3 or AlCl_3[83]. Hydrogen–halogen exchange reaction of primary, secondary and tertiary chlorides or bromides is quite efficient under these conditions (equation 40).

$$\underset{\underset{\text{CH}_3(\text{CH}_2)_9\text{CHCH}_3}{|}}{\overset{\text{Br}}{}} \xrightarrow[87\%]{\text{Bu}_3\text{SiH}, \text{AlCl}_3} \text{CH}_3(\text{CH}_2)_{10}\text{CH}_3 \qquad (40)$$

Triethylsilane reduction of alkyl halides is even better in the presence of thiols, which act as polarity reversal catalysts for hydrogen atom transfer from silane to alkyl groups (equation 41)[84]. This reagent in the presence of AlCl_3 reduces even *gem*-dichlorides to alkanes (equation 42)[85]. The powerful hydrogen donor reagent *tris*-(trimethylsilyl)silane has been introduced as a superior alternative to triethylsilane for the hydrogenolysis of

$$\text{CH}_3(\text{CH}_2)_6\text{CH}_2\text{Br} \xrightarrow[\substack{\text{di-}t\text{-butylhyponitrite} \\ \text{dodecanethiol}, 99\%}]{\text{Et}_3\text{SiH}} \text{CH}_3(\text{CH}_2)_6\text{CH}_3 \qquad (41)$$

$$\text{CH}_3(\text{CH}_2)_2\text{CCl}_2\text{CH}_3 \xrightarrow[67\%]{\text{Et}_3\text{SiH}, \text{AlCl}_3} \text{CH}_3(\text{CH}_2)_3\text{CH}_3 \qquad (42)$$

organic halides. The reaction is initiated by light and proceeds through a free radical mechanism (equation 43)[86].

$$\text{CH}_3(\text{CH}_2)_{16}\text{CH}_2\text{Cl} \xrightarrow[h\nu, 93\%]{[\text{Me}_3\text{Si}]_3\text{SiH}} \text{CH}_3(\text{CH}_2)_{16}\text{CH}_3 \qquad (43)$$

3. Low-valent metals

Dissolved alkali metals have been traditionally used for the dehalogenation of alkyl halides. The zinc–acetic acid system and metal–ammonia solutions were the oldest methods used for this purpose. They still constitute an efficient medium for dehalogenations and several convenient procedures have been introduced. Sodium in

G. Mehta and H. S. P. Rao

TABLE 8. Alkanes through dissolved metal reductions

Starting halide	Product	Reference
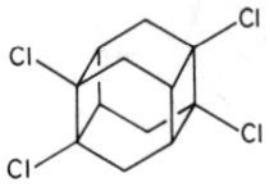	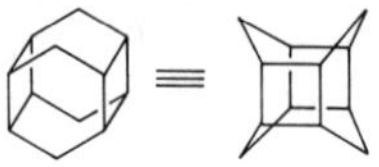	87b
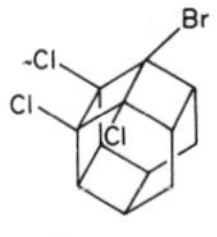		88b
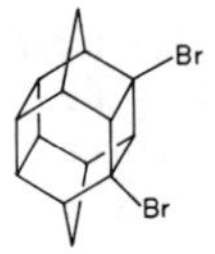	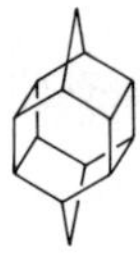	88c
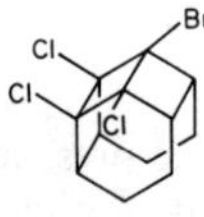	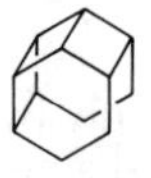	88d
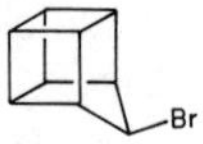	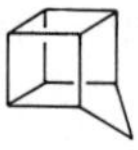	89
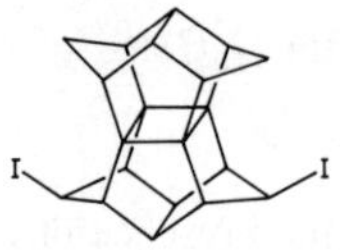	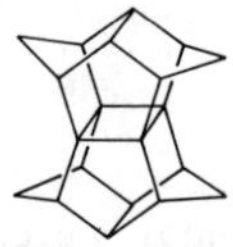	90

ethanol[87], lithium in t-butanol and THF[88,89], Na–K in t-butanol[90] are the more commonly used recipes and work well with bridgehead substituted halides. Some examples are displayed in Table 8[88–90]. In addition alkali metals in liq. NH_3 or alkylamines constitute an effective medium for dehalogenations.

Samarium iodide (SmI_2) has been introduced as an efficient reducing agent for alkyl bromides and iodides in the presence of a proton source (alcohol)[91]. Alkyl chlorides are not reduced by this reagent. It has been further observed that the SmI_2 reductions proceed exceedingly well in the presence of HMPA[92]. Thus, primary, secondary and tertiary halides are reduced to alkanes under mild conditions and in very high yields (equation 44)[92].

$$CH_3(CH_2)_8CH_2Br \xrightarrow[\text{i-PrOH, 99\%}]{\text{SmI}_2, \text{HMPA, THF}} CH_3(CH_2)_8CH_3 \qquad (44)$$

4. Metal hydrides

A variety of metal hydrides reduce alkyl halides to alkanes. Sodium borohydride has been extensively used in aprotic solvents like DMSO, DMF, diglyme and sulpholane. Reduction with this hydride in the presence of phase transfer catalysts is even more advantageous[93]. Sodium borohydride in the presence of nickel boride (Ni_2B) has been used for the reduction of polychlorinated insecticides[34]. Lithium aluminium hydride has also been employed with varying degrees of success. An example is the last step in the synthesis of iceane (equation 45)[94]. LAH in combination with $CoCl_2$ has found application in the reduction of secondary bromides[36]. Sodium cyanoborohydride in HMPA is a particularly effective system for the reduction of bromides and iodides[95]. Lithium triethylborohydride reduces a variety of alkyl halides every efficiently and rapidly to alkanes (equation 46)[96]. Lithium triethylborodeuteride can be employed for deuteriolysis with stereochemical inversion at the stereogenic centre. Copper hydride has been used for the reduction of bromo- and iodoalkanes to alkanes[97]. Magnesium hydride reduces alkyl iodides selectively to alkanes in the presence of bromides and chlorides (equation 47)[98].

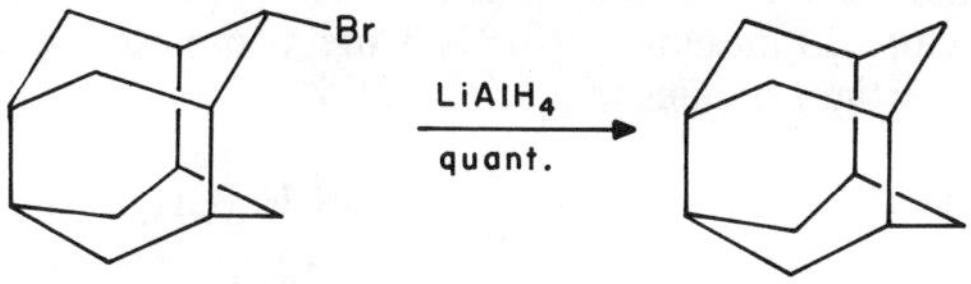

$$(45)$$

$$CH_3(CH_2)_6CH_2Cl \xrightarrow[\text{THF, 100\%}]{\text{LiEt}_3\text{BH}} CH_3(CH_2)_6CH_3 \qquad (46)$$

$$CH_3(CH_2)_8CH_2I \xrightarrow[\text{THF, 100\%}]{\text{MgH}_2} CH_3(CH_2)_8CH_3 \qquad (47)$$

Organotin hydrides have proved to be very effective reagents for the reduction of alkyl halides under homolytic conditions in the presence of radical initiators like AIBN or benzoyl peroxide. Tributyltin hydride (TBTH) is the most commonly employed reagent (equation 48)[99], but others like Bu_2SnH_2 have found utility (equation 49)[100].

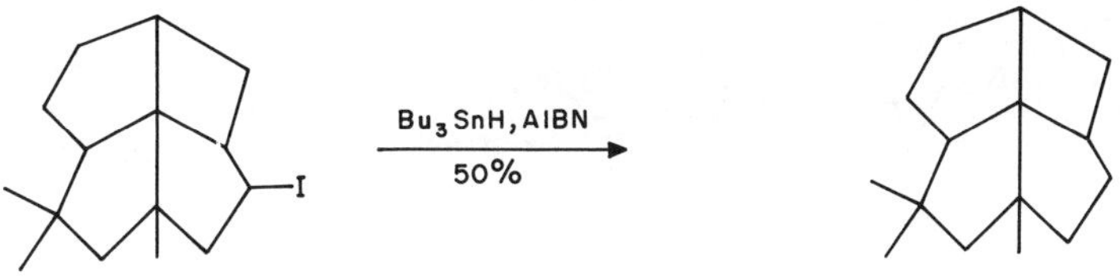

$$(48)$$

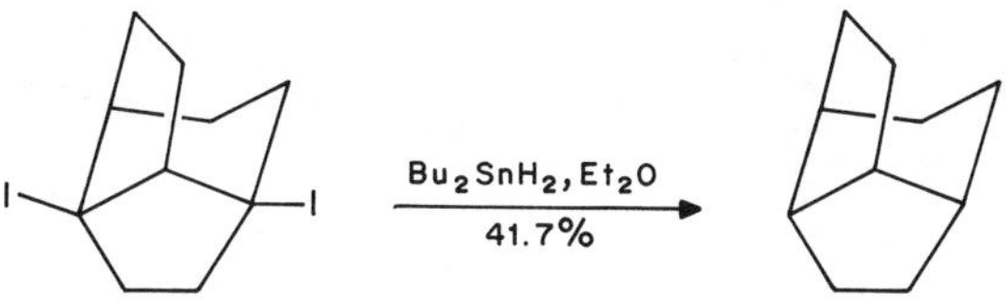

$$(49)$$

5. Electrochemical methods

Alkyl halides, excepting alkyl fluorides, can be reduced under electrochemical conditions in aprotic solvents. For example, 1-adamantyl bromide can be reduced to

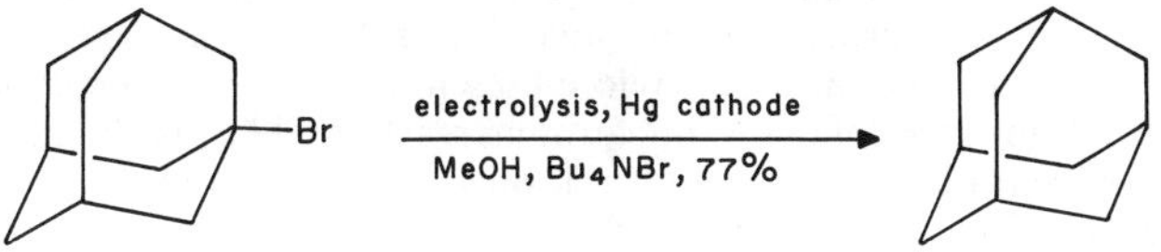

$$(50)$$

adamantane in good yield (equation 50)[101]. It has been further reported that electro-chemical hydrogenolysis of alkyl halides can be effected in the presence of aromatic hydrocarbons like naphthalene which act as electron-transfer reagents (equation 51)[102].

$$CH_3(CH_2)_4CH_2Cl \xrightarrow[\text{electrochemical, 80--100\%}]{\text{naphthalene}} CH_3(CH_2)_4CH_3 \qquad (51)$$

6. Organometallics and miscellaneous reagents

Several organometallic reagents have surfaced in recent years for the reduction of the carbon–halogen bond. A titanocene dichloride–magnesium system reduces alkyl halides to alkanes under mild conditions and in good yield (equation 52)[103]. New reagents based on 9-borabicyclo[3.3.1]nonane ("9-BBN") 'ate' complexes have been used for the efficient reduction of tertiary halides (equation 53)[104].

$$CH_3(CH_2)_7CH_2Br \xrightarrow[\text{Mg, H}_2\text{O, 87\%}]{(Cp)_2TiCl_2} CH_3(CH_2)_7CH_3 \qquad (52)$$

$$
\underset{\underset{CH_3}{|}}{\overset{\overset{Br}{|}}{\text{n-C}_4\text{H}_9\text{—C—C}_2\text{H}_5}} \xrightarrow[92\%]{\text{9-BBN, n-BuLi}} \underset{\underset{CH_3}{|}}{\overset{\overset{H}{|}}{\text{n-C}_4\text{H}_9\text{—C—C}_2\text{H}_5}} \qquad (53)
$$

Lastly, metallation of alkyl halides and protonation has been a frequently employed method for the preparation of alkanes. Examples of alkane synthesis via conversion to the Grignard reagent and protonation are legion. However, two pleasing examples have been recently reported (equations 54 and 55)[105].

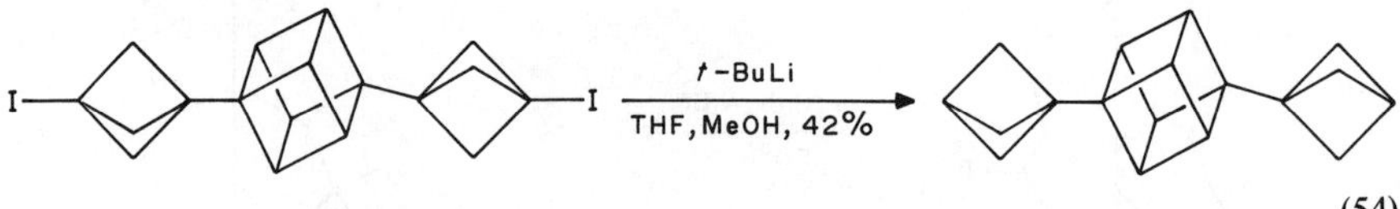

$$(54)$$

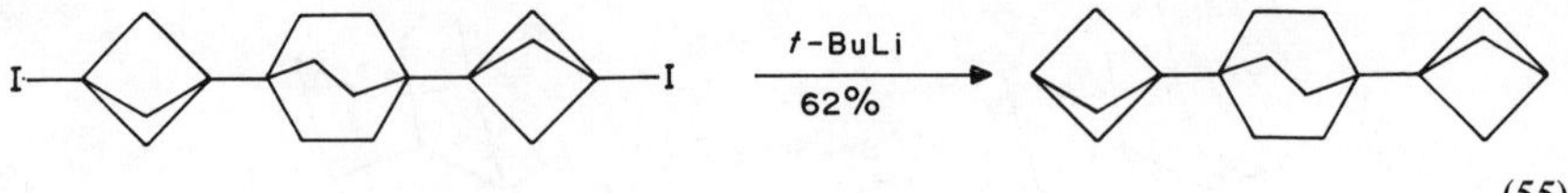

$$(55)$$

D. Aldehydes, Ketones and Related Compounds

1. General aspects

Deoxygenation of aldehydes and ketones is a step frequently encountered in organic synthesis. Its importance emanates from the fact that the carbonyl group is a very versatile handle for the purpose of key organic transformations, e.g. homologations, annulations, alkylations, etc., and is therefore present in almost every complex synthetic venture. A large number of common ring forming reactions, like Dieckmann cyclization, Thorpe–Ziegler cyclization, acyloin condensation, involve a carbonyl functionality which needs to be disposed of, once it has served its purpose, to gain access to alkanes. While the carbonyl group can be transformed to the corresponding alkane via the hydroxyl group or an alkene, the concern here will be only with methods that transform a carbonyl group to a methylene group without mediation by other functional groups.

2. Direct reduction

Catalytic hydrogenation of carbonyl compounds to alkanes is a difficult proposition under normal conditions, although limited success is attainable with aromatic ketones. However, certain enolates derived from ketones have been shown to undergo catalytic reduction to alkanes quite efficiently. For example, enol triflates of ketones are reduced over platinum oxide catalyst to alkanes (equation 56)[106]. Similarly, enol phosphates, conveniently prepared from ketones, can be quantitatively hydrogenated to alkanes (equation 57)[107].

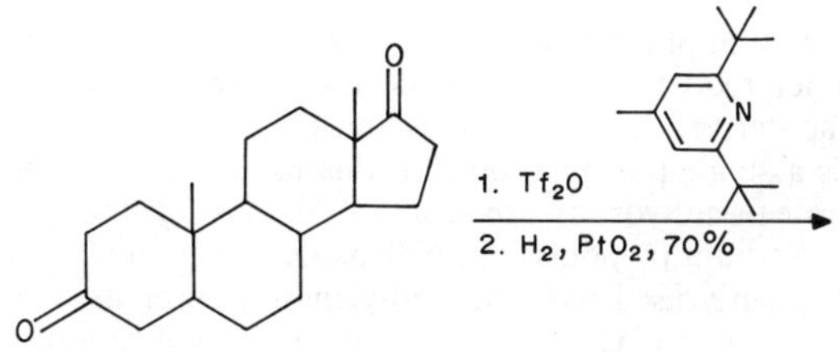

$$\text{1. Tf}_2\text{O}$$
$$\text{2. H}_2\text{, PtO}_2\text{, 70\%}$$

(56)

$$\xrightarrow[100\%]{\text{H}_2\text{, Pt—C}}$$

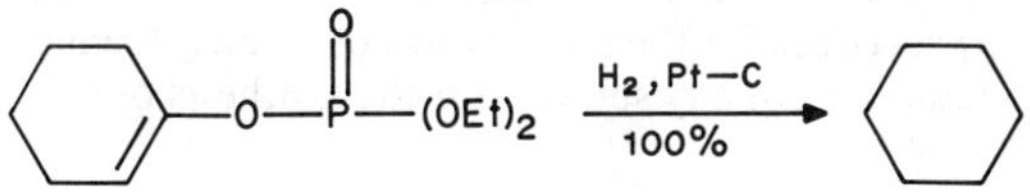

(57)

The remarkable reducing ability of systems containing an organosilicon hydride and gaseous boron trifluoride has been exploited for the direct conversion of ketones to alkanes[108]. Thus, 2-methylcyclohexanone and 2-adamantanone are efficiently transformed to methylcyclohexane and adamantane, respectively (equation 58). The procedure is general enough to be of preparative value, although it works better with aryl carbonyl compounds[108]. Diborane in the presence of BF_3 is able to deoxygenate cyclopropyl ketones (equation 59)[109].

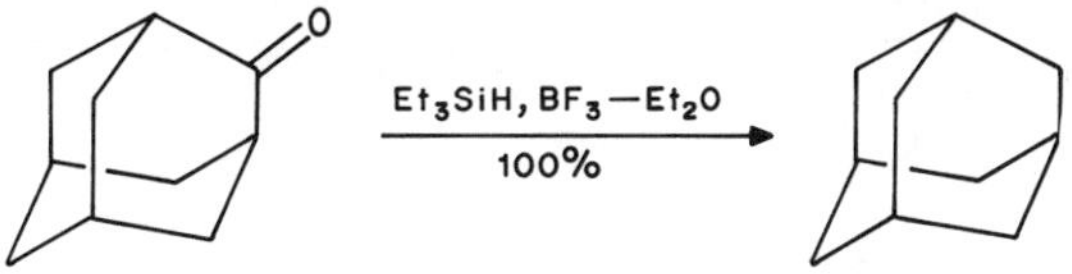

$$\xrightarrow[100\%]{\text{Et}_3\text{SiH, BF}_3\text{—Et}_2\text{O}}$$

(58)

$$\xrightarrow[\sim 80\%]{\text{B}_2\text{H}_6\text{—BF}_3}$$

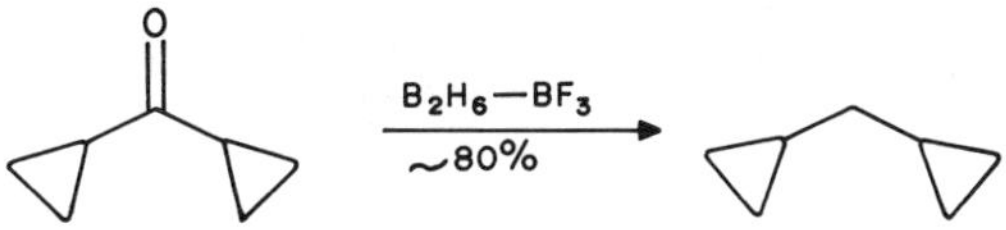

(59)

Clemmensen reduction involving reaction of ketones with amalgamated zinc in the presence of hydrochloric acid is one of the classical methods of deoxygenation[110]. It works well with aromatic ketones but its applicability to alicyclic systems is severely limited. The strongly acidic conditions and high temperature make this procedure too stringent, especially if other functional groups are present. Quite often side reactions are observed and complex mixtures of products are obtained. Nonetheless, the method has found favour in many cases (equation 60)[111].

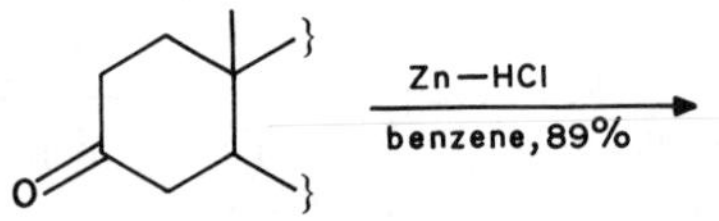

$$(60)$$

3. Reduction of hydrazones (Wolff–Kishner and Caglioti reductions)

A method of almost universal applicability for the deoxygenation of carbonyl compounds is the Wolff–Kishner reduction[11i,112]. While the earlier reductions were carried out in two steps on the derived hydrazone or semicarbazone derivatives, the Huang–Minlon modification is a single-pot operation. In this procedure, the carbonyl compound and hydrazine (hydrate or anhydrous) are heated (180–220 °C) in the presence of a base and a proton source. Sodium or potassium hydroxide, potassium-t-butoxide and other alkoxides are the frequently used bases and ethylene glycol or its oligomers are used as the solvent and proton source. Over the years, several modifications of this procedure have been used to cater to the specific needs of a given substrate. The Wolff–Kishner reaction works well with both aldehydes and ketones and remains the most routinely used procedure for the preparation of alkanes from carbonyl compounds (Table 9)[113]. This method is equally suitable for the synthesis of polycyclic and hindered alkanes.

TABLE 9. Alkanes via Wolff–Kishner reduction

Starting ketone	Product	Reference
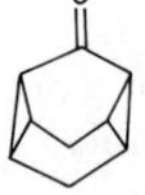	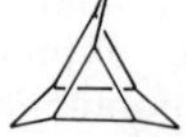	113a
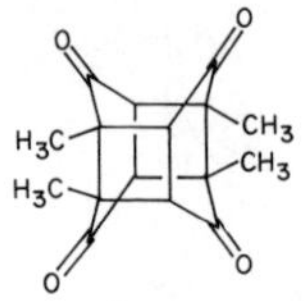	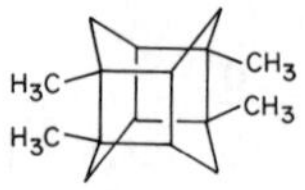	113b,c
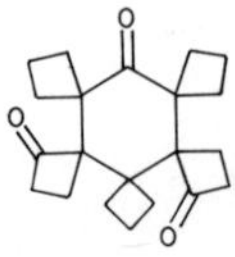	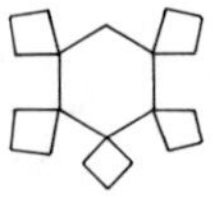	113d
		113e

(*continued*)

TABLE 9. (*continued*)

Starting ketone	Product	Reference
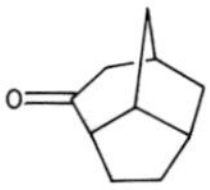	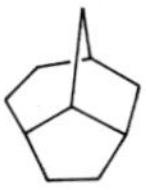	113f
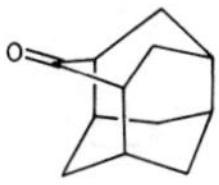	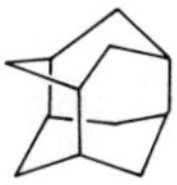	113g
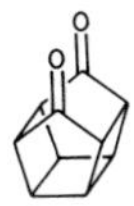	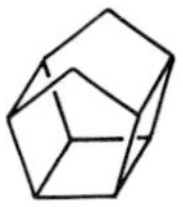	113h
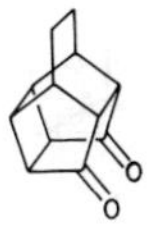		113i
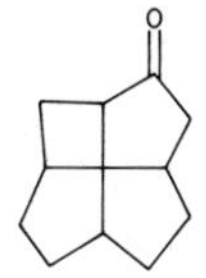	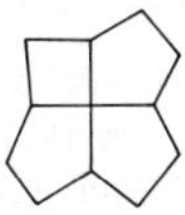	113j
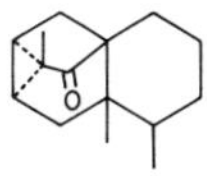	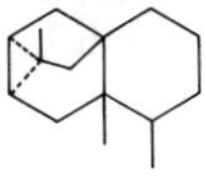	113k
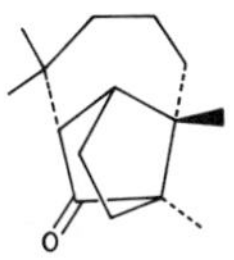	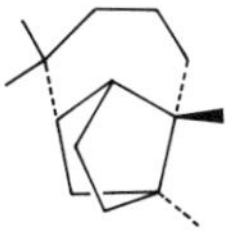	113l

A very useful variant of the hydrazone reduction is the deoxygenation of aldehydes and ketones via the hydride reduction of tosylhydrazones (Caglioti reaction)[114]. The method is mild, convenient and widely applicable. While sodium borohydride was used in the earlier procedures, considerable improvements have been achieved through the uses of sodium cyanoborohydride[115], catecholborane[116], diborane[117], *bis*-benzoyloxy borane[118] and copper borohydride[119] as reducing agents and HMPA, DMF, sulpholane, etc. as solvents. Use of the sterically crowded 2,4,6-triisopropyl tosylhydrazone derivative has greatly facilitated the reduction in some cases (equations 61–64)[119].

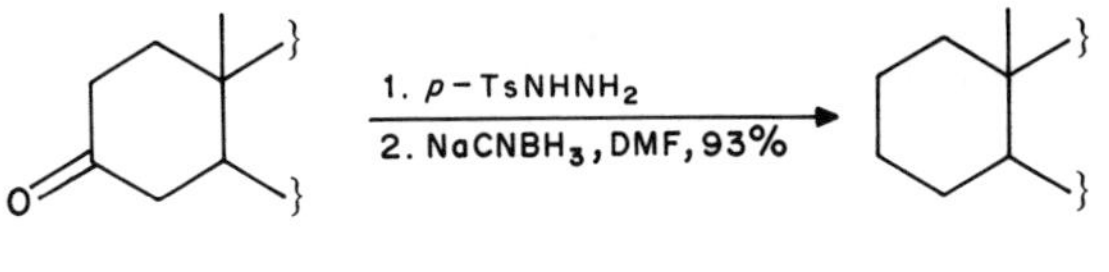

$$\text{1. } p-\text{TsNHNH}_2 \qquad \text{2. NaCNBH}_3, \text{DMF}, 93\% \qquad (61)$$

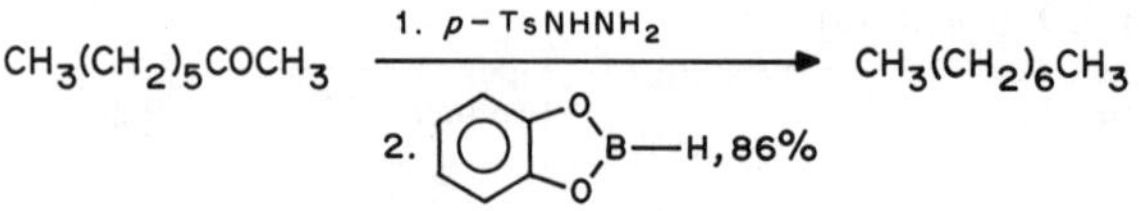

$$CH_3(CH_2)_5COCH_3 \xrightarrow[\text{2.}]{\text{1. } p-\text{TsNHNH}_2} CH_3(CH_2)_6CH_3 \quad (62)$$

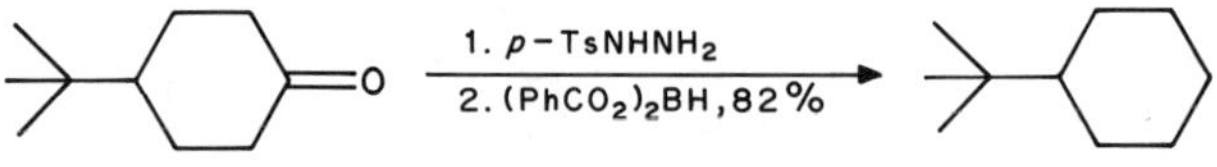

2. (benzodioxaborole) B—H, 86%

$$\xrightarrow[\text{2. } (PhCO_2)_2BH, 82\%]{\text{1. } p-\text{TsNHNH}_2} \quad (63)$$

$$\xrightarrow[\text{2. } CuBH_4(PPh_3)_2, 70\%]{\text{1. } 2,4,6-(i-Pr)_3C_6H_2SO_2NHNH_2} \quad (64)$$

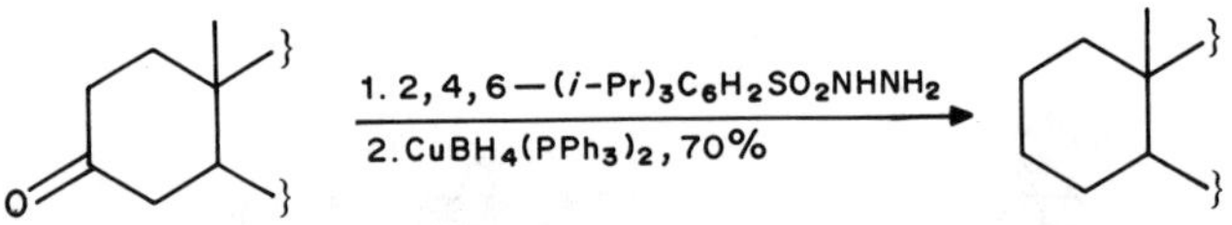

4. Thioacetalization–desulphurization (Mozingo reaction)

A popular two-step protocol for the deoxygenation of carbonyl compounds is the Mozingo reaction in which the aldehyde or ketone is transformed to a dithioacetal in the presence of a Lewis acid and then reductive desulphurization is carried out with Raney Ni[120]. The reaction is mild, efficient and particularly convenient for small-scale preparations. Ethanedithiol, propanedithiol and benzene-1,2-dithiol are employed for the preparation of dithioacetals. Some of the polycycloalkanes made by this procedure are shown in Table 10[121].

TABLE 10. Alkanes through dithioacetal desulphurization

Starting ketone	Product	Reference
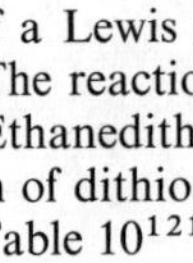	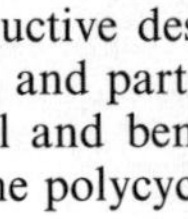	121a
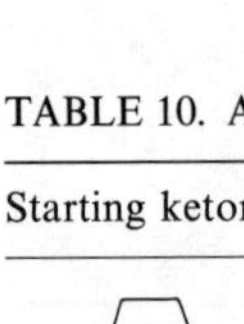	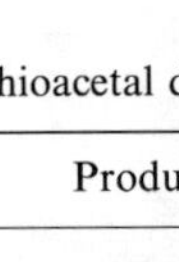	121b
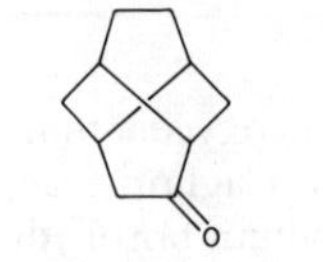	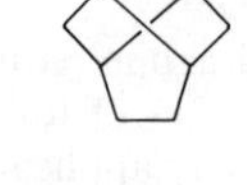	121c

Dithioacetals can also be desulphurized under radical conditions using tributyltin hydride (TBTH) (equation 65)[122] or by using metal–ammonia solutions (equation 66)[123]. Lithium aluminium hydride has also been used in some cases for reductive desulphurization. In a manner analogous to the preparation of dithioacetals, ketones can be transformed to diselenoacetals with aryl or alkyl selenol. These in turn have been reduced with Raney Ni[124] or Li–$EtNH_2$[124] and under radical conditions with TBTH or triphenyltin hydride to furnish the corresponding alkanes (equation 67)[78].

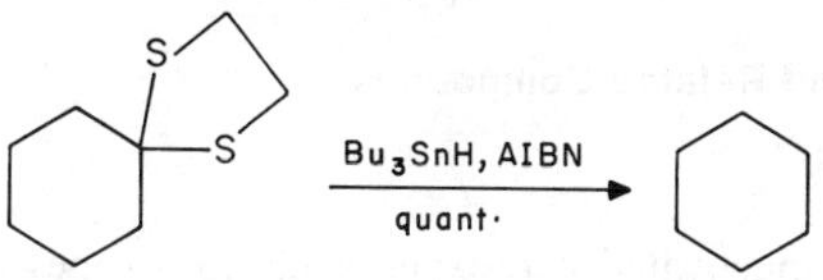

S
S
Bu_3SnH, AIBN
quant.
(65)

SH
1.
SH, PTS
$H_3C(CH_2)_7COCH_3$
2. Na—NH_3, quant.
$CH_3(CH_2)_8CH_3$
(66)

SePh
SePh
Ph_3SnH, AIBN
91%
(67)

5. Reductive alkylation

Reductive alkylation of aldehydes and ketones essentially involves replacement of oxygen with two alkyl groups to furnish alkanes. The procedure can be employed advantageously for the conversion of ketones into *gem*-dimethylated alkanes, a structural feature present in many natural products. Thus, several cyclic ketones have been transformed into *gem*-dimethylated alkanes, employing mild conditions, with dimethyltitanium dichloride (equation 68)[125]. *gem*-Dimethylation of ketones has also been achieved by heating them with $(CH_3)_3Al$ in the presence of $AlCl_3$ (equations 69 and 70)[126].

O
Me_2TiCl_2
73%
CH_3
CH_3
(68)

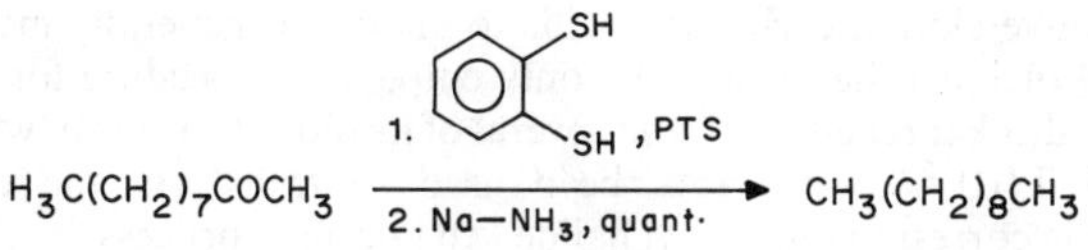

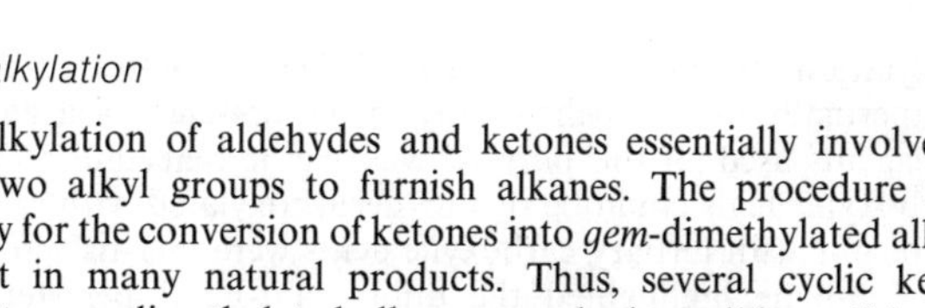

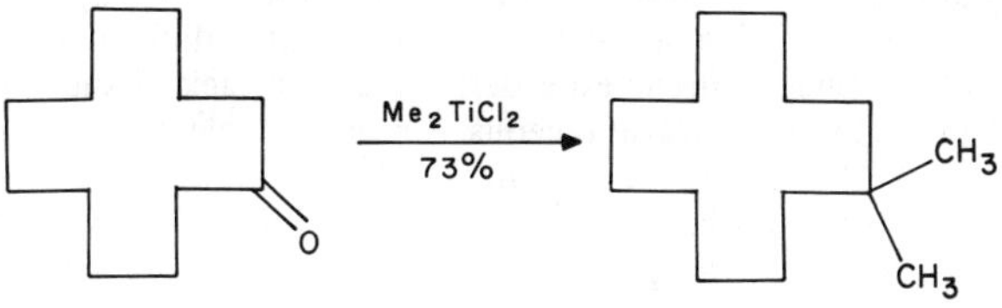

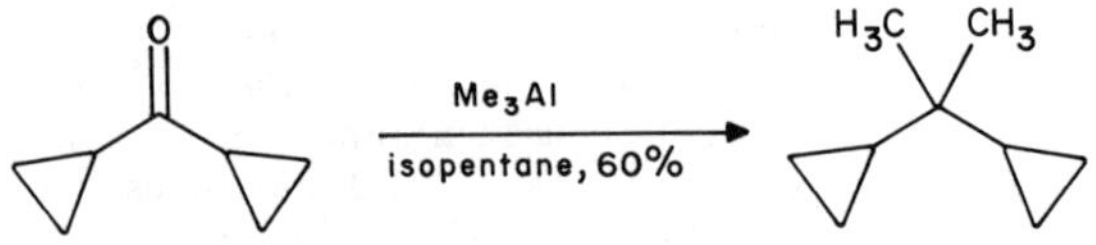

$$\text{(69)}$$

$$CH_3(CH_2)_5CHO \xrightarrow[\text{major product}]{Me_3Al} CH_3(CH_2)_5CHMe_2 \qquad (70)$$

E. Carboxylic Acids and Related Compounds

1. Indirect method

Decarboxylation of unactivated carboxylic acids to alkanes is not always easy to accomplish and therefore indirect methods are more often employed for this purpose[127]. When direct transformation to alkanes is to be carried out, suitable activation through derivatization becomes a pre-requisite in most cases.

The indirect conversion of carboxylic acids to alkanes is generally mediated through alkyl halides and olefins. The most commonly employed procedure for this purpose is the classical Hunsdiecker reaction[128] and several of its modifications in which Hg(II)[129a], lead(IV)[129b] and Tl(I)[129c] salts have been used to transform the carboxylic acid functionality to the corresponding alkyl halide. An alternate process for the Hunsdiecker reaction involves decarboxylative halogenation of N-hydroxypyridine-3-thione esters in the presence of trichloromethyl radicals[130]. The procedure can also be carried out in a single-pot operation, without the isolation of the ester. Iodoform or iodine can be used in place of the trichloromethyl radical for obtaining iodoalkanes. More recently, decarboxylative iodinations have been carried out employing t-butyl hypoiodide[131] and hypervalent iodine[132] based reagents. The most celebrated example in this context is the transformation of cubane carboxylic acids to iodocubanes[131,132]. Indirect conversion of carboxylic acids to alkanes with the same number of carbon atoms can be achieved through 1,3-benzodithiolium tetrafluoroborate salts[123a]. While these indirect methods are extremely important, they will not be dealt with here.

2. Thermal decarboxylation

Direct decarboxylations to alkanes are accomplished on activated carboxylic acid derivatives either thermally or photochemically in the presence of a hydrogen donor. One of the early reagents used for this purpose was lead tetraacetate[127]. While primary and secondary carboxylic acids could be readily decarboxylated with moderate success with this reagent, results with tertiary carboxylic acids were unsatisfactory.

The synthesis of cubane highlighted the utility of t-butyl peresters for effecting decarboxylations, particularly of bridgehead and tertiary carboxylic acids[133]. This method has been successfully used in many polycycloalkane syntheses (Table 11)[133b,134].

Several radical intermediate-based methodologies have been introduced for decarboxylations as an extension of earlier efforts on the deoxygenation of alcohols. One of the early approaches involved the TBTH reduction of the dihydrophenanthrene derivative of carboxylic acids (equation 71)[135]. However, the difficulties associated with the preparation of the phenanthrene ester derivatives and lack of success with tertiary carboxylic acids proved to be disadvantageous. A more notable discovery in this context is the demonstration that esters derived from COOH groups attached to primary, secondary and tertiary carbons and N-hydroxypyridine-2-thione undergo efficient reductive decarboxylation in the presence of TBTH in refluxing benzene or toluene

TABLE 11. Alkanes through decomposition of peresters

Starting *t*-butyl perester	Product	Reference
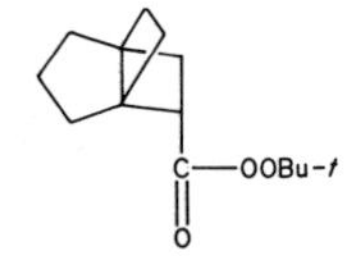	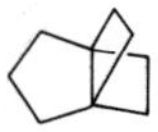	134a
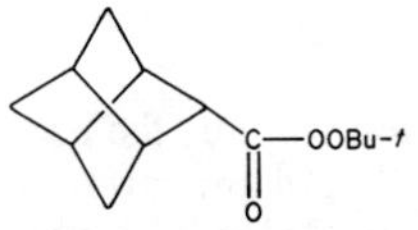	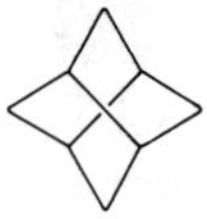	134b
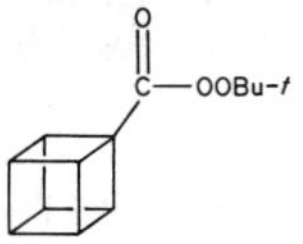	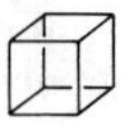	133b
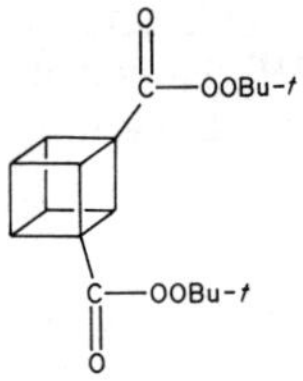	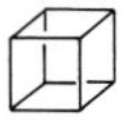	134c
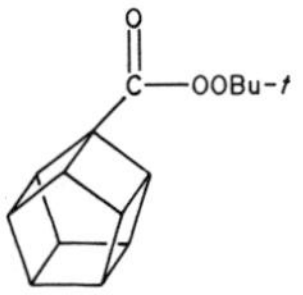		134d

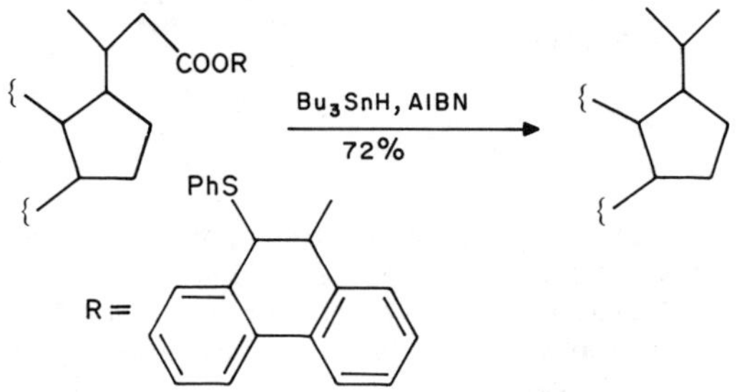

(71)

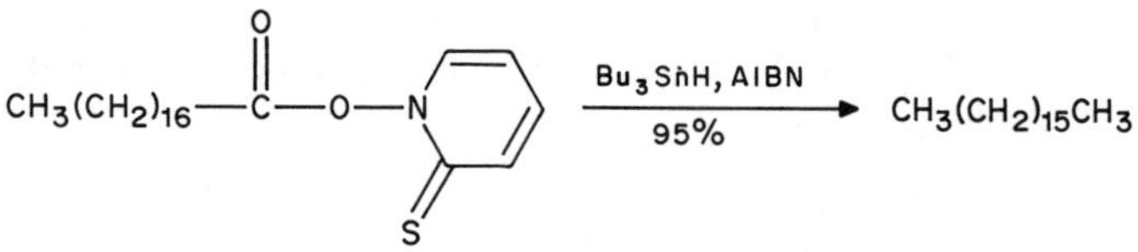

(72)

580   G. Mehta and H. S. P. Rao

(equation 72)[130b,136]. This methodology is convenient and gaining wide acceptance. Reduction of acid chlorides with tripropyltin hydride in the presence of the radical initiator di-t-butyl peroxide has also been shown to furnish alkanes in moderate yield (equation 73)[137]. Carboxylic acids can also be decarboxylated via carboselenoic acid esters with TBTH in the presence of of AIBN [138].

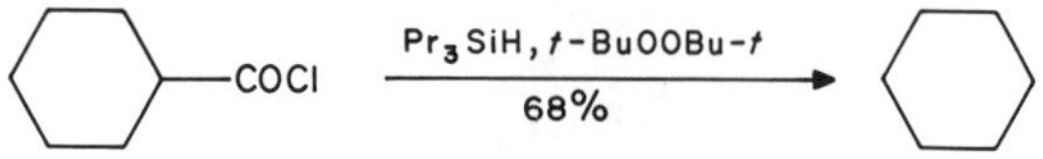

$$\langle\rangle\!-\!COCl \xrightarrow[68\%]{Pr_3SiH,\,t-BuOOBu-t} \langle\rangle \qquad (73)$$

An alternative procedure for reductive decarboxylation without the use of trialkyltin hydrides as hydrogen atom donors has been developed[136a]. Alkane carboxylic acid esters derived from N-hydroxypyridine-2-thione decomposed to alkyl radical, which can readily accept a hydrogen atom from t-BuSH (equation 74) to give alkanes. This reaction can be conveniently performed as a one-pot experiment wherein the acid chloride of an alkane carboxylic acid, the sodium salt of thiohydroxamic acid, t-BuSH and 4-dimethyl-aminopyridine (DMAP) in benzene solution are heated to reflux. This procedure works well for COOH groups attached to primary and secondary carbon atoms. Instead of N-hydroxypyridine-2-thione, one can use other thiohydroxamic acids, viz. N-hydroxy-N-methylthiobenzamide[139], 3-hydroxy-4-methylthiazole-2(3H)-thione (equation 75)[139,140] and 1-N-hydroxy-3-N-methylbenzoylenethiourea[139] for decarboxylation reactions.

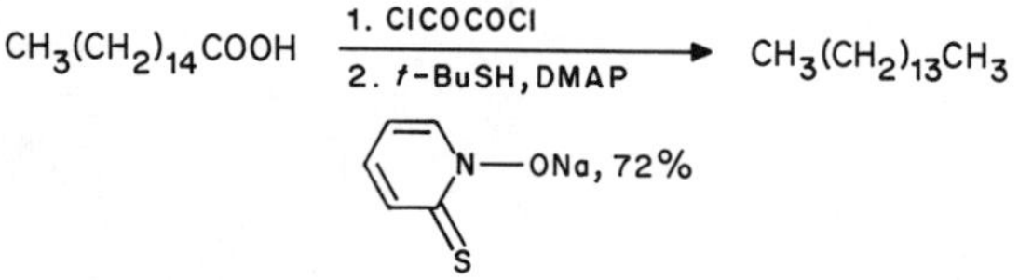

$$CH_3(CH_2)_{14}COOH \xrightarrow[\substack{2.\ t\text{-BuSH, DMAP}}]{1.\ ClCOCOCl} CH_3(CH_2)_{13}CH_3 \qquad (74)$$

N—ONa, 72%

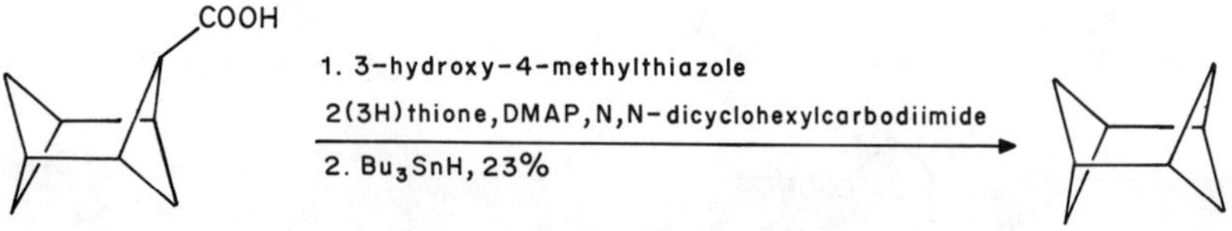

$$(75)$$

1. 3-hydroxy-4-methylthiazole
2(3H)thione, DMAP, N,N-dicyclohexylcarbodiimide
2. Bu$_3$SnH, 23%

3. Photodecarboxylation

Aliphatic carboxylic acids can be photodecarboxylated to alkanes in the presence of heteroaromatic compounds like acridine, phenanthridine, phenazine, quinoline etc. and t-BuSH (equation 76)[141]. The latter serves as the hydrogen donor. The reaction proceeds well with 1°-, 2°- and 3°-carboxylic acids. Photodecarboxylation can also be achieved via N-acyloxyphthalimides in the presence of 1,6-bis-(dimethylamino)pyrene (BDMAP) and t-BuSH in good yield (equation 77)[142]. This method is also suited for 1°-, 2°- and 3°-carboxylic acids. In a related procedure, benzophenone oxime esters of carboxylic acids have also been shown to undergo photochemical decarboxylation in the presence of t-BuSH (equation 78)[143].

$$CH_3(CH_2)_{14}COOH \xrightarrow[t\text{-BuSH, 72\%}]{hv,\,\text{acridine}} CH_3(CH_2)_{13}CH_3 \qquad (76)$$

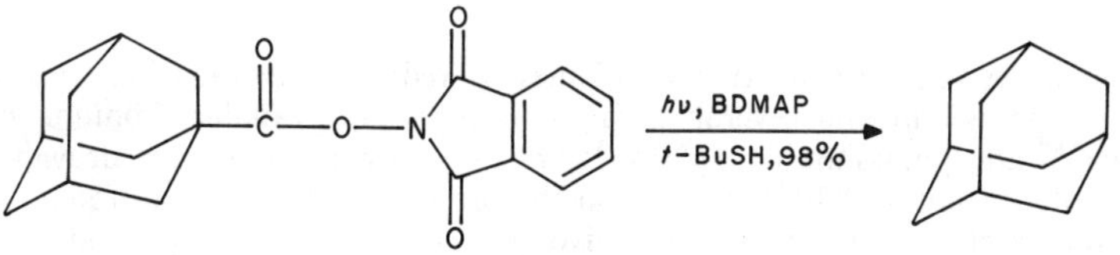

$$(77)$$

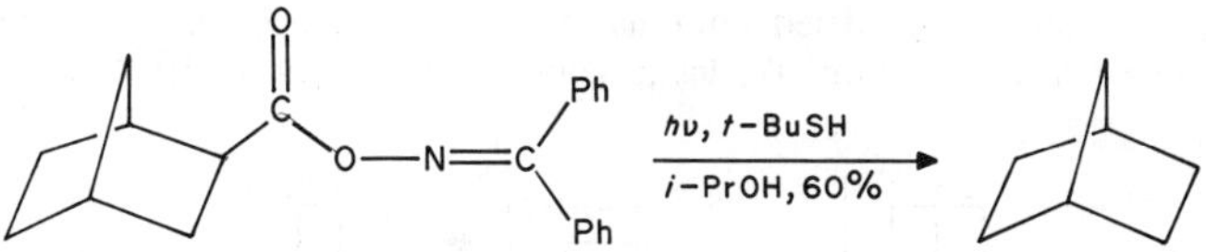

$$(78)$$

F. Miscellaneous

1. Amines and isocyanides

Direct replacement of the amino group of primary amines by hydrogen is seldom attempted, primarily because NH_2 is a poor leaving group, and consequently substitution reactions are difficult to accomplish[144]. However, reductive deamination to alkanes can be achieved indirectly through appropriate derivatization. N,N-Bis-tosylamides derived from primary amines undergo substitution reaction on carbon with sodium borohydride in HMPA to yield alkanes (equation 79)[145]. Bis-tosylation is a requirement, since with monotosyl compounds replacement of the tosyl group on nitrogen is a competing reaction. Sodium cyanoborohydride and, surprisingly, lithium triethylborohydride were found to be poor hydride donors in this reaction. Reductive deamination of primary amines to alkanes via sequentially formed hydrazo and azo compounds has also been reported, but yields in this reaction are only moderate[146].

$$CH_3(CH_2)_9 - N \underset{Ts}{\overset{Ts}{\diagdown}} \xrightarrow[91\%]{NaBH_4, HMPA} CH_3(CH_2)_8CH_3 \qquad (79)$$

Reductive deamination can be performed in reasonable yields via phenylimidoyl chlorides prepared from benzoyl derivatives of primary amines. Imidoyl chloride derivatives decompose on heating with TBTH and AIBN in xylene to yield alkanes (equation 80)[147]. Primary amines can also be reductively deaminated in high yields via isocyanide derivatives. Formylation–dehydration transforms an amino group into an isocyanide quite routinely. Reductive removal of isocyanide can be performed with sodium naphthalenide in hydrocarbon solvents[148] or with TBTH in the presence of AIBN[149]. The TBTH reduction works equally well for primary, secondary and tertiary isocyanides and yields are good (equation 81)[149].

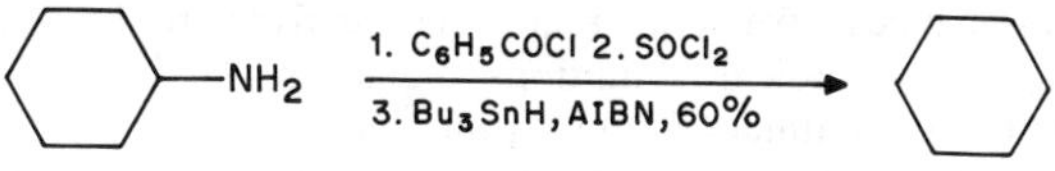

$$(80)$$

$$CH_3(CH_2)_{16} - \underset{NC}{\overset{}{\underset{|}{CH}}} - CH_3 \xrightarrow[81\%]{Bu_3SnH, AIBN} CH_3(CH_2)_{17}CH_3 \qquad (81)$$

2. Nitriles

Alkanenitriles can be converted to alkanes by reductive decyanation with metallic sodium or potassium and several useful procedures are available. Sodium in liquid ammonia[150] and potassium in HMPA[151] are efficient reagents for the decyanation of alkanenitriles (equation 82)[152]. Sodium in the presence of tris(acetylacetonato)iron III is a good combination for the reductive decyanation of primary and secondary alkanenitriles (equation 83)[153]. Highly dispersed potassium over alumina, easily prepared by melting potassium over dried alumina, converts primary, secondary and tertiary alkanenitriles to corresponding alkanes in good yield (equation 84)[154]. Recently, it has

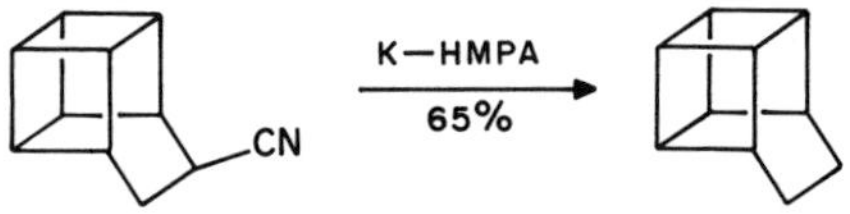

$$ \text{(82)} $$

$$ \text{CH}_3(\text{CH}_2)_6\text{CH}_2\text{CN} \xrightarrow[100\%]{\text{Na–Fe(acac)}_3} \text{CH}_3(\text{CH}_2)_6\text{CH}_3 \qquad (83) $$

$$ \underset{\underset{\text{CN}}{|}}{\text{CH}_3(\text{CH}_2)_7\text{CHCH}_2\text{CH}_3} \xrightarrow[85\%]{\text{K–Al}_2\text{O}_3} \text{CH}_3(\text{CH}_2)_9\text{CH}_3 \qquad (84) $$

been reported that toluene radical anion generated from potassium/dicyclohexyl-18-crown-6/toluene system is highly efficient for the reductive decyanation of primary, secondary and tertiary nitriles under mild conditions (equation 85)[155]. Tertiary alkanenitriles can be converted to higher alkanes by exhaustive catalytic hydrogenation on Ni–Al$_2$O$_3$ (equation 86)[156].

$$ \text{CH}_3(\text{CH}_2)_{16}\text{CH}_2\text{CN} \xrightarrow[90\%]{\text{K–dicyclohexyl-18-crown-6}} \text{CH}_3(\text{CH}_2)_{16}\text{CH}_3 \qquad (85) $$

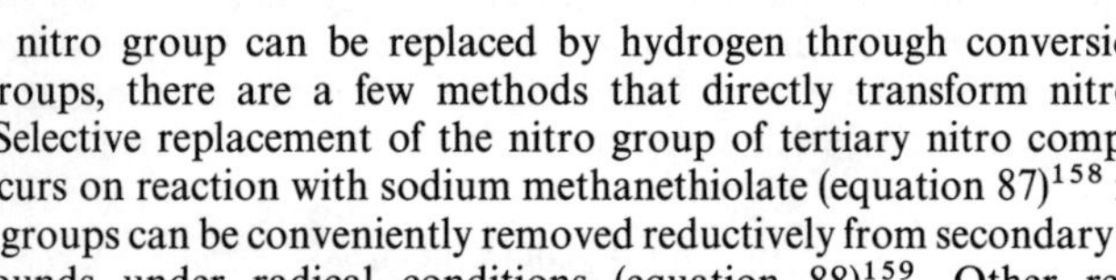

$$ \text{(86)} $$

3. Nitro compounds

While the nitro group can be replaced by hydrogen through conversion to other functional groups, there are a few methods that directly transform nitroalkanes to alkanes[157]. Selective replacement of the nitro group of tertiary nitro compounds with hydrogen occurs on reaction with sodium methanethiolate (equation 87)[158] in moderate yields. Nitro groups can be conveniently removed reductively from secondary and tertiary nitro compounds under radical conditions (equation 88)[159]. Other reagents, viz. 1-benzyl-1,4-dihydronicotinamide[160] and sodium hydrogen telluride[161], have also been reported for reduction of nitroalkanes but they are of limited synthetic utility.

$$ \underset{\underset{\text{NO}_2}{|}}{\text{Me}_3\text{CCH}_2\text{CMe}_2} \xrightarrow[55\%]{\text{CH}_3\text{SNa, HMPA}} \text{Me}_3\text{CCH}_2\text{CHMe}_2 \qquad (87) $$

$$\underset{\underset{NO_2}{|}}{Me_3CCH_2-CMe_2} \xrightarrow[75\%]{Bu_3SnH,\,AIBN} Me_3CCH_2CHMe_2 \qquad (88)$$

III. COUPLING REACTIONS

A. Alkali Metals (Wurtz Coupling)

One of the oldest carbon–carbon bond forming reactions is the coupling of organic halides with sodium to give symmetrical dimeric alkane products (Wurtz reaction). An illustrative example is the preparation of hectane from iodopentacontane (equation 89)[162]. The reaction as such is of very limited synthetic value as yields are poor and side reactions like dehalogenation and dehydrohalogenation intervene and complex mixtures of products are obtained. Nonetheless, some interesting alkanes have been synthesized through Wurtz coupling and a noteworthy example is the preparation of a highly crowded hydrocarbon 2,3-di(1-adamantyl)-2,3-dimethylbutane (equation 90)[163]. Use of pyrophoric lead has been described for Wurtz coupling and apparently better yields are obtained (equation 91)[164].

$$n\text{-}C_{50}H_{101}-I \xrightarrow[61\%]{Na} n\text{-}C_{100}H_{202} \qquad (89)$$

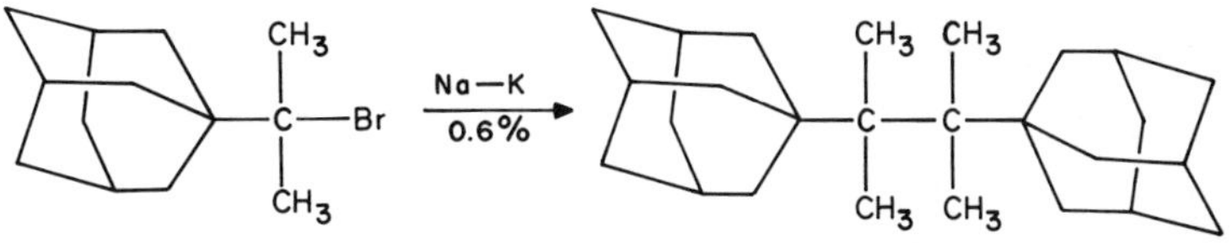

$$(90)$$

$$n\text{-}C_4H_9Br \xrightarrow[44\%]{Pb} n\text{-}C_8H_{18} \qquad (91)$$

From the synthetic point of view, intramolecular Wurtz-like coupling in dihaloalkanes has proved to be a very valuable reaction. Employing lithium, sodium or potassium or their mixed alloy, sodium naphthalenide, etc., a variety of novel, strained polycyclic alkanes of intrinsic interest have been assembled (Table 12)[165]. The use of ultrasound has been effective in some of these heterogeneous reactions. Wurtz-type coupling can also be carried out employing magnesium and alkyllithium reagents. 1,5-Diiodopentane is cyclized by t-butyllithium with great facility to the corresponding cyclopentane compound (equation 92)[160]. The reaction has considerable potential. Two pleasing illustrations are from the recently unravelled cubyl-cage chemistry (equations 93 and 94)[166,167]. These reactions have been considered to be proceeding through the dehalogenated dehydro-intermediates but result in novel coupling products.

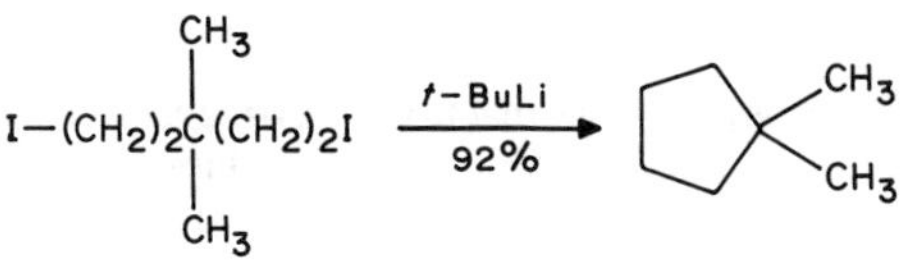

$$\underset{\underset{CH_3}{|}}{I-(CH_2)_2\overset{\overset{CH_3}{|}}{C}(CH_2)_2I} \xrightarrow[92\%]{t-BuLi} \qquad (92)$$

TABLE 12. Alkanes from intramolecular Wurtz-like coupling

Starting halide	Product	Reference
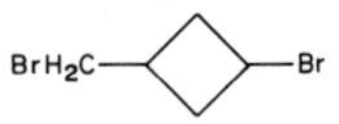		165a
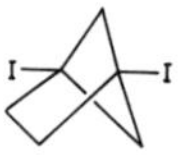		165b
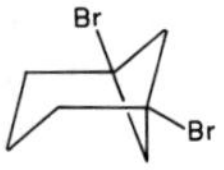		165c
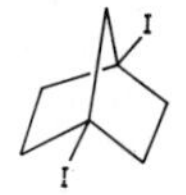	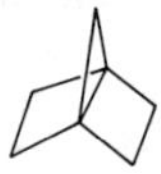	165d
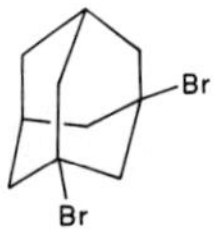	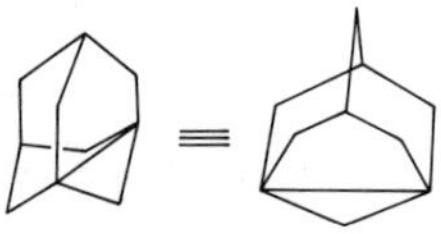	165e
		165f
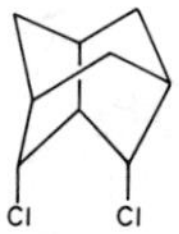	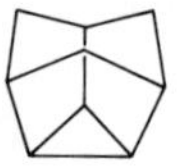	165g

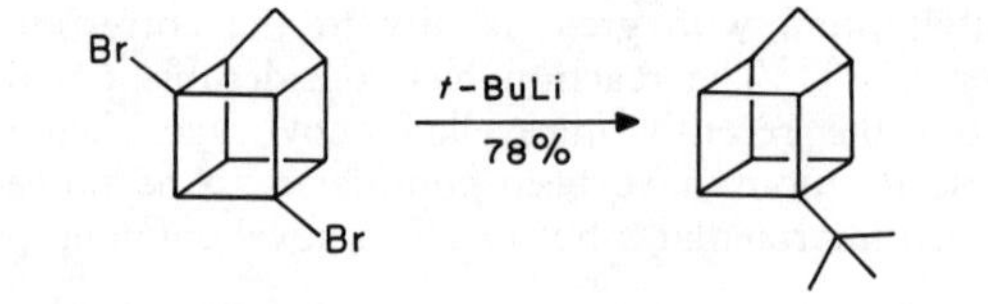

$$(93)$$

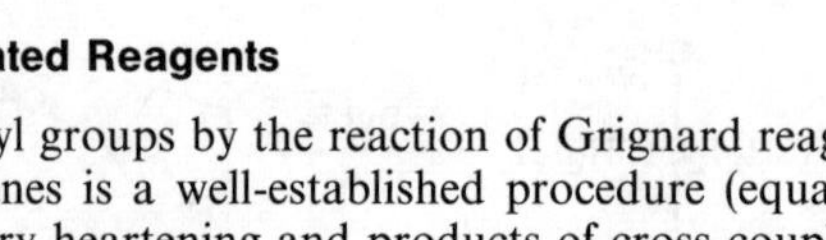

$$(94)$$

B. Grignard and Related Reagents

The coupling of alkyl groups by the reaction of Grignard reagents with alkyl halides to furnish higher alkanes is a well-established procedure (equation 95). The yields in many cases are not very heartening and products of cross coupling are also frequently encountered. However, an exception is the bridgehead coupling in an adamantane derivative (equation 96)[168]. Attempts to improve this Grignard reagent procedure has involved the use of Cu(I), Ag(I), Co(I) and Tl(I)[169]. While the yields are not too bad, formation of alkane mixtures persists as shown in equation 97[169a]. Coupling of alkylmagnesium bromides with α, ω-dibromoalkanes has been shown to furnish bis-coupled alkanes (equation 98)[170].

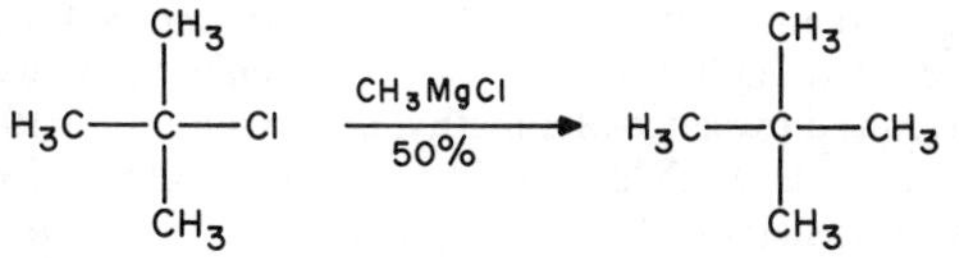

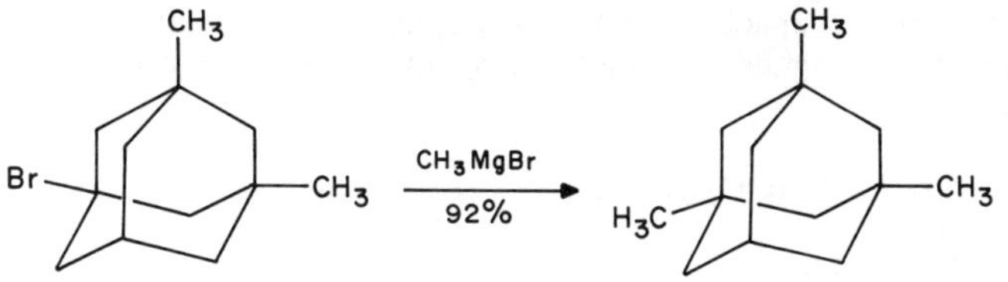

$$(CH_3)_2CHMgBr + Br(CH_2)_6Br \xrightarrow[\text{HMPA, 16\%}]{Cu(I)Br} (CH_3)_2CH(CH_2)_6CH(CH_3)_2 \qquad (98)$$

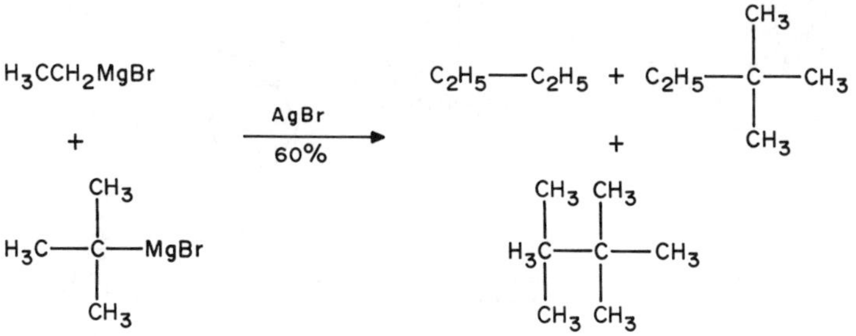

An extremely interesting example of Grignard reagent mediated alkane synthesis is the coupling reaction of 1-adamantyl-1,1-dibromo-2,2-dimethylpropane with magnesium (equation 99)[171]. The highly hindered alkane product is novel in the sense that interconversion of rotamers across the ethane moiety does not take place under ambient conditions.

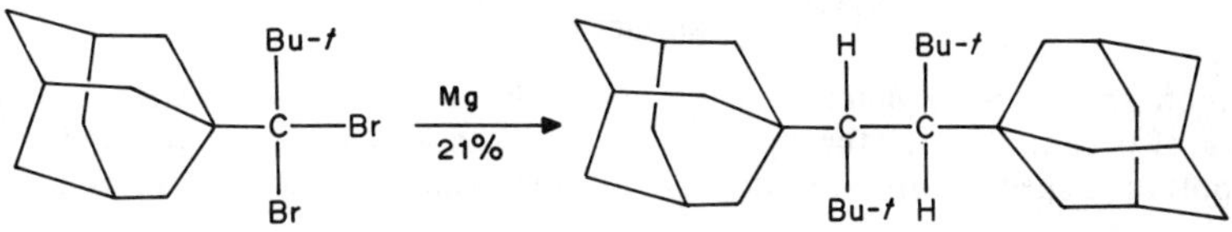

C. Transition Metal Reagents

The advent of versatile organocuprate reagents in organic synthesis has led to their increased use as effective reagents for the unsymmetrical coupling of alkyl halides[172]. It is now fairly well established that lithium dimethyl cuprate is an excellent reagent for the replacement of bromine or iodine by a methyl group in alkyl halides (equation 100)[172]. The reaction is general enough to be carried out with other LiR_2Cu reagents in which R group transfer takes place. Several reagent systems based on higher-order cuprates like Me_3CuLi_2, $R_2Cu(SCN)Li_2$, $R_2Cu(CN)Li_2$ etc. have been reported in recent years[172b]. Employing $R_2Cu(CN)Li_2$ reagent, several long-chain and branched-chain alkanes have been prepared in high yield through mixed coupling reactions. An example of butyl group transfer is shown in equation 101[173]. This reagent can be further manipulated to $RR'Cu(CN)Li_2$ and selective ligand transfer has been achieved. Thus, a butyl group can be selectively coupled in the presence of a methyl group (equation 102)[174]. Additionally, by using higher-order cuprates with 'dummy' ligands like 2-lithiothiophene, it is possible to economize on the transferring R group and also achieve selectivity as shown in equation 103[175]. Alkyl cuprates can also alkylate organomercurials, thus offering a wide variety of cross-coupled products in reasonable yield (equation 104)[176]. This is a useful procedure as organomercurials are synthetically quite accessible from other functional groups.

$$CH_3(CH_2)_8CH_2I \xrightarrow[90\%]{LiMe_2Cu} CH_3(CH_2)_9CH_3 \qquad (100)$$

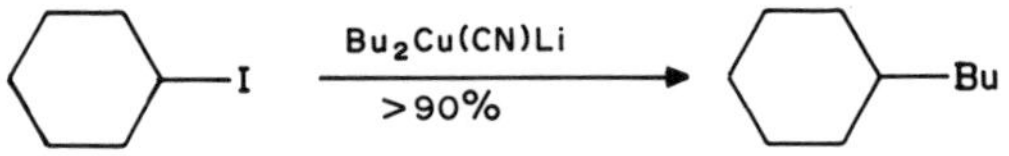

$$\text{(101)}$$

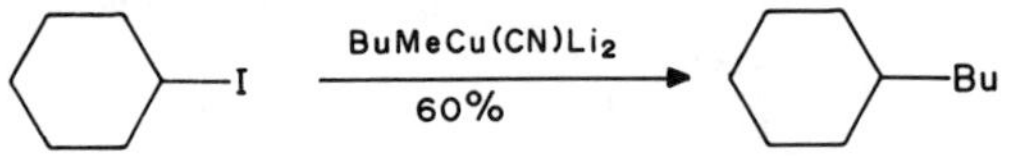

$$\text{(102)}$$

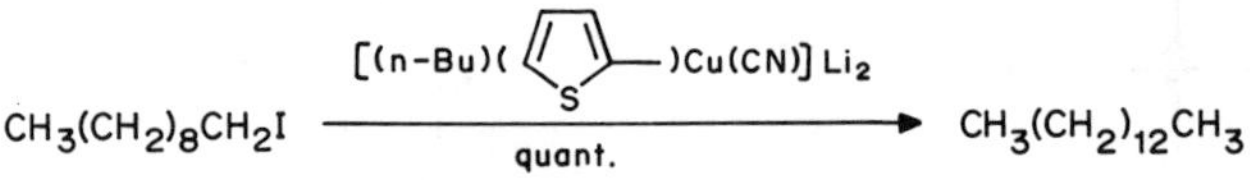

$$\text{(103)}$$

$$CH_3(CH_2)_4CH_2HgCl \xrightarrow[62\%]{Li_2Cu(CH_3)_3} CH_3(CH_2)_5CH_3 \qquad (104)$$

Palladium-based reagents have also been used in alkane synthesis through coupling reactions, though they are much more effective in the case of arenes and alkenes. An example is the preparation of long-chain alkanes, shown in equation 105[177]. Interestingly, the reaction works well for long-chain alkyl halides and the yields decrease dramatically with smaller alkanes (cf. 0.7% for iodohexane-to-dodecane conversion).

$$CH_3(CH_2)_{18}CH_2I \xrightarrow[NH_2NH_2, 74\%]{PdCl_2-NaOH} CH_3(CH_2)_{38}CH_3 \qquad (105)$$

Organotitanium and organozinc reagents have been remarkably efficient in inducing coupling of organic halides through alkyl group transfer (equations 106 and 107)[178]. An important aspect of this reaction is that dialkylated quaternary centres can be installed.

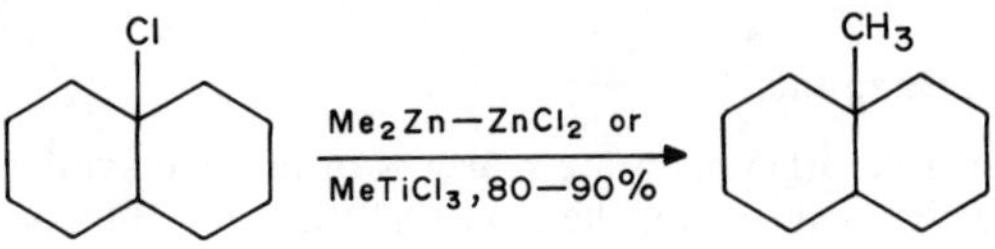

$$\text{(106)}$$

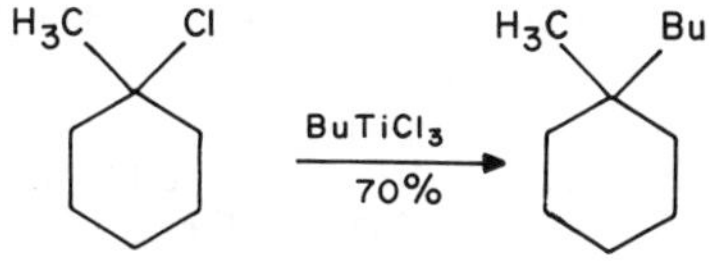

$$(107)$$

D. Organoboranes and Organoalanes

Alkane synthesis via coupling of organoboranes was discovered very early during the development of organoborane chemistry. The procedure is simple and involves the hydroboration of an alkene followed by treatment with basic silver nitrate (equations 108–110)[179]. Both acyclic and cyclic alkanes can be prepared through this procedure.

Hindered *t*-alkyl halides with lithium triethylborohydride in the presence of chromous chloride furnish highly crowded coupled products in good yield (equation 111)[180]. Alkanes obtained by hydroalumination of alkenes with LAH in the presence of titanium tetra-chloride can be dimerized in the presence of cupric acetate to furnish alkanes (equation 112)[181].

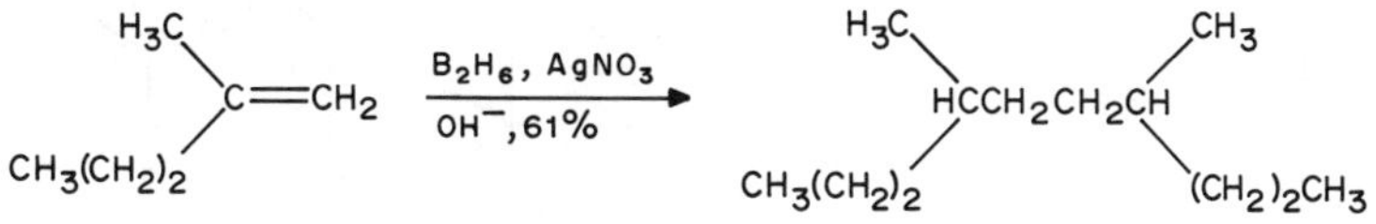

$$(108)$$

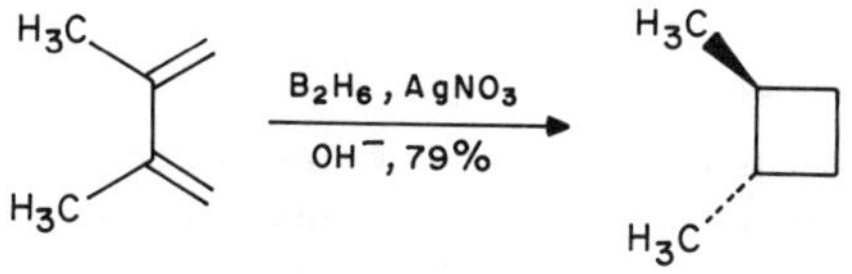

$$(109)$$

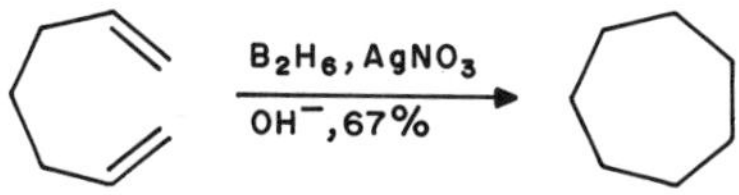

$$(110)$$

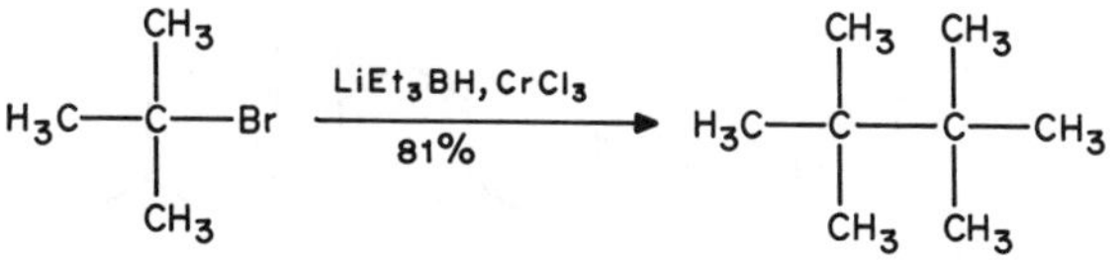

$$(111)$$

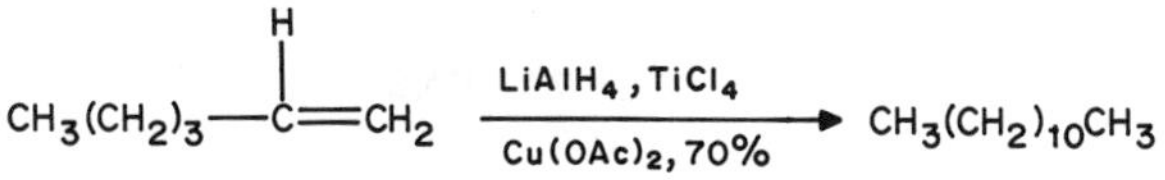

$$(112)$$

E. Kolbe Electrolysis

One of the classical reactions in organic chemistry for the synthesis of alkanes through symmetrical coupling of alkane carboxylic acids is the Kolbe reaction[182]. There are numerous examples of this reaction in the literature in which anodic oxidation of the COOH group furnishes alkyl radicals which readily dimerize. With the choice of proper reaction conditions, particularly current density, carboxylic acid concentration and the cross-section of platinum electrode, this procedure can be used for meaningful syntheses of many alkanes. Two recent examples of the synthesis of highly branched alkanes are shown in equations 113 and 114[183].

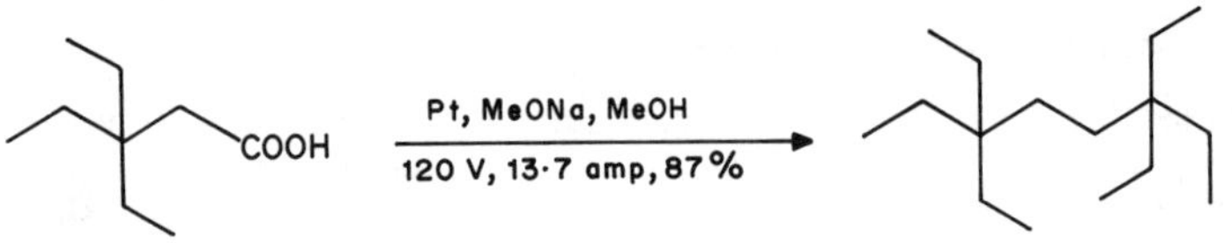

$$\text{(113)}$$

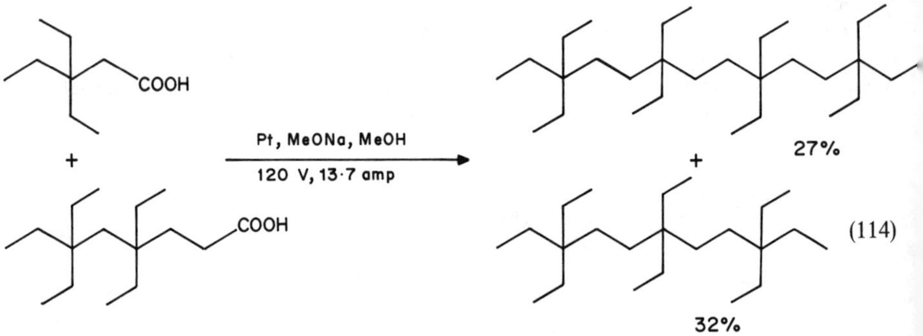

$$\text{(114)}$$

F. Photochemical Reaction

Mercury sensitized reactions of alkanes have been extensively studied in the context of C—H bond activation. It has been observed that many alkane dimerizations can be carried out photochemically on a preparative scale with high conversion and excellent yields (equations 115 and 116)[184,185].

$$\text{n-C}_{18}\text{H}_{38} \xrightarrow[\text{90—95\%}]{\text{Hg, }h\nu} \text{n-C}_{36}\text{H}_{74} \qquad \text{(115)}$$

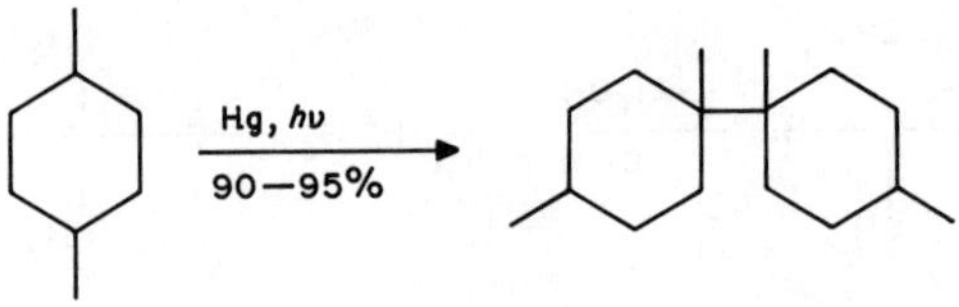

$$\text{(116)}$$

IV. CYCLIZATIONS

A. Radical Cyclization

The emergence of radical addition to olefins as a powerful C—C bond forming reaction offers many possible applications in alkane syntheses[186]. In its intramolecular mode, the reaction is extremely well suited for carboxylic ring construction with regio- and stereochemical control. A vast array of polycycloalkane skeletons present in natural products

TABLE 13. Alkanes through radical cyclizations

Radical precursor	Product	Reference
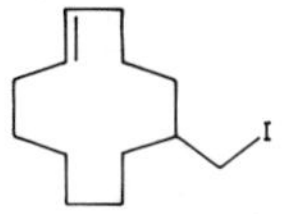	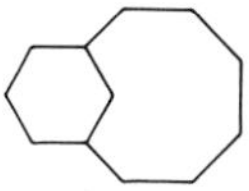	187a
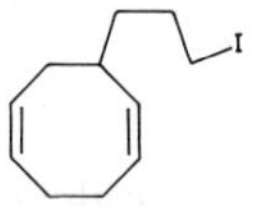	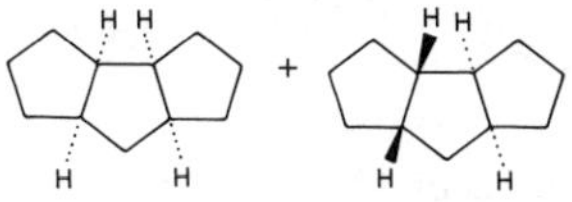	187b
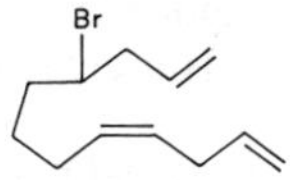	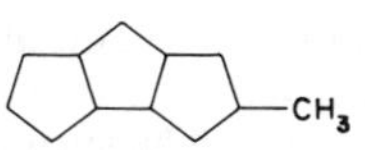	187c
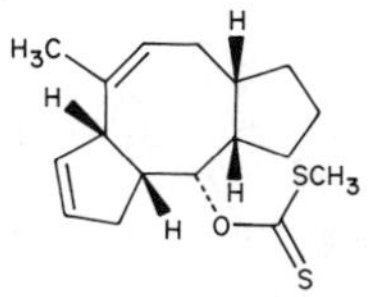	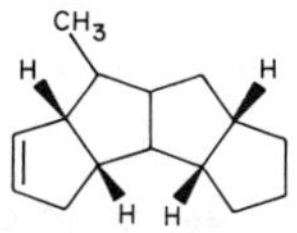	187d
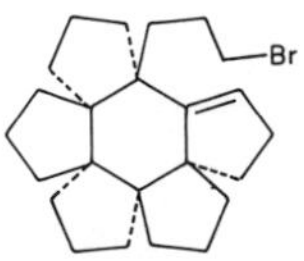	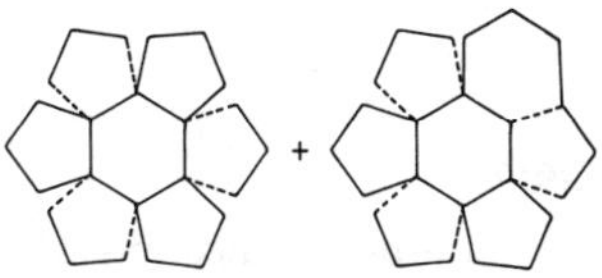	187e

have been assembled employing radical chemistry[186]. A few examples of carbocyclic ring formation, including those of tandem cyclization leading to multiple ring formation, are displayed in Table 13[187]. These radical cyclizations are carried out under mild conditions, employing a more or less standard protocol, with TBTH in the presence of AIBN under high dilution conditions. Formation of five-membered rings is generally favoured[186].

B. Organometallic Reagents

Cyclization reactions proceeding through intramolecular reaction of ene-organo-metallics are gaining ascendancy in modern organic synthesis. The method is well suited for alkane synthesis, in particular polycycloalkane syntheses. The cyclization of ene-organometallics predominantly leads to five-membered rings with good regio- and stereochemical control. The literature covering ene-organometallic cyclizations has been reviewed in 1985 and only more recent examples are presented here[188].

Alkenyl-lithiums, which can be prepared by metal–halogen interchange between an appropriate iodide and t-BuLi at $-78\,°C$, undergo cyclizations on warming to room temperature before quenching with a proton source (equation 117)[189]. This cyclization reaction has been extended for the construction of polycyclic alkanes by sequential polyolefinic cyclizations (tandem cyclizations). A propellane system could be constructed in a single step from bis-intramolecular cyclization as shown in equation 118[190].

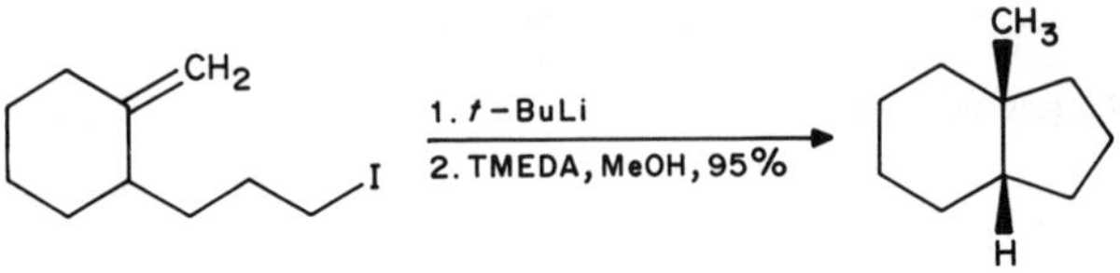

$$(117)$$

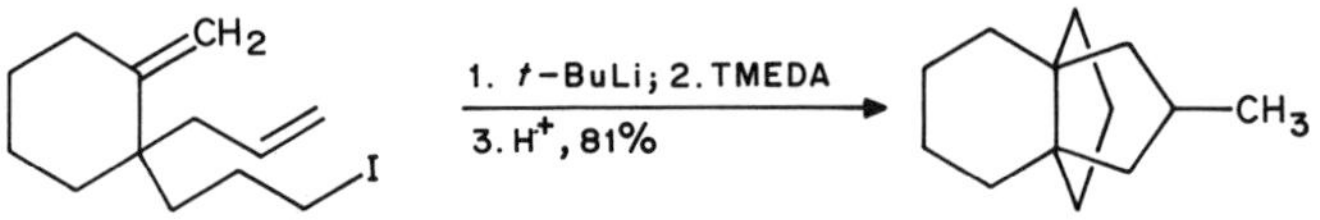

$$(118)$$

Reactions involving intramolecular insertion of alkene into early transition metal–carbon bonds, particularly to organo-Sc, organo-Zr and organo-Ti compounds, are also known.

It has been reported recently that organotitanium compounds prepared conveniently from the corresponding Grignard reagents and titanocene chloride cyclize with an appropriately positioned internal alkene in the presence of ethylaluminium chloride (equation 119)[191].

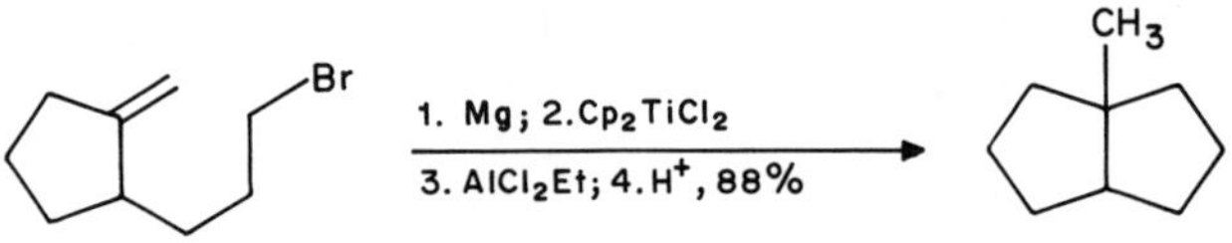

$$(119)$$

C. Solvolytic Rearrangement

Reductive solvolysis of unsaturated alcohols with double bond participation can be exploited for the synthesis of cycloalkanes. A pleasing example is the synthesis of the pentacyclic diterpene natural product trachylobane (equation 120)[192].

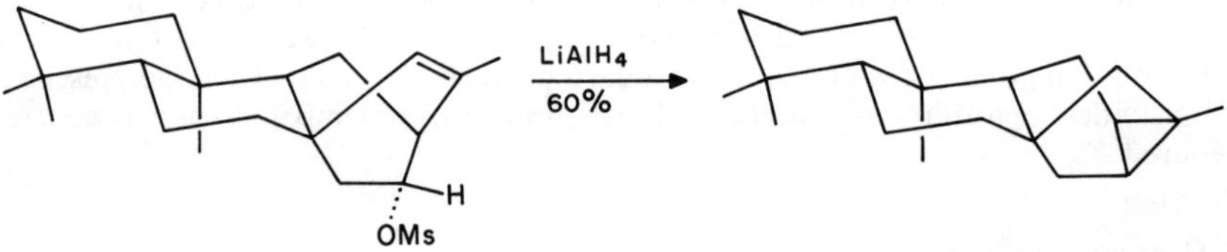

$$(120)$$

D. Photocyclization

One of the best known photochemical reactions of considerable preparative value is the [2 + 2]-photocycloaddition reaction and numerous examples are recorded in literature[193]. In its intramolecular version, the reaction has proved remarkably effective in assembling a wide array of complex polycycloalkanes. Both $\pi_s^2 + \pi_s^2$ and $\pi_s^2 + \sigma_s^2$ photocyclizations take place with considerable facility under appropriate sensitization. Conversion of norbornadiene to quadricyclane (equation 121) is one of the very well studied cases. Table 14 summarizes several other interesting examples. Intermolecular photocycloaddition reactions lead to dimeric polycycloalkanes as shown in equations 122 and 123[193-195].

TABLE 14. Alkanes through intramolecular photocycloadditions

Starting olefin	Product	Reference
		194a
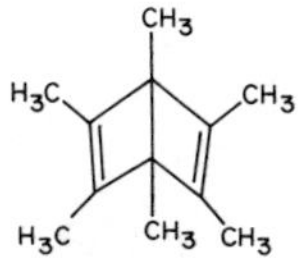	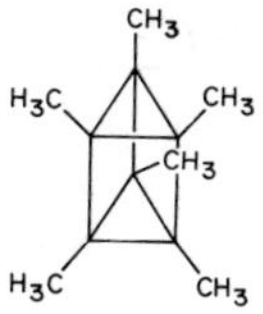	194b
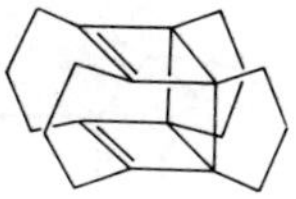	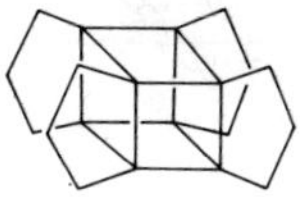	194c
	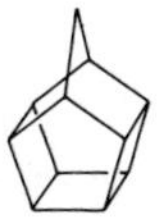	194d
	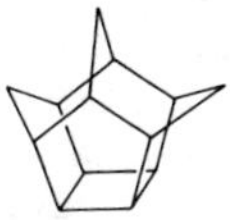	194e
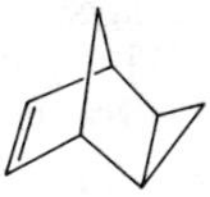	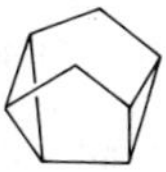	194f
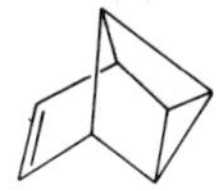	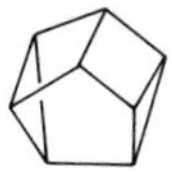	194g
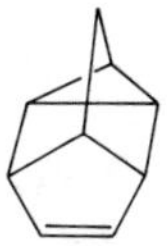	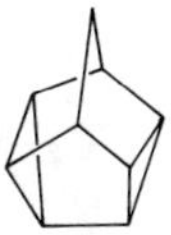	194h
	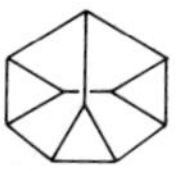	194i

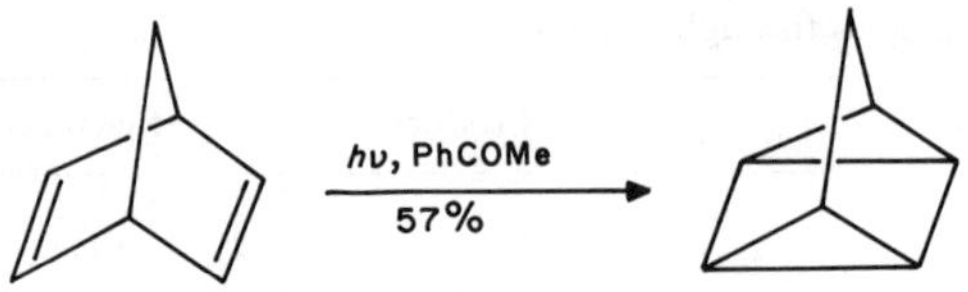

$$\xrightarrow[57\%]{h\upsilon, \text{PhCOMe}} \qquad (121)$$

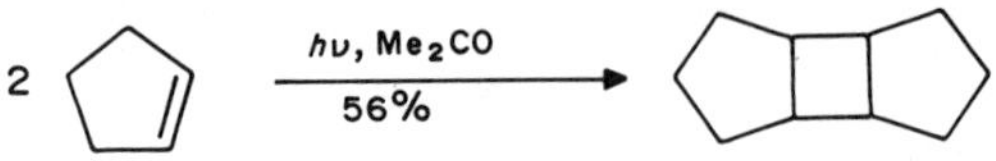

$$2 \quad \xrightarrow[56\%]{h\upsilon, \text{Me}_2\text{CO}} \qquad (122)$$

$$\xrightarrow[27\%]{h\upsilon, \text{Me}_2\text{CO}} \qquad (123)$$

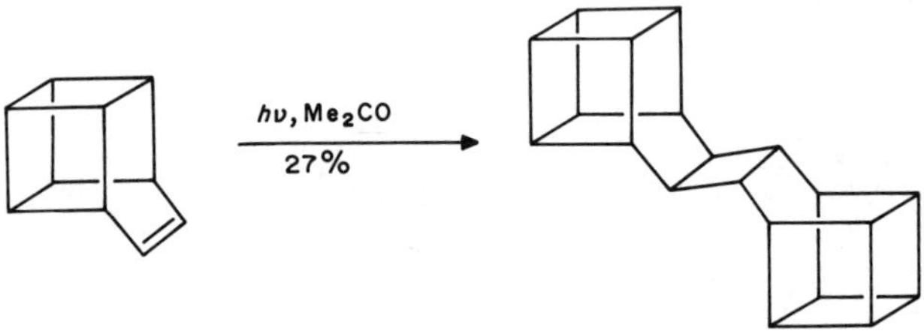

E. Dehydrocyclization

In the context of dodecahedrane synthesis, a novel stratagem was devised as the ultimate step in the synthesis. This consisted of dehydrocyclization of seco-dodecahedrane to dodecahedrane in the presence of a catalyst at elevated temperature (equation 124)[196]. Use of palladium on carbon catalyst gave a poor yield, but remarkable improvement in yield was accomplished with the Pt/Ti on an alumina catalyst system. A related and no less elegant example is the transformation of [1.1.1.1]-pagodane to dodecahedrane through an isomerization, hydrogenation and dehydrocyclization sequence (equation 125)[197]. While there are not many examples of this type of cyclization, its potential in gaining access to many interesting polycycles through their secologues remains to be fully explored.

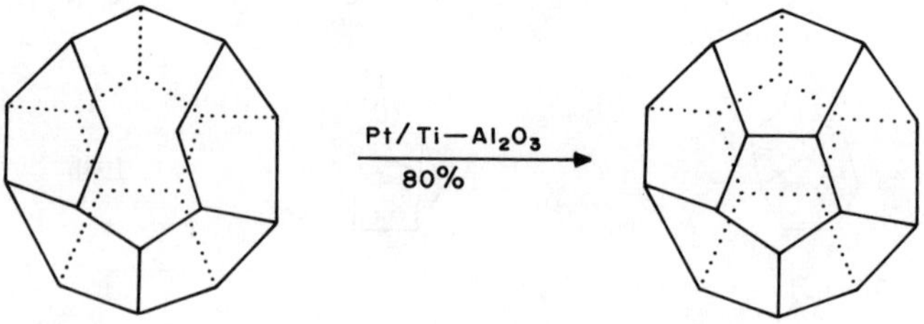

$$\xrightarrow[80\%]{\text{Pt/Ti}-\text{Al}_2\text{O}_3} \qquad (124)$$

$$\xrightarrow[8\%]{\text{H}_2, \text{Pt}-\text{Al}_2\text{O}_3} \qquad (125)$$

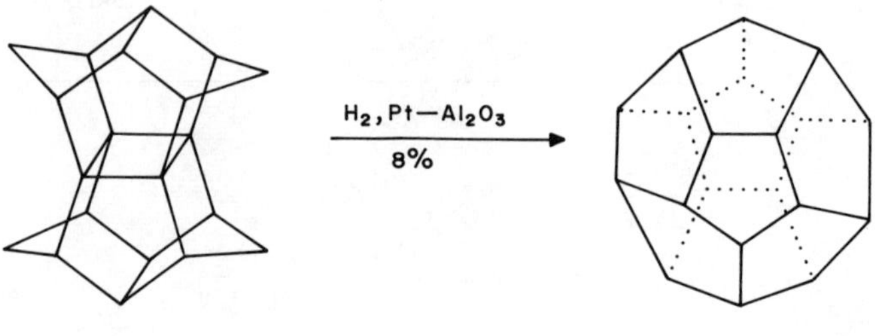

V. CARBENE INSERTION REACTIONS

Cyclopropanation of alkenes, in both inter- and intramolecular modes, constitutes an attractive route to polycycloalkanes. These cyclopropanations of alkenes are carried out through carbene or carbenoid intermediates[198]. For the direct cyclopropanation of alkenes, the methylene iodide zinc–copper couple (Simmons–Smith reaction)[199] reagent is commonly used. Several modifications of this procedure, including acceleration with ultrasound, are known[200]. A somewhat less frequently used procedure for cyclopropanation is through methylene addition from diazomethane, which can be carried out either thermally or photochemically or in the presence of metal salts, e.g. $Pd(OAc)_2$ or $Rh_2(OAc)_4$. In Table 15[201], some examples of the preparation of cyclopropane bearing cycloalkanes are summarized. Preparation of $[n]$-rotanes by this method is particularly noteworthy.

Intramolecular cyclopropanations can be carried out on alkenes bearing a suitably positioned dihalo group. Reaction of these alkenes with a strong base or organometallic reagent furnishes alkanes[198]. Such intramolecular cyclopropanations invariably lead to highly compact, small, strained polycycloalkanes. The methodology has proved useful in providing access to small-ring propellanes, particularly [1.1.1]-propellanes. Several interesting examples are shown in Table 16[202].

An important aspect of carbenes is their ability to insert into C—H bonds. Such insertions in the intramolecular mode eventuate in polycyclic frameworks. For the purpose of intramolecular insertions, carbenes are best generated through the photolysis or thermolysis of alkali metal salts of p-toluenesulphonylhydrazones. Examples of formation of polycycloalkanes through carbenoid C—H insertion are provided in Table 17[203].

TABLE 15. Alkanes through intermolecular carbene additions

Starting olefin	Product	Reference
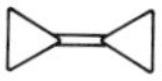		201a
		201b
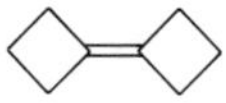		201c
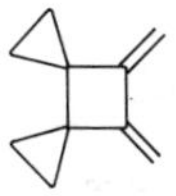	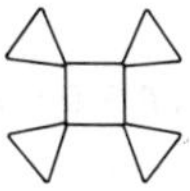	201d
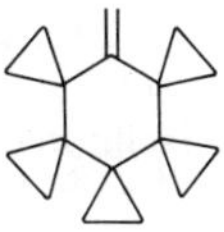	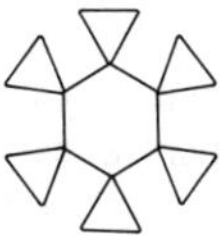	201e

TABLE 16. Alkanes through intramolecular carbene additions

Starting olefin	Product	Reference
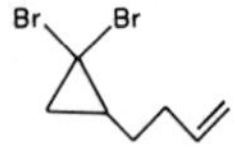 Br, Br structure with allyl group	 bicyclic structure	202a
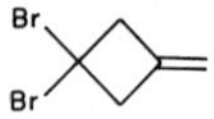 Br, Br cyclobutane with methylene	 bicyclic structure	202b
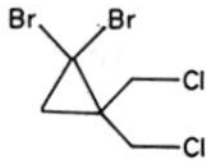 Br, Br cyclopropane with Cl, Cl	 bicyclic structure	202c
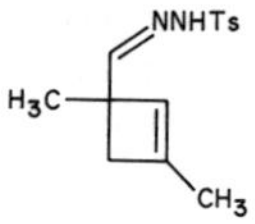 NNHTs, H₃C, CH₃ cyclobutene	 H₃C, CH₃ structure	202d
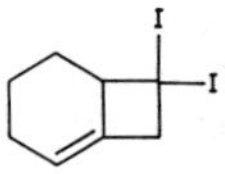 I, I bicyclic structure	 polycyclic structure	202e
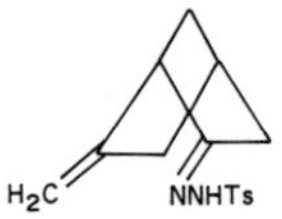 H₂C, NNHTs structure	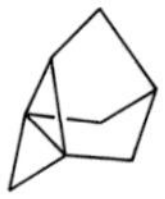 polycyclic structure	202f
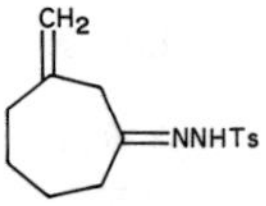 CH₂, NNHTs cycloheptane structure	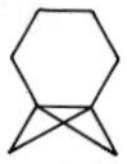 bicyclic structure	202g
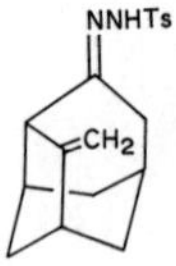 NNHTs, CH₂ adamantane structure	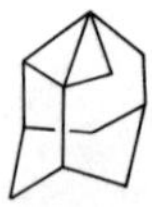 polycyclic structure	202h

VI. MOLECULAR EXTRUSION REACTIONS

Ejection of stable, neutral fragments from polycyclic molecules constitutes an elegant way of preparing cycloalkanes. The neutral species to be removed is generally dinitrogen, carbon monoxide or carbon dioxide. Heat, light and certain metal catalysts promote these reactions. In some cases, excellent yields are obtained and several preparatively useful procedures are available. Thermal extrusion reactions are facilitated by carrying out the reaction under flash vacuum pyrolysis (FVP) conditions. Examples of alkanes obtained through thermal or photochemical dinitrogen extrusion are summarized in

TABLE 17. Alkanes through intramolecular carbene insertion

Precursor	Product	Reference
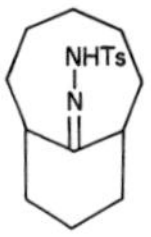 NHTs N		203a
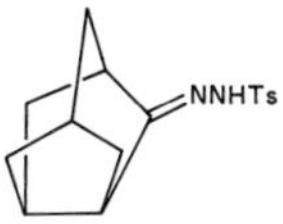 NNHTs	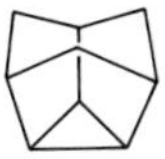	203b
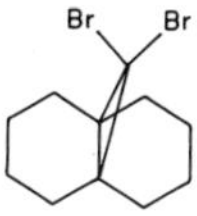 Br Br	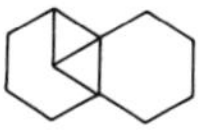	203c
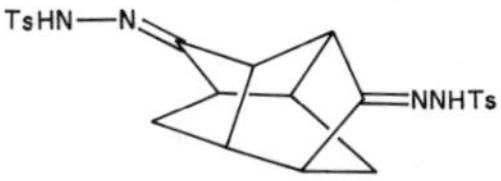 TsHN—N NNHTs	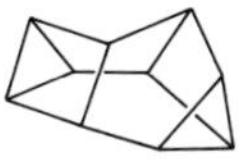	203d
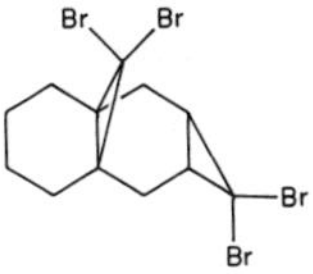 Br Br Br Br	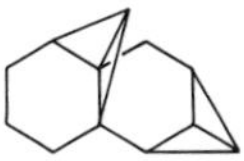	203e
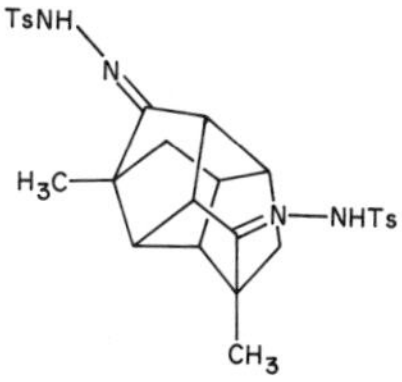 TsNH N H₃C N—NHTs CH₃	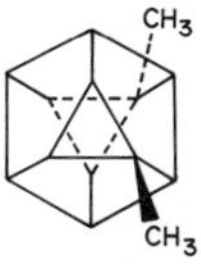 CH₃ CH₃	203f

Table 18[204]. Synthesis of [3]-prismane[204c] and [3]-peristylane[204d] are notable achievements.

Certain transition metal catalysts promote the decarbonylation of aldehydes to furnish n-alkanes (equation 126)[205]. Wilkinson's catalyst $(Ph_3P)_3RhCl$ is the one most often used for this purpose.

$$CH_3(CH_2)_5CHO \xrightarrow[78\%]{(Ph_3P)_3RhCl} CH_3(CH_2)_4CH_3 \tag{126}$$

Photodecarbonylation is a well-established reaction and its recent applications involve synthesis of the tetra-t-butyltetrahedrane compound (equation 127)[206] and [4.5]-coronane (equation 128)[207]. Three examples of carbon dioxide elimination are exhibited in equations 129–131 to furnish corresponding alkanes[208].

TABLE 18. Alkanes through nitrogen extrusion

Starting azo-compound	Product	Reference
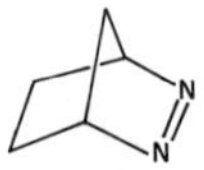		204a
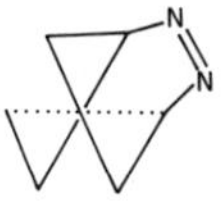		204b
	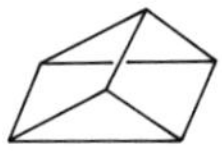	204c
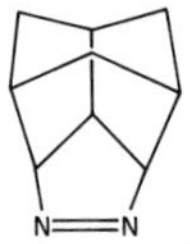		204d
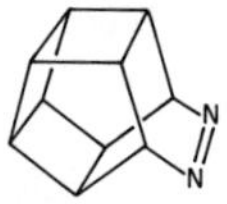	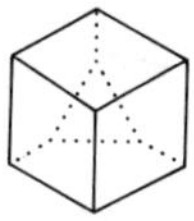	204e

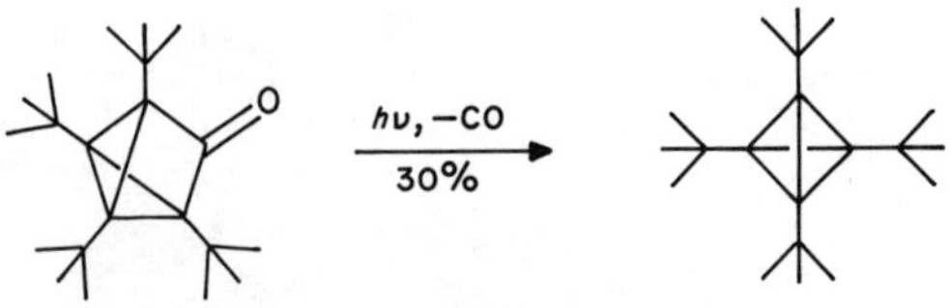

$$\xrightarrow[30\%]{h\nu,\ -CO}$$

(127)

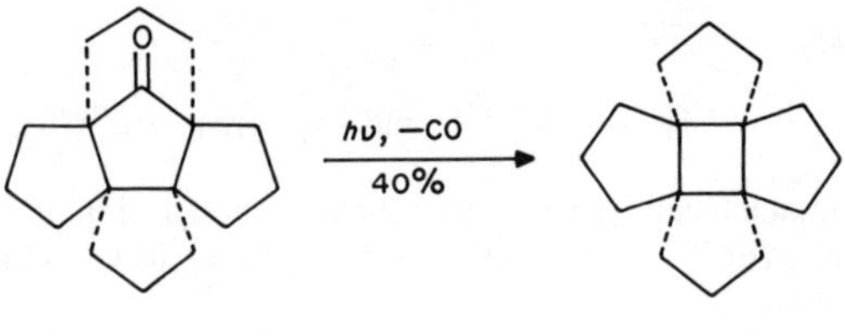

$$\xrightarrow[40\%]{h\nu,\ -CO}$$

(128)

$$\xrightarrow[25\%]{FVP,\ 700\ ^\circ C}$$

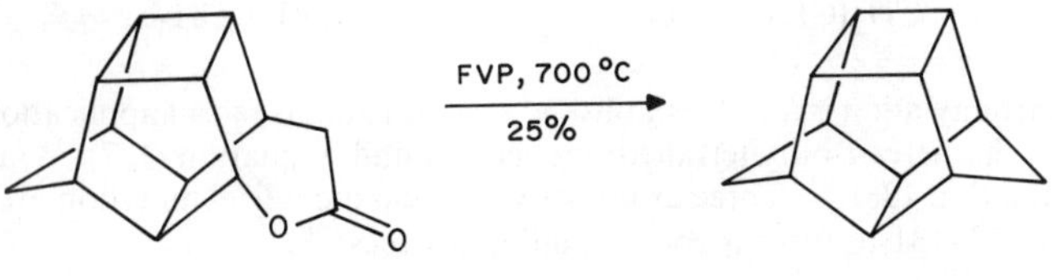

(129)

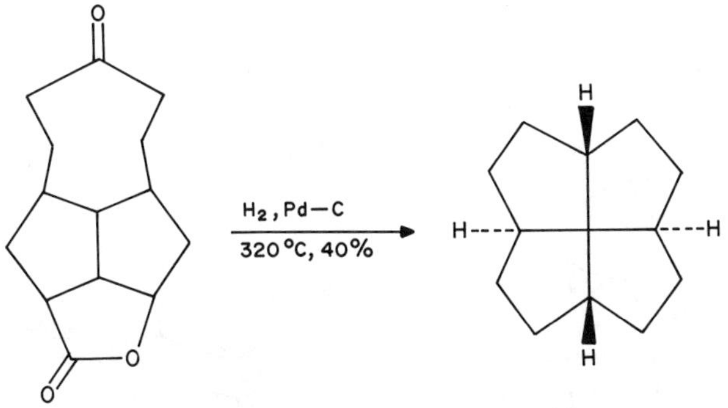

(130)

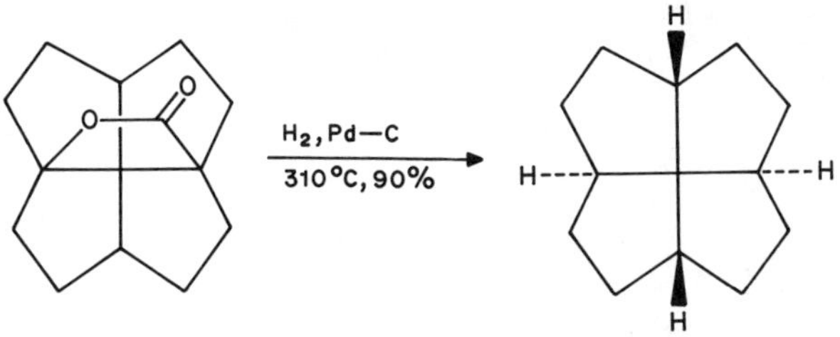

(131)

VII. REARRANGEMENTS

Rearrangement-based approaches have figured prominently in the syntheses of cycloalkanes, particularly polycycloalkanes. These rearrangements have been generally effected by Lewis acids or strong protic acids, transition metal catalysts and photoirradiation. A wide variety of novel polycycloalkanes have been synthesized and characterized during the past two decades following this approach and a few representative examples will be discussed here.

The scope of the thermodynamically driven (stabilomer approach) cycloalkane rearrangements with Lewis acids, of which tetrahydrodicyclopentadiene–adamantane transformation is a prototype, has been greatly amplified[209]. Not only have many new 'adamantane-land' polycycloalkanes been synthesized, but their rearrangements have been studied to provide new hydrocarbons. A few examples of cycloalkanes, synthesized through the stabilomer approach, are depicted in Table 19[210]. Some of these rearrangements can also be brought about thermally over a catalyst in the gas phase. Two illustrations are the synthesis of triamantane (equation 132)[211] and *anti*-tetramantane (equation 133)[212]. An intriguing example of alkene–alkane rearrangement mediated by strong acids is shown in equation 134[213]. Deep-seated methyl group migrations and a strategic C—C bond formation are its notable features. An example of carbocationic rearrangement from a cyclohexanol to methyl cyclopentane of intrinsic synthetic value is the preparation of *endo*-methyl perhydrotriquinacene (equation 135)[214].

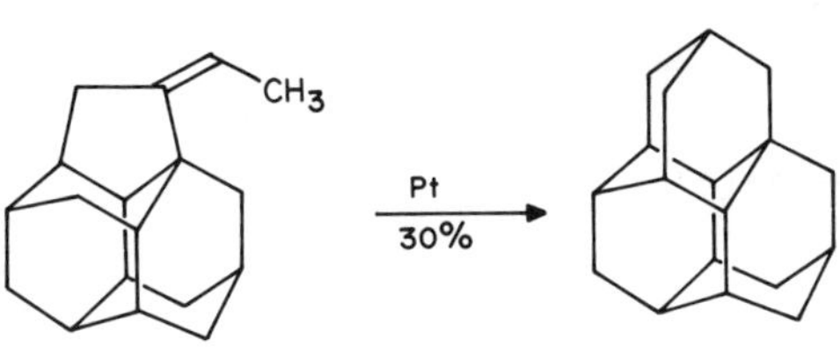

(132)

598   G. Mehta and H. S. P. Rao

TABLE 19. Alkanes through Lewis-acid catalysed rearrangements

Starting alkane	Product	Reference
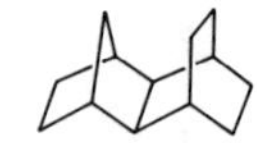	 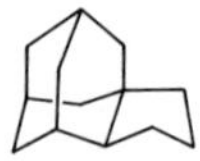	210a
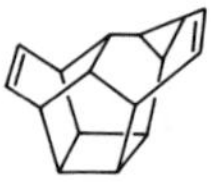	 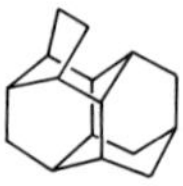	210b
	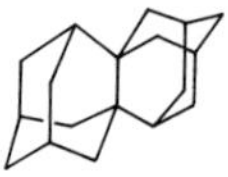	210c
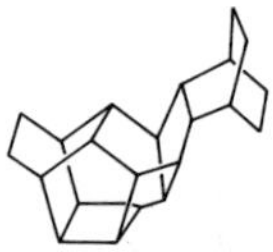	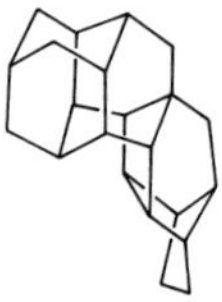	210d
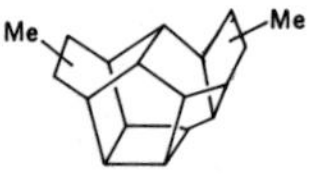	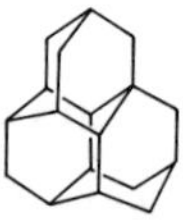	210e
	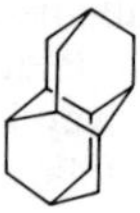	210f

TABLE 20. Alkanes through silver(I) catalysed rearrangements

Starting alkane	Product	Reference
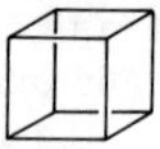	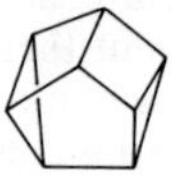	215a
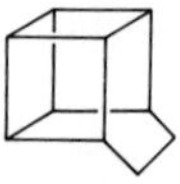	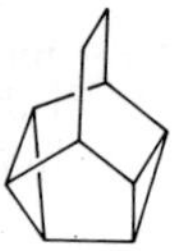	215b
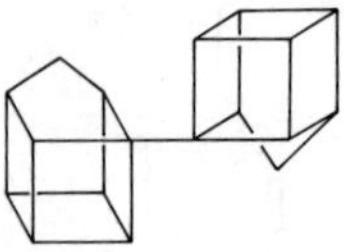	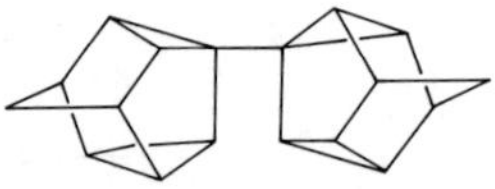	215c

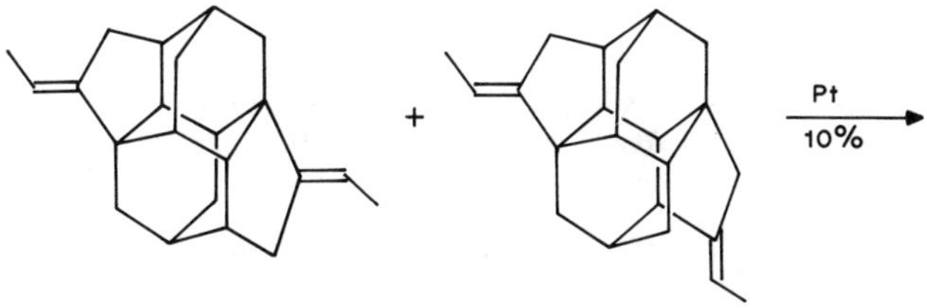

(133)

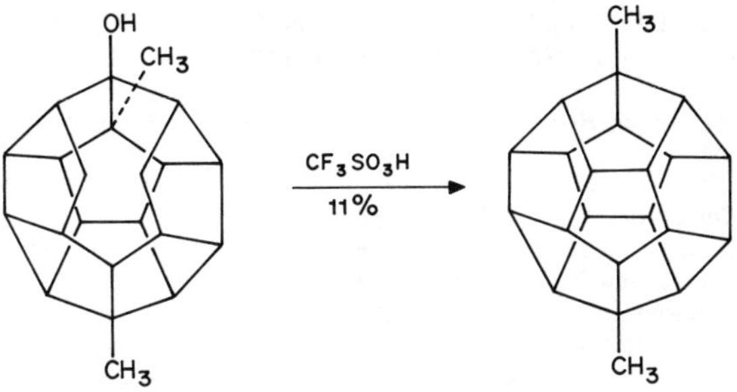

(134)

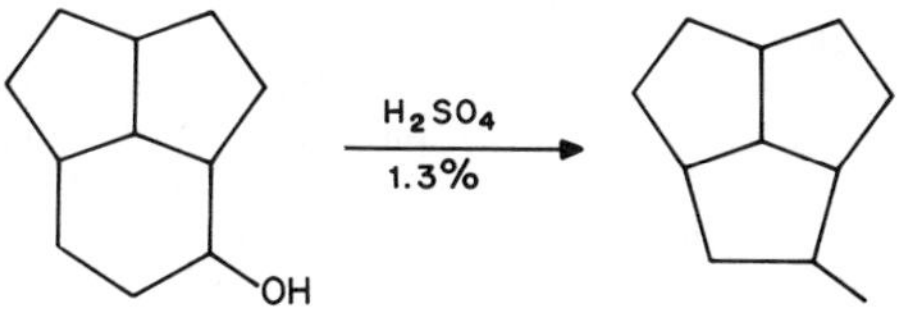

(135)

Transition metal catalysed rearrangements of polycycloalkanes, particularly those bearing multiple cyclobutane rings, have been very effective in providing entry into new hydrocarbons. Usually Ag(I) as perchlorate or tetrafluoroborate and Rh(I) as $(Ph_3P)_3RhCl$, $Rh_2(NOR)_2Cl_2$, $Rh(COD)_2Cl_2$, $Rh_2(CO)_4Cl_2$ have been employed for catalysing the rearrangements. Some of the hydrocarbons prepared through these rearrangements are shown in Table 20[215]. Examples of photochemical rearrangements, as distinct from photocycloadditions, leading to alkanes are few; see, e.g., equation 136[216].

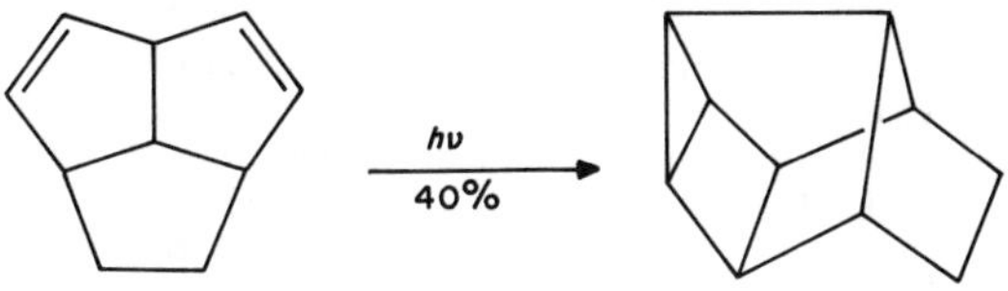

(136)

VIII. ACKNOWLEDGEMENTS

We would like to thank GM's research group and particularly Dr S. Padma for their help in the preparation of this manuscript. One of us (HSPR) appreciates financial support from CSIR and UGC and the hospitality of the School of Chemistry, University of Hyderabad, during a UGC National Associationship.

IX. REFERENCES

1. D. H. R. Barton and W. D. Ollis (Eds.), *Comprehensive Organic Chemistry, The Synthesis and Reactions of Organic Compounds*, Vols. 1–6, Pergamon Press, Oxford, 1979.
2. G. Wilkinson, F. G. A. Stone and E. W. Abel (Eds.), *Comprehensive Organometallic Chemistry, The Synthesis, Reactions and Structure of Organometallic Compounds*, Vols. 1–9, Pergamon Press, Oxford, 1982.
3. C. A. Buehler and D. E. Pearson, *Survey of Organic Synthesis*, Vols. 1 and 2, Wiley–Interscience, New York, 1970 and 1977.
4. R. C. Larock, *Comprehensive Organic Transformations. A Guide to Functional Group Preparations*, VCH, Weinheim, 1989.
5. Houben-Weyl, *Methoden der Organischen Chemie*, 4th ed., Vol. V/1a (alkanes), Vol. IV/3 (cyclopropane), Vol. IV/4 (cyclobutanes), S. Thieme, Stuttgart, 1970 and 1971.
6. W. Theilheimer, *Synthetic Methods of Organic Chemistry*, Vols. 1–37, S. Barger, New York, 1946–1983.
7. L. Fieser and M. Fieser, *Reagents in Organic Synthesis*, Vols. 1–14, Wiley–Interscience, New York, 1967–1990.
8. *Compendium of Organic Synthetic Methods*, Vols. 1–6, Wiley, New York, 1971–1980.
9. *Organic Reactions*, Vols. 1–36. Wiley, New York, 1942–1989.
10. J. March, *Advanced Organic Chemistry, Reactions, Mechanisms and Structure*, 3rd ed., Wiley–Interscience, New York, 1985.
11. (a) A. P. G. Kieboom and F. van Rantwijk, *Hydrogenation and Hydrogenolysis in Synthetic Organic Chemistry*, Delft University Press, Rotterdam, Netherlands, 1977.
    (b) P. N. Rylander, *Catalytic Hydrogenation in Organic Synthesis*, Academic Press, New York, 1979.
    (c) P. A. Gosselain, in *Catalysis, Heterogeneous and Homogeneous* (Eds. B. Delmon and G. Janner), Elsevier, Amsterdam, 1957, p. 167.
    (d) G. W. Parshall, *Homogeneous Catalysis, The Applications and Chemistry of Catalysis by Soluble Transition Metal Complexes*, Wiley–Interscience, New York, 1980.
    (e) C. Masters, *Homogeneous Transition Metal Catalysis*, Chapman and Hall, London, 1981.
    (f) M. Hudlicky, *Reductions in Organic Chemistry*, Ellis Horwood Ltd., New York, 1984, p. 33.
    (g) R. A. W. Johnston, A. H. Wilby and I. D. Entwistle, *Chem. Rev.*, **85**, 129 (1985).
    (h) P. N. Rylander, *Hydrogenation Methods*, Academic Press, New York, 1985.
    (i) R. L. Augustine (Ed.), *Reductions*, Arnold, London, 1968.
    (j) R. S. Dickson, *Homogeneous Catalysis with Compounds of Rhodium and Iridium*, D. Reidel, Dordrecht, 1985.
    (k) B. Bosnich, *Asymmetric Catalysis*, Martinus Nijhoff, Dordrecht, 1986.
    (l) Reference 2, H. B. Kagan, *Comprehensive Organometallic Chemistry*, Vol. III, p. 463.
    (m) L. H. Pignolet (Ed.), *Homogeneous Catalysis with Metal Phosphine Complexes*, Plenum, New York, 1983.
12. R. S. Mann and T. R. Lien, *J. Catal.*, **15**, 1 (1969).
13. (a) E. Igner, O. I. Paynter, D. J. Simmonds and M. C. Whiting, *J. Chem. Soc., Perkin Trans. 1*, 2447 (1987).
    (b) Z. Goren and S. E. Biali, *J. Am. Chem. Soc.*, **112**, 893 (1990).
    (c) G. Schill, N. Schweickert, H. Fritz and W. Vetter, *Angew. Chem., Int. Ed. Engl.*, **22**, 889 (1983).
    (d) I. T. Jacobson, *Chem. Scr.*, **5**, 174, 227 (1974).
    (e) L. A. Paquette, I. Itoh and W. B. Farnham, *J. Am. Chem. Soc.*, **97**, 7280 (1975); L. A. Paquette, I. Itoh and K. B. Lipkowitz, *J. Org. Chem.*, **41**, 3524 (1976).
    (f) M. Venkatachalam, G. Kubiak, J. M. Cook and U. Weiss, *Tetrahedron Lett.*, **26**, 4863 (1985).
    (g) G. Kaiser and H. Musso, *Chem. Ber.*, **118**, 2266 (1985).
    (h) A. Otterbach and H. Musso, *Chem. Ber.*, **121**, 2257 (1988).
    (i) H. Musso, *Chem. Ber.*, **108**, 337 (1975).
    (j) D. Wehle, N. Schormann and L. Fitger, *Chem. Ber.*, **121**, 2171 (1988).
14. J. M. Tour, J. P. Cooper and S. L. Pendalwar, *J. Org. Chem.*, **55**, 3452 (1990).
15. Y. Sasson and J. Blum, *J. Org. Chem.*, **40**, 1887 (1975).
16. M. E. Volpin, V. P. Kukolev, V. O. Chernyshev and I. S. Kolomnikov, *Tetrahedron Lett.*, 4435 (1971).
17. R. Grigg, T. R. B. Mitchell and N. Tongpenyai, *Synthesis*, 442 (1981).

18. K. Kochloefl, W. Liebelt and H. Knozinger, *J. Chem. Soc., Chem. Commun.*, 510 (1977).
19. T. Okano, K. Tsukiyama, H. Konishi and J. Kiji, *Chem. Lett.*, 603 (1982).
20. K. M. Ho, M.-C. Chan and T.-Y. Luh, *Tetrahedron Lett.*, **27**, 5383 (1986).
21. R. H. Crabtree, H. Felkin, T. Fillebeen-Khan and G. E. Morris, *J. Organomet. Chem.*, **168**, 183 (1979).
22. J. M. Brown and R. G. Naik, *J. Chem. Soc., Chem. Commun.*, 348 (1982).
23. (a) E. J. Corey, M. C. Desai and T. A. Engler, *J. Am. Chem. Soc.*, **107**, 4339 (1985).
    (b) D. A. Evans and M. M. Morrissey, *J. Am. Chem. Soc.*, **106**, 3866 (1984).
24. J. D. Morrison (Ed.), *Asymmetric Synthesis*, Vol. 5, Academic Press, New York, 1985.
25. (a) R. Noyori, O. Ohta, Y. Hsiao, M. Kitamura, T. Ohta and H. Takaya, *J. Am. Chem. Soc.*, **108**, 7117 (1986).
    (b) H. Takaya, T. Ohta, N. Sayo, H. Kumobayashi, S. Akutagawa, S.-I. Inoue, I. Kasahara and R. Noyori, *J. Am. Chem. Soc.*, **109**, 1596 (1987).
26. G. Brieger, T. J. Nestrick and T.-H. Fu, *J. Org. Chem.*, **44**, 876 (1979).
27. S. Hunig, H. R. Muller and W. Thier, *Angew. Chem., Int. Ed. Eng.*, **4**, 271 (1965).
28. (a) K. B. Wiberg and M. G. Matturro, *Tetrahedron Lett.*, **22**, 3481 (1981).
    (b) L. A. Paquette, R. V. C. Carr, P. Charumilind and J. F. Blount, *J. Org. Chem.*, **45**, 4922 (1980).
    (c) L. A. Paquette and D. W. Balogh, *J. Am. Chem. Soc.*, **104**, 774 (1982).
29. D. N. Kursanov, Z. N. Parnes and N. M. Loim, *Synthesis*, 633 (1974).
30. D. N. Kursanov, Z. N. Parnes, G. I. Bassova, N. M. Loim and V. I. Zdanovich, *Tetrahedron*, **23**, 2235 (1967).
31. A. N. Chekhlov, S. P. Ionov, M. V. Dodonov and S. N. Anacheriko, *Bioorg. Khim.*, **9**, 978 (1983); *Chem. Abstr.*, **99**, 176136g (1983).
32. R. M. Bullock and B. J. Rappoli, *J. Chem. Soc., Chem. Commun.*, 1447 (1989).
33. H. C. Brown and K. J. Murray, *Tetrahedron*, **42**, 5497 (1986); H. C. Brown and K. J. Murray, *J. Am. Chem. Soc.*, **81**, 4108 (1959).
34. B. Ganem and J. O. Osby, *Chem. Rev.*, **86**, 763 (1986).
35. E. C. Ashby and J. J. Lin, *J. Org. Chem.*, **43**, 2567 (1978).
36. E. C. Ashby and J. J. Lin, *Tetrahedron Lett.*, 4481 (1977).
37. F. Sato, S. Sato, M. Kodama and H. Sato, *J. Organomet. Chem.*, **142**, 71 (1977).
38. S.-K. Chung, *J. Org. Chem.*, **44**, 1014 (1979).
39. N. Satyanarayana and M. Periasamy, *Tetrahedron Lett.*, **25**, 2501 (1984).
40. A. F. Sowinski and G. M. Whitesides, *J. Org. Chem.*, **44**, 2369 (1979).
41. G. M. Whitesides and W. J. Ehmann, *J. Org. Chem.*, **35**, 3565 (1970).
42. Z. N. Parnes, G. I. Balestova, I. S. Akhrem, M. E. Volpin and D. N. Kursanov, *J. Chem. Soc., Chem. Commun.*, 748 (1980).
43. T. Yoshida, *Chem. Lett.*, 429 (1982).
44. (a) L. Fitzer, H.-J. Scheurmann, U. Klages, D. Wehle, D. S. Stephenson and G. Binsch, *Chem. Ber.*, **119**, 1144 (1986).
    (b) D. Wehle, H.-J. Scheurmann and L. Fitzer, *Chem. Ber.*, **119**, 3127 (1986).
    (c) P. K. Freeman and D. M. Balls, *J. Org. Chem.*, **32**, 2354 (1967).
    (d) D. Bosse and A. de Meijere, *Angew. Chem., Int. Ed. Engl.*, **13**, 663 (1974).
45. (a) R. Stober, H. Musso and E. Osawa, *Tetrahedron*, **42**, 1757 (1986).
    (b) E. Osawa, I. Schneider, K. J. Toyne and H. Musso, *Chem. Ber.*, **119**, 2350 (1986).
    (c) K. I. Hirao, T. Iwakuma, M. Taniguchi, E. Abe, O. Yonemitsu, T. Date and K. Kotera, *J. Chem. Soc., Chem. Commun.*, 691 (1974).
46. (a) P. Binger and U. Schuchardt, *Chem. Ber.*, **114**, 1649 (1981).
    (b) P. Binger, G. Schroth and J. McMeeking, *Angew. Chem., Int. Ed. Engl.*, **13**, 465 (1974).
    (c) P. Binger, J. McMeeking and U. Schuchardt, *Chem. Ber.*, **113**, 2372 (1980).
47. F. A. Carey and H. S. Tremper, *J. Org. Chem.*, **36**, 758 (1971).
48. M. G. Adlington, M. Orbanopoulos and J. L. Fry, *Tetrahedron Lett.*, 2955 (1976).
49. P. E. Peterson and C. Casey, *J. Org. Chem.*, **29**, 2325 (1964).
50. M. E. Krafft and W. J. Crooks III, *J. Org. Chem.*, **53**, 432 (1988).
51. (a) V. Bhaskar Rao, S. Wolff and W. C. Agosta, *Tetrahedron*, **42**, 1549 (1986).
    (b) L. A. Paquette, J. W. Fischer, A. R. Browne and C. W. Doecke, *J. Am. Chem. Soc.*, **107**, 686 (1985); L. A. Paquette, H. Kunzer, K. E. Green, O. De Lucchi, G. Licini, L. Pasquato and G. Valle, *J. Am. Chem. Soc.*, **108**, 3453 (1986).
52. F. Bohlman, J. Staffeldt and W. Skuballa, *Chem. Ber.*, **109**, 1586 (1976).

53. (a) S. Krishnamurthy and H. C. Brown, *J. Org. Chem.*, **41**, 3064 (1976).
    (b) R. W. Holder and M. G. Matturrao, *J. Org. Chem.*, **42**, 2166 (1977).
    (c) J. A. Marshall and R. A. Ruden, *J. Org. Chem.*, **36**, 594 (1971).
54. (a) Y. Fujimoto and T. Tatsuno, *Tetrahedron Lett.*, 3325 (1976).
    (b) P. Kocovsky and V. Cerny, *Collect. Czech. Chem. Commun.*, **44**, 246 (1979).
55. T. Morita, Y. Okamoto and H. Sakurai, *Synthesis*, 32 (1981).
56. C. K. Lau, C. Dufresne, P. C. Belanger, S. Pietre and J. Scheigetz, *J. Org. Chem.*, **51**, 3038 (1986).
57. (a) K. B. Wiberg, L. K. Olli, N. Golembeski and R. D. Adams, *J. Am. Chem. Soc.*, **102**, 7467 (1980).
    (b) B. Ernst and C. Ganter, *Helv. Chim. Acta*, **61**, 1107 (1978).
58. (a) R. B. Boar, L. Joukhadar, J. F. McGhie, S. C. Misra, A. G. M. Barrett, D. H. R. Barton and P. A. Pronpiou, *J. Chem. Soc., Chem. Commun.*, 698 (1978).
    (b) A. G. M. Barret, C. R. A. Godfrey, D. M. Hollinshead, P. A. Prokopiou, D. H. R. Barton, R. B. Boar, L. Joukhadar, J. F. McGhie and S. C. Misra, *J. Chem. Soc., Perkin Trans. 1*, 1501 (1981).
59. A. K. Bose and P. Mangiaracina, *Tetrahedron Lett.*, **28**, 2503 (1987).
60. (a) H. Deshayes and J.-P. Pete, *J. Chem. Soc., Chem. Commun.*, 567 (1978).
    (b) H. Deshayes and J.-P. Pete, *Can. J. Chem.*, **62**, 2063 (1984).
61. P. J. Eaton, D. R. Lauren, A. W. O'Connor and R. T. Weavers, *Aust. J. Chem.*, **34**, 1303 (1981).
62. (a) H. Sano, M. Ogata and T. Migita, *Chem. Lett.*, 77 (1986).
    (b) H. Sano, T. Takeda and T. Migita, *Chem. Lett.*, 119 (1988).
63. (a) N. C. Billingham, R. A. Jackson and F. Malek, *J. Chem. Soc., Chem. Commun.*, 344 (1977).
    (b) R. A. Jackson and F. Malek, *J. Chem. Soc., Perkin Trans. 1*, 1207 (1980).
64. D. H. R. Barton and S. W. McCombie, *J. Chem. Soc., Perkin Trans. 1*, 1574 (1975).
65. (a) W. Hartwig, *Tetrahedron*, **39**, 2609 (1983).
    (b) M. Ramaiah, *Tetrahedron*, **43**, 3541 (1987).
    (c) D. H. R. Barton and W. B. Motherwell, *Pure Appl. Chem.*, **53**, 15 (1981).
    (d) W. P. Neumann, *Synthesis*, 665 (1987).
66. (a) D. H. R. Barton and R. Subramanian, *J. Chem. Soc., Perkin Trans. 1*, 1718 (1977).
    (b) D. H. R. Barton, D. Crich, A. Lobberding and S. Z. Zard, *J. Chem. Soc., Chem. Commun.*, 646 (1985).
    (c) D. H. R. Barton, W. B. Motherwell and A. Stange, *Synthesis*, 743 (1981).
    (d) A. G. M. Barrett, P. A. Prokopiou and D. H. R. Barton, *J. Chem. Soc., Perkin Trans. 1*, 1510 (1981).
    (e) D. H. R. Barton, W. Hartwig, R. S. H. Motherwell, W. B. Motherwell and A. Stange, *Tetrahedron Lett.*, **23**, 2019 (1982).
    (f) M. J. Robins, J. S. Wilson and F. Hansske, *J. Am. Chem. Soc.*, **105**, 4059 (1983).
67. K. Nozaki, K. Oshima and K. Utimoto, *Tetrahedron Lett.*, **29**, 6125 (1988).
68. J. N. Kirwan, B. P. Roberts and C. R. Willis, *Tetrahedron Lett.*, **31**, 5093 (1990).
69. D. H. R. Barton and D. Crich, *J. Chem. Soc., Perkin Trans. 1*, 1603 (1986).
70. (a) R. E. Ireland, D. C. Muchmore and U. Hengartner, *J. Am. Chem. Soc.*, **94**, 5098 (1972).
    (b) H. J. Liu, S. P. Lee and W. H. Chan, *Can. J. Chem.*, **55**, 3797 (1977).
71. P. A. Jacobi and R. F. Frechette, *Tetrahedron Lett.*, **28**, 2937 (1987).
72. S. Becker, Y. Fort, R. Vanderesse and P. Caubere, *J. Org. Chem.*, **54**, 4848 (1989).
73. M.-C. Chan, K.-M. Cheng, M. K. Li and T.-Y. Luh, *J. Chem. Soc., Chem. Commun.*, 1610 (1985).
74. H. Alper and T. L. Prince, *Angew. Chem., Int. Ed. Engl.*, **19**, 315 (1980).
75. T. G. Back and K. Yang, *J. Chem. Soc., Chem. Commun.*, 819 (1990).
76. U. Bunz, K. Polborn, H.-U. Wagner and G. Szeimies, *Chem. Ber.*, **121**, 1785 (1988).
77. Y. Watanabe, T. Araki, Y. Ueno and T. Endo, *Tetrahedron Lett.*, **27**, 5385 (1986).
78. D. L. J. Clive, G. J. Chittattu, V. Farina, W. A. Kiel, S. M. Menchen, C. G. Russell, A. Singh, C. K. Wong and N. J. Curtis, *J. Am. Chem. Soc.*, **102**, 4438 (1980).
79. T. G. Back, V. I. Birss, M. Edwards and M. V. Krishna, *J. Org. Chem.*, **53**, 3815 (1988).
80. A. R. Pinder, *Synthesis*, 425 (1980).
81. (a) T. Ohsawa, T. Takagaki, A. Haneda and T. Oishi, *Tetrahedron Lett.*, **22**, 2583 (1981).
    (b) T. Imamoto, T. Takeyama and T. Kusumoto, *Chem. Lett.*, 1491 (1985).
82. L. Horner, L. Schlafer and H. Kammerer, *Chem. Ber.*, **92**, 1700 (1959).
83. (a) M. P. Doyle, C. C. McOsker and C. T. West, *J. Org. Chem.*, **41**, 1393 (1976).

(b) M. P. Doyle, C. T. West, S. J. Donnelly and C. C. McOsker, *J. Organomet. Chem.*, **117**, 129 (1976).

84. R. P. Allen, B. P. Roberts and C. R. Willis, *J. Chem. Soc., Chem. Commun.*, 1387 (1989).
85. Z. N. Parnes, V. S. Ramanover and M. E. Volpin, *J. Org. Chem. USSR (Engl. Trans.)*, **24**, 258 (1988); *Chem. Abstr.*, **109**, 148842r, 1988.
86. C. Chatgilialoglu, D. Griller and M. Lesage, *J. Org. Chem.*, **53**, 3641 (1988); **54**, 2492 (1989).
87. (a) B. V. Lap and M. N. Paddon-Row, *J. Org. Chem.*, **44**, 4979 (1979).
    (b) H.-M. Hutmacher, H.-G. Fritz and H. Musso, *Angew. Chem., Int. Ed. Engl.*, **14**, 80 (1975).
88. (a) P. Bruck, D. Thompson and S. Winstein, *Chem. Ind.*, 405 (1960).
    (b) G. Mehta and S. Padma, *J. Am. Chem. Soc.*, **109**, 2212 (1987).
    (c) G. Mehta and S. Padma, *J. Am. Chem. Soc.*, **109**, 7230 (1987).
    (d) P. E. Eaton and U. R. Chakraborty, *J. Am. Chem. Soc.*, **100**, 3634 (1978).
89. (a) W. G. Dauben and D. L. Whalen, *Tetrahedron Lett.*, 3743 (1966).
    (b) L. A. Paquette, J. S. Ward, R. A. Boggs and W. B. Farnham, *J. Am. Chem. Soc.*, **97**, 1101 (1975).
90. (a) W.-D. Fessner and H. Prinzbach, *Tetrahedron Lett.*, **24**, 5857 (1983).
    (b) W.-D. Fessner, G. Sedelmeier, P. R. Spurr, G. Rihs and H. Prinzbach, *J. Am. Chem. Soc.*, **109**, 4626 (1987).
91. P. Girard, J. L. Namy and H. B. Kagan, *J. Am. Chem. Soc.*, **102**, 2693 (1980).
92. J. Inanaga, M. Ishikawa and M. Yamaguchi, *Chem. Lett.*, 1485 (1987).
93. F. Rolla, *J. Org. Chem.*, **46**, 3909 (1981).
94. C. A. Cupas and L. Hodakowski, *J. Am. Chem. Soc.*, **96**, 4668 (1974).
95. R. O. Hutchins, D. Kandasamy, C. A. Maryanoff, D. Mansalamani and B. E. Maryanoff, *J. Org. Chem.*, **42**, 82 (1977).
96. S. Krishnamurthy and H. C. Brown, *J. Org. Chem.*, **48**, 3085 (1983).
97. E. C. Ashby, J.-J. Lin and A. B. Goel, *J. Org. Chem.*, **43**, 183 (1978).
98. E. C. Ashby, J.-J. Lin and A. B. Goel, *J. Org. Chem.*, **43**, 1557 (1978).
99. M. T. Crimmins and S. W. Mascarella, *Tetrahedron Lett.*, **28**, 5063 (1987).
100. J. R. Wiseman, J. J. Vanderbilt and W. M. Butler, *J. Org. Chem.*, **45**, 667 (1980).
101. (a) L. Horner and H. Roder, *Chem. Ber.*, **101**, 4179 (1968).
    (b) J. Casanova and L. Eberson, in *The Chemistry of the Carbon–Halogen Bond* (Ed. S. Patai), Wiley–Interscience, Chichester, 1978, p. 979.
102. J. W. Sease and R. C. Reed, *Tetrahedron Lett.*, 393 (1975).
103. T. R. Nelsen and J. J. Tufariello, *J. Org. Chem.*, **40**, 3159 (1975).
104. Y. Yamamoto, H. Toi, S.-I. Murahashi and I. Moritani, *J. Am. Chem. Soc.*, **97**, 2558 (1975).
105. K. Hassenruck, G. S. Murthy, V. M. Lynch and J. Michl, *J. Org. Chem.*, **55**, 1013 (1990).
106. V. B. Jigajinni and R. H. Wightman, *Tetrahedron Lett.*, **23**, 117 (1982).
107. A. Jung and R. Engel, *J. Org. Chem.*, **40**, 3652 (1975).
108. (a) J. L. Fry, M. Orfanopoulos, M. G. Adlington, W. R. Dittman, Jr. and S. B. Silverman, *J. Org. Chem.*, **43**, 374 (1978).
    (b) J. W. Larsen and L. W. Chang, *J. Org. Chem.*, **44**, 1168 (1979).
109. E. Breuer, *Tetrahedron Lett.*, 1849 (1967).
110. (a) E. Vedejs, *Org. React.*, **22**, 401 (1975).
    (b) J. G. St. C. Buchaman and P. D. Woodgate, *Quart. Rev.*, **23**, 522 (1969).
    (c) H. H. Szmant, *Angew. Chem., Int. Ed. Engl.*, **7**, 120 (1968).
    (d) E. L. Martin, *Org. React.*, **1**, 155 (1942).
111. (a) M. Toda, Y. Hirata and S. Yamamura, *J. Chem. Soc., Chem. Commun.*, 919 (1969).
    (b) M. Toda, M. Hayashi, Y. Hirata and S. Yamamura, *Bull. Chem. Soc. Jpn.*, **45**, 264 (1972).
112. D. Todd, *Org. React.*, **4**, 378 (1948).
113. (a) U. Biethan, U. V. Gizycki and H. Musso, *Tetrahedron Lett.*, 1477 (1965).
    (b) H. Hopf and H. Musso, *Chem. Ber.*, **106**, 143 (1973).
    (c) J. P. Chesick, J. D. Dunitz, U. V. Gizycki and H. Musso, *Chem. Ber.*, **106**, 150 (1973).
    (d) M. Giersig, D. Wehle, L. Fitjer, N. Schormann and W. Clegg, *Chem. Ber.*, **121**, 525 (1988).
    (e) G. Helmchen and G. Staiger, *Angew. Chem., Int. Ed. Engl.*, **16**, 116 (1977).
    (f) F. J. Jaggi, P. Buchs and C. Ganter, *Helv. Chim. Acta*, **63**, 872 (1980).
    (g) H. Tobler, R. O. Klaus and C. Ganter, *Helv. Chim. Acta*, **58**, 1455 (1975).
    (h) P. E. Eaton and D. R. Patterson, *J. Am. Chem. Soc.*, **100**, 2573 (1978).

(i) F. J. C. Martins, L. Fourie, H. J. Venter and P. L. Wessels, *Tetrahedron*, **46**, 623, 632 (1990).

(j) W. G. Dauben and D. M. Walker, *Tetrahedron Lett.*, **23**, 711 (1982).

(k) R. B. Kelly and J. Zamecnik, *J. Chem. Soc., Chem. Commun.*, 1102 (1970); R. B. Kelly, J. Zamecnik and B. A. Beckelt, *J. Chem. Soc., Chem. Commun.*, 479 (1971).

(l) J. R. Prahlad, U. R. Naik and Sukh Dev, *Tetrahedron*, **26**, 663 (1970).

114. L. Caglioti and P. Grasselli, *Chem. Ind.*, 153 (1964).

115. R. O. Hutchins, C. A. Milewski and B. E. Maryanoff, *J. Am. Chem. Soc.*, **95**, 3662 (1973).

116. G. W. Kabalka and J. H. Chandler, *Synth. Commun.*, **9**, 275 (1979).

117. S. Cacchi, L. Caglioti and G. Paolucci, *Bull. Chem. Soc. Jpn.*, **47**, 2323 (1974).

118. G. W. Kabalka and S. T. Summers, *J. Org. Chem.*, **46**, 1217 (1981).

119. C. W. J. Fleet, P. J. C. Harding and M. J. Whitcombe, *Tetrahedron Lett.*, **21**, 4031 (1980).

120. (a) G. R. Pettit and E. E. van Tamelen, *Org. React.*, **12**, 356 (1962).

(b) J. S. Pizey (Ed.), *Synthetic Reagents*, Vol. 2, Ellis Harwood Ltd., London, 1976.

121. (a) A. Belanger, J. Poupart and P. Deslongchamps, *Tetrahedron Lett.*, 2127 (1968); J. Gauthier and P. Deslongchamps, *Can. J. Chem.*, **45**, 297 (1967).

(b) W.-D. Fessner and M. Prinzbach, *Tetrahedron*, **42**, 1797 (1986).

(c) P. E. Eaton, R. H. Mueller, G. R. Carlson, D. A. Cullison, G. F. Cooper, T. C. Chou and E. P. Krebs, *J. Am. Chem. Soc.*, **99**, 2751 (1977).

122. C. G. Gutierrez, R. A. Stringham, T. Nitasaka and K. G. Glasscock, *J. Org. Chem.*, **45**, 3393 (1980).

123. (a) I. Degani and R. Fochi, *J. Chem. Soc., Perkin Trans. 1*, 1133 (1978).

(b) C. L. Bumgardner, J. R. Lever and S. T. Purrington, *Tetrahedron Lett.*, **23**, 2379 (1982).

124. M. Sevrin, D. Van Ende and A. Krief, *Tetrahedron Lett.*, 2643 (1976).

125. (a) M. T. Reetz, *Topics in Current Chemistry*, Springer-Verlag, New York, 1982, pp. 1–54.

(b) M. T. Reetz, J. Westermann and R. Steinbach, *J. Chem. Soc., Chem. Commun.*, 237 (1981).

126. A. Meisters and T. Mole, *Aust. J. Chem.*, **27**, 1655 (1974).

127. R. A. Sheldon and J. K. Kochi, *Org. React.*, **19**, 279 (1972).

128. (a) C. V. Wilson, *Org. React.*, **9**, 332 (1963).

(b) J. March, *J. Chem. Educ.*, **40**, 212 (1963).

129. (a) S. J. Cristol and W. C. Firth, *J. Org. Chem.*, **26**, 280 (1961).

(b) J. K. Kochi, *J. Am. Chem. Soc.*, **87**, 2500 (1965).

(c) A. McKillop, D. Bromley and E. C. Taylor, *J. Org. Chem.*, **34**, 1172 (1969).

130. (a) D. H. R. Barton, D. Crich and W. B. Motherwell, *Tetrahedron Lett.*, **24**, 4979 (1983).

(b) D. H. R. Barton, D. Crich and W. B. Motherwell, *Tetrahedron*, **41**, 3901 (1985).

131. (a) R. S. Abeywickrema and E. W. Della, *J. Org. Chem.*, **45**, 4226 (1980).

(b) E. W. Della and J. Tsanaktsidis, *Aust. J. Chem.*, **39**, 2061 (1986).

(c) E. W. Della and J. Tsanaktsidis, *Aust. J. Chem.*, **42**, 61, (1989).

132. R. M. Moriarty, J. S. Khosrowshahi and T. M. Dalecki, *J. Chem. Soc., Chem. Commun.*, 675 (1987).

133. (a) K. B. Wiberg, B. R. Lowry and T. H. Colby, *J. Am. Chem. Soc.*, **83**, 3998 (1961).

(b) P. E. Eaton and T. W. Cole, *J. Am. Chem. Soc.*, **86**, 315 (1964).

134. (a) P. E. Eaton and K. Nyi, *J. Am. Chem. Soc.*, **93**, 2786 (1971).

(b) R. R. Sauers, K. W. Kelley and B. R. Sickles, *J. Org. Chem.*, **37**, 537 (1972).

(c) J. C. Barborak, L. Watts and R. Pettit, *J. Am. Chem. Soc.*, **88**, 1328 (1966).

(d) P. E. Eaton, Y. S. Or, S. J. Branca and B. K. R. Shankar, *Tetrahedron*, **42**, 1621 (1986).

135. D. H. R. Barton, H. A. Dowlatshashi, W. B. Motherwell and D. Villemin, *J. Chem. Soc., Chem. Commun.*, 732 (1980).

136. D. H. R. Barton, D. Crich and W. B. Motherwell, *J. Chem. Soc., Chem. Commun.*, 939 (1983).

137. N. C. Billingham, R. A. Jackson and F. Malek, *J. Chem. Soc., Perkin Trans. 1*, 1137 (1979).

138. J. Pfenninger, C. Heuberger and W. Graf, *Helv. Chim. Acta*, **63**, 2328 (1980).

139. D. H. R. Barton and G. Kretzschmar, *Tetrahedron Lett.*, **24**, 5889 (1983).

140. A. Otterbach and H. Musso, *Angew. Chem., Int. Ed. Engl.*, **26**, 554 (1987).

141. K. Okada, K. Okubo and M. Oda, *Tetrahedron Lett.*, **30**, 6733 (1989).

142. K. Okada, K. Okamoto and M. Oda, *J. Am. Chem. Soc.*, **110**, 8736 (1988).

143. M. Hasebe and T. Tsuchiya, *Tetrahedron Lett.*, **28**, 6207 (1987).

144. E. H. White and D. J. Woodcock, in *Chemistry of the Amino Group* (Ed. S. Patai), Wiley–Interscience, London, 1968.

12. Alkanes: modern synthetic methods 605

145. R. O. Hutchins, F. Cistone, B. Goldsmith and P. Heuman, *J. Org. Chem.*, **40**, 2018 (1975).
146. G. A. Doldouras and J. Kollonitsch, *J. Am. Chem. Soc.*, **100**, 341 (1978).
147. T. Wirth and C. Ruchardt, *Chimia*, **42**, 230 (1988).
148. G. E. Niznik and H. M. Walborsky, *J. Org. Chem.*, **43**, 2396 (1978).
149. (a) D. H. R. Barton, G. Bringmann, G. Lamotte, W. B. Motherwell, R. S. H. Motherwell and A. E. A. Porter, *J. Chem. Soc., Perkin Trans. 1*, 2657 (1980).
    (b) D. H. R. Barton, G. Bringmann, G. Lamotte, R. S. H. Motherwell and W. B. Motherwell, *Tetrahedron Lett.*, 2291 (1979).
150. S. Yamada, K. Tomioka and K. Koga, *Tetrahedron Lett.*, 61 (1976).
151. T. Curvigny, M. Larcheque and H. Normant, *Bull. Soc. Chim. Fr.*, 1174 (1973).
152. G. Mehta, A. Srikrishna and S. C. Suri, *J. Org. Chem.*, **45**, 5375 (1980).
153. E. E. van Tamelen, H. Rudler and C. Bjorklund, *J. Am. Chem. Soc.*, **93**, 7113 (1971).
154. D. Savoia, E. Tagliavini, C. Trombini and A. Umani-Ronchi, *J. Org. Chem.*, **45**, 3227 (1980).
155. T. Ohsawa, T. Kobayashi, Y. Mizuguchi, T. Saitoh and T. Oishi, *Tetrahedron Lett.*, **26**, 6103 (1985).
156. J. G. Andrade, W. F. Maier, L. Zapf and P.v.R. Schleyer, *Synthesis*, 802 (1980).
157. H. Feuer and A. T. Nielsen, *Nitro compounds, Recent Advances in Synthesis & Chemistry*, VCH Publishers, 1990.
158. N. Kornblum, S. C. Carlson and R. G. Smith, *J. Am. Chem. Soc.*, **100**, 289 (1978); **101**, 647 (1979).
159. (a) N. Ono, H. Miyake, R. Tamura and A. Kaji, *Tetrahedron Lett.*, **22**, 1705 (1981).
    (b) D. D. Tanner, E. V. Blackburn and G. E. Diaz, *J. Am. Chem. Soc.*, **103**, 1557 (1981).
160. N. Ono, R. Tamura and A. Kaji, *J. Am. Chem. Soc.*, **102**, 2851 (1980); **105**, 4017 (1983).
161. H. Suzuki, K. Takaoka and A. Osuka, *Bull. Chem. Soc. Jpn.*, **58**, 1067 (1985).
162. G. Stallberg, S. Stallberg-Stenhagen and E. Stenhagen, *Acta Chem. Scand.*, **6**, 313 (1952).
163. M. A. Flamm-ter-Meer, H.-D. Beckhaus, K. Peters, H.-G. von Schnering and C. Ruchardt, *Chem. Ber.*, **118**, 4665 (1985).
164. L. Meszaros, *Tetrahedron Lett.*, 4951 (1967).
165. (a) K. B. Wiberg and D. S. Corner, *J. Am. Chem. Soc.*, **88**, 4437 (1966).
    (b) K. B. Wiberg, F. H. Walker, W. E. Pratt and J. Michl, *J. Am. Chem. Soc.*, **105**, 3638 (1983).
    (c) P. G. Gassman and G. S. Proehl, *J. Am. Chem. Soc.*, **102**, 6862 (1980).
    (d) F. H. Walker, K. B. Wiberg and J. Michl, *J. Am. Chem. Soc.*, **104**, 2056 (1982).
    (e) P. E. Pincock, J. Schmidt, W. B. Scott and E. J. Torupka, *Can. J. Chem.*, **50**, 3958 (1972).
    (f) K. B. Wiberg, W. E. Pratt and W. F. Bailey, *J. Am. Chem. Soc.*, **99**, 2297 (1977).
    (g) P. J. Garratt and J. F. White, *J. Org. Chem.*, **42**, 1733 (1977).
166. W. F. Bailey and R. P. Gagnier, *Tetrahedron Lett.*, **23**, 5123 (1982).
167. (a) P. E. Eaton and M. Maggini, *J. Am. Chem. Soc.*, **110**, 7230 (1988).
    (b) J. Schafer and G. Szeimies, *Tetrahedron Lett.*, **31**, 2263 (1990).
168. E. Osawa, Z. Majerski and P.v.R. Schleyer, *J. Org. Chem.*, **36**, 205 (1971).
169. (a) M. Tamura and J. Kochi, *J. Am. Chem. Soc.*, **93**, 1483 (1971).
    (b) M. Tamura and J. Kochi, *J. Am. Chem. Soc.*, **93**, 1485 (1971).
    (c) A. McKillop, L. F. Elsom and E. C. Taylor, *J. Am. Chem. Soc.*, **90**, 2423 (1968).
170. J. Nishimura, N. Yamada, Y. Horiuchi, E. Veda, A. Ohbayashi and A. Oku, *Bull. Chem. Soc. Jpn.*, **59**, 2035 (1986).
171. M. A. Flamm-ter-Meer, H.-D. Beckhaus, K. Peters, H.-G. von Schnering, H. Fritz and C. Ruchardt, *Chem. Ber.*, **119**, 1492 (1986).
172. (a) B. H. Lipshutz, *Synth. Lett.*, 119 (1990).
    (b) B. H. Lipshutz, R. S. Wilhelm and J. A. Kozlowski, *Tetrahedron*, **40**, 5005 (1984).
    (c) G. H. Posner, *Org. React.*, **22**, 253 (1975).
173. B. H. Lipshutz, R. S. Wilhelm, J. A. Kozlowski and D. Parker, *J. Org. Chem.*, **49**, 3928 (1984).
174. (a) B. H. Lipshutz, J. A. Kozlowski, D. A. Parker, S. L. Nguyen and K. E. McCarthy, *J. Organomet. Chem.*, **285**, 437 (1985).
    (b) B. H. Lipshutz, D. A. Parker, J. A. Kozlowski and S. L. Nguyen, *Tetrahedron Lett.*, **25**, 5959 (1984).
175. (a) B. H. Lipshutz, M. Koerner and D. A. Parker, *Tetrahedron Lett.*, **28**, 945 (1987).
    (b) B. H. Lipshutz, J. A. Kozlowski, D. A. Parker, S. L. Nguyen and K. E. McCarthy, *J. Organomet. Chem.*, **285**, 437 (1985).
176. R. C. Larock and D. R. Leach, *Tetrahedron Lett.*, **22**, 3435 (1981).

606   G. Mehta and H. S. P. Rao

177. R. Nakajima, K. Morita, and T. Hara, *Bull. Chem. Soc. Jpn.*, **54**, 3599 (1981).
178. M. T. Reetz, B. Wenderoth, R. Peter, R. Steinbach and J. Westermann, *J. Chem. Soc., Chem. Commun.*, 1202 (1980).
179. (a) H. C. Brown and C. H. Snyder, *J. Am. Chem. Soc.*, **83**, 1002 (1961).
    (b) E.-I. Negishi and M. J. Idacavage, *Org. React.* **33**, 1 (1985).
    (c) R. Murphy and R. H. Prager, *Tetrahedron Lett.*, 463 (1976).
180. R. Sustmann and R. Altevogt, *Tetrahedron Lett.*, **22**, 5167 (1981).
181. (a) F. Sato, Y. Mori and M. Sato, *Chem. Lett.*, 1337 (1978).
    (b) K. Isagawa, K. Tatsumi, H. Kosugi and Y. Otsuji, *Chem. Lett.*, 1017 (1977).
182. H.-J. Schafer, *Top. Curr. Chem.*, **152**, 90 (1990).
    (b) M. M. Baizer, *Tetrahedron*, **40**, 935 (1984).
    (c) H.-J. Schafer, *Angew. Chem., Int. Ed. Engl.*, **20**, 911 (1981).
183. N. Rabjohn and G. W. Flasch, Jr., *J. Org. Chem.*, **46**, 4082 (1981).
184. S. H. Brown and R. H. Crabtree, *J. Chem. Soc., Chem. Commun.*, 970 (1987).
185. S. H. Brown and R. H. Crabtree, *Tetrahedron Lett.*, **28**, 5599 (1987).
186. (a) D. P. Curran, *Synthesis*, 417, 489 (1988).
    (b) B. Giese, *Radicals in Organic Synthesis: Formation of Carbon–Carbon Bonds*, Pergamon Press, Oxford, 1986.
    (c) M. Ramaiah, *Tetrahedron*, **43**, 3541 (1987).
187. (a) J. D. Winkler, V. Sridar and M. G. Siegel, *Tetrahedron Lett.*, **30**, 4943 (1989).
    (b) J. D. Winkler and V. Sridar, *J. Am. Chem. Soc.*, **108**, 1708 (1986).
    (c) A. L. J. Beckwith, D. H. Roberts, C. H. Schiesser and A. Wallner, *Tetrahedron Lett.*, **26**, 3349 (1985).
    (d) L. A. Paquette, J. A. Colapret and D. R. Andrews, *J. Org. Chem.*, **50**, 201 (1985).
    (e) D. Wehle and L. Fitjer, *Angew. Chem., Int. Ed. Engl.*, **26**, 130 (1987).
188. J. L. Wardell and E. S. Paterson, in *Chemistry of the Metal–Carbon Bond* (Eds. F. R. Hartley and S. Patai), Vol. 2, Chap. 3, Wiley, Chichester, 1985.
189. W. F. Bailey, T. T. Nurmi, J. J. Patricia and W. Wang, *J. Am. Chem. Soc.*, **109**, 2442 (1987).
190. W. F. Bailey and K. Rossi, *J. Am. Chem. Soc.*, **111**, 765 (1989).
191. P. Rigollier, J. R. Young, L. A. Fowley and J. R. Stille, *J. Am. Chem. Soc.*, **112**, 9441 (1990).
192. R. B. Kelly, J. Eber and H.-K. Hung, *J. Chem. Soc., Chem. Commun.*, 689 (1973).
193. (a) M. Demuth and G. Mikhail, *Synthesis*, 145 (1989).
    (b) P. Margaretha, *Top. Curr. Chem.*, **103**, 1 (1982).
194. (a) G. D. Andrews and J. E. Baldwin, *J. Am. Chem. Soc.*, **99**, 4851 (1977); P. Dowd and H. Irangartinger, *Chem. Rev.*, **89**, 985 (1989).
    (b) D. M. Lemal and J. P. Lokensgard, *J. Am. Chem. Soc.*, **88**, 5934 (1966); W. Schafer, R. Criegee, R. Askani and H. Gruner, *Angew. Chem., Int. Ed. Engl.*, **6**, 78 (1967).
    (c) R. Gleiter and M. Karcher, *Angew. Chem., Int. Ed. Engl.*, **27**, 840 (1988).
    (d) P. E. Eaton, L. Cassar, R. A. Hudson and D. R. Hwang, *J. Org. Chem.*, **41**, 1445 (1976).
    (e) N. Skuballa, H. Musso and W. Boland, *Tetrahedron Lett.*, **31**, 497 (1990).
    (f) P. K. Freeman and D. M. Balls, *J. Org. Chem.*, **32**, 2354 (1967).
    (g) H. Prinzbach and D. Hunkler, *Angew. Chem., Int. Ed. Engl.*, **6**, 247 (1967).
    (h) E. Wiskott and P.v.R. Schleyer, *Angew. Chem., Int. Ed. Engl.*, **6**, 694 (1967).
    (i) E. L. Allred and B. R. Beck, *J. Am. Chem. Soc.*, **95**, 2393 (1973).
195. N. J. Jones, W. D. Deadman and E. LeGoff, *Tetrahedron Lett.*, 2087 (1973).
196. (a) L. A. Paquette, Y. Miyahara and C. W. Doecke, *J. Am. Chem. Soc.*, **108**, 1716 (1986).
    (b) L. A. Paquette, R. J. Ternansky, D. W. Balogh and G. Kentgen, *J. Am. Chem. Soc.*, **105**, 5446 (1983).
197. W.-D. Fessner, B. A. R. C. Murty, J. Worth, D. Hunkler, H. Fritz, H. Prinzbach, W. D. Roth, P. v. R. Schleyer, A. B. McEwen and W. F. Maier, *Angew. Chem., Int. Ed. Engl.*, **26**, 452 (1987).
198. (a) M. Jones, Jr. and R. A. Moss (Eds.), *Carbenes*, Wiley, New York, 1973.
    (b) S. Patai (Ed.), *The Chemistry of Alkenes*, Wiley–Interscience, London, 1968, pp. 633–671.
    (c) R. Huisgen, R. Grashey and J. Sauer, in *The Chemistry of Alkenes* (Ed. S. Patai), Wiley–Interscience, London 1968, pp. 755–776.
199. H. E. Simmons, T. L. Cairns, S. A. Vladuchick and C. M. Hoiness, *Org. React.*, **20**, 1 (1973).
200. (a) O. Repic and S. Vogt, *Tetrahedron Lett.*, **23**, 2729 (1982).
    (b) O. Repic, P. G. Lee and U. Giger, *Org. Prep. Proced. Int.*, **16**, 25 (1984).

201. (a) L. Fitjer and J.-M. Conia, *Angew. Chem., Int. Ed. Engl.*, **12**, 332, 334, 761 (1973); J. L. Ripoll, J. Limasset and J.-M. Conia, *Tetrahedron*, **27**, 2431 (1971); J. L. Ripoll and J.-M. Conia, *Tetrahedron Lett.*, 979 (1969).
(b) L. Fitjer, *Angew. Chem., Int. Ed. Engl.*, **15**, 762 (1976).
(c) L. K. Bee, J. Beeby, J. W. Everett and P. J. Garatt, *J. Org. Chem.*, **40**, 2212 (1975).
(d) A. de Meijere, H. Wenck and J. Kopf, *Tetrahedron*, **44**, 2427 (1988).
(e) L. Fitjer, *Angew. Chem., Int. Ed. Engl.*, **15**, 763 (1976).
202. (a) L. Skattebol, *J. Org. Chem.*, **31**, 2789 (1966).
(b) G. Szeimies, in *Strain and its Implications in Organic Synthesis* (Eds. A. de Meijere and S. Blechert), Kluwer Acad. Publishers, Dordrecht, 1989.
(c) K. Semmler, G. Szeimies and J. Belzner, *J. Am. Chem. Soc.*, **107**, 6410 (1985); P. Kaszynski and J. Michl, *J. Org. Chem.*, **53**, 4593 (1988).
(d) G. L. Closs and R. B. Larrabee, *Tetrahedron Lett.*, 287 (1965).
(e) J. Belzner and G. Szeimies, *Tetrahedron Lett.*, **28**, 3099 (1987).
(f) V. Vinkovic and Z. Majerski, *J. Am. Chem. Soc.*, **104**, 4027 (1982).
(g) K. Hirao, Y. Ohuchi and O. Yonemitsu, *J. Chem. Soc., Chem. Commun.*, 99 (1982).
(h) Z. Majerski and M. Zuanic, *J. Am. Chem. Soc.*, **109**, 3496 (1987).
203. (a) Y. Inamoto, K. Aigami, N. Takaishi, Y. Fujikura, K. Tsuchihashi and H. Ikeda, *J. Org. Chem.*, **42**, 3833 (1977).
(b) A. Nickon and G. D. Pandit, *Tetrahedron Lett.*, 3663 (1968).
(c) R. Vaidyanathaswami and D. Devaprabhakara, *Chem. Ind.*, 515 (1968).
(d) D. Bosse and A. de Meijere, *Tetrahedron Lett.*, 1155 (1977).
(e) L. A. Paquette, A. R. Browne and E. Chamot, *Angew. Chem., Int. Ed. Engl.*, **18**, 546 (1979).
(f) K.-i. Hirao, Y. Ohuchi and O. Yonemitzu, *J. Chem. Soc., Chem. Commun*, 99 (1982).
204. (a) P. G. Gassman, K. T. Mansfield, and T. J. Murphy, *J. Am. Chem. Soc.*, **91**, 1684 (1969).
(b) J. A. Jenkins, R. E. Doehner, Jr. and L. A. Paquette, *J. Am. Chem. Soc.*, **102**, 2131 (1980).
(c) T. J. Katz and N. Acton, *J. Am. Chem. Soc.*, **95**, 2738 (1973).
(d) P. J. Garratt and J. F. White, *J. Org. Chem.*, **42**, 1733 (1977).
(e) E. L. Allred and B. R. Beck, *Tetrahedron Lett.*, 437 (1974).
205. K. Ohno and J. Tsuji, *J. Am. Chem. Soc.*, **90**, 99 (1968).
206. (a) G. Rauscher, T. Clark, D. Poppinger and P. v. R. Schleyer, *Angew. Chem., Int. Ed. Engl.*, **17**, 276 (1978).
(b) G. Maier, S. Pfriem, U. Schafer and R. Matusch, *Angew. Chem., Int. Ed. Engl.*, **17**, 520 (1978).
(c) G. Maier, *Angew. Chem., Int. Ed. Engl.*, **27**, 309 (1988).
207. L. Fitjer and Q. Quabeck, *Angew. Chem., Int. Ed. Engl.*, **26**, 1023 (1987).
208. (a) A. P. Marchand and A. Wu, *J. Org. Chem.*, **51**, 1897 (1986).
(b) M. Luyten and R. Keese, *Angew. Chem., Int. Ed. Engl.*, **23**, 390 (1984).
209. S. A. Godleski, P. v. R. Schleyer, E. Osawa and W. T. Wipke, in *Progress in Physical Organic Chemistry* (Ed. R. W. Taft), Vol. 13, Wiley, New York, 1981.
210. (a) E. Osawa, Y. Tahara, A. Togashi, T. Iizuka, N. Tanaka, T. Kan, D. Farcasiu, G. J. Kent, E. M. Engler and P. v. R. Schleyer, *J. Org. Chem.*, **47**, 1923 (1982).
(b) E. Osawa, A. Furusaki, N. Hashiba, T. Matsumoto, V. Singh, Y. Tahara, E. Wiskott, M. Farcasiu, T. Iizuka, N. Tanaka, T. Kan and P. v. R. Schleyer, *J. Org. Chem.*, **45**, 2985 (1980).
(c) W. D. Graham, P. v. R. Schleyer, E. W. Hagaman and E. Wenkert, *J. Am. Chem. Soc.*, **95**, 5785 (1973).
(d) P. v. R. Schleyer, E. Osawa and M. G. B. Drew, *J. Am. Chem. Soc.*, **90**, 5034 (1968).
(e) R. Hamilton, M. A. McKervey, J. J. Rooney and J. F. Malone, *J. Chem. Soc., Chem.Commun.*, 1027 (1976).
(f) C. Cupas, V. Z. Williams, Jr., P. v. R. Schleyer and D. J. Trecker, *J. Am. Chem. Soc.*, **87**, 917 (1965).
211. W. Burns, M. A. McKervey and J. J. Rooney, *J. Chem. Soc., Chem. Commun.*, 965 (1975).
212. W. Burns, M. A. McKervey, T. R. B. Mitchell and J. J. Rooney, *J. Am. Chem. Soc.*, **100**, 906 (1978).
213. G. G. Christoph, P. Engel, R. Usha, D. W. Balogh and L. A. Paquette, *J. Am. Chem. Soc.*, **104**, 784 (1982).
214. Y. Fujikura, N. Takaishi and Y. Inamoto, *Tetrahedron*, **37**, 4465 (1981).

215. (a) L. Cassar, P. E. Eaton and J. Halpern, *J. Am. Chem. Soc.*, **92**, 6366 (1970).
(b) L. A. Paquette, R. S. Beckley and W. B. Farnham, *J. Am. Chem. Soc.*, **97**, 1089 (1975), L. A. Paquette, R. A. Boggs, W. B. Farnham and R. S. Beckley, *J. Am. Chem. Soc.*, **97**, 1112 (1975).
(c) J. Schafer, K. Polborn and G. Szeimies *Chem. Ber.*, **121**, 2263 (1988).
216. L. A. Paquette, J. D. Kramer, P. B. Lavrik and M. J. Wyvratt, *J. Org. Chem.*, **42**, 503 (1977).

CHAPTER **13**

Electrophilic reactions on alkanes

GEORGE A. OLAH and G. K. SURYA PRAKASH

Katherine B. and Donald P. Loker Hydrocarbon Research Institute and Department of Chemistry, University of Southern California, Los Angeles, CA 90089-1661, USA

I. INTRODUCTION

Before the end of the 19th century, saturated hydrocarbons (paraffins) played only a minor role in industrial chemistry. They were mainly used as a source of paraffin wax as well as for heating and lighting oils. Aromatic compounds such as benzene, toluene, phenol and naphthalene obtained from destructive distillation of coal were the main source of organic materials used in the preparation of dyestuffs, pharmaceutical products, etc. Calcium carbide-based acetylene was the key starting material for the emerging synthetic organic industry. It was the ever increasing demand for gasoline after the first world war that initiated study of isomerization and cracking reactions of petroleum

The Chemistry of Alkanes and Cycloalkanes
Edited by S. Patai and Z. Rappoport © 1992 John Wiley & Sons Ltd

fractions. After the second world war, rapid economic expansion necessitated more and more abundant and cheap sources for chemicals and this resulted in the industry switching over to petroleum-based ethylene as the main source of chemical feed-stock. One of the major difficulties that had to be overcome is the low reactivity of some of the major components of the petroleum. The lower boiling components (up to 250 °C) are mainly straight-chain saturated hydrocarbons or paraffins which, as their name indicates (parum affinis:slight reactivity), have very little reactivity. Consequently, the lower paraffins were cracked to give olefins (mainly ethylene, propylene and butylenes). The straight-chain liquid hydrocarbons have also very low octane numbers which make them less desirable as gasoline components. To transform these paraffins into useful components for gasoline and other chemical applications, they have to undergo diverse reactions such as isomerization, cracking or alkylation. These reactions, which are used on a large scale in industrial processes, necessitate acidic catalysts (at temperature around 100 °C) or noble metal catalysts (at higher temperature, 200–500 °C) capable of activating the strong covalent C—H or C—C bonds[1]. Since the early 1960s, superacids[2] are known to react with saturated hydrocarbons, even at temperatures much below 0 °C. This discovery initiated extensive studies devoted to electrophilic reactions and conversions of saturated hydrocarbons, including the parent methane. This review will encompass electrophilic reactions of alkanes using superacidic catalysts.

II. C—H (AND C—C) BOND PROTOLYSIS AND HYDROGEN–DEUTERIUM EXCHANGE

The fundamental step in the acid-catalyzed hydrocarbon conversion processes is the formation of the intermediate carbocations R^+ (equation 1). Whereas all studies involving isomerization, cracking and alkylation reactions under acidic conditions agree that a trivalent carbocation (carbenium ion) is the key intermediate, the mode of formation of this reactive species from the neutral hydrocarbon remained controversial for many years.

$$R—H \xrightarrow{\text{Acid}} R^+ \begin{cases} \rightarrow \text{Isomerization} \\ \rightarrow \text{Cracking} \\ \rightarrow \text{Alkylation–homologation} \end{cases} \qquad (1)$$

In 1946 Bloch, Pines and Schmerling[3] observed that n-butane would isomerize to isobutane under the influence of pure aluminum chloride only in the presence of HCl. They proposed that the ionization step takes place through initial protolysis of the alkane as indicated by formation of minor amounts of hydrogen in the initial stage of the reaction (equation 2). The first direct evidence of protonation of alkanes under superacid conditions was reported independently by Olah and Lukas[4] as well as Hogeveen and coworkers[5].

$$n\text{-}C_4H_{10} + HCl \xrightarrow{\text{AlCl}_3} sec\text{-}C_4H_9^+ {}^-AlCl_4 + H_2 \qquad (2)$$

When n-butane or isobutane was reacted with $HSO_3F:SbF_5$ (Magic Acid®), t-butyl cation was formed (equation 3) exclusively as evidenced by a sharp singlet at 4.5 ppm (from TMS) in the ^1H-NMR spectrum. In excess Magic Acid, the stability of the ion is remarkable and the NMR spectrum of the solution remains unchanged even after having been heated to 110 °C.

$$(CH_3)_3CH \xrightarrow[\text{room temperature}]{HSO_3F:SbF_5} (CH_3)_3C^+ SbF_5FSO_3^- + H_2 \xleftarrow{HSO_3F:SbF_5} n\text{-}C_4H_{10} \qquad (3)$$

It was shown[6] that the t-butyl cation undergoes degenerate carbon scrambling at higher temperatures. A lower limit of $E_a \sim 30\,\text{kcal mol}^{-1}$ was estimated for the scrambling process which could correspond to the energy difference between t-butyl and primary isobutyl cation (the latter being partially delocalized, 'protonated cyclopropane') involved in the isomerization process.

n-Pentane and isopentane are ionized under the same conditions to the t-amyl cation. n-Hexane and the branched C_6 isomers ionize in the same way to yield a mixture of the three tertiary hexyl ions as shown by their ^{1}H-NMR spectra (equation 4).

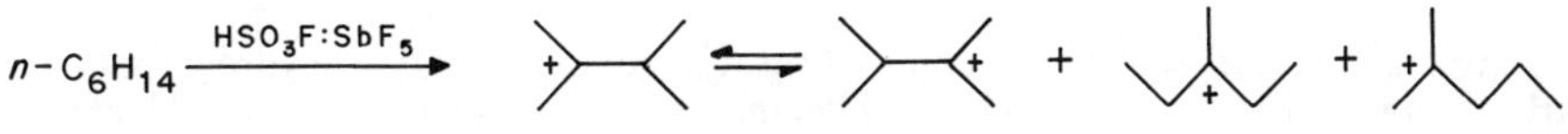

$$n-C_6H_{14} \xrightarrow{\text{HSO}_3\text{F:SbF}_5} \quad \succ\!\!\!-\!\!\!\prec \quad \rightleftharpoons \quad \succ\!\!\!-\!\!\!\prec^+ \quad + \quad \bigvee\!\!\bigvee \quad + \quad \bigvee\!\!\!\bigwedge$$

$$(4)$$

Both methylcyclopentane and cyclohexane were found to give the methylcyclopentyl ion which is stable at low temperature, in excess superacid[7]. When alkanes with seven or more carbon atoms were used, cleavage was observed with formation of the stable t-butyl cation. Even paraffin wax and polyethylene ultimately gave the t-butyl cation after complex fragmentation and ionization processes.

In compounds containing only primary hydrogen atoms such as neopentane and 2,2,3,3-tetramethylbutane, a carbon–carbon bond is broken rather than a carbon–hydrogen bond (equation 5)[8].

$$\begin{array}{c} \text{CH}_3 \\ | \\ \text{CH}_3-\overset{\displaystyle |}{\underset{\displaystyle |}{\text{C}}}-\text{CH}_3 \xrightarrow{\text{H}^+} \text{CH}_4 + (\text{CH}_3)_3\text{C}^+ \\ \text{CH}_3 \end{array} \qquad (5)$$

Hogeveen and coworkers have suggested a linear transition state for the protolytic ionization of hydrocarbons. This, however, may be the case only in sterically congested systems. Results of protolytic reactions of hydrocarbons in superacidic media were interpreted by Olah as an indication for the general electrophilic reactivity of covalent C—H and C—C single bonds of alkanes and cycloalkanes (equation 6). The reactivity is due to the σ-donor ability of a shared electron pair (of σ-bond) via two-electron, three-center bond formation. The transition state of the reactions are consequently of a three-center bound pentacoordinate carbonium ion nature. Strong indication for the mode of protolytic attack was obtained from deuterium–hydrogen exchange studies. Monodeuteromethane was reported to undergo C—D exchange without detectable side reactions in the HF:SbF$_5$ system (equation 7)[9]. d$_{12}$-Neopentane when treated with Magic Acid was also reported[10] to undergo H—D exchange before cleavage.

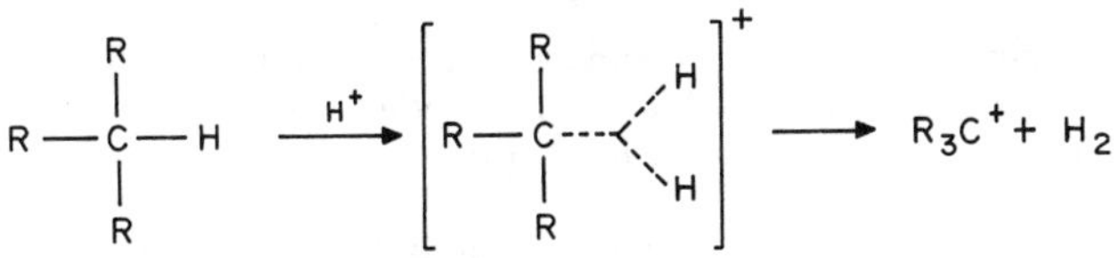

$$R-\overset{\displaystyle R}{\underset{\displaystyle R}{\overset{\displaystyle |}{\underset{\displaystyle |}{\text{C}}}}}-H \xrightarrow{\text{H}^+} \left[R-\overset{\displaystyle R}{\underset{\displaystyle R}{\overset{\displaystyle |}{\underset{\displaystyle |}{\text{C}}}}}\cdots\!\!\prec_{\text{H}}^{\text{H}} \right]^+ \longrightarrow R_3\text{C}^+ + H_2 \qquad (6)$$

$$\text{CH}_3\text{D} + \text{HF:SbF}_5 \longrightarrow \left[H_3\text{C}\cdots\!\!\prec_{\text{D}}^{\text{H}} \right]^+ \longrightarrow \text{CH}_4 + \text{DF:SbF}_5 \qquad (7)$$

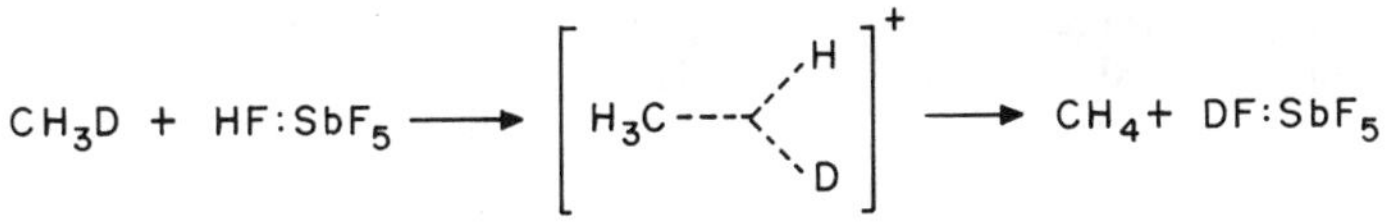

Based on the demonstration of H–D exchange of molecular hydrogen (and deuterium) in superacid solutions[11a], Olah suggested that these reactions go through trigonal isotopic H_3^+ ions in accordance with theoretical calculations and IR studies[11b].

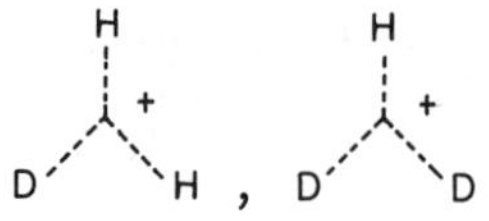

Consequently, the reverse reaction of protolytic ionization of hydrocarbons to carbenium ions, i.e. the reduction of carbenium ion by molecular hydrogen[12,13], can be considered as alkylation of H_2 by the electrophilic carbenium ion through the pentacoordinate carbonium ion. Indeed, Hogeveen has experimentally obtained support for this point by reacting stable alkyl cations in superacids with molecular hydrogen (equation 8)[12].

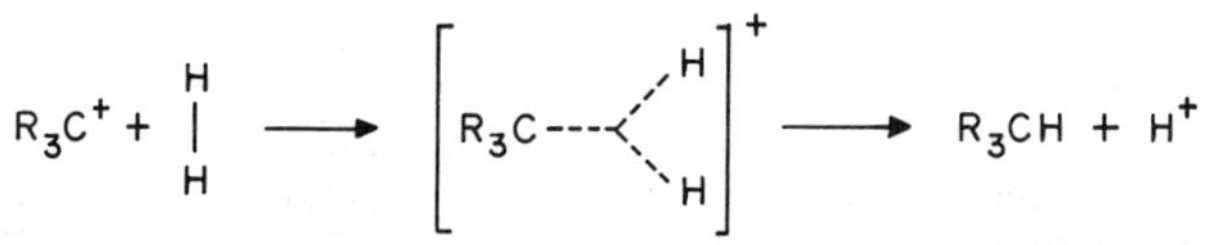

$$R_3C^+ + \begin{array}{c} H \\ | \\ H \end{array} \longrightarrow \left[R_3C \cdots \diagup^{\displaystyle H}_{\diagdown H} \right]^+ \longrightarrow R_3CH + H^+ \qquad (8)$$

Further evidence for the pentacoordinate carbonium ion mechanism of alkane protolysis was obtained in the H–D exchange reaction observed with isobutane. When isobutane is treated with deuterated superacids ($DSO_3F:SbF_5$ or $DF:SbF_5$) at low temperature ($-78\,^{\circ}C$) and atmospheric pressure, the initial hydrogen–deuterium exchange is observed only at the tertiary carbon. Ionization yields only deuterium-free t-butyl cation and HD^{14a}. Recovered isobutane from the reaction mixture shows at low temperature only methine hydrogen–deuterium exchange. This result is best explained as proceeding through a two-electron, three-center bound pentacoordinate carbonium ion (equation 9). Adamantane also undergoes facile H–D exchange at the bridgehead positions[14b].

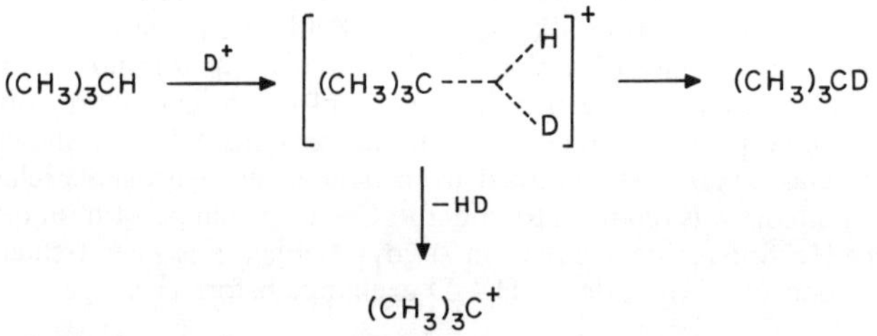

$$(CH_3)_3CH \xrightarrow{\quad D^+ \quad} \left[(CH_3)_3C \cdots \diagup^{\displaystyle H}_{\diagdown D} \right]^+ \longrightarrow (CH_3)_3CD \qquad (9)$$

$$\downarrow_{-HD}$$

$$(CH_3)_3C^+$$

The H–D exchange in isobutane in superacid media is fundamentally different from the H–D exchange observed by Otvos and coworkers[15] in D_2SO_4, who found the eventual exchange of all the nine methyl hydrogens but not the methine hydrogen (equation 10).

Otvos suggested[15] that under the reaction conditions a small amount of t-butyl cation is formed in an oxidative step, which then deprotonates to isobutylene. The reversible protonaton (deuteration) of isobutylene was responsible for the H–D exchange on the methyl hydrogens, whereas tertiary hydrogen is involved in intermolecular hydride transfer from unlabeled isobutane (at the CH position). Under superacidic conditions, where no olefin formation occurs, the reversible isobutylene protonation cannot be involved in the exchange reaction. On the other hand, a kinetic study of hydrogen

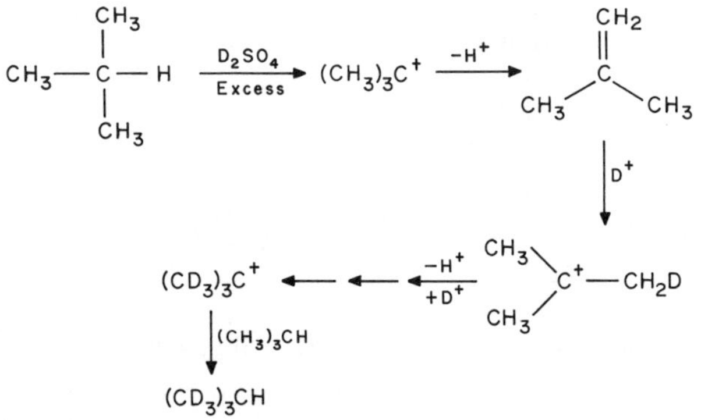

$$
\underset{\substack{|\\CH_3}}{\overset{\substack{CH_3\\|}}{CH_3-C-H}} \quad \xrightarrow[\text{Excess}]{D_2SO_4} \quad (CH_3)_3C^+ \quad \xrightarrow{-H^+} \quad \underset{CH_3}{\overset{CH_2}{\underset{\diagdown}{\overset{\parallel}{C}}}}\overset{\diagup}{CH_3}
$$

$$\Big\downarrow D^+ \tag{10}$$

$$
(CD_3)_3C^+ \longleftarrow \longleftarrow \underset{+D^+}{\overset{-H^+}{\longleftarrow}} \quad \underset{CH_3}{\overset{CH_3\diagdown}{\diagup}}C^+-CH_2D
$$

$$\Big\downarrow (CH_3)_3CH$$

$$(CD_3)_3CH$$

deuterium exchange in deuteroisobutane[16] showed that the exchange of the tertiary hydrogen was appreciably faster than the hydride abstraction by C—H protolysis.

The nucleophilic nature of the alkanes is also shown by the influence of the acidity level on their solubility. Torck and coworkers[17] have investigated the variation of the composition of the catalytic phase as a function of SbF_5 concentration in isomerization of pentane in $HF:SbF_5$.

The total amount of hydrocarbons increases from 1.6% to 14.6% in weight when the SbF_5 concentration varies from 0 to $6.8\,mol\,l^{-1}$. The amount of carbenium ions increases linearly with the SbF_5 concentration, and the solubility of the hydrocarbon itself reaches a maximum for $5\,mol\,l^{-1}$ of SbF_5. The apparent decrease in solubility of the hydrocarbon at higher SbF_5 to concentration may be due to the rapid rate of hydrocarbon protolysis as well as to change in the composition of the acid, SbF_6^- anions being transformed to $Sb_2F_{11}^-$ anions (equation 11).

$$RH + H^+Sb_2F_{11}^- \longrightarrow R^+Sb_2F_{11}^- + H_2 \tag{11}$$

One of the difficulties in understanding the carbocationic nature of acid-catalyzed transformations of alkanes via the hydride abstraction mechanism was that no stoichiometric amount of hydrogen gas evolution was observed from the reaction mixture. For this reason, a complementary mechanism was proposed involving direct hydride abstraction by the Lewis acid (equation 12)[18].

$$RH + 2SbF_5 \longrightarrow R^+SbF_6^- + SbF_3 + HF \tag{12}$$

Olah, however, has pointed out that if SbF_5 would abstract H^-, it would need to form SbF_5H^- ion involving an extremely weak Sb—H bond compared to the strong C—H bond being broken[14]. Thermodynamic calculations (equation 13) also show that the direct oxidation of alkanes by SbF_5 is not feasible. Hydrogen is generally assumed to be partially consumed in the reduction of one of the superacid components.

$$
\begin{aligned}
2HSO_3F + H_2 &\longrightarrow SO_2 + H_3O^+ + HF + SO_3F^- & \Delta H = -33\,kcal\,mol^{-1}\\
SbF_3 + H_2 &\longrightarrow SbF_2 + 2HF & \Delta H = -49\,kcal\,mol^{-1}
\end{aligned} \tag{13}
$$

The direct reduction of SbF_5 in the absence of hydrocarbons by molecular hydrogen necessitates, however, more forcing conditions (50 atm, high temperature) which suggests that the protolytic ionization of alkanes proceeds probably via solvation of protonated alkane by SbF_5 and concurrent ionization-reduction (equation 14)[14].

Culman and Sommer have reinvestigated[18b] the mechanism of SbF_5 oxidation of saturated hydrocarbons. They found efficient oxidation of isobutane to t-butyl cation by neat SbF_5 free of protic acid, at $-80\,°C$ in the presence of a proton trap like acetone (equation 15). They consequently concluded that proton is not essential for the C—H bond oxidation by SbF_5, contrary to what was previously suggested.

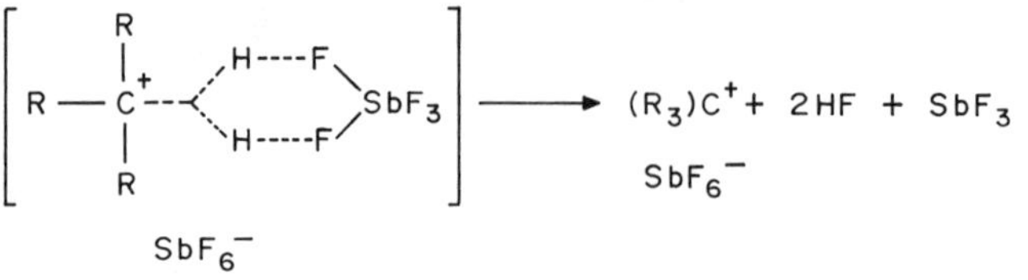

$$\left[R-\underset{\underset{R}{|}}{\overset{\overset{R}{|}}{C}}\overset{+}{-} \left\langle \begin{array}{c} H----F \\ H----F \end{array} \right\rangle SbF_3 \right] \underset{SbF_6^-}{} \longrightarrow (R_3)C^+ + 2HF + SbF_3 \qquad (14)$$

$$SbF_6^-$$

$$(CH_3)_3C\!-\!H + 3SbF_5 + (CH_3)_2C\!=\!O \longrightarrow (CH_3)_2\overset{+}{C}\bar{S}bF_6 + (CH_3)_2C\!=\!\overset{+}{O}H\bar{S}bF_6 + SbF_3 \qquad (15)$$

It must be pointed out, however, that the significant demonstration of the oxidizing ability of pure SbF_5 towards C—H bonds still produces protic acid in the system. In the absence of a proton trap, after the initial stage of the reaction, the *in situ* formed SbF_5:HF conjugate superacid will itself act as a protolytic ionizing agent. The situation is thus similar to the question of the acidic initiator in the polymerization of isobutylene. Even if Lewis acids themselves could initiate the reaction, protic acid formed in the system will provide after the initial stage conjugate Bronsted–Lewis acid to promote protic initiation. Regardless, the study of SbF_5 free of protic acid in initiating carbocationic reactions of alkanes, such as isobutane, much furthered our knowledge of the field.

In studies involving solid acid-catalyzed hydrocarbons cracking reactions using HZSM-5 zeolite, Haag and Dessau[19] were able to account nearly quantitatively for H_2 formed in the protolysis ionization step of the reaction. This is a consequence of the solid acid zeolite catalyst which is not easily reduced by the hydrogen gas evolved.

It must also be pointed out, however, that initiation of acid-catalyzed alkane transformations under oxidative conditions (chemical or electrochemical) can also involve radical cations or radical paths leading to the initial carbenium ions. In the context of our present discussion we shall not elaborate on this interesting chemistry further and limit our treatment to purely protolytic reactions.

Under strongly acidic conditions C—H bond protolysis is not the only pathway by which hydrocarbons are heterolytically cleaved. Carbon–carbon bonds can also be cleaved by protolysis[8,20] involving pentacoordinate intermediates (equation 16).

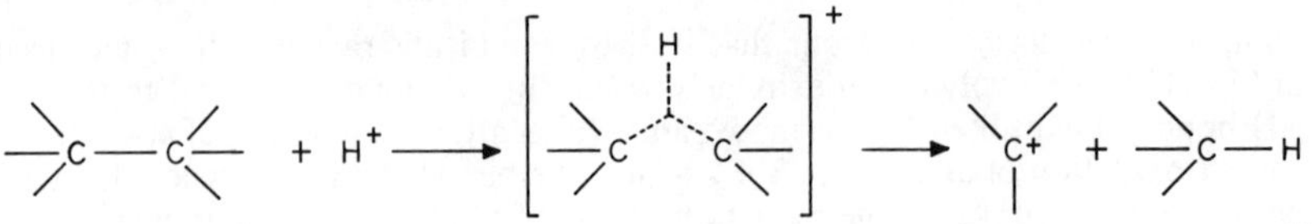

$$\overset{\diagup}{\underset{\diagdown}{C}}\!-\!\overset{\diagup}{\underset{\diagdown}{C}} + H^+ \longrightarrow \left[\overset{\diagup}{\underset{\diagdown}{C}}\cdots\overset{\overset{H}{\cdot}}{\underset{\diagdown}{C}} \right]^+ \longrightarrow \overset{\diagup}{C^+} + \overset{\diagup}{\underset{\diagdown}{C}}\!-\!H$$

$$(16)$$

Similarly, many carbon–heteroatom bonds are also cleaved[2a,21,22] under strong acid catalysis involving pentacoordinate carbon intermediates.

III. ALKANE CONVERSION BY ELECTROCHEMICAL OXIDATION IN SUPERACIDS

In 1973, Fleischmann, Plechter and coworkers[23] showed that the anodic oxidation potential of several alkanes in HSO_3F was dependent on the proton donor ability of the medium. This acidity dependence shows that there is a rapid protonation equilibrium before the electron transfer step and it is the protonated alkane that undergoes oxidation (equation 17).

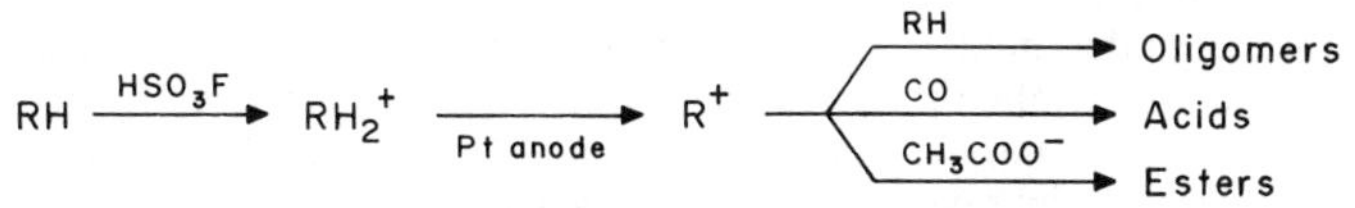

$$\text{RH} \xrightarrow{\text{HSO}_3\text{F}} \text{RH}_2^{+} \xrightarrow[\text{Pt anode}]{} \text{R}^{+}
\begin{cases}
\xrightarrow{\text{RH}} \text{Oligomers} \\
\xrightarrow{\text{CO}} \text{Acids} \\
\xrightarrow{\text{CH}_3\text{COO}^{-}} \text{Esters}
\end{cases} \quad (17)$$

More recently, the electrochemical oxidation of lower alkanes in the HF solvent system has been investigated by Devynck and coworkers over the entire pH range[24]. Classical and cyclic voltammetry show that the oxidation process depends largely on the acidity level. Isopentane (2-methylbutane, M2BH), for example, undergoes two-electron oxidation in HF:SbF$_5$ and HF:TaF$_5$ solutions (equation 18)[25].

$$\text{M2BH} - 2e^{-} \longrightarrow \text{M2B}^{+} + \text{H}^{+} \quad (18)$$

In the higher-acidity region, the intensity–potential curve shows two peaks (at 0.9 V and 1.7 V, respectively, vs the Ag/Ag + system). The first peak corresponds to the oxidation of the protonated alkane and the second to the oxidation of the alkane itself.

The chemical oxidation process in the acidic solution (equation 19) can be considered as a sum of two electrochemical reactions (equations 20 and 21).

$$\text{RH} + \text{H}^{+} \rightleftharpoons \text{R}^{+} + \text{H}_2 \quad (19)$$

$$2\text{H} + 2e^{-} \rightleftharpoons \text{H}_2(\text{E}^0 \text{ for } \text{H}^{+}/\text{H}_2) \quad (20)$$

$$\text{RH} \rightleftharpoons \text{R}^{+} + \text{H}^{+} + 2e^{-} (\text{E}^0 \text{ for } \text{H}^{+}/\text{H}_2) \quad (21)$$

The oxidation of the alkane (M2BH) by H$^+$ gives the carbocation only at pH values below 5.7. In the stronger acids it is the protonated alkane which is oxidized. At pH values higher than 5.7, oxidation of isopentane gives the alkane radical which dimerizes or oxidized in a pH-independent process (Scheme 1).

$$\text{M2BH} \xrightarrow{-e^{-}} \text{M2BH}^{\cdot +} \longrightarrow \text{M2B}^{\cdot} + \text{H}^{+}$$

$$2\text{M2B}^{\cdot} \longrightarrow (\text{M2B})_2$$

$$\text{M2B}^{\cdot} \xrightarrow{-e^{-}} \text{M2B}^{+}$$

SCHEME 1

Commeyras and coworkers[26,27] have also electrochemically oxidized alkanes (anodic oxidation). However, the reaction results in condensation and cracking of alkanes. The results are in agreement with the alkane behavior in superacid media and indicate the ease of oxidation of tertiary alkanes. However, high acidity levels are necessary for the oxidation of alkanes possessing only primary C—H bonds.

Once the alkane has been partly converted into the corresponding carbenium ion, then typical rearrangement, fragmentation, hydrogen transfer as well as alkylation reactions will occur.

IV. ISOMERIZATION OF ALKANES

Acid-catalyzed isomerization of saturated hydrocarbons was first reported in 1933 by Nenitzescu and Dragan[28]. They found that when n-hexane was refluxed with aluminum chloride, it was converted into its branched isomers. This reaction is of major economic importance as the straight-chain C_5–C_8 alkanes are the main constituents of gasoline obtained by refining of the crude oil. Because the branched alkanes have a considerably higher octane number than their linear counterparts, the combustion properties of gasoline can be substantially improved by isomerization.

The isomerization of n-butane to isobutane is of substantial importance because isobutane reacts under mild acidic conditions with olefins to give highly branched hydrocarbons in the gasoline range. A variety of useful products can be obtained from isobutane: isobutylene, t-butyl alcohol, methyl t-butyl ether and t-butyl hydroperoxide[25]. A number of methods involving solution as well as solid acid catalysts[26] have been developed to achieve isomerization of n-butane as well as other linear higher alkanes to branched isomers.

A substantial number of investigations have been devoted to this isomerization reaction and a number of reviews are available[29-31]. The isomerization is an equilibrium reaction that can be catalyzed by various strong acids. In the industrial processes, aluminum chloride and chlorinated alumina are the most widely used catalysts. Whereas these catalysts become active only at temperatures above 80–100 °C, superacids are capable of isomerizing alkanes at room temperature and below. The advantage is that lower temperatures thermodynamically favor the most branched isomers.

The electron donor character of C—H and C—C single bonds that leads to pentacoordinated carbonium ions explains the mechanism of acid-catalyzed isomerization of n-butane, as shown in Scheme 2. Carbon–carbon bond protolysis, however, can also take place giving methane, ethane and propane.

Related alkanes such as pentanes, hexanes and heptanes isomerize by similar pathways with increasing tendency towards cracking (i.e. C—C bond cleavage).

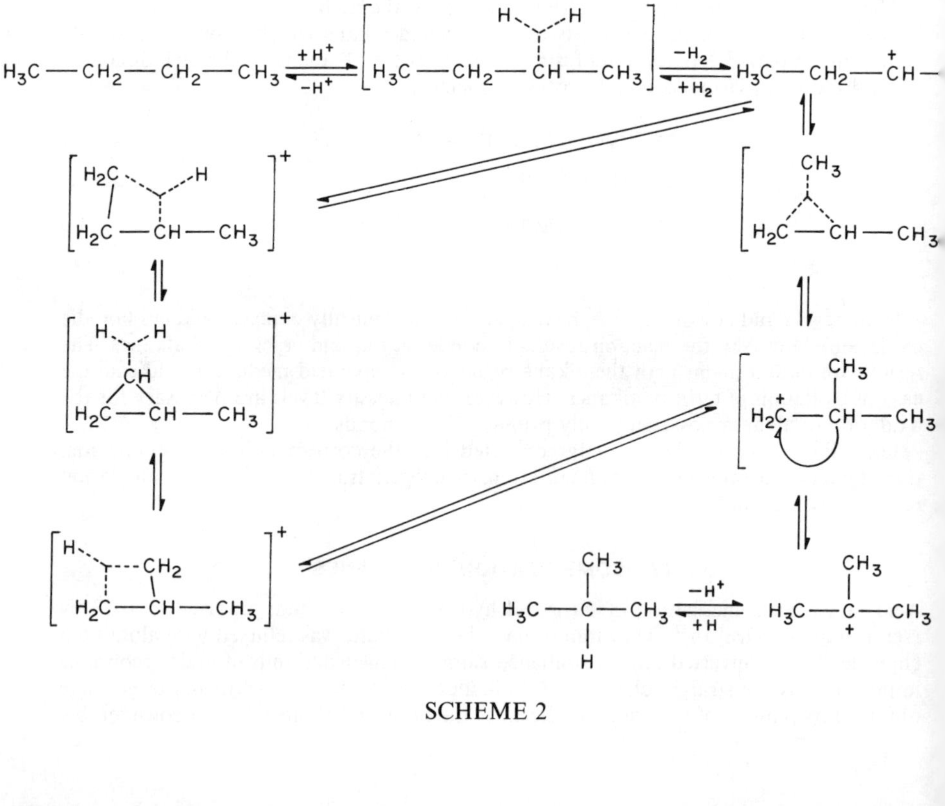

SCHEME 2

During the isomerization process of pentanes, hexanes and heptanes, cracking of the hydrocarbon is an undesirable side reaction. The discovery that cracking can be substantially suppressed by hydrogen gas under pressure was of significant importance. In our present-day understanding, the effect of hydrogen is to quench carbocationic sites through five-coordinate carbocations to the related hydrocarbons, thus decreasing the possibility of C—C bond cleavage reactions responsible for the acid-catalyzed cracking.

Isomerization of n-hexane is superacids can be depicted by the three steps shown in Scheme 3.

Step 1. Formation of the Carbenium ion:

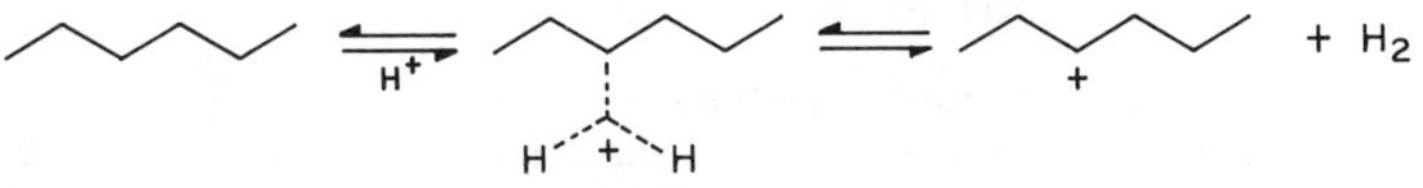

Step 2. Isomerization of the Carbenium ion via Hydride Shifts, Alkide Shifts and Protonated Cyclopropane (for the Branching Step):

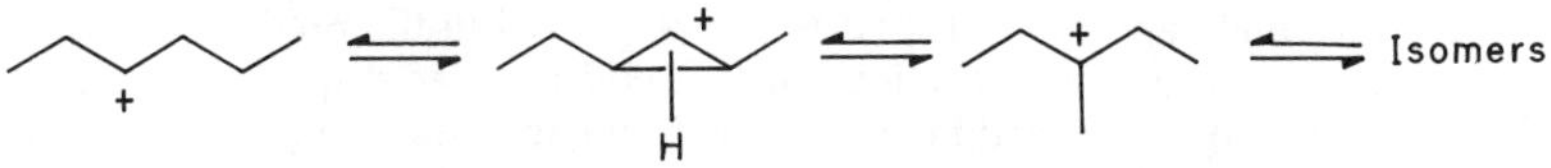

Step 3. Hydride transfer from the Alkane to the Incipient Carbenium ion:

SCHEME 3

Whereas step 1 is stoichiometric, steps 2 and 3 form a catalytic cycle involving the continuous generation of carbenium ions via hydride transfer from a new hydrocarbon molecule (step 3) and isomerization of the corresponding carbenium ion (step 2). This catalytic cycle is controlled by two kinetic and two thermodynamic parameters that can help orient the isomer distribution depending on the reaction conditions. Step 2 is kinetically controlled by the relative rates of hydrogen shifts, alkyl shifts and protonated cyclopropane formation and it is thermodynamically controlled by the relative stabilities of the secondary and tertiary ions. Step 3, however, is kinetically controlled by the hydride transfer from excess of the starting hydrocarbon and by the relative thermodynamic stability of the various hydrocarbon isomers.

For these reasons, the outcome of the reaction will be very different depending on which thermodynamic or kinetic factor will be favored. In the presence of excess hydrocarbon in equilibrium with the catalytic phase and long contact times, the thermodynamic hydrocarbon isomer distribution is attained. However, in the presence of a large excess of acid, the product will reflect the thermodynamic stability of the intermediate carbenium ions (which, of course, is different from that of hydrocarbons) if rapid hydride transfer or quenching can be achieved. Torck and coworkers[17,32] have shown that the limiting step, in the isomerization of n-hexane and n-pentane with the HF:SbF$_5$ acid catalyst, is the hydride transfer with sufficient contact in a batch reactor, as indicated by the thermodynamic isomer distribution of C$_6$ isomers.

The isomerization of n-pentane in superacids of the type R$_F$SO$_3$H:SbF$_5$ (R$_F$ = C$_n$F$_{2n+1}$) has been investigated by Commeyras and coworkers[33]. The influence of parameters such as acidity, hydrocarbon concentration, nature of the perfluoroalkyl group, total pressure, hydrogen pressure, temperature and agitation has been studied. In weaker superacids such as neat CF_3SO_3H, alkanes that have no tertiary hydrogen are isomerized only very slowly, as the acid is not strong enough to abstract hydride to form the initial carbocation. This lack of reactivity can be overcome by introducing initiator carbenium ions in the medium to start the catalytic process. For this purpose, alkenes may be added, which are directly converted into their corresponding carbenium ions by protonation, or alternatively the alkane may be electrochemically oxidized (anodic oxidation) (equation 22). Both methods are useful to initiate isomerization and cracking reactions. The latter method has been studied by Commeyras and coworkers[27].

$$RH \xrightarrow{-e} RH^{+\cdot} \xrightarrow{-H^+} R^\cdot \xrightarrow{-e} R^+ \tag{22}$$

Recently, Olah[34] has developed a method wherein natural gas liquids containing saturated straight-chain hydrocarbons can be conveniently upgraded to highly branched hydrocarbons (gasoline upgrading) using $HF:BF_3$ catalyst. The addition of a small amount of olefins, preferably butenes, increased the reaction rate. This can be readily explained by the formation of alkyl fluorides (HF addition to olefins), whereby an equilibrium concentration of cations is maintained in the system during the upgrading reaction. The gasoline upgrading process is also improved in the presence of hydrogen gas, which helps to suppress side reactions such as cracking and disproportionation and minimize the amount of hydrocarbon products entering the catalyst phase of the reaction mixture. The advantage of the above method is that the catalysts $HF:BF_3$ (being gases at ambient temperatures) can easily be recovered and recycled.

Olah has also found HSO_3F and related superacids as efficient catalysts for the isomerization of n-butane to isobutane[35].

The difficulties encountered in handling liquid superacids and the need for product separation from catalyst in batch processes have stimulated research in the isomerization of alkanes over solid superacids. The isomerization of 2-methyl- and 3-methylpentane and 2,3-dimethylbutane, using SbF_5-intercalated graphite as a catalyst, has been studied in a continuous flow system[36].

The isomerization of cycloalkanes with acid catalysts is also well known. Over SbF_5-intercalated graphite it can be achieved at room temperature without the usual ring opening and cracking reactions which occur at higher temperatures and lower acidities[37]. In the presence of excess hydrocarbon after several hours, the thermodynamic equilibrium is reached for the isomers (Scheme 4). Interconversion between cyclohexane and methylcyclopentane yields the thermodynamic equilibrium mixture. It should be mentioned, however, that the thermodynamic ratio for the neutral hydrocarbon isomerization is very different as compared with the isomerization of the corresponding ions. The large energy difference ($> 10\,kcal\,mol^{-1}$) between the secondary cyclohexyl cation and the tertiary methylcyclopentyl ion means that in solution chemistry in the presence of excess superacid only the latter can be observed[7].

Whereas the cyclohexane–methylcyclopentane isomerization involves initial formation of the cyclohexyl (methylcyclopentyl) cation (i.e. via protolysis of a C—H bond) it should be mentioned that in the acid-catalyzed isomerization of cyclohexane up to 10% hexanes are also formed and this is indicative of C—C bond protolysis.

The potential of other solid superacid catalysts, such as Lewis acid-treated metal oxides for the skeletal isomerization of hydrocarbons, has been studied in a number of cases. The reaction of butane with $SbF_5:SiO_2:TiO_2$ gave the highest conversion forming C_3, iso-C_4, iso-C_5, and traces of higher alkanes. $TiO_2:SbF_5$, on the other hand, gave the highest selectivity for skeletal isomerization of butane. With $SbF_5:Al_2O_3$, however, the conversions were very low[38,39].

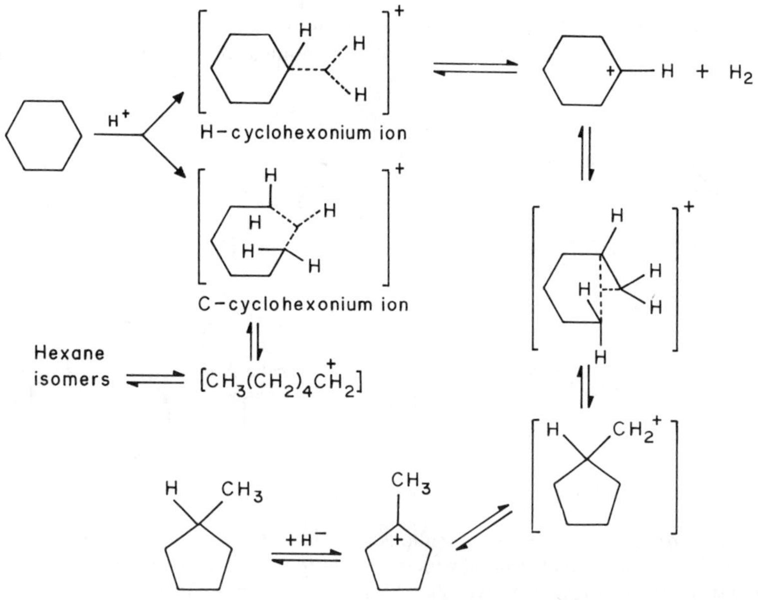

H–cyclohexonium ion

$+$ H $+$ H$_2$

C–cyclohexonium ion

Hexane
isomers $\rightleftharpoons$ $[CH_3(CH_2)_4\overset{+}{C}H_2]$

$+H^-$

SCHEME 4

Similarly, isomerization of pentane and 2-methylbutane over a number of SbF_5-treated metal oxides has been investigated. The $TiO_2:ZrO_2:SbF_5$ system was the most reactive and, at the maximum conversion, the selectivity for skeletal isomerization was found to be ca 100%[40].

A comparison of the reactivity of SbF_5-treated metal oxides with that of HSO_3F-treated catalysts showed that the former is by far the better catalyst for reaction of alkanes at room temperature, although the HSO_3F-treated catalyst showed some potential for isomerization of 1-butane[41].

$SbF_5:SiO_2:Al_2O_3$ has been used to isomerize a series of alkanes at or below room temperature. Methylcyclopentane, cyclohexane, propane, butane, 2-methylpropane and pentane all reacted at room temperature, whereas methane, ethane and 2,2-dimethylpropane could not be activated[41].

The isomerization of a large number of C_{10} hydrocarbons under strongly acidic conditions gives the unusually stable isomer adamantane. The first such isomerization was reported by Schleyer in 1957[42a]. During a study of the facile aluminum chloride-catalyzed endo:exo isomerization of tetrahydrodicyclopentadiene, difficulty was encountered with a highly crystalline material that often clogged distillation heads. This crystalline material was found to be adamantane. Adamantane can be prepared from a variety of C_{10} precursors and involves a series of hydride and alkyde shifts. The mechanism of the reaction has been reviewed in detail[42b]. Fluoroantimonic acid ($HF:SbF_5$) very effectively isomerizes tetrahydrodicyclopentadiene into adamantane[42c]. The work has been extended to other conjugate superacids[43], wherein adamantane is obtained in quantitative yields (equation 23).

More recently[44a] isomerization of polycyclic hydrocarbons $C_{4n+6}H_{4n+12}$ ($n = 1, 2, 3$) led to adamantane, diamantane and triamantane, respectively, using new generation acid systems such as $B(OSO_2CF_3)_3$, $B(OSO_2CF_3)_3 + CF_3SO_3H$ and $SbF_5:CF_3SO_3H$ (equation 24). The yields of these cage compounds substantially improved upon addition of catalytic amount of 1-bromo (chloro, fluoro)-adamantane as a source of 1-adamantyl cation. These reactions were further promoted by sonication (ultrasound treatment).

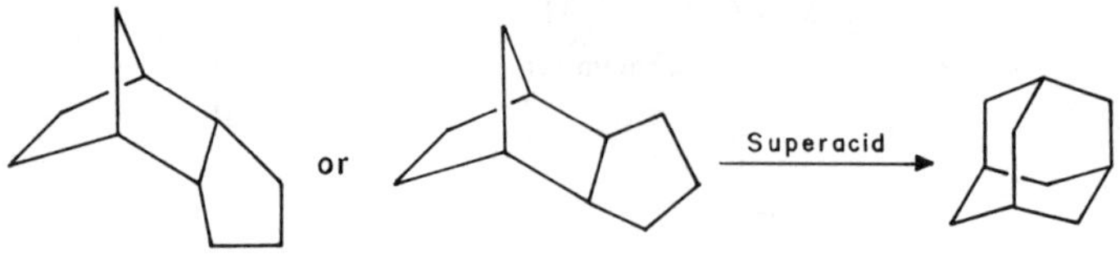

$$\text{or} \qquad \xrightarrow{\text{Superacid}} \qquad \qquad (23)$$

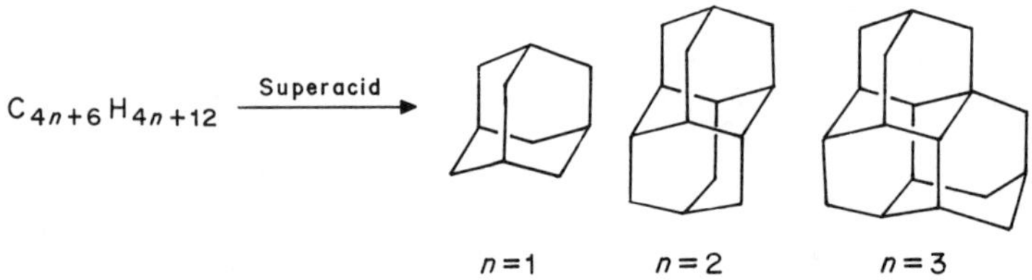

$$C_{4n+6}H_{4n+12} \xrightarrow{\text{Superacid}} \qquad \qquad (24)$$

$$n=1 \qquad\qquad n=2 \qquad\qquad n=3$$

Sodium borohydride in excess of triflic acid at low temperature provides[44b] a highly efficient reductive superacid system, which was found effective for the reductive superacid isomerization of unsaturated polycyclic precursers to cage hydrocarbons in high yields (equation 25). The key to the success of the ionic hydrogenation system appears to be due to the *in situ* formation of boron tris(triflate), a highly superacidic Lewis acid.

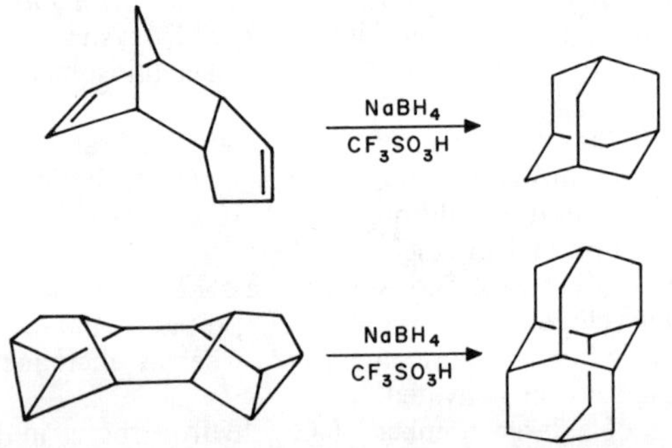

$$\xrightarrow[CF_3SO_3H]{NaBH_4}$$

$$\xrightarrow[CF_3SO_3H]{NaBH_4} \qquad\qquad (25)$$

Vol'pin and coworkers have developed significant new chemistry[45a] with $RCOX \cdot 2AlX_3$ complexes (R = alkyl, aryl; X = Cl, Br) as efficient aprotic superacids for isomerizations and transformation of alkanes and cycloalkanes. The systems also promote cleavage of C—C and C—H σ-bonds. These aprotic superacid systems were reported to be frequently more convenient and superior to protic and Lewis acid systems in their activity.

A good example of superacid catalyzed isomerization is the fascinating synthesis of 1,16-dimethyldodecahedrane from the seco-dimethyldodecahedrene (equation 26)[45b].

A novel approach to highly symmetrical pentagonal dodecahedrane which involved gas phase isomerization of [1.1.1.1]pagodane over 0.1% Pt/Al_2O_3 under an atmosphere of H_2 at 360 °C was provided by Prinzbach, Schleyer, Roth and coworkers[45c]. Dodecahedrane was obtained in 8% yield (equation 27). The isomerization probably involves carbocationic intermediates.

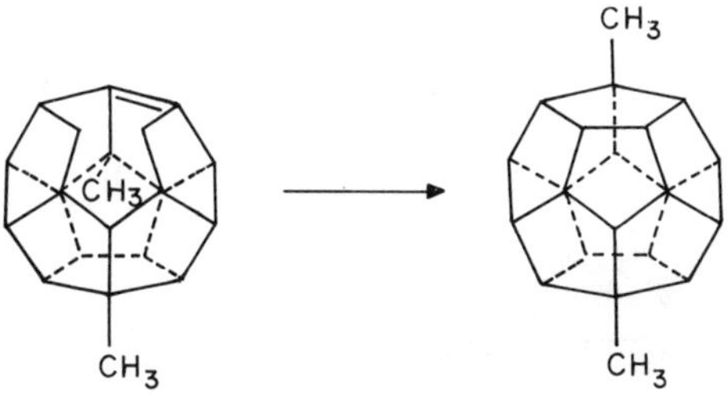

$$\text{(26)}$$

$$\text{(27)}$$

V. ACID-CATALYZED CLEAVAGE (CRACKING) REACTIONS OF ALKANES (β-CLEAVAGE vs C—C BOND PROTOLYSIS)

The reduction in molecular weight of various fractions of crude oil is an important operation in petroleum chemistry. The process is called cracking. Catalytic cracking is usually achieved by passing the hydrocarbons over a metallic or acidic catalyst, such as crystalline zeolites at about 400–600 °C. The molecular-weight reduction involves carbocationic intermediates and the mechanism is based on the β-scission of carbenium ions (equation 28).

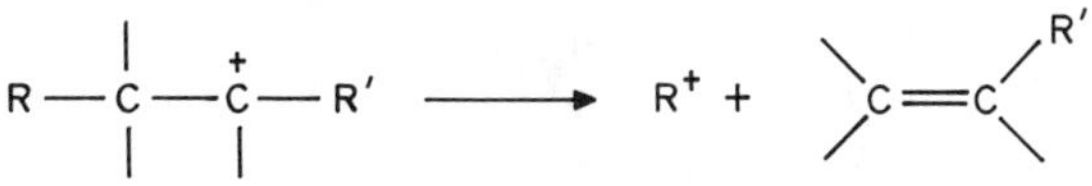

$$R-\overset{|}{\underset{|}{C}}-\overset{+}{\underset{|}{C}}-R' \longrightarrow R^+ + \overset{\diagdown}{}C=C\overset{R'}{\diagup}$$ (28)

The main goal of catalytic cracking is to upgrade higher boiling oils which, through this process, yield lower hydrocarbons in the gasoline range[46,47].

Historically, the first cracking catalyst used was aluminum trichloride. With the development of heterogeneous solid and supported acid catalysts, the use of $AlCl_3$ was soon superceded, since its activity was mainly due to the ability to bring about acid-catalyzed cleavage reactions.

The development of highly acidic superacid catalysts in the 1960s again focused attention on acid-catalyzed cracking reactions. $HSO_3F:SbF_5$, trade-named Magic Acid, derived its name due to its remarkable ability to cleave higher-molecular-weight hydrocarbons, such as paraffin wax, to lower-molecular-weight components, preferentially C_4 and other branched isomers.

As a model for cracking of alkanes, the reaction of 2-methylpentane (MP) over SbF_5-intercalated graphite has been studied in a flow system, the hydrocarbon being diluted in a hydrogen stream[36]. A careful study of the product distribution vs time on stream showed that propane was the initial cracking product whereas isobutane and isopentane (as major cracking products) appear only later.

This result can only be explained by the β-scission of the trivalent 3-methyl-4-pentyl ion as the initial step in the cracking process. Based on this and on the product distribution vs time profile, a general scheme for the isomerization and cracking process of the methylpentanes has been proposed (Scheme 5)[36].

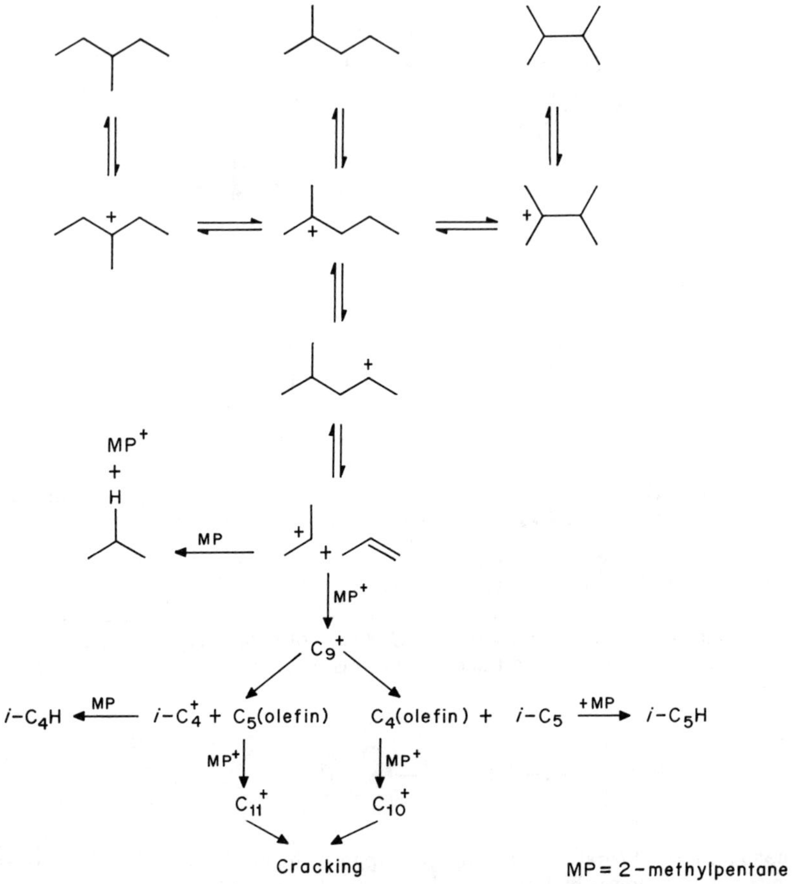
MP⁺
+
H
MP
MP⁺
C9⁺
i-C4H
MP
i-C4⁺ + C5(olefin)
C4(olefin) + i-C5
+MP
i-C5H
MP⁺
MP⁺
C11⁺
C10⁺
Cracking
MP = 2-methylpentane

SCHEME 5

The propene which is formed in the β-scission step never appears as a reaction product because it is alkylated immediately under the superacidic condition by a $C_6{}^+$ carbenium ion, forming a $C_9{}^+$ carbocation which is easily cracked to form a $C_4{}^+$ or $C_5{}^+$ ion and the corresponding C_4 or C_5 alkene. The alkenes are further alkylated by a $C_6{}^+$ carbenium ion in a cyclic process of alkylation and cracking reactions (Scheme 6). The $C_4{}^+$ or $C_5{}^+$

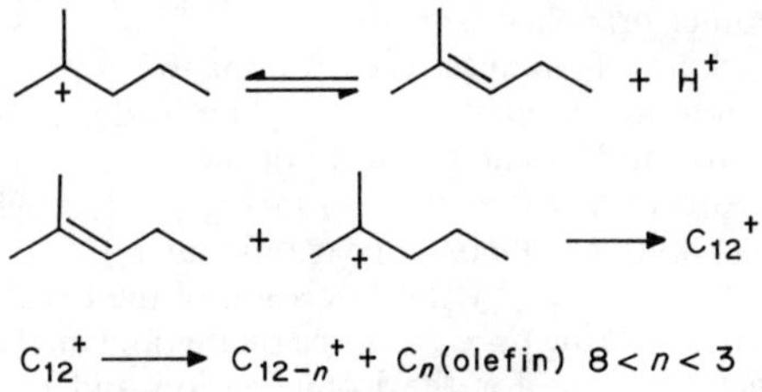
+ H⁺
C12⁺
C12⁺ ⟶ C12−n⁺ + Cn(olefin) 8 < n < 3

SCHEME 6

ions also give the corresponding alkanes (isobutane and isopentane) by hydride transfer from the starting methylpentane. The scheme, which occurs under superacidic conditions, is at variance with the scheme that was proposed for the cracking of C_6 alkanes under less acidic conditions[27].

Under superacidic conditions, it is known that the deprotonation (first type) equilibria lie too far to the left ($K = 10^{-16}$ for isobutane[16]) to make this pathway plausible. On the other hand, among the C_6 isomers, 2-methylpentane is by far the easiest to cleave by β-scission. The 2-methyl-2-pentyl ion is the only species that does not give a primary cation by this process. For this reason, this ion is the key intermediate in the isomerization–cracking reaction of C_6 alkanes.

Under superacidic conditions, β-scission is not the only pathway by which hydrocarbons are cleaved. The C—C bond can also be cleaved by protolysis (equation 29).

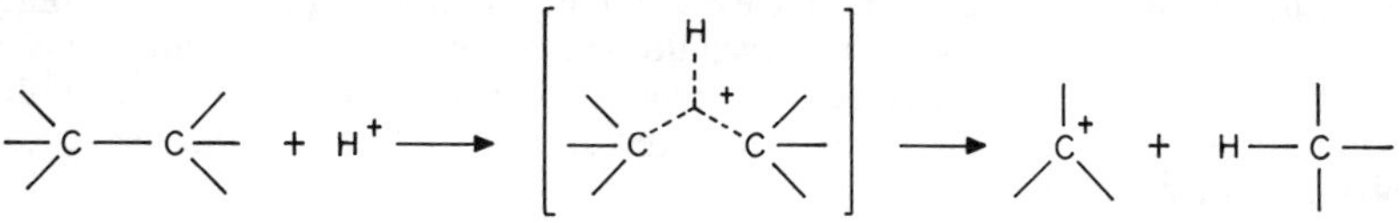

(29)

The protolysis under superacid conditions has been studied independently by Olah and coworkers[48] Hogereen and coworkers[5] and Brouwer and coworkers[49]. The carbon–carbon cleavage in neopentane yielding methane and the t-butyl ion occurs by a mechanism different from the usual β-scission of carbenium ions (equation 30).

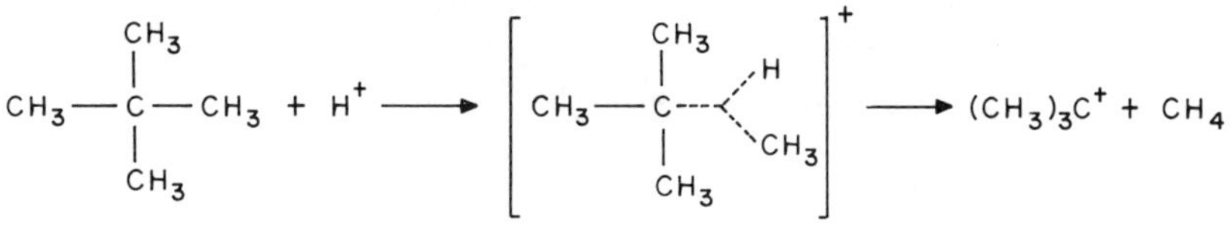

(30)

The protolysis occurs following the direct protonation of the σ-bond providing evidence for the σ-basicity of hydrocarbons. Under slightly different conditions, protolysis of a C—H bond occurs yielding rearranged t-pentyl ion (t-amyl cation, equation 31).

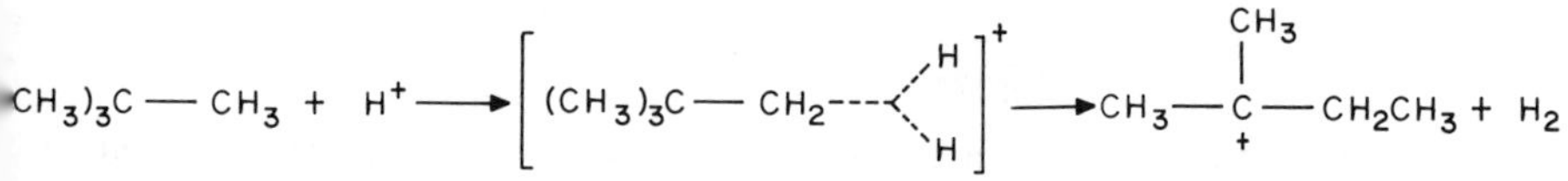

(31)

In cycloalkanes, the C—C bond cleavage leads to ring opening (equation 32)[7].

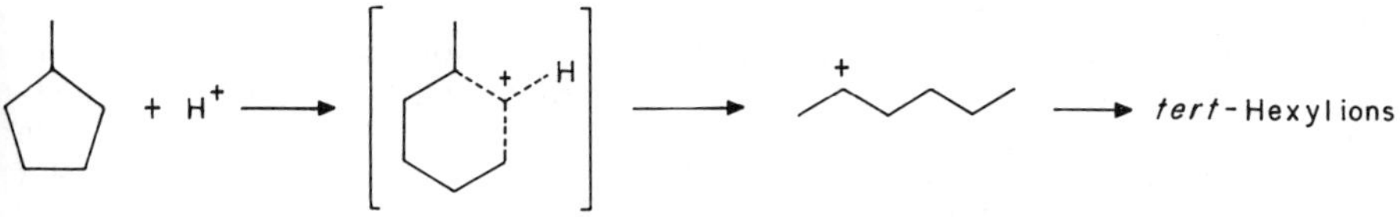

(32)

This reaction is much faster than the carbon–carbon bond cleavage in neopentane, despite the initial formation of secondary carbenium ions. Norbornane is also cleaved in a fast reaction yielding substituted cyclopentyl ions. Thus, protonation of alkanes induces cleavage of the molecule by two competitive ways: (1) protolysis of a C—H bond followed by β-scission of the carbenium ions and (2) direct protolysis of a C—C bond yielding a lower-molecular-weight alkane and a lower-molecular-weight carbenium ion. This reaction, which is of economic importance in the upgrading of higher boiling petroleum fractions to gasoline, has also been shown applicable to coal depolymerization and hydroliquefaction processes[50]. The cleavage of selected model compounds representing coal structural units in the presence of HF, BF_3, and under hydrogen pressure has been studied by Olah and coworkers[50]. Bituminous coal (Illinois No. 6) could be pyridine-solubilized to the extent of 90% by treating it with superacidic $HF:BF_3$ catalyst in the presence of hydrogen gas at 105 °C for 4 h. Under somewhat more elevated temperatures (150–170 °C), cyclohexane extractability of up to 22% and distillability of upto 28% is achieved. Addition of hydrogen donor solvents such as isopentane has been shown to improve the efficiency of coal conversion to cyclohexane-soluble products. The initial 'depolymerization' of coal involves various protolytic cleavage reactions involving those of C—C bonds.

VI. ALKYLATION OF ALKANES AND OLIGOCONDENSATION OF LOWER ALKANES

The alkylation of alkanes by olefins, from a mechanistic point of view, must be considered as the alkylation by the carbenium ion formed by the protonation of the olefin. The well-known acid-catalyzed isobutane–isobutylene reaction demonstrates the mechanism rather well (Scheme 7).

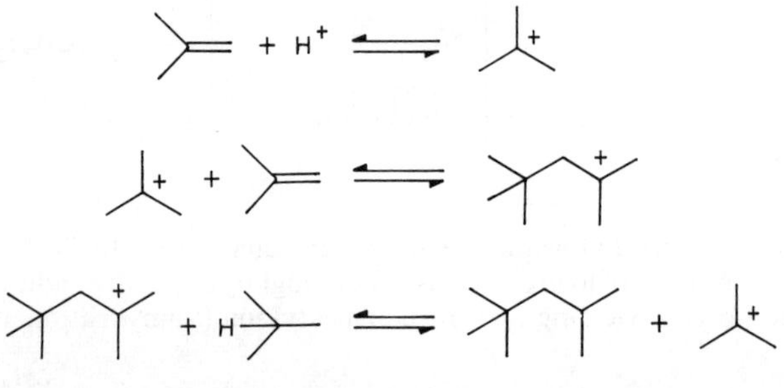

SCHEME 7

As is apparent in the last step, isobutane is not alkylated but transfers a hydride to the $C_8{}^+$ carbocation before being used up in the middle step as the electrophilic reagent (t-butyl cation). The direct alkylation of isobutane by an incipient t-butyl cation would yield 2,2,3,3-tetramethylbutane which, indeed, was observed in small amounts in the reaction of t-butyl cation with isobutane under stable ion conditions at low temperatures (vide infra).

The alkylating ability of methyl and ethyl fluoride–antimony pentafluoride complexes have been investigated by Olah and his group[51,52] showing the extraordinary reactivity of these systems. Self-condensation was observed as well as alkane alkylation. When $CH_3F:SbF_5$ was reacted with excess of CH_3F at 0 °C, at first only an exchanging complex was observed in the ^1H-NMR spectrum. After 0.5 h, the starting material was converted into the t-butyl cation (equation 33). Similar reactions were observed with the ethyl fluoride–antimony pentafluoride complex (equation 34). When the complex was treated with isobutane or isopentane direct alkylation products were observed. To improve the understanding of these alkane alkylation reactions (equation 35), Olah and his group carried out experiments involving the alkylation of the lower alkanes by stable carbenium ions under controlled superacidic stable ion conditions[48,53,54].

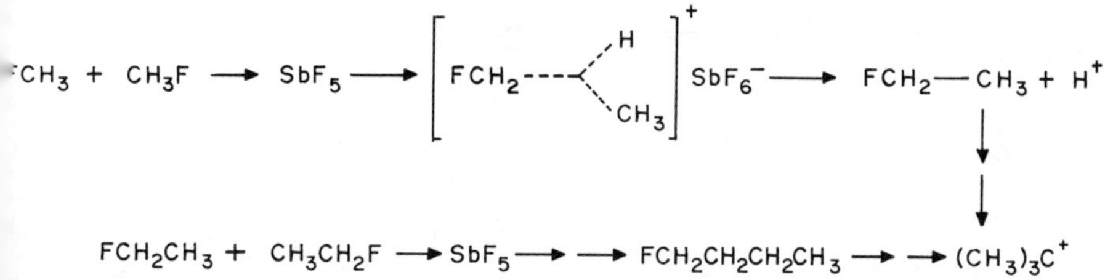

$$\cdot CH_3 + CH_3F \longrightarrow SbF_5 \longrightarrow \left[FCH_2\text{---}\underset{CH_3}{\overset{H}{\diagup}} \right]^+ SbF_6^- \longrightarrow FCH_2\text{---}CH_3 + H^+$$

$$FCH_2CH_3 + CH_3CH_2F \longrightarrow SbF_5 \longrightarrow \longrightarrow FCH_2CH_2CH_2CH_3 \longrightarrow \longrightarrow (CH_3)_3C^+$$

(33)

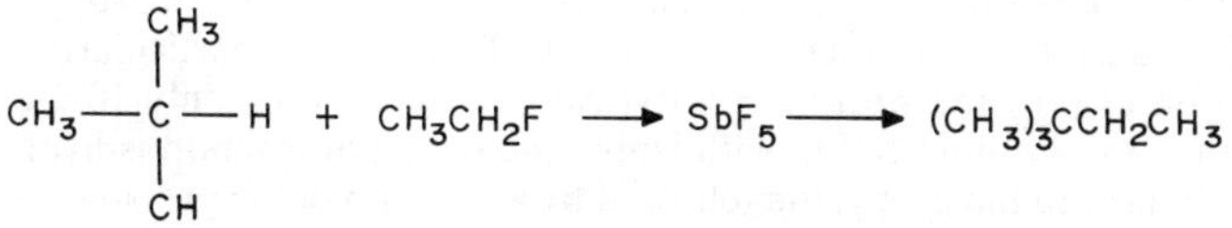

$$CH_3\text{---}\underset{\underset{CH}{|}}{\overset{\overset{CH_3}{|}}{C}}\text{---}H + CH_3CH_2F \longrightarrow SbF_5 \longrightarrow (CH_3)_3CCH_2CH_3 \quad (34)$$

$$R\text{---}H + R'^+ \longrightarrow \left[R\overset{\overset{H}{\vdots}}{\diagdown}R' \right]^+ \longrightarrow R\text{---}R' + H^+ \quad (35)$$

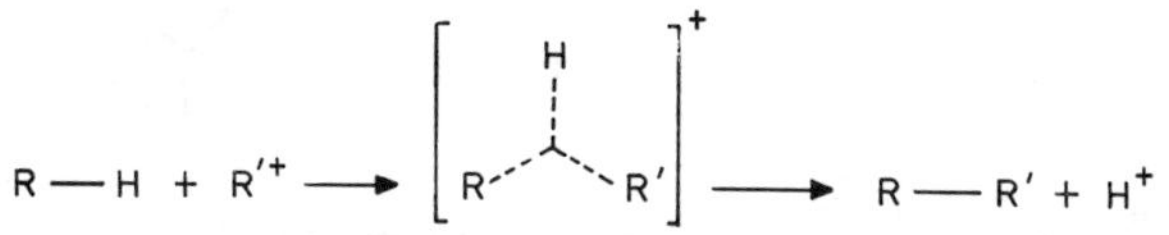

The σ-donor ability of the C—C and C—H bonds in alkanes was demonstrated by a variety of examples. The order of reactivity of single bonds was found to be: tertiary C—H > C—C > secondary C—H ≫ primary C—H, although various specific factors such as steric hindrance can influence the relative reactivities.

Typical alkylation reactions are those of propane, isobutane and n-butane by the *t*-butyl or *sec*-butyl ion. These systems are somewhat interconvertible by competing hydride transfer and rearrangement of the carbenium ions. The reactions were carried out using alkyl carbenium ion hexafluoroantimonate salts prepared from the corresponding halides and antimony pentafluoride in sulfuryl chloride fluoride solution and treating them in the same solvent with alkanes. The reagents were mixed at −78 °C, warmed up to −20 °C and quenched with ice water before analysis. The intermolecular hydride transfer between tertiary and secondary carbenium ions and alkanes is generally much faster than the alkylation reaction. Consequently, the alkylation products are also those derived from the new alkanes and carbenium ions formed in the hydride transfer reaction.

Propylation of propane by the isopropyl cation (equation 36), for example, gives a significant amount (26% of the C_6 fraction) of the primary alkylation product.

$$\diagup\!\!\!\diagdown + {}^+\diagdown\!\!\!\diagup \longrightarrow \left[\diagup\!\!\!\diagdown\overset{\overset{H}{\vdots^+}}{}\diagdown\!\!\!\diagup \right] \longrightarrow \diagup\!\!\!\diagdown\!\!\!\diagup \quad (36)$$

The C_6 isomer distribution, 2-methylpentane (28%), 3-methylpentane (14%) and n-hexane (32%) is very far from thermodynamic equilibrium, and the presence of these isomers indicates that not only isopropyl cation but also n-propyl cation are involved as intermediates (as shown by $^{13}C_2-^{13}C_1$ scrambling in the stable ion, see equation 37)[55].

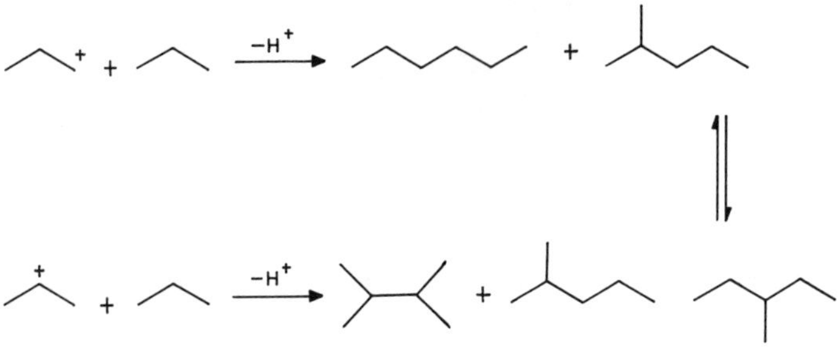

$$\text{(37)}$$

The strong competition between alkylation and hydride transfer appears in the alkylation reaction of propane by butyl cations, or of butanes by the propyl cation. The amount of C_7 alkylation products is rather low. This point is particularly emphasized in the reaction of propane with the t-butyl cation which yields only 10% of heptanes. In the interaction of propyl cation with isobutane the main reaction is hydride transfer from the isobutane to the propyl ion followed by alkylation of the propane by the propyl ions (equation 38).

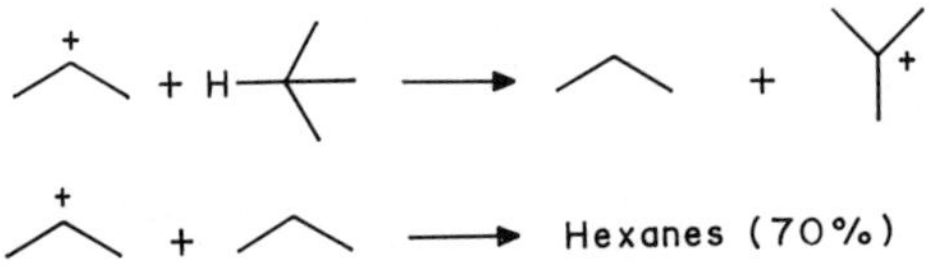

$$\text{(38)}$$

Even the alkylation of isobutane by the t-butyl cation despite the highly unfavorable steric interaction has been demonstrated[48] by the formation of small amounts of 2,2,3,3-tetramethylbutane (equation 39). This result also indicates that the related five-coordinate carbocationic transition state (or high lying intermediate) of the degenerate isobutane-t-butyl cation hydride transfer reaction is not entirely linear, despite the highly crowded nature of the system.

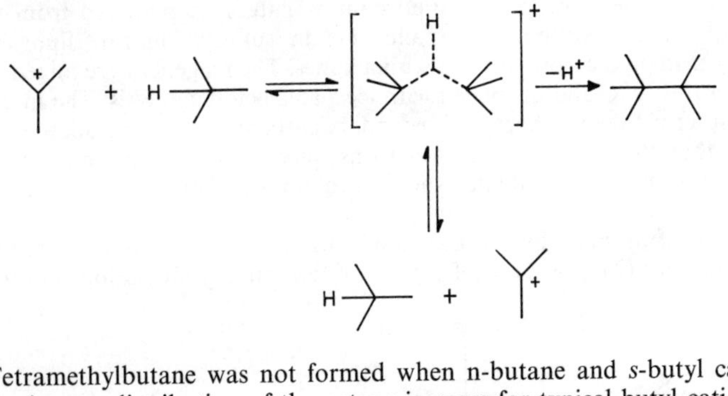

$$\text{(39)}$$

2,2,3,3-Tetramethylbutane was not formed when n-butane and s-butyl cation were reacted. The isomer distribution of the octane isomers for typical butyl cation–butane alkylations is shown in Table 1.

TABLE 1. Isomeric octane compositions obtained in typical alkylations of butanes with butyl cations

Octane	C_4—C_4			
	$(CH_3)_3CH$ $(CH_3)_3{}^+C$	$CH_3CH_2CH_2CH_3$ $(CH_3)_3{}^+C$	$(CH_3)_3CH$ $CH_3{}^+CHCH_2CH_3$	$(CH_3)_3CH$ $CH_3{}^+CHCH_2CH_3$
2,2,4-Trimethylpentane	18.0	4.0	3.8	8.5
2,2-Dimethylhexane			0.4	
2,2,3,3-Tetramethylbutane	1–2		Trace	1–2
2,5-Dimethylhexane	43.0	0.6	1.6	29.0
2,4-Dimethylhexane	7.6	Trace		6.6
2,2,3-Trimethylpentane	3.0	73.6	40.6	3.2
3,3-Dimethylhexane			12.3	7.1
2,3,4-Trimethylpentane	1.5	7.2	15.5	6.2
2,3,3-Trimethylpentane	3.6		3.8	8.8
2,3-Dimethylhexane	4.2	6.9		12.8
2-Methylheptane		Trace	10.3	6.7
3-Methylheptane	19.3	7.6	6.8	9.5
n-Octane	0.2		4.8	Trace

The superacid [CF_3SO_3H or $CF_3SO_3H:B(O_3SCF_3)_3$] catalyzed alkylation of adamantane with lower olefins (ethene, propene and butenes) was also investigated.[56] Alkyladamantanes obtained show that the reaction occurs by two pathways: (a) adamantylation of olefins by adamantyl cation formed through hydride abstraction from adamantane by alkyl cations (generated by the protonation of the olefins) and (b) direct σ-alkylation of adamantane by the alkyl cations via insertion into the bridgehead C—H bond of adamantane through a pentacoordinate carbonium ion (Scheme 8).

In order to gain understanding of the mechanism of the formation of alkyladamantanes, the reaction of adamantane with butenes is significant.

Reaction with 2-butenes gave mostly 1-n-butyladamantane, 1-sec-butyladamantane and 1-isobutyladamantane. Occasionally, trace amounts of 1-tert-butyladamantane were also formed. Isobutylene (2-methylpropene), however, consistently gave relatively good yield of 1-tert-butyladamantane along with other isomeric 1-butyladamantanes. 1-Butene gave only the isomeric butyladamantanes with only trace amounts of tert-butyladamantanes.

The formation of isomeric butyladamantanes, with the exception of tert-butyl-adamantane, is through adamantylation of olefins and carbocationic isomerization. The formation of tert-butyladamantane in the studied butylation reactions can, however, not be explained by this path. Since in control experiments attempted acid-catalyzed isomerization of isomeric 1-butyladamantanes did not give even trace amounts of 1-tert-butyladamantane, the tertiary isomer must be formed in the direct σ-tert-butylation of adamantane by tert-butyl cation through a pentacoordinate carbonium ion. The same intermediate is involved in the concomitant formation of 1-adamantyl cation via intermolecular hydrogen transfer (the major reaction). The formation of even low yields of tert-butyladamantane in the reaction is a clear indication that the pentacoordinate carbocation does not attain a linear geometry →C----H----C← (which could result only in hydrogen transfer), despite unfavorable steric interactions (equation 40). This reaction is similar to the earlier discussed reaction between tert-butyl cation and isobutane to form 2,2,3,3-tetramethylbutane. An alternate pathway (equation 41) for the formation of tert-butyladamantane through

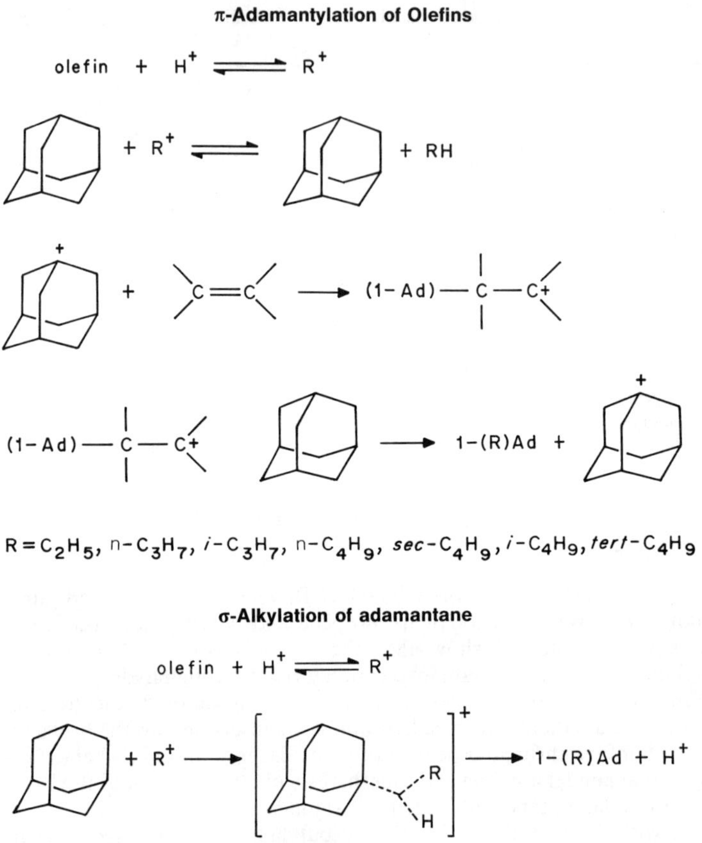

π-Adamantylation of Olefins

$$olefin \ + \ H^+ \ \rightleftharpoons \ R^+$$

$$+ \ R^+ \ \rightleftharpoons \ \ \ \ + \ RH$$

$$+ \ \ \begin{array}{c} \diagup \\ C=C \\ \diagup \end{array} \ \longrightarrow \ (1-Ad)-\overset{|}{\underset{|}{C}}-\overset{\diagup}{\underset{\diagdown}{C}}+$$

$$(1-Ad)-\overset{|}{\underset{|}{C}}-\overset{\diagup}{\underset{\diagdown}{C}}+ \ \ \ \ \ \longrightarrow \ 1-(R)Ad \ +$$

$$R = C_2H_5, \ n-C_3H_7, \ i-C_3H_7, \ n-C_4H_9, \ sec-C_4H_9, \ i-C_4H_9, \ tert-C_4H_9$$

σ-Alkylation of adamantane

$$olefin \ + \ H^+ \ \rightleftharpoons \ R^+$$

$$+ \ R^+ \ \longrightarrow \ \left[\ \ \ \begin{array}{c} \cdots R \\ \cdots H \end{array} \right]^+ \ \longrightarrow \ 1-(R)Ad \ + \ H^+$$

SCHEME 8

hydride abstraction of an intermediate 1-adamantylalkyl cation would necessitate involvement of an energetic 'primary' cation or highly distorted 'protonated cyclopropane' which is not likely under the reaction conditions.

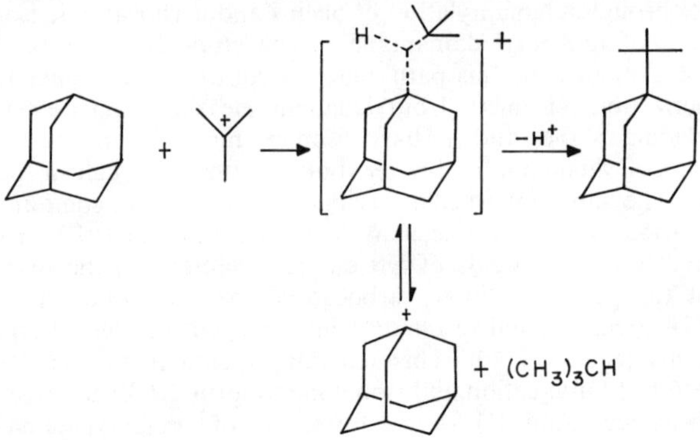

$$(40)$$

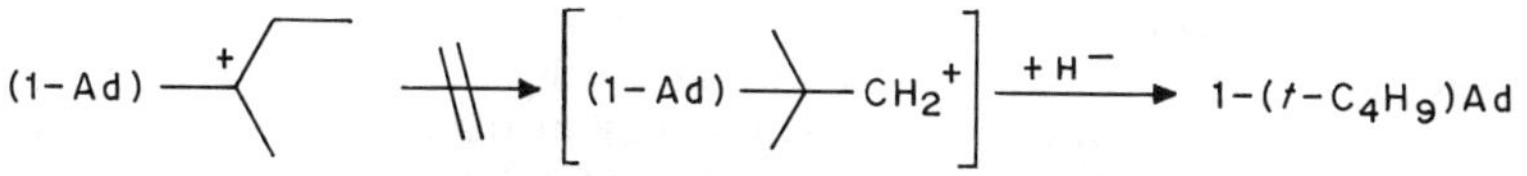

$$(1-Ad) \longrightarrow \overset{+}{\diagdown}\hspace{-0.3em}\diagup \longrightarrow \parallel \longrightarrow \left[(1-Ad) \longrightarrow \hspace{-0.3em}\diagup \hspace{-0.3em}\diagdown \hspace{-0.5em}-CH_2^+ \right] \xrightarrow{\ +H^-\ } 1-(t-C_4H_9)Ad \quad (41)$$

The observation of 1-*tert*-butyladamantane in the superacid-catalyzed reactions of adamantane with butenes provides unequivocal evidence for the σ-alkylation of adamantane by the *tert*-butyl cation. As this involves an unfavorable sterically crowded tertiary–tertiary interaction, it is reasonable to suggest that similar σ-alkylation can also be involved in less strained interactions with secondary and primary alkyl systems. Although superacid-catalyzed alkylation of adamantane with olefins occurs predominantly via adamantylation of olefins, competing direct σ-alkylation of adamantane can also occur. As the adamantane cage allows attack of the alkyl group only from the front side, the reported studies provide significant new insight into the mechanism of electrophilic reactions at saturated hydrocarbons and the nature of their carbocationic intermediates.

The protolytic oxidative condensation of methane in Magic Acid solution at 60 °C is evidenced by the formation of higher alkyl cations such as t-butyl and t-hexyl cations (equation 42). It is not necessary to assume a complete cleavage of $[CH_5]^+$ to a free, energetically unfavorable methyl cation. The carbon–carbon bond formation can indeed be visualized through reaction of the C—H bond of methane with the developing methyl cation.

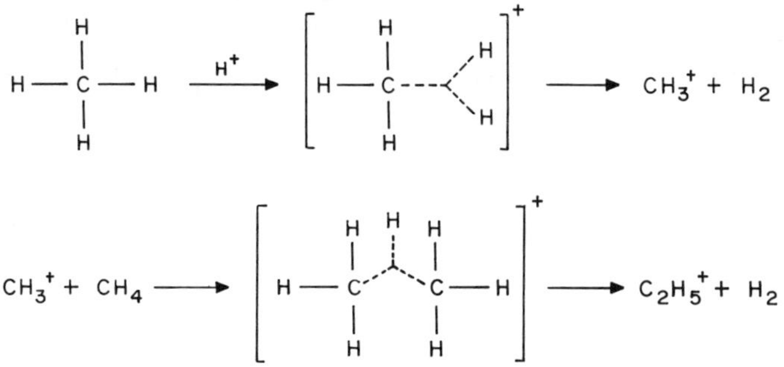

$$H-\overset{\overset{\displaystyle H}{|}}{\underset{\underset{\displaystyle H}{|}}{C}}-H \xrightarrow{\ H^+\ } \left[H-\overset{\overset{\displaystyle H}{|}}{\underset{\underset{\displaystyle H}{|}}{C}}\cdots\diagup\hspace{-0.5em}\begin{matrix}H\\ \\H\end{matrix} \right]^+ \longrightarrow CH_3^+ + H_2$$

$$(42)$$

$$CH_3^+ + CH_4 \longrightarrow \left[H-\overset{\overset{\displaystyle H}{|}}{\underset{\underset{\displaystyle H}{|}}{C}}\overset{\displaystyle H}{\cdots}\overset{\overset{\displaystyle H}{|}}{\underset{\underset{\displaystyle H}{|}}{C}}-H \right]^+ \longrightarrow C_2H_5^+ + H_2$$

Combining two methane molecules to ethane and hydrogen is (equation 43) endothermic by some 16 kcal mol^{-1}. Any condensation of methane to ethane and subsequently to higher hydrocarbons must thus overcome unfavorable thermodynamics. This can be achieved in condesation processes of an oxidative nature, where hydrogen is removed by the oxidant. In our original studies, the SbF$_5$ or FSO$_3$H component of the superacid system also acted as oxidants. The oxidative condensation of methane was subsequently studied further in more detail[57]. It was found that with added suitable oxidants such as halogens, oxygen, sulfur or selenium the superacid-catalyzed condensation of methane is feasible (equation 44). Significant practical problems, however, remain in carrying out the condensation effectively. Conversion so far achieved has been only in low yields. Due to the easy cleavage of longer-chain alkanes by the same superacids, C$_3$–C$_6$ products predominate.

$$2CH_4 \longrightarrow C_2H_6 + H_2 \quad (43)$$

$$CH_4 \xrightarrow[\text{superacid catalyst}]{\text{halogens, O}_2, S_x, Se} \text{hydrocarbons} \quad (44)$$

A further approach found useful was the use of natural gas instead of pure methane in the condensation reaction[57]. When natural gas in dehydrogenated, the C_2–C_4 alkanes it contains are converted into olefins. The resulting methane–olefin mixture can then be passed without separation through a superacid catalyst resulting in exothermic alkylative condensation (equation 45).

$$CH_4 + RCH{=}CH_3 \longrightarrow CH_2CHRCH_3 \tag{45}$$

Alkylation of methane by olefins under superacid catalysis was demonstrated both in solution chemistry under stable ion conditions[58] and in heterogeneous gas-phase alkylations over solid catalysts using a flow system[59]. Not only propylene and butylenes but also ethylene could be used as alkylating agents.

Thus alkylation of methane, ethane, propane and n-butane by the ethyl cation generated via protonation of ethylene in superacid media has been studied by Siskin[60], Sommer and coworkers[61], and Olah and coworkers, respectively[62]. The difficulty lies in generating in a controlled way a very energetic primary carbenium ion in the presence of excess methane and at the same time avoiding oligocondensation of ethylene itself. Siskin carried out the reaction of methane–ethylene (86:14) gas mixture through a 10:1 $HF:TaF_5$ solution under pressure with strong mixing (equation 46). Along with the recovered starting materials 60% of C_3 was found (propane and propylene). Propylene is formed when propane, which is substantially a better hydride donor, reacts with the ethyl cation (equation 47).

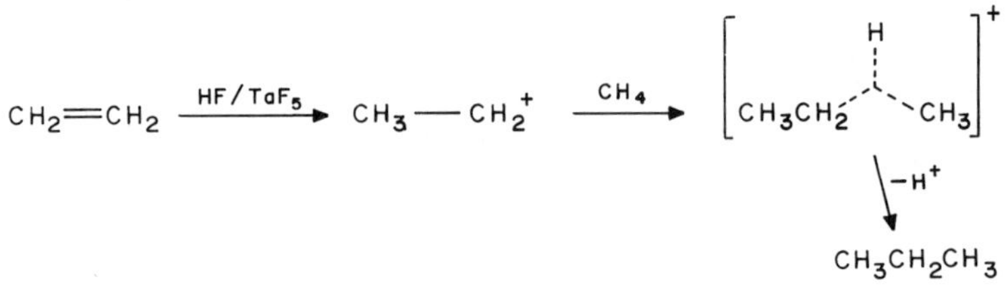

$$CH_2{=}CH_2 \xrightarrow{\text{HF/TaF}_5} CH_3{-}CH_2^+ \xrightarrow{\text{CH}_4} \left[CH_3CH_2 \overset{\overset{\displaystyle H}{\cdot\cdot\cdot}}{\diagdown} CH_3 \right]^+$$

$$\big\downarrow {-H^+} \tag{46}$$

$$CH_3CH_2CH_3$$

$$CH_3CH_2CH_3 + CH_3{-}CH_2^+ \longrightarrow CH_3{-}CH_3 + CH_3{-}\overset{+}{C}H{-}CH_3 \tag{47}$$

$$\big\downarrow {-H^+}$$

$$CH_3{-}CH{=}CH_2$$

Propane as a degradation product of polyethylene was, however, ruled out because ethylene alone under the same conditions does not give any propane. Under similar conditions but under hydrogen pressure, polyethylene reacts quantitatively to form C_3 to C_6 alkanes, 85% of which are isobutane and isopentane. These results further substantiate the direct alkane–alkylation reaction and the intermediacy of the pentacoordinate carbonium ion. Siskin also found that when ethylene was allowed to react with ethane in a flow system, n-butane was obtained as the sole product (equation 48)[60] indicating that the ethyl cation is alkylating the primary C—H bond through a five-coordinate carbonium ion.

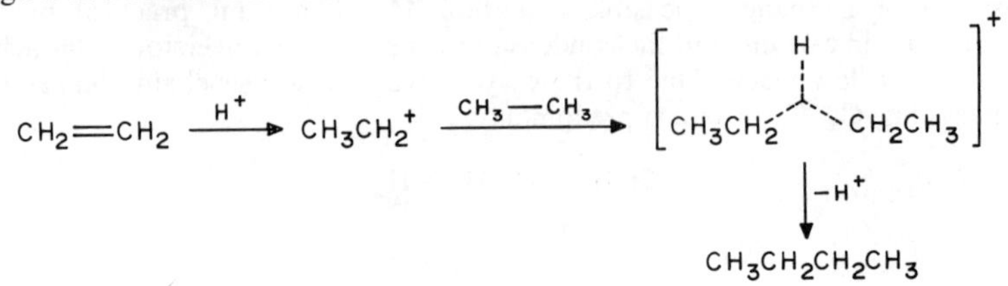

$$CH_2{=}CH_2 \xrightarrow{H^+} CH_3CH_2^+ \xrightarrow{\text{CH}_3{-}CH_3} \left[CH_3CH_2 \overset{\overset{\displaystyle H}{\cdot\cdot\cdot}}{\diagdown} CH_2CH_3 \right]^+$$

$$\big\downarrow {-H^+} \tag{48}$$

$$CH_3CH_2CH_2CH_3$$

If the ethyl cation had reacted with excess ethylene, primary 1-butyl cation would have been formed which irreversibly would rearrange to the more stable s-butyl and subsequently t-butyl cation giving isobutane as the end product.

The yield of the alkene–alkane alkylation in a homogeneous $HF:SbF_5$ system depending on the alkene:alkane ratio has been investigated by Sommer and coworkers in a batch system with short reaction times[61]. The results support direct alkylation of methane, ethane and propane by the ethyl cation and the product distribution depends on the alkene:alkane ratio.

Despite the unfavorable experimental conditions in a batch system for kinetic controlled reactions, a selectivity of 80% in n-butane was achieved through ethylation of ethane. The results show, however, that to succeed in the direct alkylation the following conditions have to be met: (1) The olefin should be totally converted to the reactive cation (incomplete protonation favors the polymerization and cracking processes); this means the use of a large excess of acid and good mixing. (2) The alkylation product must be removed from the reaction mixture before it transfers a hydride to the reactive cation, in which case the reduction of the alkene is achieved. (3) The substrate to cation hydride transfer should not be easy; for this reason the reaction shows the best yield and selectivity when methane and ethane are used.

Direct evidence for the direct ethylation of methane with ethylene was provided by Olah and his group[62] using ^{13}C-labeled methane (99.9% ^{13}C) over solid superacid catalysts such as $TaF_5:AlF_3$, TaF_5 and SbF_5: graphite. Product analyses by gas chromatography–mass spectrometry (GC–MS) are given in Table 2.

These results show a high selectivity in mono-labeled propane $^{13}CC_2H_8$ which can only arise from direct electrophilic attack of the ethyl cation on methane via pentacoordinate carbonium ion (equation 49).

$$CH_3-CH_2^+ \xrightarrow{^{13}CH_4} \left[CH_3-CH_2 \overset{\overset{H}{|}}{\cdots}{\cdot}^{13}CH_3 \right]^+ \xrightarrow{-H^+} CH_3CH_2{}^{13}CH_3 \quad (49)$$

An increase of the alkene:alkane ratio results in a significant decrease in single-labeled propane, ethylene polymerization—cracking and hydride transfer become the main reactions. This labeling experiment carried out under conditions where side reactions were negligible is indeed unequivocal proof for the direct alkylation of an alkane by a very reactive carbenium ion.

Polycondensation of alkanes over $HSO_3F:SbF_5$ has also been achieved by Roberts and Calihan[63]. Several low-molecular-weight alkanes such as methane, ethane, propane, butane and isobutane were polymerized to highly branched oily oligomers with a

TABLE 2. Ethylation of $^{13}CH_4$ with C_2H_4

$^{13}CH_4:C_2H_4$	Catalyst	Products normalized (%)[a,b]				Label content of C_3 fraction (%)	
		C_2H_6	C_3H_8	$i\text{-}C_4H_{10}$	C_2H_5F	$^{13}CC_2H_8$	C_3H_8
98.7:1.3	$TaF_5:AlF_3$	51.9	9.9	38.2		31	69
99.1:0.9	TaF_5		15.5	3.0	81.5	91	9
99.1:0.9	SbF_5:graphite	64.1	31.5		4.4	96	4

[a]All values reported are in mole percent.
[b]Excluding methane.

molecular-weight range from the molecular weight of monomers to around 700. These reactions again follow the same initial protolysis of the C—H or C—C bond which results in a very reactive carbenium ion. Similarly, the same workers[64] were also able to polycondense methane with a small amount of olefins such as ethylene, propylene, butadiene and styrene to yield oily polymethylene oligomer with a molecular weight ranging from 100 to 700.

VII. CARBOXYLATION

Alkanecarboxylic acids are readily prepared from alkenes and carbon monoxide or formic acid in strongly superacidic solutions[65]. The reaction between carbenium ions generated from alkenes and carbon monoxide affording oxocarbenium ions (acyl cations) is the key step in the well-known Koch–Haaf reaction used for the general preparation of carboxylic acids[65,66]. The conventional protic acid such as H_2SO_4 was found to be less effective for the generation of oxocarbenium ions. Subsequent investigations[67] found that the superacidic $HF:BF_3$ or CF_3SO_3H[68] are very efficient to effect this process. The volatility of the former catalyst system, however, necessitates high-pressure conditions.

The use of superacidic activation of alkanes to their related carbocations allowed the preparation of alkanecarboxylic acids from alkanes themselves with CO_2 followed by aqueous work-up.

The formation of C_6 or C_7 carboxylic acids along with some ketones was reported in the reaction of isopentane, methylcyclopentane and cyclohexane with CO in $HF:SbF_5$ at ambient temperatures and atmospheric pressures[69].

TABLE 3. Superacid-catalyzed reaction of alkanes with CO[a]

				Product distribution (%)						
	Total	C_3	C_4	C_5 acids		C_6 acids		C_7 acids		
Substrate	yield[b]	acid	acid	tert-	sec-	tert-	sec-	tert-	sec-	C_8 acids
2-Methyl	55	Some	Some	4		95		Trace		—
butane				67	33	38	62			
n-Pentane	53	3	26	14		57		Trace		—
				27	73	39	61			
2-Methyl	61	—	Some	3		3		94		—
pentane				67	33	40	60	29	71	
n-Hexane	69	2	10	18		5		65		Some
				13	87	39	61	33	67	
2,2-Dimethyl	57	—	Some	10		15		75		—
butane		.		80	20	40	60	35	65	
n-Heptane	80	Some	47	48		5		Some		Some
				90	10	38	62			
2,2,4-Tri-	102	—	—	100		—		—		—
methylpentane				100	0					
n-Octane	90	Some	6	87		7		Some		Some
				90	10	38	62			
n-Nonane	61	Some	34	31		33		2		Some
				87	13	38	62			

[a]Reaction time, 1 h. Reaction temperature 30 °C. Alkane 20 mmol, SbF_5/alkane = 2 mol ratio, HF/SbF_5 = 5 mol ratio.
[b]Based on alkane used.

Yoneda and coworkers have also found that other alkanes can also be carboxylated with CO in HF:SbF$_5$ superacid system[70]. Tertiary carbenium ions formed by protolysis of C—H bonds of branched alkanes in HF:SbF$_5$ undergo skeletal isomerization and disproportionation prior to reacting with CO in the same acid system to form oxocarboxylic acids after hydrolysis. Alkyl methyl ketones with short alkyl chain (less than C$_4$) do not react under these conditions due to the proximity of the positive charge on the protonated ketone and the developing carbenium ion[71]. The results are shown in Table 3.

When using tertiary C$_5$ or C$_6$ alkanes a considerable amount of secondary carboxylic acids[72a] are produced by the reaction of CO with secondary alkyl carbenium ions. Such cations are formed as transient intermediates by skeletal isomerization of the initially formed tertiary cations. Carboxylic acids with lower number of carbon atoms than the starting alkanes are formed from the fragment alkyl cations generated by the protolysis of C—C bonds of the straight-chain alkanes. Intermolecular hydride shift between the fragment cations and the starting alkanes gives rise to a large amount of carboxylic acids with alkyl groups of the same number of carbon atoms as the starting alkanes. For alkanes with more than C$_7$ atoms, β-scission occurs exclusively to produce C$_4$-, C$_5$- and C$_6$- carboxylic acids.

Sommer and coworkers[72b] have achieved selective carbonylation of propane in superacidic media promoted by halogen. When propane–carbon monoxide mixture (CO:propane ratio = 3) was passed through HF–SbF$_5$ in a Kel-F reactor at $-10\,°C$ with the addition of a small amount of sodium bromide (Br$^-$/Sb:0.5 mol%), the NMR spectrum indicated the formation of isopropyloxocarbenium ion in 95% yield with a total conversion of 9% of the propane. This remarkable reaction can be rationalized as in equation 50.

$$Br^- + 2H^+ \longrightarrow \left[Br\text{---}\overset{H}{\underset{H}{\diagdown}} \right]^+ \longrightarrow Br^+ + H_2$$

$$\text{(50)}$$

$$CH_3CH_2CH_3 \xrightarrow{Br^+} \left[\begin{array}{c} H\diagdown\diagup Br \\ | \\ CH_3CHCH_3 \end{array} \right]^+ \xrightarrow{CO} \underset{CH_3CHCH_3}{\overset{CO^+}{|}} + HBr$$

VIII. FORMYLATION

Whereas electrophilic formylation of aromatics with CO was studied under both the Gatterman–Koch condition and with superacid catalysis[73-78] in some detail, electrophilic formylation of saturated aliphatics remains virtually unrecognized.

Reactions of acyclic hydrocarbons of various skeletal structures with CO in superacid media were recently studied by Yoneda and coworkers[70-72] as discussed in the previous section. Products obtained were only isomeric carboxylic acids with lower number of carbon atoms than the starting alkanes. Formation of the carboxylic acids were accounted by the reactions of parent, isomerized and fragmented alkyl cations with CO to form the oxocarbenium ion intermediates (Koch–Haaf reaction) followed by their quenching with water. No formylated products in these reactions have been identified.

We have recently reported the superacid catalyzed reaction of the polycyclic cage hydrocarbon adamantane with CO (equation 51)[79]. 1-Adamantanecarboxaldehyde was obtained in various percent yields (Table 4) in different superacids along with

TABLE 4. Percentage yield of 1-adamantanecarboxaldehyde in different superacids in Freon-113[a]

Acid systems	Adamantane acid	Yield 1-Ad-CHO (%)
CF_3SO_3H	1:12	0.2
	1:12	9.1[b]
$CF_3SO_3H + B(OSO_2CF_3)_3$ (1:1)	1:3	3.4
	1:3	14.5[b]
$CF_3SO_3H + SbF_5$ (1:1)	1:3	8.2
	1:3	21.0[b]

[a] All reactions were carried out from $-78\,°C$ to room temperature.
[b] Isolated yields.

1-adamantane-carboxylic acid and 1-adamantanol (the products of the reaction of 1-adamantyl cation). The mechanism of formation of 1-adamantanecarboxaldehyde has been investigated. Formation of aldehyde product from 1-adamantanoyl cation via hydride abstraction (Scheme 9) from adamantane is only a minor pathway. The major pathway by which 1-adamantanecarboxaldehyde is formed is by direct σ-insertion of the protosolvated formyl cation into the C—H bond of adamantane at the bridgehead position (equation 52).

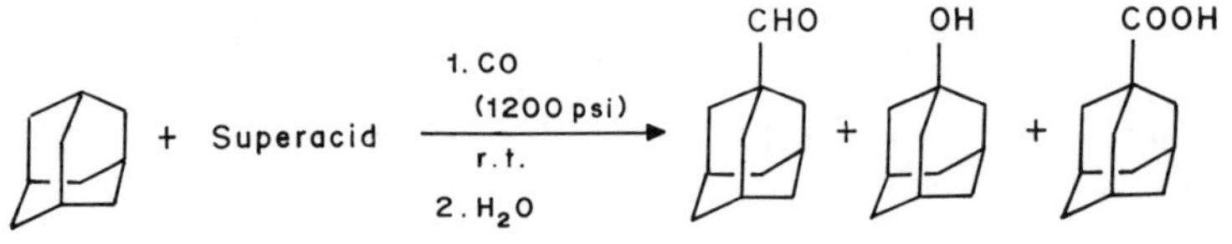

$$ (51) $$

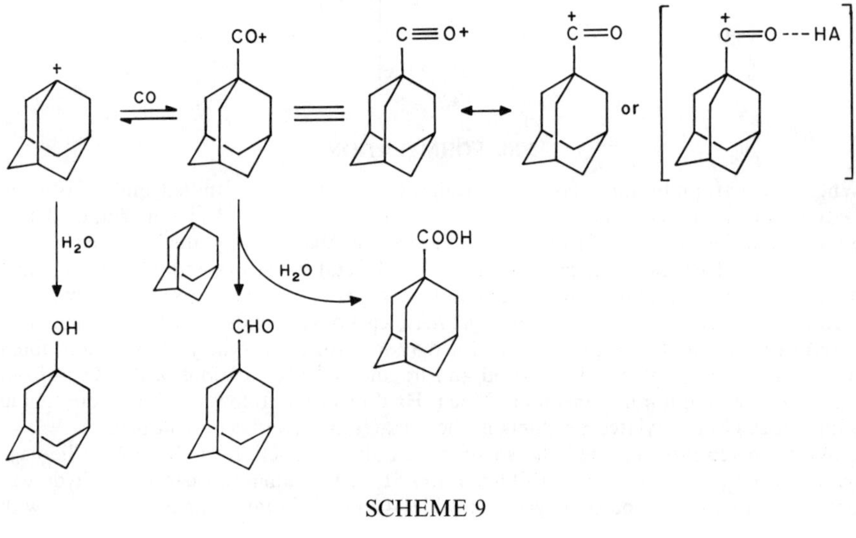

SCHEME 9

$$\text{(52)}$$

Whereas the formyl cation could not be directly observed by NMR spectroscopy[78,79], its intermediacy has been well established in aromatic formylation reactions[73-78]. In order to account for the failure to observe the formyl cation, it has been suggested that CO is protonated in acid media to generate protosolvated formyl cation[80,81], a very reactive electrophile. Protosolvation of the carbonyl oxygen allows facile deprotonation of the methine proton, thus resulting in rapid exchange via involvement of the isoformyl cation.

Insertion of protosolvated formyl cation into the C—H σ-bond has further been substantiated by carrying out the reaction of 1,3,5,7-tetradeuteroadamantane with CO in superacid media. 3,5,7-Trideutero-1-adamantanecarboxaldehyde-H and 3,5,7-trideutero-1-adamantanecarboxaldehyde-D were obtained in the ratio 94:6. This work provides the first direct evidence for the electrophilic formylation of a saturated hydrocarbon by CO under superacid catalysis.

IX. OXYFUNCTIONALIZATION

Functionalization of aliphatic hydrocarbons into their oxygenated compounds is of substantial interest. In connection with the preparation and studies of a great variety of carbocations in different superacid systems[22], electrophilic oxygenation of alkanes with ozone and hydrogen peroxide was investigated by Olah and coworkers in superacid media under typical electrophilic conditions[82]. The electrophiles are the protonated ozone (^+O_3H) or the hydrogen peroxonium ion $H_3O_2^+$. The reactions are depicted as taking place via initial electrophilic attack by the electrophiles on the σ-bonds of alkanes through the pentacoordinated carbonium ion transition state followed by proton elimination to give the desired product.

When hydrogen peroxide is protonated in superacid media, hydrogen peroxonium ion ($H_3O_2^+$) is formed. Certain peroxonium salts have been well characterized and even isolated as stable salts[83,84]. Hydrogen peroxonium ion is considered as the incipient ^+OH ion, a strong electrophile for electrophilic hydroxylation at single (σ) bonds of alkanes. The reactions are thus similar to those described previously for electrophilic protonation (protolysis), alkylation and formylation.

Superacid-catalyzed electrophilic hydroxylation of branched alkanes were carried out using $HSO_3F:SbF_5:SO_2ClF$ with various ratios of alkane and hydrogen peroxide at different temperatures[85]. Some of the results are summarized in Table 5. Protonated hydrogen peroxide inserts into the C—H bond of alkanes. The mechanism is illustrated in Scheme 10 with isobutane.

The intermediate pentacoordinate hydroxycarbonium ion, in addition to giving the insertion product, gives tert-butyl cation (by elimination of water) which is responsible for the formation of dimethylmethylcarboxonium ion through the intermediate tert-butyl hydroperoxide. Hydrolysis of the carboxonium ion gives acetone and methanol. When the reaction was carried out by passing isobutane at room temperature through a solution of Magic Acid in excess hydrogen peroxide, in addition to the insertion product, a number of other products were also formed which were rationalized as arising from hydrolysis of the carboxonium ion and from Baeyer–Villiger oxidation of acetone. The mechanism was further substantiated by independent treatment of alkane and

TABLE 5. Products of the reaction of branched-chain alkanes with H_2O_2 in HSO_3F:SbF_5:SO_2ClF solution

Alkane^a	H_2O_2 (mmol)	Temperature (C°)	Major Products
C | C—C—C | H	2	-78 -20	$(CH_3)_2C{=}O^+CH_3$ "
	4	-78 -20	$(CH_3)_2C{=}O^+CH_3$ "
	6	-78 -20	$(CH_3)_2C{=}O^+CH_3$ "
	6	$+20$	$(CH_3)_2C{=}O^+CH_3$ (trace), DAP^b (25%), CH_3OH (50%), CH_3CO—O—CH_3 (25%),
C | C—C—C—C | H	3	-78 -20	$(CH_3)_2C{=}O^+C_2H_5$ $C_2H_5(CH_3)_2C^+$
	6	-78 -20	$(CH_3)_2C{=}O^+C_2H_5$ "
C C | | C—C—C—C | | C H	4	-78	$(CH_3)_2C{=}O^+CH_3$ (50%), $(CH_3)_2C{=}O^+H$ (50%)
	6	-40	$(CH_3)_2C{=}O^+CH_3$ (50%), DAP (50%)

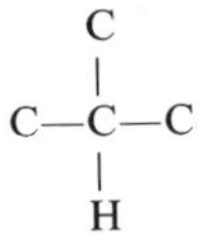

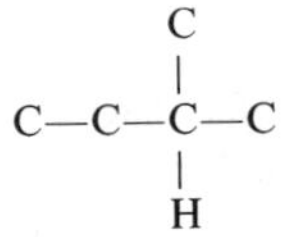

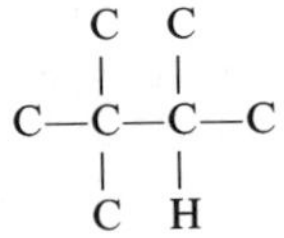

^{a}Alkane, 2 mmole; ^{b}DAP, dimeric acetone peroxide.

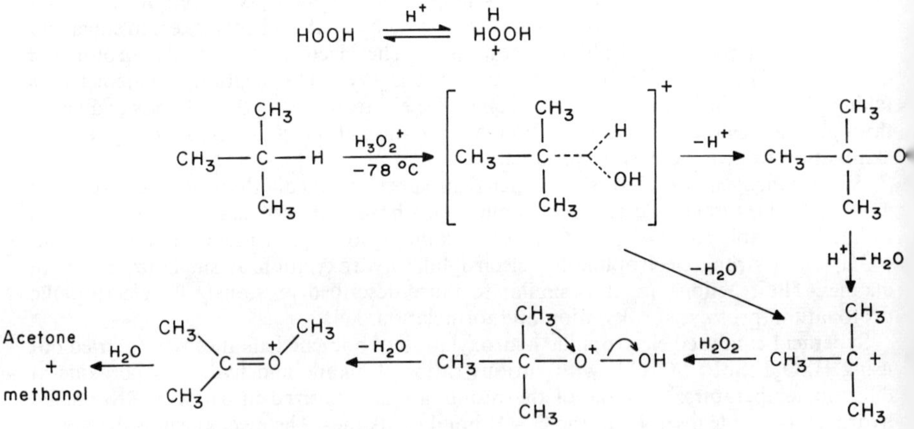

SCHEME 10

Baeyer–Villiger oxidation of several ketones with hydrogen peroxide–Magic Acid systems[86]. In the superacid-catalyzed hydroxylation of straight-chain alkanes such as ethane, propane, butane, etc. with hydrogen peroxide, related oxygenated products were identified[85].

In studies on superacid-catalyzed oxyfunctionalization of methane it has been found that hydrogen peroxide in superacidic media gives methyl alcohol with very high ($> 95\%$) selectivity[86]. Electrophilic ^+OH insertion by protonated hydrogen peroxide in the C—H bonds of methane is the indicated reaction path. The reaction is limited, however, by the use of hydrogen peroxide in the liquid phase. In the superacidic medium, methyl alcohol formed is immediately protonated to methyloxonium ion $(CH_3OH_2^+)$ and thus prevented from further oxidation. Superacid-catalyzed oxygenation of methane with ozone gives predominantly formaldehyde[87a]. The reaction is best understood as electrophilic insertion of ^+O_3H ion into the methane C—H bonds leading to a hydrotrioxide, which then eliminates hydrogen peroxide giving protonated formaldehyde (Scheme 11). The competing pathway forms protonated methyl alcohol with O_2 elimination, but this reaction is only a minor one.

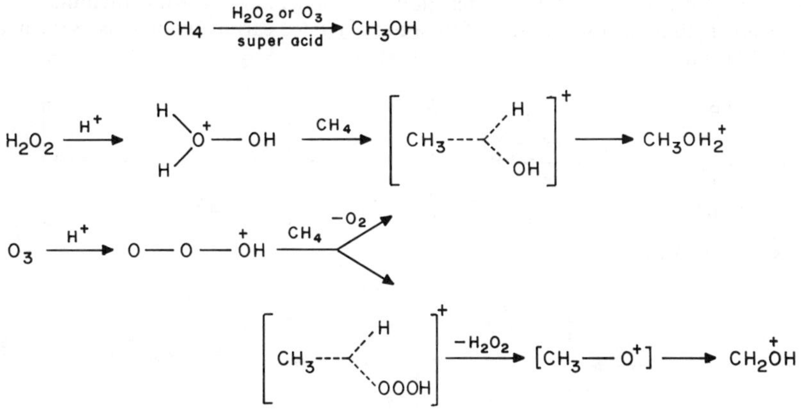

$$CH_4 \xrightarrow[\text{super acid}]{H_2O_2 \text{ or } O_3} CH_3OH$$

$$H_2O_2 \xrightarrow{H^+} \underset{H}{\overset{H}{\diagdown}}\overset{+}{O}-OH \xrightarrow{CH_4} \left[CH_3\text{---}\overset{H}{\underset{OH}{\diagup}}\right]^+ \longrightarrow CH_3OH_2^+$$

$$O_3 \xrightarrow{H^+} O-O-\overset{+}{O}H \xrightarrow[]{CH_4} \overset{-O_2}{\diagup}$$

$$\left[CH_3\text{---}\overset{H}{\underset{OOOH}{\diagup}}\right]^+ \xrightarrow{-H_2O_2} [CH_3-O^+] \longrightarrow CH_2\overset{+}{O}H$$

SCHEME 11

Electrophilic oxygenation of methane to methyl alcohol under superacidic conditions proves the high selectivity of electrophilic substitution contrasted with non-selective radical oxidation. However, as indicated, hydrogen peroxide chemistry is limited to the liquid phase and the desirable goal of achieving selective heterogeneous catalytic oxidation of methane to methyl alcohol remains elusive. Consequently, Olah and coworkers have considered combining selective halogenation of methane to methyl halides with subsequent hydrolysis to methyl alcohol (vide infra).

Bis(trimethylsilyl)peroxide/trifluoromethanesulfonic acid has been found to be an efficient electrophilic oxygenating agent for saturated hydrocarbons, such as adamantane and diamantane[87b]. With adamantane the major reaction product is 4-oxahomo-adamantane (isolated in 79% yield) obtained through C—C σ-bond insertion with very little 1-adamantanol, the C—H σ-bond insertion product (equation 53). In the case of diamantane, two isomeric oxahomodiamantanes were obtained along with two isomeric bridgehead diamantanols corresponding to C—C and C—H σ-bond insertions.

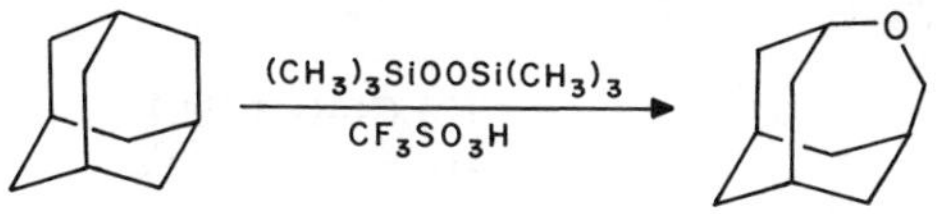

$$\text{(adamantane)} \xrightarrow[\text{CF}_3\text{SO}_3\text{H}]{(CH_3)_3SiOOSi(CH_3)_3} \text{(4-oxahomoadamantane)} \tag{53}$$

Ozone, which is a resonance hybrid of the following canonical structures (equation 54), can react as 1,3-dipole, an electrophile or a nucleophile[88]. Whereas the electrophilic nature of ozone has been established in its reactions with alkenes, alkynes, arenes, amines, sulfides, phosphines, etc.[89-93], the nucleophilic properties of ozone has not yet been well studied[94].

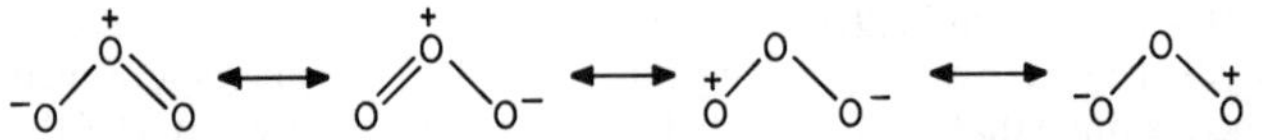

$$\begin{matrix} \overset{+}{O} \\ -O \end{matrix} \diagup\!\!\diagdown \!\! O \longleftrightarrow \begin{matrix} \overset{+}{O} \\ O \end{matrix} \diagup\!\!\diagdown \!\! O^- \longleftrightarrow \overset{+}{O} \diagup\!\!\diagdown \!\! O^- \longleftrightarrow -O \diagup\!\!\diagdown \!\! \overset{+}{O} \qquad (54)$$

In the reaction of ozone with carbenium ions[87] the nucleophilic attack of ozone has been inferred through the formation of intermediate trioxide ion leading to dialkyl alkylcarboxonium ion, similar to the acid-catalyzed rearrangement of hydroperoxides to carboxonium ions (Hock reaction) (equation 55). When a stream of oxygen containing 15% ozone was passed through a solution of isobutane in $HSO_3F{:}SbF_5{:}SO_2ClF$ solution at $-78\,^\circ C$, dimethyl methylcarboxonium ion (CH_3CO^+) was identified by ^1H and ^{13}C NMR spectroscopy[87,95]. Similar reactions of isopentane, 2,3-dimethylbutane and 2,2,3-trimethylbutane resulted in the formation of related carboxonium ions as the major product (Table 6).

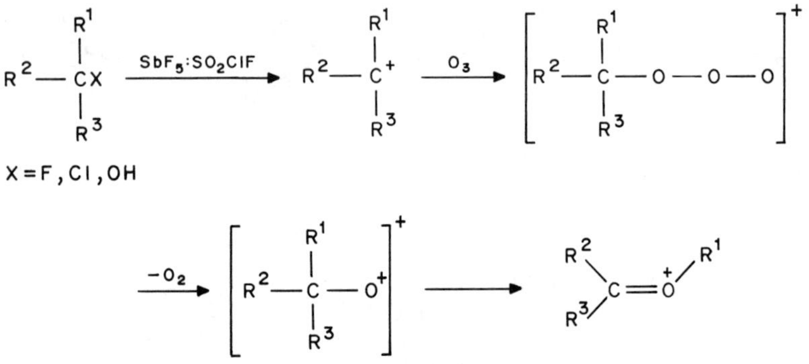

$$R^2 - \underset{\underset{R^3}{|}}{\overset{\overset{R^1}{|}}{C}}X \xrightarrow{SbF_5{:}SO_2ClF} R^2 - \underset{\underset{R^3}{|}}{\overset{\overset{R^1}{|}}{C}}{}^+ \xrightarrow{O_3} \left[R^2 - \underset{\underset{R^3}{|}}{\overset{\overset{R^1}{|}}{C}} - O - O - O \right]^+$$

$$X = F, Cl, OH$$

$$\xrightarrow{-O_2} \left[R^2 - \underset{\underset{R^3}{|}}{\overset{\overset{R^1}{|}}{C}} - \overset{+}{O} \right]^+ \longrightarrow \underset{R^3}{\overset{R^2}{\diagdown}}C = \overset{+}{\underset{}{O}} \diagup R^1 \qquad (55)$$

TABLE 6. Product of the reaction of branched alkanes with ozone in Magic Acid–SO_2ClF at $-78\,^\circ C$[87]

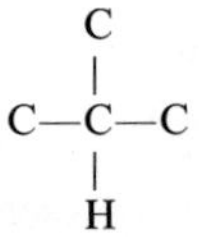 $C-\underset{\underset{H}{\mid}}{\overset{\overset{C}{\mid}}{C}}-C$	$(CH_3)_2C{=}O^+CH_3$
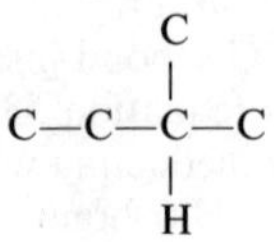 $C-C-\underset{\underset{H}{\mid}}{\overset{\overset{C}{\mid}}{C}}-C$	$(CH_3)_2C{=}O^+C_2H_5$
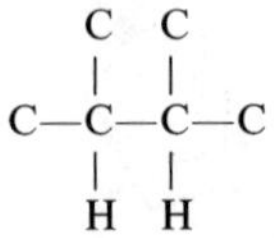 $C-\underset{\underset{H}{\mid}}{\overset{\overset{C}{\mid}}{C}}-\underset{\underset{H}{\mid}}{\overset{\overset{C}{\mid}}{C}}-C$	$(CH_3)_2C{=}O^+CH(CH_3)_2$ (60%) $(CH_3)_2C{=}O^+H$ (40%)
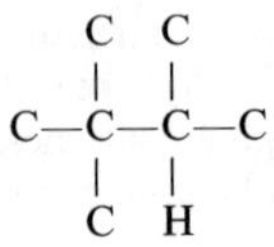 $C-\underset{\underset{C}{\mid}}{\overset{\overset{C}{\mid}}{C}}-\underset{\underset{H}{\mid}}{\overset{\overset{C}{\mid}}{C}}-C$	$(CH_3)_2C{=}O^+CH_3$ (50%) $(CH_3)_2C{=}O^+H$ (50%)

Formation of an intermediate alkylcarbenium ion which is the key step in superacid-catalyzed reaction of ozone with alkanes is considered to proceed by two mechanistic pathways as illustrated in Scheme 12. The carbenium ions subsequently undergo nucleophilic reaction with ozone as discussed previously. Reactions of ozone with alkanes giving ketones and alcohols as involved in mechanism b have been reported in several instances[96-98]. The products obtained from isobutane and isoalkanes (Table 6) are in accordance with the mechanism discussed above.

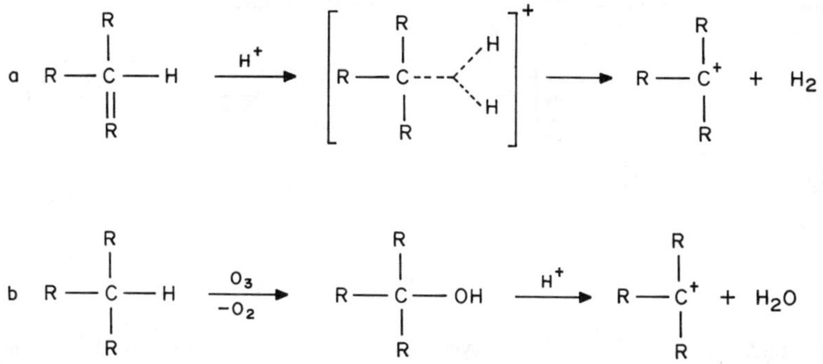

$$
\text{a} \quad R{-}\underset{\underset{R}{\|}}{\overset{\overset{R}{|}}{C}}{-}H \quad \xrightarrow{H^+} \quad \left[R{-}\underset{\underset{R}{|}}{C}{\cdots}{\overset{H}{\underset{H}{<}}} \right]^+ \quad \longrightarrow \quad R{-}\underset{\underset{R}{|}}{\overset{\overset{R}{|}}{C}}{^+} \; + \; H_2
$$

$$
\text{b} \quad R{-}\underset{\underset{R}{|}}{\overset{\overset{R}{|}}{C}}{-}H \quad \xrightarrow[-O_2]{O_3} \quad R{-}\underset{\underset{R}{|}}{\overset{\overset{R}{|}}{C}}{-}OH \quad \xrightarrow{H^+} \quad R{-}\underset{\underset{R}{|}}{\overset{\overset{R}{|}}{C}}{^+} \; + \; H_2O
$$

SCHEME 12

The rate of formation of the dimethyl methylcarboxonium ion from isobutane is considerably faster than that of the *tert*-butyl cation from isobutane in the absence of ozone under the same reaction conditions[87]. A solution of isobutane in excess Magic Acid:SO_2ClF solution gives only a trace amount of *tert*-butyl cation after standing for 5 h at $-70\,°C$. *tert*-Butyl alcohol, on the other hand, in the same acid system gives a quantitative yield of the corresponding cation. In the presence of ozone, under the same conditions it gives dimethyl methylcarboxonium ion. Isobutane does not give any oxidation product in the absence of Magic Acid under low-temperature ozonization conditions. No experimental evidence for the intermediacy of the carbenium ions could be furnished either[87]. The most probable reaction path postulated for these reactions is the electrophilic attack by protonated ozone on the alkanes resulting in a pentacoordinated transition state, from which the involved carboxonium ion is formed

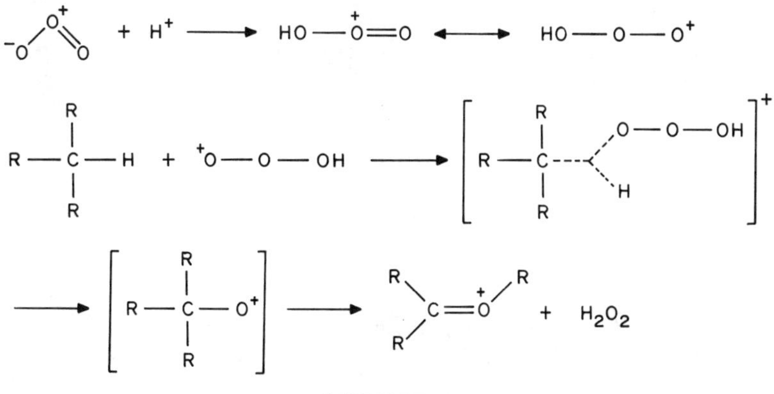

$$
{^-O}{-}\overset{+}{O}{=}\overset{O}{} \; + \; H^+ \; \longrightarrow \; HO{-}\overset{+}{O}{=}O \; \longleftrightarrow \; HO{-}O{-}O^+
$$

$$
R{-}\underset{\underset{R}{|}}{\overset{\overset{R}{|}}{C}}{-}H \; + \; {^+O}{-}O{-}OH \; \longrightarrow \; \left[R{-}\underset{\underset{R}{|}}{C}{\cdots}{\overset{O{-}O{-}OH}{\underset{H}{<}}} \right]^+
$$

$$
\longrightarrow \; \left[R{-}\underset{\underset{R}{|}}{\overset{\overset{R}{|}}{C}}{-}O^+ \right] \; \longrightarrow \; \underset{R}{\overset{R}{>}}C{=}\overset{+}{O}\overset{R}{\underset{}{}} \; + \; H_2O_2
$$

SCHEME 13

(Scheme 13). Since ozone has a strong 1,3-dipole or at least a strong polarizability[88], it is expected to be readily protonated in superacid media. Attempts to directly observe protonated ozone O_3H^+ by ^1H NMR spectroscopy[87] were rather inconclusive because of probable fast hydrogen exchange with the acid system.

Superacid-catalyzed reaction of ozone with straight-chain alkanes has also been investigated at low temperature. Magic Acid catalyzed ozonization of ethane is shown in equation 56. Reaction of methane under similar conditions was also investigated and discussed previously. Reactions of cycloalkanes have similarly been studied.[82]

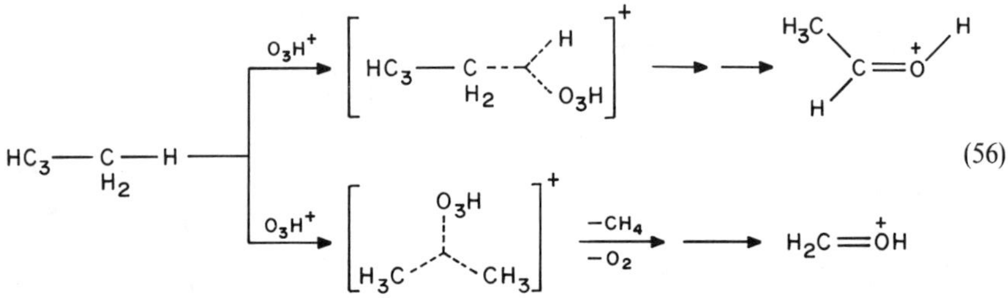

$$(56)$$

Superacid-catalyzed electrophilic oxygenation of functionalized hydrocarbons has also been achieved[82]. Oxidations of alcohols, ketones and aldehydes were carried out using protonated ozone. These reactions are illustrated in Scheme 14. In the case of carbonyl compounds, the C—H bond located farther than the γ-position seems to react with ozone in the presence of Magic Acid. The strong electron-withdrawing effect of the protonated carbonyl group is sufficient to inhibit the reaction of the C—H bonds at α-, β- and γ-positions. Superacid ($HF:SbF_5$) catalyzed oxidation with ozone has also been carried out with certain keto steroids[99a,b] bearing various substituents such as carbonyl, hydroxy and acetoxy groups.

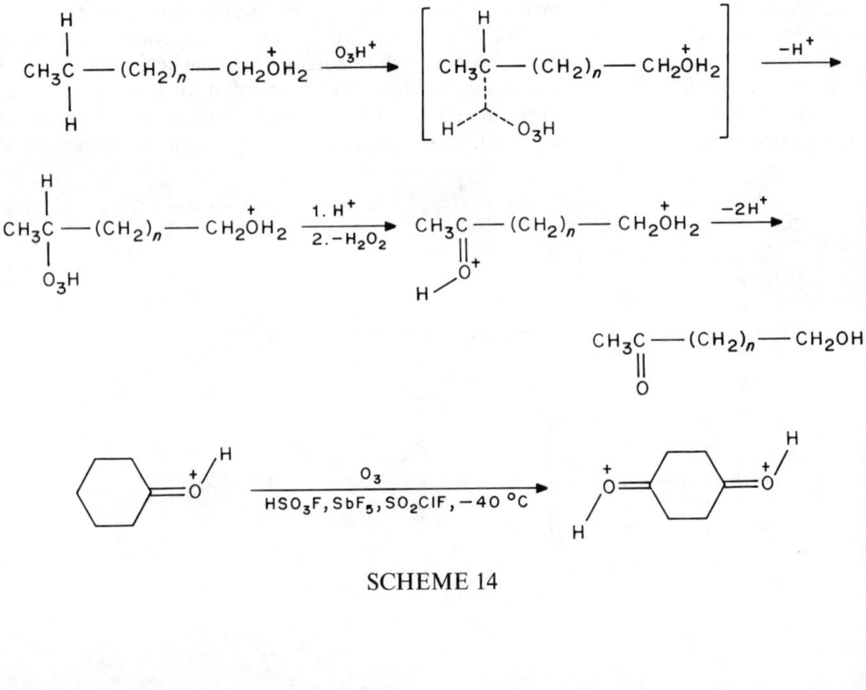

SCHEME 14

X. SULFURATION

Selective electrophilic sulfuration of alkanes and cycloalkanes to the corresponding monosulfides has been achieved[99c] by reacting alkanes with elemental sulfur and trifluoromethanesulfonic acid. Cyclopentane under relatively mild reaction conditions gives cyclopentyl sulfide in 46% isolated yield (equation 57).

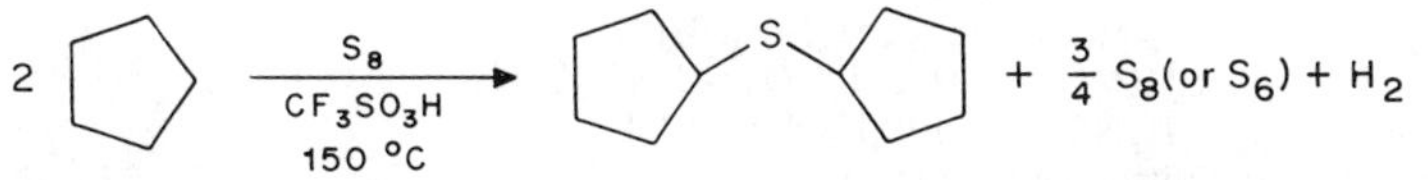

2
S_8
CF_3SO_3H
150 °C
S
+ $\frac{3}{4}$ S_8(or S_6) + H_2

(57)

This remarkable transformation can be rationalized by either of the two mechanisms A or B displayed in Scheme 15. The reaction is catalytic in the acid.

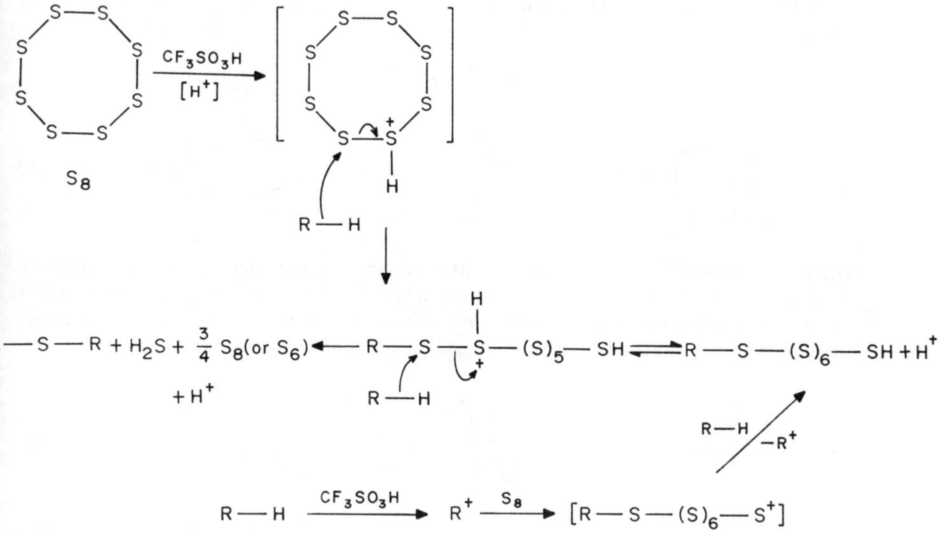

S_8
CF_3SO_3H
[H^+]
R—H
—S—R + H_2S + $\frac{3}{4}$ S_8(or S_6)
+ H^+
R—S—S—(S)$_5$—SH
R—S—(S)$_6$—SH + H^+
−R^+
R—H
CF_3SO_3H
R^+
S_8
[R—S—(S)$_6$—S^+]

SCHEME 15

XI. NITRATION AND NITROSATION

Nitronium ion is capable of nitrating not only the aromatic systems but also the aliphatic hydrocarbons. Electrophilic nitration of alkanes and cycloalkanes was carried out with $NO_2^+PF_6^-$, $NO_2^+BF_4^-$, $NO_2^+SbF_6^-$ salts in CH_2Cl_2-tetramethylenesulfone or HSO_3F solution[100a]. Some representative reactions of nitronium ion with various alkanes, cycloalkanes and polycyclic alkanes are shown in Table 7. The nitration takes place on both C—C and C—H bonds involving two-electron, three-center (2e–3c) bonded five-coordinated carbocations as depicted in equation 58 with adamantane, a cage polycycloalkane. The yield of 1-nitroadamantane obtained is excellent. Formation of 1-fluoroadamantane and 1-adamantanol as by-products in the reaction indicates that the pentacoordinate carbocation can also cleave to the 1-adamantyl cation. These results also show the non-linear nature of the ionic five-coordinate carbocationic transition state (intermediate)[100b]. Support for such an intermediate also comes from low

TABLE 7. Nitration of alkanes and cycloalkanes with $NO_2{}^+PF_6{}^-$

Hydrocarbon	Nitroalkane products and their mole ratio
Methane	CH_3NO_2
Ethane	$CH_3NO_2 > CH_3CH_2NO_2$, 2.9:1
Propane	$CH_3NO_2 > CH_3CH_2NO_2 > 2$-$NO_2C_3H_7 > 1$-$NO_2C_3H_7$, 2.8:1:0.5:0.1
Isobutane	$tert$-$NO_2C_4H_9 > CH_3NO_2$, 3:1
n-Butane	$CH_3NO_2 > CH_3CH_2NO_2 > 2$-$NO_2C_4H_9 \sim 1$-$NO_2C_4H_9$, 5:4:1.5:1
Neopentane	$CH_3NO_2 > tert$-$C_4H_9NO_2$, 3.3:1
Cyclohexane	Nitrocyclohexane
Adamantane	1-Nitroadamantane, 17.5:1

$k_H/k_D \approx 1.30$ observed for adamantane vs 1,3,5,7-tetradeuteroadamantane in the nitration reaction.

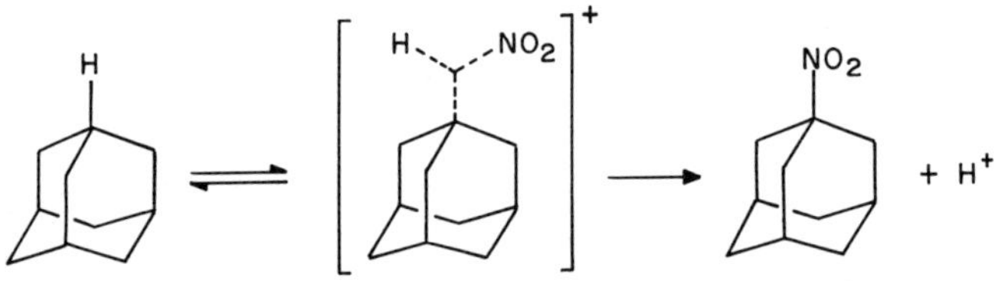

$$ (58) $$

Nitrosonium ion (NO^+) is an excellent hydride abstracting agent. Cumene undergoes hydride abstraction to provide cumyl cation, which further reacts to give condensation products[100c]. The hydride abstracting ability of NO^+ has been exploited in many organic transformations such as the Ritter reaction, ionic fluorination and so on (equation 59)[100d].

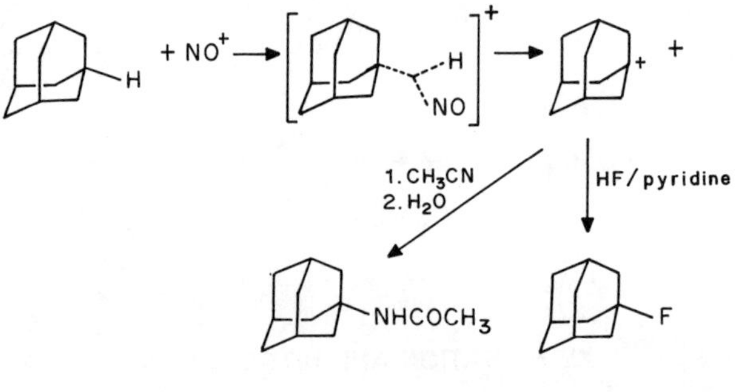

$$ (59) $$

XII. HALOGENATION

Halogenation of saturated aliphatic hydrocarbons is usually carried out by free radical processes[101]. Superacid-catalyzed ionic halogenation of alkanes has been reported by Olah and his coworkers[102]. Chlorination and chlorolysis of alkanes in the presence of SbF_5, $AlCl_3$ and $AgSbF_6$ catalysts were carried out. As a representative the reaction of methane with the Cl_2:SbF_5 system is shown in equation 60.

No methylene chloride or chloroform was observed in the reaction. Under the used stable ion conditions, dimethylchloronium ion formation (equation 61) also occurs. This is, however, a reversible process and helps to minimize competing alkylation of methane to ethane (and higher homologs) which becomes more predominant when methyl fluoride is formed via halogen exchange.

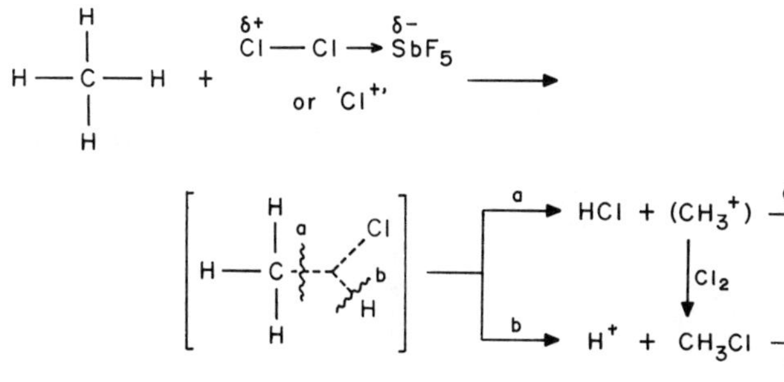

$$
\begin{array}{c}
\underset{\displaystyle H}{\overset{\displaystyle H}{H-\underset{|}{\overset{|}{C}}-H}} \;+\; \begin{array}{c} \overset{\delta+}{Cl}-Cl \to \overset{\delta-}{SbF_5} \\ \text{or } 'Cl^{+}\text{'} \end{array} \longrightarrow
\end{array}
$$

$$
\left[H-\underset{\displaystyle H}{\overset{\displaystyle H}{\underset{|}{\overset{|}{C}}}} \begin{array}{c} a \\ \diagdown \\ \diagup \\ b \end{array}\begin{array}{c} Cl \\ \\ H \end{array} \right] \longrightarrow
\begin{array}{l}
\xrightarrow{\;a\;} HCl + (CH_3^{+}) \xrightarrow{\;CH_3Cl\;} \\
\qquad\qquad\qquad\quad \downarrow Cl_2 \qquad CH_3Cl^{+}CH_3 \quad (60) \\
\xrightarrow{\;b\;} H^{+} + CH_3Cl \xrightarrow[(CH_3)]{}
\end{array}
$$

$$
CH_4 \xrightarrow[-78\,^{\circ}C]{SbF_5-Cl_2-SO_2ClF}
\left[H-\underset{\displaystyle H}{\overset{\displaystyle H}{\underset{|}{\overset{|}{C}}}} \cdots \begin{array}{c} Cl \\ \\ H \end{array} \right]^{+}
\xrightarrow{-H^{+}} CH_3Cl \underset{}{\overset{CH_3Cl}{\rightleftharpoons}} CH_3Cl^{+}CH_3
$$

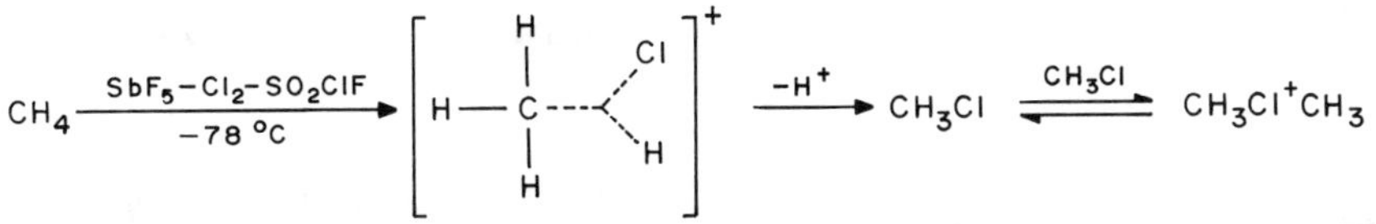

$$(61)$$

$Ag^{+}SbF_6^{+}$ catalyzed electrophilic bromination of alkanes[103] has also been carried out as shown in Scheme 16.

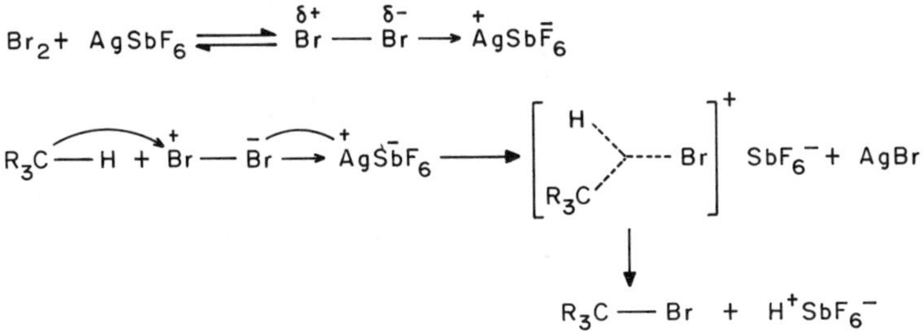

$$
Br_2 + AgSbF_6 \rightleftharpoons \overset{\delta+}{Br}-\overset{\delta-}{Br} \to \overset{+}{Ag}\overset{-}{SbF_6}
$$

$$
R_3C\overgroup{-H} + \overset{+}{Br}\overgroup{-}\overset{+}{Br} \to Ag\overset{+}{Sb}\overline{F_6} \longrightarrow
\left[\begin{array}{c} H \diagdown \\ \diagup \\ R_3C \end{array} \cdots Br \right]^{+} SbF_6^{-} + AgBr
$$

$$
\downarrow
$$

$$
R_3C - Br \;+\; H^{+}SbF_6^{-}
$$

SCHEME 16

Even electrophilic fluorination of alkanes has been reported. F_2 and fluoroxytrifluoromethane have been used to fluorinate tertiary centers in steroids and adamantanes by Barton and coworkers[104]. The electrophilic nature of a reaction involving polarized but not cationic fluorine species[105] has been invoked. Gal and Rozen[106] have carried out direct electrophilic fluorination of hydrocarbons in the presence of chloroform. F_2 appears to be strongly polarized in chloroform (hydrogen bonding with the acidic proton of $CHCl_3$). However, so far no positively charged fluorine species (fluoronium ion) is known in solution chemistry.

In extending the electrophilic halogenation (chlorination and bromination) of methane to catalytic heterogeneous gas-phase reactions, we have recently found[107] that methane can be chlorinated or brominated over various solid acid or supported platinum group metal catalysts (the latter being the heterogeneous analog of Shilov's solution chemistry) to methyl halides with high selectivity under relatively mild conditions. Table 8 summarizes some of the results.

Mechanistically, both reactions are electrophilic insertion reactions into the methane C—H bonds. In the platinum insertion reaction subsequent chlorolysis of the surface

TABLE 8. Chlorination and bromination of methane over suported acid catalysts

Catalyst	CH_4/Cl_2	Reaction temperature (°C)	$GHSV^a$ $(ml\,g^{-1}\,h^{-1})$	Conversion (%)	Product % CH_3Cl	CH_2Cl_2
10% FeO_xCl_y/Al_2O_3	1:2	250	100	16	88	12[b]
20% $TaOF_3/Al_2O_3$	1:2	235	50	14	82	6
	1:2	235	1400	15	93	7
	2:1	235	1200	13	96	4
20% $NbOF_3/Al_2O_3$	1:3	250	50	10	90	10
10% $ZrOF_2/Al_2O_3$	1:4	270	100	34	96	4
Nafion-H	1:4	185	100	18	88	12
20% GaO_xCl_y/Al_2O_3	1:2	250	100	26	90	10
20% TaF_5-Nafion-H	1:2	200	100	11	97	3
25% SbF_5-graphite	1:2	180	100	7	98	2

Catalyst	CH_4/Br_2	Reaction temperature (°C)	$GHSV^a$ $(ml\,g^{-1}\,H^{-1})$	Conversion (%)	Product (%) CH_3Br	CH_2Br_2
20% $SbOF_3/Al_2O_3$	5:1	200	100	20	99	
20% $TaOF_3/Al_2O_3$	15:1	250	50	14	99	

Supported platinum metal-catalyzed halogenation of methane

Catalyst	CH_4/X_2 (X = Cl, Br)	Reaction temperature (°C)	$GHSV^a$ $(ml\,g^{-1}\,H^{-1})$	Conversion (%)	Product (%) CH_3Cl	CH_2Cl_2
			Chlorination			
0.5% Pt/Al_2O_3	1:3	100	600	11[b]	~100	
	1:3	150	600	16[b]	92	8
	1:3	200	600	32[b]	92	8
	2:1	250	300	23[c]	98	<2
	3:1	250	300	26[c]	99	1
5% $Pd/BaSO_4$	2:1	200	600	30[c]	99	1
			Bromination			
0.5% Pt/Al_2O_3	2:1	200	300	8[c]	99	Trace

[a] Gaseous space velocity
[b] Based on methane
[c] Based on chlorine (bromine).

bound methylplatinum chloride complex regenerates the catalyst and gives methyl chloride (Scheme 17).

Concerning the electrophilic halogenation of methane it should also be pointed out that the singlet–triplet energy difference of positive halogens (as illustrated for the hypothetical X^+ ions) favor the latter[107c]. Electrophilic halogenations thus may be more complex and can involve radical ions even under conditions where conventional radical chain halogenation is basically absent.

Combining the halogenation of methane with catalytic, preferentially gas-phase hydrolysis, methyl alcohol (and dimethyl ether) can be obtained in high selectivity (Scheme 18)[107a,b]. The hydrolysis of methyl chloride with caustic was first carried out by Berthelot in the mid 1800[108,109]. This first preparation of methyl alcohol, however,

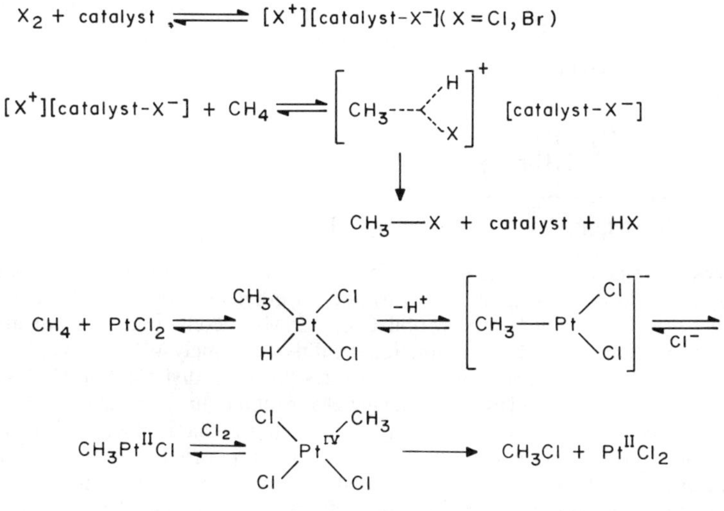

$$X_2 + \text{catalyst} \;\rightleftharpoons\; [X^+][\text{catalyst}-X^-]\,(\,X = Cl\,,\,Br\,)$$

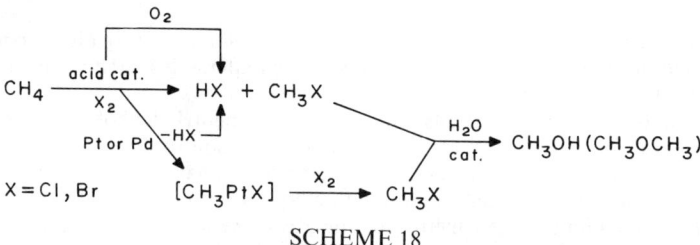

SCHEME 17

was never utilized in a practical way. Olah and coworkers carried out an extensive study of such reactions and found that methyl halides hydrolyze over alumina to methyl alcohol–dimethyl ether in yields up to 25% per pass with gaseous space velocities of up to 1500 and high turnovers[107a,b].

SCHEME 18

They also found that solid acid catalysts, such as Nafion-H, are also capable of catalyzing the hydrolysis of methyl halides and yielding dimethyl ether under the reaction temperatures of 150–170 °C (equation 62)[109b].

$$2CH_3X \xrightarrow[\text{Nafion-H}]{H_2O} CH_3OCH_3 + 2HX \tag{62}$$

Consequently halogenation of methane followed by hydrolysis is a suitable way to obtain methyl alcohol (and dimethyl ether) (equation 63).

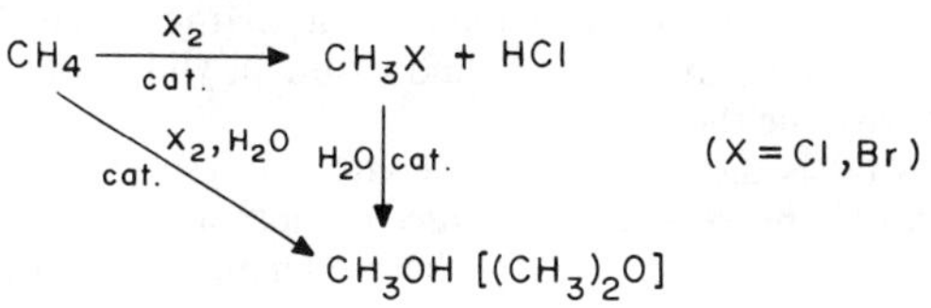

It was also possible to show that halogens (particularly bromine) together with steam can be reacted with methane over acidic catalysts in a single step producing methyl alcohol (dimethyl ether), although conversion of methane so far was only modest (equation 64)[109b].

$$CH_4 + Br_2 + H_2O \longrightarrow CH_3OH + 2HBr$$
$$2HBr + \tfrac{1}{2}O_2 \longrightarrow H_2O + Br_2 \tag{64}$$

$$\text{overall reaction: } CH_4 + \tfrac{1}{2}O_2 \longrightarrow CH_3OH$$

The feasible selective catalytic preparation of methyl halides and methyl alcohol from methane also allows subsequent condensation to ethylene and higher hydrocarbons. This can be accomplished by the Mobil Corp. ZSM-5 zeolite catalysts but also over non-shape selective acidic–basic bifunctional catalysts, namely WO_3/Al_2O_3.

Whereas the Mobil process starts with *syn* gas based methyl alcohol, Olah's studies were an extension of the previously discussed electrophilic functionalization of methane and does not involve any zeolite-type catalysts. It was found that bifunctional acidic–basic catalysts such as tungsten oxide on alumina or related supported transition metal oxides or oxyfluorides such as alumina or related supported transition metal oxides or oxyfluorides such as tantalum or zirconium oxyfluoride are capable of condensing methyl chloride, methyl alcohol (dimethyl ether), methyl mercaptan (dimethyl sulfide), primarily to ethylene (and propylene) (equation 65)[110].

$$2CH_3X \longrightarrow CH_2{=}CH_2 + 2HX$$
$$X = \text{halogen}$$
$$\text{or } 2CH_3XH \longrightarrow CH_2{=}CH_2 + 2H_2X$$
$$X = O, S \tag{65}$$

In these reactions, methane is the major by-product formed probably by competing radical reactions. However, since the overall starting material is methane this represents only the need for recycling.

According to Olah's investigations the conversion of methyl alcohol over bifunctional acidic–basic catalyst after initial acid-catalyzed dehydration to dimethyl ether involves oxonium ion formation catalyzed also by the acid functionality of the catalyst. This is followed by basic site catalyzed deprotonation to a reactive surface-bound oxonium ylide, which is then immediately methylated by excess methyl alcohol or dimethyl ether leading to the crucial $C_1 \to C_2$ conversion step. The ethyl methyl oxonium ion formed subsequently eliminates ethylene. All other hydrocarbons are derived from ethylene by known oligomerization–fragmentation chemistry. Propylene is formed via a cyclopropane intermediate. The overall reaction sequence is depicted in Scheme 19.

The intermolecular nature of the C_1–C_2 transformation step was shown by experiments using mono-^{13}C labelled dimethyl ether and analyzing the isotopic composition of the product ethylene. The intramolecular Stevens-type rearrangement under the reaction conditions was clearly ruled out.

It is not necessary to invoke a free Meerwein-type trimethyloxonium ion in the heterogeneous catalytic reaction. Lewis-type coordination complexes of dimethyl ether with the acidic catalyst sites having oxonium ion character can be involved, giving subsequently via deprotonation surface-bound oxonium ylides followed by methylation and elimination of ethylene (Scheme 20)[111,112].

Whereas Olah's studies did not involve zeolite catalysts, it is probable that in the ZSM-5 catalyzed Mobil process too no direct monomolecular dehydration of methyl alcohol to methylene is involved. This is thermodynamically not feasible even when

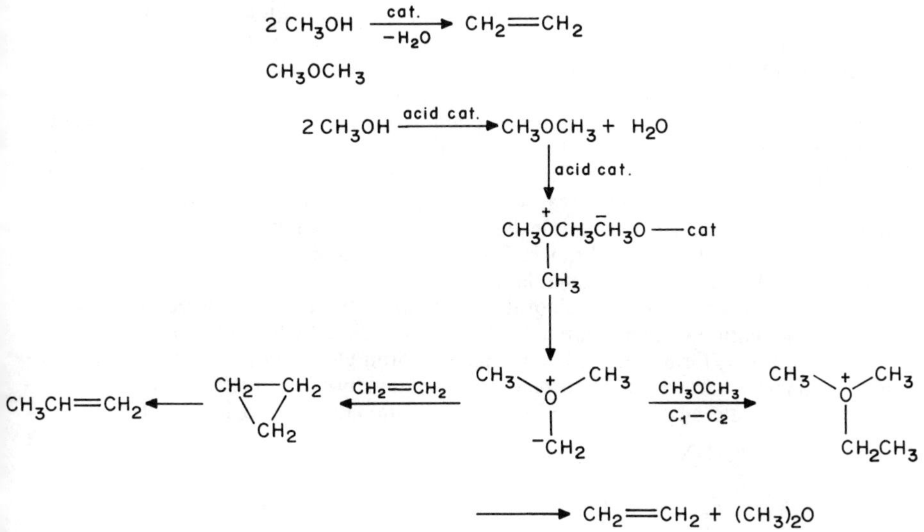

$$2\ CH_3OH \xrightarrow[-H_2O]{cat.} CH_2{=}CH_2$$

$$CH_3OCH_3$$

$$2\ CH_3OH \xrightarrow{acid\ cat.} CH_3OCH_3 + H_2O$$

$$\downarrow acid\ cat.$$

$$CH_3\overset{+}{O}CH_3\bar{C}H_3O \longrightarrow cat$$
$$|$$
$$CH_3$$

$$\longrightarrow CH_2{=}CH_2 + (CH_3)_2O$$

SCHEME 19

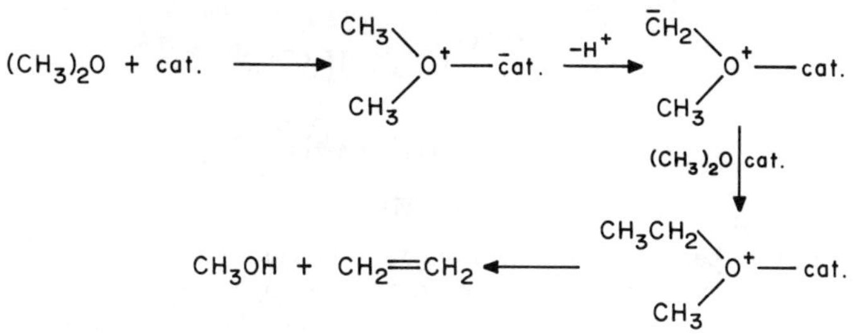

SCHEME 20

considering that surface complexation could somewhat affect the otherwise very endothermic thermodynamics (see Scheme 21).

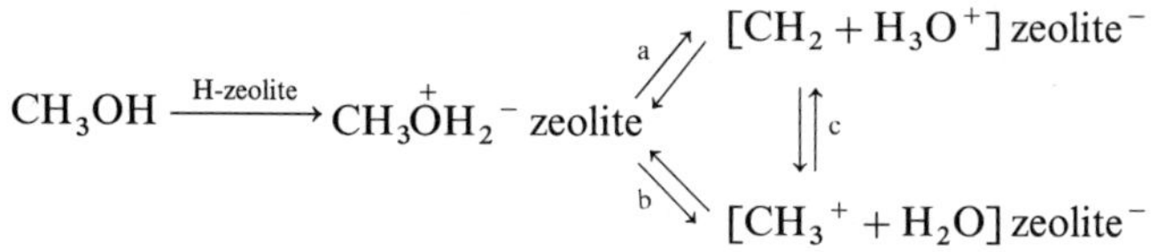

$$CH_3OH \xrightarrow{H\text{-zeolite}} CH_3\overset{+}{O}H_2{}^- zeolite \begin{array}{c} \overset{a}{\nearrow} [CH_2 + H_3O^+]\,zeolite^- \\ \Big\updownarrow c \\ \underset{b}{\searrow} [CH_3{}^+ + H_2O]\,zeolite^- \end{array}$$

$$\begin{array}{cccc} CH_3OH & \longrightarrow CH_2 & + H_2O & \\ -200.7 & +390.4 & -241.8 & \Delta H = +349.3\,kcal\,mol^{-1} \end{array}$$

$$\begin{array}{cccc} 2CH_3OH & \longrightarrow CH_3OCH_3 + H_2O & \\ 2(-200.7) & +184.1 & -241.8 & \Delta H = -24.5\,kcal\,mol^{-1} \end{array}$$

SCHEME 21

It is therefore reasonable to suggest that in the zeolite-catalyzed process too condensation proceeds via bimolecular dehydration of methyl alcohol to dimethyl ether, which subsequently is transformed via the oxonium ylide pathway and intermolecular methylation to ethyl oxonium species (the crucial C_1—C_2 bond formation step)[111]. A radical or radical ion type condensation mechanism inevitably should give also competing coupling products which are, however, not observed. Once ethylene is formed in the system it can undergo acid-catalyzed oligomerization–cleavage reactions, undergo methylene insertion or react further with methyl alcohol giving higher hydrocarbons.

With increase in the acidity of the catalyst, methyl alcohol or dimethyl ether undergo condensation to saturated hydrocarbons or aromatics, with no olefin by-products. With tantalum or niobium pentafluoride based catalyst the studies at 300 °C resulted in conversion to gasoline range branched hydrocarbons and some aromatics (30% of the product)[112]. This composition is similar to that reported with H-ZSM-5 zeolite catalyst.

The condensation of methyl chloride or bromide to ethylene proceeds by a related mechanistic path involving initial acid-catalyzed dimethylhalonium ion (or related catalyst complex) formation with subsequent proton elimination to a reactive methyl-halonium methylide, which then is readily methylated by excess methyl halide. The ethyl-halonium ion intermediate gives ethylene by β-elimination (see Scheme 22).

$$2CH_3X \longrightarrow CH_2{=}CH_2 + 2HX$$

$$CH_3X^+ \rightarrow {}^-cat. \longrightarrow [\bar{C}H_2\overset{+}{X}{-}cat.] \xrightarrow{CH_3X} CH_3CH_2\overset{+}{X}{-}cat. \xrightarrow{-HX} CH_2{=}CH_2$$

$$2CH_3X \xrightarrow{acid\,cat.} CH_3\overset{+}{X}CH_3\bar{C}l{-}cat.$$

$$\downarrow -H^+ \, base$$

$$CH_3\overset{+}{X}\bar{C}H_2 \xrightarrow[cat.]{CH_3X} CH_3\overset{+}{X}CH_2CH_3\bar{X}{-}cat.$$

$$\downarrow$$

$$CH_3X + CH_2{=}CH_2$$

SCHEME 22

Similar reaction paths can be visualized for condensation of methyl mercaptan or methylamines. It is interesting to note that Corey and Chaykovsky's well-known synthetic studies[113] with the use of dimethylsulfonium methylide mentioned the need to generate the ylide at low temperatures, which otherwise led to decomposition giving ethylene. Indeed, this reaction is similar to that involved in the higher-temperature acid–base catalyzed condensation reaction (Scheme 23).

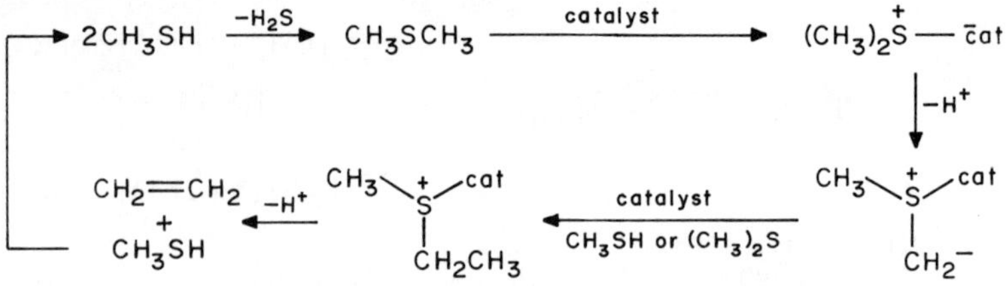

SCHEME 23

XIII. CONCLUSIONS

One of us, in the conclusion of a review in *Carbocations and Electrophilic Reactions*[48], wrote some 16 years ago on the realization of the electron donor ability of shared

σ-electron pair: 'More importantly, the concept of pentacoordinated carbonium ion formation *via* electron sharing of single bonds with electrophilic reagents in three-center bond formation promises to open up a whole new important area of chemistry. Whereas the concept of tetravalency of carbon obviously is not affected, carbon penta- (or tetra-)coordination as a general phenomenon must be recognized. Trivalent carbenium ions play a major role in electrophilic reactions of π-and n-donors, whereas pentacoordinated carbonium ions play an equally important similar role in electrophilic reactions of σ-donor saturated systems'[11]. 'The realization of the electron donor ability of shared (bonded) electron pairs (single bonds) could one day rank equal in importance with G. N. Lewis' realization[114] of the importance of the electron donor unshared (non-bonded) electron pairs. We can now not only explain the reactivity of saturated hydrocarbons and single bonds in general in electrophilic reactions, but indeed use this understanding to explore many new areas and reactions of carbocation chemistry'.

It seems that the intervening years have justified that prediction to a significant degree. The electrophilic chemistry of alkanes has rapidly expanded and has started to occupy a significant role even in the conversion of methane.

XIV. REFERENCES

1. F. Asinger, *Paraffins, Chemistry and Technology*, Pergamon Press, New York, 1965.
2. (a) For a comprehensive early review, see G. A. Olah, *Angew. Chem., Int. Ed. Engl.*, **12**, 173 (1973) and references cited therein.
   (b) For reviews on superacids, see G. A. Olah, G. K. S. Prakash and J. Sommer, *Superacids*, Wiley–Interscience, New York, 1985.
3. H. S. Bloch, H. Pines and L. Schmerling, *J. Am. Chem. Soc.*, **68**, 153 (1946).
4. (a) G. A. Olah and J. Lukas, *J. Am. Chem. Soc.*, **89**, 2227 (1967).
   (b) G. A. Olah and J. Lukas, *J. Am. Chem. Soc.*, **89**, 4739 (1967).
5. (a) A. F. Bickel, G. J. Gaasbeek, H. Hogeveen, J. M. Oelderick and J. C. Platteuw, *Chem. Commun.*, 634 (1967).
   (b) H. Hogeveen and A. F. Bickel, *Chem. Commun.*, 635 (1967).
6. G. K. S. Prakash, A. Husain and G. A. Olah, *Angew. Chem.*, **95**, 51 (1983).
7. G. A. Olah and J. Lukas, *J. Am. Chem. Soc.*, **90**, 933 (1968).
8. G. A. Olah and R. H. Schlosberg, *J. Am. Chem. Soc.*, **90**, 2726 (1968).
9. H. Hogeveen and C. J. Gaasbeek, *Recl. Trav. Chim. Pays-Bas*, **87**, 319 (1968).
10. G. A. Olah, G. Klopman and R. H. Schlosberg, *J. Am. Chem. Soc.*, **91**, 3261 (1969).
11. (a) G. A. Olah, J. Shen and R. H. Schlosberg, *J. Am. Chem. Soc.*, **92**, 3831 (1970).
    (b) T. Oka, *Phys. Rev. Lett.*, **43**, 531 (1980).
12. H. Hogeveen and A. F. Bickel, *Recl. Trav. Chim. Pays-Bas*, **86**, 1313 (1967).
13. H. Pines and N. E. Hoffman, in *Friedel–Crafts and Related Reactions*, Vol. II (Ed. G. A. Olah), Wiley–Interscience, New York, 1964, p. 1216.
14. (a) G. A. Olah, Y. Halpern, J. Shen and Y. K. Mo, *J. Am. Chem. Soc.*, **93**, 1251 (1971).
    (b) G. A. Olah, N. Trivedi and G. K. S. Prakash, unpublished results.
15. J. W. Otvos, D. P. Stevenson, C. D. Wagner and O. Beeck, *J. Am. Chem. Soc.*, **73**, 5741 (1951)
16. H. Hoffman, C. J. Gaasbeek and A. F. Bickel, *Recl. Trav. Chim. Pays-Bas*, **88**, 703 (1969).
17. R. Bonnifay, B. Torck and J. M. Hellin, *Bull. Soc. Chim. Fr.*, 808 (1977).
18. (a) J. Lucas, P. A. Kramer and A. P. Kouwenhoven, *Recl. Trav. Chim. Pays-Bas*, **92**, 44 (1973).
    (b) J-C. Culman and J. Sommer, *J. Am. Chem. Soc.*, **112**, 4057 (1990).
19. W. O., Haag and R. H. Dessau, *International Catalysis Congress* (W. Germany), 1984, II, 105.
20. For a review, see D. M. Brouwer and H. Hogeveen, *Prog. Phys. Org. Chem.*, **9**, 179 (1972).
21. E. W. Colvin, *Silicon in Organic Synthesis*, Butterworths, London, 1981.
22. W. P. Weber, *Silicon Reagents in Organic Synthesis*, Springer-Verlag, Berlin, Heidelberg, New York, 1983.
23. J. Bertram, J. P. Coleman, M. Fleischmann and D. Plechter, *J. Chem. Soc. Perkin Trans 2*, 374 (1973).
24. For a review, see P. L. Fabre, J. Devynck and B. Tremillon, *Chem. Rev.*, **82**, 591 (1982).

650   G. A. Olah and G. K. Surya Prakash

25. J. Devynck, P. L. Fabre, A. Ben Hadid and B. Tremillon, *J. Chem. Res.*, 200 (1979).
26. A. Germain, P. Ortega and A. Commeyras, *Nouv. J. Chim.*, 3, 415 (1979).
27. H. Choucroun, A. Germain, D. Brunel and A. Commeyras, *Nouv. J. Chim.*, 7, 83 (1983).
28. C. D. Nenitzescu and A. Dragan, *Ber. Dtsch. Chem. Ges.*, 66, 1892 (1933).
29. H. Pines and N. E. Joffman, in *Friedel-Crafts and Related Reactions*, Vol. 2 (Ed. G. A. Olah),
    Wiley–Interscience, New York, 1964, p. 1211.
30. C. D. Nenitzescu, in *Carbonium Ions*, Vol. II (Eds. G. A. Olah and P. v. R. Schleyer),
    Wiley–Interscience, New York, 1970, p. 490.
31. F. Asinger, in *Paraffins*, Pergamon Press, New York, 1968, p. 695.
32. (a) R. Bonnifay, B. Torck and J. M. Hellin, *Bull. Soc. Chim. Fr.*, 1057 (1977).
    (b) R. Bonnifay, B. Torck and J. M. Hellin, *Bull. Soc. Chim. Fr.*, 36 (1978).
33. D. Brunel, J. Itier, A. Commeyras, R. Phan Tan Luu and D. Mathieu, *Bull. Soc. Chim. Fr.
    II*, 249, 257 (1979).
34. G. A. Olah, U.S. Patent 4,472,268 (1984); 4,508,618 (1985).
35. G. A. Olah, U.S. Patent 4,613,723 (1986).
36. (a) F. Le Normand, F. Fajula, F. Gault and J. Sommer, *Nouv. J. Chim.* 6, 411 (1982).
    (b) F. Le Normand, F. Fajula and J. Sommer, *Nouv. J. Chim.*, 6, 291 (1982).
37. K. Laali, M. Muller and J. Sommer, *Chem. Commun.*, 1088 (1980).
38. G. A. Olah, J. Kaspi and J. Bukala, *J. Org. Chem.*, 42, 4187 (1977).
39. G. A. Olah, U.S. Patent 4,116,880 (1978).
40. G. A. Olah and J. Kaspi, *J. Org. Chem.*, 42, 3046 (1977).
41. G. A. Olah, J. R. DeMember and J. Shen, *J. Am. Chem. Soc.*, 95, 4952 (1973).
42. (a) P. v. R. Schleyer, *J. Am. Chem. Soc.*, 79, 3292 (1957).
    (b) R. C. Fort, Jr., in *Adamantane, the Chemistry of Diamond Molecules* (Ed. P. Gassman),
    Marcel Dekker, New York, 1976.
    (c) G. A. Olah and J. A. Olah, *Synthesis*, 488 (1973).
43. G. A. Olah and O. Farooq, *J. Org. Chem.*, 51, 5410 (1986).
44. (a) O. Farooq, S. M. F. Farnia, M. Stephenson and G. A. Olah, *J. Org. Chem.*, 53, 2840 (1988).
    (b) G. A. Olah, A-H. Wu, O. Farooq and G. K. S. Prakash, *J. Org. Chem.*, 54, 791 (1989).
45. (a) M. Vol'pin, I. Akhrem and A. Orlinkov, *New J. Chem.*, 13, 771 (1989).
    (b) L. Paquette and A. Balogh, *J. Am. Chem. Soc.*, 104, 774 (1982).
    (c) W.-D. Fessner, B-A. R. C. Murty, J. Wörth, D. Hunkler, H. Fritz, H. Prinzback, W. R.
    Roth, P. v. R. Schleyer, A. B. McEwen and W. F. Maier, *Angew. Chem., Int. Ed. Engl.*, 26, 452
    (1987).
46. G. C. Schuit, H. Hoog and J. Verhuis, *Recl. Trav. Chim. Pays-Bas*, 59, 793 (1940); British
    Patent 535054 (1941).
47. H. Pines and R. C. Wackher U.S. Patent 2,406,633 (H. Pines to Universal Oil); *Chem. Abstr.*,
    41, 474 (1947).
48. For a review, see G. A. Olah, *Carbocations and Electrophilic Reactions*, Verlag Chemie, Wiley,
    New York, 1974.
49. For a review, see D. M. Brouwer and H. Hogeveen, *Prog. Phys. Org. Chem.*, 9, 179 (1972).
50. (a) G. A. Olah, M. Bruce, E. H. Edelson and A. Husain, *Fuel*, 63, 1130 (1984).
    (b) G. A. Olah and A. Husian, *Fuel*, 63, 1427 (1984).
    (c) G. A. Olah, M. Bruce, E. H. Edelson and A. Husain, *Fuel*, 63, 1432 (1984).
51. G. A. Olah, J. R. DeMember and R. H. Schlosberg, *J. Am. Chem. Soc.*, 91, 2112 (1969).
52. G. A. Olah, J. R. DeMember, R. H. Schlosberg and Y. Halpern, *J. Am. Chem. Soc.*, 94, 156
    (1972).
53. D. M. Brouwer, Y. K. Mo and J. A. Olah, *J. Am. Chem. Soc.*, 95, 4939 (1973).
54. G. A. Olah, J. R. Demember and J. Shen, *J. Am. Chem. Soc.*, 95, 4952 (1973).
55. G. A. Olah and A. M. White, *J. Am. Chem. Soc.*, 91, 5801 (1969).
56. G. A. Olah, O. Farooq, V. V. Krishnamurthy, G. K. S. Prakash and K. Laali *J. Am. Chem.
    Soc.*, 107, 7541 (1985).
57. G. A. Olah, U.S. Patent 4,443,192; 4,465,893; 4,467,130 (1984); 4,513,164 (1985).
58. G. A. Olah and J. A. Olah, *J. Am. Chem. Soc.*, 93, 1256 (1971); M. Siskin, R. H. Schlosberg
    and W. P. Kocsis, in *New Strong Acid Catalyzed Alkylation and Reduction Reactions* (Eds.
    L. F. Albright and R. A. Gikdsktm), American Chemical Society Monographs, No. 55,
    Washington, D.C. (1977).
59. G. A. Olah, J. D. Felberg and K. Lammertsma, *J. Am. Chem. Soc.*, 105, 6529 (1983).

60. M. Siskin, *J. Am. Chem. Soc.*, **98**, 5413 (1976).
61. J. Sommer, M. Muller and K. Laali, *Nouv. J. Chim.*, **6**, 3 (1982).
62. G. A. Olah, J. D. Felberg and K. Lammertsma, *J. Am. Chem. Soc.*, **105**, 6529 (1983).
63. D. T. Roberts, Jr. and L. E. Calihan, *J. Macromol. Sci (Chem.)*, **A7**(8), 1629 (1973).
64. D. T. Roberts, Jr. and L. E. Calihan, *J. Macromol. Sci (Chem.)*, **A7**(8), 1641 (1973).
65. G. A. Olah and J. A. Olah, *Friedel-Crafts and Related Reactions*, Vol. 3, Part 2, Wiley–Interscience, New York, 1964, p. 1272.
66. H. Hogeveen, *Adv. Phys. Org. Chem.*, **10**, 29 (1973).
67. W. F. Gresham and G. E. Tabet, U.S. Patent 2,485,237 (1946).
68. B. L. Booth and T. A. El-Fekky, *J. Chem. Soc. Perkin Trans. 1*, 2441 (1979).
69. R. Paatz and G. Weisberger, *Chem. Ber.*, **100**, 984 (1967).
70. N. Yoneda, T. Fukuhara, Y. Takahashi and A. Suzuki, *Chem. Lett.*, 17 (1983).
71. N. Yoneda, H. Sato, T. Fukuhara, Y. Takahashi and A. Suzuki, *Chem. Lett.*, 19 (1983).
72. (a) N. Yoneda, Y. Takahashi, T. Fukuhara and A. Suzuki, *Bull. Chem. Soc. Jpn.*, **59**, 2819 (1986).
    (b) S. Delavarnne, M. Simon, M. Fauconet and J. Sommer, *J. Am. Chem. Soc.*, **111**, 383 (1989).
73. G. A. Olah, F. Pelizza, S. Kobayashi and J. A. Olah, *J. Am. Chem. Soc.*, **98**, 296 (1976).
74. S. Fujiyama, M. Takagawa and S. Kajiyama, German Patent 2,425591 (1974).
75. J. M. Delderick and A. Kwantes, British Patent 1,123,966 (1968).
76. W. F. Gresham and G. E. Tabet, U.S. Patent 21,485,237 (1949).
77. O. Farooq, Ph.D. thesis, University of Southern California, Los Angeles, CA (Dec. 1984).
78. G. A. Olah, K. Laali and O. Farooq, *J. Org. Chem.*, **50**, 1483 (1985).
79. O. Farooq, M. Marcelli, G. K. S. Prakash and G. A. Olah, *J. Am. Chem. Soc.*, **110**, 864 (1988).
80. G. A. Olah and S. Kuhn, *J. Am. Chem. Soc.*, **82**, 2380 (1960).
81. H. R. Christen, *Grundlagen des Organische Chemie*, Verlag Sauerlander-Diesterweg, Aarau–Frankfurt am Main, 1970, p. 242.
82. G. A. Olah, D. G. Parker and N. Yoneda, *Angew. Chem., Int. Ed. Engl.*, **17**, 909 (1978).
83. K. O. Christe, W. W. Wilson and E. C. Curtis, *Inorg. Chem.*, **18**, 2578 (1976).
84. G. A. Olah, A. L. Berrier and G. K. S. Prakash, *J. Am. Chem. Soc.*, **104**, 2373 (1982).
85. G. A. Olah, N. Yoneda and D. G. Parker, *J. Am. Chem. Soc.*, **99**, 483 (1977).
86. G. A. Olah, D. G. Parker, N. Yoneda and F. Pelizza, *J. Am. Chem. Soc.*, **98**, 2245 (1976).
87. (a) G. A. Olah, N. Yoneda and D. G. Parker, *J. Am. Chem. Soc.*, **98**, 5261 (1976).
    (b) G. A. Olah, T. D. Ernst, C. B. Rao and G. K. S. Prakash, *New J. Chem.*, **13**, 791 (1989).
88. R. Tambarulo, S. N. Ghosh, C. A. Barrus and W. Gordy, *J. Chem. Phys.*, **24**, 851 (1953).
89. P. D. Bartlett and M. Stiles, *J. Am. Chem. Soc.*, **77**, 2806 (1955).
90. J. P. Wibault, E. L. J. Sixma, L. W. E. Kampschidt and H. Boer, *Recl. Trav. Chim. Pays-Bas*, **69**, 1355 (1950).
91. J. P. Wibault and E. L. J. Sixma, *Recl. Trav. Chim. Pays-Bas*, **71**, 76 (1951).
92. P. S. Bailey, *Chem. Rev.*, **58**, 925 (1958).
93. L. Homer, H. Schaefer and W. Ludwig, *Chem. Ber.*, **91**, 75 (1958).
94. P. S. Bailey, J. W. Ward, R. E. Hornish and F. E. Potts, *Adv. Chem. Ser.*, **112**, 1 (1972).
95. Further oxidation products, i.e. acetylium ion and CO_2, were reported to be observed in a number of the reactions studied. Such secondary oxidations are not induced by ozone.
96. T. M. Hellman and G. A. Hamilton, *J. Am. Chem. Soc.*, **96**, 1530 (1974).
97. M. C. Whiting, A. J. N. Bolt and J. H. Parrish, *Adv. Chem. Ser.*, **77**, 4 (1968).
98. D. O. Williamson and R. J. Cvetanovic, *J. Am. Chem. Soc.*, **92**, 2949 (1970).
99. (a) J. C. Jacquesy, R. Jacquesy, L. Lamande, C. Narbonne, J. F. Patoiseau and Y. Vidal, *Nouv. J. Chim.* **6**, 589 (1982).
    (b) J. C. Jacquesy and J. F. Patoiseau, *Tetrahedron Lett.*, 1499 (1977).
    (c) G. A. Olah, Q. Wang and G. K. S. Prakash, *J. Am. Chem. Soc.*, **112**, 3698 (1990).
100. (a) G. A. Olah and H. C. Lin, *J. Am. Chem. Soc.*, **93**, 1259 (1971).
    (b) G. A. Olah and coworkers unpublished results.
    (c) G. A. Olah and N. Friedman, *J. Am. Chem. Soc.*, **88**, 5330 (1966).
    (d) G. A. Olah, *Acc. Chem. Res.*, **13**, 330 (1980).
101. M. L. Poutsma, *Methods in Free-Radical Chemistry*, Vol. II (Ed. E. S. Husyer), Marcel Dekker, New York, 1969.
102. (a) G. A. Olah and Y. K. Mo, *J. Am. Chem. Soc.*, **94**, 6864 (1972).
    (b) G. A. Olah, R. Renner, P. Schilling and Y. K. Mo, *J. Am. Chem. Soc.*, **95**, 7686 (1973).
103. G. A. Olah and P. Schilling, *J. Am. Chem. Soc.*, **95**, 7680 (1973).

652   G. A. Olah and G. K. Surya Prakash

104. D. Alker, D. H. R. Barton, R. H. Hesse, J. L. James, R. E. Markwell, M. M. Pechet, S. Rozen, T. Takeshita and H. T. Toh, *Nouv. J. Chim.* **4**, 239 (1980).
105. K. O. Christe, *J. Fluorine Chem.*, 519 (1983); K. O. Christe, *J. Fluorine Chem.*, 269 (1984).
106. C. Gal and S. Rozen, *Tetrahedron Lett.*, 449 (1984).
107. (a) G. A. Olah, B. Gupta, M. Farnia, J. D. Felberg, W. M. Ip, A. Husain, R. Karpeles, K. Lammertsma, A. K. Melhotra and N. J. Trivedi, *J. Am. Chem. Soc.*, **107**, 7097 (1985).
     (b) G. A. Olah, U.S. Patent 7,523,040 (1985) and corresponding foreign patents.
     (c) Y. Li, X. Wang, F. Jensen, K. Houk and G. A. Olah, *J. Am. Chem. Soc.*, **112**, 3922 (1990).
108. M. Berthelot, *Ann. Chim.*, **52**, 97 (1858).
109. (a) K. Weissermel and H. J. Arpe, *Industrial Organic Chemistry*, Verlag-Chemie, Weinheim, 1978, pp. 47–48.
     (b) G. A. Olah and coworkers, unpublished results.
110. G. A. Olah, H. Doggweiler, J. D. Felberg, S. Frohlich, M. J. Grdina, R. Karpeles, T. Keumi, S. Inaba, W. M. Ip, K. Lammertsma, G. Salem and D. C. Tabor, *J. Am. Chem. Soc.*, **106**, 2143 (1984); G. A. Olah, U.S. Patent 4,373,109 (1983) and corresponding foreign patents.
111 (a) G. A. Olah, G. K. S. Prakash, R. W. Ellis and J. A. Olah, *J. Chem. Soc., Chem. Commun.*, 9 (1986).
     (b) Also see G. A. Olah, *Acc. Chem. Res.*, **20**, 422 (1987).
112. G. Salem, Ph.D. thesis, University of Southern California, 1980.
113. E. J. Lewis, *J. Am. Chem. Soc.*, **38**, 762 (1916); *Valence and Structure of Atoms and Molecules*, Chemical Catalog Corp., New York, 1923.

CHAPTER **14**

Organometallic chemistry of alkane activation

ROBERT H. CRABTREE

Department of Chemistry, Yale University, 225 Prospect Street, New Haven, CT 06511, USA

I. INTRODUCTION

In this chapter, we consider the interaction of transition metal complexes with alkanes to give alkane-derived products, where organometallic species are either the final products or are suspected intermediates. In this area[1] only a few fundamentally different mechanistic reaction types appear to operate.

The Chemistry of Alkanes and Cycloalkanes
Edited by S. Patai and Z. Rappoport © 1992 John Wiley & Sons Ltd

A. Mechanistic Types

The first alkane reactions involving organometallic intermediates were investigated by Shilov. The best current understanding of the pathway involved is shown in equation 1a. The metal M appears to bind the C—H bond to form a side-on σ complex of a type discussed in more detail below. In the Shilov systems, this species loses a proton (equation 1a) to give a metal alkyl which is usually unstable and goes on to give a variety of catalytic reactions. Occasionally, however, the metal alkyl has been observed directly.

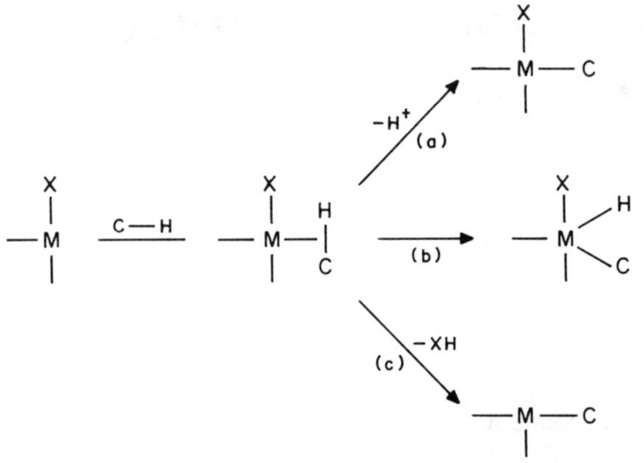

$$\tag{1}$$

In the second type of process the metal acts as a carbenoid and inserts into the C—H bond, a process generally termed 'oxidative addition' in organometallic chemistry (equation 1b). This reaction is believed to go via the same sort of alkane complex as in the Shilov system, but, instead of losing a proton, it goes instead to an alkylmetal hydride. This may be stable, in which case it is observed as the final product, or it may react further.

In the third type, predominantly seen for early metals, the C—H bond is broken by what Bercaw and coworkers[2] have called σ-bond metathesis (equation 1c). In this situation an alkyl X already present on the metal undergoes exchange with the alkane. Here too, an alkane complex is a very likely intermediate. Instead of the proton being lost as H^+, as in the Shilov case, it protonates the alkyl, a process which is promoted by its anionic character, which is in turn a result of the low electronegativity of the early metal.

Another type involves homolytic splitting of the C—H bond by a pair of metallo-radicals (equation 2). This has so far been seen in only one case, perhaps because coordinatively unsaturated paramagnetic metal complexes have only rarely been studied.

$$\tag{2}$$

We also discuss reactions of metal atoms, ions and excited metal atoms because these are ligand-free analogues of the systems mentioned above; in one case a synthetically useful method is based on this chemistry.

B. Agostic Species and Alkane Complexes

Several of the mechanisms discussed above are believed to involve C—H⋯M bridged species. A large number of compounds are now known in which a C—H bond of the

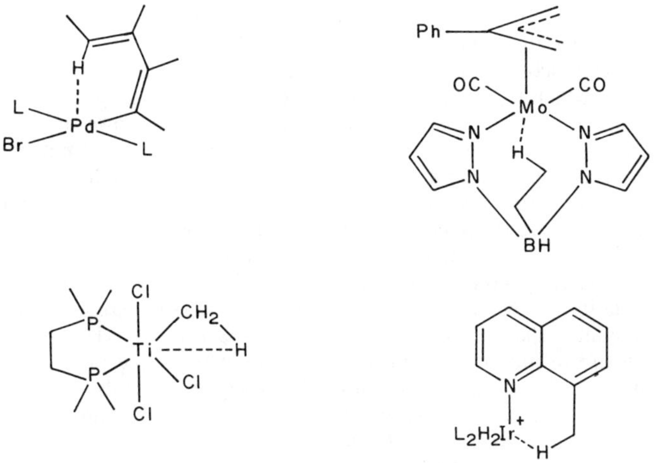

FIGURE 1. Some examples of compounds with agostic C—H bonds

ligand is chelated to the metal via a C—H···M bridge to form a so-called 'agostic' system[3]. Typical examples are shown in Figure 1. In order to form these complexes, a metal must have a 2e vacant site to accept the C—H bond and the chelate ring size must be favorable; the usual ring size is 3–6 atoms. If the C—H bond is cleaved, the oxidative addition product, a cyclometalated[4] species, is formed. The equilibria shown in equation 3 are therefore possible. Agostic binding greatly increases the proton acidity of the C—H bond. This is the reason that the alkane, which is normally an extremely weak acid, readily deprotonates in the Shilov system; the resulting alkyl anion is stabilized by the metal as a metal alkyl. So far no isolable alkane complexes have been seen. Presumably the C—H···M bond is not strong enough to hold an alkane to the metal. Alkane complexes are almost certainly involved in the reactions of alkanes, as noted in equation 1 and further discussed below.

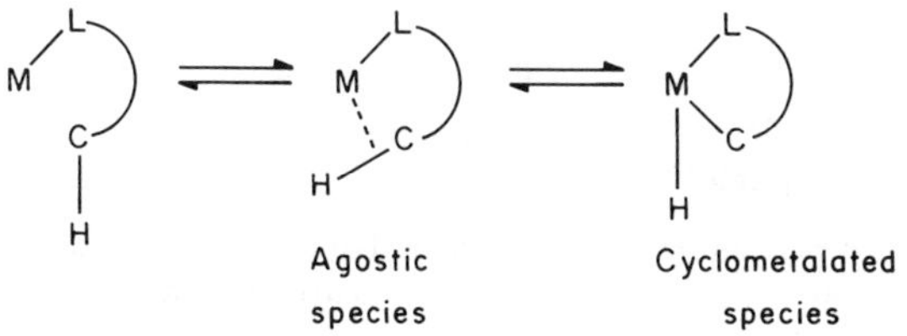

(3)

Agostic species   Cyclometalated species

C. Theoretical Aspects

Theoretical studies have recently been reviewed by Saillard[5]. The isolobal analogy between H—H and C—H is evident in the close relation between stable H_2 complexes[6] and agostic systems. In both cases, the X—H group is bound side-on to the metal without cleavage of the X—H bond. The bonding picture is similar for each: the X—H bonding pair donates into a vacant $M(d_\sigma)$ orbital and the $M(d_\pi)$ lone pairs on the metal back donate into the X—H σ^*. The metal must therefore be a good σ acid to bind the poorly basic X—H bonding pair, but a relatively poor π base, because excessive π basicity would populate the σ^* orbital of the C—H bond and so cleave the bond. H_2 complexes are much more stable than methane complexes; the latter are only proposed as reaction intermediates, while scores of H_2 complexes have now been isolated. It is therefore only the chelate effect which has so far allowed C—H $\cdots$ M species to be isolated. The more basic Si—H bond readily forms complexes without chelation, however, and these have a similar bonding scheme[7]. Some π-back bonding is helpful in strengthening agostic binding because d^0 metals have not so far given isolable H_2 complexes.

Theoretical studies are still sparse on catalytic systems of the Shilov type. It is not yet clear why Pt and Pd are so much more effective than other elements.

Oxidative addition is symmetry-allowed. The metal must be unsaturated, having sixteen valence electrons (16e) or less, and have at least two d_π electrons so that it is capable of oxidation by two units. Another way of thinking of the role of the d_π electrons is that they populate the C—H σ^* orbital and so break the C—H bond. This means that d^0 metals are unsuitable; d^2 and d^8 metals seem to be the most active. A lower activation energy is expected for the more electronegative metals and if a vacant p_π orbital is also available, as is the case for T-shaped d^8-ML_3 systems, such species are postulated as intermediates in several catalytic systems discussed in Section III.B [e.g. $RhClL_2$ and $Ir(OCOR)L_2$]. A triangular side-on transition state is found in all studies. This is less stable in the order H—H > CH_3—H > CH_3—CH_3, as a result of the directed character of the sp^3 orbital of CH_3, which is less able to bond to X and M simultaneously than is the 1s orbital of H^8. Morokuma and coworkers[8b] noted that the M—H bond is formed earlier than the M—C bond during the reaction. Experimental support for the character of the transition state comes from a study of agostic C—H $\cdots$ M bonds[9]. A calculation at the EH level has identified the main orbital interaction in C—H oxidative addition to d^8-ML_4 and $CpML^{10}$.

σ-bond metathesis has also been studied from a theoretical perspective [11,12]. A Cp_2LuH_3 species lies on the path for exchange in $Cp_2LuH + H_2$ (Cp = cyclopentadienyl). The two electrons required to form the bonds to the two hydrogen atoms in the transition state, come not from the metal as they do in an oxidative addition, but from the π electrons of the Cp. One empty orbital on the metal is required to accept the X—H bonding pair and the rate goes up as the energy of this level falls. In contrast to the situation for oxidation, d^0 metals do react, indeed they are the most active. EH calculations did not support the idea of an H_2 complex as intermediate, but that was because the Cp_2MH fragment has to bend to accommodate the incoming X—H group. Presumably if a bent Cp_2LuH were taken as the reference state, an attractive interaction with H_2 would be seen. As it is, complexation may be said to drive the bending.

D. Thermodynamic Aspects[13]

Alkanes are relatively stable species thermodynamically and so many reactions of alkanes (dehydrogenation, dehydrodimerization, carbonylation) are unfavorable under ambient conditions. This means we often need to couple some favorable process with the unfavorable alkane conversion in order to drive it. We look at the details in Section III.B, but only note here that the common appearance of photochemistry in alkane chemistry can be seen as a way to drive reactions thermodynamically and to access highly reactive transition metal fragments that are kinetically competent to react with alkanes.

The oxidative addition of H_2 is much more common than the reaction of alkane C—H bonds. Both thermodynamic and theoretical studies of oxidative addition of X—H bonds to metal complexes identify the relative weakness of M—C bonds as an important factor in the generally lower thermodynamic stability of alkyl hydrides relative to dihydrides or silyl hydrides. Halpern[14] identified a number of species where D(M—H) is about

60 kcal mol^{-1} and D(M—C) is around 30 kcal mol^{-1}; the sum is insufficient to provide a driving force for cleavage of alkane C—H bonds [D(C—H) = 95–105 kcal mol^{-1}]. On theoretical grounds, Low and Goddard[8a] have estimated the driving force for the addition of H$_2$, CH$_3$—H and CH$_3$—CH$_3$ to (H$_3$P)$_2$Pt as -16, $+9$ and $+16$ kcal mol^{-1}, respectively. Systems where alkane C—H bonds insert to give isolable adducts (Section III.A) are probably rare because few systems have M—C and M—H bonds substantially stronger than those studied by Halpern and Goddard; usually the reaction is endothermic for alkane C—H bonds. Arenes are much easier to activate by oxidative addition; this is probably related to the far higher M—Ph bond energy compared to M—CH$_3$.

Bryndza, Bercaw and coworkers[15] have shown that M—X bond strengths in Cp*Ru(PMe$_3$)$_2$X(Cp* = η^5-C$_5$Me$_5$; X = OH, C≡CPh, CH$_2$COMe, NHPh and OMe) correlate well with the X—H bond strengths for the free protonated ligand X—H. By this measure, M—H is anomalously strong (or H—H anomalously weak) for reasons that are not yet clear. In more recent work, the same group[16] has shown that the heterolytic M—PMe$_3$ bond strength in Cp*Ru(PMe$_3$)$_2$X is very dependent on the steric size of X. This result implies that we cannot assume the transferability of bond strengths in organometallic systems. If this proves to be general, we will probably have to abandon the notion of bond strengths in our discussions of the thermodynamics of organometallic processes. Unlike organic systems, organometallic complexes are often sterically very crowded as a result of the large coordination numbers of the central metal and the large cone angle of many common ligands (e.g. Cp*).

II. SHILOV CHEMISTRY

In 1967, Hodges and Garnett[17] found that PtCl$_4^{2-}$ catalyzes H/D exchange between solvent CH$_3$COOD and arenes. In 1969, Shilov and coworkers[18,19] showed that this reaction is also effective for alkanes. This surprising and striking observation attracted considerable attention. At that time and up to the late 1970s it was generally believed that alkanes were too unreactive to interact with transition metal complexes. A number of C—H bond-breaking reactions had been studied by Chatt (see next section) but none of these were to be effective for alkanes. Many observers drew the hasty conclusion that the Shilov observation was in some way misinterpreted. For example, questions were asked about the homogeneity of the solutions. Today, there is no longer any doubt that the platinum catalysts are authentically homogeneous and that Shilov's observations represent the first cases of alkane activation by low-valent metal complexes. Although the Pt complex is not itself organometallic, organometallic intermediates are important in the chemistry and so we cover them in this chapter.

A variety of Pt complexes were shown to be active[19] and the rate is suppressed by Cl$^-$ ion. Because of the reversible aquation of PtCl$_4^{2-}$ in solution, several species are present, but Pt(H$_2$O)Cl$_3^-$ and Pt(H$_2$O)$_2$Cl$_2$ were considered to be the most active catalytically, because Pt(H$_2$O)$_4^{2+}$ reacts relatively slowly. The reactivity order for different types of C—H bond is $1° > 2° > 3°$, exactly the reverse of the one usually seen for radical or electrophilic reactions. It has subsequently become clear that this is the normal order for metal-catalyzed reactions of the types discussed in Section II and III. The steric hindrance implicit in the side-on transition state is thought to be responsible. Arenes are slightly more reactive than alkanes, but only slightly.

Kinetic analysis shows that exchange is multiple, so the substrate does not dissociate until several H/D exchange steps have taken place. This was initially interpreted in terms of alkene or, for CH$_4$, carbene complexes as intermediates, but with the discovery of the stability of 'agostic' complexes[20,21] (Section I.B)[19] an alternative explanation is that the alkane remains bound to the platinum via the C—H bond as L$_3$Pt(HR). One end of an ethane molecule undergoes H/D exchange faster than the other. This was originally taken as indicating that α elimination to give $HPt=CHCH_3$ is faster than β elimination to give $HPt(CH_2=CH_2)$. Similar observations have been made in other systems[22], but an alternative explanation that should now be considered is that an ethane complex $L_3Pt(HCH_2CH_3)$ may tend to undergo an intramolecular turnstile exchange in which the Pt remains bound to the same end of the molecule, faster than it undergoes an end-for-end exchange.

As Hodges and Garnett[17] had found in the case of arenes, the addition of $PtCl_6{}^{2-}$ leads to net oxidation to give chloro-arenes or -alkanes, with reduction of Pt(IV) to Pt(II)[23a]. Methane was converted to $MeCl$[23b]. Once again, a $1° > 2° > 3°$ selectivity pattern was observed. In CF_3CO_2H, trifluoroacetate esters were also formed. For cyclohexane, benzene was a major product, perhaps as a result of dehydrochlorination of the chlorocarbon intermediates.

A catalytic variant of this system uses Cu^{2+} and air; over five turnovers of oxidation are observed[23c]. A related catalyst deposited on silica is much more active[24]. The selectivity pattern of H/D exchange and chlorination are essentially the same, and the Pt^{II}—R species formed in the C—H bond-breaking step is thought to lead to one or the other product according to conditions. In particular, the presence of an oxidant would lead to Pt^{IV}—R, which could undergo nucleophilic abstraction of R^+ from the metal by Cl^-.

An interesting development[25] was the isolation of stable aryl and alkyl platinum species from the reaction mixtures by the addition of stabilizing ligands. For example, naphthalene as substrate and NH_3 as trap give an isolable 2-naphthyl species, and methane and PPh_3 give $MePtCl_3(PPh_3)_2$ (equation 4). A Pt—Me signal has been detected by ^1H NMR in reaction mixtures containing methane; the coupling to ^{195}Pt shows that the methyl group is directly bound to a single Pt.

$$PtCl_4{}^{2-} \xrightarrow{1.\,NpH,\,2.\,NH_3} [(H_3N)Cl_4Pt(Np)]^- \qquad (4a)$$

$$PtCl_4{}^{2-} \xrightarrow{1.\,MeH,\,2.\,PPh_3} [(Ph_3P)_2Cl_3PtMe] \qquad (4b)$$
$$(Np = 2\text{-naphthyl})$$

A photochemical version of the system has also been reported[26]. In this case, irradiation of $PtCl_6{}^{2-}$ and hexane in AcOH gives a Pt(II) complex of hexene. The alkene is almost certainly formed by β elimination of the intermediate alkyl, a reaction to be discussed in more detail in the next section. With Me_2CO as substrate a stable alkyl, $[(H_3N)Cl_4Pt(CH_2COMe)]^-$ was isolated by addition of NH_3.

Labinger, Herring and Bercaw[27] have recently reinvestigated the Shilov $PtCl_4{}^{2-}/PtCl_6{}^{2-}$ system with interesting results. Water-soluble substrates, such as p-toluenesulfonic acid and ethanol, were studied. Sequential formation of the benzylic alcohol and then the aldehyde were observed from the sulfonic acid, but benzylic activation is not essential because attack also occurred at the β position of p-$EtC_6H_4SO_3H$. Ethanol gave a mixture of products including the 1,2-glycol.

Sen and coworkers[28a] have examined systems based on $Pd(OAc)_2$ in triflic acid (TfOH) at 80 °C (equation 5). $Na_2Cr_2O_7/Pd(OAc)_2/TfOH$ aromatizes cyclohexanes[28b]; for example, decalin gives naphthalene (9%) and α-tetralone (4%). Normally the functionalization product of an alkane is more reactive than the alkane itself and so only the low conversion prevents the initial product from being oxidized further. Shilov and Sen's use of triflic acid in this context means that the initial functionalization product at the alcohol oxidation level is protected as the triflate ester, which is relatively insensitive to oxidation. 1,4-Dimethylbenzene is oxidized to triflates with a 50:1 preference for oxidation at the ring rather than of the side chain, unlike the selectivity expected in a radical process. A kinetic isotope effect of 5 was measured. In certain cases the reaction can be made catalytic with $K_2S_2O_8$.

$$Pd(OAc)_2 \xrightarrow{\text{AdH, TfOH}} AdOTf + Pd(0) \qquad (5a)$$

$$Pd(OAc)_2 \xrightarrow{\text{MeH, TfOH}} CH_3OTf + Pd(0) \qquad (5b)$$

$$(Ad = 2\text{-adamantyl}; \ Tf = CF_3SO_3)$$

III. OXIDATIVE ADDITION

In significant early work, dating from 1965, Chatt and Davidson[29] found the first example of stoichiometric intermolecular C—H activation by oxidative addition with naphthalene addition to $Ru(dmpe)_2$ (equation 6). In the absence of substrate the system gave cyclometalation, an intramolecular version of C—H activation, but disfavored in this case by the small ring size of the resulting species. The structures of some of the intermediates and products were determined and the chemistry was extended to other substrates (but not to alkanes)[30-32]. Chatt and Coffey[33] found H/D exchange between C_6D_6 and $ReH_7(PPh_3)_2$ and Tebbe and Parshall[34] looked at a similar exchange in $TaH_5(dmpe)_2$ and $CpRh(C_2H_4)_2$.

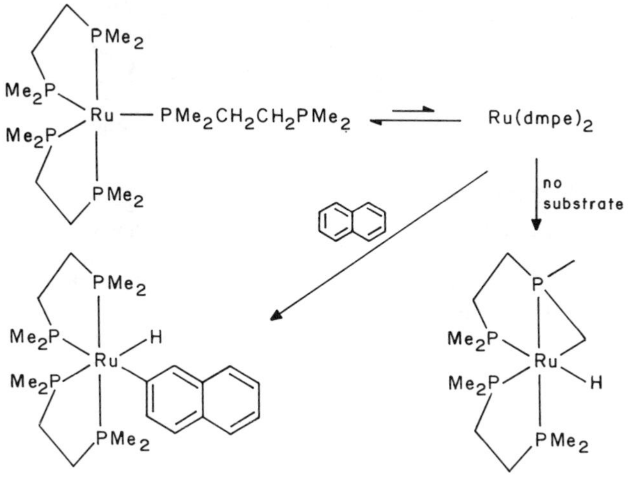

$$(dmpe = Me_2PCH_2CH_2PMe_2)$$

Gianotti and Green[35] studied the photochemistry of Cp_2WH_2 and showed that the photoexpulsion of H_2 results in the formation of the highly reactive fragment 'Cp_2W'. This reacts with a variety of arenes to give products by oxidative addition of a C—H bond; for example benzene gives $Cp_2W(Ph)H$. With $SiMe_4$, which is very weakly activated and can be regarded almost as an alkane, a dimeric product of C—H insertion was seen (equation 7). In the case of alkanes a dimer was formed, the result of C—H activation of a C—H bond of another molecule of Cp_2W (equation 8).

The first indication that metal phosphine complexes were active for catalytic reactions of alkanes came from Shilov's group[36] finding that $CoH_3(PPh_3)_3$ catalyzes H/D exchange between C_6D_6 and CH_4. These systems probably involve oxidative addition (equation 1), but they were not followed up at the time because of the greater interest in the Pt system discussed above. Since 1980, a variety of Rh, Ir, Re and Ru complexes have shown activity for alkane dehydrogenation and carbonylation. The intermediate alkyl formed in the initial oxidative addition can be trapped by β elimination to give a free alkene, or a polyenyl complex or an aldehyde by CO insertion.

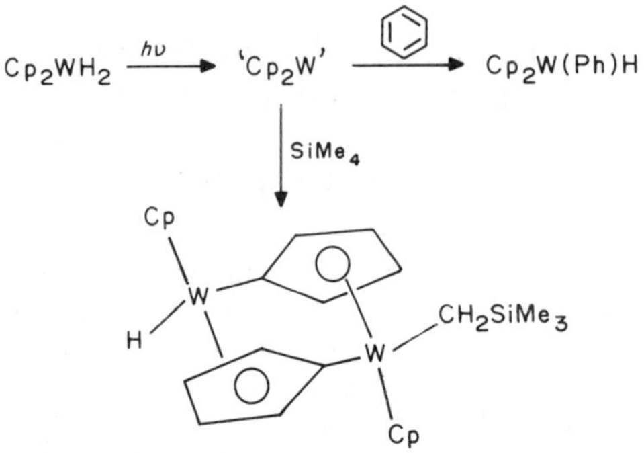

$$Cp_2WH_2 \xrightarrow{h\upsilon} {}^{\backprime}Cp_2W{}^{\backprime} \xrightarrow{\hspace{1cm}} Cp_2W(Ph)H$$

(7)

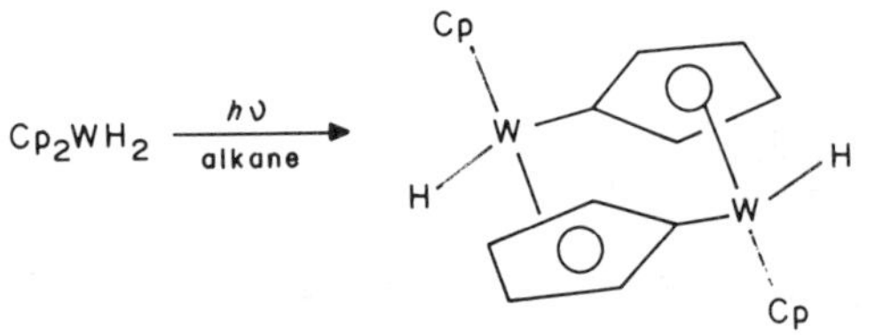

(8)

A. Stoichiometric Chemistry

Other than isotope exchange, the first reaction of metal phosphine complex with an alkane, which we reported in 1979[37a], was the dehydrogenation of cyclopentane by $[IrH_2S_2L_2]^+$ to $[CpIrHL_2]^+$ (L = PPh$_3$; S = Me$_2$CO), where the cyclopentadienyl ligand bound to iridium comes from the alkane. By using cyclopentane as substrate and providing a metal center with multiple dissociable ligands, it was possible to trap the initial alkyl hydride by successive β-elimination steps to yield the thermodynamically very stable cyclopentadienyl complexes (equation 9). Cyclooctane was dehydrogenated

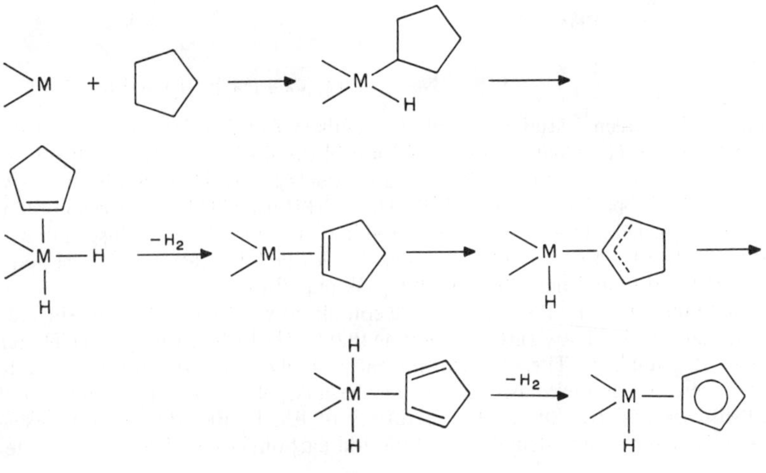

(9)

to the 1,5-cyclooctadiene complex. The hydrogen released from the alkane was transferred to the alkene t-BuCH=CH$_2$ (tbe). This is a particularly suitable hydrogen acceptor because it is bulky and therefore a poor ligand and so does not block access to the metal by alkane, as does ethylene. Tbe also lacks reactive allylic C—H bonds that might be broken by the metal[37b]. The same dehydrogenation reactions were also observed from other precursor complexes, such as [Ir(η^4-naphthalene)L$_2$]$^+$, all of which are believed to be sources of the IrL$_2^+$ fragment[37c].

Cyclohexane was insufficiently reactive to give products in the CH$_2$Cl$_2$ solvent used in the work described above, and iridium chloro complexes were formed instead, but on moving to neat alkane, several alkane-derived reaction products were seen as shown in equation 10[37d]. Particularly notable is the formation of fluorobenzene when L is $(p$-FC$_6$H$_4)_3$P. This shows that the complex degrades by P—C bond cleavage, otherwise the reaction would probably be catalytic.

$$L = (p\text{-}FC_6H_4)_3P; \; S = Me_2CO$$

(10)

Baudry, Ephritikhine and Felkin[38] found that cyclopentane could be dehydrogenated with ReH$_7$L$_2$/tbe to give CpReH$_2$L$_2$. The real importance of this system was the catalytic dehydrogenation observed with other substrates and mentioned in Section III.B.

In 1982, the groups of Bergman, Graham and Jones published the first examples of the isolation of alkyl hydride adducts from oxidative addition of alkane C—H bonds to a number of metal centers. Janowicz and Bergman[39] studied Cp*IrH$_2$L, which undergoes photoextrusion of H$_2$ on photolysis. The resulting 'CpIrL' fragment reacts with the alkane to form the product (equation 11). This intermediate is believed to be the 16e species {(η^5-Cp)IrL} (but a 'slipped' 14e structure such as {(η^3-Cp)IrL} is possible[40,41]). Thermal exchange of the adduct with other alkanes is possible at 110 °C (equation 12) so that the reaction is not exclusively photochemical. The cyclohexyl hydride has now been characterized by X-ray methods[42]. Less than 10% crossover was observed for neopentane and cyclohexane-d$_{12}$. The isotope effect for cyclohexane/cyclohexane-d$_{12}$ is $k_H k_D = 1.38$, which is consistent with the oxidative addition mechanism.

$$\text{Cp*IrH}_2\text{L} \xrightarrow{h\nu} (\text{Cp*IrL}) \xrightarrow{\text{R—H}} \text{Cp*Ir(R)(H)L}$$

(11)

$$(\text{Cp*} = \text{C}_5\text{Me}_5; \; \text{L} = \text{PMe}_3)$$

$$\text{Cp*Ir(R)(H)L} \xrightarrow{110\,°C} \{\text{Cp*IrL}\} \xrightarrow{\text{R'—H}} \text{Cp*Ir(R')(H)L}$$

(12)

The 1°:2° selectivity is 1.5:1 for propane. A mixture of isomers from n-pentane rearranges to the 1-pentyl hydride at 110 °C. This shows that the 1° product is thermodynamically the most stable, probably for steric reasons. Equilibration of the cyclohexyl and 1-pentyl derivatives showed that the Ir-pentyl bond is the stronger by 5.5 kcal mol^{-1}. This suggested that a methyl complex would be stable and, indeed, heating the cyclohexyl hydride with methane at 150 °C led to the methyl hydride. Attempted functionalization of the alkane proved disappointing, but transfer of the neopentyl group from Ir to HgCl$_2$ was possible.

The reductive elimination of cyclohexane from the cyclohexyl deuteride adduct was studied. Interestingly, a scrambling of the Ir—D with the α position of the cyclohexyl moiety was observed. This rearrangement may imply that an intermediate alkane complex is formed, which selectively undergoes a 1,1′ shift as shown in equation 13. Alternatively, a ring slip/α-elimination mechanism seems less likely.

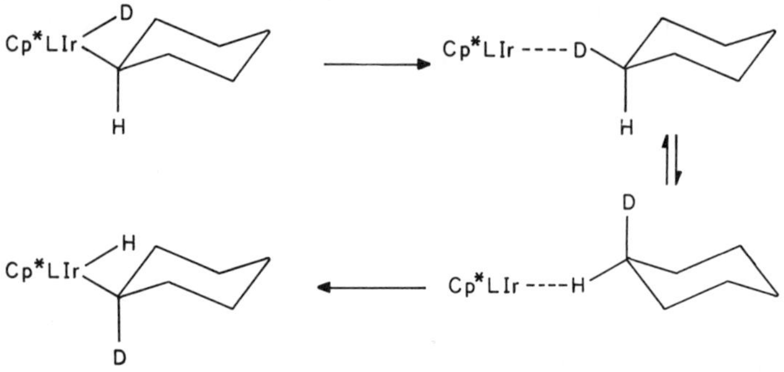

$$Cp^*LIr \cdots D \quad \rightleftharpoons \quad Cp^*LIr \cdots H \tag{13}$$

Thermochemical data[43] suggest that $D(\text{Ir—H})$ is $74 \pm 5\,\text{kcal mol}^{-1}$ and $D(\text{Ir—C})$ is $51\,\text{kcal mol}^{-1}$, considerably larger than is the case for Vaska's complex, $[\text{IrCl(CO)(PPh}_3)_2]$, which does not oxidatively add alkane C—H bonds in spite of being coordinatively unsaturated 16e species. The difference may be that Vaska's complex has the square-planar geometry favored for Ir(I) while CpIrL cannot attain this structure.

Bergman and coworkers have applied liquid Xe as a useful solvent for C—H activation. For the Cp^*IrLH_2 system, it allowed study of methane, as well as of more exotic alkanes, such as cubane and adamantane; in the latter case the 2-adamantyliridium hydride was formed on irradiation[44]. In addition, time-resolved IR spectroscopic studies were carried out in Xe and Kr which led to the proposal that alkane complexes are in equilibrium with rare gas complexes in this system (equation 14). The alkane complex is the last observed precursor before the alkyl hydride is formed[45].

$$Cp^*LIr(Xe) + C_6H_{12} \rightleftharpoons Cp^*LIr(C_6H_{12}) + Xe \tag{14}$$

The complex $Cp^*Ir(allyl)H$ in the presence of phosphine also activates C—H bonds of cyclopropane as shown in equation 15[46].

$$CP^*Ir(\eta^3\text{-allyl})H + C_3H_6 + L \rightarrow Cp^*IrL(1\text{-propyl})(C_3H_5) \tag{15}$$

Graham and coworkers[47,48] looked at the related dicarbonyl analogue, $Cp^*Ir(CO)_2$, in which photoextrusion of CO leads to very similar chemistry. Methane at 8 atm in a perfluorohexane solution also reacts to give the methyl hydride[48], a reaction which proceeds even at 12 K, illustrating the low activation energy for the C—H oxidative addition. The derivatization of the alkyl hydride with CCl_4 to give the alkyl chloro complex was particularly useful because the latter is much more kinetically stable than the former.

Tris(3,5-dimethylpyrazolyl)borate (Tp*) can replace Cp* in the system. $Tp^*Ir(CO)_2$ reacts with cyclohexane under photolysis to give the cyclohexyl hydride[49].

Jones and Feher[50] found that Cp^*RhLH_2 reacts with alkanes by photolysis at low temperature. For example, propane gave the 1-propyl hydride at $-55\,°C$ followed by isolation at $-40\,°C$. The compound was unstable on warming unless first converted to the alkyl bromo complex with bromoform. The facility of the oxidative addition and the reductive elimination allowed detailed kinetic and thermodynamic studies to be made. From a kinetic point of view, propane and benzene were found to be comparable in reactivity, but the phenyl hydride from benzene was much more stable thermodynamically, largely as a result of M—Ph being much stronger than M–alkyl bonds.

Dihapto-benzene intermediates were postulated in the benzene reaction. Periana and Bergman have also studied this system[51]. It is more selective than the Ir analogue for attacking different types of C—H bonds. Bromination led to the formation of free RBr (70–80%) from the alkylrhodium complex. The rate of alkane reaction correlates with the number of C—H bonds in the molecule, but only the 1° alkyl hydride is seen as a product. This suggests that attack takes place along the whole chain, but facile thermal rearrangement generates the most stable alkyl. The cyclopropyl deuteride shows a scrambling of the label reminiscent of that mentioned above for the iridium cyclohexyl deuteride; an intermediate alkane complex was proposed.

CpReL$_3$ (L = PMe$_3$) is also photoreactive with alkanes by loss of L[52]. The product is CpReL$_2$(R)H (R = methyl, 1-hexyl, cyclopentyl and cyclopropyl). The more sterically hindered substrate, cyclohexane, did not give an alkyl hydride; cyclometalation of the coordinated PMe$_3$ or oxidative addition of the C—H bond of the free L takes place instead. This more bulky system has a lower tolerance for steric bulk in the alkane.

Ibers, DiCosimo and Whitesides[53] showed that the bis-neopentyl platinum complex of equation 16 gave facile cyclometalation. They were surprised that the reverse reaction, which would of course be an alkane activation with neopentane as the substrate, did not take place because, thermodynamically, the bonds formed should compensate for the bonds lost. Presumably, it is the unfavorable entropic term which is the major factor in preventing the reverse reaction, but the substantial steric bulk of the neopentyl group may tend to reduce the Pt—neopentyl bond strength in a bis-neopentyl complex.

$$L_2Pt \qquad \xrightarrow{-CMe_4} \qquad L_2Pt \qquad\qquad (16)$$

Caulton and coworkers[54] showed that photolysis of ReH$_5$L$_3$ (L = PMe$_2$Ph) liberates L rather than H$_2$, and in the presence of tbe as hydrogen-acceptor, they found that cyclopentane can be dehydrogenated to CpReH$_2$L$_2$ as Felkin and coworkers[38] had previously observed thermally for ReH$_7$L$_2$.

One of the most remarkable conversions of n-alkanes was discovered in Felkin's group. In contrast to the iridium catalysts, which dehydrogenate n-alkanes to alkenes, ReH$_7$(PR$_3$)$_2$ gives stable diene complexes instead[55]. All the possible isomers of the diene are in rapid equilibrium, which can be detected by spin saturation transfer measurements in the ^{1}H NMR spectrum. The addition of P(OMe)$_3$ to this mixture induces the transfer of two of the hydrogen ligands to the diene, and essentially only the terminal monoene is formed. A possible reaction pathway is shown in equation 17. Presumably, the terminal diene complex is the most reactive and the P(OMe)$_3$ selectively liberates the terminal C=C double bond of the diene from the metal. The remaining C=C double bond is then hydrogenated. The other isomers of the diene complex convert into the reactive isomer sufficiently rapidly so that essentially all the product can be formed via the more reactive isomer.

Baker and Field[56] have reported that (dmpe)$_2$FeH$_2$ gives (dmpe)$_2$FeH(1-pentyl) on photolysis in n-pentane at −30 °C. On warming to room temperature, photolysis releases 1-pentene. The reaction is not catalytic because the iron complex reacts with the product alkene to give the very stable (dmpe)$_2$Fe(1-pentene) and (dmpe)$_2$FeH(C=CHPr-n). Hackett and Whitesides[57] have reported some stoichiometric C—H activation reactions with Pt complexes.

In interesting recent work, Sherry and Wayland[58] have shown that metalloradicals can homolyze alkane C—H bonds. Normally, two reactive metalloradicals would tend

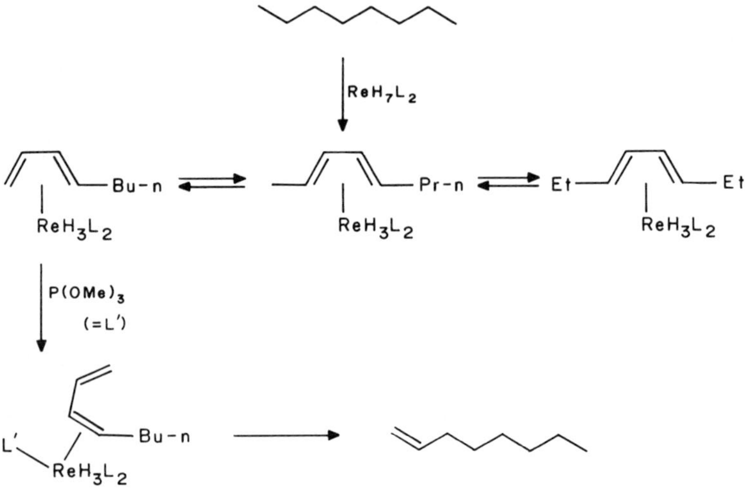

R. H. Crabtree

ReH_7L_2

$Bu-n$ $\rightleftharpoons$ $Pr-n$ $\rightleftharpoons$ Et — Et

ReH_3L_2 ReH_3L_2 ReH_3L_2

$P(OMe)_3$
$(=L')$

(17)

$Bu-n$ $\longrightarrow$

L' ReH_3L_2

to recombine, but the $\alpha,\beta,\gamma,\delta$-tetramesitylporphyrin (TMP) ligand was chosen to make this pathway sterically unfavorable (equation 18). The bond strengths for Me—Rh(TMP) and H—Rh(TMP) were estimated as 45 and 62 kcal mol^{-1}, respectively, and therefore just sufficient to break the methane C—H bond $[D(C—H) = 105$ kcal mol$^{-1}]$. This is so far the only example of a C—H activation by a two-center oxidative addition.

$$[Rh(TMP)]_2 \rightleftharpoons 2Rh(TMP) \xrightarrow{CH_4} MeRh(TMP) + HRh(TMP) \qquad (18)$$

B. Catalytic Chemistry

Two catalytic reactions have been studied in most detail, alkane dehydrogenation and alkane carbonylation. In the case of alkane dehydrogenation (equation 19), oxidative addition to the metal is followed by β elimination to liberate the alkene and leave a metal dihydride (equation 20). Catalysts that dehydrogenate alkanes to free alkenes all have additional labile ligands, which make the intermediate alkyl hydride unsaturated and so allow it to β-eliminate to give the alkene. This mechanism is illustrated in steps d–f of Figure 2 in case[59] of $IrH_2(O_2CCF_3)L_2$ in which the carboxylate group is known to be able to open and close easily.

$$RCH_2CH_3 \xrightarrow{catalyst} RCH=CH_2 + H_2 \qquad (19)$$

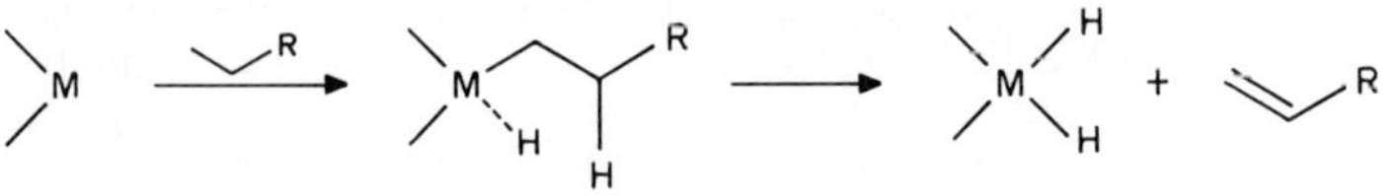

$$\text{M} \quad \longrightarrow \quad \text{M} \quad \longrightarrow \quad \text{M} \quad + \quad R \qquad (20)$$

To recycle the catalyst, the hydrogen must be removed from the metal. This is related to the fact that equation 19 is endothermic; the reverse reaction, alkene hydrogenation, normally proceeds catalytically at temperatures below 200 °C and at ambient pressure. Removing the hydrogen allows the alkane dehydrogenation to proceed by giving it a net favorable free-energy change. This can be done in a number of ways. Felkin's group studied the first method[38,55], coupling the dehydrogenation with a favorable alkene hydrogenation step (equation 19). We had introduced t-BuCH=CH$_2$ (tbe) as a hydrogen acceptor for stoichiometric alkane dehydrogenation[37], and it proved useful for the

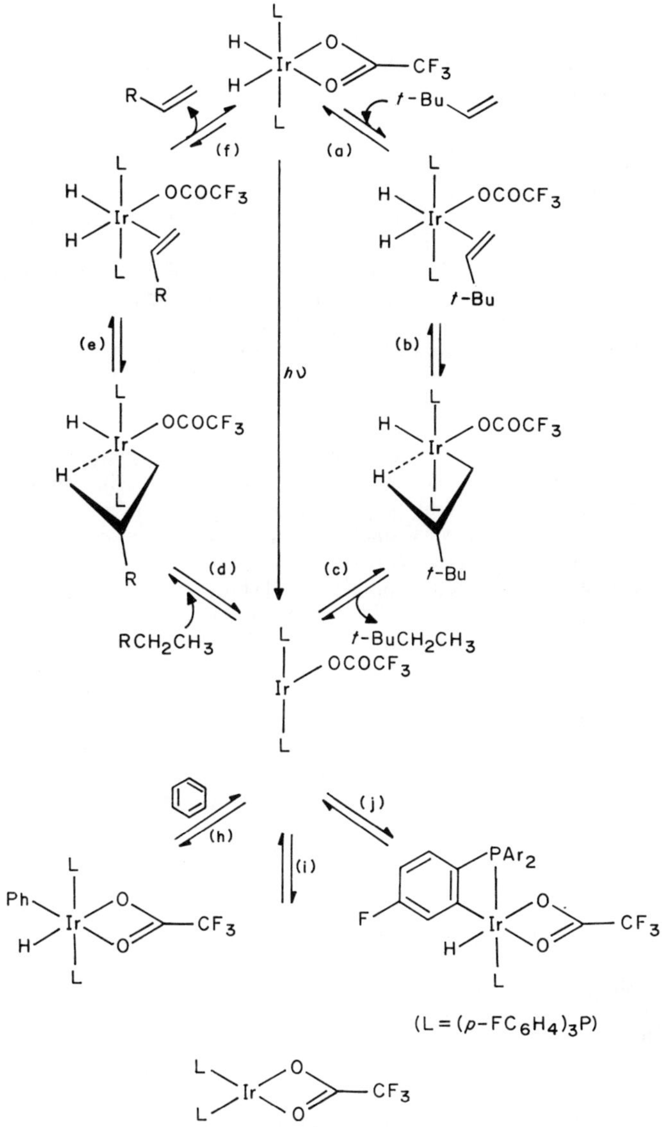

FIGURE 2. A mechanistic scheme for the thermal and photochemical dehydrogenation of RCH_2CH_3 to $RCH{=}CH_2$ by $[IrH_2(O_2CCF_3)(PR_3)_2]$. L is $(p\text{-}FC_6H_4)_3P$ for the thermal chemistry and $(C_6H_{11})_3P$ for the photochemistry

catalytic reactions. It is bulky and so does not coordinate strongly to vacant sites at the metal and it is very active in removing hydride ligands from the metal by hydrogenation to give $t\text{-}BuCH_2CH_3$. This is illustrated in steps (a – c) of Figure 2. The C—H oxidative addition step is confirmed by the isolation of the phenyl hydride when benzene is used as substrate (Figure 2, step h). Equation 21 seems to be thermodynamically downhill in almost all cases under the conditions employed. This reaction is catalyzed by several complexes $\{RuH_4[(p\text{-}FC_6H_4)_3]_3P\}^{55b}$, $[IrH_2(O_2CCF_3)(PPh_3)_2]^{59}$, $ReH_7(PPh_3)_2^{55a}$ and $IrH_5[P(Pr\text{-}i)_3]_2^{55c}$.

$$RCH_2CH_3 + t\text{-}BuCH{=}CH_2 \xrightarrow{\text{catalyst}} RCH{=}CH_2 + t\text{-}BuCH_2CH_3 \qquad (21)$$

In the first Felkin[55a] system, ReH_7L_2/tbe, a variety of cyclic alkanes could be catalytically dehydrogenated to the alkenes. Methylcyclohexane gave the most interesting selectivity information. The bulky system with $L = PCy_3$ ($Cy = C_6H_{11}$) showed selectivity for the least thermodynamically stable of the possible alkene products: methylene-cyclohexane. This indicates that attack at the 1° position is followed by a β elimination of the 3° C—H. In contrast, the smaller PEt_3 system showed preferential attack at the least hindered 3- and 4-methylene groups. In neither case was alkene isomerization an important process. The iridium system, $[IrH_2(O_2CCF_3)(PPh_3)_2]^{59}$, was less selective for the formation of methylenecyclohexane, apparently because this system catalyzes alkene isomerization more efficiently than the rhenium catalysts. By far the major product from n-hexane was trans-2-hexene.

A second way of making the endothermic alkane dehydrogenation thermodynamically favorable is to reflux the solvent to drive off the H_2. Because the solubility of a gas falls essentially to zero at the boiling point of a solvent the partial pressure of H_2, $p(H_2)$, in the solution phase is kept very small. As Saito and coworkers[60] first showed using $RhClL_3$ as catalyst, this drives the reaction, and a substantial amount of alkene can be obtained before catalyst deactivation occurs. Of course, the boiling point of the substrate (in this case cyclooctane) must be suitable because this determines the reaction temperature. In this system an excess of phosphine (3–8 moles/Rh) was beneficial for rate and catalyst stability. Up to 13 turnovers of cyclooctene were formed.

Photochemical dehydrogenation reactions can proceed directly without a hydrogen acceptor, because the reaction is driven by the energy of the photons. In some cases a gentle stream of inert gas improves the efficiency of the process by physically removing this H_2 from the system. This was first carried out in our group with $[IrH_2(O_2CCF_3)$ $(PCy_3)_2]^{59}$ as catalyst and the mechanism is illustrated in steps g, d, e and f of Figure 2. The recent extensive studies of the groups of Saito[60b] and of Tanaka[61a] have involved $RhCl(CO)L_2$ ($L = PMe_3$) as catalyst, where photoexpulsion of CO is the first step[61b]. Catalyst degradation seems to be somewhat less severe for the photochemical compared to the thermal version of the reaction and many hundreds of turnovers of product are observed in favorable cases. In the case of linear alkenes, the product mixture favors the more stable internal alkenes; but this is not simply a result of alkene isomerization activity of the catalyst[61b]. In $[IrH_2(O_2CCF_3)(PCy_3)_2]$, the presence of tbe helped preserve the kinetic products by suppressing isomerization, so, for example, methylenecyclohexane is only the major product from methylcyclohexane in the presence of tbe (equation 22). The figures in equation 22 show the number of catalytic turnovers observed.

From the point of view of catalysis the most serious limitation of metal phosphine complexes is their tendency to degrade, especially under thermal conditions. This has been studied in detail[59] in the case of $\{IrH_2(O_2CCF_3)[(p\text{-}FC_6H_4)_3P]_2\}$, where it has been shown that the main degradation pathway is P—C bond hydrogenolysis, a process that has been reviewed by Garrou[62] and treated theoretically by Hoffmann[63]. The process seems to involve oxidative addition of the P—C bond to the metal, to produce a catalytically inactive cluster, as shown schematically in equation 23. ArH has also been detected from $ReH_7(PAr_3)_2^4$ and propane from $IrH_5[P(Pr\text{-}i)_3]_2$-catalyzed[55c] alkane dehydrogenation reactions. We had always considered cyclometalation to be a catalyst deactivation pathway, but in the $IrH_2(O_2CCF_3)[(p\text{-}FC_6H_4)_3P]_2$ system, this step is reversible (Figure 2, step f) and the cyclometalation product reverts to the active catalyst

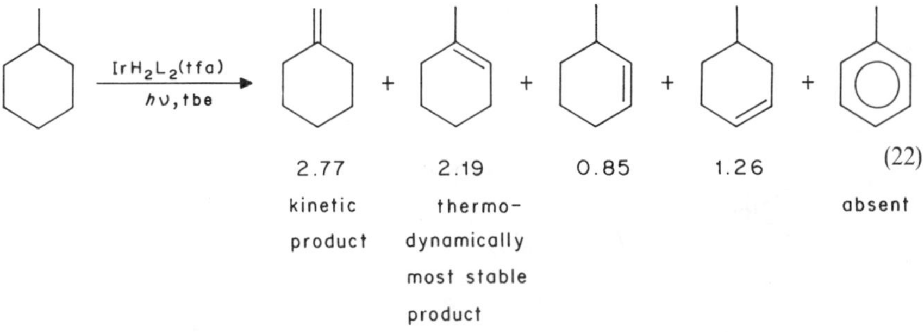

$$2.77 \qquad 2.19 \qquad 0.85 \qquad 1.26 \qquad (22)$$

kinetic   thermo-   absent

product   dynamically

most stable

product

$$L = P(C_6H_{11})_3; \; tfa = CF_3CO_2; \; tbe = t\text{-}BuCH{=}CH_2$$

on heating. In our thermal iridium system, the fall-off in catalytic activity was directly correlated with the build-up of fluorobenzene. Other deactivation pathways may also occur in other systems, and more attention needs to be given to these and to the problem of designing degradation-resistant ligands.

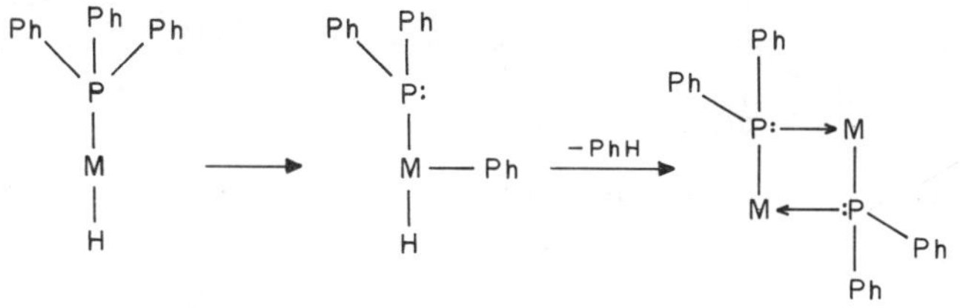

$$(23)$$

The photochemical catalysts seem to be much less sensitive to decomposition than the thermal catalysts. For $\{IrH_2(O_2CCF_3)[(C_6H_{11})_3]P\}_2/h\nu$ the production of cyclooctene from cyclooctane, while slow, is roughly linear with time (12 turnovers after 7 days, 23 turnovers after 14 days photolysis). After 7 days, 70% of the catalyst can be recovered in pure form even from the dilute solution employed. The Tanaka[61a] and Saito[60] photochemical systems have shown hundreds of turnovers and are the stablest systems to date.

Cyclooctane has proved to be the most reactive alkane of those tried for catalytic alkane dehydrogenation. The heat of dehydrogenation is unusually large for this alkane, presumably as a result of the decrease in unfavorable transannular interactions on going to the product cyclooctane. Cyclooctane is therefore a good choice for assaying new potential catalysts for alkane dehydrogenation activity. Curiously, cyclooctane is one of the least reactive in stoichiometric systems. This reactivity trend may indicate that the oxidative addition of the alkane C—H bond is not the rate-determining step in catalytic alkane dehydrogenation; instead, the β elimination may often be the rate-determining step[59,61b].

Photochemical alkane carbonylation with $RhCl(CO)(PMe_3)_2$ is also possible[61a]. This seems to operate by initial photoextrusion of CO from the catalyst, oxidative addition of the alkane C—H bond, addition of CO to the metal, followed by insertion, and then reductive elimination as shown in Figure 3. Preferential reaction at the 1° or 2° C—H bond is found. Here the initial product does not seem to isomerize, but Norrish type II photoreactions tend to degrade the aldehyde product. Moving to longer wavelengths minimizes the Norrish degradation problem, but the selectivity of the catalytic system then falls off. No more than 30 turnovers have been observed to date (e.g. equation 24)

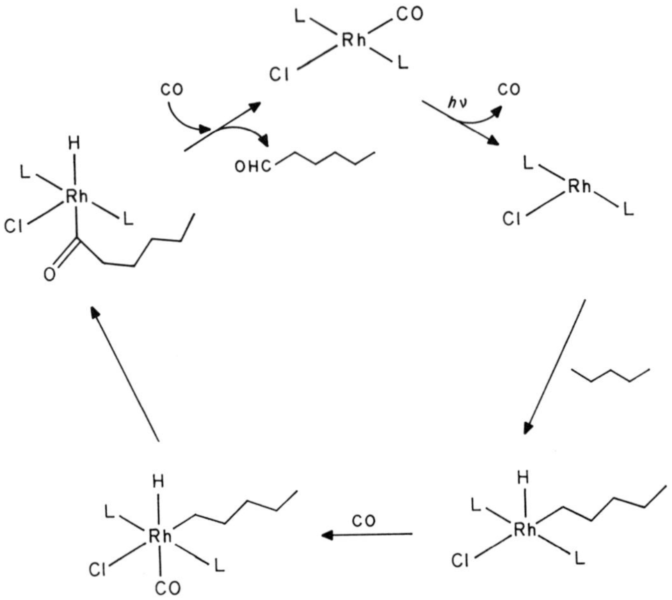

R. H. Crabtree

FIGURE 3. A mechanistic scheme for the photochemical carbonylation of n-pentane by [RhCl(CO)(PMe$_3$)$_2$]

but the system is very promising. Carbonylation of alkanes such as cyclohexane, which contain only 2° C—H bonds, is much less efficient, as is insertion of isonitriles.

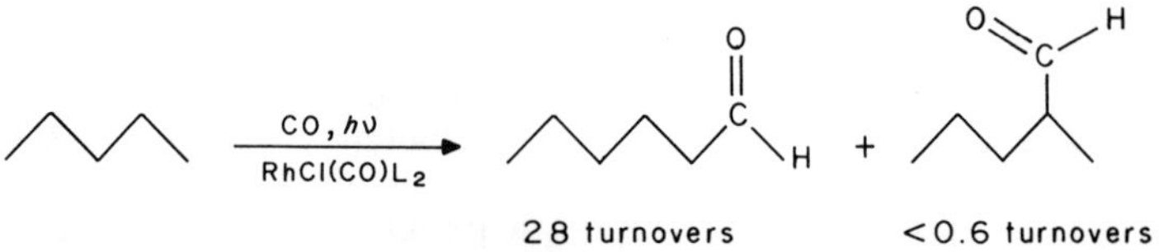

$$\qquad\qquad\qquad\qquad (24)$$

28 turnovers $\qquad$ <0.6 turnovers

In the catalytic systems, the most probable intermediate leading to the C—H bond breaking is a 14e three-coordinate T-shaped species. This would be a reasonable structure for {IrL$_2$(η^1-OCOCF$_3$)} in the system we studied, IrHL$_2$ in Felkin's iridium system as well as for {RhClL$_2$} in the Tanaka and Saito systems. The rhenium systems are unlikely to be directly analogous, but the 14e ReH$_3$L$_2$ has been proposed[55] as the key precursor from ReH$_7$L$_2$. This preference for 14e intermediates is a reflection of the feeling that simple 16e square-planar d^8 species are unlikely candidates for alkane C—H bond breaking, because, unlike 'CpIrL', so many stable examples are known. As noted above, CpIrL is probably 16e but is certainly not square planar.

The reason for the selectivity pattern appears to be the nature of the kinetic pathway for C—H oxidative addition. A Burgi–Dunitz analysis of a number of complexes containing C—H···M interactions led to the conclusion[9] that the C—H bond approaches the metal—H first with a large M—H—C angle. As the C—H bond gets closer, it pivots about the H atom so as to reduce the M—H—C angle and bring the carbon atom close to the metal. The transition state, shown as 1, appears to be late, with a short M—C distance. Substituents on the carbon atom are therefore likely to interfere with the approach to the transition state and so retard the reaction.

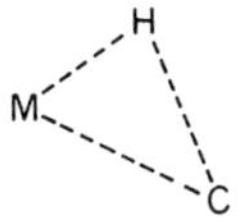

(1)

The same analysis also suggested that cyclometalation (equation 3) might have a higher barrier than the activation of an external C—H bond such as that of the substrate, because of unfavorable conformational factors in the approach of a ligand C—H bond. Jones and Feher[50] have shown that for the $CpIrH_2P(Pr-i)_3/hv/CH_3CH_2CH_3$ system, there is indeed a greater kinetic barrier for ligand C—H oxidative addition than for the addition of an alkane C—H bond. Cheney and Shaw[4b] showed that cyclometalation is strongly accelerated by steric crowding and so such crowding should be avoided if alkane activation, rather than cyclometalation, is desired. Indeed, all the successful alkane activating systems are relatively unhindered: MXL_2, ReH_3L_2, $CpIrL$.

For all the catalytic systems discussed above, radical and carbenium ion mechanisms can be excluded on several grounds, perhaps the most convincing being the $1° > 2° > 3°$ selectivity for attack at different C—H bonds. Mechanisms involving heterogeneous catalysis by metal surfaces, which might be formed by decomposition of the catalyst, have been excluded by several tests including the use of metallic mercury, which poisons platinum group metal surfaces[37e].

A detailed study[59] of the chemistry of $[IrH_2(O_2CCF_3)L_2]$ $\{1, L = [(p\text{-}FC_6H_4)_3]P\}$, which is a thermal alkane dehydrogenation catalyst using tbe as hydrogen acceptor, has given considerable insight into the mechanism of the C—H activation reactions. The addition of CO or even of such relatively weakly-binding ligands as alkenes or MeCN leads to opening of the trifluoroacetate to give the mono-adduct (equation 25). Interestingly, the CH_3CO_2 complex does not act as a catalyst and only gives the reaction in equation 25 with CO but not with MeCN or alkenes. The reaction of the trifluoroacetate with alkenes is reversible and the K_{eq} values were measured. These show the low coordinating power of tbe relative to other alkenes. This is a very useful property for a hydrogen acceptor, which must not block active sites at the metal at which the alkane is to bind.

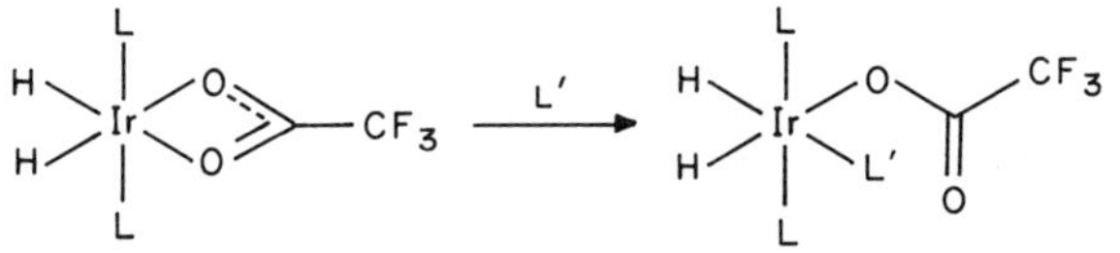

$$ \tag{25} $$

Treatment of the thermal catalyst $[IrH_2(\eta^2\text{-}OCOCF_3)L_2]$ with tbe gives $[IrH_2(tbe)$ $(\eta^1\text{-}OCOCF_3)L_2]$, which slowly converts to a species that is probably the agostic alkyl hydride shown in Figure 2. If so, the agostic character of the alkyl hydride increases the thermodynamic driving force to form the alkyl hydride and so facilitates the alkane oxidative addition. After loss of $t\text{-}BuCH_2CH_3$, the final product is an isolable ortho-metalated species, which is an equally effective catalyst precursor and so is presumably in equilibrium with the active catalyst. In benzene, the reaction of $IrH_2(O_2CCF_3)L_2$ with tbe gives an isolable phenyl hydride. This is the first case in which a catalytically active C—H activation system has also given a stable C—H oxidative addition product.

The Ir—Ph bond is unusually strong thanks to d_π-p_π interactions and the phenyl group cannot easily β-eliminate. The value of k_H/k_D for the thermal oxidative addition of benzene is 4.5. This seems to be a reasonable value for an oxidative addition, although lower isotope effects ($k_H/k_D = \sim 1.4$) were observed for cyclohexane and benzene addition to the much more reactive intermediates Cp*M(PMe$_3$)[39,50] (M = Rh, Ir). The reaction scheme shown in Figure 2 is therefore proposed for thermal alkane dehydrogenation with this catalyst. The phenyl hydride undergoes first-order decomposition at 65 °C to give an equilibrium mixture with the cyclometalation product with $K_{eq} = 3.6$ in favor of the phenyl hydride.

It has been shown[59] that the catalytic activity of IrH$_2$(O$_2$CCF$_3$)(PAr$_3$)$_2$ for alkane dehydrogenation falls as the amount of ArH rises. The formation of 0.5 mole of ArH per Ir is sufficient to deactivate the catalyst. These and other deactivation pathways require further study and general methods for slowing or preventing these processes are badly needed. Such methods might also be applicable to other homogeneous catalysts and lead to practical applications.

Grigoryan and coworkers[64] have observed H/D exchange between CD$_4$ and methyl groups bound to aluminum in Ziegler–Natta systems such as TiCl$_4$/Me$_2$AlCl at 20–50 °C. Multiple exchange is observed even at short times and Ti—CH$_2$ species were invoked. At 70 °C, Cp$_2$V catalyzes H/D exchange between D$_2$ and CH$_4$ and between D$_2$ and solvent benzene. With Ti(OBu-t)$_4$/AlEt$_3$ as catalyst, methane adds to ethylene to give as much as 15% yield of propane. With CD$_4$ and C$_2$H$_4$, d_4-propane predominates.

Catalytic H/D exchange between C$_6$D$_6$ and methane has been observed using metal phosphine catalysts. Jones has shown that CpReH$_2$(PPh$_3$)$_2$ loses phosphine on photolysis to give a species active for the isotopic exchange reaction; 68 turnovers were observed after 3 h[50]. Felkin[55] has studied thermal catalysis using IrH$_5$\{P(Pr-i)$_3$\}$_2$ activated by t-BuCH=CH$_2$. About forty turnovers were observed after 46 days at 80 °C. The thermal Ir catalysts also gave evidence for the formation of other products. Ethane, toluene and biphenyl were all observed in small and variable amounts from the C$_6$H$_6$/CH$_4$ reaction. These may have been formed by reductive elimination in an IrH$_3$L$_2$RR' species. Lin, Ma and Lu[65] have applied the IrH$_5$L$_2$/tbe system to dehdrogenation of pinane, which gives largely β-pinene.

C. Carbon–Carbon Bond Cleavage

Alkane C—C bond cleavage is a possible alternative strategy for alkane activation. So far, direct C—C bond activation by metal phosphine complexes has not yet been observed. This is probably because C—C bond cleavage is kinetically much slower than C—H bond cleavage, and is illustrated (for the reverse reaction) by the contrast between the much readier reductive elimination of alkyl hydride complexes compared to their dialkyl analogues. In addition, a direct oxidative addition of a C—C bond would put two bulky groups on the metal; this would lead to a reduced M—C bond strength. Bergman and coworkers[42] estimate that the C—C oxidative addition reaction of Cp*Ir(PMe$_3$) with C$_6$H$_{11}$—C$_6$H$_{11}$ is exothermic by 22 kcal mol^{-1} using the Ir—C$_6$H$_{11}$ bond strength derived from Cp*Ir(PMe$_3$)C$_6$H$_{11}$(H).

We have shown[66] how *gem*-dialkylcyclopentanes can be dehydrogenated to the corresponding diene complexes by [IrH$_2$(Me$_2$CO)$_2$L$_2$]SbF$_6$; these can then undergo the Green–Eilbracht[67] reaction to give alkyliridium complexes (equation 26). Extending the study to the diethyl analogues led to the unexpected formation of 1,2- and 1,3-diethyl-cyclopentadienyl iridium hydrides. A mechanism involving C—C oxidative addition, followed by migration of the alkyl back to the ring, was proposed. Formation of the aromatic cyclopentadienyl in the C—C bond cleavage step must help drive the reaction; even so, high temperatures ($\sim$ 150 °C) are needed.

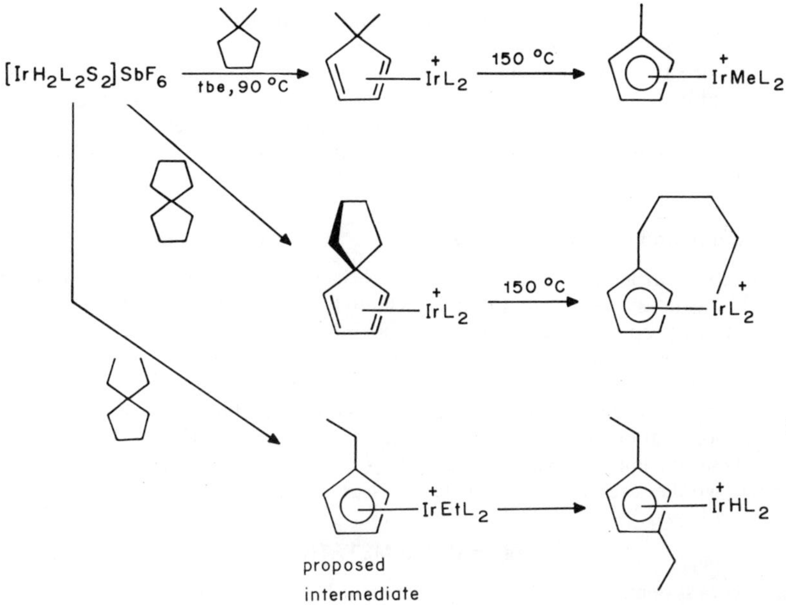

$$(26)$$

Oxidative addition of a C—C bond to the metal is much more rapid in strained hydrocarbons, thanks to the relief of strain in the transition state. The earliest report dates from 1955 (equation 27)[68].

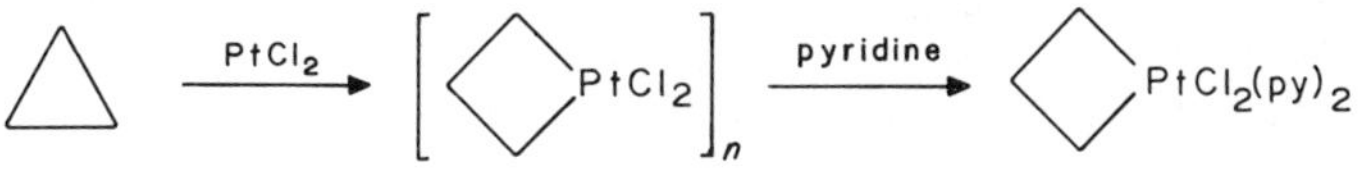

$$(27)$$

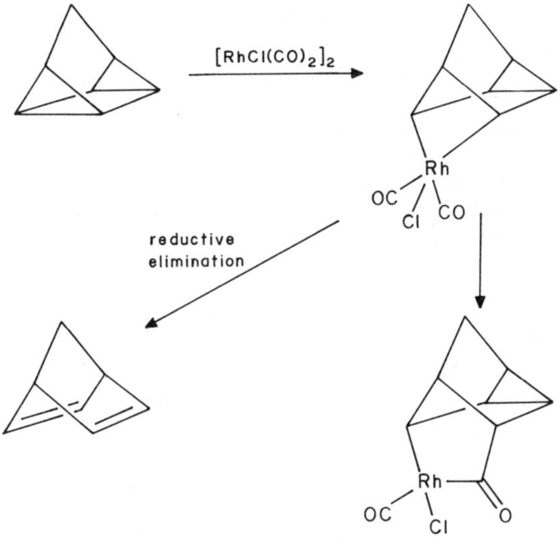

$$(28)$$

The catalytic rearrangement of strained alkanes is common[69]. Cassar and Halpern[70] showed that in the $[RhCl(CO)_2]_2$-catalyzed rearrangement of quadricyclane to norbornadiene, the admission of CO led to the formation of an acyl complex that seemed to be formed by trapping the initial oxidative addition product by a migratory insertion (equation 28). On the other hand, the Ag^+-catalyzed rearrangement of strained alkanes goes via carbonium ion intermediates which can be trapped by such nucleophiles as $MeOH^{71}$.

D. Surface-bound Organometallic Species

Yermakov, Kuznetsov and Zakharov[72] and Schwartz[73] have studied organometallic species bound to surfaces. For example, $Rh(allyl)_3$ reacts with a suitably hydroxylated silica to release propene and give an $(allyl)_2Rh(silica)$ system, which has high stoichiometric and catalytic reactivity. Methane reacts to give butenes after hydrogenation, so that the net reaction is the addition of methane to the allyl group. Methane followed by Cl_2 leads to MeCl. Physical studies suggest that the Rh is coordinated to surface oxygen atoms, an unusual coordination environment for organometallic species. The mechanism involved in this chemistry may be oxidative addition, or, perhaps more likely, Shilov chemistry. Selectivity studies are needed to decide the issue.

IV. σ-BOND METATHESIS

A number of systems operate by the σ–bond metathesis[2] route (equation 1c). The reaction of labeled methane with Cp^*_2LuMe (equation 29) was studied by Watson[749]. In spite of the high C—H bond strength ($105\,kcal\,mol^{-1}$) and the very low proton acidity of methane ($pK_a > 40$), the reaction appears to proceed very rapidly with second-order kinetics. If σ–bond metathesis is interpreted as involving a methane complex as an intermediate, the reaction becomes easier to understand, because in such a complex C—H to M dσ donation will deplete the C—H bond of electrons and increase the acidity. In the better-studied case of H_2 complexation, side-on binding can increase the proton acidity from a pK_a of ca 25 in free H_2 to values approaching zero for the most acidic of the H_2 complexes known[6]. Larger alkanes react more slowly. Comparison of the reaction rates for different types of C—H bond shows a qualitative correlation with the C—H bond acidity (rater than C—H bond strength or ionization potential), as might be expected if proton abstraction from a C—H $\cdots$ M intermediate is important.

$$Cp^*_2LuCH_3 + {}^{13}CH_4 = Cp^*_2Lu^{13}CH_3 + CH_4 \qquad (29)$$

Bercaw and coworkers[74b] have shown that a variety of hydrocarbons react with the related species Cp^*_2ScR, but the rate is not obviously dependent on the C—H acidity. The cyclometalated species $Cp^*_2Th(CH_2)_2CMe_2$ only gives the first step of equation 30 with methane at 30–60°C but goes on to the dialkyl with cyclopropane[75]. Wolczanski and coworkers[76] have shown that the Zr=NR group can activate methane. As shown in equation 31, the NR group takes up the proton liberated from the methane.

$$Cp^*_2Th(CH_2)_2CMe_2 \xrightarrow{RH} Cp^*_2Th(R)(CH_2CMe_3) \xrightarrow{RH} Cp^*_2Th(R)_2 + CMe_4 \qquad (30)$$

$$(R_3SiNH)_2Zr{=}NSiR_3 \xrightarrow{CH_4} (R_3SiNH)_3Zr{=}CH_3 \qquad (31)$$

Sec-butyl potassium is known to metalate linear alkanes at the terminal position; carboxylic acids are obtained on carboxylation[77].

V. OTHER MECHANISMS

Renneke and Hill[78a] have shown how heteropolyanions such as $H_3PW_{12}O_{40}$ can catalytically oxidize alkanes under photolytic conditions in wet CH_3CN. The reduced form of the catalyst is recycled by Pt/C-catalyzed reduction of water to H_2. The products from RH are largely RNHCOMe and the alkene. They are thought to be formed by electron transfer from the alkane to give RH$^+$, which can deprotonate to give the R· radical. This attacks the nitrile solvent to give the ketone and amide, or can be oxidized to the R$^+$ cation, which can then eliminate a proton to give the alkene. This system is particularly interesting in that such polyanions ought to have better resistance to degradation than some of the other catalysts that have been studied.

The carbene precursor, N_2CHCO_2Et, can insert into alkane C—H bonds with $Rh_2(OCOR)_4$ as catalyst. The selectivity is 7:66:27 for attack at C-1, C-2 and C-3 of n-pentane for R = CF_3 and rises to 30:61:9 when R is 9-trypticyl[78b].

VI. METAL ATOMS, IONS AND EXCITED METAL ATOMS

A. Neutral Metal Atoms in the Ground State

Matrix isolation work at 10–20 K with a variety of metal ions showed no signs of alkane reactions except for Al[79] and B. Theoretical work by Lebrilla and Maier[80] suggested that a partially filled p-shell is important for successful C—H activation, because this could lead to metal to C—H σ^* electron donation and hence to C—H cleavage. This was certainly consistent with the observed reactivity of Al and B (s^2p^1).

The second- and third-row transition metals seem to be more reactive; for example, Rh reacts with methane under matrix conditions. This higher reactivity may be a result of $D(M—H)$ being higher for these metals. Metal clusters, such as Pt_x, are also very reactive in a pulsed nozzle beam, for example, converting cyclohexane to benzene. Since Pt is known to be active as a heterogeneous catalyst for alkane reactions, this is perhaps not surprising[81]. Cluster reactions are probably responsible for the observations by the Klabunde and Skell groups[82] that Ni or Zr vapor reacts with alkane to yield metal crystallites with significant C content.

B. Metal Ions

Allison, Freas and Ridge[83] discovered in 1979 that transition metal ions in the gas phase are very reactive with alkanes. Alkanes tend to be more reactive with electrophiles than nucleophiles, so the electrophilic character of the cation may be one important factor in the success of the chemistry.

One very useful feature of gas-phase physical methods is that bond strengths can be estimated rather well in many cases. For example, $D(Fe—CH_3)$ is 37(7) kcal mol^{-1}, while $D(Fe—CH_3)^+$ is 58(2) kcal mol^{-1} [84]. It is probably the strong ion–dipole interactions that strengthen $[M—R]^+$ over $[M—R]$ and this in turn must help the thermodynamics of bond cleavage in the case of the ion. Another distinctive feature of the bond-strength data is that unlike the case of metal complexes, where the $L_nM—CH_3$ bond is much weaker than $L_nM—H$, for ions and atoms, $D(M—CH_3)$ is the same as $D(M—H)$ for any particular metal. The reason may simply be that steric effects only operate in the case of $L_nM—CH_3$. An unfortunate complication in the detailed interpretation of these data is the uncertainty about the structure of a species such as Fe—CH$_3$$^+$. Is it a simple alkyl, is it agostic, and, if the latter, how many H groups bridge? The bond strengths have been analyzed on the basis of the promotion energy from the ground state atom to the $(n-1)s^1nd^{m-1}$ state thought to be appropriate for forming the M—R species.

The first-row transition metal monocations all react with methane; some of the pathways found are illustrated in equations 32–35. The relative importance of each depends on the kinetic energy of the metal ion. For example, V$^+$ goes mostly by equation 34 at low energies and by equation 32 at high energies. The mechanism suggested is oxidative addition followed by α elimination and reductive elimination of H_2 at low energies (equation 36) or M—C bond cleavage at high energies. The predominant pathway for Fe^+ is equation 32. Spin conservation is an important limitation on the types of reaction allowed in these systems. This is believed to be the reason that the channel of equation 34 is not observed for Fe^+.

$$M^+ + CH_4 \longrightarrow MH^+ + CH_3 \cdot \tag{32}$$

$$\longrightarrow MCH_3^+ + H \tag{33}$$

$$\longrightarrow MCH_2^+ + H_2 \tag{34}$$

$$\longrightarrow MCH^+ + H_2 + H \tag{35}$$

$$M^+ + CH_4 \rightarrow H—M—CH_3^+ \longrightarrow H_2M{=}CH_2^+ \rightarrow H_2 + M{=}CH_2 \tag{36}$$

Alkane C—H and C—C bonds both react with metal ions, perhaps because the metal ions are sterically very small and $2 \times D(M—CR_3)$ is large enough to 'pay for' breaking a C—C bond (usually $85\,kcal\,mol^{-1}$) and provide some excess driving force to help overcome the kinetic barriers. The reactions of metal ions with ethane and higher alkanes therefore often lead to chain cleavage. For example, Fe^+ and propane give both $Fe^+(propene) + H_2$ as well as $Fe^+(ethylene) + MeH$. In butane, H loss occurs from the 1- and 4-positions, believed to be a result of metal attack at the central (most electron-rich) C—C bond, followed by β elimination[85]. Another example involving Fe^+ and isobutane is shown in equation 37[32].

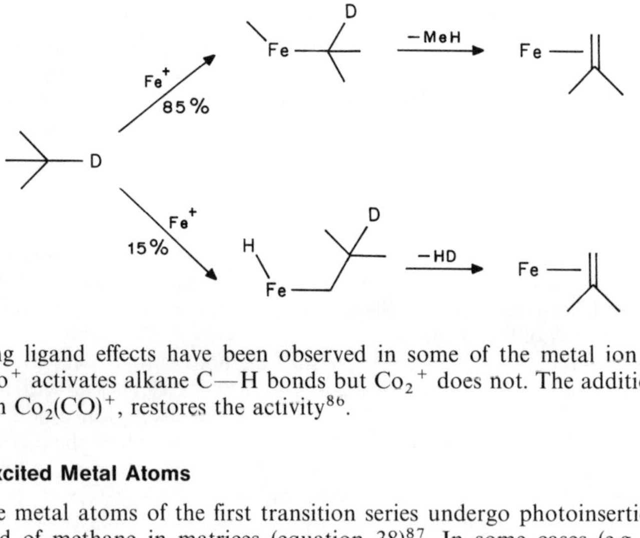

$$(37)$$

Interesting ligand effects have been observed in some of the metal ion work. For example, Co^+ activates alkane C—H bonds but Co_2^+ does not. The addition of a CO ligand, as in $Co_2(CO)^+$, restores the activity[86].

C. Photoexcited Metal Atoms

All of the metal atoms of the first transition series undergo photoinsertion into the C—H bond of methane in matrices (equation 38)[87]. In some cases (e.g. Fe, Co), a precursor to the insertion product, identified as an alkane complex, was also detected. Subsequent irradiation at a longer wavelength generally caused reductive elimination (e.g. Fe, Co). In other cases (e.g. Cu), the alkylmetal hydride fragments to an alkyl radical and an M—H group. Longer-chain alkanes also react with Fe^{88}.

$$Fe + CH_4 \underset{400\ nm}{\overset{300\ nm}{\rightleftharpoons}} H—Fe—CH_3 \tag{38}$$

Mercury photosensitization[89] has long been known to lead to alkane C—H bond cleavage. The $^3P_1$ excited state of mercury (Hg*) formed by irradiation at 254 nm follows the sequence of events shown in equations 39–43. The partially filled p shell of Hg* ($d^{10}s^1p^1$ configuration) may be important in conferring reactivity, as discussed in Section VI.A for neutral atoms. [H—Hg—R]* may be an intermediate in equation (40).

$$Hg + h\nu \longrightarrow Hg^* \qquad (39)$$

$$Hg^* + RH \longrightarrow [R—H\cdots Hg]^* \longrightarrow R\cdot + H\cdot + Hg \qquad (40)$$

$$H\cdot + RH \longrightarrow R\cdot + H_2 \qquad (41)$$

$$2R\cdot \longrightarrow R_2 + R(-H) + RH \qquad (42)$$

$$R(-H) + H\cdot \longrightarrow R\cdot \qquad (43)$$

Very recently, this reaction has been used for synthetic alkane chemistry on a multigram scale at ambient temperatures and pressures. For example, cyclooctane can be converted to the dimer in high yield even at high conversion. This would not have been the case in a solution phase radical reaction, because the product R_2 is intrinsically more reactive than the initial substrate, RH. The reason for the selectivity proved to be that the dimer has so low a vapor pressure that essentially only the original substrate, cyclooctane, is present in the vapor phase and reaction only happens in the vapor. This vapor selectivity effect allows the initial functionalization product to be isolated[90].

The observed selectivity is $3° > 2° > 1°$ as expected for a homolytic pathway and species with $4°$–$4°$ C—C bonds are very efficiently assembled, especially in the presence of H_2 which increases the selectivity because H· atoms which are formed are somewhat more selective for the weakest C—H bonds in the molecule. H atoms are also very tolerant of functional groups, so a variety of functionalized molecules (esters, epoxides, ketones etc.) can also be dehydrodimerized. Methylcyclohexane only gives 12% of the $4°$–$4°$ dehydrodimer in the absence of H_2, but in its presence enough of this dimer is formed to allow it to crystallize from the product mixture (equation 44). Alkanes can also be functionalized by cross-dimerization with other species. Equation 45 shows the results from cyclohexane and methanol. The three products are formed in approximately statistical amounts. The polarities of the three species are so different that the glycol can be removed with water and the bicyclohexyl separated by elution with pentane.

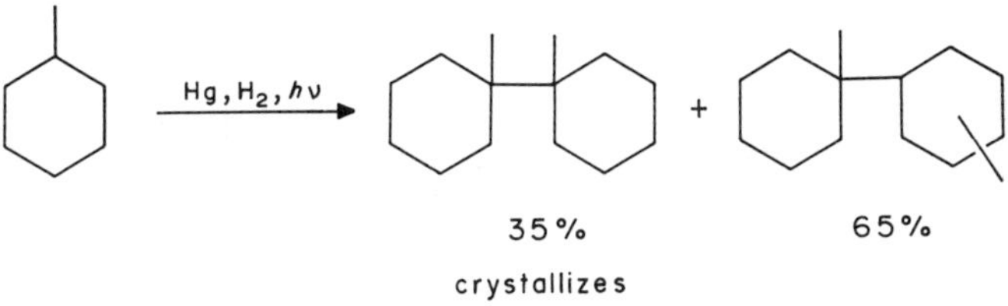

$$(44)$$

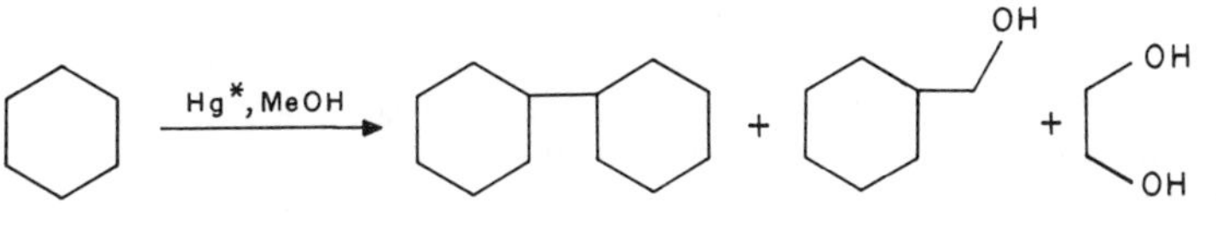

35%   65%
crystallizes

$$(45)$$

Cross-dimerizing cyclohexane with the formaldehyde trimer gives a material which yields cyclohexanecarboxaldehyde on hydrolysis (equation 46). The radicals formed by Hg* reactions of alkanes can also be efficiently trapped by SO_2 to give RSO_3H after oxidation, and by CO to give RCHO and R_2CO[91].

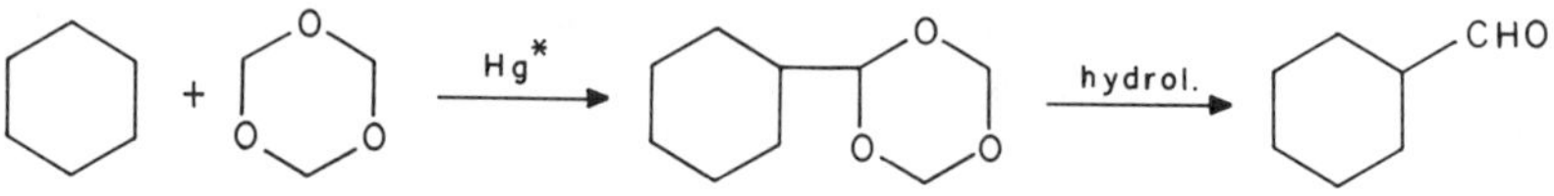

$$(46)$$

Methane and other alkanes with strong C—H bonds are not efficiently attacked by the Hg/$h\nu$ or Hg/$h\nu$/H$_2$ systems, but need a stronger H-atom abstractor, such as the oxygen atom. O atoms can be formed efficiently in the system Hg/$h\nu$/N$_2$O (equation 47). Methanol was used to trap the methyl radicals and give a functionalized product (equation 48). The labeling scheme as well as control experiments confirm the origin of the methyl group of the product ethanol[92]. Methane does react with Hg*/O$_2$ to give MeOOH in good yield, so the Hg* probably attacks O$_2$ to give O or O$_2$* rather than methane.

$$N_2O \xrightarrow{\text{Hg}/h\nu} N_2 + O \tag{47}$$

$$CD_3OH \xrightarrow{O} \cdot CD_2OH + \cdot OH \xrightarrow{CH_4} \cdot CD_2OH + \cdot CH_3 + H_2O \rightarrow CH_3CD_2OH \tag{48}$$

Green and coworkers[93] have used metal vapor synthesis (MVS) to make isolable quantities of a number of organometallic species directly from alkanes (equations 49–51). In each case the complex has the stoichiometry of the parent alkane. Equation 50 is reminiscent of some of the chemistry we saw in the homogeneous metal phosphine chemistry discussed earlier in this chapter.

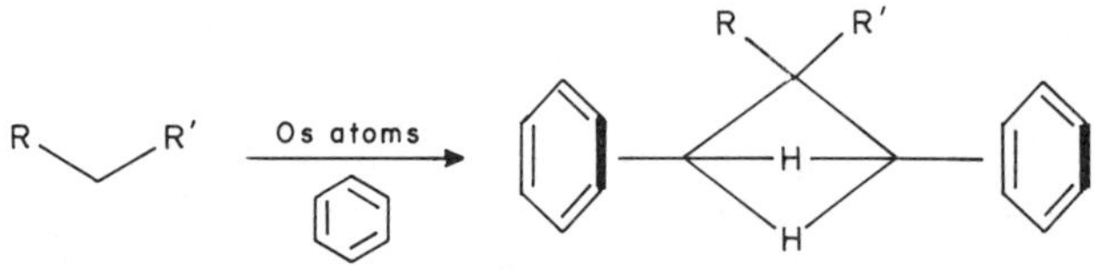

$$(49)$$

$$c\text{-}C_5H_{10} \xrightarrow{W, PMe_3} CpW(PMe_3)H_5 \tag{50}$$

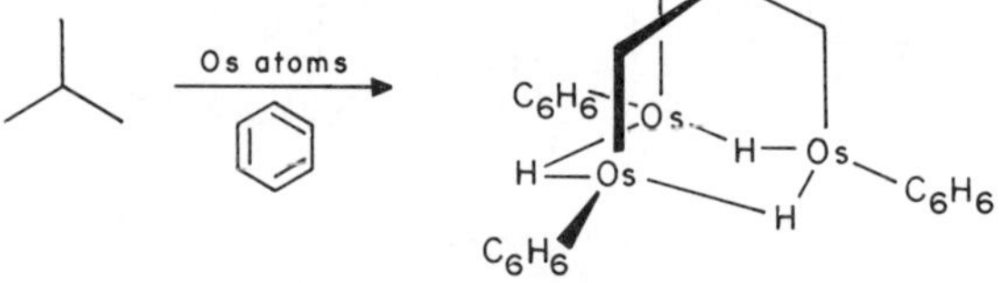

$$(51)$$

In equations 49 and 51, the metals bind the hydrogens in μ-H sites in the cluster. In the rhenium case (analogous to equation 49) the carbon fragment from the alkane is bound as a μ-carbene in the resulting cluster. This may be a reasonable analogy for the way alkane-derived fragments are bound on metal surfaces. The co-ligands, benzene and trimethylphosphine, are required to stabilize the complex. It is perhaps surprising that these ligands are not activated in the products observed. Of course, other species may have escaped detection, and the observed products are no doubt thermodynamic traps, so we cannot deduce anything about the relative reactivity of arene, phosphine and alkane C—H bonds. Photochemical processes may also occur under the MVS conditions, because the presence of a strongly glowing bead of metal near its melting point must generate a broad spectrum of radiation.

VII. ACKNOWLEDGMENTS

We thank the National Science Foundation, the Petroleum Research Fund, the Department of Energy and Exxon Corp. for funding our work in this area.

VIII. REFERENCES

1. R. H. Crabtree, *Chem. Rev.*, **85**, 245 (1985).
2. M. S. Thompson, S. M. Baxter, A. R. Bulls, B. J. Burger, M. C. Nolan, B. Santarsiero, P. Schaefer and J. E. Bercaw, *J. Am. Chem. Soc.*, **109**, 203 (1987).
3. M. Brookhart and M. L. H. Green, *J. Organomet. Chem.*, **250**, 395 (1983); R. H. Carbtree and D. G. Hamilton, *Adv. Organomet. Chem.*, **28**, 105 (1988).
4. (a) M. I. Bruce, *Angew. Chem., Int. Ed. Engl.*, **16**, 73 (1977).
   (b) A. J. Cheney and B. L. Shaw, *J. Chem. Soc., Dalton Trans.*, 754 (1952).
5. J. Y. Saillard, in *Selective Hydrocarbon Activation* (Eds. J. A. Davies, P. L. Watson, J. F. Liebman and A. Greenberg), Chap. 7, VCH, Weinheim, 1990.
6. R. H. Crabtree, *Acc. Chem. Res.*, **23**, 95 (1990); G. J. Kuba, *Acc. Chem. Res.*, **21**, 120 (1988).
7. H. Rabaa, J. Y. Saillard and U. Schubert, *J. Organomet. Chem.*, **330**, 397 (1987).
8. (a) J. J. Low and W. A. Goddard, *J. Am. Chem. Soc.*, **108**, 6115 (1986).
   (b) S. Obara, K. Kitaura and K. Morokuma, *J. Am. Chem. Soc.*, **106**, 7482 (1984).
9. R. H. Crabtree, E. M. Holt, M. E. Lavin and S. M. Morehouse, *Inorg. Chem.*, **24**, 1986 (1985).
10. J. Y. Saillard and R. Hoffmann, *J. Am. Chem. Soc.*, **106**, 2006 (1984).
11. M. Steigerwald and W. Goddard, *J. Am. Chem. Soc.*, **106**, 308 (1984).
12. S. Rabaa and R. Hoffmann, *J. Am. Chem. Soc.*, **108**, 4327 (1986).
13. J. F. Liebman and C. D. Hoff, in *Selective Hydrocarbon Activation* (Eds. J. A. Davies, P. L. Watson, J. F. Liebman and A. Greenberg), Chap. 6, VCH, Weinheim, 1990.
14. J. Halpern, *Polyhedron*, **7**, 1483 (1988).
15. H. E. Bryndza, L. K. Fong, R. A. Paciello, W. Tam and J. E. Bercaw, *J. Am. Chem. Soc.*, **109**, 1444 (1987).
16. H. E. Bryndza, P. J. Domaille, R. A. Paciello and J. E. Bercaw, *Organometallics*, **8**, 379 (1989).
17. R. L. Hodges and J. L. Garnett, *J. Phys. Chem.*, **72**, 1673 (1968).
18. N. F. Goldshleger, M. B. Tyabin, A. E. Shilov and A. A. Shteinman, *Zh. Fiz. Khim.*, **43**, 2174 (1969).
19. A. E. Shilov, *The Activation of Saturated Hydrocarbons by Transition Metal Complexes*, D. Riedel, Dordrecht, 1984.
20. F. A. Cotton and A. G. Stanislowski, *J. Am. Chem. Soc.*, **96**, 5074 (1974).
21. D. M. Roe, P. M. Bailey, K. Moseley and P. M. Maitlis, *Chem. Commun.*, 1273 (1972).
22. M. J. Burk, M. P. McGrath and R. H. Crabtree, *J. Am. Chem. Soc.*, **110**, 620 (1988).
23. (a) V. V. Eskova, A. E. Shilov and A. A. Shteinman, *Kinet. Catal.*, **13**, 534 (1972).
    (b) Y. V. Geletii and A. E. Shilov, *Kinet. Catal.*, **24**, 486 (1983).
    (c) N. F. Goldshleger, V. V. Lavrushko, A. P. Krushch and A. A. Shteinman, *Izv. Akad. Nauk SSSR, Ser. Khim.*, 2174 (1976).
24. V. P. Trateyakov, C. P. Zimtseva, E. S. Rudakov and A. S. Osetskii, *React. Kinet. Catal. Lett.*, **12**, 543 (1979).
25. V. V. Lavrushko, S. A. Lermontov and A. E. Shilov, *Kinet. Catal. Lett.*, **15**, 269 (1980).
26. G. B. Shulpin, G. V. Nizova and A. E. Shilov, *Chem. Commun.*, 761 (1983).
27. J. A. Labinger, A. H. Herring and J. E. Bercaw, *J. Am. Chem. Soc.*, **112**, 5628 (1990).
28. (a) E. Gretz, T. F. Oliver and A. Sen, *J. Am. Chem. Soc.*, **109**, 8109 (1987); A. Sen, E. Gretz, T. F. Oliver and Z. Jiang, *New J. Chem.*, **13**, 755 (1989).
    (b) L. C. Kao and A. E. Shilov, *New J. Chem.*, (1991) in press.
29. J. Chatt and J. M. Davidson, *J. Chem. Soc.*, 843 (1965).
30. F. A. Cotton, B. A. Frenz and D. L. Hunter, *Chem. Commun.*, 755 (1974).
31. C. A. Tolman, S. D. Ittel and J. P. Jesson, *J. Am. Chem. Soc.*, **100**, 4081 (1978); **101**, 1742 (1979).
32. J. Chatt and H. R. Watson, *J. Chem. Soc.*, 2545 (1962).
33. J. Chatt and R. S. Coffey, *J. Chem. Soc. (A)*, 1963 (1969).
34. F. N. Tebbe and G. W. Parshall, *J. Am. Chem. Soc.*, **92**, 5234 (1970).
35. C. Gianotti and M. L. H. Green, *Chem. Commun.*, 1114 (1972); M. L. H. Green, *Pure Appl. Chem.*, **50**, 27 (1978).

678   R. H. Crabtree

36. N. F. Gol'dschleger, M. B. Tyabin, A. E. Shilov and A. A. Shteinman, *Zh. Fiz. Khim.*, **43**, 2174 (1969).
37. (a) R. H. Crabtree, J. M. Mihelcic and J. M. Quirk, *J. Am. Chem. Soc.*, **101**, 7738 (1979).
    (b) R. H. Crabtree, M. F. Mellea, J. M. Mihelcic and J. M. Quirk, *J. Am. Chem. Soc.*, **104**, 107 (1982).
    (c) R. H. Crabtree and C. P. Parnell, *Organometallics*, **3**, 1727 (1984).
    (d) R. H. Crabtree and C. P. Parnell, *Organometallics*, **4**, 519 (1985).
38. D. Baudry, M. Ephritikhine and H. Felkin, *Chem. Commun.*, 1243 (1980).
39. A. H. Janowicz and R. G. Bergman, *J. Am. Chem. Soc.*, **104**, 352 (1982); **105**, 3929 (1983).
40. D. E. Marx and A. J. Lees, *Inorg. Chem.*, **27**, 1121 (1988).
41. The conceptual basis of organometallic chemistry, including an explanation of electron count, the η descriptor and slip, can be found in any recent text, such as R. H. Crabtree, *The Organometallic Chemistry of the Transition Metals*, Wiley, Chichester, 1988.
42. J. M. Buchanan, J. M. Stryker and R. G. Bergman, *J. Am. Chem. Soc.*, **108**, 1537 (1986).
43. S. P. Nolan, C. D. Hoff, P. O. Stoutland, L. J. Newman, J. M. Buchanan, R. G. Bergman, G. K. Yang and K. S. Peters, *J. Am. Chem. Soc.*, **109**, 3143 (1987).
44. M. B. Sponsler, B. H. Weiller, P. O. Stoutland and R. G. Bergman, *J. Am. Chem. Soc.*, **111**, 6841 (1989).
45. B. H. Weiller, E. P. Wasserman, R. G. Bergman, C. B. Moore and G. C. Pimentel, *J. Am. Chem. Soc.*, **111**, 8288 (1989).
46. W. D. McGhee and R. G. Bergman, *J. Am. Chem. Soc.*, **110**, 4246 (1988).
47. (a) J. K. Hoyano and W. A. G. Graham, *J. Am. Chem. Soc.*, **104**, 3723 (1982).
    (b) W. A. G. Graham, *J. Organomet. Chem.*, **300**, 81 (1986).
48. J. K. Hoyano, A. D. McMaster and W. A. G. Graham, *J. Am. Chem. Soc.*, **105**, 7190 (1983).
49. C. K. Ghosh and W. A. G. Graham, *J. Am. Chem. Soc.*, **109**, 4276 (1987).
50. W. D. Jones and F. J. Feher, *J. Am. Chem. Soc.*, **104**, 4240 (1982); **106**, 1650 (1984); *Acc. Chem. Res.*, **22**, 91 (1989).
51. R. A. Periana and R. G. Bergman, *J. Am. Chem. Soc.*, **106**, 7272 (1984); **108**, 7332 (1986).
52. T. T. Wenzel and R. G. Bergman, *J. Am. Chem. Soc.*, **108**, 4856 (1986).
53. J. A. Ibers, R. DiCosimo and G. M. Whitesides, *Organometallics*, **3**, 1 (1982).
54. M. A. Green, J. C. Huffman, K. G. Caulton and J. J. Ziolkowski, *J. Organomet. Chem.*, **218**, C39 (1981).
55. (a) D. Baudry, M. Ephritikhine and H. Felkin, *Chem. Commun.*, 606 (1982); 788 (1983); D. Baudry, M. Ephritikhine, H. Felkin and J. Zakrzewski, *Chem. Commun.*, 1235 (1982); *Tetrahedron Lett.*, 1283 (1984).
    (b) H. Felkin, T. Fillebeen-Khan, Y. Gault, R. Holmes-Smith and J. Zakrzewski, *Tetrahedron Lett.*, 1279 (1984).
    (c) C. J. Cameron, H. Felkin, T. Fillebeen-Khan, N. J. Forrow and E. Guittet, *Chem. Commun.*, 801 (1986).
    (d) H. Felkin, T. Fillebeen-Khan, R. Holmes-Smith and Y. Lin, *Tetrahedron Lett.*, **26**, 1999 (1985).
56. M. V. Baker and L. D. Field, *J. Am. Chem. Soc.*, **109**, 2825, 7433 (1987).
57. M. Hackett and G. M. Whitesides, *J. Am. Chem. Soc.*, **110**, 1449 (1988).
58. A. E. Sherry and B. B. Wayland, *J. Am. Chem. Soc.*, **112**, 1259 (1990).
59. M. J. Burk, R. H. Crabtree and D. V. McGrath, *Chem. Commun.*, 1829 (1985); M. J. Burk and R. H. Crabtree, *J. Am. Chem. Soc.*, **109**, 8025 (1987).
60. (a) T. Fujii and Y. Saito, *Chem. Commun.*, 757 (1990).
    (b) K. Nomura and Y. Saito, *Chem. Commun.*, 161 (1988).
    (c) K. Nomura and Y. Saito, *J. Mol. Catal.*, **54**, 57 (1989).
61. (a) M. Tanaka, *Pure Appl. Chem.*, **62**, 1147 (1990); T. Sakakura, T. Sodeyama, Y. Tokunaga and M. Tanaka, *Chem. Lett.*, 263 (1987); T. Sakakura and M. Tanaka, *Chem. Commun.*, 758 (1987); T. Sakakura, T. Sodeyama and M. Tanaka, *Chem. Lett.*, 683 (1988); *New J. Chem.*, **13**, 737 (1989); K. Sasaki, Y. Tokunaga, K. Wada and M. Tanaka, *Chem. Commun.*, 155 (1988); T. Sakakura, T. Sodeyama, Y. Tokunaga and M. Tanaka, *Chem. Lett.*, 263 (1988); T. Sakakura, T. Sodeyama and M. Tanaka, *Chem. Ind.*, 530 (1988).
    (b) J. A. Maguire, W. T. Boese and A. S. Goldman, *J. Am. Chem. Soc.*, **111**, 7088 (1989).
62. P. Garrou, *Chem. Rev.*, **85**, 171 (1985).
63. R. Hoffmann, *Helv. Chim. Acta*, **67**, 1 (1984).
64. (a) E. A. Grigoryan, F. S. Dyachkovskii and I. R. Mullagaliev, *Dokl. Akad. Nauk SSSR* **224**, 859

(1975); E. A. Grigoryan, F. S. Dyachkovskii, S. Ya. Zhuk and L. I. Vyshinskaja, *Kinet. Katal.*, **19**, 1063 (1978).

(b) N. S. Enikopopjan, Kh. R. Gyulumjan and E. A. Grigoryan, *Dokl. Akad. Nauk SSSR* **249**, 1380 (1979); E. A. Grigoryan, Kh. R. Gyulumjan, E. I. Gurtovaya, N. S. Enikopopjan and M. A. Ter-Kazarova, *Dokl. Akad. Nauk SSSR* **257**, 364 (1981).

65. Y. Lin, D. Ma and X. Lu, *J. Organomet. Chem.*, **323**, 407 (1987).
66. R. H. Crabtree, R. P. Dion, D. J. Gibboni, D. V. McGrath and E. M. Holt, *J. Am. Chem. Soc.*, **108**, 7222 (1986).
67. F. W. S. Benfield and M. L. H. Green, *J. Chem. Soc., Dalton Trans.*, 1325 (1974); P. Eilbracht, *Chem. Ber.*, **113**, 542, 1033, 1420, 2211 (1980).
68. C. H. F. Tipper, *J. Chem. Soc.*, 2043 (1955); D. M. Adams, J. Chatt, R. Guy and N. Shepard, *J. Chem. Soc.*, 738 (1961).
69. K. C. Bishop, *Chem. Rev.*, **76**, 461 (1976); H. Hogeveen and H. C. Volger, *J. Am. Chem. Soc.*, **89**, 2486 (1967).
70. L. Cassar and J. Halpern, *J. Chem. Soc. (D)*, 1082 (1971).
71. H. Hogeveen and B. J. Nusse, *Tetrahedron Lett.*, 159 (1974).
72. Y. I. Yermakov, B. N. Kuznetsov and V. Zakharov, *Catalysis by Supported Complexes*, Elsevier, Amsterdam, 1981.
73. J. Schwartz, *Acc. Chem. Res.*, **18**, 302 (1985).
74. (a) P. L. Watson, *J. Am. Chem. Soc.*, **105**, 6491 (1983).
(b) G. Parkin, E. Bunel, B. J. Burger, M. S. Trimmer, A. A. Van and J. E. Bercaw, *J. Mol. Catal.*, **41**, 21 (1987).
75. J. W. Bruno, G. M. Smith, T. J. Marks, C. K. Fair, A. J. Schultz and J. M. Williams, *J. Am. Chem. Soc.*, **109**, 203 (1987).
76. C. C. Cummins, B. M. Baxter and P. T. Wolczanski, *J. Am. Chem. Soc.*, **110**, 8731 (1988).
77. H. Mergard and F. Korte, *Monatsh. Chem.*, **98**, 763 (1967).
78. (a) R. F. Renneke and C. L. Hill, *J. Am. Chem. Soc.*, **108**, 3528 (1986); **110**, 5461 (1988); C. L. Hill, D. A. Bouchard, M. Khadkhodayan, M. M. Williamson, J. A. Schmidt and E. F. Hilinski, *J. Am. Chem. Soc.*, **110**, 5471 (1988).
(b) A. F. Noel, J. L. Costa and A. J. Hubert, *J. Mol. Catal.*, **58**, 21 (1990) and references cited therein.
79. K. J. Klabunde and Y. Tanaka, *J. Am. Chem. Soc.*, **105**, 3544 (1983).
80. C. B. Lebrilla and W. F. Maier, *Chem. Phys. Lett.*, **105**, 183 (1984).
81. D. J. Trevor, R. L. Whetton, D. M. Cox and A. Kaldor, *J. Am. Chem. Soc.*, **107**, 518 (1985).
82. S. C. Davis and K. J. Klabunde, *J. Am. Chem. Soc.*, **100**, 5973 (1978); R. J. Remick, T. A. Asunta and P. S. Skell, *J. Am. Chem. Soc.*, **101**, 1320 (1979).
83. J. Allison, R. B. Freas and D. P. Ridge, *J. Am. Chem. Soc.*, **101**, 1332 (1979).
84. P. B. Armentrout, in *Selective Hydrocarbon Activation* (Eds. J. A. Davies, P. L. Watson, J. F. Liebman and A. Greenberg), Chap. 14, VCH, Weinheim, 1990.
85. L. F. Halle, R. Houriet, M. M. Kappes, R. H. Staley and J. L. Beauchamp, *J. Am. Chem. Soc.*, **104**, 6293 (1982).
86. J. Allison, R. B. Freas and D. P. Ridge, *J. Am. Chem. Soc.*, **101**, 1332 (1979); R. B. Freas and D. P. Ridge, *J. Am. Chem. Soc.*, **102**, 7129 (1980); **106**, 825 (1984).
87. Z. H. Kafafi in *Selective Hydrocarbon Activation* (Eds. J. A. Davies, P. L. Watson, J. F. Liebman and A. Greenberg), Chap. 12, VCH, Weinheim, 1990.
88. G. A. Ozin, J. G. McCaffrey and J. M. Parnis, *Angew. Chem., Int. Ed.*, **25**, 1072 (1986).
89. R. J. Cvetanovic, *Prog. React. Kinet.*, **2**, 77 (1964).
90. (a) S. H. Brown and R. H. Crabtree, *Chem. Commun.*, 927 (1987).
(b) S. H. Brown and R. H. Crabtree, *J. Am. Chem. Soc.*, **111**, 2935 (1989).
(c) S. H. Brown and R. H. Crabtree, *J. Am. Chem. Soc.*, **111**, 2946 (1989).
(d) C. G. Boojamra, R. H. Crabtree, R. R. Ferguson and C. A. Muedas, *Tetrahedron Lett.*, **30**, 5583 (1989).
(e) C. A. Muedas, R. R. Freguson and R. H. Crabtree, *Tetrahedron Lett.*, **30**, 3389 (1989).
91. R. R. Ferguson and R. H. Crabtree, *J. Org. Chem.*, in press.
92. R. R. Ferguson and R. H. Crabtree, *New J. Chem.*, **13**, 647 (1989).
93. J. A. Bandy, F. G. N. Cloke, M. L. H. Green, D. O'Hare and K. Prout, *Chem. Commun.*, 240 (1984); 355, 356 (1985); M. L. H. Green and G. Parker, *Chem. Commun.*, 1467 (1984).

## CHAPTER 15

Rearrangements and photo-chemical reactions involving alkanes and cycloalkanes

T. OPPENLÄNDER*

F. Hoffmann-La Roche AG, CH-4002 Basle, Switzerland

G. ZANG and W. ADAM

Department of Organic Chemistry, University of Würzburg, D-8700 Würzburg, Germany

*Now at Fachhochschule Furtwangen, Abteilung Villingen–Schwenningen, Fachbereich Verfahrenstechnik, D-7730 VS-Schwenningen, Germany.

The Chemistry of Alkanes and Cycloalkanes
Edited by S. Patai and Z. Rappoport © 1992 John Wiley & Sons Ltd

I. INTRODUCTION

The present chapter deals with a heterogeneous field of organic chemistry with respect to the broad range of organic substrates involved and with respect to the application of a variety of different preparative thermal and photochemical methods. The fascinating aspect of all activities related to the chemistry of alkanes, however, is the understanding of the chemical and photochemical reactivity of hydrocarbons (C_nH_m), especially the activation of C—C and C—H bonds[1]. Furthermore, there are continuing research interests in the chemistry of strained organic molecules[2] focussing on the use of cyclopropanes in organic synthesis[3], their thermodynamic properties[4], the 'aesthetic' chemistry of 'platonic' structures like tetrahedrane[5], cubane[6] and the related strained $(CH)_8$ hydrocarbons[7] and the construction of spherical molecules like dodecahedrane[8].

Significant industrial applications of C—H activation techniques are the development of hydrotreating catalysts for the petroleum industry[9] and the ultraviolet laser ablation technique (ablative photodecomposition) which employs pulsed ArF excimer lasers ($\lambda = 193$ nm) as powerful UV light sources in material processing (microlithography) and in medicine (microsurgery, dentistry, ophthalmology)[10]. The use of tunable vacuum UV excimer lasers in the range of 100–200 nm is very promising for the exploitation of new photochemical processes in this wavelength range[11], but the broad industrial application of excimer lasers seems to be limited due to economical considerations.

This limitation may be overcome by the use of novel incoherent UV excimer light sources[12] of high intensity and of high spectral purity. They are available over a suitable wavelength range (120 nm–360 nm, Figure 1) and a variable range of geometries, i.e. planar or cylindrical discharge tubes. For example, the Xe source emits narrow-band vacuum UV radiation peaking at 172 nm (7.2 eV) characterized by a half-width of 12–14 nm and no other emissions in the wavelength region of 100 nm to 800 nm[12b]. Promising commercial applications of these lamps are photodeposition of metal films for semiconductor processing[12d,e], photoablation techniques[10] and their use in photochemistry[12], particularly waste water treatment.

Further highlights in this field of research are the investigation of the photochemistry on solid surfaces[13], e.g. semiconductors and photocatalysis[14], photochemical conversion of solar energy[15,16], reactions initiated by photoinduced electron transfer (PET)[17] and

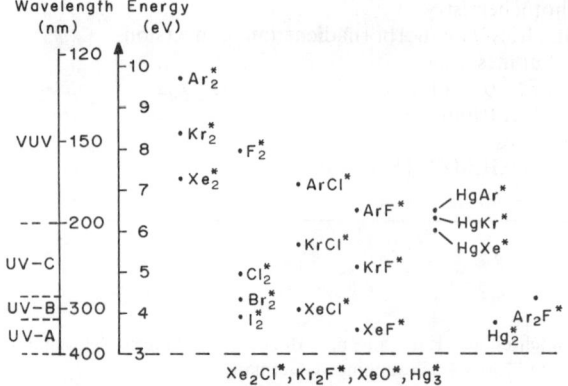

FIGURE 1. Wavelength range of potential excimer sources. Reproduced by permission of the International Union of Pure and Applied Chemistry from Reference 12b

the 185-nm photoreactivity of cycloalkanes[18]. New trends in the 185-nm photochemistry of bicyclic and tricyclic alkanes[19] are presented in Sections II.B.3 and II.B.4.

Photochemistry, by definition[20], deals with excitation energies of $\lambda \geqslant 40\,nm$ ($\leqslant 30\,eV$), whereas radiation chemistry (radiolysis) studies the photochemical effects of ionizing radiation (e.g. from radioactive sources or synchrotron radiation) such as the phenomenon of photoconductivity[21] and the formation of alkane radical cations $(RH^{\cdot +})$[22]. In this chapter we will concentrate on photochemical reactions of alkanes and cycloalkanes and their rearrangements. We mainly considered the literature since 1980.

II. REACTIONS

A. Simple Alkanes

1. Skeletal rearrangements on transition metal surfaces

Investigations which concern the mechanisms of skeletal rearrangements of saturated hydrocarbons induced by heterogeneous transition metal catalysis are of great interest for industrial applications, e.g. for petroleum reforming processes[9,23]. The developments in this field were reviewed recently by Hejtmanek[24a], and by Maire and Garin[24b], who focussed on the probable reaction mechanisms which include bond-shift and cyclic mechanisms for the skeletal isomerization of acyclic alkanes. Scheme 1 summarizes the

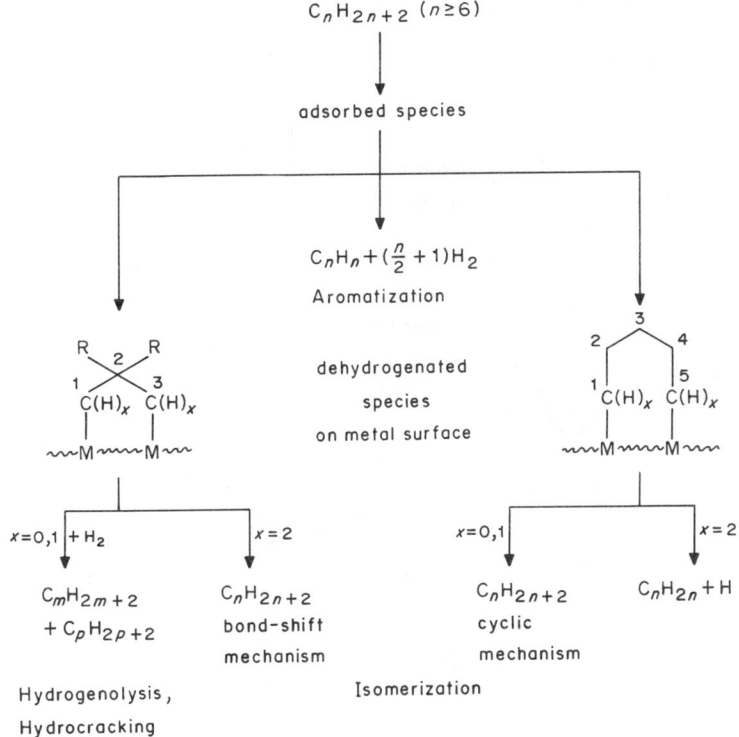

SCHEME 1. Rearrangements of hydrocarbons on metal surfaces

four main reaction pathways observed for heterogeneous catalysed reactions of saturated hydrocarbons on metal surfaces[24b,25].

The reactions include hydrogenolysis or hydrocracking (rupture of C—C bonds), isomerization, dehydrocyclization and aromatization of the C_nH_{2n+2} molecules[24,26]. They are observed with varying selectivities, which depend on the experimental conditions, e.g. temperature, pressure and the type of catalyst. A multitude of supported (Al_2O_3, SiO_2) or unsupported catalysts (e.g. monometallic transition metals[27], intermetallic compounds[28], bimetallic alloys[29], etc.) are applied to induce skeletal rearrangements of hydrocarbons. The latter, however, are the most important catalysts used in industrial practice, which allow reforming processes at low pressure and at low temperatures.

In the case of akyl substituted hydrocarbons with at least five carbon atoms in the chain (e.g. methylpentanes), the cyclic mechanism competes with the bond shift mechanism on catalysts of medium and low dispersion[25,26b], whereas n-hexane never

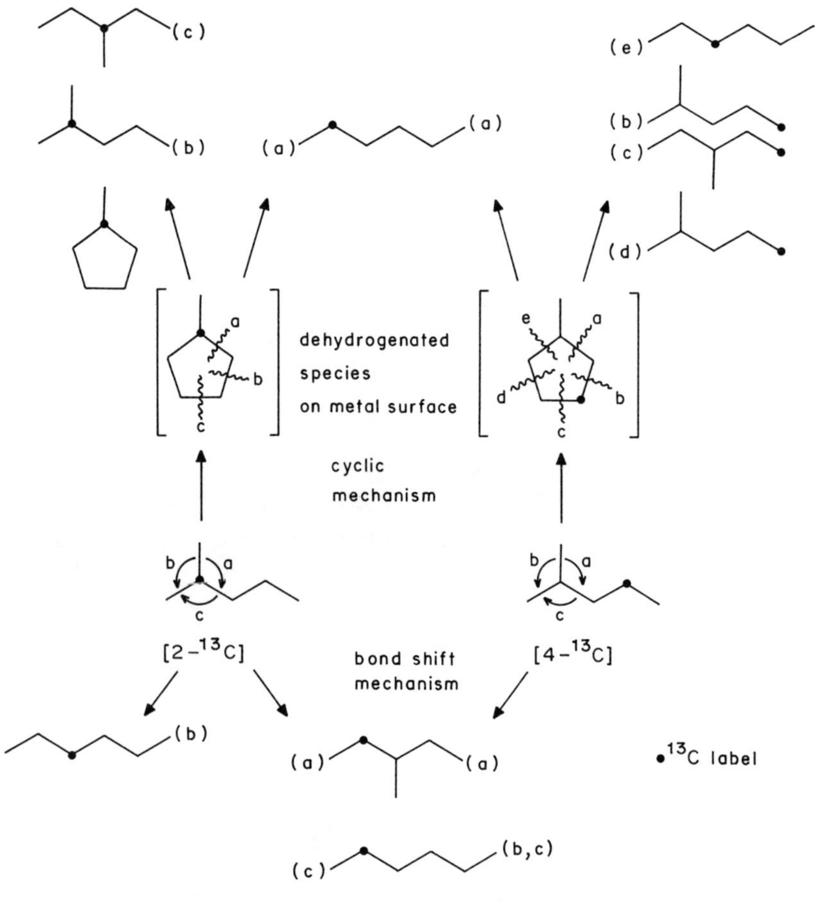

SCHEME 2. Isomerization pathways of 2-methylpentane labelled by ^{13}C at the 2- and 4-positions

participates in the cyclic pathway using Ir catalysts[29]. The evaluation of the relative contribution of bond shift versus cyclic mechanism to the overall reaction is best accomplished by using the ^{13}C tracer technique[23,24b,26b,29]. For instance, 2-methylpentane specifically ^{13}C labelled at the 2- or 4-position was used to distinguish between those two mechanisms by leading to different product mixtures (Scheme 2).

The nature of the reactive intermediates involved in hydrocracking or in skeletal rearrangements of hydrocarbons is still under active discussion. Many arguments are in favour of metalla-carbyne or metalla-carbene like complexes[24,26b]. However, several experimental results may only be explained by proposing metalla-cyclobutanes or σ-alkyl-metal species[30] as adsorbed reactive intermediates[24b], and they depend strongly on the nature of the active sites[31] (e.g. large or small size particles, single crystal metal surfaces[32]).

The ring-opening reaction of methylcyclopentane (Scheme 3) is the reverse reaction of the dehydrocyclization of methylpentanes (Scheme 2). Reactions of alkylcyclopentanes on supported and unsupported transition metal catalysts are also very important in naphtha reforming processes[28,33].

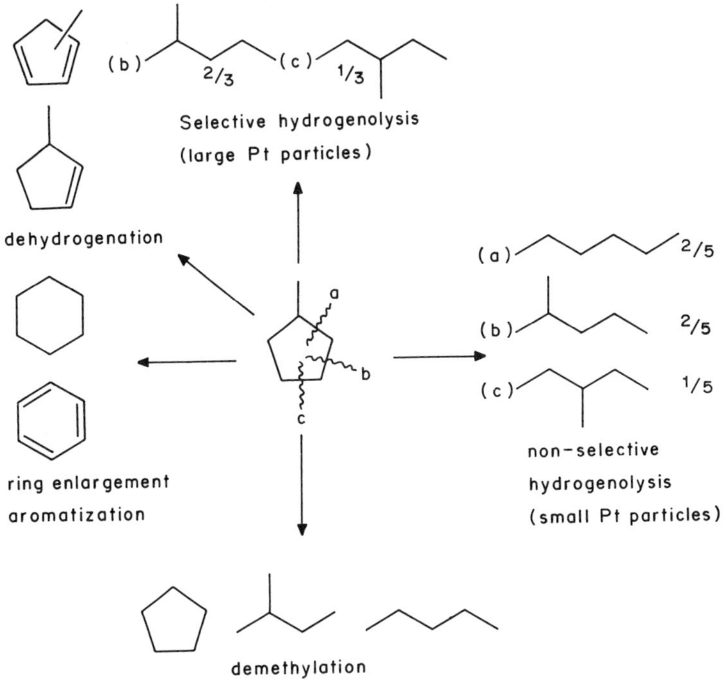

SCHEME 3. Skeletal rearrangements of methylcyclopentane on metallic catalysts

The skeletal rearrangement of methylcyclopentane on monometallic or intermetallic catalysts ($Pt^{24b}, CePd_3{}^{28}$) in a hydrogen atmosphere may lead to C—C bond rupture with formation of hexane and methylpentanes in a selective or non-selective manner depending on the particle size of the catalyst[24b]. Ring enlargement gives cyclohexane and benzene, accompanied by demethylation reactions to C_5-isomers (Scheme 3), dehydrogenation to methylcyclopentenes and methylcyclopentadienes, and successive

686 T. Oppenländer, G. Zang and W. Adam

formation of lower alkanes C_5—C_1. The intermetallic catalyst $CePd_3$ exhibits a very low activity[28] and the product distribution is different from the Pd catalysed reaction. However, the activation of $CePd_3$ with air at room temperature or at 120 °C increases the catalytic activity dramatically. This was attributed to the rearrangement of the surface generating catalytic active Pd sites. Thus, the air-activated $CePd_3$ catalyst shows the characteristic reactivity of Pd atoms in the skeletal rearrangements of methylcyclopentane, 2-methylpentane and 3-methylhexane[28].

The main goal of research in catalysis is to control the surface reactions by introducing a second metal into a supported monometallic catalyst[33]. For example, doping of Pt/Al_2O_3 with Sn gives the supported bimetallic catalyst Sn—Pt/Al_2O_3[33]. Using this catalyst, the aromatization of methylcyclopentane to benzene can be strongly suppressed compared to the undoped catalyst. The product distribution in the presence of hydrogen gas at 250 °C (methylcyclopentenes, methylcyclopentadienes, benzene) is also dependent on the type of catalyst used and the formation of ring-opened products (2-methylpentane, 3-methylpentane, hexane) under these conditions is negligible (less than 5%)[33].

On the other hand, the application of Pt black or Pt/SiO_2 as catalyst for the hydrogenolysis of the alkylcyclopentanes 1–5 leads mainly to ring-opened products and alkylcyclopentene formation, whereas aromatization is not significant. The product distribution is strongly dependent on the hydrogen gas pressure[27].

(1)   (2)   (3)   (4)   (5)

A Ni metal catalyst was recently used to produce benzene from methane under ultrahigh vacuum conditions[34a] (equation 1) by activation of CH_4 physisorbed on the Ni catalyst at 47 K by molecular beam techniques at pressures less than 10^{-4} torr. The selectivity of benzene formation under these conditions is 100% and the reaction proceeds via collision-induced dissociative chemisorption[34b]. The collision of Kr atoms with physisorbed CH_4 leads to dissociation into adsorbed H-atoms and methyl radicals; the latter dissociate to CH fragments which successively recombine to adsorbed C_2H_2. Trimerization of C_2H_2 on the metal surface yields quantitatively C_6H_6.

$$2CH_4 \xrightarrow[(Ni)]{-3H_2} [C_2H_2] \longrightarrow \bighexagon \qquad (1)$$

The aromatization of propane on supported (SiO_2/Al_2O_3) Ga, Zn or Pt catalysts is limited due to hydrocracking, dealkylation and hydrogen transfer reactions which lead to the formation of methane and ethane as the major products, and ethene and propene[35]. However, the use of a Pd membrane reactor[35] increases the yield of aromatics (e.g. benzene, toluene, C_8–C_{12} aromatics) dramatically by effective separation of the produced hydrogen gas from the reaction mixture by utilizing the H_2 permeability of Pd thin films.

An important field of industrial chemistry is the investigation of catalytic degradation methods of saturated polymers in view of recycling techniques. For example, the degradation of polypropylene on a silica–alumina catalyst[36] gives different product mixtures (gaseous, liquid and oligomeric fractions) which depend on the reaction temperature and the concentration of the catalyst.

2. Photochemistry

In contrast to the heterogeneously catalysed transformations of alkanes, the direct photochemical and radiolytic reactions with high-energy radiation in the vacuum UV are usually less selective. The reason for this non-selective photochemical behaviour is probably the excitation to different coupled excited states of the singlet manifold of the alkanes. Despite the high chemical stability of alkanes and cycloalkanes, the vacuum-UV photolysis or radiolysis of alkanes in the liquid or vapour phase leads to H_2 elimination, H radical detachment, elimination of low molecular mass alkanes and to decomposition into two-radical fragments in the primary photochemical step[37]. The final product distribution can be explained by the known reactivity of these primary radical fragments. Nevertheless, the frequency of C—H and C—C bond ruptures in alkanes by applying ionizing radiation is far from a statistical distribution and is directly correlated to the chemical structure of the substrates[38]. This structure effect in the radiolysis and photolysis of liquid alkanes is attributed to the reactivity of low-energy singlet excited states populated near the absorption onset. Therefore, branched alkanes like 2,2-dimethyl-propane and 2,2,4-trimethylpentane decompose mainly by C—C bond scission. This is in contrast to the decomposition of n-alkanes and C_5—C_{10} cycloalkanes, which yield mostly hydrogen gas and dehydrogenation products[38].

Extensive fluorescence studies[39] on alkanes reveal two decay channels for the S_1-state: the temperature independent S_1-T_n intersystem crossing (ISC) which can be accelerated in the presence of xenon (Xe heavy atom effect[40]), and the temperature-dependent singlet decomposition channel with an energy barrier of $E_A = 10$–$20\,kJ\,mol^{-1}$ [37] (equation 2). By investigation of the temperature dependence and the Xe effect on the quantum yields of the photodecomposition, the different contributions of S_1 versus T_1 reactivity can be assigned. On this basis, the primary decomposition pathways of liquid ethylcyclohexane photolysis at 7.6 eV (163 nm) can be interpreted as follows: H_2 elimination occurs from the S_1 reaction channel and is the main process, whereas H-atom detachment and ring-opening processes are a consequence of intersystem crossing to the triplet state of ethylcyclohexane. Consequently, low-temperature photolysis of hydrocarbons in solid xenon[41] reveals the formation of caged H atoms or thermally unstable Xe_nH molecules. Table 1 summarizes the quantum yields of hydrogen gas production (ΦH_2) during the photolysis of several alkanes and cycloalkanes in the liquid and/or vapour phase, which demonstrate the structural influence[42-45].

$$\text{alkane} \xrightarrow{hv} [\text{alkane}]^{S_1} \xrightarrow{E_A} \text{singlet decomposition}$$

$$\text{ISC} \downarrow \qquad\qquad\qquad\qquad (2)$$

$$[\text{alkane}]^{T_n} \longrightarrow \text{triplet decomposition}$$

A selective coupling of liquid cycloalkanes to the corresponding bicycloalkyls by 193 nm laser irradiation[46a] was recently described to proceed via homolytic C—H bond cleavage (equation 3). No olefinic products could be detected and the reaction is very inefficient at 185-nm irradiation.

$$n = 1, 2, 3, 4$$

This wavelength dependence is rationalized in terms of a one-photon absorption process at the absorption edge of the cycloalkanes (excitation to the lowest excited singlet

688 T. Oppenländer, G. Zang and W. Adam

TABLE 1. Photolysis at 7.6, 8.4 and 10.0 eV of several alkanes and cycloalkanes and the quantum yields of hydrogen gas production (ΦH_2)

Molecule	Wavelength of irradiation (nm eV^{-1})	Phase	ΦH_2	Reference
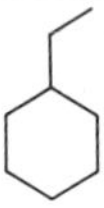	163 (7.6)	liquid	0.73	37
	163 (7.6)	liquid	0.86	38a
	147 (8.4)	vapour	0.74	42
	147 (8.4)	liquid	1.0	44
	123.6 (10.0)	vapour	0.57	42
	123.6 (10.0)	liquid	1.0	44
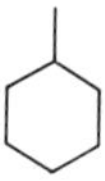	163 (7.6)	liquid	0.77	43
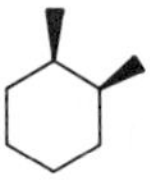	163 (7.6)	liquid	0.46	
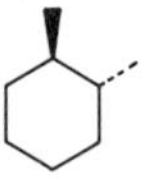	163 (7.6)	liquid	0.39	
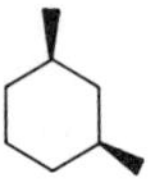	163 (7.6)	liquid	0.59	
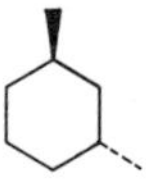	163 (7.6)	liquid	0.66	
	163 (7.6)	liquid	0.61	
	163 (7.6)	liquid	0.56	
	163 (7.6)	liquid	0.93	45
	147 (8.4)		0.73	
	123.6 (10.0)		0.65	
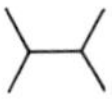	147 (8.4)	vapour	0.22	38b
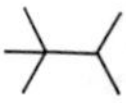	147 (8.4)	vapour	0.44	
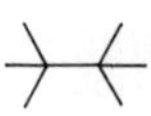	147 (8.4)	vapour	1.1	38c

state), and successive generation of the radical intermediates in high concentration by laser irradiation which favour the intermolecular dimerization. The carbene route to dicyclohexyl was unequivocally excluded by conducting a competition experiment with a 1:1 mixture of C_6H_{12}/C_6D_{12}. No formation of the intermolecular C—H insertion products $C_{12}D_{12}H_{10}$ and $C_{12}D_{10}H_{12}$ was detected by GC/mass spectrometry. This photochemical behaviour clearly contrasts the results obtained with higher-energy UV radiation[44,47], which probably proceeds through carbene formation (equation 4) to yield significant amounts of cyclohexene[47]. The combined yield of photodimers of n-pentane[46b] on 193-nm laser excitation (equation 5) is similar to that of cyclohexane (equation 3). The relative reactivity of the cycloalkanes compared to n-pentane (primary C-H coupling $= 1.0$) shows a significant dependence on ring size (Table 2)[46c].

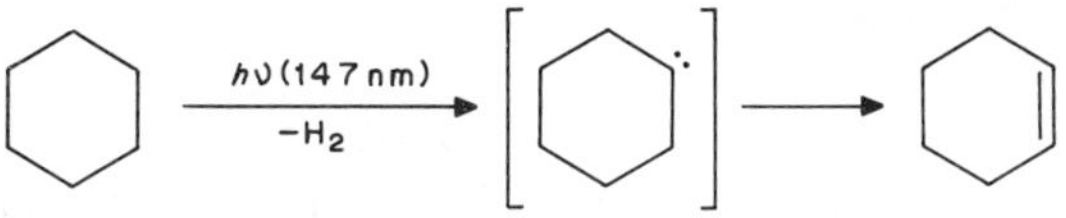

(4)

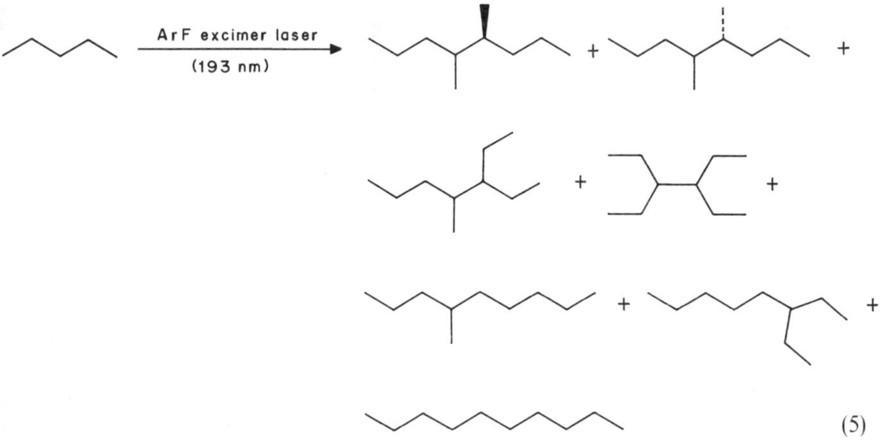

(5)

Dehydrodimerization of cyclohexane to bicyclohexyl was also achieved by mercury photosensitization[48]. However, Hg-, Cd- and Zn-photosensitized reactions of acyclic

TABLE 2. Relative photoreactivity of cycloalkanes C_nH_{2n} and n-pentane at $193\,nm$[46c]

Cycloalkane C_nH_{2n}	Relative rate of dimerization
$n = 5$	3.0
$n = 6$	3.8
$n = 7$	2.0
$n = 8$	1.6
n-pentane: $(CH)_{sec}$	5.4
$(CH)_{prim}$	1.0

alkanes (e.g. propane) are usually unselective and lead to complex mixtures of decomposition products[48c].

Carbene formation during vacuum-UV photolysis (7.6 eV) and radiolysis of cyclooctane by unimolecular H_2 elimination from a single carbon atom (1,1-elimination) was shown to be the main reaction channel of excited cyclooctane (equation 6)[49]. The carbene 6 is formed with an efficiency close to 100% in the long-wavelength photolysis of the sodium salt of the p-tosylhydrazone 7 as well as in the vacuum-UV photolysis and radiolysis of cyclooctane. This was deduced from the similar product distribution of both reactions to give cyclooctene (8), bicyclo[3.3.0]octane (9) and bicyclo[5.1.0]octane (10)[49].

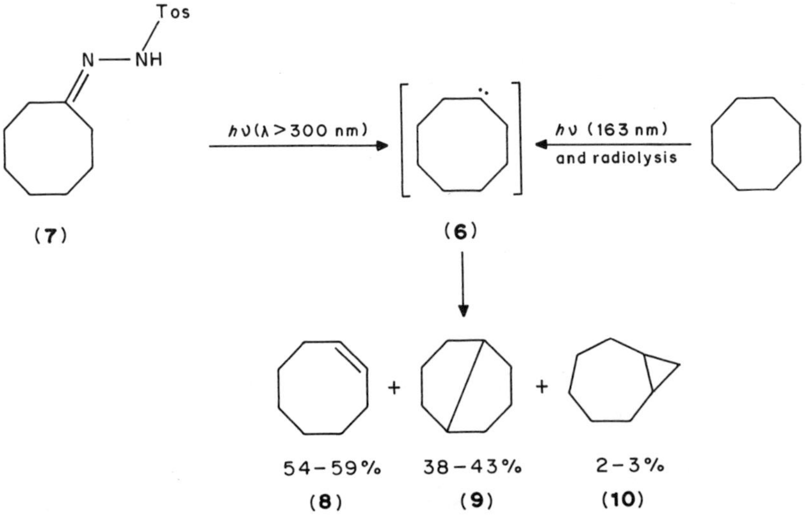

(6)

$$54-59\% \qquad 38-43\% \qquad 2-3\%$$
$$(8) \qquad (9) \qquad (10)$$

The interesting phenomenon of wavelength-dependent selectivity (cf. equations 3, 4, 6) in the vacuum-UV photolysis of organic substrates was also noticed by Inoue and coworkers[50] in the UV photolysis of 2,3-dimethylbutene. The product distribution is strongly dependent on the excitation wavelength over the narrow range of 185 nm to 228 nm.

Novel industrial applications may evolve from continuous wave (CW) CO_2 laser-driven homogeneous pyrolysis of hydrocarbons[51] (e.g. C_4H_{10}[51b], n-heptane[51c]), which seems to be advantageous over conventional pyrolysis methods[51]. For example, the CW CO_2 laser SF_6-sensitized decomposition of cyclohexane in the gas phase yields mainly ethene and 1,3-butadiene (equation 7). The product distribution is almost invariable with the extent of conversion and different from conventional pyrolysis[51a]. No acetylene is formed

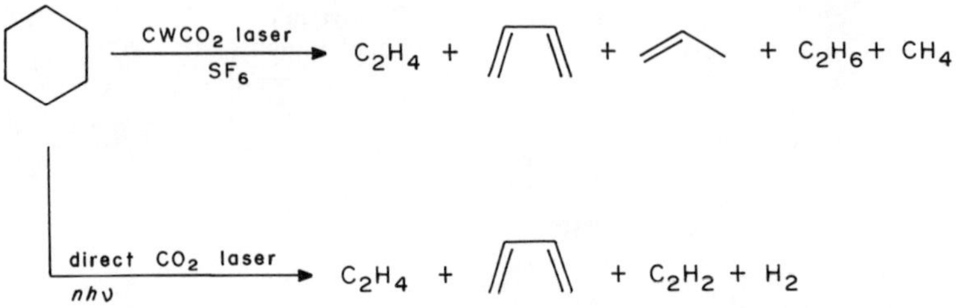

(7)

due to the absence of reactor surface effects. On the other hand, direct vibrational multi-photon ($nh\nu$) excitation of cyclohexane by using a focussed high-energy pulsed CO_2 laser (100 mJ to 1.5 J) leads to H_2, ethene and acetylene as the major products[52]; the latter results from fragmentation of primarily formed butadiene (equation 7). Under these conditions a bright luminescence is observed, indicating that high concentrations of C_2 radicals and electronically excited hydrogen atoms are present[52].

3. The adamantane rearrangement

The Lewis-acid-catalysed rearrangement of polycyclic hydrocarbons C_nH_{2n-4x} ($x = 1, 2, 3, \ldots$) into their thermodynamically more stable diamond-like isomers is usually referred to as the *adamantane rearrangement*[2a,53]. The archetype of this unique isomerization is the $AlCl_3$-catalyzed transformation of *endo*-tricyclo[5.2.1.0^{2,6}]decane (11) (tetrahydrodicyclopentadiene) into adamantane (Ad-H) (Scheme 4), which was first discovered by Schleyer[54] more than 30 years ago. This process is thermodynamically controlled and it involves carbenium ion intermediates which rearrange via successive 1,2-C bond and 1,3 hydride shifts. Numerous theoretical and experimental investigations[55,56] have established the most direct thermodynamically feasible pathway (*exo*-11 → 12 → 13 → 14 → 15 → Ad-H) for the isomerization of *endo*-11 to adamantane. This pathway is defined by the complex rearrangement graph for the interconversion of the 19 tricyclic $C_{10}H_{16}$ isomers[56]. Furthermore, the prediction of the most stable C_nH_m isomer, the so-called stabilomer[57], based on force field calculations, is consistent with experimental results for AlX_3 isomerizations. The identification of intermediates in the rearrangement of tetrahydrodicyclopentadiene (11) to adamantane is extremely difficult because of the extraordinary thermodynamic stability of adamantane, which constitutes an energy bottleneck (*endo*-11 → 12 → Ad-H)[56b]. Nevertheless, a recent renaissance[55] of research in this field led to substantial information which concerns the isomeric intermediates and the mechanisms of the $C_{10}H_{16}$ isomerizations.

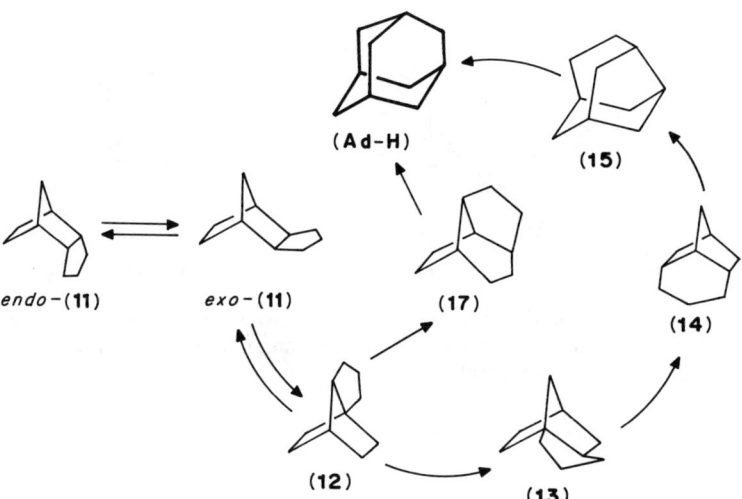

SCHEME 4. The most direct pathways in the adamantane rearrangement of tricyclo[5.2.1.0^{2,6}]decane (11)

Thus, the nitric acid catalyzed rearrangement of *endo*-**11** undoubtedly leads to the formation of the tricyclo[4.2.2.0^{1,5}]decane skeleton. This was shown by the isolation of its corresponding dinitrate **16**[58] (equation 8). Reaction 8 proves that the first stage of the adamantane rearrangement of *endo*-**11** proceeds through the intermediate **12** (Scheme 4). On the other hand, tricyclo[4.2.2.0^{1,5}]decane (**12**) itself reacts in the presence of AlBr$_3$ in CS$_2$ at 0 °C (30 min) to adamantane, *exo*-**11** and tricyclo[5.3.0.0^{4,8}]decane (**17**)[59] (equation 9). This reaction represents an entirely new pathway (*exo*-**11** → **12** → **17** → Ad-H, Scheme 4) for the formation of adamantane from **11**, which was not predicted by theoretical considerations[56]. Novel insights into the complex Lewis-acid-catalysed isomerization of the 1,2-trimethylenenorbornanes *exo*-**13** and *endo*-**13** (tricyclo[5.2.1.0^{1,5}]decanes) were achieved recently by deuterium- and ^{13}C-labelling experiments[55o,p,q]. Treatment of deuterium-labelled *exo*-**13** and *endo*-**13** with AlBr$_3$ in CS$_2$ at −60 °C for 5–15 min leads to specifically D-labelled 2-*endo*, 6-*endo*-trimethylenenorbornane **18** (tricyclo-[5.2.1.0^{3,8}]decane), which indicates the involvement of the degenerate rearrangement A ⇌ B (Scheme 5).

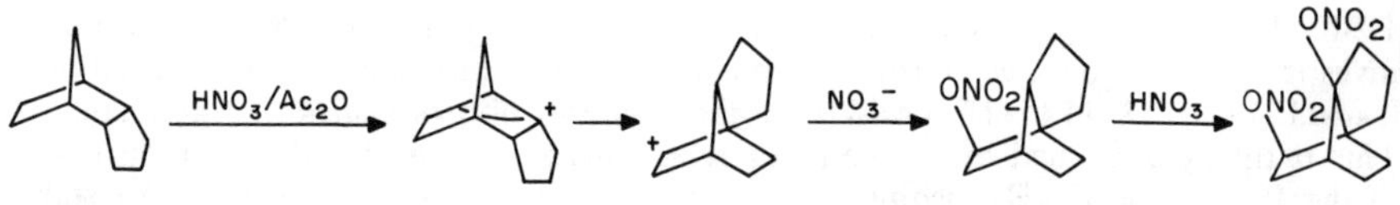

(8)

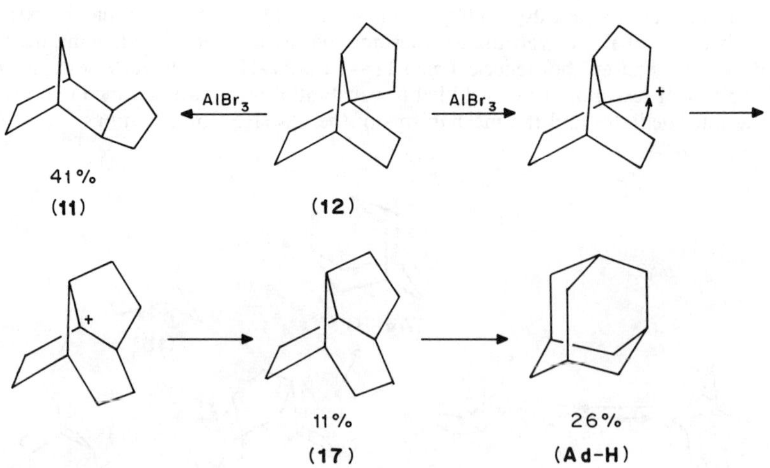

(9)

Further information on the multiple pathways of isomerizations in the 'Adamantane panorama' may be gathered from Table 3, which summarizes the known rearrangements of the C$_{10}$H$_{16}$ isomers of adamantane[54–56,58–61].

B. Strained Cycloalkanes

1. Cyclopropanes

a. Transition metal catalysed rearrangements. The Ag$^+$ ion-catalysed rearrangement of strained polycyclic molecules was first recognized in the late 1960s[62] and turned out

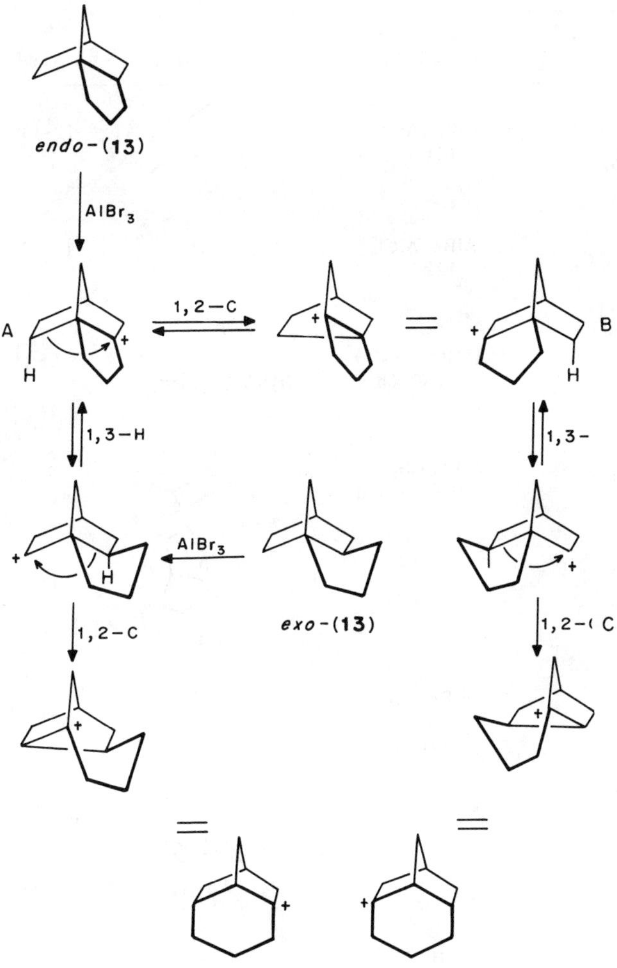

SCHEME 5. Degenerate rearrangement in the $AlBr_3$-catalysed isomerization of the 1,2-trimethylenenorbornanes *exo*- and *endo*-13 (1,2-C, 1,3-H shifts)[55o,p,q]

to be a powerful synthetic tool for obtaining a wide range of interesting carbocyclic structures[63]. For example, the synthesis of 7,7-dimethyl- and 7,7-diphenylcyclohepta-triene (19) was made readily accessible by reaction of the tetracyclo[4.1.0.0^{2,4}.0^{3,5}]-heptanes 20 with $AgClO_4$ (equation 10)[64]. This methodology utilizes benzvalene as a benzene equivalent and is superior to conventional syntheses of a variety of 7-mono- and 7-disubstituted 1,3,5-cycloheptatrienes.

Interestingly, the dehydroprotoadamantanes 21 and 22 rearrange on treatment with $AgClO_4$ in boiling benzene to the tricyclic olefins 23–25, but with different product selectivities (equation 11)[65]. Despite the fact that bicyclo[2.1.0]pentane itself is inert to

TABLE 3. The 'Adamantane panorama', which covers rearrangements of tricyclic $C_{10}H_{16}$ isomers of adamantane (**Ad-H**)

Structure	Reaction conditions	Products	Yield	Reference
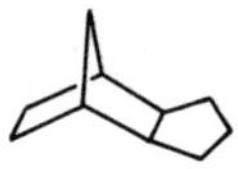	$AlBr_3/AlCl_3$ 12 h	 (**Ad-H**)	12–13%	54 55 j, r, s
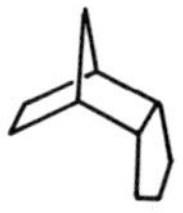	$AlBr_3/AlCl_3$ 12 h	**Ad-H**	12–13%	54, 55 j
	HNO_3/Ac_2O 10 °C, 4 h	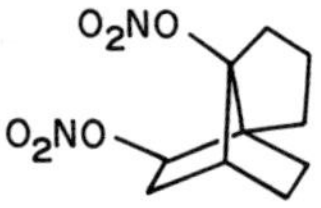 O_2NO O_2NO	65%	58
	$AlBr_3/CS_2$ 0 °C, 60 min	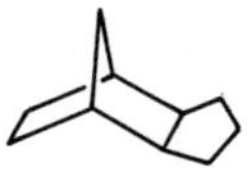	41%	59 a, b, c
		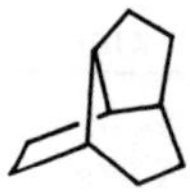	11%	
		Ad-H	26%	
	$AlBr_3/CS_2$ −60 °C, 30 min		10%	55 g, o, p 56 b
		(**Ad-H**)	not stated not stated	
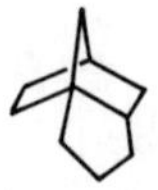	$AlBr_3/CS_2$ −60 °C, 1 min	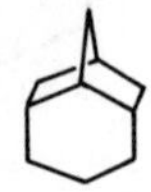	80%	55 f, g, p, q
	$AlBr_3/CS_2$ −10 °C	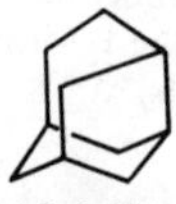	not stated	56 b
		(**Ad-H**)	not stated	
	$AlBr_3/CS_2$ 0 °C	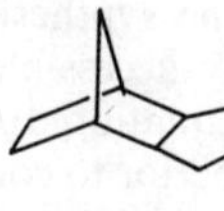	>90%	55 l
	$AlBr_3/CS_2$	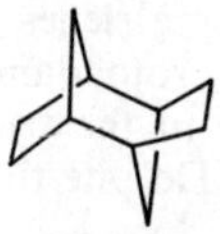	>90%	55 l

TABLE 3. (*continued*)

Structure	Reaction conditions	Products	Yield	Reference
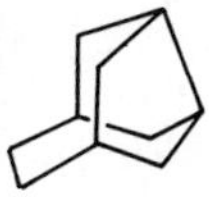	AlBr$_3$/CS$_2$	**Ad-H**	53%	55 h
	AlCl$_3$/CH$_2$Cl$_2$	**Ad-H**	90%	59 b
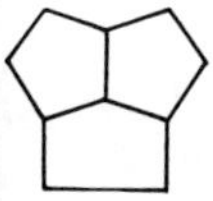	AlCl$_3$/n-hexane reflux, 45 h	**Ad-H**	91%	55 i
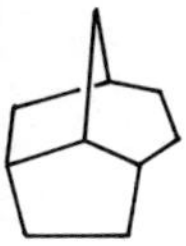	AlX$_3$	**Ad-H**	not stated	61
	AlBr$_3$/CS$_2$ −70 °C, 5 min	**Ad-H**	not stated not stated	55 n
	AlBr$_3$/CS$_2$ −75 °C	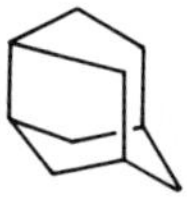 **Ad-H**	10% 60%	55 m
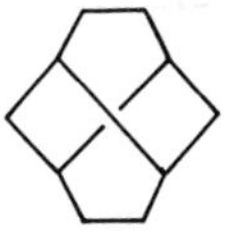	AlBr$_3$/AlCl$_3$ 70 °C, n-hexane, 2 h	**Ad-H**	85%	56 a, 60
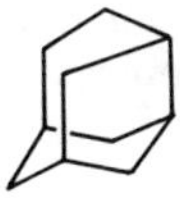	AlBr$_3$/CS$_2$ 25 °C	**Ad-H**	not stated	55 e
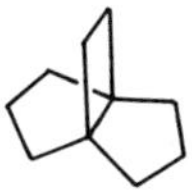	AlCl$_3$/CH$_2$Cl$_2$ 15 °C/10 h	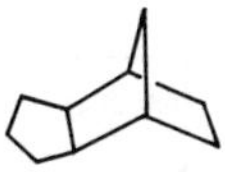 **Ad-H**	traces not stated	59 b

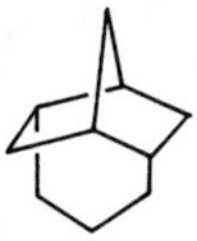

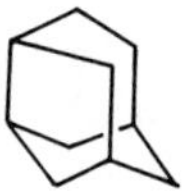

the Ag^+-induced rearrangement[66], the strained pentacyclic compound **26**, which contains a bicyclo[2.1.0]pentane structural element, leads to tetracyclo[$5.3.1.0^{2,6}.0^{3,9}$]undec-4-ene (**27**) in 85% yield (equation 12)[66a]. This result was attributed to the favourable geometry of the strained central bond in **26**, which promotes carbocationic skeletal rearrangement, and to the stabilizing interaction between silver and the rearranged carbocation **28**[66a].

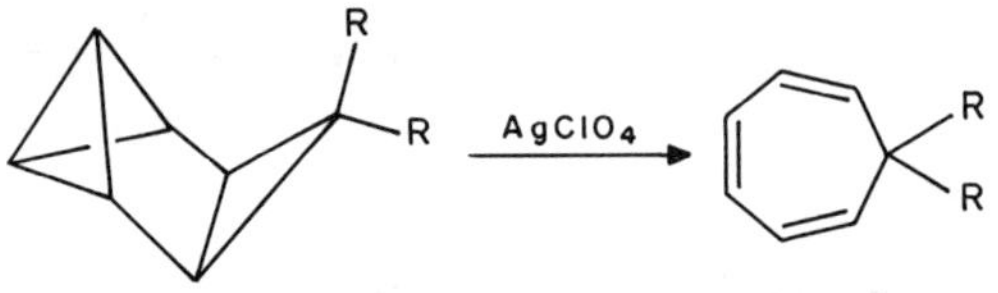

$$\text{(10)}$$

(**20a**) R = Me

(**20b**) R = Ph

(**19**)

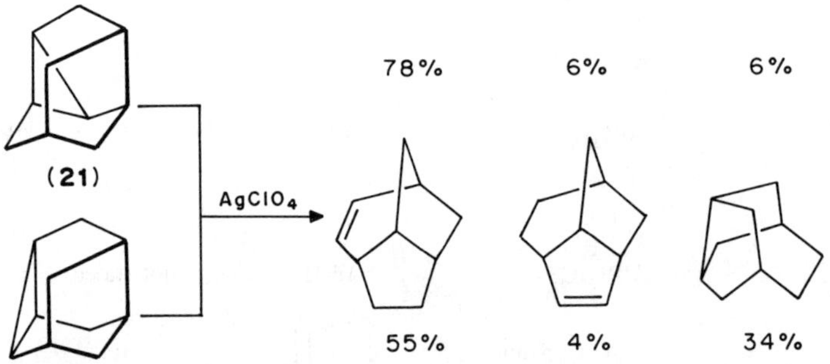

$$\text{(11)}$$

(**21**)

(**22**)

78%   6%   6%

55%   4%   34%

(**23**)   (**24**)   (**25**)

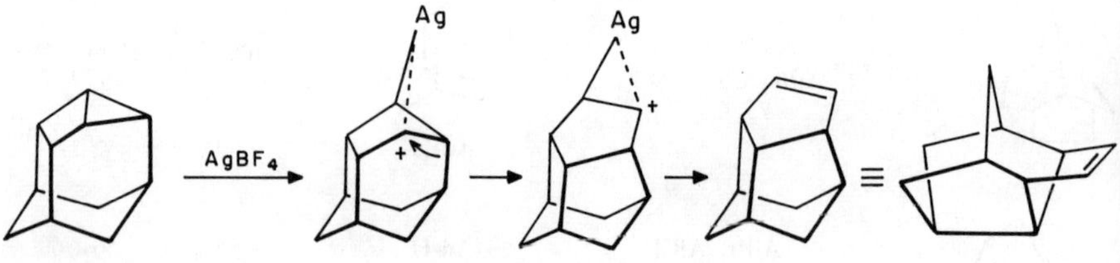

(**26**)   (**28**)   (**27**)

$$\text{(12)}$$

A mechanism was proposed which engages metal-stabilized carbenium ion intermediates like **29** and **30** in the rearrangement of alkyl-substituted cyclopropanes induced by the olefin metathesis catalysts $WCl_6/SnPh_4$ and $PhWCl_3/EtAlCl_2$[67].

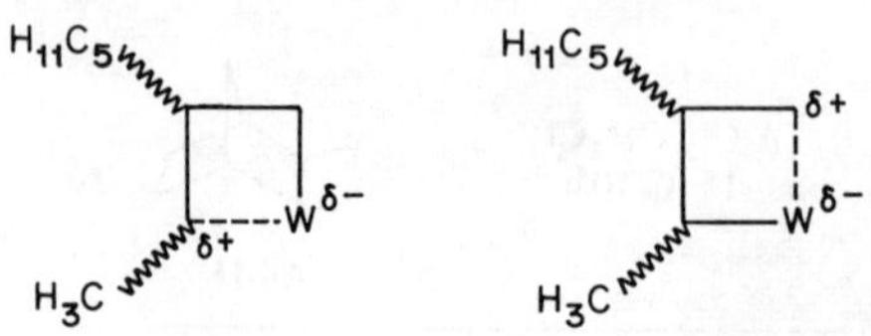

(**29**)   (**30**)

TABLE 4. Ring-opening reactions of alkyl substituted cyclopropanes with the catalysts $WCl_6/SnPh_4$ and $PhWCl_3/EtAlCl_2$.

Cyclopropane	Catalyst	Products	Yield
(cyclopropane with n-C_6H_{13})	$WCl_6/SnPh_4$	(alkene products)	$\sum$ 47% [a]
(CH_3, n-C_5H_{11} cyclopropane)	$WCl_4/SnPh_4$	(alkene products)	$\sum$ 54% [b]
(CH_3, n-C_5H_{11} cyclopropane)		(alkene products)	
(dimethyl cyclopropane)	$WCl_6/SnPh_4$	(alkene product)	60%
(bicyclic cyclopropane)	$WCl_6/SnPh_4$	(cyclohexene product)	8% [c]
(cyclopropane with n-C_6H_{13})	$PhWCl_3/EtAlCl_2$	(alkene products)	$\sum$ 19% [d]
(bicyclic cyclopropane)	$WCl_6/SnPh_4$	(cyclooctene products)	not stated
		(CH_3 substituted product)	not stated

Conversion of the cyclopropane: [a]68%; [b]98%; 45%; [d]91%.

The products of several alkyl-substituted cyclopropanes with these electrophilic catalysts are compiled in Table 4[67].

Interestingly, Z- and E-non-3-ene and Z- and E-non-2-ene could not be identified as reaction products of Z- and E-1-methyl-2-n-pentylcyclopropane. These products would correspond to the fission of the most substituted central σ-bond of the cyclopropane. Furthermore, methylene–carbene elimination leading to the lower homologous alkenes seems to be an unfavourable process with those catalysts. This is in contrast to the behaviour of bicyclo[2.1.0]pentane, which reacts in the presence of PhWCl$_3$/EtAlCl$_2$ to cyclobutene in 70% yield[67d,e].

The reactions of Pd^{2+} with cyclopropanes[68] lead to skeletal rearrangements and appear to be initiated by the heterolytic cleavage of the most substituted C—C σ-bonds of the cyclopropane moiety (equation 13). Two cleavage modes were observed which lead in the primary step to the formation of tertiary carbocations followed by successive deprotonation to the η^3-allyl complex 31 as the major pathway.

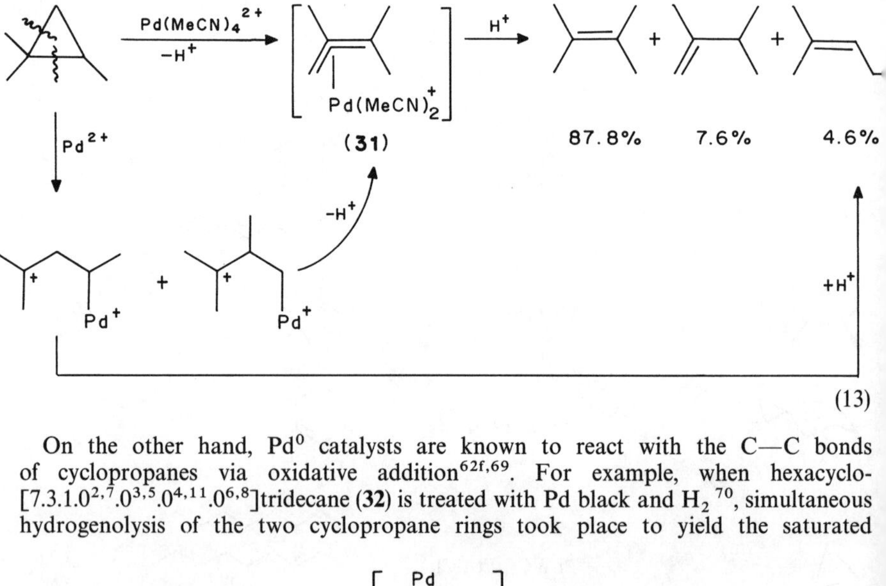

(13)

On the other hand, Pd^0 catalysts are known to react with the C—C bonds of cyclopropanes via oxidative addition[62f,69]. For example, when hexacyclo-[7.3.1.0^{2,7}.0^{3,5}.0^{4,11}.0^{6,8}]tridecane (32) is treated with Pd black and H$_2$[70], simultaneous hydrogenolysis of the two cyclopropane rings took place to yield the saturated

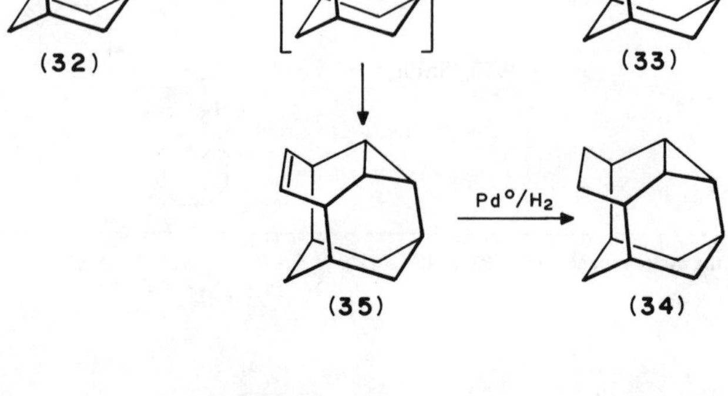

(14)

hydrocarbon 33 accompanied by the formation of 34 (equation 14). The reaction is initiated by complexation of Pd^0 to the two strained σ-bonds of the cyclopropane rings. Hydrogenation of this Pd complex before valence isomerization to the olefin 36 produces the saturated hydrocarbon tetracyclo[6.3.1.1^{2,7}.0^{4,10}]tridecane (33).

However, the Ir(I)-catalysed rearrangement of exo- and endo-tricyclo[3.2.1.0^{2,4}]-octane[71] (36) preferentially leads to the conversion of the cyclopropane moiety into an exocyclic methylene group (equation 15). The different product distribution of exo- or endo-36 was rationalized in terms of the formation of iridacyclobutane- and iridacarbene-olefin intermediates.

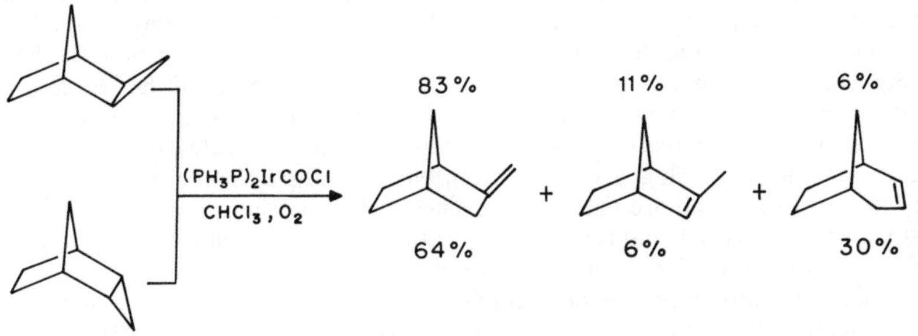

Similarly, a tetrameric platina(IV)-cyclobutane intermediate 37 was postulated in the ring homologation reaction of tetracyclo[3.3.1.0^{2,4}.0^{6,8}]nonane (38), which on treatment with Zeise's dimer $(C_2H_4PtCl_2)_2$ led cleanly to the alkene 39[72] (equation 16). The mechanism of this rearrangement was established by studies with deuterium and ^{13}C labelling experiments.

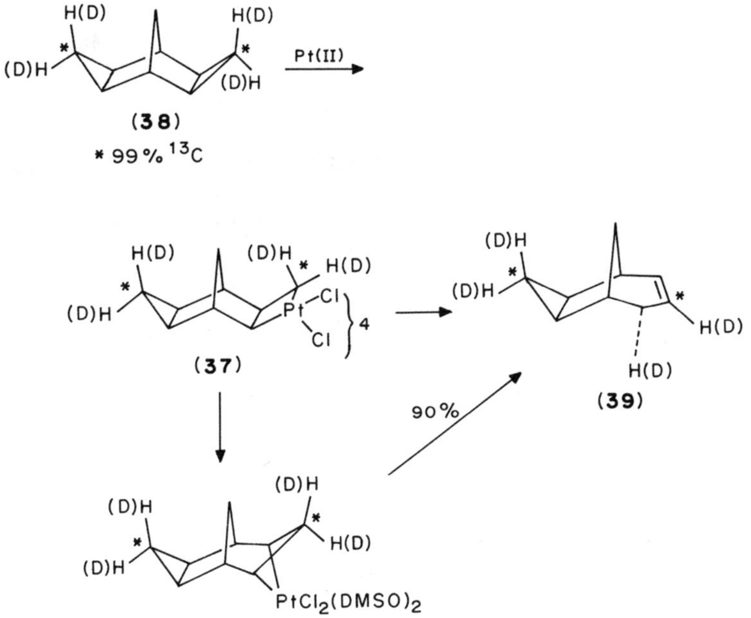

b. Photochemistry. The photochemistry of aryl- and vinylcyclopropanes was reviewed in 1979 by Hixson[73] and is characterized by the interaction of the substituent with the cyclopropyl group. On the other hand, the radiation chemistry of cyclopropane and its alkyl substituted derivatives[74] is triggered by direct ionization processes. The direct 185-nm photolysis of cyclopropanes in solution, however, seems to be a consequence of Rydberg-type excitations[18b]. The photochemistry of simple cyclopropanes in the vapour phase at 147, 163, 174 and 185 nm and their 185-nm photoreactivity in solution have been extensively reviewed recently by Steinmetz[18a] and Adam and Oppenländer[18b]. The vapour-phase photochemistry of cyclopropanes below 185 nm is dominated by carbene elimination and free radical fragmentation processes[18a], whereas the 185-nm photochemistry of cyclopropanes in solution is characterized by the competition between one- and consecutive two-bond cleavage modes of the cyclopropane moiety, the latter leading to carbene elimination[18b]. Some representative examples of the 185-nm photoreactivity of cyclopropanes in solution are listed in Table 5, which demonstrates the difference in the product selectivity compared to ground state catalytic transformations of cyclopropanes (cf. Table 4; Section II.B.1.a). Novel results of the 185-nm photoreactivity of cyclopropanes, which are part of a bicyclic or tricyclic ring system, are discussed in Section II.B.3 and II.B.4, and reveal radical cationic or zwitterionic intermediates, which might be formed via σ,3s Rydberg excitation[19].

The decomposition modes of polyatomic molecules under the influence of high-power infrared laser pulses[75] have been under active investigation (cf. Section II.A.2) and can at least qualitatively be rationalized in terms of 'hot' intermediates, which influence the product distribution[76]. For example, the infrared multiphoton (n$h\nu$) decomposition of bicyclopropyl in the gas phase by CO_2 laser irradiation[76] produces a complex reaction mixture which consists mainly of C_6H_{10} isomers (equation 17). All the products were

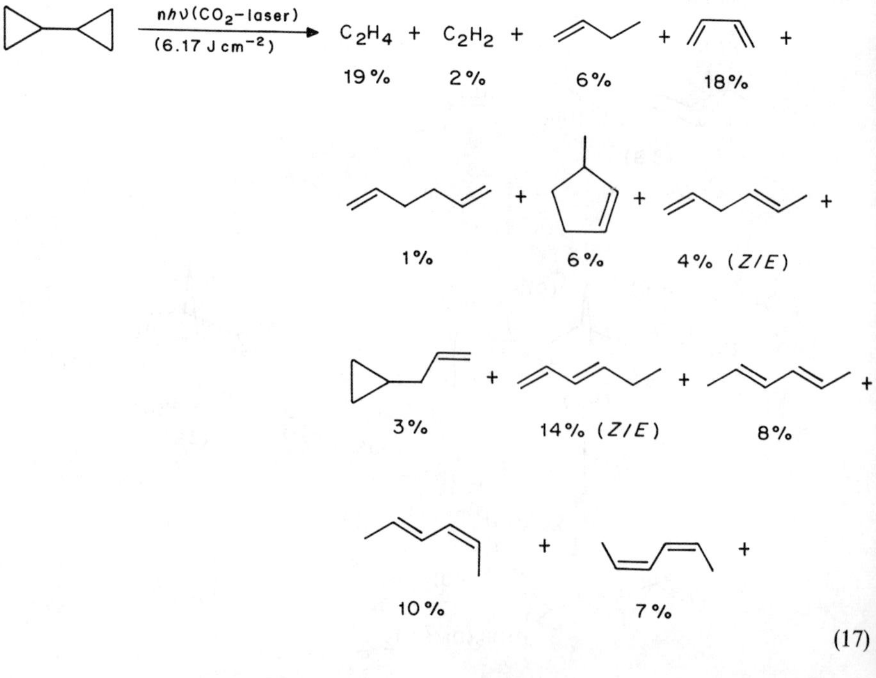

(17)

TABLE 5. 185-nm photoreactivity of some alkylcyclopropanes in solution[a]

Cyclopropane	Products	Yield (%)
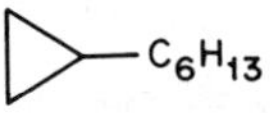 (cyclopropyl)–C$_6$H$_{13}$	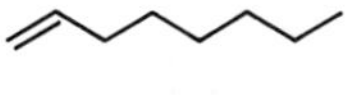 [CH$_2$=CH–(CH$_2$)$_6$–CH$_3$ terminal alkene]	13
	 [internal alkene]	17
	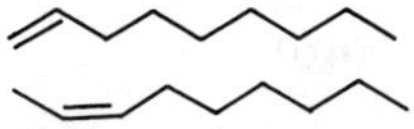 [longer alkene chain]	7
	 [branched alkene]	4
	C$_8$H$_{16}$	17
	C$_9$H$_{18}$	9
 [1,1,2-trimethylcyclopropane type structure]	 [CH$_2$=CH–CH(CH$_3$)–]	26
	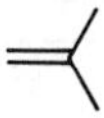 [CH$_2$=C(CH$_3$)–CH$_2$–]	24
	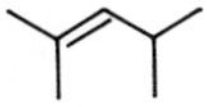 [(CH$_3$)$_2$C=CH–CH(CH$_3$)–]	23
	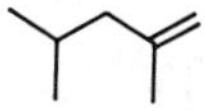 [(CH$_3$)$_2$CH–CH=C(CH$_3$)–]	27
 [spiro[2.5]octane]	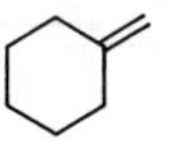 [methylenecyclohexane]	52
	 [ethylidenecyclohexene]	7
	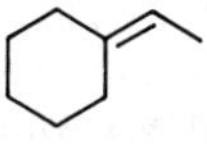 [ethylidenecyclohexane]	14
	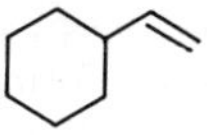 [vinylcyclohexane]	13
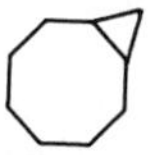 [bicyclo[5.1.0]octane type]	 [1,8-nonadiene type diene]	83
	 [cyclononene] (Z/E)	17
 [1,1,2-trimethylcyclopropane]	 [CH$_2$=CH–CH(CH$_3$)–]	29
	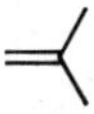 [CH$_2$=C(CH$_3$)–]	32
	C$_2$H$_4$	29
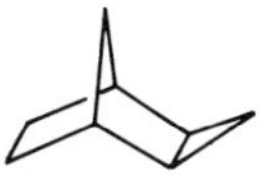 [tricyclic norbornane-fused cyclopropane]	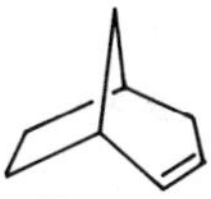 [norbornene derivative]	0.74[b]
	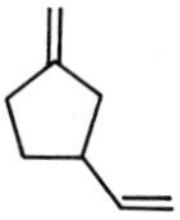 [methylenecyclopentane with vinyl]	0.35[b]

[a] Taken from Reference 18b.
[b] Quantum yield of product formation.

postulated to arise from the primary ring-opening of one cyclopropyl moiety with formation of a 1,3-diradical **40** (Scheme 6).

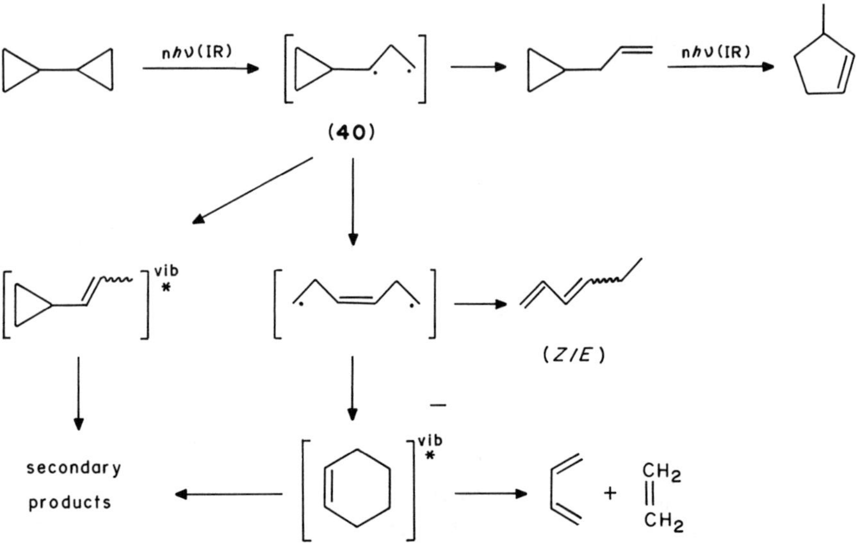

SCHEME 6. Infrared multiphoton dissociation of bicyclopropyl

Consecutive 1,2-hydrogen shifts lead to allylcyclopropane which rearranges on secondary irradiation to 3-methylcyclopentene, and to vibrationally excited 1-cyclopropylpropene. This 'hot' molecule decomposes with formation of secondary products. Furthermore, a 'hot' cyclohexene molecule was postulated to be generated via cyclopropyl–carbinyl rearrangement of the initially formed 1,3-diradical **40** to a 1,6-diradical and successive ring closure. The vibrationally 'hot' cyclohexene can undergo a *retro*-Diels–Alder reaction to produce 1,3-butadiene and ethene. A similar 'hot molecule effect' was discussed in the 185-nm photochemistry of olefins and ethers in solution[77]. On IR laser irradiation both cyclopropane moieties cleave in a statistical manner as was demonstrated by irradiation of bicyclopropyl-d_4 specifically deuterated at one ring[75b] (equation 18). From the deuterium distribution in the single-ring cleaved product

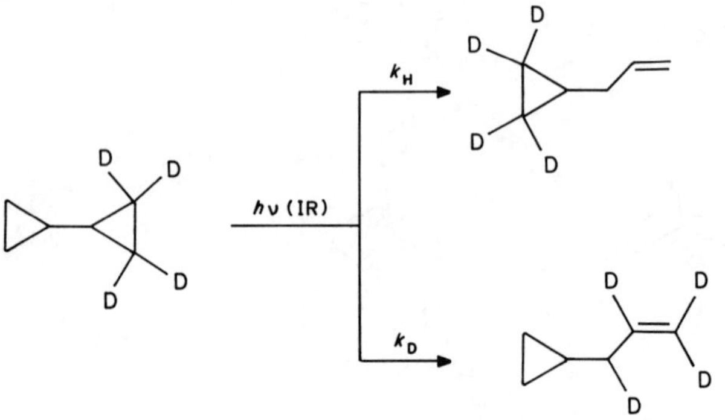

(18)

(allylcyclopropane-d_4) a branching ratio of $k_H/k_D = 1.03 \pm 0.08$ was determined. This is in agreement with the statistical prediction of random reactivity in the multiphoton infrared photochemistry of bicyclopropyl[75,76].

The SF_6-sensitized decomposition of spirohexane[78] with a continuous wave CO_2 laser (4–6 W), however, is much more selective and produces ethene, 1,3-butadiene and methylenecyclopropane and, to a minor amount, methylenecyclopentane at higher laser output (8 W) (equation 19). This indicates that the initial cleavage of the cyclobutene ring to the 1,4-diradical **41** arises from a higher-energy process. Cyclobutylidene **42** as

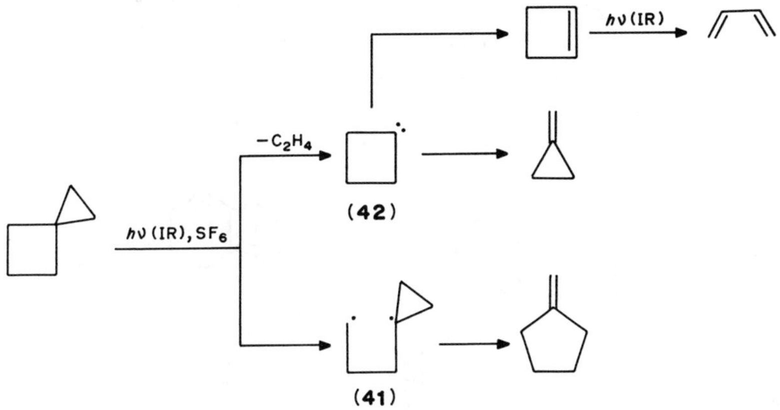

(19)

a common precursor of methylenecyclopropane and cyclobutene was also proposed in the 185-nm photochemistry of cyclobutene[79] to produce ethene, acetylene, methylenecyclopropane and 1,3-butadiene.

Direct photolysis of the unique tetramethylene-bridged prismane **43**[80] (pentacyclo[4.3.1.0^{1,6}.0^{7,9}.0^{8,10}]decane) leads cleanly to the expected tetralin **44** (equation 20).

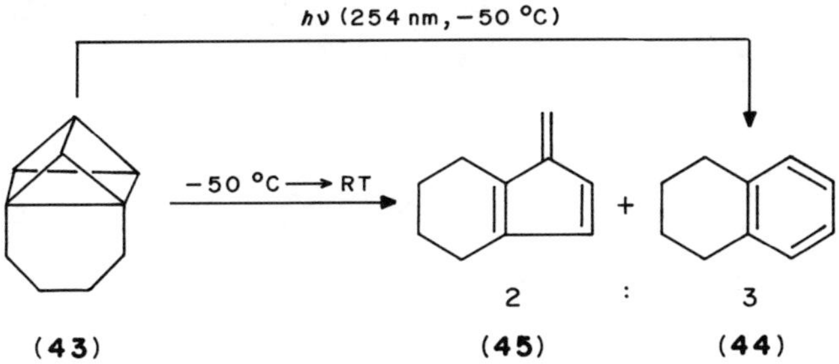

(20)

This photochemical behaviour is in accord with the well-known photoreactivity of prismane itself, which on direct excitation gives benzene[81] as the sole product. However, the direct photolysis of **43** at low temperature is accompanied by its thermal rearrangement to the fulvene **45**. The mechanism of this novel isomerization was established by deuterium labelling experiments[80] (Scheme 7) and was proposed to proceed through the intermediary benzvalene derivatives **46** and **47**. The products d-**44** and d-**45** exhibit less than 4% D scrambling.

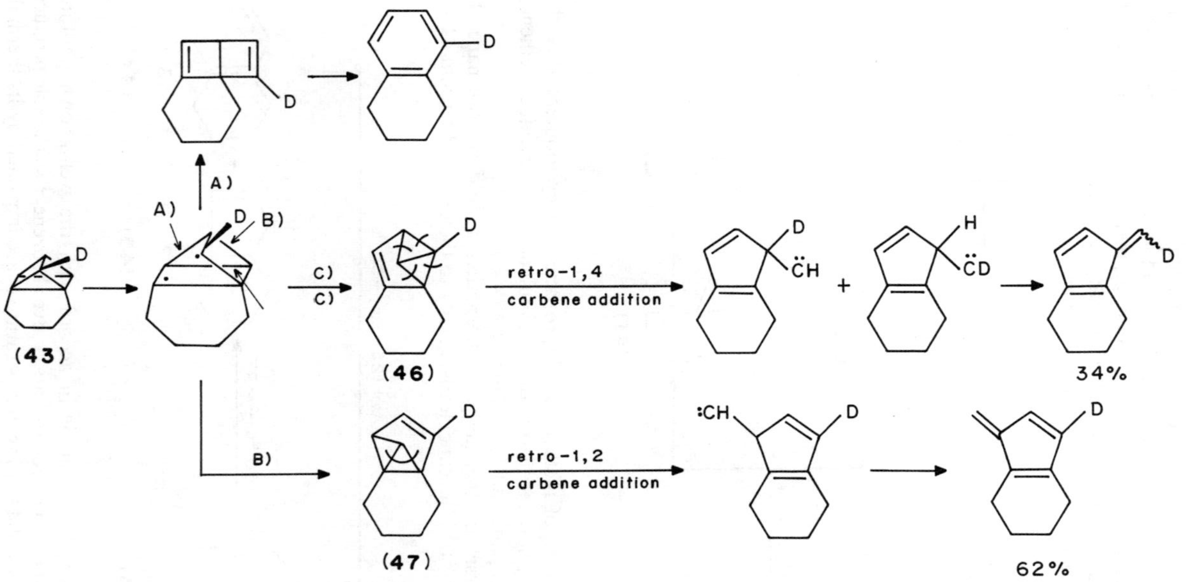

704

SCHEME 7. Proposed mechanism for the thermal rearrangement of penacyclo[4.3.1.0$^{1.6}$.10$^{7.9}$.0$^{8.10}$]decane (43)[80]

c. Quadricyclane/norbornadiene interconversion. The isomerization of norbornadiene (**NBD**: bicyclo[2.2.1]hept-2,5-diene) to quadricyclane (**Q**: tetracyclo[3.2.0.^{2,7}.0^{4,6}]heptane, equation 21) still receives much attention[82,83] as a potential solar-energy storage process.

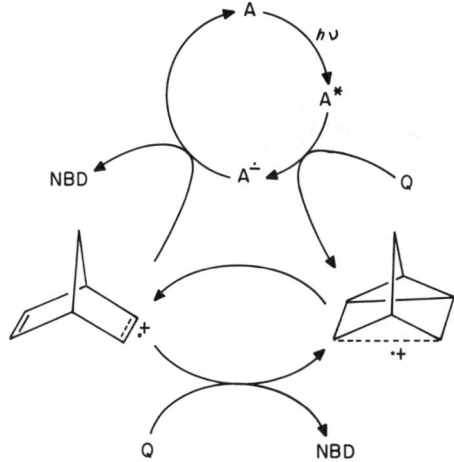

$$\text{(NBD)} \quad \xrightleftharpoons[\text{cat.}]{h\nu(\text{sens.})} \quad \text{(Q)} \qquad (21)$$

Unfortunately, **NBD** does not absorb in the visible and UV-A (400–320 nm) range of the solar spectrum. This problem may be overcome by the use of a wide variety of photosensitizers, which includes transition metal complexes[83b] and organic sensitizers like acridones[83a] and aromatic ketones[82,83c,d,e]. In this section, however, new trends in the exothermic back-isomerization **Q → NBD** are of considerable interest because in this two cyclopropane ring moieties are involved. This ring-opening reaction may be achieved by transition metal complexes[84] or by Lewis acid catalysis; the latter, however, leads to side reactions which include cationic oligomerization[84a].

More interestingly, **Q** can be converted to **NBD** by way of intermediary radical cations formed by photoinduced electron transfer (PET)[17] to suitable electron acceptors[85] (Scheme 8).

SCHEME 8. Photoinitiated catalytic cycle of the valence isomerization of **Q** to **NBD** (A = electron acceptor)

Transition metal complexes like PdCl$_2$(η^4-NBD)[85c] and [Ru(bpy)$_3$]$^{3+}$ [85a] or 9,10-dicyanoanthracene and 1,2,4,5-tetracyanobenzene[85b] have been successfully used as electron-accepting sensitizers. Reductive quenching of the excited singlet state of the electron-acceptor (A*) by **Q** leads to the formation of its radical cation Q$^{\cdot+}$ which undergoes a rapid irreversible rearrangement to the thermodynamically more stable $NBD^{\cdot+}$ radical cation (Scheme 8). $NBD^{\cdot+}$ may be reduced to **NBD** either by $A^{\cdot-}$ or by quadricyclane itself to lead to regeneration of $Q^{\cdot+}$. Consequently, **Q** re-enters the cycle, which accounts for quantum yields of **NBD** formation above 100. This catalytic cycle exhibits a significant solvent dependence in the presence of singlet or triplet sensitizers like 1-cyanonaphthalene (**1-CN**)[85b,86], and decreases the efficiency of the ring-opening process with increasing polarity of the solvent. This phenomenon was attributed to the reversible formation of solvent-separated ion pair intermediates $(1\text{-}CN^{\cdot-}/Q^{\cdot+})$ in polar solvents, whereas in non-polar solvents the isomerization proceeds through initial formation of a singlet exciplex[85b,86].

The valence isomerization of **Q**, mediated by photocatalysis on n-type semiconductors[87], seems to be a rather inefficient process. The yield of **NBD** varies with the type of semiconductor used in the order $CdS > TiO_2 \geqq ZnO$, and the nature of the solvent[87]. The quantum yields of these photoreactions are in the range of $\Phi = 10^{-4}\text{--}10^{-2}$ [87a,b].

A schematic representation of the valence isomerization $\mathbf{Q} \rightarrow \mathbf{NBD}$ at the surface of an illuminated semiconductor particle is shown in Scheme 9[87b]. Photochemical excitation of a semiconductor promotes an electron from the filled valence bond (VB) to the vacant conduction band (CB) with formation of a short-lived electron–hole pair (e^-, h^+)[88]. The

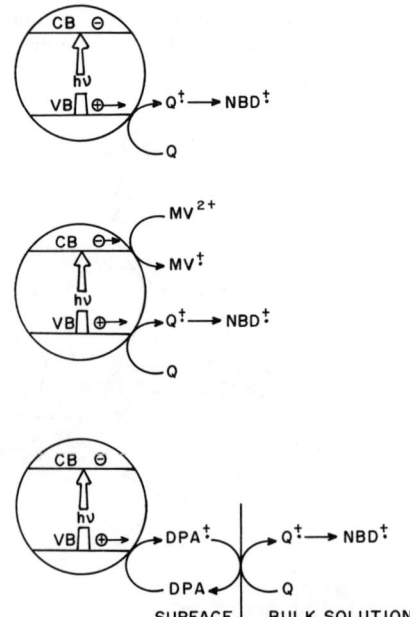

SCHEME 9. Valence isomerization of **Q** at an illuminated semiconductor particle. VB = valence band, CB = conduction band, DPA = diphenylamine, MV = methylviologen. Reprinted with permission from Ikezawa and Kutal, *J. Org. Chem.*, **52**, 3299. Copyright (1987) American Chemical Society

TABLE 6. Oxidation potentials $E_{1/2}^{ox}$ (V) vs. SCE[87a,b,d]

Compound	$E_{1/2}^{ox}$
Cds	1.6
TiO$_2$	2.3
ZnO	2.4
Q	0.91
NBD	1.54

positive holes thus formed may be reduced by an appropriate electron donor like **Q**. The decisive requirement for this redox coupling is that the oxidation potentials of the photogenerated holes are sufficiently positive to oxidize **Q**; the oxidation potentials for several n-type semiconductors are given in Table 6. Therefore, a positive hole at the semiconductor surface oxidizes **Q** to **Q**$^{·+}$, which readily isomerizes to **NBD**$^{·+}$ followed by the reduction to **NBD**. A charge transfer mechanism induced by photogenerated alkali halide colour centers of the 'missing electron' type was recently proposed by Michl and coworkers[89] for the rapid conversion of **Q** to **NBD** in a CsI matrix after illumination (equation 22). The electron deficient h-centers $I_2^{·-}$ which are formed upon irradiation of CsI are able to oxidize **Q** by electron transfer to **Q**$^{·+}$ to initiate a catalytic cycle similar to that described in Scheme 9 for the photocatalytic action of semiconductors.

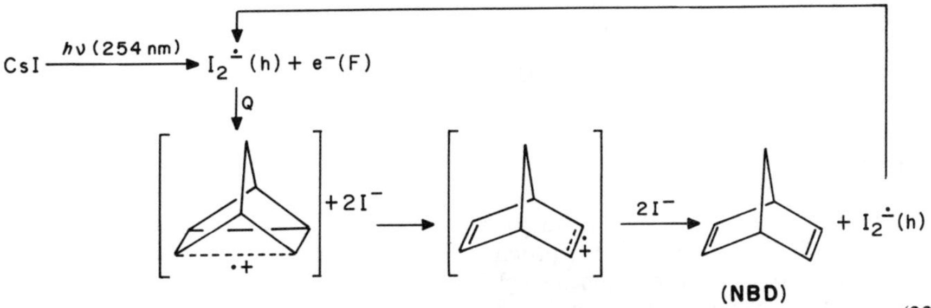

$$\text{CsI} \xrightarrow{h\nu\,(254\,nm)} I_2^{\cdot-}(h) + e^-(F)$$

(22)

In contrast to the photoinduced electron transfer reactions of quadricyclane, its direct 185-nm photolysis[90] in the liquid phase leads to **NBD** and to the formation of two side-products, namely 1,3,5-cycloheptatriene and 6-methylfulvene (Scheme 10). On the other hand, selective excitation of the CH bonds of gaseous **Q** by absorption of 598.8 nm laser photons produces a vibrationally 'hot' [NBD]^{vib*}, which fragments into cyclopentadiene, acetylene and 1,3,5-cycloheptatriene[91] (Scheme 10). These CH overtone transitions involve excitation energies greater than the activation energy for the **Q** → **NBD** isomerization to account for the 'hot' molecule effect. The photoproducts 6-methylfulvene and 1,3,5-cycloheptatriene are also formed in trace amounts in the direct 254-nm (< 5%) and triplet-sensitized (< 0.5%) photolysis of **Q**, which limits the usefulness of **NBD** $\rightleftarrows$ **Q** interconversion as an efficient solar-energy storage system[90].

An independent entry into the diradical manifold of the **NBD** $\rightleftarrows$ **Q** valence isomerization was described by Turro and coworkers[92a], who used the photochemical or thermal N$_2$ elimination from the azoalkanes 3,4-diazatricyclo[4.2.1.0^{2,5}]nona-3,7-diene (**48**) and 8,9-diazatetracyclo[4.3.0.0^{2,4}.0^{3,7}]non-8-ene (**49**) to produce the

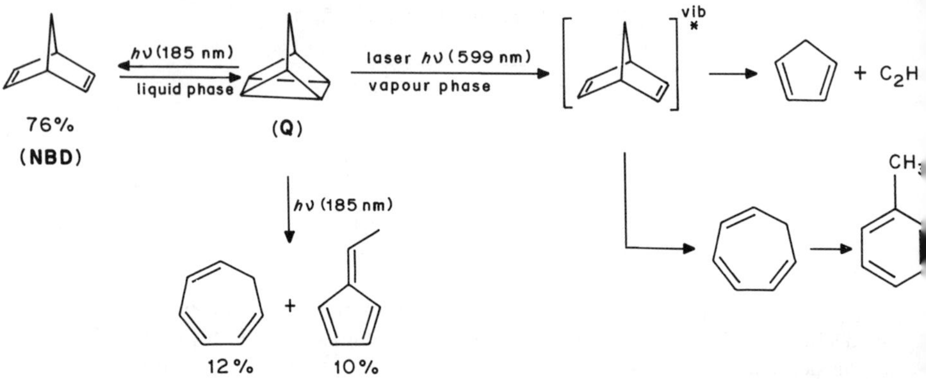

SCHEME 10. 185-nm and overtone vibrational photochemistry of quadricyclane (Q)[90,91]

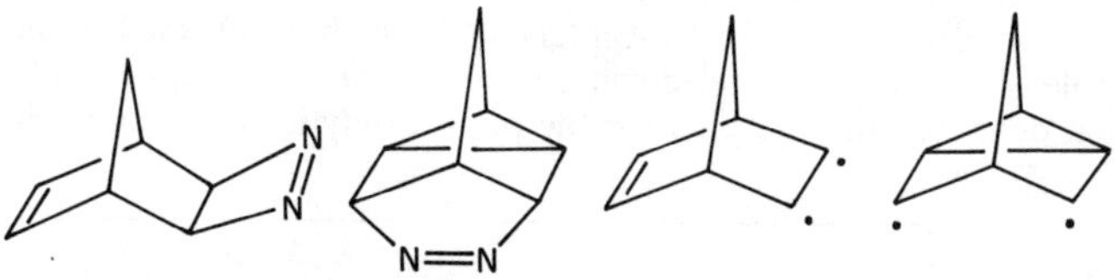

corresponding diradicals **50** and **51**. The azoalkanes **48** and **49** exhibit a distinct spin multiplicity effect, which manifests itself in a different product distribution that depends on singlet or triplet-sensitized excitation[92]. Therefore, the compounds **48** and **49** could serve as a suitable spin multiplicity probe in the direct 185-nm photolysis of azoalkanes in solution[93] to distinguish between singlet or triplet photodenitrogenation modes. The main products of direct and sensitized photolysis and thermal decomposition of the azoalkanes **48** and **49** are **NBD** and **Q** (Table 7).

Whereas the n_-,π^* photoreactivity of Z- and E-azoalkanes[94,95] is well established, the nature of higher excited states reached by 185-nm excitation (e.g. π,π^*, $n_+\pi^*$, n_+,σ^* and/or $(n_-,3s)$, $(n_-,3p)$ Rydberg transitions) is still a field of controversial discussion[18b,96]. The formation of products from higher excited states by 185-nm excitation was proposed to take place from short-lived singlet excited states ($< 10^{-10}$ s)[97] because intersystem crossing to lower triplet-excited states cannot compete[97f].

The experimental results of Table 7[92,93,98-100] are rationalized in terms of Scheme 11[93,101,102]. The quantum yields of substrate consumption (Φ_S) by 185-nm irradiation of **48** and **49** were determined applying the Z/E-cyclooctene actinometer[98]. Surprisingly, the quantum yields Φ_S^{185} are similar to those obtained on direct 350-nm excitation Φ_S^{350} (Table 7; entries 3,5 and 8,10).

The minima of the S_1 and T_1 energy surfaces which connect the lowest excited states of **NBD** and **Q** are located at the NBD and the Q side[92a,b,100]. Therefore, the recombination of the singlet diradicals **50** and **51** derived from **48** and **49** lead predominantly to **NBD** by following the S_1 surface, whereas the corresponding triplet diradicals react mainly to **Q** (Table 7; entries 2,7). Direct photolysis (singlet excitation)

TABLE 7. Photolysis and thermolysis of the azoalkanes **48** and **49**[a]

Substrate	Entry	Reaction conditions[b]	Φ_s[c]	Distribution[d] (% rel.)		Mass balance[e] ($\sum$%, absolute)	Reference
				NBD[f]	Q[f]		
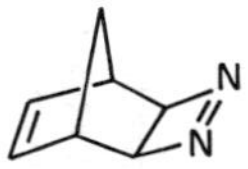 (**48**)	1	ΔT, 90–100 °C	—	>95	<5	—	92a
	2	$h\nu$, acetone	(0.40)	10	90	—	92a
	3	$h\nu$ (350 nm)[g]	(0.36)	24(29)	76(71)	90	93
	4	$h\nu$ (350 nm) piperylene[h]	—	77	23	—	93
	5	$h\nu$ (185 nm)	0.37	56	44	94	93
 (**49**)	6	ΔT, 90–100 °C	—	>95	<5	—	92a
	7	$h\nu$, acetone	(0.57)	0.3(10)	99.7(90)	—	92c
	8	$h\nu$ (350 nm)[g]	(0.44)	59(63)	41(37)	—	92c
	9	$h\nu$ (350 nm) piperylene[h]	—	53	47	—	93
	10	$h\nu$ (185 nm)	0.41	66	34	93	93

[a]The values in parentheses are also from References 92a.
[b]Photolyses in n-pentane as solvent.
[c]Quantum yield of substrate consumption[93,98].
[d]Quantitative analysis by calibrated capillary GC, experimental error 5–10%.
[e]$\sum$% = % azoalkane + %**NBD** + %**Q**.
[f]The product distribution was extrapolated to zero conversion by considering the 185-nm photoreactivity of **NBD** and **Q**.
[g]Traces of toluene (1.1%) and of 6-methylenebicyclo[3.10]hex-2-ene(7.8%) were detected; the absolute values given in Reference 92c have been normalized to 100% = %**NBD** + %**Q**.
[h]Piperylene is an effective triplet quencher, $E_T = 247\,\text{kJ}\,\text{mol}^{-1}$[99].

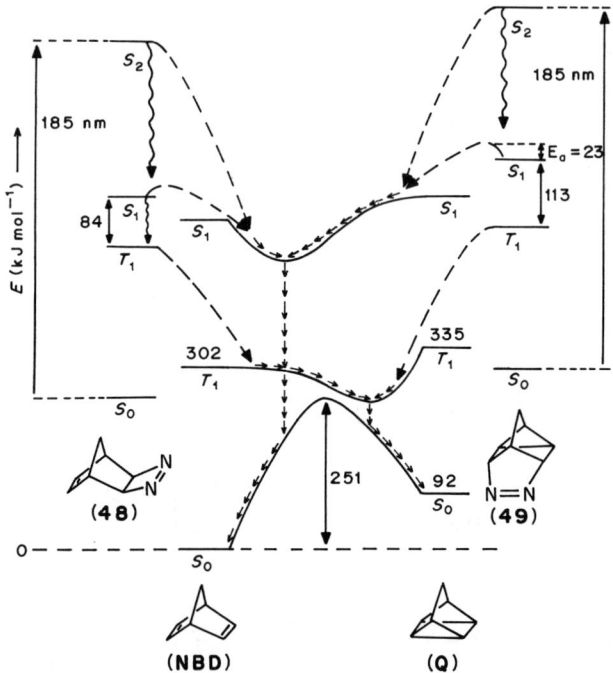

SCHEME 11. Qualitative representation of the overlapping azoalkane
NBD and Q energy surfaces[93]

of the azoalkane **48** (Table 7; entry 3), however, led to the formation of 76% **Q**, which indicates an effective S_1-T_1 intersystem crossing in the excited-state azoalkane chromophore. Consequently, in the presence of the triplet quencher piperylene (Table 7; entry 4) the singlet photoproduct **NBD** is formed in excess (77%). However, S_1-T_1 intersystem crossing seems not to be an important process in the photodenitrogenation of azoalkane **49** (Table 7; entries 8,9).

In contrast, the direct 185-nm excitation of **48** (Table 7; entry 5) preferably produces **NBD**, which indicates an ineffective $S \rightarrow T$ intersystem crossing, as is the case in the 185-nm photolysis of **49** (Table 7; entry 10). The thermolysis of both azoalkanes **48** and **49** was reported to yield **NBD** as the main product (Table 7; entries 1,6)[92]. Therefore, the 185-nm photodenitrogenation of **48** and **49** results from higher excited singlet states (Scheme 11). Furthermore, fast internal conversion (ca 10^{-13} s) may lead according to Kasha's[101] rule to the population of vibrationally excited S_1 states which exhibit enough energy to overcome the small activation barrier ($12-30$ kJ mol^{-1} [102]) of the N_2 elimination from S_1.

2. Cyclobutanes

The thermolysis of hydrocarbons which contain cyclobutane rings[103], e.g. bicyclo[2.2.0]hexane (**52**), has been studied widely with respect to its kinetics and stereo-chemistry of the ring-opening and ring-inversion reactions which reveal a multistep

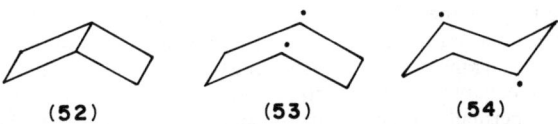

process to form cyclohexane-1,4-diyls **53** and **54**[104]. In recent years, the synthesis and the chemistry of saturated strained polycyclic cage molecules has been investigated intensively to provide understanding of the fascinating chemistry of 'platonic' structures[6,7,105] and related compounds like homocubanes[105a]. For example, the reactivity of cubane **55** (pentacyclo[4.2.0.0^{2,5}.0^{3,8}.0^{4,7}]octane) and its derivatives[6] is characterized by their extraordinary kinetic stability. One of the most intriguing aspects in this field of research is the recent approach towards the synthesis of 1,4-dehydrocubane **56**[106], which exists most likely as the ground state singlet 1,4-cubadiyl **57**[106,107] without a central diagonal C—C bond. Furthermore, the cubyl cation **58** which was proposed

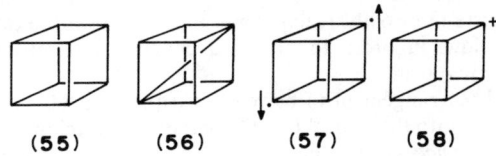

as an intermediate in a variety of solvolysis reactions of cubane derivatives seems to be a classical non-delocalized cation[108], but this is still a theme of controversial discussion.

In contrast to the vast literature that has accumulated on the thermal reactivity of alkylcyclobutanes, their photochemistry is only scarcely documented. Yet, the clean fragmentation of bicyclo[4.2.0]octane (**59**) on 185-nm irradiation[109] to octa-1,7-diene, cyclohexene and ethene (equation 23) may be attributed to a (σ,3s) Rydberg excitation[110].

$$
\text{(59)} \quad \xrightarrow[\text{n-pentane}]{h\nu\ (185\ \text{nm})} \quad \left[\ (\sigma,3s)\,\text{Ry}\ \right]^{\cdot} \longrightarrow \quad + \quad + \quad \begin{array}{c}CH_2\\ \|\\ CH_2\end{array}
$$

$$(23)$$

Alternatively, σ,σ^* excited states of simple cyclobutanes seem to be accessible on 185-nm excitation[111]. The quantum yields are $\Phi_s = 0.12$ for substrate consumption and $\Phi_p = 0.10$ and $\Phi_p = 0.03$ for the formation of octa-1,7-diene and cyclohexene[109]. Similarly, cubane (**55**)[112] decomposes on 185-nm excitation with a quantum yield of $\Phi_s = 0.08$ to give cyclooctatetraene and benzene[113] (equation 24).

$$
\textbf{55} \quad \xrightarrow[\text{n-pentane}]{h\nu\ (185\ \text{nm})} \quad + \quad \qquad (24)
$$

3. Bicyclo[2.1.0]pentanes

The physicochemical properties and rearrangements of bicyclo[2.1.0]pentanes have been reviewed in detail[2a,2b,18,114]. In this section, however, we consider molecules with

712   T. Oppenländer, G. Zang and W. Adam

the bicycloalkane itself as the functional moiety, and allow simple alkyl substitution in the polycyclic structures. With this limitation, three main activation modes are possible for the initiation of rearrangements: (a) acid and transition metal-promoted rearrangements, (b) photochemically [direct, sensitized and photoinduced electron transfer (PET) conditions] and (c) thermally induced rearrangements. In Table 8 several product studies are collected, which describe the parallel or divergent behaviour of bicyclo[2.1.0]pentanes under the different activation techniques[19,66,80,96,115-126].

The data of Table 8 reveal that the three activation modes may lead to similar (Table 8; e.g. entries 1a, c, e) or to completely different (Table 8; e.g. entries 13a, b) product distributions. Even within a series of PET experiments dramatic differences in the product distribution were observed by variation of the electron acceptors (Table 8; entry 2f). Furthermore, stereochemical factors are of considerable importance. Thus, syn-5-methylbicyclo[2.1.0]pentane easily rearranges to 1-methylcyclopentene on treatment with [Rh(CO)$_2$Cl]$_2$ (Table 8; entry 2a), whereas the anti isomer does not react even at elevated temperature (Table 8; entry 2b). Unfortunately, the activation modes for the individual substrates have not been completely examined in all cases, which limits comprehensive understanding of the reaction mechanisms. Nevertheless, several examples will be discussed below, in order to obtain a concrete description of the reaction mechanisms that are involved.

The Ni^0 catalysed rearrangement of bicyclo[2.1.0]pentane (60, cf. Table 8; entry 1b) leads to metal insertion into the central C—C bond of the bicyclopentane in the primary step (equation 25)[115b]. Successively, the trapping alkene ZCH=CHZ attacks a Ni—C

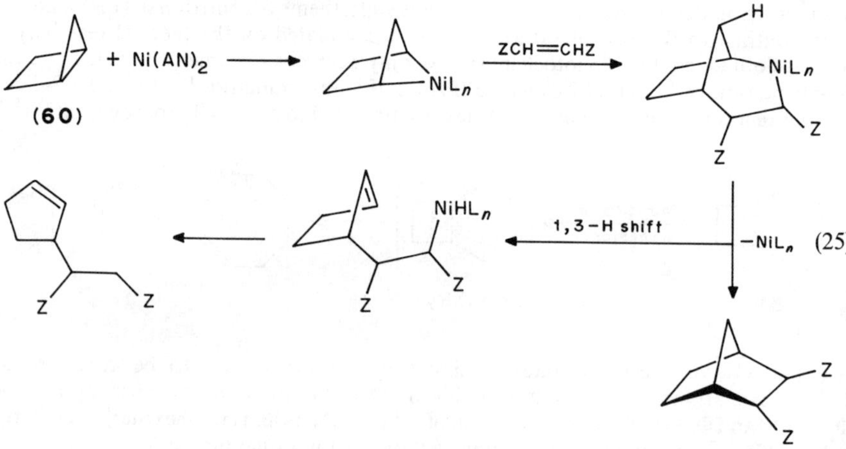

bond and the resulting intermediate is transformed in two pathways, which consist of a 1,3-H shift and catalyst elimination or immediate catalyst removal to the observed products. However, the higher homologues bicyclo[3.1.0]hexane and bicyclo[4.1.0]heptane are unreactive under the above conditions[115b]. Cyclopentene generated by Rh(I)-promoted rearrangement of selectively deuterated bicyclo[2.1.0]pentane (Table 8; entry 1a) shows total deuterium scrambling. A detailed mechanism for this observation was not suggested[115a].

Alternatively, a lateral C—C bond attack of the Rh catalyst is observed by using 3,3,6,6-tricyclo[3.1.0.0^{2,4}]hexane as substrate (Table 8; entry 11a)[120a]. Even with the strained tricyclo[3.2.0.0^{2,7}]heptane (61)—cf. Table 8; entry 12a—a central attack of the

TABLE 8. Rearrangements of bicyclo[2.1.0]pentanes

Entry	Substrate	Conditions (yield)	Relative product distribution (%)	Reference
1a		5 mol% [Rh(CO)$_2$Cl]$_2$ in CHCl$_3$, 25 °C, $t_{1/2}$: 8 h[a]	 100	115a
1b		12 mol% Ni(AN)$_2$ in CH$_2$=CHCO$_2$CH$_3$, 40 °C, 36 h, N$_2$-atmosphere, (88%) (AN: NCCH=CH$_2$)	 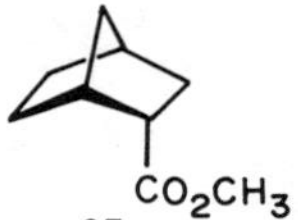 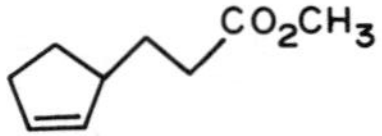 37 37 26	115b
1c	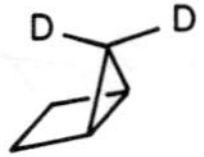	300 °C, 63 min	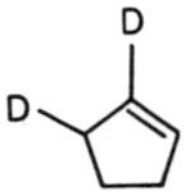 100	115c
1d		$h\nu_{185}$, n-heptane, 25 °C[b]	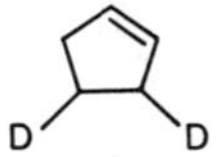  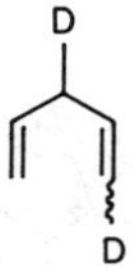 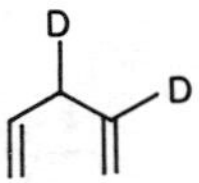 43 46 5 6	115d
1e		γ-rays, freon matrix	  $T \leqslant 100$ K $T \geqslant 100$ K	115e,f
2a		[Rh(CO)$_2$Cl]$_2$ in CDCl$_3$, 25 °C	 100	66b

(continued)

TABLE 8. (*continued*)

Entry	Substrate	Conditions (yield)	Relative product distribution (%)			Reference
2b	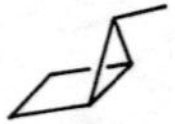	[Rh(CO)₂Cl]₂ in CDCl₃, 60 °C, 24 h	no reaction			66b
2c		HOAc 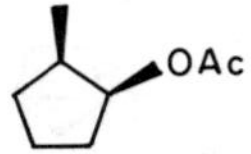 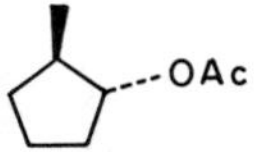 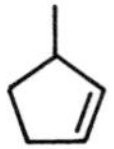	68 (OAc)	16 (OAc)	16	116a
2d		HOAc	41 (OAc)	28 (OAc)	27	116a
2d					4	116b
2e		171 °C	*syn-anti* equilibrium mixtures			116c
2f 		PET conditions, CH₂Cl₂, λ ≥ 400 nm[c]				  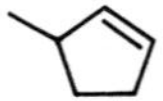 116c
		DCA	64	13	23	
		TCA	18	22	60	
		TPT	80	4	16	
		MAC	15	15	70	
2g 		PET conditions, CH₂Cl₂, λ ≥ 400 nm[c]				  116c
		DCA	40	60		
		TPT	59	41		
3a 		[Rh(CO)₂Cl]₂, CDCl₃, 25 °C	100			66b

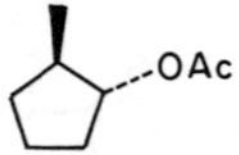

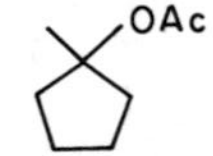

	Conditions			Ref.
3b 	[Rh(CO)₂Cl]₂, CDCl₃, 25 °C	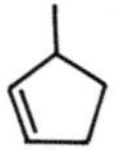 50	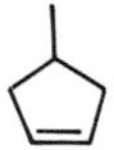 50	66b
4a	 350 °C.17 torr, (94%)	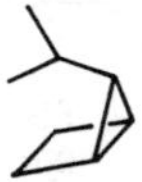 23 — (83 : 17)	64 —	 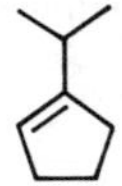 19
4b	540 °C/17 torr, (79%)		100	19
4c	hν₁₈₅₊₂₅₄, n-pentane, 25 °C conversion: 32%, mass balance: 69% conversion: 54%, mass balance:70%	35 (25 : 75^d) 39 (80 : 20^d)	26 10	13 100	others 39 51	19
5a	HOAc	20 (OAc) 71	34 (OAc) —	46 14	15	116a
5b	70 °C 116 °C 540 °C/15 torr, (80%)   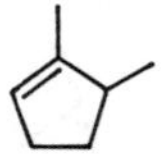		6	others 16	19
5c	hν₁₈₅₊₂₅₄, n-pentane, 25 °C, conversion: 44% mass balance: 59%	100 29	49			19

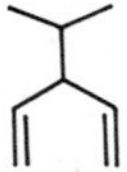

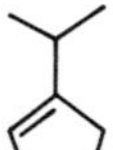

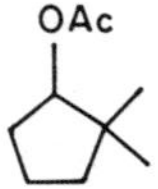

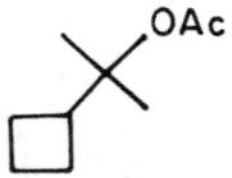

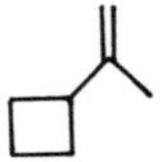

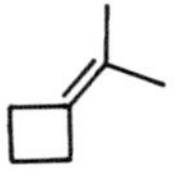

(continued)

TABLE 8. (continued)

Entry	Substrate	Conditions (yield)	Relative product distribution (%)	Reference
6a		HOAc 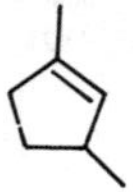 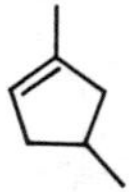 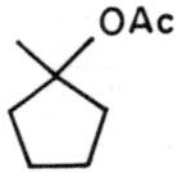	not stated (OAc structure) trace	116a
6b		540 C/17 torr, (89%)	87 13	19
			38 41	19
6c		$hv_{185+254}$, n-pentane, 25°C, conversion: 50%, mass balance: 80%, PET conditions^c, DCA, TPT	others 21	116c
6d			no reaction 100	116c
7a		170 °C   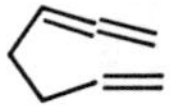	100 21	117
7b		hv_{185}, n-pentane	100 21 6	96a, 95b
			73	
8	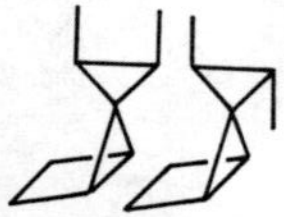	140 °C	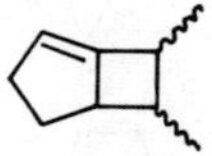 starting from each individual isomer similar cis-trans and syn-anti ratios were obtained	117

717

	ΔT			
9		19	81	
10a	117–187 °C, gas phase		 100	118
10b	135 °C, furan[e]	not stated	100	119a
				119b
10c	70 °C	not stated		119b
11a	[Rh(CO)₂Cl]₂ or CpRh(CH₂=CH₂)₂, 60 °C	40	100	120a
			60	
11b	170–200 °C/10 torr	not stated		120b
12a	1.5 n H₂SO₄ in H₂O: dioxane = 3:7, 2d, 70 °C	0.9	16.1	121a
		51.3	31.7	

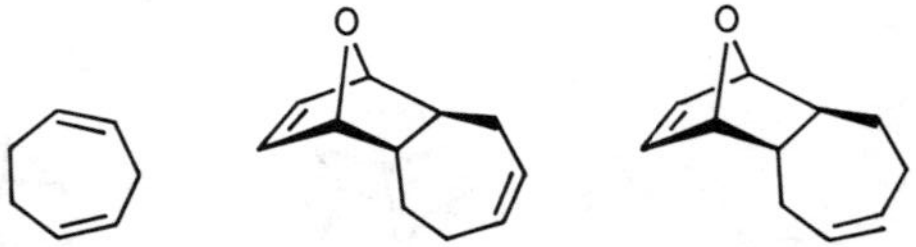

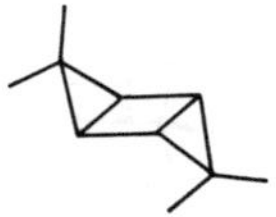

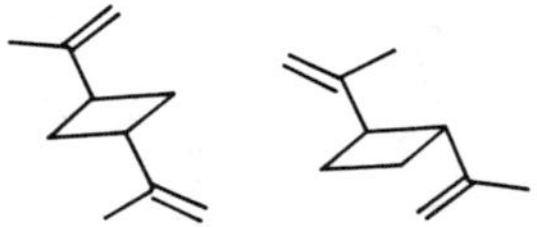

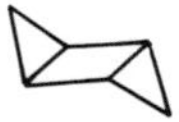

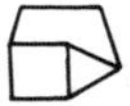

(continued)

TABLE 8. (*continued*)

Entry	Substrate	Conditions (yield)	Relative product distribution (%)					Reference
							others	121b
12b		540 °C/15 torr, conversion: 82% mass balance: 29%	100 44				16	
12c		$h\nu_{185+254}$, n-pentane, conversion: 30% mass balance: 50%		10		14	16	16	121b
13a		540 °C/15 torr, conversion: 57% mass balance: 28%	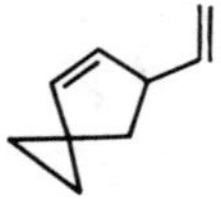	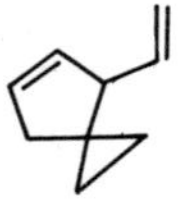		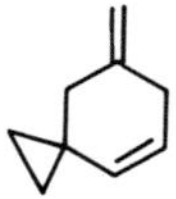	others	121b
13b		$h\nu_{185+254}$, n-pentane, conversion: 42% mass balance: 39%	44	100	12	12	32	121b
14a		540 °C/17 torr, conversion: 100% mass balance: 32%	37		others			121b
14b		$h\nu_{185+254}$, n-pentane, conversion: 42% mass balance: 39%		100 20	5	38	—	121b 121b
15a		AgBF$_4$	 100					122
15b		[Rh(CO)$_2$Cl]$_2$	100					122

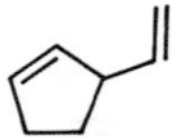

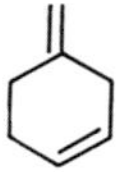

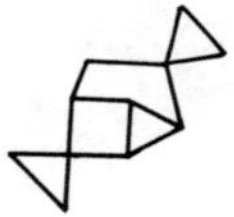

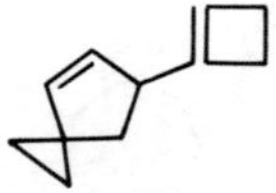

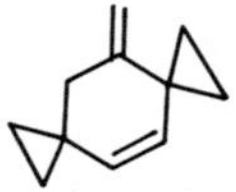

15c	0.3 N D_2SO_4 in D_2O: dioxane = 3:7, 0.5–3 h, 21 °C	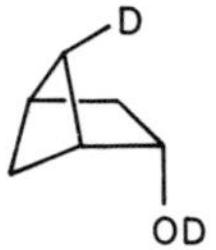 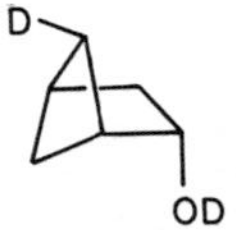  not stated	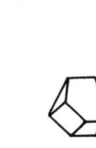 OD 19, OD 81	121a
16	180–220 °C^g 180 min.			123
17	3.6 mol% $AgBF_4$ in $CHCl_3$, 18 h, 21 °C, (85%)	(cf. equation 12)		66
18a	150 °C, benzene, 6 h (83%)	 100		124
18b	170 °C, benzene	100		124
19	21 °C, THF	cf. 18a; but higher amounts of polymeric material were generated		124
		40, 60 (cf. equation 20)		80
20a	≥125 °C	29, 71		125a

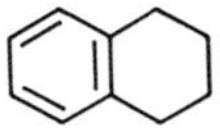

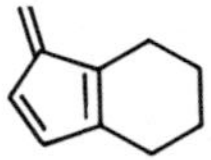

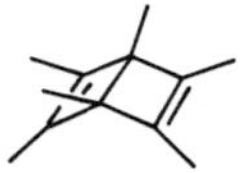

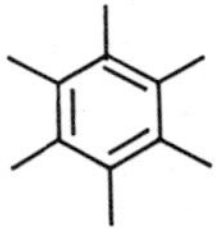

(continued)

TABLE 8. (continued)

Entry	Substrate	Conditions (yield)	Relative product distribution (%)		Reference
20b		anthraquinone, acetone, UV irradiation		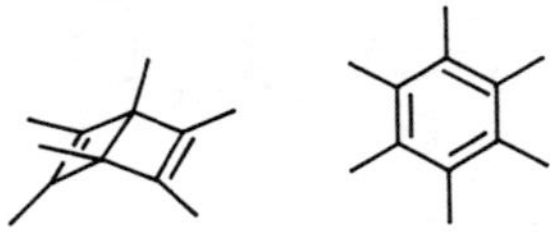 not stated	125b
20c		hv_{254}		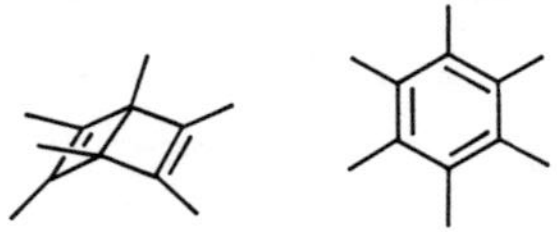 not stated	125c
21	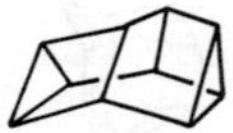	190 °C, 5 h, toluene	no reaction		126

[a] Deuterium-labelling studies with several deuterated isomers of bicyclo[2.1.0]pentane showed total scrambling of the deuterium in the cyclopentene.
[b] The exo:endo ratio of unreacted starting material was not significantly changed at the end of the photolysis.
[c] DCA = 9,10-dicyanoanthracene, TCA = tetracyanoanthracene, TPT = triphenylpyrylium tetrafluoroborate, MAC = methylacridinium tetrafluoroborate; the conversions were in the range 10–37%, the mass balances between 4% and 30%; the solutions were ca 10^{-2} M in substrate and contained 10 mol% electron acceptor; $\lambda \geqslant 400$ nm.
[d] The syn:anti ratio of unreacted starting material was not significantly changed at the end of the photolysis.
[e] The product ratio is dependent on the furan concentration.
[f] More than eighty products were detected; at 480 °C/17 torr 95% starting material was recovered.
[g] The product distribution is strongly dependent on the pressure.

proton could be excluded (equation 26)[121a].Furthermore, the entries 15a and 15b (Table 8) provide examples of 'end-on' or 'side-on' approach of different metal catalysts towards the cyclopropane moiety (equation 27)[122].

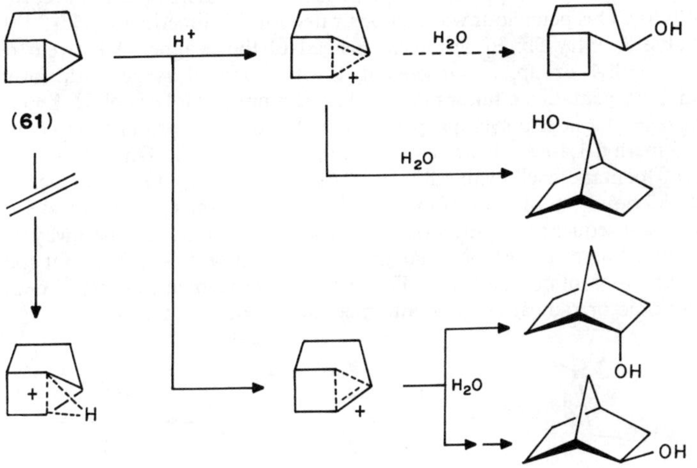

(26)

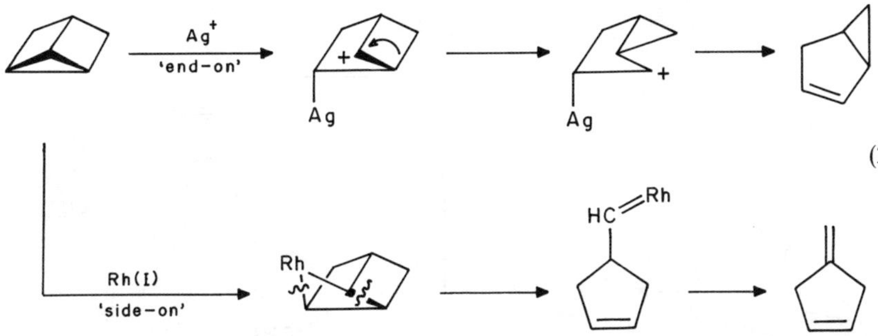

(27)

Pyrolysis of bicyclopentanes usually leads to the cleavage of the central C—C bond to give a vibrationally excited diradical, which suffers a consecutive 1,2-H shift (Table 8; entries 1c, 4b, 6b) or subsequent C—C bond cleavage (Table 8; entries 5b, 6b, 9, 10a, b, 12b, 13a, 14, 16). In other cases syn/anti isomerizations of the substrates are observed (Table 8; entries 2e, 4a, 10c). The higher annulated syn/anti-pentacyclo[5.3.0.0^{2,6}.0^{3,5}.0^{8,10}]-decanes (Table 8; entry 18) react preferably by pyrolytic cleavage of the cyclobutane unit, whereas the cyclopropane ring is not affected.

Recent efforts to elucidate the structure of the radical cation of hexamethylprismane (Table 8; entry 20b) have shown that a short-lived species (ns range) is generated in the PET reaction when anthraquinone is used as electron transfer sensitizer[125b]. Its structure is different from the radical cation of hexamethyl Dewar benzene and it rearranges to the latter and hexamethylbenzene. Both products are also generated in the pyrolysis[125a] and direct photolysis[125c] of hexamethylprismane (Table 8; entries 20a, c).

An interesting example for the kinetic stability of strained alkanes is hexacyclo-[4.4.0.0^{2,5}.0^{3,9}.0^{4,8}.0^{7,10}]decane (Table 8; entry 21). Despite its strain energy of ca

$449 \, \text{kJ mol}^{-1}$ it is thermally stable up to $200 \, °C$[126]. It seems that the structural compactness, similar to cubane[6], is responsible for this kinetic stability.

A common feature of vibrationally excited 1,3-cyclopentadiyls, generated by pyrolysis of substituted bicyclo[2.1.0]pentanes, is their lack of Wagner–Meerwein rearrangement (1,2-alkyl shifts). This behaviour was demonstrated for 5,5-dimethylbicyclo[2.1.0]pentane (**62**); cf. Table 8; entry 5b; equation 28. Instead of the Wagner–Meerwein rearrangement of a methyl group, a subsequent C—C bond cleavage with formation of 3,3-dimethyl-1,4-pentadiene takes place[19]. The 185-nm photolysis of **62**, however, leads to the cleavage of a lateral cyclopropane C—C bond in the primary step (equation 29) to yield 5-methyl-1,4-hexadiene as the major product[19]. This is in contrast to bicyclo[2.1.0]pentane itself (Table 8; entry 1d), which mainly reacts by cleavage of the central C—C bond on 185-nm irradiation[115d]. In the former case, central C—C bond rupture with subsequent formation of 3,3-dimethyl-1,4-pentadiene (pathway a, equation 29) is of subordinate importance and the corresponding 1,2-methyl shift (pathway b, equation 29) was a minor pathway. The latter rearrangement, however, demonstrates that zwitterionic or radical-cationic intermediates might be involved.

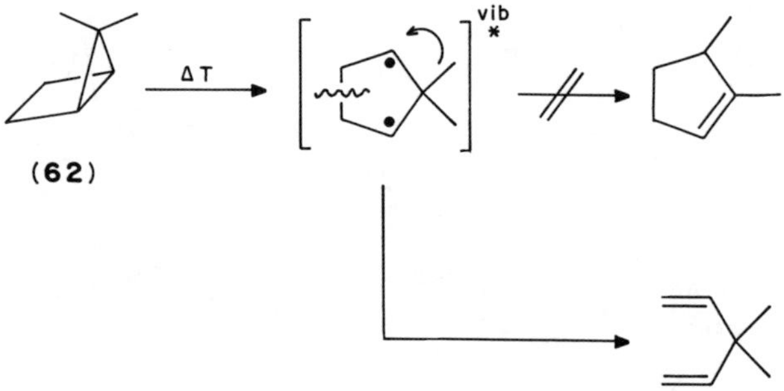

(28)

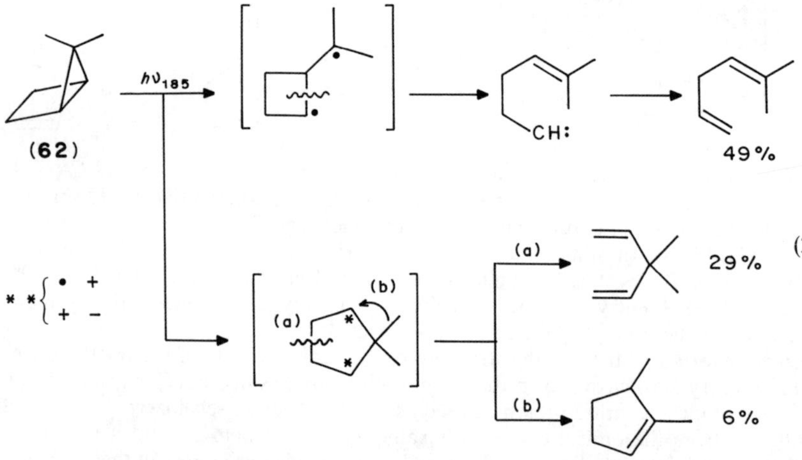

(29)

Another example of different thermal and photochemical behaviour is described by equation 30 (Table 8; entries 13a, b). Two isomeric vinylcyclopentenes are generated as major products, which depend on the activation mode[121b] (equation 30). However, if

1,2-H shifts are possible in the pyrolyses or photolyses of bicyclopentanes (Table 8; entries 1c,d,e; 4b,c; 6b,c), this reaction mode provides a considerable contribution to the rearrangements. Recent work on the 185-nm photolysis of *syn*- and *anti*-5-iso-propylbicyclo[2.1.0]pentane (**63**) shows that the 1,2-H shift is quite dependent on the stereochemistry of the starting material (Table 8; entry 4c)[19]. The different efficiencies in the formation of 1-isopropylcyclopentene demonstrate that the hydrogen shift does not occur from a common planar spacies but from a non-planar one (equation 31).

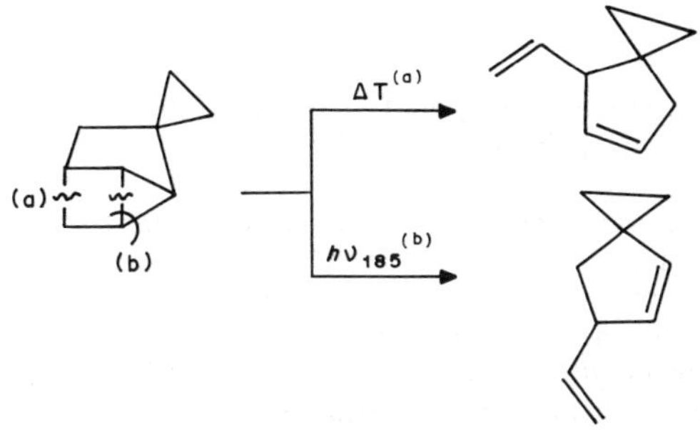

(30)

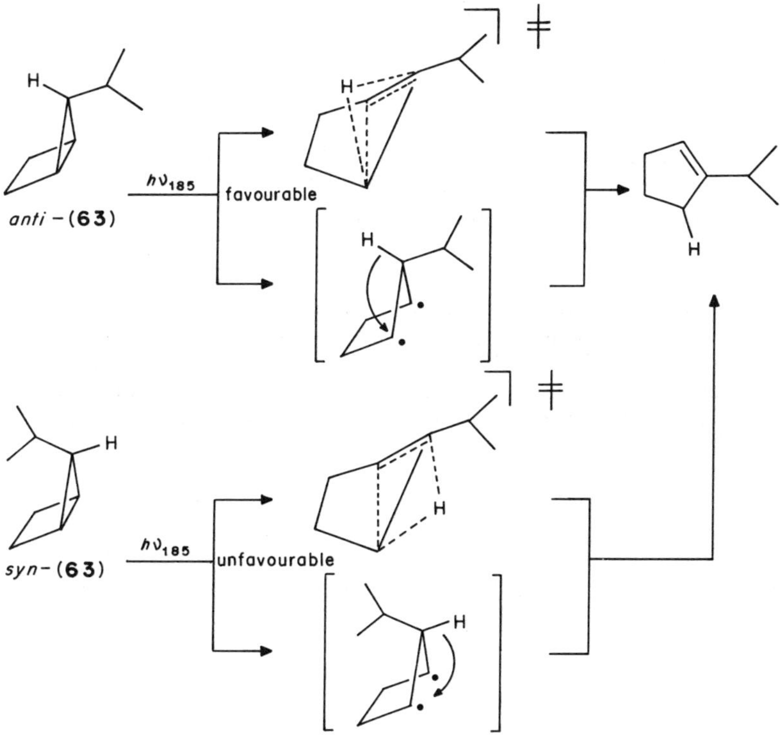

(31)

Again, the PET reaction of bicyclo[2.1.0]pentanes by employing various electron acceptors (Table 8; entries 2f,g)[116c] is strongly dependent on the stereochemistry. Thus, *syn*-5-methylbicyclo[2.1.0]pentane (Table 8; entry 2f) gives 16%–70% of 3-methylcyclopentene, which depends on the nature of the electron acceptors, whereas in the case of the *anti*-isomer (Table 8; entry 2g) this product is not generated at all. An intermediate non-planar radical cation was postulated to explain this result[116c]. Finally, the different product distributions, when applying various electron acceptors (Table 8; entries 2f,g), might be a consequence of initially formed contact ion pairs and/or solvent separated ion pairs which exhibit divergent reactivity.

4. Bicyclo[1.1.0]butanes

The rearrangement modes of bicyclo[1.1.0]butanes have been described in detail by Hoz[127]; PET experiments with bicyclo[1.1.0]butane derivatives were extensively studied by Gassman[128].

The situation concerning the transition metal catalysed rearrangement of bicyclobutanes is complex. Hoz[127] discussed at least four different parameters which affect the product distribution, namely the substitution pattern of the substrate, the type of the transition metal, its ligands and the reaction medium. For example, the divergent behaviour of 1,7-dimethyltricyclo[4.1.0.0^{2,7}]heptane (64) in the Mg^{2+}- and Ag^+-catalysed rearrangement is illustrated by equation 32[129].

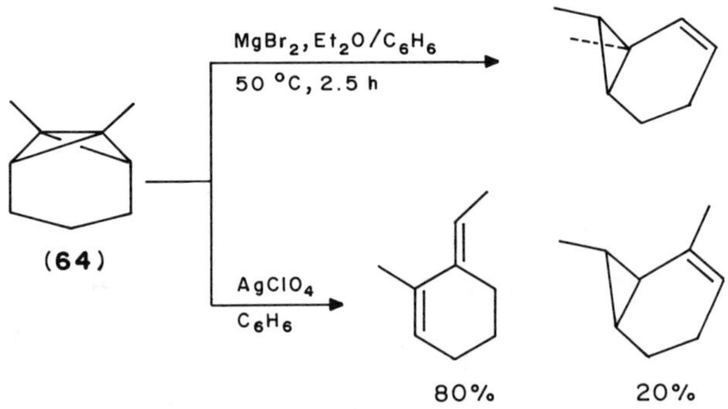

$$MgBr_2, Et_2O/C_6H_6$$
$$50\ ^\circ C, 2.5\ h$$

(32)

(64)

$$AgClO_4$$
$$C_6H_6$$

80%   20%

A prediction of the rearrangement products *a priori* was not possible, because their formation depends on too many factors. Table 9 summarizes results on the reactivity of bicyclo[1.1.0]butanes[19,79,129–136]. The Ni^0-catalysed rearrangement of bicyclo[1.1.0]-butane (Table 9; entry 1a) proceeds via a metal–carbene complex, as was demonstrated by a deuterium-labelling experiment using 65 as a mechanistic probe (equation 33)[130a].

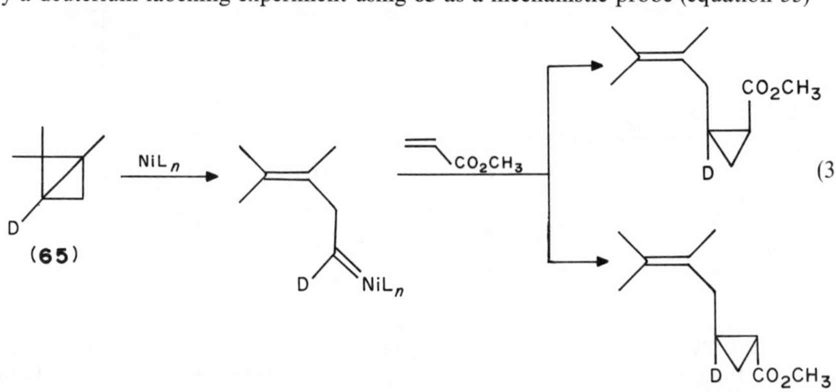

$$NiL_n$$

$$\ce{CO_2CH_3}$$

CO_2CH_3

D

(33)

(65)

D   NiL$_n$

D   CO$_2$CH$_3$

TABLE 9. Rearrangement of bicyclo[1.1.0]butanes

Entry	Substrate	Conditions (yield)	Relative product distribution (%)		Reference
1a		5 mol% Ni(COD)$_2$ in CH$_2$=CHCO$_2$CH$_3$, 0 °C, 12 h (92%)	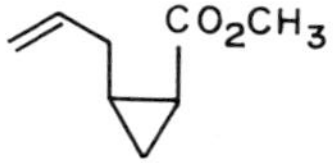 CO$_2$CH$_3$ 65	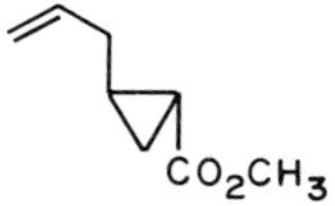 CO$_2$CH$_3$ 35	130a
1b		200 °C	 100		130b
1c		$h\nu_{185}$, n-pentane, 21 °C	 50	50	79a
1d	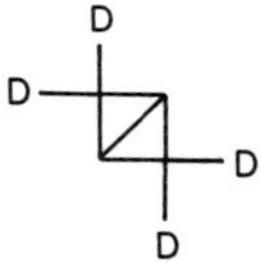	$h\nu_{185}$, n-pentane, 21 °C	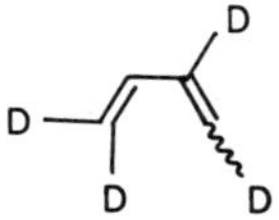 60	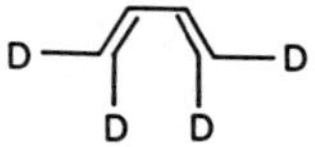 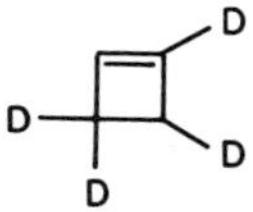 30 ... 10	79b
1e		γ-rays, CFCl$_3$ matrix, 77 K	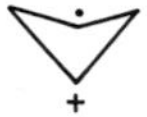 · + 100		130c
2a		AlCl$_3$	 100		131a
2b		400 °C/760 torr	 100		131b

(continued)

TABLE 9. (continued)

Entry	Substrate	Conditions (yield)	Relative product distribution (%)				Reference
2c	[structure]	$h\nu_{185+254}$, n-pentane, 21°C conversion: 42% mass balance: 91%	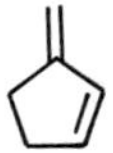 [structure] 39	 [structure] 15	 [structure] 26	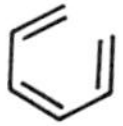 [structure] 20	19
3a	 [structure]	Lewis acid[a]	 [structure] 5–40	 [structure] 85–100	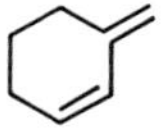 [structure] 8–98		132a–d
3b	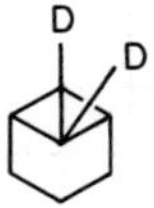 [structure, D]	430–500 °C/760 torr, (86–91%)[b]	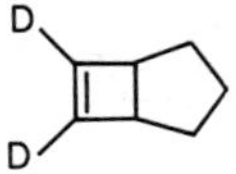 [structure, D] 6–74	 [structure, D] 26–94			129c
3c	 [structure]	$h\nu_{254}$, Hg sensitized	 [structure]	 [structure]	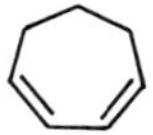 not available	 [structure]	132e
3d	[structure]	$h\nu_{185+254}$, n-pentane, 21°C conversion: 19% mass balance: 80%	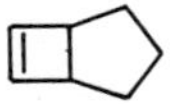 [structure] 85	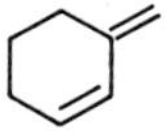 [structure] 15			19
3e	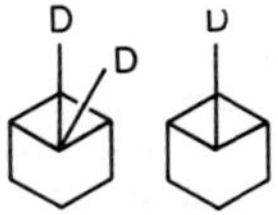 [structure, D]	$h\nu_{185+254}$, n-pentane, 21°C conversion: 72% mass balance: 30% isolated yield: 21%[c]	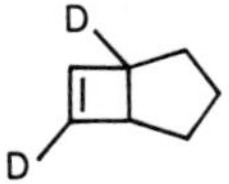 [structure, D] 33	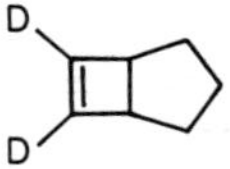 [structure, D] 17	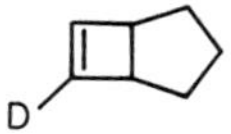 [structure, D] 33	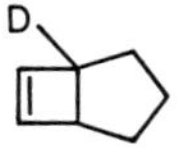 [structure, D] 17	19

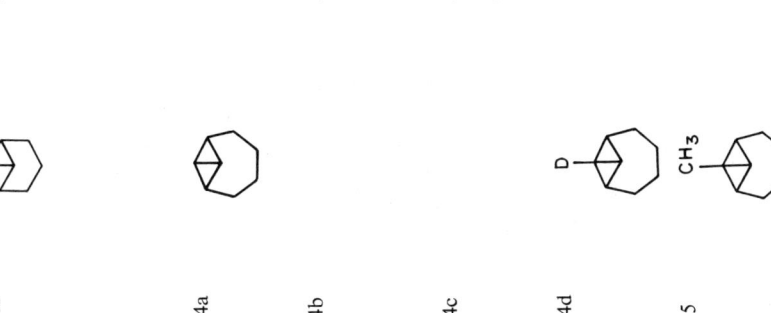

		Conditions	Products	others	Ref.
3f		DCA^d, $h\nu$ 300 nm, benzene, 21 °C			132f
4a		acid traces	100		133
4b		160 °C, [D$_6$]-benzene, $t_{1/2}$: 8 h	85	15	133, 19
4c		$h\nu_{185+254}$, n-pentane, 21 °C conversion: 33% mass balance: 64%	75	10	19
4d		$h\nu_{185+254}$, n-pentane, 3.5 h isolated yield: 13%^{e}	61	39	19
5		$h\nu_{254}$, n-pentane conversion: 100% isolated yield: 39%	100		19

(continued)

TABLE 9. (*continued*)

Entry	Substrate	Conditions (yield)	Relative product distribution (%)	Reference
6	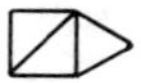	−50 °C	 100	134
7a	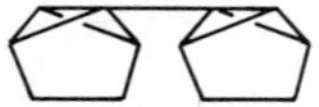	158 °C, benzene, 200 h	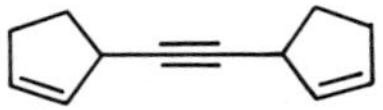 100	135
7b	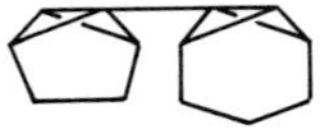	180 °C, benzene, 2 h	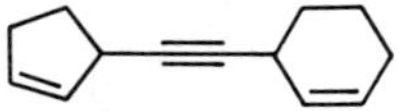 100	135
7c	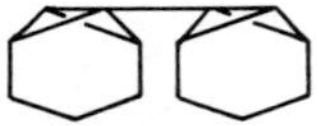	180 °C, benzene, 10 h isolated yield: 73%	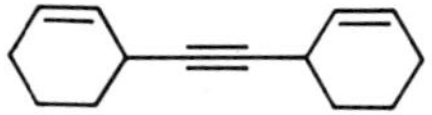 100	135
7d	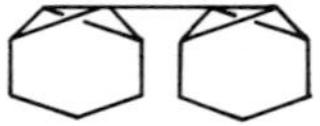	3 mol% AgBF₄	100	135
8a	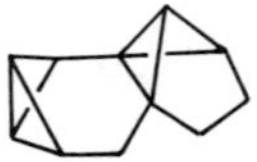	AgClO₄, benzene, (79%)	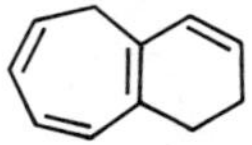 100	136a
8b	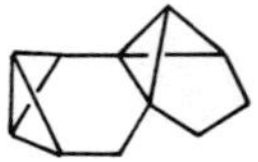	CF₃CO₂H, CHCl₃, 20 °C, (58%)	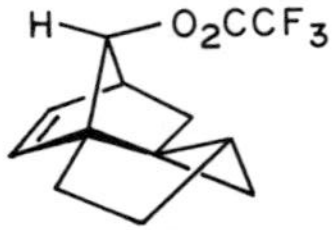 100	136b

8c

AgClO₄, 50 °C

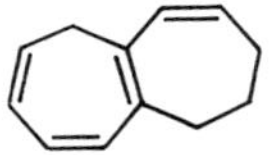

10   85

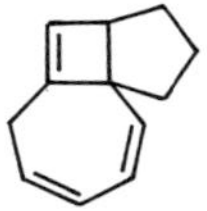

136a

8d

CF₃CO₂H, CHCl₃,
20 °C (86%)

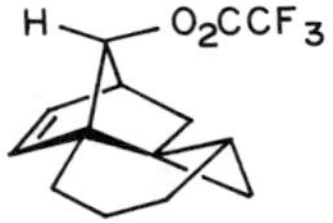

100

136b

[a] Depending on the catalyst, different yields of the three products are obtained.
[b] The generation of the products was temperature-dependent. At high tempeatures the diene dominated, and at low temperatures the bicyclobutene.
[c] The deuterated 3-methylenecyclohexene was not isolated.
[d] DCA = 9,10-dicyanoanthracene.
[e] Only the bicyclobutenes were isolated from the photolysis mixture.

The pyrolysis of bicyclo[1.1.0]butane, however (Table 9; entry 1b), occurs by way of a stepwise, stereoselective, twofold, lateral C—C bond cleavage[130b]. The 185-nm photolysis of bicyclo[1.1.0]butane was described recently by groups of Adam[79a] and of Srinivasan[79b] (Table 9; entries 1c,d). Although both groups found a different ratio of 1,3-butadiene and cyclobutene, they agreed that the dominating primary step is central C—C bond cleavage (equation 34)[79b].

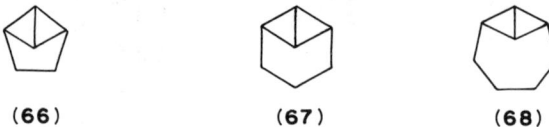

(34)

In equation 34 the primary intermediate is the 1,3-diradical, but alternatively a 1,3-zwitterion or Rydberg-like radical cation might be reached on 185-nm excitation[18a]. γ-Irradiation of bicyclo[1.1.0]butane was shown to generate a puckered 1,3-radical cation (Table 9; entry 1e)[130c]. This might be evidence for the above assumption of polar short-lived intermediates in the 185-nm photolysis of bicyclo[1.1.0]butanes. Therefore, 185-nm irradiation experiments were performed with the methylene-bridged derivatives of bicyclo[1.1.0]butane 66–68 (Table 9; entries 2c; 3d,e; 4c,d)[19] in order to examine the involvement of polar intermediates.

(66)   (67)   (68)

Deuterium-labelling experiments (Table 9; entries 3e, 4d) revealed that a Wagner–Meerwein rearrangement is responsible for the generation of the major products in the 185-nm photolysis of tricyclo[4.1.0.0^{2,7}]heptane (67) and tricyclo[5.1.0.0^{2,8}]octane (68). This rearrangement was confirmed by using the deuterated octavalane d-68 (equation 35). In fact, predominant formation of the isomer 69a with deuterium in the olefinic position (Table 9; entry 4d) reveals the existence of one or more parallel reaction channels, for example the concerted ring-opening (minor pathway, equation 35). The observation of Wagner–Meerwein chemistry, which is responsible for the formation of 69b, is experimental evidence for polar zwitterionic or radical-cationic intermediates in the

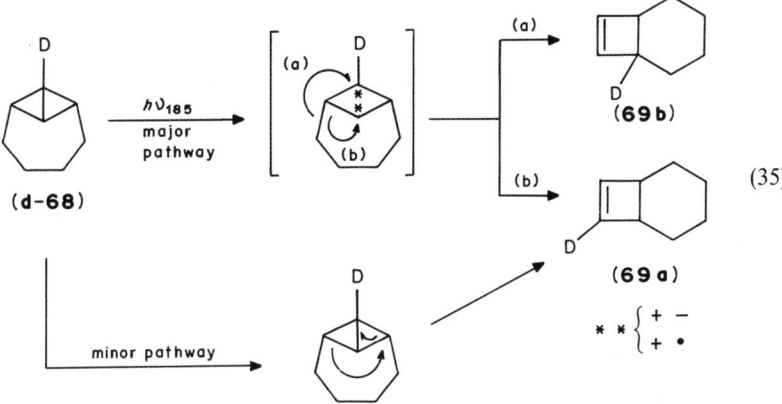

(35)

185-nm photochemistry of bicyclo[1.1.0]butanes, because 1,3-diradicals are reluctant to undergo such a rearrangement process (cf. equation 28).

Furthermore, the photochemical behaviour of the bridged bicyclobutanes **66–68** is strongly dependent on the dihedral angle Θ and thereby on the energy of the HOMO of the bicyclobutane moiety (Figure 2). The dihedral angle Θ is highest in tricyclooctane **68** (Table 9; entries 4c,d) and lowest in tricyclohexane **66**. The HOMO of tricyclooctane **68** could be assigned to the central C—C bond[137a]. Consequently, this bond cleaves on photochemical excitation (Table 9; entries 4c,d). On the other hand, the tricyclohexane

(a)   (b)

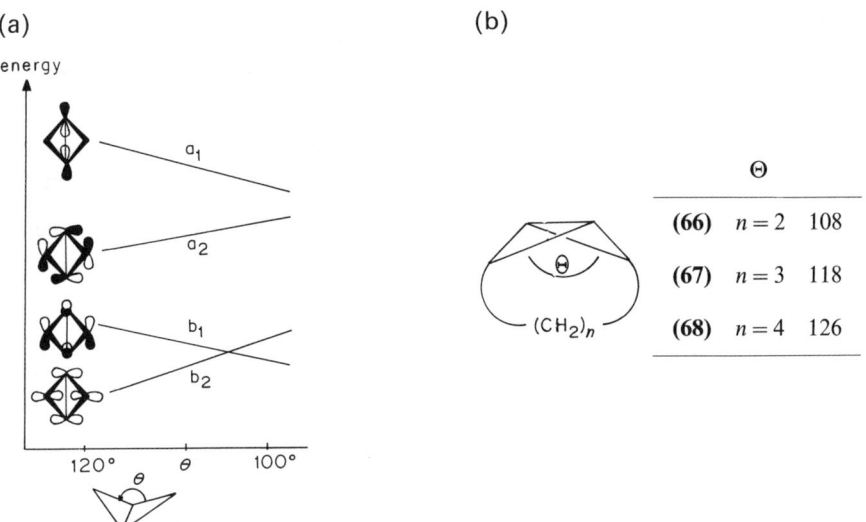

FIGURE 2. (a) Schematic representation of the valence MOs of bicyclo[1.1.0]butane as a function of the dihedral angle Θ. Reprinted with permission from Gleiter *et al.*, *J. Org. Chem.*, **50**, 5067. Copyright (1985) American Chemical Society; (b) calculated (MNDO) values of Θ[19]

66 has two occupied MOs with very similar energy (Figure 2)[137a]. One of them could be assigned to the lateral C—C bonds of the bicyclobutane moiety. This MO argument explains the dramatic change of the reaction mechanisms in the series of the bridged bicyclo[1.1.0]butanes **66–68** (Table 9; entries 2c, 3d, e, 4c, d). Lateral C—C bond cleavage is the exclusive reaction channel in the 185-nm photolysis of the tricyclohexane **66** (Table 9; entry 2c, equation 36).

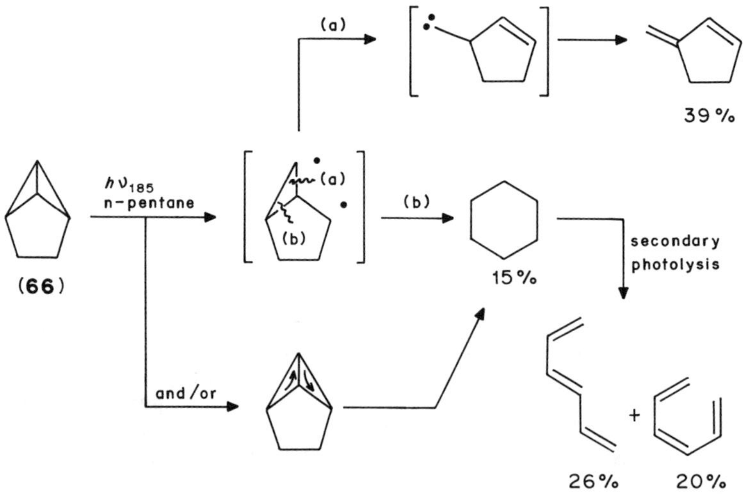

$$(36)$$

Finally, the generation of 15% 3-methylenecyclohexene (the product of lateral C—C bond cleavage) in the 185-nm irradiation of tricyclo[4.1.0.0²,⁷]heptane **67** (Table 9; entry 3d) shows that its photochemical behaviour is between the two outlined extremes. This is in accordance with the dependence of the dihedral angle Θ on the size of the methylene bridge and the energy ordering of the two highest occupied MOs of **67**, which are located between the two extremes (Figure 2). Furthermore, it was clearly demonstrated that pyrolyses and photolyses of the bridged bicyclobutanes **66–68** proceed in a different manner. This different behaviour is best evident in entry 3b (Table 9)[129c]. Pyrolysis of the deuterated tricycloheptane **d₂-67** results in the concerted ring-opening to cis/trans-1,3-cycloheptadiene **70**. Subsequently,**70** reacts to give the bicyclo[3.2.0]hept-6-ene **71** (equation 37). On the other hand, the appearance of deuterium in the bridgehead

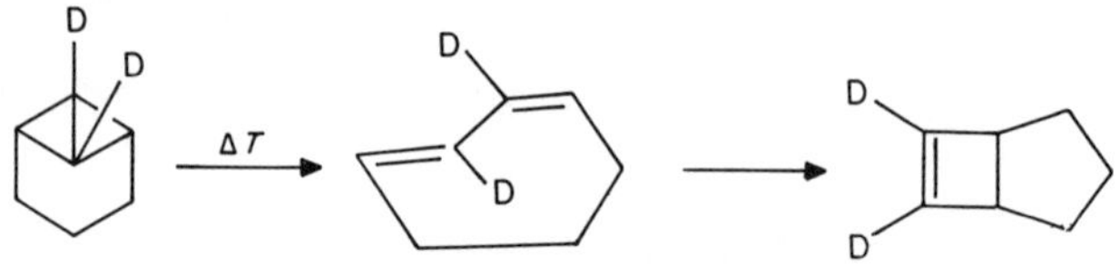

$$(37)$$

position of **71** (Table 9; entry 3e) is evidence for Wagner–Meerwein rearrangement (cf. equation 35) in the 185-nm photolysis of deuterated tricycloheptane **67**.

PET reactions of tricyclo[4.1.0.0²,⁷]heptane derivatives were carried out by Gassman and coworkers[128]. Small changes of the reaction conditions resulted in drastic changes of the product distribution or the product composition. This is also reflected by the spectral observation of radical cations which arise from tricyclo[4.1.0.0^{2,7}]heptane (**67**)[137b]. CIDNP spectroscopy of the radical cations generated with 1-CN-naphthalene in CD_3CN and CD_3COCD_3 shows central or lateral C—C bond cleavage (equation 38), which depends on the solvent used.

The mercury-sensitized photolysis of tricyclo[4.1.0.0^{2,7}]heptane (**67**)—cf. Table 9; entry 3c—probably proceeds by primary cleavage of a lateral C—C bond (equation 39).

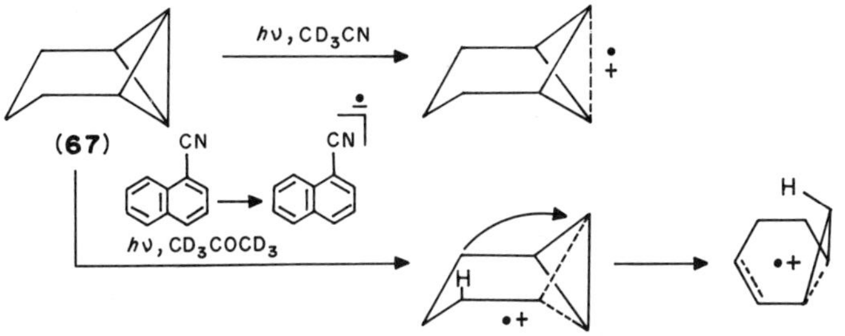

(38)

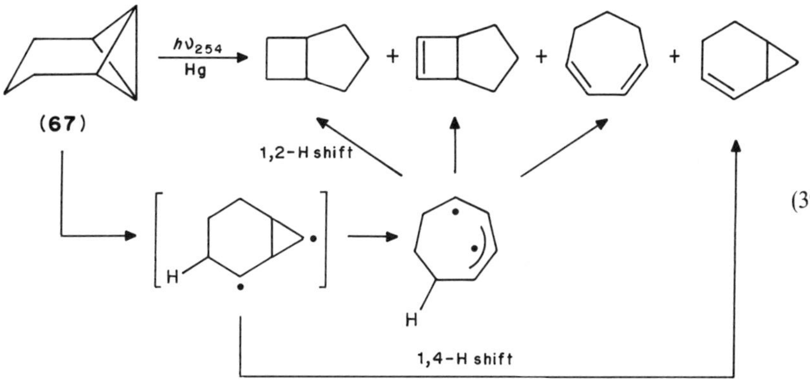

(39)

The thermolysis of dimeric bicyclobutanes (Table 9; entry 7) has been investigated by Szeimies and coworkers[135]. The reaction mechanism is not well understood. For the CF_3CO_2H-induced rearrangement of the anellated system **72** (Table 9; entries 8b,d) a mechanism was suggested, which involves the protonation of a lateral C—C bond. Subsequent Wagner–Meerwein rearrangement, generation of the non-classical carbenium ion **73**, followed by addition of the nucleophile leads to the observed products (equation 40). Finally, the systems of entry 8a and 8c (Table 9) provide one more example for the divergent behaviour of slightly different molecules in Ag$^+$-catalysed rearrangements.

5. Propellanes

The chemistry of propellanes[104] was recently reviewed by Wiberg[138a] and Ginsburg[138b]. As early as in 1971, Gassman[139] reported the transition metal catalysed rearrangement

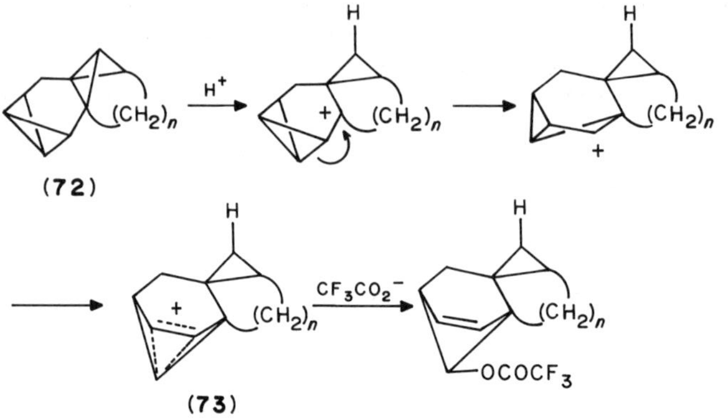

$(CH_2)_n$ (72)

(40)

$CF_3CO_2^-$ $(CH_2)_n$ $OCOCF_3$

(73)

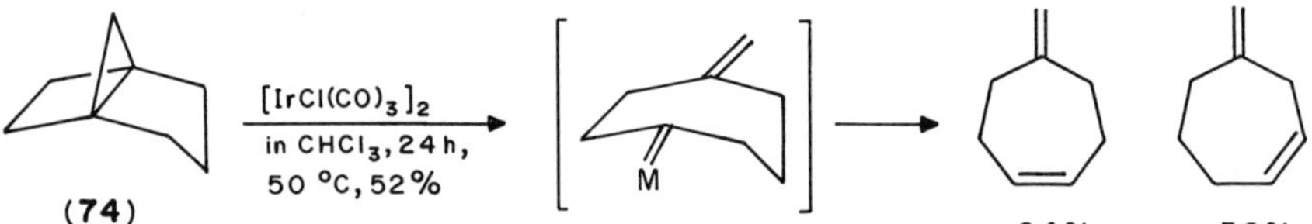

$\dfrac{[IrCl(CO)_3]_2}{in\ CHCl_3,\ 24\ h,}$ $50\ °C,\ 52\%$

(74)

M

64%   36%

(41)

of tricyclo[3.2.1.0^{1,5}]octane (74), which can be designated as a [3.2.1]propellane or a bridged bicyclo[2.1.0]pentane (equation 41).

The resulting methylenecycloheptenes are the products of an initially formed metalla–carbene complex, and the mechanism resembles that of equations 27 and 33. Blaustein and Berson[140] examined the thermal ring-opening process of dimethyl-substituted [3.2.1]propellanes (Table 10). However, the observed differences in the stereoselectivity still remain an unanswered question. It would be quite revealing to examine the 185-nm photolyses of these substrates, because the applicability of the Woodward–Hoffmann rules in the vacuum-UV range is of current interest[141].

TABLE 10. Pyrolysis of dimethyl-substituted [3.2.1]propellanes[140]

Substrate	Product distribution (%)		
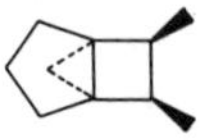	0.1	0.5	99.4
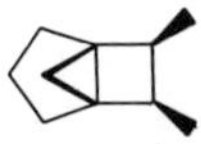	39.6	47.8	12.6
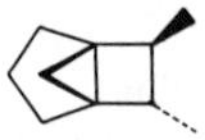	3.6	41.4	55.0

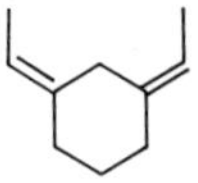

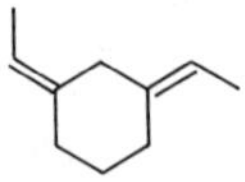

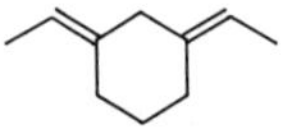

So far, considerable efforts have been made to examine thermal and acid catalysed rearrangements of propellanes. However, PET reactions and direct photolyses with high-energy photons (e.g. 185-nm excitation) have so far been neglected. It would be a challenging task for future research to fill this gap, because new information can be expected on the nature of the cleavage of the central propellane bond with its inverted carbon hybridization.

III. ACKNOWLEDGEMENTS

The authors wish to express their gratitude to Dr. W. Koch (F. Hoffmann–La Roche AG, Basle) for carrying out a *Chemical Abstracts* literature search. Furthermore, they thank Dr. H. Platsch (University of Basle) and Mr. A. Ritter (F. Hoffmann–La Roche AG, Basle) for reading and correcting the manuscript and Mr. Ph. Mühlethaler (F. Hoffmann–La Roche AG, Basle) for the skilful preparation of the original drawings. The work of the authors at Würzburg was generously supported by the Deutsche Forschungsgemeinschaft, the Fonds der Chemischen Industrie and the Stifterverband, for which we are grateful.

IV. REFERENCES

1. (a) C. L. Hill, *Activation and Functionalization of Alkanes*, Wiley, New York. 1989.
   (b) R. F. Renneke, M. Pasquali and C. L. Hill, *J. Am. Chem. Soc.*, **112**, 6585 (1990) and references cited therein.
2. (a) A. Greenberg and J. F. Liebman, *Strained Organic Molecules*, Academic Press, New York, 1978.
   (b) J. J. Gajewski, *Hydrocarbon Thermal Isomerizations*, Academic Press, New York, 1981.
   (c) L. T. Scott and M. Jones, Jr., *Chem. Rev.*, **72**, 181 (1972).
3. (a) H. N. C Wong, M.-Y. Hon, C.-W. Tse, Y.-C. Yip, J. Tanko and T. Hudlicky, *Chem. Rev.*, **89**, 165 (1989).
   (b) J. Salaün, in *The Chemistry of the Cyclopropyl Group* (Ed. Z. Rappoport), Part 2, Chap. 13, Wiley, Chichester, 1987.
4. J. F. Liebman and A. Greenberg, *Chem. Rev.*, **89**, 1225 (1989).
5. G. Maier, *Angew. Chem.*, **100**, 317 (1988).
6. G. W. Griffin and A. P. Marchand, *Chem. Rev.*, **89**, 997 (1989).
7. K. Hassenrück, H.-D. Martin and R. Walsh, *Chem. Rev.*, **89**, 1125 (1989).
8. (a) L. A. Paquette, *Chem. Rev.*, **89**, 1051 (1989).
   (b) P. E. Eaton, *Tetrahedron*, **35**, 2189 (1979).
9. (a) M. L. Occelli and R. G. Anthony (Eds.), *Hydrotreating Catalysts*, Elsevier, Amsterdam, 1989.
   (b) D. Kallo and K. M. Minachev (Eds.), *Catalysis on Zeolites*, Akademiai Kiado, Budapest, 1988.
10. (a) R. Srinivasan, B. Braren and K. G. Casey, *Pure Appl. Chem.*, **62**, 1581 (1990).
    (b) R. Srinivasan and B. Braren, *Chem. Rev.*, **89**, 1303 (1989).
11. (a) G. Hancock, *J. Photochem. Photobiol. A*, **51**, 13 (1990).
    (b) M. Brouard, *J. Photochem. Photobiol. A*, **51**, 17 (1990).
12. (a) B. Eliasson, U. Kogelschatz and H. J. Stein, *EPA Newslett*, **No. 32**, 29 (1988).
    (b) U. Kogelschatz, *Pure Appl. Chem.*, **62**, 1667 (1990).
    (c) B. Eliasson and U. Kogelschatz, *Appl. Phys., B*, **46**, 299 (1988).
    (d) H. Esram and U. Kogelschatz, *Mat. Res. Soc. Symp. Proc.*, **158**, 189 (1990).
    (e) H. Esram, J. Demny and U. Kogelschatz, *Chemtronics*, **4**, 202 (1989).
13. M. Anpo and T. Matsuura (Eds.), *Photochemistry on Solid Surfaces*, Elsevier, Amsterdam, 1989.
14. N. Serpone and E. Pelizzetti (Eds.), *Photocatalysis. Fundamentals and Applications*, Wiley, New York, 1989.
15. (a) A. Harriman, *Photochemistry*, **19**, 508 (1988).
    (b) M. Vicente and S. Esplugas, *Afinidad*, **46**, 393 (1989).
16. M. Grätzel, in Reference 17a, Part D, p. 394.

17. (a) M. A. Fox and M. Chanon (Eds.), *Photoinduced Electron Transfer*, Parts A to D, Elsevier, Amsterdam, 1988.
    (b) *Top. Curr. Chem.*, **156** (1989); **158** (1990).
    (c) J. Mattay, *Synthesis*, 233 (1989).
    (d) M. A. Fox, *Photochem. Photobiol.*, **52**, 617 (1990).
18. (a) M. G. Steinmetz, in *Organic Photochemistry* (Ed. A. Padwa), Vol. 8, Chap. 2, Marcel Dekker, New York, 1987.
    (b) W. Adam and T. Oppenländer, *Angew. Chem.*, **98**, 659 (1986); *Angew. Chem., Int. Ed. Engl.*, **25**, 661 (1986).
19. G. Zang, *Dissertation*, University of Würzburg (FRG), 1990.
20. (a) G. von Bünau and T. Wolff (Eds.), *Photochemie*, Chap. 1, Verlag Chemie, Weinheim, 1987.
    (b) cf. Reference 74.
21. (a) T. Kimura, Y. Ueda and Y. Hirai, *Can. J. Chem.*, **64**, 695 (1986).
    (b) E.-H. Böttcher and W. F. Schmidt, *J. Chem. Phys.*, **80**, 1353 (1984).
    (c) T. Funabashi, N. Okabe, T. Kimura and K. Fueki, *J. Chem. Phys.*, **75**, 1576 (1981).
22. (a) D. W. Werst, M. G. Bakker and A. D. Trifunac, *J. Am. Chem. Soc.*, **112**, 40 (1190).
    (b) M. Chanon, M. Rajzmann and F. Chanon, *Tetrahedron*, **46**, 6193 (1990).
23. S. Aeiyach, F. Garin, L. Hilaire, P. Legare and G. Maire, *J. Mol. Catal.*, **25**, 183 (1984).
24. (a) V. Hejtmanek, *Chem. Listy*, **83**, 473 (1989).
    (b) G. Maire and F. Garin, *J. Mol. Catal.*, **48**, 99 (1988).
    (c) F. Garin and G. Maire, *Acc. Chem. Res.*, **22**, 100 (1989).
25. G. Xiexian, Y. Yashu, D. Maicum, L. Huimin and L. Zhiyin, *J. Catal.*, **99**, 218 (1986).
26. (a) J. R. Anderson, *Adv. Catal.*, **25**, 1 (1976).
    (b) F. G. Gault, *Adv. Catal.*, **30**, 1 (1981).
    (c) J. K. A. Clarke and J. J. Rooney, *Adv. Catal.*, **25**, 125 (1976).
27. H. Zimmer and Z. Paal, *J. Mol. Catal.*, **51**, 261 (1989).
28. F. Le Normand, P. Girard, L. Hilaire, M. F. Ravet, G. Krill and G. Maire, *J. Catal.*, **89**, 1 (1984).
29. P. E. Puges, F. Garin, F. Weisang, P. Bernhardt, P. Girard, G. Maire, L. Guczi and Z. Schay, *J. Catal.*, **114**, 153 (1988).
30. J. K. A. Clarke, O. E. Finlayson and J. J. Rooney, *J. Chem. Soc., Chem. Commun.*, 1277 (1982).
31. J. B. F. Anderson, R. Burch and J. A. Cairns, *J. Chem. Soc., Chem. Commun.*, 1647 (1986).
32. A. Dauscher, F. Garin and G. Maire, *J. Catal.*, **105**, 233 (1987).
33. J. L. Margitfalvi, M. Hegedüs and E. Talas, *J. Mol. Catal.*, **51**, 279 (1989).
34. (a) Q. Y. Yang, A. D. Johnson, K. J. Maynard and S. T. Ceyer, *J. Am. Chem. Soc.*, **111**, 8748 (1989).
    (b) J. D. Beckerle, Q. Y. Yang, A. D. Johnson and S. T. Ceyer, *J. Chem. Phys.*, **86**, 7236 (1987).
35. S. Uemiya, T. Matsuda and E. Kikuchi, *Chem. Lett.*, 1335 (1990).
36. Y. Ishihara, H. Nanbu, C. Iwata, T. Ikemura and T. Takesue, *Bull. Chem. Soc. Jpn.*, **62**, 2981 (1989).
37. (a) G. Földiak and L. Wojnarovits, *Radiat. Phys. Chem.*, **32**, 335 (1988).
    (b) D. Phillips, *Photochemistry*, **5**, 199 (1974).
    (c) M. N. R. Ashfold, *Photochemistry*, **15**, 55 (1984).
38. (a) L. Wojnarovits and G. Földiak, *Radiat. Res.*, **91**, 638 (1982).
    (b) J. J. Doyle, S. K. Tokach, M. S. Gordon and R. D. Koob, *J. Phys. Chem.*, **86**, 3626 (1982).
    (c) S. K. Tokach and R. D. Koob, *J. Phys. Chem.*, **84**, 6 (1980).
39. (a) L. Flamigni, F. Barigelletti, S. Dellonte and G. Orlandi, *Chem. Phys. Lett*, **89**, 13 (1982).
    (b) S. Dellonte, L. Flamigni, F. Barigelletti, L. Wojnarovits and G. Orlandi, *J. Phys. Chem.*, **88**, 58 (1984).
    (c) M. A. Wickramaratchi, J. M. Presses, R. A. Holroyd and R. E. Weston, *J. Chem. Phys.*, **82**, 4745 (1985).
    (d) H. Miyasaka and N. Mataga, *Chem. Phys. Lett.*, **126**, 219 (1986).
    (e) F. P. Schwarz, D. Smith, S. G. Lias and P. Ausloos, *J. Chem. Phys.*, **75**, 3800 (1981).
40. (a) L. Wojnarovits, L. H. Luthjens, A. C. De Leng and A. Hummel, *J. Radioanal. Nucl. Chem.*, **101**, 349 (1986).
    (b) G. Orlandi, F. Barigelletti, L. Flamigni and S. Dellonte, *J. Photochem.*, **31**, 49 (1985).
41. M. Creuzburg, F. Koch and F. Wittl, *Chem. Phys. Lett.*, **156**, 387 (1989).
42. R. R. Hentz and S. J. Rzad, *J. Phys., Chem.*, **71**, 4096 (1967).

43. L. Wojnarovits, L. Kozari, C. S. Keszei and G. Földiak, *J. Photochem.*, **19**, 79 (1982).
44. J. Y. Yang, F. M. Servedio and R. A. Holroyd, *J. Chem. Phys.*, **48**, 1331 (1968).
45. P. Ausloos, S. G. Lias and R. E. Rebbert, *J. Phys. Chem.*, **85**, 2322 (1981).
46. (a) A. Ouchi, A. Yabe, Y. Inoue, A. Daino and T. Hakushi, *J. Chem. Soc., Chem. Commun.*, 1669 (1989).
   (b) A. Ouchi, A. Yabe, Y. Inoue and Y. Diano, *XIIIth IUPAC Symposium on Photochemistry*, University of Warwick, Coventry (England), July 22–28, 1990, Book of Abstracts, p. 76.
   (c) A. Ouchi, personal communication, *XIIIth IUPAC Symposium on Photochemistry*, University of Warwick, Coventry (England), July 22–28, 1990.
47. R. D. Doepker and P. Ausloos, *J. Chem. Phys.*, **42**, 3746 (1965).
48. (a) S. H. Brown and R. H. Crabtree *J. Am. Chem. Soc.*, **111**, 2946 (1989).
   (b) S. H. Brown and R. H. Crabtree, *J. Am. Chem. Soc.*, **111**, 2935 (1989).
   (c) S. Yamamoto, M. Kozasa, Y. Sueishi and N. Nishimura, *Bull. Chem. Soc. Jpn.*, **61**, 3449 (1988).
49. L. Wojnarovits, T. Szondy, E. Szekeres-Bursics and G. Földiak, *J. Photochem.*, **18**, 273 (1982).
50. Y. Inoue, T. Mukai and T. Hakushi, *Chem. Lett.*, 1665 (1983).
51. (a) J. Pola, *Collect. Czech. Chem. Commun.*, **49**, 231 (1984).
   (b) J. T. De Maleissye, F. Lempereur, C. Lalo and J. Masanet, *J. Photochem.*, **27**, 273 (1984).
   (c) J. Pola, P. Kubat, J. Vitek, M. Farkacova and A. Trka, *Collect. Czech. Chem. Commun.*, **48**, 3527 (1983).
52. M. Quinn, Z. Sabet and N. Alyassini, *J. Phys. Chem.*, **86**, 2880 (1982).
53. (a) E. Osawa, K. Aigami, N. Takaishi, Y. Inamoto, Y. Fujikura, Z. Majerski, P. von R. Schleyer, E. M. Engler and M. Farcasiu, *J. Am. Chem. Soc.*, **99**, 5361 (1977).
   (b) M. A. McKervey, *Chem. Soc. Rev.*, **3**, 479 (1974).
   (c) M. A. McKervey, *Tetrahedron*, **36**, 971 (1980).
   (d) R. C. Fort, Jr. and P. von R. Schleyer, *Chem. Rev.*, **64**, 277 (1964).
   (e) Y. Fujikura, N. Takaishi and Y. Inamoto, *Tetrahedron*, **37**, 4465 (1981).
   (f) Y. Inamoto, K. Aigami, N. Takaishi, Y. Fujikura, K. Tsuchihashi and H. Ikeda, *J. Org. Chem.* **42**, 3833 (1977).
   (g) F. S. Hollowood, M. A. McKervey, R. Hamilton and J. J. Rooney, *J. Org. Chem.*, **45**, 4954 (1980).
   (h) E. Osawa, A. Furusaki, N. Hashiba, T. Matsumoto, V. Singh, Y. Tahara, E. Wiskott, M. Farcasiu, T. Iizuka, N. Tanaka, T. Kan and P. von R. Schleyer, *J. Org. Chem.*, **45**, 2985 (1980).
   (i) K.-P. Zeller, R. Müller and R. W. Alder, *J. Chem. Soc., Perkin Trans. 2*, 1711 (1984).
   (j) K.-P. Zeller, R. Müller and R. W. Alder, *J. Chem. Soc., Chem. Commun.*, 330 (1984).
54. P. von R. Schleyer, *J. Am. Chem. Soc.*, **79**, 3292 (1957).
55. (a) R. Dutler, A. Rauk, T. S. Sorensen and S. M. Whitworth, *J. Am. Chem. Soc.*, **111**, 9024 (1989).
   (b) S. Fujita, *Tetrahedron*, **46**, 365 (1990).
   (c) R. C. Fort, Jr., *Adamantane: The Chemistry of Diamond Molecules*, Marcel Dekker, New York, 1976.
   (d) J. R. Wiseman, J. J. Vanderbilt and W. M. Butler, *J. Org. Chem.*, **45**, 667 (1980).
   (e) M. Farcasiu, E. W. Hagaman, E. Wenkert and P. von R. Schleyer, *Tetrahedron Lett.*, 1501 (1981).
   (f) F. J. Jäggi and C. Ganter, *Helv. Chim. Acta*, **63**, 866 (1980).
   (g) A. M. Klester, F. J. Jäggi and C. Ganter, *Helv. Chim. Acta*, **63**, 1294 (1980).
   (h) Z. Majerski, S. Djigas and V. Vinkovic, *J. Org. Chem.*, **44**, 4064 (1979).
   (i) L. A. Paquette, G. V. Meehan and S. J. Marshall, *J. Am. Chem. Soc.*, **91** 6779 (1969).
   (j) P. von R. Schleyer and M. M. Donaldson, *J. Am. Chem. Soc.*, **82**, 4645 (1960).
   (k) F. J. Jäggi and C. Ganter, *Helv. Chim. Acta*, **63**, 2087 (1980).
   (l) M. Brossi and C. Ganter, *Helv. Chim. Acta*, **70**, 1963 (1987).
   (m) H.-R. Känel and C. Ganter, *Helv. Chim. Acta*, **68**, 1226 (1985).
   (n) H.-R. Känel H.-G. Capraro and C. Ganter, *Helv. Chim. Acta*, **65**, 1032 (1982).
   (o) A. M. Klester and C. Ganter, *Helv. Chim. Acta*, **68**, 734 (1985).
   (p) A. M. Klester and C. Ganter, *Helv. Chim. Acta*, **68**, 104 (1985).
   (q) A. M.Klester and C. Ganter, *Helv. Chim. Acta*, **66**, 1200 (1983).
   (r) G. C. Lau and W. F. Maier, *Langmuir*, **3**, 164 (1987).
   (s) K. Honna, M. Sugimoto, N. Shimizu and K. Kurisaki, *Chem. Lett.*, 315 (1986).

738   T. Oppenländer, G. Zang and W. Adam

56. (a) H. W. Whitlock, Jr. and M. W. Siefken, *J. Am. Chem. Soc.*, **90**, 4929 (1968).
    (b) E. M. Engler, M. Farcasiu, A. Sevin, J. M. Cense and P. von R. Schleyer, *J. Am. Chem. Soc.*, **95**, 5769 (1973).
    (c) E. M. Engler, J. D. Andose and P. von R. Schleyer, *J. Am. Chem. Soc.*, **95**, 8005 (1973).
57. S. A. Godleski, P. von R. Schleyer, E. Osawa and W. T. Wipke, *Prog. Phys. Org. Chem.*, **13**, 63 (1981).
58. P. A. Krasutsky, I. R. Likhotvorik, A. L. Litvyn, A. G. Yurchenko and D. van Engen, *Tetrahedron Lett.*, **31**, 3973 (1990).
59. (a) K. L. Ghatak and C. Ganter, *Helv. Chim. Acta*, **71**, 124 (1988).
    (b) Y. Tobe, K. Terashima, Y. Sakai and Y. Odaira, *J. Am. Chem. Soc.*, **103**, 2307 (1981).
    (c) P. von R. Schleyer, P. Grubmüller, W. F. Maier, O. Vostrowsky, L. Skattebol and K. H. Holm, *Tetrahedron Lett*, **21**, 921 (1980).
60. M. Tichy, L. Kniezo and J. Hapala, *Tetrahedron Lett.*, 699 (1972).
61. S. G. Pozdnkina, O. E. Morozova and A. A. Petrov, *Neftekhimiya*, **13**, 21 (1973).
62. (a) H. Hogeveen and H. C. Volger, *J. Am. Chem. Soc.*, **89**, 2486 (1967).
    (b) W. Merk and R. Pettit, *J. Am. Chem. Soc.*, **89**, 4787, 4788 (1967).
    (c) L. A. Paquette and J. C. Stowell, *J. Am. Chem. Soc.*, **92**, 2584 (1970).
    (d) P. E. Eaton, L. Cassar, R. A. Hadson and D. R. Hwang, *J. Org. Chem.*, **41** 1445 (1976).
    (e) J. Halpern, in *Organic Synthesis via Metal Carbonyls* (Eds. I. Wender and P. Pino), Vol. II, Wiley, New York, 1977, p. 705.
    (f) K. C. Bishop III, *Chem. Rev.*, **76**, 461 (1976).
    (g) See also Reference 2a, p. 245.
63. L. A. Paquette, *Synthesis*, 1975, 347.
64. M. Christl, E. Brunn, W. R. Roth and H.-W. Lennartz, *Tetrahedron*, **45**, 2905 (1989).
65. F. J. Jäggi, P. Buchs and C. Ganter, *Helv. Chim. Acta*, **63**, 872 (1980).
66. (a) T. Katsushima, R. Yamaguchi, M. Kawanisi and E. Osawa, *Bull. Chem. Soc. Jpn.*, **53**, 3313 (1980).
    (b) K. B. Wiberg and K. C. Bishop III, *Tetrahedron Lett.*, 2727 (1973).
    (c) T. Katsushima, R. Yamaguchi, H. Kwasani and E. Osawa, *J. Chem. Soc., Chem. Commun.*, 39 (1976).
    (d) T. Katsushima, R. Yamaguchi and M. Kawanisi *Bull. Chem. Soc. Jpn.*, **55**, 3245 (1982).
67. (a) A. Uchida and K. Hata, *J. Mol. Catal.*, **15**, 111 (1982).
    (b) A. Uchida and K. Hata, *J. Chem. Soc., Dalton Trans.*, 1111 (1981).
    (c) P. G. Gassman and T. H. Johnson, *J. Am. Chem. Soc.*, **98**, 6057 (1976).
    (d) P. G. Gassman and T. H. Johnson, *J. Am. Chem. Soc.*, **99**, 622 (1977).
    (e) N. Calderon, J. P. Lawrence and E. A. Ofstead, *Adv. Organomet. Chem.*, **17**, 449 (1979).
68. A. Sen, T.-W. Lai and R. R. Thomas, *J. Organomet. Chem.*, **358**, 567 (1988).
69. (a) M. R. A. Blomberg, P. E. M. Siegbahn and J. E. Blackvall, *J. Am. Chem. Soc.*, **109**, 4450 (1987).
    (b) E.-I. Negishi, *Organometallics in Organic Synthesis*, Vol. I, Wiley, New York, 1980.
70. R. Yamaguchi, M. Ban and M. Kawanisi, *Bull. Chem. Soc. Jpn.*, **61**, 2909 (1988).
71. W. H. Campbell and P. W. Jennings, *Organometallics*, **1**, 1071 (1982).
72. J. O. Hoberg and P. W. Jennings, *J. Am. Chem. Soc.*, **112**, 4347 (1990).
73. S. S. Hixson, in *Organic Photochemistry* (Ed. A. Padwa), Vol. 4, Chap. 3, Marcel Dekker, New York, 1979.
74. Z. B. Alfassi, in *The Chemistry of the Cyclopropyl Group* (Ed. Z. Rappoport), Part 2, Chap. 14, Wiley, New York, 1987.
75. (a) W. C. Danen and J. C. Jang, in *Laser-induced Chemical Processes* (Ed. J. I. Stenfeld), Vol. 1, Chap. 2, Plenum Press, New York, 1981.
    (b) W. E. Farneth and M. W. Thomsen, *J. Phys. Chem.*, **87**, 3207 (1983).
76. W. E. Farneth and M. W. Thomsen, *J. Am. Chem. Soc.*, **105**, 1843 (1983).
77. (a) R. Srinivasan and J. A. Ors, *J. Org. Chem.*, **44**, 3426 (1979).
    (b) H. P. Schuchmann, C. von Sonntag and D. Schulte-Frohlinde, *J. Photochem.*, **3**, 267 (1974/75).
    (c) R. Srinivasan, L. S. White, A. R. Rossi and G. A. Epling, *J. Am. Chem. Soc.*, **103**, 7299 (1981).

78. E. A. Volnina, P. Kubat and J. Pola, *J. Org. Chem.*, **53**, 2612 (1988).
79. (a) W. Adam, T. Oppenländer and G. Zang, *J. Am. Chem. Soc.*, **107**, 3921 (1985).
    (b) A. F. Becknell, J. A. Berson and R. Srinivasan, *J. Am. Chem. Soc.*, **107**, 1076 (1985).
80. G. B. M. Kostermans, M. Hogenbirk, L. A. M. Turkenburg, W. H. de Wolf and F. Bickelhaupt, *J. Am. Chem. Soc.*, **109**, 2855 (1987).
81. (a) N. J. Turro, V. Ramamurthy and T. J. Katz, *Nouv. J. Chim.*, **1**, 363 (1977).
    (b) N. J. Turro, *Modern Molecular Photochemistry*, Chap. 12, Benjamin/Cummings, Menlo Park, California, 1978, p. 502.
82. A. A. Gorman, I. Hamblett and S. P. McNeeney, *Photochem. Photobiol.*, **51**, 145 (1990) and references cited therein.
83. (a) A. H. A. Tinnemans, B. den Ouden, H. J. T. Bos and A. Mackor, *Recl. Trav. Chim. Pays-Bas*, **104**, 109 (1985).
    (b) D. J. Fife, W. M. Moore and K. W. Morse, *J. Am. Chem. Soc.*, **107**, 7077 (1985).
    (c) T. Arai, T. Oguchi, T. Wakabayashi, M. Tsuchiya, Y. Nishimura, S. Oishi, H. Sakuragi and K. Tokumaru, *Bull. Chem. Soc. Jpn.*, **60**, 2937 (1987).
    (d) T. Arai, T. Wakabayashi, H. Sakuragi and K. Tokumaru, *Chem. Lett.*, 279 (1985).
    (e) S. J. Cristol and R. L. Kaufman, *J. Photochem.*, **12**, 207 (1980).
    (f) K. Maruyama, H. Tamiaki and S. Kawabata, *J. Org. Chem.*, **50**, 4742 (1985).
    (g) G. Jones II, in Reference 17a, Part A, Chap. 1.7.
84. (a) Z.-I. Yoshida, *J. Photochem.*, **29**, 27 (1985).
    (b) D. Wöhrle, H. Bohlen and H.-W. Rothkopf, *Makromol. Chem.*, **184**, 763 (1983).
85. (a) C. Kutal, C. K. Kelley and G. Ferraudi, *Inorg. Chem.*, **26**, 3258 (1987).
    (b) G. Jones II, S.-H. Chiang, W. G. Becker and J. A. Welch, *J. Phys. Chem.*, **86**, 2805 (1982).
    (c) C. K. Kelley and C. Kutal, *Organometallics*, **4**, 1351 (1985).
86. J. Santamaria, in Reference 17a, Part B, Chap. 3.1.
87. (a) N. C. Baird, A. M. Draper and P. De Mayo, *Can. J. Chem.*, **66**, 1579 (1988).
    (b) H. Ikezawa and C. Kutal, *J. Org. Chem.*, **52**, 3299 (1987).
    (c) A. M. Draper and P. De Mayo, *Tetrahedron Lett.*, **27**, 6157 (1986).
    (d) P. G. Gassman, R. Yamaguchi and G. F. Koser, *J. Org. Chem.*, **43**, 4392 (1978).
88. (a) P. Pichat an M. A. Fox, in Reference 17a, Part D, Chap. 6.1.
    (b) M. Grätzel, in Reference 17a, Part D, Chap. 6.3.
    (c) J.-M. Herrmann and P. Pichat, *J. Chem. Soc., Faraday Trans. 1*, **76**, 1138 (1980).
    (d) M. Vesely, M. Ceppan, L. Lapcik, V. Brezova and A. Blazej, *Chem. Listy*, **83**, 605 (1989).
    (e) M. A. Fox, *Acc. Chem. Res.*, **16**, 314 (1983).
    (f) M. Anpo, *Res. Chem. Intermed.*, **11**, 67 (1989).
    (g) G. L. Haller and D. E. Resasco, *Adv. Catal.*, **36**, 173 (1989).
89. E. S. Kirkor, V. M. Maloney and J. Michl, *J. Am. Chem. Soc.*, **112**, 148 (1990).
90. R. Srinivasan, T. Baum and G. Epling, *J. Chem. Soc., Chem. Commun.*, 437 (1982).
91. D. G. Lishan, K. V. Reddy, G. S. Hammond and J. E. Leonard, *J. Phys. Chem.*, **92**, 656 (1988).
92. (a) N. J. Turro, W. R. Cherry, M. F. Mirbach and M. J. Mirbach, *J. Am. Chem. Soc.*, **99**, 7388 (1977).
    (b) N. J. Turro, Ref. 81b, p. 427f.
    (c) W. Adam, M. Dörr and P. Hössel, *Angew. Chem.*, **98**, 820 (1986).
93. T. Oppenländer, *Dissertation*, University of Würzburg (FRG), 1984.
94. (a) W. Adam and O. De Lucchi, *Angew. Chem.*, **92**, 815 (1980).
    (b) P. S. Engel, *Chem. Rev.*, **80**, 99 (1980).
95. P. S. Engel and C. Steel, *Acc. Chem. Res.*, **6**, 275 (1973).
96. (a) W. Adam, T. Oppenländer and G. Zang, *J. Org. Chem.*, **50**, 3303 (1985).
    (b) M. B. Robin, *Higher Excited States of Polyatomic Molecules*, Vol. II, Academic Press, New York, 1975, p. 68.
    (c) M. B. Robin, in *The Chemistry of the Hydrazo, Azo and Azoxy Groups* (Ed. S. Patai), Part I, Wiley, New York, 1975.
    (d) H. Rau, *Angew. Chem.*, **85**, 248 (1973).
    (e) J. Michl, *Mol. Photochem.*, **4**, 243 (1972).
    (f) M. B. Robin, R. R. Hart and N. A. Kuebler, *J. Am. Chem. Soc.*, **89**, 1564 (1967).
97. (a) E. Kassab, J. T. Gleghorn and E. M. Evleth, *J. Am. Chem. Soc.*, **105**, 1746 (1983).
    (b) Y. I. Dorofeev and V. E. Skurat, *Russ. Chem. Rev.*, **51**, 527 (1982).

740   T. Oppenländer, G. Zang and W. Adam

(c) W. J. Leigh and R. Srinivasan, *J. Am. Chem. Soc.*, **104**, 4424 (1982).

(d) R. Srinivasan, T. Baum and J. A. Ors, *Tetrahedron Lett.*, **22**, 4795 (1981).

(e) R. Srinivasan and J. A. Ors, *J. Am. Chem. Soc.*, **100**, 7089 (1978).

(f) M. Julliard and M. Chanon, *Chem. Rev.*, **83**, 425 (1983).

98. W. Adam and T. Oppenländer, *Photochem. Photobiol.*, **39**, 719 (1984).

99. Reference 81b, p. 353.

100. K. Raghavachari, R. C. Haddon and H. D. Roth, *J. Am. Chem. Soc.*, **105**, 3110 (1983).

101. (a) D. O. Cowan and R. L. Drisko, *Elements of Organic Photochemistry*, Plenum Press, New York, 1976, p. 9.

(b) N. J. Turro, V. Ramamurthy, W. Cherry and W. Farneth, *Chem. Rev.*, **78**, 125 (1978).

102. (a) M. F. Mirbach, M. J. Mirbach, K.-C. Liu and N. J. Turro, *J. Photochem.*, **8**, 299 (1978).

(b) N. J. Turro, W. R. Cherry, M. J. Mirbach, M. F. Mirbach and V. Ramamurthy, *Mol. Photochem.*, **9**, 111 (1978).

103. J. A. Berson, in *Rearrangements in Ground and Excited States* (Ed. P. De Mayo), Vol. 1, Academic Press, New York, 1980.

104. K. B. Wiberg, J. J. Caringi and M. G. Matturro, *J. Am. Chem. Soc.*, **112**, 5854 (1990).

105. (a) A. P. Marchand, *Chem. Rev.*, **89**, 1011 (1989).

(b) G. W. Schriver and D. J. Gerson, *J. Am. Chem. Soc.*, **112**, 4723 (1990).

106. (a) K. Hassenrück, J. G. Radzisewski, V. Balaji, G. S. Murthy, A. J. McKinley, D. E. David, V. M. Lynch, H.-D. Martin and J. Michl, *J. Am. Chem. Soc.*, **112**, 873 (1990).

(b) P. E. Eaton and J. Tsanaktsidis, *J. Am. Chem. Soc.*, **112**, 876 (1990).

107. D. A. Hrovat and W. T. Borden, *J. Am. Chem. Soc.*, **112**, 875 (1990).

108. (a) P. E. Eaton, C.-X. Yang and Y. Xiong, *J. Am. Chem. Soc.*, **112**, 3225 (1990).

(b) D. A. Hrovat and W. T. Borden, *J. Am. Chem. Soc.*, **112**, 3227 (1990).

(c) R. M. Moriarty, S. M. Tuladhar, R. Penmasta and A. K. Awasthi, *J. Am. Chem. Soc.*, **112**, 3228 (1990).

109. W. Adam and T. Oppenländer, *Angew. Chem.*, **96**, 599 (1984), *Angew. Chem., Int. Ed. Engl.*, **23**, 641 (1984).

110. M. B. Robin, Reference 96b, Vol. 1, p. 146.

111. Reference 18a, p. 114.

112. We thank Prof. P. E. Eaton (University of Chicago) for a generous gift of cubane.

113. T. Oppenländer, *Diploma Thesis*, University of Würzburg (FRG), 1981.

114. (a) Z. Rappoport (Ed.), *The Chemistry of the Cyclopropyl Group*, Parts 1, 2, Wiley, Chichester, 1987.

(b) P. de Mayo (Ed.), *Rearrangements in Ground and Excited States*, Part 1–3, Academic Press, New York, 1980.

(c) J. F. Liebman and A. Greenberg (Eds.), *Molecular Structure and Energetics*, Vol. 3, Verlag Chemie, Weinheim, 1986.

(d) J. F. Liebman and A. Greenberg (Eds.), *Structure and Reactivity*, Vol. 7, Verlag Chemie, Weinheim, 1986.

(e) J. K. Kochi (Ed.), *Free Radicals*, Wiley, New York, 1973.

(f) C. Wentrup, *Reaktive Zwischenstufen I*, Thieme, Stuttgart, 1979, p. 110.

(g) W. T. Borden (Ed.), *Diradicals*, Wiley, New York, 1982.

115. (a) P. G. Gassman, T. J. Atkins and J. T. Lumb, *J. Am. Chem. Soc.*, **94**, 7757 (1972).

(b) R. Noyori, T. Suzuki and H. Takaya, *J. Am. Chem. Soc.*, **93**, 5896 (1971).

(c) J. E. Baldwin and G. D. Andrews, *J. Org. Chem.*, **38**, 1063 (1973).

(d) W. Adam and T. Oppenländer, *J. Am. Chem. Soc.*, **107**, 3924 (1985).

(e) K. Ushida, T. Shida and J. C. Walton, *J. Am. Chem. Soc.*, **108**, 2805(1986).

(f) F. Williams, Q.-X. Guo, T. M. Kolb and S. F. Nelsen, *J. Chem. Soc., Chem. Commun.*, 1835 (1989).

116. (a) K. B. Wiberg, S. R. Kass and K. C. Bishop III, *J. Am. Chem. Soc.*, **107**, 996 (1985).

(b) J. E. Baldwin and J. Ollerenshaw, *J. Org. Chem.*, **46**, 2116 (1981).

(c) H. Walter, *Dissertation*, University of Würzburg, in preparation.

117. W. R. Roth and K. Enderer, *Justus Liebigs Ann. Chem.*, **733**, 44 (1970).

118. P. Lahr, *Dissertation*, University of Bochum, 1972.

119. (a) H. Tanida, S. Teratake, Y. Hata and M. Watanabe, *Tetrahedron Lett.*, (1969) 5345.

(b) W. R. Roth, F.-G. Klärner, W. Grimme, H. G. Köser, R. Busch, B. Muskulus, R. Breuckmann, B. P. Scholz and H.-W. Lennartz, *Chem. Ber.*, **116**, 2717 (1983).

II. THE ENERGY DEPOSITION

A. Introduction

We may distinguish three kinds of high-energy radiation: fast-moving charged particles (electrons, protons, etc.), electromagnetic radiation (e.g. X rays, gamma rays) and neutrons. High-energy photons interact with the medium in three ways. In the photo-electric effect, which is predominant at low energies, an electron is ejected from a molecule with a kinetic energy equal to that of the photon (minus the binding energy). In the Compton effect part of the photon energy is imparted to an electron as kinetic energy and a lower-energy photon is formed. At higher energies ($E > 1.04$ MeV) photons give rise to formation of pairs of electrons and positrons. We see that the energy of the high-energy photons is converted into kinetic energy of charged particles. The charged particles then parcel out their energy in many small losses forming a track of electronic excitations and ionizations, as we shall see below.

Neutrons have no Coulomb interactions and energy loss takes place by elastic collisions with the nuclei, which may result in ejection of ionized atoms with kinetic energy. Here again the chemical effects result from the energy losses of charged particles.

High-energy charged particles lose their energy predominantly by Coulomb interaction with the electrons of the medium, in mostly small energy losses, which results in a large number of electronic excitations and ionizations along their paths. Secondary electrons are formed that cause further electronic excitations and ionizations, if their energy is sufficient, until they finally become thermalized. In this way high-energy charged particles form tracks of electronically excited molecules, positive ions (which may or may not be electronically excited) and thermalized electrons. These species subsequently undergo various reactions which eventually lead to stable products.

While the spectrum of reactive species formed initially is not largely different for different primary charged particles and different energies of these particles, the spatial distribution of these species is. While for MeV electrons in the condensed phase subsequent ionizations along the path are spaced on the order of 200 nm apart, for 1 MeV α-particles this is on the order of 0.4 nm. This difference in the density of the initially reactive species in the track has important consequences for the ensuing chemical reactions.

In the following we shall discuss the energy deposition in the tracks of high-energy charged particles. For a more extensive treatment the reader is referred to the chapters by Magee and Chatterjee in References 2 and 3 and the chapter by Mozumder in Reference 5.

B. Energy Losses of a Fast-moving Charged Particle

When a fast-moving charged particle traverses a medium, energy is transferred to the electrons of the medium. In a simple picture the interaction is described as a fast-moving charged particle that passes along a free electron at rest. Due to the Coulomb force the free electron is set in motion at the expense of the speed of the primary particle. This classical model is very good for the so-called hard collisions in which relatively large amounts of energy are transferred. It may be shown that the probability for a primary charged particle with charge ze and velocity v to transfer an energy between Q and dQ to an electron of the medium with mass m_e, when passing a unit distance through the medium, is given by

$$dP = \frac{2\pi z^2 e^4}{(4\pi\varepsilon_0)^2 m_e v^2} NZ \frac{dQ}{Q^2} \tag{1}$$

where N is the number of molecules per unit volume and Z the number of electrons per molecule, so that NZ is the electron density. We see that dP has the dimension of length^{-1}. It represents the probability of events in the interval Q to $Q + dQ$ per unit path length or the number of such events per unit path length of the primary particle. The quantity dP/NZ has a dimension (length)^2 and represents the (differential) cross section for the events per electron. We see that small losses are favoured. (From the inverse proportionality with the mass of the electron m_e we also see that relatively little energy is transferred to the nuclei of the medium).

For low values of Q, where Q is comparable to the binding energy of the electron, the picture of a collision with a free electron at rest is not any more correct. In these distant or soft collisions the effect of the varying electric fields due to the passing of the primary particle on the whole molecule has to be considered. This electric field varying in time is experienced by the molecule as electromagnetic waves with a range of frequencies v, or energies hv. These waves may now cause electronic excitation by resonant transition the same way as occurs with the absorption of light. For these soft collisions dP is given by

$$dP = \frac{2\pi z^2 e^4}{(4\pi\varepsilon_0)^2 m_e v^2} \frac{df}{dQ} \frac{N}{Q} \ln\left(\frac{2m_e v^2}{Q}\right) dQ \tag{2}$$

The quantity df/dQ, called the differential oscillator strength, is simply proportional to the optical absorption coefficient, ε. Since the logarithmic term varies only slowly with Q, we see that the probability for resonant transitions is roughly proportional to $(df/dQ)/Q$ or to $\varepsilon(hv)/hv$. The transitions described by equation 2 are called optical or resonant transitions.

The differential oscillator strength distributions have been determined in detail and over a wide range of energy for a number of molecules in the gas phase. An important feature of these df/dQ distributions is that a maximum is always found around twice the ionization potential, i.e. around 20 eV for hydrocarbons. In Figure 1 we show the

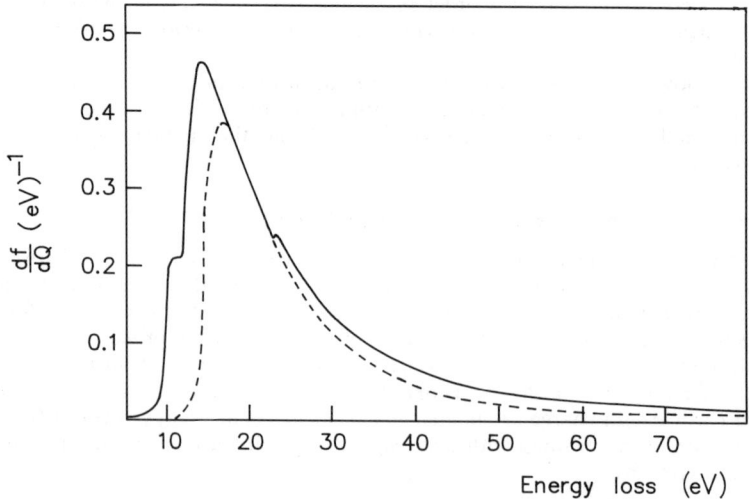

FIGURE 1. Oscillator strength distribution for methane. The dashed line indicates the oscillator strength for ionization. Reprinted with permission from Ref. 6. Copyright (1975) Pergamon Press PLC

case of gaseous methane as an example. Much less is known about the oscillator strength distributions in the condensed phase. From the few cases that have been studied, however, it appears that the distributions in the gas and the condensed phase are not grossly different.

The oscillator strength distribution for CH_4 shown in Figure 1 is a sum of contributions for different transitions. The small peak around 7–8 eV, for example, represents the transition that leads to the molecular elimination $CH_2 + H_2$. At higher energies other neutral dissociation processes are found. The ionization cross section has an onset at 12.5 eV for the process $CH_4^{+\cdot} + e^-$, however at larger energies dissociative ionizations are found (e.g. $CH_3^+ + H^\cdot + e^-$ starting at 14.4 eV). We shall discuss the various processes more extensively in Section III.

Using the two expressions in equations 1 and 2, together with the df/dQ distribution, we can now calculate the number of each energy loss that takes place along the path of the high-energy charged particle through the medium. If the contributions of the different processes to the overall df/dQ are known, the distribution of the different initial events in space can be calculated.

The total energy lost by the primary charged particle per unit path length can be obtained by integrating over all losses ($\int QdP$), using equations 1 and 2 for the hard and soft collisions, respectively, and summing the two. It may be shown that the contribution of both losses in the total energy loss of the primary particle is about equal. The energy loss per unit path, or the linear energy transfer (LET), for non-relativistic heavy particles is given by

$$-\frac{dE}{dx} = \frac{4\pi z^2 e^4}{(4\pi\varepsilon_0)^2 m_e v^2} NZ \ln \frac{2m_e v^2}{I} \tag{3}$$

where I is called the mean excitation potential, which turns out to be approximately 50 eV for saturated hydrocarbons.

We see that the LET is approximately inversely proportional to the square of the velocity v^2, and proportional to the square of the charge z^2, but that the mass of the fast-moving particle does not appear. The linear energy transfer is also called the stopping power.

For non-relativistic electrons the term under the logarithm is somewhat different $[(m_e v^2/2I)(e/2)^{1/2}]$ and for relativistic particles the expression for $-dE/dx$ is again modified.

It has been shown that the linear energy transfer is determined to a good approximation by the atomic composition of the medium only and is virtually independent of the chemical binding of the atoms. The LET divided by the electron density NZ is approximately independent of the state of aggregation.

We have seen that the energy loss distribution is heavily weighted towards small energy losses. From equation 2 and the oscillator strength distribution we find that the average loss of a fast primary particle due to resonant collisions is around 20 eV; the most probable energy loss is around 15 eV. About half of the total energy lost is in losses below 100 eV. The overall average energy loss is around 40 eV.

In Table 1 we show the number of energy losses Q in intervals of Q that appear when a 1-MeV electron slows down in a molecular medium such as a saturated hydrocarbon. For the calculation a 'typical' oscillator strength distribution has been used. The values given for the lowest losses may therefore be somewhat different in actual cases. We see the predominance of the small losses around the peak of df/dQ in the interval 10–30 eV. We have also indicated the fraction of energy lost in each interval of Q, which shows that although the large losses are few in number, they represent a non-negligible fraction of the energy.

TABLE 1. The number of primary energy losses in the track of a 1-MeV
electron slowing down in a saturated hydrocarbon[7]

Energy interval (eV)		Number of losses between Q_1 and Q_2	Fraction of total energy lost in interval Q_1 to Q_2
Q_1	Q_2		
0	10	5600	0.04
10	30	14300	0.25
30	80	4300	0.19
80	100	330	0.03
		24530	0.51
100	300	320	0.056
300	500	80	0.033
500	1000	58	0.040
1×10^3	5×10^3	46	0.093
5×10^3	1×10^4	5.8	0.040
1×10^4	1×10^5	5.2	0.134
1×10^5	5×10^5	0.5	0.093
		515	0.49
		Total: 25045	Total: 1.0

C. Spatial Distribution of Energy Losses in the Track

First we consider the spacing of the energy losses along the path of the primary
charged particle. Using the value for the LET for a given particle with a given energy
for a certain medium, we can estimate the average spacing between primary events. For
a 1-MeV electron in a medium with density $1 \, kg \, dm^{-3}$ the LET is about $0.2 \, eV \, nm^{-1}$.
With an average primary energy loss of 40 eV we find that the average spacing is
$40/0.2 = 200 \, nm$. With decreasing energy of the primary (v lower) the LET increases and
therefore the average spacing decreases. For electrons of 10 keV and 1 keV the spacings
are ca 40 nm and 5 nm, respectively.

A 1-MeV α-particle, due to the large mass and $z = 2$, has an LET approximately $500 \times$
larger than that of a 1-MeV electron. Since the energy loss distributions are not very
different, the average spacing of a primary loss will be ca $500 \times$ smaller, i.e. 0.4 nm. We
see that in this case subsequent events are very close together.

The vast majority of the primary losses involves ionization and formation of a
secondary electron, which in turn may cause further ionizations as long as the energy
is sufficiently high. As we have seen above, most primary losses involve only a few tens
of eV, which at most results in only a few more ionizations by the secondaries.
Calculations of the energy losses of these low-energy secondary electrons (i.e. a few tens
of eV) can be carried out only for a few simple molecules in the gas phase, where sufficient
knowledge is available about the nature of the transitions involved and the cross sections
for the various processes.

In the condensed phase in most cases great uncertainty remains about the processes
involved on excitation, except for the lowest levels; also the interaction of the low-energy
secondary electrons with the medium is not very well understood. It is often assumed,
however, that for consideration of the electronic transitions the assumption that the
liquid behaves as a high-density gas is not grossly in error.

Using the gas-phase cross sections for the low-energy secondary, tertiary, etc., electrons, the spatial distribution of the events in the complete track can be calculated. Let us return to the case of a 1-MeV electron in a medium of $1 \, \text{kg dm}^{-3}$. As we see from Table 1, about 98% of the losses of the 1-MeV electron is below 100 eV. The secondary electron emerging will carry a kinetic energy of at most around 90 eV, and gives rise to a few (about 4) further ionizations within a distance on the order of 1 or 2 nm. The secondary and tertiary electrons of which the energy has dropped below the lowest electronic level continue to lose energy in interactions with vibrational and rotational modes of the molecules and with phonons, until they finally become thermalized. As we shall see below, in saturated hydrocarbon liquids the thermalization distance is usually on the order of several (and in a few cases, a few tens of) nanometers.

We have found that the average spacing between primary events for a 1-MeV electron and density $1 \, \text{kg dm}^{-3}$ is ca 200 nm. We see now that, because of the large dominance of the low-energy losses, the track of a fast electron mainly consists of single ionizations and small groups of a few ionizations close together that are separated from each other by a considerable distance. Occasionally, a much larger loss appears that gives rise to a track of a secondary electron that consists of a much larger number of ions and that is more extended in space. Very large energy losses (above ca 50 keV) lead to electron tracks very similar to that of the primary, and which may be sub-divided into independent groups again. (These losses represent about 13% of the energy loss of a 1-MeV electron.)

At this point it is of interest to realize the importance of the Coulomb interaction in non-polar liquids. The Coulomb energy between single charges is given by $U = e^2/4\pi\varepsilon_0\varepsilon_r r$ and the distance r_c at which the energy U is equal to $k_B T$ with $\varepsilon_r = 2$ is equal to ca 30 nm at 300 K. We see that with the thermalization ranges for the liquids mentioned above, the thermalized electrons stay well within the range of attraction, with the result that recombination/neutralization is a predominant process in the liquid. We shall return to this in a later section. In the gas phase the situation will be different. With a density typically 100 times smaller than in the liquid, the sub-excitation electrons will mostly escape from the Coulomb field of the positive ions.

The distribution of the number of ionizations in the groups in the tracks of high-energy electrons in the gas phase has been determined from cloud-chamber experiments as well as by calculation. Results are shown in Table 2. We see that about 23% of the energy is

TABLE 2. The fraction f_n of groups containing n ion pairs in the track of a high-energy electron, and the fraction of the ions found is groups of n pairs, $F_n = nf_n/\sum nf_n$

n	$f_n{}^a$	$f_n{}^b$	$F_n{}^b$
1	0.62	0.66	0.23
2	0.20	0.19	0.13
3	0.09	0.06	0.06
4	0.04	0.031	0.044
>4		0.066	
5	0.02		
6	0.01		
>6	0.02		
5–10			0.106
11–100			0.160
>100			0.256

[a] Obtained from cloud-chamber experiments[8a,b],
[b] Calculated[8c].

deposited in single ion pairs and 47% in groups of 4 pairs and less (which corresponds to energy losses below about 100 eV).

As we have seen above, α-particles have a much higher LET than electrons with the same energy, due to the proportionality of the LET to $z^2/m_e v^2$. In the case of the 1-MeV α-particle, the spacing of the primary events was found to be 0.4 nm in the condensed phase ($d = 1 \,\text{kg dm}^{-3}$), so that all the groups of ions resulting from the small losses merge and form a cylinder with a high ionization density. Occasionally, short side tracks emerge due to larger losses. The heavy particle tracks will be more straight than electron tracks of similar energy, due to the fact that the maximum energy loss for the heavy particle in a collision with an electron is smaller (roughly m/M) than in electron–electron collisions (0.5).

For heavy particles with even higher LET, the track may be considered as a cylindrical core of high ionization density due to the low-energy losses ($< 100 \,\text{eV}$) and a cylindrical region around it (called the penumbra) with a lower ionization density due to the energetic secondary electrons. About half of the energy of the primary is spent in each of the two regions.

D. The Radiation Chemical Yield

The radiation chemical yield is defined as the number of species or events produced per unit energy absorbed by the medium. The yield is traditionally expressed in units of events per 100 eV energy absorbed and denoted as G. We shall use the symbol g when it is expressed in other units or when the units are not specified. When g is expressed in mks units (mol J^{-1}), the relation between g and G is

$$g \,(\text{in mol J}^{-1}) = 1.04 \times 10^{-7} \times G \,(\text{in events per 100 eV}) \qquad (4)$$

The yield is determined experimentally mostly by determination of a concentration c_j of a species j formed in a given amount of material, while also the average total amount of energy absorbed per unit mass, the dose D, in that material is known, with

$$c_j = g_j D d \qquad (5)$$

where d is the density of the material.

The mks unit for the dose D is J kg^{-1}, called the Gray (Gy). With g_j in mol J^{-1}, c_j in mol dm^{-3} and D in J kg^{-1}, the density is to be expressed in kg dm^{-3} in order to have a consistent set of units. (An old unit for the dose is the rad, which is equal to 0.01 Gray.)

We have seen above that we can distinguish two kinds of energy losses of the fast-moving charged particle, the optical or resonant losses and the losses in the hard collisions. The relative distribution of the various optical excitations is independent of the charge and the mass of the primary, and only weakly dependent on the velocity (see equation 2). Therefore the fast-moving secondary electrons that are created as a result of the large losses in hard collisions will give approximately the same spectrum of optical excitations as the primary particle. (For the very slow secondary electrons this does not hold.) As a first-order approximation it is often assumed that all the secondary electrons (and tertiary, etc.) give the same spectrum of optical excitations. In this way all the kinetic energy of the primary particle is converted eventually into optical excitations with the same spectrum, and this spectrum is rather independent of the initial energy of the primary. Since the energy of gamma radiation is first converted into kinetic energy of electrons, the spectrum of the resulting primary events caused by these electrons will therefore also be approximately the same again.

As a result the total number of the various primary events divided by the total energy deposited (which is the radiation chemical yield) is rather independent of the (high-energy) radiation.

While the yield of the initial events is not very different for different radiations, the yield of the end products may be very different. An important reason for this difference is the difference in the spatial distribution of the initial events. In tracks of a high LET-particle (low velocity) the intermediates are formed close together and much more reaction with each other will take place than in a low LET-particle track. We shall return to this below.

III. THE GAS PHASE

A. Introduction

A significant characteristic of the radiation chemistry in the gas phase is the importance of the fragmentation of the excited molecules and ions. Ion fragmentation competes with collisional deactivation, but also with ion–molecule reactions and charge neutralization, the relative importance of which is dependent on the pressure and on the radiation intensity (dose rate). Dissociation of neutral excited states is of course also affected by the pressure. As a result the final product distribution will in general be dependent on the pressure as well as on the dose rate.

The radiation chemistry in the gas phase is complex and mostly cannot be explained completely. In several cases, however, a fair knowledge exists, especially about ionic processes, since these (in a low-pressure gas) can be studied by mass-spectrometric techniques.

In the following we shall discuss the radiation chemistry of methane in the gas phase, which may serve as an illustration of the various aspects of the radiation chemical effects in the gas phase. We shall also consider the effect of pressure and dose rate on the competition of the various processes. Finally we shall discuss briefly the effect of pressure on the fragmentation in cyclohexane. For a more detailed study the reader is referred to the review article of György[9] and the references cited therein.

B. Primary Processes; Methane

In Figure 1 we have shown the optical oscillator strength distribution for methane in the gas phase. In this picture the contributions of neutral excitation and ionization are indicated.

The ionization onset is at 12.5 eV. We see that a substantial amount of oscillator strength for neutral excitation is found above the ionization potential, which is due to so-called superexcited states. The phenomenon of the superexcited state was in fact discovered in CH_4. Using the oscillator strength distribution for total optical absorption and assuming the yields of ionization and neutral excitation to be determined by $\int (df/dQ)\cdot(1/Q)dQ$ above and below the ionization potential, led to a calculated ratio of yields of neutral excited states and ionization much smaller than the experimental value[10]. It was therefore concluded that part of the oscillator strength for neutral excitation had to be present above the ionization potential. The phenomenon of superexcited state formation has been found to be quite general for molecular compounds, and is due to excitation of electrons in deeper-lying orbitals and double excitation. The overall probability of ionization after excitation above the ionization potential shows an isotope effect due to competition of dissociation and ionization. It is interesting to see that superexcited states are formed also at energies of several tens of eV.

There are several dissociation processes of neutral excited states[9,11]:

$$CH_4 \longrightarrow CH_3^* + H^{\cdot} \tag{6a}$$

$$\longrightarrow CH_2 + H_2 \tag{6b}$$

752   A. Hummel

$$\longrightarrow CH_2 + 2H^\bullet \qquad\qquad\qquad (6c)$$

$$\longrightarrow CH + H_2 + H^\bullet \qquad\qquad (6d)$$

$$\longrightarrow C + 2H_2 \qquad\qquad\qquad (6e)$$

All these processes take place from singlet levels above ca 8.5 eV, the level of the first excited singlet. Process 6a is believed to be the prominent decay process of the lowest triplet (at ca 6.5 eV).

At 12.5 eV we have the onset of the ionization[12]:

$$CH_4 \longrightarrow CH_4^{+\bullet} + e^- \qquad\qquad (7)$$

Several dissociative ionization processes have been observed, as is presented below, with the various onsets in parentheses[11]:

$$CH_4 \longrightarrow CH_3^+ + H^\bullet + e^- \qquad (\ 14.4\,eV) \qquad (8a)$$

$$CH_2^{+\bullet} + H_2 + e^- \qquad (\ 15.3\,eV) \qquad (8b)$$

$$CH^+ + H_2 + H^\bullet + e^- \qquad (\ 22.4\,eV) \qquad (8c)$$

$$CH^+ + 3H^\bullet + e^- \qquad (\ 23.0\,eV) \qquad (8d)$$

$$C^{+\bullet} + \cdots + e^- \qquad (\leqslant 25.2\,eV) \qquad (8e)$$

The relative abundance of the various ions has been studied by low-pressure mass-spectrometric measurement. It has been found that the relative abundance is approximately constant for electron energies above ca 100 eV. For methane the following abundances are found[13]: $CH_4^{+\bullet}$ 46%, CH_3^+ 40%, $CH_2^{+\bullet}$ 7.5%. Together with a total yield of ionization[12] of 3.6 $(100\,eV)^{-1}$ this leads to the yields of fragment ions as presented in Table 3.

At higher pressures the relative abundances of the ions change due to deactivation of excited ions before fragmentation can take place.

Yields of fragments of neutral excited states have also been obtained from final product analysis. At pressures between 100 and 700 torr 'primary' yields of CH_3, CH_2 and CH radicals of 1.4, 0.7 and ca 0.2 $(100\,eV)^{-1}$ have been obtained[13]. A total yield of neutral

TABLE 3. Yields of primary species in CH_4 in the gas phase[13]

	G value $(100\,eV)^{-1}$
$CH_4^{+\bullet}$	1.7[a]
CH_3^+	1.4[a]
$CH_2^{+\bullet}$	0.3[a]
Other	0.2
All ions	3.6
$CH_3^\bullet$	1.4[b]
CH_2	0.7[b]
$CH^\bullet$	0.2[b]
Neutral excitation	2.3

[a]From low-pressure mass-spectrometric measurements.
[b]From product analysis, 100–700 torr.

excitation of ca 2.3 $(100\,\mathrm{ev})^{-1}$ follows from these values and a ratio of yields of neutral excited states to ions of around 0.6.

Theoretical calculations have been made of the yields of ionization, singlet and triplet excitation for 100-keV electrons slowing down, using the so-called binary-encounter collision theory[11a]. Yields for the various processes have been calculated for electrons originating from three orbitals $(1t_2, 2a_1$ and $1a_1)$ with ionization potentials 13.0 eV, 23.1 eV and 290.7 eV, respectively.

It is of interest that from the total yield of processes leading to ionization of 3.6 $(100\,\mathrm{eV})^{-1}$ a yield of 0.11 $(100\,\mathrm{eV})^{-1}$ is calculated to arise from double ionization and a yield of 0.018 $(100\,\mathrm{eV})^{-1}$ is found to originate from K-shell ionization. The latter will give rise to Auger ejection of electrons, which is assumed to lead to CH_4^{6+} ions. A total yield of neutral excited states (not leading to ionization) of 4.23 $(100\,\mathrm{eV})^{-1}$ is found, of which 0.78 $(100\,\mathrm{eV})^{-1}$ is triplet. We see that a ratio of neutral excitation to ionization of $4.23/3.60 \approx 1.2$ is found, which is higher than found above from experiment.

The branching ratios for ionization and the various fragmentation processes cannot be calculated accurately. On the basis of some semi-empirical estimates yields for $CH_4^{+\cdot}$, CH_3^+ and $CH_2^{+\cdot}$ have been found of 1.63, 1.42 and 0.32 $(100\,\mathrm{eV})^{-1}$, respectively, in agreement with experiment.

C. Kinetics

As we have seen earlier, the ionization by high-energy charged particles takes place nonhomogeneously, in tracks. In the gas phase the thermalization range is very large, so that the secondary electrons escape from the Coulomb field of the positive ions. The positive ions that are formed close together will repel each other. The spatial correlation of the ions and the electrons will get lost and the charged species will become homogeneously distributed throughout the irradiated volume. Neutralization of ions and electrons will therefore take place by homogeneous second-order reaction kinetics, characterized by

$$-\frac{dc}{dt} = \alpha_e c^2 \qquad (9)$$

where c is the concentration of positive ions and electrons and α_e the second-order specific rate of recombination. With a continuous rate of production q ion pairs per cm^3 and per second, the stationary concentration will be given by

$$c = (q/\alpha_e)^{1/2} \qquad (10)$$

The average lifetime with respect to recombination is now

$$\tau_r = (1/\alpha_e c) = (1/\alpha_e q)^{1/2} \qquad (11)$$

The recombination coefficient α_e in hydrocarbon gases at pressures below a few atmospheres can be written as

$$\alpha_e = \alpha_2 + \alpha_3 c \qquad (12)$$

where α_2 and α_3 are coefficients representing the two- and three-body recombination processes, respectively. We see that α_e increases with pressure. (For neopentane, for example, $\alpha_2 = 7.0 \times 10^{-6}\,cm^3\,s^{-1}$ and $\alpha_3 = 2.5 \times 10^{-25}\,cm^6\,s^{-1}$). At high pressure ($>$ several atmospheres) the recombination is diffusion controlled and now α_e is given by

$$\alpha_e = \frac{e}{\varepsilon_0\varepsilon_r}(u_+ + u_e) \qquad (13)$$

where u_+ and u_e are the mobilities of positive ion and electron, e the electron charge and ε_0, ε_r the permittivity of vacuum and of the gas, respectively. Since the mobilities are inversely proportional to pressure, at large pressures $\alpha_e \propto p^{-1}$.

We can now make an estimate of τ_r for a given dose rate and pressure. For a typical dose rate for a gamma-irradiation experiment, dD/dt is on the order of $1\,\text{Mrad}\,\text{h}^{-1} \approx 3\,\text{Gy}\,\text{s}^{-1}$. Using $g = 4\,(100\,\text{eV})^{-1} = 4 \times 10^{-7}\,(\text{mol}\,\text{J}^{-1}) = 2.4 \times 10^{17}\,(\text{mol}\,\text{J}^{-1})$ for the yield of ion pairs, we find for a gas of molecules with a molecular mass $M = 100$ at a pressure of 1 at and a temperature $273\,\text{K}$, a rate of production of ion pairs of $q = (dD/dt)\,dg = 3.2 \times 10^{12}\,\text{cm}^{-3}\,\text{s}^{-1}$. Taking a typical value of $\alpha_e = 1 \times 10^{-5}\,\text{cm}^3\,\text{s}^{-1}$ we find $\tau_r = (1/\alpha_e q)^{1/2} = 2 \times 10^{-4}\,\text{s}$. For $p = 1$ torr we find a value of around $6 \times 10^{-3}\,\text{s}$.

In a pulse radiolysis experiment, however, the dose rates may be very much larger. With a dose rate of $3 \times 10^{11}\,\text{Gy}\,\text{s}^{-1}$ we find values of τ_r ca 3×10^5 lower than the ones given above. For $p = 1$ atmosphere we have now $\tau_r = 6 \times 10^{-10}\,\text{s}$ and for $p = 1$ torr, $\tau_r = 2 \times 10^{-8}\,\text{s}$.

The primary fragmentation after ionization by the high-energy charged particle takes place largely at a time scale of $10^{-10}\,\text{s}$ and the fragmentation pattern does not change any more drastically between 10^{-10} and $10^{-5}\,\text{s}$. Thermalization of the electrons takes place in a few times $10^{-10}\,\text{s}$ in CH_4 and C_2H_6 at one atmosphere and the thermalization time is inversely proportional to pressure; $\tau_{th} \approx 10^{-10}/p(\text{at})$ seconds.

The time between collisions of ions with the molecules is on the order of $\tau_c \approx 10^{-10}/p(\text{at})$ seconds; at $p = 1$ at, $\tau_c = 10^{-10}\,\text{s}$.

We see now that ion–molecule reactions and deactivation may start to interfere seriously with the fragmentation above ca 1 atmosphere. Also, except at very high dose rates and low pressures, ion–molecule collisions compete with neutralization, and therefore ion–molecule reactions have to be taken into consideration.

D. Reaction of the Transients in Methane

In Table 3 we have shown the yields of the 'primary species'. We have seen that $CH_4^{+\cdot}$ and CH_3^{+} are the predominantly occurring ions, and $CH_3^{\cdot}$ is the most important product of neutral excitation.

The most important ion–molecule reactions in the radiolysis of CH_4 are[9,13,15]

$$CH_4^{+\cdot} + CH_4 \longrightarrow CH_5^{+} + CH_3^{\cdot} \tag{14a}$$

$$CH_3^{+} + CH_4 \longrightarrow C_2H_5^{+} + H_2 \tag{14b}$$

$$\longrightarrow C_2H_3^{+} + 2H_2 \tag{14c}$$

The cross section for the first reaction decreases with increasing kinetic energy of the ion. The latter two reactions probably take place via the intermediate excited ion $C_2H_7^{+}$; $C_2H_3^{+}$ originates via fragmentation of excited $C_2H_5^{+}$. The ratio of $C_2H_3^{+}/C_2H_5^{+}$ is approximately 0.04 for thermal CH_3^{+} but increases dramatically with increasing kinetic energy of CH_3^{+}.

The ions CH_5^{+} and $C_2H_5^{+}$ are relatively unreactive towards CH_4 and are the major ionic species in the radiation chemistry of CH_4.

The reaction of the fragments of neutral excited states have been studied also by means of photolysis, using resonance lamps providing photons of various energies[9]. Reactions of CH_2 and $CH^{\cdot}$ with CH_4 have been observed. Both species insert into CH_4.

$$CH^{\cdot} + CH_4 \longrightarrow C_2H_5^{*\cdot} \tag{15}$$

is found; the resulting excited ethyl radical may again fragment:

$$C_2H_5^{*\cdot} \longrightarrow C_2H_4 + H. \tag{16}$$

With CH_2 the process is

$$CH_2 + CH_4 \longrightarrow C_2H_6^* \tag{17}$$

If the excited ethane molecule is not deactivated, the reaction

$$C_2H_6^* \longrightarrow 2CH_3^{\cdot} \tag{18}$$

take place.

Hot hydrogen atoms react with CH_4 to give H_2 and $CH_3^{\cdot}$ radicals:

$$H^{\cdot} + CH_4 \longrightarrow H_2 + CH_3^{\cdot} \tag{19}$$

As we mentioned above, the electrons get thermalized far away from the parent positive ion, outside the Coulomb field. Eventually, recombination of the electrons with the positive ions will take place under formation of excited species, either in the singlet or in the triplet state.

The CH_5^+ ion, on recombination, may yield either $H^{\cdot}$ or H_2[14]:

$$CH_5^+ + e^- \longrightarrow CH_4 + H^{\cdot} \tag{20a}$$

$$\longrightarrow CH_3^{\cdot} + H_2 \tag{20b}$$

The $C_2H_5^+$ ion, on recombination, will give an excited $C_2H_5^{\cdot}$ radical, which may be deactivated or fragment into $C_2H_4 + H^{\cdot}$ as mentioned earlier (equation 16).

We see now that the major 'stable' radicals remaining to react with each other are $CH_3^{\cdot}$, $C_2H_5^{\cdot}$ and $H^{\cdot}$.

In Table 4 we present yields of final products as obtained from gamma-irradiated CH_4. The yields have been found to be constant for the range of pressures of 0.12–1.2 bar. The major product, ethane, results from reaction of two $CH_3^{\cdot}$ radicals and from $CH_2 + CH_4$. Propane and butane are formed from $CH_3^{\cdot} + C_2H_5^{\cdot}$ and $2C_2H_5^{\cdot}$, respectively. There seems to be uncertainty about the yield of C_2H_4, since it may be that part of the C_2H_4 formed reacts with H atoms. The mode of formation of the polymer (with a composition approximately $C_{20}H_{40}$) is not entirely certain. It has been proposed that the ionic mechanism

$$CH_4 + C_nH_{2n+1}^+ \longrightarrow C_{n+1}H_{2n+3}^+ + H_2 \tag{21}$$

is operative[9].

It may be concluded that while the gross features of the primary processes are fairly

TABLE 4. Yields of final products in CH_4 in the gas phase at pressures 0.12–1.2 bar, Co-60 gamma irradiation[a]

Product	$G(100\,eV)^{-1}$
H_2	5.7
C_2H_6	2.2
C_2H_4	1.4
C_3H_8	0.36
n-C_4H_{10}	0.11
i-Butane	0.04
Pentanes	
Pentenes	0.03
Hexanes	
'Polymer' ($-CH_4$)	2.1

A. Hummel

well known for methane, the detailed mechanism of the product formation has not been elucidated.

E. Cyclohexane

We briefly consider cyclohexane as another example in order to illustrate the effect of density on the fragmentation. In Table 5 we show product yields obtained for pulse-irradiated (high dose rate) cyclohexane at 55 torr (a) and for gamma-irradiated cyclohexane at densities varying from 0.004 kg dm⁻³ to the liquid density (b–d).

The general trend in the product distribution is a pronounced increase of the yield of C—C break with decreasing density. In the pulsed experiments (a) with a gas pressure of 55 torr the ions have a lifetime on the order of a few nanoseconds, while the collision interval is about 0.4×10^{-9} s. Some collisional deactivation may be expected (as we shall indeed see to be the case below). The charge recombination will only interfere with slow ion–molecule reactions.

In the gamma-radiolysis experiments with 1000 torr (b), with a dose rate on the order

TABLE 5. Yields of products, in G-value units of $(100\,eV)^{-1}$, in the radiolysis of cyclohexane in the gas and liquid phase

	a room temp. 55 torr 2.5×10^{-4} kg dm⁻³ pulsed electrons	b 373 K 1000 torr 0.004 kg dm⁻³ γ-radiation	c 573 K 78 at 0.14 kg dm⁻³ γ-radiation	d room temp. 1 at 0.78 (liquid) γ-radiation
H_2		4.8	5.4	5.6
CH_4	0.75	0.36	0.13	0.01
C_2H_2	0.66	0.35	—	0.025
C_2H_4	4.62	1.7	0.37	0.10
C_2H_6	0.14	1.4	0.18	0.015
$CH_2{=}C{=}CH_2$	0.17			
$CH_3C{\equiv}CH$	0.13			
$CH_3CH{=}CH_2$	1.32	0.55	0.15	0.025
cyclo-C_3H_6	0.009			0.006
C_3H_8	0.23	0.44	0.09	0.011
1,2-C_4H_6	0.07			
1,3-C_4H_6	0.92	0.6	—	0.004
1-C_4H_8	0.53			
cis-2-C_4H_8	0.11			0.025
trans-2-C_4H_8	0.13			
i-C_4H_{10}				
n-C_4H_{10}	0.19			0.008
cyclo-C_6H_{10}	1.1	1.2	1.9	3.26
n-C_6H_{10}				0.46
CH_3-cyclo-C_5H_9				0.2
C_2H_5-cyclo-C_6H_{11}		0.3		
C_3H_7-cyclo-C_6H_{11}		0.2		
bicyclohexyl		1.0	1.1	1.81
Other C_{12}				0.1

[a]Reference 16; 80 ns pulses from a field emission source, with a total dose of 29 kGy per pulse (dose rate 0.36×10^9 kGy s⁻¹).
[b]Reference 17; gamma irradiation.
[c]Reference 18; gamma irradiation.
[d]G values from various authors, taken from Reference 19; gamma irradiation.

of $10\,\mathrm{Gy\,s^{-1}}$ we have lifetimes of the ions on the order of $10^{-4}\,\mathrm{s}$, and collision intervals shorter than $10^{-10}\,\mathrm{s}$. Here the charge recombination does not compete with ion–molecule reactions. The lifetime of the charged species is in fact so long that reaction with impurities may present a problem. The short collision interval, however, causes deactivation and decreases the fragmentation, as is observed in the product distribution.

The initial yields of the ions have been determined from gamma radiolysis in cyclohexane at a pressure of 55 torr by isotopic labelling techniques. In Table 6 we show the initial yields of the various ions together with those obtained from mass-spectrometric determinations at very low pressure. The yields are given as ion-pair yields, i.e. the yield of species per ion pair produced. The ion-pair yield M/N^+ can be converted into species produced per unit energy absorbed, by multiplication by the yield of ionization $G = 4.4$ $(100\,\mathrm{eV})^{-1}$. We see that the decomposition of the parent ion is appreciably suppressed in the gas at 55 torr, due to collisional deactivation.

The decrease in decomposition of $C_6H_{12}^{+\cdot}$ of 0.3 in the ion-pair yield is matched by an increase in the fragmentation yield of 0.35.

The fragmentation reactions are

$$(C_6H_{12}^{+\cdot}) \longrightarrow C_4H_8^{+\cdot} + C_2H_4 \tag{22a}$$

$$C_5H_9^+ + CH_3^\cdot \tag{22b}$$

$$C_3H_6^{+\cdot} + C_3H_6 \tag{22c}$$

$$C_3H_7^+ + C_3H_5^\cdot \tag{22d}$$

$$(C_4H_8^{+\cdot})^* \longrightarrow C_3H_5^+ + CH_3^\cdot \tag{23a}$$

TABLE 6. Fragmentation pattern of the ions in cyclohexane obtained mass spectrometrically (70 eV electrons) and from gamma radiolysis of cyclohexane at 55 torr[16]

	Mass spectrometer M/N^{+a}	Gamma radiolysis $p = 55$ torr	
		M/N^{+a}	$G(100\,\mathrm{eV})^{-1b}$
$C_6H_{12}^{+\cdot}$	0.16	0.46	2.0
$C_6H_{11}^+$	0.0098	$(0.14)^c$	0.62
$C_6H_{10}^{+\cdot}$		$(0.08)^c$	0.35
$C_5H_9^+$	0.049	—	—
$C_4H_8^{+\cdot}$	0.22	0.11	0.48
$C_4H_7^+$	0.077	0.044	0.19
$C_4H_6^{+\cdot}$	0.015	0.015	0.07
$C_4H_5^+$	0.011	—	—
$C_3H_7^+$	0.031	0.027	0.12
$C_3H_6^{+\cdot}$	0.068	0.036	0.16
$C_3H_5^+$	0.15	0.054	0.24
$C_3H_4^+$	0.014	0.003	0.01
$C_3H_3^+$	0.063	—	—
$C_2H_5^+$	0.028	0.014	0.06
$C_2H_4^{+\cdot}$	0.032	0.002	0.01
$C_2H_3^+$	0.067	0.043	0.19

[a] M/N^+ expressed as the fraction of the total number of ions formed.
[b] The yield in $(100\,\mathrm{eV})^{-1}$ calculated from M/N^+ using, for the total yield of ionization, $G = 4.4$ $(100\,\mathrm{eV})^{-1}$.
[c] The yields of $C_6H_{11}^+$ and $C_6H_{10}^+$ are products from secondary reactions.

$$C_4H_7^+ + H^\cdot \tag{23b}$$

$$(C_3H_6^{+\cdot})^* \longrightarrow C_3H_5^+ + H^\cdot \tag{24}$$

The ions $C_6H_{11}^+$ and $C_6H_{10}^{+\cdot}$ are formed from $C_6H_{12}^{+\cdot}$ reacting with unsaturated molecules by $H^{-\cdot}$ and H_2^- transfer.

We see now that in the gas at 55 torr the ions $C_6H_{12}^{+\cdot}$, $C_6H_{11}^+$, $C_6H_{10}^{+\cdot}$ and $C_4H_8^{+\cdot}$ with abundances 0.46, 0.14, 0.08 and 0.11, respectively, represent close to 80% of the ions.

Extensive experiments have been carried out, with gamma irradiation as well as with pulsed irradiation, with radical scavengers present in order to depress products from radical reactions, and with electron scavengers to prevent formation of products from neutralization. In this way estimates of the contribution of some of the various possible reaction pathways have been made. For a detailed discussion the reader is referred to Reference 16.

IV. LIQUIDS

A. Introduction

When comparing the early processes in the liquid and gas phases, some important differences can be observed. In the first place, in the liquid, bond dissociation of the primary excited molecules and ions takes place less frequently than in the gas, due to collisional deactivation and the cage effect. Another difference concerns what is called the charge separation. In a low-pressure gas the electrons get thermalized at a large distance away from the positive ions and they escape from the Coulomb attraction. In liquid saturated hydrocarbons the electrons get thermalized well within the Coulomb field of the positive ions and, in most liquids, the vast majority of the charges recombine at a very short time scale. The charge recombination leads to excited species that in turn may undergo dissociation into neutral fragments.

Whether or not the primary excitation and ionization processes in the liquid and gas phases are the same, we do not know. As we shall see below, the yield of ionization in the liquid appears to be somewhat larger than in the gas phase, possibly at the expense of direct excited states. Excited states formed in the primary excitation may subsequently ionize but, on the other hand, ion–electron recombination may take place at an extremely short time scale and excited states may be formed. The earliest processes are not very well understood at present.

In the following we first discuss the charge separation. Next, we consider some examples of the chemistry of the ions and of the excited states formed after recombination. Then we shall discuss the overall product formation for cyclohexane and n-pentane and indicate some general features of the decomposition of the various saturated hydrocarbons. All this work concerns the radiolysis with low LET radiation (gamma radiation, fast electrons). In the last section we give some results for radiation with different LETs for the case of cyclohexane. We also indicate how the non-homogeneity in the initial spatial distribution of the reactive species is accounted for in the (non-homogeneous) kinetics.

B. Charge Recombination and Escape from the Track

As we have seen in Section II, the slowing down of a fast electron in the liquid leads to a track of single ionizations and groups of ionizations, at a considerable distance from each other. The electrons ejected from the molecules get thermalized some distance away from the positive ion but within the Coulomb field. These thermal 'excess' electrons do not react with saturated hydrocarbon molecules, but are stable species that are more

TABLE 7. Values of G_{esc}, r_c, b, $u(-)$ and α/k for various liquids at room temperature[20,21]

	G_{esc} $(100\,eV)^{-1}$	r_c $(10^{-10}\,m)$	b^a $(10^{-10}\,m)$	$u(-)$ $(10^{-4}\,m^2\,V^{-1}\,s^{-1})$	α/k $(10^{-12}\,s)$
n-Hexane	0.13	299	67	0.08	15
Iso-octane	0.33	286	95	5.3	0.7
Neopentane	1.1	318	230	70	
Cyclohexane	0.13	297	61	0.23	2.2
cis-Decalin	0.14			0.10	13
trans-Decalin	0.13			0.013	43

^{a}The b value is the width of the gaussian distribution, used for $f(r_0)$ in equation 27 (see text).

or less trapped by potential fluctuations in the liquid. In most liquids only a minor fraction of the excess electrons and positive ions escapes recombination in the track. The yield of escape, also called free ion yield, has been measured for many liquids. Some examples are given in Table 7. We see that there are considerable differences between the liquids. We do not know the total yield of ionization in the liquid accurately, but if we take a value of 5 $(100\,eV)^{-1}$, we see that the overall probability of escape for cyclohexane is approximately $W_{esc} \simeq 0.026$ and for neopentane $W_{esc} \simeq 0.24$. The difference in escape probability reflects the difference in thermalization lengths of the electrons in the different liquids.

In order to obtain an estimate of the thermalization length it has been assumed that in the track only single pairs of oppositely charged species are present. For single ion pairs the probability of escape is given by the simple expression

$$W_{esc}(r_0) = \exp(-r_c/r_0) \tag{25}$$

Here r_0 is the initial separation and r_c the distance at which the Coulomb energy between two singly charged species is equal to $k_B T$; $r_c = e^2/4\pi\varepsilon_0\varepsilon_r k_B T$ is called the Onsager length. For $\varepsilon_r = 2$, and $T = 300\,K$, $r_c \simeq 30\,nm$ is found. We see that for $r_0 = r_c = 30\,nm$, $W_{esc} = 1/e = 0.37$.

If in the track a distribution $f(r_0)$ of initial separations is assumed, we can write for the track consisting of single pairs

$$W_{esc} = \int_0^\infty f(r_0) \exp(-r_c/r_0)\,dr_0 \tag{26}$$

or with an initial yield G_0 of single pairs

$$G_{esc} = G_0 \int_0^\infty f(r_0) \exp(-r_c/r_0)\,dr_0 \tag{27}$$

Different trial distributions have been taken for $f(r_0)$; for G_0 mostly the gas-phase value has been assumed. Comparison with the experimentally determined value of G_{esc} gives an estimate of the width of the distribution $f(r_0)$.

In Table 7 we have given the width parameters for a gaussian distribution

$$f(r_0) = (4\pi r_0^2/\pi^{3/2}b^3)\exp(-r_0^2/b^2)$$

taking for G_0 the gas-phase yield. We see that the thermalization distances differ considerably for the different liquids and that the spherical molecule neopentane has the largest value. The differences are thought to be due to differences in the efficiencies of energy loss of the electrons with energies close to thermal. An interesting correlation exists between the thermalization ranges and the mobilities of the electrons in the various hydrocarbon liquids, as is shown also in Table 7. We see that the largest mobilities are found for the liquids with the largest yields of escape and therefore the largest thermalization ranges. It may be noted that the mobilities given in this table are all orders of magnitude larger than the mobilities of molecular ions in these liquids, the latter of which have values in the range of 10^{-8} to 10^{-7} m^2 V^{-1} s^{-1}. A consequence of these large mobilities is that the recapture of the electrons by the positive ions may take place very rapidly, as we shall see below.

The single-pair treatment given above is extremely approximate since the majority of the charged species is present in multi-ion-pair groups, as we have seen above. The single-pair treatment was suggested at a time when calculation of escape probabilities for multi-ion-pair groups was not possible. Such calculations have recently been carried out using computer simulation techniques[22]. It has been shown that the ion pairs in groups of various sizes in the track of a fast electron have considerably different escape probabilities. It appears, however, that due to a compensation of errors the single-ion-pair treatment gives reasonable estimates of the width of the thermalization range distributions.

C. The Time Scale of Charge Recombination; Charge Scavenging

In saturated hydrocabon liquids most recombinations of the charged species in the track take place within a nanosecond. Direct experimental observation of a substantial yield of these charged species has not proved possible due to limitation in the time resolution of the detection techniques. We can obtain information about the lifetime (distribution) of the charged species by adding a solute to the liquid that can react with the excess electrons (or the positive species) with a known specific rate and determining the amount of reaction taking place.

Solutes in the liquid, that can react with the charged species, are called charge scavengers. An example of an electron scavenger is methyl bromide:

$$CH_3Br + e^- \longrightarrow CH_3{}^{\cdot} + Br^- \tag{28}$$

The methyl radical subsequently abstracts a hydrogen atom from a hydrocarbon molecule and CH_4 is formed. The yield of CH_4 therefore represents the electron scavenging yield. At low concentrations of CH_3Br only the yield of escaped electrons is scavenged; however, with increasing concentration an increasing fraction of the electrons will be captured before recombination has occurred.

If $G(t)$ represents the yield of the charged species as a function of time (in the absence of a scavenger), the yield of scavenging in the presence of a scavenger with concentration c is (approximately) given by

$$G_s(c) = \int_0^\infty G(t) \exp(-kct) \, kc \, dt \tag{29}$$

We see that if the yield of scavenging as a function of concentration $G_s(c)$ is known, and the specific rate of scavenging k is known, then $G(t)$, the yield of survival against time, can be determined.

It has been found that for a number of saturated hydrocarbons $G_s(c)$ can be approximately represented by the simple expression

$$G_s(c) = G_{esc} + A \frac{(\alpha c)^n}{1 + (\alpha c)^n} \tag{30}$$

where for n a value of around 0.5 is found, and for A a value of 4–5 $(100\,eV)^{-1}$. The parameter α is a measure of the efficiency of the scavenging reaction.

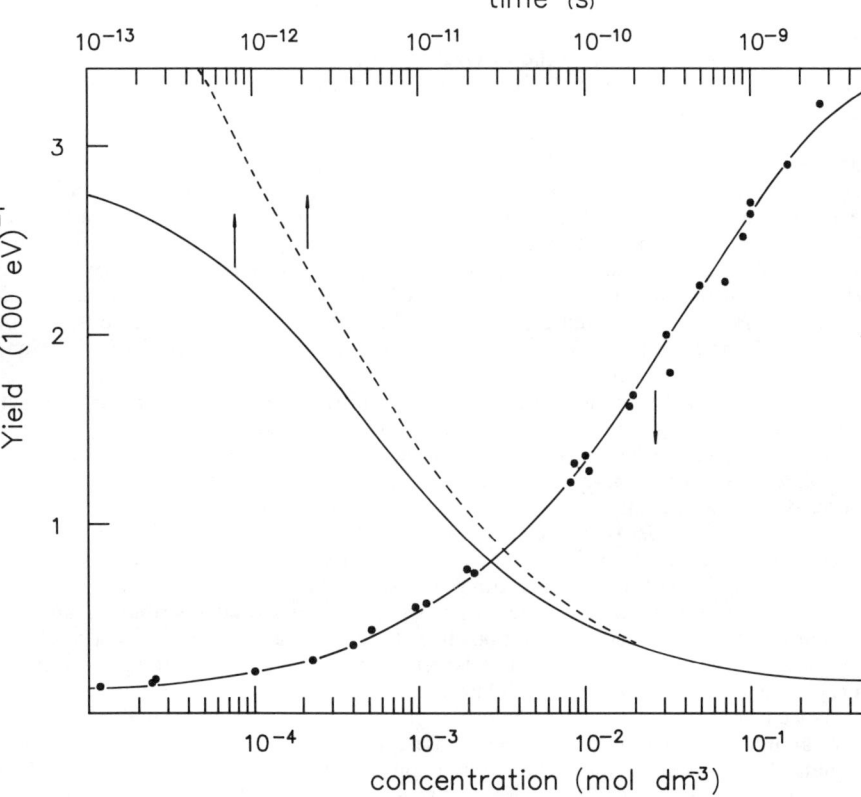

FIGURE 2. The yield of excess electrons scavenged by methyl bromide as a function of methyl bromide concentration in the gamma radiolysis of cyclohexane (lower scale). The yield of survival as a function of time for the charged species after instantaneous irradiation in cyclohexane (upper scale); the upper curve has been obtained with equation 31, the lower one with a more extensive treatment[7] (see text)

In Figure 2 we have plotted $G_s(c)$ for electron scavenging by CH_3Br in cyclohexane as an example. We see that no plateau is reached even at the highest concentrations applied, and we do not obtain a value for the total yield of electrons formed. Higher scavenger concentrations have the disadvantage that the scavenger itself absorbs energy and will decompose and correction for this effect is uncertain.

For several liquids the electron scavenging can be reasonably well represented by equation 30 using $n = 0.5$. In this case equation 29 leads to the expression

$$G(t) = G_{esc} + A \exp(kt/\alpha) \, \text{erfc} \, [(kt/\alpha)^{1/2}] \tag{31}$$

which can be approximated by

$$G(t) = G_{esc} + A \frac{2}{\pi^{1/2}} \frac{1}{(kt/\alpha)^{1/2} + (kt/\alpha + 4/\pi)^{1/2}} \tag{32}$$

The second term on the right-hand side in equation 32 decays steeply at short times;

for $kt/\alpha = 1$ this term is equal to $0.45\,A$ and at long times it approaches $A/\pi^{1/2}(kt/\alpha)^{1/2}$, which decays slowly with time.

In Figure 2 we have plotted the survival of the charged species as a function of time, $G(t)$, as obtained by equation 32 for cyclohexane (dotted curve). In order to obtain $G(t)$ the value of α/k has to be known.

Values of α have been determined for different charge scavengers from plots of $G_s(c)$ as presented in Figure 2 for CH_3Br reacting with excess electrons. Values of the specific rate k of reaction with the charged species have been obtained from pulse-radiolysis experiments, in which the decay of the concentration of charged species after a pulse of radiation due to reaction with the scavenger is followed by measuring the conductivity. For cyclohexane a value of α/k equal to 2.3×10^{-12} s is found [with a value of 5 $(100\,\text{eV})^{-1}$ for the parameter A in equation 30]. The value of α/k represents the time at which the initially recombining part of the charged species (the second term on the right-hand side of equation 32) has decayed to 45% of its initial value; for cyclohexane at $t = \alpha/k$, $G(t) = 0.13 + 5 \times 0.45 = 2.38\ (100\,\text{eV})^{-1}$.

As we mentioned, equation 29 relating $G_s(c)$ and $G(t)$ is only approximately correct. An improved calculation (using a modified version of equation 29, taking a time-dependent specific rate k as well as 'static' scavenging into consideration[32]) for cyclohexane gives the curve drawn in Figure 1. We see that this treatment leads to a faster decay than the approximative calculation. The value of α/k, however, provides a fair measure of the decay of the charged species for longer times. Values of α/k have been given in Table 7 for a few liquids. The smallest value for α/k reported is 0.7 ps for iso-octane, the largest for *trans*-decalin of 43 ps.

The lifetime distribution is determined by the initial spatial distribution and the mobilities of both the negative- and positive-charged species. Larger initial separations increase the recombination time and larger mobilities decrease it (roughly inversely proportional to the sum of the mobilities).

As we mentioned above, the mobilities of the excess electrons given in Table 7 are all several orders of magnitude larger than those of molecular ions in these liquids. In some liquids also the positive charge has been found to have a mobility far in excess of that for a molecular ion, and which is ascribed to a fast transfer of the positive charge from the parent positive ion to the neutral molecule. This phenomenon of the electron hole has been studied by charge scavenging and by pulse radiolysis[21,24].

For a more detailed treatment of the charge separation problems, see References 20–25.

D. Ions and Excited States

Neutral excited states are produced in saturated hydrocarbon liquids by direct excitation as well as by recombination of the positive ions and electrons. As we have pointed out earlier, the yield of ionization in molecular liquids is uncertain. While in the gas phase the yield of ionization can be determined by collecting the charged species in an electric field (ionization chamber), in hydrocarbon liquids this does not prove possible. Sufficiently large external fields cannot be applied in order to obtain a saturation current. In liquid noble gases (where the thermalization lengths are very large), total yields of ionization have been measured by determination of the saturation currents in a dc field, and yields have been found substantially in excess of the gas-phase values [4.2 vs 3.8 $(100\,\text{eV})^{-1}$ for argon, 4.9 vs 4.2 $(100\,\text{eV})^{-1}$ for Kr and 6.1 vs. 4.6 $(100\,\text{eV})^{-1}$ for xenon][26]. In (liquid) water, where the yield of solvated electrons has been measured spectrophotometrically at a picosecond time scale after pulsed irradiation, a yield of 4.7 $(100\,\text{eV})^{-1}$ has been found[27] while in the gas phase the yield of ionization[28] is 3.3 $(100\,\text{eV})^{-1}$. It seems likely that in saturated hydrocarbon liquids also, the yield of ionization is larger than in the gas phase. In the gas phase these yields are around 4 $(100\,eV)^{-1}$ [28]; the yields in the liquid may be estimated at around 5 $(100\,eV)^{-1}$.

Saturated hydrocarbon liquids have optical absorption edges in the far UV (*ca* 7 eV and above) and, on optical excitation, show fluorescence with a wavelength maximum around 200–230 nm (6.2–5.4 eV), which is believed to arise from the lowest excited singlet (S_1). The lifetimes of these excited states are around a nanosecond and the quantum efficiencies for fluorescence are small (< 0.02), the competing process being decomposition of the molecule. These fluorescent excited states have also been observed with high-energy radiation. Yields have been determined for compounds containing six-membered rings (which have high yields of fluorescence). The highest yields have been found for *cis*- and *trans*-decalin and bicyclohexyl [3.4, 2.8 and 3.5 $(100\,eV)^{-1}$, respectively[29]].

A substantial fraction of the fluorescent singlets in *cis*- and *trans*-decalin as well as in cyclohexane has been shown to originate from charge recombination[24].

The neutral excited states, single pairs and groups of pairs of positive ions and electrons formed initially in the track are formed in an overall singlet state, since no exchange of spin of the high-energy particle takes place. (The low-energy secondary, etc., electrons may exchange spin, but we disregard this for the moment.) It turns out that, in saturated hydrocarbon liquids, the spin state of an electron and its parent ion remains conserved for a considerable period of time (on the order of 10^{-6} s). Single pairs that recombine within that time will therefore remain singlet (geminate recombination). In multiple pair groups the situation is different, however. An electron may now recombine with another positive ion rather than its parent ion (cross recombination). In this case there is no spin correlation and the ratio of probabilities for the formation of singlets and triplets is 1/3. It can be shown that the overall probability for singlet formation for the group depends on the number of pairs initially as well as on the initial spatial distribution of the charged species. For a track of a high-energy electron in a low-yield liquid (like cyclohexane and the decalins) the overall probability for singlet formation may be estimated to be around 0.6[30]. With an estimated yield of *ca* 5 $(100\,eV)^{-1}$ for the total yield of ionization, we therefore expect a yield of *ca* 3 $(100\,eV)^{-1}$ of singlets. This is roughly what is found for the yield of S_1 for the decalins and bicyclohexyl, as mentioned above. The yields for cyclohexane and methylcyclohexane are lower, however. This may indicate that the charge recombination initially leads to a higher excited state that only partly converts to the fluorescing S_1 state or that the parent ion partly fragments.

The role of neutral excited states in radiolysis has been investigated by comparing the formation of products and intermediates in radiolysis with that in photolysis with various excitation wavelengths. We consider liquid cyclohexane.

In Figure 3 we show the quantum yields for the two predominant processes as a function of photon energy[31]:

$$c\text{-}C_6H_{12}{}^* \longrightarrow c\text{-}C_6H_{10} + H_2 \qquad (33)$$

$$c\text{-}C_6H_{12}{}^* \longrightarrow c\text{-}C_6H_{11}^{\cdot} + H^{\cdot} \qquad (34)$$

We see that the yield of H_2 elimination decreases with photon energy, while that of simple C—H break increases. At the lowest energy employed the quantum yield for H_2 elimination is approximately 0.8. It has been shown from deuterium-labelling studies that the molecular hydrogen formation takes place from one carbon atom under formation of a biradical[19].

The chemistry resulting from the two primary reactions 33 and 34 is very simple. The H atom abstracts a hydrogen atom:

$$H^{\cdot} + c\text{-}C_6H_{12} \longrightarrow H_2 + c\text{-}C_6H_{11}^{\cdot} \qquad (35)$$

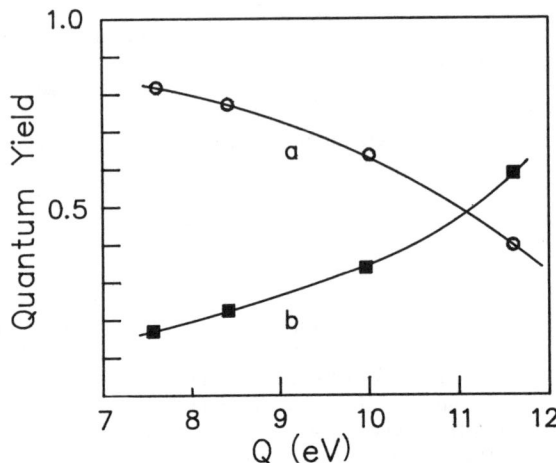

FIGURE 3. Quantum yields for the processes (a) $c\text{-}C_6H_{12} \rightarrow c\text{-}C_6H_{10} + H_2$ (○) and (b) $c\text{-}C_6H_{12} \rightarrow c\text{-}C_6H_{11}{}^{\bullet} + H^{\bullet}$ (■) in the photolysis of liquid cyclohexane at different photon energies. Reproduced by permission of the American Institute of Physics from Ref. 31

and the cyclohexyl radicals react with one another leading to either disproportionation

$$2\,c\text{-}C_6H_{11}{}^{\bullet} \longrightarrow c\text{-}C_6H_{10} + c\text{-}C_6H_{12} \qquad (36)$$

or to dimer formation

$$2\,c\text{-}C_6H_{11}{}^{\bullet} \longrightarrow (c\text{-}C_6H_{11})_2 \qquad (37)$$

The ratio of specific rates for disproportionation and combination, k_d/k_c, is 1.1 at room temperature[32]. We can deduce easily that the quantum yield for H_2 elimination ϕ_{H_2} (equation 33) is given by $\phi_{H_2} = \phi(C_6H_{10}) - 1.1 \times \phi(C_{12}H_{22})$, where ϕ (C_6H_{10}) and ϕ ($C_{12}H_{22}$) are the overall quantum yields.

In the radiolysis of liquid cyclohexane the major products are H_2, cyclohexane and the bicyclohexyl, with yields of 5.6, 3.3 and 1.8 $(100\,eV)^{-1}$, respectively[32], which suggests an important contribution of cyclohexyl radicals. This has been substantiated by radical titration experiments, where a solute is added that reacts with the radicals $R^{\bullet}$ (e.g. I_2) and the products (RI) can be measured[33]. Also, with radical scavengers present, the yields of cyclohexene and bicyclohexyl decrease with the ratio k_d/k_c for the cyclohexyl radical. When it is assumed that the dimer is formed only by the radical recombination reaction 37 [with a yield of 1.8 $(100\,eV)^{-1}$], it follows that $c\text{-}C_6H_{10}$ is formed concomitantly with a yield of $1.1 \times 1.8 = 2.0\,(100\,eV)^{-1}$. Since the total yield of $c\text{-}C_6H_{10}$ is 3.3 $(100\,eV)^{-1}$ it follows that a yield of $c\text{-}C_6H_{10}$ equal to 1.3 $(100\,eV)^{-1}$ is found that is formed in another process than disproportionation of cyclohexyl radicals. It appears that this process is molecular H_2 elimination of the S_1 state (that also fluoresces), with $\phi_{H_2} = 0.8$ as obtained from photolysis at the lowest photon energies (equation 33). The yield required for this process would be $1.3/0.8 = 1.6 \,(100 \,\text{eV})^{-1}$, which is in agreement with the yield of fluorescent singlets found.

The total yield of $c\text{-}C_6H_{11}\cdot$ radicals involved in the formation of $c\text{-}C_6H_{10}$ and $C_{12}H_{22}$ is $2 \times (2.0 + 1.8) = 7.6 \,(100 \,\text{eV})^{-1}$. If these radicals would be formed by C—H scission of excited cyclohexane (equation 34) followed by H abstraction, this would correspond to a yield of $3.8 \,(100 \,\text{eV})^{-1}$ of C—H scission. The ratio of H_2 elimination and C—H scission is now found to be $1.3/3.8 = 0.33$. In the photolysis at $7.6 \,\text{eV}$, a value of ca 4 is found, and at $11.6 \,\text{eV}$ approximately 0.6, as can be observed from Figure 3. The low value of 0.33 for the radiolysis may be an indication that highly excited states are involved. Another explanation is that a fraction of the cyclohexyl radicals originates from triplets. It has been suggested also that part of the parent positive ion decomposes, while excited[34]:

$$(C_6H_{12}^{+\cdot})^* \longrightarrow C_6H_{11}{}^+ + H\cdot \qquad (38)$$

The cyclohexyl radicals would be formed after charge recombination and by H abstraction.

Holroyd compared the photochemistry at $8.4 \,\text{eV}$ and the radiolysis of liquid n-pentane and iso-octane[35]. It was shown that the relative distribution of the yields of the different radicals originating from C—C scission is strikingly similar for the radiolysis and the photolysis, for both liquids. The total quantum yield for formation of these radicals is around 0.1. While the relative frequencies of C—C scission are rather similar, the amount of single C—H scission (and formation of the parent radical) relative to C—C scission is much larger in radiolysis. Although the similarity between the distribution of yields of radicals originating from C—C scission in photolysis and radiolysis is important evidence that excited states are the precursors, the contribution of ion fragmentation cannot be excluded. In most cases it is very difficult to distinguish experimentally between fragmentation of the excited neutral molecule and that of the parent ion. In some cases an observed difference in yields of the two parts of the molecule is evidence for ion fragmentation of the ionic fragment after neutralization [e.g. $G(CH_3\cdot) \gg G(C_4H_9\cdot)$ in neopentane[36]]. Also, the change of product yields with and without electron and positive ion scavengers has provided evidence for parent-ion fragmentation to be operative in some cases. In neopentane the fragmentation

$$\text{neo-}C_5H_{12}^{+\cdot} \longrightarrow CH_3\cdot + t\text{-}C_4H_9{}^+ \qquad (39)$$

has been shown to take place with a yield of $1.0 \,(100 \,\text{eV})^{-1}$ [36]. In iso-octane a yield of $t\text{-}C_4H_9{}^+$ of $0.4 \,(100 \,\text{eV})^{-1}$ was determined[37]. Evaluation of the contribution of parent-ion fragmentation in general remains problematic. It appears that in branched saturated hydrocarbons this process contributes significantly to the overall C—C scission.

E. Final Products

In Table 8 we show the final product distribution observed in the radiolysis with low LET radiation for cyclohexane. Such product distributions have been determined for a great number of liquids[9,19,38]. With increasing dose the products accumulate in the liquid, and reactions of the intermediates (radicals etc.) with these products become more probable, thus causing the product distribution to change. At sufficiently low doses the so-called zero-dose yields, or initial yields, can be obtained. The radiation chemical yields in principle are also dependent on the dose rate. High dose rates cause higher stationary concentrations of the reactive intermediates (radicals, ions, etc.) and shorter lifetimes due to the increased rate of the reaction of the intermediates with one another. Reaction of radicals with each other may then be favoured, e.g. compared to H-atom abstraction. This effect has been observed in a few cases[33]. In practice, however, the dose rate effects

TABLE 8. Product yields in liquid cyclohexane, irradiated with ^{60}Co gamma radiation at room temperature[a]

	$G(100\,eV)^{-1}$
Hydrogen	5.6
Cyclohexene	3.26
Methane	0.01
Ethane	0.015
Ethylene	0.10
Acetylene	0.025
Propane	0.011
Cyclopropane	0.006
Propene	0.025
n-Butane	0.008
But-1-ene + but-2-ene	0.025
Buta-1,3-diene	0.004
n-Hexane	0.08
Hex-1-ene	0.36
Hex-2 and -3-enes	0.02
Methylcyclopentane	0.20
n-Hexylcyclohexane	0.08
6-Cyclohexylhex-1-ene	0.03
Bicyclohexyl	1.81

[a]The yields of hydrogen, cyclohexene and bicyclohexyl have been taken from Reference 39, those of the other products from References 18 and 40.

appear at dose rates many orders of magnitude higher than applied with conventional gamma sources. However, on irradiation with particle beams this may play a role.

The ratio of the total yield of carbon and hydrogen atoms in the products must, of course, be equal to that in the original molecule; this is called the material balance and provides an indication about the completeness of the product spectrum. From Table 8 we find a ratio of H to C equal to 2.03 as compared with an expected value of 2 for C_6H_{12}. For the total yield of decomposition 7.9 $(100\,eV)^{-1}$ is found.

The principal products from cyclohexane are H_2, cyclohexene and bicyclohexyl with yields of 5.6, 3.2 and 1.8 $(100\,eV)^{-1}$, respectively. The yield of H_2 formation represents the yield of decomposition of cyclohexane molecules due to C—H scission (either by H_2 elimination or otherwise). The yield of C—C scission is very small. Fragment hydrocarbons account for a yield of ca 0.2 $(100\,eV)^{-1}$, corresponding to a yield of decomposed cyclohexane molecules of around 0.03 $(100\,eV)^{-1}$. Straight-chain C_6 hydrocarbons are found with a yield of 0.46 $(100\,eV)^{-1}$, methylcyclopentane has a yield of 0.2 $(100\,eV)^{-1}$ and alkylcyclohexanes have a yield of ca 0.1 $(100\,eV)^{-1}$. Altogether this corresponds to a yield of cyclohexane molecules decomposed by C—C scission of ca 0.8 $(100\,eV)^{-1}$.

The total yield of 'primary' decomposition is now $5.6 + 0.8 = 6.4$ $(100\,eV)^{-1}$. This is lower than the yield of 7.9 $(100\,eV)^{-1}$ found for the total yield of decomposition. The latter is higher due to 'secondary' reactions of intermediates with cyclohexane molecules, like hydrogen abstraction by hydrogen atoms. It should be remarked, however, that back-formation of cyclohexane also takes place, e.g. in the reaction $2\,C_6H_{11}^{\cdot} \rightarrow C_6H_{10} + C_6H_{12}$. As an example of a straight-chain alkane we show the product distribution of

TABLE 9. Product yields^a in liquid n-pentane, irradiated with ^{60}Co gamma radiation at room temperature[9]

	$G(100\,\text{eV})^{-1}$
Hydrogen	5.3
Methane	0.26
Ethane	0.61
Ethylene	0.37
Propane	0.58
Propene	0.35
n-Butane	0.10
But-1-ene	0.08
trans- and cis-But-2-ene	0.02
Pent-1-ene	1.00
trans-Pent-2-ene	1.23
cis-Pent-2-ene	0.53
2-Methylbutane	0.061
n-Hexane	0.025
2-Methylpentane	0.016
3-Methylpentane	0.0072
n-Heptane	0.050
3-Methylhexane	0.119
3-Ethylpentane	0.0573
3-Ethylpent-1-ene	0.0047
4-Methylhex-1-ene	0.0042
n-Octane	0.041
4-Methylheptane	0.0917
3-Ethylhexane	0.040
2,3-Dimethylhexane	0.011
3-Ethyl-2-methylpentane	0.014
Octanes	0.003
n-Nonane	0.0047
4-Methyloctane + ethylheptane	0.029
3-Methyloctane	0.010
3,4-Dimethylheptane + 4-Ethyl-3-methylhexane	0.03
n-Decane	0.0344
3,4-Diethylhexane + 3-Ethyl-4-methylheptane	0.183
4,5-Dimethyloctane	0.751
3-Ethyloctane + 4-Methylnonane	0.0734
Decenes	0.06

^{a}The yield of hydrogen has been obtained with a dose of 0.37×10^{19} eV g^{-1}, the yields of methane through pent-2-ene with 5×10^{19} eV g^{-1} and the remaining ones with 14.2×10^{19} eV g^{-1}.

n-pentane in Table 9. In this case we find a value of 2.43 for the ratio of H and C atoms in the products, which compares very well with the expected ratio of 2.4 for C_5H_{12}. The total yield of decomposed molecules is found to be 6.9 $(100\,\text{eV})^{-1}$.

The product spectrum of n-pentane is more complicated than that of cyclohexane. The main products are H_2, ethane, propane, pentenes and the various decanes. The decanes and a major part of the pentenes are formed by recombination and disproportionation of the parent radicals $(C_5H_{11}{}^{\bullet})$, like in cyclohexane.

The yields may be divided into yields originating from C—H scission and from C—C scission. The yield of H_2 represents the sum of the yields due to H_2 elimination and breaking of a single C—H bond. The total yield of decomposition involving C—C scission can be obtained from the yields of the various products with a carbon skeleton differing from the original molecule. For n-pentane this is found to be $G(C—C) = 1.5$ $(100\,eV)^{-1}$ (higher than in cyclohexane). Together with the yield of 5.3 $(100\,eV)^{-1}$ for decomposition with H_2 or H formation, we find a yield of 6.8 $(100\,eV)^{-1}$ for molecules undergoing 'primary' decomposition (much the same as in cyclohexane). As we have seen above, from the total yield of C and H in the products we found 6.9 $(100\,eV)^{-1}$ for the total yield of decomposition; the agreement is accidental.

For n-alkanes with different length up to C_{12} the yield of H_2 formation is about the same, as is shown in Figure 4a. For longer chains the hydrogen yield appears to drop. The yield of fragment products for alkanes is also shown in Figure 4a, which shows that also the fragment yields for n-alkanes do not change much with chain length between C_5 and C_{12}. In Figure 4b we show the H_2 yields for cycloalkanes together with the yields of fragments and open-chain products. For C_6 and larger rings the yields do not vary much, however C_3 and C_4 and to a lesser extent C_5 show large yields of fragmentation and ring opening at the expense of H_2 formation: this is obviously caused by the ring strain in these molecules.

We now turn to the branched alkanes. As an example we consider the hexane isomers. In Table 10 we show the yields of H_2 formation, together with the yields of C—C

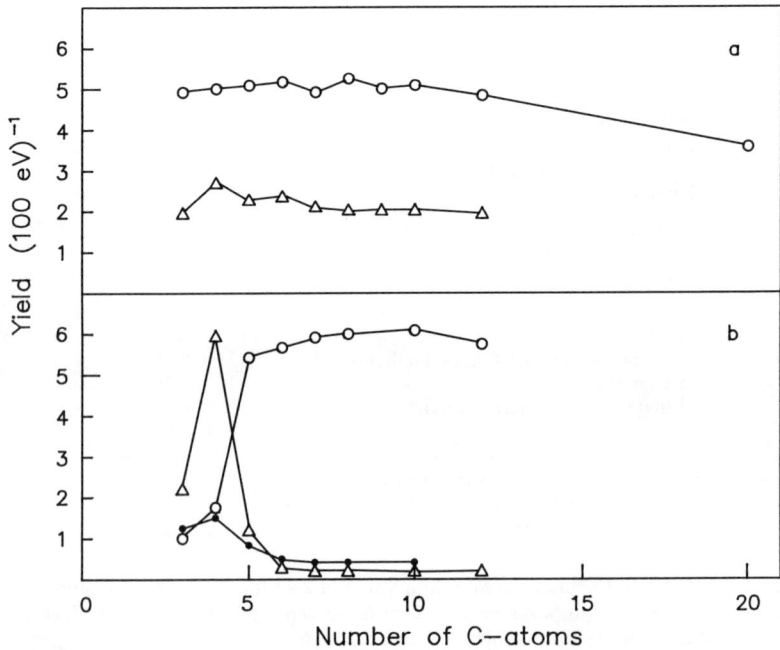

FIGURE 4. Yields of products in the gamma radiolysis of (a) n-alkanes and (b) cycloalkanes: $\bigcirc$, hydrogen, $\triangle$, fragments; $\bullet$, open-chain products from cycloalkanes. The yields for the alkanes with up to 12 C atoms have been determined at room temperature[19] and the value for C_{20} at 55 °C[41]

TABLE 10. G values of H_2 formation and C—C scission in units of $(100 \, eV)^{-1}$ in the radiolysis of different hexane isomers[a]

	$G(H_2)$	$G(C\!-\!C)$	$\dfrac{G(H_2)}{+}\\ G(C\!-\!C)$	$\dfrac{G(C\!-\!C)}{G(H_2)}$
n-Hexane	5.0	1.5	6.5	0.3
3-Methylpentane	3.4	2.4	5.8	0.7
2,3-Dimethylbutane	2.9	3.9	6.8	1.4
2,2-Dimethylbutane	2.0	5.1	7.1	2.5

scission. We see that the ratio of decomposition by C—H and C—C scission varies considerably for the different isomers, while the sum of these yields does not change much.

The yields of C—C scission have been studied extensively for a large number of compounds, and an empirical expression has been found that describes the yields of radical formation resulting from C—C scission[9]:

$$G(R) = \frac{1}{(n-1)^2}(1.0\,C_1 + 2.8\,C_2 + 8.6\,C_3 + 29\,C_4) \tag{40}$$

In this expression n represents the total number of C atoms in the original molecule; C_n represents the number of the C—C bonds in the molecule that, on scission, lead to formation of the radical R, and which have n C—C bonds adjacent to the breaking bond. For example, ethyl radicals can be formed from 3-methylhexane in two ways, either by breaking bond 2–3 or 4–5:

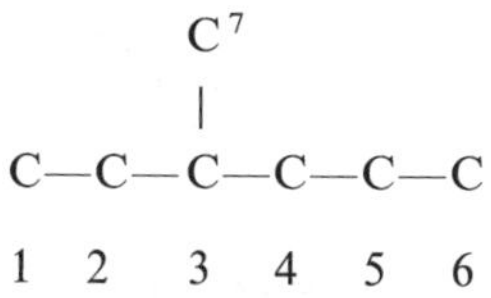

$$\begin{array}{c} C^7 \\ | \\ C\!-\!C\!-\!C\!-\!C\!-\!C\!-\!C \\ 1 \quad 2 \quad 3 \quad 4 \quad 5 \quad 6 \end{array}$$

bond 2–3 classifies as C_3 (bonds 1–2, 3–4 and 3–7 are adjacent), and there is one of these bonds, therefore $C_3 = 1$; $C_2 = 1$ (bonds 3–4 and 5–6 are adjacent); $C_1 = 0$ (applies only when R = CH_3) and $C_4 = 0$. We therefore find

$$G(C_2H_5^{\boldsymbol{\cdot}}) = \frac{1}{6^2}(2{\cdot}8 + 8{\cdot}6) = 0.32 \ (100 \, eV)^{-1}$$

We see now that in n-alkanes we have, for methyl radicals, $G(R) = 2.0/(n-1)^2$ and, for larger radicals, $G(R) = 5.6/(n-1)^2$. This shows that the ratio of the scission probability for the ultimate and the penultimate C—C bond is $2.0/5.6 = 0.36$.

An interesting correlation is found between the yield of C—C scission for a particular bond and the dissociation energy of the bond, for the various bonds in a molecule as well as in isomeric molecules[42]. This is shown in Figure 5. Each curve refers to a group of molecules with a given number (n) of C atoms. The curves decrease about exponentially with the dissociation energy in excess of that of the weakest bond. Furthermore, for each dissociation energy the yield of decomposition is lower for molecules with larger numbers of C atoms.

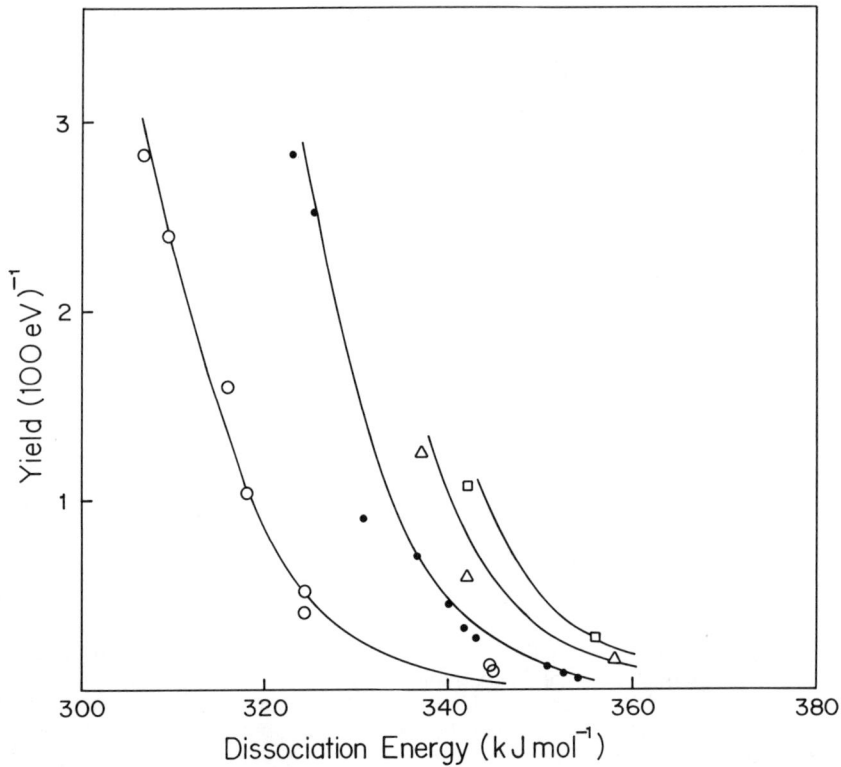

FIGURE 5. Yields of scission of C—C bonds in the gamma radiolysis of alkane isomers as a function of the dissociation energy of the bond, for molecules with a different number of C atoms: $\bigcirc$, C_8; $\bullet$, C_6; $\triangle$, C_5; $\square$, C_4. Reprinted with permission from Ref. 42. Copyright (1976) Pergamon Press PLC

F. Non-homogeneous Kinetics and LET Effects

In Section II we have discussed how the energy of the fast-moving charged particles is deposited in tracks. We have seen that subsequent energy losses along the track of a fast electron in a condensed medium are far apart ($\sim 200\,\text{nm}$). Most of the losses give rise to only one ionization, which results in a positive ion and an electron, thermalized within the Coulomb field. Also, groups of more ionizations close together are formed, with a frequency decreasing with the number of ions (Table 2).

In Section IV.B and IV.C, we have discussed the spatial separation of the thermalized electrons and the positive ions, the time scale for the recombination process and the probability of escape. We have seen that the fraction of the charges that escape and become homogeneously distributed in the liquid is small for most liquids and that most of the initial non-homogeneous recombination of the pairs and groups of pairs of oppositely charged species takes place on a time scale of 10^{-12}–10^{-10} seconds.

The non-homogeneous kinetics of single pairs of oppositely charged ions in each other's field can be treated exactly by solving the differential equation that describes the motion of one ion with respect to the other, due to diffusion and drift in the Coulomb field. Also, the scavenging probability can be calculated[20]. For multi-ion-pair groups the problem can be solved by computer simulation[22]. It should be remarked at this point that when excess electrons with a very large mobility, and therefore a very large mean free path, are involved, the existing theories are inadequate.

On recombination of the positive ions and electrons, excited molecules are formed that, in general, dissociate. In this way pairs of fragments and groups of pairs of fragments will be formed. In the case of cyclohexane this will be $C_6H_{11}^{\cdot}$ and $H^{\cdot}$, C_6H_{10} and H_2 (and some fragments resulting from C—C scission). Some of the fragments are formed with kinetic energy, notably the $H^{\cdot}$ atoms. The initial spatial distribution of these species in the regions of high concentration (called spurs) will depend on where the neutralization takes place, i.e. on the position of the positive ions at the moment of recombination. Subsequently, these species will start to carry out a diffusive motion. Part of them will encounter other species of the group, and react (non-homogeneously), the remaining part will escape from the group and become homogeneously distributed in the liquid, where they react with species from other groups.

We consider the following reactions for cyclohexane:

$$C_6H_{12}^* \longrightarrow C_6H_{11}^{\cdot}, H^{\cdot}, C_6H_{10}, H_2 \tag{41}$$

$$2\,C_6H_{11}^{\cdot} \longrightarrow (C_6H_{11})_2, C_6H_{10} + C_6H_{12} \tag{42}$$

$$H^{\cdot} + C_6H_{12} \longrightarrow H_2 + C_6H_{11}^{\cdot} \tag{43}$$

$$H^{\cdot} + C_6H_{11}^{\cdot} \longrightarrow C_6H_{12} \tag{44}$$

$$H^{\cdot} + C_6H_{10} \longrightarrow C_6H_{11}^{\cdot} \tag{45}$$

$$C_6H_{11}^{\cdot} + C_6H_{10} \longrightarrow C_{12}H_{21}^{\cdot} \tag{46}$$

A large fraction of the $H^{\cdot}$ atoms is hot initially. The hot H atoms abstract an H atom very efficiently (equation 43); thermalized H atoms [formed with a yield on the order of 1 $(100\,eV)^{-1}$ with low LET radiation] abstract much slower.

Experiments with radical scavengers (e.g. I_2) have shown that with low LET radiation ca 60% of the cyclohexyl radicals escape from the track; with increasing LET this fraction decreases. Also the yield of C_6H_{10} and $C_{12}H_{22}$ decreases, as is shown in Figure 6. The yield of H_2 first decreases somewhat with increasing LET, but then increases steeply at higher LET[39,43].

The escaped yield of $C_6H_{11}^{\cdot}$ decreases at higher LET, which is, of course, expected for a track where the spurs are formed more closely together. The decrease in the yield of C_6H_{10} and $C_{12}H_{22}$ may be explained by the increased occurrence of $H^{\cdot} + C_6H_{11}^{\cdot}$ and $C_6H_{11}^{\cdot}$ reacting with C_6H_{10} (equation 44) in the spurs. Reaction of $C_{12}H_{21}^{\cdot}$ leads to higher molecular weight products. Indeed 'polymer' yields have been observed with $G = 0.65$ $(100\,eV)^{-1}$ (on the basis of C_6 units) for a LET value of $130\,eV\,nm^{-1}$ and $G = 0.80$ $(100\,eV)^{-1}$ for $400\,eV\,nm^{-1}$ [39,43]. For values of LET below $100\,eV\,nm^{-1}$ the decrease in the H_2 yield agrees roughly with the decrease in C_6H_{10} and $C_{12}H_{22}$ and the increase in polymer yield, according to material balance.

The diffusion and reaction of the various species in the groups for initial spatial distributions representative for the different LET values have been treated theoretically by solving numerically the differential equations describing the development in time of the average local density distribution of the species in the groups:

$$\frac{\delta n_i}{\delta t} = D_i \nabla^2 n_i - \sum_j k_{ij} n_i n_j \tag{47}$$

where n_i is the density distribution in the group of species i, D_i the diffusion coefficient, k_{ij} the specific rate of reaction between species i and j, and averaging over the various groups in the track.

A. Hummel

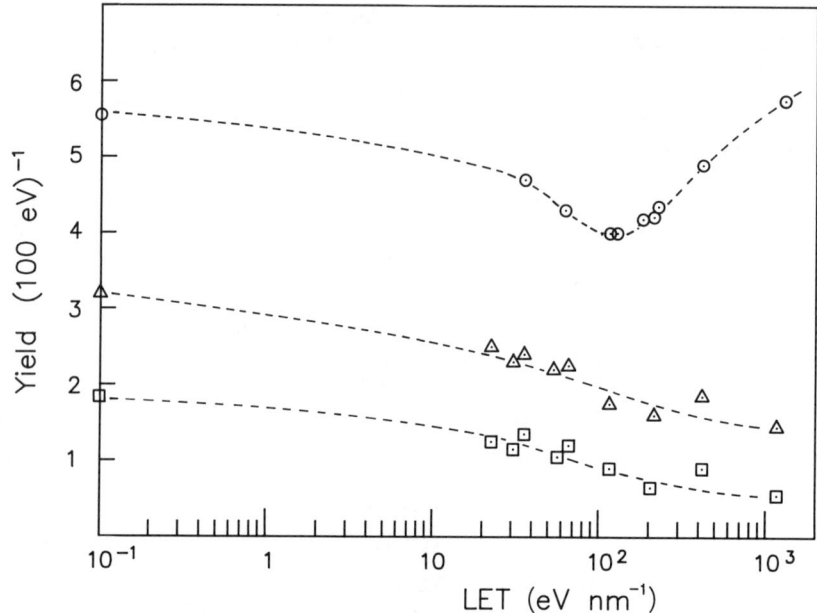

FIGURE 6. The yield of products in the radiolysis of cyclohexane as a function of LET[43];
O, H_2; △, C_6H_{10}; □, $C_{12}H_{22}$

It appears that for LET values up to $100 \, eV \, nm^{-1}$ the non-homogeneous kinetics can be rather satisfactorily described in this way[43]. At larger LET values other processes must be involved. The increase in the hydrogen yield exceeds by far the hydrogen deficiency in the polymer yield. An increase in the low molecular weight fragments, resulting from multiple C—C scission of the cyclohexane ring, has been observed. It appears that effects of hot radicals have increasing importance at the high LET values. Also, local heating in the track at the larger LET values may start to play a role.

While we have a fair degree of understanding of the radiation chemistry with low LET radiation in liquid saturated hydrocarbons, the situation is rather unclear for the chemical processes occurring in high LET charged particle tracks.

V. THE SOLID PHASE

A. Trapping of the Intermediates

An important aspect of the radiolysis in the solid is that, due to the rigidity of the matrix, the movement of reactive intermediates is often restricted to some degree. As a result certain reaction pathways involving these intermediates may be retarded or inhibited in the solid as compared to the liquid (and other processes may become more predominant in the solid). Electrons may be trapped in hydrocarbon glasses[44,45]. Yields on the order of a few tenths to one $(100 \, eV)^{-1}$ have been observed at 77 K. The trapped electrons (e_t^-) show a wide absorption spectrum in the infrared with a λ_{max} of around 1600 nm at 77 K, similar to what is observed in the liquid. The trapped electrons can also be observed by ESR. At lower temperatures the yields are somewhat larger.

At temperatures of 20–40 K the trapped electrons are found to decay by (temperature-independent) tunnelling to positive ions and radicals on a time scale of hours[46]. (If electron scavengers are present, tunnelling to these molecules also takes place.) It is interesting that the red part of the spectrum of e_t^- decays first, which shows that we deal with electrons of different trap depths.

At this point it is of interest to mention that, in the presence of solutes in the glasses, large yields of negative (and positive) ions can be obtained. In 3-methylpentane containing SF_6 and 2-methylpentene-1, a yield of 5.4 $(100 \, eV)^{-1}$ of SF_6^- has been observed[47].

It is generally considered that electron trapping in crystalline solids does not take place efficiently. In fact electron mobilities have been measured in various solid crystalline compounds. In CH_4, cyclohexane, iso-octane and neopentane mobile excess electrons have been observed[48].

Stable radical cations have been observed optically and by ESR after irradiation in matrices containing an electron scavenger[44,49]. However, ESR spectra have not been observed in the neat solids. Possible line broadening and overlap with radical spectra prevent definite conclusions about the absence of the ions from the ESR spectra. The optical absorption spectrum of the radical cation has been observed in neat 3-methyloctane at 6 K and 77 K ($\lambda_{max} = 600 \, nm$)[50] and in squalane ($C_{30}H_{62}$) at 77 K ($\lambda_{max} = 1400 \, nm$) at 100 ns after a 40-ns pulse[51].

Trapped radicals are produced in both glasses and crystalline solids and have been studied mainly by ESR[52]. In glasses the yield of radicals is typically 3–4 $(100 \, eV)^{-1}$ and in polycrystalline material 4–6 $(100 \, eV)^{-1}$. (In perdeuterated compounds this is about a factor of 1.3 lower[44].)

The radicals are formed non-homogeneously. From the time dependence of the decay of the radicals in 3-methylpentane at 77 K it was concluded that ca 60% of the radicals decays by reaction in regions of high local concentration[53]. At 77 K a small percentage of the radicals is trapped in pairs with a distance of 5–25 Å, as is evidenced by ESR experiments. At 4 K the percentage of pairs has been found to be as high as 100%[44]. The pairs may originate from H abstraction by hot H atoms or by tunnelling (see below), while the H atom may be formed by dissociation of a neutral excited molecule:

$$RH \longrightarrow R^{\bullet} + H^{\bullet} \tag{48}$$

Also, formation of the protonated molecule has been hypothesized:

$$RH^{\bullet +} + RH \longrightarrow RH_2^+ + R^{\bullet} \tag{49}$$

followed by formation of a second radical $R^{\bullet}$ by H_2 expulsion from RH_2^+ and neutralization of R^+, or by neutralization of RH_2^+ followed by H_2 expulsion[54].

A high probability for finding two radicals close together would also be expected in regions of high ionization density. Evidence for this effect being operative has been found from studies of the LET effect on radical pair formation in eicosane[55].

No trapped hydrogen atoms have been observed except in CH_4 at 5 K, where $G(H^{\bullet}) = G(CH_3^{\bullet}) = 3.3 \, (100 \, eV)^{-1}$ and $G(e_t^-) = 0$. The hydrogen atoms react either while hot or by tunnelling abstractions from C—H bonds by thermalized atoms.

Carbanions are formed by reaction of electrons with radicals: $R^{\bullet} + e^- \rightarrow R^-$. In branched alkane glasses they have an optical absorption spectrum with a λ_{max} around 240 nm ($\varepsilon \approx 6300 \, dm^3 \, mol^{-1} \, cm^{-1}$) and a long tail extending into the red[44].

B. Product Formation

Product yields are generally determined by analysing the products in the liquid obtained after melting the irradiated sample. In one technique the solid is dissolved at low temperature in a liquid containing a radical scavenger, so that the contribution of radical reactions to the products is eliminated[56].

TABLE 11. Yields of products in the gamma radiolysis of solid
n-pentane at 77 K[9,57]

	$G(100\,eV)^{-1}$
Hydrogen	5.0
Ethane	0.34
Ethylene	0.36
Propane	0.28
Propene	0.21
Pent-1-ene	1.7
trans-Pent-2-ene	1.3
cis-Pent-2-ene	0.3
n-Decane	0.16
4,5-Dimethyloctane + 3-ethyloctane	0.40
3-Ethyl-4-methylheptane + 3,4-diethylhexane	<0.01
4-Methylnonane	0.30
C_8, C_9	<0.002
C_{15}	0.01

In Table 11 we show the product yields obtained for n-pentane irradiated at 77 K (and melting before analysis). The liquid-phase results have been given in Table 9. The H_2 yield is the same in the solid and the liquid, which indicates that the yield of decomposition by C—H scission is the same. The yield of decomposition by C—C scission in the solid is approximately $0.6\,(100\,eV)^{-1}$, i.e. appreciably lower than in the liquid, where we found $1.5\,(100\,eV)^{-1}$. The lower yield of fragmentation is probably due in part to the cage effect. It is interesting to see that no CH_4 and C_4 products are found and hardly any C_6 to C_9 products. In the liquid the terminal C—C bond of the chain

TABLE 12. Yields of products in the gamma radiolysis of cyclohexane[19]

	$G(100\,eV)^{-1}$ solid, 77 K	$G(100\,eV)^{-1}$ liquid, room temperature
Hydrogen	4.53	5.6
Methane	0.002	0.001
Ethane	0.002	0.015
Ethylene	0.021	0.1
Acetylene	—	0.025
Propane	0.005	0.011
Cyclopropane	0.004	0.006
Propene	0.012	0.025
n-Butane	0.006	0.008
But-1-ene	0.009 ⎫	0.025
But-2-ene	— ⎭	
Buta-1,3-diene	0.015	0.004
Cyclohexene	2.27	3.26
n-Hexane	—	0.08
Hex-1-ene	0.19	0.36
Hex-2 and -3-enes	—	0.02
Methylcyclopentane	—	0.20
n-Hexylcyclohexane	—	0.08
6-Cyclohexylhex-1-ene	—	0.03
Bicyclohexyl	1.32	1.81

decomposes less frequently than the others; these results show that the relative frequency of breakage of the last bond is even smaller in solid n-pentane.

The experiments where the solid is dissolved at low temperature in a solvent, containing radical scavengers, have shown that in the case of n-pentane 40–70% of the dimers originates from radicals[56]. Ion–molecule reactions may contribute, however. Dimer formation via ion–molecule reactions has been evidenced for n-hexane[58]. A mechanism proposed is the direct dimerization under elimination of H_2[59]:

$$RH^{+\cdot} + RH \longrightarrow RR^{+\cdot} + H_2 \qquad (50)$$

The total yield of dimers in solid n-pentane, 0.86 $(100\,eV)^{-1}$, is less than in the liquid at room temperature, however, about the same as the yield in the liquid at $-120\,°C$, 0.9 $(100\,eV)^{-1}$. Also, the yields of n-decane in the solid and at $-120\,°C$ are not much different [0.12 and 0.085 $(100\,eV)^{-1}$, respectively][60].

The total yield of decomposition in the solid as obtained from Table 11 is 5.6 $(100\,eV)^{-1}$ as compared to 6.9 $(100\,eV)^{-1}$ in the liquid at room temperature. We see that the difference is mainly caused by the decreased yield of fragmentation in the solid as compared to the liquid.

In Table 12 we have presented the yields of products in solid cyclohexane, irradiated

TABLE 13. Yields of products in the gamma radiolysis of 3-methylpentane[9]

	$G\,(100\,eV)^{-1}$ solid, 77 K	$G\,(100\,eV)^{-1}$ liquid, room temperature
Hydrogen	3.2	3.4
Methane	0.08	0.20
Ethane	0.45	1.90
Ethylene	0.20	1.00
Propane	0.02	0.09
Isobutane	—	0.02
n-Butane	0.31	0.45
But-1-ene	0.19	0.37
But-2-enes	0.14	0.22
2-Methylbutane	0.01	0.06
2-Methyl-1-butene	0.03	0.03
n-Pentane	0.05	0.09
Pent-2-enes	0.04	0.07
2-Methylpentane	0.80	0.65
3-Methyl-1-pentene	0.40	0.27
3-Methyl-2-pentene	0.80	0.55
3-Methyl-1,3-pentadiene	0.43	0.26
Dimethylpentane	0.42	0.28
3-Methylhexane + 3-ethylpentane	0.18	0.26
2-Ethyl-3-methylpentane + 3-methylheptane + 3-ethyl-3-methylpentane	0.21	0.51
3,5-Dimethylheptane	—	0.03
4-Ethyl-3-methylhexane	0.02	0.04
5-Ethyl-3-methylheptane	—	—
3,6-Dimethyloctane + 3,4-diethylhexane	0.04	—
3,4,5-Trimethylheptane	0.07	—
3,4-Dimethyl-3-ethylhexane	0.08	0.29
2,4-Dimethyl-5-ethylheptane	0.02	0.05
3,4-Diethyl-3-methylhexane	—	0.03
C_{12}	0.68	0.98

at 77 K. Here we see that the H_2 yield in the solid is lower than in the liquid [4.5 vs 5.6 $(100\,eV)^{-1}$]. The yield of decomposition by C—C scission is *ca* 0.3 $(100\,eV)^{-1}$ in the solid, which again is lower than in the liquid [0.8 $(100\,eV^{-1})$]. The total yield of molecules decomposed is 5.0 $(100\,eV)^{-1}$ in the solid, as compared to 7.9 $(100\,eV)^{-1}$ in the liquid.

As an example of a branched hydrocarbon we give the results for solid and liquid 3-methylpentane in Table 13. The yields of C_1–C_5 fragments in the solid and liquid are 1.59 $(100\,eV)^{-1}$ and 4.7 $(100\,eV)^{-1}$, respectively, and those of C_7–C_{11} products 1.1 $(100\,eV)^{-1}$ and 1.7 $(100\,eV)^{-1}$, respectively. We see that a yield of C—C scission of *ca* 1.3 $(100\,eV)^{-1}$ and 3.2 $(100\,eV)^{-1}$ is found in the solid and the liquid, respectively. Here again we see a large difference in the yield of C—C scission between the solid and liquid (while the yield of C—H scission is not much different).

C. Polymers; Polyethylene

The radiation chemistry of polymer systems has received considerable attention, in connection with its industrial importance. Notably the radiation-induced cross-linking of polyethylene is applied on a large scale. A low degree of cross-linking of polyethylene is applied in order to regulate the viscosity of the melt for foam production. Higher degrees of cross-linking are applied for improving various properties of cable insulation, tubing material etc. A most important application is the use of the so-called memory effect of cross-linked polyethylene. With a sufficiently high degree of cross-linking the polymer, when heated above the melting point (T_m), becomes an elastomer. When stretched at a temperature above T_m and subsequently cooled down under stress, the material remains in the stretched form when the stress is removed. On heating, however, the material returns to its original unstretched form. This property of radiation cross-linked polyethylene has found a great number of applications, e.g. heat-shrinkable foil for packaging and tubes, tapes for electrical insulation etc.

In the following we shall first give some general features of the radiation chemistry of polymers and then discuss polyethylene.

One of the great problems in the interpretation of the radiation chemical effects in polymer materials is the influence of the various aspects of the structure and composition of the material. Not only the detailed chemical composition has been shown to be of importance (including the molecular weight distribution, degree of branching, occurrence of side groups, the nature of end groups, etc.), but also the detailed morphology plays an important role (the nature of the packing of the molecules in the crystallites, at the surface of the crystallites and in amorphous regions, entanglements of the chains, position of end groups, etc.). We shall illustrate some of these effects below.

An important radiation chemical effect in polymers is the production of cross-links (cl) and main-chain scissions (sc), since these have a profound effect on various macroscopic properties of the polymer material. Scission decreases the molecular weight of the polymer and cross-linking increases it. It may be shown that under conditions of random scission and cross-linking and for an initially random distribution of molecular weights, one can write for the weight-average molecular weight M_W as a function of dose[61-63]

$$\frac{1}{M_W(D)} = \frac{1}{M_W(0)} + [\tfrac{1}{2}G(sc) - 2G(cl)]D \times 10^{-10} \tag{51}$$

where $G(sc)$ and $G(cl)$ are the yields, expressed in $(100\,eV)^{-1}$, and D is the dose, in $J\,kg^{-1}$. We see that if $G(cl) = 0$, $1/M_W(D)$ increases linearly with dose, independent of the initial molecular weight $M_W(0)$. If $G(sc) = 0$, $1/M_W(D)$ decreases with dose, until $1/M_w(D) = 0$ for $2(G(cl) \times D \times 10^{-10} = 1/M_W(0)$. Since $2G(cl) \times D \times 10^{-10}$ represents the number of moles of molecules cross-linked per gram material and $1/M_W(0)$ is the number of moles of weight-average polymer molecules initially, we see that at this dose on the average each molecule contains a cross-link. This dose is called the gel dose (D_g) and at this dose a network is formed throughout the volume of the material. Experimentally, the gel dose manifests itself by the fact that the polymer material does not dissolve anymore, but remains as a gel. With increasing dose the fraction of polymer in the gel increases until it reaches a maximum, depending on $G(cl)/G(sc)$.

Information about the molecular weight distributions for $D < D_g$ can be obtained by e.g. viscosity in the melt and light scattering in solution. When $D > D_g$ the number of cross-links is determined from e.g. the sol/gel fraction, swelling or elastic moduli.

The polymer polyisobutylene $(-C(CH_3)_2-C-)_n$ presents a case where $G(cl) = 0$. This polymer is entirely amorphous (except on stretching, when it crystallizes) and the linear increase of $1/M_W(D)$ with dose has been established over a wide range of initial molecular weights. From these results it has been concluded that the scissions take place at random positions[62].

Another saturated hydrocarbon polymer that has been studied is polypropylene[64]. The isotactic form (with the methyl groups on one side of the chain) crystallizes; the atactic form has insufficient configurational order for crystallization. It has been shown that in relatively highly crystalline isotactic material $G(sc)/G(cl) \simeq 1.5 - 1.8$, while in material with low crystallinity (atactic as well as isotactic) a value of ca 0.8 is found.

We now return to polyethylene. Polyethylene may be produced with a large range of molecular weights and with varying degrees of branching. We shall be concerned with so-called linear polyethylene (with a very low number of side groups), which may have molecular weights up to 5×10^6. The polymer chains have vinyl end groups. The density of the material depends on the degree of crystallinity, which in turn depends on the mode of preparation of the sample. Samples with 60–80% crystallinity are quite common. Slow crystallization may lead to levels of crystallinity of over 90%. The crystallization takes place in flat crystallites, called lamellae, with varying thickness of the order of 10–50 nm and lateral dimensions on the order of 10 μm. The polymer chains are folded with the chain direction perpendicular to the flat surfaces of the lamellae. These surfaces therefore contain the bends of the chains, with occasional loops extending further out, and having entanglements with chains extending from other lamellae, or from amorphous regions. It appears that the chain ends are predominantly found in these fold surfaces.

The overall chemical effects in polyethylene are H_2 formation, cross-linking, main chain scission, formation of main chain unsaturation ($trans$-vinylene and diene) and disappearance of vinyl end groups[65,66]. At room temperature a yield of hydrogen of around 3.7 $(100\,eV)^{-1}$ is found. There is a considerable temperature effect; in crystalline samples at 130 °C this is 6. 3$(100\,eV)^{-1}$ and at 133 °C, in the melt, 6.2 $(100\,eV)^{-1}$.

An overall cross-link yield of ca 1–2 $(100\,eV)^{-1}$ is often found, however there is some uncertainty about the significance of this value is view of the fact that the distribution of cross-links may not be random and the theory relating the various experimental results to the number of cross-links (gel experiments, swelling, elastic moduli) may not give accurate results.

Important experiments have been carried out with isolated lamellae, also called crystal cores, of which the chain ends could be removed chemically[67]. GPC analysis has shown that the cross-links mainly occur at the folds and that the $G(cl)$ in the crystal cores is only ca 0.2 $(100\,eV)^{-1}$. The fact that cross-links are formed with such a low yield in the crystal is attributed to the fact that the carbon atoms on the adjacent chains are too far apart (> 0.41 nm) for interchain C—C bonds (0.15 nm) to be formed. It is generally accepted now that the cross-linking in polyethylene mainly takes place at the fold surfaces and in the amorphous regions.

Scission yields of around $0.2 \, (100 \, \text{eV})^{-1}$ have been estimated, however the results seem to be somewhat uncertain.

Crystal core experiments have shown that *trans*-vinylene and diene are formed throughout the crystal with G values of 2.4 and 0.5 $(100 \, \text{eV})^{-1}$, respectively. The *trans*-vinylene yield roughly agrees with the bulk yield, which appears to suggest a random process.

The mechanism of the formation of the cross-links is not clear. It has been suggested that movement of the alkyl radical site along the chain plays a role[65]. Alkyl radicals have been found to be formed by ESR in both crystalline and amorphous regions with the yields around 3 $(100 \, \text{eV})^{-1}$. Whether they can move out of the lamellae towards the folds is not entirely clear. The mechanisms for the formation of two alkyl radicals close to each other have been discussed in the foregoing section. Also, the ionic dimerization reaction[52] has been proposed for polyethylene[58].

Migration of energy and/or charge in the crystal may also play a role. The rapid migration of charge in highly crystalline ultra high molecular weight material has been shown to occur by means of microwave conductivity measurements after pulsed irradiation[69]. Whether the mobile charges are excess electrons or electron holes is not known. Migration of holes before recombination would lead to formation of excited states at positions away from the original ionization site. Also, the migration of electronic energy along the chains has been hypothesized[68].

Despite the fact that a large body of experimental results exists on the effects of high-energy radiation on polyethylene, some very basic questions concerning the mechanism of the radiation chemical processes remain unanswered.

VI. REFERENCES

1. G. Földiák (Ed.), *Radiation Chemistry of Hydrocarbons*, Elsevier, Amsterdam, 1981.
2. *Radiation Chemistry, Principles and Applications*, Farhataziz and M. A. J. Rodgers (Eds.), VCH Publ., New York, 1987.
3. G. R. Freeman (Ed.), *Kinetics of Nonhomogeneous Processes*, Wiley, New York, 1987.
4. J. H. Baxendale and F. Busi (Eds.), *The Study of Fast Processes and Transient Species by Electron Pulse Radiolysis* (Proceedings of NATO Advanced Study Institute, Capri, 1981), Reidel, Dordrecht, 1982.
5. A. Mozumder, in *Advances in Radiation Chemistry*, Vol. 1 (Eds. M. Burton and J. L. Magee), Wiley–Interscience, New York, 1969, pp. 1–102.
6. F. J. de Heer, *Int. J. Radiat. Phys. Chem.*, **7**, 137 (1975).
7. A. Hummel, unpublished results.
8. (a) W. J. Beekman, *Physica*, **15**, 327 (1949).
   (b) A. Ore and A. Larsen, *Radiat. Res.*, **21**, 331 (1964).
   (c) K. Kowari and S. Sato, *Bull. Chem. Soc. Jpn.*, **54**, 2878 (1981).
9. I. György, in *Radiation Chemistry of Hydrocarbons* (Ed. G. Földiák), Chap. 2, Elsevier, Amsterdam, 1981, pp. 61–176.
10. R. L. Platzman, *The Vortex*, **23**, no. 8, 1 (1962).
11. (a) K. Okazaki, S. Sato and S. Ohno, *Bull. Chem. Soc. Jpn.*, **49**, 174 (1976).
    (b) P. Ausloos and S. G. Lias, in *Chemical Spectroscopy and Photochemistry in the Ultra-violet* (Eds. C. Sandorfy, P. Ausloos and M. B. Robin), Reidel, Dordrecht, 1974, pp. 465–482.
    (c) P. Ausloos and S. G. Lias, *Ann. Rev. Phys. Chem.*, **22**, 85 (1971).
12. J. L. Franklin, J. G. Dillard, H. M. Rosenstock, J. T. Herron, K. Draxl and F. H. Field, *Nat. Stand. Ref. Data Ser.*, *Nat. Bur. Stand. (U.S.)*, **26**, 1 (1969).
13. G. G. Meisels, in *Fundamental Processes in Radiation Chemistry* (Ed. P. Ausloos), Chap. 6, Wiley–Interscience, New York, 1968, pp. 347–412.
14. P. Ausloos, S. G. Lias and R. Gordon, Jr., *J. Chem. Phys.*, **39**, 3341 (1963).
15. J. H. Futrell and T. O. Tiernan, in *Fundamental Processes in Radiation Chemistry* (Ed. P. Ausloos), Chap. 4, Wiley–Interscience, New York, 1968, pp. 171–280.
16. P. Ausloos, R. E. Rebbert, F. P. Schwarz and S. G. Lias, *Radiat. Phys. Chem.*, **21**, 27 (1983).

17. L. M. Theard, *J. Phys. Chem.*, **69**, 3292 (1965).
18. K. H. Jones, *J. Phys. Chem.*, **71**, 709 (1967).
19. L. Woinárowits, in *Radiation Chemistry of Hydrocarbons* (Ed. G. Földiák), chap. 3, Elsevier, Amsterdam, 1981, pp. 177–252.
20. A. Hummel, in *Kinetics of Nonhomogeneous Processes* (Ed. G. R. Freeman), Wiley, New York, 1987, pp. 215–275.
21. J. M. Warman, in *The Study of Fast Processes and Transient Species by Electron Pulse Radiolysis* (Eds. J. H. Baxendale and F. Busi), Reidel, Dordrecht, 1982, pp. 433–533.
22. (a) W. M. Bartczak and A. Hummel, *J. Phys. Chem.*, **87**, 5222 (1987).
    (b) W. M. Bartczak, M. P. de Haas and A. Hummel, *Radiat. Phys. Chem.*, **37**, 401 (1991).
23. L. H. Luthjens, H. C. de Leng, C. A. M. van den Ende and A. Hummel, in *Proc. 5th Tihany Symposium on Radiation Chemistry*, Vol. 2 (Eds. J. Dobo, P. Hedvig and R. Schiller), Akademiai Kiado, Budapest, 1983, p. 471.
24. L. H. Luthjens, H. C. de Leng, W. R. Appleton and A. Hummel, *Radiat. Phys. Chem.*, **36**, 213 (1990).
25. G. R. Freeman, in *Kinetics of Nonhomogeneous Processes* (Ed. G. R. Freeman), Wiley, New York, 1987, pp. 19–88, 277–304.
26. A. O. Allen, Yields of free ions formed in liquids by radiation, National Bureau of Standards Document NSRDS-NBS 57, p. 12 (1976).
27. C. D. Jonah, M. S. Matheson, J. R. Miller and E. J. Hart, *J. Phys. Chem.*, **80**, 1267 (1976).
28. P. Adler and H. K. Bothe, *Z. Naturforsch.*, **20a**, 1700 (1965).
29. (a) W. Rothman, F. Hirayama and S. Lipsky, *J. Chem. Phys.*, **58**, 1300 (1973).
    (b) L. Walter and S. Lipsky, *Radiat. Phys. Chem.*, **7**, 175 (1975).
    (c) L. H. Luthjens, M. P. de Haas, H. C. de Leng, A. Hummel and G. Beck, *Radiat. Phys. Chem.*, **19**, 121 (1982).
30. (a) W. M. Bartczak, M. Tachiya and A. Hummel, *Radiat. Phys. Chem.*, **36**, 195 (1990).
    (b) J. L. Magee and J.-T. J. Huang, *J. Phys. Chem.*, **76**, 3801 (1972).
    (c) B. Brocklehurst, *Nature*, **221**, 921 (1969).
    (d) T. Higashimura, K. Hirayama and K. Katsuura, *Annu. Rep. Res. Reactor, Inst. Kyoto. Univ.*, **5**, 11 (1972).
31. F. P. Schwarz, D. Smith, S. G. Lias and P. Ausloos, *J. Chem. Phys.*, **75**, 3800 (1981).
32. W. A. Cramer, in *Aspects of Hydrocarbon Radiolysis* (Eds. T. Gäumann and J. Hoigné), Academic Press, London 1968, pp. 153–212.
33. R. A. Holroyd, in *Fundamental Processes in Radiation Chemistry* (Ed. P. Ausloos), Interscience–Wiley, New York, 1968, pp. 413–514.
34. (a) T. Wada, S. Shida and Y. Hatano, *J. Phys. Chem.*, **79**, 561 (1975).
    (b) P. Ausloos, R. E. Rebbert, F. P. Schwarz and S. G. Lias, *Radiat. Phys. Chem.*, **21**, 27 (1983).
35. R. A. Holroyd, *J. Am. Chem. Soc.*, **91**, 2208 (1969).
36. R. A. Holroyd and G. W. Klein, *J. Am. Chem. Soc.*, **87**, 4983 (1965).
37. K. Tanno, T. Miyazaki, K. Shinsaka and S. Shida, *J. Phys. Chem.*, **71**, 4290 (1967).
38. T. Gäumann and H. Hoigné (Eds.), *Aspects of Hydrocarbon Radiolysis*, Academic Press, London 1968.
39. W. G. Burns and C. R. V. Reed, *Trans. Faraday Soc.*, **66**, 2159 (1970).
40. L. Wojnárovits and G. Földiák, *Radiochem. Radioanal. Lett.*, **21**, 261 (1975).
41. T. Seguchi, N. Hayakawa, N. Tamura, Y. Tabata, Y. Katsumura and N. Hayashi, *Radiat. Phys. Chem.*, **25**, 399 (1985).
42. G. Földiák, I. György and L. Wojnárovits, *Int. J. Radiat. Phys. Chem.*, **8**, 575 (1976).
43. W. G. Burns, M. J. Hopper and C. R. V. Reed, *Trans. Faraday Soc.*, **66**, 2185 (1970).
44. J. E. Willard, in *Radiation Chemistry, Principles and Applications* (Eds. Farhataziz and M. A. J. Rodgers), VCH Publ., New York, 1987, pp. 395–434.
45. L. Kevan, in *Advances in Radiation Chemistry*, Vol. 4 (Eds. M. Burton and J. L. Magee), Wiley–Interscience, New York, 1974, pp. 181–306.
46. J. Paraszcak and J. E. Willard, *J. Chem. Phys.*, **70**, 5823 (1979).
47. D. Bhattacharya and J. E. Willard, *J. Phys. Chem.*, **84**, 146 (1980).
48. T. Tezuka, H. Namba, Y. Nakamura, M. Chiba, K. Shinsaka and Y. Hatano, *Radiat. Phys. Chem.*, **21**, 197 (1983).
49. W. H. Hamill, in *Radical Ions* (Eds. J. Kaiser and L. Kevan), Wiley–Interscience, New York, 1968, pp. 321–416.

50. (a) N. V. Klassen and G. G. Teather, *J. Phys. Chem.*, **83**, 326 (1979).
    (b) J. Cygler, G. G. Teather and N. V. Klassen, *J. Phys. Chem.*, **87**, 455 (1983).
51. G. G. Teather and N. V. Klassen, *J. Phys. Chem.*, **85**, 3044 (1981).
52. R. W. Fessenden and R. H. Schuler, in *Advances in Radiation Chemistry*, Vol. 2 (Eds. M. Burton and J. L. Magee), Wiley–interscience, New York, 1970, pp. 1–176.
53. M. A. Neiss and J. E. Willard, *J. Phys. Chem.*, **79**, 783 (1975).
54. W. J. Chappas and J. Silverman, *Radiat. Phys. Chem.*, **16**, 437 (1980).
55. K. Hamanone, V. Kamansauskas, Y. Tabata and J. Silverman, *J. Chem. Phys.*, **61**, 3439 (1974).
56. P. Tilman, P. Claes and B. Tilquin, *Radiat. Phys. Chem.*, **15**, 465 (1980).
57. C. Bienfait and P. Claes, *Int. J. Radiat. Phys. Chem.*, **2**, 101 (1970).
58. (a) L. Kevan and W. F. Libby, *J. Chem. Phys.*, **39**, 1288 (1963).
    (b) L. Kevan and W. F. Libby, in *Advances in Photochemistry* (Eds. W. A. Noyes, G. S. Hammond and J. N. Pitts, Jr), Interscience, New York, 1964, pp. 183–218.
59. J. Weiss, *J. Polym. Sci.*, **29**, 425 (1958).
60. R. O. Koch, J. P. W. Houtman and W. A. Cramer, *J. Am. Chem. Soc.*, **80**, 3326 (1968).
61. A. Charlesby, *Atomic Radiation and Polymers*, Pergamon Press, Oxford, 1960.
62. A. Charlesby, in *Radiation Chemistry, Principles and Applications* (Eds. Farhataziz and M. A. J. Rodgers), VCH Publ., New York, 1987, pp. 451–475.
63. O. Saito, in *The Radiation Chemistry of Macromolecules*, Vol. 1 (Ed. M. Dole), Academic Press, London 1973, pp. 224–265.
64. D. O. Geymer, in *The Radiation Chemistry of Macromolecules*, Vol. 2 (Ed. M. Dole), Academic Press, London 1973, pp. 4–28.
65. M. Dole, in *Advances in Radiation Chemistry*, Vol. 4 (Eds. M. Burton and J. L. Magee), Wiley–Interscience, New York, 1974, pp. 307–388.
66. L. Mandelkern, in *The Radiation Chemistry of Macromolecules*, Vol. 1 (Ed. M. Dole), Academic Press, London 1972, pp. 287–334.
67. G. Ungar and A. Keller, *Polymer*, **21**, 1273 (1980).
68. R. H. Partridge, in *The Radiation Chemistry of Macromolecules*, Vol. 1 (Ed. M. Dole), Academic Press, London 1973, pp. 26–54.
69. M. P. de Haas and A. Hummel, *IEEE Trans.*, **24**, 349 (1989).

CHAPTER **17**

Electrochemical conversion of alkanes

HANS J. SCHÄFER

Organisch-Chemisches Institut der Westfälischen Wilhelms-Universität, Correns-Str. 40, 4400 Münster, Germany

I. INTRODUCTION

The regioselective functionalization of CH bonds provides the possibility to convert readily available and abundant natural compounds, such as steroids or fatty acids, into higher value products.

The Chemistry of Alkanes and Cycloalkanes
Edited by S. Patai and Z. Rappoport © 1992 John Wiley & Sons Ltd

Biological systems can perform these conversions quite specifically. For instance, in steroids nearly each carbon can be regioselectively functionalized by microbiological oxidation[1,2]. Unsaturated fatty acids can be obtained by specific aerobic desaturation of saturated fatty acids[3]. Oxidases from vertebrate animals[4,5] or yeast[5] can hydroxylate fatty acids in the ω to $(\omega - 3)$ position. Some of these reactions are used in biotechnical processes, predominantly for the preparation of otherwise difficulty accessible intermediates for the synthesis of steroid hormones[2b,c]. Besides the advantages of biotechnical processes with regard to their selectivity, they are, however, expensive due to their often low space–time yields and high demands for sterility. An interesting nonenzymatic regioselective CH oxidation is the radical-relay chlorination of steroids found by Breslow[6a]. By covalent[6b] or ionic[6c] connection of the steroid and the reagent phenyliododichloride the 9-H position is selectively oxidized. With appropriate spacer groups the 14- and 17-H positions can also be selectively functionalized[6b,7]. A disadvantage of this otherwise very elegant method is that additional steps are needed for the connection and disconnection of the reagent, which leads to a lower overall yield.

Fatty acids can be chlorinated with fair to good selectivity at the ω to $(\omega - 3)$ positions by electrostatic orientation or partial shielding. In the first case the acid is chlorinated with N,N-dialkyl-N-chloroamine in strong acid[8]. In the second case the acid is adsorbed at alumina and chlorinated at the gas–solid interphase[9].

Other oxidation reagents for nonactivated CH bonds are chromium(VI) in acetic acid[10], lead tetraacetate[11], cobalt triacetate[12], permanganate-trifluoroacetic acid[13], triethylbenzylammonium (TEBA) permanganate[14], peracids[15], P-450 analogs[16] or the Gif-system[17]. A disadvantage of these systems is their often complicated handling, high price and environmental hazards of the reagents and their products. Anodic oxidation is an interesting alternative to these conversions. The experimental procedure is simple, allows an easy scale-up and causes less environmental problems.

This review will deal with the direct and indirect anodic oxidation of unactivated CH bonds in alkanes, and of remote CH bonds in substrates with various functional groups. The conversion of CH bonds activated by vinyl, aryl, amino or alkoxy groups will not be dealt with in this chapter. However, the catalytic oxidation of alkanes in fuel cells and the oxidation of alkanes with molecular oxygen and iron catalysts are included.

II. ANODIC OXIDATION OF ALKANES

A. Anodic Substitution in Different Solvents

The unactivated CH bond in aliphatic hydrocarbons is being oxidized at potentials mostly more anodic than 2.5 V (vs SCE)[18]. This necessitates electrolytes with high anodic decomposition potentials. Because of the importance of the electrolyte in these conversions this section is organized according to the different electrolytes.

1. Fluorosulfonic acid

Fluorosulfonic acid has been frequently used as solvent in electrolysis. The half-peak potentials of alkanes are clearly visible in fluorosulfonic acid containing 0.1 M KSO_3F and 25 mM acetic acid. They range between 2.21 V (vs Pd–H_2) for propane, 1.37 V for n-octane, 1.12 V for cyclohexane and 1.00 V for isobutane[19].

The oxidation potential becomes less positive with increasing acidity of the electrolyte. This has led to the proposal of the following mechanism, whereby the alkane is protonated in a preequilibrium and the protonated form is subsequently oxidized to a carbenium ion (equation 1a–c).

This mechanism has been criticized by a French group[20-23]. In contrast to the results

$$RH + HFSO_3 \rightleftharpoons RH_2^+ + FSO_3^- \tag{1a}$$

$$RH_2^+ - e \longrightarrow RH_2^{2+} \tag{1b}$$

$$RH_2^{2+} - e \longrightarrow R^+ + 2H^+ \tag{1c}$$

in Reference 19 they found that the electrochemical oxidation of n-hexane in fluorosulfonic acid is facilitated to a small extent by increasing the concentration of the base $NaSO_3F$, namely from 1.14 V (0.1 M) to 1.09 V (1.3 M) vs perylene$^+$/perylene^{++} as reference electrode. The authors conclude, therefore, that not the protonated but the unprotonated hydrocarbon is being oxidized. Moreover, there exists a linear relationship between the halfwave potentials ($E_{1/2}$) and the ionization potentials of the hydrocarbon. Furthermore, $E_{1/2}$ shifts to less positive potentials with increasing chain length of the hydrocarbon, e.g. the oxidation potential is 2.0 V (vs perylene$^+$/perylene^{++}) for methane in $HSO_3F/0.6$ M $NaSO_3F$ at room temperature and at smooth platinum and 1.08 V for n-decane. It is proposed that the radicals form a complex with the platinum metal of the anode, with which the authors explain why the radical is 0.4 V more difficult to oxidize than the parent hydrocarbon[23].

Cyclovoltammetry of cyclopentane in fluorosulfonic acid/2 M NaF at $-60\,°C$ exhibits two irreversible peaks between 0 and 1.9 V (vs $Hg/Hg_2S_2O_6F_2$). The product of the oxidation is the cyclopentyl cation which is stable under these conditions. This behavior together with coulometry and other diagnostic criteria (e.g. the dependence of E_p on the sweep rate) indicates an ECE mechanism according to equation 2[2,23].

$$RH \xrightarrow{-e} RH^{+\cdot}$$

$$RH^{+\cdot} \xrightarrow{-H^+} R^{\cdot} \tag{2}$$

$$R^{\cdot} \xrightarrow{-e} R^+$$

The unprotonated hydrocarbon is oxidized to a radical cation, which loses a proton to form a radical that is further oxidized to a carbenium ion.

In further studies of the oxidation of cyclohexane and cyclopentane in fluorosulfonic acid, the formely proposed anodic oxidation of the protonated alkane has been reconsidered by Coleman and Pletcher[24]. Cyclovoltammetric data and the products of controlled potential electrolysis (cpe) in $FSO_3H/HOAc$ point to a mechanism, where the alkane is directly oxidized to form a carbocation according to equation 2. This undergoes, after rearrangement to the most stable carbenium ion, fast solvolysis to afford an alkyl fluorosulfonate. This is fairly stable at $-60\,°C$, but decomposes at higher temperatures to afford an olefin, which is trapped by the acetyl cation from acetic acid to afford an unsaturated ketone. The preequilibrium according to equation 1, however, cannot be fully excluded, although its role will remain uncertain until a detailed knowledge of the fluorosulfonic acid solvent system becomes available. The preparative scale anodic oxidation of cyclohexane in fluorosulfonic acid with different added acids, such as acetic acid, propionic acid or hexanoic acid, leads to a single product with a rearranged carbon skeleton, a 2-acyl-1-methylcyclopentene which is obtained in 50–60% yield (equation 3)[19,24].

Also other alkanes have been anodically converted at smooth platinum in fluorosulfonic acid/1.1 M in acetic acid to α,β-unsaturated ketones in 42–71% current yields (Table 1)[3,24].

In the absence of acetic acid, e.g. in fluorosulfonic acid/sodium fluorosulfonate, only

784 H. J. Schäfer

TABLE 1. Anodic oxidation of alkanes in fluorosulfonic acid, 1.1 M acetic acid

Alkane	Product	Current yield (%)
$(CH_3)_3CH$	$(CH_3)_2C=CHCOCH_3$	64
$n\text{-}C_5H_{12}$	$C_2H_5(CH_3)C=CHCOCH_3$	40
	$C_5H_9COCH_3$	18
$n\text{-}C_6H_{14}$	$C_6H_{11}COCH_3$	51
$n\text{-}C_7H_{16}$	$C_7H_{13}COCH_3$	56
Cyclohexane	1-Methyl-2-acetyl-1-cyclopentene	71
Cyclooctane	$C_8H_{13}COCH_3$	42

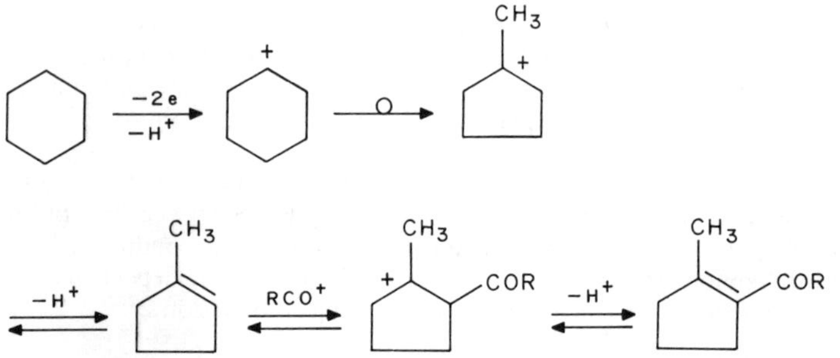

$$
\text{(3)}
$$

oligomeric products could be isolated even at $-20\,^\circ C^{24}$. Chemical conversion of the alkane competes strongly in this medium with its electrochemical oxidation. Even in the more 'basic' electrolyte: 4 M $NaSO_3F$ in fluorosulfonic acid, extensive cracking, isomerization and polymerization of n-hexane as purely chemical reactions are being observed[20] and no oxidation products are being reported.

An interesting extension of the anodic conversion in fluorosulfonic acid is the trapping of the intermediate carbenium ion with CO in an anodic Koch reaction[25] (equation 4).

$$
RH \xrightarrow[+H^+]{-2e^-} R^+ \xrightarrow{CO} RCO^+ \xrightarrow[+H^+]{H_2O} RCOOH \qquad (4)
$$

Carbon monoxide at 1 atm has no effect on the voltammetry of cyclohexane in fluorosulfonic acid. In preparative scale electrolysis in fluorosulfonic acid containing 3.0 M water the hydrocarbon cyclohexane affords in 56% current efficiency cyclohexanecarboxylic acid. n-Pentane forms at optimum conditions (1 atm CO, $-25\,^\circ C$) 74% carboxylic acids in which the alkane skeleton is largely rearranged. The product distribution is 1% 2-methylpentanoic acid, 4% 2-ethylbutanoic acid and 95% 2,2-dimethylbutanoic acid. At higher CO pressure and room temperature only 15% rearranged product is obtained, 54% 2-ethylbutanoic acid and 31% 2-methylpentanoic acid are being formed in a somewhat lower overall yield of 57%. Cyclopentane affords 80% cyclopentanecarboxylic acid.

Lower chain alkanes that resist direct anodic oxidation in fluorosulfonic acid can be indirectly converted[26–28]. For that purpose peroxodisulfuryl difluoride is generated at the anode, being either platinum gauze or vitreous carbon, by constant current electrolysis in a divided cell charged with potassium fluorosulfonate in anhydrous fluorosulfonic acid; then methane is introduced into the solution. Depending on the reaction conditions, at 45 °C selectively methyl fluorosulfonate (61% coulombic yield) and at $0\,^\circ C$ methylene bisfluorosulfonate (63%) are formed[27]. Ethane is converted to 42% ethyl fluorosulfonate (96.5% yield by NMR)[26]. Acetic acid afforded 70% methylene difluorosulfonate. Propionic acid yielded 3-fluorosulfonyloxypropionic acid (50% chemical yield), which on standing underwent elimination to acrylic acid.

The mechanism is assumed to be a radical cleavage of the peroxy bond in peroxy-disulfuryl difluoride, subsequent H abstraction of the FSO_3 radical at the alkane and oxidation of the resulting radical to a cation[28]. In carboxylic acid oxidation a subsequent rapid hydrogen shift rearranges the cation preferentially to the electrostatically least deactivated $(\omega - 1)$ position (see Section II.B).

2. Hydrogen fluoride

The anodic oxidation of alkanes in anhydrous hydrogen fluoride has been studied at various acidity levels from 'basic' medium (KF) to acidic medium (SbF_5) to establish optimum conditions for the formation of carbenium ions[29,30]. The oxidation potential depends on the structure of the hydrocarbon: methane is oxidized at 2.0 V, isopentane at 1.25 V vs Ag/Ag^+. Three cases of oxidation can be distinguished. In 'basic' medium, direct oxidation of the alkane to its radical cation occurs. In a slightly acidic medium, the first-formed radical cation disproportionates to cation, proton and alkane. The oxidation is, however, complicated by simultaneous isomerization and condensation reactions of the alkane. In strongly acidic medium, protonation of the alkane and its dissociation into a carbenium ion and molecular hydrogen occurs. In acidic medium n-pentane behaves like a tertiary alkane, which is attributed to its isomerization to isopentane. The controlled potential electrolysis in basic medium yields polymeric species.

The electrochemistry of hydrocarbons in HF/SbF_5 is mediated by Sb(V), which is a strong oxidizing agent in the superacid. It can oxidize alkanes that contain three or more carbon atoms, and the oxidation becomes easier, the longer the carbon chain; e.g. ethane is oxidized at 2.25 V (vs Ag/AgCl), propane at 1.95 V, n-butane at 1.85 V, n-hexane at $1.60\,V^{31}$. It is presumed that protonation of the alkane by the superacid facilitates its oxidation. The oxidant Sb(V) is reduced to Sb(III), which can be regenerated at the anode. This is very facile at a gold anode, possibly due to the specific adsorption of SbF_4^- on gold, while on platinum the reaction rate is considerably slower and is even undetectable at glassy carbon. Products from the alkanes are higher-molecular-weight materials.

Alkanes can be fluorinated by means of electrolysis in hydrogen fluoride. This important reaction is reviewed in Reference[32]. Electrochemical fluorination in anhydrous hydrogen fluoride (AHF), the so-called Simons process, involves the electrolysis of organic compounds (aliphatic hydrocarbons, halohydrocarbons, acid halides, esters, ethers, amines) at nickel electrodes. It mostly leads to perfluorinated compounds. It is, however, considerably accompanied by cleavage and rearrangement reactions. As mechanism, the formation of carbocations by an ECE mechanism is assumed through oxidation of the hydrocarbon by higher-valent nickel fluorides.

Partial fluorination is achieved by electrolysis in molten $KF \cdot 2HF$ in the Phillips process at porous carbon electrodes. The process is thought to involve the electrochemical generation of elemental fluorine and its radical reaction with the substrate within the pores of the anode. Hydrocarbons, acid fluorides and esters are partially fluorinated with a statistical distribution of the fluoride in the alkyl chain and in current efficiencies between 80–100%.

Anodic fluorination has also been accomplished in aprotic solvents (acetonitrile, sulfolane, nitromethane) in the presence of fluorides in the form of HF, $R_3N \cdot HF$, $R_4NF \cdot xHF$ and R_4NBF_4. Aromatic hydrocarbons (but not aliphatic ones as yet) have been oxidized by an ECE mechanism to partially fluorinated products.

3. *Trifluoroacetic acid*

In trifluoroacetic acid (0.4 M TBABF$_4$) unbranched alkanes are being oxidized in fair to good yields to the corresponding trifluoroacetates (Table 2)[33]. As mechanism, a 2e oxidation and deprotonation to an intermediate carbenium ion (according to equation 2) that undergoes solvolysis is being proposed. The isomer distribution points to a fairly unselective CH oxidation.

The oxidation potential of cyclohexane in trifluoroacetic acid/sodium trifluoroacetate is cathodically shifted from 2.57 V (vs sce) to 1.8 V in the more acidic electrolyte CF$_3$CO$_2$H/10 M FSO$_3$H. The alkane activation can also be achieved by CH$_3$SO$_3$H or CF$_3$SO$_3$H. As mechanism, the protonation of the alkane and its irreversible oxidation (according to equation 1) is being discussed. The half-peak potentials of some alkanes in CF$_3$CO$_2$H/2.5 M FSO$_3$H shift to less positive potentials with increasing chain length, namely from 2.44 V for n-pentane to 1.88 V for n-eicosane[34]. Cyclohexane yields in a preparative electrolysis in CH$_2$Cl$_2$/2.5 M CF$_3$CO$_2$H 91% cyclohexyl trifluoroacetate and 4% cyclohexyl chloride. The solvolysis products of the methylcyclopentyl cation, to which the cyclohexyl cation rearranges in more acidic media (see Section II.A.1), are not found in this more nucleophilic medium[35]. n-Decane was oxidized in dichloromethane/trifluoroacetic acid and trifluoroacetic acid/acetic acid. While in the case of n-octane a strictly statistical ratio of the isomeric 2-, 3- and 4-trifluoroacetoxy-n-octanes is reported, for n-decane the isomeric 2-, 3-, 4- and 5-trifluoroacetates are formed in relative yields of 16%, 23%, 27% and 34%, respectively. This distribution corresponds to the net positive charge distribution in the carbon skeleton of the n-decane radical cation.

Preparative electrochemical oxidation of branched alkanes in trifluoroacetic acid yields—as a function of the applied potential, which results from higher or lower current densities—mono- or bifunctionalized derivatives[36]; e.g. 2-methylpentane forms as main product 72% 2-methylpentyl-2-trifluoroacetate. The additional side products arise by oxidation at the methylene groups and rearrangement. The proposed mechanism is the oxidation of the hydrocarbon to its radical cation, deprotonation, further oxidation of the radical to a cation (according to equation 2) and solvolysis. The electrode material being Pt, Ir or Au has no specific influence on the product distribution. Furthermore 3-methylpentane, 2,2-dimethylpentane and 2,2,4-trimethylpentane have been oxidized. They undergo anodic substitution, but also partial fragmentation (see Section II.B). Product formation due to the generation of alkenes and subsequent electrophilic addition of the acidic solvent is shown to be negligible by the use of CF$_3$COOD as solvent. The relative alkane reactivities were determined by competitive electrolyses at controlled current. The relative rates are shown in Table 3.

The reaction conditions for the anodic substitution in dichloromethane/trifluoroacetic acid have been optimized for the oxidation of cyclohexane[37]. Dichloromethane appears to be the cosolvent of choice, because it sufficiently dissolves hydrocarbons and can be easily removed due to its low boiling point. Trifluoroacetic acid is a nucleophile with a fairly high resistance towards oxidation.

TABLE 2. Yields and isomer distribution of trifluoroacetates in the oxidation of alkanes in CF$_3$CO$_2$H/0.4 M TBABF$_4$

Alkane	Current yield (%)	Isomer distribution (%)		
		2-isomer	3-isomer	4-isomer
n-Octane	75	33	33	33
n-Heptane	75	45	45	10
n-Hexane	60	Not resolved		
n-Pentane	50	Not resolved		

TABLE 3. Relative rates of the anodic alkane oxidation in trifluoroacetic acid

Alkane	k_{rel}
n-Hexane	0.89
2-Methylpentane	1.00
3-Methylpentane	1.05
2,2-Dimethylpentane	1.17
n-Octane	1.22
2,3-Dimethylpentane	1.75
Methylcyclopentane	2.06
n-Decane	2.53
Cyclohexane	3.22

From the possible electrode materials gold, glassy carbon, graphite, lead dioxide on graphite and platinum, only the last one can be used, because all the other materials corrode heavily.

Different kinds of supporting electrolytes were used as $TBAClO_4$, $TBAPF_6$, $TBABF_4$, $TMABF_4$ and $TBAO_2CCF_3$. They all lead to similar yields of cyclohexyl trifluoroacetate (2) starting from 1 (equation 5). The change in temperature from -20 to $40\,°C$ had no significant influence on the yield of 2 in $CH_2Cl_2/15\%$ $CF_3CO_2H/0.05$ M $TBAPF_6$. Also, the change of the acidity from 1.2 M (10%) to 3.0 M (25%) trifluoroacetic acid did not alter the product yield. With increasing trifluoroacetic acid concentration only the side product changed gradually from chlorocyclohexyl trifluoroacetate to cyclohexyl ditrifluoroacetate.

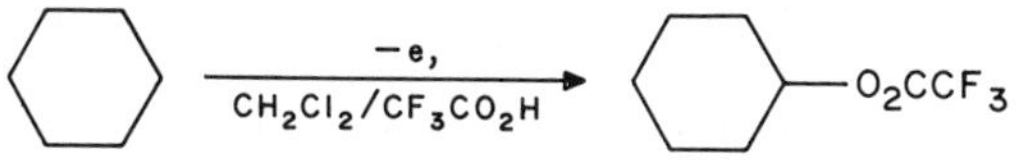

$$\text{(1)} \qquad \xrightarrow[\text{CH}_2\text{Cl}_2/\text{CF}_3\text{CO}_2\text{H}]{-e,} \qquad \text{(2)} \quad -\text{O}_2\text{CCF}_3 \qquad (5)$$

The use of freshly distilled trifluoroacetic acid, however, increased the material yield of 2 from 69 to 79% and the current yield from 43 to 63%. These improved yields could be due to a reduced water content of the acid, which otherwise unfavourably influences the reaction. The addition of $(CF_3CO)_2O$ should trap residual water in the acid. Indeed, with 4% $(CF_3CO)_2O$ in the electrolyte the material yield of 2 could be further increased to 89% and the current yield to 73%. Electrolysis of 1 in $CH_2Cl_2/20\%$ $CF_3CO_2H/4\%$ $(CF_3CO)_2O/0.05$ M $TBAPF_6$ at $0\,°C$ in an undivided cell at controlled current afforded after 1.3 current equivalents at 92% yield of 2 (gc yield); 76% cyclohexanol was isolated after hydrolysis with aqueous K_2CO_3.

The results of the oxidation of other unsubstituted cycloaliphatic compounds are summarized in Table 4, which indicates that cyclopentane and cyclohexane are oxidized in high yield and selectivity. The medium-sized rings, however, afford only low yields of the trifluoroacetates in a complex product mixture. This lack of selectivity is probably due to extensive rearrangments of the intermediates and easy follow-up oxidations of the products. In dichloromethane/acetic acid the same non-selectivity is encountered. The electrolysis of bicyclo[2.2.1]heptane (3) leads in dichloromethane/trifluoroacetic acid to exo-bicyclo[2.2.1]heptan-2-ol (4) in 83% yield as single product after hydrolysis of the trifluoroacetate (equation 6). The stereochemistry of 4 indicates that a nonclassical norbornyl cation is involved as intermediate. In dichloromethane/20% acetic acid/0.05 M $TBABF_4$ a slightly lower yield (61%) of exo-2-norbornyl acetate was obtained.

H. J. Schäfer

TABLE 4. Anodic oxidation of cycloaliphatic compounds in dichloromethane/trifluoroacetic acid

Hydrocarbon	Yield of trifluoroacetate (%)
Cyclopentane	84
Cyclohexane	92
Cyclooctane	30[a]
Cyclodecane	6[b]
Cyclododecane	9[c]

[a] 90% conversion.
[b] 30% conversion.
[c] 31% conversion.

(3)   (4)   (6)

With bicyclo[2.2.2]octane (5) in dichloromethane/acetic acid/0.05 M TBABF$_4$ a considerable amount of fluorinated products is being formed. They probably arise by reaction with TBABF$_4$. When the supporting electrolyte TBABF$_4$ was exchanged for LiClO$_4$, as expected only 6 and 7/8 were found in 7% and 79% yield, respectively (equation 7). The products can be explained by oxidation at C$_{(1)}$ and C$_{(2)}$ of 5 to the corresponding carbenium ions 9 and 10. The cation 10 rearranges to the 2-bicyclo[3.2.1]heptyl cation (11) which reacts to give an exo/endo mixture of the acetates 7/8.

These oxidations exhibit a pronounced chemo- and regioselectivity. When cyclohexyltrifluoroacetate is oxidized in CH$_2$Cl$_2$/20% CF$_3$CO$_2$H, 4% (CF$_3$CO)$_2$O and 0.05 M TBABF$_4$, the reaction proceeds slowly and affords, after 33% conversion, 11% cyclohexylditrifluoroacetate. The statistically corrected ratio of the substitution at C$_{(2)}$:C$_{(3)}$: C$_{(4)}$ is 1:3:8. This indicates the increasing inductive shielding of the CH bonds with decreasing distance from the trifluoroacetoxy group.

With methylcyclohexane the yield of trifluoroacetates (60%) is lower than with

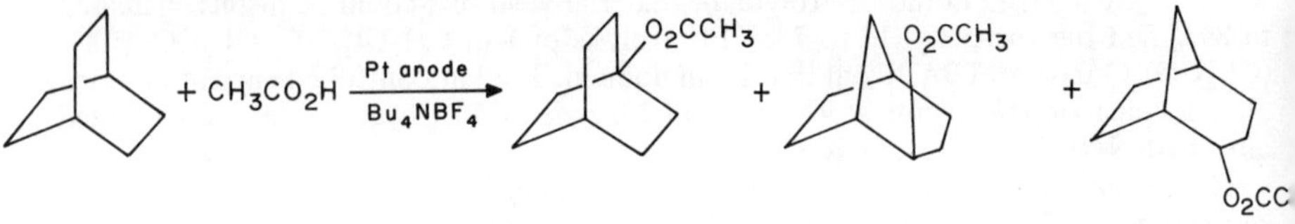

(5)   (6)   (7)   (8)

(7)

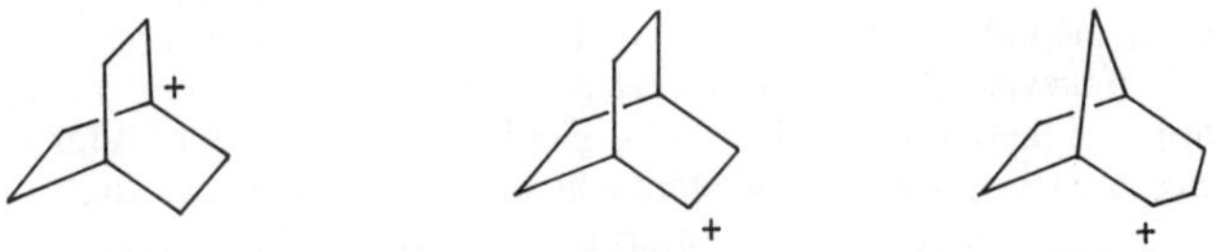

(9)   (10)   (11)

cyclohexane, presumably due to more extensive subsequent oxidations of the products. The ratio of 1-: 2-: 3-: 4-trifluoroacetoxy-1-methylcyclohexane is 11:2:2:1; it varies slightly with the kind of supporting electrolyte. This reflects the different electron densities of the CH bonds. According to MO calculations[38] $C_{(1)}$ has the highest electron density, the primary $C_{(7)}$ the lowest; the secondary carbons are between them and the density decreases from $C_{(2)}$ to $C_{(4)}$.

Chlorocyclohexane (12) yields in 0.05 M TBABF$_4$ in CH_2Cl_2/CF_3CO_2H (4:1) at 0 °C in a divided cell the products shown in equation 8. The cyclohexylchlorotrifluoroacetate (13) is formed by oxidation of chlorocyclohexane to its carbenium ion and its subsequent solvolysis. The relative yields of the 2-, 3- and 4-cyclohexylchlorotrifluoroacetates (13a–c) being 1:1.7:3.2 reflect the inductive effect of the chloro substituent, which increasingly deactivates the oxidation from the 4-H to the 2-H. 15 appears to be formed by anodic cleavage of the C—Cl bond in 12 leading to a cyclohexyl cation, that undergoes solvolysis. 14 seems to be produced by radical chlorination of chlorocyclohexane with chlorine radicals originating from the anodic cleavage of 12. This is indicated by the low selectivity for the formation of 1,2-, 1,3- and 1,4-dichlorocyclohexane being 1:1.2:1.5[39].

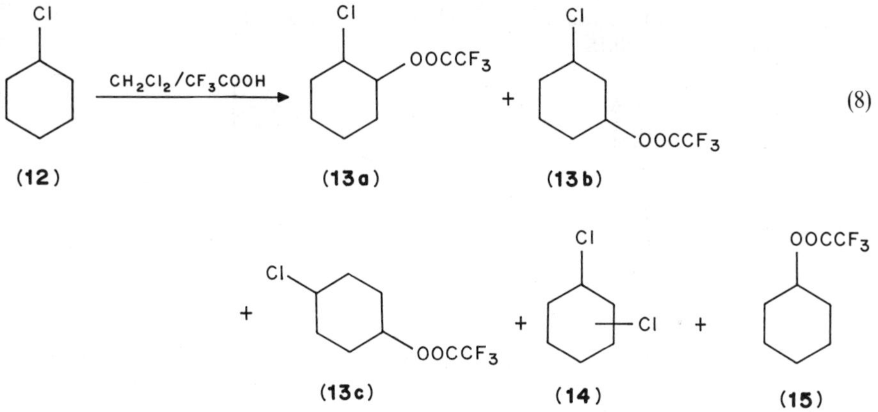

$$\text{(12)} \xrightarrow{CH_2Cl_2/CF_3COOH} \text{(13a)} + \text{(13b)} \tag{8}$$

$$+ \quad \text{(13c)} \quad + \quad \text{(14)} \quad + \quad \text{(15)}$$

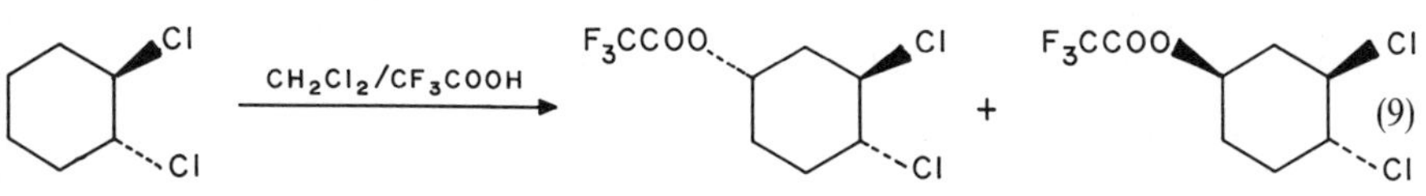

$$\text{(16)} \xrightarrow{CH_2Cl_2/CF_3COOH} \text{(17a)} + \text{(17b)} \tag{9}$$

Trans-1,2-dichlorocyclohexane (16) was electrolyzed under the same conditions as 12 to afford as main product 49% 17 (equation 9). The inductive effect of the two chlorine atoms deactivates $C_{(3)}$ and $C_{(6)}$ so strongly that the oxidation is directed mainly to $C_{(4)}$ and $C_{(5)}$. Also, the extent of cleavage of the C—Cl bond is decreased compared to the results of 12. The high selectivity and the decreased formation of side products can provide here a synthetically useful route from olefins to homoallylic alcohols via protection of the double bond, anodic oxidation, hydrolysis and deprotection.

Trans-decalin affords in dichloromethane/trifluoroacetic acid 20–26% 1- and 2-decalyl trifluoroacetates and 54% oligomers[40]. In the same solvent the anodic oxidation of the

steroid androstane proceeds unsatisfactorily. The main products are not fully characterized, high-molecular compounds, presumably oligomers.

4. Acetic acid

While in dichloromethane/trifluoroacetic acid *trans*-decalin tends to form oligomers
as major product in the less acidic electrolyte acetic acid/dichloromethane/Bu_4NBF_4,
the oligomerization can be largely suppressed, and decalyl acetates (43%) and fluorodecalins (32%) are obtained. With Bu_4NPF_6 as supporting elctrolyte, 67% decalyl acetates
(**18**) and only 10% fluorodecalins are formed. The chemoselectivity of the acetoxylation
at $C_{(4a)}$, $C_{(1)}$ and $C_{(2)}$ is 12:1:1. With Bu_4NOAc as supporting electrolyte 77% decalyl
acetate is obtained with equal selectivity but without fluorination. However,
35–40 F mol^{-1} are required for complete conversion, presumably because of a competing
Kolbe electrolysis[40,41].

Androstane is converted in glacial acetic acid $(CH_3CO_2H/CH_2Cl_2 = 3:2$, $i =$
$40\,mA\,cm^{-2}$, $T = 20\,°C$, Pt electrode) to 17% androstanyl acetate (**19**) and 38%
androstane is recovered. Remarkable is the selectivity of the oxidation: It was found
that the 6-, 7- and 12-acetates are formed in the ratio 35:1:2.5, and no other acetates
could be detected. Cholestane affords 15% cholestanyl acetates; the 6-, 7- and 12-acetates
are formed in the ratio 40:1:2.

(18)   (19)

Whereas androstane, cholestane and androstanones can be oxidized in dichloro-
methane/acetic acid and even in methanol (see Section II.A.5)[40,41], the 3- and 17-acetoxy-
androstanes as well as 3,17-diacetoxyandrostane resist CH oxidation in this solvent.
5α-Cholestan-3β-yl acetate is converted in dichloromethane/acetic acid in low yield
(5–10%) to tetratrifluoroacetoxy-5α-cholestane-3β-yl acetate. The side chain at $C_{(17)}$ is
not split off to a noticeable extent.

5. Methanol

The CH bonds of steroids surprisingly can be even oxidized in methanol as solvent[40].
The electrolysis of androstane in methanol (methanol/dichlormethane, 0.1 M $NaClO_4$,
glassy carbon electrode) proceeds very selectively. At 53% conversion, 14% 6-methoxy-
androstane (**20a**) and 27% 6-(methoxymethoxy)androstane (**20b**) are formed (equation
10). Similarly, after 50% conversion cholestane forms 15% 6-methoxycholestane and 25%
6-(methoxymethoxy)cholestane. These findings reveal a surprising and remarkable
selectivity. Of the 32 hydrogen atoms in androstane and the 48 in cholestane, predominantly only those on $C_{(6)}$ are oxidized. Acetates or methyl ethers originating from
attack at the reactive tertiary hydrogen atoms on $C_{(5)}$, $C_{(9)}$ and $C_{(14)}$ and, in the case
of cholestane, also those on $C_{(17)}$, $C_{(20)}$ and $C_{(25)}$ are not found. An explanation for this
would seem premature, but the selectivity could be effected by an association of the
steroid molecules at the electrode surface. It appears that the androstane or cholestane

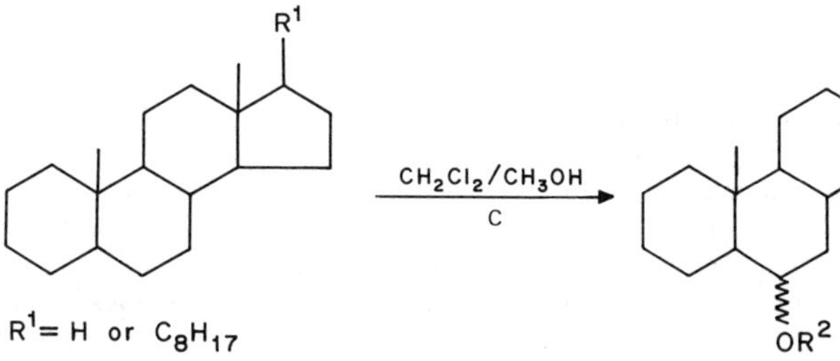

$$\begin{array}{ccc}
 & R^1 & R^2 \\
\hline
a & H & OCH_3 \\
b & H & OCH_2OCH_3 \\
\hline
\end{array}$$

molecules are stapled perpendicular to the plane of the electrode. Thereby the tertiary hydrogen atoms would be shielded within the stack, while on the front and rear side of the stack the hydrogen atoms on $C_{(6)}$, $C_{(7)}$ and $C_{(12)}$ would be exposed to the electrode for the electron transfer.

The methoxymethyl ethers **20b** could be formed by a follow-up oxidation of the methyl ethers **20a**[42]. That the less reactive but obviously more easily accessible methyl group is hereby oxidized, and not the methine hydrogen, could be a further indication of the formation of stacked associates.

The electrolysis of ketosteroids in dichloromethane/methanol does not lead to the oxidation of unactivated CH bonds. With the 3-ketosteroids 5α-androstanone and 5α-cholestan-3-one as substrates besides the acid-catalyzed ketalization of the keto group only chlorination α to the carbonyl group occurs. The anodic oxidation of the 17-keto-steroids 5α-androstan-17-one, 3β-acetoxy-5α-androstan-17-one and 5α-androstan-3,17-dione (**21**) leads to the opening of the D-ring with the formation of 13,17-secosteroids, which form lactones under the weakly acidic conditions of the electrolysis and work-up (equation 11)[41].

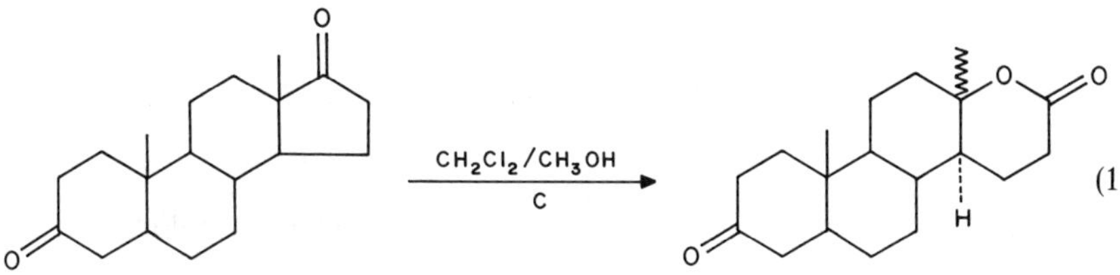

Thus the anode is a cheap and easily employable reagent for the oxidation of CH bonds of alkanes in trifluoroacetic acid, in acetic acid and in methanol. The chemo-selectivity of the anode compares favorably with other reagents; R^t, the ratio of the rate of attack at tertiary and secondary CH bonds, is 12 for decalin[43]. The regioselectivity can be controlled by inductive effects. The steroid hydrocarbons androstane and cholestane are remarkably regioselectively oxidized at $C_{(6)}$.

6. Acetonitrile

Acetonitrile has a very anodic decomposition potential. Exchange of the perchlorate anion against tetrafluoroborate or hexafluorophosphate shifts the anodic decomposition potential (1 mA cm^{-2}) of acetonitrile from 2.48 to 2.91 or 3.02 V vs Ag/Ag^+, respectively. In acetonitrile/TEABF$_4$ some half-wave potentials have been determined to be 3.0 V for i-pentane, 3.01 V for 2-methylpentane, 3.28 V for 2,2-dimethyl-butane and > 3.4 V for n-octane, n-heptane and n-hexane[44]. The half-wave potentials of some alkanes have also been obtained by the use of ultramicroelectrodes in acetonitrile without any supporting electrolyte: the values range from 3.87 V for methane to 3.5 V for n-heptane vs Ag/Ag^+. These values correlate well with the ionization potentials of the alkanes[45]. Controlled potential electrolyses were carried out on saturated solutions of four alkanes in acetonitrile 0.1 M TBABF$_4$ to yield acetamidoalkanes (Table 5)[33].

Voltammetry and coulometry indicate a 2e oxidation to a carbenium ion that subsequently reacts with the nitrogen of the acetonitrile in a Ritter reaction (equation 12).

$$RH \xrightarrow{-e} RH^{\overset{+}{\cdot}} \xrightarrow[-e]{-H^+} R^+ \xrightarrow{CH_3C\equiv N} RN\overset{+}{=}\overset{}{C}CH_3 \xrightarrow[-H^+]{H_2O} RNHCOCH_3$$

$$(12)$$

The oxidation of substituted adamantanes has been studied in detail in acetonitrile by the groups of Miller[46,47] and Mellor[48-50]. Preparative scale electrolyses of several substituted adamantanes were performed by cpe in a divided cell in acetonitrile-lithium perchlorate at 2.5 V (vs Ag/Ag^+). The products obtained are shown in equation 13 and in Table 6[47]. The $E_{p/2}$ values of the adamantanes range from 2.56 V (X = CO$_2$Me) to

TABLE 5. Yields and isomer distribution of acetamidoalkanes at cpe of alkanes at a Pt-anode in CH$_3$CN/0.1 M TBABF$_4$

		Isomer distribution (%)		
Alkane	Yield (%)	2-isomer	3-isomer	4-isomer
n-Octane	45	33	35	32
n-Heptane	48	45	45	10
n-Hexane	45	55	45	
n-Pentane	50	Not resolved		

TABLE 6. Oxidation products of substituted adamantanes 22 in acetonitrile

	Product (% yield)		
Substituent X in 22	path (a)	(equation 13)	path (b)
H	27 (90)		
Br			27 (89)
Cl	25, X = Cl (91)		
F	25, X = F (65)		
CH$_3$	25, X = CH$_3$ (91)		
CH$_2$OH			27 (37)
COCH$_3$			27 (44)
CO$_2$CH$_3$	25, X = CO$_2$CH$_3$ (64)		
CN	25, X = CN (41)		
OH			27 (41)
OCH$_3$			27 (58)

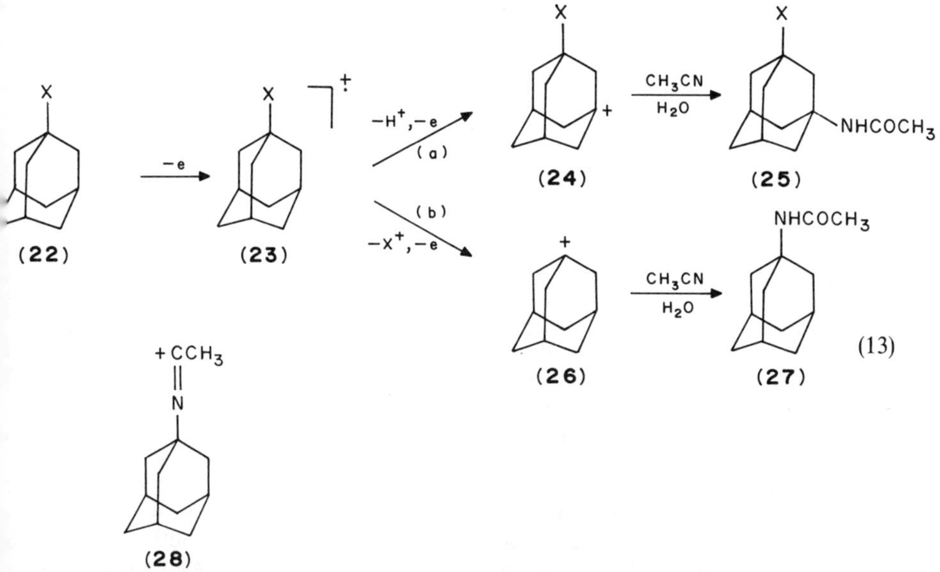

X

X

X

(22)
(23)

−e

−H⁺,−e
(a)

(24)

+

CH₃CN
H₂O

(25)

NHCOCH₃

(b)
−X⁺,−e

(26)

+

CH₃CN
H₂O

NHCOCH₃

(27)

(13)

+CCH₃
‖
N

(28)

1.90 V (X = NHCOCH$_3$) vs Ag/Ag$^+$. The peak heights in cyclovoltammetry correspond to $n = 2$.

Two general reaction types are observed: substitution of acetamide for hydrogen (equation 13, path a) and substitution of acetamide for the functional group (equation 13, path b). The isolation of high yields of 27 is only possible because the nonoxidizable precursor 28 is actually the stable product and the amides 27 or 25 result from aqueous work-up. 28 could be verified by quenching it with either D$_2$O or methanol.

A larger-scale electrolysis of 22 (X = H) could also be performed in an undivided cell providing readily gram amounts of 27. This reaction has therefore synthetic interest as a route to pharmacologically useful substituted 1-adamantylamines. Adamantyl chloride and fluoride lead to substitution of nitrogen for hydrogen at C$_{(3)}$ (path a), whilst with the bromide 22 (X = Br) 27 was formed in high yield, indicating that bromide competes favorably with hydrogen as leaving group. In each of the three adamantyl halides initial electron transfer involves an orbital from the adamantyl moiety and the substitution of bromine accounts for the weak carbon–bromine bond, whilst substitution of chlorine and fluorine is prevented by their stronger bonds. In carbon-substituted adamantanes the CO$_2$$^-$, CH$_2$OH and COCH$_3$ groups are cleaved, whilst the CH$_3$, CH$_2$OCOCH$_3$, CO$_2$CH$_3$ and CN are not. All this chemistry can be rationalized from the stability of the fragments formed by the fragmentation of the carbon-substituted bond.

The cleavages of 1-acetyl-, 1-bromo- and 1-hydroxymethyladamantane are of great interest. Each of them mimics a mass spectral result and they indicate the feasibility of preparative-scale mass spectrometry.

In a series of alkyl-substituted adamantanes the influence of the carbon substituents upon the competing pathways of substitution and fragmentation have been investigated (Table 7)[48,50]. Good coulombic and product yields are being obtained. At first, loss of an electron from the adamantyl portion is to be expected in each case. Fragmentation is observed as the dominating pathway of the radical cation when loss of a tertiary fragment can occur (equation 13, path b); no fragmentation is observed when a primary fragment might leave (equation 13, path a).

In addition a large number of substituted adamantanes has been investigated by

TABLE 7. Distribution of products from anodic oxidation
of alkyl-substituted adamantanes in acetonitrile

Compound 22	Product (% yield)	
X	Substitution (25)	Fragmentation (27)
H	74	0
Et	77	0
i-Pr	75	9
t-Bu	7	62

cyclovoltammetry. Their $E_{p/2}$ values range around 2.37 to 2.40 V vs Ag/Ag$^+$ in acetonitrile/TBABF$_4$. For comparison the corresponding values of other cycloalkanes have been determined, e.g. for diamantane: 2.14 V, bicyclo[3.3.1]nonane: 2.58 V, methylcyclohexane: 2.85 V and cis/trans-decalin: 2.57 V. No evidence of a cathodic wave was observed even at − 78 °C. The peak heights indicate a 2e wave and a diffusion-controlled oxidation. The results suggest that the adamantanes are being oxidized by an ECE process to an adamantyl cation via an intermediate delocalized radical cation. The radical ion can undergo deprotonation or fragmentation; the balance between the two processes is determined by the nature of the fragmenting group. In acetonitrile 1-t-butyladamantane gives mainly fragmentation, while in contrast 1-ethyladamantane gives no fragmentation and 1-isopropyladamantane fragments only to a small extent. A difference of temperature of ca 50 °C hardly affects the product distribution.

The comparison of the oxidation of a series of substituted adamantanes (t-butyl, methyl, adamantyl) by lead(IV), cobalt(III) and manganese(III) trifluoroacetates with the corresponding anodic oxidations in acetonitrile or trifluoroacetic acid shows that the electrochemical oxidation proceeds via a radical cation intermediate, whilst the metal salts form products by CH abstraction[49]. The preparative result is that, in anodic oxidation, preferentially products of fragmentation of the intermediate radical cation are found, whilst with the metal salts hydrogen substitution is being obtained preferentially.

7. Sulfur dioxide

In liquid SO$_2$ and CsAsF$_6$ as supporting electrolyte the oxidation of saturated hydrocarbons could be studied up to potentials of 6.0 V vs sce by using platinum ultramicroelectrodes. Irreversible voltammetric waves were found for the compounds methane to n-octane that were studied[51]. The anodic peak currents suggest n-values around 2; a bulk electrolysis consumed 2 faradays per mole of alkane. The primary products in dilute solutions (1–10 mM) were shorter chain hydrocarbons. Electrolysis of more concentrated solutions led, via reaction of the oxidation products with starting material, to longer-chain hydrocarbons.

The correlation of anodic peak potentials (at − 70 °C: methane 5.1 V to n-octane 3.97 V vs sce) with gas-phase ionization potentials and calculated HOMO energies shows that useful data can be obtained even at potentials exceeding 5.0 V vs sce. Because of the low solubility of CsAsF$_6$ appreciable solution resistance is observed. This presents difficulties when large electrodes are employed, but ultramicroelectrodes create no problems.

Cyclopentane afforded in 0.05 M TBAClO$_4$ and SO$_2$/CF$_3$CO$_2$H (6:1) at − 25 °C at a platinum anode 62% cyclopentyl trifluoroacetate[37]. For chlorocyclohexane the chlorination and dechlorination are reduced in comparison to the electrolyte

CH_2Cl_2/CF_3CO_2H. The selectivity for the formation of 2-, 3- and 4-**13** (equation 8) with 1:3.3:9.1 is much more pronounced than in CH_2Cl_2/CF_3CO_2H. This could be partially due to the lower reaction temperature.

With 1,2-dichlorocyclohexane the same products are obtained as in CH_2Cl_2/CF_3CO_2H albeit in lower yields. The same is true for methylcyclohexane. Unpleasant in these electrolyses, that are conducted in an undivided cell, is the formation of sulfur by cathodic reduction of sulfur dioxide, which partially covers the anode and cathode as a sticky coat.

8. 2,2,2-Trifluoroethanol

As the strongly acidic electrolyte CH_2Cl_2/CF_3CO_2H is prohibitive for the oxidation of acid-sensitive substrates, trifluoroethanol is a possible alternative. It is less acidic (pH = 4.5), has a fairly high anodic decomposition potential (1.9 V vs $Ag/AgNO_3$) and a low boiling point (73 °C, 760 torr). At 0 °C in 0.05 M $TBABF_4$ and 2,2,2-trifluoroethanol as electrolyte, cyclopentane afforded 71% (gc yield), 56% (isolated yield) cyclopentyl trifluoroethyl ether[37]. With chlorocyclohexane the chlorination and dechlorination are less pronounced compared to the electrolyte CH_2Cl_2/CF_3CO_2H. The yield of chloro-2,2,2-trifluoroethoxycyclohexane is, with 46%, twice as high as that of the corresponding product **13** in CH_2Cl_2/CF_3CO_2H.

9. Other solvents

Other solvents that have high anodic decomposition potentials (above 2.5 V vs sce) are nitromethane[52], nitroethane[52], propylene carbonate[52], sulfolane[52], dichloromethane/methanesulfonic acid[34] or trifluoromethanesulfonic acid[34,35]. Except for the latter two[34,53] there are no reports on their use as solvents for alkane oxidation.

B. Oxidation of Remote C—H Bonds in Ketones, Carboxylic Acids and Steroids

Ketones that lack α-branching are oxidized at 2.2 to 2.3 V vs Ag/Ag^+ in acetonitrile/0.1 N lithium perchlorate at a platinum anode to form acetamidoketones[54,55]. In general the mass balance is 60–80%. The products obtained are shown in Table 8. In 2-hexanone, 2-heptanone and 4-heptanone the acetamido group is introduced into the penultimate ($\omega - 1$) position of the carbon chain. 2-Octanone led to ($\omega - 1$)-,

TABLE 8. Remote oxidation of ketones in acetonitrile/lithium perchlorate

Ketone	Product (% yield)
2-Hexanone	5-Acetamido-2-hexanone (40)
2-Heptanone	6-Acetamido-2-heptanone (30)
2-Octanone	7-Acetamido-2-octanone (21)
	6-Acetamido-2-octanone (30)
	5-Acetamido-2-octanone (7)
4-Heptanone	2-Acetamido-4-heptanone (31)
2-Methyl-4-pentanone	2-Acetamido-2-methyl-4-pentanone (20)
	4-Acetamido-2-hexanone (20)
2,6-Dimethyl-4-heptanone	2-Acetamido-2,6-dimethyl-4-heptanone (20)
	6-Acetamido-2-methyl-4-octanone (20)
2,5-Dimethyl-3-hexanone	5-Acetamido-2-methyl-3-heptanone (25)
	5-Acetamido-2,5-dimethyl-3-hexanone (35)
4,4-Dimethyl-2-pentanone	4-Acetamido-4-methyl-2-hexanone (50)

$(\omega - 2)$- and to a small amount of $(\omega - 3)$-acetamidoketones. With branched ketones two major products are obtained (equation 14).

An intramolecular mechanism which explains the products is shown in equation 15. At first the ketocarbonyl group is oxidized to its radical cation. The $E_{p/2}$ values fit a crude correlation with the corresponding ionization potentials. Then a γ-hydrogen abstraction occurs, which is analogous to that found in mass spectrometry and photochemistry. With the electrolysis of a ketone in the presence of 2,3-dimethyl propane it was shown that the intramolecular abstraction of hydrogen is more rapid than an inter-molecular reaction with a good hydrogen donor. The rearranged product is formed by intramolecular H-abstraction at the methyl group, which has been proven by appropriate deuterium substitution (equation 15).

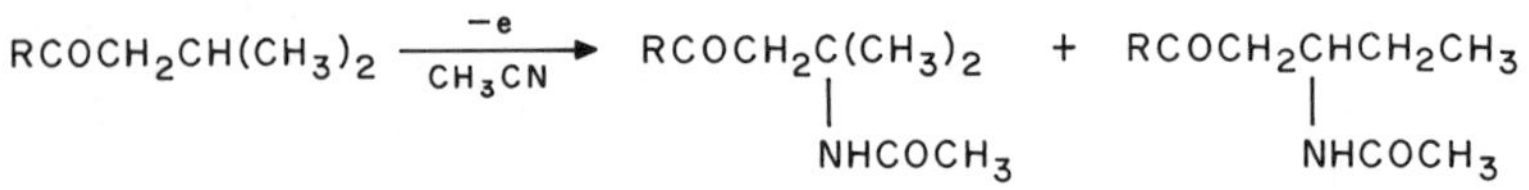

$$
RCOCH_2CH(CH_3)_2 \xrightarrow[CH_3CN]{-e} \underset{\underset{NHCOCH_3}{|}}{RCOCH_2C(CH_3)_2} + \underset{\underset{NHCOCH_3}{|}}{RCOCH_2CHCH_2CH_3} \quad (14)
$$

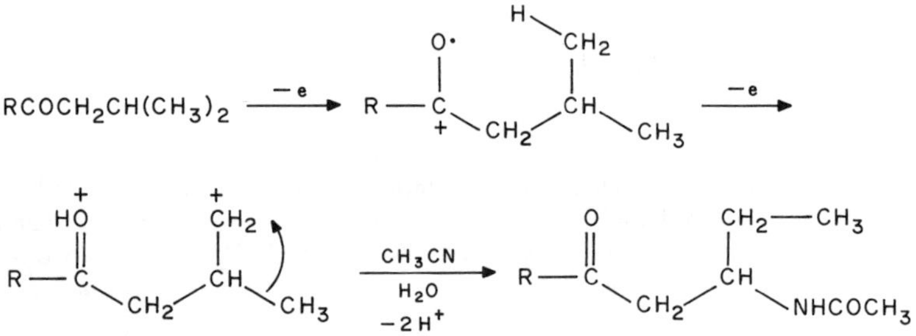

$$
RCOCH_2CH(CH_3)_2 \xrightarrow{-e} \quad (15)
$$

In dichloromethane/trifluoroacetic acid, fatty acids are preferentially trifluoroaceto-xylated at the $(\omega - 2)$- to $(\omega - 4)$-CH bonds; no attack at C-2 to C-6 is observed, or, if at all, only to a small extent (Table 9, equation 16)[40].

The remarkable regioselectivity could be due to the inductive effect of the carboxylic group, which can be protonated in the more acidic region at the anode surface. In

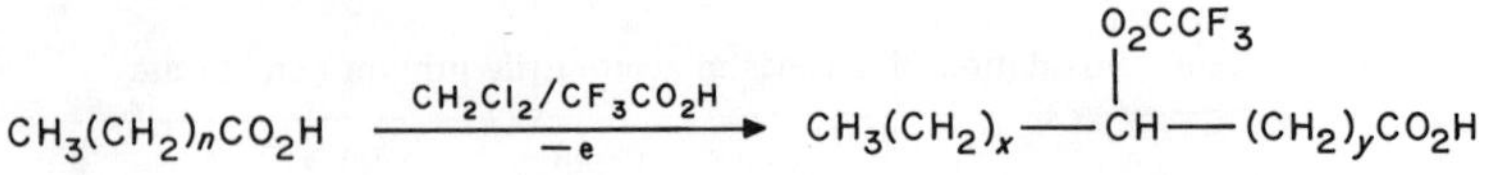

$$
\underset{(29)}{CH_3(CH_2)_nCO_2H} \xrightarrow[-e]{CH_2Cl_2/CF_3CO_2H} \underset{(30)}{CH_3(CH_2)_x\overset{\overset{O_2CCF_3}{|}}{-}CH-(CH_2)_yCO_2H} \quad (16)
$$

TABLE 9. Anodic oxidation of fatty acids **29** to trifluoroacetoxycarboxylic acids **30** in dichloromethane/trifluoroacetic acid

n in **29**	**30**[a] Yield (%)	Relative isomer distribution in **30**							rec. **29** Yield (%)
		y 5	6	7	8	9	10	11	
12	40	3	7	19	28	20	16	7	27
10	15	12	23	26	29	10	—	—	42
8	3	39	46	15	—	—	—	—	71

[a]For the determination of the isomeric structures **30** has been oxidized to the corresponding oxo acid.

hydrogen abstractions from protonated substrates by radical cations, the methylene group farthest removed from the protonated position is preferentially attacked[8,56]. In the anodic oxidation of the CH bond the transition state is undoubtedly more positively charged than that in the radical abstraction and consequently is much more sensitive towards inductive effects. Skeletal rearrangements as in the oxidation in fluorosulfonic acid[57] (see below) do not appear to occur.

The cyclovoltammograms of a series of carboxylic acids in fluorosulfuric acid–potassium fluorosulfate (1.0 M) show irreversible oxidation peaks that are diffusion controlled[57]. No peaks for the reduction of stable intermediates can be detected even at rapid potential scan rates. The oxidation potentials range from 1.95 to 1.64 V vs Pd/H$_2$. They shift cathodically with increasing length of the alkyl chain. The acids are oxidized less readily than the corresponding alkanes due to the influence of the positive charge on the protonated carboxyl groups.

Preparative scale electrolysis gave different products depending on whether the electrolyte was quenched immediately after the electrolysis or not. In the first case the major products of heptanoic acid, 4-methylpentanoic acid, 5-methylhexanoic acid or octanoic acid were always γ- to δ-lactones with overall yields of 40–70% (equation 17a). After being some time in the acidic electrolyte the lactones are converted to α,β-unsaturated ketones by a chemical follow-up reaction (equation 17b).

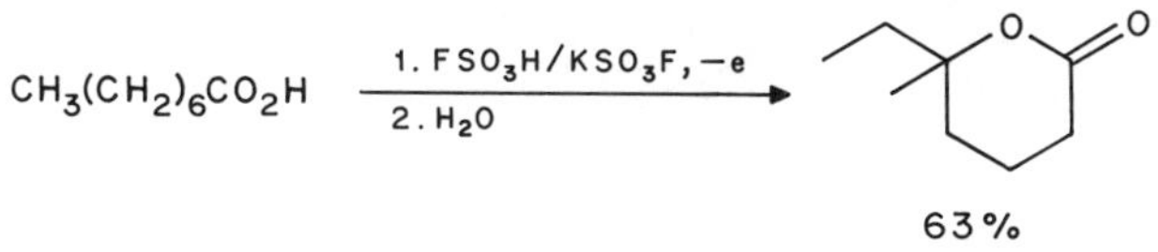

$$CH_3(CH_2)_6CO_2H \xrightarrow[\text{2 . H}_2\text{O}]{\text{1. FSO}_3\text{H/KSO}_3\text{F, } -e} \qquad (17a)$$

63%

$$CH_3(CH_2)_6CO_2H \xrightarrow[\substack{\text{2. 22 h in the electrolyte}\\ \text{3. H}_2\text{O}}]{\text{1. FSO}_3\text{H/KSO}_3\text{F, } -e} \quad + \quad \qquad (17b)$$

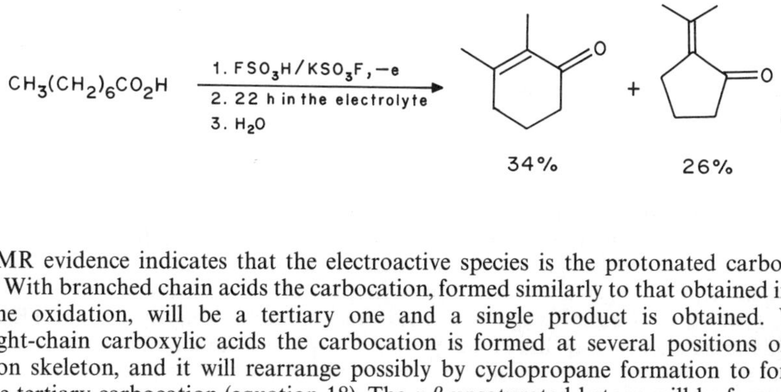

34%   26%

NMR evidence indicates that the electroactive species is the protonated carboxylic acid. With branched chain acids the carbocation, formed similarly to that obtained in the alkane oxidation, will be a tertiary one and a single product is obtained. With straight-chain carboxylic acids the carbocation is formed at several positions of the carbon skeleton, and it will rearrange possibly by cyclopropane formation to form a stable tertiary carbocation (equation 18). The α,β-unsaturated ketone will be formed by slow loss of the HSO$_3$F group with the formation of a double bond and subsequent cyclization of the protonated acid.

Whilst medium-chain-length carboxylic acids (six or more carbon atoms) can be directly oxidized in anhydrous fluorosulfuric acid containing potassium fluorosulfate, those of lower molecular weight are not oxidized before the oxidation of the fluorosulfate anion. For their conversion, solutions of peroxydisulfuryl difluoride, which is known to react also with short-chain alkanes[26–28], are prepared by constant-current electrolysis of potassium fluorosulfate and added to the carboxylic acids in fluorosulfonic acid. γ- and δ-Lactones are obtained in yields between 7 and 50% (equation 19)[58].

In all cases the products isolated were consistent with a mechanism where the initial step is the cleavage of a secondary CH bond remote from the carboxyl group. Ring

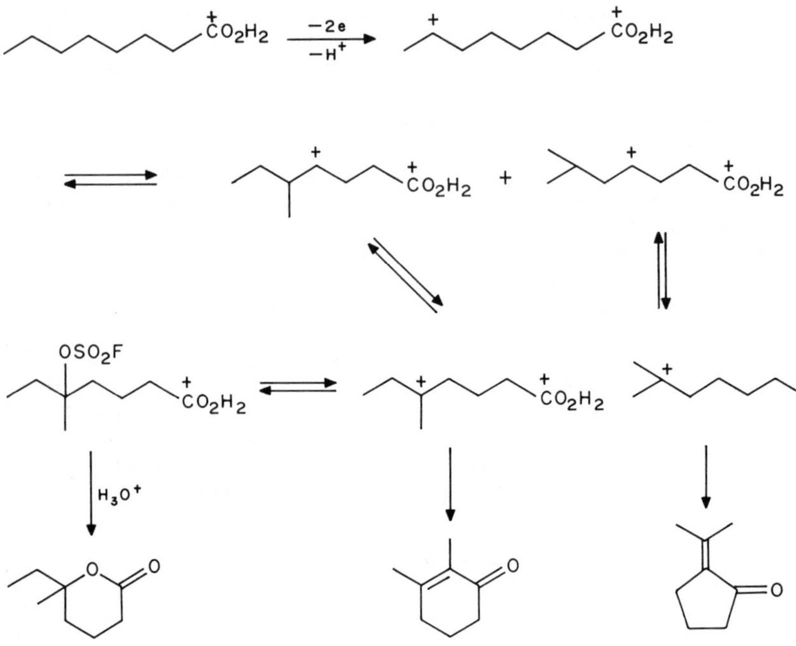

$$\text{(18)}$$

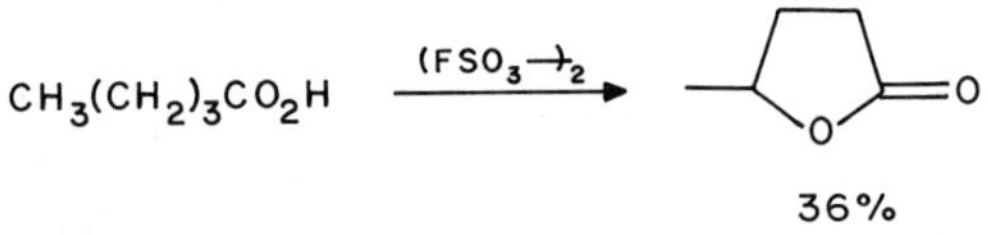

$$CH_3(CH_2)_3CO_2H \xrightarrow{(FSO_3\rightarrow)_2} \underset{36\%}{\boxed{}}=O \qquad (19)$$

closure occurs probably during work-up. The products isolated are similar to those obtained by direct oxidation of the carboxylic acids. The yields of lactones by the indirect method are, however, as good or better than those obtained by the direct oxidation. On standing, the longer-chain acids again afforded cyclic unsaturated ketones. Acyclic esters have been oxidized by cpe in a divided cell in acetonitrile/LiClO$_4$ (or TEABF$_4$) at a platinum anode. Ethyl butanoate yielded 70% of 3-acetamidobutanoate, and a high selectivity for the $(\omega - 1)$ position was also found for other esters (Table 10)[59].

TABLE 10. Remote oxidation of esters in acetonitrile

Ester	Acetamido ester (% yield)
Ethyl butanoate	Ethyl 3-acetamidobutanoate (70)
Methyl hexanoate	Methyl 5-acetamidohexanoate (42)
	Methyl 4-acetamidohexanoate (11)
Methyl nonanoate	Methyl 8-acetamidononanoate (25)
Methyl 3-methylbutanoate	Methyl 3-acetamido-3-methylbutanoate (68)
Methyl 2-methylpropanoate	Methyl 2-acetamido-2-methylpropanoate (58)
	Methyl 3-acetamidobutanoate (4)

In an undivided cell the yields are reduced substantially. Methyl hexanoate gave $(\omega - 1)$ and also some $(\omega - 2)$ products. The selectivity of the reaction is reminiscent of the ester chlorination with N-chloro amines[8,60]. In the anodic oxidation, however, carbocations seem to be involved, being formed most remote from the ester function. In suitable cases carbocation rearrangements are observed.

Remote oxidation has also been achieved with a covalently attached mediator. Anodic oxidation of 5α-cholestan-3α-yl esters with a carbonyl or m-iodophenyl group at their 3α-substituent gave in acetonitrile 0.1 M $LiClO_4$ at cpe the 6α-acetamidated cholestanyl esters in 7 to 26% yields together with esters carrying two or more acetamide functions[61]. Cyclovoltammetry indicates that the oxidation of the carbonyl compounds takes place at 2.3 to 2.7 V (vs $AgNO_3$). The results suggest that a radical cation is formed first by oxidation of the carbonyl or iodophenyl groups at the 3α-substituent which abstracts intramolecularly an equatorial 6α-hydrogen (equation 20)

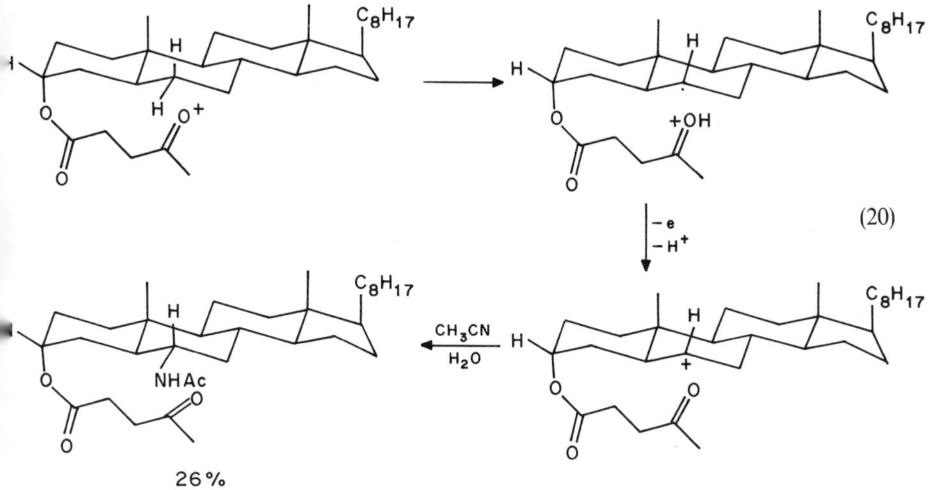

$$(20)$$

26%

C. C—C Bond Cleavage in Hydrocarbons, Cyclopropanes and Other Strained Cyclic Hydrocarbons

By analogy with their behavior in mass spectrometry, branched hydrocarbons are being cleaved upon oxidation in acetonitrile/$TEABF_4$ at $-45\,°C$. Cpe of different branched hydrocarbons affords the acetamides of the fragments. These are formed by cleavage of the initial radical cation at the C—C bond between the secondary and tertiary carbon to afford, after a second electron transfer, carbocations which react with the acetonitrile (Table 11 and equation 21)[62]. n-Octane, however, leads to the anodic substitution of the 2-, 3- and 4-H.

In cyclopropanes a C—C bond can be cleaved by way of anodic oxidation. The electrochemical oxidation of bicyclo[4.1.0]heptane (31) and bicyclo[3.1.0]hexane gave products in which the cyclopropane ring was opened. Anodic oxidation was conducted at a carbon rod anode in methanol/TEATos. The products obtained from 31 are given in equation 22[63]. The products found are completely different from those of the acidic hydrolysis of the cyclopropanes. Whilst in the methanolysis of 31 in methanol/toluene-sulfonic acid the selectivity of external/internal bond cleavage is 88/12, that of anodic oxidation is 0/100. Also, in the cleavage with $Pb(OAc)_4$ or $Tl(OAc)_3$ the external bond

TABLE 11. Anodic oxidation of branched hydrocarbons in acetonitrile/TEABF$_4$

Hydrocarbon	Products
2,3-Dimethylbutane	2-Propylacetamide
2,2,4-Trimethylpentane	t-Butylacetamide, 2-propylacetamide, 2-butylacetamide
2,2-Dimethylbutane	t-Butylacetamide, t-pentylacetamide, isopentylacetamide
t-Butylcyclohexane	t-Butylacetamide, cyclohexylacetamide
n-Octane	2-Octylacetamide, 3-octylacetamide, 4-octylacetamide

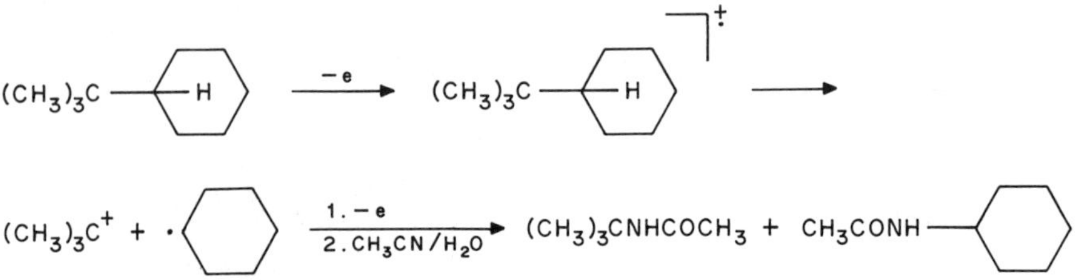

(21)

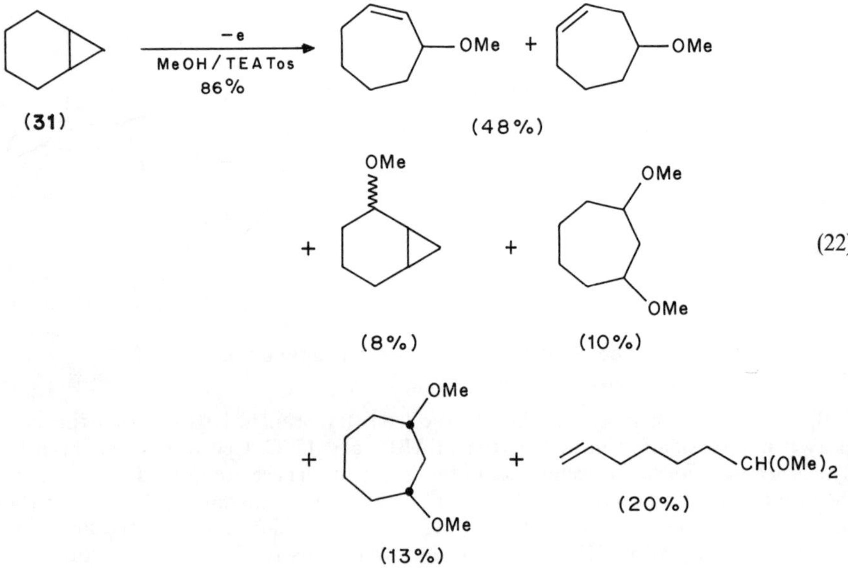

(22)

cleavage predominates with 71–91/29–9. The mechanism is assumed to be a direct electron transfer from the cyclopropane, leading by internal bond cleavage to the products. Attack of an oxidizing agent or of acid, on the other hand, induces a mainly external bond cleavage.

Anodic oxidation of polyalkyl-substituted cyclopropanes and spiroalkanes in methanol/TEATos gave monomethoxy and dimethoxy products in yields between 6 and 71%[64]. These results from the cleavage of the most highly substituted carbon–carbon bond. The oxidation potentials, that were measured in acetonitrile/TEABF$_4$, range between 2.05 V and > 2.5 V vs sce. Depending on the degree of alkyl substitution, the potential is cathodically shifted about 0.3 V per alkyl group. In the acid-catalyzed methanolysis, which corresponds to an electrophilic cleavage, the proton attacks at the least hindered side of the molecule in contrast to the anodic cleavage. The anodic cleavage of substituted cyclopropanes is reported elsewhere[65,66].

The half-wave potentials of polycyclic hydrocarbons as quadricyclane (**32**), **33**, cubane (**34**), **35** and others range between 0.91 V and 2.12 V vs sce[67]. The half-wave potentials correlate well with the correponding ionization potentials, which means that the electron is being transferred from the HOMO of the hydrocarbon. In the anodic oxidation of [1.1.0]-bicyclobutanes the 1,3-bond is oxidatively cleaved, which is in accord with theoretical predictions. Cpe of **33** affords in 65% yield a product **36** being formed by cleavage of the 1,3-bond of a radical cation, methanolysis, rearrangement and further methanolysis (equation 23). Cpe of another bicyclobutane affords dimethoxycyclobutanes and acyclic acetals, whose formation can be similarly accounted for by a cyclobutyl cation as intermediate and its further deprotonation or rearrangement.

(32)	(33)	(34)	(35)
$E_{1/2}$: 0.91 V (sce)	1.50 V	1.73 V	1.91 V

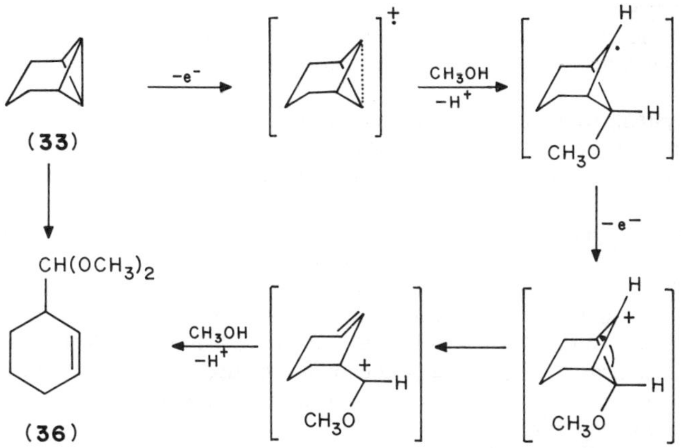

(23)

The ease of oxidation of a strained hydrocarbon is a good measure of the relative energies of the HOMOs. Thus the ability of strained hydrocarbons to quench naphthalene fluorescence, which also reflects the energy of the HOMO, correlates reasonably well with their oxidation potentials.

Tricyclene (**37**) has been oxidized—after careful optimization of the reaction conditions—in acetic acid/Et$_3$N to the Nojigiku alcohol **38** in 77% yield, again with cleavage of a cyclopropane C—C bond (equation 24). The reaction was also conducted in a 2.25 kg scale to afford **38** from impure **37** mixed with alkenes. Due to the higher oxidation potentials of the alkenes, **38** was formed in 65% yield without conversion of the alkenes[68].

$$\text{(37)} \xrightarrow[\text{2. }^-\text{OH}]{\text{1. }-2e, AcOH/4\,F/mol} HO\cdots \quad \text{(38)}$$

(37)   (38)

D. Indirect Anodic Oxidation

Saturated hydrocarbons have been partially oxidized to ketones by indirect oxidation with electrogenerated NO_3 radicals in t-butanol/water/HNO_3, which was saturated with oxygen[69]. A statistical H abstraction at the methylene groups was found. The following mechanism is proposed: (a) oxidation of NO_3^- to NO_3 radicals, (b) H abstraction of the NO_3 radical, (c) reaction of the alkyl radical with oxygen to form peroxy radicals, which decay to ketones.

Cyclohexane and norbornane can be oxidized indirectly in trifluoroacetic acid/acetic acid with $Co(OAc)_2$ as mediator in a divided cell at a platinum anode. Cyclohexanone and 2-norbornanone were obtained in 47–52% yields, accompanied by minor amounts of the corresponding alcohols and acetates[70]. n-Octane yielded 15% of the octanone in the distribution 2- (55%), 3- (22%) and 4-octanone (20%). Without cobalt(II) acetate the corresponding alcohols and acetates are obtained in similar yields.

III. OXIDATION OF ALKANES BY CATHODIC REDUCTION OF OXYGEN IN THE PRESENCE OF CATALYSTS

Saturated hydrocarbons can be oxidized by higher-valent Fe- and Mn-oxide complexes, which are formed by reduction of oxygen in the presence of the corresponding metal complexes. For that purpose chemical reductants can be exchanged beneficially for the cathode.

The Gif system, which consists of triplet oxygen, acetic acid, pyridine, zinc and an iron catalyst, oxidizes saturated hydrocarbons mainly to ketones and gives minor amounts of aldehydes. Tertiary hydrogen is only substituted in exceptional cases. With the Gif–Orsay II system[71], in which zinc is replaced by the cathode [divided cell, cpe at -0.6 to $-0.7\,V$ vs sce, trifluoroacetic acid, pyridine, $Fe_3O(OAc)_6Pyr_{3.5}$], adamantane is converted in 3.8% coulombic yield; the ratio of attack at a secondary: tertiary CH bond (C^2/C^3 ratio) is 15.0. Comparable conversions were carried out with cyclododecane to afford 21% oxidation with a ratio of alcohol:ketone = 1:14. $Trans$-decalin yielded 22% product, consisting of 0.6% 9-ol, 0.9% 1-ol, 9.0% 1-on 0.65% 2-ol and 11% 2-one. A radical mechanism for this conversion can be excluded since for the cobalt-catalyzed radical oxidation of $trans$-decalin the C^2/C^3 ratio is 0.13, which is far removed from 36 found with the Gif system.

The Gif–Orsay II system has been further improved in yields and simplified with respect to the reaction conditions[72]. Cyclododecane, adamantane and cyclooctane are oxidized in 17–30 mmolar amounts and coulombic yields of around 30% with oxygen, pyridine, trifluoroacetic acid and an iron catalyst in an undivided cell. Cyclododecane afforded a 18.9% yield of alcohol and ketone in a ratio of 0.14, adamantane a 14% coulombic yield, cyclooctane and cyclohexane coulombic yields up to 48%. The C^2/C^3 ratio with adamantane reached values up to 40. With paraquat or 4,4'-bipyridyl as electron transfer reagent and under otherwise the same conditions, a coulombic yield of 49% cyclohexanol and cyclohexanone (1:7.88) was obtained. In this process oxygen is reduced to superoxide, which oxidizes iron(II) to an active iron catalyst that is able to react with the hydrocarbon.

The chemical (Gif system) and the electrochemical conversion (Gif–Orsay system) have been compared in the oxidation of six saturated hydrocarbons (cyclohexane, 3-ethylpentane, methylcyclopentane, cis- and trans-decalin and adamantane)[73]. The results obtained for pyridine, acetone and pyridine–acetone were similar for both systems. Total or partial replacement of pyridine for acetone affects the selectivity for the secondary position and lowers the ratio ketone:secondary alcohol. The formation of the same ratio of cis- and trans-decal-9-ol from either cis- or trans-decalin indicates that tertiary alcohols result from a mechanism essentially radical in nature. The C^2/C^3 ratio between 6.5 and 32.7 rules out a radical mechanism for the formation of ketones and secondary alcohols. Ratios of 0.14 and 0.4 were reported for radical-type oxidations of adamantane and cis-decalin. Partial replacement of pyridine by methanol, ethanol or i-propanol results in diminished yields and a lower selectivity. Acetone gives comparable yields; however, the C^2/C^3 ratio drops to 0.2–10.7.

In developing and optimizing the Gif–Orsay system the components of the Gif system were studied by cyclovoltammetry[74]. This led to the following conclusions: (i) in pyridine the reduction of dioxygen is catalyzed by acids, which with increasing strength increase the reduction current and shift the potential to less negative values; (ii) in the absence of acid the reduction of dioxygen is also catalyzed by ferrous salts; (iii) the lowest negative potential in concentrated acid solution corresponds to the reduction of dioxygen; (iv) in wet pyridine Fe(III) reacts to polyhydroxy Fe(III) complexes.

The first Gif–Orsay system (Pt-anode, Hg-cathode, cpe, $TEABF_4$ in pyridine–acetic acid, divided cell) converted adamantane to 3.5% 1-adamantanol, 0.7% 2-adamantanol and 3.0% adamantone. Replacement of acetic acid by trifluoroacetic acid and its continuous addition to maintain a constant acidity improved the coulombic yield and selectivity. The oxidation of adamantane yielded 18% product with a C^2/C^3 ratio of 8.5, this one of cyclodecane 21.1% product, and this one of trans-decalin 22.2% product, with a C^2/C^3 ratio of 36. The results come close to those obtained in the chemical system. The third system

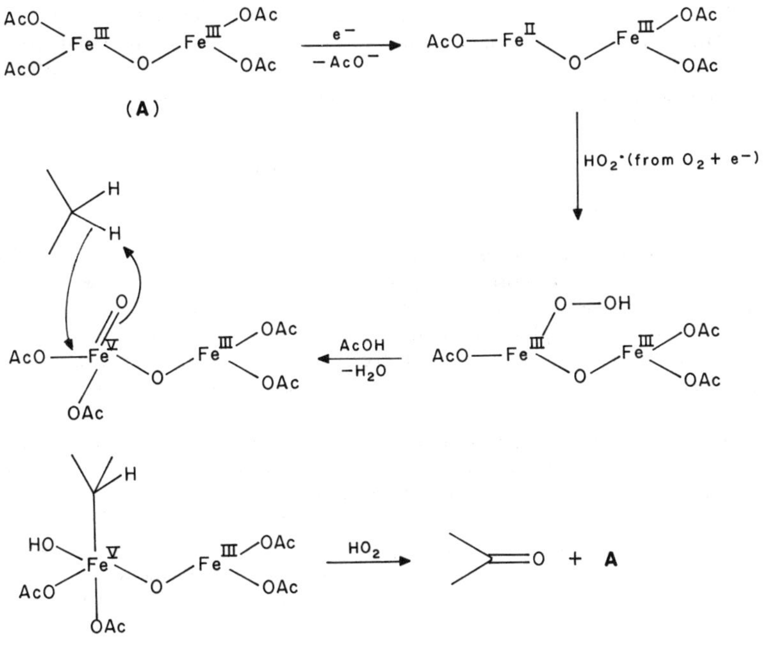

(25)

used an undivided cell at cpe; cyclododecane was oxidized with similar yields as in the Gif–Orsay II system. In the fourth system an undivided cell, a higher substrate concentration and controlled current density were used. Adamantane, cyclododecane, cyclooctane and cyclohexane were oxidized at partial conversion with good coulombic yields and high selectivity. As mechanism, both Fe^V=O- and binuclear complexes are assumed as reactive intermediates (equation 25).

Another system uses manganese-porphyrin as catalyst[75]. Hydrocarbons in acetonitrile/water containing lithium perchlorate, catalytic amounts of manganese tetraphenylporphyrin chloride, imidazole and acetic acid were electrolyzed in a divided cell at a platinum gauze working electrode with continuous oxygen bubbling. Cyclooctane was converted to cyclooctanol and cyclooctanone in a ratio of 5:15 and 7.5% coulombic yield, ethylbenzene to alcohol and ketone in the ratio 67:33 and adamantane to 1- and 2-adamantanol in a ratio of 85:15 each in 3% coulombic yield. The mechanism proposed is conversion of the manganese porphyrin to a highly oxidizing $Mn(V)$=O intermediate.

Also, with iron porphyrin deposited at the cathode, alkanes have been catalytically oxidized to ketones and alcohols in an electrochemical cell in the presence of oxygen[76]. An oxidation mechanism similar to that of a P-450 oxidation is assumed in this case.

In a much less sophisticated system hydrocarbons are being oxidized by OH radicals at the cathode[77]. This occurs by simultaneous reduction of Fe(III) and oxygen at cpe. The OH radicals are generated by a Fenton reaction from Fe(II) and cathodically formed hydrogen peroxide. Linear alkanes from C_5 to C_{10} are being oxidized to ketones as the only products. The yields decrease with increasing number of carbon atoms and with the Fe(III) concentration.

IV. OXIDATION OF HYDROCARBONS IN FUEL CELLS

In fuel cells hydrocarbons are dehydrogenated by active catalysts. The main purpose of these cells is the direct conversion of combustion energy into electrical energy. At the fuel cell cathode the hydrocarbon is in must cases oxidized to carbon dioxide (equation 26a), because the intermediates are more easily oxidized than the starting hydrocarbon. Only in few cases are substitution, dehydrogenation or coupling products of the hydrocarbon obtained. At the fuel cell anode oxygen is reduced to water (equation 26b).

$$C_nH_{2n+2} + 2nH_2O - (6n + 2)e \rightarrow nCO_2 + (6n + 2)H^+ \tag{26a}$$

$$(6n + 2)e + (6n + 2)H^+ + (3n + 1)O_2 \rightarrow (3n + 1)H_2O \tag{26b}$$

Most important in these processes is an active cathode. In most cases finely divided platinum is used for this purpose. Unsupported platinum, particularly the one reduced from Adams catalyst, and carbon-supported platinum electrocatalysts are effective in reducing the amount of platinum required for a given amount of power from a hydrocarbon fuel cell. A propane–oxygen fuel cell, which uses 10 mg Pt per cm^2 supported on carbon can deliver about 35 mW cm^{-2} using a hydrofluoric acid electrolyte at 105 °C[78].

A low-temperature aqueous fuel cell, that employs conducting porous teflon electrodes (Pt-black imbedded in teflon powder), shows excellent performance (current densities up to 200 mA cm^{-2}) in acidic and basic electrolytes[79].

In a fuel cell, whose cathode consisted of different metal chlorides supported on graphite and teflon, cyclohexane was oxidized at 298 K. $SmCl_3$ proved to be the best catalyst. Cyclohexane formed both cyclohexanol and cyclohexanone in a current efficiency of 7%. The mechanism may involve generation of active oxygen on the cathode[80].

The electrolyte also plays an important role in the cell performance. For the propane oxidation in a fuel cell the monohydrate of trifluoromethanesulfonic acid as electrolyte had, at a platinum electrode at 135 °C, a limiting current which was more than 1000% greater than that of phosphoric acid[81].

Hydrocarbons from ethane to dodecane are being oxidized at Raney platinum in 3 N sulfuric acid at 100 °C at 500 mV vs NHE (Normal Hydrogen Electrode) with current densities between 10mA cm^{-2} (dodecane) and 150 mV cm^{-2} (propane) to carbon dioxide. The rate of the alkane oxidation depends strongly on the length of the alkyl chain and on the temperature. A maximum rate is achieved for ethane and propane. The following mechanism is being proposed: the hydrocarbon in chemisorbed at the platinum, dehydrogenated and cleaved at the C—C bond, the radicals react with water until carbon dioxide is being formed. The cell and the reaction conditions as well as the preparation of the electrode are described[82-84].

Variation in the rate of anodic oxidation of normal saturated hydrocarbons with the number of carbon atoms parallels the variation in the rate of diffusion of the hydrocarbon through the electrolyte. This holds for different electrolytes (CsF/HF, H_3PO_4, HF, H_2SO_4) and electrodes (Pt-black, Raney-Pt). The rates are low for methane, highest for ethane and propane and then gradually decrease[85].

The effects of molecular structure, electrolyte and temperature on the rate and extent of adsorption and electrochemical oxidation of hydrocarbons (alkanes, alkenes, alkynes) have been reviewed[86].

The selective dehydrogenation of short-chain hydrocarbons can be achieved at higher temperatures. Ethane is dehydrogenated at 700 °C at a fuel cell current of 252 mA cm^{-2} in the presence of oxygen to ethene with 10.6% conversion and 96.9% selectivity[87]. Propane is converted in good yield to propene in a lithium iodide melt at 465 °C at 1–2 A between carbon electrodes and 12 V cell voltage[88].

The dehydrodimerization of short-chain hydrocarbons has also been achieved. The oxidative coupling of methane at the electrode Ag/Bi_2O_3 and a solid electrolyte yielded at 700 °C and 2% conversion, in 72% selectivity ethane and in 18% selectivity ethylene[89]. In another case Y_2O_3/ZrO_2 was used as solid electrolyte for the oxidative coupling of methane to ethane on $BaCO_3$ deposited on a gold anode. High pressures for CH_4 and O_2 are applied; the optimal temperature is 800 °C[90]. Furthermore, methane is dehydrodimerized at 760 °C in 11% efficiency to a mixture of ethane (58%), ethylene (37%) and acetylene (4%). The fuel cell has the general configuration: CH_4, Pt/Sm_2O_3/$La_{0.89}Sr_{0.1}MnO_3$/yttria stabilized zirconia/$La_{0.89}Sr_{0.1}MnO_3$/Pt, air[91]. Methane is also oxidized mainly to ethane and ethylene with the solid electrolyte Y_2O_3/ZrO_2 and a number of different cathode materials (e.g. Ag, Ni, Cu, Bi, Pt) and anode materials (e.g. Ag, Bi, Mn, V) which are blended with alkali and alkaline-earth metal oxides. The reaction occurs at 0.5–10 atm and 500–900 °C[92].

Once this solid oxide fuel cell anode electrocatalysis becomes mechanistically understood and optimized, it may provide a route for the simultaneous electrochemical synthesis of C_2 hydrocarbon species and generation of electrical energy from CH_4.

V. REFERENCES

1. G. S. Fonken and R. A. Johnson, *Chemical Oxidations with Microorganisms*, M. Dekker Inc., New York, 1972, p. 1.
2. (a) K. Kieslich, *Microbial Transformations of Non-Steroid Cyclic Compounds*, G. Thieme, Stuttgart, 1976.
   (b) K. Kieslich, In *Biotechnologie* (Ed. K. Dohmen), Metzler, Stuttgart, 1983.
   (c) K. Kieslich, *Bull. Soc. Chim. Fr.*, 9 (1980).
3. (a) N. J. Russell, *Trends Biochem. Sci.*, 108 (1984).

(b) H. Baumann, M. Bühler, H. Fochem, F. Hirsinger, H. Zoebelein and J. Falbe, *Angew. Chem.*, **100**, 41 (1988); *Angew. Chem., Int. Ed. Engl.*, **27**, 41 (1988).
4. Y. Miura, *Lipids*, **16**, 721 (1981).
5. P. P. Ho and A. J. Fulco, *Biochim. Biophys. Acta*, **431**, 249 (1976).
6. (a) R. Breslow, R. J. Corcoran and B. B. Snider, *J. Am. Chem. Soc.*, **96**, 6971 (1974).
   (b) R. Breslow, *Acc. Chem. Res.*, **13**, 170 (1980).
   (c) R. Breslow, *Tetrahedron Lett.*, **24**, 5039 (1983).
7. (a) P. Welzel, K. Hobert, A. Ponty, D. Neunert and H. Klein, *Tetrahedron*, **41**, 4509 (1985).
   (b) R. Breslow and D. Heyer, *J. Am. Chem. Soc.*, **104**, 2045 (1982).
   (c) R. Breslow, R. J. Corcoran, B. B. Snider, R. J. Doll, P. L. Khanna and R. Kaleya, *J. Am. Chem. Soc.*, **99**, 905 (1977).
8. (a) N. C. Deno, W. E. Billups, R. Fishbein, C. Pierson, R. Whalen and J. Wyckoff, *J. Am. Chem. Soc.*, **93**, 438 (1971).
   (b) E. Cramer and H. J. Schäfer, *Fat Sci. Technol.*, **90**, 351 (1988).
   (c) E. Cramer, Ph.D. Thesis Münster, 1989.
9. (a) C. Eden and Z. Shaked, *Isr. J. Chem.*, **13**, 1 (1975).
   (b) L. Hinkamp, M. S. Thesis, 1990, Münster.
10. K. B. Wiberg, *J. Am. Chem. Soc.*, **83**, 423 (1961).
11. S. R. Jones and J. M. Mellor, *J. Chem. Soc.*, *Perkin Trans. 1*, 2576 (1976).
12. J. Hanotier, Ph. Camerman and M. Hanotier-Bridoux, *J. Chem. Soc.*, *Perkin Trans. 2*, 2247 (1972).
13. R. Stewart and U. A. Spitzer, *Can. J. Chem.*, **56**, 1273 (1978).
14. H. J. Schmidt and H. J. Schäfer, *Angew. Chem.*, **91**, 77 (1979); *Angew. Chem., Int. Ed. Engl.*, **18**, 68 (1979).
15. (a) W. Müller and H. J. Schneider, *Angew. Chem.*, **91**, 438 (1979); *Angew. Chem., Int. Ed. Engl.*, **18**, 407 (1979).
   (b) N. C. Deno, E. J. Jedziniak, L. A. Messer, M. D. Meyer, S. G. Stroud and E. S. Tomeszko, *Tetrahedron*, **33**, 2503 (1977).
   (c) A. Rotman and Y. Mazur, *J. Am. Chem. Soc.*, **94**, 6628 (1972).
16. D. Mansuy, in *Activation and Functionalization of Alkanes* (Ed. C. L. Hill), Wiley, New York, 1989, p. 195.
17. D. H. R. Barton and N. Ozbalik, in Reference 16, p. 281.
18. (a) C. K. Mann and K. K. Barnes, *Electrochemical Reactions in Non-aqueous Systems*, M. Dekker, New York, 1970.
   (b) H. Siegermann, in *Technique of Electroorganic Synthesis, Part II* (Ed. N. L. Weinberg), Wiley, New York, 1975.
   (c) J. Perichon, M. Herlem, F. Bobilliart, A. Thiebault and K. Nyberg, in *Encyclopedia of Electrochemistry of the Elements, Vol. VI* (Eds. A. J. Bard and H. Lund), M. Dekker, New York, 1978.
   (d) L. Eberson, J. H. P. Utley and O. Hammerich, in *Organic Electrochemistry* (Eds. H. Lund and M. M. Baizer), M. Dekker, New York, 1991, p. 515.
19. J. Bertram, J. P. Coleman, M. Fleischmann and D. Pletcher, *J. Chem. Soc.*, *Perkin Trans. 2*, 374 (1973).
20. F. Bobilliart, A. Thiebault and M. Herlem, *C. R. Acad. Sci. Paris, Ser. C*, **278**, 1485 (1974).
21. C. Pitti, F. Bobilliart, A. Thiebault and M. Herlem, *Anal. Lett.*, **8**, 241 (1975). *Chem. Abstr.*, **83**, 34722r (1975).
22. S. Pitti, M. Herlem and J. Jordan, *Tetrahedron Lett.*, 3221 (1976).
23. Ch. Pitti, M. Cerles, A. Thiebault and M. Herlem, *J. Electroanal. Chem.*, **126**, 163 (1981).
24. J. P. Coleman and D. Pletcher, *J. Electroanal. Chem.*, **87**, 111 (1978).
25. J. P. Coleman, *J. Electrochem. Soc.*, **128**, 2561 (1981)
26. D. Pletcher and C. Z. Smith, *Chem. Ind.*, 371 (1976).
27. J. P. Coleman and D. Pletcher, *Tetrahedron Lett.*, 147 (1974).
28. C. J. Myall and D. Pletcher, *J. Electroanal. Chem.*, **85**, 371 (1976).
29. P. L. Fabre, J. Devynck and B. Tremillon, *Tetrahedron*, **38**, 2697 (1982).
30. J. Devynck, P. L. Fabre, A. Ben Hadid and B. Tremillion, *J. Chem. Res. (M)*, 2469 (1979).
31. G. Brilmyer and R. Jasinski, *J. Electrochem. Soc.*, **129**, 1950 (1982).
32. (a) W. V. Childs, L. Christensen, F. W. Klink and C. F. Kolpin, in *Organic Electrochemistry*, 3rd ed. (Eds. H. Lund and M. M. Baizer), M. Dekker, New York, 1991, p. 1103.

(b) I. N. Rozhkow, in *Organic Electrochemistry*, 2nd ed. (Eds. M. M. Baizer and H. Lund), M. Dekker, New York, 1983, p. 805.
33. D. B. Clark, M. Fleischmann and D. Pletcher, *J. Chem. Soc., Perkin Trans. 2*, 1578 (1973).
34. H. P. Fritz and T. Würminghausen, *J. Electroanal. Chem.*, **54**, 181 (1974).
35. H. P. Fritz and T. Würminghausen, *J. Chem. Soc., Perkin Trans. 1*, 610 (1976).
36. H. P. Fritz and T. Würminghausen, *Z. Naturforsch.*, **32b**, 241 (1977).
37. H. J. Schäfer, E. Cramer, A. Hembrock and G. Zimmermann, *M. M. Baizer Memorial Book*, M. Dekker Inc., New York, 1991, in press.
38. H. Preuss, *Z. Naturforsch.*, **26a**, 1020 (1971).
39. (a) D. S. Ashton and J. M. Tedder, *J. Chem. Soc. (B)*, 1031 (1970).
    (b) D. S. Ashton, J. M. Tedder, J. C. Walton, A. Nechvatal and I. K. Stoddart, *J. Chem. Soc., Perkin Trans 1*, 846 (1973).
40. (a) A. Hembrock and H. J. Schäfer, *Angew. Chem.*, **97**, 1048 (1985); *Angew. Chem., Int. Ed. Engl.*, **24**, 1055 (1985).
    (b) A. Hembrock and H. J. Schäfer, *Proceeding of the 1st International Symposium on Electroorganic Synthesis* (Ed. S. Torii), Kodansha, Tokyo, 1987, p. 121.
41. A. Hembrock, Ph.D. Thesis, University of Münster, 1989.
42. T. Shono and Y. Matsumura, *J. Am. Chem. Soc.*, **91**, 2803 (1969).
43. The oxidation of decalin at 20 °C with chlorine in benzene or in CS_2 or with $C_6H_5ICl_2$ in benzene proceeds with $R_s^t = 6.3$, 10.5 and 8.3, respectively. F. Kämper, Ph.D. Thesis, University of Münster, 1979.
44. M. Fleischmann and D. Pletcher, *Tetrahedron Lett.*, 6255 (1968).
45. J. Cassidy, S. B. Khoo, S. Pons and M. Fleischmann, *J. Phys. Chem.*, **89**, 3933 (1985).
46. V. R. Koch and L. L. Miller, *Tetrahedron Lett.*, 693 (1973).
47. V. R. Koch and L. L. Miller, *J. Am. Chem. Soc.*, **95**, 8631 (1973).
48. G. J. Edwards, S. R. Jones and J. M. Mellor, *J. Chem. Soc., Chem. Commun.*, 816 (1975).
49. S. R. Jones and J. M. Mellor, *J. Chem. Soc., Chem. Commun.*, 385 (1976).
50. G. J. Edwards, S. R. Jones and J. M. Mellor, *J. Chem. Soc., Perkin Trans. 2*, 505 (1977).
51. E. Garcia and A. J. Bard, *J. Electrochem. Soc.*, **137**, 2752 (1990).
52. (a) A. T. Kuhn and J. G. Sunderland, *Electrochim. Acta*, **18**, 119 (1973).
    (b) D. B. Clark, M. Fleischmann and D. Pletcher, *J. Electroanal. Chem.*, **42**, 133 (1973).
53. B. Carre, P. L. Fabre and J. Devynck, *Bull. Soc. Chim. Fr.*, 255 (1987).
54. J. Y. Becker, L. R. Byrd and L. L. Miller, *J. Am. Chem. Soc.*, **96**, 4718 (1974).
55. J. Y. Becker, L. R. Byrd, L. L. Miller and Y.-H. So, *J. Am. Chem. Soc.*, **97**, 853 (1975).
56. F. Kämper, H. J. Schäfer and H. Luftmann, *Angew. Chem.*, **88**, 334 (1976); *Angew. Chem., Int. Ed. Engl.*, **15**, 306 (1976).
57. D. Pletcher and C. Z. Smith, *J. Chem. Soc., Perkin Trans. 1*, 948 (1975).
58. Ch. J. Myall, D. Pletcher and C. Z. Smith, *J. Chem. Soc., Perkin Trans. 1*, 2035 (1976).
59. L. L. Miller and V. Ramachandran, *J. Org. Chem.*, **39**, 369 (1974).
60. F. Minisci, R. Galli, A. Galli and R. Bernardi, *Tetrahedron Lett.*, 2207 (1967).
61. M. Tokuda, T. Yamashita and H. Suginome, in *Recent Advances in Electroorganic Synthesis* (Ed. S. Torii), Elsevier, Kodanska, Tokyo, 1987.
62. T. M. Siegel and L. L. Miller, *J. Chem. Soc., Chem. Commun.*, 341 (1974).
63. T. Shono, Y. Matsumura and Y. Nakagawa, *J. Org. Chem.*, **36**, 1771 (1971).
64. T. Shono and Y. Matsumura, *Bull. Chem. Soc. Jpn.*, **48**, 2861 (1975).
65. (a) M. Klehr and H. J. Schäfer, *Angew. Chem., Int. Ed. Engl.*, **14**, 247 (1975).
    (b) S. Torii, T. Okomoto and N. Ueno, *J. Chem. Soc., Chem. Commun.*, 293 (1978).
66. (a) S. Torii, T. Inokuchi and N. Takahashi, *J. Org. Chem.*, **43**, 5020 (1978).
    (b) A. J. Baggaley, R. Brettle and J. R. Sutton, *J. Chem. Soc., Perkin Trans 1*, 1055 (1975).
67. (a) P. G. Gassmann and R. Yamaguchi, *J. Am. Chem. Soc.*, **101**, 1308 (1979).
    (b) P. G. Gassmann and R. Yamaguchi, *Tetrahedron*, **38**, 1113 (1982).
68. T. Uchida, Y. Matsubara, I. Nishiguchi, T. Hirashima, T. Ohnishi and K. Kanehira, *J. Org. Chem.*, **55**, 2938 (1990).
69. R. Tomat and A. Rigo, *J. Appl. Electrochem.*, **16**, 8 (1986).
70. I. Nishiguchi, R. Shundou, A. Ikeda, Y. Matsubara and T. Hirashima, 2nd International IUPAC Symposium of Organic Chemistry: Technological Perspectives, Baden-Baden, FRG, 14.–19.4.91, Poster 54.

71. G. Balavoine, D. H. R. Barton, J. Boivin, A. Gref, N. Ozbalik and H. Riviere, *Tetrahedron Lett.*, **27**, 2849 (1986).
72. G. Balavoine, D. H. R. Barton, J. Boivin, A. Gref, N. Ozbalik and H. Riviere, *J. Chem. Soc., Chem. Commun.*, 1727 (1986).
73. G. Balavoine, D. H. R. Barton, J. Boivin, A. Gref, P. Le Coupanec, N. Ozbalik, J. A. X. Pestana and H. Riviere, *Tetrahedron*, **44**, 1091 (1988).
74. G. Balavoine, D. H. R. Barton, J. Boivin, A. Gref, I. Hallerey, N. Ozbalik, J. A. Pestana and H. Riviere, *New J. Chem.*, **14**, 175 (1990).
75. P. Leduc, P. Battioni, J. F. Bartoli and D. Mansuy, *Tetrahedron Lett.*, **29**, 205 (1988).
76. A. M. Khenkin and A. E. Shilov, *React. Kinet. Catal. Lett.*, **33**, 125 (1987); *Chem. Abstr.*, **108**, 37015h (1987).
77. R. Tomat and A. Rigo, *J. Appl. Electrochem.*, **15**, 167 (1985); *Chem. Abstr.*, **102**, 174903 (1985).
78. E. J. Cairns and E. J. Mc Inerney, *J. Electrochem. Soc.*, **114**, 980 (1967).
79. L. W. Niedrach and H. R. Alford, *J. Electrochem. Soc.*, **112**, 117 (1965).
80. K. Otsuka, I. Yamanaka and K. Hosakawa, *Nature*, **345**, 697 (1990).
81. A. A. Adams and H. J. Barger, Jr., *J. Electrochem. Soc.*, **121**, 987 (1974).
82. H. Binder, A. Köhling, H. Krupp, K. Richter and G. Sandstede, *J. Electrochem. Soc.*, **112**, 355 (1965).
83. G. Sandstede, *Chem. -Ing.-Tech.*, **38**, 676 (1966).
84. G. Sandstede, *Chem. -Ing.-Tech.*, **37**, 632 (1965).
85. E. J. Cairns, *Science*, **155**, 1245 (1967).
86. E. J. Cairns, *Adv. Electrochem. Electrochem. Eng.*, **8**, 337 (1971); *Chem. Abstr.*, **75**, 104358n (1971).
87. T. I. Mazanec and T. L. Cable, Brit. UK Pat. 2203446 A; *Chem. Abstr.*, **110**, 135899k (1989).
88. S. G. Mill, US Pat. 3321386; *Chem. Abstr.*, **67**, 45781e (1967).
89. K. Otsuka and A. Morikawa, Jap. Pat. 61030688 A2; *Chem. Abstr.*, **105**, 114383u (1986).
90. K. Otsuka, K. Suga and I. Yamanaka, *Catal. Today*, **6**, 587 (1990); *Chem. Abstr.*, **113**, 118105t (1990).
91. N. U. Pujare and A. F. Sammells, *J. Electrochem. Soc.*, **135**, 10, 2544 (1988).
92. T. I. Mazanec and T. L. Cable, US Pat. 4802958 A; *Chem. Abstr.*, **110**, 215208z (1989).

Fast tritium/hydrogen exchange (equation 7) preceding the pyrolytic decomposition has been utilized to determine the specific activity of tritiated water added to AcONa/NaOH mixture[4]. [^{14}C]methane with molar activity greater than 1850 MBq mmol^{-1} (50 mCi mmol^{-1}), free of other gases and volatile products, has been prepared[6] in a modification of the method of Hutchins and coworkers[7] (equation 8). The intermediate diborane formed in this reaction has been trapped quantitatively as an adduct with 1,5-cyclooctadiene yielding the non-volatile 9-borabicyclononane (equation 8a).

$$CH_3COONa + NaOT \rightleftharpoons CH_2TCOONa + NaOH \qquad (7)$$

$$^{14}CO_2 \xrightarrow[\text{2. HI conc. (58\%)}]{\text{1. LiAlH}_4} {}^{14}CH_3I \xrightarrow[\text{and 1,5-cyclooctadiene, stirring overnight}]{\text{NaBH}_4 \text{ in MeOCH}_2CH_2OCH_2CH_2OMe} {}^{14}CH_4 + NaI + BH_3$$

82% yield,
400 mCi

radiochemical
purity > 99%   (8)

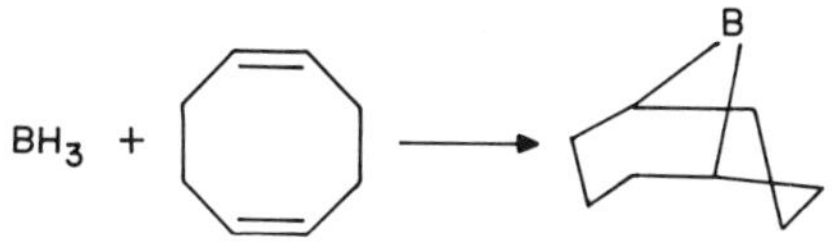

3. Synthesis and chromatographic separation of ^{11}C-methane

^{11}C Radiopharmaceuticals of very high specific radioactivity are needed for 'in vivo' studies of specific receptor binding. The maximum specific activity achieved in the course of synthesis of various ^{11}C-labelled compounds had been of the order of 1–2 Ci μmol^{-1} after taking all precautions concerning pollution with atmospheric $^{12}CO_2$ at all stages of the synthesis[8]. The maximum possible specific radioactivity of $^{11}CH_4$ equals[9] 8.38×10^8 Ci g^{-1} or 76.1 Ci μmol^{-1}. In a preliminary study[8] it has been shown that $^{13}CH_4/^{12}CH_4$, $^{12}CH_3D/^{12}CH_4$, $^{12}CD_3H/^{12}CH_4$ and $^{12}CD_4/^{12}CH_4$ isotopic derivatives can be separated at $-206 \pm 1\,^\circ$C using a 100-m-long solft-glass capillary tube. ^{11}C methane with a specific activity of 40 Ci μmol^{-1} has been obtained[10a] by chromatography using a capillary (300-m-long soft-glass columns) in the presence of 10% N_2 in helium, performed at $-207\,^\circ$C to $-208\,^\circ$C. The starting $^{11}CH_4$ had been produced in ^{14}N(p, α)^{11}C nuclear reaction by irradiating ($N_2 + 5\%$ hydrogen) with 20 MeV protons. The non-reactive methane has been transformed quantitatively into more reactive methyl iodide, methanol, hydrogen cyanide, CNH and formaldehyde without isotopic dilution for rapid labelling of radiopharmaceuticals. CD_3CH_3 has been prepared[10b] from methyl-$^2H_3$ benzenesulphonate with dilithium methyl 2-thienylcyanocuprate.

B. Synthesis of Isotopically Labelled Propanes

1. Synthesis of ^{13}C-labelled propanes

Propane-1-^{13}C and 2-methylpropane-1-^{13}C have been obtained[1f] according to equations 9 and 10:

$$Et^{13}COOH \xrightarrow{\text{LiAlH}_4} Et^{13}CH_2OH \xrightarrow{I_2} Et^{13}CH_2I \xrightarrow[\text{couple}]{\text{zinc–copper}} Et^{13}CH_3 \qquad (9)$$

$$\text{Me}_2\text{CO} \xrightarrow{^{13}\text{CH}_3\text{MgI}} \underset{\underset{\text{OMgI}}{|}}{^{13}\text{CH}_3\text{CMe}_2} \xrightarrow{\text{HI}} \underset{\underset{\text{I}}{|}}{^{13}\text{CH}_3\text{CHMe}_2} \xrightarrow{\Delta}$$

$$[^{13}\text{C}]\text{H}_3\text{C}{=}[^{13}\text{C}]\text{H}_2 \xrightarrow{\underset{\text{Ni}}{\text{H}_2}} [^{13}\text{C}]\text{H}_3\text{CHMe}_2 \qquad (10)$$
$$\underset{|}{}$$
$$\text{CH}_3$$

2. Synthesis of propane-1-d and 2-methylpropane-1-d

These compounds have been prepared[11] from 1-bromopropane and 2-methyl-1-bromopropane via the corresponding Grignard reagent (equation 11). The appearance of the C—D stretching band at $2180\,\text{cm}^{-1}$ in the IR spectrum of 1 has been observed.

$$\text{RBr} \xrightarrow[\text{THF}]{\text{Mg}} \text{RMgBr} \xrightarrow{\text{D}_2\text{O}} \text{RD}$$

$$\begin{aligned}&\textbf{1a: } R = \text{n-Pr}\\&\textbf{1b: } R = i\text{-Bu}\end{aligned} \qquad (11)$$

3. Synthesis of propane-1,2-^{14}C

Propane-1,2-^{14}C, specific activity of about $110\,\text{mCi mmol}^{-1}$, has been synthesized from doubly labelled acetylene in reaction series shown in equation 12. This synthesis[12] included preparation of doubly labelled acetylene from barium carbonate-^{14}C according to Cox and Warne[13], nearly quantitative hydrogenation of acetylene to ethylene, addition of hydroiodic acid to the latter to form iodoethane, preparation of ethylmagnesium iodide followed by carbonation to yield propionic acid and reduction o n-propanol with 70–75% yield. The latter yielded a tosylate which was finally reduced to ^{14}C doubly labelled propane 2 with sodium borohydride, and purified by gas chromatography on alumina or silica. Its specific activity was close to the maximal possible specific activity of $^{14}\text{C}_2$-acetylene (i.e. $124.9\,\text{mCi mmol}^{-1}$).

$$\text{Ba}^{14}\text{CO}_3 \xrightarrow{\text{Ba}} \text{Ba}^{14}\text{C}_2 \xrightarrow{\text{HCl}} {}^{14}\text{C}_2\text{H}_2 \xrightarrow[\text{RT, overnight}]{\text{Cr}^{2+},\,\text{H}^+}$$

sp. act. $50–60\,\text{mCi mmol}^{-1}$

$${}^{14}\text{C}_2\text{H}_4 \xrightarrow[150-160\,°\text{C, 1 h}]{\text{HI(sat. water solution)}} {}^{14}\text{C}_2\text{H}_5\text{I} \xrightarrow{\text{Mg}} {}^{14}\text{C}_2\text{H}_5\text{MgI}$$

$$\xrightarrow[\text{dry ice temp.}]{\text{CO}_2,\,\text{Et}_2\text{O}} {}^{14}\text{C}_2\text{H}_5\text{COOH} \xrightarrow[\text{ether, 6 h refl.}]{\text{LiAlH}_4} {}^{14}\text{C}_2\text{H}_5\text{CH}_2\text{OH}$$

$$70–75\%\text{ yield}$$

$$\xrightarrow[\text{NaOH, stir 4h at 5°C}]{p\text{-MeC}_6\text{H}_4\text{SO}_3\text{H/ether}} {}^{14}\text{C}_2\text{H}_5\text{CH}_2\text{OTs} \xrightarrow[\text{under He, 85°C, 3 h}]{\text{NaBH}_4\text{(excess)}}$$

$$80\%\text{ yield}$$

$$\begin{aligned}&{}^{14}\text{CH}_3{}^{14}\text{CH}_2\text{CH}_3 \qquad (12)\\&\qquad\quad\textbf{(2)}\end{aligned}$$

$$60\%\text{ yield, sp. act. } 110\,\text{mCi mmol}^{-1}$$

C. Synthesis of ^{14}C-Labelled Pentane, Heptane and Other ^{14}C- and ^{11}C-Containing Alkanes

1. Synthesis of 2,2,4-trimethylpentane-$^{14}C_2$

This compound (isooctane-$^{14}C_2$), ^{14}C-labelled at both methyl groups of the isopropyl moiety, has been prepared[14] according to equation 13. 3,3-Dimethylbutyric acid, synthesized from vinylidene dichloride, has been utilized as one of the starting materials. The mixture of isomeric diisobutenes, obtained by direct dehydration of the tertiary carbinol-^{14}C intermediate, has been hydrogenated according to Reference 15, to yield **3**.

$$H_2C{=}CCl_2 + Me_3COH + H_2O \xrightarrow[\substack{0-5\,^\circ C,\ 2h\ (-HCl)}]{94\%\ H_2SO_4,\ BF_3\ \text{etherate}} Me_3CCH_2COOH$$

$$\xrightarrow[\substack{10\,h\ reflux,\\ water\ separator}]{EtOH,\ H_2SO_4} Me_3CCH_2COOEt \xrightarrow{2^{14}CH_3MgI,\ Et_2O,\ 1\ h\ reflux}$$

$$Me_3CCH_2C(^{14}CH_3)_2OH \xrightarrow{I_2,\ 1\ h\ reflux}$$

$$Me_3CCH_2C(^{14}CH_3){=}^{14}CH_2 + Me_3CCH{=}C(^{14}CH_3)_2$$

$$\xrightarrow[\substack{EtOH\ solution\ of\ sodium\ borohydride/\\ conc.\ HCl,\ RT,\ 1\ h\ stirring\ work\ up}]{2H/methanolic\ hexachloroplatinic\ acid;\ 1M} Me_3CCH_2CH(^{14}CH_3)_2 \qquad (13)$$
$$\textbf{(3)}$$

b.p. 99–100 °C, 55% chem. and radiochem.
yield relative to diisobutene mixture

2. Synthesis of ^{14}C-labelled heptanes

a. Synthesis of [1-^{14}C]- and [2-^{14}C]-n-heptane. [1-^{14}C]-n-Heptane has been prepared[16] in five steps (equation 14). The first four steps are well documented in the periodical literature and monographs[1,5]. In the last step [1-^{14}C]-n-$C_6H_{13}{}^{14}CH_2Br$ has been added dropwise to lithium aluminium hydride and lithium hydride in tetrahydrofuran, the content refluxed for several hours, cooled, diluted and acidified [2.5 M sulphuric acid) and the product **4** was worked up by microdistillation. **4** has been used to study the structure and composition of micelles formed by dinonylnaphthalenesulphonic acid in n-heptane in the presence of water[17].

$$C_6H_{13}Br \xrightarrow{Mg} C_6H_{13}MgBr \xrightarrow{^{14}CO_2} C_6H_{13}{}^{14}CO_2MgBr \xrightarrow{HCl}$$

$$C_6H_{13}{}^{14}CO_2H \xrightarrow{LiAlH_4} C_6H_{13}{}^{14}CH_2OH \xrightarrow{HBr,\ H_2SO_4}$$

$$C_6H_{13}{}^{14}CH_2Br \xrightarrow[THF]{LiAlH_4,\ LiH} C_6H_{13}{}^{14}CH_3 \qquad (14)$$
$$\textbf{(4)}$$
b.p. 98.4 °C,
90% yield in last step

n-Heptane-1-^{14}C has also been prepared by Pines and coworkers[18,19] by dehydrating n-heptanol-1-^{14}C and subsequent hydrogenation at 150 °C at 100 atm. Mitchell[20] prepared **4** in a similar manner and applied it to study the mechanism of the dehydrocyclization of n-heptane to toluene when 29% of the ^{14}C was found in the methyl group of toluene. Pines and Chen[18] investigated aromatization of n-heptane-1-^{14}C over chromia-alumina and suggested a mechanism involving five-, six- and seven-membered ring intermediates.

Balaban and coworkers[21] obtained **4** and (2-^{14}C)-n-heptane according to equations 15 and 16, using ^{14}CH$_3$I as the source of ^{14}C in the former case. In the latter case, ^{14}CO$_2$ was the source of the isotope.

$$^{14}CH_3I \xrightarrow{Mg} {}^{14}CH_3MgI \xrightarrow[\text{2 h reflux}]{CdCl_2/\text{ether}} (^{14}CH_3)_2Cd \xrightarrow[\text{benzene, 15°C}]{n-C_5H_{11}COCl} \qquad (15)$$

$$n-C_5H_{11}CO^{14}CH_3 \xrightarrow[\substack{\text{Na in diethylene} \\ \text{glycol, 20 h reflux}}]{N_2H_4,} n-C_6H_{13}{}^{14}CH_3$$
$$\textbf{(5)}\ 67\% \text{ yield} \qquad\qquad\qquad \textbf{(4)}$$
$$\text{sp. act. } 14\,\mu\text{Ci mmol}^{-1}$$

$$n-C_5H_{11}Br \xrightarrow[\text{ether, reflux}]{Mg, I} n-C_5H_{11}MgBr \xrightarrow[-20\,°C,\,5\,h\,\text{stirring}]{^{14}CO_2,\,3\,mCi}$$

$$n-C_5H_{11}{}^{14}COOH \xrightarrow[\text{3 h reflux}]{SOCl_2,\,\text{excess}} n-C_5H_{11}{}^{14}COCl \xrightarrow{Me_2Cd}$$

$$\qquad\textbf{(6)} \qquad\quad 151\text{--}153\,°C,\,56\%\text{ yield}$$

$$n-C_5H_{11}{}^{14}COMe \xrightarrow{N_2H_4} n-C_5H_{11}{}^{14}CH_2Me \qquad (16)$$
$$\textbf{(7)}$$
$$25\%\text{ yield, overall, chem. and radiochem.}$$

b. Synthesis of n-heptane (4-^{14}C). This compound, **8**, has been synthesized by reacting n-propylmagnesium bromide with ethyl formate carboxyl-^{14}C. The 4-heptanol-4-^{14}C obtained was acetylated, pyrolyzed to n-heptene and hydrogenated over a platinum catalyst[22-24] (equation 17).

$$H^{14}COOH \xrightarrow{EtOH} H^{14}CO_2Et \xrightarrow{n-PrMgBr} H^{14}C{\overset{\displaystyle OEt}{\underset{\displaystyle OMgBr}{\Big\langle}}}\!\!-Pr-n \xrightarrow{n-PrMgBr}$$

$$H^{14}C{\overset{\displaystyle OH}{\underset{\displaystyle Pr-n}{\Big\langle}}}\!\!-Pr-n \xrightarrow[\substack{2.\,400\,°C \\ 3.\,H_2,\,PtO_2}]{1.\,AcOH} n-Pr-{}^{14}CH_2Pr-n \qquad (17)$$
$$\textbf{(8)}$$

c. 1,4-Dimethylcyclohexane-7-^{14}C. This has been obtained[23,24] from 4-methylcyclohexanone with ^{14}CH$_3$MgI and hydrogenating the product at 100°C and 150 atm over Raney Ni. 1,2-Dimethylcyclohexane-7-^{14}C has been prepared similarly from 2-methylcyclohexanone and ^{14}CH$_3$MgI.

Cyclic hydrocarbons are produced as intermediates in the course of synthesis of ^{14}C-labelled benzene via carboxyl-labelled acids[25] (equation 18).

3. Synthesis of ^{14}C-ring-labelled adamantane derivatives

a. 2-Methyladamantane-2-^{14}C. This has been prepared[26] according to equation 19. Ring expansion of adamantanone **9** with ^{14}C-labelled diazomethane yielded the 4-homoadamantanone **10** which was converted to the diazoketone **11**, followed by photolytic

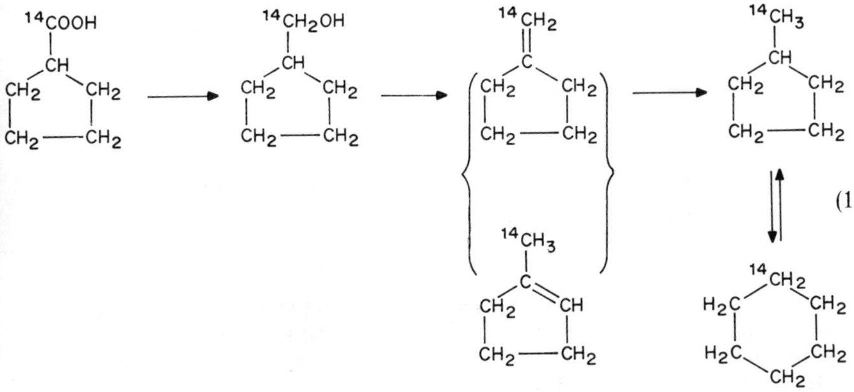

(18)

Wolf ring contraction to 2-adamantanecarboxylic acid-2-^{14}C (**12**) and reduction of the latter resulted in 2-methyladamantane (2-^{14}C), **13** (m.p. of white waxy crystals 144–146 °C, purity greater than 99%, specific activity 8.52 nCi per mgC, 5% yield from ^{14}CH$_2$N$_2$).

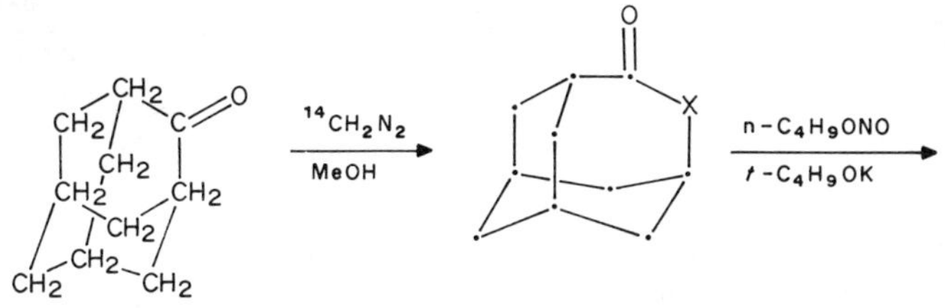

(**9**)

(**10**) 49.8% yield
from ^{14}CH$_2$N$_2$

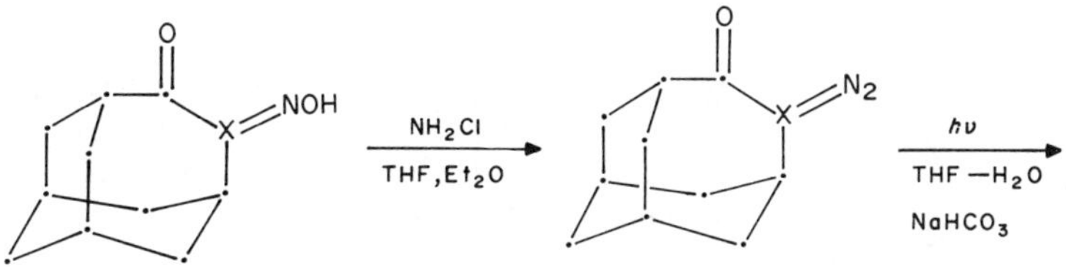

(**11**)

(19)

(·) denotes C, CH or CH$_2$; (X) denotes ^{14}CH$_2$, ^{14}CH or ^{14}C

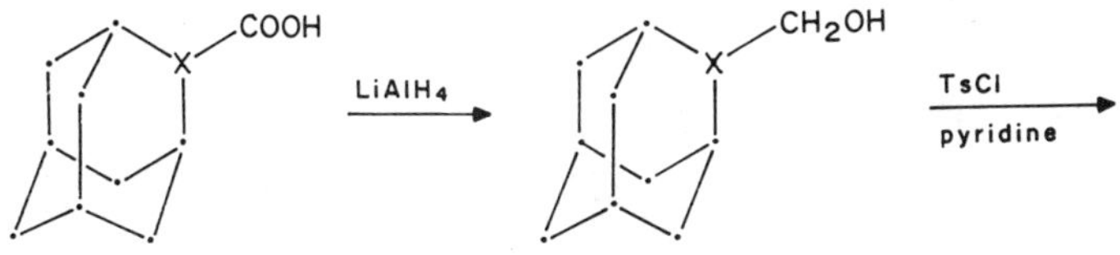

(**12**) 11.5% yield from ^{14}CH$_2$N$_2$

23.1% overall yield from **10**

78.5% yield

(2-adamantyl-2-^{14}C)methanol

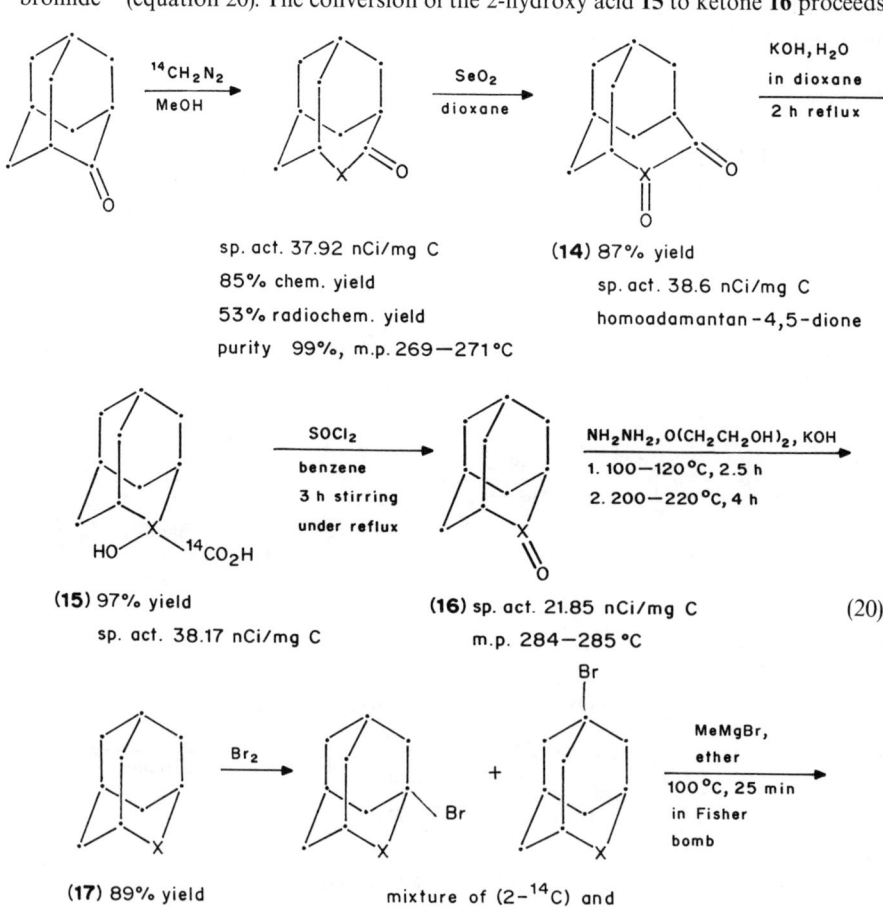

94% yield   (13) 63.5% yield from the tosylate

(2-adamantyl-2-^{14}C)carbinyl

tosylate

b. *1-Methyladamantane-2 or -4-^{14}C*. These have been prepared[27] by ring expansion of adamantanone with diazomethane-^{14}C, oxidation of the 4-homoadamantanone-4-^{14}C to homoadamantane diketone **14**, benzylic acid rearrangement of **14** to the hydroxy acid, **15**, a novel conversion of **15** with SOCl$_2$ to 2-adamantanone-2-^{14}C (**16**) with overall 66% yield, Wolff–Kishner reduction of **16** to adamantane-2-^{14}C (**17**) and conversion of **17** into **18a** and **18b** by treatment of **17** first with bromine and then with methylmagnesium bromide[28] (equation 20). The conversion of the 2-hydroxy acid **15** to ketone **16** proceeds

sp. act. 37.92 nCi/mg C   (14) 87% yield

85% chem. yield   sp. act. 38.6 nCi/mg C

53% radiochem. yield   homoadamantan-4,5-dione

purity 99%, m.p. 269—271°C

(15) 97% yield   (16) sp. act. 21.85 nCi/mg C   (20)

sp. act. 38.17 nCi/mg C   m.p. 284—285°C

(17) 89% yield   mixture of (2-^{14}C) and

sp. act. 21.80 nCi/mg C   (4-^{14}C) isomers, 92% yield

m.p. 268—269°C

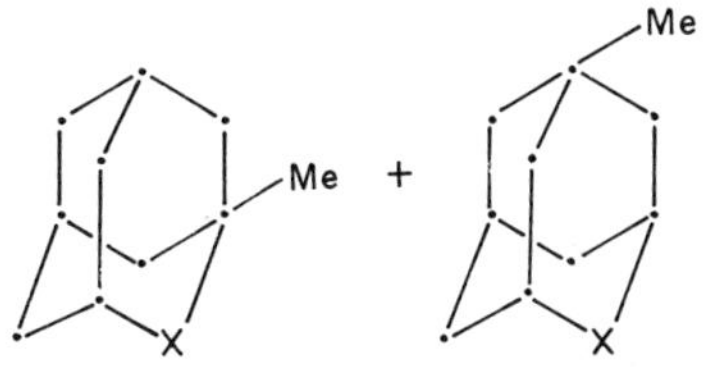

(18a)   **(18b)**
sp. act.
3.52 nCi/mg C
82% yield

probably through the cyclic intermediate **19**, which decomposes according to equation 21. The rearrangement of adamantane-2-^{14}C to adamantane-1-^{14}C taking place[29] in

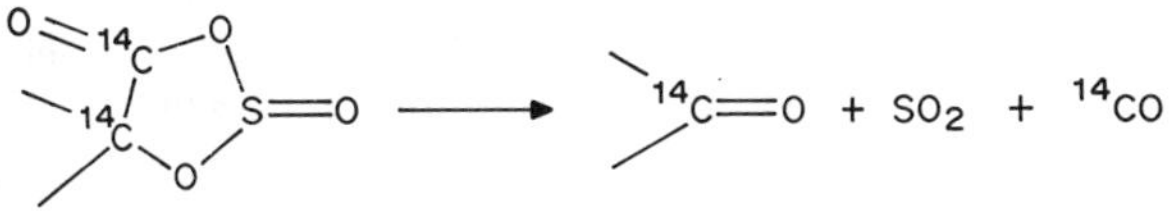

$$^{14}C{=}O + SO_2 + {}^{14}CO \qquad (21)$$

(19)

AlBr$_3$/CS$_2$ at 110 °C (after 8 h) and the corresponding rearrangement of 2-methyladamantane-2-^{14}C studied in AlBr$_3$/CS$_2$ at 250 °C during 45 h[30] have been briefly reviewed by Muccino[31].

4. Synthesis of n-octadecane [1-^{14}C] and n-hexadecane [1-^{14}C]

These compounds have been prepared[32] according to equation 22. In a similar manner Me(CH$_2$)$_{14}$^{14}CH$_2$OH (m.p. 47–49 °C, y. 92–95%), Me(CH$_2$)$_{14}$^{14}CH$_2$I (m.p. 18–20 °C) and

$$Me(CH_2)_{16}{}^{14}CO_2H \xrightarrow[\text{RT, 100 h stirring}]{\text{LiAlH}_4/\text{Et}_2\text{O}} Me(CH_2)_{16}{}^{14}CH_2OH$$

93% yield, m.p. 56–58 °C

$$\xrightarrow[\text{180 °C, 1 h stirring}]{\text{I, red P}} Me(CH_2)_{16}{}^{14}CH_3I \xrightarrow[\text{100 °C, 30 min stirring}]{\text{Et}_2\text{O, Zn, AcOH sat. with HCl}}$$

$$Me(CH_2)_{16}{}^{14}CH_3 \qquad (22)$$
85–95% yield
m.p. 27–29 °C

Me(CH$_2$)$_{14}$^{14}CH$_3$ (76%, b.p. $_{14}$: 156–158 °C) have been obtained starting from palmitic acid (1-^{14}C), Me(CH$_2$)$_{14}$^{14}CO$_2$H. Reduction of cetyl iodide with lithium aluminium hydride yields -n-hexadecane in higher (95%) yield[33] (equation 23):

$$4RX + LiAlH_4 \longrightarrow 4RH + LiAlX_4 \qquad (23)$$

5. Synthesis of hydrocarbons labelled with ^{14}C at 1,2 and 3 positions

These ^{14}C-labelled hydrocarbons have been synthesized[34,35] mostly according to the schemes shown in equations 24–26. The Wittig reaction is also suitable for preparing[35] alkenes and alkanes labelled with ^{14}C in various positions (equations 27 and 28). ^{14}C-labelled alkenes have been reduced to the corresponding alkanes. The yields of

$$^{14}CO_2 \longrightarrow R^{14}COOH \longrightarrow R^{14}CH_2OH \longrightarrow R^{14}CH_2X \longrightarrow R^{14}CH_3 \qquad (24)$$

$$R^{14}CH_2X \xrightarrow{K^{14}CN} R^{14}CH_2{}^{14}CN \longrightarrow$$

$$R^{14}CH_2{}^{14}CH_2OH \longrightarrow R^{14}CH_2{}^{14}CH_2X \longrightarrow R^{14}CH_2{}^{14}CH_3 \qquad (25)$$

$$R^{14}CH_2X \longrightarrow R^{14}CH_2MgX \xrightarrow{(CH_2)_2O} R^{14}CH_2(CH_2)_2OH$$

$$R^{14}CH_2(CH_2)_2OH \longrightarrow R^{14}CH_2(CH_2)_2X \longrightarrow R^{14}CH_2CH_2CH_3 \qquad (26)$$

$$R^{1\ 14}CH_2X \longrightarrow [R^{1\ 14}CH_2PPh_3]X \longrightarrow$$
$$R^{1\,14}\overset{-}{C}H\overset{+}{P}Ph_3 \xrightarrow{R^2CHO} R^1 - {}^{14}CH{=}CHR^2 \qquad (27)$$

$$R^{1\ 14}COOH \longrightarrow R^{1\ 14}CHO \xrightarrow{RCH\overset{+}{P}Ph_3} R^{1\ 14}CH{=}CHR \qquad (28)$$

hydrocarbons in the Wittig reaction is rather low (10%) but it is nevertheless useful for the preparation of alkenes possessing a ^{14}C label and double bond in the same position. The starting alkyl halides (specific activity $2000\,GBq\,mol^{-1}$) have been synthesized according to equation 29. ^{14}C-labelled aldehydes have been produced from ^{14}C-labelled carboxylic acids through the imidazolides reduced in turn[36] with $LiAlH_4$ (equation 30).

$$3\,ROH + PX_3 \longrightarrow HPO(OH)_2 + 3\,RX \qquad (29)$$
$$40\text{--}85\% \text{ yield}$$
$$\text{chemical purity} > 98\%$$

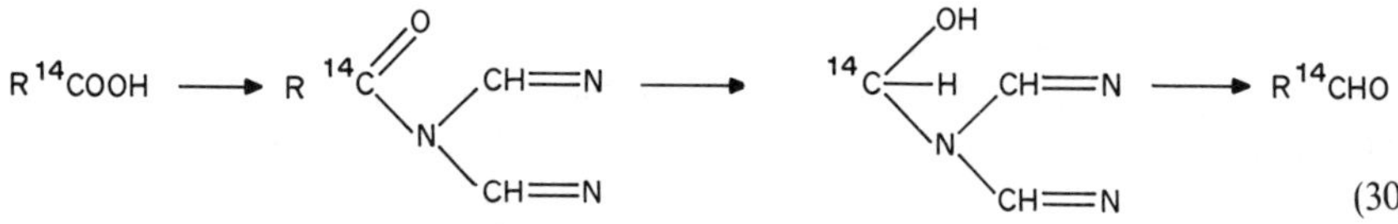

$$(30)$$

6. Synthesis of methyl-^{11}C-labelled hydrocarbons

[1-^{11}C]pentane, [1-^{11}C]nonane, [1-^{11}C]undecane, 2-[^{11}C]methylnaphthalene and [^{11}C]methylbenzene have been obtained by selectively coupling ^{11}C methyl iodide with the appropriate lithium organocuprates[37-39] (equation 31). The lithium dialkylcuprates

$$R_2CuLi \xrightarrow[\text{2. }H^+,\,H_2O]{\text{1. }[^{11}C]H_3I/\text{ether}} R[^{11}C]H_3 \qquad (31)$$

$$R = n\text{-}C_4H_9,\ n\text{-}C_8H_{17},\ n\text{-}C_{10}H_{21}$$

have been prepared[38,39] by reacting cuprous iodide with alkyllithium and the [^{11}C]methyl iodide from [^{11}C]carbon dioxide[40], produced in $^{14}N(p,\alpha)^{11}C$ nuclear reaction. The starting key compounds [^{11}C]O, [^{11}C]O$_2$, H[^{11}C]N[41], [^{11}C]OCl$_2$[42] and H[^{11}C]HO[43,44] are frequently obtained[41-46] directly at the target in the course of nuclear syntheses of [^{11}C]. In the [1-^{11}C]pentane synthesis metal–halogen exchange took place and the by-product [^{11}C]H$_4$ in 20% yield has been produced. The coupling reaction is very slow at low temperatures ($-40\,^\circ$C). The specific activity of the [^{11}C]methyl iodide used was of the order of $10\text{--}30\,mCi\,\mu mol^{-1}$. (The maximal possible specific activity of ^{11}C is $9 \times 10^9\,Ci\,mol^{-1}$.)

[1-^{11}C]nonane

$$\text{CuI/ether} + 2\,C_8H_{17}\text{Li/hexane} \longrightarrow (C_8H_{17})_2\text{CuLi} \xrightarrow[\text{2. hydrolysis}]{\text{1. } [^{11}\text{C]H}_3\text{I}} C_8H_{17}[^{11}\text{C]H}_3$$

$$\tag{31a}$$

[1-^{11}C]undecane. Lithium didecylcuprate, prepared by reacting copper(I) iodide in ether with decyllithium in hexane, has been coupled with [^{11}C]H$_3$I. The above [^{11}C]-labelled aliphatic and aromatic hydrocarbons have been used to study their effects on man, since they are common components in air pollutants. Aliphatic saturated and unsaturated hydrocarbons act also as pheromones and defensive secretions among some social insects such as ants and bees. The ^{11}C-labelled compounds have been applied for *in vivo* studies[45,46].

7. Radiochemical synthesis of ^{11}C-labelled hydrocarbons

Direct interaction of recoil ^{11}C atoms or ^{11}CH fragments, obtained in the ^{12}C(γ, n)^{11}C process with various C$_6$ hydrocarbons, resulted in the formation of ^{11}C-labelled methane, acetylene, ethane + ethylene, propane + propylene and 1,3-butadiene[47a]. Methane-^{11}C and acetylene-^{11}C have been produced in the radiochemical reaction between ^{11}C and liquid benzene, with the yield ratio ($\overset{*}{C}H_4/\overset{*}{C}_2H_2$) being in the range of 0.02 to 0.03.

Synchrotron irradiation of 2,2-dimethylbutane yielded ^{11}C-labelled methane, ethane, ethylene, acetylene, propane, propylene and butadiene. The yields of methane and ethylene + ethane relative to acetylene, extrapolated to zero dose, where 0.12 ± 0.03 and 0.49 ± 0.00, respectively. Iodine does not affect the purely hot atom reactions producing ^{11}CH$_4$ and $\overset{*}{C}_2$ compounds. The yield ratios to acetylene extrapolated to zero dose were found to be 0.15 ± 0.05 for propylene, 0.1 ± 0.05 for propane and 0.1 ± 0.05 for butadiene.

Synchrotron irradiation of 2-methylpentane produced ^{11}C-labelled methane, ethane + ethylene, acetylene, propane + propylene and 1,3-butadiene. The yield ratios of methane and ethane + ethylene to acetylene were 0.16 ± 0.01 and 0.40 ± 0.04, respectively, while the yield ratios for propane + propylene and butane were 0.097 ± 0.015 and 0.15 ± 0.02, respectively.

Six synchrotron irradiations of n-hexane[47b] with initial iodine concentration of 0.0061 mole fraction yielded ^{11}C-labelled methane, ethane + ethylene, acetylene, propane + propylene and 1,3-butadiene. Yields of methane and ethylene + ethane relative to acetylene were 0.19 ± 0.01 and 0.25 ± 0.02, respectively. The yield ratio was 0.059 ± 0.008 for propane + propylene and 0.14 ± 0.03 for butadiene. The (C$_2$H$_4$ + C$_2$H$_6$)/C$_2$H$_2$ yield ratios observed in the case of propane[48a] and isobutane[49a] were equal to 0.48 and 0.47, respectively. The above observations fit a mechanism which involves the insertion of an energetic ^{11}C or ^{11}CH fragment into a C—H bond, and formation of an intermediate decomposing into two-carbon or three-carbon stable products. Thermal reactions of the ^{11}C group occur mainly with surrounding molecules, since the concentration of the radicals is small and iodine as scavenger cannot be used to distinguish between hot and thermal reactions. The relative experimental yields of CH$_4$, C$_2$H$_4$ + C$_2$H$_6$ and C$_3$H$_6$ extrapolated to zero dose are therefore independent of the iodine concentration. In benzene and in cyclohexane all the C—H bonds are identical and only one kind of intermediate is formed, designated 'C$_6$H$_5$CH:' and 'C$_6$H$_{11}$CH:'. In n-hexane, 2-methylpentane and 2,2-dimethylbutane different intermediates are possible depending on the particular C—H bond attacked. The yields of ^{11}C-labelled products increase in going from benzene to cyclohexane to n-hexane, parallelling the hydrogen availability on carbon. The number of hydrogen-rich Me sites increases with increase of branching in the hexane isomers and the observed $(C_2H_4 + C_2H_6)/C_2H_2$ ratio increases in the sequence: benzene < cyclohexane < n-hexane < 2-methylpentane < 2,2-dimethylbutane. The above insertion is to a great extent a statistical process with little chemical selectivity. The methane yield from several hydrocarbons is relatively constant, which implies that $^{11}CH_4$ is produced in the hot region by a series of abstraction reactions as the ^{11}C fragments cool down.

Irradiation of gaseous NH_3 with 10-MeV protons (using a 60-in. cyclotron) resulted in ^{11}C production by a $^{14}N(p, \alpha)$ ^{11}C nuclear process, and in direct synthesis of $^{11}CH_4$ and $^{11}CH_3NH_2$ (major reaction products)[48b].

D. Synthesis of Dodecane-1,12-$^{13}C_2$ and Hexadeuteriododecane

Dodecane-1,12-^{13}C has been synthesized[49b] by the procedure shown in equation 32. 1,10-Dibromodecane treated with cyanide-^{13}C in aqueous ethanol[50] gave the dinitrile which was hydrolyzed *in situ* to the acid, **20**, and in turn reduced with a borane–dimethyl sulphide complex[51,52] to 1,12-dodecanediol-1,12-^{13}C in 90% yield. The bis-tosylate of **20** has then been reduced to the required dodecane (1,12-^{13}C) in 30% yield, enriched with ^{13}C up to 88.9%.

$$BrCH_2(CH_2)_8CH_2Br \xrightarrow[\substack{1.\ EtOH/H_2O,\ 5\,h\ reflux \\ 2.\ HCl(conc.),\ 12\,h\ reflux}]{Na^{13}CN} HOO^{13}C(CH_2)_{10}{}^{13}COOH$$

(**20**) 86% yield,
m.p. 127–128 °C

$$\xrightarrow[RT,\ 8\,h]{(Me_2S\cdots BH_3)/THF,\ 25\,°C} HO^{13}CH_2(CH_2)_{10}{}^{13}CH_2OH \xrightarrow[py,\ 0\,°C,\ 24\,h]{p\text{-}MeC_6H_4SO_2Cl}$$

90% yield,
m.p. 81–83 °C

$$TsO^{13}CH_2(CH_2)_{10}{}^{13}CH_2OTs \xrightarrow[2\ days\ reflux]{LiAlH_4/ether} {}^{13}CH_3(CH_2)_{10}{}^{13}CH_3 \qquad (32)$$

80% yield, m.p. 90–91 °C

(**21**) 212–216 °C, b.p. at 760
mole fractions: ^{13}C—2.2%,
$^{13}C_1$—19.4%, $^{13}C_2$—78.4%

The reduction of a 1,6-dicarboxylic acid dimethyl ester to the corresponding hydrocarbon had been described[1]. In a similar manner hexadeuteriododecane (**22**) has been prepared[49b] (equation 33). The final LAD reduction yielded dodecane-1,1,1,12,12,12-$^2H_6$ (**22**) which contained, according to mass spectra: $^2H_0$—0%, $^2H_1$—0%, $^2H_2$—0.9%, $^2H_3$—0.9%, $^2H_4$—0.9%, $^2H_5$—6.3%, $^2H_6$—90.7%.

$$HOOC(CH_2)_{10}COOH \xrightarrow{MeOH/HCl} MeO_2C(CH_2)_{10}CO_2Me \xrightarrow{LiAlD_4}$$

$$HOCD_2(CH_2)_{10}CD_2OH \xrightarrow{TsOH} TsOCD_2(CH_2)_{10}CD_2OTs$$

$$\xrightarrow[ether,\ 8\,h\ reflux]{LiAlD_4} CD_3(CH_2)_{10}CD_3$$

(**22**)   (33)

E. Syntheses of Deuteriated Alkanes

1. Synthesis of n-decane-d_{22}

The synthesis of n-decane-d_{22}, needed for neutron diffusion studies, has been elaborated[53] by optimization of a patented process[54–56] based on H/D isotopic exchange over Pd/C catalyst. In a typical experiment 142 g of decane and 20 g of Pd on carbon (5%) were used. After 995 h at 165 °C, deuteriation of 99.4–99.7% has been reached as measured by MS, NMR and densimetry.

2. Synthesis of 1,1-dideuterio- and 2,2-dideuterio-4-phenylbutane

a. Synthesis of 2,2-dideuterio-4-phenylbutane (**23**). This has been undertaken to study the fragmentation of deuteriated alkylchromium compounds. **23** has been prepared[58] using LAD in the reduction steps (equation 34). The product **23** contained 2,2-dideuterio-4-phenylbutane (96.3%), 2-deuterio-4-phenylbutane (3.5%) and a trideuterio species (0.2%).

$$PhCH_2CH_2CO_2Et \xrightarrow[\text{2 h reflux}]{LiAlD_4} PhCH_2CH_2CD_2OH \xrightarrow[-5\,°C\ to\ 20\,°C]{PBr_3}$$

$$PhCH_2CH_2CD_2Br \xrightarrow[\text{2. } CO_2,\ -70\,°C]{1.\ Mg} PhCH_2CH_2CD_2COOH \xrightarrow{LiAlH_4}$$

$$\text{m.p. 48–49 °C}$$

$$PhCH_2CH_2CD_2CH_2OH \xrightarrow{PBr_3} PhCH_2CH_2CD_2CH_2Br$$

$$\xrightarrow[\text{2. } H_2O]{1.\ Mg/THF} PhCH_2CH_2CD_2CH_3 \qquad (34)$$

$$\text{(23) 99.2\% chem. purity}$$

$$Ph(CH_2)_3CO_2Et \xrightarrow{LiAlD_4} Ph(CH_2)_3CD_2OH$$

$$\xrightarrow{PBr_3} Ph(CH_2)_3CD_2Br \xrightarrow[\text{2. } H_2O]{1.\ Mg/THF} Ph(CH_2)_3CD_2H \qquad (35)$$

$$\text{(24) 99.4\% chem. purity}$$

b. 1,1-Dideuterio-4-phenylbutane (**24**). This has been obtained similarly in the reaction sequence shown in equation 35. The product **24** contained 1,1-dideuterio-4-phenylbutane (97.2%) and monodeuterio-4-phenylbutane (2.8%). No H/D exchange took place during the formation or hydrolysis of the organomagnesium halide intermediate compounds or during age chromatographic separations of the hydrocarbons obtained, nor in the reaction of deuterated alkylmagnesium bromides with $CrCl_3(THF)_3$ (equation 36). In the course of methanolysis of solvated trialkylchromium compounds **25** with oxygen-free methanol at $-40\,°C$, the 2,2-dideuterio- and 1,1-dideuterio-4-phenylbutanes have been obtained (as shown by gas chromatographic analysis followed by mass spectrometry and NMR spectroscopy).

$$3\,RMgX + CrCl_3(THF)_3 \xrightarrow[-70\,°C\ to\ -40\,°C]{THF} R_3Cr(THF)_n \qquad (36)$$
$$\text{(25)}$$

$$R = PhCH_2CH_2CD_2CH_2 \text{ or } Ph(CH_2)_3CD_2$$

3. Synthesis of dideuteriomethylene hydrocarbons by copper catalysed reaction of Grignard reagents

Hydrocarbons containing a single specific methylene group labelled with deuterium have been needed for a vibrational study of chain conformations and organizational changes occurring during phase transitions[59]. Dilithium tetrachlorocuprate has been found to catalyse effectively the coupling[60] of Grignard reagents with alkyl halides and long-chain alkyl *p*-toluenesulphonates[61]. Lithium dimethyl cuprate, Me_2CuLi, has also been used for preparation of deuteriated hydrocarbons[62,63]. Deuteriated toluenesulphonates of the desired chain length are easily prepared from esters of fatty acids by $LiAlD_4$ treatment followed by tosylation in pyridine. Equation 37 has been applied[64] to synthesize $n\text{-}C_5H_{11}CD_2CH_3$, $n\text{-}C_{10}H_{21}CD_2CH_2CH_3$, $n\text{-}C_{19}H_{39}CD_2CH_3$, $n\text{-}C_{17}H_{35}CD_2C_3H_7$, $n\text{-}C_{15}H_{31}CD_2C_5H_{11}\text{-}n$ and $n\text{-}C_{10}H_{21}CD_2C_{10}H_{21}\text{-}n$. The yields of heneicosanes deuteriated in the 2-, 4-, 6- and 11-positions were 85%, 74%, 50% and 83%, respectively.

$$RCO_2CH_3 + LiAlD_4 \xrightarrow{\text{RT, 2h stirring}} RCD_2OH \xrightarrow[T < 15\,°C]{\text{TsCl, Py}}$$

$$RCD_2OTs \xrightarrow[\text{6h stir at RT}]{R^1MgX/THF/,\ Li_2CuCl_4/THF} RCD_2R^1 \tag{37}$$

$$R^1 = Me,\ Et,\ n\text{-}C_3H_7,\ n\text{-}C_5H_{11}\ \text{or}\ n\text{-}C_{10}H_{21}$$

4. Deuteriation of symmetrical diphenylethane and trans-stilbene

1,2-Diphenylethane; hexestrol or 3,4-bis(*p*-hydroxyphenylhexane-n (**26**); *erythro-* and *threo*-4,4′-dihydroxy-α-ethyl-α′-methyl-1,2-diphenylethanes have been deuterium and tritium labelled by catalytic exchange with isotopic water in the presence of platinum oxide pre-reduced with sodium borohydride[65]. Both aromatic hydrogens and alkyl hydrogens were exchanged. Methoxy groups on the aromatic rings create the steric hindrance diminishing the hydrogen exchange in adjacent positions. Aromatic hydrogen

$$HOC_6H_4CHEt\text{---}CHEtC_6H_4OH$$

(**26**)

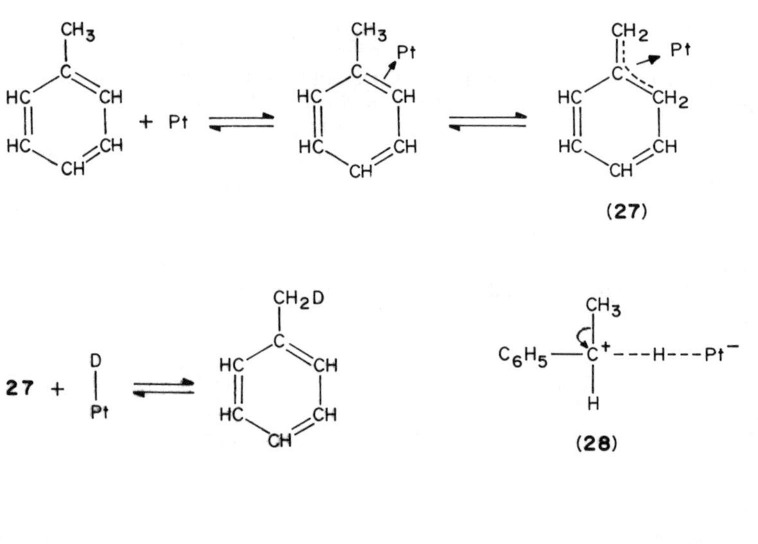

(**27**)

(37a)

(**28**)

exchange proceeds according to a π-complex adsorption mechanism[65]. An allylic mechanism (equation 37a) proposed to explain the deuteriation in the methyl group, and an electrophilic radical abstraction in intermediate **28** contribute to the overall hydrogen exchange observed in these synthetic estrogens. The extensive exchange of the substituents on the 1,2-carbon atoms of 1,2-diphenylethane is attributed to rotation about the single bond joining these atoms. At certain stages of the rotation, the steric hindrance caused by the aromatic rings disappears and the aliphatic group can be bonded to the catalyst. In 1,2-diphenylethane the aliphatic hydrogens underwent 90.8% deuteriation. The intramolecular distribution of deuterium both in labelled *trans*-stilbene and in 1,2-diphenylethane derivatives has been presented[65].

5. Deuteriated hydrocarbons used in gas-chromatographic studies

The following deuteriated paraffins have been prepared by exchange in D_2O or with D_2 over Ni, or by standard chemical methods, and used[66] for determination of gas-chromatographic retention indices I_r:

1. Hexane-1,2,3-d_7, hexane-d_{14}, 2-bromohexane-1,2,3-d_6, 2-C_2H_5-hexane-d_{13}, 2-C_2D_5-hexane.
2. Cyclohexane-1,2,3-d_6, cyclohexane-d_{12}, methylcyclohexane-d_3, methylcyllohexane-d_{14}.
3. Heptane-1-d_3, heptane-1,2,3-d_7, heptane-1,2,3,4-d_9, heptane-1,2,3,4,5,6-d_{13}, heptane-d_{16}, 2-methylheptane-d_{18}.
4. Octane-1,2-d_5, octane-1,2,3-d_7, octane-1,2,3,4-d_9, octane-1,2,3,4,5,6-d_{13}, octane-d_{18}, 3-octanone-2,4-d_4, 2-octanone-1,3-d_5, 3-bromooctane-2,3,4-d_5.
5. Dodecane-d_{26}.

The temperature dependences of the retention indices, I_R, have been calculated with the use of equation 38:

$$I_R = a + bT \qquad (38)$$

The temperature dependences of the differences (Equation 39) and of their ratios ($\Delta I/D$), where D is the number of deuterium atoms within the molecule have been calculated according to equation 40, where

$$\Delta I = I_{R(\text{protiated compd.})} - I_{R(\text{deuteriated compd.})} \qquad (39)$$

$$\Delta I_R = A - B/T^2 = A(1 - T_1^2/T^2) \qquad (40)$$

T_1 is an "inversion temperature". Below this temperature the non-deuteriated compound is eluted first. The deuteriated compounds always have smaller temperature coefficients of the retention indices than the protiated ones. There exists therefore a crossing temperature. The values of $\Delta I_R/D$ have been found to be constant for one class of substances and a particular stationary phase (for instance, $\Delta I_R/D$ amounts to 0.73 ± 0.06 using **MBMA** (= m-bis(phenoxyphenoxy)benzene + Apiezon grease L) for deuteriated isomeric dodecanes, nonanes and octanes). The value of α equals zero if we neglect the anharmonicity of low-frequency vibrations caused by expansion of the crystal lattice with temperature and the dependence of the frequency shift of compounds in condensed phases on the temperature[67-69a]. Van Hook[69b-d] simplifies the Bigeleisen relation (equation 41) while discussing the temperature dependence of gas-chromatographic isotopic separations by taking $\alpha = 0$ (α has generally a small value) and introducing temperature-dependent force constants[69c,d].

$$\ln(P_1/P_2) = \alpha - \beta/T + \gamma/T^2 \qquad (41)$$

The difference ΔI has also been expressed in the form of equation 42 and the coefficients α, β and γ were tabulated for hexane-d$_7$, heptane-d$_{16}$, cyclohexane-d$_6$, cyclohexano-d$_{17}$, methylcyclohexane-d$_3$ and -d$_{14}$ and for benzene-d$_6$. Equation 42 has also been transformed for $\alpha = 0$ into straight-line dependence

$$\Delta I = \alpha - \beta/T + \gamma/T^2 \tag{42}$$

$$T \cdot \Delta I_2 = A' - B'/T = A'(1 - T_1'/T) \tag{43}$$

where T_1' is the inversion temperature.

The gas-chromatographic results have been used to determine the differences between the thermodynamic quantities of deuterated[70a] and protiated hydrocarbons with the help of the equation

$$10^{-2}\Delta I_R \cdot b = (-\Delta H_D + \Delta H_H)/2 \cdot 3RT + (\Delta S_D - \Delta S_H)/2.3R \tag{44}$$

where $b = \log(V_{n+1}/V_n)$. Here V_n is the specific retention volume of the standard paraffin having n carbon atoms; ΔI_R is taken as a positive quantity. For hexane-d$_7$ and hexane-d$_{14}$ the $(\Delta H_H - \Delta H_D)$ values are equal to 84 ± 4 cal mol^{-1} and 167 ± 7 cal mol^{-1}, while the $(\Delta S_H - \Delta S_D)$ values are 0.18 ± 0.01 cal deg^{-1} mol^{-1} and 0.37 ± 0.02 cal deg^{-1} mol^{-1}, respectively. For heptane-d$_{16}$ the corresponding values are 181 ± 8 cal mol^{-1} and 0.39 ± 0.02 cal deg^{-1} mol^{-1} (with squalane as the stationary phase). Thermodynamic quantities calculated for other deuterated hydrocarbons with the use of equation 44 are of similar magnitudes. For benzene-d$_6$, $(\Delta H_H - \Delta H_D) = 55 \pm 2$ cal mol^{-1} and $(\Delta S_H - \Delta S_D) = 0.12 \pm 0.01$ cal deg^{-1} mol^{-1} (also for squalane). The thermodynamic quantities estimated for MBMA differ remarkably from those estimated for the squalane stationary phase[66] for all hydrocarbons investigated.

Hydrogen isotope exchanges of organic compounds including hydrocarbons in dilute acid at elevated temperatures were systematically investigated by Werstiuk and coworkers[70-72]. The procedures described can be applied for perdeuteriation of a variety of organic compounds, as well as for specific deuteriations, for instance, of alkylated derivatives of benzene, since exchange of the aromatic hydrogens is much faster than that of the aliphatic hydrogens.

F. Synthesis of Tritium-labelled Long-chain Alkanes

1. Synthesis of n-hexadecane-1-T

Tritium- and carbon-14-labelled n-hexadecane is frequently used for efficiency determinations in liquid scintillation counting. n-Hexadecane-1,2(n)-T is usually prepared by catalytic reduction of n-hexadec-1-ene with tritium gas[73]. Tritiated n-hexadecane has also been obtained by exchange between tritiated water and n-hexadecane in the presence of a Co–Mo–S catalyst[74]. Reduction of cetyl iodide, $CH_3(CH_2)_{15}I$, with Zn in AcOH has been found to be rather difficult[75]. However, by reduction of cetyl iodide, cetyl bromide, sodium-n-hexadecyl sulphate and n-hexadecyl p-toluenesulphonate with LiAlH$_4$[76] about 97% yields are obtained when the molar ratio of RX to LiAlH$_4$ is $1:0.75$, and yields of 100% have been achieved with a $1:1$ molar ratio of the reactants. Thus, n-hexadecane-1-T has been synthesized by reduction of cetyl bromide with LiAlT$_4$ (5 mCi) in THF (reflux, 26 h) yielding n-hexadecane-1-T, $CH_3(CH_2)_{14}CH_2T$, with a specific activity of $14.0 \mu Ci\,ml^{-1}$.

2. Synthesis of tritium-labelled triacontane $(C_{30}H_{62})$, hexacosane $(C_{26}H_{54})$ and docosane $(C_{22}H_{46})$ of high specific activity

These tritiated hydrocarbons, needed to study hydrocarbon utilization by microorganisms, have been prepared by simultaneous Kolbe[77] electrolysis of mixtures of T-labelled sodium palmitate (**29**) and sodium laureate (**30**) (equation 45). **29** and **30** have been tritium-labelled[77] by exchange reaction with tritiated water, giving a mixture of **29** and **30** with a specific activity of $0.04\,Ci\,mmol^{-1}$. After the Kolbe electrolysis the hydrocarbon mixture had a specific activity of $0.02\,Ci\,mmol^{-1}$, containing **31**, **33** and **32** in 1:1:2 ratio, as expected. In the Kolbe electrolysis of the carboxylate, coupling of two hydrocarbon radicals (equation 46) yields the desired hydrocarbon R_2. If two different acids are used, three hydrocarbons are obtained in the statistically expected proportions (equation 47).

$$4\,C_{15}\overset{*}{H}_{31}CO_2Na + 4\,C_{11}\overset{*}{H}_{23}CO_2Na \xrightarrow[\substack{100\,V,\,0.9\,A,\,RT,\,ca\,4\,h \\ \text{platinum electrodes}}]{CH_3OH/CH_3ONa}$$

$$\text{(29)} \qquad\qquad\qquad \text{(30)}$$

$$C_{22}\overset{*}{H}_{46} + C_{30}\overset{*}{H}_{62} + 2\,C_{26}\overset{*}{H}_{54} + 8\,CO_2 + 8e \quad (45)$$

$$\text{(33)} \qquad\quad \text{(31)} \qquad\quad \text{(32)}$$

The asterisk * denotes a tritium label

$$2\,RCO_2^- \xrightarrow{\text{electrolysis}} R\text{—}R + 2\,CO_2 + 2e \quad (46)$$

$$4\,R^1CO_2^- + 4\,R^2CO_2^- \longrightarrow R^1\text{—}R^1 + 2\,R^1\text{—}R^2 + R^2\text{—}R^2 + 8\,CO_2 + 8e \quad (47)$$

3. Synthesis of tritium-labelled dotriacontane with high specific activity

[14]C-Labelled dotriacontane has been used[78] to study the retention of tobacco smoke in the respiratory system of smoke-exposed laboratory animals. High-resolution autoradiography studies within the respiratory tract needed tritium-labelled dotriacontane (**34**) with a specific activity of $0.5\,Ci\,mmol^{-1}$. This compound has been prepared[79] by synthesizing the diacetylene **35** and by its reduction to dotriacontane-15,15,16,16,17,17,18,18-^{3}H, **34** (equation 48). Sodium acetylide with myristyl bromide yielded hexadec-1-yne **36**. Oxidative coupling of **36** using cupric acetate gave the diyne **35** which, upon reduction with tritium gas in the presence of Adams catalyst, yielded **34**, with the specific activity of $0.50\,Ci\,mmol^{-1}$ as required.

$$HC\equiv CH \xrightarrow[\text{in liq. NH}_3,\,4\,h]{1.\,Fe(NO_3)_3\,cat.,\,2.\,Na} HC\equiv CNa$$

$$\xrightarrow[65\,°C\text{ under }N_2,\,3\,h]{n\text{-}C_{14}H_{29}Br,\,DMF} CH_3(CH_2)_{13}C\equiv CH + NaBr \quad (48a)$$

$$\text{(36) reddish brown}$$

$$2 \times 36 \xrightarrow[\substack{MeOH/pyr/Et_2O \\ 4\,h\text{ reflux}}]{CuOAc\text{ anhydr.}} Me(CH_2)_{13}C\equiv C\text{—}C\equiv C(CH_2)_{13}Me$$

$$\text{(35) platelets, m.p. }47\,°C$$

$$\xrightarrow[\text{dioxane}]{\overset{*}{H}_2(5\,Ci\text{ of tritium})/Pt\text{ in MeOH}} Me(CH_2)_{13}(C\overset{*}{H}_2)_4(CH_2)_{13}Me \quad (48b)$$

$$\text{(34)}$$

while platelets, m.p. 69.7°C, chem. purity 99% confirmed by mass spectrometry and gas–liquid chromatography; 1.02 Ci total activity of crude product

4. Use of recoil energy for synthesis of ^3H-labelled hydrocarbons

Highly energetic tritium atoms used for labelling of solid and liquid organic materials are obtained in the nuclear reaction $^6Li(n, \alpha)T$. Irradiation of $LiCO_3$ with neutrons initiates the double reaction (n–t–n). Tritium produced in reaction $^6Li(n, \alpha)T$ yields ^{18}F with oxygen-16. The fluorine-18 $(T_{1/2} = 112\,min)$ emits positrons. Irradiation of $LiBO_2$ yields, besides T, also ^{13}N: $^{10}B(\alpha, n)^{13}N$. Tritium atoms used for irradiation of gaseous organic targets are generated in the (n, p) reaction with helium-3, $^3He(n, p)^3H$. Irradiation of 1 g of Li of natural isotopic composition in the 10^{12} neutrons $cm^{-2}s^{-1}$ flux during 1 h yields 1 μCi of ^3H. In the course of irradiation of the mixture of the lithium salt with the organic target, about 30–50% of the tritium atoms are stabilized in the form of molecular hydrogen, 10–50% of ^3H replace the hydrogen in the parent molecule and 5–24% stabilize as the products of destruction of the parent molecules or new, more complex, products[80]. For instance, during the irradiation of cyclohexane with hot tritium atoms, besides tritium-labelled cyclohexane, various unsaturated compounds shown in equation 49 are also obtained[81a]. Irradiation of aliphatic chain hydrocarbons with hot tritium atoms

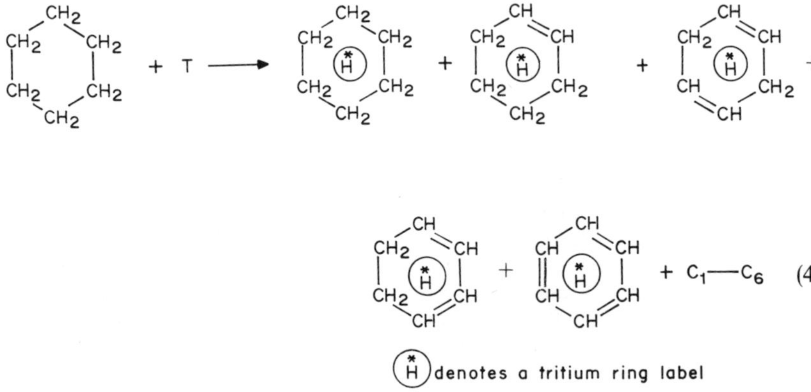

$$\text{(cyclohexane)} + T \longrightarrow \text{(tritiated cyclohexane, } \overset{*}{H}) + (\overset{*}{H}) + (\overset{*}{H}) +$$

$$(\overset{*}{H}) + (\overset{*}{H}) + C_1\!-\!C_6 \qquad (49)$$

$\overset{*}{H}$ denotes a tritium ring label

results in the formation of tritium-labelled parent molecules[81b] and of hydrocarbons with shorter aliphatic chain (equation 50). The yield in the hot tritium atom reactions is not sensitive to the temperature. For instance, the yield of HT, CH_3T and CH_3CH_2T obtained in the reaction of hot T with CH_4 was equal to 61%, 30.6% and 2.5%, respectively at room temperature and these values changed only slightly to 61.2%, 31.2% and 2.1% at 200 °C. On the other hand additives which change the average energy of hot tritium atoms very markedly the yields of the products. For instance, addition of helium to cyclopropane diminished the yield of C_3H_5T from 22% to 11%. However, only changes in the average energy of tritium atoms in the 1–100 eV interval are important, and the yields are not sensitive to the initial energies of the hot tritium atoms. Thus, the yield of CH_3T obtained in the irradiation of methane with $T(E_0 = 2.7\,MeV)$ atoms generated in the $^6Li(n, \alpha)T$ reaction equals 30.5%, and the yield was similar (31.8%) when T atoms $(E_0 = 0.2\,MeV)$ produced in the $^3He(n, p)T$ nuclear reaction were used.

$$CH_3(CH_2)_2CH_3 + T \longrightarrow CH_3CH_2CHTCH_3 \longrightarrow CH_3CH_2CH_2T$$
$$\longrightarrow CH_3CH_2T \longrightarrow CH_3T \qquad (50)$$

Radiolysis of the parent molecules is smaller when a heterogeneous paste consisting of Li_2CO_3 (or other lithium derivatives) mixed with the compound to be labelled is neutron-irradiated. In this case α-particles and tritium atoms leave the solid and enter the liquid

60 (tritium concentrates in CH_2T^+ ion). About one-third of the excited $C\overset{*}{H}_5{}^+$ ions from tritiation of CH_4 with $^3He^3H^+$ are stabilized by collision in methane at 1 atm.; the rest dissociate to produce methyl ions. The reaction of tritiated ethyl ions with traces of water is probably responsible for production of tritiated ethylene (equation 63) in the absence of added propane.

$$C_2\overset{*}{H}_5{}^+ + H_2O \longrightarrow C_2\overset{*}{H}_4 + \overset{*}{H}_3O^+ \qquad (63)$$

The tritium exchange between $^3H_2$ gas and CH_4 gas at room temperature, both alone and in the presence of He, Ne, Ar, or Kr, Xe, I_2 or NO added, has been studied by Pratt and Wolfgang[94b] and the rate R of the formation of radioactive methane was found to be $R = 8 \times 10^7 a^{3/2}$ for unscavenged experiments and $R = 9.2 \times 10^6 a + 7.8 \times 10^5 a^2$ for scavenged samples, where a is the activity of $^3H_2$ in $mCi\,cm^{-3}$. The a-term is attributed to the processes $^3H_2 \longrightarrow {}^3He^3H^+ + e$ and $^3He^3H^+ + CH_4 \longrightarrow CH_3{}^3H$, the $a^{3/2}$-term to the $CH_4{}^+ + {}^3H_2 \longrightarrow CH_3{}^3H$ reaction and the a^2-term to the exchange with radiation-activated $^3H_2$ or to reactions involving radicals or excited states. Besides tritium-labelled methane, labelled ethane, ethylene, propane, i-C_4H_{10}, n-C_4H_{10}, MeI and higher hydrocarbons have also been isolated.

2. Reactions of $^3He^3H^+$ ions with ethane

Tritiated methane and ethane are produced in 14% and 34–36% yields, respectively, in the reaction of $^3He^3H^+$ ions both with pure ethane as well as with ethane containing 2 mol% of oxygen[92b]. The low yield of tritiated ethane can be explained by the formation of excited $C_2H_6{}^3H^+$ ions which decompose into labelled ethyl ions (without any appreciable tritium isotope effect)[95] (equations 64a–c) or by the simultaneous exothermic hydride ion transfer reaction (equation 65) competing with the tritiation of ethane (equation 64a), since in the reaction (equation 66) of the stabilized tritiated ion $C_2\overset{*}{H}_7{}^+$ no activity is lost as hydrogen tritide.

$$^3He^3H^+ + C_2H_6 \longrightarrow (C_2\overset{*}{H}_7)^+_{exc.} + {}^3He \qquad (64a)$$

$$(C_2\overset{*}{H}_7)^+_{exc.} \longrightarrow C_2\overset{*}{H}_5{}^+ + \overset{*}{H}_2 \qquad (64b)$$

$$C_2\overset{*}{H}_5{}^+ + C_2H_6 \longrightarrow C_2\overset{*}{H}_6 + C_2H_5{}^+ \qquad (64c)$$

$$C_2H_6 + {}^3He^3H^+ \longrightarrow C_2H_5{}^+ + He + {}^3H^1H \qquad (65)$$

$$C_2\overset{*}{H}_7{}^+ + C_2H_6 \longrightarrow C_2\overset{*}{H}_6 + C_2H_7{}^+ \qquad (66)$$

The excited tritiated $(C_2\overset{*}{H}_7)^+_{exc.}$ ions have enough excitation energy to dissociate through other reaction pathways energetically less favourable than that in equation 64b, leading to the observed tritiated methane formation (equation 67):

$$(C_2\overset{*}{H}_7)^+_{exc.} \longrightarrow C\overset{*}{H}_3{}^+ + C\overset{*}{H}_4 \qquad (67)$$

$$C\overset{*}{H}_3{}^+ + C_2H_6 \longrightarrow C\overset{*}{H}_4 + C_2H_5{}^+$$

3. Reactions of $^3He^3H^+$ ions with propane

Tritiated methane (13.9%) and propane (14.6%) are the major products in the reaction of $^3He^3H^+$ ions with propane. Smaller amounts of ethane (4.7%), ethylene (4.5%) and traces of butane have also been produced. The above yields are insensitive to the addition of 2 mol% of oxygen as a radical scavenger and to a decrease in pressure from 760 to 20 torr (except for an increase in the methane yield from 14 to 18%). These results have been explained[96] by equations 68–71.

$$^3He^3H^+ + C_3H_8 \longrightarrow {}^3He + (C_3\overset{*}{H}_9)^+_{exc.} \qquad (68)$$

$$100\%$$

$$(C_3\overset{*}{H}_9)^+_{exc.}
\begin{cases}
\xrightarrow[A]{20\%} \begin{cases} C_2H_5^+ + C\overset{*}{H}_4(15\%) & (a) \\ C_2\overset{*}{H}_5^+ (5\%) + CH_4 & (b) \end{cases} \\[2em]
\xrightarrow[B]{80\%} \begin{cases} C_3\overset{*}{H}_7^+ (15\%) + H_2 & (a) \\ C_3H_7^+ + \overset{*}{H}_2 (65\%) & (b) \end{cases}
\end{cases}$$

(69)

(70)

$$C_3\overset{*}{H}_7^+ + C_3H_8 \xrightarrow{a} C_3H_7^+ + C_3\overset{*}{H}_8 (15\%)$$

(71)

$$C_2\overset{*}{H}_5^+ + C_3H_8 \xrightarrow{b} C_3H_7^+ + C_2\overset{*}{H}_6$$

The combined yield of all identified tritiated products accounted for only 30–40% of the $^3He^3H^+$ ions produced, owing to the probable formation of large amounts of hydrogen tritide which has been difficult to determine due to its presence as impurity in the molecular tritium. In addition, no long-lived $C_3\overset{*}{H}_9^+$ intermediate seems to be formed in the initial protonation step. Reaction 71a is probably the single channel which leads to the formation of tritiated propane. The high yield of tritiated methane in reaction 69a and the low yield of tritiated ethyl radical in reaction 69b indicate the non-statistical distribution of tritium within the excited $(C_3\overset{*}{H}_9)^+_{exc.}$ ion 37 (equation 72).

$$(C\overset{*}{H}_3C\overset{*}{H}_2C\overset{*}{H}_4)^+_{exc.} \longrightarrow (CH_3CH_2)^+ + C\overset{*}{H}_4 \qquad (72)$$
$$(37)$$

The product yields given by the authors[96] indicate that hot tritium atoms insert preferentially into *terminal* methyl groups of propane and the time between ^3H insertion and decomposition of 37 is insufficient for full equilibration of tritium within the excited ion 37. This 'tritium effect', that is, the preferential presence of a tritium atom in one of the two terminal groups, should stimulate tritiated methane formation by rupture of the methylene carbon–$C\overset{*}{H}_4$ bond due to electronic rearrangements and not due only to the mass effect of tritium. Similarly, the preferential $(65/80)\cdot 100\% = 81.25\%$ localization of tritium in $\overset{*}{H}_2$ in the reaction shown in equation 70a also indicates the non-equilibrium nature of this decomposition process.

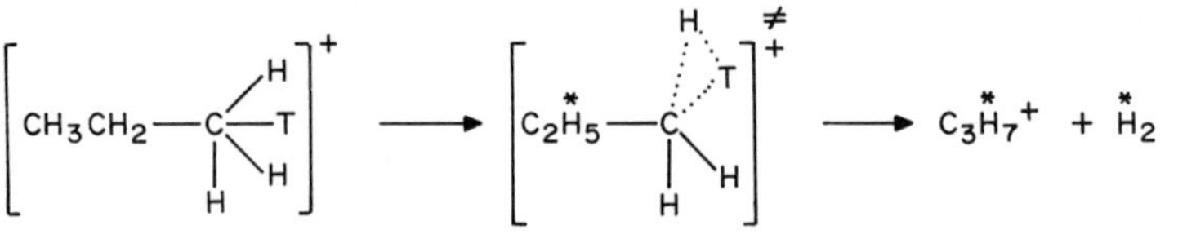

$$\left[CH_3CH_2{-}C\overset{H}{\underset{H}{\diagup}}{-}T \right]^+ \longrightarrow \left[C_2\overset{*}{H}_5{-}C\overset{H}{\underset{H}{\diagdown}}{\cdots}T \right]^{\neq +} \longrightarrow C_3\overset{*}{H}_7^+ + \overset{*}{H}_2 \qquad (70a)$$

The 'tritium effect' is not the simple result of the higher mass of tritium relative to ^1_1H, which lowers the frequency of carbon–hydrogen vibrations, but rather the effect of the local excitation of the electronic structure around the terminal carbon of propane caused by collisions and 40% fragment according to equation 79b. The remaining 35% dissociate preferential tritiated methane formation (equation 72a). Due to its higher mass the tritium atom is located more closely to the terminal carbon than the light hydrogens and participates less frequently than ^1_1H atoms in the 'migration exchange' with the C_2H_5 radical. But this is only a secondary mass effect, and the stimulation of the electronic rearrangement around the terminal carbon caused by insertion of the tritium ion is the principal reason for the preferential CH_3T formation. Also, the stronger and more stable CH_3—T bond ($= 104\ kcal\ mol^{-1}$) is formed in larger abundance in this process than the weaker C_2H_5—T bond ($= 95$–$85\ kcal\ mol^{-1}$).

$$[CH_3CH_2-\overset{+}{C}H_3T]_{exc.} \longrightarrow \left[CH_3\overset{+}{C}H_2 \cdots \overset{\overset{\displaystyle H\ \ H}{\displaystyle |\diagup}}{\underset{\underset{\displaystyle T}{\displaystyle |}}{C}}-H \right]^* \longrightarrow CH_3\overset{+}{C}H_2 + CH_3T \quad (72a)$$

The results of earlier studies[97] of a system containing propane gas and T_2(2.2 Ci) have been summarized in the form of linear equations 73–75 of the type $L(X) = a + b[T_2]$, where a and b are constants and $[T]$ denotes the total β-activity in the sample; $L(X)$, the number of tritium atoms incorporated per T decay, increases linearly with increasing tritium concentration $[T_2]$ in the 0–100 Ci interval.

$$L(CH_4-T) = (0.11 \pm 0.01) + (0.020 \pm 0.002)[T_2] \quad (73)$$

$$L(C_2H_6-T) = (0.020 \pm 0.003) + (0.011 \pm 0.001)[T_2] \quad (74)$$

$$L(C_3H_8-T) = (0.15 \pm 0.01) + (0.022 \pm 0.001)[T_2] \quad (75)$$

The rate of formation of tritiated products in the propane–tritium system at a total β-particle activity of 2.2 Ci depends linearly on time (20–100 h period). The yields decreased to about half when T_2 was replaced by HT. Irradiation of the propane–T_2 system with an external ^{60}Co γ- source (1×10^4 Ci) has increased the $L(X)$ values due to an increase in the rate of electron production. The tritium-labelling process is initiated by two reactive species: the $^3He^3H^+$ ion (so-called 'decay labelling') and the electron (β-labelling)[98,99]. Nitric oxide reduced the $L(X)$ values.

4. Reactions of $^3He^3H^+$ ions with cyclopropane and with 1,2-dimethylcyclopropane

a. Gas-phase reactions of $^3He^3H^+$ ions with cyclopropane. These yield tritium-labelled methane (22%), ethylene (15.2%) and cyclopropane[96] (18.8%) as the major products. Ethane (4%), propane (2%) and propylene (7.1%) are the minor ones. The relative yields of main products appear to remain constant upon decrease of the cyclopropane pressure in the system from 760 to 20 torr and by addition of O_2. The total yield of the products corresponds to 70% of the radioactivity of the generated $^3He^3H^+$ ions. Excited protonated cyclopropane rings are produced in the first reaction step (equation 76). They dissociate partly or are stabilized by collision (equation 77). The stabilized ions transfer protons to cyclopropane molecules (equation 78) forming tritiated C_3H_6 hydrocarbons. Tritiated cyclopropane is the major product with only relatively minor rearrangement of cyclic $C_3H_7^+$ ions to a linear structure. About 25% of the excited protonated ions are stabilized by collisions and 40% fragment according to equation 79b. The remaining 35% dissociate to form a hydrogen molecule and a $C_3H_5^+$ ion (equation 79c).

$$c\text{-}C_3H_6 + {}^3He^3H^+ \longrightarrow (C_3\overset{*}{H}_7)^+_{exc.} + {}^3He \quad (76)$$

$$(C_3\overset{*}{H}_7)_{exc.} \overset{+M}{\longrightarrow} C_3\overset{*}{H}_7{}^+ \quad (77)$$

$$C_3\overset{*}{H}_7{}^+ + c\text{-}C_3H_6 \longrightarrow C_3\overset{*}{H}_6 + C_3H_7{}^+ \quad (78)$$

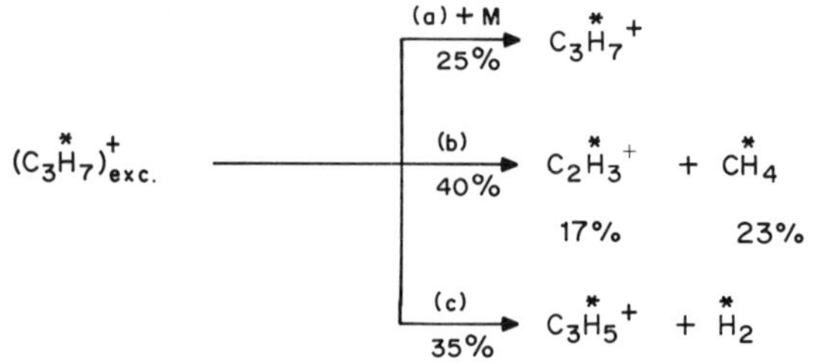

$$(C_3\overset{*}{H}_7)^+_{exc.} \quad \underset{}{\xrightarrow{}} \quad \begin{cases} \overset{(a)+M}{\underset{25\%}{\longrightarrow}} \; C_3\overset{*}{H}_7{}^+ \\[2mm] \overset{(b)}{\underset{40\%}{\longrightarrow}} \; C_2\overset{*}{H}_3{}^+ \; + \; \overset{*}{C}H_4 \\[1mm] \qquad\;\; 17\% \qquad\quad 23\% \\[2mm] \overset{(c)}{\underset{35\%}{\longrightarrow}} \; C_3\overset{*}{H}_5{}^+ \; + \; \overset{*}{H}_2 \end{cases} \tag{79}$$

Hydride transfer (equation 80) from cyclopropane to $C_2\overset{*}{H}_3{}^+$ excited ions, formed in reaction 79b, is one of the possible routes of the tritiated ethylene formations ($\Delta H^{24}_{(80)} \cong -58\,\text{kcal mol}^{-1}$).

$$C_2\overset{*}{H}_3{}^+ \; + \; c\text{-}C_3H_6 \; \longrightarrow \; C_2\overset{*}{H}_4 \; + \; C_3H_5{}^+ \tag{80}$$

b. Reactions of helium tritide ions with cis- and trans-1,2-dimethylcyclopropane[100]. Storage of 2 mCi of $^3H_2$ mixed with an organic substrate at 300 torr (in the presence of 2 mol% O_2 during 176 to 246 days in sealed ampoules) resulted in the formation, in the case of cis-1,2-dimethylcyclopropane, of tritiated methane (9.3%), cis-1,2-dimethylcyclo-propane (8.3%), 2-methyl-2-butene (8.3%) and trans-2-pentene (2.4%), the total yield of the organic products being 28.3%. Using trans-1,2-dimethylcyclopropane the combined yield of the labelled organic products was only 20.7%, containing methane (6.7%), trans-1,2-dimethylcyclopropane (8.3%) and an unknown (probably tritiated cyclopentene, 5.7%). The low total yields of tritiated products in both systems indicate that besides the reaction

$$^3He^3H^+ + RH \longrightarrow RH^3H^+ + {}^3He \tag{81}$$

$$^3He^3H^+ + RH \longrightarrow R^+ + H^3H + {}^3He \tag{82}$$

which lead to tritiated hydrocarbons, hydride-ion abstraction (equation 82) also takes place, yielding hydrogen tritide as the only labelled product. The different total yields show that the competition between reactions 81 and 82 depends on structural factors. The retention of configuration after the $^3He^3H^+$ attack rules out any intermediacy of linear pentyl ions and provides additional evidence for the occurrence of gaseous protonated cyclopropanes 38 (equation 83).

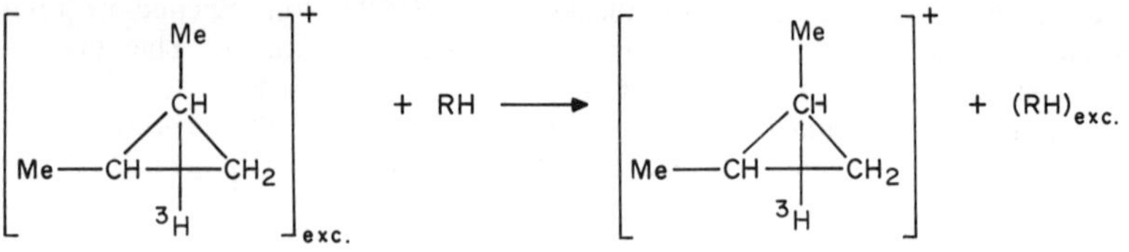

$$\left[\begin{array}{c} Me \\ | \\ CH \\ Me\text{---}CH\!\!\overset{\triangle}{\underset{}{\text{---}}}\!\!CH_2 \\ {}^3H \end{array}\right]^+_{exc.} + RH \longrightarrow \left[\begin{array}{c} Me \\ | \\ CH \\ Me\text{---}CH\!\!\overset{\triangle}{\underset{}{\text{---}}}\!\!CH_2 \\ {}^3H \end{array}\right]^+ + (RH)_{exc.} \tag{83}$$

(38)

Formation of tritiated methane in both systems and the absence of $C_4H_7{}^3H$ hydrocarbons in the products indicate that production of methane involves the attack of $^3He^3H^+$ on a methyl group of the substrate followed by fast dissociation of the methyl-tritiated intermediate (equation 84). The butenyl ions, $C_4H_7{}^+$, are transformed by hydride-ion abstraction from the substrate to butenes. Methane formation according to a

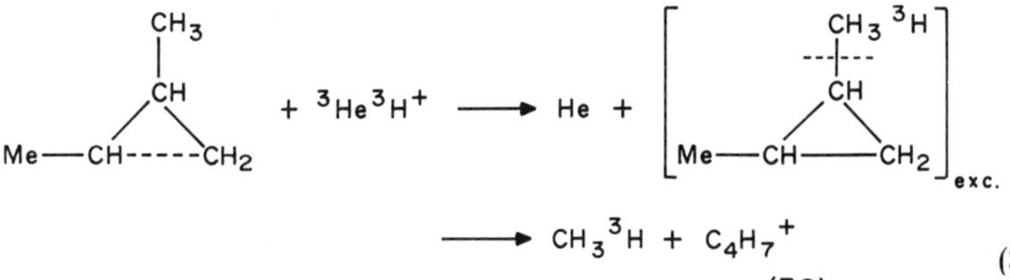

$$\text{(structures and reaction)} \quad (84)$$

$$\longrightarrow CH_3{}^3H + C_4H_7{}^+$$
$$(39)$$

protolytic cleavage mechanism has also been observed in the case of alkanes dissolved in $HFSO_3–SbF_5$ solutions[101]. The products of cis- and trans-dimethylcyclopropane trition-ation, apart from tritiated methane and tritiated parent, are significantly different. The cis isomer yields two linear amylenes via cleavage of the cyclopropane ring (equation 85) and subsequent proton transfer from these ions to molecules of the substrate. No labelled amylenes have been found in the products of the reaction of trans-1,2-dimethylcyclopropane with helium tritide ions.

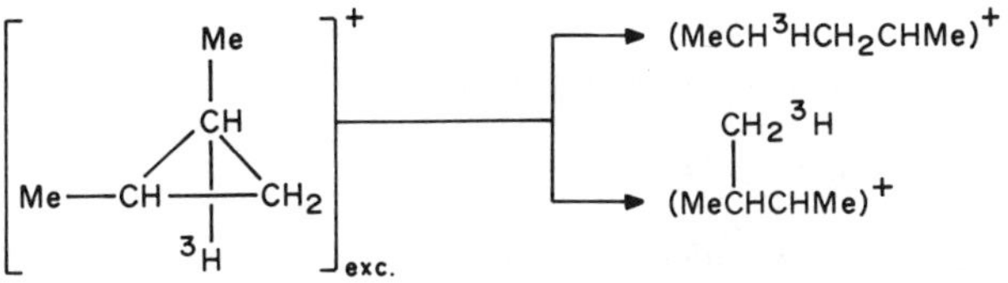

$$\text{(structures and reaction)} \quad (85)$$

$$\longrightarrow (MeCH^3HCH_2CHMe)^+$$
$$CH_2{}^3H$$
$$\longrightarrow (MeCHCHMe)^+$$

5. Reactions of $^3He^3H^+$ ions with n-butane, i-butane and cyclobutane

a. Reactions of helium tritide ions with n-butane[96]. Tritiated methane (15%), tritiated butane (15%) tritiated ethane (6%), ethylene (3.4% and 2.5% yields in the absence and presence of 2% O_2, respectively) and propane (2.3% both in the absence and presence of oxygen) are obtained in this reaction. The attack on n-butane by the $^3He^3H^+$ ions (equation 86) forms the excited protonated cations that fragment and react subsequently according to equations 87 and 88.

$$n\text{-}C_4H_{10} + {}^3He^3H^+ \longrightarrow (C_4\overset{*}{H}_{11})^+_{exc.} + {}^3He \qquad (86)$$

$$(C_4\overset{*}{H}_{11})^+_{exc.} \quad
\begin{cases}
\xrightarrow[78\%]{(a)} C_4\overset{*}{H}_9{}^+ (14\%) + \overset{*}{H}_2(64\%) \\
\xrightarrow[6\%]{(b)} C_2\overset{*}{H}_5{}^+ + C_2\overset{*}{H}_6 \\
\xrightarrow[16\%]{(c)} C_3\overset{*}{H}_7{}^+ + C\overset{*}{H}_4 (14\%)
\end{cases} \Bigg\rbrace 6\% \qquad (87)$$

$$n\text{-}C_4H_{10} + C_2\overset{*}{H}_5{}^+ \longrightarrow C_4H_9{}^+ + C_2\overset{*}{H}_6$$

$$n\text{-}C_4H_{10} + C_3\overset{*}{H}_7{}^+ \xrightarrow{(a)} C_4H_9{}^+ + C_3\overset{*}{H}_8 \ (2\%) \qquad (88)$$

$$n\text{-}C_4H_{10} + C_4\overset{*}{H}_9{}^+ \xrightarrow{(b)} C_4H_9{}^+ + n\text{-}C_4\overset{*}{H}_{10}{}^+ \ (14\%)$$

The distribution of tritium among the different radioactive species isolated indicates that the life time of the excited $(C_4\overset{*}{H}_{11})^+$ species is too short for the attaintment of the equilibrium distribution of the label within the tritiated cation.

b. *Reaction of* $^3He^3H^+$ *ions with isobutane.* Tritiated isobutane[96] (ca 18%), methane (25.3% with pure i-C_4H_{10} and 23.4% in the presence of 2% O_2), ethane (2.2 and 1.7% respectively), ethylene (3.9 and 3.5%) and propane (3.4 and 3.2) are the products. Equations 89 and 90 illustrate the distribution of tritium label in the products. Most of the tritium activity (90%) is contained in the methyl group[102] of the labelled i-C_4H_{10}[103]. The attack at the branched carbon is probably followed by elimination of $^1H^3H$ (equation 89b).

$$^3He^3H^+ + i\text{-}C_4H_{10} \longrightarrow {}^3He + (C_4\overset{*}{H}_{11})^+_{exc.}$$
$$100\%$$

$$(C_4\overset{*}{H}_{11})^+_{exc.} \quad \begin{cases} \xrightarrow[28\%]{(a)} C_3\overset{*}{H}_7{}^+ (3\%) + C\overset{*}{H}_4 (25\%) \\ \xrightarrow[72\%]{(b)} C_4\overset{*}{H}_9{}^+ (18\%) + \overset{*}{H}_2 (54\%) \end{cases} \tag{89}$$

$$i\text{-}C_4H_{10} + C_3\overset{*}{H}_7{}^+ \longrightarrow C_4H_9{}^+ \rightarrow C_3\overset{*}{H}_8 (3\%)$$

$$i\text{-}C_4H_{10} + C_4\overset{*}{H}_9{}^+ \longrightarrow C_4H_9{}^+ + i\text{-}C_4\overset{*}{H}_{10} (18\%) \tag{90}$$

The schemes in equations 86–90 do not involve stabilized protonated ions such as were postulated in the cyclopropane reaction in which evidence for cyclopropanium ion formation has been found.

c. *Reactions of* $^3He^3H^+$ *ions with cyclobutane.* These were unaffected by the presence of added oxygen. The following tritiated hydrocarbons have been[104] isolated in the reaction (at 400 torr): *trans*-2-butene (12.1%), ethylene (11.1%), propylene (8.4%), cyclobutane (6.9%), ethane (6.1%) and methane (1.3%). The total yield of organic tritiated products represents 45.9% of the reacting $^3He^3H^+$ ions. As before, tritiation of cyclobutane (equation 91) is the first step in the formation of labelled products, followed by collisional stabilizations and reactions of the stabilized cation (equations 92) which retains partly a cyclic structure, or undergoes fragmentations (equations 93). The yield of tritiated ethylene is about twice as high as the yield of tritiated ethane[104].

$$^3He^3H^+ + c\text{-}C_4H_8 \longrightarrow (C_4\overset{*}{H}_9)^+_{exc.} + {}^3He \tag{91}$$

$$(C_4\overset{*}{H}_9)^+_{exc.} + M \longrightarrow C_4\overset{*}{H}_9{}^+ + M_{exc.}$$

$$(C_4\overset{*}{H}_9)^+ + C_4H_8 \longrightarrow C_4H_9{}^+ + C_4\overset{*}{H}_8 \begin{cases} \text{cyclobutane} \\ + \\ \textit{trans}\text{-2-butene} \end{cases} \tag{92}$$

$$c\text{-}C_4\overset{*}{H}_9{}^+ \longrightarrow \text{secondary-}C_4\overset{*}{H}_9{}^+ \longrightarrow \textit{trans}\text{-2-butene}$$

$$(C_4\overset{*}{H}_9)^+_{exc.} \longrightarrow C\overset{*}{H}_4 + C_3\overset{*}{H}_5{}^+ \xrightarrow{c\text{-}C_4H_8} C_3\overset{*}{H}_6 + C_4H_7{}^+$$

$$(C_4\overset{*}{H}_9)^+_{exc.} \longrightarrow C_2\overset{*}{H}_4 + C_2\overset{*}{H}_5{}^+ \xrightarrow{c\text{-}C_4H_8} C_2\overset{*}{H}_6 + C_4H_7{}^+ \tag{93}$$

d. *Reaction of fast* T_2 *and* T_2^+ *with n-butane.* The translational energy dependences of the probability of reaction between n-butane and accelerated T_2^+ ions or T_2 molecules have been determined by efficient collection of the tritiated n-butane molecules and measuring their radioactivity by a radio gas-chromatographic method[105]. $T_2{}^+$ ions have been accelerated to discrete energies ranging from 5 to 100 eV in a specially designed 'chemical accelerator'[105]. The energy-selected tritium ions either collided with a crossed beam of the organic molecules or were transformed to T_2 of the same translational energy using D_2 and then reacted with the n-butane. The threshold of 6.0 ± 0.6 eV has been found for hydrogen displacement between neutral energetic T_2 and n-butane. The probability of this reaction increases with increase in the translational energy of T_2 and reaches a rather flat plateau extending between 30 eV and 100 eV. The threshold energy of 6.0 eV is close to the endothermicity (5.0 eV) of reaction 94.

$$T_2 + \text{n-}C_4H_{10} \longrightarrow \text{n-}C_4H_9T + T + H \qquad (94)$$

A similar Walden inversion mechanism has been proposed for displacement of H atoms in methane and ethane by fast T atoms[106,107]. The reaction cross section for the formation of n-C_4H_9T as a function of translational energy of T_2^+ decreases rapidly with increase in the energy of the ion from 5 eV to 20 eV, then remains relatively constant up to 40 eV and finally shows a broad maximum centered at about 60 eV. The shape of the observed experimental yield curve has been rationalized by suggesting that there are at least two mechanisms for the reaction of T_2^+ with butane. Below ca 20 eV T_2^+ energy, the most likely mode is tritiation followed by loss of H_2 and hydride abstraction (equation 95). At higher energies, collisional dissociation of the butane by fast T_2^+, followed by reaction of the butyl radical with molecular tritium, has been suggested (equation 96).

$$T_2^+ + \text{n-}C_4H_{10} \longrightarrow C_4H_{10}T^+ + T$$
$$C_4H_{10}T^+ \longrightarrow C_4H_8T^+ + H_2 \qquad (95)$$
$$C_4H_8T^+ + H^-(\text{wall}) \longrightarrow \text{n-}C_4H_9T$$
$$T_2^+ + \text{n-}C_4H_{10} \longrightarrow C_4H_9 + H + T^+ + T$$
$$C_4H_9 + T_2 \longrightarrow C_4H_9T + T \qquad (96)$$

These experimental observations have been compared with theoretical considerations of Porter[108], Menzinger and Wolfgang[109] and Porter and Kunt[110] concerning the reacting systems consisting of T_2 (or T^+) and solid cyclohexane. A similar shape of reaction probability curves for both organic targets has been obtained.

6. Reactions of $^3He^3H^+$ ions with cyclopentane and cyclohexane

a. Reactions with cyclopentane. The total yield of tritiated[104] products formed from cyclopentane (+ 2 mol% O_2) with $^3He^3H^+$ ions at 100 torr amounts to 34.2% of the total tritium activity present in the helium tritide (cyclopentane—11.1%, trans-2-pentene—4.5%, propane—4.6%, ethylene—4.1%, propylene—3.4%, ethane—2.6%, unknown product, probably pentadiene—3.9%, methane—not determined). Cyclopentane and its rearrangement product, trans-2-pentene, arise in reaction 97.

$$^3He^3H^+ + \text{c-}C_5H_{10} \longrightarrow (C_5\overset{*}{H}_{11})^+_{exc.} + He$$
$$(C_5\overset{*}{H}_{11})^+_{exc.} + M \longrightarrow C_5\overset{*}{H}_{11}^+ + M_{exc.} \qquad (97)$$
$$C_5\overset{*}{H}_{11}^+ + \text{c-}C_5H_{10} \longrightarrow C_5\overset{*}{H}_{10} + C_5H_{11}^+$$

Ethane and ethylene are formed in fragmentation processes (equation 98) and propane and propylene in hydride ion transfer reactions (equation 99).

$$(C_5\overset{*}{H}_{11})^+_{exc.} \longrightarrow C_3\overset{*}{H}_5^+ + C_2\overset{*}{H}_6 \qquad (98)$$
$$(C_5\overset{*}{H}_{11})^+_{exc.} \longrightarrow C_3\overset{*}{H}_7^+ + C_2\overset{*}{H}_4$$
$$C_3\overset{*}{H}_5^+ + \text{c-}C_5H_{10} \longrightarrow C_3\overset{*}{H}_6 + C_5H_9^+ \qquad (99)$$
$$C_3\overset{*}{H}_7^+ + \text{c-}C_5H_{10} \longrightarrow C_3\overset{*}{H}_8 + C_5H_9^+$$

b. Reactions of $^3He^3H^+$ with cyclohexane + 2 mol% O_2 at 100 torr. Tritiated cyclohexane has been produced in 9.7 ± 0.9% yield in this reaction[104]. The other products are produced in much smaller yields (trans-2-hexene 3.1%, trans-2-pentene 1.3%, propylene 3.0%, propane 2.7%, ethylene 1.6%, ethane 1.2%, methane—not determined). Formation of the tritiated cyclohexane (equation 100) requires one to assume the existence of the cyclic protonated intermediate cyclohexanium ion, **40**, Isomerization of **40** and subsequent proton transfer lead only to one of the possible hexenes, *trans*-2-hexene.

$$^3He^3H^+ + c\text{-}C_6H_{12} \longrightarrow (C_6\overset{*}{H}_{13})^+_{exc.} + He$$
$$[(\mathbf{40})]$$
$$(C_6\overset{*}{H}_{13})^+_{exc.} + M \longrightarrow C_6\overset{*}{H}_{13}{}^+ + M_{exc.} \tag{100}$$
$$C_6\overset{*}{H}_{13}{}^+ + c\text{-}C_6H_{12} \longrightarrow C_6\overset{*}{H}_{12} + C_6H_{11}{}^+$$

$C_1\text{–}C_5$ hydrocarbons arise from the fragmentation of the excited carbonium ion (equation 101):

$$(C_6\overset{*}{H}_{13})^+_{exc.} \longrightarrow C\overset{*}{H}_4 + C_5\overset{*}{H}_9{}^+$$
$$(C_6\overset{*}{H}_{13})^+_{exc.} \longrightarrow C_2\overset{*}{H}_6 + C_4\overset{*}{H}_7{}^+ \tag{101}$$
$$(C_6\overset{*}{H}_{13})^+_{exc.} \longrightarrow C_2\overset{*}{H}_4 + C_4\overset{*}{H}_9{}^+$$
$$(C_6\overset{*}{H}_{13})^+_{exc.} \longrightarrow C_3\overset{*}{H}_6 + C_3\overset{*}{H}_7{}^+$$

and in the exothermic hydride ion transfer reactions 102:

$$C_5\overset{*}{H}_9{}^+ + c\text{-}C_6H_{12} \longrightarrow C_5\overset{*}{H}_{10} + C_6H_{11}{}^+ \tag{102}$$
$$C_3\overset{*}{H}_7{}^+ + c\text{-}C_6H_{12} \longrightarrow C_3\overset{*}{H}_8 + C_6H_{11}{}^+$$

c. Concluding remarks concerning reactions of c-alkanes with helium tritide. After reaction of cycloalkanes with $^3He^3H^+$ part of the tritiated carbonium ions retains a cyclic structure. Normal tritiated alkanes are not formed in the reactions of the corresponding cycloalkanes. This means that the linear alkyl ions formed from the isomerization of protonated cycloalkanes react according to equation 103:

$$n\text{-}C_n\overset{*}{H}^+_{2n+1} + c\text{-}C_nH_{2n} \longrightarrow n\text{-}C_n\overset{*}{H}_{2n} + C_nH^+_{2n+1} \tag{103}$$

and not according to equation 104:

$$n\text{-}C_n\overset{*}{H}^+_{2n+1} + c\text{-}C_nH_{2n} \longrightarrow n\text{-}C_n\overset{*}{H}_{2n+2} + C_nH^+_{2n-1} \tag{104}$$

The manner in which $^3He^+$ is bound in the cycloalkanium ion is not known precisely. Combined yields of the isolated organic products diminish from 67% with cyclopropane to 46% with cyclobutane, to 34% with cyclopentane and to 23% with cyclohexane. This indicates that the abstraction process

$$^3He^3H^+ + RH \longrightarrow R^+ + H^3H + {}^3He \tag{105}$$

prevails over the tritiation process (equation 106) with increasing number of carbon atoms in the ring. No-polymeric compounds have been detected in these studies.

$$^3He^3H^+ + RH \longrightarrow RH^3H^+ + {}^3He \tag{106}$$

7. Reactions of $^3HeT^+$ ions with gaseous bicyclo[n.1.0]alkanes

The gas-phase reactions of $^3HeT^+$ with bicyclo[2.1.0]pentane, bicyclo[3.1.0]hexane and bicyclo[4.1.0]heptane (**41**) have been investigated[111] and the formation of tritiated hydrocarbons retaining the bicyclic structure of the starting substrate has been observed. This finding indicates that the gaseous bicycloalkylium ions are formed via exothermic trition transfer from the $^3HeT^+$ to the bicyclic substrate (equation 107). Due to the exothermicity of this electrophilic attack, a large fraction of the bicyclic organic ions formed do not undergo stabilization by collisions in the 100–300 torr pressure range investigated and decompose or isomerize by cleavage of one or more C—C bonds.

$$^3He^3H^+ + bicyclo[n,1,0]C_mH_{2m-2} \longrightarrow ^3He + (bicyclo[n.1.0]C_mH_{2m-2}{}^3H)^+_{exc.}$$
$$+ M$$

$$(bicyclo[n.1.0]C_mH_{2m-2}{}^3H)^+ + M_{exc.} + fragments \qquad (107)$$
$$\downarrow s$$
$$bicyclo[n.1.0]C_mH_{2m-3}{}^3H + SH^+$$

In the case of **41** (experimental conditions: **41**—100 torr, oxygen—10 torr, $^3H_2$—2 mCi, volume of the pyrex vessel 500 ml, storage period 250–263 days, temperature 90 °C), besides tritiated **41** (6.2%), tritiated methane (15.2%), ethene (1.9%), propene (14.2%), cis-but-2-ene (2.3%) and 2-methylbut-2-ene + trans-pent-2-ene (8.1%) have been isolated. In the case of reaction of $^3He^3H^+$ with bicyclo[2.1.0]pentane (**42**), the yields of the tritiated products were: **42**—3.8%, methane—16.1%, ethene—0.5%, propene—12.2%, 2-methyl-but-2-ene—1.4%, cyclopentene—5.7%, other $(C_4),(C_5)$ products—5%. $^3He^3H^+$ with bicyclo[3.1.0]hexane (**43**) gave **43** (1.7%), methane (12.4%), propene (4.0%), trans-pent-2-ene (2.9%),1-methylcyclopentene (0.4%), unknown $(C_4, 5.2\%)$.

The above results indicate that the fragmentation of the bicycloalkylium ions is very sensitive to specific structural factors. The formation of tritiated cyclopentene (the major C_5 product from gas-phase tritonation of **42**) proceeds according to equation 108, which involves the initial cleavage of the cyclopropane ring.

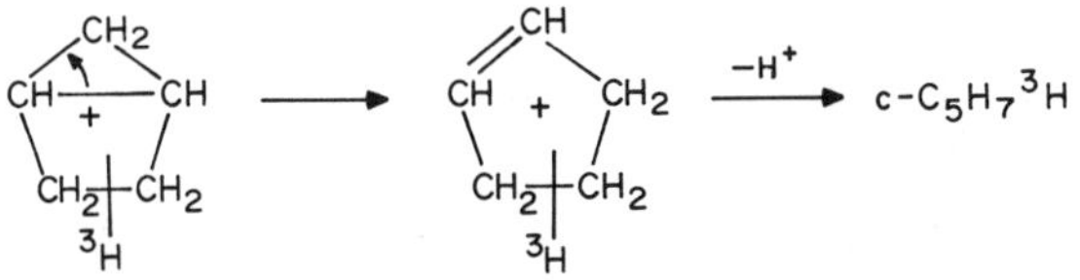

$$(108)$$

No tritiated cyclohexene has been found in the reaction of $^3He^3H^+$ with bicyclo[3.1.0]hexane and, contrary to liquid-phase reactions only a low yield of 1-methylcyclopentene has been obtained. Gas-phase electrophilic reactions of $^3He^3H^+$ with gaseous alkanes and cycloalkanes have been reviewed by Cacace[112].

8. Reactions of tritium atoms generated in the $^3He(n,p)^3H$ nuclear reaction with hydrocarbons

Recoil tritium atoms, produced in the $^3He(n,p)$ ^3H reaction, were studied with different hydrocarbons including[113a] alkanes $(C_2H_6, n-C_3H_8, n-C_4H_{10}, CMe_4, CH_3CHMe_2, n-C_5H_{12}, n-C_6H_{14})$, alkenes $(CH_2=CH—CH_2CH_3, CH_2=CHCHMe_2, CH_2=CHCH_3,$ trans- and cis-$CH_3CH=CHCH_3, CH_2=CMe_2)$ and an alkyne, $CH\equiv CCH_3$ in hexaflu-orocyclobutene, and the normalized yields of $CH_3{}^3H$ formed by the attack of hot tritium on terminal carbon–carbon single bonds have been determined (equation 109). These

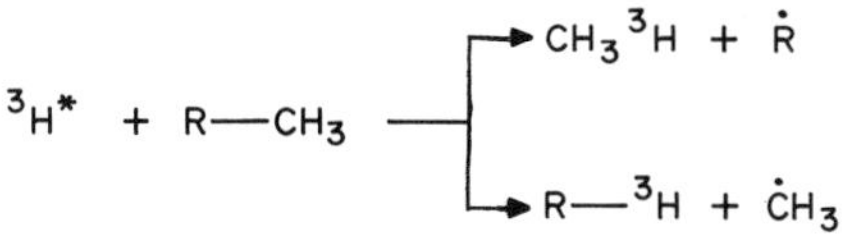

$$(109)$$

* denotes a 'hot' atom

yields have been found to correlate inversly with the R—CH_3 bond dissociation energies. A suggestion[113] has been made that the reaction of recoil tritium atoms at C—C single bonds is faster than the vibrational displacement of the allyl radical. The time for this type of interaction has been assumed to be less than 2–5×10^{-14} s.

Linear (44) and triangular (45) collisional complexes have been proposed for the $CH_3{}^3H$ formation process. The cleavage of the R—CH_3 bond should be the rate-determining step for the Me abstraction in the linear complex 44 while the triangular

(44)   (45)

complex 45 implies that the electron-overlapping process and the C—T bond formation should be rate-determining. The dependence of the $CH_3{}^3H$ yields on $D(R$—$CH_3)$ and the minor role which the electron density plays in determining the yields of tritiated methane support the structure 44 for the 'transition state complex' of the Me^3H formation. A ^{14}C methyl kinetic isotope effect study should help to formulate correctly the mechanism of these reactions.

9. Physico-chemical studies of hot tritium atom reactions

a. The replacement of an H atom by a hot tritium atom is facilitated by the presence of a high electron density in the C—H bond under attack[114]. This view has been supported by the linear correlation between the relative yields of R^i—^3H in reaction $^3H^* + R^{i1}H \longrightarrow R^{i3}H + {}^1H$, and the NMR proton chemical shift (ppm) for a series of substituted halomethanes and alkanes: CH_4 (100% yield; $\delta_{NMR} = 0$ ppm) > alkanes > CH_3Br > CH_3Cl > CH_3F > CH_2Cl_2 > CHF_3 (55% yields, $\delta_{NMR} = -6.2$). The replacement of H by T is also more probable in Me groups present in Me_3Si in which the electron density is higher by a factor 1,18 than in Me_4C, in good agreement with the above linear correlation[115].

b. Complete ($> 99\%$) retention of configuration at the asymmetric[116] carbon atoms has been found in the O_2-scavenged gas-phase replacement of hydrogen by energetic tritium in *dl-* and *meso-*$(CHFCl)_2$. I_2-scavenged liquid samples showed measurable isomerization on irradiation of a LiF solution with a neutron flux of approximate 1×10^{12} n cm^{-2} s^{-1}. In gas-phase experiments ^3He has been used as the source of recoil tritium. In unscavenged gas-phase experiments and in liquid-phase experiments with I_2 as scavenger, the formation of 'inversion' products caused by radical combination reactions has been observed. The possibility of negligible inversion of configuration in the case of gas-phase T for H substitution in CH_4 leading to $CH_3{}^3H$ formation has been considered[116].

c. The effect of all 'chemical' factors, such as[117] electronegativity, electron density and bond energy, upon the yields of the hot tritium for hydrogen substitutions in various halomethane and alkane molecules has also been studied by Rowland and coworkers[117]. They concluded that the probability of the T/H reaction, i.e. $MeX \rightarrow CH_2{}^3HX$ formation, decreases with increasing electronegativity of the substituent X (measured, as before, by the NMR shift). The view has been expressed that highly electronegative substituents such as F might serve as a sink for electron density developed by the departure of the H atom, leaving less electron density in the vicinity of the incipient C—^{3}H bond and minimizing the probability of this new bond formation. The hypothesis that halogen substituents influenced sterically the T for H substitution has been rejected. The relative yields of $CH_3{}^3H$ formation in the reaction of hot tritium with CH_3X increased by a factor of 7.5 from CH_3F to CH_3I. The bond dissociation energy of the C—X bond is the most important one among the factors controlling the yields in these T/H substitution reactions. This means that when tritium atom has an excess of kinetic energy, geometrical and physical factors such as atomic masses and sizes, bond angles, angle of approach at the moment of collision, etc., are not of great importance. (The cumulative steric alkyl substituent effects in the T/H substitution reactions are between 0 and 10% per alkyl group.) Rowland[117] warns strongly that in the evaluation of primary yields using the observed product yields, it is necessary to consider the secondary isomerization or decomposition of internally excited molecules during the substitution process.

10. Isotope effects in the substitution of 2.8 eV tritium atoms with methane

The intermolecular[118] isotope effect in reactions 110 and 111 for 2.8 eV tritium atoms generated by photolysis of TBr at 1849 Å, as measured by the ratio $[(CH_3T/CH_4)/(CD_3T/CD_4)]$, has been found to equal 7.2. The ratio for the intramolecular competition for the T for H versus T for D substitution in partially deuteriated methanes (equations 112 and 113) is approximately unity, as measured semi-quantitatively by the ratio of product yields, $[CHD_2T]/[CH_2DT] = 1.06 \pm 0.1$.

$$T^* + CH_4 \longrightarrow CH_3T + {}^1H \tag{110}$$

$$T^* + CD_4 \longrightarrow CD_3T + D \tag{111}$$

$$T^* + CH_2D_2 \longrightarrow CHD_2T + H \tag{112}$$

$$T^* + CH_2D_2 \longrightarrow CH_2DT + D \tag{113}$$

The above results have been classified as 'primary replacement isotope effect of 1.6 ± 0.2 favouring the substitution of H over D' and as 'secondary isotope effect of 1.6 ± 0.2 per H/D favouring bond formation to methyl group with more H atoms in it'[118].

The *per bond* abstraction yields have also been determined and found to be 1.4–2.0 times higher for HT than for DT formation. Experimental difficulties encountered in the determinations of HT abstraction yields did not allow a more detailed study of this effect. The probability of tritium substitution into the methane molecule by 2.8 eV T atoms increases with the decrease in the number of D atoms in the methane, as shown by the relative yields for this process: $CH_4 (7.2 \pm 0.2)$, $CH_3D (5.6 \pm 0.3)$, $CH_2D_2 (3.1 \pm 0.3)$ and $CD_4 (1.0)$.

Little preference has been found for replacement of H or D atoms. In the case of CH_2D_2 target molecule (products CHD_2T or CH_2DT) the ratio of product yields, T for H or T for D, equals 1.06 ± 0.1. For the CHD_3 target molecule (CD_3T or CHD_2T formed) the T for H or T for D ratio has been found to be 0.4 ± 0.1 (per bond, 1.2 ± 0.3). The primary replacement isotope effect for very high energy tritium atoms (192,000 eV) from the $^3He(n,p)^3H$ process and substituting into CHX_3 versus CDX_3 molecules (where $X = CH_3$[119], CD_3[119] or F[120]) equals 1.30 ± 0.05, indicating also the preference for H replacement. No detailed analysis has been given by Rowland and coworkers, but several ideas aimed at understanding the substitution processes have been discussed.

The 'billiard ball' atom–atom collisions favouring the easier replacement of D do not explain the observed experimental deuterium isotope effects in substitution reactions with hot tritium. The linear structures, 'T—^{i}H—R', of the transition states have also been rejected. An attempt has been made to rationalize the experimental findings by invoking the participation in equations 110–113 of '*pseudo-complexes*' containing five hydrogen substituents near the central C atom, including the energetic tritium atom, which participate actively in the motion along the reaction coordinate. The electron densities between the C atom and the five substituents in these complexes are different from that in the normal C—H bonds in methane. A suggestion has been made that the reaction is concerted, involving simultaneous formation of the C—T bond and cleavage of the C—H or C—D bond, but 3-centred calculations are inadequate for the description of these processes. More elaborate calculations taking into account all five hydrogen substituents should explain the data and establish the correct structure of the transient complex[121]. The authors[118] reject the idea that the displacement is a fast direct, localized event occurring on a time scale comparable to a bond vibration (ca 10^{-14} s) and express the view that the five-bonded carbon system exists for several C—H vibrations 3–7×10^{-14} s) and substantial distribution, albeit not full internal equilibration, of the energy occurs during its life-time.

The lack of a large intramolecular deuterium isotope effect in the reaction of hot tritium with partially deuteriated methanes has been explained by Rowland and coworkers[118] by postulating the existence of a very large secondary deuterium isotope effect which counterbalances the normal primary replacement isotope effect.

Assuming the intermediacy of 'five-bonded carbon pseudo-complexes' the small but preferred formation of HT over that of DT is understood if we accept that the hydrogen abstraction by tritium atoms proceeds through a *bent triangular* transition state which gives low values for the kinetic isotope effects (equations 114 and 115). Formation of HT

$$T + \begin{array}{c} H \\ \diagdown \\ \diagup \\ H \end{array} C \begin{array}{c} D \\ \diagup \\ \diagdown \\ D \end{array} \longrightarrow \left[\begin{array}{c} H \\ \diagdown \\ T—C \\ \diagup \\ H \end{array} \begin{array}{c} D \\ \diagup \\ \diagdown \\ D \end{array} \right] \longrightarrow \left[\begin{array}{c} H \\ \diagdown \\ T\cdots C \\ \diagup \\ H \end{array} \begin{array}{c} D \\ \diagup \\ \diagdown \\ D \end{array} \right]^{\neq} \longrightarrow HT + \overset{.}{C}D_2H$$

$$(114)$$

and DT by a free radical mechanism with a linear symmetrical transition state, $[T \cdots H(D) \cdots R]^{\neq}$, would be accompanied by a larger deuterium kinetic isotope effect.

The normal deuterium isotope effect observed in equations 110 and 111 might be caused by large zero-point energy contributions of the C—H/C—D bonds weakened substantially or broken in the transition state (with retention of configuration) of reaction paths shown in equations 116 and 117, and by some contribution to the total yield of H_2 or D_2

$$T + \begin{array}{c} H \\ \diagdown \\ \diagup \\ D \end{array} C \begin{array}{c} H \\ \diagup \\ \diagdown \\ D \end{array} \longrightarrow \left[\begin{array}{c} H \\ \diagdown \\ T—C—D \\ \diagup \\ H \quad | \\ \quad D \end{array} \right] \longrightarrow \left[\begin{array}{c} H \\ \diagdown \\ T\cdots C \\ \diagup \\ \quad | \\ \quad D \end{array} \begin{array}{c} H \\ \diagup \\ \diagdown \\ D \end{array} \right]^{\neq} \longrightarrow TD + \overset{.}{C}H_2D$$

$$(115)$$

$$T + CH_4 \longrightarrow \left[\begin{array}{c} H \\ \diagdown \\ T—C \\ \diagup \\ H \end{array} \begin{array}{c} H \\ \diagup \\ \diagdown \\ H \end{array} \right] \longrightarrow \left[\begin{array}{c} H \\ \diagdown \\ T—C \\ \diagup \\ H \end{array} \begin{array}{c} H \\ \diagup \\ \diagdown \\ H \end{array} \right]^{\neq} \longrightarrow TCH_3 + H^{.}$$

$$(116)$$

$$T + CD_4 \longrightarrow \left[\begin{array}{c} D \\ \diagdown \\ T—C \\ \diagup \\ D \end{array} \begin{array}{c} D \\ \diagup \\ \diagdown \\ D \end{array} \right] \longrightarrow \left[\begin{array}{c} D \\ \diagdown \\ T—C \\ \diagup \\ D \end{array} \begin{array}{c} D \\ \diagup \\ \diagdown \\ D \end{array} \right]^{\neq} \longrightarrow TCD_3 + D^{.}$$

$$(117)$$

by the reaction paths shown in equations 118 and 119, accompanied by a smaller deuterium isotope effect. Assuming that the reactions 112 and 113 proceed through the transition states shown in equation 120, a very large secondary deuterium isotope effect had to be invoked by Rowland[118] to explain the observed, very negligible intramolecular deuterium isotope effect in these reactions. However, if isotopic hydrogen molecules are

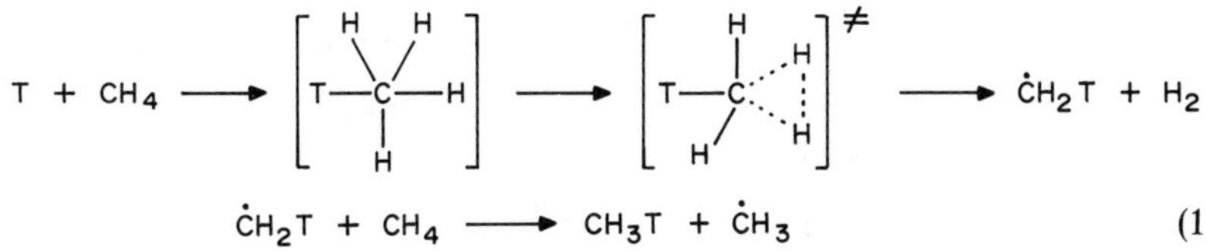

$$T + CH_4 \longrightarrow \left[\begin{array}{c} H \quad H \\ T-C-H \\ H \end{array} \right] \longrightarrow \left[\begin{array}{c} H \\ T-C \overset{H}{\underset{H}{\cdots}} \end{array} \right]^{\neq} \longrightarrow \dot{C}H_2T + H_2$$

$$\dot{C}H_2T + CH_4 \longrightarrow CH_3T + \dot{C}H_3 \tag{118}$$

$$T + CD_4 \longrightarrow \left[\begin{array}{c} D \quad D \\ T-C-D \\ D \end{array} \right] \longrightarrow \left[\begin{array}{c} D \\ T-C \overset{D}{\underset{D}{\cdots}} \end{array} \right]^{\neq} \longrightarrow \dot{C}D_2T + D_2$$

$$\dot{C}D_2T + CD_4 \longrightarrow CD_3T + \dot{C}D_3 \tag{119}$$

$$T^* + CH_2D_2 \longrightarrow \left[\begin{array}{c} H \quad D \\ T-C \\ H \quad D \end{array} \right]$$

$$\longrightarrow \left[\begin{array}{c} H \quad D \\ T-C \\ H \quad D \end{array} \right]^{\neq} \longrightarrow THCD_2 + H^{\cdot}$$

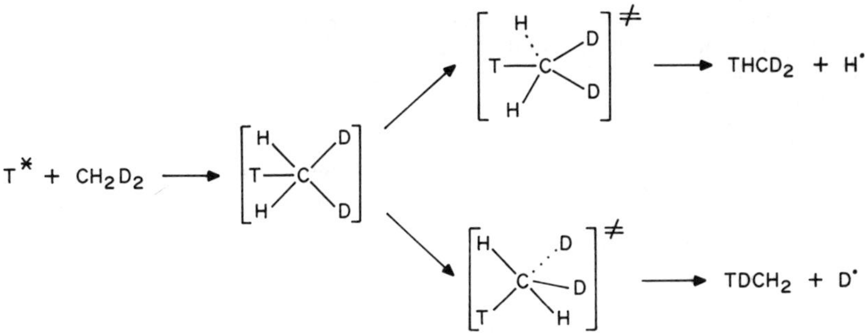

$$\tag{120}$$

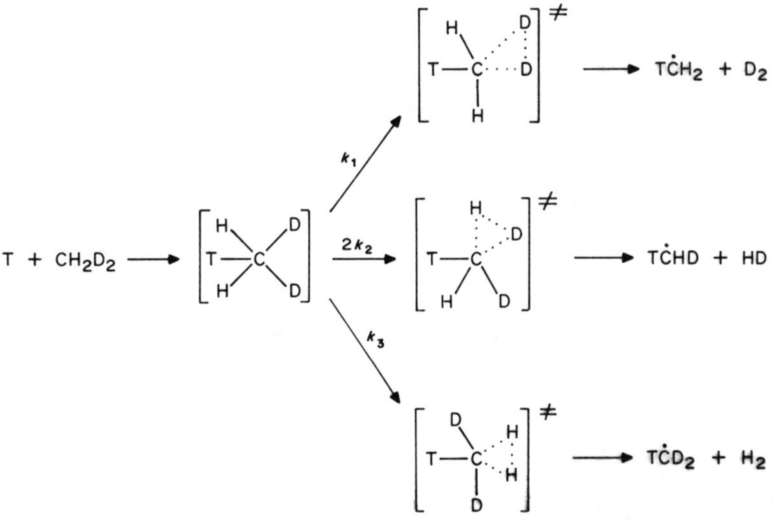

$$\longrightarrow \left[\begin{array}{c} H \quad D \\ C-D \\ T \quad H \end{array} \right]^{\neq} \longrightarrow TDCH_2 + D^{\cdot}$$

$$\longrightarrow \left[\begin{array}{c} H \quad D \\ T-C \cdots D \\ H \end{array} \right]^{\neq} \longrightarrow T\dot{C}H_2 + D_2 \tag{121}$$

$$T + CH_2D_2 \longrightarrow \left[\begin{array}{c} H \quad D \\ T-C \\ H \quad D \end{array} \right] \overset{k_1}{\underset{k_3}{\overset{2k_2}{\longrightarrow}}} \left[\begin{array}{c} H \\ T-C \cdots D \\ H \quad D \end{array} \right]^{\neq} \longrightarrow T\dot{C}HD + HD \tag{122}$$

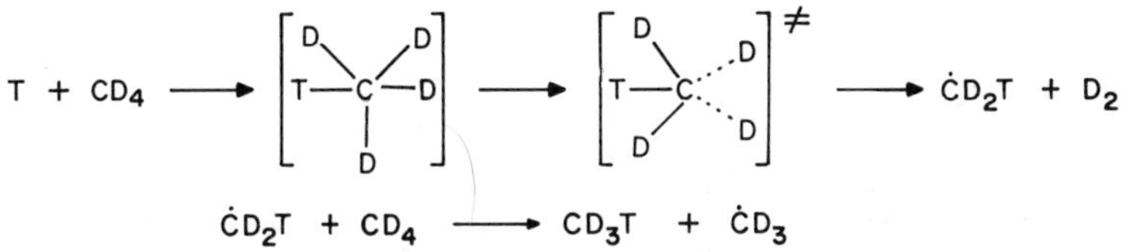

$$\longrightarrow \left[\begin{array}{c} D \quad H \\ T-C \cdots H \\ D \end{array} \right]^{\neq} \longrightarrow T\dot{C}D_2 + H_2 \tag{123}$$

produced directly as shown in equations 121–123a, then there is no need to postulate this unusually large secondary deuterium isotope effect.

$$T\dot{C}H_2 + CH_2D_2 \begin{cases} \nearrow TCH_3 + H\dot{C}D_2 \\ \\ \searrow TDCH_2 + D\dot{C}H_2 \end{cases} \tag{121a}$$

$$T\dot{C}DH + CH_2D_2 \begin{cases} \nearrow TDCH_2 + H\dot{C}D_2 \\ \\ \searrow THCD_2 + D\dot{C}H_2 \end{cases} \tag{12a}$$

$$T\dot{C}D_2 + CH_2D_2 \begin{cases} \nearrow THCD_2 + H\dot{C}D_2 \\ \\ \searrow TCD_3 + D\dot{C}H_2 \end{cases} \tag{123a}$$

(In terms of nonequilibrium interaction conditions the lack of the intramolecular deuterium isotope effect in the 'hot' tritium atoms reaction with partial deuteriated methane, CH_xD_{4-x}, implies that the greater chance of encounter between fast moving tritium and deuterium and transfer to it of the excess kinetic energy rather than to the fast vibrating hydrogen-1 atom, counterbalance the opposite acting zero-point energy effect which manifests itself in chemical reactions taking place under thermal equilibrium conditions.)

C. Chemical Consequences of Tritium Decay in Tritium Multilabelled Hydrocarbons

1. Nuclear decay of methane — $^3H_4$

a. Fully tritiated methane–$^3H_4$, synthesized according to equation 124, has been used[122] to help the identification of the neutral stable products formed in the ionic reactions initiated by the β-decay of tritium atoms contained in methane molecules at atmospheric pressure. In a gaseous mixture of C^3H_4 in methane (specific activity ~ 0.2 mCi mmol^{-1}) + 2% oxygen (to scavenge thermal radicals), stored in glass ampoules at atmospheric pressure and room temperature for 30 days, the radiochemical yields of products have been found to be: CHT_3 (5%), hydrogen (28.70%), ethylene (10.90%), ethane (1%), propane (0.8%), propylene (0.02%), n-butane (0.3%), the total product activity being less than 50% of the calculated activity of the decay fragments. In samples containing, in addition to methane, 3% C_3H_8 and 2% O_2, after 30 days the yields of CH^3H_3, hydrogen and ethane were found to be 5%, 34% and 49%, respectively; ethylene, propane, propylene and n-butane have been produced in 0.02% radiochemical yield only.

$$T_2 \xrightarrow[500\,°C]{CuO} T_2O \xrightarrow[150\,°C]{Al_4C_3} CT_4 \tag{124}$$

In the system comprising 84% C_3H_8, 14% CH_4 and 2% O_2 the major labelled products were CH^3H_3 (74%), hydrogen (16.60% radiochemical yield) and ethane (7%); tritium-labelled ethylene and propane have been produced in 0.2% yield, propylene and n-butane in 0.06% yield only. The effect of self-β-radiolysis of the sample has been found to be insignificant. The above data have been interpreted as indicating that in the system containing 98% CH_4 and 2% O_2 the major products, tritiated hydrogen and ethylene, are produced in reactions 125, where asterisk * indicates labelled species containing an unknown number of ^3H atoms. Oxygen present in the system scavenges the hydrogen atoms and, in its presence, only a minor fraction of the ethyl ions form ethylene (equation 126a) at atmospheric pressure:

$$CH_3^{*+} \longrightarrow C^+ + H_2^* + H^*$$
$$CH_3^{*+} \longrightarrow CH^{*+} + H_2^*$$
$$CH_3^{*+} \longrightarrow CH_2^{*+} + H^* \qquad (125)$$
$$CH_2^{*+} + CH_4 \longrightarrow C_2H_4^{*+} + H_2^*$$
$$CH_2^{*+} + CH_4 \longrightarrow C_2H_3^{*+} + H + H_2^*$$
$$C_2H_5^{*+} + e \longrightarrow C_2H_4^* + H^* \qquad (126a)$$

The low yield of CH^3H_3 in methane indicates that the hydride ion transfer (reaction 126b)

$$C^3H_3^+ + CH_4 \longrightarrow CH^3H_3 + CH_3^+ \qquad (126b)$$

occurring in the mass spectrometer is slow in comparison with reaction 126c:

$$C^3H_3^+ + CH_4 \longrightarrow C_2H_5^{*+} + H_2^* \qquad (126c)$$

The primary fragmentation pattern from the decay of tritium in tritiated methane, $CH_3^{3}H$ (equation 127), proceeding in the mass spectrometer at low pressure consists of the following charged fragments[123]: H^+ (2.40% abundance), H_2^+ (0.14%), $\{(^3He^+), H_3^+\}$ (0.12%), $^{12}C^+$ (4.9%), $\{^{12}CH^+, (^{13}C^+)\}$ (4%), CH_2^+ (4.90%), CH_3^+ (82.00 ± 1.5%).

$$CH_3^{3}H \xrightarrow{\ -\beta^-\ } [\overset{\frown}{CH_3^{3}He^+}] \longrightarrow [CH_3^{+3}He] \longrightarrow CH_3^+ + {}^3He \qquad (127)$$

Mass spectrometric and theoretical[83] evidence shows that about 80% of the original ions have little or no excitation energy and produce stable methyl ions after the loss of ^3He. The remaining 20% are in highly excited states (to about 20 eV) and undergo further fragmentation producing the secondary ions listed above. In the presence of excess of propane the fragmentation of $CH_3^{3}He^+$ at 760 torr is similar to the fragmentation observed at low pressure in the mass spectrometer. The methyl ions undergo a hydride ion transfer[124] (equation 128)

$$C^3H_3^+ + C_3H_8 \longrightarrow CH^3H_3 + C_3H_7^+ \qquad (128)$$

or react with methane yielding ethyl ions (equations 129a and 129b).

$$C^3H_3^+ + CH_4 \longrightarrow C_2H_5^{*+} + H_2^* \qquad (129a)$$
$$C_2H_5^{*+} + C_3H_8 \longrightarrow C_2H_6^* + C_3H_7^+ \qquad (129b)$$

In the system $[CH_4(14\%) + C_3H_8(84\%) + O_2(2\%)]$, CH^3H_3 and $C_2H_6^*$ are produced in 74% and 7.0%, respectively. In the system containing $CH_4(95\%)$, $C_3H_8(3\%)$ and $O_2(2\%)$, the C^3H_4 β-decay produces H_2^* with 34% yield and ethane with a 49.0% yield through reactions (129a) and (129b). The low yield of CH^3H_3 indicates that reaction (128) is not significant and ethane is produced according to equation (129b).

b. Gas-phase reactions of tritiated methyl cations. Gas-phase reactions of methyl cations (generated in the β-decay of tritiated hydrocarbons) with molecules of water, alcohols (MeOH, EtOH), hydrogen halides, ethyl halides, benzene, toluene, xylenes, ethylene and (ethylene + H_2O) have been investigated by Nefedov and coworkers[125–130]. Oxocations such as those at the top of page 848 are formed in the reactions of free carbocations with water and alcohols. Further transformations of these cations occur with departure of R^+ or H^+. Departure of R^+ and radioactive methanol formation is the preferable

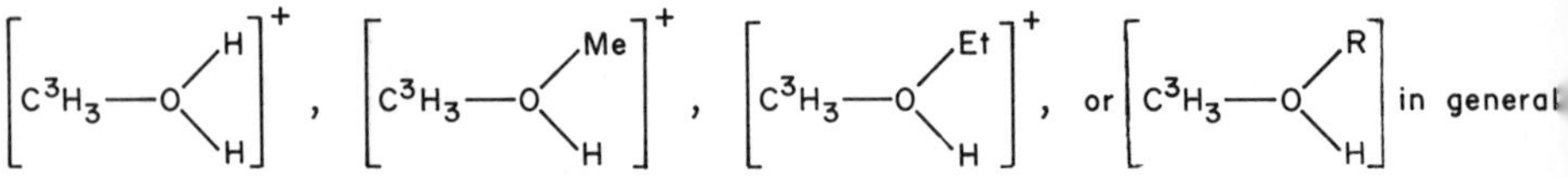

$$\left[C^3H_3-O\diagdown{}_H^{+H} \right]^+ , \left[C^3H_3-O\diagdown{}_H^{+Me} \right]^+ , \left[C^3H_3-O\diagdown{}_H^{+Et} \right]^+ , \text{ or } \left[C^3H_3-O\diagdown{}_H^{+R} \right] \text{ in general}$$

exothermic process. The heats of formation (expressed in $kcal\,mol^{-1}$) have been found to be 0 (MeOH), 24 (EtOH), 27 (n-C_3H_7OH), 28 (n-C_4H_9OH), 35 (i-C_4H_9OH), 44 (2-C_4H_9OH) and 46 (i-C_3H_7OH), respectively. The yields of radioactive methanol were 86, 81, 85 and 94% for the four alcohols and 100% for the last three. The endothermic formation of ethers (departure of H^+) has not been observed, but the less endothermic synthesis of alcohols containing one carbon atom more than the substrate has been found to take place. Thus in the reactions with MeOH, EtOH, PrOH and BuOH, respectively, the formation of ethanol (10%), i-C_3H_7OH (16%), secondary-C_4H_9OH (10%) and secondary-C_5H_{11}OH (6%) takes place. This implies that the rearrangement of the intermediate oxo-complex consisting in migration of the C^3H_3 group from oxygen to the nearest carbon atom and accompanied by proton abstraction is energetically a more favourable process than abstraction of the proton bound to oxygen. Reactions[127] of methyl cations $^+C^3H_3$, generated as before in β-decay of C^3H_4, with hydrogen halides ($HX = HCl$, HBr and HI) yield the corresponding C^3H_3X compounds. The reactions of cation $^+C^3H_3$ with ethyl halogenides (RX = EtCl, EtBr, EtI) yield (equation 130) the corresponding methyl halides (C^3H_3X; $X = Cl$ yield $15 \pm 1\%$; $X =$ Br yield $37 \pm 1\%$; $X =$ I yield 100%) and also other radioactive compounds such as EtCl, i-C_3H_7Cl and n-C_3H_7Cl in the case of the [EtCl–C^3H_4] system, EtBr and i-PrBr in the case of the [EtBr–C^3H_4] system, but only MeI in the [EtI–C^3H_4] system (iodine being a better nucleophile than bromine and chlorine). The gas-phase reactions of $^+C^3H_3$ cations, produced from C^3H_4, with ethylene and with [ethylene + water] were also studied[130]. The methyl cations with the ethylene medium initiate reaction 131, which is a cationic polymerization process:

$$EtX + {}^+C^3H_3 \longrightarrow Et—\overset{+}{X}—C^3H_3 \longrightarrow Et^+ + C^3H_3X \qquad (130)$$

$${}^+C^3H_3 + CH_2{=}CH_2 \longrightarrow C^3H_3—CH_2CH_2{}^+ \xrightarrow{C_2H_4} C^3H_3(CH_2)_3CH_2{}^+, \text{ etc.} \quad (131)$$

Non-volatile tritium-labelled compounds have been produced, but the degree of polymerization has been found to be not very high since the intermediate carbocations participate also in other reactions which lead to shortening of the chain length (equation 132). Reaction 132c seems to be the most probable. No simple saturated aliphatic hydrocarbon has been formed in the system '$^+C^3H_3$'—'$CH_2{=}CH_2$'. In a mixture consisting of $^+C^3H_3$, $CH_2{=}CH_2$ and H_2O no n-alcohols containing more than two carbon atoms have been detected. This means that the isomerization of the carbocations formed is a faster process than their reactions with water or with ethylene. The reaction

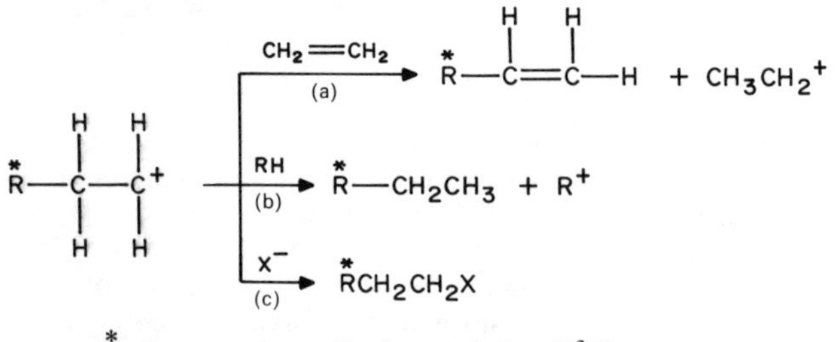

$$\overset{*}{R} \text{ denotes the radical containing } C^3H_3 \text{ group}$$

scheme in equation 133 cannot explain the observed formation of tritiated ethanol. Equation 134 has been suggested as the probable route for the observed $C^3H_3CH_2OH$ formation.

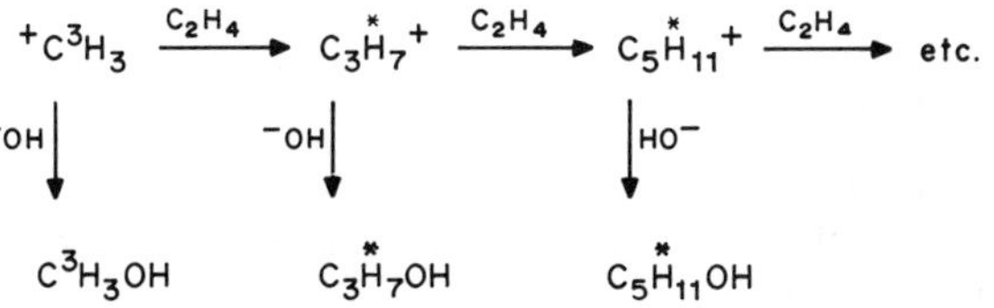

$$^+C^3H_3 \xrightarrow{C_2H_4} C_3\overset{*}{H}_7{}^+ \xrightarrow{C_2H_4} C_5\overset{*}{H}_{11}{}^+ \xrightarrow{C_2H_4} \text{etc.}$$

$$\downarrow^{-OH} \qquad \downarrow^{-OH} \qquad \downarrow^{HO^-} \qquad\qquad (133)$$

$$C^3H_3OH \qquad C_3\overset{*}{H}_7OH \qquad C_5\overset{*}{H}_{11}OH$$

$$CH_2{=}CH_2 + {}^+C^3H_3 \longrightarrow C^3H_3CH_2CH_2{}^+ \longrightarrow \underset{CH_3}{\overset{C^3H_3}{>}}CH^+$$

$$\xrightarrow{H_2O} \underset{CH_3}{\overset{C^3H_3}{>}}CH{-}\overset{+}{O}\overset{H}{\underset{H}{\big<}} \longrightarrow C^3H_3CH_2OH + {}^+CH_3$$

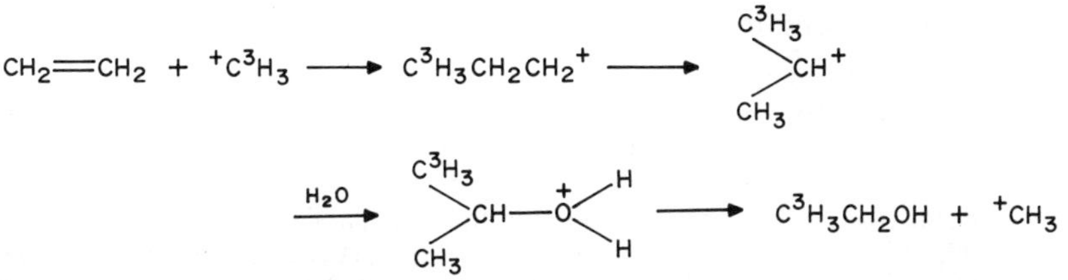

$$(134)$$

In the reaction of free $^+C^3H_3$ cations with a gaseous ethylene–water (1:1) mixture, the relative yields of the tritiated products were as follows: MeOH—82.5 ± 8.0%, EtOH—4.1 ± 0.8%, i-C_3H_7OH—4.1%, $tert$-$C_5H_{11}OH$—3.4 ± 0.3%, 1-$C_5H_{11}OH$—3.4%, and two unknown products in 3.4 ± 0.8% and 2.5% yields. In the reaction of $^+C^3H_3$ with 5:1 gaseous mixture of ethylene and water, the corresponding relative yields have been found to be 18.9%, 21.2%, 5.7%, 23.6%, 18.5%, 2.5% and 9.6%.

2. Reactions promoted by the decay of tritium atoms contained in ethane

[1,2-$^3H_2$]Ethane, obtained by reacting ethylene with $^3H_2$ gas over a chromium catalyst at $-7\,^\circ$C[131,132], has been allowed to decay either as above or in the presence of 0.5% O_2, at atmospheric pressure. The labelled ethyl ions formed in about 80% of the β-transitions undergo a hydride ion transfer reaction (equation 135) with inactive ethane molecules.

$$C_2H_4T^+ + C_2H_6 \longrightarrow C_2H_5T + C_2H_5{}^+ \qquad (135)$$

The other identified labelled products (produced with the following radiochemical yields: $^1H^3H$—11.6%, methane—3.0%, ethylene—5.5%, propane—1.6%, n-butane—2.7%) originate from fragmentation processes that take place in about 20% of the decay events and from the reactions of the fragment ions. $^3H^1H$ is produced in reactions 136. Methane and ethylene are formed in reactions 137 and also in equations 138 and 139.

$$(C_2H_4{}^3H^3He)^+_{exc.} \longrightarrow C_2\overset{*}{H}_3{}^+ + {}^3He + \overset{*}{H}_2$$

$$\longrightarrow C_2\overset{*}{H}_2{}^+ + {}^3He + \overset{*}{H}_2 + \overset{*}{H}$$

$$\longrightarrow C_2\overset{*}{H}{}^+ + {}^3He + 2\overset{*}{H}_2 \qquad (136)$$

$$\longrightarrow C_2{}^+ + {}^3He + 2\overset{*}{H}_2 + \overset{*}{H}$$

$$C_2\overset{*}{H}_3{}^+ + C_2H_6 \longrightarrow C_2\overset{*}{H}_4 + C_2H_5{}^+ \qquad (137)$$

$$C_2\overset{*}{H}_3{}^+ + C_2H_6 \longrightarrow C\overset{*}{H}_4 + C_2\overset{*}{H}_5{}^+$$

$$C_2\overset{*}{H}_2{}^+ + C_2H_6 \longrightarrow C_2\overset{*}{H}_4 + C_2\overset{*}{H}_4{}^+ \qquad (138)$$

$$C_2\overset{*}{H}{}^+ + C_2H_6 \longrightarrow C\overset{*}{H}_4 + C_3\overset{*}{H}_3{}^+ \qquad (139)$$

The suggestion has been made[131] that tritiated butane, C_4H_9T, is formed in ethane (2.7% yield) and in propane (1% yield) by reactions of tritiated fragment ions that give rise to C_3 and C_4 species. But it is also possible[131] that in the presence of a third body ethyl ions could react with ethane at high pressure to produce C_4 ions competitively with the reaction shown in equation 135.

Quite satisfactory agreement between the experimental and theoretical yields, expected from ion–molecule reactions, has been obtained assuming a lack of large isotope effects in the fragmentation of the primary decay species. Wexler and Hess[133] have determined mass-spectrometrically the decay-induced fragmentation pattern of a series of tritiated organic molecules including C_2H_5T.

3. Chemical consequences of β-decay in [1,2-$^3H_2$] propane

The use of multi-labelled alkanes has been extended for reactions of the carbocations produced in the decay of propane-1,2-$^3H_2$ both in gaseous and in liquid phase[134]. Low-pressure, decay-induced fragmentation of propane-1-^3H and propane-2-^3H has been investigated by Wexler, Anderson and Singer[135] by mass spectrometry, and the yields of $C_3H_7^+$ ions in the two reactions have been found to be 56% and 41%. In contrast, much higher yields (72–83%) of initial fragments were found in the decay of CH_3^3H, $C_2H_5^3H$, $C_6H_5^3H$ and four monotritiated toluenes. The $C_3H_7^+$ ions undergo facile decomposition into $C_3H_5^+$ ions and H_2 with an activation energy of 4–10 kcal mol^{-1} for primary and 30–40 kcal mol^{-1} for secondary propyl ions. Propane-1,2-$^3H_2$ has been obtained by reacting propylene (0.033 mmol) with tritium gas (4 Ci) over a chromium oxide–gel catalyst at $-12\,^\circ$C for 36 h (equation 140).

$$CH_3CH{=}CH_2 \xrightarrow{\ T_2\ } CH_3CHT{-}CH_2T \tag{140}$$

The crude labelled propane has been purified by preparative gas chromatography and diluted with inactive propane to the desired specific activity (ca 1 mCi mmol^{-1}). Preliminary investigations have shown that the contribution of self-radiolytic processes caused by β-particles produced in the decay of $C_3H_6^3H_2$ during storage (R.T., 1 atm., 4 weeks) of samples having a specific activity less than 0.5 mCi mmol^{-1} is negligible. The decomposition has also been studied in pure liquid propane and also in the presence of 2% O_2, at $-130\,^\circ$C during 4 weeks. In gaseous propane the total tritium activity found in products other than propane was only 34.9%. It was assumed that the decay-induced fragmentation processes do not depend on the pressure (as with methane–$^3H_4$ and ethane–$^3H_2$), and that the fraction of $C_3H_6^3H^+$ ions escaping dissociation is higher at 760 torr than at low pressure (10^{-5} torr) and the dissociation process (equation 141) is

$$C_3H_6^3H^+ = C_3\overset{*}{H}_5^+ + \overset{*}{H}_2 \tag{141}$$

$$C_3H_4^3H^+ + C_3H_8 \longrightarrow C_3H_5^3H + C_3H_7^+ \tag{142}$$

largely prevented at 1 atm. by collisional deactivation processes. The yield of propene formed in the hydride ion-transfer process (equation 142) is only one-third of the value expected from low-pressure fragmentation of an equimolecular mixture of propane-1-^3H and propane-2-^3H following β-decay[135]. The formation of tritiated products, such as $^1H^3H$, Me^3H, Et^3H, $C_2H_3^3H$, C_2H^3H, $C_3H_5^3H$, $C_3H_3^3H$, besides $C_3H_7^3H$, suggest that only 70–80% of organic ions are produced by the nuclear β-decay in states of low excitation energy and are stabilized in collisions at 760 torr. The remaining 20–30% of the β-transitions produces organic ions in higher excited states (up to 20 eV) which fragment into smaller ions and yield finally tritiated products other than $C_3H_7^3H$. In the liquid phase (at $-130\,^\circ$C) the tritiated propylene is the major product observed, while the yield of other tritiated products is sharply reduced. This suggests that the shorter time required for collision in the liquid phase tends to stabilize all the daughter $C_3H_5{}^3H^+$ ions formed in lower excitation states, including part of those with higher energies which are dissociating in the gas phase at 1 atm. These stabilized ions react with C_3H_8 (equation 143) yielding tritiated propane.

$$C_3H_6{}^3H^+ + C_3H_8 \longrightarrow C_3H_7{}^3H + C_3H_7{}^+ \tag{143}$$

The highly energetic $C_3H_6{}^3H^+$ dissociates according to equation 144:

$$C_3H_6{}^3H^+ \longrightarrow C_3\overset{*}{H}_5{}^+ + \overset{*}{H}_2 \tag{144}$$

The $C_3H_4{}^3H^+$ ions reacting with inactive propane give tritiated propylene, $C_3H_5{}^3H$, through a hydride ion transfer process (equation 145):

$$C_3H_4{}^3H^+ + C_3H_8 \longrightarrow C_3H_5{}^3H + C_3H_7{}^+ \tag{145}$$

The yields of tritiated methane, acetylene and ethylene in liquid propane are much lower than in the gas phase.

4. Chemical consequences of β-decay in [1,2-³H₂]cyclopentane

[1,2-$^3H_2$]Cyclopentane, obtained by reacting cyclopentene with $^3H_2$ over a Cr_2O_3 catalyst at $-10\,°C$ during 24 h, has been allowed to decay during two months in pure cyclopentane at 700 torr or in cyclopentane with 2% O_2 at pressures ranging from 10 to 700 torr[136]. Over 90% of the β-transitions yield monotritiated cyclopentane (46) as the

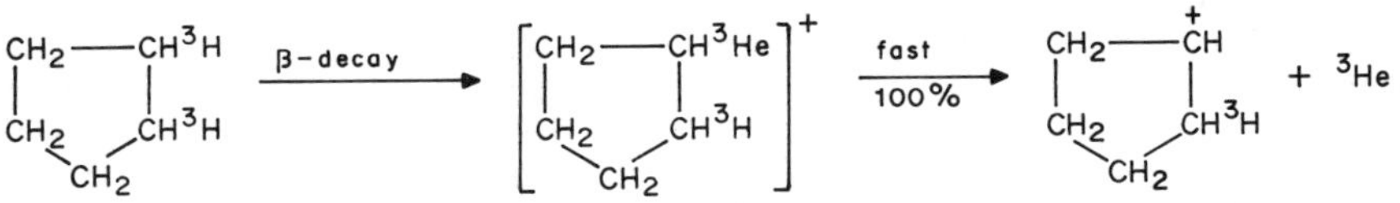

(47)   (146)

major labelled product (equations 146 and 147). The $C_5H_8{}^3H^+$ daughter ions 47, formed in the ground state or with an excitation energy insufficient to cause further dissociation, abstract a hydride ion from inactive cyclopentane yielding 46 (equation 147) which could not be detected by radio-GLC, since the available columns did not allow the separation of the undecayed cyclo-$C_5H_8{}^3H_2$ from the minor ($\leqslant 1\%$) cyclo-$C_5H_9{}^3H$ component.

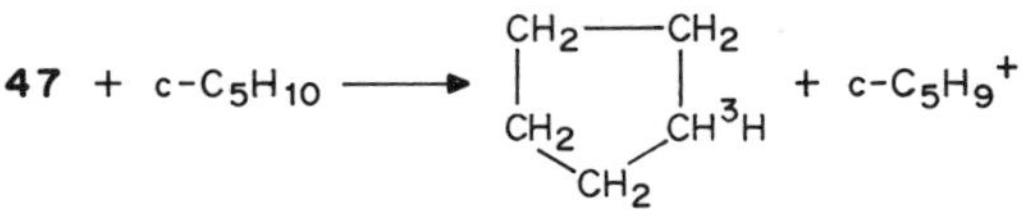

(46)

A small fraction of the ions receive in the β-transition sufficient excitation energy to cause isomerization and fragmentation of the parent molecule to tritiated $C_1–C_5$ hydrocarbons. The yields in the case of cyclopentane at 700 torr have been: MeT—1.6%, Et$^3H_3$—0.4%, $C_2{}^1H^3H$—about 0.1%, n-Pr^3H—0.5%, n-Bu^3H—0.7%, [$^3H_1$]pent-1-ene—3.5%, combined yields of the tritiated products—6.7%.

Equation 148 is the likely route to the major linear by-product 48.

Formation of tritiated propane and acetylene requires the rupture of two C—C bonds in cyclopentyl ions with considerable excitation energy.

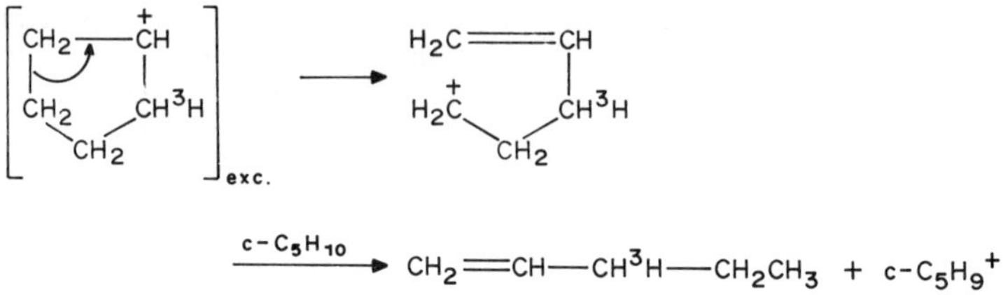

$$
\left[
\begin{array}{c}
CH_2 \overset{+}{-}CH \\
| \\
CH_2 \quad CH^3H \\
\diagdown CH_2
\end{array}
\right]_{exc.}
\longrightarrow
\begin{array}{c}
H_2C \!=\!=\! CH \\
\overset{+}{H_2C} \diagdown CH^3H \\
CH_2
\end{array}
$$

$$
\xrightarrow{c-C_5H_{10}} CH_2\!=\!=\!CH\!-\!CH^3H\!-\!CH_2CH_3 + c\text{-}C_5H_9{}^+ \qquad (148)
$$

(48)

D. Chemical Effects of β-Decay of ^{14}C Studied by Double Isotopic Labelling

1. β-Decay of ethane doubly labelled with ^{14}C

The fate of the nitrogen-14 resulting from the β-decay of carbon-14 in labelled ethane molecules has been investigated. Doubly labelled ethane, $^{14}CH_3{}^{14}CH_3$, was obtained according to equation 149 and was used to observe the rate of formation of singly labelled $^{14}CH_3NH_2$[137].

$$
Ba^{14}CO_3 \xrightarrow{H_2SO_4} {}^{14}CO_2 \xrightarrow{Ba} Ba_2C \xrightarrow{H_2O} {}^{14}C_2H_2 \xrightarrow[Pd]{H_2} {}^{14}C_2H_6 \qquad (149)
$$

To the gaseous $^{14}C_2H_6$, a known amount of CH_3NH_2 carrier was added, and the $^{14}CH_3NH_2$ diluted with $^{12}CH_3NH_2$ was purified by reacting it with phenyl isothiocyanate (equation 150), followed by repeated crystallization of the ^{14}C-methyl phenyl thiourea, 49, up to constant specific activity. Precise specific activity determinations of 49 revealed that the fraction of the labelled ethane molecules which do not dissociate during the $^{14}C \longrightarrow {}^{14}N$ decay, but become methyl amine molecules, equals 47% in agreement with theoretical expectations[138]. The maximum energy of β-particles emitted by ^{14}C is relatively low, namely 0.154 MeV; the maximum recoil energy of the daughter ^{14}N is also low, 7.0 eV. 59% of this recoil energy (4.1 eV) will go into internal motions of the $^{14}CH_3{}^{14}NH_3{}^+$ molecule. A recoil energy of 3.56 eV is required to supply an energy in vibration and rotation equal to the C—N bond energy of 2.1 eV[137] (this energy can be provided in the emission of β-particles of 80 keV energy). However, the β-emitting isotopes have an average energy $\langle \beta^- \rangle$ per disintegration of ^{14}C which equals 49.5 keV only[10]. Thus, in about 80% of the decay events the β-particles have less energy than required to cause carbon–nitrogen bond rupture. Probably, electronic excitation of the $MeNH_3{}^+$ cation directly, or in the course of its transformation to $MeNH_2$, results in additional ruptures of C—N bonds since the observed bond dissociation versus non-dissociation ratio reaches about the 1:1 level.

$$
Ph\!-\!N\!=\!C\!=\!S + {}^{14}CH_3NH_2 \longrightarrow PhNHC\overset{\displaystyle S}{\overset{\|}{}}NH\!-\!{}^{14}CH_3 \qquad (150)
$$

m.p. 115 °C (49)

E. Reactions of ^{11}C Hot Atoms

The reactions of ^{14}C and ^{11}C recoil atoms with simple aliphatic and alicyclic hydrocarbons investigated in gaseous and liquid phase have been reviewed by Stöcklin[89,139]. Acetylene-^{11}C has been the most abundant product[140] when the $^{12}C(\gamma,n)^{11}C$ nuclear reaction took place in benzene, n-hexane, 2-methylpentane or 2,2-dimethylbutane targets. Ethylene-^{11}C and ethane-^{11}C became increasingly important when branching in the target molecules increased. The hot reaction products are the result of insertion of ^{11}C atoms and the moieties ^{11}CH and $^{11}CH_2$ into C—H bonds followed by bond ruptures in the resulting intermediates. Decrease of the fragmentation products and increase of stabilization products has been found in the condensed phase. The energetic radicals do not distinguish between various C—H compounds.

The recoil chemistry of carbon-11 in liquid $C_5–C_7$ hydrocarbons has been investigated by Clark[141]. The recoil atoms have been produced in the $^{12}(\gamma,n)^{11}C$ reaction. The study of the product yields of ethane, ethylene and acetylene (as well as of methane) from different target molecules has been the main concern in this work. Iodine affected the yields of all the volatile products except acetylene. Acetylene-^{11}C, the principal products in all the hydrocarbons investigated, is produced in hot reactions with naked ^{11}C atoms. Insertion of recoil fragments into C—H and C=C bonds leads to various ^{11}C-labelled hydrocarbons.

IV. ISOTOPIC HYDROCARBONS IN BIOLOGICAL AND NATURAL SYSTEMS

The study of the mechanism of the oxidation and assimilation of *iso*-paraffins and paraffins by microorganisms is needed for the solution of two practical problems: protein synthesis by microorganisms utilizing the by-products of the petroleum industry as a source of carbon and of energy[142–144] and decontamination of the marine environment from oil pollution[145,146]. The elucidation of the role and fate of hydrocarbons in living organisms has also been a general concern of biologically oriented chemists. These researches are described in the following sections.

B. Hydrocarbon Oxidations by Bacterial Enzyme Systems

1. Octane[1-^{14}C] oxidation by Pseudomonad bacterium strain

A cell-free, soluble enzyme preparation of a Pseudomonad bacterium strain, isolated from soil[147a], has been found to oxidize octane to n-octanol and octanoic acid. Pyridine nucleotide, oxygen and Fe^{++} ions are involved in this bioconversion. The reaction has been studied at 28 °C with a system consisting of buffer, octane[1-[4]C] (specific activity 5×10^4 cpm μmol^{-1}) dissolved in EtOH or acetone, enzyme and suitable cofactors. The reaction could be stopped by the addition of dilute H_2SO_4. Suitable work-up of the system yielded unreacted ^{14}C-octane, ^{14}C-octanol and ^{14}C-octanoate, respectively. The sum of the radioactivity found in octanol and octanoate served as a measure of the hydrocarbon-oxidizing activity of the enzyme system since the intermediates (hydroperoxide, octanal) do not accumulate.

A linear relationship has been found between the amount of enzyme employed and the [1-^{14}C]octanol and [1-^{14}C]octanoate produced. The products have also been linearly dependent on time. The oxidase system could be separated into two distinct enzyme fractions. One of them is stabilized by ascorbate, the second requires Fe^{+2} or Fe^{+3} ions for activity. The oxidation is the result of a direct attack by an (activated oxygen) and its mechanism in the investigated cell-free enzyme preparations is in accord with the one found in intact bacteria. DPNH (or DPN) and Fe^{+2} (or Fe^{+3}) ions participate in the initial oxidative attack at the methyl group (rather than at a methylene group) of paraffins having seven or more carbon atoms and produce primary alcohols with the same carbon skeleton. A suggestion has been made [147b–e] that flavoproteins participate in enzymatic reactions involving pyridine nucleotides and oxygen.

2. Microsomal hydroxylation of [1-^{14}C]decane

The mechanism of oxidation of decane has been studied[148a] using [1-^{14}C]decane. The oxidation of decane to decanol, decanoic acid and decamethylene glycol by mouse liver microsomes required NADPH and O_2. This indicates that the process is initiated by the hydroxylation of decane to decanol. The Michaelis constant for this reaction was found to be 0.5 mM and the optimum pH was about 8.1. The reaction has been strongly inhibited by CO and t-butyl isocyanide. Evidence has been presented indicating that the enzymic systems for decane hydroxylation and for decanoate ω-oxidation are not identical. Organometallic oxidation catalysts have been reviewed by Speier[148b.]

B. Metabolism of Isotopically Labelled Paraffins

1. Metabolism of deuteriated n-hexadecane

n-Hexadecane containing an excess of deuterium, fed to rats for a period of 9 days, has been found to be absorbed from the gastrointestinal tract and partially deposited in the lipid tissue[149]. The absorbed deuteriated hexadecane has been oxidized to fatty acid in the body, largely in the liver. 7.8.9.10-Tetradeuterio hexadecane used in this bioexperiment has been prepared by Kolbe electrolysis of potassium α,β-dideuterio pelargonate $CH_3(CH_2)_5(CHD)_2COOK$, which in turn has been obtained by addition of deuterium to the double bond of nonen-2-oic, $CH_3(CH_2)_5CH{=}CHCOOH$, while the latter has been prepared by a malonic acid synthesis[150]. No deuterium exchange of α-hydrogens of the acid and the hydrogen of the solvent has been noticed in the conditions of the Kolbe synthesis.

2. Oxidation of [1-^{14}C]n-hexadecane

a. The oxidation of [1-^{14}C]n-hexadecane to palmitic acid, $C_{15}H_{31}$ $^{14}COOH$, by subcellular fractions of the intestinal mucosa of guinea pig has been investigated by Mitchell and Hübscher[151,152a]. Nonlinear kinetics have been observed. Maximum formation of palmitic acid occurred when both nucleotides involved in the oxidation and hydroxylation, i.e. NAD and NADP + glucose-6-phosphate, were added to the incubation system. A 42% reduction in the amount of palmitic acid formed has been found when the air phase was replaced by carbon monoxide. Only labelled n-hexadecanol had been isolated among the possible lobelled products.

A search for other labelled fatty acids has been unsuccessful. Paraffins are minor components of the dietary lipids and are absorbed from the small intestine. The presence of an enzyme system transforming alkanes to fatty acids prevents the harmful accumulation of paraffins in the body. It has been suggested[152a–g] that this oxidation proceeds according to two simultaneously operating mechanisms. In the one, the alkane is first hydroxylated and then dehydrogenated to aldehyde and in turn oxidized to the acid. In the second mechanism, n-1-alkene is the primary product which is subsequently hydrated and dehydrogenated to aldehyde (and acid). However, the identification of the second intermediate, n-hexadecene, has been tentative because of analytical difficulties encountered in the course of separation of n-hexadecane and n-hexadecene. The problem of the correct mechanisms of oxidation in biosystems can be solved by studying the kinetic deuterium isotope effects in the oxidation of [1-D_3]- and [1,2-D_5]-n-hexadecanes as well as ^{14}C kinetic isotope effects in the oxidation of [1-^{14}C] and [2-^{14}C]-n-hexadecanes to corresponding palmitic acids.

b. 1-^{14}C hexadecane and 1-^{14}C octadecane are converted directly to the corresponding fatty acids of the same chain length in the rumen of goat, rat and chicken as well as in liver homogenates from goat and rat[153]. Ichinara, Kusunose and Kusunose[154] have found that the microsomal fraction of mouse liver contains an enzyme system requiring NADPH and oxygen which hydroxylates long-chain aliphatic hydrocarbons[154]. The same authors[155] investigated also the oxidation of n-hexadecane by a mouse liver misrosomal fraction[156]. Radioactive scan of a thin-layer chromatogram of the oxidation products of 1-^{14}C hexadecane showed a main peak corresponding to the [1-^{14}C] substrate and two additional peaks corresponding to ^{14}C-palmitic acid and to ^{14}C-cetyl (n-hexadecyl) alcohol, the products of hydroxylation, obtained in 10–45% yield. Carbon monoxide strongly inhibited this reaction, suggesting the involvement of cytochrome P-450[157]. Bergstrom and collaborators[158] have found that administration of 2,2-dimethyl[1-^{14}C]stearic acid, $CH_3(CH_2)_{15}C(Me)_2{}^{14}COOH$, to the rat led to excretion of 90% of the administered label in the urine as 2,2-dimethyl[1-^{14}C]adipic acid, $HOO^{12}C(CH_2)_3C(Me)_2{}^{14}COOH$. This has been used as evidence for the ω-oxidation mechanism. Weeny[159] administered to rats a 2% solution of [1-^{14}C]n-hexadecane in olive oil and found that, after 15 h, 20–30% of the radioactivity appeared in the lymph lipids as fatty acid. It has been concluded that the oxidation of the alkane occurred in the epithelial cells of the intestinal mucosa.

3. The metabolism of [1-^{14}C]octadecane

n-Alkanes have been identified in bovine and human cardiac muscle and in animal brain[160,161]. The metabolism of [1-^{14}C]octadecane in rats has been studied by Popovic[162]. A paraffin emulsion containing 10–20 μCi (specific activity 21 mCi mmol^{-1}) of [1-^{14}C]n-octadecane has been administered to rats both orally and intravenously. The presence of the $^{14}CO_2$ in the air expired showed that the rats are able to absorb and metabolise n-octadecane. The radioactivity has been present in all tissues investigated (liver, spleen, heart, kidney, lung, brain). Most of the ^{14}C activity has been found in the liver. After intravenous administration of [1-^{14}C]-n-octadecane, the spleen was found to contain extremely high radioactivity. Liver lipid analysis showed that n-octadecane incorporates into fatty acids.

4. The metabolism of [8-^{14}C]n-heptadecane

The metabolism of [8-^{14}C]n-heptadecane, widespread in the marine food chain and one of the components of n-paraffins used for the growing of yeast, has been studied in rats by Tulliez and Bories[163]. About 65% of the ^{14}C-labelled heptadecane has been used by the rats as an energy source and completely metabolized to $^{14}CO_2$. There was no hydrocarbon in the urine. The radioactivity in the faeces has been entirely contained in heptadecane. About 7% of the absorbed ^{14}C-heptadecane has been stored while the rest has been ω-oxidized to ^{14}C-heptadecanoic acid, which underwent in turn the normal fatty acid degradation pathway and contributed to the synthesis of lipids and non-lipids. The heptadecanoic and heptadecenoic acids are the most strongly labelled acids. The presence of $C_{(13)}$ and $C_{(15)}$ fatty acids indicates that the heptadecanoic acid undergoes also the usual β-oxidation. The presence of ^{14}C-labelled squalene, $C_{30}H_{50}$, and cholesterol (which needed ^{14}C-acetate and ^{14}C-acetyl CoA for their synthesis) indicate that (8-^{14}C)heptadecanoic acid had been thoroughly metabolized. The even distribution of radioactivity in the fatty acids of the phospholipids implies that heptadecane did not interfere with the biochemical mechanisms of these functional lipids.

5. Metabolism of 2-[^{14}C-methyl]n-hexadecane and [1-^{14}C]octadecane by yeast C. guilliermondii

The utilization of 2-methylhexadecane-methyl-^{14}C (50) and octadecane[1-^{14}C] (51) in various combinations by the yeast *Candida guilliermondii* has been investigated by

Davidov and coworkers[144]. The kinetics and the rate of yeast assimilation of $CH_3CH(^{14}CH_3)$—$(CH_2)_{13}CH_3$ (50) and $CH_3(CH_2)_{16}{}^{14}CH_3$ (51) dissolved in various combinations in the inert component pristane, 2,6,10,14-tetramethylpentadecane (52), and containing *C. guilliermondii*, has been studied[144]. 52 does not undergo degradation but enlarges the volume of the hydrocarbon phase to 12–13% of the total volume and creates conditions imitating the process of cultivation of yeast on petroleum. ^{14}C from 50 has been transformed into $^{14}CO_2$ indicating complete degradation of 50 by *C. guilliermondii*. The rate of oxidation of the branched 50 was 5–6 times lower than that of the n-alkane 51. 5.5% of the carbon of the biomass originates from carbons of 50, which incorporates mainly into lipidic fractions(triglycerides, diglycerides, phospholipids and free fatty acids) and water-soluble products. Cell lipids incorporate only 15-methylhexadecanoic acid formed as a result of oxidation at the unbranched end of the aliphatic chain. It has been noted that significant quantities of the ^{14}C-label of 50 accumulate in the amines and non-proteinogenic amino acids which are released in appreciable amounts from the cell into the extracellular culture liquid. 85% of the radioactive label of the cellular organic acids and the culture liquid has been found in isovaleric acid. Equation 151 has been proposed for the metabolic pathway of 50 in *C. guilliermondii*.

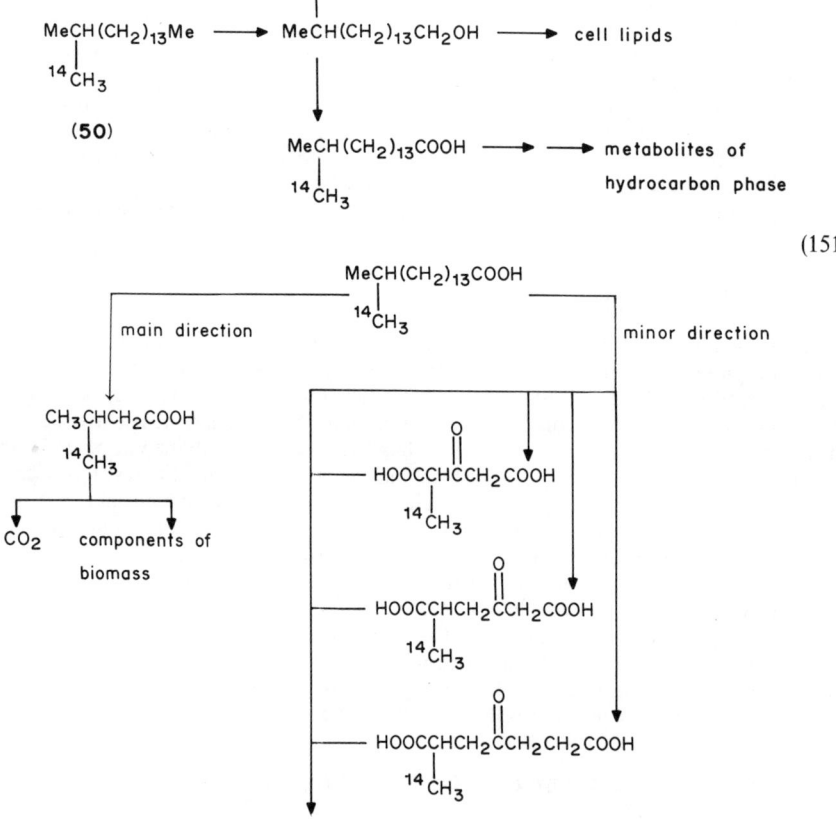

(151)

6. Hydrocarbons, fatty acids and drug hydroxylation catalysed by liver microsomal enzyme systems

Liver microsomes in the presence of NADPH and molecular oxygen catalyse the hydroxylation of various lipid-soluble drugs (equation 152). A hypothesis has been

$$RH + NADPH + H^+ + O_2 \longrightarrow ROH + NADP^+ + H_2O \qquad (152)$$

proposed that these microsomal hydroxylations proceed by way of a common enzyme system which involves the NADPH–cytochrome c reductase and the cytochrome P-450. This hypothesis has been tested[164] by studying the oxidation of [^{14}C]testosterone, heptane and ω-oxidation of lauric acid, $Me(CH_2)_{10}COOH$, in rat liver microsomes. Both laurate and heptane stimulated the rate of NADPH oxidation catalysed by liver microsomes isolated from rats. Heptyl alcohol from heptane and 12-hydroxylauric acid from lauric acid have been isolated and identified besides various hydroxylation products of testosterone (2β-hydroxy-, 6α-hydroxy-, 7α-hydroxy- and 16α-hydroxytestosterone). The inhibition in the presence of CO has been of about 30% (carbon monoxide interacts with cytochrome P-450 forming an enzymatically inactive complex); 10–40% inhibition has been observed when pure nitrogen has been used as the gas phase, and 40% CO in the gas phase caused a complete inhibition of the stimulation of the rate of NADPH oxidation by all four substrates tested (aminopyrine, testosterone, heptane and laurate). However, in the absence of any added substrate only one-third of the overall NADPH disappearance catalysed by the microsomes was cancelled when 40% CO was added to the gas phase. This indicates that cytochrome P-450 is involved only in part of the reactions in which NADPH participates. Obviously further experimental work is needed to elucidate this microsomal hydroxylating enzyme system. A mechanism for the hydroxylation of alkanes and fatty acids and also for that of the demethylation of various drugs (equation 153) by the liver microsomal cytochrome P-450 system (Scheme 1), has

$$R_2NCH_3 + NADPH + H^+ + O_2 \longrightarrow R_2NH + CH_2O + NADP^+ + H_2O \quad (153)$$

been proposed as a working hypothesis by Coon and coworkers[165] showing the possible role of the superoxide radical ion ($O_2^{\cdot-} \rightleftharpoons O_4^{2-}$). It is possible that cytochrome P-450 generates the intermediate 'superoxide' in the presence of molecular oxygen. In the final step, e, the attack of superoxide on the substrate, accompanied by the uptake of a second electron, yields the hydroxylated substrate and water (equation 154).

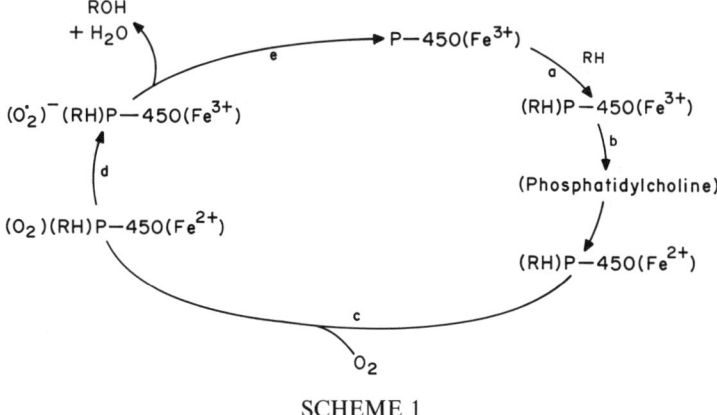

SCHEME 1

$$RH + 2O_2^{\cdot-} + 2H^+ \longrightarrow ROH + H_2O + O_2 \qquad (154)$$

Oxo-bridged binuclear iron centres in oxygen transport, oxygen reduction and oxygenation have been studied and reviewed by Sanders-Loer and coworkers[166,167]. Oxidations and reductions in general, including reported reductions and oxidations of biological interest, have been reviewed critically by Fleet[168].

C. Isotopic Environmental Bio-geological Studies

The recent general availability of precise McKinney–Nier-type isotopic ratio mass spectrometers allows one to measure[169-172,172a,172b,172c] the natural variations in stable isotope concentrations of organic and inorganic matter, including hydrocarbons. The results are expressed as stable hydrogen (δD) and carbon ($\delta^{13}C$) isotope ratios, where δ is defined by

$$\delta(\text{‰}) = [(R_{sample}/R_{standard}) - 1] \times 1000$$

(R is D/H or $^{13}C/^{12}C$, and the standards are Vienna Standard Mean Ocean Water[172] and Peedee Belemnite[170], respectively, for δD and $\delta^{13}C$).

These data have been reviewed recently[173,174] by Tan for inorganic carbon isotopes dissolved in marine and estuarine (Saguenay Fjord and St. Lawrence Estuary) environments and by Anati and Gat[175a] and Pierre[175b] for restricted marine basins and marginal sea environments. Sackett[176] wrote a special chapter on stable carbon isotope studies of organic matter in the marine environment and found that stable carbon isotope composition may serve as a good indicator for organic pollutions in marine sediments and as a technique for marine pollution studies in general[177]. The stable hydrogen and carbon isotope composition of sedimentary methane stirred from the Florida Everglades has also been investigated[178,179], but the mean values $\delta^{13}C = -61.7$‰ (standard deviation 3.6‰, $n = 51$) did not differ significantly from the average $\delta^{13}C$ of all natural sources, which is -58.3‰. The mean δD of Everglades sedimentary methane ($\delta D = -293$‰, standard deviation 14‰, $n = 50$) has been found to be slightly enriched in deuterium in comparison with $\delta D = -360 \pm 30$‰ corresponding to methane from all sources. This difference has been explained as the result of partial bacterial oxidation of methane flux evolving from sediments through the marine water to the atmosphere. The ratio of the rate constants (k_H/k_D) for the reactions of CH_4 and CH_3D with OH radicals has been estimated[180] to be 1.5 at 416 K in pulse radiolysis of water vapours. Mean sedimentary methane values may serve as a rough estimate of the average of methane emitted to the atmosphere from the Everglades system.

The isotope composition of gases taken from hermetically sealed holes drilled in rocks of the ijolite–urtit complex[181] of the Khibiny Massif (Kola Penisula), consisting of C_2H_6, C_3H_8, isobutene, butane, isopentane, pentane, C_6H_{14}, H_2, O_2, N_2, Ar, He, CO and CO_2, is characterized by the ratios $^{13}C/^{12}C$ equal to $1.1160\%–1.11482\%$ in ijolites and $1.1160\%–1.11482\%$ in urtites ($^{13}C/^{12}C \cong 0.0112$). Bitumens had a $^{13}C/^{12}C$ percentage ratio equal to 1.0905%. Carbon of hydrocarbon gases from gas and oil is, as a rule, lighter than carbon of petroleum. However, the carbon of hydrocarbon gases from rocks of the Khibiny Massif is heavier than carbon from bitumens from the same massif. This implies that hydrocarbon gases have been formed there at higher temperatures than the bitumens (in line with the hypothesis that bitumens in the Khibiny Massif have inorganic origin).

The experimentally determined $\delta^{13}C$ differences between CH_4 and C_2H_6 and CH_4 and C_3H_6 of natural gases have been found to be most important for the prospecting for petroleum[182]. The $\delta^{13}C$ values of CH_4 vary with the origin and depth of its location and with the CO_2 content. Mass spectrometric determinations of the $^{13}C/^{12}C$ ratios in individual hydrocarbons separated from Italian natural gases by gas chromatography and oxidized to CO_2 led Tongiorgi and his coworkers[183] to the conclusion that while the two-fold bacterial and chemical mechanism of CH_4 origin cannot be discarded, isotopic fractionation taking place in the course of migration of these gases is predominant and responsible for the observed variations in the $^{13}C/^{12}C$ ratios. The structure and ^{13}C contents of individual alkanes in bat guano found in the Carlsbad region, New Mexico[184] have been found to be related to the photosynthetic pathways of local plants and the feeding habits of insects, which constitute a major percentage of the bat's diet. Two isotopically distinct groups of branched alkanes, derived from two chemotaxonomically distinct insect populations having different feeding habits, have been identified (one population of insects feeds on crops and the other on native vegetation).

V. CHEMICAL STUDIES WITH LABELLED HYDROCARBONS

A. Isotopic Tracer Studies with Alkanes and Cycloalkanes

1. ^{14}C study of the dehydrogenation of butane/butene mixtures

Dehydrogenation of hydrocarbons is a most important industrial process. Rubber production from inorganic matter is one of the related examples to which the isotopic kinetic method[185,185a] has been applied[186]. ^{14}C-labelled butane and butene used in this study[186] have been obtained according to equation 155. Various labelled 1:1 mixtures of

$$Ba^{14}CO_3 \xrightarrow{H_2SO_4} {}^{14}CO_2 \xrightarrow{LiAlH_4} {}^{14}CH_3OH \xrightarrow{PI_3} {}^{14}CH_3I \xrightarrow{Mg}$$

$$^{14}CH_3MgI \xrightarrow{C_3H_5Br} {}^{14}CH_3CH_2CH{=}CH_2 \xrightarrow{Ni, H_2} {}^{14}CH_3(CH_2)_2CH_3$$

$$(155)$$

butane and butene have been used in this investigation. Specific activities of butane, butene and butadiene and also of carbon dioxide have been determined in the post-reaction mixture at different 'times of contact' with the chromia catalyst at 635 °C ('contact time' is defined as the reciprocal of the rate of passing the reactant mixture over the catalyst). In the experiments with ^{14}C-butene and unlabelled butane the specific activities of butene and butadiene have been practically the same. However, using ^{14}C-butane with unlabelled butene, the specific activity of butadiene has been found to be smaller than that of butene. This fits equation 156, where ω_1, ω_2 and ω_3 are reaction velocities, α, β and γ are specific

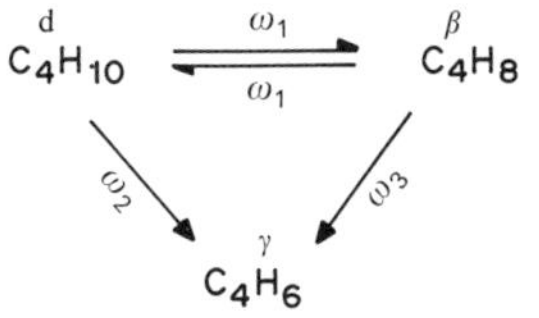

$$\begin{array}{c} \overset{d}{C_4H_{10}} \underset{\omega_1}{\overset{\omega_1}{\rightleftharpoons}} \overset{\beta}{C_4H_8} \\ {}_{\omega_2}\diagdown \quad \diagup {}_{\omega_3} \\ \underset{C_4H_6}{\overset{\gamma}{}} \end{array} \qquad (156)$$

activities of butane, butene and butadiene, respectively, and where $\omega_2 \ll \omega_3$. This means that butadiene is produced by dehydrogenation of butene, and practically no direct transformation of butane into butadiene takes place. $\omega_2:\omega_3 \cong 1:25$ when chromia oxides on alumina are used as catalyst, and $\omega_1:\omega_2 \cong 20:1$. The rates of transfer of the ^{14}C label from butane to butene and from butene to butane have been found to be similar (equation 157),

$$CH_3CH_2CH_2CH_3 + {}^{14}CH_3CH_2CH{=}CH_2{=}CH_2 \longrightarrow$$
$$CH_3CH_2CH{=}CH_2 + {}^{14}CH_3CH_2CH_2CH_3 \quad (157)$$
$${}^{14}CH_3CH_2CH_2CH_3 + CH_3CH_2CH{=}CH_2 \longrightarrow$$
$$CH_3CH_2CH_2CH_3 + \sim \tfrac{1}{2}{}^{14}CH_3CH_2CH{=}CH_2 + \sim \tfrac{1}{2}CH_3CH_2CH_2CH{=}{}^{14}CH_2$$
$$(157a)$$

as the rate constants for both the direct and reverse direction have been found to be nearly equal (negligible ^{14}C isotope effect). More than 90% of the butadiene is produced through butene. Simultaneously with the dehydrogenation of butene, dehydrogenation of butane and hydrogenation of butene also take place.

2. *^{14}C- and ^{2}H-tracer studies of the mechanism of dehydrocyclization of paraffins*

This problem has been extensively reviewed[187,188]. In this section we summarize the results obtained by Kazanskii and coworkers[189-192] and by Paál and Tétényi[193-195]. The problem of intermediates formed over oxide catalysts has been studied by using ^{14}C-labelled compounds[189-192]. The mechanism of C_6-dehydrocyclization has been investigated using n-hexane mixed with various amounts of ^{14}C-labelled cyclohexane. No decrease in the specific activity of cyclohexane has been observed, meaning that the latter is not an intermediate in the process. n-Hexane is transformed into benzene by-passing the stage of cyclohexane formation[190]. When a mixture of hexane and hexene-1-[1-^{14}C] was used, the radioactivity of hexene decreased in the course of the reaction[191] and labelled hexadienes were also found, showing that the formation of benzene proceeds via hexene-1. Similarly it has been shown that hexadiene-[^{14}C] does not cyclo-isomerize to cyclohexene, since cyclohexene acquires no radioactivity when a mixture of hexadiene-^{14}C and cyclohexane is subjected to aromatization. Using a mixture of hexadienes-[^{14}C] with hexatriene, the ^{14}C label has been found both in the cyclohexadienes and in benzene, leading to the conclusion that benzene formation proceeds[191] according to equation 158:

$$\text{paraffin} \longrightarrow \text{olefins} \longrightarrow \text{dienes} \longrightarrow \text{trienes} \longrightarrow \text{cyclodienes} \longrightarrow \text{benzene} \quad (158)$$

Thus dehydrocyclization of hexane and of other n-paraffins over oxide catalysts occurs through dehydrogenation steps up to formation of a triene, which first yields a cyclodiene and then undergoes dehydrogenation to aromatics. Chromia on alumina, molybdenum and vanadium oxides on alumina act according to the above catalytic pathway. Paál and Tétényi[193-195] have investigated the dehydrocyclization of hydrocarbons over metallic Ni, Pt and over mixed Ni/Al_2O_3 and Pt/Al_2O_3 catalysts. The ^{14}C-labelled hydrocarbons have been obtained[196,197] according to equation 159:

$$CH_3(CH_2)_4MgBr \xrightarrow{{}^{14}CO_2} CH_3(CH_2)_4{}^{14}COOH \xrightarrow{LiAlH_4} CH_3(CH_2)_4{}^{14}CH_2OH$$

$$\xrightarrow[-H_2O]{Al_2O_3,\,370°C,\,under\,N_2} CH_3(CH_2)_3CH{=}{}^{14}CH_2 \xrightarrow[185°C]{H_2/Ni} CH_3(CH_2)_4{}^{14}CH_3 \quad (159)$$
$$\textbf{(53)}$$

specific activity $0.56\,mCi\,g^{-1}$

The reaction sequence hexane $\longrightarrow$ hexenes $\longrightarrow$ hexadienes $\longrightarrow$ benzene has been verified using the radioactive tracer technique. A mixture of radioactive n-hexane-[1-^{14}C] and inactive hexene-1 has been subjected to dehydrocyclization in the presence of Pt and Ni catalysts used either as free metals or on alumina carriers. In each experiment the increase in the specific activity of the hexene and hexadiene fraction has been observed. These unsaturated hydrocarbons are therefore intermediates in the dehydrocyclization of hexane. The above conclusion has been subsequently corroborated by studying the dehydrocyclization of a mixture of radioactive n-hexene-1-[1-^{14}C] and inactive cyclohexane[194]. No sign of transfer of the radioactivity between hexene and cyclohexane has been found. No detectable transfer of radioactivity from n-hexane or from n-hexene to cyclohexane has been observed in conversions of various mixtures of open-chain and cyclic hydrocarbons, where one of the compounds has been radioactive. The dehydrogenation of cyclohexane and of n-hexane have been found to be parallel reactions[195] converging only in some of the final reaction steps (equation 160).

$$n\text{-hexane} \longrightarrow n\text{-hexenes} \longrightarrow n\text{-hexadienes} \longrightarrow (n\text{-hexatriene})$$
$$\searrow \text{benzene}$$
$$\text{cyclohexane} \longrightarrow \text{cyclohexene} \longrightarrow \text{cyclohexadiene} \nearrow \qquad (160)$$

Ring closure probably occurs through cyclohexadiene[198], but the elucidation of the exact mechanism of ring closure needs further isotope studies. The initial steps, that is the dissociative adsorption in the dehydrogenation of cyclohexane and in the D/H exchange reactions taking place between cyclohexane and deuterium on metal surfaces, are similar[199,200]. Cyclohexane, cyclohexene and benzene have been found to contain tritium activity when a catalyst covered with tritium has been used in the dehydrocyclization experiments. No hexane $\longrightarrow$ cyclohexane reaction occurs.

H/D exchange data between gaseous and solid states have been used[201] to explain the mechanism of the catalytic action of hydrides of transition metals of group IV–VI(Ti, Zr, Hf, V and Cr) in the conversion of hydrocarbons and to formulate equation 161 illustrating

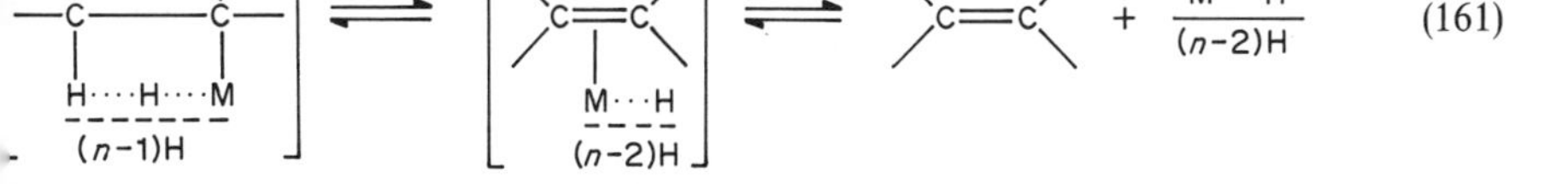

$$(161)$$

the hydrocarbon interaction with the active site of the catalyst. Hydrogen loss from the hydride crystal lattice takes place during the hydrogenation–dehydrogenation process. The distance between metal atoms in hydrides is longer than in hydrogen-free metals and the latter do not accelerate the reaction. The effective length of the M—H bond varies with temperature. The hydrogen atom bound to the reactive site (M—H) differs from the 'free hydrogen atom'.

Whilst discussing the mechanisms of catalytic hydrogenation and hydrogenolysis of methylbenzenes on platinum at 65–110 °C, it was pointed out[202] that the methyl groups exchange hydrogen to deuterium easier than the ring hydrogens.

3. Catalytic reactions of propene[1-^{14}C] on a Y-type decationized molecular sieve

The mechanism of heterogeneous reactions of propylene on Y-type decationized molecular sieve catalysts at about or below 200 °C yielding saturated hydrocarbons such as propane, 2-methylpropane (isobutane), 2-methylbutane and 2-methylpentane has been studied[203] using propene-[1-^{14}C]. The specific activity of propane has been found to be the same as the initial radioactivity of propene. The specific activity of 2-methylpentane has been twice as high as the initial specific activity of propene. The specific activities of isobutane and of isopentane relative to that of propene have been 1.1 and 1.6, respectively. These data indicate that propane is not a decomposition product of 2-methylpentyl cations but is produced by hydrogenation of the isopropyl cations formed on Brönsted-type acidic sites of the catalyst (equation 162), probably by intermolecular hydride ion abstraction. The yield of propane increased in the presence of compounds rich in weakly bound hydrogen such as tetralin, etc., and also when the catalyst surface has been covered with 'carbonaceous deposits' from which atomic hydrogen could be split off according to a radical process. 2-Methylpentane is formed from isopropyl cations with adsorbed propene molecules (equation 163) followed by intermolecular hydrogen transfer.

$$CH_3CH{=}^{14}CH_2 + \{H^+Y^-\} \longrightarrow i\text{-}\overset{*}{C}_3H_7{}^+Y^- \tag{162}$$

$$\{i\text{-}\overset{*}{C}_3H_7{}^+Y^-\} + (\overset{*}{C}_3H_6)_{ads.} \longrightarrow \{\overset{*}{C}_6H_{13}{}^+Y^-\} \tag{163}$$

$$* \text{ denotes a } {}^{14}C \text{ label}$$

The 2-methylpentyl cations undergo isomerization, carbon–carbon bond cleavage and intramolecular hydrogen transfer leading to isobutane and isopentane (equations 164). No methane or ethane has been found in the reaction products, neither have any hydrocarbons containing more than six carbon atoms been isolated.

$$CH_3\overset{+}{C}CH_2CH_2CH_3 \quad\quad CH_3CHCH_3 + (C_2H_3{}^+)$$
$$| \qquad\qquad\qquad \longrightarrow \qquad |$$
$$CH_3 \qquad\qquad\qquad\qquad CH_3$$
specific activity ≈ 2 | relative specific activity ≈ 1.1 $\quad\quad$ (164)

$$CH_3CH\overset{+}{C}HCH_2CH_3 \quad\quad CH_3CHC_2H_5 + (CH^+)$$
$$| \qquad\qquad\qquad \longrightarrow \qquad |$$
$$CH_3 \qquad\qquad\qquad\qquad CH_3$$
specific activity $\cong 2$ relative | relative specific activity $\cong 1.6$

The specific radioactivity of the carbonaceous deposits has not been determined. The intramolecular distribution of the ${}^{14}C$ label in isobutane and in isopentanes has not been studied.

4. ${}^{13}C$ tracer study of the degradation of 3-methylpentane on a super-acid catalyst

Doubly ${}^{13}C$-labelled hexane[204], [1,5-${}^{13}C_2$]3-methylpentane, ${}^{13}CH_3CH_2CH(Me)\cdot CH_2{}^{13}CH_3$(54) containing 78% dilabelled in positions 1 and 5, 21% monolabelled in position 1 and 1% of unlabelled molecules has been used to study the mechanism of degradation of branched hexanes[205] over a catalyst consisting of antimony pentafluoride inserted into graphite ('$C_{6.5}SbF_5$'), at $0\,^{\circ}C$. The distribution of the ${}^{13}C$ label in the 3-methylpentane samples and in the paraffins of lower and higher molecular weight has been determined mass spectrometrically. The isotopic distribution in the parent ion has been unaffected but the isomerization to yield $MeCH(Me)CH_2CH_2Me$, $MeCH_2CH(Me)CH_2Me$ and $Me_2CHCHMe_2$ is a fast intramolecular process leading to isotopic equilibration of all six carbon atoms. The isotopic distribution of the initially produced propane agrees with the hypothesis that it is produced via direct β-scission of the carbenium ion shown in equation 165 followed by intermolecular hydride transfer. However, at the end of the run, propane is obtained involving a polymerization mechanism. Isobutane and isopentane are also produced by a depolymerization–polymerization process. No primary ${}^{13}C$-isotope effect has been noted in the fragmentation of ^{13}C-labelled branched hexanes at $0\,°C$. All studies of the conversion of ^{13}C-labelled **54** over SbF_5 graphite have been rationalized by the scheme in equation 166.

$$CH_3CH(CH_3)CH_2\overset{+}{C}HCH_3 \longrightarrow \longrightarrow CH_2{=}CHCH_3 + CH_3CH_2CH_3 \qquad (165)$$

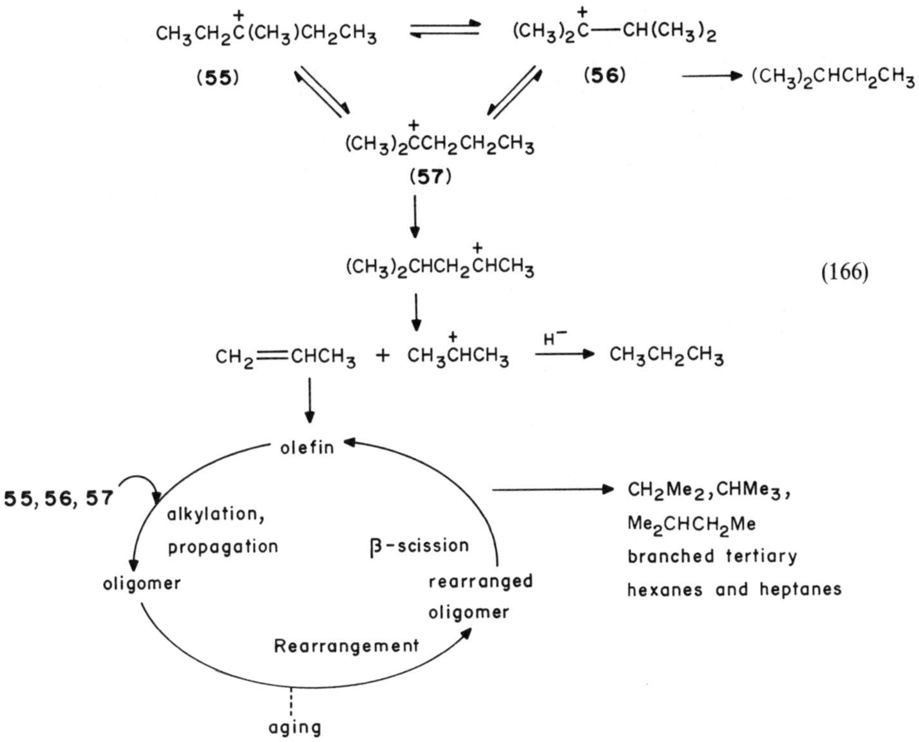

$$(166)$$

5. *Use of perdeuteriopropene to study the isomerization and exchange of olefins on iron catalysts*

The use of perdeuteriopropene, C_3D_6 (containing 5.3% of C_3D_5H), permitted study[206] of the isomerization and exchange processes in olefins without interference from H/D exchange reactions. In all cases adsorbed deuterium atoms produced either by adsorption of D_2 or by dissociative adsorption of perdeuterioolefins on the catalyst have been found to be the reacting species. The same technique has also been applied to study at $-37\,°C$ the exchange and isomerization of ethylene, propene, butenes, pentenes and 2-methylbutenes on iron films prepared by evaporation under high vacuum of a pure iron wire.

Large differences have been observed between H/D exchange rates of α- and β-olefins. In the weakly adsorbed α-olefins the vinylic hydrogen atoms are easily exchanged. Thus in ethylene all four hydrogen atoms underwent facile H/D exchange; in 1-pentene and propene a definite break between d_3 and d_4 in deuterium distribution patterns has been found. (5.7% of D_3 molecules but only 0.75% of D_4 molecules of 1-pentene, $CH_3CH_2CH_2CH{=}CH_2$, have been detected after 6 min of H/D exchange of this olefin in the presence of C_3D_6 on an iron film at $-37\,°C$; in the case of H/D exchange with C_2H_4

the corresponding numbers for D_3 and D_4 molecules are 34.0% and 59.2% after 15 min exchange).

The H/D exchange of β-olefins is extremely slow. Various mechanisms of double-bond migration, taking place on iron films at $-37\,°C$ in the presence of C_3D_6, in α- and β-olefins have been studied and interpreted.* α-Olefins yield poorly exchanged cis and trans isomers, e.g. large amounts of d_0 cis- and trans-butenes. β-Olefins yield highly deuteriated α-olefins with maxima in their deuterium paterns. Thus, pronounced maxima at d_8 and d_5 were found in 1-butenes and 1-pentenes obtained from cis-butene and cis-pentene, respectively. These, as well as the maximum at d_4 in 3-methyl-1-butene obtained from 2-methyl-2-butene, indicate that the dissociative adsorption of the olefin at a vinylic carbon, accompanied by a complete H/D exchange of all adjacent allylic hydrogen atoms followed by intramolecular rearrangement, is the primary act of the $\beta-\alpha$ isomerization (equation 167). The electron attractive character of the metal 'M' in the carbon–metal bond

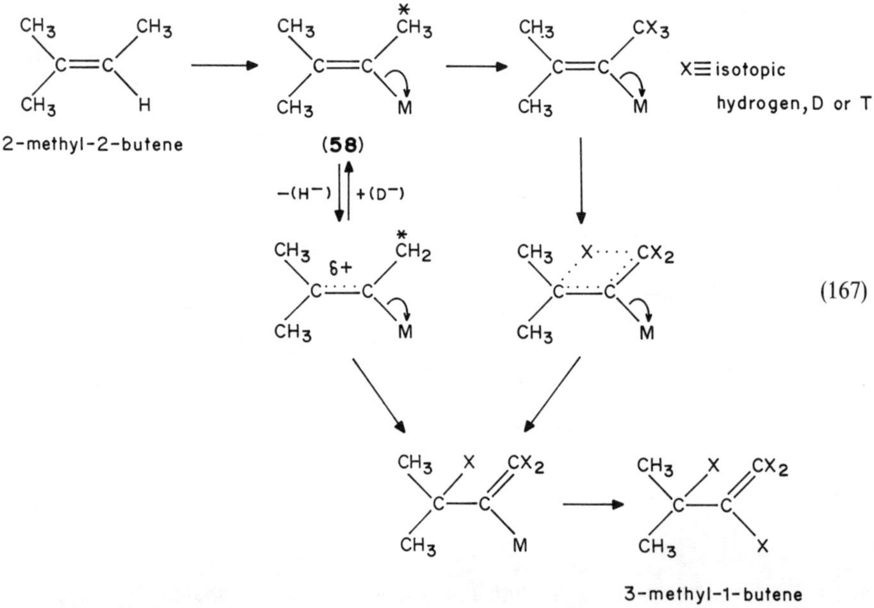

$$(167)$$

in structure **58** loosens the adjacent allylic C—H bonds and allows multiple H/D exchange, hydrogen migration and 'intramolecular rearrangement'. The formation of d_1 molecules from cis-butene and cis-pentene proceeds by addition of a deuterium atom to the double bond, followed by free rotation around the $C_{(2)}$—$C_{(3)}$ bond and dehydrogenation to the adsorbed trans olefin. The presence of d_0 isomers implies the possibility of a direct cis–trans process without formation or breaking of any C—H bond. The authors[206] observed a deuterium kinetic isotope effect of 2.3 for the double-bond migration.

6. Catalytic hydropolymerization of ethylene in the presence of ^{14}CO

A detailed review (containing 163 references) of catalytic reactions leading to the synthesis of hydrocarbons and alcohols from ^{14}CO and hydrogen has been presented by

---

*α- and β-olefins are adsorbed with different strength on two different adsorption sites on the surface. Intramolecular hydrogen shifts are responsible for isomerizations and exchanges.

catalyse the exchange process. Complexes of Pt(0) are also inactive and even at 100 °C no deuterium exchange occurs in the presence of $Pt(PPh_3)_3$ or $PtCO(HCl)_2$ after 10 hours[220].

$$MD_3 \rightleftharpoons MD + D_2$$

$$MD + RH \rightleftharpoons \overset{\overset{\displaystyle D}{\displaystyle |}}{\underset{\underset{\displaystyle H}{\displaystyle |}}{M}}-R \rightleftharpoons DR + MH \qquad (182)$$

The rate of deuterium exchange in ethane in the presence of $K_2[PtCl_4]$ has been found to be proportional to the first power of C_2H_6, 0.62 order with respect to the catalyst concentration and independent of the ionic strength. Polydeuterio ethanes have been found besides EtD in the initial stages of the reactions[220]. This indicates that more than 1 D atom is transferred to ethane in the elementary act of exchange. Equation 183 has been suggested to explain the formation of mono- and polydeuterioethanes.

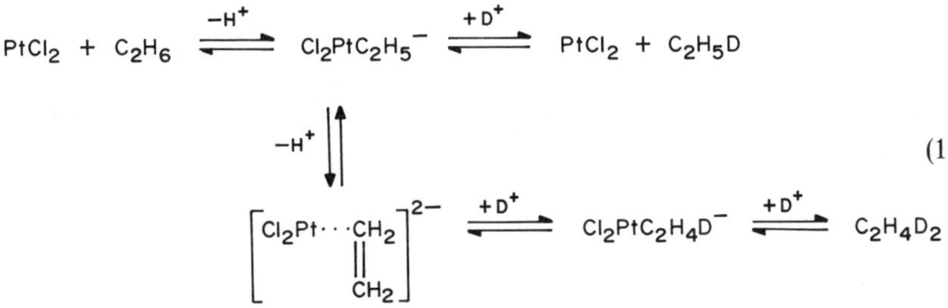

$$PtCl_2 + C_2H_6 \underset{-H^+}{\overset{}{\rightleftharpoons}} Cl_2PtC_2H_5^- \underset{}{\overset{+D^+}{\rightleftharpoons}} PtCl_2 + C_2H_5D$$

$$-H^+ \updownarrow$$

$$\begin{bmatrix} Cl_2Pt\cdots CH_2 \\ \quad\quad\, \| \\ \quad\quad\, CH_2 \end{bmatrix}^{2-} \underset{}{\overset{+D^+}{\rightleftharpoons}} Cl_2PtC_2H_4D^- \underset{}{\overset{+D^+}{\rightleftharpoons}} C_2H_4D_2 \qquad (183)$$

The isotope effect $k_{CH_4}/k_{CD_4} = 3.0 \pm 0.5$ observed in CD_4–H_2O and CH_4–D_2O exchange experiments indicates that the C—H bond rupture is rate-determining. The rates of hydrogen exchange in CH_4, C_2H_6 and AcOH are found to be in the ratio 1:5:0.5. This indicates that electrophilic attack of the platinum compound at electrons of the C—H bond is the most effective one. The activation energy of deuterium exchange in ethane deduced from the data in the 80–100 °C interval equals 18.6 kcal mol^{-1}. The rate of deuterium exchange is 30 times faster in $D_2O/AcOD$ (1:1) than in D_2O only. The above data indicate that the hydrocarbons are activated by entering into the coordination sphere of the central metal atom and that the metal catalyst should have electron-accepting properties (equation 184).

$$[Pt(II)] + RH \rightleftharpoons \left[Pt(IV) \overset{\displaystyle R}{\underset{\displaystyle H}{<}} \right] \qquad (184)$$

Pt(IV) and Pt(0) do not catalyse the deuterium exchange owing to saturation of the coordination sphere and to weakening of the electron-accepting properties. The electron accepting properties and the catalytic activity in complexes $K_2[PtX_4]$ diminish in the series X: Cl > Br > I and in the series $PtCl_2 > PtCl_3^- > PtCl_4^{2-}$ of chloride complexes which are in equilibrium with $K_2[PtCl_4]$ in solution. $PtCl_2$ is the most active species in solution. In the presence of excess Cl$^-$ ions the equilibrium (equation 185) shifts to left.

$$PtCl_4^{2-} \rightleftharpoons PtCl_3^- + Cl^-$$
$$\qquad (185)$$
$$PtCl_3^- \rightleftharpoons PtCl_2 + Cl^-$$

Acetic acid plays the role of a weak chelating agent of $PtCl_2$ (**59**) and prevents the formation of dimers having low catalytic activity (equation 186). The reactions in equation 187 have been proposed as the route[219] leading to EtD formation in $AcOD/D_2O$ solutions.

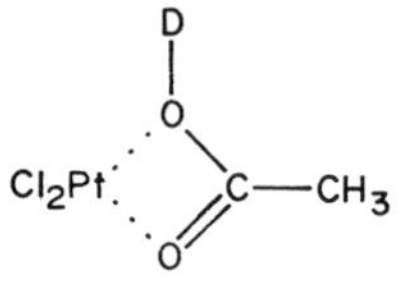

(59)

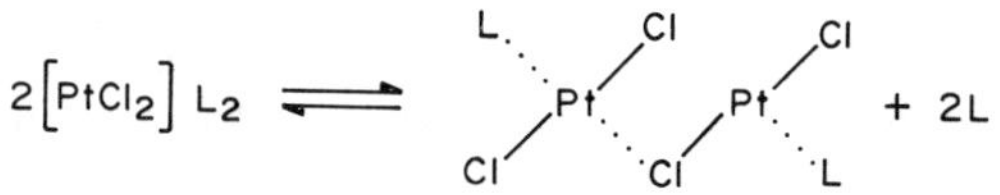

$$2\left[PtCl_2\right] L_2 \rightleftharpoons \quad\quad + 2L \quad\quad (186)$$

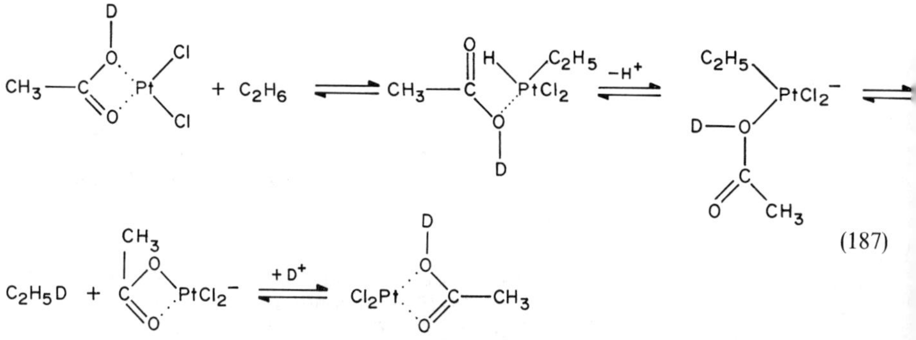

(187)

The R—H bond rupture followed by the formation of the metal–alkyl bond Pt—R is the rate-determining step in the activation of saturated hydrocarbons by metal complexes. The exchange rate decreases with the increase of branching in the hydrocarbon molecules. The scheme in equation 188 illustrates the simultaneous formation of organic chlorides and unsaturated hydrocarbons besides the deuteration of saturated hydrocarbons (methane, ethane, propane, n-butane, i-butane, n-hexane, c-hexane, toluene in 0.1–0.7 M water solutions of H_2PtCl_6, containing 5% K_2PtCl_4 at 120 °C in sealed ampoules[221,222].

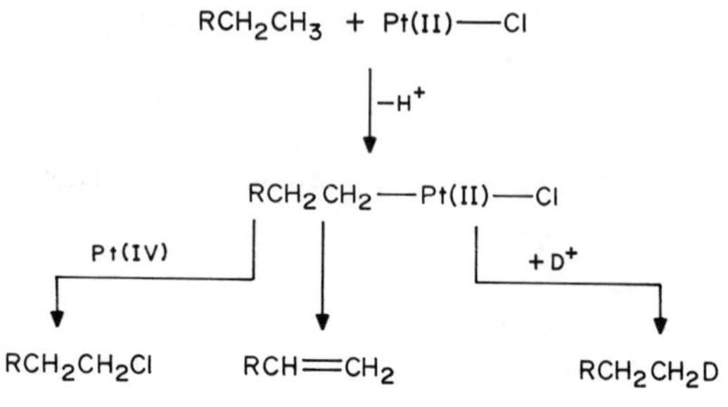

(188)

Deuteriation of methane and simultaneous reaction of K_2PtX_4 ($X = Cl, Br, CN$) with RHgBr ($R = Me, Et, Bu$) carried out in $D_2O/AcOD$ (1:1), 0.02 M Pt(II), 0.02 M DCl at 100 °C is illustrated[223] in equation 189. The same distribution of deuteriated methanes

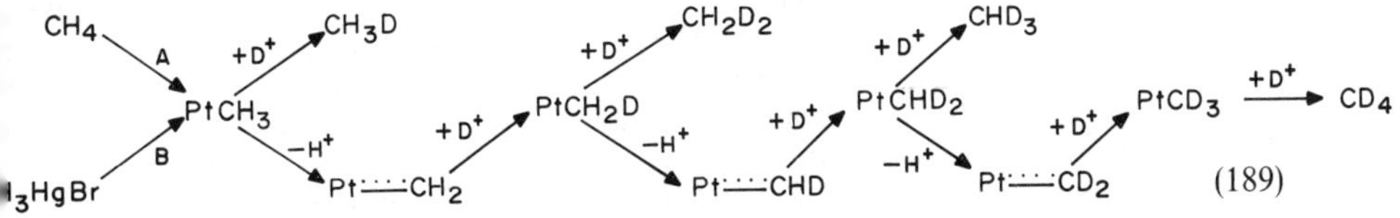

(189)

obtained both in direct H/D exchange of CH_4 catalysed by K_2PtCl_4 and in the reaction of MeHgBr with K_2PtCl_4 convincingly indicates the formation of Pt—R compounds in the course of homogeneous activation of saturated hydrocarbons. Both exchanges proceed through the same intermediate compound RPt(II)X. All reactions of saturated hydrocarbons including H/D exchanges in solutions of transition metal complexes have been reviewed by Shilov and Shteinman's group[224]: Earlier deuteriation studies including saturated hydrocarbons have been reviewed by Agova[225]. Systematic studies of isotopically labelled aliphatic and aromatic hydrocarbon ligand exchanges in different ^{14}C-labelled organometallic complexes have been carried out by Korshunov, Batalov and their coworkers[226-228]. Deuteriation of hydrocarbons has also been investigated using a deuteriated phosphoric acid–boron trifluoride complex[229]. Much information on radiation-induced methods of tritium labelling[230], on photochemical methods of labelling including ^{14}C-diazomethane (equation 190)[231] and on catalytic exchange methods of hydrogen isotope labelling[232] has been published recently.

$$CH_2N_2 \xrightarrow{h\nu} CH_2:$$

$$CH_3(CH_2)_3CH_3 + {}^{14}CH_2N_2 \longrightarrow {}^{14}CH_3(CH_2)_4CH_3 + {}^{14}CH_3CH(CH_3)(CH_2)_2Me$$
$$+ CH_3CH_2CH({}^{14}CH_3)CH_2CH_3 \qquad (190)$$

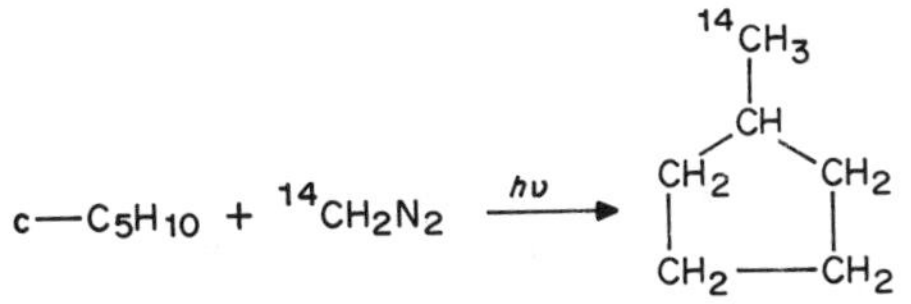

(190a)

2. Interaction of isotopic methanes with dimethylcadmium

In the course of investigations on the activation of isotopic methanes by organometallic compounds, the non-monotonic deuterium isotope effect on the inhibition of the free-radical auto-oxidation by dimethyl cadmium, $(CH_3)_2Cd$, in n-decane has been discovered[233-235]. The inhibitory power of the isotopic methanes increases in the order $CD_4 < CH_4 < CH_2D_2 < CH_3D < CD_3H$, where the inhibitory effect of CD_3H is about an order of magnitude higher than that of CH_4. The induction time $\tau_{ind.}$ (in minutes) depends linearly on the pressure of CH_4 (or CD_3H) (equations 191 and 192).

$$\tau_{ind.} = 0.136 P_{CH_4} + 28.43 \text{ min} \qquad \text{correl. coefficient } 0.999 \qquad (191)$$
$$(P_{CH_4} \text{ interval } 0\text{--}230 \text{ mm Hg})$$

$$\tau_{ind.} = 0.809 P_{CD_3H} + 32.65 \text{ min} \qquad \text{correl. coefficient } 0.974 \qquad (192)$$
$$(P_{CD_3H} \text{ interval } 0\text{--}50 \text{ mm Hg})$$

(experimental conditions: C_0 (of Me_2Cd) $= 3 \times 10^{-2}\,mol\,l^{-1}$, $t = 50\,°C$, $P_{O_2} = 260\,mm\,Hg$ over the reaction mixture)

Under similar experimental conditions, but in the absence of oxygen, H/D exchange in the systems $(CH_3)_2Cd-CD_3H$ and $(C_2H_5)_2Cd-CD_3H$ has been noted. These observations indicate that the interaction of methane with dialkylcadmium compounds $(Et_2Cd,\ Bu_2Cd,\ etc.)$ and with intermediates takes place in the thermolysis and in auto-oxidations and that the organocadmium compounds activate methane selectively. Similar conclusions were drawn from an earlier exchange study of methane and monodeuteriomethane with atomic deuterium[236-238] which showed that D/H exchange proceeds by hydrogen abstraction $(D + CH_4 = \dot{C}H_3 + HD)$ followed by fast exchange of the methyl radical with D atoms prior to stabilization as methane, and that C—H bond rupture is 20% more probable in CH_3D than in CH_4.

3. Carbon acid–D_2 exchanges over solid MgO

CH/D_2 exchanges in hydrocarbons over thermally activated MgO at 250–500 °C have been investigated [239,240] and the enthalpies of activation $\Delta H^{\neq}$ (given below in parentheses) deduced from experimental Arrhenius activation energies E_a (equation 193) were

$$\Delta H^{\neq} = E_a - mRT \qquad (193)$$

($m = 2$ for the bimolecular transition state,

$R = 1.987\,kcal\,mol^{-1}$, T in K is the reaction temperature)

determined for ethane (the 300–450 °C temperature interval was studied, $\Delta H^{\neq} = 6.9 \pm 0.1\,kcal\,mol^{-1}$), propane (350–450 °C, 17.1 ± 0.3), cyclopropane (250–400 °C, 10.3 ± 0.1), i-butane (350–500 °C, 16.2 ± 0.3), cyclobutane (300–400 °C, 13.2 ± 0.3), neopentane (CMe_4, 250–420 °C, 16.3 ± 0.1), cyclopentane (300–450 °C, 13.8 ± 0.2) and cyclohexane (300–400 °C, 14.3 ± 0.1). These values were found to correlate linearly (equation 194) with gas-phase

$$\Delta H^{\neq} = 0.38\Delta H_a^{\circ} - 144.7 \qquad \text{correl. coefficient } 0.9998 \qquad (194)$$

acidities (defined as standard enthalpies ΔH_a° for the reaction: $R—H_{(g)} \rightarrow R_{(g)}^{-} + H_{(g)}^{+}$). A suggestion has been made that gas-phase acidities (Bronsted acidities in the absence of solvent)[241] for other carbon acids, ΔH_a°) can be estimated by carrying out CH/D_2 exchange over a solid base such as magnesium oxide and measuring accurately E_a and using equations 193 and 194. The k_H/k_D ratio has been estimated to be about 1.6. It has been assumed reasonably that D_2 and H_2 dissociate extremely rapidly, that diffusion of RH and D_2 through the pores is also rapid and that the heterolytic dissociation of RH is the rate-determining step. Structure **60** has been proposed for the activated complex. The charge on the carbanion R^- is localized on the carbon orbital of the C—H bond. The carbon–hydrogen bond is elongated but the other internuclear distances and angles in the transition state remain unchanged. Besides the thermodynamic cycle the detailed reaction coordinate diagram for heterogeneous H/D exchange has been presented and its physical interpretation given.

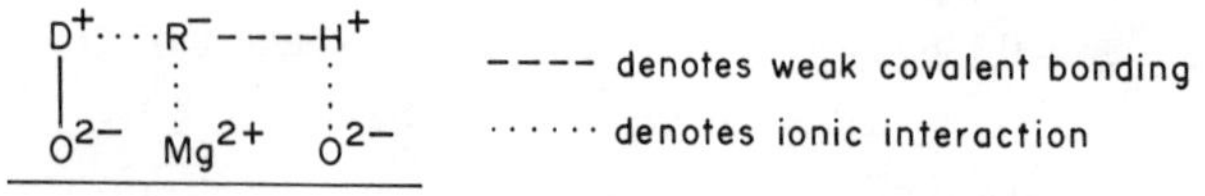

     ‐ ‐ ‐ ‐ denotes weak covalent bonding

     · · · · · · denotes ionic interaction

(60)

C. Isotope Effect Studies with Alkanes and Cycloalkanes

1. Hydrogen kinetic isotope effects in cyclopropane–propene and related rearrangements

Hydrogen and carbon isotope effects are the most sensitive means for studying the elementary acts in hydrocarbon reactions. Unfortunately, precise determinations of isotopic compositions deal mainly with hydrocarbons produced in natural conditions (Section IV.C). No systematic isotope effect studies for light hydrocarbon production or decomposition were carried out in controlled laboratory conditions, except some deuterium isotope effect determinations carried out on rearrangements of cycloalkanes, reviewed in this section.

a. *Tritium isotope effect in the thermal isomerization of cyclopropane-t_1.* Cyclopropane, first synthesized by Freund[242] in 1882 (equation 195), has been the object of extensive isotope studies[243]. It isomerizes thermally to propene in a homogeneous and unimolecular[24] (equation 196).

$$BrCH_2CH_2CH_2Br + 2Na = CH_2\!\!\underset{\diagdown\,\diagup}{\overset{\diagup}{}}\!\!CH_2 + 2NaBr \tag{195}$$

$$CH_2$$

$$\log k = 15.17 - (65000 \pm 2000)/2.3RT \tag{196}$$

Chambers and Kistiakovsky[244] suggested two possible reaction paths (equation 197)

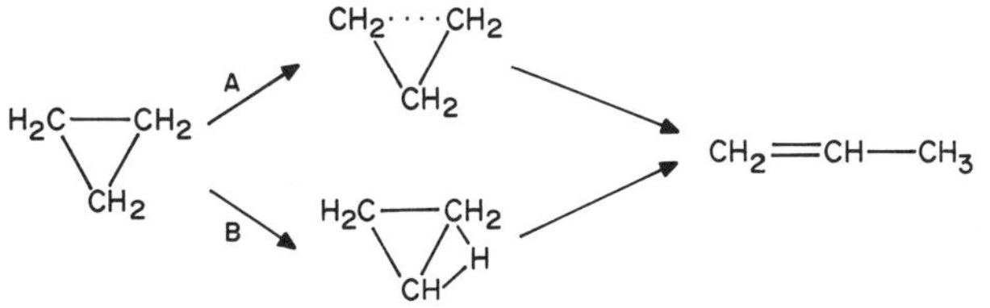

$$(197)$$

leading to propene. In path A a trimethylene biradical is produced in the rate-determining step and rearranges to propene. In path B, primary breaking of a C—H bond and hydrogen migration take place, leading to unsaturation. Tritium, deuterium and ^{13}C kinetic isotope effects as well as labelling of the ring by methyl substituents have been used to solve this reaction and other theoretical problems of unimolecular reactions[245-247]. Tritium isotope effects in the thermal isomerization of cyclopropane-t_1, produced in reaction 198, have been examined by Weston[245,247] and Lindquist and

$$CH_2\!\!=\!\!CH\!\!-\!\!CH_2Br + \overset{*}{H}Br \longrightarrow BrCH_2C\overset{*}{H}_2CH_2Br$$

$$\xrightarrow[\text{abs. EtOH, reflux}]{\text{Zn dust,}} C_3\overset{*}{H}_6\,(77\% \text{ yield, sp. act. } 3\,mCi\,mol^{-1}) \tag{198}$$

Rollefson[246] by comparing[248a] the tritium specific activity S_f of the unreacted labelled cyclopropane from the post-reaction mixture with the specific activity S_0 of cyclopropane-t_1 used as the starting material or with the specific activity of propenes produced at a given fraction of completion (f) of the reaction (equation 199):

$$\varepsilon = (1 - k^*/k) = -\log(S_f/S_0)/\log(1-f) \tag{199}$$

where k and k^* are the rate constants for the isomerization of normal and labelled cyclopropane.

The tritium isotope effect depends on temperature[246] (at a pressure of 200 mm Hg) in the temperature interval 447–555 °C according to equation 200:

$$\varepsilon = k/k^* = 0.63 \pm 0.02 \exp((825 \pm 60)/RT) \tag{200}$$

$$\varepsilon \text{ (at } 450\,^\circ\text{C)} = 1.1187$$

or, according to equation 201, at the same pressure and in the temperature interval 406–492 °C[247]:

$$\varepsilon = k/k^* = 0.86 \pm 0.06 \exp((385 \pm 95)/RT) \tag{201a}$$

$$= 0.89 \pm 0.06 \exp((332 \pm 105)/RT) \tag{201b}$$

$$\varepsilon \text{(at } 450\,^\circ\text{C)} = 1.1243$$

[Equations 200 and 201 have been obtained in different laboratories (see References 246 and 247). Equations 201a and 201b are two experimental dependences from the same laboratory.]

The k/k^* values are lowered as the pressure decreases and become equal to unity at about 1 mm Hg. The conversions of k/k^* values to the k_H/k_T isotope effects defined by equations 202 are of limited value in view of the lack of data concerning secondary tritium isotope effects in this system.

$$\text{cyclo-C}_3\text{H}_6 \longrightarrow \text{CH}_3\text{—CH==CH}_2 \qquad 6k_1$$

$$\text{cyclo-C}_3\text{H}_5\text{T} \xrightarrow{\text{H shift}} \text{CH}_2\text{T—CH==CH}_2 \qquad 2k_2$$

$$\xrightarrow{\text{T shift}} \text{CH}_2\text{T—CH==CH}_2 \qquad k_3 \tag{202}$$

$$\xrightarrow{\text{H shift}} \text{CH}_3\text{—CT==CH}_2 \qquad k_4$$

$$\xrightarrow{\text{H shift}} \text{CH}_3\text{—CH==CHT} \qquad 2k_5$$

$$k/k^* = 6k_1/(2k_2 + k_3 + k_4 + 2k_5)$$

Assuming that $k_1 = k_2 = k_4 = k_5$, designating k_1/k_3 by k_H/k_T and neglecting the secondary tritium isotope effect, one obtains equation 203:

$$k/k^* = 6/[5 + (k_T/k_H)] \quad \text{or} \quad k_H/k_T = \frac{\varepsilon}{6 - 5\varepsilon} \tag{203}$$

By substituting the value $\varepsilon = 1.1243$ (at 450 °C) into equation 203 one obtains $k_H/k_T = 2.970$.

Tritium specific activity determinations are usually carried out with larger uncertainties than mass spectrometric deuterium determinations. Calculations of k_H/k_T ratios with the use of equation 203 are therefore not very reliable. Nevertheless, the data obtained in the high-temperature region indicate clearly that the 'critical coordinate' for reaction 197 is the C—H rather than the C—C distance and support the Slater assumption[248b] based on theoretical considerations concerning unimolecular decomposition of cyclopropane.

b. Deuterium isotope effects. The first deuterium isotope effect studies with cyclo-propane d_2 were carried out by Schlag and Rabinovitch[249]. Unlabelled cyclopropane molecules undergo isomerization 1.22 times faster than cyclopropane-d_2 at 718 K.

TABLE 1. Pressure variation of deuterium isotope effect in cyclopropane isomerization at 755 K

P (cm)		10^2	10	1	10^{-1}	10^{-2}	10^{-3}	0
$(k_H/k_D)_{exp.}$	—		1.96	1.88	1.70	1.50	1.30	—	
$(k_H/k_D)_{calc.}$		2.10	2.08	2.00	1.80	1.50	1.19	0.83	0.25

Neglecting secondary deuterium isotope effects, this value gives $k_H/k_D = 2.18$. The deuterium isotope effect in the thermal copyrolysis of cyclo-C_3H_6 and cyclo-C_3D_6 in the temperature interval 407–514 °C has been determined by Blades[250]. At about 60 cm Hg the k_H/k_D ratio depends on the absolute temperature according to equation 204:

$$k_H/k_D = 0.82 \exp(1300/RT) \tag{204}$$

$(k_H/k_D)_{482\,°C} = 1.9504$, $(k_H/k_D)_{100.5\,°C} = 4.7244$, $(k_H/k_D)_{25\,°C} = 7.3609$, $(k_H/k_D)_{0\,°C} = 8.9883$

The k_H/k_D isotope effect is pressure-dependent and decreases from 1.98 at 76 cm Hg to 1.35 at 0.0178 cm Hg for temperature 482 °C. This decrease has been confirmed by subsequent studies and by the inversion of the $k_{(C_3H_6)}/k_{(C_3D_6)}$ ratio found to occur at about 10^{-2} mm Hg in agreement with theoretical expectations[251,252] (Table 1, line 3). Below this pressure the wall effects caused the $k_{(C_3H_6)}/k_{(C_3D_6)}$ ratio to be constant and equal to about 0.8, down to a pressure of 10^{-4} (at 510 °C).

The high-pressure results are comparable with the deuterium isotope effect observed in the decomposition of ethyl and ethyl-D_5 acetates[253] ($k_H/k_D = 2.23$ at 445 °C is comparable with $k_H/k_D = 2.04$ with cyclopropanes). The activation energy difference of 1300 cal mol^{-1} found by Blades[250] is slightly above the zero-point energy difference of 1150 cal mol^{-1}, corresponding to C—H/C—D oscillators. This implies that the hydrogen isotope effect observed in this reaction is a primary effect and considerable carbon–hydrogen bond weakening takes place during activation in the transition state. A suggestion has been made that the hydrogen atom is weakly bonded both to the original and the terminal carbon atoms involved in the migration. The deuterium isotope effect of 1.37 at 449 °C (activation energy difference of 1400 cal mol^{-1}) observed in the decomposition of cyclobutane to ethylene[254] has been interpreted similarly by assuming C—H bond weakening in the transition state.

Setser and Rabinovitch have investigated also the unimolecular thermal geometrical and structural isomerization of methyl cyclopropane[255] and determined the deuterium isotope effects in the structural isomerization of 1,2-dideuterio-3-methylcyclopropane (61). (Geometrical and structural isomerization rates differ by a factor of twenty.)

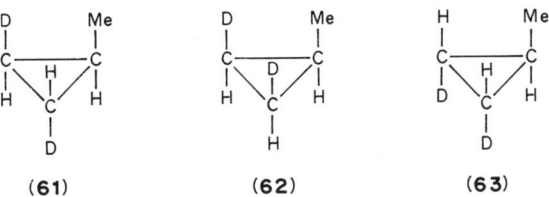

(61)   (62)   (63)

The nearly identical rate expressions for the geometrical isomerization of 61 to yield 62 and 63 (at a pressure of 1.1 cm and temperature range 380–420 °C) are given by equations 205 and 206.

$$k_{(62)} = 10^{15.3 \pm 0.4} \exp - ((60,400 \pm 900)/RT) \tag{205}$$

$$k_{(63)} = 10^{15.4 \pm 0.5} \exp - ((60,700 \pm 1700)/RT) \tag{206}$$

The thermal structural isomerization rate constants for methylcyclopropane and methylcyclopropane-D_2 to yield butenes depend on the temperature according to equations 207 and 208 in the 420–475 °C temperature range.

$$k_{(struct.)} = 10^{14.16} \exp(-62,400/RT) \tag{207}$$

$$k_{D_2(struct.)} = 10^{14.43} \exp(-62,300/RT) \tag{208}$$

where

$$k_{(struct.)}/k_{D_2(struct.)} = 1,5136 \exp(-100/RT) \tag{209}$$

$k_{(struct.)}/k_{D_2(struct.)}$ at 420 °C = 1.4075, $k_{(struct.)}/k_{D_2(struct.)}$ at 450 °C = 1.4118, $k_{(struct.)}/k_{D_2(struct.)}$ at 475 °C = 1.4150.

The precision of these determinations has not been high enough to reveal the very small positive temperature coefficient of the deuterium isotope effect in only partially (D_2) deuteriated 3-methylcyclopropanes. In the perdeuteriocyclopropane pyrolysis a much larger positive temperature coefficient for k_{H_6}/k_{D_6} equal to 1300 cal mol^{-1} has been noted by Blades[250]. Setser and Ralinovitch[255] deduced also the secondary isotope effect for the methylcyclopropane to isobutene isomerization $k_{iH}/k_{iH}^* = 1.10 \pm 0.03$ and estimated the isotope effect per H or D atom as defined by equations 210 and 211: $k_{tH}/k_{tD}^* = 1.59$, $k_{cH}/k_{cD}^* = 1.41$ and $k_{bH}/k_{bD}^* = 2.36$. These values are comparable to $k_H/k_D = 2.18$ and 1.96 determined for cyclopropane-D_2 and cyclopropane-D_6, respectively, near the high-pressure region.

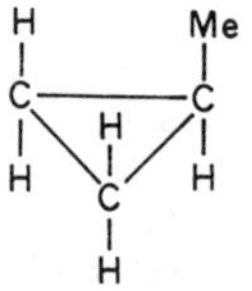

$$\xrightarrow{k_{iH}} \text{isobutane}$$

$$\xrightarrow{k_{tH}} trans\text{-butene-2} \tag{210}$$

$$\xrightarrow{k_{cH}} cis\text{-butene-2}$$

$$\xrightarrow{k_{bH}} \text{butene-1}$$

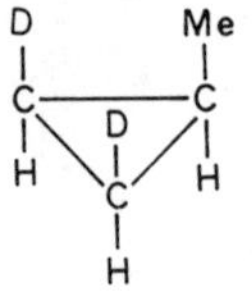

$$\xrightarrow{k_{iH}^*} CH_2DC(Me)CHD$$

$$\xrightarrow{k_{tH}^*, k_{cH}'} cis,trans-CH_2DCD{=}CHCH_3 \tag{211}$$

$$\xrightarrow{k_{tD}^*, k_{cD}'} cis,trans-CHD_2CH{=}CHCH_3$$

$$\xrightarrow{k_{bH}^*} CHD{=}CDCH_2CH_3$$

$$\xrightarrow{k_{bD}^*} CHD{=}CHCHDCH_3$$

Thus the secondary and primary deuterium isotope effects determined in this study also indicate that entropy and zero-point energy factors associated with breaking of the C—D bond and migration in the activated complexes are important for the structural isomerization to yield butene-1 and butene-2, and reaction path B (equation 197) must be included in the mechanistic considerations concerning the cyclopropane isomerization. But the higher activation energy for isobutane formation ($Q = 64.3$ kcal mol^{-1}) than that for butene-2 or butene-1 ($Q = 62.0 \pm 0.6$ kcal mol^{-1}) indicates also that the rupture of the particular C—C bond postulated in mechanism A of equation 197 is also important in the formation of the transition state.

The authors[255] suggest a concerted isomerization mechanism with the transition state having the structure **64**, but are inclined also to suppose that the structural and

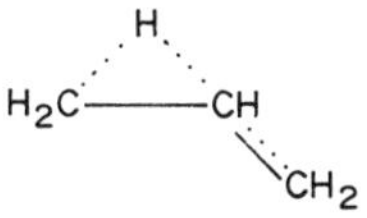

H
H_2C———CH
CH_2

(64)

geometrical isomerizations proceed according to a common reaction scheme (equation 212) presented in the review chapter by Carpenter[243] which involves the intermediacy of a geometrically equilibrated biradical. ^{13}C kinetic isotope effect studies[247,261] have been undertaken to clarify this problem, since skeletal ring vibrations are insensitive to deuterium substitutions.

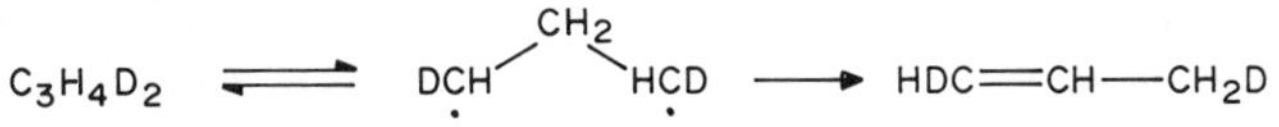

$$C_3H_4D_2 \rightleftharpoons \overset{CH_2}{DCH \diagup \diagdown HCD} \longrightarrow HDC = CH - CH_2D \qquad (212)$$

c. Rotational kinetic deuterium isotope effects. Double deuteration[256] of the terminal carbon of the vinyl group caused a 1.125 ± 0.04-fold decrease in the rate constant for cyclopentene formation in the rearrangement (equation 213) of (R, R)-*trans*-2-methyl-1-(3,3-dimethyl-1-buten-1,1-D_2-2-yl)cyclopropane **65** and 1.11 ± 0.04 increase in the diene formation. These values have been interpreted as caused by a rotational isotope effect disfavouring the formation of deuteriated cyclopentene and favouring that of the deuteriated diene.

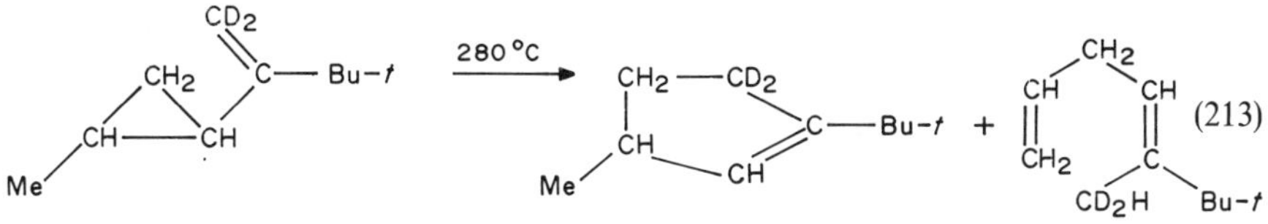

(65)

A rotational kinetic deuterium isotope effect,[257] $k_{H_2}/k_{D_2} = 1.31 \pm 0.04$, has been found[243] in the thermal structural rearrangement of *cis*- and trans-2,3-dimethyl-1-dideuteriomethylenecyclopropane, **65t** and **65c**, to **66** but little effect was found in the geometrical isomerization of **65t** to **65c**.

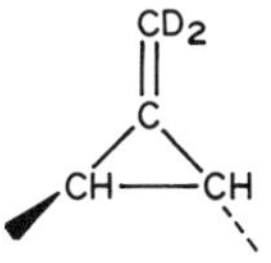

(65t) 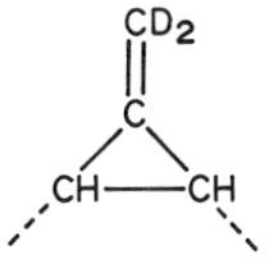   **(65c)** 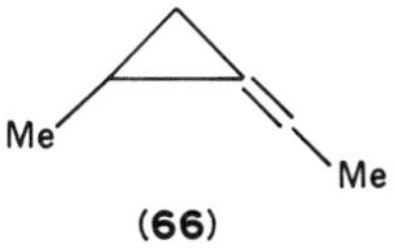   **(66)**

(2-methylethylidenecyclopropane)

Ratio k_{H_2}/k_{D_2} equal to 1.08 ± 0.02 was observed in the interconversion of 2-methyl-6-dideuteriomethylenebicyclo[3.1.0]hexane, **67**, into **68** (equation 214). This suggests that the *exo* methylene rotation occurs in a fast step after rate-determining formation of a planar trimethylenemethane biradical **69**[243].

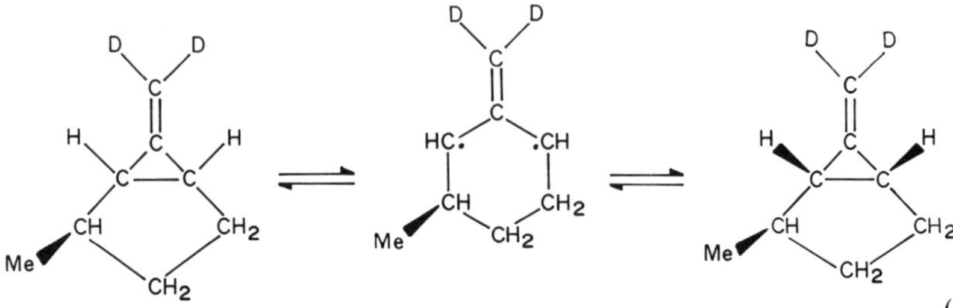

(214)

(67)   (69)   (68)

d. Deuterium isotope effects in pyrazoline decomposition. Secondary deuterium kinetic isotope effects on the thermolysis of two α-deuterated 1-pyrazolines (**70, 71**) and two β-deuteriated 1-pyrazolines (**72, 73**) at 229.40 ± 05 °C (or corrected to 105 °C) have been

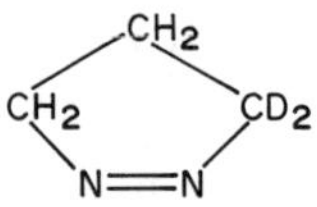

(70)

$k_H/k_D = 1.19 \pm 0.01 (= 1.26 \pm 0.01$ at 105 °C)

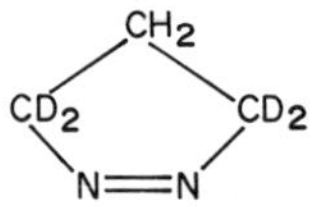

(71)

$1.40 \pm 0.02 \ (1.55 \pm 0.02)$

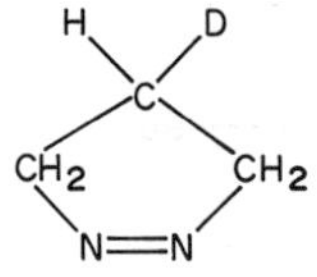

(72)

$1.05 \pm 0.01 (1.06 \pm 0.01)$

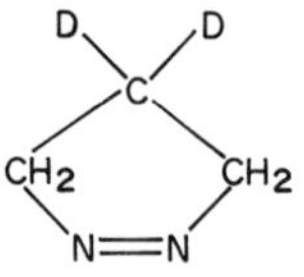

(73)

$1.12 \pm 0.01 \ (1.14 \pm 0.01)$

found[258]. These indicated values have been interpreted as supporting a mechanism involving a symmetrical transition state and two synchronous C—N bond cleavages (equation 215).

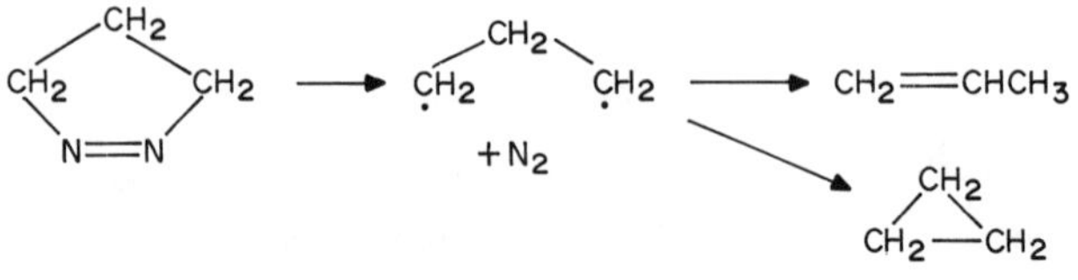

(215)

A thermal deazotization study[259] of 4-methylpyrazoline-4-d in the temperature interval 170–290 °C has revealed that inctroduction of deuterium at the 4 position decreased the first-order decomposition rate constants by a factor of 1.06 ± 0.03 at 241.75 °C and gave an isotope effect $k_H/k_D = 1.80$ in the product-determining step leading to cyclopropane formation. The results have been interpreted as indicating that the formation of the isobutylene diradical **74** (equation 216) is the rate-determining step of the decomposition while hydrogen migration versus ring closure in this intermediate determines the ratio of the products. ^{13}C kinetic isotope effect studies should provide conclusive evidence supporting this reaction scheme.

$$(216)$$

2. Carbon-13 kinetic isotope effect studies

a. The first ^{13}C kinetic isotope effect study on the structural isomerization of cyclopropane of natural isotopic composition to propene (equation 217) was carried out by Weston[247]. This was found (at 492 °C and one atm.) to give $k_{(^{12}C_3H_6)}/k_{(^{12}C_2{}^{13}CH_6)} =$ 1.0072 ± 0.0006. This value should provide indirect information about ring relaxation in the transition state (T.S.) of the gas-phase isomerization. While the non-skeletal vibrations in cyclopropane are very sensitive to deuterium substitution, they are only very little affected by carbon isotopy. On the other hand, skeletal vibrations are strongly affected by substitution of carbon-13 for carbon-12. ^{13}C studies should therefore throw light on the degree of change of these vibrations in the transition state of the rate-determining step of the cyclopropane isomerization. The results have been interpreted in terms of physical and isotopic concepts[248,260] presented by the theoretical equation 217

$$k/k^* = (m^{*\ddagger}/m^{\ddagger})^{1/2}[1 + \sum G(u_i)\Delta u_i - \sum G(u_i^{\ddagger})\Delta u_i^{\ddagger}]$$ (217)

where k corresponds to the lighter molecule and k^* to the heavier molecule.

Two extreme situations have been considered. In the first, corresponding to the reaction path B in equation 197, it has been assumed that ^{13}C—H bond rupture and subsequent hydrogen migration are the rate-determining steps. In this case $(m^{*\ddagger}/m^{\ddagger})^{1/2}$, the temperature-independent term, was calculated[247] using a three-centre reaction model $^{13}C\cdots H\cdots{}^{12}C$ in which one C—H bond breaks and a new one is formed, and was found to be 0.0010. The temperature-dependent contribution in the square brackets of equation 217 should in this case be equal to 0.0062. If $\omega_{(^{12}C-H)}$ and $\omega_{(^{13}C-H)}$ are assumed to equal 2985 cm^{-1} and 2976.06 cm^{-1} correspondingly, one finds that $G(u)\Delta u = 0.00546$ at 492 °C. The agreement is quite satisfactory.

The second extreme reaction path A in equation 197 involves carbon–carbon bonds in the rate-determining step (a single bond is breaking, a double bond is formed). Three-centre formulation gives[247] in this case the value 0.0128 for $(m^{*\neq}/m^{\neq})^{1/2}$. Weston[247] concluded that 'some tightening of bonds' in the transition state, lowering the ^{13}C isotope effect, takes place. Increase of the bonding in the transition state should lead to $\sum G(u_i^{\neq})\Delta u_i^{\neq} - \sum G(u_i)\Delta u_i = 0.0055$. This is more than what corresponds to a change of two single C—C bonds to one double bond as given by equation 218,

$$[G(u_{^{13}C=^{12}C}^{\neq})\Delta u_i - 2G(u_{^{13}C-^{12}C})\Delta u_i] = 0.00243$$ (218)

calculated taking $\omega_{(^{12}C-^{12}C)} = 993$ cm^{-1}, $\omega_{(^{12}C-^{12}C)} = 973.656$ cm^{-1}, $\omega_{(^{13}C=^{12}C)} = 1588.442$ cm^{-1} and $\omega_{(^{12}C=^{12}C)} = 1620$ cm^{-1}.

No agreement is obtained between experimental and calculated k/k^* ratios at

$492\,°C$ when considering the 'statistical factor' relating to the experimental k/k^* value with more explicit isotopic modes of reactions defined by Scheme 2 with product-like transition states. In this case the temperature-dependent factor (T.D.F.) equals

$$\tfrac{1}{3}(3 - 0.00243 - 0.0002011 - 0.00223) = 0.99838$$

(calculated from the expression in the square brackets of equation 217, using the frequencies ω mentioned above). Multiplication[247] by the temperature-independent factor (T.I.F.) 1.0128 does not reproduce the experimental k/k^* value at $492\,°C$.

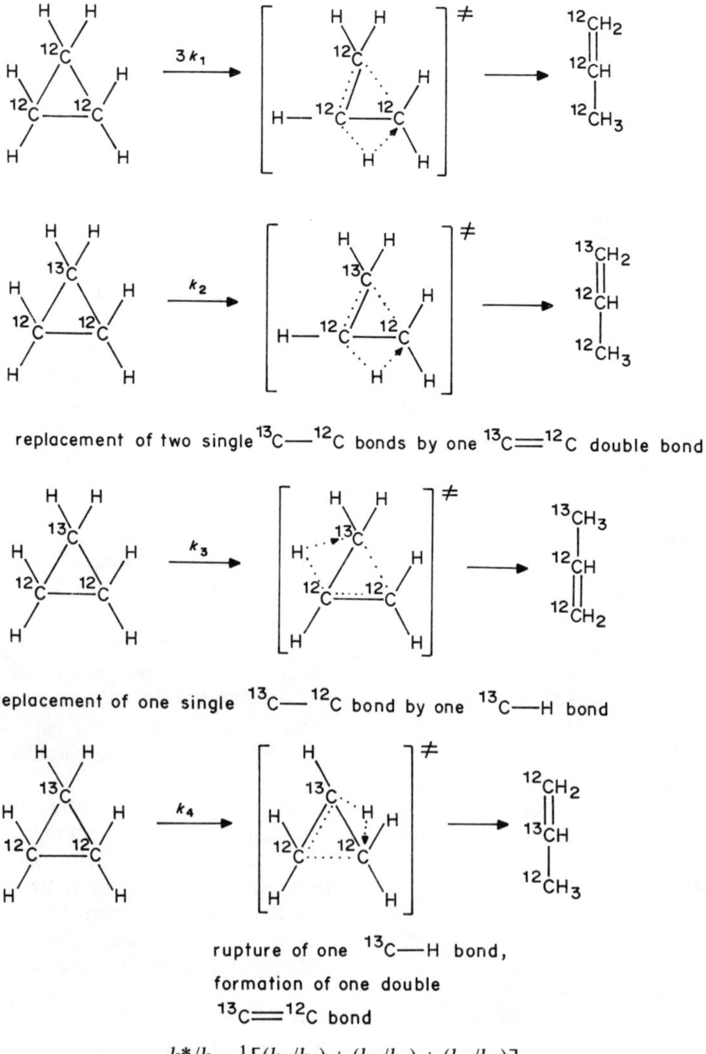

(218)

replacement of two single $^{13}C—^{12}C$ bonds by one $^{13}C{=}^{12}C$ double bond

replacement of one single $^{13}C—^{12}C$ bond by one $^{13}C—H$ bond

rupture of one $^{13}C—H$ bond,
formation of one double
$^{13}C{=}^{12}C$ bond

$$k^*/k = \tfrac{1}{3}[(k_2/k_1) + (k_3/k_1) + (k_4/k_1)]$$

SCHEME 2

Product-like transition states for each isotopic reaction lead to $(1 - k_2/k_1)$, $(1 - k_3/k_1)$ and $(1 - k_4/k_1)$ values, which give a calculated $k^*_{^{12}C \; ^{13}CH}{}_{)}/k_{(^{12}C_3H_6)}$ ratio not equal to the experimentally determined k^*/k ratio. Assuming that the transition state for the isomerization of cyclopropane to propene resembles the trimethylene biradical **75**, in

(75)

which one C—H bond is broken and the practically free hydrogen moves towards one of the terminal methylene carbons, one obtains that at 492 °C the mean T.D.F. equals $\frac{1}{3}(3 + 0.005257 + 0.005257 + 0.005458) = 1.005324$. This value should be multiplied by T.I.F. = 1.00184 to reproduce the experimentally determined value (1.0072) for the ^{13}C isotope effect. Inclusion of the tunnel corrections for both the C—C and C—H isotopic bond ruptures should bring the T.D.F. even closer to the experimental ^{13}C kinetic isotope effect. It should be noted also that the biradical-like structure **75** of the transition complex predicts the correct direction of changes of the $^{13}C(k/k^*)$ isotope effect with temperature (k/k^* rises with decrease of temperature, T.D.F. = 0.0173 and 0.02461 at 100 °C and 25 °C, respectively). The structures of transition complexes, such as for instance **64**, in which the double bond is highly developed, lead to a slightly inverse temperature dependence of the ^{13}C kinetic isotrope effect, in disagreement with later-found[261] positive values of the temperature coefficient ($\Delta Q > 0$) in the expression $k/k^* = $ T.I.F. $\times \exp \Delta Q/RT$.

b. A more detailed ^{13}C kinetic isotope effect study of the isomerization of cyclopropane was carried out by Sims and Yankwich[261]. The effects of temperature and pressure on the first-order rate constants k_s for structural isomerization of cyclopropane as well as the same effects on the ^{13}C isotopic rate constant ratio $k_{str.}/k^*_{str.}$ for the structural isomerization (equations 219 and 220) have been investigated[261,262].

$$\text{cyclopropane}(^{12}C_3H_6) \xrightarrow{k_s} \text{propylene}(^{12}C_3H_6) \tag{219}$$

$$\text{cyclopropane}(^{12}C_2{}^{13}CH_6) \xrightarrow{k^*_s} \text{propylene}(^{12}C_2{}^{13}CH_6) \tag{220}$$

The high-pressure rate constant k_s obtained by extrapolation to $1/p^{1/2} = 0$ of plots k_s^{-1} vs $p^{-1/2}$ depends on the temperature according to equation 221,

$$\log(k_s^\infty) = (15.39 \pm 0.01) - [(65679 \pm 308)/2.303RT] \tag{221}$$

which agrees well with earlier results[263-266]. The ^{13}C isotope effect is relatively small, nevertheless the high precision ($\pm 0.003\%$) of modern mass-spectrometric determinations of the isotopic ratios $R_i = {}^{13}C^{16}O_2/{}^{12}C^{16}O_2$ when measuring the isotopic composition of CO_2 derived from the combustion[169-172c] permitted one also to determine the pressure dependence. The latter has been found to be similar to that observed in the case of deuterium isotope effects[250]. Below about 100 mm Hg pressure the ^{13}C isotope effect decreases approximately linearly with $\log p$. The high-pressure limiting values of k_s/k^*_s are 1.0080 (at 513.8 °C) and 1.0090 (at 450.3 °C). The results corresponding to about 1 atm pressure are expressed by equation 222.

$$k_s/k^*_s = (0.995 \pm 0.001)\exp[(19.1 \pm 2.0)/RT] \tag{222}$$

This dependence suggests that considerable ring relaxation parallels the hydrogen bridging process. The k_s/k_s^* ratio at 25 °C, calculated from equation 222, equals 1.025–1.030, i.e. quite close to the temperature-dependent factor of ~ 1.0246 at 25 °C calculated assuming a transition complex of a diradical structure **75**. Thus the positive temperature coefficient in equation 222 for k_s/k_s^* clearly indicates that product-like transition states with an already developed double bond must be rejected. Such product-like transition states would imply the increase in k_s/k_s^* values with temperature, in disagreement with the experimentally established relation in equation 222.

Additional information concerning the structures of the transition states for the reactions of Scheme 2 should be provided by intramolecular ^{13}C measurements on the structural isomerization, particularly including studies which will indicate the degree of new ^{13}C—H bond formation, as well as on the degree of H—^{13}C bond rupture in the transition state.

Assuming that $k_2 = k_3$ in Scheme 2 and neglecting the isotope effect in the ^{13}C—H bond rupture, Sims and Yankwich[261] recalculated the measured k/k^* values into $^{12}k/^{13}k$ isotope effects with the aid of the relation

$$^{12}k/^{13}k = 2(k/k^*)/[3 - (k/k^*)]$$

and obtained the temperature dependence given in equation 223. These new values and their temperature dependence indicate a considerable ring relaxation in the transition state and have been reproduced in subsequent theoretical calculations by making certain changes of normal vibrations[262] in going from reactants to transition state [ring deformation frequencies are affected very much by carbon isotope substitution and the values $\omega = 883 \, cm^{-1}$ (C_3H_6) and $\omega = 854 \, cm^{-1}$ ($^{13}C_3H_6$) have been used in the calculation].

Computations of k_{12C}/k_{13C} using the equation

$$(k_{12C}/k_{13C}) = 2(k/k^*)/[3 - (k/k^*) \times (k_4/k_1)]$$

result in lower absolute values of carbon-13 kinetic isotope effects and lower the value of the temperature coefficient, since k_4/k_1 is temperature-dependent.

$$
\begin{aligned}
^{12}k/^{13}k &= 0.995 \exp(27 \pm 2)/RT \qquad\qquad (223)\\
&= 1.012 \quad \text{(at 513.8 °C)}\\
&= 1.014 \quad \text{(at 450.3 °C)}\\
&= 1.032 \quad \text{(at 100.5 °C)}\\
&= 1.041 \quad \text{(at 25 °C)}
\end{aligned}
$$

3. Isotopic oxidation studies of alkanes

The isotope effect studies of the oxidation of aliphatic chains will be covered by us more extensively in a chapter in Supplement B2 of this series ^{13}C and tritium isotope effects in the oxidation of isotopic methanes over CuO to carbon dioxide and water[267,268] have shown that the C—H bond rupture is the rate-determining step in this process, average $k_{(^{12}CH_4)}/k_{(^{13}CH_4)} = 1.0074 \pm 0.0010$ at 700 °C, average $k_{(CH_4)}/k_{(CH_3{}^3H)} = 1.18 \pm 0.05$, $k_{(C-H)}/k_{(C-T)} \cong 2.6$ at 700–750 °C neglecting secondary tritium isotope effects and back $^3H/^1H$ exchange interferences. Isotope effects in diffusion processes are characterized by different values of ^{13}C and ^3H isotope effects. The ^{13}C kinetic isotope effect in the oxidation of cyclopropane to carbon dioxide and water over CuO has been found to be normal[269]: $k_{(^{12}C_3H_6)}/k_{(^{12}C_2{}^{13}CH_6)} = 1.027 \pm 0.005$ at 500 °C and 1.017 ± 0.005 at 650 °C.

The kinetic deuterium isotope effects at 25 °C in the oxidation of cyclohexane $(c-C_6H_{12}/c-C_6D_{12})$ and t-butane $(CH_3)_3CH/(CH_3)_3CD$ with CrO_3 in 45–60% H_2SO_4 have been found[270] to be 4.7 ± 0.4. k_H/k_D for 3-ethylpentane[271] and diphenylmethane[272] are 2.5 and 6.4, respectively. The value $k_0^{(H_2O)}[HMnO_4]/k_0^{(D_2O)}[DMnO_4] = 0.62$ for i-butane oxidation[273] has been found to be in agreement with the general rule that acid–base catalysed reactions proceed faster in D_2O than in H_2O because of the smaller value of pK_a and higher concentration of the deuteriated reagent. A conclusion has also been drawn that oxidations of alkanes with CrO_3, MnO_4^- and $HMnO_4$[274], showing a kinetic isotope effect for deuterium of about 4, proceed through a linear or cyclic 'late' transition state (equation 224) with an oxygen bridge.

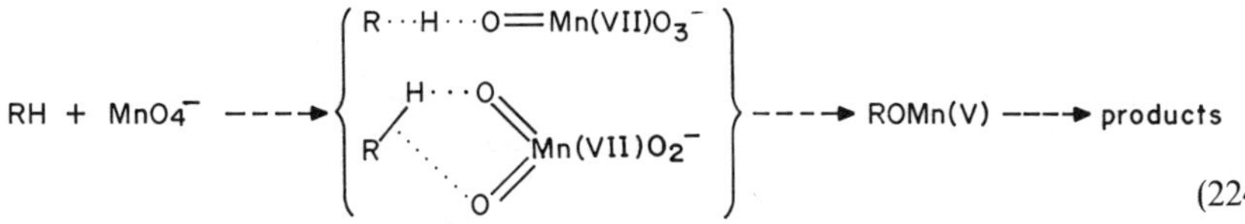

$$RH + MnO_4^- \dashrightarrow \left\{ \begin{array}{l} R\cdots H \cdots O = Mn(VII)O_3^- \\ \begin{array}{c} H \cdots O \\ R \diagdown \diagup \diagdown Mn(VII)O_2^- \\ \diagup O \end{array} \end{array} \right\} \dashrightarrow ROMn(V) \longrightarrow products$$

$$(224)$$

4. Free-radical oxidation of [1–^{14}C]butane with oxygen

The theory of 'destructive oxidation' of hydrocarbons with oxygen has been tested[185c] by oxidizing [1–^{14}C]butane with oxygen. According to this theory the oxidation of butane should proceed according to equations 225.

$$CH_3CH_2CH_2{}^{14}CH_3 + O_2 \begin{array}{c} \nearrow H_2O + CH_3CH_2CH_2{}^{14}C{\diagup\!\!\!\diagdown}{}^O_H \\ \searrow H_2O + {}^{14}CH_3CH_2CH_2C{\diagup\!\!\!\diagdown}{}^O_H \end{array}$$

$$CH_3CH_2CH_2{}^{14}C{\diagup\!\!\!\diagdown}{}^O_H + O_2 \longrightarrow H_2O + {}^{14}CO + CH_3CH_2C{\diagup\!\!\!\diagdown}{}^O_H \qquad (225)$$

$${}^{14}CH_3CH_2CH_2C{\diagup\!\!\!\diagdown}{}^O_H + O_2 \longrightarrow {}^{14}CH_3CH_2C{\diagup\!\!\!\diagdown}{}^O_H + CO + H_2O$$

$${}^{14}CH_3CH_2C{\diagup\!\!\!\diagdown}{}^O_H + O_2 \longrightarrow CO + H_2O + {}^{14}CH_3C{\diagup\!\!\!\diagdown}{}^O_H$$

$${}^{14}CH_3C{\diagup\!\!\!\diagdown}{}^O_H + O_2 \longrightarrow H{}^{14}C{\diagup\!\!\!\diagdown}{}^O_H + CO + H_2O$$

Determinations of the ^{14}C-labelled products isolated at different oxidation times[185c] showed that HCHO is produced not only from the terminal methyl carbons but also from middle carbons. Similar results have been obtained for acetaldehyde, 82% of which originates from the terminal carbons (1–2, 3–4), 18% from the middle carbons (2–3) of the aliphatic chain. This implies that the theory of 'destructive oxidation' does not fit the results in this case. Moreover, the specific activity of carbon dioxide has been found to be lower than that of carbon monoxide, indicating that CO_2 is not only the product of carbon monoxide oxidation.

The rates of formation and the rates of oxidation of acetaldehyde at different stages of butane oxidation process have also been studied[185d] by introducing at different reaction times ^{14}C—AcOH and measuring its specific activity vs reaction time by Neiman's method[185a,b]. Labelled acetic acid was isolated when ^{14}C—AcOH was added to the reacting mixture at the end of the 'cold flame stage' of the butane oxidation.

By introducing $[^{14}C]O$ into butane–oxygen mixtures simultaneously with a small quantity of carbon dioxide of natural isotopic composition and determining the changes in the specific activities of carbon monoxide and carbon dioxide vs the butane oxidation times, it has been shown that no more than 1.3–5.3% of CO_2 is produced directly from CO oxidation[185e]. A suggestion has been made that carbon dioxide originates from the decomposition of $RCOOO^{\bullet}$ radicals, formed in the course of butane oxidation, which split into CO_2 and $RO^{\bullet}$ radicals, and that CO is oxidized partly to CO_2 in the free radical reactions (see equation 226).

$$CO + {}^{\bullet}OH \rightarrow CO_2 + H^{\bullet}, \quad H^{\bullet} + O_2 \rightarrow HO_2^{\bullet}, \quad :CO + HO_2^{\bullet} \rightarrow CO_2 + {}^{\bullet}OH \quad (226)$$

The mechanism of the low-temperature oxidation of methane has also been studied following the kinetic isotopic method[185f,g]. By adding known portions of $H[^{14}C]HO$ and CO to the methane–air mixture and measuring the changes with time of the specific activities of formaldehyde and carbon monoxide in the mixture, it has been established that formaldehyde is the unique precursor of carbon monoxide and the oxidation of methane proceeds according to equation 227. A suggestion has been made that

$$^{14}CH_4 \xrightarrow{\ O_2\ } H^{14}CHO \xrightarrow{\ O_2\ } {}^{14}CO \xrightarrow{\ O_2\ } {}^{14}CO_2 \quad (227)$$

formaldehyde results from isomerization of methylperoxy radicals (equations 228).

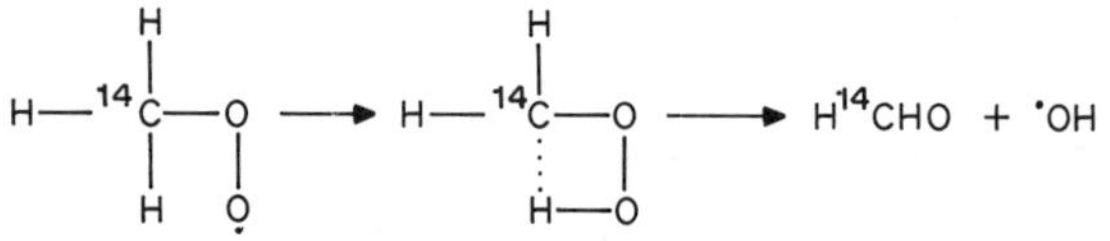

$$(228a)$$

$$CH_3OO^{\bullet} + {}^{14}CH_4 \longrightarrow CH_3COOH + {}^{14\bullet}CH_3 \quad (228b)$$

Applications of the kinetic isotopic method to the study of gas-phase oxidations of alkanes as well as liquid-phase oxidation of cyclohexane (equation 229) are presented in a monograph by Neiman and Gal[185h].

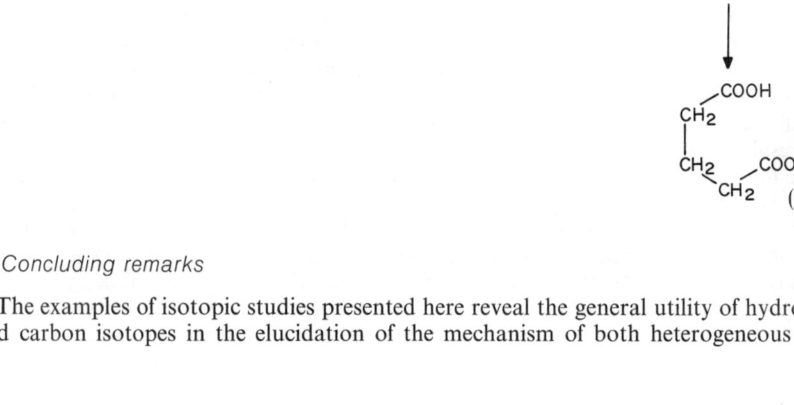

$$(229)$$

5. Concluding remarks

The examples of isotopic studies presented here reveal the general utility of hydrogen and carbon isotopes in the elucidation of the mechanism of both heterogeneous and homogeneous hydrocarbon reactions. It has been possible to establish the nature of intermediates in dehydration–cyclization processes and in the slow methane oxidation, as well as in reactions of other alkanes with oxygen in general. Hydrogen exchange studies with hydrocarbons permitted one to solve the problem of the reversibility of certain steps in heterogeneous catalytic processes and to obtain information concerning the structure of the species adsorbed on the surfaces of the catalysts. A correlation between gas-phase acidities and experimental enthalpies of activation characterizing the H/D exchange reactions over solid base catalysts has been established.

Isotopic tracer studies are complex, time consuming and expensive, hence their use with hydrocarbons is rather limited. Even the simplest alkanes have not yet been investigated in all their aspects, and the number of studies aiming at the elucidation of the elementary steps in hydrocarbon reactions is by far smaller still. Only the reactions of cyclopropane have been probed satisfactorily with both hydrogen and carbon kinetic isotope effect techniques. Experimental difficulties and the necessity of achieving very high precision in the kinetic and isotopic determinations are the main reasons for the scarcity of tracer and isotope effect investigations, and the few experimental studies of hydrocarbon reactions have not been accorded full theoretical and calculational treatment. Particularly, the problems of secondary deuterium isotope effects[275] and of intramolecular deuterium isotope effects in hot radiochemical reactions and in Boltzman reactions involving isotopic methane require detailed theoretical and experimental reinvestigation. Some of the ideas presented by us as explanations for the observed lack of intramolecular deuterium isotope effects in hot and free radical reactions with deuteriated methanes are suggestions but not necessarily the solutions of these problems.

VI. ACKNOWLEDGEMENTS

Preparation of this chapter was possible due to support by The Faculty of Chemistry of The Jagiellonian University in Cracow and by the University of Warsaw. M.Z. thanks also his mother for offering her house and services in Witeradow enabling uninterrupted work on this chapter during the summer. We also extend our gratitude to many scientists (Professors Alexandrov, Korshunov, Tishchenko, Carpenter, Klabunde, Sims, Longinelli and many others cited in the text and references) for providing us with reprints or manuscripts of their papers for helpful suggestions and hospitality. Discussions with Mgr Papiernik-Zielińska and Dr. R. Kański on biological and radiochemical problems covered in this chapter are also acknowledged.

VII. REFERENCES

1. (a) T. Ito, T. Watanabe, T. Tashiro, M. Kawasaki, K. Toi and H. Kobayashi, Room-temperature activation and oxidation of methane over magnesium oxides, in *Acid–Base Catalysis* (Eds. K. Tanabe, H. Hattori, T. Yamaguchi and T. Tanaka), Kodansa Ltd, Tokyo, 1989, pp. 483–490.
    (b) A. Machocki, A. Denis, T. Borowiecki and J. Barcicki, Oxidative coupling of CH_4 over Na_2SO_4-doped CaO catalyst, Communication No. II K 26, presented during Annual Meeting of the Polish Chemical Society, Szezecin, September 5–8, 1990, p. 178.
    (c) T. Ito, T. Watanabe, T. Tashiro and K. Toi, *J. Chem. Soc., Faraday Trans. 1*, **85**, 2381 (1989).
    (d) T. Tashiro, T. Ito and K. Toi, *J. Chem. Soc., Faraday Trans.* 1, **86**, 1139 (1990).
    (e) C-H. Lin, J-X. Wang and J. H. Lunsford, *J. Catal.*, **111**, 302 (1988).
    (f) K. D. Campbell, H. Zhang and J. H. Lunsford, *J. Phys. Chem.*, **92**, 750 (1988).
    (g) K. D. Campbell, E. Morales and J. H. Lunsford, *J. Am. Chem. Soc.*, **109**, 7900 (1987).
    (h) C-H. Lin, T. Ito, J-X. Wang and J. H. Lunsford, *J. Am. Chem. Soc.*, **109**, 4808 (1987).
    (i) J-X. Wang and J. H. Lunsford, *J. Phys. Chem.*, **90**, 5883 (1986).
    (j) J-X. Wang and J. H. Lunsford, *J. Phys. Chem.*, **90**, 3890 (1986).

(k) T. Ito, J-X. Wang, C-H. Lin and J. H. Lunsford, *J. Am. Chem. Soc.*, **107**, 5062 (1985).

(l) M. I. Kabachnik, *Metalloorganic Compounds and Radicals* (collection of chapters from The Chemistry Institute, USSR Acad. Sci., Gorky, dedicated to G. A. Razuvayev), Science, Moscow, 1985.

(m) Yu. A. Alexandrov, Achievements in the autooxidation of organometallic compounds, chapter with 59 references in Reference 11.

(n) G. A. Abakumov, Complexes of metals with free radical ligands, review chapter with 69 references in Reference 11.

(o) G. A. Domrachev, The problems of stability of organometallic compounds in the processes of their synthesis and decompositions, review chapter with 96 references in Reference 11.

(p) M. R. Leonov, G. V. Solov'yova and S. V. Patrikeev, *Metalloorganic Chemistry* (*Acad. Sci. USSR*), **3**, 343 (1990).

(q) V. A. Yablokov, A. V. Dozorov, N. I. Podolskaya, E. I. Makarov and V. A. Gonina, *J. Gen. Chem. USSR*, **58**, 2132 (1988).

(t) A. Murray and D. L. Williams, *Organic Synthesis with Isotopes*, Part I, Interscience, London, 1958, pp. 801–803.

2. M. Alei, Jr., *J. Labelled Compd. Radiopharm.*, **17**, 115 (1980).
3. P. Povinec, *Int. J. Appl. Radiat. Isot.*, **26**, 465 (1975).
4. M. Zieliński, *Nukleonika*, **13**, 1011 (1968).
5. M. Zieliński, in the *Chemistry of Carboxylic Acids and Esters* (Ed. S. Patai), Chap. 10, Interscience, London, 1969, p. 462.
6. T. Elbert, Proceedings of the 12th Radiochemical Conference, Mariánské Lázné, May 7–11, 1990 (Ed. Z. Prâŝil), p. 21.
7. R. D. Hutchins, D. Hoke, J. Keogh and D. Koharski, *Tetrahedron Lett.*, 3495 (1969).
8. G. Berger, C. Prenant, J. Sastre and D. Comar, *J. Labelled Compd. Radiopharm.*, **14**, 1486 (1982).
9. G. Berger, C. Prenant, J. Sastre and D. Comar, *J. Labelled Compd. Radiopharm.*, **21**, 1294 (1984).
10. (a) E. Browne and R. B. Firestone, in *Table of Radioactive Isotopes* (Ed. V. S. Shirley) Wiley, New York, (1986)

    (b) A. L. Bailey and G. S. Bates *Labelled Compd. Radiopharm.*, **25**, 1267 (1988).
11. L. L. Braun and R. L. Law, *J. Chem. Educ.*, **58**, 79 (1981).
12. C. Raadschelders-Buijze, *Radiochem. Radioanal. Lett.*, **7**, 351 (1971).
13. J. D. Cox and R. J. Warne, *J. Chem. Soc.*, 1893 (1951).
14. I. Bally and A. T. Balaban, *Rev. Roum. Chim.*, **20**, 1471 (1975).
15. C. M. Brown and H. C. Brown, *J. Org. Chem.*, **31**, 3989 (1966).
16. G. G. David, D. F. C. Morris and E. L. Short, *J. Labelled Compd. Radiopharm.*, **28**, 603 (1981).
17. D. F. C. Morris, *J. Colloid Interface Sci.*, **51**, 52 (1975).
18. H. Pines and C. T. Chen, *J. Org. Chem.*, **26**, 1057 (1961).
19. H. Pines and A. W. Shaw, *J. Am. Chem. Soc.*, **79**, 1474 (1957).
20. J. J. Mitchell, *J. Am. Chem. Soc.*, **80**, 5848 (1958).
21. I. Bally, E. Gard, E. Ciornei, M. Biltz and A. T. Balaban, *J. Labelled Compd.*, **11**, 63 (1975).
22. R. E. McMahon, *Ind. Eng. Chem.*, **47**, 844 (1955).
23. J. A. Feighan and B. H. Davis, *J. Catal.*, **4**, 594 (1965).
24. L. I. Shevlyakova, M. A. Lur'e, A. V. Vysotskii and V. G. Lipovich, *Izv. Nauchno.-Issled. Inst. Nefte-Uglekhim. Sin. Irkutsk. Univ.*, **9**, 125 (1967). *Chem. Abstr.*, **72**, 132135u (1970).
25. *The Radiochemical Manual*, second edition, The Radiochemical Centre, Amersham, 1966, references 1–12 to Table C1 therein.
26. Z. Majerski, A. P. Wolf and P. v. R. Schleyer, *J. Labelled Compd.*, **6**, 179 (1970).
27. S. H. Liggero, Z. Majerski, P.v.R. Schleyer, A. P. Wolf and C. S. Redvanly, *J. Labelled Compd.*, **7**, 3 (1971).
28. E. Ösawa, Z. Majerski and P. v. R. Schleyer, *J. Org. Chem.*, **36**, 205 (1971).
29. Z. Majerski, S. H. Liggero, P. v. R. Schleyer and A. P. Wolf, *J. Chem. Soc. (D)*, *Chem. Commun.*, 1596 (1970).
30. Z. Majerski, P. v. R. Schleyer and A. P. Wolf, *J. Am. Chem. Soc.*, **92**, 5731 (1970).
31. R. R. Muccino, *Organic Synthesis With Carbon-14*, Wiley, New York, 1983, p. 254.
32. V. A. Klimashevskaya, I. S. Sedletskaya and V. S. Volkova, *Tr. Gos. Inst. Prikl. Khim.*, No. 52, 53 (1964); *Chem. Abstr.*, **63**, 13053h (1965).

18. Syntheses and uses of isotopically labelled alkanes and cycloalkanes   887

33. R. F. Nystrom and W. G. Brown, *J. Am. Chem. Soc.*, **70**, 3738 (1948).
34. M. Bubner and L. Schmidt, *Synthese Kohlenstoff-14-markirter Verbindungen*, Leipzig, 1966.
35. I. Gütert, M. Bubner, Ch. Gorner and P. Jander, Synthesis of hydrocarbons labelled with ¹⁴C, in *Organic Compounds Labelled With Radioactive Isotopes*, Part 1 (Ed. Central Sci. Atom. Inform. Inst.), Moscow, 1982, pp. 67–75.
36. L. Pichat, J. C. Levron and J. P. Guermont, *Bull. Soc. Chim. Fr.*, 1200 (1969).
37. B. Langström and S. Sjöberg, *J. Labelled Compd. Radiopharm.*, **18**, 671 (1981).
38. E. J. Corey and G. H. Posner, *J. Am. Chem. Soc.*, **90**, 5615 (1968).
39. G. M. Whitesides, W. F. Fischer Jr., J. San Filippo, R. W. Bashe and H. O. House, *J. Am. Chem. Soc.*, **91**, 4871 (1969).
40. C. Marazano, M. Maziere, G. Berger and D. Comar, *Int. J. Appl. Radiat. isot.*, **28**, 49 (1977).
41. D. R. Christmann, R. D. Finn, K. I. Karlstrom and A. P. Wolf, *Int. J. Appl. Radiat Isot.*, **26**, 435 (1975).
42. D. Roeda, C. Crouzel and B. van Zanten, *Radiochem. Radioanal. Lett.*, **33**, 175 (1978).
43. D. R. Christman, E. J. Crawford, M. Friedkin and A. P. Wolf, *Proc. Natl. Acad. Sci. USA*, **69**, 98 (1972).
44. M. G. Straatmann and M. J. Welch, *J. Nucl. Med.*, **16**, 425 (1975).
45. G. Bergström and J. Lofqvist, *J. Insect Physiol.*, **19**, 877 (1973).
46. R. H. Whittaker and P. P. Fenny, *Science*, **171**, 757 (1971).
47. (a) E. P. Rack, C. E. Lang and A. F. Voigt, *J. Chem. Phys.*, **38**, 1211 (1963).
    (b) C. E. Lang and A. F. Voigt, *J. Phys. Chem.*, **65**, 1542 (1961).
48. (a) C. MacKay and R. Wolfgang, *J. Am. Chem. Soc.*, **83**, 2399 (1961).
    (b) F. Cacace and A. P. Wolf, *J. Am. Chem. Soc.*, **84**, 3202 (1962).
49. (a) C. MacKay and R. Wolfgang, *Radiochim, Acta*, **1**, 42 (1961).
    (b) G. J. Shaw and G. W. A. Milne, *J. Labelled Compd. Radiopharm.*, **12**, 557 (1976).
50. F. B. Latore, N. Green and W. A. Gersdoft, *J. Am. Chem. Soc.*, **70**, 3707 (1948).
51. C. F. Lane, *J. Org. Chem.*, **39**, 1437 (1974).
52. Aldrich Fine Chemicals, Catalogue 1988/1989, p. 204.
53. P. Bouchet, R. Lazaro and J. Rouviere, *J. Labelled Compd. Radiopharm.*, **18**, 1071 (1981).
54. J. G. Atkinson, M. O. Luke and R. Stuart, *Can. J. Chem.*, **45**, 1511 (1967).
55. J. G. Atkinson, M. O. Luke and R. Stuart, *Chem. Commun.*, 474 (1967).
56. (a) J. G. Atkinson and M. O. Luke, French patent 1560953 (1969); *Chem. Abstr.*, **71**, 162862 (1969).
    (b) J. G. Atkinson, M. O. Luke and R. S. Stuart, *Chem. Commun.*, 283 (1969).
57. R. P. A. Sneeden and H. H. Zeiss, *J. Labelled Compd.*, **5**, 54 (1969).
58. D. G. Hill, W. A. Judge, P. S. Skell, S. W. Kantor and C. R. Hauser, *J. Am. Chem. Soc.*, **74**, 5599 (1952).
59. R. G. Snyder, M. Maroncelli, S. P. Qi and H. L. Strauss, *Science*, **214**, 188 (1981).
60. M. Tamura and J. Kochi, *Synthesis*, 303 (1971).
61. G. Fouquet and M. Schlosser, *Angew. Chem., Int. Ed. Engl.*, **13**, 82 (1974).
62. G. J. Shaw, *J. Labelled Compd. Radiopharm.*, **18**, 1641 (1981).
63. R. T. Morrison and R. N. Boyd, *Organic Chemistry*, Polish second edition, PWN, Warszawa, 1990, p. 126.
64. C. A. Elliger, *J. Labelled Compd. Radiopharm.*, **20**, 135 (1983).
65. P. J. Claringbold, J. L. Garnett and J. H. O'Keefe, *J. Labelled Compd.*, **5**, 21 (1969).
66. T. Gäumann and R. Bonzo, *Helv. Chim. Acta*, **56**, 1165 (1973).
67. J. Bigeleisen, *J. Chem. Phys.*, **34**, 1485 (1961).
68. T. Ishida and J. Bigeleisen, *J. Chem. Phys.*, **49**, 5498 (1968).
69. (a) M. Zieliński, *Isotope Effects in Chemistry*, Polish Sci. Publ., Warsaw, 1979, p. 138.
    (b) W. A. Van Hook, *J. Chromatogr. Sci.*, **10**, 191 (1972).
    (c) G. Janco and W. A. Van Hook, *Chem. Rev.*, **74**, 689 (1974).
    (d) T. C. Chan and W. A. Van Hook, *J. Chem. Thermodyn.*, **7**, 1119 (1975).
70. (a) R. V. Golovnya an Yu. N. Arsenyev, *Chromatographia*, **4**, 250 (1971).
    (b) N. H. Werstiuk and G. Timmins, *Can. J. Chem.*, **59**, 3218 (1981).
71. N. H. Werstiuk and G. Timmins, *Can. J. Chem.*, **63**, 530 (1985); **64**, 1072 (1986); **64**, 1072 (1986); **66**, 2309 (1988).

72. N. H. Werstiuk, in *Labelled Compounds* (Part A) (Eds. J. R. Jones and E. Buncel), Elsevier, Amsterdam, 1987, pp. 124–155.
73. E. A. Evans, *Tritium and its Compounds*, 2nd edn., Butterworths, London, 1974, p. 323.
74. B. E. Gordon and J. A. Van Klavern, *Int. J. Appl. Radiat. Isot.*, **13**, 103 (1962).
75. P. C. Carey and J. C. Smith, *J. Chem. Soc.*, 346 (1933).
76. P. Adriaens, S. Asselberghs, L. Dumon and B. Meesschaert, *J. Labelled Compd. Radiopharm.*, **16**, 785 (1979).
77. G. Ayrey and J. M. Lynch, *J. Labelled Compd.*, **8**, 175 (1972).
78. B. R. Davis, T. H. Houseman and H. R. Roderick, *Beitr. Tabakforsch.*, **7**, 138 (1973).
79. T. H. Houseman, S. H. Binns and K. Phillips, *J. Labelled Compd. Radiopharm.*, **14**, 163 (1978).
80. B. G. Dzantiev, Use of recoil energy of atoms for the synthesis of labelled compounds, in *Metody Polucheniya Radioaktivn. Preparatov*, Sb. Statei (USSR), 1962, pp. 75–88; *Chem. Abstr.*, **59**, 1446 (1963).
81. (a) A. N. Nesmeyanov, B. G. Dzantiev, V. V. Pozdeev and Yu. M. Rumyancev, Synthesis of tritium labelled cyclic hydrocarbons with the use of tritium recoil atoms (in Russian), Lecture RICC/321 presented during The International Conference on Applications of Radioactive Isotopes in Physical Sciences and in Industry, Copenhagen, 1960.
    (b) M. Henchman, D. Urch and R. Wolfgang, *Can. J. Chem.*, **38**, 1722 (1960).
    (c) F. F. Farley and B. S. Gordon, World Pet. Congr., Proc. 5th, N.Y., **10**, 47 (1959); *Chem. Abstr.*, **56**, 10444i (1962).
82. (a) M. Zieliński, in *Selected Topics of the Contemporary Nuclear Chemistry and Radiochemistry* (Ed. B. Waligóra), The Jagiellonian University Press, Cracow 1992/93 (in press).
    (b) R. Serber and H. Snyder, *Phys. Rev.*, **87**, 152 (1952).
    (c) S. Wexler, in *Action Chimiques et Biologiques des Radiation* (Ed. M. Haissinsky), Huitième Serie, Masson et Cie, 1965, pp. 107–241.
83. A. Migdal, *J. Phys. USSR (Moscow)*, **4**, 449 (1941).
84. S. Wexler and D. C. Hess, *J. Phys. Chem.*, **62**, 1382 (1958).
85. T. A. Carlson, *J. Chem. Phys.*, **32**, 1234 (1960).
86. C. A. Raadschelders-Buijze, C. L. Ross and P. Ross, *Chem. Phys.*, **1**, 468 (1973).
87. S. Wexler, *J. Inorg. Nucl. Chem.*, **10**, 8 (1959).
88. A. P. Wolf, Labeling of organic compounds by recoil methods, *Annu. Rev. Nucl. Sci.*, **10**, 259 (1960).
89. G. Stöcklin, *Chemie heisser Atome-Chemische Reaktionen als Folge von Kernprozessen*, Verlag Chemie, GmGH, Weinheim/Bergstr., 1969.
90. F. S. Rowland, Gas phase reactions of atomic tritium, fluorine and chlorine, IAEA-615/9, in *Hot Atom Chemistry Status Report*, Proceedings of a Panel, Vienna, 13–17 May, 1975, IAEA, Austria, December 1975, pp. 139–160.
91. G. R. Choppin and J. Rydberg, *Nuclear Chemistry*, Pergamon Press, Oxford, 1980.
92. (a) F. Cacace and S. Caronna, *J. Am. Chem. Soc.*, **89**, 6848 (1967).
    (b) F. Cacace, R. Cipollini and G. Ciranni, *J. Am. Chem. Soc.*, **90**, 1122 (1968).
93. P. L. Gant and K. Yang, *J. Chem. Phys.*, **30**, 1108 (1959).
94. (a) T. H. Pratt and R. Wolfgang, *J. Am. Chem. Soc.*, **83**, 10 (1961).
    (b) T. H. Pratt and R. Wolfgang, *Radioisotopes Phys. Ind.*, Proc. Conf. Use, Copenhagen, **3**, 159 (1962); *Chem. Abstr.*, **58**, 953 h (1963).
95. V. Aquilanti and G. G. Volpi, *J. Chem. Phys.*, **44**, 2307 (1966).
96. F. Cacace, M. Caroselli, R. Cipollini and G. Ciranni, *J. Am. Chem. Soc.*, **90**, 2222 (1968).
97. K. Yang and P. L. Gant, *J. Phys. Chem.*, **66**, 1619 (1962).
98. K. Yang and P. L. Gant, *J. Chem. Phys.*, **31**, 1589 (1959).
99. M. Cantwell, *Phys. Rev.*, **101**, 1747 (1956).
100. F. Cacace, A. Guarino and M. Speranza, *J. Am. Chem. Soc.*, **93**, 1088 (1971).
101. G. A. Olah, G. Klopman and R. H. Schlosberg, *J. Am. Chem. Soc.*, **91**, 3261 (1969).
102. J. W. Otvos, D. P. Stevenson, C. D. Wagner and O. Beek, *J. Am. Chem. Soc.*, **73**, 5741 (1951).
103. R. Wolfgang, *J. Chem. Phys.*, **40**, 3730 (1964).
104. F. Cacace, A. Guarino and E. Possagno, *J. Am. Chem. Soc.*, **91**, 3131 (1969).
105. J. W. Beatty and S. Wexler, *J. Phys. Chem.*, **75**, 2417 (1971).
106. D. L. Bunker and M. D. Pattengill, *J. Chem. Phys.*, **53**, 3041 (1970).
107. P. J. Kuntz, E. M. Nemeth, J. C. Polanyi and W. H. Wong, *J. Chem. Phys.*, **52**, 4654 (1970).

108. R. N. Porter, *J. Chem. Phys.*, **45**, 2284 (1966).
109. M. Menzinger and R. Wolf gang, *J. Chem. Phys.*, **50**, 2991 (1969).
110. R. N. Porter and S. Kunt, *J. Chem. Phys.*, **52**, 3240 (1970).
111. F. Cacace, A. Guarino and M. Speranza, *J. Chem. Soc., Perkin Trans. 2*, 66 (1973).
112. F. Cacace, *Adv. Phys. Org. Chem.*, **8**, 79 (1970).
113. (a) J. L. Williams, S. H. Daniel and Y. N. Tang, *J. Phys. Chem.*, **77**, 2464 (1973).
    (b) F.S. Rowland, *J. Am. Chem. Soc.*, **90**, 3584 (1968).
114. F. S. Rowland, E. K. C. Lee and Y. N. Tang, *J. Phys. Chem.*, **73**, 4024 (1969).
115. T. Tominaga, A. Hosaka and F. S. Rowland, *J. Phys. Chem.*, **73**, 465 (1969).
116. G. F. Palino and F. S. Rowland, *J. Phys. Chem.*, **75**, 1299 (1971).
117. Y-N. Tang, E. K. C. Lee, E. Tachikawa and F. S. Rowland, *J. Phys. Chem.*, **75**, 1290 (1971).
118. C. C. Chou and F. S. Rowland, *J. Phys. Chem.*, **75**, 1283 (1971).
119. T. Smail and F. S. Rowland, *J. Phys. Chem.*, **74**, 456 (1970).
120. T. Smail and F. S. Rowland, *J. Phys. Chem.*, **74**, 1859 (1970).
121. E. K. C. Lee and F. S. Rowland, *J. Am. Chem. Soc.*, **85**, 2907 (1963).
122. F. Cacace, C. Ciranni and A. Guarino, *J. Am. Chem. Soc.*, **88**, 2903 (1966).
123. A. H. Snell and F. Pleasonton, *J. Phys. Chem.*, **62**, 1377 (1958).
124. F. W. Lampe and F. H. Field, *Tetrahedron*, **7**, 189 (1959).
125. V. D. Nefedov, E. N. Sinotova, G. P. Akulov and V. A. Syreyshchikov, *Radiochemistry (USSR)*, **10**, 602 (1968).
126. V. D. Nefedov, E. N. Sinotova, G. P. Akulov and V. A. Syreyshchikov, *Radiochemistry (USSR)*, **10**, 600 (1968).
127. V. D. Nefedov, E. N. Sinotova and G. P. Akulov, *Radiochemistry (USSR)*, **10**, 609 (1968).
128. V. D. Nefedov, E. N. Sinotova, G. P. Akulov and V. P. Sass, *Radiochemistry (USSR)*, **10**, 761 (1968).
129. V. D. Nefedov, E. N. Sinotova, G. P. Akulov and M. V. Korsakov, *J. Gen. Radiochemistry*, **6**, 1214 (1970).
130. V. D. Nefedov, E. N. Sinotova, G. P. Akulov and M. V. Korsakov, *Radiochemistry*, **15**, 286 (1973).
131. B. Aliprandi, F. Cacace and A. Guarino, *J. Chem. Soc. (B)*, 519 (1967).
132. R. L. Burwell, Jr., A. B. Littlewood, M. Cardow, G. Pass and C. T. H. Stoddart, *J. Am. Chem. Soc.*, **82**, 6272 (1960).
133. S. Wexler and D. C. Hess, *J. Phys. Chem.*, **62**, 1382 (1958); M. Cantwell, *Phys. Rev.*, **101**, 1747 (1956).
134. F. Cacace, M. Caroselli and A. Guarino, *J. Am. Chem. Soc.*, **89**, 4584 (1967).
135. S. Wexler, G. R. Anderson and L. A. Singer, *J. Chem. Phys.*, **32**, 417 (1960).
136. L. Babernics and F. Cacace, *J. Chem. Soc. (B)*, 2313 (1971).
137. R. L. Wolfgang, R. C. Anderson and R. W. Dodson, *J. Chem. Phys.*, **24**, 16 (1956).
138. M. Wolfsberg, *J. Chem. Phys.*, **24**, 24 (1956).
139. G. Stöcklin, Chemistry of nucleogeneic carbon atoms (Kernforschungs anlage juelich, Germany), AEC Accession No. 5570, Rep. No. Jul-228-RC.
140. E. P. Rack, C. E. Lang and A. F. Voigt, *J. Chem. Phys.*, **38**, 1211 (1963).
141. D. E. Clark U.S. At. Energy Comm. IS-T-23, 93 pp (1965); *Chem. Abstr.*, **63**, 17838f (1965).
142. M.P. Pirnic, Microbial oxidation of methyl-branched alkanes, *Crit. Rev. Microbiol.*, **5**, 413 (1977).
143. E. J. McKenna, Microbial metabolism of normal and branched alkanes, in *Degradation of Synthetic Organic Molecules in the Biosphere*, Natl. Acad. Sci, Washington D.C., 1972, p. 73.
144. E. R. Davidov, Yu. I. Sokolov, N. F. Demanova and A. D. Golobov, *Prikl. Biokhim. Mikrobiol.*, **17**, 328 (1981).
145. R. R. Colwell and J. D. Walker, Ecological aspects of microbiological degradation of petroleum in the marine environment, *Crit. Rev. Microbiol.*, **5**, 423 (1977).
146. W. M. Sackett, in NATO Adv. Study Inst Ser., *Strategies and Advanced Techniques for Marine Pollution Studies: Mediterranean Sea*, Vol. 69 (Eds. C. S. Giam and H. J. M. Don), Springer-Verlag, Berlin, Heidelberg, 1986, pp. 289–301.
147. (a) R. K. Gholson, J. N. Baptist and M. J. Coon, *Biochemistry*, **2**, 1155 (1963).
    (b) M. Katagiri, S. Yamamoto and O. Hayaishi, *J. Biol. Chem.*, **237**, 2413 (1962).
    (c) A. J. Fulco and K. Bloch, *Biochim. Biophys. Acta*, **63**, 545 (1962).

(d) H. E. Conrad, R. DuBus and I. C. Gunsalus, *Biochim. Biophys. Res. Commun.*, **6**, 293 (1961).

(e) M. Kusunose, E. Kusunose and M. J. Coon, unpublished results.

148. (a) K. Ichihara, E. Kusunose and M. Kusunose, *Biochim. Biophys. Acta*, **176**, 713 (1969).

(b) G. Speier, in *The Chemistry of the Metal–Carbon Bond*, Vol. 5 (Ed. F. R. Hartley), Chap. 5, Wiley, Chichester, 1989, pp. 147–198.

149. D. W. Stetten, Jr, *J. Biochem.*, **147**, 327 (1943).

150. V. J. Harding and C. Weizmann, *J. Chem. Soc.*, **97**, 299 (1910).

151. M. P. Mitchell and G. Hübscher, *Biochem. J.*, **103**, 23P (1967).

152. (a) M. P. Mitchell and G. Hübscher, *Eur. J. Biochem.*, 7, 90 (1968).

(b) M. Kusunose, E. Kusunose and M. J. Coon, *J. Biol. Chem.*, **239**, 1374 (1964).

(c) A. J. Peterson, D. Basu and M. J. Coon, *J. Biol. Chem.*, **241**, 5162 (1966).

(d) J. Chouteau, E. Azoulay and J. C. Senez, *Bull. Soc. Chim. Biol.*, **241**, 5162 (1966).

(e) J. Chouteau, E. Azoulay, and J. C. Senez, *Nature*, **194**, 576 (1962).

(f) E. Azoulay and J. C. Senez, *Biochim. Biophys. Acta*, 77, 554 (1963).

(g) F. Wagner, W. Zahn and V. Bühring, *Angew. Chem.*, 79, 314 (1967).

153. R. D. McCarthy, *Biochim. Biophys. Acta*, **84**, 74 (1964).

154. K. Ichinara, E. Kusunose and M. Kusunose, *J. Japan. Biochem. Soc.*, **40**, 464 (1968).

155. M. Kusunose, K. Ichinara and E. Kusunose, *Biochim. Biophys. Acta*, **176**, 679 (1969).

156. R. K. Gholson, J. N. Baptist and M. J. Coon, *Biochemistry*, **2**, 1155 (1963); R. K. Gholson and M. J. Coon, *Abstracts, Am. Chem. Soc., 138th Meeting*, New York, 1960.

157. D. Y. Cooper, S. Levin, S. Narasimhulu, O. Rosenthal and R. W. Estabrook, *Science*, **147**, 400 (1965).

158. S. Bergstrom, B. Borgstrom, N. Tryding and G. Westoo, *Biochem. J.*, **58**, 604 (1954).

159. D. J. McWeeny, Ph.D. Thesis, University of Birmingham (1957).

160. C. E. Cain, O. F. Bell, Jr., H. B. White, Jr. L. L. Sulya and R. R. Smith, *Biochim. Biophys. Acta*, **144**, 493 (1967).

161. H. Dannenberg and R. Richter, *Z. Physiol. Chem.*, **349**, 565 (1968).

162. M. Popovic, *FEBS Lett.*, **12**, 49 (1970).

163. J. E. Tulliez and G. F. Bories, *Lipids*, **13**, 103 (1978).

164. M. L. Das, S. Ororenius and L. Ernster, *Eur. J. Biochem.*, **4**, 519 (1968).

165. M. J. Coon, H. W. Srobel, A. P. Autor, J. Heidema and W. Duppel, In *Biological Hydroxylation Mechanism* (Eds. G. S. Boyd and R. M. S. Smellie), Academic Press, London and New York, 1972, pp. 45–54.

166. J. Sanders-Loehr, in *Oxidases and Related Redox Systems*, Alan R. Liss, Inc., New York, 1988, pp. 193–209.

167. J. Sanders-Loehr, W. D. Wheeler, A. K. Shiemke, B. A. Averill and T. M. Loehr, *J. Am. Chem. Soc.*, **111**, 8084 (1989).

168. C. W. J. Fleet, in *Organic Reaction Mechanisms 1985* (Eds. A. C. Knipe and W. E. Watts), Chapter 5, Wiley, London, 1987, pp. 169–222.

169. C. R. McKinney, J. M. McCrea, S. Epstein, H. A. Allen and H. C. Urey, *Rev. Sci. Instrum.*, **21**, 724 (1950).

170. H. Craig, *Geochim. Cosmochim. Acta*, **3**, 53 (1953).

171. H. Craig and L. L. Gordon, in *Stable Isotopes in Oceanographic Studies and Paleotemperatures-Spoleto* (Ed. E. Tongiorgi), CNR, Pisa, 1965, pp. 9–130.

172. R. Gonfiantini, *Nature*, **271**, 534 (1978).

(a) A. Longinelli and S. Nuti, *Earth and Planetary Science Letters*, **5**, 13 (1968).

(b) D. D' Angela and A. Longinelli, *Chemical Geology (Isotope Geoscience Section)*, **86**, 75 (1990).

(c) J. Oesselmann, "On the development of automatic sample preparation devices", Finnigan MAT GmbH, D-2800 Bremen and Textronica AG, CH-9463, Oberriet/Swizerland (1991).

173. F. C. Tan, in *Handbook of Environmental Isotope Geochemistry*, Vol. 3, *The Marine Environment, A* (Eds. P. Fritz and J. Ch. Fontes), Chap. 5, Elsevier, Amsterdam, 1989, pp. 171–190.

174. F. C. Tan and P. M. Strain, in *Canadian Bulletin of Fisheries and Aquatic Sciences*. No. 220. *Chemical Oceanography in the Gulf of St. Lawrence* (Ed. P. M. Strain), Chap. V, pp. 59–77 (1990).

175. (a) D. A. Anati and J. R. Gat, Chap. 2 in Reference 173, pp. 29–73.

(b) C. Pierre, Chap. 8 in Reference 173, pp. 257–315.

176. W. M. Sackett, Chap. 4 in Reference 173, pp. 139–169.
177. E. S. Van Vleet, W. M. Sackett, S. B. Reinhardt an M. E. Mangini, Distribution, sources and fates of floating oil residues in the Eastern Gulf of Mexico, *Mar. Pollut. Bull.*, **15**, 106 (1984); and Reference 146.
178. R. A. Burke, Jr. and T. R. Barber, *Global Biogeochemical Cycles*, **2**, 329 (1988).
179. T. R. Barber and R. A. Burke, Jr., *Global Biogeochemical Cycles*, **2**, 411 (1988).
180. S. Gordon, *Int. J. Chem. Kinet.*, **7**, Symp. 1, 289 (1975).
181. V. S. Lebedev and I. A. Petersil'e, *Dokl. Akad. Nauk SSSR*, **158**, 1102 (1964).
182. S. M. Katchenkov, *Tr. Vses. Neft. nauchno-Issled. Geologorazved. Inst.* (USSR), **355**, 127 (1974); *Chem. Abstr.*, **84**, 7262 (1976).
183. U. Colombo, F. Gazzarrini, G. Sironi, R. Gonfiantini and E. Tongiorgi, *Nature*, **205**, 1303 (1965).
184. D. J. Des Marais, J. M. Mitchell, W. G. Meinschein and J. M. Hayes, *Geochim. Cosmochim. Acta*, **44**, 2075 (1980).
185. (a) M. B. Neiman, *Z. Phys. Chem.*, **28**, 1235 (1954).
     (b) M. Zieliński and M. Kańska, in *The Chemistry of Quinonoid Compounds*, Vol. II (Eds. S. Patai and Z. Rappoport), Chap. 19, Wiley, Chichester, 1988, pp. 1167–1171.
     (c) M. B. Neiman and G. I. Feklisov, *Dokl. Akad. Nauk SSSR.*, **91**, 877 (1953).
     (d) M. B. Neiman and G. I. Feklisov, *Dokl. Akad. Nauk SSSR*, **91**, 1137 (1953).
     (e) A. F. Lukovnikov and M. B. Neiman, *Z. Phys. Chem.*, **29**, 1410 (1955).
     (f) I. N. Antonova, V. A. Kuzmin, R. I. Moshkina, A. B. Nalbandian, M. B. Neiman and G. I. Feklisov, *Izv. Akad. Nauk SSSR, Otd. Khim. Nauk*, 789 (1955).
     (g) A. B. Nalbandian, *Dokl. Akact. Nauk SSSR*, **60**, 607 (1948); *Z. Phys. Chem.*, **22**, 1443 (1948).
     (h) M. B. Neiman and D. Gal, *The Kinetic Isotope Method and Its Application*, Akademiai Kiadó, Budapest, 1971.
186. A. A. Balandin, M. B. Neiman, O. K. Bogdanova, G. V. Isagulyanc, A. P. Shcheglova and E. I. Popov, *Izv. Akad. Nauk SSSR, Otd. Khim. Nauk*, 157 (1957).
187. V. G. Lipovich, O. I. Shmidt, M. A. Lurie and I. V. Kalechits, *I.N.U.S. of the University of Irkutzk*, **9**, 33 (1967).
188. Yu. I. Derbentser, G. V. Isagulyants, *Usp. Khim.*, **38**, 1597 (1969) and references cited therein.
189. B. A. Kazanskii, G. V. Isagulyants, M. I. Rozengart, Yu. D. Dubinskii and L. I. Kovalenko, Vth International Congress on Catalysis, Palm Beach, Florida, USA, 1972, preprint 94.
190. B. A. Kazanskii, G. V. Isagulyants, M. I. Rozengart, Yu. I. Debrentsev and Yu. G. Dubinskii, *Dokl. Akad. Nauk SSSR*, **191**, 600 (1970).
191. B. A. Kazanskii, M. I. Rozengart, G. V. Isagulyants, Yu. I. Derbentser, L. I. Kovalenko, A. V. Greish and Yu. G. Dubinskii, *Dokl. Akad. Nauk SSSR*, **197**, 1085 (1971).
192. B. A. Kazanskii, Symposium on the Mechanism of Hydrocarbon Reactions, 5–7 June, 1973, Siofok, Hungary (Eds. F. Marta and D. Kallo), Akademiai Kiado, Budapest, 1975, pp. 15–47.
193. Z. Paál and P. Tétényi, *Acta Chim. Acad. Sci. Hung.*, **54**, 175 (1967).
194. Z. Paál and P. Tétényi, *Acta Chim. Acad. Sci. Hung.*, **55**, 273 (1968).
195. Z. Paál and P. Tétényi, *Acta Chim. Acad. Sci. Hung.*, **58**, 105 (1968).
196. V. I. Komarewsky, S. C. Uhlick and M. J. Murray, *J. Am. Chem. Soc.*, **67**, 557 (1945).
197. E. O. Weinman, I. L. Chaikoff and W. G. Dauben, *J. Biol. Chem.*, **184**, 735 (1950).
198. R. D. Mullineaux and J. H. Raley, *J. Am. Chem. Soc.*, **85**, 3178 (1963).
199. J. R. Anderson and C. Kemball, *Proc. Roy. Soc. London*, **A226**, 472 (1954).
200. P. Tétényi, *Acta Chim. Acad. Sci. Hung.*, **40**, 157 (1964).
201. V. V. Lunin, G. V. Lisihkin, Y. M. Bondarev and L. K. Denisov, in Symposium on the Mechanism of Hydrocarbon Reactions, 5–7 June, 1973, Siofok, Hungary (Eds. F. Marta and D. Kallo), Akademiai Kiado, Budapest, 1975, pp. 71–79.
202. G. Lietz and J. Völter, in Reference 201, pp. 151–161.
203. O. Örhalmi and P. Fejes, in Reference 201, pp. 457–468.
204. M. Daage and F. Fajula, *J. Catal.*, **81**, 394 (1983).
205. F. Le Normand and F. Fajula, in *Catalysis by Acids and Bases* (Eds. B. Imelik, C. Naccache, G. Coudurier, Y. Ben Taarit and J. C. Vedrine); Elsevier, Amsterdam, 1985, pp. 325–334.
206. R. Touroude and F. G. Gault, in Reference 201, pp. 215–225.
207. Ya. T. Eidus, *Usp. Khim.*, **36**, 824 (1967).
208. Yu. I. Debrencey and G. V. Isagulyants, *Usp. Khim.*, **38**, 1597 (1969).

209. G. V. Isagulyand, N. I. Ershov, Yu. I. Debrencev and Ya. T. Eidus, *Izv. Akad. Nauk SSSR*, *Ser. Khim.*, 1234 (1968).
210. N. I. Ershov, Ya. T. Eidu and I. V. Guseva, *Dokl. Akad. Nauk SSSR*, **167**, 583 (1966).
211. J. R. H. Ross, M. C. F. Steel and A. Zeini-Isfahani, in Reference 201, pp. 201–214.
212. M. Zieliński, unpublished results.
213. G. B. Skinner, C. S. Ronald and S. K. Davis, *J. Phys. Chem.*, **75**, 1 (1971).
214. S. W. Benson and G. R. Haugen, *J. Phys. Chem.*, **71**, 1735 (1967); *Chem. Abstr.*, **66**, 64789 m; **67**, 11022t. (1967)
215. P. Ausloos, in Reference 201, pp. 603–624.
216. V. L. Talroze and A. K. Lyubimova, *Dokl. Akad. Nauk SSSR*, **86**, 909 (1952).
217. T. Gaümann and A. Ruf, in Reference 201, pp. 647–658.
218. A. E. Shilov and A. A. Shteinman, in Reference 201, pp. 479–485 and references cited therein; *Chem. Abstr.*, **84**, 42653e (1976).
219. N. F. Goldschleger, M. B. Tjabin, A. E. Shilov and A. A. Shteinman, *Zh. Fiz. Khim.*, **43**, 2174 (1969).
220. M. B. Tjabin, A. E. Shilov and A. A. Shteinman, *Dokl. Akad. Nauk SSSR*, **198**, 380 (1971).
221. N. F. Goldschleger, V. V. Eskova, A. E. Shilov and A. A. Shteinman, *Zh. Fiz. Khim.*, **46**, 1353 (1972).
222. R. J. Hodges, D. E. Webster and P. B. Wells, *Chem. Commun.*, 462 (1971).
223. N. F. Goldschleger, I. I. Moiseev, M. L. Khidekel and A. A. Shteinman, *Dolk. Akad. Nauk SSSR*, **206**, 106 (1972).
224. N. F. Goldschleger, M. B. Tjabin, A. E. Shilov and A. A. Shteinman, *Zh. Fiz. Khim.*, **43**, 2174 (1968); also Reference 218.
225. M. Agova, *Khim. Ind. (Sofia)*, **35**, 23 (1963).
226. I. A. Korshunov and A. P. Batalov, *J. Gen. Chem. USSR*, **29**, 4048 (1959).
227. A. P. Batalov, *Dokl. Akad. Nauk SSSR*, **291**, 139 (1986).
228. A. V. Severin and A. P. Batalov, *Radiokhimiya*, **30**, 537 (1988).
229. N. Makabe, S. Yokoyama, M. Itoh and G. Takeyu (Hokkaido Daigaku Kogakubu Kenkyu Hokoku, **62**, 77 (1971).), *Chem. Abstr.*, **79**, 136638c (1973).
230. C. T. Peng, in *Isotopes in the Physical and Biomedical Sciences*, Vol. I, *Labelled Compounds* (Part A) (Eds. E. Buncel and J. R. Jones), Chap. 2, Elsevier, Amsterdam, 1987, pp. 8–51 and references cited therein.
231. M. Yoshida and S. Nakayama, Chap. 3 in Reference 230, pp. 52–85.
232. J. L. Garnet, *Catal. Rev.*, **5**, 229 (1971); also J. L. Garnett and M. A. Long, Chap. 4 in Reference 230, pp. 88–121.
233. Yu. A. Alexandrov, N. V. Kuznetzova, S. A. Lebedev, E. A. Grigorian and A. F. Shestakov, *Dokl. Akad. Nauk SSSR* (in press) (personal communication, see also *Chem. Abstr.*, **113**, 191531 h (1990)).
234. Yu. A. Alexandrov, S. A. Lebedev and N. V. Kuznetzova, *Zh. Obsch. Khim.*, **56**, 1667 (1986).
235. Yu. A. Alexandrov, *Liquid Phase Autooxidation of Organometallic Compounds*, Science, Moscow, 1978, p. 41.
236. D. W. Coillet and G. M. Harris, *J. Am. Chem. Soc.*, **75**, 1486 (1953).
237. G. M. Harris, *J. Phys. Chem.*, **56**, 891 (1952).
238. S. Z. Roginski, *Theoretical Principles of the Isotopic Methods of Study of Chemical Reactions*, Izd. AN SSSR, Moscow, 1956, p. 48.
239. M. F. Hoq and K. J. Klabunde, in *Acid–Base Catalysis* (Eds. K. Tanabe, H. Hattori, T. Yamaguchi and T. Tanaka), Kodanska Ltd, 1989, pp. 105–121. (Ref 1a).
240. M. F. Hoq and K. J. Klabunde, *J. Am. Chem. Soc.*, **108**, 2114 (1986).
241. C. H. DePuy and V. M. Bierbaum, *J. Am. Chem. Soc.*, **106**, 4051 (1984).
242. I. Z. Siemion, August Freund (1835–1892). On the hundredth anniversary of the cyclopropane synthesis, in *Wiadomości Chemiczne*, **37**, 509 (1983), and reference 14 in this review: "Ueber Trimethylen", *Sitzungsber.*, **86**, 359 (1882); *Monatsh. Chem.*, **3**, 625 (1882); *J. prakt. Chem.*, **26**, 367 (1882).
243. B. K. Carpenter, in *The Chemistry of the Cyclopropyl Group* (Ed. Z. Rappoport), Chap. 17, Wiley, Chichester, 1987, pp. 1027–1082.
244. T. S. Chambers and G. B. Kistiakovsky, *J. Am. Chem. Soc.*, **56**, 399 (1934).
245. R. E. Weston, *J. Chem. Phys.*, **23**, 988 (1955).
246. R. H. Lindquist and G. K. Rollefson, *J. Chem. Phys.*, **24** 725 (1956).

247. R. E. Weston, *J. Chem. Phys.*, **26**, 975 (1957).
248. (a) M. Zieliński, in *The Chemistry of Quinonoid Compounds* (Eds. S. Patai and Z. Rappoport), Chap. 12, Wiley, London, 1974, pp. 619–627.
    (b) N. B. Slater, *Proc. Roy. Soc. London*, **A218**, 224 (1953).
249. E. W. Schlag and B. S. Rabinovitch, *J. Am. Chem. Soc.*, **82**, 5996 (1960).
250. A. T. Blades, *Can. J. Chem.*, **39**, 1401 (1961).
251. B. S. Rabinovitch, P. W. Gilderson and A. T. Blades, *J. Am. Chem. Soc.*, **86**, 2994 (1964).
252. B. S. Rabinovitch, D. W. Setser and F. W. Schneider, *Can. J. Chem.*, **39**, 2609 (1961).
253. A. T. Blades and P. W. Gilderson, *Can. J. Chem.*, **38**, 1407 (1960).
254. J. Langrish and O. Pritchard, *J. Phys. Chem.*, **62**, 761 (1958).
255. D. W. Setser and B. S. Rabinovitch, *J. Am. Chem. Soc.*, **86**, 564 (1964).
256. J. J. Gajewski and J. M. Warner, *J. Am. Chem. Soc.*, **106**, 802 (1984).
257. J. J. Gajewski and S. K. Chou, *J. Am. Chem. Soc.*, **99**, 5696 (1977).
258. B. H. Al-Sader and R. J. Crawford, *Can. J. Chem.*, **46**, 3301 (1968).
259. R. J. Crawford, R. J. Dummel and A. Mishra, *J. Am. Chem. Soc.*, **87**, 3023 (1965).
260. J. Bigeleisen, *J. Chem. Phys.*, **17**, 675 (1949).
261. L. B. Sims and P. E. Yankwich, *J. Phys. Chem.*, **71**, 3459 (1967).
262. L. B. Sims, H. P. E. Sachse and E. Mei, Report on research: *Vibrational Assignments and Descriptions of Normal Modes of Vibration of Cyclopropane Molecules*, Department of Chemistry, University of Arkansas, Fayetteville, 1972.
263. E. W. Schlag and B. S. Rabinovitch, *J. Am. Chem. Soc.*, **82**, 5996 (1960).
264. W. E. Falconer, T. F. Hunter and A. F. Trotman-Dickenson, *J. Chem. Soc.*, 609 (1961).
265. H. O. Pritchard, R. G. Sowden and A. F. Trotman-Dickenson, *Proc. Roy. Soc. London*, **A217**, 563 (1953).
266. H. S. Johnston and J. R. White, *J. Chem. Phys.*, **22**, 1969 (1954).
267. M. Zieliński, *Nucl. Appl.*, **2**, 51 (1966).
268. M. Zieliński, *Naukleonika*, **11**, 807 (1966).
269. L. B. Sims, *Kinetic Isotope Effects and Mechanism of Thermal Reactions of Small Ring Compounds*, Report on Research No. 320-G, Michigan State University (1968).
270. E. S. Rudakov, N. A. Tishchenko and A. I. Lucky, *Dokl. Akad. Nauk SSSR*, **252**, 893 (1980).
271. K. B. Wiberg and G. Foster, *J. Am. Chem. Soc.*, **83**, 423 (1961).
272. I. Necsoiu, A. T. Balaban I. Pascarus, E. Sliam, M. Elian and C. D. Nenitroscer, *Tetrahedron*, **19**, 1133 (1963).
273. I. S. Rudakov, N. A. Tishchenko and L. K. Volkova, *Kinetica i Kataliz*, **27**, 1101 (1986).
274. I. S. Rudarkov and L. K. Volkova, *Kinetica i Kataliz*, **24**, 542 (1983).
275. D. H. Su, D. Maurice, D. G. Truhlar, *J. Am. Chem. Soc.*, **112**, 6206 (1990).

## CHAPTER 19

Natural occurrence, biochemistry and toxicology

DEREK V. BANTHORPE

Department of Chemistry, University College London, 20 Gordon Street, London WC1H OAJ, England

The Chemistry of Alkanes and Cycloalkanes
Edited by S. Patai and Z. Rappoport © 1992 John Wiley & Sons Ltd

I. INTRODUCTION

Vast quantities of numerous alkanes and cycloalkanes together with other hydrocarbons occur in natural gas and petroleum deposits as well as in shales and related marine and terrestrial sediments. Crude oils and their head-space gases may contain virtually the entire range (i.e. all carbon numbers) from C_1 to C_{70} or C_{78} compounds[1-3], and the literature on the nature, occurrence and origin of these hydrocarbons is voluminous (cf Section II.A). In contrast, the existence of such compounds in living organisms is almost entirely ignored in compilations and reviews on natural products and secondary metabolites, although they seem to be almost universally distributed, albeit at low levels, in the Animal and Plant Kingdoms and often occur in micro-organisms. A range (not continuous) up to C_{62} compounds has been reported from such sources[4], although the main distribution falls from about C_{21} to C_{35}. An excellent summary discusses the alkanes of all types that occur amongst the components of the waxes of bacteria, algae, plants and animals[5], and the occurrence, metabolism and pathology of n-alkanes in living organisms has been reviewed[6]. Up to 1990 at least 2000 papers have appeared containing information concerning the distribution of alkanes in the living world.

Numerous quantitative analyses of petroleum deposits have been made, but few studies on the occurrence of the alkanes in living organisms record quantitative data (e.g. as % wt–wet wt) or survey genera or families. Many of the latter studies record one-off, apparently unrelated and undiscussed observations tucked away in lists of other secondary metabolites.

Almost always GC–MS has been the chosen tool for identification of alkanes from geological or living sources, and such methods in general are outlined in preceding chapters. More recently, the tandem mass-spectrometric (MS–MS) technique has been applied to the effluent from GC columns[7] and adaptions of the standard GC–MS methods have been applied for the analysis of triterpanes and steranes (C_{30} compounds and the like) that often occur in petroleum and are valuable geological markers[8]. Supercritical fluid chromatography linked to MS will undoubtedly also become a widely used technique for separation and analysis of petroleum and natural waxes[9], as will thermospray HPLC–MS methods.

Considerable caution is needed in assigning hydrocarbons to be natural products of the living world owing to the widespread, possibly global, pollution generated by the mining, refining and transporting activities of the petroleum industry and the ubiquity

TABLE 1. Nomenclature of n- and branched alkanes

No. of C atoms (n-chain) or branching-type		Name
(n-C)	10–19	Decane to nonadecane[a]
(n-C)	20	Eicosane
(n-C)	21	Heneicosane
(n-C)	22, 23, $\cdots$, 29	Docosane, tricosane, $\cdots$, nonacosane
(n-C)	30	Triacontane
(n-C)	31	Hentriacontane
(n-C)	32, $\cdots$ etc.	Dotriacontane, $\cdots$, etc.
(n-C)	40, 50 etc.	Tetracontane, pentacontane, etc.
		iso- $\equiv$ (2-methylalkyl-)
		anteiso $\equiv$ (3-methylalkyl-)

[a] n-Alkanes higher than (and including) $C_{17}H_{36}$ are solids at normal room temperatures, whilst the range C_5H_{12} to $C_{16}H_{34}$ are liquids. Chain branching produces an increase in volatility.

of petroleum-based products: thus, extraneous hydrocarbons can turn up in virtually all environments, e.g. in rain[10], and can enter the food chain as well as blanketing (albeit usually at very low levels) animal and plant materials. At present it seems an act of faith that this pollution is regarded as negligible in its long-term biological effects.

An outline of the perhaps unfamiliar nomenclature often applied to the higher alkanes (and widely used in the literature) is appended in Table 1. In this chapter the more esoteric names will be ignored in favour of references to C_n alkanes or C_n compounds.

II. NATURAL OCCURRENCE

A. Natural Gas, Petroleum, Shale Oils

Methane accumulates in coal seams as fire damp formed by anaerobic bacterial decomposition of organic material such as cellulose, and similarly is generated as marsh gas in swamps and water-logged vicinities. It is also liberated by bacterial decomposition of animal and bird excrement (e.g. cow manure and bat guano) and tissue[11], and can be detected (ca. $1 \mu g l^{-1}$) in marine surface waters, presumably mainly formed from decomposition of zoo- and phyto-plankton[12]. Surface waters also contain higher alkanes (C_{12} to C_{24}) at levels up to a total of $400 \mu g l^{-1}$, and similar patterns of compounds are found in deep-sea sediments[12].

Natural gas is present in enormous quantities in the earth's crust and is usually associated with oil deposits. It is chiefly composed of C_5 and C_6 hydrocarbons (mainly n-alkanes) but the range extends to C_{12} as is shown by recent analyses of gas from Canadian and US well-heads[13].

Petroleums contain three main classes of hydrocarbons: alkanes (paraffins, especially prevalent in US petroleums), naphthenes (cycloalkanes) and aromatics, the proportions of which vary with the geographical source. A classical analysis of a US crude oil made over 30 years ago led to the characterization of 175 components including all the n-alkanes up to $C_{33}H_{68}$, and indicated that the crude comprised about 30% n-alkanes and 15% branched-chain alkanes and naphthenes based on recoveries[14]. The data are summarized in Table 2, which indicates the conventional names of the fractions. One of the significant conclusions is that, although petroleum is a complex mixture, it is composed of an insignificant proportion of the number of compounds that are theoretically possible. Thus, for hydrocarbons with C_{13}, C_{20} and C_{30} the number of possible isomers is 802, 366 and 319 and greater than 4×10^9, respectively; this count excludes stereoisomers, and if the latter are included the numbers become astronomical! More recent analyses, such as that of Arabian crudes[15], reveal broadly similar patterns to those obtained for the US material, although recoveries of identified products as a proportion of the total are now more quantitative owing to the more sophisticated methods of separation and analysis.

TABLE 2. Summary of components of a representative US petroleum

Fraction	Gas	Gasoline	Kerosene	Light oil[a]	Heavy oil	Lubricating oil
Boiling point (°C)	<40	40–180	180–230	230–305	305–405	405–515
n-C_xH_{2x+2}	C_1–C_5	C_6–C_{10}	C_{11}, C_{12}	C_{12}–C_{17}	C_{18}–C_{25}	C_{26}–C_{38}
% in petroleum	4	33	13	19	15	10
No. of compounds	7	101	37	12	10	8
% accounted for	100	82	38	30	21	10
% of original petroleum	4	28	5	7	3	0.9

$a \equiv$ Diesel oil

898   D. V. Banthorpe

TABLE 3. Isoprenoid hydrocarbons from geological sources

Name	Structure	Source
(1) Pristane (C$_{19}$)	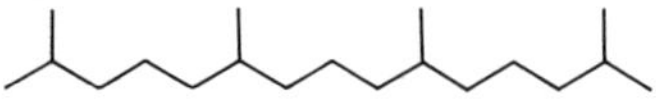	Recent and ancient marine sediments; petroleum
(2) Phytane (C$_{20}$)	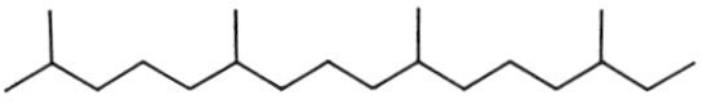	Petroleum; shales
(3) Cholestane (C$_{27}$)	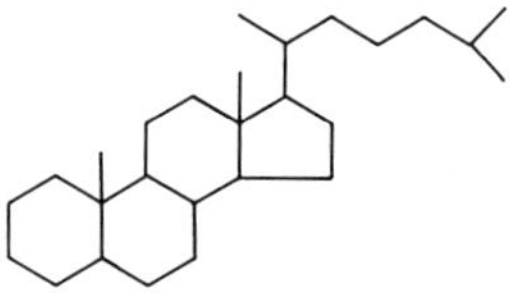	Shales, marine sediments
(4) Hopane (C$_{30}$)	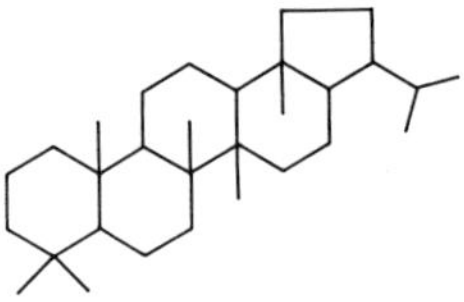	Ancient sediments
(5) Ergosterane (C$_{28}$) 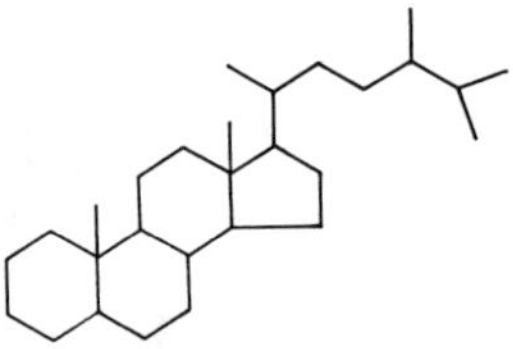		(6) Stigmasterane (C$_{29}$) 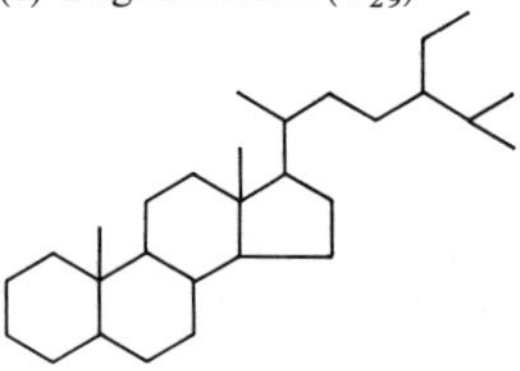

The occurrence of cycloalkanes, especially n-alkylated cyclopentanes and cyclohexanes, in petroleum has been reviewed in detail[16]. These types of hydrocarbons are widespread in petroleums but are rarely found in living organisms (cf Sections II.B–II.E). An especially interesting class of compounds that is not widespread but does occur in some petroleums[17-20] and also in some coals[17] are the isoprenoid hydrocarbons. These are chemical fossils, derived from isoprenoids ($\equiv$ terpenoids) of biological origin that have such stable structures as to be little altered by deposition and accumulation into petroleum or shale oil over recent geological times and, of course, their existence provides evidence for the biological origin, at least in part, of petroleum[17,21,22].

Some isoprenoid hydrocarbons are shown in Table 3. Pristane (1) and phytane (2) are probably the most widely distributed and are obviously derived from the diterpenoid phytol (C$_{20}$) that is a component of the chlorophylls common to all green plants. Analogous compounds with C$_{14}$ to C$_{21}$ skeletons also have been characterized and all betray their terpenoid ancestry by the presence of the characteristic isopentane (2-methylbutane) sub-unit. The C$_{17}$ compound is believed to be degraded from squalene (a triterpenoid) and is rare[23], whereas its more common C$_{21}$ relative was probably derived from breakdown of carotane that in turn was formed from carotenoids (C$_{40}$ compounds) that comprise the pigments of higher plants and many micro-organisms. Steranes and triterpanes, that are undoubtedly produced by geological modification of steroids and triterpenoids of plant and algal origin, are found in recent shales, some ancient sediments and a few petroleums; examples are cholestane (3) and hopane (4). Many related skeletal types occur, e.g. ergosterane (5, C$_{28}$) and stigmasterane (6, C$_{29}$), which have obvious relationships with ergosterol from yeasts and stigmasterol, the major steroid of green leaves[19,20,24,25].

In addition to numerous technical papers concerning the components of the gasoline, kerosene and various oil fractions of different petroleums, a great deal of work has involved the analysis of the ratio of n-alkanes with odd or even numbers of carbon atoms ever since the notable difference in this ratio between oils from different sediments and reservoirs was first emphasized[26]. The incentive for these studies was the long controversy concerning the origin of petroleum (cf Section III.A). It was well known (cf. Section II.B and II.E) that higher plants and many micro-organisms predominantly produce n-alkanes with odd-carbon chains, and it was presumed that these were derived from decarboxylation of the natural series of fatty acids with even-carbon numbers. Thus, determination of the odd–even ratio in sediments and oils of different known geological ages was thought to provide insight to the validity of the rival biogenic and geochemical theories for the origin of petroleum. In order to provide quantification, various criteria were derived to measure the n-alkane distribution, e.g. the carbon preference index (CPI) or the odd–even preference (OEP), and these parameters were determined for recent and ancient sediments and for crude oils and used to assess the source beds of the petroleums[27,28]. The general consensus was that there was a marked preference for odd-number n-alkanes in recent oil-bearing sediments (ranging from the first few thousand years of deposition up to several tens of thousands of years) but older sediments contained oil with a more even distribution for the odd and even types. A simplified summary of such analyses is in Figure 1. The significance of these observations will be outlined in Section III.A.

Considerable effort has been applied to the analysis of the hydrocarbon content of non-oil-bearing sediments and rocks. Recent non-reservoir sediments may contain 30 to 60 ppm hydrocarbons, whilst the levels in non-reservoir ancient sediments may increase to 300 ppm, probably due to seepage (up) from reservoir sediments over the course of geological time[16]. Studies have centred on three broad, ill-defined classes of material: soil waxes—the organic material extracted by ether or benzene–methanol mixtures from very recent deposits (e.g. soils), bitumens—the material similarly extracted from sedimentary fossiliferous rocks or shales; and kerogens—the organic material in shales and older sediments that is insoluble in the commonly used petroleum solvents and is probably derived from lignin.

Numerous soil waxes have been analysed[29]. The hydrocarbon pattern characteristic of petroleum is not observed and typically the distribution of n-alkanes falls within the

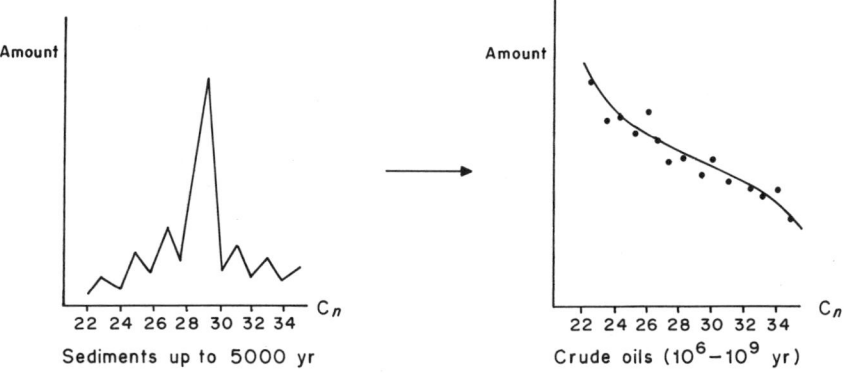

FIGURE 1. Variation of odd–even carbon number for n-alkanes from deposits of different ages

range C_{19} to C_{36} with up to 90% being odd-carbon compounds, especially C_{29}, C_{31} and $C_{33}{}^{30-32}$. This pattern is characteristic of that of the waxes of higher plants. Alkane levels were particularly high (ca 12 ppm) in podzol-type soils that also had high (ca 3%) levels of organic matter in general[31]. The patterns have been considered as indicating a mixed marine and terrestrial (plant and algae) origin of the extracts[32]. Long-chain methyl ketones also occur in some soils and may be directly derived from alkanes by microbial oxidation (cf Section IV.A) or may similarly originate from fatty acids by β-oxidation followed by decarboxylation[33]. Bitumenous extracts of ozocerite from East Europe and the USA yielded solid n-alkanes containing both odd and even carbon numbers in The range C_{17} to $C_{24}{}^{34}$, whereas extracts from shales and deep sediments such as gypsum gave mainly odd-carbon n-alkanes (C_{29} and C_{31}) and isoprenoid hydrocarbons characteristic of terrestrial higher plants and algae. Amongst the isoprenoid compounds, phytane (2), pristane (1), pregnanes, norcholestanes, gammaceranes, α- and β-hopanes (4), α- and β-norhopanes and cholestanes were present derived from di- and tri-terpenoids of higher plants[35,36]. Isoprenoids such as pristane and the hopanes have also been characterized from recent lake sediments. $17\beta(H)$-Hopane does not occur in petroleum and was thought to be here derived from bacterial degradation of hop-22(29)-ene of plant origin, whereas $17\alpha(H)$-norhopane in the lake peat arose from bacterial decomposition of spaghnum moss[37]. The hydrocarbon content of other marine sediments correlated with the content of organic carbon and the presence of the alkanes was taken as an early indication of the capacity of these deposits to generate petroleum[38].

Kerogens from various locations have been investigated as being sources of those bed fossil oils that ultimately, after geochemical modification, could generate petroleum[39]. Such material usually yields long-chain hydrocarbons on heating under conditions that could simulate such geochemical processes[40].

B. Higher Plants

The term 'plant wax' is used colloquially to describe the mixture of lipids covering the external surfaces of plants. This is more accurately described as cuticular wax as there are other types. The term cutin is also used.

1. Cuticular wax

This essentially complete sealing agent for the surfaces of most leaves, fruits, seeds and other external organs acts as a mechanical barrier against wear and tear, reduces water loss and prevents invasion by fungi and other micro-organisms. The ubiquity of wax coverings in the living world bears testimony to the biological importance of the material. Such waxes typically consist of a mixture of relatively simple hydrocarbons, wax esters, fatty acids, ketones, fatty alcohols etc., usually with more than 17 carbon atoms, which can be easily removed without damaging the underlying membranes by dipping in organic solvents. The hydrocarbons in common with the other wax components are enzymically formed in the epidermal cells and extruded by an ill-defined mechanism to the tissue surface. Most fruits, e.g. apples, grapes, and pears, show a more or less pronounced waxy fluorescence due to light scattering at the surface of the lipid layers. In many plants up to 50% of the total wax may be composed of alkanes, e.g. the commercially available candelilla wax contains ca 46% n-$C_{31}H_{64}$. Detailed reviews have appeared on the botanical aspects and chemical composition of these cuticular waxes[41-46].

The distribution of types and sizes of alkanes in the cuticular waxes of higher plants can be summarized, albeit over-simplified, as in Table 4. Typically 5 to 100 mg of wax per $1000\,cm^2$ of leaf occurs, and the yield of total alkane may vary from less than 1% in the commercial carnauba wax or that from pine needles to over 90% of the wax from *Solandra grandiflora* (Solanaceae). Much of the data on which Table 4 is based has been excellently discussed and collated[44].

n-Alkanes are the most widespread components of cuticular wax and occur in virtually all plants in a pattern dominated by a few odd-carbon compounds (especially the C_{29} and C_{31} compounds) although even-carbon homologues do occur. Wax from apple and pea contained about 33% n-alkane and this fraction was composed mainly of the n-C_{29} compound (>97% of total) or its n-C_{31} homologues (>90%), respectively[47]. However, a more usual situation is in grape varieties where n-alkanes comprise *ca* 1.2% of the wax and the pattern of the former is C_{29} (22%), C_{27} (20), C_{31} (15) and C_{25} (17)[48]. When the n-alkane content of the wax is low (*ca* 1% or less), as in *Pinus* species and other gymnosperms, the predominance of odd-number carbon chains over even tends to disappear although no example is known in which n-alkanes with an even number of carbons become dominant either in total or individually. When even-numbered compounds do form appreciable proportions of the n-alkane fraction, those with C_{28} and C_{30} skeletons tend to predominate[49]. No relationship could be discerned between the chain length of the n-alkanes and those of the fatty acids that co-occur (as esters) in the wax of a range of *Aloe* species[49].

Important studies have been reported on the cuticular waxes of the Rosaceae[50], the Cruciferae[51], *Citrus* species[52,53] and *Nicotiana* species[54,55] and many other citations are available[44]. The proportions and patterns of alkanes in the waxes from plants within the same genus or family tend to be similar, but there are no known correlations between the results for different genera and families except those reported in Table 4.

Branched-chain alkanes are common in cuticular wax although much less widely distributed than n-alkanes. They comprise a major proportion (nearly 50%) of the alkane fraction from some *Nicotiana* and *Geonium* species. Generally, the iso- and anteiso-alkanes (cf Table 1) with odd and even numbers of carbon atoms, respectively, are the main examples although compounds with internal and/or multiple methyl branching and even cycloalkanes have been reported as minor components[44,50-52]. iso-Alkanes with even-carbon numbers (e.g. C_{32} and C_{34} compounds) occur in wax from rose and lavender petals, respectively[52], and iso and anteiso homologues lower than the range listed in Table 4 (e.g. $C_{21}-C_{23}$ compounds) have been found in lilac blossom[52].

It is clear that the pattern of alkanes may not be invariant throughout the life of the

TABLE 4. Alkanes of cuticular wax from higher plants

Type	Main range	Main components	Source
n-Alkanes	$C_{25}-C_{35}$	C_{29}, C_{31} (less C_{27}, C_{33})	Most plants
iso-Alkanes	$C_{25}-C_{35}$	C_{27}, C_{29}, C_{31}, C_{33}	Less widespread; occurs in blossom (rose, lavender) etc.
anteiso-Alkanes	$C_{24}-C_{36}$	C_{28}, C_{30}, C_{32}	Petals, algae
Internally branched alkanes	$C_{14}-C_{34}$	?	Rare: petals
Cycloalkanes	$C_{18}-C_{33}$	?	Seed oils; <1% of total alkanes, rare
(Isoprenoid)-cycloalkanes	$C_{19}-C_{30}$	C_{19}, C_{20}	Rare: apple, podocarpaceae

plant. Thus the chain length of n-alkanes increased with age for both *Solandra* and *Abies* species[49,53]—in the latter the predominant alkane was the n-C_{23} compound in the first season followed by the n-C_{29} homologue in the second year. Also the ratio of n-C_{31} alkane to anteiso-C_{30} compound varied dramatically with leaf age and position on the stalk for *Nicotiana* species[55]. In addition, greenhouse studies on cabbage varieties revealed a marked dependence of wax composition on intensity of illumination and humidity[54], although the composition for *Halimium* species was unaffected by variations in soil or climate[55]. Much more study of these aspects of wax metabolism is needed as the interspecific and intergeneric variations in proportions and patterns of components observed may reflect differing environmental conditions.

In addition to the obvious biological functions of cuticular waxes listed at the head of this section, a further role may lie in the control of insect infestation. Thus host selection by phytophagous insects in their early (infective) stages is often partially or totally dependent on wax composition[56]. Consequently, the variation in wax composition for different plant species may represent divergent evolution providing the plants with chemical defences by altering or modifying the (presumably tactile) properties of target compounds. The n-alkane fraction and especially n-$C_{32}H_{66}$ isolated from *Vicia faba* (broad bean) had very significant insect-attracting properties and may represent such a target molecule that is exploited by predators to locate their food plant[57].

The patterns of alkanes in higher plants also appear to apply to the surface waxes of mosses and liverworts, except for the less prominent dominance of a few alkanes in each species and a slightly higher content of the even-chain alkanes. For ferns, C_{17} and C_{29} compounds usually predominate, whereas in multicellular algae the n-C_{15} and n-C_{17} alkanes or sometimes alkenes or internally branched alkanes are the main components[44-46]. For convenience of comparison, the algae are further considered in Section II.E.

Alkenes occur as minor components of the waxes of higher plants and generally have the same chain lengths and predominance of odd carbon chains over even as in the accompanying n-alkanes.

2. Suberin-associated waxes

Suberin impregnates the walls of cork cells and in layers covers bark, tubers, roots, wound peridorm and bundle sheaths of monocotyledons to provide a protective mantle impervious to liquids and gases. The associated waxes are not as well studied as those from cutin but do exhibit certain periodicities. The hydrocarbons have a broader distribution of chain lengths than the cuticular material, a predominance of shorter carbon chains and a higher proportion of even-length chains. In addition, no single alkane predominates; usually several are present in similar proportions[58-60].

3. Intracellular and internal waxes

Although cuticular and suberin-associated waxes are primarily external coverings, they also surround specialized structures such as seeds and embryos or line organs such as nectaries or the juice sacs of citrus fruits. Up to half of the dry weight of the cytoplasmic (plasma) membrane of plant cells is lipid and this also contains hydrocarbons of uncertain nature[58]. Very little is known of the intracellular waxes although certain seeds[61], algae and marine zooplankton[58] use such hydrocarbon-containing materials as energy sources. Heartwood waxes—whether of inter- or intra-cellular origin is unclear—of *Gottifereae* and *Myristicaceae* species contain n-alkanes with similar proportions of odd- and even-carbon chains[62,63] and patterns quite unlike those in the external cutin are also found for internal waxes of leaves of several species[49].

4. Essential oils

Ethane has been detected in the aroma components of apple; methane, 3-methylhexane, n-hexane and cyclohexane occur similarly in orange, whereas strawberries evolve isopentane, n-hexane, cyclohexane, 2-methylpentane and methylcyclopentane[64]. Lower alkanes also occur in numerous essential oils. Thus n-heptane comprises over 90% of the oil from *Pinus jefferii* needles[65] and is a component of many other *Pinus* oils[66], whilst gum turpentines from the same genus yield n-heptane, n-nonane and n-undecane[67]. Leaf oil from *Ruta graveolens* is very unusual in containing 2-undecanone, 2-nonanone and 2-nonyl acetate (total *ca* 66%)[68]: these compounds are presumably formed by *in vivo* oxidation of the corresponding n-alkanes. Most essential oils are obtained by steam distillation of foliage and higher alkanes are not recorded as being components. Solvent extraction (e.g. petroleum ether, benzene) of flowerheads of plant species of many genera frequently yields extracts (concretes) containing large amounts of higher ($> C_{15}$) alkanes: these arise from the deposits of cuticular wax on the blossoms. Thus n-C_{27} alkane together with close homologues were found in extracts from Neroli and other[69] genera. Surprisingly, steam distillation of petals sometimes also yields products containing much wax of largely hydrocarbon nature: analysis of many *Rosa* species yielded oils containing n-C_{21} alkane (and possibly n-C_{19} and n-C_{20}) in 5–43% of the total[70].

Commercially available plant oils that have importance in the aroma and flavour industries often are of uncertain provenance and, even if derived from the nominal source, may contain components of both foliage and flowerheads extracted by unspecified methods. Table 5 records the pattern of n-alkanes typical of a selection of such oils[71]. When the oil contains relatively large quantities of alkanes, the pattern of odd-carbon chains, as is found in cuticular wax, predominates, although compounds from n-C_{29} to n-C_{31} (that is the optimum range for the latter) were not detected. Even-chain alkanes appeared to predominate when small amounts of hydrocarbon were present in the extract: this recalls the situation for cuticular waxes mentioned in Section II.B.1. Because of the uncertain and probably non-quantitative extraction precedures at source, these results cannot be held to indicate that the even-chain alkanes were predominant in any of the parent plant tissues.

5. Taxonomy

The hydrocarbon composition of cuticular waxes is usually much simpler than that of the co-occurring fatty acids, and the ease of assay of the former fraction by GC–MS

TABLE 5. n-Alkanes in commercially available essential oils

Oil	n-Alkane (ppt)[b]
Elderflower (English; concrete)	C_{19}(146); C_{21}(114); C_{23}(37); $+ C_{20}$, C_{12}, C_{15}
Otto Rose (Bulgarian)	C_{19}(109); C_{21}(46); C_{17}(16); $+ C_{20}$, C_{15}, C_{23}
Rose Absolute (Egyptian)[a]	C_{19}(151); C_{21}(45); C_{17}(27); $+ C_{21}$, C_{18}, C_{15}, C_{16}, C_{14}
Rose Essence (Turkish)	C_{19}(32); C_{21}(13); C_{17}(7); C_{23}(3); $+ C_{20}$
Mimosa Absolute[a]	C_{19}(19); C_{17}(6); $+ C_{21}$, C_{23}, C_{15}, C_{18}
Jasmine Absolute (English)	C_{22}(2); $+ C_{19}$
Citronella (Sri Lankan)	C_{12}(1); $+ C_{13}$, C_{11}, C_{14}
Citronella (Argentinian)	C_{10}(1); C_{12}(1); C_{13}(1); $+ C_{11}$

[a]Absolute indicates that the oil has been extracted into layers of paraffin wax. Other oils are solvent extracted or steam distilled.
[b]ppt ≡ part per thousand.

has led to exploration of the use of the alkanes as taxonomic markers[72]. Such applications seem uncertain. It is evident that the wax components are under overall genetic control but that variable factors associated with age and environment may be superimposed on the pattern in some examples—although in others the genetically imposed pattern appears to be very stable and is little affected by external influences[49]. Findings such that cuticular waxes from stem, leaves and fruit of grape species differ considerably in alkane pattern[73] as well as the environmental effects previously mentioned (Section II.B.1) cast doubt on any widespread taxonomic value of the approach[74]. Nevertheless, many chemosystematic studies have been based on the pattern of alkanes in cuticular wax, e.g. for *Abies*[53], *Pisum*[75], *Brassica*[76] and *Arbutus*[77] species. Other investigations include the use of the n-C_{33} alkane as a taxonomic marker for maize[78], for n-alkanes in inheritance studies of guayule (*Parthenium argentatum*) and its hybrids[79] and the distinction between *Sinapsis* and *Brassica* (wild mustard and rape) species by analysis of the surface waxes[80]. The whole subject has been extensively reviewed[43,81-83].

6. Tissue culture

Numerous studies of secondary metabolism in plant cell cultures have been reported but few have mentioned alkane accumulation, although probably it is widespread. The alkane pattern in cell suspensions of *Euphorbia* species was similar to that in leaves (mainly n-C_{28}, n-C_{29} alkanes) and recoveries were 5 to 60% (in different cell lines) of that obtained from the parent tissue[84], and similar products (n-C_{29}, n-C_{31} predominant) were obtained for cultures of guayule[85]. In contrast, cell lines from *Pogostemon* species yielded alkanes (n-C_{15} and n-C_{17}) of lower chain length than the major (n-C_{31}) alkane of the cutin of the parent at levels only *ca* 10% that in the latter[86]. Callus from a *Rosmarinus* species similarly yielded n-C_{15}, n-C_{16} and n-C_{17} whereas the parent leaf tissue mainly accumulated a n-C_{29} alkane[87].

GC-monitoring of the surface wax from callus and cell suspensions of eighteen members of the *Lavandula, Rosa, Pinus, Tanacetum, Artemisia, Jasminum* and *Salvia* genera revealed the occurrence of n-alkanes in the optimum lines at levels 2 to 85% those in the cuticular waxes of the respective parent leaves. Both odd- and even-chain n-alkanes were present in the range n-C_{14} to n-C_{31}. Except in one case, the odd-chain alkanes predominated and the n-C_{29} compound was the mode[88]. The initial explants in these (and in previously described) studies were unavoidably mixtures of epidermal and internal leaf cells, and as these types are known to yield different patterns of n-alkanes (cf Section II.B.1 and Reference 49) the variations in alkane types and levels may reflect morphological differences in the resulting stable (for up to 3 yr and 50 subcultures) cell lines. However, regeneration of plantlets from callus of *Tanacetum* and *Artemisia* species resulted in a pattern of surface wax identical with that of the field-grown plants[88]. Light-requiring suspensions of a *Ruta* species yielded oil containing undecan-2-one as main product as does the parent plant[89]. Consequently both alkane synthesizing and oxidizing enzymes were expressed in these cell lines.

C. Insects

The arthropods, especially the insects, exhibit probably the greatest ability to synthesize and utilize alkanes of any class of the Animal Kingdom. Their external surfaces are covered with cuticular waxes that provide a barrier which is impervious to water and prevents invasion by micro-organisms as well as providing general protection. This barrier may also effect (both positively and negatively) the penetration of insecticides. The waxes contain a wealth of hydrocarbons with n-alkanes sometimes predominating. Over one-hundred hydrocarbons have also been isolated and characterized from internal tissue and from exocrine glands and organs. The anatomy and physiology and chemical composition of arthropods have been comprehensively reviewed[90-92].

The differences in hydrocarbon patterns in surface waxes and in the components of interior tissue are illustrated by analysis of pupae of *Manduca* (tobacco hornworm). n-Alkanes only comprised *ca* 3% of the hydrocarbon fraction of the cuticular wax, the balance being unsaturated compounds. In contrast, internal tissues (fat bodies, muscle, gut) contained the same carbon spectrum (C_{21} to C_{41}) as in the wax but now branched alkanes made up the bulk of the hydrocarbon fraction (*ca* 80%), followed by n-alkanes (9%) with the residue being unsaturated compounds[93]. The proportion of n-alkanes in the hydrocarbon fraction from cuticular wax of a *Bombyx* silkworm fell from 95 to 35% on passage from the larval to the pupal stage[94] and similar results have been found for *Trichoplusia* (cabbage looper)[95] and *Drosophilia* (fruitfly) species[96]. However, it is likely that the cuticular wax has a more stable composition over the adult life of most insects and is only synthesized at (low) rates sufficient to replace that lost by wear and tear[97]. The site of synthesis has been demonstrated to be in the cuticle in a cockroach species: no hydrocarbon synthesis occurred in preparations from fat bodies[98].

More vigorous synthesis is likely in glandular tissue where hydrocarbons are lost as pheromonal and defensive secretions. It is generally considered that the composition of cuticular lipids is not correlated with the diet of insects, although the latter may have some ecological significance as indicated by recent work on insect–insect predation[99]. Nevertheless, it has frequently been observed that different insect species and their parasites which live in association have similar cuticular waxes that often also resemble the wax from the food plant[100].

Examples of the alkanes from the cutin of a variety of insects are shown in Table 6. The conclusion from these and numerous other studies, e.g. on *Blattella* (cockroach)[109,110], *Locusta*[111] and *Attagenus* (beetle) species[112] is that cuticular insect waxes resemble in general the surface waxes of plants in that, although both odd- and even-chain n-alkanes occur, the former predominate and the mode is in the range n-C_{27} to n-C_{31}. However, iso and especially anteiso and internally branched compounds are much more common in insect waxes. Monomethylated n-alkanes are often the main component of the entire wax and the majority of the internally branched monomethyl compounds are of *even* total carbon number with the substituent being sited at an odd-numbered carbon atom, usually at C-11, C-13 or C-15. Such compounds with carbon ranges C_{24}–C_{28} or C_{24}–C_{50} occur in cockroach[109,113,114] and grasshopper[115] species and the range C_{26}–C_{38} is

TABLE 6. Alkanes of cuticular wax of insects

Genus	Name	Types, range[a]	Main	References
Apis	Bee	n-(96%) Int-(C_{26}–C_{30})	n-C_{31} 13-Me-C_{27}	101–103
Himatismus	Beetle	n-(42%; C_{18}–C_{33}) anteiso (4%; C_{28}–C_{30}) Int. (15%; C_{26}–C_{30})	n-C_{25}, C_{27}	104
Eurychora	Beetle	n-(98%; C_{25}–C_{35})	n-C_{31}	106
Sitophilus	Weevil	n-(30%; C_{16}–C_{39})	n-C_{27}, C_{29}, C_{31}	105
Drosicha	Scale insect	n-(13%; C_{27}–C_{33}) iso-(4%; C_{27}–C_{33})	n-C_{28}, C_{29} 2-Me-C_{28}	107
Periplanata	Cockroach	Int (59%; C_{26}–C_{28}) anteiso (14%) n-(25%)	n-C_{25}, C_{27}, C_{29} 13-Me-C_{25}	96, 108

[a]Int = internally branched alkane; percentage figures refer to the proportion of the component in the wax.

D. V. Banthorpe

frequent in other insects[116]. In some, a series of isomers with the methyl located on *all* odd-numbered carbons has been reported. In contrast to the complex mixture found in many insects, certain *Periplanata* cockroaches[117] contain 13-methyl-C_{27} as essentially the only alkanes in the wax. Mono- di- and tri-methylated n-alkanes also have been found in cuticular wax[95,118] and the 9,13-dimethyl-C_{27} compound is the main component from *Icanthoscelides* weevils[119]; *Manduca* tobacco hornworms contain oxygenated n-alkanes with the functionalization either at C-11 or at C-11 and C-12 for C_{26} and C_{28} compounds, respectively[120].

The Dufours glands of ants are veritable hydrocarbon factories and produce large amounts of short-chain alkanes that are secreted in admixture with the formic acid-based venom to act as potent alarm and aggregating pheromones[92,121]. The main components are n-C_9, n-C_{10}, n-C_{11}, 3-Me-C_9, 3-Me-C_{11} and 5-Me-C_{11} compounds with the n-C_{11} alkane predominating[92,122–126] but wide variations occur and *Iridomyrex* ants synthesize mainly odd-chain compounds in the range C_{23} to C_{31}[127]. Other insects produce similar defensive secretions: the n-C_{13} alkane is the most common and can comprise up to half of the two-phase system, but n-C_{15} is widespread and n-C_{19} is also common, and the sequence can go up to n-C_{31}[92,122]. Relevant studies were published on *Edessa* (stinkbug)[128], Chrysopa (lacewing)[91] and Drusilla (beetle)[129] species. Different ant species form aldehyde or ketone derivatives of the n-C_{11} to n-C_{15} alkanes, but the patterns are similar and the alarm signals seem to be intergeneric as well as interspecific[130].

Despite their apparent chemical unreactivity and lack of wide variation in physical properties, specific alkanes apparently play an important role in mediating other pheromonal and social behaviour of insects. Generally, homologues considerably higher than those present in defensive secretions are utilized. Branched alkanes, e.g. 15-methyl and 15,19-dimethyl-C_{33} compounds, are sex hormones of a *Stomoxys* stable fly[131], and the 3,11-dimethyl-C_{29} alkane is the female sex contact pheromone of *Blattella* cockroaches[132]. Similar branched alkanes comprise the sex pheromones of a *Musca* housefly species[133] and the contact mating pheromone for *Colcoides* biting midges[134]. In contrast, n-alkanes (n-C_{24}, C_{25}) are the copulation releasers of *Orgyia* moths[135] whereas n-C_{21}, n-C_{23} and n-C_{25} compounds are marking pheromones for various bees, wasps and aphids[136]. Cuticular hydrocarbons can also act as recognition signals, e.g. for *Formica* ants[137] and *Amblyomma* tick species[138]. Termites produce caste and species–specific mixtures of cuticular hydrocarbons that allow recognition, but termitophiles associated with the colonies have evolved identical surface coatings that allow them to penetrate the colony as wolves in sheeps' clothing. Such chemical mimicry is shown by a *Staphylemid* beetle that to this end secretes 9-methyl-C_{25}, n-C_{25} and n-C_{26} alkanes[139].

D. Higher Animals

Remarkably little is known concerning the nature and distribution of alkanes in mammals. Usually less than 1%, if any, of mammalian waxes are hydrocarbons (compared with up to 90% of insect cutin) and when alkanes do occur in waxes and in fats the pattern is very complex[140]. This may be the result of the accumulation in the fat tissues of alkanes from a variety of dietary sources[141,142]. Egg, duck and chicken fat contain predominantly C_{20} to C_{23} alkanes whereas lard and meat fats generally include small amounts of the C_{25} to C_{31} compounds[143]. Bovine tissues (e.g. liver, heart) contain n-C_{12} to n-C_{31} alkanes and phytane[144]. The surface lipids of human skin contain up to 1% alkanes in a range C_{15}–C_{35} which shows widespread variations between subjects. ^{14}C-dating for these alkanes showed an age *ca* 30,000 years consistent with contamination with petroleum products from the environment[145].

Nevertheless, there is no doubt that mammals can synthesize alkanes *de novo*. The saturated long-chain fatty acids that are the necessary precursors (cf Sections III.B and III.C) are formed in extracts from rat liver and brain[146] and cell-free preparations from the sciatic nerve of rabbit synthesize C_{19}–C_{33} alkanes from fatty acids by the conventional chain-elongation route (see Section III.B)[147,148]. The quantities and profiles of n-alkanes differ in the sciatic nerves of normal mice from those of trembling or quaking mutants in a manner related to levels of myelination, and clearly play some fundamental physiological role[149]. Long-chain di-α,ω-diols (C_{29}–C_{34}) in the form of esters occur in lipids secreted from bovine and human meibomian glands: such compounds had previously been isolated from plant cutin but it was presumed they were derived from n-alkanes generated in situ in the animals[150,151].

The uropygial gland (a modified sebaceous gland responsible for the surface coat of oil) of grebes secreted a lipid fraction containing ca 40% alkanes, mainly the n-C_{21}, n-C_{23}, n-C_{25} and n-C_{27} compounds[152], and long-chain diols (see above) were isolated from the same gland of chicken[153].

n-C_{17} and n-C_{18} alkanes seem to be the most common alkanes in fish oils[154,155] and pristane (cf Table 1) also occurs[156]. Levels are higher in marine fish (e.g. hearring, cod liver) than in freshwater fish oils and presumably represent ingested plankton. Arctic fish contain up to 50 mg/100 g dry weight of alkanes, mainly n- and branched (C_{10}–C_{26}) but also iso (C_{14}–C_{20})[157]. The n-C_9 to n-C_{13} alkanes of sea mullet are metabolized with half-lives of ca 18 days compared with up to 1200 days for chlorinated hydrocarbons[158]. Analysis may often be invalidated by pollutants: petroleum hydrocarbons have been traced in the food chain from mussels to crabs where they accumulate in specific tissues, such as liver or gonads[159].

E. Micro-organisms

The lipids of prokaryotes have been reviewed[160]. Unlike higher plants, bacteria (with the exception of mycobacteria) do not usually contain significant amounts of waxes in their surface layers. Lipid droplets are infrequently seen internally—in Acinetobacter species these may contain up to 12% alkanes mainly as the n-C_{16} compound[58]. The main interest surrounding the alkane composition has been its use in differentiating species of micrococci (in which hydrocarbons may contain up to 20% of total lipid) from staphylococci, which lack hydrocarbons. Two Sarcina species were found to contain the same hydrocarbon pattern and have been reclassified on the basis of this and other criteria as a single Micrococcus species[58].

Fungal cell walls usually contain appreciable quantities of aliphatic hydrocarbons with chain lengths similar to those in higher plants or algae and the components are presumed to play similar functional roles[58]. Thus yeasts contain both straight-chain and branched alkanes and the predominant chain length is in the range C_{16} to C_{39}[161,162]. The dangers of contamination by extraneous hydrocarbons during assay of alkanes from fungi have been stressed[163]. Alkanes in the range C_{18} to C_{35} occur in fungal spores with the n-C_{27}, n-C_{29} and n-C_{31} compounds predominating. Components with even-numbered carbon chains are present in very small concentrations (much less than in mycelia) and branched-chain isomers are restricted to a few species. These findings have found use in taxonomy[162].

Analysis of alkanes of bacteria, fungi and algae made in the course of attempts to determine the source materials for petroleums[164–166] are summarized in Table 7. Typically, the algae contained 0.075% alkane per dry weight, chlorella 0.03%, yeast 0.005% and bacteria less than 0.005%. More recent studies have suggested that blue green algae more usually have n-alkanes (especially n-C_{17}) predominating rather than branched-chain alkanes as shown[167]. The non-photosynthetic organisms differ from the rest in producing alkanes with carbon chains considerably in excess of that of the longest

908   D. V. Banthorpe

TABLE 7. Alkane composition of micro-organisms

Type	Nature[a]	Main Components
Chlorella	P	n-C_{15} to n-C_{19}
Green algae	P	n-C_{16} to n-C_{21}; pristane; phytane
Blue green algae	P	7-Me-C_{17}; 8-Me-C_{18}
Purple bacteria	P	n-C_{17} + pristane + phytane
Bacteria	N	n-C_{18}; n-C_{24} to n-C_{29} (less)
Yeast	N	n-C_{22} to n-C_{24}; C_{26}-C_{29} (less)

[a]P = non-photosynthetic and N = non-photosynthetic organism.

fatty acid (stearic acid, C_{18}) that is constructed by the conventional fatty acid synthetases. In contrast, the photosynthetic organisms accumulate pristane and phytane that are markers derived from the metabolism of phytol—the ubiquitous component of chlorophyll.

III. FORMATION AND BIOGENESIS

A. Origin of Petroleum

There has been great debate concerning the abiogenic or biogenic origin of petroleum[168-172]. The former view dates back to Mendeleev and posits hydrocarbons being formed by reaction of metal carbides with steam at great depths. The Fischer–Tropsch process for the formation of alkane mixtures by catalytic reduction of carbon monoxide (made by the water gas reaction) with hydrogen was suggested as a model[172-174]. One problem with this theory is the huge number of isomers (ca 366,320 for C_{20} compounds) all of similar stabilities and presumably about equally probable of occurrence that become possible as the carbon number increases. In the event, very few indeed of these have been detected in crude petroleums and these few are types that occur in biological systems[175]. A modified view is that oil in surface deposits is conceded to be of biological origin whereas the larger quantities in deeper sediments are inorganic nature. Such a 'bimodal' theory has been imaginatively argued[174]. Most geologists now accept a biogenic origin for petroleum[16,176]. Recent deposits may contain liquid oils derived from the lipids of plankton and other organic material, but these differ from petroleums[177] and are transformed, usually in deeply buried sediments at temperatures of ca 200 °C, into the final pattern of products[178,179]. The kinetic and thermodynamic stabilities of the modified hydrocarbons can be estimated (from data on commercial cracking processes) to be such as to curtail interconversions far below any statistical distribution of products and so allow the tell-tale characteristics of biological origin to be discerned[166].

Direct evidence for the biogenic theory is provided by: (a) the almost complete predominance of methyl substitution in branched chain products[176], (b) observation of ^{13}C-^{12}C ratios in petroleum compounds characteristic of biological materials[180]; (c) the occurrence of 'terpenoid' alkanes such as pristane and phytane (Table 1), characteristic of the existence of photosynthetic organisms[181] and also other modified terpenoids (e.g. carotane, ergostane) that are undoubtedly derived from living material[16]; and (d) the observation of optical activity in some compounds (e.g. gammacerane, hopanes, steranes) from petroleum and shales[182-184], which have the same absolute configuration as the notional parent natural materials[185,186]. Stages in the modification of relics of living

TABLE 8. Variation in n-alkane pattern in different geological sources[a]

Sediment	ca Age (yr)	Alkane pattern (predominant)
River shale (USA)	6×10^7	n-C_{17}–C_{19}; n-C_{29}–C_{31} biomodal distribution odd C dominance; phytane; pristane
Ancient shale and chert (USA; Africa)	10^9	n-C_{18}–n-C_{31}; no odd–even dominance n-C_{29}–n-C_{31} predominant; pristane etc.
Very old sedimentary rocks (Sudan)	4×10^9	continuous distribution n-C_{18} to n-C_{33} spikes at n-C_{17} and n-C_{26}

[a]Data retrieved from Reference 166.

organisms (characterized by n-alkanes with an odd carbon number predominance) to the patterns typically found in petroleum are illustrated by some data in Table 8; also cf. Figure 1 in Section II.A.

Three particularly influential studies have summarized and filled in details of the biogenic theory. It has been emphasized that most producing horizons are much shallower than the source beds and upwards (vertical) migration of the components of petroleum through faults and fractures is common. However, neither a fixed depth nor a universally applicable scheme of transformations can be assigned as petroleum formation is governed by (a) the amount of organic material, (b) the thermal history of the beds and (c) the catalysts (e.g. acid clays) available[187]. Another interpretation attempts to trace the pattern of maturation and transformation of the various types of kerogen and considers that the modification of the pattern of hydrocarbons from that characteristic of the biological source material mostly occurs in sediments at ca 1000 m deep, whereas at ca 2000 m, cracking to light alkane chains with the production of much methane is common[188]. More recently the nature of the thermocracking process that yields light n-alkanes and alkenes has been discussed together with the equilibration and reactions of the latter (in the presence of acidic rocks) to yield mono- and dimethyl-branched alkanes[189].

Model laboratory systems have been set up for the cracking and rearrangements of n-fatty acids and alkanes[190], steroids[191] and kerogen[192] that demonstrate that the odd carbon preference for n-alkanes originating from biological source material could be rapidly abolished in oil-bearing strata at ca 200 °C. These systems also account for the formation of cyclic and aromatic components of petroleum. Similar transformations could occur in sediments at much lower temperatures over the course of geological time, especially if organic materials were absorbed on the acidic clays and shales[179]. Microbiological alteration of petroleum deposits in the producing beds has been claimed[193,194], but this phenomenon has been dismissed as negligible[195] especially at the estimated temperatures of many such sites[187].

B. Biogenesis in Higher Plants

1. General

The biogenesis of n-alkanes in plants presents two points of considerable and novel interest: firstly, chain construction. Alkyl chains are built up in plants, animals and micro-organisms by sequential condensation of C_2 units of acetate to yield fatty acid and related polyketides, but the fatty acid synthetase complexes involved rarely, if at all, catalyse chain elongation beyond C_{18} compounds: thus typically stearic acid $CH_3(CH_2)_{16}COOH$ is the end product although palmitic acid (C_{16}) is usually the predominant product. However, most plants produce n-alkanes in the range C_{18} to C_{31} or even greater chain lengths and both odd- and even-carbon compounds occur although the former almost always largely predominate. If we presume that the n-alkanes are derived from fatty acids or their biogenetic equivalents, and there seems no reasonable alternative, then the long-chain alkanes must be produced either by

(a) coupling of two smaller (C_{18} or less) sub-units, or
(b) chain elongation beyond the normal C_{18} stage.

And if the latter occurs, either the complete even-carbon series could be formed in sequence *or* the synthetase could take the chain straight through to (say) the C_{32} compound on a conveyor belt before release, whereupon the lower (C_{20}–C_{30}) fatty acids could be formed by β-oxidation and subsequent decarboxylation (to remove 2 carbon atoms in each cycle). In the event, nearly all the evidence (see later) favours the sequential chain elongation route.

Secondly, decarboxylation. The available routes to this involve decarboxylation or decarbonylation of the long-chain fatty acid or its aldehyde, respectively; this would yield the predominant series of odd-carbon n-alkanes (Figure 2; route A). The occurrence of the alternative even-carbon alkanes would necessitate an additional α-oxidation step (route B).

When studies of the subject were initiated, both the putative extended-chain elongation and the reduction steps were not known in nature, but the above is the situation as now perceived. Other mechanisms are possible: thus various even and odd n-alkanes could be produced by modification of aldehydes formed by cleavage of a double bond situated internally in a long-chain precursor; odd-carbon fatty acids and thus smaller even-carbon n-alkanes could result from chain elongation using a propionate (C_3) unit as a 'starter' rather than acetate; and n-fatty acids could be directly reduced to n-alkanes without loss of carbon via the intermediacy of aldehydes and alcohols. However, no evidence has been uncovered for any of these processes, although admittedly work in the field is not extensive.

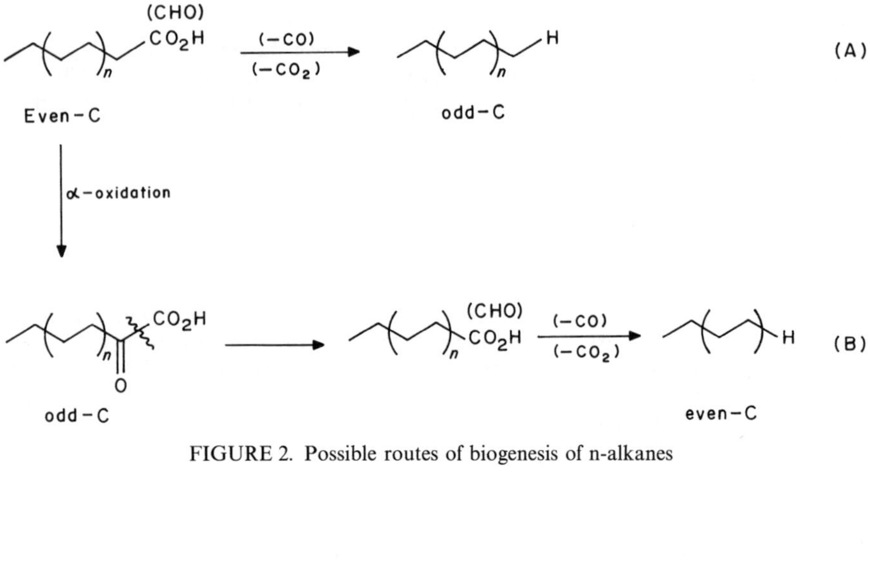

FIGURE 2. Possible routes of biogenesis of n-alkanes

2. n-Alkanes

Tracer investigations of the formation of n-alkanes (n-C_{25} to n-C_{35}) and the co-occurring iso, anteiso and branched alkanes in a *Nicotiana* species were held to support the head to head condensation of small (< C_{18}) units to form the long chains[196]. But this is now generally rejected in view of the weight of evidence supporting the elongation–decarboxylation/decarbonylation pathway[197,198].

In eucaryotes, at least, fatty acids higher than palmitic (C_{16})—the almost universal major product of the fatty acid synthetase complex—appear to be formed in this latter route by enzymes situated on the cytosolic face of the membrane of the endoplasmic reticulum (opposite the face constraining the synthetase)[199] and the hydrocarbons subsequently generated on the reticulum lumen are probably excreted via vesicles which fuse with the plasma membrane. For the study of these systems cell-free extracts are often used in which the membrane is fragmented into closed vessels (microsome fraction). Tracer work using chopped leaves[200] or microsomes[199] from pea, cabbage and brocolli indicated that chain elongation by the usual malonate pathway of addition of acetate units occurs, e.g. as in Figure 3. Later work using cell-free extracts from the epidermal cells where alkanes are generated also yielded long-chain acids (> C_{18}) when malonyl-coenzyme A and NADPH were added. Often the chain length of the fatty acids produced *in vitro* did not match that of the n-alkanes produced *in vivo*. As the epidermis generates many classes of lipids, each with its characteristic chain-length distribution, some partial separation may have been achieved[198].

Enzyme extracts such as produced very long chain acids did not resemble extracts of fatty acid synthetase. No acyl carrier protein (ACP) was involved[201]; chlorophenyl-dimethylurea which inhibits the incorporation of acetate into normal-range fatty acids had no effect on n-alkane biosynthesis, and trichloroacetate at low concentration strongly suppressed alkane formation yet did not effect normal fatty acid synthesis[198]. *In vivo*, the absence of light, which inhibits the incorporation of acetate into normal fatty acids, had little effect on the biosynthesis of alkanes provided that the C_{18} and C_{16} fatty acids were available as starters[198]. Until they are solubilized, resolved and purified the nature of these microsomal elongation enzymes cannot be elucidated. No work has been reported on higher plants, but a chain-elongating enzyme has been purified from the mycocerosic acid synthetase complex of *Mycobacterium tuberculosis*: here 2,4,6,8-tetramethyl-n-C_{28} acid and related homologues are generated by elongation of C_{16} and C_{18} acids with malonyl-CoA as source of C_2 units and with NADPH[202].

Extracts from tea leaves in the presence of oxygen and ascorbate converted fatty acids (n-C_{18} to n-C_{32}) into n-alkanes containing 2 carbons less: thus α-oxidation must have preceded final loss of carbon. However, the aldehydes prepared from the C_{18} and C_{24} acids were straightforwardly decarbonylated to the acids with one carbon atom less. Tea leaves *in vivo* produced the odd-number series of n-alkanes and it was presumed the latter route here predominated[203,204]. Thus the implication was that the final step in n-alkane formation was a decarbonylation rather than a decarboxylation and studies using particulate preparations from peas have confirmed this. The mechanism is obscure, but tracer studies have shown that the conversion, RCHO → RH + CO, involves retention of the aldehydic hydrogens, and it has also been demonstrated that a metal ion is implicated (effect of chelating agents)[205]. This type of mechanism is consistent with

$$CH_3(CH_2)_{14}CO_2H \xrightarrow{+(7 \times C_2)} CH_3(CH_2)_{28}CO_2H \longrightarrow CH_3(CH_2)_{27}CH_3$$

Palmitic acid (C_{16})   (C_{30})   n-Nonacosane (C_{29})

FIGURE 3. Elongation–decarboxylation pathway to n-alkanes

D. V. Banthorpe

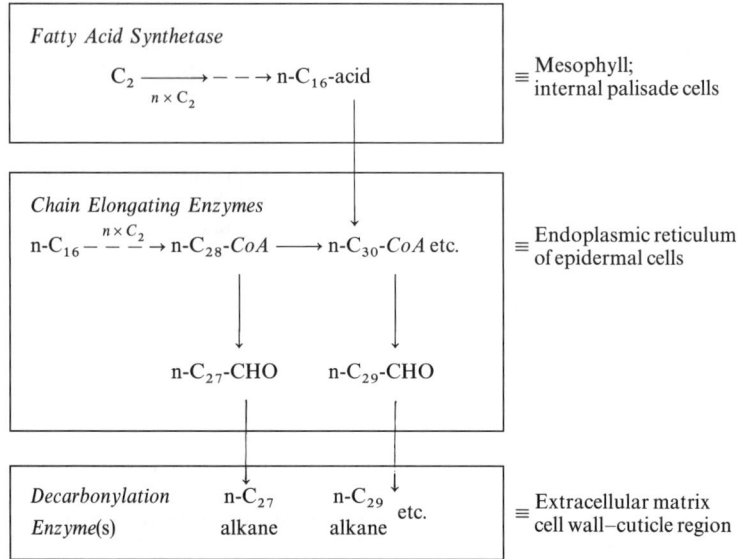

FIGURE 4. Scheme for chain elongation and decarbonylation in n-alkane biosynthesis (CoA = coenzyme A)

observations of decarboxylation of fatty acids and aldehydes promoted by complexes of Ru and Rh[206] and by demonstrations that such porphyrin complexes could catalyse decarbonylation at near-ambient temperatures[198,205].

Current views on the compartmentation of the enzyme systems that promote chain elongation and decarbonylation are represented in Figure 4. This scheme is based on electron microscopy and on studies of enzyme location and inhibition[198,205]. The chain length of alkanes formed probably depends on the aldehydes available rather than on the specificity of the decarbonylation: thus crude extracts of leaves generated n-alkanes from $n\text{-}C_{16}$ to $n\text{-}C_{32}$ fatty acids with no particular preference for substrate. The restricted ranges of n-alkanes observed *in vivo* presumably result from the specificity of products from the fatty-acid elongation system that produce the substrate for decarbonylation.

The widespread occurrence, albeit in lesser amounts, of n-alkane with even rather than odd numbers of carbons implies the existence of α-oxidation (cf Figure 2, and

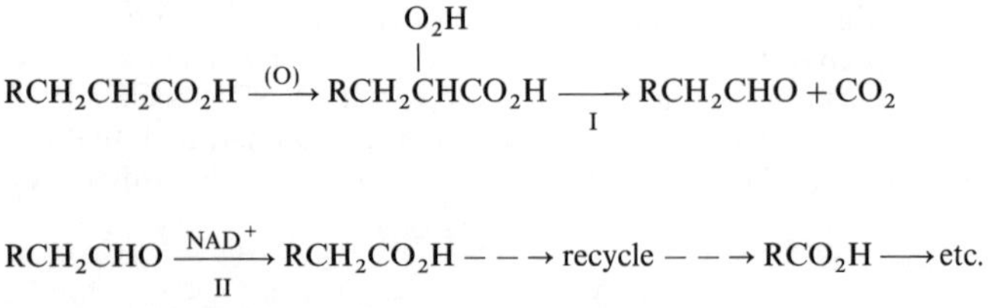

$$RCH_2CH_2CO_2H \xrightarrow{(O)} RCH_2\overset{\overset{\displaystyle O_2H}{|}}{C}HCO_2H \xrightarrow{} RCH_2CHO + CO_2$$
$$\text{I}$$

$$RCH_2CHO \xrightarrow{NAD^+} RCH_2CO_2H \dashrightarrow recycle \dashrightarrow RCO_2H \longrightarrow etc.$$
$$\text{II}$$

I. Long-chain fatty acid peroxidase
II. Aldehyde dehydrogenase

FIGURE 5. α-Oxidation of fatty acids

References 203 and 204). This route was demonstrated using extracts from germinating peanuts with only NAD^+ as added confactor and H_2O_2 was implicated as oxygen donor[207,208]. Leaf systems in contrast utilize molecular O_2[209]. The process, which can recycle, is shown in Figure 5. As well as producing odd-carbon fatty acids wherever they are needed for metabolic purposes, α-oxidation can displace the reading frame by one position whenever β-oxidation of a fatty acid is prevented by a methyl or other blocking group at the β-position[210].

3. Branched and cyclic alkanes

The reasons for the presence of significant amounts of branched alkanes in some plants but not in others is not clear. Early studies showed that the amino acids valine, threonine, isoleucine and leucine were readily incorporated into the branched alkanes of *Nicotiana* species[211] and use of valine or isoleucine as starter unit, as in Figure 6, is consistent with later tracer work on incorporations into iso- and anteiso-alkanes[198]. In plants, as well as in *Mollusca* and *Arthropoda*, the majority of the iso-alkanes have an odd number of carbons whilst the majority of the anteiso-alkanes have an even carbon number, and this follows from the scheme outlined. The methyl substituent in internally branched alkanes is introduced by transfer from S-adenosyl methionine to an unsaturated fatty acid (cf Figure 6) in blue green algae and a similar situation may occur for plants[198]. Use of oleic acid (C_{18}; see Figure 6) as the 'normal' fatty acid that is subjected to chain elongation would result in the ultimate formation of 9- or 10-methylated alkanes, which are types that frequently occur. Feeding of ^{14}C-labelled isoleucine or isobutyric acid to *Brassica* species that do not normally produce branched alkanes led to labelled iso- and anteiso-branched fatty acids ($< C_{18}$): neither very long chain fatty acids nor branched alkanes were formed[198]. This suggests that the non-acceptance by the decarbonylation system of short-chain precursors, rather than lack of availability of branched-chain precursors, precludes the formation of the alkanes although compartmentation effects could play a

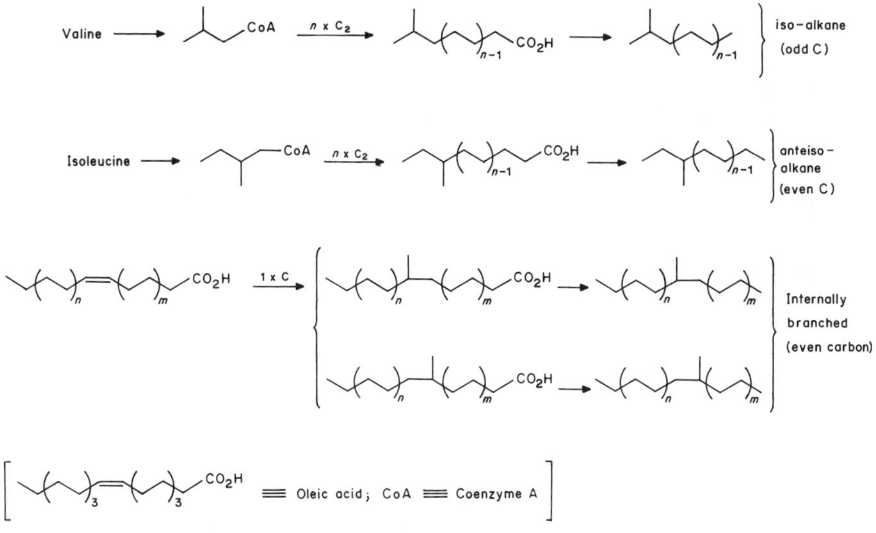

FIGURE 6. Biosynthesis of branched-chain alkanes

914   D. V. Banthorpe

major role. Little is known about the biogenesis of the cycloalkanes that sometimes
occur as minor components of plant waxes: it is possible they may be derived by cyclization of dicarboxylic acid derivatives or diene derivatives of the alkane precursors or
indeed shikimic acid may be involved. The possible routes to these and other minor
alkanes and their derivatives has been summarized[198].

C. Biogenesis in Other Organisms

Less detailed tracer studies have been made on a variety of organisms other than higher
plants and, except for some observations on bacteria (see later), the chain-elongation
mechanism to the higher alkanes seems consistent with all results. The most extensive
work has concerned the synthesis of n-C_{21} to n-C_{27} alkanes in microsomal preparation
from the uropygial gland of grebe[212]. Fatty acids and their corresponding aldehydes
were efficiently converted into n-alkanes in the absence of oxygen and the effect of
inhibitors and chelating agents suggested a decarbonylation mechanism as in plants.
However, there were differences. 1-^{3}H-octadecanal was converted into n-heptadecane
with exchange of tracer with water and the resolution of this discrepancy must await
the isolation and purification of the appropriate enzymes. Extracts of mammalian sciatic
nerve[213,214] and rat brain[215] also synthesized alkanes (n-C_{19} to C_{33}) by the chain-
elongation pathway.

The latter pathway also appears to be present in several insect species[216-218] and
both ^{13}C and ^{14}C studies revealed that iso- and anteiso-alkanes were formed with
incorporation of valine and isoleucine in *Musca* (housefly) and *Blattella* (cockroach)
species[219,220]. However, the situation seems not necessarily always to be similar in higher
plants. ^{14}C-Propionate incorporated during chain elongation serves as the donor for
the methyl of both the anteiso- and internally branched alkanes (e.g. 3-methyl-C_{25} etc.)
that predominate in *Periplanata* (cockroach) species[221,222] and the 3-methyl branch is
added just prior to chain completion[222]. In contrast, in another cockroach species
(*Blattella*), this group is introduced early (as in plants)[220].

Blue green algae (*Nostoc*, *Anacystis*) species form n-heptadecane from stearate
by decarboxylation rather than by decarbonylation[223,224] and branched hydrocarbons
with methyl at C-7 or C-8 were formed by group transfer from S-adenosyl
methionine[223,225].

The sole evidence for construction of alkane chains by coupling of two moieties rather
than by sequential chain elongation comes from work on bacteria. Extracts of *Sarcina*
species synthesized even-chain n-alkanes under conditions where α-oxidation was
supposedly inhibited by imidazole and head-to-head condensation of two fatty acids
($< C_{18}$) was proposed[226,227]. Corynomycolic acid (15-COOH, 6-OH-n-C_{31}) from
Corynebacterium species was considered to be formed by similar coupling of two
molecules of palmitic acid[228], and the pathway was suggested for formation of alkanes
of *Sarcina* species with iso and anteiso branching at both ends of the molecule[229]. The
status of these claims is difficult to assess: similar couplings were thought to occur in
higher plants, but the view was discounted on further study (see Section III.B).

IV. METABOLISM

A. By Micro-organisms

Oxidation and further functionalization of alkanes by microbes has assumed great
importance in respect of facets of the petroleum industry. Thus such metabolism can
exploit petroleum components of low commercial value to form cheap carbon sources

TABLE 9. Main genera of micro-organisms members of which can metabolize alkanes

Bacteria	Fungi	Moulds
Pseudomonas	Candida	Botrytris
Brevibacterium	Mycotorula	Penicillium
Achromobacterium	Trorulopsis	
Corynebacterium		
Cellulomonas		
Micrococcus		
Rhodococcus		
Mycobacterium		
Nocardia		
Streptomyces		

for the fermentation industry or to generate biomass for use as edible protein, especially as animal fodder. Also the utilization of alkanes as a sole carbon source by some micro-organisms has led to blockage of fuel leads of cars and planes by burgeoning cultures and corrosion of fuel tanks and lines by acidic metabolites. Numerous reviews summarize the microbial metabolism of alkanes and related hydrocarbons[230-238].

Genera (mostly closely related) that encompass the most potent micro-organisms that can oxidize alkanes are listed in Table 9. A few of the species occur in oil deposits and can utilize gaseous or liquid hydrocarbons as sole carbon source[238], but most inhabit moorland or acidic soils and the underlying shales where they utilize soil waxes as well as a variety of other apparently unpromising substrates[239,240]. n-Alkanes seen to be generally utilized most effectively of the hydrocarbons, although this may reflect the situation that pure samples of branched alkanes have not easily been available for laboratory screening. A few select species of micro-organisms only metabolize methane, but there is a fairly clear distinction between these that grow on $n-C_{10}$ and lower alkanes and those that prefer the $n-C_{10}$ to $n-C_{20}$ range. Few grow well on homologues greater than $n-C_{20}$[238]. There are an increasing number of examples where an organism grown on an n-alkane as sole carbon source is enabled partially to metabolize another alkane which alone is unable to support growth.

The usual sequence of metabolism is terminal oxidation to the alcohol and thence to the acid, followed by β-oxidation to clip off acetate units as energy sources or as building units for the construction of other metabolites. Fatty acids ($>C_{14}$) produced by the β-oxidation pathway can be incorporated directly into the microbial lipids. Some organisms oxidize n-alkanes at C-2 to yield methyl ketones, many of which are important compounds in the aroma or flavour industry and in some such as Myobacterium species, oxidation at both C-1 and C-2 cooccurs[237]. Small amounts of acids are usually secreted into the medium where, in a few documented cases, they inhibit growth. Branched alkanes (e.g. 2-methyl-C_{11} or pristane) are converted into dicarboxylic acids by α,ω-oxidation in a pathway that can be inducible[241]. Acyclic terpenoid hydrocarbons, e.g. pristane or farnesane, may also be degraded by this route[242] but cycloalkanes are rarely attacked[237]. All attempts to isolate micro-organisms that could oxidize cyclohexane failed, but marine muds did contain species which could accept this substrate to yield cyclohexanone that could then be further metabolized by a plethora of organisms in situ[243].

The induction of alkane hydroxylases by unoxidized alkanes (such as $n-C_7$) has been demonstrated in a Pseudomonas species[241] and cytochrome P450—a principal oxygenase of the substate—was induced in a *Candida* species after two hours[244]. In recent studies the regulatory locus for hydroxylase activity in a *Pseudomonas* species has been explored[245]; and a genomic DNA clone encoding isocitrate lyase (a key enzyme of the glycoxylate cycle), which is a peroxisomal enzyme of the yeast *Candida* that converts n-alkane into citric and iso-citric acids, was isolated with a DNA probe from the gene library of the yeast[246].

Yeasts, especially *Candida* species, mostly grow well on even-carbon numbered n-alkanes (n-C_{10} to C_{18}) but much less so on the odd-carbon homologues[247]. Both sets of compounds can be metabolized to yield citric and other hydroxy acids in excellent (up to 50%) yields[248-250]. These latter presumably accumulate by flooding of the Krebs cycle with C_2 units derived from breakdown of the hydrocarbon. Yeasts that metabolize petroleum fractions have also been extensively exploited to generate biomass to manufacture 'single cell protein' for animal foodstuffs. One ton of n-alkane yielded roughly the same mass of protein-vitamin conjugate when adapted species were used and the process was economically viable[251-255]. Such petroleum-metabolizing lines can be immobilized on clay or sand[256] or can be encapsulated[257].

Marine bacteria and yeast may ameliorate the effects of the apparently inevitable oil spillages at sea[258]. Hundreds of such organisms have been isolated from the Arctic that can metabolize diesel oils down to 10 °C and bacterial communities that can degrade n-chain and branched alkanes have been characterized[259,260].

B. By Animals and Plants

Appreciable quantities of alkanes are ingested by animals as constituents of plant material and also by man as artificial additives (e.g. mineral oils) of foodstuffs or as components of natural or artificially produced oils and fats. Very little is known about the details of the metabolic turnover of such compounds. As in micro-organisms, it is likely that alkanes are detoxified in animals and plants by a mixed-function oxidase which forms the alcohols that may be further metabolized or excreted in the urine as conjugates, such as glucoronides or sulphates. Most of the compounds reported were characterized (GC–MS) after extraction from urine fractions that had been incubated with the appropriate enzyme to cleave the conjugate[261-263]. Some biotransformations of alkanes by animal systems (mainly *in vivo*) are summarized in Figure 7.

A few facts are also known concerning the metabolism of alkanes by plants. Cuticular waxes typically contain secondary alcohols, ketones and β-diketones in addition to alkanes and the former have the skeletons as in Figure 8, where functionalization is near the centre of the molecule and the chain lengths of the alcohols and ketones are closely similar to those of the major n-alkanes present in any particular species. Tracer evidence indicates that the secondary alcohol and ketone are formed in sequence by oxidation of the corresponding n-alkane by a mixed function oxidase which is inhibited by chelating agents. Thus in *Brassica* species, the 14- or 15-hydroxy- and oxo-derivatives of the n-C_{29} alkane were thus formed both *in vivo* and in cell-free extracts[200,274].

The β-diketones and their less-common hydroxylated derivatives (α- and γ-ketols) are not common but have been found to be the major components of the cuticular wax of a few species[198]. Unlike the alcohols and ketones these do not have chain lengths corresponding to the predominant n-alkanes of the species. Genetic studies with barley mutants indicate that the hydrocarbons and the β-diketones are generated by two distinct enzyme systems and the routes can be differentially inhibited in tissue slices. There is also tracer evidence that the β-diketones are not derived from the n-alkanes. The mechanism of formation of these derivatives has not been elucidated, but it is possible that the keto groups are preserved during the chain elongation as shown in Figure 9. Additional hydroxy groups are presumably introduced later by a mixed-function oxidase similar to that responsible for the formation of the monoketones[198,275].

References

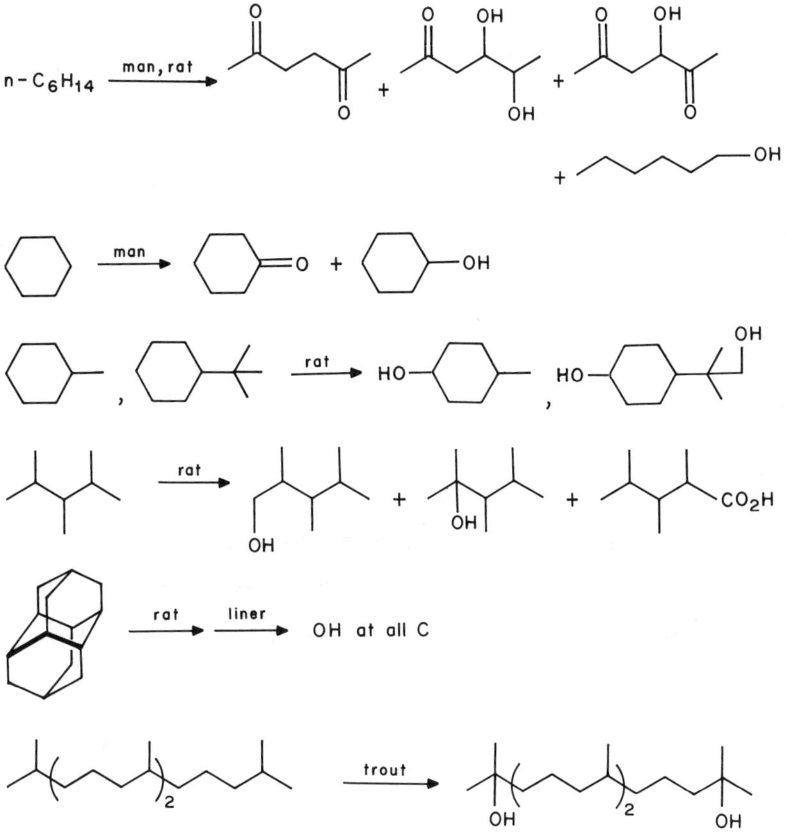

264–266

264

267–269

270, 271

272

273

FIGURE 7. Biotransformation of some alkanes

Type	Usual C range
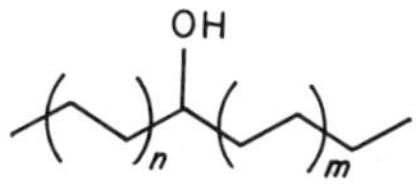	C_{20}–C_{33}
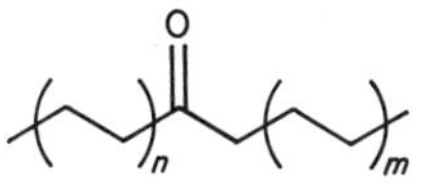	C_{24}–C_{33}
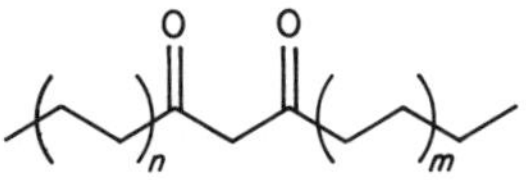	C_{31}–C_{33}

FIGURE 8. Oxygenated components of plant waxes

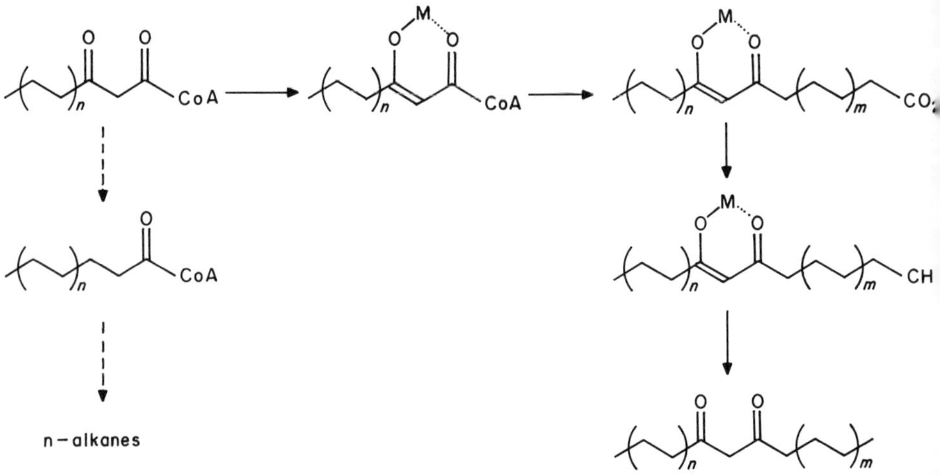

n – alkanes

FIGURE 9. Biogenetic route to β-diketones in plant waxes (CoA $\equiv$ coenzyme A, M $\equiv$ metal or H)

V. PHARMACOLOGY AND TOXICOLOGY

On account of their widespread occurrence both as pollutants and as components of (mainly vegetable) foodstuffs, the potential toxicology of the alkanes has been subject to severe scrutiny. Their general toxicity[276,277] and neurotoxicity[278] have been reviewed.

The lower alkanes are much less toxic than their chlorinated derivatives. Methane and ethane are simple asphyxiants not lethal to mice except at high $(680\,\mathrm{mg\,l^{-1}})$ concentrations[279]. The C_5 to C_8 alkanes all cause depression, dizziness and uncoordination when inhaled at the 5000 ppm level and may irritate the respiratory system with narcotic effects at higher concentrations[280,281]. The lower hydrocarbons are also anaesthetics with a cut-off in potency which correlates with the decrease in absorption into the lipid bilayer of nerve cells as the homology increases. This cut-off occurs for n-alkanes at C_9 to C_{12} in different species[282,283]. Inhalation of even small amounts of the lower alkanes can cause cyanosis, hypopyrexia and tachycordia leading to a chemical pneumonitis, and oral ingestion can cause hepatic damage. As a consequence, ingestions larger than $1\,\mathrm{g\,kg^{-1}}$ body weight should be vomited and a $10\,\mathrm{g}$ dose could be potentially fatal[284].

These consequences may be due to physical damage to membranes by the hydrocarbons, but the major effect may result from metabolism to form potentially toxic derivatives. On either count not all alkanes may contribute to the symptoms. Thus n-hexane on ingestion causes polyneuropathy in humans and experimental animals and is much more toxic than the n-C_5 and n-C_7 homologues. The potency may be due to formation of the 2,5-dione which can couple with pyrrole rings in proteins[264,285,286]; t-butylcyclohexane is similarly oxidized in rat to give derivatives hydroxylated in the side chain that can cause renal damage[269]. The n-C_{14} alkane has a much greater irritant effect in rat than its near homologues[287,288]; and inhalation by rat of the n-C_9 to n-C_{13} alkanes at close to air saturation revealed only the first homologue to be appreciably toxic[289]. The lower hydrocarbons comprise a high proportion (typically 12%) of air pollutants. 2-Methylpentane is a mucus coagulant[290] but the lower alkanes per se are probably not a significant health hazard. However, they are oxidized to form the highly irritating components of photochemical smogs[284].

Alkanes found in mammalian tissue have usually been considered of exogenous origin (e.g. from diet) and have been assigned neither a normal nor a pathological function, but this opinion may have to change. Recently, the n-alkane component of human epidermal tissues (ca 6% total lipid) has been shown to increase strikingly (up to 25%) in some scaling diseases, and all the evidence is that the enhanced level of alkane is not of exogeneous origin. The implication is that the compounds play a role in the pathology of the disease[291]. Alkane levels are also increased abnormally in the aorta of atherosclerotic rabbits[292]. Myelin of both the central and peripheral nervous systems contains remarkably high levels of long-chain fatty acids ($> C_{18}$) and their corresponding n-alkanes, which probably affect the physical and chemical properties of the sheath[293,294]. Impaired synthesis or incorporation of alkanes during myelin assembly could be the cause of such defective mylenisation in 'quaking' mutant mice[295]. Brain myelin of mouse accumulates especially high quantities of alkanes whilst miochondria, synaphosomes and microsomes possess very little[296].

Aliphatic hydrocarbons including alkanes are widely distributed in the internal organs and non-neural tissue of mammals at low level, usually 3 to 60 μg g^{-1}, such as to suggest exogenous origin[297-301]. Vaseline and mineral oils, and by implication higher alkanes have been repeatedly cleared as potentially toxic materials. However, oil droplets with all the characteristics of mineral oils can accumulate and cause lesions in spleen and liver[302] and it is possible that some alkanes at least have insidious effects. Recently a novel, sudden fatal pathological condition has been discovered that is associated with the visceral accumulation of long-chain n-alkanes (n-C_{29} to n-C_{33}) in a human patient. Plant cuticular waxes as source were implicated by this alkane pattern and indeed the patient had indulged in gross consumption of apples (ca 1 kg day^{-1} over 18 years[303]!).

VI. REFERENCES

1. W. A. Gruse and D. R. Stevens, Chemistry and Technology of Petroleum, McGraw-Hill, New York, 1975.
2. A. J. W. Headlee and R. E. McClelland, Ind. Eng. Chem., 43, 2547 (1951).
3. M. O. Denekas, F. T. Coulson, J. W. Moore and C. G. Dodd, Ind. Eng. Chem., 43, 1165 (1951).
4. C. B. Cooks, G. W. Jamieson and L. S. Ciekeszko, Am. Assoc. Pet. Geol. Bull., 49, 301 (1965).
5. P. E. Kolattukudy, Chemistry and Biochemistry of Natural Waxes, Elsevier, Amsterdam, 1976.
6. D. E. Lester, Prog. Food Nutr. Sci., 3, 1 (1979).
7. J. F. Carter, J. Rechka and N. Robinson, Biomed. Environ. Mass Spectrom., 18, 939 (1989).
8. B. J. Kimble, J. R. Maxwell, R. P. Philp and G. Eglinton, Chem. Geol., 14, 173 (1974).
9. S. B. Hawthorne and D. J. Miller, J. Chromatogr., 388, 397 (1987).
10. D. I. Welch and C. D. Watts, Int. J. Environ. Anal. Chem., 38, 185 (1990).
11. R. A. Mah, D. M. Ward, L. Baresi and T. L. Glass, Ann. Rev. Microbiol., 31, 309 (1977).
12. J. R. Riley and R. Chester, Introduction to Marine Chemistry, Academic, London, 1971, p. 189.
13. B. J. Moore and S. Sigler, Bur. Mines Inf. Circl. (U.S.A.), 1C, 9188 (1988); Chem. Abstr., 109, 152704 (1988).
14. D. Rossini, J. Chem. Educ., 37, 554 (1960).
15. S. A. Peg, F. Mahmud and D. K. Al Harbi, Fuel Sci. Technol. Int., 8, 125 (1990); Chem. Abstr., 112, 59315, (1990).
16. W. G. Meinschein, in Organic Geochemistry (Eds. G. Eglinton and M. J. J. Murphy), Longmans, London, 1969, p. 330.
17. J. D. Brooks, K. Gould and J. W. Smith, Nature (London), 222, 257 (1969).
18. I. R. Hills, G. W. Smith and E. V. Whitehead, J. Inst. Petr., 56, 127 (1970).
19. A. A. Ensminger, A. van Dorsselar, C. Spyckerelle, P. Albrecht and G. Ourisson, Adv. Org. Geochem., 245 (1973).
20. E. V. Whitehead, Adv. Org. Geochem., 225 (1973).
21. J. R. Maxwell, C. T. Pilliger and G. Eglinton, Quart. Rev. Chem. Soc. (London), 25, 571 (1971).
22. J. E. Gallegos, Anal. Chem., 43, 1151 (1971).

23. E. D. McCarthy and M. Calvin, *Tetrahedron*, **23**, 2609 (1967).
24. W. Henderson, V. Wollrab and G. Eglinton, *Adv. Org. Geochem.*, 203 (1969).
25. P. C. Anderson, P. M. Gardner, E. V. Whitehead, D. E. Anders and W. E. Robinson, *Geochim. Cosmochim. Acta*, **33**, 1304 (1969).
26. E. D. Evans, G. S. Kenny, W. G. Meinschein and E. E. Bray, *Anal. Chem.*, **29**, 1858 (1957).
27. E. E. Bray and E. D. Evans, *Geochim. Cosmochim. Acta*, **22**, 2 (1961).
28. R. S. Scanlon and J. E. Smith, *Geochim. Cosmochim. Acta*, **34**, 611 (1970).
29. H. Yonetani and P. Terashima, *Sekiyu Gijutsu Kyotaishi*, **49**, 141 (1984); *Chem. Abstr.*, **102**, 23554 (1985).
30. R. I. Morrison and W. Rick, *J. Sci. Food Agric.*, **18**, 351 (1967).
31. R. De Borger and Y. van Elsen, *Rev. Agric. (Brussels)*, **27**, 1493 (1974).
32. K. Pihlaja, E. Malinski and E. L. Poutanen, *Org. Geochem.*, **15**, 321 (1990).
33. A. C. Chibnall and S. H. Piper, *Biochem. J.*, **28**, 2209 (1934).
34. J. H. Butler, D. T. Downing, R. J. Swaby, *Aust. J. Chem.*, **17**, 817 (1964).
35. M. L. John (Ed.), *Organic Marine Geochemistry*, ACS Symposium Series 305, *Am. Chem. Soc.*, Washington D. C., 1986, p. 22.
36. A. V. Botello and E. F. Mandelli, *Bull. Environ. Contam. Toxicol.*, **19**, 162 (1978).
37. M. M. Quirk, R. L. Patience, J. R. Maxwell and D. E. Wheatley, *Pergamon Ser. Environ. Sci.*, **3**, 23 (1980); *Chem. Abstr.*, **93**, 209902 (1980).
38. J. M. Hunt, *Nature (London)*, **254**, 411 (1975).
39. R. J. Cordell, *Am. Assoc. Pet. Geol. Bull.*, **56**, 2029 (1972).
40. R. W. C. Wyckoff, *Biochemistry of Animal Fossils*, Scientechnica, Bristol, 1972, p. 118.
41. P. E. Kolattukudy, *Science*, **159**, 498 (1968).
42. C. Eglinton and R. J. Hamilton, *Science*, **156**, 1322 (1967).
43. P. Mazliak, *Prog. Phytochem.*, **49**, 1 (1968).
44. P. E. Kolattukudy, in *Biochemistry of Plants* (Eds. P. K. Stumpf and E. E. Cohn), Vol. 4, Academic Press, New York, 1980, p. 580.
45. E. A. Baker, in *The Plant Cuticle* (Eds. D. F. Cutler, K. L. Alvin and C. E. Price), Academic Press, London, 1982, p. 139.
46. G. Eglinton and R. J. Hamilton, in *Chemical Plant Taxonomy* (Ed. T. Swain), Academic Press, London, 1963, p. 187.
47. P. Mazliak, *Rev. Gen. Bot.*, **70**, 43 (1963).
48. F. Radler and D. H. S. Horn, *Aust. J. Chem.*, **18**, 1059 (1965).
49. G. A. Herbin and P. A. Robins, *Phytochemistry*, **8**, 1985 (1969).
50. P. E. Kolattukudy, *Plant Physiol.*, **43**, 375 (1968).
51. T. Kaneda, *Biochemistry*, **6**, 2023 (1967).
52. V. Wollrab, M. Streibl and F. Sorm, *Chem. Ind. (London)*, 1872 (1967).
53. J. Neubeller, *Gartenbauwissenschaft*, **55**, (1990); *Chem. Abstr.*, **113**, 168992 (1990).
54. E. Sutter, *Can. J. Bot.*, **62**, 74 (1984).
55. P. G. Guelz, M. Rosinski and C. Eich, *Z. Pflanzenphysiol.*, **107**, 281 (1982).
56. F. Klingauf, K. Noecker-Wenzel and V. Roëttger, *Z. Pflanzenkr. Pflanzenschutz*, **85**, 228 (1978); *Chem. Abstr.*, **89**, 143768 (1978).
57. K. Noecker-Wenzel, W. Klein and F. Klingauf, *Tetrahedron Lett.*, 4409 (1971).
58. J. L. Harwood and N. J. Russell, *Lipids of Plants and Microbes*, Allen and Unwin, London, 1984.
59. B. B. Dean and P. E. Kolattukudy, *Plant Physiol.*, **59**, 48 (1977).
60. K. E. Espelic, N. Z. Sadek and P. E. Kolattukudy, *Planta*, **148**, 468 (1980).
61. A. R. S. Kartha and S. P. Singh, *Chem. Ind. (London)*, 1347 (1969).
62. R. E. Grice, H. D. Locksley and F. Scheinmann, *Nature (London)*, **218**, 892 (1968).
63. W. Cocker, T. B. H. McMurry and M. S. Njamila, *J. Chem. Soc.*, 1692 (1965).
64. H. E. Nursten, in *Biochemistry of Fruits and their Products* (Ed. A. C. Hulme), Vol. 1, Academic Press, London, 1970, p. 245.
65. W. Sandermann and W. Scheers, *Chem. Ber.*, **93**, 2266 (1960).
66. E. Guenther, *The Essential Oils*, Vol. 2, Van Nostrand–Reinhold, New York, 1946, p. 7.
67. W. Haagen-Smit, P. Redemann and N. V. Mirov, *J. Am. Chem. Soc.*, **69**, 2014 (1947).
68. K. Yaacob, C. M. Abdullah and D. Joulain, *J. Essent. Oil Res.*, **1**, 203 (1989).
69. F. Guildemeister and F. Hoffman, *Die Anterischen Öle*, Vol. 3, Akademie Verlag, Berlin, 1960, pp. 242, 243, 256.

256. S. H. Omar and H. J. Rehm, *Appl. Microbiol. Biotechnol.*, **28**, 103 (1988).
257. S. Fukui, *J. Am. Oil Chem. Soc.*, **65**, 96 (1988).
258. K. Lee, C. S. Wong, W. J. Cretney, F. A. Whitney, T. R. Parsons C. M. Lalli and J. Wu, *Microb. Ecol.*, **11**, 337 (1985).
259. T. V. Koronelli, V. V. Ilinskii, S. G. Dermacheva, T. Komarova, A. N. Belyaeva, Z. O. Fileppora and B. V. Rozinkov, *Izv. Akad. Nauk SSSR, Ser. Biol.*, 581 (1989).
260. J. F. Rontani and G. Guisti, *Mar. Chem.*, **20**, 197 (1986).
261. E. Hodgson and W. Daulemann, in *Introduction to Biochemical Toxicology* (Eds. E. Hodgson and S. E. Guthrie), Blackwell, Oxford, 1980, p. 78.
262. R. Deutsch-Wenzel, *Ber. Dtsch. Wiss. Ges. Erdoel Erdgas Kohle*, 174 (1987).
263. D. R. Hawkins (Ed.), *Biotransformations*, Royal Society of Chemistry, London, 1988, Vol. 1; 1989, Vol. 2.
264. M. Governa, R. Calisti, G. Coppa, G. Tagliarento, A. Colombri and W. Troni, *J. Toxicol. Environ. Health*, **20**, 219 (1987).
265. N. Fedtke and H. M. Bolt, *Biomed. Environ. Mass. Spectrom.*, **14**, 563 (1987).
266. N. Fedtke and H. M. Bolt, *Arch. Toxicol.*, **61**, 131 (1987).
267. M. J. Parnell, G. M. Henningsen, C. J. Hixson, K. O. Yu, G. A. McDonald and M. P. Serve, *Chemosphere*, **17**, 1321 (1988).
268. G. M. Henningsen, R. A. Sollomon, K. O. Yu, I. Lopez, J. Roberts and M. P. Serve, *Toxicol. Environ. Health*, **24**, 19 (1988).
269. G. M. Henningsen, K. O. Yu, R. O. Sollomon, M. J. Ferry, I. Lopez, J. Roberts and M. P. Serve, *Toxicol. Lett.*, **39**, 313 (1987).
270. M. Charbonneau, B. A. Lock, J. Strasser, M. G. Cox. M. J. Turner and J. S. Burr, *Toxicol. Appl. Pharmacol.*, **91**, 171 (1987).
271. C. T. Olson, D. W. Hobson, K. O. Yu and M. P. Serve, *Toxicol. Lett.*, **37**, 199 (1987).
272. P. Hodek. P. J. Janscak, P. Anzenbacher, J. Burkhard, J. Janku and L. Vodicka, *Xerobiotica*, **18**, 1109 (1988).
273. A. M. Lebon, J. P. Cravedi and J. E. Tuilliez, *Chemosphere*, **17**, 1063 (1988).
274. P. E. Kolattukudy and T-Y. J. Liu, *Biochem. Biophys. Res. Commun.*, **41**, 1369 (1970).
275. J. D. Mikkelsen and P. von Weltstein-Knowles, *Arch. Biochem. Biophys.*, **188**, 172 (1978).
276. L. S. Mullin, A. W. Ader, W. C. Daughtney, D. Z. Frost and M. R. Greenwood, *J. Appl. Toxicol.*, **10**, 135 (1990).
277. J. W. Griffin, *Neurobehav. Toxicol. Teratol.*, **3**, 437 (1981).
278. J. L. O'Donohue, *Neurotoxic Ind. Commer. Chem.*, **2**, 61 (1985).
279. B. B. Shugaev, *Arch. Environ. Health*, **18**, 878 (1969).
280. E. Hodgson, R. B. Mailman and J. E. Chambers, *Dictionary of Toxicology*, Macmillan, London, 1988, p. 20.
281. G. D. Muir (Ed.), *Hazards in the Chemical Laboratory*, 2nd ed., Chemical Society, London, 1972.
282. D. A. Haydon, B. M. Hendry, S. R. Levinson and J. Requera, *Nature (London)*, **268**, 356 (1977).
283. D. A. Haydon, B. M. Hendry, S. R. Levinson and J. Requera, *Biochim. Biophys. Acta*, **470**, 17 (1977).
284. J. Doull C. D. Klassen and M. O. Amdur (Eds.), *Cassareti and Doull's Toxicology*, 2nd ed., Macmillan, New York, 1990, p. 168.
285. Y. Takeuchi, Y. Ono, N. Hisanaga, J. Kitoh and Y. Sugiura, *Br. J. Ind. Med.*, **37**, 241 (1980).
286. E. Hodgson, R. P. Mailman and J. E. Chambers, *Dictionary of Toxicology*, Macmillan, London, 1988, p. 20.
287. M. Cejka, *Schmierstoffe Schmierungstech.*, 27 (1969); *Chem. Abstr.*, **71**, 104946 (1969).
288. S. J. Moloney and J. J. Teal, *Arch. Dermatol. Res.*, **280**, 375 (1988).
289. A. G. Nilsen, O. A. Haugen, K. Zahlsen, J. Halgunset, A. Helseth, H. Aarset and I. Eide, *Pharmacol. Toxicol. (Copenhagen)*, **62**, 259 (1988).
290. C. E. Searle, *Chemical Carcinogens*, ACS Monograph, Washington, DC, 1976, p. 339.
291. M. L. Williams and B. M. Elias, *Biochem. Biophys. Res. Commun.*, **107**, 322 (1982).
292. R. B. Ramsey, R. T. Aexel, S. Jain and H. J. Nicholas, *Atherosclerosis*, **15**, 301 (1972).
293. E. Rouser, *Adv. Lipid Res.*, **10**, 331 (1972).
294. S. G. Kayser and S. Patton, *Biochem. Biophys. Res. Commun.*, **41**, 1572 (1970).
295. J. M. Bourre, F. Boiron, C. Cassagre, O. Dumont, F. Le Terrier, H. Metzger and J. Viret, *Neurochem. Pathol.*, **4**, 29 (1986).

296. J. M. Bourre, C. Cassagre, S. Larrouguere-Regnier and D. Darriet, *J. Neurochem.*, **20**, 325 (1977).
297. D. T. Downing, in *Chemistry and Biochemistry of Natural Waxes* (Ed. P. E. Kolattukudy), Elsevier, Amsterdam, 1976, p. 17.
298. H. J. Nicholas and K. J. Bambaugh, *Biochim. Biophys. Acta*, **98**, 372 (1965).
299. B. Nagy, V. F. Modzeleski and W. M. Scott, *Biochem. J.*, **114**, 645 (1969).
300. E. L. Bandurski and B. Nagy, *Lipids*, **10**, 67 (1975).
301. L. Schlunegger, *Biochem. Biophys. Acta*, **260**, 339 (1972).
302. J. K. Bostnott and J. Margolis, *Johns Hopkins Med. J.*, **127**, 65 (1970); *Chem. Abstr.*, **74**, 11534 (1971).
303. S. Salvayre, A. Negre, F. Rocchiccioli, C. Duboucher, A. Maret, C. Vieu, A, Lageron, J. Polonovskii and L. Douste-Blazy, *Biochem. Biophys. Acta*, **958**, 477 (1988).

CHAPTER **20**

Inverted and planar carbon

WILLIAM C. AGOSTA

The Rockefeller University, New York, New York 10021, USA

The Chemistry of Alkanes and Cycloalkanes
Edited by S. Patai and Z. Rappoport © 1992 John Wiley & Sons Ltd

I. INTRODUCTION, SCOPE AND NOMENCLATURE

The effects of angle strain in organic molecules have attracted interest and study for more than a century. The first notable success in this area was Perkin's syntheses of cyclobutane-[1] and cyclopropanecarboxylic[2] acids in 1883 and 1884, when it was generally presumed that such small rings could not exist. At the time, Perkin was a twenty-two-year-old student in Baeyer's laboratory in Munich. Ignoring the collective advice of Baeyer, Emil Fischer and Viktor Meyer, he adapted the newly developed malonic ester synthesis to his needs, prepared these novel cycloalkane derivatives and opened a new field of investigation. This led directly to Baeyer's strain theory[3], which ascribed the unusual reactivity of small carbocyclic rings to the necessary departure from tetrahedral bonding of the carbon atoms involved. The concept of strain has since grown far more complex[4,5], but the theory that Baeyer presented in two and a half small pages in 1885 has provided the qualitative basis for a hundred years of inquiry into the distortion of bond angles. During this time organic chemists have contrived numerous ways of deforming the ordinary tetrahedral sp^3, planar sp^2 and linear sp bonds of the carbon atom in order to assess the effects of such changes on molecular properties. Theory has increasingly provided both understanding of these effects and also guidance in devising systems for future investigation, while the growing power of synthesis has made ever more structural types available for experimental examination.

All this activity has led to a large body of information which, along with investigations into other types of strain, has been the subject of excellent reviews[4,5]. In the present chapter we wish to focus attention specifically on theoretical and experimental studies of molecules containing flattened or inverted carbon. 'Flattened carbon' will denote a carbon atom with attached nuclei W, X, Y and Z, having angle deformations that would lead, in the limit, to a planar molecular fragment CWXYZ. Calculations reveal that the energy difference between square-planar and tetrahedral methane is considerably greater than typical carbon–hydrogen and carbon–carbon bond strengths[6], so that the limiting case of a molecule containing planar carbon with these ordinary bonds is not experimentally likely. In terms of symmetry coordinates such deformations can be analyzed into two types of motion[7-9]. The first is compression deformation, in which one pair of opposite angles, as θ_{WX} and θ_{YZ} in 1, is opened at the expense of the other four angles. If both θ_{WX} and θ_{YZ} open to 180°, a square-planar fragment is formed. The second type of motion is a twist deformation, in which one pair of opposite angles remains invariant, a second pair opens up and the third closes. This can be visualized as a rotation of, for example, the CWX portion of CWXYZ relative to CYZ in 2. In the limit, a rotation of 90° furnishes a planar fragment. Both compression and twisting can of course occur in a single system, although *ab initio* calculations on spiropentane show bending to be energetically more economical than twisting[10].

By 'inverted carbon' we will mean a fragment CWXYZ in which all four bonds lie in the same half of a sphere centered on the carbon atom C, for example, 3[11]. Equivalent definitions are that the plane determined by W, X and Y passes between C and Z; or that C lies outside the tetrahedron determined by W, X, Y and Z. Numerous compounds containing inverted carbon are known.

Before turning to the types of carbon skeletons in which these distortions are imposed on tetrahedral carbon, we should note that an alternative approach to such geometry is through use of other sorts of bonds associated with different hybridization at carbon. Details lie outside the scope of this discussion, but calculations indicate that in carbon bonded to two lithium atoms there should be a great increase in stability of a planar arrangement; the unknown systems 4–6 are predicted to the planar[8]. More recently, an X-ray structure of the tetramer of (2,6-dimethoxyphenyl)lithium has provided the first experimental evidence for a planar tetracoordinate carbon containing a CLi_2 fragment[12]. Earlier organometallic compounds of interest here include a carborane for which an

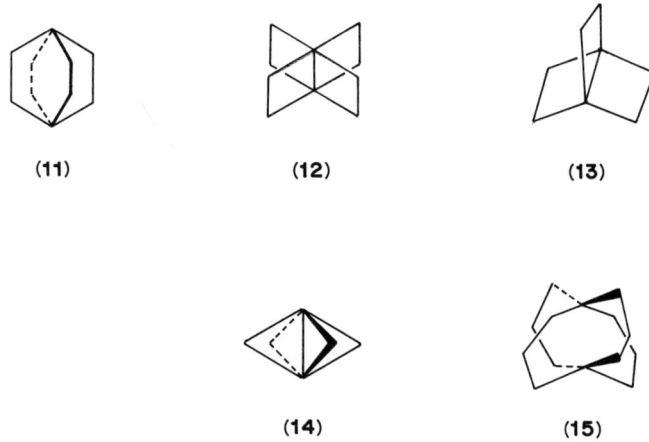

(11)   (12)   (13)

(14)   (15)

bridgehead carbon atoms and a strain energy of 456 kcal mol^{-1} [37]. Molecular mechanics calculations for [3.3.3.3]paddlane yield a structure with only somewhat flattened bridgehead atoms, but extensive nonbonded interactions between the three-carbon chains, as shown in 15[38].

C. Fenestranes

There has been considerable discussion of the possibility of stabilizing a planar carbon atom by placing it at the center of an annulene that would be aromatic if planar. Hoffmann and coworkers originally suggested that the unsaturated [5.6.5.6]- and [5.5.6.6]fenestranes 16 and 17 were promising systems in this regard, and that the corresponding [4.4.4.4], [4.5.4.5] and [5.5.5.5] systems 18, 19 and 20 would be unstable[30]. MNDO calculations support these conclusions for 18–20, but suggest that 16 also lacks stability[39]; MINDO/3[40] predicts instability for 16, 17 and 20[41]. It has been proposed, however, that 20 should be stabilized and flattened owing to interaction between the energetically unfavorable highest occupied molecular orbital (HOMO) at the central atom and the lowest unoccupied molecular orbital (LUMO) of the peripheral twelve π-electron system. Related calculations also indicated that the delocalization energy in homoconjugated 21 is nearly as great as that in 20 and suggested that these hydrocarbons should be attractive goals for synthesis[42].

For saturated fenestranes *ab initio* molecular orbital calculations give acceptable estimates of energies and structures[43]. Semiempirical methods have also proved quite useful, but molecular mechanics calculations for fenestranes are less accurate. Experience indicates that with the usual force constants molecular mechanics tends to overestimate the energy required for large deformation of bond angles, introducing the possibility of sizable errors both in absolute strain energy and in molecular geometry[44,45]. These empirical calculations are very convenient, however, for estimating energy differences between homologous or isomeric systems.

Ab initio calculations give strain energies in the range of 148–157 kcal mol^{-1} for the unknown $(1\alpha, 2\beta, 5\alpha, 6\beta)$[3.5.3.5]fenestrane (22) and its $1\alpha, 2\beta, 5\beta, 6\beta$ and $1\alpha, 2\alpha, 5\alpha, 6\alpha$ diastereomers[10]. This study included similar calculations on the two diastereomeric [3.3.5]fenestranes. The structure calculated for the known[10,46] $3\alpha, 6\beta$ isomer 24,

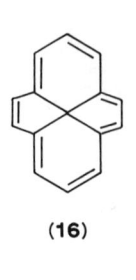

(16)

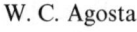
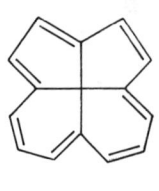

(17)

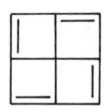

(18)

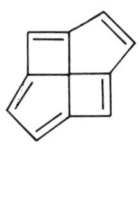

(19)

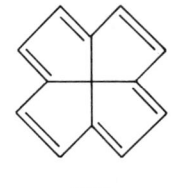

(20)

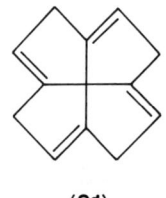

(21)

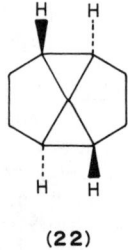

(22)

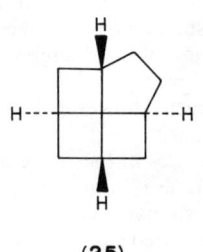

(23)

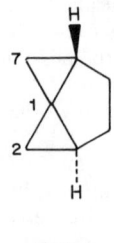

(24)

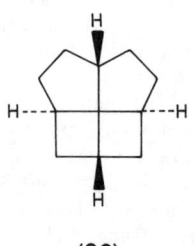

(25)

(26)

preparation of which is mentioned below, agrees well with the values obtained experimentally by electron diffraction[47]. The calculated strain energy of **24** is 80 kcal mol^{-1}, and the unknown $3\alpha,6\alpha$ isomer is predicted to have an energy 46 kcal mol^{-1} greater. *Ab initio* calculations on $(1\alpha,5\alpha,7\beta)$-[4.4.4.4]fenestrane (**23**) yield a strain energy of 160 kcal mol^{-1}, as well as several other interesting conclusions[48,49]. The carbon–carbon bond to the central atom is unusually short, 1.480 Å; the external carbon–carbon bond is rather long, 1.600 Å; and the bond angle across the central atom is 130.4°. The very short bond seems to be the result of poor directionality, leading to markedly bent bonds.

Fenestranes larger than the [4.4.4.4] system have not been the subject of published *ab initio* calculations, but there are MNDO studies of several of these hydrocarbons[9,19,38,50]. For [4.4.4.5]fenestrane (**25**) and its [4.4.5.5] homologue **26**, the calculated angles across the quaternary atom agree within 1° or better with experimental results. MNDO calculations for the unknown [3.5.3.5]fenestrane provide the bond angles across the central atom as 127° and 67°[38], which may be compared with the *ab initio* results of 123.8° and 72.4°[10]; for [4.4.4.4]fenestrane MNDO gives 132° for the central angle[38], while the *ab initio* structure has 130.4°[48,49]. Using data from crystal structures and MNDO calculations for a wide variety of spiro systems, including the embedded spiro systems of several fenestranes, Keese and coworkers have classified these systems according to the amounts of compression and twisting contributing to bond deformation at the spiro atom[8,9,51]. Plots of these data show that among the systems examined fenestranes are unique in that flattening at the central carbon atom occurs almost solely through compression with little or no twisting.

D. Propellanes

There has been intense theoretical interest in the smallest member of the series, [1.1.1]propellane (**27**), and most of the physical properties of this surprisingly stable compound were correctly predicted from *ab initio* calculations performed before its synthesis[52–54]. The central bond has been the object of much attention, particularly since earlier studies suggested that it had no charge density[55]. An attractive current description is that this is a rather fat bond with charge spread out at the midpoint and having high local charge density at the bridgehead carbon atoms. The charge density is about 80% of that in the central bond of butane. This comes from a detailed analysis of the charge topology[56] of **27**[57], and it is supported by other calculations[58,59]. However, there is another detailed theoretical description of the bonding in **27** suggesting that the HOMO is nonbonding and not concerned with the $C_{(1)}$—$C_{(3)}$ bond. In this picture $C_{(1)}$ and $C_{(3)}$ are held together by a 'σ-bridged π-bond', consisting of three-center, two-electron orbitals that use $2p\pi$ orbitals on $C_{(1)}$ and $C_{(3)}$ and sp^2 lone pairs on the three methylene bridges[60]. Some support for this view comes from the photoelectron spectrum of **27**, where little change in structure accompanies the lowest energy ionization, although this also could be the consequence of the rigid cage structure of **27**[61]. However, later work using this theoretical approach, but with correlated wave functions, did indicate bonding between the orbitals on $C_{(1)}$ and $C_{(3)}$[62].

Ab initio calculations on [2.1.1]propellane (**28**) and [2.2.1]propellane (**29**) suggest that the central bond in these hydrocarbons is quite weak[54,57,63]. This is in keeping with the observation that, unlike **27**, both **28** and **29** are stable only below 50 K in argon matrix[64,65].

Inclusion of d orbitals in *ab initio* calculations is necessary in order to achieve consistent relative energies for [4.2.1]- and smaller propellanes[54]. Calculated structures are available for six propellanes, and the resulting *ab initio* energies have been converted into estimated enthalpies of formation (± 2 kcal mol^{-1}) using a set of group equivalents[66]. For the three

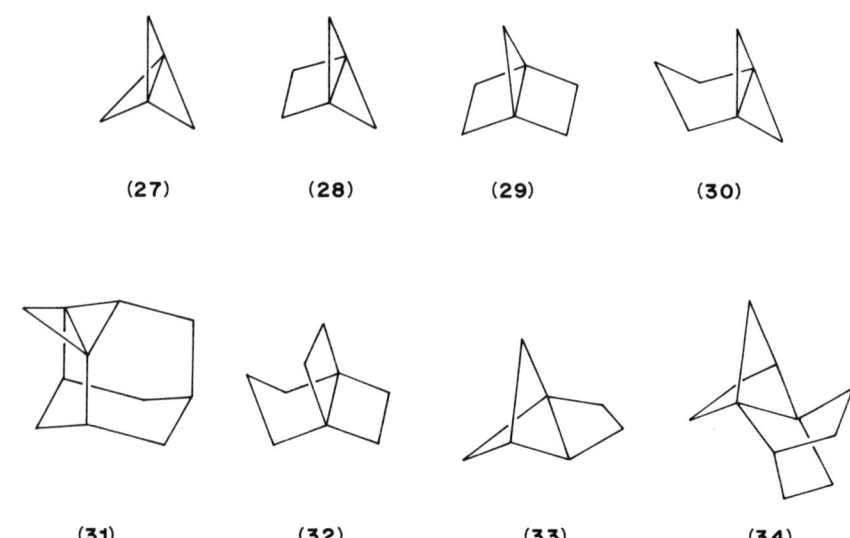

smallest, [1.1.1]- (27), [2.1.1]- (28) and [2.2.1]propellane (29), strain energies were estimated to be nearly constant at 102, 104 and 102 kcal mol⁻¹, respectively. For 27 the calculated structure and enthalpy of formation are in good agreement with experimental data[67].

MNDO appears to provide satisfactory geometry for derivatives of [3.1.1]propellane (30), and HAM/3[68] calculations based on MNDO geometry reproduce reasonably well the experimental ionization energies of 31, which is a [3.1.1]propellane[69]. Molecular mechanics calculations are useful here only for less distorted systems, such as [3.2.2]propellane (32) and larger systems that do not contain inverted carbon[70].

E. Bicyclo[1.1.0]butanes

Extended Hückel calculations on 10 are in substantial agreement with the structure derived from microwave spectroscopic studies[21]. Extensive restricted Hartree–Fock calculations have been carried out on the inversion of substituted bicyclo[1.1.0]butanes and on the energetics of variation of the flap angle[23,71]. The results indicate that the ring system is surprisingly flexible, with little energy required for variations of up to 20° in the flap angle. These calculations also reveal that inversion of the system involves non-least-motion movement of the substituents at $C_{(1)}$ and $C_{(3)}$.

From infrared and Raman spectra of bicyclobutane and appropriate deuterated isomers a new vibrational assignment was made with the help of the spectrum calculated using the 6-31G* basis set[72]. A normal coordinate analysis furnished atomic polar tensors and related properties. The results were compared with similar data for cyclopropane and [1.1.1]propellane.

F. Other Carbon Skeletons

Inverted carbon atoms can be incorporated into many other ring systems of lower symmetry than those discussed above. Molecular mechanics calculations have been reported for several such hydrocarbons, along with the opinion that the results indicate that these compounds should be attractive synthetic targets[73]. A characteristic difficulty with such calculations is that they do not allow for energetic changes involved in the rehybridization that necessarily occurs in these distorted systems. For two of these hydrocarbons, **33** and **34**, MNDO calculations, which have their own limitations, suggest strain energies that may be prohibitively high. The MNDO enthalpy of formation for **33** is 108 kcal mol^{-1}, and that for **34** is 101 kcal mol^{-1} [38]. For comparison, MNDO yields 37 kcal mol^{-1} as the enthalpy of formation of [4.4.4.5]fenestrane (**25**), which is isomeric with **34**[38].

III. PREPARATIVE CHEMISTRY AND PROPERTIES

Our purpose here is to review the physical properties indicative of flattening or inversion of carbon in appropriate compounds and also to provide a brief synopsis of the synthetic approaches employed in their preparation.

A. Paddlanes

1. Complex structures

Less progress has been made in preparation of strained paddlanes than in the other areas under discussion. A number of compounds formally represented by **7** or heteroatom derivatives of **7** are known, including some, such as **35**[74], **36**[75] and **37**[76], that were prepared without specific interest in inverted carbon. Others result from exploration of routes that might be adapted to highly strained systems but have led so far to compounds with conformational restrictions but not great angle strain. One of these is **39**, which is available on Diels–Alder addition of dicyanoacetylene to furanophane **38**[77]. The proton NMR

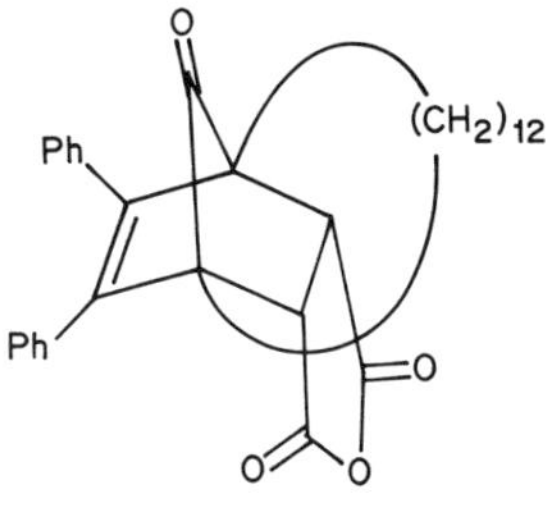

(35) 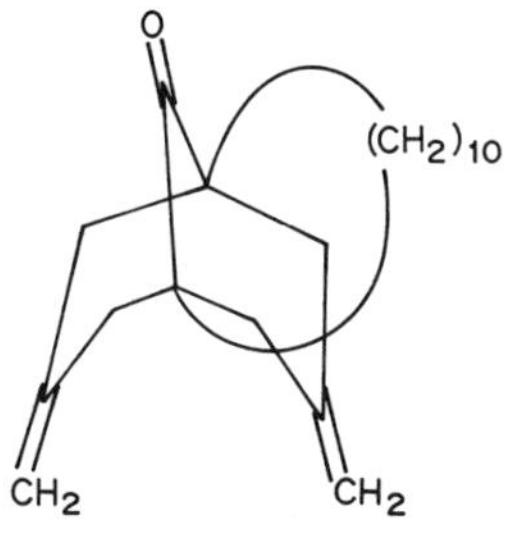   (36) 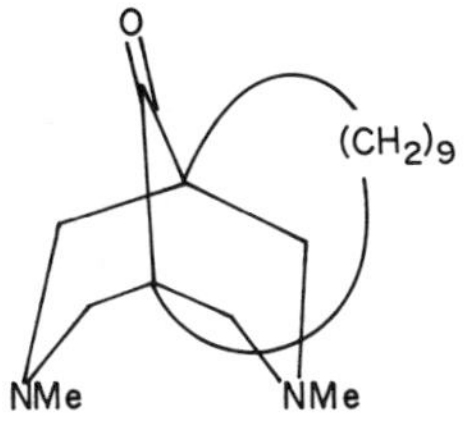   (37)

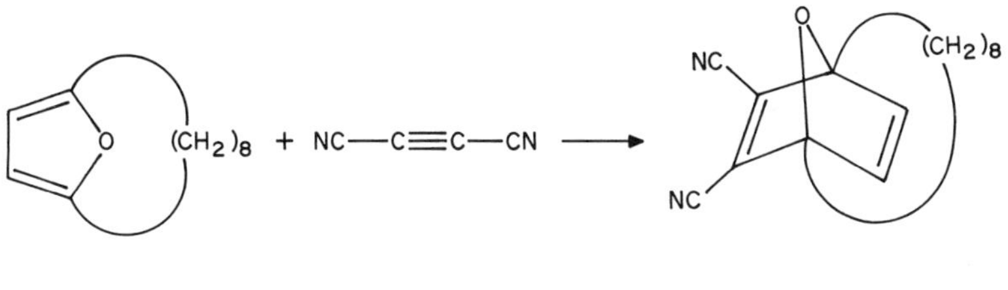

(38)   (39)

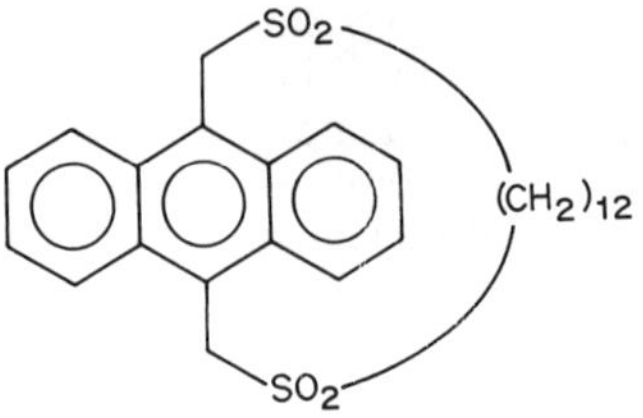

(40)

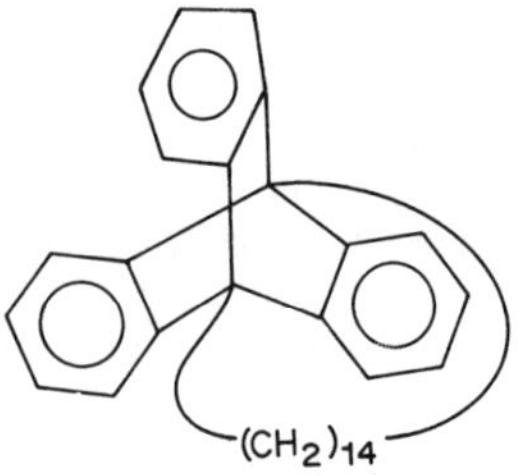

(41)

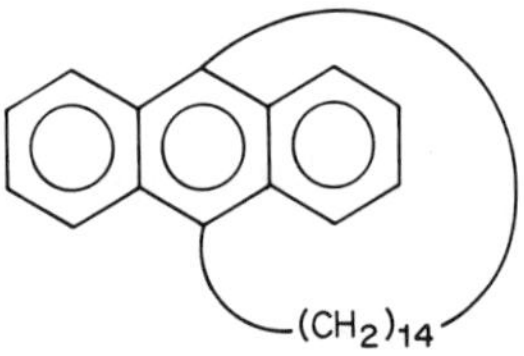

(42)

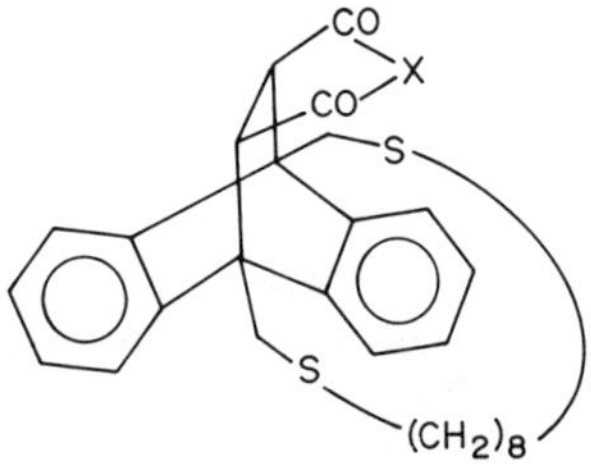

(43) X=O
(44) X=NPh

spectrum of **39** shows evidence for two conformers at $-100\,°C$. The interesting triptycene derivative **41** results from benzyne addition to the bridged anthracene bis-sulfone **40**, followed by pyrolytic elimination of SO_2[78]. Both spectroscopic evidence and models indicate that the polymethylene bridge is fixed between two of the aromatic rings and cannot rotate about the triptycene skeleton. Reversal of the order of the synthetic steps was unsuccessful, as benzyne failed to add to the 9,10-polymethyleneanthracene **42**. Maleic anhydride and N-phenylmaleimide did add, however, to a related bis-sulfide to furnish **43** and **44**, in which movement of the chain is also restricted. Wiberg and O'Donnell investigated the addition of dicyanoacetylene to the [n](1,4)naphthalenophanes **45**, where n is 8, 9, 10 and 14[79]. In all cases the major product was **46**, but with n = 14 a small amount of **47** was obtained. There was no evidence of significant distortion in this latter compound. Efforts by Warner and coworkers to close the fourth ring in dibromides **48** and **49**, where n = 0 and 1, led instead to dimeric bis-sulfides such as **50**[80].

2. Simple [n.2.2.2]paddlanediones

A most interesting series of paddlanes are three compounds prepared by Eaton and Leipzig[81]. Reaction of bicyclo[2.2.2]octane-1,4-dicarboxaldehyde with lithium metal and 1,12-dibromododecane furnished the paddlanedione **51**. Wolff rearrangement of **51** by way of the bis α-diazoketone **52** gave the expected bis ketene **53**, which was converted to the new, lower homolog paddlanedione **54** on ozonolysis in the presence of propionaldehyde. Repetition of this sequence furnished the third member of the series, the [10.2.2.2]paddlanedione **55**. There is a noteworthy progression of conformational

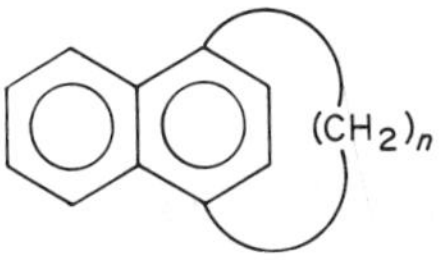

(45) 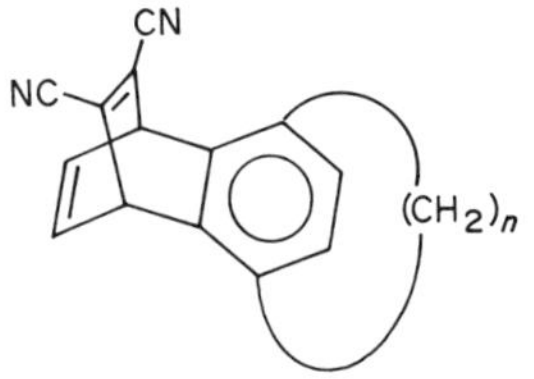   (46) 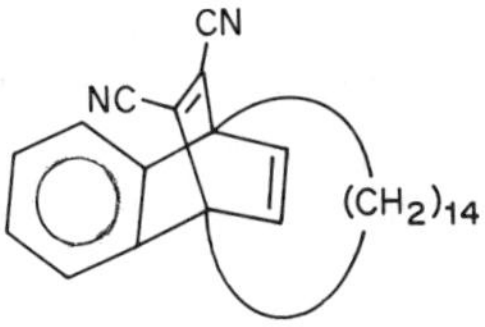   (47)

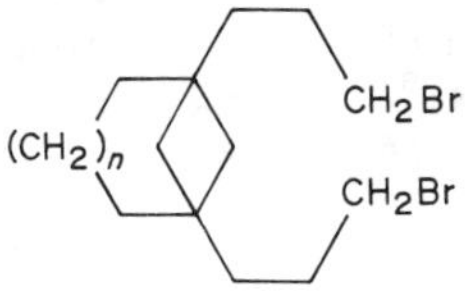

(48) 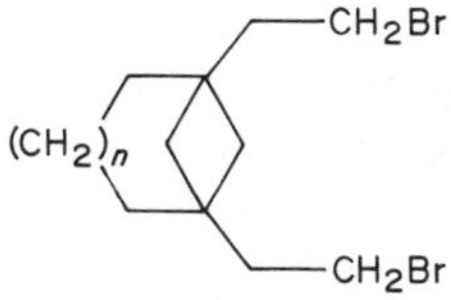   (49) 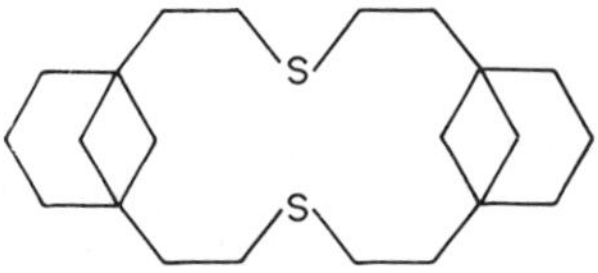   (50)

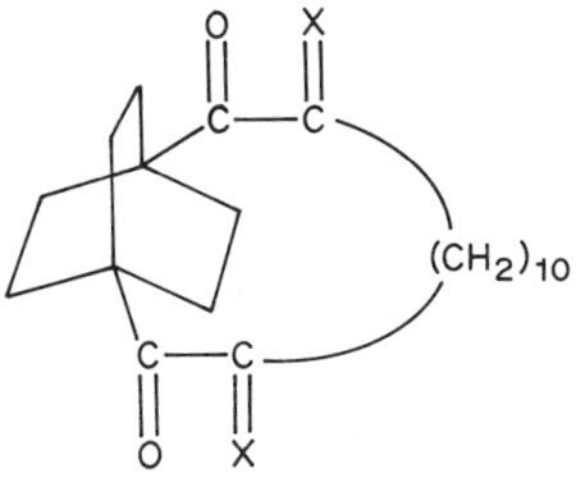

(51) X=H₂   (53) 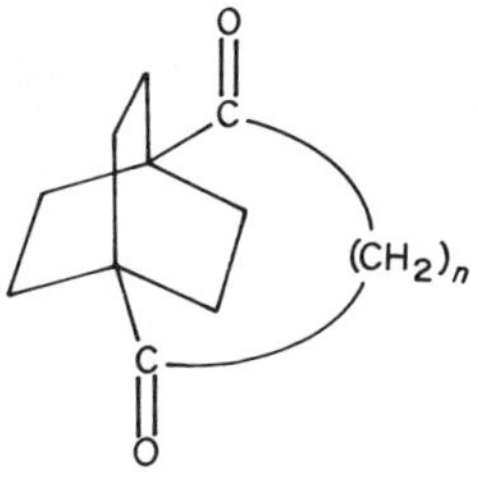   (54) n=10
(52) X=N₂   (55) n=8

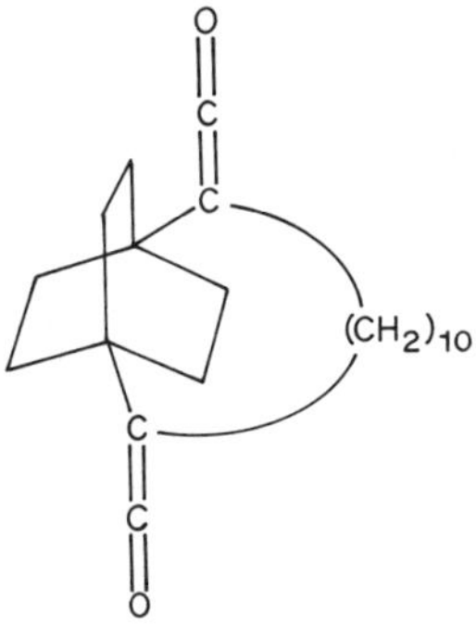

restriction in the series **51, 54, 55**, NMR spectra indicate that motion of the polymethylene chain relative to the [2.2.2]bicyclooctane unit is free in **51**, somewhat restricted and temperature-dependent in **54** and impossible in **55**. This third compound is also the lowest member of the homologous series that can be constructed with space-filling models and, of course, it should be possible to attempt additional ring contractions from **55** in an effort to reach still lower members of this series. Molecular mechanics calculations suggest that, already in **55**, there is some distortion of bond angles.

B. Fenestranes

The material in this section is simply organized by decreasing ring size of the fenestranes of specific interest in the investigation under discussion. Owing to the use of ring contractions in preparation of many of these compounds, more than one fenestrane ring system may be encountered in a particular section.

1. [5.5.5.7]Fenestranes

These large ring fenestranes are of interest because the only reported naturally occurring fenestrane is the diterpene lauren-1-ene (**56**). Since the structure was determined by crystallographic methods, bond distances and angles are available for the derived bromide **57**. Angles across the central atom are 117.9° $[C_{(1)}-C_{(8)}-C_{(7)}]$ and 118.9° $[C_{(4)}-C_{(8)}-C_{(9)}]$, reflecting a small measure of flattening, and bonds to $C_{(8)}$ are 1.527–1.609 Å, indicating no bond shortening in this system.

Three quite different syntheses have been reported for **56**[82-84]. In the first of these[82] the previously known ketol **58** was elaborated into keto aldehyde **59**; this cyclized to yield **60**, and introduction of the final methyl group and stepwise conversion of the carbonyl group to a double bond gave lauren-1-ene (**56**). The second synthesis[83] proceeds through intramolecular photochemical cycloaddition in the key intermediate **61** with formation of **62**. Conversion to **63** and then reduction, first with sodium in ammonia and then by hydrogenation of the side-chain double bond, furnished **64**, and this was converted from ester into aldehyde and cyclized to **60**. Once again, although by a different route, this was converted to the natural product. The third synthesis[84], which is due to Wender and coworkers, is the shortest and most efficient, and is conceptually remarkable. It proceeds through intermediate **66**, which is prepared by alkylation of lactone **65** with homoprenyl iodide. This is reduced with hydride to the corresponding lactol, and then, in the key step,

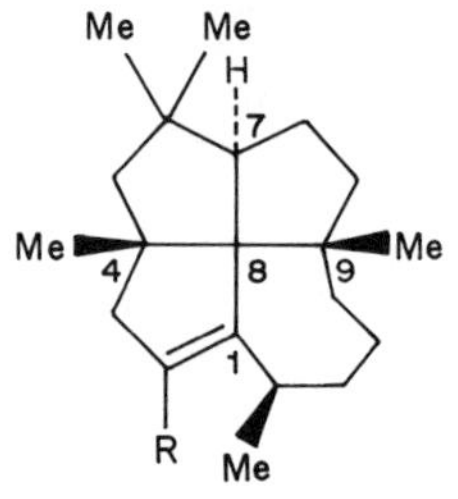

(56) R=H   **(58)**
(57) R=Br

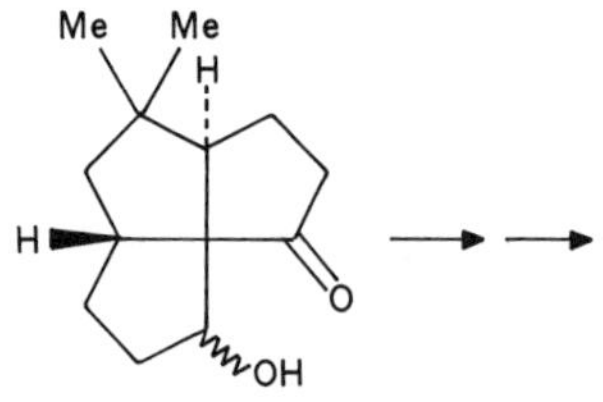

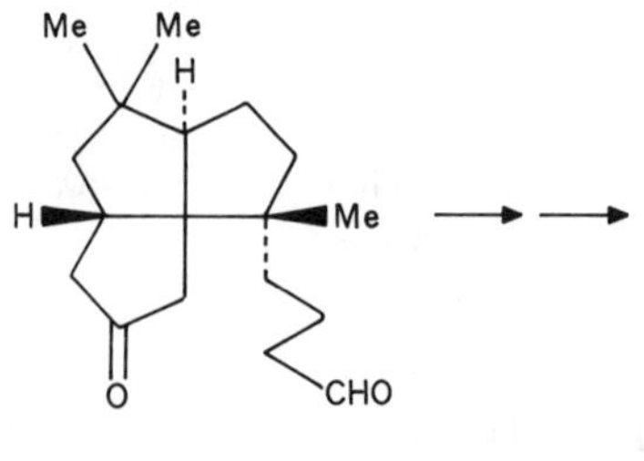

(59) 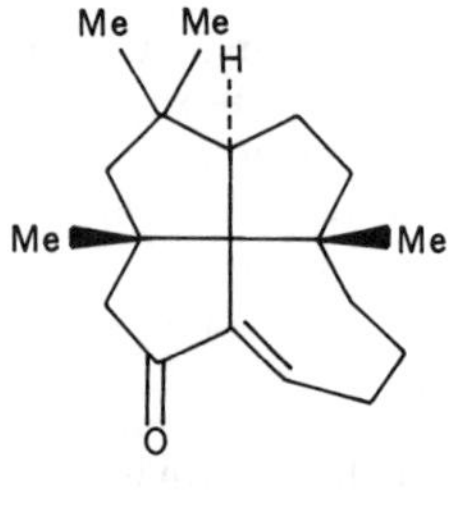   **(60)**

intramolecular photocycloaddition of the side chain of the lactol to the aromatic ring gives **67**. Unlike previous cases of this cycloaddition[85] (see Section III.B.2, for example), only a single vinylcyclopropane isomer was detected in this reaction, a result attributed to the greater strain in the alternative product. Reductive cleavage of the $C_{(3)}$—$C_{(5)}$ bond of the cyclopropane with lithium in methylamine and concomitant reduction of the lactol yielded diol **68**. Deoxygenation of **68** then furnished lauren-1-ene (**56**).

2. [5.5.5.5]Fenestranes

There has been greater synthetic activity in this series of fenestranes than in any other. The first simple member of the series to be prepared was tetraketone **69**[86]. This synthesis by Cook, Weiss and their coworkers is the prototype of several studies by these investigators. In the case at hand, **70** could be cyclized to **69** in good yield in hot cumene-diglyme containing a large quantity of 1-naphthalenesulfonic acid. An X-ray study confirmed the structure of **69** and showed that in the crystal the molecule exists in a chiral

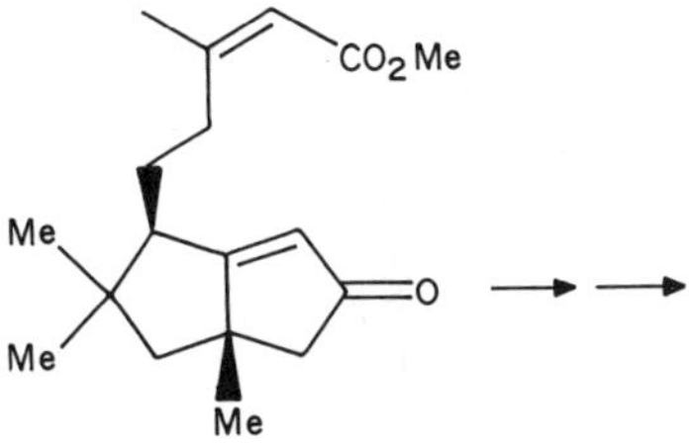

(61)

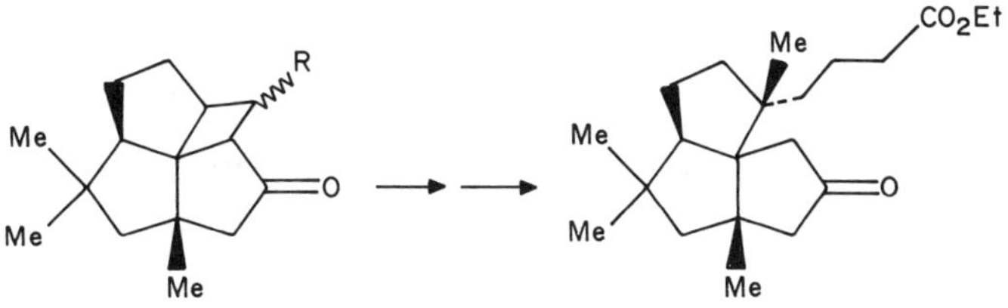

(62) R=CO₂Me
(63) R=CH=CHCO₂Et

(64)

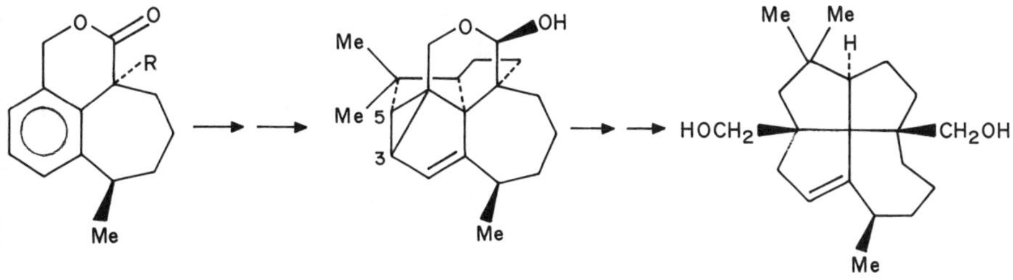

(65) R=H
(66) R=CH₂CH₂CH=CMe₂

(67)

(68)

conformation with each crystal composed of a single enantiomer. As one would expect, in this [5.5.5.5] system there is less flattening at the central atom $[C_{(13)}]$ than in fenestranes with four-membered rings to be discussed below. Angle $C_{(1)}$—$C_{(13)}$—$C_{(7)}$ is 117.5° and $C_{(4)}$—$C_{(13)}$—$C_{(10)}$ is 115.1°; bond lengths at $C_{(13)}$ range from 1.527 Å $[C_{(7)}$—$C_{(13)}]$ to 1.568 Å $[C_{(10)}$—$C_{(13)}]$. The stereochemistry of 69 $(1\alpha,4\beta,7\alpha,10\beta)$ is such that the approach of a nucleophile from either side of the molecule to one of the carbonyl groups is sterically equivalent to the hindered *endo* approach in a *cis*-bicyclo[3.3.0]octanone. This unavoidable steric interaction nicely rationalizes the relative stability of 69 to base-catalyzed ring opening of its two β-dicarbonyl groupings[87]. Treatment with methoxide leads to slow, regiospecific cleavage, forming only 71 and none of its spiro-fused isomer. Finally, reduction of 69 with diborane in tetrahydrofuran gave a diastereomeric mixture of tetrahydroxy compounds that was dehydrated by hot hexamethylphosphoramide to provide $(1\alpha,4\beta,7\alpha,10\beta)$-2,5,8,11-[5.5.5.5]fenestratetraene (72) along with a smaller amount of the isomeric bridgehead alkene 73[88]. Proton and carbon nuclear magnetic resonance spectra of 72 are consistent with its expected D_{2d} symmetry. These two hydrocarbons are the most highly unsaturated fenestranes presently known.

Keese has explored quite different routes to [5.5.5.5]fenestranes. The first of these proceeds from lactone 74, which is available in several steps from 1,5-cyclooctadiene. This lactone was elaborated into 75, and this diester underwent Dieckmann cyclization to give, after decarboxylation, cyclooctanone 76. Photolysis of the potassium salt of the derived tosylhydrazone then gave a carbene that inserted in the nearer tertiary carbon–hydrogen bond, furnishing the pentacyclic fenestrane lactone 77. In an unusual transformation 77 was converted to (13s,1α,4β,7α,10β)-[5.5.5.5]fenestrane (78) on heating at 310 °C for 4.5 hours in the presence of palladium-on-carbon and hydrogen[89]. Without added hydrogen, 77 is essentially stable under these conditions; with hydrogen present, the product is a single hydrocarbon accompanied by *ca* 10% starting material. This hydrocarbon had spectroscopic properties appropriate for a [5.5.5.5]fenestrane, and its stereochemistry was assigned as depicted in 78 on the basis of its ^{13}C-NMR spectrum. This consisted of a three-line pattern compatible only with the symmetry of a 1α,4β,7α,10β fenestrane. In principle, there are two such isomers, 78 (13s) and the contorted compound with inverted stereochemistry (13r) at the central carbon atom. Since the calculated strain energy of the latter species is *ca* 182 kcal mol^{-1}, Keese concluded that the reaction conditions leading to this product assured that it was the relatively strain-free 78 rather than its highly unstable isomer.

A second route to 78 was first attempted from 79[90,91]. Unlike 75 above, 79 did not cyclize in the Dieckmann reaction, so it was converted to the corresponding dinitrile 80 for Ziegler–Thorpe ring closure. This cyclization was successful, and exposure of the resulting enamino nitrile to acid hydrolysis and then Jones oxidation yielded ketolactone 81. In this case the carbene insertion failed. These differences in the fates of closely related carbenes presumably reflect changes in dynamic conformation as a function of specific structure[92]. In contrast to the carbene insertion, the thermal reaction in the presence of palladium-on-carbon used previously on 77 did succeed with 81, giving directly a modest yield of 78.

In an attempt to reach an isomeric [5.5.5.5]fenestrane, Luyten and Keese prepared epimeric ketolactone 82 from dicyclopentadiene. Here decomposition of the tosylhydrazone salt led only to olefin 83. Once again the high-temperature palladium and hydrogen process was effective, but the product was 78 rather than the desired $(1\alpha,4\alpha,7\alpha,10\beta)$-[5.5.5.5]fenestrane[51]. According to molecular mechanics calculations this latter unknown isomer should be about 13 kcal mol^{-1} more strained than 78[38].

Keese and coworkers have also used the intermolecular arene–alkene photocycloaddition reaction as a key step in another route to [5.5.5.5]fenestranes[93]. Irradiation of 84 gave a mixture of products from which the desired adduct 85 could be isolated. Structure and stereochemistry of this tricyclic ester are supported both by earlier experience with the

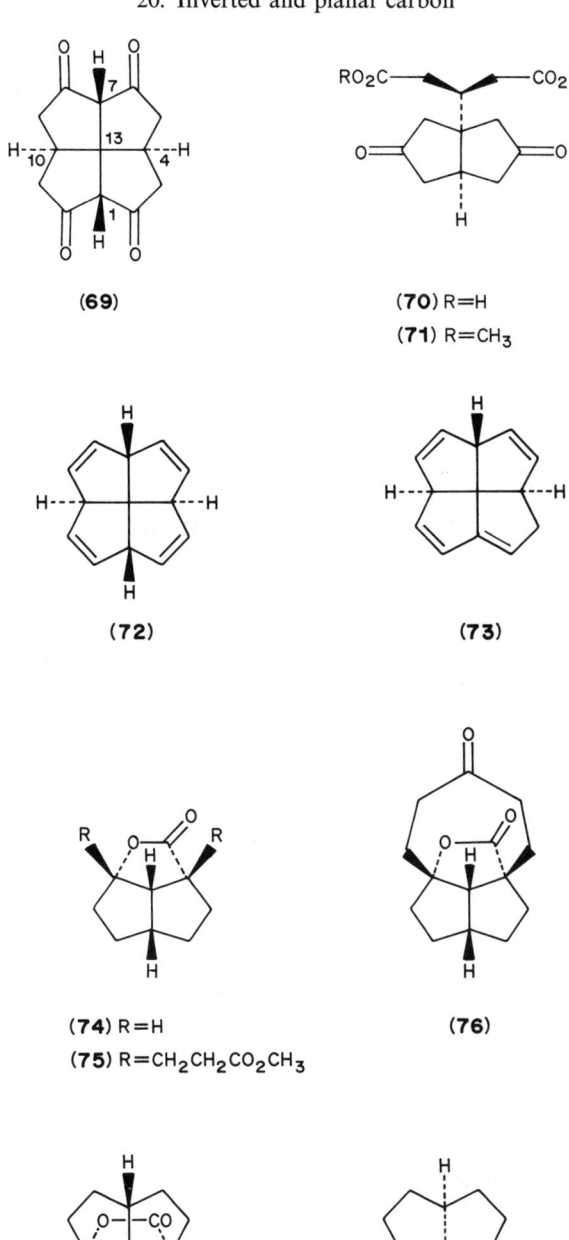

(69)

(70) R=H
(71) R=CH₃

(72)

(73)

(74) R=H
(75) R=CH₂CH₂CO₂CH₃

(76)

(77)

(78)

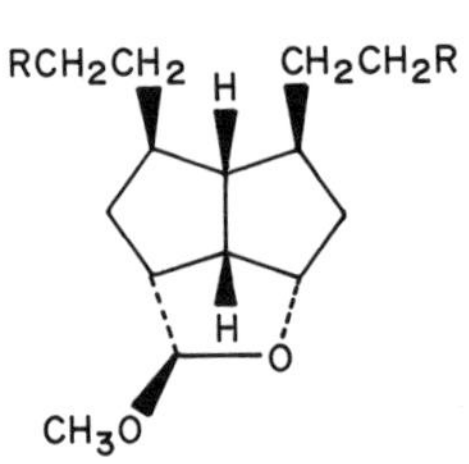

(79) R=CO₂CH₃
(80) R=CN

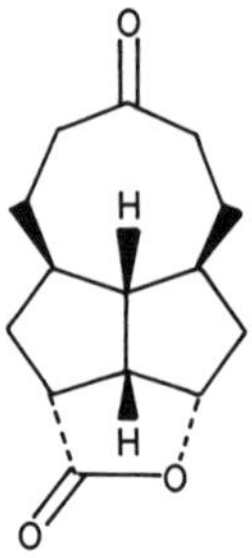

(81)

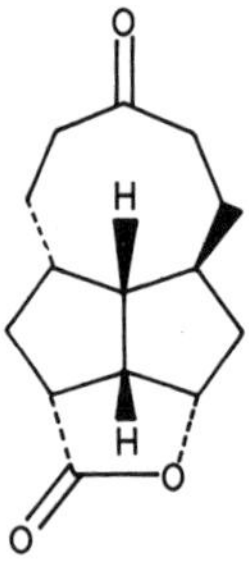

(82)

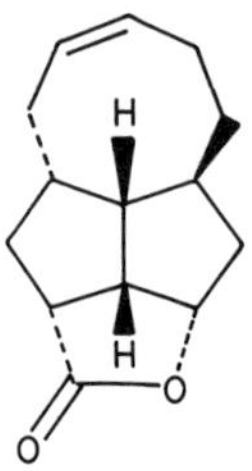

(83)

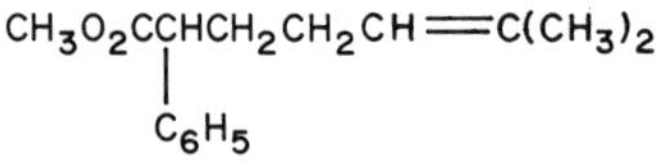

(84)

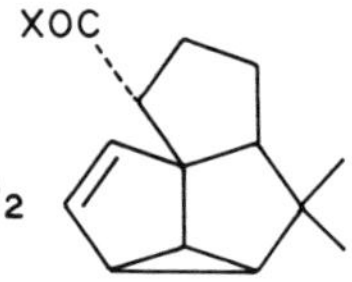

(85) X=CH₃O
(86) X=N₂CH

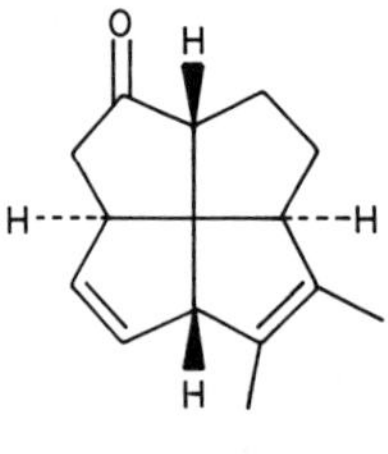

(87)

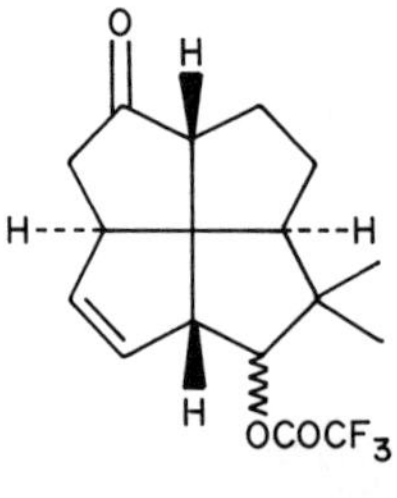

(88)

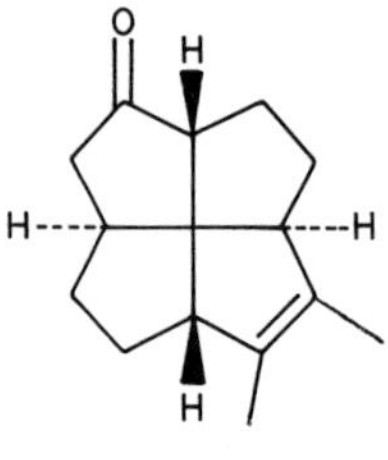

(89)

photochemistry of 5-phenyl-1-pentenes[85] and by NMR spectra. The corresponding diazoketone **86** was prepared by way of the mixed anhydride with isovaleric acid, and on treatment with trifluoroacetic acid this lost nitrogen and closed to a mixture (2.2:1) of fenestradienone **87** and trifluoroacetoxyfenestrenone **88**. The structures of these products were established by proton and ¹³C NMR spectra. Hydrogenation of **87** over palladium-on-carbon in methanol selectively reduced only the less substituted double bond and furnished **89**.

A totally different approach has led to the symmetrical tetrabenzo-[5.5.5.5]fenestratetraene **90**[94], which is available in gram amounts in nine steps from 1,3-indanedione (**91**) and dibenzylideneacetone (**92**)[95]. The first key reaction in this synthesis by Kuck and Bögge[94] is the Friedel–Crafts cyclization of triol **93** to **94**, catalyzed by phosphoric acid. Alcohol **94** is converted to **95** by way of Favorskii ring contraction of the derived ketone, followed by decarboxylation. Cycloaddition of **95** with tetrachloro-thiophene S,S-dioxide (**96**)[96] furnishes **97**, which is reduced by sodium and alcohol to **90**. The X-ray structure of **90** indicates envelope-like bending of each cyclopentene ring, so that the molecular symmetry is S_4 rather than D_{2d}. The flattening of the central quaternary carbon is similar to that of **69** with angle $C_{(1)}$—$C_{(13)}$—$C_{(7)}$ equal to $116.5°$.

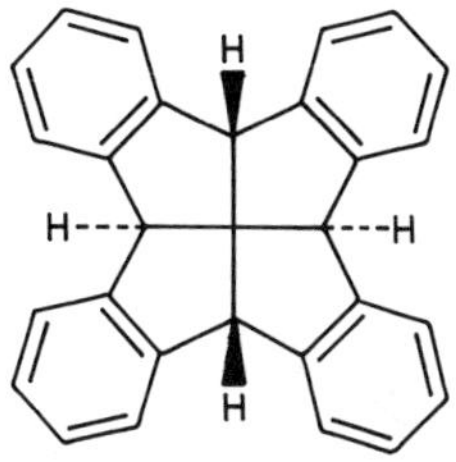

(90) 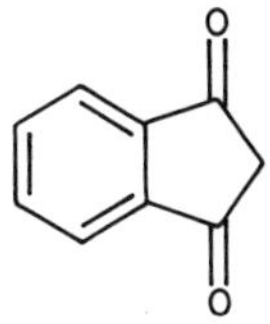   (91) 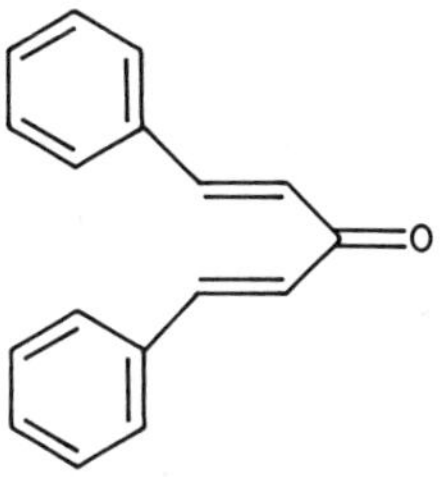   (92)

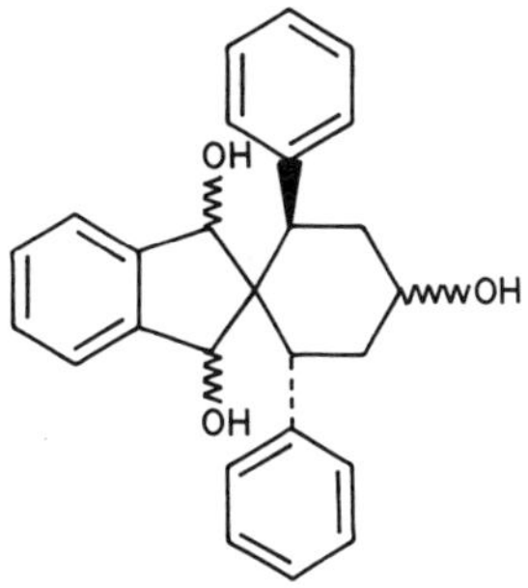

(93) 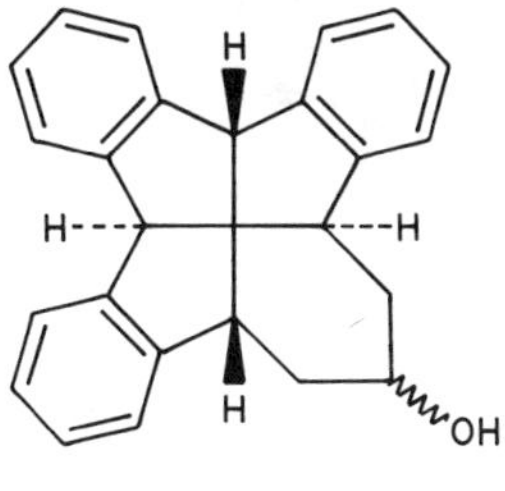   (94)

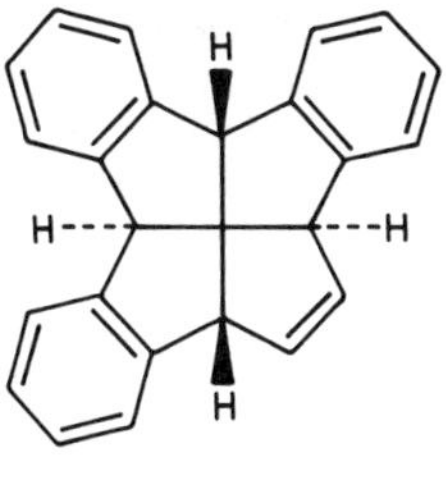

(95) 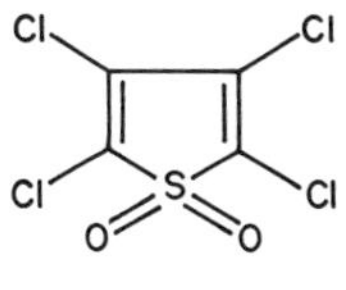   (96) 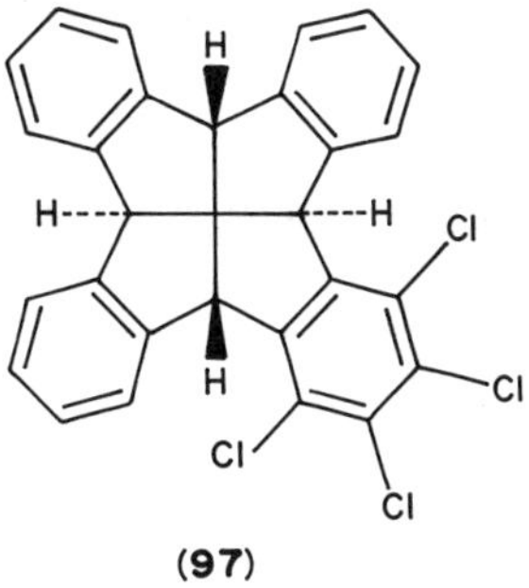   (97)

W. C. Agosta

3. [4.4.5.5]Fenestranes

Several $(1\alpha,3\beta,6\alpha,9\beta)$-[4.4.5.5]fenestranes are known, including the parent hydrocarbon[97], but the only compound for which structural information is available is keto amide **98**, which was prepared by Dauben and his coworkers from **99**[19,98]. Irradiation of **99** yielded the [4.5.5.5]fenestrane **100**; ring contraction by way of the α-diazo ketone gave **101**, and removal of the silyl protecting group and oxidation afforded two keto esters. These were separated, and the major isomer was converted to **98** for X-ray analysis. The resulting structure revealed considerable flattening at $C_{(11)}$; angle $C_{(1)}$—$C_{(11)}$—$C_{(6)}$ is 128.2° and $C_{(3)}$—$C_{(11)}$—$C_{(9)}$ is 123.0°. These observed values are in good agreement with those from MNDO calculations[9,19,50].

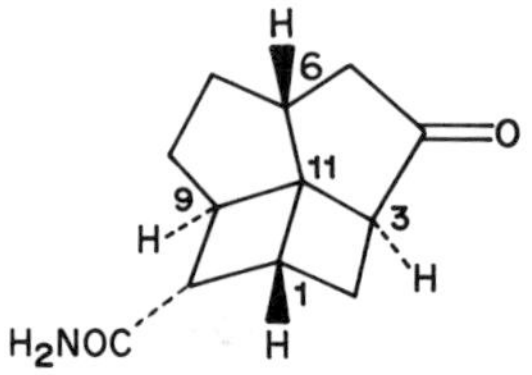

(98) 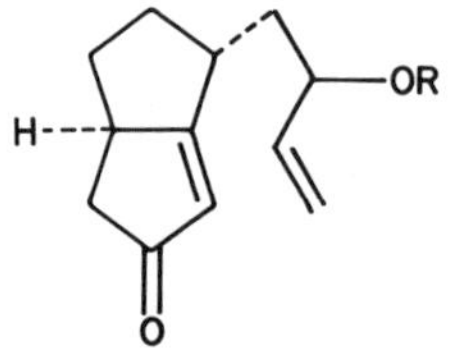   **(99)**

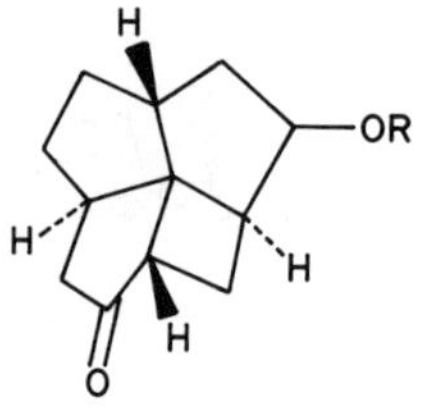

(100) 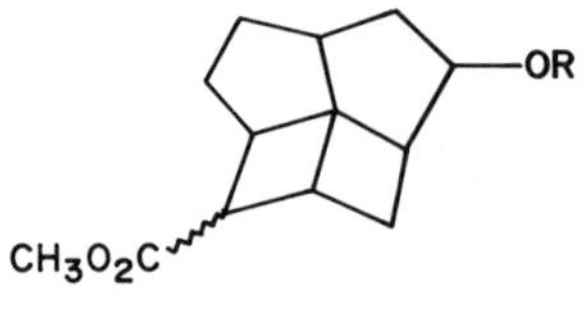   **(101)**

$R = Si(CH_3)_2C(CH_3)_3$

4. [4.4.4.5]Fenestranes

The [4.4.4.5]fenestranes are currently the tetracyclic fenestranes of smallest ring size and most flattened central carbon atom that have been synthesized and studied experimentally. Representatives of this series are accessible from keto ester **102**[99]. Photochemical [2 + 2] cyclization of **102** led regiospecifically to **103**, and this was converted to diazoketone **104** after formation of the ethylene ketal. The keto carbene formed on exposure of **104** to rhodium acetate cleanly inserted into the only readily accessible carbon–hydrogen bond, yielding the [4.4.5.5]fenestrane **105**. Removal of the carbonyl group of **105** and deketalization gave **106**. This was converted to the α-diazoketone, and photochemical Wolff rearrangement in methanol then yielded the [4.4.4.5]fenestrane methyl ester **107**. This compound was thermally reasonably stable, surviving gas chromatography at 130 °C for a half hour with only slight decomposition. The derived p-bromoanilide **108** furnished crystals suitable for X-ray analysis.

There are several noteworthy features in the crystallographic studies on **108**. The angles $C_{(1)}$—$C_{(10)}$—$C_{(6)}$ and $C_{(3)}$—$C_{(10)}$—$C_{(8)}$, which reflect the enforced flattening at $C_{(10)}$, are 128.3° and 129.2°, respectively. Two of the bond distances involving $C_{(10)}$ are quite short,

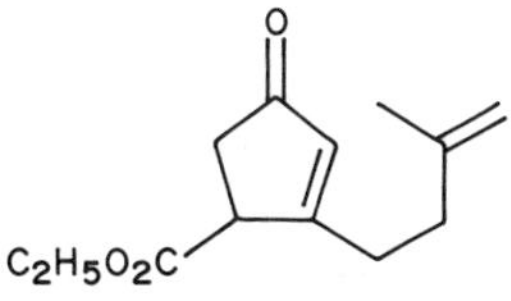

(102)

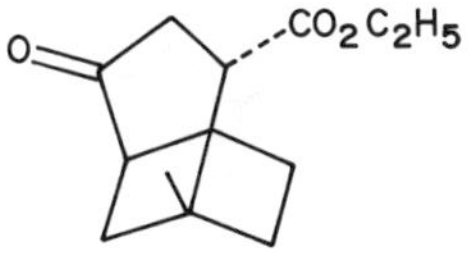

(103)

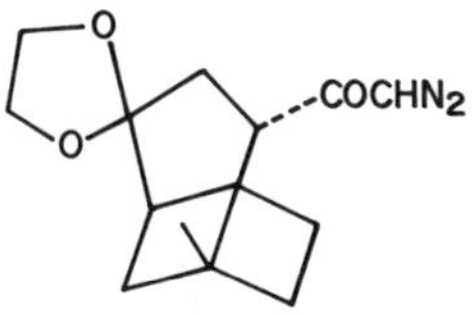

(104)

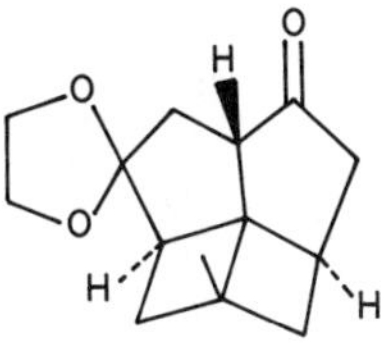

(105)

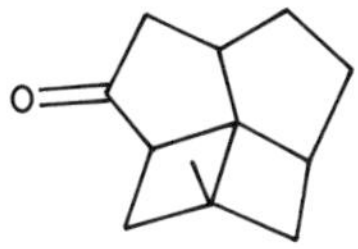

(106)

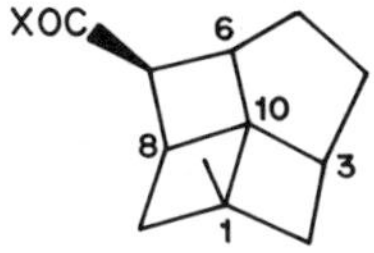

(107) X=CH₃O

(108) X= p-BrC₆H₄NH

with both $C_{(3)}$—$C_{(10)}$ and $C_{(6)}$—$C_{(10)}$ only 1.49 Å. The perimeter bonds for the cyclobutane rings are correspondingly lengthened to an average value of 1.574 Å. These values are in quite good agreement with those given by MNDO calculations[9,19,50]. The shortened bonds to $C_{(10)}$ are also in line with the *ab initio* calculations for [4.4.4.4]fenestrane (23) discussed above, although in 108 the two shortest bonds to $C_{(10)}$ are associated with the five-membered ring.

The chemical effects of these skeletal distortions appear in the thermal and photochemical behavior of the [4.4.4.5]fenestrane ring system[100]. The observed reactions reveal a significant role in these processes for the short, weak bonds to the central carbon atom. On heating to 100 °C in benzene keto ester 109, which was prepared from keto ketal 105, undergoes isomerization to two products, 110 (15%) and 111 (45%). The formation of 110 appears to be a symmetry allowed [$_\pi 2_s + _\pi 2_s + _\pi 2_s$] thermal cycloreversion[101] proceeding directly from 109 to the product, as shown in the structure with arrows. The fenestrane skeleton imposes on the six reacting carbon atoms [$C_{(1)}$ and $C_{(6)}$ to $C_{(10)}$] a rigid boat cyclohexane geometry that is ideal for such concerted fragmentation[101].

Structure 111 for the second thermolysis product was substantiated by an X-ray crystallographic study. There is evidence that the mechanism of this singular transformation combines homolytic and heterolytic steps. Homolysis of the $C_{(8)}$—$C_{(10)}$ bond in 109 gives 112, which fragments to 113, and under catalysis by adventitious acid this undergoes a 1,2 alkyl shift to form 111. In keeping with this mechanism, thermolysis of 109 in benzene at 100 °C as before, but in a stirred solution containing solid sodium bicarbonate, yielded 110 but no 111, and when the thermolysis was carried out in methanol rather than benzene as solvent, the product was 114, the compound expected[102] from conjugate addition of methanol to 113. The initial major thermal product from 109 then is 113, which does not survive the reaction conditions but can be trapped as 111 or 114. We see then that the thermal rearrangements of 109 to 110 and 113 involve rupture of either $C_{(1)}$—$C_{(10)}$ or $C_{(8)}$—$C_{(10)}$, the two short central bonds expected to be weakest in the [4.4.4.5]fenestrane ring system[44,103].

The photochemistry of 109 probes reactions that can be initiated by electronic excitation of the ketone carbonyl. Irradiation of 109 through Pyrex ($\lambda > 280$ nm) in

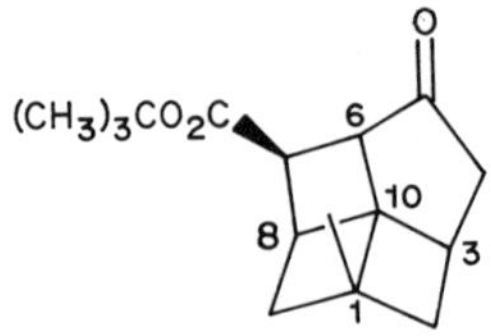

(109)

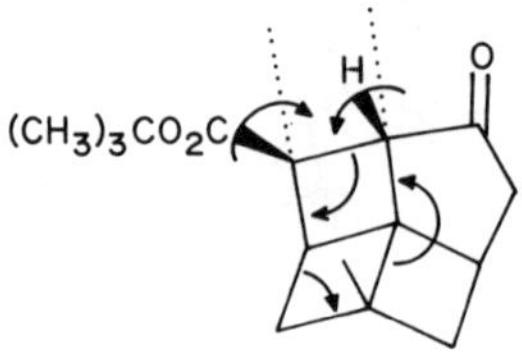

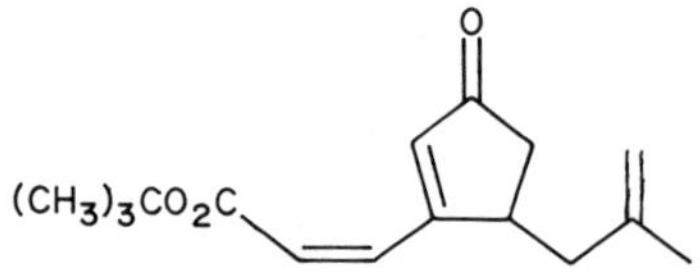

(110)

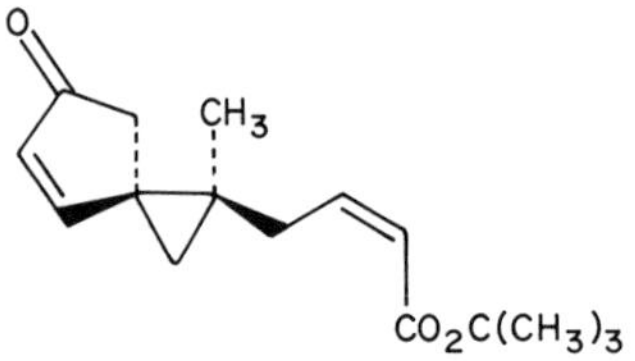

(111)

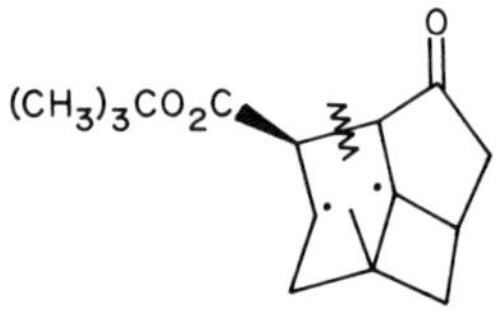

(112)

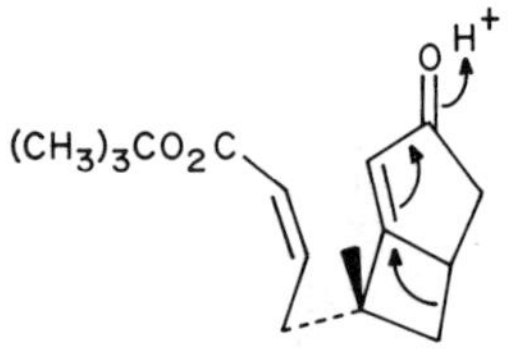

(113)

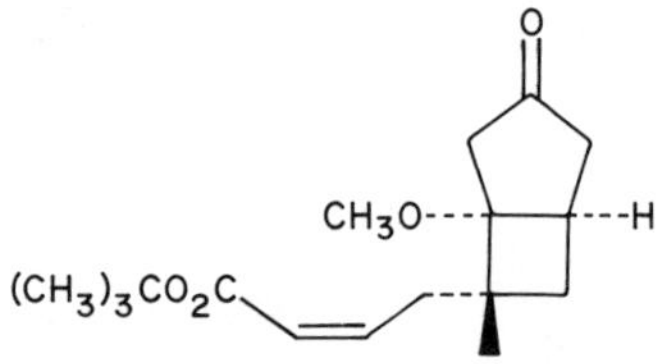

(114)

benzene containing *ca* 7% methanol led to three isolated products, the *Z* and *E* unsaturated esters **114** and **115** (*ca* 5% each) and aldehyde **116** (36%). Absorption of light leads to an nπ^* state of the ketone carbonyl and then β cleavage of the $C_{(6)}$—$C_{(7)}$ bond to form biradical **117**; after conformational relaxation this can cleave to both **113** and **118**. Subsequent Michael addition of methanol yields **114** and **115**. Such β cleavage is not unusual in the photochemistry of cyclopropyl ketones, where its occurrence is attributed to strain. The strain inherent in **109** apparently is sufficient to permit β cleavage of a cyclobutane, although simple acylcyclobutanes do not behave in this fashion.

Formation of **116** requires a cleavage of **109** toward the adjacent four-membered ring (cf **119**), subsequent homolysis of the $C_{(1)}$—$C_{(10)}$ bond (cf **120**) and then hydrogen transfer. This mechanistic sequence raises an interesting point. Typically, type I cleavage of cyclobutyl ketones takes place preferentially away from the four-membered ring or else gives both possible products, as formation of a cyclobutyl radical is somewhat disfavored[104]. Cleavage of **109** toward the cyclobutane suggests that the bond suffering initial cleavage here [$C_{(5)}$—$C_{(6)}$] is relatively weak. In these photochemical reactions we see that short, weak bonds to the central carbon atom cleave in the second step, following light-promoted α or β cleavage of the ketone.

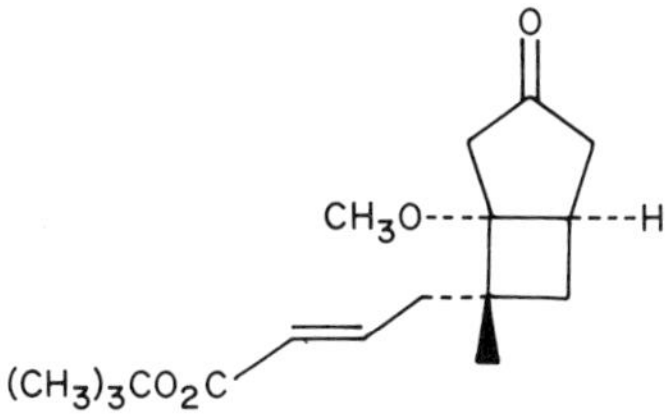

(115)

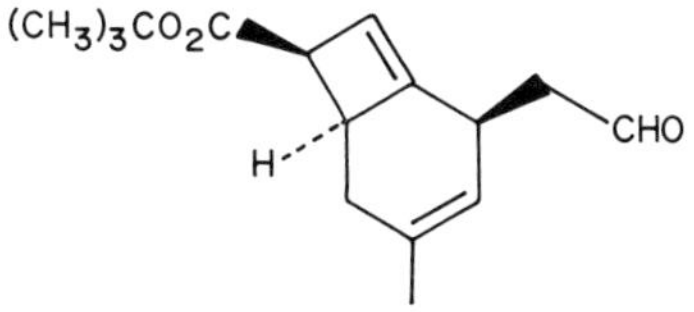

(116)

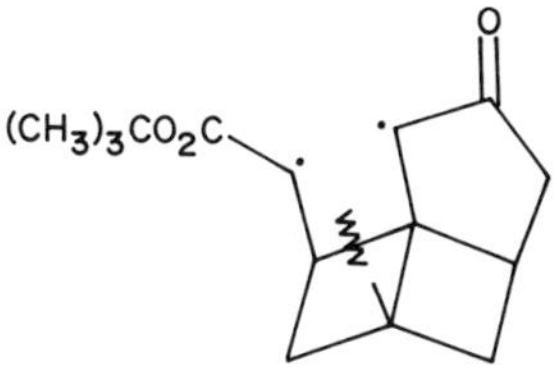

(117)

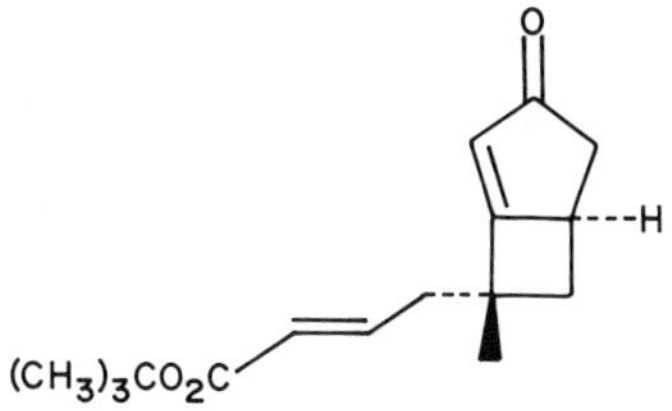

(118)

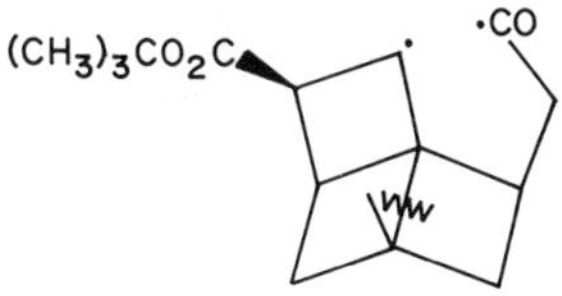

(119)

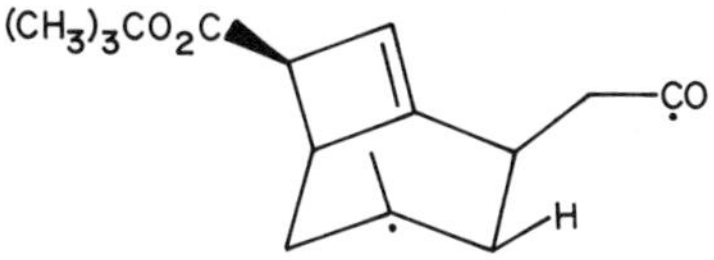

(120)

C. Propellanes

1. General

We discuss preparation and properties of relevant propellanes according to ring size. For the larger ring systems frequently there has been an X-ray crystallographic study, showing that the system does contain inverted carbon. Typically these compounds undergo rapid free radical reactions with opening of the central propellane bond, and such transformations are not specifically mentioned in all cases.

2. [4.2.1]Propellanes

Elimination of HCl from 1-cholorquadricyclane (121) leads to transient olefin 122, which can be captured by anthracene to furnish the highly substituted [4.2.1]propellane 123. The X-ray structure of 123 indicates that the bridgehead carbon atoms of the propellane system are very slightly inverted; the plane of the three starred substituents of the nearer bridgehead passes 0.05 Å behind the bridgehead. The necessary rehybridization is apparently sufficient to lead to typical propellane reactions, and 123 adds thiophenol readily to form 124[105]. Related adducts form with substituted anthracenes. The parent [4.2.1]propellane is also known and is little more reactive than bicyclo[2.1.0]hexane from which it is formally derived[106].

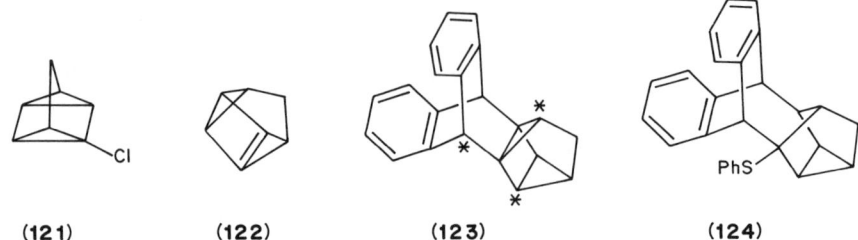

(121)   **(122)**   **(123)**   **(124)**

3. [4.1.1]Propellanes

The parent [4.1.1]propellane (**125**) is one of the products of pyrolysis of the sodium salt of tosylhydrazone **126**[107]. This carbene insertion gives a mixture of methylenecyclo-heptenes and the desired propellane. While **125** has been characterized spectroscopically, it is quite unstable. Its solutions are stabilized by addition of a radical inhibitor.

Szeimies and his coworkers have prepared a number of more stable and structurally more complex [4.1.1]propellanes using two related approaches. The bridgehead positions of bicyclo[1.1.0]butanes are somewhat acidic, so that **127** can be selectively lithiated and

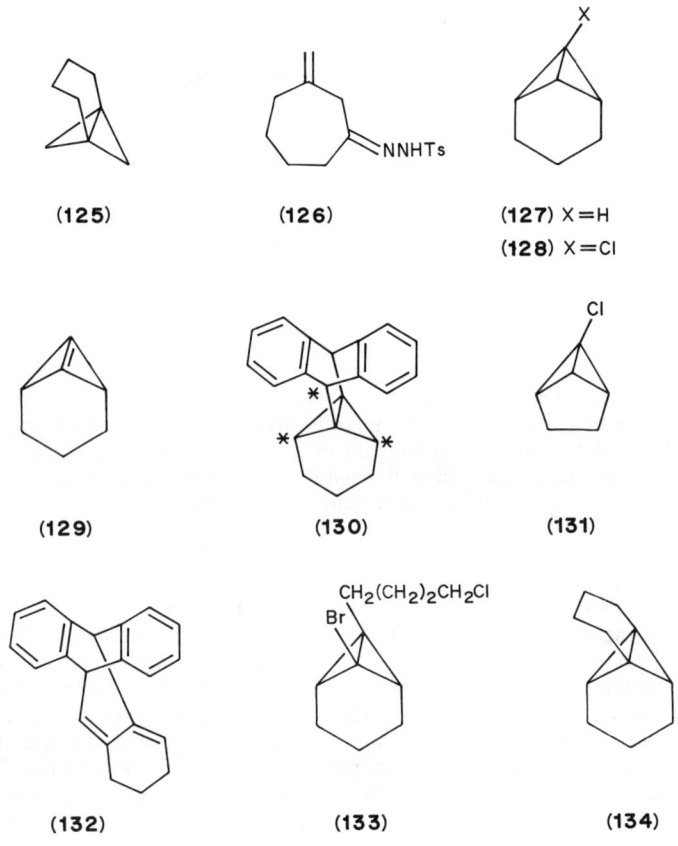

(125)   **(126)**   **(127)** X=H
                                              (128) X=Cl

(129)   **(130)**   **(131)**

(132)   **(133)**   **(134)**

then converted to, for example, **128**. The first approach is similar to that described above for preparation of **123**. Reaction of **128** with strong base leads to the very reactive olefin **129**, which is trapped in the presence of added anthracene as the propellane **130**[108]. Similar reactions occur with **128** and various other anthracenes substituted at $C_{(9)}$ or both $C_{(9)}$ and $C_{(10)}$, and analogous sequences have also been carried out with other bridged bicyclo[1.1.0]butanes, such as **131**[109,110]. X-ray crystallographic studies show that the three starred substituents of **130** define a plane that passes between the two bridgehead atoms 0.31 Å behind the near bridgehead[111]. The same extent of inversion is present in the homologous adduct from anthracene and **131**[109]. These various [4.1.1]propellanes undergo a general thermal rearrangement that involves cleavage of the central propellane bond. Thermolysis of **130**, for example, at 180 °C gives **132**[112].

In Szeimies's second propellane synthesis **127** was converted to **133** by way of organometallic intermediates. Treatment with butyllithium then gave **134**, which is stable to storage in the cold[113].

A second intramolecular carbene addition similar to that used for **125** has provided another type of complex [4.1.1]propellane. Pyrolysis of the dry sodium salt of **135** furnished the rather stable **136** in 75% yield[114]. This hydrocarbon was recovered

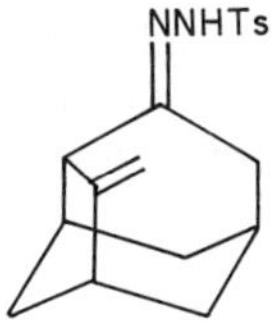

(135) 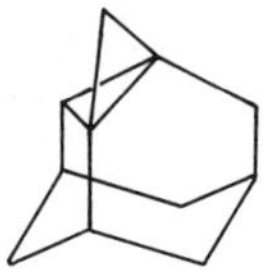   (136) 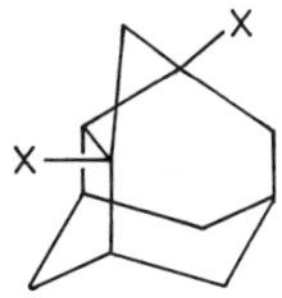   (137) X = SMe
                              (138) X = H

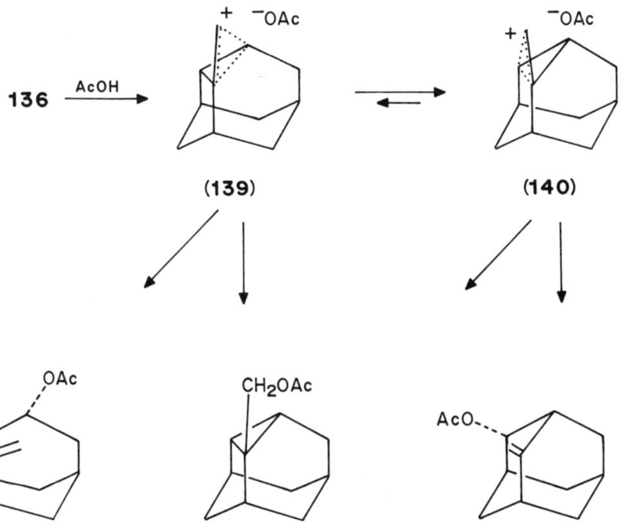

136 $\xrightarrow{\text{AcOH}}$

(139)   (140)

(141)   (142)   (143)

unchanged after 24 hours at 80 °C and is inert toward nucleophiles at room temperature. It reacts rapidly, however, with electrophiles and free radicals. Dimethyl disulfide, for example, adds instantaneously to yield 137. Reduction of 136 by lithium in ethylamine–hexane cleaves the central propellane bond in an electron transfer process to give 80% of 138. The two inverted carbons of 136 are structurally nonequivalent, and for this reason the behavior of this propellane on protonation is of some interest. Glacial acetic acid adds rapidly and essentially quantitatively to 136 to furnish a mixture of 141 (15%), 142 (25%) and 143 (60%). Majerski and Zuniac have suggested that these products arise from the rapidly equilibrating bicyclobutonium ions 139 and 140[114]. The observation that products from protonation of only one of the two inverted carbons were isolated in this reaction was taken as evidence for a difference in electron density at these two sites.

4. [3.3.1]Propellanes

Although [3.3.1]propellane itself is not expected to have an inverted carbon atom, one of the earliest known compounds that does is its derivative 1,3-dehydroadamantane (144). This hydrocarbon is prepared by debromination of 1,3-dibromoadamantane (145) with sodium–potassium dispersion[115]; it is unchanged after three days at 100 °C in degassed heptane, but the cyclopropane ring is quite reactive with nucleophiles and molecular oxygen, even at room temperature. In aqueous acid 144 rapidly gives 1-adamantanol, and treatment with bromine in heptane yields 145. Bromine in ether at − 70 °C reacts with 144 to form a yellow precipitate formulated as the oxonium tribromide 146. On warming and isolation this yields 145 and the bromo ether 147[115].

5. [3.2.1]Propellanes

The parent [3.2.1]propellane 148 is available through carbene addition to 1(5)-bicyclo[3.2.0]heptene 150[7,15,116,117]. This was the first compound with inverted carbon

(144)

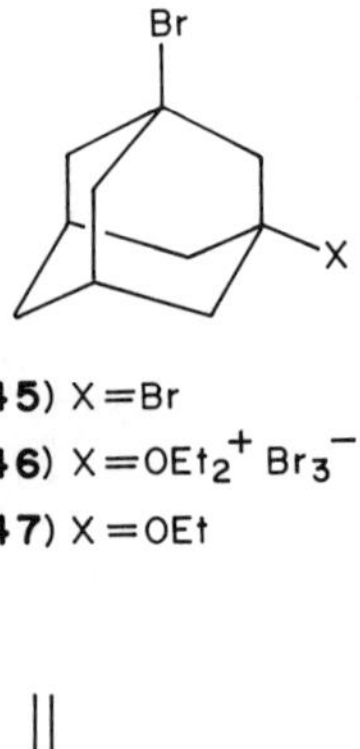

(145) X=Br
(146) X=OEt₂⁺ Br₃⁻
(147) X=OEt

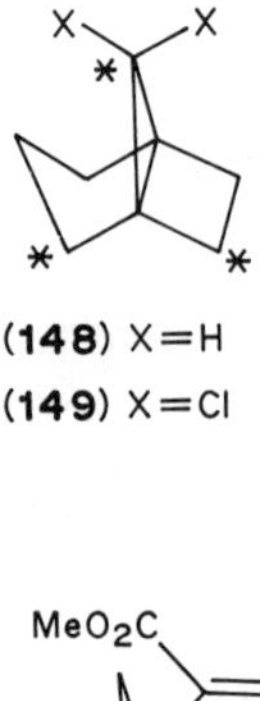

(148) X=H
(149) X=Cl

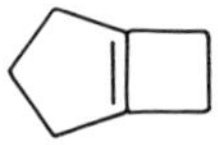

(150)

(151)

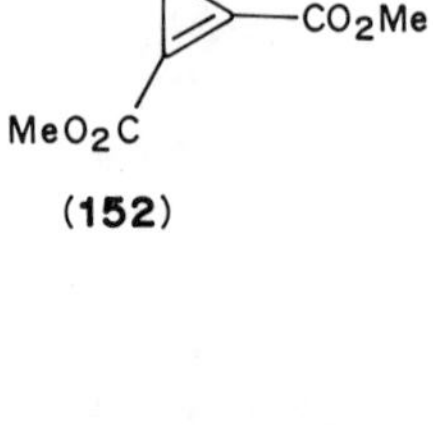

(152)

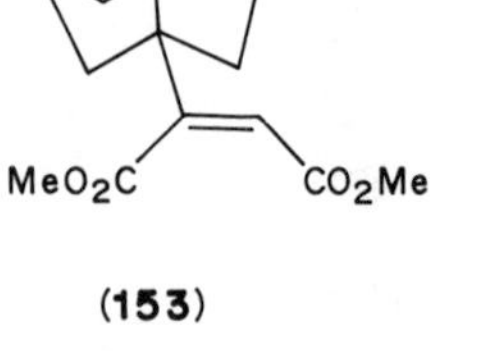

(153)

to be prepared, and the X-ray structure of **149**, which is the product of addition of dichlorocarbene to **150**, indicated that the plane of the starred carbon atoms passes 0.1 Å behind the nearer bridgehead[11]. While **148** is thermally stable to 320 °C, where it fragments to 1,3-dimethylenecyclohexane (**151**), it reacts rapidly at room temperature with acetic acid, oxygen and free radicals. Dimethyl acetylenedicarboxylate adds at 25 °C to form **152** and **153**[118]. Similar reactions take place with the tetracyclic [3.2.1]propellane **154**[118].

Several heterocyclic [3.2.1]propellanes are available through addition of furans and 1,2,3-trimethylisoindole (**155**) to **122**. Furan, for example, yields **156**[105].

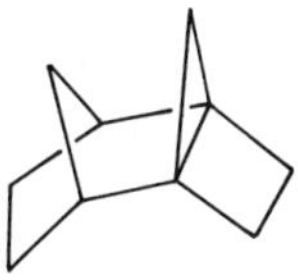

(**154**) 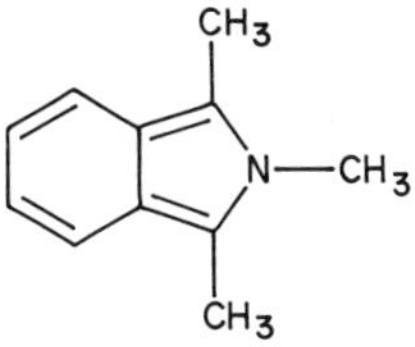   (**155**) 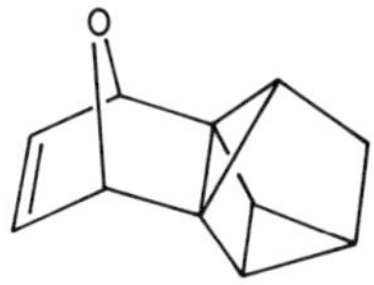   (**156**)

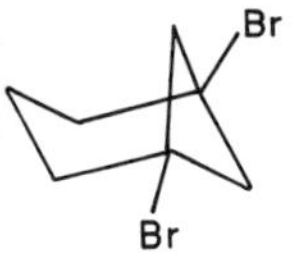

(**157**)    (**158**)    (**159**) 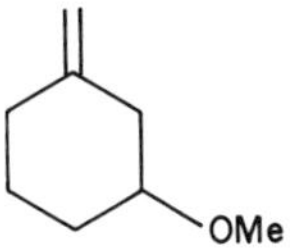   (**160**)

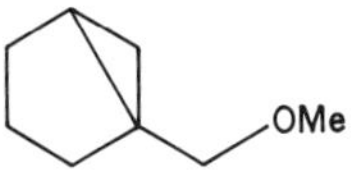

(**161**) 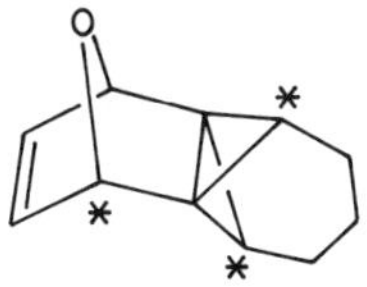   (**162**) 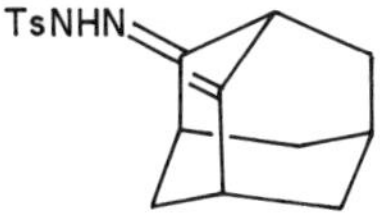   (**163**) 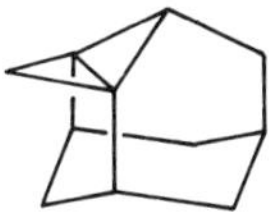   (**164**)

6. [3.1.1]Propellanes

Treatment of the bridgehead dibromide **157** with sodium in refluxing triglyme (216 °C), with direct distillation of the product into a cold trap, furnishes 75% of the parent [3.1.1]propellane (**158**)[119]. This compound is stable in toluene solution at room temperature but polymerizes in the absence of solvent. It reacts at − 78 °C with methanol to give largely a mixture of the three isomeric ethers **159**, **160** and **161**.

Numerous [3.1.1]propellanes have been prepared through trapping of transient bicyclo[1.1.0]butenes such as **129** with furans, cyclopentadienes and isoindole (**155**)[108,110,112,120]. The furan adduct with **129** is **162**, the X-ray structure of which reveals that the plane of the starred atoms passes 0.40 Å behind the bridgehead atom[121].

Using the intramolecular carbene insertion route, Mlinarić-Majerski and Majerski have prepared two [3.1.1]propellanes. Pyrolysis of the sodium salt of **163** provides **164** in 70% yield[122]. This hydrocarbon decomposes only slowly at room temperature in benzene, but reacts rapidly in radical reactions with methanol and with oxygen[122,123]. With acetic acid, **164** gives products similar to those formed from the related [4.1.1]propellane **136**. Majerski has suggested that these reactions reflect high electron density at the back side of the bridgehead atoms and diradicaloid character at these centers[122]. The photoelectron spectrum of **164** has been interpreted as indicating bonding character in the HOMO and, as noted above, MNDO calculations support this conclusion[69].

Vincović and Majerski prepared a second [3.1.1]propellane **165** from pyrolysis of **166**[124]. This system is more strained than the parent **158**, decomposing in 30 minutes at room temperature in benzene. It reacts with HCl at $-80\,^{\circ}$C to yield only the nortricyclylmethyl chloride **167**. Although this appears to involve external opening, rather than cleavage of the central propellane bond, protonation of the central bond to give **168** followed by cyclobutyl to cyclopropylcarbinyl rearrangement can also provide **167**.

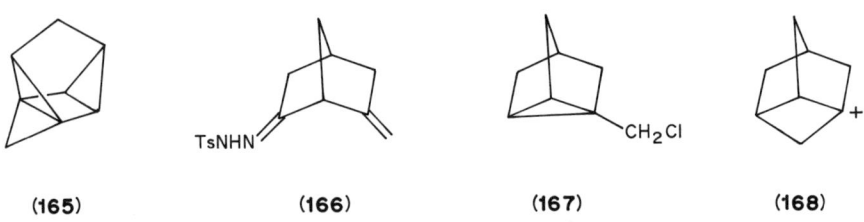

| (165) | (166) | (167) | (168) |

7. [2.2.1]Propellanes

[2.2.1]Propellane (**169**) is much less stable than the larger ring systems discussed above. Its preparation involves dehalogenation of 1,4-diiodonorbornane (**170**) by potassium in the gas phase, with direct trapping of the product at 20 K in an argon matrix[65]. In line with calculations mentioned above[57,63], the central bond appears to be quite weak; upon warming to 50 K and softening of the argon matrix, **169** polymerizes. It reacts with bromine to form **171**.

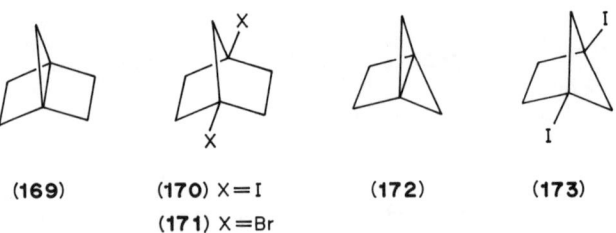

| (169) | (170) X = I | (172) | (173) |
| | (171) X = Br | | |

8. [2.1.1]Propellanes

Preparation and properties of [2.1.1]propellane (**172**) are analogous to those of **169** just described. In this case 1,4-diiodobicyclo[2.1.1]hexane (**173**) served as precursor to the hydrocarbon[64].

The dehalogenation approach used successfully for preparation of **134** from **133** was

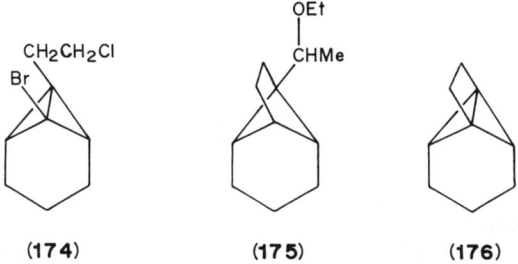

(174)   (175)   (176)

also applied to **174**. Reaction of this intermediate with butyllithium in ether led to **175**, presumably by way of the desired [2.1.1]propellane **176**, which then added solvent in a radical reaction.

9. [1.1.1]Propellanes

The original preparation of [1.1.1]propellane (**177**) involved Hunsdiecker degradation of bicyclo[1.1.0]pentane-1,3-dicarboxylic acid (**178**) to the dibromide **179** and then treatment of **179** with an alkyllithium[63]. Unfortunately, **178** was only accessible with difficulty[125], and [1.1.1]propellane remained effectively unavailable until Szeimies and coworkers reported a remarkable two-step synthesis in 1985[126]. The commercially

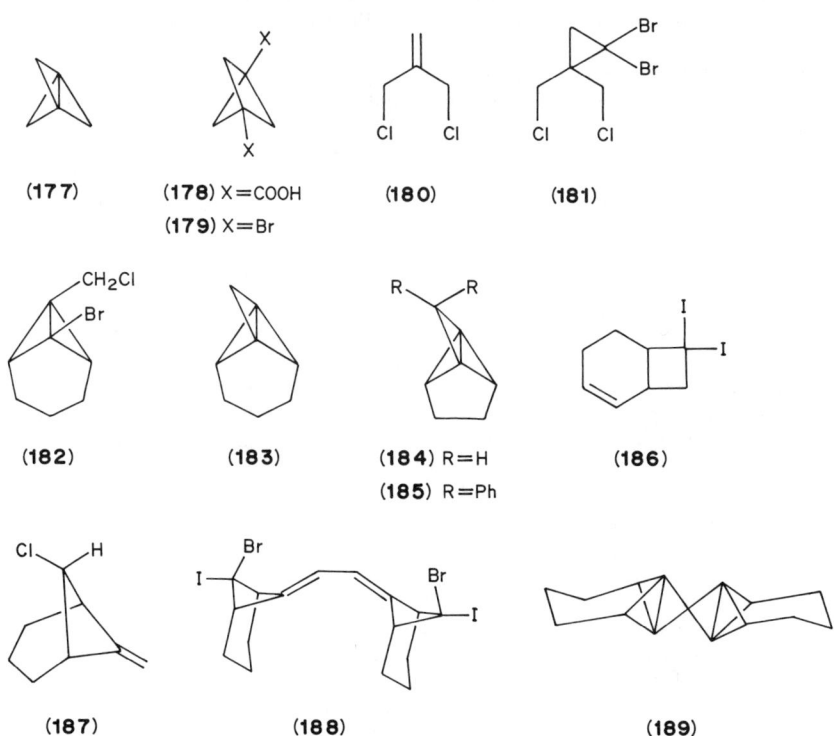

(177)   (178) X=COOH   (180)   (181)
               (179) X=Br

(182)   (183)   (184) R=H   (186)
                              (185) R=Ph

(187)   (188)   (189)

available dichloride **180** adds dibromocarbene to form the expected cyclopropane **181** in moderate yield, and treatment of **181** with methyllithium gives [1.1.1]propellane (**177**) in good yield. This synthesis has made **177** available in multigram quantities and, as we note below, the propellane now provides a convenient starting material for preparation of many bicyclo[1.1.1]penanes, including **178**. Related syntheses by Szeimies's group yielded tetracyclic derivatives of **177**. Reaction of **182** with methyllithium furnished the trimethylene-bridged propellane **183**, and a similar approach led to lower homologue **184**[127,128]. In addition, a carbene route to [1.1.1]propellanes was developed. For example, treatment of either diiodide **186** or the chloro compound **187** with methyllithium provides an alternative route to **183**[128,129], and similar treatment of **188** gives the dimeric propellane **189**[130].

The structure of **177** has been determined by a neutron diffraction study[131] and confirmed by analysis of the vibrational spectrum[67]. The bridgehead to methylene [$C_{(1)}$—$C_{(2)}$] distance is 1.525 Å, while the bridgehead to bridgehead [$C_{(11)}$—$C_{(3)}$] bond length is 1.596 Å; the $C_{(2)}$—$C_{(1)}$—$C_{(4)}$ bond angle is 95.1°. These results may be compared with those from the X-ray structure of the more complex **185**[128]. There are two molecules per unit cell, so that two sets of data emerge from the study. The bridgehead to methylene distance is 1.534 (1.526) Å, the bridgehead to bridgehead distance is 1.592 (1.586) Å and the methylene–bridgehead–methylene bond angles are 89.3° (89.7°) for the symmetric pair and 98.0° (98.1°) for the angle involving the diphenylmethylene group. For all [1.1.1]propellanes for which structural information is available, the central bridgehead–bridgehead bond is between 1.577 and 1.601 Å[128,132]. An interesting point to emerge from X-ray studies on **183** and **184** at 81 K is that there appears to be no accumulation of charge between the bridgehead nuclei, but there is charge density outside each bridgehead carbon atom[132]. This provides a nice explanation for the ease of electrophilic attack on the [1.1.1]propellane system.

The photoelectron spectrum of **177** has a narrow first band with an intense 0,0 vibrational component, strongly suggesting that there is little change in equilibrium geometry on passing from the hydrocarbon to the radical cation[61]. As mentioned earlier, this could reflect failure of the HOMO to provide significant $C_{(1)}$—$C_{(3)}$ bonding, or it may simply be a consequence of the rigid structure. The ionization potential (-9.74 eV) lies between that of cyclopropane (-10.9 eV) and bicyclobutane (-9.25 eV). The $C_{(1)}$—$C_{(2)}$ coupling constant in the ^{13}C NMR spectrum of **177** is 9.9 Hz, and the C—H coupling constant is 163.7 Hz. From these data estimates of the hybridization at each bond have been made[133]. This approach[134] leads to a value of $sp^{0.5}$ at each bridgehead for the orbital forming the central bond.

Thermolysis of [1.1.1]propellane at 114 °C in a gas flow system yields methylenecyclobutene (**191**), presumably by way of carbene **190** and 1,2 shift of hydrogen[63]. At 430 °C in a second study the only product was dimethylenecyclopropane, suggested to result from a symmetry-allowed [$_{\sigma}2_s + _{\sigma}2_a$] rearrangement[127]. Similar thermal reaction of **183** at 370 °C leads to **192** and **193**, which were thought to arise from secondary rearrangement of the unstable primary product analogous to **192**, which is **194**. Formation of **195** from **184** can be rationalized in parallel fashion by way of **196** and **197**[127].

Exposure of **177** to AgBF$_4$, Rh(I) or complexes of Pd(II) yields largely the symmetrical dimer **198**. A small amount of methylenecyclobutene (**191**) is observed in some cases, and **191** becomes the major product with Pt(0) and Ir(I) complexes[135]. Reactions of [1.1.1]propellane **177** with electron-deficient alkenes and alkynes also occur readily[135]. Acetylene dicarboxylic ester (**199**), for example, yields a 1:1 adduct **201** and two 2:1 adducts **203** and **204**. These are most easily understood through biradical mechanisms; addition of one of the propellane central bonding electrons to **199** forms **205**, which can then collapse to **201** with one bond cleavage. Reaction of a second molecule of **177** with **201** leads to **206**, and simple reorganization then furnishes **203** and **204**. The same types of

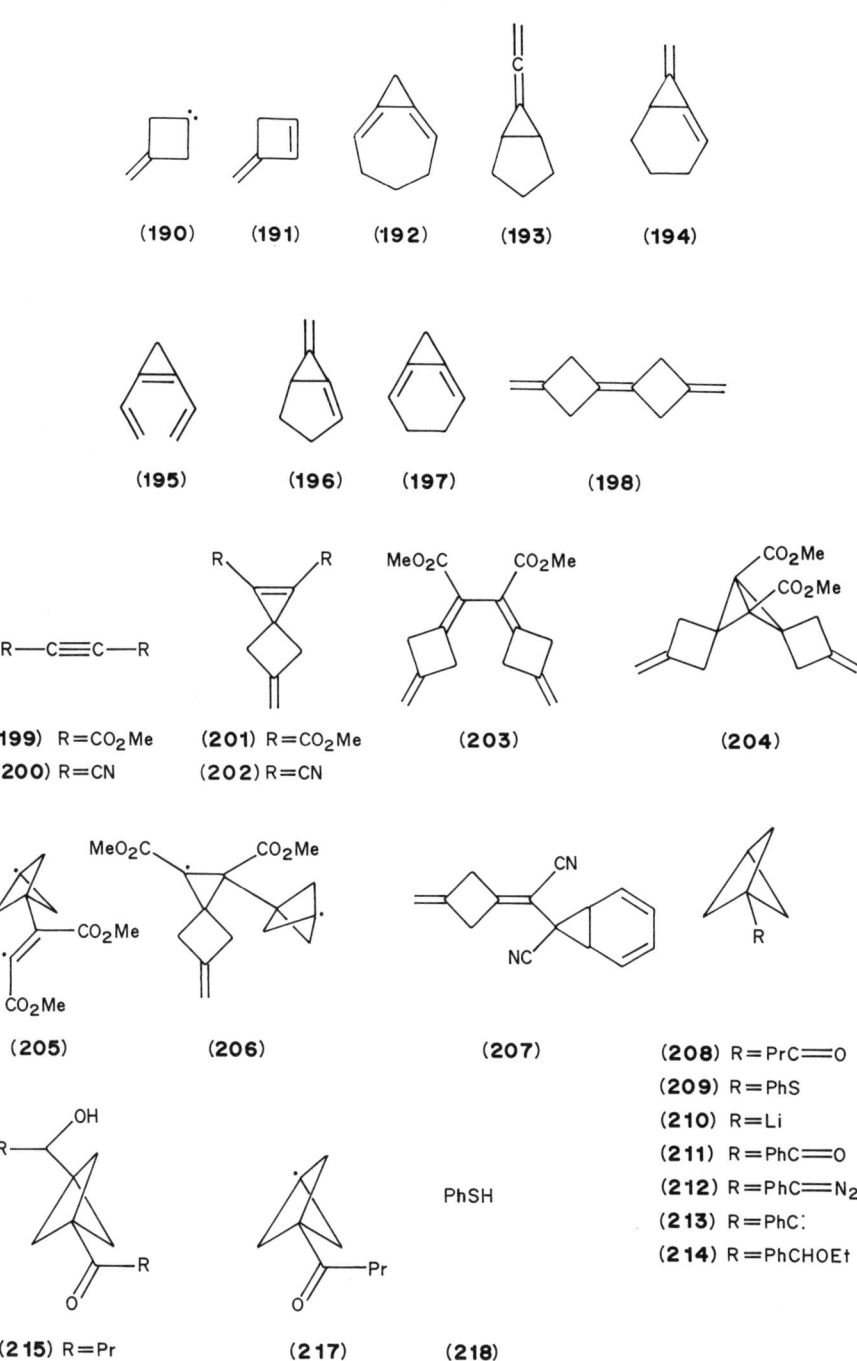

(190)   (191)   (192)   (193)   (194)

(195)   (196)   (197)   (198)

R—C≡C—R

(199) R=CO₂Me   (201) R=CO₂Me   (203)   (204)
(200) R=CN   (202) R=CN

(205)   (206)   (207)   (208) R=PrC=O
                          (209) R=PhS
                          (210) R=Li
                          (211) R=PhC=O
                          (212) R=PhC=N₂
                          (213) R=PhC:
                          (214) R=PhCHOEt

PhSH

(215) R=Pr   (217)   (218)
(216) R=Me

$$\left[t\text{-Bu} \longleftarrow \bigcirc \!\!-\!\! \bigcirc \longrightarrow \text{Bu-}t \right]^{-\cdot} \text{Li}^+ \qquad \text{EtO} \!\!\diagup\!\!\triangle \qquad \text{Ph} \!\!\diagup\!\!\triangle$$

(219)   (220)   (221)

products are formed with dicyanoacetylene (200) in methylene chloride, but if the solvent is benzene the 1:1 adduct 202 is surprisingly attacked by solvent to form 207[135].

The free radical additions common to small propellanes take place particularly readily with 177. Aldehydes add spontaneously and give both 1:1 and 1:2 adducts[135]. Butyraldehyde, for example, furnishes a 1:2 mixture of ketone 208 and ketol 215 by way of the bridgehead radical 217. Radical addition to an aldehyde carbonyl in successful competition with abstraction of the aldehydic hydrogen is exceptional. Here it occurs preferentially, and both acetaldehyde and benzaldehyde yield only ketol adducts. From acetaldehyde this ketol is 216, and haloform oxidation of this adduct with iodine in base provides a convenient route to dicarboxylic acid 178. The formerly difficult-of-access precursor to propellane 177 is thus now readily available in two steps from the propellane itself.

Thiophenol (218) also adds spontaneously to 177 to give the expected phenyl thioether 209[135]. Cleavage of thioethers with lithium 4,4'-di-tert-butylbiphenyl (219) is a useful method for preparing organolithium compounds[136], and this reaction has been employed here as a route to 1-lithiobicyclo[1.1.1]pentane (210)[135]. This in turn offers synthetically useful access to a variety of 1-substituted bicyclo[1.1.1]pentanes through the host of well-known transformations of alkyllithiums. One particularly noteworthy set of reactions made possible by the availability of 210 involves preparation of the phenyl ketone 211 and its conversion to the tosylhydrazone. This permitted examination of the behavior of diazo compound 212, which is formed transiently on decomposition of the tosylhydrazone in ethanolic ethoxide. Under the conditions employed 212 should lose nitrogen and form carbene 213. The observed products from this reaction are ethyl ether 214, explainable through insertion of 213 into solvent, and rearranged ether 220. Formation of 220 is nicely explained by way of a bond shift in carbene 213 to yield the anti-Bredt, bridgehead olefin 2-phenyl-1-bicyclo[2.1.1]hexene (221), which should add ethanol rapidly to give 220[135].

A related transformation that promises to be equally useful synthetically is the reaction of 219 directly with 177 to furnish the dilithio derivative 222 of bicyclo[1.1.1]pentane[137].

The bridgehead radicals formed on addition of free radicals to 177 can also attack a second molecule of the propellane. Depending on the rate constants of the competitive processes involved, this can lead to dimeric, trimeric and higher oligomeric species with various number of bicyclo[1.1.1]pentane units[138,139]. For example, treatment of 177 with dimethyl disulfide leads to a small amount of 223, among other products[138]. Molecules containing n bicyclopentane units joined by (n − 1) bridgehead bonds have been called [n]staffanes[139]. Both these oligomers of bicyclopentane[139,140] and the conceptually related oligomers in which bicyclopentanes are connected by acetylenic units[137,141] can be prepared from [1.1.1]propellanes, and both are of interest as providing a series of rigid molecules of known dimensions.

D. Bicyclo[1.1.0]butanes

The chemistry of bicyclo[1.1.0]butanes is well developed, with particular attention over the last two decades paid to reactivity of the central bond with metal ions and complexes. Earlier work has been reviewed[28,142], and here we shall emphasize recent studies in this area. As can be seen from the discussion of propellanes above, in recent years several bicyclobutanes have been prepared as intermediates in the synthesis of complex [n.1.1]propellanes.

Two types of reactivity of bicyclobutanes with electron-deficient alkenes and alkynes have been known for many years[143]. These are an ene-like reaction and a cycloaddition-like reaction. The first is typified by the room-temperature addition of dicyanoacetylene (200) to 3-methylbicyclobutane-1-carbonitrile (224) to furnish 225 as the major product[144], and the second by the reaction between 224 and acrylonitrile at elevated temperature to give the bicyclo[2.1.1]hexane 226[145]. More recently a third mode of reaction has been described[135,146], for which the prototype is addition of tricyclo[3.1.1.0^{1,7}]heptane ('Moore's hydrocarbon', 227) to tetracyanoethylene (228) to form cyclopropane 229[146]. It is noteworthy that 227 is incapable of undergoing either the ene-type reaction, which would produce an olefin violating Bredt's Rule, or the

(222)  (223) 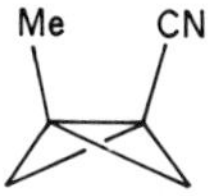 (224)

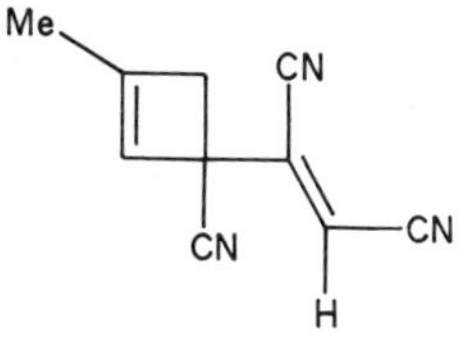

(225) 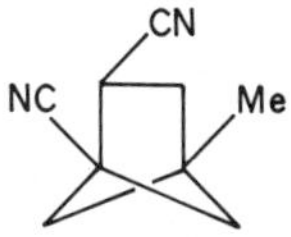 (226)  (227) 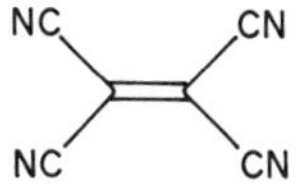 (228)

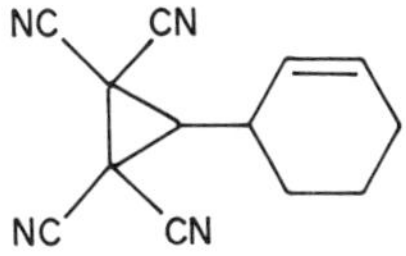

(229)  (230) 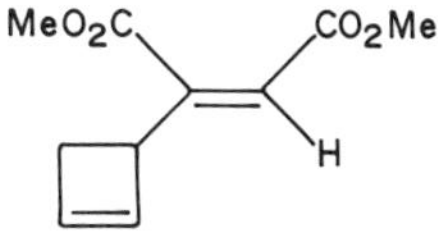 (231)

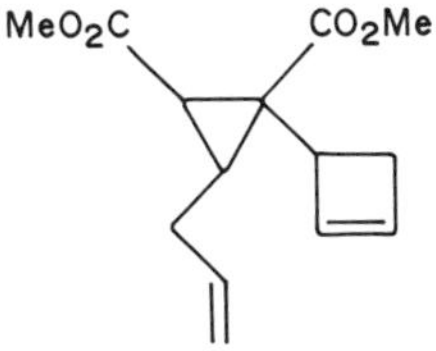

(232) 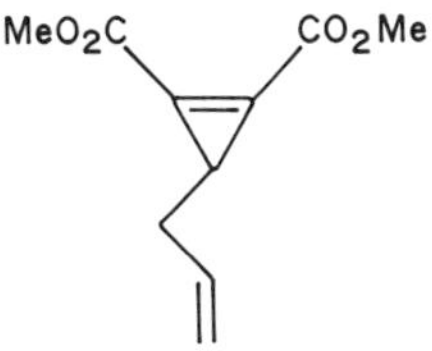 (233) 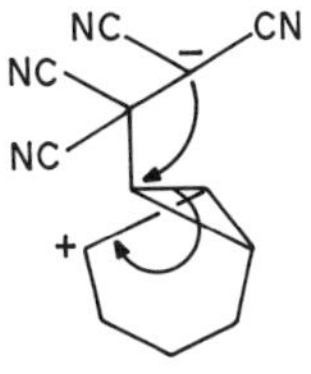 (234)

958   W. C. Agosta

cycloaddition-type process, owing to steric hindrance by the trimethylene bridge to the expected[147] mode of addition. However, bicyclobutane (230) itself has now been seen to give adducts of this third type in some cases[135]. While addition of acetylene dicarboxylic ester (199) to 230 yields only ene product 231 in hydrocarbon solvent, this is accompanied by increasing amounts of a 2:1 adduct 232 in increasingly polar solvent. In acetonitrile or methanol the ratio of 231 to 232 is 3:2. Formation of 232 is most readily understood as the result of an ene-type reaction of a second molecule of bicyclobutane (230) with the initially formed third-type adduct 233; kinetic examination of the reaction supports this assumption. Similar transformations occur with dicyanoacetylene (200). Earlier mechanistic work on these modes of addition to bicyclobutanes has been summarized[143]. It appears that the ene-type reaction is generally concerted and that the cycloaddition-type reaction usually has a biradical intermediate. The new, third mode of addition appears to involve ionic intermediates in view of its increasing importance in polar solvents[135]. Christl and coworkers proposed earlier that reaction between 227 and 228 proceeds through zwitterion 234, which then collapses to product as shown[146].

IV. CONCLUSION

Interest in planar and inverted carbon has been driven by curiosity concerning the chemical and physical effects of increasing distortion of the tetrahedral carbon atom. As we have seen, this has led to the preparation of a wonderful variety of new ring systems, and there is now structural information for many of these, documenting quantitatively the extent of flattening or degree of inversion at the distorted center. Of equal importance, and perhaps of greater interest to more organic chemists, there is now ample evidence that these bonding distortions do indeed lead to hitherto unobserved types of chemical behavior, offering new transformations and also access to novel or systems otherwise difficult of access. There remain unsolved problems. There are flattened or inverted systems that have not yet yielded to synthesis, and there are other systems not yet incorporated into structures providing the practical, kinetic stability that permits chemical and physical examination of the distorted center at leisure. All this said, it has been a rather successful and stimulating chapter in organic chemistry, so far.

V. ACKNOWLEDGMENT

It is a pleasure to thank Dr. S. T. Waddell for convenient access to his *Dissertation*[135a], which provides an excellent review of the chemistry of [1.1.1]propellanes and bicyclo[1.1.0]butanes, as well as a considerable contribution to their chemistry.

VI. REFERENCES

1. W. H. Perkin, Jr., *Ber. Deut. Chem. Gesell.*, **16**, 1787 (1883).
2. W. H. Perkin, Jr., *Ber. Deut. Chem. Gesell.*, **17**, 54, 323 (1884).
3. A. Baeyer, *Ber. Deut. Chem. Gesell*, **18**, 2277 (1885).
4. A. Greenberg and J. F. Liebman, *Strained Organic Molecules*, Academic Press, New york, 1978.
5. K. B. Wiberg, *Angew. Chem.*, **98**, 312 (1986); *Angew. Chem., Int. Ed. Engl.*, **25**, 312 (1986).
6. J. B. Collins, J. D. Dill, E. D. Jemmis, Y. Apeloig, P. v. R. Schleyer, R. Seeger and J. A. Pople, *J. Am. Chem. Soc.*, **98**, 5419 (1976).
7. K. B. Wiberg and G. B. Ellison, *Tetrahedron*, **30**, 1573 (1974).
8. W. Luef, R. Keese and H.-B. Bürgi, *Helv. Chim. Acta*, **70**, 534 (1987).
9. W. Luef and R. Keese, *Helv. Chim. Acta*, **70**, 543 (1987).
10. K. B. Wiberg, *J. Org. Chem.*, **50**, 5285 (1985).
11. K. B. Wiberg, G. B. Burgmaier, K.-W. Shen, S. J. La Placa, W. C. Hamilton and M. D. Newton, *J. Am. Chem. Soc.*, **94**, 7402 (1972).

12. S. Harder, J. Boersma, L. Brandsma, A. van Hateren, J. A. Kanters, W. Bauer and P. v. R. Schleyer, *J. Am. Chem. Soc.*, **110**, 7802 (1988); H. Dietrich, W. Mahdi and W. J. Storch, *J. Organomet. Chem.*, **349**, 1 (1988). See also in this regard F. A. Cotton and M. Millar, *J. Am. Chem. Soc.*, **99**, 7886 (1977).
13. M. L. Thompson and R. N. Grimes, *J. Am. Chem. Soc.*, **93**, 6677 (1971).
14. E. H. Braye, L. E. Dahl, W. Hubel and D. L. Wampler, *J. Am. Chem. Soc.*, **84**, 4633 (1962).
15. K. B. Wiberg, J. E. Hiatt and G. Burgmaier, *Tetrahedron Lett.*, 5855 (1968).
16. V. Georgian and M. Saltzman, *Tetrahedron Lett.*, 4315 (1972).
17. A. Nickon and E. F. Silversmith, *Organic Chemistry: The Name Game*, Pergamon Press, New York, 1987.
18. E. H. Hahn, H. Bohm and D. Ginsburg, *Tetrahedron Lett.*, 507 (1973).
19. B. R. Venepalli and W. C. Agosta, *Chem. Rev.*, **87**, 399 (1987).
20. J. Altman, E. Babad, J. Itzchaki and D. Ginsburg, *Tetrahedron Suppl. No. 8 (Part 1)*, 279 (1966).
21. K. W. Cox, M. D. Harmony, G. Nelson and K. B. Wiberg, *J. Chem. Phys.*, **50**, 1976 (1969).
22. P. L. Johnson and J. P. Schaefer, *J. Org. Chem.*, **37**, 2762 (1972).
23. P. G. Gassman, M. L. Greenlee, D. A. Dixon, S. Richtsmeier and J. Z. Gougoutas, *J. Am. Chem. Soc.*, **105**, 5865 (1983).
24. D. Ginsburg, *Propellanes*, Verlag Chemie, Weinheim, 1975; *Propellanes: Structure and Reactions, Sequels I and II*, Technion, Haifa, 1981, 1985; D. Ginsburg, in *The Chemistry of the Cyclopropyl Group* (Ed. Z. Rappoport), Wiley, Chichester, 1987, pp. 1193–1221.
25. K. B. Wiberg, *Acc. Chem. Res.*, **17**, 379 (1984).
26. K. B. Wiberg, *Chem. Rev.*, **89**, 975 (1989).
27. K. Krohn, *Nachr. Chem. Tech. Lab.*, **35**, 264 (1987).
28. K. B. Wiberg, *Rec. Chem. Prog.*, **26**, 143 (1965); S. Hoz in *The Chemistry of the Cyclopropyl Group*, (ed. Z. Rappoport), Wiley, Chichester, 1987, pp. 1121–1192.
29. M. D. Newton, in *Modern Theoretical Chemistry*, Vol. 4 (Ed. H. F. Schaefer), Plenum Press, New York, 1977, p. 223; N. D. Epiotis, *Unified Valence Bond Theory and Electronic Structure*, Springer-Verlag, New York, 1983.
30. R. Hoffmann, R. W. Alder and C. F. Wilcox, Jr., *J. Am. Chem. Soc.*, **92**, 4992 (1970); R. Hoffmann, *Pure Appl. Chem.*, **28**, 181 (1971).
31. S. Durmaz, J. N. Murrell and J. B. Pedlay, *J. Chem. Soc., Chem. Commun.*, 933 (1972); J. N. Murrell, J. B. Pedlay and S. Durmaz, *J. Chem. Soc., Faraday Trans. 2*, **69**, 1370 (1973).
32. D. C. Crans and J. P. Snyder, *J. Am. Chem. Soc.*, **102**, 7153 (1980).
33. M.-B. Krogh-Jespersen, J. Chandrasekhar, E.-U. Würthwein, J. B. Collins and P. v. R. Schleyer, *J. Am. Chem. Soc.*, **102**, 2263 (1980); H. Monkhorst, *J. Chem. Soc., Chem. Commun.*, 1111 (1968); G. Olah and G. Klopman, *Chem. Phys. Lett.*, **11**, 604 (1971); V. I. Minkin, R. M. Minyaev and I. I. Zacharov, *J. Chem. Soc., Chem. Commun.*, 213 (1977); W. A. Lathan, W. J. Hehre, L. A. Curtiss and J. A. Pople, *J. Am. Chem. Soc.*, **93**, 6377 (1971).
34. K. B. Wiberg, G. B. Ellison and J. J. Wendoloski, *J. Am. Chem. Soc.*, **98**, 1212 (1976).
35. M. J. S. Dewar and W. Thiel, *J. Am. Chem. Soc.*, **99**, 4899, 4907 (1977).
36. E.-U. Würthwein, J. Chandrasekhar, E. D. Jemmis and P. v. R. Schleyer, *Tetrahedron Lett.*, **22**, 843 (1981).
37. K. B. Wiberg, *Tetrahedron Lett.*, **26**, 5967 (1985).
38. W. C. Agosta, unpublished results.
39. J. Chandrasekhar, E.-U. Würthwein and P. v. R. Schleyer, *Tetrahedron*, **37**, 921 (1981).
40. R. C. Bingham, M. J. S. Dewar and D. H. Ho, *J. Am. Chem. Soc.*, **97**, 1285, 1294 (1975).
41. M. C. Böhm, R. Gleiter and P. Schang, *Tetrahedron Lett.*, 2575 (1979).
42. R. Keese, A. Pfenninger and A. Roesle, *Helv. Chim. Acta*, **62**, 326 (1979).
43. J. A. Pople, *Mod. Theor. Chem.*, **4**, 1 (1977).
44. K. B. Wiberg, L. K. Olli, N. Golembeski and R. D. Adams, *J. Am. Chem. Soc.*, **102**, 7467 (1980).
45. P. Gund and T. M. Gund, *J. Am. Chem. Soc.*, **103**, 4458 (1981).
46. L. Skattebøl, *J. Org. Chem.*, **31**, 2789 (1966).
47. Z. Smith, B. Andersen and S. Bunce, *Acta Chem. Scand.*, **31A**, 557 (1977); H. M. Frey, R. G. Hopkins and L. Skattebøl, *J. Chem. Soc. (B)*, 539 (1971).
48. K. B. Wiberg and J. J. Wendoloski, *J. Am. Chem. Soc.*, **104**, 5679 (1982).
49. J. M. Schulman, M. L. Sabio and R. L. Disch, *J. Am. Chem. Soc.*, **105**, 743 (1983).
50. W. Luef and R. Keese, unpublished results.
51. M. Luyten and R. Keese, *Tetrahedron*, **42**, 1687 (1986).

960   W. C. Agosta

52. M. D. Newton and J. M. Schulman, *J. Am. Chem. Soc.*, **94**, 773 (1972).
53. W.-D. Stohrer and R. Hoffmann, *J. Am. Chem. Soc.*, **94**, 779 (1972).
54. K. B. Wiberg, *J. Am. Chem. Soc.*, **105**, 1227 (1983).
55. P. Chackrabarti, P. Seiler, J. D. Dunitz, W.-D. Schlüter and G. Szeimies, *J. Am. Chem. Soc.*, **103**, 7378 (1981).
56. R. F. W. Bader, *Acc. Chem Res.*, **18**, 9 (1985).
57. K. B. Wiberg, R. F. W. Bader and C. D. H. Lau, *J. Am. Chem. Soc.*, **109**, 985, 1001 (1987).
58. A. B. Pierini, H. F. Reale and J. A. Medrano, *J. Mol. Struct. Theochem*, **33**, 109 (1986).
59. D. Feller and E. R. Davidson, *J. Am. Chem. Soc.*, **109**, 4133 (1987).
60. J. E. Jackson and L. C. Allen, *J. Am. Chem. Soc.*, **106**, 591 (1984).
61. E. Honegger, H. Huber, E. Heilbronner, W. P. Dailey and K. B. Wiberg, *J. Am. Chem. Soc.*, **107**, 7172 (1985).
62. R. F. Messmer and P. A. Schultz, *J. Am. Chem. Soc.*, **108**, 7407 (1986).
63. K. B. Wiberg and F. H. Walker, *J. Am. Chem. Soc.*, **104**, 5239 (1982).
64. K. B. Wiberg, F. H. Walker, W. E. Pratt and J. Michl, *J. Am. Chem. Soc.*, **105**, 3638 (1983).
65. F. H. Walker, K. B. Wiberg and J. Michl, *J. Am. Chem. Soc.*, **104**, 2056 (1982).
66. K. B. Wiberg, *J. Comput. Chem.*, **5**, 197 (1984).
67. K. B. Wiberg, W. P. Dailey, F. H. Walker, S. T. Waddell, L. S. Crocker and M. Newton, *J. Am. Chem. Soc.*, **107**, 7247 (1985).
68. L. Åsbrink, C. Fridh and C. Lindholm, *Chem. Phys. Lett.*, **52**, 63, 69, 72 (1977).
69. M. Eckert-Maksić, K. Mlinarić-Majerski and Z. Majerski, *J. Org. Chem.*, **52**, 2098 (1987).
70. O. Ermer, R. Gerdil and J. D. Dunitz, *Helv. Chim. Acta*, **54**, 2476 (1971); H. Dodziuk, *J. Comput. Chem.*, **5**, 571 (1984); N. L. Allinger, Y. H. Yuh and J.-H. Lii, *J. Am. Chem. Soc.*, **111**, 8551 (1989).
71. M. N. Paddon-Row, K. N. Houk, P. Dowd, P. Garner and R. Schappert, *Tetrahedron Lett.*, **22**, 4799 (1981).
72. K. B. Wiberg, S. T. Waddell and R. E. Rosenberg, *J. Am. Chem. Soc.*, **112**, 2184 (1990).
73. H. Dodziuk, *Tetrahedron*, **44**, 2951 (1988).
74. C. F. H. Allen and J. A. Van Allan, *J. Org. Chem.*, **18**, 882 (1953).
75. T. Mori, K. Kimoto, M. Kawanisi and H. Nozaki, *Tetrahedron Lett.*, 3653 (1969).
76. C. W. Thornber, *J. Chem. Soc., Chem. Commun.*, 238 (1973).
77. R. Helder and H. Wynberg, *Tetrahedron Lett.*, 4321 (1973).
78. F. Vögtle and P. K. T. Mew, *Angew. Chem.*, **90**, 58 (1978); *Angew. Chem., Int. Ed. Engl.*, **17**, 60 (1978).
79. K. B. Wiberg and M. J. O'Donnell, *J. Am. Chem. Soc.*, **101**, 6660 (1979).
80. P. Warner, B.-L. Chen, C. A. Bronski, B. A. Karcher and R. A. Jacobson, *Tetrahedron Lett.*, **22**, 375 (1981).
81. P. E. Eaton and B. D. Leipzig, *J. Am. Chem. Soc.*, **105**, 1656 (1983).
82. T. Tsunoda, M. Amaike, U. S. F. Tambunan, Y. Fujise and S. Ito, *Tetrahedron Lett.*, **28**, 2537 (1987).
83. M. T. Crimmins and L. D. Gould, *J. Am. Chem. Soc.*, **109**, 6199 (1987).
84. P. A. Wender, T. W. von Geldern and B. H. Levine, *J. Am. Chem. Soc.*, **110**, 4858 (1988).
85. H. Morrison, *Acc. Chem. Res.*, **12**, 383 (1979); P. A. Wender and G. B. Dreyer, *Tetrahedron*, **37**, 4445 (1981); *J. Am. Chem. Soc.*, **104**, 5805 (1982); P. A. Wender and J. J. Howbert, *Tetrahedron Lett.*, **23**, 3983 (1982).
86. R. Mitschka, J. M. Cook and U. Weiss, *J. Am. Chem. Soc.*, **100**, 3973 (1978); R. Mitschka, J. Oehldrich, K. Takahashi, J. M. Cook, U. Weiss and J. V. Silverton, *Tetrahedron*, **37**, 4521 (1981).
87. W. C. Han, K. Takahashi, J. M. Cook, U. Weiss and J. V. Silverton, *J. Am. Chem. Soc.*, **104**, 318 (1982).
88. M. N. Deshpande, M. Jawdosiuk, G. Kubiak, M. Venkatachalam, U. Weiss and J. M. Cook, *J. Am. Chem. Soc.*, **107**, 4786 (1985); M. Venkatachalam, M. N. Deshpande, M. Jawdosiuk, G. Kubiak, S. Wehrli, J. M. Cook and U. Weiss, *Tetrahedron*, **42**, 1597 (1986). See also in this regard M. N. Deshpande, S. Wehrli, M. Jawdosiuk, J. T. Guy, Jr., D. W. Bennett, J. M. Cook, M. R. Depp and U. Weiss, *J. Org. Chem.*, **51**, 2436 (1986).
89. R. Keese, *Nachr. Chem. Tech. Lab.*, **30**, 844 (1982); M. Luyten and R. Keese, *Angew. Chem.*, **96**, 358 (1984); *Angew. Chem., Int. Ed. Engl.*, **23**, 390 (1984).
90. H. Schori, B. B. Patil and R. Keese, *Tetrahedron*, **37**, 4457 (1981).
91. M. Luyten and R. Keese, *Helv. Chim. Acta*, **67**, 2242 (1984).

92. S. D. Burke and P. A. Grieco, *Org. React.*, **26**, 361 (1979).
93. J. Mani, J.-H. Cho, R. R. Astik, E. Stamm, P. Bigler, V. Meyer and R. Keese, *Helv. Chim. Acta*, **67**, 1930 (1984); J. Mani and R. Keese, *Tetrahedron*, **41**, 5697 (1985).
94. D. Kuck and H. Bögge, *J. Am. Chem. Soc.*, **108**, 8107 (1986).
95. D. Kuck and A. Schuster, *Angew. Chem.*, **100**, 1222 (1988); *Angew. Chem., Int. Ed. Engl.*, **27**, 1201 (1988).
96. M. S. Raasch, *J. Org. Chem.*, **45**, 856 (1980).
97. V. B. Rao, S. Wolff and W. C. Agosta, *Tetrahedron*, **42**, 1549 (1986).
98. W. G. Dauben, J. Pesti and C. H. Cummins, unpublished results.
99. V. B. Rao, C. F. George, S. Wolff and W. C. Agosta, *J. Am. Chem. Soc.*, **107**, 5732 (1985).
100. S. Wolff, B. R. Venepalli, C. F. George and W. C. Agosta, *J. Am. Chem. Soc.*, **110**, 6785 (1988).
101. R. B. Woodward and R. Hoffmann, *Angew. Chem.*, **81**, 797 (1969); *Angew. Chem., Int. Ed. Engl.*, **8**, 781 (1969).
102. H. O. House, M. B. DeTar and D. VanDerveer, *J. Org. Chem.*, **44**, 3793 (1979).
103. S. Wolff and W. C. Agosta, *J. Chem. Soc., Chem. Commun.*, 118 (1981); S. Wolff and W. C. Agosta, *J. Org. Chem.*, **46**, 4821 (1981).
104. A. G. Fallis, *Can. J. Chem.*, **53**, 1657 (1975); P. D. Hobbs and P. D. Magnus, *J. Am. Chem. Soc.*, **98**, 4594 (1976).
105. O. Baumgärtel, J. Harnisch, G. Szeimies, M. Van Meerssche, G. Germain and J.-P. Declercq, *Chem. Ber.*, **116**, 2205 (1983).
106. P. Warner and R. LaRose, *Tetrahedron Lett.*, 2141 (1972).
107. D. P. G. Hamon and V. C. Trenerry, *J. Am. Chem. Soc.*, **103**, 4962 (1981).
108. U. Szeimies-Seebach, A. Schöffer, R. Römer and G. Szeimies, *Chem. Ber.*, **114**, 1767 (1981).
109. U. Szeimies-Seebach, J. Harnisch, G. Szeimies, M. Van Meerssche, G. Germain and J. P. Declercq, *Angew. Chem., Int. Ed. Engl.*, **17**, 848 (1978).
110. A.-D. Schlüter, H. Harnisch, J. Harnisch, U. Szeimies-Seebach and G. Szeimies, *Chem. Ber.*, **118**, 3513 (1985).
111. J.-P. Declercq, G. Germain and M. Van Meerssche, *Acta Crystallogr., Sect. B*, **34**, 3472 (1978).
112. K.-D. Baumgart, H. Harnisch, U. Szeimies-Seebach and G. Szeimies, *Chem. Ber.*, **118**, 2883 (1985).
113. J. Morf and G. Szeimies, *Tetrahedron Lett.*, **44**, 5363 (1986).
114. Z. Majerski and M. Zuniac, *J. Am. Chem. Soc.*, **109**, 3496 (1987).
115. R. E. Pincock and E. J. Torupka, *J. Am. Chem. Soc.*, **91**, 4593 (1969); R. E. Pincock, J. Schmidt, W. B. Scott and E. J. Torupka, *Can. J. Chem.*, **50**, 3958 (1972).
116. K. B. Wiberg and G. Burgmaier, *Tetrahedron Lett.*, 317 (1969); *J. Am. Chem. Soc.*, **94**, 7396 (1972).
117. P. G. Gassman, A. Topp and J. W. Keller, *Tetrahedron Lett.*, 1093 (1969).
118. D. H. Aue and R. N. Reynolds, *J. Org. Chem.*, **39**, 2315 (1974).
119. P. G. Gassman and G. S. Proehl, *J. Am. Chem. Soc.*, **102**, 6862 (1980).
120. H.-G. Zoch. A.-D. Schlüter and G. Szeimies, *Tetrahedron Lett.*, **22**, 3835 (1981).
121. U. Szeimies-Seebach, G. Szeimies, M. Van Meerssche, G. Germain and J.-P. Declercq, *Nouv. J. Chim.*, **3**, 357 (1979).
122. K. Mlinarić-Majerski and Z. Majerski, *J. Am. Chem. Soc.*, **102**, 1418 (1980); **105**, 7389 (1983).
123. K. Mlinarić-Majerski, Z. Majerski, B. Rakvin and Z. Veksli, *J. Org. Chem.*, **54**, 545 (1989); Z. Majerski and K. Mlinarić-Majerski, *J. Org. Chem.*, **51**, 3219 (1986).
124. V. Vinković and Z. Majerski, *J. Am. Chem. Soc.*, **104**, 4027 (1982).
125. D. E. Applequist, T. L. Renkin and J. W. Wheeler, *J. Org. Chem.*, **47**, 4985 (1982).
126. K. Semmler, G. Szeimies and J. Belzner, *J. Am. Chem. Soc.*, **107**, 6410, (1985); J. Belzner, U. Bunz, K. Semmler, G. Szeimies, K. Opitz and A.-D. Schlüter, *Chem. Ber.*, **122**, 397 (1989).
127. J. Belzner and G. Szeimies, *Tetrahedron Lett.*, **27**, 5839 (1986).
128. J. Belzner, G. Gareiss, K. Polborn, W. Schmid, K. Semmler and G. Szeimies, *Chem. Ber.*, **122**, 1509 (1989).
129. J. Belzner and G. Szeimies, *Tetrahedron Lett.*, **28**, 3099 (1987).
130. G. Kottirsch, K. Polborn and G. Szeimies, *J. Am. Chem. Soc.*, **110**, 5588 (1988).
131. L. Hedberg and K. Hedberg, *J. Am. Chem. Soc.*, **107**, 7257 (1985).
132. P. Seiler, J. Belzner, U. Bunz and G. Szeimies, *Helv. Chim. Acta*, **71**, 2100 (1988).
133. R. M. Jarret and L. Cusumano, *Tetrahedron Lett.*, **31**, 171 (1990).

134. E. W. Della and P. E. Pigou, *J. Am. Chem. Soc.*, **106**, 1085 (1984); K. Frei and H. J. Bernstein, *J. Chem. Phys.*, **38**, 1216 (1963).
135. (a) S. T. Waddell, *Dissertation*, Yale University, 1988.
     (b) K. B. Wiberg and S. T. Waddell, *J. Am. Chem. Soc.*, **112**, 2194 (1990).
136. P. K. Freeman and L. L. Hutchinson, *Tetrahedron Lett.*, 1849 (1976); *J. Org. Chem.*, **45**, 1924 (1980); C. Rucker, *Tetrahedron Lett.*, **25**, 4349 (1984).
137. U. Bunz and G. Szeimies, *Tetrahedron Lett.*, **31**, 651 (1990).
138. U. Bunz, K. Polborn, H.-U. Wagner and G. Szeimies, *Chem. Ber.*, **121**, 1785 (1988).
139. P. Kaszynski and J. Michl, *J. Am. Chem. Soc.*, **110**, 5225 (1988).
140. R. Gleiter, K.-H. Pfeifer, G. Szeimies and U. Bunz, *Angew. Chem.*, **102**, 418 (1990); *Angew. Chem., Int. Ed. Engl.*, **29**, 413 (1990).
141. U. Bunz and G. Szeimies, *Tetrahedron Lett.*, **30**, 2087 (1989).
142. L. A. Paquette, *Acc. Chem. Res.*, **4**, 280 (1971); P. G. Gassman, *Acc. Chem. Res.*, **4**, 128 (1971).
143. M. Pomerantz, R. N. Wilke, G. W. Gruber and U. Roy, *J. Am. Chem. Soc.*, **94**, 2752 (1972).
144. P. G. Gassman and K. Mansfield, *J. Am. Chem. Soc.*, **90**, 1517 (1968).
145. E. P. Blanchard and A. Cairncross, *J. Am. Chem. Soc.*, **88**, 487, 496 (1966).
146. M. Christl, R. Lange, C. Herzog, R. Stangl, K. Peters, E. M. Peters and H. G. von Schnering, *Angew. Chem.*, **97**, 595 (1985); *Angew. Chem., Int. Ed. Engl.*, **24**, 611 (1985).
147. P. G. Gassman, K. Mansfield and T. J. Murphy, *J. Am. Chem. Soc.*, **90**, 4746 (1968); *J. Am. Chem. Soc.*, **91**, 1684 (1969).

## CHAPTER 21

Radical reactions of alkanes and cycloalkanes

GLEN A. RUSSELL

Department of Chemistry, Iowa State University, Ames, Iowa 50011, USA

I. GENERAL INTRODUCTION

Attack of an atom or free radical upon alkanes generally involves attack on hydrogen atoms via a more or less linear transition state (**1**). Attack at carbon in acyclic alkanes is excluded by a number of experimental observations including the absence of rupture of carbon–carbon bonds. Reactions proceeding via transition state **2** are excluded by the observed primary deuterium isotope effects which vary considerably with the nature of the attacking radical or atom. Moreover, when $X^{\cdot} = D^{\cdot}$, no hydrogen–deuterium exchange is observed by EPR spectroscopy for the radicals formed by attack of $D^{\cdot}$ upon

The Chemistry of Alkanes and Cycloalkanes
Edited by S. Patai and Z. Rappoport © 1992 John Wiley & Sons Ltd

methylene groups[1].

$$RH + X^{\cdot} \rightarrow [R\text{---}H\text{---}X]^{\ddagger} \rightarrow R^{\cdot} + HX \tag{1}$$

(1)

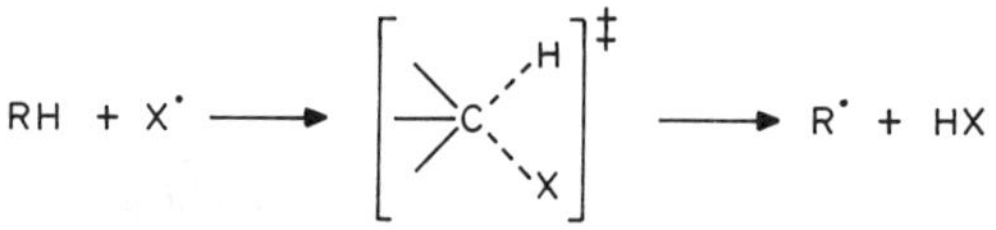

$$RH + X^{\cdot} \longrightarrow \left[\begin{array}{c} \\ \diagdown \\ \diagup C \diagdown \\ \end{array} \substack{,,H \\ \diagdown \\ X} \right]^{\ddagger} \longrightarrow R^{\cdot} + HX \tag{2}$$

(2)

With cyclopropanes, radical attack can lead either to the cyclopropyl radical via 1 or to the ring-opened 3-substituted propyl radical via 2. The relief in ring strain as well as the high bond strength of cyclopropane carbon–hydrgen bonds[2] ($\sim 106\,\text{kcal}\,\text{mol}^{-1}$) are the primary considerations in these competitive processes. With reactive radicals (as judged by the overall ΔH for reaction 1) such as t-BuO$^{\cdot}$, Cl$^{\cdot}$ or 3,3-dimethyl-N-glutarimidyl radical, hydrogen atom abstraction from the cyclopropane predominates[3,4], whereas for less reactive species such as Br$^{\cdot}$ (where ΔH for reaction 1 is highly endothermic) only the ring-opening process is observed[5,6]. With more highly strained polycyclic compounds, such as [1.1.1]propellane, the efficiency of radical attack leading to ring opening increases. Thus reaction of propellane with BrCN forms the telomers 3 with $n = 1$–3 (X = CN, Y = Br)[7] and even higher telomers ($n = 4$ or 5) or polymers are formed with benzoyl peroxide initiated reactions leading to 3 with X = MeO$_2$C, HO$_2$C, (EtO$_2$C)$_2$CH, (EtO$_2$C)$_3$C, (NC)$_2$CH, (EtO$_2$C)$_2$C(Ph), MeO$_2$CCH(OMe), MeO$_2$CCH(CN), (EtO)$_2$P(O) and Y = H[8]. Telomers with X = Y = Br or MeC(O)S are formed upon free radical reactions with Br$_2$ or MeC(O)SSC(O)Me[8] and even hydrogen will add in a free radical manner to give 3, X = Y = H[9].

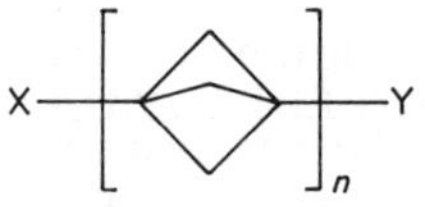

$$X\text{---}\left[\diamondsuit \right]_n\text{---}Y$$

(3)

Hydrogen atom abstraction reactions of alkanes (reaction 1) can be observed in chain reactions such as photohalogenation (Scheme 1), autoxidation (Scheme 2) or in nonchain processes such as shown in Scheme 3. In addition to the bimolecular trapping reactions of R$^{\cdot}$ as typified in Schemes 1–3 the intermediate alkyl radicals can undergo competing unimolecular reactions such as ring opening in the cyclopropylcarbinyl radical (reaction 3) or the rupture of an unstrained carbon–carbon bond in alkane pyrolysis. In the absence of a trapping reagent, internal hydrogen atom transfers can occur in alkyl radicals (reaction 4) while adduct radicals arising from alkyl radicals often are observed to participate in an internal 1,5-hydrogen atom transfer, e.g. reactions 5 and 6[10,11]. Because of the possibility of intramolecular or intermolecular hydrogen atom transfers, the observed radical trapping products are not necessarily a quantitative measured of the point of initial attack upon the alkane. The presence of radical rearrangements involving hydrogen atom migrations can in principle be detected by varying the concentration of the radical trapping reagent or by using appropriately deuterium-labeled alkanes[12].

$$X_2 \xrightarrow{hv} 2\,X^{\bullet}$$

initiator $\rightarrow R^{\bullet}$

$$X^{\bullet} + RH \rightarrow HX + R^{\bullet}$$

$$R^{\bullet} + O_2 \rightarrow ROO^{\bullet}$$

$$R^{\bullet} + X_2 \rightarrow RX + X^{\bullet}$$

$$ROO^{\bullet} + RH \rightarrow ROOH + R^{\bullet}$$

$$2\,X^{\bullet} \xrightarrow{3rd\ body} X_2$$

$$2\,ROO^{\bullet} \rightarrow nonradical\ products$$

SCHEME 1   SCHEME 2

$$PhN{=}NCPh_3 \xrightarrow{\Delta} Ph^{\bullet} + N_2 + Ph_3C^{\bullet}$$

$$Ph^{\bullet} + RH \rightarrow PhH + R^{\bullet}$$

$$R^{\bullet} + Ph_3C^{\bullet} \rightarrow Ph_3CR$$

SCHEME 3

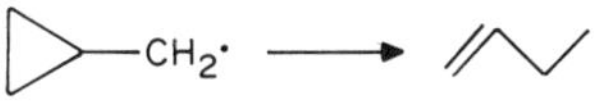

(3)

$$CH_3CH_2CH_2CH_2CH_2^{\bullet} \xrightarrow{E_{act}\,=\,23\,kcal\,mol^{-1}} CH_3\overset{\bullet}{C}HCH_2CH_2CH_3 \quad (4)$$

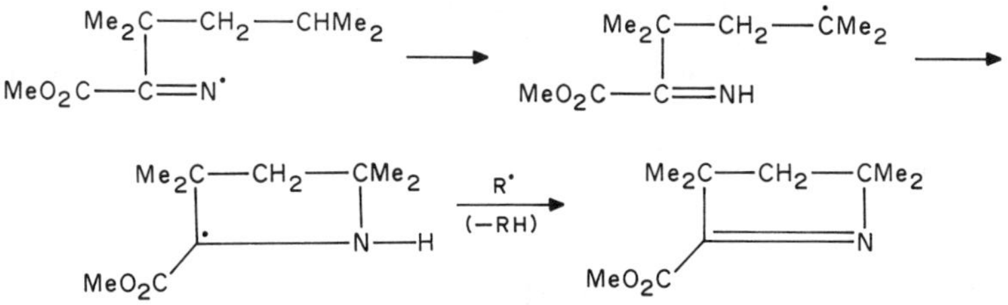

(5)

$$Me_2CHCH_2CHMe_2 + O_2 \xrightarrow{115-120\,^{\circ}C} Me_2C(OOH)CH_2C(OOH)Me_2\,(95\%) \quad (6)$$

The attack of an atom or radical upon an alkane will depend on the energy content of the attacking radical with the rate constant increasing and selectivity decreasing with an increase in temperature. It is usually assumed that the attacking radical species is thermally equilibrated. However, for the more reactive attacking species such as Cl$^{\bullet}$, only a few molecular collisions need occur before a carbon–hydrogen bond is broken. If the attacking species has been generated with excess kinetic energy, e.g. in Scheme 1 where the reaction of R$^{\bullet}$ with Cl$_2$ is highly exothermic, it is not certain that thermal equilibration can compete with hydrogen abstraction, particularly for reactions involving undiluted substrates. Some observations concerning primary deuterium isotope effects in gas-phase chlorination suggest that such hot atom effects may indeed be important. Primary hydrogen–deuterium isotope rate ratios are quite sensitive to the reactivity of the attacking radical as measured by ΔH for reaction 1. Thus, the vapor-phase reaction of Cl$^{\bullet}$ with CH$_4$ has an energy of activation for hydrogen abstraction of about 2.5 kcal mol^{-1}[13] and $k(H)/k(D)$ for the photochlorination of CH$_2$D$_2$ is observed to be 14 at -23, 12 at 0 and 8.2 at 52 °C[14]. Absolute rate measurements for the reaction of CH$_4$ and CD$_4$ with an excess of photochemically generated chlorine atoms in the absence of molecular chlorine lead to similar values, $k(H)/k(D) = 12–13$ at 25 °C decreasing to about 2 at 500 °C with $\Delta E_a = 1.8$ kcal mol^{-1}.[15] On the other hand, the energy of activation for attack of Cl· upon the 3°-hydrogen atom of isobutane is no greater than 0.3 kcal mol^{-1} in the gas phase[16] and probably very close to 0. The photochlorination of Me$_3$CH and Me$_3$CD yields $k(H)/k(D)$ as only 1.4 in the liquid phase at -15 °C (based on the ratios of t-BuCl/i-BuCl and t-BuCl/Me$_2$CDCH$_2$Cl formed[12]). The gas-phase competitive chlorination of C$_2$H$_6$ and C$_2$D$_6$ yields values of $k(H)/k(D) = 2.6$–2.8 at 27 °C[17,18]. A critical survey of literature data places the energy of activation for Cl· attack upon ethane to be in the range of 0.1–0.2 kcal mol^{-1}[113]. Surprisingly, the absolute rate constants for the reaction of Cl· with C$_2$H$_6$ and C$_2$D$_6$ in the absence of a chain chlorination reaction leads to much higher values of $k(H)/k(D)$ of ~ 6 at 22 °C[19]. Although the competitive chlorination should lead to the more accurate measurement, the discrepancy is too large to be ignored. One possibility is that, in chain reactions involving molecular chlorine, the chlorine atom carries excess kinetic energy from the exothermic (~ 25 kcal mol^{-1}) reaction of C$_2$H$_5$· with Cl$_2$. The chlorine atom is essentially hot and in a low energy of activation process yields a $k(H)/k(D)$ more consistent with a much higher temperature. Vapor-phase competitive photochlorinations in the presence of an inert diluent gas generally give a higher selectivity than liquid-phase chlorinations of neat hydrocarbons, possibly reflecting the thermalization of the chlorine atom in the presence of inert diluents.

An additional complication of chain reactions such as photochlorination in the liquid phase is that, not only is the chlorine atom generated with excess kinetic energy from the propagation reaction, but the Cl· is generated in close proximity to the RCl molecule formed by attack of R· upon Cl$_2$. Although the excess kinetic energy should help the chlorine atom escape from the solvent cage, there is a high probability of a second hydrogen atom abstraction reaction occurring with the RCl molecule in the solvent cage wall unless the cage wall contains molecules which are themselves quite reactive to chlorine atoms[20,21]. This leads to the production of di- and trichlorination products when photochlorinations are performed at low alkane concentrations in solvents which are inert to chlorine atom interaction.

II. REACTIVITY OF HYDROGEN ATOMS IN BRANCHED CHAIN ALKANES

A. Bond Dissociation Energies

The reactivity of carbon–hydrogen bonds towards a given free radical is a function of the bond dissociation energy, since this controls the value of ΔH for reaction 1. Complexities in measuring the enthalpy change in reaction 7, and particularly in measuring the energy of activation for the reaction of R· with HI, have led to widely quoted values of 2°- and 3°-alkyl–hydrogen bond dissociation energies 3–5 kcal mol^{-1} lower than the best current estimates based on dissociation–recombination equilibria measured for reactions 8 and 9[22]. A compilation of 'best values' of bond dissociation energies (b.d.e.) based on dissociation–recombination equilibria and upon a reinvestigation of the energy of activation of the reaction of R· with HI is given in Table 1[23,24].

$$I· + RH \rightleftharpoons HI + R· \tag{7}$$

$$R—R' \rightleftharpoons R· + R'' \tag{8}$$

$$R· \rightleftharpoons \text{alkene} + H· \text{ (or CH}_3·) \tag{9}$$

The bond dissociation energies of a variety of H—X species are given in Table 2. These values when combined with the data of Table 1 define the enthalpy change of reaction 1.

TABLE 1. Bond dissociation energies (b.d.e.) for R—H at 25 °C

R	ΔH_f^0 of R'(25 °C), kJ mol^{-1}	Bond dissociation energy	
		kcal mol^{-1}	kJ mol^{-1}
Me	146 ± 1	105	439
Et	118.5 ± 1.7	101	420
Pr	100 ± 2	101	422
i-Pr	89 ± 3	98	408
sec-Bu	67 ± 3	99	411
t-Bu	48.6 ± 1.7	96	401

TABLE 2. Bond dissociation energies (b.d.e.) of H—X at 25 °C

X	b.d.e. (kcal mol^{-1})
F	135
OH	118
CN	111
NH$_2$	108
CF$_3$	106
C$_6$H$_5$	105
H	104
Cl	103
t-BuO	102
CCl$_3$	96
t-BuOO	90
Br	87
NF$_2$	76
I	71

B. Relative Reactivities of 1°-, 2°- and 3°-Hydrogen Atoms

1. Methane and ethane

The high bond dissociation energy for a C—H bond of methane greatly reduces the reactivity towards radical attack and methane is quite unreactive compared with other simple alkanes. Cyclopropane hydrogen atoms are similarly unreactive as are the methylene hydrogens in many bicycloalkanes containing a one-carbon bridge. Only the most reactive of radicals will abstract a hydrogen atom from methane at ordinary temperatures. Table 3 gives a compilation of data from gas-phase kinetic studies using critical evaluations of literature data where available[25]. As shown in Table 3, there is a general correlation between the energy of activation for attack on methane and the selectivity when methane and ethane are compared. Evans and Polanyi demonstrated that the simple relationship $E_a = \alpha(\Delta H°) + \beta$ holds in many radical methathetical or transfer reactions such as reaction 1[26,27]. The value of α, of course, should be less than 1[28]. In the Evans–Polanyi equation the factor α can be associated with the amount of bond breaking in the transition state. The values of α calculated from the energies of activation for attack of halogen atoms on methane and ethane vary from 0.2 for F' to 0.6 for Cl'. However, with Br' and I' values of $\alpha > 1$ are observed, since the experimental

TABLE 3. Relative reactivities of CH_4 and C_2H_6 at 25 °C

Attacking radical	$E_a(CH_4)$	$\Delta H°(CH_4)$	$k(C_2H_6)/k(CH_4)$ molecule	per H
F˙	1.2–1.9	− 35	3	1:2
Cl˙	2.5	+ 2	560	1:370
Br˙	18	+ 18	3150	1:2100
I˙	34–35	+ 34	1.2×10^6	1:800,000
HO˙	5	− 13	39	1:26
NC˙	1–8	− 6	29	1:19
H˙	12	+ 1	42	1:28
CF_3˙	10–12	− 1	110	1:73
CD_3˙	14	0	∼ 450	∼ 1:300
CCl_3˙	22	+ 9	∼ 355^a	∼ 1:250

^{a} Assuming a constant A value and reported E_a values.

values of $E_a(CH_4) - E_a(C_2H_6)$ exceed the difference in bond dissociation energies (4 kcal mol^{-1}) with $\Delta E_a = 4.8$ for Br˙ and 7.5 for I˙. Certainly with these unreactive species participating in highly endothermic reactions there must be a considerable change in the nature of the transition state when the substrate is changed from methane to ethane. The reaction of alkyl radicals with HI is now postulated to involve an intermediate such as R---İ---H[23] and, by microscopic reversibility, a similar intermediate may be involved in the attack of I˙ upon a carbon–hydrogen bond. (The overall energies of activation for reaction of t-Bu˙ with HI and HBr are reported to be about − 1.5 kcal mol^{-1})[29].

For the carbon-centered radicals CF_3˙, CD_3˙ and CCl_3˙ one would expect an increase in selectivity in the order listed. Values of $\Delta E_a(CH_4$—$C_2H_6)$ are roughly constant at 4 ± 0.2 kcal mol^{-1} and are approximately equal to the change in the b.d.e., although at 164 °C, CF_3˙ is approximately 100 times more reactive than CD_3˙ or CH_3˙ towards both CH_4 and C_2H_6[30].

2. Higher alkanes

The reactivity of methyl groups in alkanes towards radical attack is reasonably constant towards Cl˙ at 25 °C in the gas phase; the reactivities per methyl group relative to ethane vary from 1.2–1.5 in propane, butane, isobutane and neopentane[31]. Intramolecular competition studies can more precisely determine small differences in reactivity than intermolecular competitions or absolute rate measurements. The gas-phase chlorination of $Me_3CCH_2CH_2CH_2Me$ at 101 °C is reported to yield (among other isomers) a 2:1 ratio of $ClCH_2CMe_2CH_2CH_2CH_2Me$ and $Me_3CCH_2CH_2CH_2$-CH_2Cl[32]. This would seem to indicate a much lower reactivity of t-butyl methyl groups compared with the methyl group in a n-alkane. However, in this system the low yield of $ClCH_2CMe_2CH_2CH_2CH_2Me$ is accompanied by an unexpectedly high yield of $Me_3CCH_2CH_2CH(Cl)Me$ (29.6% of the monochlorinated products) when compared with $Me_3CCH_2CH(Cl)CH_2Me$ (21.9%). This suggests the possibility that the chlorination products are not a true measure of the point of Cl˙ attack and that ˙$CH_2CMe_2CH_2CH_2CH_2Me$ may undergo a facile 1,5-hydrogen atom migration to form $Me_3CCH_2CH_2ĊHMe$ in a process which competes with the reaction of Cl_2 with the 1°-alkyl radical.

Direct competition of butane and neopentane with F˙ at room temperature indicates that the four methyl groups of neopentane are 1.8 times as reactive as the two methyl

21. Radical reactions of alkanes and cycloalkanes 969

groups of butane in a reaction with zero energy of activation[33]. In a competition
between ethane and propane the methyl group reactivities showed a slight preference
(35%) for attack on propane[34]. Towards CD_3· the rate constants for attack on
ethane:neopentane:hexamethylethane are in the ratio 2:3.2:5.2, i.e. very close to the
statistical ratio of 2:4:6[35]. In Table 4 some typical $1°:2°:3°$ rate ratios are summarized,
based on intramolecular competition with the assumptions that all methyl groups have
the same reactivity towards a given radical. Where possible, the values of $k(2°)/k(1°)$ and
$k(3°)/k(1°)$ are based on data for n-alkanes, isobutane and 2,3-dimethylbutane.

Table 4 presents a fairly consistent correlation of reactivity and selectivity. The value
of α in the Evans–Polanyi equation increases as the strength of the new bond (HX in

TABLE 4. Relative reactivities (per atom)

Attacking radical	Conditions	$k(2°)/k(1°)$	$k(3°)/k(1°)$	References
F·	25 °C, gas	1.1–1.4	1.4	33, 34, 36, 37
Cl·	25 °C, gas	3.7–3.8	5	31
Cl·	25 °C, liquid	2.2–3.0	3.5–4.5	12,38
'Cl''	25 °C, 4 M C_6H_6	5	20	39
'Cl''	25 °C, 12 M CS_2	25	225	39
H·	25 °C, liquid	6–7	~90	40
H·	480 °C, gas	6–6.5	26	41
D·	35 °C, liquid	5	42	42
HO·	480 °C, gas	3.3	8.6	41
t-BuO·	40 °C, liquid	12	50	43
Ph·	60 °C, liquid	9	44	44
p-$O_2NC_6H_4$·	60 °C, liquid	11	50	45
CH_3·	77 °C, gas	20	215	46
CD_3·	164 °C, gas	10	100	30
CF_3·	184 °C, gas	10	60	47
CCl_3·	150 °C, gas	70	~1300	48
Br·	150 °C, gas	80	~1600	49,50
NF_2·	100 °C, gas	8–30	~700	51
I·	25 °C, gas	~1850	~210,000	52
PhICl·	40 °C, liquid	21	370	53
Cl_3CS·	0 °C, liquid	29–33	110–120	54
ClO·	95 °C, gas	22	71	55
$R_3CC(CO_2Me){=}N$·	98 °C, liquid	~12	~55	10
	25 °C, liquid	$k(1°) > k(2°) > k(3°)^a$		56
$PhCO_2$·	100 °C, liquid	$k(3°)/k(2°) = 17$–22		57
SO_2Cl·	55 °C, liquid	> 10		58
	25 °C, liquid	$k(3°)/k(2°) = 16$–20		59
t-BuOO·	50 °C, liquid	See Table 9, $k(3°)/k(2°) = 10$–20		60

[a]$k(1°$ at C-4$) > k(2°)$ in 2,2-dimethylbutane and $k(1°) > k(3°)$ in 2-methylbutane.
[b]Cytochrome P-450 model formed by oxygen atom transfer to chloro(5,10,15,20-tetra-o-tolylporphyri-
nato)iron(III). With adamantane $k(3°)/k(2°)$ varies with the aromatic substituent on the porphyrin ring from 11
to 20 to 4 for the tetramethyl, tetraphenyl and tetra-o-tolyl derivatives[60].

TABLE 5. Values of $k(2°)/k(1°)$ on a per-hydrogen basis for heptane photochlorination with Cl_2

	Position				
Conditions	2	3	4	Average	Reference
Neat, 20 °C	3.0	2.9	2.8	2.9	62[a]
Neat, 80 °C	2.1	1.9	1.8	2.0	63
Gas, 100 °C				2.4	64
Gas, 260 °C				2.0	64
C_6H_6 soln., 20 °C	8.1	9.9	9.1	8.9	65
CS_2 soln., 20 °C	27	35	31	30	66
Cl_2C=$C(Cl)C(Cl)$=CCl_2					
soln., 80 °C	2.7	2.4	2.0	2.4	64

[a]Average of several experimental reports.

reaction 1) being formed decreases. Thus, for $CH_3^{\cdot}$ and $CF_3^{\cdot}$ α is ~ 0.5 while α increases to 0.86 for $Br^{\cdot}$, 0.90 for $NF_2^{\cdot}$ and 0.97 for $I^{\cdot}$[51]. The E_a value for the gas-phase reaction of chlorine atoms with 1° carbon–hydrogen bonds is no more than $0.2\,kcal\,mol^{-1}$[13] and is essentially zero for the reaction of 2° and 3° carbon–hydrogen bonds[25]. Nevertheless, there is an appreciable 1°:2°:3° selectivity which is somewhat greater in the gas phase than in neat hydrocarbon solution. A careful study of the photochlorination of hexane in the gas phase led to the equation[61]

$$k(2°)/k(1°) = (2.2 \pm 0.6)\exp\left[(0.214 \pm 0.127)/RT\right]$$

On the other hand, the selectivities in solution seem to be mainly determined by energy of activation differences. Thus, in the liquid phase hexane yields the equation

$$k(2°)/k(1°) = (0.8 \pm 0.2)\exp\left[(0.597 \pm 0.020)/RT\right]$$

A variety of data for heptane are collected in Table 5.

There is no generally accepted explanation for the differences in selectivities for gas- and liquid-phase photochlorinations. As mentioned earlier, in the liquid phase a chlorine atom formed by the reaction $R^{\cdot} + Cl_2 \rightarrow RCl + R^{\cdot}$ has a good chance of reacting with a molecule in the wall of solvent cage before the chlorine atom can diffuse out of the cage. The $Cl^{\cdot}$ may also possess excess kinetic energy (from ΔH of the reaction in which it was formed) and this energy can be shared with the cage ensemble and raise the effective temperature of the cage considerably above the ambient temperature of the solution. This would lead to a lower selectivity if indeed there is an appreciable E_a value for hydrogen atom abstraction. The temperature effects routinely observed in liquid-phase chlorinations do suggest a higher ΔE_a value than in the gas phase, but this may reflect barriers to the orientation of the hydrocarbon molecules (in the cage walls surrounding the $Cl^{\cdot}$) to form the nearly linear transition state depicted in 1.

3. Solvent effects and complexed chlorine atoms

The appreciable effect of certain solvents in liquid-phase photochlorination reactions is the result of the formation of complexes between the extremely electronegative $Cl^{\cdot}$ and the solvent. These can be σ-complexes such as $CS_2Cl^{\cdot}$, $SO_2Cl^{\cdot}$, $PhICl^{\cdot}$, $C_5H_5NCl^{\cdot}$ or π-complexes as observed for most aromatic molecules[67–69]. As expected, the selectivity increases with the concentration of the complexing agent. Table 6 presents data for the relative reactivity of the 3°- and 1°-hydrogen atoms in 2,3-dimethylbutane (DMB) and the 2°- and 1°-hydrogen atoms in heptane in CS_2 and PhH solutions.

TABLE 6. Solvent effects in the photochlorination of 2,3-dimethylbutane[68,70] and n-heptane[66,70] at 20–25 °C

	DMB	n-Heptane $[k(2°)/k(1°)]^a$		
Solvent	$k(3°)/k(1°)^a$	C-2	C-3	C-4
Neat	4.2	3.0	2.9	2.8
CS_2, 2 M	15			
CS_2, 8 M	105			
CS_2, 12 M	225	27	35	31
PhH, 2 M	11			
PhH, 4 M	20			
PhH, 8 M	49	8.1	9.6	9.1

^{a} Per hydrogen atom.

The π-complex of a chlorine atom with an aromatic molecule can approach the selectivity of a bromine atom. The observed selectivity of a chlorine atom in the presence of a complexing agent also depends upon the substrate concentration[70], because the rate of the reaction of Cl· with the substrate and the complexing agent are of the same magnitude[21]. Scheme 4 presents the absolute rate constants that have been deduced by a study of the chlorination of DMB by the laser flash photolysis technique[21].

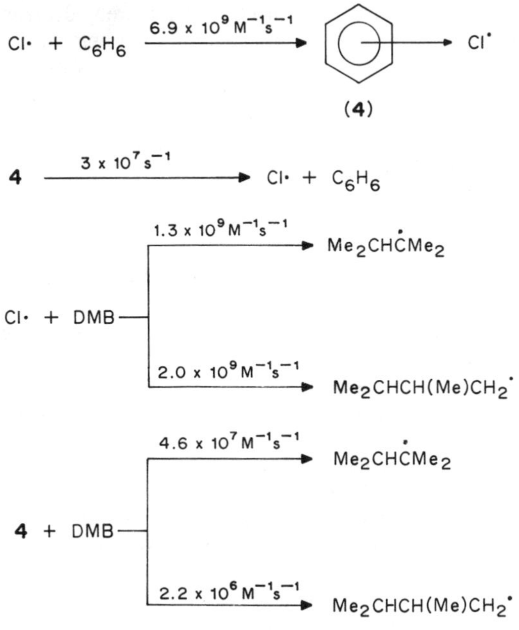

SCHEME 4. Chlorination of 2,3-dimethylbutane (DMB) in benzene solution

The complex of 4 gives a $k(3°)/k(1°)$ selectivity of 126, considerably greater than the selectivity of ~ 50 observed in 8 M benzene where both the free chlorine atom and the complexed atom abstract hydrogen atoms from the alkane.

The selectivity of the chlorine atom in complexing aromatic solvents increases with a decrease in the ionization potential of the solvent up to the point where the π-complex is so unreactive that most of the reaction occurs via the free Cl˙. Thus, selectivity for attack on DMB at 4 M aromatic concentration increases from $PhNO_2$ to $PhCF_3$ to PhCl to PhF to PhH and $1,3\text{-}Me_2C_6H_4$. However, making the aromatic molecule a better complexing agent, e.g. $1,3,5\text{-}Me_3C_6H_3$, $1,2,4,5\text{-}Me_4C_6H_2$ or Me_5C_6H, results in a decrease in selectivity because a higher fraction of the hydrogen atom abstraction occurs by attack of the uncomplexed chlorine atom[71]. Pyridine forms a complex with Cl˙, which is known to be a σ-complex with the chlorine attached to the nitrogen atom and in the plane of the pyridine ring[72]. Although the structure of this complex is quite different from the arene π-complex 4, photochlorination of DMB in CCl_4 with 2 M pyridine gave about the same ratio of 3° to 1° attack as observed with 2 M $1,4\text{-}(t\text{-}Bu)_2C_6H_4$ forming the 3°- and 1°-alkyl chlorides in a 8.0–8.5:1 ratio at room temperature[71]. Under these conditions 2 M PhI gives a 3°/1° ratio of chlorides of 23, $k(3°)/k(1°) = 138$.

Photochlorination of low concentrations of substrates such as DMB in inert solvents leads to high percentages of di- and trichlorination products even at low conversions of the substrate[20]. The polychlorination is drastically reduced by the presence of complexing solvents. Reaction in an inert solvent of an alkyl radical with molecular chlorine forms a molecule of the alkyl chloride and a chlorine atom with excess kinetic energy. Under these conditions there is a high probability that the Cl˙ will react with abstraction of a hydrogen atom from the alkyl chloride in the wall of the cage of molecules surrounding the chlorine atom before the chlorine atom can diffuse out of the solvent cage. In the presence of solutes such as benzene, the cage walls will contain molecules which readily complex the chlorine atoms, thereby decreasing their reactivity and allowing them to escape from the molecule of alkyl chloride and thus decreasing the amount of polychlorination. Quantitative measurements of the effects of different aromatic solvents upon the ratio of mono- to polychlorinated neopentane in $Cl_2C(F)C(Cl)F_2$ has provided a measurement of the relative rates of complexation of Cl˙ by aromatic solvents[73]. As would be expected for a process involving diffusion out of a cage, viscosity of the solvent can have an effect on the ratio of mono- to polychlorination products[74].

'Complexed chlorine atoms' can be generated directly by alkyl radical attack upon species such as $PhICl_2$, SO_2Cl_2 or PCl_5. In the case of PhICl˙ little dissociation into PhI and free Cl˙ occurs and high selectivities are observed (Table 4). With SO_2Cl_2, dissociation to the free chlorine atom occurs readily. With reactive alkanes at low temperatures (e.g. 55 °C), SO_2Cl_2 chlorinations are considerably more selective than photochlorinations with Cl_2[58]. Thus, for DMB at 55 °C a $k(3°)/k(1°)$ reactivity of 10 is observed for SO_2Cl_2 vs 3.7 for Cl_2. Addition of C_6H_6 gives rise to the benzene-complexed chlorine atom and, if the SO_2 is allowed to escape, essentially the same selectivities are observed in chlorination by the two reagents. At higher temperatures, or with hydrocarbons with a low reactivity, there are only minor differences between the product ratios observed with SO_2Cl_2 and Cl_2[75]. Photochlorination of octane by PCl_5/Bz_2O_2 at ~ 100 °C, or with Cl_2/PCl_3, gives a very high selectivity (approximately equivalent to that observed in CCl_3˙ attack), indicating that the species PCl_4˙ is capable of abstracting a hydrogen atom[66].

The radical Cl_3CSO_2˙ (formed by alkyl radical attack on Cl_3SO_2Cl) can dissociate into SO_2 and CCl_3˙[76]. In Table 7 the selectivities observed are identical for the benzoyl peroxide (Bz_2O_2) initiated chlorination of n-octane with CCl_4 or CCl_3SO_2Cl at 98 °C, suggesting that complete dissociation of CCl_3SO_2˙ to CCl_3˙ occurred[66]. From Table 7 it is also obvious that N-chlorosuccinimide is reacting via a chlorine atom chain because of the facile reaction with HCl (reaction 10).

TABLE 7. Chorination of n-octane at 98 °C[66]

| Reagents[a] | % of monochlorides $^{A}CH_3$—$^{B}CH_2$—$^{C}CH_2$—$^{D}CH_2$—C_4H_9 | | | |
	H^A	H^B	H^C	H^D
t-BuOCl, Bz$_2$O$_2$	8	40	28	25
CCl$_4$, Bz$_2$O$_2$	2	42	30	26
Cl$_3$CSO$_2$Cl, Bz$_2$O$_2$	2	42	30	26
Cl$_3$SCl, hv	5	32	33	31
PCl$_5$, Bz$_2$O$_2$	2	28	35	35
NCS, Bz$_2$O$_2$	15	31	29	25
Cl$_2$, hv, 20 °C	14	30	28	27

[a]Bz$_2$O$_2$ = benzoyl peroxide; NCS = N-chlorosuccinimide.

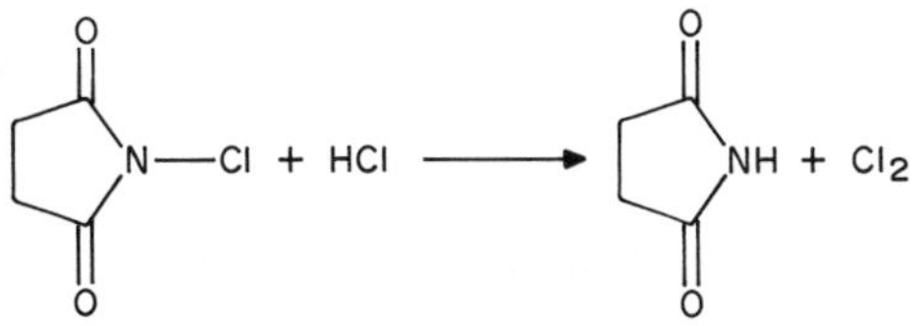

(10)

4. Steric effects in radical attack upon alkanes

The values of $k(2°)/k(1°)$ and $k(3°)/k(1°)$ of Table 4 have been derived mainly from the simplest of alkanes. With highly branched alkanes steric effects become apparent even with the reactive Cl˙. With sterically hindered radicals such as the amine radical cation derived from N-chloro-2,2,6,6-tetramethylpiperidinium ion, the reactivity of the 2-position of pentane is 5 times that of the more hindered 3-position[56]. In general, methylene hydrogen atoms near the end of a long alkane chain are more reactive towards radical attack than methylene groups nearer to the center of the chain; e.g., see Table 7.

With 2,3,4-trimethylpentane the two 3°-hydrogen atoms are quite different in reactivity towards Cl˙ and even the two kinds of methyl groups are differentiated by the hindered RSO$_2$N(t-Bu)˙ radicals (Table 8)[75]. With the sterically hindered amino radicals the 3°-hydrogen atom at the 3-position is about twice as reactive as the 3°-hydrogen at the

TABLE 8. Chlorination of 2,3,4-trimethylpentane at 85 °C

| Chlorinating system | % of monochlorides [(Me^A)$_2$CH^B]$_2$CH^{C}Me^D | | | |
	H^A	H^B	H^C	H^D
SO$_2$Cl$_2$, Bz$_2$O$_2$	40	24	26	10
MeSO$_2$N(t-Bu)Cl, Bz$_2$O$_2$	52	19	19	9
PhSO$_2$N(t-Bu)Cl, Bz$_2$O$_2$	64	15	13	8
Cl$_3$CSO$_2$Cl, Bz$_2$O$_2$	5	71	24	—

2-position, possibly because of relief in strain in forming radical **5**. However, the methylene hydrogen atoms in 2,2,4,4-tetramethylpentane do not have an increased reactivity towards Cl·, suggesting there is little relief in strain in forming **6**.

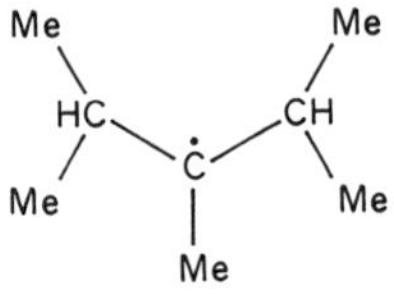

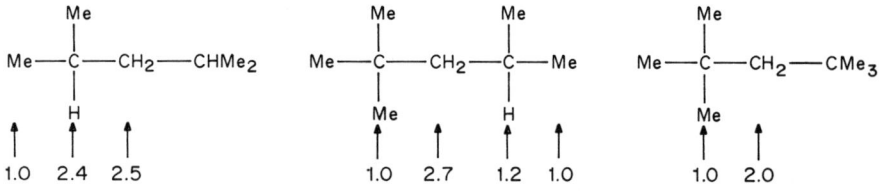

(5) 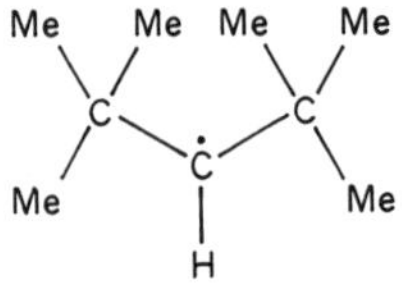   (6)

Photochlorination at 40 °C of 2,4-dimethylpentane and 2,2,4-trimethylpentane actually give examples where the 2°-hydrogen atoms at C-3 are more reactive than the 3°-hydrogen atoms at C-2 (Scheme 5)[77]. For 2,4-dimethylpentane and 2,2,4-trimethylpentane the 2°-hydrogen atoms have a normal reactivity relative to the 1°-hydrogen atoms whereas the reactivity of the 3°-hydrogen atoms are greatly reduced by the presence of isobutyl or neopentyl substituents at the 3°-position. These substituents apparently obstruct the approach of Cl· to the 3°-position.

$$
\begin{array}{ccc}
\underset{\uparrow\ \ \uparrow\ \ \uparrow}{\text{Me}-\overset{\displaystyle\text{Me}}{\underset{\displaystyle\text{H}}{\text{C}}}-\text{CH}_2-\text{CHMe}_2} &
\text{Me}-\overset{\displaystyle\text{Me}}{\underset{\displaystyle\text{Me}}{\text{C}}}-\text{CH}_2-\overset{\displaystyle\text{Me}}{\underset{\displaystyle\text{H}}{\text{C}}}-\text{Me} &
\text{Me}-\overset{\displaystyle\text{Me}}{\underset{\displaystyle\text{Me}}{\text{C}}}-\text{CH}_2-\text{CMe}_3
\end{array}
$$

    1.0   2.4   2.5   1.0   2.7   1.2   1.0   1.0   2.0

SCHEME 5. Relative reactivities per hydrogen atom in photochlorination at 40 °C[77]

The rather approximate values of the 1°:2°:3° rate ratios of Table 4 are further illustrated by the reported selectivities for chlorine atom attack on 2-methylheptane at 20 °C (Scheme 6)[75]. Again an isobutyl substituent appears to reduce greatly the reactivity of the methylene hydrogens at C-4.

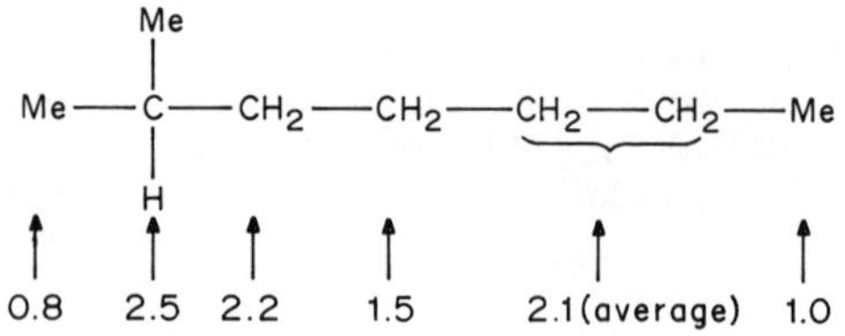

$$
\text{Me}-\overset{\displaystyle\text{Me}}{\underset{\displaystyle\text{H}}{\text{C}}}-\text{CH}_2-\text{CH}_2-\text{CH}_2-\text{CH}_2-\text{Me}
$$

    0.8   2.5   2.2   1.5   2.1(average)   1.0

SCHEME 6. Relative reactivities of hydrogen atoms in 2-methylheptane towards Cl· at 20 °C[75]

The reaction of t-BuOO· with a series of alkanes in the liquid phase at 30 °C has been studied by the AIBN-initiated autoxidation of the alkane in the presence of 1–2 M t-BuOOH[60]. Trapping of the alkyl radical by oxygen leads to a peroxy radical which readily undergoes hydrogen atom transfer with t-BuOOH to regenerate the hydrogen-abstracting species t-BuOO·. The results of co-autoxidation with 3-methyl-pentane are given in Table 9, where absolute rate constants (per hydrogen atom) have been calculated from product analysis and a rate constant for tertiary attack on 3-methylpentane of $7 \times 10^{-3}\ \text{M}^{-1}\ \text{s}^{-1}$.

TABLE 9. Reactivity of carbon–hydrogen bonds towards t-BuOO$\cdot$ at 30 °C[60]

Hydrocarbon	$k(3°)^a$	$k(2°)^a$
Me$_2$CHCH$_2$Me	7×10^{-3}	1.4×10^{-4}
Me$_2$CHCH$_2$CH$_2$Me	7×10^{-3}	3.3×10^{-4} (C-4)
		1.0×10^{-4} (C-3)
MeCH$_2$CH(Me)CH$_2$Me	7×10^{-3}	
MeCH$_2$CH$_2$CH$_2$CH$_2$Me	—	3.5×10^{-4} (C-3)
		4.4×10^{-4} (C-2)
Me$_2$CHCHMe$_2$	9.3×10^{-3}	—
Me$_3$CCHMe$_2$	$13 \ \times 10^{-3}$	—
Me$_2$CHCH(Me)CH$_2$Me	8.9×10^{-3} (C-3)	—
	5.5×10^{-3} (C-2)	—
Me$_2$CHCH$_2$CHMe$_2$	2.5×10^{-3}	—
Me$_3$CCH(Me)CH$_2$Me	3.2×10^{-3}	—
Me$_3$CCH$_2$CHMe$_2$	1.5×10^{-3}	—
Cyclohexane	—	2.6×10^{-4}
Cyclopentane	—	8.7×10^{-4}
Methylcyclohexane	10×10^{-3}	—
Methylcyclopentane	21×10^{-3}	—

^{a}Per hydrogen atom in M^{-1} s^{-1}.

The most reactive secondary hydrogen atoms are nearly as reactive as the least reactive tertiary hydrogen atoms, although in general a value of $k(3°)/k(2°)$ equal to 10–20 is observed. The low reactivity of the 3°-hydrogen atoms in Me$_2$CHCH$_2$CHMe$_2$ or Me$_3$CCH$_2$CHMe$_2$ is consistent with the chlorination results (Schemes 5 and 6) and reflects hindrance to radical attack as a 3°-carbon center containing an isobutyl or neopentyl substituent.

The E_a value for reaction of t-BuOO$\cdot$ with 3-methylpentane is 1603 ± 0.07 for 3°-attack and 17.4 ± 0.8 kcal mol^{-1} for 2°-attack. Surprisingly, the intermolecular hydrogen–deuterium primary isotope effect is 24–28 (as measured with methylcyclohexane-d$_{14}$ or 3-methylpentane-d$_{14}$) and is much greater than the theoretical limit of ~ 10 for complete loss of the zero-point energies for carbon–hydrogen bonds in the transition state[60].

5. Reactivity of cycloalkanes

The methylene groups in cycloalkanes show only a small difference in relative reactivity towards Cl$\cdot$ for the C$_4$–C$_8$ cycles (Table 10)[3,69]. These differences are accenturated

TABLE 10. Relative reactivities (per methylene group) of cycloalkanes towards Cl$\cdot$

Cycloalkane	Solvent, temperature		
	neat, 40 °C[69]	12 M CS$_2$, 40 °C[69]	5 M CCl$_4$, 0 °C[3]
C$_3$			0.05
C$_4$			0.84
C$_5$	1.0	1.1	0.95
C$_6$	1.0	1.0	1.0
C$_7$	1.1	2.0	
C$_8$	1.6	3.8	

by using the more selective complexed chlorine atom in $12\,M\ CS_2$[69]. Cyclopropane is, of course, much less reactive than the other cycloalkanes because of the high carbon–hydrogen bond strength.

Towards Br· at 60 °C the relative reactivities of cyclopentane and cyclohexane hydrogen atoms are $\sim 9:1$[78] while towards t-BuOO· at 30 °C (Table 9) the reactivity difference is 2–3.

6. Radical reactions of spiroalkanes

The radicals **7** ($n = 1$–4) are formed by t-BuO· attack upon the hydrocarbons in the cavity of an EPR spectrometer[79,80]. Since **7** is a cyclopropylcarbinyl radical, ring opening to **8** can be expected. However, the spiro[2.2]pentyl radical (**7**, $n = 1$) did not undergo ring opening up to 110 °C. The vapor-phase chlorination of spiropentane produces a mixture of products, but none appears to be derived from radical **8** ($n = 1$) or from the cyclopropyl radical **9** (Scheme 7)[81]. Instead, the products are derived from radicals **7** ($n = 1$), **10** and its cyclopropylcarbinyl ring-opening product **11**. Lower temperatures favored the formation of **10** over **7** and reduced the amount of **10** converted to **11**. In CCl_4 solution at 20 °C the higher concentration of Cl_2 completely supressed the conversion of **10** to **11**[82]. Photobromination or photoiodination of spiropentane at 20 °C produced the dibromide or diiodide expected from radical **10** with Br or I in place of Cl[79]. Although spiropentane is not more reactive than cyclopropane in hydrogen abstraction by t-BuO· or CF_3·, it is > 100 times as reactive as cyclopropane in S_H2 attack at carbon by Br·[79].

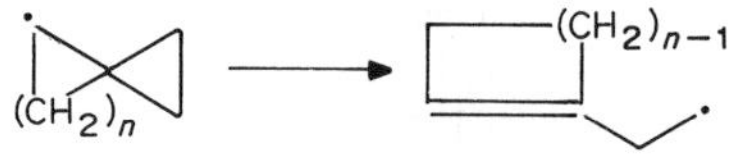

$$(11)$$

(7)   **(8)**

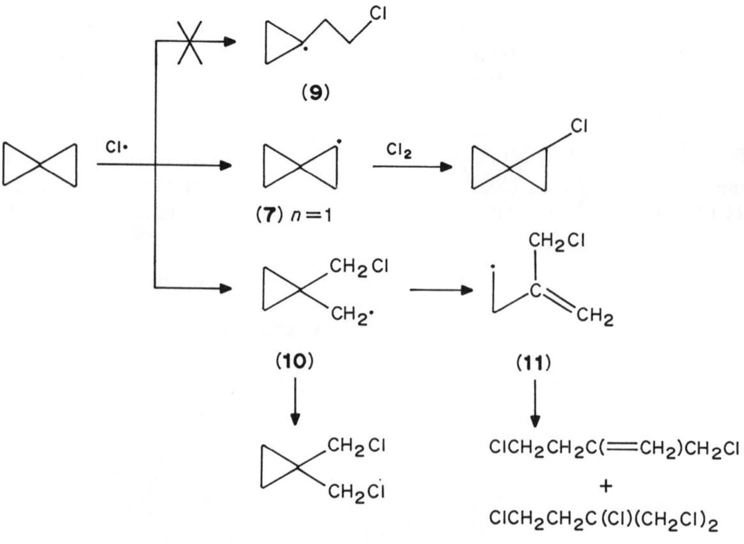

SCHEME 7. Chlorination products of spiropentane[81]

Both radicals **7** and **10** are cyclopropylcarbinyl radicals and are stabilized relative to simple alkyl radicals or to a cyclopropyl radical such as **9**. The cyclopropylcarbinyl radical (**12**) exists in a preferred bisected conformation with an energy barrier for C—C bond rotation (interconversion of H_A and H_B) of $2.8\,\text{kcal mol}^{-1}$ reflecting a cyclopropylcarbinyl resonance stabilization of $2.5\,\text{kcal mol}^{-1}$[183].

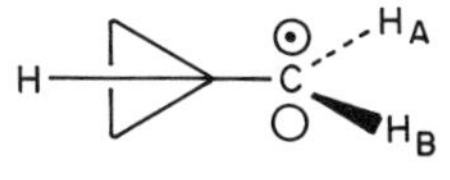

(12)

The rate constant for ring opening of **7** ($n = 1$) cannot be greater than $10^4\,\text{s}^{-1}$ at $25\,°\text{C}$ ($E_a > 12\,\text{kcal mol}^{-1}$)[80]. However, the radicals with $n = 2$–4 undergo facile ring opening. With $n = 2$ the ring opening occurs $\sim 1/10$ as fast as for **12** while with $n = 3$ or 4 reaction 11 occurs more rapidly than reaction 3. Table 11 summarizes data based on EPR measurements[80,83].

Photobromination of spirohexane produces only the products of S_H2 attack of Br· upon the cyclopropyl carbon atom (Scheme 8). However, in contrast to spiropentane, both the cycloalkylcarbinyl and cycloalkyl radicals are formed with the latter species predominating[80].

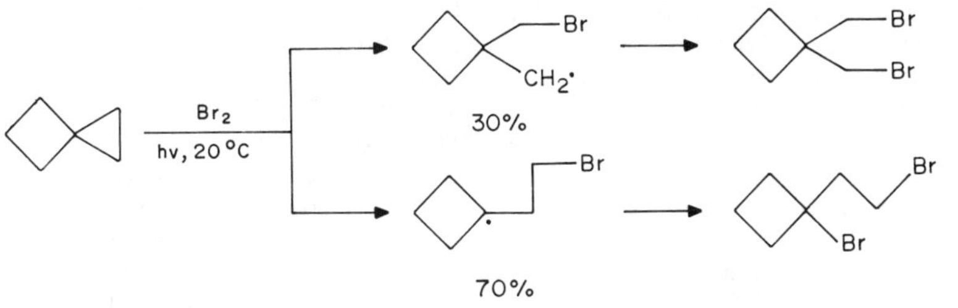

SCHEME 8. Photobromination of spirohexane

In photochlorination of spirohexane, hydrogen atom abstraction from the cyclobutane ring greatly predominated over reactions of the cyclopropyl ring (Scheme 9)[84.]

TABLE 11. Cyclopropylcarbinyl radical ring opening[80,83]

System	$k(25\,°\text{C})(\text{s}^{-1})$	$E_a(\text{kcal mol}^{-1})$
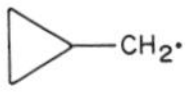 ▷—CH₂·	1.3×10^8	5.9
7, $n = 1$	$< 10^4$	> 12
7, $n = 2$	1×10^7	7.2
7, $n = 3$	$> 10^{8.5}$	< 6.5
7, $n = 4$	$> 10^{8.5}$	< 6.5
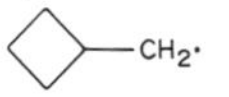 ◇—CH₂·	4.7×10^3	12.1

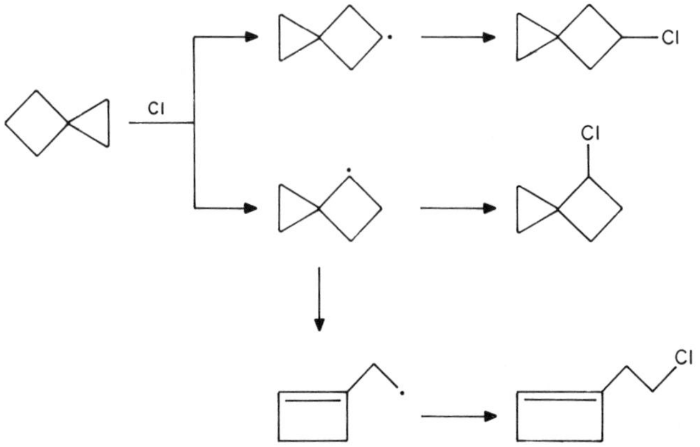

SCHEME 9. Photochlorination of spirohexane[84]

7. Radical reactions of bicyclo[m.n.0]alkanes

Iodination of bicyclo[1.1.0]butane leads to 1,3-diiodocyclobutane, but chlorination or bromination led to complex product mixtures[85]. Bicyclo[2.1.0]pentane reacts with Cl· or Br· to form products derived from the 3-halocyclopentyl radical as a result of reaction 12[86-88]. Trichloromethyl, t-butoxy or N-succinimidyl radicals preferentially extract a hydrogen from the 2-position to yield a cyclopropylcarbinyl-type radical which undergoes ring opening with a rate constant of $2.4 \times 10^9 \, s^{-1}$ at $37 \, °C$[88,89]. Bicyclo[2.1.0]pentane reacts with cytochrome P-450 to produce a mixture of ~ 7 parts of the unrearranged alcohol to 1 part of the ring-opened alcohol, whereas methylcyclopropane yields only cyclopropylmethanol[90]. These results are consistent with an oxygen rebound mechanism for P-450 oxidations of saturated hydrocarbons which can occur with retention of stereochemistry (Scheme 10), with k(rebound) $\approx 2 \times 10^{10} \, s^{-1}$[89]. Interestingly, bicyclo[2.1.0]pentane reacts with P-450 by hydrogen abstraction and rebound only at the endo face[90].

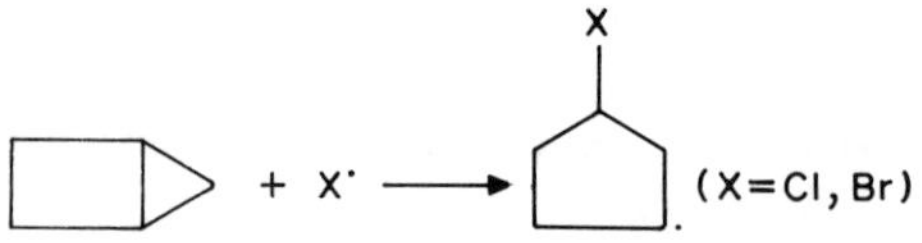

(12)

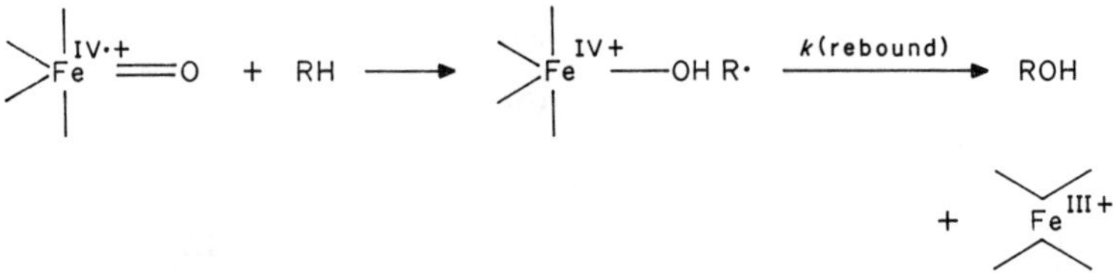

SCHEME 10. Oxygen rebound mechanism for cytochrome P-450 oxidation

Oxidation of cis-decahydronaphthalene with chloro(5,10,15,20-tetraarylporphyrinato)iron(III) in the presence of the oxygen transfer agent PhIO, also gives a stereospecific

TABLE 12. Decomposition of 9-carbo-t-butylperoxydecalins in the presence of O_2 at 50 °C in 1,2-dimethoxyethane[91]

Perester	O_2 pressure (atm)	trans-9-Decalol/cis-9-decalol[a]
cis	1	85/15
trans	1	91/9
cis	146	61/39
cis	340	46/54
cis	545	30/70

[a] After LiAlH$_4$ reduction.

hydroxylation expected for a rebound mechanism with $k(3°)/k(2°) = 16$. cis-9-Decalol is formed as the major product and insignificant amounts of trans-9-decalol are detected[59]. Generation of the cis- and trans-9-decalyl radicals in the presence of oxygen at 50 °C by the decomposition of the 9-carbo-t-butylperoxydecalins leads to hydroperoxides which can be reduced to cis- and trans-9-decalols by LiAlH$_4$[91]. Table 12 presents data indicating the difficulty of trapping the cis-9-decalyl radical before it isomerizes to the more stable trans form.

Somewhat surprisingly, the cis-9-decalyl radical can be efficiently trapped by 0.5 M perbenzoic acid at 100 °C[57]. Reaction of perbenzoic acid with decalin is presumed to proceed according to Scheme 11. Table 13 summarizes pertinent data for this reaction.

$$R^{\cdot} + PhC(O)OOH \longrightarrow ROH + PhCO_2^{\cdot}$$

$$PhCO_2^{\cdot} + RH \longrightarrow PhCO_2H + R^{\cdot}$$

or

$$PhCO_2^{\cdot} \longrightarrow Ph^{\cdot} \xrightarrow{RH} PhH + R^{\cdot}$$

SCHEME 11.

Ring opening in radicals 13 depends upon the value of n (Scheme 12)[92]. Radical 13 ($n = 0$) can be observed by EPR at low temperatures but 13 ($n = 1$) rearranges to the 3-cyclopentyl radical too rapidly to be observed. Although at low temperatures 13 with $n = 2$–5 rearranges preferentially to the cycloalkenylmethyl radicals, chemical reactions at higher temperature indicate that both decomposition pathways occur for $n = 2$. Photochlorination of bicyclo[3.1.0]hexane with t-BuOCl at 70–85 °C produces only the

TABLE 13. 9-Decalols produced in the reactions of PhCO$_3$H with decalins at 100 °C[57]

PhCO$_3$H (M)	trans-9-Decalol/cis-9-decalol
cis-decalin	
0.1	36/54
0.2	16/84
0.5	3/97
trans-decalin	
0.1	97/3
0.2	(95/5)
0.5	95/5

unrearranged 2- and 3-chloro isomers (Scheme 13)[93]. However, chloroformylation with ClCOCOCl at 70–100 °C yields products derived from both the 4-cyclohexenyl and the 2-cyclopentenylmethyl radical. The chloroformylation reaction involves the radical sequence in Scheme 14[94–97]. Oxalyl chloride is apparently a much poorer radical trap than *t*-BuOCl. Thus the photochemical reaction of oxalyl chloride with bicyclo[3.1.0]-hexane leads to both rearranged and unrearranged products as shown in reaction 13[93]. Bicyclo[3.1.0]hexane undergoes photobromination to yield products from Br⋅ attack on the cyclopropyl carbon atoms (Scheme 15). Although bicyclo[2.1.0]pentane reacted only via path (a) of Scheme 15[88], bicyclo[3.1.0]hexane reacts by paths (a), (b) and (c) with path (a) predominating[93]. However, for bicyclo[4.1.0]heptane paths (b) and/or (c) predominate over (a), although the major reaction product appears to be formed by hydrogen abstraction at C-2[92]. Photochlorination of bicyclo[4.1.0]heptane in the vapor phase with low Cl_2 concentrations at 114 °C produces as the major product the ring-opened product expected from Scheme 12 with $n = 3$, but *syn* and *anti*-3-chloronorcarane were also formed[98]. With *t*-BuOCl at 25 °C the 2-norcaranyl radical was largely trapped before ring opening occurred giving 14 and 15 in a ratio of ≈ 1:2. With *t*-BuOCl the 3-chloronorcaranes are also formed in a 3:1 ratio of *anti* to *syn* isomers.

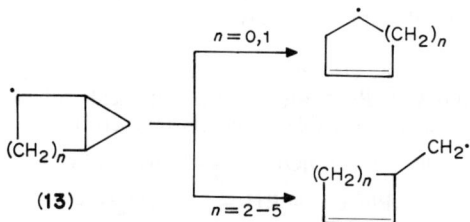

SCHEME 12. β-Elimination in bicyclo[n.1.0]-alkyl radicals generated by *t*-BuO⋅ on attack in the cavity of an EPR spectrometer[92]

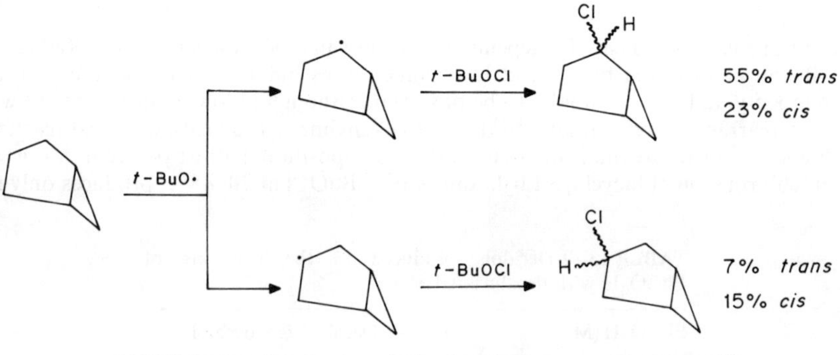

SCHEME 13. Chlorination of bicyclo[3.1.0]hexane with *t*-BuOCl[93]

$$ClCOCOCl \xrightarrow{h\nu} 2ClCO\cdot$$
$$ClCO\cdot \longrightarrow Cl\cdot + CO$$
$$RH + Cl\cdot \longrightarrow R\cdot + HCl$$
$$R\cdot + ClCOCOCl \longrightarrow RCOCl + ClCO\cdot$$

SCHEME 14. Free radical chloroformylation[94–97]

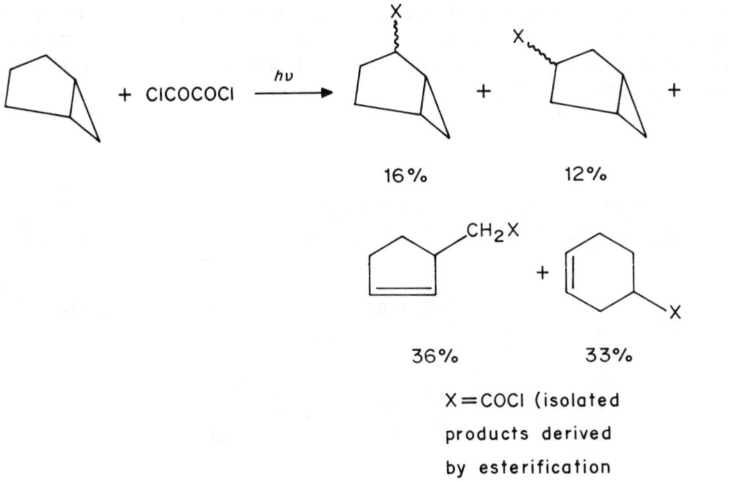

16%   12%

CH$_2$X

(13)

36%   33%

X=COCl (isolated
products derived
by esterification

(a)

(CH$_2$)$_n$—CH
Br

(b)

CH$_2$·
(CH$_2$)$_n$—Br
H

(c)

CH$_2$Br
(CH$_2$)$_n$·

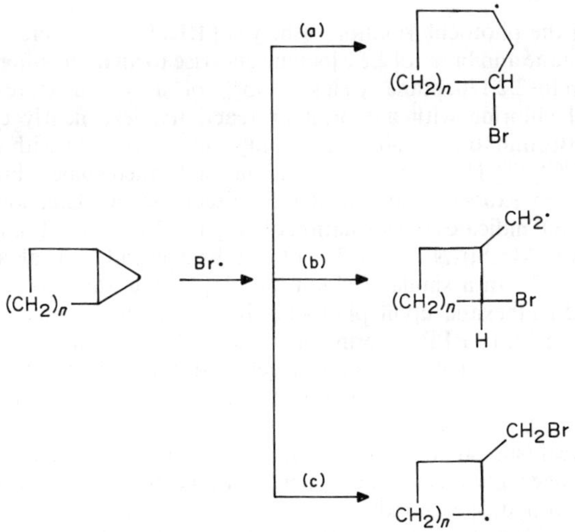

SCHEME 15. Bromine atom attack on bicyclo[n.1.0]alkanes

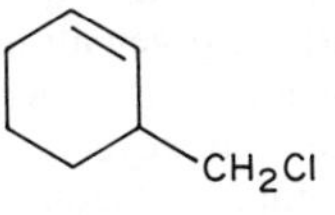

(14) 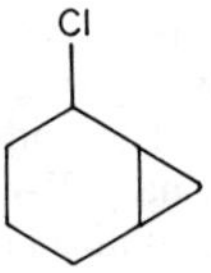   (15)

Photochlorination of bicyclo[2.2.0]hexane in the vapor phase produces **16, 17** and **18** in a 2:1:1 ratio (reaction 14)[99]. Compound **17** is apparently formed at low chlorine concentrations via the cyclobutylcarbinyl ring opening of reaction 15. In bicyclo[2.2.0]-

hexane the bridgehead hydrogens appear to be slightly more reactive ($\sim 30\%$) on a per hydrogen basis than the methylene hydrogen (assuming that Cl˙ can attack from both the *exo* and *endo* face). Further examples of bridgehead reactivity will be considered in the following section.

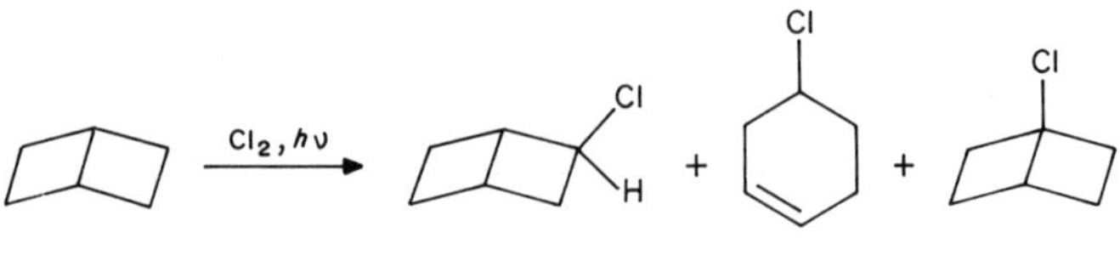

$$\hspace{5cm} (14)$$

                    (16)   **(17)**   **(18)**

$$\hspace{5cm} (15)$$

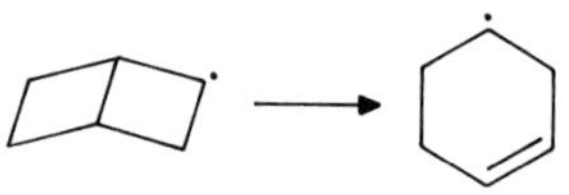

8. Reactivity of bridgehead hydrogen atoms

Products from the photochlorination of bicyclo[1.1.1]pentane, bicyclo[2.2.1]hexane, bicyclo[2.2.1]heptane and bicyclo[2.2.2]octane give rise to an interesting series of relative reactivities. Bicyclo[2.2.1]heptane yields $\sim 95\%$ of a 3:1 mixture of 2-*exo*- and 2-*endo*-norbornyl chloride with a molecular reactivity very nearly equal to that of cyclohexane[100]. Bromination at 80 °C gave only the 2°-bromide with *exo/endo* = 3 for Br_2 and 5 for $BrCCl_3$[100]. However, NBS bromination formed some 1-bromonorbornane ($\sim 9\%$) and gave an *exo/endo* ratio of 2:1[78]. Direct radical detection by EPR upon reaction with *t*-BuO˙ indicates the formation of only the 2-norbornyl radical[83]. However, the more reactive $(Me_3Si)_2N$˙ gives 3% of the bridgehead radical and 97% of the 2-norbornyl radical[83]. In a similar fashion bicyclo[2.1.1]hexane gives mainly ($\sim 95\%$) 2-chlorobicyclo[2.1.1]hexane upon photochlorination with a molecular reactivity 0.8 that of cyclohexane[101]. In EPR experiments none of the bridgehead radical is observed with *t*-BuO˙ but 29% bridgehead attack is detected with $(Me_3Si)_2N$˙[83]. For both the [2.2.1] and [2.1.1] systems very little if any attack on the methylene bridge is observed. Because of the C—C—C bond angle at this bridge, radical attack would lead to a nonplanar σ radical (similar to a cyclopropyl radical) and a reactivity comparable to a cyclopropyl hydrogen is expected, i.e. a per-hydrogen reactivity towards Cl˙ of ~ 0.1 that of a 1°-carbon hydrogen bond.

Both bicyclo[2.2.2]octane and bicyclo[1.1.1]pentane give about equal amounts of Cl˙ attack at the bridgehead positions and in the ethylene or methylene bridges[102,103]. However, bicyclo[2.2.2]octane has a molecular reactivity similar to cyclohexane whereas bicyclo[1.1.1]pentane has a very low reactivity. In bicyclo[2.2.2]octane the 3°(bridgehead)/2° reactivity per hydrogen atom is ~ 6 and k(3°-bridgehead)/k(cyclohexane) per hydrogen atom is ~ 3[102]. The bridgehead hydrogen atoms are as reactive or perhaps even more reactive than ordinary 3°-hydrogen atoms, indicating that the nonplanar (σ) bridgehead radical is not destabilized relative to an acyclic 3°-alkyl radical. On the other hand, the 3°-bridgehead radicals from bicyclo[2.2.1]heptane and bicyclo[2.1.1]hexane are much less readily formed than acyclic 3°-alkyl radicals.

The bridgehead radical formed from bicyclo[1.1.1]pentane has an unusually large coupling to the other bridgehead hydrogen atom of ~ 70 gauss **(19)**[104]. The hybrid orbital containing the unpaired electron forms three W-plan arrangements with the other bridgehead radical, resulting in unusual stability for **19** compared with the bridgehead radicals from bicyclo[2.1.1]hexane or bicyclo[2.2.1]heptane where the corresponding bridgehead radicals have EPR coupling to the other bridgehead hydrogen atom of 2.5 and 2.3 gauss, respectively[105,106].

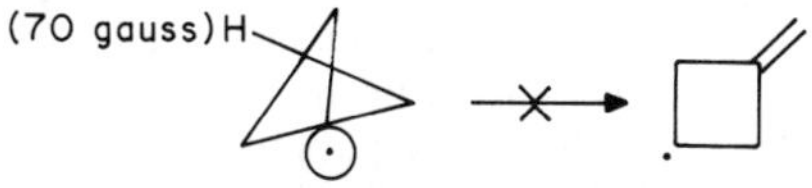

$$(16)$$

(19)   **(20)**

Although chlorine atoms attack bicyclo[1.1.1]pentane to give about equal parts of the bridgehead radical and the 2°-radical from attack at one of the three methylene groups, EPR experiments at $-33\,°C$ indicate that t-BuO˙ attack yields only the bridgehead radical[83]. Furthermore, the bridgehead radical is stable to 37 °C and the rearranged radical **20** is not observed by EPR spectroscopy; the rate constant for the conversion of **19** to **20** must be less than $10^{-3}\,s^{-1}$.

Attack of Cl˙ upon adamantane forms both the 2°- and 3°-chloroadamantanes with $k(3°)/k(2°) = 1.9^{107}$. However, for tricyclo[3.3.0.0$^{2.6}$] octane, only chlorine atom attack at the methylene groups is observed (reaction 17)[108].

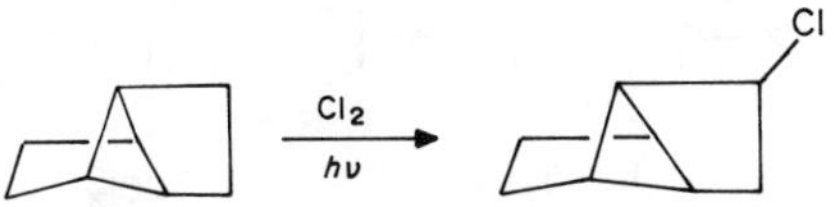

$$(17)$$

III. RADICAL ATTACK ON CYCLOPROPANES

Cyclopropanes can be attacked by radicals at either hydrogen or carbon (Scheme 16). Only very reactive radicals such as Cl˙, CF$_3$˙, t-BuO˙ or imidyl radicals can abstract cyclopropane hydrogen atoms[3,4,78,82] while with less reactive species such as Br˙ or I˙ only S_H2 substitution at carbon is observed[5,6,79,85].

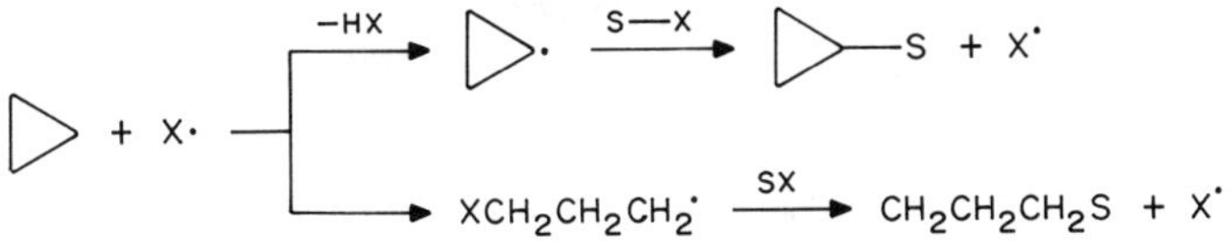

SCHEME 16. Competing reactions of cyclopropanes

Vapor-phase chlorination of cyclopropane at $\sim 100\,°C$ gives a good yield of chlorocyclopropane essentially free of products derived from the allyl radical[109]. Ring opening of the cyclopropyl radical to the allyl radical, although highly exothermic, has a high energy of activation (reaction 18)[110]. However, at lower temperatures in solution the major reaction product is 1,3-dichlorocyclopropane. Chlorination of cyclopropane with t-BuOCl gives excellent yields of chlorocyclopropane[3], and bromination with N-bromosuccinimide or N-bromo-3,3-dimethylglutarimide gives excellent yields of bromocyclopropane[4,78]*.

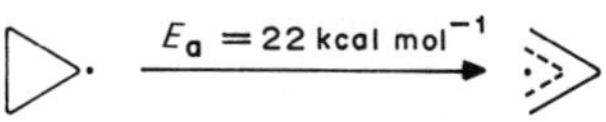

$$(18)$$

*For further discussion of cyclopropyl radical chemistry see G. Boche and H. M. Walborsky, in *The Chemistry of the Cyclopropyl Group* (Ed. Z. Rappoport), Chap. 12, Wiley, Chichester, 1987.

With alkylcyclopropanes, reaction with Cl· can lead to a variety of products including those resulting from the cyclopropylcarbinyl ring opening. However, the gas-phase chlorination of bicyclopropyl yielded the unrearranged 3°-chloride in good yield (reaction 19)[111]. The gas-phase photochlorination of methylcyclopropane yields as major products cyclopropylcarbinyl chloride ($\sim 35\%$), homoallyl chloride ($\sim 35\%$) and minor amounts of 1,3-dichlorobutane, 1,3-dichloro-2-methylpropane and products of substitution in the ring[112,113]. In CCl_4 solution at 0–60 °C the photochlorination of methylcyclopropane forms cyclopropcarbinyl chloride (56%), 1,3-dichlorobutane (7.3%) and 1,3-dichloro-2-methylpropane[3]. In the liquid phase the high concentration of Cl_2 prevents the cyclopropylcarbinyl radical from undergoing ring opening. The S_H2 substitution at carbon occurs only at the unsubstituted carbon atoms, but there is little control over the regiochemistry of the ring opening. Photochlorination of 1,1-dichlorocyclopropane forms $ClCH_2CH_2CCl_3$ rather than $ClCH_2CCl_2CH_2Cl$, indicating attack of Cl· at the least hindered carbon and ring opening to form the most stable radical ($ClCH_2CH_2CCl_2\cdot$)[114]. Chlorination of cis- and trans-1,1-dichloro-2,3-dideuteriocyclopropanes demonstrated that Cl· attack at carbon proceeds with inversion of configuration (Scheme 17)[114].

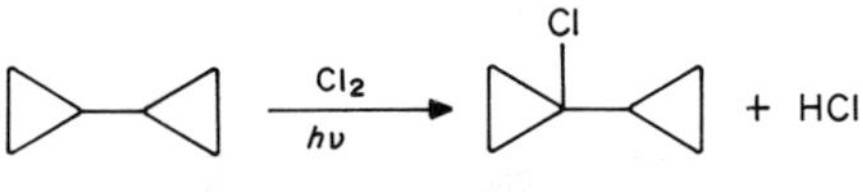

$$\text{(19)}$$

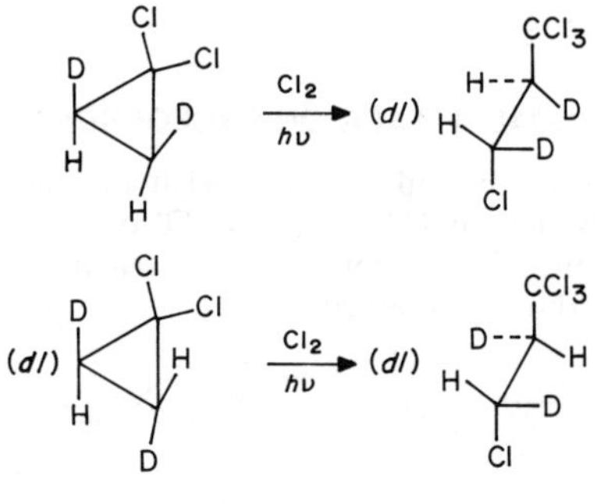

SCHEME 17. S_H2 attack of Cl· on cyclopropanes[114]

Photochlorination of nortricyclene at 25 °C in 1,1,2-trichloroethylene gave about equal amounts of products resulting from hydrogen atom abstraction (21, 22) and S_H2 substitution (23); see Scheme 18[115]. The predominance of chlorine in the exo-positions of 23 is consistent with an inversion mechanism for S_H2 attack of Cl·.

Photobrominations are more selective than photochlorinations and essentially only the ring-opened products are observed[6]. Furthermore, these ring openings are highly regioselective, producing only products derived from the most stable radical formed by Br· attack at the last substituted carbon atom. Table 14 summarizes some pertinent results[6].

The increase in the rate of the attack of Br· upon the methylcyclopropanes may indicate that complexes of Br· and the cyclopropane are involved (reaction 20)[6].

The stereochemistry of bromine atom attack upon the cyclopropyl carbon—carbon bond is clearly one of inversion. Thus, 2,4-dehydroadamantane reacts to form a,e-2,4-dibromoadamantane (66%), e,e-2,4-dibromoadamantane and none of the a,a-isomer (reaction 21)[6].

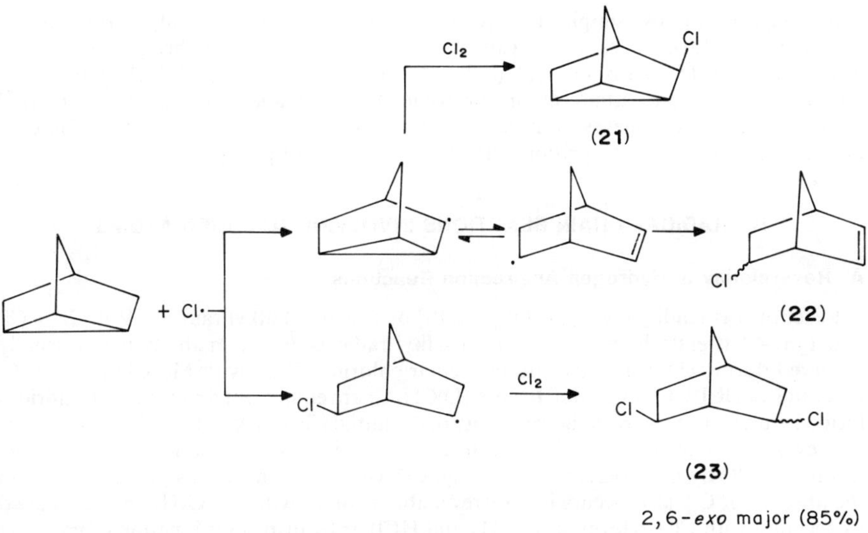

SCHEME 18. Photochlorination of nortricyclene[116]

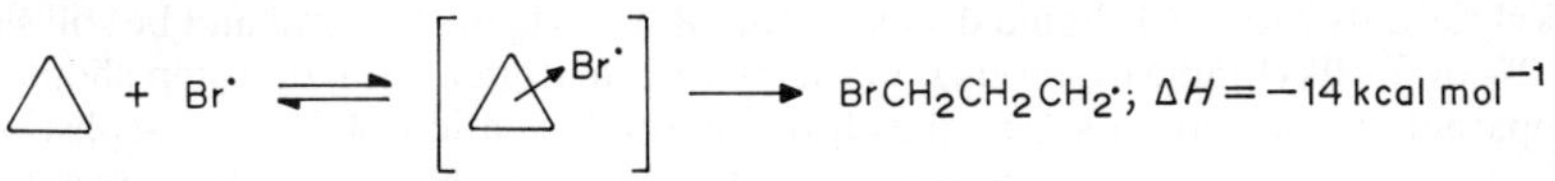

$$BrCH_2CH_2CH_2^{\cdot}; \quad \Delta H = -14 \text{ kcal mol}^{-1}$$

(20)

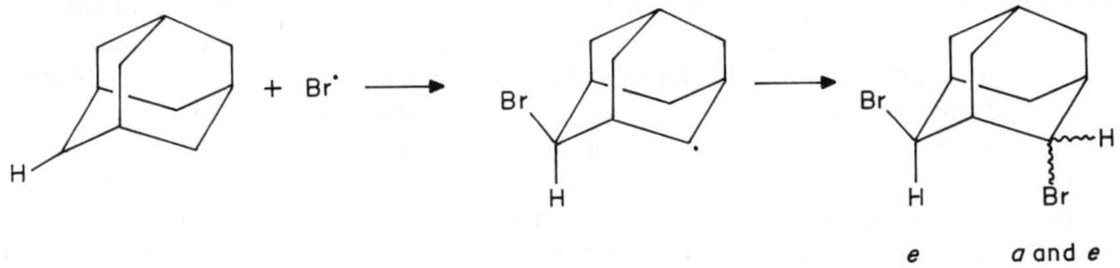

(21)

TABLE 14. Photobromination products of alkylcyclopropanes at $-78\,^\circ$C in CH_2Cl_2[6]

Substrate	Product	$k(\text{rel})^a$
MeC$_3$H$_5$	MeCH(Br)CH$_2$CH$_2$Br	390
1,1-Me$_2$C$_3$H$_4$	BrCH$_2$CH$_2$C(Br)Me$_2$	18,500
trans-1,2-Me$_2$C$_3$H$_4$ } cis-1,2-Me$_2$C$_3$H$_4$ }	MeCH(Br)CH(Me)CH$_2$Br^b	8650
1,1,2-Me$_3$C$_3$H$_3$	BrCH$_2$CMe$_2$CH(Br)Me (7%) and BrCH$_2$CH(Me)C(Br)Me$_2$ (83%)	61,000

^{a}Rel. to c-C$_3$H$_6$ = 1.00.
^{b}Mixture of diastereomers.

Although Br$^{\bullet}$ attacks simple alkylcyclopropanes at the least substituted carbon atom, with bicyclo[n.1.0]alkanes attack can be at the more substituted bridgehead carbon atom [Scheme 15, paths (a) and (b)][90,94]. In the case of bicyclo[1.1.0]butane or bicyclo[2.1.0]pentane, attack at the bridgehead carbon and opening of the internal cyclopropane bond relieves considerable strain. However, with bicyclo[3.1.0]hexane attack of Br$^{\bullet}$ still seems to predominate at the bridgehead position[94].

IV. RADICAL CHAIN REACTIONS INVOLVING HALOGEN ATOMS

A. Reversibility of Hydrogen Abstraction Reactions

Reaction 1 is readily reversible for $X^{\bullet} = Br^{\bullet}$ or $I^{\bullet}$ with all alkyl radicals. With $X^{\bullet} = Cl^{\bullet}$ reaction -1 is endothermic for 2°- or 3°-alkyl radicals and reversibility is not usually observed during chlorinations with molecular chlorine. Thus, with Me_3CD at $-15\,°C$, the ratios of DCl/HCl and t-$BuCl/Me_2C(D)CH_2Cl$ are equivalent and the alkyl chlorides formed are a true measure of the selectivity of the initial $Cl^{\bullet}$ attack[12]. In such chlorinations the reversal of reaction 1 would have to compete with the very rapid and exothermic trapping of $R^{\bullet}$ by Cl_2. Reactions of atomic $Cl^{\bullet}$ with C_2H_6 in the gas phase at 25 °C in the absence of Cl_2 also occurs in an irreversible manner with the $C_2H_5^{\bullet}$ being trapped by reaction with $Cl^{\bullet}$ (to form only C_2H_4 and HCl) or by disproportionation or coupling with another ethyl radical[116]. When chlorine atoms are generated in a system not containing molecular Cl_2, or other good alkyl radical trap, there is a better chance of observing the reversal of the hydrogen atom abstraction step. Since the rate of reaction of alkyl radicals with HCl should decrease from $R^{\bullet} = CH_3^{\bullet}$ to 1°-alkyl$^{\bullet}$ and be still slower with 2°- or 3°-alkyl radicals, reversal of the hydrogen atom abstraction step should lead to apparent chlorine atom selectivities favoring the formation of 2°- or 3°-substitution products in alkanes. Such appears to be the case in the free radical reaction of chloroolefins with alkanes in the presence of peroxides (Scheme 19)[117,118]. The isomer distribution in the chlorovinylation of hexane differs considerably from that observed in the photochlorination reaction with Cl_2 (Table 15)[118]. The selectivities observed with different vinyl chlorides are distinctly different than the distribution of products observed in chlorination with molecular Cl_2, and have been ascribed to the reversal of the hydrogen abstraction reaction leading to a predominance of the more stable 2°-alkyl radicals and in particular to the 2-hexyl radical[119]. Of course, the observed selectivities may be in part controlled by the relative rates of trapping of the alkyl radicals by the chloroalkanes, but this would not explain the preference for 2°-alkyl substitution. The percentage of n-$C_6H_{13}C(Cl){=}CCl_2$ formed in the reaction between hexane and $CCl_2{=}CCl_2$ drastically decreases as the reaction proceeds (and HCl concentration increases) or when HCl is added to the reaction mixture. For example, at 0.2–0.3% conversion, n-$C_6H_{13}C(Cl){=}Cl_2$ decreases from 9.6% of the chlorovinylation products to 2.4% when 6 mol% of HCl is added[119]. Reaction of hexane and perdeuteriocyclohexane with $CCl_2{=}CCl_2$ give both of the expected trichlorovinylation products but with extensive hydrogen–deuterium exchange in the unreacted alkanes.

$$RH + Cl^{\bullet} \rightleftharpoons R^{\bullet} + HCl$$

$$R^{\bullet} + CCl_2{=}CCl_2 \longrightarrow RCCl_2CCl_2^{\bullet}$$

$$RCCl_2CCl_2^{\bullet} \longrightarrow RC(Cl){=}CCl_2 + Cl^{\bullet}$$

SCHEME 19. Chlorovinylation of alkanes[117]

Similar observations have been made in the photocyanation of butane with ClCN (Scheme 20)[120]. The extrapolated ratio of $MeCH(CN)CH_2Me/MeCH_2CH_2CH_2CN$ at

TABLE 15. Relative reactivities per hydrogen atom in hexane[118]

| Reagent | CH_3^A—CH_2^B—CH_2^C—C_3H_7 | | |
	H^A	H^B	H^C
$Cl_2, h\nu, 65\,°C$	0.35	1.15	1.0
cis-$C_2H_2Cl_2, 125\,°C^a$	0.06	2.0	1.0
$C_2HCl_3, 125\,°C^b$	0.04	1.7	1.0
$C_2Cl_4, 125\,°C^c$	tr	4.2	1.0

^{a}To give RCH=CHCl.
^{b}To give RCH=CCl$_2$.
^{c}To give RC(Cl)=CCl$_2$.

0% reaction was 2/1, consistent with the selectivity of a chlorine atom of $k(2°)/k(1°) = 3$. However, as HCl built up in the solution the $2°/1°$ product increased to 24/1 at 1% reaction and to 168/1 at 5% reaction[120]. Extensive hydrogen–deuterium exchange was observed for the cyanation of a mixture of c-C_6H_{12} and c-C_6D_{12} and addition of DCl resulted in the incorporation of D in unreacted alkanes such as cyclohexane or 2,3-dimethylbutane. The Bz_2O_2-initiated reaction of ClCN with 2,3-dimethylbutane gave only the 3°-cyanide ($Me_2CHC(CN)Me_2$) in 95% yield, although irreversible attack of $Cl^{•}$ upon $Me_2CHCHMe_2$ actually produces more of the 1°-alkyl radical (~ 12 parts 1° to 8 parts 3°). Reaction of methyl cyanoformate ($CH_3OC(=O)CN$) with $Me_2CHCHMe_2$ followed a pathway similar to ClCN (Scheme 20) with β-elimination of $CH_3CO_2^{•}$ from the adduct radical followed by decarboxylation and irreversible hydrogen atom abstraction by $Me^{•}$ to yield up to 77% of $Me_2CHC(CN)Me_2$[10]. With cyanogen (NCCN) 2,3-dimethylbutane in the presence of Bz_2O_2 produced $Me_2CHCMe_2C(CN)=NH$ in >70% yield from the addition of $Me_2CHCMe_2^{•}$ to cyanogen followed by intermolecular hydrogen atom transfer (reaction 22)[10].

$$Cl^{•} + MeCH_2CH_2Me \rightleftharpoons HCl + n\text{-}Bu^{•} + sec\text{-}Bu^{•}$$

$$R^{•} + ClCN \rightarrow RC(Cl)=N^{•}$$

$$RC(Cl)=N^{•} \rightarrow RC\equiv N + Cl^{•}$$

SCHEME 20. Photocyanation with ClCN[120]

$$Me_2CHCMe_2C(CN)=N^{•} + RH \rightarrow R^{•} + Me_2CHCMe_2C(CN)=NH \qquad (22)$$

Another example of the reversal of the hydrogen atom abstraction reaction with $X^{•} = Cl^{•}$ is furnished by the reaction of HCl with Bz_2O_2 in CH_3CN at 98 °C in the presence of an alkane[121]. The reaction is presumed to yield Cl_2 by reactions 23 and 24 and chlorination of the alkane is observed. Although the nature of the radical which attacks the alkane is uncertain, the reaction rate seems to be controlled by a slow reaction 23 followed by a fast reaction 24. With 0.3 mol% Bz_2O_2 the observed $k(2°)/k(1°)$ for butane was 400 decreasing to 60 at 3.5 mol% Bz_2O_2, where the rate of formation and steady-state concentration of Cl_2 would be much higher[121]. For isobutane, the values of $k(3°)/k(1°)$ calculated from the ratio of the observed t-BuCl and i-BuCl decreased from 12,000 at 0.3 mol% Bz_2O_2 to 900 at 3.5 mol% Bz_2O_2. Of course, the relative reactivities for $Cl^{•}$ attack in the absence of a reaction between $R^{•}$ and HCl (i.e. in the presence of significant amounts of Cl_2) are much lower, $k(3°):k(2°):k(1°) = 4:3:1$ at 98 °C.

$$HCl + PhC(O)OC(O)Ph \rightarrow PhCO_2H + PhC(O)OCl \qquad (23)$$

$$PhC(O)OCl + HCl \rightarrow PhCO_2H + Cl_2 \qquad (24)$$

Because reversal in the hydrogen atom abstraction step can be observed in the absence of a good trapping agent for the alkyl radical, attention should be given to the nature of the hydrogen abstracting species for chlorinations using reagents other than Cl_2 or SO_2Cl_2, for example with reagents such as PCl_5 or $RSO_2N(t\text{-}Bu)Cl$ (Tables 7 and 8)[66,75]. The reversal of reaction 1 with $X^{\cdot} = Br^{\cdot}$ should occur more readily than for $X^{\cdot} = Cl^{\cdot}$. Thus, even in the presence of molecular bromine reversal can be observed (as reflected in the composition of the bromination product) if HBr is not rapidly removed from the system[122,123]. The reaction of alkyl radicals with HBr is, of course, well known in the peroxide-initiated addition of HBr to alkenes. By use of Br_2 for the reaction solvent the reversal can be effectively eliminated. Reversal of hydrogen abstraction with $X^{\cdot} = Br^{\cdot}$ may even occur in the solvent cage before $R^{\cdot}$ and HBr have diffused apart. Although trapping of $R^{\cdot}$ that has escaped the solvent cage by HBr can be effectively eliminated by removal of the HBr, for example by using a large excess of an epoxide or of NBS to trap HBr[124,125], prevention of the cage reaction between $R^{\cdot}$ and HBr is more difficult unless the trapping agent, e.g. Br_2, is part of the cage wall and can effectively compete with HBr. Viscosity of the solvent will have an important effect on the lifetime of the cage and, with a lower viscosity, the probability is increased for $R^{\cdot}$ to escape from the molecule of HBr formed in the hydrogen atom abstraction reaction. Important viscosity effects have been demonstrated in the bromination of substituted toluenes even when good trapping agents for HBr (ethylene oxide, NBS) are employed[126]. Since viscosity varies with temperature, this introduces an uncertainty in the interpretation of temperature effects on relative reactivities observed in photobromination reactions. Cage return also explains the grossly different kinetic isotope effects observed in gas and liquid-phase photobrominations of $PhCH_2D$: $k(H)/k(D) = 4.6$ in CCl_4 solution at 77 °C[123]; $k(H)/k(D) = 6.5$ vapor phase at 121 °C (8.2 calculated at 77 °C)[127]. On the other hand, photochlorination of $PhCH_2D$ in CCl_4, both in solution and in the vapor phase, at 70 °C gives essentially identical values of $k(H)/k(D) = 1.99$ and 2.08, respectively[128].

Competitive bromination of $c\text{-}C_6H_{12}$ and $c\text{-}C_6D_{12}$ in the gas and liquid phases at 21 °C gave values of $k(H)/k(D) = 5.4$ (vapor) and 4.3 (liquid, 0.07–3.6 M Br_2). However, at 10–18 M Br_2, $k(H)/k(D)$ was identical to the gas-phase result[129]. External exchange of $R^{\cdot}$ and HBr is apparently prevented by a bromine concentration of 0.07 M or greater, but exchange within the solvent cage still occurs up to very high Br_2 concentration. At 12–18 M, bromine molecules are an important part of the cage walls and the cage return is effectively quenched.

Photobrominations by Cl_3CBr can occur by mixed radical chains (Schemes 21 and 22)[122], further complicated by reversal of the hydrogen atom abstraction step in Scheme 22[119]. Photobromination of toluene, ethylbenzene and cumene at high Br_2 concentration with rapid removal of the HBr gives, for this benzylic series, $k(1°):k(2°):k(3°) = 1:17:37$ at 40 °C[122]. A similar result is observed for photobromination with NBS[130]. With $BrCCl_3$, a relative reactivity series of 1:50:260 is observed which suggests the occurrence of Scheme 21 with a greater selectivity for $CCl_3^{\cdot}$ than for $Br^{\cdot}$[122]. However, Scheme 22 is more susceptible to reversal of the hydrogen atom abstraction step than is bromination with molecular Br_2 because of the lower reactivity of Cl_3CBr towards $R^{\cdot}$. When the CCl_3Br brominations are performed in the presence of efficient HBr traps such as K_2CO_3 or ethylene oxide, a relative reactivity series towards $CCl_3^{\cdot}$ attack of 1:10:29 is observed and $CCl_3^{\cdot}$ is found to be slightly less selective than $Br^{\cdot}$[124,131]. In the absence of such HBr trapping agents the photolysis of $BrCCl_3$ leads to the bromine atom chain of Scheme 22, but reversal of the hydrogen atom abstraction step dominates over the bromine atom transfer from $BrCCl_3$. Photolysis of $BrCCl_3$ with bicyclo[2.1.0]pentane yields a complex set of reaction products apparently formed by $Br^{\cdot}$ attack (to yield the 3-bromocyclopentyl radical) and hydrogen atom abstraction by $CCl_3^{\cdot}$ (to yield the 3-cyclopentenyl radical)[88]. Chlorinations with CCl_4 do not give any evidence for a competing chlorine atom chain, presumably because HCl is much less reactive than HBr towards CCl_3·[119].

$$R^· + Cl_3CBr \longrightarrow RBr + CCl_3^·$$

$$CCl_3^· + RH \longrightarrow HCCl_3 + R^·$$

$$Br^· + RH \rightleftharpoons HBr + R^·$$

$$R^· + BrCCl_3 \longrightarrow RBr + CCl_3^·$$

$$CCl_3^· + HBr \longrightarrow HCCl_3 + Br^·$$

SCHEME 21   SCHEME 22

B. Halogenations Involving Mixed Radical Chains

Alkane halogenations often occur by processes which do not involve halogen atom attack on the alkane (Scheme 23). Among the Z groups that have been previously mentioned, e.g. in Table 4, are t-BuO, CCl_3, ClO, CCl_3S, PCl_4 and PhICl. Similar reactions are observed with $Z = (CH_3)_2NH^{+}$[132-134] or Et_3N^{+}[134] and alkane hydroxylations will occur with Et_2NHOH^{+} or Et_3NOH^{+} in the presence of Fe(II) in CF_3CO_2H[135]. With $Z = SO_2Cl$ hydrogen abstraction can occur by both Cl· and SO_2Cl· but SO_2Cl· attack is observed only for the more reactive alkanes[59]. A similar situation exists for Cl_3CSO_2·$\rightleftharpoons CCl_3^· + SO_2$[136]. (Attack of R· upon ZX to yield RZ and a halogen atom to continue the chain can also occur, for example with ZX = ClCOCOCl, vinyl halides or NCCl; see Schemes 14, 19 and 20.)

$$Z^· + RH \rightarrow R^· + HZ$$
$$R^· + Z - X \rightarrow R - Z + X^·$$

SCHEME 23. (X = halogen and Z not a halogen)

Mixed radical chains, where the attacking species are both Z· and X·, are sometimes observed when HX can readily generate X_2 from reaction with the reagent ZX. Such processes have been identified in the reactions of t-BuOCl, NBS and Cl_2O. On the other hand, N-chlorosuccinimide seems to react always by a chlorine atom chain by virtue of reaction 10.

With Cl_2O in the gas phase the $k(1°):k(2°):k(3°)$ reactivities (Table 4) exclude Cl· as the chain carrying species and are consistent with hydrogen abstraction by ClO· followed by the reaction of R· with Cl_2O to form RCl (reaction 25)[55]. In CCl_4 a much different selectivity is observed, i.e. $k(1°):k(2°):(3°) = 1:11.5:24$, and the stoichiometry is closer to equation 26. This suggests that in solution reactions 27 and 28 occur and that a chlorine atom chain competes with hydrogen atom abstraction by ClO·. With HOCl in aqueous solution at 40 °C the reaction goes completely by the chlorine atom chain with $k(1°):k(2°):k(3°)$ the same as observed for photochlorination with Cl_2[137].

$$Cl_2O + RH \longrightarrow RCl + HOCl \tag{25}$$

$$2RH + Cl_2O \longrightarrow 2RCl + H_2O \tag{26}$$

$$Cl_2O + HCl \longrightarrow HOCl + Cl_2 \tag{27}$$

$$HOCl + HCl \longrightarrow H_2O + Cl_2 \tag{28}$$

Photochemical reactions of t-BuOCl generally involve hydrogen atom abstraction by t-BuO·, a radical which is considerably more selective than Cl·[138]. In the reaction of t-BuOCl with 2,3-dimethylbutane at 64 °C, there is a report of a selectivity identical to that of Cl· upon photolysis but a higher selectivity with initiation by Bz_2O_2[66]. Photolysis of t-BuOCl may lead to chlorine atoms, hydrogen chloride and molecular chlorine whereas Bz_2O_2 leads to the t-BuO· radical chain. Reaction 29 will shift t-BuOCl chlorination to a chlorine atom chain reaction. The occurrence of a chlorine atom chain will result in reduced selectivity and in an increase in the ratio of t-BuOH/Me$_2$CO formed in the competitive reactions of t-BuO$^\cdot$ (Scheme 24).

$$t\text{-BuOCl} + \text{HCl} \longrightarrow t\text{-BuOH} + \text{Cl}_2 \tag{29}$$

$$t\text{-BuO}^\cdot \begin{cases} \xrightarrow{\quad k_d \quad} \text{Me}_2\text{C}=\text{O} + \text{Me}^\cdot \\ \xrightarrow{\quad k_a[\text{RH}] \quad} t\text{-BuOH} + \text{R}^\cdot \end{cases}$$

SCHEME 24. Competitive reactions of t-BuO$^\cdot$

With aralkyl hydrocarbons there is considerable evidence that t-BuOCl prefers to react by a chlorine atom chain[139]. The chlorine atom chain also occurs more readily for PhCH$_2$CMe$_2$OCl which, at 40 °C with hv or AIBN initiation, yields $k(3°):k(1°) = 6$–7 for DMB whereas t-BuOCl gives a value of 44 and Cl$_2$ a value of 4.5. With toluene the k_a/k_d ratios (Scheme 24) provide a diagnostic probe to the nature of the hydrogen-abstracting species (Table 16). Surprisingly, t-BuOBr reacts by a t-BuO$^\cdot$ chain under conditions where t-BuOCl gives a mixture of Cl$^\cdot$ and t-BuO$^\cdot$ chains (Table 16). The presence of an alkene [e.g. CH(Cl)=CCl$_2$ in Table 16] in the photoinitiated reaction of t-BuOCl traps Cl$^\cdot$ and/or Cl$_2$ and allows the t-BuO$^\cdot$ chain to dominate.

Direct measurements of the relative reactivities of c-C$_6$H$_{12}$ and PhCH$_3$ (by disappearance of the reactants) and indirect measurements (from the k_a/k_d ratios observed in two separate experiments) gave the results summarized in Table 17[139]. Again the presence of CH(Cl)=CH$_2$Cl$_2$ stopped the chlorine atom chain for t-BuOCl and again t-BuOBr appeared to react with no interference from a bromine atom chain.

TABLE 16. Reaction of t-BuO$^\cdot$ with toluene in CCl$_4$ at 70 °C[139]

Radical precursor	Conditions	k_a/k_d
t-BuON=NOBu-t	thermal	2.2
t-BuOBr	hv	2.0
t-BuOCl	hv	~ 10
t-BuOCl	hv plus 2% Cl$_2$	150–450
t-BuOCl	AIBN[a]	2.4
t-BuOCl	hv plus 0.1 M C$_2$HCl$_3$	2.5

[a] AIBN = azobisisobutyronitrile.

TABLE 17. Relative reactivities of cyclohexane and toluene at 40 °C[139]

Reagent	Solvent	Indirect[a]	Direct[b]
t-BuOCl	PhCl	1.8	6.5
t-BuOCl	C$_2$HCl$_3$	6.0	6.2
t-BuOBr	PhCl	4.7	4.2
t-BuOBr	C$_2$HCl$_3$	5.4	4.3
[t-BuOC(O)$\cdot$]$_2$	PhCl	5.3	5.2
[t-BuON=]$_2$	PhCl	6.0	5.7

[a] From k_a/k_d ratios.
[b] From disappearance of hydrocarbons.

With hydrocarbons which yield a resonance-stabilized benzylic radical, apparently the nature of the hydrogen atom abstracting species is determined by the competition between the benzyl radical and the possible chlorine atom donors. The reaction of $PhCH_2^{\cdot}$ with t-BuOCl is apparently slow enough that the benzyl radical will preferentially react with traces of Cl_2 regenerated by reaction 29. With the more reactive t-BuOBr, apparently the benzyl radical reacts readily to generate t-BuO$^{\cdot}$ and a bromine atom chain does not ensue. A bromine atom chain is also less likely to be observed than a $Cl^{\cdot}$ or BuO$^{\cdot}$ chain because Br$^{\cdot}$ is a less reactive hydrogen atom abstracting agent.

The results observed with t-BuOCl have considerable similarity to halogenations with N-haloimides and, in particular, with N-halosuccinimides. With the N-chloro compounds a chlorine atom chain is observed with all substrates[140]. The N-bromoimides are more reactive towards carbon radicals. For example, with Pr$^{\cdot}$ NBS is 7.3 times as reactive as NCS[141] and 1°-alkyl radicals are reported to react with N-bromoimides or Br_2 at a diffusion-controlled rate, $k \sim (1.3–2.2) \times 10^{10} \, M^{-1} s^{-1}$ at 15 °C in CH_2Cl_2[4], although one report indicates that Br_2 is much more reactive than NBS towards 3°-alkyl radicals[142]. With reactive hydrocarbons, particularly with aralkyl hydrocarbons, NBS reacts by a bromine atom chain (Scheme 25)[143–145]. However, with less reactive hydrocarbons such as butane or neopentane which yield a more reactive alkyl radical, NBS reacts by an imidyl radical chain. The imidyl radical is relatively unselective and gives relative reactivities closer to Cl$^{\cdot}$ than to Br$^{\cdot}$[146]. Reaction of NBS with butane in the presence of C_2H_4 gives $k(2°)/k(1°) \cong 4.6$ vs ~ 3 for Cl$^{\cdot}$[146].

$$Br^{\cdot} + PhCH_3 \longrightarrow PhCH_2^{\cdot} + HBr$$

$$HBr + NBS \longrightarrow Br_2 + succinimide$$

$$PhCH_2^{\cdot} + Br_2 \longrightarrow PhCH_2Br + Br^{\cdot}$$

SCHEME 25. NBS bromination of toluene

Again, the nature of the chain is apparently controlled by the reaction of the alkyl radical with the halogenating agent. With N-chloroimides, alkyl or benzyl radicals search out traces of molecular chlorine from reaction 10 and a chlorine atom chain predominates. With the more reactive NBS, 1°-alkyl radicals can readily abstract a bromine atom and an imidyl chain results, particularly at high NBS concentrations and in the presence of alkenes, to trap Br$^{\cdot}$ and/or Br_2. However, as the radical becomes less reactive, such as a benzyl radical, the radical reacts more slowly with NBS but readily with traces of Br_2, and the bromine atom chain dominates. With radicals of intermediate reactivity such as cycloalkyl, both chains can occur simultaneously[147,148]. Again the presence of an alkene traps Br$^{\cdot}$ and/or Br_2 and allows the imidyl radical chain to dominate. Thus, the relative reactivities of c-C_5H_{10}:c-C_6H_{12} are 9.2 towards Br$^{\cdot}$ (bromination in liquid Br_2) but only 0.82 towards the N-succinimidyl radical (NBS/CH_2Cl_2/CH_2=CH_2)[147]. With bicyclo[2.1.0]pentane NBS reacts to give about equal amounts of the products expected from bromine atom attack (1,3-dibromocyclopentane) and succinimidyl radical hydrogen atom abstraction (to yield 4-bromocyclopentene)[88]. The occurrence of competing bromine atom and N-succinimidyl radical chains has led to considerable confusion in the literature[146,149].

The N-haloimide halogenations are also controlled partially by the fact that Cl$^{\cdot}$ or the N-succinimidyl radical are much more reactive than Br$^{\cdot}$ in hydrogen abstracting reactions and, towards a hydrocarbon of low reactivity such as neopentane, a bromine atom chain would be quite ineffective. With mixtures of NBS and Cl_2, halogenation occurs to form the alkyl bromides but with the selectivity expected for chlorine atom attack[150]. Apparently ClBr is formed and reacts with the alkyl radical to form RBr and a chlorine atom. A similar situation exists for the bromination of alkanes using a mixture of Br_2 and Cl_2, at least for unreactive hydrocarbons[151]. In such a halogenation system the predominant species present will be BrCl and Br$^\cdot$ because of the following equilibria[152]:

$$Cl^\cdot + Br_2 \rightleftharpoons BrCl + Br^\cdot \qquad k_e = 1.6 \times 10^5$$
$$Br^\cdot + Cl_2 \rightleftharpoons BrCl + Cl^\cdot \qquad k_e = 1.3 \times 10^{-4}$$

With a hydrocarbon of low reactivity, hydrogen atom abstraction by Cl$^\cdot$ dominates but, with a more reactive hydrocarbon, Br$^\cdot$ can be the major hydrogen atom abstracting species. Thus, with $PhCH_2CH_3$ $k(\alpha)/k(\beta)$ per hydrogen atom is 116, characteristic of Br$^\cdot$ attack rather than attack by Cl$^\cdot$ or the complexed chlorine atom; the ratio of the α-halo compounds is about 10:1 in favor of the bromide[152].

Chain reactions involving halogen atoms are observed in several oxidation processes involving molecular oxygen. One example is the HBr-promoted vapor-phase autoxidation of hydrocarbons such as isobutane[153,154]. Reaction of a 1:1 mixture of isobutane and O_2 in the presence of 8 mol% of HBr at 163 °C gives up to 70% of t-butyl hydroperoxide by the mechanism of Scheme 26. Straight-chain hydrocarbons are oxidized to ketones[155] and ethane to acetic acid[156], products arising from further reactions of initially formed hydroperoxides. The side chains of aralkyl hydrocarbons are autoxidized in the presence of HBr in the liquid phase[157]. The two-step propagation sequence of Scheme 26 occurs more readily than the one-step reaction involving attack of t-BuOO$^\cdot$ upon Me_3CH. The energy of activation for Br$^\cdot$ attack on a 3°-carbon–hydrogen bond of 7.5 kcal mol^{-1} [25] is considerably lower than the value of ~ 16 kcal mol^{-1} for t-BuOO$^\cdot$ [60]. Furthermore, the rapid conversion of t-BuOO$^\cdot$ to Br$^\cdot$ changes the termination mechanism. In particular, the formation of Br_2 by the coupling of two bromine atoms is reversible at 163 °C whereas the t-BuOO$^\cdot$ termination process would be irreversible. The net effect is that the HBr-promoted reaction of Scheme 26 occurs more readily than autoxidation in the absence of HBr even though the two propagation steps have the same overall thermodynamics as the direct reaction of t-BuOO$^\cdot$ with Me_2CH.

$$Br^\cdot + Me_3CH \longrightarrow HBr + Me_3C^\cdot$$
$$Me_3C^\cdot + O_2 \longrightarrow Me_3COO^\cdot$$
$$Me_3COO^\cdot + HBr \longrightarrow Me_3COOH + Br^\cdot$$

SCHEME 26. HBr-promoted autoxidation of isobutane

The reaction of alkanes with a mixture of SO_2 and O_2 also occurs more readily than simple autoxidation of the hydrocarbon (reaction 30)[158]. Here the radical $RSO_2OO^\cdot$, formed as shown in Scheme 27, is apparently less prone to termination reactions than a simple alkylperoxy radical[159].

$$RH + SO_2 + O_2 \longrightarrow RSO_2OOH \qquad (30)$$
$$R^\cdot + SO_2 \longrightarrow RSO_2^\cdot$$
$$RSO_2^\cdot \longrightarrow RSO_2OO^\cdot$$
$$RSO_2OO^\cdot + RH \longrightarrow RSO_2OOH + R^\cdot$$
$$RSO_2OOH + H_2O \longrightarrow RSO_3H + H_2O_2$$

SCHEME 27. Sulfoxidation of hydrocarbons

Another intriguing reaction of molecular oxygen that involves a chlorine atom as the hydrogen abstraction species is the chlorophosphonation of alkanes (reaction 31)[160-163]. Reaction 31 and the competing reaction 32 spontaneously form chlorine atoms and proceed readily at $-78\,^{\circ}\mathrm{C}$ with alkanes with a low reactivity such as ethane[163]. Evidence that reaction 31 involves Cl$^{\cdot}$ attack on the alkane includes the observation that the isomer distribution in the chlorophosphonation of butane is the same as in photochlorination. Scheme 28 summarizes the more important reactions involved in the overall reaction 31[164]. The relative reactivity of $R^{\cdot} = c$-$C_6H_{11}^{\cdot}$ towards PCl_3 and O_2 is about 0.5[164]. This leads to Scheme 28 at high PCl_3 or low oxygen concentrations. At low PCl_3 or high oxygen concentrations alkyl phosphorodichloridates are formed (reaction 33). As shown in Scheme 29, trapping of $R^{\cdot}$ by O_2 leads to a peroxy/alkoxy radical chain. In competition with Schemes 28 and 29 are reactions involving the trapping of Cl$^{\cdot}$ by PCl_3 with the regeneration of a Cl$^{\cdot}$ (Scheme 30). At low oxygen pressures the relative reactivities of PCl_3 and c-C_6H_{12} towards Cl$^{\cdot}$ are measured to be ~ 0.36. The chlorophosphonation reaction gives very poor yields with aralkyl hydrocarbons such as $PhCH_3$ or Ph_2CH_2, while Ph_3CH fails to react[165,166]. With more stable alkyl radicals the intermediate $RPCl_3O^{\cdot}$ can decompose to yield $R^{\cdot}$ and $POCl_3$ and the substitution process is thereby sabotaged. The reaction occurs for other chlorophosphines such as $EtPCl_2$ (reaction 34)[167].

$$RH + 2PCl_3 + O_2 \longrightarrow RPOCl_2 + POCl_3 + HCl \tag{31}$$

$$PCl_3 + \tfrac{1}{2}O_2 \longrightarrow POCl_3 \tag{32}$$

$$Cl^{\cdot} + RH \longrightarrow HCl + R^{\cdot}$$

$$R^{\cdot} + PCl_3 \longrightarrow PRCl_3^{\cdot}$$

$$RPCl_3^{\cdot} + O_2 \longrightarrow RPCl_3OO^{\cdot}$$

$$RPCl_3OO^{\cdot} + PCl_3 \longrightarrow RPCl_3OOPCl_3^{\cdot}$$

$$RPCl_3OOPCl_3^{\cdot} \longrightarrow RPCl_3O^{\cdot} + POCl_3$$

$$RPCl_3O^{\cdot} \longrightarrow RPOCl_2 + Cl^{\cdot}$$

SCHEME 28. Chlorophosphonation mechanism

$$RH + PCl_3 + O_2 \longrightarrow ROH + POCl_3 \longrightarrow ROPOCl_2 + HCl \tag{33}$$

$$R^{\cdot} + O_2 \longrightarrow ROO^{\cdot}$$

$$ROO^{\cdot} + PCl_3 \longrightarrow ROOPCl_3^{\cdot}$$

$$ROOPCl_3^{\cdot} \longrightarrow RO^{\cdot} + POCl_3$$

$$RO^{\cdot} + RH \longrightarrow ROH + R^{\cdot}$$

SCHEME 29. Alkylphosphorodichloridation mechanism

$$Cl^{\cdot} + PCl_3 \longrightarrow PCl_4^{\cdot}$$

$$PCl_4^{\cdot} + O_2 \longrightarrow PCl_4OO^{\cdot}$$

$$PCl_4OO^{\cdot} + PCl_3 \longrightarrow PCl_4OOPCl_3^{\cdot}$$

$$PCl_4OOPCl_3^{\cdot} \longrightarrow PCl_4O^{\cdot} + POCl_3$$

$$PCl_4O^{\cdot} \longrightarrow POCl_3 + Cl^{\cdot}$$

SCHEME 30. Autoxidation of PCl_3

$$EtPCl_2 + c\text{-}C_6H_{12} + O_2 \rightarrow c\text{-}C_6H_{11}P(Et)(Cl)O \tag{34}$$

The trapping of peroxy radicals by PCl_3 to yield a phosphoranyl radical is known to occur readily[168]. In the case of Ph_3P, reaction with t-$BuOO^{\cdot}$ occurs with an energy of activation of only $3\,\mathrm{kcal\,mol^{-1}}$[169], much less than the energy of activation for attack

994   G. A. Russell

of t-BuOO$^{\bullet}$ upon 2°- or 3°-carbon hydrogen bonds, i.e. ~ 16–17 kcal mol^{-1} [60]. One might expect the reaction of alkanes with PCl_3 to form $RPCl_2$ and HCl to occur readily by Scheme 31. The reaction has been observed with methane or ethane, but only in low yields at temperatures of 575–600 °C [170]. The absence of a facile reaction may reflect the absence of an initiation reaction, and in fact the reaction is reported to occur more readily in the presence of traces of oxygen.

$$R^{\bullet} + PCl_3 \rightleftharpoons RPCl_3^{\bullet}$$

$$RPCl_3^{\bullet} \longrightarrow RPCl_2 + Cl^{\bullet}$$

$$Cl^{\bullet} + RH \longrightarrow R^{\bullet} + HCl$$

SCHEME 31. Free radical alkylation of PCl_3

V. REFERENCES

1. T. Cole and H. Heller, *J. Chem. Phys.*, **42**, 1668 (1965).
2. D. F. McMillen and D. M. Golden, *Annu. Rev. Phys. Chem.*, **35**, 493 (1980).
3. C. Walling and P. S. Fredricks, *J. Am. Chem. Soc.*, **84**, 3326 (1962).
4. J. M. Tanko, P. S. Skell and S. J. Seshadri, *J. Am. Chem. Soc.*, **110**, 3221 (1988).
5. G. G. Maynes and D. E. Applequist, *J. Am. Chem. Soc.*, **95**, 856 (1973).
6. K. J. Shea and P. S. Skell, *J. Am. Chem. Soc.*, **95**, 6728 (1973).
7. K. B. Wiberg and F. M. Walker, *J. Am. Chem. Soc.*, **104**, 5239 (1982).
8. P. Kaszynski and J. Michl, *J. Am. Chem. Soc.*, **110**, 5225 (1988).
9. G. S. Murthy, K. Hassenrück, V. M. Lynch and J. Michl, *J. Am. Chem. Soc.*, **111**, 7262 (1989).
10. D. D. Tanner and P. M. Rahimi, *J. Org. Chem.*, **44**, 1674 (1979).
11. F. F. Rust, *J. Am. Chem. Soc.*, **79**, 4000 (1957).
12. H. C. Brown and G. A. Russell, *J. Am. Chem. Soc.*, **74**, 3995 (1952).
13. R. T. Watson, *J. Phys. Chem. Ref. Data*, **6**, 871 (1977).
14. K. B. Wiberg and E. L. Motell, *Tetrahedron*, **19**, 2009 (1963).
15. M. A. A. Clyne and R. F. Walker, *J. Chem. Soc., Faraday Trans. 1*, **69**, 1547 (1973).
16. P. Cadman, A. W. Kirk and A. F. Trotman-Dickenson, *J. Chem. Soc. Faraday Trans. 1*, **72**, 1027 (1976).
17. G. Chiltz, R. Eckling, P. Goldfinger, G. Huybrechts, H. J. Johnston, L. Meyers and G. Verkeke, *J. Chem. Phys.*, **38**, 1053 (1963).
18. E. Tschuikow-Roux, J. Niedzielski and F. Farah, *Can. J. Chem.*, **63**, 1093 (1985).
19. S. S. Parmar and S. W. Benson, *J. Am. Chem. Soc.*, **111**, 57 (1989).
20. P. S. Skell and H. N. Baxter III, *J. Am. Chem. Soc.*, **107**, 3823 (1985).
21. K. U. Ingold, J. Lusztyk and K. D. Raner, *Acc. Chem. Res.*, **23**, 219 (1990).
22. W. Tsang, *J. Am. Chem. Soc.*, **107**, 2872 (1985).
23. J. A. Seetula, J. J. Russell and D. Gutman, *J. Am. Chem. Soc.*, **112**, 1347 (1990).
24. J. A. Seetula and D. Gutman, *J. Phys. Chem.*, **94**, 7529 (1990).
25. J. A. Kerr and S. J. Moss (Eds.), *Handbook of Bimolecular and Termolecular Gas Phase Reaction*, Vol. 1, CRC Press Inc., Boca Raton, Florida, 1981.
26. M. G. Evans and M. Polanyi, *Trans. Faraday Soc.*, **34**, 11 (1938).
27. E. T. Butler and M. Polanyi, *Trans. Faraday Soc.*, **39**, 19 (1943).
28. A. F. Trotman-Dickenson, *Gas Kinetics*, Butterworths, London, 1955.
29. D. Gutman, *Acc. Chem. Res.*, **23**, 375 (1990).
30. P. Gray, A. A. Herod and A. Jones, *Chem. Rev.*, **71**, 247 (1971).
31. J. H. Knox and R. L. Nelson, *Trans. Faraday Soc.*, **55**, 937 (1959).
32. V. R. Descei, A. Nechvatal and J. M. Tedder, *J. Chem. Soc. (B)*, 387 (1970).
33. R. Foon and N. A. McAskill, *Trans. Faraday Soc.*, **65**, 3005 (1969).
34. R. Foon and G. P. Reid, *Trans. Faraday Soc.*, **67**, 3513 (1971).
35. A. F. Trotman-Dickenson, J. P. Birchard and E. W. R. Steacie, *J. Chem. Phys.*, **19**, 163 (1951).
36. G. C. Fettis, J. H. Knox and A. F. Trotman-Dickenson, *J. Chem. Soc.*, 1064 (1960).
37. P. C. Anson, P. S. Fredricks and J. M. Tedder, *J. Chem. Soc.*, 918 (1959).

38. I. Galiba, J. M. Tedder and J. C. Walton, *J. Chem. Soc. (B)*, 609 (1966).
39. G. A. Russell, *J. Am. Chem. Soc.*, **80**, 4987 (1958).
40. T. J. Hardwick, *J. Phys. Chem.*, **65**, 101 (1961).
41. R. B. Baker, R. R. Baldwin and R. W. Walker, *Trans. Faraday Soc.*, **66**, 2812 (1970).
42. W. A. Pryor and J. P. Stanley, *J. Am. Chem. Soc.*, **93**, 1412 (1971).
43. C. Walling, *Pure Appl. Chem.*, **15**, 69 (1967).
44. R. F. Bridger and G. A. Russell, *J. Am. Chem. Soc.*, **85**, 3754 (1963).
45. W. A. Pryor, K. Smith, J. T. Echols and D. L. Fuller, *J. Org. Chem.*, **37**, 1753 (1972).
46. J. A. Kerr and M. J. Parsonage, *Evaluated Kinetic Data on Gas Phase Hydrogen Transfer Reactions of Methyl Radicals*, Butterworths, London, 1976.
47. M. H. Arican, E. Potter and D. A. Whytock, *J. Chem. Soc., Faraday Trans. 1*, **69**, 184 (1973).
48. H. W. Sidebottom, J. M. Tedder and J. C. Walton, *Int. J. Chem. Kinet.*, **4**, 249 (1972).
49. J. M. Tedder, *Q. Rev. (London)*, **14**, 336 (1960).
50. W. A. Thaler, *Methods in Free-Radical Chemistry*, **2**, 189 (1969).
51. P. Cadman, C. Dodwell, A. J. White and A. F. Trotman-Dickenson, *J. Chem. Soc. (A)*, 2967 (1971).
52. J. H. Knox and R. G. Musgrave, *Trans. Faraday Soc.*, **63**, 2201 (1967).
53. D. D. Tanner and P. B. Van Bostelen, *J. Org. Chem.*, **32**, 1517 (1967).
54. H. Kloosterziel, *Recl. Trav. Chim. Pays-Bas*, **82**, 497 (1963).
55. R. Shaw, *J. Chem. Soc. (B)*, 513 (1968).
56. N. C. Deno, D. G. Pohl and H. J. Spinelli, *Biorg. Chem.*, **3**, 66 (1974).
57. J. Fossey, D. Lefort, M. Massoudi, J.-Y. Nedelec and J. Surba, *Can. J. Chem.*, **63**, 678 (1985).
58. G. A. Russell, *J. Am. Chem. Soc.*, **80**, 5002 (1958).
59. J. T. Groves and T. E. Nemo, *J. Am. Chem. Soc.*, **105**, 6243 (1983).
60. J. H. B. Chenier, S.-B. Tong and T. A. Howard, *Can. J. Chem.*, **56**, 3047 (1978).
61. I. Galiba, J. M. Tedder and J. C. Walton, *J. Chem. Soc.*, 918 (1959).
62. M. L. Poutsma, *Methods in Free-Radical Chemistry*, **1**, 79 (1969).
63. B. Blouri, C. Cerceau and G. Lanchec, *Bull. Soc. Chim. Fr.*, 304 (1964).
64. G. Lanchec, *Chem. Ind. (Paris)*, **94**, 46 (1965).
65. P. Smit and H. J. den Hertog, *Recl. Trav. Chim. Pays-Bas*, **83**, 891 (1964).
66. B. Fell and L. H. Krug, *Chem. Ber.* **98**, 2871 (1965).
67. G. A. Russell, *J. Am. Chem. Soc.*, **79**, 2977 (1957).
68. G. A. Russell, *J. Am. Chem. Soc.*, **80**, 4897 (1958).
69. G. A. Russell, *J. Am. Chem. Soc.*, **80**, 4997 (1958).
70. P. S. Skell, H. N. Baxter III and G. K. Taylor, *J. Am. Chem. Soc.*, **105**, 20 (1983).
71. K. D. Raner, J. Lusztyk and K. U. Ingold, *J. Am. Chem. Soc.*, **111**, 3652 (1989).
72. R. Breslow, M. Brandl, J. Hunger, N. Turro, K. Cassidy, K. Krogh-Jesperson and J. D. Westbrook, *J. Am. Chem. Soc.*, **109**, 7204 (1987).
73. J. M. Tanko and F. E. Anderson III, *J. Am. Chem. Soc.*, **110**, 3525 (1988).
74. D. D. Tanner, H. Oumar-Mahamat, C. P. Meintzer, E. C. Tsai, T. T. Lu and D. Yang, *J. Am. Chem. Soc.*, **113**, 5397 (1991).
75. A. E. Fuller and W. J. Higgenbottom, *J. Chem. Soc.*, 3228 (1965).
76. E. S. Huyser, *J. Am. Chem. Soc.*, **82**, 5246 (1960).
77. G. A. Russell and P. G. Hafley, *J. Org. Chem.*, **31**, 1869 (1966).
78. J. G. Tranham and Y-S. Lee, *J. Am. Chem. Soc.*, **96**, 3590 (1974).
79. A. J. Kennedy, J. C. Walton and K. U. Ingold, *J. Chem. Soc., Perkin Trans. 2*, 751 (1982).
80. C. Roberts and J. C. Walton, *J. Chem. Soc., Perkin Trans. 2*, 841 (1985).
81. D. E. Applequist, G. F. Fanta and B. W. Henrikson, *J. Am. Chem. Soc.*, **82**, 2368 (1960).
82. S. H. Jones and E. Whittle, *Int. J. Chem. Kinet.*, **2**, 479 (1970).
83. F. MacCorquodale and J. C. Walton, *J. Chem. Soc., Faraday Trans. 1*, **84**, 3233 (1988).
84. D. E. Applequist and J. A. Landgrebe, *J. Am. Chem. Soc.*, **86**, 1543 (1964).
85. K. B. Wiberg, G. M. Lampman, R. P. Ciula, D. J. Connor, P. Schertier and J. Lavanish, *Tetrahedron*, **21**, 2749 (1965).
86. R. S. Boikess and M. D. Mackay, *Tetrahedron Lett.*, 5991 (1968).
87. R. S. Boikess and M. D. Mackay, *J. Org. Chem.*, **36**, 901 (1971).
88. C. Jamieson, J. C. Walton and K. U. Ingold, *J. Chem. Soc., Perkin Trans. 2*, 1366 (1980).
89. V. W. Bowry, J. Lusztyk and K. U. Ingold, *J. Am. Chem. Soc.*, **111**, 1927 (1989).

90. P. R. Ortis de Montellano and R. A. Stearns, *J. Am. Chem. Soc.*, **109**, 3415 (1987).
91. P. D. Bartlett, R. E. Pincock, J. R. Rolstan, W. G. Schindel and L. A. Singer, *J. Am. Chem. Soc.*, **87**, 2590 (1965).
92. C. Roberts and J. C. Walton, *J. Chem. Soc., Perkin Trans. 2*, 879 (1983).
93. P. K. Freeman, F. A. Raymond, J. C. Sutton and W. R. Kindley, *J. Org. Chem.*, **33**, 1448 (1968).
94. M. S. Kharasch and H. C. Brown, *J. Am. Chem. Soc.*, **62**, 454 (1940).
95. M. S. Kharasch and H. C. Brown, *J. Am. Chem. Soc.*, **64**, 329 (1942).
96. M. S. Kharasch, S. S. Kane and H. C. Brown, *J. Am. Chem. Soc.*, **64**, 333 (1942).
97. M. S. Kharasch, S. S. Kane and H. C. Brown, *J. Am. Chem. Soc.*, **64**, 1621 (1942).
98. R. S. Boikess, M. Mackay and D. Blithe, *Tetrahedron Lett.*, 401 (1971).
99. R. Srinivasan and F. I. Sonntag, *Tetrahedron Lett.*, 603 (1967).
100. E. C. Kooyman and G. C. Vegter, *Tetrahedron*, **4**, 382 (1958).
101. R. Srinivasan and F. J. Sonntag, *J. Am. Chem. Soc.*, **89**, 407 (1967).
102. A. F. Bickel, J. Knotnerus, E. C. Kooyman and G. C. Vegter, *Tetrahedron*, **9**, 230 (1960).
103. K. B. Wiberg and V. Z. Williams Jr., *J. Am. Chem. Soc.*, **89**, 3373 (1967).
104. B. Maillard and J. C. Walton, *J. Chem. Soc., Chem. Commun.*, 900 (1983).
105. T. Kawamura, M. Matsunaga and T. Yonezawa, *J. Am. Chem. Soc.*, **97**, 3234 (1975).
106. T. Kawamura and T. Yonezawa, *J. Chem. Soc., Chem. Commun.*, 948 (1976).
107. G. W. Smith and H. D. Williams, *J. Org. Chem.*, **26**, 2207 (1961).
108. J. Meinwald and B. E. Kaplan, *J. Am. Chem. Soc.*, **89**, 2611 (1967).
109. J. D. Roberts and P. H. Dirstine, *J. Am. Chem. Soc.*, **67**, 1281 (1945).
110. J. A. Kerr, H. Smith and A. F. Trotman-Dickenson, *J. Chem. Soc. (A)*, 1400 (1972).
111. J. A. Landgrebe and L. W. Becker, *J. Am. Chem. Soc.*, **89**, 2505 (1967).
112. J. D. Roberts and R. H. Mazur, *J. Am. Chem. Soc.*, **73**, 2509 (1951).
113. E. Renk, P. R. Schafer, W. H. Graham, R. H. Mazur and J. D. Roberts, *J. Am. Chem. Soc.*, **83**, 1987 (1961).
114. J. H. Incremna and C. J. Upton, *J. Am. Chem. Soc.*, **94**, 301 (1972).
115. M. L. Poutsma, *J. Am. Chem. Soc.*, **87**, 4293 (1965).
116. O. Dobis and S. W. Benson, *J. Am. Chem. Soc.*, **112**, 1023 (1990).
117. L. Schmerling and J. P. West, *J. Am. Chem. Soc.*, **71**, 2015 (1949).
118. F. F. Rust and C. S. Bell, *J. Am. Chem. Soc.*, **92**, 5530 (1970).
119. D. D. Tanner, S. C. Lewis and N. Wada, *J. Am. Chem. Soc.*, **94**, 7034 (1972).
120. D. D. Tanner and N. J. Bunce, *J. Am. Chem. Soc.*, **91**, 3028 (1969).
121. N. J. Bunce and D. D. Tanner, *J. Am. Chem. Soc.*, **91**, 6096 (1969).
122. G. A. Russell and C. DeBoer, *J. Am. Chem. Soc.*, **85**, 3136 (1963).
123. K. B. Wiberg and L. H. Slaugh, *J. Am. Chem. Soc.*, **80**, 3033 (1958).
124. D. D. Tanner and N. Wada, *J. Am. Chem. Soc.*, **97**, 2190 (1975).
125. D. D. Tanner, T. C.-S. Ruo, H. Takiguchi, A. Guillaume, D. W. Reed, B. P. Setiloane, S. L. Tan and C. P. Meintzer, *J. Org. Chem.*, **48**, 2743 (1983).
126. D. D. Tanner, C. P. Meintzer, E. C. Tsai and H. Oumar-Mahamat, *J. Am. Chem. Soc.*, **112**, 7369 (1990).
127. R. B. Timmons, J. de Guzman and R. E. Vanerin, *J. Am. Chem. Soc.*, **90**, 5996 (1968).
128. C. Walling and B. Miller, *J. Am. Chem. Soc.*, **79**, 4181 (1957).
129. D. D. Tanner, T. Ochiai and T. Pace, *J. Am. Chem. Soc.*, **97**, 6162 (1975).
130. G. A. Russell and K. M. Desmond, *J. Am. Chem. Soc.*, **85**, 3139 (1963).
131. D. D. Tanner, R. J. Arhart, E. V. Blackburn, N. C. Das and N. Wada, *J. Am. Chem. Soc.*, **96**, 829 (1974).
132. F. Minisci, G. P. Gardini and F. Bertini, *Can. J. Chem.*, **48**, 544 (1970).
133. J. Spanswick and K. U. Ingold, *Can. J. Chem.*, **48**, 546 (1970).
134. N. C. Deno, K. Eisenhardt, R. Fishbein, C. Pierson, D. Pohl and H. Spinelli, *XXIIIrd. Int. Cong. of Pure and Appl. Chem., Boston, 1971*, **4**, 155 (1972).
135. N. C. Deno and D. G. Pohl, *J. Am. Chem. Soc.*, **96**, 6680, (1974).
136. E. S. Huyser, H. Schimke and R. L. Burham, *J. Org. Chem.*, **28**, 2141 (1963).
137. D. D. Tanner and N. Nychka, *J. Am. Chem. Soc.*, **89**, 121 (1967).
138. C. Walling and B. B. Jacknow, *J. Am. Chem. Soc.*, **82**, 6108, 6113 (1960).

139. C. Walling and J. A. McGuinness, *J. Am. Chem. Soc.*, **91**, 2053 (1969).
140. J. Adam, P. A. Grosselain and P. Goldfinger, *Bull. Soc. Chim. Belg.*, **65**, 533 (1956).
141. A. G. Davies, B. P. Roberts and N. M. Smith, *J. Chem. Soc., Perkin Trans. 2*, 2221 (1972).
142. P. S. Skell, D. L. Tuleen and P. D. Readio, *J. Am. Chem. Soc.*, **85**, 2850 (1963).
143. G. A. Russell, C. DeBoer and K. M. Desmond, *J. Am. Chem. Soc.*, **85**, 365 (1963).
144. R. E. Pearson and J. C. Martin, *J. Am. Chem. Soc.*, **85**, 354, 3142 (1963).
145. C. Walling, A. L. Rieger and D. D. Tanner, *J. Am. Chem. Soc.*, **85**, 3129 (1963).
146. P. S. Skell, V. Lüning, D. S. McBain and J. M. Tanko, *J. Am. Chem. Soc.*, **108**, 121 (1986).
147. D. D. Tanner, C. P. Meintzer and S. L. Tan, *J. Org. Chem.*, **50**, 1534 (1985).
148. D. D. Tanner, T. C.-S. Ruo, H. Takiguchi, A. Guillaume, D. W. Reed, B. P. Setiloane, S. L. Tan and C. P. Meintzer, *J. Org. Chem.*, **48**, 2743 (1983).
149. D. D. Tanner, D. W. Reed, S. L. Tan, C. P. Meintzer, C. Walling and A. Sopchik, *J. Am. Chem. Soc.*, **107**, 6576 (1985).
150. C. Walling, G. M. El-Taliawi and A. Sopchik, *J. Org. Chem.*, **51**, 736 (1986).
151. J. L. Speier, *J. Am. Chem. Soc.*, **73**, 826 (1951).
152. P. S. Skell, H. N. Baxter III and J. M. Tanko, *Tetrahedron Lett.*, **27**, 5181 (1986).
153. F. F. Rust and W. E. Vaughan, *Ind. Eng. Chem.*, **41**, 2595 (1949).
154. E. R. Bell, F. H. Dickey, J. H. Raley, F. F. Rust and W. E. Vaughan, *Ind. Eng. Chem.*, **41**, 2597 (1949).
155. P. J. Nawrocki, J. H. Raley, F. F. Rust and W. E. Vaughan, *Ind. Eng. Chem.*, **41**, 2604 (1949).
156. E. R. Bell, G. E. Irish, J. H. Raley, F. F. Rust and W. E. Vaughan, *Ind. Eng. Chem.*, **41**, 2609 (1949).
157. B. Barnett, E. R. Bell, F. H. Dickey, F. F. Rust and W. E. Vaughan, *Ind. Eng. Chem.*, **41**, 2612 (1949).
158. R. Graf, *Justus Liebigs Ann. Chem.*, **578**, 50 (1952).
159. G. A. Russell, *J. Am. Chem. Soc.*, **79**, 3871 (1957).
160. J. O. Clayton and W. L. Jensen, *J. Am. Chem. Soc.*, **70**, 3880 (1948).
161. L. Z. Soborovskii, Y. M. Zinov'ev and M. A. Englin, *Dokl. Akad. Nauk SSSR*, **67**, 293 (1949).
162. R. Graf, *Chem. Ber.*, **85**, 9 (1952).
163. A. F. Isbell and F. T. Wadsworth, *J. Am. Chem. Soc.*, **78**, 6042 (1956).
164. F. R. Mayo, L. J. Durham and K. S. Griggs, *J. Am. Chem. Soc.*, **85**, 3156 (1963).
165. W. L. Jensen and C. R. Noller, *J. Am. Chem. Soc.*, **71**, 2384 (1949).
166. P. Lesfauries and P. Rumpf, *Bull. Soc. Chim. Fr.*, 542 (1950).
167. L. Z. Soborovskii and Y. M. Zinov'ev, *Zh. Obshch. Khim.*, **24**, 516 (1954).
168. A. G. Davies and B. P. Roberts, *Angew. Chem., Int. Ed. Engl.*, **10**, 738 (1971).
169. J. A. Howard, *Free Radicals*, Vol. II (Ed J. Kochi), Wiley, New York, 1973, p. 31.
170. J. A. Pianfetti and L. D. Quin, *J. Am. Chem. Soc.*, **84**, 851 (1962).

Author index

This author index is designed to enable the reader to locate an author's name and work with the aid of the reference numbers appearing in the text. The page numbers are printed in normal type in ascending numerical order, followed by the reference numbers in parentheses. The numbers in *italics* refer to the pages on which the references are actually listed.

Martin, R.O. 913 (208), *923*
Martins, F.J.C. 574, 575 (113i), *604*
Maruyama, K. 705 (83f), *739*
Marx, D.E. 661 (40), *678*
Marx, J.N. (135), *184*
Maryanoff, B.E. 571 (95), 575 (115), *603, 604*
Maryanoff, C.A. 571 (95), *603*
Marzio, A. di 430 (181), *453*
Masamune, S. 85–87 (40), *94*, 540 (86), *549*
Masanet, J. 690 (51b), *737*
Mascarella, S.W. 571 (99), *603*
Mason, J. 352, 355 (11), *389*
Mason, R. 540 (78), *549*
Mason, S.F. 152 (81, 89), 163, 164 (112), *183, 184*
Massoudi, M. 969, 979 (57), *995*
Masters, C. 555–557 (11e), *600*
Mastryukov, V.S. 125 (193), 127 (193, 206, 210), *132, 133*, 379 (172), 380, 382 (179), *392*
Mataga, N. 687 (39d), *736*
Mateescu, G.D. 478 (97, 102), *527*
Matheson, M.S. 762 (27), *779*
Mathias, A. 400 (29), *450*
Mathieu, D. 617 (33), *650*
Matsubara, Y. 801 (68), 802 (70), *807*
Matsuda, K. 407 (90, 91), *451*
Matsuda, T. 686 (35), *736*
Matsumoto, T. 103, 112, 114, 116, 117 (63), *129*, 597, 598 (210b), *607*, 691 (53h), *737*
Matsumura, Y. 791 (42), 799 (63), 800 (64), *807*
Matsunaga, M. 983 (105), *996*
Matsura, K. 425 (167), *453*
Matsuura, T. 682 (13), *735*
Mattay, J. 682, 705 (17c), *736*
Matthews, C.S. 235 (42), 237 (49), *286*
Matthews, D.E. 341 (110), *349*
Matthews, R.S. 361 (43), *390*
Mattice, W.L. 128 (211), *133*
Matturro, M.G. 558 (28a), 562 (53b), *601, 602*, 711, 733 (104), *740*
Matusch, R. 364 (56), *390*, 595 (206b), *607*
Maunder, C.M. 109 (93), *130*
Maurice, D. 885 (275), *893*
Maxwell, J.R. 294, 295 (10), *346*, 896 (8), 898 (21), 900 (37), 908 (183), 909 (185, 186), *919, 920, 923*
Mayer, B. 479 (109), 507 (153), *527, 528*
Mayer, G. 478, 516, 517 (106), *527*
Maynard, K.J. 686 (34a), *736*
Mayne, C.L. 388 (228), *393*
Maynes, G.G. 964, 983 (5), *994*
Mayo, F.R. 993 (164), *997*
Mayo, P.de 711 (114b), *740*

Mays, R. 321 (61), *348*
Mazanec, T.I. 805 (87, 92), *808*
Maziere, M. 820 (40), *887*
Mazliak, P. 900 (43), 901 (47), 904 (43), *920*
Mazur, R.H. 984 (112, 113), *996*
Mazur, Y. 782 (15c), *806*
McAdoo, D.J. 444 (228), *454*
McAffee, A.M. 538 (61), *549*
McAskill, N.A. 969 (33), *994*
McBain, D.S. 991 (146), *997*
McBride, B.J. 231 (27), *286*
McCaffrey, J.G. 674 (88), *679*
McCarthy, E.D. 898 (23), 907 (165), *920, 923*
McCarthy, K.E. 586 (174a, 175b), *605*
McCarthy, R.D. 854 (153), *890*
McClelland, R.E. 895 (2), *919*
McClusky, J.V. 724, 728 (134), *741*
McCombie, S.W. 565, 566 (64), *602*
McCormick, A. 321, 323 (64), *348*
McCrea, J.M. 858, 881 (169), *890*
McDaniel, C.A. 906 (139), *922*
McDonald, G.A. 917 (267), *925*
McDonald, R.A. 228 (23), 259 (60), *285, 287*
McEwen, A.B. 592 (197), *606*, 620 (45c), *650*
McEwen, C.N. 433 (191–195), *453*
McFadden, W.H. 411 (120), 444 (227), *452, 454*
McFarland, C.W. 478 (102), *527*
McGhee, W.D. 662 (46), *678*
McGhie, J.F. 564 (58a, 58b), *602*
McGlynn, S.P. 479 (121), *527*
McGrath, D.V. 664, 666, 667, 669 (59), 670 (59, 66), *678, 679*
McGrath, M.P. 85, 87 (41), *94*, 658 (22), *677*
McGuinness, J.A. 990 (139), *997*
McInerney, E.J. 804 (78), *808*
McInnes, A.G. 914 (224), *924*
McIver, R.T. 543 (140), 544 (146), *550, 551*
McKean, D.C. 89, 90 (55), *94*
McKee, M.L. 460, 507 (11), *524*
McKeever, L.D. 542 (124), *550*
McKenna, E.J. 853 (143), *889*, 915 (234), *924*
McKervey, M.A. 116 (115), *130*, 368 (69), *390*, 597 (210e, 211, 212), 598 (210e), *607*, 691 (53b, 53c, 53g), *737*
McKillop, A. 578 (129c), 585 (169c), *604, 605*
McKinley, A.J. 711 (106a), *740*
McKinney, C.R. 858, 881 (169), *890*
McLafferty, F.W. 332 (90), *348*, 412 (125), 424 (159, 163), 434, 444 (163), 447 (238), *452–454*
McLafferty, W.J. 123 (182), *132*
McLean, A.D. 532, 537, 538, 547 (33), *548*
McLuckey, S.A. 405 (59), *451*
McMahon, R.E. 816 (22), *886*

Author index

Author index

Index compiled by K. Raven

Subject index

Index compiled by P. Raven